Errata and Corrigenda No. 7 (Vol. 1-13) Jan. 1979, have been entered

Errata and Corrigenda No. 8 (Vols. 1-13), January 1980, have been entered in this volume.

Errata and Corrigenda No 9 (Vols. 1-13) 4-16-81 have been entered in this volume.

THERMAL RADIATIVE PROPERTIES

Coatings

THERMOPHYSICAL PROPERTIES OF MATTER

The TPRC Data Series

A Comprehensive Compilation of Data by the
Thermophysical Properties Research Center (TPRC), Purdue University

Y. S. Touloukian, Series Editor
C. Y. Ho, Series Technical Editor

New data on thermophysical properties are being constantly accumulated at TPRC. Contact TPRC and use its interim updating services for the most current information.

THERMOPHYSICAL PROPERTIES OF MATTER
VOLUME 9

THERMAL RADIATIVE PROPERTIES
Coatings

Y. S. Touloukian

Director
Thermophysical Properties Research Center
and
Distinguished Atkins Professor of Engineering
School of Mechanical Engineering
Purdue University
and
Visiting Professor of Mechanical Engineering
Auburn University

D. P. DeWitt

Deputy Director and
Associate Senior Researcher
Thermophysical Properties Research Center
and
Associate Professor of Mechanical Engineering
Purdue University

R. S. Hernicz

Assistant Senior Researcher
Thermophysical Properties Research Center
Purdue University

IFI/PLENUM • NEW YORK-WASHINGTON • 1972

Library of Congress Catalog Card Number 73-129616

ISBN (13-Volume Set) 0-306-67020-8
ISBN (Volume 9) 0-306-67029-1

IFI/Plenum Data Corporation is a subsidiary of
Plenum Publishing Corporation
227 West 17th Street, New York, N.Y. 10011

Distributed in Europe by Heyden & Son, Ltd.
Spectrum House, Alderton Crescent
London NW4 3XX, England

Printed in the United States of America

"In this work, when it shall be found that much is omitted, let it not be forgotten that much likewise is performed..."

SAMUEL JOHNSON, A.M.

From last paragraph of Preface to his two-volume *Dictionary of the English Language*, Vol. I, page 5, 1755, London, Printed by Strahan.

Foreword

In 1957, the Thermophysical Properties Research Center (TPRC) of Purdue University, under the leadership of its founder, Professor Y. S. Touloukian, began to develop a coordinated experimental, theoretical, and literature review program covering a set of properties of great importance to science and technology. Over the years, this program has grown steadily, producing bibliographies, data compilations and recommendations, experimental measurements, and other output. The series of volumes for which these remarks constitute a foreword is one of these many important products. These volumes are a monumental accomplishment in themselves, requiring for their production the combined knowledge and skills of dozens of dedicated specialists. The Thermophysical Properties Research Center deserves the gratitude of every scientist and engineer who uses these compiled data.

The individual nontechnical citizen of the United States has a stake in this work also, for much of the science and technology that contributes to his well-being relies on the use of these data. Indeed, recognition of this importance is indicated by a mere reading of the list of the financial sponsors of the Thermophysical Properties Research Center; leaders of the technical industry of the United States and agencies of the Federal Government are well represented.

Experimental measurements made in a laboratory have many potential applications. They might be used, for example, to check a theory, or to help design a chemical manufacturing plant, or to compute the characteristics of a heat exchanger in a nuclear power plant. The progress of science and technology demands that results be published in the open literature so that others may use them. Fortunately for progress, the useful data in any single field are not scattered throughout the tens of thousands of technical journals published throughout the world. In most fields, fifty percent of the useful work appears in no more than thirty or forty journals. However, in the case of TPRC, its field is so broad that about 100 journals are required to yield fifty percent. But that other fifty percent! It is scattered through more than 3500 journals and other documents, often items not readily identifiable or obtainable. Nearly 50,000 references are now in the files.

Thus, the man who wants to use existing data, rather than make new measurements himself, faces a long and costly task if he wants to assure himself that he has found all the relevant results. More often than not, a search for data stops after one or two results are found—or after the searcher decides he has spent enough time looking. Now with the appearance of these volumes, the scientist or engineer who needs these kinds of data can consider himself very fortunate. He has a single source to turn to; thousands of hours of search time will be saved, innumerable repetitions of measurements will be avoided, and several billions of dollars of investment in research work will have been preserved.

However, the task is not ended with the generation of these volumes. A critical evaluation of much of the data is still needed. Why are discrepant results obtained by different experimentalists? What undetected sources of systematic error may affect some or even all measurements? What value can be derived as a "recommended" figure from the various conflicting values that may be reported? These questions are difficult to answer, requiring the most sophisticated judgment of a specialist in the field. While a number of the volumes in this Series do contain critically evaluated and recommended data, these are still in the minority. The data are now being more intensively evaluated by the staff of TPRC as an integral part of the effort of the National Standard Reference Data System (NSRDS). The task of the National Standard Reference Data System is to organize and operate a comprehensive program to prepare compilations of critically evaluated data on the properties of substances. The NSRDS is administered by the National Bureau of Standards under a directive from the Federal Council for Science and Technology, augmented by special legislation of the Congress of the United States. TPRC is one of the national resources participating in the National Standard Reference Data System in a united effort to satisfy the needs of the technical community for readily accessible, critically evaluated data.

As a representative of the NBS Office of Standard Reference Data, I want to congratulate Professor Touloukian and his colleagues on the accomplishments represented by this Series of reference data books. Scientists and engineers the world over are indebted to them. The task ahead is still an awesome one and I urge the nation's private industries and all concerned Federal agencies to participate in fulfilling this national need of assuring the availability of standard numerical reference data for science and technology.

EDWARD L. BRADY
Associate Director for Information Programs
National Bureau of Standards

Preface

Thermophysical Properties of Matter, the TPRC Data Series, is the culmination of twelve years of pioneering effort in the generation of tables of numerical data for science and technology. It constitutes the restructuring, accompanied by extensive revision and expansion of coverage, of the original *TPRC Data Book*, first released in 1960 in loose-leaf format, 11″ × 17″ in size, and issued in June and December annually in the form of supplements. The original loose-leaf *Data Book* was organized in three volumes: (1) metallic elements and alloys, (2) nonmetallic elements, compounds, and mixtures which are solid at N.T.P., and (3) nonmetallic elements, compounds, and mixtures which are liquid or gaseous at N.T.P. Within each volume, each property constituted a chapter.

Because of the vast proportions the *Data Book* began to assume over the years of its growth and the greatly increased effort necessary in its maintenance by the user, it was decided in 1967 to change from the loose-leaf format to a conventional publication. Thus, the December 1966 supplement of the original *Data Book* was the last supplement disseminated by TPRC.

While the manifold physical, logistic, and economic advantages of the bound volume over the loose-leaf oversize format are obvious and welcome to all who have used the unwieldy original volumes, the assumption that this work will no longer be kept on a current basis because of its bound format would not be correct. Fully recognizing the need of many important research and development programs which require the latest available information, TPRC has instituted a *Data Update Plan* enabling the subscriber to inquire, by telephone if necessary, for specific information and receive, in many instances, same-day response on any new data processed or revision of published data since the latest edition. In this context, the TPRC Data Series departs drastically from the conventional handbook and giant multivolume classical works, which are no longer adequate media for the dissemination of numerical data of science and technology without a continuing activity on contemporary coverage. The loose-leaf arrangements of many works fully recognize this fact and attempt to develop a combination of bound volumes and loose-leaf supplement arrangements as the work becomes increasingly large. TPRC's *Data Update Plan* is indeed unique in this sense since it maintains the contents of the TPRC Data Series current and live on a day-to-day basis between editions. In this spirit, I strongly urge all purchasers of these volumes to complete in detail and return the *Volume Registration Certificate* which accompanies each volume in order to assure themselves of the continuous receipt of annual listing of corrigenda during the life of the edition.

The TPRC Data Series consists initially of 13 independent volumes. The initial seven were published in 1970. Volumes 8, 9, and 10 are planned for 1972 and Volumes 11, 12, and 13 for 1973. It is also contemplated that subsequent to the first edition, each volume will be revised, updated, and reissued in a new edition approximately every fifth year. The organization of the TPRC Data Series makes each volume a self-contained entity available individually without the need to purchase the entire Series.

The coverage of the specific thermophysical properties represented by this Series constitutes the most comprehensive and authoritative collection of numerical data of its kind for science and technology.

Whenever possible, a uniform format has been used in all volumes, except when variations in presentation were necessitated by the nature of the property or the physical state concerned. In spite of the wealth of data reported in these volumes, it should be recognized that all volumes are not of the same degree of completeness. However, as additional data are processed at TPRC on a continuing basis, subsequent editions will become increasingly more complete and up to date. Each volume in the Series basically comprises three sections, consisting of a text, the body of numerical data with source references, and a material index.

The aim of the textual material is to provide a complementary or supporting role to the body of numerical data rather than to present a treatise on the subject of the property. The user will find a basic theoretical treatment, a comprehensive presentation of selected works which constitute reviews, or compendia of empirical relations useful in estimation of the property when there exists a paucity of data or when data are completely lacking. Established major experimental techniques are also briefly reviewed.

The body of data is the core of each volume and is presented in both graphical and tabular formats for convenience of the user. Every single point of numerical data is fully referenced as to its original source and no secondary sources of information are used in data extraction. In general, it has not been possible to critically scrutinize all the original data presented in these volumes, except to eliminate perpetuation of gross errors. However, in a significant number of cases, such as for the properties of liquids and gases and the thermal conductivity of all the elements, the task of full evaluation, synthesis, and correlation has been completed. It is hoped that in subsequent editions of this continuing work, not only new information will be reported but the critical evaluation will be extended to increasingly broader classes of materials and properties.

The third and final major section of each volume is the Material Index. This is the key to the volume, enabling the user to exercise full freedom of access to its contents by any choice of substance name or detailed alloy and mixture composition, trade name, synonym, etc. Of particular interest here is the fact that in the case of those properties which are reported in separate companion volumes, the Material Index in each of the volumes also reports the contents of the other companion volumes.* The sets of companion volumes are as follows:

Thermal conductivity:	Volumes 1, 2, 3
Specific heat:	Volumes 4, 5, 6
Radiative properties:	Volumes 7, 8, 9
Thermal expansion:	Volumes 12, 13

The ultimate aims and functions of TPRC's Data Tables Division are to extract, evaluate, reconcile, correlate, and synthesize all available data for the thermophysical properties of materials with the result of obtaining internally consistent sets of property values, termed the "recommended reference values." In such work, gaps in the data often occur, for ranges of temperature, composition, etc. Whenever feasible, various techniques are used to fill in such missing information, ranging from empirical procedures to detailed theoretical calculations. Such studies are resulting in valuable new estimation methods being developed which have made it possible to estimate values for substances and/or physical conditions presently unmeasured or not amenable to laboratory investigation. Depending on the available information for a particular property and substance, the end product may vary from simple tabulations of isolated values to detailed tabulations with generating equations, plots showing the concordance of the different values, and, in some cases, over a range of parameters presently unexplored in the laboratory.

The TPRC Data Series constitutes a permanent and valuable contribution to science and technology. These constantly growing volumes are invaluable sources of data to engineers and scientists, sources in which a wealth of information heretofore unknown or not readily available has been made accessible. We look forward to continued improvement of both format and contents so that TPRC may serve the scientific and technological community with ever-increasing excellence in the years to come. In this connection, the staff of TPRC is most anxious to receive comments, suggestions, and criticisms from all users of these volumes. An increasing number of colleagues are making available at the earliest possible moment reprints of their papers and reports as well as pertinent information on the more obscure publications. I wish to renew my earnest request that this procedure become a universal practice since it will prove to be most helpful in making TPRC's continuing effort more complete and up to date.

It is indeed a pleasure to acknowledge with gratitude the multisource financial assistance received from over fifty of TPRC's sponsors which has made the continued generation of these tables possible. In particular, I wish to single out the sustained major support being received from the Air Force Materials Laboratory–Air Force Systems Command, the Office of Standard Reference Data–National Bureau of Standards, and the Office of Advanced Research and Technology–National Aeronautics and Space Administration. TPRC is indeed proud to have been designated as a National Information Analysis Center for the Department of Defense as well as a component of the National

*For the first edition of the Series, this arrangement was not feasible for Volumes 7 and 8 due to the sequence and the schedule of their publication. This situation will be resolved in subsequent editions.

Standard Reference Data System under the cognizance of the National Bureau of Standards.

While the preparation and continued maintenance of this work is the responsibility of TPRC's Data Tables Division, it would not have been possible without the direct input of TPRC's Scientific Documentation Division and, to a lesser degree, the Theoretical and Experimental Research Divisions. The authors of the various volumes are the senior staff members in responsible charge of the work. It should be clearly understood, however, that many have contributed over the years and their contributions are specifically acknowledged in each volume. I wish to take this opportunity to personally thank those members of the staff, research assistants, graduate research assistants, and supporting graphics and technical typing personnel without whose diligent and painstaking efforts this work could not have materialized.

Y. S. Touloukian

Director
Thermophysical Properties Research Center
Distinguished Atkins Professor of Engineering

Purdue University
Lafayette, Indiana
May 1972

Introduction to Volume 9

This volume of *Thermophysical Properties of Matter*, the TPRC Data Series, follows the general format of the volumes on other properties except where departures and innovations were found necessary for the effective presentation of thermal radiative properties.

The volume consists of three major sections: the front text material together with its bibliography, the main body of numerical data and its references, and the material index.

The text material is intended to assume a role complementary to the main body of numerical data, the presentation of which is the primary purpose of this volume. It is felt that a moderately detailed discussion of the theoretical nature of the properties under consideration together with an overview of predictive procedures and recognized experimental techniques will be appropriate in a major reference work of this kind. Common to each of the companion Volumes 7, 8, and 9 are sections dealing with the physics of thermal radiation—definition of terms, basic concepts, and useful relations—and with experimental techniques suitable for all classes of materials. In addition, Volume 7 discusses the classical electron theory explaining metallic behavior and surface characterization, a very special concern for the metallics. Volume 8 treats the theoretical behavior of the simple dielectric and semitransparent materials, with and without scattering, as well as some experimental techniques particularly suited for measurements on the nonmetallics. Volume 9 includes theoretical discussions on coating systems and treats the problems relating to environmental influences. Sufficient detail is given to make the text material self-contained, but it has been the practice to provide extensive reference citations to lead the interested reader to the literature for a more comprehensive treatment.

The main body of the volume consists of the presentation of numerical data compiled over the years in a most comprehensive manner on coatings for all applications particularly thermal control. In the context of this volume, a *coating* is a system consisting of a layer (or layers) of any substance(s) upon a substrate; they are further classified in three categories entitled pigmented, contact, and conversion coatings, as is discussed in the introduction to the *Numerical Data* section. The extraction of all data directly from their original sources ensures freedom from errors of transcription. Furthermore, some gross errors appearing in the original source documents have been corrected. The organization and presentation of the data together with other pertinent information in the use of the tables and figures are discussed in detail in the introduction to the *Numerical Data* section.

In addition to the original data presentation in the *Numerical Data* section, the Analyzed Data Graphs will give the user an evaluative review of the data. This analysis work is first a filtering process; it identifies data which are felt to be reliably or typically identified with the materials and gives the user a good feeling of "relief" from the "spaghetti" type of presentation shown in the original or archival data graphs. The analyzed curves are based, in some instances, on experiences of the research team as well as the original data sources. This analysis work is an innovative feature of the radiative properties volumes and should not be considered as recommended reference values identified in other volumes of this Series; the work is intended to make the best reliable data available in a convenient form to the thermal designer.

As stated earlier, all data have been obtained from their original sources and each data set is so referenced. TPRC has in its files all documents cited in this volume. Those that cannot readily be obtained elsewhere are available from TPRC in microfiche form.

The volumes on the thermal radiative properties have grown out of activities supported principally by the National Aeronautics and Space Administration, Office of Advanced Research and Technology (NSR-005-037), under the monitorship of Mr.

Conrad Mook, and by the Jet Propulsion Laboratory, California Institute of Technology (NAS 7-100), under the monitorship of Mr. W. F. Carroll. We wish to acknowledge the benefit of extensive discussions with Mr. D. W. Gates, Space Sciences Laboratory, NASA-MSFC, and Mr. Carroll, Structures, Dynamics, and Materials Section, JPL, who have served in an active advisory capacity throughout the progress of work on Volumes 7, 8, and 9 from the outset.

Over the past five years, many graduate students, research assistants, and technical staff have contributed to the preparation of these volumes for varying periods under the authors' supervision. In chronological order of their association with TPRC, we wish to acknowledge the contributions of I. M. Yeyinmen, B. Compani-Tabrizi, M. C. Muinzer, P. Sioshansi, J. J. Hsia, C. K. Hsieh, R. L. Jones, K. F. Sohn, S. L. Miller, and D. L. Gudgel.

The authors acknowledge the assistance of Mr. Joseph C. Richmond, Institute for Basic Standards, National Bureau of Standards, who provided valuable suggestions relating to nomenclature and subproperty classification; many of the ideas in the text portion of the Volumes 7, 8, and 9 are a result of his direct contributions.

Inherent in the character of this work is the fact that in the preparation of this volume we have drawn most heavily upon the scientific literature and feel a debt of gratitude to the authors of the referenced articles. While their often discordant results have caused us much difficulty in reconciling their findings, we consider this to be our challenge and our contribution to negative entropy of information, as an effort is made to create from the randomly distributed data a condensed, more orderly state.

While this volume is primarily intended as a reference work for the designer, researcher, experimentalist, and theoretician, the teacher at the graduate level may also use it as a teaching tool to point out to his students the topography of the state of knowledge on the thermal radiative properties of coatings. We believe there is also much food for reflection by the specialist and the academician concerning the meaning of "original" investigation and its "information content."

The authors and their contributing associates are keenly aware of the possibility of many weaknesses in a work of this scope. We hope that we will not be judged too harshly and that we will receive the benefit of suggestions regarding references omitted, additional material groups needing more detailed treatment, improvements in presentation or in recommended values, and, most important, any inadvertent errors. If the *Volume Registration Certificate* accompanying this volume is returned, the reader will assure himself of receiving annually a list of corrigenda as possible errors come to our attention.

Lafayette, Indiana
May 1972

Y. S. TOULOUKIAN
D. P. DEWITT
R. S. HERNICZ

Contents

Numerical Data

Material Index

GROUPING OF MATERIALS AND LIST OF FIGURES AND TABLES

1. PIGMENTED COATINGS

A. Metallic Pigmented Coatings

B. Nonmetallic Pigmented Coatings

Note: Figure number with "A" indicates analyzed data graph.
* No figure

1. PIGMENTED COATINGS (continued)

B. Nonmetallic Pigmented Coatings (continued)

Note: Figure number with "A" indicates analyzed data graph.
* No figure

1. PIGMENTED COATINGS (continued)

B. Nonmetallic Pigmented Coatings (continued)

Note: Figure number with "A" indicates analyzed data graph.
* No figure

1. PIGMENTED COATINGS (continued)

B. Nonmetallic Pigmented Coatings (continued)

Note: Figure number with "A" indicates analyzed data graph.
* No figure

1. PIGMENTED COATINGS (continued)

B. Nonmetallic Pigmented Coatings (continued)

C. Enamels (Fused Vitreous Porcelain)

D. Trade Names

Note: Figure number with "A" indicates analyzed data graph.
* No figure

1. PIGMENTED COATINGS (continued)

D. Trade Names (continued)

Note: Figure number with "A" indicates analyzed data graph.
* No figure

1. PIGMENTED COATINGS (continued)

D. Trade Names (continued)

Note: Figure number with "A" indicates analyzed data graph.
* No figure

2. CONTACT COATINGS (continued)

A. Metallic Contact Coatings (continued)

Note: Figure number with "A" indicates analyzed data graph.
* No figure

2. CONTACT COATINGS (continued)

A. Metallic Contact Coatings (continued)

---

* No figure

2. CONTACT COATINGS (continued)

A. Metallic Contact Coatings (continued)

B. Nonmetallic Inorganic Contact Coatings

* No figure

2. CONTACT COATINGS (continued)

B. Nonmetallic Inorganic Contact Coatings (continued)

* No figure

2. CONTACT COATINGS (continued)

B. Nonmetallic Inorganic Contact Coatings (continued)

* No figure

2. CONTACT COATINGS (continued)

B. Nonmetallic Inorganic Contact Coatings (continued)

C. Second Surface Mirrors

Note: Figure number with "A" indicates analyzed data graph.
* No figure

2. CONTACT COATINGS (continued)

C. Second Surface Mirrors (continued)

D. Anti-Reflection Coatings

E. Resin Coatings

Note: Figure number with "A" indicates analyzed data graph.
* No figure

2. CONTACT COATINGS (continued)

E. Resin Coatings (continued)

F. Metallic Black Coatings

3. CONVERSION COATINGS

A. Anodized Coatings

Note: Figure number with "A" indicates analyzed data graph.
* No figure

3. CONVERSION COATINGS (continued)

A. Anodized Coatings (continued)

B. Oxidized Coatings

---

* No figure

3. CONVERSION COATINGS (continued)

C. Miscellaneous

* No figure

Theory, Estimation, and Measurement

Notation

a	Absorption coefficient
c	Speed of light in a medium
c_0	Speed of light in a vacuum
c_1	Planck's first radiation constant, $2\pi c_0^2 h$
c_2	Planck's second radiation constant, $c_0 h/k$
C_p	Specific heat (constant pressure)
d	Thickness of film
dA	Elemental area on radiating surface
e	Base of natural logarithms
E	Irradiance; Electric field strength
g	Solid angle of cone
h	Planck constant
H	Magnetic field strength
I	Radiant intensity
j	Imaginary unit; Volume spectral emissive power
J	Current density
k	Index of absorption; Boltzmann constant
K	Diffuse absorption coefficient
K^*	Complex dielectric constant
K'	Relative permittivity or real dielectric constant, real part of K^*
K''	Relative loss factor, imaginary part of K^*
L	Radiance
l	Mean free path
m	Electronic mass
M	Sample mass; Radiant exitance
n	Index of refraction
n^*	Complex index of refraction, $n - jk$
N	Number density of free electrons
P	Degree of polarization
q	Electronic charge
Q	Radiant energy
r	Electrical resistivity; Fresnel reflectivity coefficient
R	Reflectance factor; (Fresnel) Reflectivity
s	Solar (wavelength distribution); Scattering coefficient
S	Diffuse scattering coefficient
t	Time; Total (blackbody wavelength distribution)
T	Temperature (absolute); Internal transmittance or transmissivity
v	Volume
v_f	Fermi velocity
W	Radiant density
x	Distance
Z^*	Intrinsic impedance
α	Absorptance; Absorptivity; Plane half angle of right circular cone; Attenuation factor, real part of γ
α_p	Polarizability
β	Radiance factor; Phase factor, imaginary part of γ
γ	Complex propagation factor
δ	Penetration depth
ϵ	Emittance; Emissivity
ϵ_0	Permittivity of free space
ϵ^*	Complex permittivity
ϵ'	Real permittivity, real part of ϵ^*
ϵ''	Loss factor, imaginary part of ϵ^*
θ	Angle between surface normal and direction of incident flux, zenith angle, or co-latitude
θ'	Angle between surface normal and direction of reflected or emitted flux
λ	Wavelength
$\bar{\lambda}$	Integrated (wavelength distribution)
μ_0	Permeability of free space
μ^*	Complex permeability
ν	Frequency
π	Constant 3.14159 $\cdots$
ρ	Reflectance; Reflectivity
σ	Stefan–Boltzmann constant; electrical conductivity
τ	Transmittance; relaxation time
ϕ	Azimuthal angle of incident flux
ϕ'	Azimuthal angle of reflected or emitted flux
Φ	Radiant flux
ω	Solid angle of incident beam; $2\pi \times$ frequency
ω'	Solid angle of reflected or emitted beam

Subscripts

s	Perpendicular polarized component; Solar (spectral distribution)	b	Blackbody conditions
p	Parallel polarized component	λ	Spectral concentration
		t	Total (blackbody wavelength distribution)

Thermal Radiative Properties of Coatings

1. INTRODUCTION

Radiation is one of the three fundamental modes of heat transfer, the others being conduction and convection. Radiation differs from the other modes in two important respects: first, no medium is required for transport of energy by radiation, and second, the rate of heat dissipation by radiation varies approximately as the fourth power of the absolute temperature, while that by the other modes varies approximately as the first power of temperature. For these reasons, radiation becomes the dominant mode of heat transfer at high temperatures and in the absence of an atmosphere.

The thermal radiative properties—emittance, reflectance, absorptance, and transmittance—are the parameters which are descriptive of the energy transported by the radiation mode. The properties can be prescribed in greater detail to account for the spectral or wavelength conditions and the geometrical or directional conditions in which the radiant energy interacts with the solid. This interaction can be phenomenologically described by other properties as well, such as the optical constants, complex dielectric constant, or propagation factor, each of which is especially convenient for studying various aspects of the interaction. If the designer is to make effective use of radiative properties data, it is helpful that he be aware of the various thermal radiative and optical properties describing the basic mechanisms of the radiant energy–matter interaction which models the radiative transport process.

There is a marked contrast between the radiative properties of metallic and nonmetallic solids. The text of the companion Volume 7 [167] of this Series shows that understanding of the basic mechanism of interaction between radiant energy and metallic solids is reasonably well developed. The behavior of the metallic solid is adequately described by the free electron models, which indeed are only approximate but provide simple and useful tools to the designer. The more sophisticated theories, while not useful as yet for the prediction of numerical values from structural parameters, do provide a means for evaluation of test data and a basis for developing empirical relations to meet specific conditions. The theory of nonmetallic behavior, the subject of the text in companion Volume 8 [168], is less well developed. The simplest model ascribes the nonmetallic behavior due to a combination of several types of free electrons and electrons bound to the lattice. The theory is quite useful for basic understanding of behavior but not tractable for direct computation of property values. The problem is further complicated by transparency, scattering phenomena, and temperature gradients within the solid, which usually can only be treated in a gross or oversimplified manner.

The magnitude of the radiative properties of the *metallic* solid is determined to a large extent by the surface condition; due to the large extinction coefficient, radiant energy will not travel more than a few hundred angstroms into the metal before being totally absorbed. As a result, surface roughness, oxide layer formations, structural defects due to mechanical stresses, etc., can be dominating influences on the property variations as a function of wavelength and temperature; however, most important is the dependence of the properties upon the environmental conditions. The *nonmetallic* or dielectric materials are known to be less sensitive to surface conditions; the absorption and emission processes are "bulk" or "volume" phenomena as a significant surface layer of the material is responsible for the behavior. This is a consequence of appreciable transparency of the nonmetallic solid to thermal radiation. A *coating*, as defined for usage in this volume, consists of layer(s) of any substance upon a substrate. The optical behavior of the coating can be metallic or nonmetallic depending upon the optical properties and thickness of the coating layer(s) and of the optical properties of the substrate. The metallic–nonmetallic characteristics may also be wavelength dependent, as, for example, the optical solar reflector (see Fig. 7). As suspected, the coatings as a class are more difficult to treat theoretically; it is the added complexity that makes coatings especially attractive since their optical properties can be tailored to meet application requirements.

The principal differences between developments in the metallic and nonmetallic properties are twofold: (1) the lack of theoretical tools and simplified models for nonmetallic solids such as are available for metallic materials and (2) the contributions of the transparency of nonmetallic solids giving rise to "volume" effects rather than the "surface" effects which dominate the behavior of metallic solids.

The difficulties of characterizing data—unambiguously relating the measured property data to the conditions of the specimen—and of understanding environmental influences have frequently required the designer to measure the desired property of the actual material (metallic and nonmetallic) as it will be used in the environment of the application. Much of such effort can be reduced through proper use of the extensive compendia presented in this volume; also through the availability of such a bulk of data, it is likely that attention to the characterization of materials (surfaces) will increase.

The purpose of this text material is to expose the user to some of the pitfalls and limitations, as well as advantages, in the use of existing data reported in the literature. The text follows along lines similar to that of the Series companion volumes; the first seven sections are general in character, dealing with basic concepts, laws, definitions, and relations which are applicable to nonmetallic and metallic solids and to coatings.

The exposition begins with Section 2 briefly defining the terms—processes, things, quantities, properties, and modifiers—used in discussing thermal radiation phenomena. Following this, Section 3 presents a more rigorous definition of the properties and the notion of their dependency upon wavelength, temperature, polarization, and geometric directions. Section 4 reviews the physics of thermal radiation; then Section 5 discusses the interaction of radiant energy with materials. The interrelations between the properties—as described by geometric and wavelength modifiers—need to be understood in order for the user to synthesize fragments of available data. Section 6 treats the subproperty descriptors, and Section 7 discusses the important concepts in their interrelationships.

Section 8 summarizes briefly theoretical models of the more important types of coatings amenable to analyses. Thin metallic films at room and moderate temperatures and in the infrared can be treated by the classical free electron models, while for subnormal or cryogenic temperatures the complications of the anomalous skin effect are introduced. Multilayer films—with layers and substrates being simple dielectrics or absorbing media, which can be optically thick or thin (interference occurs)—have been extensively studied as they represent an important class of coatings. There is less understanding of the optical performance of diffuse coatings such as paints and particulate systems, where both surface and volume scattering mechanisms are significant. The reader should also find the text portions of the two Series companion volumes useful for developing some background on theoretical models of the radiative properties for coating systems.

The problems of measuring the radiative properties of coatings are first discussed in Section 9, which is a merging of the texts presented in the companion volumes. Many methods are equally appropriate to any type of solid or surface. The primary difficulty arises when the sample must be heated or cooled and if the material thermal conductivity is low (nonmetallics); then substantial temperature gradients can exist and there is ambiguity in identifying a surface temperature. The methods discussed in this section should introduce the reader to the more conventional approaches used by the experimentalist. It is not the aim here to evaluate and recommend techniques but rather to briefly review the limitations and capability of the various techniques.

The depth of presentation is aimed at the reader with a background in the physics of thermal radiation. For his benefit, an attempt is made to provide general references, usually standard texts or extensive review articles, which are intended to complement the necessarily brief treatment of the subject in this volume. The development of thermal radiation studies over the past years can be followed through the conference proceedings on thermal radiation, starting as an informal one in 1958, to the annual AIAA meetings in most recent years [10, 11, 15, 27, 75, 77, 90, 110, 138, 148]. The data compilations of references [65, 162, 164, 165] are most useful supplements for property coverage on materials not contained in the data sections of the Series volumes. The TPRC's *Thermophysical Properties Research Literature Retrieval Guide* [166] permits rapid access to the literature on the thermal radiative properties—and eight other thermophysical properties—of all classes of materials. Through these references, the reader will have access to a vital portion of knowledge created in the field of thermal radiation and a better appreciation of the technical problems involved in the application of the basic principles to actual design situations.

2. DEFINITION OF TERMS [74]

A. Processes

Radiation. The process by which radiant energy is emitted by a body; also the process by which energy is transferred in the form of radiant energy.

Reflection. The process by which radiant energy incident on a surface or medium leaves that surface or medium from the incident side.

Transmission. The process by which radiant energy incident on a surface or medium leaves that surface or medium on a side other than the incident side.

Absorption. The process by which radiant energy is converted into another form of energy.

Refraction. The process by which a beam of radiant energy, on transmission through the interface between two media of different index of refraction, is deviated toward the normal to the interface in the medium of higher index of refraction.

Propagation. The process or processes by which radiant energy is transferred from one point to another in space.

B. Things

Radiator. A source of radiant energy.

Thermal Radiator. A radiator that emits thermal radiant energy, as a consequence of its temperature only.

Blackbody. A body or surface that absorbs all of the radiant energy incident upon it, and emits the maximum possible amount of thermal radiant energy at each frequency for a body at its temperature.

Reflector. A body that reflects incident radiant energy.

Transmitter. A body that transmits incident radiant energy.

Absorber. A body that absorbs incident radiant energy.

Retroreflector. A reflector that reflects incident radiant energy in directions close to the direction of incidence.

C. Quantities

Radiant energy, Q. Energy in the form of electromagnetic waves or photons. Joules, ergs, or kilowatt-hours.

Thermal Radiant Energy, Q. Radiant energy that is emitted by a thermal radiator.

Radiant Density, W. $W = dQ/dv$. Radiant energy per unit volume. Joule per cubic meter, erg per cubic centimeter.

Radiant Flux, Φ. $\Phi = dQ/dt$. Time rate of flow of radiant energy. Erg per second, watt.

Radiant Intensity, I. $I = d\Phi/d\omega$. Flux per unit solid angle from a source. Watt per steradian.

Radiance, L. $L = d^2\Phi/d\omega dA \cos\theta$. Flux propagated in a given direction, per unit solid angle about that direction and per unit area projected normal to the direction.

Exitance, M. $M = d\Phi/dA$. Flux per unit area leaving a surface.

Irradiance, E. $E = d\Phi/dA$. Flux per unit area incident on a surface.

D. Properties

*Reflectance,** ρ. The ratio of reflected flux to incident flux.

*Absorptance,** α. The ratio of absorbed flux to incident flux.

Transmittance, τ. The ratio of transmitted flux to incident flux.

Internal Transmittance, T. The ratio of the radiant flux reaching the exit surface to the flux which leaves the entry surface of a transparent body.

*Emittance,** ϵ. The ratio of the radiant exitance of a body at a given temperature to that of a blackbody radiator at the same temperature.

Reflectivity, ρ, ρ_∞. The reflectance of a specimen that has an optically smooth surface and is thick enough to be opaque.

Absorptivity, α, α_∞. The absorptance of a specimen that has an optically smooth surface and is thick enough to be opaque.

Emissivity, ϵ, ϵ_∞. The emittance of a specimen that has an optically smooth surface and is thick enough to be opaque.

Reflectance Factor, R. The ratio of the flux reflected by a specimen under specified conditions of irradiation and viewing to that reflected by the ideal completely reflecting, perfectly diffusing surface, identically irradiated and viewed.

**Note:* Properties ending in "ance" are properties of real specimens, regardless of thickness or surface condition. Properties ending in "ivity" are instrinsic properties of the material of which the specimen is composed, and can only be approached by values measured on real specimens that have clean optically smooth surfaces and are opaque.

Radiance Factor, β. The ratio of the reflected radiance of a specimen in a given direction under specified conditions of irradiation to that of the ideal completely reflecting, perfectly diffusing surface, identically irradiated.

E. Modifiers

Spectral. For a property, at a given wavelength, designated by (λ) following the symbol for the property. For a quantity, spectral concentration, designated by the subscript λ, such as $\Phi_\lambda = d\Phi/d\lambda$.

*Total.** Refers to blackbody wavelength distribution. For a quantity, the spectral quantity is integrated and designated by the subscript t. For the property emittance, which is a ratio, the numerator and denominator are integrated separately and designated by the symbol (t) following the property symbol.

Solar. Having the spectral distribution of solar energy, or integrated over the solar spectrum. It is designated by the symbol (s) following the property symbol.

Integrated.† Having a spectral distribution prescribed by integration over some specified portion of the spectrum; designated by the symbol ($\bar{\lambda}$) following the symbol for the property and necessarily some comments regarding the spectrum must be given.

Directional. In a given direction. The direction is completely specified by two angles, θ and ϕ; θ is the angle between the specified direction and the normal to the surface, and ϕ is the azimuth of the specified direction measured from some fiducial mark on the specimen. For quantities, direction is denoted by the subscripts θ, ϕ (ϕ is only required when the surface structure has lay). For properties, the symbols θ, ϕ indicating direction are enclosed in parentheses following the symbol for the property as ($\theta, \phi; \theta', \phi'$); those indicating the incident direction first, followed by a semicolon, then the primed symbols indicating the direction of the reflected or transmitted rays.

Normal. A special case of directional where $\theta = 0°$; in the context of this volume, this modifier includes conditions where $\theta < 15°$, see Section 6.B.a for further detail.

Angular. A more general case of directional where $\theta > 0°$, that is, for cases other than normal; in the context of this volume, this modifier includes $\theta \geq 15°$, see Section 6.B.a for further detail.

*Conical.** Over a finite solid angle smaller than a hemisphere. Both the size and direction of the solid angle must be specified. If the angle is a right circular cone, the direction is the axis of the cone and is designated by the symbols θ, ϕ as above, and the size is designated by the plane half angle of the cone, α. If the solid angle is not a right circular cone, it must be described in detail in the text and is designated by the symbol g. As above, primed symbols are used to indicate reflected or transmitted beams.

*Hemispherical.** Over a complete hemisphere, designated by the symbol 2π replacing the θ, ϕ, g or θ' ϕ', g' in parentheses following the symbol for the property or quantity.

Specular. In the direction of mirror reflection. Under these conditions, $\theta' = \theta$ and $\phi' = \phi + 180°$.

Diffuse. Applied to a reflector or transmitter; reflecting or transmitting in all directions over a hemisphere. Applied to incident radiant energy, incident from angles over a hemisphere.

Perfectly Diffuse. With equal radiance in all directions from a surface.

*Frequently the modifier *total* is used to include any wavelength distribution including blackbody; in this volume it refers only to blackbody conditions and as such is applicable only to the property emittance.

†This modifier is not in widespread usage but is most convenient for sub-property classification purposes in this volume.

**Note:* Unless otherwise indicated, it is assumed that the incident radiance is uniform over the specified solid angle for conical or hemispherical irradiation. No such assumption is made for emitted, reflected or transmitted radiance.

3. THERMAL RADIATIVE PROPERTIES†

A. Interrelationships of Properties

All matter is continually emitting radiant energy as a result of the thermal vibration of the particles (electrons, ions, atoms, and molecules) of which it is composed. This process is called thermal radiation, and the radiant energy so emitted is called thermal radiant energy.

Each solid body is not only continually emitting thermal radiant energy, but it is also continually being bombarded by radiant energy from its surroundings, some of which is absorbed. The net rate of heat transfer by radiation to or from the body is equal to the difference in the rates of emission and absorption. Hence, the properties of the body that

†References for general background review are [132, 147, 152, 153, 157, 174].

influence these rates are called thermal radiative properties.

When a body is irradiated, part of the incident radiant energy is reflected, part is absorbed, and the rest is transmitted. Nothing else can happen to it. The incident flux Φ_i is equal to the sum of the reflected flux Φ_r, the absorbed flux Φ_a, and the transmitted flux Φ_t:

$$\Phi_i = \Phi_r + \Phi_a + \Phi_t \tag{1}$$

The reflectance ρ is the ratio of reflected flux to incident flux; the absorptance α is the ratio of absorbed flux to incident flux; and the transmittance τ is the ratio of transmitted flux to incident flux. Dividing both sides of equation (1) by Φ_i gives

$$1 = \rho + \alpha + \tau \tag{2}$$

For opaque materials, $\tau = 0$, hence for such materials

$$\rho + \alpha = 1 \qquad (\tau = 0) \tag{3}$$

Kirchhoff's law states that the absorptance is equal to the emittance

$$\alpha = \epsilon \tag{4}$$

Thus, for an opaque material

$$\rho + \epsilon = 1 \tag{5}$$

and the thermal radiative properties of an opaque body are fully described by either the reflectance or the emittance. However, there are certain restrictions that apply to equations (2) through (5) which will be discussed later.

B. Blackbody Radiation

A blackbody radiator absorbs all radiant energy incident upon it and emits the maximum possible amount of flux per unit area at any given wavelength or wavelength interval for any body at its temperature.

The only true blackbody radiator that exists is a completely enclosed cavity with opaque walls at a uniform temperature. All real materials reflect part of the radiant energy incident upon them and emit less radiant energy than a blackbody radiator at the same temperature. Nevertheless, the concept of a blackbody radiator is indispensable to a discussion of thermal radiation processes. The radiant exitance M (the flux emitted per unit area) of a blackbody radiator is given by the Stefan–Boltzmann equation

$$M_t = \sigma T^4 \tag{6}$$

in units of watts per square meter, where σ is the Stefan–Boltzmann constant, $5.6696 \times 10^{-8}\ \mathrm{W \cdot m^{-2} \cdot K^{-4}}$ and T is temperature in kelvins [124]. The spectral, or wavelength, distribution of this flux is given by the Planck equation

$$M_\lambda = c_1 \lambda^{-5} [e^{c_2/\lambda T} - 1]^{-1} \tag{7}$$

in which M_λ is the spectral exitance in watts per square meter and meter wavelength interval, c_1 is the first radiation constant, $3.7418 \times 10^{-16}\ \mathrm{W \cdot m^2}$, c_2 is the second radiation constant, $1.4388 \times 10^{-2}\ \mathrm{m \cdot K}$, and e is the base of natural logarithms [124]. The geometric distribution of this radiant exitance is given by Lambert's cosine law

$$I_\theta = I_0 \cos \theta \tag{8}$$

in which I_θ is the directional intensity of a plane source in the direction θ (measured from the normal to the surface) and I_0 is the intensity of the source in a direction normal to its surface.

While the radiation laws expressed in equations (6), (7), and (8) apply only to blackbody radiators,* they can be applied to real surfaces by using the emittance as a proportionality factor. For instance, the exitance M of a real specimen is given by

$$M = \epsilon(2\pi; t)\sigma T^4 \tag{9}$$

where $\epsilon(2\pi; t)$ is the hemispherical total emittance of the specimen at temperature T, and the spectral exitance M_λ is given by

$$M_\lambda = \epsilon(2\pi; \lambda) c_1 \lambda^{-5} [e^{c_2/\lambda T} - 1]^{-1} \tag{10}$$

where $\epsilon(2\pi; \lambda)$ is the hemispherical spectral emittance of the specimen at wavelength λ and temperature T. The directional radiance L_θ, where θ is the angle from the given direction to the normal to the surface, is given by

$$L_\theta = \epsilon(\theta; t)\sigma \pi^{-1} T^4 \tag{11}$$

where $\epsilon(\theta; t)$ is the total directional emittance in direction θ at temperature T. The directional spectral radiance, $L_{\theta,\lambda}$, of a specimen is given by

$$L_{\theta,\lambda} = \epsilon(\theta; \lambda) c_1 \pi^{-1} \lambda^{-5} [e^{c_2/\lambda T} - 1]^{-1} \tag{12}$$

where $\epsilon(\theta, \lambda)$ is the directional spectral emittance of the specimen at wavelength λ and temperature T in the direction θ.

*There is a further restriction. These equations apply rigorously only to the case where the blackbody is emitting into a vacuum. Whem emitting into a medium of index of refraction greater than unity, the emitted flux is increased by n^2, where n is the index of refraction of the medium. The increase in emitted energy when radiating into air ($n = 1.0003$ approx.) is too small to be detected in ordinary measurements.

Equations (9), (10), (11), and (12) suggest that the emittance of a specimen may change with wavelength, angle of incidence, and temperature. This is indeed the case. All thermal radiative properties vary with wavelength, direction (measured from the normal to the surface) of the incident or exitent radiant energy, temperature of the specimen, the degree of polarization, and, for polarized incident or exitent flux, with the angle between the plane of polarization and the plane of incidence. The plane of incidence is the plane defined by the direction of the incident ray and the normal to the surface. The modifiers defined in the Definitions of Terms are used to indicate the conditions under which the properties or quantities were evaluated.

The variation of the thermal radiative properties with temperature, wavelength, and geometric conditions (including polarization) of irradiation and viewing pose certain restrictions on equations (2), (3), (4), and (5). For equations (2) and (3) to be valid, α, ρ, and τ must be evaluated under the same conditions, which means that the temperature of the specimen must be the same, and the spectral composition, direction, solid angle, and degree and direction of polarization of the incident radiant energy must be identical, and all of the reflected and transmitted radiant energy must be measured.

C. Kirchhoff's Law

Kirchhoff's law, equation (4), is derived for the condition that the specimen is irradiated in a blackbody cavity with walls at the same temperature as the specimen, which means that the specimen is uniformly irradiated over a hemisphere with unpolarized radiant energy having the spectral distribution of that of a blackbody radiator at the temperature of the specimen. However, it can be proved that equation (4) is also valid for the two conditions: (1) any solid angle less than a hemisphere if the direction and solid angle of the incident beam for the absorption evaluation is identical to the direction and solid angle (but opposite in sense) of the emitted beam for the emittance evaluation, and (2) for plane-polarized radiant energy with the plane of polarization at any given angle to the plane of measurement, provided that it is the same for the incident radiant energy for the absorption evaluation and the emitted radiant energy for the emittance evaluation. Even with these modifications, equation (4) applies strictly only provided the spectral composition of the incident radiant energy for the absorptance is that of blackbody radiant energy at the temperature of the specimen. This would appear to impose a severe restriction on the general applicability of equation (5). However, it can also be shown that equation (4) applies to any small wavelength band, as well as to total blackbody radiant energy. The properties of reflectance, absorptance, and transmittance do not vary with the amount of incident radiant energy until very high flux densities are reached. Within the narrow wavelength band used in measuring spectral thermal radiative properties the spectral distribution of radiant energy from almost any thermal source is approximately the same as that from a blackbody radiator at the temperature of the specimen. Also, polarization effects are completely absent for normally incident radiant energy and are negligible at angles near the normal. Hence equations (4) and (5) can be considered valid for normal spectral properties and can be used to convert normal hemispherical reflectance to normal emittance with but little error.

4. PHYSICS OF THERMAL RADIATION

A. The Nature of Radiant Energy

Radiant energy can be treated in two ways, as electromagnetic waves, or as photons. In both forms it travels in straight lines at the speed of light, which, in vacuum, is a fundamental constant of nature c_0, with a value of 2.997925×10^{10} cm sec^{-1}.

Photons are particles of zero rest mass, each of which contains or consists of a fixed amount of energy, called a quantum. Quantum mechanical treatment of photons is the most convenient way of studying the generation and interaction of radiant energy with matter on a micro scale, but interactions on a macro scale are handled more readily by wave mechanics, in which radiant energy is considered as waves.

Electromagnetic waves are characterized by frequency or wavelength and amplitude. Frequency ν and wavelength λ are related by

$$c = \nu\lambda \tag{13}$$

The energy content of a wave is the square of the amplitude.

The frequency ν is constant, regardless of the medium through which the wave is propagated, but the speed and wavelength change with the index of refraction n of the medium

$$n = c_0/c_n = \lambda_0/\lambda_n \tag{14}$$

where λ_n and c_n are the wavelength and speed in the medium of index of refraction n, and λ_0 and c_0 are the corresponding values in vacuum.

Electromagnetic waves consist of magnetic (H) and electric (E) field vectors which successively oscillate in directions perpendicular to the direction of travel and to each other; this is a characteristic of transverse waves as described by the Maxwell equations governing electromagnetic wave phenomena. The plane of polarization of the wave is defined as the plane containing the electric field vector. If the waves are oriented with their planes of polarization parallel, the beam is *plane polarized.* If the planes are randomly oriented, the wave is termed *unpolarized.* The terms *circularly* or *elliptically* polarized identify other conditions of amplitude and oscillation direction of the electric field. Most practical radiant energy sources and beams within optical systems—particularly after reflectance from a plane surface—are partially polarized; that is, there is a partial, but not complete, preferred orientation of the planes of polarization of the waves making up the beam.

The plane of polarization of such a partially polarized beam is taken as the plane of polarization for which the intensity of the beam is maximum. The degree of polarization P of such a beam is measured as

$$P = \frac{I_p - I_s}{I_p + I_s} \tag{15}$$

where I_p is the intensity of the beam when polarized parallel to its plane of polarization, and I_s is the intensity when polarized normal to its plane of polarization.

The frequency, and hence the wavelength, of electromagnetic radiation theoretically can vary from zero to infinity. Thermal radiant energy, however, is generally restricted to the wavelength range of 0.1 μm to 1 mm. This overall range has been broken down into seven subranges [74], as follows:

Range	*Designation*	*Wavelength Range,* μm
Ultraviolet	UVC	0.1–0.28
	UVB	0.28–0.315
	UVA	0.315–0.400
Visible	VIS	0.380–0.780
Infrared	IRA	0.78–1.4
	IRB	1.4–3
	IRC	3–1000

B. Basic Laws

The Planck equation, (7), derived by a quantum-mechanical approach, is the basis for all thermal radiation measurements. All of the other important relationships follow from this equation:

$$M_\lambda = c_1\lambda^{-5}[e^{c_2/\lambda T} - 1]^{-1} \tag{7}$$

The radiation constants c_1 and c_2 are related to more fundamental constants as follows:

$$c_1 = 2\pi c_0{}^2 h$$
$$c_2 = c_0 h/k$$

where c_0 is the speed of light in vacuum, h is the Planck constant, $6.6262 \times 10^{-34}\,\text{J}\cdot\text{sec}$, and k is the Boltzmann constant, $1.3806 \times 10^{-23}\,\text{J}\cdot\text{K}^{-1}$.

Since for a blackbody radiator, $L = M/\pi$, equation (7) can be rewritten to give spectral radiance:

$$L_\lambda = c_1\pi^{-1}\lambda^{-5}[e^{c_2/\lambda T} - 1]^{-1} \tag{16}$$

An inspection of equation (16) shows that the right-hand side can be reduced to a single variable, λT, by multiplying both sides of the equation by T^{-5}:

$$L_\lambda T^{-5} = c_1\pi^{-1}(\lambda T)^{-5}[e^{c_2/\lambda T} - 1]^{-1} \tag{17}$$

Equation (17) shows that the shape of the spectral distribution of the radiance of a blackbody radiator is a function of λT and not of λ and T separately. This is one form of Wien's displacement law, which states that the peak of the spectral distribution curve occurs at a wavelength that is inversely proportional to the absolute temperature. This is shown in Fig. 1. Setting $dL/d\lambda$ equal to zero and solving for $(\lambda T)_{\text{max}}$ gives

$$(\lambda T)_{\text{max}} = 2898\ \mu\text{m}\cdot\text{K} \tag{18}$$

which is the Wien displacement equation.

Two other equations were developed prior to the Planck equation and are important not only for their historical significance but also because they are useful approximations in certain wavelength ranges. Both were developed by classical wave mechanics. The first is known as Wien's law, and in its original form was expressed as

$$L_\lambda = F(\lambda)e^{-f(\lambda)/T} \tag{19}$$

in which $F(\lambda)$ and $f(\lambda)$ were unknown functions of λ. It was later found that the true equation is

$$L_\lambda = c_1\pi^{-1}\lambda^{-5}e^{-c_2/\lambda T} \tag{20}$$

Note that it is identical to the Planck equation except for the minus one in the denominator. This

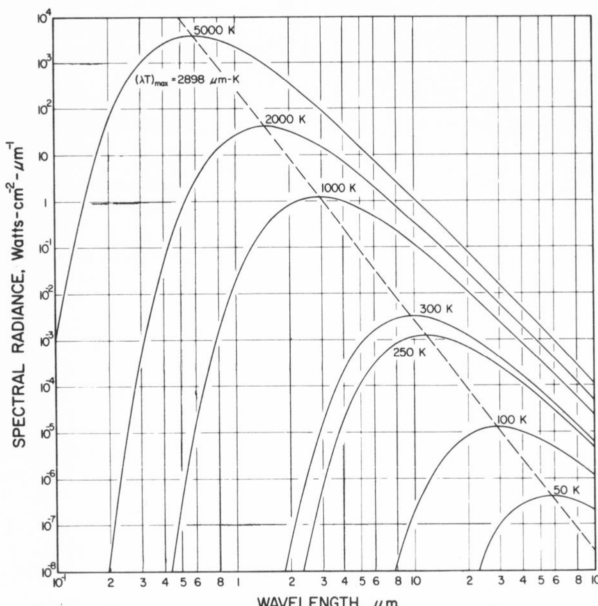

Fig. 1. The Planck distribution law, spectral radiance of blackbody radiation as a function of temperature and wavelength.

equation is a useful approximation at short wavelengths, and is much used in optical pyrometry. It gives values of L_λ that are too low at long wavelengths. The error is less than 1 percent when λT is less than 2898 μm·K. The second important relation is the Rayleigh–Jeans law, and is expressed as

$$L_\lambda = \frac{c_1 T}{c_2 \pi \lambda^4} \tag{21}$$

Equation (21) is valid at very long wavelengths where $\lambda T > 10{,}000$ μm·K with but small error.

If equation (7) is integrated over all wavelengths,

$$M = \int_0^\infty \frac{c_1 \lambda^{-5}\, d\lambda}{e^{c_2/\lambda T} - 1} = \frac{\pi^4 c_1}{15 c_2{}^4} T^4 = \sigma T^4 \tag{22}$$

which is equation (6). Thus, σ, the Stefan–Boltzmann constant, is expressed in terms of other constants as

$$\sigma = \frac{\pi^4 c_1}{15 c_2{}^4} = \frac{2k^4 \pi^5}{15 c_0{}^2 h^3} \tag{23}$$

The fraction of the total exitance of a blackbody at temperature T that is emitted within the wavelength interval 0 to λ, may be obtained as follows. The exitance in the wavelength interval will be designated as $M(0 - \lambda;\, T)$ and is given by

$$M(0 - \lambda; T) = \int_0^\lambda M_\lambda(T) d\lambda = \int_0^\lambda \frac{c_1 \lambda^{-5}\, d\lambda}{e^{c_2/\lambda T} - 1}$$

$$= \int_0^\lambda \frac{T^5 c_1 (\lambda T)^{-5}\, d(\lambda T)}{e^{c_2/\lambda T} - 1}$$

$$= \int_0^{\lambda T} \frac{T^4 c_1 (\lambda T)^{-5}\, d(\lambda T)}{e^{c_2/\lambda T} - 1}$$

The total exitance is $M_t = \sigma T^4$, thus

$$M(0 - \lambda; T)/M_t = \int_0^{\lambda T} \frac{T^4 c_1 (\lambda T)^{-5}\, d(\lambda T)}{e^{c_2/\lambda T} - 1} \Big/ \sigma T^4$$

$$= \frac{1}{\sigma} \int_0^{\lambda T} \frac{c_1 (\lambda T)^{-5}\, d(\lambda T)}{e^{c_2/\lambda T} - 1} \tag{24}$$

Substituting M_λ within the integral for its equivalent, $c_1 \lambda^{-5}/(e^{c_2/\lambda T} - 1)$, gives

$$M(0 - \lambda; T)/M_t = \frac{1}{\sigma T^5} \int_0^{\lambda T} M_{\lambda T}\, d(\lambda T) \tag{25}$$

Since for blackbody $M = \pi L$, then equation (25) is equally valid if L is substituted for M wherever it occurs.

The ratio $L_b(0 - \lambda T)/L_t$ or $M_b(0 - \lambda T)/M_t$ is given in Table I for various values of λT.* The fraction of the total exitance or radiance of a blackbody at temperature T, occurring in any wavelength interval λ_1 to λ_2 ($\lambda_2 > \lambda_1$) may be obtained by subtracting the value from the table for $\lambda_1 T$ from that for $\lambda_2 T$. The value for the exitance $M([\lambda_1 - \lambda_2]T)$ may be obtained by multiplying the resulting value by σT^5, and the radiance $L([\lambda_1 - \lambda_2]T)$, by multiplying the value by $\sigma T^5/\pi$.

Figure 2A is a plot of the ratio $L_b(0 - \lambda T)/\sigma T^4$, dotted line, and $L_{b,\lambda}(\lambda, T)/\sigma T^5$, solid line, both plotted as a function of λT.

Another useful relationship is the fractional change in L_λ or M_λ produced by a fractional change in the temperature T of a blackbody radiator. Differentiating equation (16) with respect to T gives

$$\frac{dL_\lambda}{L_\lambda} \Big/ \frac{dT}{T} = (c_2/\lambda T) \frac{e^{c_2/\lambda T}}{e^{c_2/\lambda T} - 1} \tag{26}$$

This ratio is plotted as a function of λT in Fig. 2B, dotted line, and shows that the change in spectral radiance with temperature increases with decrease in λT, particularly at values of λT below 2898 μm·K. The term on the right of equation (26) can be simplified to $c_2/\lambda T$ for use at $\lambda T < 2898$ μm·K with an error of considerably less than 1 percent. Again

*See also [131] for more detailed tables, which, however, were computed with the older set of radiation constants, $c_1 = 3.7413 \times 10^{-16}$ W·m² and $c_2 = 1.4380 \times 10^{-2}$ m·K.

Table I. Blackbody Radiation Functions

λT, μm·K	$L_{b,\lambda}(\lambda, T)/\sigma T^5$, $m^{-1}\cdot K^{-1}\cdot sr^{-1}$		$\pi L_b\,(0-\lambda T)/\sigma T^4$		$L_{b,\lambda}(\lambda, T)/(L_{b,\lambda})_{\lambda T\,=\,2898}$	
	p	q	p	q	p	q
200	0.375195	−21	0.341796	−26	0.519451	−23
400	0.490424	−7	0.186468	−11	0.678981	−9
600	0.104056	−2	0.929299	−7	0.144064	−4
800	0.991183	−1	0.164351	−4	0.137277	−2
1000	0.118508	+1	0.320780	−3	0.164072	−1
1200	0.523935	+1	0.213431	−2	0.725376	−1
1400	0.134411	+2	0.779084	−2	0.186089	0
1600	0.249128	+2	0.197204	−1	0.344913	0
1800	0.375563	+2	0.393449	−1	0.519959	0
2000	0.493422	+2	0.667347	−1	0.683133	0
2200	0.589636	+2	0.100897	0	0.816338	0
2400	0.658848	+2	0.140268	0	0.912161	0
2600	0.701271	+2	0.183135	0	0.970894	0
2800	0.720216	+2	0.227908	0	0.997123	0
2898	0.722294	+2	0.250126	0	0.100000	+1
3000	0.720229	+2	0.273252	0	0.997142	0
3200	0.705948	+2	0.318124	0	0.977369	0
3400	0.681517	+2	0.361760	0	0.943546	0
3600	0.650369	+2	0.403633	0	0.900422	0
3800	0.615199	+2	0.443411	0	0.851730	0
4000	0.578040	+2	0.480907	0	0.800283	0
4200	0.540370	+2	0.516046	0	0.748131	0
4400	0.503231	+2	0.548830	0	0.696712	0
4600	0.467321	+2	0.579316	0	0.646996	0
4800	0.433089	+2	0.607597	0	0.599602	0
5000	0.400794	+2	0.633786	0	0.554890	0
5200	0.370562	+2	0.658011	0	0.513035	0
5400	0.342428	+2	0.680402	0	0.474084	0
5600	0.316361	+2	0.701090	0	0.437994	0
5800	0.292287	+2	0.720203	0	0.404664	0
6000	0.270108	+2	0.737864	0	0.373958	0
6200	0.249710	+2	0.754187	0	0.345718	0
6400	0.230973	+2	0.769282	0	0.319777	0
6600	0.213775	+2	0.783248	0	0.295967	0
6800	0.197997	+2	0.796180	0	0.274123	0
7000	0.183524	+2	0.808160	0	0.254085	0
7200	0.170247	+2	0.819270	0	0.235703	0
7400	0.158065	+2	0.829580	0	0.218837	0
7600	0.146883	+2	0.839157	0	0.203356	0
7800	0.136614	+2	0.848060	0	0.189139	0
8000	0.127177	+2	0.856344	0	0.176075	0
8500	0.106766	+2	0.874666	0	0.147816	0
9000	0.901414	+1	0.890090	0	0.124798	0
9500	0.765296	+1	0.903147	0	0.105953	0
10000	0.653243	+1	0.914263	0	0.904400	−1
10500	0.560490	+1	0.923775	0	0.775987	−1
11000	0.483294	+1	0.931956	0	0.669110	−1
11500	0.418701	+1	0.939027	0	0.579683	−1
12000	0.364373	+1	0.945167	0	0.504466	−1
13000	0.279441	+1	0.955210	0	0.386880	−1

Table I (continued)

λT, μm·K	$L_{b,\lambda}(\lambda, T)/\sigma T^5$, $m^{-1} \cdot K^{-1} \cdot sr^{-1}$		$\pi L_b\ (0-\lambda T)/\sigma T^4$		$L_{b,\lambda}(\lambda, T)/(L_{b,\lambda})_{\lambda T\,=\,2898}$	
	p	q	p	q	p	q
14000	0.217628	+1	0.962970	0	0.301301	−1
15000	0.171855	+1	0.969056	0	0.237930	−1
16000	0.137421	+1	0.973890	0	0.190257	−1
18000	0.908187	0	0.980939	0	0.125736	−1
20000	0.623273	0	0.985683	0	0.862908	−2
25000	0.276458	0	0.992299	0	0.382750	−2
30000	0.140461	0	0.995427	0	0.194465	−2
40000	0.473862	−1	0.998057	0	0.656052	−3
50000	0.201592	−1	0.999045	0	0.279100	−3
75000	0.418572	−2	0.999807	0	0.579503	−4
100000	0.135744	−2	1.000000	0	0.187934	−4

Note: Value in each case is $p \times 10^q$. The radiation constants used for computation (1969) of tabular values are $c_1 = 3.7415 \times 10^{-16}$ W·m² and $c_2 = 1.43879 \times 10^{-2}$ m·K.

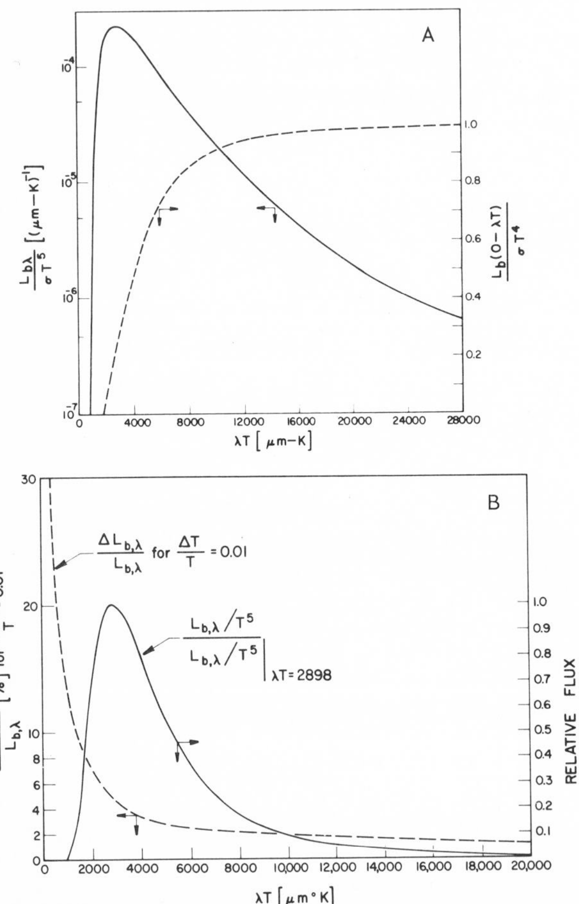

Fig. 2. (A) Relative spectral radiance of blackbody radiation as a function of λT, solid curve; and fraction of spectral radiance in the wavelength interval 0–λT, dashed curve. (B) Relative spectral radiance of blackbody radiation as a function of λT, solid curve; and percentage increase in spectral radiance produced by a 1 percent increase in absolute temperature as a function of λT, dashed curve.

the equation applies to either spectral radiance L_λ or spectral exitance M_λ.

The value of the right side of equation (26) can be used for the exponent x in the relationship

$$L_\lambda \sim T^x \tag{27}$$

which is a good approximation when the change in temperature is a small fraction of T. Again M_λ can be substituted for L_λ in equation (27). The value of x is 5.0 at 2898 μm·K, and varies as $1/\lambda T$, hence is about 10 at 1500 μm·K, 15 at 1000 μm·K, and 2.5 at 6000 μm·K.

The relationship shown in equation (26) and plotted in Fig. 2B is extremely important to remember in making measurements of L_λ by direct comparison of the radiance of a hot specimen to that of a blackbody radiator at the same temperature; it shows the effect of a small temperature difference on the accuracy of the results obtained.

5. INTERACTION OF RADIANT ENERGY WITH MATTER

A. Wave Behavior [16, 53]

When an electromagnetic wave in vacuum is incident on the plane surface of an optically homogeneous specimen, interaction of the wave with the material of the specimen will occur. The electrical and magnetic properties of the specimen will be different from those of the vacuum, and, as a result, there may be a change in the direction of propagation of the wave, its velocity, amplitude, wavelength, and phase, and it may be separated into two portions, one reflected and one transmitted. The transmitted portion will be partially or totally absorbed. The only property of the wave that never changes is its frequency.

Similar changes in the wave will occur whenever it is incident on an interface between two media of different properties. The changes can be computed from the properties of the material, or the differences in properties on the two sides of the interface, and from the direction of propagation of the wave relative to the interface and the direction of its plane of polarization relative to the plane containing the direction of incidence and the normal to the interface at the point of incidence.

B. Optical Properties [59]

The Maxwell equations describe the change in a wave as it crosses an interface and propagates into the material:

$$E = E_0 \exp(j\omega t - \gamma x) \tag{28}$$

where

$$\gamma = \alpha + j\beta \tag{29}$$

Equation (28) indicates that the electric field vector E_0 at the interface where $x = 0$ (x is the distance from the interface in the material) is attenuated on penetrating into the material by an amount α and that a phase change β has occurred in crossing the interface. Thus, two parameters (or one complex one) of the material are needed to define the changes in amplitude and phase which occur in crossing the interface.

Many different sets of parameters, called optical properties in this discussion, can be found in the literature, each having some merit in interpreting the interaction of a wave with a material. Table II presents several of the more commonly used optical properties, together with a brief description and summary of their interrelationship.

C. Thermal Radiative Properties

The optical properties describe the interaction of an electromagnetic wave with matter in terms of phase and amplitude, while the thermal radiative properties describe the energy transfer during the interaction. It is obvious that the two types of properties, optical and thermal radiative, are related. In some cases the relationships are simple.

One situation in which the relation is not simple is that for the general case of a wave incident on an interface. By solving the Maxwell equations for the boundary conditions, the Fresnel relations for specular reflection can be derived. The specular reflectance at the interface (fraction of incident flux reflected in the direction of mirror reflectance) is given as

$$\rho_s(\theta) = \frac{a^2 + b^2 - 2a\cos\theta + \cos^2\theta}{a^2 + b^2 + 2a\cos\theta + \cos^2\theta} \tag{30}$$

$$\rho_p(\theta) = \rho_s(\theta)\frac{a^2 + b^2 - 2a\sin\theta\tan\theta + \sin^2\theta\tan^2\theta}{a^2 + b^2 + 2a\sin\theta\tan\theta + \sin^2\theta\tan^2\theta} \tag{31}$$

where

$$2a^2 = [(n^2 - k^2 - \sin^2\theta)^2 + 4n^2k^2]^{1/2} + (n^2 - k^2 - \sin^2\theta) \tag{32}$$

$$2b^2 = [(n^2 - k^2 - \sin^2\theta)^2 + 4n^2k^2]^{1/2} - (n^2 - k^2 - \sin^2\theta) \tag{33}$$

Table II. Parameters and Relations Descriptive of Electromagnetic Wave–Material Interaction

Complex permittivity	$\epsilon^* = \epsilon' - j\epsilon''$	ϵ'—Real permittivity ϵ''—Loss factor
Complex permeability	$\mu^* = \mu_0$	Valid assumption for almost all applications.
Complex propagation factor	$\gamma = \alpha + j\beta$	α—Attenuation factor β—Phase factor
Complex dielectric constant	$K^* = \dfrac{\epsilon^*}{\epsilon_0} = K' - jK''$	K'— Relative permittivity or real dielectric constant K''— Relative loss factor
Complex index of refraction	$n^* = n - jk$	n—Index of refraction, ratio of phase velocities k—Index of absorption
Intrinsic impedance	$Z^* = \sqrt{\dfrac{\mu^*}{\epsilon^*}}$	Z^*—Ratio of electric to magnetic field vectors
Normal spectral reflectivity	$\rho(0;\lambda) = \dfrac{(n-1)^2 + k^2}{(n+1)^2 + k^2}$	Ratio of squares of the electric field vector of reflected and incident waves
$\gamma = j\omega(\epsilon^*\mu^*)^{1/2}$	$n^* = \sqrt{K^*}$	$\epsilon^* = \epsilon_0\left(1 - j\dfrac{1}{\omega r}\right)$

The angle θ is the angle between the incident ray and the normal to the interface, ρ_s is the reflectance for plane-polarized incident radiant energy with its plane of polarization normal to the plane of incidence (the plane containing the incident ray and the normal to the interface at the point of incidence), and ρ_p is the reflectance for plane-polarized incident radiant energy with its plane of polarization parallel to the plane of incidence.

If the incident radiant energy is completely unpolarized, it can be shown that

$$\rho(\theta) = \tfrac{1}{2}[\rho_s(\theta) + \rho_p(\theta)] \tag{34}$$

The Fresnel equations, (30) and (31), have been expressed in terms of n and k, but the relations are found in various forms in the literature. The simplest case occurs for normal incidence ($\theta = 0$), where the equations reduce to

$$\rho_p(0) = \rho_s(0) \tag{35}$$

and

$$a = n \qquad b = k \tag{36}$$

Hence, for radiant energy incident from vacuum or a medium of index of refraction of 1,

$$\rho(0) = \frac{(n-1)^2 + k^2}{(n+1)^2 + k^2} \tag{37}$$

The above relations express the energy transfer in terms of electromagnetic wave theory for specular reflection, and no specification of the material, other than its optical properties n and k, is necessary. As indicated previously, the reflected, absorbed, and transmitted flux sum to the incident flux, hence

$$\rho + \tau + \alpha = 1 \tag{2}$$

6. DETAILED DISCUSSION OF THERMAL RADIATIVE PROPERTIES

A. Primary Properties

The primary properties—emissivity, reflectivity, absorptivity and transmissivity—have been previously introduced. The suffix *ivity* denotes the property of the ideal material—optically smooth and homogeneous. For metallic materials that are neither smooth nor homogeneous, the radiative parameters are not unique properties of the bulk material, but rather are properties of the surface. In these more frequent situations the properties are denoted by the suffix *ance* and hereafter the properties are referred to and defined as follows:

ϵ, emittance —Ratio of the radiant exitance of the specimen to that emitted by a blackbody radiator at the same temperature and under the same geometric and wavelength conditions.

ρ, reflectance —Ratio of the reflected radiant flux to the incident flux.

α, absorptance —Ratio of absorbed radiant flux to the incident flux.

τ, transmittance—Ratio of some specified portion of the transmitted flux to the incident radiant flux.

For each of these primary properties, it is necessary to specify the geometric and wavelength conditions to which the particular property corresponds. Unfortunately, there is no generally accepted convention on the choice of symbols or terminology to describe these conditions. After due consideration and consultation, a nomenclature system suitable to classification needs in the organization of this compendium was adopted. To assist the reader, the system used in this volume will be fully explained and then related when appropriate to the various terminologies found in the literature [74, 89, 117, 169].

B. Subproperty Descriptors

a. Geometric Descriptors

Figure 3 shows the general case of reflection at a surface and indicates the necessary geometric parameters required to fully describe the incident and reflected fluxes. The beams representing the incident or viewed flux are described by the zenith angles for θ or θ' and by the beam solid angles ω or ω'. The longitudinal angles ϕ and ϕ' relate the axes of the

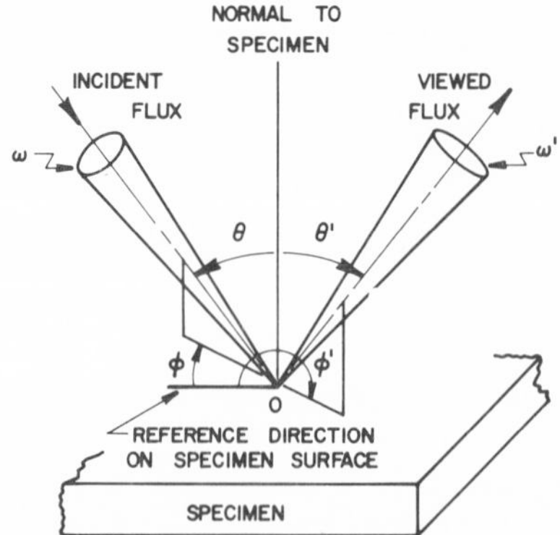

Fig. 3. Geometric parameters descriptive of reflection from a surface. θ is the zenith angle, or co-latitude, in degrees; ϕ is the azimuthal angle, or longitude, in degrees; ω is the beam solid angle, in steradians; and the symbol $'$ refers to viewing conditions.

beams to each other and some reference line on the specimen; as a practical matter very few measurements so specify this angular descriptor. It is the convention in this volume to distinguish three sets of conditions as follows:

Normal —Conditions for incidence and/or viewing through a solid angle ω or ω', normal to the specimen; that is θ or $\theta' < 15°$

Angular —Conditions for incidence and/or viewing through a solid angle ω or ω' at some direction specified by θ or $\theta' \geq 15°$

Hemispherical—Conditions for incidence and/or viewing of flux over a hemispherical region; that is ω or $\omega' = 2\pi$

The descriptors normal and angular do not fully describe the geometric conditions; ω and/or ω' and θ and/or θ' must be provided to fully specify the geometry.

It has been suggested that other descriptors be used to indicate the two extreme conditions for ω or ω'. If the incident or viewed beam is parallel, then ω or $\omega' = 0$; this condition can be approximated in practice where ω or ω' is so small that slight increases have no influence. In this case, the term *directional* is used and when ω or ω' is not negligible, the term *conical* is applied. As a practical matter for categorization of subproperties, only the terms *normal*, *angular*, and *hemispherical* are separately distinguished in the subsequent data tables. Whenever information is given on ω or ω', it is reported in an appropriate column. To a great many thermophysicists details on ω or ω' may not be essential, but they should be fully aware that such information is available.

For the subproperties of emittance and absorptance, only one geometric descriptor is required to designate the conditions of viewing and incidence respectively. For the subproperties of reflectance and transmittance, two geometric descriptors are required since both incidence and viewing conditions need to be specified. While the three descriptors selected for categorization are not fully descriptive, they give good practical resolution from a classification and retrieval viewpoint.

b. Wavelength Descriptors

These descriptors indicate spectral or wavelength conditions for which the subproperties are specified. Each of the four prime properties needs to be characterized by the wavelength conditions of the radiant flux—emitted, incident, reflected, or transmitted. The following terminology is used in this volume to refer to the common conditions:

Spectral—For a very narrow band of wavelengths, also referred to as monochromatic; no maximum band width criteria has been established by convention; symbolically denoted by λ.

Total —Refers to blackbody wavelength distribution; this descriptor is applicable only to emittance; symbolically denoted by t. The temperature of the blackbody should be given, since spectral distribution varies with temperature.

The properties, being dimensionless ratios, cannot be integrated. The term *integrated*, when applied to such ratios, means that the numerator and denominator are integrated separately, then divided. An "integrated" ratio can also be thought of as a weighted average, where the weighting factor is the spectral distribution of the source, or blackbody in the case of emittance.

Integrated—Relative to some specified wavelength distribution of the irradiating source or some specified broad band; for such subproperties it is necessary that details of the spectral characteristics of the source be fully presented; symbolically denoted by $\bar{\lambda}$.

Solar —Having the wavelength distribution of the sun; the solar distribution can be either natural or artificial—lamps, arcs, etc., in which case the nature of the source needs to be specified; symbolically denoted by s.

In a later section, interrelationships between the various wavelength conditions will be discussed with a view to exposing the reader to methods of computing the subproperties from fragments of other subproperty data.

c. Symbolic Representation

In subsequent discussion the various subproperties are expressed according to the following convention. The symbols for the four primary properties—ϵ, ρ, α, and τ—have already been presented. The geometric (incidence and viewing condition) and wavelength descriptors, in that same order, are symbolically represented within the parentheses being separated by semicolons. The most general case would be (using reflectance as an example):

$$\rho(\theta, \phi, \omega; \theta', \phi', \omega'; \lambda)$$

where the wavelength descriptors used in this text are λ, t, $\bar{\lambda}$, and s indicating spectral conditions previously defined.

As a practical matter not all the designations are always needed and many are omitted for convenience sake; usually ϕ and ϕ' are not used and, of course, for emittance and absorptance, the incidence and viewing geometry symbols, respectively, are not applicable.

C. Reflectance Factor

The *reflectance factor R* is defined as the ratio of the flux reflected by a specimen under specified geometric conditions of irradiation and viewing to that reflected by the ideal completely reflecting, perfectly diffusing surface, under identical conditions of irradiation and viewing [89]. A perfectly diffusing surface is defined as a surface whose reflected radiance is independent of direction of viewing and hence obeys Lambert's law.

Reflectance factor is relatively easy to measure because comparatively simple equipment can be used. Integrating sphere reflectometers are frequently used for these measurements. While the ideal completely reflecting, perfectly diffusing surface does not exist, several materials can be used that are reasonably close approximations to such a surface. Examples are freshly smoked magnesium oxide, magnesium carbonate, high-purity barium sulphate, sodium chloride, flowers of sulphur, and some good white paints [3, 64, 102]. For those conditions where the specimen is diffusely irradiated over a hemisphere, a high-quality first-surface mirror can be used as the reference standard. When integrating sphere reflectometers are used, the sphere coating itself may be used as the reference.

Integrating sphere reflectometers are generally restricted to operation over the solar range (0.25 to 2.5 μm) where conventional sphere coatings such as smoked magnesium oxide and barium sulphate have high reflectance. Operation has been extended farther into the infrared by use of flowers of sulfur and sodium chloride sphere coatings.

Reflectometers such as the hohlraum reflectometer, where the specimen is irradiated over a hemisphere, actually measure reflectance factor. This also applies to the ellipsoidal mirror, paraboloid mirror, and Coblentz hemisphere reflectometers when used in the inverse mode, so that the specimen is irradiated over a hemisphere.

The primary distinction between true reflectance and reflectance factor measurements is that in true reflectance measurements the incident flux is measured directly, then the reflected flux is measured directly, and the reported reflectance is the ratio of the second measured value to the first. In reflectance factor measurements either (1) the incident flux is measured indirectly by measuring the flux reflected by a diffusely reflecting standard, or (2) the radiance of the flux incident on the specimen is measured over a small solid angle, either by viewing the source directly as in a hohlraum, or by viewing it indirectly by use of a mirror standard, as in the ellipsoidal mirror, paraboloid mirror, or Coblentz hemisphere instruments.

Reflectance factor data are useful to the thermophysicist, because for a diffuse reflector the reflectance factor measured under any given geometric conditions of irradiation and viewing is a good estimate of the reflectance of the specimen measured under the same conditions of irradiation, but with hemispherical viewing. The error in the estimate is zero if the specimen is a perfectly diffuse reflector, and increases with both the departure of the specimen from a perfectly diffuse reflector and with the variation of the solid angle of viewing from a hemisphere.

The fundamental relationship by the use of which the various types of reflectance and reflectance factor can be related to each other is [48]

$$\frac{\rho(\theta_1, \phi_1, \omega_1; \theta_2, \phi_2, \omega_2)}{\cos\theta_2\omega_2} = \frac{\rho(\theta_2, \phi_2, \omega_2; \theta_1, \phi_1, \omega_1)}{\cos\theta_1\omega_1} \tag{38}$$

where the bidirectional reflectance, $\rho(\theta, \phi, d\omega; \theta', \phi', d\omega')$ is defined as

$$\rho(\theta, \phi, d\omega; \theta', \phi', d\omega') = \frac{L'(\theta', \phi')\cos\theta'\,d\omega'}{L(\theta, \phi)\cos\theta\,d\omega} \tag{39}$$

where L is the radiance of the beam incident on the specimen over the elementary solid angle $d\omega$ from direction θ, ϕ, and L' is the reflected radiance of the specimen over the elementary solid angle $d\omega'$ in the direction θ', ϕ'. The geometry for equations (38) and (39) is shown in Fig. 3.

The only restriction on the general application of equation (38) to solid angles of any size is that L, the incident radiance, must not vary with angle over the solid angle of incidence in either case.

By the use of equation (38) it can be shown that the following relationships hold:

$$R(2\pi; \theta', \omega') = R(\theta, \omega; 2\pi) \quad (40)$$

where $\theta = \theta'$ and $\omega = \omega'$.

$$\rho(2\pi; 2\pi) = R(2\pi; 2\pi) \quad (41)$$

$$\rho(\theta, \omega; 2\pi) = R(\theta, \omega; 2\pi) \quad (42)$$

where θ and ω are the same for ρ and R.

$$R(\theta_1, \omega_1; \theta_2, \omega_2) = R(\theta_2, \omega_2; \theta_1, \omega_1) \quad (43)$$

$$\rho(\theta, \omega; \theta', \omega') = \frac{\omega' \cos\theta'}{\pi} R(\theta, \omega; \theta', \omega') \quad (44)$$

where for large solid angles, $\omega' \cos\theta' = \int_{\omega'} \cos\theta' \cdot \sin\theta' \, d\theta' \, d\phi'$.

7. INTERRELATIONSHIPS BETWEEN PROPERTIES

A. Geometric Relations

In many cases the data required for a specific application will not be available in the literature. In some of these cases, it is possible to obtain the desired data from other subproperties for which data may be available for the same or a similar material, by judicious use of the relationships previously mentioned and realistic estimates of the geometric distribution of the reflected or emitted radiant energy.

The geometric distribution of the radiant energy emitted or reflected from the surface of a nonmetal is nearly constant over the hemisphere. The emittance of a typical nonconductor, for either spectral or total wavelength conditions, is constant to an angle of incidence of nearly 70° and then drops rapidly toward grazing conditions. The distribution is in general not affected by the surface roughness, since, as will be discussed in some detail later, the radiative behavior is dependent upon conditions at a plane below the surface, which is usually at some appreciable depth [52, 139]. The common notion that nonmetals are diffuse emitters is reasonable and follows directly from the simple theory of dielectric materials and the Fresnel relations. In contrast, the behavior of metals is quite the opposite. The radiance of a smooth metal surface increases with angle from the normal to a maximum somewhere less than 90° and then drops to zero at grazing conditions. Only for angles less than 15° from the normal is it reasonable to assume that the geometric distribution of emitted or reflected flux is constant. As the metallic surface is roughened, the distribution changes markedly until it resembles a diffuse (not totally Lambertian) emitter or reflector. The high sensitivity of the radiative properties to surface effects is a characteristic of metals, but is not a serious consideration for nonmetals.

The two extremes for geometric distribution of reflected flux are specular and diffuse reflection; these are the useful limits for the treatment of real materials. The equations below can be applied for converting data from one set of geometric parameters to another. Additional detail is given in references [157] and [80]. Reference [89] treats the reflectance and reflectance factor subproperties.

From Kirchhoff's law [equations (4) and (5)]

$$\epsilon = \alpha = 1 - \rho \quad (45)$$

Equation (45) is restricted by the geometric and wavelength distribution of the reflected and emitted radiant energy. Considering the geometric distribution only, for opaque specimens

$$\alpha(\theta, \omega) = 1 - \rho(\theta, \omega; 2\pi) \quad (46)$$

where θ, ω are the same for α and ρ, and

$$\epsilon(\theta', \omega') = \alpha(\theta, \omega) \quad (47)$$

where $\theta = \theta'$ and $\omega = \omega'$. Equation (46) was derived on the basis of conservation of energy. Incident radiant energy that is not reflected must be absorbed, and equation (47) is a statement of Kirchhoff's law. Equations (46) and (47) can be used to convert one type of data (subproperty) to another. If normal emittance data are not available, for instance, normal absorptance or normal hemispherical reflectance can be used to compute the desired values.

In the classification system used in the data section of this volume, reflectance and transmittance subproperties are grouped (geometrically) by common incidence conditions. The relations which can be used to compute one subproperty of reflectance from another for different geometric conditions are as follows:

For perfectly diffuse reflecting surfaces

$$\rho(\theta, \omega; 2\pi) = \text{constant, for all } \theta \quad (48)$$

$$\rho(\theta, \omega; 2\pi) = \frac{\pi}{\omega' \cos\theta'} \rho(\theta, \omega; \theta', \omega') \quad (49)$$

where θ and ω are the same on both sides of the equation.

$$R(\theta_1, \omega_1; \theta_1', \omega_1') = R(\theta_2, \omega_2; \theta_2', \omega_2') \quad (50)$$

where θ_1 and θ_2, θ_1' and θ_2', ω_1 and ω_2, ω_1' and ω_2' can have any values.

For perfectly specular reflecting surfaces

$$\rho(\theta, \phi, \omega; 2\pi) = \rho(\theta, \phi, \omega; \theta', \phi', \omega') \quad (51)$$

if and only if $\theta' = \theta, \phi' = \phi + \pi$, and $\omega' \geq \omega$.

To compute one subproperty of emittance from another, for different geometric conditions, the following equation may be used:

$$\epsilon(2\pi) = \frac{\int_{\theta=0}^{\pi/2} \int_{\phi=0}^{2\pi} L_b \epsilon(\theta', \phi', \omega') \cos\theta' \sin\theta' \, d\theta' \, d\phi'}{\pi L_b} \quad (52)$$

This equation is sometimes written

$$\epsilon(2\pi) = \frac{1}{\pi} \int_{\Omega} \epsilon(\theta', \omega') \cos\theta' \, d\omega' \quad (53)$$

For a perfectly diffuse emitter, which may be approximated by the typical nonconductor or a roughened metal surface,

$$\epsilon(2\pi) = \epsilon(\theta', \omega') = \epsilon(0, \omega') \quad (54)$$

As a matter of convenience and practicality in much of the development for theoretical relations, little attention has been given to specification of the solid angles of illumination or viewing (ω or ω'). Two cases appear in the literature distinguishing the importance of the relative size of the solid angle. The term *conical* is applied when the solid angle is sufficiently large that there is a variation of radiance across the front of the solid angle. In the limit when the solid angle is 2π, the term *hemispherical* has been used. The term *directional* is applied when the solid angle is so small that there is insignificant variation of the radiance within this solid angle. This concept is carried through the literature for bidirectional reflectance [$\rho(\theta, \omega; \theta', \omega')$ where $\omega = \omega' \simeq 0$] and directional emittance [$\epsilon(\theta', \omega')$ where $\omega' \simeq 0$].

More detailed considerations of angular or geometric relations for the reflectance (and transmittance) properties utilize the concept of a distribution function [109]. This function permits computation of any of the various geometric subproperties of reflectance. There are not too many instances where such a function is available in the literature and normally there are insufficient fragments of subproperty data available to permit one to formulate it.

B. Wavelength Relations

The four specific wavelength conditions identified in the data section of this volume have been defined earlier as: spectral (λ), total (t), integrated ($\bar{\lambda}$), and solar (s). Straightforward methods can be used to obtain one subproperty from another, although in some instances the calculations can be laborious.

The interrelation between *spectral* and *total* emittance which follows directly from their definitions is

$$\epsilon(t) = \frac{\int_0^\infty \epsilon(\lambda) M_{b,\lambda}(\lambda, T) \, d\lambda}{\int_0^\infty M_{b,\lambda}(\lambda, T) \, d\lambda} \quad (55)$$

If the function $\epsilon(\lambda)$ can be simply prescribed, the integration can be performed by a numerical technique on a high-speed computer. In practice, it is convenient to redefine the limits of integration to correspond to the wavelength limits between which some specific fraction of the total energy is found.

The Planck blackbody function indicates the distribution of energy throughout the spectrum. From such considerations, Fig. 4 is derived with the cross-hatched region indicating the wavelength band encompassing 99 percent of the flux emitted by a blackbody source at the specified temperature. For example, 99 percent of the energy emitted by a blackbody at 1000 K is in the wavelength band 2 to 28 μm. Hence, for this temperature, the integration limits of equation (55) above can be simplified with small error in the result.

In the process of computing total emittance data from spectral values, it is essential to identify the wavelength band over which most of the radiant energy is emitted. To do so permits one to assess the validity of a very common assumption used in heat-transfer calculations; namely, that the surface in question is a grey body which has no spectral variation in emittance. Another situation arises when it is necessary to evaluate the possible effect extrapolation of limited spectral data may have. An example of this arises when considering metallic conductors in which the spectral emittance in the visible and near infrared (say to 1 μm) is quite variable, but in the longer wavelength region is monotonic of a form approximated by the Hagen–Rubens relation. The spectral character of nonmetallic solids is usually not as simple, and hence care should be taken that all absorption bands are identified before making extrapolations.

The computations by integration of equation (55) can be done by finite wavelength intervals

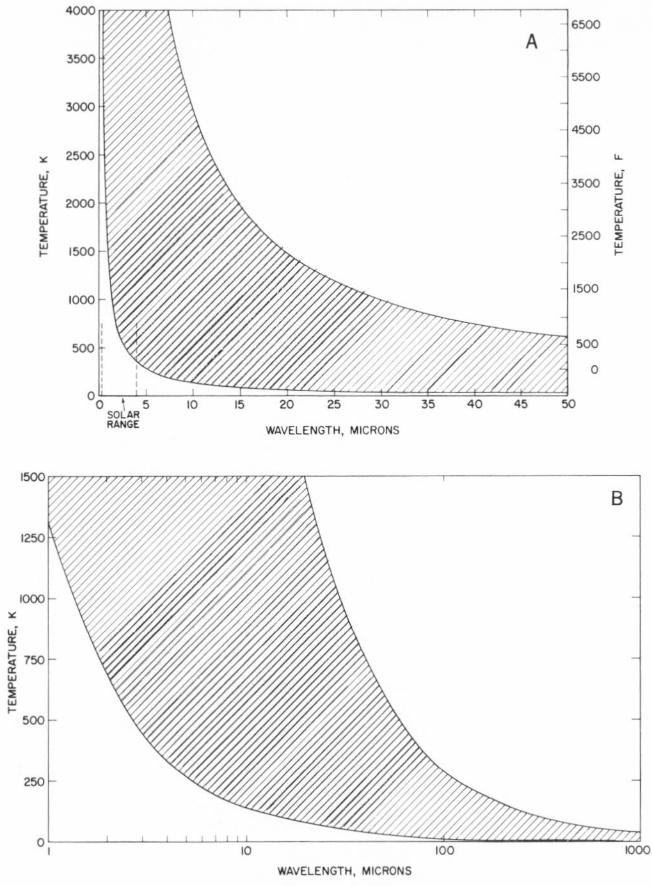

Fig. 4. Wavelength band containing 99% of the flux emitted by a blackbody at the indicated temperature: (A) for high temperatures and (B) for low temperatures.

$$\epsilon(t) = \frac{\sum_{\lambda_1}^{\lambda_2} \epsilon(\lambda) M_{b,\lambda}(\lambda, T)\Delta\lambda}{\sum_{\lambda_1}^{\lambda_2} M_{b,\lambda}(\lambda, T)\Delta\lambda} \tag{56}$$

where the limits λ_1 and λ_2 are selected to include substantially all of the flux emitted by a blackbody radiator at the temperature T.

In the Weighted Ordinate Method, uniform values of $\Delta\lambda$ are used, and each value of $\epsilon(\lambda)$ must be weighted by a factor proportional to $L_b(\lambda, T)$, the flux emitted per unit area, unit solid angle, and, within the wavelength interval λ and $(\lambda + \Delta\lambda)$, by a blackbody at the test temperature T. In the Selected Ordinate Method, the area under the blackbody spectral distribution curve is divided vertically into X units of equal area. The X selected ordinates are then the X median wavelengths of the X areas. In this case, $\Delta\lambda$ varies, but the product $L(\lambda, T)\Delta\lambda$ is held constant and for $X = 100$ this is 0.01. Under these conditions then

$$\epsilon(t) \simeq 0.01 \sum_{\lambda_1}^{\lambda_{100}} \epsilon(\lambda) \tag{57}$$

A high degree of precision in the computation of $\epsilon(t)$ can be attained using either the Weighted or Selected Ordinate Method by taking a large enough number of ordinates. With any given number of ordinates, the computation error will be smaller by the Selected Ordinate Method than by the Weighted Ordinate Method. Since the spectral emittance curves for solids do not normally have sharp peaks or valleys, the 100-Selected Ordinate Method gives values of adequate precision for most applications.

The wavelengths representing the 100 selected ordinates for the temperature range 600 to 1400 K are presented in Table III [140]. This table along with equation (55) should be most useful for com-

Table III. Wavelengths for Computation of Total Emittance from Spectral Data by the 100 Selected Ordinate Method [140]

		λ, μm, at various values of T								
%*	λT, μm·K	600 K	700 K	800 K	900 K	1000 K	1100 K	1200 K	1300 K	1400 K
0.5	1322	2.203	1.889	1.652	1.469	1.322	1.202	1.102	1.017	0.944
1.5	1534	2.557	2.191	1.918	1.704	1.534	1.395	1.278	1.180	1.096
2.5	1662	2.770	2.374	2.078	1.847	1.622	1.511	1.385	1.278	1.187
3.5	1762	2.937	2.517	2.202	1.958	1.762	1.602	1.468	1.355	1.258
4.5	1846	3.077	2.637	2.308	2.051	1.846	1.678	1.538	1.420	1.318
5.5	1920	3.200	2.743	2.400	2.133	1.920	1.745	1.600	1.477	1.371
6.5	1989	3.315	2.841	2.486	2.210	1.989	1.808	1.657	1.530	1.424
7.5	2052	3.420	2.931	2.565	2.280	2.052	1.865	1.710	1.578	1.466
8.5	2111	3.518	3.016	2.639	2.346	2.111	1.919	1.759	1.624	1.508
9.5	2168	3.613	3.097	2.710	2.409	2.168	1.971	1.806	1.668	1.549
10.5	2222	3.703	3.174	2.778	2.469	2.222	2.020	1.852	1.709	1.587
11.5	2274	3.790	3.249	2.842	2.527	2.274	2.067	1.895	1.749	1.624
12.5	2325	3.875	3.321	2.906	2.583	2.325	2.114	1.937	1.788	1.661
13.5	2374	3.957	3.391	2.968	2.638	2.374	2.158	1.978	1.826	1.696
14.5	2423	4.038	3.461	3.029	2.692	2.423	2.203	2.019	1.864	1.731
15.5	2470	4.117	3.529	3.088	2.744	2.470	2.245	2.058	1.900	1.764
16.5	2517	4.195	3.596	3.146	2.797	2.517	2.288	2.097	1.936	1.798
17.5	2563	4.271	3.662	3.204	2.848	2.563	2.330	2.136	1.972	1.831
18.5	2609	4.348	3.727	3.261	2.899	2.609	2.372	2.174	2.007	1.864
19.5	2654	4.423	3.792	3.318	2.949	2.654	2.413	2.212	2.042	1.896
20.5	2698	4.496	3.854	3.372	2.998	2.698	2.453	2.248	2.075	1.927
21.5	2743	4.571	3.919	3.429	3.048	2.743	2.494	2.286	2.110	1.959
22.5	2887	4.645	3.982	3.484	3.097	2.787	2.534	2.322	2.144	1.991
23.5	2831	4.718	4.044	3.539	3.146	2.831	2.574	2.359	2.178	2.022
24.5	2876	4.793	4.109	3.595	3.196	2.876	2.615	2.397	2.212	2.054
25.5	2920	4.866	4.172	2.650	3.244	2.920	2.655	2.433	2.246	2.086
26.5	2964	4.940	4.234	3.705	3.293	2.964	2.695	2.470	2.280	2.117
27.5	3008	5.013	4.297	3.760	3.342	3.008	2.735	2.507	2.314	2.149
28.5	3052	5.086	4.360	3.815	3.391	3.052	2.775	2.543	2.348	2.180
29.5	3097	5.161	4.424	3.871	3.441	3.097	2.815	2.581	2.382	2.212
30.5	3141	5.235	4.487	3.926	3.490	3.141	2.855	2.617	2.416	2.244
31.5	3186	5.310	4.552	3.982	3.540	3.186	2.896	2.655	2.451	2.275
32.5	3231	5.385	4.616	4.039	3.590	3.231	2.937	2.692	2.485	2.308
33.5	3277	5.461	4.682	4.096	3.641	3.277	2.979	2.731	2.521	2.341
34.5	3323	5.538	4.747	4.154	3.692	3.323	3.021	2.769	2.556	2.374
35.5	3369	5.615	4.813	4.211	3.743	3.369	3.063	2.807	2.592	2.406
36.5	3415	5.691	4.879	4.269	3.794	3.415	3.105	2.846	2.627	2.439
37.5	3462	5.770	4.946	4.328	3.847	3.462	3.147	2.885	2.663	2.473
38.5	3510	5.850	5.014	4.388	3.900	3.510	3.191	2.925	2.700	2.507
39.5	3558	5.930	5.083	4.448	3.953	3.558	3.235	2.965	2.737	2.542
40.5	3607	6.011	5.152	4.509	4.008	3.607	3.279	3.006	2.775	2.576
41.5	3656	6.093	5.223	4.570	4.062	3.656	3.324	3.047	2.812	2.611
42.5	3706	6.176	5.294	4.632	4.118	3.706	3.369	3.088	2.851	2.647
43.5	3757	6.261	5.367	4.696	4.174	3.757	3.415	3.131	2.890	2.684
44.5	3809	6.348	5.441	4.761	4.232	3.809	3.463	3.174	2.930	2.721
45.5	3861	6.435	5.516	4.826	4.290	3.861	3.510	3.217	2.970	2.758
46.5	3914	6.523	5.592	4.892	4.349	3.914	3.558	3.262	3.011	2.796
47.5	3968	6.613	5.669	4.960	4.409	3.968	3.607	3.307	3.052	2.834
48.5	4023	6.705	5.747	5.029	4.470	4.023	3.657	3.352	3.095	2.874
49.5	4079	6.798	5.827	5.099	4.532	4.079	3.708	3.399	3.138	2.914
50.5	4136	6.893	5.909	5.170	4.596	4.136	3.750	3.447	3.182	2.954
51.5	4194	6.990	5.992	5.242	4.660	4.194	3.813	3.495	3.226	2.996
52.5	4254	7.090	6.077	5.318	4.727	4.254	3.867	3.545	3.272	3.039
53.5	4314	7.190	6.163	5.392	4.793	4.314	3.922	3.595	3.318	3.081
54.5	4377	7.295	6.253	5.471	4.863	4.377	3.979	3.647	3.367	3.126

Table III (continued)

%*	λT, μm·K	λ, μm, at various values of T: 600 K	700 K	800 K	900 K	1000 K	1100 K	1200 K	1300 K	1400 K
55.5	4440	7.400	6.343	5.550	4.933	4.440	4.036	3.700	3.415	3.171
56.5	4505	7.508	6.436	5.631	5.006	4.505	4.095	3.754	3.465	3.218
57.5	4572	7.620	6.532	5.715	5.080	4.572	4.156	3.810	3.517	3.266
58.5	4640	7.733	6.629	5.800	5.156	4.640	4.218	3.867	3.569	3.314
59.5	4710	7.856	6.729	5.888	5.233	4.710	4.282	3.925	3.623	3.364
60.5	4782	7.970	6.832	5.978	5.313	4.782	4.347	3.985	3.678	3.416
61.5	4856	8.093	6.937	6.070	5.396	4.856	4.415	4.047	3.735	3.469
62.5	4932	8.220	7.046	6.165	5.480	4.932	4.484	4.110	3.794	3.523
63.5	5010	8.350	7.157	6.262	5.567	5.010	4.555	4.175	3.854	3.579
64.5	5091	8.485	7.273	6.364	5.657	5.091	4.628	4.242	3.916	3.636
65.5	5175	8.625	7.393	6.469	5.750	5.175	4.705	4.312	3.981	3.696
66.5	5262	8.770	7.517	6.578	5.847	5.262	4.784	4.385	4.048	3.759
67.5	5351	8.918	7.644	6.689	5.945	5.351	4.865	4.459	4.116	3.822
68.5	5444	9.073	7.777	6.805	6.049	5.444	4.945	4.536	4.188	3.889
69.5	5541	9.235	7.916	6.926	6.157	5.541	5.037	4.617	4.262	3.958
70.5	5641	9.401	8.059	7.051	6.268	5.641	5.128	4.701	4.339	4.029
71.5	5745	9.575	8.207	7.181	6.383	5.745	5.223	4.787	4.419	4.105
72.5	5854	9.756	8.363	7.318	6.504	5.854	5.322	4.878	4.503	4.181
73.5	5968	9.946	8.526	7.460	6.631	5.968	5.425	4.923	4.591	4.263
74.5	6087	10.145	8.696	7.609	6.783	6.087	5.533	5.072	4.682	4.348
75.5	6212	10.353	8.874	7.765	6.902	6.212	5.647	5.176	4.778	4.437
76.5	6343	10.571	9.062	7.929	7.048	6.343	5.766	5.286	4.879	4.531
77.5	6482	10.803	9.260	8.102	7.202	6.482	5.893	5.401	4.986	4.630
78.5	6628	11.046	9.469	8.285	7.364	~~4.628~~ 6.628	6.025	5.523	5.098	4.734
79.5	6783	11.304	9.690	8.479	7.537	6.783	6.166	5.652	5.218	4.845
80.5	6948	11.580	9.926	8.685	7.720	6.948	6.316	5.790	5.345	4.963
81.5	7123	11.871	10.176	8.904	7.914	7.123	6.475	5.936	5.479	5.088
82.5	7311	12.185	10.444	9.139	8.123	7.311	6.646	6.092	5.624	5.222
83.5	7514	12.523	10.735	9.392	8.349	7.514	6.831	6.261	5.780	5.367
84.5	7732	12.886	11.046	9.665	8.591	7.732	7.029	6.443	5.948	5.523
85.5	7969	13.281	11.385	9.961	8.854	7.969	7.245	6.641	6.130	5.692
86.5	8228	13.712	11.755	10.285	9.142	8.228	7.480	6.856	6.329	5.877
87.5	8513	14.188	12.162	10.641	9.459	8.513	7.739	7.094	6.548	6.081
88.5	8829	14.714	12.613	11.036	9.810	8.829	8.026	7.357	6.691	6.306
89.5	9183	15.304	13.119	11.479	10.203	9.183	8.348	7.652	7.064	6.559
90.5	9583	15.971	13.690	11.979	10.648	9.583	8.712	7.986	7.371	6.845
91.5	10042	16.746	14.346	12.552	11.158	10.042	9.129	8.386	7.726	7.173
92.5	10577	17.628	15.110	13.221	11.752	10.577	9.615	8.813	8.136	7.555
93.5	11215	18.524	16.022	14.019	12.461	11.215	10.195	9.345	8.627	8.011
94.5	11996	19.993	17.137	14.995	13.329	11.996	10.878	9.996	9.228	8.569
95.5	12990	21.649	18.558	16.238	14.433	12.990	11.809	10.825	9.992	9.278
96.5	14327	23.877	20.468	17.909	15.919	14.327	13.025	11.939	10.021	10.233
97.5	16295	27.157	23.279	20.369	18.105	16.295	14.814	13.579	12.535	11.639
98.5	19724	32.872	28.178	24.655	21.915	19.724	17.931	16.436	15.172	14.088
99.5	29372	48.951	41.961	36.715	32.635	29.372	26.702	24.476	22.594	20.980

*The wavelength λ is chosen so that the indicated percentage of blackbody radiation occurs at wavelengths shorter than the indicated wavelength.

computing total emittance from spectral emittance data.

The spectral subproperties of reflectance, absorptance, and transmittance are used to find the corresponding *integrated* values according to the relation (shown for absorptance)

$$\alpha(\bar{\lambda}) = \frac{\int_a^b \alpha(\lambda) E(\lambda)\, d\lambda}{\int_a^b E(\lambda)\, d\lambda} \qquad (58)$$

where $E(\lambda)$ is the spectral irradiance of the illuminating source. The integration is performed over the wavelength interval or band denoted by the limits of a and b.

Should the source be a blackbody at some temperature T, the absorptance can be evaluated by use of an equation similar to equation (55). A special case of the integrated subproperty is where the illuminating source is that of the solar disc, natural or artificial. The solar absorptance is normally specified for the geometric condition of normal or angular incidence and can be determined from spectral data in the following manner.

$$\alpha(\theta; s) = \frac{\int_0^\infty \alpha(\theta; \lambda) E_s(\lambda)\, d\lambda}{\int_0^\infty E_s(\lambda)\, d\lambda} \tag{59}$$

where $E_s(\lambda)$ denotes the solar irradiance, which has been tabulated in several references [88, 156]. Frequently, the solar absorptance is determined from room temperature reflectance data

$$\alpha(\theta; \lambda) = 1 - \rho(\theta; 2\pi; \lambda) \tag{60}$$

using Kirchoff's law and integrating over the solar spectrum.

8. THEORY OF COATING PROPERTIES

A. General Considerations

A coating is a system consisting of a layer (or layers) of any substance(s) upon a substrate. For the purposes of this work, three separate categories have been identified: *pigmented coatings*, including vitreous enamels and paints, are mixtures of a pigment and vehicle applied to a substrate; *contact coatings* are formed by a layer or layers of a substance coated on a substrate without chemical reaction occurring between the coating material and the substrate; and *conversion* or *diffusion coatings* are layers of a compound, or mixture of compounds formed by the chemical reaction of the substrate with another material. Coatings have characteristics—optical and physical—which are not necessarily like the properties of any single component of the system; by proper combination of components, a system or composite can be generated which has the superior application qualities of each without their respective disadvantages. Examples are the refractory metal coated systems. The base metal (molybdenum, for example) has superior high-temperature strength but poor oxidation resistance characteristics; the coating (flame-sprayed molybdenum disulfide) protects the substrate from oxidation giving the system favorable high-temperature application qualities. The optical behavior of the coating system may not always be determined by the outer layer or surface, but will be dependent upon the properties of the intermediate layer or layers and even in many instances the substrate materials, into which radiant flux can penetrate. There are a few systems—particularly metallic and multilayer dielectric films—which can be analytically treated wherein the optical properties of the individual components can be used to determine the optical properties of the system. The class of thermal control coatings are tailored to provide desired optical properties in order that the surface shall have the proper ratio of absorbed power from the solar source (or other extraterrestrial source) to emitted power; hence the equilibrium temperature of the surface can be controlled within its application environment by proper design of the coating. In subsequent paragraphs of this section, the major classes of coatings—primarily for thermal control—will be identified and described in order to introduce the wide variety of coatings being used and some of the principles upon which such coatings are designed [133].

The parameter in common use to describe the performance of thermal control coatings is the ratio of the normal solar absorptance, $\alpha(0; s)$ to the hemispherical total emittance, $\epsilon(2\pi; t)$, which for

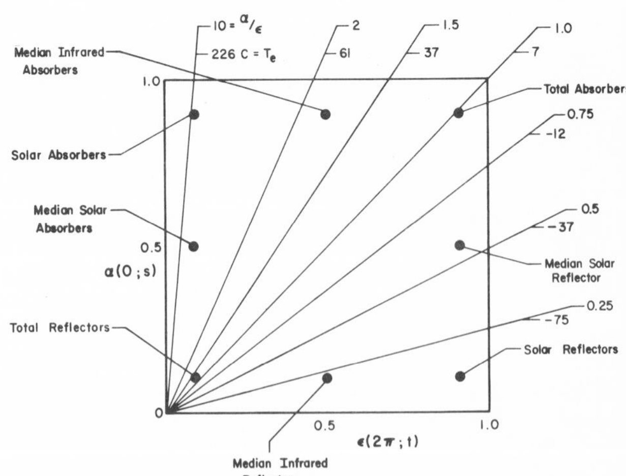

Fig. 5. Characteristics of basic types of thermal control surfaces showing $\alpha(0; s)/\epsilon(2\pi; t)$ ratios and equilibrium temperature, T_e, of a coated isothermal sphere at 1 AU [133].

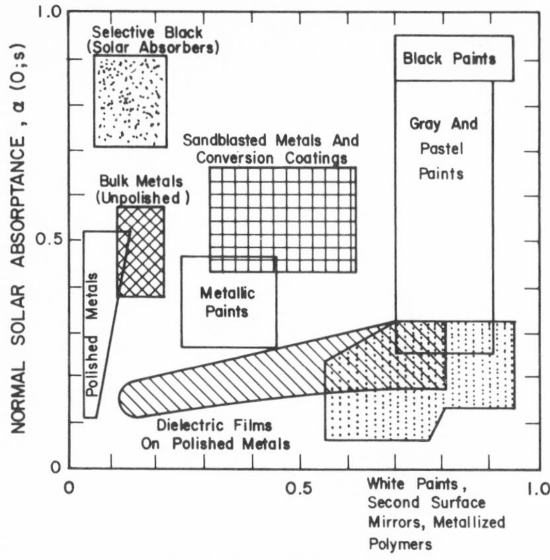

Fig. 6. Range of $\alpha(0; s)$ and $\epsilon(2\pi; t)$ covered by available thermal control coatings and surfaces [76].

brevity will be referred to simply as α/ϵ. It should be remembered that the individual values of α and/or ϵ are equally important in the consideration of thermal control surface performance. Figure 5 shows the basic types of surfaces as a plot of hemispherical total emittance $\epsilon(2\pi; t)$ *vs* normal solar absorptance $\alpha(0; s)$ and lines of constant α/ϵ. For each of these lines, two values are labeled as follows: (1) $T_e(C)$ is the equilibrium temperature for an isothermal sphere at 1 AU,* and (2) the ratio $\alpha(s)/\epsilon(t)$. Figure 6 shows the types of coatings or the methods which can be used to obtain the various types of surfaces displaying the characteristics demonstrated in the previous figure. The feature of these coatings which explains their performance is the spectral reflectivity, which, as shown in the last section, is used to compute solar absorptance and total emittance by integration techniques using the proper weighting function. By way of example of spectral behavior, Fig. 7 shows several coatings displaying a number of especially interesting features making them attractive for use as thermal control surfaces. The normal spectral reflectivity is plotted logarithmically *vs* wavelength and the two scales on the upper abscissa are the percentage increment of energy falling below any wavelength for solar and blackbody conditions. The first of these scales represents the percentage of solar irradiance $E_\lambda(s)$ associated with wavelengths shorter than that indicated; this shows that 98 percent of the solar spectrum is within the region 0.28 to 3.9 μm. The second of the upper scales represents the percentage of blackbody (300 K) radiation associated with wavelengths shorter than that indicated; the region 4.8 to 77 μm encompasses 98 percent of the total energy within this blackbody spectrum. Obviously, these are the spectral regions of interest in which measurement techniques have been developed. As will be seen, there are some limitations in adequately obtaining data over the entire region, but these are not serious. Typical metallic reflection is displayed by the spectra of evaporated aluminum, Curve A; bulk materials will, in general, have lower reflectance values very much dependent upon surface treatment and texture. The thickness of aluminum oxide layers overcoated on aluminum primarily influences the infrared character, as shown by Curves *B*, *C*, and *D*; the effect of the layers 0.5 μm, 1.0 μm, and 1.5 μm thickness is to increase the hemispherical total emittance from an uncoated value of 0.03 to 0.12, 0.23, and 0.40, respectively [68]. This gives some indication of the large variations in the optical characteristics that can be caused by thin film overcoating techniques. Curve *E* shows the spectra of an optical solar reflector, OSR, which in this case is a second surface mirror of 2000 Å thickness silver film over a silica substrate; the solar absorptance is 0.047 and the hemispherical total emittance 0.74, that is, $\alpha(s)/\epsilon(t) = 0.064$ [63]. Quite another class of coatings consists of the paint systems of which zinc oxide (S-13G, ZnO in methyl silicone binder) is typical; Curves *E* and *F* show the reflectance for the conditions before and after ultraviolet exposure (800 ESU), respectively [180]. The solar absorptance before UV exposure was 0.17 and the change in absorptance $\Delta\alpha(s) = 0.01$ indicating that this is a relatively stable coating. An example of severe degradation is the coating of TiO_2 pigment in an epoxy binder (see Fig. 109, page 280, of the *Numerical Data* section) having an initial value of $\alpha(s) = 0.225$ and a change $\Delta\alpha(s) = 0.182$ after 509 ESH. The most stable of all the coatings is Z-93, ZnO pigment in potassium silicate binder (see Fig. 130A, page 391, of the *Numerical Data* section) which has a change $\Delta\alpha(s) = 0.00$ after 2500 ESH. The resistance of a coating to environmental effects is one of the primary considerations in the selection of a system and, as a result, there is an extremely large number of measurements on such effects in the literature.

*1 AU is the mean distance between the sun and earth, 1.496×10^8 km; the solar constant at 1 AU is 0.140 watt cm^{-2} [156].

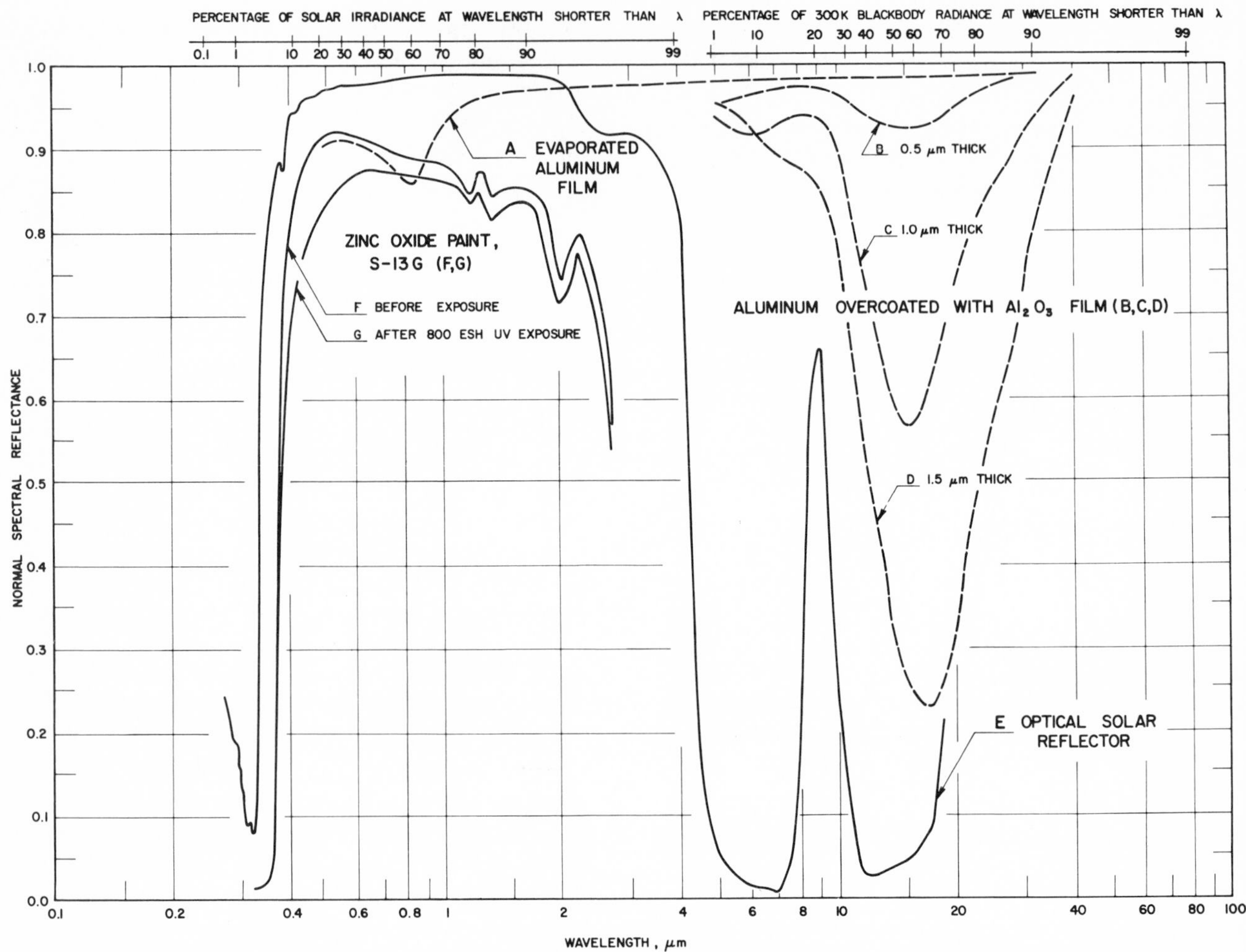

Fig. 7. Typical spectral characteristics of selected coatings: *A*, evaporated aluminum, *B*, *C*, and *D*, evaporated aluminum overcoated with aluminum oxide films of 0.5 μm, 1.0 μm, and 1.5 μm thickness, respectively: *E*, optical solar reflector, 2000 Å silver over silica substrate; *F*, *G*, zinc oxide white paint before and after ultraviolet exposure, to 800 ESA, respectively.

B. Thin Metallic Films

Metallic films of the order of 2000–3000 Å thickness will be opaque and exhibit all the characteristics of the nearly ideal bulk metal. Films formed by vapor deposition under controlled vacuum conditions have a higher reflectivity than a bulk surface which has been polished mechanically for reason of structural damage caused by deformation; however, using electropolishing techniques it is possible to achieve nearly the same reflectivity of bulk and film surfaces. It is appreciated that the well-developed and understood processes of film deposition to generate highly reproducible surfaces on nearly any substrate makes their use attractive in a wide range of applications.

The uses for thin films are primarily in the range of room to subnormal temperatures; because of their oxidation at higher temperatures, other types of films are more useful as protective or thermal control surfaces. The major portion of the data that have been generated on these metallic coatings is in the aforementioned temperature region primarily because of their utility as high reflectors in insulating systems (dewars, and the like) and additionally because of the theoretical interest in having (spectral) data as a function of temperature to determine electronic structural parameters. There is an appreciable amount of data generated for the purpose of studying key materials rather than for determining the performances of the material in a thermal environment. Such data, in general, are obtained on well-characterized samples, generated under carefully controlled conditions, and are most useful to the thermophysicist in his better understanding of the material.

In the subsequent discussion of this section, the basis for theoretical treatment of thin films will be briefly presented. The approach will be limited to the room and subnormal temperature range; a discussion of the higher temperature range has already been presented in companion Volume 7. In this work, the dependency of emittance as a function of temperature, wavelength, and angle of exitance was discussed within the framework of the simple free electron theory. This approach yields tractable relations suited for engineering estimations; refined relations including influences due to several types of free electrons and/or resonance (bound electron) effects are available for correlation of available data. A review of this discussion would be useful background for better understanding of the following presentation.

a. Drude Theory

The earliest attempts to predict the optical properties of metals were begun by Drude and have been described in many classical reference texts [151, 175]. They have also been the subject of recent investigations [6, 8]. The Drude theory is based upon the Maxwell wave equation expressed in Gaussian form as

$$\nabla^2\bar{E} + \frac{\omega^2}{c^2}\bar{E} - j\frac{4\pi\omega}{c^2}\bar{J} = 0 \tag{61}$$

where the electric field vector $\bar{E}$ of frequency ω and the current density vector J are related by the complex conductivity

$$\bar{J} = (j\omega\alpha_p + \sigma)\bar{E} \tag{62}$$

The polarizability, α_p, and conductivity, σ, are both frequency-dependent parameters of the material which must be determined from some model of the electronic structure and are not calculable from electromagnetic theory. Both the polarizability and conductivity can be expressed as the sum of contributions from various mechanisms: intraband, interband, lattice absorption, localized states, etc. The real part of the complex conductivity, σ, is related to the energy absorbed by the material and its various components will be nonzero only in frequency intervals where absorption from the particular mechanism occurs. On the other hand, the components of the polarizability can have nonzero terms outside the frequency interval in which absorption takes place. This causes difficulty for the more complex materials in separating the effects due to any one mechanism.

The most significant of the mechanisms which will depend greatly upon the frequency (wavelength) region are described as interband (change in energy state of an electron) and intraband (change in energy level of an electron) transitions. There is a low-frequency limit below which intraband transitions do not take place; this is clearly seen in reflectance spectra as the absorption edge, which is in the neighborhood of 1 μm for most metals, although the influence will extend beyond this region. The effect of intraband transitions predominates at the lower frequencies (longer wavelengths) and is the primary mechanism responsible for the infrared behavior of metals; this is fortunately the more important wavelength region for heat transfer studies and the only region for which the theory provides tractable solutions.

The classical treatment is to simultaneously solve Maxwell's equations and the Boltzmann transport equation for a free electron gas which describes the time and spatial changes in the velocity distribution. A key assumption in the derivation is that the depth of penetration of the field is large enough that the contributions due to changes in the velocity distribution over this distance can be neglected. This is equivalent to the classical explanation that the electric field which the electron experiences is spatially uniform during the time between collisions. For the case of only intraband contributions, the optical constants are expressed as

$$n^2 - k^2 = 1 - 4\pi\alpha_{p,\text{intra}} + 4\pi\alpha_{p,0}$$

$$= 1 - \frac{4\pi\sigma_0}{\tau(\omega^2 + \tau^{-2})} + 4\pi\alpha_{p,0} \tag{63}$$

$$nk = \frac{2\pi\sigma_{\text{intra}}}{\omega} = \frac{2\pi\sigma_0}{\omega^2\tau^2(\omega^2 + \tau^{-2})} \tag{64}$$

where σ_0 is the dc electrical conductivity and τ is the relaxation time. The term $\alpha_{p,0}$ accounts for any contributions to α_p made by other absorption mechanisms outside the specified frequency range. Bennett [6, 8] has developed a simplified explicit expression for relating the reflectivity in terms of the two basic Drude parameters σ_0 and τ. The normal incidence reflectivity was given as a simple function of n and k in Eq. (37); for the conditions where $n > 1$ and $k > 1$, typical of most metals in the infrared, an excellent approximation for the normal (specular) reflectance is

$$\rho(0; \lambda) = \exp[-4n/(n^2 + k^2)] \tag{65}$$

Substitution of the Drude relations, Eqs. (63) and (64), into (65) gives

$$\rho(0;\lambda) = \exp\{-(2\omega/\pi\sigma_0)^{1/2}[(\omega^2\tau^2+1)^{1/2}-\omega\tau]^{1/2} \times [1+g(\alpha_{p,0})]\} \quad (66)$$

where

$$g(\alpha_{p,0}) = \alpha_{p,0}[\omega(\omega^2\tau^2+1)^{1/2}+2\omega^2\tau]/2\sigma_0 \quad (67)$$

The latter is only the first-order term of an expansion; however, errors in $\rho(0;\lambda)$ or n and k are rather small and can be neglected for calculations on most metals. For silver, $g(\alpha_{p,0}) = 0.057$, giving an error of 0.0002 in $\rho(0;\lambda)$ at 1 μm and becomes even smaller at longer wavelengths [8].

For the condition of $\omega\tau \ll 1$, the reflectance can be expressed as a function of only the direct electrical conductivity

$$\rho(0;\lambda) = 1-(2\omega/\pi\sigma_0)^{1/2} \quad (68)$$

This is the well known Hagen–Rubens relation applicable only at longer wavelengths in the infrared as will be shown in a subsequent illustration on the experimental verification of the Drude theory. In the higher frequency range where $\omega\tau \approx 1$, the two parameters σ_0 and τ are necessary to characterize the behavior of the metal. The electronic relaxation time can be approximated from the Lorentz–Sommerfeld relation

$$\tau = m^*\sigma_0/Ne^2 \quad (69)$$

where m^* is the effective mass of the electron, determined from specific heat data or band calculations, and is nearly the free electronic mass for many metals. The number density, N, most conveniently determined from Hall coefficient measurements, can be approximately calculated using the valence of the metal as an estimate of the number of free electrons per atom. Hence, we see that the reflectance can be calculated from nonoptical parameters for regions where the Drude theory is valid.

Additional useful relations for predicting the angular variation of spectral reflectance for the two degrees of polarization are

$$\rho(\theta;\lambda)_s = 1-\frac{4(30\lambda/r)^{1/2}\cos\theta}{60(\lambda/r)+2(2\lambda/r)^{1/2}\cos\theta+\cos^2\theta} \quad (70)$$

$$\rho(\theta;\lambda)_p = 1-\frac{4(30\lambda/r)^{1/2}}{60(\lambda/r)\cos^2\theta+2(30\lambda/r)^{1/2}\cos\theta+1} \quad (71)$$

The wavelength dependence is that of the Hagen–Rubens relation resulting from the long wavelength simplification of the Drude theory. The angular dependence and polarization are prescribed by the Fresnel relation, Eqs. (30) and (31), which is a direct consequence of electromagnetic wave theory.

Figure 8 shows the results of a recent study [6] regarding the validity of the Drude theory; the infrared reflectance of carefully prepared ultrahigh-vacuum-deposited aluminum, gold, and silver films

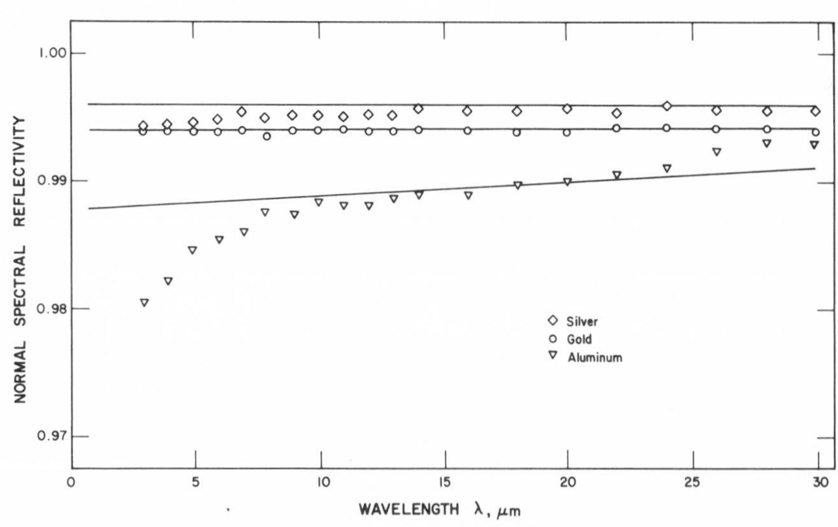

Fig. 8. The infrared reflectance of silver, gold, and aluminum high-vacuum evaporated films compared with the simple Drude model [6]. See Table IV for parameters.

Table IV. Values of the Drude Theory Parameters for Several Metals at Room Temperature

	Free electrons per atom	Conductivity, σ_0 ($\text{ohm}^{-1}\ \text{m}^{-1}$)	Relaxation time, τ (sec)
Aluminum [6]	2.7	3.53×10^3	0.801×10^{-14}
Copper [5]	0.41	1.71×10^3	1.41×10^{-14}
Gold [6]	1.0	4.09×10^3	2.46×10^{-14}
Nickel [5]	0.09	1.13×10^2	0.50×10^{-14}
Silver [6]	1.09	5.39×10^3	2.99×10^{-14}

are shown to be in excellent agreement with the theoretical predictions. For the short wavelength region, particularly noticeable for aluminum, the absorption edge or interband transitions begin to have an influence. As will be pointed out, the sample must be very carefully prepared if the simple Drude model is to be applicable in the infrared beyond, say, 10 μm. The Drude parameters for the above films and for several other simple materials are shown in Table IV. These values can only be used for illustrative purposes since they are dependent upon film thickness, evaporation rate, and substrate smoothness, to name a few. For commercially available films, one should expect significant variation in σ_0 and τ and use bulk values for estimation only with caution.

To this point, we have considered only intraband transitions of one type of electron in our treatment of the Drude theory while quantum theory admits there may be several types responsible for the absorption mechanism. Roberts [144] has extended the simple theory to include this effect (several types of free electrons) and also includes contributions due to interband transitions (several types of bound electrons). The resulting expressions for the optical constants are

$$n^2 - k^2 = 1 + \sum_m \frac{K_{om}\lambda^2(\lambda^2 - \lambda_{sm}^2)}{(\lambda^2 - \lambda_{sm}^2)^2 + (\delta_m \lambda_{sm} \lambda)^2} - \frac{\lambda^2}{2\pi c \epsilon_0} \sum_n \frac{\sigma_n \lambda_n}{\lambda_n^2 + \lambda^2} \tag{72}$$

$$2nk = -\sum_m \frac{K_{om}\lambda^2(\delta_m \lambda_{sm} \lambda)}{(\lambda^2 - \lambda_{sm}^2)^2 + (\delta_m \lambda_{sm} \lambda)^2} + \frac{\lambda^2}{2\pi c \epsilon_0} \sum_n \frac{\sigma_n \lambda}{\lambda_n^2 + \lambda^2} \tag{73}$$

The first terms (summation over m types) describe contributions of the electrons bound to the lattice which are characterized by three parameters: restoring force constant, damping factor, and the number density.† The second term describes the free electron contributions and has the same form as the Drude theory relations of Eqs. (63) and (64) except the contribution is summed over different types of electrons. The theory was developed for correlation of high-temperature optical constants of metals. As an example, the transition metal nickel has a complex electronic structure displaying inflection points as a function of wavelength; the films of Fig. 8 are much simpler in structure. In this instance, the material parameters of the above relations were determined by fitting the theory to the data; hence the parameters are determined optically rather than from an independent set of measurements, and the theory has its primary utility as a correlation tool. Despite the large number of parameters, the correlation is not an arbitrary curve fitting operation, as there are certain physical constraints that the relation parameters must satisfy. Seban [150] studied the Roberts model to describe the emittance of the transition metals iron, nickel, and platinum in the spectral region 2 to 15 μm at elevated temperatures; several types of free electron carriers and one resonance term are required in this correlation. Edwards and deVolo [42] studied the influence of several types of free electron carriers for some twenty metals in the infrared. An appealing aspect of their work is the resulting accurate analytical expression, of reasonable tractability with parameters of physical significance, for emittance as a function of wavelength. While the fundamental basis for application of the Drude theory to real surfaces may be suspect, the approach seems justified on the grounds of the best approximation available.

b. Anomalous Skin Effect

In the Drude theory described above resulting in the Eqs. (63) and (64), a basic assumption is that the electron experiences a uniform spatial electrical field between collisions with the lattice. This assumption is satisfied when the amplitude penetration depth, δ, of the incident electric field (light) is much greater than the mean free path, l, of the conduction electrons, That is,

$$\delta / l \gg 2/(1 + \omega^2 \tau^2) \tag{74}$$

†The classical explanation for the Roberts parameters can be found in Reference [140], page 58.

where $l = \tau v_f$ and v_f is the Fermi velocity and

$$\delta = Re\left[\frac{(2\pi\sigma_0\omega)^{1/2}(1+j)}{c(1+j\omega\tau)^{1/2}}\right]^{-1} = \frac{\lambda}{2\pi k} \tag{75}$$

At low frequencies (long wavelengths) where $\omega\tau \ll 1$, Eq. (74) requires that $\delta \gg 2l$; in this long frequency range l is nearly constant, while δ is proportional to $1/\omega^{1/2}$ and hence the assumption is satisfied for nearly any reasonable value of τ and σ_0. For high frequencies (shorter wavelengths) where $\omega\tau \gg 1$, the Drude theory is applicable, according to Eq. (74), when $\delta/l \gg 2/\omega^2\tau^2$; in this region δ will be independent of frequency and l is constant, satisfying the necessary inequality for sufficiently high frequencies. However, contributions from other mechanisms than the interband transitions as included in Eqs. (63) and (64) must be considered as the spectral region approaches the absorption edge. The frequency (spectral) region between the two limits of applicability of the Drude theory, where the amplitude penetration depth and mean free path are of comparable magnitude, is called the anomalous skin effect region [155]. In this region, the spatial variation of the electric field within the metal is more complicated than that described by the simple exponential decaying field of the Drude theory. An additional surface parameter p is introduced by the theory of Reuter and Sondheimer [136] which is related to the probability that the free electrons are specularly reflected upon striking the inner surface of the metal.

Figure 9 shows, for the case of silver [8], the comparison of the Drude theory (assuming bulk conductivity and one free electron per atom) with the anomalous skin effect modifications for the two cases of specular reflection, $p = 1$, and of diffuse reflection, $p = 0$, of the electrons at the inner surface. The comparison cannot be made below 1 μm as the effect of absorption by interband transitions will begin to appear. For the long wavelength region, the three models converge and it is especially interesting to note that this is in the neighborhood of 100 μm; however, it should be noted that the reflectance scale is very much enlarged and to see differences, state-of-the-art measurements are needed. When computing absorptances, $1 - \rho(0;\lambda)$, these differences become appreciable on a percentage basis. The degree of surface roughness determines the value of the parameter p and the evidence indicates that only supersmooth surfaces will permit specular reflection with $p = 1$. For the case of silver [8] the roughness must be less than 45 Å rms, a condition which can only be achieved by special sample preparation techniques. This study was the first to show that the transition from specular to diffuse electron reflection is a rather abrupt affair. For most well prepared films with routine commercial finishes, it is more likely that diffuse reflection will be the case.

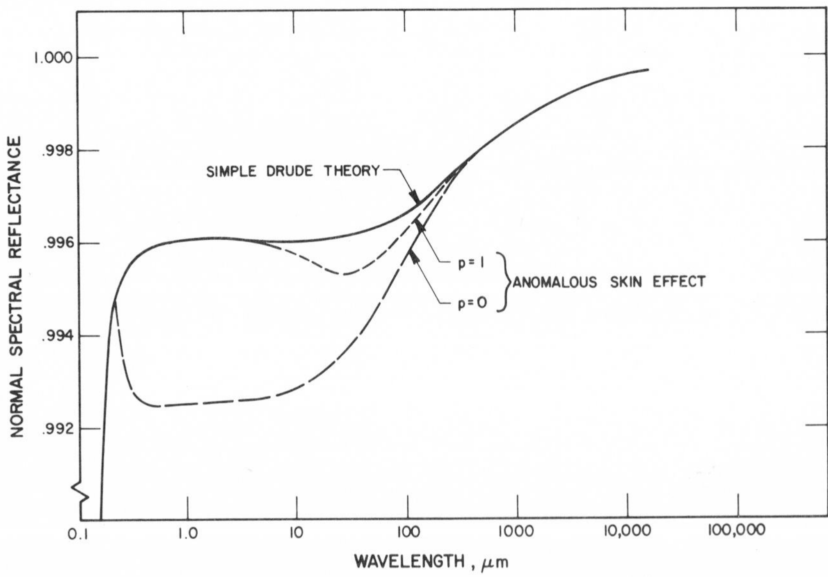

Fig. 9. Reflectance of silver calculated by the anomalous skin effect (dashed curves for diffuse and specular electronic reflection conditions) and the simple Drude theory assuming bulk dc conductivity and 1 free electron per atom [8].

Tractable solutions for the limiting cases of specular and diffuse electronic reflections have been developed by Dingle [35]. The solutions are expressed in series form for the surface impedance, Z^*:

$$Z^* = -(4\pi j\omega\mu l/c_0^2)\,[E(0)/E'(0)] \tag{76}$$

where $E(0)$ is the electric field of the incident electromagnetic wave and $E'(0)$ the electric field of the wave in the material just beneath the surface; for all practical purposes, the magnetic permeability is assumed to be unity. From the relations of Table II, the optical constants n and k can be determined as

$$n^* = (\epsilon_0)^{1/2}\,(Z^*)^{-1} \tag{77}$$

from which the normal spectral reflectance can be determined.

For the case of specular reflection, $p = 1$, the frequency dependence of the ratio of the absorptance computed by the anomalous skin effect, α_0, to that computed by the standard Drude theory, α_D, is shown in Fig. 10 as a function of $\omega\tau$, where τ is the relaxation time. A family of three curves is shown for values 4, 10^2, and 10^4 of the dimensionless parameters $\bar{\sigma} = (\tfrac{3}{2}l^2/\delta^2)/\omega\tau$, which includes the electronic mean free path, l, and the classical skin depth, δ [136]. For the materials silver, gold, and aluminum at room temperature represented in Fig. 8, the dimensionless parameter $\bar{\sigma}$ has values of 4.244, 1.850 and 0.995, respectively; for these conductors at room temperature, the "anomalous" behavior can only be detected with very careful measurements [8], but within the wavelength region of the measurements of Fig. 8, behavior is explained by the Drude theory which coincides with the anomalous skin theory for specular reflection $p = 1$. The "anomalous" behavior becomes more prominent at lower temperatures and long wavelengths. For the case of silver at liquid helium temperature, the dimensionless parameter $\bar{\sigma}$ has a value 10^6 beyond the range of that shown in Fig. 10; the peak of the absorptance ratio—the location of the maximum difference between the anomalous and the standard theory predictions—occurs at $\omega\tau \simeq 3.0$, corresponding to approximately 70 μm. At these lower temperatures and long wavelengths, the effect is pronounced and the classical Drude theory is no longer a reasonable approximation to experimental values. At such temperatures, the inequality of Eq. (74) is no longer satisfied primarily because of the extremely long mean free path. However, at these cryogenic temperatures other effects, including photon-multiple photon processes and sample size effects on electrical conductivity, can be important and must be considered.

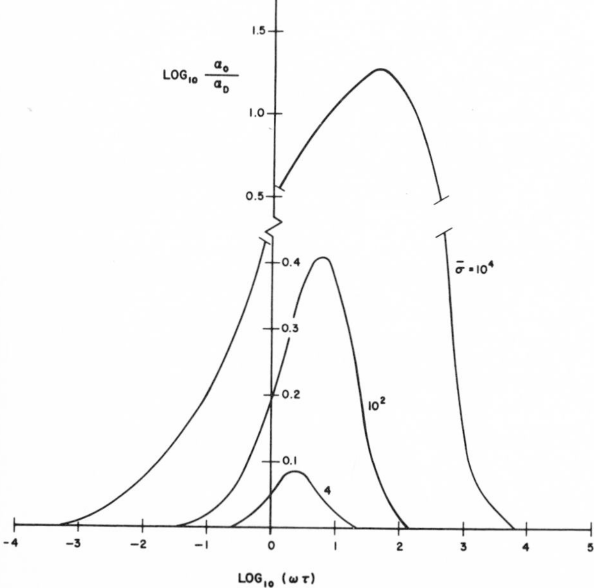

Fig. 10. Frequency dependence of the absorptivity ratio according to the anomalous skin effect ($p = 1$), α_0, to the classical Drude theory, α_D [136].

For the case of diffuse reflection by the electrons at the inner face of the surface where $p = 0$, Dingle [35] has given simplified analytical relations and tabulated values for the near and far infrared regions (see his Table III). The absorptance is determined from tabulated values of dimensionless conductivity and frequency parameters which include the two basic Drude parameters σ_0 and τ. The long dashed curve of Fig. 9 for the room temperature reflectance of silver has been computed using the Dingle values. Dingle also treats the more general case of oblique incidence in which the electric field will have components in two directions for which the surface impedance will be different. The criteria that the difference is negligible is that

$$(\nu/e)\,(m/N)^{1/2} \ll 1 \tag{78}$$

which is easily satisfied by most metals. Simple analytical expressions, primarily for restrictive frequency regions (values of $\omega\tau$), are available for computation of hemispherical reflectance properties.

C. Multilayer Films

The application of a semitransparent coating or film to a substrate can significantly alter the reflectance of the substrate; this phenomenon gives the possibility of generating systems with optical characteristics to meet special applications. From an optical viewpoint there is a clear distinction between thick and thin films where in the latter interference effects predominate, as will be subsequently discussed. There is a large body of knowledge on the optical properties of thin films—usually referred to as *thin film optics*—since there is considerable flexibility and latitude possible in designing or controlling optical parameters. Thin film optics is based upon the solution of the classical electromagnetic wave equations according to the boundary conditions that are created in considering n-layer systems. The computational techniques are not simple especially when attempting to determine what layer properties (thickness and index of refraction) are necessary to achieve an overall desired performance. For the more sophisticated systems such as narrow band interference filters, broad band filters, and antireflection coatings, the reader can find considerable coverage of such topics in the current research literature [69, 72] and classical texts on the subject.

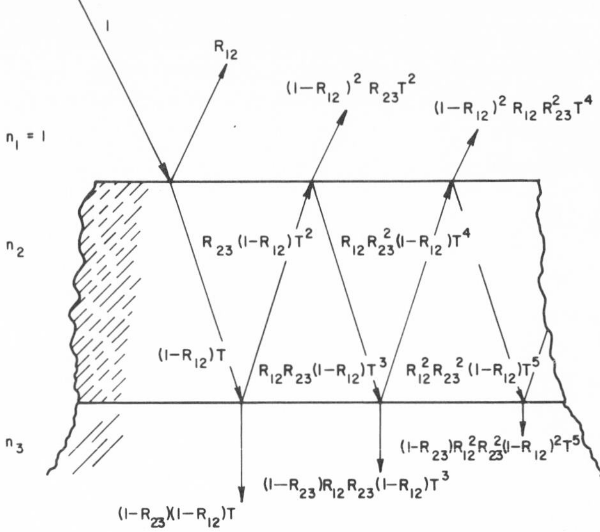

$$\rho = R_{12} + (1-R_{12})^2 R_{23} T^2 + (1-R_{12})^2 R_{12} R_{23}^2 T^4 + \ldots\ldots$$

$$\rho = R_{12} + \frac{(1-R_{12})^2 R_{23} T^2}{1 - R_{12} R_{23} T^2}$$

Fig. 11. Multiple reflections at normal incidence in a semitransparent dielectric thick film, n_2, on a dielectric substrate, n_2.

a. Thick Dielectric Films

The optical behavior of a thick film is determined by consideration of only wave amplitude changes across the boundaries; that is, the effects of interference between the reflected rays within the film layer and the incident rays are negligible. Consider the simple system of Fig. 11 which depicts a ray in vacuum or air incident upon a dielectric layer of index of refraction n_2 and thickness d_1 upon a second dielectric, the substrate in this instance, with an index of refraction n_3. The reflectance of the system, assuming that there will be no reflections from the backside of the substrate, can be determined by tracing the incident ray through the multiple reflections and transmission paths as indicated on the figure. From earlier developments for the nonabsorbing dielectric media, the reflectances R_{12} and R_{23}, due to index of refraction changes at the two interfaces 1–2 and 2–3, respectively, are

$$R_{12} = \left(\frac{n_2 - 1}{n_2 + 1}\right)^2 \tag{79}$$

and

$$R_{23} = \left(\frac{n_3 - n_2}{n_3 + n_2}\right)^2 \tag{80}$$

for the especially simple case of normal incidence; for the general incidence conditions on metallic surfaces, the Fresnel relations of Eqs. (30) and (31) can be used in which the index of refraction must be considered as a complex quantity to account for absorption effects. It should be noticed that in this treatment as well as that following, effects due to scattering are not considered. To determine the reflectance of the system shown in Fig. 11, the individual components of the incident ray returning to the incident medium are summed to give

$$\rho = R_{12} + \frac{(1 - R_{12})^2 R_{23} T^2}{1 - R_{12} R_{23} T^2} \tag{81}$$

For the special case where $n_3 = n_1 = 1$, then $R_{12} = R_{23}$ and the reflectance relation reduces to the form of the semitransparent slab treated in Volume 8, Section 8B. If it is assumed the absorption of the film is negligible (T, the internal transmittance, is unity) then the reflectance of the dielectric film and substrate system can be expressed as

$$\rho = 1 - \frac{4 n_3 n_2}{(n_2{}^2 + n_3)(n_3 + 1)} \tag{82}$$

With this simple relation, now it is easy to see how the film properties affect the system reflectance. The condition for minimized reflectance is found by differentiation ($d\rho/dn_2 = 0$, and also showing the second derivative is positive) to be

$$n_2 = \sqrt{n_3} \tag{83}$$

The ratio of the reflectance of the substrate to the system with this condition is

$$\frac{R_{31}}{\rho(n_2 = \sqrt{n_3})} = 1 + \frac{2\sqrt{n_3}}{n_3 + 1} \tag{84}$$

and for the case of glass where $n_3 = 1.5$, the ratio is approximately 2; that is, for the best case, the reflectance of glass can be halved (from 4% to 2%) by the application of the thick film acting as an antireflection coating. The consequences of such a coating to the surfaces of optical devices containing many elements (30 refracting surfaces, for example) can well be appreciated in which the overall transmittance increases from 21.5% to 45.4% as a result of this coating.

b. Thin Dielectric Films

In thick films the multiple reflection process does not include any interference phenomena. The more interesting case—so far as the capability to design coatings or multilayer films with desired optical characteristics—is the optically thin film, where the thickness is of the order of the wavelength of the radiant flux. The optical behavior and the influence of a thin film over a substrate is dependent not on the thickness alone but on the optical thickness which is defined as the product nd, where n is the refractive index and d the thickness of the film. The wavelength of a monochromatic beam is $1/n$ times less in the film than in vacuum, and hence there will be n times more waves in the film than in vacuum for the same distance interval. The interference phenomena is explained in terms of the number of waves in the film. Based upon visual perception of interference effects, an empirical upper limit on the optical thickness is $nd < 2.5\lambda$; for dielectrics where indices of refraction can be sizable (like oxide films, $n = 2$ to 3 and germanium $n \approx 4$) the optical thickness and the actual dimensional thickness can be quite different.

Consider the theoretical model for a transparent homogeneous dielectric film of index of refraction n_1 and thickness d over a smooth metallic substrate which has a complex index of refraction, $n_2^* = (n_2 - jk_2)$ accounting for the dispersion influence represented by the absorption coefficient [170]. Figure 11 pictorially represents the multiple reflections that occur with an incident ray of the plane wave striking the film at an angle of incidence θ_1. A portion of the wave incident upon the system is reflected at the film–vacuum interface and the balance is transmitted (without absorption in this treatment) through the film reaching the film-metallic interface. At this interface there is a phase change in the reflection process as a consequence of the dispersive behavior of the metal substrate. This wave is reflected back toward the vacuum side and is partially reflected, with no phase change, on the inside of the film–vacuum interface and partially transmitted out into the vacuum. The resulting multiple reflections with phase changes at the metallic substrate cause the waves emerging from the film to have a phase which is different than the incident wave, and interference, constructive or destructive, can occur. At the vacuum–film interface, the first reflection is determined from the Fresnel relation for a simple dielectric

$$R_{12,p}^{1/2} = r_{12,p} = \frac{n_2 \cos\theta_1 - n_1 \cos\theta_2}{n_2 \cos\theta_1 + n_1 \cos\theta_2} \tag{85}$$

$$R_{12,s}^{1/2} = r_{12,s} = \frac{n_1 \cos\theta_1 - n_2 \cos\theta_2}{n_1 \cos\theta_2 + n_2 \cos\theta_2} \tag{86}$$

where R_{12} is the reflectivity and r_{12} is amplitude of the wave (frequently referred to as the Fresnel reflection coefficient) and the subscripts identify the polarization states, s and p, denoting perpendicular and parallel components. At the interface between the film and the substrate, there is a change in phase of the wave, ϕ_{23}, reduction in the wave amplitude, r_{23} given by the relation

$$r_{23,p}e^{j\phi_{23,p}} = \frac{n_3^* \cos\theta_2 - n_2 \cos\theta_3}{n_3^* \cos\theta_2 + n_2 \cos\theta_3} \tag{87}$$

$$r_{23,s}e^{j\phi_{23,s}} = \frac{n_2 \cos\theta_2 - n_3^* \cos\theta_3}{n_2 \cos\theta_2 + n_3^* \cos\theta_3} \tag{88}$$

The angles θ_1, θ_2, and θ_3 are related by Snell's law, giving $n_1 \sin\theta_1 = n_2 \sin\theta_2 = n_3^* \sin\theta_3$.

The magnitude of the interference effect from successively reflected waves differing in phase is determined by following a plane wave through the multiple reflection process. Using the Fresnel coefficients for the appropriate interfaces, it has been shown that the reflectance of the system—film and substrate—is given as

$$\rho = \frac{r_{12}^2 + r_{23}^2 + 2r_{12}r_{23}\cos(\beta - \phi_{23})}{1 + r_{12}^2 r_{23}^2 + 2r_{12}r_{23}\cos(\beta - \phi_{23})} \tag{89}$$

where the reflectance and the Fresnel coefficients should be subscripted, s or p, for perpendicular or parallel polarization components. In this relation β denotes the phase difference between successively reflected waves because of the differences in their path lengths ($4\pi d(n_2/\lambda_0)\cos\theta_2$, where λ_0 is the wavelength in vacuum and d the film thickness) and ϕ_{23} subscripted for the s or p conditions denotes the phase difference between successively reflected waves due to reflections from the metal substrate and hence would depend upon the optical constant n_3^*. For the case where the substrate is a dielectric, the phase difference is only due to β and the resulting simplifications would serve to explain the performance of thin films for use as antireflection coatings over glass optical elements, as was covered by example in the discussion of thick films. It should be noted that the reflectance of the system will be periodic in $\beta - \phi_{23}$ and hence the application of a film to achieve a certain effect in one wavelength region may create quite another effect elsewhere. For example, since an antireflection coating is $\lambda/4$ thick at only one wavelength where its performance is optimized, its effectiveness decreases gradually with departure from this wavelength. The approach then called for is to consider multifilm layers, in which case the relations become even more complicated [85, 170].

The above model is limited to cases where the film can be considered a perfectly transparent dielectric; that is, there is no absorption in the film itself. There are many instances where this is an adequate assumption, but the more general case of the previous model is to include the imaginary part of the index of refraction of the film, that is, $n_2^* = n_2 - jk_2$. The model has been developed for the two polarization conditions [145, 146], from which Fig. 12 shows the remarkable effect only a slight amount of absorption causes. The figure represents the reflectance for the conditions of $n_1 = 1.0$, $n_2 = 2.0 - jk_2$, and $n_3^* = 1.21 - 6.15j$; the absorption coefficient is variously assumed to be $k_2 = 0.000$, 0.002, and 0.004. As a further example of the effect of surface film, mention can be made of the classical example of aluminum oxide on aluminum shown earlier as Curves B, C, and D of Fig. 7. The function of this coating, by way of example, is to reduce reflectance in the infrared without adversely effecting reflectance in the visible or solar region; the result is a low and controllable α/ϵ ratio, inasmuch as the solar absorptance is low (because visible reflectance is high) and the total emittance is higher. The use of MgF_2 and CaF_2 for use as antireflective coatings in the ultraviolet are further examples of the use of films to modify, improve, or otherwise control surface optical properties.

A great amount of the theoretical effort devoted to thin film optics concerns the development of techniques to measure film thickness and properties and substrate properties. Of very real concern are the

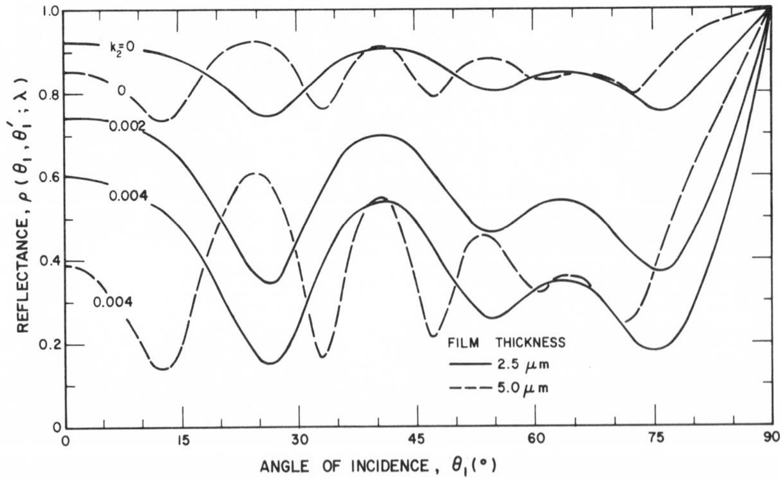

Fig. 12. The effect of absorption on specular reflectivity for parallel plane polarized light of a thin film (thickness 2.5 and 5.0 μm, $n_2^* = 1.21 - jk_2$) on a metallic substrate ($n_3^* = 1.21 - 6.15j$) at 0.63 μm [146].

effects due to slight films—naturally occurring ones particularly—on carefully prepared surfaces undergoing ellipsometry measurements or similar examination. Such studies are widely represented in the literature. Less effort has been given to the effect of thin films on the heat transfer from surfaces. This is especially important at lower temperatures, in the vacuum environment, where the heat transfer mode is radiation. A recent study [32] treats the effects of thin films of mylar, oil, and oxides on the hemispherical total absorptance of aluminum, gold, and copper in the cryogenic region. When $nd/\lambda_{max} < 0.1$ (where λ_{max} is at the peak of blackbody curve for the temperature considered), the films have virtually no effect, while for the condition ≈ 0.3, the effect was maximized.

D. Scattering Phenomena

When the phenomena of reflection, transmission, refraction, and diffraction are no longer separable, we say scattering occurs. Consider the material—in coating or bulk form—to consist of particles imbedded in a matrix; if the wavelength of the incident light is much less than the particle size, regular reflection at elementary mirrors inclined on some statistical basis at all angles as predicted by geometric optics is an acceptable approach. When the wavelength approaches that of the particle size, scattering occurs in which reflection, absorption, refraction, and diffraction mechanisms must be considered. The simplest of all such theories is the Rayleigh scattering model applicable to dilute gases, fogs, and aerosols where absorption effects are minimal and the molecule acts as a single dipole excited by the incident light. The theory predicts that the intensity of the scattered light is inversely proportional to the fourth power of the wavelength and that scattering is not isotropic but proportional to $(1 - \cos^2 \theta_s)$, where θ_s is the angle between the incident and observed directions.

When the wavelength of the incident light is comparable to the particle dimensions, the electrons within the particle will be excited with differing phase and, since the secondary waves emitted are coherent, interference can occur. Hence in certain directions, scattering can be completely diminished and the angular distribution will be quite different from Rayleigh scattering. The Mie theory treats the rigorous theory of single scattering of a wave at a spherical particle of any size and optical properties—dielectric or absorbing. For more tightly packed particle layers where multiple scattering occurs, the theory is much more complex and must take into account particle size and optical properties, packing density, and matrix optical properties.

Particularly important for the practical cases of particulate coating layers is to determine whether the scattering process is due primarily to surface or volume effects. The distinction between bulk and surface scattering processes will be subsequently discussed, but by far the simplest to deal with are the nearly perfect diffusers such as smoked magnesium oxide and barium sulfate, which have been extensively studied in the visible. The radiative property of greatest interest for diffusing materials is the spectral bidirectional reflectance as shown in the polar representation of Fig. 13. The test for diffuseness is conformity to Lambert's law in which the reflection indicatrix appears as a circle. Deviations from this normally occur at higher angles of incidence and for materials which exhibit appreciable absorption. A recent study on barium sulfate [12] has shown that perfect diffuseness for large angles of incidence approaching 60° can be obtained by proper preparation techniques. For larger angles of incidence the regular reflection component increases; with improperly prepared samples, one would expect this effect to appear at smaller incidence angles. The indicatrices in the infrared region for loosely and densely packed germanium powder for 45° inci-

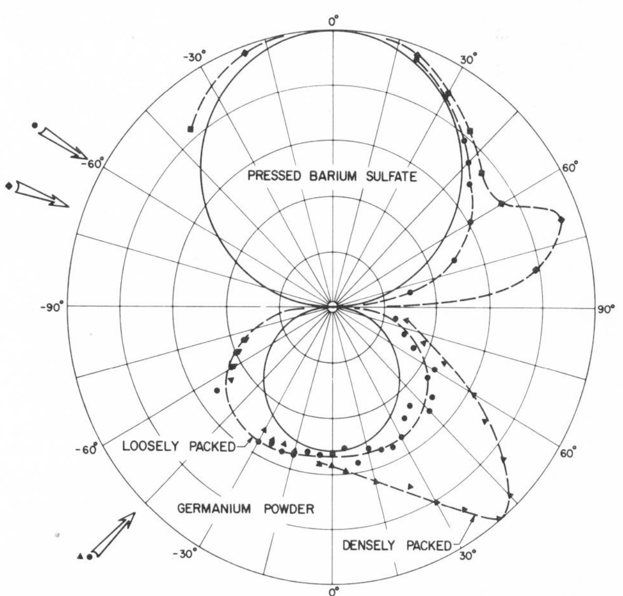

Fig. 13. Reflection indicatrices of (1) pressed barium sulfate for incident angles of 60° (●) and 70° (▲) at 0.52 μm [12], (2) powdered germanium for 45° incidence in loose packed (●) and densely packed (▲) conditions in the 4- to 9-μm band [2], and (3) perfect diffusers, solid curves.

dence angle are shown in the lower half of Fig. 13 [2]. The regular component dominates the reflection process for densely packed conditions, but even for the best packing conditions (loose), this surface is only a fair approximation to a perfect diffuser. This effect is partially a consequence of surface scattering resulting from the high index of refraction of the germanium particles; for the barium sulfate, volume scattering is the more prominent mechanism, giving uniform scattering characteristics. One of the more interesting characteristics of a complete or perfect diffuser is its ability to completely depolarize incident flux. For moderate angles of incidence this was thought in general to be the case, but recent work has shown polarization effects, as well as back and forward scattering effects, are present with good diffusers [25]. The amount of directional reflectance data found in the literature is limited and for the most part, confined to the visible, although techniques for infrared measurements have been developed as well. The bulk of the visible wavelength region data, particularly as relates to paint technology, is concerned with gloss and color interpretation.

Nearly perfect diffuse reflection is a volume or bulk rather than a surface effect and is theoretically described by classical treatment. Consider a specimen of very large area, so that edge effects may be neglected, that is irradiated diffusely over its entire area. Under steady-state conditions, the incident flux not reflected at the surface will penetrate into the specimen diminishing with the depth of penetration; the amount of the diminution at any point along a plane parallel to the surface will be a function only of the distance of the plane from the surface. Under these conditions a one-dimensional analysis can be used. All flux lost by lateral scattering will be compensated by an equal gain through similar scattering from the surrounding area.

The flux is separated into two portions: I, traveling outward from the interior of the specimen, and J, traveling inward from the surface, both traveling normal to the plane of the surface. Both are attenuated by scattering and absorption. A diffuse absorption coefficient, K, is defined by equating $KI\,dx$ to the reduction in I by absorption within an infinitesimal layer dx. It should be noted that the diffuse absorption coefficient, K, is exactly twice the absorption coefficient, a, referred to previously. A diffuse backscattering coefficient S is similarly defined by equating $SI\,dx$ to the flux scattered backwards from I and added to J in the layer dx; S is twice the scattering coefficient, s, defined in an analogous way to the absorption coefficient, a. Within the distance dx the flux I will be diminished by absorption and backscattering, and reinforced by the flux backscattered by J.
Hence we can write:

$$dI/dx = -(K + S)\,I + SJ \tag{90}$$

and

$$dJ/dx = (K + S)\,J - SI \tag{91}$$

These are the basic differential equations used by Kubelka [103], Hamaker [66], and others, and treated in several texts [98, 172]. The solution to these two differential equations can be easily obtained by a direct integration technique applying the boundary conditions at $x = 0$, where $(J/I)_{x=0} = R_g$, the reflectance of the background and at $x = d$, $(J/I)_{x=d} = R$, reflectance of the sample. For the special case where $d \to \infty$, the result can be expressed as

$$\frac{K}{S} = \frac{(1 - R_\infty)^2}{2R_\infty} \tag{92}$$

where R_∞ is the diffuse reflectance for the infinitely thick coating. This equation is the Kubelka–Monk function showing the dependence of the reflectance on only the ratio of the absorption and scattering coefficients. Another simplifying case, but of some practical significance, occurs when the background is an ideal nonreflecting surface [100] giving tractable expressions for interpretation. For the case of finite thickness, the reflectance, R, is a function of R_∞ (that is, the ratio K/S), R_g, and the product Sd, sometimes referred to as the scattering power. The numerous relations which follow from this theory—the Kubelka–Monk theory—have been conveniently summarized for the reader's study in References [98 (p. 120) and 172 (p. 60)]. These references also contain further discussion on the application of the theory to paints, coatings, and diffusers, including a summary of typical scattering parameter values. The constraint of the theory is the neglect of any surface reflection; that is, the reflection process is primarily a volume effect. Furthermore, the theory does not account for any wavelength dependence except to the extent any of the parameters are variable. The diffuse reflectance, expressed using the fundamental parameters, is

$$\rho = \frac{(1 - \beta)^2\,(e^{\gamma d} - e^{-\gamma d})}{(1 + \beta)^2\,e^{\gamma d} - (1 - \beta)^2\,e^{-\gamma d}} \tag{93}$$

$$\tau = \frac{4\beta}{(1+\beta)^2 e^{\gamma d} - (1-\beta)^2 e^{-\gamma d}} \tag{94}$$

where

$$\beta = [K/(K + 2S)]^{1/2} \tag{95}$$

$$\gamma = [K(K + 2S)]^{1/2} \tag{96}$$

Note that these expressions satisfy the extreme case for a nonscattering or simple absorbing material where $S = 0$, then $\gamma = K$ or a, for this case, giving $\tau = e^{-ad}$. The above forms are similar to those developed in our treatment of scattering phenomena in the text of companion Volume 8. The theory has been extended to include effects of external specular reflectance of the specimen and internal specular reflection at both surfaces of the specimen. Such work was found useful for interpretation of reflectance properties of unglazed ceramics and powders [95, 137].

Experimentally the Kubelka–Monk theory fails for intense scattering materials which have a large absorption coefficient. This is attributed primarily to the fact that there is increased reflection at the surface (or interface between the environment and coating) and that such reflection is specular in nature. The reflection process occurring at the front surface is dependent upon the indices of refraction n, and absorption, k, the latter being related to the absorption coefficient by $a = (4\pi/\lambda)k$. As a increases there is an increase in the specular reflection at the surface according to Fresnel relations which govern this process. The volume reflectance, on the other hand, has an exponential dependence upon a. Hence there will be some marked changes in the relative amounts of reflected flux which is surface or volume reflected. It is postulated that the total reflectance is the sum of the volume and surface components, giving a theoretical approach found useful in explaining certain observable effects [171]. When the particle grain size of a coating is decreased, the specular rays at the surface undergo more multiple reflections since there are more interfaces; therefore, the surface reflectance decreases with decreasing grain size. The effect of decreasing grain size on the volume reflection process is the opposite for as the particles become smaller, the incident radiation meets more interfaces but has passed through less absorbing media; hence more flux is reflected [44].

The Kubelka–Monk theory is a "two-flux theory" based upon diffuse forward and backward flux scattering. This approach assumes isotropic, diffuse distribution of flux within the medium while the four-flux and multiflux models [122] with increased complexity do consider the angular distribution effects. The reflection coefficients of the internal surfaces are functions of angle and the absorption and scattering processes will be angular dependent. As K/S increases, there is an increased departure of the flux from being diffusely internally scattered, which explains the failure of the two-flux approach for intense absorbing, scattering materials. While the theory of the multiflux model is tractable and computations are made possible with computer assistance, there is a lack of well characterized reflectance measurements (total or directional) for extensive evaluation of the theory. The particularly useful feature of the two-flux Kubelka–Monk theory is that effective scattering and absorption coefficients can be computed easily from simple directional observations [22]. Hence this theory has proved adequate to handily explain the performance of many scattering materials.

E. Space Environment Effects

The effect of the space environment on the thermal radiative properties of materials used for thermal control applications has received considerable attention in the research literature. While considerable evidence has been collected on the magnitude of property changes, the mechanisms for UV, proton, and electron exposure for many classes of materials are not fully understood. In addition to the coating having stable or predicted degradation of its radiative properties in the space environment, the coating must be relatively easy to apply on large areas and easy to maintain or repair. In this section, only the highlights of environmental effects will be noted. For further detail the reader is referred to a recent summary and handbook [19] or to the several recent studies which are typical of current interest and activity [58, 101, 134].

The principal problem in temperature control concerns the change in the $\alpha(0, s)/\epsilon(2\pi; t)$ ratio (normal solar absorptance to hemispherical total emittance) of a coating due to space environment effects causing degradation. The natural effects of primary importance are (1) electromagnetic solar radiation; (2) particle radiation due to protons: Van Allen and solar wind; and due to electrons: Van Allen and auroral; and (3) to a lesser extent, physical impacts due to atmospheric particles and micrometeorites. Artificial environmental effects of a persistent or transient nature must be considered including nuclear effects from detonations or space borne sources. One of the more significant effects is the contamination arising from rocket plume exhaust and from outgassing or vaporization of spacecraft materials, particularly hydrocarbons; such effects are especially important for optical surfaces where performance can be altered by only very thin layers [57, 67]. A second major area where artificial environmental effects are important occurs in vacuum systems, especially solar simulators, where frost, such as CO_2, and hydrocarbon films deposit and build up on cold surfaces [111].

The environment which affects coatings most seriously is solar radiation, particularly UV; especially susceptible are the metal oxide paints. The solar spectrum lies within 0.3 to 4.0 μm having approximately 1% beyond each of these limits [156]. The primary effect of IR radiation is heating or thermal agitation of the coating molecules; in general, there is insufficient energy per quantum to break ordinary chemical bonds or initiate chemical reactions. However, UV and soft X-ray components have sufficient quantum energy to initiate reactions in organic coatings, and such reactions proceed at higher rates resulting from increased temperatures caused by IR absorption. The influence of high photon energy radiation on metallic systems is negligible.

Considerable effort has been devoted to determining the effect of solar irradiation, particularly with appreciable energies in the lower UV region, and the combination of UV irradiation *and* vacuum on thermal control coatings, both organic and inorganic. In the early 1960's, the first approach was to separately expose samples to solar irradiation, vacuum, or other effects and then measure the reflectance under normal ambient conditions. Much of this information was found to be in poor agreement with actual flight data and felt to be unreliable. It was found [115] that the damage occurring to the irradiated samples had healed when the samples were returned to the air environment. This follows from a simple point of view of the damage mechanism where the effect of the irradiation (solar, electron, or proton) is the migration of oxygen ions from the coating causing the bleaching effect and, upon return to the atmosphere, the oxygen is replaced and the reflectance returns to the pre-exposed value. This effect led to *in situ* procedures in which all the measurements and exposure testing is done under hard vacuum conditions; the curves 112, 144, and 115 of Fig. 124*A*(2), page 311 of the *Numerical Data* section, illustrate this effect.

Particle radiation has a similar adverse effect on coatings causing permanent reflectance change in a similar manner to UV irradiation. The better coatings can withstand fluences of the order of 10^{15} to 10^{16} e/cm^2 at energies of 145 keV. Specular surfaces and leafing aluminum–silicone coatings are, in general, relatively resistant to reflectance degradation for energies less than 50 keV. Diffuse coatings, or paints, excepting leafing aluminum, will suffer severe in-air recoverable degradation in the infrared region and substantial visible region reflectance losses which are less recoverable or bleachable upon re-exposure to air. The wavelength region and extent of damage is highly dependent upon the binder; coatings with methyl silicone binders sustain greatest reflectance degradation in the IR, while coatings with potassium silicate binders suffer largest electron-induced reflectance degradation in the visible. Tests on various coatings have shown that irradiation rate effects from 50 keV electrons are evident over a wide range of fluxes (4×10^8 to 1.7×10^{12} e/cm^2 sec) and fluences (10^{13} to 8×10^{15} e/cm^2). Damage by electron exposure at low temperatures (77 K) is generally less severe than at room temperature. The combination of UV or solar and electron damage is generally more severe than the sum of the damage caused by the individual factors [19].

Exposure to proton sources, primarily the Van Allen and solar-wind space environment, can substantially degrade reflectance of coatings for fluences greater than $3 \times 10^{15} p/cm^2$ and energies 3–468 keV. Damage has been found in some instances to be greater at lower temperatures 77 K than at room temperature. The sequential combination of proton and UV irradiation is only slightly more damaging than UV alone and the UV tends to bleach proton irradiation damage.

As a consequence of the need to study thermal control coatings *in situ* and to determine their degradation due to the various environmental effects, the experimental apparatus to obtain useful data for the thermal designer is quite complex. Such apparatus referred to as combined effects facilities allow for sequential or combined exposure to usually UV, proton, electron, or perhaps other effects and measurement of the spectral reflectance. Since temperature of the sample during exposure is an important damage parameter, provisions are made to cool or otherwise control sample temperature within the environmental chamber. For solar degradation studies, the most useful wavelength range for reflectance would be 0.28 to 4.0 μm; in practice this range can be covered by integrating sphere reflectometers similar to the methods described in Section 9.B.b. for which practical sphere wall coatings like magnesium oxide and barium sulfate limit the reflectance range from 0.28 to 2.5 μm. Typical facilities are described in References [20, 159, 179]. For the reader interested in general study on damage mechanisms and degradation, the handbook of Reference [19] will provide an excellent overview and the symposium series on space thermophysics problems will provide a comprehensive coverage of the most important developments.

In the search for stable diffuse thermal control coatings which satisfy simple application requirements, zinc oxide systems have been given the most serious study. One of the earliest formulations, designated as S-13, consisted of a high-purity zinc oxide (New Jersey Zinc Co. SP500) in a dimethyl silicone binder (GE, RTV-602). UV degradation, measured in laboratory or flight data, was more extensive than desired. An improved formulation more resistant to UV irradiation in a vacuum environment was developed; designated as S-13G it consisted of potassium silicate protected encapsulated zinc oxide in RTV-602 silicone binder. The most stable of the zinc oxide systems is designated as Z93; the degradation of three zinc oxide coatings is illustrated on Fig. 24A(2), page 311 of the *Numerical Data* section. The work to develop these coatings requires detailed attention to preparation procedures and Reference [178] summarizes these experiences, which have spanned nearly eight years.

For further information on the thermal control problem, the reader is encouraged to consult Reference [19] and the AIAA Symposium series cited in the Introduction.

9. METHODS OF MEASUREMENT

A. Basic Techniques

Many experimental techniques for measuring the thermal radiative properties of materials have been described in the literature. It is obviously impossible to describe all of the methods in detail here. Instead, the general principles involved will be discussed briefly, and a few methods will be outlined. The reader is referred to the original references for detailed descriptions of the methods.

In selecting a method for evaluating a given property, it is well to keep in mind the equations previously discussed, which can be used to compute one property from another. It frequently happens that the desired property data can be obtained more easily by computation from data obtained by direct measurement of another property, than by direct measurement of the desired property.

Methods of evaluating radiant flux fall into two general categories, calorimetric and radiometric. In calorimetric techniques the radiant flux absorbed or emitted by a sample is evaluated in terms of the heat lost or gained by the sample. In radiometric techniques, the radiant flux is measured directly. In general, calorimetric techniques tend to be relatively simple and free from systematic error, but of relatively low precision. Radiometric techniques, on the other hand, tend to require more elaborate equipment, and, while capable of relatively high precision, are subject to systematic errors that may be difficult to evaluate.

Calorimetric techniques are used to measure absorptance and emittance; they are not suitable for direct measurement of reflectance. In essence, a sample is placed in an environment where all heat transfer to or from it is by radiation, or is in a form such that it can be evaluated directly in power units, such as power input to a direct electrically heated sample, or heat loss in terms of the rate of boiling of a liquid such as water or nitrogen. If a steady-state condition is achieved, the heat input can be equated to the heat output, and the net radiant heat transfer can be related to the temperature of the sample. If the temperature of the sample is changing, the net heat transfer to or from the sample can be related to the temperature and rate of temperature change. The desired thermal radiative properties can be computed from the measured rate of radiant heat transfer, the geometry of the system, and the temperature of the sample and its surroundings.

In radiometric techniques the emitted incident and/or reflected radiant flux is measured directly and the desired property is computed as the ratio of measured fluxes. Rather elaborate optical systems may be required to collect the radiant flux from the source and focus it onto the sample for reflectance measurements and to collect the flux reflected or emitted by the specimen and focus it onto the detector for reflectance and emittance measurements. If spectral measurements are made, a monochromator must be included in the optical path. The major errors in radiometric determinations arise from flux losses in the optical system.

The physical properties of nonmetallic materials—their low thermal conductivity and high total emittance—give rise to thermal gradients which create serious problems in the measurement and use of their thermal radiative properties. These materials in general tend to be somewhat translucent and hence emit and absorb radiant energy within a surface layer of appreciable thickness. The thermal gradients tend to be normal to the emitting or absorbing surface, and frequently even are nonlinear very close to the surface. For such a condition—a large gradient across a radiating or absorbing layer of appreciable thickness—it is not possible to define a unique temperature, and hence the emittance, defined as the ratio of the flux emitted by the sample to that emitted by a blackbody radiator *at the same temperature*, is difficult to evaluate. For such a nonisothermal condition it is necessary to establish an "effective temperature," defined as that of an isothermal sample which emits at the same rate as the nonisothermal one. This is not a totally satisfying solution for several reasons, the principal one being that the resulting emittance value has significance only for the peculiar details of the test and is not a fundamental property of the material. A similar situation can exist with low-thermal-conducting coatings deposited on a substrate; if there is appreciable heat transfer through the system, the temperatures of the substrate and coating can be quite different. Because of these considerations, all measurements of emittance of nonmetallics or coated systems should be made when thermal gradients are as small as possible.

B. Reflectance Techniques [41]

The more important techniques for measuring reflectance are summarized in Table V. Only methods that measure essentially all of the reflected radiant energy are included in the table, since

Table V. Primary Reflectance Techniques

Type of reflectometer	Property	Wavelength range, μm	Temperature range, K	References	Remarks
Heated Cavity (*Hohlraum*)	$\rho(0; 2\pi)$	2–35	RT–900	56	A versatile infrared instrument; in widespread use.
Integrating Spheres					
Substitution/comparative type	$R(0; 2\pi)$	0.3–2.6	RT	61, 83	Sample mounted at sphere surface; technique used in many commercial spectrophotometers; may be sensitive to sample texture.
Absolute	$\rho(\theta; 2\pi)$	0.2–2.6	RT	43	Sample mounted at sphere center; suitable for sample of arbitrary reflection distribution function.
Laser source	$R(0; 2\pi)$	0.63, 1.15	RT–2500	97	Laser used as conventional source.
Integrating Mirrors					
Coblentz hemisphere	$\rho(\theta; 2\pi)$	1–15	RT–600	29, 86	Sample and detector or source are located at conjugate focal points of a hemispherical mirror; aberrations are a serious source of error.
Paraboloidal	$\rho(\theta; 2\pi)$	2–100	RT	36, 125	Suitable for samples of arbitrary reflection distribution function.
Ellipsoidal	$\rho(\theta; 2\pi)$	1–15	RT–400	40	Aberrations are reduced by using an ellipsoidal mirror with true focal points instead of a hemispherical mirror.
Multiple Reflection					
Strong technique	$\rho(0; 0)$	0.3–35	RT	54, 161	For specular reflectors only.
Bennett–Koehler modification	$\rho(0; 0)$	0.2–35	RT	7	Errors minimized by unique optical design, accuracy state of the art ± 0.001 unit.

methods that measure only a portion of the reflected radiant energy have only limited application. Goniometric reflectance methods, which measure the geometric distribution of the reflected radiant energy are described in a separate section.

Five separate techniques are described in the following sections; they are: (a) heated cavity reflectometers, (b) integrating sphere reflectometers, (c) integrating mirror reflectometers, (d) specular reflectometers, and (e) gonioreflectometers.

a. Heated Cavity Reflectometers

The Gier–Dunkle reflectometer [56] is sketched in Figs. 14A and B. The sample forms part of the wall of a blackbody cavity and is irradiated over a

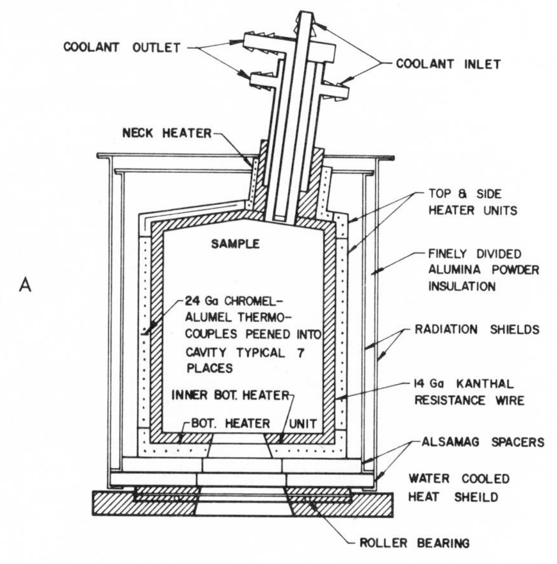

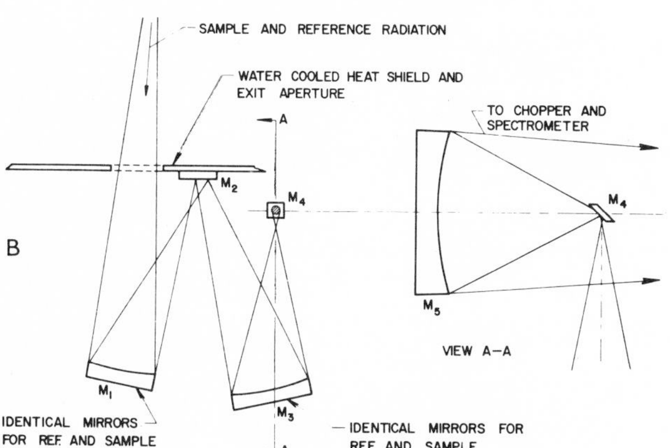

Fig. 14. Schematic of the heated cavity (Hohlraum) reflectometer—the cavity and specimen arrangement. (A) Cavity and sample arrangement. (B) Auxiliary optics—sample and reference beam viewing.

hemisphere by blackbody radiation from the hot cavity walls. Images of the sample and a spot on the cavity wall are alternately focused by the optical system onto the entrance slit of the monochromator. The output of the monochromator is the ratio of the radiances in the two beams, which is a measure of the hemispherical directional reflectance factor $R(2\pi;\theta',\omega')$ of the specimen.

The sample is usually viewed at an angle of about 4° from the normal, so that the specular component is included in the measured value. If desired, the sample can be viewed normally, so that the specular component is not included in the measured value.

The double-beam monochromator frequently used with the hohlraum reflectometer is a prism instrument with a thermocouple detector that covers the wavelength range of 1 to 35 μm. The wavelength range is set primarily by the energy available for measurement and the sensitivity of the detector.

Because $R(2\pi;\theta',\omega') = \rho(\theta,\omega;2\pi)$, the results are frequently reported as the directional hemispherical reflectance and are reported as such in the data compilation of this volume, with a footnote to the effect that $R(2\pi;\theta',\omega')$ was actually measured.

There are many sources of error in heated cavity reflectance measurements which have been discussed in detail in the literature [36, 78, 160]. The principal sources of error arise due to nonuniform irradiance of the sample over the hemisphere as a result of the aperture and thermal gradients in the cavity, emission from the sample due to heating, and failure of the radiance of the spot viewed on the cavity wall to represent the average radiance of the cavity walls, again due to thermal gradients in the cavity.

The hohlraum reflectometer irradiates the sample over nearly a complete hemisphere with nearly uniform radiance, hence the measured values are nearly independent of the geometric distribution of flux reflected by the sample. The instrument can be used to measure both specularly and diffusely reflecting samples. It is not recommended that measurements be made on heated samples since it is not possible to separate the radiant energy reflected and emitted by the sample.

Several versions of the heated cavity reflectometer have been built and the reader is referred to the literature for details. One major advance [37] has been to place the water-cooled specimen at the center of the cavity, away from the walls, so that the angle of viewing can be varied from normal to near grazing. This modification has the further advantage that cooling of the cavity walls by the cooling water for the specimen is virtually eliminated, hence thermal gradients within the cavity are usually much smaller than for the type of cavity shown in Fig. 14A.

b. Integrating Sphere Reflectometers

Integrating sphere reflectometers are used in greater numbers than any other type, and many different commercial instruments of this type are available. The reflectometer uses an integrating sphere to collect the radiant energy reflected by a sample. The reflected flux is distributed uniformly over the surface of the sphere, where it can be sampled by a detector. The sphere has an inside surface that has a high uniform reflectance and is a near-perfect diffuse reflector. There are several apertures in the sphere, one for the incident beam, one for the detector, and one for the sample. There may also be one for the comparison standard.

The theory of the integrating sphere is based on two fundamental laws of radiation: (1) the flux received by an elemental area from a point source is inversely proportional to the square of the distance from the source and directly proportional to the cosine of the angle between the normal to the surface and the direction of incidence, and (2) the flux reflected by a perfect diffuser follows the cosine distribution law, which means that the flux per unit solid angle reflected from a unit surface area in a given direction is proportional to the cosine of the angle between the normal to the surface and the given direction. Thus, for a sphere having an inner surface of uniform perfectly diffuse reflectance, the flux reflected by an area on the sphere wall is uniformly distributed over the surface of the sphere.

Integrating sphere instruments can be operated in the direct or indirect mode. In the direct mode the sample is irradiated directly, and the detector views an area on the sphere wall. A similar reading is then taken on a comparison standard of known reflectance under the same conditions. The measured directional-hemispherical reflectance factor $R(\theta, \omega; 2\pi)$ is then the ratio of the sample reading to the standard reading times the known reflectance of the standard.

In the indirect mode the beam from the source is first incident on the sphere wall, and the sample is irradiated uniformly over the hemisphere. The detector views the sample directly. A similar reading is then taken on a comparison standard of known reflectance under the same conditions. The measured hemispherical-directional reflectance factor $R(2\pi; \theta', \omega')$ is then the ratio of the sample reading to the standard reading times the known reflectance of the standard. The measured reflectance factors $R(\theta, \omega; 2\pi)$ and $R(2\pi; \theta', \omega')$ are each equal to the directional hemispherical reflectance $\rho(\theta, \omega; 2\pi)$ and are frequently reported as such.

The sample may be removed and replaced by the standard (substitution method), or there may be separate sample and standard apertures which are alternately irradiated or viewed (comparison method). Practically all double-beam recording instruments use the comparison method. In this method the sample and standard apertures should be, and usually are, located symmetrically with respect to the entrance aperture, the detector aperture, and the area of sphere wall viewed by the detector or irradiated by the source.

Errors which may be significant for some specimens can be avoided if there is an internal shield in the sphere which prevents the radiant energy reflected by this specimen, in direct-mode operation, from falling directly onto the area of sphere wall viewed by the detector, or, in indirect-mode operation, that prevents the radiant energy reflected by the sphere wall on the first reflection from falling on the specimen.

Integrating sphere reflectometers frequently include a monochromator, which usually is located in the incident beam between the source and the sphere in instruments operated in the direct mode, and between the sphere and the detector in instruments operated in the indirect mode.

The wavelength range over which an integrating sphere reflectometer can be used is determined by the reflectance characteristics of the sphere walls. For the most commonly used wall coatings, magnesium oxide and barium sulfate [119], the spectral range is about 0.25 to 2.5 μm. There are many reports [3, 118] in the literature on the techniques of preparing smoked magnesium oxide, and its reflectance properties. A recent report [64] indicates that high-purity barium sulfate may be superior to smoked magnesium oxide as a sphere coating. Other materials that have been used at longer wavelengths include flowers of sulfur [102], sprayed sodium chloride [97], and coated roughened glass [39]. Such materials have not found extensive use in integrating sphere reflectometers, but have been used in flux-averaging spheres in the infrared [38].

Comparison standards of reflectance are required for most integrating sphere reflectometers. The working standard that is used routinely is usually a piece of white vitrolite glass or white porcelain enamel that has been calibrated against freshly smoked magnesium oxide, MgO [158]. The measured relative reflectance (actually the reflectance factor) may be reported relative to MgO or as "absolute" reflectance, which is the reflectance relative to MgO times the reflectance of MgO. In many cases, the values have been reported simply as reflectance, with no further designation. A second standard frequently used is a block of magnesium carbonate, $MgCO_3$, the surface of which is scraped from time to time to provide a fresh layer of material. Both MgO and $MgCO_3$ have reflectances greater than 0.95 over most of the wavelength range where they are used. Hence, for most engineering purposes, no serious error is introduced if the reflectance of the reference standard is considered to be 1.0 instead of 0.95 to 0.99, as it will be in the visible and near infrared. However, the reflectance of MgO and $MgCO_3$ is low in the ultraviolet, and significant errors can be introduced if it is taken as 1.0 in the wavelength range of 0.25 to 0.4 μm.

Goebel *et al.* [60] have shown how to obtain reflectance of a coating directly through the use of an auxiliary sphere with an integrating sphere reflectometer. Kneissl and Richmond [97] have shown that, in an integrating sphere in which the area viewed by the detector is shielded from the specimen, the ratio of the signal when the specimen is irradiated to that when an area on the sphere wall not in the area viewed by the detector and not shielded from that area is irradiated is the absolute reflectance of the specimen, except for small errors due to flux losses out the apertures in the sphere.

Errors in integrating sphere reflectometers are due to (1) flux losses out the apertures of the sphere, (2) direct impingement of radiant energy reflected by the specimen or standard onto the area viewed by the detector, (3) direct impingement of radiant energy reflected by the sample onto the detector, and (4) variations in the reflectance of the sphere coating and deviations of the sphere coating from a perfectly diffuse reflector. Error 1 can be minimized by reducing the ratio of the total area of the apertures to the area of the sphere wall. Errors 2 and 3 can be virtually eliminated by use of suitable radiation shields in the sphere. Error 4 can be reduced by care in the choice and application of the sphere coating. For well designed spheres, carefully maintained, the errors should not exceed 2 or 3 percent. If the sample and comparison standard have the same geometric distribution of reflected radiant energy, the flux losses cancel out, and there is no error. The error for any particular sample is a function of the geometric distribution of reflected radiant energy for the sample.

The integrating sphere developed by Edwards *et al.* [43] is unusual in that it is designed to measure directional-hemispherical reflectance at angles of incidence from 0° (normal) to 80° from the normal. This instrument is illustrated in Figs. 15A and B. The specimen is irradiated with monochromatic radiant energy through port *E*, and the detector is located at port *C*. For absolute reflectance measurements the sphere wall, of coated magnesium oxide, is used as the reference. This is accomplished by translation and rotation of the sphere so that the incident beam misses the sample and strikes the sphere wall as shown in Fig. 15C. For relative measurements, a calibrated standard is attached to the back of the specimen holder as shown in the figures.

A recent development is the use of a laser-source integrating sphere reflectometer [96, 97] to measure directional-hemispherical reflectance of samples at temperatures up to 2500 K. The instrument is used at wavelengths of 0.6328, 1.15, and 3.39 μm. The high-intensity narrow-bandwidth chopped source, a spike filter in front of the detector that transmits only at the laser wavelength, and synchronous amplification of the signal from the detector combine to produce a good signal-to-noise ratio in the presence of the large amount of radiant energy emitted by the hot specimen.

While not an integrating sphere reflectometer, the technique of McNicholas [114] and Martin [116] gives data equivalent to those obtained with such an instrument. In this type reflectometer, the specimen is located at the center of curvature of a hemispherical source and is viewed directionally through a small hole in the hemispherical source. Hemispherical illumination of the sample is achieved by distributed light sources about the exterior of the translucent hemisphere.

c. *Integrating Mirror Reflectometers*

An alternative to the integrating sphere is to replace the sphere with an integrating mirror. There are both advantages and disadvantages to this approach. Mirrors in general have high reflectance in the infrared and are much more efficient than integrating spheres, hence are highly advantageous in the infrared where energy is low. On the other hand, it is physically impossible to collect the radiant energy, reflected by a sample into a hemisphere, into a parallel beam of small cross section. For this

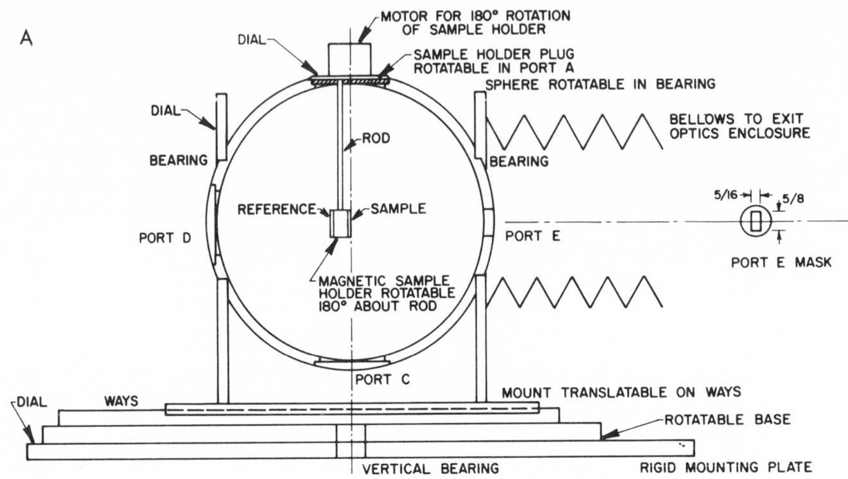

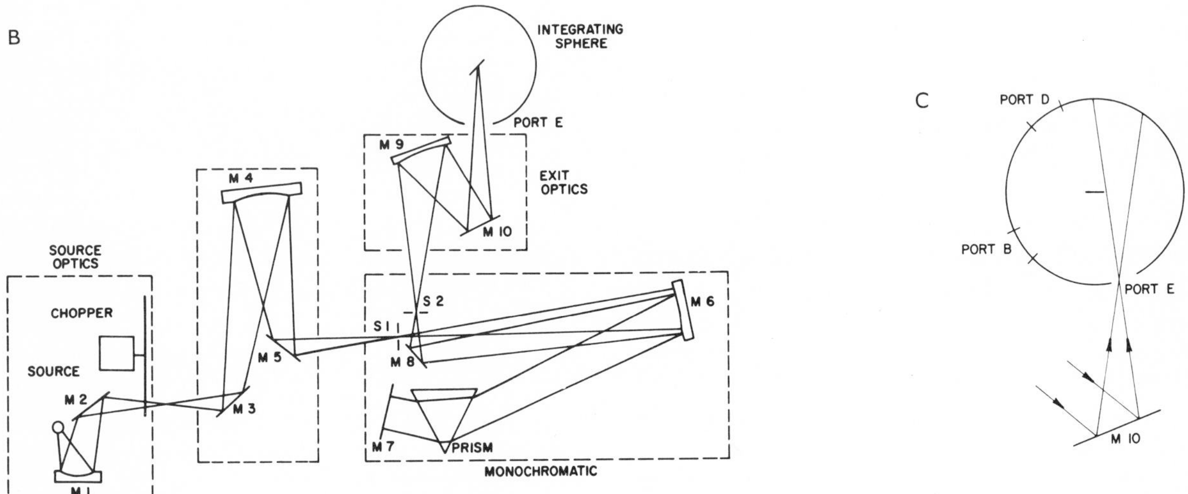

Fig. 15. Integrating sphere reflectometer, after Edwards *et al.* [43]. (A) The sphere. (B) Auxiliary optics. (C) Top view in the reference position for absolute reflectance measurements.

reason, an integrating mirror reflectometer requires either a large area detector, a detector that is equally responsive to energy incident on it from different directions within a large solid angle, or a flux-averaging device [39] in front of the detector.

Three types of integrating mirrors have been used: hemispherical, ellipsoidal, and paraboloidal.*

(i) *Hemispherical Mirror Reflectometers.* The hemispherical or Coblentz [29] mirror reflectometer was one of the earliest reflectance techniques developed. A diagram of the instrument is shown in Fig. 16. The operation of hemispherical mirror reflectometers is based on the fact that radiant energy leaving a point source in the plane of the edge of the hemisphere will be focused into a small image surrounding a second point in the same plane, diametrically opposite the source, and at the same distance as the source from the center of the hemisphere [17]. These two points are called conjugate focal points. The size of the image is determined by spherical aberrations, which for any hemisphere increase with the distance from the source to the center.

In the Coblenz instrument the sample is centered

*This discussion does not include the arc-image-type apparatus [128, 176] which employ extensive integrating mirrors. Such apparatus is most useful for measurements on nonmetallic materials and for elevated temperatures in the 2500 to 3700 K range.

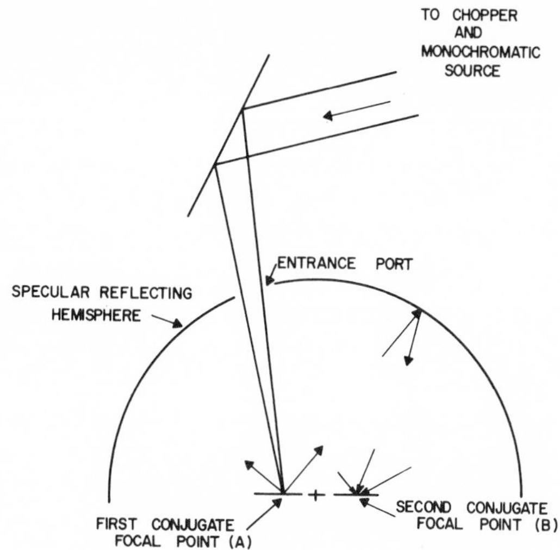

Fig. 16. The Coblentz hemisphere.

on one conjugate focal point, and the detector is centered on the other. The two conjugate focal points selected are close to the center, and thus to each other. The sample is irradiated directionally through a small hole in the hemisphere. An enlarged image of the irradiated area of the sample is focused onto the detector. The detector can be moved to the first focal point to measure the incident flux directly, or a standard of known reflectance can be measured to evaluate the incident flux. The ratio of the reflected flux to the measured incident flux, corrected for system losses, is the directional hemispherical reflectance $\rho(\theta; 2\pi)$. When a standard of known reflectance is used to evaluate the incident flux, the instrument measures the normal hemispherical reflectance factor $R(\theta; 2\pi)$.

Major losses in the system are the absorptance of the mirror, flux lost out the entrance aperture, atmospheric absorption, and flux that misses the detector due to aberrations. Additional errors are due to failure of the detector to sense equally flux incident from different angles (angular sensitivity) and on different portions of the sensitive area (area sensitivity).

Janssen and Torborg [86] modified the original Coblentz instrument as illustrated in Fig. 17. A diffuser at one conjugate focal point is irradiated through a hole in the mirror, and the sample at the second conjugate focal point is irradiated hemispherically by the reflected flux from the diffuser. The specimen is viewed directionally near normal through a second hole in the mirror. This modification permits heating of the sample without affecting the detector, which could not be done with the original Coblentz instrument. A comparison standard is required to evaluate the incident flux. The instrument operated over the wavelength range of 0.4 to 20 μm and at sample temperatures up to 300 C.

White [173] designed an automatic recording double-beam instrument employing a hemispherical reflector. A Nichrome wire source is located at the first focal point, and the sample at the second. The source and specimen are viewed through a single

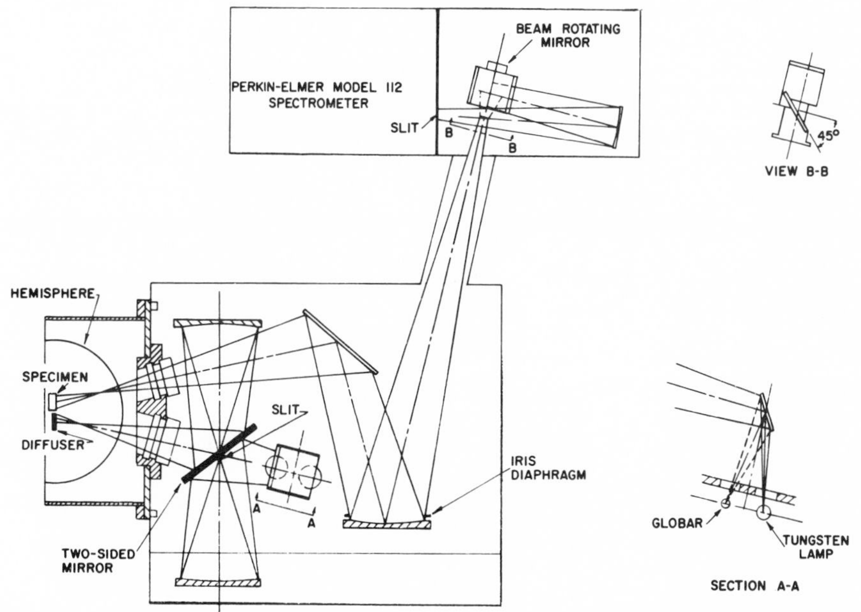

Fig. 17. The Janssen–Torborg [86] hemispherical reflectometer.

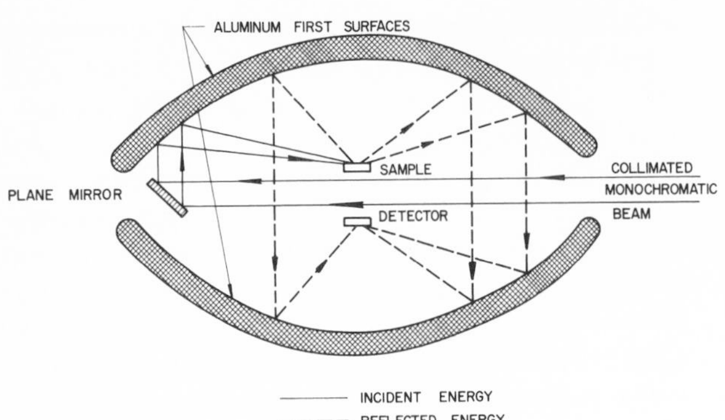

Fig. 18. The paraboloid reflectometer.

large hole in the hemisphere, by optical systems that focus images on the entrance slits for the two beams of a double-beam spectrometer. The wavelength range is 2.5 to 22.5 μm. A unique internal chopper is provided, so that measurements can be made on heated specimens at temperatures up to 1000 C. Measurements have also been made on specimens at temperatures as low as -196 C [91].

Birkebak and Hartnett [14] extended the hemisphere to about three-fourths of a sphere, which permitted the sample and detector to be tilted so that the reflectance could be evaluated for different angles of incidence. Kozyrev and Vershinin [99] used baffles within the hemisphere to measure the flux reflected into preselected solid angles. Other variations of the hemispherical mirror instrument are described in references [33, 149].

(ii) *Paraboloidal Mirror Reflectometers.* The paraboloidal mirror reflectometer [36] is shown in Fig. 18. The basic principle of the instrument is the same as for the hemispherical mirror reflectometer. The movable mirror between the two paraboloids permits the sample to be irradiated at any desired angle. Neher and Edwards [125] used off-axis paraboloids, as shown in Fig. 19. In this case the sample is irradiated hemispherically by a large-area source, and is viewed directionally. The mirrors designated *FM* in the figure are filter mirrors that reflect only in the infrared and absorb at wavelengths below 2 μm, thus reducing errors due to stray radiant energy. The instrument operates in the 2 to 100 μm range.

(iii) *Ellipsoidal Mirror Reflectometers.* Aberrations in the optical system are reduced when the hemispherical mirror is replaced with a prolate ellipsoidal mirror, which has true focal points. If an ellipsoid of low eccentricity is cut along a plane through the major axis, one half can be used to replace the hemisphere in a hemispherical mirror instrument, with reduced aberrations in the image and improved operation. Neu [127] and Heinisch [73] have constructed instruments of this type.

If a prolate ellipsoid of appreciable eccentricity is cut in a plane normal to the major axis that contains one focal point, the smaller portion of the ellipsoid can be used in an instrument of the configuration shown in Fig. 20 [40]. In this case, the sample is located at the first focal point, and irradiated by a small chopped monochromatic beam through a hole in the mirror. An enlarged image of the irradiated area of the sample is formed at the second focal point where the detector is located. Because the image is blurred, a flux-averaging device [39],

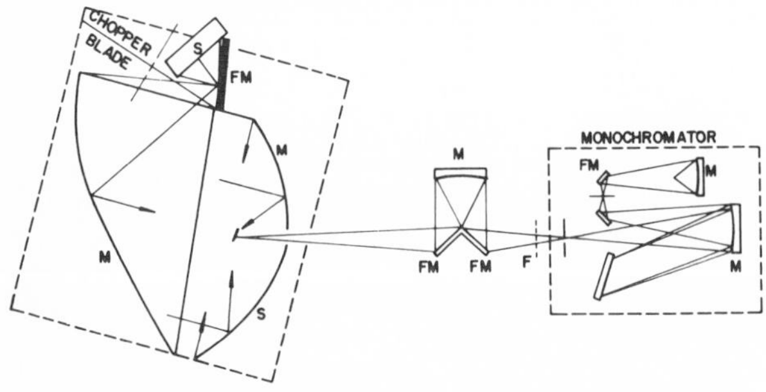

Fig. 19. Neher–Edwards [125] paraboloidal reflectometer.

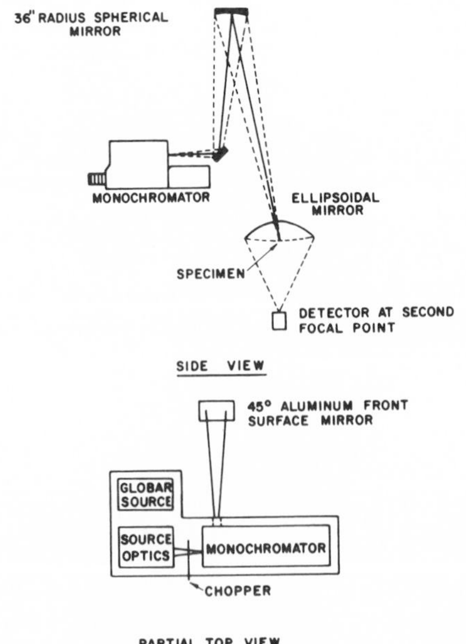

Fig. 20. Ellipsoidal mirror reflectometer.

usually an integrating sphere coated with mu sulfur or sodium chloride, is used in front of the detector. Shields can be placed in the first focal plane to confine the measured flux to that reflected by the specimen into any desired solid angle. Corrections can thus be made for all flux losses on the basis of the actual geometric distribution of reflected radiant energy from the sample being measured, and the final error can be reduced to less than one percent. A high-quality mirror of known reflectance is used as the reference standard. The useful range of the instrument used in this manner is about 0.4 to 8.0 μm with the long-wavelength limit being imposed by the low efficiency of the averaging sphere. Detector scanning devices [141] can be used to extend the wavelength range to 15 μm.

An improved version of this instrument [141] employs a large area source surrounding the second focal point, which irradiates a sample at the first focal point uniformly over a hemisphere. The specimen is viewed through a small hole in the mirror, and the reflected beam is passed through a monochromator in front of the detector. When used in this mode, satisfactory operation was attained at wavelengths out to beyond 30 μm.

The ellipsoidal mirror reflectometer measures directional-hemispherical reflectance factor $R(\theta, \omega; 2\pi)$ when used in the direct mode (detector at second focal point) and hemispherical-directional reflectance factor $R(2\pi; \theta', \omega')$ when used in the inverse mode (source at the second focal point).

d. Specular Reflectometers

For high-quality mirrors the multiple-reflection technique [47, 54, 161] is simple and accurate. Figure 21 shows a schematic of the optical path. The reflectance is computed as the square root of the ratio of two measured fluxes, hence the uncertainty in the reflectance is about half that in the measured ratio. The apparatus of Kelsall [93], using a small-diameter, highly collimated beam, achieves very high measurement sensitivity resulting from multiple reflections in the range 11 to 39. Very accurate alignment of the mirrors, and optically flat mirrors, are required to maintain the same beam geometry for the two configurations. The major source of error is due to slight displacement of the reflected beam on the detector, since for most detectors the output varies with the position of the incident beam on the detector. Flux-averaging devices in front of the detector can help to minimize this error [7].

Bennett and Koehler [7] have improved the optical system to reduce the errors due to slight shifts in the alignment of the sample mirror, or slight deviations of the sample from optical flatness. Accuracies of $\pm$ 0.001 reflectance units over the wavelength range of 0.45 to 22.5 μm have been reported [79].

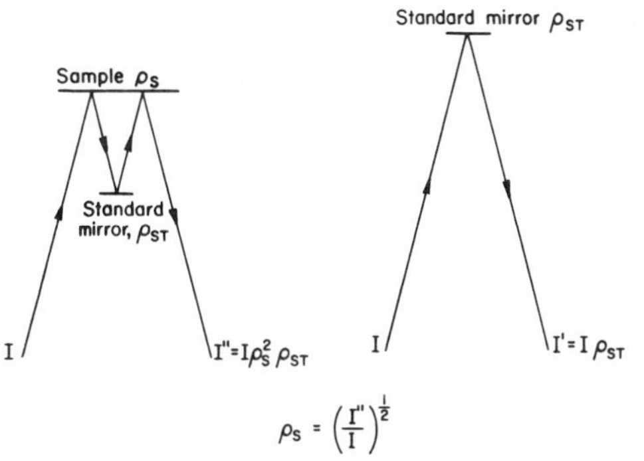

Fig. 21. The Strong [161] multiple-reflection technique.

e. Gonioreflectometers

A gonioreflectometer measures the bidirectional reflectance as a function of angle of reflectance for any given angle of incidence. Several instruments [97] have been constructed to measure bidirectional reflectance in the plane of incidence. More elaborate instruments [18, 31, 120] have been developed to measure bidirectional reflectance over the entire hemisphere. The latter instruments permit scanning over nearly the entire hemisphere for any given direction of incidence, and the angle of incidence can be varied from normal to near grazing. In addition, the measurements can be made with monochromatic incident radiant energy over the wavelength range of about 0.5 to 1.0 μm. Reference [123] describes an infrared technique for measuring the angular scattering of transmitted flux of polycrystalline materials.

C. Radiometric Emittance Techniques

Thermal emittance may be measured directly by measuring the ratio of the radiance of a sample to that of a blackbody radiator at the same temperature, and under the same geometric and wavelength conditions of viewing. The measured property is then the directional emittance of the sample, either total, if all wavelengths are measured at once with a nonselective detector, or spectral, if the measurement is confined to a narrow spectral band.

a. General Considerations

The comparison blackbody may be either an integral blackbody cavity, whose walls are formed by the sample [4, 34, 81, 95, 135, 163], or a separate blackbody controlled at the temperature of the sample [62, 140]. The integral blackbody is preferred when working at very high temperatures, where temperature measurement and control may be difficult, or at very short wavelengths, where extremely close temperature control is required for accurate measurements. The blackbody cavity should be sufficiently deep to appear "black" and yet be isothermal; these are difficult requirements to satisfy in many nonmetallic samples. The isothermal requirement can be better satisfied by a shallow cavity, which then requires a correction for deviation from an emittance of unity according to the extensive studies of Kelly and Moore *et al.* [92, 121].

The integral blackbody technique is particularly suitable when measuring samples that are heated in a vacuum or in controlled atmospheres, where they must be viewed through a window. The transmittance of the window may be changed as a result of condensation on its inner surface of material volatilized from the hot specimen. Such a change in transmittance can cause serious errors, which are compensated when both the sample and comparison blackbody are viewed through the same window. The separate blackbody technique is most accurate at temperatures below 1800 K, where temperature measurement and control by use of conventional thermocouples present no serious problems.

Spectral emittance is usually desired at all wavelengths where significant amounts of radiant energy are emitted. This range was shown as a function of temperature in Fig. 4. As a practical matter, most measurements are made in the 1 to 15 μm range, with a few being extended out to 30 or 35 μm. Measurements at wavelengths below about 1 μm are difficult because of temperature control problems; measurements at long wavelengths are difficult because of the low energy available for measurement.

The change in temperature required to change the spectral radiance of a blackbody radiator by 0.5 percent is given in Table VI for several wavelengths and temperatures. These data indicate the degree of temperature control that is required for accurate emittance measurements, particularly by the separate blackbody technique.

b. Separate Blackbody Technique

An example of the separate blackbody technique, described in reference [84], is shown in Fig. 22. The sample, in the form of a flat strip, is heated by passing a current through it and is enclosed in a water-cooled shield to reduce thermal gradients due to convective cooling. Two identical laboratory blackbodies are used, designed to have an effective emittance of 0.998 or better. One blackbody always is used as the source for the comparison beam of the spectrometer; either the hot sample or the reference blackbody can be used as the source for the sample beam of the spectrometer.

In operation, the comparison blackbody is brought to the desired temperature by manual adjustment of the power input. The reference blackbody is placed in position to serve as the source for the sample beam of the spectrometer, and its temperature is controlled to be the same as that of the comparison blackbody by means of a differential thermocouple, one junction of which is in each blackbody. The ratio of the signals from the two beams, plotted as a function of wavelength, is referred to as the "100 percent line." The reference blackbody is then replaced by the specimen furnace, and the sample is controlled at the same temperature

Table VI. Allowable Temperature Variation for a $\pm 1/2\%$ Variation of Blackbody Radiance

	Temperature differences, ΔT (K)					
T (K)	$\lambda = 0.4\ \mu$m	$\lambda = 0.5\ \mu$m	$\lambda = 0.75\ \mu$m	$\lambda = 1.0\ \mu$m	$\lambda = 5\ \mu$m	$\lambda = 15\ \mu$m
500	0.035	0.043	0.065	0.087	0.433	1.112
1000	0.139	0.174	0.261	0.348	1.640	3.215
1500	0.313	0.391	0.586	0.782	3.335	5.538
2000	0.556	0.695	1.042	1.389	5.302	7.949
2500	0.869	1.086	1.628	2.164	7.425	10.389
3000	1.251	1.564	2.343	3.102	9.639	12.848

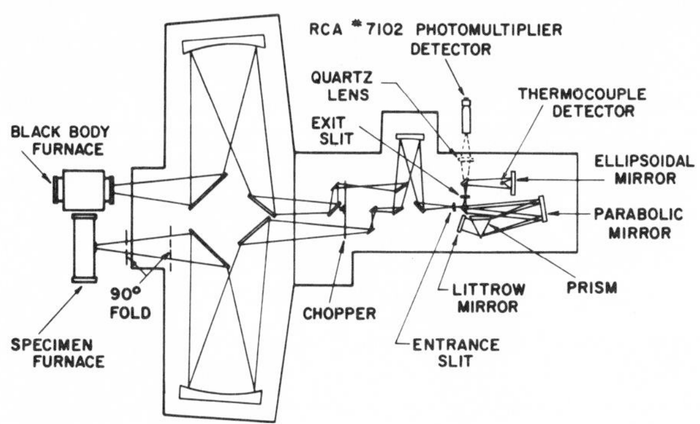

Fig. 22. Schematic of apparatus for measurement of normal spectral emittance [143].

as the comparison blackbody, again by means of a differential thermocouple, one junction of which is in the blackbody cavity, and the other being formed by separate wire welds to the sample. The ratio of the signals from the two beams, plotted as a function of wavelength, is referred to as the "sample line." The specimen beam is then blocked near the specimen furnace, and the ratio of the signals from the two beams is referred to as the "zero line." The spectral emittance at any wavelength, $\epsilon(0; \lambda)$ is then computed as

$$\epsilon(0; \lambda) = \frac{S - Z}{H - Z} \tag{97}$$

where S is the height of the sample line, H the height of the 100 percent line, and Z the height of the zero line at wavelength λ. Standards of emittance [140] have been calibrated by this technique, and are available from the Office of Standard Reference Materials, National Bureau of Standards, Washington, D.C. 20234.

c. Integral Blackbody Technique for Conductors

The integral blackbody approach is illustrated in Fig. 23. A thin-walled tubular specimen is heated by passing a current through it. The small hole in the wall of the tube approximates a blackbody at the temperature of the wall if the tube walls are thin and of uniform thickness and the hole diameter is small compared to the diameter of the tube [34]. Thermal gradients along the tube are virtually eliminated by proper adjustment of the power input to the end heaters. Larrabee [105, 106] has generated very-high-precision emittance measurements on tungsten with this approach although Kunz [104] has raised a question regarding the manner of viewing to obtain the "zero" reading.

Normal total emittance can be measured by the techniques described above if a detector is used that has essentially the same spectral response over the wave-length range of emitted radiant energy, or if the sample is a graybody emitter [108]. This technique has been used [1] to study the variation of total directional emittance with angle of viewing. Normal total emittance data are frequently computed from normal spectral data, as has been discussed previously.

d. Rotating Cylindrical Sample Technique

The alternate heating and viewing of a moving sample is an approach for reducing and controlling the thermal gradients. The sample is enclosed in a furnace where it is heated then moved in front of a cooled port where it is viewed briefly. It may also involve alternate heating by radiation and viewing of a stationary sample, where the periods of heating and viewing are controlled by rotating shutters. Movable specimens have been tested in the form of a rotating disc [28, 94, 113, 154], a rotating cylinder [26, 46], or as discs mounted on an oscillating beam [9]. Stationary specimens have been heated in an arc-image furnace [30, 45, 128, 176] and in a solar furnace [107].

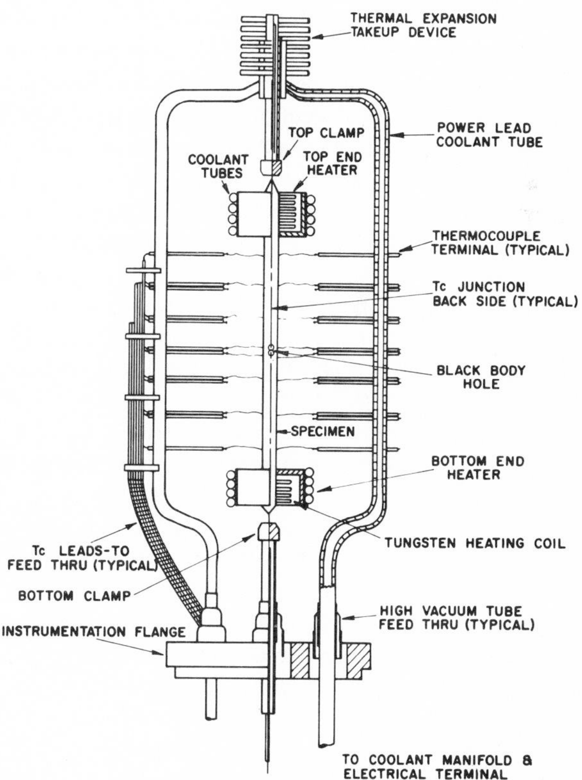

Fig. 23. Typical arrangement of integral blackbody technique for normal spectral emittance measurements, by direct electrical heating of sample.

In the case of a moving specimen that is alternately heated and cooled, the specimen will reach a steady state temperature, and if the motion is rapid enough that the viewing time is short, the temperature fluctuation of the viewed surface will be so small that it can be neglected.

Photon conduction does not generate a thermal gradient in a material. The photons pass through the material without affecting it in any way until they are absorbed. As a result, a specimen that is heated by radiation tends to have a small thermal gradient normal to the heated surface. The radiation is absorbed within the surface layer from which the emitted radiation originates, and after the specimen has reached a steady state condition, there is little phonon conduction within the surface layer. If the time periods during which the specimen is alternately heated and viewed are brief, the temperature fluctuation of the specimen will be small. Usually these periods are only a fraction of a second. With both moving specimens and radiantly heated stationary specimens the thermal gradients normal to the surface can be reduced to a point where they do not introduce a significant error into the measurement.

The rotating cylinder method used at the National Bureau of Standards [26] will be described briefly and should serve to illustrate the basis for the radiometric techniques of emittance determinations. The specimen was a hollow cylinder 25 mm (1 in.) in diameter and 25 mm (1 in.) high, with an outer surface ground to be round to about 0.05 mm (0.002 in.). A cross-sectional view of the furnace is shown in Fig. 24. The specimen was mounted on top of an alumina pedestal which revolved inside a platinum-wound resistance furnace. The winding of the furnace was designed so that there was no axial thermal gradient along the specimen. The specimen revolved in front of a water-cooled viewing port. A theoretical analysis [130], confirmed by temperature measurements at a point near the surface of a rotating specimen, indicated (1) that the temperature drop at a point on the surface of the rotating specimen as it passed in front of the viewing port was inversely proportional to the speed of rotation of the specimen, and was only about 2 K at a speed of 50 rpm, and (2) that the temperature measured by a stationary, radiation-shielded thermocouple located at the center of the hollow rotating specimen was the same, within about 1 K, as that measured by the thermocouple imbedded near the surface of the rotating specimen.

The flux from the hot specimen was focused on the entrance slit of one beam of a double-beam infrared spectrometer, and the flux from a laboratory blackbody furnace was focused on the entrance slit of the second beam of the spectrometer. The blackbody furnace was controlled to the same temperature as the specimen. The spectrometer automatically scanned over the spectral range of 1 to 15 μm, and plotted the ratio of the signals from the two beams, which after correction for spectrometer errors, was the normal spectral emittance of the specimen.

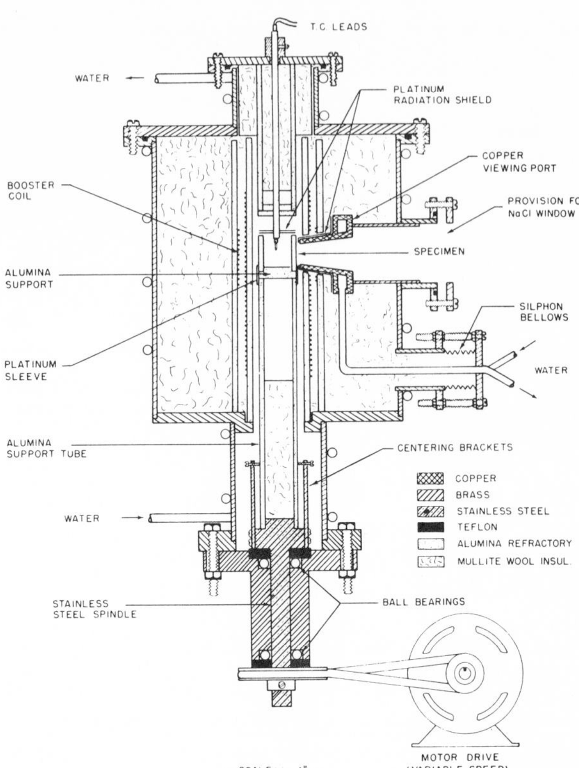

Fig. 24. Cross-sectional view of the rotating specimen furnace used in measuring normal spectral emittance of nonmetals [26].

D. Calorimetric Techniques

In calorimetric techniques the emittance or absorptance is measured in terms of the heat lost or gained by the specimen through radiant heat transfer. The ratio of the observed rate of heat transfer to the computed rate of radiant heat transfer for a blackbody surface under the same conditions gives the emittance or absorptance. The technique is only suitable for total emittance, and usually hemispherical total emittance, but can be used for absorptance for incident radiant energy of any desired spectral distribution, including narrow spectral bands.

One common factor in all calorimetric methods is that the sample is thermally isolated from its surroundings, so that essentially all heat transfer to or from it is by radiation. The rate of heat transfer is evaluated in terms of the rate of temperature change of the sample, or the power input or output to or from the specimen is in a form such that it can be accurately evaluated. This assures that when the sample reaches an equilibrium temperature, the radiant heat transfer can be equated to the heat transfer by other modes.

a. Measurements above 500 K

The simplest calorimetric technique for measuring hemispherical total emittance is the hot-filament method [177]. The sample, in the form of a long filament, is heated in vacuum by passing a current through it. If the wire is of uniform cross section and resistance, a section at the center will quickly come to a uniform equilibrium temperature. A small length of this central portion is instrumented with thermocouples, which can also serve as potential leads. After the filament has come to equilibrium, its temperature, the current flowing through it, and the potential drop across the central section are measured. The total hemispherical emittance is computed from the following equation

$$\epsilon(2\pi; t) = VI/A\sigma(T_1^4 - T_2^4) \qquad (98)$$

where $\epsilon(2\pi; t)$ is the hemispherical total emittance, V is the voltage drop across the central section, I is the current flowing in the filament, A is the surface area of the center section, σ is the Stefan–Boltzmann constant, T_1 is the temperature of the central section, and T_2 is the temperature of the surroundings. Several simplifying assumptions were made in deriving equation (98), which are discussed in the literature. The same general method has been used for metal strip samples [142] with the addition of a water-cooled chamber, and provision for expansion of the heated specimen, and even guard heaters for the ends of the sample. Figure 25 is a drawing of one such apparatus. The overall accuracy is dependent upon the magnitude of $T_1 - T_2$, and the relative magnitude of the heat losses by conduction in the atmosphere of the chamber, through thermocouple leads and the ends of the sample, and the assumptions made in deriving equation (98). With careful work, accuracy on the order of ± 2 percent is easily attained.

Several error analyses of the method [112, 126] have been published. Measurements to 1800 K are relatively simple and reliable, with the use of conventional noble metal thermocouples. Measurements have been made to 3000 K with the use of refractory metal thermocouples, with a significant loss in accuracy due to the instability of the thermocouples.

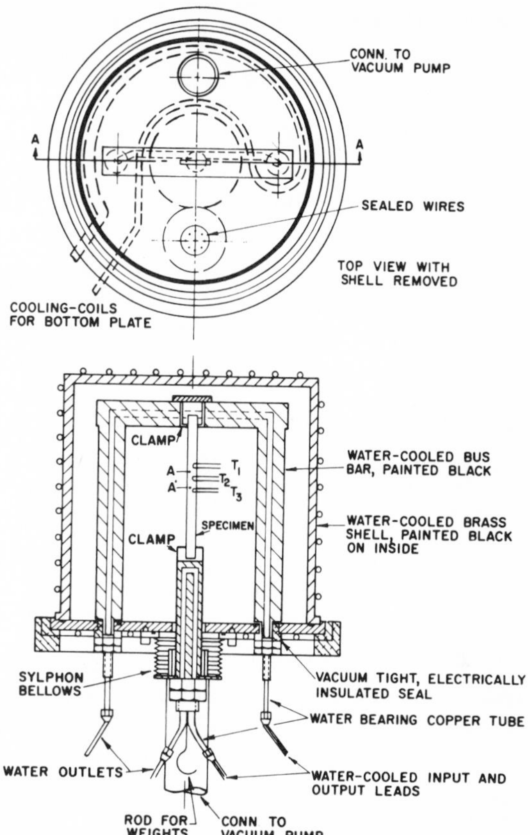

Fig. 25. Typical arrangement of calorimetric technique for hemispherical total emittance measurements in the moderate-to-high-temperature region by direct electrical heating of sample.

b. Measurements in the Range 270 *to* 500 K

The hot-filament method can be used at lower temperatures by reducing T_2. This is commonly done by liquid nitrogen cooling and extends the useful temperature range downwards to about 270 K [129]. Heat losses by conduction become a much larger fraction of the total heat loss at these temperatures, and more stringent precautions must be taken to reduce them. Also longer times are required to reach thermal equilibrium.

In this temperature range the heat input is frequently by radiation [51, 55]. The sample may be heated by a beam from a solar simulator, and the $\alpha(s)/\epsilon(2\pi; t)$ ratio computed from the equilibrium temperature attained by use of the equation [21]

$$\alpha(s)/\epsilon(2\pi; t) = A_2\sigma(T_1^4 - T_2^4)/EA_1 \qquad (99)$$

or during heating by use of the equation

$$\alpha(s) = [\epsilon(2\pi; t) A_2 \sigma(T_1^4 - T_2^4) + MC_p \, dT_1/dt]/EA_1 \tag{100}$$

where $\alpha(s)$ is the solar absorptance, $\epsilon(2\pi; t)$ is the hemispherical total emittance, A_2 is the total surface area of the sample, σ is the Stefan–Boltzmann constant, T_1 is the temperature of the sample, T_2 is the temperature of the chamber walls, E is the irradiance on the sample, A_1 is the irradiated area of the sample, M is the mass of the sample, C_p is the specific heat of the sample, and dT_1/dt is the rate of temperature change of the sample. The source is then turned off, and the sample is allowed to cool. The hemispherical total emittance can then be computed by the equation

$$\epsilon(2\pi; t) = \frac{MC_p \, dT_1/dt}{\sigma A_2(T_1^4 - T_2^4)} \tag{101}$$

Jack [82] has developed an alternate approach in which the same properties are determined from temperature phase and amplitude observations of a sample subjected to cyclic (sinusoidal) solar irradiation.

c. Measurements below 270 K

For measurements at temperatures below 270 K the chamber walls may be cooled by liquid helium [21, 87] and the $\alpha(s)/\epsilon(2\pi; t)$ ratio measured by the procedures outlined above. For steady-state measurements, the heat input may be electrical [23, 49, 50]. Figure 26 shows a helium-cooled chamber used for measurements at temperatures in the 20 to 300 K range, in which electrical heating is employed. The hemispherical total emittance can be computed from the net heat transfer rate between two parallel plates held at different temperatures. Figure 27 is one such type of apparatus [70, 71] in which two samples of the same material are used. The lower sample, 8 in. in diameter, is electrically heated and maintained at about 290 K, and the upper sample, 4 in. in diameter, is cooled by the liquid nitrogen to about 80 K. The net heat transfer rate is measured by the rate of boil-off of the liquid nitrogen from the inner dewar. The emittance, ϵ, is computed from

$$\Phi = \sigma \frac{\epsilon}{2 - \epsilon} A(T_1^4 - T_2^4) \tag{102}$$

where Φ is the measured heat-transfer rate, A is the area of the cold (upper) sample, T_1 is the temperature of the hot sample, and T_2 is the temperature of the cold sample. The measured emittance, ϵ, is an

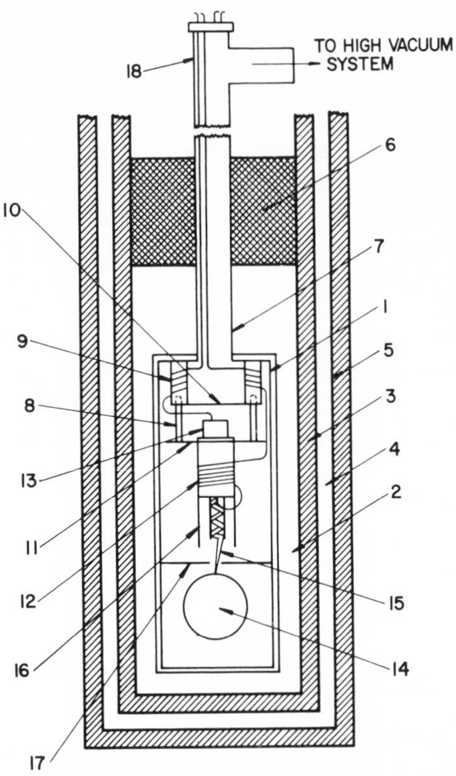

Fig. 26. Concentric sphere technique for low-temperature hemispherical total emittance [23]: 1) Sample container, 2) liquid helium space, 3) super-insulated dewar, 4) liquid nitrogen space, 5) dewar, 6) heat exchanger, 7) support tube, 8) nylon studs, 9) copper posts, 10) copper plate, 11) copper plate, 12) copper block, 13) germanium resistance thermometer, 14) sample, 15) differential thermocouple, 16) radiation shield, 17) radiation shield, 18) hermetic seal.

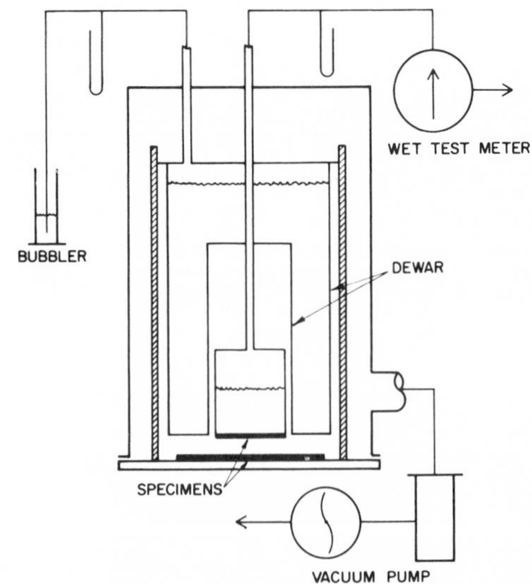

Fig. 27. Flat-plate technique for averaged hemispherical total emittance, low-temperature region [71].

average total hemispherical emittance for the sample material at the two temperatures. The factor $\epsilon/(2 - \epsilon)$ in the equation arises from the geometrical view factor between infinite parallel plates. This condition is closely approximated by the small separation between the plates, $\frac{1}{4}$ in., and the larger size of the lower plate. Caren [24] describes an instrument for the range 10 to 300 K of a similar nature.

Biondi [13] measured the spectral absorptance and reflectance of specularly reflecting samples at 4.2 K. The sample and a black absorber are thermally bonded to copper stages of appreciable thermal mass, which are in turn thermally connected to the liquid helium sink through heat leads of the proper thermal conductivity to give the stage a thermal time constant of about 10. The black absorber is positioned to receive the radiant energy specularly reflected by the sample. The stages can also be electrically heated. The sample is irradiated with a beam of monochromatic radiant energy of known irradiance and allowed to come to thermal equilibrium. The temperature of the copper stages is then measured, as well as that of the liquid helium sink. The source is then turned off, and the copper stages are heated electrically to the same temperature observed during irradiation. The radiant flux absorbed by the sample and black absorber is then equated to the electrical power required to maintain each stage at its respective temperature, and the reflectance and absorptance of the specimen is computed.

References to Text

1. Abbott, G. L., "Total Normal and Total Hemispherical Emittance of Polished Metals," in *Proc. Symp. on Measurement of Thermal Radiative Properties of Solids* (J. C. Richmond, Editor), Dayton, Ohio (1962), NASA-SP-31, 293–306, 1963.
2. Agnew, J. T. and McQuistan, "Experiments Concerning Infrared Diffuse Reflectance Standards in the Range 0.8 to 20 Microns," *J. Opt. Soc. Amer.*, **43**, 999–1007, 1953.
3. Anon., "Preparation and Reference White Reflectance Standards, ASTM Description E259-66," ASTM Standards, General Test Methods, Part 30, 803–5, May 1967.
4. Askwyth, W. H., Yahes, R. J., House, R. D., and Mikk, G., "Determination of Emissivity of Materials," Vol. I, NASA-CR-56496, 1962; Vol. II, NASA-CR-56497, 1963; Vol. III, NASA-CR-56498, 1964.
5. Beattie, J. R. and Conn, G. K. T., "Optical Constants of Metals in the Infrared–Conductivity of Silver, Copper, and Nickel," *Phil. Mag.*, **46**, 989–1001, ~~1965.~~ 1955
6. Bennett, H. E. and Bennett, J. M., "Validity of the Drude Theory for Silver, Gold, and Aluminum in the Infrared," from *Optical Properties and Electronic Structure of Metals and Alloys* (F. Abeles, Editor), North-Holland, Amsterdam, 175–88, 1966.
7. Bennett, H. E. and Koehler, W. F., "Precision Measurement of Absolute Specular Reflectance with Minimized Systematic Errors," *J. Opt. Soc. Am.*, **50**, 1–6, 1960.
8. Bennett, H. E., et al., "Verification of the Anomalous Skin Effect Theory for Silver in the Infrared," *Phys. Rev.*, **165**, 755–65, 1968.
9. Betz, H. T., Olson, O. H., Schurin, B. D., and Morris, J. C., "Determination of Emissivity and Reflectivity Data on Aircraft Structural Materials, Part II, Techniques for Measurement of Total Normal Emissivity, Normal Spectral Emissivity, Solar Absorptivity and Presentation of Results," WADC-TR-56-222, Pt. II, 184 pp., 1957.
10. Bevans, J. T. (Editor), "Thermal Design Principles of Spacecraft and Entry Bodies," from *Progr. Astron. Aeron.*, Academic Press, **21**, 855 pp., 1969.
11. Bevans, J. T. (Editor), "Thermophysics: Applications to Thermal Design of Spacecraft," from *Progr. Astron. Aeron.*, Academic Press, **23**, 580 pp., 1970.
12. Billmeyer, F. W., Jr., Lewis, D. L., and Davidson, J. G., "Goniophotometry of Pressed Barium Sulphate," *Color Engineering*, May/June, 31–6, 1971.
13. Biondi, A., "Optical Absorption of Copper and Silver at 4.2 K," *Phys. Rev.*, **102**, 964–7, 1956.
14. Birkebak, R. C. and Hartnett, J. P., "Measurements of the Total Absorptivity for Solar Radiation of Several Engineering Materials," *Trans. ASME*, **80**, 373–8, 1958.
15. Blau, H. and Fischer, H. (Editors), *Radiative Transfer from Solid Materials*, MacMillan and Co., 257 pp., 1962.
16. Born, W. and Wolf, E., *Principles of Optics*, Pergamon Press, 3rd Edition, 1965.
17. Brandenberg, W. M., "Focusing Properties of Hemispherical and Ellipsoidal Mirror Reflectometers," *J. Opt. Soc. Am.*, **54**, 1235–7, 1964.
18. Brandenberg, W. M. and Neu, J. T., "Unidirectional Reflectance of Imperfectly Diffuse Surfaces," *J. Opt. Soc. Am.*, **56**(1), 97–103, 1966.
19. Broadway, N. J., "Thermal Control Coatings," in *Radiation Effects Design Handbook*, Section 2, NASA-CR-1786, 201 pp., June 1971.
20. Brown, R. R., Fogdall, L. B., and Cannady, S. S., "Electron-Ultraviolet Radiation Effects on Thermal Control Coatings," from *Progr. Astron. Aeron.*, Academic Press, **21**, 697–724, 1969.
21. Butler, C. P. and Jenkins, R. J., "Space Chamber Emittance Measurements," in *Proc. Symp. on Measurement of Thermal Radiation Properties of Solids*, J. C. Richmond, Editor), Dayton, Ohio (1962), NASA-SP-31, 39–43, 1963.
22. Caldwell, B., "Kubelka–Munk Coefficients from Transmittance," *Opt. Soc. Am.*, **58**, 755–58, 1968.
23. Caren, R. P., "Cryogenic Emittance Measurements," in *Proc. Symp. on Measurement of Thermal Radiation Properties of Solids* (J. C. Richmond, Editor), Dayton, Ohio (1962), NASA-SP-31, 45–7, 1963.
24. Caren, R. P., "Low-Temperature Emittance Determinations," in *Progr. Astron. Aeron.*, Academic Press, **18**, 61–73, 1966.
25. Carmer, D. C. and Bair, M., "Some Polarization Characteristics of Magnesium Oxide and Other Diffuse Reflectors," *Appl. Optics*, **8**(8), 1597–1605, 1969.
26. Clark, H. E. and Moore, D. G., "A Rotating Cylinder Method for Measuring Normal Spectral Emittance of Ceramic Oxide Specimens from 1200 to 1600 K," *J. Res. Natl. Bur. Stand.*, **70A**(5), 393–415, 1966.
27. Clauss, F. T. (Editor), *First Symposium—Surface Effects on Spacecraft Materials*, Wiley and Sons, 404 pp., 1960.
28. Clayton, W. A., "A 500 to 4500 F Thermal Radiation Test Facility for Transparent Materials," in *Proc. Symp. on Measurement of Thermal Radiation Properties of Solids* (J. C. Richmond, Editor), Dayton, Ohio (1962), NASA-SP-31, 445–60, 1963.
29. Coblentz, W. W., "The Diffuse Reflecting Power of Various Substances," *Bull. Natl. Bur. Stand.*, **9**, 283–325, 1913.
30. Comstock, D. F., Jr., "A Radiation Technique for Determining the Emittance of Refractory Oxides," in *Proc.*

Symp. on Measurement of Thermal Radiation Properties of Solids (J. C. Richmond, Editor), Dayton, Ohio (1962), NASA-SP-31, 461–8, 1963.

31. Comstock, D. F., Jr., A. D. Little Rept. to Jet Propulsion Lab., Contract No. 950867, Subcontract NAS-7-100, March 1966.
32. Cravalho, E. G. and Drazen, E. L. C., "Effect of Thin Surface Films on the Radiative Properties of Metal Surfaces," in *Progr. Astron. Aeron.*, Academic Press, **23**, 363–83, 1970.
33. Derksen, W. L. and Monahan, T. I., "Automatic Recording Reflectometer for Measuring Diffuse Reflectance in the Visible and Infrared Regions," *J. Opt. Soc. Am.*, **42**, 263–5, 1952.
34. DeVos, J. S., "A New Determination of the Emissivity of Tungsten Ribbon," *Physica*, **20**, 690–714, 1954.
35. Dingle, R. B., "Anomalous Skin Effect and the Reflectivity of Metals I," *Physica*, **19**, 311–64, 1953.
36. Dunkle, R. V., "Spectral Reflectance Measurements," in *Surface Effects on Spacecraft Materials* (F. J. Clauss, Editor), Wiley and Sons, 117–37, 1960.
37. Dunkle, R. V., Edwards, D. K., Gier, J. T., Nelson, K. E., and Roddick, R. D., "Heated Cavity Reflectometer for Angular Reflectance Measurement," in *Progr. International Res. on Thermodynamic and Transport Properties*, ASME, 541–62, 1962.
38. Dunn, S. T., "Application of Sulphur Coatings to Integrating Spheres," *Appl. Opt.*, **4**(4), p. 377, 1965.
39. Dunn, S. T., "Flux Averaging Devices for the Infrared," NBS-TN-279, 44 pp., 1965.
40. Dunn, S. T., Richmond, J. C., and Wiebelt, J. A., "Ellipsoidal Mirror Reflectometer," *J. Res. Natl. Bur. Stand.*, **70C**(2), 75–88, 1966.
41. Dunn, S. T., Richmond, J. C., and Parmer, J. F., "Survey of Infrared Techniques and Computational Methods in Radiant Heat Transfer," *J. Spacecr. and Rockets*, **3**, 961–75, 1966.
42. Edwards, D. K. and deVolo, B. N., "Useful Approximations for the Spectral and Total Emissivity of Smooth Bare Metals," in *Advances in Thermophysical Properties at Extreme Temperatures and Pressures* (S. Gratch, Editor), Am. Soc. Mech. Eng., 174–88, 1965.
43. Edwards, D. K., Gier, J. T., Nelson, E. D., and Roddick, R. D., "Integrating Sphere for Imperfectly Diffuse Samples," *Appl. Opt.*, **51**, 1279–88, 1961.
44. Emslie, A. G., "Theory of Diffuse Spectral Reflectance of a Thick Layer of Absorbing and Scattering Particles," in *Progr. Astron. Aeron.*, Academic Press, **18**, 281–90, 1966.
45. Evans, R. J., Clayton, W. A., and Fries, M., "A Very Rapid 3000 F Technique for Measuring Emittance of Opaque Solid Materials," in *Proc. Symp. on Measurement of Thermal Radiation of Properties of Solids* (J. C. Richmond, Editor), Dayton, Ohio (1962), NASA-SP-31, 483–88, 1963.
46. Folweiler, R. C., "Thermal Radiation Characteristics of Transparent, Semi-Transparent and Translucent Material Under Non-Isothermal Conditions," WADC-ASD-TDR-62-719, 115 pp., 1964.
47. Fowler, P., "Far Infrared Absorptance of Gold," National Technical Information Services, U.S. Govt., 57 pp., 1960. [AD 418 456]
48. Fragstein, C. V., "On the Formulation of Kirchhoff's Law and Its Use for a Suitable Definition of Diffuse Reflection Factors," *Optik*, **12**, 60–70, 1955.
49. Fulk, M. M., Reynolds, M. M., and Park, O. E., "Thermal Radiation Absorption by Metals," in *Proc. 1954 Cryogenics Engr. Conf.*, NBS-R-3517, 151–7, 1955. [AD 125 047].
50. Fulk, M. M., Reynolds, M. M., and Park, O. E., "Thermal Radiation Absorption by Metals," in *Advances in Cryogenic Engineering*, Plenum Press, **1**, 224–9, 1960.
51. Fussell, W. B., Triolo, J. J., and Henninger, J. H., "A Dynamic Thermal Vacuum Technique for Measuring the Solar Absorption and Thermal Emittance of Spacecraft Coatings," in *Proc. Symp. on Measurement of Thermal Radiation Properties of Solids* (J. C. Richmond, Editor), Dayton, Ohio (1962), NASA-SP-31, 83–101, 1963.
52. Gannon, R. E. and Linder, B., "Effect of Surface Roughness and Porosity on Emittance of Alumina," *J. Am. Ceram. Soc.*, **47**(11), 592–3, 1964.
53. Garbundy, M., *Optical Physics*, Academic Press, 466 pp., 1965.
54. Gates, D. M., Shaw, C. C., and Beaumont, D., "Infrared Reflectance of Evaporated Metal Films," *J. Opt. Soc. Am.*, **48**(2), 88–9, 1958.
55. Gaumer, R. E. and Stewart, J. V., "Calorimetric Determination of Infrared Emittance and the α_s/ϵ Ratio," in *Proc. Symp. on Measurement of Thermal Radiation Properties of Solids* (J. C. Richmond, Editor), Dayton, Ohio (1962), NASA-SP-31, 127–33, 1963.
56. Gier, J. T., Dunkle, R. V., and Bevans, J. T., "Measurement of Absolute Spectral Reflectivity from 1.0 to 15 Microns," *J. Opt. Soc. Am.*, **44**, 558–62, 1954.
57. Gillette, R. B., "Cleaning of Contaminated Spacecraft Surfaces Using Reactive Gas Plasmas," from AIAA 6th Thermophysics Conference, Paper No. 71-463, April 1971.
58. Gilligan, J. E. and Brzuskiewicz, J., "A Theoretical and Experimental Study of Light Scattering in Thermal Control Materials," in *Progr. Astron. Aeron.*, Academic Press, **24**, 69–92, 1971.
59. Given, M. P., "Optical Properties of Metals," in *Solid State Physics, Advances in Research and Applications*, Academic Press, **6**, 313–52, 1958.
60. Goebel, D. G., Caldwell, B. P., and Hammond, H. K., III, "Use of an Auxiliary Sphere with a Spectroreflectometer to Obtain Absolute Reflectance," *J. Opt. Soc. Am.*, **56**, 783–8, 1966.
61. Goebel, D. G., "Generalized Integrating Sphere Theory," *Appl. Opt.*, **6**, 125–8, 1967.
62. Gravina, A., Bastian, R., and Dyer, J., "Instrumentation for Emittance Measurements in the 400 to 1800 F Temperature Range," in *Proc. Symp. Measurement of Thermal Radiation Properties of Solids* (J. C. Richmond, Editor), Dayton, Ohio (1962), NASA-SP-31, 329–36, 1963.
63. Greenberg, S. A., Vance, D. A., and Streed, E. R., "Low Solar Absorptance Surfaces with Controlled Emittance: A Second Generation of Thermal Control Coatings," in *Progr. Astron. Aeron.*, Academic Press, **20**, 297–309, 1967.

64. Grum, F. and Luckey, G. W., "Optical Sphere Paint and a Working Standard of Reflectance," *Appl. Opt.*, **7**(11), 2289–94, 1968.
65. Gubareff, G. G., Janssen, J. E., and Torborg, R. H., *Thermal Radiation Properties Survey*, Minneapolis-Honeywell Regulator Co., 2nd Edition, 293 pp., 1960.
66. Hamaker, H. C., "Radiation and Heat Transfer in Light Scattering Materials, "*Philips Research Repts.*, **2**, 55–67, 103–11, 112–25, 420–5, 1947.
67. Hass, G. and Hunter, W. R., "Laboratory Experiments to Study Surface Contamination and Degradation of Optical Coatings and Materials in Simulated Space Environments," *Appl. Opt.*, **9**(9), 2101–10, 1970.
68. Hass, G., Ramsey, J. B., Triolo, J. T., and Albright, H. T., "Solar Absorptance and Thermal Emittance of Aluminum Coated with Surface Films of Evaporated Aluminum Oxide," in *Progr. Astron Aeron.*, Academic Press, **18**, 47–60, 1966.
69. Hass, G. and Thun, R. E. (General Editors), "Physics of Thin Films," in *Advances in Research and Development*, Academic Press, 6 Volumes, 1963–1971.
70. Haury, G. L., "An Apparatus for the Measurement of Total Hemispherical Emissivity and Thermal Conductivity between Ambient and Liquid Nitrogen Temperatures," ASD-TDR-63-146, 16 pp., 1960. [AD 411 140]
71. Haury, G. L., "An Apparatus for Measuring Total Hemispherical Emittance between Ambient and Liquid Nitrogen Temperatures," in *Proc. Symp. on Measurement of Thermal Radiation of Properties of Solids* (J. C. Richmond, Editor), Dayton, Ohio (1962), NASA-SP-31, 51–4, 1953.
72. Heavens, O., *Optical Properties of Thin Films*, Oxford Press, 261 pp., 1955.
73. Heinisch, R. P., Bradac, F. J., and Perlick, D. B., "On the Fabrication and Evaluation of an Integrating Hemiellipsoid," *Appl. Opt.*, **9**(2), 483–9, 1970.
74. Heller, G. B. (Editor), "Electromagnetic Radiation Definitions," in *Progr. Astron. Aeron.*, Academic Press, **20**, 947–61, 1967.
75. Heller, G. B. (Editor), "Thermophysics and Temperature Control of Spacecraft and Entry Vehicles," in *Progr. Astron. Aeron.*, Academic Press, **18**, 867 pp., 1966.
76. Heller, G. B., "The Status of Thermophysics as a Multidiscipline Area in Astronautics and Aeronautics," in *Progr. Astron. Aeron.*, Academic Press, **24**, 3–25, 1970.
77. Heller, G. B. (Editor), "Thermophysics of Spacecraft and Planetary Bodies, Radiation Properties of Solids and Electromagnetic Radiation Environment in Space," in *Progr. Astron. Aeron.*, Academic Press, **20**, 975 pp., 1967.
78. Hembach, R. J., Hemmerdinger, L., and Katz, A. J., "Heated Cavity Reflectometer Modifications," in *Proc. Symp. on Measurement of Thermal Radiation Properties of Solids* (J. C. Richmond, Editor), Dayton, Ohio (1962), NASA-SP-31, 153–67, 1963.
79. Hernicz, R. S., "Design and Performance Evaluation of a High Accuracy Reflectometer for Use in the Near Ultraviolet to Far Infrared Region of the Spectrum," M.S. Thesis, School of Mechanical Engineering, Purdue University, 121 pp., June 1970.
80. Hottel, H. C. and Sarofim, A. F., *Radiative Transfer*, McGraw-Hill, 1968.
81. House, R. D., Lyons, G. J., and Askwyth, W. H., "Measurement of Spectral Normal Emittance of Materials under Simulated Spacecraft Powerplant Operation Conditions," in *Proc. Symp. Measurement of Thermal Radiation Properties of Solids* (J. C. Richmond, Editor), Dayton, Ohio (1962), NASA-SP-31, 343–55, 1963.
82. Jack, J. R., "Technique for Measuring Absorptance and Emittance by Using Cyclic Incident Radiation," *AIAA J.*, **5**(9), 1603–6, 1967.
83. Jacquez, J. A. and Kuppenheim, H. F., "Theory of the Integrating Sphere," *J. Opt. Soc. Am.*, **45**, 460–70, 1954.
84. Jakob, M., *Heat Transfer*, Academic Press, 51–2, 1949.
85. Jamieson, J. A., McFee, R. H., Plass, G. N., Grube, R. H., and Richards, R. G., *Infrared Physics and Engineering*, McGraw-Hill, 673 pp., 1963.
86. Janssen, J. E. and Torborg, R. H., "Measurement of Spectral Reflectance Using an Integrating Hemisphere," in *Proc. Symp. on Measurement of Thermal Radiation Properties of Solids* (J. C. Richmond, Editor), Dayton, Ohio (1962), NASA-SP-31, 169–82, 1963.
87. Jenkins, R. J., Butler, C. P., and Parker, W. J., "Total Hemispherical Emittance Measurements over the Temperature Range 77 to 300 K," USNRDL-TR-663, 57 pp., 1963. [AD 419 067]
88. Johnson, F. S., "The Solar Constant," *J. Meteorol.*, **11**, p. 431, 1954.
89. Judd, D. B., "Terms, Definitions and Symbols in Reflectometry," *J. Opt. Soc. Am.*, **57**(4), 445–52, 1967.
90. Katzoff, S. (Editor), *Symposium on Thermal Radiation of Solids*, San Francisco (1964), NASA-SP-55, 620 pp., 1965.
91. Keegan, H. J. and Weidner, V. R., "Infrared Spectral Reflectance of Frost," *J. Opt. Soc. Am.*, **56**(4), 523–4, 1966.
92. Kelly, F. J., "On Kirchhoff's Law and Its Generalized Application to Absorption and Emission by Cavities," *J. Res. Natl. Bur. Stand.*, **69B**(3), 165–71, 1965.
93. Kelsall, O., "Absolute Specular Reflectance Measurements of Highly Reflecting Optical Coatings at 10.6μ," *Appl. Opt.*, **9**(1), 85–90, 1970.
94. Kjelby, A. S., "Emittance Measurement Capability for Temperatures up to 3000 F," in *Proc. Symp. on Measurement of Thermal Radiation Properties of Solids* (J. C. Richmond, Editor), Dayton, Ohio (1962), NASA-SP-31, 499–503, 1963.
95. Klein, J. D., "Heat Transfer by Radiation in Powders," MIT, Ph.D. Dissertation, 1960.
96. Kneissl, G. J., Richmond J. C., and Wiebelt, J. A., "A Laser Source Integrating Sphere for the Measurement of Directional Hemispherical Reflectance at High Temperature," in *Progr. Astron. Aeron.*, Academic Press, **20**, 177–202, 1967.
97. Kneissl, G. J. and Richmond, J. C., "A Laser Source Integrating Sphere Reflectometer," NBS-TN-439, 1968.
98. Kortum, G., *Reflectance Spectroscopy*, Springer Verlag, 366 pp., 1969.
99. Kozyrev, B. P. and Vershinin, O. E., "Determination of Spectral Coefficients of Diffuse Reflection of Infrared Radiation from Blackened Surfaces," *Opt. Spectry.*, **6**, 345–50, 1959.
100. Krewinghaus, A. B., "Infrared Reflectance of Paints," *Appl. Optics*, **8**(4), 807–12, 1969.

101. Kroes, R. L., et al., "Effects of Ultraviolet Irradiation on Zinc Oxide," in *Progr. Astron. Aeron.*, Academic Press, **24**, 29–60, 1971.
102. Kronstein, M., Krauschaar, R. J., and Deacle, R. E., "Sulphur as a Standard of Reflectance in the Infrared," *J. Opt. Soc. Am.*, **53**, 458–65, 1963.
103. Kubelka, P., "New Contributions to the Optics of Intensely Light-Scattering Material," *J. Opt. Soc. Am.*, **38**(5), 448–57, 1948.
104. Kunz, H., "Prüfen technischer Strahlungspyrometere," VDI-Berichte, **112**, 37–46, 1966.
105. Larrabee, R. D., "The Spectral Emissivity and Optical Properties of Tungsten," Techn. Rept. 328, Res. Lab. of Electronics, MIT, April 1957. [AD 156 602]
106. Larrabee, R. D., "Spectral Emissivity of Tungsten," *J. Opt. Soc. Am.*, **49**, 619–25, 1959.
107. Laszio, T. S., Gannon, R. E., and Sheehan, P. J., "Emittance Measurements of Solids Above 2000 C," in *Proc. Symp. on Thermal Radiation of Solids* (S. Katzoff, Editor), San Francisco (1964), NASA-SP-55, 277–86, 1965.
108. Limperis, T., Szeles, D. M., and Wolfe, W. L., "The Measurement of Total Normal Emittance of Three Nuclear Reactor Materials," in *Proc. Symp. on Measurement of Thermal Radiative Properties of Solids* (J. C. Richmond, Editor), Dayton, Ohio (1962), NASA-SP-31, 357–64, 1963.
109. Love, T. J. and Francis, R. E., "Experimental Determination of Reflectance Function for Type 302 Stainless Steel," in *Progr. Astron. Aeron.*, Academic Press, **20**, 115–35, 1967.
110. Lucas, J. W. (Editor), "Heat Transfer and Spacecraft Thermal Control." in *Progr. Astron. Aeron.*, Academic Press, **24**, p. 427, 1971.
111. McCullough, B. A., Wood, B. E., Smith, A. M., and Birkebak, R. C., "A Vacuum Integrating Sphere for In-Situ Reflectance Measurements at 77 K from 0.5 to 1.0 μ," in *Progr. Astron. Aeron.*, Academic Press, **20**, 137–50, 1967.
112. McElroy, D. L. and Kollie, T. G., "The Total Hemispherical Emittance of Platinum, Columbium-1% Zirconium, and Polished and Oxidized INOR-8 in the Range 100 to 1200 C," in *Proc. Symp. Measurement of Thermal Radiation Properties of Solids* (J. C. Richmond, Editor), Dayton, Ohio (1962), NASA-SP-31, 365–79, 1963.
113. McMahon, H. O., "Thermal Radiation Characteristics of Some Glasses," *J. Am. Ceram. Soc.*, **34**(3), 91–6, 1951.
114. McNicholas, H. J., "Absolute Methods in Reflectometry," *J. Res. Natl. Bur. Stand.*, **1**, 29–73, 1928.
115. MacMillan, H. F., Sklensky, A. F., and McKellar, L. A., "Apparatus for Spectral Bidirectional Reflectance Measurements During Ultraviolet Irradiation in Vacuum," in *Progr. Astron. Aeron.*, Academic Press, **18**, 129–49, 1966.
116. Martin, W. E., "Hemispherical Spectral Reflectance of Solids," in *Proc. Symp. on Measurement of Thermal Radiation Properties of Solids* (J. C. Richmond, Editor), Dayton, Ohio (1962), NASA-SP-31, 183–92, 1963.
117. Meyer-Arendt, J. R., "Radiometry and Photometry; Units and Conversion Factors," *J. Appl. Opt.*, **7**(10), 2081–4, 1968.
118. Middleton, W. E. K. and Sanders, C. L., "The Absolute Spectral Diffuse Reflectance of Magnesium Oxide," *J. Opt. Soc. Am.*, **41**, 419–24, 1951.
119. Middleton, W. E. K. and Sanders, C. L., "An Improved Sphere Paint," *Illum. Engr.*, **48**, 254–6, 1953.
120. Miller, E. R. and VunKannon, R. S., "Development and Use of a Bidirectional Spectro-reflectometer," in *Progr. Astron. Aeron.*, Academic Press, **20**, 219–33, 1967.
121. Moore, D. G., "Investigation of Shallow Reference Cavities for High Temperature Emittance Measurements," in *Proc. Symp. on Measurement of Thermal Radiation Properties of Solids* (J. C. Richmond, Editor), Dayton, Ohio (1962), NASA-SP-31, 515–26, 1963.
122. Mudgett, P. S. and Richards, L. W., "Multiple Scattering Calculations for Technology," *Appl. Opt.*, **10**(7), 1485–1502, 1971.
123. Munis, R. H. and Finkel, M. W., "Goniometric Measurements of Infrared Transmitting Materials," *Appl. Opt.*, **7**(10), 2001–4, 1968.
124. National Bureau of Standards, *General Physical Constants*, NBS Special Publication No. 344, 1971.
125. Neher, R. T. and Edwards, D. K., "Far Infrared Reflectometer for Imperfectly Diffuse Specimens," *Appl. Opt.*, **4**, 775–80, 1965.
126. Nelson, K. E. and Bevans, J. T., "Errors of the Calorimetric Method of Total Emittance Measurements," in *Proc. Symp. on Measurement of Thermal Radiation Properties of Solids* (J. C. Richmond, Editor), Dayton, Ohio (1962), NASA-SP-31, 55–65, 1963.
127. Neu, J. T., "Design Fabrication and Performance of an Ellipsoidal Spectroreflectometer," NASA-CR-73193, 1968.
128. Null, M. R. and Lozier, W. W., "Measurement of Reflectance and Emittance at High Temperature With a Carbon Arc Image Furnace," in *Proc. Symp. on Measurement of Thermal Radiation Properties of Solids* (J. C. Richmond, Editor), Dayton, Ohio (1962), NASA-SP-31, 535–9, 1963.
129. Nyland, T. W., "Apparatus for the Measurement of Hemispherical Emittance from 270 to 650 K," in *Proc. Symp. on Measurement of Thermal Radiation Properties of Solids* (J. C. Richmond, Editor), Dayton, Ohio (1962), NASA-SP-31, 393–401, 1963.
130. Peavy, B. A. and Eubanks, A. G., "Periodic Heat Flow in a Hollow Cylinder Rotating in a Furnace with a Viewing Port," in *Proc. Symp. on Measurement of Thermal Radiation Properties of Solids* (J. C. Richmond, Editor), Dayton, Ohio (1962), NASA-SP-31, 553–63, 1963.
131. Pivovonsky, M. and Nagel, M. R., *Tables of Blackbody Radiation Functions*, MacMillan Co., N.Y., 1962.
132. Planck, M., *The Theory of Heat Radiation*, Dover Publications, 224 pp., 1959.
133. Plunkett, J. D., *NASA Contributions to the Technology of Inorganic Coatings*, NASA-SP-5014, 260 pp., November 1964.
134. Progar, D. J. and Wade, W. R., "Vacuum and Ultraviolet Radiation Effects on Binders and Pigments for Spacecraft Thermal Control Coatings," NASA-TN-D-6546, November 1971.
135. Reithof, T. R. and DeSantis, V. J., "Techniques of Measuring Normal Spectral Emissivity of Conductive

Refractory Compounds at High Temperatures," in *Proc. Symp. on Measurement of Thermal Radiation Properties of Solids* (J. C. Richmond, Editor), Dayton, Ohio (1962), NASA-SP-31, 565–84, 1963.

136. Reuter, G. E. H. and Sondheimer, E. H., "The Theory of the Anomalous Skin Effect in Metals," *Proc. Royal Soc.* (London), **A195**, 336–64, 1948.

137. Richmond, J. C., "Relation of Emittance to Other Optical Properties," *J. Res. Natl. Bur. Stand.*, **67C**(3), 217–26, 1963.

138. Richmond, J. C. (Editor), *Proc. Symp. on Measurement of Thermal Radiation Properties of Solids*, Dayton, Ohio (1962), NASA-SP-31, 587 pp., 1963.

139. Richmond, J. C., "Effect of Surface Roughness on Emittance of Nonmetals," *J. Opt. Soc. Am.*, **56**(2), 253–4, 1966.

140. Richmond, J. C., Dunn, S. T., DeWitt, D. P., and Hayes, W. D., Jr., "Procedures for Precise Determination of Thermal Radiation Properties," NBS-TN-267, 62 pp., 1965.

141. Richmond, J. C. and Geist, J. C., "Infrared Reflectance Measurements," NBS-R-10071, 95 pp., 1969.

142. Richmond, J. C. and Harrison, W. N., "Equipment and Procedures for Evaluation of Total Hemispherical Emittance," *Am. Ceram. Soc. Bull.*, **39**, 668–73, 1960.

143. Richmond, J. C., Harrison, W. N. and Shorten, F. J., "An Approach to Thermal Emittance Standards," in *Proc. Symp. on Measurement of Thermal Radiation Properties of Solids* (J. C. Richmond, Editor), Dayton, Ohio (1962), NASA-SP-31, 402–23, 1963.

144. Roberts, S., "Interpretation of the Optical Properties of Metal Surfaces," *Phys. Rev.*, **100**, p. 1667, 1955.

145. Ruiz-Urbieta, M., Sparrow, E. M., and Eckert, E. R. G., "Methods for Determining Film Thickness and Optical Constants of Films and Substrates," *J. Opt. Soc. Am.*, **61**(3), 351–59, 1971.

146. Ruiz-Urbieta, M., Sparrow, E. M., and Eckert, E. R. G., "Use of Parallel Polarized Radiation in Determinations of Optical Constants and Thickness of Films," *Int. J. Heat Mass Transfer*, **15**, 169–72, 1972.

147. Rutger, G. A. W., "Temperature Radiation of Solids," in *Handbuch der Physik*, **26**, 9, 1958.

148. Sagamore Ordnance Materials Research 5th Conference —Materials in Space Environment, Syracuse University Research Institute, September 1958.

149. Sanderson, J. A., "The Diffuse Spectral Reflectance of Paints in the Near Infrared," *J. Opt. Soc. Am.*, **37**, 771–7, 1947.

150. Seban, R. A., "The Emissivity of Transition Metals in the Infrared," *J. Heat Transfer*, **C87**, 173–6, 1965.

151. Seitz, F., *The Modern Theory of Solids*, McGraw-Hill, 629–42, 1940.

152. Siegel, R. and Howell, J. R., *Thermal Radiation Heat Transfer*, McGraw-Hill, 600 pp., 1971.

153. Siegel, R. and Howell, J. R., "Thermal Radiation Heat Transfer: Blackbody, Electromagnetic Theory and Materials Properties," NASA-SP-164(1), 194 pp., 1968.

154. Slemp, W. S. and Wade, W. R., "A Method for Measuring the Spectral Normal Emittance in Air of a Variety of Materials Having Stable Emittance Characteristics," in *Proc. Symp. on Measurement of Thermal Radiation Properties of Solids* (J. C. Richmond, Editor), Dayton, Ohio (1962), NASA-SP-31, 433–9, 1963.

155. Sokolov, A. V., *Optical Properties of Metals*, American Elsevier, 472 pp., 1967.

156. "Solar Electromagnetic Radiation," NASA-SP-8005, 1965; see also Olson, O. H., "Selected Ordinates for Solar Absorptivity Calculations," *Appl. Optics*, **2**(1), 109–10, 1963.

157. Sparrow, E. M. and Cess, R. D., *Radiation Heat Transfer*, Brooks/Cole Publishing Co., 1966.

158. "Standards for Checking the Calibration of Spectrophotometers (200 to 1000 μm)," NBS-Letter-Circular-1017.

159. Streed, E. R., "An Experimental Study of the Combined Space Environmental Effects on a Zinc-Oxide/Potassium-Silicate Coating," in *Progr. Astron. Aeron.*, Academic Press, **20**, 237–64, 1967.

160. Streed, E. R., McKellar, L. A., Rollings, R., Jr., and Smith, C. A., "Errors Associated with Hohlraum Radiation Characteristics Determinations," in *Proc. Symp. on Measurement of Thermal Radiation Properties of Solids* (J. C. Richmond, Editor), Dayton, Ohio (1962), NASA-SP-31, 237–52, 1963.

161. Strong, J., *Procedures in Experimental Physics*, Prentice-Hass, 376 pp., 1938.

162. Svet, D. Ya., *Thermal Radiation (Metals, Semiconductors, Ceramics, Partly Transparent Bodies, and Films)*, New York, Consultants Bureau, 98 pp., 1965.

163. Taylor, R. E., Davis, F. E., and Powell, R. W., "Direct Heating Methods for Measuring Thermal Conductivity of Solids at High Temperatures," *High Temperatures—High Pressures*, **1**, 663–73, 1969.

164. Tingwaldt, A. O. and Kunz, H., "Optische Temperaturmessung (Pyrometrie)," Band VI/4a, 47–147, Landolt-Börnstein, 1967.

165. Touloukian, Y. S. (Editor), *Thermophysical Properties of High Temperature Solid Materials*, 6 Volumes, MacMillan Co., New York, 1967.

166. Touloukian, Y. S., Gerritsen, J. K., and Moore, N. Y. (Editors), *Thermophysical Properties Research Literature Retrieval Guide*, Plenum Press, 3 Volumes, 1967.

167. Touloukian, Y. S. and DeWitt, D. P., *Thermal Radiative Properties—Metallic Elements and Alloys*, Volume 7 of *Thermophysical Properties of Matter—The TPRC Data Series*, Plenum, New York, 1644 pp., 1970.

168. Touloukian, Y. S. and DeWitt, D. P., *Thermal Radiative Properties—Nonmetallic Solids*, Volume 8 of *Thermophysical Properties of Matter—The TPRC Data Series*, Plenum, New York, 1890 pp., 1972.

169. USA Standard—Nomenclature and Definitions for Illuminating Engineering, PP-16-USAS Z7.1–1967; Revision of Z7.1–1942 UBC 653.014, 8:62132. See also Vocabulaire Intl. Eclairage, 3rd Edition, Publication CIE No. 17 (E-1.1), 1970.

170. Vasicek, A., *Optics of Thin Films*, North-Holland Publishing Co., Amsterdam, 403 pp., 1960.

171. Vincent, R. K. and Hunt, G. R., "Infrared Reflectance from Mat Surfaces," *Appl. Optics*, **7**(1), 53–9, 1968.

172. Wendlandt, W. W. and Hecht, H. G., *Reflectance Spectroscopy*, Interscience, 298 pp., 1966.

173. White, J. U., "New Method for Measuring Diffuse Reflectance in the Infrared," *J. Opt. Soc. Am.*, **54**, 1332–7, 1964.

174. Wiebelt, J. A., *Engineering Radiation Heat Transfer*, Holt, Rinehart and Winston Publishing Co., 1966.

175. Wilson, A. H., *The Theory of Metals*, Cambridge University Press, London, Second Edition, 1954.

176. Wilson, R. G., "Hemispherical Spectral Emittance of Ablation Chars, Carbon and Zirconia (to 3700 K)," in *Proc. Symp. on Thermal Radiation of Solids* (Katzoff, S., Editor), San Francisco (1964), NASA-SP-55, 259–75, 1965.

177. Worthing, A. G., "Temperature Radiation Emissivities and Emittances," in *Temperature, Its Measurement and Control in Science and Industry*, Reinholt Publishing Corp., 1164–87, 1941.

178. Zerlaut, G. A., "Investigation of Environmental Effects on Coatings for Thermal Control of Large Space Vehicles," IIT Research Institute, NASA (MSFC) Contract No. NAS8-5379, Rept. No. IITRI-06002-97, 355 pp., October 1971.

179. Zerlaut, G. A. and Courtney, W. J., "Space-Simulation Facility for In-Situ Reflectance Measurements," in *Progr. Astron. Aeron.*, Academic Press, **20**, 349–68, 1967.

180. Zerlaut, G. A., Rogers, F. O., and Noble, G., "The Development of S-13G Type Thermal Control Coatings," in *Progr. Astron. Aeron.*, Academic Press, **21**, 741–66, 1969.

Numerical Data

Data Presentation and Related General Information

1. SCOPE OF COVERAGE

Included in this volume are data on 116 material groups of pigmented coatings, 143 material groups of contact coatings, and 29 material groups of conversion coatings. Materials within each group are arranged in alphabetical order by name, as listed in the *Grouping of Materials and List of Figures and Tables* in the front of the volume. In all, this volume reports 5269 sets of data on 1161 materials, which are identified by generic and/or trade names in the *Material Index* at the end of the volume. For the user's convenience, the *Materials Index* also lists the materials contained in Volume 7 and 8.

These data were obtained by systematically examining over 879 documents on radiative properties (journals, reports, theses, etc.) dated from 1970 back to the early 1900's. To strike a compromise between completeness of coverage, broad representation of available data, and economy of time and effort, it was necessary to limit the references to those containing useful data. Only 295 references are included in this volume as data sources, and 85 percent of these have been published since 1960. The documents were collected, organized, and coded by TPRC's Scientific Documentation Division, which scans, on a continuing basis, the world literature through abstracting journals and a large number of primary technical and scientific journals.

Since the major area of interest in thermal radiative properties is on materials in their solid state, the data presented cover only the temperature range from near 0 K to the melting point of the materials presented.

The four primary properties covered in this work, as defined in an earlier section, are emittance, reflectance, absorptance, and transmittance. The suffix *ance* is used to indicate that the radiative

Table VII. Subproperty Designation

Emittance

Hemispherical total emittance
Normal total emittance
Angular total emittance

Hemispherical spectral emittance
Normal spectral emittance
Angular spectral emittance

*Reflectance**

Hemispherical integrated reflectance
Normal integrated reflectance
Angular integrated reflectance

Hemispherical spectral reflectance
Normal spectral reflectance
Angular spectral reflectance

Hemispherical solar reflectance
Normal solar reflectance
Angular solar reflectance

Absorptance

Hemispherical integrated absorptance
Normal integrated absorptance
Angular integrated absorptance

Hemispherical spectral absorptance
Normal spectral absorptance
Angular spectral absorptance

Hemispherical solar absorptance
Normal solar absorptance
Angular solar absorptance

*Transmittance**

Hemispherical integrated transmittance
Normal integrated transmittance
Angular integrated transmittance

Hemispherical spectral transmittance
Normal spectral transmittance
Angular spectral transmittance

Hemispherical solar transmittance
Normal solar transmittance
Angular solar transmittance
Absorptance-to-emittance ratio

*The geometry descriptors refer to the conditions of the incident radiant flux.

parameters are not unique properties of the bulk material, but rather depend strongly on the character of the surface. This contrasts with the use of the suffix *ivity*, which denotes the property of the ideal, optically smooth, and homogeneous material. Application of the geometric and wavelength descriptors to these four primary properties results in the 34 subproperties listed in Table VII.

In addition to the 34 subproperties, we have also included data on the change in normal spectral reflectance and the change in normal solar absorptance. These properties, usually the consequence of environmental effects such as UV and proton exposure, are the difference between the *final value* (exposed) of reflectance/absorptance and the *initial value* (pre-exposed) of reflectance/absorptance. This change is presented as a function of wavelength, UV exposure, flight time, or other pertinent variables. There are some cases, however, where it is more convenient to represent such changes on the *initial* minus the *final* value. We present these data as an increase, with a comment in the Specification Table to the effect that a positive change in reflectance/absorptance from the initial (pre-exposed) condition represents a decrease in reflectance/absorptance. This volume also includes data on the ratio of normal solar absorptance to hemispherical total emittance when such data were taken from actual satellite flights.

2. PRESENTATION OF DATA

For each material, subproperty data are separately presented in graphical and tabular formats accompanied by a table presenting details of the test conditions and specimen preparation and characterization.

The format for the presentation of the thermal radiative properties is designed specifically to supply the reader with the aspects of the properties in a comprehensive yet concise form. Each presentation consists of four sections*: Analyzed Data Graph, Original Data Plot, Specification Table, and Data Table, in that same order.

The Analyzed Data Graph presents a new and powerful approach to increasing the effectiveness of literature data. It is an evaluative review identifying and "recommending" reliable or typical data for various surface or environmental conditions. These analyzed graphs were generated to give the user an indication of trends characteristic of the material under various conditions. As such, these graphs should not be viewed as being "recommended" values in the purest sense, but only as giving characteristic behavior of the material.

The Original Data Plot is a graphical representation which presents most of the tabulated data. In overcrowded figures some of the data which are repetitive in nature are omitted. Occasionally comments on the test conditions are included to add clarity to the presentation.

The Specification Table gives the most important information concerning a set of data: the curve number for correlating the information on the Specification Table with that of the figure and Data Table, the reference number corresponding to the number given in the listed references, the year of the publication from which the data were extracted, independent variable range (temperature or wavelength), geometric descriptors (θ, θ', ω, ω'), and the error (%) reported by the author.

The "Composition (weight percent), Specification, and Remarks" column of the Specification Table provides all the available pertinent information about the specimen and test conditions. The presentation is standardized in the following order:

(1) trade name of coating if applicable
(2) generic name of coating
(3) thickness of coating
(4) substrate
(5) composition of coating (weight percent)
(6) application technique
(7) condition of coating and substrate
(8) measurement conditions important to data (environment, etc.)
(9) properties of coating
(10) other pertinent information given by author
(11) author's sample designation

In cases where a reference standard is used in the measurement—the data are measured relative to a standard—this is indicated under Item (10). An example of such a statement is: "data measured relative to MgO." No mention is made of a reference standard if the author converts the data into "absolute" values.

Following the Specification Table is the Data Table, a tabular presentation of all the property values shown or not shown on the figure and described in the Specification Table.

Table VII lists the grouping of the various subproperties that are presented in this volume. It

*In certain cases, where there exists only a small amount of data, the graphical presentation is omitted.

shows the 34 subproperties which are classified for organizational and retrieval purposes. The amount of existing data for some of these subproperties is quite small, but there are good reasons to present the data using this generalized scheme. First, the clarity of presentation is better by not grouping together data which logically are unrelated. Also, this scheme lends itself especially well to the systematic updating and expansion contemplated in this continuing work.

3. CLASSIFICATION OF MATERIALS

This volume is classified into three major sections—pigmented coatings, contact coatings, and conversion coatings. The definitions of these coatings are as follows:

Pigmented Coatings—a mixture of pigment and binder applied to a substrate. Classified alphabetically by pigment.

Contact Coatings—a layer, or layers, of a substance coated on a substrate without a chemical reaction occurring between the coating material and the substrate. Classified alphabetically by coating material itself.

Conversion Coatings—a layer of a compound, or mixture of compounds, formed by the chemical reaction of the substrate with another material. Classified alphabetically by substrate.

The material index gives an alphabetical listing by trade name, generic name, and class of all materials contained in Volume 9 as well as in Volumes 7 and 8. The pigmented coatings are listed by both pigment and binder, while the contact coatings are listed by both coating and substrate. The conversion coatings are listed by substrate and coating process.

4. SYMBOLS AND ABBREVIATIONS USED IN THE FIGURES AND TABLES

e	Electrons
ESH	Equivalent sun hours
p	Protons
PVC or P-VC	Pigment-to-volume concentration
SH	Sun hours
UV	Ultraviolet
ϵ	Emittance
ρ	Reflectance
α	Absorptance
τ	Transmittance
λ	Wavelength, μm
T	Temperature, K
M.P.	Melting Point
θ, θ'	Zenith angle, degrees
ϕ, ϕ'	Azimuthal angle, degrees
ω, ω'	Solid angle, steradians
′ (prime)	Viewing conditions
$>$	Greater than
$<$	Less than
$\approx$	Approximately
③	Curve number
[4]	Single data point number

5. CONVENTION FOR BIBLIOGRAPHIC CITATIONS

For the following types of documents the bibliographic information is cited in the sequences given below.

Journal Articles

a. Author(s)—The names and initials of all authors are given. The last name is written first, followed by initials.
b. Title of article.
c. Journal name—The abbreviated name of the journal as used in *Chemical Abstracts* is given.
d. Series, volume, and number—If the series is designated by a letter, no comma is used between the letter for series and the numeral for volume, and they are underlined together. If the series is also designated by a numeral, a comma is used between the numeral for series and the numeral for volume, and only the numeral representing volume is underlined. No comma is used between the numerals representing volume and number. The numeral for number is enclosed in parentheses.
e. Pages—The inclusive page numbers of the article are given.
f. Year—the year of publication.

Reports

a. Author(s)
b. Title of report
c. Name of the responsible organization—The name of the organization that executed the research, if different, is not given
d. Report, or bulletin, circular, technical note, etc.
e. Number
f. Part
g. Pages
h. Year
i. ASTIA's AD number—This is given in square brackets whenever available.

Books

a. Author(s)
b. Title
c. Volume
d. Edition
e. Publisher
f. Place
g. Pages
h. Year

Theses

a. Author
b. Title
c. University
d. Degree—Ph.D. (or Sc.D., M.S., etc.) Thesis
e. Pages
f. Year

6. CRYSTAL STRUCTURES, TRANSITION TEMPERATURES, AND OTHER PERTINENT PHYSICAL CONSTANTS OF THE ELEMENTS

Table XI of Volume 7 (pp. 52a–58a) contains information on the crystal structures, transition temperatures, and certain pertinent physical constants of all the elements—metallic and nonmetallic. No attempt was made to critically evaluate the temperatures and constants given in that table and they should not be considered recommended values.

7. CONVERSION FACTORS

Wavelength

1 micrometer (μm) = 10^{-6} meter (m)
= 10^{-4} centimeter (cm)
= 10^{-3} millimeter (mm)
= 10^{3} nanometers (nm)
= 10^{4} angstroms (Å)
= 3.937×10^{-5} inch
= 10^{4}/wave number (cm^{-1})
= 1.24/photon energy (eV)
= 2.998×10^{14}/frequency (cps)

Temperature

$$C = K - 273.2$$
$$F = (9/5)(K - 273.2) + 32$$
$$R = (9/5) K$$

where

C = degrees Centigrade
F = degrees Fahrenheit
R = degrees Rankine
K = kelvins

Numerical Data on Thermal Radiative Properties of Coatings

1. PIGMENTED COATINGS

A. Metallic Pigmented Coatings

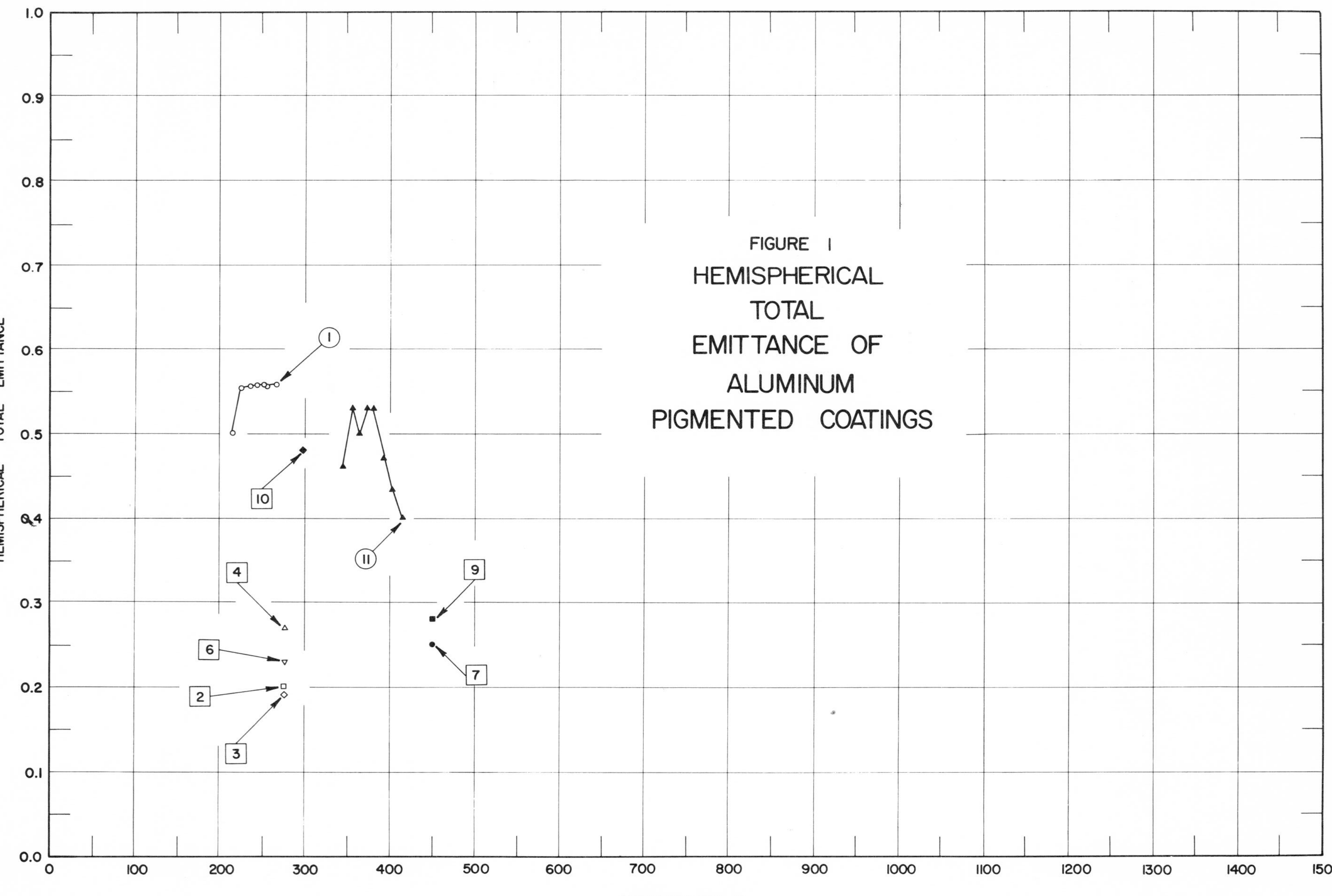
FIGURE 1
HEMISPHERICAL
TOTAL
EMITTANCE OF
ALUMINUM
PIGMENTED COATINGS
HEMISPHERICAL TOTAL EMITTANCE
TEMPERATURE, K
1.0
0.9
0.8
0.7
0.6
0.5
0.4
0.3
0.2
0.1
0.0
0
100
200
300
400
500
600
700
800
900
1000
1100
1200
1300
1400
1500
1
10
11
4
9
6
7
2
3

SPECIFICATION TABLE NO. 1 HEMISPHERICAL TOTAL EMITTANCE OF ALUMINUM PIGMENTED COATINGS

Curve No.	Ref. No.	Year	Temperature Range, K	Reported Error, %	Composition (weight percent), Specifications and Remarks
1	48	1963	216-269		Aluminum in nitrocellulose binder; aluminum alloy substrate; measured in vacuum (10^{-8} mm Hg).
2	49	1961	278	10	Leafing aluminum in silicone binder; Fuller aluminum silicone; data is avg of several measurements.
3	49	1961	278		Aluminum in silicone binder; magnesium alloy substrate. [Authors' designation: Specimen No. 1]
4	49	1961	278		Above specimen and conditions except exposed to ultraviolet radiation in vacuum (10^{-6} to 8 x 10^{-6} mm Hg) from an argon-filled AH-6 high pressure Hg-arc lamp for 80 hrs; slight cracking observed.
5*	49	1961	278		Similar to curve 3 specimen and conditions. [Authors' designation: Specimen No. 2]
6	49	1961	278		Above specimen and conditions except exposed to ultraviolet radiation in vacuum (10^{-6} to 8 x 10^{-6} mm Hg) from an argon-filled AH-6 high pressure Hg-arc lamp for 80 hrs; slight cracking observed.
7	15	1965	450		Leafing aluminum in silicone binder; Fuller aluminum silicone paint No. 171-A-152; measured in vacuum (10^{-6} mm Hg); exposed for 24 hrs at 450 K; avg of 8 specimens.
8*	15	1965	450		Leafing aluminum in silicone binder; Fuller aluminum silicone paint No. 171-A-152; weathered for 24 hrs (sprayed with distilled water for 10 min at each hr of test interval; continuous exposure to a carbon arc which simulates earth sunlight); avg of 8 specimens.
9	15	1965	450		Leafing aluminum in silicone binder; Fuller aluminum silicone paint No. 171-A-152; as received; avg of 8 specimens.
10	50	1965	~298		Aluminum in acrylic binder; LMSC nonleafing aluminum paint.
11	43	1962	346-415	±3	Aluminum powder in nitrocellulose binder (0.076 mm thick); measured in vacuum (10^{-8} mm Hg); data extracted from smooth curve.

* Not shown on plot.

DATA TABLE NO. 1 HEMISPHERICAL TOTAL EMITTANCE OF ALUMINUM PIGMENTED COATINGS

[Temperature, T, K; Emittance, ϵ]

T	ϵ	T	ϵ
CURVE 1		CURVE 11	
216	0.50	346	0.460
224	0.53	356	0.530
235	0.55	363	0.500
243	0.57	371	0.530
251	0.58	380	0.530
255	0.575	393	0.470
269	0.59	403	0.435
		415	0.400
CURVE 2			
278	0.20		
CURVE 3			
278	0.19		
CURVE 4			
278	0.27		
CURVE 5*			
278	0.20		
CURVE 6			
278	0.23		
CURVE 7			
450	0.25		
CURVE 8*			
450	0.25		
CURVE 9			
450	0.28		
CURVE 10			
298	0.48		

* Not shown on plot

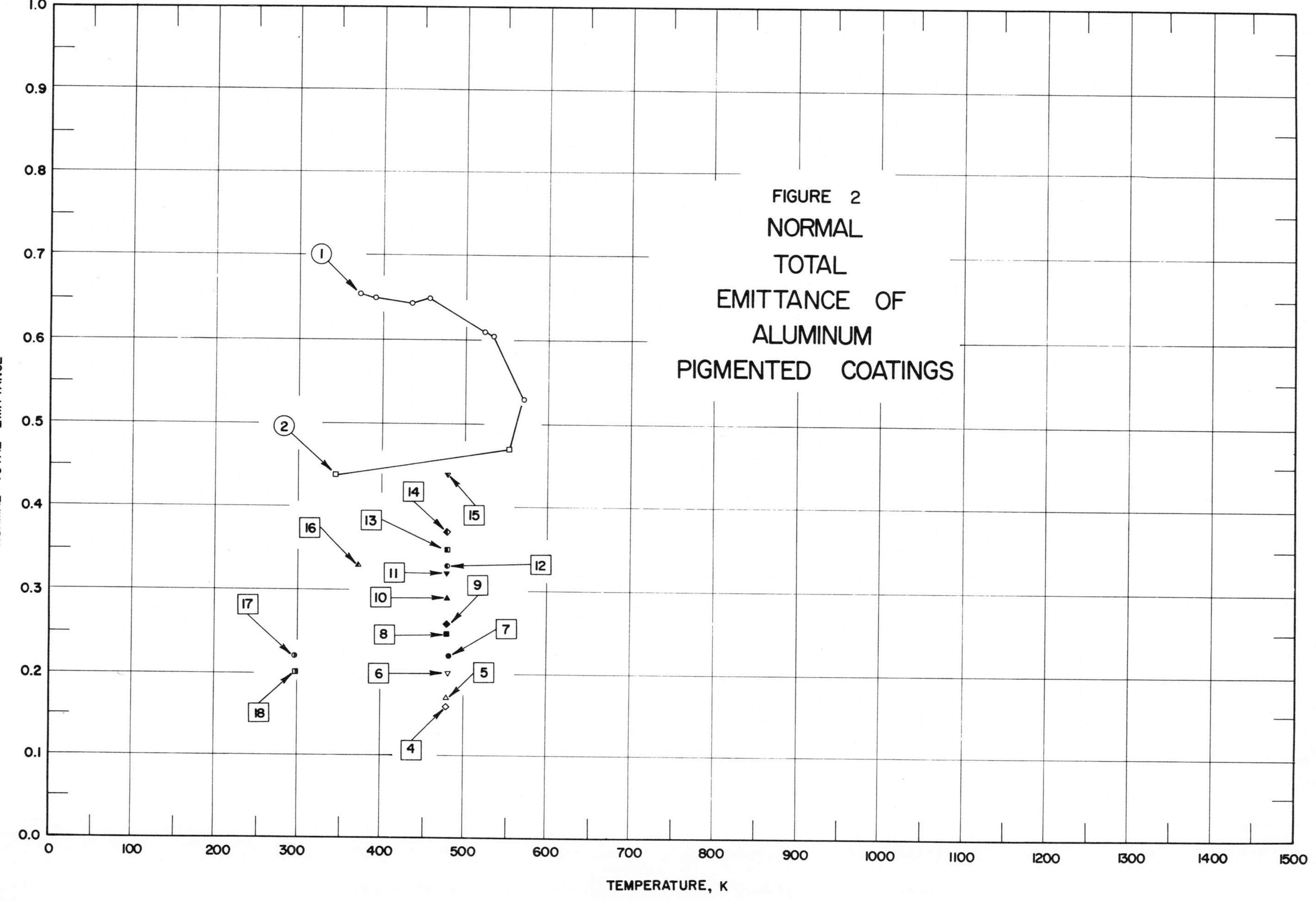

FIGURE 2 NORMAL TOTAL EMITTANCE OF ALUMINUM PIGMENTED COATINGS

SPECIFICATION TABLE NO. 2 NORMAL TOTAL EMITTANCE OF ALUMINUM PIGMENTED COATINGS

Curve No.	Ref. No.	Year	Temperature Range, K	Geometry θ'	Reported Error, %	Composition (weight percent), Specifications and Remarks
1	59	1948	372-572	$\sim 0^\circ$		Aluminum in lacquer 1234 binder; 75-ST Alclad substrate of nominal composition: 5.6 Zn, 2.5 Mg, 1.6 Cu, 0.30 Cr, Al balance; lacquer burned off at approx 536 K.
2	59	1948	572-348	$\sim 0^\circ$		Above specimen and conditions; residue of aluminum flakes on 75-ST Alclad surface; cooling.
3*	60	1954	111-593	$\sim 0^\circ$		Aluminum paint in silicone (2 coatings), Inconel substrate; measured in argon (10^{-3} mm Hg). [Author's designation: Sample 1]
4	61	1960	478	$\sim 0^\circ$		Trial Aluminum No. 18270 paint from Trial Chemical Co.; HAE anodic coating and HK-31A magnesium sand casting substrates; coating system subjected to environmental tests of heat aging, salt spray, humidity, and hydraulic oil resistance. [Authors' designation: Panel No. 446]
5	61	1960	478	$\sim 0^\circ$		Trial Aluminum No. 18270 paint from Trial Chemical Co.; Products Techniques PT-201 epoxy, HAE anodic coating, and HK-31A magnesium sand casting substrates; coating system subjected to environmental tests of heat aging, salt spray, humidity, and hydraulic oil resistance. [Authors' designation: Panel No. 445]
6	61	1960	478	$\sim 0^\circ$		Sherwin-Williams Silverbrite No. 55 paint; HAE anodic coating and HK-31A magnesium sand casting substrates; coating system subjected to environmental tests of heat aging, salt spray, humidity, and hydraulic oil resistance. [Authors' designation: Panel No. 440]
7	61	1960	478	$\sim 0^\circ$		Trial Aluminum No. 18270 paint from Trial Chemical Co.; Products Techniques PT-201 epoxy, HAE anodic coating, and HK-31A magnesium sand casting substrates; coating system subjected to environmental tests of heat aging, salt spray, humidity, and hydraulic oil resistance. [Authors' designation: Panel No. 448]
8	61	1960	478	$\sim 0^\circ$		Sherwin-Williams Silverbrite No. 55 paint; Bradley Paint Co. Silicone No. 37-3-4, Dow 17 anodic coating, and HK-31A magnesium sand casting substrates; coating system subjected to environmental tests of heat aging, salt spray, humidity, and hydraulic oil resistance. [Authors' designation: Panel No. 414]
9	61	1960	478	$\sim 0^\circ$		Sherwin-Williams Silverbrite No. 55 paint; HAE anodic coating and HK-31A magnesium sand casting substrates; coating system subjected to environmental tests of heat aging, salt spray, humidity, and hydraulic oil resistance. [Authors' designation: Panel No. 441]
10	61	1960	478	$\sim 0^\circ$		Sherwin-Williams Silverbrite No. 55 paint; Products Techniques PT-201 epoxy, Dow 17 anodic coating, and HK-31A magnesium sand casting substrates; coating system subjected to environmental tests of heat aging, salt spray, humidity, and hydraulic oil resistance. [Authors' designation: Panel No. 405]
11	61	1960	478	$\sim 0^\circ$		Sherwin-Williams Silverbrite No. 55 paint; Dow 17 anodic coating and HK-31A magnesium sand casting substrates; coating system subjected to environmental tests of heat aging, salt spray, humidity, and hydraulic oil resistance. [Authors' designation: Panel No. 408]

* Not shown on plot

SPECIFICATION TABLE NO. 2 NORMAL TOTAL EMITTANCE OF ALUMINUM PIGMENTED COATINGS (continued)

Curve No.	Ref. No.	Year	Temperature Range, K	Geometry θ'	Reported Error, %	Composition (weight percent), Specifications and Remarks
12	61	1960	478	$\sim 0°$		Sherwin-Williams Silverbrite No. 55 paint; Dow 17 anodic coating and HK-31A magnesium sand casting substrates; coating system subjected to environmental tests of heat aging, salt spray, humidity, and hydraulic oil resistance. [Authors' designation: Panel No. 400]
13	61	1960	478	$\sim 0°$		Sherwin-Williams Silverbrite No. 55 paint; Products Techniques PT-201 epoxy, Dow 17 anodic coating, and HK-31A magnesium sand casting substrates; coating system subjected to environmental tests of heat aging, salt spray, humidity, and hydraulic oil resistance. [Authors' designation: Panel No. 410]
14	61	1960	478	$\sim 0°$		Sherwin-Williams Silverbrite No. 55 paint; Products Techniques PT-201 epoxy, HAE anodic coating, and HK-31A magnesium sand casting substrates; coating system subjected to environmental tests of heat aging, salt spray, humidity, and hydraulic oil resistance. [Authors' designation: Panel No. 445]
15	61	1960	478	$\sim 0°$		Sherwin-Williams Silverbrite No. 55 paint; Products Techniques PT-201 epoxy, Dow 17 anodic coating, and HK-31A magnesium sand casting substrates; coating system subjected to environmental tests of heat aging, salt spray, humidity, and hydraulic oil resistance. [Authors' designation: Panel No. 406]
16	7	1961	373	$\sim 0°$		Aluminum (leafing pigment) in silicone binder; measured in vacuum ($\leq 10^{-5}$ mm Hg).
17	19	1965	298	$0°$		Al leafing paint, Al substrate; reported error ± 0.02 unit.
18	19	1965	298	$0°$		Above specimen and conditions except computed from 1-R (2π, $0°$).

DATA TABLE NO. 2 NORMAL TOTAL EMITTANCE OF ALUMINUM PIGMENTED COATINGS

[Temperature, T, K; Emittance, ϵ]

T	ϵ
CURVE 1	
372	0.655
390	0.650
434	0.645
454	0.650
521	0.610
532	0.605
572	0.530
CURVE 2	
572	0.530*
554	0.470
348	0.440
CURVE 3*	
209	0.322
362	0.415
531	0.301
590	0.262
209	0.247
111	0.239
427	0.292
583	0.267
113	0.261
593	0.271
113	0.230
CURVE 4	
478	0.16
CURVE 5	
478	0.17
CURVE 6	
478	0.20
CURVE 7	
478	0.22
CURVE 8	
478	0.25
CURVE 9	
478	0.26
CURVE 10	
478	0.29
CURVE 11	
478	0.32
CURVE 12	
478	0.33
CURVE 13	
478	0.35
CURVE 14	
478	0.37
CURVE 15	
478	0.44
CURVE 16	
373	0.33
CURVE 17	
298	0.22
CURVE 18	
298	0.20

* Not shown on plot

SPECIFICATION TABLE NO. 3 ANGULAR INTEGRATED REFLECTANCE OF ALUMINUM PIGMENTED COATINGS

Curve No.	Ref. No.	Year	Temperature, K	Angular Range, °	Geometry θ	θ'	ω'	Reported Error, %	Composition (weight percent), Specifications, and Remarks
1*	70	1953	~298	15-75	θ	θ			Leafing aluminum (3XD polished powder) in medium oil-length binder; mineral spirits solvent, non-volatile content 50%; pigment concentration 1.5 lbs gal^{-1}; spray application; data measured with a Hunter Hazemeter; $\theta = \theta'$.
2*	70	1953	~298	15-75	θ	θ			Leafing aluminum (3A paste) in medium oil-length binder; mineral spirits solvent, non-volatile content 50%; pigment concentration 2 lbs gal^{-1}; spray application; data measured with a Hunter Hazemeter; $\theta = \theta'$.
3*	70	1953	~298	15-75	θ	θ			Leafing aluminum (# 5 paste) in medium oil-length binder; mineral spirits solvent, non-volatile content 50%; pigment concentration 2 lbs gal^{-1}; spray application; data measured with a Hunter Hazemeter; $\theta = \theta'$.
4*	70	1953	~298	15-75	θ	θ			Leafing aluminum (5XD polished powder) in medium oil-length binder; mineral spirits solvent, non-volatile content 50%; pigment concentration 1.5 lbs gal^{-1}; spray application; data measured with a Hunter Hazemeter; $\theta = \theta'$.
5*	70	1953	~298	15-75	θ	θ			Leafing aluminum (# 10 paste) in medium oil-length binder; mineral spirits solvent, non-volatile content 50%; pigment concentration 1 lb gal^{-1}; spray application; data measured with a Hunter Hazemeter; $\theta = \theta'$.
6*	70	1953	~298	15-75	θ	θ			Leafing aluminum (# 20 paste) in medium oil-length binder; mineral spirits solvent, non-volatile content 50%; pigment concentration 1 lb gal^{-1}; spray application; data measured with a Hunter Hazemeter; $\theta = \theta'$.
7*	70	1953	~298	15-75	θ	θ			Leafing aluminum (# 30 paste) in medium oil-length binder; mineral spirits solvent, non-volatile content 50%; pigment concentration 1 lb gal^{-1}; spray application; data measured with a Hunter Hazemeter; $\theta = \theta'$.
8*	70	1953	~298	15-75	θ	θ			Leafing aluminum (30XD polished powder) in medium oil-length binder; mineral spirits solvent, non-volatile content 50%; pigment concentration 0.75 lbs gal^{-1}; spray application; data measured with a Hunter Hazemeter; $\theta = \theta'$.
9*	70	1953	~298	15-75	θ	θ			Leafing aluminum (# 30A paste) in medium oil-length binder; mineral spirits solvent, non-volatile content 50%; pigment concentration 1 lb gal^{-1}; spray application; data measured with a Hunter Hazemeter; $\theta = \theta'$.
10*	70	1953	~298	15-75	θ	θ			Leafing aluminum (# 37 paste) in medium oil-length binder; mineral spirits solvent, non-volatile content 50%; pigment concentration 1 lb gal^{-1}; spray application; data measured with a Hunter Hazemeter; $\theta = \theta'$.
11*	70	1953	~298	15-75	θ	θ			Leafing aluminum (# 40 paste) in medium oil-length binder; mineral spirits solvent, non-volatile content 50%; pigment concentration 0.75 lbs gal^{-1}; spray application; data measured with a Hunter Hazemeter; $\theta = \theta'$.
12*	70	1953	~298	15-75	θ	θ			Leafing aluminum (# 42 paste) in medium oil-length binder; mineral spirits solvent, non-volatile content 50%; pigment concentration 0.75 lbs gal^{-1}; spray application; data measured with a Hunter Hazemeter; $\theta = \theta'$.
13*	70	1953	~298	15-75	θ	θ			Leafing aluminum (# 47 paste) in medium oil-length binder; mineral spirits solvent, non-volatile content 50%; pigment concentration 0.75 lbs gal^{-1}; spray application; data measured with a Hunter Hazemeter; $\theta = \theta'$.

* No plot given

SPECIFICATION TABLE NO. 3 ANGULAR INTEGRATED REFLECTANCE OF ALUMINUM PIGMENTED COATINGS (continued)

Curve No.	Ref. No.	Year	Temperature, K	Angular Range, °	Geometry θ	Geometry θ'	Geometry ω'	Reported Error, %	Composition (weight percent), Specifications, and Remarks
14*	70	1953	~298	15-75	θ	θ			Leafing aluminum (4XD polished powder) in medium oil-length binder; mineral spirits solvent, non-volatile content 50%; pigment concentration 0.5 lbs gal^{-1}; spray application; data measured with a Hunter Hazemeter; $\theta = \theta'$.
15*	70	1953	~298	15-75	θ	θ			Leafing aluminum (# 50 paste) in medium oil-length binder; mineral spirit solvent, non-volatile content 50%; pigment concentration 0.75 lbs gal^{-1}; spray application; data measured with a Hunter Hazemeter; $\theta = \theta'$.

DATA TABLE NO. 3 ANGULAR INTEGRATED REFLECTANCE OF ALUMINUM PIGMENTED COATINGS

[Angle, *°; Reflectance, ρ; Temperature, T, K]

θ	ρ	θ	ρ	θ	ρ
CURVE 1*		CURVE 6*		CURVE 11*	
T ~ 298		T ~ 298		T ~ 298	
15	0.020	15	0.0490	15	0.055
30	0.0230	30	0.057	30	0.066
45	0.0275	45	0.0740	45	0.090
60	0.0385	60	0.116	60	0.145
75	0.0640	75	0.230	75	0.283
CURVE 2*		CURVE 7*		CURVE 12*	
T ~ 298		T ~ 298		T ~ 298	
15	0.0185	15	0.0660	15	0.066
30	0.0210	30	0.077	30	0.080
45	0.0255	45	0.1020	45	0.107
60	0.0375	60	0.154	60	0.168
75	0.0715	75	0.285	75	0.311
CURVE 3*		CURVE 8*		CURVE 13*	
T ~ 298		T ~ 298		T ~ 298	
15	0.0190	15	0.0480	15	0.075
30	0.0215	30	0.055	30	0.090
45	0.0260	45	0.0710	45	0.122
60	0.0375	60	0.110	60	0.185
75	0.070	75	0.200	75	0.321
CURVE 4*		CURVE 9*		CURVE 14*	
T ~ 298		T ~ 298		T ~ 298	
15	0.0230	15	0.0520	15	0.158
30	0.0260	30	0.060	30	0.185
45	0.0315	45	0.079	45	0.237
60	0.0445	60	0.121	60	0.325
75	0.0755	75	0.240	75	0.466
CURVE 5*		CURVE 10*		CURVE 15*	
T ~ 298		T ~ 298		T ~ 298	
15	0.0510	15	0.092	15	0.076
30	0.060	30	0.109	30	0.090
45	0.0780	45	0.143	45	0.123
60	0.120	60	0.247	60	0.190
75	0.230	75	0.357	75	0.335

* No plot given

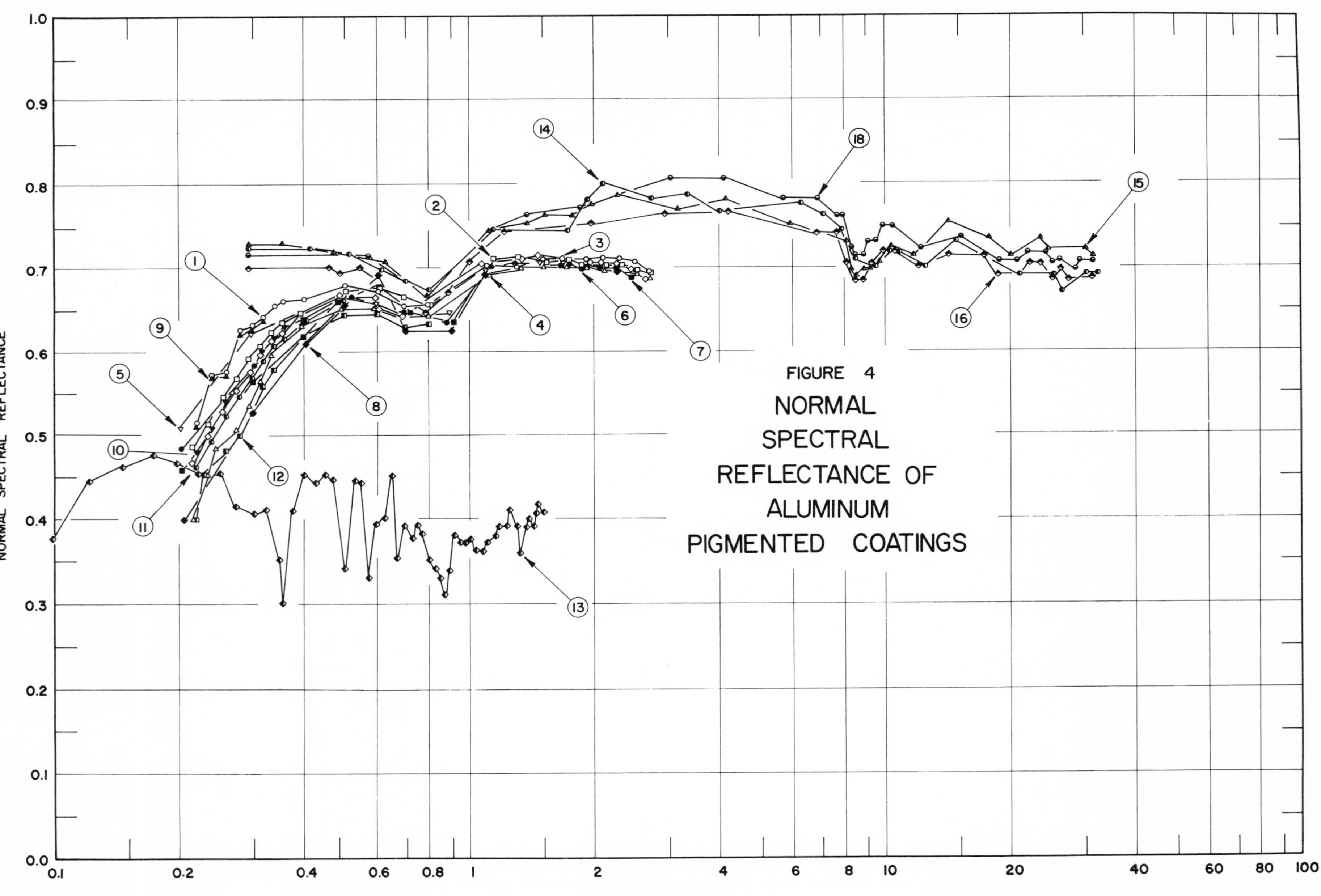
FIGURE 4
NORMAL
SPECTRAL
REFLECTANCE OF
ALUMINUM
PIGMENTED COATINGS
NORMAL SPECTRAL REFLECTANCE
WAVELENGTH, μm
1.0
0.9
0.8
0.7
0.6
0.5
0.4
0.3
0.2
0.1
0.0
0.1
0.2
0.4
0.6
0.8
1
2
4
6
8
10
20
40
60
80
100
1
2
3
4
5
6
7
8
9
10
11
12
13
14
15
16
18

Curve No.	Ref. No.	Year	Temperature, K	Wavelength Range, μm	Geometry θ	θ'	ω'	Reported Error, %	Composition (weight percent), Specifications, and Remarks
1	1	1960	358	0.221-2.706	$\sim 0^\circ$		2π		Leafing aluminum flake in silicone binder (0.0508 mm thick) on anodized aluminum alloy 24S-T substrate; measured relative to MgO.
2	1	1960	358	0.216-2.740	$\sim 0^\circ$		2π		Above specimen and conditions except exposed in vacuum (10^{-5} mm Hg) to UV radiation from a G.E. UA-3 lamp for 20 hrs.
3	1	1960	358	0.216-2.686	$\sim 0^\circ$		2π		Curve 1 specimen and conditions except exposed in vacuum (10^{-5} mm Hg) to UV radiation from a G.E. UA-3 lamp for 42 hrs.
4	1	1960	358	0.217-2.712	$\sim 0^\circ$		2π		Curve 1 specimen and conditions except exposed in vacuum (10^{-5} mm Hg) to UV radiation from a G.E. UA-3 lamp for 100 hrs.
5	7	1961	298	0.202-2.696	$\sim 0^\circ$		2π		Leafing aluminum flake in silicone resin binder (0.0508 mm thick) on anodized aluminum alloy 24S-T substrate; measured relative to MgO.
6	7	1961	298	0.202-2.719	$\sim 0^\circ$		2π		Above specimen and conditions except exposed in vacuum (10^{-5} mm Hg) to UV radiation from a G.E. UA-3 lamp for 20 hrs.
7	7	1961	298	0.202-2.675	$\sim 0^\circ$		2π		Curve 5 specimen and conditions except exposed in vacuum (10^{-5} mm Hg) to UV radiation from a G.E. UA-3 lamp for 42 hrs.
8	7	1961	298	0.206-2.694	$\sim 0^\circ$		2π		Curve 5 specimen and conditions except exposed in vacuum (10^{-5} mm Hg) to UV radiation from a G.E. UA-3 lamp for 100 hrs.
9	7	1961	298	0.220-0.800	$\sim 0^\circ$		2π		Curve 5 specimen and conditions.
10	7	1961	298	0.220-0.805	$\sim 0^\circ$		2π		Curve 6 specimen and conditions.
11	7	1961	298	0.220-0.800	$\sim 0^\circ$		2π		Curve 7 specimen and conditions.
12	7	1961	298	0.219-0.804	$\sim 0^\circ$		2π		Curve 8 specimen and conditions.
13	62	1949	$\sim$298	1.00-14.99	$\sim 0^\circ$		2π	5	Dutch Boy quick drying enamel (aluminum paint); data extracted from smooth curve; converted from R (2π, $\sim 0^\circ$).
14	72	1963	310	0.295-32.2	$\sim 0^\circ$		2π		W.P. Fuller 172-A-1 silicone-aluminum filled paint; aluminum in silicone binder ($\sim$0.0838 mm thick) on 2024 aluminum substrate; data extracted from smooth curve; property measured in air soon after sample preparation; property is representative value for 2 samples.
15	72	1963	310	0.295-31.7	$\sim 0^\circ$		2π		Similar to above specimen and conditions except sample stored in dry chamber for several days, then in nitrogen for several days; property is avg value of 3 samples.
16	72	1969	310	0.295-31.8	$\sim 0^\circ$		2π		W.P. Fuller 172-A-1 silicone-aluminum filled paint; aluminum in silicone binder ($\sim$0.0838 mm thick) on 2024 aluminum substrate; exposed to gamma radiation (dose $\sim 1.7 \times 10^{10}$ ergs gm^{-1} C^{-1}) in vacuum ($\sim 10^{-6}$ mm Hg) maintained by diffusion pump; sample stored in nitrogen for several days before measuring property in air. [Author's designation: Sample No. 1]
17*	72	1963	310	0.295-31.8	$\sim 0^\circ$		2π		Similar to above specimen and conditions. [Author's designation: Sample No. 2]
18	72	1963	310	0.295-31.5	$\sim 0^\circ$		2π		Similar to above specimen and conditions. [Author's designation: Sample No. 3]

DATA TABLE NO. 4 NORMAL SPECTRAL REFLECTANCE OF ALUMINUM PIGMENTED COATINGS

[Wavelength, λ, μm; Reflectance, ρ; Temperature, T, K]

CURVE 1
T = 358

λ	ρ
0.221	0.514
0.241	0.570
0.260	0.574
0.281	0.627
0.300	0.632
0.320	0.642
0.341	0.653
0.359	0.659
0.400	0.662
0.502	0.678
0.600	0.671
0.696	0.652
0.797	0.654
1.115	0.702
1.303	0.700
1.510	0.711
1.711	0.711
1.913	0.711
2.109	0.713
2.309	0.713
2.512	0.708
2.706	0.696

CURVE 2
T = 358

λ	ρ
0.216	0.485
0.236	0.513
0.257	0.545
0.276	0.568
0.296	0.591
0.315	0.608
0.335	0.624
0.357	0.635
0.394	0.647
0.502	0.672
0.602	0.675
0.697	0.664
0.802	0.654
1.140	0.712
1.311	0.714
1.539	0.707
1.738	0.710
1.890	0.704

CURVE 2 (cont.)

λ	ρ
2.138	0.701
2.332	0.703
2.529	0.697
2.740	0.695

CURVE 3
T = 358

λ	ρ
0.216	0.466
0.235	0.498
0.255	0.527
0.276	0.553
0.297	0.575
0.315	0.596
0.335	0.614
0.356	0.623
0.399	0.646*
0.492	0.668
0.592	0.665
0.704	0.650
0.796	0.641
1.088	0.705
1.336	0.713
1.484	0.715
1.685	0.713
1.944	0.707
2.090	0.706
2.290	0.706
2.490	0.697
2.686	0.686

CURVE 4
T = 358

λ	ρ
0.217	0.399
0.234	0.452
0.256	0.484
0.276	0.506
0.296	0.534
0.315	0.563
0.334	0.583
0.355	0.598
0.395	0.632
0.495	0.650
0.597	0.650

CURVE 4 (cont.)

λ	ρ
0.697	0.642*
0.799	0.643
1.114	0.691
1.320	0.700
1.507	0.700
1.709	0.701
1.918	0.700
2.112	0.696
2.317	0.696
2.512	0.694
2.712	0.687

CURVE 5
T = 298

λ	ρ
0.202	0.508
0.299	0.624
0.398	0.647*
0.500	0.668
0.598	0.655
0.697	0.640*
0.894	0.646
1.097	0.697
1.292	0.697
1.499	0.708
1.696	0.708
1.899	0.708*
2.098	0.712*
2.300	0.712*
2.498	0.704*
2.696	0.694*

CURVE 6
T = 298

λ	ρ
0.202	0.482
0.302	0.582
0.399	0.634
0.522	0.665
0.601	0.666*
0.692	0.647
0.883	0.634
1.120	0.701
1.294	0.708
1.524	0.705*

CURVE 6 (cont.)

λ	ρ
1.721	0.705
1.873	0.699
2.113	0.702*
2.321	0.702*
2.516	0.699*
2.719	0.693*

CURVE 7
T = 298

λ	ρ
0.202	0.458
0.303	0.564
0.398	0.619
0.484	0.659
0.605	0.679
0.724	0.647
0.914	0.636
1.075	0.701*
1.320	0.710*
1.479	0.705*
1.675	0.705
1.933	0.701
2.077	0.701
2.280	0.699
2.470	0.697
2.675	0.688

CURVE 8
T = 298

λ	ρ
0.206	0.398
0.301	0.525
0.404	0.610
0.502	0.656
0.604	0.692
0.701	0.626
0.902	0.626
1.096	0.694
1.311	0.699*
1.501	0.699*
1.702	0.700*
1.903	0.697*
2.103	0.697*
2.293	0.695
2.497	0.690*
2.694	0.684*

CURVE 9
T = 298

λ	ρ
0.220	0.509
0.240	0.568
0.259	0.570
0.279	0.622
0.299	0.627
0.318	0.638
0.339	0.651
0.359	0.656
0.400	0.654
0.501	0.667
0.601	0.659
0.700	0.644
0.800	0.644

CURVE 10
T = 298

λ	ρ
0.220	0.479
0.240	0.508
0.259	0.539
0.280	0.559
0.300	0.585*
0.319	0.602
0.340	0.617
0.359	0.630
0.399	0.638
0.504	0.666*
0.604	0.666*
0.701	0.656*
0.805	0.646*

CURVE 11
T = 298

λ	ρ
0.220	0.460
0.239	0.491
0.260	0.522
0.279	0.547
0.300	0.567
0.318	0.589
0.339	0.608
0.359	0.617
0.406	0.639
0.497	0.660

CURVE 11 (cont.)

λ	ρ
0.596	0.658*
0.706	0.643*
0.800	0.635*

CURVE 12
T = 298

λ	ρ
0.219	0.399
0.239	0.450
0.259	0.480
0.279	0.498
0.299	0.526*
0.318	0.557
0.339	0.578
0.359	0.593
0.399	0.618*
0.500	0.643
0.600	0.642
0.700	0.629
0.804	0.633

CURVE 13

λ	ρ
1.00	0.376
1.23	0.446
1.48	0.461
1.75	0.475
1.98	0.466
2.24	0.451
2.52	0.451
2.75	0.414
3.00	0.405
3.26	0.410
3.49	0.350
3.54	0.300
3.77	0.410
4.00	0.452
4.25	0.441
4.53	0.451
4.77	0.445
5.01	0.340
5.26	0.444
5.50	0.441
5.75	0.330
5.99	0.392

CURVE 13 (cont.)

λ	ρ
6.22	0.401
6.49	0.450
6.75	0.351
6.99	0.390
7.25	0.376
7.50	0.391
7.74	0.380
7.97	0.350
8.25	0.341
8.49	0.330
8.74	0.309
8.98	0.339
9.24	0.378
9.51	0.370
9.76	0.370
10.00	0.373
10.25	0.361
10.50	0.362*
10.74	0.360
11.01	0.370
11.27	0.372*
11.52	0.379
11.74	0.391
12.01	0.390*
12.27	0.391
12.51	0.410
12.74	0.401*
13.00	0.390
13.26	0.359
13.50	0.369*
13.75	0.390
13.99	0.400
14.25	0.390
14.52	0.408
14.77	0.414
14.99	0.408

CURVE 14
T = 310

λ	ρ
0.295	0.724
0.416	0.724
0.514	0.719
0.617	0.699
0.700	0.684

CURVE 14 (cont.)

λ	ρ
0.800	0.672
1.13	0.748
1.71	0.748
1.93	0.782
2.10	0.800
2.75	0.783
3.34	0.787
4.00	0.768
6.22	0.779
7.13	0.764
7.91	0.746
8.64	0.710
9.51	0.700
10.2	0.719
10.8	0.719
12.5	0.700
14.9	0.732
21.0	0.690
25.5	0.690
26.8	0.670
30.2	0.691
32.2	0.691

CURVE 15
T = 310

λ	ρ
0.295	0.730
0.356	0.730
0.476	0.720
0.621	0.708
0.796	0.666
1.11	0.748
1.39	0.753
1.52	0.763
1.77	0.763
1.99	0.778
2.28	0.788
3.19	0.771
4.14	0.783
5.96	0.751
8.01	0.731
8.36	0.698
8.57	0.690
8.85	0.697
9.20	0.697

* Not shown on plot

DATA TABLE NO. 4 NORMAL SPECTRAL REFLECTANCE OF ALUMINUM PIGMENTED COATINGS (continued)

λ	ρ
CURVE 15 (cont.) T = 310	
9.48	0.704*
9.75	0.704
10.4	0.724
11.8	0.715
14.2	0.754
17.9	0.737
20.0	0.714
23.6	0.736
24.9	0.721
30.1	0.721
31.7	0.713
CURVE 16 T = 310	
0.295	0.700
0.460	0.700
0.496	0.695
0.541	0.700
0.791	0.643
0.898	0.670
1.00	0.708
1.21	0.744
1.99	0.752
2.98	0.765
4.20	0.769
6.97	0.742
7.62	0.742
8.22	0.704
8.47	0.684
8.92	0.684
9.25	0.702
9.64	0.705*
9.91	0.720
10.4	0.720
12.1	0.700
14.1	0.713
17.3	0.713
18.9	0.690
21.1	0.690*
22.3	0.702
23.6	0.702
25.1	0.688
26.6	0.698
27.7	0.686
31.8	0.688

λ	ρ
CURVE 17* T = 310	
0.295	0.700
0.460	0.700
0.496	0.695
0.541	0.700
0.791	0.643
0.898	0.670
1.00	0.708
1.21	0.744
1.99	0.752
2.98	0.765
4.20	0.769
6.97	0.742
7.62	0.742
8.22	0.704
8.47	0.684
8.92	0.684
9.25	0.702
9.64	0.705
9.91	0.720
10.4	0.720
12.1	0.700
14.1	0.713
17.3	0.713
18.9	0.690
21.1	0.690
22.3	0.702
23.6	0.702
25.1	0.688
26.6	0.698
27.7	0.686
31.8	0.688
CURVE 18 T = 310	
0.295	0.717
0.570	0.713
0.793	0.664*
1.12	0.748
1.38	0.764
1.85	0.772
3.04	0.808
4.14	0.808
5.67	0.783
6.95	0.783

λ	ρ
CURVE 18 (cont.)	
7.61	0.762
7.98	0.762
8.38	0.723
8.55	0.714
8.91	0.714
9.19	0.731
9.55	0.731
9.91	0.750
10.3	0.750
12.1	0.722
15.1	0.739
18.8	0.708
20.6	0.708
22.0	0.718
24.2	0.718
25.3	0.704
26.3	0.708
28.2	0.698
29.4	0.706
31.5	0.706

* Not shown on plot

FIGURE 5A

ANALYZED CHANGE IN NORMAL SPECTRAL REFLECTANCE OF LEAFING ALUMINUM PIGMENTED COATINGS

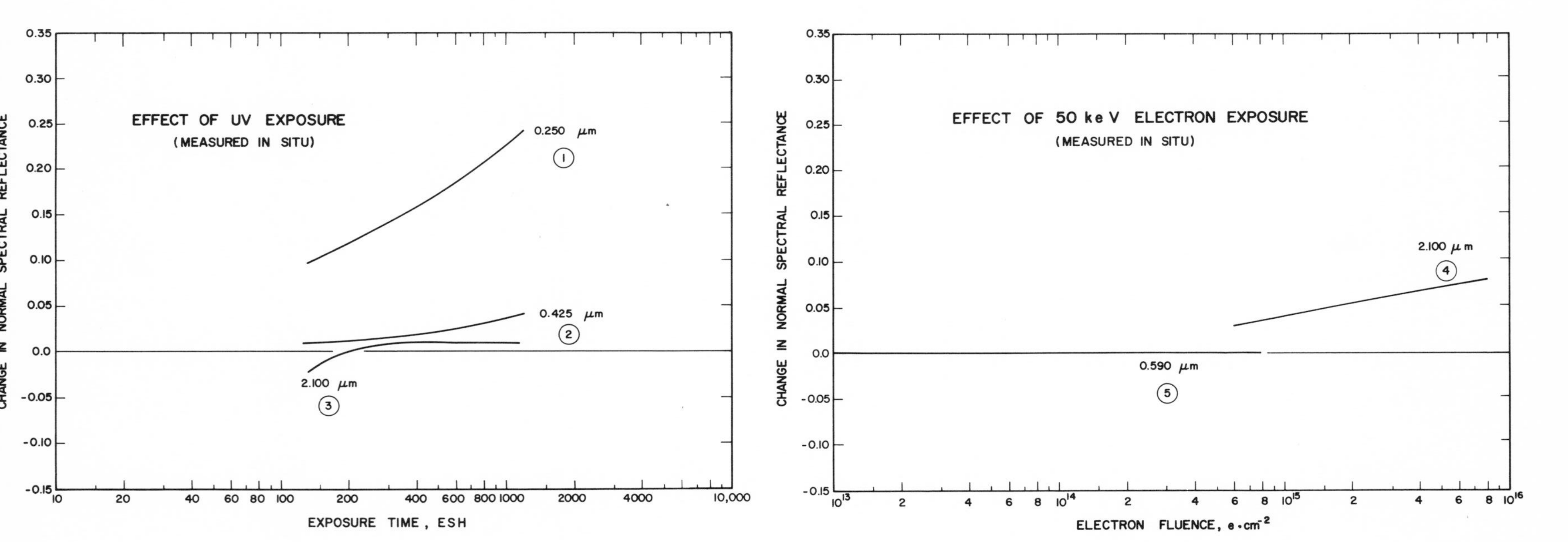

SPECIFICATION TABLE NO. 5 CHANGE IN NORMAL SPECTRAL REFLECTANCE OF ALUMINUM PIGMENTED COATINGS

Curve No.	Ref. No.	Year	Temperature, K	Wavelength Range, μm	Geometry θ	θ'	ω'	Reported Error, %	Composition (weight percent), Specifications, and Remarks
1	46	1969	295	0.250	~0°		2π		Fine leafing aluminum in Dow 805 and 806A phenylated silicone binder (0.0763 mm thick); exposed to UV radiation (4.7-UV sun rate) in vacuum (10^{-8} mm Hg); vacuum maintained by ion pump; ESH is variable; reflectance measured in situ; positive change indicates decrease in reflectance from preirradiation, in air reflectance. [Authors' designation: I]
2	46	1969	295	0.425	~0°		2π		Above specimen and conditions.
3	46	1969	295	2.100	~0°		2π		Above specimen and conditions.
4	46	1969	295	2.100	~0°		2π		Fine leafing aluminum in Dow 805 and 806A phenylated silicone binder (0.0763 mm thick); exposed to 50 keV electrons (4×10^{8}-1.7×10^{12} e cm^{-2} sec^{-1}) in vacuum (10^{-8} mm Hg); vacuum maintained by ion pump; electron fluence (e cm^{-2}) is variable; reflectance measured in situ; positive change indicates decrease in reflectance from preirradiation, in air reflectance. [Authors' designation: I]
5	46	1969	295	0.590	~0°		2π		Above specimen and conditions.

DATA TABLE NO. 5 CHANGE IN NORMAL SPECTRAL REFLECTANCE OF ALUMINUM PIGMENTED COATINGS

[Wavelength, λ, μm; Reflectance, ρ; Temperature, T, K]

ESH	$\Delta\rho$
CURVE 1 $\lambda = 0.250$ T = 295	
135	0.10
250	0.13
490	0.17
1130	0.24
CURVE 2 $\lambda = 0.425$ T = 295	
135	0.01*
250	0.01
490	0.02
770	0.03
1130	0.04

ESH	$\Delta\rho$
CURVE 3 $\lambda = 2.100$ T = 295	
135	-0.02
250	0.01*
490	0.01
1130	0.01

Electron fluence	$\Delta\rho$
CURVE 4 $\lambda = 2.100$ T = 295	
6×10^{14}	0.03
8×10^{15}	0.08
CURVE 5 $\lambda = 0.590$ T = 295	
1×10^{13}	0.0
2×10^{14}	0.0
6×10^{14}	0.0
8×10^{14}	0.0

* Not shown on plot

SPECIFICATION TABLE NO. 6 NORMAL SOLAR REFLECTANCE OF ALUMINUM PIGMENTED COATINGS

Curve No.	Ref. No.	Year	Temperature Range, K	Geometry θ	θ'	ω'	Reported Error, %	Composition (weight percent), Specifications and Remarks
1*	60	1954	311	~0°		2π		Aluminum in silicone binder; heated to 421 K; measured in air; calculated from solar absorptivity.
2*	60	1954	311	~0°		2π		Above specimen and conditions except reheated to 594 K.

DATA TABLE NO. 6 NORMAL SOLAR REFLECTANCE OF ALUMINUM PIGMENTED COATINGS

[Temperature, T, K; Reflectance, ρ]

T	ρ
CURVE 1*	
311	0.535
CURVE 2*	
311	0.695

* No plot given.

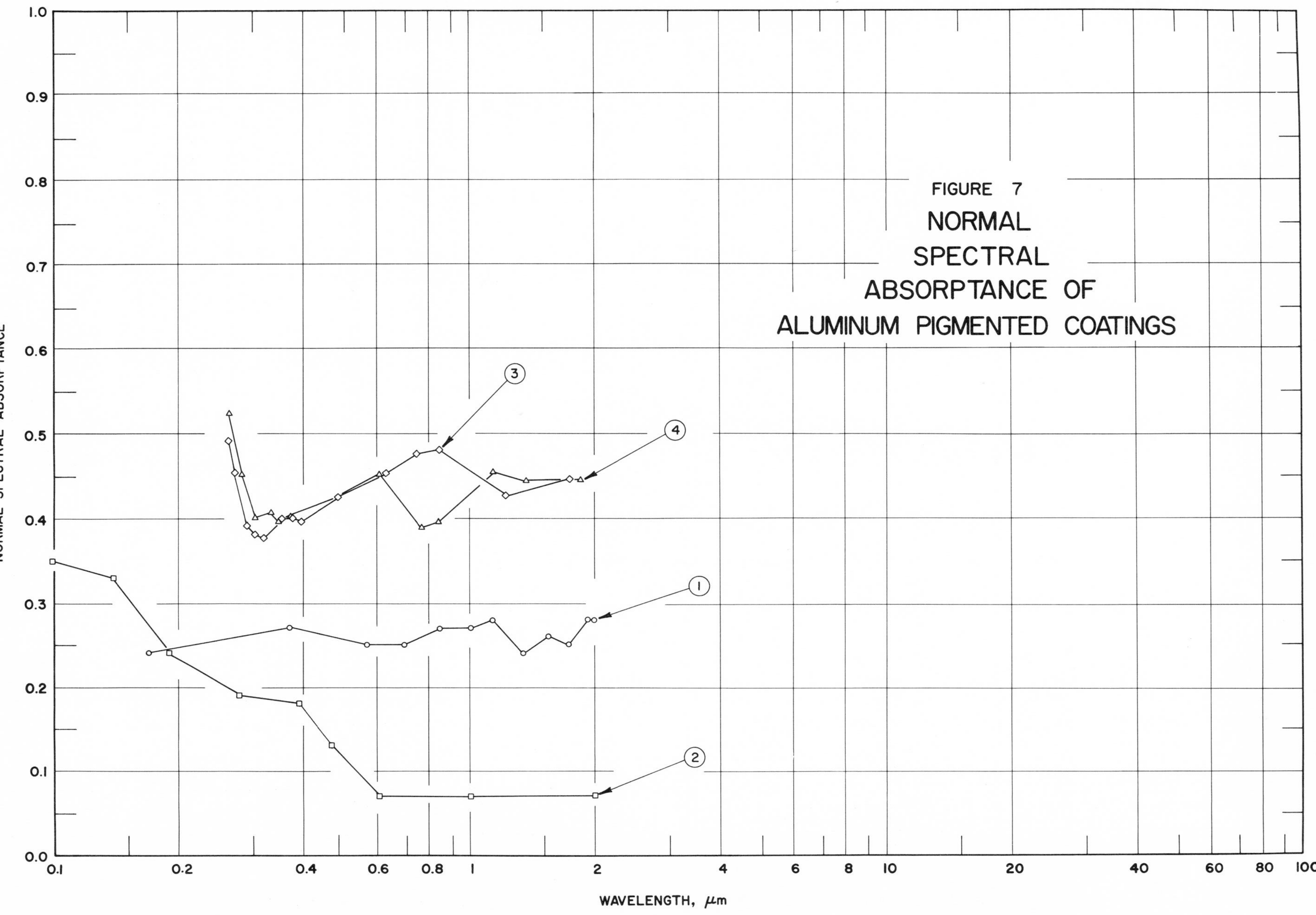
FIGURE 7
NORMAL
SPECTRAL
ABSORPTANCE OF
ALUMINUM PIGMENTED COATINGS
3
4
1
2
NORMAL SPECTRAL ABSORPTANCE
1.0
0.9
0.8
0.7
0.6
0.5
0.4
0.3
0.2
0.1
0.0
0.1
0.2
0.4
0.6
0.8
1
2
4
6
8
10
20
40
60
80
100
WAVELENGTH, μm

SPECIFICATION TABLE NO. 7 NORMAL SPECTRAL ABSORPTANCE OF ALUMINUM PIGMENTED COATINGS

Curve No.	Ref. No.	Year	Temperature, K	Wavelength Range, μm	Geometry θ	Reported Error, %	Composition (weight percent), Specifications, and Remarks
1	66	1962	298	1.7-20.0	$\sim 0^\circ$		Data extracted from smooth curve.
2	66	1962	298	0.8-20.0	$\sim 0^\circ$		Polished; data extracted from smooth curve.
3	50	1965	$\sim$298	0.266-1.851	$\sim 0^\circ$		LMSC nonleafing aluminum in acrylic binder; data extracted from smooth curve.
4	50	1965	$\sim$298	0.267-1.851	$\sim 0^\circ$		Similar to above specimen and conditions except irradiated in vacuum at 291 K with 1 Mev electrons to a total dose of 10^{16} e-cm^{-2}.

DATA TABLE NO. 7 NORMAL SPECTRAL ABSORPTANCE OF ALUMINUM PIGMENTED COATINGS

[Wavelength, λ, μm; Absorptance, α; Temperature, T, K]

λ	α	λ	α	λ	α
CURVE 1 T = 298		CURVE 2 (cont.)		CURVE 4 T $\sim$298	
1.7	0.24	6.3	0.07	0.267	0.524
3.7	0.27	10.0	0.07	0.283	0.456
5.7	0.25	20.0	0.07	0.307	0.402
7.0	0.25	CURVE 3 T $\sim$298		0.334	0.408
8.5	0.27			0.350	0.399
10.1	0.27			0.371	0.402
11.4	0.28	0.266	0.491	0.399	0.398*
13.5	0.24	0.274	0.455	0.482	0.426*
15.6	0.26	0.291	0.393	0.605	0.451
17.3	0.25	0.304	0.380	0.770	0.490
19.2	0.28	0.318	0.378	0.844	0.496
20.0	0.28	0.356	0.400	1.138	0.457
		0.372	0.401	1.363	0.445
CURVE 2 T = 298		0.399	0.398	1.851	0.447
		0.482	0.426		
		0.636	0.454		
0.8	0.35	0.752	0.478		
1.4	0.33	0.849	0.481		
1.9	0.24	1.216	0.428		
2.8	0.19	1.851	0.447		
3.9	0.18				
4.7	0.13				

* Not shown on plot

SPECIFICATION TABLE NO. 8 NORMAL SOLAR ABSORPTANCE OF ALUMINUM PIGMENTED COATINGS

Curve No.	Ref. No.	Year	Temperature Range, K	Geometry θ	Reported Error, %	Composition (weight percent), Specifications and Remarks
1*	60	1954	311	$\sim 0^\circ$		Aluminum in silicone binder; substrate unknown; heated to 422 K; measured in air at sea level.
2*	60	1954	311	$\sim 0^\circ$		Above specimen and conditions except reheated to 594 K.
3*	49	1961	278	$\sim 0^\circ$		Fuller aluminum silicone; leafing aluminum in silicone binder; substrate unknown; measured for extraterrestrial conditions.
4*	49	1961	278	$\sim 0^\circ$		Aluminum in silicone binder; Mg alloy substrate; exposed to ultraviolet radiation in vacuum (10^{-6} - 8 x 10^{-6} mm Hg) from an argon filled A-H6 high pressure Hg-arc lamp for 80 hrs; slight crackling observed. [Authors' designation: Specimen No. 1]
5*	49	1961	278	$\sim 0^\circ$		Above specimen and conditions before exposure.
6*	49	1961	278	$\sim 0^\circ$		Similar to Curve 4 specimen and conditions. [Authors' designation: Specimen No. 2]
7*	49	1961	278	$\sim 0^\circ$		Above specimen and conditions before exposure.
8*	15	1965	450	$\sim 0^\circ$		Fuller aluminum silicone paint No. 171-A-152; leafing aluminum in silicone binder; measured in vacuum (10^{-6} mm Hg); exposed for 24 hrs at 450 K; avg of 8 specimens.
9*	15	1965	450	$\sim 0^\circ$		Fuller aluminum silicone paint No. 171-A-152; leafing aluminum in silicone binder; weathered for 24 hrs (sprayed with distilled water for 10 min at each hr of test interval, continuous exposure to a carbon arc which simulates earth sunlight); avg of 8 specimens.
10*	15	1965	450	$\sim 0^\circ$		Fuller aluminum silicone paint No. 171-A-152; leafing aluminum in silicone binder; avg of 8 specimens.
11*	50	1965	~298	$\sim 0^\circ$		LMSC nonleafing aluminum in flat acrylic binder; computed from spectral reflectance data.
12*	50	1965	~298	$\sim 0^\circ$		Similar to above specimen and conditions except irradiated in vacuum at 291 K with 1 Mev electrons to a total dose of 10^{16} e cm^{-2}.

* No plot given.

DATA TABLE NO. 8 NORMAL SOLAR ABSORPTANCE OF ALUMINUM PIGMENTED COATINGS

[Temperature, T, K; Absorptance, α]

T	α
CURVE 1*	
311	0.47
CURVE 2*	
311	0.31
CURVE 3*	
278	0.23
CURVE 4*	
278	0.33
CURVE 5*	
278	0.22
CURVE 6*	
278	0.30
CURVE 7*	
278	0.21
CURVE 8*	
450	0.31
CURVE 9*	
450	0.32
CURVE 10*	
450	0.32
CURVE 11*	
298	0.44
CURVE 12*	
298	0.45

* No plot given.

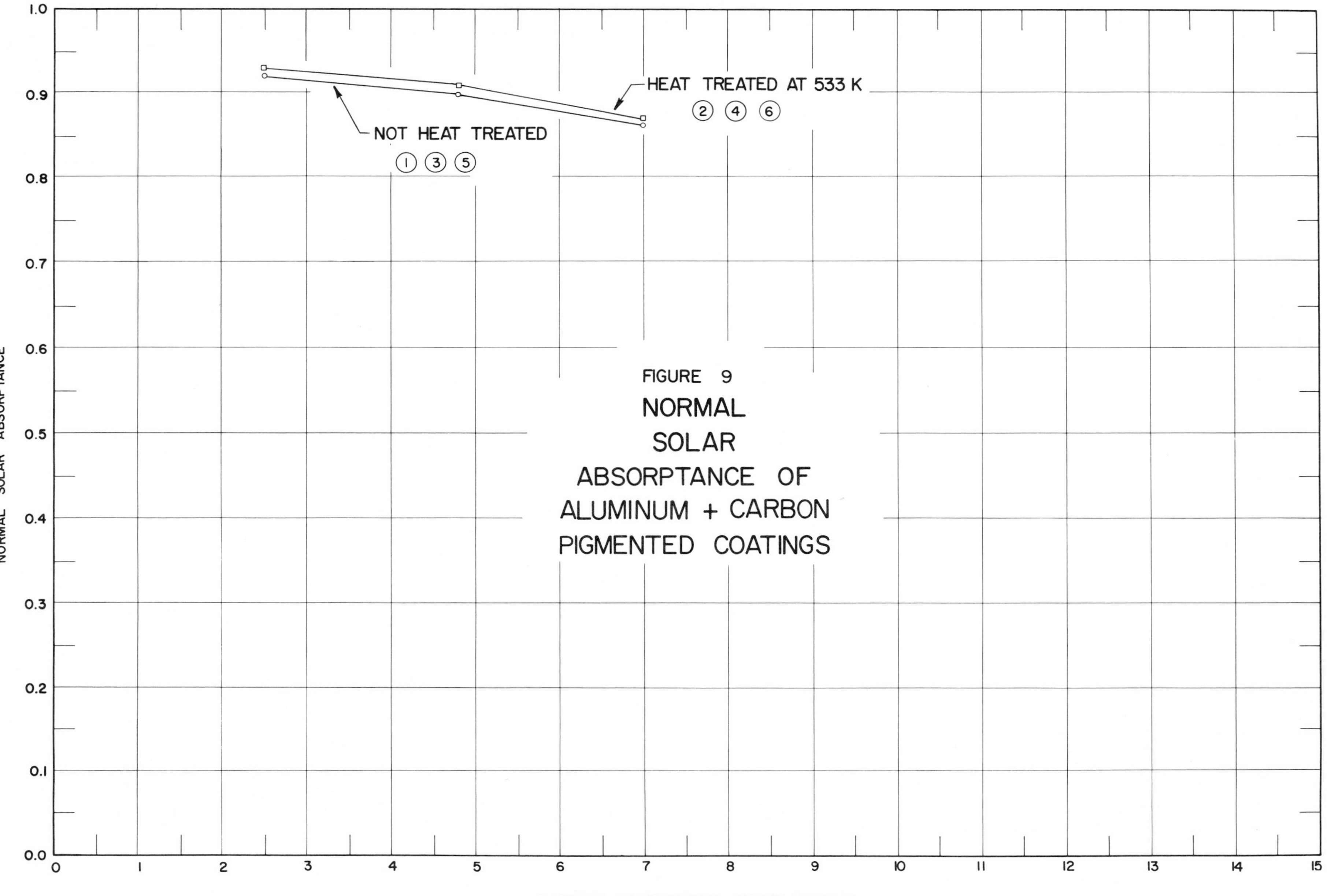

FIGURE 9 NORMAL SOLAR ABSORPTANCE OF ALUMINUM + CARBON PIGMENTED COATINGS

SPECIFICATION TABLE NO. 9 NORMAL SOLAR ABSORPTANCE OF ALUMINUM + CARBON PIGMENTED COATINGS

Curve No.	Ref. No.	Year	Temperature Range, K	Geometry θ	Reported Error, %	Composition (weight percent), Specifications and Remarks
1	77	1962	~298	~0°		Composition (wet): 2.5 aluminum, 2.5 carbon black, 11.4 silaceous inerts, 30.6 phthalic alkyd, balance organic solvents; film thickness 0.064 ± 0.013 mm; prepared by mixing black enamel (MIL-E-5557, Type III) with leafing aluminum paste (Federal Specification TT-A-468, Type II, Class A); applied by standard spraying techniques; substrate is 2024 aluminum alloy sandblasted and etched, then coated with MIL-P-8585 zinc chromate primer 0.013 mm thick; computed from spectral reflectance data.
2	77	1962	~298	~0°		Above specimen and conditions except heated to 533 K in vacuum (10^{-5} mm Hg) on a 2-hr cycle.
3	77	1962	~298	~0°		Similar to Curve 1 specimen and conditions except composition (wet): 4.8 aluminum, 2.4 carbon black, 11.1 silaceous inerts, 29.9 phthalic alkyd, balance organic solvents.
4	77	1962	~298	~0°		Above specimen and conditions except heated to 533 K in vacuum (10^{-5} mm Hg) on a 2-hr cycle.
5	77	1962	~298	~0°		Similar to Curve 1 specimen and conditions except composition (wet): 7.0 aluminum, 2.3 carbon black, 10.9 siliceous inerts, 29.2 phthalic alkyd, balance organic solvents.
6	77	1962	~298	~0°		Above specimen and conditions except heated to 533 K in vacuum (10^{-5} mm Hg) on a 2-hr cycle.

DATA TABLE NO. 9 NORMAL SOLAR ABSORPTANCE OF ALUMINUM + CARBON PIGMENTED COATINGS

[Temperature, T, K; Absorptance, α]

T	α	T	α
CURVE 1		CURVE 5	
298	0.922	298	0.862
CURVE 2		CURVE 6	
298	0.930	298	0.871
CURVE 3			
298	0.898		
CURVE 4			
298	0.907		

SPECIFICATION TABLE NO. 10 HEMISPHERICAL TOTAL EMITTANCE OF GOLD PIGMENTED COATINGS

Curve No.	Ref. No.	Year	Temperature Range, K	Reported Error, %	Composition (weight percent), Specifications and Remarks
1*	43	1962	384-477	±3	Leafing gold (6 μm thick) in resin binder; copper substrate; measured in air vacuum (10^{-3} mm Hg); data extracted from smooth curve.

DATA TABLE NO. 10 HEMISPHERICAL TOTAL EMITTANCE OF GOLD PIGMENTED COATINGS

[Temperature, T, K; Emittance, ε]

T	ε
CURVE 1*	
384	0.130
393	0.120
413	0.120
433	0.140
453	0.135
473	0.120
477	0.115

* No plot given.

SPECIFICATION TABLE NO. 11 HEMISPHERICAL TOTAL EMITTANCE OF ZIRCONIUM PIGMENTED COATINGS

Curve No.	Ref. No.	Year	Temperature Range, K	Reported Error, %	Composition (weight percent), Specifications and Remarks
1*	71	1962	221-688		Zirconium in sodium silicate binder; developed by LMSC.
2*	71	1962	298		Similar to above specimen and conditions.
3*	71	1962	298		Similar to above specimen and conditions.

DATA TABLE NO. 11 HEMISPHERICAL TOTAL EMITTANCE OF ZIRCONIUM PIGMENTED COATINGS

[Temperature, T, K; Emittance, ϵ]

T	ϵ
CURVE 1*	
221	0.861
427	0.861
468	0.842
559	0.756
688	0.561
CURVE 2*	
298	0.856
CURVE 3*	
298	0.84

* No plot given.

1. PIGMENTED COATINGS (continued)

B. Nonmetallic Pigmented Coatings

SPECIFICATION TABLE NO. 12 NORMAL TOTAL EMITTANCE OF ALUMINUM OXIDE PIGMENTED COATINGS

Curve No.	Ref. No.	Year	Temperature Range, K	Geometry θ'	Reported Error, %	Composition (weight percent), Specifications and Remarks
1*	73	1962	589-1255	~0°		Aluminum oxide paint (0.025 mm thick) on stably oxidized Inconel substrate; heated in quiescent air at 1255 K for 30 min; 2 other different samples of thickness 0.051 and 0.070 mm gave the same results.
2*	73	1962	1255	~0°		Aluminum oxide paint (0.025 mm thick) on stably oxidized Inconel substrate; 2 other different samples of thickness 0.051 and 0.070 mm gave the same results.
3*	73	1962	1255	~0°		Above specimen and conditions except heated at 1255 K for 15 min.
4*	73	1962	1255	~0°		Curve 2 specimen and conditions except heated at 1255 for 30 min.
5*	19	1965	298	0°		Al_2O_3 in potassium silicate binder; Al substrate; reported error ±0.02 emittance units.
6*	19	1965	298	0°		Above specimen and conditions except computed from 1-R (2π, 0°).
7*	68	1969	300	~0°		55 Al_2O_3, 30 ZnO and 15 TiO_2 in potassium silicate binder (~0.089 mm thick) on 2024 aluminum substrate (~1.59 mm thick); emittance calculated from reflectance; property measured in air. [Authors' designation: Al_2O_3/PS-7]
8*	68	1969	300	~0°		Similar to above specimen and conditions except exposed to vacuum (10^{-8} mm Hg) for 300 hr; vacuum maintained by ion pump; property measured in air after vacuum exposure.
9*	68	1969	300	~0°		Similar to Curve 7 specimen and conditions except exposed to UV radiation in vacuum (10^{-5} mm Hg) for 300 hr; vacuum maintained by oil-diffusion pump; H_2 gas UV source; property measured in air after exposure.
10*	68	1969	300	~0°		Similar to above specimen and conditions except used He gas UV source.

* No plot given

DATA TABLE NO. 12 NORMAL TOTAL EMITTANCE OF ALUMINUM OXIDE PIGMENTED COATINGS

[Temperature, T, K; Emittance, ϵ]

T	ϵ
CURVE 1*	
589	0.84
700	0.82
811	0.81
922	0.82
1033	0.82
1144	0.83
1255	0.83
CURVE 2*	
1255	0.78
CURVE 3*	
1255	0.84
CURVE 4*	
1255	0.84
CURVE 5*	
298	0.89
CURVE 6*	
298	0.91
CURVE 7*	
300	0.86
CURVE 8*	
300	0.86
CURVE 9*	
300	0.86
CURVE 10*	
300	0.85

* No plot given

SPECIFICATION TABLE NO. 13 NORMAL SPECTRAL EMITTANCE OF ALUMINUM OXIDE PIGMENTED COATINGS

Curve No.	Ref. No.	Year	Temperature, K	Wavelength Range, μm	Geometry θ'	Reported Error, %	Composition (weight percent), Specifications, and Remarks
1*	74	1959	306	1.67-21.00	~0°		Al_2O_3 in phosphoric acid cement binder; data extracted from smooth curve; property converted from R (2π, ~0°).

DATA TABLE NO. 13 NORMAL SPECTRAL EMITTANCE OF ALUMINUM OXIDE PIGMENTED COATINGS

[Wavelength, λ, μm; Emittance, ∈; Temperature, T, K]

λ	∈	λ	∈
CURVE 1* T = 306		CURVE 1 (cont.)	
		11.82	0.828
1.67	0.568	12.04	0.828
1.85	0.550	12.41	0.849
2.10	0.550	12.73	0.816
2.35	0.564	12.93	0.850
2.57	0.600	13.13	0.861
2.65	0.703	14.22	0.825
2.84	0.788	15.74	0.742
3.20	0.860	16.08	0.742
3.67	0.902	16.78	0.695
4.04	0.912	17.10	0.695
4.57	0.912	17.39	0.676
5.36	0.889	17.61	0.676
5.76	0.889	18.14	0.713
6.34	0.912	18.98	0.755
6.96	0.921	19.62	0.771
7.46	0.921	19.83	0.778
8.48	0.901	20.45	0.778
8.99	0.881	20.78	0.770
10.24	0.881	21.00	0.758
10.64	0.873		
11.20	0.846		
11.60	0.846		

* No plot given

FIGURE 14A

ANALYZED NORMAL SPECTRAL REFLECTANCE OF ALUMINUM OXIDE PIGMENTED COATINGS

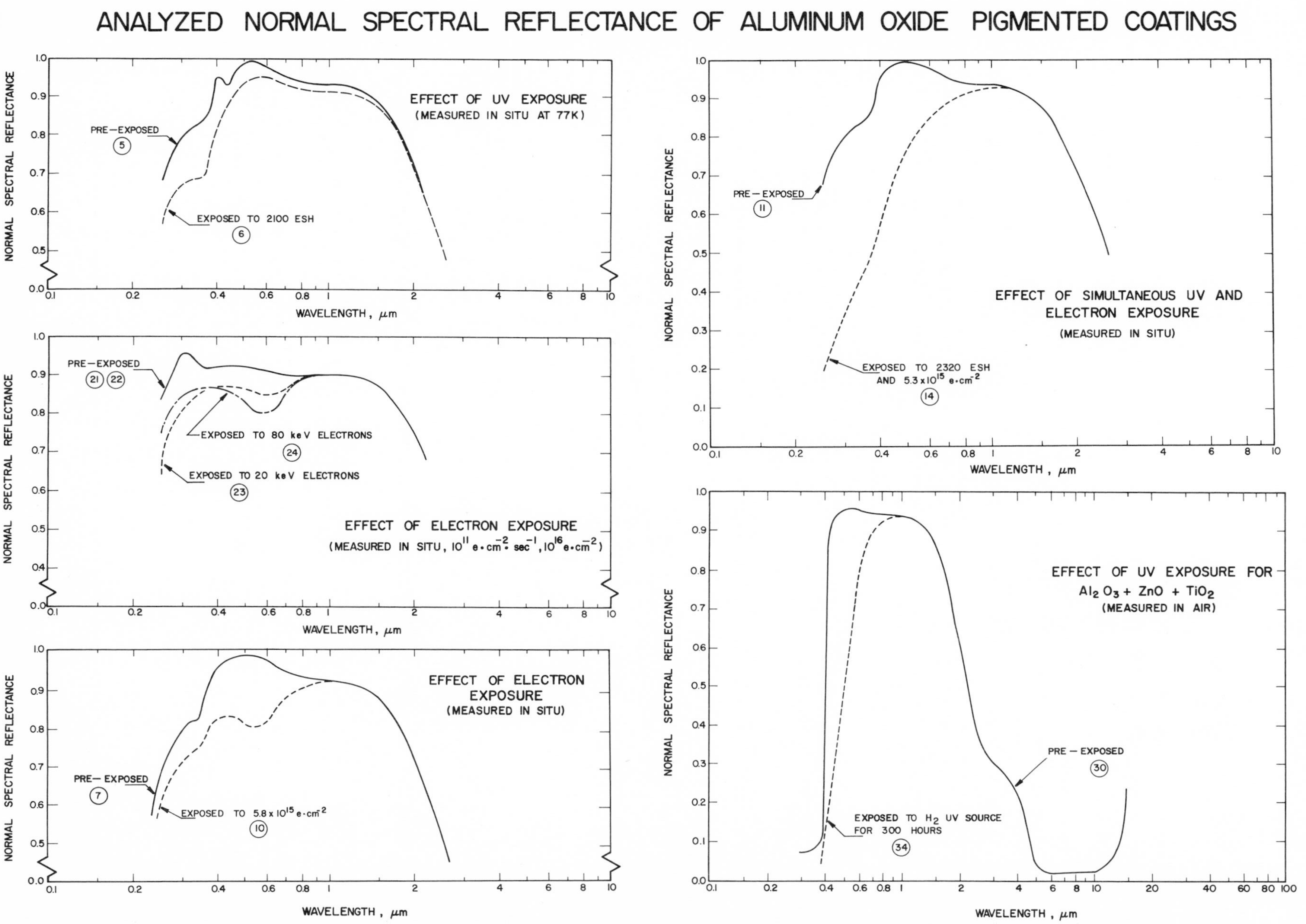

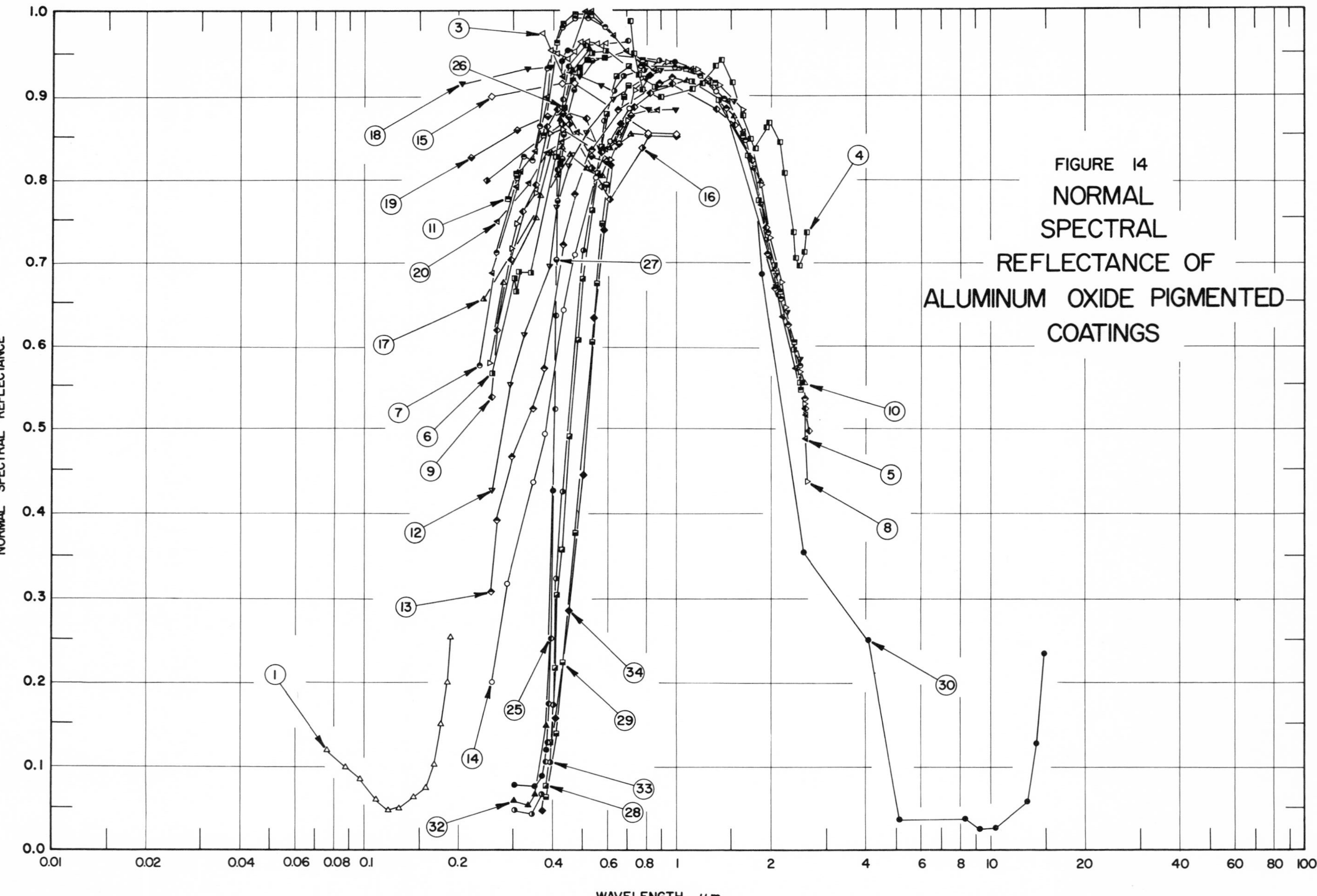
FIGURE 14
NORMAL
SPECTRAL
REFLECTANCE OF
ALUMINUM OXIDE PIGMENTED
COATINGS
NORMAL SPECTRAL REFLECTANCE
WAVELENGTH, μm
1.0
0.9
0.8
0.7
0.6
0.5
0.4
0.3
0.2
0.1
0.0
0.01
0.02
0.04
0.06
0.08
0.1
0.2
0.4
0.6
0.8
1
2
4
6
8
10
20
40
60
80
100
1
3
4
5
6
7
8
9
10
11
12
13
14
15
16
17
18
19
20
25
26
27
28
29
30
32
33
34

SPECIFICATION TABLE NO. 14 NORMAL SPECTRAL REFLECTANCE OF ALUMINUM OXIDE PIGMENTED COATINGS

Curve No.	Ref. No.	Year	Temperature, K	Wavelength Range, μm	Geometry θ	θ'	ω'	Reported Error, %	Composition (weight percent), Specifications, and Remarks
1	30	1965	298	0.0754-0.1883	<15°		2π		Al_2O_3 pigmented paint; measured in vacuum.
2*	17	1968	~298	0.220-2.500	~0°		2π		CM-145 Al_2O_3 paint; data extracted from smooth curve.
3	3	1960	~298	0.375-0.698	~0°		2π		Al_2O_3(0.5 μm polishing alumina, Buehler Ltd.), duPont RC 7007 alkyd-melamine binder; pigment concentration 85%; sprayed onto mild steel substrate; data extracted from smooth curve; measured relative to MgO.
4	3	1960	~298	0.704-2.604	~0°		2π		Similar to above specimen and conditions.
5	75	1969	77	0.256-2.594	~0°		2π		Al_2O_3 in potassium silicate binder; exposed to vacuum (5×10^{-9}-2×10^{-7} mm Hg) maintained with diffusion pump; reflectance calculated from R(2π, 0°); property measured in situ; data extracted from smooth curve.
6	75	1969	77	0.255-2.587	~0°		2π		Above specimen and conditions except exposed to 6 sun intensity UV radiation in vacuum for 350 SH.
7	75	1969	77	0.233-2.612	~0°		2π		Similar to curve 5 specimen and conditions.
8	75	1969	77	0.250-2.602	~0°		2π		Above specimen and conditions except exposed to electron radiation (8.6×10^{10}-1.6×10^{12} e cm^{-2} sec^{-1}) in vacuum; total dose 2.3×10^{15} e cm^{-2}.
9	75	1969	77	0.255-2.628	~0°		2π		Above specimen and conditions except total dose 1.2×10^{16} e cm^{-2}.
10	75	1969	77	0.250-2.628	~0°		2π		Above specimen and conditions except total dose 5.8×10^{15} e cm^{-2}.
11	75	1969	77	0.256-2.600	~0°		2π		Similar to curve 5 specimen and conditions.
12	75	1969	77	0.255-2.603	~0°		2π		Above specimen and conditions except simultaneous exposure to 1.8×10^{15} e cm^{-2} (3.5×10^{10} e cm^{-2} sec^{-1}) and 110 SH of UV at 8 suns in vacuum.
13	75	1969	77	0.251-2.600	~0°		2π		Above specimen and conditions except 3.2×10^{15} e cm^{-2} and 190 SH.
14	75	1969	77	0.256-2.600	~0°		2π		Above specimen and conditions except 5.3×10^{15} e cm^{-2} and 290 SH.
15	46	1969	295	0.256-1.000	~0°		2π		Linde α-phase Al_2O_3 in potassium silicate (0.28 mm thick); exposed to vacuum (10^{-7} mm Hg) maintained by ion pump; reflectance measured in situ; data extracted from smooth curve. [Authors' designation: D]
16	46	1969	295	0.247-1.000	~0°		2π		Above specimen and conditions except exposed to 50 keV electrons (5×10^{14} e cm^{-2}) in vacuum; reflectance measured in situ after electron exposure.
17	46	1969	295	0.240-1.000	~0°		2π		Above specimen and conditions except exposed to UV radiation (~5 equivalent sun rate in UV) in vacuum after electron exposure; ESH 18; reflectance measured in situ after UV exposure.
18	46	1969	295	0.207-1.000	~0°		2π		Linde α-phase Al_2O_3 in potassium silicate (0.28 mm thick); exposed to vacuum (10^{-7} mm Hg) maintained by ion pump; reflectance measured in situ; data extracted from smooth curve. [Authors' designation: D]
19	46	1969	295	0.220-1.000	~0°		2π		Above specimen and conditions except exposed to 50 keV electrons (5×10^{14} e cm^{-2} over 4.1 hrs with 90% during first hr) and UV radiation (4.4 equivalent sun rate in UV) in vacuum; reflectance measured in situ after 1 hr of exposure.

SPECIFICATION TABLE NO. 14 NORMAL SPECTRAL REFLECTANCE OF ALUMINUM OXIDE PIGMENTED COATINGS (continued)

Curve No.	Ref. No.	Year	Temperature, K	Wavelength Range, μm	Geometry θ	θ'	ω'	Reported Error, %	Composition (weight percent), Specifications, and Remarks
20	46	1969	295	0.267-1.000	~0°		2π		Above specimen and conditions except reflectance measured in situ at end of exposure.
21	45	1970	298	0.249-2.200	~0°		2π		Al_2O_3 in potassium silicate binder on aluminum substrate; reflectance measured in air; data extracted from smooth curve. [Authors' designation: G]
22	45	1970	298	0.250-2.200	~0°		2π		Similar to above specimen and conditions.
23	45	1970	281	0.251-2.200	~0°		2π		Above specimen and conditions except exposed to 20 keV electrons in dark in vacuum (10^{-8} mm Hg); vacuum maintained by ion pump; flux 1 x 10^{10}-5 x 10^{11} e cm^{-2} sec^{-1}; substrate held at 281 ± 2 K; reflectance measured in situ after exposure to 10^{16} e cm^{-2}.
24	45	1970	281	0.251-2.200	~0°		2π		Similar to above specimen and conditions except exposed to 80 keV electrons.
25	68	1969	300	0.3800-0.7000	~0°		2π		55 Al_2O_3, 30 ZnO and 15 TiO_2 in potassium silicate binder (~0.089 mm thick) on 2024 aluminum substrate (~1.59 mm thick); reflectance measured in air; data extracted from smooth curve. [Authors' designation: Al_2O_3/PS-7]
26	68	1969	300	0.3800-0.7000	~0°		2π		Similar to above specimen and conditions except exposed to vacuum (10^{-8} mm Hg) for 300 hrs; vacuum maintained by ion pump; reflectance measured in air after vacuum exposure.
27	68	1969	300	0.3800-0.7000	~0°		2π		Similar to curve 25 specimen and conditions except exposed to vacuum (10^{-5} mm Hg) for 300 hrs; vacuum maintained by oil-diffusion pump; reflectance measured in air after vacuum exposure.
28	68	1969	300	0.3800-0.7000	~0°		2π		Similar to above specimen and conditions except exposed to UV radiation in vacuum for 300 hrs; H_2 gas UV source.
29	68	1969	300	0.3800-0.7000	~0°		2π		Similar to above specimen and conditions except He gas UV source.
30	68	1969	300	0.30-14.95	~0°		2π		Similar to curve 25 specimen and conditions.
31	68	1969	300	0.30-14.95	~0°		2π		Similar to curve 26 specimen and conditions.
32	68	1969	300	0.30-14.95	~0°		2π		Similar to curve 27 specimen and conditions.
33	68	1969	300	0.30-14.95	~0°		2π		Similar to curve 28 specimen and conditions.
34	68	1969	300	0.30-14.95	~0°		2π		Similar to curve 29 specimen and conditions.
35*	27	1964	~298	0.440-0.600	~0°		2π		C-35 $Al_2O_3 \cdot 3H_2O$ (trihydrated aluminum oxide from Alcoa) in PS-7 potassium silicate binder; PBR 4.30; solids content 63.9%; aluminum substrate abraded with No. 60 Aloxite cloth; sample cured at 413 K for 18 hrs. [Authors' designation: Sample C2]
36*	27	1964	~298	0.440-0.600	~0°		2π		Above specimen and conditions except irradiated in vacuum (10^{-6} mm Hg) with 200 ESH at solar factor of 3 suns.

* Not shown on plot

DATA TABLE NO. 14 NORMAL SPECTRAL REFLECTANCE OF ALUMINUM OXIDE PIGMENTED COATINGS

[Wavelength, λ, μm; Reflectance, ρ; Temperature, T, K]

CURVE 1
T = 298

λ	ρ
0.0754	0.118
0.0877	0.100
0.0969	0.084
0.1080	0.060
0.1184	0.047
0.1294	0.050
0.1417	0.061
0.1574	0.073
0.1685	0.101
0.1758	0.150
0.1817	0.200
0.1883	0.255

CURVE 2*
T ~ 298

λ	ρ
0.220	0.554
0.250	0.664
0.275	0.811
0.288	0.911
0.307	0.930
0.365	0.932
0.394	0.940
0.413	0.937
0.438	0.924
0.507	0.920
0.613	0.897
0.769	0.880
0.849	0.871
0.974	0.874
1.095	0.889
1.140	0.887
1.238	0.871
1.296	0.874
1.331	0.862
1.379	0.819
1.475	0.778
1.576	0.747
1.639	0.712
1.797	0.576
1.904	0.432
1.933	0.423
2.080	0.409
2.170	0.385

CURVE 2 (cont.)*

λ	ρ
2.297	0.338
2.406	0.288
2.500	0.244

CURVE 3
T ~ 298

λ	ρ
0.375	0.976
0.394	0.959
0.413	0.952
0.426	0.946*
0.441	0.946
0.455	0.948*
0.471	0.955
0.492	0.967
0.518	0.967
0.559	0.964
0.669	0.961
0.698	0.964

CURVE 4
T ~ 298

λ	ρ
0.704	0.990
0.727	0.952
0.751	0.929
0.776	0.909
0.818	0.901*
0.888	0.900
1.037	0.905*
1.136	0.910
1.216	0.917*
1.337	0.939
1.401	0.945
1.446	0.939*
1.520	0.919
1.620	0.878
1.664	0.866*
1.726	0.850
1.803	0.837
1.852	0.844*
1.918	0.861
1.979	0.868
2.050	0.866*
2.121	0.845

CURVE 4 (cont.)

λ	ρ
2.206	0.807
2.317	0.739
2.375	0.706
2.411	0.697
2.460	0.699*
2.521	0.711
2.604	0.738

CURVE 5
T = 77

λ	ρ
0.256	0.687
0.303	0.793
0.316	0.810
0.356	0.833
0.387	0.900
0.398	0.958*
0.438	0.926
0.505	1.000
0.532	1.000
0.630	0.974
0.689	0.955
0.784	0.940
0.851	0.934
1.111	0.934
1.286	0.918
1.476	0.892
1.642	0.856
1.791	0.811
1.886	0.770
1.996	0.706
2.077	0.671
2.186	0.636
2.393	0.571
2.557	0.516
2.594	0.487

CURVE 6
T = 77

λ	ρ
0.255	0.567
0.302	0.666
0.300	0.682
0.310	0.687
0.340	0.687

CURVE 6 (cont.)

λ	ρ
0.361	0.696*
0.409	0.828
0.430	0.879*
0.484	0.929
0.535	0.951
0.561	0.954*
0.594	0.954
0.747	0.923*
0.850	0.911
1.011	0.916
1.187	0.916*
1.388	0.896
1.548	0.869*
1.701	0.829
1.848	0.775
2.014	0.696
2.077	0.671*
2.186	0.636*
2.353	0.595
2.456	0.555
2.587	0.487*

CURVE 7
T = 77

λ	ρ
0.233	0.577
0.264	0.712
0.301	0.801
0.323	0.829
0.343	0.822
0.366	0.864*
0.385	0.934
0.399	0.957*
0.431	0.981
0.463	0.992
0.489	0.994*
0.519	0.994
0.587	0.984
0.760	0.937
0.999	0.937
1.200	0.926
1.425	0.897
1.688	0.848*
1.795	0.815
1.943	0.737

CURVE 7 (cont.)

λ	ρ
2.040	0.692
2.153	0.655
2.390	0.575
2.419	0.575*
2.491	0.557
2.568	0.517
2.612	0.471

CURVE 8
T = 77

λ	ρ
0.250	0.580
0.292	0.719
0.306	0.748
0.352	0.784
0.390	0.856
0.433	0.860
0.609	0.821
0.695	0.870
0.728	0.907
0.776	0.937*
0.922	0.941
1.188	0.933
1.397	0.912
1.612	0.881
1.745	0.848
1.882	0.792
1.999	0.730
2.084	0.694*
2.138	0.675
2.202	0.645*
2.380	0.597
2.469	0.568*
2.544	0.528
2.584	0.484*
2.602	0.439

CURVE 9
T = 77

λ	ρ
0.255	0.538
0.262	0.620
0.292	0.704
0.320	0.761
0.355	0.795

CURVE 9 (cont.)

λ	ρ
0.382	0.861
0.399	0.885
0.413	0.885
0.448	0.871
0.512	0.871
0.529	0.865*
0.568	0.831
0.611	0.821*
0.697	0.872*
0.739	0.889
0.837	0.905
0.969	0.916
1.060	0.916
1.368	0.885
1.555	0.864
1.614	0.864*
1.744	0.823
1.910	0.742
2.040	0.692
2.153	0.655*
2.426	0.572
2.517	0.554*
2.591	0.523*
2.628	0.499

CURVE 10
T = 77

λ	ρ
0.250	0.580*
0.278	0.677
0.295	0.704*
0.351	0.755
0.365	0.781
0.396	0.831
0.409	0.839*
0.433	0.839
0.505	0.813
0.565	0.805
0.656	0.858
0.739	0.889
0.837	0.905*
0.969	0.916*
1.098	0.920
1.239	0.920
1.540	0.877

CURVE 10 (cont.)

λ	ρ
1.696	0.850
1.847	0.793*
1.998	0.707
2.040	0.692*
2.153	0.655*
2.426	0.572*
2.517	0.554
2.591	0.523*
2.628	0.499*

CURVE 11
T = 77

λ	ρ
0.256	0.689*
0.289	0.777
0.304	0.804
0.353	0.836*
0.372	0.858
0.406	0.965
0.428	0.985
0.469	0.996
0.539	0.996
0.775	0.943
0.861	0.934*
1.093	0.934
1.321	0.916
1.482	0.890*
1.647	0.855
1.770	0.817*
1.933	0.743*
2.005	0.702*
2.112	0.664
2.329	0.603
2.484	0.548
2.553	0.517*
2.600	0.494*

CURVE 12
T = 77

λ	ρ
0.255	0.427
0.290	0.553
0.322	0.613
0.390	0.699
0.411	0.768

CURVE 12 (cont.)

λ	ρ
0.455	0.817
0.513	0.856
0.628	0.899
0.782	0.924
0.886	0.934
1.093	0.934*
1.340	0.919*
1.513	0.894
1.677	0.856
1.792	0.820*
1.980	0.728
2.087	0.684
2.231	0.640
2.447	0.583
2.510	0.552
2.603	0.499*

CURVE 13
T = 77

λ	ρ
0.251	0.309
0.262	0.391
0.293	0.467
0.346	0.522
0.373	0.572
0.423	0.723
0.463	0.782
0.534	0.835
0.658	0.885
0.756	0.906*
0.890	0.919
0.971	0.923
1.172	0.913*
1.454	0.885
1.618	0.856*
1.777	0.810
1.979	0.710
2.094	0.667
2.258	0.625
2.399	0.601*
2.540	0.536
2.600	0.494*

* Not shown on plot

DATA TABLE NO. 14 NORMAL SPECTRAL REFLECTANCE OF ALUMINUM OXIDE PIGMENTED COATINGS (continued)

λ	ρ
CURVE 14 T = 77	
0.256	0.200
0.281	0.316
0.344	0.437
0.378	0.492
0.432	0.645
0.467	0.710
0.541	0.802
0.605	0.849
0.705	0.885
0.865	0.916
0.955	0.923
1.205	0.923
1.361	0.905
1.454	0.885*
1.618	0.856*
1.777	0.810*
1.992	0.709*
2.092	0.675*
2.239	0.628*
2.393	0.589
2.481	0.557*
2.553	0.517*
2.600	0.494*
CURVE 15 T = 295	
0.256	0.900
0.421	0.917
0.475	0.917
0.810	0.855
1.000	0.855
CURVE 16 T = 295	
0.247	0.800
0.375	0.855*
0.429	0.866
0.453	0.866
0.525	0.813
0.564	0.791
0.610	0.778
0.727	0.837
0.810	0.855
1.000	0.855
CURVE 17 T = 295	
0.240	0.656
0.414	0.806
0.455	0.830
0.606	0.837
0.718	0.854
1.000	0.855*
CURVE 18 T = 295	
0.207	0.917
0.336	0.934
0.396	0.936
0.486	0.926
0.564	0.912
0.809	0.884
1.000	0.884
CURVE 19 T = 295	
0.220	0.827
0.306	0.860
0.386	0.878
0.427	0.880
0.450	0.876
0.538	0.828
0.592	0.821
0.606	0.821*
0.652	0.843
0.719	0.879
0.800	0.884*
1.000	0.884*
CURVE 20 T = 295	
0.267	0.750
0.333	0.799
0.386	0.831
0.427	0.846
0.446	0.851*
0.480	0.857
0.586	0.834
0.632	0.849
0.707	0.881*
CURVE 20 (cont.)	
0.768	0.884
1.000	0.884
CURVE 21 T = 298	
0.249	0.849
0.262	0.858
0.292	0.947
0.308	0.962
0.343	0.927
0.354	0.931
0.362	0.911
0.386	0.939
0.409	0.930
0.444	0.926
0.455	0.932
0.486	0.932
0.581	0.916
0.696	0.909
0.743	0.899
1.376	0.899
1.499	0.890
1.590	0.875
1.696	0.845
1.809	0.797
1.897	0.775
2.013	0.738
2.102	0.721
2.200	0.684
CURVE 22 T = 298	
0.250	0.834
0.261	0.849
0.288	0.936
0.306	0.952
0.352	0.918
0.396	0.922
0.426	0.912
0.492	0.912
0.663	0.889
0.712	0.889
0.743	0.876
1.400	0.879
1.514	0.876
CURVE 22 (cont.)	
1.648	0.846
1.746	0.822
1.813	0.788
1.896	0.771
2.028	0.731
2.135	0.703
2.200	0.680
CURVE 23 T = 281	
0.251	0.677
0.263	0.693
0.295	0.804
0.350	0.855
0.384	0.866
0.412	0.870
0.494	0.870
0.578	0.849
0.613	0.849
0.724	0.876
1.400	0.879
1.514	0.876
1.648	0.846
1.746	0.822
1.813	0.788
1.896	0.771
2.028	0.731
2.135	0.703
2.200	0.680
CURVE 24 T = 281	
0.251	0.747
0.265	0.776
0.284	0.832
0.300	0.850
0.343	0.864
0.364	0.872
0.417	0.859
0.486	0.833
0.545	0.810
0.568	0.800
0.602	0.800
0.656	0.817
0.699	0.846
CURVE 24 (cont.)	
0.717	0.862
0.737	0.876
0.784	0.888
0.986	0.893
1.011	0.899
1.376	0.899
1.499	0.890
1.590	0.875
1.696	0.845
1.809	0.797
1.897	0.775
2.013	0.738
2.102	0.721
2.200	0.684
CURVE 25 T = 300	
0.3800	0.104
0.3835	0.128
0.3875	0.173
0.3924	0.251
0.3969	0.357*
0.4018	0.523
0.4068	0.638
0.4173	0.812
0.4251	0.868
0.4318	0.898
0.4406	0.918
0.4556	0.937
0.4782	0.953
0.5096	0.961
0.7000	0.966
CURVE 26 T = 300	
0.3800	0.104*
0.3835	0.128*
0.3875	0.173*
0.3924	0.251*
0.3969	0.357*
0.4018	0.523*
0.4068	0.638*
0.4143	0.764
0.4207	0.818
0.4304	0.861*
CURVE 26 (cont.)	
0.4398	0.889
0.4638	0.920
0.4824	0.935
0.5189	0.945
0.5800	0.949
0.7000	0.957
CURVE 27 T = 300	
0.3800	0.104*
0.3835	0.128*
0.3875	0.173*
0.3924	0.251*
0.3969	0.357*
0.4018	0.523*
0.4068	0.638*
0.4108	0.705
0.4181	0.775
0.4250	0.824
0.4315	0.857
0.4405	0.879
0.4651	0.909
0.4930	0.929
0.5385	0.943
0.5800	0.949*
0.7000	0.957*
CURVE 28 T = 300	
0.3800	0.073
0.3901	0.128*
0.4021	0.217
0.4131	0.304
0.4227	0.357
0.4504	0.490
0.4801	0.608
0.4996	0.680
0.5273	0.764
0.5447	0.806
0.5617	0.839*
0.5912	0.880
0.6406	0.925
0.6654	0.934*
0.7000	0.939
CURVE 29 T = 300	
0.3800	0.061
0.3856	0.068*
0.4093	0.138
0.4328	0.221
0.4746	0.378
0.5338	0.605
0.5540	0.673
0.5787	0.748
0.5985	0.793
0.6300	0.848*
0.6570	0.882*
0.6769	0.900
0.7000	0.912
CURVE 30 T = 300	
0.30	0.077
0.35	0.075
0.37	0.086
0.38	0.117
0.40	0.426
0.41	0.813*
0.42	0.917*
0.43	0.942
0.45	0.955
0.99	0.942
1.20	0.930
1.43	0.888
1.89	0.685
2.51	0.353
4.01	0.250
5.02	0.035
8.22	0.035
9.22	0.023
10.04	0.023
13.04	0.055
14.00	0.122
14.95	0.231
CURVE 31* T = 300	
0.30	0.077
0.35	0.075
0.37	0.086
CURVE 31 (cont.)*	
0.38	0.117
0.40	0.426
0.41	0.813
0.42	0.874
0.44	0.929
0.49	0.955
0.99	0.942
1.20	0.930
1.43	0.888
1.89	0.685
2.51	0.353
4.01	0.250
5.02	0.035
8.22	0.035
9.22	0.023
10.04	0.023
13.04	0.055
14.00	0.122
14.95	0.231
CURVE 32 T = 300	
0.30	0.059
0.33	0.052
0.35	0.065
0.38	0.149
0.40	0.426*
0.41	0.813*
0.42	0.874
0.44	0.913*
0.46	0.935*
0.52	0.954
0.99	0.942*
1.20	0.930*
1.43	0.888*
1.89	0.685*
2.51	0.353*
4.01	0.250*
5.02	0.035*
8.22	0.035*
9.22	0.023*
10.04	0.023*
13.04	0.055*
14.00	0.122*
14.95	0.231

* Not shown on plot

DATA TABLE NO. 14 NORMAL SPECTRAL REFLECTANCE OF ALUMINUM OXIDE PIGMENTED COATINGS (continued)

λ	ρ
CURVE 33	
T = 300	
0.30	0.049
0.34	0.042
0.37	0.065*
0.39	0.104*
0.40	0.171
0.41	0.323
0.43	0.425
0.50	0.715
0.54	0.814
0.58	0.870
0.63	0.906
0.67	0.924
0.78	0.935
0.88	0.942
0.99	0.942*
1.20	0.930*
1.43	0.888*
1.89	0.685*
2.51	0.353*
4.01	0.250*
5.02	0.035*
8.22	0.035*
9.22	0.023*
10.04	0.023*
13.04	0.055*
14.00	0.122*
14.95	0.231*
CURVE 34	
T = 300	
0.30	0.049*
0.34	0.042*
0.37	0.048
0.39	0.066*
0.41	0.156
0.45	0.284
0.50	0.445
0.54	0.634
0.58	0.740*
0.62	0.817
0.66	0.867
0.69	0.889*
0.76	0.914
0.82	0.927
0.92	0.940*
CURVE 34 (cont.)	
0.99	0.942*
1.20	0.930*
1.43	0.888*
1.89	0.685*
2.51	0.353*
4.01	0.250*
5.02	0.035*
8.22	0.035*
9.22	0.023*
10.04	0.023*
13.04	0.055*
14.00	0.122*
14.95	0.231*
CURVE 35*	
T ~ 298	
0.44	0.755
0.60	0.755
CURVE 36*	
T ~ 298	
0.44	0.645
0.60	0.725

* Not shown on plot

FIGURE 15A

ANALYZED CHANGE IN NORMAL SPECTRAL REFLECTANCE OF ALUMINUM OXIDE PIGMENTED COATINGS

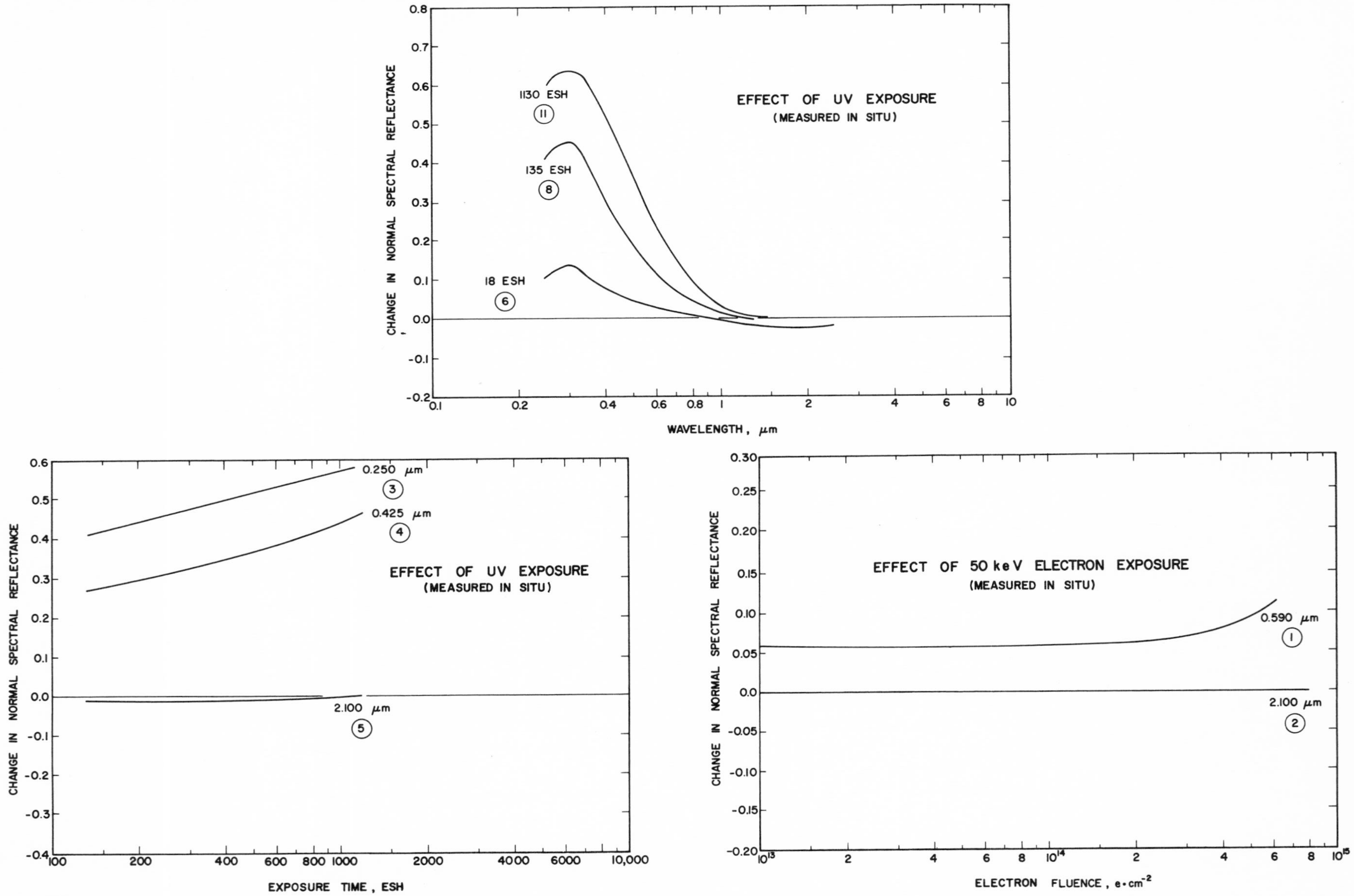

EFFECT OF UV EXPOSURE
(MEASURED IN SITU)

1130 ESH (11)
135 ESH (8)
18 ESH (6)

CHANGE IN NORMAL SPECTRAL REFLECTANCE

0.8, 0.7, 0.6, 0.5, 0.4, 0.3, 0.2, 0.1, 0.0, -0.1, -0.2

0.1, 0.2, 0.4, 0.6, 0.8, 1, 2, 4, 6, 8, 10

WAVELENGTH, μm

EFFECT OF UV EXPOSURE
(MEASURED IN SITU)

0.250 μm (3)
0.425 μm (4)
2.100 μm (5)

CHANGE IN NORMAL SPECTRAL REFLECTANCE

0.6, 0.5, 0.4, 0.3, 0.2, 0.1, 0.0, -0.1, -0.2, -0.3, -0.4

100, 200, 400, 600, 800, 1000, 2000, 4000, 6000, 10,000

EXPOSURE TIME, ESH

EFFECT OF 50 keV ELECTRON EXPOSURE
(MEASURED IN SITU)

0.590 μm (1)
2.100 μm (2)

CHANGE IN NORMAL SPECTRAL REFLECTANCE

0.30, 0.25, 0.20, 0.15, 0.10, 0.05, 0.0, -0.05, -0.10, -0.15, -0.20

10^{13}, 2, 4, 6, 8, 10^{14}, 2, 4, 6, 8, 10^{15}

ELECTRON FLUENCE, $e \cdot cm^{-2}$

SPECIFICATION TABLE NO. 15 CHANGE IN NORMAL SPECTRAL REFLECTANCE OF ALUMINUM OXIDE PIGMENTED COATINGS

Curve No.	Ref. No.	Year	Temperature, K	Wavelength Range, μm	Geometry θ	θ'	ω'	Reported Error, %	Composition (weight percent), Specifications, and Remarks
1	46	1969	295	0. 590	~0°		2π		Linde α-phase Al_2O_3 in potassium silicate binder (0. 28 mm thick); exposed to 50 keV electrons (4 x 10^9-1. 7 x 10^{12} e cm^{-2} sec^{-1}) in vacuum (10^{-9} mm Hg); vacuum maintained by ion pump; electron fluence (e cm^{-2}) is variable; reflectance measured in situ; positive change indicates decrease in reflectance from preirradiation in air, reflectance. [Authors' designation: D]
2	46	1969	295	2. 100	~0°		2π		Above specimen and conditions.
3	46	1969	295	0. 250	~0°		2π		Linde α-phase Al_2O_3 in potassium silicate binder; exposed to UV radiation (4. 7 UV sun rate) in vacuum (10^{-8} mm Hg); vacuum maintained by ion pump; ESH is variable; reflectance measured in situ; positive change indicates decrease in reflectance from preirradiation, in air reflectance. [Authors' designation: D]
4	46	1969	295	0. 425	~0°		2π		Above specimen and conditions.
5	46	1969	295	2. 100	~0°		2π		Above specimen and conditions.
6	46	1969	295	0. 250-2. 450	~0°		2π		Linde α-phase Al_2O_3 in potassium silicate binder (0. 28 mm thick); exposed to UV radiation (with rates equivalent to 5 suns in UV) in vacuum (10^{-7} mm Hg) for 18 ESH; vacuum maintained by ion pump; reflectance measured in situ; positive change indicates decrease in reflectance from preirradiation, in air reflectance; data extracted from smooth curve. [Authors' designation: D]
7	46	1969	295	0. 250-0. 708	~0°		2π		Above specimen and conditions except ESH 53.
8	46	1969	295	0. 250-1. 083	~0°		2π		Above specimen and conditions except ESH 135.
9	46	1969	295	0. 250-1. 420	~0°		2π		Above specimen and conditions except ESH 250.
10	46	1969	295	0. 250-1. 115	~0°		2π		Above specimen and conditions except ESH 490.
11	46	1969	295	0. 250-1. 573	~0°		2π		Above specimen and conditions except ESH 1130.

DATA TABLE NO. 15 CHANGE IN NORMAL SPECTRAL REFLECTANCE OF ALUMINUM OXIDE PIGMENTED COATINGS

[Wavelength, λ, μm; Reflectance, ρ; Temperature, T, K]

CURVE 1, $\lambda = 0.590$, T = 295

Electron fluence	Δρ
1×10^{13}	0.06
2×10^{14}	0.06
6×10^{14}	0.11

CURVE 2, $\lambda = 2.100$, T = 295

Electron fluence	Δρ
1×10^{13}	0.0
2×10^{14}	0.0
6×10^{14}	0.0
8×10^{14}	0.0

CURVE 3, $\lambda = 0.250$, T = 295

ESH	Δρ
135	0.41
250	0.46
490	0.51
1130	0.58

CURVE 4, $\lambda = 0.425$, T = 295

ESH	Δρ
135	0.27
250	0.31
490	0.37
770	0.42
1130	0.45

CURVE 5, $\lambda = 2.100$, T = 295

ESH	Δρ
135	-0.01
250	-0.01
490	-0.01
1130	0.0

CURVE 6, T = 295

λ	Δρ
0.250	0.109
0.293	0.139
0.303	0.143
0.364	0.098
0.456	0.057
0.803	0.011
1.202	-0.013
1.509	0.000
1.723	-0.008
1.916	-0.018
2.450	-0.016

CURVE 7, T = 295

λ	Δρ
0.250	0.164
0.284	0.215
0.300	0.220
0.309	0.220
0.363	0.163
0.433	0.120
0.520	0.082
0.708	0.023

CURVE 8, T = 295

λ	Δρ
0.250	0.415
0.297	0.467
0.390	0.326
0.501	0.170
0.597	0.102
0.907	0.027
0.946	0.020
1.083	0.005

CURVE 9, T = 295

λ	Δρ
0.250	0.478
0.305	0.521
0.369	0.391
0.448	0.284

CURVE 9 (cont.)

λ	Δρ
0.531	0.192
0.616	0.137
0.708	0.094
0.857	0.045
0.959	0.025
1.038	0.017
1.275	0.005
1.420	0.000

CURVE 10, T = 295

λ	Δρ
0.250	0.533
0.303	0.581
0.488	0.275
0.543	0.216
0.687	0.109
0.853	0.057
0.953	0.032
1.115	0.014

CURVE 11, T = 295

λ	Δρ
0.250	0.600
0.300	0.650
0.372	0.540
0.434	0.467
0.499	0.350
0.527	0.316
0.652	0.194
0.704	0.138
0.746	0.118
0.869	0.070
0.943	0.045
1.039	0.033
1.202	0.016
1.345	0.009
1.573	0.000

SPECIFICATION TABLE NO. 16 NORMAL SPECTRAL ABSORPTANCE OF ALUMINUM OXIDE PIGMENTED COATINGS

Curve No.	Ref. No.	Year	Temperature, K	Wavelength Range, μm	Geometry θ	Reported Error, %	Composition (weight percent), Specifications, and Remarks
1*	76	1966	~298	0.300-0.700	~0°		Alucer MC Al_2O_3 pigment in PS7 potassium silicate binder; PBR 5.37; ground in ball mill for ~15 hrs; aluminum substrate; surface sandblasted and cleaned; air brush application (5 coats). [Author's designation: Batch 2]

DATA TABLE NO. 16 NORMAL SPECTRAL ABSORPTANCE OF ALUMINUM OXIDE PIGMENTED COATINGS

[Wavelength, λ, μm; Absorptance, α; Temperature, T, K]

λ	α
CURVE 1*	
0.300	0.050
0.350	0.050
0.400	0.041
0.450	0.033
0.500	0.020
0.550	0.020
0.600	0.022
0.650	0.028
0.700	0.031

* No plot given

SPECIFICATION TABLE NO. 17   NORMAL SOLAR ABSORPTANCE OF ALUMINUM OXIDE PIGMENTED COATINGS

Curve No.	Ref. No.	Year	Temperature Range, K	Geometry θ	Reported Error, %	Composition (weight percent), Specifications and Remarks
1*	17	1968	298	$\sim 0°$		CM-145 Al_2O_3 inorganic paint; incident energy is earth Albedo approximated by the solar spectrum truncated below 0.35 μm; sample supplied by ESRO II Project Group.
2*	17	1968	298	$\sim 0°$		CM-145 Al_2O_3 inorganic paint; sample supplied by ESRO II Project Group.
3*	68	1969	300	$\sim 0°$		55 Al_2O_3, 30 ZnO and 15 TiO_2 in potassium silicate binder ($\sim$0.089 mm thick) on 2024 aluminum substrate ($\sim$1.59 mm thick); absorptance calculated from reflectance; property measured in air. [Authors' designation: Al_2O_3/PS-7]
4*	68	1969	300	$\sim 0°$		Similar to above specimen and conditions except exposed to vacuum (10^{-8} mm Hg) for 300 hr; vacuum maintained by ion pump; property measured in air after vacuum exposure.
5*	68	1969	300	$\sim 0°$		Similar to Curve 3 specimen and conditions except exposed to vacuum (10^{-5} mm Hg) for 300 hr; vacuum maintained by oil-diffusion pump; property measured in air after vacuum exposure.
6*	68	1969	300	$\sim 0°$		Similar to above specimen and conditions except exposed to UV radiation in vacuum for 300 hr; H_2 gas UV source.
7*	68	1969	300	$\sim 0°$		Similar to above specimen and conditions except He gas UV source.
8*	45	1969	298	$\sim 0°$		Al_2O_3 in potassium silicate binder on aluminum substrate; absorptance calculated from normal spectral reflectance; property measured in air. [Authors' designation: D3]
9*	45	1969	281	$\sim 0°$		Above specimen and conditions except exposed to 20 keV electrons (flux 1 x 10^{10} - 5 x 10^{11} $e \cdot cm^{-2} \cdot sec^{-1}$) in dark in vacuum ($10^{-8}$ mm Hg); vacuum maintained by ion pump; substrate held at 281 ± 2 K; property measured in situ after exposure to $10^{16} e \cdot cm^{-2}$.
10*	45	1969	298	$\sim 0°$		Similar to Curve 8 specimen and conditions.
11*	45	1969	281	$\sim 0°$		Above specimen and conditions except exposed to 80 keV electrons (flux 1 x 10^{10} - 5 x 10^{11} e cm^{-2} sec^{-1}) in dark in vacuum (10^{-8} mm Hg); vacuum maintained by ion pump; substrate held at 281 ± 2 K; property measured in situ after exposure to 10^{16} e cm^{-2}.
12*	75	1969	77	$\sim 0°$		Aluminum oxide in potassium silicate binder; exposed to vacuum (5 x 10^{-9} - 2 x 10^{-7} mm Hg); vacuum maintained with diffusion pump; absorptance calculated from R (2π, 0); property measured in situ.
13*	75	1969	77	$\sim 0°$		Above specimen and conditions except exposed to 350 SH at 6 sun intensity in vacuum.
14*	75	1969	77	$\sim 0°$		Similar to Curve 12 specimen and conditions except exposed to 5.8 x 10^{15} e cm^{-2} at 8.6 x 10^{10} - 1.6 x 10^{12} e cm^{-2} sec^{-1} in vacuum.
15*	75	1969	77	$\sim 0°$		Similar to Curve 12 specimen and conditions except simultaneous exposure to 5.8 x 10^{15} e cm^{-2} at 3.5 x 10^{10} e cm^{-2} sec^{-1} and 350 SH at 8 suns in vacuum.

* No plot given

SPECIFICATION TABLE NO. 17 NORMAL SOLAR ABSORPTANCE OF ALUMINUM OXIDE PIGMENTED COATINGS (continued)

Curve No.	Ref. No.	Year	Temperature Range, K	Geometry θ'	Reported Error, %	Composition (weight percent), Specifications and Remarks
16*	27	1964	~298	~0°		C-35 $Al_2O_3 \cdot 3H_2O$ (trihydrated aluminum oxide from Alcoa) in PS7 potassium silicate binder; PBR 4.30; solids content 63.9%; aluminum substrate abraded with No. 60 Aloxite cloth; sample cured at 413 K for 18 hrs. [Authors' designation: Sample C2]
17*	27	1964	~298	~0°		Above specimen and conditions except irradiated in vacuum (10^{-6} mm Hg) with 200 ESH at solar factor of 3 suns.

* No plot given

DATA TABLE NO. 17 NORMAL SOLAR ABSORPTANCE OF ALUMINUM OXIDE PIGMENTED COATINGS

[Temperature, T, K; Absorptance, α]

T	α	T	α
CURVE 1*		CURVE 13*	
298	0.12	77	0.16
CURVE 2*		CURVE 14*	
298	0.16	77	0.19
CURVE 3*		CURVE 15*	
300	0.22	77	0.24
CURVE 4*		CURVE 16*	
300	0.26	298	0.035
CURVE 5*		CURVE 17*	
300	0.25	298	0.371
CURVE 6*			
300	0.32		
CURVE 7*			
300	0.38		
CURVE 8*			
298	0.09		
CURVE 9*			
281	0.13		
CURVE 10*			
298	0.11		
CURVE 11*			
281	0.14		
CURVE 12*			
77	0.11		

* No plot given

SPECIFICATION TABLE NO. 18 ABSORPTANCE TO EMITTANCE RATIO OF ALUMINUM OXIDE PIGMENTED COATINGS

Curve No.	Ref. No.	Year	Temperature Range, K	Reported Error, %	Composition (weight percent), Specifications and Remarks
1*	58	1967	~298	15.2	Al_2O_3 in K_2O-3.3 SiO_2 binder on aluminum (6061 Al alloy) substrate; substrate has phosphate Al surface; property deduced from reflectance; laboratory data measured 6 mo. before flight of ATS-1.
2*	58	1967	unknown	8.3	Above specimen and conditions except property deduced from temp of substrate from "initial" point flight data (48 hrs after launch) of ATS-1.

DATA TABLE NO. 18 ABSORPTANCE TO EMITTANCE RATIO OF ALUMINUM OXIDE PIGMENTED COATINGS

[Temperature, T, K; Absorptance to Emittance Ratio, α/ϵ]

T	α/ϵ
CURVE 1*	
~298	0.14
CURVE 2*	
unknown	0.21

* No plot given

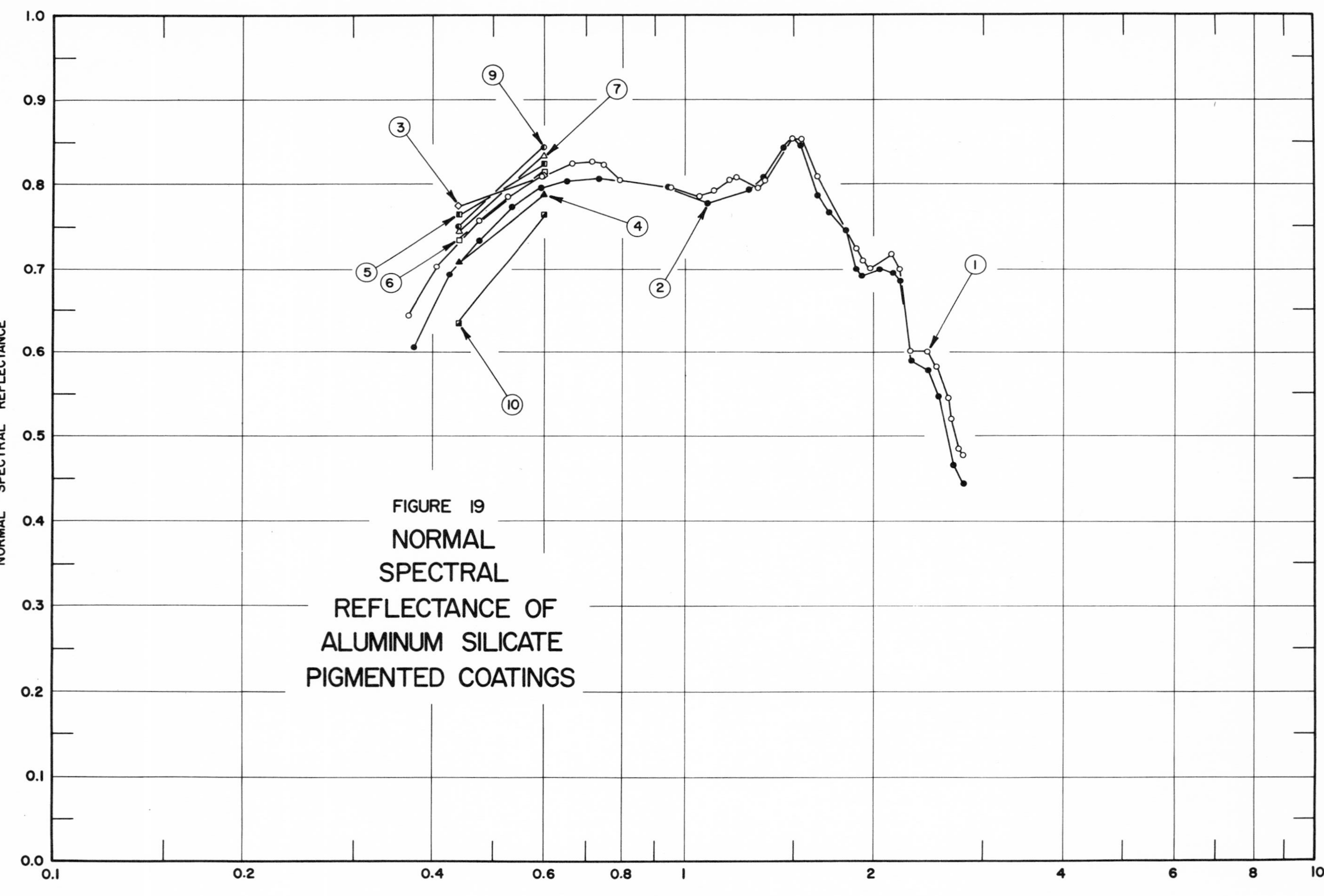

FIGURE 19 NORMAL SPECTRAL REFLECTANCE OF ALUMINUM SILICATE PIGMENTED COATINGS

SPECIFICATION TABLE NO. 19 NORMAL SPECTRAL REFLECTANCE OF ALUMINUM SILICATE PIGMENTED COATINGS

Curve No.	Ref. No.	Year	Temperature, K	Wavelength Range, μm	Geometry θ	θ'	ω'	Reported Error, %	Composition (weight percent), Specifications, and Remarks
1	20	1963	~298	0.367-2.762	~0°		2π		Molochite No. 6 aluminum silicate in PS-7 potassium silicate binder; PBR 4.30; solids content 56.9%; paint ball milled; aluminum substrate grit blasted with 40 mesh silicon carbide; air brush application; sample cured at 413 K for 18 hrs; data extracted from smooth curve. [Authors' designation: Sample C36]
2	20	1963	~298	0.374-2.774	~0°		2π		Above specimen and conditions except exposed in vacuum (10^{-5} mm Hg) to 300 ESH at solar factor of 3 suns.
3	27	1964	~298	0.44-0.60	~0°		2π		Molochite SF aluminum silicate in PS-7 potassium silicate binder; PBR 4.30; solids content 56.9%; aluminum substrate abraded with No. 60 Aloxite cloth; sample cured at 413 K for 18 hrs. [Authors' designation: Sample C34]
4	27	1964	~298	0.44-0.60	~0°		2π		Above specimen and conditions except irradiated in vacuum (10^{-6} mm Hg) with 300 ESH at solar factor of 3 suns.
5	27	1964	~298	0.44-0.60	~0°		2π		Molochite No. 6 aluminum silicate in PS-7 potassium silicate binder; PBR 4.30; solids content 56.9%; aluminum substrate abraded with No. 60 Aloxite cloth; sample cured at 413 K for 18 hrs. [Authors' designation: Sample C36]
6	27	1964	~298	0.44-0.60	~0°		2π		Above specimen and conditions except irradiated in vacuum (10^{-6} mm Hg) with 300 ESH at solar factor of 3 suns.
7	27	1964	~298	0.44-0.60	~0°		2π		Molochite No. 6 aluminum silicate (HCl leached) in PS-7 potassium silicate binder; PBR 4.30; solids content 61.7%; aluminum substrate abraded with No. 60 Aloxite cloth; sample cured at 413 K for 18 hrs. [Authors' designation: Sample C39]
8*	27	1964	~298	0.44-0.60	~0°		2π		Above specimen and conditions except irradiated in vacuum (10^{-6} mm Hg) with 250 ESH at solar factor of 2.50 suns.
9	27	1964	~298	0.44-0.60	~0°		2π		Molochite No. 6 aluminum silicate (HCl leached) in PS-7 potassium silicate binder; PBR 4.30; solids content 61.7%; aluminum substrate abraded with No. 60 Aloxite cloth; sample cured at 413 K for 18 hrs. [Authors' designation: Sample C42]
10	27	1964	~298	0.44-0.60	~0°		2π		Above specimen and conditions except irradiated in vacuum (10^{-6} mm Hg) with 2100 ESH at solar factor of 10 suns.

* Not shown on plot

DATA TABLE NO. 19 NORMAL SPECTRAL REFLECTANCE OF ALUMINUM SILICATE PIGMENTED COATINGS

[Wavelength, λ, μm; Reflectance, ρ; Temperature, T, K]

CURVE 1
T ~ 298

λ	ρ
0.367	0.644
0.404	0.701
0.471	0.759
0.528	0.788
0.598	0.810
0.664	0.823
0.715	0.826
0.749	0.822
0.796	0.805
0.951	0.797
1.068	0.788
1.130	0.792
1.180	0.804
1.214	0.809
1.289	0.798*
1.319	0.797
1.355	0.805
1.457	0.851*
1.495	0.856
1.532	0.853
1.632	0.810
1.768	0.779*
1.883	0.724
1.928	0.710
1.975	0.708
2.061	0.719*
2.124	0.719
2.168	0.712*
2.204	0.700
2.249	0.628*
2.295	0.600
2.432	0.600
2.512	0.582
2.611	0.545
2.652	0.520
2.685	0.492*
2.711	0.484
2.737	0.479*
2.762	0.479

CURVE 2
T ~ 298

λ	ρ
0.374	0.606
0.426	0.693
0.473	0.737
0.531	0.772
0.597	0.796
0.654	0.802
0.737	0.807
0.793	0.807*
0.946	0.798
1.096	0.779
1.276	0.794
1.341	0.809
1.431	0.846
1.495	0.857*
1.537	0.848
1.647	0.788
1.702	0.768
1.811	0.749
1.885	0.701
1.921	0.691
1.971	0.695*
1.997	0.700*
2.048	0.700
2.081	0.695*
2.161	0.695
2.200	0.685
2.249	0.623*
2.308	0.590
2.436	0.579
2.531	0.547
2.599	0.538*
2.698	0.464
2.741	0.443*
2.774	0.442

CURVE 3
T ~ 298

λ	ρ
0.44	0.775
0.60	0.810

CURVE 4
T ~ 298

λ	ρ
0.44	0.710
0.60	0.790

CURVE 5
T ~ 298

λ	ρ
0.44	0.765
0.60	0.825

CURVE 6
T ~ 298

λ	ρ
0.44	0.735
0.60	0.815

CURVE 7
T ~ 298

λ	ρ
0.44	0.745
0.60	0.835

CURVE 8*
T ~ 298

λ	ρ
0.44	0.745
0.60	0.835

CURVE 9
T ~ 298

λ	ρ
0.44	0.750
0.60	0.845

CURVE 10
T ~ 298

λ	ρ
0.44	0.635
0.60	0.765

* Not shown on plot

SPECIFICATION TABLE NO. 20 NORMAL SOLAR ABSORPTANCE OF ALUMINUM SILICATE PIGMENTED COATINGS

Curve No.	Ref. No.	Year	Temperature Range, K	Geometry θ	Reported Error, %	Composition (weight percent), Specifications and Remarks
1*	27	1964	~298	~0°		Molochite SF aluminum silicate in PS7 potassium silicate binder; PBR 4.30; solids content 56.9%; aluminum substrate abraded with No. 60 Aloxite cloth; sample cured at 413 K for 18 hrs. [Authors' designation: Sample C34]
2*	27	1964	~298	~0°		Above specimen and conditions except irradiated in vacuum (10^{-6} mm Hg) with 300 ESH at solar factor of 3 suns.
3*	27	1964	~298	~0°		Molochite No. 6 aluminum silicate in PS7 potassium silicate binder; PBR 4.30; solids content 56.9%; aluminum substrate abraded with No. 60 Aloxite cloth; sample cured at 413 K for 18 hrs. [Authors' designation: Sample C36]
4*	27	1964	~298	~0°		Above specimen and conditions except irradiated in vacuum (10^{-6} mm Hg) with 300 ESH at solar factor of 3 suns.

DATA TABLE NO. 20 NORMAL SOLAR ABSORPTANCE OF ALUMINUM SILICATE PIGMENTED COATINGS

[Temperature, T, K; Absorptance, α]

T	α
CURVE 1*	
298	0.251
CURVE 2*	
298	0.281
CURVE 3*	
298	0.243
CURVE 4*	
298	0.260

* No plot given

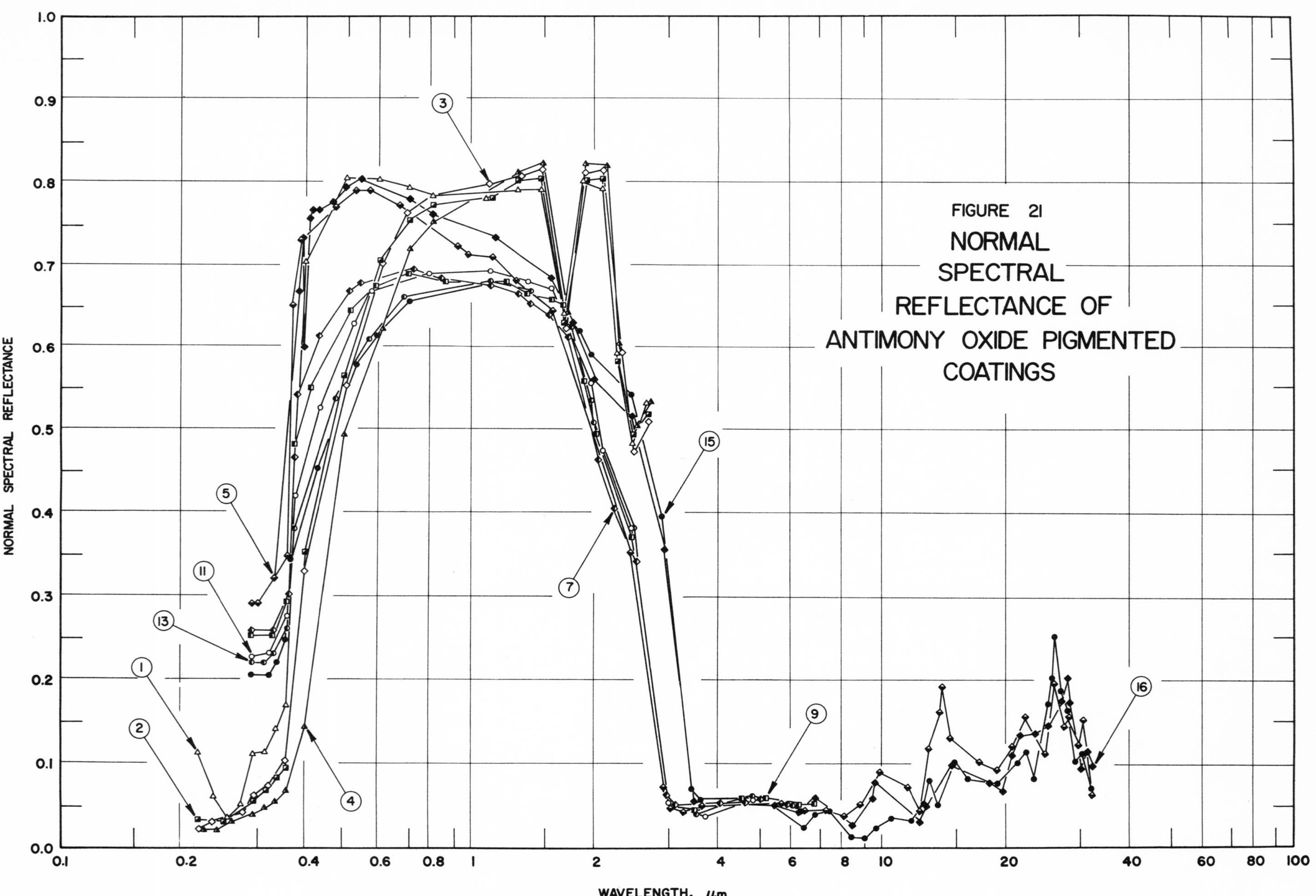

FIGURE 21 NORMAL SPECTRAL REFLECTANCE OF ANTIMONY OXIDE PIGMENTED COATINGS

SPECIFICATION TABLE NO. 21 NORMAL SPECTRAL REFLECTANCE OF ANTIMONY OXIDE PIGMENTED COATINGS

Curve No.	Ref. No.	Year	Temperature, K	Wavelength Range, μm	Geometry θ	θ'	ω'	Reported Error, %	Composition (weight percent), Specifications, and Remarks
1	1	1960	358	0.219-2.697	~0°		2π		Antimony oxide in silicone (0.051 mm thick); 30 antimony oxide; anodized aluminum alloy 24S-T substrate; measured in vacuum (10^{-5} mm Hg); measured relative to MgO.
2	1	1960	358	0.219-2.701	~0°		2π		Above specimen and conditions except exposed to UV radiation for 20 hrs.
3	1	1960	358	0.220-2.701	~0°		2π		Above specimen and conditions except exposed to UV radiation for 60 hrs.
4	1	1960	358	0.225-2.717	~0°		2π		Above specimen and conditions except exposed to UV radiation for 100 hrs.
5	84	1964	~298	0.298-32.28	~0°		2π		Sb_2O_3 in potassium silicate binder; data extracted from smooth curve; property converted from R (2π, 0°) which was measured relative to MgO for λ 0.3-2.0 μm and relative to NiO for 1.5-32.0 μm. [Authors' designation: Control]
6*	84	1964	~298	0.298-5.83	~0°		2π		Similar to above specimen and conditions. [Authors' designation: Sample 31]
7	84	1964	~298	0.298-5.83	~0°		2π		Above specimen and conditions except exposed to UV radiation for 100 ESH from AH-6 lamp (lamp-sample distance 6.35 cm) in vacuum (~5 x 10^{-7} mm Hg); vacuum maintained by oil diffusion pump.
8*	84	1964	~298	0.298-6.99	~0°		2π		Similar to curve 5 specimen and conditions. [Authors' designation: Sample 32]
9	84	1964	~298	0.298-6.99	~0°		2π		Above specimen and conditions except exposed to UV radiation for 250 ESH from AH-6 lamp (lamp-sample distance 6.35 cm) in vacuum (~5 x 10^{-7} mm Hg); vacuum maintained by oil diffusion pump.
10*	84	1964	~298	0.298-6.95	~0°		2π		Similar to curve 5 specimen and conditions. [Authors' designation: Sample 34]
11	84	1964	~298	0.298-6.95	~0°		2π		Above specimen and conditions except exposed to UV radiation for 400 ESH from AH-6 lamp (lamp-sample distance 6.35 cm) in vacuum (~5 x 10^{-7} mm Hg); vacuum maintained by oil diffusion pump.
12*	84	1964	~298	0.298-6.95	~0°		2π		Similar to curve 5 specimen and conditions. [Authors' designation: Sample 35]
13	84	1964	~298	0.298-6.95	~0°		2π		Above specimen and conditions except exposed to UV radiation for 650 ESH from AH-6 lamp (lamp-sample distance 6.35 cm) in vacuum (~5 x 10^{-7} mm Hg); vacuum maintained by oil diffusion pump.
14*	84	1964	~298	0.298-32.28	~0°		2π		Similar to curve 5 specimen and conditions. [Authors' designation: Sample 36]
15	84	1964	~298	0.298-32.28	~0°		2π		Above specimen and conditions except exposed to UV radiation for 1000 ESH from AH-6 lamp (lamp-sample distance 6.35 cm) in vacuum (~5 x 10^{-7} mm Hg); vacuum maintained by oil diffusion pump.
16	84	1964	~298	0.298-32.35	~0°		2π		Curve 5 specimen and conditions except measured after all other samples had been irradiated for 1000 ESH.

* Not shown on plot

DATA TABLE NO. 21 NORMAL SPECTRAL REFLECTANCE OF ANTIMONY OXIDE PIGMENTED COATINGS

[Wavelength, λ, μm; Reflectance, ρ; Temperature, T, K]

λ	ρ
CURVE 1	
T = 358	
0.219	0.116
0.239	0.062
0.259	0.037
0.279	0.053
0.299	0.112
0.319	0.115
0.338	0.142
0.359	0.170
0.399	0.706
0.500	0.807
0.600	0.805
0.701	0.796
0.802	0.788
1.099	0.785
1.301	0.794
1.496	0.795
1.697	0.641
1.895	0.804
2.091	0.795
2.289	0.595
2.497	0.485
2.697	0.533
CURVE 2	
T = 358	
0.219	0.036
0.254	0.033
0.298	0.059
0.319	0.070
0.340	0.084
0.359	0.096
0.398	0.355
0.499	0.567
0.600	0.708
0.702	0.757
0.802	0.777
1.130	0.784
1.302	0.804
1.492	0.806
1.697	0.631
1.919	0.803
2.096	0.806
2.292	0.585
CURVE 2 (cont.)	
2.495	0.496
2.701	0.520
CURVE 3	
T = 358	
0.220	0.024
0.237	0.033
0.280	0.043
0.300	0.062
0.321	0.072
0.338	0.089*
0.357	0.105
0.399	0.332
0.500	0.554
0.604	0.703
0.700	0.766
0.805	0.778*
1.110	0.800
1.327	0.810
1.506	0.817
1.704	0.624
1.909	0.814
2.101	0.817
2.324	0.595
2.501	0.473
2.701	0.510
CURVE 4	
T = 358	
0.225	0.023
0.242	0.023
0.263	0.032
0.277	0.034*
0.298	0.041
0.319	0.050
0.338	0.057
0.357	0.070
0.398	0.145
0.498	0.495
0.601	0.625
0.702	0.721
0.803	0.755
1.303	0.814
1.503	0.825
CURVE 4 (cont.)	
1.717	0.641
1.907	0.825
2.133	0.821
2.303	0.607
2.514	0.503
2.717	0.531
CURVE 5	
T ~ 298	
0.298	0.293
0.305	0.293
0.336	0.323
0.360	0.350
0.370	0.651
0.389	0.734
0.405	0.763*
0.430	0.769*
0.474	0.771
0.487	0.781*
0.526	0.793
0.567	0.793
0.679	0.777
0.931	0.726
0.995	0.717
1.12	0.712
1.18	0.706*
1.29	0.683
1.59	0.646
1.69	0.631*
2.46	0.353
2.96	0.075
3.08	0.050
3.30	0.045
3.68	0.052
5.79	0.055
6.53	0.049
7.27	0.049
8.09	0.040
8.93	0.054
9.97	0.092
11.53	0.075
12.02	0.063*
12.35	0.047
13.00	0.120
CURVE 5 (cont.)	
13.83	0.162
14.06	0.193
14.25	0.159*
14.65	0.131
15.24	0.116*
17.10	0.105
19.01	0.095
20.89	0.121
22.33	0.159
23.17	0.153*
23.76	0.132*
25.00	0.114
25.70	0.182*
26.18	0.199
26.60	0.186*
27.16	0.157*
27.79	0.146
28.64	0.157
29.30	0.123*
30.06	0.123
30.40	0.140*
31.04	0.153
32.28	0.066
CURVE 6*	
T ~ 298	
0.298	0.296
0.331	0.308
0.362	0.332
0.372	0.468
0.376	0.626
0.389	0.705
0.401	0.745
0.435	0.749
0.474	0.749
0.504	0.760
0.570	0.762
0.709	0.727
0.857	0.694
1.04	0.672
1.20	0.662
1.36	0.640
1.52	0.615
1.59	0.615
CURVE 6 (cont.)*	
1.70	0.596
2.24	0.408
2.51	0.344
2.55	0.320
3.00	0.065
3.11	0.054
4.05	0.053
4.69	0.056
5.14	0.060
5.83	0.054
CURVE 7	
T ~ 298	
0.298	0.260
0.331	0.260
0.362	0.303
0.372	0.468
0.380	0.541
0.426	0.616
0.505	0.670
0.540	0.680
0.647	0.696*
0.724	0.698
0.841	0.686
1.12	0.677
1.30	0.667
1.40	0.653
1.57	0.641
1.71	0.615
2.05	0.464
2.24	0.408
2.51	0.344
2.55	0.320*
3.00	0.065
3.11	0.054
4.05	0.053
4.69	0.056
5.14	0.060
5.83	0.054
CURVE 8*	
T ~ 298	
0.298	0.299
0.331	0.315
0.349	0.328
0.359	0.342
0.376	0.663
0.389	0.730
0.400	0.766
0.453	0.770
0.507	0.770
0.601	0.756
0.714	0.727
0.796	0.699
0.899	0.694
0.997	0.680
1.19	0.680
1.42	0.649
1.58	0.641
1.69	0.625
1.77	0.602
2.02	0.498
2.48	0.371
3.01	0.052
3.52	0.049
4.60	0,061
5.23	0.061
6.39	0.054
6.99	0.056
CURVE 9	
T ~ 298	
0.298	0.256
0.331	0.256
0.360	0.296
0.377	0.483
0.410	0.551
0.514	0.648
0.598	0.678
0.704	0.691
0.869	0.682
1.22	0.682
1.38	0.667
1.58	0.660
1.69	0.652
CURVE 9 (cont.)	
1.90	0.560
2.02	0.498
2.48	0.371
3.01	0.052*
3.52	0.049
4.60	0.061
5.23	0.061
6.39	0.054
6.99	0.056
CURVE 10*	
T ~ 298	
0.298	0.290
0.333	0.298
0.360	0.337
0.370	0.463
0.375	0.660
0.399	0.754
0.437	0.761
0.516	0.761
0.578	0.753
0.696	0.729
0.829	0.700
0.995	0.683
1.10	0.680
1.34	0.657
1.69	0.619
2.10	0.476
2.48	0.383
3.02	0.054
3.27	0.046
3.71	0.040
4.92	0.060
5.23	0.060
5.66	0.052
6.95	0.056
CURVE 11	
T ~ 298	
0.298	0.230
0.328	0.234
0.342	0.262*
0.360	0.278
CURVE 11 (cont.)	
0.379	0.421
0.431	0.529
0.521	0.631
0.579	0.670
0.717	0.698*
0.798	0.691
1.11	0.693
1.39	0.681
1.58	0.671
1.69	0.651*
1.96	0.557
2.02	0.500*
2.10	0.476
2.48	0.383
3.02	0.054
3.27	0.046*
3.71	0.040
4.92	0.060
5.23	0.060*
5.66	0.052*
6.95	0.056*
CURVE 12*	
T ~ 298	
0.298	0.301
0.339	0.311
0.360	0.351
0.369	0.531
0.375	0.672
0.402	0.754
0.462	0.754
0.508	0.760
0.532	0.755
0.562	0.755
0.895	0.683
1.04	0.683
1.32	0.664
1.59	0.637
1.71	0.621
1.83	0.583
2.00	0.510
2.50	0.382
3.01	0.053
3.52	0.044

* Not shown on plot

DATA TABLE NO. 21 NORMAL SPECTRAL REFLECTANCE OF ANTIMONY OXIDE PIGMENTED COATINGS (continued)

λ	ρ
CURVE 12 (cont.)*	
T ~ 298	
4.94	0.063
6.01	0.053
6.39	0.054
6.95	0.059
CURVE 13	
T ~ 298	
0.298	0.223
0.316	0.223
0.332	0.233
0.360	0.262
0.373	0.381
0.477	0.539
0.564	0.611
0.696	0.664
1.11	0.681
1.40	0.670
1.57	0.659*
1.73	0.629
1.99	0.537
2.00	0.510
2.50	0.382
3.01	0.053*
3.52	0.044
4.94	0.063
6.01	0.053
6.39	0.054*
6.95	0.059*
CURVE 14*	
T ~ 298	
0.298	0.289
0.316	0.289
0.343	0.307
0.355	0.327
0.367	0.559
0.367	0.651
0.382	0.703
0.387	0.748
0.403	0.755
0.476	0.757
0.568	0.750
0.663	0.734
CURVE 14 (cont.)*	
0.829	0.695
1.04	0.679
1.01	0.672
1.18	0.672
1.40	0.645
1.52	0.640
1.67	0.611
2.48	0.362
2.90	0.077
2.99	0.054
3.37	0.042
5.03	0.057
5.74	0.047
6.51	0.027
6.96	0.042
7.41	0.048
8.79	0.029
10.35	0.065
10.93	0.065
12.50	0.053
13.03	0.139
13.93	0.151
15.10	0.114
16.10	0.099
19.86	0.091
23.12	0.151
24.54	0.115
26.42	0.156
27.47	0.176
28.24	0.204
28.57	0.169
29.92	0.119
31.18	0.148
32.28	0.104
CURVE 15	
T ~ 298	
0.298	0.207
0.324	0.207
0.342	0.222
0.358	0.250
0.369	0.346
0.425	0.454
0.529	0.580
0.598	0.619
CURVE 15 (cont.)	
0.704	0.658
1.11	0.678*
1.38	0.666*
1.70	0.640*
1.85	0.621
1.99	0.594
2.46	0.544
2.91	0.399
3.45	0.072
3.61	0.060
5.08	0.060*
6.51	0.027
6.96	0.042
7.44	0.048
8.45	0.017
9.01	0.013
9.77	0.026
10.47	0.039
11.93	0.036
12.64	0.055
12.91	0.066*
13.00	0.082
13.96	0.053
15.03	0.104
16.10	0.084
19.01	0.080
21.62	0.102
22.23	0.116
23.28	0.084
25.46	0.171
26.06	0.204
26.18	0.254
27.16	0.137
28.11	0.165
29.85	0.105
30.97	0.113
32.28	0.072
CURVE 16	
T ~ 298	
0.298	0.293*
0.305	0.293*
0.336	0.323*
0.360	0.350*
0.394	0.601
CURVE 16 (cont.)	
0.386	0.670
0.392	0.737
0.405	0.760
0.418	0.770
0.433	0.770
0.461	0.780
0.500	0.799
0.542	0.806
0.709	0.784
0.812	0.763
1.15	0.736
1.58	0.686
1.78	0.631
2.00	0.561
2.47	0.519
2.98	0.358
3.50	0.059
3.67	0.053*
5.22	0.061*
5.54	0.052
6.39	0.046
6.91	0.061
8.51	0.030
9.52	0.061
9.68	0.080
9.97	0.092*
11.50	0.075*
12.38	0.032
12.91	0.052
13.00	0.120*
13.83	0.162*
14.06	0.193*
14.25	0.163*
14.65	0.131*
14.99	0.100
18.49	0.080
19.09	0.070
20.84	0.112
21.62	0.137
23.93	0.139
25.29	0.148
27.16	0.177
28.11	0.204
28.64	0.174
30.06	0.098
31.84	0.118
32.35	0.10[illegible]

* Not shown on plot

SPECIFICATION TABLE NO. 22 NORMAL INTEGRATED ABSORPTANCE OF ANTIMONY OXIDE PIGMENTED COATINGS

Curve No.	Ref. No.	Year	Temperature Range, K	Geometry θ	Reported Error, %	Composition (weight percent), Specifications and Remarks
1*	23	1964	~298	~0°		Antimony trioxide in potassium silicate binder; prepared by Hughes Aircraft Co.; computed from spectral reflectance data; integrated between 0.40 and 0.70 μm.

DATA TABLE NO. 22 NORMAL INTEGRATED ABSORPTANCE OF ANTIMONY OXIDE PIGMENTED COATINGS

[Temperature, T, K; Absorptance, α]

T	α
CURVE 1*	
298	0.14

* No plot given

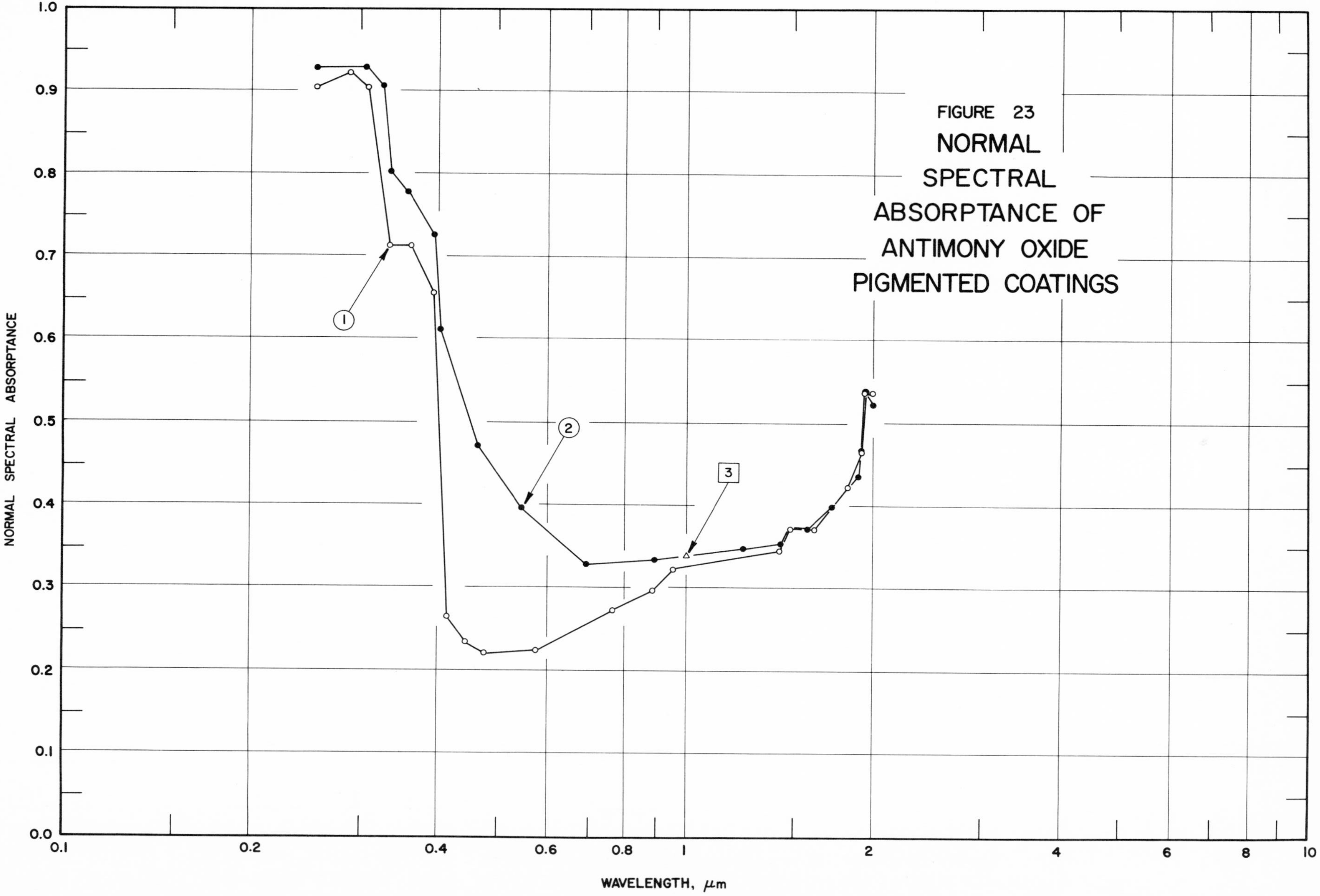

FIGURE 23 NORMAL SPECTRAL ABSORPTANCE OF ANTIMONY OXIDE PIGMENTED COATINGS

SPECIFICATION TABLE NO. 23 NORMAL SPECTRAL ABSORPTANCE OF ANTIMONY OXIDE PIGMENTED COATINGS

Curve No.	Ref. No.	Year	Temperature, K	Wavelength Range, μm	Geometry θ	Reported Error, %	Composition (weight percent), Specifications, and Remarks
1	26	1964	~298	0.253-2.007	~0°		Antimony trioxide in potassium silicate binder; prepared by Hughes Aircraft Co.; aluminum substrate; data extracted from smooth curve.
2	26	1964	~298	0.253-2.005	~0°		Above specimen and conditions except exposed in vacuum (8×10^{-7} mm Hg) to 100 hrs UV radiation at solar factor of 3 suns; sample temp during exposure, 287 K.
3	23	1964	~298	1.0	~0°		Antimony trioxide in potassium silicate binder; prepared by Hughes Aircraft Co.

DATA TABLE NO. 23 NORMAL SPECTRAL ABSORPTANCE OF ANTIMONY OXIDE PIGMENTED COATINGS

[Wavelength, λ, μm; Absorptance, α; Temperature, T, K]

λ	α
CURVE 1	**T ~ 298**
0.253	0.905
0.289	0.921
0.308	0.908
0.332	0.712
0.360	0.712
0.391	0.655
0.417	0.266
0.444	0.235
0.477	0.221
0.577	0.226
0.765	0.274
0.881	0.297
0.955	0.322
1.418	0.347
1.476	0.371
1.611	0.371
1.833	0.425
1.920	0.469
CURVE 1 (cont.)	
1.949	0.538
2.007	0.538
CURVE 2	**T ~ 298**
0.253	0.930
0.302	0.930
0.325	0.909
0.334	0.803
0.358	0.779
0.393	0.729
0.401	0.612
0.461	0.471
0.545	0.398
0.693	0.330
0.893	0.335
1.241	0.350
1.423	0.353
CURVE 2 (cont.)	
1.477	0.371*
1.582	0.371
1.737	0.400
1.847	0.421*
1.902	0.438
1.921	0.470
1.955	0.540
2.002	0.523
CURVE 3	**T ~ 298**
1.0	0.34

* Not shown on plot

SPECIFICATION TABLE NO. 24 NORMAL SOLAR ABSORPTANCE OF ANTIMONY OXIDE PIGMENTED COATINGS

Curve No.	Ref. No.	Year	Temperature Range, K	Geometry θ	Reported Error, %	Composition (weight percent), Specifications and Remarks
1*	28	1964	~298	~0°		Antimony trioxide in potassium silicate binder; prepared by Hughes Aircraft Co.; computed from spectral reflectance data.
2*	28	1964	~298	~0°		Similar to above specimen and conditions.
3*	28	1964	~298	~0°		Similar to above specimen and conditions except gamma-irradiated in vacuum (10^{-6} mm Hg) at 320 K to a total dose of 77 Megarads (dose rate, ~1.5 Megarads-hr^{-1}, Cobalt-60 source); measured within one hr after exposure to air.
4*	28	1964	~298	~0°		Similar to above specimen and conditions.
5*	28	1964	~298	~0°		Similar to above specimen and conditions except irradiated to a total does of 385 Megarads.
6*	28	1964	~298	~0°		Similar to Curve 1 specimen and conditions except ultraviolet irradiated in vacuum ($< 10^{-6}$ mm Hg) at 273-288 K for 300 equivalent sun hrs (~3 sun intensity, General Electric BH-6 lamp source); measured within one hr after exposure to air.
7*	28	1964	~298	~0°		Similar to above specimen and conditions.
8*	23	1964	~298	~0°		Antimony trioxide in potassium silicate binder; prepared by Hughes Aircraft Co.; computed from spectral reflectance data obtained with a Gier-Dunkle integrating sphere.
9*	23	1964	~298	~0°		Similar to above specimen and conditions.
10*	23	1964	~298	~0°		Similar to above specimen and conditions except reflectance obtained with a Gier-Dunkle Solar Reflectometer.
11*	23	1964	~298	~0°		Similar to above specimen and conditions except reflectance obtained with a Cary 14 spectrophotometer and integrating sphere.
12*	23	1964	~298	~0°		Similar to above specimen and conditions.
13*	23	1964	~298	~0°		Similar to above specimen and conditions except reflectance obtained with a G. E. and P-E spectrophotometer and integrating sphere.
14*	23	1964	~298	~0°		Similar to above specimen and conditions except reflectance obtained with a Beckman DK-2A spectrophotometer and integrating sphere.
15*	23	1964	~298	~0°		Similar to above specimen and conditions.
16*	23	1964	~298	~0°		Similar to above specimen and conditions.
17*	23	1964	~298	~0°		Similar to above specimen and conditions except reflectance obtained with a Bausch and Lomb 505 spectrophotometer and integrating sphere.
18*	52	1965	193-263	~0°		Antimony trioxide in potassium silicate binder; measured in earth orbit on OSO-II.

* No plot given

DATA TABLE NO. 24 NORMAL SOLAR ABSORPTANCE OF ANTIMONY OXIDE PIGMENTED COATINGS

[Temperature, T, K; Absorptance, α]

T	α
CURVE 1*	
298	0.26
CURVE 2*	
298	0.26
CURVE 3*	
298	0.36
CURVE 4*	
298	0.31
CURVE 5*	
298	0.52
CURVE 6*	
298	0.34
CURVE 7*	
298	0.34
CURVE 8*	
298	0.28
CURVE 9*	
298	0.27
CURVE 10*	
298	0.29
CURVE 11*	
298	0.30
CURVE 12*	
298	0.28
CURVE 13*	
298	0.26
CURVE 14*	
298	0.25
CURVE 15*	
298	0.27
CURVE 16*	
298	0.26
CURVE 17*	
298	0.25
CURVE 18*	
193	0.275
263	0.275

* No plot given

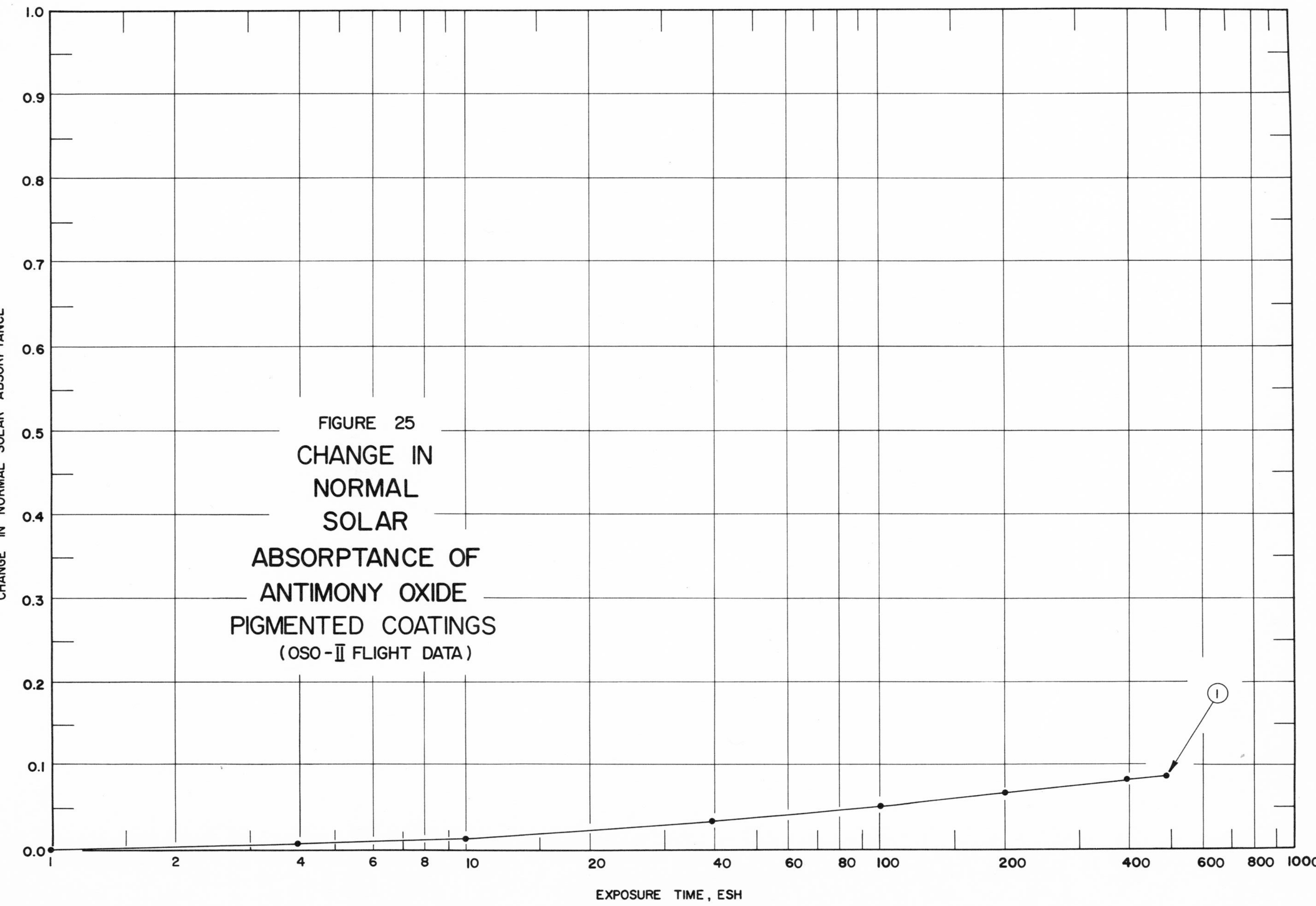

FIGURE 25 CHANGE IN NORMAL SOLAR ABSORPTANCE OF ANTIMONY OXIDE PIGMENTED COATINGS (OSO-II FLIGHT DATA)

SPECIFICATION TABLE NO. 25 CHANGE IN NORMAL SOLAR ABSORPTANCE OF ANTIMONY OXIDE PIGMENTED COATINGS

Curve No.	Ref. No.	Year	Temperature Range, K	Geometry θ	Reported Error, %	Composition (weight percent), Specifications and Remarks
1	52	1966	~298	~0°		Antimony trioxide in potassium silicate binder; data taken from OSO-II flight experiment; variable is equivalent length of time exposed to the sun.

DATA TABLE NO. 25 CHANGE IN NORMAL SOLAR ABSORPTANCE OF ANTIMONY OXIDE PIGMENTED COATINGS

[Temperature, T, K; Absorptance, α]

ESH	$\Delta\alpha$
CURVE 1 T ~ 298	
1.00	0.000
3.98	0.007
10.0	0.013
39.8	0.034
100	0.051
200	0.068
398	0.084
495	0.088

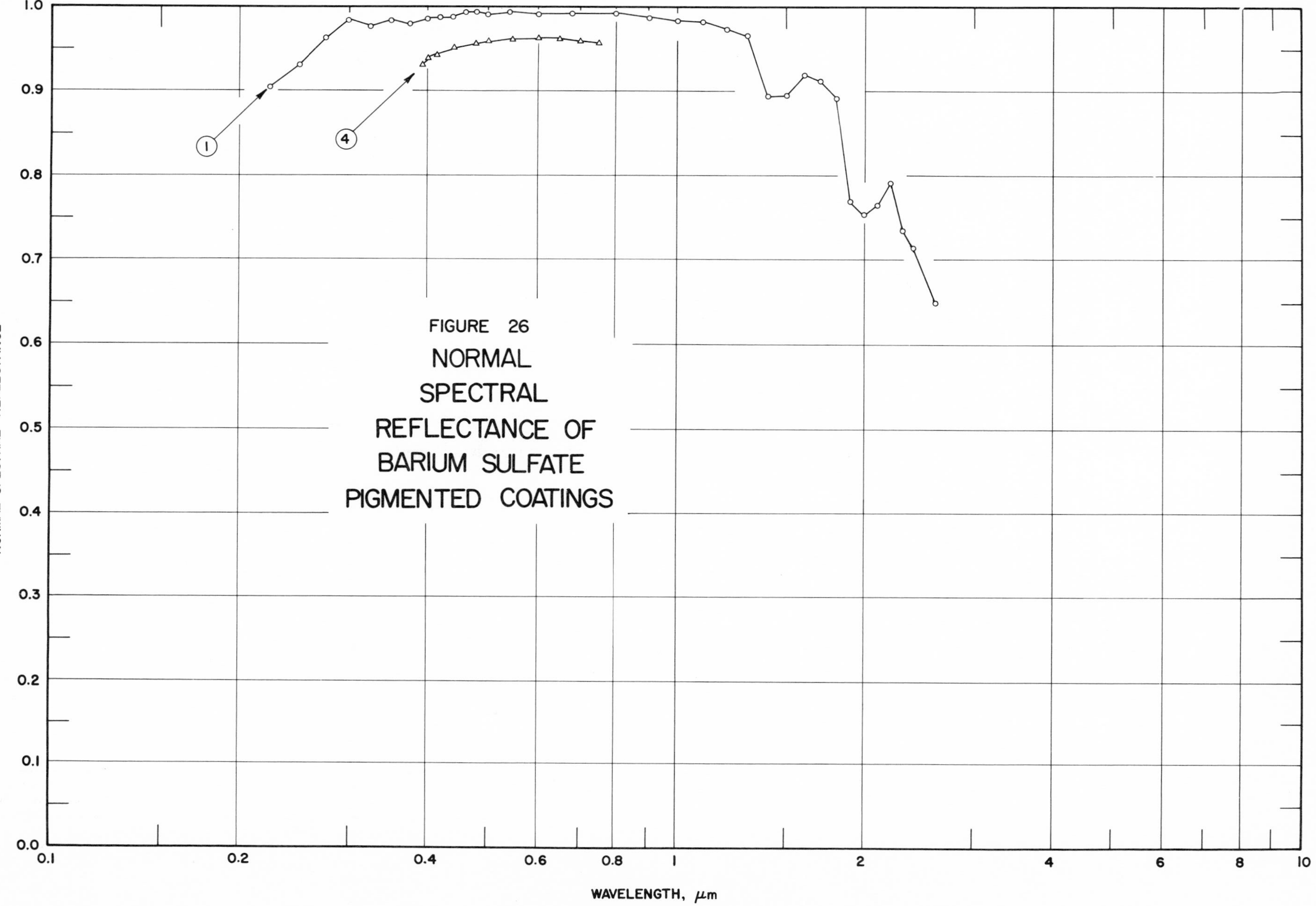

FIGURE 26 NORMAL SPECTRAL REFLECTANCE OF BARIUM SULFATE PIGMENTED COATINGS

SPECIFICATION TABLE NO. 26 NORMAL SPECTRAL REFLECTANCE OF BARIUM SULFATE PIGMENTED COATINGS

Curve No.	Ref. No.	Year	Temperature, K	Wavelength Range, μm	Geometry θ	θ'	ω'	Reported Error, %	Composition (weight percent), Specifications, and Remarks
1	87	1968	~298	0.225-2.500	~0°		2π		$BaSO_4$ (specially purified grade) in purified polyvinyl alcohol binder (1.0 mm thick); PBR 100 : 1; poured onto flat aluminum substrate, air dried, and scraped with glass edge to assure flatness; avg of 15 samples.
2*	87	1968	~298	0.200-0.800	~0°		2π		$BaSO_4$ (specially purified grade) in purified polyvinyl alcohol binder (1.0 mm thick); PBR 100; poured onto flat aluminum substrate, air dried, scraped with glass edge to assure flatness; data extracted from smooth curve; measured relative to MgO.
3*	87	1968	~298	0.213-0.800	~0°		2π		Eastman White Reflectance paint, Distillation Products Industries (1.0 μm thick); $BaSO_4$ in polyvinyl alcohol binder; PBR 100 : 1; data extracted from smooth curve; measured relative to $BaSO_4$ pressed powder.
4	88	1967	~298	0.391-0.750	~0°		2π	0.1	$BaSO_4$ in carboxy-methyl-cellulose binder (0.7 mm thick); % by weight with respect to pigment - 1st coat: 3 binder, 300 distilled H_2O, 11th and final coat: 0.0 binder, 196 distilled H_2O; spray application; specimen measured was the inner coating of a 5 meter integrating sphere; paint surface aged 10 mo. before measurement; data extracted from smooth curve.

DATA TABLE NO. 26 NORMAL SPECTRAL REFLECTANCE OF BARIUM SULFATE PIGMENTED COATINGS

[Wavelength, λ, μm; Reflectance, ρ; Temperature, T, K]

λ	ρ	λ	ρ	λ	ρ	λ	ρ	λ	ρ	λ	ρ
CURVE 1		CURVE 1 (cont.)		CURVE 1 (cont.)		CURVE 2 (cont.)*		CURVE 3 (cont.)*		CURVE 4 (cont.)	
T ~ 298											
		0.600	0.992	2.000	0.754	0.300	1.021	0.332	0.998	0.600	0.962
0.225	0.905	0.620	0.992*	2.100	0.765	0.332	1.011	0.402	1.001	0.650	0.962
0.250	0.932	0.640	0.992*	2.200	0.792	0.399	1.006	0.544	0.997	0.700	0.960
0.275	0.962	0.660	0.992*	2.300	0.735	0.466	1.005	0.723	0.994	0.750	0.959
0.300	0.983	0.680	0.992	2.400	0.713	0.585	1.011	0.800	0.995		
0.325	0.978	0.700	0.992*	2.500	0.650	0.721	1.013				
0.350	0.982	0.800	0.992			0.800	1.015	CURVE 4			
0.375	0.980	0.900	0.989	CURVE 2*				T ~ 298			
0.400	0.986	1.000	0.983	T ~ 298		CURVE 3*					
0.420	0.988	1.100	0.982			T ~ 298		0.391	0.934		
0.440	0.989	1.200	0.973	0.200	1.074			0.400	0.940		
0.460	0.993	1.300	0.966	0.208	1.078	0.213	1.008	0.411	0.944		
0.480	0.993	1.400	0.895	0.215	1.069	0.216	0.983	0.420	0.947*		
0.500	0.992	1.500	0.896	0.225	1.041	0.221	0.974	0.440	0.953		
0.520	0.992*	1.600	0.920	0.233	1.023	0.228	0.968	0.459	0.956*		
0.540	0.993	1.700	0.913	0.241	1.014	0.251	0.991	0.479	0.958		
0.560	0.992*	1.800	0.892	0.250	1.009	0.264	0.991	0.500	0.960		
0.580	0.992*	1.900	0.770	0.264	1.009	0.275	0.987	0.549	0.962		

* Not shown on plot

SPECIFICATION TABLE NO. 27 HEMISPHERICAL TOTAL EMITTANCE OF BARIUM TITANATE PIGMENTED COATINGS

Curve No.	Ref. No.	Year	Temperature Range, K	Reported Error, %	Composition (weight percent), Specifications and Remarks
1*	81	1963	435-826		FCE-11 BaO · TiO_2 (Continental Coating Corp); aluminum phosphate binder (0.178 mm thick); 99 Nb-1 Zr substrate; measured in vacuum ($<2.3 \times 10^{-6}$ mm Hg); heating cycle. [Author's designation: Run No. 1]
2*	81	1963	487-826		Above specimen and conditions; cooling cycle.
3*	49	1961	278		Barium titanate in silicone resin binder; Dow 17 treated Mg alloy substrate; exposed to ultraviolet radiation in vacuum (10^{-6}-8×10^{-6} mm Hg) from an argon-filled A-H6 high pressure Hg arc lamp for 127 hrs; vacuum maintained by VacIon pump.

DATA TABLE NO. 27 HEMISPHERICAL TOTAL EMITTANCE OF BARIUM TITANATE PIGMENTED COATINGS

[Temperature, T, K; Emittance, ϵ]

T	ϵ
CURVE 1*	
435	0.853
558	0.727
657	0.741
770	0.662
826	0.588
CURVE 2*	
826	0.588
713	0.639
599	0.700
487	0.771
CURVE 3*	
278	0.84

* No plot given

SPECIFICATION TABLE NO. 28 NORMAL SOLAR ABSORPTANCE OF BARIUM TITANATE PIGMENTED COATINGS

Curve No.	Ref. No.	Year	Temperature Range, K	Geometry θ	Reported Error, %	Composition (weight percent), Specifications and Remarks
1*	49	1961	278	~0°		36.8 barium titanate in 63.2 silicone resin binder; Dow 17 treated Mg alloy substrate; exposed in vacuum (10^{-6}–8×10^{-6} mm Hg) to UV radiation from an argon-filled A-H6 high pressure Hg arc lamp for 127 hrs; vacuum maintained by VacIon pump.

DATA TABLE NO. 28 NORMAL SOLAR ABSORPTANCE OF BARIUM TITANATE PIGMENTED COATINGS

[Temperature, T, K; Absorptance, α]

T	α
CURVE 1*	
278	0.34

* No plot given

SPECIFICATION TABLE NO. 29 HEMISPHERICAL TOTAL EMITTANCE OF BORON CARBIDE PIGMENTED COATINGS

Curve No.	Ref. No.	Year	Temperature Range, K	Reported Error, %	Composition (weight percent), Specifications and Remarks
1*	82	1962	396-1230	<± 2.5	Boron carbide in Synar (colloidal silica) binder (0.076 mm thick); molybdenum substrate; measured in vacuum; data extracted from smooth curve.

DATA TABLE NO. 29 HEMISPHERICAL TOTAL EMITTANCE OF BORON CARBIDE PIGMENTED COATINGS

[Temperature, T, K; Emittance, ∈]

T	∈
CURVE 1*	
396	0.765
648	0.767
753	0.769
993	0.782
1102	0.790
1193	0.798
1230	0.800

* No plot given

SPECIFICATION TABLE NO. 30 HEMISPHERICAL TOTAL EMITTANCE OF BORON NITRIDE PIGMENTED COATINGS

Curve No.	Ref. No.	Year	Temperature Range, K	Reported Error, %	Composition (weight percent), Specifications and Remarks
1*	15	1965	533-644		Boron nitride in potassium silicate binder (0.051 mm thick).

DATA TABLE NO. 30 HEMISPHERICAL TOTAL EMITTANCE OF BORON NITRIDE PIGMENTED COATINGS

[Temperature, T, K; Emittance, ∈]

T	∈
CURVE 1*	
533	0.83
588	0.80
644	0.78

* No plot given.

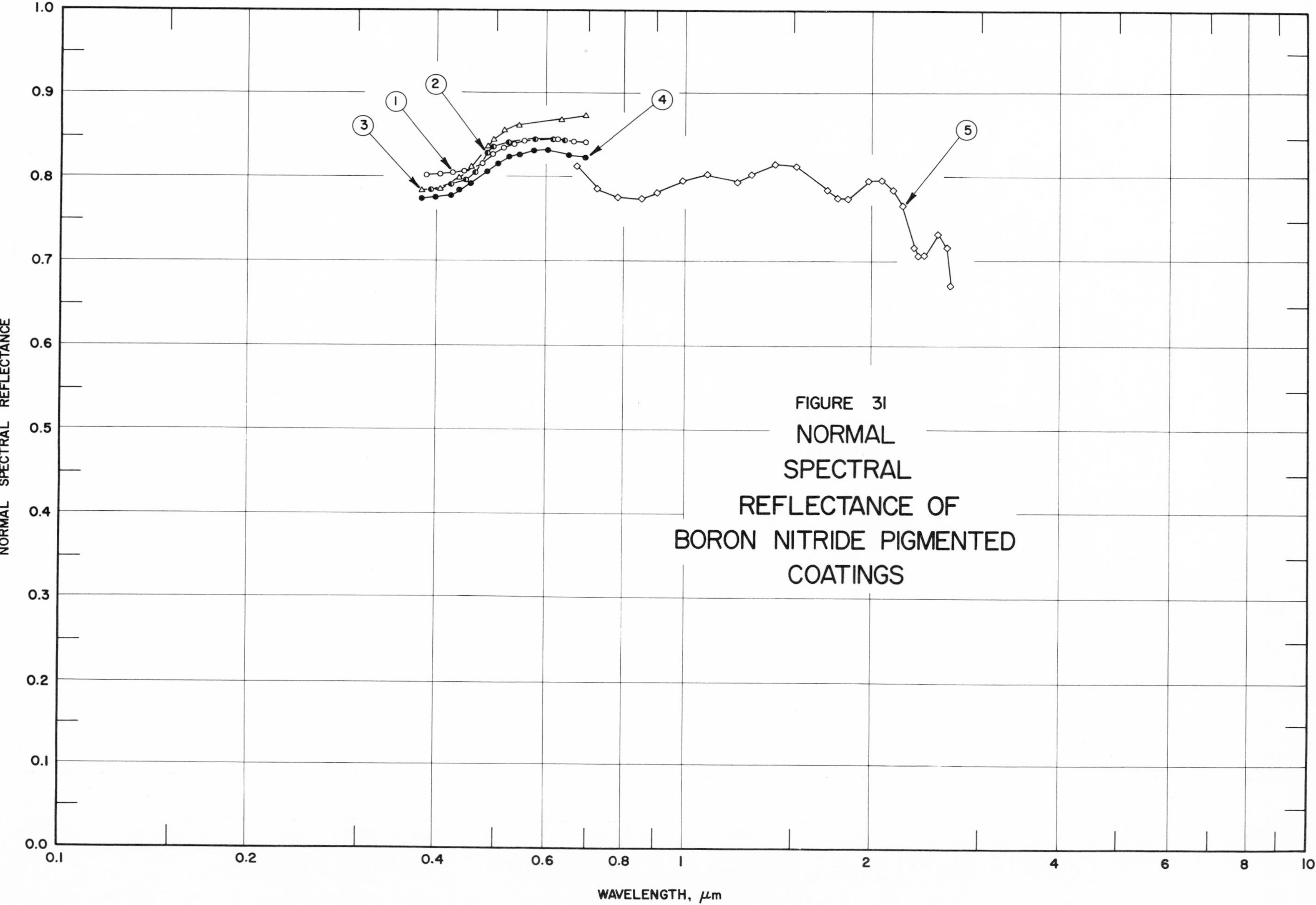

FIGURE 31 NORMAL SPECTRAL REFLECTANCE OF BORON NITRIDE PIGMENTED COATINGS

SPECIFICATION TABLE NO. 31 NORMAL SPECTRAL REFLECTANCE OF BORON NITRIDE PIGMENTED COATINGS

Curve No.	Ref. No.	Year	Temperature, K	Wavelength Range, μm	Geometry θ θ'	ω'	Reported Error, %	Composition (weight percent), Specifications, and Remarks
1	3	1960	~298	0.387-0.693	~0°	2π		Boron nitride (white hexagonal from National Carbon) in Dow Corning 806A silicone binder (0.076 mm thick); coating weight 7.3×10^{-2} kg m^{-2}; pigment concentration, 60% of solids content; voids concentration 43%; mild steel substrate; sprayed; data extracted from smooth curve; measured relative to MgO.
2	3	1960	~298	0.391-0.693	~0°	2π		Similar to curve 1 specimen and conditions except 0.102 mm thick; voids concentration 58%.
3	3	1960	~298	0.380-0.699	~0°	2π		Similar to curve 1 specimen and conditions except 0.127 mm thick; coating weight 11.97×10^{-2} kg m^{-2}; voids concentration 44%.
4	3	1960	~298	0.380-0.698	~0°	2π		Similar to curve 1 specimen and conditions except 0.122 mm thick; coating weight 10.94×10^{-2} kg m^{-2}; voids concentration 46%.
5	3	1960	~298	0.675-2.690	~0°	2π		Similar to curve 1 specimen and conditions except 0.102 mm thick; coating weight 7.36×10^{-2} kg m^{-2}.

DATA TABLE NO. 31 NORMAL SPECTRAL REFLECTANCE OF BORON NITRIDE PIGMENTED COATINGS

[Wavelength, λ, μm; Reflectance, ρ; Temperature, T, K]

λ	ρ
CURVE 1 T ~ 298	
0.387	0.801
0.407	0.802
0.426	0.805
0.443	0.808
0.458	0.813*
0.473	0.819
0.491	0.828
0.512	0.835
0.531	0.840
0.555	0.844
0.577	0.846*
0.628	0.847
0.665	0.845
0.693	0.843

λ	ρ
CURVE 2 T ~ 298	
0.391	0.785
0.423	0.791
0.449	0.799
0.461	0.806
0.474	0.820*
0.486	0.830
0.496	0.836
0.521	0.841
0.555	0.844*
0.576	0.846
0.617	0.847
0.646	0.846
0.668	0.845*
0.693	0.843*

λ	ρ
CURVE 3 T ~ 298	
0.380	0.784
0.394	0.783*
0.407	0.786
0.421	0.791*
0.436	0.800
0.454	0.813
0.482	0.838
0.496	0.848
0.517	0.857
0.541	0.862
0.571	0.866*
0.639	0.870
0.672	0.872*
0.699	0.875

λ	ρ
CURVE 4 T ~ 298	
0.380	0.773
0.400	0.775
0.421	0.778
0.438	0.784
0.455	0.792
0.482	0.807
0.501	0.818
0.522	0.825
0.544	0.829
0.573	0.832
0.606	0.832
0.652	0.829
0.698	0.826

λ	ρ
CURVE 5 T ~ 298	
0.675	0.815
0.726	0.789
0.787	0.776
0.852	0.774
0.904	0.781
0.995	0.797
1.081	0.803
1.225	0.795
1.293	0.802
1.409	0.817
1.511	0.814
1.700	0.786
1.763	0.776
1.837	0.776
1.997	0.799
2.096	0.799

λ	ρ
CURVE 5 (cont.)	
2.167	0.787
2.232	0.767
2.340	0.719
2.397	0.706
2.432	0.709
2.564	0.732
2.598	0.732*
2.633	0.719
2.690	0.672

* Not shown on plot

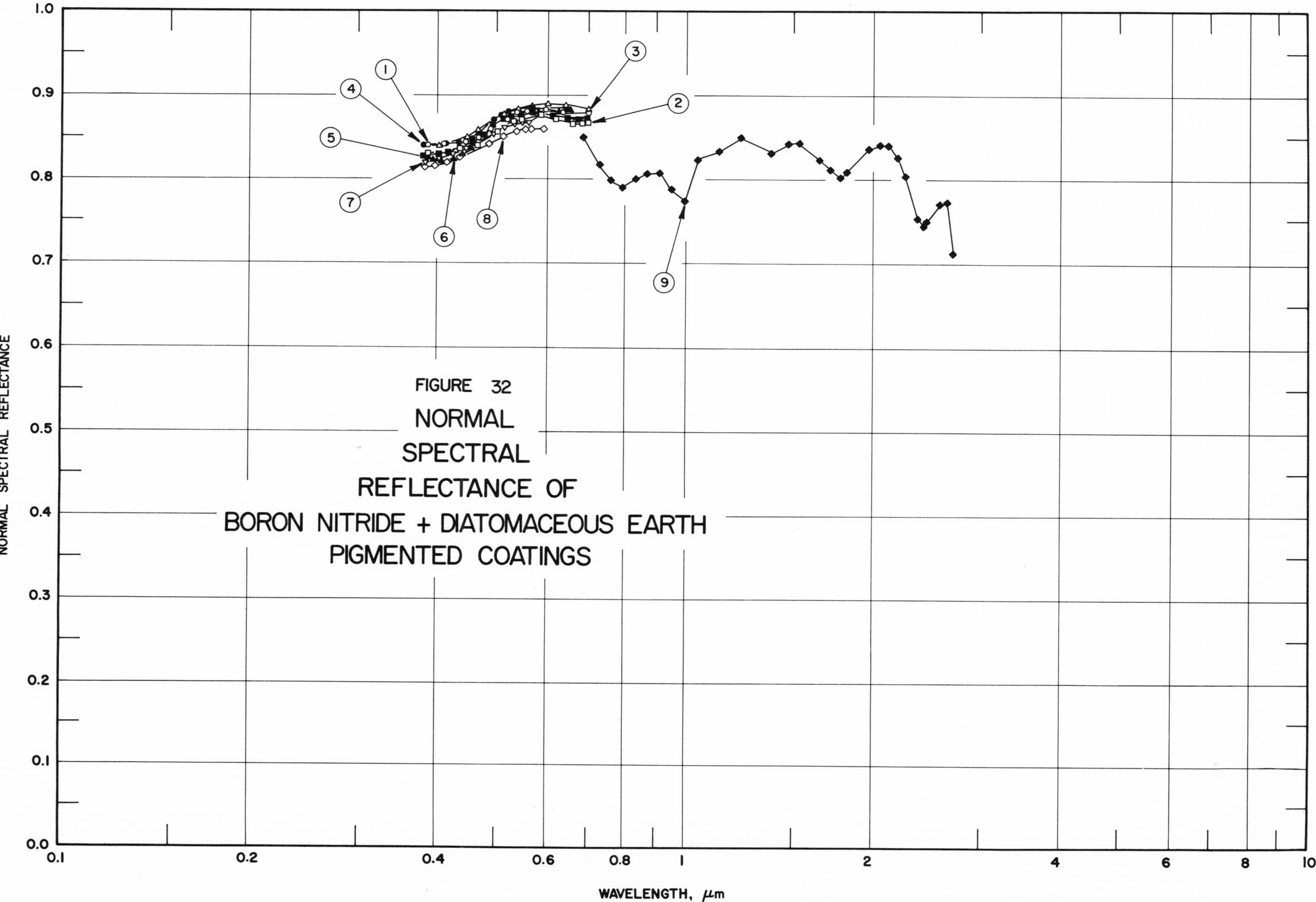

FIGURE 32
NORMAL
SPECTRAL
REFLECTANCE OF
BORON NITRIDE + DIATOMACEOUS EARTH
PIGMENTED COATINGS

SPECIFICATION TABLE NO. 32 NORMAL SPECTRAL REFLECTANCE OF BORON NITRIDE + DIATOMACEOUS EARTH PIGMENTED COATINGS

Curve No.	Ref. No.	Year	Temperature, K	Wavelength Range, μm	Geometry θ	θ'	ω'	Reported Error, %	Composition (weight percent), Specifications, and Remarks
1	3	1960	~298	0.385-0.698	~0°		2π		90 BN, National Carbon, and 10 Johns Manville Micro-Cell "C" diatomaceous earth in Dow Corning 806A silicone binder (0.109 mm thick, 0.76 x 10^{-2} g cm^{-2}) on mild steel substrate; pigment 60% of solids content; voids concentration 58%; sprayed; data extracted from smooth curve; measured relative to MgO.
2	3	1960	~298	0.386-0.698	~0°		2π		Similar to above specimen and conditions except coating 0.0965 mm thick, 0.71 x 10^{-2} g cm^{-2}; voids concentration 55%.
3	3	1960	~298	0.380-0.699	~0°		2π		Similar to curve 1 specimen and conditions except coating 0.119 mm thick, 0.827 x 10^{-2} g cm^{-2}; voids concentration 57%.
4	3	1960	~298	0.380-0.695	~0°		2π		Similar to curve 1 specimen and conditions except coating 0.0991 mm thick, 0.746 x 10^{-2} g cm^{-2}; voids concentration 54%.
5	3	1960	~298	0.380-0.694	~0°		2π		Similar to curve 1 specimen and conditions except coating 0.0889 mm thick, 0.635 x 10^{-2} g cm^{-2}; voids concentration 57%.
6	3	1960	~298	0.384-0.699	~0°		2π		80 BN, National Carbon, and 20 Johns Manville Micro-Cell "C" diatomaceous earth in Dow Corning 806A silicone binder (0.127 mm thick, 1.015 x 10^{-2} g cm^{-2}) on mild steel substrate; pigment 60% of solids content; voids concentration 52%; sprayed; data extracted from smooth curve; measured relative to MgO.
7	3	1960	~298	0.382-0.695	~0°		2π		Similar to curve 6 specimen and conditions except coating 0.094 mm thick, 0.70 x 10^{-2} g cm^{-2}; voids concentration 55%.
8	3	1960	~298	0.382-0.690	~0°		2π		Similar to curve 6 specimen and conditions except coating 0.0584 mm thick, 0.624 x 10^{-2} g cm^{-2}; voids concentration 36%.
9	3	1960	~298	0.685-2.691	~0°		2π		90 BN, National Carbon, and 10 Johns Manville Micro-Cell "C" diatomaceous earth in Dow Corning 806A silicone (0.0965 mm thick, 0.712 x 10^{-2} g cm^{-2}) on mild steel substrate; pigment 55% of solids content; sprayed; data extracted from smooth curve; measured relative to MgO.

DATA TABLE NO. 32 NORMAL SPECTRAL REFLECTANCE OF BORON NITRIDE + DIATOMACEOUS EARTH PIGMENTED COATINGS

[Wavelength, λ, μm; Reflectance, ρ; Temperature, T, K]

λ	ρ
CURVE 1 $T \sim 298$	
0.385	0.842
0.409	0.843
0.442	0.846
0.463	0.852
0.487	0.862
0.515	0.874
0.537	0.880
0.556	0.882
0.597	0.883
0.636	0.881
0.698	0.880
CURVE 2 $T \sim 298$	
0.386	0.832
0.434	0.837
0.448	0.839
0.463	0.842
0.500	0.859
0.530	0.870
0.543	0.873
0.562	0.877
0.587	0.877
0.618	0.874
0.657	0.869
0.681	0.869
0.698	0.870
CURVE 3 $T \sim 298$	
0.380	0.843
0.403	0.842
0.433	0.845
0.447	0.851
0.466	0.861
0.490	0.872
0.512	0.879
0.537	0.885
0.568	0.889
0.600	0.891
0.644	0.890
0.699	0.885
CURVE 4 $T \sim 298$	
0.380	0.840
0.414	0.842
0.433	0.844
0.445	0.848*
0.455	0.852
0.467	0.858
0.492	0.872
0.507	0.879
0.520	0.882
0.534	0.885*
0.567	0.886
0.646	0.886
0.695	0.884*
CURVE 5 $T \sim 298$	
0.380	0.830
0.401	0.831
0.419	0.834
0.434	0.838*
0.454	0.846
0.476	0.856
0.493	0.866
0.510	0.872
0.522	0.876
0.545	0.879
0.576	0.880
0.611	0.878
0.649	0.875
0.672	0.874
0.694	0.875
CURVE 6 $T \sim 298$	
0.384	0.828
0.392	0.823
0.405	0.822
0.424	0.826
0.439	0.831
0.453	0.839
0.479	0.856*
0.498	0.865*
0.514	0.870*
CURVE 6 (cont.)	
0.541	0.876*
0.576	0.880
0.624	0.882
0.654	0.884
0.699	0.889*
CURVE 7 $T \sim 298$	
0.382	0.821
0.403	0.825
0.426	0.831
0.449	0.839
0.493	0.857
0.514	0.865
0.531	0.869
0.549	0.870
0.567	0.871
0.658	0.871
0.681	0.872
0.695	0.873*
CURVE 8 $T \sim 298$	
0.382	0.817
0.398	0.818
0.415	0.821
0.433	0.827
0.484	0.845
0.510	0.852
0.538	0.858
0.555	0.860
0.663	0.860
0.690	0.861
CURVE 9 $T \sim 298$	
0.685	0.852
0.729	0.818
0.758	0.799
0.795	0.791
0.833	0.800
0.876	0.809
0.908	0.809
CURVE 9 (cont.)	
0.950	0.789
0.992	0.774
1.042	0.774
1.132	0.783
1.230	0.803
1.382	0.834
1.461	0.843
1.537	0.843
1.642	0.826
1.712	0.813
1.770	0.806
1.819	0.810
1.970	0.837
2.032	0.842
2.116	0.842
2.199	0.828
2.263	0.806
2.376	0.752
2.402	0.745
2.446	0.750
2.560	0.770
2.609	0.773
2.691	0.712

* Not shown on plot

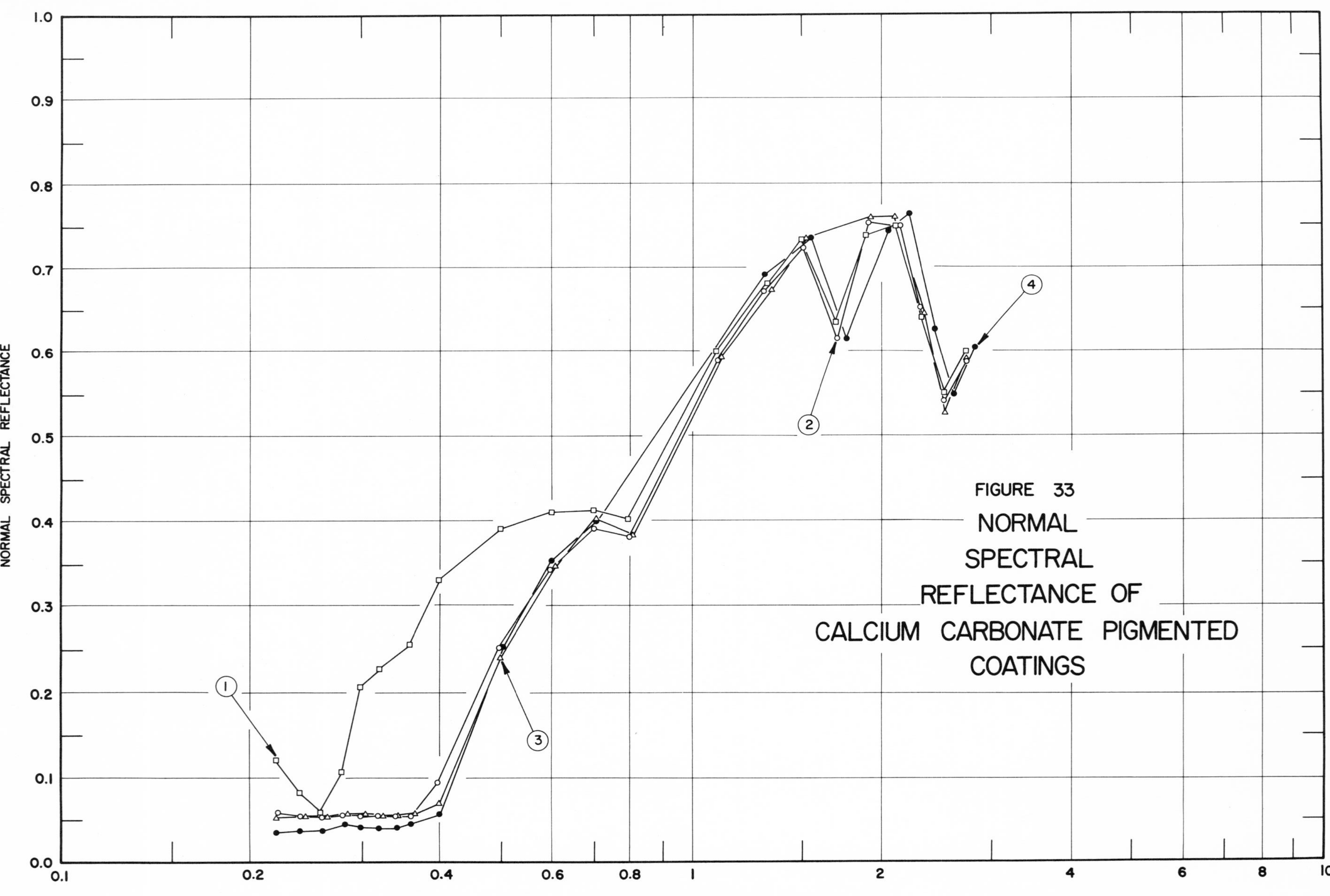

FIGURE 33 NORMAL SPECTRAL REFLECTANCE OF CALCIUM CARBONATE PIGMENTED COATINGS

SPECIFICATION TABLE NO. 33 NORMAL SPECTRAL REFLECTANCE OF CALCIUM CARBONATE PIGMENTED COATINGS

Curve No.	Ref. No.	Year	Temperature, K	Wavelength Range, μm	Geometry θ	θ'	ω'	Reported Error, %	Composition (weight percent), Specifications, and Remarks
1	1	1960	298	0.219-2.705	~0°		2π		Calcium carbonate in silicone binder (0.0508 mm thick) on anodized aluminum 24S-T substrate; measured relative to MgO.
2	1	1960	298	0.221-2.707	~0°		2π		Above specimen and conditions except exposed in vacuum (10^{-5} mm Hg) at 358 K to UV radiation from a G. E. Type UA-3 lamp for 20 hrs.
3	1	1960	298	0.219-2.728	~0°		2π		Above specimen and conditions except exposed to UV radiation for a total of 60 hrs.
4	1	1960	298	0.220-2.818	~0°		2π		Above specimen and conditions except exposed to UV radiation for a total of 100 hrs.

DATA TABLE NO. 33 NORMAL SPECTRAL REFLECTANCE OF CALCIUM CARBONATE PIGMENTED COATINGS

[Wavelength, λ, μm; Reflectance, ρ; Temperature, T, K]

CURVE 1
T = 298

λ	ρ
0.219	0.121
0.239	0.082
0.258	0.062
0.279	0.108
0.299	0.207
0.320	0.228
0.340	0.239*
0.359	0.258
0.399	0.332
0.498	0.391
0.599	0.412
0.698	0.413
0.798	0.402
1.095	0.601
1.302	0.681
1.492	0.735
1.697	0.633

CURVE 1 (cont.)

λ	ρ
1.897	0.739
2.105	0.750
2.304	0.640
2.506	0.551
2.705	0.600

CURVE 2
T = 298

λ	ρ
0.221	0.060
0.240	0.054
0.260	0.054
0.281	0.057
0.298	0.056
0.319	0.055
0.338	0.055
0.358	0.056
0.397	0.096

CURVE 2 (cont.)

λ	ρ
0.497	0.252
0.598	0.345
0.699	0.392
0.799	0.381
1.096	0.590
1.302	0.672
1.504	0.724
1.701	0.617
1.903	0.754
2.130	0.750
2.300	0.651
2.508	0.541
2.707	0.589

CURVE 3
T = 298

λ	ρ
0.219	0.052
0.244	0.054
0.266	0.054
0.284	0.059
0.303	0.057
0.324	0.055
0.341	0.055
0.363	0.059
0.399	0.071
0.498	0.240
0.604	0.348
0.705	0.404
0.804	0.384
1.125	0.594
1.332	0.673
1.523	0.736
1.903	0.762

CURVE 3 (cont.)

λ	ρ
2.100	0.762
2.338	0.646
2.510	0.529
2.728	0.591

CURVE 4
T = 298

λ	ρ
0.220	0.035
0.239	0.038
0.260	0.038
0.282	0.045
0.299	0.042
0.320	0.042
0.341	0.042
0.360	0.046
0.400	0.059
0.503	0.253

CURVE 4 (cont.)

λ	ρ
0.600	0.355
0.710	0.406
1.310	0.694
1.542	0.735
1.755	0.618
2.031	0.746
2.218	0.764
2.420	0.627
2.616	0.550
2.818	0.605

* Not shown on plot

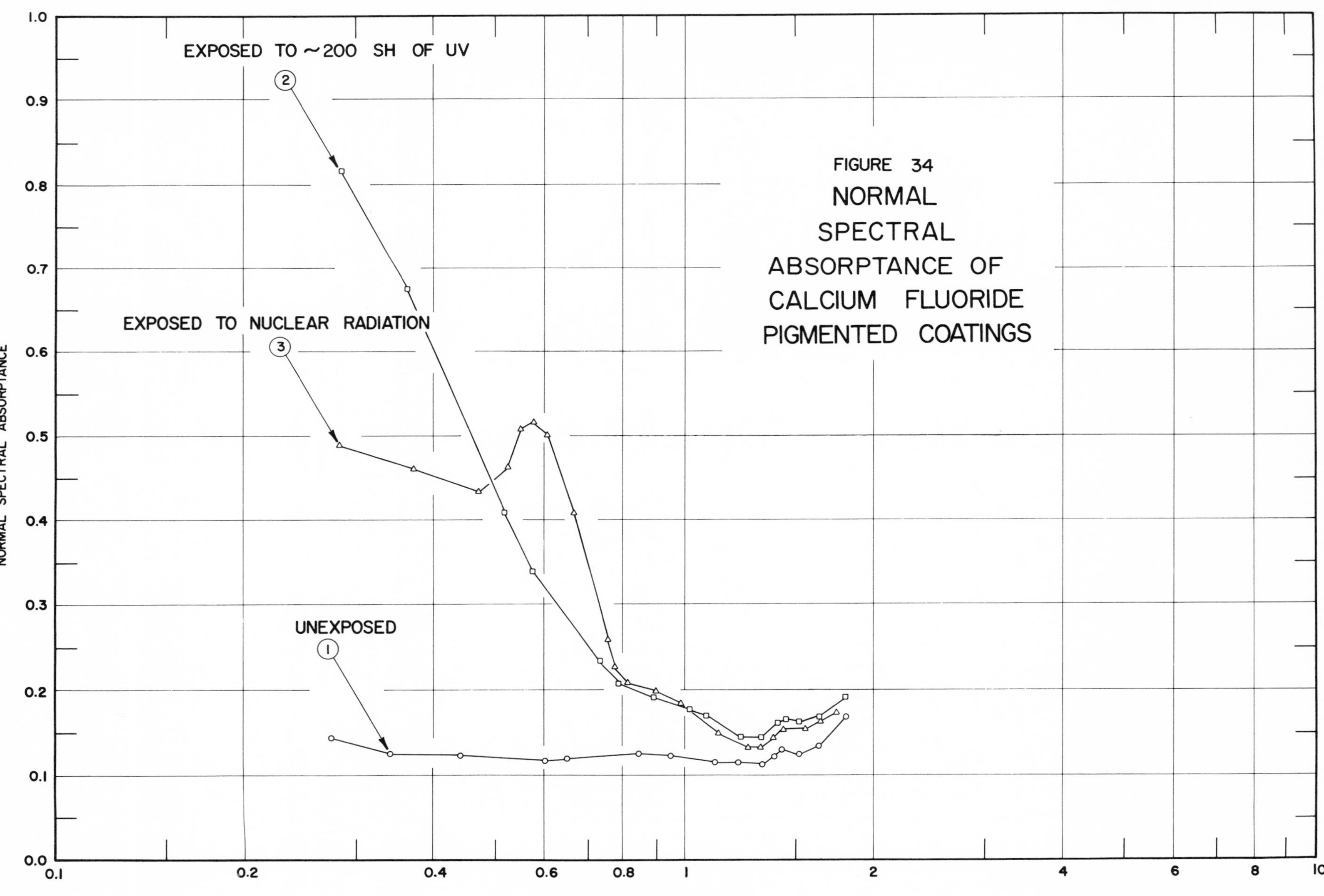
EXPOSED TO ~200 SH OF UV
2
EXPOSED TO NUCLEAR RADIATION
3
UNEXPOSED
1
FIGURE 34
NORMAL
SPECTRAL
ABSORPTANCE OF
CALCIUM FLUORIDE
PIGMENTED COATINGS
NORMAL SPECTRAL ABSORPTANCE
1.0
0.9
0.8
0.7
0.6
0.5
0.4
0.3
0.2
0.1
0.0
0.1
0.2
0.4
0.6
0.8
1
2
4
6
8
10
WAVELENGTH, μm

SPECIFICATION TABLE NO. 34 NORMAL SPECTRAL ABSORPTANCE OF CALCIUM FLUORIDE PIGMENTED COATINGS

Curve No.	Ref. No.	Year	Temperature, K	Wavelength Range, μm	Geometry θ	Reported Error, %	Composition (weight percent), Specifications, and Remarks
1	24	1965	298	0.274-1.800	~0°		CaF_2 in sodium silicate binder; data extracted from smooth curve.
2	24	1965	298	0.287-1.800	~0°		Similar to above specimen and conditions except exposed to approx 200 sun hrs of ultraviolet radiation.
3	24	1965	298	0.284-1.800	~0°		Similar to curve 1 specimen and conditions except exposed to a nuclear-radiation dose of approx 10^8 R of gamma and 5 x 10^{14} neutrons ($E \geq 2.9$ MeV) per cm^2.

DATA TABLE NO. 34 NORMAL SPECTRAL ABSORPTANCE OF CALCIUM FLUORIDE PIGMENTED COATINGS

[Wavelength, λ, μm; Absorptance, α; Temperature, T, K]

λ	α	λ	α	λ	α
CURVE 1		CURVE 2		CURVE 3	
T = 298		T = 298		T = 298	
0.274	0.146	0.287	0.818	0.284	0.490
0.341	0.128	0.362	0.675	0.372	0.463
0.442	0.123	0.517	0.412	0.473	0.435
0.601	0.118	0.575	0.340	0.526	0.465
0.652	0.120	0.733	0.234	0.552	0.509
0.846	0.126	0.791	0.209	0.578	0.518
0.949	0.124	0.898	0.191	0.605	0.502
1.114	0.118	1.024	0.178	0.668	0.410
1.226	0.118	1.077	0.172	0.755	0.260
1.338	0.114	1.241	0.148	0.773	0.227
1.394	0.125	1.333	0.145	0.811	0.211
1.434	0.131	1.404	0.162	0.900	0.199
1.511	0.125	1.448	0.166	0.981	0.185
1.641	0.136	1.516	0.164	1.137	0.150
1.800	0.169	1.647	0.170	1.257	0.133
		1.800	0.193	1.321	0.133
				1.390	0.146
				1.438	0.155
				1.550	0.155
				1.644	0.165
				1.751	0.174
				1.800	0.193*

*Not shown on plot

SPECIFICATION TABLE NO. 35 NORMAL SPECTRAL REFLECTANCE OF CALCIUM METASILICATE PIGMENTED COATINGS

Curve No.	Ref. No.	Year	Temperature, K	Wavelength Range, μm	Geometry θ	Geometry θ'	Geometry ω'	Reported Error, %	Composition (weight percent), Specifications, and Remarks
1*	27	1964	~298	0.44-0.60	~0°		2π		$CaSiO_3$ (Wallastonite) in PS-7 potassium silicate binder; PBR 4.30; solids content 62.8%; aluminum substrate abraded with No. 60 Aloxite cloth; sample cured at 413 K for 18 hrs. [Authors' designation: Sample C3]
2*	27	1964	~298	0.44-0.60	~0°		2π		Above specimen and conditions except irradiated in vacuum (10^{-6} mm Hg) with 180 ESH at solar factor of 3 suns.

DATA TABLE NO. 35 NORMAL SPECTRAL REFLECTANCE OF CALCIUM METASILICATE PIGMENTED COATINGS

[Wavelength, λ, μm; Reflectance, ρ; Temperature, T, K]

λ	ρ
CURVE 1*	T ~ 298
0.44	0.785
0.60	0.835
CURVE 2*	T ~ 298
0.44	0.525
0.60	0.715

* No plot given

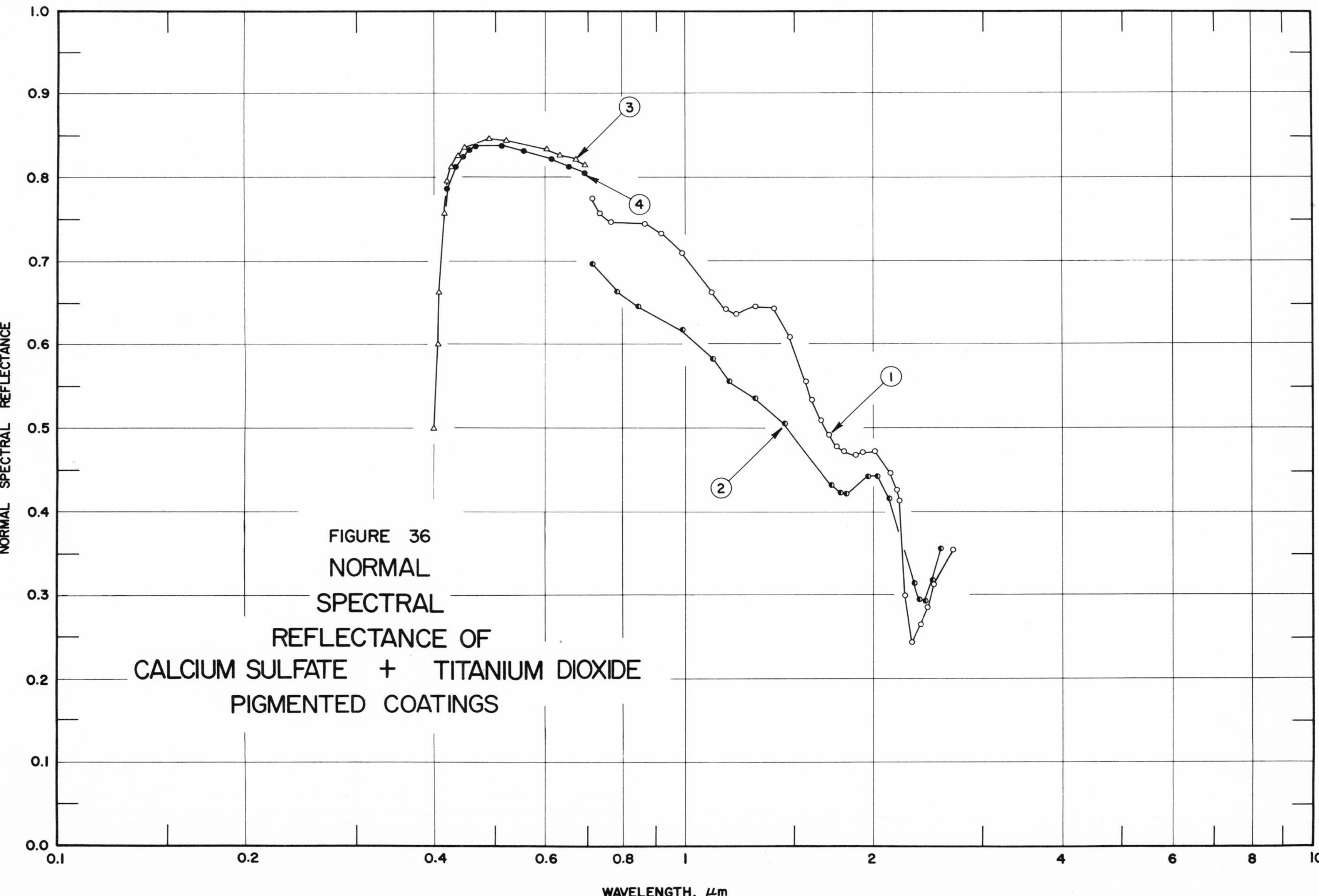

FIGURE 36 NORMAL SPECTRAL REFLECTANCE OF CALCIUM SULFATE + TITANIUM DIOXIDE PIGMENTED COATINGS

SPECIFICATION TABLE NO. 36 NORMAL SPECTRAL REFLECTANCE OF CALCIUM SULFATE + TITANIUM DIOXIDE PIGMENTED COATINGS

Curve No.	Ref. No.	Year	Temperature, K	Wavelength Range, μm	Geometry θ	θ'	ω'	Reported Error, %	Composition (weight percent), Specifications, and Remarks
1	3	1960	~298	0.714-2.696	~0°		2π		Titanox RC (70 $CaSO_4$ and 30 TiO_2) in DuPont RC 7007 alkyd-melamine resin binder; mild steel substrate; coating weight 0.098 kg m^{-2}; pigment concentration 40% of solids content; sprayed; data extracted from smooth curve; measured relative to MgO.
2	3	1960	~298	0.711-2.587	~0°		2π		Titanox C-50 (50 $CaSO_4$ and 50 TiO_2) in DuPont RC-7007 alkyd-melamine resin binder; mild steel substrate; coating weight 0.107 kg m^{-2}; pigment concentration 40% of solids content; sprayed; data extracted from smooth curve; measured relative to MgO.
3	3	1960	~298	0.400-0.695	~0°		2π		Titanox C-50 (50 $CaSO_4$ and 50 TiO_2) in DuPont RC-7007 alkyd-melamine resin binder (0.0762 mm thick); mild steel substrate; pigment 40% of solids content; voids concentration 2.6%; sprayed; data extracted from smooth curve; measured relative to MgO.
4	3	1960	~298	0.400-0.697	~0°		2π		Similar to curve 3 specimen and conditions except coating thickness 0.0711 mm; voids concentration 2.4%.

DATA TABLE NO. 36 NORMAL SPECTRAL REFLECTANCE OF CALCIUM SULFATE + TITANIUM DIOXIDE PIGMENTED COATINGS

[Wavelength, λ, μm; Reflectance, ρ; Temperature, T, K]

λ	ρ	λ	ρ	λ	ρ	λ	ρ	λ	ρ	λ	ρ
CURVE 1 T ~ 298		CURVE 1 (cont.)		CURVE 2 T ~ 298		CURVE 2 (cont.)		CURVE 3 (cont.)		CURVE 4 (cont.)	
		1.805	0.472			2.501	0.318	0.637	0.829	0.616	0.822
0.714	0.775	1.871	0.470	0.711	0.696	2.587	0.356	0.672	0.821	0.658	0.813
0.734	0.759	1.930	0.472	0.781	0.663			0.695	0.816	0.697	0.803
0.768	0.749	2.010	0.473	0.844	0.646	CURVE 3 T ~ 298					
0.866	0.744	2.070	0.467*	0.993	0.619			CURVE 4 T ~ 298			
0.914	0.735	2.133	0.449	1.111	0.583						
0.990	0.710	2.186	0.429	1.188	0.557	0.400	0.500				
1.111	0.662	2.207	0.414	1.304	0.533	0.403	0.600	0.400	0.500*		
1.176	0.641	2.252	0.300	1.458	0.504	0.407	0.664	0.403	0.600*		
1.217	0.638	2.282	0.253*	1.727	0.432	0.415	0.759	0.407	0.664*		
1.300	0.646	2.302	0.243	1.771	0.424	0.420	0.796	0.415	0.759*		
1.394	0.643	2.341	0.246*	1.819	0.422	0.426	0.813	0.420	0.787		
1.471	0.610	2.393	0.266	1.967	0.441	0.436	0.829	0.431	0.812		
1.560	0.555	2.442	0.288	2.030	0.441	0.448	0.838	0.443	0.826		
1.601	0.533	2.509	0.312	2.117	0.417	0.461	0.844*	0.453	0.832		
1.653	0.510	2.696	0.354	2.322	0.314	0.490	0.848	0.466	0.837		
1.703	0.493			2.374	0.296	0.520	0.847	0.511	0.837		
1.752	0.480			2.438	0.292	0.608	0.834	0.558	0.832		

* Not shown on plot

SPECIFICATION TABLE NO. 37 HEMISPHERICAL TOTAL EMITTANCE OF CALCIUM TITANATE PIGMENTED COATINGS

Curve No.	Ref. No.	Year	Temperature Range, K	Reported Error, %	Composition (weight percent), Specifications and Remarks
1*	81	1963	421-1088		Calcium titanate in aluminum phosphate binder (0.102 mm thick); Nb-1 Zr substrate; measured in vacuum (6.0×10^{-7} to 2.2×10^{-6} mm Hg); temp measured with thermocouple; heating cycle. [Author's designation: Run No. 1]
2*	81	1963	1088-810		Above specimen and conditions; cooling cycle.
3*	118	1962	423-1025		Calcium titanate, Metco, Inc., in aluminum phosphate binder (0.127 mm thick); Nb-1 Zr substrate; medium grit texture; measured in vacuum ($<5.0 \times 10^{-6}$ mm Hg).

DATA TABLE NO. 37 HEMISPHERICAL TOTAL EMITTANCE OF CALCIUM TITANATE PIGMENTED COATINGS

[Temperature, T, K; Emittance, ϵ]

T	ϵ	T	ϵ
CURVE 1*		CURVE 3*	
421	0.975	423	0.883
533	0.933	531	0.838
644	0.884	644	0.793
755	0.799	756	0.687
810	0.750	814	0.634
866	0.716	818	0.621
922	0.657	866	0.608
977	0.596	920	0.607
1033	0.576	976	0.616
1088	0.613	1025	0.605
CURVE 2*			
1088	0.613		
978	0.632		
810	0.681		

* No plot given

SPECIFICATION TABLE NO. 38 NORMAL SPECTRAL EMITTANCE OF CALCIUM TITANATE PIGMENTED COATINGS

Curve No.	Ref. No.	Year	Temperature, K	Wavelength Range, μm	Geometry θ'	Reported Error, %	Composition (weight percent), Specifications, and Remarks
1*	118	1962	755	1.34-12.36	~0°		Calcium titanate, Metco, Inc., in aluminum phosphate binder (0.127 mm thick); Nb-1 Zr substrate; medium grit texture.

DATA TABLE NO. 38 NORMAL SPECTRAL EMITTANCE OF CALCIUM TITANATE PIGMENTED COATINGS

[Wavelength, λ, μm; Emittance, ∈; Temperature, T, K]

λ	∈
CURVE 1* T = 755	
1.34	0.413
2.16	0.361
3.05	0.349
3.87	0.369
4.23	0.488
5.19	0.640
5.95	0.738
6.66	0.825
7.13	0.893
8.15	0.979
9.09	0.980
10.28	0.982
11.03	0.995
12.36	0.992

* No plot given

SPECIFICATION TABLE NO. 39 HEMISPHERICAL TOTAL EMITTANCE OF CARBON PIGMENTED COATINGS

Curve No.	Ref. No.	Year	Temperature Range, K	Reported Error, %	Composition (weight percent), Specifications and Remarks
1*	82	1962	422-1019	<± 2.5	Acetylene black in xylol binder (Acheson Colloid Co., Dag EC 1652); 310 stainless steel substrate; measured in vacuum; data extracted from smooth curve.
2*	49	1961	278	10	Black Kemacryl lacquer, M49BC12; carbon black in acrylic resin binder; substrate unknown; value avg of several determinations.
3*	49	1961	278		Black Kemacryl lacquer, M49BC12; carbon black in acrylic resin binder on Mg alloy substrate; substrate Dow 17 treated.
4*	49	1961	278		Above specimen and conditions except exposed to ultraviolet radiation in vacuum (10^{-6}-8 x 10^{-6} mm Hg) from an argon-filled A-H6 high pressure Hg arc lamp for 80 hrs.
5*	49	1961	278	10	Fuller flat black silicone; carbon in silicone binder; value avg of several determinations.

DATA TABLE NO. 39 HEMISPHERICAL TOTAL EMITTANCE OF CARBON PIGMENTED COATINGS

[Temperature, T, K; Emittance, ∈]

T	∈	T	∈
CURVE 1*		CURVE 5*	
422	0.878	278	0.81
606	0.892		
737	0.900		
837	0.903		
1019	0.905		
CURVE 2*			
278	0.83		
CURVE 3*			
278	0.81		
CURVE 4*			
278	0.79		

* No plot given

SPECIFICATION TABLE NO. 40 NORMAL TOTAL EMITTANCE OF CARBON PIGMENTED COATINGS

Curve No.	Ref. No.	Year	Temperature Range, K	Geometry θ'	Reported Error, %	Composition (weight percent), Specifications and Remarks
1*	83	1963	332-366	~0°		Carbon black in unknown binder (0.051 mm thick); measured in vacuum (1.6×10^{-4} mm Hg).
2*	19	1965	298	0°		Carbon black; methyl silicone binder; Al substrate; reported error ± 0.02 unit.
3*	19	1965	298	0°		Above specimen and conditions except computed from 1 - R (2π, 0°).
4*	89	1958	366-672	~0°		North American Aviation Paint No. 277.2.1; lampblack in DC807 silicone alkyd binder; 321 stainless steel substrate; air dried. [Author's designation: Specimen 5]
5*	19	1965	298	0°		Lampblack in epoxy binder; Cat-a-Lac black paint; aluminum substrate; reported error ± 0.02.
6*	19	1965	298	0°		Above specimen and conditions except computed from 1 - R(2π, 0°).

DATA TABLE NO. 40 NORMAL TOTAL EMITTANCE OF CARBON PIGMENTED COATINGS

[Temperature, T, K; Emittance, ϵ]

T	ϵ	T	ϵ
CURVE 1*		CURVE 4 (cont.)*	
332	0.763	600	0.866
350	0.761	672	0.862
366	0.760		
		CURVE 5*	
CURVE 2*		298	0.89
298	0.86		
		CURVE 6*	
CURVE 3*		298	0.89
298	0.88		
CURVE 4*			
366	0.942		
444	0.940		
522	0.930		

* No plot given

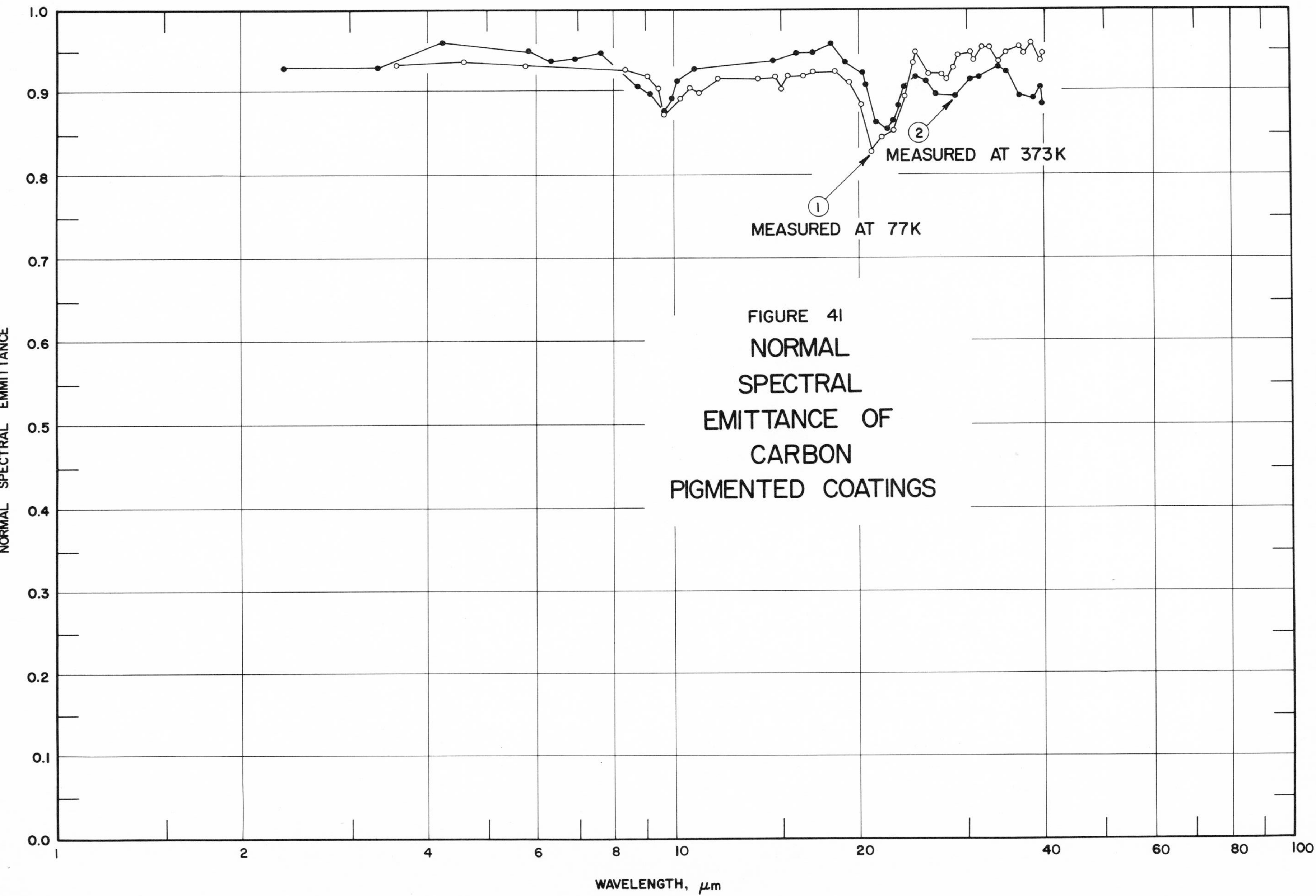

FIGURE 41
NORMAL SPECTRAL EMITTANCE OF CARBON PIGMENTED COATINGS

SPECIFICATION TABLE NO. 41 NORMAL SPECTRAL EMITTANCE OF CARBON PIGMENTED COATINGS

Curve No.	Ref. No.	Year	Temperature, K	Wavelength Range, μm	Geometry θ'	Reported Error, %	Composition (weight percent), Specifications, and Remarks
1	93	1965	77	3.58-39.81	~0°		Lampblack in epoxy binder (~0.089 mm thick); Cat-A-Lac Black paint; data extracted from smooth curve.
2	93	1965	373	2.35-39.81	~0°		Similar to above specimen and conditions.

DATA TABLE NO. 41 NORMAL SPECTRAL EMITTANCE OF CARBON PIGMENTED COATINGS

[Wavelength, λ, μm; Emittance, ϵ; Temperature, T, K]

λ	ϵ	λ	ϵ	λ	ϵ	λ	ϵ	λ	ϵ	λ	ϵ
CURVE 1 T = 77		CURVE 1 (cont.)		CURVE 1 (cont.)		CURVE 2 (cont.)		CURVE 2 (cont.)		CURVE 2 (cont.)	
		18.47	0.922	32.61	0.951	6.38	0.938	21.75	0.854*	39.27	0.901
3.58	0.931	19.43	0.910	33.68	0.937	6.94	0.940	22.37	0.854	39.81	0.886
4.60	0.936	20.17	0.884	34.20	0.937*	7.63	0.948	22.88	0.864		
5.79	0.931	21.00	0.826	34.99	0.947	8.79	0.904	23.26	0.881		
8.36	0.926	21.94	0.845	36.27	0.952	9.15	0.898	23.69	0.905		
9.03	0.919	22.98	0.853	37.00	0.947	9.62	0.875	24.17	0.916*		
9.41	0.902	23.84	0.895	37.66	0.949*	9.95	0.892	24.86	0.919		
9.69	0.872	24.52	0.935	38.07	0.958	10.26	0.913	25.97	0.913		
10.21	0.891	24.92	0.947	38.98	0.946	10.71	0.927	26.96	0.897		
10.67	0.902	26.08	0.920	39.81	0.939	14.55	0.937	28.91	0.895		
11.01	0.899	27.13	0.920			15.82	0.947	30.28	0.911		
11.80	0.915	27.86	0.915	CURVE 2		16.86	0.947	31.48	0.916		
13.73	0.915	28.57	0.928	T = 373		18.00	0.957	32.53	0.929*		
14.68	0.918	29.01	0.942			19.06	0.935	33.71	0.930		
15.00	0.901	29.49	0.948*	2.35	0.930	20.20	0.921	34.99	0.925		
15.44	0.919	30.18	0.948	3.33	0.930	20.51	0.907	36.29	0.895		
16.23	0.919	30.68	0.939	4.24	0.960	20.85	0.879*	38.19	0.891		
16.98	0.922	31.71	0.951	5.81	0.950	21.33	0.862	38.81	0.901*		

* Not shown on plot

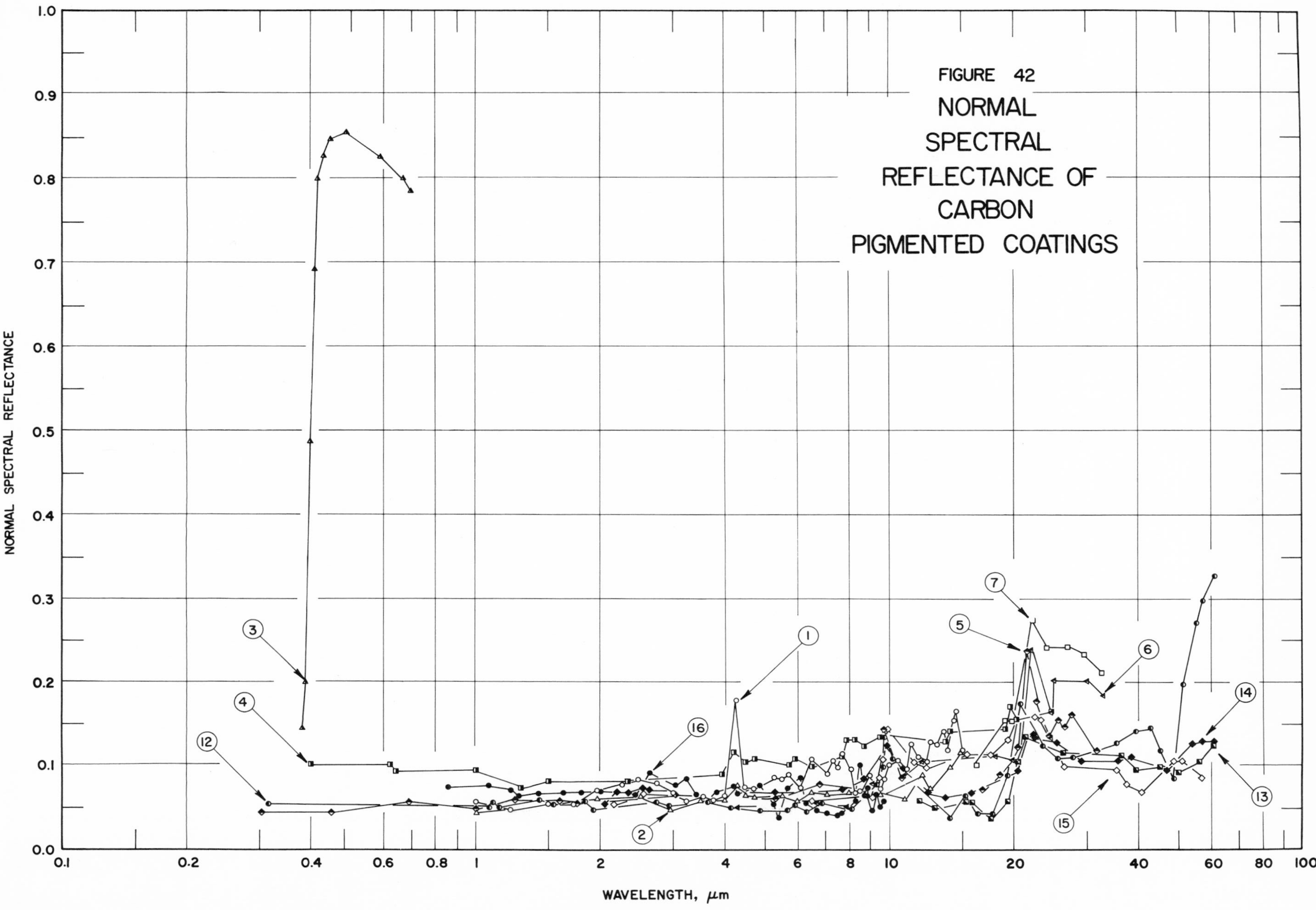

FIGURE 42 NORMAL SPECTRAL REFLECTANCE OF CARBON PIGMENTED COATINGS

SPECIFICATION TABLE NO. 42 NORMAL SPECTRAL REFLECTANCE OF CARBON PIGMENTED COATINGS

Curve No.	Ref. No.	Year	Temperature, K	Wavelength Range, μm	Geometry θ	θ'	ω'	Reported Error, %	Composition (weight percent), Specifications, and Remarks
1	62	1949	~298	1.00-15.00	0°		2π	5	Lampblack in turpentine binder; copper substrate; data extracted from smooth curve.
2	62	1949	~298	1.00-14.99	0°		2π	5	Lampblack in Decoret binder; copper substrate; data extracted from smooth curve.
3	90	1966	~298	0.380-0.700	~0°		2π		Carbon black in nitrocellulose binder; heavy coated chart paper substrate (Morest Opacity Panel Form 018); applied with Bird applicator over white background; data extracted from smooth curve.
4	89	1958	~298	0.40-24.9	~0°	~0°			North American Aviation paint No. 277.2.1; lampblack in DC807 silicone alkyd binder; 321 stainless steel substrate; air dried; data extracted from smooth curve. [Author's designation: Specimen 5]
5	72	1963	310	0.301-31.6	~0°		2π		Carbon black in acrylic binder (0.051 mm thick); Sherwin-Williams pretreatment primer P40GC1 Kemacryl Lacquer (Black No. M49 BC12); 2024 aluminum substrate; property measured in air soon after sample preparation; property is representative value (2 samples). [Author's designation: Acrylic Black]
6	72	1963	310	0.295-32.7	~0°		2π		Similar to above specimen and conditions except sample stored in dry chamber several days, in nitrogen several days, then property measured in air. [Author's designation: Acrylic Black, Sample No. 4]
7	72	1963	310	0.295-33.0	~0°		2π		Similar to above specimen and conditions. [Author's designation: Acrylic Black, Sample No. 5]
8*	72	1963	310	0.295-32.7	~0°		2π		Similar to above specimen and conditions. [Author's designation: Acrylic Black, Sample No. 6]
9*	72	1963	310	0.295-32.7	~0°		2π		Similar to curve 5 specimen and conditions except exposed to gamma radiation (dose ~1.7 x 10^{10} ergs gm^{-1} C^{-1}) in vacuum (~10^{-6} mm Hg) maintained by diffusion pump; sample stored in nitrogen for several days, then property measured in air. [Author's designation: Acrylic Black, Sample No. 1]
10*	72	1963	310	0.295-32.7	~0°		2π		Similar to above specimen and conditions. [Author's designation: Acrylic Black, Sample No. 2]
11*	72	1963	310	0.295-32.7	~0°		2π		Similar to above specimen and conditions. [Author's designation: Acrylic Black, Sample No. 3]
12	91	1966	~298	0.318-61.0	~20°		2π		Lampblack in epoxy binder (25 μm thick); Cat-A-Lac epoxy black paint (Finch Paint and Chemical Co.); aluminum substrate; 3 apparatuses used in the spectral ranges 0.33-2.5 μm (θ=20°), 1.5-23 μm (θ=25°), 20-61 μm (θ=17°).
13	91	1966	~298	11.8-60.9	~20°		2π		Similar to above specimen and conditions except thickness 75 μm.
14	91	1966	~298	2.06-61.2	~20°		2π		Similar to above specimen and conditions except thickness 250 μm.
15	91	1966	~298	2.18-56.2	~20°		2π		Similar to above specimen and conditions except thickness 750 μm.
16	92	1967	77	0.86-9.73	20°		2π		Lampblack in epoxy binder; Cat-A-Lac Black paint; measured in vacuum (10^{-6} mm Hg); converted from R (2π, 20°); measured relative to NaCl.

* Not shown on plot

DATA TABLE NO. 42 NORMAL SPECTRAL REFLECTANCE OF CARBON PIGMENTED COATINGS

[Wavelength, λ, μm; Reflectance, ρ; Temperature, T, K]

λ	ρ
CURVE 1	
T ~ 298	
1.00	0.058
1.22	0.048
1.50	0.053
1.73	0.053
1.98	0.070
2.27	0.076
2.49	0.081
2.74	0.078
3.00	0.067
3.21	0.058
3.53	0.061
3.75	0.059
4.00	0.062
4.25	0.179
4.49	0.073
4.74	0.071
5.01	0.072*
5.29	0.084
5.51	0.081
5.77	0.087
6.02	0.070
6.26	0.074*
6.53	0.107
6.77	0.095*
7.03	0.090
7.26	0.105
7.51	0.098
7.77	0.113
8.03	0.094
8.28	0.067
8.52	0.066
8.76	0.064*
9.00	0.087
9.26	0.073*
9.54	0.071
9.76	0.083
10.00	0.100
10.26	0.105*
10.50	0.105
10.76	0.097*
11.01	0.095
11.27	0.124
11.55	0.110*
11.76	0.110
12.01	0.107*

λ	ρ
CURVE 1 (cont.)	
12.29	0.103
12.53	0.127
12.79	0.126*
13.03	0.125
13.31	0.133*
13.55	0.140
13.80	0.119
14.05	0.132*
14.28	0.151
14.55	0.164
14.78	0.142*
15.00	0.119
CURVE 2	
T ~ 298	
1.00	0.043
1.49	0.056*
1.99	0.060
2.50	0.061
2.99	0.049
3.51	0.058
4.01	0.059
4.25	0.075
4.51	0.063
4.76	0.061
5.02	0.065*
5.52	0.063
6.00	0.058
6.48	0.068
7.01	0.066
7.51	0.068*
8.00	0.067
8.50	0.065*
8.98	0.063*
9.51	0.063
9.98	0.069*
10.48	0.067*
10.99	0.070
11.51	0.080*
12.01	0.089
12.54	0.072
13.03	0.085*
13.53	0.084*
14.00	0.099
14.51	0.104*
14.99	0.114

λ	ρ
CURVE 3	
T ~ 298	
0.380	0.144
0.389	0.200
0.400	0.487
0.410	0.693
0.417	0.800
0.427	0.828
0.434	0.839*
0.453	0.849
0.494	0.853
0.588	0.826
0.666	0.800
0.700	0.787
CURVE 4	
T ~ 298	
0.40	0.100
0.62	0.100
0.64	0.091
1.00	0.092
1.28	0.071
1.50	0.080
2.31	0.080
2.48	0.089*
3.97	0.089
4.20	0.115
4.51	0.101
4.78	0.108
5.00	0.100*
5.73	0.100
5.96	0.108
6.30	0.099*
6.58	0.099
6.75	0.110*
7.71	0.110
7.98	0.130
8.22	0.130
8.44	0.121*
8.71	0.121
8.95	0.131*
9.44	0.131
9.74	0.139
10.1	0.129*
13.7	0.129
14.0	0.140

λ	ρ
CURVE 4 (cont.)	
19.0	0.142
19.4	0.170
19.8	0.152*
20.1	0.153
20.3	0.161*
24.9	0.164
CURVE 5	
T = 310	
0.301	0.044
0.450	0.044
0.695	0.055
1.00	0.047
1.23	0.058
1.60	0.058
1.75	0.054
3.03	0.064
3.98	0.060*
5.35	0.060
6.80	0.077
8.35	0.068*
9.41	0.069
9.74	0.141
10.6	0.082
12.6	0.070*
16.9	0.070
18.5	0.088
20.4	0.121
21.8	0.236
22.6	0.176
24.4	0.133
25.7	0.153
26.7	0.145
28.3	0.160
31.6	0.119
CURVE 6	
T = 310	
0.295	0.050
4.38	0.050
6.71	0.059
7.92	0.052
8.79	0.079
9.18	0.079

λ	ρ
CURVE 6 (cont.)	
9.48	0.108
9.97	0.142
10.8	0.089
12.0	0.081
18.6	0.110
20.7	0.103
22.2	0.239
24.7	0.166
25.2	0.201
30.1	0.201
32.7	0.183
CURVE 7	
T = 310	
0.295	0.050*
4.38	0.050*
6.71	0.059*
7.92	0.052*
8.79	0.079*
9.18	0.079*
9.48	0.108*
9.97	0.142*
10.8	0.089*
12.0	0.081*
16.4	0.100
19.1	0.154
20.1	0.151
22.0	0.272
24.0	0.241
27.1	0.241
29.3	0.235
33.0	0.213
CURVE 8*	
T = 310	
0.295	0.050
4.38	0.050
6.71	0.059
7.92	0.052
8.79	0.079
9.18	0.079
9.48	0.108
9.97	0.142
10.8	0.089

λ	ρ
CURVE 8 (cont.)*	
12.0	0.081
18.6	0.110
20.7	0.103
22.2	0.239
24.7	0.166
25.2	0.201
30.1	0.201
32.7	0.183
CURVE 9*	
T = 310	
0.295	0.050
4.38	0.050
6.71	0.059
7.92	0.052
8.79	0.079
9.18	0.079
9.48	0.108
9.97	0.142
10.8	0.089
12.0	0.081
18.6	0.110
20.7	0.103
22.2	0.239
24.7	0.166
25.2	0.201
30.1	0.201
32.7	0.183
CURVE 10*	
T = 310	
0.295	0.050
4.38	0.050
6.71	0.059
7.92	0.052
8.79	0.079
9.18	0.079
9.48	0.108
9.97	0.142
10.8	0.089
12.0	0.081
18.6	0.110
20.7	0.103
22.2	0.239

λ	ρ
CURVE 10 (cont.)*	
24.7	0.166
25.2	0.201
30.1	0.201
32.7	0.183
CURVE 11*	
T = 310	
0.295	0.050
4.38	0.050
6.71	0.059
7.92	0.052
8.79	0.079
9.18	0.079
9.48	0.108
9.97	0.142
10.8	0.089
12.0	0.081
18.6	0.110
20.7	0.103
22.2	0.239
24.7	0.166
25.2	0.201
30.1	0.201
32.7	0.183
CURVE 12	
T ~ 298	
0.318	0.052
1.08	0.050
1.10	0.055
1.13	0.050
1.16	0.058*
1.43	0.058
1.53	0.052
1.84	0.056
1.91	0.049
2.73	0.056
2.94	0.051
3.62	0.056
4.17	0.050
4.86	0.048
5.61	0.048
5.97	0.051
6.38	0.046

λ	ρ
CURVE 12 (cont.)	
6.83	0.055
8.10	0.049
9.26	0.076
9.61	0.098
9.74	0.133
10.2	0.107
10.6	0.081*
12.3	0.047*
14.0	0.039
15.1	0.062
16.3	0.041
17.9	0.041
19.4	0.087
20.9	0.172
23.3	0.121
25.5	0.108
27.7	0.110
31.7	0.119*
35.1	0.125
39.5	0.140
42.9	0.142
45.1	0.116
48.6	0.081
51.6	0.198
55.4	0.270
57.5	0.299
61.0	0.326
CURVE 13	
T ~ 298	
11.8	0.057
12.9	0.050
15.2	0.056
15.9	0.056
17.5	0.037
19.1	0.057
20.2	0.102
21.4	0.132
26.2	0.113
36.1	0.111
39.4	0.093
45.3	0.097
50.3	0.090
56.4	0.102
60.9	0.122

* Not shown on plot

DATA TABLE NO. 42 NORMAL SPECTRAL REFLECTANCE OF CARBON PIGMENTED COATINGS (continued)

λ	ρ
CURVE 14	
T ~ 298	
2.06	0.054
2.20	0.067
2.34	0.067
2.51	0.071
2.62	0.070
5.34	0.068
5.55	0.066*
7.62	0.070
8.62	0.084
9.90	0.122
10.7	0.095
12.4	0.067
13.6	0.061
15.9	0.067
20.2	0.093
21.6	0.132*
22.5	0.131
22.5	0.138
25.2	0.129
27.6	0.110*
29.0	0.104
35.9	0.104
38.1	0.110
40.9	0.108*
46.9	0.092
54.0	0.125
57.6	0.129
61.2	0.129
CURVE 15	
T ~ 298	
2.18	0.054
2.55	0.063
6.59	0.058
8.29	0.067*
9.01	0.061
9.41	0.084
9.63	0.108
9.81	0.142
11.5	0.102
12.3	0.097
15.3	0.112
17.6	0.112
19.3	0.130
20.3	0.153*
22.7	0.158
CURVE 15 (cont.)	
23.1	0.152
23.4	0.116*
26.1	0.099
35.6	0.094
37.1	0.077
40.4	0.069
48.8	0.103
51.8	0.103
56.2	0.083
CURVE 16	
T = 77	
0.86	0.074
1.06	0.074
1.21	0.070
1.27	0.062
1.41	0.066
1.61	0.066
1.80	0.066
2.00	0.069
2.20	0.069*
2.41	0.063
2.61	0.090
2.71	0.080*
3.01	0.075
3.22	0.081
3.42	0.064
3.61	0.055*
3.82	0.067
3.95	0.067*
4.20	0.073
4.33	0.066
4.50	0.068*
4.71	0.075*
4.81	0.075
5.06	0.075*
5.24	0.051
5.40	0.039
5.56	0.068*
5.70	0.072
5.85	0.072*
6.00	0.072*
6.14	0.084
6.30	0.055
6.41	0.054*
6.55	0.052*
6.69	0.047
CURVE 16 (cont.)	
6.82	0.045*
6.95	0.045*
7.06	0.045
7.21	0.045*
7.31	0.040*
7.41	0.040*
7.50	0.040*
7.59	0.040
7.70	0.043
7.85	0.043*
7.97	0.043*
8.15	0.049*
8.25	0.057
8.33	0.066*
8.41	0.066*
8.51	0.100
8.61	0.062
8.72	0.062*
8.82	0.062*
8.92	0.062
8.99	0.072
9.07	0.059*
9.16	0.048
9.22	0.063
9.32	0.060*
9.41	0.050
9.48	0.064*
9.56	0.058*
9.64	0.056*
9.73	0.057

* Not shown on plot

SPECIFICATION TABLE NO. 43 NORMAL SOLAR ABSORPTANCE OF CARBON PIGMENTED COATINGS

Curve No.	Ref. No.	Year	Temperature Range, K	Geometry θ	Reported Error, %	Composition (weight percent), Specifications and Remarks
1*	49	1961	278	~0°	10	Carbon black in acrylic resin binder; Kemacryl black lacquer (M49BC12); substrate unknown; value is avg of several determinations.
2*	49	1961	278	~0°		Carbon in silicone binder; Fuller flat black silicone; value is avg of several determinations.
3*	49	1961	278	~0°		Carbon black in acrylic resin binder; black Kemacryl lacquer (M49BC12); Mg alloy substrate; substrate Dow 17 treated.
4*	49	1961	278	~0°		Above specimen and conditions except exposed in vacuum (10^{-6}–8 x 10^{-6} mm Hg) to UV radiation from an argon-filled A-H6 high pressure Hg arc lamp for 80 hrs.
5*	17	1968	298	~0°		Lampblack in epoxy binder; Cat-A-Lac black paint (Finch Paint and Chemical Co.) supplied by ESRO I Project Group.
6*	77	1962	~298	~0°		Carbon black in phthalic alkyd binder (0.064 mm thick); composition (wet): 2.5 carbon black, 11.7 silaceous inerts, 31.4 phthalic alkyd, balance organic solvents; applied by standard spraying techniques; zinc chromate primer on sandblasted and etched 2024 aluminum substrates; computed from spectral reflectance data.
7*	77	1962	~298	~0°		Above specimen and conditions except heated to 533 K in vacuum (10^{-5} mm Hg) on a 2-hr cycle.

DATA TABLE NO. 43 NORMAL SOLAR ABSORPTANCE OF CARBON PIGMENTED COATINGS

[Temperature, T, K; Absorptance, α]

T	α	T	α
CURVE 1*		CURVE 5*	
278	0.94	298	0.96
CURVE 2*		CURVE 6*	
278	0.89	298	0.955
CURVE 3*		CURVE 7*	
278	0.94	298	0.958
CURVE 4*			
278	0.92		

* No plot given

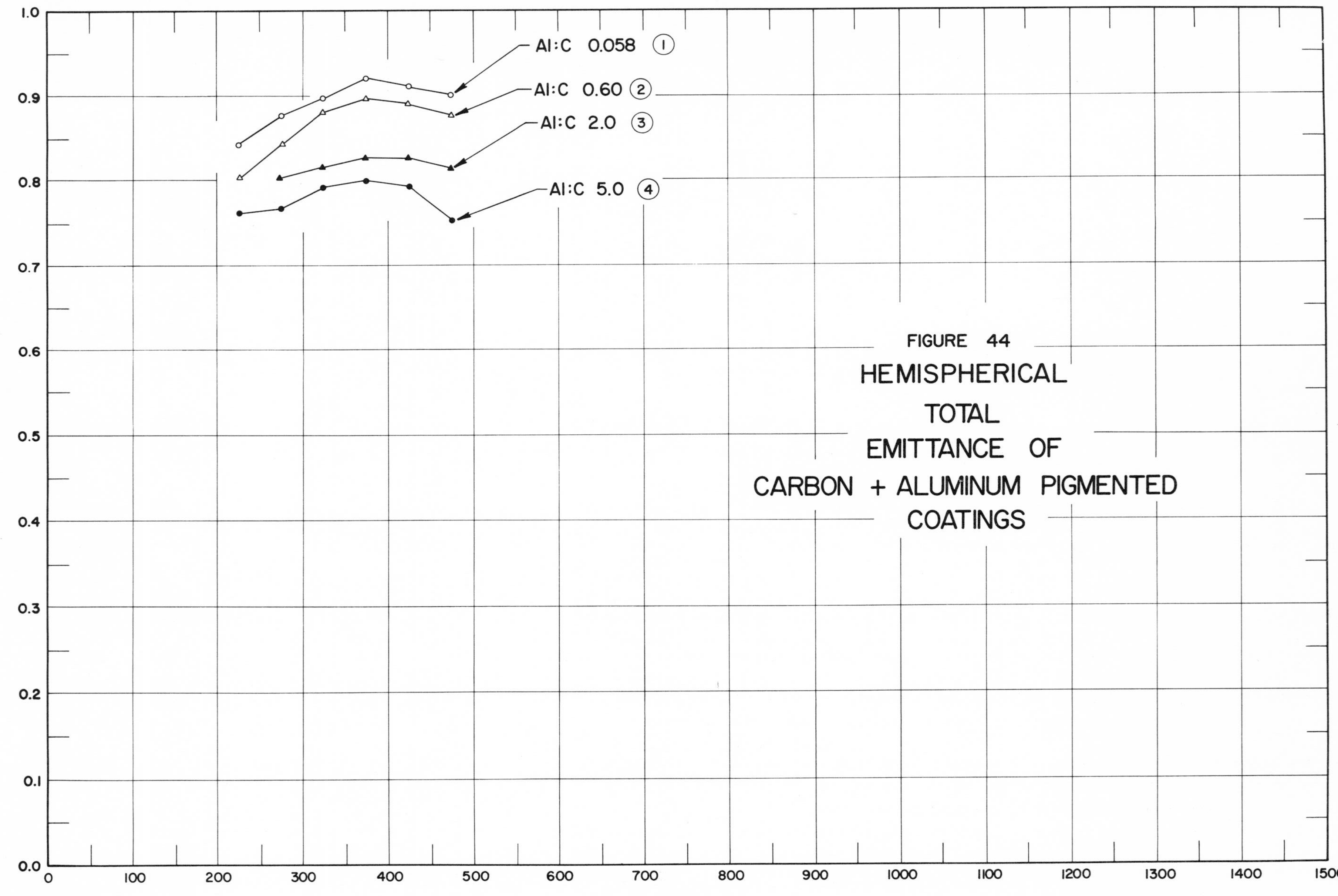

FIGURE 44 HEMISPHERICAL TOTAL EMITTANCE OF CARBON + ALUMINUM PIGMENTED COATINGS

SPECIFICATION TABLE NO. 44 HEMISPHERICAL TOTAL EMITTANCE OF CARBON + ALUMINUM PIGMENTED COATINGS

Curve No.	Ref. No.	Year	Temperature Range, K	Reported Error, %	Composition (weight percent), Specifications and Remarks
1	94	1963	228-473		Carbon black (25 Columbia Carbon Raven 11 beads and 75 Binney and Smiths Superb beads) and leafing aluminum in phthalic alkyd binder; composition wet: 2.5 carbon and 31.4 phthalic alkyd; aluminum/carbon black ratio 0.058; zinc chromate on roughened 2024 aluminum substrate; spray application; air cured. [Author's designation: Coating No. 3]
2	94	1963	229-474		Similar to curve 1 specimen and conditions except aluminum/carbon black ratio 0.60. [Author's designation: Coating No. 6]
3	94	1963	273-474		Similar to curve 1 specimen and conditions except aluminum/carbon black ratio 2.0. [Author's designation: Coating No. 9]
4	94	1963	228-474		Similar to curve 1 specimen and conditions except aluminum/carbon black ratio 5.0. [Author's designation: Coating No. 12]

DATA TABLE NO. 44 HEMISPHERICAL TOTAL EMITTANCE OF CARBON + ALUMINUM PIGMENTED COATINGS

[Temperature, T, K; Emittance, ∈]

T	∈	T	∈
CURVE 1		CURVE 3	
228	0.841	273	0.802
274	0.876	324	0.817
323	0.899	374	0.828
373	0.920	423	0.828
424	0.910	474	0.814
473	0.900		
CURVE 2		CURVE 4	
229	0.802	228	0.762
274	0.842	273	0.767
324	0.880	324	0.791
373	0.896	374	0.800
424	0.890	424	0.792
474	0.878	474	0.751

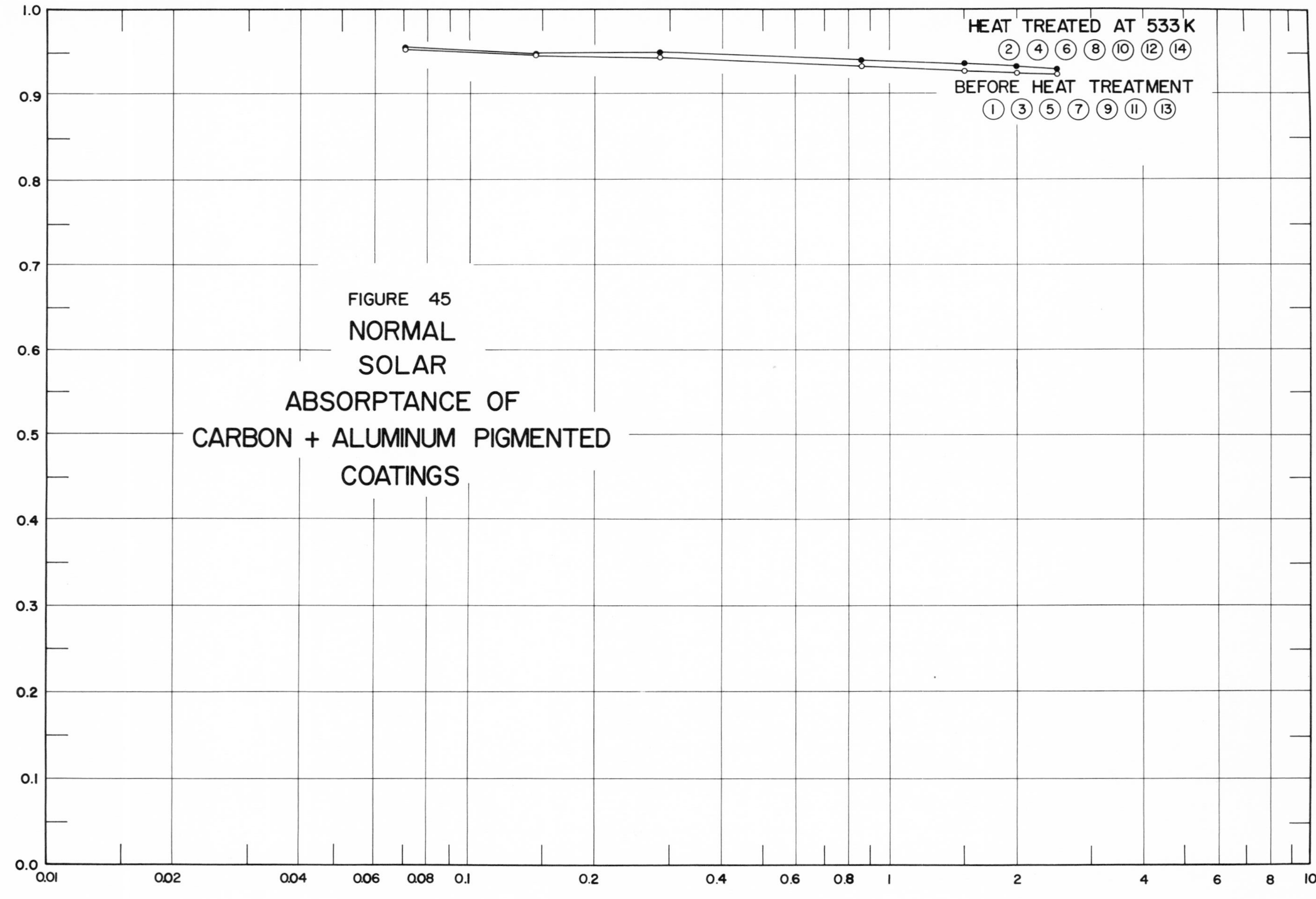

FIGURE 45
NORMAL
SOLAR
ABSORPTANCE OF
CARBON + ALUMINUM PIGMENTED
COATINGS

SPECIFICATION TABLE NO. 45 NORMAL SOLAR ABSORPTANCE OF CARBON + ALUMINUM PIGMENTED COATINGS

Curve No.	Ref. No.	Year	Temperature Range, K	Geometry θ	Reported Error, %	Composition (weight percent), Specifications and Remarks
1	77	1962	~298	~0°		Carbon and leafing aluminum in phthalic alkyd binder (0.064 mm thick); composition (wet): 2.5 carbon black, 0.287 leafing aluminum, 11.7 silaceous inerts, 37.3 phthalic alkyd, balance organic solvents; applied by standard spraying techniques; zinc chromate on sandblasted and etched 2024 aluminum; computed from spectral reflectance data.
2	77	1962	~298	~0°		Above specimen and conditions except heated to 533 K in vacuum (10^{-5} mm Hg) on a 2 hr cycle.
3	77	1962	~298	~0°		Similar to curve 1 specimen and conditions except composition (wet): 2.5 carbon black, 0.860 leafing aluminum, 11.6 silaceous inerts, 31.2 phthalic alkyd, balance organic solvents.
4	77	1962	~298	~0°		Above specimen and conditions except heated to 533 K in vacuum (10^{-5} mm Hg) on a 2 hr cycle.
5	77	1962	~298	~0°		Similar to curve 1 specimen and conditions except composition (wet): 2.5 carbon black, 1.5 leafing aluminum, 11.5 silaceous inerts, 30.9 phthalic alkyd, balance organic solvents.
6	77	1962	~298	~0°		Above specimen and conditions except heated to 533 K in vacuum (10^{-5} mm Hg) on a 2 hr cycle.
7	77	1962	~298	~0°		Similar to curve 1 specimen and conditions except composition (wet): 2.5 carbon black, 2.0 leafing aluminum, 11.5 silaceous inerts, 30.8 phthalic alkyd, balance organic solvents.
8	77	1962	~298	~0°		Above specimen and conditions except heated to 533 K in vacuum (10^{-5} mm Hg) on a 2 hr cycle.
9	77	1962	~298	~0°		Similar to curve 1 specimen and conditions except composition (wet): 2.5 carbon black, 2.5 leafing aluminum, 11.4 silaceous inerts, 30.6 phthalic alkyd, balance organic solvents.
10	77	1962	~298	~0°		Above specimen and conditions except heated to 533 K in vacuum (10^{-5} mm Hg) on a 2 hr cycle.
11	77	1962	~298	~0°		Carbon and leafing aluminum in phthalic alkyd binder (0.064 mm thick); composition (wet): 2.5 carbon black, 11.7 silaceous inerts, 31.4 phthalic alkyd, 0.07 leafing aluminum, balance organic solvents; sandblasted and etched 2024 aluminum substrate coated with zinc chromate primer; computed from reflectance.
12	77	1962	~298	~0°		Above specimen and conditions except heated to 533 K in vacuum (10^{-5} mm Hg) on a 2 hr cycle.
13	77	1962	~298	~0°		Similar to curve 1 specimen and conditions except composition (wet): 2.5 carbon black, 11.7 silaceous inerts, 31.4 phthalic alkyd, 0.145 leafing aluminum, balance organic solvents.
14	77	1962	~298	~0°		Above specimen and conditions except heated to 533 K in vacuum (10^{-5} mm Hg) on a 2 hr cycle.

DATA TABLE NO. 45 NORMAL SOLAR ABSORPTANCE OF CARBON + ALUMINUM PIGMENTED COATINGS

[Temperature, T, K; Absorptance, α]

T	α	T	α
CURVE 1		CURVE 13	
298	0.943	298	0.947
CURVE 2		CURVE 14	
298	0.950	298	0.949
CURVE 3			
298	0.932		
CURVE 4			
298	0.940		
CURVE 5			
298	0.929		
CURVE 6			
298	0.936		
CURVE 7			
298	0.925		
CURVE 8			
298	0.931		
CURVE 9			
298	0.922		
CURVE 10			
298	0.930		
CURVE 11			
298	0.951		
CURVE 12			
298	0.952		

SPECIFICATION TABLE NO. 46 NORMAL TOTAL EMITTANCE OF CHINA CLAY PIGMENTED COATINGS

Curve No.	Ref. No.	Year	Temperature Range, K	Geometry θ'	Reported Error, %	Composition (weight percent), Specifications and Remarks
1*	56	1969	311	~0°		Hughes H-10; calcinted china clay in RTV-602 silicone resin binder; property calculated from reflectance; lab data taken on sample to be tested on Lunar Orbiter V.

DATA TABLE NO. 46 NORMAL TOTAL EMITTANCE OF CHINA CLAY PIGMENTED COATINGS

[Temperature, T, K; Emittance, ϵ]

T	ϵ
CURVE 1*	
311	0.860

* No plot given

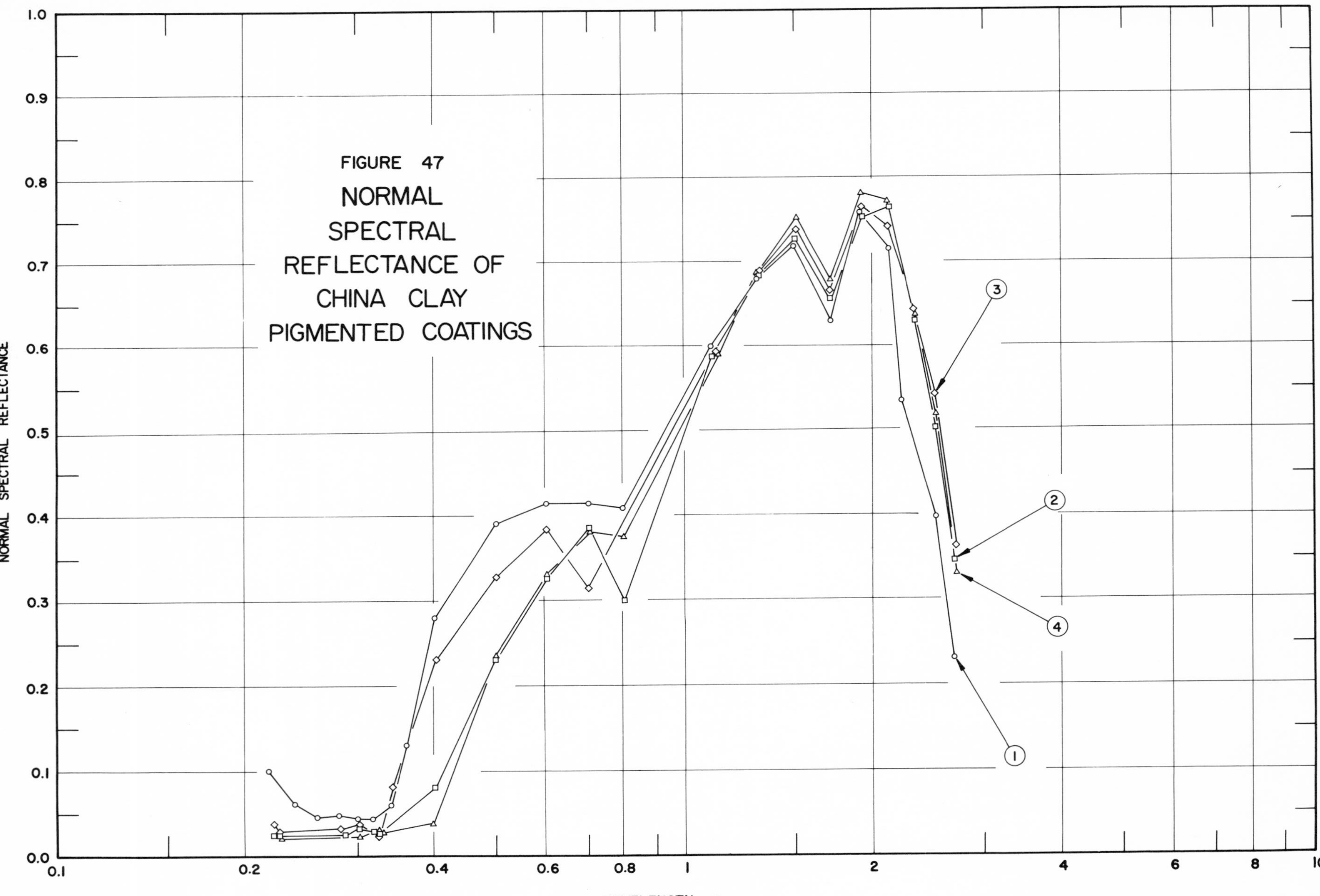

FIGURE 47 NORMAL SPECTRAL REFLECTANCE OF CHINA CLAY PIGMENTED COATINGS

SPECIFICATION TABLE NO. 47 NORMAL SPECTRAL REFLECTANCE OF CHINA CLAY PIGMENTED COATINGS

Curve No.	Ref. No.	Year	Temperature, K	Wavelength Range, μm	Geometry θ	θ'	ω'	Reported Error, %	Composition (weight percent), Specifications, and Remarks
1	1	1960	358	0.219-2.695	~0°		2 π		China clay in silicone binder (0.051 mm thick); 30% china clay; anodized aluminum alloy 24 S-T substrate; measured relative to MgO.
2	1	1960	358	0.221-2.699	~0°		2 π		Above specimen and conditions except exposed to UV radiation for 60 hrs with G.E. type UA-3 lamp in vacuum (10^{-5} mm Hg).
3	1	1960	358	0.221-2.700	~0°		2 π		Above specimen and conditions except exposed to UV radiation for 20 additional hours.
4	1	1960	358	0.225-2.700	~0°		2 π		Above specimen and conditions except exposed to UV radiation for 100 additional hours.

DATA TABLE NO. 47 NORMAL SPECTRAL REFLECTANCE OF CHINA CLAY PIGMENTED COATINGS

[Wavelength, λ, μm; Reflectance, ρ; Temperature, T, K]

λ	ρ	λ	ρ	λ	ρ	λ	ρ
CURVE 1 T = 358		CURVE 2 T = 358		CURVE 3 T = 358		CURVE 4 T = 358	
0.219	0.101	0.221	0.024	0.221	0.038	0.225	0.024*
0.240	0.063	0.226	0.024	0.225	0.027	0.282	0.021
0.260	0.044	0.287	0.025	0.281	0.032	0.306	0.023
0.282	0.048	0.305	0.033	0.301	0.036	0.324	0.030
0.302	0.046	0.319	0.029	0.324	0.024	0.330	0.027
0.319	0.046	0.326	0.028	0.401	0.082	0.399	0.037
0.340	0.062	0.400	0.081	0.501	0.233	0.505	0.233
0.361	0.133	0.500	0.234	0.602	0.323	0.603	0.333
0.400	0.283	0.606	0.327	0.703	0.385	0.704	0.381
0.501	0.392	0.708	0.386	0.800	0.315	0.800	0.376
0.602	0.414	0.801	0.300	1.115	0.595	1.127	0.593
0.701	0.416	1.101	0.590	1.314	0.691	1.319	0.691
0.800	0.408	1.320	0.684	1.502	0.740	1.503	0.755
1.101	0.600	1.500	0.730	1.706	0.668	1.710	0.680
1.301	0.683	1.706	0.658	1.912	0.768	1.908	0.783
1.495	0.720	1.932	0.757	2.108	0.749	2.110	0.775
1.701	0.632	2.107	0.765	2.301	0.645	2.331	0.640
1.907	0.759	2.306	0.633	2.501	0.546	2.502	0.519
2.108	0.718	2.501	0.505	2.700	0.365	2.700	0.333
2.296	0.535	2.699	0.348				
2.501	0.399						
2.695	0.233						

* Not shown on plot

SPECIFICATION TABLE NO. 48 NORMAL SOLAR ABSORPTANCE OF CHINA CLAY PIGMENTED COATINGS

Curve No.	Ref. No.	Year	Temperature Range, K	Geometry θ	Reported Error, %	Composition (weight percent), Specifications and Remarks
1*	56	1969	298	~0°		Hughes H-10; calcined china clay in RTV-602 silicone resin binder; property calculated from reflectance; lab data taken on sample to be tested on Lunar Orbiter V.

DATA TABLE NO. 48 NORMAL SOLAR ABSORPTANCE OF CHINA CLAY PIGMENTED COATINGS

[Temperature, T, K; Absorptance, α]

T	α
CURVE 1*	
298	0.147

* No plot given

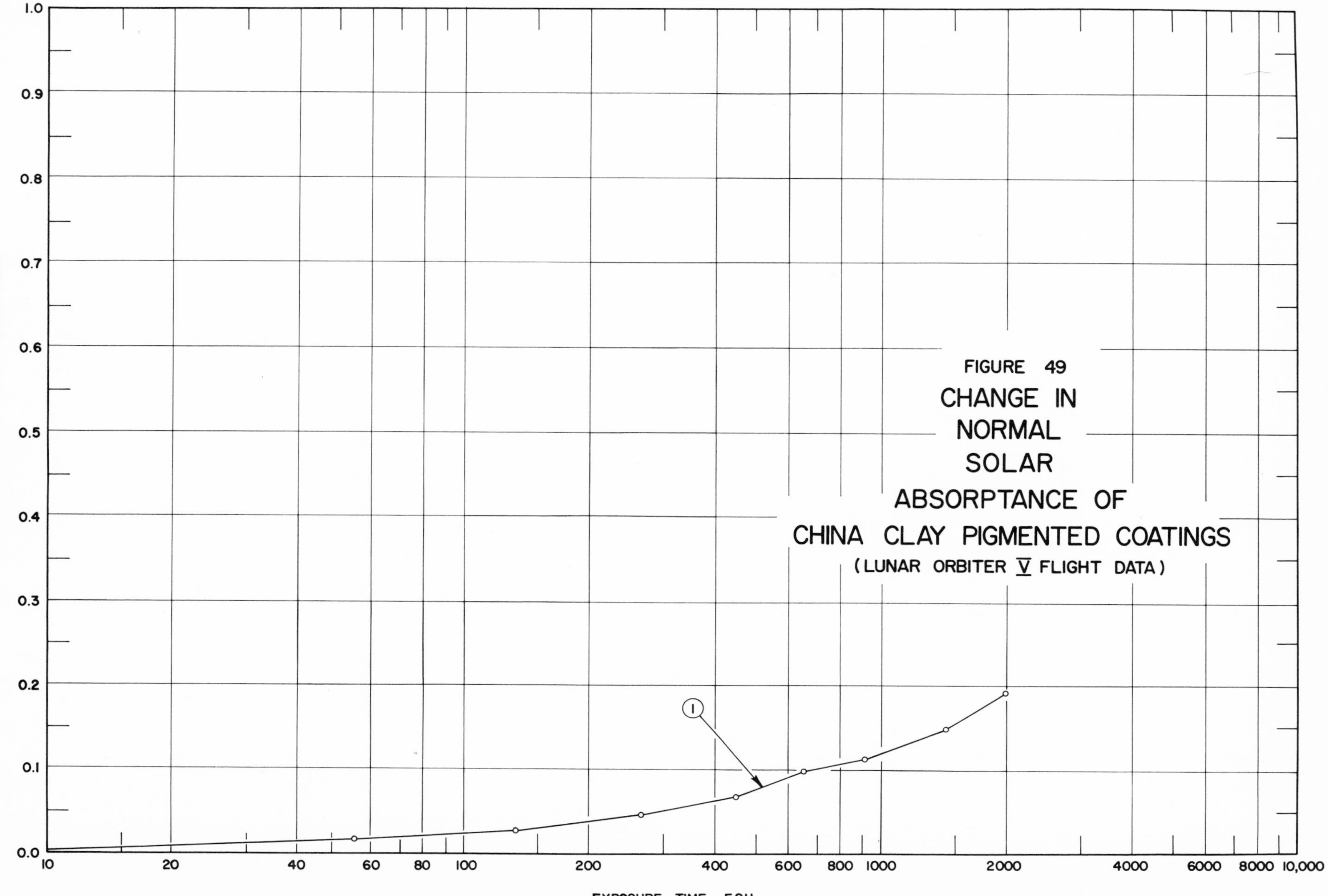
FIGURE 49
CHANGE IN NORMAL SOLAR ABSORPTANCE OF CHINA CLAY PIGMENTED COATINGS
(LUNAR ORBITER V FLIGHT DATA)
1
CHANGE IN NORMAL SOLAR ABSORPTANCE
1.0
0.9
0.8
0.7
0.6
0.5
0.4
0.3
0.2
0.1
0.0
10
20
40
60
80
100
200
400
600
800
1000
2000
4000
6000
8000
10,000
EXPOSURE TIME, ESH

SPECIFICATION TABLE NO. 49 CHANGE IN NORMAL SOLAR ABSORPTANCE OF CHINA CLAY PIGMENTED COATINGS

Curve No.	Ref. No.	Year	Temperature Range, K	Geometry θ	Reported Error, %	Composition (weight percent), Specifications and Remarks
1	56	1969	250-300	~0°		Hughes H-10; calcined china clay in G. E. RTV-602 silicone resin binder; property calculated from temp of sample; in-flight data of Lunar Orbiter V; ESH is variable; data extracted from smooth curve.

DATA TABLE NO. 49 CHANGE IN NORMAL SOLAR ABSORPTANCE OF CHINA CLAY PIGMENTED COATINGS

[Temperature, T, K; Absorptance, α]

ESH	$\Delta\alpha$
CURVE 1 T = 250-300	
0	0.0030*
54	0.0160
131	0.0297
262	0.0480
442	0.0695
647	0.0894
903	0.1120
1405	0.1496
1994	0.1907

* Not shown on plot

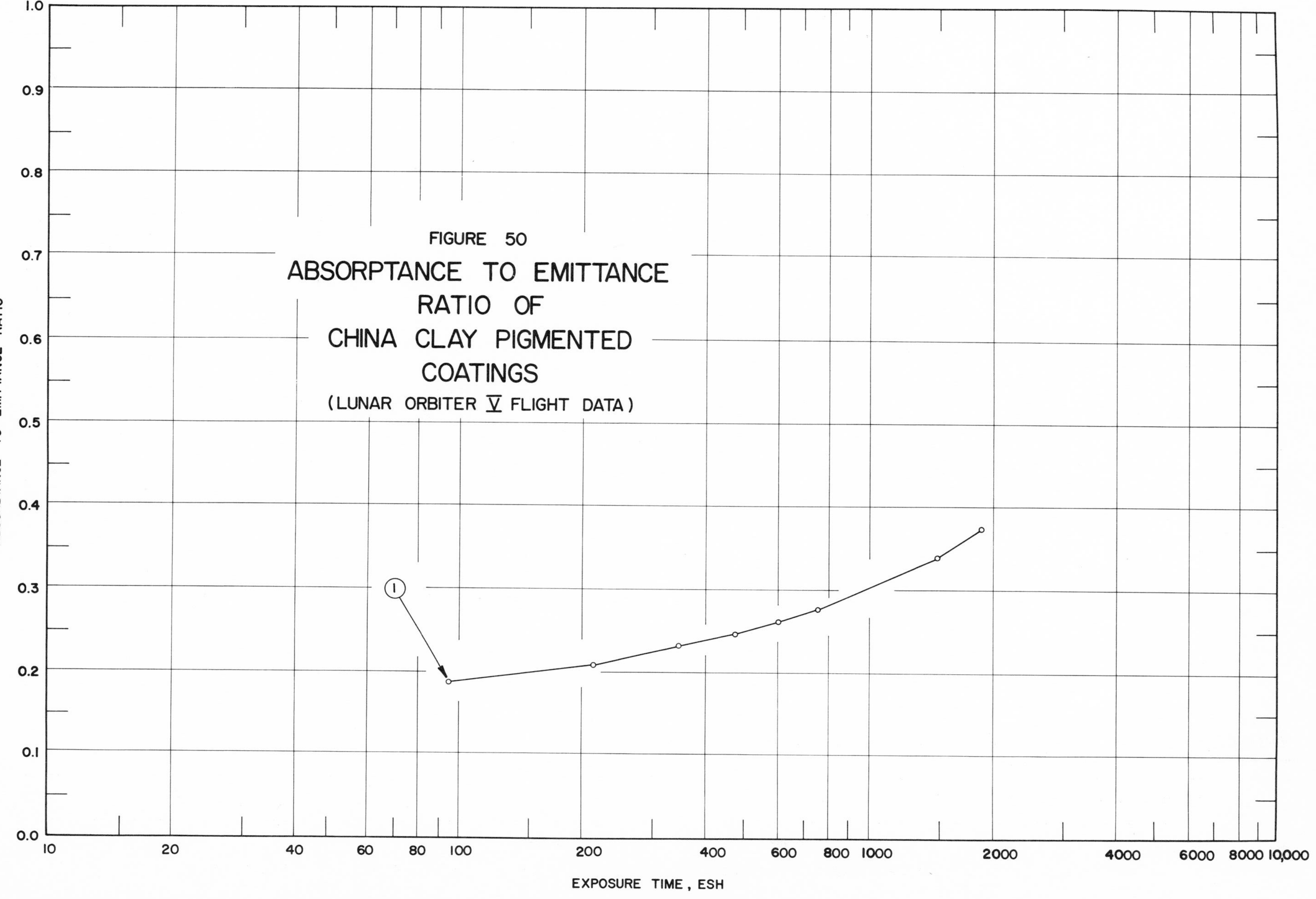
FIGURE 50
ABSORPTANCE TO EMITTANCE
RATIO OF
CHINA CLAY PIGMENTED
COATINGS
(LUNAR ORBITER V FLIGHT DATA)
ABSORBTANCE TO EMITTANCE RATIO
1.0
0.9
0.8
0.7
0.6
0.5
0.4
0.3
0.2
0.1
0.0
10
20
40
60
80
100
200
400
600
800
1000
2000
4000
6000
8000
10,000
EXPOSURE TIME, ESH
1

SPECIFICATION TABLE NO. 50 ABSORPTANCE TO EMITTANCE RATIO OF CHINA CLAY PIGMENTED COATINGS

Curve No.	Ref. No.	Year	Temperature Range, K	Reported Error, %	Composition (weight percent), Specifications and Remarks
1	56	1969	250-300		Hughes H-10; calcined china clay in G. E. RTV-602 silicone resin binder; property calculated from temp of sample; in-flight data of Lunar Orbiter V; ESH is variable; data extracted from smooth curve.

DATA TABLE NO. 50 ABSORPTANCE TO EMITTANCE RATIO OF CHINA CLAY PIGMENTED COATINGS

[Temperature, T, K; Absorptance, α; Emittance, ϵ]

ESH	α/ϵ
CURVE 1	
T = 250-300	
0	0.161*
95	0.187
212	0.210
347	0.231
468	0.247
600	0.262
759	0.279
1448	0.340
1892	0.377

* Not shown on plot

SPECIFICATION TABLE NO. 51 HEMISPHERICAL TOTAL EMITTANCE OF CHROMIUM OXIDE + IRON OXIDE + NICKEL OXIDE PIGMENTED COATINGS

Curve No.	Ref. No.	Year	Temperature Range, K	Reported Error, %	Composition (weight percent), Specifications and Remarks
1*	67	1967	1750		Cr_2O_3, Fe_3O_4 and NiO in synar binder; D-36 columbium substrate.

DATA TABLE NO. 51 HEMISPHERICAL TOTAL EMITTANCE OF CHROMIUM OXIDE + IRON OXIDE + NICKEL OXIDE PIGMENTED COATINGS

[Temperature, T, K; Emittance, ϵ]

T	ϵ
CURVE 1*	
1750	0.88

* No plot given

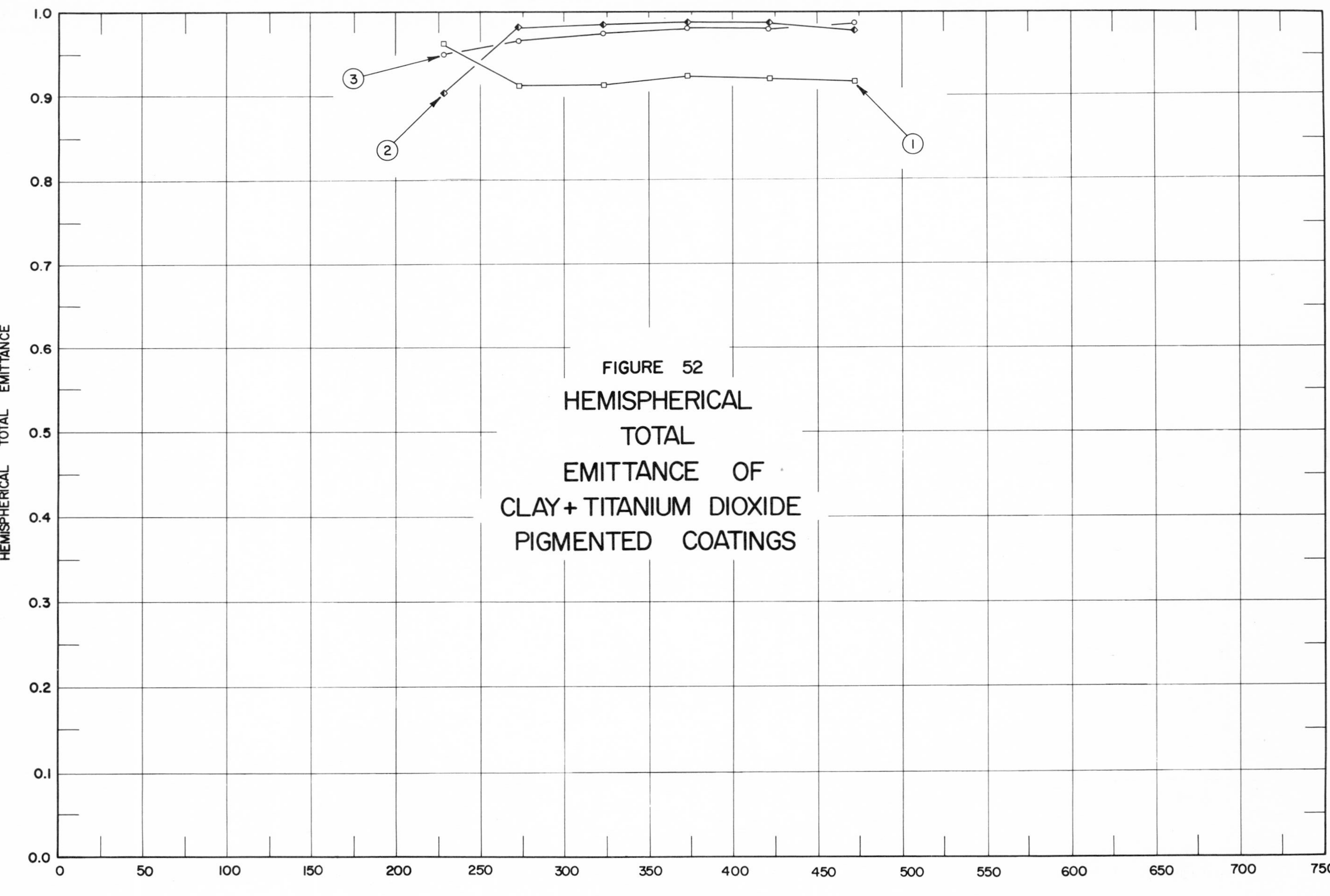

FIGURE 52
HEMISPHERICAL
TOTAL
EMITTANCE OF
CLAY + TITANIUM DIOXIDE
PIGMENTED COATINGS

SPECIFICATION TABLE NO. 52 HEMISPHERICAL TOTAL EMITTANCE OF CLAY + TITANIUM DIOXIDE PIGMENTED COATINGS

Curve No.	Ref. No.	Year	Temperature Range, K	Reported Error, %	Composition (weight percent), Specifications and Remarks
1	94	1963	229-473		26.1 rutile TiO_2, 30.0 calcined clay (Southern Clay Al Sil Ate W), and 21.6 organic solvents in 22.3 Plaskon ST-873 silicone-alkyd binder mixed with black base enamel (2.5 carbon black, 11.7 silaceous inerts, and 54.4 thinners and driers in 31.4 phthalic alkyd binder) (~0.0635 mm thick); zinc chromate (0.0127 mm thick) and 2024 aluminum alloy substrates; carbon black to TiO_2 ratio by weight 9.2×10^{-4}; aluminum substrate roughened then primed with zinc chromate primer (MIL-P-8585); spray application; air cured at ~298 K; carbon black: 25 Columbia Carbon's Raven II beads and 7.5 Binney and Smiths Superb beads. [Authors' designation: silicone-alkyd/black enamel coating No. 3]
2	94	1963	229-473		Similar to above specimen and conditions except carbon black/TiO_2 ratio 5.5×10^{-3}. [Authors' designation: silicone-alkyd/black enamel coating No. 6]
3	94	1963	227-473		Similar to above specimen and conditions except carbon black/TiO_2 ratio 1.8×10^{-2}. [Authors' designation: silicone-alkyd/black enamel coating No. 9]

DATA TABLE NO. 52 HEMISPHERICAL TOTAL EMITTANCE OF CLAY + TITANIUM DIOXIDE PIGMENTED COATINGS

[Temperature, T, K; Emittance, ϵ]

T	ϵ	T	ϵ
CURVE 1		CURVE 3	
229	0.962	227	0.950
274	0.912	274	0.965
324	0.912	324	0.975
374	0.922	373	0.980
423	0.920	422	0.980
473	0.916	473	0.986
CURVE 2			
229	0.903		
274	0.981		
323	0.984		
373	0.986		
422	0.985		
473	0.977		

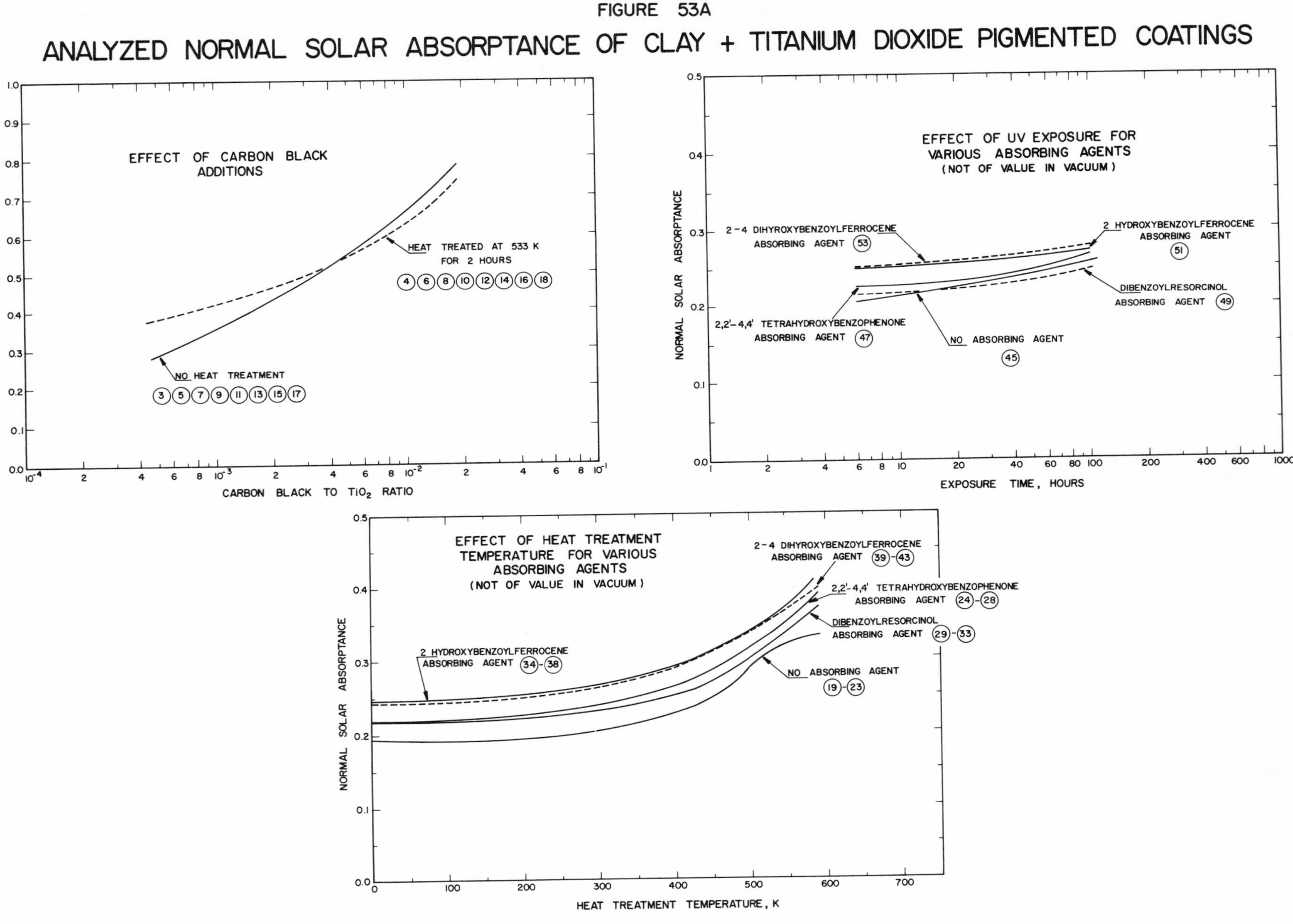

FIGURE 53A

ANALYZED NORMAL SOLAR ABSORPTANCE OF CLAY + TITANIUM DIOXIDE PIGMENTED COATINGS

SPECIFICATION TABLE NO. 53 NORMAL SOLAR ABSORPTANCE OF CLAY + TITANIUM DIOXIDE PIGMENTED COATINGS

Curve No.	Ref. No.	Year	Temperature Range, K	Geometry θ	Reported Error, %	Composition (weight percent), Specifications and Remarks
1	77	1962	~298	~0°		26.1 rutile TiO_2, 30.0 calcined clay (southern clay Al Sil Ate W), and 21.6 organic solvents in 22.3 Plaskon ST-873 silicone - alkyd binder (~0.0635 mm thick) on zinc chromate (0.0127 mm thick) and 2024 aluminum alloy substrates; aluminum substrate roughened, then primed with zinc chromate primer (MIL-P-8585); spray application; air cured at ~298 K; computed from spectral reflectance data. [Authors' designation: silicone alkyd/black enamel coating No. 1]
2	77	1962	~298	~0°		Above specimen and conditions except heat treated to 533 K in a 2-hr cycle in vacuum (10^{-5} mm Hg); vacuum maintained by diffusion pump.
3	77	1962	~298	~0°		26.1 rutile TiO_2, 30.0 calcined clay (southern clay Al Sil Ate W), and 21.6 organic solvents in 22.3 Plaskon ST-873 silicone - alkyd binder mixed with black base enamel (2.5 carbon black, 11.7 silaceous inerts, and 54.4 thinners and driers in 31.4 phthalic alkyd binder) (~0.0635 mm thick) on zinc chromate (0.0127 mm thick) and 2024 aluminum alloy substrates; carbon black to TiO_2 ratio by weight 4.6 x 10^{-4}; aluminum substrate roughened, then primed with zinc chromate primer (MIL-P-8585); spray application; air cured at ~298 K; carbon black - 25% Columbia Carbon's Raven II beads and 75% Binney and Smiths Superb beads; computed from spectral reflectance data. [Authors' designation: silicone alkyd/black enamel coating No. 2]
4	77	1962	~298	~0°		Above specimen and conditions except heat treated to 533 K in a 2-hr cycle in vacuum (10^{-5} mm Hg); vacuum maintained by diffusion pump.
5	77	1962	~298	~0°		Similar to curve 3 specimen and conditions except ratio of carbon black to TiO_2 is 9.2 x 10^{-4}. [Authors' designation: silicone-alkyd/black enamel coating No. 3]
6	77	1962	~298	~0°		Above specimen and conditions except heat treated to 533 K in a 2-hr cycle in vacuum (10^{-5} mm Hg); vacuum maintained by diffusion pump.
7	77	1962	~298	~0°		Similar to curve 3 specimen and conditions except ratio of carbon black to TiO_2 is 1.8 x 10^{-3}. [Authors' designation: silicone-alkyd/black enamel coating No. 4]
8	77	1962	~298	~0°		Above specimen and conditions except heat treated to 533 K in a 2-hr cycle in vacuum (10^{-5} mm Hg); vacuum maintained by diffusion pump.
9	77	1962	~298	~0°		Similar to curve 3 specimen and conditions except ratio of carbon black to TiO_2 is 3.7 x 10^{-3}. [Authors' designation: silicone-alkyd/black enamel coating No. 5]
10	77	1962	~298	~0°		Above specimen and conditions except heat treated to 533 K in a 2-hr cycle in vacuum (10^{-5} mm Hg); vacuum maintained by diffusion pump.
11	77	1962	~298	~0°		Similar to curve 3 specimen and conditions except ratio of carbon black to TiO_2 is 5.5 x 10^{-3}. [Authors' designation: silicone-alkyd/black enamel coating No. 6]
12	77	1962	~298	~0°		Above specimen and conditions except heat treated to 533 K in a 2-hr cycle in vacuum (10^{-5} mm Hg); vacuum maintained by diffusion pump.
13	77	1962	~298	~0°		Similar to curve 3 specimen and conditions except ratio of carbon black to TiO_2 is 7.4 x 10^{-4}. [Authors' designation: silicone-alkyd/black enamel coating No. 7]
14	77	1962	~298	~0°		Above specimen and conditions except heat treated to 533 K in a 2-hr cycle in vacuum (10^{-5} mm Hg); vacuum maintained by diffusion pump.

SPECIFICATION TABLE NO. 53 NORMAL SOLAR ABSORPTANCE OF CLAY + TITANIUM DIOXIDE PIGMENTED COATINGS (continued)

Curve No.	Ref. No.	Year	Temperature Range, K	Geometry θ	Reported Error, %	Composition (weight percent), Specifications and Remarks
15	77	1962	~298	~0°		Similar to curve 3 specimen and conditions except ratio of carbon black to TiO_2 is 9.2×10^{-3}. [Authors' designation: silicone-alkyd/black enamel coating No. 8]
16	77	1962	~298	~0°		Above specimen and conditions except heat treated to 533 K in a 2-hr cycle in vacuum (10^{-5} mm Hg); vacuum maintained by diffusion pump.
17	77	1962	~298	~0°		Similar to curve 3 specimen and conditions except ratio of carbon black to TiO_2 is 1.8×10^{-2}. [Authors' designation: silicone-alkyd/black enamel coating No. 9]
18	77	1932	~298	~0°		Above specimen and conditions except heat treated to 533 K in a 2-hr cycle in vacuum (10^{-5} mm Hg); vacuum maintained by diffusion pump.
19	77	1962	~298	~0°		26.1 rutile TiO_2, 30.0 calcined clay (southern clay Al Sil Ate W), and 21.6 organic solvents in 22.3 Plaskon ST-873 silicone-alkyd binder (~0.0635 mm thick) on zinc chromate (0.0127 mm thick) and 2024 aluminum alloy substrates; aluminum substrate roughened then primed with zinc chromate primer (MIL-P-8585); Fischer-Payne Dipcoater application; air cured at 298 K; computed from spectral reflectance data.
20	77	1962	~298	~0°		Similar to above specimen and conditions except heat treated to 422 K in a 24-hr cycle in vacuum (10^{-5} mm Hg); vacuum maintained by diffusion pump.
21	77	1962	~298	~0°		Similar to above specimen and conditions except heat treated to 478 K in a 24-hr cycle in vacuum (10^{-5} mm Hg); vacuum maintained by diffusion pump.
22	77	1962	~298	~0°		Similar to above specimen and conditions except heat treated to 533 K in a 24-hr cycle in vacuum (10^{-5} mm Hg); vacuum maintained by diffusion pump.
23	77	1962	~298	~0°		Similar to above specimen and conditions except heat treated to 589 K in a 24-hr cycle in vacuum (10^{-5} mm Hg); vacuum maintained by diffusion pump.
24	77	1962	~298	~0°		Similar to curve 19 specimen and conditions except 2,2'-4,4'-tetrahydroxybenzophenone (D-SO) absorbing agent added to the coating formulation at 1% by weight of resin content.
25	77	1962	~298	~0°		Similar to above specimen and conditions except heat treated to 422 K in a 24-hr cycle in vacuum (10^{-5} mm Hg); vacuum maintained by diffusion pump.
26	77	1962	~298	~0°		Similar to above specimen and conditions except heat treated to 478 K in a 24-hr cycle in vacuum (10^{-5} mm Hg); vacuum maintained by diffusion pump.
27	77	1962	~298	~0°		Similar to above specimen and conditions except heat treated to 533 K in a 24-hr cycle in vacuum (10^{-5} mm Hg); vacuum maintained by diffusion pump.
28	77	1962	~298	~0°		Similar to above specimen and conditions except heat treated to 589 K in a 24-hr cycle in vacuum (10^{-5} mm Hg); vacuum maintained by diffusion pump.
29	77	1962	~298	~0°		Similar to curve 19 specimen and conditions except dibenzoylresorcinol (DBR) absorbing agent added to the coating formulation at 1% by weight of resin content.
30	77	1962	~298	~0°		Similar to above specimen and conditions except heat treated to 422 K in a 24-hr cycle in vacuum (10^{-5} mm Hg); vacuum maintained by diffusion pump.
31	77	1962	~298	~0°		Similar to above specimen and conditions except heat treated to 478 K in a 24-hr cycle in vacuum (10^{-5} mm Hg); vacuum maintained by diffusion pump.
32	77	1962	~298	~0°		Similar to above specimen and conditions except heat treated to 533 K in a 24-hr cycle in vacuum (10^{-5} mm Hg); vacuum maintained by diffusion pump.

SPECIFICATION TABLE NO. 53 NORMAL SOLAR ABSORPTANCE OF CLAY + TITANIUM DIOXIDE PIGMENTED COATINGS (continued)

Curve No.	Ref. No.	Year	Temperature Range, K	Geometry θ	Reported Error, %	Composition (weight percent), Specifications and Remarks
33	77	1962	~298	~0°		Similar to above specimen and conditions except heat treated to 589 K in a 24-hr cycle in vacuum (10^{-5} mm Hg); vacuum maintained by diffusion pump.
34	77	1962	~298	~0°		Similar to curve 19 specimen and conditions except 2-hydroxybenzoylferrocene (HBF) absorbing agent added to the coating formulation at 1% by weight of resin content.
35	77	1962	~298	~0°		Similar to above specimen and conditions except heat treated to 422 K in a 24-hr cycle in vacuum (10^{-5} mm Hg); vacuum maintained by diffusion pump.
36	77	1962	~298	~0°		Similar to above specimen and conditions except heat treated to 478 K in a 24-hr cycle in vacuum (10^{-5} mm Hg); vacuum maintained by diffusion pump.
37	77	1962	~298	~0°		Similar to above specimen and conditions except heat treated to 533 K in a 24-hr cycle in vacuum (10^{-5} mm Hg); vacuum maintained by diffusion pump.
38	77	1962	~298	~0°		Similar to above specimen and conditions except heat treated to 589 K in a 24-hr cycle in vacuum (10^{-5} mm Hg); vacuum maintained by diffusion pump.
39	77	1962	~298	~0°		Similar to curve 19 specimen and conditions except 2-4-dihydroxybenzoylferrocene (DHBF) absorbing agent added to the coating formulation at 1% by weight of resin content.
40	77	1962	~298	~0°		Similar to above specimen and conditions except heat treated to 422 K in a 24-hr cycle in vacuum (10^{-5} mm Hg); vacuum maintained by diffusion pump.
41	77	1962	~298	~0°		Similar to above specimen and conditions except heat treated to 478 K in a 24-hr cycle in vacuum (10^{-5} mm Hg); vacuum maintained by diffusion pump.
42	77	1962	~298	~0°		Similar to above specimen and conditions except heat treated to 533 K in a 24-hr cycle in vacuum (10^{-5} mm Hg); vacuum maintained by diffusion pump.
43	77	1962	~298	~0°		Similar to above specimen and conditions except heat treated to 589 K in a 24-hr cycle in vacuum (10^{-5} mm Hg); vacuum maintained by diffusion pump.
44	77	1962	~298	~0°		Similar to curve 19 specimen and conditions.
45	77	1962	~298	~0°		Above specimen and conditions except exposed to ultraviolet irradiation from G. E. B-H6 lamp located 15.24 cm from sample; solar factor ~5 suns; exposure time (hrs) is variable.
46	77	1962	~298	~0°		Similar to curve 19 specimen and conditions except 2, 2'-4, 4'-tetrahydroxybenzophenone (D-SO) absorbing agent added to the coating formulation at 1% by weight of resin content.
47	77	1962	~298	~0°		Above specimen and conditions except exposed to ultraviolet irradiation from G. E. B-H6 lamp located 15.24 cm from sample; solar factor ~5 suns; exposure time (hrs) is variable.
48	77	1962	~298	~0°		Similar to curve 19 specimen and conditions except dibenzoylresorcinol (DBR) absorbing agent added to the coating formulation at 1% by weight of resin content.
49	77	1962	~298	~0°		Above specimen and conditions except exposed to ultraviolet irradiation from G. E. B-H6 lamp located 15.24 cm from sample; solar factor ~5 suns; exposure time (hrs) is variable.
50	77	1962	~298	~0°		Similar to curve 19 specimen and conditions except 2-hydroxybenzoylferrocene (HBF) absorbing agent added to the coating formulation at 1% by weight of resin content.

SPECIFICATION TABLE NO. 53 NORMAL SOLAR ABSORPTANCE OF CLAY + TITANIUM DIOXIDE PIGMENTED COATINGS (continued)

Curve No.	Ref. No.	Year	Temperature Range, K	Geometry θ	Reported Error, %	Composition (weight percent), Specifications and Remarks
51	77	1962	~298	~0°		Above specimen and conditions except exposed to ultraviolet irradiation from G. E. B-H6 lamp located 15.24 cm from sample; solar factor ~5 suns; exposure time (hrs) is variable.
52	77	1962	~298	~0°		Similar to curve 19 specimen and conditions except 2-4-dihydroxybenzoylferrocene absorbing agent added to the coating formulation at 1% by weight of resin content.
53	77	1962	~298	~0°		Above specimen and conditions except exposed to ultraviolet irradiation from G. E. B-H6 lamp located 15.24 cm from sample; solar factor ~5 suns; exposure time (hrs) is variable.

DATA TABLE NO. 53 NORMAL SOLAR ABSORPTANCE OF CLAY + TITANIUM DIOXIDE PIGMENTED COATINGS

[Temperature, T, K; Absorptance, α]

T	α
CURVE 1*	
298	0.191
CURVE 2*	
298	0.298
CURVE 3*	
298	0.289
CURVE 4*	
298	0.376
CURVE 5*	
298	0.346
CURVE 6*	
298	0.401
CURVE 7*	
298	0.421
CURVE 8*	
298	0.456
CURVE 9*	
298	0.498
CURVE 10*	
298	0.526
CURVE 11*	
298	0.570
CURVE 12*	
298	0.555

T	α
CURVE 13*	
298	0.617
CURVE 14*	
298	0.594
CURVE 15*	
298	0.651
CURVE 16*	
298	0.612
CURVE 17*	
298	0.772
CURVE 18*	
298	0.731
CURVE 19*	
298	0.193
CURVE 20*	
298	0.238
CURVE 21*	
298	0.266
CURVE 22*	
298	0.313
CURVE 23*	
298	0.331
CURVE 24*	
298	0.219

T	α
CURVE 25*	
298	0.273
CURVE 26*	
298	0.303
CURVE 27*	
298	0.339
CURVE 28*	
298	0.386
CURVE 29*	
298	0.219
CURVE 30*	
298	0.257
CURVE 31*	
298	0.286
CURVE 32*	
298	0.325
CURVE 33*	
298	0.368
CURVE 34*	
298	0.246
CURVE 35*	
298	0.301
CURVE 36*	
298	0.327

T	α
CURVE 37*	
298	0.361
CURVE 38*	
298	0.408
CURVE 39*	
298	0.244
CURVE 40*	
298	0.297
CURVE 41*	
298	0.323
CURVE 42*	
298	0.360
CURVE 43*	
298	0.396
CURVE 44*	
298	0.193
Exp. Time	α
CURVE 45*	
6	0.205
24	0.227
54	0.243
104	0.257
T	α
CURVE 46*	
298	0.219

Exp. Time	α
CURVE 47*	
6	0.226
24	0.233
54	0.249
104	0.267
T	α
CURVE 48*	
298	0.205
Exp. Time	α
CURVE 49*	
6	0.215
24	0.220
54	0.234
104	0.247
T	α
CURVE 50*	
298	0.246
Exp. Time	α
CURVE 51*	
6	0.250
24	0.255
54	0.265
104	0.272
T	α
CURVE 52*	
298	0.244

Exp. Time	α
CURVE 53*	
6	0.253
24	0.257
54	0.267
104	0.278

* No plot given

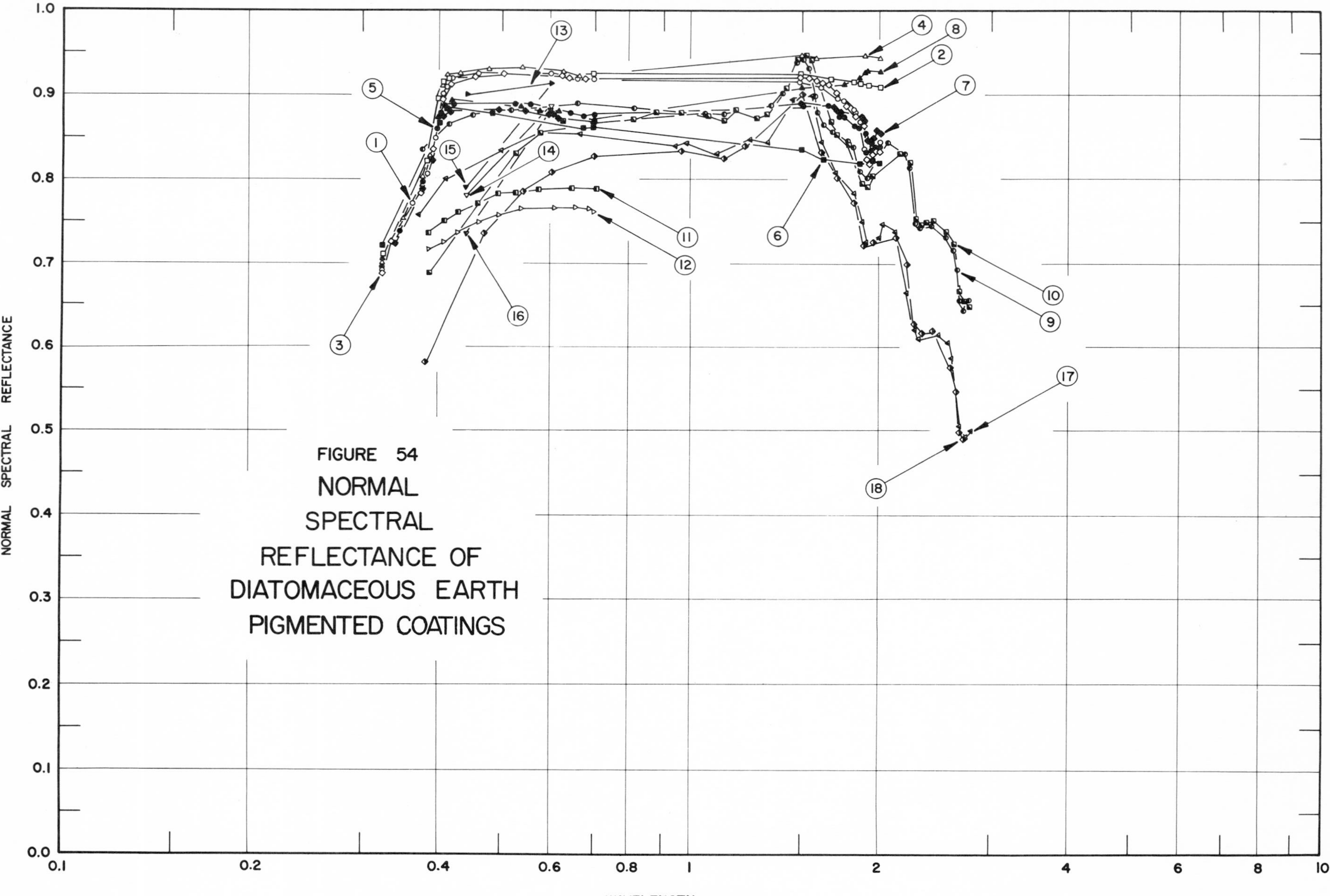

FIGURE 54 NORMAL SPECTRAL REFLECTANCE OF DIATOMACEOUS EARTH PIGMENTED COATINGS

SPECIFICATION TABLE NO. 54 NORMAL SPECTRAL REFLECTANCE OF DIATOMACEOUS EARTH PIGMENTED COATINGS

Curve No.	Ref. No.	Year	Temperature, K	Wavelength Range, μm	Geometry θ	θ'	ω'	Reported Error, %	Composition (weight percent), Specifications, and Remarks
1	35	1964	298	0.325-2.0	~0°		2π		Diatomaceous earth in potassium silicate binder (0.203 mm thick); air dried, no heat treatment; data taken from smooth curve.
2	35	1964	298	0.325-2.0	~0°		2π		Above specimen and conditions except heat treated at 773 K for 2 hrs.
3	35	1964	298	0.325-2.0	~0°		2π		Similar to curve 1 specimen and conditions.
4	35	1964	298	0.325-2.0	~0°		2π		Above specimen and conditions except heat treated at 423 K for 2 hrs.
5	35	1964	298	0.325-2.0	~0°		2π		Diatomaceous earth in potassium silicate binder (0.102 mm thick); air dried, no heat treatment; data taken from smooth curve.
6	35	1964	298	0.325-2.0	~0°		2π		Above specimen and conditions except heat treated at 773 K for 2 hrs.
7	35	1964	298	0.325-2.0	~0°		2π		Similar to curve 5 specimen and conditions.
8	35	1964	298	0.325-2.0	~0°		2π		Above specimen and conditions except heat treated at 423 K for 60 hrs.
9	20	1963	~298	0.375-2.760	~0°		2π		Dicalite WB-5 (from Great Lakes Carbon); PS7 potassium silicate binder; PBR 2.13; PVC 30.4%; paint ball-milled; aluminum substrate grit blasted with 40 mesh silicon carbide; air brush application; cured at 413 K for 18 hrs; data extracted from smooth curve. [Authors' designation: Sample C23]
10	20	1963	~298	0.385-2.765	~0°		2π		Above specimen and conditions except exposed in vacuum (10^{-5} mm Hg) to 300 ESH at solar factor of 3 suns.
11	3	1960	~298	0.383-0.701	~0°		2π		Johns-Mansville Micro-Cell "C" diatomaceous earth in Dow Corning 806A silicone binder (0.145 mm thick); pigment concentration 60% of solids content; voids concentration, 40%; sprayed onto mild steel substrate; data extracted from smooth curve; measured relative to MgO.
12	3	1960	~298	0.385-0.701	~0°		2π		Similar to curve 11 specimen and conditions except coating thickness 0.140 mm; voids concentration 47%.
13	27	1964	~298	0.440-0.600	~0°		2π		Dicalite WB-5 in PS7 potassium silicate binder; PBR 2.13; solids content 30.4%; aluminum substrate abraded with No. 60 Aloxite cloth; sample cured at 413 K for 18 hrs. [Authors' designation: Sample C23]
14	27	1964	~298	0.440-0.600	~0°		2π		Above specimen and conditions except irradiated in vacuum (10^{-6} mm Hg) with 300 ESH at solar factor of 3 suns.
15	27	1964	~298	0.440-0.600	~0°		2π		Dicalite WB-5 in PS7 potassium silicate binder; PBR 4.30; solids content 26.5%; aluminum substrate abraded with No. 60 Aloxite cloth; sample cured at 413 K for 18 hrs. [Authors' designation: Sample C25]
16	27	1964	~298	0.440-0.600	~0°		2π		Above specimen and conditions except irradiated in vacuum (10^{-6} mm Hg) with 300 ESH at solar factor of 3 suns.
17	29	1962	298	0.37-2.76	0°		2π		Dicalite WB-5 (94.57 SiO_2, 2.35 Na_2O, 1.70 Al_2O_3, 0.90 Fe_2O_3, 0.25 MgO, 0.10 CaO, 0.07 TiO_2, 0.05 V_2O_5) in potassium silicate vehicle; pigment binder ratio 1.07:1; 6061-T6 Al plate (grit blasted with 40 mesh SiC) substrate; sprayed by air-brush; air dried overnight and then 395-415 K heat-cured for 24 hrs; data extracted from smooth curve "2-10-3". [Author's designation: Sample 2-10-3]
18	29	1962	298	0.38-2.73	0°		2π		Above specimen and conditions except exposed to simulated solar radiation at 4 suns for 67 hrs.

DATA TABLE NO. 54 NORMAL SPECTRAL REFLECTANCE OF DIATOMACEOUS EARTH PIGMENTED COATINGS

[Wavelength, λ, μm; Reflectance, ρ; Temperature, T, K]

λ	ρ
CURVE 1 T = 298	
0.325	0.693
0.362	0.772
0.381	0.805
0.391	0.849
0.404	0.901
0.411	0.912
0.422	0.920
0.460	0.925
0.597	0.926
0.661	0.920
0.700	0.919
1.49	0.917
1.61	0.911
1.71	0.897
1.74	0.894
1.81	0.882
1.83	0.881
1.87	0.870
1.88	0.859
1.89	0.837
1.91	0.830
1.93	0.839
1.97	0.843
2.0	0.844
CURVE 2 T = 298	
0.325	0.711
0.382	0.821
0.398	0.897
0.407	0.915
0.414	0.919
0.460	0.925*
0.700	0.926
1.49	0.926
1.66	0.919
1.75	0.917
1.80	0.915
1.92	0.914
2.0	0.911
CURVE 3 T = 298	
0.325	0.688
0.339	0.724
0.373	0.784
0.390	0.838
0.402	0.899
0.407	0.909*
0.415	0.913
0.452	0.921
0.501	0.927
0.627	0.925
0.639	0.922
0.677	0.920
0.700	0.924*
1.49	0.921
1.53	0.921
1.59	0.918
1.61	0.913
1.65	0.913
1.68	0.905
1.70	0.903
1.77	0.887
1.84	0.876
1.88	0.863
1.90	0.825
1.91	0.820
1.94	0.831
1.96	0.835
2.0	0.834
CURVE 4 T = 298	
0.325	0.700
0.350	0.754
0.374	0.804
0.383	0.828
0.402	0.911
0.410	0.921
0.429	0.927
0.477	0.931
0.539	0.933
0.628	0.928
0.664	0.925
0.700	0.927*
CURVE 4 (cont.)	
1.50	0.947
1.58	0.945
1.88	0.947
2.0	0.944
CURVE 5 T = 298	
0.325	0.695*
0.343	0.737
0.378	0.797
0.394	0.860
0.401	0.879
0.408	0.887
0.420	0.889
0.526	0.890
0.555	0.889
0.642	0.879
0.675	0.875
0.700	0.877
1.50	0.889
1.65	0.888
1.73	0.875
1.75	0.875
1.81	0.866
1.86	0.864
1.88	0.857
1.90	0.831
1.91	0.824*
1.93	0.836
1.96	0.838
1.98	0.841*
2.0	0.837
CURVE 6 T = 298	
0.325	0.721
0.381	0.820*
0.396	0.868
0.401	0.878*
0.411	0.883
0.482	0.877
0.676	0.861
0.700	0.863
CURVE 6 (cont.)	
1.50	0.834
1.62	0.824
1.85	0.820
1.90	0.824*
1.94	0.823
2.0	0.820
CURVE 7 T = 298	
0.325	0.691*
0.340	0.722
0.375	0.785
0.387	0.823
0.399	0.871*
0.404	0.876
0.417	0.880
0.495	0.883
0.547	0.882
0.592	0.877
0.603	0.878
0.617	0.873
0.700	0.868
1.50	0.890
1.67	0.888
1.70	0.884
1.73	0.884
1.83	0.876*
1.87	0.876
1.88	0.870
1.90	0.848
1.91	0.844
1.94	0.852
1.98	0.859
2.0	0.857
CURVE 8 T = 298	
0.325	0.707
0.382	0.823*
0.396	0.878
0.406	0.889*
0.419	0.894
0.539	0.887
CURVE 8 (cont.)	
0.573	0.882
0.617	0.881
0.680	0.874*
0.699	0.873
1.50	0.911
1.75	0.914
1.84	0.920
1.87	0.929
1.90	0.930
2.0	0.929
CURVE 9 T ~ 298	
0.375	0.835
0.413	0.865
0.450	0.876
0.518	0.883
0.660	0.890
0.812	0.885
0.864	0.879
1.050	0.879
1.133	0.875
1.196	0.885
1.342	0.888
1.400	0.903
1.473	0.940
1.503	0.944
1.543	0.934
1.572	0.899
1.598	0.880
1.630	0.866
1.687	0.857
1.778	0.849
1.814	0.839
1.851	0.812
1.901	0.803
2.001	0.841*
2.043	0.845
2.198	0.831
2.227	0.816
2.296	0.749
2.321	0.744
2.414	0.747
2.557	0.732
CURVE 9 (cont.)	
2.608	0.716
2.640	0.696
2.672	0.656
2.709	0.642
2.734	0.642*
2.760	0.655
CURVE 10 T ~ 298	
0.385	0.688
0.459	0.772
0.528	0.833
0.577	0.856
0.625	0.870
0.705	0.877*
0.808	0.872
0.882	0.880
0.960	0.881
1.057	0.878
1.133	0.872
1.181	0.883
1.277	0.874
1.324	0.878
1.409	0.908
1.478	0.944
1.513	0.949
1.546	0.943
1.575	0.914*
1.663	0.870
1.702	0.857
1.786	0.847
1.816	0.838*
1.879	0.796
1.909	0.792
1.948	0.805
1.991	0.838*
2.028	0.844*
2.187	0.832
2.220	0.821
2.288	0.755
2.324	0.745
2.384	0.750
2.431	0.753
2.543	0.737
CURVE 10 (cont.)	
2.601	0.723
2.684	0.668
2.714	0.655
2.762	0.649
CURVE 11 T ~ 298	
0.383	0.735
0.408	0.751
0.429	0.762
0.459	0.772*
0.494	0.781
0.529	0.786
0.571	0.789
0.646	0.790
0.701	0.789
CURVE 12 T ~ 298	
0.385	0.712
0.407	0.728
0.429	0.737
0.461	0.750
0.495	0.758
0.539	0.765
0.607	0.768
0.657	0.769
0.689	0.765
0.701	0.763
CURVE 13 T ~ 298	
0.440	0.900
0.600	0.915
CURVE 14 T ~ 298	
0.440	0.770
0.600	0.885
CURVE 15 T ~ 298	
0.440	0.890
0.600	0.915*
CURVE 16 T ~ 298	
0.440	0.735
0.600	0.880
CURVE 17 T = 298	
0.37	0.759
0.41	0.800
0.50	0.838
0.58	0.854*
0.67	0.855
0.94	0.840
0.98	0.846
1.10	0.831
1.24	0.849
1.33	0.848
1.45	0.895
1.50	0.914*
1.55	0.900
1.61	0.848
1.70	0.810
1.82	0.785
1.86	0.750
1.90	0.727
1.99	0.731
2.09	0.748
2.13	0.738
2.23	0.667
2.28	0.621
2.32	0.610
2.47	0.614
2.54	0.605
2.60	0.588
2.68	0.503
2.72	0.494
2.76	0.500

*Not shown on plot

DATA TABLE NO. 54 NORMAL SPECTRAL REFLECTANCE OF DIATOMACEOUS EARTH PIGMENTED COATINGS

λ	ρ
CURVE 18	
T = 298	
0.38	0.581
0.47	0.734
0.54	0.785
0.60	0.809
0.70	0.828
0.96	0.835
1.13	0.826
1.22	0.841
1.33	0.843*
1.50	0.903
1.62	0.831
1.73	0.802
1.82	0.774
1.89	0.721
1.95	0.727
2.03	0.748*
2.11	0.734
2.20	0.700
2.28	0.627
2.33	0.619
2.43	0.620
2.49	0.594
2.59	0.576
2.64	0.547
2.68	0.498
2.71	0.490
2.73	0.491*

* Not shown on plot

SPECIFICATION TABLE NO. 55 NORMAL SOLAR ABSORPTANCE OF DIATOMACEOUS EARTH PIGMENTED COATINGS

Curve No.	Ref. No.	Year	Temperature Range, K	Geometry θ	Reported Error, %	Composition (weight percent), Specifications and Remarks
1*	35	1964	298	$\sim 0^\circ$		Diatomaceous earth in PS7 potassium silicate binder; pigment/binder ratio 2.15; air dried, no heat treatment; measured in vacuum (10^{-7} mm Hg). [Authors' designation: Sample H-13-14]
2*	35	1964	298	$\sim 0^\circ$		Above specimen and conditions except exposed to 2005 ESH at solar factor of 10.7 suns.
3*	35	1964	298	$\sim 0^\circ$		Similar to curve 1 specimen and conditions except heat treated at 423 K for 66 hrs. [Authors' designation: Sample H-13-16]
4*	35	1964	298	$\sim 0^\circ$		Above specimen and conditions except exposed to 2005 ESH at solar factor of 10.7 suns.
5*	35	1964	298	$\sim 0^\circ$		Similar to curve 1 specimen and conditions except heat treated at 773 K for 2 hrs. [Authors' designation: Sample H-13-15]
6*	35	1964	298	$\sim 0^\circ$		Above specimen and conditions except exposed to 2005 ESH at solar factor of 10.7 suns.

DATA TABLE NO. 55 NORMAL SOLAR ABSORPTANCE OF DIATOMACEOUS EARTH PIGMENTED COATINGS

[Temperature, T, K; Absorptance, α]

T	α
CURVE 1*	
298	0.172
CURVE 2*	
298	0.229
CURVE 3*	
298	0.170
CURVE 4*	
298	0.228
CURVE 5*	
298	0.163
CURVE 6*	
298	0.200

* No plot given

SPECIFICATION TABLE NO. 56 HEMISPHERICAL TOTAL EMITTANCE OF IRON OXIDE PIGMENTED COATINGS

Curve No.	Ref. No.	Year	Temperature Range, K	Reported Error, %	Composition (weight percent), Specifications and Remarks
1*	67	1967	1750		Pyromark paint, Fe_3O_4 + other oxides in silicone binder on TZM molybdenum substrate.

DATA TABLE NO. 56 HEMISPHERICAL TOTAL EMITTANCE OF IRON OXIDE PIGMENTED COATINGS

[Temperature, T, K; Emittance, ϵ]

T	ϵ
CURVE 1*	
1750	0.92

* No plot given

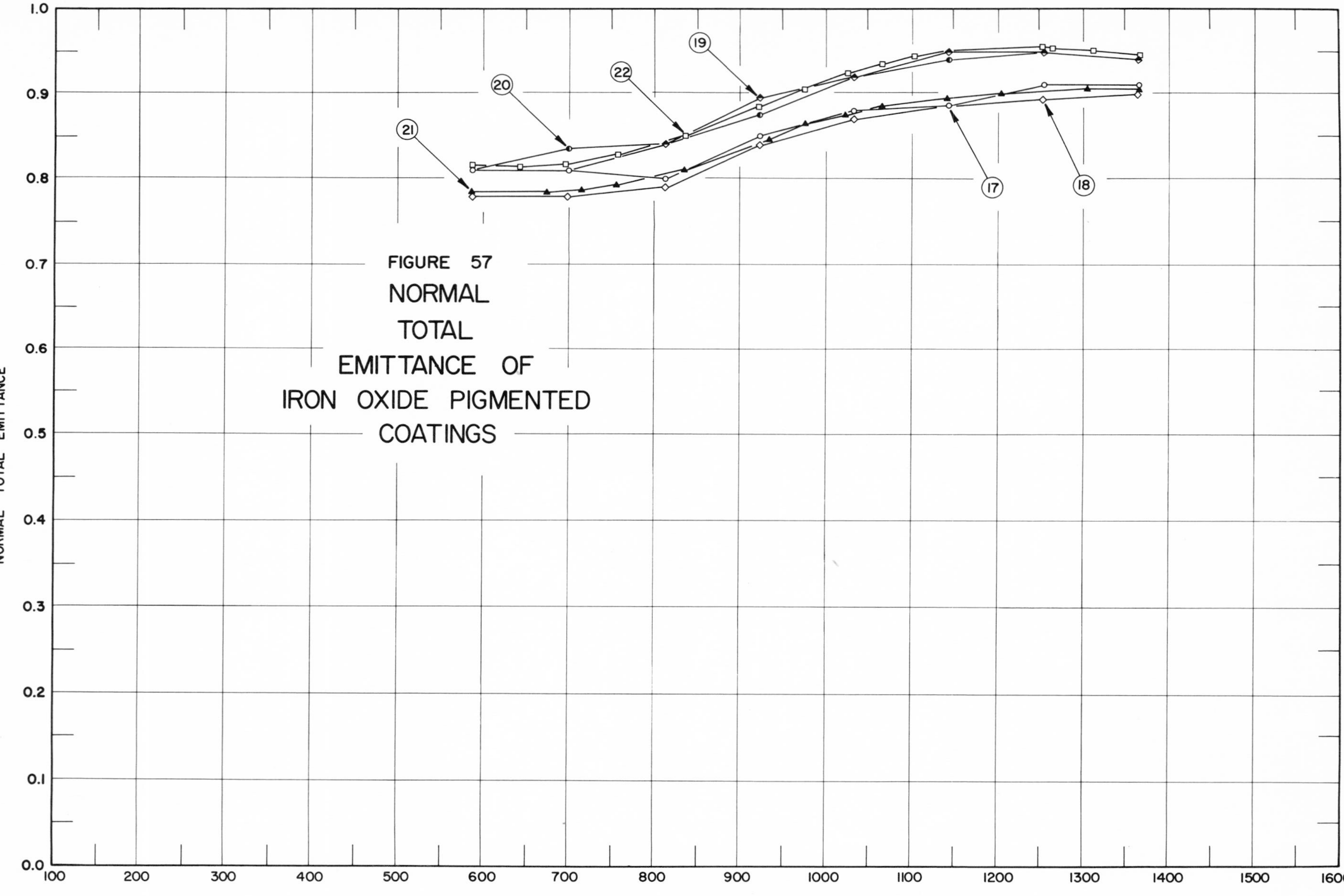

FIGURE 57 NORMAL TOTAL EMITTANCE OF IRON OXIDE PIGMENTED COATINGS

SPECIFICATION TABLE NO. 57 NORMAL TOTAL EMITTANCE OF IRON OXIDE PIGMENTED COATINGS

Curve No.	Ref. No.	Year	Temperature Range, K	Geometry θ'	Reported Error, %	Composition (weight percent), Specifications and Remarks
1*	73	1962	1367	~0°		Pyromark paint, Fe_3O_4 + other oxides in silicone binder, (0.0254 mm thick) on Inconel substrate; substrate polished; air dried at ~298 K. [Authors' designation: Specimen 1]
2*	73	1962	1367	~0°		Above specimen and conditions except measured after heating in air at 1367 K for 5 min.
3*	73	1962	1367	~0°		Above specimen and conditions except measured after heating in air at 1367 K for a total of 10 min.
4*	73	1962	1367	~0°		Above specimen and conditions except measured after heating in air at 1367 K for a total of 15 min.
5*	73	1962	1367	~0°		Similar to curve 1 specimen and conditions. [Authors' designation: Specimen 2]
6*	73	1962	1367	~0°		Above specimen and conditions except measured after heating in air at 1367 K for 5 min.
7*	73	1962	1367	~0°		Above specimen and conditions except measured after heating in air at 1367 K for a total of 10 min.
8*	73	1962	1367	~0°		Above specimen and conditions except measured after heating in air at 1367 K for a total of 15 min.
9*	73	1962	1367	~0°		Pyromark paint, Fe_3O_4 + other oxides in silicone binder, (0.0254 mm thick) on as-rolled stainless steel 321 substrate; air dried at ~298 K. [Authors' designation: Specimen 1]
10*	73	1962	1367	~0°		Above specimen and conditions except heated at 1367 K for 5 min in air.
11*	73	1962	1367	~0°		Above specimen and conditions except heated at 1367 K for 10 min in air.
12*	73	1962	1367	~0°		Above specimen and conditions except heated at 1367 K for 15 min in air.
13*	73	1962	1367	~0°		Pyromark paint, Fe_3O_4 + other oxides in silicone binder, (0.0254 mm thick) on as-rolled stainless steel 321 substrate; air dried at ~298 K. [Authors' designation: Sample 1]
14*	73	1962	1367	~0°		Above specimen and conditions except heated at 1367 K for 5 min in air.
15*	73	1962	1367	~0°		Above specimen and conditions except heated at 1367 K for 10 min in air.
16*	73	1962	1367	~0°		Above specimen and conditions except heated at 1367 K for 15 min in air.
17	73	1962	589-1367	~0°		Pyromark paint, Fe_3O_4 + other oxides in silicone binder, (0.0254 mm thick) on as-rolled stainless steel 321 substrate; air dried at ~298 K. [Authors' designation: Specimen 1]
18	73	1962	589-1367	~0°		Similar to above specimen and conditions. [Authors' designation: Specimen 2]
19	73	1962	589-1367	~0°		Pyromark paint, Fe_3O_4 + other oxides in silicone binder, (0.0254 mm thick) on Inconel substrate; substrate polished; coating air dried at ~298 K. [Authors' designation: Sample 1]
20	73	1962	589-1367	~0°		Similar to above specimen and conditions. [Authors' designation: Specimen 2]
21	95	1962	589-1367	~0°		Pyromark paint, Fe_3O_4 + other oxides in silicone binder, (~0.0254 mm thick) on stainless steel 321 substrate; substrate solvent-cleaned, as-rolled; air dried at ~298 K for 24 hrs; data extracted from smooth curve.
22	95	1962	589-1367	~0°		Similar to above specimen and conditions except on Inconel substrate; substrate solvent-cleaned and polished.

* Not shown on plot.

DATA TABLE NO. 57 NORMAL TOTAL EMITTANCE OF IRON OXIDE PIGMENTED COATINGS

[Temperature, T, K; Emittance, ∈]

T	∈	T	∈	T	∈
CURVE 1*		CURVE 13*		CURVE 20	
1367	0.94	1367	0.89	589	0.810*
				700	0.835
CURVE 2*		CURVE 14*		811	0.840*
				922	0.875
1367	0.94	1367	0.89	1033	0.920*
				1144	0.940
CURVE 3*		CURVE 15*		1255	0.950*
				1367	0.940*
1367	0.94	1367	0.89		
				CURVE 21	
CURVE 4*		CURVE 16*			
				589	0.784
1367	0.94	1367	0.90	675	0.784
				715	0.786
CURVE 5*		CURVE 17		755	0.792
				832	0.810
1367	0.94	589	0.810	933	0.848
		700	0.810	978	0.864
CURVE 6*		811	0.800	1021	0.875
		922	0.850	1069	0.885
1367	0.94	1033	0.880	1141	0.894
		1144	0.885	1203	0.900
CURVE 7*		1255	0.910	1304	0.906
		1367	0.910	1367	0.906
1367	0.94				
		CURVE 18		CURVE 22	
CURVE 8*					
		589	0.780	589	0.813
1367	0.94	700	0.780	642	0.811
		811	0.790	699	0.817
CURVE 9*		922	0.840	759	0.829
		1033	0.870	836	0.850
1367	0.88	1144	0.885*	921	0.882
		1255	0.894	978	0.907
CURVE 10*		1367	0.900	1029	0.922
				1066	0.934
1367	0.89	CURVE 19		1101	0.942
				1143	0.948*
CURVE 11*		589	0.810*	1215	0.952
		700	0.810*	1265	0.952
1367	0.91	811	0.840	1311	0.950
		922	0.896	1367	0.945
CURVE 12*		1033	0.920		
		1144	0.950		
1367	0.91	1255	0.950		
		1367	0.940		

* Not shown on plot.

SPECIFICATION TABLE NO. 58 NORMAL SPECTRAL EMITTANCE OF IRON OXIDE PIGMENTED COATINGS

Curve No.	Ref. No.	Year	Temperature, K	Wavelength Range, μm	Geometry θ'	Reported Error, %	Composition (weight percent), Specifications, and Remarks
1*	96	1963	323.2	3.58-20.12	0°		Pyromark Standard Black paint, Fe_3O_4 + other oxides in silicone binder on aluminum substrate. [Authors' designation: Sample BEC-1]

DATA TABLE NO. 58 NORMAL SPECTRAL EMITTANCE OF IRON OXIDE PIGMENTED COATINGS

[Wavelength, λ, μm; Emittance, ϵ; Temperature, T, K]

λ	ϵ	λ	ϵ	λ	ϵ
CURVE 1*		CURVE 1 (cont.)*		CURVE 1 (cont.)*	
T = 323.2		8.82	0.966	19.07	0.949
3.58	0.733	9.21	0.934	19.59	0.941
3.75	0.671	9.54	0.918	20.12	0.922
3.82	0.654	10.08	0.918		
4.24	0.645	10.41	0.938		
4.52	0.653	11.39	0.946		
4.78	0.714	11.71	0.964		
5.01	0.746	12.05	0.970		
5.46	0.845	12.47	0.971		
5.75	0.922	13.03	0.978		
6.10	0.933	13.57	0.971		
6.34	0.933	14.16	0.969		
6.59	0.911	15.11	0.988		
6.87	0.945	16.00	0.982		
7.16	0.939	16.56	0.960		
7.48	0.949	17.20	0.941		
7.87	0.957	17.83	0.938		
8.25	0.970	18.41	0.948		
8.57	0.974	18.84	0.945		

* No plot given.

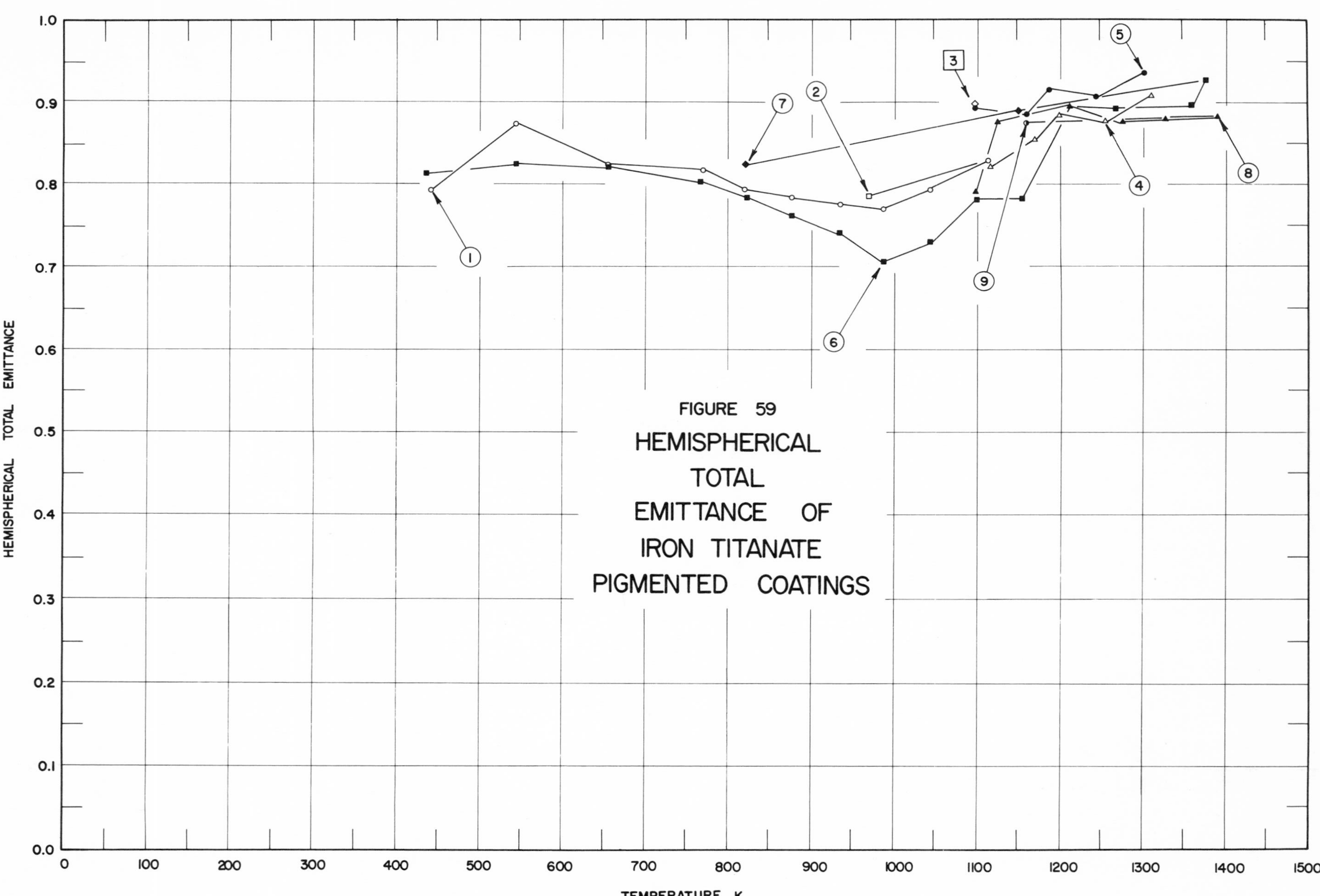

FIGURE 59
HEMISPHERICAL TOTAL EMITTANCE OF IRON TITANATE PIGMENTED COATINGS

SPECIFICATION TABLE NO. 59 HEMISPHERICAL TOTAL EMITTANCE OF IRON TITANATE PIGMENTED COATINGS

Curve No.	Ref. No.	Year	Temperature Range, K	Reported Error, %	Composition (weight percent), Specifications and Remarks
1	81	1963	441-1119		Iron titanate in aluminum phosphate binder (0.1016 mm thick) on Nb-1 Zr substrate; measured in vacuum ($<7.8 \times 10^{-6}$ mm Hg); temp measured with thermocouple; Run No. 1, heating cycle.
2	81	1963	1119-971		Above specimen and conditions; cooling cycle.
3	81	1963	1097		Curve 1 specimen and conditions except temp measured with optical pyrometer.
4	81	1963	1117-1310		Curve 1 specimen and conditions; Run No. 2.
5	81	1963	1097-1301		Above specimen and conditions except temp measured with optical pyrometer.
6	81	1963	434-1377		Iron titanate + alumina in aluminum phosphate binder (0.1016 mm thick) on Nb-1 Zr substrate; measured in vacuum ($<4.5 \times 10^{-6}$ mm Hg); temp measured with thermocouple; Run No. 1, heating cycle.
7	81	1963	1377-823		Above specimen and conditions; cooling cycle.
8	81	1963	1097-1391		Curve 6 specimen and conditions except temp measured with optical pyrometer.
9	81	1963	1391-1159		Above specimen and conditions; cooling cycle.

DATA TABLE NO. 59 HEMISPHERICAL TOTAL EMITTANCE OF IRON TITANATE PIGMENTED COATINGS

[Temperature, T, K; Emittance, ϵ]

T	ϵ	T	ϵ
CURVE 1		CURVE 6 (cont.)	
441	0.790	989	0.708
545	0.873	1046	0.730
656	0.822	1100	0.781
770	0.818	1155	0.782
826	0.794	1211	0.897*
879	0.785	1269	0.893
935	0.775	1360	0.895
988	0.770	1377	0.926
1044	0.791		
1119	0.829	CURVE 7	
CURVE 2		1377	0.926*
		1155	0.888
1119	0.829*	823	0.825
971	0.784		
		CURVE 8	
CURVE 3			
		1097	0.790
1097	0.898	1124	0.874
		1212	0.895*
CURVE 4		1276	0.875
		1328	0.879
1117	0.830	1391	0.888
1171	0.853		
1199	0.882	CURVE 9	
1255	0.876		
1310	0.908	1391	0.888*
		1159	0.876
CURVE 5			
1097	0.894		
1160	0.885		
1188	0.915		
1244	0.906		
1301	0.934		
CURVE 6			
434	0.813		
546	0.827		
656	0.821		
707	0.803		
822	0.783		
878	0.761		
934	0.740		

* Not shown on plot

FIGURE 60A

ANALYZED NORMAL SPECTRAL REFLECTANCE OF LANTHANUM OXIDE PIGMENTED COATINGS

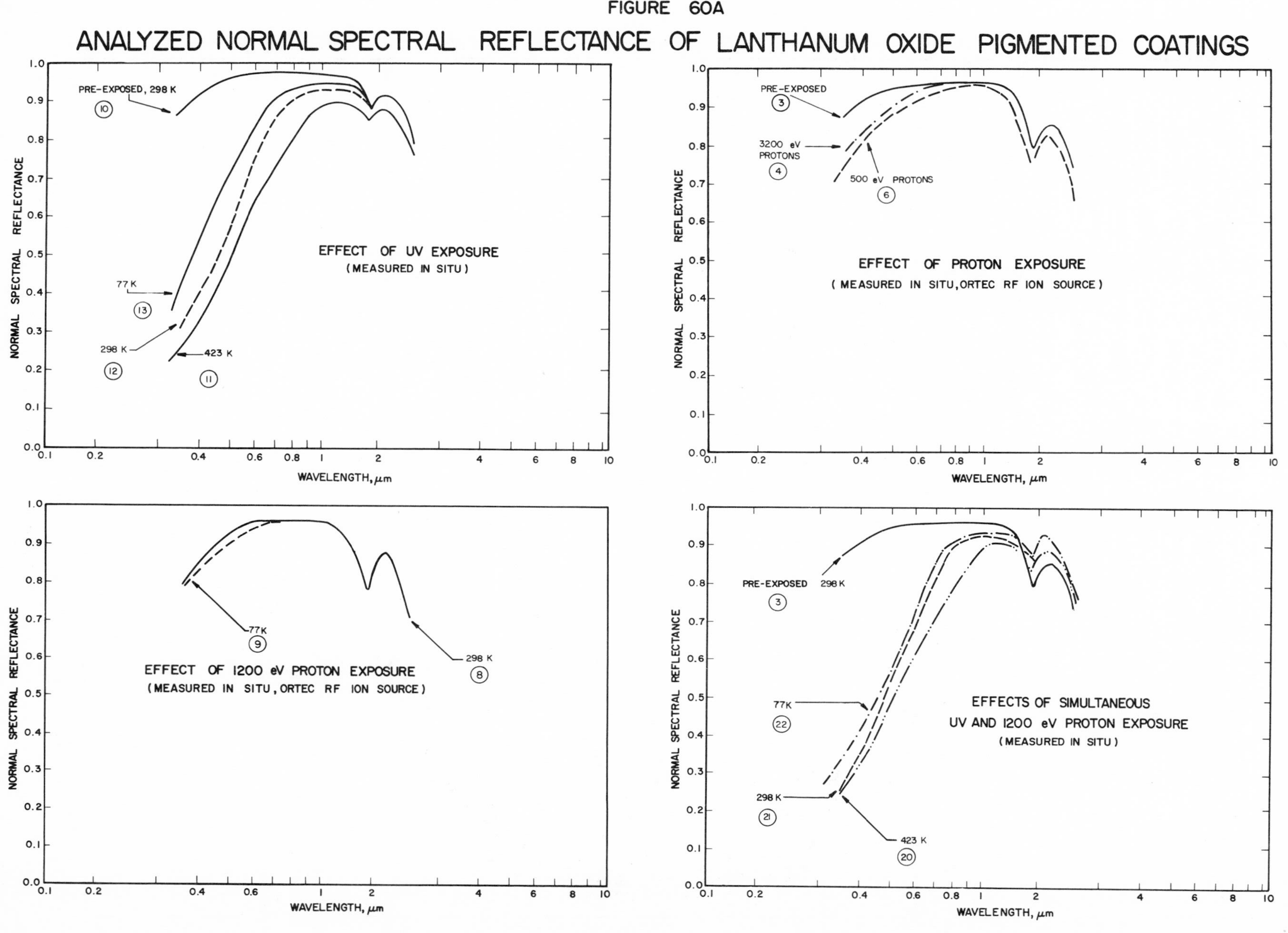

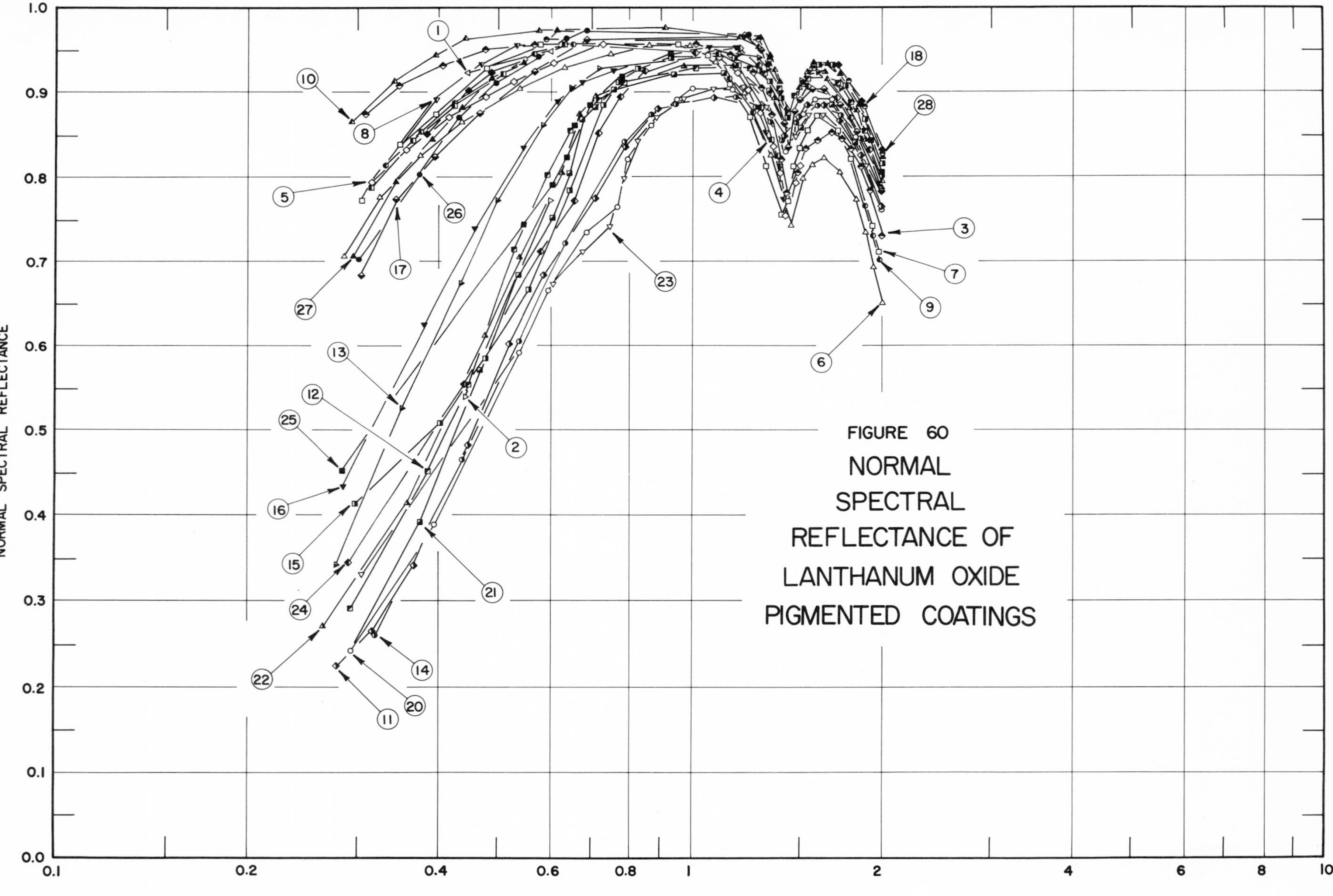

FIGURE 60 NORMAL SPECTRAL REFLECTANCE OF LANTHANUM OXIDE PIGMENTED COATINGS

SPECIFICATION TABLE NO. 60 NORMAL SPECTRAL REFLECTANCE OF LANTHANUM OXIDE PIGMENTED COATINGS

Curve No.	Ref. No.	Year	Temperature, K	Wavelength Range, μm	Geometry θ	θ'	ω'	Reported Error, %	Composition (weight percent), Specifications, and Remarks
1	27	1964	~298	0.44-0.60	~0°		2π		La_2O_3 in PS-7 potassium silicate binder; PBR 4.30; solids content 56.9%; aluminum substrate abraded with No. 60 Aloxite cloth; sample cured at 413 K for 18 hrs. [Authors' designation: Sample C4]
2	27	1964	~298	0.44-0.60	~0°		2π		Above specimen and conditions except irradiated in vacuum (10^{-6} mm Hg) with 180 ESH at solar factor of 3 suns.
3	39	1970	298	0.307-2.005	~0°		2π		La_2O_3 in K_2SiO_3 binder; 53.4 La_2O_3; ball milled for 2 hrs, then compacted; property measured in air; data extracted from smooth curve.
4	39	1970	298	0.313-2.005	~0°		2π		Similar to above specimen and conditions except exposed to 3200 eV protons from Ortec rf ion source in vacuum (5 x 10^{-7} mm Hg); property measured in situ.
5	39	1970	298	0.313-2.005	~0°		2π		Similar to above specimen and conditions except proton energy 1200 eV.
6	39	1970	298	0.285-2.004	~0°		2π		Similar to above specimen and conditions except proton energy 500 eV.
7	39	1970	423	0.304-1.987	~0°		2π		Similar to above specimen and conditions except proton energy 1200 eV.
8	39	1970	298	0.304-1.987	~0°		2π		Similar to above specimen and conditions.
9	39	1970	77	0.304-1.981	~0°		2π		Similar to above specimen and conditions.
10	39	1970	298	0.291-2.000	~0°		2π		Similar to curve 3 specimen and conditions.
11	39	1970	423	0.277-2.000	~0°		2π		Above specimen and conditions except exposed to UV source with 4 sun intensity in vacuum (5 x 10^{-7} mm Hg); property measured in situ.
12	39	1970	298	0.292-2.000	~0°		2π		Similar to above specimen and conditions.
13	39	1970	77	0.277-2.000	~0°		2π		Similar to above specimen and conditions.
14	39	1970	423	0.318-2.000	~0°		2π		Similar to above specimen and conditions except exposed to UV radiation for $\lambda < 0.4 \mu m$ with Corning filter 7-54 in vacuum (5 x 10^{-7} mm Hg).
15	39	1970	298	0.298-2.000	~0°		2π		Similar to above specimen and conditions.
16	39	1970	77	0.282-2.000	~0°		2π		Similar to above specimen and conditions.
17	39	1970	423	0.302-2.000	~0°		2π		Similar to above specimen and conditions except exposed to UV radiation for $\lambda > 0.4 \mu m$ with Corning filter 3-73 in vacuum (5 x 10^{-7} mm Hg).
18	39	1970	298	0.590-2.000	~0°		2π		Similar to above specimen and conditions.
19	39	1970	77	0.349-2.000	~0°		2π		Similar to above specimen and conditions.
20	39	1970	423	0.292-2.000	~0°		2π		Similar to curve 3 specimen and conditions except exposed to 1200 eV proton source and UV source in vacuum (5 x 10^{-7} mm Hg); property measured in situ.
21	39	1970	298	0.292-2.000	~0°		2π		Similar to above specimen and conditions.
22	39	1970	77	0.262-2.000	~0°		2π		Similar to above specimen and conditions.
23	39	1970	423	0.304-2.000	~0°		2π		Similar to above specimen and conditions except exposed to UV source for $\lambda < 0.4 \mu m$ with Corning filter 7-54.

SPECIFICATION TABLE NO. 60 NORMAL SPECTRAL REFLECTANCE OF LANTHANUM OXIDE PIGMENTED COATINGS (continued)

Curve No.	Ref. No.	Year	Temperature, K	Wavelength Range, μm	Geometry θ	θ'	ω'	Reported Error, %	Composition (weight percent), Specifications, and Remarks
24	39	1970	298	0.290-2.000	~0°		2π		Similar to above specimen and conditions.
25	39	1970	77	0.283-2.000	~0°		2π		Similar to above specimen and conditions.
26	39	1970	423	0.300-2.000	~0°		2π		Similar to above specimen and conditions except exposed to UV source for $\lambda > 0.4$ μm with Corning filter 3-73.
27	39	1970	298	0.294-2.000	~0°		2π		Similar to above specimen and conditions.
28	39	1970	77	0.294-2.000	~0°		2π		Similar to above specimen and conditions.

DATA TABLE NO. 60 NORMAL SPECTRAL REFLECTANCE OF LANTHANUM OXIDE PIGMENTED COATINGS

[Wavelength, λ, μm; Reflectance, ρ; Temperature, T, K]

λ	ρ
CURVE 1	
T ~ 298	
0.44	0.925
0.60	0.950
CURVE 2	
T ~ 298	
0.44	0.540
0.60	0.775
CURVE 3	
T = 298	
0.307	0.875
0.349	0.910
0.405	0.934
0.471	0.951
0.566	0.958
1.010	0.958
1.151	0.946
1.258	0.924
1.296	0.907
1.341	0.875*
1.386	0.821
1.412	0.781*
1.441	0.781*
1.474	0.808
1.523	0.834*
1.585	0.848
1.668	0.855
1.734	0.849*
1.793	0.834*
1.852	0.813
1.919	0.784
2.005	0.731
CURVE 4	
T = 298	
0.313	0.790
0.362	0.846
0.425	0.886
0.507	0.923
0.630	0.958
1.010	0.958*
CURVE 4 (cont.)	
1.102	0.946
1.188	0.931
1.297	0.883
1.349	0.847
1.378	0.822*
1.412	0.781*
1.441	0.781*
1.474	0.808*
1.523	0.834*
1.585	0.848*
1.668	0.855*
1.734	0.849*
1.793	0.834*
1.852	0.813*
1.919	0.784*
2.005	0.731*
CURVE 5	
T = 298	
0.313	0.794*
0.355	0.833
0.417	0.872
0.475	0.896
0.527	0.914
0.602	0.936
0.722	0.958*
1.010	0.958*
1.115	0.942*
1.180	0.924
1.230	0.905
1.288	0.879*
1.357	0.838
1.379	0.808
1.410	0.756
1.441	0.756*
1.466	0.795*
1.492	0.815
1.523	0.834*
1.585	0.848*
1.668	0.855*
1.734	0.849*
1.793	0.834*
1.852	0.813*
1.919	0.784*
2.005	0.731*
CURVE 6	
T = 298	
0.285	0.708
0.322	0.779
0.375	0.827
0.438	0.869
0.539	0.907
0.630	0.931
0.743	0.948
0.855	0.955
0.990	0.955
1.094	0.937
1.211	0.900
1.277	0.873*
1.349	0.828
1.381	0.801*
1.410	0.756*
1.445	0.743
1.464	0.769*
1.506	0.800
1.557	0.816*
1.627	0.822
1.721	0.807
1.812	0.775
1.877	0.737
1.946	0.695
2.004	0.651
CURVE 7	
T = 423	
0.304	0.773
0.349	0.840
0.395	0.878
0.478	0.930
0.579	0.957
0.950	0.957
1.062	0.945
1.159	0.912
1.246	0.872
1.328	0.815
1.394	0.756
1.429	0.774*
1.459	0.813
1.489	0.837*
1.532	0.859
CURVE 7 (cont.)	
1.574	0.873*
1.635	0.873*
1.707	0.858*
1.794	0.822
1.943	0.742
1.987	0.714
CURVE 8	
T = 298	
0.304	0.773*
0.349	0.840*
0.399	0.892
0.469	0.933
0.530	0.957
0.950	0.957*
1.062	0.945*
1.215	0.901*
1.319	0.853
1.371	0.813
1.405	0.774
1.429	0.774*
1.459	0.813*
1.489	0.837*
1.532	0.859*
1.574	0.873*
1.635	0.873*
1.707	0.858*
1.794	0.822*
1.943	0.742*
1.987	0.714*
CURVE 9	
T = 77	
0.304	0.773*
0.330	0.815
0.373	0.854
0.425	0.889*
0.507	0.925*
0.569	0.946
0.650	0.957
0.950	0.957*
1.062	0.945*
1.154	0.918*
CURVE 9 (cont.)	
1.257	0.876
1.361	0.810*
1.391	0.781*
1.429	0.774*
1.459	0.813*
1.489	0.837*
1.532	0.859*
1.574	0.873*
1.635	0.873*
1.707	0.858*
1.794	0.822*
1.891	0.768
1.944	0.732
1.981	0.701
CURVE 10	
T = 298	
0.291	0.866
0.340	0.914
0.397	0.945
0.440	0.962
0.573	0.974
0.901	0.977
1.198	0.967
1.277	0.953
1.357	0.928
1.397	0.901*
1.410	0.882
1.442	0.883*
1.453	0.898*
1.489	0.908
1.541	0.919*
1.649	0.919
1.707	0.912*
1.822	0.880
2.000	0.796
CURVE 11	
T = 423	
0.277	0.228
0.314	0.268
0.368	0.342
0.445	0.482
CURVE 11 (cont.)	
0.518	0.602
0.588	0.685
0.707	0.778
0.790	0.839
0.835	0.866*
0.885	0.881
1.089	0.895
1.281	0.887*
1.359	0.874*
1.403	0.850
1.463	0.850*
1.480	0.873
1.529	0.887*
1.622	0.887
1.725	0.867*
1.852	0.829
2.000	0.767
CURVE 12	
T = 298	
0.292	0.292
0.387	0.453
0.446	0.556
0.528	0.715
0.594	0.804
0.648	0.857
0.712	0.892
0.786	0.912
1.015	0.930
1.242	0.930
1.319	0.921*
1.374	0.906
1.411	0.878*
1.444	0.878*
1.458	0.899
1.489	0.908*
1.537	0.917
1.619	0.919*
1.705	0.911
1.787	0.890
1.881	0.856
2.000	0.796*
CURVE 13	
T = 77	
0.277	0.343
0.350	0.529
0.433	0.675
0.499	0.774
0.581	0.861
0.646	0.909
0.712	0.930
1.197	0.946
1.271	0.942
1.332	0.928
1.379	0.907*
1.411	0.878*
1.444	0.878*
1.458	0.899*
1.492	0.916
1.572	0.924
1.667	0.922*
1.732	0.906
1.787	0.890*
1.881	0.856*
2.000	0.796*
CURVE 14	
T = 423	
0.318	0.262
0.437	0.468
0.536	0.609
0.635	0.722
0.702	0.782*
0.784	0.841
0.862	0.874
0.941	0.889
1.074	0.896*
1.179	0.896
1.264	0.884
1.345	0.867
1.399	0.845
1.463	0.866
1.511	0.882*
1.572	0.888
1.646	0.888*
1.739	0.874
1.850	0.848
2.000	0.800
CURVE 15	
T = 298	
0.298	0.416
0.401	0.510
0.474	0.588
0.553	0.669
0.646	0.786
0.728	0.889
0.765	0.912
0.821	0.930
0.925	0.941
1.172	0.941
1.270	0.929
1.341	0.913
1.386	0.896
1.417	0.871
1.449	0.879*
1.458	0.896*
1.474	0.909*
1.502	0.919*
1.582	0.933
1.663	0.933
1.756	0.910
1.887	0.870
2.000	0.819
CURVE 16	
T = 77	
0.282	0.437
0.380	0.628
0.458	0.740
0.546	0.838
0.615	0.891
0.671	0.913
0.757	0.929
1.065	0.952
1.188	0.952
1.275	0.941*
1.349	0.922*
1.386	0.900
1.417	0.871*
1.449	0.879*
1.458	0.896*
1.474	0.909*
1.502	0.919*

* Not shown on plot

DATA TABLE NO. 60 NORMAL SPECTRAL REFLECTANCE OF LANTHANUM OXIDE PIGMENTED COATINGS (continued)

CURVE 16 (cont.) T = 77

λ	ρ
1.582	0.933*
1.633	0.933*
1.756	0.913*
1.887	0.870*
2.000	0.819*

CURVE 17 T = 423

λ	ρ
0.302	0.684
0.344	0.775
0.397	0.828
0.464	0.877
0.566	0.927
0.684	0.962
1.219	0.969
1.280	0.951
1.364	0.904*
1.402	0.865
1.421	0.838
1.454	0.880
1.486	0.893
1.556	0.905
1.627	0.905
1.719	0.887
1.831	0.853
2.000	0.782

CURVE 18 T = 298

λ	ρ
0.590	0.962
1.219	0.969*
1.273	0.962
1.327	0.940
1.380	0.909*
1.417	0.873*
1.454	0.880*
1.503	0.911
1.561	0.930
1.661	0.932*
1.762	0.915
1.861	0.885
2.000	0.830

CURVE 19* T = 77

λ	ρ
0.349	0.779
0.393	0.838
0.470	0.903
0.555	0.947
0.590	0.962
1.219	0.969
1.273	0.962
1.327	0.940
1.380	0.909
1.417	0.873
1.454	0.880
1.503	0.911
1.561	0.930
1.661	0.932
1.762	0.915
1.861	0.891
2.000	0.843

CURVE 20 T = 423

λ	ρ
0.292	0.243
0.394	0.390
0.535	0.594
0.595	0.668
0.687	0.737
0.766	0.766
0.797	0.825
0.863	0.862
0.948	0.893
1.000	0.905
1.162	0.905
1.297	0.885*
1.397	0.866*
1.419	0.831
1.476	0.843*
1.501	0.873
1.560	0.893*
1.661	0.893
1.759	0.874*
1.831	0.850
2.000	0.763

CURVE 21 T = 298

λ	ρ
0.292	0.243*
0.378	0.393
0.464	0.573
0.533	0.684
0.603	0.753
0.644	0.807
0.674	0.870*
0.704	0.886
0.756	0.903
0.939	0.923
1.137	0.923
1.336	0.901*
1.403	0.869*
1.419	0.831*
1.476	0.843*
1.491	0.860
1.530	0.879*
1.576	0.886*
1.661	0.886*
1.718	0.875*
1.827	0.842*
2.000	0.762*

CURVE 22 T = 77

λ	ρ
0.262	0.272
0.359	0.417
0.472	0.612
0.539	0.708
0.601	0.773*
0.624	0.806
0.664	0.874
0.710	0.898
0.771	0.915*
0.970	0.932
1.177	0.932
1.279	0.920*
1.358	0.896
1.412	0.864*
1.450	0.873*
1.459	0.894*
1.483	0.910*
1.529	0.926

CURVE 22 (cont.)

λ	ρ
1.611	0.926
1.679	0.913*
1.774	0.886
1.892	0.844
2.000	0.789

CURVE 23 T = 423

λ	ρ
0.304	0.331
0.535	0.594*
0.611	0.672
0.675	0.711
0.745	0.742
0.787	0.800
0.828	0.846
0.889	0.874
0.975	0.894
1.071	0.905
1.212	0.905
1.314	0.883
1.415	0.850*
1.463	0.850
1.548	0.877
1.628	0.877
1.735	0.858*
1.874	0.812
2.000	0.760*

CURVE 24 T = 298

λ	ρ
0.290	0.348
0.440	0.556
0.580	0.714
0.654	0.772
0.717	0.856
0.772	0.899
0.843	0.928
1.005	0.948
1.180	0.951
1.272	0.934
1.366	0.901*
1.408	0.871*
1.455	0.871*

CURVE 24 (cont.)

λ	ρ
1.474	0.892*
1.547	0.909
1.624	0.909*
1.705	0.895
1.799	0.870
1.912	0.831
2.000	0.796

CURVE 25 T = 77

λ	ρ
0.283	0.451
0.549	0.749
0.604	0.792
0.638	0.824
0.657	0.863
0.694	0.888
0.780	0.920
0.927	0.946
1.065	0.952*
1.181	0.951*
1.272	0.939*
1.345	0.919*
1.408	0.871*
1.455	0.871*
1.474	0.892*
1.547	0.909*
1.619	0.909*
1.731	0.895*
1.911	0.843
2.000	0.805

CURVE 26 T = 423

λ	ρ
0.300	0.701
0.371	0.805
0.430	0.871
0.492	0.911
0.572	0.941
0.637	0.965
0.685	0.974
1.239	0.970
1.289	0.951*
1.339	0.933

CURVE 26 (cont.)

λ	ρ
1.416	0.876*
1.444	0.876*
1.473	0.895*
1.513	0.905*
1.605	0.905*
1.693	0.889*
1.800	0.857*
2.000	0.769*

CURVE 27 T = 298

λ	ρ
0.294	0.707
0.341	0.798
0.390	0.848
0.435	0.884
0.486	0.919
0.547	0.939
0.613	0.974
1.239	0.970*
1.288	0.963
1.332	0.943
1.377	0.914
1.416	0.876*
1.444	0.876*
1.478	0.909*
1.562	0.935
1.641	0.935
1.718	0.922
1.837	0.887
2.000	0.827

CURVE 28 T = 77

λ	ρ
0.294	0.707*
0.341	0.798*
0.383	0.852
0.449	0.903
0.487	0.923
0.547	0.939*
0.613	0.974*
1.239	0.970*
1.288	0.963*
1.332	0.943*

CURVE 28 (cont.)

λ	ρ
1.377	0.914*
1.416	0.876*
1.444	0.876*
1.478	0.909*
1.562	0.935*
1.641	0.935*
1.703	0.932
1.851	0.891
2.000	0.832

* Not shown on plot

FIGURE 6IA

ANALYZED NORMAL SOLAR ABSORPTANCE OF LANTHANUM OXIDE PIGMENTED COATINGS

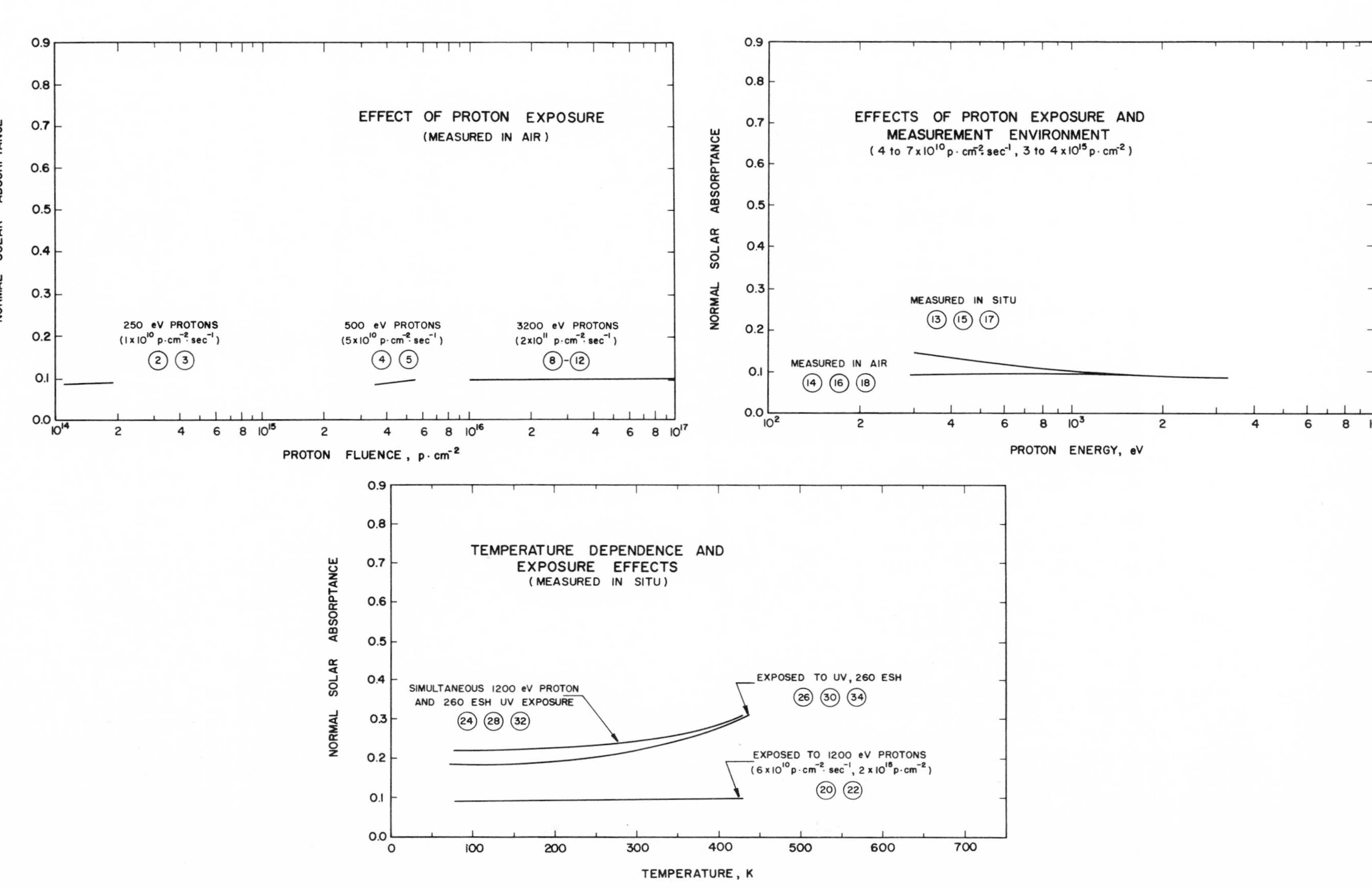

SPECIFICATION TABLE NO. 61 NORMAL SOLAR ABSORPTANCE OF LANTHANUM OXIDE PIGMENTED COATINGS

Curve No.	Ref. No.	Year	Temperature Range, K	Geometry θ	Reported Error, %	Composition (weight percent), Specifications and Remarks
1	39	1969	298	$\sim 0°$		La_2O_3 in K_2SiO_3 binder; 53.4 La_2O_3; compacted after 2 hrs ball milling; absorptance calculated from reflectance; property measured in air.
2	39	1969	298	$\sim 0°$		Above specimen and conditions except exposed to 250 ev protons from Ortec rf ion source in vacuum (5×10^{-7} mm Hg); vacuum maintained by diffusion pump; fluence 1.1×10^{14} p cm^{-2}; flux 9×10^{9} p cm^{-2} sec^{-1}.
3	39	1969	298	$\sim 0°$		Similar to above specimen and conditions except flux 1.3×10^{10} p cm^{-2} sec^{-1}; fluence 1.9×10^{14} p cm^{-2}.
4	39	1969	298	$\sim 0°$		Similar to above specimen and conditions except proton energy 500 ev; flux 5×10^{10} p cm^{-2} sec^{-1}; fluence 3.5×10^{15} p cm^{-2}.
5	39	1969	298	$\sim 0°$		Similar to above specimen and conditions except flux 4.5×10^{10} p cm^{-2} sec^{-1}; fluence 5.5×10^{15} p cm^{-2}.
6	39	1969	298	$\sim 0°$		Similar to above specimen and conditions except proton energy 1200 ev; flux 2.1×10^{11} p cm^{-2} sec^{-1}; fluence 3.2×10^{16} p cm^{-2}.
7	39	1969	298	$\sim 0°$		Similar to above specimen and conditions except flux 1.9×10^{11} p cm^{-2} sec^{-1}; fluence 5.5×10^{16} p cm^{-2}.
8	39	1969	298	$\sim 0°$		Similar to above specimen and conditions except proton energy 3200 ev; flux 2×10^{11} p cm^{-2} sec^{-1}; fluence 1×10^{16} p cm^{-2}.
9	39	1969	298	$\sim 0°$		Similar to above specimen and conditions except flux 2.4×10^{11} p cm^{-2} sec^{-1}; fluence 2.2×10^{16} p cm^{-2}.
10	39	1969	298	$\sim 0°$		Similar to above specimen and conditions except flux 2.1×10^{11} p cm^{-2} sec^{-1}; fluence 2.6×10^{16} p cm^{-2}.
11	39	1969	298	$\sim 0°$		Similar to above specimen and conditions except flux 2.5×10^{11} p cm^{-2} sec^{-1}; fluence 3.1×10^{16} p cm^{-2}.
12	39	1969	298	$\sim 0°$		Similar to above specimen and conditions except flux 2.6×10^{11} p cm^{-2} sec^{-1}; fluence 1×10^{17} p cm^{-2}.
13	39	1969	298	$\sim 0°$		Similar to curve 4 specimen and conditions except flux 3.7×10^{10} p cm^{-2} sec^{-1}; fluence 4.2×10^{15} p cm^{-2}; property measured in situ.
14	39	1969	298	$\sim 0°$		Above specimen and conditions except final measurement in air.
15	39	1969	298	$\sim 0°$		Similar to curve 8 specimen and conditions except flux 7.0×10^{10} p cm^{-2} sec^{-1}; fluence 3.4×10^{15} p cm^{-2}; property measured in situ (5×10^{-7} mm Hg).
16	39	1969	298	$\sim 0°$		Above specimen and conditions except final measurement in air.
17	39	1969	298	$\sim 0°$		Similar to curve 6 specimen and conditions except flux 6.4×10^{10} p cm^{-2} sec^{-1}; fluence 2.8×10^{15} p cm^{-2}; property measured in situ (5×10^{-7} mm Hg).
18	39	1969	298	$\sim 0°$		Above specimen and conditions except final measurement in air.
19	39	1969	298	$\sim 0°$		Similar to curve 1 specimen and conditions.

SPECIFICATION TABLE NO. 61 NORMAL SOLAR ABSORPTANCE OF LANTHANUM OXIDE PIGMENTED COATINGS (continued)

Curve No.	Ref. No.	Year	Temperature Range, K	Geometry θ	Reported Error, %	Composition (weight percent), Specifications and Remarks
20	39	1969	423	$\sim 0^\circ$		Above specimen and conditions except flux 6.1×10^{10} p cm^{-2} sec^{-1}; fluence 2.1×10^{15} p cm^{-2}; proton energy 1200 ev; property measured in situ (5×10^{-7} mm Hg).
21	39	1969	298	$\sim 0^\circ$		Above specimen and conditions except final measurement in air.
22	39	1969	77	$\sim 0^\circ$		Similar to curve 6 specimen and conditions except flux 6.5×10^{10} p cm^{-2} sec^{-1}; fluence 2.1×10^{15} p cm^{-2}; property measured in situ (5×10^{-7} mm Hg).
23	39	1969	298	$\sim 0^\circ$		Similar to curve 1 specimen and conditions.
24	39	1969	77	$\sim 0^\circ$		Above specimen and conditions except exposed to 1200 ev protons (flux 2.0×10^{10} p cm^{-2} sec^{-1}, fluence 4.6×10^{15} p cm^{-2}) and to UV (254 ESH) in vacuum (5×10^{-7} mm Hg); vacuum maintained by diffusion pump; property measured in situ.
25	39	1969	298	$\sim 0^\circ$		Above specimen and conditions except final measurement in air.
26	39	1969	77	$\sim 0^\circ$		Similar to curve 23 specimen and conditions except exposed to UV source in vacuum (5×10^{-7} mm Hg); vacuum maintained by diffusion pump; ESH 254; property measured in situ.
27	39	1969	298	$\sim 0^\circ$		Above specimen and conditions except final measurement in air.
28	39	1969	298	$\sim 0^\circ$		Similar to curve 24 specimen and conditions except flux 1.6×10^{10} p cm^{-2} sec^{-1}; fluence 3.7×10^{15} p cm^{-2}; ESH 260.
29	39	1969	298	$\sim 0^\circ$		Above specimen and conditions except final measurement in air.
30	39	1969	298	$\sim 0^\circ$		Similar to curve 26 specimen and conditions except ESH 260.
31	39	1969	298	$\sim 0^\circ$		Above specimen and conditions except final measurement in air.
32	39	1969	423	$\sim 0^\circ$		Similar to curve 28 specimen and conditions except flux 2×10^{10} p cm^{-2} sec^{-1}; fluence 4.7×10^{15} p cm^{-2}.
33	39	1969	298	$\sim 0^\circ$		Above specimen and conditions except final measurement in air.
34	39	1969	423	$\sim 0^\circ$		Similar to curve 30 specimen and conditions.
35	39	1969	298	$\sim 0^\circ$		Above specimen and conditions except final measurement in air.
36	39	1969	77	$\sim 0^\circ$		Similar to curve 28 specimen and conditions except exposed to UV radiation for $\lambda < .4$ μm with Corning filter 7-54; ESH 230; flux 2.2×10^{10} p cm^{-2} sec^{-1}; fluence 4.5×10^{15} p cm^{-2}.
37	39	1969	298	$\sim 0^\circ$		Above specimen and conditions except final measurement in air.
38	39	1969	77	$\sim 0^\circ$		Similar to curve 26 specimen and conditions except exposed to UV radiation for $\lambda < .4$ μm with Corning filter 7-54 in vacuum (5×10^{-7} mm Hg); vacuum maintained by diffusion pump; ESH 230.
39	39	1969	298	$\sim 0^\circ$		Above specimen and conditions except final measurement in air.
40	39	1969	298	$\sim 0^\circ$		Similar to curve 36 specimen and conditions except flux 1.8×10^{10} p cm^{-2} sec^{-1}; fluence 4.4×10^{15} p cm^{-2}; ESH 270.

SPECIFICATION TABLE NO. 61 NORMAL SOLAR ABSORPTANCE OF LANTHANUM OXIDE PIGMENTED COATINGS (continued)

Curve No.	Ref. No.	Year	Temperature Range, K	Geometry θ	Reported Error, %	Composition (weight percent), Specifications and Remarks
41	39	1969	298	$\sim 0^\circ$		Above specimen and conditions except final measurement in air.
42	39	1969	298	$\sim 0^\circ$		Similar to curve 38 specimen and conditions except ESH 270.
43	39	1969	298	$\sim 0^\circ$		Above specimen and conditions except final measurement in air.
44	39	1969	423	$\sim 0^\circ$		Similar to curve 36 specimen and conditions except flux 2.1 x 10^{10} p cm^{-2} sec^{-1}; fluence 4.7 x 10^{15} p cm^{-2}; ESH 248.
45	39	1969	298	$\sim 0^\circ$		Above specimen and conditions except final measurement in air.
46	39	1969	423	$\sim 0^\circ$		Similar to curve 38 specimen and conditions except ESH 248.
47	39	1969	298	$\sim 0^\circ$		Above specimen and conditions except final measurement in air.
48	39	1969	77	$\sim 0^\circ$		Similar to curve 28 specimen and conditions except exposed to UV radiation for $\lambda > .4$ μm with Corning filter 3-73; ESH 254; flux 2.0 x 10^{10} p cm^{-2} sec^{-1}; fluence 4.6 x 10^{15} p cm^{-2}.
49	39	1969	298	$\sim 0^\circ$		Above specimen and conditions except final measurement in air.
50	39	1969	77	$\sim 0^\circ$		Similar to curve 26 specimen and conditions except exposed to UV radiation for $\lambda > .4$ μm with Corning filter 3-73; ESH 254.
51	39	1969	298	$\sim 0^\circ$		Above specimen and conditions except final measurement in air.
52	39	1969	298	$\sim 0^\circ$		Similar to curve 48 specimen and conditions except flux 1.9 x 10^{10} p cm^{-2} sec^{-1}; fluence 4.4 x 10^{15}; ESH 260; final measurement in air.
53	39	1969	298	$\sim 0^\circ$		Similar to curve 50 specimen and conditions except ESH 260; final measurement in air.
54	39	1969	423	$\sim 0^\circ$		Similar to curve 48 specimen and conditions except ESH 230; flux 2.4 x 10^{10} p cm^{-2} sec^{-1}; fluence 4.9 x 10^{15} p cm^{-2}.
55	39	1969	298	$\sim 0^\circ$		Above specimen and conditions except final measurement in air.
56	39	1969	423	$\sim 0^\circ$		Similar to curve 50 specimen and conditions except ESH 230.
57	39	1969	298	$\sim 0^\circ$		Above specimen and conditions except final measurement in air.

DATA TABLE NO. 61 NORMAL SOLAR ABSORPTANCE OF LANTHANUM OXIDE PIGMENTED COATINGS

[Temperature, T, K; Absorptance, α]

T	α	T	α	T	α	T	α	T	α
CURVE 1		CURVE 13		CURVE 25		CURVE 37		CURVE 49	
298	0.083	298	0.150	298	0.164	298	0.126	298	0.078
CURVE 2		CURVE 14		CURVE 26		CURVE 38		CURVE 50	
298	0.086	298	0.094	77	0.189	77	0.143	77	0.078
CURVE 3		CURVE 15		CURVE 27		CURVE 39		CURVE 51	
298	0.090	298	0.088	298	0.141	298	0.114	298	0.070
CURVE 4		CURVE 16		CURVE 28		CURVE 40		CURVE 52	
298	0.086	298	0.087	298	0.240	298	0.220	298	0.080
CURVE 5		CURVE 17		CURVE 29		CURVE 41		CURVE 53	
298	0.095	298	0.100	298	0.195	298	0.180	298	0.086
CURVE 6		CURVE 18		CURVE 30		CURVE 42		CURVE 54	
298	0.097	298	0.097	298	0.217	298	0.220	423	0.092
CURVE 7		CURVE 19		CURVE 31		CURVE 43		CURVE 55	
298	0.098	298	0.080	298	0.173	298	0.199	298	0.081
CURVE 8		CURVE 20		CURVE 32		CURVE 44		CURVE 56	
298	0.096	423	0.096	423	0.309	423	0.307	423	0.089
CURVE 9		CURVE 21		CURVE 33		CURVE 45		CURVE 57	
298	0.098	298	0.079	298	0.238	298	0.246	298	0.076
CURVE 10		CURVE 22		CURVE 34		CURVE 46			
298	0.095	77	0.090	423	0.298	423	0.304		
CURVE 11		CURVE 23		CURVE 35		CURVE 47			
298	0.096	298	0.060	298	0.235	298	0.261		
CURVE 12		CURVE 24		CURVE 36		CURVE 48			
298	0.097	77	0.220	77	0.176	77	0.080		

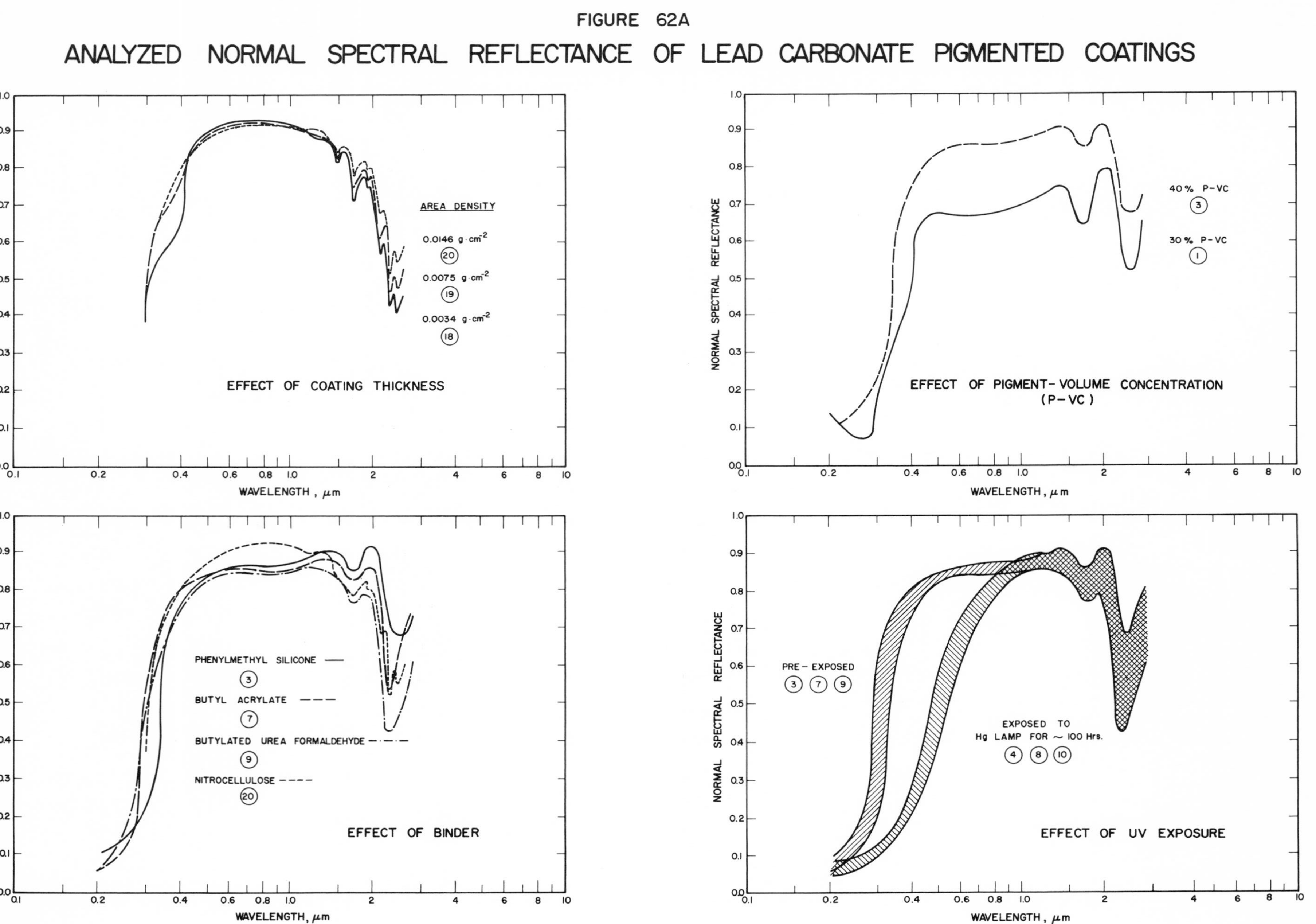

FIGURE 62A

ANALYZED NORMAL SPECTRAL REFLECTANCE OF LEAD CARBONATE PIGMENTED COATINGS

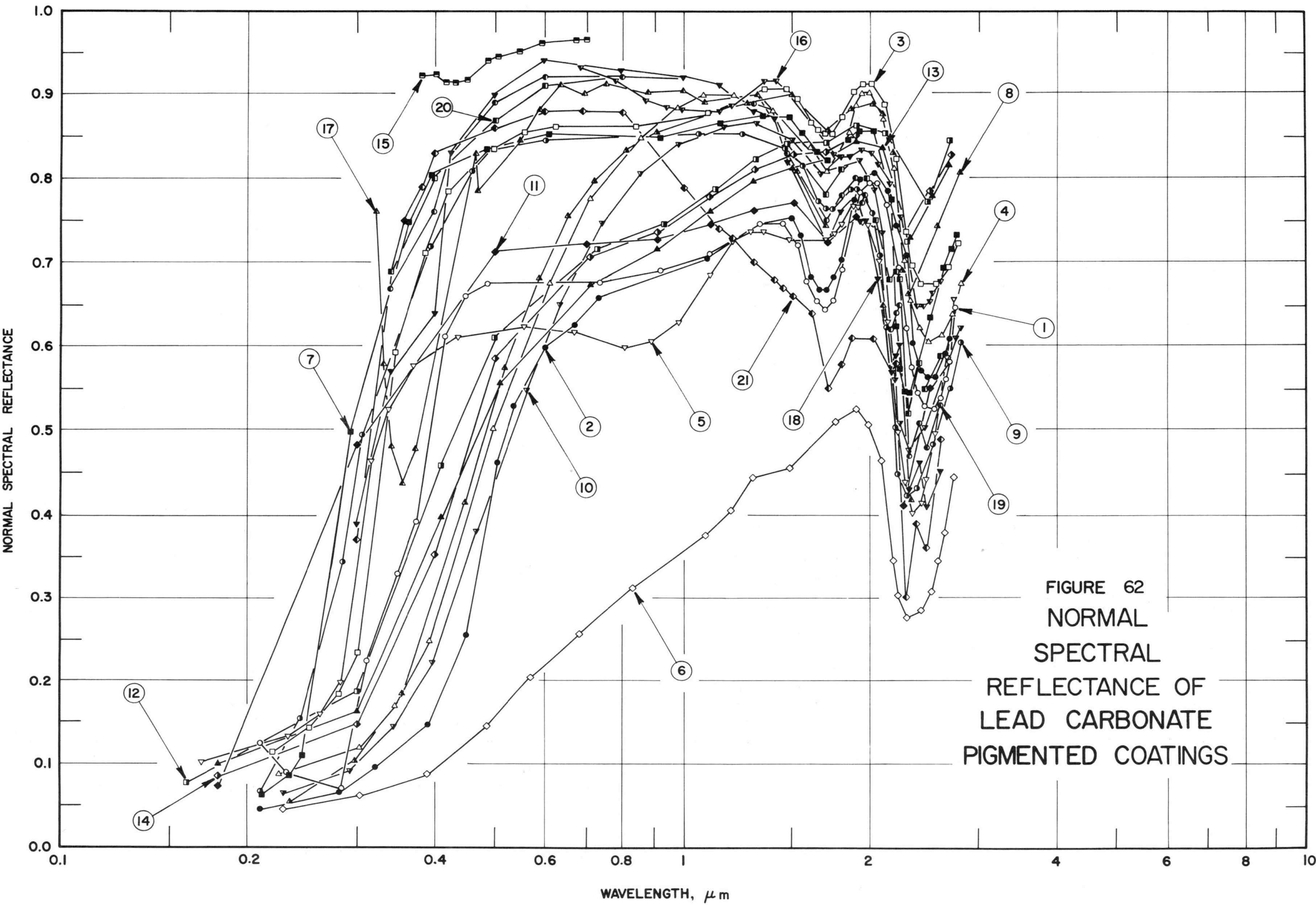

FIGURE 62 NORMAL SPECTRAL REFLECTANCE OF LEAD CARBONATE PIGMENTED COATINGS

SPECIFICATION TABLE NO. 62 NORMAL SPECTRAL REFLECTANCE OF LEAD CARBONATE PIGMENTED COATINGS

Curve No.	Ref. No.	Year	Temperature K	Wavelength Range, μm	Geometry θ	θ'	ω'	Reported Error, %	Composition (weight percent), Specifications, and Remarks
1	1	1960	~358	0.210-2.74	~0°		2π		Basic white lead carbonate in phenylmethyl silicone binder (5.08 mm thick); anodized 24S-T aluminum substrate; pigment-volume ratio: 30%; data extracted from smooth curve; measured relative to MgO.
2	1	1960	~358	0.210-2.74	~0°		2π		Above specimen and conditions except UV exposed for 100 hrs to G.E. UA-3 mercury lamp in vacuum (10^{-5} mm Hg); intensity of lamp about 10 times intensity of sun in UV; distance from lamp to source: 8.5 cm.
3	1	1960	~358	0.219-2.77	~0°		2π		Basic white lead carbonate in phenylmethyl silicone binder (5.08 mm thick); anodized 24S-T aluminum substrate; pigment-volume ratio: 40%; data extracted from smooth curve; measured relative to MgO.
4	1	1960	~358	0.226-2.80	~0°		2π		Above specimen and conditions except UV exposed for 105 hrs to G.E. UA-3 mercury lamp in vacuum (10^{-5} mm Hg); intensity of lamp about 10 times intensity of sun in UV; distance from lamp to source: 8.5 cm.
5	1	1960	~358	0.169-2.73	~0°		2π		Basic white lead carbonate in butylated melamine formaldehyde binder (5.08 mm thick); anodized 24 S-T aluminum substrate; pigment-volume ratio: 20%; data extracted from smooth curve; measured relative to MgO.
6	1	1960	~358	0.229-2.73	~0°		2π		Above specimen and conditions except UV exposed for 104 hrs to G.E. UA-3 mercury lamp in vacuum (10^{-5} mm Hg); intensity of lamp about 10 times intensity of sun in UV; distance from lamp to source: 8.5 cm.
7	1	1960	~358	0.211-2.76	~0°		2π		Basic white lead carbonate in butyl acrylate copolymer binder (5.08 mm thick); anodized 24 S-T aluminum substrate; pigment-volume ratio: 40%; data extracted from smooth curve; measured relative to MgO.
8	1	1960	~358	0.234-2.79	~0°		2π		Above specimen and conditions except UV exposed for 105 hrs to G.E. UA-3 mercury lamp in vacuum (10^{-5} mm Hg); intensity of lamp about 10 times intensity of sun in UV; distrance from lamp to source: 8.5 cm.
9	1	1960	~358	0.210-2.80	~0°		2π		Basic white lead carbonate in butylated urea formaldehyde binder (5.08 mm thick); anodized 24 S-T aluminum substrate; pigment-volume ratio: 40%; data extracted from smooth curve; measured relative to MgO.
10	1	1960	~358	0.228-2.80	~0°		2π		Above specimen and conditions except UV exposed for 105 hrs to G.E. UA-3 mercury lamp in vacuum (10^{-5} mm Hg); intensity of lamp about 10 times intensity of sun in UV; distance from lamp to source: 8.5 cm.
11	2	1962	339	0.18-2.69	~0°		2π		Basic white lead carbonate in Acryloid A10 binder (5.08 mm thick); substrate unknown; pigment-volume ratio: 30%; measured relative to MgO.
12	2	1962	339	0.16-2.69	~0°		2π		Above specimen and conditions except UV exposed for 20 hrs to mercury lamp in vacuum (10^{-5} mm Hg).
13	2	1962	339	0.18-2.68	~0°		2π		Curve 11 specimen and conditions except UV exposed for 45 hrs to mercury lamp in vacuum (10^{-5} mm Hg).
14	2	1962	339	0.18-2.70	~0°		2π		Curve 11 specimen and conditions except UV exposed to mercury lamp in vacuum (10^{-5} mm Hg) for 45 hrs and equipment breakdown.

SPECIFICATION TABLE NO. 62 NORMAL SPECTRAL REFLECTANCE OF LEAD CARBONATE PIGMENTED COATINGS (continued)

Curve No.	Ref. No.	Year	Temperature K	Wavelength Range, μm	Geometry θ θ'	ω'	Reported Error, %	Composition (weight percent), Specifications, and Remarks
15	3	1960	~298	0.381-0.700	~0°	2π		Basic $PbCO_3$ (reagent grade) in alkyd-melamine binder (DuPont RC 7007); sprayed onto mild steel substrate; pigment concentration: 85%; data extracted from smooth curve; measured relative to MgO.
16	3	1960	~298	0.689-2.60	~0°	2π		Similar to curve 15 specimen and conditions.
17	4	1961	~298	0.322-2.33	~0°	2π		Basic white lead in alkyd binder; substrate unknown; converted from R $(2\pi, 0)$; measured relative to MgO.
18	4	1961	~298	0.300-2.60	~0°	2π		Basic white lead in nitrocellulose binder; rutile TiO_2 in nitrocellulose and aluminum substrates; weight of paint: 0.0034 g cm^{-2}; converted from R $(2\pi, 0)$; measured relative to MgO; data extracted from smooth curve.
19	4	1961	~298	0.300-2.60	~0°	2π		Similar to above specimen and conditions except weight of paint 0.0075 g cm^{-2}.
20	4	1961	~298	0.300-2.60	~0°	2π		Similar to above specimen and conditions except weight of paint 0.0146 g cm^{-2}.
21	4	1961	~298	0.300-2.60	~0°	2π		Basic white lead in acrylic binder; substrate unknown; converted from R $(2\pi, 0)$; measured relative to MgO; data extracted from smooth curve.

DATA TABLE NO. 62 NORMAL SPECTRAL REFLECTANCE OF LEAD CARBONATE PIGMENTED COATINGS

[Wavelength, λ, μm; Reflectance, ρ; Temperature, T, K]

λ	ρ
CURVE 1 T ~ 358	
0.210	0.126
0.231	0.090
0.284	0.070
0.312	0.225
0.349	0.330
0.374	0.393
0.416	0.613
0.448	0.661
0.489	0.677
0.737	0.677
0.918	0.691
1.11	0.710
1.33	0.747
1.45	0.747
1.53	0.720
1.58	0.678
1.64	0.653
1.69	0.644
1.75	0.654
1.81	0.692
1.89	0.762
1.93	0.782
2.00	0.794
2.06	0.794
2.14	0.768
2.23	0.694
2.29	0.622
2.34	0.576
2.39	0.546
2.45	0.529
2.54	0.525
2.60	0.539
2.65	0.561
2.68	0.582
2.74	0.648
CURVE 2 T ~ 358	
0.210	0.047
0.281	0.067
0.321	0.094
0.390	0.147
0.448	0.255
0.503	0.461
0.536	0.529
CURVE 2 (cont.)	
0.601	0.598
0.673	0.627
0.832	0.659
1.09	0.707
1.33	0.747*
1.50	0.754
1.55	0.732
1.62	0.684
1.66	0.669
1.70	0.669
1.75	0.684
1.79	0.704
1.88	0.775
1.93	0.799
2.04	0.807
2.13	0.785
2.20	0.743
2.29	0.659
2.35	0.604
2.42	0.572
2.48	0.564
2.55	0.565
2.64	0.593
2.69	0.611
2.74	0.648*
CURVE 3 T ~ 358	
0.219	0.113
0.252	0.142
0.281	0.182
0.301	0.235
0.346	0.593
0.386	0.711
0.421	0.785
0.498	0.836
0.559	0.854
0.625	0.861
0.839	0.861
1.09	0.877
1.36	0.905
1.47	0.905
1.53	0.895
1.61	0.867
1.65	0.858
CURVE 3 (cont.)	
1.69	0.852
1.74	0.852
1.82	0.872
1.89	0.902
1.95	0.912
2.01	0.912
2.12	0.889
2.20	0.823
2.28	0.737
2.34	0.697
2.42	0.676
2.56	0.676
2.68	0.696
2.77	0.723
CURVE 4 T ~ 358	
0.226	0.088
0.303	0.119
0.345	0.169
0.392	0.249
0.498	0.501
0.610	0.675
0.709	0.776
0.855	0.848
0.979	0.883
1.08	0.898
1.21	0.898
1.39	0.880
1.60	0.836
1.71	0.809
1.78	0.823
1.85	0.855
1.90	0.888
1.94	0.900
2.00	0.900
2.10	0.871
2.18	0.813
2.27	0.701
2.32	0.655
2.40	0.622
2.48	0.605
2.62	0.614
2.71	0.639
2.80	0.676
CURVE 5 T ~ 358	
0.169	0.101
0.233	0.133
0.261	0.159
0.282	0.198
0.316	0.463
0.336	0.524
0.370	0.577
0.435	0.612
0.557	0.624
0.673	0.618
0.805	0.598
0.889	0.606
0.981	0.630
1.11	0.686
1.20	0.724
1.28	0.738
1.35	0.738
1.48	0.728
1.70	0.728
1.76	0.734
1.80	0.745
1.87	0.765
1.93	0.765
1.99	0.743
2.07	0.700
2.14	0.629
2.21	0.498
2.27	0.439
2.34	0.402
2.42	0.415
2.47	0.441
2.55	0.496
2.66	0.586
2.73	0.658
CURVE 6 T ~ 358	
0.229	0.047
0.304	0.062
0.389	0.088
0.485	0.144
0.570	0.204
0.681	0.257
0.830	0.312
CURVE 6 (cont.)	
1.08	0.376
1.19	0.404
1.29	0.445
1.48	0.456
1.76	0.510
1.90	0.524
1.99	0.508
2.08	0.463
2.17	0.346
2.22	0.301
2.29	0.276
2.42	0.286
2.51	0.309
2.57	0.344
2.64	0.379
2.73	0.444
CURVE 7 T ~ 358	
0.211	0.062
0.233	0.087
0.246	0.110
0.294	0.498
0.363	0.748
0.396	0.803
0.485	0.835
0.609	0.852
0.917	0.848
1.15	0.867
1.34	0.874
1.48	0.874
1.56	0.855
1.64	0.831
1.72	0.822
1.84	0.847
1.93	0.857
2.03	0.857
2.16	0.787
2.20	0.637
2.23	0.574
2.28	0.547
2.32	0.545
2.50	0.636
2.62	0.694
2.70	0.715
2.76	0.733
CURVE 8 T ~ 358	
0.234	0.055
0.297	0.103
0.352	0.183
0.447	0.416
0.517	0.575
0.587	0.681
0.654	0.755
0.720	0.799
0.811	0.833
0.907	0.855
1.27	0.890
1.50	0.898
1.61	0.865*
1.67	0.854*
1.72	0.856
1.87	0.882
2.03	0.889
2.10	0.877
2.18	0.830
2.25	0.692
2.30	0.662
2.34	0.662*
2.57	0.744
2.79	0.807
CURVE 9 T ~ 358	
0.210	0.065
0.244	0.153
0.285	0.344
0.306	0.496
0.394	0.719
0.459	0.809
0.500	0.835*
0.602	0.846
1.07	0.853
1.25	0.853
1.46	0.835
1.56	0.817
1.65	0.772
1.70	0.764
1.74	0.764
1.86	0.788
1.92	0.788
2.01	0.759
CURVE 9 (cont.)	
2.08	0.709
2.13	0.627
2.19	0.504
2.22	0.449
2.29	0.423
2.37	0.432
2.53	0.483
2.69	0.550
2.80	0.604
CURVE 10 T ~ 358	
0.228	0.065
0.293	0.091
0.342	0.142
0.398	0.223
0.468	0.379
0.561	0.548
0.634	0.650
0.740	0.746
0.853	0.806
0.983	0.840
1.17	0.862
1.32	0.865
1.50	0.845
1.67	0.806
1.93	0.820
2.04	0.788
2.10	0.735
2.19	0.560
2.23	0.509
2.30	0.476
2.46	0.502
2.65	0.559*
2.76	0.609
2.80	0.622
CURVE 11 T = 339	
0.18	0.074
0.30	0.483
0.50	0.712
0.70	0.720
0.91	0.729
1.11	0.745
CURVE 11 (cont.)	
1.31	0.761
1.51	0.770
1.71	0.723
1.90	0.757
2.09	0.733*
2.27	0.412
2.50	0.550
2.69	0.581*
CURVE 12 T = 339	
0.16	0.079
0.30	0.188
0.41	0.458
0.50	0.612
0.73	0.717
0.93	0.747
1.13	0.788
1.32	0.823
1.51	0.844*
1.71	0.842
1.90	0.864
2.13	0.856
2.29	0.723
2.49	0.771
2.69	0.846
CURVE 13 T = 339	
0.18	0.100
0.30	0.162
0.41	0.398
0.51	0.558
0.71	0.672
0.91	0.714
1.11	0.761
1.30	0.797
1.50	0.814
1.70	0.822*
1.90	0.844
2.10	0.838
2.33	0.730
2.52	0.779
2.68	0.817

*Not shown on plot

λ	ρ
CURVE 14	
T = 339	
0.18	0.087
0.30	0.147
0.40	0.352
0.50	0.582
0.71	0.707
0.91	0.738
1.11	0.779
1.31	0.812
1.51	0.829
1.70	0.831
1.90	0.852
2.11	0.851*
2.31	0.737*
2.50	0.786
2.70	0.829
CURVE 15	
T ~ 298	
0.381	0.922
0.402	0.923
0.418	0.916
0.433	0.914
0.451	0.918
0.488	0.940
0.505	0.945
0.548	0.952
0.595	0.960
0.675	0.965
0.700	0.965
CURVE 16	
T ~ 298	
0.689	0.932
0.783	0.917
0.874	0.893
0.941	0.885
0.999	0.882
1.14	0.880
1.20	0.884
1.35	0.916
1.40	0.914
1.50	0.897*
1.70	0.839*
1.74	0.829
CURVE 16 (cont.)	
1.80	0.824
1.87	0.826
1.94	0.832
2.01	0.830
2.08	0.817
2.16	0.793
2.22	0.754
2.39	0.649
2.45	0.649
2.49	0.653
2.54	0.664
2.60	0.676
CURVE 17	
T ~ 298	
0.322	0.760
0.331	0.588
0.340	0.481
0.355	0.437
0.372	0.479
0.465	0.830
0.468	0.784
0.546	0.845
0.589	0.882
0.637	0.912
0.695	0.899
0.752	0.913
0.879	0.902
1.00	0.902
1.08	0.890
1.32	0.899
1.53	0.810
1.71	0.743
1.98	0.743*
2.10	0.650
2.33	0.419
CURVE 18	
T ~ 298	
0.300	0.39
0.340	0.57
0.400	0.64
0.425	0.83
0.500	0.90
0.600	0.94
CURVE 18 (cont.)	
0.800	0.93
1.00	0.92
1.15	0.91
1.21	0.89*
1.31	0.88
1.40	0.87
1.47	0.82
1.70	0.72*
1.80	0.76
1.90	0.78*
1.94	0.75
1.97	0.75
2.06	0.68
2.16	0.57
2.20	0.59
2.23	0.60
2.31	0.43
2.40	0.46
2.47	0.41
2.60	0.45
CURVE 19	
T ~ 298	
0.300	0.39*
0.340	0.67
0.400	0.76
0.425	0.83*
0.500	0.89
0.600	0.92
0.800	0.92
1.00	0.92*
1.15	0.91*
1.21	0.89*
1.31	0.89
1.40	0.88*
1.47	0.83
1.70	0.75
1.80	0.78
1.90	0.80
1.94	0.77
1.97	0.78
2.06	0.71*
2.16	0.62
2.20	0.64
2.23	0.65
2.31	0.47
CURVE 19 (cont.)	
2.40	0.51
2.47	0.84
2.60	0.53
CURVE 20	
T ~ 298	
0.300	0.39*
0.340	0.69
0.400	0.80
0.425	0.83*
0.500	0.87
0.600	0.91
0.800	0.92*
1.00	0.92*
1.15	0.91*
1.21	0.89*
1.31	0.90*
1.40	0.88*
1.47	0.84
1.70	0.78
1.80	0.81
1.90	0.82*
1.94	0.80*
1.97	0.80
2.06	0.75
2.16	0.68
2.20	0.69
2.23	0.68
2.31	0.52
2.40	0.58
2.47	0.55
2.60	0.59
CURVE 21	
T ~ 298	
0.300	0.37
0.340	0.69*
0.357	0.75
0.381	0.79
0.400	0.83
0.500	0.86
0.600	0.88
0.690	0.88
0.800	0.88
1.00	0.79
CURVE 21 (cont.)	
1.14	0.74
1.20	0.73
1.31	0.70
1.40	0.68
1.44	0.67
1.51	0.66
1.62	0.64
1.72	0.55
1.80	0.58
1.87	0.61
2.03	0.61
2.14	0.57*
2.20	0.58
2.29	0.30
2.38	0.39
2.46	0.36
2.60	0.49

*Not shown on plot

SPECIFICATION TABLE NO. 63 HEMISPHERICAL TOTAL EMITTANCE OF LITHIUM ALUMINUM SILICATE PIGMENTED COATINGS

Curve No.	Ref. No.	Year	Temperature Range, K	Reported Error, %	Composition (weight percent), Specifications and Remarks
1*	50	1965	~298		LMSC Lithafrax ($Li_2O + Al_2O_3 + 8\ SiO_2$) in Na_2SiO_3 binder; pigment/binder ratio 4:1; impurities: 0.1 Fe_2O_3, 0.4 K_2O, 0.2 Na_2O, 0.011 Fe; cured at 472 K.

DATA TABLE NO. 63 HEMISPHERICAL TOTAL EMITTANCE OF LITHIUM ALUMINUM SILICATE PIGMENTED COATING

[Temperature, T, K; Emittance, ϵ]

T	ϵ
CURVE 1*	
298	0.87

* No plot given

SPECIFICATION TABLE NO. 64 NORMAL TOTAL EMITTANCE OF LITHIUM ALUMINUM SILICATE PIGMENTED COATINGS

Curve No.	Ref. No.	Year	Temperature Range, K	Geometry θ'	Reported Error, %	Composition (weight percent), Specifications and Remarks
1*	18	1964	296-811	$\sim 0^{\circ}$		Li-Al-SiO_3 in K_2SiO_3 binder ($\sim$0.1015 mm thick); 6061-T6 aluminum substrate; air brush application; calculated from spectral reflectance.

DATA TABLE NO. 64 NORMAL TOTAL EMITTANCE OF LITHIUM ALUMINUM SILICATE PIGMENTED COATING

[Temperature, T, K; Emittance, ϵ]

T	ϵ
CURVE 1*	
296	0.907
480	0.914
648	0.873
811	0.808

* No plot given

SPECIFICATION TABLE NO. 65 NORMAL SPECTRAL REFLECTANCE OF LITHIUM ALUMINUM SILICATE PIGMENTED COATINGS

Curve No.	Ref. No.	Year	Temperature, K	Wavelength Range, μm	Geometry θ	θ'	ω	Reported Error, %	Composition (weight percent), Specifications, and Remarks
1*	27	1964	~298	0.44-0.60	~0°		2π		$LiAlSiO_4$ (from Foote Mineral Co.) in PS7 potassium silicate binder; PBR 4.30; solids content 64.4%; aluminum substrate abraded with No. 60 Aloxite cloth; sample cured at 413 K for 18 hrs. [Authors' designation: Sample C5]
2*	27	1964	~298	0.44-0.60	~0°		2π		Above specimen and conditions except irradiated in vacuum (10^{-6} mm Hg) with 250 ESH at solar factor of 2.5 suns.
3*	27	1964	~298	0.44-0.60	~0°		2π		Lithafrax in PS7 potassium silicate binder; PBR 4.30; solids content 64.4%; aluminum substrate abraded with No. 60 Aloxite cloth; sample cured at 413 K for 18 hrs. [Authors' designation: Sample C7]
4*	27	1964	~298	0.44-0.60	~0°		2π		Above specimen and conditions except irradiated in vacuum (10^{-6} mm Hg) with 2100 ESH at solar factor of 10 suns.

DATA TABLE NO. 65 NORMAL SPECTRAL REFLECTANCE OF LITHIUM ALUMINUM SILICATE PIGMENTED COATING

[Wavelength, λ, μm; Reflectance, ρ; Temperature, T, K]

λ	ρ
CURVE 1*	
0.44	0.590
0.60	0.645
CURVE 2*	
0.44	0.520
0.60	0.625
CURVE 3*	
0.44	0.850
0.60	0.860
CURVE 4*	
0.44	0.435
0.60	0.605

* No plot given

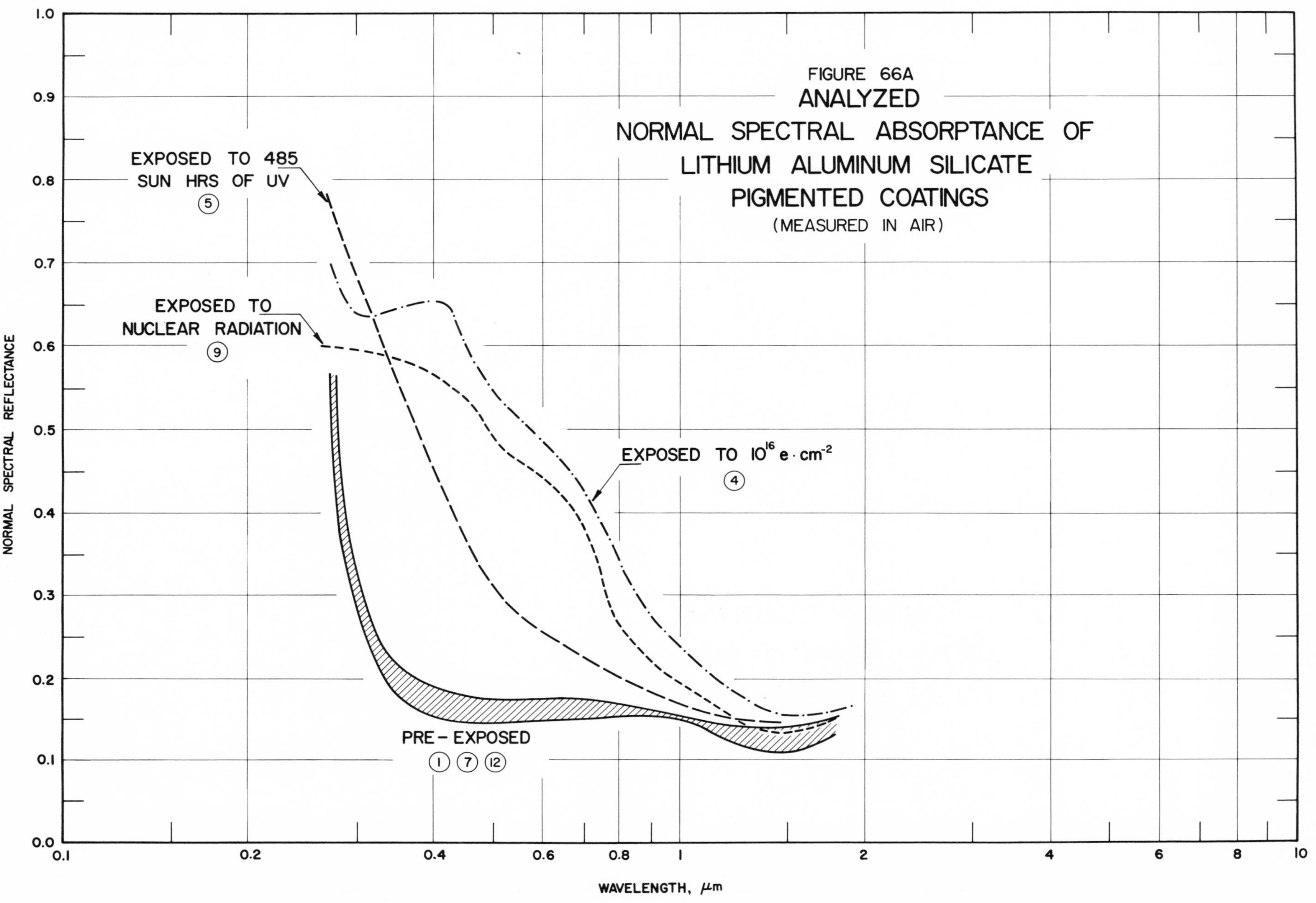

FIGURE 66A
ANALYZED
NORMAL SPECTRAL ABSORPTANCE OF
LITHIUM ALUMINUM SILICATE
PIGMENTED COATINGS
(MEASURED IN AIR)
EXPOSED TO 485
SUN HRS OF UV
(5)
EXPOSED TO
NUCLEAR RADIATION
(9)
EXPOSED TO 10^{16} e · cm^{-2}
(4)
PRE – EXPOSED
(1) (7) (12)
NORMAL SPECTRAL REFLECTANCE
WAVELENGTH, μm
1.0
0.9
0.8
0.7
0.6
0.5
0.4
0.3
0.2
0.1
0.0
0.1
0.2
0.4
0.6
0.8
1
2
4
6
8
10

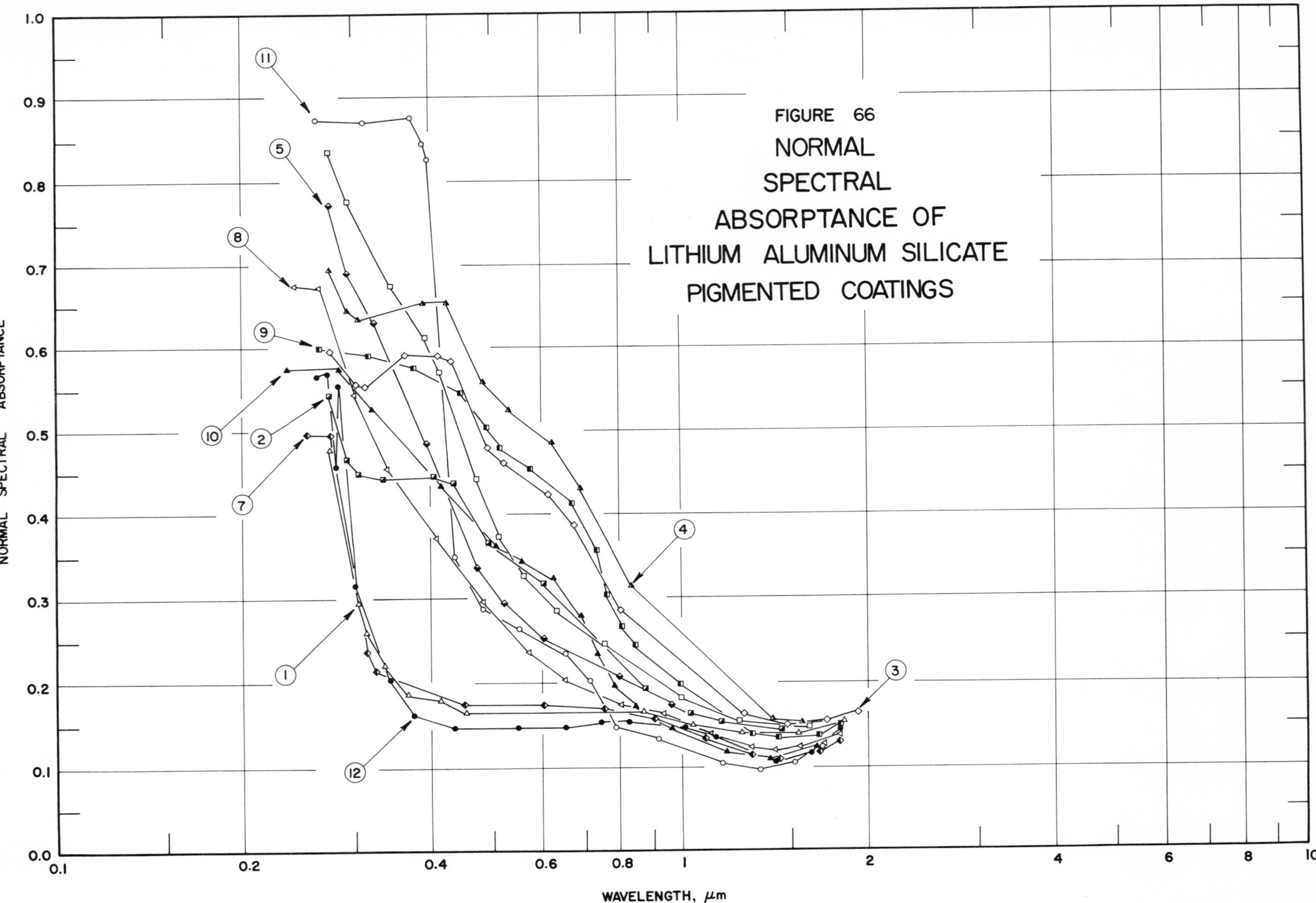

FIGURE 66 NORMAL SPECTRAL ABSORPTANCE OF LITHIUM ALUMINUM SILICATE PIGMENTED COATINGS

SPECIFICATION TABLE NO. 66 NORMAL SPECTRAL ABSORPTANCE OF LITHIUM ALUMINUM SILICATE PIGMENTED COATINGS

Curve No.	Ref. No.	Year	Temperature, K	Wavelength Range, μm	Geometry θ	Reported Error, %	Composition (weight percent), Specifications, and Remarks
1	50	1965	~298	0.278-1.882	~0°		LMSC Lithafrax ($Li_2O + Al_2O_3 + 8SiO_2$) in Na_2SiO_3 binder; pigment/binder ratio 4:1; impurities 0.1 Fe_2O_3, 0.4 K_2O, 0.2 Na_2O, 0.011 Fe; cured at 472 K; data extracted from smooth curve.
2	50	1965	~298	0.277-1.454	~0°		Similar to above specimen and conditions except irradiated in vacuum at 291 K with 1 MeV electrons to a total dose of 10^{14} e cm^{-2}.
3	50	1965	~298	0.278-1.939	~0°		Similar to above specimen and conditions except irradiated to a total dose of 10^{15} e cm^{-2}.
4	50	1965	~298	0.277-1.939	~0°		Similar to above specimen and conditions except irradiated to a total dose of 10^{16} e cm^{-2}.
5	50	1965	~298	0.278-1.454	~0°		Similar to above specimen and conditions except irradiated in vacuum at 291 K with ultraviolet radiation for 485 sun hrs.
6	50	1965	~298	0.276-1.600	~0°		Similar to above specimen and conditions except irradiated in vacuum at 291 K with 1 MeV electrons to a total dose of 10^{16} e cm^{-2} and additionally with ultraviolet radiation for 485 sun hrs.
7	24	1965	298	0.254-1.800	~0°		Lithafrax in sodium silicate binder; data extracted from smooth curve.
8	24	1965	298	0.244-1.800	~0°		Similar to above specimen and conditions except exposed to ~200 sun hrs of ultraviolet irradiation.
9	24	1965	298	0.268-1.800	~0°		Similar to curve 7 specimen and conditions except exposed to a nuclear-radiation dose of approx 10^8 R of gamma and 5 x 10^{14} neutrons cm^{-2} ($E \geq 2.9$ MeV).
10	24	1965	298	0.238-1.800	~0°		Similar to above specimen and conditions except exposed to nuclear-radiation dose of approx 10^8 R of gamma and 5 x 10^{14} neutrons cm^{-2} ($E \geq 2.9$ MeV).
11	24	1965	298	0.263-1.800	~0°		Similar to above specimen and conditions except exposed to a nuclear-radiation dose of approx 10^8 R of gamma and 5 x 10^{14} neutrons cm^{-2} ($E \geq 2.9$ MeV) at 77 K.
12	24	1965	298	0.263-1.800	~0°		Lithafrax in sodium silicate binder; data extracted from smooth curve.

DATA TABLE NO. 66 NORMAL SPECTRAL ABSORPTANCE OF LITHIUM ALUMINUM SILICATE PIGMENTED COATINGS

[Wavelength, λ, μm; Absorptance, α; Temperature, T, K]

CURVE 1
T ~ 298

λ	α
0.278	0.479
0.306	0.295
0.314	0.260
0.337	0.221
0.369	0.189
0.414	0.170
0.453	0.166
0.870	0.166
1.049	0.150
1.254	0.140
1.542	0.140
1.882	0.151

CURVE 2
T ~ 298

λ	α
0.277	0.542
0.295	0.466
0.309	0.450
0.337	0.442
0.405	0.446
0.432	0.438
0.496	0.366
0.606	0.318
0.876	0.193
1.032	0.164
1.163	0.151
1.454	0.144

CURVE 3
T ~ 298

λ	α
0.278	0.598
0.304	0.559
0.317	0.554
0.365	0.591
0.411	0.591
0.433	0.584
0.498	0.480
0.524	0.461
0.618	0.424
0.680	0.388
0.805	0.288
1.267	0.163
1.471	0.150

CURVE 3 (cont.)

λ	α
1.729	0.155
1.939	0.163

CURVE 4
T ~ 298

λ	α
0.277	0.697
0.296	0.645
0.309	0.633
0.390	0.652
0.424	0.652
0.489	0.560
0.531	0.525
0.627	0.489
0.699	0.431
0.836	0.316
1.406	0.156
1.561	0.151
1.729	0.155*
1.939	0.163*

CURVE 5
T ~ 298

λ	α
0.278	0.771
0.297	0.692
0.327	0.631
0.395	0.487
0.434	0.438*
0.476	0.338
0.527	0.294
0.601	0.251
0.800	0.209
0.962	0.175
1.032	0.164*
1.163	0.151*
1.454	0.144*

CURVE 6
T ~ 298

λ	α
0.276	0.837
0.298	0.777
0.348	0.676
0.394	0.611
0.418	0.571

CURVE 6 (cont.)

λ	α
0.474	0.442
0.514	0.374
0.561	0.326
0.631	0.286
0.753	0.248
1.000	0.181
1.243	0.153
1.454	0.144*
1.600	0.149

CURVE 7
T = 298

λ	α
0.254	0.497
0.277	0.496
0.318	0.239
0.329	0.215
0.365	0.189*
0.454	0.175
0.602	0.175
0.756	0.170
0.901	0.159
1.092	0.134
1.305	0.115
1.447	0.110
1.669	0.118
1.800	0.130

CURVE 8
T = 298

λ	α
0.244	0.675
0.268	0.673
0.301	0.544
0.341	0.455
0.410	0.372
0.487	0.298
0.571	0.238
0.652	0.203
0.806	0.175
0.934	0.163
1.117	0.140
1.304	0.123
1.413	0.120
1.559	0.123
1.701	0.128
1.800	0.138

CURVE 9
T = 298

λ	α
0.268	0.600
0.320	0.593
0.378	0.579
0.449	0.547
0.492	0.503
0.517	0.480
0.578	0.455
0.672	0.412
0.736	0.356
0.768	0.301
0.803	0.266
0.845	0.244
1.000	0.199
1.160	0.156*
1.305	0.139
1.431	0.135
1.665	0.137
1.800	0.150

CURVE 10
T = 298

λ	α
0.238	0.576
0.285	0.576
0.323	0.526
0.415	0.436
0.509	0.363
0.560	0.344
0.627	0.325
0.698	0.280
0.738	0.233
0.783	0.197
0.845	0.173
0.963	0.147
1.182	0.118
1.395	0.110
1.478	0.111*
1.652	0.124
1.800	0.151*

CURVE 11
T = 298

λ	α
0.263	0.876
0.315	0.872

CURVE 11 (cont.)

λ	α
0.372	0.877
0.390	0.848
0.398	0.829
0.435	0.350
0.484	0.288
0.552	0.264
0.656	0.234
0.718	0.201
0.790	0.148
0.914	0.133
1.167	0.102
1.343	0.098
1.510	0.103
1.697	0.126
1.800	0.149

CURVE 12
T = 298

λ	α
0.263	0.568
0.273	0.570
0.285	0.556
0.281	0.458
0.301	0.317
0.342	0.202
0.375	0.161
0.433	0.147
0.550	0.149
0.656	0.148
0.743	0.152
0.829	0.152
1.010	0.147
1.141	0.136
1.317	0.114*
1.438	0.106
1.624	0.116
1.800	0.142

* Not shown on plot

FIGURE 67A

ANALYZED NORMAL SOLAR ABSORPTANCE OF LITHIUM ALUMINUM SILICATE PIGMENTED COATINGS

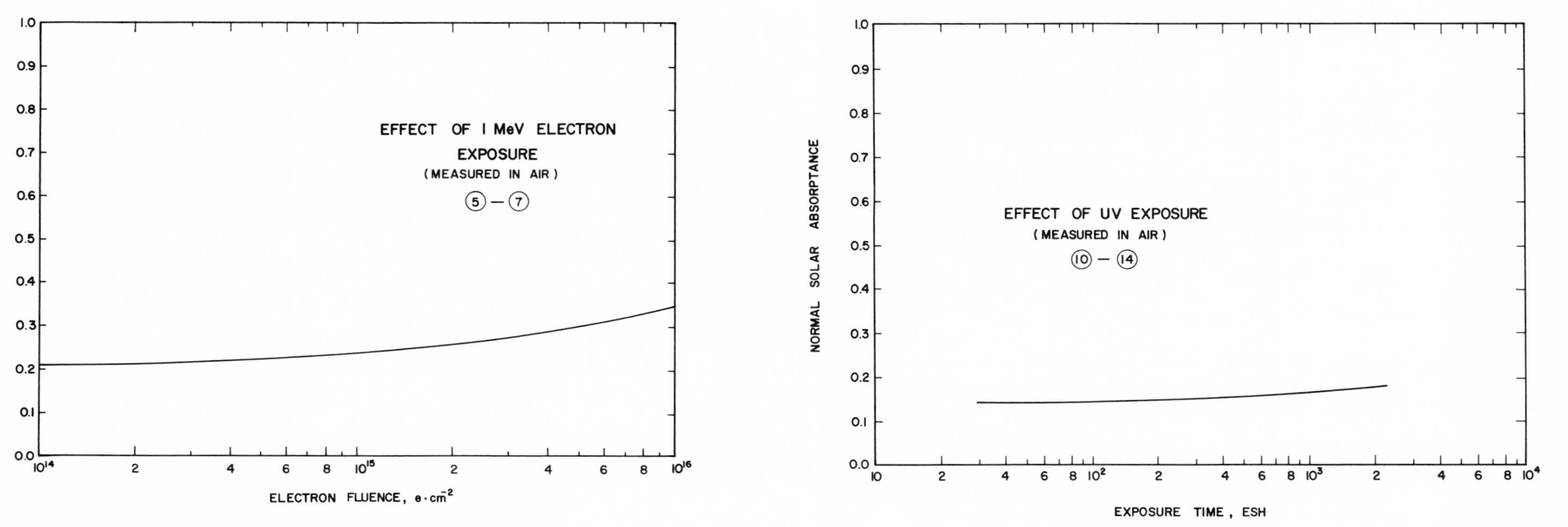

SPECIFICATION TABLE NO. 67 NORMAL SOLAR ABSORPTANCE OF LITHIUM ALUMINUM SILICATE PIGMENTED COATINGS

Curve No.	Ref. No.	Year	Temperature Range, K	Geometry θ	Reported Error, %	Composition (weight percent), Specifications and Remarks
1	52	1965	193-263	$\sim 0°$		Synthetic spodumene in sodium silicate; measured in earth orbit on OSO-II; data is an avg value over the indicated temp range.
2	50	1965	~ 298	$\sim 0°$		LMSC Lithafrax (composition: $Li_2O + Al_2O_3 + 8\ SiO_2$) in Na_2SiO_3 binder; pigment to binder ratio 4:1; impurities; 0.1 Fe_2o_3, 0.4 K_2O, 0.2 Na_2O and cured at 472 K; computed from spectral reflectance measured in air.
3	50	1965	~ 298	$\sim 0°$		Similar to above specimen and conditions except irradiated in vacuum at 291 K with 1 Mev electrons to a total dose of 10^{16} e cm^{-2}, and additionally with ultraviolet radiation for 485 sun hrs.
4	50	1965	~ 298	$\sim 0°$		Similar to curve 2 specimen and conditions except irradiated in vacuum at 291 K with ultraviolet radiation for 485 sun hrs.
5	50	1965	~ 298	$\sim 0°$		Similar to above specimen and conditions except irradiated with 1 Mev electrons at 155 K to a total dose of 10^{14} e cm^{-2}.
6	50	1965	~ 298	$\sim 0°$		Similar to curve 2 specimen and conditions except irradiated at 422 K to a total dose of 10^{15} e cm^{-2}.
7	50	1965	~ 298	$\sim 0°$		Similar to above specimen and conditions except irradiated to a total dose of 10^{16} e cm^{-2}.
8	50	1965	~ 298	$\sim 0°$		Similar to above specimen and conditions except irradiated to a total dose of 10^{15} e cm^{-2}.
9	50	1965	~ 298	$\sim 0°$		Similar to above specimen and conditions except irradiated in vacuum at 291 K with 1 Mev electrons to a total dose of 10^{14} e cm^{-2}.
10	78	1961	298	$\sim 0°$		$LiAlSiO_4$ in Na_2SiO_3 binder (0.127 mm thick); aluminum substrate; pigment/volume ratio 4:1; sprayed; baked at 473 K for 2 hrs; exposed to UV radiation for 30 sun hrs in vacuum (10^{-7} mm Hg); data extracted from smooth curve; calculated from reflectance measured in air.
11	78	1961	298	$\sim 0°$		Above specimen and conditions except with 500 sun hrs total irradiation.
12	78	1961	298	$\sim 0°$		Above specimen and conditions except with 1000 sun hrs total irradiation.
13	78	1961	298	$\sim 0°$		Above specimen and conditions except with 1500 sun hrs total irradiation.
14	78	1961	298	$\sim 0°$		Above specimen and conditions except with 2200 sun hrs total irradiation.

DATA TABLE NO. 67 NORMAL SOLAR ABSORPTANCE OF LITHIUM ALUMINUM SILICATE PIGMENTED COATINGS

[Temperature, T, K; Absorptance, α]

T	α	T	α
CURVE 1		CURVE 12	
193	0.179	298	0.167
263	0.179		
CURVE 2		CURVE 13	
298	0.15	298	0.170
CURVE 3		CURVE 14	
298	0.25	298	0.17
CURVE 4			
298	0.21		
CURVE 5			
298	0.21		
CURVE 6			
298	0.24		
CURVE 7			
298	0.35		
298	0.34		
CURVE 8			
298	0.30		
CURVE 9			
298	0.22		
CURVE 10			
298	0.142		
CURVE 11			
298	0.158		

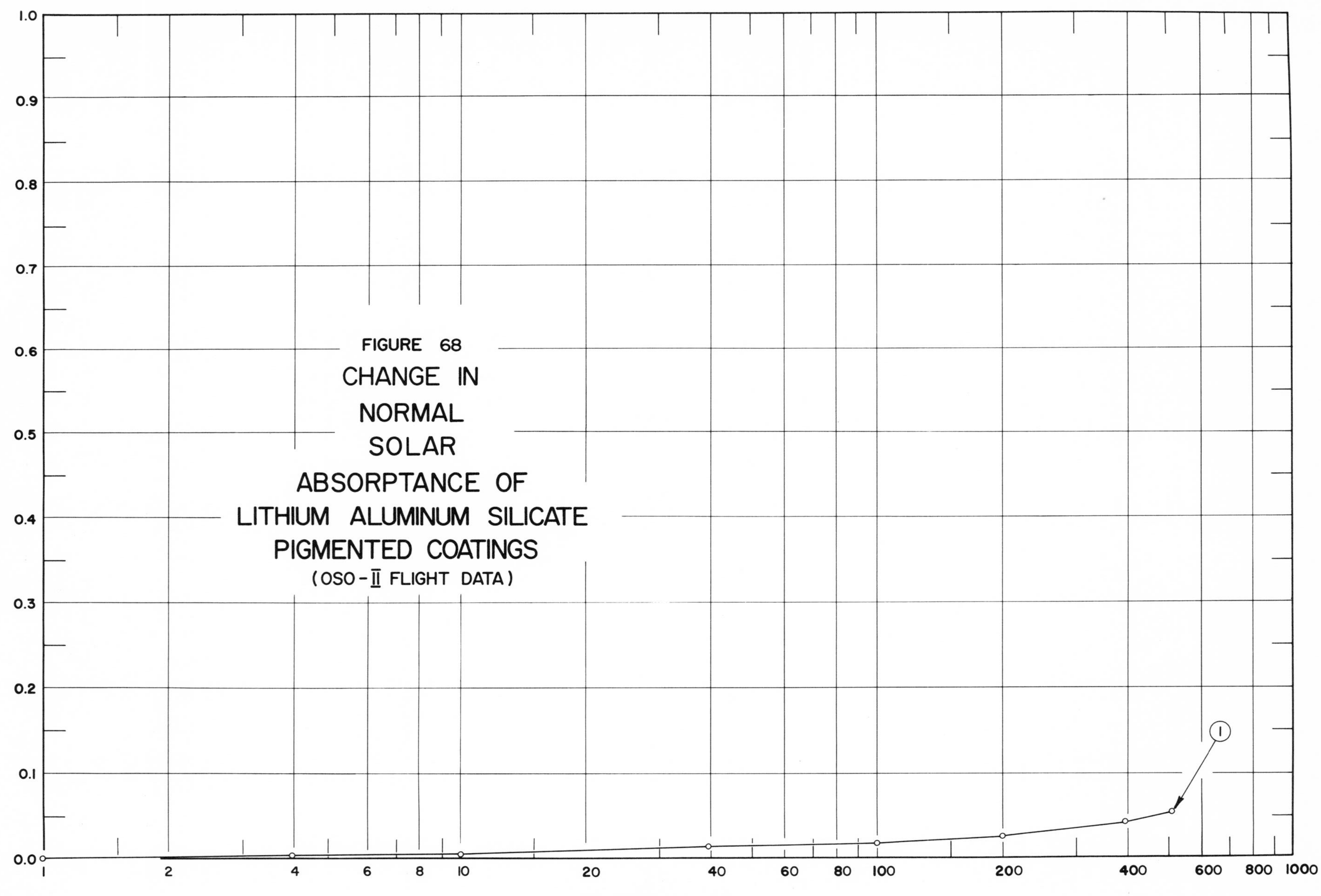

FIGURE 68
CHANGE IN NORMAL SOLAR ABSORPTANCE OF LITHIUM ALUMINUM SILICATE PIGMENTED COATINGS
(OSO-II FLIGHT DATA)

SPECIFICATION TABLE NO. 68 CHANGE IN NORMAL SOLAR ABSORPTANCE OF LITHIUM ALUMINUM SILICATE PIGMENTED COATINGS

Curve No.	Ref. No.	Year	Temperature Range, K	Geometry θ	Reported Error, %	Composition (weight percent), Specifications and Remarks
1	52	1966	~298	~0°		Synthetic spodumene in sodium silicate binder; data taken from OSO-II flight experiment; variable is equivalent length of time exposed to the sun.

DATA TABLE NO. 68 CHANGE IN NORMAL SOLAR ABSORPTANCE OF LITHIUM ALUMINUM SILICATE PIGMENTED COATINGS

[Temperature, T, K; Absorptance, α]

ESH	$\Delta\alpha$
CURVE 1	
T ~ 298	
1.00	0.000
3.98	0.002
10.0	0.005
39.8	0.012
100	0.019
200	0.026
398	0.043
513	0.055

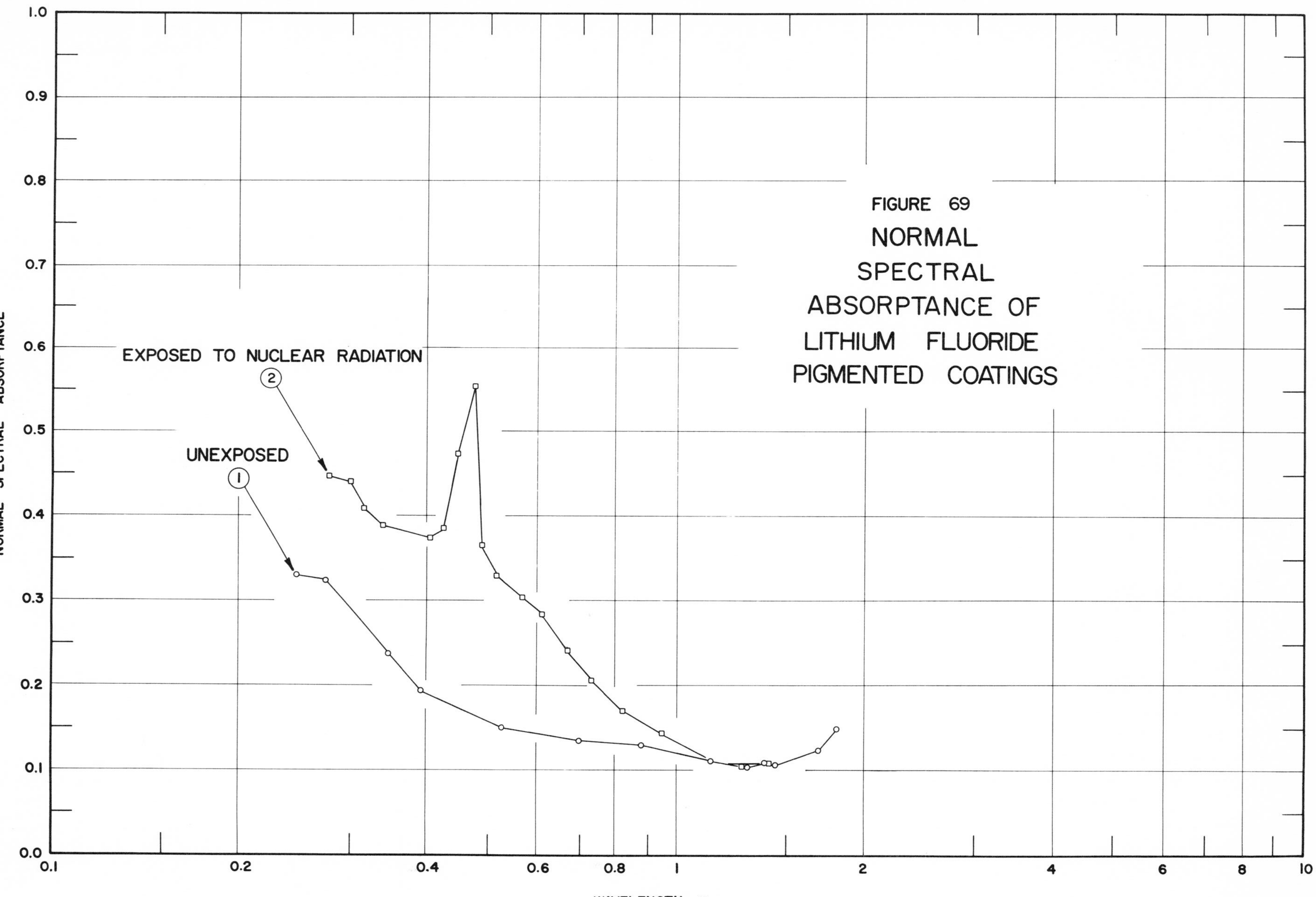
FIGURE 69
NORMAL
SPECTRAL
ABSORPTANCE OF
LITHIUM FLUORIDE
PIGMENTED COATINGS
EXPOSED TO NUCLEAR RADIATION
2
UNEXPOSED
1
NORMAL SPECTRAL ABSORPTANCE
1.0
0.9
0.8
0.7
0.6
0.5
0.4
0.3
0.2
0.1
0.0
0.1
0.2
0.4
0.6
0.8
1
2
4
6
8
10
WAVELENGTH, μm

SPECIFICATION TABLE NO. 69 NORMAL SPECTRAL ABSORPTANCE OF LITHIUM FLUORIDE PIGMENTED COATINGS

Curve No.	Ref. No.	Year	Temperature K	Wavelength Range, μm	Geometry θ	Reported Error, %	Composition (weight percent), Specifications, and Remarks
1	24	1965	298	0.246-1.800	~0°		LiF in sodium silicate binder; data extracted from smooth curve.
2	24	1965	298	0.278-1.400	~0°		Similar to above specimen and conditions except exposed to approx 10^8 R of gamma and 5×10^{14} neutrons ($E \geq 2.9$ MeV) per cm^2.

DATA TABLE NO. 69 NORMAL SPECTRAL ABSORPTANCE OF LITHIUM FLUORIDE PIGMENTED COATINGS

[Wavelength, λ, μm; Absorptance, α; Temperature, T, K]

λ	α	λ	α
CURVE 1		CURVE 2	
T = 298		T = 298	
0.246	0.332	0.278	0.450
0.274	0.326	0.300	0.443
0.348	0.240	0.315	0.413
0.389	0.195	0.339	0.391
0.525	0.152	0.403	0.376
0.697	0.137	0.423	0.389
0.875	0.131	0.442	0.477
1.128	0.114	0.475	0.558
1.296	0.105	0.487	0.369
1.380	0.111	0.514	0.332
1.446	0.109	0.562	0.308
1.664	0.126	0.607	0.286
1.800	0.152	0.667	0.244
		0.729	0.208
		0.819	0.172
		0.940	0.146
		1.132	0.120*
		1.281	0.108
		1.400	0.111

*Not shown on plot

SPECIFICATION TABLE NO. 70 NORMAL INTEGRATED REFLECTANCE OF MAGNESIUM CARBONATE PIGMENTED COATINGS

Curve No.	Ref. No.	Year	Temperature Range, K	Geometry θ	θ'	ω'	Reported Error, %	Composition (weight percent), Specifications and Remarks
1*	13	1940	298	0°		2π		Magnesium carbonate in 60 ethylcellulose plus 40 Dow 7 binder; pigment volume 40%; milled; measured over the wavelength band 0.28 to 0.32 μm. [Authors' designation: No. G-10]
2*	13	1940	298	0°		2π		Above specimen and conditions except exposed to ultraviolet radiation (one ft from a GE S-1 sunlamp) for 7 days.
3*	13	1940	298	0°		2π		Above specimen and conditions except exposed to ultraviolet radiation (one ft from a GE S-1 sunlamp) for 15 days.
4*	13	1940	298	0°		2π		Above specimen and conditions except exposed to ultraviolet radiation (one ft from a GE S-1 sunlamp) for one mo.
5*	13	1940	298	0°		2π		Above specimen and conditions except exposed to ultraviolet radiation (one ft from a GE S-1 sunlamp) for two mo.
6*	13	1940	298	0°		2π		Above specimen and conditions except exposed to ultraviolet radiation (one ft from a GE S-1 sunlamp) for four mo.
7*	13	1940	298	0°		2π		Above specimen and conditions except exposed to ultraviolet radiation (one ft from a GE S-1 sunlamp) for eight mo.
8*	13	1940	298	0°		2π		Similar to curve 1 specimen and conditions.
9*	13	1940	298	0°		2π		Above specimen and conditions except aged in air for 85 days.

* No plot given

DATA TABLE NO. 70 NORMAL INTEGRATED REFLECTANCE OF MAGNESIUM CARBONATE PIGMENTED COATINGS

[Temperature, T, K; Reflectance, ρ]

T	ρ
CURVE 1*	
298	0.316
CURVE 2*	
298	0.320
CURVE 3*	
298	0.270
CURVE 4*	
298	0.180
CURVE 5*	
298	0.230
CURVE 6*	
298	0.440
CURVE 7*	
298	0.590
CURVE 8*	
298	0.316
CURVE 9*	
298	0.300

* No plot given

SPECIFICATION TABLE NO. 71 NORMAL INTEGRATED REFLECTANCE OF MAGNESIUM OXIDE PIGMENTED COATINGS

Curve No.	Ref. No.	Year	Temperature Range, K	Geometry θ	θ'	ω'	Reported Error, %	Composition (weight percent), Specifications and Remarks
1*	13	1940	298	0°		2π		Magnesium Oxide in 60 ethylcellulose plus 40 Dow 7 binder; pigment volume 40%; milled; measured over the wavelength band 0.28 to 0.32 μm. [Authors' designation: No. G-9]
2*	13	1940	298	0°		2π		Above specimen and conditions except exposed to the ultraviolet radiation (1 ft from a GE S-1 sunlamp) for seven days.
3*	13	1940	298	0°		2π		Above specimen and conditions except exposed to the ultraviolet radiation (1 ft from a GE S-1 sunlamp) for 15 days.
4*	13	1940	298	0°		2π		Above specimen and conditions except exposed to the ultraviolet radiation (1 ft from a GE S-1 sunlamp) for one mo.
5*	13	1940	298	0°		2π		Above specimen and conditions except exposed to the ultraviolet radiation (1 ft from a GE S-1 sunlamp) for two mos.
6*	13	1940	298	0°		2π		Above specimen and conditions except exposed to the ultraviolet radiation (1 ft from a GE S-1 sunlamp) for four mos.
7*	13	1940	298	0°		2π		Above specimen and conditions except exposed to the ultraviolet radiation (1 ft from a GE S-1 sunlamp) for eight mos.
8*	13	1940	298	0°		2π		Similar to curve 1 specimen and conditions.
9*	13	1940	298	0°		2π		Above specimen and conditions except aged in air for 85 days.

* No plot given

DATA TABLE NO. 71 NORMAL INTEGRATED REFLECTANCE OF MAGNESIUM OXIDE PIGMENTED COATINGS

[Temperature, T, K; Reflectance, ρ]

T	ρ
CURVE 1*	
298	0.473
CURVE 2*	
298	0.440
CURVE 3*	
298	0.360
CURVE 4*	
298	0.245
CURVE 5*	
298	0.130
CURVE 6*	
298	0.160
CURVE 7*	
298	0.350
CURVE 8*	
298	0.473
CURVE 9*	
298	0.425

* No plot given

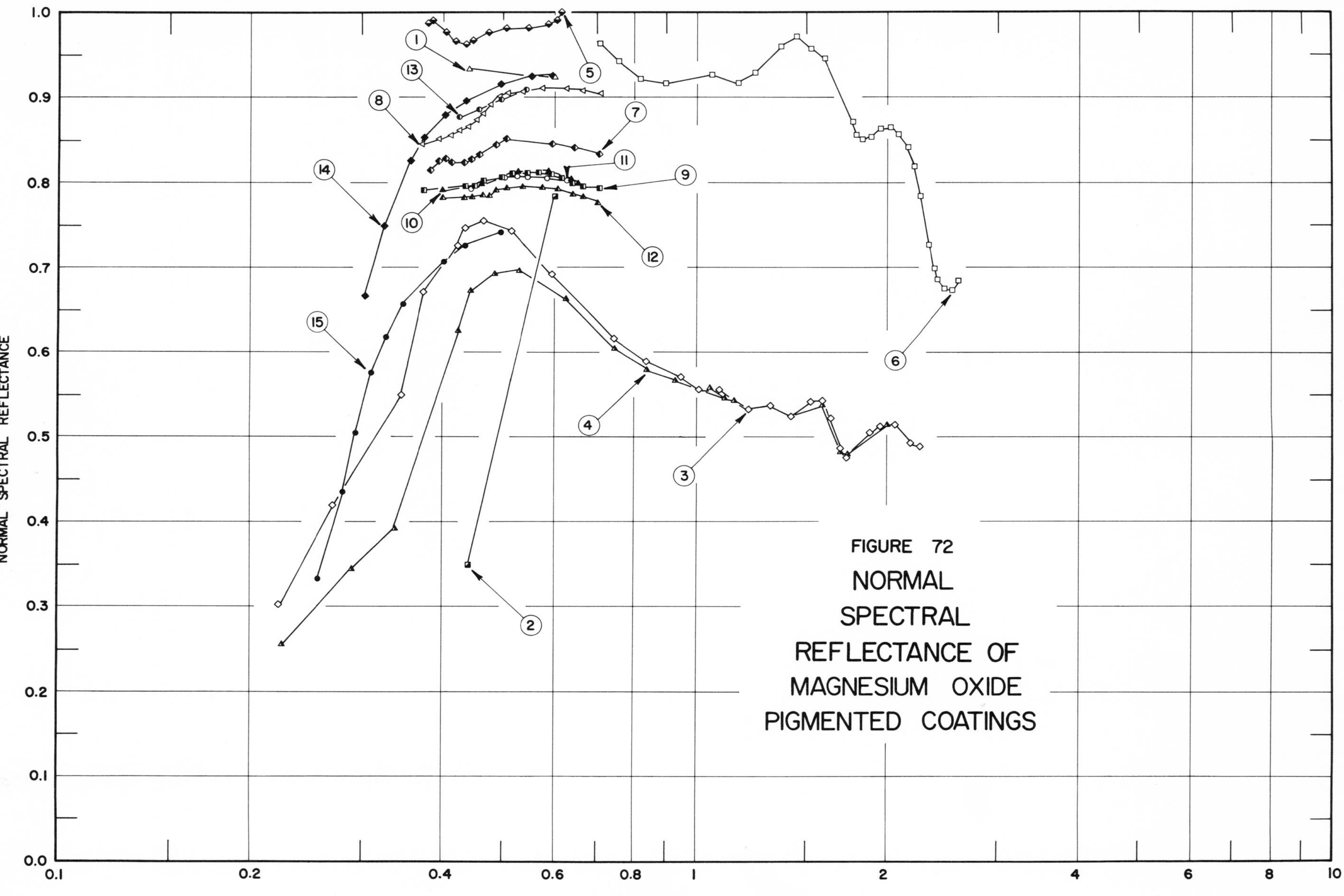

FIGURE 72 NORMAL SPECTRAL REFLECTANCE OF MAGNESIUM OXIDE PIGMENTED COATINGS

SPECIFICATION TABLE NO. 72 NORMAL SPECTRAL REFLECTANCE OF MAGNESIUM OXIDE PIGMENTED COATINGS

Curve No.	Ref. No.	Year	Temperature, K	Wavelength Range, μm	Geometry θ	θ'	ω'	Reported Error, %	Composition (weight percent), Specifications, and Remarks
1	27	1966	~298	0.44-0.60	~0°		2π		MgO in Leonite 201-S silicone-epoxy modified acrylic resin binder; aluminum substrate; PBR 0.44; paint ground in porcelain jar mill for 16 hrs. [Authors' designation: Sample P20]
2	27	1964	~298	0.44-0.60	~0°		2π		Above specimen and conditions except irradiated in vacuum (10^{-6} mm Hg) with 108 ESH at solar factor of 4 suns.
3	79	1967	~298	0.221-2.245	12.5°		2π		MgO in acrylic binder (supplied by Elastomers and Coatings Branch, Air Force Materials Lab, Wright-Patterson AFB); painted on stainless steel disk; measured in vacuum ($2-5 \times 10^{-7}$ mm Hg); data extracted from smooth curve; converted from reflectance factor. [Authors' designation: Sample No. 9-4]
4	79	1967	~298	0.225-2.245	12.5°		2π		Above specimen and conditions except irradiated in a vacuum ($\sim 4 \times 10^{-6}$ mm Hg) with protons of nominal energy 3 keV; total dose, 10^{16} p cm^{-2}; measured in situ.
5	3	1960	~298	0.380-0.613	~0°		2π		MgO (reagent grade; particle size 0.1 μm) in duPont RC 7007 alkyd-melamine binder; pigment concentration 85%; sprayed onto mild steel substrate; data extracted from smooth curve; measured relative to MgO.
6	3	1960	~298	0.703-2.596	~0°		2π		Similar to above specimen and conditions.
7	3	1960	~298	0.383-0.701	~0°		2π		MgO (reagent grade) in Dow Corning 806A silicone binder (0.079 mm thick); pigment concentration, 60% of solids content; voids concentration, 13%; sprayed onto mild steel substrate; data extracted from smooth curve; measured relative to MgO.
8	3	1960	~298	0.372-0.710	~0°		2π		MgO (reagent grade) in Dow Corning 806A silicone binder (0.213 mm thick); pigment concentration, 60% of solids content; voids concentration, 37.5%;sprayed onto mild steel substrate; data extracted from smooth curve; measured relative to MgO.
9	3	1960	~298	0.374-0.707	~0°		2π		Similar to curve 8 specimen and conditions except 0.061 mm thick and voids concentration 26.3%.
10	3	1960	~298	0.400-0.700	~0°		2π		MgO (reagent grade) in Dow Corning 806A silicone binder (0.061 mm thick and weight 0.854 g cm^{-2}); pigment concentration, 60% of solids content; voids concentration, 26.3%; sprayed onto mild steel substrate; data extracted from smooth curve; measured relative to MgO.
11	3	1960	~298	0.400-0.701	~0°		2π		Similar to above specimen and conditions except 0.071 mm thick, weight 0.937 g cm^{-2} and voids concentration 30.7%.
12	3	1960	~298	0.400-0.700	~0°		2π		Similar to above specimen and conditions except 0.069 mm thick, weight 0.918 g cm^{-2} and voids concentration 30.0%.
13	80	1966	~298	0.4260-0.5400	0°		2π		MgO in B-72 acryloid binder; 30.3 MgO and 69.7 B-72; measured relative to MgO. [Author's designation: Reflector No. VI]
14	80	1966	~298	0.3024-0.5972	0°		2π		Above specimen and conditions except converted from R (0°, 2π).
15	80	1966	~298	0.2565-0.4956	45°	0°			Curve 13 specimen and conditions except geometry of data is for reflectance factor; data extracted from smooth curve.

DATA TABLE NO. 72 NORMAL SPECTRAL REFLECTANCE OF MAGNESIUM OXIDE PIGMENTED COATINGS

[Wavelength, λ, μm; Reflectance, ρ; Temperature, T, K]

CURVE 1, T ~ 298

λ	ρ
0.44	0.935
0.60	0.925

CURVE 2, T ~ 298

λ	ρ
0.44	0.350
0.60	0.785

CURVE 3, T ~ 298

λ	ρ
0.221	0.301
0.270	0.420
0.346	0.550
0.373	0.671
0.421	0.726
0.433	0.749
0.462	0.755
0.511	0.743
0.592	0.693
0.749	0.619
0.835	0.590
0.947	0.572
1.012	0.559
1.089	0.559
1.212	0.532
1.312	0.538
1.413	0.525
1.510	0.541
1.577	0.541
1.638	0.522
1.693	0.489
1.738	0.479
1.871	0.506
1.956	0.514
2.060	0.515
2.191	0.493
2.245	0.480

CURVE 4, T ~ 298

λ	ρ
0.225	0.256
0.290	0.346
0.339	0.391
0.422	0.627
0.444	0.671
0.481	0.693
0.528	0.698
0.624	0.661
0.742	0.605
0.837	0.580
0.923	0.569
1.056	0.560
1.115	0.547
1.152	0.543
1.212	0.533*
1.312	0.538*
1.413	0.525*
1.499	0.538
1.570	0.538
1.692	0.485
1.748	0.480
2.002	0.515
2.066	0.515*
2.191	0.493*
2.245	0.480*

CURVE 5, T ~ 298

λ	ρ
0.380	0.986
0.389	0.991
0.406	0.978
0.420	0.969
0.436	0.964
0.449	0.967
0.473	0.977
0.502	0.981
0.546	0.981
0.588	0.987
0.602	0.992
0.613	1.000

CURVE 6, T ~ 298

λ	ρ
0.703	0.962
0.753	0.942
0.817	0.921
0.896	0.919
1.067	0.927
1.174	0.918
1.249	0.930
1.366	0.960
1.441	0.971
1.523	0.969
1.593	0.949
1.761	0.871
1.797	0.858
1.830	0.852
1.883	0.855
1.954	0.864
2.027	0.864
2.097	0.857
2.156	0.841
2.204	0.820
2.257	0.785
2.325	0.728
2.366	0.703
2.408	0.687
2.462	0.676
2.525	0.675
2.558	0.678*
2.596	0.684

CURVE 7, T ~ 298

λ	ρ
0.383	0.817
0.395	0.826
0.405	0.829
0.414	0.825
0.431	0.824
0.443	0.827
0.459	0.833
0.487	0.847
0.503	0.852
0.544	0.852
0.593	0.849
0.647	0.842
0.701	0.833

CURVE 8, T ~ 298

λ	ρ
0.372	0.847
0.395	0.853
0.412	0.858
0.426	0.862
0.440	0.868
0.451	0.874
0.462	0.881
0.478	0.894
0.491	0.901
0.506	0.906
0.531	0.909
0.573	0.911
0.627	0.911
0.666	0.910
0.710	0.906

CURVE 9, T ~ 298

λ	ρ
0.374	0.791
0.434	0.797
0.450	0.798
0.464	0.801
0.497	0.808
0.516	0.811
0.541	0.812
0.567	0.812
0.592	0.810
0.614	0.806
0.640	0.800
0.663	0.798
0.707	0.794

CURVE 10, T ~ 298

λ	ρ
0.400	0.794
0.460	0.800
0.510	0.812*
0.526	0.813
0.546	0.814*
0.587	0.812
0.613	0.809*
0.637	0.803

CURVE 10 (cont.)

λ	ρ
0.655	0.800
0.700	0.797

CURVE 11, T ~ 298

λ	ρ
0.400	0.791*
0.444	0.794
0.462	0.798
0.502	0.806
0.521	0.809
0.542	0.809
0.585	0.807
0.627	0.802
0.664	0.798*
0.701	0.793*

CURVE 12, T ~ 298

λ	ρ
0.400	0.781
0.432	0.781
0.449	0.783
0.461	0.785
0.472	0.788
0.487	0.792
0.503	0.795
0.532	0.796
0.571	0.795
0.608	0.792
0.640	0.789
0.667	0.784
0.700	0.778

CURVE 13, T ~ 298

λ	ρ
0.4260	0.879
0.4570	0.888
0.4950	0.900
0.5400	0.91

CURVE 14, T ~ 298

λ	ρ
0.3024	0.669
0.3270	0.750
0.3583	0.829
0.3753	0.853
0.4045	0.880
0.4391	0.898
0.4993	0.918
0.5514	0.923
0.5972	0.923

CURVE 15, T ~ 298

λ	ρ
0.2565	0.332
0.2806	0.434
0.2934	0.503
0.3106	0.576
0.3282	0.620
0.3498	0.659
0.4011	0.709
0.4342	0.729
0.4956	0.741

* Not shown on plot

SPECIFICATION TABLE NO. 73 NORMAL SPECTRAL ABSORPTANCE OF MAGNESIUM OXIDE PIGMENTED COATINGS

Curve No.	Ref. No.	Year	Temperature, K	Wavelength Range, μm	Geometry θ	Reported Error, %	Composition (weight percent), Specifications, and Remarks
1*	76	1966	~298	0.300-0.700	~0°		MgO (analytical reagent grade) in GE SR-112 binder; aluminum substrate; spray application (2 coats).

DATA TABLE NO. 73 NORMAL SPECTRAL ABSORPTANCE OF MAGNESIUM OXIDE PIGMENTED COATINGS

[Wavelength, λ, μm; Absorptance, α; Temperature, T, K]

λ	α
CURVE 1*	
T ~298	
0.300	0.139
0.350	0.113
0.400	0.091
0.450	0.080
0.500	0.062
0.550	0.054
0.600	0.059
0.650	0.064
0.700	0.071

* No plot given

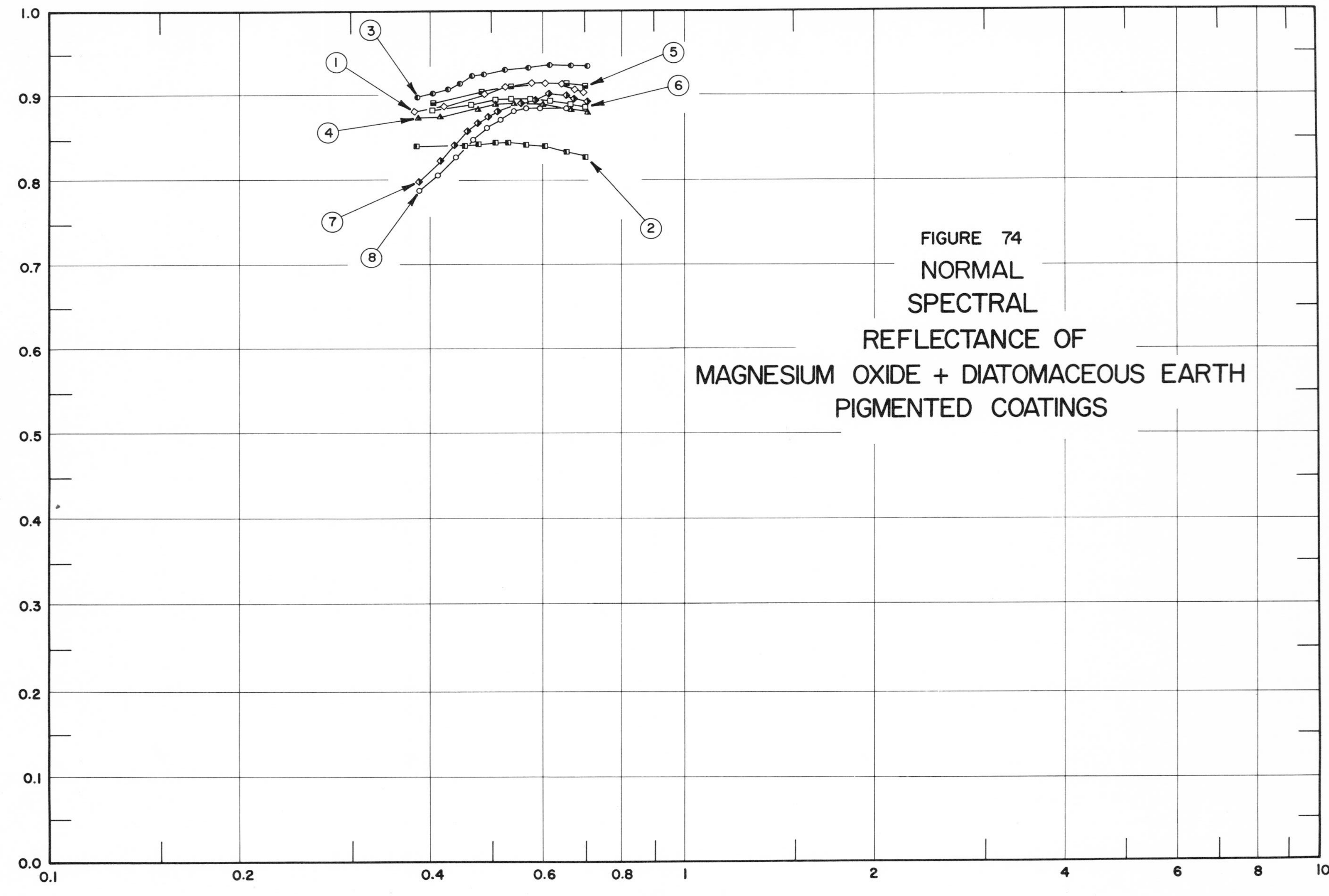

FIGURE 74 NORMAL SPECTRAL REFLECTANCE OF MAGNESIUM OXIDE + DIATOMACEOUS EARTH PIGMENTED COATINGS

SPECIFICATION TABLE NO. 74 NORMAL SPECTRAL REFLECTANCE OF MAGNESIUM OXIDE + DIATOMACEOUS EARTH PIGMENTED COATINGS

Curve No.	Ref. No.	Year	Temperature, K	Wavelength Range, μm	Geometry θ	Geometry θ'	Geometry ω'	Reported Error, %	Composition (weight percent), Specifications, and Remarks
1	3	1960	~298	0.379-0.698	~0°		2π		90 MgO (reagent grade) and 10 diatomaceous earth (Johns Manville Microcell "C") in Dow Corning 806A silicone binder (0.130 mm thick); pigment concentration, 60% of solids content; voids concentration, 46.5%; sprayed onto mild steel substrate; data extracted from smooth curve; measured relative to MgO.
2	3	1960	~298	0.380-0.700	~0°		2π		Similar to above specimen and conditions except 0.051 mm thick; voids concentration 24.0%.
3	3	1960	~298	0.381-0.702	~0°		2π		Similar to curve 1 specimen and conditions except 80 MgO and 20 diatomaceous earth; thickness 0.127 mm; voids concentration 39.8%.
4	3	1960	~298	0.381-0.701	~0°		2π		Similar to curve 3 specimen and conditions except 0.053 mm thick; voids concentration 34.3%.
5	3	1960	~298	0.401-0.700	~0°		2π		80 MgO (reagent grade) and 20 diatomaceous earth (Johns Manville MicroCell "C") in Dow Corning 806A silicone binder (0.079 mm thick); coating weight 0.781×10^{-2} to 0.942×10^{-2} g cm^{-2}; pigment concentration, 60% of solids content; voids concentration 30.7%; sprayed onto mild steel substrate; data extracted from smooth curve; measured relative to MgO.
6	3	1960	~298	0.401-0.700	~0°		2π		Similar to above specimen and conditions except 0.064 mm thick; voids concentration 26.0%.
7	3	1960	~298	0.383-0.701	~0°		2π		80 MgO (reagent grade) and 20 diatomaceous earth (Microcell "C") in Dow Corning 806A silicone binder (0.122 mm thick); pigment concentration, 60% of solids content; voids concentration 46%; sprayed onto mild steel substrate; data extracted from smooth curve; measured relative to MgO.
8	3	1960	~298	0.383-0.701	~0°		2π		Similar to above specimen and conditions except 70 MgO and 30 diatomaceous earth (0.203 mm thick); voids concentration 69%.

DATA TABLE NO. 74 NORMAL SPECTRAL REFLECTANCE OF MAGNESIUM OXIDE + DIATOMACEOUS EARTH PIGMENTED COATINGS

[Wavelength, λ, μm; Reflectance, ρ; Temperature, T, K]

λ	ρ
CURVE 1	
T ~ 298	
0.379	0.884
0.399	0.886*
0.420	0.890
0.484	0.904
0.523	0.911
0.573	0.916
0.609	0.917
0.641	0.915
0.676	0.910
0.698	0.906
CURVE 2	
T ~ 298	
0.380	0.841
0.451	0.844
0.479	0.847
0.503	0.848
0.530	0.848
0.565	0.846
0.605	0.842
0.652	0.837
0.700	0.830
CURVE 3	
T ~ 298	
0.381	0.900
0.404	0.904
0.428	0.910
0.443	0.916
0.467	0.923
0.487	0.927
0.523	0.931
0.570	0.934
0.614	0.937
0.665	0.937
0.702	0.937
CURVE 4	
T ~ 298	
0.381	0.876
0.415	0.879
0.475	0.888
0.506	0.891
0.548	0.892
0.600	0.891
0.662	0.886
0.701	0.881
CURVE 5	
T ~ 298	
0.401	0.894
0.482	0.906
0.538	0.912
0.575	0.915*
0.605	0.916*
0.651	0.916
0.700	0.912
CURVE 6	
T ~ 298	
0.401	0.886
0.463	0.891
0.507	0.897
0.536	0.898
0.573	0.898
0.617	0.896
0.666	0.892
0.700	0.889
CURVE 7	
T ~ 298	
0.383	0.800
0.413	0.826
0.438	0.845
0.459	0.860
0.475	0.870
0.491	0.878
0.510	0.884
0.558	0.893
0.589	0.899
0.613	0.902
0.653	0.901
0.676	0.899
0.701	0.894
CURVE 8	
T ~ 298	
0.383	0.790
0.410	0.809
0.439	0.830
0.467	0.850
0.490	0.864
0.513	0.873
0.540	0.882
0.564	0.887
0.593	0.889
0.657	0.889
0.701	0.886*

* Not shown on plot

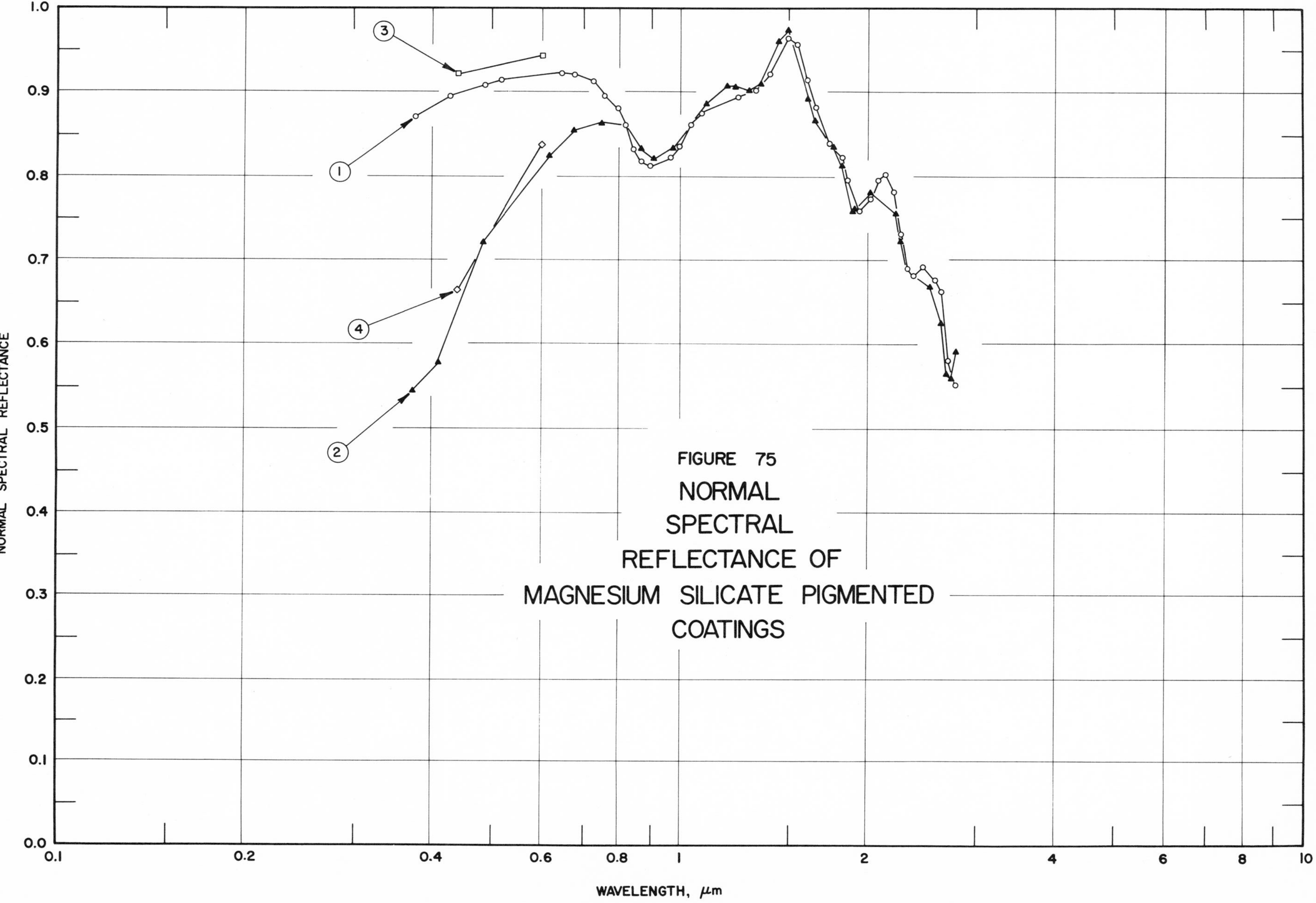

FIGURE 75 NORMAL SPECTRAL REFLECTANCE OF MAGNESIUM SILICATE PIGMENTED COATINGS

SPECIFICATION TABLE NO. 75 NORMAL SPECTRAL REFLECTANCE OF MAGNESIUM SILICATE PIGMENTED COATINGS

Curve No.	Ref. No.	Year	Temperature, K	Wavelength Range, μm	Geometry θ	Geometry θ' ω'	Reported Error, %	Composition (weight percent), Specifications, and Remarks
1	20	1963	~298	0.377-2.770	~0°	2π		Magnesium silicate in PS-7 potassium silicate binder; PBR 4.30; solids content 56.9%; paint ball milled; aluminum substrate grit blasted with 40 mesh silicon carbide; brush application; sample cured at 413 K for 18 hrs; data extracted from smooth curve. [Authors' designation: Sample C9]
2	20	1963	~298	0.372-2.766	~0°	2π		Above specimen and conditions except exposed in vacuum (10^{-5} mm Hg) to 200 ESH at solar factor of 3 suns.
3	27	1964	~298	0.44-0.60	~0°	2π		$MgSiO_3$ in PS-7 potassium silicate binder; PBR 4.30; solids content 56.9%; aluminum substrate abraded with No. 60 Aloxite cloth; sample cured at 413 K for 18 hrs. [Authors' designation: Sample C9]
4	27	1964	~298	0.44-0.60	~0°	2π		Above specimen and conditions except exposed in vacuum (10^{-6} mm Hg) to 200 ESH at solar factor of 3 suns.

DATA TABLE NO. 75 NORMAL SPECTRAL REFLECTANCE OF MAGNESIUM SILICATE PIGMENTED COATINGS

[Wavelength, λ, μm; Reflectance, ρ; Temperature, T, K]

λ	ρ	λ	ρ	λ	ρ	λ	ρ	λ	ρ	λ	ρ
CURVE 1		CURVE 1 (cont.)		CURVE 1 (cont.)		CURVE 2		CURVE 2 (cont.)		CURVE 2 (cont.)	
T ~ 298						T ~ 298					
		1.079	0.878	2.135	0.802			1.533	0.964*	2.626	0.628
0.377	0.871	1.241	0.898	2.201	0.781	0.372	0.548	1.604	0.895	2.676	0.587
0.429	0.897	1.320	0.902	2.258	0.731	0.410	0.580	1.656	0.870	2.689	0.569
0.488	0.910	1.397	0.923	2.282	0.707*	0.485	0.721	1.699	0.849*	2.719	0.561
0.517	0.915	1.471	0.962*	2.310	0.691	0.617	0.828	1.764	0,836	2.746	0.569*
0.647	0.923	1.499	0.966	2.338	0.684*	0.676	0.857	1.814	0.826*	2.766	0.594
0.678	0.923	1.546	0.958	2.376	0.683	0.741	0.863	1.836	0.815		
0.721	0.915	1.605	0.919	2.459	0.694	0.813	0.863*	1.847	0.799*	CURVE 3	
0.758	0.899	1.651	0.885	2.524	0.690*	0.865	0.836	1.863	0.774*	T ~ 298	
0.796	0.882	1.746	0.841	2.578	0.679	0.901	0.823	1.892	0.760		
0.818	0.861	1.814	0.825	2.615	0.665	0.970	0.835	1.922	0.761	0.44	0.925
0.840	0.834	1.837	0.814*	2.642	0.643*	1.108	0.890	2.013	0.782	0.60	0.945
0.863	0.820	1.852	0.799	2.666	0.611*	1.195	0.910	2.210	0.759		
0.892	0.814	1.875	0.774*	2.700	0.582	1.237	0.910	2.259	0.722	CURVE 4	
0.922	0.816*	1.913	0.765*	2.727	0.566*	1.292	0.904	2.305	0.692*	T ~ 298	
0.967	0.824	1.947	0.760	2.770	0.553	1.352	0.912	2.352	0.680*		
0.998	0.837	2.016	0.772			1.445	0.963	2.523	0.670	0.44	0.665
1.046	0.862	2.090	0.799			1.493	0.975	2.566	0.656*	0.60	0.840

* Not shown on plot

SPECIFICATION TABLE NO. 76 NORMAL SOLAR ABSORPTANCE OF MAGNESIUM SILICATE PIGMENTED COATINGS

Curve No.	Ref. No.	Year	Temperature Range, K	Geometry θ	Reported Error, %	Composition (weight percent), Specifications and Remarks
1*	27	1964	~298	~0°		$MgSiO_3$ in PS7 potassium silicate binder; PBR 4.30; solids content 56.9%; aluminum substrate abraded with No. 60 Aloxite cloth; sample cured at 413 K for 18 hrs. [Authors' designation: Sample C9]
2*	27	1964	~298	~0°		Above specimen and conditions except irradiated in vacuum (10^{-6} mm Hg) with 200 ESH at solar factor of 3 suns.

DATA TABLE NO. 76 NORMAL SOLAR ABSORPTANCE OF MAGNESIUM SILICATE PIGMENTED COATINGS

[Temperature, T, K; Absorptance, α]

T	α
CURVE 1*	
298	0.130
CURVE 2*	
298	0.219

* No plot given

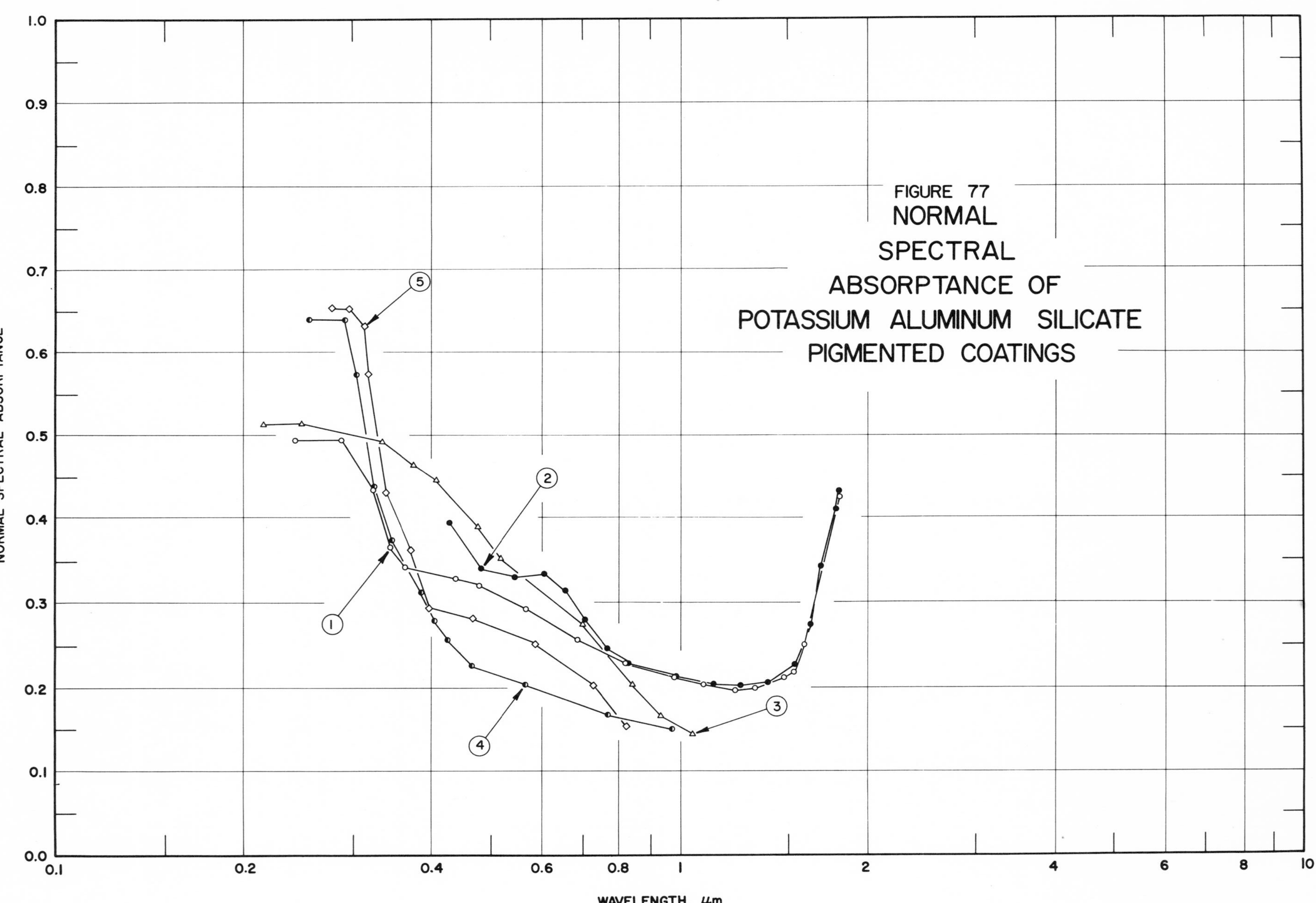

FIGURE 77 NORMAL SPECTRAL ABSORPTANCE OF POTASSIUM ALUMINUM SILICATE PIGMENTED COATINGS

SPECIFICATION TABLE NO. 77 NORMAL SPECTRAL ABSORPTANCE OF POTASSIUM ALUMINUM SILICATE PIGMENTED COATINGS

Curve No.	Ref. No.	Year	Temperature K	Wavelength Range, μm	Geometry θ	Reported Error, %	Composition (weight percent), Specifications, and Remarks
1	24	1965	298	0.241-1.800	~0°		Potassium aluminum silicate in sodium silicate binder; specimen exposed to a nuclear-radiation dose of approx 10^8 R of gamma and 5 x 10^{14} neutrons ($E \geq 2.9$ MeV) per cm^2; data extracted from smooth curve.
2	24	1965	298	0.273-1.800	~0°		Similar to above specimen and conditions except exposed at 77 K.
3	24	1965	298	0.215-1.049	~0°		Similar to curve 1 specimen and conditions.
4	24	1965	298	0.255-0.969	~0°		Potassium aluminum silicate in sodium silicate binder; specimen exposed to approx 200 sun hrs of ultraviolet irradiation; data extracted from smooth curve.
5	24	1965	298	0.277-0.816	~0°		Potassium aluminum silicate in sodium silicate binder; specimen exposed to a nuclear radiation dose of approx 10^8 R of gamma and 5 x 10^{14} neutrons ($E \geq 2.9$ MeV) per cm^2 and 500 sun hrs of ultraviolet irradiation; data extracted from smooth curve.

DATA TABLE NO. 77 NORMAL SPECTRAL ABSORPTANCE OF POTASSIUM ALUMINUM SILICATE PIGMENTED COATINGS

[Wavelength, λ, μm; Absorptance, α; Temperature, T, K]

λ	α	λ	α	λ	α	λ	α	λ	α
CURVE 1 T = 298		CURVE 2 T = 298		CURVE 3 T = 298		CURVE 4 T = 298		CURVE 5 T = 298	
0.241	0.494	0.273	0.829	0.215	0.513	0.255	0.640	0.277	0.654
0.288	0.495	0.330	0.825	0.247	0.515	0.290	0.640	0.295	0.653
0.321	0.435	0.366	0.835	0.332	0.492	0.302	0.574	0.312	0.631
0.344	0.367	0.397	0.818	0.373	0.463	0.322	0.438	0.317	0.575
0.362	0.344	0.429	0.395	0.407	0.447	0.346	0.375	0.339	0.431
0.439	0.330	0.480	0.341	0.474	0.390	0.383	0.313	0.370	0.364
0.478	0.321	0.546	0.331	0.517	0.353	0.402	0.280	0.399	0.295
0.563	0.295	0.602	0.335	0.699	0.276	0.423	0.257	0.465	0.282
0.685	0.258	0.655	0.316	0.836	0.203	0.465	0.228	0.582	0.253
0.815	0.230	0.708	0.281	0.926	0.168	0.563	0.202	0.722	0.201
0.975	0.212	0.766	0.249	1.049	0.145	0.765	0.169	0.816	0.153
1.089	0.202	0.825	0.230			0.969	0.150		
1.234	0.198	0.980	0.214						
1.316	0.200	1.136	0.203						
1.465	0.211	1.253	0.202						
1.519	0.220	1.370	0.207						
1.574	0.251	1.522	0.227						
1.800	0.427	1.619	0.275						
		1.698	0.345						
		1.770	0.412						
		1.800	0.436						

SPECIFICATION TABLE NO. 78 NORMAL TOTAL EMITTANCE OF POTASSIUM TITANATE PIGMENTED COATINGS

Curve No.	Ref. No.	Year	Temperature Range, K	Geometry θ'	Reported Error, %	Composition (weight percent), Specifications and Remarks
1*	83	1963	358	$\sim 0°$		Potassium titanate in phosphate binder; potassium titanate porcelain enamel substrate; measured in vacuum (1.6×10^{-4} mm Hg).

DATA TABLE NO. 78 NORMAL TOTAL EMITTANCE OF POTASSIUM TITANATE PIGMENTED COATINGS

[Temperature, T, K; Emittance, ϵ]

T	ϵ
CURVE 1*	
358	0.868

* No plot given

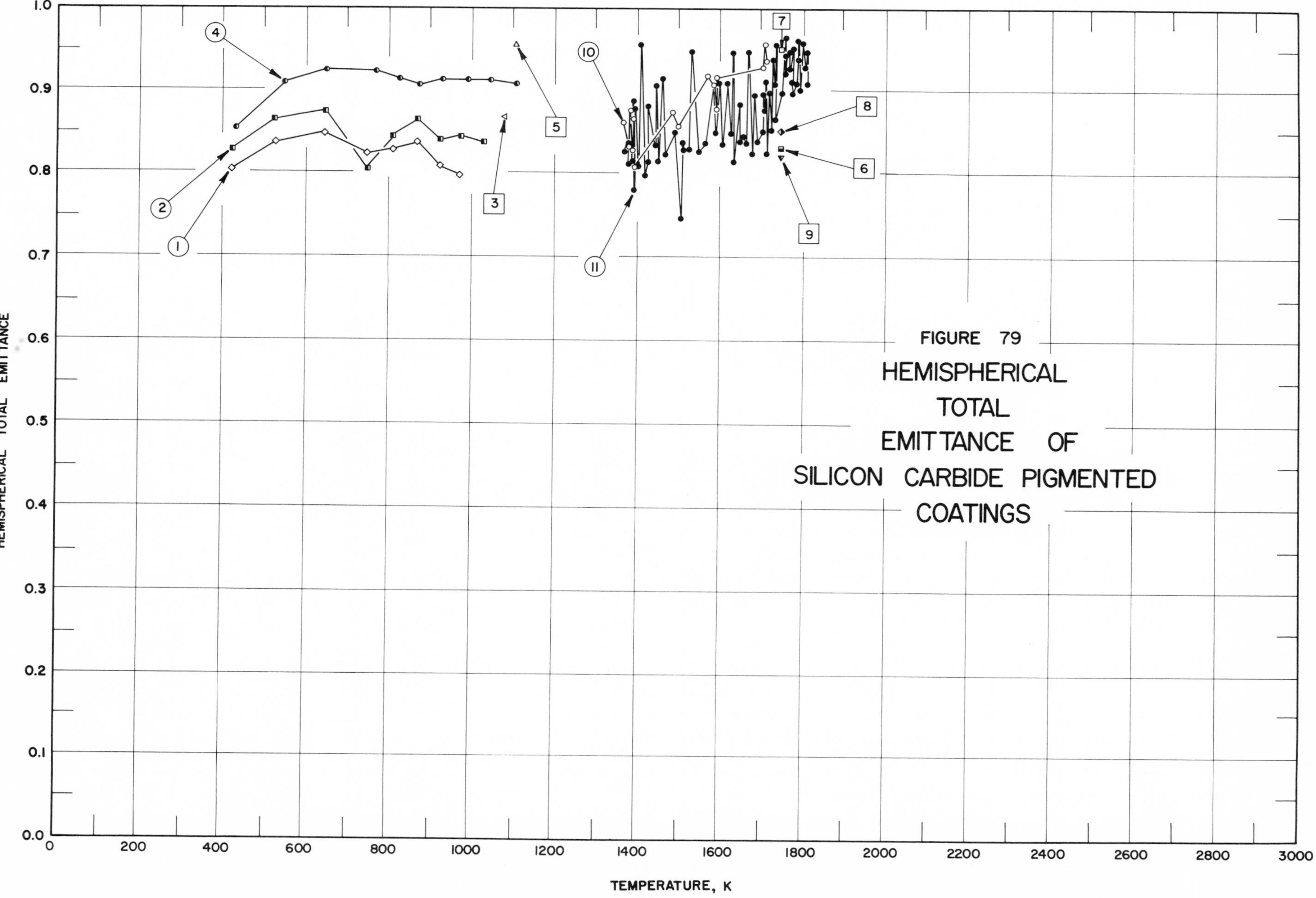

FIGURE 79 HEMISPHERICAL TOTAL EMITTANCE OF SILICON CARBIDE PIGMENTED COATINGS

SPECIFICATION TABLE NO. 79 HEMISPHERICAL TOTAL EMITTANCE OF SILICON CARBIDE PIGMENTED COATINGS

Curve No.	Ref. No.	Year	Temperature Range, K	Reported Error, %	Composition (weight percent), Specifications and Remarks
1	81	1963	422-977		Silicon carbide in aluminum phosphate binder (0.178 mm thick); Nb-1Zr substrate; measured in vacuum ($<3.8 \times 10^{-6}$ mm Hg); temp measured with thermocouple. [Author's designation: Run No. 1]
2	81	1963	422-1034		Similar to above specimen and conditions except 0.102 mm thick and vacuum $<3.4 \times 10^{-6}$ mm Hg.
3	81	1963	1080		Above specimen and conditions except temp measured with optical pyrometer.
4	81	1963	438-1115		Similar to curve 1 specimen and conditions except 0.127 mm thick and vacuum $<9.8 \times 10^{-6}$ mm Hg.
5	81	1963	1101		Above specimen and conditions except temp measured with optical pyrometer.
6	67	1967	1750		SiC in Synar and water binder; D-36 niobium substrate; cured at high temp; measured in air.
7	67	1967	1750		SiC in Synar binder; D-36 niobium substrate; cured at high temp; measured in air.
8	67	1967	1750		SiC in Synar and water binder; TZM molybdenum substrate; cured at high temp; measured in air.
9	67	1967	1750		SiC in DC20 binder; TZM molybdenum substrate; cured at high temp; measured in air.
10	67	1967	1368-1716		SiC in silica binder; disilicide on D-36 niobium substrate; cured at high temp; measured in air.
11	67	1967	1375-1810		Similar to above specimen and conditions except measured at a reduced pressure with sonic velocity air passed over sample.

DATA TABLE NO. 79 HEMISPHERICAL TOTAL EMITTANCE OF SILICON CARBIDE PIGMENTED COATINGS

[Temperature, T, K; Emittance, ∈]

T	∈
CURVE 1	
422	0. 802
533	0. 836
644	0. 846
755	0. 823
810	0. 828
866	0. 839
922	0. 809
977	0. 796
CURVE 2	
422	0. 829
533	0. 863
644	0. 872
755	0. 803
810	0. 844
866	0. 863
922	0. 840
977	0. 845
1034	0. 839
CURVE 3	
1080	0. 867
CURVE 4	
438	0. 851
545	0. 908
657	0. 923
768	0. 921
822	0. 911
879	0. 906
933	0. 912
989	0. 911
1044	0. 911
1115	0. 907
CURVE 5	
1101	0. 956
CURVE 6	
1750	0. 83
CURVE 7	
1750	0. 95
CURVE 8	
1750	0. 85
CURVE 9	
1750	0. 82
CURVE 10	
1368	0. 860
1376	0. 855*
1380	0. 832
1382	0. 874
1384	0. 829
1387	0. 864
1388	0. 849*
1388	0. 807
1488	0. 871
1500	0. 869*
1500	0. 856
1508	0. 858*
1575	0. 918
1592	0. 906
1597	0. 917
1597	0. 896*
1597	0. 879
1703	0. 928
1710	0. 926*
1711	0. 956
1711	0. 942*
1716	0. 934
CURVE 11	
1375	0. 825
1380	0. 834
1380	0. 810
1393	0. 812
1395	0. 889
1396	0. 780
1398	0. 825*
1399	0. 878
1403	0. 809
CURVE 11 (cont.)	
1406	0. 904*
1413	0. 955
1417	0. 815*
1420	0. 824*
1420	0. 799
1425	0. 828*
1431	0. 812
1434	0. 880
1435	0. 840*
1443	0. 831
1447	0. 904
1447	0. 821*
1450	0. 845*
1454	0. 813
1460	0. 912
1463	0. 839*
1472	0. 839*
1475	0. 821
1490	0. 836*
1499	0. 848
1501	0. 826*
1504	0. 746
1511	0. 815*
1515	0. 838
1530	0. 827
1534	0. 850*
1535	0. 940*
1537	0. 947
1538	0. 867*
1545	0. 844*
1550	0. 826
1556	0. 827*
1564	0. 835
1569	0. 826*
1576	0. 835*
1583	0. 905
1583	0. 848
1595	0. 864*
1596	0. 850*
1600	0. 909
1600	0. 877*
1612	0. 860*
1612	0. 833
1615	0. 850*
1620	0. 906
CURVE 11 (cont.)	
1627	0. 848
1629	0. 944
1634	0. 844*
1634	0. 812
1640	0. 829*
1646	0. 874*
1648	0. 881
1648	0. 838
1659	0. 843
1663	0. 835
1669	0. 947
1680	0. 876*
1680	0. 822
1693	0. 895
1695	0. 866*
1697	0. 881*
1699	0. 837
1702	0. 850
1703	0. 897
1704	0. 876
1716	0. 910
1717	0. 828
1720	0. 899
1723	0. 851
1725	0. 938
1727	0. 918*
1727	0. 906
1737	0. 955
1737	0. 866
1749	0. 955*
1751	0. 934*
1754	0. 895
1759	0. 942*
1760	0. 964
1760	0. 920*
1762	0. 932*
1764	0. 944
1767	0. 925
1769	0. 950
1774	0. 924*
1775	0. 910
1778	0. 894
1782	0. 906
1784	0. 918*
1785	0. 938
1787	0. 945*
CURVE 11 (cont.)	
1790	0. 960
1792	0. 915*
1792	0. 925*
1795	0. 900
1796	0. 946*
1804	0. 958
1805	0. 927
1807	0. 946
1808	0. 915*
1810	0. 907

* Not shown on plot

SPECIFICATION TABLE NO. 80 HEMISPHERICAL TOTAL EMITTANCE OF SILICON CARBIDE + X PIGMENTED COATINGS

Curve No.	Ref. No.	Year	Temperature Range, K	Reported Error, %	Composition (weight percent), Specifications and Remarks
1*	82	1962	533-981	< ±2.5	SiC and SiO_2 in aluminum phosphate binder (0.076 mm thick); aluminum substrate; measured in vacuum (10^{-8} mm Hg); data extracted from smooth curve.
2*	67	1967	1750		SiC and UO_2 in Synar and water binder; D-36 niobium substrate; cured at high temp; measured in air.
3*	67	1967	1750		SiC and talc in Ludox binder; D-36 niobium substrate; cured at high temp; measured in air.
4*	67	1967	1750		SiC in DC806A (contains titanium carbide) binder; TZM molybdenum substrate; cured at high temp; measured in air.

DATA TABLE NO. 80 HEMISPHERICAL TOTAL EMITTANCE OF SILICON CARBIDE + X PIGMENTED COATINGS

[Temperature, T, K; Emittance, ϵ]

T	ϵ
CURVE 1*	
533	0.876
588	0.881
685	0.882
801	0.879
883	0.876
943	0.870
981	0.865
CURVE 2*	
1750	0.77
CURVE 3*	
1750	0.72
CURVE 4*	
1750	0.89

* No plot given

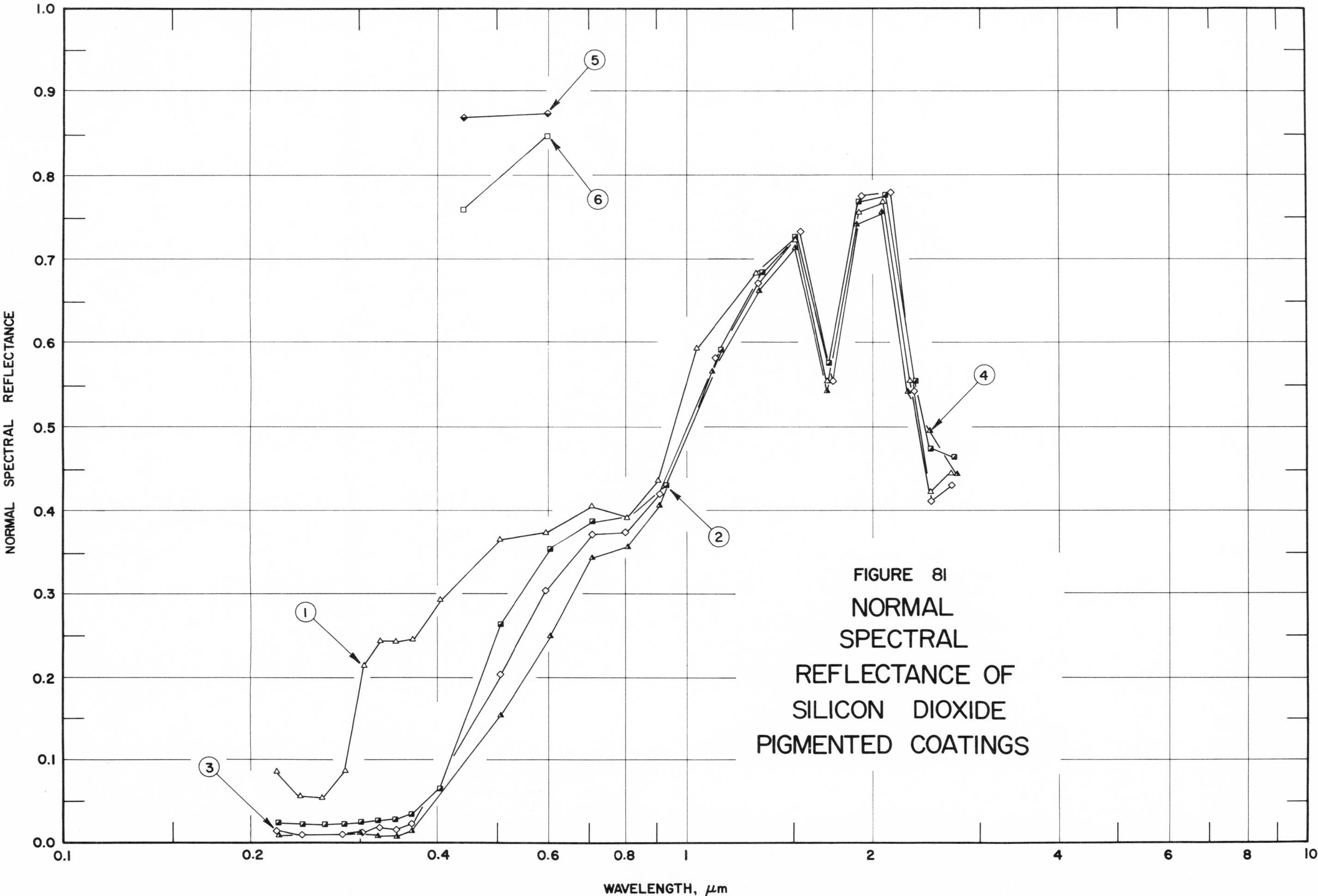
FIGURE 81
NORMAL
SPECTRAL
REFLECTANCE OF
SILICON DIOXIDE
PIGMENTED COATINGS
NORMAL SPECTRAL REFLECTANCE
WAVELENGTH, μm
0.0
0.1
0.2
0.3
0.4
0.5
0.6
0.7
0.8
0.9
1.0
0.1
0.2
0.4
0.6
0.8
1
2
4
6
8
10
1
2
3
4
5
6

SPECIFICATION TABLE NO. 81 NORMAL SPECTRAL REFLECTANCE OF SILICON DIOXIDE PIGMENTED COATINGS

Curve No.	Ref. No.	Year	Temperature, K	Wavelength Range, μm	Geometry θ θ'	ω'	Reported Error, %	Composition (weight percent), Specifications, and Remarks
1	1	1960	358	0.220-2.697	~0°	2π		SiO_2 in silicone (0.051 mm thick); pigment concentration by weight 20%; anodized aluminum alloy 24 substrate; measured in vacuum (10^{-5} mm Hg); measured relative to MgO.
2	1	1960	358	0.221-2.700	~0°	2π		Above specimen and conditions except exposed to UV radiation for 6 hrs from G. E. type UA-3 lamp.
3	1	1960	358	0.220-2.690	~0°	2π		Above specimen and conditions except exposed to UV radiation for 40 hrs.
4	1	1960	358	0.221-2.722	~0°	2π		Above specimen and conditions except exposed to UV radiation for 104 hrs.
5	27	1964	~298	0.44-0.60	~0°	2π		Fused quartz powder (mfgr., General Electric) in PS-7 potassium silicate binder; PBR 4.30; solids content 62.8%; aluminum substrate abraded with No. 60 Aloxite cloth; sample cured at 413 K for 18 hrs. [Authors' designation: Sample C11]
6	27	1964	~298	0.44-0.60	~0°	2π		Above specimen and conditions except irradiated in vacuum (10^{-6} mm Hg) with 300 ESH at solar factor of 3 suns.

DATA TABLE NO. 81 NORMAL SPECTRAL REFLECTANCE OF SILICON DIOXIDE PIGMENTED COATINGS

[Wavelength, λ, μm; Reflectance, ρ; Temperature, T, K]

CURVE 1, T = 358

λ	ρ
0.220	0.086
0.240	0.056
0.260	0.053
0.281	0.085
0.301	0.214
0.321	0.241
0.341	0.243
0.363	0.245
0.401	0.294
0.501	0.356
0.599	0.374
0.701	0.406
0.801	0.392
0.900	0.438
1.104	0.595
1.305	0.684
1.502	0.722
1.698	0.556
1.905	0.756
2.096	0.770
2.301	0.556
2.491	0.424
2.697	0.447

CURVE 2, T = 358

λ	ρ
0.221	0.023
0.242	0.022
0.262	0.022
0.281	0.022
0.300	0.025
0.320	0.026
0.342	0.028
0.361	0.033
0.401	0.064
0.501	0.263
0.601	0.354
0.702	0.389
0.804	0.391*
0.928	0.431
1.140	0.593
1.333	0.686
1.507	0.738
1.702	0.579
1.900	0.770
2.106	0.778
2.329	0.555
2.499	0.476
2.700	0.467

CURVE 3, T = 358

λ	ρ
0.220	0.014
0.241	0.010
0.263	0.010*
0.280	0.010
0.301	0.012
0.321	0.018
0.341	0.016
0.361	0.021
0.501	0.203
0.599	0.304
0.701	0.371
0.800	0.374
0.905	0.421
1.123	0.582
1.320	0.671
1.530	0.732
1.725	0.552
1.935	0.775
2.133	0.780
2.323	0.542
2.497	0.412
2.690	0.431

CURVE 4, T = 358

λ	ρ
0.221	0.009
0.300	0.011
0.320	0.009
0.341	0.009
0.362	0.013
0.501	0.152
0.601	0.250
0.701	0.343
0.801	0.359
0.905	0.407
1.106	0.567
1.310	0.661
1.504	0.714
1.695	0.542
1.898	0.741
2.077	0.758
2.277	0.542
2.498	0.497
2.722	0.444

CURVE 5, T ~ 298

λ	ρ
0.44	0.870
0.60	0.875

CURVE 6, T ~ 298

λ	ρ
0.44	0.760
0.60	0.850

* Not shown on plot

SPECIFICATION TABLE NO. 82 NORMAL SOLAR ABSORPTANCE OF SILICON DIOXIDE PIGMENTED COATINGS

Curve No.	Ref. No.	Year	Temperature Range, K	Geometry θ	Reported Error, %	Composition (weight percent), Specifications and Remarks
1*	27	1964	~298	~0°		Fused quartz powder in PS7 potassium silicate binder; PBR 4.30; solids content 62.8%; aluminum substrate abraded with No. 60 Aloxite cloth; sample cured at 413 K for 18 hrs. [Authors' designation: Sample C11]
2*	27	1964	~298	~0°		Above specimen and conditions except irradiated in vacuum (10^{-6} mm Hg) with 300 ESH at solar factor of 3 suns.
3*	27	1964	~298	~0°		Dicalite WB-5 in PS7 potassium silicate binder; PBR 2.13; solids content 30.4%; aluminum substrate abraded with No. 60 Aloxite cloth; sample cured at 413 K for 18 hrs. [Authors' designation: Sample C23]
4*	27	1964	~298	~0°		Above specimen and conditions except irradiated in vacuum (10^{-6} mm Hg) with 300 ESH at solar factor of 3 suns.
5*	27	1964	~298	~0°		Dicalite WB-5 in PS7 potassium silicate binder; PBR 4.30; solids content 26.5%; aluminum substrate abraded with No. 60 Aloxite cloth; sample cured at 413 K for 18 hrs. [Authors' designation: Sample C25]
6*	27	1964	~298	~0°		Above specimen and conditions except irradiated in vacuum (10^{-6} mm Hg) with 300 ESH at solar factor of 3 suns.

DATA TABLE NO. 82 NORMAL SOLAR ABSORPTANCE OF SILICON DIOXIDE PIGMENTED COATINGS

[Temperature, T, K; Absorptance, α]

T	α	T	α
CURVE 1*		CURVE 6*	
298	0.177	298	0.186
CURVE 2*			
298	0.221		
CURVE 3*			
298	0.136		
CURVE 4*			
298	0.173		
CURVE 5*			
298	0.128		

* No plot given

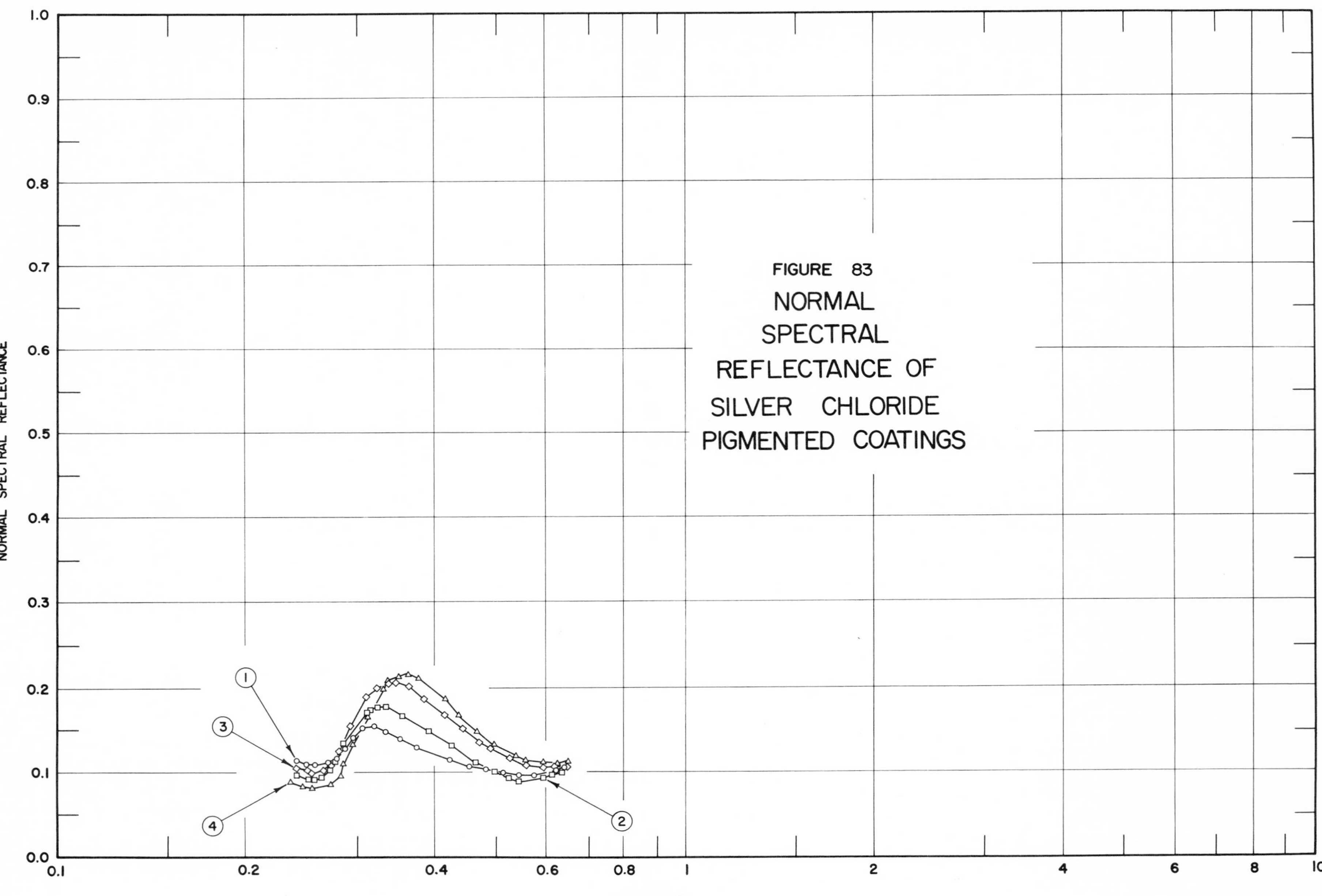

FIGURE 83 NORMAL SPECTRAL REFLECTANCE OF SILVER CHLORIDE PIGMENTED COATINGS

SPECIFICATION TABLE NO. 83 NORMAL SPECTRAL REFLECTANCE OF SILVER CHLORIDE PIGMENTED COATINGS

Curve No.	Ref. No.	Year	Temperature, K	Wavelength Range, μm	Geometry θ θ' ω'	Reported Error, %	Composition (weight percent), Specifications, and Remarks
1	42	1963	298	0.2415-0.6405	~0° ~0°		Silver chloride suspended in gelatin (concentration 0.062 mole); quartz plate substrate; data extracted from smooth curve; measured relative to $MgCO_3$.
2	42	1963	298	0.2414-0.6385	~0° ~0°		Silver chloride suspended in gelatin (concentration 0.123 mole); quartz plate substrate; data extracted from smooth curve; measured relative to $MgCO_3$.
3	42	1963	298	0.2417-0.6522	~0° ~0°		Silver chloride suspended in gelatin (concentration 0.185 mole); quartz plate substrate; data extracted from smooth curve; measured relative to $MgCO_3$.
4	42	1963	298	0.2360-0.6494	~0° ~0°		Silver chloride suspended in gelatin (concentration 0.246 mole); quartz plate substrate; data extracted from smooth curve; measured relative to $MgCO_3$.

DATA TABLE NO. 83 NORMAL SPECTRAL REFLECTANCE OF SILVER CHLORIDE PIGMENTED COATINGS

[Wavelength, λ, μm; Reflectance, ρ; Temperature, T, K]

λ	ρ	λ	ρ	λ	ρ	λ	ρ
CURVE 1 T = 298		CURVE 2 T = 298		CURVE 3 T = 298		CURVE 4 T = 298	
0.2415	0.115	0.2414	0.099	0.2417	0.107	0.2360	0.090
0.2509	0.111	0.2524	0.094	0.2508	0.101	0.2484	0.085
0.2571	0.110	0.2569	0.094	0.2578	0.100	0.2564	0.084
0.2706	0.112	0.2646	0.096	0.2676	0.103	0.2710	0.087
0.2800	0.119	0.2719	0.104	0.2725	0.109	0.2819	0.097
0.2895	0.130	0.2781	0.116	0.2807	0.127	0.2869	0.111
0.2989	0.142	0.2861	0.134	0.2948	0.158	0.2979	0.136
0.3093	0.152	0.3101	0.172	0.3122	0.190	0.3131	0.167
0.3204	0.156	0.3146	0.175	0.3220	0.200	0.3303	0.199
0.3358	0.150	0.3245	0.179	0.3377	0.206	0.3391	0.208
0.3537	0.141	0.3340	0.179	0.3472	0.207	0.3553	0.215
0.3757	0.132	0.3584	0.169	0.3640	0.203	0.3633	0.216
0.4218	0.118	0.3934	0.150	0.3878	0.188	0.3760	0.213
0.4566	0.109	0.4273	0.133	0.4164	0.170	0.4165	0.188
0.4845	0.105	0.4664	0.114	0.4445	0.154	0.4395	0.169
0.5129	0.101	0.4963	0.102	0.4721	0.139	0.4680	0.150
0.5472	0.099	0.5279	0.095	0.4912	0.130	0.4979	0.135
0.5771	0.099	0.5573	0.093	0.5297	0.118	0.5361	0.121
0.6216	0.104	0.5931	0.095	0.5624	0.110	0.5582	0.115
0.6405	0.108	0.6182	0.098	0.5983	0.106	0.5915	0.112
		0.6385	0.103	0.6193	0.106	0.6228	0.111
				0.6522	0.107	0.6494	0.112

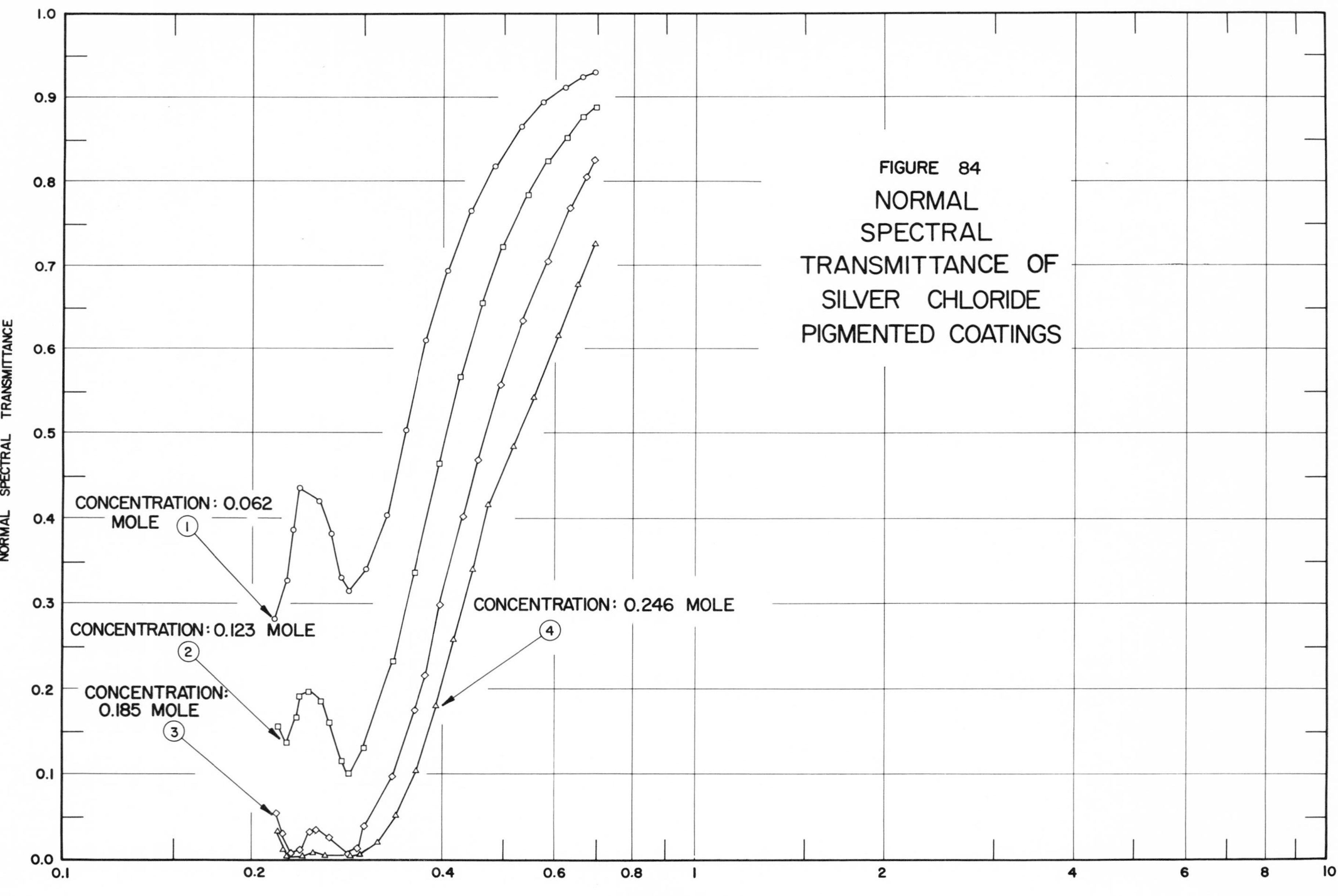

FIGURE 84 NORMAL SPECTRAL TRANSMITTANCE OF SILVER CHLORIDE PIGMENTED COATINGS

SPECIFICATION TABLE NO. 84 NORMAL SPECTRAL TRANSMITTANCE OF SILVER CHLORIDE PIGMENTED COATINGS

Curve No.	Ref. No.	Year	Temperature, K	Wavelength Range, μm	Geometry θ θ' ω'	Reported Error, %	Composition (weight percent), Specifications, and Remarks
1	42	1963	298	0.2195-0.6967	~0° ~0°		Silver chloride suspended in gelatin (concentration 0.062 mole); quartz plate substrate; data extracted from smooth curve.
2	42	1963	298	0.2210-0.6975	~0° ~0°		Silver chloride suspended in gelatin (concentration 0.123 mole); quartz plate substrate; data extracted from smooth curve.
3	42	1963	298	0.2193-0.6967	~0° ~0°		Silver chloride suspended in gelatin (concentration 0.185 mole); quartz plate substrate; data extracted from smooth curve.
4	42	1963	298	0.2191-0.6964	~0° ~0°		Silver chloride suspended in gelatin (concentration 0.246 mole); quartz plate substrate; data extracted from smooth curve.

DATA TABLE NO. 84 NORMAL SPECTRAL TRANSMITTANCE OF SILVER CHLORIDE PIGMENTED COATINGS

[Wavelength, λ, μm; Transmittance, τ; Temperature, T, K]

λ	τ	λ	τ	λ	τ	λ	τ
CURVE 1 T = 298		CURVE 2 T = 298		CURVE 3 T = 298		CURVE 4 T = 298	
0.2195	0.281	0.2210	0.156	0.2193	0.054	0.2191	0.034
0.2276	0.327	0.2271	0.138	0.2230	0.030	0.2244	0.011
0.2311	0.387	0.2353	0.167	0.2307	0.009	0.2295	0.003
0.2398	0.439	0.2398	0.192	0.2367	0.012	0.2408	0.004
0.2551	0.421	0.2465	0.197	0.2492	0.034	0.2501	0.009
0.2675	0.383	0.2586	0.185	0.2547	0.035	0.2601	0.004
0.2779	0.332	0.2662	0.162	0.2679	0.026	0.2806	0.004
0.2853	0.314	0.2772	0.115	0.2837	0.009	0.2992	0.008
0.3041	0.342	0.2861	0.100	0.2949	0.016	0.3186	0.021
0.3272	0.406	0.3023	0.131	0.3064	0.040	0.3392	0.051
0.3522	0.502	0.3347	0.233	0.3344	0.097	0.3648	0.105
0.3796	0.612	0.3635	0.339	0.3643	0.175	0.3910	0.181
0.4075	0.695	0.3976	0.468	0.3764	0.215	0.4175	0.261
0.4461	0.767	0.4285	0.568	0.3988	0.300	0.4471	0.341
0.4865	0.820	0.4630	0.658	0.4302	0.403	0.4792	0.417
0.5325	0.865	0.4999	0.725	0.4548	0.470	0.5161	0.485
0.5780	0.894	0.5459	0.785	0.4944	0.558	0.5550	0.545
0.6252	0.913	0.5871	0.825	0.5362	0.635	0.6091	0.618
0.6677	0.925	0.6299	0.854	0.5870	0.708	0.6556	0.679
0.6967	0.931	0.6689	0.877	0.6364	0.770	0.6964	0.729
		0.6975	0.891	0.6737	0.809		
				0.6967	0.826		

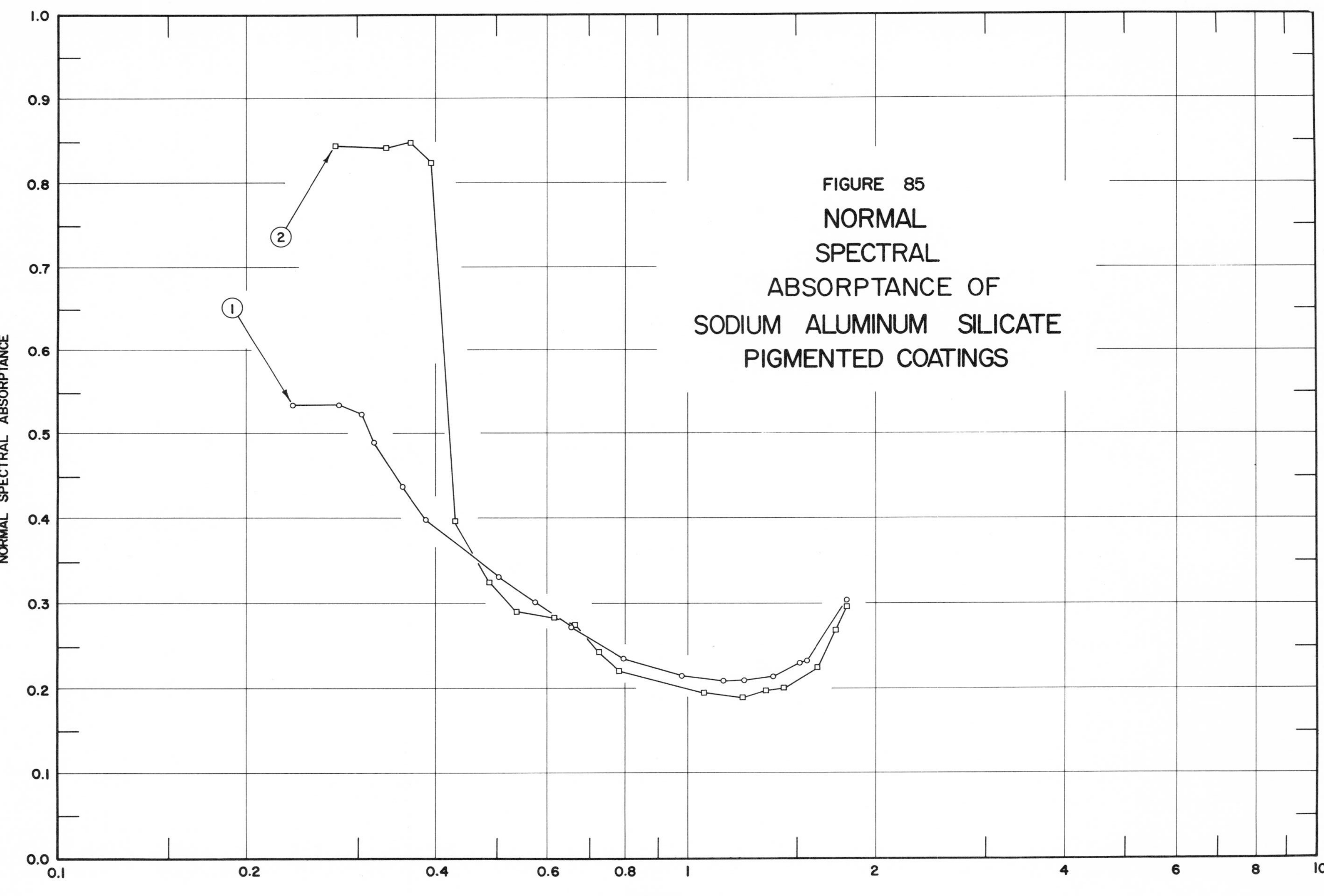

FIGURE 85 NORMAL SPECTRAL ABSORPTANCE OF SODIUM ALUMINUM SILICATE PIGMENTED COATINGS

SPECIFICATION TABLE NO. 85 NORMAL SPECTRAL ABSORPTANCE OF SODIUM ALUMINUM SILICATE PIGMENTED COATINGS

Curve No.	Ref. No.	Year	Temperature, K	Wavelength Range, μm	Geometry θ	Reported Error, %	Composition (weight percent), Specifications, and Remarks
1	24	1965	298	0.237-1.800	~0°		Sodium aluminum silicate in sodium silicate binder; specimen exposed to a nuclear-radiation dose of approx 10^8 R of gamma and 5 x 10^{14} neutrons (E ≥ 2.9 Mev) per cm^2; data extracted from smooth curve; calculated from spectral reflectance $\rho(\sim 0°, 2\pi)$.
2	24	1965	298	0.278-1.800	~0°		Similar to above specimen and conditions except exposed at 77 K.

DATA TABLE NO. 85 NORMAL SPECTRAL ABSORPTANCE OF SODIUM ALUMINUM SILICATE PIGMENTED COATINGS

[Wavelength, λ, μm; Absorptance, α; Temperature, T, K]

λ	α	λ	α
CURVE 1 T = 298		CURVE 2 T = 298	
0.237	0.535	0.278	0.848
0.281	0.535	0.333	0.845
0.305	0.525	0.367	0.851
0.320	0.491	0.395	0.828
0.354	0.439	0.430	0.397
0.385	0.400	0.487	0.327
0.508	0.334	0.539	0.292
0.573	0.303	0.616	0.286
0.656	0.272	0.669	0.277
0.798	0.238	0.725	0.243
0.979	0.216	0.783	0.223
1.149	0.210	1.067	0.197
1.243	0.210	1.227	0.191
1.371	0.215	1.340	0.198
1.502	0.231	1.425	0.202
1.566	0.234	1.603	0.228
1.800	0.305	1.739	0.270
		1.800	0.299

SPECIFICATION TABLE NO. 86 HEMISPHERICAL TOTAL EMITTANCE OF SPINEL PIGMENTED COATINGS

Curve No.	Ref. No.	Year	Temperature Range, K	Reported Error, %	Composition (weight percent), Specifications and Remarks
1*	15	1965	533-644		Cr-Co-Ni spinel in aluminum phosphate binder (0.0508 mm thick); substrate unknown.
2*	82	1962	422-1075	<±2.5	$NiO \cdot Cr_2O_3$ spinel and SiO_2 in aluminum phosphate binder (~0.0762 mm thick); stainless steel substrate; measured in vacuum (10^{-8} mm Hg); data extracted from smooth curve.

DATA TABLE NO. 86 HEMISPHERICAL TOTAL EMITTANCE OF SPINEL PIGMENTED COATINGS

[Temperature, T, K; Emittance, ϵ]

T	ϵ
CURVE 1*	
533	0.86
588	0.86
644	0.86
CURVE 2*	
422	0.881
1075	0.884

* No plot given

SPECIFICATION TABLE NO. 87 NORMAL SOLAR ABSORPTANCE OF SPINEL PIGMENTED COATINGS

Curve No.	Ref. No.	Year	Temperature Range, K	Geometry θ	Reported Error, %	Composition (weight percent), Specifications and Remarks
1*	86	1964	~298	~0°		Magnesium aluminate spinel in potassium silicate binder; PBR 2.15; cured by air-drying; heat treated 16 hrs at 773 K; substrate unknown; calculated from reflectance. [Authors' designation: Test V-55, Sample No. 7006]
2*	86	1964	~298	~0°		Above specimen and conditions except exposed to 690 ESH of UV radiation in vacuum ($< 10^{-6}$ mm Hg); avg solar factor of 9.4 suns.
3*	86	1964	~298	~0°		Similar to curve 1 specimen and conditions except an additional 2 hr heat treatment at 773 K. [Authors' designation: Test V-55, Sample No. 7007]
4*	86	1964	~298	~0°		Above specimen and conditions except exposed to 690 ESH of UV radiation in vacuum ($< 10^{-6}$ mm Hg); avg solar factor of 9.4 suns.

DATA TABLE NO. 87 NORMAL SOLAR ABSORPTANCE OF SPINEL PIGMENTED COATINGS

[Temperature, T, K; Absorptance, α]

T	α
CURVE 1*	
298	0.137
CURVE 2*	
298	0.282
CURVE 3*	
298	0.135
CURVE 4*	
298	0.300

* No plot given

SPECIFICATION TABLE NO. 88 HEMISPHERICAL TOTAL EMITTANCE OF STRONTIUM MOLYBDATE PIGMENTED COATINGS

Curve No.	Ref. No.	Year	Temperature Range, K	Reported Error, %	Composition (weight percent), Specifications and Remarks
1*	85	1969	158-398		$SrMoO_4$ in PS-7 potassium silicate binder (~0.139 mm thick); Irridite 14 and 2024 clad aluminum substrates; PBR 3; aluminum substrate cleaned and etched with an alcohol-phosphoric acid mixture; 6 min Irridite 14 conversion coating on aluminum substrate; $SrMoO_4$/K-Silicate spray applied with lamp grade nitrogen; cured at 423 K and air dried; property measured in vacuum by calorimetric method; data extracted from smooth curve. [Author's designation: B-5 $SrMoO_4$/K-Silicate, NRDL-RTD-81-6]

DATA TABLE NO. 88 HEMISPHERICAL TOTAL EMITTANCE OF STRONTIUM MOLYBDATE PIGMENTED COATINGS

[Temperature, T, K; Emittance, ϵ]

T	ϵ
CURVE 1*	
158	0.850
279	0.842
398	0.835

* No plot given.

SPECIFICATION TABLE NO. 89 NORMAL TOTAL EMITTANCE OF STRONTIUM MOLYBDATE PIGMENTED COATINGS

Curve No.	Ref. No.	Year	Temperature Range, K	Geometry θ	Reported Error, %	Composition (weight percent), Specifications and Remarks
1*	68	1969	300	$\sim 0°$		Strontium molybdate in acrylic binder ($\sim$0.305 mm thick); 2024 aluminum substrate ($\sim$1.59 mm thick); emittance calculated from reflectance; property measured in air. [Authors' designation: $SrMoO_4$/acrylic]
2*	68	1969	300	$\sim 0°$		Similar to curve 1 specimen and conditions except exposed to UV radiation in vacuum (10^{-5} mm Hg) for 300 hr; vacuum maintained by oil-diffusion pump; H_2 gas UV source; property measured in air after exposure.
3*	68	1969	300	$\sim 0°$		Similar to above specimen and conditions except He gas UV source.

DATA TABLE NO. 89 NORMAL TOTAL EMITTANCE OF STRONTIUM MOLYBDATE PIGMENTED COATINGS

[Temperature, T, K; Emittance, ϵ]

T	ϵ
CURVE 1*	
300	0.83
CURVE 2*	
300	0.83
CURVE 3*	
300	0.82

* No plot given

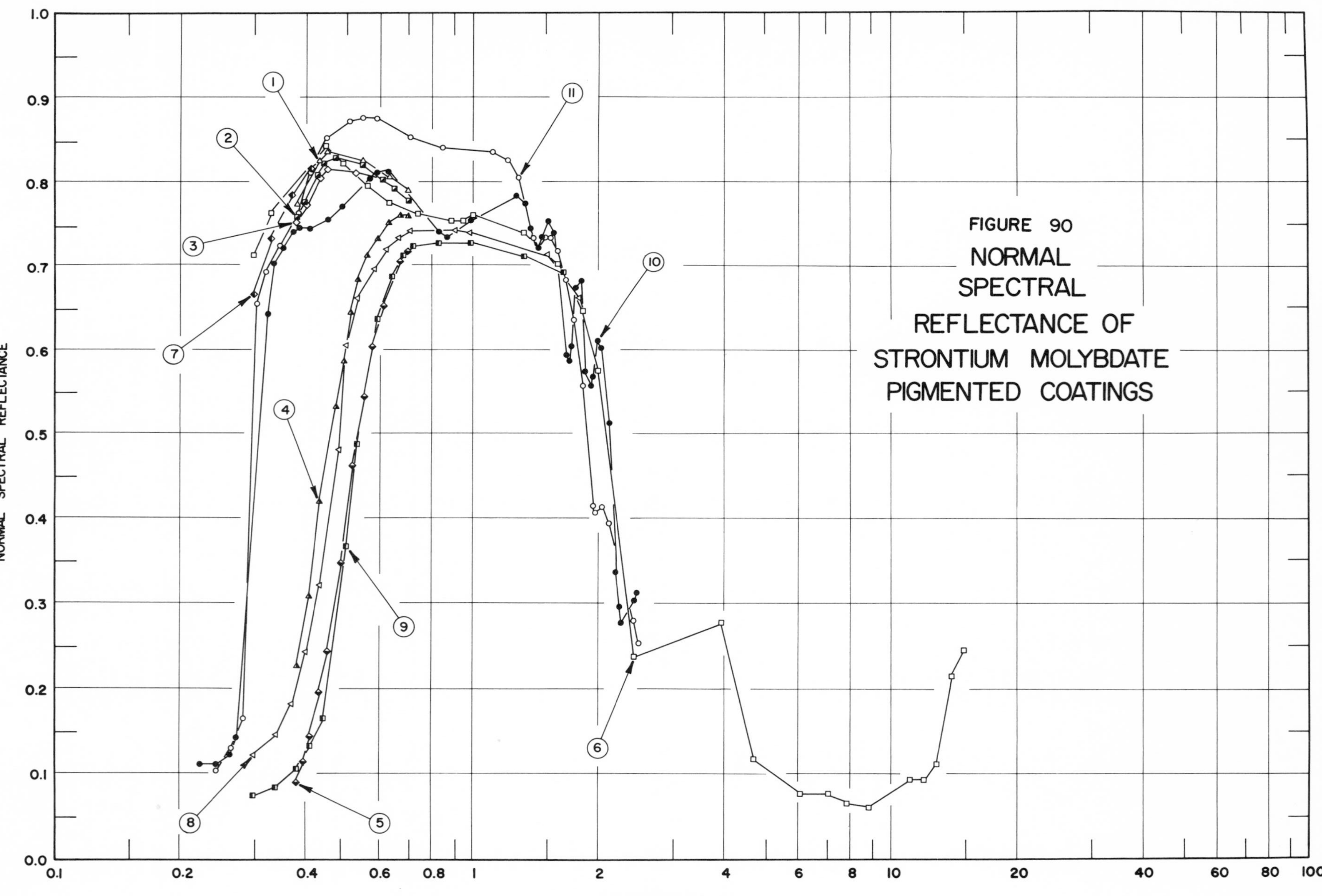

FIGURE 90 NORMAL SPECTRAL REFLECTANCE OF STRONTIUM MOLYBDATE PIGMENTED COATINGS

SPECIFICATION TABLE NO. 90 NORMAL SPECTRAL REFLECTANCE OF STRONTIUM MOLYBDATE PIGMENTED COATINGS

Curve No.	Ref. No.	Year	Temperature, K	Wavelength Range, μm	Geometry θ	Geometry θ'	Geometry ω'	Reported Error, %	Composition (weight percent), Specifications, and Remarks
1	53	1969	300	0.3800-0.7000	~0°		2π		Strontium molybdate in acrylic binder (~0.305 mm thick); 2024 aluminum substrate (~1.59 mm thick); reflectance measured in air; data extracted from smooth curve. [Authors' designation: $SrMoO_4$/acrylic]
2	53	1969	300	0.3800-0.7000	~0°		2π		Similar to above specimen and conditions except exposed to vacuum (10^{-8} mm Hg) for 300 hrs; vacuum maintained by ion pump; reflectance measured in air after vacuum exposure.
3	53	1969	300	0.3800-0.7000	~0°		2π		Similar to curve 1 specimen and conditions except exposed to vacuum (10^{-5} mm Hg) for 300 hrs; vacuum maintained by oil-diffusion pump; reflectance measured in air after vacuum exposure.
4	53	1969	300	0.3800-0.7000	~0°		2π		Similar to above specimen and conditions except exposed to UV radiation in vacuum for 300 hrs; H_2 gas UV source.
5	53	1969	300	0.3800-0.7000	~0°		2π		Similar to above specimen and conditions except He gas UV source.
6	53	1969	300	0.30-15.02	~0°		2π		Similar to curve 1 specimen and conditions.
7	53	1969	300	0.30-15.02	~0°		2π		Similar to curve 3 specimen and conditions.
8	53	1969	300	0.30-15.02	~0°		2π		Similar to curve 4 specimen and conditions.
9	53	1969	300	0.30-15.02	~0°		2π		Similar to curve 5 specimen and conditions.
10	85	1969	~298	0.225-2.489	~0°		2π		$SrMoO_4$ in B-44 acrylic (Rohm and Haas) binder (~0.266 mm thick); Irridite 14 and 2024 clad aluminum substrates; PBR 1; aluminum substrate cleaned and etched with an alcohol-phosphoric acid mixture; 6 min Irridite 14 conversion coating on aluminum substrate; $SrMoO_4$/acrylic spray applied with filtered air, then air dried; data extracted from smooth curve. [Author's designation: B-1 $SrMoO_4$/acrylic]
11	85	1969	~298	0.246-2.501	~0°		2π		$SrMoO_4$ in PS-7 potassium silicate binder (~0.139 mm thick); Irridite 14 and 2024 clad aluminum substrates; PBR 3; aluminum substrate cleaned and etched with an alcohol-phosphoric acid mixture; 6 min Irridite 14 conversion coating on aluminum substrate; $SrMoO_4$/K-silicate spray applied with lamp grade nitrogen; cured at 423 K and air dried; data extracted from smooth curve. [Author's designation: B-5 $SrMoO_4$/K-silicate]

DATA TABLE NO. 90 NORMAL SPECTRAL REFLECTANCE OF STRONTIUM MOLYBDATE PIGMENTED COATINGS

[Wavelength, λ, μm; Reflectance, ρ; Temperature, T, K]

λ	ρ
CURVE 1 T = 300	
0.3800	0.776
0.4305	0.827
0.4468	0.838
0.5431	0.829
0.6390	0.809
0.7000	0.792
CURVE 2 T = 300	
0.3800	0.760
0.3999	0.779
0.4228	0.810
0.4406	0.824
0.4706	0.830
0.5429	0.821
0.6063	0.804
0.6565	0.794
0.7000	0.780
CURVE 3 T = 300	
0.3800	0.755
0.4001	0.776
0.4327	0.806
0.4554	0.817
0.5213	0.812
0.6063	0.804*
0.6565	0.794*
0.7000	0.780*
CURVE 4 T = 300	
0.3800	0.226
0.4037	0.308
0.4355	0.420
0.4686	0.531
0.4895	0.588
0.5152	0.647
0.5377	0.686
0.5601	0.714
0.5881	0.736
0.6285	0.755
0.6795	0.762
0.7000	0.761
CURVE 5 T = 300	
0.3800	0.090
0.3953	0.115
0.4104	0.144
0.4328	0.199
0.4503	0.245
0.4812	0.347
0.5191	0.461
0.5506	0.545
0.5795	0.606
0.6100	0.652
0.6421	0.686*
0.6709	0.706
0.7000	0.720
CURVE 6 T = 300	
0.30	0.715
0.33	0.766
0.44	0.846
0.49	0.823
0.56	0.799
0.63	0.779
0.74	0.765
0.88	0.758
0.95	0.758
1.00	0.763
1.32	0.742
1.60	0.703
1.84	0.649
2.00	0.575
2.42	0.237
3.98	0.279
4.77	0.119
6.01	0.077
7.08	0.077
7.83	0.066
8.90	0.062
11.11	0.093
12.01	0.093
12.93	0.111
14.00	0.216
15.02	0.247
CURVE 7 T = 300	
0.30	0.667
0.33	0.735
0.37	0.787
0.41	0.819
0.45	0.838*
0.49	0.823
0.56	0.799
0.63	0.779
0.74	0.765
0.88	0.758
0.95	0.758
1.00	0.763
1.32	0.742
1.60	0.703
1.84	0.649
2.00	0.575
2.42	0.237
3.98	0.279
4.77	0.119
6.01	0.077
7.08	0.077
7.83	0.066
8.90	0.062
11.11	0.093
12.01	0.093
12.93	0.111
14.00	0.216
15.02	0.247
CURVE 8 T = 300	
0.30	0.122
0.34	0.146
0.37	0.181
0.40	0.243
0.43	0.322
0.48	0.481
0.50	0.606
0.53	0.661
0.58	0.698
0.62	0.722
0.67	0.737
0.71	0.745
0.91	0.745
0.99	0.744
1.51	0.716
1.80	0.661
1.84	0.649*
2.00	0.575*
2.42	0.237*
3.98	0.279*
4.77	0.119*
6.01	0.077*
7.08	0.077*
7.83	0.066*
8.90	0.062*
11.11	0.093*
12.01	0.093*
12.93	0.111*
14.00	0.216*
15.02	0.247*
CURVE 9 T = 300	
0.30	0.073
0.34	0.083
0.38	0.106
0.41	0.131
0.44	0.163
0.50	0.367
0.53	0.489
0.55	0.559*
0.59	0.639
0.64	0.689
0.68	0.713
0.72	0.724
0.83	0.730
0.99	0.730
1.34	0.713
1.66	0.692
1.84	0.649*
2.00	0.575*
2.42	0.237*
3.98	0.279*
4.77	0.119*
6.01	0.077*
7.08	0.077*
7.83	0.066*
8.90	0.062*
11.11	0.093*
12.01	0.093*
12.93	0.111*
14.00	0.216*
15.02	0.247*
CURVE 10 T ~ 298	
0.225	0.111
0.246	0.111
0.264	0.122
0.274	0.141
0.327	0.644
0.338	0.703
0.351	0.723
0.371	0.743
0.388	0.749
0.412	0.749
0.452	0.758
0.495	0.772
0.573	0.805
0.592	0.812
0.627	0.812
0.826	0.743
0.850	0.739*
0.869	0.739
0.997	0.758
1.271	0.787
1.307	0.787*
1.338	0.778
1.389	0.749
1.415	0.727*
1.429	0.724*
1.444	0.724
1.480	0.737
1.524	0.755
1.550	0.755*
1.574	0.741
1.670	0.596
1.686	0.588*
1.701	0.588
1.720	0.604
1.772	0.675
1.791	0.682*
1.812	0.682
1.836	0.668*
1.851	0.645*
1.870	0.574
1.885	0.562*
1.913	0.559
1.936	0.569
1.988	0.606*
2.001	0.612
2.025	0.612*
2.049	0.602
2.090	0.567*
2.131	0.511
2.201	0.336
2.226	0.298
2.260	0.277*
2.295	0.277
2.421	0.302
2.489	0.312
CURVE 11 T ~ 298	
0.246	0.102
0.266	0.130
0.284	0.165
0.307	0.655
0.320	0.692
0.348	0.726
0.383	0.763
0,404	0.815
0.450	0.852
0.504	0.872
0.541	0.877
0.594	0.877
0.701	0.855
0.853	0.842
1.117	0.839
1.214	0.829
1.297	0.809
1.358	0.779*
1.407	0.734
1.438	0.726*
1.503	0.736
1.540	0.736
1.609	0.720
1.688	0.683
1.751	0.638
1.836	0.559
1.950	0.415
1.965	0.406
2.019	0.412*
2.059	0.412
2.121	0.394
2.413	0.280
2.501	0.253

* Not shown on plot

FIGURE 9IA

ANALYZED NORMAL SOLAR ABSORPTANCE OF STRONTIUM MOLYBDATE PIGMENTED COATINGS

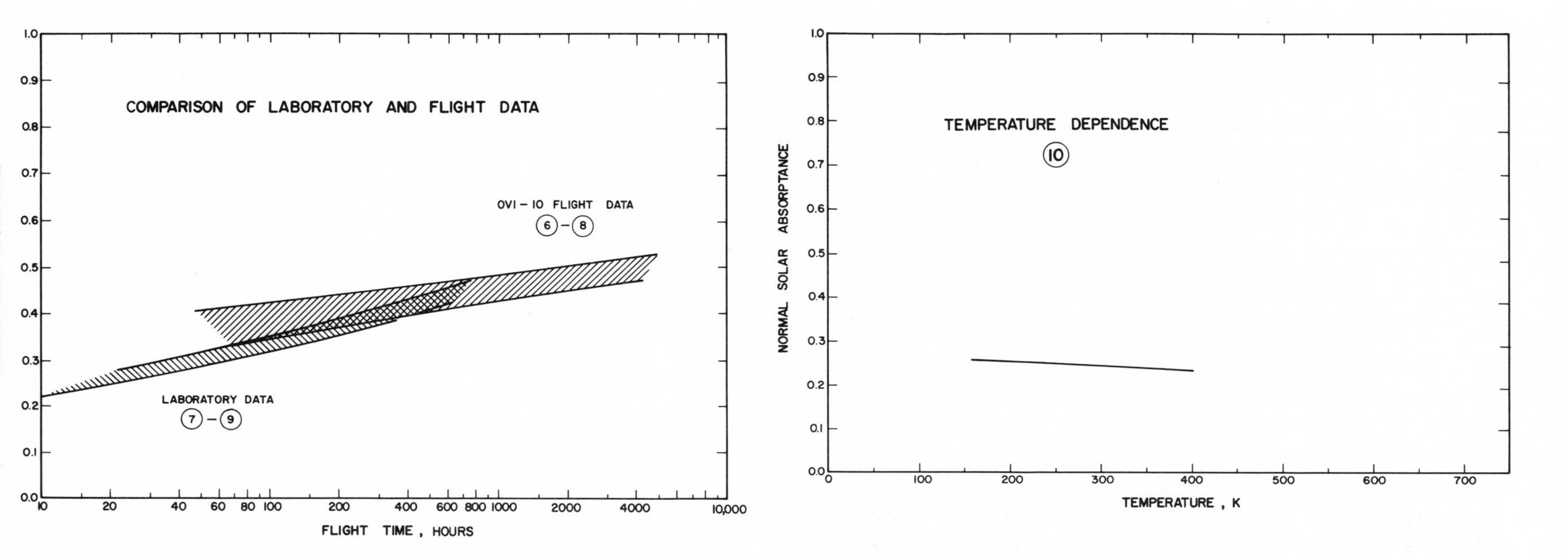

SPECIFICATION TABLE NO. 91 NORMAL SOLAR ABSORPTANCE OF STRONTIUM MOLYBDATE PIGMENTED COATINGS

Curve No.	Ref. No.	Year	Temperature Range, K	Geometry θ	Reported Error, %	Composition (weight percent), Specifications and Remarks
1	68	1969	300	~0°		Strontium molybdate in acrylic binder (~0.305 mm thick) on 2024 aluminum substrate (~1.59 mm thick); absorptance calculated from reflectance; property measured in air. [Authors' designation: $SrMoO_4$/acrylic]
2	68	1969	300	~0°		Similar to above specimen and conditions except exposed to vacuum (10^{-8} mm Hg) for 300 hrs; vacuum maintained by ion pump; property measured in air after vacuum exposure.
3	68	1969	300	~0°		Similar to curve 1 specimen and conditions except exposed to vacuum (10^{-5} mm Hg) for 300 hrs; vacuum maintained by oil-diffusion pump; property measured in air after vacuum exposure.
4	68	1969	300	~0°		Similar to above specimen and conditions except exposed to UV radiation in vacuum for 300 hrs; H_2 gas UV source.
5	68	1969	300	~0°		Similar to above specimen and conditions except He gas UV source.
6	85	1969	~300	~0°		$SrMoO_4$ in B-44 acrylic (Rohm and Haas) binder (~0.266 mm thick); Irridite 14 and 2024 clad aluminum substrates; PBR 1; aluminum substrate cleaned and etched with an alcohol-phosphoric acid mixture; 6 min Irridite 14 conversion coating on aluminum substrate; $SrMoO_4$/acrylic spray applied with filtered air, then air dried; data extracted from smooth curve; property calculated by calorimetric method from in-flight temp data of OVI-10; satellite flight time (hrs) is variable; actual UV ESH is from 0.25 to 0.375 of flight time. [Author's designation: B-1 $SrMoO_4$/acrylic]
7	85	1969	~300	~0°		Similar to above specimen and conditions except exposed to UV in vacuum; property converted from spectral reflectance measured in the lab in situ; actual UV ESH is one third of flight time.
8	85	1969	~300	~0°		$SrMoO_4$ in PS-7 potassium silicate binder (~0.139 mm thick) on Irridite 14 and 2024 clad aluminum substrates; PBR 3; aluminum substrate cleaned and etched with an alcohol-phosphoric acid mixture; 6 min Irridite 14 conversion coating on aluminum substrate; $SrMoO_4$/K-Silicate spray applied with lamp grade nitrogen; cured at 423 K and air dried; data extracted from smooth curve; property calculated by calorimetric method from in-flight temp data of OVI-10; satellite flight time (hrs) is variable; actual UV ESH is from 0.25 to 0.375 of flight time. [Author's designation: B-5 $SrMoO_4$/K-Silicate]
9	85	1969	~300	~0°		Similar to above specimen and conditions except exposed to UV in vacuum; property converted from spectral reflectance measured in the lab in situ; actual UV ESH is one third of flight time.
10	85	1969	156-400	~0°		$SrMoO_4$ in PS-7 potassium silicate binder (~0.139 mm thick); Irridite 14 and 2024 clad aluminum substrates; PBR 3; aluminum substrate cleaned and etched with an alcohol-phosphoric acid mixture; 6 min Irridite 14 conversion coating on aluminum substrate; $SrMoO_4$/K-Silicate spray applied with lamp grade nitrogen; cured at 423 K and air dried; property measured in vacuum by calorimetric method; data extracted from smooth curve. [Author's designation: B-5 $SrMoO_4$/K-Silicate, NRDL-RTD-81-6]

DATA TABLE NO. 91 NORMAL SOLAR ABSORPTANCE OF STRONTIUM MOLYBDATE PIGMENTED COATINGS

[Temperature, T, K; Absorptance, α]

T	α
CURVE 1	
300	0.28
CURVE 2	
300	0.29
CURVE 3	
300	0.30
CURVE 4	
300	0.44
CURVE 5	
300	0.50

Flight time	α
CURVE 6 ($T \sim 300$)	
47.1	0.405
497	0.465
4850	0.527
CURVE 7 ($T \sim 300$)	
22.8	0.283
141	0.361
743	0.474
CURVE 8 ($T \sim 300$)	
69.7	0.336
545	0.399
4090	0.474
CURVE 9 ($T \sim 300$)	
10.0	0.221
76.0	0.306
593	0.425

T	α
CURVE 10	
156	0.260
279	0.247
400	0.232

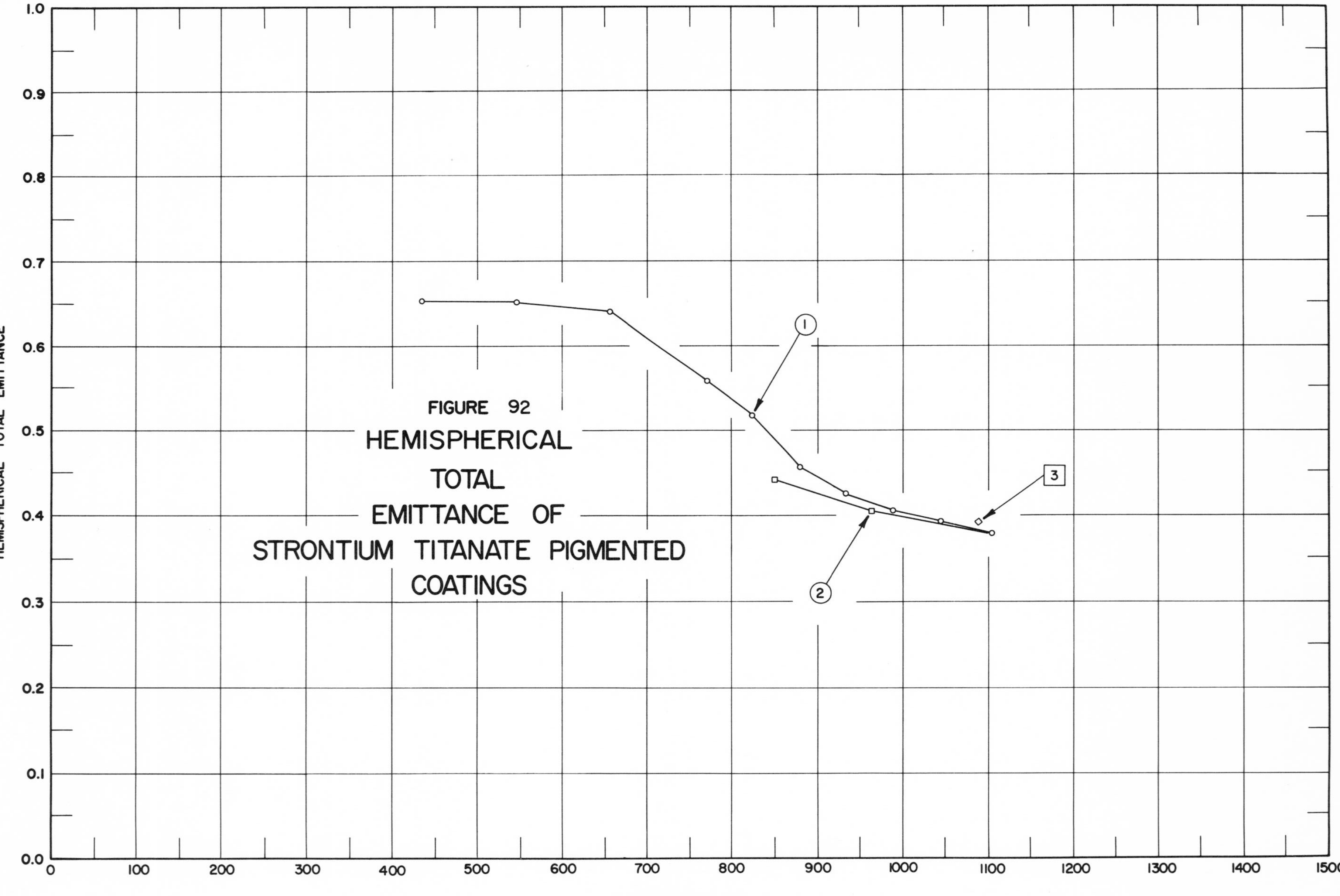

FIGURE 92 HEMISPHERICAL TOTAL EMITTANCE OF STRONTIUM TITANATE PIGMENTED COATINGS

SPECIFICATION TABLE NO. 92 HEMISPHERICAL TOTAL EMITTANCE OF STRONTIUM TITANATE PIGMENTED COATINGS

Curve No.	Ref. No.	Year	Temperature Range, K	Reported Error, %	Composition (weight percent), Specifications and Remarks
1	81	1963	434-1102		Strontium titanate (Plasmadyne Corp) in aluminum phosphate binder (0.254 mm thick) on AISI-310 stainless steel substrate; measured in vacuum (<8.3 x 10^{-6} mm Hg); temp measured with thermocouple; Run No. 1, heating cycle.
2	81	1963	1102-850		Above specimen and conditions; cooling cycle.
3	81	1963	1093		Curve 1 specimen and conditions except temp measured with optical pyrometer.

DATA TABLE NO. 92 HEMISPHERICAL TOTAL EMITTANCE OF STRONTIUM TITANATE PIGMENTED COATINGS

[Temperature, T, K; Emittance, ϵ]

T	ϵ
CURVE 1	
434	0.653
547	0.652
658	0.642
770	0.558
827	0.518
880	0.456
935	0.426
990	0.406
1045	0.394
1102	0.379
CURVE 2	
1102	0.379*
962	0.405
850	0.443
CURVE 3	
1093	0.391

* Not shown on plot

SPECIFICATION TABLE NO. 93 HEMISPHERICAL TOTAL EMITTANCE OF STRONTIUM ZIRCONATE PIGMENTED COATINGS

Curve No.	Ref. No.	Year	Temperature Range, K	Reported Error, %	Composition (weight percent), Specifications and Remarks
1*	49	1961	278		Strontium zirconate in silicone resin binder; Dow 17 treated Mg alloy substrate; exposed in vacuum (10^{-6} - 8 x 10^{-6} mm Hg) to UV radiation from an argon filled A-H6 high pressure Hg arc lamp for 127 hrs.

DATA TABLE NO. 93 HEMISPHERICAL TOTAL EMITTANCE OF STRONTIUM ZIRCONATE PIGMENTED COATINGS

[Temperature, T, K; Emittance, ϵ]

T	ϵ
CURVE 1*	
278	0.85

* No plot given

SPECIFICATION TABLE NO. 94 NORMAL SOLAR ABSORPTANCE OF STRONTIUM ZIRCONATE PIGMENTED COATINGS

Curve No.	Ref. No.	Year	Temperature Range, K	Geometry θ	Reported Error, %	Composition (weight percent), Specifications and Remarks
1*	49	1961	278	$\sim 0^\circ$		Strontium zirconate in silicone resin binder; Dow 17 treated Mg alloy substrate; exposed in vacuum (10^{-6} - 8 x 10^{-6} mm Hg) to UV radiation from an argon filled A-H6 high pressure Hg arc lamp for 127 hrs.

DATA TABLE NO. 94 NORMAL SOLAR ABSORPTANCE OF STRONTIUM ZIRCONATE PIGMENTED COATINGS

[Temperature, T, K; Absorptance, α]

T	α
CURVE 1*	
278	0.47

* No plot given

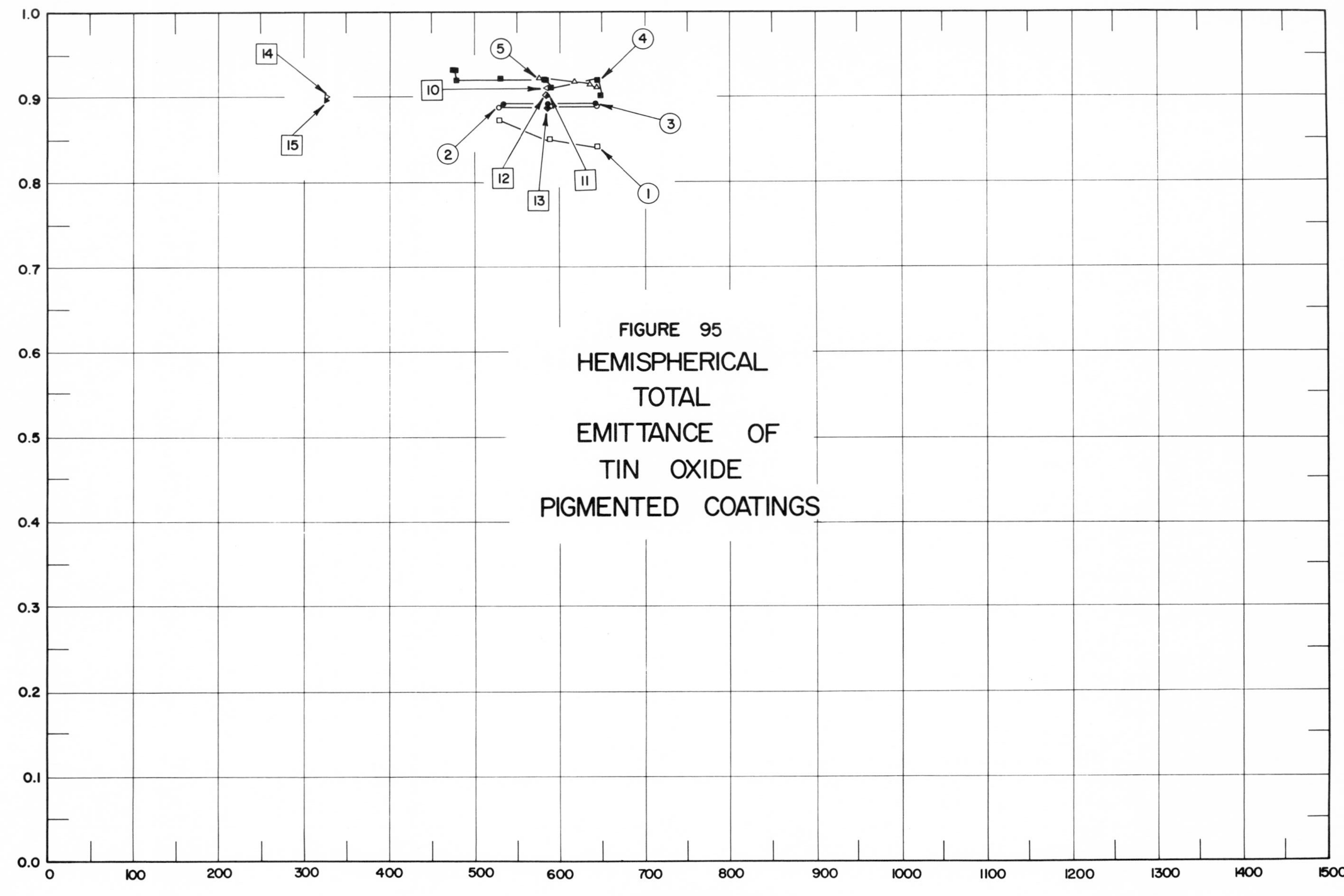

FIGURE 95
HEMISPHERICAL TOTAL EMITTANCE OF TIN OXIDE PIGMENTED COATINGS
HEMISPHERICAL TOTAL EMITTANCE
TEMPERATURE, K
1.0 0.9 0.8 0.7 0.6 0.5 0.4 0.3 0.2 0.1 0.0
0 100 200 300 400 500 600 700 800 900 1000 1100 1200 1300 1400 1500
14
15
5
10
2
12
13
11
4
3
1

SPECIFICATION TABLE NO. 95 HEMISPHERICAL TOTAL EMITTANCE OF TIN OXIDE PIGMENTED COATINGS

Curve No.	Ref. No.	Year	Temperature Range, K	Reported Error, %	Composition (weight percent), Specifications and Remarks
1	15	1965	533-644		SnO_2 in potassium silicate binder (0.0762 mm thick); substrate unknown.
2	15	1965	533-644		SnO_2 in aluminum phosphate binder (0.0762 mm thick); substrate unknown.
3	15	1965	533-644		AI 93; SnO_2 in aluminum phosphate binder (0.0762 mm thick) on Cr-Co-Ni spinel in aluminum phosphate binder substrate.
4	15	1965	477-649		AI 93; SnO_2 in aluminum phosphate binder on Cr-Co-Ni spinel in aluminum phosphate substrate; composite coating thickness 0.0762 mm; measured in vacuum (1.0×10^{-5} - 4.0×10^{-7} mm Hg).
5	15	1965	477-644		Similar to the above specimen and conditions.
6*	15	1965	~589		AI 93 (tailored, adaptable to copper substrate); SnO_2 in aluminum phosphate binder; maintained in vacuum (10^{-5} mm Hg) at 589 K; time (hrs), at indicated temp, is variable.
7*	15	1965	~589		AI 93 (tailored, adaptable to aluminum substrates); SnO_2 in aluminum phosphate binder; maintained in vacuum (10^{-5} mm Hg) at 589 K; measured in situ; time (hrs), at indicated temp, is variable.
8*	15	1965	584-589		AI 93 (tailored, applicable to copper substrate); SnO_2 in aluminum phosphate binder (0.0206 mm thick) on Cr-Co-Ni spinel in aluminum phosphate binder (0.0381 mm thick) and vitreous ceramic (0.0206 mm thick) substrates; irradiated in vacuum (10^{-5} mm Hg) to a neutron flux leveled up to 3.0×10^{18} nvt and a max gamma exposure of 1.9×10^{10} R over a period of 13 days; measured in situ. [Author's designation: Specimen No. 1]
9*	15	1965	~589		Above specimen and conditions; neutron flux, nvt, is variable.
10	53	1966	588		SnO_2 in aluminum phosphate binder (0.0152 mm thick); measured in vacuum (10^{-5} mm Hg).
11	53	1966	588		Above specimen and conditions except exposed to fast neutron irradiation of 3.0×10^{18} neutrons cm^{-2} and gamma irradiation of 1.9×10^{10} R in vacuum.
12	53	1966	588		SnO_2 in aluminum phosphate binder (0.0152 mm thick).
13	53	1966	588		Above specimen and conditions except exposed in air to 3.0×10^{18} e cm^{-2} at 1.25 MeV.
14	53	1966	329		SnO_2 in aluminum phosphate binder (0.0152 mm thick).
15	53	1966	329		Above specimen and conditions except exposed in air to 3.0×10^{18} e cm^{-2} at 1.25 MeV.

* Not shown on plot

DATA TABLE NO. 95 HEMISPHERICAL TOTAL EMITTANCE OF TIN OXIDE PIGMENTED COATINGS

[Temperature, T, K; Emittance, ϵ]

CURVE 1

T	ϵ
533	0.87
588	0.85
644	0.84

CURVE 2

T	ϵ
533	0.88
588	0.88
644	0.88

CURVE 3

T	ϵ
533	0.91
588	0.91
644	0.91

CURVE 4

T	ϵ
477	0.93
478	0.93
479	0.92
533	0.92
587	0.92
588	0.92
590	0.91
644	0.92
646	0.92
649	0.90

CURVE 5

T	ϵ
477	0.930
577	0.923
619	0.919
633	0.915
644	0.911

CURVE 6*
T ~ 589

Time	ϵ
24	0.91
1000	0.91
2000	0.91
3000	0.91

CURVE 6 (cont.)*

Time	ϵ
4000	0.93
5900	0.93

CURVE 7*
T ~ 589

Time	ϵ
24	0.90
1000	0.90
2000	0.90
3000	0.92
4900	0.92

CURVE 8*

T	ϵ
589	0.93
586	0.93
586	0.93
587	0.94
586	0.94
587	0.94
586	0.94
588	0.94
587	0.94
588	0.92
587	0.93
588	0.94
588	0.93
588	0.94
588	0.93
588	0.94
586	0.93
587	0.93
588	0.93
588	0.93
584	0.94

CURVE 9*
T ~ 589

Neutron flux	ϵ
0	0.93
4.3×10^{16}	0.93
3.2×10^{17}	0.93
4.5×10^{17}	0.94
5.8×10^{17}	0.94
7.1×10^{17}	0.94
8.4×10^{17}	0.94
9.7×10^{17}	0.94
1.1×10^{18}	0.94
1.2×10^{18}	0.92
1.3×10^{18}	0.93
1.5×10^{18}	0.94
1.6×10^{18}	0.93
1.8×10^{18}	0.94
1.9×10^{18}	0.93
2.0×10^{18}	0.94
2.3×10^{18}	0.93
2.5×10^{18}	0.93
2.8×10^{18}	0.93
3.0×10^{18}	0.93
3.0×10^{18}	0.94

CURVE 10*

T	ϵ
588	0.91

CURVE 11

T	ϵ
588	0.90

CURVE 12

T	ϵ
588	0.90

CURVE 13

T	ϵ
588	0.88

CURVE 14

T	ϵ
329	0.90

CURVE 15

T	ϵ
329	0.89

* Not shown on plot

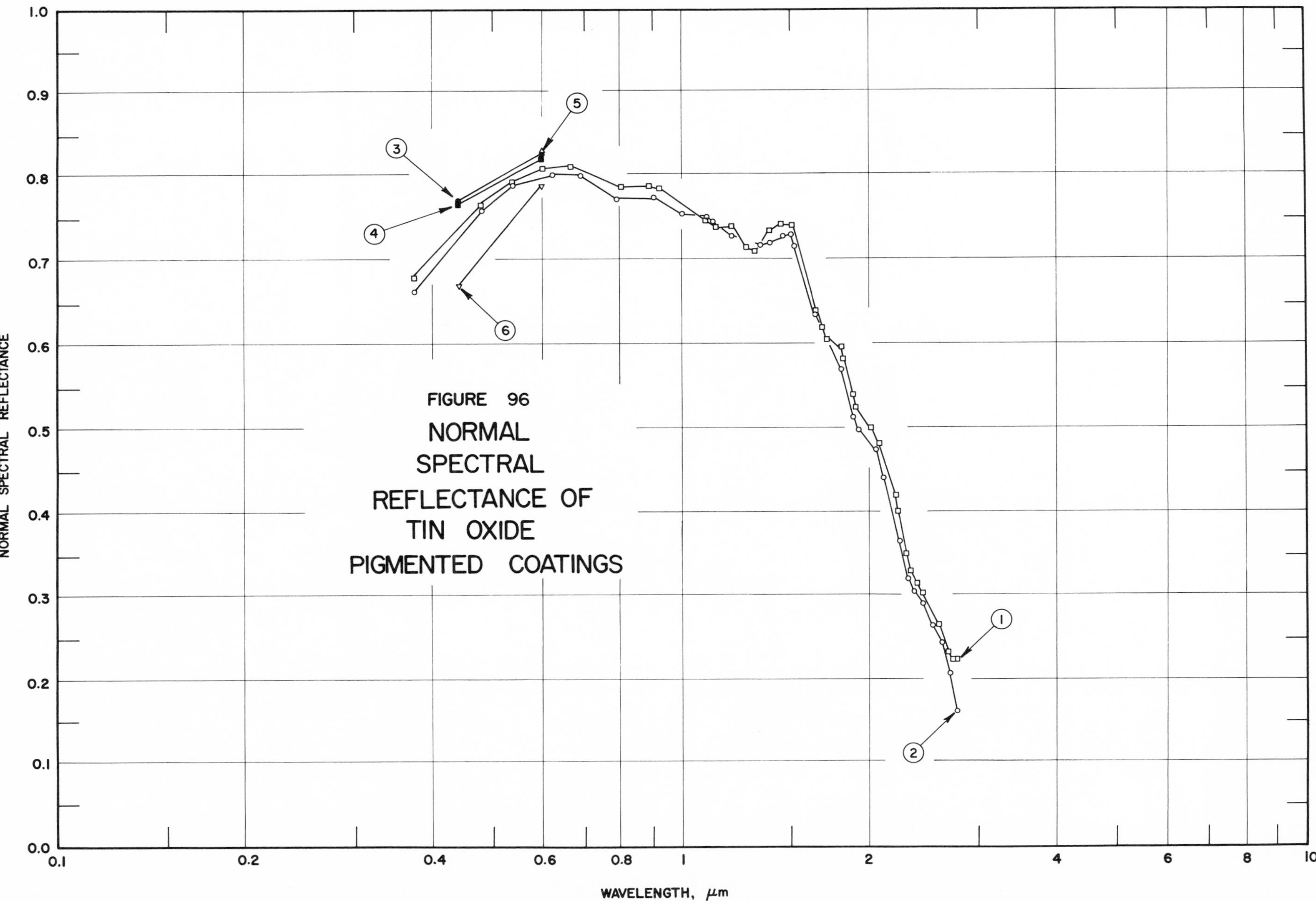
FIGURE 96
NORMAL
SPECTRAL
REFLECTANCE OF
TIN OXIDE
PIGMENTED COATINGS
NORMAL SPECTRAL REFLECTANCE
WAVELENGTH, μm
1.0
0.9
0.8
0.7
0.6
0.5
0.4
0.3
0.2
0.1
0.0
0.1
0.2
0.4
0.6
0.8
1
2
4
6
8
10
1
2
3
4
5
6

SPECIFICATION TABLE NO. 96 NORMAL SPECTRAL REFLECTANCE OF TIN OXIDE PIGMENTED COATINGS

Curve No.	Ref. No.	Year	Temperature K	Wavelength Range, μm	Geometry θ	θ'	ω'	Reported Error, %	Composition (weight percent), Specifications, and Remarks
1	20	1963	~298	0.375-2.768	~0°		2π		Tin oxide in PS-7 potassium silicate binder on aluminum substrate; PBR 4.30; solids content 61.7%; paint ball milled; substrate grit blasted with 40 mesh silicone carbide; air brush application; sample cured at 413 K for 18 hrs; data extracted from smooth curve. [Authors' designation: Sample C13]
2	20	1963	~298	0.375-2.762	~0°		2π		Above specimen and conditions except exposed in vacuum (10^{-5} mm Hg) to 300 ESH of UV radiation at solar factor of 3 suns.
3	27	1964	~298	0.44-0.60	~0°		2π		SnO_2 in PS-7 potassium silicate binder on aluminum substrate; PBR 4.30; solids content 61.7%; substrate abraded with No. 60 Aloxite cloth; sample cured at 413 K for 18 hrs. [Authors' designation: Sample C13]
4	27	1964	~298	0.44-0.60	~0°		2π		Above specimen and conditions except irradiated in vacuum (10^{-6} mm Hg) with 300 ESH of UV radiation at solar factor of 3 suns.
5	27	1964	~298	0.44-0.60	~0°		2π		SnO_2 in PS-7 potassium silicate binder on aluminum substrate; PBR 4.30; solids content 61.7%; substrate abraded with No. 60 Aloxite cloth; sample cured at 413 K for 18 hrs. [Authors' designation: Sample C17]
6	27	1964	~298	0.44-0.60	~0°		2π		Above specimen and conditions except irradiated in vacuum (10^{-6} mm Hg) with 2100 ESH of UV radiation at solar factor of 10 suns.

DATA TABLE NO. 96 NORMAL SPECTRAL REFLECTANCE OF TIN OXIDE PIGMENTED COATINGS

[Wavelength, λ, μm; Reflectance, ρ; Temperature, T, K]

CURVE 1
T ~ 298

λ	ρ
0.375	0.679
0.479	0.766
0.536	0.792
0.601	0.808
0.669	0.810
0.804	0.785
0.891	0.787
0.923	0.784
1.080	0.747
1.148	0.739
1.205	0.739
1.279	0.715
1.324	0.710
1.397	0.732
1.450	0.740
1.510	0.740
1.643	0.640
1.683	0.620
1.726	0.609
1.805	0.597
1.829	0.584
1.883	0.540
1.918	0.524
2.032	0.500
2.088	0.482
2.206	0.420
2.228	0.402
2.290	0.351
2.318	0.331
2.380	0.315
2.448	0.303
2.597	0.268
2.680	0.232
2.710	0.226
2.768	0.226

CURVE 2
T ~ 298

λ	ρ
0.375	0.663
0.481	0.756
0.536	0.786
0.625	0.800
0.694	0.800
0.791	0.771
0.905	0.773
1.001	0.755
1.100	0.751
1.138	0.744
1.201	0.726
1.343	0.715
1.384	0.719
1.466	0.728
1.502	0.728
1.527	0.715
1.647	0.633
1.805	0.571
1.888	0.514
1.936	0.498
2.040	0.475
2.113	0.440
2.246	0.367
2.309	0.320
2.378	0.305
2.452	0.292
2.536	0.265
2.619	0.245
2.694	0.209
2.762	0.162

CURVE 3
T ~ 298

λ	ρ
0.44	0.770
0.60	0.825

C **CURVE 4**
T ~ 298

λ	ρ
0.44	0.765
0.60	0.820

CURVE 5
T ~ 298

λ	ρ
0.44	0.765
0.60	0.830

CURVE 6
T ~ 298

λ	ρ
0.44	0.670
0.60	0.785

SPECIFICATION TABLE NO. 97 NORMAL SOLAR ABSORPTANCE OF TIN OXIDE PIGMENTED COATINGS

Curve No.	Ref. No.	Year	Temperature Range, K	Geometry θ	Reported Error, %	Composition (weight percent), Specifications and Remarks
1*	53	1966	298	~0°		SnO_2 in aluminum phosphate binder (0.0381 mm thick); measured in vacuum (10^{-6} mm Hg).
2*	53	1966	298	~0°		Above specimen and conditions except exposed to 1.0 x 10^{16} protons cm^{-2} at 150 keV in vacuum.
3*	53	1966	298	~0°		SnO_2 in aluminum phosphate binder (0.0381 mm thick); measured in vacuum (10^{-6} mm Hg).
4*	53	1966	298	~0°		Above specimen and conditions except exposed to 1.0 x 10^{15} protons cm^{-2} at 150 keV in vacuum.
5*	52	1965	193-263	~0°		SnO_2 in silicate binder on chrome-nickel spinel in phosphate binder substrate; measured in earth orbit on OSO-II; property avg over temp range.
6*	27	1964	~298	~0°		SnO_2 in PS7 potassium silicate binder on aluminum substrate; PBR 4.30; solids content 61.7%; substrate abraded with No. 60 Aloxite cloth; sample cured at 413 K for 18 hrs. [Authors' designation: Sample C 13]
7*	27	1964	~298	~0°		Above specimen and conditions except irradiated in vacuum (10^{-6} mm Hg) with 300 ESH of UV radiation at solar factor of 3 suns.

DATA TABLE NO. 97 NORMAL SOLAR ABSORPTANCE OF TIN OXIDE PIGMENTED COATINGS

[Temperature, T, K; Absorptance, α]

T	α	T	α
CURVE 1*		CURVE 6*	
298	0.35	298	0.264
CURVE 2*		CURVE 7*	
298	0.50	298	0.278
CURVE 3*			
298	0.35		
CURVE 4*			
298	0.42		
CURVE 5*			
193	0.235		
263	0.235		

* No plot given

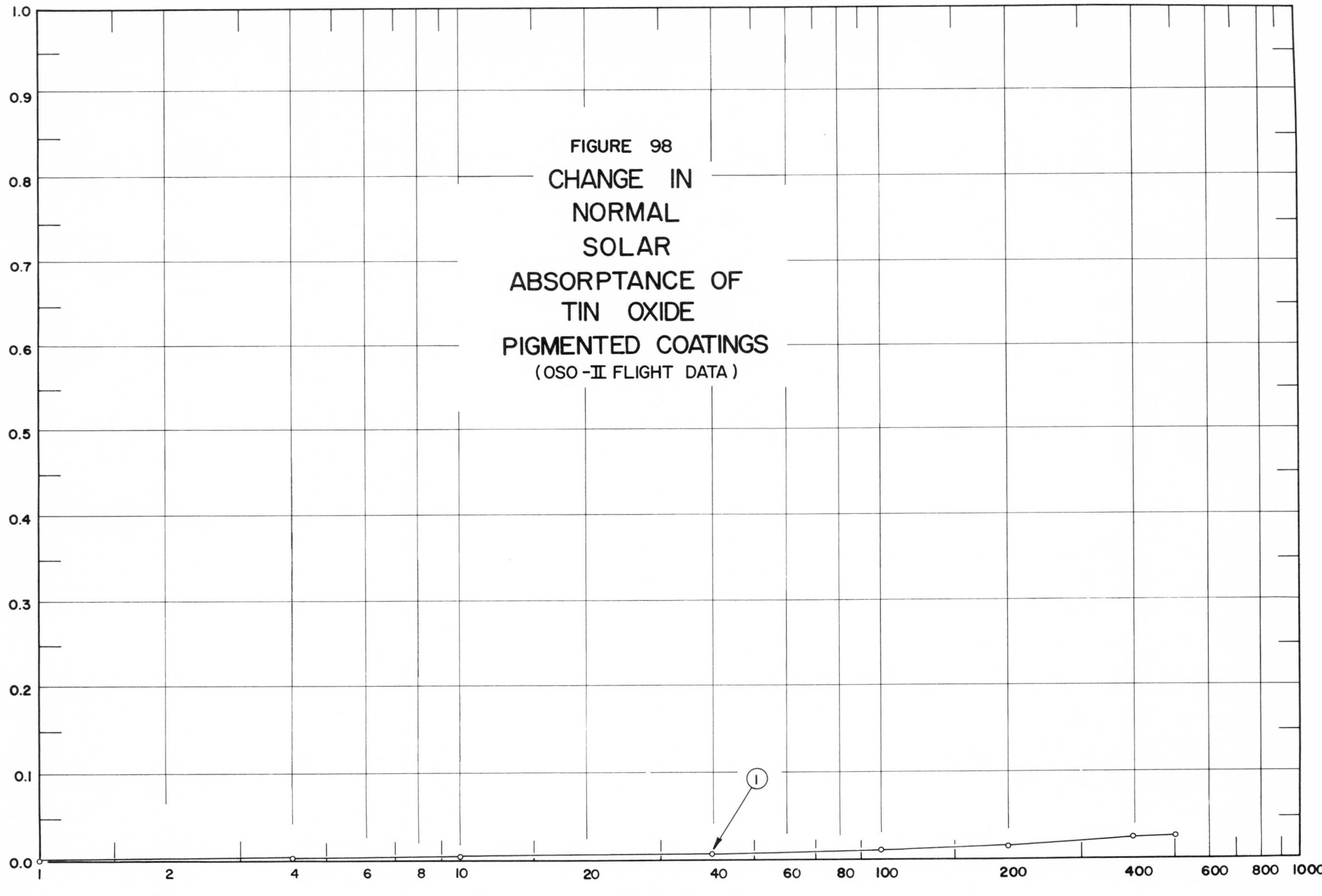

FIGURE 98
CHANGE IN NORMAL SOLAR ABSORPTANCE OF TIN OXIDE PIGMENTED COATINGS
(OSO-II FLIGHT DATA)

SPECIFICATION TABLE NO. 98 CHANGE IN NORMAL SOLAR ABSORPTANCE OF TIN OXIDE PIGMENTED COATINGS

Curve No.	Ref. No.	Year	Temperature Range, K	Geometry θ	Reported Error, %	Composition (weight percent), Specifications and Remarks
1	52	1966	~298	~0°		Tin dioxide in silicate binder on chrome-nickel spinel in phosphate binder substrate; data taken from OSO-II flight experiment; variable is equivalent length of time exposed to the sun.

DATA TABLE NO. 98 CHANGE IN NORMAL SOLAR ABSORPTANCE OF TIN OXIDE PIGMENTED COATINGS

[Temperature, T, K; Absorptance, α]

ESH	$\Delta\alpha$
CURVE 1	
T ~ 298	
1.00	0.000
3.98	0.002
10.0	0.003
39.8	0.007
100	0.011
200	0.015
398	0.025
501	0.026

SPECIFICATION TABLE NO. 99 HEMISPHERICAL TOTAL EMITTANCE OF TITANIUM CARBIDE PIGMENTED COATINGS

Curve No.	Ref. No.	Year	Temperature Range, K	Reported Error, %	Composition (weight percent), Specifications and Remarks
1*	67	1967	1371-1712		TiC in silica binder on disilicide and TZM molybdenum substrates; measured in air.
2*	67	1967	1369-1919		Similar to above specimen and conditions except measured at reduced pressure with sonic velocity air passed over sample.

DATA TABLE NO. 99 HEMISPHERICAL TOTAL EMITTANCE OF TITANIUM CARBIDE PIGMENTED COATINGS

[Temperature, T, K; Emittance, ϵ]

T	ϵ	T	ϵ	T	ϵ	T	ϵ	T	ϵ	T	ϵ	T	ϵ	T	ϵ
CURVE 1*		CURVE 2 (cont.)		CURVE 2 (cont.)		CURVE 2 (cont.)		CURVE 2 (cont.)		CURVE 2 (cont.)		CURVE 2 (cont.)		CURVE 2 (cont.)	
1371	0.787	1385	0.891	1474	0.803	1618	0.821	1688	0.799	1753	0.885	1858	0.935	1897	0.909
1387	0.787	1387	0.876	1527	0.854	1624	0.866	1691	0.891	1753	0.870	1865	0.897	1898	0.936
1430	0.849	1409	0.878	1527	0.869	1629	0.878	1691	0.868	1755	0.828	1869	0.910	1900	0.947
1525	0.798	1414	0.907	1531	0.879	1629	0.855	1692	0.858	1757	0.850	1870	0.949	1900	0.887
1595	0.809	1415	0.836	1533	0.868	1630	0.849	1693	0.791	1759	0.885	1872	0.914	1902	0.976
1663	0.817	1419	0.826	1535	0.805	1633	0.904	1695	0.868	1762	0.860	1874	0.985	1903	0.966
1710	0.836	1421	0.808	1537	0.839	1634	0.877	1698	0.789	1788	0.914	1875	0.945	1904	0.950
1712	0.855	1422	0.848	1539	0.855	1635	0.893	1699	0.860	1788	0.855	1882	0.945	1906	0.930
		1424	0.825	1577	0.878	1635	0.865	1707	0.839	1796	0.963	1882	0.897	1906	0.908
CURVE 2*		1457	0.873	1583	0.898	1638	0.898	1709	0.805	1797	0.915	1887	0.903	1907	0.896
		1465	0.819	1583	0.849	1639	0.857	1732	0.843	1818	0.965	1888	0.925	1908	0.941
1369	0.778	1465	0.823	1592	0.878	1640	0.878	1742	0.880	1822	0.894	1891	0.937	1909	0.919
1370	0.879	1466	0.866	1595	0.860	1656	0.785	1742	0.936	1825	0.964	1892	0.978	1912	0.899
1370	0.824	1466	0.797	1603	0.867	1668	0.888	1748	0.876	1835	0.966	1892	0.954	1914	0.966
1371	0.834	1469	0.873	1607	0.883	1673	0.878	1748	0.838	1835	0.892	1894	0.904	1914	0.909
1371	0.809	1469	0.846	1608	0.857	1675	0.851	1749	0.864	1840	0.995	1895	0.927	1917	0.948
1373	0.800	1472	0.877	1612	0.864	1679	0.815	1751	0.905	1846	0.944	1895	0.917	1918	0.968
1380	0.869	1473	0.836	1612	0.819	1682	0.882	1751	0.816	1853	0.955	1897	0.960	1918	0.931
1381	0.812	1474	0.895	1617	0.867	1685	0.897	1752	0.842	1855	0.954			1918	0.914
														1919	0.900

* No plot given

SPECIFICATION TABLE NO. 100 HEMISPHERICAL TOTAL EMITTANCE OF TITANIUM HYDRIDE PIGMENTED COATINGS

Curve No.	Ref. No.	Year	Temperature Range, K	Reported Error, %	Composition (weight percent), Specifications and Remarks
1*	97	1961	773-1473	± 2.5	Titanium hydride in necoloidine (200-300 mesh) binder (6 x 10^{-3} gm cm^{-2}) on molybdenum ribbon substrate; sintered at 1573 K in vacuum; measured in vacuum (<5 x 10^{-6} mm Hg); data extracted from smooth curve.

DATA TABLE NO. 100 HEMISPHERICAL TOTAL EMITTANCE OF TITANIUM HYDRIDE PIGMENTED COATINGS

[Temperature, T, K; Emittance, ϵ]

T	ϵ
CURVE 1*	
773	0.610
873	0.630
973	0.655
1073	0.680
1173	0.700
1273	0.715
1373	0.730
1473	0.745

* No plot given

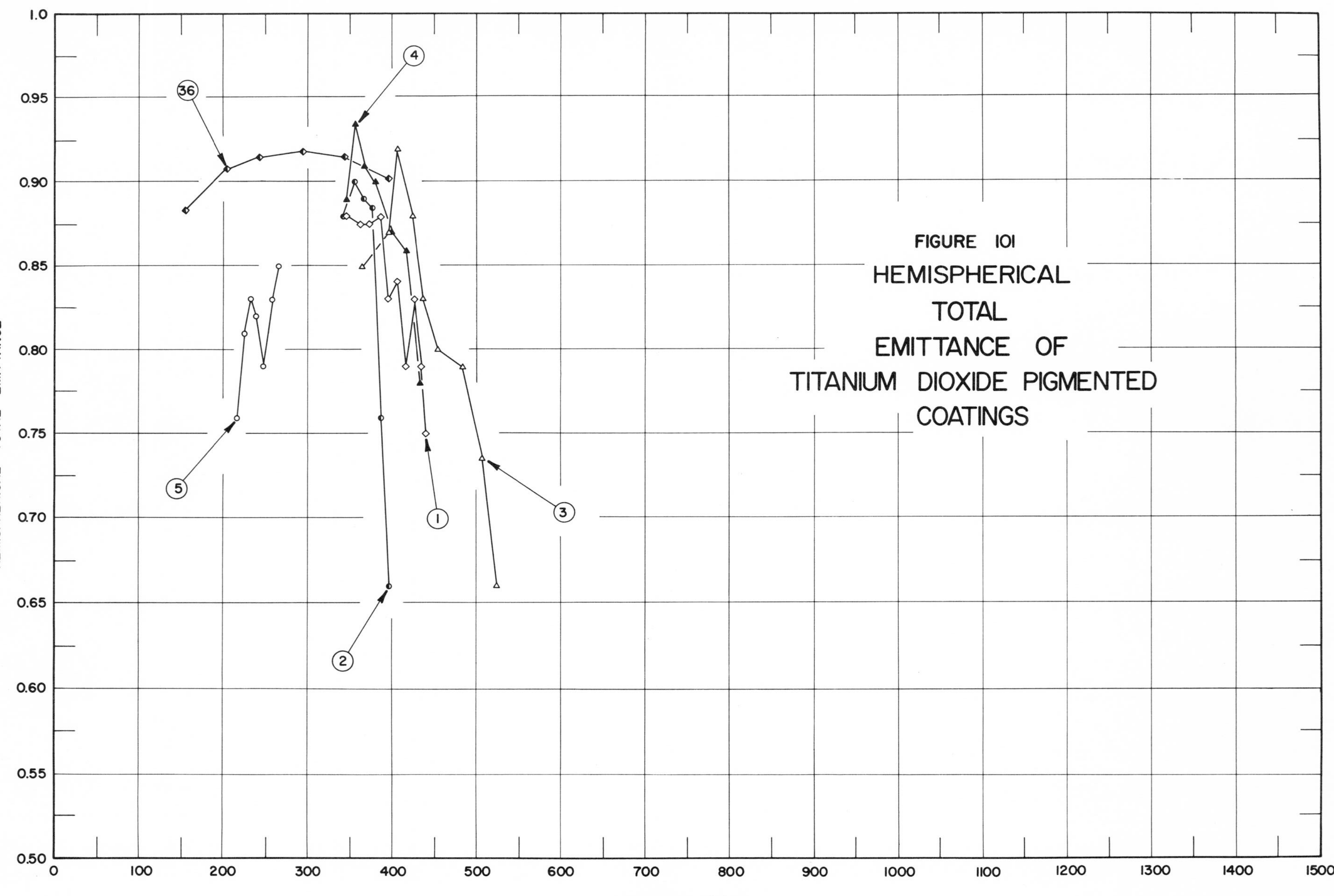

FIGURE 101 HEMISPHERICAL TOTAL EMITTANCE OF TITANIUM DIOXIDE PIGMENTED COATINGS

SPECIFICATION TABLE NO. 101 HEMISPHERICAL TOTAL EMITTANCE OF TITANIUM DIOXIDE PIGMENTED COATINGS

Curve No.	Ref. No.	Year	Temperature Range, K	Reported Error, %	Composition (weight percent), Specifications and Remarks
1	43	1962	349-440	±3	TiO_2 in epoxide binder (~0.0762 mm thick); measured in vacuum (10^{-3} mm Hg); data extracted from smooth curve.
2	43	1962	346-397	±3	TiO_2 in nitrocellulose binder (~0.0762 mm thick); measured in vacuum (10^{-3} mm Hg); data extracted from smooth curve.
3	43	1962	365-523	±3	TiO_2 in acrylic binder (~0.0762 mm thick); measured in vacuum (10^{-3} mm Hg); data extracted from smooth curve.
4	43	1962	349-431	±3	Similar to above specimen and conditions except aged at 423 K for 2000 hrs.
5	48	1963	218-267	±3	TiO_2 in epoxide binder on aluminum alloy substrate; measured in vacuum (10^{-3}mm Hg).
6*	50	1965	~298		LMSC/Dow Corning thermatrol paint; TiO_2 (rutile) in silicone binder.
7*	50	1965	~298		White Skyspar enamel, A. Brown Co., A423 Color SA9185; TiO_2 in epoxy binder.
8*	50	1965	~298		Fuller 517-W-7 gloss white paint; TiO_2 in silicone-modified alkyd binder; cured at 513 K.
9*	67	1967	1750		TiO_2 in DC806A (contains titanium carbide) binder on TZM molybdenum substrate.
10*	49	1961	278		White Skyspar enamel (A423-SA9185, A. Brown Co.); TiO_2, untinted, in epoxy resin binder on Mg alloy substrate.
11*	49	1961	278		Above specimen and conditions except exposed in vacuum (10^{-6}-8 x 10^{-6} mm Hg) to UV radiation from an argon filled A-H6 high pressure Hg arc lamp for 100 hrs.
12*	49	1961	278		White Skyspar enamel (SA-8818 A423-SA8818, A. Brown Co.); TiO_2, tinted, in epoxy resin binder on Dow 15 treated Mg alloy substrate.
13*	49	1961	278		Above specimen and conditions except exposed in vacuum (10^{-6}-8 x 10^{-6} mm Hg) to UV radiation from an argon filled A-H6 high pressure Hg arc lamp for 46 hrs.
14*	49	1961	278		White Skyspar enamel (SA-8818 A423-SA8818, A. Brown Co.); TiO_2, tinted, in epoxy resin binder on Dow 17 treated Mg alloy substrate.
15*	49	1961	278		Above specimen and conditions except exposed in vacuum (10^{-6}-8 x 10^{-6} mm Hg) to UV radiation from an argon filled A-H6 high pressure Hg arc lamp for 80 hrs.
16*	49	1961	278		Similar to curve 14 specimen and conditions.
17*	49	1961	278		White Skyspar enamel (A423-SA9185, A. Brown Co.); TiO_2, untinted, in epoxy resin binder on Dow 15 treated Mg alloy substrate.
18*	49	1961	278		Above specimen and conditions except exposed in vacuum (10^{-6}-8 x 10^{-6} mm Hg) to UV radiation from an argon filled A-H6 high pressure Hg arc lamp for 12 hrs.
19*	49	1961	278		White Skyspar enamel (A423-SA9185, A. Brown Co.); TiO_2, untinted, in epoxy resin binder on Dow 17 treated Mg alloy substrate.
20*	49	1961	278		Above specimen and conditions except exposed in vacuum (10^{-6}-8 x 10^{-6} mm Hg) to UV radiation from an argon filled A-H6 high pressure Hg arc lamp for 26 hrs.

* Not shown on plot

SPECIFICATION TABLE NO. 101 HEMISPHERICAL TOTAL EMITTANCE OF TITANIUM DIOXIDE PIGMENTED COATINGS (continued)

Curve No.	Ref. No.	Year	Temperature Range, K	Reported Error, %	Composition (weight percent), Specifications and Remarks
21*	49	1961	278		Fuller gloss white silicone 517-W-1; TiO_2 in silicone binder; substrate unknown; value avg of several determinations.
22*	49	1961	278		Fuller gloss white silicone 517-W-1; TiO_2 in silicone binder on Mg alloy substrate.
23*	49	1961	278		Fuller gloss white silicone 517-W-1; TiO_2 in silicone binder on Dow 15 treated Mg alloy substrate.
24*	49	1961	278		Above specimen and conditions except exposed in vacuum (10^{-6}-8 x 10^{-6} mm Hg) to UV radiation from an argon filled A-H6 high pressure Hg arc lamp for 46 hrs; vacuum maintained by VacIon pump.
25*	49	1961	278		Fuller gloss white silicone 517-W-1; TiO_2 in silicone binder on Mg alloy substrate.
26*	49	1961	278		Above specimen and conditions except exposed in vacuum (10^{-6}-8 x 10^{-6} mm Hg) to UV radiation from an argon filled A-H6 high pressure Hg arc lamp for 12 hrs; vacuum maintained by VacIon pump.
27*	49	1961	278		Fuller gloss white silicone 517-W-1; TiO_2 in silicone on Mg alloy substrate.
28*	49	1961	278		Above specimen and conditions except exposed in vacuum (10^{-6}-8 x 10^{-6} mm Hg) to UV radiation from an argon filled A-H6 high pressure Hg arc lamp for 100 hrs; vacuum maintained by VacIon pump.
29*	49	1961	278		Fuller gloss white silicone 517-W-1; TiO_2 in silicone binder on Dow 15 treated Mg alloy substrate.
30*	49	1961	278		Above specimen and conditions except exposed in vacuum (10^{-6}-8 x 10^{-6} mm Hg) to UV radiation from an argon filled A-H6 high pressure Hg arc lamp for 26 hrs; vacuum maintained by VacIon pump.
31*	49	1961	278	10	White Skyspar enamel, (B, A-423-SA9185); TiO_2, untinted, in epoxy resin binder; substrate unknown; value avg of several determinations.
32*	49	1961	278		Sicon white 7X1153; TiO_2 in silicone binder on Dow 15 treated Mg alloy substrate.
33*	49	1961	278		Above specimen and conditions except exposed in vacuum (10^{-6}-8 x 10^{-6} mm Hg) to UV radiation from an argon filled A-H6 high pressure Hg arc lamp for 12 hrs; vacuum maintained by VacIon pump.
34*	49	1961	278		Sicon white 7X1153; TiO_2 in silicone binder on Dow 15 treated Mg alloy substrate.
35*	49	1961	278		Above specimen and conditions except exposed in vacuum (10^{-6}-8 x 10^{-6}) to UV radiation from an argon filled A-H6 high pressure Hg arc lamp for 100 hrs; vacuum maintained by VacIon pump.
36	85	1969	157-399		PV-100, rutile TiO_2 in silicone-alkyd binder (0.152± 0.00763 mm thick) on zinc chromate primer (0.0152± 0.00254 mm thick) and 2024 aluminum substrates; aluminum substrate cleaned and etched with an alcohol-phosphoric acid mixture, then primed; PV-100 spray applied and air dried; property measured in vacuum by calorimetric method; data extracted from smooth curve. [Author's designation: A-1 PV-100, NRDL-RTD-81-2]
37*	85	1969	157-399		Similar to above specimen and conditions. [Author's designation: B-4 PV-100, NRDL-RTD-81-2]

* Not shown on plot

DATA TABLE NO. 101 HEMISPHERICAL TOTAL EMITTANCE OF TITANIUM DIOXIDE PIGMENTED COATINGS

[Temperature, T, K; Emittance, ϵ]

T	ϵ
CURVE 1	
349	0.880
363	0.875
373	0.875
386	0.880
396	0.830
406	0.840
417	0.790
427	0.830
433	0.790
440	0.750
CURVE 2	
346	0.880
356	0.900
369	0.890
376	0.885
387	0.760
397	0.660
CURVE 3	
365	0.850
398	0.870
408	0.920
423	0.880
437	0.830
452	0.800
482	0.790
509	0.735
523	0.660
CURVE 4	
349	0.890
358	0.935
367	0.910
380	0.900
400	0.870
417	0.860
431	0.780
CURVE 5	
218	0.76
225	0.81
CURVE 5 (cont.)	
232	0.83
239	0.82
247	0.79
254	0.83
267	0.85
CURVE 6*	
298	0.86
CURVE 7*	
298	0.91
CURVE 8*	
298	0.90
CURVE 9*	
1750	0.70
CURVE 10*	
278	0.86
CURVE 11*	
278	0.82
CURVE 12*	
278	0.85
CURVE 13*	
278	0.87
CURVE 14*	
278	0.85
CURVE 15*	
278	0.82
CURVE 16*	
278	0.82
CURVE 17*	
278	0.86
CURVE 18*	
278	0.84
CURVE 19*	
278	0.83
CURVE 20*	
278	0.85
CURVE 21*	
278	0.81
CURVE 22*	
278	0.83
CURVE 23*	
278	0.83
CURVE 24*	
278	0.82
CURVE 25*	
278	0.81
CURVE 26*	
278	0.83
CURVE 27*	
278	0.81
CURVE 28*	
278	0.84
CURVE 29*	
278	0.81
CURVE 30*	
278	0.78
CURVE 31*	
278	0.86
CURVE 32*	
278	0.83
CURVE 33*	
278	0.83
CURVE 34*	
278	0.84
CURVE 35*	
278	0.83
CURVE 36	
157	0.884
205	0.907
242	0.915
296	0.918
345	0.915
399	0.901
CURVE 37*	
157	0.884
205	0.907
242	0.915
296	0.918
345	0.915
399	0.901

* Not shown on plot

SPECIFICATION TABLE NO. 102 NORMAL TOTAL EMITTANCE OF TITANIUM DIOXIDE PIGMENTED COATINGS

Curve No.	Ref. No.	Year	Temperature Range, K	Geometry θ'	Reported Error, %	Composition (weight percent), Specifications and Remarks
1*	19	1965	298	0°		TiO_2 in potassium silicate vehicle on aluminum substrate; reported error ± 0. 02 unit.
2*	19	1965	298	0°		Above specimen and conditions except property computed from 1 - R(2π, 0°).
3*	19	1965	298	0°		TiO_2 in potassium silicate (boron treated) binder on aluminum substrate; reported error ± 0. 002 unit.
4*	19	1965	298	0°		Above specimen and conditions except property computed from 1 - R(2π, 0°).
5*	68	1969	300	~0°		Titanium dioxide in silicone alkyd binder (~0. 152 mm thick) on 2024 aluminum substrate (~1. 59 mm thick); emittance computed from reflectance; property measured in air. [Authors' designation: PV-100]
6*	68	1969	300	~0°		Similar to above specimen and conditions except exposed to vacuum (10^{-8} mm Hg) for 300 hrs; vacuum maintained by ion pump; property measured in air after vacuum exposure.
7*	68	1969	300	~0°		Similar to curve 5 specimen and conditions except exposed to UV radiation in vacuum (10^{-5} mm Hg) for 300 hrs; vacuum maintained by oil-diffusion pump; H_2 gas UV source; property measured in air after exposure.
8*	68	1969	300	~0°		Similar to above specimen and conditions except He gas UV source.
9*	68	1969	300	~0°		TiO_2 in potassium silicate binder (~0. 1015 mm thick) on 2024 aluminum substrate (~1. 59 mm thick); emittance computed from reflectance; property measured in air. [Authors' designation: TiO_2/PS-7]
10*	68	1969	300	~0°		Similar to above specimen and conditions except exposed to vacuum (10^{-8} mm Hg) for 300 hrs; vacuum maintained by ion pump; property measured in air after vacuum exposure.
11*	68	1969	300	~0°		Similar to curve 9 specimen and conditions except exposed to UV radiation in vacuum (10^{-5} mm Hg) for 300 hrs; vacuum maintained by oil-diffusion pump; H_2 gas UV source; property measured in air after exposure.
12*	68	1969	300	~0°		Similar to above specimen and conditions except He gas UV source.
13*	57	1969	311	~0°		Hughes H-2; Cabot RF-1 titanium dioxide in PS-7 potassium silicate binder; substrate unknown; property computed from reflectance; lab data taken on sample to be tested on Lunar Orbiter IV.
14*	83	1963	360	~0°		A-31-45 paint; anatase TiO_2; unknown binder; unknown substrate; measured in vacuum (1. 6 x 10^{-4} mm Hg).

* No plot given

DATA TABLE NO. 102 NORMAL TOTAL EMITTANCE OF TITANIUM DIOXIDE PIGMENTED COATINGS

[Temperature, T, K; Emittance, ϵ]

T	ϵ	T	ϵ
CURVE 1*		CURVE 13*	
298	0.87	311	0.876
CURVE 2*		CURVE 14*	
298	0.89	360	0.910
CURVE 3*			
298	0.90		
CURVE 4*			
298	0.88		
CURVE 5*			
300	0.84		
CURVE 6*			
300	0.84		
CURVE 7*			
300	0.84		
CURVE 8*			
300	0.84		
CURVE 9*			
300	0.86		
CURVE 10*			
300	0.86		
CURVE 11*			
300	0.86		
CURVE 12*			
300	0.86		

* No plot given

FIGURE 103A (I)

ANALYZED NORMAL SPECTRAL REFLECTANCE OF TITANIUM DIOXIDE PIGMENTED COATINGS

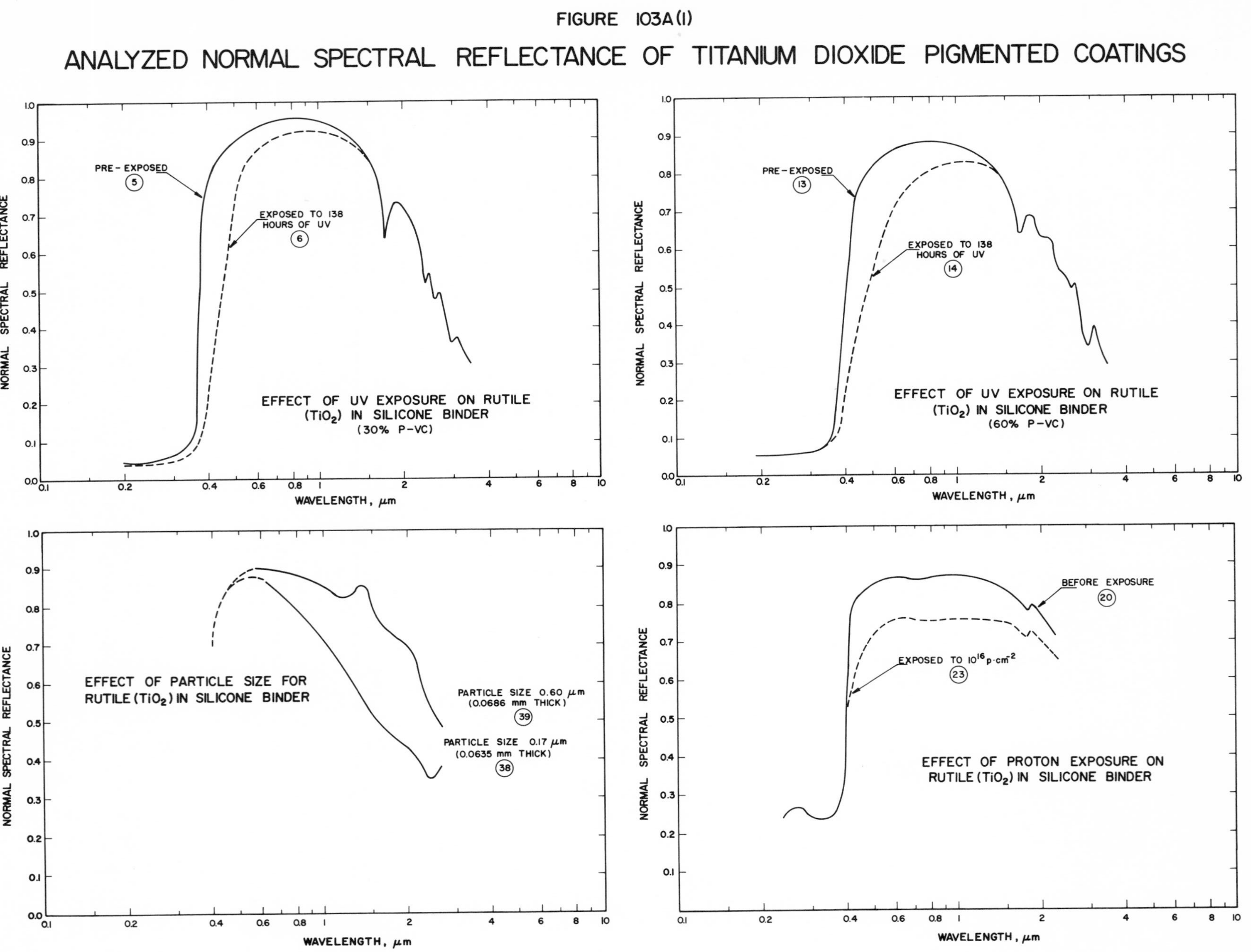

FIGURE 103A (2)

ANALYZED NORMAL SPECTRAL REFLECTANCE OF TITANIUM DIOXIDE PIGMENTED COATINGS

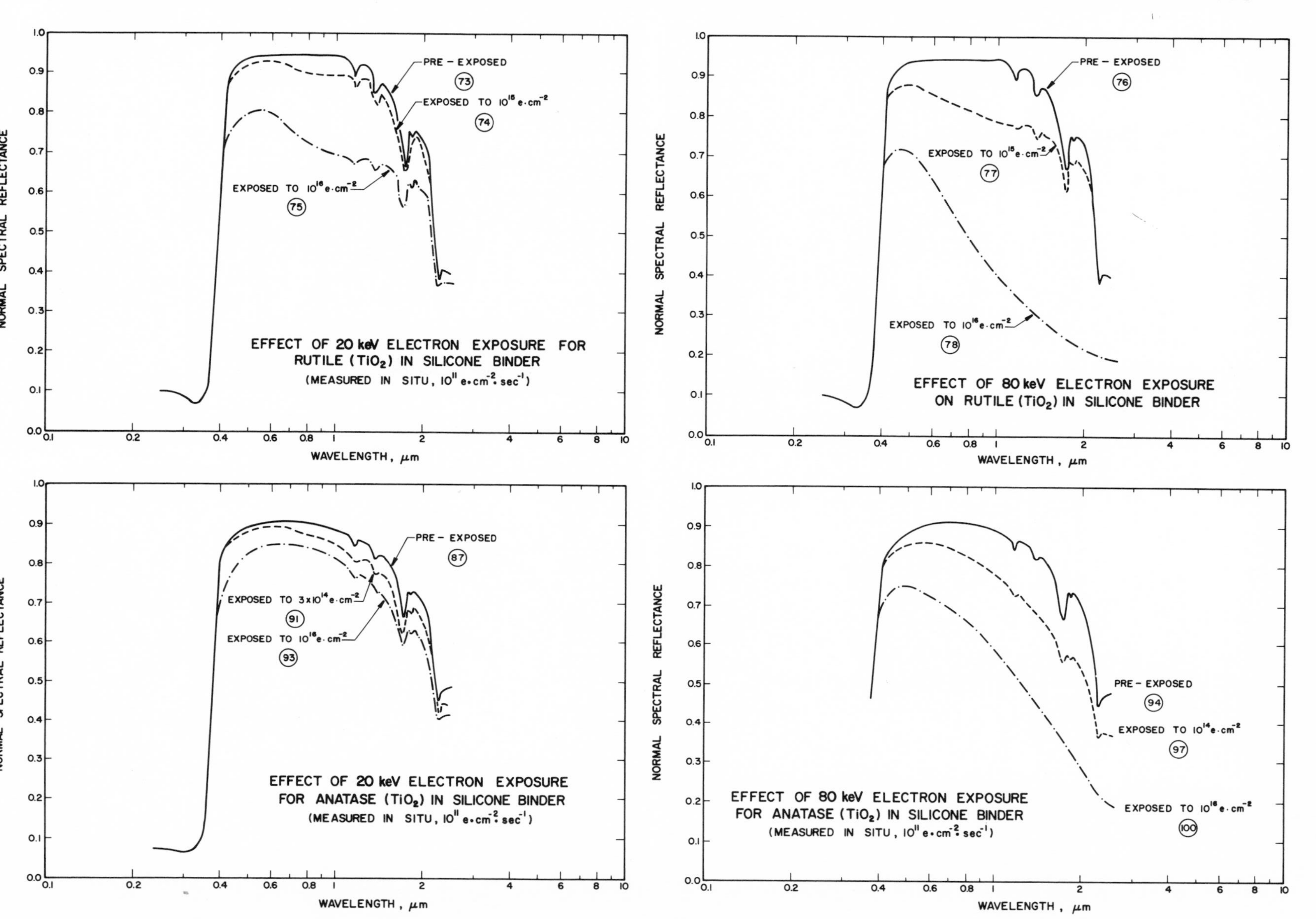

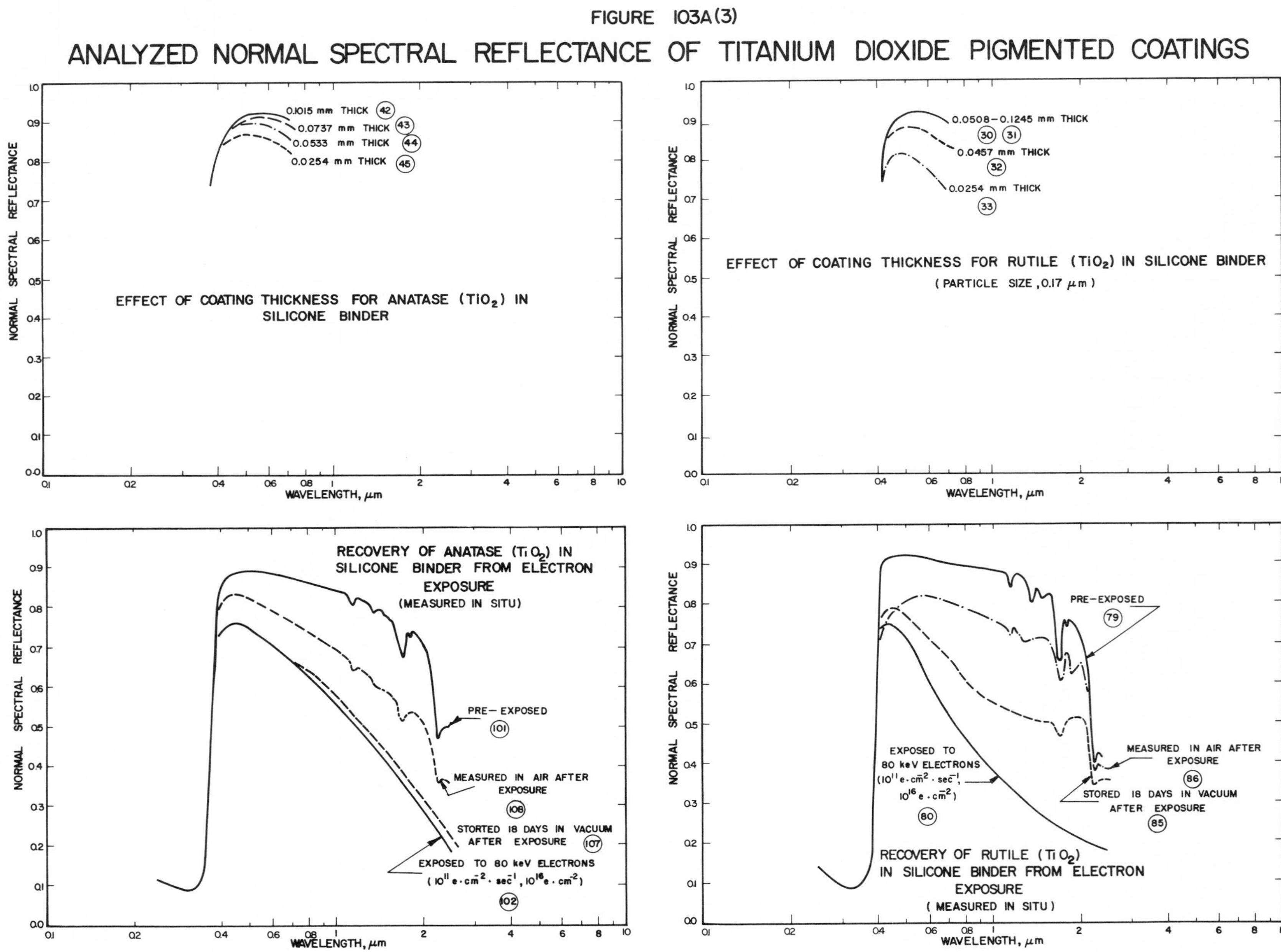
FIGURE 103A(3)
ANALYZED NORMAL SPECTRAL REFLECTANCE OF TITANIUM DIOXIDE PIGMENTED COATINGS
EFFECT OF COATING THICKNESS FOR ANATASE (TiO2) IN SILICONE BINDER
0.1015 mm THICK (42)
0.0737 mm THICK (43)
0.0533 mm THICK (44)
0.0254 mm THICK (45)
EFFECT OF COATING THICKNESS FOR RUTILE (TiO2) IN SILICONE BINDER
(PARTICLE SIZE, 0.17 μm)
0.0508-0.1245 mm THICK (30) (31)
0.0457 mm THICK (32)
0.0254 mm THICK (33)
RECOVERY OF ANATASE (TiO2) IN SILICONE BINDER FROM ELECTRON EXPOSURE
(MEASURED IN SITU)
PRE-EXPOSED (101)
MEASURED IN AIR AFTER EXPOSURE (108)
STORTED 18 DAYS IN VACUUM AFTER EXPOSURE (107)
EXPOSED TO 80 keV ELECTRONS (10^11 e·cm^-2·sec^-1, 10^16 e·cm^-2) (102)
RECOVERY OF RUTILE (TiO2) IN SILICONE BINDER FROM ELECTRON EXPOSURE
(MEASURED IN SITU)
PRE-EXPOSED (79)
EXPOSED TO 80 keV ELECTRONS (10^11 e·cm^-2·sec^-1, 10^16 e·cm^-2) (80)
MEASURED IN AIR AFTER EXPOSURE (86)
STORED 18 DAYS IN VACUUM AFTER EXPOSURE (85)
NORMAL SPECTRAL REFLECTANCE
WAVELENGTH, μm

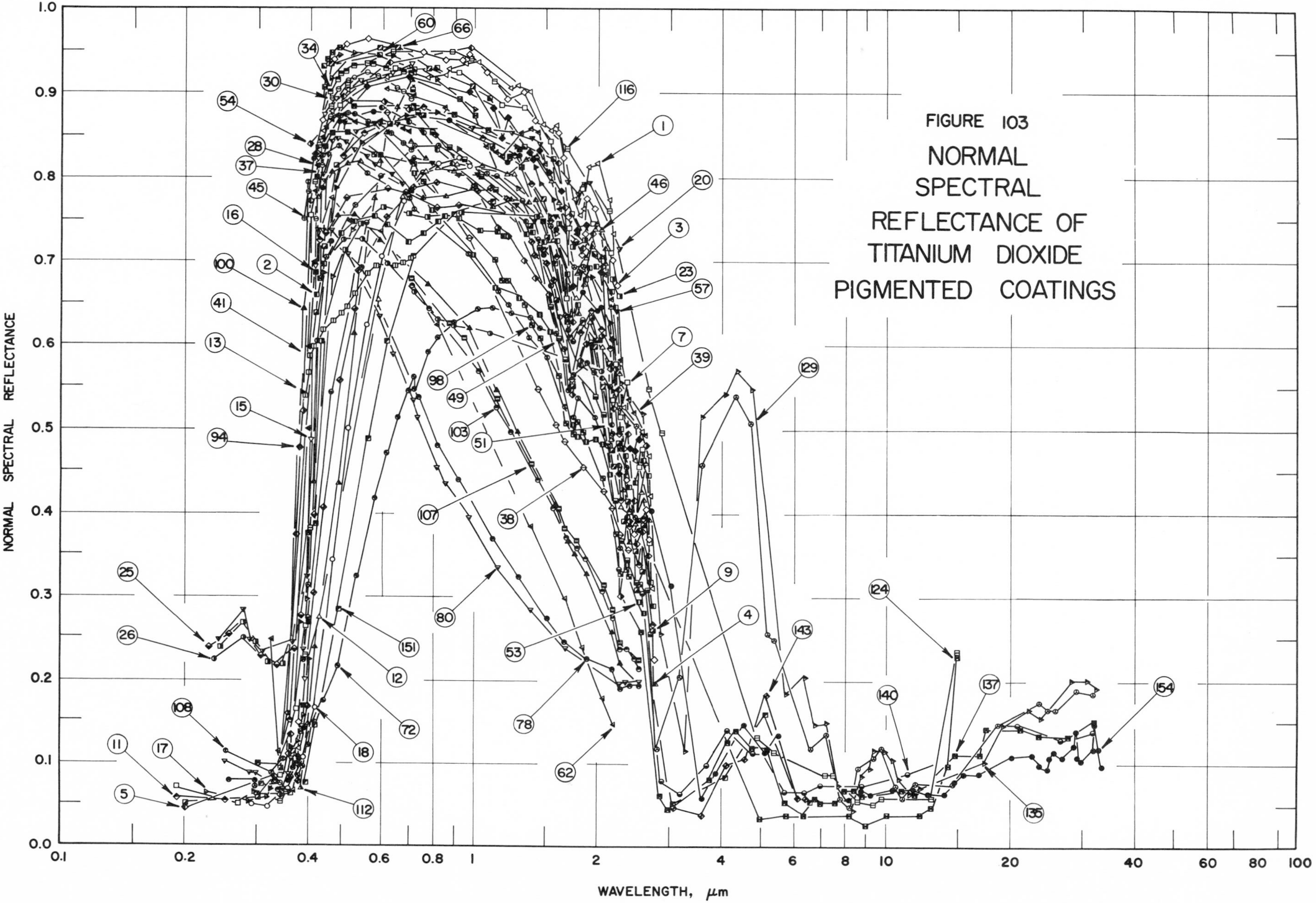

FIGURE 103 NORMAL SPECTRAL REFLECTANCE OF TITANIUM DIOXIDE PIGMENTED COATINGS

SPECIFICATION TABLE NO. 103 NORMAL SPECTRAL REFLECTANCE OF TITANIUM DIOXIDE PIGMENTED COATINGS

Curve No.	Ref. No.	Year	Temperature, K	Wavelength Range, μm	Geometry θ	θ'	ω'	Reported Error, %	Composition (weight percent), Specifications, and Remarks
1	20	1963	~298	0.377-2.704	~0°		2π		Titanox RA-10 rutile TiO_2 in G. E. LTV-602 binder on aluminum substrate; PBR 1.80; PVC 30%; air brush application; data extracted from smooth curve; measured relative to MgO; corrected for reflectance of MgO. [Authors' designation: Sample 28]
2	20	1963	~298	0.377-2.704	~0°		2π		Above specimen and conditions except exposed in vacuum (10^{-5} mm Hg) to 1850 ESH with G. E. AH-6 UV source; vacuum maintained by ion pump.
3	20	1963	~298	0.384-2.713	~0°		2π		TiPure R-900-1 rutile TiO_2 in G. E. LTV-602 binder on aluminum substrate; PBR 2.27; PVC 35%; air brush application; data extracted from smooth curve; measured relative to MgO; corrected for reflectance of MgO. [Authors' designation: Sample S-32]
4	20	1963	~298	0.384-2.711	~0°		2π		Above specimen and conditions except exposed in vacuum (10^{-5} mm Hg) to 1650 ESH; G. E. AH-6 lamp UV source; vacuum maintained by ion pump.
5	2	1962	337	0.20-2.70	~0°		2π		Rutile TiO_2 in silicone binder (0.0508 mm thick); PVC 30%; substrate unknown; data extracted from smooth curve; measured relative to MgO; corrected for reflectance of MgO.
6*	2	1962	337	0.20-2.70	~0°		2π		Rutile TiO_2 in silicone binder (0.0508 mm thick); substrate unknown; PVC 30%; exposed to UV (UA-3 lamp at 8.5 cm) in vacuum (10^{-5} mm Hg) for 138 hrs; data extracted from smooth curve; measured relative to MgO; corrected for reflectance of MgO.
7	2	1962	~333	0.19-2.70	~0°		2π		Rutile TiO_2 in silicone binder (0.0508 mm thick); substrate unknown; PVC 40%; data extracted from smooth curve; measured relative to MgO; corrected for reflectance of MgO.
8*	2	1962	~333	0.19-2.70	~0°		2π		Rutile TiO_2 in silicone binder (0.0508 mm thick); substrate unknown; PVC 40%; exposed to UV (UA-3 lamp at 8.5 cm) in vacuum (10^{-5} mm Hg) for 100 hrs; data extracted from smooth curve; measured relative to MgO; corrected for reflectance of MgO.
9	2	1962	330	0.20-2.70	~0°		2π		Rutile TiO_2 in silicone binder (0.0508 mm thick); substrate unknown; PVC 45%; data extracted from smooth curve; measured relative to MgO; corrected for reflectance of MgO.
10*	2	1962	330	0.20-2.70	~0°		2π		Rutile TiO_2 in silicone binder (0.0508 mm thick); substrate unknown; PVC 45%; exposed to UV (UA-3 lamp at 8.5 cm) in vacuum (10^{-5} mm Hg) for 138 hrs; data extracted from smooth curve; measured relative to MgO; corrected for reflectance of MgO.
11	2	1962	336	0.19-2.70	~0°		2π		Rutile TiO_2 in silicone binder (0.0508 mm thick); substrate unknown; PVC 50%; data extracted from smooth curve; measured relative to MgO; corrected for reflectance of MgO.
12	2	1962	336	0.19-2.70	~0°		2π		Rutile TiO_2 in silicone binder (0.0508 mm thick); substrate unknown; PVC 50%; exposed to UV (UA-3 lamp at 8.5 cm) in vacuum (10^{-5} mm Hg) for 138 hrs; data extracted from smooth curve; measured relative to MgO; corrected for reflectance of MgO.

* Not shown on plot

SPECIFICATION TABLE NO. 103 NORMAL SPECTRAL REFLECTANCE OF TITANIUM DIOXIDE PIGMENTED COATINGS (continued)

Curve No.	Ref. No.	Year	Temperature, K	Wavelength Range, μm	Geometry θ θ'	ω'	Reported Error, %	Composition (weight percent), Specifications, and Remarks
13	2	1962	337	0.19-2.70	~0°	2π		Rutile TiO_2 in silicone binder (0.0508 mm thick); substrate unknown; PVC 60%; data extracted from smooth curve; measured relative to MgO; corrected for reflectance of MgO.
14*	2	1962	337	0.19-2.70	~0°	2π		Rutile TiO_2 in silicon binder (0.0508 mm thick); substrate unknown; exposed to UV (UA-3 lamp at 8.5 cm) in vacuum (10^{-5} mm Hg) for 138 hrs; data extracted from smooth curve; measured relative to MgO; corrected for reflectance of MgO.
15	90	1966	~298	0.380-0.700	~0°	2π		Rutile TiO_2 in nitrocellulose lacquer binder (0.0381 mm thick) on heavy coated paper chart (Morest Opacity Panel Form 018) substrate; applied with Bird applicator over white background; data extracted from smooth curve.
16	90	1966	~298	0.380-0.700	~0°	2π		Similar to above specimen and conditions except applied over a black background (Morest Opacity Panel Form 018).
17	28	1964	~298	0.222-1.003	~0°	2π		Skyspar A423, color SA9185, untinted, manufactured by Andrew Brown Co.; rutile TiO_2 in epoxy binder; substrate unknown; data extracted from smooth curve.
18	28	1964	~298	0.222-0.976	~0°	2π		Similar to above specimen and conditions except ultraviolet irradiated in vacuum ($<10^{-6}$ mm Hg) at 273-288 K for 300 equivalent sun hrs (~3-sun intensity, G. E. BH-6 lamp source); measured within 1 hr after exposure to air.
19*	28	1964	~298	0.222-1.003	~0°	2π		Similar to curve 17 specimen and conditions except gamma-irradiated in vacuum (10^{-6} mm Hg) at 320 K to a total dose of 385 Megarads (dose rate ~1.5 Megarads hr^{-1}, Cobalt-60 source); measured within 1 hr after exposure to air.
20	79	1967	~298	0.240-2.248	12.5°	2π		PV-100; TiO_2 in silicone alkyd binder on stainless steel substrate; measured in vacuum (2-5 x 10^{-7} mm Hg); data extracted from smooth curve; converted from reflectance factor. [Authors' designation: Sample No. 9-2]
21*	79	1967	~298	0.241-2.243	12.5°	2π		Above specimen and conditions except irradiated in vacuum (~4 x 10^{-6} mm Hg) with protons of nominal energy 3 keV; total dose 10^{14} p cm^{-2}; measured in situ after irradiation.
22	79	1967	~298	0.241-2.248	12.5°	2π		Above specimen and conditions except irradiated additionally to a total dose of 10^{15} p cm^{-2}; measured in situ after irradiation.
23	79	1967	~298	0.241-2.238	12.5°	2π		Above specimen and conditions except irradiated additionally to a total dose of 10^{16} p cm^{-2}; measured in situ after irradiation.
24*	79	1967	~298	0.229-2.263	12.5°	2π		Similar to curve 20 specimen and conditions. [Authors' designation: Sample No. 8-2]
25	79	1967	~298	0.228-2.272	12.5°	2π		Above specimen and conditions except irradiated in vacuum (~4 x 10^{-6} mm Hg) with electrons of energy 145 keV; total dose 1.3 x 10^{15} e cm^{-2}; measured in situ after irradiation.
26	79	1967	~298	0.234-2.311	12.5°	2π		Above specimen and conditions except irradiated additionally to a total dose of 1.3 x 10^{16} e cm^{-2}; measured in situ after irradiation.
27*	79	1967	~298	0.234-2.237	12.5°	2π		Above specimen and conditions except irradiated additionally to a total dose of 4 x 10^{16} e cm^{-2}; measured in situ after irradiation.
28	3	1960	~298	0.395-0.698	~0°	2π		DuPont R-510 pure rutile TiO_2 in DuPont RC 7007 alkyd melamine resin binder (0.0585 mm thick, 11.94 x 10^{-2} kg m^{-2}) on mild steel substrate; pigment 40% of solids content; sprayed; data extracted from smooth curve; measured relative to MgO.

* Not shown on plot

SPECIFICATION TABLE NO. 103 NORMAL SPECTRAL REFLECTANCE OF TITANIUM DIOXIDE PIGMENTED COATINGS (continued)

Curve No.	Ref. No.	Year	Temperature, K	Wavelength Range, μm	Geometry θ	Geometry θ'	Geometry ω'	Reported Error, %	Composition (weight percent), Specifications, and Remarks
29*	3	1960	~298	0.395-0.695	~0°		2π		DuPont R-100 pure rutile TiO_2 in DuPont RC 7007 alkyd melamine resin binder (0.0686 mm thick, 11.01 x 10^{-2} kg m^{-2}) on mild steel substrate; pigment 40% of solids content; sprayed; data extracted from smooth curve; measured relative to MgO.
30	3	1960	~298	0.413-0.700	~0°		2π		Rutile TiO_2 (New Jersey Zinc Co. MLW-42-43, mean particle size 0.17 μm) in Dow Corning 806A silicone binder (0.1245 mm thick) on mild steel substrate; pigment 60% of solids content; sprayed; data extracted from smooth curve; measured relative to MgO.
31*	3	1960	~298	0.413-0.700	~0°		2π		Similar to above specimen and conditions except coating 0.0508 mm thick.
32*	3	1960	~298	0.413-0.700	~0°		2π		Similar to curve 30 specimen and conditions except coating 0.0457 mm thick.
33*	3	1960	~298	0.427-0.640	~0°		2π		Similar to curve 30 specimen and conditions except coating 0.0254 mm thick.
34	3	1960	~298	0.413-0.700	~0°		2π		Rutile TiO_2 (New Jersey Zinc Co. 8902-96-B) (mean particle size 0.60 μm) in Dow Corning 806A silicone binder (0.0686 mm thick) on mild steel substrate; pigment 60% of solids content; sprayed; data extracted from smooth curve; measured relative to MgO.
35*	3	1960	~298	0.414-0.700	~0°		2π		Similar to above specimen and conditions except coating 0.0381 mm thick.
36*	3	1960	~298	0.414-0.700	~0°		2π		Similar to curve 34 specimen and conditions except coating 0.0254 mm thick.
37	3	1960	~298	0.413-0.700	~0°		2π		Similar to curve 34 specimen and conditions except coating 0.0203 mm thick.
38	3	1960	~298	0.616-2.600	~0°		2π		Similar to curve 34 specimen and conditions except mean particle size 0.17 μm; coating 0.0635 mm thick.
39	3	1960	~298	0.617-2.600	~0°		2π		Similar to curve 34 specimen and conditions.
40*	3	1960	~298	0.388-0.698	~0°		2π		Titanox A-WD TiO_2 (anatase 98% pure, electrostatically screened, particle size >0.5 μm) in Dow Corning 806A silicone binder on mild steel substrate; pigment 60% of solids content; sprayed; data extracted from smooth curve; measured relative to MgO.
41	3	1960	~298	0.388-0.700	~0°		2π		Similar to above specimen and conditions except pigment particle size <0.5 μm.
42*	3	1960	~298	0.387-0.700	~0°		2π		Titanox A-WD TiO_2 (98% pure anatase) in Dow Corning 806A silicone binder (0.1015 mm thick) on mild steel substrate; pigment 60% of solids content; sprayed; data extracted from smooth curve; measured relative to MgO.
43*	3	1960	~298	0.387-0.700	~0°		2π		Similar to above specimen and conditions except coating thickness 0.0737 mm.
44*	3	1960	~298	0.387-0.700	~0°		2π		Similar to curve 42 specimen and conditions except coating thickness 0.0533 mm.
45	3	1960	~298	0.387-0.700	~0°		2π		Similar to curve 42 specimen and conditions except coating thickness 0.0254 mm.
46	3	1960	~298	0.700-2.600	~0°		2π		DuPont 510 rutile TiO_2 in DuPont RC-7007 alkyd melamine resin binder (9.3 x 10^3 g cm^{-2}) on aluminum foil substrate; pigment 40% of solids content; sprayed; data extracted from smooth curve; measured relative to MgO.

* Not shown on plot

SPECIFICATION TABLE NO. 103 NORMAL SPECTRAL REFLECTANCE OF TITANIUM DIOXIDE PIGMENTED COATINGS (continued)

Curve No.	Ref. No.	Year	Temperature, K	Wavelength Range, μm	Geometry θ	θ'	ω'	Reported Error, %	Composition (weight percent), Specifications, and Remarks
47*	3	1960	~298	0.700-2.600	~0°		2π		Similar to above specimen and conditions except coating (8.75×10^3 g cm^{-2}) on aluminum paint substrate.
48*	3	1960	~298	0.700-2.600	~0°		2π		Similar to curve 46 specimen and conditions except coating (8.95×10^3 g cm^{-2}) on mild steel substrate.
49	3	1960	~298	0.702-2.682	~0°		2π		Titanox AMO anatase TiO_2 in Dow Corning 806A silicone binder (0.0940 mm thick, 16.5×10^3 g cm^{-2}) on mild steel substrate; pigment 60% of solids content; sprayed; data extracted from smooth curve; measured relative to MgO.
50*	3	1960	~298	0.700-2.682	~0°		2π		Similar to above specimen and conditions except coating 0.0991 mm thick, 17.6×10^3 g cm^{-2}.
51	3	1960	~298	0.597-2.589	~0°		2π		DuPont R-100 rutile TiO_2 in DuPont RC 7007 alkyd melamine resin binder (0.0610 mm thick, 12.25×10^3 g cm^{-2}) on mild steel substrate; pigment 60% of solids content; sprayed; data extracted from smooth curve; measured relative to MgO.
52*	3	1960	~298	0.709-2.590	~0°		2π		Similar to above specimen and conditions except coating 0.0305 mm thick, 6.77×10^3 g cm^{-2}, pigment 40% of solids content.
53	3	1960	~298	0.716-2.588	~0°		2π		Titanox AMO, anatase TiO_2 in DuPont RC 7007 alkyd melamine resin binder (0.0330 mm thick, 7.27×10^3 g cm^{-2}) on mild steel substrate; pigment 40% of solids content; sprayed; data extracted from smooth curve; measured relative to MgO.
54	4	1961	~298	0.300-2.60	~0°		2π		Anatase TiO_2 in acrylic binder; substrate unknown; converted from R (2π, 0°); measured relative to MgO; data extracted from smooth curve.
55*	4	1961	~298	0.337-2.32	~0°		2π		Anatase TiO_2 in alkyd binder; substrate unknown; converted from R (2π, 0°); measured relative to MgO.
56*	4	1961	~298	0.338-2.35	~0°		2π		Rutile TiO_2 in alkyd binder; substrate unknown; converted from R (2π, 0°); measured relative to MgO.
57	4	1961	~298	0.379-2.201	~0°		2π		Rutile TiO_2 in nitrocellulose binder on white primer and aluminum substrates; weight of paint 3.39×10^3 g cm^{-2}; converted from R (2π, 0°); measured relative to MgO; data extracted from smooth curve.
58*	4	1961	~298	0.379-2.186	~0°		2π		Similar to above specimen and conditions except weight of paint 7.05×10^{-3} g cm^{-2}.
59*	4	1961	~298	0.322-2.205	~0°		2π		Similar to above specimen and conditions except weight of paint 10.5×10^{-3} g cm^{-2}.
60	4	1961	~298	0.379-2.189	~0°		2π		Similar to above specimen and conditions except weight of paint 13.2×10^{-3} g cm^{-2}.
61*	4	1961	~298	0.379-2.202	~0°		2π		Similar to above specimen and conditions except weight of paint 16.8×10^{-3} g cm^{-2}.
62	4	1961	~298	0.322-2.198	~0°		2π		Rutile TiO_2 in nitrocellulose binder on black primer and aluminum substrates; weight of paint 3.73×10^{-3} g cm^{-2}; converted from R (2π, 0°); measured relative to MgO; data extracted from smooth curve.
63*	4	1961	~298	0.322-2.205	~0°		2π		Similar to above specimen and conditions except weight of paint 7.13×10^{-3} g cm^{-2}.
64*	4	1961	~298	0.364-2.205	~0°		2π		Similar to above specimen and conditions except weight of paint 10.15×10^{-3} g cm^{-2}.
65*	4	1961	~298	0.364-2.200	~0°		2π		Similar to above specimen and conditions except weight of paint 13.4×10^{-3} g cm^{-2}.

* Not shown on plot

SPECIFICATION TABLE NO. 103 NORMAL SPECTRAL REFLECTANCE OF TITANIUM DIOXIDE PIGMENTED COATINGS (continued)

Curve No.	Ref. No.	Year	Temperature, K	Wavelength Range, μm	Geometry θ	θ'	ω'	Reported Error, %	Composition (weight percent), Specifications, and Remarks
66	4	1961	~298	0.364-2.198	~0°		2π		Similar to above specimen and conditions except weight of paint 16.95 x 10^{-3} g cm^{-2}.
67*	38	1966	~298	0.361-2.40	~0°		2π		Thermatrol ZA-100 thermal control coating, Dow Corning Corp.; TiO_2 in silicone binder; substrate unknown; data extracted from smooth curve.
68*	38	1966	~298	0.361-2.40	~0°		2π		Above specimen and conditions except exposed to UV in vacuum (<8 x 10^{-6} mm Hg) for 1330 ESH at solar factor of 20; measured 24 hrs after air pressure raised to 0.3 mm Hg.
69*	37	1967	293	0.360-2.50	~0°		2π		TiO_2 in Dow Corning XR6-3488 methyl silicone binder; substrate unknown; PBR 1.5; measured in vacuum (<4 x 10^{-7} mm Hg); data extracted from smooth curve; measured relative to MgO; corrected for reflectance of MgO. [Authors' designation: Sample 52-P]
70*	37	1967	293	0.360-2.50	~0°		2π		Above specimen and conditions except exposed to UV (from 4 G. E. UA-11 and 2 UA-3 lamps) in situ for 1130 ESH at solar factor of 4-5.
71*	37	1967	293	0.360-2.50	~0°		2π		Pyromark; TiO_2 in methyl phenyl silicone binder; substrate unknown; measured in vacuum (<4 x 10^{-7} mm Hg); data extracted from smooth curve; measured relative to MgO; corrected for reflectance of MgO. [Authors' designation: Sample 49 Pyromark]
72	37	1967	293	0.360-2.50	~0°		2π		Above specimen and conditions except exposed to UV (from 4 G. E. UA-11 and 2 UA-3 lamps) in situ for 1130 ESH at solar factor of 4-5.
73*	45	1970	298	0.250-2.495	~0°		2π		Rutile TiO_2 in methyl silicone binder on aluminum substrate; reflectance measured in air; data extracted from smooth curve. [Authors' designation: O]
74*	45	1970	281	0.250-2.495	~0°		2π		Above specimen and conditions except exposed to 20 keV electrons in dark in vacuum (10^{-8} mm Hg); vacuum maintained by ion pump; flux 1 x 10^{10}-5 x 10^{11} e cm^{-2} sec^{-1}; substrate held at 281 ± 2 K; reflectance measured in situ after 1 x 10^{15} e cm^{-2}.
75*	45	1970	281	0.250-2.497	~0°		2π		Above specimen and conditions except reflectance measured after 10^{16} e cm^{-2}.
76*	45	1970	281	0.256-2.505	~0°		2π		Similar to above specimen and conditions except exposed to 80 keV electrons; reflectance measured after 3 x 10^{14} e cm^{-2}.
77*	45	1970	281	0.256-2.505	~0°		2π		Above specimen and conditions except reflectance measured after 1 x 10^{15} e cm^{-2}.
78	45	1970	281	0.256-2.505	~0°		2π		Above specimen and conditions except reflectance measured after 1 x 10^{16} e cm^{-2}.
79*	45	1970	298	0.250-2.500	~0°		2π		Similar to curve 73 specimen and conditions.
80	45	1970	281	0.250-2.500	~0°		2π		Above specimen and conditions except exposed to 80 keV electrons in dark in vacuum (10^{-8} mm Hg); substrate held at 281 ± 2 K; reflectance measured in situ after 10^{16} e cm^{-2}.
81*	45	1970	281	0.250-0.706	~0°		2π		Above specimen and conditions except reflectance measured in situ 2 days after end of 10^{16} e cm^{-2} exposure.
82*	45	1970	281	0.250-0.706	~0°		2π		Above specimen and conditions except reflectance measured 3 days after end of 10^{16} e cm^{-2} exposure.
83*	45	1970	281	0.250-0.706	~0°		2π		Above specimen and conditions except reflectance measured 4 days after end of 10^{16} e cm^{-2} exposure.
84*	45	1970	281	0.250-0.706	~0°		2π		Above specimen and conditions except reflectance measured 7 days after end of 10^{16} e cm^{-2} exposure.

* Not shown on plot

Curve No.	Ref. No.	Year	Temperature, K	Wavelength Range, μm	Geometry θ	θ'	ω'	Reported Error, %	Composition (weight percent), Specifications, and Remarks
85*	45	1970	281	0.250-2.500	~0°		2π		Above specimen and conditions except reflectance measured 18 days after end of 10^{16} e cm^{-2} exposure.
86*	45	1970	298	0.250-2.500	~0°		2π		Above specimen and conditions except reflectance measured in air after 10^{16} e cm^{-2} test.
87*	45	1970	298	0.250-2.504	~0°		2π		Anatase TiO_2 in methyl silicone binder on aluminum substrate; reflectance measured in air; data extracted from smooth curve. [Authors' designation: L_1]
88*	45	1970	281	0.250-2.504	~0°		2π		Above specimen and conditions except exposed to 20 keV electrons in dark in vacuum (10^{-8} mm Hg); vacuum maintained by ion pump; flux 1 x 10^{10}-5 x 10^{11} e cm^{-2} sec^{-1}; substrate held at 281 ± 2 K; reflectance measured in situ after 1 x 10^{13} e cm^{-2}.
89*	45	1970	281	0.250-2.504	~0°		2π		Above specimen and conditions except reflectance measured after 5 x 10^{13} e cm^{-2}.
90*	45	1970	281	0.250-2.504	~0°		2π		Above specimen and conditions except reflectance measured after 1 x 10^{14} e cm^{-2}.
91*	45	1970	281	0.250-2.504	~0°		2π		Above specimen and conditions except reflectance measured after 3 x 10^{14} e cm^{-2}.
92*	45	1970	281	0.250-2.504	~0°		2π		Above specimen and conditions except reflectance measured after 1 x 10^{15} e cm^{-2}.
93*	45	1970	281	0.250-2.504	~0°		2π		Above specimen and conditions except reflectance measured after 1 x 10^{16} e cm^{-2}.
94	45	1970	298	0.378-2.505	~0°		2π		Similar to curve 87 specimen and conditions.
95*	45	1970	281	0.378-2.505	~0°		2π		Above specimen and conditions except exposed to 80 keV electrons in dark in vacuum (10^{-8} mm Hg); vacuum maintained by ion pump; flux 1 x 10^{10}-5 x 10^{11} e cm^{-2} sec^{-1}; substrate held at 281 ± 2 K; reflectance measured in situ after 1 x 10^{13} e cm^{-2}.
96*	45	1970	281	0.378-2.505	~0°		2π		Above specimen and conditions except reflectance measured after 5 x 10^{13} e cm^{-2}.
97*	45	1970	281	0.378-2.505	~0°		2π		Above specimen and conditions except reflectance measured after 1 x 10^{14} e cm^{-2}.
98	45	1970	281	0.378-2.505	~0°		2π		Above specimen and conditions except reflectance measured after 3 x 10^{14} e cm^{-2}.
99*	45	1970	281	0.378-2.505	~0°		2π		Above specimen and conditions except reflectance measured after 1 x 10^{15} e cm^{-2}; α_s = 0.33.
100	45	1970	281	0.378-2.505	~0°		2π		Above specimen and conditions except reflectance measured after 1 x 10^{16} e cm^{-2}.
101*	45	1970	298	0.250-2.459	~0°		2π		Similar to curve 87 specimen and conditions.
102*	45	1970	281	0.250-2.500	~0°		2π		Above specimen and conditions except exposed to 80 keV electrons in dark in vacuum (10^{-8} mm Hg); vacuum maintained by ion pump; flux 1 x 10^{10}-5 x 10^{11} e cm^{-2} sec^{-1}; substrate held at 281 ± 2 K; reflectance measured in situ after end of 10^{16} e cm^{-2} exposure.
103	45	1970	281	0.700-2.500	~0°		2π		Above specimen and conditions except reflectance measured 2 days after end of 10^{16} e cm^{-2} exposure.
104*	45	1970	281	0.700-2.500	~0°		2π		Above specimen and conditions except reflectance measured 3 days after end of 10^{16} e cm^{-2} exposure.
105*	45	1970	281	0.700-2.500	~0°		2π		Above specimen and conditions except reflectance measured 4 days after end of 10^{16} e cm^{-2} exposure.
106*	45	1970	281	0.700-2.500	~0°		2π		Similar to above specimen and conditions except reflectance measured 7 days after end of 10^{16} e cm^{-2} exposure.
107	45	1970	281	0.700-2.500	~0°		2π		Similar to above specimen and conditions except reflectance measured 18 days after end of 10^{16} e cm^{-2} exposure.
108	45	1970	281	0.700-2.500	~0°		2π		Above specimen and conditions except reference measured in air after 10^{16} e cm^{-2} test.

* Not shown on plot.

SPECIFICATION TABLE NO. 103 NORMAL SPECTRAL REFLECTANCE OF TITANIUM DIOXIDE PIGMENTED COATINGS (continued)

Curve No.	Ref. No.	Year	Temperature, K	Wavelength Range, μm	Geometry θ	θ'	ω'	Reported Error, %	Composition (weight percent), Specifications, and Remarks
109*	68	1969	300	0.380-0.700	~0°		2π		Titanium dioxide in silicone alkyd binder (~0.152 mm thick) on 2024 aluminum substrate (~1.59 mm thick); reflectance measured in air; data extracted from smooth curve. [Authors' designation: PV-100]
110*	68	1969	300	0.380-0.700	~0°		2π		Similar to above specimen and conditions except exposed to vacuum (10^{-8} mm Hg) for 300 hrs; vacuum maintained by ion pump; reflectance measured in air after vacuum exposure.
111*	68	1969	300	0.380-0.700	~0°		2π		Similar to curve 109 specimen and conditions except exposed to vacuum (10^{-5} mm Hg) for 300 hrs; vacuum maintained by oil-diffusion pump; reflectance measured in air after vacuum exposure.
112	68	1969	300	0.380-0.700	~0°		2π		Similar to above specimen and conditions except exposed to UV radiation in vacuum for 300 hrs; H_2 gas UV source.
113*	68	1969	300	0.380-0.700	~0°		2π		Similar to above specimen and conditions except He gas UV source.
114*	68	1969	300	0.300-14.92	~0°		2π		Similar to curve 109 specimen and conditions.
115*	68	1969	300	0.300-14.92	~0°		2π		Similar to curve 110 specimen and conditions.
116	68	1969	300	0.300-14.92	~0°		2π		Similar to curve 111 specimen and conditions.
117*	68	1969	300	0.300-14.92	~0°		2π		Similar to curve 112 specimen and conditions.
118*	68	1969	300	0.300-14.92	~0°		2π		Similar to curve 113 specimen and conditions.
119*	68	1969	300	0.380-0.700	~0°		2π		TiO_2 in potassium silicate binder (~0.1015 mm thick) on 2024 aluminum substrate (~1.59 mm thick); reflectance measured in air; data extracted from smooth curve. [Authors' designation: TiO_2/PS-7]
120*	68	1969	300	0.380-0.700	~0°		2π		Similar to above specimen and conditions except exposed to vacuum (10^{-8} mm Hg) for 300 hrs; vacuum maintained by ion pump; reflectance measured in air after vacuum exposure.
121*	68	1969	300	0.380-0.700	~0°		2π		Similar to curve 119 specimen and conditions except exposed to vacuum (10^{-5} mm Hg) for 300 hrs; vacuum maintained by oil-diffusion pump; reflectance measured in air after vacuum exposure.
122*	68	1969	300	0.380-0.700	~0°		2π		Similar to above specimen and conditions except exposed to UV radiation in vacuum for 300 hrs; H_2 gas UV source.
123*	68	1969	300	0.380-0.700	~0°		2π		Similar to above specimen and conditions except He gas UV source.
124	68	1969	300	0.300-14.99	~0°		2π		Similar to curve 119 specimen and conditions.
125*	68	1969	300	0.300-14.99	~0°		2π		Similar to curve 120 specimen and conditions.
126*	68	1969	300	0.300-14.99	~0°		2π		Similar to curve 121 specimen and conditions.
127*	68	1969	300	0.300-14.99	~0°		2π		Similar to curve 122 specimen and conditions.
128*	68	1969	300	0.310-14.99	~0°		2π		Similar to curve 123 specimen and conditions.

* Not shown on plot

SPECIFICATION TABLE NO. 103 NORMAL SPECTRAL REFLECTANCE OF TITANIUM DIOXIDE PIGMENTED COATINGS (continued)

Curve No.	Ref. No.	Year	Temperature, K	Wavelength Range, μm	Geometry θ	Geometry θ'	Geometry ω'	Reported Error, %	Composition (weight percent), Specifications, and Remarks
129	72	1963	310	0.295-31.1	~0°		2π		W. P. Fuller 517-W-1 glossy white silicone paint; TiO_2 in silicone binder (~0.0915 mm thick) on 2024 aluminum substrate; data extracted from smooth curve; property measured in air soon after sample preparation; property is representative value (2 samples). [Author's designation: Silicone White]
130*	72	1963	310	0.295-31.6	~0°		2π		Similar to above specimen and conditions except sample stored in dry chamber for several days then in nitrogen for several days, then preoperty measured in air. [Author's designation: Silicone White, Sample No. 4]
131*	72	1963	310	0.295-31.6	~0°		2π		Similar to above specimen and conditions. [Author's designation: Silicone White, Sample No. 5]
132*	72	1963	310	0.295-31.6	~0°		2π		Similar to above specimen and conditions. [Author's designation: Silicone White, Sample No. 6]
133*	72	1963	310	0.295-32.0	~0°		2π		W. P. Fuller 517-W-1 glossy white silicone paint; TiO_2 in silicone binder (~0.0915 mm thick) on 2024 aluminum substrate; exposed to gamma radiation (dose ~1.7×10^{10} ergs g^{-1} C^{-1}) in vacuum (~10^{-6} mm Hg) maintained by diffusion pump; sample stored in nitrogen for several days, then property measured in air. [Author's designation: Silicone White, Sample No. 1]
134*	72	1963	310	0.295-32.0	~0°		2π		Similar to above specimen and conditions. [Author's designation: Silicone White, Sample No. 2]
135	72	1963	310	0.295-32.0	~0°		2π		Similar to above specimen and conditions. [Author's designation: Silicone White, Sample No. 3]
136*	72	1963	310	0.295-31.7	~0°		2π		Skyspar A-423 SA9185; TiO_2, untinted, in epoxy resin binder on epoxy primer SA9184 (~0.137 mm thick) and 2024 aluminum substrate; data extracted from smooth curve; property measured in air soon after sample preparation; property is representative value (2 samples). [Author's designation: Epoxy White]
137	72	1963	310	0.295-31.9	~0°		2π		Skyspar A-423 SA9185; TiO_2, untinted, in epoxy resin binder on epoxy primer SA9185 (~0.137 mm thick) and 2024 aluminum substrate; data extracted from smooth curve; sample stored in dry chamber several days, then in nitrogen several days, then property measured in air. [Author's designation: Epoxy White, Sample No. 4]
138*	72	1963	310	0.295-31.9	~0°		2π		Similar to above specimen and conditions. [Author's designation: Epoxy White, Sample No. 5]
139*	72	1963	310	0.295-31.9	~0°		2π		Similar to above specimen and conditions. [Author's designation: Epoxy White, Sample No. 6]
140	72	1963	310	0.295-31.4	~0°		2π		Skyspar A-423 SA9185; TiO_2, untinted, in epoxy resin binder on epoxy primer SA9184 (~0.137 mm thick) and 2024 aluminum substrate; exposed to gamma radiation (dose ~1.7×10^{10} ergs g^{-1} C^{-1}) in vacuum (~10^{-6} mm Hg) maintained by diffusion pump; sample stored in nitrogen for several days, then property measured in air; property is avg value of 3 samples. [Author's designation: Epoxy White]
141*	84	1964	~298	0.298-32.80	~0°		2π		Skyspar paint; TiO_2 in epoxy binder; data extracted from smooth curve; property converted from R $(2\pi, 0°)$ which was measured relative to MgO from 0.3 to 2.0 μm and relative to NiO from 0.5 to 32 μm. [Authors' designation: Control]
142*	84	1964	~298	0.298-6.29	~0°		2π		Similar to above specimen and conditions. [Authors' designation: Sample I]

* Not shown on plot

SPECIFICATION TABLE NO. 103 NORMAL SPECTRAL REFLECTANCE OF TITANIUM DIOXIDE PIGMENTED COATINGS (continued)

Curve No.	Ref. No.	Year	Temperature, K	Wavelength Range, μm	Geometry θ θ'	ω'	Reported Error, %	Composition (weight percent), Specifications, and Remarks
143	84	1964	~298	0.298-6.29	~0°	2π		Above specimen and conditions except exposed to UV radiation from AH-6 lamp (lamp-sample distance 6.35 cm) in vacuum (~5 x 10^{-7} mm Hg); vacuum maintained by oil diffusion pump; ESH 100.
144*	84	1964	~298	0.298-6.29	~0°	2π		Above specimen and conditions except ESH 250.
145*	84	1964	~298	0.298-6.29	~0°	2π		Above specimen and conditions except ESH 400.
146*	84	1964	~298	0.298-6.29	~0°	2π		Above specimen and conditions except ESH 650.
147*	84	1964	~298	0.298-6.29	~0°	2π		Above specimen and conditions except ESH 1000.
148*	84	1964	~298	0.298-6.57	~0°	2π		Similar to curve 141 specimen and conditions. [Authors' designation: Sample II]
149*	84	1964	~298	0.307-6.57	~0°	2π		Above specimen and conditions except exposed to UV radiation from AH-6 lamp (lamp-sample distance 6.35 cm) in vacuum (~5 x 10^{-7} mm Hg); vacuum maintained by oil diffusion pump; ESH 100.
150*	84	1964	~298	0.307-6.57	~0°	2π		Above specimen and conditions except ESH 250.
151	84	1964	~298	0.307-6.57	~0°	2π		Above specimen and conditions except ESH 400.
152*	84	1964	~298	0.307-6.57	~0°	2π		Above specimen and conditions except ESH 650.
153*	84	1964	~298	0.307-6.57	~0°	2π		Above specimen and conditions except ESH 1000.
154	84	1964	~298	0.298-32.8	~0°	2π		Curve 141 specimen and conditions measured after other samples had been irradiated for 1000 ESH of UV.
155*	85	1969	~298	0.250-2.497	~0°	2π		Rutile TiO_2 (DuPont R-960) in RTV-602 (General Electric) dimethyl siloxane; PBR 2; baked in air at 423 K for 1 hr; data extracted from smooth curve. [Author's designation: A-5 78- B2]

* Not shown on plot

DATA TABLE NO. 103 NORMAL SPECTRAL REFLECTANCE OF TITANIUM DIOXIDE PIGMENTED COATINGS

[Wavelength, λ, μm; Reflectance, ρ; Temperature, T, K]

λ	ρ
CURVE 1 T ~ 298	
0.377	0.116
0.420	0.833
0.429	0.856
0.453	0.878
0.493	0.901
0.540	0.910
0.612	0.917
0.710	0.921
0.778	0.917
0.847	0.920
0.871	0.928
0.955	0.932
1.038	0.934
1.226	0.904
1.279	0.907
1.362	0.900
1.437	0.878*
1.477	0.863
1.545	0.857
1.594	0.861
1.742	0.773
1.800	0.755
1.830	0.761
1.901	0.814
2.000	0.819
2.115	0.773
2.196	0.733
2.269	0.577
2.300	0.525
2.344	0.492*
2.394	0.465*
2.454	0.461
2.565	0.497
2.614	0.489*
2.643	0.469
2.676	0.420
2.704	0.310
CURVE 2 T ~ 298	
0.377	0.116*
0.416	0.659
0.426	0.695
0.466	0.732
CURVE 2 (cont.)	
0.511	0.748
0.574	0.752
0.612	0.743
0.691	0.721
0.755	0.733
0.801	0.748
0.852	0.752
0.914	0.750
1.014	0.738
1.130	0.736
1.168	0.730
1.333	0.731
1.399	0.735
1.491	0.722
1.595	0.723
1.692	0.683
1.795	0.649
1.846	0.654*
1.881	0.677
1.903	0.691
1.970	0.692
2.151	0.662
2.204	0.644
2.292	0.500
2.316	0.469
2.389	0.430
2.450	0.435
2.584	0.461
2.626	0.444
2.678	0.317
2.704	0.289
CURVE 3 T ~ 298	
0.384	0.149
0.405	0.436*
0.421	0.867
0.459	0.939
0.485	0.959
0.545	0.963
0.747	0.948
0.859	0.938
0.943	0.938
0.972	0.944
1.076	0.923
CURVE 3 (cont.)	
1.207	0.892
1.254	0.897*
1.316	0.900
1.375	0.879*
1.402	0.869
1.441	0.864*
1.492	0.864*
1.524	0.854*
1.603	0.851
1.641	0.823
1.698	0.751
1.750	0.723*
1.810	0.709
1.847	0.734
1.860	0.761*
1.887	0.774
1.965	0.762
2.014	0.741*
2.111	0.731*
2.180	0.700
2.212	0.670
2.270	0.497*
2.313	0.437
2.364	0.399*
2.413	0.373
2.602	0.404
2.639	0.393*
2.659	0.369
2.713	0.225
CURVE 4 T ~ 298	
0.384	0.149*
0.405	0.436
0.419	0.776
0.457	0.807
0.498	0.819
0.525	0.820
0.744	0.776
0.852	0.779
0.985	0.788
1.103	0.765
1.213	0.774
1.352	0.770
1.402	0.773
CURVE 4 (cont.)	
1.609	0.769*
1.745	0.655
1.787	0.639
1.820	0.643*
1.843	0.671
1.862	0.699*
1.918	0.718
1.961	0.707*
2.014	0.705
2.088	0.715
2.157	0.694*
2.202	0.664*
2.268	0.476
2.306	0.427
2.393	0.365
2.472	0.376*
2.563	0.410*
2.599	0.415
2.633	0.390
2.649	0.360*
2.711	0.195
CURVE 5 T = 337	
0.20	0.046
0.33	0.081
0.36	0.132
0.37	0.371
0.40	0.771
0.48	0.895
0.62	0.935
0.98	0.952
1.28	0.907*
1.60	0.819
1.65	0.764
1.71	0.629
1.81	0.714
1.91	0.732
2.04	0.692
2.13	0.667*
2.19	0.524
2.23	0.546
2.30	0.479
2.35	0.494
2.38	0.441*
CURVE 5 (cont.)	
2.43	0.413
2.48	0.361
2.55	0.377
2.61	0.348
2.70	0.314*
CURVE 6* T = 337	
0.20	0.039
0.32	0.056
0.38	0.124
0.43	0.406
0.50	0.617
0.64	0.848
0.74	0.913
0.94	0.931
1.23	0.900
1.39	0.876
1.60	0.815
1.66	0.743
1.71	0.629
1.81	0.717
1.93	0.732
2.13	0.669
2.19	0.524
2.23	0.546
2.30	0.480
2.35	0.494
2.38	0.439
2.43	0.412
2.49	0.362
2.55	0.376
2.61	0.342
2.70	0.312
CURVE 7 T ~ 333	
0.19	0.070
0.27	0.050
0.34	0.083
0.37	0.164
0.39	0.261
0.40	0.754
0.47	0.886
CURVE 7 (cont.)	
0.64	0.927
0.91	0.923
1.17	0.885
1.31	0.883
1.38	0.852*
1.53	0.842
1.58	0.814*
1.66	0.655
1.70	0.740
1.75	0.753*
1.81	0.789
1.89	0.776*
1.94	0.745
2.08	0.734*
2.13	0.710
2.16	0.532
2.20	0.576
2.26	0.512
2.32	0.554
2.45	0.360
2.50	0.454
2.55	0.476
2.64	0.364
2.70	0.312*
CURVE 8* T ~ 333	
0.19	0.070
0.27	0.051
0.36	0.103
0.42	0.211
0.46	0.437
0.52	0.713
0.59	0.836
0.68	0.895
0.84	0.923
1.15	0.885
1.33	0.881
1.38	0.852
1.55	0.841
1.59	0.801
1.66	0.654
1.71	0.739
1.76	0.756
1.81	0.787
CURVE 8 (cont.)*	
1.89	0.772
1.94	0.745
2.10	0.729
2.16	0.532
2.21	0.574
2.26	0.513
2.33	0.552
2.46	0.360
2.50	0.449
2.54	0.473
2.63	0.363
2.70	0.316
CURVE 9 T = 330	
0.20	0.050
0.30	0.058
0.34	0.092
0.41	0.639
0.43	0.716
0.46	0.813
0.62	0.862
0.75	0.871
1.10	0.838
1.30	0.811
1.50	0.728
1.59	0.652
1.71	0.550
1.91	0.632
2.00	0.605
2.11	0.577
2.18	0.474
2.21	0.482
2.30	0.390
2.35	0.413
2.38	0.371*
2.43	0.352
2.49	0.308
2.53	0.330
2.58	0.386
2.70	0.259
CURVE 10* T = 330	
0.20	0.041
0.31	0.058
0.37	0.118
0.52	0.566
0.66	0.779
0.74	0.825
0.93	0.850
1.15	0.815
1.35	0.789
1.53	0.703
1.61	0.626
1.70	0.549
1.92	0.631
2.01	0.603
2.10	0.582
2.18	0.473
2.21	0.484
2.30	0.391
2.35	0.413
2.39	0.371
2.43	0.352
2.49	0.309
2.53	0.330
2.58	0.385
2.70	0.257
CURVE 11 T = 336	
0.19	0.057
0.25	0.053
0.33	0.078
0.36	0.150
0.42	0.724
0.48	0.825
0.54	0.852
0.75	0.882
1.05	0.849
1.25	0.829
1.38	0.791
1.51	0.749
1.57	0.704
1.71	0.558
1.80	0.612
1.92	0.640

* Not shown on plot

DATA TABLE NO. 103 NORMAL SPECTRAL REFLECTANCE OF TITANIUM DIOXIDE PIGMENTED COATINGS (continued)

λ	ρ
CURVE 11 (cont.)	
T = 336	
2.02	0.617
2.11	0.595
2.19	0.473*
2.23	0.495
2.29	0.415
2.35	0.415*
2.39	0.383
2.50	0.338
2.58	0.393*
2.70	0.265*
CURVE 12	
T = 336	
0.19	0.056*
0.25	0.053*
0.34	0.086*
0.42	0.273
0.58	0.654
0.67	0.779
0.76	0.808
0.90	0.816
1.25	0.803
1.45	0.755
1.57	0.700*
1.71	0.556*
1.90	0.629
2.03	0.599
2.12	0.489
2.16	0.464*
2.23	0.462*
2.29	0.409
2.36	0.400
2.50	0.337*
2.58	0.393*
2.70	0.265*
CURVE 13	
T = 337	
0.19	0.057*
0.32	0.069
0.36	0.116
0.38	0.269
0.38	0.541
0.43	0.685

λ	ρ
CURVE 13 (cont.)	
0.46	0.789
0.53	0.841
0.68	0.873
0.81	0.877
1.26	0.837
1.30	0.819*
1.40	0.813
1.44	0.792
1.55	0.775
1.59	0.704*
1.66	0.685*
1.70	0.633
1.80	0.687
1.92	0.680
1.96	0.641*
2.11	0.629
2.17	0.543
2.24	0.521*
2.29	0.494*
2.33	0.504
2.39	0.400*
2.47	0.339*
2.55	0.391*
2.70	0.297*
CURVE 14*	
T = 337	
0.19	0.052
0.33	0.068
0.38	0.117
0.45	0.413
0.60	0.701
0.69	0.787
0.88	0.819
1.26	0.811
1.37	0.815
1.44	0.790
1.56	0.769
1.60	0.698
1.66	0.684
1.70	0.634
1.83	0.687
1.93	0.675
1.96	0.644
2.11	0.628
2.16	0.543

λ	ρ
CURVE 14 (cont.)*	
2.24	0.523
2.28	0.493
2.34	0.504
2.39	0.391
2.46	0.339
2.56	0.392
2.70	0.296
CURVE 15	
T ~ 298	
0.380	0.144
0.389	0.200
0.400	0.487
0.410	0.693
0.418	0.822
0.421	0.843
0.426	0.859*
0.433	0.873
0.460	0.892
0.486	0.904*
0.528	0.913
0.639	0.924
0.700	0.932
CURVE 16	
T ~ 298	
0.380	0.142*
0.385	0.169
0.394	0.266
0.411	0.688
0.417	0.800
0.432	0.839
0.453	0.852
0.495	0.852
0.591	0.827
0.700	0.785
CURVE 17	
T ~ 298	
0.222	0.061
0.284	0.052
0.315	0.057
0.348	0.083*
0.372	0.124

λ	ρ
CURVE 17 (cont.)	
0.382	0.155
0.391	0.224
0.418	0.832
0.427	0.878*
0.448	0.918
0.474	0.943
0.505	0.948
0.633	0.942
0.817	0.915
1.003	0.880
CURVE 18	
T ~ 298	
0.222	0.061*
0.290	0.050
0.318	0.047
0.352	0.069
0.384	0.106
0.411	0.166
0.459	0.342
0.495	0.500
0.541	0.621
0.598	0.703
0.674	0.773
0.741	0.806
0.817	0.819
0.915	0.821
0.976	0.814
CURVE 19*	
T ~ 298	
0.222	0.061
0.284	0.052
0.315	0.057
0.338	0.073
0.351	0.084
0.383	0.137
0.402	0.214
0.420	0.280
0.446	0.326
0.471	0.366
0.540	0.539
0.576	0.616
0.622	0.665
0.682	0.699

λ	ρ
CURVE 19 (cont.)*	
0.749	0.713
0.823	0.703
0.922	0.677
1.003	0.668
CURVE 20	
T ~ 298	
0.240	0.248
0.278	0.284
0.289	0.249
0.306	0.231
0.362	0.242
0.393	0.323
0.419	0.738
0.424	0.738
0.435	0.796
0.519	0.853
0.581	0.863
0.646	0.862
0.740	0.856
0.981	0.859
1.227	0.853
1.381	0.848
1.449	0.840
1.523	0.833*
1.696	0.795
1.790	0.780
1.854	0.791
1.900	0.793
2.248	0.711
CURVE 21*	
T ~ 298	
0.241	0.237
0.275	0.269
0.296	0.243
0.318	0.220
0.344	0.219
0.362	0.242
0.393	0.323
0.419	0.719
0.437	0.768
0.470	0.812
0.536	0.847
0.636	0.849

λ	ρ
CURVE 21 (cont.)*	
0.800	0.841
0.857	0.854
0.902	0.851
1.122	0.846
1.336	0.835
1.407	0.837
1.546	0.815
1.649	0.799
1.755	0.771
1.830	0.768
1.878	0.774
2.243	0.700
CURVE 22*	
T ~ 298	
0.241	0.237
0.275	0.269
0.296	0.243
0.318	0.220
0.344	0.219
0.362	0.242
0.393	0.323
0.418	0.738
0.424	0.738
0.435	0.796
0.502	0.845
0.606	0.863
0.753	0.855
0.863	0.868
0.936	0.868
1.005	0.857
1.356	0.847
1.428	0.840
1.454	0.844
1.508	0.849
1.576	0.838
1.644	0.813
1.706	0.793
1.790	0.780
1.854	0.791
1.900	0.793
2.248	0.711

λ	ρ
CURVE 23	
T ~ 298	
0.241	0.237
0.275	0.269
0.296	0.243
0.318	0.220
0.344	0.219
0.362	0.242*
0.393	0.323*
0.426	0.602
0.483	0.712
0.531	0.746
0.600	0.764
0.684	0.758
0.765	0.751
0.921	0.758
1.414	0.752
1.463	0.749*
1.528	0.750*
1.600	0.745
1.715	0.717
1.771	0.717*
1.856	0.728
2.174	0.676
2.238	0.659
CURVE 24*	
T ~ 298	
0.229	0.250
0.241	0.245
0.271	0.274
0.298	0.241
0.343	0.235
0.381	0.278
0.402	0.398
0.438	0.799
0.450	0.832
0.617	0.903
0.694	0.896
0.794	0.896
0.817	0.888
0.878	0.889
0.932	0.900
1.311	0.878
1.446	0.873
1.636	0.847
1.726	0.816

λ	ρ
CURVE 24 (cont.)*	
1.796	0.809
1.875	0.818
2.200	0.756
2.263	0.721
CURVE 25	
T ~ 298	
0.228	0.239
0.254	0.251
0.276	0.265*
0.301	0.229
0.334	0.219
0.367	0.236
0.380	0.277
0.402	0.398
0.421	0.684*
0.434	0.736
0.490	0.777
0.618	0.787
0.680	0.781
0.739	0.766*
0.813	0.764
0.918	0.768
1.479	0.755*
1.627	0.741*
1.763	0.722
1.849	0.732*
1.933	0.726*
2.034	0.707*
2.185	0.696*
2.245	0.673*
2.272	0.657*
CURVE 26	
T ~ 298	
0.234	0.222
0.276	0.250
0.305	0.227*
0.331	0.219*
0.378	0.274*
0.402	0.398*
0.416	0.602
0.434	0.692*
0.485	0.728
0.531	0.728

* Not shown on plot

DATA TABLE NO. 103 NORMAL SPECTRAL REFLECTANCE OF TITANIUM DIOXIDE PIGMENTED COATINGS (continued)

CURVE 26 (cont.) T ~ 298

λ	ρ
0.611	0.697
0.711	0.661
0.808	0.630
0.898	0.622
1.091	0.612
2.018	0.567
2.192	0.543*
2.246	0.547*
2.311	0.557*

CURVE 27* T ~ 298

λ	ρ
0.234	0.222
0.276	0.250
0.305	0.227
0.331	0.219
0.378	0.274
0.402	0.398
0.416	0.602
0.431	0.649
0.489	0.703
0.526	0.709
0.606	0.692
0.720	0.646
0.831	0.618
0.930	0.604
1.348	0.578
1.430	0.568
1.548	0.568
1.729	0.560
1.811	0.557
1.874	0.561
1.950	0.556
2.016	0.553
2.184	0.558
2.237	0.555

CURVE 28 T ~ 298

λ	ρ
0.395	0.500
0.398	0.599
0.405	0.699
0.413	0.784
0.417	0.815

CURVE 28 (cont.)

λ	ρ
0.422	0.833*
0.429	0.847
0.439	0.860
0.450	0.869
0.461	0.872
0.486	0.873
0.517	0.865
0.543	0.862
0.567	0.862*
0.599	0.866
0.635	0.864*
0.676	0.859
0.698	0.855

CURVE 29* T ~ 298

λ	ρ
0.395	0.500
0.398	0.599
0.405	0.699
0.413	0.784
0.418	0.814
0.425	0.831
0.437	0.848
0.448	0.856
0.460	0.861
0.473	0.864
0.493	0.862
0.536	0.853
0.564	0.850
0.615	0.847
0.666	0.832
0.680	0.829
0.695	0.827

CURVE 30 T ~ 298

λ	ρ
0.413	0.750
0.417	0.782*
0.423	0.814
0.431	0.839*
0.443	0.871
0.457	0.893
0.465	0.902
0.477	0.909*
0.491	0.914

CURVE 30 (cont.)

λ	ρ
0.509	0.918
0.525	0.921*
0.548	0.922
0.567	0.921*
0.585	0.920
0.601	0.917*
0.642	0.906
0.667	0.900
0.700	0.892

CURVE 31* T ~ 298

λ	ρ
0.413	0.750
0.417	0.782
0.423	0.814
0.432	0.833
0.439	0.850
0.450	0.865
0.462	0.881
0.476	0.891
0.487	0.899
0.501	0.905
0.515	0.910
0.537	0.913
0.557	0.914
0.582	0.913
0.597	0.910
0.623	0.903
0.656	0.893
0.700	0.882

CURVE 32* T ~ 298

λ	ρ
0.413	0.750
0.417	0.782
0.423	0.799
0.427	0.813
0.435	0.832
0.444	0.848
0.454	0.862
0.465	0.871
0.476	0.877
0.490	0.881
0.506	0.885
0.523	0.886

CURVE 32 (cont.)*

λ	ρ
0.537	0.886
0.549	0.885
0.562	0.882
0.573	0.878
0.595	0.869
0.614	0.861
0.639	0.852
0.668	0.843
0.700	0.833

CURVE 33* T ~ 298

λ	ρ
0.427	0.750
0.432	0.780
0.440	0.797
0.448	0.808
0.455	0.814
0.467	0.817
0.477	0.817
0.489	0.816
0.505	0.814
0.528	0.807
0.549	0.800
0.569	0.792
0.583	0.785
0.609	0.772
0.624	0.761
0.640	0.750

CURVE 34 T ~ 298

λ	ρ
0.413	0.750*
0.419	0.822*
0.423	0.845*
0.432	0.875*
0.443	0.900
0.451	0.909*
0.461	0.915
0.472	0.921
0.487	0.926
0.504	0.930
0.519	0.931*
0.568	0.931
0.611	0.929
0.673	0.925
0.700	0.923

CURVE 35* T ~ 298

λ	ρ
0.414	0.750
0.418	0.805
0.427	0.859
0.434	0.879
0.445	0.890
0.461	0.899
0.475	0.904
0.493	0.907
0.511	0.907
0.542	0.905
0.581	0.900
0.626	0.893
0.675	0.883
0.700	0.878

CURVE 36* T ~ 298

λ	ρ
0.414	0.750
0.418	0.796
0.427	0.830
0.448	0.871
0.455	0.877
0.470	0.881
0.484	0.881
0.515	0.876
0.586	0.862
0.654	0.850
0.700	0.842

CURVE 37 T ~ 298

λ	ρ
0.413	0.750*
0.430	0.803
0.441	0.827
0.449	0.841
0.454	0.844*
0.464	0.846
0.476	0.846*
0.488	0.843
0.515	0.834
0.541	0.826
0.574	0.816
0.628	0.801
0.666	0.792
0.700	0.785*

CURVE 38 T ~ 298

λ	ρ
0.616	0.877
0.979	0.728
1.424	0.548
1.482	0.526*
1.564	0.502
1.636	0.485
1.725	0.467*
1.829	0.453
1.978	0.437*
2.061	0.425
2.156	0.406
2.277	0.372
2.321	0.360*
2.361	0.355
2.440	0.355*
2.505	0.364*
2.600	0.380*

CURVE 39 T ~ 298

λ	ρ
0.617	0.903
0.719	0.896
0.841	0.881
1.078	0.838
1.123	0.830
1.174	0.827
1.220	0.829*
1.257	0.835*
1.304	0.853
1.342	0.859*
1.385	0.859
1.420	0.853
1.473	0.819
1.505	0.799*
1.556	0.775*
1.604	0.761
1.672	0.746*
1.763	0.734
1.895	0.719*
1.989	0.704
2.064	0.683
2.137	0.653
2.267	0.580
2.311	0.556*
2.364	0.533

CURVE 39 (cont.)

λ	ρ
2.415	0.519
2.484	0.505*
2.539	0.499*
2.600	0.492

CURVE 40* T ~ 298

λ	ρ
0.388	0.633
0.392	0.663
0.397	0.683
0.404	0.703
0.416	0.726
0.429	0.744
0.443	0.757
0.458	0.766
0.475	0.771
0.527	0.779
0.586	0.785
0.638	0.786
0.669	0.785
0.698	0.781

CURVE 41 T ~ 298

λ	ρ
0.388	0.539
0.392	0.565
0.395	0.584
0.401	0.598
0.412	0.611*
0.427	0.619
0.458	0.629
0.473	0.637
0.490	0.646
0.510	0.660
0.531	0.675*
0.551	0.685
0.567	0.691*
0.589	0.694
0.640	0.694
0.666	0.697
0.700	0.701

CURVE 42* T ~ 298

λ	ρ
0.387	0.750
0.396	0.783
0.409	0.822
0.424	0.853
0.437	0.877
0.443	0.885
0.453	0.894
0.475	0.909
0.489	0.916
0.503	0.921
0.518	0.923
0.537	0.924
0.602	0.921
0.652	0.916
0.700	0.911

CURVE 43* T ~ 298

λ	ρ
0.387	0.750
0.396	0.783
0.409	0.822
0.424	0.853
0.437	0.877
0.445	0.884
0.456	0.893
0.468	0.899
0.480	0.905
0.498	0.909
0.515	0.913
0.535	0.915
0.552	0.915
0.600	0.911
0.700	0.898

CURVE 44* T ~ 298

λ	ρ
0.387	0.750
0.396	0.783
0.409	0.822
0.424	0.853
0.440	0.877
0.446	0.883
0.461	0.892
0.475	0.897

* Not shown on plot

DATA TABLE NO. 103 NORMAL SPECTRAL REFLECTANCE OF TITANIUM DIOXIDE PIGMENTED COATINGS (continued)

CURVE 44 (cont.)*
T ~ 298

λ	ρ
0.489	0.900
0.506	0.902
0.520	0.903
0.540	0.901
0.565	0.898
0.593	0.893
0.632	0.885
0.668	0.876
0.700	0.868

CURVE 45
T ~ 298

λ	ρ
0.387	0.750
0.396	0.783
0.410	0.822
0.421	0.841*
0.436	0.856*
0.448	0.863*
0.459	0.868*
0.475	0.873
0.489	0.875
0.511	0.875
0.530	0.872*
0.559	0.865
0.624	0.851
0.668	0.840
0.700	0.832

CURVE 46
T ~ 298

λ	ρ
0.700	0.842
0.760	0.823
0.813	0.812
0.888	0.809
1.046	0.816
1.135	0.806*
1.208	0.805
1.261	0.812*
1.343	0.821
1.423	0.824
1.476	0.817*
1.547	0.796
1.608	0.770
1.751	0.696*

CURVE 46 (cont.)

λ	ρ
1.778	0.686
1.817	0.688*
1.863	0.701
1.955	0.739
1.985	0.744
2.015	0.738
2.087	0.697
2.158	0.643
2.242	0.564
2.315	0.476*
2.372	0.402
2.405	0.386*
2.436	0.387
2.516	0.423
2.600	0.460

CURVE 47*
T ~ 298

λ	ρ
0.700	0.842
0.760	0.823
0.813	0.811
0.862	0.802
0.992	0.799
1.055	0.788
1.153	0.764
1.213	0.754
1.376	0.742
1.457	0.723
1.605	0.658
1.726	0.596
1.806	0.549
1.867	0.551
2.003	0.565
2.037	0.565
2.086	0.549
2.173	0.502
2.261	0.436
2.314	0.378
2.377	0.334
2.414	0.328
2.500	0.344
2.600	0.363

CURVE 48*
T ~ 298

λ	ρ
0.700	0.842
0.760	0.823
0.810	0.808
0.862	0.799
0.983	0.780
1.057	0.764
1.100	0.753
1.162	0.737
1.254	0.726
1.333	0.714
1.415	0.697
1.521	0.667
1.620	0.630
1.676	0.606
1.749	0.573
1.780	0.560
1.794	0.549
1.837	0.532
1.932	0.497
1.990	0.481
2.017	0.477
2.163	0.475
2.195	0.467
2.220	0.454
2.414	0.328
2.500	0.344
2.600	0.363

CURVE 49
T ~ 298

λ	ρ
0.702	0.903
0.729	0.876
0.782	0.851
0.869	0.837
0.941	0.821
1.046	0.800
1.127	0.785
1.370	0.741
1.440	0.716
1.538	0.645
1.602	0.608
1.659	0.581
1.730	0.565*
1.819	0.552*
1.968	0.538

CURVE 49 (cont.)

λ	ρ
2.007	0.531*
2.042	0.517
2.204	0.413
2.290	0.367
2.342	0.341
2.388	0.327*
2.485	0.318
2.588	0.317*
2.621	0.308
2.637	0.293*
2.682	0.255

CURVE 50*
T ~ 298

λ	ρ
0.700	0.886
0.725	0.860
0.761	0.845
0.801	0.834
0.868	0.827
0.976	0.827
1.013	0.823
1.118	0.776
1.171	0.755
1.204	0.749
1.354	0.780
1.400	0.782
1.439	0.759
1.491	0.712
1.549	0.646
1.602	0.608
1.666	0.581
1.756	0.560
1.916	0.518
2.023	0.486
2.101	0.447
2.265	0.356
2.331	0.322
2.394	0.301
2.444	0.297
2.570	0.292
2.608	0.285
2.637	0.260
2.682	0.198

CURVE 51
T ~ 298

λ	ρ
0.597	0.882
0.692	0.868
0.757	0.852
0.807	0.831
0.880	0.816
1.145	0.782*
1.327	0.749
1.389	0.731*
1.519	0.670
1.602	0.633
1.747	0.586*
1.801	0.575
1.906	0.565*
2.010	0.560
2.053	0.549*
2.139	0.504
2.223	0.456
2.291	0.419*
2.361	0.395
2.422	0.384*
2.512	0.379
2.589	0.379*

CURVE 52*
T ~ 298

λ	ρ
0.709	0.816
0.808	0.780
1.138	0.681
1.225	0.670
1.399	0.663
1.492	0.638
1.748	0.541
1.845	0.527
1.988	0.522
2.065	0.506
2.185	0.461
2.333	0.392
2.407	0.369
2.500	0.353
2.590	0.343

CURVE 53
T ~ 298

λ	ρ
0.716	0.803
0.751	0.775*
0.797	0.760
0.998	0.707
1.126	0.668
1.169	0.659*
1.280	0.655
1.372	0.650*
1.447	0.636
1.518	0.614
1.725	0.504
1.781	0.490
1.869	0.485
1.986	0.489
2.035	0.482
2.140	0.444
2.347	0.337
2.431	0.308*
2.504	0.292
2.588	0.280

CURVE 54
T ~ 298

λ	ρ
0.300	0.06
0.340	0.06
0.357	0.11
0.381	0.52
0.400	0.84
0.500	0.89
0.600	0.89
0.690	0.88
0.800	0.85
1.00	0.79
1.14	0.74
1.20	0.73
1.31	0.70
1.40	0.68
1.44	0.67*
1.51	0.66
1.62	0.64*
1.72	0.54
1.80	0.58
1.87	0.60
2.03	0.61
2.14	0.57

CURVE 54 (cont.)

λ	ρ
2.20	0.57
2.29	0.30
2.38	0.39
2.46	0.36*
2.60	0.48

CURVE 55*
T ~ 298

λ	ρ
0.337	0.588
0.348	0.358
0.361	0.253
0.374	0.374
0.381	0.515
0.392	0.653
0.398	0.741
0.411	0.819
0.474	0.880
0.537	0.880
0.595	0.895
0.646	0.919
0.700	0.923
0.766	0.925
0.896	0.907
1.01	0.896
1.10	0.879
1.31	0.855
1.56	0.782
1.74	0.662
1.91	0.689
2.08	0.596
2.32	0.318

CURVE 56*
T ~ 298

λ	ρ
0.338	0.566
0.343	0.332
0.359	0.174
0.372	0.132
0.382	0.147
0.391	0.208
0.398	0.361
0.411	0.667
0.473	0.896
0.473	0.917
0.541	0.917

CURVE 56 (cont.)*

λ	ρ
0.596	0.925
0.648	0.942
0.702	0.942
0.772	0.952
0.901	0.944
1.01	0.942
1.11	0.930
1.32	0.909
1.58	0.831
1.75	0.744
1.92	0.744
2.09	0.659
2.35	0.380

CURVE 57
T ~ 298

λ	ρ
0.379	0.100
0.429	0.874*
0.481	0.863
0.571	0.863
0.700	0.819
0.782	0.819
0.910	0.801
1.020	0.801
1.286	0.769
1.404	0.748
1.589	0.751
1.802	0.709*
1.986	0.706*
2.201	0.640

CURVE 58*
T ~ 298

λ	ρ
0.379	0.100
0.429	0.874
0.481	0.909
0.581	0.909
0.707	0.878
0.973	0.848
1.156	0.832
1.396	0.793
1.573	0.795
1.647	0.779
1.702	0.742
1.743	0.724

* Not shown on plot

CURVE 58 (cont.)*
T ~ 298

λ	ρ
1.832	0.718
2.002	0.713
2.186	0.624

CURVE 59*
T ~ 298

λ	ρ
0.322	0.024
0.335	0.180
0.335	0.101
0.379	0.100
0.429	0.874
0.593	0.936
0.684	0.908
0.729	0.908
0.783	0.916
0.905	0.897
1.004	0.897
1.299	0.831
1.399	0.812
1.493	0.810
1.593	0.816
1.714	0.764
1.824	0.731
1.895	0.722
1.980	0.716
2.122	0.646
2.205	0.583

CURVE 60
T ~ 298

λ	ρ
0.379	0.100*
0.429	0.874*
0.436	0.930
0.584	0.943
0.635	0.943*
0.715	0.925
0.814	0.929
1.093	0.893
1.328	0.836
1.426	0.822*
1.566	0.822
1.714	0.754
1.815	0.720*

CURVE 60 (cont.)

λ	ρ
1.977	0.709
2.189	0.575*

CURVE 61*
T ~ 298

λ	ρ
0.379	0.100
0.436	0.951
0.590	0.960
0.685	0.949
0.771	0.954
1.235	0.889
1.341	0.850
1.445	0.841
1.494	0.846
1.588	0.846
1.747	0.768
1.848	0.737
1.992	0.717
2.127	0.638
2.202	0.585

CURVE 62
T ~ 298

λ	ρ
0.322	0.248
0.336	0.112
0.364	0.065
0.451	0.776
0.500	0.749
0.595	0.736
0.805	0.625
1.371	0.381
1.658	0.299
1.817	0.237
2.046	0.179
2.198	0.146

CURVE 63*
T ~ 298

λ	ρ
0.322	0.427
0.339	0.227
0.357	0.197
0.381	0.124
0.451	0.888

CURVE 63 (cont.)*

λ	ρ
0.517	0.876
0.578	0.886
0.748	0.839
0.927	0.784
0.993	0.736
1.257	0.669
1.630	0.537
1.841	0.423
2.071	0.349
2.205	0.293

CURVE 64*
T ~ 298

λ	ρ
0.364	0.065
0.453	0.918
0.528	0.911
0.607	0.921
0.886	0.853
1.022	0.829
1.352	0.703
1.592	0.639
1.780	0.533
2.048	0.442
2.205	0.362

CURVE 65*
T ~ 298

λ	ρ
0.364	0.065
0.454	0.928
0.590	0.931
0.634	0.936
0.703	0.917
0.800	0.910
0.905	0.882
1.015	0.871
1.287	0.775
1.621	0.689
1.812	0.584
2.049	0.502
2.200	0.416

CURVE 66
T ~ 298

λ	ρ
0.364	0.065*
0.454	0.934
0.581	0.951
0.649	0.951
0.774	0.928
0.998	0.907
1.482	0.773
1.614	0.745*
1.787	0.653*
1.954	0.598
2.043	0.550
2.120	0.494*
2.198	0.456

CURVE 67*
T ~ 298

λ	ρ
0.361	0.049
0.371	0.049
0.379	0.070
0.418	0.848
0.425	0.884
0.442	0.914
0.458	0.929
0.486	0.935
0.914	0.947
1.06	0.947
1.14	0.934
1.17	0.890
1.20	0.946
1.25	0.926
1.28	0.938
1.31	0.941
1.35	0.926
1.39	0.890
1.43	0.907
1.44	0.913
1.51	0.913
1.54	0.904
1.61	0.904
1.66	0.870
1.67	0.730
1.68	0.736
1.70	0.719
1.72	0.824

CURVE 67 (cont.)*

λ	ρ
1.74	0.720
1.75	0.750
1.76	0.819
1.78	0.819
1.79	0.874
1.83	0.832
1.86	0.848
1.87	0.886
1.99	0.846
2.03	0.832
2.11	0.829
2.16	0.779
2.25	0.537
2.28	0.447
2.31	0.605
2.33	0.618
2.34	0.542
2.36	0.498
2.37	0.510
2.38	0.566
2.40	0.566

CURVE 68*
T ~ 298

λ	ρ
0.361	0.049
0.371	0.049
0.379	0.070
0.411	0.678
0.427	0.795
0.440	0.827
0.469	0.856
0.546	0.892
0.648	0.909
0.911	0.912
1.05	0.932
1.13	0.914
1.15	0.900
1.17	0.880
1.19	0.913
1.21	0.920
1.24	0.910
1.29	0.927
1.34	0.911
1.38	0.879
1.41	0.879

CURVE 68 (cont.)*

λ	ρ
1.44	0.898
1.49	0.908
1.54	0.886
1.56	0.886
1.60	0.910
1.61	0.904
1.66	0.870
1.67	0.730
1.68	0.736
1.70	0.719
1.72	0.824
1.74	0.720
1.75	0.750
1.76	0.819
1.78	0.819
1.79	0.874
1.83	0.832
1.85	0.844
1.88	0.873
1.99	0.833
2.10	0.823
2.16	0.777
2.25	0.537
2.28	0.447
2.31	0.605
2.33	0.618
2.34	0.542
2.36	0.498
2.37	0.510
2.38	0.566
2.40	0.566

CURVE 69*
T = 293

λ	ρ
0.360	0.074
0.367	0.084
0.372	0.097
0.379	0.122
0.383	0.153
0.386	0.182
0.390	0.236
0.399	0.406
0.408	0.644
0.413	0.733
0.420	0.777

CURVE 69 (cont.)*

λ	ρ
0.426	0.794
0.445	0.802
0.502	0.820
0.568	0.829
0.613	0.838
0.671	0.842
0.710	0.842
0.710	0.829
0.736	0.842
0.766	0.848
0.998	0.843
1.12	0.840
1.16	0.830
1.19	0.810
1.21	0.816
1.23	0.831
1.29	0.831
1.31	0.826
1.35	0.824
1.41	0.782
1.43	0.785
1.44	0.795
1.50	0.803
1.54	0.791
1.56	0.789
1.61	0.796
1.64	0.789
1.66	0.771
1.70	0.696
1.73	0.678
1.75	0.672
1.77	0.687
1.81	0.742
1.83	0.747
1.86	0.747
1.89	0.754
1.91	0.754
1.99	0.735
2.07	0.737
2.11	0.732
2.17	0.686
2.22	0.623
2.28	0.501
2.30	0.470
2.32	0.459
2.40	0.468

CURVE 69 (cont.)*

λ	ρ
2.45	0.470
2.48	0.466
2.50	0.466

CURVE 70*
T = 293

λ	ρ
0.360	0.080
0.364	0.085
0.372	0.102
0.378	0.126
0.382	0.147
0.386	0.190
0.392	0.283
0.403	0.491
0.409	0.614
0.414	0.662
0.419	0.693
0.428	0.709
0.468	0.745
0.508	0.768
0.541	0.782
0.582	0.791
0.634	0.799
0.710	0.802
0.710	0.782
0.763	0.801
0.897	0.803
0.971	0.809
1.13	0.809
1.16	0.801
1.19	0.782
1.24	0.807
1.34	0.804
1.37	0.798
1.41	0.765
1.45	0.783
1.50	0.787
1.55	0.775
1.61	0.784
1.65	0.772
1.68	0.734
1.71	0.684
1.72	0.674
1.76	0.666
1.79	0.705

* Not shown on plot

DATA TABLE NO. 103 NORMAL SPECTRAL REFLECTANCE OF TITANIUM DIOXIDE PIGMENTED COATINGS (continued)

λ	ρ
CURVE 70 (cont.)* T = 293	
1.81	0.728
1.83	0.742
1.86	0.738
1.90	0.750
1.93	0.747
1.97	0.734
2.03	0.729
2.09	0.731
2.11	0.723
2.18	0.667
2.21	0.620
2.27	0.504
2.28	0.473
2.31	0.453
2.35	0.463
2.41	0.466
2.50	0.464
CURVE 71* T = 293	
0.360	0.084
0.369	0.093
0.376	0.108
0.380	0.122
0.384	0.142
0.389	0.195
0.397	0.325
0.409	0.596
0.416	0.688
0.420	0.712
0.429	0.729
0.455	0.752
0.499	0.774
0.532	0.778
0.646	0.778
0.710	0.772
0.710	0.760
0.760	0.768
0.841	0.753
0.938	0.742
1.00	0.739
1.09	0.723
1.10	0.723
1.13	0.712
1.22	0.703

λ	ρ
CURVE 71 (cont.)*	
1.24	0.703
1.36	0.682
1.42	0.662
1.57	0.648
1.63	0.632
1.65	0.604
1.67	0.577
1.69	0.569
1.71	0.584
1.77	0.602
1.85	0.605
1.90	0.599
1.95	0.586
2.08	0.581
2.11	0.573
2.15	0.528
2.16	0.516
2.19	0.509
2.22	0.509
2.26	0.498
2.30	0.468
2.40	0.429
2.44	0.409
2.47	0.399
2.50	0.399
CURVE 72 T = 293	
0.360	0.080
0.371	0.082
0.382	0.094
0.395	0.120
0.408	0.148
0.429	0.174
0.461	0.215
0.519	0.321
0.572	0.417
0.608	0.470
0.642	0.511
0.681	0.544
0.710	0.560
0.710	0.546
0.764	0.590
0.813	0.609
0.883	0.626
1.01	0.641

λ	ρ
CURVE 72 (cont.)	
1.10	0.643
1.23	0.639
1.37	0.631
1.43	0.620
1.59	0.612
1.63	0.595
1.67	0.545*
1.69	0.545
1.71	0.562
1.79	0.578*
1.88	0.579
1.94	0.567
2.08	0.562*
2.11	0.550
2.15	0.506*
2.18	0.495
2.23	0.495*
2.31	0.452
2.39	0.425*
2.45	0.394
2.47	0.391*
2.50	0.390
CURVE 73* T = 298	
0.250	0.100
0.300	0.087
0.334	0.071
0.356	0.090
0.377	0.164
0.413	0.840
0.428	0.904
0.502	0.936
0.708	0.951
0.742	0.946
1.064	0.937
1.113	0.930
1.141	0.916
1.157	0.893
1.197	0.921
1.240	0.921
1.302	0.912
1.356	0.850
1.370	0.848
1.419	0.873
1.447	0.870

λ	ρ
CURVE 73 (cont.)*	
1.510	0.845
1.565	0.848
1.624	0.802
1.639	0.775
1.664	0.704
1.701	0.684
1.717	0.661
1.781	0.756
1.818	0.740
1.841	0.753
1.859	0.753
1.967	0.714
2.036	0.714
2.092	0.687
2.199	0.494
2.235	0.407
2.255	0.386
2.302	0.410
2.321	0.413
2.351	0.404
2.417	0.404
2.446	0.397
2.495	0.401
CURVE 74* T = 281	
0.250	0.082
0.306	0.082
0.334	0.071
0.356	0.090
0.377	0.164
0.413	0.840
0.430	0.901
0.509	0.922
0.694	0.922
0.719	0.908
1.037	0.894
1.104	0.892
1.141	0.881
1.160	0.861
1.200	0.886
1.298	0.881
1.324	0.865
1.349	0.830
1.373	0.821
1.414	0.848

λ	ρ
CURVE 74 (cont.)*	
1.435	0.848
1.493	0.828
1.562	0.828
1.617	0.792
1.662	0.694
1.699	0.676
1.716	0.656
1.777	0.738
1.815	0.726
1.844	0.743
1.861	0.743
1.968	0.703
2.042	0.703
2.086	0.684
2.118	0.644
2.199	0.494
2.235	0.407
2.255	0.386
2.302	0.410
2.321	0.413
2.351	0.404
2.417	0.404
2.446	0.397
2.495	0.401
CURVE 75* T = 281	
0.250	0.082
0.306	0.082
0.334	0.071
0.356	0.090
0.377	0.164
0.408	0.699
0.441	0.770
0.474	0.802
0.524	0.809
0.697	0.764
0.748	0.734
0.897	0.707
1.053	0.690
1.101	0.690
1.152	0.673
1.204	0.688
1.247	0.689
1.318	0.683
1.358	0.659

λ	ρ
CURVE 75 (cont.)*	
1.402	0.671
1.446	0.675
1.487	0.666
1.515	0.666
1.534	0.668
1.571	0.668
1.612	0.649
1.654	0.579
1.672	0.579
1.712	0.566
1.743	0.607
1.772	0.625
1.813	0.618
1.832	0.629
1.854	0.634
1.949	0.611
2.031	0.611
2.079	0.599
2.237	0.367
2.255	0.360
2.295	0.380
2.316	0.384
2.345	0.376
2.400	0.379
2.446	0.372
2.497	0.378
CURVE 76* T = 281	
0.256	0.087
0.311	0.081
0.336	0.066
0.364	0.091
0.387	0.221
0.415	0.841
0.426	0.891
0.473	0.907
0.705	0.901
0.781	0.883
0.989	0.877
1.081	0.877
1.125	0.870
1.161	0.848
1.213	0.873
1.289	0.870
1.327	0.863

λ	ρ
CURVE 76 (cont.)*	
1.360	0.822
1.383	0.814
1.421	0.839
1.453	0.839
1.500	0.819
1.522	0.819
1.563	0.823
1.613	0.797
1.638	0.755
1.657	0.686
1.697	0.674
1.723	0.650
1.753	0.709
1.778	0.740
1.816	0.724
1.837	0.739
1.868	0.739
1.882	0.726
1.960	0.700
2.041	0.701
2.080	0.687
2.192	0.509
2.242	0.399
2.264	0.386
2.314	0.405
2.365	0.400
2.505	0.396
CURVE 77* T = 281	
0.256	0.087
0.311	0.081
0.336	0.066
0.364	0.091
0.387	0.221
0.426	0.879
0.509	0.882
0.779	0.823
0.904	0.801
1.042	0.789
1.143	0.783
1.166	0.770
1.183	0.770
1.215	0.783
1.327	0.779
1.359	0.755

λ	ρ
CURVE 77 (cont.)*	
1.385	0.747
1.427	0.766
1.517	0.749
1.562	0.753
1.619	0.734
1.641	0.713
1.660	0.668
1.679	0.638
1.700	0.638
1.721	0.617
1.753	0.660
1.783	0.690
1.798	0.690
1.817	0.682
1.861	0.696
1.964	0.668
2.041	0.668
2.085	0.652
2.120	0.622
2.189	0.502
2.237	0.395
2.261	0.376
2.320	0.399
2.350	0.393
2.418	0.393
2.448	0.386
2.505	0.391
CURVE 78 T = 281	
0.256	0.078
0.299	0.078
0.336	0.066
0.364	0.091
0.387	0.221
0.414	0.688*
0.429	0.715*
0.459	0.724*
0.535	0.691
0.703	0.560*
0.728	0.536
0.813	0.480
0.907	0.438
1.101	0.368
1.274	0.321
1.504	0.273

* Not shown on plot

λ	ρ
CURVE 78 (cont.) T = 281	
1.656	0.244
1.880	0.225
2.150	0.212
2.260	0.189
2.383	0.191
2.505	0.191
CURVE 79* T = 298	
0.250	0.136
0.337	0.083
0.356	0.088
0.385	0.234
0.410	0.848
0.429	0.918
0.450	0.907
0.490	0.928
0.583	0.915
0.699	0.910
0.764	0.896
1.097	0.883
1.120	0.876
1.160	0.844
1.177	0.864
1.210	0.874
1.297	0.865
1.319	0.853
1.364	0.802
1.414	0.832
1.448	0.832
1.499	0.814
1.520	0.814
1.557	0.823
1.617	0.787
1.652	0.677
1.673	0.685
1.693	0.668
1.716	0.661
1.738	0.711
1.775	0.757
1.797	0.742
1.811	0.742
1.835	0.759
1.857	0.754
CURVE 79 (cont.)*	
1.877	0.743
1.937	0.727
2.030	0.730
2.068	0.714
2.116	0.660
2.231	0.408
2.257	0.396
2.294	0.425
2.320	0.425
2.340	0.415
2.379	0.418
2.435	0.404
2.476	0.414
2.500	0.414
CURVE 80 T = 281	
0.250	0.100
0.287	0.087
0.299	0.087
0.326	0.069*
0.342	0.069
0.354	0.082
0.360	0.105
0.385	0.235
0.407	0.698*
0.427	0.738*
0.450	0.738
0.498	0.709
0.582	0.633
0.632	0.586
0.711	0.531
0.721	0.511
0.804	0.453
0.845	0.433
0.963	0.392
1.148	0.333
1.365	0.282
1.662	0.236
2.139	0.218*
2.252	0.192
2.314	0.198
2.500	0.198
CURVE 81* T = 281	
0.250	0.100
0.287	0.087
0.299	0.087
0.326	0.069
0.342	0.069
0.354	0.082
0.360	0.105
0.385	0.235
0.403	0.711
0.418	0.736
0.439	0.743
0.469	0.737
0.540	0.683
0.636	0.593
0.706	0.542
CURVE 82* T = 281	
0.250	0.100
0.287	0.087
0.299	0.087
0.326	0.069
0.342	0.069
0.354	0.082
0.360	0.105
0.385	0.235
0.403	0.711
0.418	0.736
0.439	0.743
0.469	0.737
0.540	0.683
0.636	0.593
0.706	0.542
CURVE 83* T = 281	
0.250	0.100
0.287	0.087
0.299	0.087
0.326	0.069
0.342	0.069
0.354	0.082
0.360	0.105
0.385	0.235
CURVE 83 (cont.)*	
0.411	0.747
0.440	0.756
0.471	0.747
0.670	0.592
0.706	0.567
CURVE 84* T = 281	
0.250	0.100
0.287	0.087
0.299	0.087
0.326	0.069
0.342	0.069
0.354	0.082
0.360	0.105
0.385	0.235
0.415	0.762
0.442	0.776
0.486	0.769
0.639	0.662
0.706	0.609
CURVE 85* T = 281	
0.250	0.100
0.287	0.087
0.299	0.087
0.326	0.069
0.342	0.069
0.354	0.082
0.360	0.105
0.385	0.235
0.415	0.779
0.431	0.788
0.474	0.788
0.665	0.682
0.711	0.665
0.721	0.648
0.773	0.622
0.838	0.596
0.940	0.569
1.194	0.523
1.369	0.504
1.613	0.505
1.654	0.473
CURVE 85 (cont.)*	
1.681	0.473
1.697	0.468
1.717	0.468
1.780	0.505
1.821	0.505
1.856	0.513
1.966	0.513
2.030	0.517
2.075	0.514
2.106	0.502
2.169	0.441
2.234	0.348
2.256	0.336
2.299	0.356
2.396	0.356
2.438	0.352
2.470	0.356
2.500	0.356
CURVE 86* T = 298	
0.250	0.136
0.337	0.083
0.356	0.088
0.385	0.234
0.421	0.853
0.443	0.865
0.476	0.867
0.716	0.803
0.810	0.777
1.007	0.754
1.134	0.744
1.160	0.729
1.198	0.741
1.318	0.741
1.380	0.708
1.458	0.718
1.515	0.712
1.557	0.719
1.600	0.713
1.626	0.692
1.656	0.632
1.670	0.619
1.695	0.616
1.712	0.604
1.779	0.675
CURVE 86 (cont.)*	
1.816	0.668
1.834	0.671
1.874	0.632
1.891	0.625
2.019	0.653
2.053	0.653
2.083	0.640
2.150	0.563
2.231	0.408
2.255	0.379
2.314	0.396
2.419	0.389
2.440	0.385
2.500	0.385
CURVE 87* T = 298	
0.250	0.077
0.337	0.071
0.360	0.158
0.393	0.756
0.405	0.825
0.437	0.868
0.491	0.894
0.575	0.909
0.687	0.914
0.727	0.917
0.802	0.909
0.896	0.898
1.001	0.891
1.090	0.884
1.127	0.874
1.161	0.847
1.171	0.847
1.203	0.864
1.236	0.864
1.318	0.853
1.384	0.819
1.404	0.826
1.450	0.826
1.515	0.801
1.579	0.801
1.622	0.770
1.636	0.756
1.660	0.698
1.682	0.684
CURVE 87 (cont.)*	
1.701	0.681
1.716	0.665
1.782	0.733
1.821	0.724
1.858	0.736
1.887	0.736
1.999	0.702
2.072	0.684
2.162	0.626
2.229	0.503
2.258	0.450
2.280	0.454
2.297	0.473
2.318	0.481
2.349	0.476
2.401	0.484
2.443	0.476
2.504	0.489
CURVE 88* T = 281	
0.250	0.077
0.337	0.071
0.360	0.158
0.393	0.756
0.405	0.825
0.437	0.868
0.491	0.894
0.575	0.909
0.687	0.914
0.708	0.914
0.782	0.901
0.908	0.892
0.990	0.886
1.085	0.875
1.126	0.864
1.168	0.840
1.200	0.854
1.252	0.854
1.312	0.845
1.391	0.810
1.416	0.818
1.438	0.818
1.521	0.793
1.544	0.794
1.582	0.791
CURVE 88 (cont.)*	
1.634	0.753
1.666	0.685
1.685	0.672
1.705	0.672
1.724	0.658
1.784	0.722
1.862	0.727
1.885	0.727
1.959	0.701
2.064	0.676
2.137	0.623
2.166	0.608
2.229	0.490
2.259	0.441
2.277	0.447
2.308	0.469
2.334	0.469
2.348	0.466
2.399	0.471
2.442	0.464
2.504	0.477
CURVE 89* T = 281	
0.250	0.077
0.337	0.071
0.360	0.158
0.393	0.756
0.405	0.825
0.437	0.868
0.491	0.894
0.575	0.909
0.687	0.914
0.708	0.914
0.782	0.901
0.908	0.892
1.076	0.872
1.128	0.860
1.161	0.824
1.196	0.848
1.229	0.850
1.327	0.832
1.361	0.808
1.391	0.808
1.414	0.814
1.524	0.788

* Not shown on plot

DATA TABLE NO. 103 NORMAL SPECTRAL REFLECTANCE OF TITANIUM DIOXIDE PIGMENTED COATINGS (continued)

λ	ρ
CURVE 89 (cont.)* T = 281	
1.573	0.786
1.634	0.753
1.666	0.685
1.685	0.672
1.705	0.672
1.724	0.658
1.784	0.722
1.820	0.708
1.861	0.723
1.959	0.694
2.039	0.677
2.090	0.650
2.123	0.625
2.147	0.609
2.229	0.490
2.259	0.441
2.277	0.447
2.308	0.469
2.334	0.469
2.348	0.466
2.399	0.471
2.442	0.464
2.504	0.475
CURVE 90* T = 281	
0.250	0.074
0.337	0.069
0.360	0.158
0.393	0.756
0.405	0.825
0.437	0.868
0.491	0.894
0.711	0.904
0.807	0.888
0.914	0.880
1.076	0.861
1.143	0.848
1.168	0.823
1.188	0.823
1.203	0.838
1.230	0.841
1.268	0.832
1.314	0.827
1.383	0.792

λ	ρ
CURVE 90 (cont.)*	
1.413	0.798
1.458	0.796
1.523	0.772
1.576	0.772
1.611	0.753
1.632	0.740
1.660	0.680
1.682	0.656
1.706	0.656
1.723	0.643
1.768	0.696
1.783	0.703
1.808	0.703
1.821	0.696
1.854	0.703
1.898	0.703
2.041	0.665
2.093	0.646
2.140	0.605
2.153	0.602
2.255	0.434
2.296	0.444
2.323	0.457
2.360	0.455
2.399	0.457
2.442	0.451
2.504	0.462
CURVE 91* T = 281	
0.250	0.074
0.337	0.069
0.360	0.158
0.393	0.756
0.405	0.825
0.437	0.868
0.491	0.887
0.568	0.893
0.712	0.897
0.728	0.885
0.881	0.871
0.959	0.863
1.082	0.846
1.112	0.840
1.144	0.826
1.164	0.808
1.208	0.824

λ	ρ
CURVE 91 (cont.)*	
1.282	0.816
1.314	0.808
1.360	0.784
1.380	0.776
1.422	0.782
1.476	0.772
1.530	0.755
1.565	0.755
1.602	0.743
1.643	0.705
1.661	0.654
1.700	0.642
1.717	0.629
1.780	0.688
1.817	0.679
1.853	0.693
2.000	0.655
2.081	0.636
2.155	0.590
2.202	0.529
2.260	0.422
2.285	0.429
2.296	0.441
2.323	0.449
2.358	0.444
2.397	0.447
2.440	0.442
2.504	0.452
CURVE 92* T = 281	
0.250	0.071
0.337	0.067
0.360	0.158
0.393	0.756
0.405	0.825
0.434	0.849
0.469	0.869
0.516	0.875
0.705	0.875
0.732	0.869
0.952	0.847
1.058	0.829
1.145	0.808
1.162	0.787
1.205	0.797

λ	ρ
CURVE 92 (cont.)*	
1.284	0.788
1.322	0.777
1.380	0.750
1.426	0.755
1.533	0.729
1.567	0.729
1.630	0.696
1.645	0.680
1.658	0.642
1.678	0.624
1.702	0.624
1.718	0.610
1.738	0.623
1.781	0.660
1.823	0.652
1.864	0.662
1.999	0.627
2.071	0.609
2.158	0.563
2.201	0.522
2.252	0.421
2.264	0.409
2.281	0.410
2.323	0.428
2.422	0.428
2.464	0.423
2.504	0.431
CURVE 93* T = 281	
0.250	0.071
0.337	0.067
0.360	0.158
0.388	0.624
0.407	0.722
0.466	0.823
0.499	0.845
0.715	0.856
0.766	0.845
0.925	0.833
1.096	0.798
1.147	0.783
1.166	0.766
1.220	0.774
1.328	0.756
1.388	0.730

λ	ρ
CURVE 93 (cont.)*	
1.434	0.731
1.522	0.703
1.551	0.703
1.620	0.681
1.641	0.657
1.659	0.618
1.717	0.592
1.744	0.619
1.773	0.633
1.792	0.633
1.820	0.624
1.864	0.632
1.973	0.604
2.074	0.582
2.135	0.552
2.183	0.519
2.220	0.454
2.238	0.418
2.259	0.399
2.288	0.407
2.326	0.421
2.353	0.418
2.373	0.418
2.400	0.420
2.447	0.412
2.504	0.418
CURVE 94 T = 298	
0.378	0.478
0.412	0.829
0.467	0.885*
0.542	0.907
0.695	0.920
0.978	0.900
1.119	0.881
1.165	0.848
1.204	0.871
1.321	0.855*
1.359	0.830
1.382	0.822
1.405	0.830*
1.446	0.831
1.517	0.808
1.570	0.808*
1.619	0.784

λ	ρ
CURVE 94 (cont.)	
1.646	0.745*
1.663	0.689*
1.691	0.686*
1.722	0.670
1.761	0.723*
1.782	0.738
1.822	0.726*
1.862	0.743
1.909	0.734*
2.002	0.701*
2.065	0.685
2.174	0.607
2.232	0.475*
2.259	0.448
2.288	0.463*
2.297	0.472*
2.319	0.478*
2.341	0.472*
2.398	0.479
2.451	0.473
2.505	0.488
CURVE 95* T = 281	
0.378	0.478
0.412	0.829
0.470	0.878
0.646	0.892
0.852	0.877
1.085	0.837
1.142	0.822
1.174	0.802
1.201	0.812
1.272	0.802
1.301	0.802
1.385	0.768
1.435	0.768
1.517	0.745
1.592	0.741
1.645	0.697
1.682	0.637
1.704	0.637
1.723	0.620
1.758	0.662
1.788	0.673
1.824	0.665

λ	ρ
CURVE 95 (cont.)*	
1.860	0.676
2.078	0.616
2.157	0.569
2.198	0.529
2.240	0.442
2.265	0.418
2.290	0.426
2.319	0.439
2.403	0.439
2.456	0.430
2.505	0.438
CURVE 96* T = 281	
0.378	0.478
0.412	0.829
0.471	0.868
0.517	0.872
0.718	0.862
0.935	0.831
1.117	0.788
1.150	0.776
1.168	0.761
1.200	0.768
1.323	0.745
1.388	0.717
1.452	0.712
1.526	0.689
1.593	0.680
1.629	0.659
1.646	0.642
1.671	0.598
1.705	0.593
1.722	0.580
1.760	0.607
1.787	0.618
1.826	0.607
1.864	0.613
2.062	0.561
2.175	0.505
2.266	0.389
2.311	0.404
2.465	0.397
2.505	0.398

λ	ρ
CURVE 97* T = 281	
0.378	0.478
0.412	0.821
0.471	0.868
0.509	0.866
0.604	0.855
0.705	0.849
0.757	0.832
0.790	0.832
0.982	0.796
1.133	0.759
1.171	0.730
1.200	0.736
1.326	0.704
1.391	0.681
1.428	0.681
1.535	0.648
1.603	0.635
1.647	0.600
1.660	0.567
1.706	0.559
1.726	0.550
1.751	0.569
1.785	0.576
1.821	0.567
1.855	0.571
2.068	0.520
2.183	0.464
2.243	0.385
2.265	0.367
2.315	0.380
2.349	0.376
2.413	0.376
2.454	0.367
2.505	0.371
CURVE 98 T = 281	
0.378	0.478*
0.412	0.829*
0.430	0.836*
0.473	0.849*
0.573	0.828
0.709	0.817
0.748	0.795
0.845	0.779*

* Not shown on plot

DATA TABLE NO. 103 NORMAL SPECTRAL REFLECTANCE OF TITANIUM DIOXIDE PIGMENTED COATINGS (continued)

CURVE 98 (cont.)
T = 281

λ	ρ
1.128	0.700
1.167	0.676
1.203	0.676
1.311	0.648
1.377	0.622
1.449	0.607
1.603	0.561
1.644	0.539*
1.661	0.509
1.723	0.493
1.766	0.507
1.826	0.497
1.866	0.497*
2.089	0.440
2.164	0.411*
2.209	0.374
2.261	0.327
2.318	0.332
2.346	0.329*
2.399	0.324
2.446	0.316*
2.505	0.316

CURVE 99*
T = 281

λ	ρ
0.378	0.478
0.412	0.804
0.461	0.834
0.506	0.829
0.585	0.799
0.721	0.781
0.766	0.765
0.938	0.727
1.152	0.655
1.170	0.642
1.200	0.637
1.469	0.559
1.642	0.501
1.684	0.466
1.706	0.466
1.723	0.456
1.788	0.460
1.908	0.439
2.123	0.379
2.193	0.356

CURVE 99 (cont.)*

λ	ρ
2.267	0.298
2.340	0.298
2.446	0.285
2.505	0.282

CURVE 100
T = 281

λ	ρ
0.378	0.478*
0.388	0.642
0.405	0.700
0.462	0.755
0.510	0.755
0.593	0.725
0.713	0.698*
0.774	0.671
0.961	0.620
1.139	0.546
1.279	0.499
1.561	0.410
1.671	0.365
1.731	0.350
1.875	0.329
2.193	0.258
2.267	0.220
2.313	0.219*
2.505	0.197*

CURVE 101*
T = 298

λ	ρ
0.250	0.113
0.321	0.085
0.343	0.101
0.353	0.159
0.391	0.795
0.401	0.864
0.469	0.898
0.713	0.879
0.980	0.848
1.106	0.836
1.139	0.822
1.159	0.802
1.198	0.823
1.318	0.809
1.356	0.789
1.379	0.782

CURVE 101 (cont.)*

λ	ρ
1.399	0.790
1.439	0.794
1.497	0.778
1.542	0.778
1.561	0.781
1.601	0.765
1.621	0.751
1.637	0.730
1.655	0.676
1.682	0.676
1.715	0.665
1.750	0.715
1.777	0.736
1.816	0.728
1.830	0.739
1.855	0.745
1.904	0.743
1.942	0.727
2.045	0.707
2.118	0.660
2.176	0.623
2.235	0.489
2.256	0.464
2.294	0.488
2.318	0.498
2.340	0.492
2.396	0.498
2.432	0.494
2.459	0.494
2.500	0.508

CURVE 102*
T = 281

λ	ρ
0.250	0.089
0.299	0.073
0.324	0.070
0.339	0.082
0.353	0.159
0.390	0.730
0.438	0.760
0.468	0.763
0.508	0.750
0.583	0.704
0.709	0.667
0.718	0.657
0.761	0.639
0.790	0.635

CURVE 102 (cont.)*

λ	ρ
1.012	0.562
1.150	0.506
1.367	0.440
1.622	0.375
1.664	0.358
1.716	0.347
1.766	0.347
1.836	0.334
1.880	0.330
2.189	0.262
2.253	0.226
2.313	0.227
2.392	0.217
2.500	0.200

CURVE 103
T = 281

λ	ρ
0.700	0.670
0.774	0.642
1.011	0.568
1.130	0.523
1.218	0.496
1.421	0.438
1.551	0.405
1.626	0.386*
1.668	0.370
1.712	0.358*
1.768	0.358
1.858	0.343
2.038	0.306
2.184	0.272
2.256	0.235
2.320	0.235
2.414	0.223
2.500	0.213

CURVE 104*
T = 281

λ	ρ
0.700	0.670
0.774	0.642
1.011	0.568
1.130	0.523
1.218	0.496
1.421	0.438
1.551	0.405

CURVE 104 (cont.)*

λ	ρ
1.626	0.386
1.668	0.370
1.712	0.358
1.768	0.358
1.858	0.343
2.038	0.306
2.184	0.272
2.256	0.235
2.320	0.235
2.414	0.223
2.500	0.213

CURVE 105*
T = 281

λ	ρ
0.700	0.673
0.774	0.645
1.011	0.573
1.230	0.498
1.388	0.448
1.476	0.426
1.589	0.400
1.635	0.384
1.651	0.374
1.716	0.362
1.777	0.362
1.807	0.351
1.858	0.345
2.037	0.312
2.184	0.280
2.256	0.241
2.320	0.238
2.414	0.225
2.500	0.219

CURVE 106*
T = 281

λ	ρ
0.700	0.673
0.774	0.645
1.011	0.573
1.230	0.498
1.388	0.448
1.476	0.426
1.589	0.400
1.635	0.384
1.651	0.374

CURVE 106 (cont.)*

λ	ρ
1.716	0.362
1.777	0.362
1.807	0.351
1.858	0.345
2.037	0.312
2.184	0.280
2.256	0.241
2.320	0.238
2.414	0.225
2.500	0.219

CURVE 107
T = 281

λ	ρ
0.700	0.678
0.870	0.626
0.940	0.607
1.147	0.532
1.388	0.459
1.601	0.403
1.637	0.394*
1.654	0.380
1.724	0.368*
1.776	0.368
2.065	0.314
2.185	0.284
2.255	0.244
2.324	0.242*
2.441	0.227*
2.500	0.223

CURVE 108
T = 298

λ	ρ
0.250	0.113
0.321	0.085
0.343	0.101
0.353	0.159
0.391	0.795
0.429	0.827
0.466	0.837
0.548	0.803
0.696	0.774
0.715	0.761
0.974	0.709
1.141	0.662
1.159	0.647*

CURVE 108 (cont.)

λ	ρ
1.203	0.649
1.329	0.623*
1.359	0.608
1.445	0.601*
1.501	0.585
1.574	0.576*
1.618	0.563*
1.641	0.546*
1.656	0.525
1.714	0.512
1.737	0.527*
1.781	0.537
1.815	0.532*
1.860	0.534*
1.977	0.512
2.101	0.480
2.167	0.449*
2.235	0.373*
2.256	0.358
2.320	0.365*
2.395	0.365*
2.445	0.358*
2.500	0.358*

CURVE 109*
T = 300

λ	ρ
0.380	0.149
0.386	0.211
0.390	0.267
0.395	0.355
0.403	0.557
0.405	0.643
0.409	0.718
0.417	0.808
0.423	0.851
0.428	0.877
0.435	0.892
0.442	0.901
0.479	0.920
0.537	0.939
0.585	0.950
0.643	0.956
0.700	0.954

CURVE 110*
T = 300

λ	ρ
0.380	0.149
0.386	0.211
0.390	0.267
0.395	0.355
0.403	0.557
0.405	0.643
0.409	0.718
0.417	0.808
0.423	0.833
0.430	0.858
0.437	0.875
0.445	0.883
0.493	0.911
0.517	0.923
0.566	0.938
0.627	0.949
0.700	0.952

CURVE 111*
T = 300

λ	ρ
0.380	0.116
0.384	0.155
0.389	0.210
0.395	0.305
0.401	0.437
0.407	0.611
0.412	0.697
0.417	0.766
0.423	0.804
0.430	0.830
0.440	0.846
0.468	0.876
0.510	0.906
0.544	0.923
0.601	0.942
0.627	0.949
0.700	0.952

CURVE 112
T = 300

λ	ρ
0.380	0.069
0.388	0.098*
0.398	0.146
0.415	0.238

* Not shown on plot

DATA TABLE NO. 103 NORMAL SPECTRAL REFLECTANCE OF TITANIUM DIOXIDE PIGMENTED COATINGS (continued)

CURVE 112 (cont.)
T = 300

λ	ρ
0.463	0.435
0.509	0.612
0.534	0.692*
0.566	0.778
0.587	0.820
0.612	0.852
0.644	0.883
0.700	0.915

CURVE 113*
T = 300

λ	ρ
0.380	0.058
0.393	0.071
0.402	0.089
0.423	0.134
0.443	0.189
0.469	0.267
0.491	0.341
0.530	0.476
0.565	0.593
0.596	0.689
0.623	0.752
0.643	0.792
0.665	0.826
0.700	0.866

CURVE 114*
T = 300

λ	ρ
0.300	0.129
0.350	0.106
0.360	0.109
0.390	0.229
0.400	0.380
0.410	0.894
0.430	0.920
0.470	0.941
0.560	0.947
0.880	0.947
0.980	0.940
1.11	0.911
1.69	0.835
2.11	0.760
2.63	0.548
2.85	0.495

CURVE 114 (cont.)*

λ	ρ
4.82	0.130
5.37	0.112
7.02	0.085
7.40	0.085
8.51	0.054
9.28	0.050
9.73	0.057
12.90	0.057
14.92	0.232

CURVE 115*
T = 300

λ	ρ
0.300	0.129
0.350	0.106
0.360	0.109
0.390	0.229
0.400	0.380
0.410	0.865
0.440	0.905
0.460	0.921
0.530	0.947
0.880	0.947
0.980	0.940
1.11	0.911
1.69	0.835
2.11	0.760
2.63	0.548
2.85	0.495
4.82	0.130
5.37	0.112
7.02	0.085
7.40	0.085
8.51	0.054
9.28	0.050
9.73	0.057
12.90	0.057
14.92	0.232

CURVE 116
T = 300

λ	ρ
0.300	0.061*
0.340	0.054
0.350	0.062
0.360	0.109*
0.390	0.229

CURVE 116 (cont.)

λ	ρ
0.400	0.380
0.410	0.797
0.420	0.838
0.450	0.881
0.490	0.909
0.550	0.934
0.630	0.947
0.880	0.947
0.980	0.940
1.11	0.911
1.69	0.835
2.11	0.760
2.63	0.548
2.85	0.495
4.82	0.130
5.37	0.112
7.02	0.085
7.40	0.085
8.51	0.054
9.28	0.050
9.73	0.057
12.90	0.057
14.92	0.232

CURVE 117*
T = 300

λ	ρ
0.300	0.089
0.350	0.071
0.370	0.082
0.390	0.146
0.420	0.305
0.520	0.666
0.580	0.803
0.620	0.863
0.670	0.898
0.720	0.921
0.770	0.937
0.820	0.947
0.880	0.947
0.980	0.940
1.11	0.911
1.69	0.835
2.11	0.760
2.63	0.548
2.85	0.495
4.82	0.130

CURVE 117 (cont.)*

λ	ρ
5.37	0.112
7.02	0.085
7.40	0.085
8.51	0.054
9.28	0.050
9.73	0.057
12.90	0.057
14.92	0.232

CURVE 118*
T = 300

λ	ρ
0.300	0.076
0.340	0.063
0.360	0.063
0.380	0.080
0.420	0.150
0.460	0.255
0.490	0.360
0.620	0.722
0.670	0.817
0.740	0.889
0.790	0.919
0.830	0.933
0.880	0.947
0.980	0.940
1.11	0.911
1.69	0.835
2.11	0.760
2.63	0.548
2.85	0.495
4.82	0.130
5.37	0.112
7.02	0.085
7.40	0.085
8.51	0.054
9.28	0.050
9.73	0.057
12.90	0.057
14.92	0.232

CURVE 119*
T = 300

λ	ρ
0.380	0.136
0.384	0.169
0.390	0.224

CURVE 119 (cont.)*

λ	ρ
0.395	0.298
0.403	0.478
0.412	0.717
0.420	0.896
0.425	0.920
0.432	0.942
0.442	0.954
0.467	0.960
0.485	0.960
0.505	0.948
0.615	0.928
0.700	0.924

CURVE 120*
T = 300

λ	ρ
0.380	0.136
0.384	0.169
0.390	0.224
0.395	0.298
0.403	0.478
0.412	0.717
0.421	0.896
0.426	0.912
0.434	0.932
0.445	0.944
0.472	0.951
0.505	0.948
0.615	0.928
0.700	0.924

CURVE 121*
T = 300

λ	ρ
0.380	0.136
0.384	0.169
0.390	0.224
0.395	0.298
0.403	0.478
0.412	0.717
0.420	0.881
0.426	0.901
0.432	0.915
0.442	0.928
0.463	0.939
0.497	0.949
0.615	0.928
0.700	0.924

CURVE 122*
T = 300

λ	ρ
0.380	0.096
0.387	0.132
0.395	0.183
0.400	0.214
0.410	0.316
0.424	0.416
0.439	0.491
0.468	0.611
0.484	0.664
0.504	0.729
0.535	0.802
0.550	0.829
0.577	0.857
0.601	0.874
0.687	0.895
0.700	0.901

CURVE 123*
T = 300

λ	ρ
0.380	0.076
0.390	0.100
0.397	0.125
0.402	0.150
0.411	0.199
0.452	0.362
0.484	0.482
0.517	0.599
0.551	0.696
0.571	0.749
0.595	0.791
0.618	0.823
0.648	0.853
0.700	0.878

CURVE 124
T = 300

λ	ρ
0.300	0.099
0.340	0.084*
0.370	0.098
0.380	0.136
0.410	0.386
0.420	0.679
0.430	0.908
0.440	0.937
0.450	0.946

CURVE 124 (cont.)

λ	ρ
0.470	0.951
0.920	0.891
1.00	0.872
4.96	0.031
5.69	0.038
8.07	0.038
8.88	0.026
10.07	0.039
12.05	0.039
12.99	0.048
14.00	0.096
14.99	0.228

CURVE 125*
T = 300

λ	ρ
0.300	0.099
0.340	0.084
0.370	0.098
0.380	0.136
0.410	0.386
0.420	0.679
0.430	0.908
0.450	0.934
0.470	0.943
0.510	0.945
0.920	0.891
1.00	0.872
4.96	0.031
5.69	0.038
8.07	0.038
8.88	0.026
10.07	0.039
12.05	0.039
12.99	0.048
14.00	0.096
14.99	0.228

CURVE 126*
T = 300

λ	ρ
0.300	0.099
0.340	0.084
0.370	0.098
0.380	0.136
0.410	0.386
0.420	0.679
0.430	0.908

CURVE 126 (cont.)*

λ	ρ
0.440	0.937
0.450	0.946
0.470	0.951
0.920	0.891
1.00	0.872
4.96	0.031
5.69	0.038
8.07	0.038
8.88	0.026
10.07	0.039
12.05	0.039
12.99	0.048
14.00	0.096
14.99	0.228

CURVE 127*
T = 300

λ	ρ
0.300	0.067
0.340	0.055
0.360	0.065
0.390	0.096
0.410	0.304
0.460	0.544
0.500	0.701
0.550	0.802
0.590	0.850
0.630	0.878
0.680	0.893
0.770	0.893
1.00	0.870
4.96	0.031
5.69	0.038
8.07	0.038
8.88	0.026
10.07	0.039
12.05	0.039
12.99	0.048
14.00	0.096
14.99	0.228

CURVE 128*
T = 300

λ	ρ
0.310	0.068
0.380	0.061
0.390	0.077
0.510	0.531

* Not shown on plot

DATA TABLE NO. 103 NORMAL SPECTRAL REFLECTANCE OF TITANIUM DIOXIDE PIGMENTED COATINGS (continued)

CURVE 128 (cont.)* T = 300

λ	ρ
0.570	0.713
0.600	0.786
0.650	0.839
0.700	0.869
0.770	0.875
0.880	0.875
1.00	0.865
4.96	0.031
5.69	0.038
8.07	0.038
8.88	0.026
10.07	0.039
12.05	0.039
12.99	0.048
14.00	0.096
14.99	0.228

CURVE 129 T = 310

λ	ρ
0.295	0.070
0.340	0.074
0.398	0.311
0.439	0.702
0.518	0.767
0.704	0.836
0.810	0.837
1.06	0.802
1.51	0.744
1.71	0.692
1.84	0.692
2.01	0.719*
2.26	0.523
2.48	0.503
2.79	0.118
3.12	0.203
3.59	0.459
4.29	0.539
4.64	0.508
5.17	0.252
5.39	0.249
6.59	0.117
7.01	0.134
7.62	0.058
8.16	0.048
8.47	0.094

CURVE 129 (cont.)

λ	ρ
9.26	0.106
9.68	0.116
10.5	0.072
10.9	0.059
11.8	0.075
14.3	0.075
18.5	0.146
23.2	0.172
24.4	0.164
25.6	0.164
28.8	0.189
31.1	0.184

CURVE 130* T = 310

λ	ρ
0.295	0.069
0.346	0.069
0.399	0.359
0.449	0.725
0.510	0.772
0.717	0.855
1.00	0.836
1.29	0.786
1.53	0.767
1.60	0.749
1.70	0.691
1.92	0.749
2.06	0.731
2.26	0.548
2.54	0.504
2.86	0.200
2.98	0.154
3.04	0.124
3.71	0.480
4.23	0.522
4.37	0.570
4.68	0.540
5.22	0.296
5.45	0.261
6.65	0.120
7.12	0.150
7.58	0.060
7.74	0.060
8.12	0.032
8.66	0.074
9.16	0.080

CURVE 130 (cont.)*

λ	ρ
9.57	0.113
11.4	0.051
11.9	0.064
14.4	0.074
16.1	0.066
18.1	0.125
21.7	0.175
24.4	0.155
27.1	0.184
28.8	0.169
31.6	0.164

CURVE 131* T = 310

λ	ρ
0.295	0.069
0.346	0.069
0.399	0.359
0.449	0.725
0.510	0.772
0.717	0.855
1.00	0.836
1.29	0.786
1.53	0.767
1.60	0.749
1.70	0.691
1.92	0.749
2.06	0.731
2.26	0.548
2.54	0.504
2.86	0.200
2.98	0.154
3.06	0.112
3.77	0.444
4.02	0.478
4.28	0.505
4.64	0.476
5.19	0.223
5.66	0.204
6.63	0.097
7.01	0.109
7.46	0.060
7.74	0.060
8.12	0.032
8.66	0.074
9.16	0.080
9.57	0.113

CURVE 131 (cont.)*

λ	ρ
11.4	0.051
11.9	0.064
14.4	0.074
16.1	0.066
18.1	0.125
21.7	0.175
24.4	0.155
27.1	0.184
28.8	0.169
31.6	0.164

CURVE 132* T = 310

λ	ρ
0.295	0.069
0.346	0.069
0.399	0.359
0.449	0.725
0.510	0.772
0.717	0.855
1.00	0.836
1.29	0.786
1.53	0.767
1.60	0.749
1.70	0.691
1.92	0.749
2.06	0.731
2.26	0.548
2.54	0.504
2.86	0.200
2.98	0.154
3.04	0.124
3.69	0.482
4.12	0.537
4.25	0.579
4.76	0.551
5.21	0.308
5.61	0.299
6.54	0.146
7.16	0.162
7.58	0.060
7.74	0.060
8.12	0.032
8.66	0.074
9.16	0.080
9.57	0.113
11.4	0.051

CURVE 132 (cont.)*

λ	ρ
11.9	0.064
14.4	0.074
16.1	0.066
18.1	0.125
21.7	0.175
24.4	0.155
27.1	0.184
28.8	0.169
31.6	0.164

CURVE 133* T = 310

λ	ρ
0.295	0.064
0.340	0.064
0.398	0.298
0.457	0.680
0.501	0.739
0.699	0.833
0.903	0.833
1.28	0.788
1.51	0.752
1.59	0.731
1.70	0.667
1.82	0.708
1.93	0.729
2.01	0.704
2.29	0.553
2.58	0.519
2.82	0.253
3.24	0.115
3.67	0.481
3.99	0.506
4.34	0.535
4.64	0.503
5.78	0.157
6.39	0.202
6.71	0.146
7.16	0.149
7.60	0.072
8.29	0.044
8.70	0.085
9.14	0.093
9.39	0.115
9.84	0.115
10.2	0.103
10.6	0.080

CURVE 133 (cont.)*

λ	ρ
10.9	0.080
11.4	0.062
12.0	0.070
17.1	0.101
19.0	0.145
22.0	0.163
23.7	0.152
28.0	0.200
30.3	0.200
32.0	0.190

CURVE 134* T = 310

λ	ρ
0.295	0.064
0.340	0.064
0.398	0.298
0.457	0.680
0.501	0.739
0.699	0.833
0.903	0.833
1.28	0.788
1.51	0.752
1.59	0.731
1.70	0.667
1.82	0.708
1.93	0.729
2.01	0.704
2.29	0.553
2.58	0.519
2.82	0.253
3.24	0.115
3.58	0.513
4.01	0.611
4.30	0.625
4.65	0.598
5.80	0.205
6.39	0.202
6.71	0.146
7.16	0.149
7.60	0.072
8.29	0.044
8.70	0.085
9.14	0.093
9.39	0.115
9.84	0.115
10.2	0.103

CURVE 134 (cont.)*

λ	ρ
10.6	0.080
10.9	0.080
11.4	0.062
12.0	0.070
17.1	0.101
19.0	0.145
22.0	0.163
23.7	0.152
28.0	0.200
30.3	0.200
32.0	0.190

CURVE 135 T = 310

λ	ρ
0.295	0.064
0.340	0.064*
0.398	0.298
0.457	0.680
0.501	0.739
0.699	0.833*
0.903	0.833
1.28	0.788
1.51	0.752
1.59	0.731
1.70	0.667
1.82	0.708*
1.93	0.729*
2.01	0.704*
2.29	0.553
2.58	0.519
2.82	0.253
3.24	0.115
3.58	0.513
4.09	0.540
4.37	0.568
4.67	0.546
5.74	0.183
6.39	0.202
6.71	0.146
7.16	0.149
7.60	0.072
8.29	0.044
8.70	0.085
9.14	0.093
9.39	0.115
9.84	0.115

CURVE 135 (cont.)

λ	ρ
10.2	0.103
10.6	0.080
10.9	0.080*
11.4	0.062
12.0	0.070
17.1	0.101
19.0	0.145
22.0	0.163
23.7	0.152
28.0	0.200
30.3	0.200
32.0	0.190

CURVE 136* T = 310

λ	ρ
0.295	0.067
0.344	0.067
0.399	0.296
0.425	0.810
0.505	0.873
0.597	0.891
0.799	0.882
1.15	0.842
1.33	0.814
1.51	0.752
1.64	0.669
1.81	0.647
2.03	0.686
2.60	0.341
2.93	0.302
3.41	0.087
3.62	0.071
4.54	0.137
5.19	0.102
5.19	0.089
5.52	0.076
6.71	0.078
8.95	0.067
10.7	0.080
12.3	0.070
17.4	0.070
19.6	0.114
21.1	0.114
22.0	0.118
23.5	0.099
26.2	0.100

* Not shown on plot

DATA TABLE NO. 103 NORMAL SPECTRAL REFLECTANCE OF TITANIUM DIOXIDE PIGMENTED COATINGS (continued)

CURVE 136 (cont.)*
T = 310

λ	ρ
28.5	0.138
31.7	0.100

CURVE 137
T = 310

λ	ρ
0.295	0.082*
0.341	0.082*
0.399	0.376
0.443	0.858*
0.511	0.883
0.717	0.883
0.803	0.874*
0.899	0.874
1.08	0.856
1.21	0.825
1.28	0.835
1.40	0.771*
1.48	0.748
1.57	0.729*
1.69	0.629
1.78	0.648*
1.99	0.604*
2.25	0.301*
2.51	0.257
2.81	0.060
2.99	0.045
3.70	0.080
4.14	0.124
4.29	0.139
4.68	0.119
5.17	0.119
5.63	0.051
6.29	0.039
6.74	0.057
6.96	0.051
7.55	0.051
7.97	0.068
8.37	0.068
8.70	0.060
10.9	0.069
11.5	0.062
14.5	0.110
16.6	0.110
17.3	0.140
21.0	0.140

CURVE 137 (cont.)

λ	ρ
23.2	0.131
27.1	0.131
31.9	0.150

CURVE 138*
T = 310

λ	ρ
0.295	0.082
0.341	0.082
0.399	0.376
0.443	0.858
0.511	0.883
0.717	0.883
0.803	0.874
0.899	0.874
1.08	0.856
1.21	0.825
1.28	0.835
1.40	0.771
1.48	0.748
1.57	0.729
1.69	0.629
1.78	0.648
1.99	0.604
2.25	0.350
2.55	0.309
3.00	0.078
4.36	0.140
4.68	0.119
5.17	0.119
5.53	0.098
5.75	0.069
6.02	0.060
6.23	0.060
6.66	0.070
7.11	0.070
7.63	0.061
7.88	0.082
8.45	0.075
9.14	0.075
9.97	0.080
11.9	0.080
13.2	0.099
14.5	0.110
16.6	0.110
17.3	0.140
21.0	0.140

CURVE 138 (cont.)*

λ	ρ
23.2	0.131
27.1	0.131
31.9	0.150

CURVE 139*
T = 310

λ	ρ
0.295	0.082
0.341	0.082
0.399	0.376
0.443	0.858
0.511	0.883
0.717	0.883
0.803	0.874
0.899	0.874
1.08	0.856
1.21	0.825
1.28	0.835
1.40	0.771
1.48	0.748
1.57	0.729
1.69	0.629
1.78	0.648
1.99	0.604
2.25	0.350
2.55	0.309
3.00	0.078
4.36	0.140
4.68	0.119
5.17	0.119
5.53	0.098
5.78	0.074
6.23	0.072
6.63	0.086
6.96	0.086
7.44	0.077
7.88	0.091
8.37	0.091
8.91	0.084
9.70	0.092
11.1	0.092
12.9	0.110
16.6	0.110
17.3	0.140
21.0	0.140
23.2	0.131
27.1	0.131
31.9	0.150

CURVE 140
T = 310

λ	ρ
0.295	0.080*
0.339	0.080*
0.399	0.271
0.445	0.541
0.542	0.746
0.587	0.801
0.707	0.821
0.881	0.816*
1.12	0.800
1.21	0.772*
1.30	0.772
1.42	0.726
1.59	0.702*
1.72	0.601
1.82	0.630
1.91	0.601
2.02	0.597*
2.24	0.331
2.54	0.304
2.84	0.079
3.14	0.062
3.65	0.097
4.12	0.139
4.78	0.112
5.26	0.112*
5.61	0.065
6.35	0.065
6.91	0.072
8.72	0.072
11.2	0.087
14.8	0.112*
16.3	0.112*
18.7	0.144*
20.7	0.144
26.1	0.127
31.4	0.138

CURVE 141*
T ~ 298

λ	ρ
0.298	0.090
0.335	0.079
0.355	0.097
0.381	0.195
0.399	0.266
0.423	0.789

CURVE 141 (cont.)*

λ	ρ
0.441	0.865
0.511	0.886
0.547	0.893
0.695	0.884
0.783	0.884
1.19	0.843
1.29	0.822
1.40	0.774
1.48	0.708
1.59	0.739
1.69	0.618
1.76	0.647
1.81	0.661
1.91	0.639
1.99	0.631
2.99	0.060
3.51	0.055
3.89	0.086
4.50	0.145
5.03	0.112
5.45	0.131
6.03	0.059
7.88	0.064
8.27	0.071
8.51	0.083
8.99	0.075
9.46	0.075
10.00	0.084
10.49	0.084
10.99	0.093
11.56	0.086
14.96	0.113
16.51	0.109
17.33	0.109
18.53	0.114
20.04	0.108
22.38	0.110
23.12	0.096
24.26	0.091
24.66	0.106
25.29	0.113
26.36	0.107
28.18	0.120
28.37	0.137
28.90	0.108
29.64	0.102
30.61	0.120

CURVE 141 (cont.)*

λ	ρ
31.33	0.148
32.21	0.117
32.80	0.096

CURVE 142*
T ~ 298

λ	ρ
0.298	0.079
0.349	0.073
0.369	0.121
0.394	0.328
0.400	0.427
0.417	0.832
0.430	0.843
0.448	0.872
0.472	0.872
0.492	0.878
0.570	0.878
0.798	0.858
1.10	0.820
1.19	0.788
1.32	0.769
1.45	0.725
1.58	0.702
1.66	0.635
1.71	0.621
1.75	0.633
1.86	0.636
1.97	0.615
3.02	0.048
3.54	0.038
4.04	0.099
4.57	0.105
5.05	0.181
6.09	0.058
6.29	0.058

CURVE 143
T ~ 298

λ	ρ
0.298	0.061*
0.350	0.061*
0.375	0.077
0.399	0.167
0.412	0.302
0.429	0.406
0.474	0.557

CURVE 143 (cont.)

λ	ρ
0.511	0.641
0.517	0.662*
0.520	0.689
0.615	0.753
0.701	0.789
0.803	0.793
1.01	0.785
1.09	0.789
1.20	0.774
1.29	0.766*
1.38	0.722
1.40	0.707
1.50	0.699
1.55	0.676
1.62	0.662*
1.66	0.635
1.71	0.621*
1.75	0.628*
1.86	0.628*
2.00	0.602*
3.02	0.048
3.54	0.038
4.04	0.099
4.57	0.105
5.05	0.181
6.09	0.058
6.29	0.058

CURVE 144*
T ~ 298

λ	ρ
0.298	0.061
0.350	0.061
0.375	0.077
0.390	0.097
0.403	0.163
0.423	0.284
0.444	0.348
0.471	0.439
0.496	0.500
0.538	0.617
0.623	0.709
0.680	0.746
0.792	0.779
0.916	0.781
1.12	0.770
1.19	0.760

CURVE 144 (cont.)*

λ	ρ
1.31	0.756
1.35	0.728
1.45	0.698
1.50	0.699
1.55	0.676
1.62	0.662
1.66	0.635
1.71	0.621
1.75	0.628
1.86	0.628
2.00	0.602
3.02	0.048
3.54	0.038
4.04	0.099
4.57	0.105
5.05	0.181
6.09	0.058
6.29	0.058

CURVE 145*
T ~ 298

λ	ρ
0.298	0.061
0.350	0.061
0.375	0.077
0.390	0.097
0.415	0.149
0.469	0.286
0.557	0.505
0.656	0.653
0.706	0.704
0.778	0.726
0.895	0.753
1.06	0.760
1.19	0.742
1.33	0.742
1.37	0.732
1.39	0.704
1.50	0.704
1.61	0.682
1.64	0.662
1.66	0.635
1.71	0.621
1.75	0.633
1.86	0.636
1.97	0.615
3.02	0.048

* Not shown on plot

DATA TABLE NO. 103 NORMAL SPECTRAL REFLECTANCE OF TITANIUM DIOXIDE PIGMENTED COATINGS (continued)

CURVE 145 (cont.)* T ~ 298

λ	ρ
3.54	0.038
4.04	0.099
4.57	0.105
5.05	0.181
6.09	0.058
6.29	0.058

CURVE 146* T ~ 298

λ	ρ
0.298	0.061
0.350	0.061
0.374	0.066
0.399	0.099
0.456	0.206
0.520	0.348
0.613	0.534
0.712	0.678
0.783	0.715
0.877	0.739
0.995	0.751
1.11	0.757
1.19	0.744
1.30	0.744
1.37	0.732
1.39	0.704
1.50	0.704
1.61	0.682
1.64	0.662
1.66	0.635
1.71	0.621
1.75	0.633
1.86	0.636
1.97	0.615
3.02	0.048
3.54	0.038
4.04	0.099
4.57	0.105
5.05	0.181
6.09	0.058
6.29	0.058

CURVE 147* T ~ 298

λ	ρ
0.298	0.061
0.357	0.061
0.381	0.065
0.404	0.080
0.439	0.138
0.497	0.256
0.544	0.362
0.699	0.645
0.805	0.750
0.895	0.774
1.09	0.789
1.28	0.769
1.39	0.731
1.55	0.692
1.61	0.682
1.64	0.662
1.66	0.635
1.71	0.621
1.75	0.633
1.86	0.636
1.97	0.615
3.02	0.048
3.54	0.038
4.04	0.099
4.57	0.105
5.05	0.181
6.09	0.058
6.29	0.058

CURVE 148* T ~ 298

λ	ρ
0.298	0.092
0.350	0.083
0.376	0.144
0.399	0.300
0.425	0.789
0.430	0.834
0.445	0.851
0.450	0.870
0.477	0.872
0.508	0.879
0.594	0.879
0.798	0.862
0.920	0.841
1.10	0.827

CURVE 148 (cont.)*

λ	ρ
1.20	0.792
1.29	0.790
1.41	0.741
1.60	0.703
1.70	0.624
1.81	0.639
2.00	0.601
3.02	0.050
4.02	0.081
5.01	0.159
6.01	0.055
6.57	0.055

CURVE 149* T ~ 298

λ	ρ
0.307	0.072
0.343	0.064
0.378	0.081
0.393	0.129
0.405	0.199
0.426	0.380
0.484	0.579
0.554	0.718
0.712	0.800
0.950	0.800
1.00	0.790
1.10	0.793
1.21	0.779
1.30	0.773
1.41	0.725
1.48	0.715
1.52	0.696
1.57	0.696
1.61	0.701
1.70	0.624
1.81	0.639
2.00	0.601
3.02	0.050
4.02	0.081
5.01	0.159
6.01	0.055
6.57	0.055

CURVE 150* T ~ 298

λ	ρ
0.307	0.072
0.343	0.064
0.378	0.081
0.393	0.129
0.405	0.199
0.441	0.333
0.480	0.462
0.515	0.552
0.535	0.612
0.594	0.670
0.616	0.681
0.674	0.749
0.749	0.773
0.857	0.785
0.955	0.780
1.16	0.777
1.27	0.765
1.40	0.719
1.49	0.697
1.56	0.697
1.60	0.686
1.67	0.622
1.80	0.634
2.00	0.601
3.02	0.050
4.02	0.081
5.01	0.159
6.01	0.055
6.57	0.055

CURVE 151 T ~ 298

λ	ρ
0.307	0.072
0.343	0.064*
0.360	0.064
0.390	0.075
0.412	0.146
0.469	0.283
0.547	0.487
0.615	0.602
0.701	0.704
0.899	0.757
1.10	0.770
1.23	0.751*
1.30	0.751*

CURVE 151 (cont.)

λ	ρ
1.45	0.711
1.49	0.697*
1.56	0.697
1.60	0.686*
1.65	0.638*
1.67	0.638*
1.70	0.607
1.81	0.628
2.00	0.601
3.02	0.050
4.02	0.081
5.01	0.159
6.01	0.055*
6.57	0.055

CURVE 152* T ~ 298

λ	ρ
0.307	0.072
0.343	0.064
0.360	0.064
0.390	0.075
0.424	0.139
0.466	0.224
0.518	0.344
0.605	0.531
0.701	0.671
0.811	0.738
0.899	0.758
1.10	0.770
1.21	0.750
1.30	0.751
1.49	0.697
1.56	0.697
1.60	0.686
1.65	0.638
1.67	0.638
1.70	0.607
1.81	0.628
2.00	0.601
3.02	0.050
4.02	0.081
5.01	0.159
6.01	0.055
6.57	0.055

CURVE 153* T ~ 298

λ	ρ
0.307	0.072
0.343	0.064
0.360	0.064
0.390	0.075
0.404	0.087
0.456	0.170
0.512	0.279
0.558	0.401
0.599	0.487
0.696	0.644
0.798	0.753
0.907	0.785
1.10	0.793
1.20	0.778
1.30	0.772
1.41	0.723
1.48	0.716
1.51	0.703
1.60	0.703
1.70	0.624
1.80	0.644
2.00	0.601
3.02	0.050
4.02	0.081
5.01	0.159
6.01	0.055
6.57	0.055

CURVE 154 T ~ 298

λ	ρ
0.298	0.082*
0.326	0.082*
0.357	0.079
0.377	0.106
0.387	0.140
0.399	0.266*
0.423	0.789
0.433	0.827*
0.472	0.866
0.562	0.877
0.698	0.877*
0.863	0.869
0.891	0.871*
1.09	0.853*
1.20	0.820

CURVE 154 (cont.)

λ	ρ
1.29	0.827
1.40	0.774*
1.48	0.708
1.59	0.739
1.69	0.618
1.76	0.647*
1.81	0.661
1.91	0.639*
1.99	0.640
1.99	0.631*
2.01	0.645
2.67	0.402
3.00	0.314
3.56	0.058
3.89	0.086
4.50	0.145
5.03	0.112
5.45	0.131
6.03	0.059*
7.76	0.059*
8.12	0.056
8.62	0.065
9.18	0.061
10.04	0.069
11.66	0.068
12.07	0.062
13.33	0.062
14.62	0.077
15.27	0.086
16.78	0.086
20.04	0.108
22.38	0.110
23.12	0.096
24.26	0.091
24.66	0.106
25.29	0.113
26.36	0.107
28.18	0.120
28.37	0.137
28.90	0.108
29.64	0.102
31.26	0.117
31.33	0.148
32.21	0.117
32.80	0.096

CURVE 155* T ~ 298

λ	ρ
0.250	0.052
0.433	0.052
0.455	0.063
0.465	0.074
0.477	0.904
0.488	0.923
0.512	0.933
0.913	0.925
1.112	0.910
1.149	0.887
1.169	0.880
1.183	0.882
1.210	0.890
1.231	0.900
1.251	0.900
1.311	0.861
1.351	0.841
1.382	0.828
1.438	0.837
1.489	0.837
1.551	0.826
1.604	0.805
1.645	0.772
1.678	0.727
1.711	0.698
1.735	0.698
1.760	0.706
1.785	0.731
1.801	0.736
1.827	0.735
1.866	0.724
1.933	0.680
1.951	0.675
1.971	0.675
1.995	0.687
2.021	0.695
2.041	0.695
2.065	0.685
2.129	0.614
2.184	0.530
2.214	0.475
2.249	0.441
2.288	0.418
2.316	0.404
2.360	0.400
2.497	0.390

* Not shown on plot

FIGURE 104A

ANALYZED CHANGE IN NORMAL SPECTRAL REFLECTANCE OF TITANIUM DIOXIDE PIGMENTED COATINGS

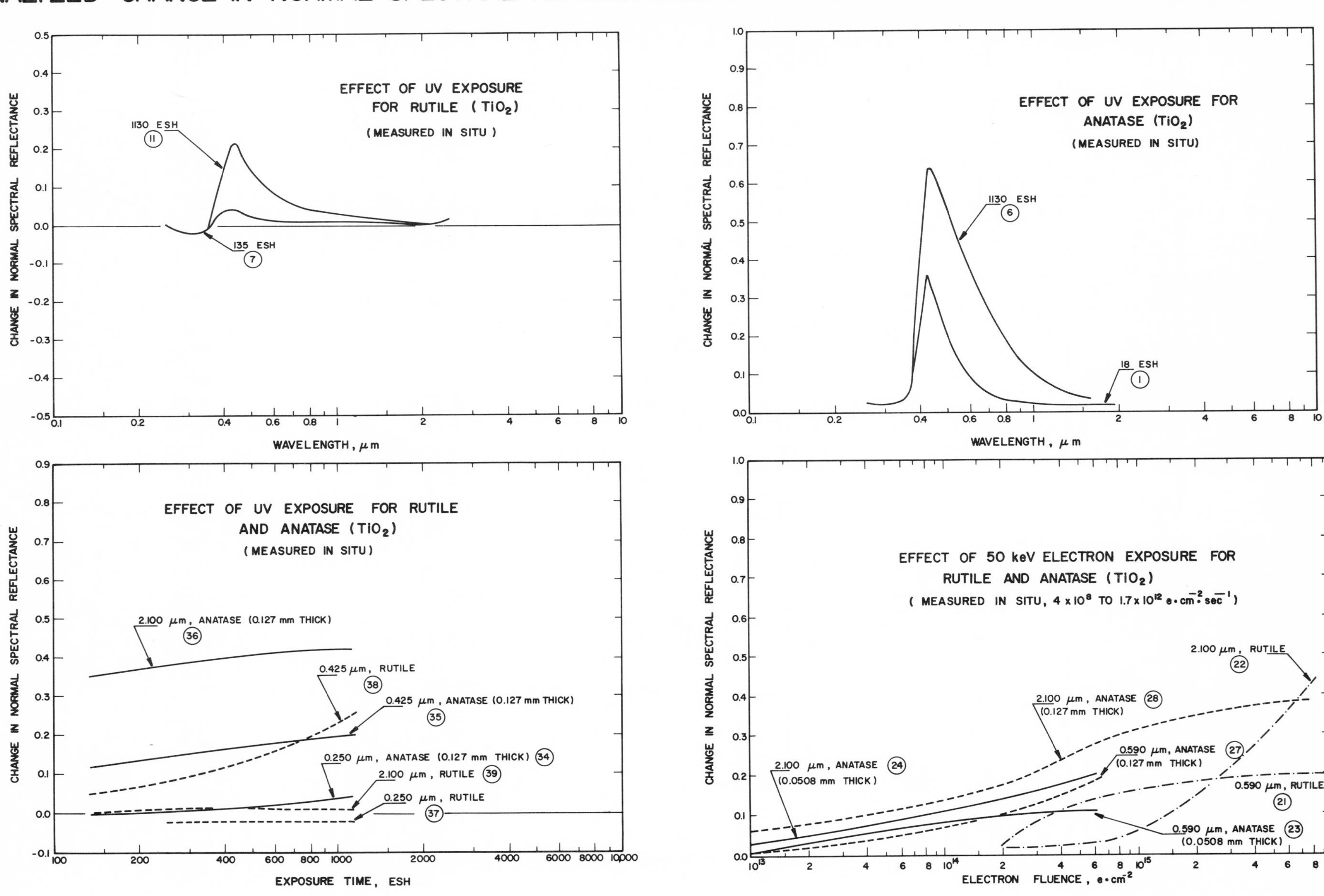

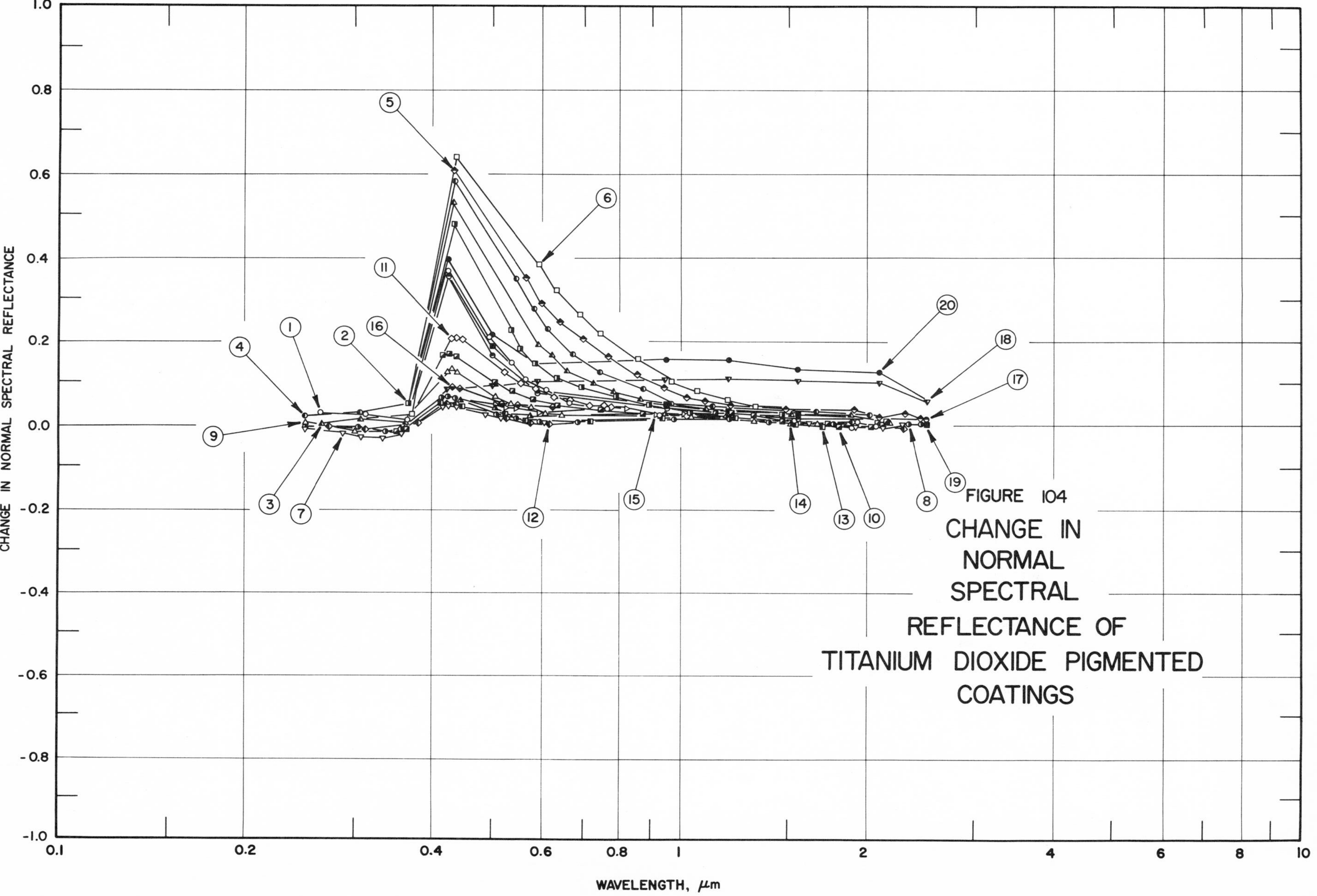
FIGURE 104
CHANGE IN NORMAL SPECTRAL REFLECTANCE OF TITANIUM DIOXIDE PIGMENTED COATINGS
CHANGE IN NORMAL SPECTRAL REFLECTANCE
WAVELENGTH, μm
1.0
0.8
0.6
0.4
0.2
0.0
-0.2
-0.4
-0.6
-0.8
-1.0
0.1
0.2
0.4
0.6
0.8
1
2
4
6
8
10
1
2
3
4
5
6
7
8
9
10
11
12
13
14
15
16
17
18
19
20

SPECIFICATION TABLE NO. 104 CHANGE IN NORMAL SPECTRAL REFLECTANCE OF TITANIUM DIOXIDE PIGMENTED COATINGS

Curve No.	Ref. No.	Year	Temperature K	Wavelength Range, μm	Geometry θ	θ'	ω'	Reported Error, %	Composition (weight percent), Specifications, and Remarks
1	46	1969	295	0.265-2.064	~0°		2π		Anatase TiO_2 in methyl phenyl silicone binder (OAO Pyromark); substrate unknown; exposed to UV radiation (with UV rates equivalent to 5 suns) in vacuum (10^{-7}mm Hg); ESH 18; vacuum maintained by ion pump; reflectance measured in situ; positive change indicates decrease in reflectance from preirradiation, in air reflectance; data extracted from smooth curve. [Authors' designation: Y]
2	46	1969	295	0.367-1.488	~0°		2π		Above specimen and conditions except ESH 53.
3	46	1969	295	0.264-2.085	~0°		2π		Above specimen and conditions except ESH 135.
4	46	1969	295	0.250-2.029	~0°		2π		Above specimen and conditions except ESH 250.
5	46	1969	295	0.367-2.450	~0°		2π		Above specimen and conditions except ESH 490.
6	46	1969	295	0.264-2.450	~0°		2π		Above specimen and conditions except ESH 1130.
7	46	1969	295	0.250-2.450	~0°		2π		Rutile TiO_2 in RTV 602 methyl silicone binder; substrate unknown; exposed to UV radiation (with UV rates equivalent to 5 suns) in vacuum (10^{-7} mm Hg); ESH 135; vacuum maintained by ion pump; reflectance measured in situ; positive change indicates decrease in reflectance from preirradiation, in air reflectance; data extracted from smooth curve. [Authors' designation: O]
8	46	1969	295	0.304-2.450	~0°		2π		Above specimen and conditions except ESH 250.
9	46	1969	295	0.250-2.450	~0°		2π		Above specimen and conditions except ESH 490.
10	46	1969	295	0.304-2.450	~0°		2π		Above specimen and conditions except ESH 770.
11	46	1969	295	0.304-2.450	~0°		2π		Above specimen and conditions except ESH 1130.
12	46	1969	295	0.271-2.450	~0°		2π		Rutile TiO_2 in XR6-3488 methyl silicone binder; substrate unknown; exposed to UV radiation (with UV rates equivalent to 5 suns) in vacuum (10^{-7} mm Hg); ESH 135; vacuum maintained by ion pump; reflectance measured in situ; positive change indicates decrease in reflectance from preirradiation, in air reflectance; data extracted from smooth curve. [Authors' designation: P]
13	46	1969	295	0.271-2.450	~0°		2π		Above specimen and conditions except ESH 250.
14	46	1969	295	0.250-2.450	~0°		2π		Above specimen and conditions except ESH 490.
15	46	1969	295	0.271-2.450	~0°		2π		Above specimen and conditions except ESH 770.
16	46	1969	295	0.271-2.450	~0°		2π		Above specimen and conditions except ESH 1130.
17	46	1969	295	0.425-2.50	~0°		2π		Anatase TiO_2 in methyl phenyl silicone binder (OAO Pyromark); substrate unknown; exposed to UV radiation (4.4-UV sun rate) in vacuum (10^{-8} mm Hg); vacuum maintained by ion pump; ESH 18; reflectance measured in situ; positive change indicates decrease in reflectance from preirradiation, in air reflectance. [Authors' designation: Y]
18	46	1969	295	0.425-2.50	~0°		2π		Similar to above specimen and conditions except exposed to 50 keV electrons in vacuum; electron fluence 5×10^{14} e cm^{-2}.

SPECIFICATION TABLE NO. 104 CHANGE IN NORMAL SPECTRAL REFLECTANCE OF TITANIUM DIOXIDE PIGMENTED COATINGS (continued)

Curve No.	Ref. No.	Year	Temperature K	Wavelength Range, μm	Geometry θ	θ'	ω'	Reported Error, %	Composition (weight percent), Specifications, and Remarks
19	46	1969	295	0.425-2.50	$\sim 0^\circ$		2π		Similar to curve 17 specimen and conditions except consecutive exposure to UV radiation (4.4 - UV sun rate) then to 50 keV electrons in vacuum; ESH 18; electron fluence 5 x 10^{14} e cm^{-2}.
20	46	1969	295	0.425-2.50	$\sim 0^\circ$		2π		Similar to curve 17 specimen and conditions except simultaneous exposure to UV radiation (4.4 UV sun rate) and to 50 keV electrons in vacuum; total exposure time 4.1 hrs; ESH 18; electron fluence 5 x 10^{14} e cm^{-2} with 90% occurring during the first hr of UV exposure.
21*	46	1969	295	0.590	$\sim 0^\circ$		2π		Rutile TiO_2 in XR6-3488 methyl silicone binder; substrate unknown; exposed to 50 keV electrons (4 x 10^8 - 1.7 x 10^{12} e cm^{-2} sec^{-1}) in vacuum (10^{-8} mm Hg); vacuum maintained by ion pump; electron fluence (e cm^{-2}) is variable; reflectance measured in situ; positive change indicates decrease in reflectance from preirradiation, in air reflectance. [Authors' designation: P]
22*	46	1969	295	2.100	$\sim 0^\circ$		2π		Above specimen and conditions.
23*	46	1969	295	0.590	$\sim 0^\circ$		2π		Anatase TiO_2 in methyl silicone binder (0.0508 mm thick) on Cat-A-Lac white primer (0.0508 mm thick) substrate; final substrate unknown; exposed to 50 keV electrons (4 x 10^8 - 1.7 x 10^{12} e cm^{-2} sec^{-1}) in vacuum (10^{-8} mm Hg); vacuum maintained by ion pump; electron fluence (e cm^{-2}) is variable; reflectance measured in situ; positive change indicates decrease in reflectance from preirradiation, in air reflectance. [Authors' designation: A]
24*	46	1969	295	2.100	$\sim 0^\circ$		2π		Above specimen and conditions.
25*	46	1969	295	0.590	$\sim 0^\circ$		2π		Anatase TiO_2 in methyl phenyl silicone binder (OAO Pyromak) ; substrate unknown; exposed to 50 keV electrons (4 x 10^8 - 1.7 x 10^{12} e cm^{-2} sec^{-1}) in vacuum (10^{-8} mm Hg); vacuum maintained by ion pump; electron fluence (e cm^{-2}) is variable; reflectance measured in situ; positive change indicates decrease in reflectance from preirradiation, in air reflectance. [Authors' designation: Y]
26*	46	1969	295	2.100	$\sim 0^\circ$		2π		Above specimen and conditions.
27*	46	1969	295	0.590	$\sim 0^\circ$		2π		Anatase TiO_2 in methyl silicone binder (0.127 mm thick) on Cat-A-Lac white primer (0.0508 mm thick) substrate; final substrate unknown; exposed to 50 keV electrons (4 x 10^8 - 1.7 x 10^{12} e cm^{-2} sec^{-1}) in vacuum (10^{-8} mm Hg); vacuum maintained by ion pump; electron fluence (e cm^{-2}) is variable; reflectance measured in situ; positive change indicates decrease in reflectance from preirradiation, in air reflectance. [Authors' designation: L]
28*	46	1969	295	2.100	$\sim 0^\circ$		2π		Above specimen and conditions.
29*	46	1969	295	0.590	$\sim 0^\circ$		2π		Rutile TiO_2 in RTV602 methyl silicone binder; substrate unknown; exposed to 50 keV electrons (4 x 10^8 - 1.7 x 10^{12} e cm^{-2} sec^{-1}) in vacuum (10^{-8}mm Hg); vacuum maintained by ion pump; electron fluence (e cm^{-2}) is variable; reflectance measured in situ; positive change indicates decrease in reflectance from preirradiation, in air reflectance. [Authors' designation: O]

* Not shown on plot

SPECIFICATION TABLE NO. 104 CHANGE IN NORMAL SPECTRAL REFLECTANCE OF TITANIUM DIOXIDE PIGMENTED COATINGS (continued)

Curve No.	Ref. No.	Year	Temperature K	Wavelength Range, μm	Geometry θ	θ'	ω'	Reported Error, %	Composition (weight percent), Specifications, and Remarks
30*	46	1969	295	2.100	$\sim 0°$		2π		Above specimen and conditions.
31*	46	1969	295	0.250	$\sim 0°$		2π		Anatase TiO_2 in methyl phenyl silicone binder (OAO Pyromark); substrate unknown; exposed to UV radiation (4.7 UV sun rate) in vacuum (10^{-8} mm Hg); vacuum maintained by ion pump; ESH is variable; reflectance measured in situ; positive change indicates decrease in reflectance from preirradiation, in air reflectance. [Authors' designation: Y]
32*	46	1969	295	0.425	$\sim 0°$		2π		Above specimen and conditions.
33*	46	1969	295	2.100	$\sim 0°$		2π		Above specimen and conditions.
34*	46	1969	295	0.250	$\sim 0°$		2π		Anatase TiO_2 in methyl silicone binder (0.127 mm thick) on Cat-A-Lac white primer (0.0508 mm thick) substrate; final substrate unknown; exposed to UV radiation (4.7 UV sun rate) in vacuum (10^{-8} mm Hg); vacuum maintained by ion pump; ESH is variable; reflectance measured in situ; positive change indicates decrease in reflectance from preirradiation, in air reflectance. [Authors' designation: L]
35*	46	1969	295	0.425	$\sim 0°$		2π		Above specimen and conditions.
36*	46	1969	295	2.100	$\sim 0°$		2π		Above specimen and conditions.
37*	46	1969	295	0.250	$\sim 0°$		2π		Rutile TiO_2 in RTV602 methyl silicone binder; substrate unknown; exposed to UV radiation (4.7 UV sun rate) in vacuum (10^{-8} mm Hg); vacuum maintained by ion pump; ESH is variable; reflectance measured in situ; positive change indicates decrease in reflectance from preirradiation, in air reflectance. [Authors' designation: O]
38*	46	1969	295	0.425	$\sim 0°$		2π		Above specimen and conditions.
39*	46	1969	295	2.100	$\sim 0°$		2π		Above specimen and conditions.
40*	46	1969	295	0.250	$\sim 0°$		2π		Rutile TiO_2 in XR6-3488 methyl silicone binder; substrate unknown; exposed to UV radiation (4.7 UV sun rate) in vacuum (10^{-8} mm Hg); vacuum maintained by ion pump; ESH is variable; reflectance measured in situ; positive change indicates decrease in reflectance by preirradiation, in air reflectance. [Authors' designation: P]
41*	46	1969	295	0.425	$\sim 0°$		2π		Above specimen and conditions.
42*	46	1969	295	2.100	$\sim 0°$		2π		Above specimen and conditions.

* Not shown on plot

DATA TABLE NO. 104 CHANGE IN NORMAL SPECTRAL REFLECTANCE OF TITANIUM DIOXIDE PIGMENTED COATINGS

[Wavelength, λ, μm; Reflectance, ρ; Temperature, T, K]

λ	Δρ
CURVE 1 T = 295	
0.265	0.034
0.312	0.026
0.364	0.017
0.425	0.364
0.498	0.212
0.537	0.150
0.567	0.116
0.610	0.083
0.668	0.058
0.766	0.040
0.921	0.029
1.54	0.018
1.95	0.016
2.06	0.000
CURVE 2 T = 295	
0.367	0.058
0.435	0.481
0.537	0.228
0.559	0.182
0.587	0.149
0.635	0.118
0.696	0.090
0.791	0.066
0.894	0.047
1.04	0.037
1.49	0.025
CURVE 3 T = 295	
0.264	0.013
0.308	0.019
0.362	0.011
0.431	0.539
0.591	0.197
0.624	0.168
0.660	0.138
0.728	0.102
0.784	0.082
0.865	0.063
1.11	0.041
CURVE 3 (cont.)	
1.52	0.029*
1.89	0.029
2.09	0.014
CURVE 4 T = 295	
0.250	0.021
0.306	0.035
0.367	0.058*
0.435	0.583
0.549	0.356
0.581	0.280
0.611	0.236
0.674	0.175
0.746	0.132
0.871	0.086
0.970	0.058
1.13	0.047
1.67	0.038
2.03	0.026
CURVE 5 T = 295	
0.367	0.058*
0.434	0.614
0.568	0.358
0.600	0.296
0.642	0.249
0.700	0.203
0.770	0.163
0.856	0.123
0.943	0.092
1.03	0.074
1.12	0.060
1.21	0.054
1.49	0.040
1.92	0.040
2.10	0.022
2.32	0.035
2.45	0.021
CURVE 6 T = 295	
0.264	0.013*
0.370	0.027
0.436	0.643
0.591	0.386
0.632	0.328
0.690	0.269
0.746	0.223
0.851	0.162
0.964	0.113
1.07	0.084
1.20	0.061
1.33	0.050
1.49	0.040*
1.92	0.040*
2.10	0.022*
2.32	0.035*
2.45	0.021*
CURVE 7 T = 295	
0.250	-0.003
0.289	-0.018
0.307	-0.022
0.331	-0.022
0.356	-0.018
0.416	0.044
0.424	0.046
0.438	0.044
0.488	0.026
0.514	0.020
0.602	0.012
1.61	0.010
1.93	0.000
2.13	0.000
2.29	0.004
2.45	0.013*
CURVE 8 T = 295	
0.304	0.000
0.338	-0.012
0.355	-0.009
CURVE 8 (cont.)	
0.413	0.071
0.422	0.074
0.435	0.070
0.506	0.032
0.542	0.020
0.595	0.013
0.684	0.013
0.977	0.020
1.20	0.020
1.38	0.016
1.77	0.009
1.91	0.000
1.94	0.000*
2.18	0.013
2.45	0.013*
CURVE 9 T = 295	
0.250	0.007
0.300	-0.014
0.336	-0.016*
0.357	-0.007*
0.422	0.138
0.430	0.140
0.439	0.136
0.502	0.076
0.533	0.057
0.574	0.042
0.608	0.034
0.649	0.030
1.04	0.021
1.33	0.018
1.66	0.012
2.45	0.013*
CURVE 10 T = 295	
0.304	0.000*
0.350	-0.007
0.362	-0.004
0.416	0.174
0.426	0.178
0.437	0.174
CURVE 10 (cont.)	
0.502	0.105
0.536	0.084
0.580	0.064
0.639	0.050
0.742	0.040
0.870	0.034
1.04	0.034*
1.14	0.038
1.21	0.038
1.32	0.031
1.46	0.018
1.54	0.013*
1.74	0.011
1.80	0.000
2.02	0.000
2.08	0.007
2.18	0.013
2.45	0.013*
CURVE 11 T = 295	
0.304	0.000*
0.350	-0.007*
0.362	-0.004*
0.430	0.210
0.438	0.214
0.447	0.209
0.522	0.128
0.556	0.103
0.587	0.087
0.628	0.070
0.667	0.061
0.715	0.052
0.774	0.045
0.873	0.037*
0.925	0.034*
1.04	0.034
1.14	0.038
1.21	0.038*
1.32	0.031*
1.46	0.018*
1.54	0.013*
1.74	0.011*
1.80	0.000*
CURVE 11 (cont.)	
2.02	0.000*
2.08	0.007*
2.18	0.013*
2.45	0.013*
CURVE 12 T = 295	
0.271	0.000
0.311	-0.006
0.362	0.000
0.380	0.012
0.423	0.048
0.437	0.047
0.529	0.020
0.573	0.013
0.619	0.012
1.91	0.013
2.11	0.000*
2.29	0.000
2.45	0.014*
CURVE 13 T = 295	
0.271	0.000*
0.311	-0.005*
0.358	-0.005*
0.412	0.051
0.424	0.054*
0.446	0.052
0.522	0.026
0.566	0.016
0.711	0.016
0.930	0.021
1.23	0.022
1.41	0.017*
1.55	0.014
1.71	0.000
1.89	0.013
2.11	0.000*
2.29	0.000*
2.45	0.014*
CURVE 14 T = 295	
0.250	0.000
0.309	-0.014*
0.357	-0.009*
0.420	0.062*
0.429	0.064*
0.444	0.062
0.503	0.042
0.525	0.036*
0.573	0.032
0.948	0.032
1.19	0.023
1.52	0.013
1.90	0.013*
2.11	0.000*
2.29	0.000*
2.45	0.014*
CURVE 15 T = 295	
0.271	0.000*
0.311	-0.006*
0.362	0.000*
0.415	0.073*
0.439	0.075*
0.505	0.050
0.523	0.044
0.545	0.041
0.593	0.038
0.720	0.041
0.914	0.032
1.03	0.030
1.19	0.023*
1.52	0.013*
1.90	0.013*
2.11	0.000*
2.29	0.000*
2.45	0.014*
CURVE 16 T = 295	
0.271	0.000*
0.311	-0.006*
CURVE 16 (cont.)	
0.362	0.000*
0.430	0.095
0.442	0.093
0.518	0.058
0.547	0.051
0.576	0.048*
0.624	0.046
0.712	0.050*
0.752	0.050
0.947	0.032*
1.03	0.030*
1.19	0.023*
1.52	0.013*
1.90	0.013*
2.11	0.000*
2.29	0.000*
2.45	0.014*
CURVE 17 T = 295	
0.425	0.36
0.500	0.17
0.590	0.08
0.950	0.04
1.20	0.03*
1.55	0.02
2.10	0.02*
2.50	0.02
CURVE 18 T = 295	
0.425	0.09
0.500	0.10
0.590	0.12
0.950	0.18
1.20	0.19
1.55	0.17
2.10	0.12
2.50	0.06

* Not shown on plot

DATA TABLE NO. 104 CHANGE IN NORMAL SPECTRAL REFLECTANCE OF TITANIUM DIOXIDE PIGMENTED COATINGS (continued)

CURVE 19
T = 295

λ	Δρ
0.425	0.36*
0.500	0.19
0.590	0.09*
0.950	0.05
1.20	0.04
1.55	0.03
2.10	0.02*
2.50	0.01

CURVE 20
T = 295

λ	Δρ
0.425	0.40
0.500	0.22
0.590	0.15*
0.950	0.16
1.20	0.16
1.55	0.14
2.10	0.13
2.50	0.06*

CURVE 21*
λ = 0.590
T = 295

electron fluence	Δρ
2 x 10^{14}	0.03
6 x 10^{14}	0.14
8 x 10^{15}	0.20

CURVE 22*
λ = 2.100
T = 295

electron fluence	Δρ
2 x 10^{14}	0.02
6 x 10^{14}	0.03
8 x 10^{15}	0.43

CURVE 23*
λ = 0.590
T = 295

electron fluence	Δρ
1 x 10^{13}	0.01
2 x 10^{14}	0.10
6 x 10^{14}	0.11

CURVE 24*
λ = 2.100
T = 295

electron fluence	Δρ
1 x 10^{13}	0.03
2 x 10^{14}	0.14
6 x 10^{14}	0.20

CURVE 25*
λ = 0.590
T = 295

electron fluence	Δρ
2 x 10^{14}	0.04
6 x 10^{14}	0.12

CURVE 26*
λ = 2.100
T = 295

electron fluence	Δρ
2 x 10^{14}	0.02
6 x 10^{14}	0.13

CURVE 27*
λ = 0.590
T = 295

electron fluence	Δρ
1 x 10^{13}	0.01
2 x 10^{14}	0.14
6 x 10^{14}	0.19

CURVE 28*
λ = 2.100
T = 295

electron fluence	Δρ
1 x 10^{13}	0.06
2 x 10^{14}	0.17
6 x 10^{14}	0.28
8 x 10^{15}	0.39

CURVE 29*
λ = 0.590
T = 295

electron fluence	Δρ
2 x 10^{14}	0.02
6 x 10^{14}	0.03

CURVE 30*
λ = 2.100
T = 295

electron fluence	Δρ
2 x 10^{14}	0.01
6 x 10^{14}	0.02
8 x 10^{15}	0.37

CURVE 31*
λ = 0.250
T = 295

ESH	Δρ
135	0.02
250	0.02
490	0.01
1130	0.03

CURVE 32*
λ = 0.425
T = 295

ESH	Δρ
135	0.58
250	0.60
490	0.64
1130	0.67

CURVE 33*
λ = 2.100
T = 295

ESH	Δρ
135	0.01
250	0.01
490	0.01
1130	0.02

CURVE 34*
λ = 0.250
T = 295

ESH	Δρ
135	0.0
250	0.0
490	0.02
1130	0.04

CURVE 35*
λ = 0.425
T = 295

ESH	Δρ
135	0.12
250	0.14
490	0.17
1130	0.20

CURVE 36*
λ = 2.100
T = 295

ESH	Δρ
135	0.35
250	0.38
490	0.41
1130	0.42

CURVE 37*
λ = 0.250
T = 295

ESH	Δρ
250	-0.02
490	-0.02
1130	-0.02

CURVE 38*
λ = 0.425
T = 295

ESH	Δρ
135	0.05
250	0.08
490	0.14
770	0.19
1130	0.25

CURVE 39*
λ = 2.100
T = 295

ESH	Δρ
135	0.0
250	0.01
490	0.01
1130	0.01

CURVE 40*
λ = 0.250
T = 295

ESH	Δρ
250	-0.02
490	-0.02
1130	-0.01

CURVE 41*
λ = 0.425
T = 295

ESH	Δρ
135	0.05
250	0.06
490	0.07
770	0.08
1130	0.10

CURVE 42*
λ = 2.100
T = 295

ESH	Δρ
135	0.0
250	0.0
490	0.01
1130	0.01

* Not shown on plot

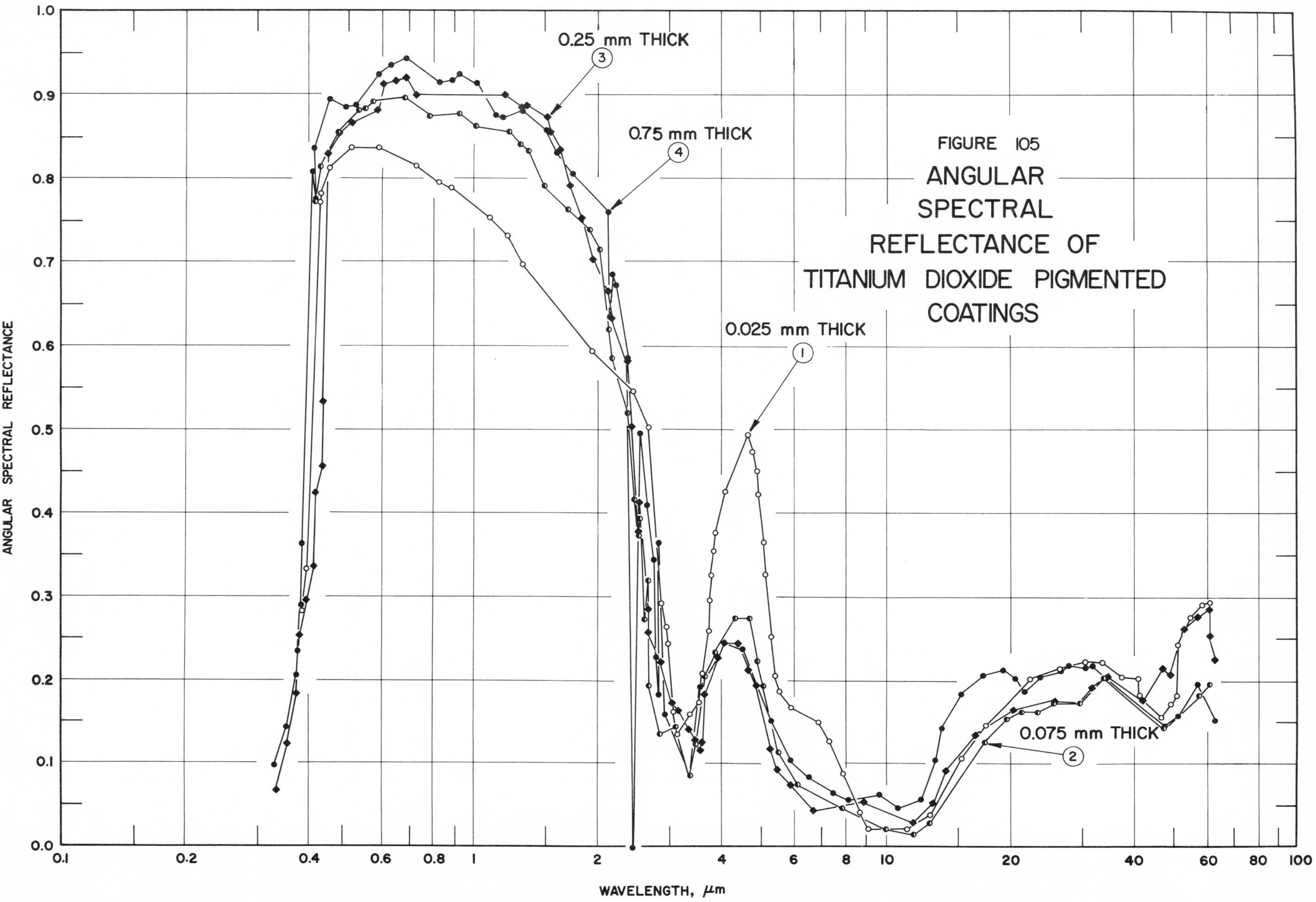

FIGURE 105 ANGULAR SPECTRAL REFLECTANCE OF TITANIUM DIOXIDE PIGMENTED COATINGS

SPECIFICATION TABLE NO. 105 ANGULAR SPECTRAL REFLECTANCE OF TITANIUM DIOXIDE PIGMENTED COATINGS

Curve No.	Ref. No.	Year	Temperature K	Wavelength Range, μm	Geometry θ	Geometry θ'	Geometry ω'	Reported Error, %	Composition (weight percent), Specifications, and Remarks
1	91	1966	~298	0.387-61.2	~20°		2π		PV-100; TiO_2 in silicone alkyd binder (0.025 mm thick) on aluminum substrate; 3 instruments used: for 0.33-2.5 μm $\theta = 20°$, for 1.5-23 μm $\theta = 25°$, for 20-61 μm $\theta = 17°$.
2	91	1966	~298	0.418-60.3	~20°		2π		Similar to above specimen and conditions except 0.075 mm thick.
3	91	1966	~298	0.333-62.6	~20°		2π		Similar to above specimen and conditions except 0.25 mm thick.
4	91	1966	~298	0.330-62.2	~20°		2π		Similar to above specimen and conditions except 0.75 mm thick.

DATA TABLE NO. 105 ANGULAR SPECTRAL REFLECTANCE OF TITANIUM DIOXIDE PIGMENTED COATINGS

[Wavelength, λ, μm; Reflectance, ρ; Temperature, T, K]

λ	ρ
CURVE 1	**T ~ 298**
0.387	0.284
0.399	0.335
0.427	0.772
0.427	0.783
0.450	0.814
0.502	0.839
0.599	0.839
0.731	0.817
0.826	0.796
0.893	0.790
1.10	0.752
1.21	0.731
1.32	0.699
1.95	0.595
2.45	0.546
2.68	0.502
2.89	0.294
2.94	0.266
2.97	0.245
3.06	0.162
3.11	0.136
3.36	0.160
3.52	0.174
3.60	0.210
3.71	0.260
3.71	0.296
3.77	0.329
3.81	0.356
3.83	0.379
4.10	0.428
4.61	0.496
4.79	0.474
4.86	0.451
4.84	0.423
5.05	0.366
5.17	0.329
5.34	0.253
5.40	0.209
5.58	0.187
5.90	0.169
6.98	0.150
7.36	0.127
7.97	0.089
CURVE 1 (cont.)	
8.64	0.042
9.09	0.021
11.3	0.022
12.9	0.040
15.2	0.108
17.5	0.149
22.4	0.203
26.5	0.218
30.2	0.225
33.1	0.223
37.4	0.206
41.0	0.204
41.7	0.184
46.5	0.158
49.0	0.171
50.4	0.182
51.0	0.245
54.9	0.279
58.6	0.291
61.2	0.295
CURVE 2	**T ~ 298**
0.418	0.771
0.429	0.814
0.471	0.856
0.527	0.881
0.561	0.891
0.680	0.898
0.553	0.885
0.781	0.875
0.935	0.877
1.01	0.862
1.22	0.856
1.30	0.842
1.36	0.835
1.50	0.793
1.70	0.763
1.92	0.740
2.02	0.715
2.13	0.620
2.19	0.586
2.39	0.520
CURVE 2 (cont.)	
2.48	0.418
2.54	0.393
2.51	0.373
2.66	0.320
2.60	0.272
2.67	0.194
2.69	0.194*
2.83	0.135
3.10	0.145
3.35	0.086
3.47	0.124
3.61	0.209
3.87	0.234
4.30	0.275
4.65	0.275
4.90	0.223
5.04	0.194
5.49	0.113
6.16	0.075
7.81	0.049
10.0	0.022
11.7	0.017
12.8	0.030
17.4	0.126
19.8	0.154
21.2	0.161
23.4	0.161
25.6	0.174
29.5	0.174
33.8	0.204
47.9	0.141
57.1	0.183
60.3	0.199
CURVE 3	**T ~ 298**
0.333	0.068
0.356	0.122
0.371	0.183
0.380	0.254
0.397	0.295
0.410	0.337
0.417	0.425
CURVE 3 (cont.)	
0.425	0.456
0.430	0.534
0.447	0.830
0.479	0.853
0.510	0.868
0.582	0.882
0.606	0.912
0.656	0.919
0.691	0.920
0.736	0.900
1.20	0.900
1.31	0.887
1.35	0.889
1.51	0.875
1.55	0.856
1.63	0.839
1.71	0.793
1.85	0.752
1.97	0.701
2.12	0.666
2.17	0.632
2.37	0.581
2.43	0.505
2.52	0.411
2.54	0.395*
2.51	0.377
2.68	0.285
2.68	0.258
2.87	0.223
3.01	0.171
3.14	0.165
3.31	0.140
3.48	0.129
3.54	0.117
3.60	0.126
3.63	0.184
3.92	0.228
4.06	0.245
4.40	0.245
4.67	0.214
4.85	0.195
5.23	0.119
5.53	0.091
5.97	0.074
6.76	0.043
CURVE 3 (cont.)	
8.85	0.055
11.6	0.030
13.0	0.053
14.0	0.091
16.5	0.135
20.4	0.166
25.8	0.175
29.4	0.172*
31.9	0.192
34.3	0.207
42.0	0.178
46.7	0.216
48.6	0.210
53.0	0.261
57.4	0.277
61.0	0.286
61.5	0.269*
61.2	0.254
62.6	0.228
CURVE 4	**T ~ 298**
0.330	0.097
0.352	0.141
0.374	0.208
0.378	0.235
0.381	0.290
0.389	0.362
0.415	0.773
0.412	0.809
0.415	0.837
0.455	0.895
0.492	0.888
0.520	0.889
0.592	0.923
0.638	0.935
0.693	0.941
0.835	0.915
0.893	0.918
0.939	0.923
1.04	0.914
1.14	0.877
1.19	0.874
CURVE 4 (cont.)	
1.32	0.881
1.51	0.858
1.60	0.831
1.75	0.806
2.11	0.760
2.19	0.685
2.22	0.672
2.17	0.633
2.37	0.583
2.43	0.000
2.55	0.497
2.64	0.410
2.81	0.365
2.76	0.344
2.79	0.229
2.81	0.183
2.91	0.159
3.36	0.084*
3.54	0.194
4.04	0.243
4.54	0.238
5.22	0.151
5.90	0.103
6.47	0.084
7.49	0.065
8.16	0.057
9.70	0.063
10.6	0.049
12.1	0.058
13.1	0.104
13.6	0.142
15.2	0.184
17.1	0.209
19.4	0.214
20.7	0.204
21.7	0.188
23.7	0.206
26.6	0.215
27.8	0.220
30.1	0.217
31.6	0.220
47.4	0.144
50.8	0.159
56.3	0.197
62.2	0.151

* Not shown on plot

SPECIFICATION TABLE NO. 106 NORMAL INTEGRATED ABSORPTANCE OF TITANIUM DIOXIDE PIGMENTED COATINGS

Curve No.	Ref. No.	Year	Temperature Range, K	Geometry θ	Reported Error, %	Composition (weight percent), Specifications and Remarks
1*	23	1964	~298	~0°		Skyspar A423, color SA 9185 untinted, manufactured by Andrew Brown Co.; rutile TiO_2 in an epoxy binder; substrate unknown; computed from spectral reflectance data; integrated between 0.40 and 0.70 μm.
2*	23	1964	~298	~0°		Rutile TiO_2 in silicone binder; substrate unknown; computed from spectral reflectance data; integrated between 0.40 and 0.70 μm. [IIT Research Institute designation: TC-50-19]

DATA TABLE NO. 106 NORMAL INTEGRATED ABSORPTANCE OF TITANIUM DIOXIDE PIGMENTED COATINGS

[Temperature, T, K; Absorptance, α]

T	α
CURVE 1*	
298	0.08
CURVE 2*	
298	0.11

* No plot given

FIGURE 107A

ANALYZED NORMAL SPECTRAL ABSORPTANCE OF TITANIUM DIOXIDE PIGMENTED COATINGS

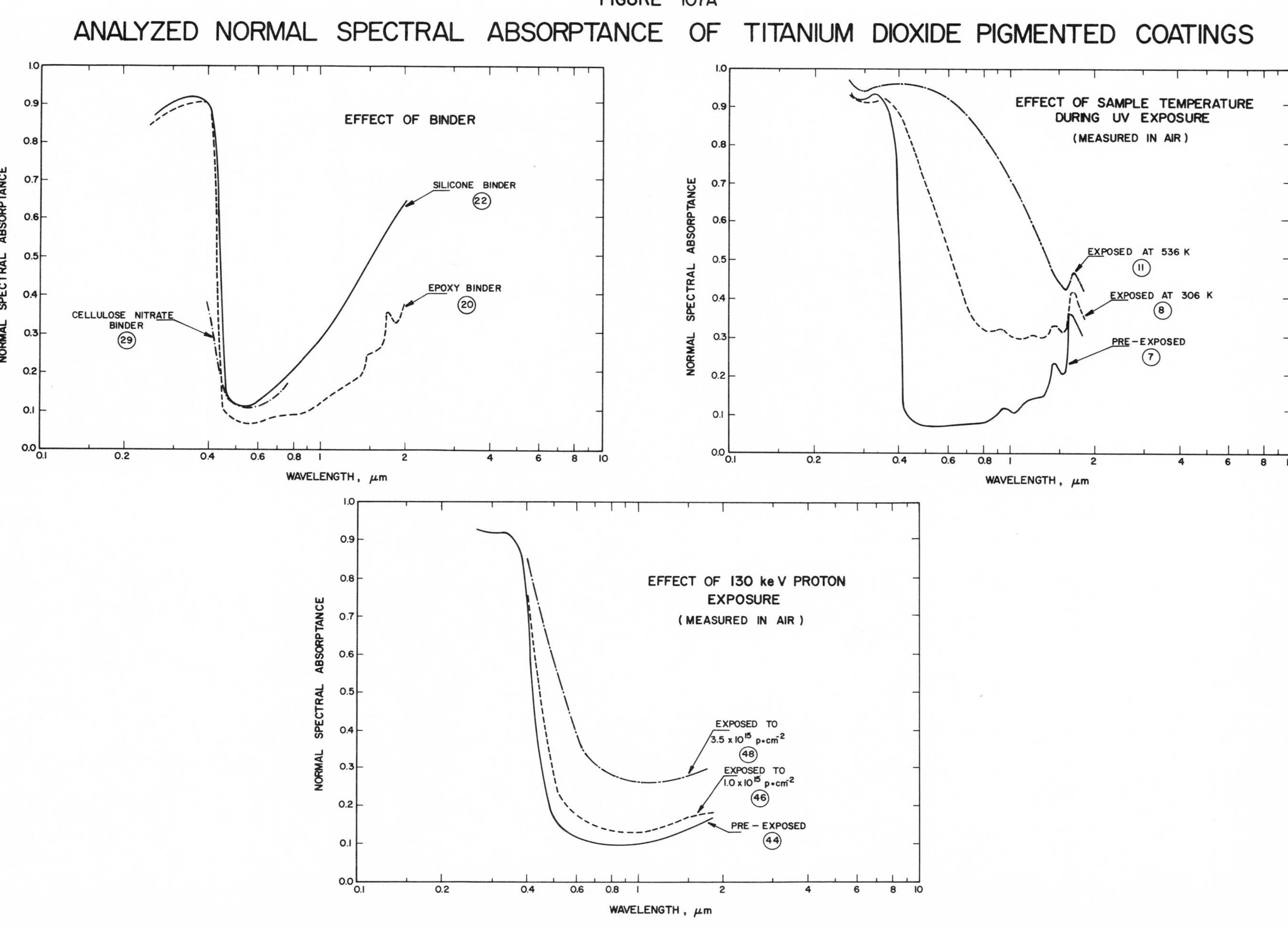

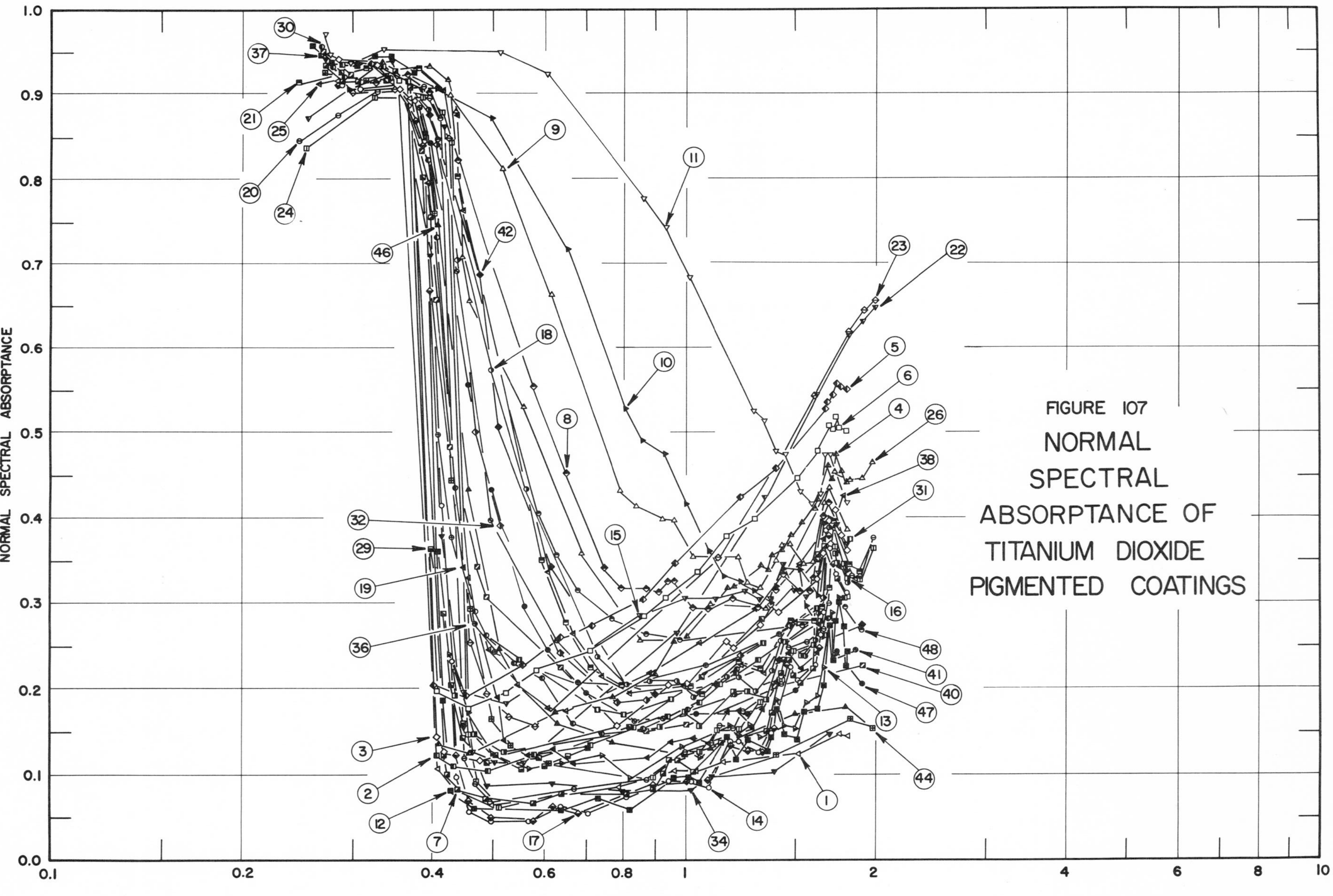

FIGURE 107 NORMAL SPECTRAL ABSORPTANCE OF TITANIUM DIOXIDE PIGMENTED COATINGS

SPECIFICATION TABLE NO. 107 NORMAL SPECTRAL ABSORPTANCE OF TITANIUM DIOXIDE PIGMENTED COATINGS

Curve No.	Ref. No.	Year	Temperature, K	Wavelength Range, μm	Geometry θ	Reported Error, %	Composition (weight percent), Specifications, and Remarks
1	24	1965	298	0.284-1.800	~0°		TiO_2 in sodium silicate binder; substrate unknown; data extracted from smooth curve.
2	24	1965	298	0.289-1.800	~0°		Similar to above specimen and conditions except exposed to a nuclear-radiation dose of approx 10^8 R of gamma and 5 x 10^{14} neutrons ($E \geq 2.9$ MeV) cm^{-2}.
3	24	1965	298	0.284-1.800	~0°		TiO_2 in silicone binder; substrate unknown; data extracted from smooth curve.
4	24	1965	298	0.284-1.800	~0°		Similar to above specimen and conditions except exposed to a nuclear-radiation dose of approx 10^8 R of gamma and 5 x 10^{14} neutrons ($E \geq 2.9$ MeV) cm^{-2}.
5	24	1965	298	0.281-1.800	~0°		Similar to curve 3 specimen and conditions except exposed to approx 500 sun hrs of ultraviolet irradiation and a nuclear radiation dose of approx 10^8 R of gamma and 5 x 10^{14} neutrons ($E \geq 2.9$ MeV) cm^{-2}.
6	24	1965	298	0.285-1.799	~0°		Similar to curve 3 specimen and conditions except exposed to a nuclear-radiation dose of approx 10^8 R of gamma and 5 x 10^{14} neutrons ($E \geq 2.9$ MeV) cm^{-2} and to low intensity ultraviolet radiation.
7	25	1964	~298	0.270-1.800	~0°		White skyspar enamel (A423, color SA9185); TiO_2 in epoxy binder; substrate unknown; data extracted from smooth curve; converted from spectral reflectance $\rho(\sim 0°, 2\pi)$.
8	25	1964	~298	0.270-1.800	~0°		Similar to above specimen and conditions except exposed to 10 suns UV radiation from a 1 KW A-H6 (PEK Labs Type C) mercury-argon lamp in vacuum (10^{-6}-10^{-7} mm Hg) for 50 hrs; sample maintained at 306 K during exposure.
9	25	1964	~298	0.270-1.800	~0°		Similar to above specimen and conditions except sample maintained at 383 K during exposure.
10	25	1964	~298	0.270-1.800	~0°		Similar to curve 8 specimen and conditions except sample maintained at 450 K during exposure.
11	25	1964	~298	0.270-1.800	~0°		Similar to curve 8 specimen and conditions except sample maintained at 536 K during exposure.
12	25	1964	~298	0.259-1.800	~0°		TiO_2 in polymethylvinyl siloxane binder (LMSC/Dow Corning); substrate unknown; data extracted from smooth curve; converted from spectral reflectance $\rho(\sim 0°, 2\pi)$.
13	25	1964	~298	0.259-1.800	~0°		Similar to above specimen and conditions except exposed to UV radiation from a 1 KW A-H6 (PEK Labs Type C) mercury-argon lamp (99.1 mm from sample) in vacuum (10^{-6}-10^{-7} mm Hg) for 100 hrs.
14	25	1964	~298	0.287-1.800	~0°		White Skyspar enamel (A423, color SA9185); TiO_2 in epoxy binder; substrate unknown; data extracted from smooth curve; converted from spectral reflectance $\rho(\sim 0°, 2\pi)$.
15	25	1964	~298	0.287-1.800	~0°		Similar to above specimen and conditions except exposed to UV radiation from a 1 KW A-H6 (PEK Labs Type C) mercury-argon lamp with a Corning 0-54 filter (190.5 mm from sample) in vacuum (10^{-6}-10^{-7} mm Hg) for 100 hrs.
16	25	1964	~298	0.287-1.800	~0°		Above specimen and conditions except after UV exposure sample exposed to fluorescent lamp (photons of energies less than 3.9 eV) (12.7 mm from sample) in air for 120 hrs.
17	25	1964	~298	0.282-1.800	~0°		White Skyspar enamel (A423, color SA9185); TiO_2 in epoxy binder; substrate unknown; data extracted from smooth curve; converted from spectral reflectance $\rho(\sim 0°, 2\pi)$.

SPECIFICATION TABLE NO. 107 NORMAL SPECTRAL ABSORPTANCE OF TITANIUM DIOXIDE PIGMENTED COATINGS (continued)

Curve No.	Ref. No.	Year	Temperature, K	Wavelength Range, μm	Geometry θ	Reported Error, %	Composition (weight percent), Specifications, and Remarks
18	25	1964	~298	0.282-1.800	~0°		Above specimen and conditions except exposed to UV radiation from a 1 KW A-H6 (PEK Labs Type C) mercury-argon lamp (190.5 mm from sample) in vacuum (10^{-6}-10^{-7} mm Hg) for 100 hrs.
19	25	1964	~298	0.282-1.800	~0°		Above specimen and conditions except after UV exposure sample exposed to fluorescent lamp (photons of energies less than 3.9 eV) (12.7 mm from sample) in air for 120 hrs.
20	26	1964	~298	0.246-1.994	~0°		Skyspar A-423; TiO_2 in epoxy resin (Shell Chemical Co. No. 1001) binder on aluminum substrate; data extracted from smooth curve; converted from spectral reflectance $\rho(\sim 0°, 2\pi)$.
21	26	1964	~298	0.246-1.994	~0°		Above specimen and conditions except exposed to UV radiation from a G. E. B-H6 lamp at solar factor of 3 suns in vacuum (8 x 10^{-7} mm Hg) for 100 hrs; vacuum maintained by ion pump; measured in air within 4 hrs of UV exposure.
22	26	1964	~298	0.254-2.000	~0°		TiO_2 in LTV-602 silicone resin binder on aluminum substrate; data extracted from smooth curve; converted from spectral reflectance $\rho(\sim 0°, 2\pi)$.
23	26	1964	~298	0.254-2.000	~0°		Similar to above specimen and conditions except exposed to UV radiation from a G. E. B-H6 lamp at solar factor of 3 suns in vacuum (8 x 10^{-7} mm Hg) for 100 hrs; vacuum maintained by ion pump; measured in air within 4 hrs of UV exposure.
24	26	1964	~298	0.252-1.985	~0°		Skyspar A-423; TiO_2 in epoxy resin (Shell Chemical Co. No. 1001) binder on aluminum substrate; data extracted from smooth curve; converted from spectral reflectance $\rho(\sim 0°, 2\pi)$.
25	26	1964	~298	0.264-1.985	~0°		Similar to above specimen and conditions except exposed to UV radiation from a G. E. B-H6 lamp at solar factor of 3 suns in vacuum (10^{-7} mm Hg) for 300 ESH; measured in air within 4 hrs after exposure.
26	26	1964	~298	0.264-1.970	~0°		Similar to curve 24 specimen and conditions except exposed to gamma radiation from a cobalt-60 source (A. E. C. L. Gammacell) at a dose rate of 1.6 megarad hr^{-1} and total radiation of 385 megarads in vacuum (5 x 10^{-7} mm Hg); sample outgassed for ~12 hrs at 5 x 10^{-7} mm Hg, then sealed in Pyrex tube, then irradiated; Gammacell ambient temp 320 K; measured within 1 hr after Pyrex tube opened.
27*	23	1964	~298	1.0	~0°		Rutile TiO_2 in silicone binder; substrate unknown. [IIT Research Institute designation: TC-50-19]
28*	23	1964	~298	1.0	~0°		Skyspar A423, color SA9185 untinted; rutile TiO_2 in an epoxy binder; substrate unknown.
29	76	1966	~298	0.399-0.700	~0°	~3	Howe and French E4709, flat white lacquer; rutile TiO_2 in cellulose nitrate binder; substrate unknown; sprayed (3 coats).
30	50	1965	~298	0.268-1.77	~0°		Fuller 517-W-1 gloss white paint; TiO_2 in silicone-modified alkyd binder; substrate unknown; cured at 513 K; data extracted from smooth curve.
31	50	1965	~298	0.268-1.82	~0°		Similar to above specimen and conditions except irradiated in vacuum at 291 K with 1 MeV electrons to a total dose of 10^{16} e cm^{-2}.
32	50	1965	~298	0.268-1.80	~0°		Similar to curve 30 specimen and conditions except irradiated in vacuum at 291 K with UV radiation for 485 sun hrs.

* Not shown on plot

SPECIFICATION TABLE NO. 107 NORMAL SPECTRAL ABSORPTANCE OF TITANIUM DIOXIDE PIGMENTED COATINGS (continued)

Curve No.	Ref. No.	Year	Temperature, K	Wavelength Range, μm	Geometry θ	Reported Error, %	Composition (weight percent), Specifications, and Remarks
33*	50	1965	~298	0.268-1.72	~0°		Similar to curve 30 specimen and conditions except irradiated in vacuum at 291 K with 1 MeV electrons to a total dose of 10^{16} e cm^{-2}, and additionally with UV radiation for 485 sun hrs.
34	50	1965	~298	0.270-1.699	~0°		LMSC/Dow Corning thermatrol paint; rutile TiO_2 in silicone binder; substrate unknown; data extracted from smooth curve.
35*	50	1965	~298	0.270-1.722	~0°		Similar to above specimen and conditions except irradiated in vacuum at 291 K with 1 MeV electrons to a total dose of 10^{16} e cm^{-2}.
36	50	1965	~298	0.270-1.746	~0°		Similar to curve 34 specimen and conditions except irradiated in vacuum at 291 K with UV radiation for 485 sun hrs.
37	50	1965	~298	0.268-1.722	~0°		White Skyspar enamel, A. Brown A423 color SA9185; TiO_2 in epoxy binder; substrate unknown; data extracted from smooth curve.
38	50	1965	~298	0.268-1.771	~0°		Similar to above specimen and conditions except irradiated in vacuum at 291 K with 1 MeV electrons to a total dose of 10^{16} e cm^{-2}.
39*	50	1965	~298	0.268-1.722	~0°		Similar to curve 37 specimen and conditions except irradiated in vacuum at 291 K with UV radiation for 485 sun hrs.
40	117	1965	~298	0.269-1.908	~0°		Fuller 517-W-1 gloss white paint; TiO_2 in silicone-modified-alkyd binder; substrate unknown; cured at 513 K; data extracted from smooth curve.
41	117	1965	~298	0.269-1.851	~0°		Similar to above specimen and conditions except irradiated in vacuum at 291 K with 130 keV protons to a total dose of 1.3 x 10^{15} p cm^{-2}.
42	117	1965	~298	0.269-1.908	~0°		Similar to above specimen and conditions except irradiated to a total dose of 2.0 x 10^{15} p cm^{-2}.
43*	117	1965	~298	0.269-1.908	~0°		Similar to above specimen and conditions except irradiated to a total dose of 3.2 x 10^{15} p cm^{-2}.
44	117	1965	~298	0.270-1.968	~0°		LMSC/Dow Corning Thermatrol paint; rutile TiO_2 in silicone binder; substrate unknown; data extracted from smooth curve.
45*	117	1965	~298	0.270-1.968	~0°		Similar to above specimen and conditions except irradiated in vacuum at 291 K with 130 keV protons to a total dose of 0.6 x 10^{15} p cm^{-2}.
46	117	1965	~298	0.270-1.968	~0°		Similar to above specimen and conditions except irradiated to a total dose of 1.0 x 10^{15} p cm^{-2}.
47	117	1965	~298	0.270-1.908	~0°		Similar to above specimen and conditions except irradiated to a total dose of 1.5 x 10^{15} p cm^{-2}.
48	117	1965	~298	0.270-1.908	~0°		Similar to above specimen and conditions except irradiated to a total dose of 3.5 x 10^{15} p cm^{-2}.

* Not shown on plot

DATA TABLE NO. 107 NORMAL SPECTRAL ABSORPTANCE OF TITANIUM DIOXIDE PIGMENTED COATINGS

[Wavelength, λ, μm; Absorptance, α; Temperature, T, K]

λ	α
CURVE 1 T = 298	
0.284	0.918
0.345	0.916
0.368	0.899
0.407	0.108
0.427	0.086
0.493	0.065
0.789	0.085
0.955	0.102
1.035	0.101
1.151	0.113
1.504	0.124
1.750	0.147
1.800	0.143
CURVE 2 T = 298	
0.289	0.937
0.333	0.935
0.363	0.917
0.404	0.122
0.430	0.110
0.488	0.105
0.575	0.124
0.701	0.134
0.825	0.155
0.930	0.167
1.050	0.184
1.133	0.203
1.205	0.219
1.308	0.232
1.784	0.303
1.800	0.305
CURVE 3 T = 298	
0.284	0.942
0.328	0.937
0.354	0.909
0.405	0.145
0.418	0.127
0.472	0.119
0.780	0.157
0.937	0.201
CURVE 3 (cont.)	
1.004	0.206
1.163	0.254
1.190	0.249
1.318	0.272
1.367	0.297
1.406	0.289
1.611	0.327
1.659	0.357
1.688	0.394
1.703	0.367
1.735	0.408
1.760	0.377
1.800	0.360
CURVE 4 T = 298	
0.284	0.943*
0.332	0.937
0.351	0.910
0.408	0.137
0.452	0.126
0.620	0.172
0.767	0.190
0.891	0.218
0.956	0.254
1.007	0.260
1.177	0.317
1.193	0.318
1.323	0.342
1.352	0.338
1.379	0.355
1.427	0.360
1.477	0.379
1.533	0.392
1.635	0.420
1.686	0.460
1.701	0.444
1.727	0.471
1.762	0.453
1.800	0.440
CURVE 5 T = 298	
0.281	0.942*
0.320	0.939
0.353	0.909*
0.400	0.203
0.450	0.193
0.552	0.229
0.634	0.260
0.710	0.273
0.851	0.303
0.960	0.348
1.225	0.425
1.391	0.458
1.661	0.526
1.673	0.535
1.710	0.542
1.741	0.557
1.769	0.551
1.800	0.550
CURVE 6 T = 298	
0.285	0.942*
0.333	0.935*
0.351	0.915
0.405	0.199
0.454	0.177
0.520	0.196
0.580	0.221
0.705	0.245
0.860	0.285
0.921	0.304
1.047	0.336
1.169	0.377
1.283	0.398
1.509	0.446
1.638	0.478
1.686	0.508
1.711	0.502
1.729	0.517
1.753	0.502
1.799	0.500
CURVE 7 T ~ 298	
0.270	0.936
0.286	0.917
0.305	0.917
0.325	0.938*
0.349	0.930
0.367	0.910
0.397	0.756
0.415	0.125
0.420	0.100
0.437	0.083
0.482	0.071
0.571	0.069
0.636	0.079
0.803	0.079
0.886	0.084
0.914	0.100
0.932	0.117
0.961	0.117
1.036	0.102
1.144	0.137
1.340	0.150
1.366	0.171
1.410	0.219
1.436	0.233
1.459	0.233
1.488	0.215
1.526	0.207
1.599	0.222
1.615	0.327*
1.636	0.358
1.652	0.362
1.742	0.329
1.800	0.301*
CURVE 8 T ~ 298	
0.270	0.936*
0.286	0.916*
0.325	0.916
0.363	0.924
0.394	0.904
0.439	0.824
0.579	0.554
0.650	0.451
CURVE 8 (cont.)	
0.744	0.340
0.795	0.317
0.864	0.317
0.904	0.311
0.936	0.325
0.952	0.325
1.025	0.293
1.083	0.293
1.180	0.303
1.331	0.293
1.387	0.313
1.431	0.332
1.446	0.332
1.510	0.312
1.579	0.312
1.607	0.327
1.652	0.400
1.680	0.418
1.800	0.360*
CURVE 9 T ~ 298	
0.270	0.936*
0.286	0.916*
0.325	0.916*
0.363	0.924*
0.423	0.900
0.514	0.812
0.612	0.662
0.790	0.431
0.831	0.413
0.919	0.399
0.958	0.399
1.039	0.354
1.211	0.354
1.259	0.324
1.314	0.319
1.380	0.350
1.411	0.365
1.462	0.367
1.544	0.344
1.590	0.347
1.655	0.418
1.681	0.436
1.800	0.384
CURVE 10 T ~ 298	
0.270	0.936*
0.295	0.909
0.407	0.909
0.498	0.872
0.655	0.717
0.803	0.529
0.853	0.490
0.923	0.473
1.003	0.417
1.081	0.360
1.155	0.330
1.224	0.323
1.280	0.311
1.317	0.293*
1.363	0.299
1.396	0.319
1.430	0.339
1.526	0.339
1.557	0.295
1.599	0.280
1.629	0.309
1.660	0.392
1.681	0.399
1.800	0.337
CURVE 11 T ~ 298	
0.270	0.971
0.275	0.949
0.297	0.939
0.333	0.954
0.510	0.950
0.602	0.923
0.859	0.777
0.930	0.743
1.013	0.682
1.285	0.522
1.337	0.511
1.387	0.478
1.440	0.473
1.524	0.430
1.589	0.417
1.641	0.429
1.665	0.472
CURVE 11 (cont.)	
1.693	0.472
1.800	0.418
CURVE 12 T ~ 298	
0.259	0.959
0.277	0.939
0.302	0.935
0.323	0.947
0.344	0.947
0.372	0.926
0.388	0.856
0.408	0.360
0.414	0.188
0.428	0.083
0.462	0.062
0.634	0.060
0.723	0.073
0.812	0.060
0.958	0.099
1.012	0.099
1.050	0.091
1.176	0.142
1.206	0.118
1.256	0.140
1.350	0.128
1.365	0.142
1.389	0.175
1.430	0.147
1.500	0.140
1.545	0.171
1.617	0.177
1.658	0.201
1.698	0.273
1.714	0.231
1.734	0.279
1.751	0.300
1.775	0.271
1.786	0.226
1.800	0.243
CURVE 13 T ~ 298	
0.259	0.959*
0.277	0.939*
0.302	0.935*
0.323	0.947*
0.344	0.947*
0.372	0.926*
0.388	0.856*
0.408	0.360*
0.424	0.240
0.453	0.171
0.497	0.125
0.563	0.107
0.639	0.113
0.736	0.121
0.818	0.098
0.980	0.131
1.050	0.120
1.172	0.157
1.215	0.146
1.268	0.158
1.316	0.142
1.351	0.143*
1.386	0.187
1.432	0.156
1.494	0.152
1.553	0.182
1.616	0.190
1.659	0.224
1.703	0.288
1.717	0.237
1.735	0.285*
1.759	0.303
1.785	0.227*
1.800	0.244*
CURVE 14 T ~ 298	
0.287	0.913
0.308	0.909
0.352	0.912*
0.374	0.894
0.387	0.848
0.412	0.415
0.428	0.231

* Not shown on plot

DATA TABLE NO. 107 NORMAL SPECTRAL ABSORPTANCE OF TITANIUM DIOXIDE PIGMENTED COATINGS (continued)

CURVE 14 (cont.)
$T \sim 298$

λ	α
0.433	0.099
0.458	0.059
0.495	0.046
0.563	0.046
0.631	0.062
0.700	0.054
0.802	0.077
0.933	0.092
1.029	0.092
1.082	0.086
1.176	0.139
1.217	0.139
1.301	0.126
1.364	0.149
1.424	0.217
1.462	0.224
1.508	0.208
1.601	0.225
1.634	0.261
1.650	0.289
1.666	0.381
1.686	0.362
1.705	0.366*
1.739	0.331
1.800	0.310

CURVE 15
$T \sim 298$

λ	α
0.287	0.913*
0.308	0.909*
0.352	0.912*
0.374	0.894*
0.387	0.848*
0.412	0.415*
0.428	0.231*
0.448	0.195
0.520	0.181
0.849	0.285
0.994	0.305
1.113	0.304
1.246	0.312
1.314	0.301
1.366	0.301
1.439	0.343
1.499	0.314

CURVE 15 (cont.)

λ	α
1.558	0.307
1.629	0.334
1.668	0.416*
1.696	0.416*
1.749	0.390
1.800	0.366

CURVE 16
$T \sim 298$

λ	α
0.287	0.913*
0.308	0.909*
0.352	0.912*
0.374	0.894*
0.387	0.848*
0.412	0.415*
0.428	0.231*
0.433	0.193
0.466	0.149
0.517	0.127
0.908	0.156
0.995	0.171
1.059	0.157
1.197	0.196
1.252	0.196
1.347	0.186
1.433	0.252
1.536	0.239
1.594	0.250
1.634	0.278
1.673	0.392*
1.737	0.357
1.749	0.346
1.774	0.346
1.800	0.329

CURVE 17
$T \sim 298$

λ	α
0.282	0.913
0.348	0.913*
0.376	0.870
0.384	0.841
0.394	0.669
0.433	0.121
0.453	0.070
0.492	0.050

CURVE 17 (cont.)

λ	α
0.571	0.047
0.617	0.064
0.677	0.056
0.885	0.087
1.002	0.093
1.079	0.093
1.170	0.135
1.215	0.139*
1.258	0.128
1.325	0.128
1.371	0.151
1.418	0.213*
1.456	0.226
1.497	0.213*
1.592	0.222*
1.638	0.249
1.671	0.376
1.692	0.370*
1.713	0.342
1.755	0.327*
1.800	0.311*

CURVE 18
$T \sim 298$

λ	α
0.282	0.911*
0.363	0.912*
0.393	0.881
0.492	0.573
0.560	0.433
0.721	0.238
0.803	0.203
0.878	0.208
0.962	0.208
1.038	0.190
1.163	0.211
1.224	0.211
1.298	0.196
1.366	0.220
1.437	0.255
1.548	0.237
1.602	0.254
1.631	0.289
1.671	0.388
1.700	0.387*
1.720	0.354*
1.750	0.341*
1.800	0.324

CURVE 19
$T \sim 298$

λ	α
0.282	0.913*
0.348	0.913*
0.376	0.870*
0.384	0.841*
0.394	0.669*
0.447	0.341
0.501	0.190
0.517	0.140
0.562	0.122
0.617	0.130
0.665	0.114
0.868	0.120
0.945	0.139
1.032	0.141
1.072	0.136
1.174	0.171
1.211	0.171
1.274	0.154
1.320	0.176
1.387	0.209
1.452	0.240
1.536	0.223
1.610	0.240
1.634	0.273
1.657	0.372
1.671	0.388*
1.700	0.387*
1.720	0.354*
1.750	0.341
1.800	0.324*

CURVE 20
$T \sim 298$

λ	α
0.246	0.849
0.282	0.877
0.328	0.901
0.386	0.910
0.410	0.872
0.449	0.120
0.465	0.090
0.491	0.068
0.572	0.068*
0.664	0.083
0.863	0.093
0.998	0.127

CURVE 20 (cont.)

λ	α
1.080	0.137
1.138	0.156
1.177	0.156*
1.189	0.151
1.212	0.155*
1.238	0.171
1.279	0.178*
1.376	0.185*
1.417	0.205
1.455	0.249
1.552	0.252
1.655	0.267
1.678	0.298
1.704	0.376*
1.795	0.344*
1.830	0.330
1.882	0.330
1.994	0.375

CURVE 21
$T \sim 298$

λ	α
0.246	0.915
0.314	0.932
0.380	0.932
0.422	0.900*
0.439	0.805
0.593	0.350
0.648	0.277
0.709	0.225
0.798	0.201
1.015	0.202
1.064	0.192
1.213	0.230
1.317	0.221
1.469	0.279
1.597	0.279
1.647	0.294
1.683	0.317
1.714	0.395
1.734	0.395*
1.822	0.342
1.872	0.334
1.994	0.375*

CURVE 22
$T \sim 298$

λ	α
0.254	0.871
0.294	0.905*
0.339	0.924
0.383	0.909
0.417	0.862
0.460	0.127
0.500	0.115
0.558	0.115
0.967	0.265
1.341	0.424
1.613	0.542
1.824	0.612
1.915	0.630
2.000	0.646

CURVE 23
$T \sim 298$

λ	α
0.254	0.871*
0.299	0.903
0.349	0.909
0.394	0.899
0.423	0.850
0.459	0.253
0.485	0.195
0.524	0.169
0.579	0.156
1.129	0.351
1.600	0.543
1.819	0.617
1.920	0.642
2.000	0.655

CURVE 24
$T \sim 298$

λ	α
0.252	0.839
0.324	0.899
0.386	0.899
0.411	0.880
0.426	0.846
0.452	0.149
0.466	0.092
0.509	0.062
0.634	0.079*
0.888	0.097

CURVE 24 (cont.)

λ	α
0.992	0.127*
1.054	0.127
1.137	0.152
1.214	0.152
1.316	0.183
1.385	0.183*
1.429	0.210
1.478	0.242
1.575	0.250*
1.658	0.261
1.687	0.314*
1.719	0.365
1.794	0.329*
1.886	0.329
1.985	0.362

CURVE 25
$T \sim 298$

λ	α
0.264	0.914
0.380	0.931*
0.415	0.908
0.436	0.877
0.445	0.763
0.608	0.337
0.662	0.263
0.734	0.214
0.813	0.195
0.991	0.201
1.053	0.195
1.128	0.206*
1.232	0.228
1.332	0.219
1.488	0.277
1.605	0.277*
1.663	0.295*
1.689	0.319*
1.729	0.386
1.859	0.329
1.909	0.336*
1.985	0.362*

CURVE 26
$T \sim 298$

λ	α
0.264	0.914*
0.395	0.933

CURVE 26 (cont.)

λ	α
0.421	0.919
0.434	0.880
0.445	0.709
0.458	0.655
0.553	0.530
0.684	0.359
0.841	0.256
0.952	0.256*
1.091	0.296
1.340	0.292
1.604	0.410
1.660	0.420*
1.722	0.454
1.815	0.441
1.908	0.446
1.970	0.463

CURVE 27*
$T \sim 298$

λ	α
1.0	0.50

CURVE 28*
$T \sim 298$

λ	α
1.0	0.14

CURVE 29
$T \sim 298$

λ	α
0.399	0.364
0.449	0.160
0.500	0.123
0.550	0.107
0.600	0.110
0.650	0.121
0.700	0.131

CURVE 30
$T \sim 298$

λ	α
0.268	0.956
0.277	0.934
0.290	0.920
0.341	0.920
0.355	0.908*
0.367	0.889

* Not shown on plot

DATA TABLE NO. 107 NORMAL SPECTRAL ABSORPTANCE OF TITANIUM DIOXIDE PIGMENTED COATINGS (continued)

CURVE 30 (cont.) T ~ 298

λ	α
0.392	0.823
0.406	0.731
0.431	0.436
0.446	0.341*
0.468	0.290
0.488	0.262
0.539	0.230
0.593	0.211
0.663	0.160
0.738	0.147
0.855	0.152
1.00	0.174
1.20	0.220
1.68	0.280
1.77	0.329*

CURVE 31 T ~ 298

λ	α
0.268	0.956*
0.277	0.934*
0.290	0.920*
0.341	0.920*
0.355	0.908*
0.367	0.889*
0.392	0.823*
0.406	0.731*
0.431	0.436*
0.446	0.341*
0.468	0.290*
0.488	0.262*
0.544	0.233
0.729	0.176
0.800	0.170
0.946	0.189
1.33	0.252
1.61	0.291
1.82	0.373

CURVE 32 T ~ 298

λ	α
0.268	0.956*
0.277	0.934*
0.290	0.920*
0.347	0.920*

CURVE 32 (cont.)

λ	α
0.375	0.897*
0.405	0.841
0.439	0.705
0.468	0.500
0.510	0.390
0.633	0.241
0.780	0.184
0.867	0.187
1.22	0.239
1.46	0.272
1.65	0.304
1.80	0.349*

CURVE 33* T ~ 298

λ	α
0.268	0.959
0.279	0.930
0.293	0.922
0.339	0.922
0.371	0.910
0.391	0.892
0.401	0.872
0.466	0.614
0.544	0.405
0.626	0.289
0.725	0.229
0.790	0.208
0.946	0.200
1.28	0.250
1.48	0.281
1.72	0.332

CURVE 34 T ~ 298

λ	α
0.270	0.941
0.297	0.937*
0.343	0.940
0.357	0.934*
0.371	0.897*
0.383	0.844*
0.399	0.710
0.412	0.379
0.428	0.240*
0.448	0.157
0.488	0.089

CURVE 34 (cont.)

λ	α
0.611	0.090
0.795	0.081
1.016	0.083
1.107	0.097
1.378	0.102
1.699	0.149

CURVE 35* T ~ 298

λ	α
0.270	0.941
0.297	0.937
0.343	0.940
0.357	0.934
0.371	0.897
0.383	0.844
0.399	0.710
0.412	0.379
0.428	0.240
0.444	0.180
0.486	0.107
0.502	0.099
0.747	0.121
1.016	0.128
1.138	0.145
1.512	0.151
1.722	0.185

CURVE 36 T ~ 298

λ	α
0.270	0.941*
0.297	0.937*
0.343	0.940*
0.357	0.934*
0.371	0.897*
0.383	0.844*
0.399	0.710*
0.408	0.499
0.429	0.376
0.464	0.275
0.500	0.244
0.674	0.207
0.790	0.203*
1.078	0.228
1.409	0.261
1.746	0.299*

CURVE 37 T ~ 298

λ	α
0.268	0.946
0.287	0.927
0.308	0.917*
0.337	0.919
0.354	0.915
0.367	0.887*
0.385	0.802
0.418	0.289
0.425	0.203
0.443	0.143
0.454	0.126*
0.482	0.114
0.582	0.120
0.816	0.153
1.060	0.211
1.333	0.279
1.610	0.350
1.722	0.406*

CURVE 38 T ~ 298

λ	α
0.268	0.946*
0.287	0.927*
0.308	0.917*
0.337	0.919*
0.354	0.915*
0.367	0.887*
0.391	0.799
0.432	0.432*
0.453	0.330
0.490	0.248
0.541	0.203
0.608	0.178
0.649	0.174
0.761	0.190*
0.873	0.219
1.051	0.253
1.292	0.291
1.512	0.342
1.676	0.384*
1.771	0.426

CURVE 39* T ~ 298

λ	α
0.268	0.946
0.287	0.927
0.308	0.917
0.376	0.917
0.400	0.908
0.426	0.875
0.579	0.549
0.689	0.372
0.770	0.313
0.867	0.288
1.170	0.277
1.409	0.317
1.676	0.384
1.722	0.406

CURVE 40 T ~ 298

λ	α
0.269	0.951
0.284	0.928*
0.298	0.925
0.338	0.925*
0.353	0.914*
0.384	0.835
0.404	0.657
0.425	0.484
0.470	0.342
0.488	0.307
0.633	0.225
0.785	0.156*
0.861	0.150
1.000	0.159
1.378	0.204
1.908	0.228

CURVE 41 T ~ 298

λ	α
0.269	0.951*
0.284	0.928*
0.298	0.925*
0.346	0.925*
0.380	0.886
0.438	0.690
0.492	0.398
0.626	0.256

CURVE 41 (cont.)

λ	α
0.827	0.161
0.919	0.163
1.192	0.195
1.851	0.246

CURVE 42 T ~ 298

λ	α
0.269	0.951*
0.284	0.928*
0.298	0.925*
0.351	0.924*
0.397	0.878
0.440	0.803*
0.475	0.688
0.508	0.509
0.611	0.342
0.810	0.203*
0.892	0.194
1.138	0.204
1.425	0.231
1.908	0.274

CURVE 43* T ~ 298

λ	α
0.269	0.951
0.284	0.928
0.298	0.925
0.346	0.925
0.380	0.911
0.418	0.880
0.444	0.839
0.484	0.723
0.539	0.517
0.770	0.247
0.816	0.222
1.016	0.218
1.192	0.220
1.459	0.245
1.653	0.269
1.908	0.298

CURVE 44 T ~ 298

λ	α
0.270	0.927
0.289	0.918*

CURVE 44 (cont.)

λ	α
0.311	0.917
0.355	0.917*
0.371	0.891*
0.386	0.852
0.401	0.760
0.429	0.445
0.456	0.292
0.492	0.165
0.530	0.134
0.605	0.113
0.925	0.093*
1.088	0.096*
1.393	0.121
1.824	0.163
1.968	0.152

CURVE 45* T ~ 298

λ	α
0.270	0.927
0.289	0.918
0.311	0.917
0.355	0.917
0.371	0.891
0.386	0.852
0.401	0.760
0.429	0.445
0.456	0.292
0.498	0.198
0.541	0.171
0.838	0.108
0.886	0.103
1.042	0.107
1.292	0.127
1.746	0.170
1.968	0.152

CURVE 46 T ~ 298

λ	α
0.270	0.927*
0.289	0.918*
0.311	0.917*
0.355	0.917*
0.371	0.891*
0.386	0.852*
0.409	0.749

CURVE 46 (cont.)

λ	α
0.454	0.433
0.506	0.249
0.549	0.204*
0.629	0.160
0.765	0.139
0.992	0.125
1.097	0.130
1.494	0.170
1.771	0.178
1.968	0.152*

CURVE 47 T ~ 298

λ	α
0.270	0.927*
0.289	0.918*
0.311	0.917*
0.357	0.917*
0.378	0.892*
0.397	0.844
0.454	0.556
0.492	0.433
0.559	0.296
0.608	0.245
0.697	0.196
0.821	0.160*
1.042	0.170
1.253	0.170
1.494	0.199
1.746	0.243
1.908	0.204

CURVE 48 T ~ 298

λ	α
0.270	0.927*
0.289	0.918*
0.311	0.917*
0.357	0.917*
0.378	0.892*
0.403	0.847
0.585	0.404
0.623	0.356
0.678	0.314
0.765	0.280
0.867	0.262
0.976	0.256
1.531	0.272
1.797	0.294
1.908	0.269

* Not shown on plot

FIGURE 108A (I)

ANALYZED NORMAL SOLAR ABSORPTANCE OF TITANIUM DIOXIDE PIGMENTED COATINGS

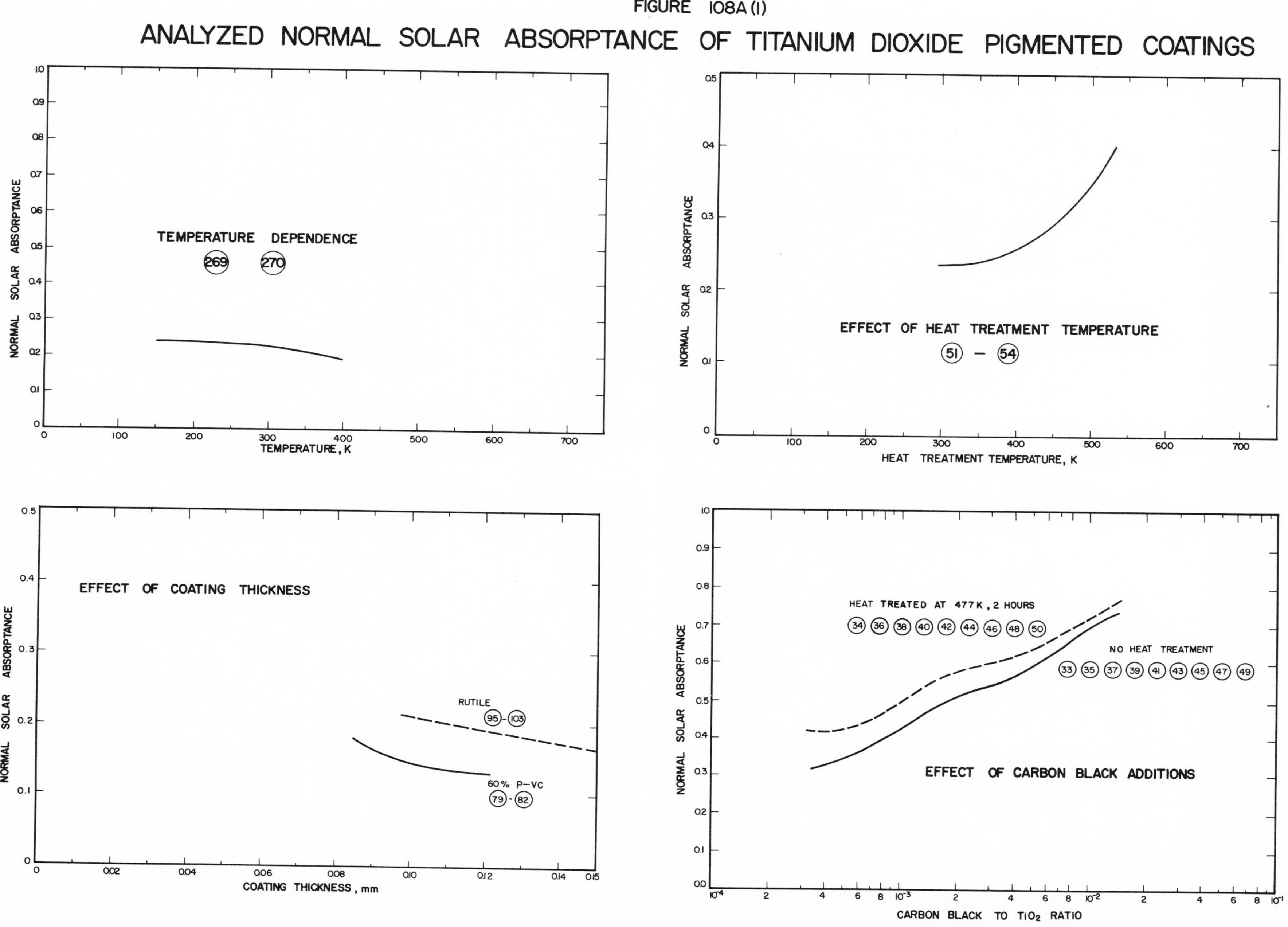

FIGURE 108A (2)

ANALYZED NORMAL SOLAR ABSORPTANCE OF TITANIUM DIOXIDE PIGMENTED COATINGS

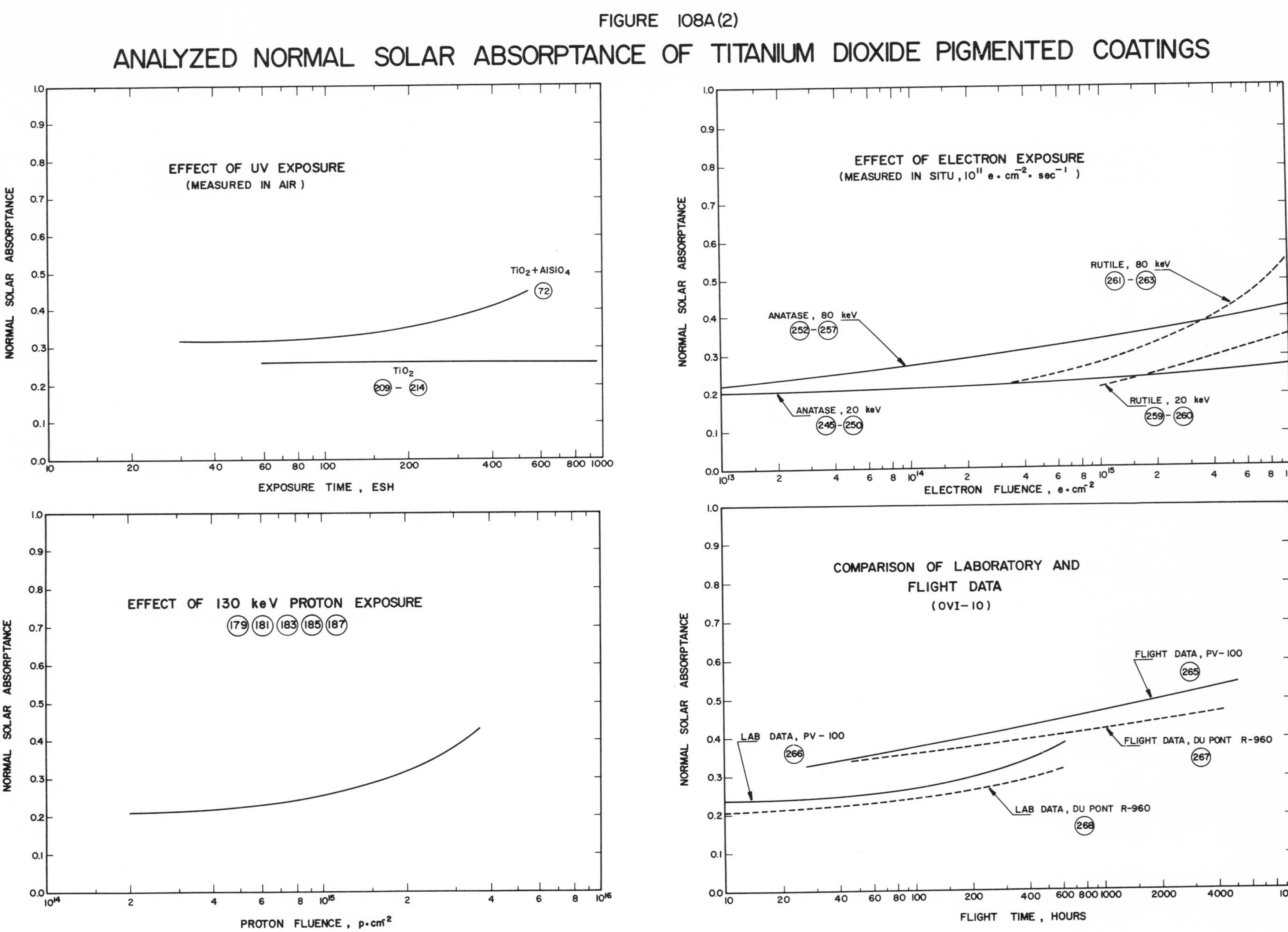

SPECIFICATION TABLE NO. 108 NORMAL SOLAR ABSORPTANCE OF TITANIUM DIOXIDE PIGMENTED COATINGS

Curve No.	Ref. No.	Year	Temperature Range, K	Geometry θ	Reported Error, %	Composition (weight percent), Specifications and Remarks
1	49	1961	278	0°		Fuller gloss white silicone 517-W-1; TiO_2 in silicone resin binder on Dow 15 treated Mg alloy substrate.
2	49	1961	278	0°		Above specimen and conditions except exposed in vacuum (10^{-6}-8 x 10^{-6} mm Hg) to UV radiation from an argon-filled A-H6 high pressure Hg arc lamp for 26 hrs; vacuum maintained by VacIon pump.
3	49	1961	278	0°		Fuller Gloss White silicone 517-W-1; TiO_2 in silicone resin binder on Mg alloy substrate.
4	49	1961	278	0°		Above specimen and conditions except exposed in vacuum (10^{-6}-8 x 10^{-6} mm Hg) to UV radiation from an argon-filled A-H6 high pressure Hg arc lamp for 100 hrs; vacuum maintained by VacIon pump.
5	49	1961	278	0°		Fuller gloss white silicone 517-W-1; TiO_2 in silicone resin binder on Mg alloy substrate.
6	49	1961	278	0°		Above specimen and conditions except exposed in vacuum (10^{-6}-8 x 10^{-6} mm Hg) to UV radiation from an argon-filled A-H6 high pressure Hg arc lamp for 26 hrs; vacuum maintained by VacIon pump.
7	49	1961	278	0°		Similar to curve 1 specimen and conditions.
8	49	1961	278	0°		Above specimen and conditions except exposed in vacuum (10^{-6}-8 x 10^{-6} mm Hg) to UV radiation from an argon-filled A-H6 high pressure Hg arc lamp for 26 hrs; vacuum maintained by VacIon pump.
9	49	1961	278	0°		Fuller gloss white silicone 517-W-1; TiO_2 in silicone resin binder on Dow 15 treated Mg alloy substrate.
10	49	1961	278	0°		Above specimen and conditions except exposed in vacuum (10^{-6}-8 x 10^{-6} mm Hg) to UV radiation from an argon-filled A-H6 high pressure Hg arc lamp for 46 hrs; vacuum maintained by VacIon pump.
11	49	1961	278	0°		Sicon white 7X1153; TiO_2, tinted, in silicone binder on Dow 15 treated Mg alloy substrate.
12	49	1961	278	0°		Above specimen and conditions except exposed in vacuum (10^{-6}-8 x 10^{-6} mm Hg) to UV radiation from an argon-filled A-H6 high pressure Hg arc lamp for 12 hrs; vacuum maintained by VacIon pump.
13	49	1961	278	0°		White Skyspar enamel, B, A423-SA9185; TiO_2, untinted, in epoxy resin binder on Mg alloy substrate.
14	49	1961	278	0°		Above specimen and conditions except exposed in vacuum (10^{-6}-8 x 10^{-6} mm Hg) to UV radiation from an argon-filled A-H6 high pressure Hg arc lamp for 100 hrs; vacuum maintained by VacIon pump.
15	49	1961	278	0°		White Skyspar enamel, B, A423-SA9185; TiO_2, untinted, in epoxy resin binder on Dow 15 treated Mg alloy substrate.
16	49	1961	278	0°		Above specimen and conditions except exposed in vacuum (10^{-6}-8 x 10^{-6} mm Hg) to UV radiation from an argon-filled A-H6 high pressure Hg arc lamp for 12 hrs; vacuum maintained by VacIon pump.
17	49	1961	278	0°		White Skyspar enamel, A, A423-SA8818; TiO_2, tinted, in epoxy resin binder on Dow 15 treated Mg alloy substrate.
18	49	1961	278	0°		Above specimen and conditions except exposed in vacuum (10^{-6}-8 x 10^{-6} mm Hg) to UV radiation from an argon-filled A-H6 high pressure Hg arc lamp for 46 hrs; vacuum maintained by VacIon pump.
19	49	1961	278	0°		White Skyspar enamel, B, A423-SA9185; TiO_2, untinted, in epoxy resin binder; substrate unknown; value avg of several determinations.
20	49	1961	278	0°		White Skyspar enamel, A, A423-SA8818; TiO_2, tinted, in epoxy resin binder on Dow 17 treated Mg alloy substrate.
21	49	1961	278	0°		Above specimen and conditions except exposed in vacuum (10^{-6}-8 x 10^{-6} mm Hg) to UV radiation from an argon-filled A-H6 high pressure Hg arc lamp for 80 hrs; vacuum maintained by VacIon pump.

SPECIFICATION TABLE NO. 108 NORMAL SOLAR ABSORPTANCE OF TITANIUM DIOXIDE PIGMENTED COATINGS (continued)

Curve No.	Ref. No.	Year	Temperature Range, K	Geometry θ	Reported Error, %	Composition (weight percent), Specifications and Remarks
22	49	1961	278	0°		Similar to curve 1 specimen and conditions.
23	49	1961	278	0°		White Skyspar enamel, B, A423-SA9185; TiO_2, untinted, in epoxy resin binder on Dow 17 treated Mg alloy substrate.
24	49	1961	278	0°		Above specimen and conditions except exposed in vacuum (10^{-6}-8 x 10^{-6} mm Hg) to UV radiation from an argon-filled A-H6 high pressure Hg arc lamp for 26 hrs; vacuum maintained by VacIon pump.
25	49	1961	278	0°		Sicon White 7X1153; TiO_2, tinted, in silicone binder on Dow 15 treated Mg alloy substrate.
26	49	1961	278	0°		Above specimen and conditions except exposed in vacuum (10^{-6}-8 x 10^{-6} mm Hg) to UV radiation from an argon-filled A-H6 high pressure Hg arc lamp for 100 hrs; vacuum maintained by VacIon pump.
27	20	1963	~298	~0°		TiO_2 (Titanox RA-10) in G. E. LTV-602 methyl silicone resin binder on aluminum substrate; PBR 1. 80; PVC 30%; spray application; computed from spectral reflectance data. [Authors' designation: S-28]
28	20	1963	~298	~0°		Above specimen and conditions except exposed to 1850 ESH of UV at solar factor of 10. 1 suns.
29	20	1963	~298	~0°		TiO_2 (TiPure R-900-1) in G. E. LTV-602 methyl silicone resin binder on aluminum substrate; PVC 35%; spray application; computed from spectral reflectance data. [Authors' designation: S-32]
30	20	1963	~298	~0°		Above specimen and conditions except exposed to 1650 ESH of UV at solar factor of 9 suns.
31	77	1962	~298	~0°		36. 1 TiO_2 [(RA)-Type III (TT-T-425)], 5. 0 stearated aluminum silicate, and 27. 2 organic solvents in 31. 7 polyester resin (Multron R-4, R-10, R-22) binder (~0. 0635 mm thick) on zinc chromate (0. 0127 mm thick) and 2024 aluminum alloy substrates; aluminum substrate roughened,then primed with zinc chromate primer (MIL-P-8585); spray application; air cured at ~298 K; computed from spectral reflectance data. [Authors' designation: polyurethane/black enamel coating No. 1]
32	77	1962	~298	~0°		Above specimen and conditions except heat treated to 477 K in a 2-hr cycle in vacuum (10^{-5} mm Hg); vacuum maintained by diffusion pump.
33	77	1962	~298	~0°		36. 1 TiO_2 [(RA)-Type III (TT-T-425)], 5. 0 stearated aluminum silicate, and 27. 2 organic solvents in 31. 7 polyester resin (Multron R-4, R-10, R-22) binder mixed with black base enamel (2. 5 carbon black, 11. 7 silaceous inerts, and 54. 4 thinners and driers in 31. 4 phthalic alkyd binder) (~0. 0635 mm thick) on zinc chromate (0. 0127 mm thick) and 2024 aluminum alloy substrates; carbon black to TiO_2 ratio by weight 3. 4 x 10^{-4}; aluminum substrate roughened,then primed with zinc chromate primer (MIL-P-8585); spray application; air cured at ~298 K; carbon black - 25 Columbia Carbon's Raven II beads and 75 Binney and Smiths Superb beads; computed from spectral reflectance data. [Authors' designation: polyurethane/black enamel coating No. 2]
34	77	1962	~298	~0°		Above specimen and conditions except heat treated to 477 K in a 2-hr cycle in vacuum (10^{-5} mm Hg); vacuum maintained by diffusion pump.

SPECIFICATION TABLE NO. 108 NORMAL SOLAR ABSORPTANCE OF TITANIUM DIOXIDE PIGMENTED COATINGS (continued)

Curve No.	Ref. No.	Year	Temperature Range, K	Geometry θ	Reported Error, %	Composition (weight percent), Specifications and Remarks
35	77	1962	~298	~0°		Similar to curve 33 specimen and conditions except ratio of carbon black to TiO_2 6.8 x 10^{-4}. [Authors' designation: polyurethane/black enamel coating No. 3]
36	77	1962	~298	~0°		Above specimen and conditions except heat treated to 477 K in a 2-hr cycle in vacuum (10^{-5} mm Hg); vacuum maintained by diffusion pump.
37	77	1962	~298	~0°		Similar to curve 33 specimen and conditions except ratio of carbon black to TiO_2 1.1 x 10^{-3}. [Authors' designation: polyurethane/black enamel coating No. 4]
38	77	1962	~298	~0°		Above specimen and conditions except heat treated to 477 K in a 2-hr cycle in vacuum (10^{-5} mm Hg); vacuum maintained by diffusion pump.
39	77	1962	~298	~0°		Similar to curve 33 specimen and conditions except ratio of carbon black to TiO_2 1.4 x 10^{-3}. [Authors' designation: polyurethane/black enamel coating No. 5]
40	77	1962	~298	~0°		Above specimen and conditions except heat treated to 477 K in a 2-hr cycle in vacuum (10^{-5} mm Hg); vacuum maintained by diffusion pump.
41	77	1962	~298	~0°		Similar to curve 33 specimen and conditions except ratio of carbon black to TiO_2 2.8 x 10^{-3}. [Authors' designation: polyurethane/black enamel coating No. 6]
42	77	1962	~298	~0°		Above specimen and conditions except heat treated to 477 K in a 2-hr cycle in vacuum (10^{-5} mm Hg); vacuum maintained by diffusion pump.
43	77	1962	~298	~0°		Similar to curve 33 specimen and conditions except ratio of carbon black to TiO_2 5.6 x 10^{-3}. [Authors' designation: polyurethane/black enamel coating No. 7]
44	77	1962	~298	~0°		Above specimen and conditions except heat treated to 477 K in a 2-hr cycle in vacuum (10^{-5} mm Hg); vacuum maintained by diffusion pump.
45	77	1962	~298	~0°		Similar to curve 33 specimen and conditions except ratio of carbon black to TiO_2 8.3 x 10^{-3}. [Authors' designation: polurethane/black enamel coating No. 8]
46	77	1962	~298	~0°		Above specimen and conditions except heat treated to 477 K in a 2-hr cycle in vacuum (10^{-5} mm Hg); vacuum maintained by diffusion pump.
47	77	1962	~298	~0°		Similar to curve 33 specimen and conditions except ratio of carbon black to TiO_2 1.0 x 10^{-2}. [Authors' designation: polyurethane/black enamel coating No. 9]
48	77	1962	~298	~0°		Above specimen and conditions except heat treated to 477 K in a 2-hr cycle in vacuum (10^{-5} mm Hg); vacuum maintained by diffusion pump.
49	77	1962	~298	~0°		Similar to curve 33 specimen and conditions except ratio of carbon black to TiO_2 1.4 x 10^{-2}. [Authors' designation: polyurethane/black enamel coating No. 10]
50	77	1962	~298	~0°		Above specimen and conditions except heat treated to 477 K in a 2-hr cycle in vacuum (10^{-5} mm Hg); vacuum maintained by diffusion pump.
51	77	1962	~298	~0°		36.1 TiO_2 [(RA)-Type III (TT-T-425)], 5.0 stearated aluminum silicate, and 27.2 organic solvents in 31.7 polyester resin (Multron R-4, R-10, R-22) binder (~0.0635 mm thick) on zinc chromate (0.0127 mm thick) and 2024 aluminum alloy substrates; aluminum substrate roughened,then primed with zinc chromate primer (MIL-P-8585); Fischer-Payne Dipcoater application; air cured at ~298 K; computed from spectral reflectance data.
52	77	1962	~298	~0°		Similar to above specimen and conditions except heat treated to 422 K in a 24-hr cycle in vacuum (10^{-5} mm Hg); vacuum maintained by diffusion pump.

SPECIFICATION TABLE NO. 108 NORMAL SOLAR ABSORPTANCE OF TITANIUM DIOXIDE PIGMENTED COATINGS (continued)

Curve No.	Ref. No.	Year	Temperature Range, K	Geometry θ	Reported Error, %	Composition (weight percent), Specifications and Remarks
53	77	1962	~298	~0°		Similar to above specimen and conditions except heat treated to 478 K in a 24-hr cycle in vacuum (10^{-5} mm Hg); vacuum maintained by diffusion pump.
54	77	1962	~298	~0°		Similar to above specimen and conditions except heat treated to 533 K in a 24-hr cycle in vacuum (10^{-5} mm Hg); vacuum maintained by diffusion pump.
55	77	1962	~298	~0°		Similar to curve 51 specimen and conditions except 2, 2'-4, 4'-tetrahydroxybenzophenone absorbing agent added to the coating formulation at 1% by weight of resin content.
56	77	1962	~298	~0°		Similar to above specimen and conditions except heat treated to 422 K in a 24-hr cycle in vacuum (10^{-5} mm Hg); vacuum maintained by diffusion pump.
57	77	1962	~298	~0°		Similar to above specimen and conditions except heat treated to 478 K in a 24-hr cycle in vacuum (10^{-5} mm Hg); vacuum maintained by diffusion pump.
58	77	1962	~298	~0°		Similar to above specimen and conditions except heat treated to 533 K in a 24-hr cycle in vacuum (10^{-5} mm Hg); vacuum maintained by diffusion pump.
59	77	1962	~298	~0°		Similar to curve 51 specimen and conditions except dibenzoylresorcinol (DBR) absorbing agent added to the coating formulation at 1% by weight of resin content.
60	77	1962	~298	~0°		Similar to above specimen and conditions except heat treated to 422 K in a 24-hr cycle in vacuum (10^{-5} mm Hg); vacuum maintained by diffusion pump.
61	77	1962	~298	~0°		Similar to above specimen and conditions except heat treated to 478 K in a 24-hr cycle in vacuum (10^{-5} mm Hg); vacuum maintained by diffusion pump.
62	77	1962	~298	~0°		Similar to above specimen and conditions except heat treated to 533 K in a 24-hr cycle in vacuum (10^{-5} mm Hg); vacuum maintained by diffusion pump.
63	77	1962	~298	~0°		Similar to curve 51 specimen and conditions except 2-hydroxybenzoylferrocene (HBF) absorbing agent added to the coating formulation at 1% by weight of resin content.
64	77	1962	~298	~0°		Similar to above specimen and conditions except heat treated to 422 K in a 24-hr cycle in vacuum (10^{-5} mm Hg); vacuum maintained by diffusion pump.
65	77	1962	~298	~0°		Similar to above specimen and conditions except heat treated to 478 K in a 24-hr cycle in vacuum (10^{-5} mm Hg); vacuum maintained by diffusion pump.
66	77	1962	~298	~0°		Similar to above specimen and conditions except heat treated to 533 K in a 24-hr cycle in vacuum (10^{-5} mm Hg); vacuum maintained by diffusion pump.
67	77	1962	~298	~0°		Similar to curve 51 specimen and conditions except 2, 4-dihydroxybenzoylferrocene (DHBF) absorbing agent added to the coating formulation at 1% by weight of resin content.
68	77	1962	~298	~0°		Similar to above specimen and conditions except heat treated to 422 K in a 24-hr cycle in vacuum (10^{-5} mm Hg); vacuum maintained by diffusion pump.
69	77	1962	~298	~0°		Similar to above specimen and conditions except heat treated to 478 K in a 24-hr cycle in vacuum (10^{-5} mm Hg); vacuum maintained by diffusion pump.
70	77	1962	~298	~0°		Similar to above specimen and conditions except heat treated to 533 K in a 24-hr cycle in vacuum (10^{-5} mm Hg); vacuum maintained by diffusion pump.

SPECIFICATION TABLE NO. 108 NORMAL SOLAR ABSORPTANCE OF TITANIUM DIOXIDE PIGMENTED COATINGS (continued)

Curve No.	Ref. No.	Year	Temperature Range, K	Geometry θ	Reported Error, %	Composition (weight percent), Specifications and Remarks
71	77	1962	~298	~0°		Similar to curve 51 specimen and conditions.
72	77	1962	~298	~0°		Above specimen and conditions except exposed to UV radiation from G. E. B-H6 lamp 15. 24 cm from sample; solar factor ~5 suns; exposure time (hrs) is variable.
73	77	1962	~298	~0°		Similar to curve 51 specimen and conditions except dibenzoylresorcinol (DBR) absorbing agent added to the coating formulation at 1% by weight of resin content.
74	77	1962	~298	~0°		Above specimen and conditions except exposed to UV radiation from G. E. B-H6 lamp 15. 24 cm from sample; solar factor ~5 suns; exposure time (hrs) is variable.
75	77	1962	~298	~0°		Similar to curve 51 specimen and conditions except 2 hydroxybenzoylferrocene absorbing agent added to the coating formulation at 1% by weight of resin content.
76	77	1962	~298	~0°		Above specimen and conditions except exposed to UV radiation from G. E. B-H6 lamp 15. 24 cm from sample; solar factor ~5 suns; exposure time (hrs) is variable.
77	101	1964	~298	~0°		TiO_2, American Cyanamid Co. Unitane OR640, in melamine modified polyvinyl butyral (90 Shawinigan Resins Corp Butvar B-98 polyvinyl butyral resin and 10 American Cyanamid Co. Cymel 300 hexa-methoxymethyl melamine) binder (0. 039 mm thick) on quartz substrate; PVC 30%; applied with Bird Film applicator; calculated from spectral reflectance data. [Authors' designation: Sample 251]
78	101	1964	~298	~0°		Similar to curve 77 specimen and conditions except coating thickness 0. 089 mm. [Authors' designation: Sample 271]
79	101	1964	~298	~0°		Similar to curve 77 specimen and conditions except coating thickness 0. 097 mm; PVC 60%. [Authors' designation: Sample 298]
80	101	1964	~298	~0°		Similar to curve 77 specimen and conditions except coating thickness 0. 120 mm; PVC 60%. [Authors' designation: Sample 350]
81	101	1964	~298	~0°		Similar to curve 77 specimen and conditions except coating thickness 0. 113 mm; PVC 60%; aluminum substrate. [Authors' designation: Sample 349]
82	101	1964	~298	~0°		Similar to curve 77 specimen and conditions except coating thickness 0. 085 mm; PVC 60%; aluminum substrate. [Authors' designation: Sample 393]
83	25	1964	~298	~0°		Fuller Gloss White Silicone (517-W-1); TiO_2 in silicone binder; substrate unknown; data extracted from smooth curve; calculated from spectral reflectance $\rho(\sim 0°, 2\pi)$.
84	25	1964	~298	~0°		Similar to above specimen and conditions except exposed to 10 suns UV radiation from a 1 KW A-H6 (PEK Labs Type C) mercury-argon lamp in vacuum (10^{-6}-10^{-7} mm Hg) for 50 hrs; sample maintained at 304 K during exposure.
85	25	1964	~298	~0°		Similar to above specimen and conditions except sample maintained at 311 K during exposure.
86	25	1964	~298	~0°		Similar to above specimen and conditions except sample maintained at 381 K during exposure.
87	25	1964	~298	~0°		Similar to above specimen and conditions except sample maintained at 456 K during exposure.
88	25	1964	~298	~0°		Similar to above specimen and conditions except sample maintained at 544 K during exposure.
89	25	1964	~298	~0°		White Skyspar enamel (A 423, color SA 9185); TiO_2 in epoxy binder; substrate unknown; data extracted from smooth curve; calculated from spectral reflectance $\rho(\sim 0°, 2\pi)$.

SPECIFICATION TABLE NO. 108 NORMAL SOLAR ABSORPTANCE OF TITANIUM DIOXIDE PIGMENTED COATINGS (continued)

Curve No.	Ref. No.	Year	Temperature Range, K	Geometry θ	Reported Error, %	Composition (weight percent), Specifications and Remarks
90	25	1964	~298	~0°		Similar to above specimen and conditions except exposed to 10 suns UV radiation from a 1 KW A-H6 (PEK Labs Type C) mercury-argon lamp in vacuum (10^{-6}-10^{-7} mm Hg) for 50 hrs; sample maintained at 306 K during exposure.
91	25	1964	~298	~0°		Similar to above specimen and conditions except sample maintained at 311 K during exposure.
92	25	1964	~298	~0°		Similar to above specimen and conditions except sample maintained at 383 K during exposure.
93	25	1964	~298	~0°		Similar to above specimen and conditions except sample maintained at 450 K during exposure.
94	25	1964	~298	~0°		Similar to above specimen and conditions except sample maintained at 383 K during exposure.
95	28	1964	~298	~0°		Skyspar A423, color SA9185 untinted; rutile TiO_2 in epoxy binder (0.0978 mm thick); substrate unknown; calculated from spectral reflectance data.
96	28	1964	~298	~0°		Similar to above specimen and conditions.
97	28	1964	~298	~0°		Similar to above specimen and conditions except coating thickness 0.107 mm.
98	28	1964	~298	~0°		Similar to above specimen and conditions except coating thickness 0.114 mm.
99	28	1964	~298	~0°		Similar to above specimen and conditions except coating thickness 0.129 mm.
100	28	1964	~298	~0°		Similar to above specimen and conditions except coating thickness 0.135 mm.
101	28	1964	~298	~0°		Similar to above specimen and conditions except coating thickness 0.136 mm.
102	28	1964	~298	~0°		Similar to above specimen and conditions except coating thickness 0.147 mm.
103	28	1964	~298	~0°		Similar to above specimen and conditions except coating thickness 0.155 mm.
104	28	1964	~298	~0°		Skyspar A423, color SA 9185 untinted, manufactured by Andrew Brown Co.; rutile TiO_2 in an epoxy binder; substrate unknown; calculated from spectral reflectance data.
105	28	1964	~298	~0°		Similar to above specimen and conditions.
106	28	1964	~298	~0°		Similar to above specimen and conditions except gamma-irradiated in vacuum (10^{-6} mm Hg) at 320 K to a total dose of 77 Megarads (dose rate ~1.5 Megarads hr^{-1}, Cobalt-60 source); measured within 1 hr after exposure to air.
107	28	1964	~298	~0°		Similar to above specimen and conditions.
108	28	1964	~298	~0°		Similar to above specimen and conditions except irradiated to a total dose of 385 Megarads.
109	28	1964	~298	~0°		Similar to curve 104 specimen and conditions except UV radiated in vacuum ($<10^{-6}$ mm Hg) at 273-288 K for 300 equivalent sun hrs (3-sun intensity, G. E. BH-6 lamp source); measured within 1 hr after exposure to air.
110	28	1964	~298	~0°		Similar to above specimen and conditions.
111	28	1964	~298	~0°		Rutile TiO_2 in silicone binder; substrate unknown; computed from spectral reflectance data. [IIT Research Institute designation: TC-50-19]
112	28	1964	~298	~0°		Similar to above specimen and conditions.

SPECIFICATION TABLE NO. 108 NORMAL SOLAR ABSORPTANCE OF TITANIUM DIOXIDE PIGMENTED COATINGS (continued)

Curve No.	Ref. No.	Year	Temperature Range, K	Geometry θ	Reported Error, %	Composition (weight percent), Specifications and Remarks
113	28	1964	~298	~0°		Similar to above specimen and conditions except gamma-irradiated in vacuum (10^{-6} mm Hg) at 320 K to a total dose of 77 Megarads(dose rate, ~1. 5 Megarads hr^{-1}, Cobalt-60 source); measured within 1 hr after exposure to air.
114	28	1964	~298	~0°		Similar to above specimen and conditions.
115	28	1964	~298	~0°		Similar to above specimen and conditions except irradiated to a total dose of 385 Megarads.
116	28	1964	~298	~0°		Similar to curve 111 specimen and conditions except ultraviolet irradiated in vacuum ($<10^{-6}$ mm Hg) at 273-288 K for 300 equivalent sun hrs (~3-sun intensity, G. E. BH-6 lamp source); measured within 1 hr after exposure to air.
117	28	1964	~298	~0°		Similar to above specimen and conditions.
118	23	1964	~298	~0°		Rutile TiO_2 in silicone binder; substrate unknown; computed from spectral reflectance data obtained with a Gier-Dunkle integrating sphere. [IIT Research Institute designation: TC-50-19]
119	23	1964	~298	~0°		Similar to above specimen and conditions except reflectance obtained with a Gier-Dunkle Solar Reflectometer.
120	23	1964	~298	~0°		Similar to above specimen and conditions except reflectance obtained with a Cary 14 spectrophotometer and integrating sphere.
121	23	1964	~298	~0°		Similar to above specimen and conditions.
122	23	1964	~298	~0°		Similar to above specimen and conditions except reflectance obtained with a G. E. and P. E. spectrophotometer and integrating sphere.
123	23	1964	~298	~0°		Similar to above specimen and conditions except reflectance obtained with a Beckman DK-2A spectrophotometer and integrating sphere.
124	23	1964	~298	~0°		Similar to above specimen and conditions.
125	23	1964	~298	~0°		Similar to above specimen and conditions except reflectance obtained with a Bausch and Lomb 505 spectrophotometer and integrating sphere.
126	23	1964	~298	~0°		Skyspar A423, color SA 9185 untinted, manufactured by Andrew Brown Co.; rutile TiO_2 in an epoxy binder; substrate unknown; computed from spectral reflectance data obtained with a Gier-Dunkle integrating sphere.
127	23	1964	~298	~0°		Similar to above specimen and conditions.
128	23	1964	~298	~0°		Similar to above specimen and conditions except reflectance obtained with a Gier-Dunkle Solar Reflectometer.
129	23	1964	~298	~0°		Similar to above specimen and conditions except reflectance obtained with a Cary 14 spectrophotometer and integrating sphere.
130	23	1964	~298	~0°		Similar to above specimen and conditions.
131	23	1964	~298	~0°		Similar to above specimen and conditions except reflectance obtained with a G. E. and P. E. spectrophotometer and integrating sphere.
132	23	1964	~298	~0°		Similar to above specimen and conditions except reflectance obtained with a Beckman DK-2A spectrophotometer and integrating sphere.
133	23	1964	~298	~0°		Similar to above specimen and conditions.

SPECIFICATION TABLE NO. 108 NORMAL SOLAR ABSORPTANCE OF TITANIUM DIOXIDE PIGMENTED COATINGS (continued)

Curve No.	Ref. No.	Year	Temperature Range, K	Geometry θ	Reported Error, %	Composition (weight percent), Specifications and Remarks
134	23	1964	~298	~0°		Similar to above specimen and conditions.
135	23	1964	~298	~0°		Similar to above specimen and conditions except reflectance obtained with a Bausch and Lomb 505 spectrophotometer and integrating sphere.
136	50	1965	~298	~0°		White Skyspar enamel, A. Brown A423 color SA 9185; TiO_2 in epoxy binder; substrate unknown; computed from spectral reflectance data.
137	50	1965	~298	~0°		Similar to above specimen and conditions except irradiated in vacuum ($\leq 10^{-6}$ mm Hg) at 289 K with 1 MeV electrons, to a total dose of 10^{14} e cm^{-2}.
138	50	1965	~298	~0°		Similar to above specimen and conditions except irradiated to a total dose of 10^{15} e cm^{-2}.
139	50	1965	~298	~0°		Similar to above specimen and conditions except irradiated to a total dose of 10^{16} e cm^{-2}; maintained in vacuum at 155 K for 12 hrs after irradiation.
140	50	1965	~298	~0°		Similar to above specimen and conditions.
141	50	1965	~298	~0°		Similar to curve 137 specimen and conditions except irradiated at 422 K to a total dose of 10^{15} e cm^{-2}.
142	50	1965	~298	~0°		Similar to above specimen and conditions except irradiated at 155 K to a total dose of 10^{15} e cm^{-2}; maintained in vacuum at 155 K for 12 hrs after irradiation.
143	50	1965	~298	~0°		Similar to curve 136 specimen and conditions except irradiated in vacuum (10^{-7} mm Hg) at 289 K with UV radiation for 485 sun hrs.
144	50	1965	~298	~0°		Similar to above specimen and conditions except irradiated in vacuum at 291 K with 1 MeV electrons, to a total dose of 10^{16} e cm^{-2}, and additionally with UV radiation for 485 sun hrs.
145	50	1965	~298	~0°		Fuller 577-W-1 gloss white paint; TiO_2 in silicone-modified-alkyd binder; substrate unknown; cured at 513 K; computed from spectral reflectance data.
146	50	1965	~298	~0°		Similar to above specimen and conditions except irradiated in vacuum ($\leq 10^{-6}$ mm Hg) at 289 K with 1 MeV electrons to a total dose of 10^{16} e cm^{-2}; maintained in vacuum at 155 K for 12 hrs after irradiation.
147	50	1965	~298	~0°		Similar to above specimen and conditions.
148	50	1965	~298	~0°		Similar to curve 145 specimen and conditions except irradiated in vacuum ($\leq 10^{-6}$ mm Hg) at 422 K with 1 MeV electrons to a total dose of 10^{15} e cm^{-2}.
149	50	1965	~298	~0°		Similar to above specimen and conditions except irradiated at 155 K to a total dose of 10^{15} e cm^{-2}; maintained in vacuum at 155 K for 12 hrs after irradiation.
150	50	1965	~298	~0°		Similar to curve 145 specimen and conditions except irradiated in vacuum (10^{-7} mm Hg) at 289 K with UV radiation for 485 sun hrs.
151	50	1965	~298	~0°		Similar to curve 145 specimen and conditions except irradiated in vacuum ($\leq 10^{-6}$ mm Hg) at 289 K with 1 MeV electrons to a total dose of 10^{16} e cm^{-2} and additionally with UV radiation for 485 sun hrs.
152	50	1965	~298	~0°		LMSC/Dow Corning thermatrol paint; TiO_2 (rutile) in silicone binder; computed from spectral reflectance data.

SPECIFICATION TABLE NO. 108 NORMAL SOLAR ABSORPTANCE OF TITANIUM DIOXIDE PIGMENTED COATINGS (continued)

Curve No.	Ref. No.	Year	Temperature Range, K	Geometry θ	Reported Error, %	Composition (weight percent), Specifications and Remarks
153	50	1965	~298	~0°		Similar to above specimen and conditions except irradiated in vacuum ($\leq 10^{-6}$ mm Hg) at 289 K with 1 MeV electrons to a total dose of 10^{14} e cm^{-2}.
154	50	1965	~298	~0°		Similar to above specimen and conditions except irradiated to a total dose of 10^{15} e cm^{-2}.
155	50	1965	~298	~0°		Similar to above specimen and conditions except irradiated to a total dose of 10^{16} e cm^{-2}; maintained in vacuum at 155 K for 12 hrs after irradiation.
156	50	1965	~298	~0°		Similar to above specimen and conditions.
157	50	1965	~298	~0°		Similar to curve 153 specimen and conditions except irradiated at 422 K to a total dose of 10^{15} e cm^{-2}.
158	50	1965	~298	~0°		Similar to above specimen and conditions except irradiated at 155 K to a total dose of 10^{15} e cm^{-2}; maintained in vacuum at 155 K for 12 hrs after irradiation.
159	50	1965	~298	~0°		Similar to curve 152 specimen and conditions except irradiated in vacuum (10^{-7} mm Hg) at 289 K with UV radiation for 485 sun hrs.
160	117	1965	~298	~0°		Fuller 517-W-1 gloss white paint; TiO_2 in silicone-modified-alkyd binder; cured at 513 K; computed from spectral reflectance data. [Authors' designation: Sample No. 3]
161	117	1965	~298	~0°		Above specimen and conditions except irradiated in vacuum with 130 keV protons to a total dose of $>5 \times 10^{15}$ p cm^{-2}.
162	117	1965	~298	~0°		Similar to curve 160 specimen and conditions. [Author's designation: Sample No. 4]
163	117	1965	~298	~0°		Above specimen and conditions except irradiated in vacuum with 130 keV protons to a total dose of 3.2×10^{15} p cm^{-2}.
164	117	1965	~298	~0°		Similar to curve 160 specimen and conditions. [Author's designation: Sample No. 15]
165	117	1965	~298	~0°		Above specimen and conditions except irradiated in vacuum with 130 keV protons to a total dose of 2.0×10^{15} p cm^{-2}.
166	117	1965	~298	~0°		Similar to curve 160 specimen and conditions. [Author's designation: Sample No. 16].
167	117	1965	~298	~0°		Above specimen and conditions except irradiated in vacuum with 130 keV protons to a total dose of 1.3×10^{15} p cm^{-2}.
168	117	1965	~298	~0°		Similar to curve 160 specimen and conditions. [Author's designation: Sample No. 25]
169	117	1965	~298	~0°		Above specimen and conditions except irradiated in vacuum with 130 keV protons to a total dose of 0.6×10^{15} p cm^{-2}.
170	117	1965	~298	~0°		Similar to curve 160 specimen and conditions. [Author's designation: Sample No. 26]
171	117	1965	~298	~0°		Above specimen and conditions except irradiated in vacuum with 130 keV protons to a total dose of 0.3×10^{15} p cm^{-2}.
172	117	1965	~298	~0°		LMSC/Dow Corning Thermatrol paint; TiO_2 (rutile) in silicone binder; computed from spectral reflectance data. [Author's designation: Sample No. 5]
173	117	1965	~298	~0°		Above specimen and conditions except irradiated in vacuum with 130 keV protons to a total dose of 3.5×10^{15} p cm^{-2}.
174	117	1965	~298	~0°		Similar to curve 172 specimen and conditions. [Author's designation: Sample No. 6]

SPECIFICATION TABLE NO. 108 NORMAL SOLAR ABSORPTANCE OF TITANIUM DIOXIDE PIGMENTED COATINGS (continued)

Curve No.	Ref. No.	Year	Temperature Range, K	Geometry θ	Reported Error, %	Composition (weight percent), Specifications and Remarks
175	117	1965	~298	~0°		Above specimen and conditions except irradiated in vacuum with 130 keV protons to a total dose of 1.5×10^{15} p cm^{-2}.
176	117	1965	~298	~0°		Similar to curve 172 specimen and conditions. [Author's designation: Sample No. 12]
177	117	1965	~298	~0°		Above specimen and conditions except irradiated in vacuum with 130 keV protons to a total dose of 1.0×10^{15} p cm^{-2}.
178	117	1965	~298	~0°		Similar to curve 172 specimen and conditions. [Author's designation: Sample No. 17]
179	117	1965	~298	~0°		Above specimen and conditions except irradiated in vacuum with 130 keV protons to a total dose of 3.5×10^{15} p cm^{-2}.
180	117	1965	~298	~0°		Similar to curve 172 specimen and conditions. [Author's designation: Sample No. 18]
181	117	1965	~298	~0°		Above specimen and conditions except irradiated in vacuum with 130 keV protons to a total dose of 1.5×10^{15} p cm^{-2}.
182	117	1965	~298	~0°		Similar to curve 172 specimen and conditions. [Author's designation: Sample No. 24]
183	117	1965	~298	~0°		Above specimen and conditions except irradiated in vacuum with 130 keV protons to a total dose of 1.0×10^{15} p cm^{-2}.
184	117	1965	~298	~0°		Similar to curve 172 specimen and conditions. [Author's designation: Sample No. 27]
185	117	1965	~298	~0°		Above specimen and conditions except irradiated in vacuum with 130 keV protons to a total dose of 0.6×10^{15} p cm^{-2}.
186	117	1965	~298	~0°		Similar to curve 172 specimen and conditions. [Author's designation: Sample No. 28]
187	117	1965	~298	~0°		Above specimen and conditions except irradiated in vacuum with 130 keV protons to a total dose of 0.2×10^{15} p cm^{-2}.
188	52	1965	193-263	~0°		Titanium dioxide in epoxy binder; measured in earth orbit on OSO-II.
189	52	1965	193-263	~0°		TiO_2 in silicone binder (adhesive backed elastomeric film); measured in earth orbit on OSO-II.
190	55	1968	297	~0°		Pyromark standard white, Tempil Corp; TiO_2 in Dow Corning methyl-phenyl silicone binder. [Author's designation: Sample 29]
191	55	1968	225	~0°		Above specimen and conditions except measured in vacuum ($< 10^{-6}$ mm Hg).
192	55	1968	225	~0°		Above specimen and conditions except exposed in vacuum ($<10^{-6}$ mm Hg) to 5×10^{13} e cm^{-2} at 3.0 keV (flux density 6×10^{12} e cm^{-2}) and 5×10^{14} p cm^{-2} at 2.5 keV (flux density 8×10^{12} p cm^{-2} sec^{-1}); measured in situ.
193	55	1968	225	~0°		Above specimen and conditions; measured 6 days later.
194	55	1968	225	~0°		Above specimen and conditions; measured 8 days later.
195	55	1968	225	~0°		Above specimen and conditions; measured 10 days later.
196	55	1968	225	~0°		Above specimen and conditions; measured 11 days later.
197	55	1968	297	~0°		Above specimen and conditions except measured in air 8 days later.
198	55	1968	297	~0°		Similar to curve 190 specimen and conditions. [Author's designation: Sample 4]

SPECIFICATION TABLE NO. 108 NORMAL SOLAR ABSORPTANCE OF TITANIUM DIOXIDE PIGMENTED COATINGS (continued)

Curve No.	Ref. No.	Year	Temperature Range, K	Geometry θ	Reported Error, %	Composition (weight percent), Specifications and Remarks
199	55	1968	225	~0°		Above specimen and conditions except measured in vacuum ($<10^{-6}$ mm Hg).
200	55	1968	225	~0°		Above specimen and conditions except exposed in vacuum ($<10^{-6}$ mm Hg) to 61 ESH of UV at solar factor of 1 sun; measured in situ.
201	55	1968	225	~0°		Above specimen and conditions except exposed to a total of 200 ESH of UV radiation and additionally to 5×10^{13} e cm^{-2} at 3.0 keV (flux density 6×10^{13} e cm^{-2} sec^{-1}) and 5×10^{14} p cm^{-2} at 2.5 keV (flux density 8×10^{12} p cm^{-2} sec^{-1}).
202	55	1968	225	~0°		Above specimen and conditions except exposed to a total of 335 ESH of UV.
203	55	1968	225	~0°		Above specimen and conditions except exposed to a total of 500 ESH of UV.
204	55	1968	225	~0°		Above specimen and conditions except exposed to a total of 714 ESH of UV.
205	55	1968	225	~0°		Above specimen and conditions except exposed to a total of 945 ESH of UV.
206	55	1968	297	~0°		Above specimen and conditions except measured in air 8 days later.
207	55	1968	297	~0°		Goddard 78-2B white; DuPont R-960 TiO_2 in RTV-602 resin binder; PBR 2.0; toluene solvent. [Author's designation: Sample 11]
208	55	1968	225	~0°		Above specimen and conditions except measured in air.
209	55	1968	225	~0°		Above specimen and conditions except exposed in vacuum ($<10^{-6}$ mm Hg) to 61 ESH of UV radiation at solar factor 1 sun.
210	55	1968	225	~0°		Above specimen and conditions except exposed to a total of 200 ESH of UV.
211	55	1968	225	~0°		Above specimen and conditions except exposed to a total of 335 ESH of UV.
212	55	1968	225	~0°		Above specimen and conditions except exposed to a total of 500 ESH of UV.
213	55	1968	225	~0°		Above specimen and conditions except exposed to a total of 714 ESH of UV.
214	55	1968	225	~0°		Above specimen and conditions except exposed to a total of 945 ESH of UV.
215	55	1968	297	~0°		Above specimen and conditions except measured in air 8 days later.
216	55	1968	297	~0°		Similar to curve 207 specimen and conditions. [Author's designation: Sample 27]
217	55	1968	225	~0°		Above specimen and conditions except measured in vacuum ($<10^{-6}$ mm Hg).
218	55	1968	225	~0°		Above specimen and conditions except exposed in vacuum ($<10^{-6}$ mm Hg) to 5×10^{13} e cm^{-2} at 3.0 keV (flux density 6×10^{12} e cm^{-2} sec^{-1}) and 5×10^{14} p cm^{-2} at 2.5 keV (flux density 8×10^{12} p cm^{-2} sec^{-1}), measured in situ.
219	55	1968	225	~0°		Above specimen and conditions; measured 6 days later.
220	55	1968	225	~0°		Above specimen and conditions; measured 8 days later.
221	55	1968	225	~0°		Above specimen and conditions; measured 10 days later.
222	55	1968	225	~0°		Above specimen and conditions; measured 11 days later.
223	55	1968	225	~0°		Above specimen and conditions except measured in air 8 days later.

SPECIFICATION TABLE NO. 108 NORMAL SOLAR ABSORPTANCE OF TITANIUM DIOXIDE PIGMENTED COATINGS (continued)

Curve No.	Ref. No.	Year	Temperature Range, K	Geometry θ	Reported Error, %	Composition (weight percent), Specifications and Remarks
224	55	1968	297	~0°		Similar to curve 207 specimen and conditions. [Author's designation: Sample 2]
225	55	1968	225	~0°		Above specimen and conditions except measured in vacuum ($<10^{-6}$ mm Hg).
226	55	1968	225	~0°		Above specimen and conditions except exposed in vacuum ($<10^{-6}$ mm Hg) to 61 ESH of UV radiation at solar factor 1 sun; measured in situ.
227	55	1968	225	~0°		Above specimen and conditions except exposed to a total of 200 ESH of UV radiation and additionally to 5×10^{13} e cm^{-2} at 3.0 keV (flux density 6×10^{12} e cm^{-2} sec^{-1}) and 5×10^{14} protons at 2.5 keV (flux density 8×10^{12} p cm^{-2} sec^{-1}).
228	55	1968	225	~0°		Above specimen and conditions except exposed to a total of 335 ESH of UV.
229	55	1968	225	~0°		Above specimen and conditions except exposed to a total of 500 ESH of UV.
230	55	1968	225	~0°		Above specimen and conditions except exposed to a total of 714 ESH of UV.
231	55	1968	225	~0°		Above specimen and conditions except exposed to a total of 945 ESH of UV.
232	55	1968	297	~0°		Above specimen and conditions except measured in air 8 days later.
233	17	1968	298	~0°		CM-146 white paint; TiO_2 pigmented inorganic paint; sample supplied by ESRO II Project Group.
234	68	1969	300	~0°		Titanium dioxide in silicone alkyd binder (~0.152 mm thick) on 2024 aluminum substrate (~1.59 mm thick); absorptance calculated from reflectance; property measured in air. [Authors' designation: PV-100]
235	68	1969	300	~0°		Similar to above specimen and conditions except exposed to vacuum (10^{-8} mm Hg) for 300 hrs; vacuum maintained by ion pump; property measured in air after vacuum exposure.
236	68	1969	300	~0°		Similar to curve 234 specimen and conditions except exposed to vacuum (10^{-5} mm Hg) for 300 hrs; vacuum maintained by oil-diffusion pump; property measured in air after vacuum exposure.
237	68	1969	300	~0°		Similar to above specimen and conditions except exposed to UV radiation in vacuum for 300 hrs; H_2 gas UV source.
238	68	1969	300	~0°		Similar to above specimen and conditions except He gas UV source.
239	68	1969	300	~0°		TiO_2 in potassium silicate binder (~0.1015 mm thick) on 2024 aluminum substrate (~1.59 mm thick); absorptance calculated from reflectance; property measured in air. [Authors' designation: TiO_2/PS-7]
240	68	1969	300	~0°		Similar to above specimen and conditions except exposed to vacuum (10^{-8} mm Hg) for 300 hrs; vacuum maintained by ion pump; property measured in air after vacuum exposure.
241	68	1969	300	~0°		Similar to curve 239 specimen and conditions except exposed to vacuum (10^{-5} mm Hg) for 300 hrs; vacuum maintained by oil-diffusion pump; property measured in air after vacuum exposure.
242	68	1969	300	~0°		Similar to above specimen and conditions except exposed to UV radiation in vacuum for 300 hrs; H_2 gas UV source.
243	68	1969	300	~0°		Similar to above specimen and conditions except He gas UV source.
244	45	1969	298	~0°		TiO_2 (anatase) in methyl silicone binder on aluminum substrate; absorptance calculated from normal spectral reflectance; property measured in air. [Authors' designation: L_1]

SPECIFICATION TABLE NO. 108 NORMAL SOLAR ABSORPTANCE OF TITANIUM DIOXIDE PIGMENTED COATINGS (continued)

Curve No.	Ref. No.	Year	Temperature Range, K	Geometry θ	Reported Error, %	Composition (weight percent), Specifications and Remarks
245	45	1969	281	~0°		Above specimen and conditions except exposed to 20 keV electrons (flux 1×10^{10}-5×10^{11} e cm^{-2} sec^{-1}) in dark in vacuum (10^{-8} mm Hg); vacuum maintained by ion pump; substrate held at 281 ± 2 K; property measured in situ after 1×10^{13} e cm^{-2}.
246	45	1969	281	~0°		Above specimen and conditions except property measured in situ after 5×10^{13} e cm^{-2}.
247	45	1969	281	~0°		Above specimen and conditions except property measured in situ after 1×10^{14} e cm^{-2}.
248	45	1969	281	~0°		Above specimen and conditions except property measured in situ after 3×10^{14} e cm^{-2}.
249	45	1969	281	~0°		Above specimen and conditions except property measured in situ after 1×10^{15} e cm^{-2}.
250	45	1969	281	~0°		Above specimen and conditions except property measured in situ after 1×10^{16} e cm^{-2}.
251	45	1969	298	~0°		Similar to curve 244 specimen and conditions.
252	45	1969	281	~0°		Above specimen and conditions except exposed to 80 keV electrons (flux 1×10^{10}-5×10^{11} e cm^{-2} sec^{-1}) in dark in vacuum (10^{-8} mm Hg); vacuum maintained by ion pump; substrate held at 281 ± 2 K; property measured in situ after 1×10^{13} e cm^{-2}.
253	45	1969	281	~0°		Above specimen and conditions except property measured in situ after 5×10^{13} e cm^{-2}.
254	45	1969	281	~0°		Above specimen and conditions except property measured in situ after 1×10^{14} e cm^{-2}.
255	45	1969	281	~0°		Above specimen and conditions except property measured in situ after 3×10^{14} e cm^{-2}.
256	45	1969	281	~0°		Above specimen and conditions except property measured in situ after 1×10^{15} e cm^{-2}.
257	45	1969	281	~0°		Above specimen and conditions except property measured in situ after 1×10^{16} e cm^{-2}.
258	45	1969	298	~0°		TiO_2 (rutile) in methyl silicone binder on aluminum substrate; absorptance calculated from normal spectral reflectance; property measured in air. [Authors' designation: O]
259	45	1969	281	~0°		Above specimen and conditions except exposed to 20 keV electrons (flux 1×10^{10}-5×10^{11} e cm^{-2} sec^{-1}) in dark in vacuum (10^{-8} mm Hg); vacuum maintained by ion pump; substrate held at 281 ± 2 K; property measured in situ after 10^{15} e cm^{-2}.
260	45	1969	281	~0°		Above specimen and conditions except property measured after 10^{16} e cm^{-2}.
261	45	1969	281	~0°		Similar to curve 259 specimen and conditions except exposed to 80 keV electrons; property measured in situ after 3×10^{14} e cm^{-2}.
262	45	1969	281	~0°		Above specimen and conditions except property measured in situ after 10^{15} e cm^{-2}.
263	45	1969	281	~0°		Above specimen and conditions except property measured in situ after 10^{16} e cm^{-2}.
264	56	1969	298	~0°		Hughes H-2; Cabot RF-1 TiO_2 in PS-7 potassium silicate binder; property calculated from reflectance; lab data taken on sample to be tested on Lunar Orbiter IV.
265*	85	1969	Unknown	~0°		PV-100; rutile TiO_2 in silicone-alkyd binder (0.152 ± 0.00763 mm thick) on zinc chromate primer (0.0152 ± 0.00254 mm thick) and 2024 aluminum substrates; aluminum substrate cleaned and etched with an alcohol-phosphoric acid mixture, then primed; PV-100 spray applied and air dried; data extracted from smooth curve; property calculated by calorimetric method from in-flight temp data of OVI-10; satellite flight time (hrs) is variable; actual UV ESH is from 0.25 to 0.375 of flight time. [Author's designation: A-1 PV-100]

* Not shown on plot

SPECIFICATION TABLE NO. 108 NORMAL SOLAR ABSORPTANCE OF TITANIUM DIOXIDE PIGMENTED COATINGS (continued)

Curve No.	Ref. No.	Year	Temperature Range, K	Geometry θ	Reported Error, %	Composition (weight percent), Specifications and Remarks
266*	85	1969	Unknown	~0°		Similar to above specimen and conditions except exposed to UV in vacuum; property calculated from spectral reflectance measured in situ; actual UV ESH is one third of flight time.
267*	85	1969	Unknown	~0°		Rutile TiO_2 (Dupont R-960) in RTV-602 (General Electric) dimethyl siloxane; PBR 2; baked in air at 423 K for 1 hr; data extracted from smooth curve; property calculated by calorimetric method from in-flight temp data of OVI-10; satellite flight time (hrs) is variable; actual UV ESH is from 0.25 to 0.375 of flight time. [Author's designation: A-5 78-B2]
268*	85	1969	Unknown	~0°		Similar to above specimen and conditions except exposed to UV in vacuum; property calculated from spectral reflectance measured in situ; actual UV ESH is one third of flight time.
269	85	1969	157-393	~0°		PV-100; rutile TiO_2 in silicone-alkyd binder (0.152 ± 0.00763 mm thick) on zinc chromate primer (0.0152 ± 0.00254 mm thick) and 2024 aluminum substrates; aluminum substrate cleaned and etched with an alcohol-phosphoric acid mixture, then primed; PV-100 spray applied and air dried; data extracted from smooth curve; property measured in vacuum by calorimetric method. [Author's designation: A-1 PV-100, NRDL-RTD-81-2]
270	85	1969	157-393	~0°		Similar to above specimen and conditions. [Author's designation: B-4 PV-100, NRDL-RTD-81-2]

* Not shown on plot

DATA TABLE NO. 108 NORMAL SOLAR ABSORPTANCE OF TITANIUM DIOXIDE PIGMENTED COATINGS

[Temperature, T, K; Absorptance, α]

Curve	T	α
CURVE 1	278	0.30
CURVE 2	278	0.34
CURVE 3	278	0.30
CURVE 4	278	0.33
CURVE 5	278	0.33
CURVE 6	278	0.35
CURVE 7	278	0.27
CURVE 8	278	0.30
CURVE 9	278	0.290
CURVE 10	278	0.293
CURVE 11	278	0.25
CURVE 12	278	0.33

Curve	T	α
CURVE 13	278	0.26
CURVE 14	278	0.36
CURVE 15	278	0.26
CURVE 16	278	0.31
CURVE 17	278	0.25
CURVE 18	278	0.37
CURVE 19	278	0.26
CURVE 20	278	0.23
CURVE 21	278	0.39
CURVE 22	278	0.22
CURVE 23	278	0.24
CURVE 24	278	0.37

Curve	T	α
CURVE 25	278	0.26
CURVE 26	278	0.37
CURVE 27	298	0.197
CURVE 28	298	0.347
CURVE 29	298	0.179
CURVE 30	298	0
CURVE 31	298	0.230
CURVE 32	298	0.342
CURVE 33	298	0.319
CURVE 34	298	0.420
CURVE 35	298	0.374
CURVE 36	298	0.447

Curve	T	α
CURVE 37	298	0.437
CURVE 38	298	0.511
CURVE 39	298	0.476
CURVE 40	298	0.543
CURVE 41	298	0.536
CURVE 42	298	0.595
CURVE 43	298	0.604
CURVE 44	298	0.650
CURVE 45	298	0.668
CURVE 46	298	0.693
CURVE 47	298	0.701
CURVE 48	298	0.727

Curve	T	α
CURVE 49	298	0.730
CURVE 50	298	0.764
CURVE 51	298	0.237
CURVE 52	298	0.274
CURVE 53	298	0.305
CURVE 54	298	0.403
CURVE 55	298	0.255
CURVE 56	298	0.299
CURVE 57	298	0.336
CURVE 58	298	0.448
CURVE 59	298	0.248
CURVE 60	298	0.290

Curve	T	α
CURVE 61	298	0.316
CURVE 62	298	0.424
CURVE 63	298	0.310
CURVE 64	298	0.347
CURVE 65	298	0.371
CURVE 66	298	0.445
CURVE 67	298	0.319
CURVE 68	298	0.361
CURVE 69	298	0.392
CURVE 70	298	0.465
CURVE 71	298	0.237

Exp. time	α
CURVE 72 (T ~ 298)	
6	0.314
24	0.333
54	0.364
104	0.438

T	α
CURVE 73	
298	0.248

Exp. time	α
CURVE 74 (T ~ 298)	
6	0.304
24	0.325
54	0.351
104	0.395

T	α
CURVE 75	
298	0.310

Exp. time	α
CURVE 76 (T ~ 298)	
6	0.347
24	0.369
54	0.388
104	0.413

T	α
CURVE 77	
298	0.155

Curve	T	α
CURVE 78	298	0.198
CURVE 79	298	0.149
CURVE 80	298	0.133
CURVE 81	298	0.135
CURVE 82	298	0.180
CURVE 83	298	0.21
CURVE 84	298	0.34
CURVE 85	298	0.34
CURVE 86	298	0.42
CURVE 87	298	0.41
CURVE 88	298	0.52
CURVE 89	298	0.20

DATA TABLE NO. 108 NORMAL SOLAR ABSORPTANCE OF TITANIUM DIOXIDE PIGMENTED COATINGS (continued)

T	α	T	α	T	α	T	α	T	α	T	α	T	α	T	α
CURVE 90		CURVE 102		CURVE 114		CURVE 126		CURVE 138		CURVE 150		CURVE 162		CURVE 174	
298	0.50	298	0.183	298	0.33	298	0.21	298	0.28	298	0.35	298	0.29	298	0.19
CURVE 91		CURVE 103		CURVE 115		CURVE 127		CURVE 139		CURVE 151		CURVE 163		CURVE 175	
298	0.53	298	0.162	298	0.43	298	0.22	298	0.32	298	0.39	298	0.42	298	0.32
CURVE 92		CURVE 104		CURVE 116		CURVE 128		CURVE 140		CURVE 152		CURVE 164		CURVE 176	
298	0.58	298	0.20	298	0.28	298	0.22	298	0.32	298	0.17	298	0.29	298	0.19
CURVE 93		CURVE 105		CURVE 117		CURVE 129		CURVE 141		CURVE 153		CURVE 165		CURVE 177	
298	0.62	298	0.20	298	0.28	298	0.22	298	0.32	298	0.17	298	0.38	298	0.25
CURVE 94		CURVE 106		CURVE 118		CURVE 130		CURVE 142		CURVE 154		CURVE 166		CURVE 178	
298	0.78	298	0.42	298	0.30	298	0.21	298	0.26	298	0.18	298	0.29	298	0.19
CURVE 95		CURVE 107		CURVE 119		CURVE 131		CURVE 143		CURVE 155		CURVE 167		CURVE 179	
298	0.212	298	0.38	298	0.27	298	0.23	298	0.49	298	0.22	298	0.33	298	0.42
CURVE 96		CURVE 108		CURVE 120		CURVE 132		CURVE 144		CURVE 156		CURVE 168		CURVE 180	
298	0.217	298	0.47	298	0.31	298	0.16	298	0.49	298	0.22	298	0.29	298	0.19
CURVE 97		CURVE 109		CURVE 121		CURVE 133		CURVE 145		CURVE 157		CURVE 169		CURVE 181	
298	0.205	298	0.39	298	0.27	298	0.26	298	0.29	298	0.19	298	0.30	298	0.32
CURVE 98		CURVE 110		CURVE 122		CURVE 134		CURVE 146		CURVE 158		CURVE 170		CURVE 182	
298	0.198	298	0.39	298	0.30	298	0.21	298	0.29	298	0.19	298	0.29	298	0.19
CURVE 99		CURVE 111		CURVE 123		CURVE 135		CURVE 147		CURVE 159		CURVE 171		CURVE 183	
298	0.199	298	0.25	298	0.33	298	0.19	298	0.30	298	0.19	298	0.29	298	0.25
CURVE 100		CURVE 112		CURVE 124		CURVE 136		CURVE 148		CURVE 160		CURVE 172		CURVE 184	
298	0.183	298	0.25	298	0.28	298	0.25	298	0.30	298	0.29	298	0.19	298	0.19
CURVE 101		CURVE 113		CURVE 125		CURVE 137		CURVE 149		CURVE 161		CURVE 173		CURVE 185	
298	0.183	298	0.37	298	0.26	298	0.26	298	0.30	298	0.54	298	0.42	298	0.22

DATA TABLE NO. 108 NORMAL SOLAR ABSORPTANCE OF TITANIUM DIOXIDE PIGMENTED COATINGS (continued)

T	α
CURVE 186	
298	0.19
CURVE 187	
298	0.21
CURVE 188	
193	0.225
263	0.225
CURVE 189	
193	0.189
263	0.189
CURVE 190	
297	0.24
CURVE 191	
225	0.24
CURVE 192	
225	0.26
CURVE 193	
225	0.28
CURVE 194	
225	0.25
CURVE 195	
225	0.26
CURVE 196	
225	0.26
CURVE 197	
297	0.23

T	α
CURVE 198	
297	0.24
CURVE 199	
225	0.24
CURVE 200	
225	0.28
CURVE 201	
225	0.31
CURVE 202	
225	0.32
CURVE 203	
225	0.32
CURVE 204	
225	0.32
CURVE 205	
225	0.32
CURVE 206	
297	0.27
CURVE 207	
297	0.22
CURVE 208	
225	0.23
CURVE 209	
225	0.26

T	α
CURVE 210	
225	0.26
CURVE 211	
225	0.24
CURVE 212	
225	0.27
CURVE 213	
225	0.20
CURVE 214	
225	0.26
CURVE 215	
297	0.25
CURVE 216	
297	0.21
CURVE 217	
225	0.23
CURVE 218	
225	0.25
CURVE 219	
225	0.23
CURVE 220	
225	0.22
CURVE 221	
225	0.27

T	α
CURVE 222	
225	0.23
CURVE 223	
225	0.23
CURVE 224	
297	0.24
CURVE 225	
225	0.23
CURVE 226	
225	0.25
CURVE 227	
225	0.25
CURVE 228	
225	0.26
CURVE 229	
225	0.25
CURVE 230	
225	0.25
CURVE 231	
225	0.25
CURVE 232	
297	0.24
CURVE 233	
298	0.20

T	α
CURVE 234	
300	0.21
CURVE 235	
300	0.23
CURVE 236	
300	0.25
CURVE 237	
300	0.33
CURVE 238	
300	0.39
CURVE 239	
300	0.26
CURVE 240	
300	0.30
CURVE 241	
300	0.30
CURVE 242	
300	0.36
CURVE 243	
300	0.42
CURVE 244	
298	0.20
CURVE 245	
281	0.20

T	α
CURVE 246	
281	0.21
CURVE 247	
281	0.21
CURVE 248	
281	0.22
CURVE 249	
281	0.24
CURVE 250	
281	0.26
CURVE 251	
298	0.19
CURVE 252	
281	0.22
CURVE 253	
281	0.25
CURVE 254	
281	0.27
CURVE 255	
281	0.31
CURVE 256	
281	0.33
CURVE 257	
281	0.42

T	α
CURVE 258	
298	0.19
CURVE 259	
281	0.21
CURVE 260	
281	0.34
CURVE 261	
281	0.22
CURVE 262	
281	0.27
CURVE 263	
281	0.54
CURVE 264	
298	0.178
Flight time	α
CURVE 265* T unknown	
27.4	0.327
391	0.424
5220	0.539
CURVE 266* T unknown	
10.0	0.233
46.6	0.252
79.2	0.264
139	0.277
230	0.299
360	0.326
470	0.346
600	0.373

Flight time	α
CURVE 267* T unknown	
47.2	0.340
466	0.395
4400	0.461
CURVE 268* T unknown	
10.0	0.209
50.1	0.225
77.8	0.231
130	0.242
207	0.256
358	0.282
485	0.298
589	0.316
T	α
CURVE 269	
157	0.241
219	0.239
286	0.227
338	0.213
393	0.194
CURVE 270	
157	0.241
219	0.239
286	0.227
338	0.213
393	0.194

* Not shown on plot

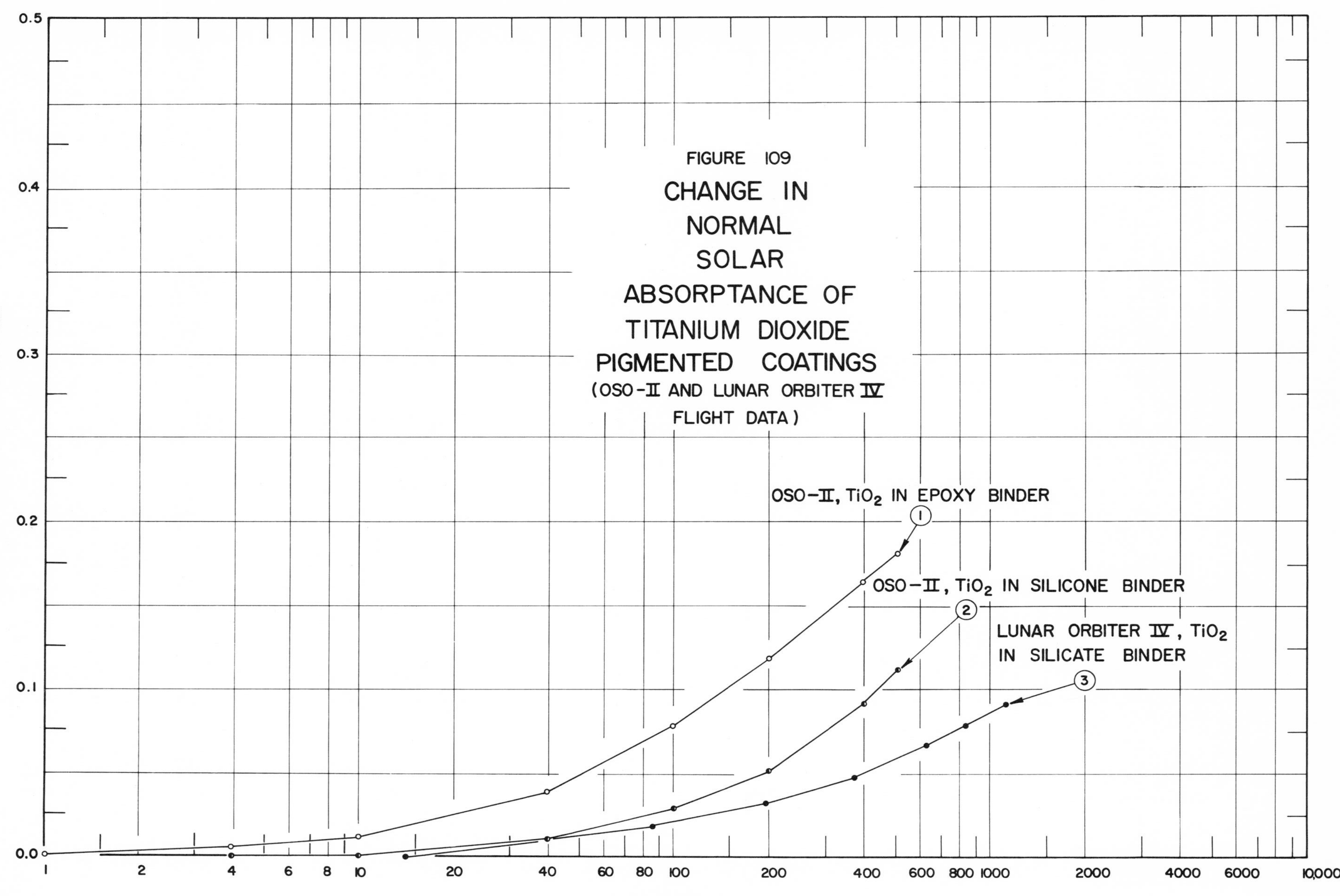

FIGURE 109 CHANGE IN NORMAL SOLAR ABSORPTANCE OF TITANIUM DIOXIDE PIGMENTED COATINGS (OSO-II AND LUNAR ORBITER IV FLIGHT DATA)

SPECIFICATION TABLE NO. 109 CHANGE IN NORMAL SOLAR ABSORPTANCE OF TITANIUM DIOXIDE PIGMENTED COATINGS

Curve No.	Ref. No.	Year	Temperature Range, K	Geometry θ	Reported Error, %	Composition (weight percent), Specifications and Remarks
1	52	1966	~298	~0°		TiO_2 in epoxy binder; data taken from OSO-II flight experiment; variable is equivalent length of time exposed to the sun.
2	52	1966	~298	~0°		TiO_2 in silicone binder (adhesive-backed elastomeric film); data taken from OSO-II flight experiment; variable is equivalent length of time exposed to the sun.
3	56	1969	269-297	~0°		Hughes H-2; Cabot RF-1 titanium dioxide in PS-7 potassium silicate binder; property calculated from temp of sample; in-flight data of Lunar Orbiter IV; ESH is variable; data extracted from smooth curve.

DATA TABLE NO. 109 CHANGE IN NORMAL SOLAR ABSORPTANCE OF TITANIUM DIOXIDE PIGMENTED COATINGS

[Temperature, T, K; Absorptance, α]

ESH	$\Delta\alpha$	ESH	$\Delta\alpha$
CURVE 1 T ~ 298		CURVE 2 (cont.)	
		200	0.053
1.00	0.000	398	0.094
3.98	0.005	504	0.113
10.0	0.012		
39.8	0.037	CURVE 3 T = 269-297	
100	0.077		
200	0.119		
398	0.165	14	0.000
509	0.182	86	0.018
		199	0.032
CURVE 2 T ~ 298		378	0.048
		630	0.066
		840	0.079
1.00	0.000	1122	0.092
3.98	0.000		
10.0	0.000		
39.8	0.010		
100	0.028		

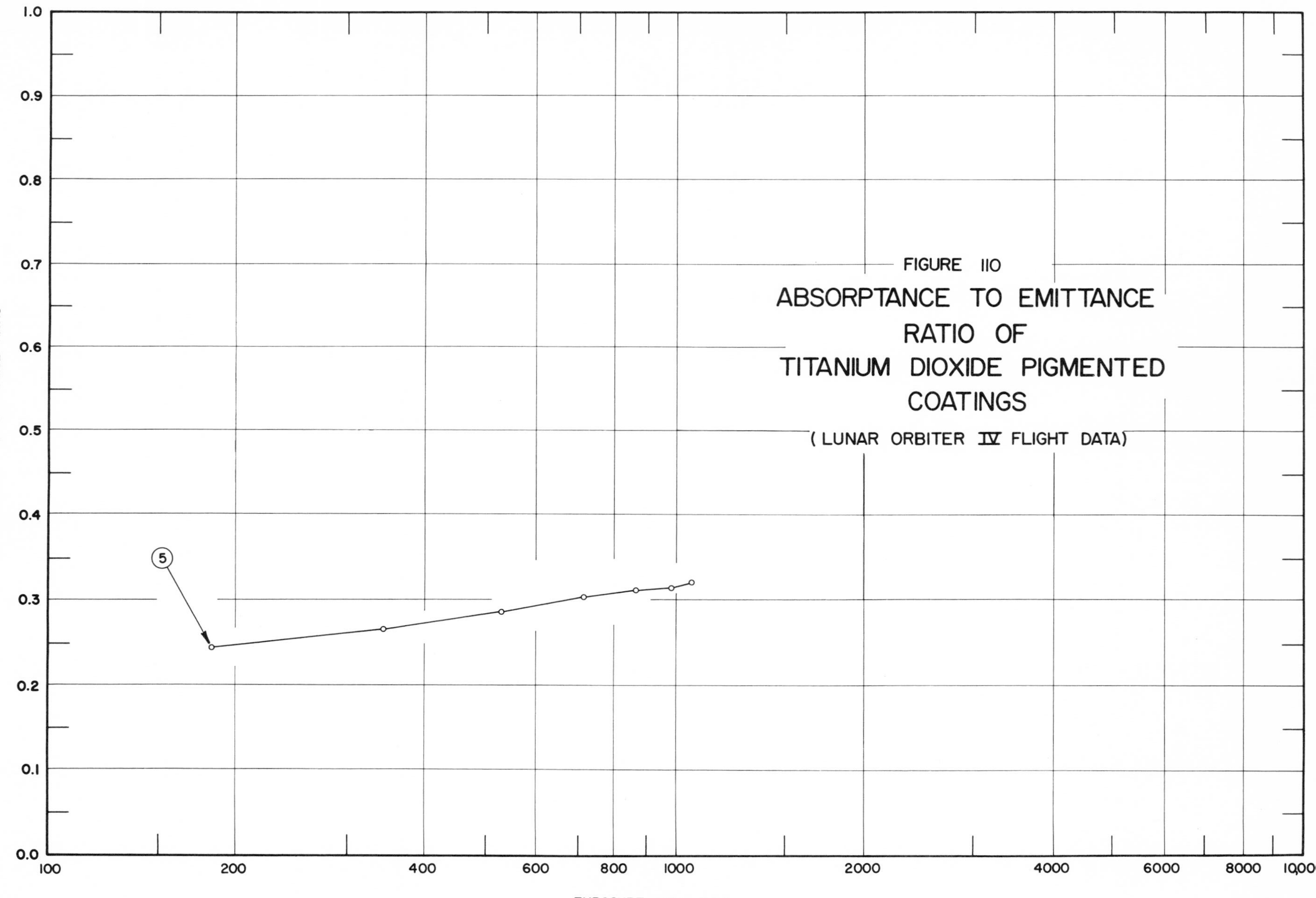
FIGURE 110
ABSORPTANCE TO EMITTANCE
RATIO OF
TITANIUM DIOXIDE PIGMENTED
COATINGS
(LUNAR ORBITER IV FLIGHT DATA)
ABSORPTANCE TO EMITTANCE RATIO
1.0
0.9
0.8
0.7
0.6
0.5
0.4
0.3
0.2
0.1
0.0
5
100
200
400
600
800
1000
2000
4000
6000
8000
10,000
EXPOSURE TIME, ESH

SPECIFICATION TABLE NO. 110 ABSORPTANCE TO EMITTANCE RATIO OF TITANIUM DIOXIDE PIGMENTED COATINGS

Curve No.	Ref. No.	Year	Temperature Range, K	Reported Error, %	Composition (weight percent), Specifications and Remarks
1*	58	1967	~298	11.2	TiO_2 and Al_2O_3 in K_2O-3.3 SiO_2 binder on 6061 aluminum alloy substrate; substrate has phosphate Al surface; property calculated from reflectance; laboratory data measured 6 mo before flight of ATS-1.
2*	58	1967	unknown	7.0	Above specimen and conditions except property calculated from temp of substrate from "initial" point flight data (48 hrs after launch) of ATS-1.
3*	58	1967	~298	9.4	Dow Corning Q92-090; TiO_2 (anatase AMO) in methyl silicone binder on aluminum (6061 Al alloy) substrate; aluminum substrate treated with Cat-a-lac white prime coat; property calculated from reflectance; lab data measured 6 mo before flight of ATS-1.
4*	58	1967	unknown	5.2	Above specimen and conditions except property calculated from temp of substrate from "initial" point flight data (48 hrs after launch) of ATS-1.
5	56	1969	269-297		Hughes H-2; Cabot RF-1 TiO_2 in PS-7 potassium silicate binder; property calculated from temp of sample from in-flight data of Lunar Orbiter IV; ESH is variable; data extracted from smooth curve.

DATA TABLE NO. 110 ABSORPTANCE TO EMITTANCE RATIO OF TITANIUM DIOXIDE PIGMENTED COATINGS

[Temperature, T, K; Absorptance, α; Emittance, ϵ]

T	α/ϵ
CURVE 1*	
298	0.20
CURVE 2*	
unknown	0.23
CURVE 3*	
298	0.22
CURVE 4*	
unknown	0.32

ESH	α/ϵ
CURVE 5 T = 269-297	
0	0.215*
184	0.245
342	0.268
530	0.286
719	0.302
863	0.311
986	0.316
1106	0.321

* Not shown on plot

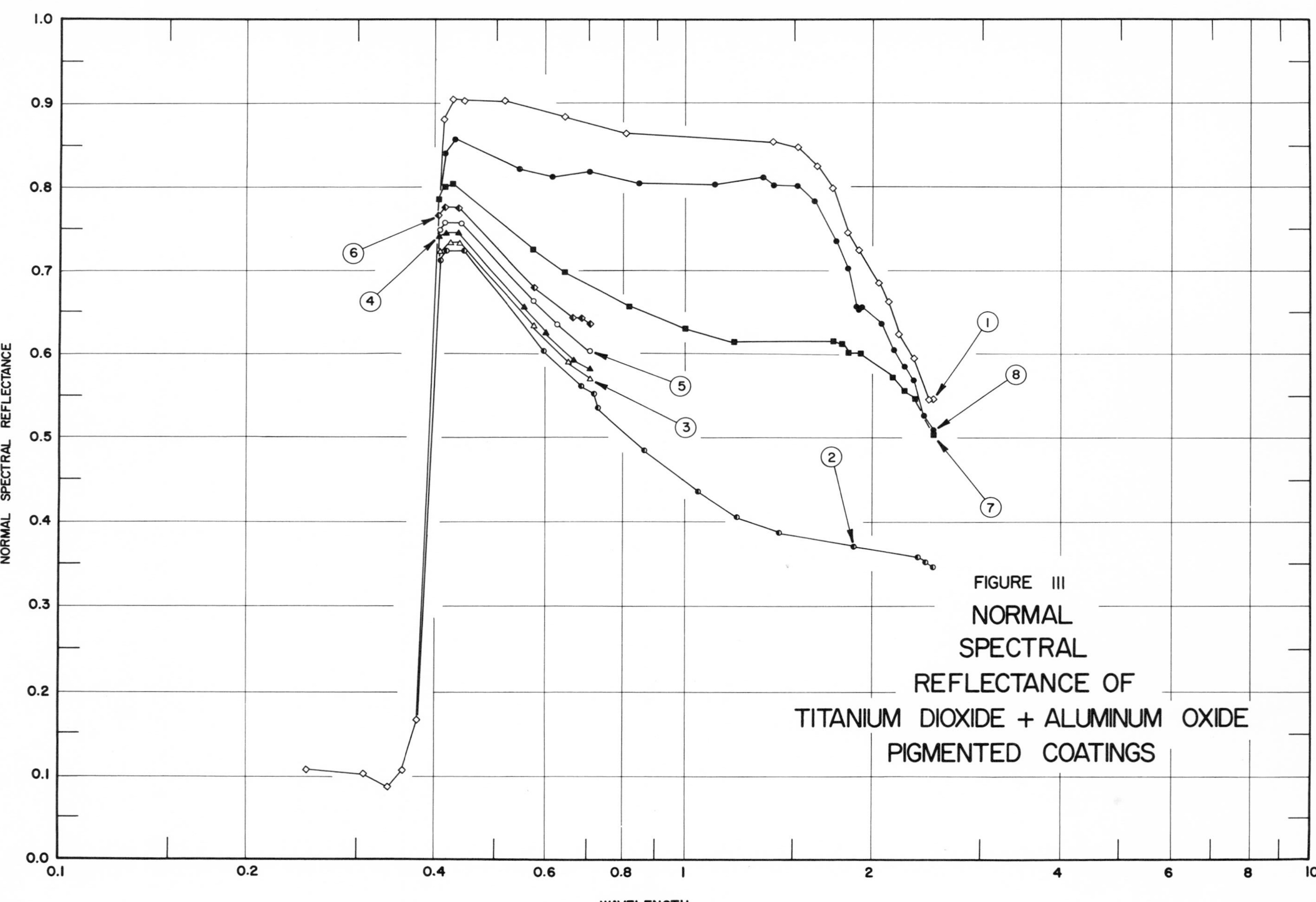
FIGURE III
NORMAL
SPECTRAL
REFLECTANCE OF
TITANIUM DIOXIDE + ALUMINUM OXIDE
PIGMENTED COATINGS
NORMAL SPECTRAL REFLECTANCE
WAVELENGTH, μm
0.0
0.1
0.2
0.3
0.4
0.5
0.6
0.7
0.8
0.9
1.0
0.1
0.2
0.4
0.6
0.8
1
2
4
6
8
10
1
2
3
4
5
6
7
8

SPECIFICATION TABLE NO. 111 NORMAL SPECTRAL REFLECTANCE OF TITANIUM DIOXIDE + ALUMINUM OXIDE PIGMENTED COATINGS

Curve No.	Ref. No.	Year	Temperature, K	Wavelength Range, μm	Geometry θ	Geometry θ'	Geometry ω'	Reported Error, %	Composition (weight percent), Specifications, and Remarks
1	45	1970	298	0.250-2.500	~0°		2π		TiO_2 and Al_2O_3 in potassium silicate binder on aluminum substrate; reflectance measured in air; data extracted from smooth curve. [Authors' designation: E_3]
2	45	1970	281	0.250-2.500	~0°		2π		Above specimen and conditions except exposed to 80 keV electrons in dark in vacuum (10^{-8} mm Hg); flux 1 x 10^{10} - 5 x 10^{11} e cm^{-2} sec^{-1}; aluminum substrate held at 281 ± 2 K; reflectance measured in situ after 10^{16} e cm^{-2}.
3	45	1970	281	0.250-0.706	~0°		2π		Above specimen and conditions except reflectance measured 2 days after end of exposure to 10^{16} e cm^{-2}.
4	45	1970	281	0.250-0.706	~0°		2π		Above specimen and conditions except reflectance measured 3 days after end of exposure to 10^{16} e cm^{-2}.
5	45	1970	281	0.250-0.706	~0°		2π		Above specimen and conditions except reflectance measured 4 days after end of exposure to 10^{16} e cm^{-2}.
6	45	1970	281	0.250-0.706	~0°		2π		Above specimen and conditions except reflectance measured 7 days after end of exposure to 10^{16} e cm^{-2}.
7	45	1970	281	0.250-2.500	~0°		2π		Above specimen and conditions except reflectance measured 18 days after end of exposure to 10^{16} e cm^{-2}.
8	45	1970	298	0.250-2.500	~0°		2π		Above specimen and conditions except reflectance measured in air after end of exposure to 10^{16} e cm^{-2} test.

DATA TABLE NO. 111 NORMAL SPECTRAL REFLECTANCE OF TITANIUM DIOXIDE + ALUMINUM OXIDE PIGMENTED COATINGS

[Wavelength, λ, μm; Reflectance, ρ; Temperature, T, K]

λ	ρ
CURVE 1 T = 298	
0.250	0.107
0.309	0.101
0.336	0.089
0.354	0.107
0.373	0.167
0.411	0.883
0.426	0.908
0.441	0.905
0.516	0.905
0.646	0.888
0.804	0.868
1.399	0.858
1.515	0.850
1.627	0.827
1.722	0.800
1.832	0.747
1.909	0.727
2.038	0.687
2.122	0.661
2.204	0.625
2.325	0.596
2.482	0.549
2.500	0.549
CURVE 2 T = 281	
0.250	0.107*
0.309	0.101*
0.336	0.089*
0.354	0.107*
0.373	0.167*
0.407	0.712
0.418	0.723
0.444	0.723
0.592	0.605
0.681	0.563
0.712	0.553
0.727	0.536
0.860	0.484
1.051	0.438
1.215	0.407
1.410	0.386
1.868	0.370
CURVE 2 (cont.)	
2.362	0.359
2.432	0.352
2.500	0.349
CURVE 3 T = 281	
0.250	0.107*
0.309	0.101*
0.336	0.089*
0.354	0.107*
0.373	0.167*
0.407	0.726
0.423	0.737
0.438	0.737
0.572	0.635
0.650	0.592
0.706	0.572
CURVE 4 T = 281	
0.250	0.107*
0.309	0.101*
0.336	0.089*
0.354	0.107*
0.373	0.167*
0.407	0.741
0.418	0.747
0.432	0.747
0.559	0.657
0.600	0.626
0.667	0.594
0.706	0.583
CURVE 5 T = 281	
0.250	0.107*
0.309	0.101*
0.336	0.089*
0.354	0.107*
0.373	0.167*
0.407	0.750
0.414	0.759
CURVE 5 (cont.)	
0.440	0.759
0.571	0.663
0.626	0.636
0.706	0.603
CURVE 6 T = 281	
0.250	0.107*
0.309	0.101*
0.336	0.089*
0.354	0.107*
0.373	0.167*
0.407	0.769
0.418	0.777
0.439	0.777
0.579	0.680
0.669	0.645
0.687	0.645
0.706	0.638
CURVE 7 T = 281	
0.250	0.107*
0.309	0.101*
0.336	0.089*
0.354	0.107*
0.373	0.167*
0.407	0.788
0.414	0.801
0.426	0.806
0.571	0.727
0.641	0.700
0.812	0.657
1.007	0.632
1.200	0.617
1.731	0.619
1.793	0.613
1.827	0.601
1.919	0.601
2.143	0.574
2.221	0.557
2.323	0.548
2.500	0.502
CURVE 8 T = 298	
0.250	0.107*
0.309	0.101*
0.336	0.089*
0.354	0.107*
0.373	0.167*
0.418	0.842
0.430	0.860
0.545	0.823
0.615	0.815
0.703	0.820
0.845	0.806
1.122	0.806
1.346	0.813
1.399	0.802
1.510	0.802
1.620	0.785
1.754	0.737
1.834	0.703
1.881	0.657
1.900	0.654
1.928	0.658
2.076	0.638
2.165	0.609
2.226	0.587
2.318	0.570
2.418	0.527
2.500	0.510

* Not shown on plot

SPECIFICATION TABLE NO. 112 CHANGE IN NORMAL SPECTRAL REFLECTANCE OF TITANIUM DIOXIDE + ALUMINUM OXIDE PIGMENTED COATINGS

Curve No.	Ref. No.	Year	Temperature, K	Wavelength Range, μm	Geometry θ	Geometry θ'	Geometry ω'	Reported Error, %	Composition (weight percent), Specifications, and Remarks
1*	46	1969	295	0.590	~0°		2π		Rutile TiO_2 (Cabot, Inc.) and Al_2O_3 in potassium silicate binder (0.1015 mm thick); substrate unknown; exposed to 50 keV electrons (4×10^8-1.7×10^{12} e cm^{-2} sec^{-1}) in vacuum (10^{-8} mm Hg); vacuum maintained by ion pump; electron fluence (e cm^{-2}) is variable; reflectance measured in situ; positive change indicates decrease in reflectance from preirradiation, in air reflectance. [Authors' designation: E]
2*	46	1969	295	2.100	~0°		2π		Above specimen and conditions.

DATA TABLE NO. 112 CHANGE IN NORMAL SPECTRAL REFLECTANCE OF TITANIUM DIOXIDE + ALUMINUM OXIDE PIGMENTED COATINGS

[Wavelength, λ, μm; Reflectance, ρ; Temperature, T, K]

Electron fluence	$\Delta\rho$
CURVE 1*	
$\lambda = 0.590$	
T = 295	
1×10^{13}	0.10
2×10^{14}	0.14
6×10^{14}	0.22
CURVE 2*	
$\lambda = 2.100$	
T = 295	
1×10^{13}	0.0
2×10^{14}	0.0
6×10^{14}	0.0
8×10^{14}	0.0

*No plot given

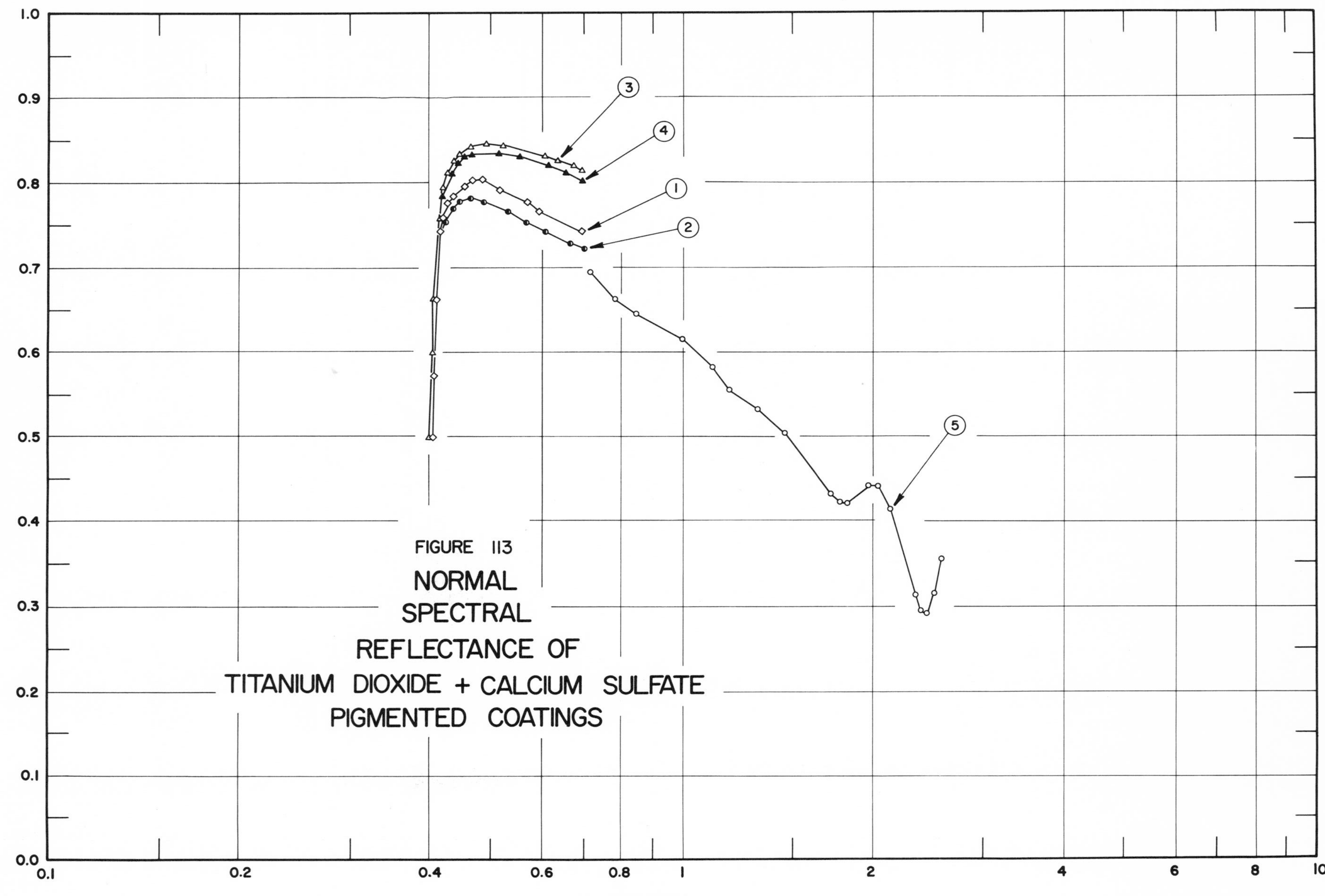

FIGURE 113
NORMAL
SPECTRAL
REFLECTANCE OF
TITANIUM DIOXIDE + CALCIUM SULFATE
PIGMENTED COATINGS

SPECIFICATION TABLE NO. 113 NORMAL SPECTRAL REFLECTANCE OF TITANIUM DIOXIDE + CALCIUM SULFATE PIGMENTED COATINGS

Curve No.	Ref. No.	Year	Temperature, K	Wavelength Range, μm	Geometry θ θ'	ω'	Reported Error, %	Composition (weight percent), Specifications, and Remarks
1	3	1960	~298	0.405-0.699	~0°	2π		Titanox RC, 70 rutile TiO_2, 30 $CaSO_4$, in DuPont RC7007 alkyd melamine resin binder (0.0762 mm thick, 1.22 x 10^{-2} g cm^{-2}) on mild steel substrate; pigment 40% of solids content; sprayed; data extracted from smooth curve; measured relative to MgO.
2	3	1960	~298	0.405-0.700	~0°	2π		Similar to above specimen and conditions except coating 0.066 mm thick, 1.05 x 10^{-2} g cm^{-2}.
3	3	1960	~298	0.400-0.695	~0°	2π		Titanox C-50, 50 TiO_2, 50 $CaSO_4$, in DuPont RC7007 alkyd-melamine resin binder (0.0762 mm thick, 1.25 x 10^{-2} g cm^{-2}) on mild steel substrate; pigment 40% of solids content; sprayed; data extracted from smooth curve; measured relative to MgO.
4	3	1960	~298	0.400-0.697	~0°	2π		Similar to above specimen and conditions except coating 0.0711 mm thick, 1.15 x 10^{-2} g cm^{-2}.
5	3	1960	~298	0.711-2.587	~0°	2π		Titanox C-50, 50 TiO_2 and 50 $CaSO_4$, in DuPont RC7007 alkyd-melamine resin binder on mild steel substrate; coating weight 1.07 x 10^{-2} g cm^{-2}; pigment concentration 40% of solids content; sprayed; data extracted from smooth curve; measured relative to MgO.

DATA TABLE NO. 113 NORMAL SPECTRAL REFLECTANCE OF TITANIUM DIOXIDE + CALCIUM SULFATE PIGMENTED COATINGS

[Wavelength, λ, μm; Reflectance, ρ; Temperature, T, K]

λ	ρ	λ	ρ	λ	ρ	λ	ρ	λ	ρ	λ	ρ
CURVE 1 T ~ 298		CURVE 2 T ~ 298		CURVE 3 T ~ 298		CURVE 4 T ~ 298		CURVE 5 T ~ 298		CURVE 5 (cont.)	
0.405	0.500	0.405	0.500*	0.400	0.500	0.400	0.500*	0.711	0.696	2.438	0.292
0.406	0.574	0.406	0.574*	0.403	0.600	0.403	0.600*	0.781	0.663	2.501	0.318
0.410	0.663	0.410	0.663*	0.407	0.664	0.407	0.664*	0.844	0.646	2.587	0.356
0.417	0.743	0.418	0.741*	0.415	0.759	0.415	0.759*	0.993	0.619		
0.420	0.760	0.422	0.755	0.420	0.796	0.420	0.787	1.111	0.583		
0.426	0.777	0.432	0.770	0.426	0.813	0.431	0.812	1.188	0.557		
0.435	0.786	0.445	0.779	0.436	0.829	0.443	0.826	1.304	0.533		
0.454	0.798	0.461	0.782	0.448	0.838	0.453	0.832	1.458	0.504		
0.468	0.803	0.485	0.779	0.461	0.844	0.466	0.837	1.727	0.432		
0.484	0.804	0.530	0.766	0.490	0.848	0.511	0.837	1.771	0.424		
0.515	0.794	0.567	0.754	0.520	0.847	0.558	0.832	1.819	0.422		
0.570	0.777	0.609	0.743	0.608	0.834	0.616	0.822	1.967	0.441		
0.593	0.769	0.663	0.730	0.637	0.829	0.658	0.813	2.030	0.441		
0.699	0.743	0.700	0.723	0.672	0.821	0.697	0.803	2.117	0.417		
				0.695	0.816			2.322	0.314		
								2.374	0.296		

* Not shown on plot

SPECIFICATION TABLE NO. 114 HEMISPHERICAL TOTAL EMITTANCE OF TITANIUM DIOIXDE + TALC PIGMENTED COATINGS

Curve No.	Ref. No.	Year	Temperature Range, K	Reported Error, %	Composition (weight percent), Specifications and Remarks
1*	50	1965	~298		White Kemacryl lacquer, Sherwin-Williams M49WC17; 50 TiO_2 and 50 talc in acrylic binder; substrate unknown.
2*	49	1961	278	10	Kemacryl lacquer white M49WC17; 50 TiO_2 and 50 talc in acrylic resin binder; substrate unknown; value avg of several determinations.
3*	49	1961	278		White Kemacryl lacquer (M49WC17, Sherwin-Williams Co.); 50 TiO_2 and 50 talc in acrylic resin binder on Dow 17 treated Mg alloy substrate.
4*	49	1961	278		Above specimen and conditions except exposed in vacuum (10^{-6}-8 x 10^{-6} mm Hg) to UV radiation from an argon-filled A-H6 high pressure Hg arc lamp for 26 hrs.
5*	49	1961	278		Similar to curve 3 specimen and conditions.
6*	49	1961	278		Above specimen and conditions except exposed in vacuum (10^{-6}-8 x 10^{-6} mm Hg) to UV radiation from an argon-filled A-H6 high pressure Hg arc lamp for 26 hrs.
7*	49	1961	278		Similar to curve 3 specimen and conditions.
8*	49	1961	278		Above specimen and conditions except exposed in vacuum (10^{-6}-8 x 10^{-6} mm Hg) to UV radiation from an argon-filled A-H6 high pressure Hg arc lamp for 80 hrs.
9*	49	1961	278		Similar to curve 3 specimen and conditions.
10*	49	1961	278		Above specimen and conditions except exposed in vacuum (10^{-6}-8 x 10^{-6} mm Hg) to UV radiation from an argon-filled A-H6 high pressure Hg arc lamp for 100 hrs.

* No plot given

DATA TABLE NO. 114 HEMISPHERICAL TOTAL EMITTANCE OF TITANIUM DIOXIDE + TALC PIGMENTED COATINGS

[Temperature, T, K; Emittance, ∈]

T	∈
CURVE 1*	
298	0.86
CURVE 2*	
278	0.75
CURVE 3*	
278	0.73
CURVE 4*	
278	0.75
CURVE 5*	
278	0.73
CURVE 6*	
278	0.76
CURVE 7*	
278	0.73
CURVE 8*	
278	0.78
CURVE 9*	
278	0.74
CURVE 10*	
278	0.76

* No plot given

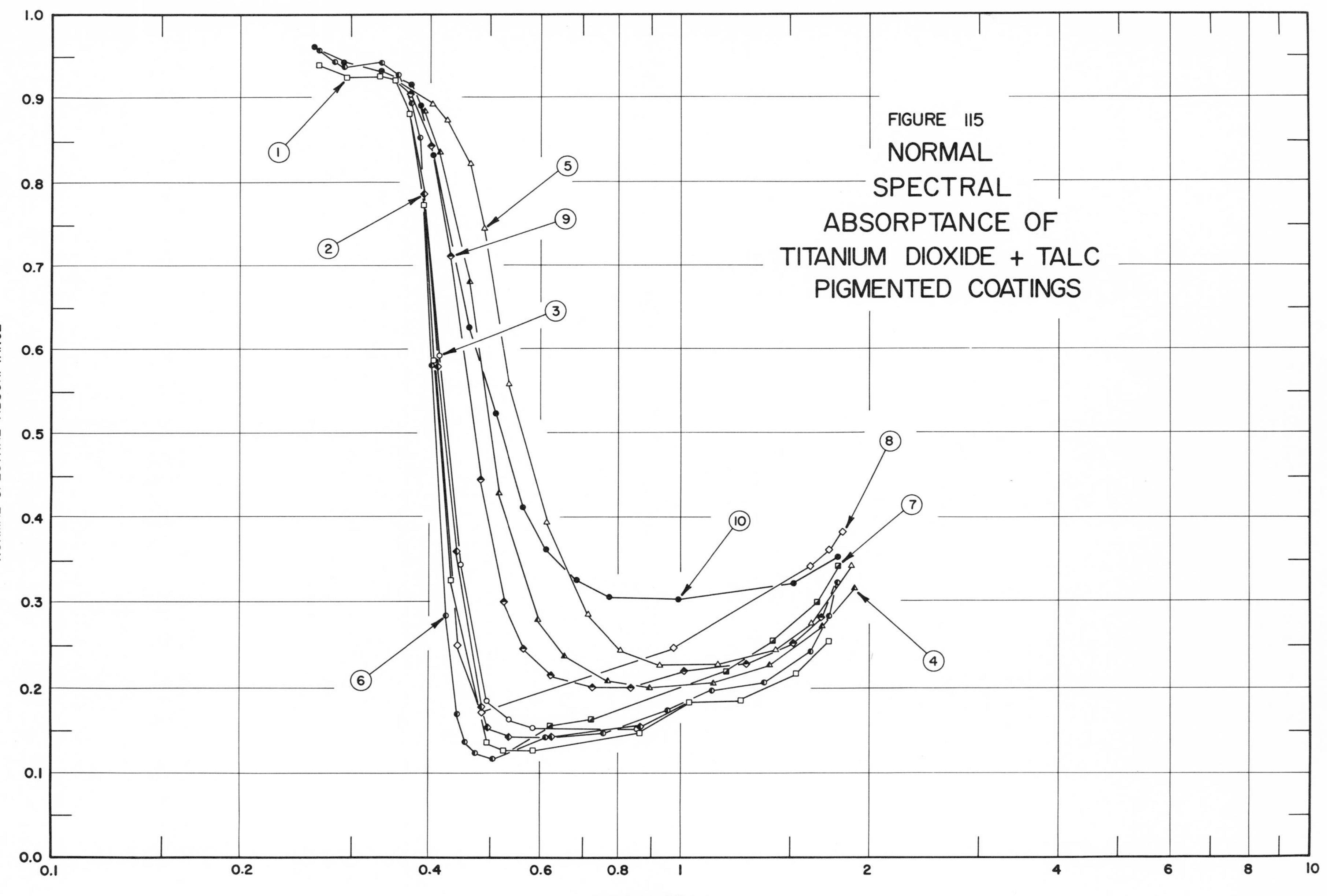
FIGURE 115
NORMAL
SPECTRAL
ABSORPTANCE OF
TITANIUM DIOXIDE + TALC
PIGMENTED COATINGS
NORMAL SPECTRAL ABSORPTANCE
WAVELENGTH, μm
0.0
0.1
0.2
0.3
0.4
0.5
0.6
0.7
0.8
0.9
1.0
0.1
0.2
0.4
0.6
0.8
1
2
4
6
8
10
1
2
3
4
5
6
7
8
9
10

SPECIFICATION TABLE NO. 115 NORMAL SPECTRAL ABSORPTANCE OF TITANIUM DIOXIDE + TALC PIGMENTED COATINGS

Curve No.	Ref. No.	Year	Temperature, K	Wavelength Range, μm	Geometry θ	Reported Error, %	Composition (weight percent), Specifications, and Remarks
1	117	1965	~298	0.269-1.746	~0°		White Kemacryl lacquer, Sherwin-Williams M49WC17; 50 TiO_2 + 50 Talc in acrylic binder; substrate unknown; data extracted from smooth curve.
2	117	1965	~298	0.269-1.746	~0°		Similar to above specimen and conditions except irradiated in vacuum with 130 keV protons to a total dose of 0.5 x 10^{15} p cm^{-2}.
3	117	1965	~298	0.269-1.746	~0°		Similar to above specimen and conditions except irradiated to a total dose of 1.1 x 10^{15} p cm^{-2}.
4	117	1965	~298	0.269-1.908	~0°		Similar to above specimen and conditions except irradiated to a total dose of 2.0 x 10^{15} p cm^{-2}.
5	117	1965	~298	0.269-1.879	~0°		Similar to above specimen and conditions except irradiated to a total dose of 3.9 x 10^{15} p cm^{-2}.
6	50	1965	~298	0.268-1.771	~0°		White Kemacryl lacquer, Sherwin-Williams M49WC17; 50 TiO_2 and 50 talc in acrylic binder; data extracted from smooth curve.
7	50	1965	~298	0.268-1.797	~0°		Similar to above specimen and conditions except irradiated in vacuum at 291 K with 1 MeV electrons to a total dose of 10^{15} e cm^{-2}.
8	50	1965	~298	0.268-1.82	~0°		Similar to above specimen and conditions except irradiated to a total dose of 10^{16} e cm^{-2}.
9	50	1965	~298	0.268-1.771	~0°		Similar to curve 6 specimen and conditions except irradiated with UV radiation for 485 sun hrs.
10	50	1965	~298	0.263-1.797	~0°		Similar to curve 6 specimen and conditions except irradiated in vacuum at 291 K with 1 MeV electrons to a total dose of 10^{16} e cm^{-2} and additionally with UV radiation for 485 sun hrs.

DATA TABLE NO. 115 NORMAL SPECTRAL ABSORPTANCE OF TITANIUM DIOXIDE + TALC PIGMENTED COATINGS

[Wavelength, λ, μm; Absorptance, α, Temperature, T, K]

λ	α
CURVE 1	
T ~ 298	
0.269	0.940
0.297	0.926
0.335	0.928
0.352	0.923
0.372	0.881
0.391	0.773
0.407	0.588
0.431	0.327
0.496	0.137
0.525	0.127
0.582	0.127
0.861	0.149
1.033	0.182
1.253	0.186
1.531	0.218
1.746	0.255
CURVE 2	
T ~ 298	
0.269	0.940*
0.297	0.926*
0.335	0.928*
0.352	0.923*
0.372	0.882*
0.392	0.789
0.412	0.580
0.441	0.360
0.488	0.179
0.498	0.152
0.539	0.142
0.629	0.142
0.867	0.153
1.033	0.182*
1.253	0.186*
1.531	0.218*
1.746	0.255*
CURVE 3	
T ~ 298	
0.269	0.940*
0.297	0.926*
0.335	0.928*
0.352	0.923*
CURVE 3 (cont.)	
0.372	0.883*
0.397	0.789*
0.415	0.593
0.450	0.345
0.496	0.184
0.537	0.162
0.585	0.153
0.855	0.152
1.033	0.182*
1.253	0.186*
1.531	0.218*
1.746	0.255*
CURVE 4	
T ~ 298	
0.269	0.940*
0.297	0.926*
0.335	0.928*
0.354	0.926*
0.392	0.889
0.418	0.839
0.461	0.680
0.519	0.430
0.599	0.280
0.656	0.239
0.770	0.209
0.899	0.200
1.138	0.206
1.393	0.229
1.699	0.271
1.908	0.318
CURVE 5	
T ~ 298	
0.269	0.940*
0.297	0.926*
0.335	0.928*
0.354	0.926*
0.407	0.896
0.428	0.876
0.463	0.823
0.490	0.748
0.537	0.560
0.611	0.395
CURVE 5 (cont.)	
0.713	0.286
0.805	0.244
0.925	0.229
1.159	0.229
1.425	0.245
1.610	0.275
1.879	0.342
CURVE 6	
T ~ 298	
0.268	0.958
0.281	0.944
0.294	0.939
0.337	0.941
0.357	0.927
0.375	0.895
0.387	0.853
0.403	0.583
0.426	0.285
0.444	0.170
0.456	0.137
0.473	0.122
0.504	0.118
0.610	0.141
0.756	0.148
0.953	0.176
1.127	0.197
1.363	0.206
1.610	0.241
1.746	0.283
1.771	0.325
CURVE 7	
T ~ 298	
0.268	0.958*
0.281	0.944*
0.294	0.939*
0.337	0.941*
0.357	0.927*
0.375	0.895*
0.387	0.853*
0.403	0.583*
0.426	0.285*
0.444	0.170*
0.456	0.137*
CURVE 7 (cont.)	
0.473	0.122*
0.504	0.118*
0.620	0.156
0.721	0.164
1.181	0.220
1.409	0.255
1.653	0.300
1.797	0.343
CURVE 8	
T ~ 298	
0.268	0.958*
0.281	0.944*
0.294	0.939*
0.337	0.941*
0.357	0.927*
0.375	0.895*
0.387	0.853*
0.407	0.583*
0.445	0.250
0.466	0.199*
0.488	0.171
0.976	0.249
1.61	0.343
1.74	0.362
1.82	0.383
CURVE 9	
T ~ 298	
0.268	0.958*
0.281	0.944*
0.299	0.936*
0.351	0.929*
0.374	0.907
0.401	0.846
0.433	0.711
0.482	0.448
0.527	0.300
0.566	0.248
0.626	0.214
0.729	0.200
0.837	0.200
1.016	0.220
1.278	0.229
CURVE 9 (cont.)	
1.512	0.251
1.676	0.282
1.771	0.325*
CURVE 10	
T ~ 298	
0.263	0.960
0.294	0.942
0.339	0.933
0.374	0.918
0.389	0.892
0.407	0.832
0.461	0.628
0.510	0.524
0.561	0.412
0.611	0.361
0.689	0.326
0.779	0.306
0.991	0.302
1.512	0.321
1.797	0.354

* Not shown on plot

SPECIFICATION TABLE NO. 116 NORMAL SOLAR ABSORPTANCE OF TITANIUM DIOXIDE + TALC PIGMENTED COATINGS

Curve No.	Ref. No.	Year	Temperature Range, K	Geometry θ	Reported Error, %	Composition (weight percent), Specifications and Remarks
1*	49	1961	278	0°	10	White Kemacryl lacquer M49WC17; 50 TiO_2 and 50 talc in acrylic resin binder; substrate unknown; value avg of several determinations.
2*	49	1961	278	0°		White Kemacryl lacquer M49WC17; 50 TiO_2 and 50 talc in acrylic resin binder on Dow 17 treated Mg alloy substrate.
3*	49	1961	278	0°		Above specimen and conditions except exposed in vacuum (10^{-6}–8 x 10^{-6} mm Hg) to UV radiation from an argon-filled A-H6 high pressure Hg arc lamp for 80 hrs; vacuum maintained by VacIon pump.
4*	49	1961	278	0°		White Kemacryl lacquer M49WC17; 50 TiO_2 and 50 talc in acrylic resin binder on Dow 17 treated Mg alloy substrate.
5*	49	1961	278	0°		Above specimen and conditions except exposed in vacuum (10^{-6}–8 x 10^{-6} mm Hg) to UV radiation from an argon-filled A-H6 high pressure Hg arc lamp for 26 hrs; vacuum maintained by VacIon pump.
6*	49	1961	278	0°		White Kemacryl lacquer M49WC17; 50 TiO_2 and 50 talc in acrylic resin binder on Dow 17 treated Mg alloy substrate.
7*	49	1961	278	0°		Above specimen and conditions except exposed in vacuum (10^{-6}–8 x 10^{-6} mm Hg) to UV radiation from an argon-filled A-H6 high pressure Hg arc lamp for 100 hrs; vacuum maintained by VacIon pump.
8*	49	1961	278	0°		Similar to curve 6 specimen and conditions.
9*	49	1961	278	0°		Above specimen and conditions except exposed in vacuum (10^{-6}–8 x 10^{-6} mm Hg) to UV radiation from an argon-filled A-H6 high pressure Hg arc lamp for 100 hrs; vacuum maintained by VacIon pump.
10*	49	1961	278	0°		White Kemacryl lacquer M49WC17; 50 TiO_2 and 50 talc in acrylic resin binder on Dow 17 treated Mg alloy substrate.
11*	49	1961	278	0°		Above specimen and conditions except exposed in vacuum (10^{-6}–8 x 10^{-6} mm Hg) to UV radiation from an argon-filled A-H6 high pressure Hg arc lamp for 26 hrs; vacuum maintained by VacIon pump.
12*	50	1965	~298	~0°		White Kemacryl lacquer, Sherwin-Williams M49WC17; 50 TiO_2 and 50 talc in acrylic binder; substrate unknown; computed from spectral reflectance data.
13*	50	1965	~298	~0°		Similar to above specimen and conditions except irradiated in vacuum ($\leq 10^{-6}$ mm Hg) at 289 K with 1 MeV electrons to a total dose of 10^{14} e cm^{-2}.
14*	50	1965	~298	~0°		Similar to above specimen and conditions except irradiated to a total dose of 10^{15} e cm^{-2}.
15*	50	1965	~298	~0°		Similar to above specimen and conditions except irradiated to a total dose of 10^{16} e cm^{-2}; maintained in vacuum at 155 K for 12 hrs after irradiation.
16*	50	1965	~298	~0°		Similar to above specimen and conditions.
17*	50	1965	~298	~0°		Similar to above specimen and conditions.
18*	50	1965	~298	~0°		Similar to above specimen and conditions.
19*	50	1965	~298	~0°		Similar to above specimen and conditions.
20*	50	1965	~298	~0°		Similar to above specimen and conditions.
21*	50	1965	~298	~0°		Similar to curve 13 specimen and conditions except irradiated at 422 K to a total dose of 10^{15} e cm^{-2}.

* No plot given.

SPECIFICATION TABLE NO. 116 NORMAL SOLAR ABSORPTANCE OF TITANIUM DIOXIDE + TALC PIGMENTED COATINGS (continued)

Curve No.	Ref. No.	Year	Temperature Range, K	Geometry θ	Reported Error, %	Composition (weight percent), Specifications and Remarks
22*	50	1965	~298	~0°		Similar to above specimen and conditions except irradiated at 155 K to a total dose of 10^{15} e cm^{-2}; maintained in vacuum at 155 K for 12 hrs after irradiation.
23*	50	1965	~298	~0°		Similar to curve 12 specimen and conditions except irradiated in vacuum (10^{-7} mm Hg) at 289 K with ultraviolet radiation for 485 sun hrs.
24*	50	1965	~298	~0°		Similar to curve 12 specimen and conditions except irradiated in vacuum ($\leq 10^{-6}$ mm Hg) at 289 K with 1 MeV electrons to a total dose of 10^{16} e cm^{-2} and additionally with ultraviolet radiation for 485 sun hrs.
25*	117	1965	~298	~0°		White Kemacryl lacquer, Sherwin-Williams M49WC17; 50 TiO_2 and 50 talc in acrylic binder; computed from spectral reflectance data. [Authors' designation: Sample No. 1]
26*	117	1965	~298	~0°		Above specimen and conditions except irradiated in vacuum with 130 keV protons to a total dose of 3.9×10^{15} p cm^{-2}.
27*	117	1965	~298	~0°		Similar to curve 25 specimen and conditions. [Author's designation: Sample No. 2]
28*	117	1965	~298	~0°		Above specimen and conditions except irradiated in vacuum with 130 keV protons to a total dose of 2.0×10^{15} p cm^{-2}.
29*	117	1965	~298	~0°		Similar to curve 25 specimen and conditions. [Author's designation: Sample No. 11]
30*	117	1965	~298	~0°		Above specimen and conditions except irradiated in vacuum with 130 keV protons to a total dose of 2.0×10^{15} p cm^{-2}.
31*	117	1965	~298	~0°		Similar to curve 25 specimen and conditions. [Author's designation: Sample No. 13]
32*	117	1965	~298	~0°		Above specimen and conditions except irradiated in vacuum with 130 keV protons to a total dose of 3.9×10^{15} p cm^{-2}.
33*	117	1965	~298	~0°		Similar to curve 25 specimen and conditions. [Author's designation: Sample No. 14]
34*	117	1965	~298	~0°		Above specimen and conditions except irradiated in vacuum with 130 keV protons to a total dose of 2.0×10^{15} p cm^{-2}.
35*	117	1965	~298	~0°		Similar to curve 25 specimen and conditions. [Author's designation: Sample No. 23]
36*	117	1965	~298	~0°		Above specimen and conditions except irradiated in vacuum with 130 keV protons to a total dose of 2.0×10^{15} p cm^{-2}.
37*	117	1965	~298	~0°		Similar to curve 25 specimen and conditions. [Author's designation: Sample No. 29]
38*	117	1965	~298	~0°		Above specimen and conditions except irradiated in vacuum with 130 keV protons to a total dose of 1.1×10^{15} p cm^{-2}.
39*	117	1965	~298	~0°		Similar to curve 25 specimen and conditions. [Author's designation: Sample No. 30]
40*	117	1965	~298	~0°		Above specimen and conditions except irradiated in vacuum with 130 keV protons to a total dose of 0.5×10^{15} p cm^{-2}.

* No plot given.

DATA TABLE NO. 116 NORMAL SOLAR ABSORPTANCE OF TITANIUM DIOXIDE + TALC PIGMENTED COATINGS

[Temperature, T, K; Absorptance, α]

T	α	T	α	T	α	T	α
CURVE 1*		CURVE 13*		CURVE 25*		CURVE 37*	
278	0.26	298	0.26	298	0.24	298	0.24
CURVE 2*		CURVE 14*		CURVE 26*		CURVE 38*	
278	0.26	298	0.26	298	0.44	298	0.26
CURVE 3*		CURVE 15*		CURVE 27*		CURVE 39*	
278	0.33	298	0.30	298	0.24	298	0.24
CURVE 4*		CURVE 16*		CURVE 28*		CURVE 40*	
278	0.27	298	0.29	298	0.33	298	0.25
CURVE 5*		CURVE 17*		CURVE 29*			
278	0.35	298	0.28	298	0.24		
CURVE 6*		CURVE 18*		CURVE 30*			
278	0.26	298	0.30	298	0.35		
CURVE 7*		CURVE 19*		CURVE 31*			
278	0.32	298	0.31	298	0.24		
CURVE 8*		CURVE 20*		CURVE 32*			
278	0.26	298	0.30	298	0.44		
CURVE 9*		CURVE 21*		CURVE 33*			
278	0.35	298	0.23	298	0.24		
CURVE 10*		CURVE 22*		CURVE 34*			
278	0.27	298	0.28	298	0.35		
CURVE 11*		CURVE 23*		CURVE 35*			
278	0.32	298	0.35	298	0.24		
CURVE 12*		CURVE 24*		CURVE 36*			
298	0.24	298	0.44	298	0.37		

* No plot given.

SPECIFICATION TABLE NO. 117 HEMISPHERICAL TOTAL EMITTANCE OF TITANIUM PYROPHOSPHATE PIGMENTED COATINGS

Curve No.	Ref. No.	Year	Temperature Range, K	Reported Error, %	Composition (weight percent), Specifications and Remarks
1*	43	1962	343-402	± 3.0	Titanium pyrophosphate in nitrocellulose binder (~0.0762 mm thick); data extracted from smooth curve; measured in vacuum (10^{-3} mm Hg).

DATA TABLE NO. 117 HEMISPHERICAL TOTAL EMITTANCE OF TITANIUM PYROPHOSPHATE PIGMENTED COATINGS

[Temperature, T, K; Emittance, ϵ]

T	ϵ
CURVE 1*	
343	0.960
352	0.890
359	0.905
368	0.865
381	0.830
390	0.770
402	0.660

* No plot given

SPECIFICATION TABLE NO. 118 NORMAL TOTAL EMITTANCE OF ZINC CHROMATE PIGMENTED COATINGS

Curve No.	Ref. No.	Year	Temperature Range, K	Geometry θ'	Reported Error, %	Composition (weight percent), Specifications and Remarks
1*	98	1940	340-430	~0°		Zinc chromate in unknown binder (0. 002-0. 005 mm thick) on 24 S-T Alclad (4. 5 Cu, 1. 5 Mg, 0. 6 Mn, balance Al) substrate.

DATA TABLE NO. 118 NORMAL TOTAL EMITTANCE OF ZINC CHROMATE PIGMENTED COATINGS

[Temperature, T, K; Emittance, ϵ]

T	ϵ	T	ϵ
CURVE 1*		CURVE 1 (cont.)*	
340	0. 445	420	0. 604
340	0. 463	428	0. 556
341	0. 503	429	0. 554
348	0. 532	430	0. 547
349	0. 500		
350	0. 515		
364	0. 546		
365	0. 556		
367	0. 544		
377	0. 533		
377	0. 550		
378	0. 556		
388	0. 564		
393	0. 544		
393	0. 565		
405	0. 557		
405	0. 554		
407	0. 566		
419	0. 611		

* No plot given

SPECIFICATION TABLE NO. 119 NORMAL SPECTRAL REFLECTANCE OF ZINC CHROMATE PIGMENTED COATINGS

Curve No.	Ref. No.	Year	Temperature, K	Wavelength Range, μm	Geometry θ	θ'	ω'	Reported Error, %	Composition (weight percent), Specifications, and Remarks
1*	99	1966	~300	2.01-14.46	~0°		2π	<± 2	Zinc chromate primer, no. 960, manufactured by Rust-Oleum Corp; zinc chromate primer (~0.0381 mm thick) on AISI 316 stainless steel (1.27 mm thick) substrate; sprayed; air dried for a min.of 24 hrs; converted from R (2π, 0°). [Authors' designation: specimen No. 11]
2*	99	1966	~300	2.23-14.40	~0°		2π	<± 2	Rust-Oleum Corp zinc chromate primer (15 parts) and Glidden Co. missle black (1 part) (~0.0381 mm thick) on AISI 316 stainless steel (1.27 mm thick) substrate; sprayed; air dried for a min.of 24 hrs; converted from R (2π, 0°). [Authors' designation: specimen No. 12]

DATA TABLE NO. 119 NORMAL SPECTRAL REFLECTANCE OF ZINC CHROMATE PIGMENTED COATINGS

[Wavelength, λ, μm; Reflectance, ρ; Temperature, T, K]

λ	ρ
CURVE 1*	**T ~ 300**
2.01	0.407
2.12	0.372
2.59	0.399
3.14	0.054
3.21	0.070
2.08	0.091
3.49	0.046
3.72	0.156
4.01	0.166
4.41	0.261
4.55	0.244
5.21	0.220
5.74	0.056
6.07	0.083
6.39	0.051
6.70	0.102
6.97	0.050
7.54	0.051
CURVE 1 (cont.)*	
8.47	0.042
9.35	0.040
10.14	0.042
10.96	0.040
11.53	0.064
12.06	0.048
12.53	0.061
13.23	0.054
13.84	0.052
14.46	0.066
CURVE 2*	**T ~ 300**
2.23	0.054
3.23	0.069
3.64	0.106
4.33	0.190
5.02	0.181
CURVE 2 (cont.)*	
5.77	0.195
6.38	0.079
7.48	0.077
8.44	0.078
9.33	0.082
10.03	0.098
10.88	0.092
11.47	0.107
12.16	0.090
12.74	0.108
13.38	0.102
13.90	0.102
14.40	0.109

* No plot given

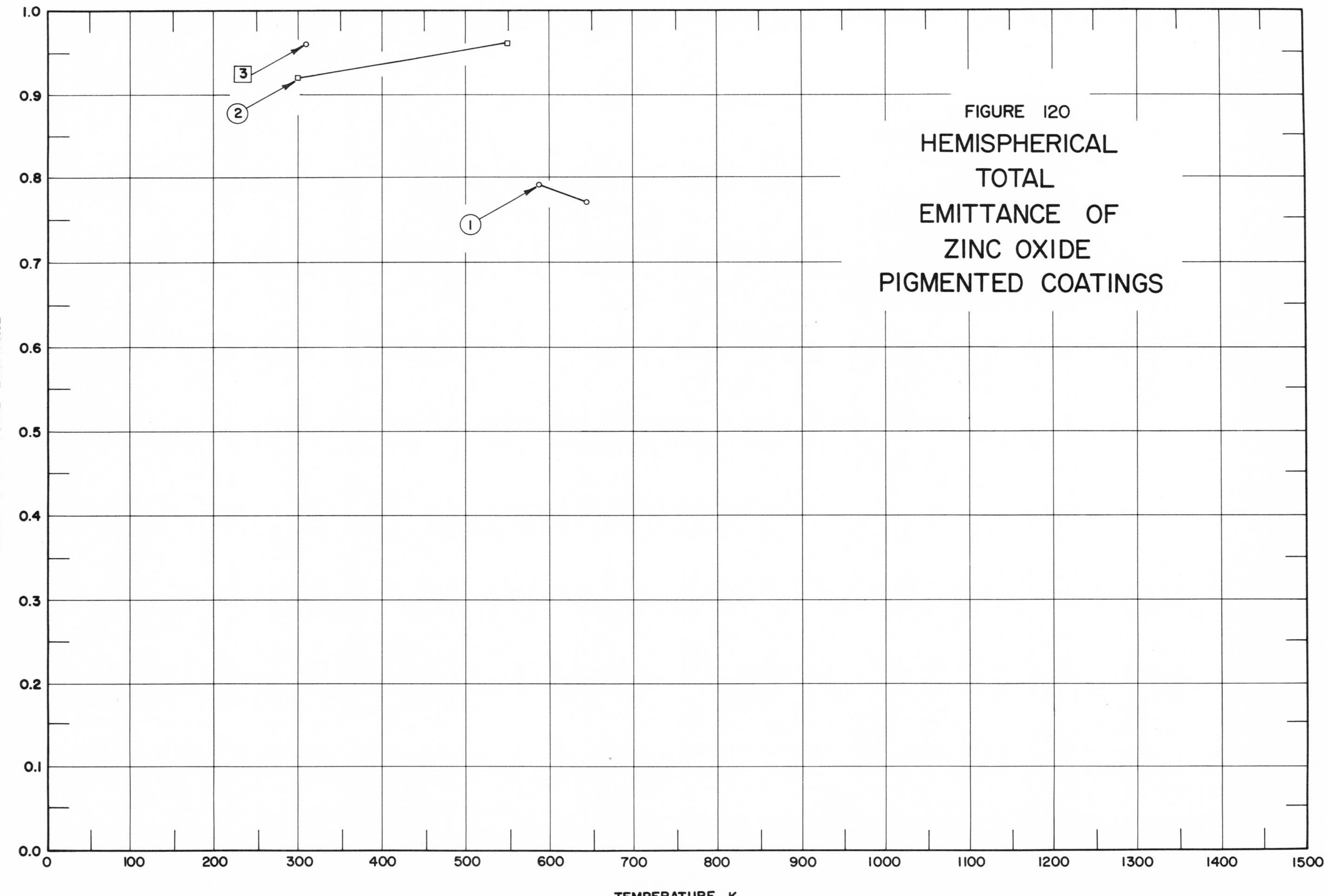

FIGURE 120
HEMISPHERICAL TOTAL EMITTANCE OF ZINC OXIDE PIGMENTED COATINGS
HEMISPHERICAL TOTAL EMITTANCE
TEMPERATURE, K
1.0
0.9
0.8
0.7
0.6
0.5
0.4
0.3
0.2
0.1
0.0
0
100
200
300
400
500
600
700
800
900
1000
1100
1200
1300
1400
1500
1
2
3

SPECIFICATION TABLE NO. 120 HEMISPHERICAL TOTAL EMITTANCE OF ZINC OXIDE PIGMENTED COATINGS

Curve No.	Ref. No.	Year	Temperature Range, K	Reported Error, %	Composition (weight percent), Specifications and Remarks
1	15	1965	588-644		ZnO in potassium silicate binder (0.051 mm thick).
2	16	1965	300-550	± 5	ZnO in methyl silicone binder (0.127 mm thick); measured in vacuum (10^{-7} mm Hg).
3	17	1968	309.8		Z-93 paint.

DATA TABLE NO. 120 HEMISPHERICAL TOTAL EMITTANCE OF ZINC OXIDE PIGMENTED COATINGS

[Temperature, T, K; Emittance, ϵ]

T	ϵ
CURVE 1	
588	0.79
644	0.77
CURVE 2	
300	0.92
550	0.96
CURVE 3	
309.8	0.96

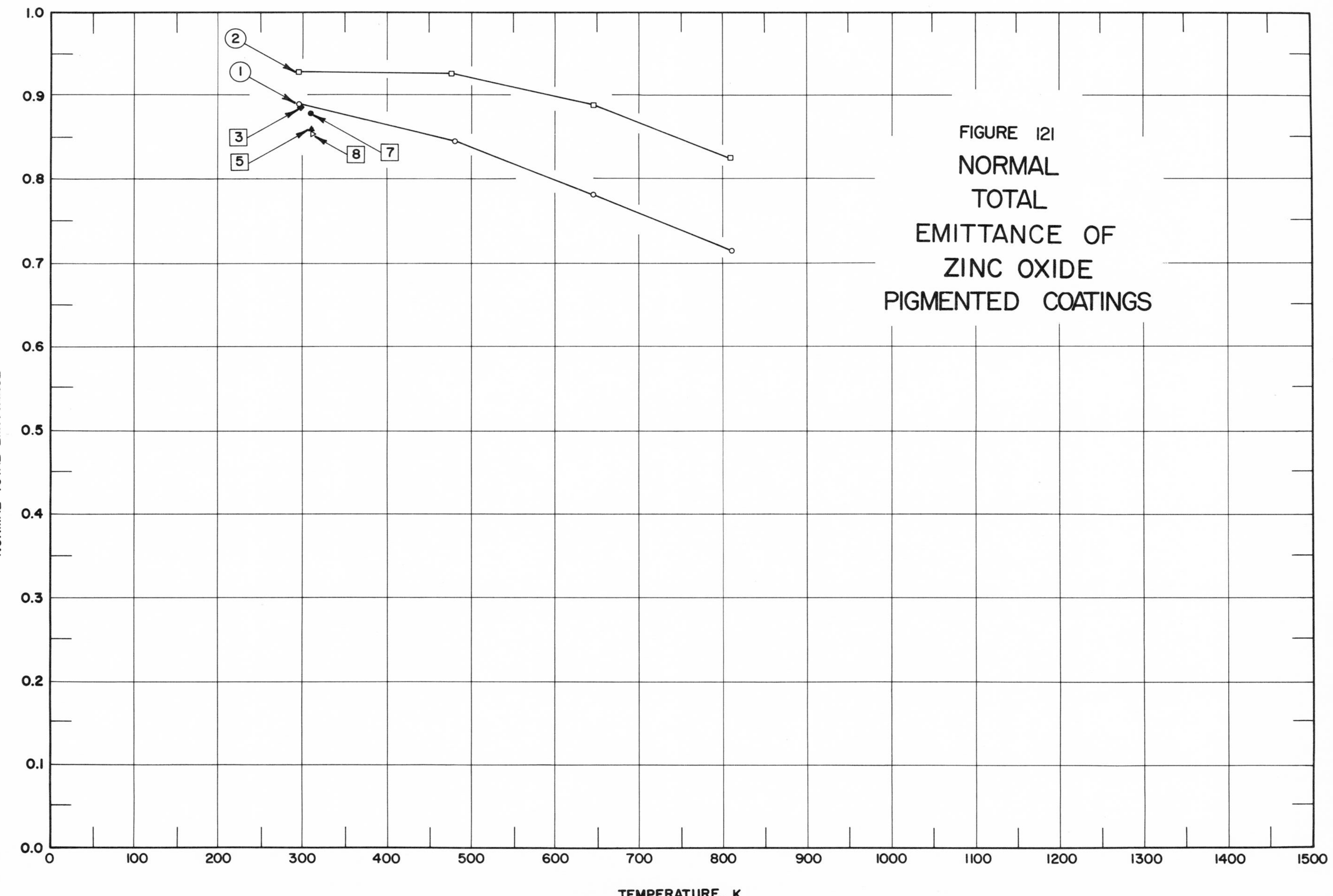

FIGURE 121 NORMAL TOTAL EMITTANCE OF ZINC OXIDE PIGMENTED COATINGS

SPECIFICATION TABLE NO. 121 NORMAL TOTAL EMITTANCE OF ZINC OXIDE PIGMENTED COATINGS

Curve No.	Ref. No.	Year	Temperature Range, K	Geometry θ'	Reported Error, %	Composition (weight percent), Specifications and Remarks
1	18	1964	296-811	$\sim 0^{\circ}$		ZnO in methyl silicone binder (0.102 mm thick); 6061-T6 aluminum substrate; applied with an air brush; calculated from spectral reflectance.
2	18	1964	295-810	$\sim 0^{\circ}$		ZnO in K_2SiO_3 binder (0.102 mm thick); 6061-T6 aluminum substrate; applied with an air brush; calculated from spectral reflectance.
3	19	1965	298	$\sim 0^{\circ}$		ZnO in potassium silicate binder; Al substrate; reported error: ±0.02 emittance units.
4	19	1965	298	$\sim 0^{\circ}$		Above specimen and conditions except computed from 1-R (2π, 0°).
5	56	1969	311	$\sim 0^{\circ}$		S-13G over B-1056; ZnO in methyl silicone binder (0.0508 mm thick) over ZnO in RTV-602 silicone resin binder substrate (0.254 mm thick); property calculated from reflectance; lab data taken on sample to be used on Lunar Orbiter IV.
6	56	1969	311	$\sim 0^{\circ}$		Similar to above specimen and conditions except sample tc be tested on Lunar Orbiter V.
7	56	1969	311	$\sim 0^{\circ}$		S-13G; ZnO in methyl silicone binder (0.254 mm thick); property calculated from reflectance; lab data taken on sample tab tested on Lunar Orbiter IV.
8	56	1969	311	$\sim 0^{\circ}$		B-1060; SP-500 ZnO in silicone binder (0.264 mm thick); ZnO silicate treated; property calculated from reflectance; lab data taken on sample to be tested on Lunar Orbiter IV.
9	56	1969	311	$\sim 0^{\circ}$		Z-93; SP-500 ZnO in PS-7 potassium silicate binder; property calculated from from reflectance; lab data taken on sample to be tested on Lunar Orbiter V.

DATA TABLE NO. 121 NORMAL TOTAL EMITTANCE OF ZINC OXIDE PIGMENTED COATINGS

[Temperature, T, K; Emittance, ϵ]

T	ϵ
CURVE 1	
296	0.890
481	0.846
646	0.781
811	0.715
CURVE 2	
295	0.929
478	0.926
646	0.889
810	0.826
CURVE 3	
298	0.88
CURVE 4	
298	0.89*
CURVE 5	
311	0.360
CURVE 6	
311	0.860*
CURVE 7	
311	0.879
CURVE 8	
311	0.855
CURVE 9	
311	0.880*

* Not shown on plot

SPECIFICATION TABLE NO. 122 HEMISPHERICAL SPECTRAL REFLECTANCE OF ZINC OXIDE PIGMENTED COATINGS

Curve No.	Ref. No.	Year	Temperature K	Wavelength, μm	Angular Range	Geometry ω	Geometry θ'	Reported Error, %	Composition (weight percent), Specifications and Remarks
1*	21	1966	298	0.533	0-80	2π	θ'		SP-500 ZnO in potassium silicate binder; data extracted from smooth curve.

DATA TABLE NO. 122 HEMISPHERICAL SPECTRAL REFLECTANCE OF ZINC OXIDE PIGMENTED COATINGS

[Angle, θ', °; Reflectance, ρ; Temperature, T, K; Wavelength, λ, μm]

θ'	ρ
CURVE 1*	
T = 298	
0	0.909
10	0.909
20	0.910
30	0.911
40	0.913
50	0.915
60	0.919
70	0.925
80	0.933

* No plot given

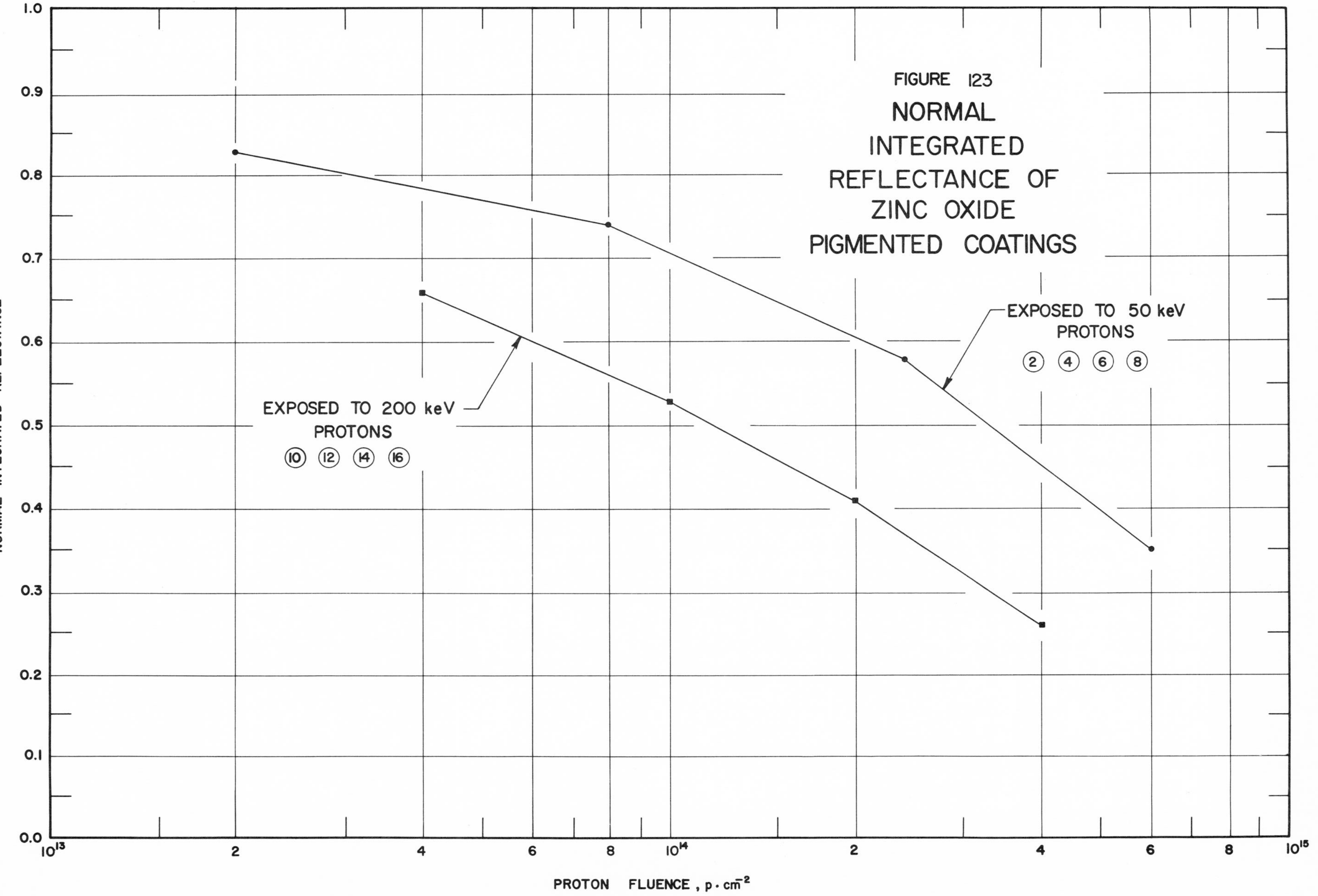

FIGURE 123 NORMAL INTEGRATED REFLECTANCE OF ZINC OXIDE PIGMENTED COATINGS

SPECIFICATION TABLE NO. 123 NORMAL INTEGRATED REFLECTANCE OF ZINC OXIDE PIGMENTED COATINGS

Curve No.	Ref. No.	Year	Temperature Range, K	Geometry θ	θ'	ω'	Reported Error, %	Composition (weight percent), Specifications and Remarks
1	22	1965	~298	~0°		2π		SP-500 ZnO in Dow Corning Q90016 methyl silicone binder (0.076 mm thick); spray application; integrated over wavelength range 0.400-0.433 μm.
2	22	1965	~298	~0°		2π		Above specimen and conditions except irradiated in vacuum with 50 keV protons to a total dose of 2 x 10^{13} p cm^{-2}.
3	22	1965	~298	~0°		2π		Similar to curve 1 specimen and conditions.
4	22	1965	~298	~0°		2π		Above specimen and conditions except irradiated in vacuum with 50 keV protons to a total dose of 8 x 10^{13} p cm^{-2}.
5	22	1965	~298	~0°		2π		Similar to curve 1 specimen and conditions.
6	22	1965	~298	~0°		2π		Above specimen and conditions except irradiated in vacuum with 50 keV protons to a total dose of 2.4 x 10^{14} p cm^{-2}.
7	22	1965	~298	~0°		2π		Similar to curve 1 specimen and conditions.
8	22	1965	~298	~0°		2π		Above specimen and conditions except irradiated in vacuum with 50 keV protons to a total dose of 6.0 x 10^{14} p cm^{-2}.
9	22	1965	~298	~0°		2π		Similar to curve 1 specimen and conditions.
10	22	1965	~298	~0°		2π		Above specimen and conditions except irradiated in vacuum with 200 keV protons to a total dose of 4 x 10^{13} p cm^{-2}.
11	22	1965	~298	~0°		2π		Similar to curve 1 specimen and conditions.
12	22	1965	~298	~0°		2π		Above specimen and conditions except irradiated in vacuum with 200 keV protons to a total dose of 1 x 10^{14} p cm^{-2}.
13	22	1965	~298	~0°		2π		Similar to curve 1 specimen and conditions.
14	22	1965	~298	~0°		2π		Above specimen and conditions except irradiated in vacuum with 200 keV protons to a total dose of 2 x 10^{14} p cm^{-2}.
15	22	1965	~298	~0°		2π		Similar to curve 1 specimen and conditions.
16	22	1965	~298	~0°		2π		Above specimen and conditions except irradiated in vacuum with 200 keV protons to a total dose of 4 x 10^{14} p cm^{-2}.

DATA TABLE NO. 123 NORMAL INTEGRATED REFLECTANCE OF ZINC OXIDE PIGMENTED COATINGS

[Temperature, T, K; Reflectance, ρ]

T	ρ	T	ρ
CURVE 1		CURVE 13	
298	0.885	298	0.890
CURVE 2		CURVE 14	
298	0.83	298	0.41
CURVE 3		CURVE 15	
298	0.885	298	0.890
CURVE 4		CURVE 16	
298	0.74	298	0.26
CURVE 5			
298	0.885		
CURVE 6			
298	0.58		
CURVE 7			
298	0.885		
CURVE 8			
298	0.35		
CURVE 9			
298	0.890		
CURVE 10			
298	0.66		
CURVE 11			
298	0.890		
CURVE 12			
298	0.53		

FIGURE 124A (I)

ANALYZED NORMAL SPECTRAL REFLECTANCE OF ZINC OXIDE PIGMENTED COATINGS

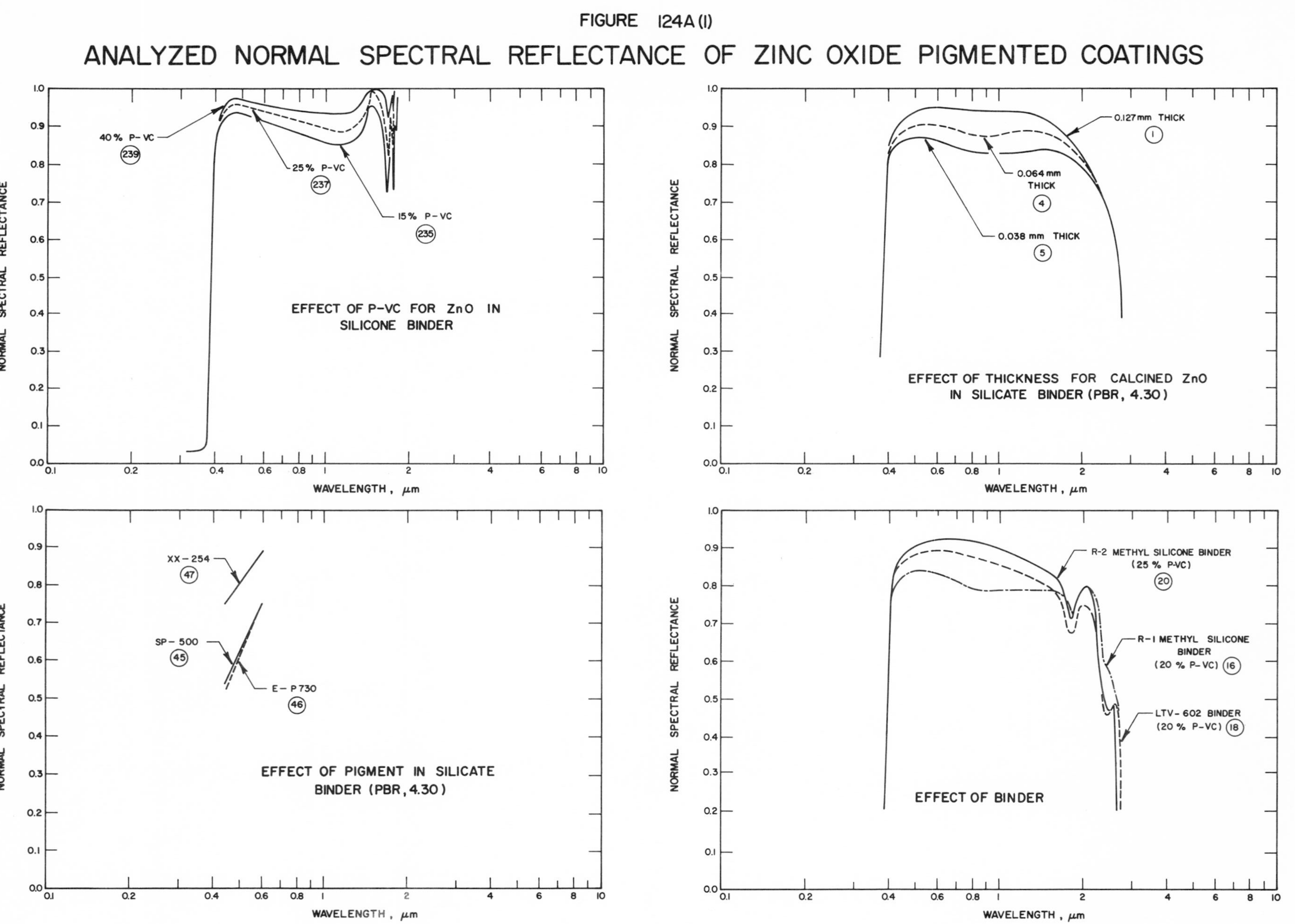

FIGURE 124A (2)

ANALYZED NORMAL SPECTRAL REFLECTANCE OF ZINC OXIDE PIGMENTED COATINGS

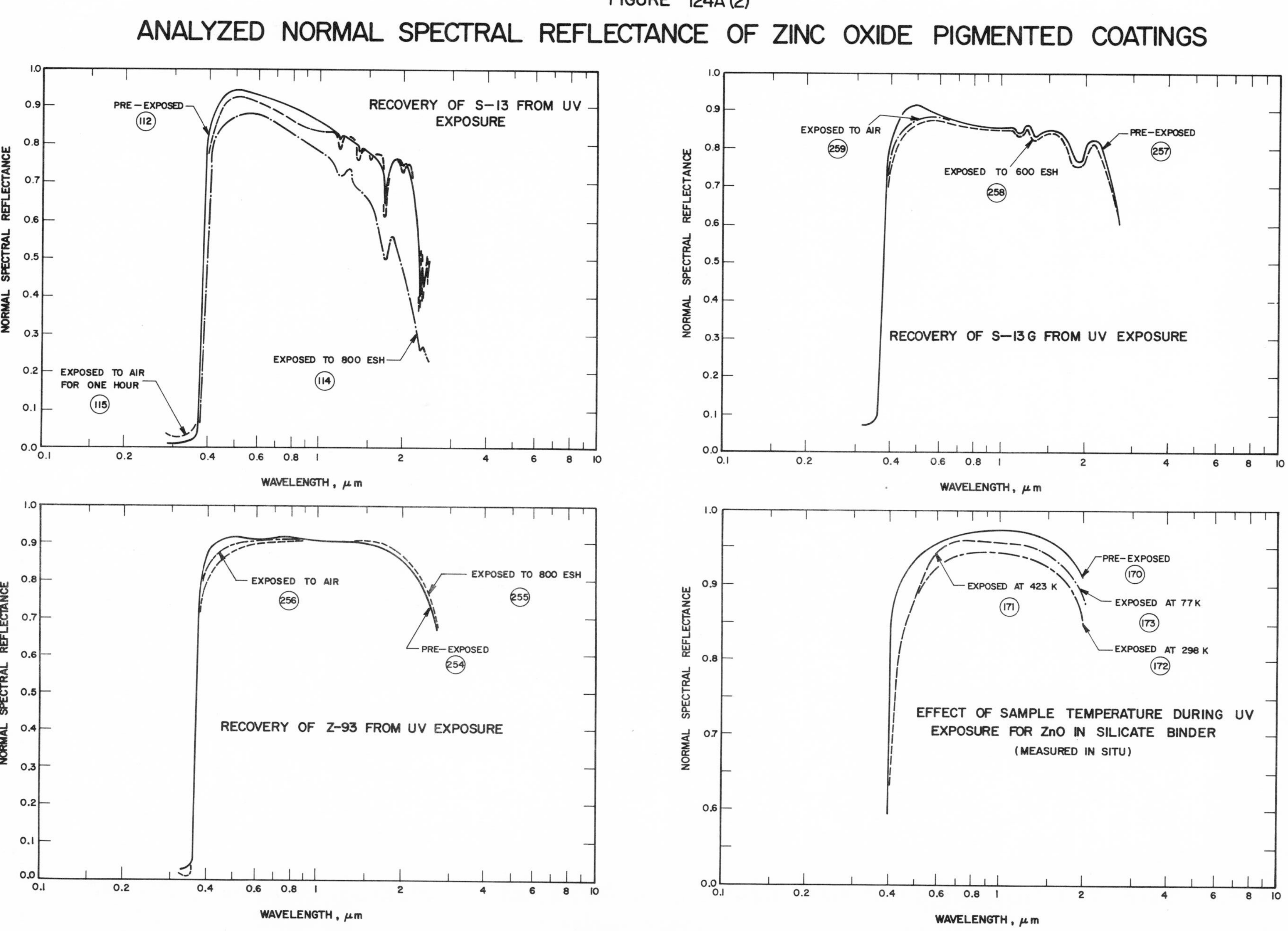

FIGURE 124A (3)

ANALYZED NORMAL SPECTRAL REFLECTANCE OF ZINC OXIDE PIGMENTED COATINGS

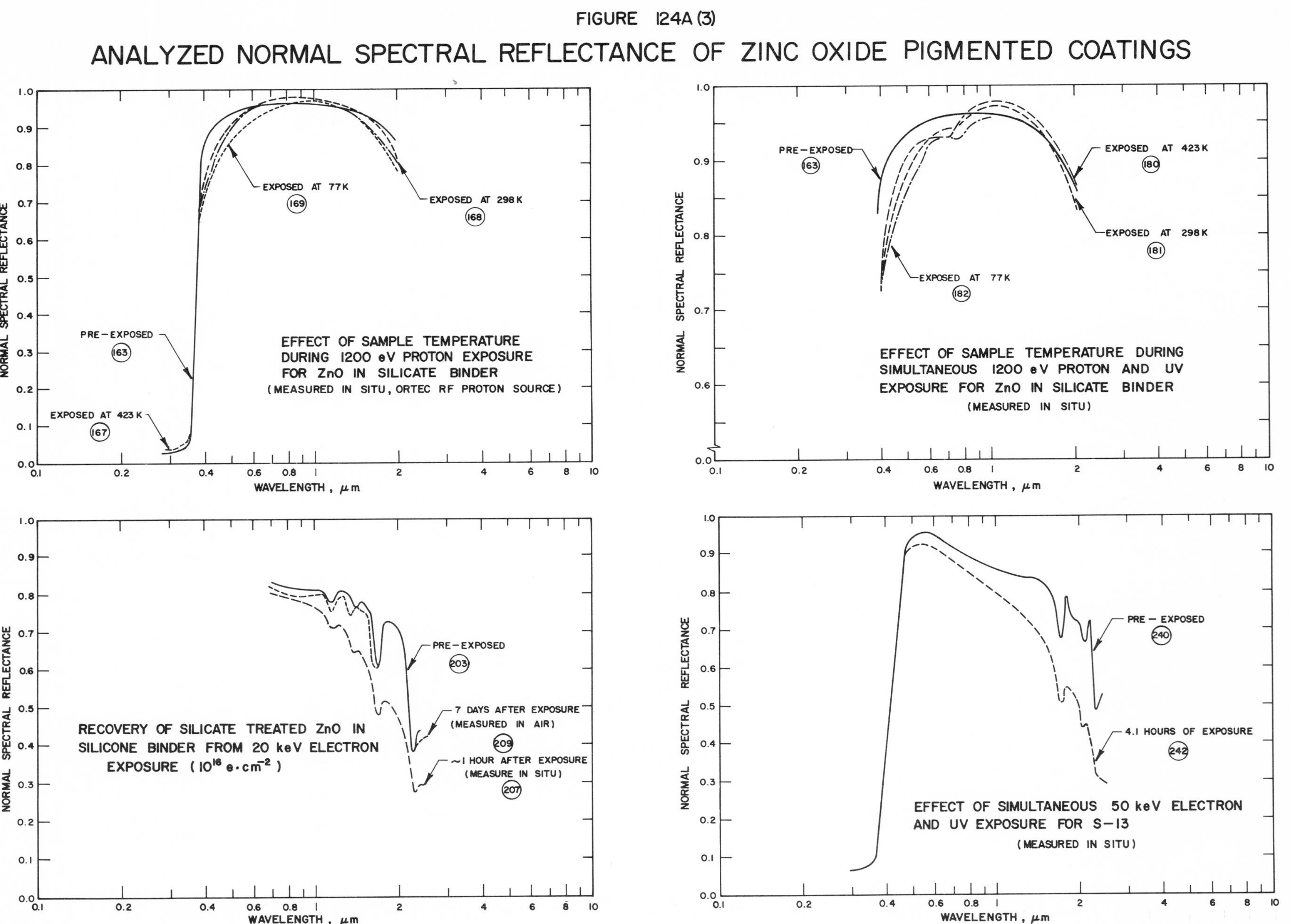

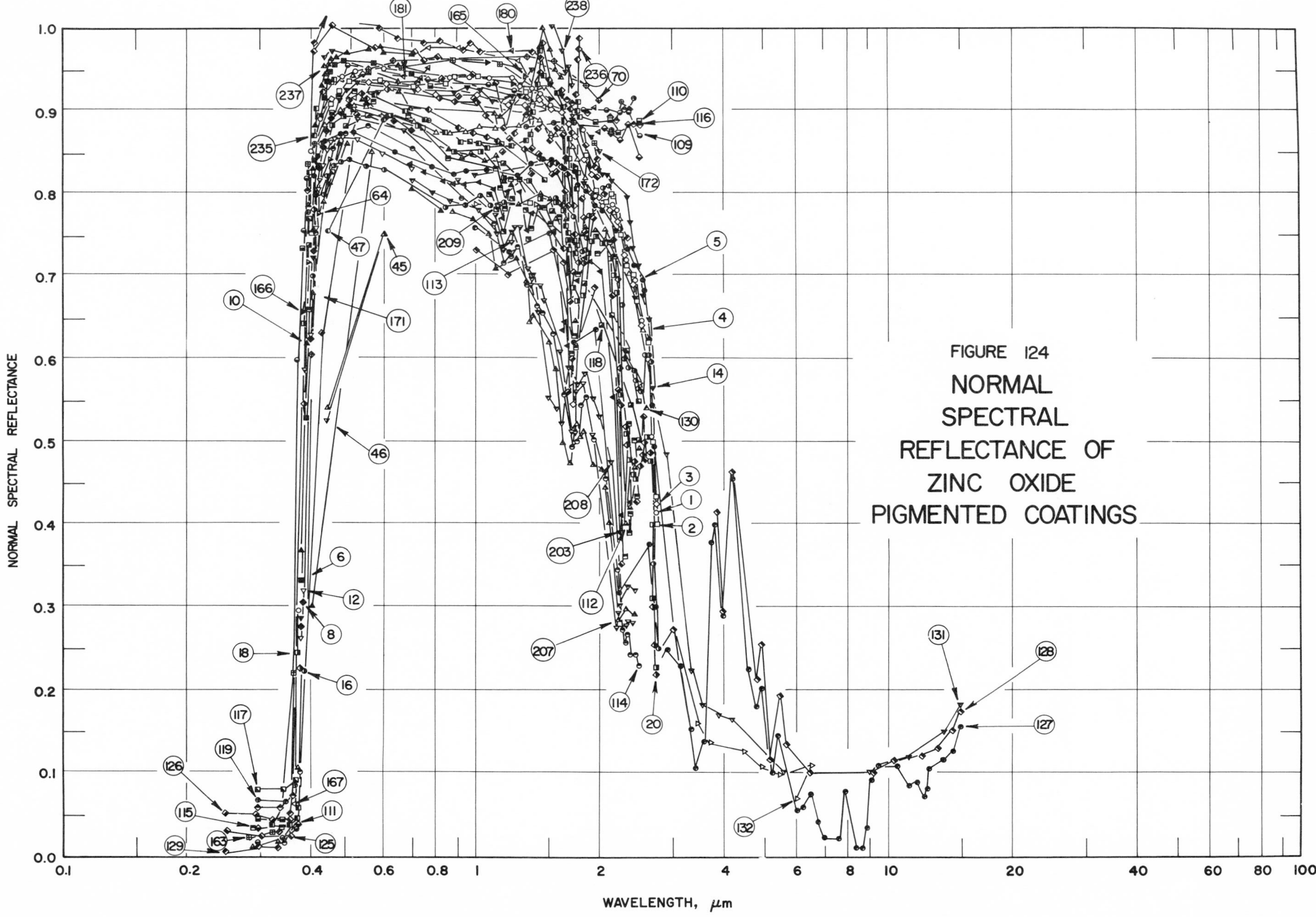
FIGURE 124
NORMAL
SPECTRAL
REFLECTANCE OF
ZINC OXIDE
PIGMENTED COATINGS
NORMAL SPECTRAL REFLECTANCE
WAVELENGTH, μm

SPECIFICATION TABLE NO. 124 NORMAL SPECTRAL REFLECTANCE OF ZINC OXIDE PIGMENTED COATINGS

Curve No.	Ref. No.	Year	Temperature, K	Wavelength Range, μm	Geometry θ θ'	ω'	Reported Error, %	Composition (weight percent), Specifications, and Remarks
1	20	1963	~298	0.372-2.762	~0°	2π		Calcined SP-500 ZnO in PS-7 potassium silicate binder (0.127 mm thick); PBR 4.30; solids content 46.3%; aluminum substrate abraded with No. 60 Aloxite cloth; airbrush application; data extracted from smooth curve.
2	20	1963	~298	0.372-2.757	~0°	2π		Similar to curve 1 specimen and conditions except 0.102 mm thick.
3	20	1963	~298	0.372-2.757	~0°	2π		Similar to curve 1 specimen and conditions except 0.089 mm thick.
4	20	1963	~298	0.372-2.757	~0°	2π		Similar to curve 1 specimen and conditions except 0.064 mm thick.
5	20	1963	~298	0.372-2.762	~0°	2π		Similar to curve 1 specimen and conditions except 0.038 mm thick.
6	20	1963	~298	0.381-0.700	~0°	2π		SP-500 in LTV-602 silicone binder; PBR 5.00; aluminum substrate; data extracted from smooth curve. [Authors' designation: Sample P-4]
7*	20	1963	~298	0.380-0.700	~0°	2π		Above specimen and conditions except exposed in vacuum (10^{-5} mm Hg) to 108 ESH.
8	20	1963	~298	0.381-0.700	~0°	2π		SP-500 ZnO in DuPont Viton B copolymer binder; PBR 4.0; aluminum substrate; data extracted from smooth curve. [Authors' designation: Sample P-10]
9*	20	1963	~298	0.381-0.700	~0°	2π		Above specimen and conditions except exposed in vacuum (10^{-5} mm Hg) to 108 ESH.
10	20	1963	~298	0.380-0.700	~0°	2π		SP-500 ZnO in 3M Kel-F8213 copolymer acetone binder; PBR 5.0; aluminum substrate; data extracted from smooth curve. [Authors' designation: Sample P-14].
11*	20	1963	~298	0.380-0.700	~0°	2π		Above specimen and conditions except exposed in vacuum (10^{-5} mm Hg) to 108 ESH.
12	20	1963	~298	0.380-0.700	~0°	2π		SP-500 ZnO in Leonite 201-S silicone epoxy modified acrylic resin; PBR 0.67; aluminum substrate; data extracted from smooth curve. [Authors' designation: Sample P-19]
13*	20	1963	~298	0.380-0.700	~0°	2π		Above specimen and conditions except exposed in vacuum (10^{-5} mm Hg) to 108 ESH.
14	20	1963	~298	0.380-2.706	~0°	2π		SP-500 pigment; calcined 16 hrs at 973 C; PBR 4.30; solids content 56.9%; paint ball-milled; aluminum substrate abraded with No. 60 Aloxite cloth; airbrush application; cured by air drying; data extracted from smooth curve. [Authors' designation: Sample Z93]
15*	20	1963	~298	0.369-2.708	~0°	2π		Above specimen and conditions except exposed in vacuum (10^{-5} mm Hg) to 4170 ESH at solar factor of 10.6 suns.
16	20	1963	~298	0.387-2.767	~0°	2π		SP-500 ZnO in R-1 experimental methyl silicone binder; PBR 1.35; PVC 25%; aluminum substrate; airbrush application; data extracted from smooth curve. [Authors' designation: Sample S-4]
17*	20	1963	~298	0.387-2.768	~0°	2π		Above specimen and conditions except exposed in vacuum (10^{-5} mm Hg) to 1460 ESH.
18	20	1963	~298	0.371-2.745	~0°	2π		SP-500 ZnO in G. E. LTV-602 binder; PBR 1.40; PVC 20%; catalyst, G. E. SRC-04; solvent, toluene; aluminum substrate; airbrush application; data extracted from smooth curve. [Authors' designation: Sample S-7]
19*	20	1963	~298	0.371-2.745	~0°	2π		Above specimen and conditions except exposed in vacuum (10^{-5} mm Hg) to 1460 ESH.
20	20	1963	~298	0.380-2.757	~0°	2π		SP-500 ZnO in R-2 experimental methyl silicone binder; Me/Si ratio 1.46; PBR 1.70; PVC 25%; aluminum substrate; airbrush application; data extracted from smooth curve. [Authors' designation: Sample S-8]

*Not shown on plot

Curve No.	Ref. No.	Year	Temperature, K	Wavelength Range, μm	Geometry θ θ'	ω'	Reported Error, %	Composition (weight percent), Specifications, and Remarks
21*	20	1963	~298	0.380-2.749	~0°	2π		Above specimen and conditions except exposed in vacuum (10^{-5} mm Hg) to 1460 ESH.
22*	20	1963	~298	0.380-2.763	~0°	2π		SP-500 ZnO in R-5 experimental methyl silicone binder; Me/Si ratio 1.38; PBR 1.63; PVC 25%; data extracted from smooth curve. [Authors' designation: Sample S-11]
23*	20	1963	~298	0.380-2.753	~0°	2π		Above specimen and conditions except exposed in vacuum (10^{-5} mm Hg) to 1460 ESH.
24*	20	1963	~298	0.380-2.757	~0°	2π		SP-500 ZnO in R-7 experimental methyl silicone binder; Me/Si ratio 1.33; PBR 1.64; PVC 25%; aluminum substrate; airbrush application; data extracted from smooth curve. [Authors' designation: Sample S-16]
25*	20	1963	~298	0.380-2.754	~0°	2π		Above specimen and conditions except exposed in vacuum (10^{-5} mm Hg) to 1600 ESH.
26*	20	1963	~298	0.377-2.705	~0°	2π		SP-500 ZnO in R-5 experimental methyl silicone binder; PBR 2.68; PVC 35%; aluminum substrate; data extracted from smooth curve. [Authors' designation: Sample S-19]
27*	20	1963	~298	0.377-2.705	~0°	2π		Above specimen and conditions except exposed in vacuum (10^{-5} mm Hg) to 1200 ESH.
28*	20	1963	~298	0.377-2.705	~0°	2π		Above specimen and conditions except exposed in vacuum (10^{-5} mm Hg) to 1850 ESH.
29*	20	1963	~298	0.386-2.708	~0°	2π		SP-500 ZnO in G. E. LTV-602 binder; PBR 3.73; PVC 40%; aluminum substrate; airbrush application; data extracted from smooth curve. [Authors' designation: Sample S-26]
30*	20	1963	~298	0.386-2.709	~0°	2π		Above specimen and conditions except exposed in vacuum (10^{-5} mm Hg) to 1200 ESH.
31*	20	1963	~298	0.376-2.704	~0°	2π		SP-500 ZnO in R-8 experimental methyl silicone binder; PBR <1.38; PVC 40%; aluminum substrate; airbrush application; data extracted from smooth curve. [Authors' designation: Sample S-31]
32*	20	1963	~298	0.376-2.702	~0°	2π		Similar to above specimen and conditions except heat-treated for 1 hr at 533 K and exposed to 1780 ESH.
33*	20	1963	~298	0.381-2.708	~0°	2π		SP-500 ZnO in R-8 experimental methyl silicone binder; PBR <1.38; PVC 40%; aluminum substrate; airbrush application; data extracted from smooth curve. [Authors' designation: Sample S-31]
34*	20	1963	~298	0.381-2.704	~0°	2π		Above specimen and conditions except exposed in vacuum (10^{-5} mm Hg) to 1780 ESH.
35*	27	1964	~298	0.440-0.600	~0°	2π		SP-500 ZnO in PS-7 potassium silicate binder; PBR 4.30; solids content 46.3%; aluminum substrate abraded with No. 60 Aloxite cloth; sample cured at 413 K for 18 hrs. [Authors' designation: Sample Z5]
36*	27	1964	~298	0.440-0.600	~0°	2π		Above specimen and conditions except irradiated in vacuum (10^{-6} mm Hg) for 200 ESH at solar factor of 3 suns.
37*	27	1964	~298	0.440-0.600	~0°	2π		SP-500 ZnO in PS-7 potassium silicate binder; PBR 4.30; solids content 56.9%; aluminum substrate abraded with No. 60 Aloxite cloth; sample cured for 18 hrs at 413 K. [Authors' designation: Sample Z8]

*Not shown on plot

SPECIFICATION TABLE NO. 124 NORMAL SPECTRAL REFLECTANCE OF ZINC OXIDE PIGMENTED COATINGS (continued)

Curve No.	Ref. No.	Year	Temperature, K	Wavelength Range, μm	Geometry θ	θ'	ω	Reported Error, %	Composition (weight percent), Specifications, and Remarks
38*	27	1964	~298	0.440-0.600	~0°		2π		Above specimen and conditions except irradiated in vacuum (10^{-6} mm Hg) for 200 ESH at solar factor of 3 suns.
39*	27	1964	~298	0.440-0.600	~0°		2π		SP-500 ZnO in PS-7 potassium silicate binder; PBR 2.13; solids content 51.9%; aluminum substrate abraded with No. 60 Aloxite cloth; sample cured at 413 K for 18 hrs. [Authors' designation: Sample Z9]
40*	27	1964	~298	0.440-0.600	~0°		2π		Above specimen and conditions except irradiated in vacuum (10^{-6} mm Hg) for 200 ESH at solar factor of 3 suns.
41*	27	1964	~298	0.440-0.600	~0°		2π		SP-500 ZnO in PS-7 potassium silicate binder; PBR 4.30; solids content 46.3%; aluminum substrate abraded with No. 60 Aloxite cloth; sample cured at 413 K for 18 hrs. [Authors' designation: Sample Z10]
42*	27	1964	~298	0.440-0.600	~0°		2π		Above specimen and conditions except irradiated in vacuum (10^{-6} mm Hg) for 300 ESH at solar factor of 3 suns.
43*	27	1964	~298	0.440-0.600	~0°		2π		SP-500 ZnO in PS-7 potassium silicate binder; PBR 4.30; solids content 46.3%; aluminum substrate abraded with No. 60 Aloxite cloth; sample cured at 413 K for 18 hrs. [Authors' designation: Sample Z27]
44*	27	1964	~298	0.440-0.600	~0°		2π		SP-500 ZnO in PS-7 potassium silicate binder; PBR 3.22; solids content 51.3%; aluminum substrate abraded with No. 60 Aloxite cloth; sample cured at 413 K for 18 hrs. [Authors' designation: Sample Z29]
45	27	1964	~298	0.440-0.600	~0°		2π		SP-500 ZnO (calcined at 1073 K for 12 hrs) in PS-7 potassium silicate binder; PBR 4.30; solids content 64.4%; aluminum substrate abraded with No. 60 Aloxite cloth; sample cured at 413 K for 18 hrs. [Authors' designation: Sample Z35]
46	27	1964	~298	0.440-0.600	~0°		2π		E-P730 ZnO in PS-7 potassium silicate binder; PBR 4.30; solids content 64.4%; aluminum substrate abraded with No. 60 Aloxite cloth; sample cured at 413 K for 18 hrs. [Authors' designation: Sample Z39]
47	27	1964	~298	0.440-0.600	~0°		2π		XX254 ZnO in PS-7 potassium silicate binder; PBR 4.30; solids content 73.0%; aluminum substrate abraded with No. 60 Aloxite cloth; sample cured at 413 K for 18 hrs. [Authors' designation: Sample Z42]
48*	27	1964	~298	0.38-2.71	~0°		2π		SP-500 ZnO in potassium silicate binder; PBR 4.30; solids content 56.9%; aluminum substrate; data extracted from smooth curve. [Authors' designation: Sample Z93]
49*	27	1964	~298	0.38-2.71	~0°		2π		Similar to above specimen and conditions except exposed to 4170 ESH in vacuum (10^{-6} mm Hg).
50*	27	1964	~298	0.38-2.70	~0°		2π		SP-500 ZnO (calcined) in PS-7 potassium silicate binder; PBR 4.30; spray application; aluminum substrate abraded with No. 60 Aloxite cloth; cured by air drying; data extracted from smooth curve.
51*	27	1964	~298	0.38-2.71	~0°		2π		Similar to above specimen and conditions except irradiated in vacuum (10^{-6} mm Hg) for 4170 ESH.
52*	27	1964	~298	0.440-0.600	~0°		2π		SP-500 ZnO in G. E. LTV-602 polydimethyl siloxane binder; PVC 20%; aluminum substrate; paint ground in porcelain jar mill for 16 hrs. [Authors' designation: Sample S-7]

*Not shown on plot

SPECIFICATION TABLE NO. 124 NORMAL SPECTRAL REFLECTANCE OF ZINC OXIDE PIGMENTED COATINGS (continued)

Curve No.	Ref. No.	Year	Temperature, K	Wavelength Range, μm	Geometry θ	θ'	ω'	Reported Error, %	Composition (weight percent), Specifications, and Remarks
53*	27	1964	~298	0.440-0.600	~0°		2π		SP-500 ZnO in Dow Corning 806A methyl phenyl silicone binder; PBR 0.7; aluminum substrate; paint ground in porcelain jar mill for 16 hrs. [Authors' designation: Sample P-1]
54*	27	1964	~298	0.440-0.600	~0°		2π		Above specimen and conditions except irradiated in vacuum (10^{-6} mm Hg) for 108 ESH at solar factor of 4 suns.
55*	27	1964	~298	0.440-0.600	~0°		2π		SP-500 ZnO in G. E. SE551 methyl-phenyl silicone; PBR 3.4; aluminum substrate; paint ground in porcelain jar mill for 16 hrs. [Authors' designation: Sample P-2]
56*	27	1964	~298	0.440-0.600	~0°		2π		Above specimen and conditions except irradiated in vacuum (10^{-6} mm Hg) for 500 ESH at solar factor of 4 suns.
57*	27	1964	~298	0.440-0.600	~0°		2π		SP-500 ZnO in G. E. RTV-11 polydimethyl siloxane polymer binder; PBR 1.0; aluminum substrate; paint ground in porcelain jar mill for 16 hrs. [Authors' designation: Sample P-3]
58*	27	1964	~298	0.440-0.600	~0°		2π		Above specimen and conditions except irradiated in vacuum (10^{-6} mm Hg) for 500 ESH at solar factor of 4 suns.
59*	27	1964	~298	0.440-0.600	~0°		2π		SP-500 ZnO in G. E. LTV-602 silicone potting compound; PBR 5.0; aluminum substrate; paint ground in porcelain jar mill for 16 hrs. [Authors' designation: Sample P-4]
60*	27	1964	~298	0.440-0.600	~0°		2π		Above specimen and conditions except irradiated in vacuum (10^{-6} mm Hg) for 500 ESH at solar factor of 4 suns.
61*	27	1964	~298	0.440-0.600	~0°		2π		SP-500 ZnO in DuPont Teflon TFE 852-202 polytetrafluoroethylene binder; PBR 0.67; aluminum substrate; paint ground in porcelain jar mill for 16 hrs. [Authors' designation: Sample P-7]
62*	27	1964	~298	0.440-0.600	~0°		2π		Above specimen and conditions except irradiated in vacuum (10^{-6} mm Hg) for 108 ESH at solar factor of 4 suns.
63*	27	1964	~298	0.440-0.600	~0°		2π		SP-500 ZnO in DuPont Teflon FEP 120 copolymer resin binder; PBR 0.4; aluminum substrate; paint ground in porcelain jar mill for 16 hrs. [Authors' designation: Sample P-8]
64*	27	1964	~298	0.440-0.600	~0°		2π		Above specimen and conditions except irradiated in vacuum (10^{-6} mm Hg) for 314 ESH at solar factor of 4 suns.
65*	27	1964	~298	0.440-0.600	~0°		2π		SP-500 ZnO in DuPont Viton B gum binder; PBR 4.0; aluminum substrate; paint ground in porcelain jar mill for 16 hrs. [Authors' designation: Sample P-10]
66*	27	1964	~298	0.440-0.600	~0°		2π		Above specimen and conditions except irradiated in vacuum (10^{-6} mm Hg) for 500 ESH at solar factor of 4 suns.
67*	27	1964	~298	0.440-0.600	~0°		2π		SP-500 ZnO in 3M Kel-F800 copolymer resin binder; PBR 0.5; aluminum substrate; paint ground in porcelain jar mill for 16 hrs. [Authors' designation: Sample P-11]
68*	27	1964	~298	0.440-0.600	~0°		2π		Above specimen and conditions except irradiated in vacuum (10^{-6} mm Hg) for 108 ESH at solar factor of 4 suns.

*Not shown on plot

SPECIFICATION TABLE NO. 124 NORMAL SPECTRAL REFLECTANCE OF ZINC OXIDE PIGMENTED COATINGS (continued)

Curve No.	Ref. No.	Year	Temperature, K	Wavelength Range, μm	Geometry θ θ'	ω'	Reported Error, %	Composition (weight percent), Specifications, and Remarks
69*	27	1964	~298	0.440-0.600	~0°	2π		SP-500 ZnO in 3M Kel-F 8213 copolymer latex binder; PBR 1.5; aluminum substrate; paint ground in porcelain jar mill for 16 hrs. [Authors' designation: Sample P-13]
70*	27	1964	~298	0.440-0.600	~0°	2π		Above specimen and conditions except irradiated in vacuum (10^{-6} mm Hg) for 108 ESH at solar factor of 4 suns.
71*	27	1964	~298	0.440-0.600	~0°	2π		SP-500 ZnO in 3M Kel-F 8213 copolymer latex binder; PBR 5.0; aluminum substrate; paint ground in porcelain jar mill for 16 hrs. [Authors' designation: Sample P-14]
72*	27	1964	~298	0.440-0.600	~0°	2π		Above specimen and conditions except irradiated in vacuum (10^{-6} mm Hg) for 500 ESH at solar factor of 4 suns.
73*	27	1964	~298	0.440-0.600	~0°	2π		SP-500 ZnO in Lenonite 201-S silicone-epoxy modified acrylic resin binder; PBR 0.67; aluminum substrate; paint ground in porcelain jar mill for 16 hrs. [Authors' designation: Sample P-19]
74*	27	1964	~298	0.440-0.600	~0°	2π		Above specimen and conditions except irradiated in vacuum (10^{-6} mm Hg) for 108 ESH at solar factor of 4 suns.
75*	28	1964	~298	0.382-0.741	~0°	2π		S-13; zinc oxide in silicone binder; data extracted from smooth curve.
76*	28	1964	~298	0.385-0.743	~0°	2π		Similar to above specimen and conditions except ultraviolet irradiated in vacuum ($<10^{-6}$ mm Hg) at 273-288 K for 720 ESH (~3-sun intensity, General Electric BH-6 lamp source); measured within 1 hr after exposure to air.
77*	28	1964	~298	0.381-0.741	~0°	2π		Similar to above specimen and conditions except additionally irradiated for 5.9 ESH in air.
78*	28	1964	~298	0.381-0.743	~0°	2π		Similar to curve 76 specimen and conditions except additionally irradiated for 5.9 ESH in nitrogen.
79*	29	1962	298	0.380-0.700	0°	2π		SP-500 ZnO in potassium silicate binder; 6061-T6Al plate (grit-blasted with 40 mesh SiC) substrate; sprayed by airbrush; air dried overnight and then 395-415 K heat-cured for 24 hrs; data extracted from smooth curve. [Authors' designation: Sample 1-9-5]
80*	29	1962	298	0.380-0.700	0°	2π		Above specimen and conditions except exposed to simulated solar radiation at 1.5 suns for 50 hrs.
81*	29	1962	298	0.380-0.700	0°	2π		Above specimen and conditions except exposed to simulated solar radiation at 1.5 suns for 50 hrs and then 4 suns for 60 hrs.
82*	29	1962	298	0.380-0.700	0°	2π		Above specimen and conditions except exposed to simulated solar radiation at 1.5 suns for 50 hrs and then 4 suns for 60 hrs and finally 4 suns for 67 hrs.
83*	29	1962	298	0.38-2.75	0°	2π		SP-500 ZnO in potassium silicate binder; 6061-T6 Al plate (grit-blasted with 40 mesh SiC) substrate; applied by airbrush; air dried overnight and then 395-415 K heat-cured for 24 hrs; data extracted from smooth curve. [Authors' designation: Sample 2-13-06]
84*	29	1962	298	0.40-2.76	0°	2π		Above specimen and conditions except exposed to simulated solar radiation at 4 suns for 67 hrs.

*Not shown on plot

Curve No.	Ref. No.	Year	Temperature, K	Wavelength Range, μm	Geometry θ	θ'	ω'	Reported Error, %	Composition (weight percent), Specifications, and Remarks
85*	29	1962	298	0.380-0.700	0°		2π		ZnO in Leonite 201-S binder; pigment to binder ratio 0.67; 6061-T6 Al plate (grit-blasted with 40 mesh SiC) substrate; sprayed; cured at 398 K for 24 hrs; data extracted from smooth curve. [Author's designation: Sample 405-2B]
86*	29	1962	298	0.380-0.700	0°		2π		Above specimen and conditions except exposed to simulated solar radiation at 4 suns for 18.5 hrs.
87*	29	1962	298	0.380-0.700	0°		2π		Above specimen and conditions except exposed to simulated solar radiation at 4 suns for 78.5 hrs.
88*	29	1962	298	0.380-0.700	0°		2π		ZnO in Leonite 201-S binder; pigment to binder ratio 0.67; 6061-T6 Al plate (grit-blasted with 40 mesh SiC) substrate; sprayed; cured at 298 K and then cured at 398 K for 30 min. [Author's designation: Sample 416-1]
89*	29	1962	298	0.380-0.700	0°		2π		Above specimen and conditions except exposed to simulated solar radiation at 4 suns for 27 hrs.
90*	29	1962	298	0.380-0.700	0°		2π		ZnO in Viton B binder; pigment to binder ratio 4; 6061-T6 Al plate (grit-blasted with 40 mesh SiC) substrate; sprayed; cured at 374 K for 24 hrs and then at 400 K for 24 hrs. [Author's designation: Sample 181]
91*	29	1962	298	0.380-0.700	0°		2π		Above specimen and conditions except exposed to simulated solar radiation at 4 suns for 27 hrs.
92*	29	1962	298	0.380-0.700	0°		2π		ZnO in silicone RTV-11 binder; pigment to binder ratio 1; 6061-T6 Al plate (grit-blasted with 40 mesh SiC) substrate; sprayed; cured at 298 K for 24 hrs and then at 422 K for 24 hrs. [Author's designation: Sample 192]
93*	29	1962	298	0.380-0.700	0°		2π		Above specimen and conditions except exposed to simulated solar radiation at 4 suns for 27 hrs.
94*	29	1962	298	0.380-0.700	0°		2π		ZnO in KEL-F 8213 (Ketone dispersion) binder; pigment to binder ratio 5; 6061-T6 Al plate (grit-blasted with 40 mesh SiC) substrate; sprayed; cured at 298 K for 24 hrs and then at 422 K for 16 hrs. [Author's designation: Sample 194]
95*	29	1962	298	0.380-0.700	0°		2π		Above specimen and conditions except exposed to simulated solar radiation at 4 suns for 27 hrs.
96*	29	1962	298	0.380-0.700	0°		2π		ZnO in silicone LTV-602 binder; pigment to binder ratio 5; 6061-T6 Al plate (grit-blasted with 40 mesh SiC) substrate; sprayed; cured at 298 K for 16 hrs, then at 393 K for 4 hrs, and then at 423 K for 4 hrs. [Author's designation: Sample 197]
97*	29	1962	298	0.380-0.700	0°		2π		Above specimen and conditions except exposed to simulated solar radiation at 4 suns for 27 hrs.
98*	29	1962	298	0.380-0.700	0°		2π		ZnO in silicone LTV-602 binder; pigment to binder ratio 5; 6061-T6 Al plate (grit-blasted with 40 mesh SiC) substrate; sprayed; cured at 298 K for 16 hrs, then at 393 K for 4 hrs, and then at 423 K for 4 hrs. [Author's designation: Sample 198]
99*	29	1962	298	0.380-0.700	0°		2π		Above specimen and conditions except exposed to simulated solar radiation at 4 suns for 124 hrs.

*Not shown on plot

SPECIFICATION TABLE NO. 124 NORMAL SPECTRAL REFLECTANCE OF ZINC OXIDE PIGMENTED COATINGS (continued)

Curve No.	Ref. No.	Year	Temperature, K	Wavelength Range, μm	Geometry θ	θ'	ω'	Reported Error, %	Composition (weight percent), Specifications, and Remarks
100	30	1965	298	0.0700-0.1888	<15°		2π		Measured in vacuum.
101*	31	1965	~300	0.250-0.360	~0°		2π		S-13; ZnO in LTV-602 binder; sprayed onto a 1.27 mm aluminum substrate; measured in vacuum (10^{-6} mm Hg); data extracted from smooth curve; measured relative to MgO. [Authors' designation: Sample No. 227]
102*	31	1965	~300	0.30-2.60	~0°		2π		Above specimen and conditions.
103*	31	1965	~300	0.250-0.360	~0°		2π		Similar to the above specimen and conditions except exposed to 6.1 x 10^{15} protons cm^{-2}. [Authors' designation: Sample No. 216]
104*	31	1965	~300	0.30-2.60	~0°		2π		Above specimen and conditions.
105*	31	1965	~300	1.00-15.00	~0°		2π		S-13; ZnO in LTV-602 binder; sprayed onto a 1.27 mm aluminum substrate; measured in vacuum (10^{-6} mm Hg); data extracted from smooth curve. [Authors' designation: Sample No. 214]
106*	31	1965	~300	1.00-15.00	~0°		2π		Similar to the above specimen and conditions except exposed to 6 x 10^{15} protons cm^{-2}. [Authors' designation: Sample No. 216]
107*	31	1965	~300	1.00-15.00	~0°		2π		Z-93; ZnO in potassium silicate binder; sprayed onto a 1.27 mm aluminum substrate; measured in vacuum (10^{-6} mm Hg); data extracted from smooth curve. [Authors' designation: Sample No. 227]
108*	31	1965	~300	1.00-6.57	~0°		2π		Similar to the above specimen and conditions except exposed to 1.0 x 10^{16} protons cm^{-2}. [Authors' designation: Sample No. 232]
109	32	1967	~298	0.30-2.50	7°		2π		ZnO treated with potassium silicate; prepared by spraying a "water mull" of silicate-zinc oxide on a warmed aluminum substrate; measured in situ after 17 hrs in vacuum (<10^{-6} mm Hg); data extracted from smooth curve.
110	32	1967	~298	0.30-2.50	7°		2π		Above specimen and conditions except ultraviolet irradiated in vacuum with 6-8 sun intensity for 200 ESH; measured in situ.
111	32	1967	~298	0.30-2.50	7°		2π		Above specimen and conditions except exposed to air for about 1 hr; measured in air.
112	32	1967	~298	0.291-2.50	7°		2π		S-13; SP-500 ZnO in silicone binder; aluminum substrate; measured in situ after 17 hrs in vacuum (<10^{-6} mm Hg); data extracted from smooth curve.
113	32	1967	~298	0.291-2.50	7°		2π		Above specimen and conditions except ultraviolet irradiated in vacuum with 6-8 sun intensity for 200 ESH; measured in situ.
114	32	1967	~298	0.291-2.500	7°		2π		Above specimen and conditions except irradiated additionally to a total exposure of 800 ESH; measured in situ.
115	32	1967	~298	0.291-2.500	7°		2π		Above specimen and conditions except exposed to air for about 1 hr; measured in air.
116	32	1967	~298	0.30-2.500	7°		2π		SP-500 ZnO; prepared by spraying a "water mull" of ZnO on a warmed aluminum substrate; measured in situ after 17 hrs in vacuum (<10^{-6} mm Hg); data extracted from smooth curve.
117	32	1967	~298	0.30-2.50	7°		2π		Above specimen and conditions except ultraviolet irradiated in vacuum with 6-8 sun intensity for 200 ESH; measured in situ.

*Not shown on plot

SPECIFICATION TABLE NO. 124 NORMAL SPECTRAL REFLECTANCE OF ZINC OXIDE PIGMENTED COATINGS (continued)

Curve No.	Ref. No.	Year	Temperature, K	Wavelength Range, μm	Geometry θ	θ'	ω'	Reported Error, %	Composition (weight percent), Specifications, and Remarks
118	32	1967	~298	0.30-2.50	7°		2π		Above specimen and conditions except irradiated additionally to a total exposure of 800 ESH; measured in situ.
119	32	1967	~298	0.30-2.50	7°		2π		Above specimen and conditions except exposed to air for about 1 hr; measured in air.
120*	22	1965	~298	0.300-0.700	~0°		2π		SP-500 ZnO in Dow Corning Q90016 methyl silicone binder (0.076 mm thick); xylene solvent; spray application; data extracted from smooth curve; measured relative to MgO.
121*	22	1965	~298	0.300-0.700	~0°		2π		Similar to above specimen and conditions except irradiated at 300-325 K in vacuum (10^{-5} mm Hg) with 200 keV protons at a flux of 2.0 x 10^{11} p cm^{-2} sec^{-1} to a total dose of 4 x 10^{13} p cm^{-2}.
122*	22	1965	~298	0.300-0.700	~0°		2π		Similar to above specimen and conditions except irradiated to a total dose of 1 x 10^{14} p cm^{-2}.
123*	22	1965	~298	0.300-0.700	~0°		2π		Similar to above specimen and conditions except irradiated to a total dose of 2 x 10^{14} p cm^{-2}.
124*	22	1965	~298	0.300-0.700	~0°		2π		Similar to above specimen and conditions except irradiated to a total dose of 4 x 10^{14} p cm^{-2}.
125	33	1965	~298	0.250-2.600	5°		2π		S-13; ZnO in LTV-602 silicone binder; sprayed on an aluminum substrate; data extracted from smooth curve; measured relative to MgO. [Authors' designation: Sample No. 215]
126	33	1965	~298	0.250-2.600	5°		2π		Similar to above specimen and conditions except irradiated in vacuum at 300 K with 8.2 keV protons to a total dose of 6.1 x 10^{15} p cm^{-2}. [Authors' designation: Sample No. 216]
127	33	1965	~298	1.0-15.0	~0°		2π		Similar to curve 125 specimen and conditions except not referenced to MgO. [Authors' designation: Sample No. 214]
128	33	1965	~298	1.0-15.0	~0°		2π		Curve 126 specimen and conditions not referenced to MgO.
129	33	1965	~298	0.250-2.602	5°		2π		Z-93; ZnO in potassium silicate binder; sprayed on an aluminum substrate; data extracted from smooth curve; measured relative to MgO. [Authors' designation: Sample No. 233]
130	33	1965	~298	0.250-2.602	5°		2π		Similar to above specimen and conditions except irradiated in vacuum at 300 K with 7.7 keV protons to a total dose of 1 x 10^{16} p cm^{-2}. [Authors' designation: Sample No. 232]
131	33	1965	~298	1.0-15.0	~0°		2π		Similar to curve 129 specimen and conditions except not referenced to MgO. [Authors' designation: Sample No. 227]
132	33	1965	~298	1.00-6.59	~0°		2π		Curve 130 specimen and conditions except not referenced to MgO.
133*	34	1964	298	0.300-0.469	~0°		2π		S-13; ZnO in LTV-602 methyl silicate binder; data extracted from smooth curve.
134*	34	1964	~298	0.325-2.007	~0°		2π		S-13; ZnO in LTV-602 methyl silicone binder (0.127 mm thick); PVC 30%; film peeled off stainless steel substrate after curing and fixed backside up to aluminum substrate; data extracted from smooth curve.

*Not shown on plot

SPECIFICATION TABLE NO. 124 NORMAL SPECTRAL REFLECTANCE OF ZINC OXIDE PIGMENTED COATINGS (continued)

Curve No.	Ref. No.	Year	Temperature, K	Wavelength Range, μm	Geometry θ θ'	ω'	Reported Error, %	Composition (weight percent), Specifications, and Remarks
135*	34	1964	298	0.325-2.0	~0°	2π		SP-500 ZnO in SR80 binder (0.203 mm thick); PVC 30%; surface abraded down to 0.170 mm; data extracted from smooth curve.
136*	34	1964	298	0.325-2.0	~0°	2π		Similar to above specimen and conditions except prepared as 4 coats; film thickness 0.203 mm abraded down to 0.172 mm.
137*	34	1964	298	0.325-2.0	~0°	2π		S-13; ZnO in LTV-602 silicone binder (0.127 mm thick); PVC 30%; film free and detached from aluminum substrate; data extracted from smooth curve.
138*	34	1964	298	0.325-2.0	~0°	2π		S-13; SP-500 ZnO in LTV silicone resin (0.127 mm thick); PVC 30%; clean, unpolished aluminum substrate; data extracted from smooth curve.
139*	34	1964	298	0.325-2.0	~0°	2π		S-13; SP-500 ZnO in LTV-602 binder; PVC 30%; data extracted from smooth curve.
140*	34	1964	~298	0.325-2.0	~0°	2π		Similar to above specimen and conditions except pigment ground 54 hrs (borundum fortified porcelain mill).
141*	34	1964	298	0.325-2.0	~0°	2π		ZnO in silicate binder; measured relative to MgO. [Authors' designation: Sample 1-A]
142*	34	1964	298	0.325-2.0	~0°	2π		ZnO in silicate binder; measured relative to MgO. [Authors' designation: Sample 3-A]
143*	35	1964	298	0.325-0.7; 1.5-2.0	0°	2π		ZnO in potassium silicate binder (0.127 mm thick); applied on abraded aluminum substrate; air dried, no heat-treatment; data extracted from smooth curve.
144*	35	1964	298	0.325-0.7; 1.5-2.0	0°	2π		Above specimen and conditions except heat-treated at 773 K for 2 hrs.
145*	35	1964	298	0.2-2.0	~0°	2π		SP-500 ZnO in LTV-602 silicone binder; data extracted from smooth curve.
146*	35	1964	298	0.2-2.0	~0°	2π		Similar to above specimen and conditions except pigment calcined (913 K for 16 hrs).
147*	17	1968	~298	0.250-2.498	~0°	2π		Z-93; data extracted from smooth curve.
148*	36	1966	~298	0.326-0.700	~0°	2π		S-13; SP-500 ZnO in G. E. RTV-602 silicone binder; G. E. SRC05 catalyst; toluene solvent; measured in vacuum; data extracted from smooth curve.
149*	36	1966	~298	0.701-2.70	~0°	2π		Similar to curve 148 specimen and conditions except exposed in situ to 1200 ESH; measured in situ.
150*	36	1966	~298	0.701-2.70	~0°	2π		Similar to curve 149 specimen and conditions except vacuum chamber vented to the air.
151*	36	1966	~298	0.325-0.700	~0°	2π		Z-93; SP-500 ZnO in PS7 potassium silicate binder; PBR 4.30; measured in vacuum; data extracted from smooth curve.
152*	36	1966	~298	0.700-2.70	~0°	2π		Above specimen and conditions except exposed in situ to 200 ESH; measured in situ.
153*	36	1966	~298	0.325-0.700	~0°	2π		Owens-Illinois type 650 glass resin paint; measured in vacuum; data extracted from smooth curve.
154*	36	1966	~298	0.700-2.70	~0°	2π		Above specimen and conditions except exposed in situ to 500 ESH; measured in situ.
155*	36	1966	~298	0.700-2.70	~0°	2π		S-13G; SP-500 ZnO in G. E. RTV-602 silicone binder; measured in vacuum; data extracted from smooth curve.
156*	36	1966	~298	0.700-2.70	~0°	2π		Above specimen and conditions except exposed in situ to 160 ESH; measured in situ.

*Not shown on plot

SPECIFICATION TABLE NO. 124 NORMAL SPECTRAL REFLECTANCE OF ZINC OXIDE PIGMENTED COATINGS (continued)

Curve No.	Ref. No.	Year	Temperature, K	Wavelength Range, μm	Geometry θ θ'	ω'	Reported Error, %	Composition (weight percent), Specifications, and Remarks
157*	36	1966	~298	0.697-2.70	~0°	2π		S-13H; SP-500 ZnO in G. E. RTV-602 silicone binder; PVC 40%; measured in vacuum; data extracted from smooth curve.
158*	36	1966	~298	0.697-2.70	~0°	2π		Above specimen and conditions except exposed in situ to 150 ESH; measured in situ.
159*	37	1967	293	0.360-2.50	~0°	2π		S-13G; SP-500 ZnO in LTV-602 binder (0.254-0.305 mm thick); measured in vacuum (<4 x 10^{-7} mm Hg); data extracted from smooth curve; measured relative to MgO. [Authors' designation: Sample 54-M7]
160*	37	1967	293	0.360-2.50	~0°	2π		Above specimen and conditions except exposed in situ to 1130 ESH at solar factor of 4-5; measured in situ.
161*	38	1966	~298	0.357-2.40	~0°	2π		S-13; SP-500 ZnO in LTV-602 silicone binder; data extracted from smooth curve.
162*	38	1966	~298	0.357-2.40	~0°	2π		Above specimen and conditions except exposed to 1330 ESH at solar factor of 20 suns in vacuum (<8 x 10^{-6} mm Hg); measured 24 hrs after air-pressure raised to 0.3 mm Hg.
163	39	1970	298	0.287-1.975	~0°	2π		SP-500 ZnO in K_2SiO_3 binder; 53.4 ZnO; ball-milled for 2 hrs, then compacted; measured in air; data extracted from smooth curve.
164	39	1970	298	0.278-1.979	~0°	2π		Above specimen and conditions except exposed to 3200 eV protons from Ortec rf ion source in vacuum (5 x 10^{-7} mm Hg); measured in situ.
165*	39	1970	298	0.287-1.966	~0°	2π		Similar to above specimen and conditions except proton energy 1200 eV.
166	39	1970	298	0.287-1.989	~0°	2π		Similar to above specimen and conditions except proton energy 500 eV.
167	39	1970	423	0.290-1.922	~0°	2π		Similar to above specimen and conditions except proton energy 1200 eV.
168*	39	1970	298	0.290-1.991	~0°	2π		Similar to above specimen and conditions.
169*	39	1970	77	0.290-1.991	~0°	2π		Similar to above specimen and conditions.
170	39	1970	298	0.401-2.000	~0°	2π		Similar to curve 163 specimen and conditions.
171	39	1970	423	0.401-1.174	~0°	2π		Above specimen and conditions except exposed to UV source with 4 sun intensity in vacuum (5 x 10^{-7} mm Hg); property measured in situ.
172	39	1970	298	0.401-2.000	~0°	2π		Similar to above specimen and conditions.
173	39	1970	77	0.655-2.000	~0°	2π		Similar to above specimen and conditions.
174*	39	1970	423	0.553-2.000	~0°	2π		Similar to above specimen and conditions except exposed to UV radiation for $\lambda < 0.4$ μm with Corning filter 7-54 in vacuum (5 x 10^{-7} mm Hg).
175*	39	1970	298	0.392-2.000	~0°	2π		Similar to above specimen and conditions.
176*	39	1970	77	0.392-2.000	~0°	2π		Similar to above specimen and conditions.
177*	39	1970	423	0.393-2.000	~0°	2π		Above specimen and conditions except exposed to UV radiation for $\lambda > 0.4$ μm with Corning filter 3-73 in vacuum (5 x 10^{-7} mm Hg).
178*	39	1970	298	0.393-2.000	~0°	2π		Similar to above specimen and conditions.
179*	39	1970	77	0.393-2.000	~0°	2π		Similar to above specimen and conditions.

*Not shown on plot

SPECIFICATION TABLE NO. 124 NORMAL SPECTRAL REFLECTANCE OF ZINC OXIDE PIGMENTED COATINGS (continued)

Curve No.	Ref. No.	Year	Temperature, K	Wavelength Range, μm	Geometry θ	θ'	ω'	Reported Error, %	Composition (weight percent), Specifications, and Remarks
180	39	1970	423	0.400-2.000	~0°		2π		Similar to curve 163 specimen and conditions except exposed to 1200 eV protons source and UV source in vacuum (5 x 10^{-7} mm Hg); measured in situ.
181	39	1970	298	0.400-2.000	~0°		2π		Similar to above specimen and conditions.
182	39	1970	77	0.400-2.000	~0°		2π		Similar to above specimen and conditions.
183*	39	1970	423	0.385-2.000	~0°		2π		Similar to above specimen and conditions except exposed to UV source for $\lambda<0.4$ μm with Corning filter 7-54.
184*	39	1970	298	0.385-2.000	~0°		2π		Similar to above specimen and conditions.
185*	39	1970	77	0.385-2.000	~0°		2π		Similar to above specimen and conditions.
186*	39	1970	423	0.386-2.000	~0°		2π		Similar to above specimen and conditions except exposed to UV source for $\lambda>0.4$ μm with Corning filter 3-73.
187*	39	1970	298	0.386-2.000	~0°		2π		Similar to above specimen and conditions.
188*	39	1970	77	0.386-2.000	~0°		2π		Similar to above specimen and conditions.
189*	45	1970	298	0.250-2.504	~0°		2π		Silicate treated ZnO in methyl silicone binder; aluminum substrate; reflectance measured in air. [Authors' designation: 101-7]
190*	45	1970	281	0.250-2.504	~0°		2π		Above specimen and conditions except exposed to 20 keV electrons in dark in vacuum (10^{-8} mm Hg); vacuum maintained by ion pump; flux 1 x 10^{10} to 5 x 10^{11} e cm^{-2} sec^{-1}; substrate held at 281 ± 2 K; reflectance measured in situ after exposure to 1 x 10^{13} e cm^{-2}.
191*	45	1970	281	0.250-2.504	~0°		2π		Above specimen and conditions except reflectance measured after exposure to 5 x 10^{13} e cm^{-2}.
192*	45	1970	281	0.250-2.504	~0°		2π		Above specimen and conditions except reflectance measured after exposure to 1 x 10^{14} e cm^{-2}.
193*	45	1970	281	0.250-2.507	~0°		2π		Above specimen and conditions except reflectance measured after exposure to 3 x 10^{14} e cm^{-2}.
194*	45	1970	281	0.250-2.504	~0°		2π		Above specimen and conditions except reflectance measured after exposure to 1 x 10^{15} e cm^{-2}.
195*	45	1970	281	0.250-2.504	~0°		2π		Above specimen and conditions except reflectance measured after exposure to 1 x 10^{16} e cm^{-2}.
196*	45	1970	298	0.385-2.505	~0°		2π		Similar to curve 189 specimen and conditions.
197*	45	1970	281	0.385-2.505	~0°		2π		Above specimen and conditions except exposed to 80 keV electrons in dark in vacuum (10^{-8} mm Hg); vacuum maintained by ion pump; flux 1 x 10^{10} to 5 x 10^{11} e cm^{-2} sec^{-1}; substrate held at 281 ± 2 K; reflectance measured in situ after 1 x 10^{13} e cm^{-2}.
198*	45	1970	281	0.385-2.505	~0°		2π		Above specimen and conditions except reflectance measured after exposure to 5 x 10^{13} e cm^{-2}.
199*	45	1970	281	0.385-2.505	~0°		2π		Above specimen and conditions except reflectance measured after exposure to 1 x 10^{14} e cm^{-2}.

*Not shown on plot

SPECIFICATION TABLE NO. 124 NORMAL SPECTRAL REFLECTANCE OF ZINC OXIDE PIGMENTED COATINGS (continued)

Curve No.	Ref. No.	Year	Temperature, K	Wavelength Range, μm	Geometry θ	Geometry θ' ω'	Reported Error, %	Composition (weight percent), Specifications, and Remarks
200*	45	1970	281	0.385-2.505	~0°	2π		Above specimen and conditions except reflectance measured after exposure to 3 x 10^{14} e cm^{-2}.
201*	45	1970	281	0.385-2.505	~0°	2π		Above specimen and conditions except reflectance measured after exposure to 1 x 10^{15} e cm^{-2}.
202*	45	1970	281	0.385-2.505	~0°	2π		Above specimen and conditions except reflectance measured after exposure to 1 x 10^{16} e cm^{-2}.
203	45	1970	298	0.700-2.496	~0°	2π		Similar to curve 189 specimen and conditions.
204*	45	1970	281	0.700-2.490	~0°	2π		Above specimen and conditions except exposed to 20 keV electrons in dark in vacuum (10^{-8} mm Hg); vacuum maintained by ion pump; reflectance measured in situ 3 hrs 52 min after end of exposure to 10^{15} e cm^{-2}.
205*	45	1970	281	0.700-2.490	~0°	2π		Above specimen and conditions except reflectance measured in vacuum 54 hrs after electron exposure.
206*	45	1970	281	0.700-2.494	~0°	2π		Above specimen and conditions except reflectance measured 123 hrs after exposure to electrons.
207	45	1970	281	0.700-2.493	~0°	2π		Similar to curve 204 specimen and conditions except reflectance measured 1 hr 5 min after end of exposure to 10^{16} e cm^{-2}.
208	45	1970	281	0.700-2.493	~0°	2π		Above specimen and conditions except reflectance measured 50 hrs after end of exposure to 10^{16} e cm^{-2}.
209	45	1970	298	0.700-2.493	~0°	2π		Above specimen and conditions except reflectance measured in air 5 days after dry air backfill (7 days after end of exposure to 10^{16} e cm^{-2}).
210*	45	1970	298	0.250-2.500	~0°	2π		Similar to curve 203 specimen and conditions.
211*	45	1970	281	0.250-2.500	~0°	2π		Above specimen and conditions except exposed to 80 keV electrons in dark in vacuum (10^{-8} mm Hg); vacuum maintained by ion pump; flux 1 x 10^{10} to 5 x 10^{11} e cm^{-2} sec^{-1}; substrate held at 281 ± 2 K; reflectance measured in situ after exposure to 10^{16} e cm^{-2}.
212*	45	1970	281	0.700-2.500	~0°	2π		Above specimen and conditions except reflectance measured 2 days after end of exposure to 10^{16} e cm^{-2}.
213*	45	1970	281	0.700-2.500	~0°	2π		Above specimen and conditions except reflectance measured 3 days after end of exposure to 10^{16} e cm^{-2}.
214*	45	1970	281	0.700-2.500	~0°	2π		Above specimen and conditions except reflectance measured 4 days after end of exposure to 10^{16} e cm^{-2}.
215*	45	1970	281	0.700-2.500	~0°	2π		Above specimen and conditions except reflectance measured 7 days after end of exposure to 10^{16} e cm^{-2}.
216*	45	1970	281	0.700-2.500	~0°	2π		Above specimen and conditions except reflectance measured 18 days after end of exposure to 10^{16} e cm^{-2}.
217*	45	1970	298	0.250-2.500	~0°	2π		Above specimen and conditions except reflectance measured in air after above tests had been made.

*Not shown on plot

SPECIFICATION TABLE NO. 124 NORMAL SPECTRAL REFLECTANCE OF ZINC OXIDE PIGMENTED COATINGS (continued)

Curve No.	Ref. No.	Year	Temperature, K	Wavelength Range, μm	Geometry θ	θ'	ω'	Reported Error, %	Composition (weight percent), Specifications, and Remarks
218*	45	1970	298	0.703-2.500	~0°		2π		ZnO in methyl silicone binder; aluminum substrate; reflectance measured in air. [Authors' designation: S-13G]
219*	45	1970	281	0.703-2.500	~0°		2π		Above specimen and conditions except exposed to 20 keV electrons in dark in vacuum (10^{-8} mm Hg); vacuum maintained by ion pump; flux 1 x 10^{10} to 5 x 10^{11} e cm^{-2} sec^{-1}; substrate held at 281 ± 2 K; reflectance measured in situ 4 min after exposure to 1 x 10^{15} e cm^{-2}.
220*	45	1970	281	0.704-2.500	~0°		2π		Above specimen and conditions except reflectance measured 49 min after end of exposure to 1 x 10^{15} e cm^{-2}.
221*	45	1970	281	0.704-2.500	~0°		2π		Above specimen and conditions except reflectance measured 1 hr 39 min after exposure to 1 x 10^{15} e cm^{-2}.
222*	45	1970	281	0.704-2.500	~0°		2π		Above specimen and conditions except reflectance measured 4 hrs 7 min after end of exposure to 1 x 10^{15} e cm^{-2}.
223*	45	1970	281	0.704-2.500	~0°		2π		Above specimen and conditions except reflectance measured 6 hrs 55 min after end of exposure to 10^{15} e cm^{-2}.
224*	45	1970	281	0.704-2.500	~0°		2π		Above specimen and conditions except reflectance measured 23 hrs 4 min after end of exposure to 10^{15} e cm^{-2}.
225*	45	1970	281	0.704-2.500	~0°		2π		Above specimen and conditions except reflectance measured 53 hrs 41 min after end of exposure to 10^{15} e cm^{-2}.
226*	45	1970	281	0.704-2.500	~0°		2π		Above specimen and conditions except reflectance measured 122 hrs 52 min after end of exposure to 10^{15} e cm^{-2}.
227*	45	1970	298	0.250-2.500	~0°		2π		Similar to curve 218 specimen and conditions.
228*	45	1970	281	0.250-2.500	~0°		2π		Above specimen and conditions except exposed to 80 keV electrons in dark in vacuum (10^{-8} mm Hg); vacuum maintained by ion pump; flux 1 x 10^{10} to 5 x 10^{11} e cm^{-2} sec^{-1}; substrate held at 281 ± 2 K; reflectance measured in situ after exposure to 10^{16} e cm^{-2}.
229*	45	1970	281	0.700-2.500	~0°		2π		Above specimen and conditions except reflectance measured 2 days after end of exposure to 10^{16} e cm^{-2}.
230*	45	1970	281	0.700-2.500	~0°		2π		Above specimen and conditions except reflectance measured 3 days after end of exposure to 10^{16} e cm^{-2}.
231*	45	1970	281	0.700-2.500	~0°		2π		Above specimen and conditions except reflectance measured 4 days after end of exposure to 10^{16} e cm^{-2}.
232*	45	1970	281	0.700-2.500	~0°		2π		Above specimen and conditions except reflectance measured 7 days after end of exposure to 10^{16} e cm^{-2}.
233*	45	1970	281	0.700-2.500	~0°		2π		Above specimen and conditions except reflectance measured 18 days after end of exposure to 10^{16} e cm^{-2}.
234*	45	1970	298	0.250-2.500	~0°		2π		Above specimen and conditions except reflectance measured in air after above tests had been made.

*Not shown on plot

Curve No.	Ref. No.	Year	Temperature, K	Wavelength Range, μm	Geometry θ θ'	ω'	Reported Error, %	Composition (weight percent), Specifications, and Remarks
235	44	1963	~298	0.321-1.800	~0°	2π		SP-500 ZnO in G. E. LTV-602 methyl silicone binder (~0.170 mm thick); 6061 aluminum substrate; PVC 15%; data is avg for several specimens; data extracted from smooth curve; measured relative to magnesium carbonate.
236	44	1963	~298	0.321-1.790	~0°	2π		Similar to above specimen and conditions except PVC 20%.
237	44	1963	~298	0.321-1.790	~0°	2π		Similar to above specimen and conditions except PVC 25%.
238	44	1963	~298	0.321-1.784	~0°	2π		Similar to above specimen and conditions except PVC 35%.
239*	44	1963	~298	0.321-1.769	~0°	2π		Similar to above specimen and conditions except PVC 40%.
240	46	1969	295	0.300-2.470	~0°	2π		ZnO in methyl silicone binder (0.228 mm thick) over S54044 primer; exposed to vacuum (10^{-7} mm Hg); vacuum maintained by ion pump; reflectance measured in situ; data extracted from smooth curve. [Authors' designation: S-13]
241	46	1969	295	0.300-2.470	~0°	2π		Above specimen and conditons except exposed to 50 keV electrons (5×10^{14} e cm^{-2} over 4.1 hrs with 90% during first hr) and UV radiation (4.4 equivalent sun rate in UV) in vacuum; reflectance measured in situ after 1 hr of exposure.
242	46	1969	295	0.300-2.470	~0°	2π		Above specimen and conditions except reflectance measured in situ after 4.1 hrs of exposure.
243*	47	1967	298	0.250-2.749	~0°	2π		SP-500 ZnO in PS-7 K_2SiO_3 binder (0.178 mm thick); ball-milled for 2 hrs; aluminum alloy 6061 substrate; cured in oven for 15 min at 400 K; exposed to vacuum (~10^{-8} mm Hg); vacuum maintained by an ion pump and a sublimation pump; reflectance measured in air immediately after exposure; data extracted from smooth curve.
244*	47	1967	298	0.250-2.749	~0°	2π		Similar to above specimen and conditions except exposed to 10-100 keV protons (3×10^{9}-5.5×10^{11} p cm^{-2} sec^{-1}) in vacuum (~10^{-8} mm Hg); proton exposure 2×10^{15} p cm^{-2}; reflectance measured in air immediately after exposure to protons.
245*	47	1967	298	0.250-2.749	~0°	2π		Above specimen and conditions except exposed to solar spectrum simulated with a mercury-xenon 5 kw short-arc lamp in vacuum after proton exposure; reflectance measured in air immediately after solar spectrum exposure.
246*	47	1967	298	0.250-2.749	~0°	2π		Similar to curve 243 specimen and conditions except exposed simultaneously to 10-100 keV protons (3×10^{9}-5.5×10^{11} p cm^{-2} sec^{-1}) and to solar spectrum simulated with a mercury-xenon 5 kw short-arc lamp in vacuum (~10^{-8} mm Hg); reflectance measured in air immediately after exposure.
247*	18	1964	298	0.255-1.600	~0°	2π		ZnO in K_2SiO_3 binder; 6061-T6 aluminum substrate.
248*	18	1964	394	0.376-1.600	~0°	2π		Above specimen and conditions except exposed to 235 ESH of ultraviolet radiation.
249*	18	1964	394	0.359-1.602	~0°	2π		Curve 247 specimen and conditions except exposed to 2300 ESH of ultraviolet radiation.
250*	18	1964	530	0.304-1.601	~0°	2π		Curve 247 specimen and conditions except exposed to 7000 ESH of ultraviolet radiation.

*Not shown on plot

SPECIFICATION TABLE NO. 124 NORMAL SPECTRAL REFLECTANCE OF ZINC OXIDE PIGMENTED COATINGS (continued)

Curve No.	Ref. No.	Year	Temperature, K	Wavelength Range, μm	Geometry θ	θ'	ω'	Reported Error, %	Composition (weight percent), Specifications, and Remarks
251*	63	1969	~298	0.325-2.60	~0°		2π		S-13 ; ZnO in polydimethyl siloxane (RTV-602) binder; exposed to vacuum; reflectance measured in situ.
252*	63	1969	~298	0.325-2.60	~0°		2π		Above specimen and conditions except exposed to UV radiation (from a G. E. AH-6 lamp) in vacuum; ESH 800.
253*	63	1969	~298	0.325-2.60	~0°		2π		Above specimen and conditions except measured in air after UV exposure.
254*	63	1969	~298	0.325-2.681	~0°		2π		Z-93; SP-500 ZnO in potassium silicate binder; exposed to vacuum; reflectance measured in situ.
255*	63	1969	~298	0.325-2.681	~0°		2π		Above specimen and conditions except exposed to UV radiation (from a G. E. AH-6 lamp) in vacuum; ESH 800.
256*	63	1969	~298	0.325-2.681	~0°		2π		Above specimen and conditions except measured in air after UV exposure.
257*	63	1969	~298	0.325-2.620	~0°		2π		S-13G; ZnO in methyl silicone binder; prepared from 16-hr-sweated pigment; only toluene used as solvent; exposed to vacuum; reflectance measured in situ.
258*	63	1969	~298	0.325-2.60	~0°		2π		Above specimen and conditions except exposed to UV radiation (from a G. E. AH-6 lamp) in vacuum; ESH 600.
259*	63	1969	~298	0.325-2.60	~0°		2π		Above specimen and conditions except measured in air after UV exposure.
260*	63	1969	~298	0.325-2.727	~0°		2π		S-13G; ZnO in methyl silicone binder; prepared from sweated pigment, calcined for 16 hrs at 923 K; exposed to vacuum; reflectance measured in situ.
261*	63	1969	~298	0.325-2.701	~0°		2π		Above specimen and conditions except exposed to UV radiation (from a G. E. AH-6 lamp) in vacuum; ESH 600.
262*	63	1969	~298	0.325-2.695	~0°		2π		Above specimen and conditions except measured in air after UV exposure.
263*	63	1969	~298	0.325-2.694	~0°		2π		S-13G; ZnO in methyl silicone binder; prepared from sweated pigment; toluene and petroleum ether solvent; exposed to vacuum; reflectance measured in situ.
264*	63	1969	~298	0.325-2.690	~0°		2π		Above specimen and conditions except exposed to UV radiation (from G. E. AH-6 lamp) in vacuum; ESH 800.
265*	63	1969	~298	0.325-2.694	~0°		2π		Above specimen and conditions except measured in air after UV exposure.
266*	63	1969	~298	0.325-2.778	~0°		2π		S-13G; ZnO in methyl silicone binder; ZnO silicate treated; pigment was only sifted prior to wet-grinding; paint grind time 3 hrs; exposed to vacuum; reflectance measured in situ.
267*	63	1969	~298	0.325-2.798	~0°		2π		Above specimen and conditions except exposed to UV radiation (from a G. E. AH-6 lamp) in vacuum; ESH 1400.
268*	63	1969	~298	0.325-2.778	~0°		2π		Above specimen and conditions except measured in air after UV exposure.
269*	63	1969	~298	0.334-2.595	~0°		2π		Similar to curve 266 specimen and conditions except pigment was unsifted and unground prior to wet-grinding; paint grind time 4 hrs.
270*	63	1969	~298	0.334-2.595	~0°		2π		Above specimen and conditions except exposed to UV radiation (from a G. E. AH-6 lamp) in vacuum; ESH 1400.

*Not shown on plot

SPECIFICATION TABLE NO. 124 NORMAL SPECTRAL REFLECTANCE OF ZINC OXIDE PIGMENTED COATINGS (continued)

Curve No.	Ref. No.	Year	Temperature, K	Wavelength Range, μm	Geometry θ	θ'	ω'	Reported Error, %	Composition (weight percent), Specifications, and Remarks
271*	63	1969	~298	0.335-2.595	~0°		2π		Above specimen and conditions except measured in air after UV exposure.
272*	63	1969	~298	0.332-2.279	~0°		2π		Similar to curve 266 specimen and conditions except pigment dry-ground 30 min; paint grind time 3 hrs.
273*	63	1969	~298	0.332-2.279	~0°		2π		Above specimen and conditions except exposed to UV radiation (from G. E. AH-6 lamp) in vacuum; ESH 1400.
274*	63	1969	~298	0.332-2.279	~0°		2π		Above specimen and conditions except measured in air after UV exposure.
275*	63	1969	~298	0.325-2.436	~0°		2π		Similar to curve 266 specimen and conditions except pigment hand-mulled prior to wet-grinding; paint grind time 3 hrs.
276*	63	1969	~298	0.325-2.488	~0°		2π		Above specimen and conditions except exposed to UV radiation (from a G. E. AH-6 lamp) in vacuum; ESH 1400.
277*	63	1969	~298	0.325-2.455	~0°		2π		Above specimen and conditions except measured in air after UV exposure.
278*	63	1969	~298	0.328-2.655	~0°		2π		Similar to curve 266 specimen and conditions except pigment remulled from first-hand mulling; paint grind time 5 hrs.
279*	63	1969	~298	0.325-2.628	~0°		2π		Above specimen and conditions except exposed to UV radiation (from a G. E. AH-6 lamp) in vacuum; ESH 1400.
280*	63	1969	~298	0.328-2.628	~0°		2π		Above specimen and conditions except measured in air after UV exposure.
281*	63	1969	~298	0.327-2.733	~0°		2π		S-13G; ZnO in methyl silicone binder; ZnO silicate treated; paint prepared from sweated pigment neutralized with formic acid and calcined for 16 hrs at 923 K; exposed to vacuum; reflectance measured in situ.
282*	63	1969	~298	0.327-2.651	~0°		2π		Above specimen and conditions except exposed to UV radiation (from a G. E. AH-6 lamp) in vacuum; ESH 600.
283*	63	1969	~298	0.327-2.645	~0°		2π		Above specimen and conditions except measured in air after UV exposure.
284*	63	1969	~298	0.328-2.716	~0°		2π		Similar to curve 281 specimen and conditions except pigment neutralized with sodium acid phosphate.
285*	63	1969	~298	0.328-2.636	~0°		2π		Above specimen and conditions except exposed to UV radiation (from a G. E. AH-6 lamp) in vacuum; ESH 600.
286*	63	1969	~298	0.328-2.636	~0°		2π		Above specimen and conditions except measured in air after UV exposure.
287*	63	1969	~298	0.332-2.728	~0°		2π		ZnO in Owens-Illinois 650 resin binder; ZnO silicate treated; PVC 32%; pigment slurried for 15 min with 50 g of a 3% solution of NaH_2PO_4, filtered, redispersed in 50 g of 3% NaH_2PO_4, slurried 15 min, filtered, redispersed in 50 g of distilled water, filtered, and dried for 18 hrs at 372 K; exposed to vacuum; reflectance measured in situ.
288*	63	1969	~298	0.332-2.728	~0°		2π		Above specimen and conditions except exposed to UV radiation (from a G. E. AH-6 lamp) in vacuum; ESH 300.
289*	63	1969	~298	0.332-2.728	~0°		2π		Above specimen and conditions except ESH 550.

*Not shown on plot

SPECIFICATION TABLE NO. 124 NORMAL SPECTRAL REFLECTANCE OF ZINC OXIDE PIGMENTED COATINGS (continued)

Curve No.	Ref. No.	Year	Temperature, K	Wavelength Range, μm	Geometry θ	θ'	ω'	Reported Error, %	Composition (weight percent), Specifications, and Remarks
290*	63	1969	~298	0.332-2.728	~0°		2π		Above specimen and conditions except ESH 1200.
291*	63	1969	~298	0.331-2.728	~0°		2π		Above specimen and conditions except measured in air after UV exposure.

*Not shown on plot

DATA TABLE NO. 124 NORMAL SPECTRAL REFLECTANCE OF ZINC OXIDE PIGMENTED COATINGS

[Wavelength, λ, μm; Reflectance, ρ; Temperature, T, K]

λ	ρ
CURVE 1 T ~ 298	
0.372	0.295
0.400	0.852
0.420	0.891
0.452	0.927
0.479	0.942
0.514	0.949
0.600	0.951
0.871	0.940
0.970	0.941
1.085	0.940
1.222	0.941
1.275	0.930
1.315	0.923
1.394	0.928
1.446	0.921
1.510	0.907
1.679	0.902
1.743	0.894
1.809	0.878
1.878	0.846
1.982	0.798
2.022	0.788
2.154	0.789
2.207	0.782
2.290	0.744
2.357	0.722
2.476	0.696
2.569	0.647
2.618	0.602
2.693	0.507
2.762	0.414
CURVE 2 T ~ 298	
0.372	0.295*
0.403	0.853*
0.445	0.908
0.466	0.924
0.499	0.936
0.584	0.937
0.713	0.930
0.843	0.933
0.886	0.941
1.002	0.942
CURVE 2 (cont.)	
1.261	0.931*
1.315	0.924
1.399	0.927*
1.447	0.921*
1.513	0.908*
1.682	0.902*
1.720	0.898
1.878	0.849
1.969	0.805
2.010	0.790
2.066	0.791
2.121	0.802
2.147	0.799
2.249	0.751
2.355	0.716
2.422	0.701
2.488	0.670
2.558	0.647*
2.603	0.622
2.700	0.501
2.734	0.434
2.757	0.400
CURVE 3 T ~ 298	
0.372	0.295*
0.403	0.853*
0.422	0.891*
0.446	0.914
0.468	0.925*
0.494	0.932*
0.542	0.934
0.605	0.932
0.723	0.928
0.800	0.920
1.001	0.920
1.200	0.909
1.303	0.907
1.404	0.914
1.522	0.903
1.606	0.897
1.693	0.885
1.742	0.871
1.806	0.844
1.915	0.826
CURVE 3 (cont.)	
1.986	0.797
2.031	0.789*
2.114	0.783
2.173	0.777
2.215	0.769
2.308	0.726
2.411	0.690
2.545	0.641
2.604	0.599*
2.680	0.500*
2.735	0.425
2.757	0.400*
CURVE 4 T ~ 298	
0.372	0.295*
0.403	0.853*
0.431	0.884
0.470	0.900
0.502	0.907
0.541	0.906
0.710	0.888
0.802	0.877
0.873	0.875
0.948	0.878
1.017	0.883
1.101	0.881
1.290	0.884
1.386	0.893
1.434	0.890
1.504	0.875
1.545	0.873
1.613	0.880
1.667	0.873
1.711	0.859
1.769	0.843
1.894	0.821
1.973	0.800*
2.013	0.791*
2.111	0.784*
2.181	0.779
2.213	0.770
2.251	0.750
2.305	0.721*
2.339	0.711*
CURVE 4 (cont.)	
2.376	0.706
2.545	0.639
2.601	0.601*
2.681	0.500*
2.735	0.425*
2.757	0.400*
CURVE 5 T ~ 298	
0.372	0.295*
0.406	0.822
0.416	0.847
0.433	0.864
0.472	0.874
0.507	0.875
0.640	0.851
0.781	0.835
0.884	0.827
0.956	0.830
1.073	0.832
1.202	0.828
1.285	0.824
1.379	0.838
1.446	0.844
1.526	0.844
1.647	0.832
1.733	0.829
1.795	0.813
1.856	0.804
1.921	0.797
1.963	0.788
2.026	0.783
2.067	0.785*
2.131	0.782*
2.176	0.777*
2.241	0.757
2.344	0.736
2.476	0.714
2.526	0.697
2.559	0.683
2.620	0.624
2.676	0.545
2.702	0.496
2.762	0.414*
CURVE 6 T ~ 298	
0.381	0.331
0.390	0.528
0.400	0.752
0.408	0.842
0.418	0.892
0.426	0.918
0.436	0.938
0.445	0.947*
0.457	0.955
0.479	0.961
0.564	0.959
0.700	0.945
CURVE 7* T ~ 298	
0.380	0.314
0.391	0.520
0.403	0.744
0.410	0.838
0.419	0.881
0.429	0.908
0.440	0.929
0.459	0.943
0.481	0.948
0.525	0.947
0.611	0.942
0.700	0.929
CURVE 8 T ~ 298	
0.380	0.276
0.385	0.306
0.399	0.625
0.405	0.731
0.411	0.799
0.418	0.832
0.426	0.857
0.436	0.877
0.448	0.890
0.462	0.898
0.492	0.904
0.544	0.906
0.602	0.894
CURVE 8 (cont.)	
0.633	0.888
0.700	0.879
CURVE 9* T ~ 298	
0.381	0.248
0.389	0.366
0.399	0.553
0.405	0.618
0.415	0.672
0.430	0.722
0.443	0.750
0.465	0.784
0.498	0.817
0.541	0.845
0.570	0.856
0.606	0.864
0.664	0.867
0.700	0.866
CURVE 10 T ~ 298	
0.380	0.368
0.392	0.620
0.398	0.753
0.404	0.832
0.410	0.884
0.422	0.930
0.429	0.945
0.439	0.957
0.460	0.966
0.490	0.971
0.551	0.974
0.593	0.977
0.599	0.972
0.700	0.968
CURVE 11* T ~ 298	
0.380	0.321
0.387	0.406
0.398	0.618
0.405	0.702
CURVE 11 (cont.)*	
0.414	0.761
0.422	0.792
0.438	0.825
0.467	0.861
0.499	0.890
0.547	0.920
0.582	0.932
0.641	0.941
0.700	0.951
CURVE 12 T ~ 298	
0.380	0.261
0.385	0.318
0.400	0.627
0.410	0.759
0.417	0.804
0.426	0.833
0.435	0.853
0.444	0.863
0.451	0.868
0.479	0.871
0.519	0.867
0.592	0.849
0.700	0.832
CURVE 13* T ~ 298	
0.380	0.212
0.385	0.242
0.390	0.313
0.397	0.439
0.409	0.571
0.425	0.672
0.435	0.704
0.458	0.754
0.479	0.783
0.504	0.805
0.531	0.819
0.579	0.827
0.661	0.826
0.679	0.817
0.700	0.816
CURVE 14 T ~ 298	
0.380	0.285
0.408	0.723
0.414	0.832
0.435	0.881*
0.463	0.913
0.482	0.925
0.514	0.934
0.713	0.929
0.890	0.922
0.957	0.928
1.010	0.925
1.097	0.910
1.149	0.911
1.207	0.919
1.269	0.919
1.524	0.932
1.614	0.919
1.729	0.890
1.798	0.881
1.902	0.870
1.993	0.830
2.020	0.82[illegible]
2.096	0.826
2.134	0.818
2.201	0.796
2.318	0.766
2.396	0.733
2.472	0.713
2.595	0.679
2.634	0.649
2.706	0.565
CURVE 15* T ~ 298	
0.369	0.252
0.380	0.285
0.415	0.688
0.415	0.791
0.439	0.841
0.481	0.879
0.579	0.899
0.689	0.910
0.805	0.912
0.874	0.924

*Not shown on plot

DATA TABLE NO. 124 NORMAL SPECTRAL REFLECTANCE OF ZINC OXIDE PIGMENTED COATINGS (continued)

λ	ρ
CURVE 15 (cont.)* T ~ 298	
0.921	0.927
0.976	0.927
1.061	0.917
1.135	0.920
1.263	0.933
1.307	0.931
1.481	0.947
1.523	0.944
1.601	0.923
1.691	0.915
1.800	0.913
1.902	0.897
2.005	0.866
2.102	0.865
2.198	0.843
2.396	0.789
2.509	0.770
2.619	0.722
2.708	0.651
CURVE 16 T ~ 298	
0.387	0.223
0.400	0.700
0.426	0.799
0.438	0.822
0.457	0.836
0.472	0.840
0.497	0.842
0.544	0.835
0.600	0.830
0.856	0.785
0.944	0.787
1.005	0.799
1.084	0.787
1.174	0.792
1.404	0.785
1.581	0.796
1.710	0.774
1.802	0.731
1.834	0.732
1.865	0.749
1.892	0.775
1.968	0.798*
2.030	0.809

λ	ρ
CURVE 16 (cont.)	
2.120	0.813
2.166	0.798*
2.223	0.759
2.276	0.665
2.302	0.602
2.334	0.591
2.407	0.588
2.457	0.575
2.485	0.566
2.528	0.561
2.600	0.607
2.622	0.607
2.638	0.598
2.660	0.563*
2.706	0.352
2.730	0.301
2.767	0.250
CURVE 17* T ~ 298	
0.387	0.223
0.391	0.546
0.411	0.574
0.445	0.702
0.472	0.743
0.546	0.779
0.629	0.797
0.714	0.804
0.808	0.822
0.838	0.822
1.046	0.842
1.196	0.841
1.311	0.857
1.386	0.851
1.593	0.868
1.643	0.861
1.706	0.823
1.765	0.794
1.804	0.787
1.848	0.797
1.920	0.843
1.996	0.847
2.120	0.846
2.211	0.803
2.271	0.694
2.309	0.633

λ	ρ
CURVE 17 (cont.)*	
2.372	0.593
2.416	0.579
2.495	0.579
2.534	0.587
2.564	0.606
2.598	0.627
2.641	0.615
2.657	0.588
2.683	0.456
2.706	0.352
2.730	0.301
2.767	0.250
CURVE 18 T ~ 298	
0.371	0.245
0.386	0.672
0.411	0.838*
0.425	0.866*
0.450	0.885
0.480	0.895
0.619	0.894
0.728	0.882
0.889	0.853
1.016	0.853
1.168	0.821
1.354	0.809
1.441	0.795
1.589	0.792
1.636	0.781
1.695	0.745
1.746	0.687
1.780	0.671
1.838	0.679
1.880	0.733
1.905	0.745
1.979	0.751
2.027	0.742
2.049	0.741*
2.138	0.723
2.182	0.700
2.209	0.681
2.278	0.557
2.314	0.510
2.359	0.479
2.409	0.461

λ	ρ
CURVE 18 (cont.)	
2.459	0.456
2.603	0.507
2.629	0.498*
2.653	0.477
2.676	0.401
2.689	0.311
2.745	0.227
CURVE 19* T ~ 298	
0.371	0.245
0.395	0.722
0.412	0.749
0.462	0.815
0.526	0.841
0.606	0.855
0.701	0.855
0.891	0.826
1.017	0.829
1.149	0.797
1.329	0.796
1.416	0.777
1.589	0.766
1.712	0.699
1.741	0.669
1.775	0.652
1.798	0.652
1.877	0.691
1.943	0.691
2.107	0.661
2.193	0.631
2.221	0.600
2.255	0.546
2.327	0.458
2.368	0.433
2.429	0.417
2.577	0.448
2.592	0.442
2.605	0.444
2.633	0.420
2.653	0.354
2.667	0.310
2.745	0.227

λ	ρ
CURVE 20 T ~ 298	
0.380	0.225
0.385	0.548
0.403	0.781
0.425	0.860
0.443	0.885*
0.495	0.913
0.532	0.921
0.611	0.927
0.819	0.917
0.897	0.911
1.001	0.911
1.059	0.896
1.333	0.855
1.443	0.844
1.508	0.839
1.575	0.841
1.616	0.836
1.646	0.825
1.726	0.774*
1.796	0.719
1.819	0.719
1.844	0.734*
1.897	0.775*
1.957	0.794*
1.999	0.800*
2.066	0.800
2.134	0.771*
2.181	0.746
2.217	0.699
2.254	0.590
2.287	0.546
2.308	0.519
2.362	0.498
2.447	0.478
2.517	0.470
2.549	0.486
2.584	0.500
2.611	0.500*
2.630	0.488
2.665	0.401*
2.694	0.301
2.719	0.255
2.757	0.220

λ	ρ
CURVE 21* T ~ 298	
0.380	0.225
0.385	0.548
0.395	0.697
0.471	0.838
0.508	0.867
0.654	0.900
0.717	0.905
0.777	0.905
0.865	0.896
1.028	0.891
1.180	0.849
1.230	0.845
1.285	0.852
1.326	0.845
1.429	0.817
1.560	0.810
1.607	0.795
1.667	0.762
1.758	0.693
1.789	0.682
1.825	0.682
1.854	0.698
1.888	0.722
1.943	0.727
2.089	0.712
2.159	0.687
2.229	0.641
2.278	0.549
2.308	0.500
2.363	0.465
2.427	0.443
2.494	0.431
2.555	0.441
2.589	0.453
2.622	0.430
2.679	0.347
2.702	0.276
2.714	0.241
2.749	0.200
CURVE 22* T ~ 298	
0.380	0.235
0.392	0.677
0.411	0.807

λ	ρ
CURVE 22 (cont.)*	
0.461	0.888
0.490	0.895
0.706	0.874
0.844	0.847
0.928	0.829
1.119	0.827
1.229	0.803
1.329	0.803
1.396	0.786
1.473	0.786
1.536	0.789
1.583	0.781
1.734	0.712
1.793	0.688
1.827	0.688
1.861	0.701
1.900	0.740
2.000	0.749
2.204	0.688
2.226	0.659
2.276	0.546
2.304	0.517
2.351	0.498
2.440	0.474
2.526	0.472
2.550	0.488
2.599	0.502
2.634	0.502
2.673	0.447
2.696	0.304
2.713	0.268
2.736	0.242
2.763	0.221
CURVE 23* T ~ 298	
0.380	0.235
0.392	0.677
0.420	0.784
0.435	0.812
0.482	0.846
0.559	0.860
0.661	0.861
0.807	0.854
0.929	0.830
1.041	0.826

λ	ρ
CURVE 23 (cont.)*	
1.121	0.811
1.181	0.796
1.341	0.773
1.441	0.754
1.696	0.744
1.737	0.721
1.769	0.662
1.798	0.645
1.847	0.645
1.914	0.658
2.005	0.658
2.101	0.645
2.184	0.613
2.226	0.582
2.284	0.491
2.338	0.447
2.387	0.422
2.455	0.414
2.600	0.419
2.652	0.395
2.682	0.265
2.700	0.242
2.753	0.200
CURVE 24* T ~ 298	
0.380	0.214
0.406	0.677
0.439	0.810
0.474	0.840
0.539	0.857
0,646	0.857
0.767	0.837
0.833	0.815
0.866	0.796
0.914	0.802
1.000	0.843
1.017	0.844
1.081	0.813
1.184	0.789
1.212	0.789
1.429	0.753
1.467	0.751
1.544	0.756
1.663	0.728
1.699	0.722

*Not shown on plot

DATA TABLE NO. 124 NORMAL SPECTRAL REFLECTANCE OF ZINC OXIDE PIGMENTED COATINGS (continued)

λ	ρ
CURVE 24 (cont.)*	
T ~ 298	
1.782	0.674
1.833	0.676
1.879	0.692
1.905	0.707
2.030	0.720
2.106	0.713
2.158	0.698
2.203	0.676
2.235	0.648
2.257	0.600
2.315	0.540
2.393	0.501
2.480	0.474
2.538	0.481
2.559	0.511
2.577	0.520
2.599	0.527
2.626	0.521
2.642	0.508
2.673	0.401
2.699	0.358
2.757	0.324
CURVE 25*	
T ~ 298	
0.380	0.214
0.411	0.698
0.469	0.812
0.512	0.834
0.552	0.844
0.597	0.849
0.673	0.841
0.790	0.822
0.938	0.810
0.994	0.808
1.037	0.804
1.072	0.804
1.183	0.784
1.303	0.767
1.402	0.783
1.515	0.751
1.602	0.751
1.728	0.700
1.804	0.700
1.850	0.711
CURVE 25 (cont.)*	
1.919	0.740
1.959	0.745
2.001	0.745
2.069	0.716
2.177	0.668
2.274	0.626
2.394	0.531
2.417	0.523
2.458	0.525
2.502	0.548
2.567	0.548
2.617	0.540
2.651	0.513
2.714	0.419
2.754	0.331
CURVE 26*	
T ~ 298	
0.377	0.247
0.411	0.728
0.438	0.819
0.479	0.867
0.537	0.886
0.601	0.895
0.666	0.895
0.808	0.874
0.909	0.867
1.014	0.867
1.087	0.858
1.146	0.842
1.187	0.834
1.225	0.833
1.290	0.845
1.348	0.838
1.501	0.799
1.574	0.799
1.612	0.801
1.724	0.753
1.764	0.743
1.810	0.737
1.910	0.768
1.955	0.773
2.020	0.774
2.077	0.767
2.147	0.747
2.199	0.717
CURVE 26 (cont.)*	
2.220	0.689
2.256	0.588
2.280	0.561
2.321	0.541
2.374	0.533
2.552	0.537
2.599	0.528
2.638	0.499
2.705	0.335
CURVE 27*	
T ~ 298	
0.377	0.247
0.408	0.699
0.424	0.775
0.481	0.853
0.515	0.870
0.561	0.880
0.600	0.886
0.722	0.885
0.790	0.881
0.892	0.872
0.955	0.864
1.000	0.859
1.105	0.861
1.187	0.856
1.243	0.839
1.285	0.826
1.310	0.821
1.392	0.836
1.413	0.836
1.519	0.804
1.564	0.792
1.604	0.790
1.750	0.735
1.802	0.729
1.858	0.740
1.941	0.763
2.006	0.760
2.090	0.724
2.193	0.691
2.237	0.608
2.256	0.589
2.272	0.560
2.302	0.530
2.360	0.509
CURVE 27 (cont.)*	
2.403	0.501
2.511	0.507
2.601	0.512
2.620	0.497
2.674	0.416
2.705	0.335
CURVE 28*	
T ~ 298	
0.377	0.247
0.407	0.688
0.474	0.817
0.522	0.865
0.611	0.881
0.757	0.876
0.827	0.866
0.953	0.860
1.016	0.863
1.072	0.847
1.120	0.847
1.189	0.834
1.220	0.832
1.245	0.838
1.304	0.822
1.336	0.816
1.416	0.816
1.447	0.808
1.501	0.788
1.617	0.771
1.699	0.739
1.747	0.727
1.801	0.723
1.941	0.731
1.988	0.722
2.019	0.719
2.200	0.662
2.224	0.647
2.231	0.601
2.257	0.561
2.325	0.516
2.377	0.494
2.418	0.489
2.495	0.491
2.579	0.502
2.605	0.491
2.671	0.403
CURVE 28 (cont.)*	
2.690	0.376
2.705	0.335
CURVE 29*	
T ~ 298	
0.386	0.259
0.401	0.712
0.421	0.801
0.433	0.854
0.473	0.910
0.601	0.943
0.651	0.948
0.709	0.948
0.780	0.938
0.864	0.936
0.896	0.941
1.127	0.937
1.164	0.943
1.228	0.939
1.301	0.929
1.409	0.944
1.632	0.914
1.780	0.865
1.810	0.864
1.856	0.877
1.876	0.894
1.922	0.905
2.049	0.901
2.109	0.888
2.199	0.875
2.227	0.858
2.326	0.742
2.373	0.723
2.414	0.717
2.563	0.729
2.597	0.724
2.626	0.707
2.681	0.642
2.708	0.590
CURVE 30*	
T ~ 298	
0.386	0.259
0.401	0.712
0.426	0.804
CURVE 30 (cont.)*	
0.470	0.884
0.501	0.903
0.543	0.917
0.638	0.940
0.711	0.940
0.796	0.931
0.894	0.941
0.934	0.942
0.950	0.946
1.011	0.948
1.075	0.934
1.154	0.928
1.212	0.935
1.246	0.935
1.297	0.930
1.371	0.936
1.433	0.936
1.498	0.907
1.548	0.902
1.608	0.905
1.713	0.864
1.780	0.849
1.824	0.848
1.907	0.873
1.975	0.876
1.989	0.872
2.012	0.872
2.133	0.835
2.201	0.818
2.232	0.792
2.279	0.730
2.379	0.666
2.403	0.658
2.505	0.676
2.570	0.676
2.614	0.660
2.647	0.631
2.709	0.548
CURVE 31*	
T ~ 298	
0.376	0.263
0.405	0.754
0.449	0.843
0.493	0.867
0.543	0.867
CURVE 31 (cont.)*	
0.601	0.857
0.686	0.847
0.776	0.822
0.841	0.817
0.888	0.821
1.011	0.800
1.115	0.783
1.218	0.784
1.320	0.770
1.410	0.770
1.460	0.757
1.517	0.744
1.600	0.745
1.701	0.703
1.763	0.693
1.816	0.698
1.953	0.724
2.008	0.718
2.067	0.707
2.197	0.664
2.254	0.550
2.280	0.527
2.321	0.514
2.436	0.512
2.504	0.527
2.584	0.536
2.623	0.529
2.683	0.415
2.704	0.348
CURVE 32*	
T ~ 298	
0.376	0.263
0.401	0.712
0.453	0.826
0.485	0.851
0.600	0.856
0.693	0.845
0.763	0.823
0.831	0.815
0.892	0.822
0.976	0.810
1.155	0.782
1.319	0.764
1.551	0.730
1.618	0.714
CURVE 32 (cont.)*	
1.722	0.671
1.801	0.662
1.931	0.688
2.055	0.663
2.119	0.638
2.200	0.623
2.218	0.577
2.266	0.508
2.302	0.486
2.368	0.470
2.412	0.464
2.502	0.463
2.574	0.468
2.616	0.460
2.702	0.320
CURVE 33*	
T ~ 298	
0.381	0.251
0.408	0.635
0.447	0.813
0.480	0.860
0.521	0.873
0.628	0.861
0.705	0.848
0.767	0.822
0.832	0.808
0.958	0.804
1.087	0.790
1.132	0.786
1.318	0.755
1.438	0.761
1.500	0.744
1.538	0.727
1.641	0.700
1.727	0.700
1.802	0.677
1.822	0.677
1.886	0.696
1.942	0.706
2.012	0.698
2.067	0.689
2.128	0.684
2.201	0.661
2.216	0.647
2.249	0.564

*Not shown on plot

DATA TABLE NO. 124 NORMAL SPECTRAL REFLECTANCE OF ZINC OXIDE PIGMENTED COATINGS (continued)

λ	ρ
CURVE 33 (cont.)* T ~ 298	
2.273	0.524
2.303	0.495
2.337	0.489
2.423	0.497
2.495	0.493
2.583	0.515
2.604	0.513
2.630	0.500
2.684	0.402
2.708	0.327
CURVE 34* T ~ 298	
0.381	0.251
0.408	0.635
0.422	0.717
0.454	0.797
0.499	0.833
0.571	0.840
0.743	0.829
0.914	0.807
0.944	0.807
1.070	0.790
1.213	0.754
1.298	0.742
1.408	0.740
1.515	0.712
1.632	0.691
1.672	0.666
1.737	0.640
1.812	0.631
2.004	0.630
2.115	0.592
2.199	0.570
2.284	0.455
2.327	0.435
2.441	0.429
2.584	0.429
2.624	0.410
2.704	0.294
CURVE 35* T ~ 298	
0.44	0.960
0.60	0.985
CURVE 36* T ~ 298	
0.44	0.935
0.60	0.975
CURVE 37* T ~ 298	
0.44	0.945
0.60	0.980
CURVE 38* T ~ 298	
0.44	0.905
0.60	0.960
CURVE 39* T ~ 298	
0.44	0.885
0.60	0.895
CURVE 40* T ~ 298	
0.44	0.865
0.60	0.885
CURVE 41* T ~ 298	
0.44	0.940
0.60	0.970
CURVE 42* T ~ 298	
0.44	0.930
0.60	0.980
CURVE 43* T ~ 298	
0.44	0.910
0.60	0.945
CURVE 44* T ~ 298	
0.44	0.895
0.60	0.925
CURVE 45 T ~ 298	
0.44	0.540
0.60	0.750
CURVE 46 T ~ 298	
0.44	0.525
0.60	0.750*
CURVE 47 T ~ 298	
0.44	0.755
0.60	0.890
CURVE 48* T ~ 298	
0.38	0.300
0.46	0.800
0.49	0.920
0.57	0.940
0.71	0.930
0.89	0.920
0.98	0.925
1.13	0.910
1.40	0.925
1.54	0.935
1.80	0.885
1.94	0.855
2.01	0.830
2.11	0.820
2.31	0.770
CURVE 48 (cont.)*	
2.43	0.725
2.53	0.700
2.60	0.670
2.64	0.640
2.71	0.555
CURVE 49* T ~ 298	
0.38	0.250
0.42	0.800
0.49	0.870
0.57	0.900
0.71	0.910
0.79	0.910
0.89	0.920
0.98	0.925
1.12	0.920
1.34	0.950
1.47	0.945
1.52	0.945
1.56	0.935
1.63	0.920
1.80	0.910
1.86	0.905
2.01	0.865
2.14	0.860
2.37	0.800
2.54	0.760
2.60	0.725
2.71	0.650
CURVE 50* T ~ 298	
0.38	0.301
0.43	0.841
0.48	0.903
0.55	0.932
0.64	0.934
0.89	0.923
1.06	0.923
1.13	0.912
1.54	0.930
1.94	0.848
2.02	0.822
CURVE 50 (cont.)*	
2.12	0.817
2.31	0.770
2.40	0.729
2.56	0.684
2.62	0.656
2.70	0.561
CURVE 51* T ~ 298	
0.38	0.253
0.42	0.794
0.50	0.879
0.61	0.904
0.81	0.909
0.92	0.922
1.10	0.913
1.32	0.929
1.50	0.945
1.66	0.916
1.89	0.900
2.02	0.861
2.13	0.861
2.43	0.782
2.56	0.753
2.71	0.644
CURVE 52* T ~ 298	
0.44	0.875
0.60	0.925
CURVE 53* T ~ 298	
0.44	0.890
0.60	0.875
CURVE 54* T ~ 298	
0.44	0.560
0.60	0.835
CURVE 55* T ~ 298	
0.44	0.885
0.60	0.945
CURVE 56* T ~ 298	
0.440	0.840
0.600	0.935
CURVE 57* T ~ 298	
0.44	0.910
0.60	0.920
CURVE 58* T ~ 298	
0.440	0.880
0.600	0.915
CURVE 59* T ~ 298	
0.44	0.910
0.60	0.935
CURVE 60* T ~ 298	
0.44	0.900
0.60	0.935
CURVE 61* T ~ 298	
0.44	0.840
0.60	0.915
CURVE 62* T ~ 298	
0.44	0.460
0.60	0.745
CURVE 63* T ~ 298	
0.44	0.842
0.60	0.773
CURVE 64* T ~ 298	
0.44	0.524
0.60	0.675
CURVE 65* T ~ 298	
0.440	0.885
0.600	0.920
CURVE 66* T ~ 298	
0.44	0.765
0.60	0.890
CURVE 67* T ~ 298	
0.44	0.840
0.60	0.776
CURVE 68* T ~ 298	
0.440	0.648
0.600	0.727
CURVE 69* T ~ 298	
0.44	0.870
0.60	0.875
CURVE 70* T ~ 298	
0.440	0.520
0.600	0.725
CURVE 71* T ~ 298	
0.440	0.970
0.600	0.980
CURVE 72* T ~ 298	
0.44	0.805
0.60	0.950
CURVE 73* T ~ 298	
0.44	0.860
0.60	0.850
CURVE 74* T ~ 298	
0.440	0.350
0.600	0.785
CURVE 75* T ~ 298	
0.382	0.500
0.395	0.661
0.404	0.784
0.415	0.863
0.427	0.894
0.448	0.920
0.472	0.935
0.499	0.943
0.520	0.943
0.585	0.922
0.667	0.896
0.724	0.886
0.741	0.886
CURVE 76* T ~ 298	
0.385	0.500
0.421	0.662
0.439	0.749
0.464	0.800

*Not shown on plot

DATA TABLE NO. 124 NORMAL SPECTRAL REFLECTANCE OF ZINC OXIDE PIGMENTED COATINGS (continued)

λ	ρ
CURVE 76 (cont.)*	
T ~ 298	
0.495	0.837
0.529	0.859
0.578	0.874
0.612	0.875
0.670	0.863
0.724	0.864
0.743	0.866
CURVE 77*	
T ~ 298	
0.381	0.500
0.397	0.661
0.414	0.782
0.431	0.838
0.448	0.872
0.469	0.894
0.499	0.912
0.532	0.917
0.572	0.912
0.617	0.895
0.691	0.881
0.741	0.880
CURVE 78*	
T ~ 298	
0.381	0.500
0.386	0.524
0.400	0.584
0.406	0.615
0.420	0.684
0.430	0.726
0.441	0.762
0.459	0.803
0.489	0.840
0.528	0.866
0.558	0.877
0.598	0.881
0.620	0.877
0,664	0.863
0.710	0.863
0.743	0.866
CURVE 79*	
T = 298	
0.380	0.317
0.383	0.345
0.398	0.743
0.403	0.822
0.406	0.859
0.422	0.926
0.428	0.944
0.444	0.963
0.463	0.972
0.510	0.980
0.561	0.982
0,629	0.977
0.700	0.977
CURVE 80*	
T = 298	
0.380	0.317
0.383	0.345
0.398	0.731
0.401	0.772
0.406	0.846
0.412	0.884
0.423	0.925
0.432	0.944
0.449	0.961
0.483	0.976
0.564	0.980
0.630	0.975
0.700	0.974
CURVE 81*	
T = 298	
0.380	0.317
0.383	0.346
0.388	0.427
0.399	0.704
0.406	0.829
0.414	0.882
0.431	0.934
0.441	0.945
0.459	0.959
0.502	0.970
0.562	0.976
CURVE 81 (cont.)*	
0.625	0.973
0.700	0.973
CURVE 82*	
T = 298	
0.380	0.310
0.386	0.345
0.407	0.738
0.415	0.833
0.427	0.888
0.434	0.906
0.469	0.938
0.477	0.939
0.492	0.946
0.510	0.949
0.523	0.955
0.563	0.958
0.610	0.961
0.700	0.961
CURVE 83*	
T = 298	
0.38	0.296
0.40	0.804
0.43	0.912
0.51	0.951
0.64	0.945
0.77	0.929
1.17	0.915
1.29	0.901
1.37	0.911
1.41	0.932
1.45	0.954
1.50	0.963
1.68	0.868
1.74	0.862
1.82	0.861
1.85	0.836
1.88	0.804
1.93	0.792
2.07	0.823
2.12	0.827
2.20	0.806
2.25	0.738
CURVE 83 (cont.)*	
2.30	0.721
2.36	0.711
2.48	0.700
2.60	0.673
2.62	0.649
2.67	0.580
2.69	0.569
2.71	0.571
2.75	0.594
CURVE 84*	
T = 298	
0.40	0.621
0.43	0.890
0.47	0.920
0.55	0.934
0.82	0.916
0.87	0.917
1.00	0.930
1.10	0.891
1.21	0.913
1.34	0.914
1.42	0.936
1.47	0.962
1.50	0.965
1.54	0.955
1.67	0.872
1.73	0.862
1.80	0.863
1.84	0.855
1.89	0.812
1.94	0.820
2.00	0.864
2.16	0.849
2.21	0.831
2.29	0.750
2.32	0.742
2.48	0.739
2.58	0.714
2.66	0.650
2.69	0.617
2.71	0.614
2.76	0.619
CURVE 85*	
T = 298	
0.380	0.330
0.385	0.409
0.395	0.674
0.400	0.771
0.404	0.820
0.416	0.875
0.428	0.908
0.449	0.929
0.525	0.965
0.562	0.967
0.700	0.967
CURVE 86*	
T = 298	
0.380	0.259
0.389	0.417
0.402	0.572
0.411	0.638
0.427	0.712
0.443	0.763
0.467	0.819
0.488	0.857
0.520	0.898
0.561	0.923
0.593	0.934
0.646	0.943
0.700	0.944
CURVE 87*	
T = 298	
0.380	0.182
0.385	0.222
0.390	0.294
0.404	0.383
0.427	0.483
0.473	0.647
0.500	0.727
0.521	0.778
0.547	0.825
0.575	0.860
0.611	0.892
0.643	0.909
0.659	0.916
CURVE 87 (cont.)*	
0.679	0.924
0.700	0.927
CURVE 88*	
T = 298	
0.380	0.260
0.385	0.318
0.396	0.552
0.407	0.729
0.412	0.779
0.420	0.813
0.430	0.843
0.441	0.858
0.453	0.869
0.492	0.870
0.542	0.862
0.642	0.842
0.700	0.835
CURVE 89*	
T = 298	
0.380	0.211
0.386	0.252
0.392	0.334
0.397	0.451
0.409	0.569
0.420	0.642
0.431	0.694
0.436	0.704
0.442	0.715
0.460	0.756
0.480	0.785
0.504	0.807
0.530	0.819
0.575	0.828
0.659	0.827
0.680	0.819
0.700	0.818
CURVE 90*	
T = 298	
0.380	0.274
0.383	0.287
CURVE 90 (cont.)*	
0.389	0.385
0.398	0.593
0.407	0.741
0.414	0.808
0.420	0.838
0.428	0.862
0.440	0.883
0.453	0.895
0.472	0.903
0.506	0.909
0.551	0.908
0.606	0.898
0.680	0.887
0.700	0.886
CURVE 91*	
T = 298	
0.380	0.248
0.385	0.296
0.402	0.583
0.410	0.648
0.420	0.689
0.433	0.729
0.449	0.760
0.479	0.800
0.515	0.833
0.558	0.855
0.602	0.868
0.658	0.872
0.700	0.871
CURVE 92*	
T = 298	
0.380	0.258
0.385	0.300
0.399	0.628
0.408	0.793
0.414	0.837
0.422	0.872
0.433	0.901
0.444	0.917
0.461	0.927
0.504	0.930
0.575	0.926
CURVE 92 (cont.)*	
0.663	0.915
0.700	0.909
CURVE 93*	
T = 298	
0.380	0.258
0.385	0.300
0.399	0.624
0.407	0.724
0.415	0.782
0.422	0.815
0.431	0.842
0.446	0.868
0.466	0.888
0.486	0.902
0.535	0.914
0.579	0.916
0.651	0.911
0.700	0.903
CURVE 94*	
T = 298	
0.380	0.368
0.389	0.514
0.399	0.751
0.406	0.845
0.411	0.884
0.419	0.918
0.429	0.944
0.442	0.959
0.477	0.971
0.594	0.978
0.600	0.975
0.700	0.971
CURVE 95*	
T = 298	
0.380	0.319
0.388	0.420
0.399	0.621
0.406	0.700
0.412	0.748
0.422	0.790

*Not shown on plot

DATA TABLE NO. 124 NORMAL SPECTRAL REFLECTANCE OF ZINC OXIDE PIGMENTED COATINGS (continued)

λ	ρ
CURVE 95 (cont.)*	**T = 298**
0.438	0.825
0.466	0.861
0.498	0.890
0.547	0.918
0.574	0.931
0.700	0.954
CURVE 96*	**T = 298**
0.380	0.320
0.389	0.483
0.395	0.639
0.404	0.793
0.411	0.858
0.418	0.896
0.423	0.916
0.434	0.938
0.444	0.951
0.458	0.960
0.478	0.965
0.595	0.961
0.700	0.953
CURVE 97*	**T = 298**
0.380	0.312
0.382	0.320
0.392	0.500
0.399	0.661
0.408	0.800
0.412	0.842
0.416	0.876
0.424	0.905
0.436	0.927
0.446	0.940
0.459	0.948
0.494	0.954
0.600	0.949
0.700	0.939
CURVE 98*	**T = 298**
0.380	0.319
0.384	0.338
0.388	0.416
0.398	0.636
0.404	0.746
0.410	0.816
0.419	0.872
0.431	0.906
0.448	0.918
0.469	0.933
0.489	0.940
0.531	0.936
0.600	0.936
0.616	0.930
0.650	0.930
0.689	0.920
0.700	0.919
CURVE 99*	**T = 298**
0.380	0.319
0.384	0.338
0.388	0.416
0.398	0.636
0.402	0.700
0.412	0.807
0.420	0.863
0.430	0.891
0.438	0.899
0.465	0.924
0.491	0.937
0.603	0.936
0.616	0.930
0.650	0.930
0.688	0.927
0.700	0.927
CURVE 100	**T = 298**
0.0700	0.076
0.0800	0.072
0.0899	0.068
0.0998	0.067
0.1100	0.068
CURVE 100 (cont.)	
0.1201	0.068
0.1306	0.070
0.1401	0.068
0.1525	0.062
0.1624	0.050
0.1722	0.066
0.1801	0.102
0.1888	0.131
CURVE 101*	**T ~ 300**
0.250	0.030
0.300	0.021
0.340	0.023
0.344	0.020
0.350	0.024
0.360	0.024
CURVE 102*	**T ~ 300**
0.30	0.025
0.34	0.046
0.40	0.969
0.44	1.029
0.55	0.975
0.66	0.969
0.72	0.971
0.75	0.960
0.80	0.965
0.85	0.954
0.93	0.950
1.00	0.940
1.12	0.931
1.18	0.885
1.20	0.911
1.24	0.929
1.27	0.916
1.33	0.909
1.39	0.854
1.42	0.878
1.46	0.888
1.53	0.867
1.58	0.882
1.62	0.866
1.65	0.820
CURVE 102 (cont.)*	
1.68	0.686
1.76	0.669
1.80	0.822
1.85	0.794
1.88	0.857
1.95	0.843
2.00	0.820
2.08	0.823
2.12	0.802
2.15	0.761
2.22	0.609
2.31	0.350
2.35	0.489
2.39	0.409
2.44	0.474
2.48	0.427
2.52	0.509
2.56	0.530
2.60	0.470
CURVE 103*	**T ~ 300**
0.250	0.049
0.287	0.052
0.320	0.041
0.360	0.041
CURVE 104*	**T ~ 300**
0.30	0.045
0.35	0.045
0.43	0.624
0.48	0.803
0.53	0.904
0.60	0.936
0.72	0.940
0.84	0.938
0.92	0.916
0.99	0.916
1.04	0.901
1.12	0.897
1.18	0.853
1.22	0.884
1.32	0.869
1.39	0.824
CURVE 104 (cont.)*	
1.43	0.846
1.48	0.847
1.53	0.824
1.59	0.828
1.65	0.808
1.68	0.695
1.69	0.662
1.72	0.703
1.74	0.646
1.78	0.766
1.82	0.766
1.85	0.756
1.91	0.797
2.00	0.761
2.12	0.748
2.22	0.609
2.31	0.350
2.35	0.489
2.39	0.409
2.44	0.474
2.48	0.427
2.52	0.509
2.56	0.530
2.60	0.470
CURVE 105*	**T ~ 300**
1.00	0.763
1.25	0.734
1.52	0.755
1.64	0.753
1.75	0.622
1.92	0.688
2.06	0.671
2.26	0.316
2.55	0.372
2.68	0.373
2.79	0.250
3.05	0.241
3.25	0.203
3.37	0.165
3.47	0.108
3.65	0.141
3.74	0.383
3.87	0.410
4.03	0.291
CURVE 105 (cont.)*	
4.21	0.456
4.36	0.457
4.53	0.257
4.66	0.215
4.78	0.180
5.00	0.202
5.10	0.111
5.27	0.100
5.52	0.144
6.01	0.057
6.20	0.059
6.51	0.076
6.90	0.037
7.08	0.031
7.35	0.036
7.68	0.032
7.92	0.077
8.20	0.036
8.50	0.011
8.78	0.019
8.97	0.044
9.12	0.090
9.42	0.107
10.00	0.111
11.04	0.084
11.51	0.088
12.02	0.072
12.32	0.080
12.41	0.107
13.83	0.118
14.34	0.127
15.00	0.156
CURVE 106*	**T ~ 300**
1.00	0.733
1.24	0.703
1.59	0.731
1.75	0.622
1.92	0.688
2.06	0.671
2.26	0.316
2.55	0.372
2.68	0.373
2.79	0.250
3.05	0.241
CURVE 106 (cont.)*	
3.25	0.203
3.37	0.165
3.47	0.108
3.65	0.141
3.74	0.383
3.87	0.410
4.03	0.291
4.21	0.456
4.39	0.457
4.67	0.261
4.70	0.217
4.86	0.218
4.99	0.254
5.24	0.119
5.55	0.196
5.68	0.140
6.02	0.121
6.37	0.102
9.18	0.101
9.54	0.111
10.37	0.116
11.72	0.116
12.69	0.127
13.36	0.133
14.43	0.154
15.00	0.176
CURVE 107*	**T ~ 300**
1.00	0.830
1.33	0.840
1.63	0.829
1.83	0.803
2.31	0.695
2.57	0.619
2.65	0.564
2.90	0.365
3.15	0.288
3.43	0.202
3.58	0.181
3.85	0.172
4.22	0.164
5.67	0.104
9.06	0.101
10.62	0.120
11.54	0.122
CURVE 107 (cont.)*	
13.80	0.147
15.00	0.186
CURVE 108*	**T ~ 300**
1.00	0.788
1.35	0.786
2.52	0.531
2.84	0.351
2.93	0.294
3.36	0.175
3.59	0.144
4.02	0.128
4.55	0.125
5.30	0.100
5.53	0.100
5.93	0.072
6.19	0.076
6.57	0.108
CURVE 109	**T ~ 298**
0.300	0.019
0.352	0.021
0.371	0.036
0.371	0.600
0.382	0.755
0.396	0.819
0.424	0.886*
0,445	0.931*
0.499	0.956
0.868	0.936*
1.089	0.936
1.157	0.924
1.763	0.913*
1.828	0.901
2.094	0.901
2.190	0.884
2.273	0.899
2.377	0.899
2.500	0.872

*Not shown on plot

DATA TABLE NO. 124 NORMAL SPECTRAL REFLECTANCE OF ZINC OXIDE PIGMENTED COATINGS (continued)

CURVE 110
T ~ 298

λ	ρ
0.300	0.044
0.346	0.044
0.371	0.074*
0.371	0.600*
0.380	0.736
0.443	0.898
0.477	0.929*
0.557	0.946
0.797	0.930
1.753	0.899*
1.925	0.889
2.135	0.890
2.207	0.874
2.500	0.889

CURVE 111
T ~ 298

λ	ρ
0.300	0.035
0.371	0.043
0.371	0.600*
0.380	0.736*
0.429	0.884*
0.471	0.933*
0.529	0.947
0.770	0.940
1.285	0.906*
2.009	0.880
2.134	0.876
2.202	0.865
2.305	0.886
2.364	0.886*
2.500	0.845

CURVE 112
T ~ 298

λ	ρ
0.291	0.012
0.349	0.018
0.370	0.041*
0.378	0.104
0.395	0.787
0.429	0.901*
0.459	0.930*
0.501	0.943
0.547	0.934*

CURVE 112 (cont.)

λ	ρ
0.657	0.924
0.867	0.855
0.927	0.846
0.978	0.851
1.052	0.843
1.135	0.848
1.180	0.819
1.213	0.814
1.249	0.828
1.338	0.819
1.382	0.787
1.477	0.802
1.523	0.794
1.572	0.770
1.652	0.770
1.704	0.724*
1.732	0.647
1.801	0.750
1.850	0.761
1.963	0.757
2.038	0.745
2.077	0.757
2.142	0.721*
2.234	0.549
2.303	0.403
2.363	0.480
2.381	0.420
2.419	0.467
2.453	0.478*
2.474	0.454*
2.500	0.493

CURVE 113
T ~ 298

λ	ρ
0.291	0.012*
0.349	0.018*
0.370	0.041*
0.378	0.104*
0.390	0.587
0.420	0.816
0.454	0.878
0.489	0.896*
0.669	0.876
0.907	0.818
1.127	0.776
1.188	0.734

CURVE 113 (cont.)

λ	ρ
1.220	0.743
1.259	0.761
1.280	0.761
1.351	0.709
1.388	0.689
1.432	0.689
1.497	0.672
1.621	0.611
1.726	0.513
1.744	0.513
1.793	0.562
1.837	0.572
1.977	0.509
2.114	0.477
2.303	0.276
2.340	0.293
2.403	0.277
2.432	0.284
2.500	0.283

CURVE 114
T ~ 298

λ	ρ
0.291	0.012*
0.349	0.018
0.370	0.041*
0.378	0.104
0.390	0.587*
0.412	0.754
0.457	0.855
0.486	0.876
0.556	0.881
0.701	0.856
0.934	0.795
1.110	0.765
1.182	0.718
1.220	0.726
1.274	0.739
1.346	0.690
1.413	0.666
1.478	0.658
1.550	0.631
1.650	0.558
1.726	0.495
1.764	0.503
1.814	0.546
1.844	0.551

CURVE 114 (cont.)

λ	ρ
1.942	0.505
2.080	0.456
2.202	0.346
2.277	0.272
2.303	0.259
2.344	0.269
2.399	0.244
2.441	0.244
2.500	0.230

CURVE 115
T ~ 298

λ	ρ
0.291	0.036
0.344	0.036
0.373	0.045*
0.378	0.104*
0.394	0.660
0.419	0.843*
0.444	0.896*
0.466	0.917
0.505	0.926
0.564	0.914
0.688	0.897
0.860	0.846
1.132	0.832
1.167	0.809
1.180	0.784
1.230	0.820
1.294	0.820
1.361	0.790
1.379	0.759
1.439	0.787*
1.488	0.787
1.534	0.767
1.600	0.773
1.658	0.773
1.695	0.745*
1.717	0.609
1.727	0.630
1.735	0.615
1.772	0.703
1.825	0.752*
1.944	0.756*
1.996	0.729
2.118	0.744
2.153	0.726

CURVE 115 (cont.)

λ	ρ
2.302	0.361
2.355	0.522
2.388	0.390
2.399	0.414
2.407	0.459*
2.441	0.506
2.467	0.435
2.500	0.483*

CURVE 116
T ~ 298

λ	ρ
0.300	0.057
0.339	0.057
0.365	0.072
0.396	0.804
0.423	0.889*
0.455	0.917*
0.486	0.926*
0.777	0.902*
1.339	0.898
1.549	0.908*
1.607	0.904*
1.695	0.915
1.752	0.912
1.834	0.899
2.123	0.894
2.199	0.890
2.300	0.904
2.358	0.904
2.445	0.885
2.500	0.885

CURVE 117
T ~ 298

λ	ρ
0.300	0.079
0.346	0.079
0.369	0.090
0.396	0.771
0.420	0.844*
0.451	0.878
0.504	0.901
0.745	0.881
0.979	0.866
1.157	0.863
1.498	0.819

CURVE 117 (cont.)

λ	ρ
1.850	0.734*
2.110	0.655
2.316	0.611
2.500	0.550

CURVE 118
T ~ 298

λ	ρ
0.300	0.057*
0.339	0.057*
0.365	0.072*
0.396	0.742
0.412	0.815*
0.453	0.869*
0.482	0.884
0.546	0.891
0.857	0.856*
1.182	0.837
1.532	0.781
1.864	0.693
2.043	0.642
2.238	0.596*
2.392	0.545
2.500	0.502

CURVE 119
T ~ 298

λ	ρ
0.300	0.067
0.351	0.066
0.368	0.086
0.393	0.779
0.414	0.861
0.451	0.898*
0.494	0.911*
0.777	0.890
0.839	0.871*
1.079	0.871*
1.320	0.881
1.597	0.887
1.682	0.906*
1.751	0.888
2.177	0.879
2.243	0.898*
2.282	0.913
2.432	0.899
2.500	0.917

CURVE 120*
T ~ 298

λ	ρ
0.300	0.022
0.328	0.028
0.355	0.023
0.365	0.041
0.370	0.070
0.375	0.176
0.380	0.443
0.382	0.712
0.387	0.811
0.397	0.892
0.405	0.945
0.411	0.972
0.421	0.999
0.430	1.007
0.503	1.014
0.537	0.994
0.568	0.970
0.599	0.947
0.629	0.939
0.700	0.926

CURVE 121*
T ~ 298

λ	ρ
0.300	0.022
0.328	0.028
0.355	0.023
0.365	0.041
0.370	0.070
0.375	0.176
0.387	0.621
0.393	0.689
0.402	0.740
0.423	0.835
0.437	0.884
0.452	0.926
0.466	0.955
0.482	0.974
0.503	0.986
0.542	0.982
0.568	0.971
0.606	0.949
0.665	0.931
0.700	0.926

CURVE 122
T ~ 298

λ	ρ
0.300	0.034
0.361	0.034
0.367	0.042
0.373	0.080
0.380	0.156
0.391	0.379
0.396	0.465
0.405	0.529
0.422	0.613
0.448	0.721
0.469	0.791
0.488	0.847
0.512	0.893
0.530	0.911
0.552	0.921
0.580	0.926
0.609	0.923
0.652	0.912
0.700	0.892

CURVE 123
T ~ 298

λ	ρ
0.300	0.044
0.363	0.031
0.372	0.049
0.386	0.162
0.389	0.221
0.395	0.320
0.401	0.388
0.412	0.465
0.429	0.539
0.449	0.612
0.480	0.738
0.496	0.791
0.517	0.848
0.536	0.888
0.553	0.904
0.569	0.912
0.599	0.914
0.634	0.905
0.700	0.889

*Not shown on plot

DATA TABLE NO. 124 NORMAL SPECTRAL REFLECTANCE OF ZINC OXIDE PIGMENTED COATINGS (continued)

CURVE 124, T ~ 298

λ	ρ
0.300	0.053
0.353	0.038
0.368	0.028
0.375	0.042
0.380	0.076
0.390	0.153
0.398	0.207
0.416	0.289
0.446	0.411
0.481	0.591
0.501	0.674
0.524	0.754
0.541	0.797
0.557	0.827
0.583	0.863
0.608	0.883
0.638	0.893
0.673	0.894
0.700	0.886

CURVE 125, T ~ 298

λ	ρ
0.250	0.030
0.308	0.025
0.360	0.025
0.360	0.044
0.404	0.977
0.445	1.030
0.564	0.975
0.719	0.971
0.757	0.961
0.806	0.966
0.993	0.943
1.116	0.935
1.172	0.880
1.203	0.918*
1.235	0.929*
1.260	0.917*
1.322	0.917
1.387	0.858
1.420	0.882
1.458	0.890*
1.523	0.867
1.568	0.885
1.615	0.875
1.651	0.806
1.681	0.688
1.746	0.666
1.800	0.826*
1.840	0.794
1.880	0.860
1.995	0.822
2.076	0.825*
2.129	0.787*
2.184	0.679
2.292	0.352
2.343	0.494*
2.388	0.410*
2.435	0.478*
2.478	0.427
2.513	0.485*
2.549	0.532
2.600	0.479

CURVE 126, T ~ 298

λ	ρ
0.250	0.051
0.298	0.051
0.329	0.044
0.351	0.044
0.438	0.632
0.517	0.893
0.591	0.935*
0.838	0.940
0.926	0.919
0.987	0.919
1.111	0.899*
1.174	0.853
1.222	0.887
1.327	0.868*
1.396	0.827
1.453	0.849
1.538	0.825
1.591	0.829
1.653	0.804
1.688	0.660*
1.710	0.705
1.734	0.646*
1.768	0.715*
1.804	0.771
1.845	0.758
1.905	0.800
1.978	0.771
2.093	0.754*
2.159	0.679*
2.223	0.562
2.292	0.352*
2.343	0.494*
2.388	0.410*
2.435	0.478*
2.478	0.427*
2.520	0.507*
2.577	0.507*
2.600	0.479*

CURVE 127, T ~ 298

λ	ρ
1.00	0.761
1.27	0.732
1.52	0.755
1.73	0.621
1.95	0.687
2.24	0.318
2.61	0.376
2.79	0.250*
2.94	0.250
3.17	0.230
3.35	0.153
3.43	0.107
3.62	0.138
3.71	0.378
3.83	0.400
4.00	0.289
4.26	0.456
4.62	0.229
4.79	0.181
4.97	0.204
5.12	0.101*
5.25	0.101
5.49	0.145
6.01	0.057
6.22	0.061
6.51	0.075
6.82	0.043
7.10	0.034
7.66	0.034
7.90	0.079
8.42	0.012
8.71	0.012
8.93	0.038
9.18	0.094
9.57	0.110
10.16	0.110
11.07	0.086
11.50	0.089
12.05	0.072
12.30	0.081
12.42	0.106
13.77	0.119
14.32	0.127
15.00	0.154

CURVE 128, T ~ 298

λ	ρ
1.00	0.733
1.22	0.701
1.56	0.734
1.66	0.716
1.73	0.621*
1.95	0.687
2.24	0.318*
2.61	0.376*
2.79	0.250*
3.07	0.271
3.35	0.153*
3.43	0.107*
3.62	0.138*
3.71	0.378*
3.86	0.416
4.00	0.296
4.29	0.463
4.81	0.218
4.99	0.255
5.21	0.117
5.54	0.196
5.73	0.136
6.52	0.101
9.22	0.101
9.57	0.110*
10.34	0.117
12.01	0.123
13.37	0.133
14.42	0.155
15.00	0.177

CURVE 129, T ~ 298

λ	ρ
0.250	0.004
0.300	0.013
0.335	0.011
0.338	0.030
0.360	0.050
0.382	0.300*
0.411	0.984
0.430	1.031
0.465	1.035
0.509	1.003
0.590	1.000
0.647	0.987
0.748	0.983
0.820	0.985
0.937	0.976
0.977	0.981
1.021	0.974
1.394	0.974
1.424	0.965
1.502	0.965*
1.639	0.948*
1.756	0.914*
1.862	0.873*
1.934	0.834
1.961	0.845
2.282	0.766*
2.387	0.749
2.458	0.712*
2.602	0.683

CURVE 130, T ~ 298

λ	ρ
0.250	0.004*
0.300	0.013*
0.335	0.011*
0.338	0.030*
0.360	0.050*
0.405	0.300
0.560	0.854
0.582	0.913*
0.651	0.954
0.841	0.954
1.393	0.910
1.411	0.895*
1.483	0.895*
1.612	0.864*
1.747	0.817*
1.852	0.773*
1.926	0.719*
1.966	0.725*
2.068	0.707*
2.451	0.591*
2.602	0.541

CURVE 131, T ~ 298

λ	ρ
1.000	0.828
1.50	0.838*
1.85	0.817
2.45	0.674
2.67	0.598*
2.92	0.385
3.36	0.222
3.57	0.183
3.90	0.170
4.18	0.166
5.60	0.103
9.09	0.102
10.45	0.119*
11.31	0.119
13.87	0.150
15.00	0.184

CURVE 132, T ~ 298

λ	ρ
1.00	0.788
1.41	0.788
2.47	0.569
2.66	0.482*
3.02	0.270*
3.48	0.159
3.78	0.138
4.53	0.127
4.94	0.109
5.56	0.099
6.02	0.070
6.59	0.110

CURVE 133*, T = 298

λ	ρ
0.300	0.050
0.352	0.044
0.365	0.040
0.370	0.042
0.375	0.047
0.377	0.058
0.379	0.072
0.380	0.091
0.392	0.619
0.398	0.837
0.401	0.872
0.406	0.907
0.409	0.924
0.417	0.946
0.423	0.957
0.428	0.966
0.438	0.976
0.447	0.981
0.459	0.982
0.469	0.980

CURVE 134*, T ~ 298

λ	ρ
0.305	0.014
0.352	0.015
0.372	0.019
0.401	0.879
0.422	0.962
0.433	0.985
0.450	0.993
0.493	0.989
0.611	0.948
0.684	0.930
0.815	0.912
0.931	0.903
1.120	0.903
1.146	0.896
1.157	0.885
1.175	0.844
1.188	0.891
1.193	0.898
1.211	0.902
1.603	0.908
1.622	0.904
1.639	0.892
1.652	0.857
1.664	0.702
1.674	0.731
1.679	0.701
1.705	0.848
1.722	0.675
1.758	0.855
1.771	0.821
1.790	0.884
1.810	0.908
1.824	0.853
1.836	0.860
1.844	0.855
1.851	0.867
1.859	0.914
1.887	0.925
1.910	0.928
1.966	0.922
2.007	0.923

CURVE 135*, T = 298

λ	ρ
0.307	0.046
0.358	0.046
0.376	0.052
0.409	0.834
0.440	0.915
0.472	0.938
0.498	0.945
0.633	0.946
0.817	0.930
1.017	0.902
1.110	0.892
1.142	0.879
1.165	0.852
1.210	0.893
1.274	0.897
1.317	0.891
1.385	0.890
1.510	0.874
1.559	0.876
1.614	0.884
1.648	0.871
1.676	0.834
1.694	0.738
1.701	0.767

*Not shown on plot

λ	ρ
CURVE 135 (cont.)*	
T = 298	
1.713	0.706
1.728	0.813
1.744	0.688
1.764	0.822
1.775	0.798
1.798	0.849
1.812	0.859
1.840	0.845
1.881	0.872
1.916	0.875
1.956	0.870
1.989	0.855
CURVE 136*	
T = 298	
0.307	0.046
0.358	0.046
0.376	0.052
0.409	0.834
0.437	0.909
0.457	0.938
0.494	0.950
0.541	0.955
0.705	0.947
0.962	0.924
1.115	0.904
1.149	0.895
1.166	0.864
1.201	0.903
1.226	0.910
1.326	0.910
1.458	0.902
1.523	0.892
1.563	0.894
1.617	0.897
1.660	0.891
1.694	0.738
1.701	0.767
1.713	0.706
1.728	0.813
1.744	0.688
1.764	0.822
1.775	0.809
1.793	0.861
1.807	0.870

λ	ρ
CURVE 136 (cont.)*	
1.824	0.866
1.836	0.852
1.859	0.861
1.875	0.879
1.901	0.886
1.943	0.888
1.967	0.883
1.994	0.868
CURVE 137*	
T = 298	
0.309	0.045
0.341	0.045
0.375	0.051
0.407	0.920
0.426	0.950
0.444	0.973
0.493	0.977
0.764	0.922
0.827	0.907
0.885	0.899
1.044	0.886
1.128	0.867
1.156	0.852
1.171	0.840
1.182	0.818
1.185	0.838
1.206	0.860
1.270	0.858
1.401	0.832
1.450	0.825
1.522	0.805
1.571	0.803
1.619	0.791
1.642	0.774
1.657	0.754
1.683	0.614
1.694	0.646
1.704	0.606
1.719	0.662
1.734	0.606
1.772	0.708
1.777	0.699
1.791	0.734
1.811	0.746
1.821	0.715

λ	ρ
CURVE 137 (cont.)*	
1.827	0.731
1.838	0.711
1.852	0.749
1.898	0.753
1.950	0.752
1.973	0.746
1.999	0.732
2.055	0.724
CURVE 138*	
T = 298	
0.309	0.045
0.341	0.045
0.375	0.051
0.407	0.920
0.426	0.950
0.444	0.973
0.493	0.977
0.626	0.957
0.766	0.937
0.875	0.915
1.105	0.913
1.142	0.902
1.159	0.883
1.175	0.857
1.185	0.901
1.198	0.911
1.238	0.917
1.437	0.919
1.615	0.910
1.640	0.901
1.655	0.885
1.671	0.847
1.690	0.665
1.699	0.728
1.711	0.657
1.726	0.815
1.738	0.644
1.761	0.819
1.770	0.779
1.780	0.839
1.806	0.876
1.812	0.832
1.821	0.860
1.832	0.832
1.841	0.872

λ	ρ
CURVE 138 (cont.)*	
1.865	0.895
1.906	0.911
1.954	0.898
1.975	0.881
2.011	0.871
2.049	0.868
CURVE 139*	
T = 298	
0.318	0.049
0.373	0.050
0.405	0.934
0.416	0.980
0.435	0.993
0.477	0.996
0.572	0.978
0.667	0.963
0.809	0.936
0.876	0.922
1.083	0.924
1.122	0.917
1.149	0.901
1.174	0.847
1.177	0.904
1.188	0.918
1.220	0.925
1.266	0.923
1.300	0.926
1.343	0.922
1.403	0.928
1.610	0.916
1.633	0.911
1.659	0.890
1.673	0.856
1.687	0.723
1.691	0.700
1.705	0.738
1.711	0.685
1.734	0.825
1.737	0.720
1.746	0.683
1.774	0.826
1.776	0.791
1.813	0.887
1.820	0.830
1.828	0.859

λ	ρ
CURVE 139 (cont.)*	
1.834	0.835
1.841	0.868
1.863	0.894
1.892	0.908
1.920	0.911
1.946	0.895
1.984	0.884
2.013	0.881
CURVE 140*	
T ~ 298	
0.313	0.042
0.334	0.037
0.361	0.041
0.373	0.051
0.409	0.716
0.429	0.839
0.442	0.869
0.464	0.899
0.495	0.917
0.533	0.929
0.617	0.926
0.869	0.905
0.942	0.899
1.123	0.899
1.146	0.886
1.173	0.834
1.193	0.899
1.214	0.905
1.300	0.903
1.363	0.906
1.507	0.910
1.542	0.900
1.579	0.902
1.611	0.906
1.634	0.901
1.643	0.893
1.673	0.726
1.689	0.672
1.700	0.696
1.708	0.661
1.723	0.707
1.745	0.648
1.768	0.790
1.775	0.792
1.813	0.887

λ	ρ
CURVE 140 (cont.)*	
1.820	0.830
1.828	0.859
1.834	0.835
1.841	0.868
1.863	0.894
1.892	0.908
1.920	0.911
1.946	0.895
2.015	0.870
CURVE 141*	
T = 298	
0.325	0.020
0.350	0.020
0.375	0.040
0.400	0.790
0.450	0.960
0.500	0.965
0.600	0.960
0.700	0.950
0.800	0.940
0.900	0.935
1.0	0.930
1.2	0.935
1.4	0.960
1.6	0.905
1.8	0.850
2.0	0.770
CURVE 142*	
T = 298	
0.325	0.030
0.350	0.030
0.375	0.050
0.400	0.795
0.450	0.965
0.500	0.980
0.600	0.980
0.700	0.970
0.800	0.965
0.900	0.960
1.0	0.955
1.2	0.960
1.4	0.970
1.6	0.915

λ	ρ
CURVE 142 (cont.)*	
1.8	0.850
2.0	0.770
CURVE 143*	
T = 298	
0.397	0.700
0.411	0.799
0.420	0.865
0.435	0.902
0.447	0.926
0.460	0.939
0.495	0.958
0.522	0.965
0.560	0.967
0.643	0.970
0.680	0.968
0.700	0.969
1.499	0.957
1.557	0.953
1.640	0.933
1.735	0.913
1.776	0.905
1.801	0.904
1.853	0.899
1.867	0.894
1.881	0.883
1.906	0.841
1.914	0.839
1.928	0.844
1.951	0.853
1.982	0.854
CURVE 144*	
T = 298	
0.397	0.700
0.404	0.810
0.420	0.890
0.441	0.950
0.450	0.964
0.480	0.974
0.505	0.977
0.558	0.980
0.683	0.980
0.699	0.983
1.500	0.990

λ	ρ
CURVE 144 (cont.)*	
1.573	0.986
1.651	0.974
1.777	0.968
1.820	0.971
1.870	0.983
1.910	0.968
1.939	0.979
1.969	0.965
2.000	0.960
CURVE 145*	
T = 298	
0.305	0.056
0.341	0.061
0.366	0.069
0.379	0.086
0.382	0.123
0.400	0.741
0.417	0.881
0.435	0.915
0.459	0.943
0.520	0.952
0.557	0.949
0.859	0.951
0.978	0.924
1.146	0.906
1.170	0.897
1.184	0.878
1.200	0.913
1.495	0.900
1.593	0.882
1.626	0.866
1.647	0.839
1.677	0.748
1.701	0.861
1.744	0.737
1.761	0.860
1.795	0.890
1.821	0.861
1.878	0.906
2.000	0.878

*Not shown on plot

DATA TABLE NO. 124 NORMAL SPECTRAL REFLECTANCE OF ZINC OXIDE PIGMENTED COATINGS (continued)

λ	ρ
CURVE 146*	
T = 298	
0.305	0.056
0.341	0.061
0.366	0.069
0.379	0.086
0.382	0.123
0.400	0.741
0.417	0.881
0.435	0.915
0.459	0.943
0.520	0.952
0.557	0.949
0.859	0.951
1.068	0.939
1.144	0.939
1.181	0.887
1.200	0.947
1.437	0.941
1.571	0.946
1.600	0.925
1.632	0.879
1.677	0.748
1.701	0.861
1.744	0.737
1.752	0.823
1.786	0.947
1.804	0.909
1.821	0.901
1.827	0.907
1.838	0.901
1.841	0.957
1.872	0.964
2.000	0.944
CURVE 147*	
T ~ 298	
0.250	0.020
0.323	0.021
0.349	0.039
0.416	0.743
0.439	0.805
0.479	0.827
0.563	0.839
0.723	0.839
0.859	0.833

λ	ρ
CURVE 147 (cont.)*	
0.946	0.829
0.996	0.826
1.070	0.825
1.207	0.807
1.400	0.797
1.512	0.785
1.683	0.752
1.792	0.738
1.881	0.661
1.914	0.646
2.023	0.648
2.068	0.638
2.183	0.588
2.235	0.574
2.272	0.558
2.393	0.543
2.498	0.514
CURVE 148*	
T ~ 298	
0.326	0.072
0.343	0.075
0.355	0.073
0.364	0.071
0.366	0.075
0.371	0.080
0.373	0.089
0.375	0.113
0.380	0.216
0.384	0.382
0.386	0.404
0.386	0.432
0.387	0.438
0.387	0.500
0.388	0.523
0.391	0.560
0.394	0.600
0.399	0.624
0.403	0.655
0.407	0.673
0.411	0.689
0.417	0.707
0.426	0.729
0.429	0.729
0.436	0.741

λ	ρ
CURVE 148 (cont.)*	
0.449	0.749
0.463	0.751
0.467	0.748
0.470	0.753
0.519	0.745
0.528	0.743
0.539	0.746
0.552	0.741
0.554	0.737
0.567	0.740
0.586	0.740
0.645	0.738
0.674	0.736
0.679	0.733
0.693	0.735
0.697	0.736
0.700	0.734
CURVE 149*	
T ~ 298	
0.701	0.735
0.759	0.730
0.823	0.716
0.903	0.704
0.938	0.696
0.950	0.696
0.984	0.687
0.991	0.680
1.02	0.679
1.05	0.669
1.12	0.654
1.12	0.645
1.15	0.633
1.16	0.622
1.19	0.624
1.25	0.610
1.31	0.591
1.33	0.575
1.36	0.526
1.37	0.521
1.38	0.523
1.38	0.532
1.38	0.529
1.40	0.543
1.41	0.546

λ	ρ
CURVE 149 (cont.)*	
1.43	0.546
1.52	0.512
1.54	0.512
1.55	0.503
1.56	0.504
1.60	0.485
1.64	0.466
1.67	0.532
1.68	0.416
1.69	0.415
1.73	0.401
1.75	0.414
1.78	0.416
1.82	0.404
1.84	0.401
1.86	0.401
1.88	0.403
1.90	0.392
1.92	0.392
1.92	0.384
1.96	0.373
2.00	0.362
2.10	0.327
2.13	0.311
2.16	0.261
2.18	0.234
2.23	0.195
2.25	0.195
2.28	0.205
2.31	0.229
2.35	0.221
2.41	0.219
2.43	0.206
2.46	0.192
2.53	0.170
2.60	0.132
2.65	0.095
2.70	0.079
CURVE 150*	
T ~ 298	
0.701	0.745
0.748	0.737
0.829	0.729
0.916	0.720

λ	ρ
CURVE 150 (cont.)*	
0.953	0.714
1.11	0.697
1.13	0.691
1.15	0.680
1.18	0.674
1.23	0.684
1.28	0.681
1.29	0.679
1.31	0.680
1.33	0.672
1.34	0.651
1.35	0.635
1.36	0.627
1.38	0.624
1.40	0.636
1.42	0.670
1.44	0.670
1.49	0.663
1.56	0.659
1.57	0.659
1.61	0.652
1.63	0.633
1.68	0.553
1.70	0.543
1.73	0.541
1.75	0.550
1.78	0.606
1.79	0.613
1.80	0.611
1.81	0.614
1.83	0.614
1.88	0.640
1.97	0.620
2.03	0.611
2.09	0.601
2.11	0.590
2.19	0.395
2.23	0.331
2.25	0.290
2.28	0.273
2.29	0.297
2.30	0.335
2.33	0.373
2.34	0.369
2.37	0.372
2.40	0.368

λ	ρ
CURVE 150 (cont.)*	
2.42	0.368
2.44	0.359
2.47	0.343
2.51	0.336
2.54	0.326
2.57	0.300
2.62	0.207
2.70	0.114
CURVE 151*	
T ~ 298	
0.325	0.063
0.342	0.063
0.356	0.065
0.362	0.069
0.369	0.076
0.371	0.085
0.374	0.103
0.375	0.134
0.378	0.178
0.383	0.341
0.388	0.413
0.391	0.461
0.395	0.498
0.403	0.549
0.412	0.581
0.418	0.597
0.432	0.619
0.446	0.624
0.454	0.627
0.459	0.630
0.462	0.628
0.470	0.633
0.482	0.632
0.489	0.635
0.507	0.633
0.511	0.637
0.529	0.632
0.558	0.634
0.598	0.635
0.608	0.639
0.649	0.640
0.686	0.638
0.692	0.638
0.693	0.636
0.700	0.634

λ	ρ
CURVE 152*	
T ~ 298	
0.700	0.678
0.950	0.680
1.23	0.679
1.26	0.682
1.48	0.678
1.53	0.671
1.54	0.665
1.61	0.663
1.68	0.653
1.77	0.637
1.82	0.629
1.89	0.616
1.96	0.606
2.00	0.601
2.17	0.566
2.23	0.548
2.33	0.538
2.40	0.527
2.48	0.509
2.60	0.493
2.70	0.514
CURVE 153*	
T ~ 298	
0.325	0.117
0.364	0.118
0.368	0.125
0.372	0.142
0.374	0.168
0.380	0.325
0.385	0.463
0.387	0.519
0.389	0.551
0.393	0.590
0.403	0.650
0.411	0.677
0.419	0.698
0.425	0.701
0.436	0.712
0.446	0.716
0.473	0.716
0.506	0.705
0.510	0.700
0.526	0.700
0.543	0.688

λ	ρ
CURVE 153 (cont.)*	
0.554	0.692
0.575	0.685
0.593	0.687
0.598	0.685
0.601	0.688
0.604	0.688
0.606	0.683
0.632	0.682
0.658	0.683
0.674	0.679
0.681	0.679
0.688	0.675
0.700	0.674
CURVE 154*	
T ~ 298	
0.700	0.670
0.756	0.662
0.846	0.639
0.881	0.630
0.903	0.630
1.00	0.610
1.03	0.603
1.07	0.600
1.09	0.591
1.10	0.591
1.15	0.569
1.22	0.570
1.24	0.560
1.28	0.557
1.34	0.537
1.36	0.529
1.37	0.520
1.40	0.519
1.44	0.511
1.48	0.498
1.52	0.486
1.59	0.475
1.63	0.464
1.67	0.434
1.71	0.418
1.73	0.426
1.76	0.432
1.85	0.423
1.89	0.416
1.93	0.406

* Not shown on plot

DATA TABLE NO. 124 NORMAL SPECTRAL REFLECTANCE OF ZINC OXIDE PIGMENTED COATINGS (continued)

λ	ρ
CURVE 154 (cont.)*	
T ~ 298	
2.00	0.391
2.06	0.378
2.10	0.364
2.15	0.339
2.19	0.310
2.22	0.284
2.24	0.248
2.26	0.241
2.32	0.285
2.34	0.295
2.38	0.296
2.40	0.291
2.44	0.287
2.47	0.292
2.52	0.311
2.57	0.303
2.64	0.258
2.70	0.222
CURVE 155*	
T ~ 298	
0.700	0.715
1.01	0.714
1.04	0.716
1.08	0.715
1.11	0.709
1.14	0.689
1.17	0.664
1.19	0.691
1.21	0.700
1.27	0.699
1.29	0.691
1.32	0.691
1.36	0.659
1.37	0.649
1.38	0.644
1.40	0.656
1.41	0.664
1.43	0.667
1.44	0.673
1.48	0.662
1.50	0.649
1.56	0.649
1.56	0.654

λ	ρ
CURVE 155 (cont.)*	
1.58	0.654
1.60	0.647
1.60	0.648
1.65	0.608
1.66	0.573
1.67	0.517
1.68	0.511
1.70	0.531
1.72	0.509
1.73	0.491
1.73	0.491
1.74	0.522
1.76	0.548
1.77	0.548
1.78	0.568
1.80	0.571
1.81	0.567
1.82	0.551
1.83	0.551
1.86	0.566
1.88	0.552
1.90	0.530
1.92	0.531
1.98	0.513
1.99	0.515
2.00	0.513
2.06	0.510
2.10	0.492
2.14	0.445
2.21	0.369
2.25	0.293
2.25	0.277
2.27	0.272
2.27	0.283
2.28	0.289
2.30	0.341
2.32	0.341
2.35	0.301
2.40	0.325
2.41	0.320
2.45	0.291
2.48	0.310
2.52	0.314
2.55	0.307
2.59	0.256
2.62	0.214
2.70	0.129

λ	ρ
CURVE 156*	
T ~ 298	
0.700	0.700
0.798	0.697
0.805	0.703
0.820	0.703
0.828	0.700
0.840	0.702
0.879	0.703
0.896	0.698
0.921	0.700
0.921	0.705
0.939	0.699
1.01	0.696
1.04	0.698
1.08	0.693
1.13	0.680
1.16	0.676
1.22	0.681
1.26	0.677
1.30	0.669
1.33	0.659
1.37	0.639
1.41	0.648
1.43	0.646
1.47	0.644
1.52	0.628
1.56	0.628
1.58	0.623
1.61	0.614
1.63	0.607
1.65	0.592
1.66	0.566
1.68	0.553
1.69	0.554
1.70	0.554
1.72	0.535
1.74	0.546
1.75	0.561
1.76	0.568
1.77	0.572
1.78	0.579
1.82	0.568
1.85	0.569
1.88	0.574
1.90	0.569
1.95	0.556
1.97	0.549
2.02	0.544

λ	ρ
CURVE 156 (cont.)*	
2.10	0.526
2.13	0.508
2.18	0.475
2.26	0.399
2.29	0.362
2.29	0.374
2.30	0.385
2.32	0.428
2.37	0.391
2.42	0.410
2.46	0.386
2.52	0.403
2.56	0.392
2.61	0.341
2.64	0.281
2.66	0.265
2.66	0.238
2.70	0.206
CURVE 157*	
T ~ 298	
0.697	0.486
0.767	0.504
0.796	0.509
0.828	0.509
0.857	0.520
0.898	0.525
0.939	0.527
0.961	0.530
0.978	0.528
1.02	0.530
1.08	0.526
1.10	0.526
1.11	0.522
1.16	0.511
1.19	0.522
1.22	0.522
1.24	0.520
1.28	0.518
1.33	0.508
1.35	0.501
1.38	0.498
1.40	0.501
1.45	0.496
1.49	0.486
1.51	0.486

λ	ρ
CURVE 157 (cont.)*	
1.58	0.478
1.62	0.469
1.64	0.454
1.65	0.439
1.67	0.430
1.69	0.426
1.71	0.426
1.73	0.434
1.76	0.460
1.79	0.466
1.83	0.471
1.85	0.467
1.93	0.469
1.95	0.477
1.97	0.475
2.15	0.487
2.17	0.482
2.20	0.459
2.23	0.411
2.26	0.388
2.31	0.423
2.32	0.420
2.35	0.418
2.41	0.443
2.46	0.455
2.49	0.474
2.51	0.485
2.52	0.490
2.53	0.490
2.58	0.454
2.63	0.419
2.70	0.404
CURVE 158*	
T ~ 298	
0.697	0.481
0.724	0.481
0.762	0.491
0.821	0.500
0.912	0.503
0.979	0.508
1.05	0.506
1.10	0.501
1.13	0.498
1.16	0.490
1.18	0.494
1.18	0.499

λ	ρ
CURVE 158 (cont.)*	
1.19	0.501
1.20	0.498
1.23	0.501
1.27	0.498
1.30	0.489
1.31	0.489
1.37	0.474
1.41	0.474
1.43	0.476
1.47	0.463
1.48	0.465
1.52	0.458
1.58	0.454
1.63	0.437
1.66	0.417
1.67	0.405
1.68	0.400
1.72	0.403
1.74	0.413
1.76	0.424
1.77	0.438
1.82	0.441
1.86	0.449
2.08	0.456
2.16	0.449
2.19	0.440
2.21	0.419
2.21	0.404
2.25	0.370
2.26	0.367
2.28	0.377
2.31	0.398
2.35	0.393
2.41	0.416
2.46	0.426
2.48	0.438
2.50	0.458
2.54	0.458
2.59	0.425
2.63	0.397
2.65	0.392
2.70	0.393

λ	ρ
CURVE 159*	
T = 293	
0.360	0.045
0.366	0.045
0.372	0.056
0.374	0.069
0.376	0.109
0.379	0.173
0.383	0.335
0.386	0.449
0.390	0.560
0.394	0.608
0.398	0.643
0.403	0.666
0.410	0.694
0.419	0.724
0.435	0.746
0.479	0.772
0.523	0.795
0.559	0.799
0.604	0.806
0.710	0.800
0.710	0.787
0.727	0.787
0.760	0.799
0.807	0.795
0.983	0.790
1.13	0.779
1.16	0.764
1.17	0.753
1.18	0.749
1.21	0.770
1.24	0.775
1.34	0.767
1.38	0.740
1.40	0.734
1.43	0.751
1.47	0.751
1.53	0.734
1.58	0.744
1.61	0.739
1.63	0.730
1.65	0.702
1.69	0.621
1.73	0.600
1.75	0.600
1.80	0.680
1.89	0.682
1.93	0.670

λ	ρ
CURVE 159 (cont.)*	
1.96	0.663
2.08	0.662
2.11	0.658
2.16	0.630
2.20	0.589
2.23	0.536
2.27	0.436
2.30	0.396
2.32	0.386
2.40	0.399
2.50	0.399
CURVE 160*	
T = 293	
0.360	0.049
0.370	0.049
0.373	0.062
0.375	0.081
0.378	0.125
0.383	0.268
0.385	0.313
0.388	0.349
0.392	0.384
0.400	0.431
0.410	0.466
0.435	0.537
0.458	0.586
0.485	0.636
0.515	0.680
0.554	0.722
0.584	0.746
0.627	0.767
0.658	0.778
0.690	0.784
0.710	0.784
0.710	0.757
0.759	0.780
0.983	0.777
1.12	0.769
1.14	0.761
1.18	0.734
1.20	0.753
1.22	0.761
1.33	0.750
1.35	0.740
1.39	0.715
1.46	0.730

* Not shown on plot

DATA TABLE NO. 124 NORMAL SPECTRAL REFLECTANCE OF ZINC OXIDE PIGMENTED COATINGS (continued)

λ	ρ
CURVE 160 (cont.)*	
T = 293	
1.51	0.715
1.56	0.712
1.58	0.715
1.61	0.712
1.64	0.692
1.66	0.661
1.69	0.595
1.71	0.584
1.73	0.573
1.75	0.573
1.80	0.644
1.82	0.644
1.84	0.641
1.87	0.644
1.90	0.644
1.95	0.627
2.00	0.612
2.07	0.611
2.11	0.601
2.15	0.577
2.19	0.544
2.21	0.522
2.28	0.387
2.29	0.365
2.31	0.357
2.36	0.367
2.38	0.367
2.40	0.370
2.47	0.368
2.50	0.371
CURVE 161*	
T ~298	
0.357	0.038
0.368	0.054
0.374	0.066
0.385	0.554
0.394	0.723
0.404	0.824
0.418	0.880
0.437	0.908
0.459	0.923
0.503	0.932
0.575	0.932

λ	ρ
CURVE 161 (cont.)*	
0.714	0.908
0.946	0.875
1.12	0.852
1.16	0.837
1.17	0.802
1.19	0.843
1.29	0.838
1.34	0.830
1.38	0.796
1.41	0.811
1.46	0.815
1.53	0.796
1.59	0.813
1.62	0.801
1.65	0.795
1.66	0.753
1.67	0.649
1.68	0.653
1.69	0.631
1.71	0.743
1.73	0.606
1.75	0.740
1.77	0.729
1.80	0.800
1.84	0.751
1.85	0.788
1.86	0.799
1.93	0.799
1.98	0.778
2.08	0.778
2.14	0.746
2.19	0.650
2.22	0.541
2.24	0.422
2.26	0.351
2.28	0.327
2.29	0.335
2.31	0.500
2.33	0.548
2.34	0.469
2.35	0.438
2.36	0.429
2.37	0.438
2.38	0.490
2.40	0.489

λ	ρ
CURVE 162*	
T ~298	
0.357	0.038
0.368	0.054
0.374	0.066
0.385	0.445
0.392	0.544
0.400	0.677
0.407	0.731
0.421	0.787
0.456	0.856
0.482	0.887
0.518	0.902
0.714	0.908
0.946	0.875
1.12	0.852
1.16	0.837
1.17	0.802
1.19	0.843
1.29	0.838
1.34	0.830
1.38	0.796
1.41	0.811
1.46	0.815
1.53	0.796
1.59	0.813
1.62	0.801
1.65	0.795
1.66	0.753
1.67	0.649
1.68	0.653
1.69	0.631
1.71	0.743
1.73	0.606
1.75	0.740
1.77	0.729
1.80	0.800
1.84	0.751
1.85	0.788
1.86	0.799
1.93	0.784
1.98	0.767
2.09	0.767
2.15	0.721
2.19	0.650
2.22	0.541
2.24	0.422

λ	ρ
CURVE 162 (cont.)*	
2.26	0.351
2.28	0.327
2.29	0.335
2.31	0.500
2.33	0.548
2.34	0.469
2.35	0.438
2.36	0.429
2.37	0.438
2.38	0.490
2.40	0.489
CURVE 163	
T = 298	
0.287	0.023
0.326	0.029
0.352	0.062*
0.366	0.222
0.394	0.838
0.413	0.885*
0.443	0.911*
0.493	0.933*
0.661	0.957
0.882	0.964
1.159	0.957
1.457	0.940*
1.742	0.911*
1.975	0.868
CURVE 164	
T = 298	
0.287	0.023*
0.326	0.029*
0.352	0.062*
0.366	0.222*
0.389	0.719
0.416	0.778
0.446	0.841
0.478	0.888*
0.535	0.925*
0.602	0.951*
0.661	0.957*
0.882	0.964*
1.159	0.957*

λ	ρ
CURVE 164 (cont.)	
1.457	0.940*
1.628	0.915
1.866	0.860*
1.979	0.824*
CURVE 165	
T = 298	
0.287	0.023*
0.326	0.029*
0.352	0.062*
0.366	0.222*
0.389	0.719*
0.416	0.778*
0.446	0.841*
0.478	0.888*
0.535	0.925*
0.602	0.951*
0.661	0.957*
0.882	0.964*
1.159	0.957*
1.321	0.947
1.531	0.911*
1.724	0.875*
1.966	0.823*
CURVE 166	
T = 298	
0.287	0.023*
0.326	0.029*
0.352	0.062*
0.366	0.222*
0.386	0.654
0.404	0.728*
0.441	0.812
0.473	0.863
0.525	0.907*
0.573	0.936*
0.661	0.957*
0.882	0.964*
1.083	0.958
1.276	0.934
1.743	0.846
1.931	0.798*
1.989	0.776*

λ	ρ
CURVE 167	
T = 423	
0.290	0.036*
0.339	0.036*
0.367	0.061
0.393	0.739*
0.414	0.799*
0.456	0.864*
0.510	0.916
0.539	0.938*
0.575	0.960*
0.786	0.978
0.913	0.978
1.211	0.961
1.478	0.937
1.701	0.905
1.901	0.854*
1.992	0.822*
CURVE 168*	
T = 298	
0.290	0.036
0.339	0.036
0.367	0.061
0.392	0.664
0.441	0.809
0.490	0.879
0.536	0.925
0.578	0.949
0.641	0.966
0.786	0.978
0.913	0.978
1.211	0.961
1.391	0.938
1.649	0.889
1.991	0.812
CURVE 169*	
T = 77	
0.290	0.036
0.339	0.036
0.367	0.061
0.392	0.664
0.441	0.809
0.483	0.845

λ	ρ
CURVE 169 (cont.)*	
0.542	0.891
0.626	0.925
0.737	0.951
0.872	0.966
0.968	0.971
1.201	0.955
1.415	0.929
1.673	0.876
1.991	0.791
CURVE 170	
T = 298	
0.401	0.607
0.401	0.794*
0.407	0.845*
0.434	0.888*
0.464	0.917*
0.501	0.939*
0.549	0.952
0.633	0.962
1.329	0.968
1.531	0.961
1.862	0.931
2.000	0.914
CURVE 171	
T = 423	
0.401	0.605*
0.401	0.680
0.433	0.794
0.481	0.868
0.535	0.915
0.596	0.940*
0.655	0.956*
1.174	0.956*
CURVE 172	
T = 298	
0.401	0.605*
0.401	0.680*
0.433	0.794*
0.481	0.862*
0.517	0.890

λ	ρ
CURVE 172 (cont.)	
0.580	0.923
0.646	0.936*
1.090	0.942*
1.479	0.929
1.656	0.915*
1.801	0.897*
2.000	0.852
CURVE 173	
T = 77	
0.655	0.956
1.174	0.956
1.499	0.942
1.705	0.927
1.920	0.894
2.000	0.878
CURVE 174*	
T = 423	
0.553	0.882
0.609	0.910
0.671	0.930
0.752	0.946
0.829	0.967
1.218	0.967
1.367	0.948
1.644	0.919
2.000	0.859
CURVE 175*	
T = 298	
0.392	0.409
0.400	0.543
0.428	0.660
0.461	0.749
0.507	0.816
0.560	0.868
0.634	0.907
0.712	0.934
0.866	0.959
1.150	0.959
1.265	0.949
1.436	0.928

* Not shown on plot

DATA TABLE NO. 124   NORMAL SPECTRAL REFLECTANCE OF ZINC OXIDE PIGMENTED COATINGS (continued)

λ	ρ
CURVE 175* (cont.) T = 298	
1.644	0.899
1.844	0.865
2.000	0.836
CURVE 176* T = 77	
0.392	0.409
0.400	0.600
0.438	0.727
0.466	0.792
0.519	0.856
0.581	0.909
0.681	0.939
0.797	0.967
1.218	0.967
1.370	0.946
1.531	0.924
1.737	0.895
2.000	0.853
CURVE 177* T = 423	
0.393	0.700
0.395	0.762
0.413	0.841
0.454	0.898
0.501	0.921
0.550	0.934
0.629	0.937
0.685	0.948
0.830	0.965
1.200	0.965
1.418	0.950
1.622	0.931
1.747	0.921
1.904	0.905
2.000	0.894
CURVE 178* T = 298	
0.393	0.700
0.395	0.762
0.413	0.841

λ	ρ
CURVE 178 (cont.)	
0.454	0.898
0.504	0.929
0.563	0.941
0.686	0.947
0.830	0.965
1.200	0.965
1.418	0.950
1.622	0.931
1.747	0.921
1.904	0.905
2.000	0.894
CURVE 179* T = 77	
0.393	0.700
0.395	0.762
0.413	0.841
0.454	0.898
0.504	0.929
0.563	0.941
0.686	0.947
0.830	0.965
1.200	0.965
1.418	0.950
1.622	0.931
1.747	0.921
1.904	0.905
2.000	0.894
CURVE 180 T = 423	
0.400	0.723*
0.425	0.814*
0.454	0.863*
0.490	0.903*
0.524	0.922*
0.605	0.930*
0.724	0.930
0.770	0.952
0.833	0.964*
0.924	0.974*
1.228	0.974
1.405	0.961
1.664	0.929
1.832	0.904
2.000	0.875

λ	ρ
CURVE 181 T = 298	
0.400	0.723*
0.428	0.805*
0.481	0.866*
0.541	0.912*
0.601	0.932*
0.668	0.942
0.753	0.942
0.932	0.967*
1.214	0.967*
1.434	0.943*
1.627	0.924*
1.766	0.906*
1.918	0.869*
2.000	0.845*
CURVE 182 T = 77	
0.400	0.723
0.436	0.787
0.503	0.860
0.546	0.890
0.615	0.921
0.645	0.930
0.738	0.930
0.789	0.928
0.861	0.950
1.160	0.954
1.334	0.950
1.434	0.943
1.627	0.924
1.758	0.899
1.862	0.878
2.000	0.840
CURVE 183* T = 423	
0.385	0.405
0.389	0.523
0.439	0.721
0.481	0.807
0.528	0.861
0.591	0.904
0.644	0.931
0.716	0.953

λ	ρ
CURVE 183 (cont.)	
0.790	0.961
0.883	0.970
1.208	0.970
1.383	0.954
1.555	0.937
1.759	0.904
2.000	0.862
CURVE 184* T = 298	
0.385	0.405
0.389	0.523
0.439	0.721
0.473	0.778
0.501	0.810
0.544	0.836
0.589	0.887
0.639	0.924
0.690	0.941
0.775	0.970
1.174	0.970
1.342	0.954
1.541	0.931
1.733	0.899
2.000	0.845
CURVE 185* T = 77	
0.385	0.405
0.389	0.523
0.428	0.673
0.456	0.734
0.510	0.796
0.580	0.844
0.639	0.858
0.676	0.886
0.759	0.918
0.877	0.942
1.016	0.955
1.201	0.955
1.361	0.944
1.522	0.925
1.726	0.893
2.000	0.845

λ	ρ
CURVE 186* T = 423	
0.386	0.600
0.388	0.692
0.442	0.802
0.489	0.861
0.541	0.904
0.590	0.935
0.653	0.951
0.912	0.971
1.208	0.971
1.306	0.963
1.471	0.941
1.603	0.921
1.830	0.875
2.000	0.842
CURVE 187* T = 298	
0.386	0.600
0.388	0.692
0.438	0.810
0.480	0.876
0.547	0.924
0.624	0.954
0.668	0.964
0.899	0.971
1.208	0.971
1.408	0.958
1.734	0.906
2.000	0.853
CURVE 188* T = 77	
0.386	0.600
0.387	0.637
0.420	0.735
0.459	0.805
0.521	0.869
0.601	0.918
0.698	0.942
0.925	0.965
1.209	0.965
1.332	0.953
1.514	0.930
1.666	0.902
1.824	0.875
2.000	0.840

λ	ρ
CURVE 189* T = 298	
0.250	0.111
0.305	0.090
0.329	0.081
0.358	0.081
0.370	0.170
0.396	0.796
0.409	0.826
0.436	0.843
0.495	0.877
0.571	0.887
0.706	0.892
0.759	0.884
1.096	0.874
1.147	0.853
1.163	0.831
1.201	0.866
1.304	0.860
1.378	0.811
1.406	0.828
1.435	0.837
1.524	0.810
1.567	0.816
1.605	0.799
1.635	0.762
1.658	0.680
1.693	0.651
1.710	0.646
1.722	0.627
1.756	0.708
1.784	0.736
1.822	0.719
1.844	0.735
1.862	0.731
1.970	0.700
2.059	0.694
2.105	0.670
2.159	0.602
2.224	0.453
2.257	0.378
2.292	0.401
2.302	0.413
2.325	0.417
2.350	0.409
2.404	0.416
2.442	0.410
2.504	0.426

λ	ρ
CURVE 190* T = 281	
0.250	0.093
0.305	0.078
0.329	0.069
0.358	0.069
0.370	0.170
0.396	0.796
0.409	0.826
0.436	0.843
0.495	0.877
0.571	0.887
0.706	0.892
0.759	0.884
1.096	0.874
1.147	0.853
1.163	0.831
1.201	0.866
1.304	0.860
1.378	0.811
1.406	0.828
1.435	0.837
1.524	0.810
1.567	0.816
1.605	0.799
1.635	0.762
1.658	0.680
1.693	0.651
1.710	0.646
1.722	0.627
1.756	0.708
1.784	0.736
1.822	0.714
1.844	0.724
1.862	0.731
1.970	0.700
2.059	0.694
2.105	0.670
2.159	0.602
2.224	0.453
2.257	0.378
2.292	0.401
2.302	0.404
2.325	0.412
2.350	0.409
2.404	0.416
2.442	0.410
2.504	0.424

λ	ρ
CURVE 191* T = 281	
0.250	0.093
0.305	0.078
0.329	0.069
0.358	0.069
0.370	0.170
0.396	0.796
0.409	0.826
0.436	0.842
0.491	0.872
0.539	0.881
0.707	0.892
1.102	0.870
1.147	0.853
1.163	0.832
1.201	0.866
1.307	0.860
1.381	0.808
1.423	0.830
1.458	0.828
1.518	0.803
1.540	0.809
1.584	0.809
1.621	0.775
1.659	0.678
1.700	0.643
1.722	0.625
1.773	0.710
1.784	0.726
1.821	0.710
1.847	0.722
1.872	0.722
1.945	0.692
2.004	0.684
2.062	0.680
2.123	0.636
2.163	0.581
2.227	0.429
2.259	0.373
2.305	0.403
2.332	0.403
2.349	0.397
2.370	0.397
2.410	0.404
2.442	0.399
2.504	0.417

* Not shown on plot

DATA TABLE NO. 124 NORMAL SPECTRAL REFLECTANCE OF ZINC OXIDE PIGMENTED COATINGS (continued)

λ	ρ
CURVE 192*	
T = 281	
0.250	0.093
0.305	0.078
0.329	0.069
0.358	0.069
0.370	0.170
0.396	0.796
0.409	0.826
0.436	0.840
0.476	0.859
0.504	0.874
0.539	0.881
0.708	0.892
1.096	0.868
1.145	0.850
1.164	0.822
1.186	0.827
1.206	0.851
1.243	0.855
1.314	0.844
1.385	0.793
1.435	0.814
1.468	0.814
1.528	0.788
1.568	0.794
1.607	0.775
1.639	0.735
1.663	0.656
1.682	0.630
1.704	0.633
1.724	0.600
1.784	0.699
1.796	0.699
1.824	0.687
0.853	0.699
1.922	0.682
2.005	0.660
2.064	0.655
2.107	0.626
2.157	0.574
2.205	0.479
2.251	0.377
2.265	0.363
2.285	0.367
2.300	0.383
2.327	0.390
2.357	0.382

λ	ρ
CURVE 192 (cont.)	
2.414	0.388
2.445	0.383
2.504	0.396
CURVE 193*	
T = 281	
0.250	0.090
0.330	0.068
0.359	0.068
0.370	0.170
0.396	0.796
0.409	0.826
0.436	0.840
0.476	0.859
0.504	0.874
0.558	0.879
0.704	0.883
0.848	0.869
0.878	0.863
0.949	0.863
1.008	0.854
1.112	0.845
1.145	0.825
1.169	0.802
1.214	0.831
1.310	0.817
1.359	0.777
1.384	0.765
1.425	0.778
1.473	0.768
1.524	0.744
1.565	0.744
1.618	0.715
1.643	0.678
1.668	0.600
1.685	0.590
1.705	0.585
1.721	0.567
1.760	0.628
1.783	0.641
1.822	0.626
1.840	0.634
1.863	0.636
1.946	0.604
2.010	0.591
2.049	0.584

λ	ρ
CURVE 193 (cont.)	
2.103	0.556
2.167	0.496
2.242	0.350
2.266	0.329
2.299	0.348
2.324	0.353
2.349	0.346
2.419	0.350
2.445	0.346
2.507	0.356
CURVE 194*	
T = 281	
0.250	0.090
0.330	0.068
0.359	0.068
0.370	0.170
0.396	0.796
0.409	0.826
0.436	0.840
0.476	0.859
0.504	0.874
0.558	0.879
0.704	0.883
0.724	0.872
0.858	0.859
0.968	0.843
1.058	0.828
1.104	0.815
1.141	0.799
1.169	0.773
1.213	0.788
1.324	0.756
1.389	0.702
1.406	0.707
1.433	0.707
1.518	0.665
1.578	0.651
1.630	0.612
1.667	0.542
1.683	0.520
1.704	0.520
1.723	0.503
1.760	0.534
1.782	0.540
1.821	0.526

λ	ρ
CURVE 194 (cont.)	
1.866	0.526
1.976	0.487
2.043	0.473
2.127	0.436
2.197	0.370
2.247	0.296
2.262	0.283
2.319	0.299
2.379	0.294
2.428	0.294
2.463	0.292
2.504	0.294
CURVE 195*	
T = 281	
0.250	0.069
0.354	0.059
0.359	0.069
0.370	0.170
0.389	0.557
0.400	0.628
0.431	0.722
0.497	0.830
0.530	0.848
0.620	0.868
0.701	0.871
0.749	0.861
0.787	0.861
0.890	0.846
1.001	0.831
1.099	0.812
1.146	0.786
1.165	0.763
1.214	0.781
1.308	0.756
1.379	0.700
1.415	0.704
1.468	0.690
1.514	0.662
1.583	0.643
1.629	0.602
1.662	0.527
1.722	0.495
1.782	0.540
1.821	0.521
1.846	0.524

λ	ρ
CURVE 195 (cont.)	
1.865	0.524
1.951	0.491
2.025	0.470
2.099	0.446
2.187	0.380
2.247	0.296
2.262	0.283
2.319	0.294
2.428	0.294
2.463	0.292
2.504	0.294
CURVE 196*	
T = 298	
0.385	0.476
0.400	0.792
0.421	0.831
0.470	0.840
0.487	0.869
0.536	0.881
0.699	0.890
0.731	0.883
0.819	0.883
0.851	0.879
0.968	0.879
0.996	0.872
1.095	0.872
1.128	0.865
1.162	0.828
1.200	0.863
1.283	0.863
1.327	0.853
1.382	0.811
1.398	0.827
1.444	0.837
1.487	0.825
1.520	0.810
1.551	0.818
1.575	0.818
1.613	0.790
1.640	0.741
1.663	0.664
1.702	0.648
1.720	0.633
1.763	0.715
1.780	0.733

λ	ρ
CURVE 196 (cont.)	
1.819	0.721
1.841	0.735
1.861	0.735
1.882	0.725
1.958	0.703
2.043	0.703
2.107	0.675
2.165	0.592
2.237	0.401
2.261	0.379
2.286	0.393
2.303	0.407
2.322	0.413
2.349	0.403
2.401	0.414
2.446	0.409
2.505	0.428
CURVE 197*	
T = 281	
0.385	0.476
0.400	0.792
0.421	0.831
0.470	0.840
0.487	0.869
0.536	0.881
0.699	0.890
0.946	0.875
1.044	0.868
1.124	0.861
1.162	0.828
1.188	0.834
1.200	0.854
1.251	0.858
1.265	0.855
1.289	0.855
1.328	0.845
1.428	0.824
1.465	0.824
1.504	0.800
1.540	0.800
1.563	0.806
1.612	0.785
1.640	0.741
1.663	0.664
1.680	0.642

λ	ρ
CURVE 197 (cont.)	
1.700	0.642
1.720	0.616
1.781	0.716
1.824	0.701
1.843	0.714
1.864	0.719
1.966	0.685
2.045	0.682
2.081	0.671
2.146	0.612
2.201	0.509
2.237	0.401
2.265	0.367
2.293	0.384
2.306	0.395
2.326	0.400
2.344	0.395
2.400	0.400
2.445	0.395
2.505	0.410
CURVE 198*	
T = 281	
0.385	0.476
0.400	0.792
0.421	0.831
0.470	0.840
0.487	0.869
0.536	0.879
0.640	0.883
0.826	0.872
0.847	0.865
0.905	0.860
0.921	0.866
1.043	0.850
1.120	0.839
1.145	0.826
1.163	0.803
1.200	0.826
1.264	0.821
1.322	0.810
1.381	0.762
1.401	0.772
1.420	0.778
1.440	0.778
1.521	0.746

λ	ρ
CURVE 198 (cont.)	
1.565	0.746
1.605	0.730
1.644	0.681
1.665	0.606
1.707	0.590
1.720	0.568
1.784	0.644
1.824	0.628
1.858	0.635
1.965	0.596
2.049	0.577
2.108	0.552
2.154	0.503
2.243	0.342
2.263	0.323
2.305	0.341
2.328	0.343
2.362	0.336
2.395	0.338
2.450	0.335
2.505	0.341
CURVE 199*	
T = 281	
0.385	0.476
0.400	0.792
0.421	0.831
0.470	0.840
0.487	0.869
0.536	0.879
0.707	0.879
0.743	0.864
0.895	0.847
1.068	0.824
1.116	0.813
1.140	0.800
1.165	0.773
1.201	0.791
1.249	0.787
1.328	0.761
1.366	0.726
1.381	0.717
1.424	0.724
1.488	0.702
1.524	0.682
1.570	0.676

* Not shown on plot

DATA TABLE NO. 124 NORMAL SPECTRAL REFLECTANCE OF ZINC OXIDE PIGMENTED COATINGS (continued)

CURVE 199* (cont.)
T = 281

λ	ρ
1.606	0.657
1.644	0.610
1.663	0.550
1.705	0.534
1.720	0.513
1.758	0.550
1.783	0.562
1.824	0.542
1.842	0.545
1.866	0.543
1.978	0.499
2.067	0.472
2.128	0.435
2.201	0.368
2.244	0.291
2.266	0.275
2.305	0.287
2.325	0.287
2.373	0.281
2.412	0.281
2.442	0.275
2.505	0.278

CURVE 200*
T = 281

λ	ρ
0.385	0.476
0.400	0.792
0.412	0.828
0.439	0.846
0.493	0.862
0.613	0.862
0.710	0.852
0.724	0.836
0.919	0.790
1.101	0.725
1.166	0.669
1.205	0.669
1.284	0.637
1.386	0.557
1.452	0.535
1.519	0.488
1.570	0.467
1.626	0.428
1.661	0.375
1.723	0.345

CURVE 200 (cont.)

λ	ρ
1.743	0.351
1.774	0.351
1.829	0.329
1.938	0.293
2.114	0.245
2.180	0.214
2.261	0.165
2.350	0.167
2.459	0.158
2.505	0.158

CURVE 201*
T = 281

λ	ρ
0.385	0.476
0.400	0.792
0.423	0.825
0.444	0.822
0.483	0.849
0.540	0.846
0.706	0.814
0.766	0.784
0.901	0.727
1.061	0.639
1.141	0.580
1.173	0.549
1.208	0.532
1.386	0.403
1.523	0.321
1.637	0.265
1.678	0.236
1.732	0.216
1.852	0.187
2.002	0.158
2.183	0.132
2.261	0.112
2.339	0.112
2.505	0.104

CURVE 202*
T = 281

λ	ρ
0.385	0.476
0.400	0.626
0.485	0.753
0.517	0.779
0.559	0.793

CURVE 202 (cont.)

λ	ρ
0.615	0.800
0.691	0.795
0.763	0.759
0.903	0.701
1.026	0.638
1.127	0.569
1.169	0.525
1.211	0.509
1.368	0.398
1.466	0.340
1.535	0.295
1.640	0.245
1.692	0.218
1.796	0.186
2.000	0.141
2.104	0.124
2.242	0.100
2.505	0.092

CURVE 203*
T = 298

λ	ρ
0.700	0.837
0.753	0.828
0.841	0.817
0.883	0.814
0.946	0.814
1.034	0.816
1.105	0.808
1.151	0.771
1.158	0.776*
1.197	0.806
1.218	0.813*
1.248	0.813*
1.292	0.804
1.308	0.796*
1.347	0.758
1.360	0.763*
1.381	0.778
1.409	0.782*
1.582	0.758
1.604	0.751*
1.649	0.634
1.659	0.641*
1.672	0.645
1.690	0.617
1.698	0.611*

CURVE 203 (cont.)

λ	ρ
1.709	0.622*
1.746	0.696
1.757	0.714*
1.779	0.727*
1.801	0.711*
1.825	0.725
1.858	0.731*
1.900	0.718
1.956	0.710
2.033	0.707
2.056	0.701*
2.095	0.674*
2.163	0.556
2.234	0.390
2.242	0.384*
2.274	0.413
2.299	0.426*
2.411	0.426
2.496	0.432*

CURVE 204*
T = 281

λ	ρ
0.700	0.821
0.792	0.799
0.878	0.787
1.027	0.787
1.070	0.783
1.098	0.774
1.140	0.717
1.152	0.713
1.196	0.734
1.246	0.725
1.304	0.704
1.351	0.658
1.394	0.666
1.585	0.604
1.611	0.585
1.649	0.507
1.677	0.505
1.697	0.483
1.758	0.529
1.766	0.534
1.781	0.534
1.802	0.520
1.837	0.526
1.971	0.484

CURVE 204 (cont.)

λ	ρ
2.017	0.473
2.061	0.463
2.114	0.434
2.229	0.294
2.243	0.285
2.300	0.302
2.457	0.300
2.490	0.298

CURVE 205*
T = 381

λ	ρ
0.700	0.831
0.790	0.813
0.923	0.804
1.053	0.791
1.095	0.784
1.128	0.761
1.151	0.739
1.206	0.767
1.233	0.767
1.292	0.756
1.352	0.701
1.404	0.716
1.531	0.687
1.573	0.682
1.607	0.665
1.636	0.596
1.647	0.558
1.659	0.561
1.673	0.561
1.692	0.550
1.700	0.537
1.732	0.568
1.763	0.607
1.771	0.611
1.809	0.597
1.854	0.606
1.983	0.561
2.047	0.553
2.080	0.538
2.126	0.503
2.178	0.427
2.237	0.325
2.249	0.319
2.295	0.336
2.490	0.336

CURVE 206*
T = 281

λ	ρ
0.700	0.831
0.792	0.815
0.856	0.808
1.033	0.800
1.095	0.792
1.122	0.775
1.150	0.748
1.200	0.778
1.248	0.778
1.295	0.765
1.356	0.718
1.384	0.730
1.407	0.732
1.492	0.712
1.530	0.710
1.592	0.693
1.632	0.638
1.652	0.573
1.671	0.581
1.681	0.581
1.701	0.554
1.718	0.572
1.772	0.635
1.802	0.621
1.817	0.630
1.840	0.637
1.950	0.601
1.995	0.588
2.060	0.577
2.099	0.552
2.166	0.470
2.239	0.327
2.252	0.327
2.297	0.351
2.494	0.350

CURVE 207
T = 381

λ	ρ
0.700	0.807
0.824	0.782
0.893	0.782
0.994	0.770
1.082	0.753
1.138	0.709
1.153	0.703*

CURVE 207 (cont.)

λ	ρ
1.175	0.708*
1.206	0.723*
1.226	0.723
1.312	0.692
1.348	0.646
1.363	0.646*
1.399	0.652
1.417	0.652*
1.502	0.621
1.595	0.588
1.648	0.499
1.677	0.496*
1.701	0.476
1.739	0.514
1.768	0.521
1.802	0.506
1.846	0.514
1.965	0.473
2.022	0.467
2.078	0.446
2.141	0.402
2.234	0.285
2.253	0.277*
2.298	0.297*
2.338	0.297
2.412	0.291
2.493	0.291*

CURVE 208
T = 281

λ	ρ
0.700	0.819
0.790	0.802
0.875	0.794
0.915	0.794*
1.013	0.781
1.076	0.776
1.119	0.752
1.151	0.731*
1.204	0.754
1.227	0.754*
1.295	0.734*
1.343	0.689*
1.371	0.697

CURVE 208 (cont.)

λ	ρ
1.381	0.702
1.408	0.702*
1.560	0.657
1.599	0.647*
1.641	0.553
1.652	0.543*
1.668	0.547*
1.689	0.540
1.702	0.521
1.763	0.586*
1.774	0.590
1.789	0.585*
1.800	0.572
1.813	0.572*
1.852	0.582
1.930	0.553
2.013	0.531*
2.057	0.520*
2.128	0.475
2.213	0.349*
2.241	0.303
2.254	0.303*
2.302	0.326*
0.349	0.326
2.438	0.320*
2.493	0.320

CURVE 209
T = 298

λ	ρ
0.700	0.819*
0.790	0.802*
1.094	0.800
1.120	0.783
1.151	0.757
1.170	0.767*
1.192	0.788*
1.228	0.795
1.265	0.795*
1.306	0.783
1.355	0.745
1.397	0.762
1.502	0.754
1.577	0.752*
1.601	0.746*
1.633	0.642*
1.649	0.620*

* Not shown on plot

DATA TABLE NO. 124 NORMAL SPECTRAL REFLECTANCE OF ZINC OXIDE PIGMENTED COATINGS (continued)

λ	ρ
CURVE 209 (cont.)	
T = 298	
1.669	0.630*
1.686	0.613*
1.699	0.608*
1.715	0.620*
1.749	0.683*
1.778	0.710*
1.799	0.700*
1.824	0.707*
1.849	0.716*
1.888	0.705*
2.044	0.688*
2.103	0.663*
2.185	0.520*
2.221	0.417*
2.233	0.393*
2.244	0.381*
2.267	0.386*
2.292	0.408*
2.309	0.412*
2.404	0.411*
2.493	0.421*
CURVE 210*	
T = 298	
0.250	0.137
0.324	0.085
0.355	0.082
0.373	0.089
0.396	0.795
0.405	0.826
0.413	0.845
0.422	0.845
0.442	0.836
0.474	0.842
0.487	0.865
0.510	0.869
0.763	0.836
0.864	0.827
1.040	0.825
1.106	0.821
1.125	0.816
1.165	0.780
1.203	0.815
1.303	0.811
1.326	0.806
CURVE 210 (cont.)	
1.354	0.782
1.380	0.768
1.402	0.785
1.440	0.796
1.504	0.778
1.522	0.778
1.562	0.789
1.579	0.789
1.617	0.767
1.662	0.650
1.697	0.638
1.719	0.625
1.783	0.731
1.818	0.719
1.839	0.737
1.862	0.737
1.879	0.729
1.961	0.713
2.042	0.718
2.101	0.692
2.168	0.605
2.241	0.404
2.261	0.389
2.283	0.404
2.301	0.421
2.323	0.428
2.339	0.421
2.400	0.427
2.442	0.424
2.466	0.427
2.500	0.442
CURVE 211*	
T = 281	
0.250	0.086
0.285	0.073
0.350	0.061
0.373	0.072
0.393	0.639
0.445	0.713
0.508	0.769
0.540	0.777
0.637	0.766
0.700	0.757
0.763	0.719
0.884	0.664
CURVE 211 (cont.)	
0.904	0.662
1.013	0.603
1.133	0.526
1.184	0.486
1.222	0.471
1.380	0.367
1.494	0.302
1.643	0.236
1.721	0.201
1.800	0.183
2.002	0.142
2.123	0.126
2.209	0.116
2.258	0.104
2.500	0.095
CURVE 212*	
T = 281	
0.700	0.757
0.783	0.729
0.885	0.682
0.905	0.679
1.010	0.627
1.085	0.583
1.130	0.552
1.163	0.521
1.221	0.500
1.322	0.438
1.400	0.387
1.583	0.293
1.670	0.243
1.728	0.222
1.789	0.213
1.808	0.201
1.942	0.169
2.089	0.140
2.206	0.123
2.262	0.109
2.414	0.103
2.500	0.100
CURVE 213*	
T = 281	
0.700	0.757
0.775	0.733
CURVE 213 (cont.)	
0.952	0.664
1.048	0.613
1.146	0.552
1.165	0.529
1.201	0.519
1.331	0.440
1.378	0.405
1.534	0.322
1.647	0.265
1.720	0.227
1.747	0.227
1.801	0.211
2.002	0.161
2.206	0.123
2.263	0.109
2.402	0.105
2.500	0.102
CURVE 214*	
T = 281	
0.700	0.766
0.836	0.720
0.877	0.700
0.907	0.695
1.028	0.642
1.107	0.598
1.165	0.548
1.205	0.538
1.394	0.421
1.464	0.386
1.607	0.310
1.668	0.267
1.721	0.244
1.803	0.229
1.998	0.173
2.188	0.141
2.254	0.120
2.285	0.115
2.500	0.111
CURVE 215*	
T = 281	
0.700	0.771
0.834	0.733
0.885	0.706
CURVE 215 (cont.)	
0.914	0.700
1.002	0.664
1.119	0.604
1.163	0.564
1.201	0.557
1.327	0.485
1.386	0.440
1.438	0.421
1.492	0.387
1.638	0.309
1.660	0.287
1.721	0.260
1.766	0.257
1.821	0.235
2.004	0.187
2.186	0.148
2.258	0.124
2.292	0.124
2.500	0.111
CURVE 216*	
T = 281	
0.700	0.786
0.945	0.710
1.101	0.644
1.142	0.617
1.161	0.594
1.201	0.588
1.347	0.503
1.384	0.477
1.410	0.470
1.551	0.398
1.639	0.347
1.659	0.319
1.719	0.289
1.749	0.289
1.805	0.278
1.826	0.264
1.862	0.259
2.032	0.205
2.167	0.171
2.259	0.138
2.500	0.123
CURVE 217*	
T = 298	
0.250	0.137
0.324	0.085
0.355	0.082
0.373	0.089
0.391	0.701
0.406	0.774
0.427	0.808
0.460	0.830
0.503	0.843
0.688	0.835
0.760	0.823
0.871	0.814
1.106	0.808
1.145	0.789
1.162	0.764
1.181	0.788
1.207	0.800
1.303	0.800
1.329	0.786
1.362	0.754
1.384	0.754
1.404	0.767
1.423	0.776
1.466	0.776
1.504	0.760
1.536	0.764
1.564	0.769
1.604	0.756
1.621	0.741
1.644	0.695
1.664	0.634
1.685	0.631
1.723	0.612
1.742	0.654
1.781	0.709
1.815	0.700
1.840	0.710
1.870	0.700
1.913	0.689
1.949	0.686
2.039	0.691
2.068	0.682
2.109	0.663
2.146	0.621
2.193	0.528
2.241	0.404
CURVE 217 (cont.)	
2.264	0.382
2.295	0.401
2.319	0.408
2.378	0.408
2.445	0.411
2.500	0.421
CURVE 218*	
T = 298	
0.703	0.855
0.799	0.833
0.951	0.829
1.102	0.815
1.125	0.806
1.153	0.785
1.179	0.803
1.204	0.814
1.286	0.810
1.359	0.762
1.369	0.762
1.380	0.769
1.398	0.786
1.421	0.786
1.476	0.774
1.499	0.771
1.542	0.775
1.588	0.766
1.610	0.741
1.646	0.645
1.651	0.636
1.661	0.641
1.666	0.650
1.675	0.653
1.679	0.649
1.693	0.616
1.700	0.611
1.709	0.616
1.754	0.704
1.766	0.718
1.778	0.723
1.792	0.719
1.795	0.709
1.800	0.705
1.817	0.709
1.847	0.724
1.855	0.724
CURVE 218 (cont.)	
1.898	0.702
1.937	0.695
2.051	0.693
2.075	0.681
2.115	0.651
2.155	0.591
2.225	0.433
2.244	0.399
2.259	0.388
2.284	0.415
2.292	0.427
2.311	0.435
2.341	0.435
2.402	0.426
2.452	0.426
2.500	0.433
CURVE 219*	
T = 281	
0.703	0.804
0.803	0.770
0.930	0.730
1.053	0.675
1.097	0.654
1.155	0.601
1.200	0.596
1.306	0.540
1.357	0.496
1.431	0.469
1.494	0.436
1.570	0.407
1.606	0.398
1.653	0.342
1.675	0.340
1.700	0.325
1.734	0.343
1.756	0.351
1.773	0.351
1.809	0.340
1.851	0.341
1.953	0.325
2.034	0.321
2.089	0.309
2.150	0.285
2.227	0.219
2.250	0.210

* Not shown on plot

DATA TABLE NO. 124 NORMAL SPECTRAL REFLECTANCE OF ZINC OXIDE PIGMENTED COATINGS (continued)

λ	ρ
CURVE 219* (cont.)	
T = 281	
2.293	0.223
2.324	0.226
2.451	0.224
2.500	0.227
CURVE 220*	
T = 281	
0.704	0.806
0.807	0.772
0.939	0.735
1.046	0.689
1.102	0.662
1.154	0.612
1.207	0.609
1.299	0.558
1.335	0.529
1.373	0.508
1.415	0.492
1.503	0.448
1.600	0.412
1.648	0.352
1.671	0.349
1.699	0.334
1.749	0.356
1.757	0.359
1.777	0.359
1.808	0.349
1.867	0.348
1.940	0.337
2.039	0.324
2.095	0.315
2.170	0.280
2.237	0.220
2.247	0.215
2.259	0.213
2.293	0.223
2.324	0.226
2.451	0.224
2.500	0.227
CURVE 221*	
T = 281	
0.704	0.811
0.793	0.782

λ	ρ
CURVE 221 (cont.)	
0.885	0.755
0.986	0.723
1.074	0.681
1.100	0.671
1.153	0.620
1.164	0.618
1.201	0.619
1.240	0.602
1.321	0.551
1.369	0.516
1.403	0.507
1.476	0.475
1.540	0.446
1.603	0.420
1.652	0.359
1.676	0.356
1.700	0.340
1.748	0.361
1.763	0.364
1.784	0.364
1.805	0.353
1.858	0.353
1.921	0.340
1.999	0.330
2.056	0.328
2.103	0.316
2.178	0.278
2.241	0.220
2.258	0.216
2.294	0.225
2.324	0.227
2.452	0.225
2.500	0.227
CURVE 222*	
T = 281	
0.704	0.813
0.816	0.779
0.907	0.753
1.032	0.710
1.106	0.679
1.158	0.629
1.205	0.632
1.298	0.587
1.363	0.542
1.457	0.494

λ	ρ
CURVE 222 (cont.)	
1.554	0.450
1.602	0.433
1.639	0.394
1.653	0.367
1.677	0.362
1.700	0.349
1.722	0.358
1.744	0.370
1.774	0.374
1.814	0.362
1.848	0.362
1.924	0.348
2.000	0.336
2.063	0.331
2.125	0.307
2.178	0.278
2.241	0.220
2.258	0.216
2.294	0.227
2.324	0.229
2.452	0.226
2.500	0.227
CURVE 223*	
T = 281	
0.704	0.813
0.806	0.782
0.899	0.761
1.024	0.719
1.094	0.691
1.157	0.647
1.156	0.644
1.321	0.582
1.371	0.545
1.507	0.482
1.610	0.437
1.653	0.375
1.666	0.375
1.701	0.356
1.749	0.378
1.771	0.382
1.807	0.370
1.852	0.370
1.916	0.352
2.062	0.334
2.105	0.319

λ	ρ
CURVE 223 (cont.)	
2.164	0.290
2.230	0.234
2.257	0.219
2.309	0.241
2.405	0.240
2.500	0.233
CURVE 224*	
T = 281	
0.704	0.819
0.803	0.789
0.913	0.767
1.060	0.726
1.135	0.682
1.150	0.666
1.175	0.666
1.202	0.670
1.271	0.643
1.331	0.602
1.359	0.577
1.397	0.575
1.536	0.506
1.602	0.481
1.650	0.406
1.672	0.406
1.701	0.386
1.734	0.404
1.767	0.413
1.800	0.401
1.833	0.401
1.941	0.372
2.043	0.354
2.081	0.346
2.143	0.316
2.212	0.256
2.236	0.232
2.257	0.225
2.309	0.241
2.500	0.233
CURVE 225*	
T = 281	
0.704	0.829
0.788	0.802
0.903	0.784

λ	ρ
CURVE 225 (cont.)	
1.059	0.755
1.098	0.742
1.154	0.691
1.202	0.703
1.291	0.674
1.353	0.628
1.377	0.624
1.568	0.544
1.597	0.535
1.659	0.454
1.701	0.430
1.766	0.451
1.778	0.451
1.802	0.439
1.861	0.435
1.907	0.416
1.993	0.395
2.073	0.378
2.116	0.361
2.172	0.324
2.244	0.243
2.262	0.239
2.311	0.252
2.379	0.251
2.500	0.242
CURVE 226*	
T = 281	
0.704	0.834
0.811	0.803
0.922	0.788
1.060	0.762
1.107	0.748
1.150	0.704
1.204	0.720
1.303	0.691
1.367	0.641
1.410	0.641
1.557	0.579
1.595	0.567
1.615	0.546
1.654	0.468
1.671	0.471
1.699	0.446
1.727	0.472
1.772	0.489

λ	ρ
CURVE 226 (cont.)	
1.793	0.481
1.803	0.472
1.832	0.472
1.950	0.433
2.059	0.410
2.111	0.389
2.214	0.288
2.253	0.248
2.284	0.262
2.312	0.267
2.372	0.267
2.436	0.259
2.500	0.259
CURVE 227*	
T = 298	
0.250	0.114
0.315	0.075
0.352	0.068
0.372	0.079
0.400	0.797
0.414	0.854
0.427	0.877
0.505	0.897
0.720	0.867
0.875	0.839
0.951	0.839
1.108	0.826
1.139	0.810
1.163	0.786
1.191	0.809
1.208	0.818
1.289	0.811
1.329	0.804
1.365	0.772
1.384	0.772
1.406	0.787
1.446	0.789
1.508	0.772
1.565	0.780
1.624	0.752
1.644	0.703
1.659	0.649
1.686	0.625
1.721	0.625
1.737	0.667

λ	ρ
CURVE 227 (cont.)	
1.784	0.717
1.824	0.705
1.839	0.717
1.862	0.714
1.882	0.699
1.962	0.685
2.042	0.690
2.083	0.677
2.118	0.653
2.184	0.555
2.241	0.404
2.257	0.391
2.299	0.428
2.317	0.428
2.345	0.421
2.401	0.427
2.440	0.421
2.467	0.427
2.500	0.437
CURVE 228*	
T = 281	
0.250	0.072
0.288	0.063
0.355	0.056
0.372	0.065
0.392	0.598
0.486	0.709
0.514	0.727
0.557	0.729
0.715	0.691
0.720	0.668
0.948	0.510
1.151	0.339
1.295	0.243
1.415	0.184
1.536	0.145
1.659	0.117
1.774	0.103
2.102	0.085
2.500	0.074

λ	ρ
CURVE 229*	
T = 281	
0.700	0.733
0.847	0.659
0.886	0.632
0.907	0.628
1.048	0.532
1.123	0.474
1.160	0.429
1.200	0.414
1.307	0.332
1.399	0.280
1.490	0.231
1.644	0.166
1.729	0.145
1.932	0.113
2.129	0.094
2.241	0.084
2.500	0.080
CURVE 230*	
T = 281	
0.700	0.748
0.910	0.640
1.066	0.535
1.160	0.458
1.248	0.401
1.368	0.317
1.506	0.241
1.674	0.174
1.845	0.135
2.002	0.112
2.129	0.097
2.241	0.084
2.500	0.080

* Not shown on plot

DATA TABLE NO. 124 NORMAL SPECTRAL REFLECTANCE OF ZINC OXIDE PIGMENTED COATINGS (continued)

λ	ρ
CURVE 231*	
T = 281	
0.700	0.765
0.914	0.666
1.055	0.582
1.131	0.527
1.163	0.494
1.208	0.470
1.396	0.341
1.489	0.289
1.652	0.206
1.722	0.183
1.837	0.155
2.002	0.124
2.128	0.107
2.171	0.107
2.243	0.095
2.500	0.085
CURVE 232*	
T = 281	
0.700	0.773
0.874	0.700
0.883	0.690
0.910	0.687
1.071	0.589
1.139	0.542
1.165	0.517
1.211	0.500
1.385	0.374
1.460	0.336
1.648	0.233
1.732	0.196
1.762	0.194
1.837	0.170
1.999	0.135
2.140	0.115
2.243	0.095
2.500	0.085
CURVE 233*	
T = 281	
0.700	0.777
0.969	0.670
1.143	0.561
1.159	0.538
CURVE 233 (cont.)	
1.204	0.523
1.371	0.405
1.521	0.318
1.642	0.260
1.656	0.241
1.717	0.215
1.754	0.210
1.929	0.165
2.013	0.145
2.193	0.117
2.241	0.103
2.439	0.090
2.500	0.090
CURVE 234*	
T = 298	
0.250	0.114
0.315	0.075
0.352	0.068
0.372	0.079
0.400	0.734
0.432	0.820
0.466	0.860
0.495	0.871
0.641	0.861
0.786	0.836
0.847	0.832
0.886	0.826
0.966	0.826
1.064	0.820
1.112	0.811
1.145	0.799
1.163	0.774
1.200	0.801
1.329	0.795
1.365	0.762
1.382	0.756
1.425	0.774
1.485	0.768
1.510	0.757
1.548	0.757
1.564	0.763
1.586	0.761
1.628	0.738
1.665	0.637
1.682	0.648
CURVE 234 (cont.)	
1.722	0.600
1.756	0.666
1.780	0.693
1.822	0.684
1.844	0.690
1.864	0.683
1.881	0.664
1.905	0.655
1.986	0.655
2.045	0.660
2.095	0.647
2.158	0.584
2.184	0.546
2.241	0.404
2.261	0.381
2.285	0.389
2.314	0.405
2.444	0.405
2.500	0.411
CURVE 235	
T ~ 298	
0.321	0.038
0.357	0.038
0.372	0.046
0.375	0.058
0.397	0.828
0.404	0.871
0.417	0.903
0.445	0.935
0.452	0.939
0.467	0.939*
0.565	0.918
0.673	0.900
0.758	0.891
0.851	0.875
0.923	0.864
1.035	0.863
1.069	0.857
1.206	0.860
1.329	0.867
1.354	0.873
1.388	0.899
1.407	0.933
1.426	0.974
1.447	0.974*
CURVE 235 (cont.)	
1.479	0.944
1.512	0.934
1.591	0.923
1.644	0.906
1.659	0.886*
1.667	0.764
1.673	0.751
1.677	0.734
1.681	0.755
1.685	0.846
1.690	0.869
1.694	0.882*
1.698	0.882*
1.703	0.863*
1.712	0.746
1.720	0.730
1.725	0.740
1.730	0.809
1.738	0.829
1.740	0.900
1.746	0.912
1.750	0.896
1.755	0.889*
1.760	0.896*
1.765	0.924*
1.776	0.956*
1.785	0.972*
1.800	0.979*
CURVE 236	
T ~ 298	
0.321	0.038*
0.357	0.038*
0.372	0.046*
0.375	0.058*
0.397	0.828*
0.403	0.867*
0.421	0.927
0.440	0.946*
0.453	0.949*
0.474	0.949*
0.553	0.928*
0.645	0.914
0.771	0.903
0.911	0.881
0.986	0.874
CURVE 236 (cont.)	
1.087	0.870
1.135	0.868
1.269	0.876
1.299	0.885*
1.352	0.887*
1.373	0.903
1.396	0.926*
1.409	0.949
1.415	0.970*
1.427	0.985
1.436	0.985
1.532	0.939
1.642	0.927
1.648	0.919*
1.659	0.789
1.664	0.735
1.672	0.735*
1.678	0.847
1.686	0.873*
1.692	0.904*
1.700	0.904
1.708	0.875
1.715	0.787
1.719	0.753
1.725	0.806*
1.731	0.834*
1.733	0.883*
1.738	0.898*
1.741	0.920
1.748	0.925*
1.752	0.920*
1.755	0.902*
1.760	0.932*
1.767	0.964
1.777	0.977
1.790	0.988
CURVE 237	
T ~ 298	
0.321	0.038*
0.357	0.038*
0.372	0.046*
0.375	0.058*
0.397	0.828*
0.403	0.867*
0.433	0.953
CURVE 237 (cont.)	
0.445	0.964
0.479	0.964*
0.649	0.934
0.861	0.909
0.897	0.901
1.106	0.898
1.138	0.894
1.318	0.901*
1.363	0.906*
1.383	0.928*
1.409	0.978
1.429	1.000
1.458	1.000
1.511	0.975
1.585	0.958
1.636	0.942
1.654	0.928*
1.665	0.886*
1.671	0.826*
1.677	0.879*
1.694	0.921*
1.709	0.908
1.718	0.801
1.730	0.890
1.738	0.917
1.742	0.936
1.745	0.941
1.750	0.941
1.751	0.929
1.756	0.927
1.767	0.964*
1.777	0.977*
1.790	0.988*
CURVE 238	
T ~ 298	
0.321	0.038*
0.357	0.038*
0.372	0.046*
0.375	0.058*
0.397	0.828*
0.403	0.867*
0.439	0.965
0.448	0.972
0.483	0.972*
0.598	0.950*
CURVE 238 (cont.)	
0.731	0.940*
0.873	0.919*
1.088	0.912*
1.205	0.912
1.297	0.917*
1.340	0.922*
1.378	0.950*
1.427	1.000*
1.502	1.000
1.608	0.972
1.644	0.968*
1.657	0.949
1.668	0.900*
1.668	0.878*
1.672	0.874*
1.682	0.939*
1.687	0.951*
1.699	0.954
1.712	0.930
1.721	0.900*
1.725	0.865*
1.728	0.912*
1.744	0.968*
1.748	0.973*
1.756	0.956*
1.768	0.967*
1.769	0.978*
1.774	0.984*
1.784	0.990*
CURVE 239*	
T ~ 298	
0.321	0.038
0.357	0.038
0.372	0.046
0.375	0.058
0.397	0.828
0.403	0.867
0.439	0.965
0.456	0.977
0.479	0.980
0.614	0.958
0.727	0.949
0.920	0.943
1.192	0.933
1.268	0.938
CURVE 239 (cont.)*	
1.330	0.947
1.415	1.000
1.604	1.000
1.634	0.986
1.660	0.986
1.667	0.981
1.671	0.922
1.675	0.930
1.679	0.975
1.686	0.985
1.708	0.977
1.722	0.918
1.734	0.968
1.745	0.988
1.749	0.988
1.756	0.980
1.769	1.000
CURVE 240	
T = 295	
0.300	0.064
0.366	0.076
0.430	0.566
0.480	0.928
0.537	0.953
0.580	0.953
0.751	0.890
1.232	0.833
1.378	0.833
1.615	0.791
1.708	0.673
1.798	0.785
1.875	0.82[illegible]
1.996	0.823
2.108	0.768
2.148	0.723
2.213	0.486
2.276	0.461
2.470	0.528

* Not shown on plot

DATA TABLE NO. 124 NORMAL SPECTRAL REFLECTANCE OF ZINC OXIDE PIGMENTED COATINGS (continued)

CURVE 241
T = 295

λ	ρ
0.300	0.064
0.366	0.076
0.430	0.566
0.470	0.904
0.499	0.926
0.554	0.926
1.148	0.742
1.535	0.602
1.708	0.477
1.812	0.515
2.006	0.466
2.109	0.421
2.303	0.267
2.391	0.277
2.470	0.289

CURVE 242
T = 295

λ	ρ
0.300	0.064
0.366	0.076
0.430	0.566
0.470	0.904
0.499	0.926
0.554	0.926
0.830	0.842
1.187	0.757
1.510	0.660
1.571	0.633
1.682	0.518
1.706	0.508
1.791	0.547
1.849	0.546
2.044	0.491
2.166	0.422
2.295	0.311
2.470	0.295

CURVE 243*
T = 298

λ	ρ
0.250	0.038
0.298	0.044
0.325	0.027
0.349	0.027
0.362	0.048

CURVE 243 (cont.)

λ	ρ
0.369	0.093
0.396	0.668
0.408	0.774
0.434	0.857
0.455	0.883
0.504	0.902
0.813	0.920
0.971	0.910
1.210	0.910
1.348	0.918
1.836	0.880
1.962	0.846
2.148	0.846
2.211	0.826
2.332	0.805
2.528	0.795
2.749	0.789

CURVE 244*
T = 298

λ	ρ
0.250	0.038
0.298	0.044
0.325	0.027
0.349	0.027
0.362	0.045
0.370	0.075
0.396	0.528
0.405	0.602
0.416	0.656
0.436	0.733
0.477	0.822
0.500	0.862
0.536	0.891
0.575	0.907
0.629	0.919
0.813	0.920
0.971	0.910
1.210	0.910
1.348	0.918
1.836	0.880
1.962	0.846
2.148	0.846
2.211	0.826
2.332	0.805
2.528	0.795
2.749	0.789

CURVE 245*
T = 298

λ	ρ
0.250	0.038
0.298	0.044
0.325	0.027
0.349	0.027
0.362	0.048
0.369	0.093
0.392	0.486
0.403	0.603
0.420	0.694
0.458	0.796
0.495	0.850
0.562	0.895
0.727	0.920
0.813	0.920
0.971	0.910
1.210	0.910
1.348	0.918
1.836	0.880
1.962	0.846
2.148	0.846
2.211	0.826
2.332	0.805
2.528	0.788
2.749	0.785

CURVE 246*
T = 298

λ	ρ
0.250	0.024
0.342	0.024
0.359	0.034
0.368	0.066
0.390	0.487
0.405	0.656
0.420	0.732
0.442	0.790
0.465	0.825
0.501	0.864
0.529	0.888
0.572	0.906
0.629	0.919
0.813	0.920
0.971	0.910
1.210	0.910
1.348	0.918
1.836	0.880

CURVE 246 (cont.)

λ	ρ
1.962	0.846
2.148	0.846
2.211	0.826
2.332	0.805
2.528	0.795
2.749	0.789

CURVE 247*
T = 298

λ	ρ
0.255	0.048
0.334	0.047
0.374	0.118
0.379	0.277
0.391	0.600
0.400	0.743
0.428	0.873
0.503	0.938
0.551	0.938
0.614	0.934
0.701	0.925
0.800	0.906
0.902	0.926
0.999	0.919
1.060	0.906
1.196	0.878
1.348	0.859
1.401	0.932
1.501	0.932
1.600	0.901

CURVE 248*
T = 294

λ	ρ
0.376	0.072
0.395	0.526
0.504	0.895
0.597	0.916
0.701	0.912
0.803	0.892
0.902	0.911
1.001	0.903
1.101	0.886
1.200	0.859
1.299	0.850
1.361	0.841
1.404	0.916

CURVE 248 (cont.)

λ	ρ
1.502	0.919
1.600	0.884

CURVE 249*
T = 394

λ	ρ
0.359	0.040
0.402	0.699
0.502	0.871
0.597	0.901
0.703	0.897
0.801	0.881
0.903	0.900
1.000	0.903
1.199	0.858
1.298	0.832
1.438	0.916
1.501	0.902
1.602	0.865

CURVE 250*
T = 530

λ	ρ
0.304	0.039
0.400	0.547
0.496	0.649
0.594	0.682
0.694	0.714
0.802	0.776
0.900	0.770
0.996	0.776
1.096	0.798
1.198	0.799
1.298	0.800
1.400	0.806
1.498	0.806
1.601	0.794

CURVE 251*
T ~ 298

λ	ρ
0.325	0.040
0.349	0.048
0.360	0.069
0.369	0.561
0.384	0.803
0.394	0.840

CURVE 251 (cont.)

λ	ρ
0.411	0.867
0.433	0.893
0.464	0.913
0.487	0.918
0.525	0.923
0.630	0.907
0.776	0.889
0.985	0.878
1.046	0.872
1.123	0.847
1.169	0.822
1.191	0.838
1.223	0.847
1.255	0.847
1.295	0.835
1.328	0.796
1.353	0.808
1.449	0.815
1.559	0.815
1.707	0.802
1.929	0.628
1.971	0.622
2.091	0.737
2.143	0.758
2.198	0.765
2.283	0.758
2.452	0.678
2.535	0.554
2.600	0.367

CURVE 252*
T ~ 298

λ	ρ
0.325	0.040
0.349	0.048
0.360	0.069
0.365	0.332
0.386	0.657
0.402	0.755
0.420	0.805
0.439	0.837
0.470	0.861
0.497	0.872
0.525	0.876
0.630	0.870
0.776	0.850
0.881	0.830

CURVE 252 (cont.)

λ	ρ
0.985	0.790
1.101	0.741
1.158	0.705
1.182	0.715
1.271	0.687
1.330	0.625
1.408	0.625
1.697	0.544
1.949	0.435
2.012	0.440
2.324	0.334
2.487	0.245
2.600	0.151

CURVE 253*
T = 298

λ	ρ
0.325	0.074
0.340	0.075
0.360	0.100
0.369	0.561
0.375	0.690
0.393	0.787
0.420	0.840
0.443	0.873
0.468	0.889
0.493	0.898
0.525	0.900
0.630	0.888
0.776	0.874
0.985	0.871
1.049	0.862
1.120	0.835
1.158	0.805
1.189	0.808
1.211	0.837
1.226	0.840
1.256	0.840
1.291	0.824
1.315	0.791
1.331	0.783
1.363	0.788
1.436	0.802
1.525	0.807
1.652	0.802
1.836	0.691
1.892	0.621

CURVE 253 (cont.)

λ	ρ
1.978	0.609
2.061	0.628
2.184	0.738
2.226	0.746
2.287	0.739
2.467	0.617
2.535	0.538
2.600	0.367

CURVE 254*
T ~ 298

λ	ρ
0.325	0.030
0.342	0.030
0.350	0.037
0.356	0.048
0.359	0.066
0.374	0.733
0.382	0.799
0.389	0.835
0.399	0.863
0.411	0.882
0.427	0.898
0.443	0.904
0.480	0.914
0.525	0.914
0.630	0.908
0.776	0.916
0.985	0.905
1.340	0.900
1.569	0.896
1.873	0.872
2.098	0.846
2.289	0.812
2.452	0.772
2.576	0.727
2.681	0.679

* Not shown on plot

DATA TABLE NO. 124 NORMAL SPECTRAL REFLECTANCE OF ZINC OXIDE PIGMENTED COATINGS (continued)

λ	ρ
CURVE 255* T ~ 298	
0.325	0.030
0.342	0.030
0.350	0.037
0.356	0.048
0.359	0.066
0.373	0.618
0.380	0.707
0.390	0.770
0.402	0.799
0.425	0.832
0.446	0.851
0.480	0.872
0.525	0.886
0.630	0.896
0.776	0.898
0.985	0.905
1.340	0.904
1.636	0.900
1.895	0.880
2.214	0.840
2.550	0.767
2.681	0.699
CURVE 256* T ~ 298	
0.325	0.015
0.339	0.011
0.351	0.025
0.356	0.045
0.359	0.064
0.375	0.729
0.381	0.785
0.397	0.825
0.413	0.850
0.435	0.874
0.457	0.886
0.525	0.900
0.630	0.905
0.776	0.908
0.985	0.908
1.340	0.900
1.569	0.896
1.873	0.872
2.098	0.846
2.289	0.812

λ	ρ
CURVE 256 (cont.)*	
2.452	0.772
2.576	0.727
2.681	0.679
CURVE 257* T ~ 298	
0.325	0.075
0.349	0.075
0.355	0.083
0.361	0.093
0.365	0.110
0.369	0.289
0.386	0.695
0.396	0.800
0.401	0.824
0.410	0.849
0.428	0.875
0.444	0.894
0.464	0.906
0.484	0.913
0.507	0.913
0.525	0.908
0.630	0.883
0.776	0.861
0.985	0.856
1.053	0.857
1.109	0.858
1.157	0.842
1.185	0.843
1.210	0.852
1.225	0.861
1.251	0.860
1.308	0.834
1.322	0.834
1.517	0.852
1.634	0.844
1.816	0.779
1.874	0.766
1.948	0.777
2.121	0.827
2.179	0.828
2.253	0.815
2.546	0.702
2.620	0.632

λ	ρ
CURVE 258* T ~ 298	
0.325	0.075
0.349	0.075
0.355	0.083
0.361	0.093
0.365	0.110
0.369	0.289
0.383	0.613
0.401	0.738
0.415	0.805
0.432	0.826
0.464	0.850
0.503	0.866
0.525	0.870
0.630	0.870
0.776	0.855
0.985	0.847
1.096	0.847
1.162	0.837
1.214	0.848
1.236	0.852
1.285	0.825
1.323	0.825
1.478	0.844
1.601	0.837
1.773	0.769
1.841	0.758
1.952	0.764
2.166	0.813
2.381	0.749
2.600	0.632
CURVE 259* T ~ 298	
0.325	0.075
0.349	0.075
0.355	0.083
0.361	0.093
0.365	0.110
0.369	0.289
0.383	0.613
0.404	0.767
0.425	0.830
0.441	0.850
0.470	0.863
0.525	0.879

λ	ρ
CURVE 259 (cont.)*	
0.630	0.879
0.776	0.855
0.985	0.847
1.096	0.847
1.162	0.837
1.214	0.848
1.236	0.852
1.285	0.825
1.323	0.825
1.478	0.844
1.601	0.837
1.773	0.769
1.841	0.758
1.952	0.764
2.166	0.813
2.381	0.749
2.600	0.632
CURVE 260* T ~ 298	
0.325	0.014
0.339	0.014
0.349	0.024
0.357	0.033
0.362	0.052
0.373	0.548
0.386	0.760
0.397	0.802
0.409	0.827
0.426	0.851
0.439	0.867
0.465	0.884
0.492	0.892
0.525	0.896
0.630	0.888
0.776	0.878
0.985	0.870
1.181	0.844
1.237	0.855
1.262	0.855
1.335	0.831
1.528	0.847
1.678	0.845
1.790	0.836
2.029	0.747
2.197	0.794

λ	ρ
CURVE 260 (cont.)*	
2.295	0.801
2.498	0.752
2.662	0.647
2.727	0.571
CURVE 261* T ~ 298	
0.325	0.014
0.339	0.014
0.349	0.024
0.357	0.033
0.362	0.052
0.373	0.548
0.379	0.662
0.396	0.730
0.413	0.767
0.437	0.802
0.483	0.838
0.525	0.856
0.630	0.868
0.776	0.864
0.985	0.849
1.095	0.840
1.194	0.821
1.241	0.833
1.271	0.833
1.340	0.801
1.593	0.815
1.671	0.810
1.869	0.755
2.013	0.706
2.108	0.708
2.186	0.737
2.275	0.753
2.350	0.743
2.573	0.615
2.701	0.536
CURVE 262* T ~ 298	
0.325	0.014
0.339	0.014
0.349	0.024
0.357	0.033
0.362	0.052

λ	ρ
CURVE 262 (cont.)*	
0.373	0.548
0.382	0.705
0.403	0.770
0.423	0.807
0.450	0.839
0.484	0.858
0.525	0.872
0.630	0.872
0.776	0.868
0.985	0.860
1.160	0.836
1.204	0.836
1.237	0.841
1.277	0.841
1.341	0.817
1.554	0.837
1.662	0.837
1.823	0.804
2.007	0.735
2.068	0.735
2.216	0.785
2.259	0.791
2.492	0.714
2.695	0.578
CURVE 263* T ~ 298	
0.325	0.015
0.339	0.018
0.353	0.033
0.361	0.060
0.380	0.676
0.389	0.765
0.402	0.821
0.413	0.849
0.430	0.873
0.448	0.890
0.469	0.908
0.491	0.918
0.525	0.920
0.630	0.911
0.776	0.892
0.985	0.883
1.049	0.877
1.185	0.848
1.202	0.861

λ	ρ
CURVE 263 (cont.)*	
1.224	0.872
1.265	0.872
1.293	0.862
1.340	0.842
1.460	0.852
1.513	0.854
1.651	0.853
1.761	0.845
1.872	0.819
1.963	0.752
2.007	0.745
2.045	0.754
2.152	0.790
2.222	0.798
2.460	0.742
2.631	0.642
2.694	0.569
CURVE 264* T ~ 298	
0.325	0.015
0.339	0.018
0.353	0.033
0.361	0.060
0.374	0.427
0.388	0.643
0.404	0.712
0.435	0.776
0.474	0.819
0.525	0.849
0.630	0.876
0.776	0.872
0.985	0.866
1.082	0.859
1.187	0.835
1.221	0.848
1.247	0.855
1.266	0.855
1.289	0.848
1.323	0.816
1.397	0.823
1.530	0.836
1.628	0.836
1.718	0.822
1.891	0.742
1.929	0.727

λ	ρ
CURVE 264 (cont.)*	
1.999	0.717
2.111	0.736
2.222	0.778
2.460	0.690
2.581	0.618
2.690	0.540
CURVE 265 T = 298	
0.325	0.015
0.339	0.018
0.353	0.033
0.361	0.060
0.379	0.648
0.399	0.735
0.417	0.781
0.438	0.816
0.465	0.848
0.487	0.865
0.525	0.880
0.630	0.890
0.776	0.884
0.985	0.875
1.073	0.865
1.188	0.843
1.227	0.860
1.272	0.860
1.319	0.836
1.340	0.831
1.524	0.844
1.659	0.845
1.822	0.816
1.948	0.740
2.013	0.732
2.076	0.742
2.136	0.779
2.195	0.791
2.331	0.777
2.447	0.744
2.631	0.642
2.694	0.569

*Not shown on plot

DATA TABLE NO. 124 NORMAL SPECTRAL REFLECTANCE OF ZINC OXIDE PIGMENTED COATINGS (continued)

CURVE 266* T ~ 298

λ	ρ
0.325	0.009
0.339	0.023
0.351	0.047
0.362	0.086
0.373	0.248
0.382	0.601
0.394	0.739
0.401	0.792
0.408	0.809
0.417	0.823
0.431	0.833
0.454	0.843
0.501	0.843
0.525	0.837
0.630	0.821
0.776	0.801
0.985	0.774
1.102	0.769
1.193	0.755
1.209	0.743
1.249	0.754
1.293	0.754
1.340	0.740
1.516	0.710
1.614	0.722
1.754	0.727
1.849	0.728
2.015	0.714
2.150	0.678
2.235	0.640
2.314	0.659
2.385	0.666
2.484	0.672
2.687	0.598
2.778	0.517

CURVE 267* T ~ 298

λ	ρ
0.325	0.009
0.339	0.023
0.351	0.047
0.362	0.086
0.373	0.248

CURVE 267 (cont.)*

λ	ρ
0.379	0.471
0.390	0.579
0.407	0.659
0.430	0.733
0.450	0.776
0.468	0.797
0.486	0.806
0.513	0.806
0.525	0.803
0.630	0.803
0.776	0.784
0.985	0.774
1.140	0.772
1.209	0.755
1.251	0.765
1.292	0.765
1.411	0.747
1.575	0.730
1.734	0.738
1.856	0.737
2.009	0.728
2.150	0.698
2.250	0.661
2.360	0.676
2.434	0.681
2.523	0.681
2.640	0.662
2.739	0.613
2.798	0.545

CURVE 268* T ~ 298

λ	ρ
0.325	0.009
0.339	0.023
0.351	0.047
0.362	0.086
0.373	0.248
0.382	0.601
0.414	0.734
0.438	0.796
0.466	0.816
0.485	0.820
0.525	0.820
0.630	0.811

CURVE 268 (cont.)*

λ	ρ
0.776	0.794
0.985	0.774
1.102	0.769
1.193	0.755
1.209	0.743
1.249	0.754
1.293	0.754
1.340	0.740
1.516	0.710
1.614	0.722
1.754	0.727
1.849	0.728
2.015	0.714
2.150	0.678
2.235	0.640
2.314	0.659
2.385	0.666
2.484	0.672
2.687	0.598
2.778	0.517

CURVE 269* T ~ 298

λ	ρ
0.334	0.010
0.353	0.010
0.373	0.024
0.380	0.035
0.383	0.286
0.395	0.625
0.407	0.781
0.422	0.821
0.434	0.847
0.447	0.870
0.464	0.886
0.485	0.898
0.501	0.904
0.531	0.904
0.572	0.895
0.621	0.873
0.630	0.865
0.730	0.856
0.776	0.856
0.881	0.851
0.985	0.832

CURVE 269 (cont.)*

λ	ρ
1.059	0.822
1.122	0.798
1.153	0.809
1.186	0.809
1.228	0.795
1.265	0.768
1.289	0.777
1.324	0.777
1.449	0.769
1.593	0.748
1.710	0.679
1.746	0.672
1.796	0.678
1.901	0.714
2.008	0.714
2.210	0.670
2.595	0.513

CURVE 270* T ~ 298

λ	ρ
0.334	0.010
0.353	0.010
0.373	0.024
0.380	0.035
0.383	0.286
0.395	0.625
0.407	0.674
0.428	0.722
0.448	0.759
0.473	0.793
0.495	0.812
0.524	0.830
0.559	0.843
0.595	0.852
0.630	0.856
0.776	0.856
0.881	0.851
0.985	0.832
1.059	0.822
1.122	0.798
1.153	0.809
1.186	0.809
1.228	0.795
1.265	0.768

CURVE 270 (cont.)*

λ	ρ
1.289	0.777
1.324	0.777
1.449	0.769
1.593	0.748
1.710	0.679
1.746	0.672
1.796	0.678
1.901	0.714
2.008	0.714
2.210	0.670
2.595	0.513

CURVE 271* T ~ 298

λ	ρ
0.334	0.010
0.353	0.010
0.373	0.024
0.380	0.035
0.383	0.286
0.395	0.625
0.399	0.676
0.421	0.731
0.441	0.780
0.468	0.815
0.493	0.835
0.526	0.851
0.574	0.863
0.630	0.863
0.730	0.856
0.776	0.856
0.881	0.851
0.985	0.832
1.059	0.822
1.122	0.798
1.153	0.809
1.228	0.795
1.265	0.768
1.289	0.777
1.324	0.777
1.449	0.769
1.593	0.748
1.710	0.679
1.746	0.672
1.796	0.678

CURVE 271 (cont.)*

λ	ρ
1.901	0.714
2.008	0.714
2.210	0.670
2.595	0.513

CURVE 272* T ~ 298

λ	ρ
0.332	0.013
0.352	0.023
0.362	0.033
0.367	0.049
0.385	0.535
0.409	0.768
0.418	0.803
0.431	0.828
0.445	0.849
0.468	0.871
0.497	0.884
0.516	0.888
0.551	0.888
0.598	0.875
0.630	0.860
0.776	0.846
0.985	0.819
1.069	0.802
1.114	0.783
1.162	0.795
1.196	0.795
1.224	0.785
1.249	0.763
1.287	0.780
1.305	0.780
1.338	0.759
1.548	0.675
1.606	0.670
1.714	0.707
1.789	0.720
1.866	0.720
1.960	0.712
2.279	0.505

CURVE 273* T ~ 298

λ	ρ
0.332	0.013
0.352	0.023
0.362	0.033
0.367	0.049
0.385	0.535
0.418	0.659
0.432	0.722
0.446	0.759
0.464	0.788
0.482	0.802
0.509	0.817
0.554	0.829
0.630	0.833
0.776	0.822
0.985	0.802
1.052	0.792
1.105	0.767
1.124	0.767
1.158	0.782
1.186	0.782
1.247	0.754
1.267	0.754
1.292	0.762
1.323	0.748
1.461	0.689
1.562	0.655
1.663	0.670
1.758	0.694
1.822	0.694
2.071	0.612
2.192	0.567
2.279	0.505

CURVE 274* T ~ 298

λ	ρ
0.332	0.013
0.352	0.023
0.362	0.033
0.367	0.049
0.385	0.535
0.401	0.666
0.420	0.727
0.436	0.772

CURVE 274 (cont.)*

λ	ρ
0.459	0.805
0.489	0.826
0.515	0.837
0.561	0.842
0.630	0.842
0.776	0.830
0.985	0.809
1.059	0.799
1.111	0.775
1.161	0.787
1.196	0.787
1.246	0.759
1.307	0.775
1.345	0.751
1.548	0.675
1.606	0.670
1.714	0.707
1.830	0.709
2.033	0.661
2.126	0.606
2.279	0.505

CURVE 275* T ~ 298

λ	ρ
0.325	0.022
0.340	0.010
0.353	0.019
0.361	0.041
0.368	0.099
0.381	0.679
0.387	0.761
0.398	0.824
0.409	0.857
0.428	0.886
0.445	0.898
0.461	0.906
0.478	0.908
0.513	0.909
0.573	0.896
0.630	0.872
0.776	0.833
0.985	0.800
1.210	0.782
1.282	0.786

* Not shown on plot

DATA TABLE NO. 124 NORMAL SPECTRAL REFLECTANCE OF ZINC OXIDE PIGMENTED COATINGS (continued)

λ	ρ
CURVE 275 (cont.)*	
T ~ 298	
1.327	0.775
1.436	0.758
1.598	0.773
1.704	0.774
1.833	0.761
1.895	0.726
1.911	0.697
1.954	0.697
2.115	0.734
2.257	0.684
2.398	0.547
2.436	0.524
CURVE 276*	
T ~ 298	
0.325	0.022
0.340	0.010
0.353	0.019
0.361	0.041
0.368	0.099
0.375	0.492
0.387	0.578
0.405	0.658
0.420	0.694
0.445	0.738
0.481	0.774
0.512	0.793
0.554	0.813
0.630	0.824
0.776	0.814
0.985	0.789
1.228	0.773
1.282	0.776
1.428	0.755
1.530	0.754
1.642	0.764
1.730	0.759
1.859	0.716
1.916	0.678
2.105	0.736
2.272	0.687
2.377	0.606
2.455	0.541
CURVE 277*	
T ~ 298	
0.325	0.022
0.340	0.010
0.353	0.019
0.361	0.041
0.368	0.099
0.378	0.571
0.397	0.667
0.411	0.704
0.434	0.751
0.464	0.793
0.495	0.814
0.540	0.837
0.582	0.845
0.630	0.852
0.703	0.852
0.776	0.833
0.985	0.800
1.205	0.787
1.247	0.791
1.284	0.791
1.420	0.767
1.593	0.782
1.704	0.781
1.836	0.763
1.895	0.726
1.911	0.697
1.954	0.697
2.058	0.738
2.115	0.747
2.310	0.691
2.455	0.541
CURVE 278*	
T ~ 298	
0.328	0.020
0.339	0.020
0.356	0.040
0.368	0.481
0.378	0.669
0.397	0.768
0.409	0.801
0.420	0.815
0.475	0.842
0.512	0.852
0.570	0.852
CURVE 278 (cont.)*	
0.630	0.842
0.703	0.833
0.776	0.833
0.881	0.822
0.985	0.802
1.107	0.772
1.195	0.735
1.216	0.723
1.239	0.735
1.278	0.750
1.327	0.750
1.411	0.738
1.455	0.705
1.515	0.701
1.581	0.716
1.648	0.722
1.728	0.726
1.826	0.719
1.952	0.666
1.985	0.612
2.008	0.606
2.063	0.614
2.159	0.654
2.246	0.670
2.319	0.670
2.383	0.665
2.563	0.549
2.655	0.441
CURVE 279*	
T ~ 298	
0.325	0.032
0.345	0.046
0.357	0.079
0.368	0.444
0.383	0.516
0.410	0.595
0.425	0.634
0.444	0.664
0.526	0.739
0.589	0.780
0.630	0.793
0.703	0.797
0.776	0.791
0.881	0.781
0.985	0.762
CURVE 279 (cont.)*	
1.117	0.737
1.219	0.709
1.307	0.732
1.332	0.729
1.455	0.687
1.673	0.704
1.782	0.695
1.853	0.642
1.959	0.604
2.034	0.604
2.077	0.614
2.104	0.629
2.236	0.649
2.297	0.649
2.415	0.612
2.628	0.441
CURVE 280*	
T ~ 298	
0.328	0.020
0.339	0.020
0.356	0.040
0.368	0.481
0.396	0.617
0.424	0.662
0.488	0.744
0.543	0.797
0.590	0.817
0.630	0.821
0.703	0.816
0.776	0.805
0.881	0.797
0.985	0.779
1.100	0.754
1.197	0.721
1.239	0.720
1.307	0.738
1.325	0.738
1.384	0.726
1.487	0.699
1.615	0.710
1.727	0.716
1.831	0.701
1.976	0.622
2.017	0.605
2.063	0.611
CURVE 280 (cont.)*	
2.105	0.630
2.215	0.634
2.322	0.622
2.427	0.601
2.628	0.441
CURVE 281*	
T ~ 298	
0.327	0.028
0.340	0.028
0.353	0.045
0.357	0.062
0.369	0.506
0.385	0.727
0.393	0.789
0.402	0.817
0.414	0.836
0.431	0.861
0.448	0.874
0.471	0.885
0.496	0.892
0.572	0.892
0.630	0.886
0.776	0.866
0.985	0.845
1.090	0.839
1.145	0.819
1.177	0.800
1.232	0.818
1.244	0.820
1.274	0.820
1.307	0.805
1.340	0.777
1.434	0.792
1.539	0.799
1.685	0.791
2.029	0.680
2.167	0.724
2.227	0.732
2.326	0.725
2.432	0.681
2.733	0.485
CURVE 282*	
T ~ 298	
0.327	0.028
CURVE 282 (cont.)*	
0.340	0.028
0.353	0.045
0.357	0.062
0.369	0.506
0.375	0.626
0.394	0.686
0.411	0.722
0.442	0.765
0.471	0.795
0.507	0.819
0.547	0.835
0.585	0.842
0.630	0.847
0.776	0.840
0.881	0.824
0.985	0.817
1.075	0.806
1.121	0.795
1.180	0.770
1.228	0.776
1.264	0.771
1.331	0.733
1.473	0.728
1.731	0.686
1.956	0.598
2.022	0.593
2.170	0.605
2.280	0.594
2.425	0.523
2.651	0.372
CURVE 283*	
T ~ 298	
0.327	0.028
0.340	0.028
0.353	0.045
0.357	0.062
0.369	0.506
0.377	0.670
0.402	0.746
0.431	0.791
0.456	0.819
0.489	0.841
0.543	0.865
0.585	0.872
0.630	0.872
0.776	0.853
CURVE 283 (cont.)*	
0.985	0.837
1.104	0.824
1.177	0.792
1.221	0.806
1.270	0.812
1.303	0.797
1.340	0.768
1.539	0.790
1.711	0.765
1.916	0.684
2.011	0.671
2.254	0.717
2.392	0.686
2.645	0.498
CURVE 284*	
T ~ 298	
0.328	0.033
0.337	0.028
0.355	0.043
0.361	0.066
0.379	0.682
0.388	0.758
0.403	0.809
0.416	0.835
0.436	0.855
0.456	0.873
0.477	0.885
0.499	0.893
0.535	0.896
0.582	0.896
0.630	0.888
0.776	0.871
0.985	0.855
1.069	0.855
1.171	0.838
1.230	0.851
1.270	0.851
1.323	0.831
1.500	0.843
1.625	0.842
1.755	0.833
1.873	0.803
2.004	0.747
2.212	0.790
2.269	0.797
2.324	0.791
CURVE 284 (cont.)*	
2.590	0.667
2.716	0.588
CURVE 285*	
T ~ 298	
0.328	0.033
0.337	0.028
0.355	0.043
0.361	0.066
0.371	0.327
0.385	0.663
0.399	0.737
0.412	0.770
0.439	0.804
0.468	0.828
0.499	0.845
0.551	0.858
0.598	0.862
0.630	0.862
0.776	0.848
0.985	0.838
1.124	0.834
1.164	0.821
1.229	0.835
1.266	0.835
1.324	0.813
1.367	0.811
1.519	0.821
1.652	0.816
1.756	0.805
1.955	0.726
2.013	0.718
2.084	0.723
2.217	0.777
2.276	0.781
2.555	0.659
2.636	0.598
CURVE 286*	
T ~ 298	
0.328	0.033
0.337	0.028
0.355	0.043
0.361	0.066
0.379	0.682
0.390	0.736

*Not shown on plot

DATA TABLE NO. 124 NORMAL SPECTRAL REFLECTANCE OF ZINC OXIDE PIGMENTED COATINGS (continued)

λ	ρ
CURVE 286 (cont.)*	
T ∼ 298	
0.407	0.781
0.428	0.811
0.450	0.833
0.476	0.850
0.525	0.863
0.587	0.870
0.630	0.870
0.776	0.856
0.985	0.846
1.076	0.846
1.170	0.831
1.230	0.843
1.258	0.843
1.320	0.826
1.589	0.833
1.726	0.827
1.937	0.734
2.024	0.734
2.113	0.759
2.217	0.777
2.276	0.781
2.555	0.659
2.636	0.598
CURVE 287*	
T ∼ 298	
0.332	0.038
0.346	0.038
0.354	0.048
0.368	0.196
0.380	0.625
0.392	0.826
0.401	0.874
0.409	0.897
0.420	0.917
0.425	0.924
0.438	0.924
0.463	0.909
0.495	0.913
0.537	0.901
0.579	0.893
0.630	0.885
0.776	0.843
0.985	0.807
CURVE 287 (cont.)*	
1.106	0.794
1.159	0.779
1.182	0.770
1.202	0.780
1.226	0.788
1.265	0.788
1.299	0.773
1.331	0.731
1.466	0.754
1.551	0.761
1.719	0.761
1.902	0.749
1.987	0.732
2.060	0.684
2.113	0.690
2.238	0.723
2.329	0.737
2.393	0.736
2.491	0.717
2.606	0.647
2.728	0.474
CURVE 288*	
T ∼ 298	
0.332	0.038
0.346	0.038
0.354	0.048
0.368	0.196
0.380	0.625
0.391	0.805
0.404	0.865
0.415	0.891
0.420	0.904
0.437	0.913
0.454	0.910
0.535	0.891
0.584	0.887
0.630	0.875
0.703	0.869
0.776	0.845
0.868	0.835
0.985	0.811
1.112	0.805
1.147	0.799
1.178	0.781
CURVE 288 (cont.)*	
1.230	0.795
1.266	0.795
1.311	0.775
1.332	0.746
1.514	0.768
1.603	0.773
1.721	0.773
1.916	0.758
2.047	0.705
2.095	0.705
2.239	0.732
2.389	0.747
2.466	0.741
2.610	0.661
2.728	0.474
CURVE 289*	
T ∼ 298	
0.332	0.038
0.346	0.038
0.354	0.048
0.368	0.196
0.380	0.625
0.396	0.815
0.406	0.863
0.417	0.890
0.424	0.902
0.457	0.906
0.557	0.881
0.588	0.881
0.630	0.868
0.664	0.865
0.698	0.867
0.776	0.840
0.861	0.816
0.985	0.803
1.106	0.791
1.159	0.779
1.182	0.770
1.202	0.780
1.226	0.788
1.265	0.788
1.299	0.773
1.331	0.731
1.466	0.754
CURVE 289 (cont.)*	
1.551	0.761
1.719	0.761
1.902	0.749
1.987	0.732
2.060	0.684
2.113	0.690
2.238	0.723
2.329	0.737
2.393	0.736
2.491	0.717
2.606	0.647
2.728	0.474
CURVE 290*	
T ∼ 298	
0.332	0.031
0.343	0.031
0.354	0.044
0.368	0.196
0.380	0.625
0.392	0.826
0.401	0.874
0.409	0.897
0.420	0.917
0.425	0.924
0.456	0.922
0.510	0.912
0.579	0.893
0.630	0.885
0.776	0.843
0.985	0.807
1.106	0.794
1.159	0.779
1.182	0.770
1.202	0.780
1.226	0.788
1.265	0.788
1.299	0.773
1.331	0.731
1.466	0.754
1.551	0.761
1.719	0.761
1.902	0.749
1.987	0.732
2.060	0.684
CURVE 290 (cont.)*	
2.113	0.690
2.238	0.723
2.329	0.737
2.393	0.736
2.491	0.717
2.606	0.647
2.728	0.474
CURVE 291*	
T ∼ 298	
0.331	0.022
0.338	0.023
0.354	0.038
0.361	0.056
0.363	0.077
0.363	0.096
0.361	0.109
0.368	0.196
0.380	0.625
0.392	0.826
0.401	0.871
0.409	0.895
0.420	0.913
0.425	0.921
0.438	0.921
0.463	0.909
0.495	0.913
0.517	0.898
0.568	0.894
0.609	0.885
0.616	0.892
0.624	0.892
0.642	0.877
0.701	0.877
0.776	0.848
0.830	0.848
0.933	0.827
0.985	0.821
1.140	0.811
1.179	0.788
1.224	0.799
1.257	0.802
1.292	0.797
1.313	0.787
1.340	0.761
CURVE 291 (cont.)*	
1.667	0.778
1.724	0.778
1.921	0.766
1.995	0.753
2.085	0.718
2.135	0.718
2.264	0.746
2.355	0.751
2.493	0.742
2.559	0.727
2.649	0.623
2.728	0.474

* Not shown on plot

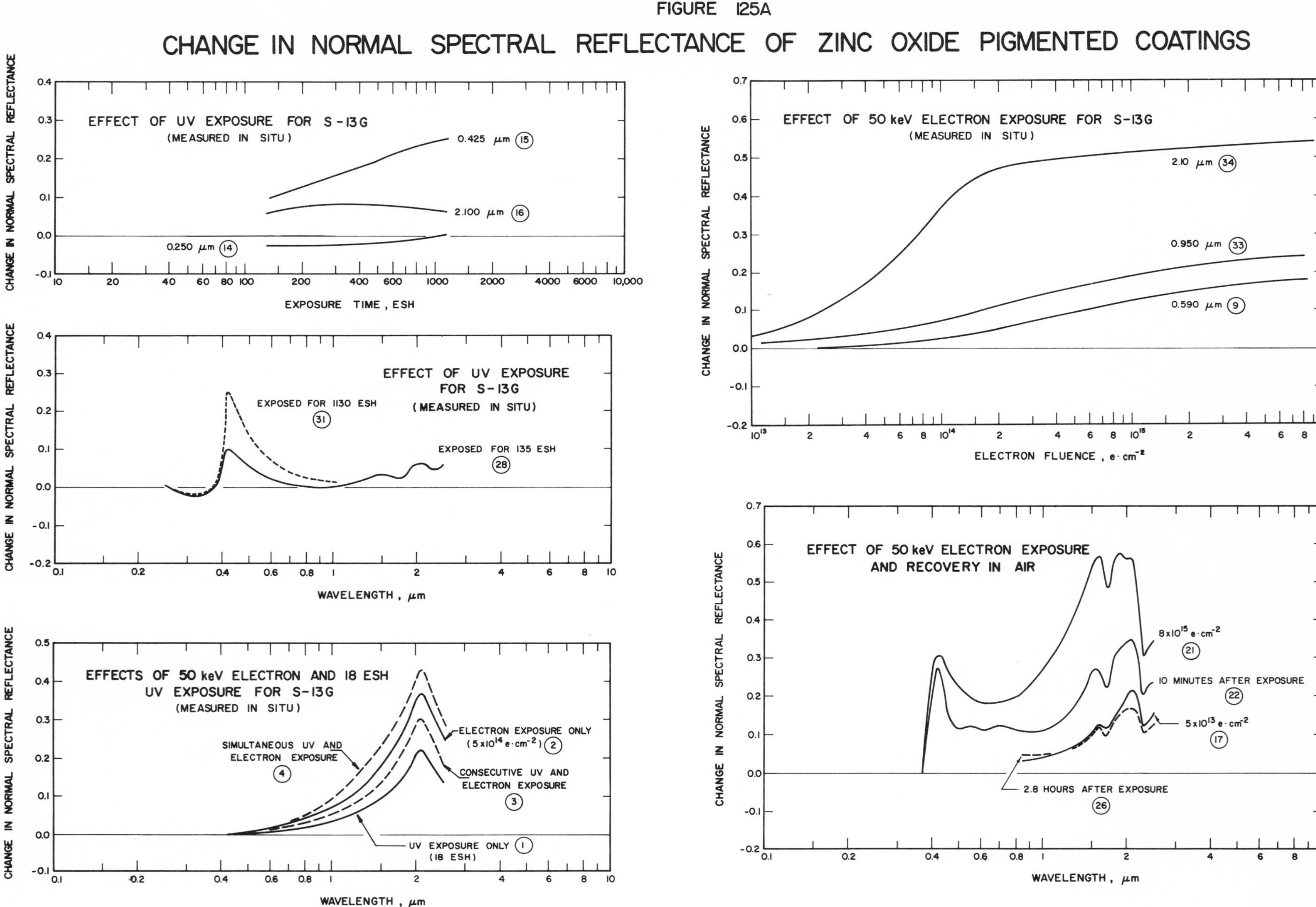

FIGURE 125A
CHANGE IN NORMAL SPECTRAL REFLECTANCE OF ZINC OXIDE PIGMENTED COATINGS
EFFECT OF UV EXPOSURE FOR S-13G
(MEASURED IN SITU)
0.425 μm (15)
2.100 μm (16)
0.250 μm (14)
EXPOSURE TIME, ESH
CHANGE IN NORMAL SPECTRAL REFLECTANCE
EFFECT OF UV EXPOSURE FOR S-13G
(MEASURED IN SITU)
EXPOSED FOR 1130 ESH (31)
EXPOSED FOR 135 ESH (28)
WAVELENGTH, μm
EFFECTS OF 50 keV ELECTRON AND 18 ESH UV EXPOSURE FOR S-13G
(MEASURED IN SITU)
SIMULTANEOUS UV AND ELECTRON EXPOSURE (4)
ELECTRON EXPOSURE ONLY (5×10^{14} e·cm^{-2}) (2)
CONSECUTIVE UV AND ELECTRON EXPOSURE (3)
UV EXPOSURE ONLY (1) (18 ESH)
EFFECT OF 50 keV ELECTRON EXPOSURE FOR S-13G
(MEASURED IN SITU)
2.10 μm (34)
0.950 μm (33)
0.590 μm (9)
ELECTRON FLUENCE, e·cm^{-2}
EFFECT OF 50 keV ELECTRON EXPOSURE AND RECOVERY IN AIR
8×10^{15} e·cm^{-2} (21)
10 MINUTES AFTER EXPOSURE (22)
5×10^{13} e·cm^{-2} (17)
2.8 HOURS AFTER EXPOSURE (26)

SPECIFICATION TABLE NO. 125 CHANGE IN NORMAL SPECTRAL REFLECTANCE OF ZINC OXIDE PIGMENTED COATINGS

Curve No.	Ref. No.	Year	Temperature, K	Wavelength Range, μm	Geometry θ	θ'	ω'	Reported Error, %	Composition (weight percent), Specifications, and Remarks
1	46	1969	295	0.425-2.500	$\sim 0^\circ$		2π		ZnO in methyl silicone binder (0.228 mm thick) over S54044 primer; exposed to UV radiation (4.4-UV sun rate) in vacuum (10^{-8} mm Hg); vacuum maintained by ion pump; ESH 18; reflectance measured in situ; positive change indicates decrease in reflectance from preirradiation, in air reflectance. [Authors' designation: S-13]
2	46	1969	295	0.425-2.500	$\sim 0^\circ$		2π		Similar to above specimen and conditions except exposed to 50 keV electrons only in vacuum; electron fluence 5 x 10^{14} e cm^{-2}.
3	46	1969	295	0.425-2.500	$\sim 0^\circ$		2π		Similar to curve 1 specimen and conditions except consecutive exposure to UV radiation (4.4-UV sun rate) then to 50 keV electrons in vacuum; ESH 18; electron fluence 5 x 10^{14} e cm^{-2}.
4	46	1969	295	0.425-2.500	$\sim 0^\circ$		2π		Similar to curve 1 specimen and conditions except simultaneous exposure to UV radiation (4.4-UV sun rate) and to 50 keV electrons in vacuum; total exposure time 4.1 hours; ESH 18; electron fluence 5 x 10^{14} e cm^{-2} with 90% occurring during the first hour of EV exposure.
5*	46	1969	295	0.590	$\sim 0^\circ$		2π		ZnO in methyl silicone binder (.254 mm thick) over .0508 mm thick Cat-A-Lac white primer; exposed to 50 keV electrons (4 x 10^8 - 1.7 x 10^{12} e cm^{-2} sec^{-1}) in vacuum (10^{-8} mm Hg); vacuum maintained by ion pump; electron fluence (e cm^{-2}) is variable; reflectance measured in situ; positive change indicates decrease in reflectance from preirradiation, in air reflectance. [Authors' designation: C]
6*	46	1969	295	2.100	$\sim 0^\circ$		2π		Above specimen and conditions.
7*	46	1969	295	2.100	$\sim 0^\circ$		2π		ZnO in methyl silicone binder (.228 mm thick) over S54044 primer; exposed to 50 keV electrons (4 x 10^8 - 1.7 x 10^{12} e cm^{-2} sec^{-1}) in vacuum (10^{-8} mm Hg); vacuum maintained by ion pump; electron fluence (e cm^{-2}) is variable; reflectance measured in situ; positive change indicates decrease in reflectance from preirradiation, in air reflectance. [Authors' designation: S-13]
8*	46	1969	295	0.590	$\sim 0^\circ$		2π		Above specimen and conditions.
9	46	1969	295	0.590	$\sim 0^\circ$		2π		ZnO in methyl silicone binder (0.28 mm thick) over S54044 primer; exposed to 50 keV electrons (4 x 10^8 - 1.7 x 10^{12} e cm^{-2} sec^{-1}) in vacuum (10^{-8} mm Hg); vacuum maintained by ion pump; electron fluence (e cm^{-2}) is variable; reflectance measured in situ; positive change indicates decrease in reflectance from preirradiation, in air reflectance. [Authors' designation: S-13G]
10*	46	1969	295	2.100	$\sim 0^\circ$		2π		Above specimen and conditions.
11*	46	1969	295	0.250	$\sim 0^\circ$		2π		ZnO in methyl silicone binder (.228 mm thick) over S54044 primer; exposed to UV radiation (4.7 UV sun rate) in vacuum (10^{-8} mm Hg); vacuum maintained by ion pump; ESH is variable; reflectance measured in situ; positive change indicates decrease in reflectance from preirradiation, in air reflectance. [Authors' designation: S-13]
12*	46	1969	295	0.425	$\sim 0^\circ$		2π		Above specimen and conditions.

* Not shown on plot

SPECIFICATION TABLE NO. 125 CHANGE IN NORMAL SPECTRAL REFLECTANCE OF ZINC OXIDE PIGMENTED COATINGS (continued)

Curve No.	Ref. No.	Year	Temperature, K	Wavelength Range, μm	Geometry θ	θ'	ω'	Reported Error, %	Composition (weight percent), Specifications, and Remarks
13*	46	1969	295	2.100	$\sim 0^\circ$		2π		Above specimen and conditions.
14	46	1969	295	0.250	$\sim 0^\circ$		2π		ZnO in methyl silicone binder (.28 mm thick) over S54044 primer; exposed to UV radiation (4.7 UV sun rate) in vacuum (10^{-8} mm Hg); vacuum maintained by ion pump; ESH is variable; reflectance measured in situ; positive change indicates decrease in reflectance from preirradiation, in air reflectance. [Authors' designation: S-13G]
15	46	1969	295	0.425	$\sim 0^\circ$		2π		Above specimen and conditions.
16	46	1969	295	2.100	$\sim 0^\circ$		2π		Above specimen and conditions.
17	46	1969	295	0.581-2.500	$\sim 0^\circ$		2π		ZnO in methyl silicone binder (0.28 mm thick); exposed to 50 keV electrons in vacuum (10^{-7} mm Hg); vacuum maintained by ion pump; reflectance measured in situ after 5 x 10^{13} e cm^{-2}; positive change indicates decrease in reflectance from preirradiation, in air reflectance; data extracted from smooth curve. [Authors' designation: S-13G]
18*	46	1969	295	0.820-2.500	$\sim 0^\circ$		2π		Above specimen and conditions except reflectance measured after 1 x 10^{14} e cm^{-2}.
19*	46	1969	295	0.370-0.700	$\sim 0^\circ$		2π		Above specimen and conditions except reflectance measured after 2 x 10^{14} e cm^{-2}.
20*	46	1969	295	0.370-2.500	$\sim 0^\circ$		2π		Above specimen and conditions except reflectance measured after 5.5 x 10^{14} e cm^{-2}.
21	46	1969	295	0.370-2.500	$\sim 0^\circ$		2π		Above specimen and conditions except reflectance measured after 8 x 10^{15} e cm^{-2}.
22	46	1969	295	0.370-2.500	$\sim 0^\circ$		2π		Above specimen and conditions except reflectance measured in air 10 minutes after end of exposure to 8 x 10^{15} (50 keV) e cm^{-2}.
23*	46	1969	298	0.860-2.500	$\sim 0^\circ$		2π		Above specimen and conditions except reflectance measured 57 minutes after end.
24*	46	1969	298	0.370-2.500	$\sim 0^\circ$		2π		Above specimen and conditions except reflectance measured 1 hour 11 minutes after end.
25*	46	1969	298	0.868-2.500	$\sim 0^\circ$		2π		Above specimen and conditions except reflectance measured 1 hour 39 minutes after end.
26	46	1969	298	0.869-2.500	$\sim 0^\circ$		2π		Above specimen and conditions except reflectance measured 2 hours and 48 minutes after end.
27*	46	1969	298	0.370-0.700	$\sim 0^\circ$		2π		Above specimen and conditions except reflectance measured 3 hours and 4 minutes after end.
28	46	1969	295	0.250-2.450	$\sim 0^\circ$		2π		ZnO in methyl silicone binder (0.28 mm thick); exposed to UV radiation (with rates equivalent to 5 suns in UV) in vacuum (10^{-7} mm Hg); ESH 135; vacuum maintained by ion pump; reflectance measured in situ; positive change indicates decrease in reflectance from preirradiation, in air reflectance; data extracted from smooth curve. [Authors' designation: S-13G]
29*	46	1969	295	0.376-2.450	$\sim 0^\circ$		2π		Above specimen and conditions except ESH 350.
30*	46	1969	295	0.250-2.450	$\sim 0^\circ$		2π		Above specimen and conditions except ESH 490.
31	46	1969	295	0.250-0.999	$\sim 0^\circ$		2π		Above specimen and conditions except ESH 1130.

* Not shown on plot

SPECIFICATION TABLE NO. 125 CHANGE IN NORMAL SPECTRAL REFLECTANCE OF ZINC OXIDE PIGMENTED COATINGS (continued)

Curve No.	Ref. No.	Year	Temperature, K	Wavelength Range, μm	Geometry θ	θ'	ω'	Reported Error, %	Composition (weight percent), Specifications, and Remarks
32*	46	1969	295	0.590	$\sim 0°$		2π		ZnO and Al_2O_3 in potassium silicate binder (0.127 mm thick); exposed to 50 keV electrons in vacuum (10^{-7} mm Hg); vacuum maintained by ion pump: electron fluence (e cm^{-2}) is variable; reflectance measured in situ; positive change indicates decrease in reflectance as compared to preirradiation, in air reflectance; data extracted from smooth curve. [Authors' designation: F]
33	46	1969	295	0.950	$\sim 0°$		2π		ZnO in methyl silicone binder (0.28 mm thick); exposed to 50 keV electrons in vacuum (10^{-7} mm Hg); vacuum maintained by ion pump; electron fluence (e cm^{-2}) is variable; reflectance measured in situ; positive change indicates decrease in reflectance as compared to preirradiation, in air reflectance; data extracted from smooth curve. [Authors' designation: S-13G]
34	46	1969	295	2.10	$\sim 0°$		2π		Above specimen and conditions.

* Not shown on plot

DATA TABLE NO. 125 CHANGE IN NORMAL SPECTRAL REFLECTANCE OF ZINC OXIDE PIGMENTED COATINGS

[Wavelength, λ, μm; Reflectance, ρ; Temperature, T, K]

CURVE 1 (T = 295)

λ	Δρ
0.425	0.01
0.590	0.01
0.950	0.03
1.200	0.06
1.550	0.10
2.100	0.22
2.500	0.14

CURVE 2 (T = 295)

λ	Δρ
0.425	0
0.590	0.02
0.950	0.06
1.200	0.11
1.550	0.20
2.100	0.37
2.500	0.26

CURVE 3 (T = 295)

λ	Δρ
0.425	0*
0.590	0.02*
0.950	0.04
1.200	0.07
1.550	0.15
2.100	0.30
2.500	0.19

CURVE 4 (T = 295)

λ	Δρ
0.425	0*
0.590	0.02*
0.950	0.07
1.200	0.12
1.550	0.24
2.100	0.43
2.500	0.30

CURVE 5* (λ = 590, T = 295)

Electron Fluence	Δρ
1 x 10^{13}	0
2 x 10^{14}	0
6 x 10^{14}	0.03

CURVE 6* (λ = 2.100, T = 295)

Electron Fluence	Δρ
2 x 10^{14}	0.13
6 x 10^{14}	0.19

CURVE 7* (λ = 2.100, T = 295)

Electron Fluence	Δρ
1 x 10^{13}	0.06
2 x 10^{14}	0.30
6 x 10^{14}	0.33
8 x 10^{15}	0.45

CURVE 8* (λ = 0.590, T = 295)

Electron Fluence	Δρ
1 x 10^{13}	0
2 x 10^{14}	0
6 x 10^{14}	0
8 x 10^{14}	0

CURVE 9 (λ = 0.590, T = 295)

Electron Fluence	Δρ
1 x 10^{13}	0
2 x 10^{14}	0.05
6 x 10^{14}	0.10
8 x 10^{15}	0.18

CURVE 10* (λ = 2.100, T = 295)

Electron Fluence	Δρ
1 x 10^{13}	0.03
2 x 10^{14}	0.48
6 x 10^{14}	0.51
8 x 10^{15}	0.55

CURVE 11* (λ = 0.250, T = 295)

ESH	Δρ
135	-0.10
250	-0.10
490	-0.09
1130	-0.06

CURVE 12* (λ = 0.425, T = 295)

ESH	Δρ
135	0.05
250	0.08
490	0.14
770	0.20
1130	0.26

CURVE 13* (λ = 2.100, T = 295)

ESH	Δρ
135	0.30
250	0.34
490	0.37
1130	0.40

CURVE 14 (λ = 0.250, T = 295)

ESH	Δρ
135	-0.03
250	-0.02
490	-0.02
1130	0

CURVE 15 (λ = 0.425, T = 295)

ESH	Δρ
135	0.10
250	0.14
490	0.19
770	0.23
1130	0.25

CURVE 16 (λ = 2.100, T = 295)

ESH	Δρ
135	0.06
250	0.08
490	0.08
1130	0.06

CURVE 17 (T = 295)

λ	Δρ
0.851	0.032
1.116	0.054
1.256	0.062
1.340	0.079
1.590	0.125
1.698	0.116
1.787	0.148
1.860	0.167
1.997	0.185
2.082	0.215
2.099	0.215
2.204	0.191
2.286	0.121
2.306	0.121
2.500	0.157

CURVE 18* (T = 295)

λ	Δρ
0.820	0.059
0.977	0.076
1.059	0.101
1.116	0.122
1.217	0.145
1.369	0.198
1.496	0.239
1.552	0.271
1.579	0.283
1.605	0.283
1.681	0.249
1.696	0.249
1.798	0.315
1.861	0.339
1.931	0.351
1.993	0.351
2.067	0.370
2.099	0.370
2.157	0.352
2.223	0.293
2.284	0.216
2.301	0.207
2.330	0.207
2.393	0.232
2.500	0.238

CURVE 19* (T = 295)

λ	Δρ
0.370	0.000
0.445	0.035
0.503	0.030
0.548	0.056
0.612	0.056
0.700	0.070

CURVE 20* (T = 295)

λ	Δρ
0.370	0.000
0.419	0.080
0.430	0.080
0.467	0.062
0.488	0.062
0.555	0.082
0.636	0.115
0.700	0.127
0.820	0.151
0.900	0.173
0.999	0.168
1.134	0.238
1.251	0.301
1.351	0.358
1.466	0.412
1.593	0.489
1.627	0.479
1.682	0.419
1.709	0.419
1.784	0.486
1.839	0.509
1.904	0.518
1.987	0.505
2.103	0.505
2.166	0.461
2.234	0.366
2.263	0.314
2.293	0.287
2.349	0.287
2.500	0.308

CURVE 21 (T = 295)

λ	Δρ
0.370	0.000
0.396	0.266
0.405	0.294
0.418	0.304
0.437	0.301
0.454	0.261
0.500	0.226
0.601	0.177
0.700	0.200
0.830	0.198
0.898	0.234
1.000	0.269
1.039	0.286
1.078	0.318
1.173	0.353
1.220	0.380
1.264	0.414
1.294	0.448
1.409	0.474
1.529	0.550
1.577	0.567
1.607	0.567
1.630	0.552
1.672	0.488
1.689	0.482
1.713	0.482
1.795	0.553
1.833	0.569
1.871	0.575
1.898	0.575
1.998	0.559
2.093	0.562
2.159	0.521
2.209	0.465
2.252	0.370
2.299	0.304
2.319	0.304
2.372	0.323
2.432	0.337
2.500	0.346

CURVE 22 (T = 298)

λ	Δρ
0.370	0.000
0.418	0.275
0.437	0.241
0.446	0.170
0.491	0.115
0.511	0.115
0.534	0.120
0.568	0.120
0.606	0.111
0.677	0.120
0.867	0.107
0.956	0.107
1.286	0.164
1.411	0.199
1.475	0.251
1.538	0.270
1.575	0.270
1.627	0.250
1.690	0.221
1.818	0.307
1.903	0.328
2.007	0.334
2.089	0.349
2.149	0.331
2.231	0.279
2.263	0.226
2.289	0.204
2.315	0.204
2.389	0.227
2.428	0.232
2.500	0.232

CURVE 23* (T = 298)

λ	Δρ
0.860	0.096
0.989	0.096
1.236	0.119
1.577	0.178
1.641	0.170
1.694	0.157
1.810	0.213
1.823	0.226
1.856	0.235
1.918	0.235
2.109	0.261
2.160	0.252
2.220	0.216
2.292	0.163
2.323	0.150
2.360	0.144
2.415	0.144
2.458	0.157
2.500	0.171

CURVE 24* (T = 298)

λ	Δρ
0.370	0.000
0.418	0.275
0.437	0.241
0.446	0.170
0.491	0.115
0.511	0.115
0.534	0.120
0.568	0.120
0.606	0.111
0.677	0.120
0.867	0.107
0.956	0.107
1.286	0.164
1.411	0.199
1.475	0.251
1.538	0.270
1.575	0.270
1.627	0.250
1.690	0.221
1.818	0.307
1.903	0.328
2.007	0.334
2.089	0.349
2.149	0.331

* Not shown on plot

DATA TABLE NO. 125 CHANGE IN NORMAL SPECTRAL REFLECTANCE OF ZINC OXIDE PIGMENTED COATINGS (continued)

CURVE 24* (cont.)
T = 298

λ	Δρ
2.231	0.279
2.263	0.226
2.289	0.204
2.315	0.204
2.389	0.227
2.428	0.232
2.500	0.232

CURVE 25*
T = 298

λ	Δρ
0.868	0.068
0.936	0.064
0.989	0.064
1.042	0.073
1.102	0.073
1.141	0.069
1.222	0.075
1.350	0.100
1.430	0.106
1.554	0.133
1.572	0.138
1.612	0.138
1.701	0.115
1.823	0.169
2.050	0.203
2.140	0.203
2.172	0.197
2.292	0.118
2.355	0.124
2.500	0.150

CURVE 26
T = 298

λ	Δρ
0.869	0.046
0.979	0.042
1.094	0.051
1.166	0.054
1.355	0.075
1.429	0.081
1.590	0.119
1.690	0.097
1.820	0.140
1.864	0.150

CURVE 26 (cont.)

λ	Δρ
2.004	0.164
2.196	0.166
2.310	0.108
2.500	0.126

CURVE 27*

λ	Δρ
0.370	0.000
0.421	0.161
0.497	0.072
0.566	0.087
0.591	0.069
0.619	0.069
0.700	0.083

CURVE 28
T = 295

λ	Δρ
0.250	0.005
0.300	-0.021
0.327	-0.024
0.353	-0.019
0.422	0.101
0.506	0.051
0.589	0.022
0.956	0.000
1.338	0.021
1.546	0.032
1.713	0.022
1.912	0.054
2.099	0.063
2.302	0.043
2.450	0.054

CURVE 29*
T = 295

λ	Δρ
0.376	0.034
0.426	0.143
0.523	0.065
0.597	0.033
0.957	0.011
1.375	0.031
1.559	0.041
1.713	0.043
2.077	0.083

CURVE 29 (cont.)

λ	Δρ
2.105	0.083
2.289	0.045
2.450	0.063

CURVE 30*
T = 295

λ	Δρ
0.250	0.015
0.306	-0.019
0.330	-0.019
0.353	-0.013
0.417	0.188
0.502	0.097
0.591	0.042
0.950	0.014
0.984	0.012
1.375	0.031
1.559	0.041
1.714	0.043
1.888	0.061
2.098	0.073
2.289	0.045
2.450	0.063

CURVE 31
T = 295

λ	Δρ
0.250	0.005
0.300	-0.019
0.330	-0.019
0.353	-0.013
0.418	0.250
0.466	0.185
0.513	0.134
0.564	0.088
0.592	0.069
0.618	0.059
0.654	0.049
0.712	0.038
0.820	0.026
0.945	0.019
0.989	0.012
0.999	0.011

CURVE 32*
λ = 0.590
T = 295

Electron Fluence	Δρ
$.113 \times 10^{14}$	0.065
$.409 \times 10^{14}$	0.095
$.175 \times 10^{15}$	0.136
$.787 \times 10^{16}$	0.326

CURVE 33
λ = .950 μm
T = 295

Electron Fluence	Δρ
1.18×10^{13}	0.013
4.04×10^{13}	0.037
8.17×10^{13}	0.062
1.43×10^{14}	0.090
2.42×10^{14}	0.121
4.31×10^{14}	0.153
7.29×10^{14}	0.177
1.20×10^{15}	0.198
2.58×10^{15}	0.219
7.87×10^{15}	0.239

CURVE 34
λ = 2.10 μm
T = 295

Electron Fluence	Δρ
1.00×10^{13}	0.029
3.07×10^{13}	0.124
4.52×10^{13}	0.192
6.41×10^{13}	0.262
9.26×10^{13}	0.348
1.20×10^{14}	0.407
1.86×10^{14}	0.472
6.73×10^{14}	0.504
7.87×10^{15}	0.542

* Not shown on plot

SPECIFICATION TABLE NO. 126 NORMAL INTEGRATED ABSORPTANCE OF ZINC OXIDE PIGMENTED COATING

Curve No.	Ref. No.	Year	Temperature Range, K	Geometry θ	Reported Error, %	Composition (weight percent), Specifications and Remarks
1*	23	1964	~298	~0°		IIT Research Institute designation: 441-2; zinc oxide pigment in potassium silicate; computed from spectral reflectance data; integrated between 0.40 and 0.70 μ.
2*	17	1968	298	~0°		Z-93; SP-500 ZnO in potassium silicate binder (0.089 mm thick); aluminum substrate; measured with earth albedo as source, approximated by solar spectrum truncated below 0.35 μm; sample supplied by ESRO I Project Group.

DATA TABLE NO. 126 NORMAL INTEGRATED ABSORPTANCE OF ZINC OXIDE PIGMENTED COATING

[Temperature, T, K; Absorptance, α]

T	α
CURVE 1*	
298	0.02
CURVE 2*	
298	0.26

* No plot given

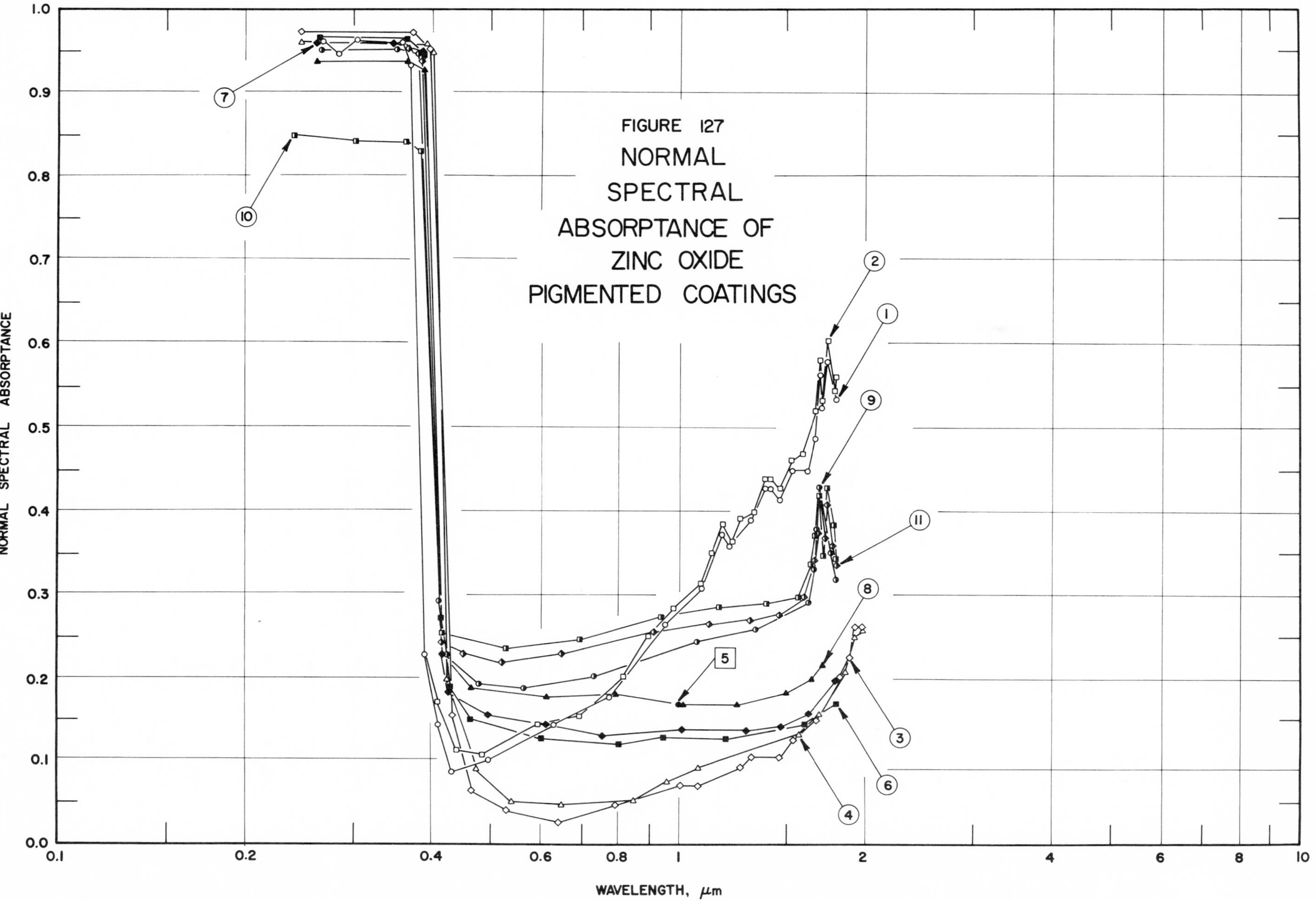

FIGURE 127 NORMAL SPECTRAL ABSORPTANCE OF ZINC OXIDE PIGMENTED COATINGS

SPECIFICATION TABLE NO. 127 NORMAL SPECTRAL ABSORPTANCE OF ZINC OXIDE PIGMENTED COATINGS

Curve No.	Ref. No.	Year	Temperature, K	Wavelength Range, μm	Geometry θ	Reported Error, %	Composition (weight percent), Specifications, and Remarks
1	25	1964	~298	0.269-1.8	~0°		ZnO in polymethylvinyl siloxane binder (LMSC/Dow-Corning); substrate unknown; data extracted from smooth curve; converted from spectral reflectance. [Authors' designation: ZnO/Silicone]
2	25	1964	~298	0.269-1.8	~0°		Similar to above specimen and conditions except exposed to UV radiation from a 1 kW A-H6 (PEK Labs Type C) mercury-argon lamp (at 99.1 mm from sample) in vacuum (10^{-6} - 10^{-7} mm Hg) for 100 hrs.
3	26	1964	~298	0.247-1.997	~0°		ZnO in potassium silicate; aluminum substrate; data extracted from smooth curve.
4	26	1964	~298	0.247-1.997	~0°		Above specimen and conditions except exposed in vacuum (8 x 10^{-7} mm Hg) to 100 hrs UV radiation at solar factor 3 suns; sample temp during exposure 288 K.
5	23	1964	~298	1.0	~0°		ZnO in potassium silicate binder; IIT Research Inst. designation: 441-2.
6	24	1965	298	0.264-1.800	~0°		ZnO in sodium silicate binder; data extracted from smooth curve.
7	24	1965	298	0.263-1.799	~0°		Similar to above specimen and conditions except exposed to approx 200 sun hrs of ultraviolet radiation.
8	24	1965	298	0.263-1.800	~0°		Similar to curve 6 specimen and conditions except exposed to nuclear radiation of approx 10^8R of gamma and 5 x 10^{14} neutrons ($E \geq 2.9$ Mev) per cm^2 and 500 sun hrs of ultraviolet radiation.
9	24	1965	298	0.267-1.800	~0°		ZnO in Dow Corning Q90090 methyl silicone binder; data extracted from smooth curve.
10	24	1965	298	0.241-1.800	~0°		Similar to above specimens and conditions except exposed to nuclear radiation of approx 10^8R of gamma and 5 x 10^{14} neutrons ($E \geq 2.9$ Mev) per cm^2.
11	24	1965	298	0.268-1.800	~0°		Similar to above specimen and conditions except exposed to nuclear radiation of approx 10^8R of gamma and 5 x 10^{14} neutrons ($E \geq 2.9$ Mev) per cm^2 and 500 sun hrs of ultraviolet radiation.

DATA TABLE NO. 127 NORMAL SPECTRAL ABSORPTANCE OF ZINC OXIDE PIGMENTED COATINGS

[Wavelength, λ, μm; Absorptance, α; Temperature, T, K]

CURVE 1
T ~298

λ	α
0.269	0.963
0.285	0.949
0.306	0.967
0.360	0.962
0.372	0.938
0.391	0.229
0.411	0.144
0.434	0.088
0.498	0.101
0.631	0.146
0.779	0.179
0.956	0.268
1.091	0.310
1.183	0.374
1.210	0.360
1.317	0.391
1.382	0.430
1.416	0.430
1.461	0.418
1.534	0.451
1.621	0.451
1.669	0.491
1.694	0.567
1.707	0.526
1.745	0.582
1.782	0.536*
1.800	0.536

CURVE 2
T ~298

λ	α
0.269	0.963*
0.285	0.949*
0.306	0.967*
0.360	0.962*
0.372	0.938*
0.391	0.229*
0.411	0.173
0.442	0.113
0.486	0.108
0.595	0.145
0.694	0.155
0.816	0.204
0.896	0.251

CURVE 2 (cont.)

λ	α
0.980	0.286
1.079	0.315
1.138	0.353
1.177	0.387
1.219	0.367
1.260	0.395
1.331	0.401
1.386	0.441
1.418	0.441
1.465	0.430
1.523	0.465
1.589	0.471
1.664	0.523
1.698	0.584
1.716	0.536
1.750	0.608
1.782	0.548
1.800	0.564

CURVE 3
T ~298

λ	α
0.247	0.976
0.374	0.976
0.400	0.956
0.433	0.154
0.467	0.065
0.530	0.042
0.642	0.029
0.791	0.049
1.005	0.071
1.079	0.071
1.253	0.096
1.318	0.106
1.451	0.106
1.544	0.129
1.665	0.150
1.824	0.204
1.890	0.229
1.942	0.264
1.997	0.266

CURVE 4
T ~298

λ	α
0.247	0.963
0.394	0.963
0.406	0.950
0.429	0.200
0.475	0.093
0.540	0.051
0.650	0.047
0.844	0.053
0.954	0.077
1.066	0.094
1.572	0.137
1.678	0.158
1.853	0.211
1.928	0.255
1.997	0.262

CURVE 5
T ~298

λ	α
1.0	0.17

CURVE 6
T = 298

λ	α
0.264	0.968
0.366	0.968
0.389	0.949
0.416	0.274
0.430	0.193
0.465	0.151
0.602	0.129
0.804	0.122
0.945	0.129
1.191	0.128
1.599	0.144
1.800	0.173

CURVE 7
T = 298

λ	α
0.263	0.963
0.348	0.963
0.389	0.951
0.418	0.232

CURVE 7 (cont.)

λ	α
0.429	0.187
0.497	0.157
0.612	0.145
0.754	0.133
1.011	0.138
1.289	0.138
1.462	0.143
1.639	0.160
1.799	0.200

CURVE 8
T = 298

λ	α
0.263	0.940
0.367	0.940
0.389	0.931
0.419	0.230*
0.463	0.191
0.613	0.180
0.796	0.183
1.026	0.172
1.248	0.170
1.499	0.186
1.648	0.202
1.706	0.218
1.800	0.245*

CURVE 9
T = 298

λ	α
0.267	0.955
0.352	0.955
0.382	0.950
0.414	0.297
0.422	0.230
0.479	0.197
0.564	0.191
0.732	0.205
1.075	0.246
1.340	0.262
1.612	0.294
1.663	0.334
1.677	0.382
1.696	0.433
1.756	0.354
1.800	0.322

CURVE 10
T = 298

λ	α
0.241	0.853
0.304	0.848
0.365	0.848
0.399	0.833
0.419	0.256
0.530	0.239
0.699	0.249
0.934	0.276
1.164	0.288
1.381	0.293
1.556	0.299
1.637	0.339
1.658	0.372
1.693	0.423
1.728	0.349
1.744	0.432
1.770	0.386
1.800	0.345

CURVE 11
T = 298

λ	α
0.268	0.955*
0.368	0.955
0.388	0.942
0.418	0.245
0.453	0.231
0.523	0.222
0.651	0.233
0.913	0.259
1.131	0.269
1.316	0.273
1.456	0.278
1.593	0.300
1.658	0.344
1.675	0.377
1.696	0.421*
1.723	0.371
1.741	0.413
1.769	0.362
1.800	0.337

* Not shown on plot

FIGURE 128A

ANALYZED CHANGE IN NORMAL SPECTRAL ABSORPTANCE OF ZINC OXIDE PIGMENTED COATINGS

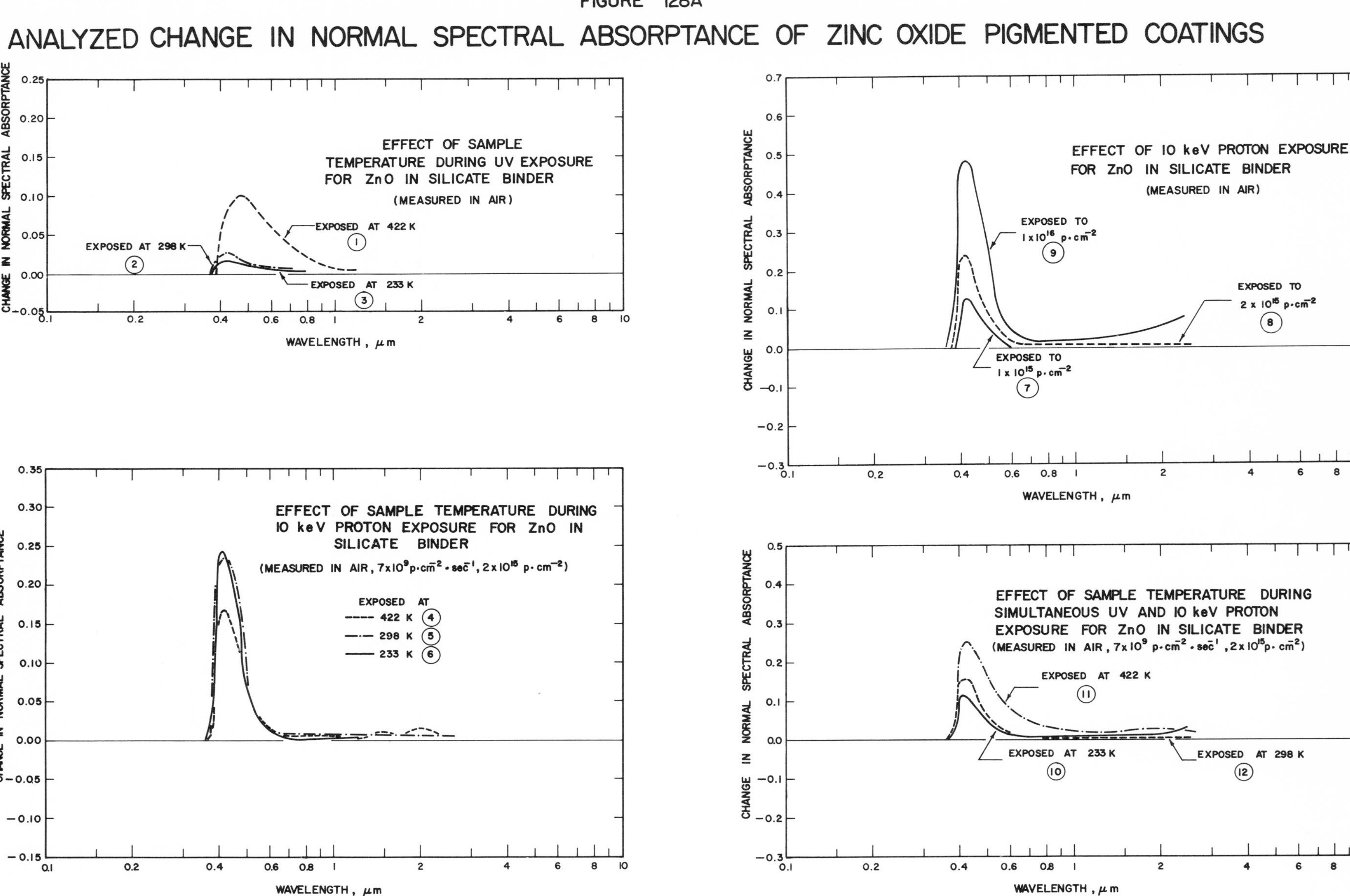

SPECIFICATION TABLE NO. 128 CHANGE IN NORMAL SPECTRAL ABSORPTANCE OF ZINC OXIDE PIGMENTED COATINGS

Curve No.	Ref. No.	Year	Temperature, K	Wavelength Range, μm	Geometry θ	Reported Error, %	Composition (weight percent), Specifications, and Remarks
1	47	1967	298	0.384-1.162	$\sim 0^\circ$		SP500 ZnO in PS-7 K_2SiO_3 binder (0.178 mm thick); ball milled for 2 hours; aluminum alloy (6061) substrate; cured in oven for 15 minutes at 400 K; exposed to near UV and EUV for 75 hours to 10 equivalent suns (with a Mercury-Xenon 5-kw short-arc lamp) in vacuum ($\sim 10^{-8}$ mm Hg) at 422 K; vacuum maintained with an ion pump and a sublimation pump; absorptance calculated from reflectance; property measured in air immediately after exposure; positive change indicates increase in absorptance from preirradiation conditions.
2	47	1967	298	0.371-1.025	$\sim 0^\circ$		Similar to above specimen and conditions except exposed at 298 K.
3	47	1967	298	0.381-0.776	$\sim 0^\circ$		Similar to above specimen and conditions except exposed at 233 K.
4	47	1967	298	0.369-2.300	$\sim 0^\circ$		SP500 ZnO in PS-7 K_2SiO_3 binder (0.178 mm thick); ball milled for 2 hours; aluminum alloy (6061) substrate; cured in oven for 15 minutes at 400 K; exposed to 10 keV protons (7.4 x 10^9 p cm^{-2} sec^{-1}) in vacuum (10^{-8} mm Hg) at 422 K; proton exposure 2 x 10^{15} p cm^{-2}; vacuum maintained by an ion pump and a sublimation pump; absorptance calculated from reflectance; property measured in air immediately after exposure; positive change indicates increase in absorptance from preirradiation conditions.
5	47	1967	298	0.369-2.48	$\sim 0^\circ$		Similar to above specimen and conditions except exposed at 298 K.
6	47	1967	298	0.368-1.202	$\sim 0^\circ$		Similar to above specimen and conditions except exposed at 233 K.
7	47	1967	298	0.380-0.599	$\sim 0^\circ$		Similar to curve 4 specimen and conditions except proton flux 5.5 x 10^{11} p cm^{-2} sec^{-1}; proton exposure 1 x 10^{15} p cm^{-2}; exposed at 298 K.
8	47	1967	298	0.377-2.480	$\sim 0^\circ$		Above specimen and conditions except proton exposure 2 x 10^{15} p cm^{-2}.
9	47	1967	298	0.372-2.380	$\sim 0^\circ$		Similar to above specimen and conditions except proton exposure 1 x 10^{16} p cm^{-2}.
10	47	1967	298	0.375-2.450	$\sim 0^\circ$		Similar to curve 4 specimen and conditions except exposed simultaneously to 10 keV protons (7.4 x 10^9 p $cm^{-2}sec^{-1}$) and 75 hours of UV and EUV at 10 ES in vacuum at 233 K; proton exposure 2 x 10^{15} p cm^{-2}.
11	47	1967	298	0.371-2.480	$\sim 0^\circ$		Similar to above specimen and conditions except exposed at 422 K.
12	47	1967	298	0.374-2.480	$\sim 0^\circ$		Similar to above specimen and conditions except exposed at 298 K.
13*	47	1967	298	0.376-2.480	$\sim 0^\circ$		Similar to curve 4 specimen and conditions except exposed simultaneously to 75 hours of UV and EUV at 10 ES and protons at start of UV exposure (5.5 x 10^{11} p cm^{-2} sec^{-1}) in vacuum at 298 K; proton exposure 2 x 10^{15} p cm^{-2}.
14*	47	1967	298	0.380-2.480	$\sim 0^\circ$		Similar to above specimen and conditions except exposed to protons continuously during UV exposure (7.4 x 10^9 p cm^{-2} sec^{-1}).
15*	47	1967	298	0.380-2.480	$\sim 0^\circ$		Similar to above specimen and conditions except exposed to protons during final part of UV exposure (5.5 x 10^{11} p cm^{-2} sec^{-1}).
16*	47	1967	298	0.376-2.398	$\sim 0^\circ$		Similar to curve 13 specimen and conditions except exposed at 422 K.
17*	47	1967	298	0.376-2.403	$\sim 0^\circ$		Similar to curve 14 specimen and conditions except exposed at 422 K.

* Not shown on plot

SPECIFICATION TABLE NO. 128 CHANGE IN NORMAL SPECTRAL ABSORPTANCE OF ZINC OXIDE PIGMENTED COATINGS (continued)

Curve No.	Ref. No.	Year	Temperature, K	Wavelength Range, μm	Geometry θ	Reported Error, %	Composition (weight percent), Specifications, and Remarks
18*	47	1967	298	0.370-0.658	$\sim 0^\circ$		Similar to curve 4 specimen and conditions except exposed to 100 keV protons (5.5×10^{11} p cm^{-2} sec^{-1}) in vacuum at 298 K.
19*	47	1967	298	0.379-2.283	$\sim 0^\circ$		Similar to curve 13 specimen and conditions except 100 keV protons.
20*	47	1967	298	0.393-2.366	$\sim 0^\circ$		Similar to curve 14 specimen and conditions except 100 keV protons.
21*	47	1967	298	0.369-0.703	$\sim 0^\circ$		Similar to curve 13 specimen and conditions except 10 keV protons.

* Not shown on plot

DATA TABLE NO. 128 CHANGE IN NORMAL SPECTRAL ABSORPTANCE OF ZINC OXIDE PIGMENTED COATINGS

[Wavelength, λ, μm; Absorptance, α; Temperature, T, K]

λ	Δα
CURVE 1	
T = 298	
0.384	0.014
0.406	0.060
0.417	0.085
0.439	0.098
0.474	0.100
0.508	0.091
0.606	0.056
0.629	0.054
0.640	0.054
0.678	0.039
0.720	0.029
0.775	0.022
0.894	0.013
1.017	0.007
1.162	0.005
CURVE 2	
T = 298	
0.371	0.004
0.379	0.010
0.410	0.027
0.426	0.027
0.471	0.017
0.500	0.013
0.728	0.007
1.025	0.005
CURVE 3	
T = 298	
0.381	0.006
0.397	0.011
0.410	0.014
0.443	0.016
0.478	0.013
0.501	0.012
0.560	0.007
0.776	0.004
CURVE 4	
T = 298	
0.369	0.000
0.379	0.011
0.382	0.023
0.383	0.044
0.393	0.120
0.402	0.150
0.415	0.166
0.423	0.168
0.432	0.162
0.451	0.143
0.475	0.110
0.494	0.075
0.521	0.048
0.548	0.029
0.588	0.013
0.623	0.008
1.049	0.006
1.230	0.000
1.433	0.012
1.729	0.006
2.029	0.014
2.300	0.008
CURVE 5	
T = 298	
0.369	0.000
0.374	0.012
0.380	0.035
0.382	0.103
0.392	0.182
0.407	0.222
0.415	0.238
0.426	0.238
0.435	0.229
0.454	0.198
0.480	0.120
0.499	0.089
0.530	0.051
0.563	0.024
0.616	0.009
0.660	0.006
2.480	0.006
CURVE 6	
T = 298	
0.368	0.009
0.374	0.013
0.379	0.028
0.388	0.083
0.391	0.178
0.397	0.211
0.403	0.231
0.413	0.245
0.428	0.231
0.447	0.208
0.456	0.191
0.467	0.173
0.473	0.151
0.478	0.122
0.486	0.097
0.502	0.069
0.525	0.045
0.551	0.028
0.610	0.009
0.644	0.006
0.653	0.002
0.731	0.000
1.202	0.002
CURVE 7	
T = 298	
0.380	0.000
0.390	0.030
0.396	0.056
0.404	0.098
0.413	0.122
0.418	0.127
0.432	0.128
0.480	0.063
0.510	0.035
0.541	0.020
0.573	0.010
0.599	0.000
CURVE 8	
T = 298	
0.377	0.000
0.380	0.027
0.383	0.068
0.385	0.145
0.392	0.192
0.404	0.229
0.412	0.241
0.424	0.241
0.447	0.194
0.469	0.141
0.492	0.092
0.508	0.067
0.541	0.037
0.576	0.017
0.610	0.006
0.842	0.007
1.020	0.004
2.480	0.006
CURVE 9	
T = 298	
0.372	0.000
0.380	0.028
0.387	0.269
0.396	0.413
0.403	0.465
0.412	0.480
0.420	0.487
0.429	0.484
0.451	0.432
0.478	0.315
0.518	0.156
0.534	0.114
0.559	0.077
0.601	0.043
0.658	0.023
0.789	0.014
0.925	0.015
1.136	0.020
1.431	0.030
1.864	0.048
2.123	0.060
2.380	0.073
CURVE 10	
T = 298	
0.375	0.006
0.380	0.015
0.394	0.085
0.399	0.102
0.407	0.113
0.414	0.117
0.428	0.109
0.447	0.093
0.489	0.050
0.510	0.038
0.533	0.026
0.565	0.016
0.601	0.011
0.635	0.012
0.682	0.005
0.722	0.003
0.791	0.000
0.902	0.003
1.420	0.012
1.569	0.012
1.653	0.011
1.875	0.011
2.101	0.014
2.279	0.021
2.450	0.028
CURVE 11	
T = 298	
0.371	0.003
0.377	0.007
0.380	0.011
0.392	0.166
0.399	0.208
0.410	0.238
0.425	0.255
0.447	0.238
0.469	0.214
0.520	0.141
0.548	0.115
0.569	0.098
0.604	0.081
0.654	0.062
0.713	0.047
0.782	0.034
0.873	0.024
1.048	0.014
1.102	0.015
1.198	0.015
1.337	0.013
1.441	0.013
1.595	0.022
1.693	0.024
1.839	0.021
2.029	0.020
2.183	0.027
2.266	0.027
2.344	0.022
2.480	0.020
CURVE 12	
T = 298	
0.374	0.004
0.378	0.010
0.387	0.083
0.395	0.121
0.404	0.142
0.413	0.156
0.420	0.160
0.427	0.160
0.434	0.157
0.454	0.128
0.492	0.068
0.514	0.050
0.546	0.032
0.590	0.017
0.623	0.010
0.750	0.008
1.187	0.007
2.480	0.005
CURVE 13*	
T = 298	
0.376	0.000
0.381	0.011
0.392	0.082
0.400	0.142
0.412	0.181
0.417	0.186
0.426	0.186
0.442	0.173
0.466	0.138
0.495	0.088
0.511	0.065
0.537	0.044
0.564	0.031
0.615	0.016
0.654	0.009
0.687	0.007
1.881	0.005
2.480	0.004
CURVE 14*	
T = 298	
0.380	0.000
0.383	0.022
0.389	0.055
0.399	0.108
0.409	0.141
0.418	0.159
0.423	0.163
0.443	0.149
0.460	0.130
0.491	0.083
0.519	0.057
0.540	0.042
0.573	0.027
0.610	0.017
0.695	0.011
1.642	0.008
2.480	0.004
CURVE 15*	
T = 298	
0.380	0.000
0.383	0.022
0.389	0.055
0.399	0.102
0.418	0.147
0.426	0.147
0.458	0.119
0.484	0.085
0.506	0.057
0.526	0.038
0.547	0.025
0.566	0.018
0.596	0.012
0.651	0.010
1.061	0.010
1.642	0.008
2.480	0.004
CURVE 16*	
T = 298	
0.376	0.006
0.381	0.016
0.384	0.027
0.388	0.055
0.390	0.132
0.400	0.181
0.416	0.209
0.420	0.209
0.438	0.197
0.463	0.173
0.513	0.099
0.547	0.071
0.573	0.056
0.608	0.046
0.764	0.023
0.908	0.011
1.079	0.006
1.344	0.005
2.398	0.006
CURVE 17*	
T = 298	
0.376	0.006
0.381	0.016
0.384	0.027
0.388	0.055
0.390	0.132
0.400	0.212
0.410	0.239
0.420	0.253
0.438	0.242
0.462	0.217
0.481	0.186

*Not shown on plot

DATA TABLE NO. 128 CHANGE IN NORMAL SPECTRAL ABSORPTANCE OF ZINC OXIDE PIGMENTED COATINGS (continued)

λ	Δα
CURVE 17* (cont.) T = 298	
0.497	0.161
0.523	0.128
0.557	0.102
0.661	0.060
0.720	0.045
0.786	0.034
0.875	0.024
0.993	0.018
1.233	0.015
1.368	0.013
1.495	0.018
1.604	0.023
1.729	0.025
1.802	0.020
1.898	0.020
2.006	0.018
2.210	0.028
2.403	0.020
CURVE 18* T = 298	
0.370	0.000
0.378	0.029
0.382	0.067
0.386	0.165
0.392	0.211
0.408	0.266
0.415	0.280
0.419	0.282
0.439	0.264
0.462	0.231
0.477	0.195
0.487	0.150
0.502	0.115
0.536	0.068
0.556	0.045
0.586	0.024
0.631	0.012
0.658	0.010
CURVE 19* T = 298	
0.379	0.022
0.388	0.114

λ	Δα
CURVE 19 (cont.)	
0.394	0.197
0.404	0.246
0.411	0.261
0.422	0.271
0.441	0.255
0.456	0.236
0.476	0.188
0.487	0.153
0.488	0.113
0.521	0.088
0.553	0.056
0.576	0.035
0.603	0.023
0.645	0.012
0.708	0.009
0.805	0.011
0.897	0.011
1.004	0.008
1.377	0.013
1.651	0.016
2.283	0.016
CURVE 20* T = 298	
0.393	0.173
0.400	0.207
0.414	0.234
0.426	0.242
0.439	0.233
0.456	0.214
0.467	0.195
0.483	0.152
0.507	0.106
0.549	0.066
0.588	0.041
0.626	0.029
0.679	0.018
0.738	0.012
0.856	0.008
1.109	0.008
2.254	0.014
2.366	0.019

λ	Δα
CURVE 21* T = 298	
0.369	0.000
0.376	0.016
0.381	0.045
0.384	0.145
0.401	0.216
0.410	0.239
0.417	0.239
0.424	0.236
0.435	0.223
0.486	0.107
0.521	0.057
0.543	0.034
0.570	0.019
0.603	0.010
0.703	0.006

*Not shown on plot

FIGURE 129A

ANALYZED NORMAL SOLAR ABSORPTANCE OF ZINC OXIDE PIGMENTED COATINGS

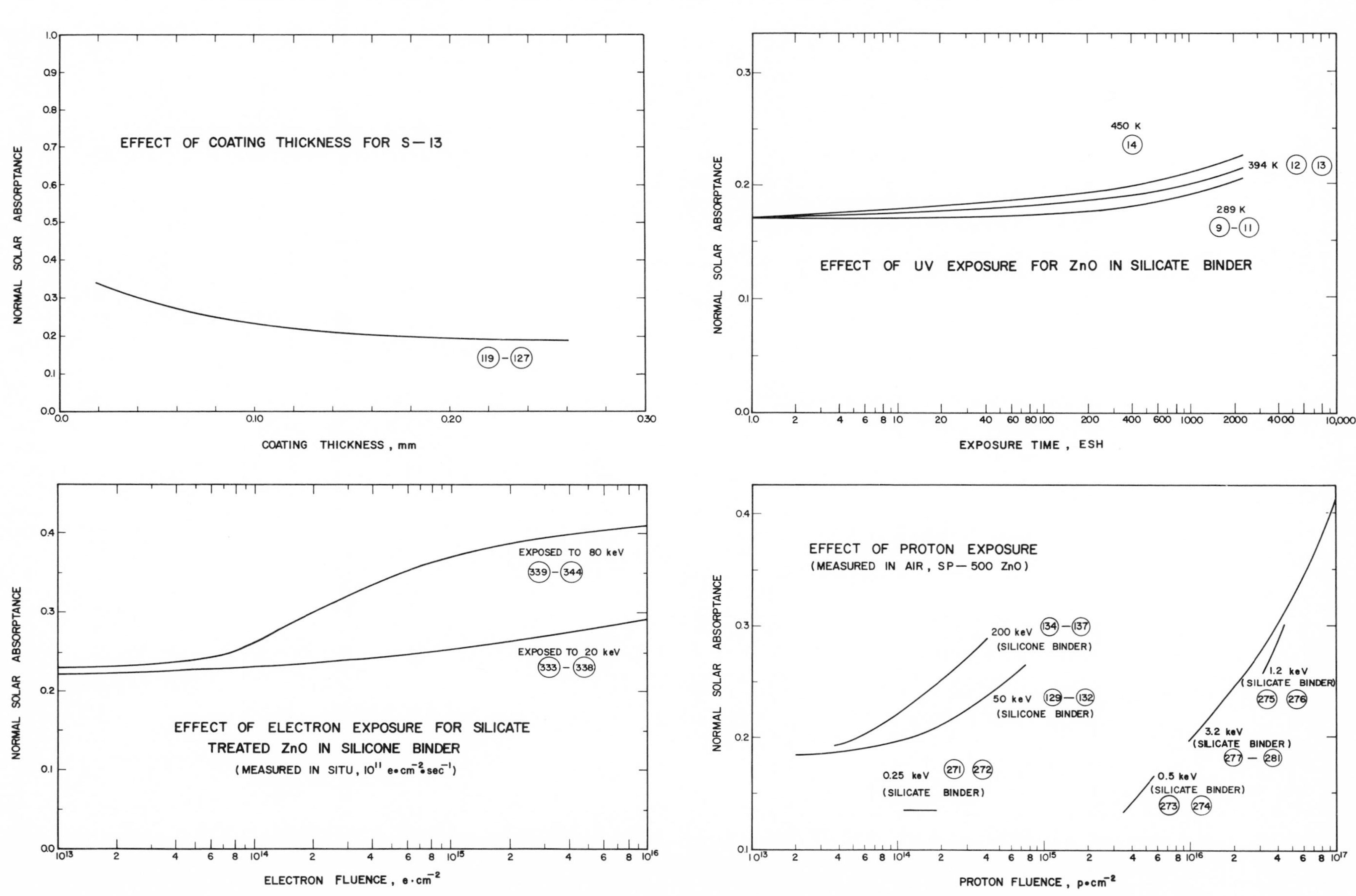

SPECIFICATION TABLE NO. 129 NORMAL SOLAR ABSORPTANCE OF ZINC OXIDE PIGMENTED COATINGS

Curve No.	Ref. No.	Year	Temperature Range, K	Geometry θ	Reported Error, %	Composition (weight percent), Specifications and Remarks
1*	27	1964	~298	~0°		SP-500 ZnO in PS-7 potassium silicate binder; PBR 4.30; solids content 46.3; aluminum substrate abraded with No. 60 Aloxite cloth; sample cured at 413 K for 18 hrs. [Authors' designation: Sample Z-5]
2*	27	1964	~298	~0°		Above specimen and conditions except irradiated in vacuum (10^{-6} mm Hg) with 200 ESH at solar factor of 3 suns.
3*	27	1964	~298	~0°		SP-500 ZnO in PS-7 potassium silicate binder; PBR 4.30; solids content 56.9; aluminum substrate abraded with No. 60 Aloxite cloth; sample cured at 413 K for 18 hrs. [Authors' designation: Sample Z-8]
4*	27	1964	~298	~0°		Above specimen and conditions except irradiated in vacuum (10^{-6} mm Hg) with 200 ESH at solar factor of 3 suns.
5*	27	1964	~298	~0°		SP-500 ZnO in PS-7 potassium silicate binder; PBR 2.13; solids content 51.9; aluminum substrate abraded with No. 60 Aloxite cloth; sample cured at 413 K for 18 hrs. [Authors' designation: Sample Z-9]
6*	27	1964	~298	~0°		Above specimen and conditions except irradiated in vacuum (10^{-6} mm Hg) with 200 ESH at solar factor of 3 suns.
7*	27	1964	~298	~0°		SP-500 ZnO in PS-7 potassium silicate binder; PBR 4.30; solids content 46.3; aluminum substrate abraded with No. 60 Aloxite cloth; sample cured at 413 K for 18 hrs. [Authors' designation: Sample Z-10]
8*	27	1964	~298	~0°		Above specimen and conditions except irradiated in vacuum (10^{-6} mm Hg) with 300 ESH at solar factor of 3 suns.
9	18	1964	289	~0°		ZnO in potassium silicate binder; specimen exposed to 1 ESH of ultraviolet radiation; calculated from reflectance.
10	18	1964	289	~0°		Above specimen and conditions except exposed to 235 ESH of ultraviolet radiation.
11	18	1964	289	~0°		Above specimen and conditions except exposed to 2300 ESH of ultraviolet radiation.
12	18	1964	289	~0°		ZnO in potassium silicate binder; specimen exposed to 235 ESH of ultraviolet radiation; calculated from reflectance.
13	18	1964	289	~0°		Above specimen and conditions except exposed to 2300 ESH of ultraviolet radiation.
14	18	1964	289	~0°		ZnO in potassium silicate binder; specimen exposed to 2300 ESH of ultraviolet radiation; calculated from reflectance.
15*	28	1964	~298	~0°		ZnO in potassium silicate binder; calculated from reflectance. [Authors' designation: IITRI 441-2]
16*	28	1964	~298	~0°		Similar to above specimen and conditions.
17*	28	1964	~298	~0°		Similar to above specimen and conditions except gamma-irradiated in vacuum (10^{-6} mm Hg) at 320 K to a total dose of 77 Megarads (dose rate ~1.5 Megarads hr^{-1}, cobalt-60 source); measured within 1 hr after exposure to air.

*Not shown on plot

SPECIFICATION TABLE NO. 129 NORMAL SOLAR ABSORPTANCE OF ZINC OXIDE PIGMENTED COATINGS (continued)

Curve No.	Ref. No.	Year	Temperature Range, K	Geometry θ	Reported Error, %	Composition (weight percent), Specifications and Remarks
18*	28	1964	~298	~0°		Similar to above specimen and conditions.
19*	28	1964	~298	~0°		Similar to above specimen and conditions except irradiated to a total dose of 385 Megarads.
20*	28	1964	~298	~0°		Similar to curve 15 specimen and conditions except ultraviolet irradiated in vacuum ($<10^{-6}$ mm Hg) at 273-288 K for 300 ESH (~3-sun intensity, General Electric BH-6 lamp source); measured within 1 hr after exposure to air.
21*	28	1964	~298	~0°		Similar to above specimen and conditions.
22*	28	1964	~298	~0°		S-13; ZnO in silicone binder; calculated from reflectance.
23*	28	1964	~298	~0°		Similar to above specimen and conditions.
24*	28	1964	~298	~0°		Similar to above specimen and conditions except gamma-irradiated in vacuum (10^{-6} mm Hg) at 320 K to a total dose of 77 Megarads (dose rate ~1.5 Megarads hr^{-1}, cobalt-60 source); measured within 1 hr after exposure to air.
25*	28	1964	~298	~0°		Similar to above specimen and conditions.
26*	28	1964	~298	~0°		Similar to above specimen and conditions except irradiated to a total dose of 385 Megarads.
27*	28	1964	~298	~0°		Similar to curve 22 specimen and conditions except ultraviolet irradiated in vacuum ($<10^{-6}$ mm Hg) at 273-288 K for 300 ESH (~3-sun intensity, General Electric BH-6 lamp source); measured within 1 hr after exposure to air.
28*	28	1964	~298	~0°		Similar to above specimen and conditions.
29*	28	1964	~298	~0°		S-13; ZnO in silicone binder (0.104 mm thick); calculated from reflectance.
30*	28	1964	~298	~0°		Above specimen and conditions except variable is film thikness in mm.
31*	23	1964	~298	~0°		ZnO in potassium silicate binder; calculated from reflectance; data measured with a Gier-Dunkle integrating sphere. [Authors' designation: IITRI 441-2]
32*	23	1964	~298	~0°		Similar to above specimen and conditions except reflectance obtained with a Gier-Dunkle Solar Reflectometer.
33*	23	1964	~298	~0°		Similar to above specimen and conditions except reflectance obtained with a Cary 14 spectrophotometer and integrating sphere.
34*	23	1964	~298	~0°		Similar to above specimen and conditions.
35*	23	1964	~298	~0°		Similar to above specimen and conditions except reflectance obtained with a G. E. and P. E. spectrophotometer and integrating sphere.
36*	23	1964	~298	~0°		Similar to above specimen and conditions except reflectance obtained with a Beckman DK-2A spectrophotometer and integrating sphere.
37*	23	1964	~298	~0°		Similar to above specimen and conditions.
38*	23	1964	~298	~0°		Similar to above specimen and conditions.
39*	23	1964	~298	~0°		Similar to above specimen and conditions except reflectance obtained with a Bausch and Lomb 505 spectrophotometer and integrating sphere.

*Not shown on plot

SPECIFICATION TABLE NO. 129 NORMAL SOLAR ABSORPTANCE OF ZINC OXIDE PIGMENTED COATINGS (continued)

Curve No.	Ref. No.	Year	Temperature Range, K	Geometry θ	Reported Error, %	Composition (weight percent), Specifications and Remarks
40*	16	1965	300-550	0°	± 5	ZnO in methyl silicone binder (0.127 mm thick); measured in vacuum (10^{-7} mm Hg).
41*	51	1965	298	~0°		Z-93; ZnO in potassium silicate binder; substrate abraded with No. 60 Aloxite cloth; PBR 4.30. [Authors' designation: Sample 6001]
42*	51	1965	298	~0°		Above specimen and conditions except heat-treated (418 K for 16 hrs); time lapse, 1 day.
43*	51	1965	298	~0°		Above specimen and conditions except heat-treated (773 K for 2 hrs); time lapse, 6 hrs.
44*	51	1965	298	~0°		Similar to curve 41 specimen and conditions except after time lapse of 3 mo.
45*	51	1965	298	~0°		Z-93; ZnO in potassium silicate binder; abraded magnesium silicate substrate; PBR 4.30. [Authors' designation: Sample 6003]
46*	51	1965	298	~0°		Similar to above specimen and conditions except heat-treated (418 K for 16 hrs); time lapse, 1 day.
47*	51	1965	298	~0°		Similar to curve 45 specimen and conditions except heat-treated (973 K for 2 hrs); time lapse, 1 day.
48*	51	1965	298	~0°		Similar to curve 45 specimen and conditions except heat-treated (1073 K for 2 hrs); time lapse, 6 hrs.
49*	51	1965	298	~0°		Similar to curve 45 specimen and conditions except after time lapse of 3 mo.
50*	51	1965	298	~0°		Z-93; ZnO in potassium silicate binder; abraded alumina substrate; PBR 4.30. [Authors' designation: Sample 6004]
51*	51	1965	298	~0°		Similar to above specimen and conditions except heat-treated (418 K for 16 hrs); time lapse, 1 day.
52*	51	1965	298	~0°		Similar to curve 50 specimen and conditions except heat-treated (973 K for 2 hrs); time lapse, 1 day.
53*	51	1965	298	~0°		Similar to curve 50 specimen and conditions except heat-treated (1073 K for 2 hrs); time lapse, 6 hrs.
54*	51	1965	298	~0°		Similar to curve 50 specimen and conditions except after time lapse of 3 mo.
55*	51	1965	298	~0°		Z-93; ZnO in potassium silicate binder; PBR 4.30; measured in vacuum (<3.0 x 10^{-5} mm Hg). [Authors' designation: Sample 6005]
56*	51	1965	298	~0°		Above specimen and conditions except exposed to 1000 ESH at solar factor of 11.7 suns.
57*	51	1965	279	~0°		Z-93; ZnO in potassium silicate binder; PBR 4.3; air dried at 100% relative humidity; measured in vacuum (<10^{-6} mm Hg). [Authors' designation: Sample 7071]
58*	51	1965	279	~0°		Above specimen and conditions except exposed to 2160 ESH at solar factor of 10.0 suns.
59*	51	1965	279	~0°		Similar to curve 57 specimen and conditions except air dried at 0% relative humidity. [Authors' designation: Sample 7072]
60*	51	1965	279	~0°		Above specimen and conditions except exposed to 2160 ESH at solar factor of 10.0 suns.
61*	51	1965	279	~0°		Similar to curve 57 specimen and conditions except air dried at 35% relative humidity. [Authors' designation: Sample 7073]

*Not shown on plot

SPECIFICATION TABLE NO. 129 NORMAL SOLAR ABSORPTANCE OF ZINC OXIDE PIGMENTED COATINGS (continued)

Curve No.	Ref. No.	Year	Temperature Range, K	Geometry θ	Reported Error, %	Composition (weight percent), Specifications and Remarks
62*	51	1965	279	~0°		Above specimen and conditions except exposed to 2160 ESH at solar factor of 10.0 suns.
63*	51	1965	279	~0°		S-13; calcined (923 K for 16 hrs) SP-500 ZnO in RTV methyl silicone binder (0.127 mm thick); n-Butanol solvent; PVC 30%; measured in vacuum (10^{-7} mm Hg).
64*	51	1965	279	~0°		Above specimen and conditions except exposed to 2100 ESH at solar factor of 10.1 suns.
65*	51	1965	279	~0°		Similar to curve 63 specimen and conditions except paint formulated using isooctane solvent.
66*	51	1965	279	~0°		Above specimen and conditions except exposed to 2100 ESH at solar factor of 10.1 suns.
67*	51	1965	279	~0°		Similar to curve 63 specimen and conditions except paint formulated using toluene solvent.
68*	51	1965	279	~0°		Above specimen and conditions except exposed to 2100 ESH at solar factor of 10.1 suns.
69*	51	1965	278	~0°		Z-93; SP-500 ZnO in potassium silicate binder; PBR 4.30; measured in vacuum (9×10^{-7} mm Hg). [Authors' designation: Sample 7088]
70*	51	1965	278	~0°		Above specimen and conditions except exposed to 1020 ESH at solar factor of 3.9 suns.
71*	51	1965	278	~0°		Similar to curve 69 specimen and conditions. [Authors' designation: Sample 7089]
72*	51	1965	278	~0°		S-13; SP-500 ZnO in RTV-602 binder; measured in vacuum (3.0×10^{-6} mm Hg). [Authors' designation: Sample 5068-1]
73*	51	1965	278	~0°		Above specimen and conditions except exposed to 1020 ESH at solar factor of 3.9 suns.
74*	51	1965	278	~0°		Similar to curve 72 specimen and conditions. [Authors' designation: Sample 5068-2]
75*	51	1965	278	~0°		Above specimen and conditions except exposed to 1250 ESH at solar factor of 14.4 suns.
76*	51	1965	278	~0°		S-33; SP-500 ZnO in methyl silicone binder (Me/Si ratio 1.4); 30% PVC; measured in vacuum (6×10^{-7} mm Hg). [Authors' designation: Sample 5012-1C]
77*	51	1965	278	~0°		Above specimen and conditions except exposed to 1020 ESH at solar factor of 3.9 suns.
78*	51	1965	278	~0°		Similar to curve 76 specimen and conditions.
79*	51	1965	278	~0°		Similar to above specimen and conditions except exposed to 1250 ESH at solar factor of 14.4 suns.
80*	51	1965	279	~0°		Z-93; calcined SP-500 ZnO in potassium silicate binder; PBR 4.30; spray application; measured in vacuum ($<10^{-6}$ mm Hg); fresh preparation. [Authors' designation: Sample 7135]
81*	51	1965	279	~0°		Above specimen and conditions except exposed to 300 ESH at solar factor of 4.8 suns.
82*	51	1965	279	~0°		Similar to curve 80 specimen and conditions except measured in vacuum (5×10^{-6} mm Hg). [Authors' designation: Sample 7140]
83*	51	1965	279	~0°		Above specimen and conditions except exposed to 4500 ESH at solar factor of 6.1 suns.
84*	51	1965	279	~0°		Similar to curve 80 specimen and conditions except measured in vacuum (3.5×10^{-6} mm Hg). [Authors' designation: Sample 7130]
85*	51	1965	279	~0°		Above specimen and conditions except exposed to 7900 ESH at solar factor of 5.8 suns.

*Not shown on plot

SPECIFICATION TABLE NO. 129 NORMAL SOLAR ABSORPTANCE OF ZINC OXIDE PIGMENTED COATINGS (continued)

Curve No.	Ref. No.	Year	Temperature Range, K	Geometry θ	Reported Error, %	Composition (weight percent), Specifications and Remarks
86*	51	1965	279	~0°		S-33; SP-500 ZnO in experimental R-9 methyl silicate resin binder (Me/Si ratio 1.4); n-butanol solvent; spray application; measured in vacuum ($<10^{-6}$ mm Hg). [Authors' designation: Sample 5012-C3-2]
87*	51	1965	279	~0°		Above specimen and conditions except exposed to 300 ESH at solar factor of 4.8 suns.
88*	51	1965	279	~0°		Similar to curve 86 specimen and conditions except measured in vacuum (5×10^{-6} mm Hg). [Authors' designation: Sample 5012-C3-3]
89*	51	1965	279	~0°		Above specimen and conditions except exposed to 4500 ESH at solar factor of 6.1 suns.
90*	51	1965	279	~0°		Similar to curve 86 specimen and conditions except measured in vacuum (3.5×10^{-6} mm Hg). [Authors' designation: Sample 5012-C3]
91*	51	1965	279	~0°		Above specimen and conditions except exposed to 7900 ESH at solar factor of 5.8 suns.
92*	51	1965	279	~0°		S-34; calcined SP-500 ZnO in RTV-602 binder; 40% PVC; n-butanol, solvent; catalyst, GE SRC-05 (1 drop/20 g); spray application; measured in vacuum ($<10^{-6}$ mm Hg). [Authors' designation: Sample 5096-2]
93*	51	1965	279	~0°		Above specimen and conditions except exposed to 300 ESH at solar factor of 4.8 suns.
94*	51	1965	279	~0°		Z-94; calcined SP-500 ZnO in potassium silicate binder; PBR 6.45; spray application; measured in vacuum ($<10^{-6}$ mm Hg); preparation aged 5 mo. [Authors' designation: Sample 7138]
95*	51	1965	279	~0°		Above specimen and conditions except exposed to 300 ESH at solar factor of 4.8 suns.
96*	51	1965	279	~0°		Similar to curve 94 specimen and conditions except measured at pressure of 5×10^{-6} mm Hg. [Authors' designation: Sample 7143]
97*	51	1965	279	~0°		Above specimen and conditions except exposed to 4500 ESH at solar factor of 6.1 suns.
98*	51	1965	279	~0°		Similar to curve 94 specimen and conditions except measured at pressure of 3.5×10^{-6} mm Hg. [Authors' designation: Sample 7133]
99*	51	1965	279	~0°		Above specimen and conditions except exposed to 7900 ESH at solar factor of 5.8 suns.
100*	51	1965	279	~0°		Z-93; calcined SP-500 ZnO in potassium silicate binder; PBR 4.30; spray application; measured in vacuum ($<10^{-6}$ mm Hg); preparation aged 4 mo. [Authors' designation: Sample 7137]
101*	51	1965	279	~0°		Above specimen and conditions except heat-treated at 773 K for 2 hrs. [Authors' designation: Sample 7137 HT]
102*	51	1965	279	~0°		Above specimen and conditions except exposed to 300 ESH at solar factor of 4.8 suns.
103*	51	1965	279	~0°		Similar to curve 100 except measured at pressure of 5×10^{-6} mm Hg. [Authors' designation: Sample 7142]
104*	51	1965	279	~0°		Above specimen and conditions except heat-treated at 773 K for 2 hrs. [Authors' designation: Sample 7142 HT]
105*	51	1965	279	~0°		Above specimen and conditions except exposed to 4500 ESH at solar factor of 6.1 suns.

*Not shown on plot

SPECIFICATION TABLE NO. 129 NORMAL SOLAR ABSORPTANCE OF ZINC OXIDE PIGMENTED COATINGS (continued)

Curve No.	Ref. No.	Year	Temperature Range, K	Geometry θ	Reported Error, %	Composition (weight percent), Specifications and Remarks
106*	51	1965	279	~0°		Similar to curve 100 specimen and conditions except measured at pressure of 3.5 x 10^{-6} mm Hg. [Authors' designation: Sample 7132]
107*	51	1965	279	~0°		Above specimen and conditions except heat-treated at 773 K for 2 hrs. [Authors' designation: Sample 7132HT]
108*	51	1965	279	~0°		Above specimen and conditions except exposed to 7900 ESH at solar factor of 5.8 suns.
109*	51	1965	279	~0°		Z-93; calcined SP-500 ZnO in potassium silicate binder; PBR 4.30; spray application; measured in vacuum (<10^{-6} mm Hg); preparation aged 4 mo. [Authors' designation: Sample 7131]
110*	51	1965	279	~0°		Above specimen and conditions except exposed to 300 ESH at solar factor of 4.8 suns.
111*	51	1965	279	~0°		Similar to curve 109 specimen and conditions except measured in vacuum (5 x 10^{-6} mm Hg). [Authors' designation: Sample 7141]
112*	51	1965	279	~0°		Above specimen and conditions except exposed to 4500 ESH at solar factor of 6.1 suns.
113*	51	1965	279	~0°		Similar to curve 109 specimen and conditions except measured at pressure of 3.5 x 10^{-6} mm Hg. [Authors' designation: Sample 7131]
114*	51	1965	278	~0°		Above specimen and conditions except exposed to 7900 ESH at solar factor of 5.8 suns.
115*	51	1965	279	~0°		S-34; calcined SP-500 ZnO in RTV-602 binder; 40% PVC; n-butanol solvent; catalyst GE SRC-05 (1 drop/20 g); spray application; measured in vacuum (5 x 10^{-6} mm Hg). [Authors' designation: Sample 5096-3]
116*	51	1965	279	~0°		Above specimen and conditions except exposed to 4500 ESH at solar factor of 6.1 suns.
117*	51	1965	279	~0°		S-34; calcined SP-500 ZnO in RTV-602 binder; 40% PVC; n-butanol solvent; SRC-05 catalyst (1 drop/20 g). [Authors' designation: Sample 5096-1]
118*	51	1965	279	~0°		Above specimen and conditions except exposed to 7900 ESH at solar factor of 5.8 suns.
119	51	1965	298	~0°		S-13; SP-500 ZnO in RTV-602 binder (0.025 mm thick); SRC-05 catalyst; toluene solvent; spray application to primed surface; air dried.
120	51	1965	298	~0°		Similar to above specimen and conditions except 0.051 mm thick.
121	51	1965	298	~0°		Similar to above specimen and conditions except 0.076 mm thick.
122	51	1965	298	~0°		Similar to above specimen and conditions except 0.102 mm thick.
123	51	1965	298	~0°		Similar to above specimen and conditions except 0.127 mm thick.
124	51	1965	298	~0°		Similar to above specimen and conditions except 0.152 mm thick.
125	51	1965	298	~0°		Similar to above specimen and conditions except 0.203 mm thick.
126	51	1965	298	~0°		Similar to above specimen and conditions except 0.229 mm thick.
127	51	1965	298	~0°		Similar to above specimen and conditions except 0.254 mm thick.
128	22	1965	~298	~0°		SP-500 ZnO in Q90016 methyl silicone binder (0.076 mm thick); xylene solvent; spray application; calculated from reflectance.

*Not shown on plot

SPECIFICATION TABLE NO. 129 NORMAL SOLAR ABSORPTANCE OF ZINC OXIDE PIGMENTED COATINGS (continued)

Curve No.	Ref. No.	Year	Temperature Range, K	Geometry θ	Reported Error, %	Composition (weight percent), Specifications and Remarks
129	22	1965	~298	~0°		Similar to above specimen and conditions except irradiated in vacuum with 50 keV protons to a total dose of 2×10^{13} p cm^{-2}.
130	22	1965	~298	~0°		Similar to above specimen and conditions except irradiated to a total dose of 8×10^{13} p cm^{-2}.
131	22	1965	~298	~0°		Similar to above specimen and conditions except irradiated to a total dose of 2.4×10^{14} p cm^{-2}.
132	22	1965	~298	~0°		Similar to above specimen and conditions except irradiated to a total dose of 6.0×10^{14} p cm^{-2}.
133*	22	1965	~298	~0°		Similar to curve 128 specimen and conditions.
134	22	1965	~298	~0°		Similar to above specimen and conditions except irradiated in vacuum with 200 keV protons to a total dose of 4×10^{13} p cm^{-2}.
135	22	1965	~298	~0°		Similar to above specimen and conditions except irradiated to a total dose of 1×10^{14} p cm^{-2}.
136	22	1965	~298	~0°		Similar to above specimen and conditions except irradiated to a total dose of 2×10^{14} p cm^{-2}.
137	22	1965	~298	~0°		Similar to above specimen and conditions except irradiated to a total dose of 4×10^{14} p cm^{-2}.
138*	52	1965	193-263	~0°		ZnO in S-83-14 resin binder; toluene thinner; measured in earth orbit on OSO-II; data is avg over indicated temp range.
139*	52	1965	193-263	~0°		ZnO in silicone resin binder; xylene thinner; measured in earth orbit on OSO-II; data is avg over indicated temp range.
140*	52	1965	193-263	~0°		Z-52; AZO 55 LO ZnO in PS 7 potassium silicate binder; water thinner; measured in earth orbit on OSO-II; data is avg over indicated temp range.
141*	53	1966	298	~0°		ZnO in potassium silicate binder (0.127 to 0.152 mm thick); computed from reflectance measured relative to MgO.
142*	53	1966	298	~0°		Above specimen and conditions except exposed to proton irradiation of 1.2×10^{16} p cm^{-2} at 150 keV in vacuum (10^{-6} mm Hg).
143*	53	1966	323	~0°		ZnO in potassium silicate binder (0.127 to 0.152 mm thick); computed from reflectance measured relative to MgO.
144*	53	1966	323	~0°		Above specimen and conditions except exposed to proton irradiation of 1.0×10^{16} p cm^{-2} at 150 keV in vacuum (10^{-6} mm Hg).
145*	53	1966	298	~0°		ZnO in potassium silicate binder (0.127 to 0.152 mm thick); computed from reflectance measured relative to MgO.
146*	53	1966	298	~0°		Above specimen and conditions except exposed to proton irradiation of 1.0×10^{16} p cm^{-2} at 20 keV in vacuum (10^{-6} mm Hg).
147*	53	1966	298	~0°		ZnO in potassium silicate binder (0.127 to 0.152 mm thick); computed from reflectance measured relative to MgO.
148*	53	1966	298	~0°		Above specimen and conditions except exposed to proton irradiation of 1.0×10^{15} p cm^{-2} at 150 keV in vacuum (10^{-6} mm Hg).
149*	53	1966	298	~0°		ZnO in potassium silicate binder (0.127 to 0.152 mm thick); computed from reflectance measured relative to MgO.

*Not shown on plot

SPECIFICATION TABLE NO. 129 NORMAL SOLAR ABSORPTANCE OF ZINC OXIDE PIGMENTED COATINGS (continued)

Curve No.	Ref. No.	Year	Temperature Range, K	Geometry θ	Reported Error, %	Composition (weight percent), Specifications and Remarks
150*	53	1966	298	~0°		Above specimen and conditions except exposed to proton irradiation of 1.0 x 10^{14} p cm^{-2} at 150 keV in vacuum (10^{-6} mm Hg).
151*	53	1966	298	~0°		ZnO in potassium silicate binder (0.127 to 0.152 mm thick); computed from reflectance measured relative to MgO.
152*	53	1966	298	~0°		Above specimen and conditions except exposed to electron irradiation of 1.0 x 10^{16} e cm^{-2} at 2 MeV in vacuum (10^{-6} mm Hg).
153*	53	1966	298	~0°		ZnO in potassium silicate binder (0.127 to 0.152 mm thick); computed from reflectance measured relative to MgO.
154*	53	1966	298	~0°		Above specimen and conditions except exposed to electron irradiation of 1.0 x 10^{15} e cm^{-2} at 2 MeV in vacuum (10^{-6} mm Hg).
155*	53	1966	298	~0°		ZnO in potassium silicate binder (0.127 to 0.152 mm thick); computed from reflectance measured relative to MgO.
156*	53	1966	298	~0°		Above specimen and conditions except exposed to electron irradiation of 1.0 x 10^{14} e cm^{-2} at 2 MeV in vacuum (10^{-6} mm Hg).
157*	54	1965	298	~0°		S-13; SP-500 ZnO in LTV-602 binder; SRC 05 catalyst; measured in simulated space environment. [Authors' designation: Sample 5082]
158*	54	1965	298	~0°		Above specimen and conditions except exposed to 2000 ESH.
159*	54	1965	298	~0°		Similar to curve 157 specimen and conditions except prepared with Shell H2 catalyst. [Authors' designation: Sample 5080]
160*	54	1965	298	~0°		Above specimen and conditions except exposed to 2000 ESH.
161*	54	1965	298	~0°		OI No. 650; SP-500 ZnO in Owens Illinois glass resin base; PVC 35%; measured in vacuum. [Authors' designation: Sample 5066 (lot 1025)]
162*	54	1965	298	~0°		Above specimen and conditions except exposed to 2000 ESH.
163*	54	1965	298	~0°		Similar to curve 161 specimen and conditions. [Authors' designation: Sample 5076-2 (lot 276)]
164*	54	1965	298	~0°		Above specimen and conditions except exposed to 2000 ESH.
165*	55	1968	225	~0°		Grumman RTV-602 white; SP-500 ZnO (240 pbw) in RTV-602 (100 pbw) binder; toluene solvent (70 pbw); measured in vacuum (<10^{-6} mm Hg). [Author's designation: Sample 28]
166*	55	1968	225	~0°		Above specimen and conditions except measured in air.
167*	55	1968	225	~0°		Curve 165 specimen and conditions except exposed to corpuscular radiation of 5 x 10^{13} e cm^{-2} at 3.0 keV (flux density 6 x 10^{12} e cm^{-2} sec^{-1}) and 5 x 10^{14} p cm^{-2} at 2.5 keV (flux density 8 x 10^{12} p cm^{-2} sec^{-1}).
168*	55	1968	225	~0°		Above specimen and conditions.
169*	55	1968	225	~0°		Above specimen and conditions.
170*	55	1968	297	~0°		Above specimen and conditions except measured in air.

*Not shown on plot

SPECIFICATION TABLE NO. 129 NORMAL SOLAR ABSORPTANCE OF ZINC OXIDE PIGMENTED COATINGS (continued)

Curve No.	Ref. No.	Year	Temperature Range, K	Geometry θ	Reported Error, %	Composition (weight percent), Specifications and Remarks
171*	55	1968	225	~0°		Similar to curve 165 specimen and conditions. [Authors' designation: Sample 3]
172*	55	1968	297	~0°		Above specimen and conditions except measured in air.
173*	55	1968	225	~0°		Curve 171 specimen and conditions except exposed to 200 ESH at solar factor of 1 sun and corpuscular radiation similar to curve 167.
174*	55	1968	225	~0°		Above specimen and conditions except exposed to 500 ESH.
175*	55	1968	225	~0°		Above specimen and conditions except exposed to 945 ESH.
176*	55	1968	297	~0°		Above specimen and conditions except measured in air.
177*	17	1968	298	~0°		Z-93; SP-500 ZnO in potassium silicate binder; aluminum substrate; supplied by IITRI.
178*	17	1968	298	~0°		Similar to above specimen and conditions except supplied by ESRO I Project Group; coating thickness 0.089 mm.
179*	17	1968	298	~0°		Similar to above specimen and conditions except soiled.
180*	17	1968	298	~0°		CM-147; ZnO pigmented inorganic paint; sample supplied by ESRO II Project Group.
181*	20	1963	~298	~0°		S-4; SP-500 ZnO in R-1 experimental methyl silicone binder; Me/Si ratio 1.29; PBR 1.35; PVC 25%; aluminum substrate; spray application; computed from spectral reflectance data.
182*	20	1963	~298	~0°		Above specimen and conditions except exposed to 1460 ESH at solar factor of 9 suns.
183*	20	1963	~298	~0°		S-16; SP-500 ZnO in R-7 experimental methyl silicone binder; Me/Si ratio 1.33; PBR 1.64; PVC 25%; aluminum substrate; spray application; computed from spectral reflectance data.
184*	20	1963	~298	~0°		Above specimen and conditions except exposed to 615 ESH at solar factor of 9 suns.
185*	20	1963	~298	~0°		S-16; SP-500 ZnO in R-7 experimental methyl silicone binder; Me/Si ratio 1.33; PBR 1.64; PVC 25%; aluminum substrate; spray application; computed from spectral reflectance data.
186*	20	1963	~298	~0°		Above specimen and conditions except exposed to 1600 ESH at solar factor of 11 suns.
187*	20	1963	~298	~0°		S-11; SP-500 ZnO in R-5 experimental methyl silicone binder; Me/Si ratio 1.38; PBR 1.63; PVC 25%; aluminum substrate; spray application; computed from spectral reflectance data.
188*	20	1963	~298	~0°		Above specimen and conditions except exposed to 1460 ESH at solar factor of 9 suns.
189*	20	1963	~298	~0°		S-8; SP-500 ZnO in R-2 experimental methyl silicone binder; Me/Si ratio 1.46; PBR 1.70; PVC 25%; aluminum substrate; spray application; computed from spectral reflectance data.
190*	20	1963	~298	~0°		Above specimen and conditions except exposed to 1460 ESH at solar factor of 9 suns.
191*	20	1963	~298	~0°		S-16; SP-500 ZnO in R-7 experimental methyl silicone binder; Me/Si ratio 1.33; PBR 1.64; PVC 25%; aluminum substrate; spray application; computed from spectral reflectance data.
192*	20	1963	~298	~0°		Above specimen and conditions except exposed to 615 ESH at solar factor of 8.9 suns.

*Not shown on plot

SPECIFICATION TABLE NO. 129 NORMAL SOLAR ABSORPTANCE OF ZINC OXIDE PIGMENTED COATINGS (continued)

Curve No.	Ref. No.	Year	Temperature Range, K	Geometry θ	Reported Error, %	Composition (weight percent), Specifications and Remarks
193*	20	1963	~298	~0°		S-16; SP-500 ZnO in R-7 experimental methyl silicone binder; Me/Si ratio 1.33; PBR 1.64; PVC 25%; aluminum substrate; spray application; computed from spectral reflectance data.
194*	20	1963	~298	~0°		Above specimen and conditions except exposed to 1600 ESH at solar factor of 11 suns.
195*	20	1963	~298	~0°		S-16; SP-500 ZnO in R-7 experimental methyl silicone binder; Me/Si ratio 1.33; PBR 1.64; PVC 25%; aluminum substrate; spray application; computed from spectral reflectance data.
196*	20	1963	~298	~0°		Above specimen and conditions except exposed to 1700 ESH at solar factor of 10.7 suns.
197*	20	1963	~298	~0°		S-20; SP-500 ZnO in R-7 experimental methyl silicone binder; Me/Si ratio 1.33; PBR 2.10; PVC 30%; aluminum substrate; spray application; computed from spectral reflectance data.
198*	20	1963	~298	~0°		Above specimen and conditions except exposed to 1600 ESH at solar factor of 11 suns.
199*	20	1963	~298	~0°		S-11; SP-500 ZnO in R-5 experimental methyl silicone binder; Me/Si ratio 1.38; PBR 1.63; PVC 25%; aluminum substrate; spray application; computed from spectral reflectance data.
200*	20	1963	~298	~0°		Above specimen and conditions except exposed to 1460 ESH at solar factor of 9 suns.
201*	20	1963	~298	~0°		S-15; SP-500 ZnO in R-5 experimental methyl silicone binder; Me/Si ratio 1.38; PBR 2.12; PVC 30%; aluminum substrate; spray application; computed from spectral reflectance data.
202*	20	1963	~298	~0°		Above specimen and conditions except exposed to 1700 ESH at solar factor of 10.7 suns.
203*	20	1963	~298	~0°		S-19; SP-500 ZnO in R-5 experimental methyl silicone binder; Me/Si ratio 1.38; PBR 2.68; PVC 35%; aluminum substrate; spray application; computed from spectral reflectance data.
204*	20	1963	~298	~0°		Above specimen and conditions except exposed to 630 ESH at solar factor of 9.1 suns.
205*	20	1963	~298	~0°		S-19; SP-500 ZnO in R-5 experimental methyl silicone binder; Me/Si ratio 1.38; PBR 2.68; PVC 35%; aluminum substrate; spray application; computed from spectral reflectance data.
206*	20	1963	~298	~0°		Above specimen and conditions except exposed to 1200 ESH at solar factor of 8.7 suns.
207*	20	1963	~298	~0°		S-19; SP-500 ZnO in R-5 experimental methyl silicone binder; Me/Si ratio 1.38; PBR 2.68; PVC 35%; aluminum substrate; spray application; computed from spectral reflectance data.
208*	20	1963	~298	~0°		Above specimen and conditions except exposed to 1850 ESH at solar factor of 10.1 suns.
209*	20	1963	~298	~0°		S-21; pigment, 1:1 mixture SP-500 and E-P414 ZnO in R-5 experimental methyl silicone binder; Me/Si ratio 1.38; PBR 2.68; PVC 35%; aluminum substrate; spray application; computed from spectral reflectance data.
210*	20	1963	~298	~0°		Above specimen and conditions except exposed to 660 ESH at solar factor of 9.6 suns.

*Not shown on plot

SPECIFICATION TABLE NO. 129 NORMAL SOLAR ABSORPTANCE OF ZINC OXIDE PIGMENTED COATINGS (continued)

Curve No.	Ref. No.	Year	Temperature Range, K	Geometry θ	Reported Error, %	Composition (weight percent), Specifications and Remarks
211*	20	1963	~298	~0°		S-7; SP-500 ZnO in G. E. LTV-602 methyl silicone binder; PBR 1.40; PVC 20%; aluminum substrate; spray application; computed from spectral reflectance data.
212*	20	1963	~298	~0°		Above specimen and conditions except exposed to 1460 ESH at solar factor of 9 suns.
213*	20	1963	~298	~0°		S-7; SP-500 ZnO in G. E. LTV-602 methyl silicone binder; PBR 1.87; PVC 25%; aluminum substrate; spray application; computed from spectral reflectance data.
214*	20	1963	~298	~0°		Above specimen and conditions except exposed to 1460 ESH at solar factor of 9 suns.
215*	20	1963	~298	~0°		S-13; SP-500 ZnO in G. E. LTV-602 methyl silicone binder; PBR 2.40; PVC 30%; aluminum substrate; spray application; computed from spectral reflectance data.
216*	20	1963	~298	~0°		Above specimen and conditions except exposed to 1460 ESH at solar factor of 9 suns.
217*	20	1963	~298	~0°		S-27; SP-500 ZnO in G. E. LTV-602 methyl silicone binder; PBR 3.04; PVC 35%; aluminum substrate; spray application; computed from spectral reflectance data.
218*	20	1963	~298	~0°		Above specimen and conditions except exposed to 1600 ESH at solar factor of 10.2 suns.
219*	20	1963	~298	~0°		S-26; SP-500 ZnO in G. E. LTV-602 methyl silicone binder; PBR 3.73; PVC 40%; aluminum substrate; spray application; computed from spectral reflectance data.
220*	20	1963	~298	~0°		Above specimen and conditions except exposed to 1200 ESH at solar factor of 8.7 suns.
221*	20	1963	~298	~0°		S-11; SP-500 ZnO in R-5 experimental methyl silicone binder; Me/Si ratio 1.38; PBR 1.63; PVC 25%; aluminum substrate; spray application; computed from spectral reflectance data.
222*	20	1963	~298	~0°		Above specimen and conditions except exposed to 1460 ESH at solar factor of 9 suns.
223*	20	1963	~298	~0°		S-15; SP-500 ZnO in R-5 experimental methyl silicone binder; Me/Si ratio 1.38; PBR 2.12; PVC 30%; aluminum substrate; spray application; computed from spectral reflectance data.
224*	20	1963	~298	~0°		Above specimen and conditions except exposed to 1700 ESH at solar factor of 10.7 suns.
225*	20	1963	~298	~0°		S-19; SP-500 ZnO in R-5 experimental methyl silicone binder; Me/Si ratio 1.38; PBR 2.68; PVC 35%; aluminum substrate; spray application; computed from spectral reflectance data.
226*	20	1963	~298	~0°		Above specimen and conditions except exposed to 1200 ESH at solar factor of 8.7 suns.
227*	20	1963	~298	~0°		S-19; SP-500 ZnO in R-5 experimental methyl silicone binder; Me/Si ratio 1.38; PBR 2.68; PVC 35%; aluminum substrate; spray application; computed from spectral reflectance data.
228*	20	1963	~298	~0°		Above specimen and conditions except exposed to 1850 ESH at solar factor of 10.1 suns.
229*	20	1963	~298	~0°		S-22; SP-500 ZnO in R-5A experimental methyl silicone binder; Me/Si ratio <1.38; PBR 2.13; PVC 35%; aluminum substrate; spray application; computed from spectral reflectance data.
230*	20	1963	~298	~0°		Above specimen and conditions except exposed to 620 ESH at solar factor of 9 suns.

*Not shown on plot

SPECIFICATION TABLE NO. 129 NORMAL SOLAR ABSORPTANCE OF ZINC OXIDE PIGMENTED COATINGS (continued)

Curve No.	Ref. No.	Year	Temperature Range, K	Geometry θ	Reported Error, %	Composition (weight percent), Specifications and Remarks
231*	20	1963	~298	~0°		S-23; SP-500 ZnO in R-5B experimental methyl silicone binder; PBR 2.55; PVC 35%; aluminum substrate; spray application; computed from spectral reflectance data.
232*	20	1963	~298	~0°		Above specimen and conditions except exposed to 1850 ESH at solar factor of 10.1 suns.
233*	20	1963	~298	~0°		S-23; SP-500 ZnO in R-5B experimental methyl silicone binder; PBR 2.55; PVC 35%; aluminum substrate; spray application; computed from spectral reflectance data.
234*	20	1963	~298	~0°		Above specimen and conditions except exposed to 1140 ESH at solar factor of 8.3 suns.
235*	20	1963	~298	~0°		S-24; SP-500 ZnO in R-5B experimental methyl silicone binder; PBR 2.55; PVC 35%; aluminum substrate; spray application; computed from spectral reflectance data.
236*	20	1963	~298	~0°		Above specimen and conditions except exposed to 1140 ESH at solar factor of 8.3 suns.
237*	20	1963	~298	~0°		S-25; SP-500 ZnO in R-5A experimental methyl silicone binder; Me/Si ratio <1.38; PBR 2.13; PVC 35%; aluminum substrate; spray application; computed from spectral reflectance data.
238*	20	1963	~298	~0°		Above specimen and conditions except exposed to 1960 ESH at solar factor of 14.2 suns.
239*	20	1963	~298	~0°		S-25; SP-500 ZnO in R-5A experimental methyl silicone binder; Me/Si ratio <1.38; PBR 2.13; PVC 35%; aluminum substrate; spray application; computed from spectral reflectance data.
240*	20	1963	~298	~0°		Above specimen and conditions except exposed to 1850 ESH at solar factor of 10.1 suns.
241*	20	1963	~298	~0°		S-25; SP-500 ZnO in R-5A experimental methyl silicone binder; Me/Si ratio <1.38; PBR 2.13; PVC 35%; aluminum substrate; spray application; computed from spectral reflectance data.
242*	20	1963	~298	~0°		Above specimen and conditions except exposed to 1580 ESH at solar factor of 11.4 suns.
243*	20	1963	~298	~0°		S-29; SP-500 ZnO in R-8 experimental methyl silicone binder; Me/Si ratio <1.38; PBR 2.64; PVC 35%; aluminum substrate; spray application; computed from spectral reflectance data.
244*	20	1963	~298	~0°		Above specimen and conditions except exposed to 1600 ESH at solar factor of 10.2 suns.
245*	20	1963	~298	~0°		S-29; SP-500 ZnO in R-8 experimental methyl silicone binder; Me/Si ratio <1.38; PBR 2.64; PVC 35%; aluminum substrate; spray application; computed from spectral reflectance data.
246*	20	1963	~298	~0°		Above specimen and conditions except exposed to 1000 ESH at solar factor of 10.5 suns.
247*	20	1963	~298	~0°		S-31; SP-500 ZnO in R-8 experimental methyl silicone binder; Me/Si ratio <1.38; PBR 3.16; PVC 40%; aluminum substrate; spray application; computed from spectral reflectance data.
248*	20	1963	~298	~0°		Above specimen and conditions except exposed to 1780 ESH at solar factor of 9.4 suns.
249*	20	1963	~298	~0°		S-31; SP-500 ZnO in R-8 experimental methyl silicone binder; Me/Si ratio <1.38; PBR 3.16; PVC 40%; aluminum substrate; spray application; computed from spectral reflectance data.
250*	20	1963	~298	~0°		Above specimen and conditions except exposed to 1600 ESH at solar factor of 9 suns.
251*	20	1963	~298	~0°		S-31; SP-500 ZnO in R-8 experimental methyl silicone binder; Me/Si ratio <1.38; PBR 3.16; PVC 40%; aluminum substrate; spray application; heat-treated 1 hr at 533 K; computed from spectral reflectance data.
252*	20	1963	~298	~0°		Above specimen and conditions except exposed to 1780 ESH at solar factor of 9.4 suns.

*Not shown on plot

SPECIFICATION TABLE NO. 129 NORMAL SOLAR ABSORPTANCE OF ZINC OXIDE PIGMENTED COATINGS (continued)

Curve No.	Ref. No.	Year	Temperature Range, K	Geometry θ	Reported Error, %	Composition (weight percent), Specifications and Remarks
253*	20	1963	~298	~0°		S-30; SP-500 ZnO in G. E. 81932 methyl silicone binder; PBR 1.45; PVC 25%; aluminum substrate; spray application; computed from spectral reflectance data.
254*	20	1963	~298	~0°		Above specimen and conditions except exposed to 1600 ESH at solar factor of 10.2 suns.
255*	20	1963	~298	~0°		S-18; SP-500 ZnO in G. E. SR-80 methyl binder; PBR 1.45; PVC 25%; aluminum substrate; spray application; computed from spectral reflectance data.
256*	20	1963	~298	~0°		Above specimen and conditions except exposed to 1600 ESH at solar factor of 11 suns.
257*	20	1963	~298	~0°		S-33; SP-500 ZnO in R-9 experimental methyl silicone binder; Me/Si ratio <1.38; PBR 3.16; PVC 40%; aluminum substrate; spray application; cured 1 hr at 422 K and 1 hr at 533 K; computed from spectral reflectance data.
258*	20	1963	~298	~0°		Above specimen and conditions except exposed to 4170 ESH at solar factor of 10.6 suns.
259*	20	1963	~298	~0°		S-33; SP-500 ZnO in R-9 experimental methyl silicone binder; Me/Si ratio <1.38; PBR 3.16; PVC 40%; aluminum substrate; spray application; cured 1 hr at 422 K; computed from spectral reflectance data.
260*	20	1963	~298	~0°		Above specimen and conditions except exposed to 4170 ESH at solar factor of 10.6 suns.
261*	20	1963	~298	~0°		S-31; SP-500 ZnO in R-8 experimental methyl silicone binder; Me/Si ratio <1.38; PBR 3.16; PVC 40%; aluminum substrate; spray application; cured 1 hr at 422 K; computed from spectral reflectance data.
262*	20	1963	~298	~0°		Above specimen and conditions except exposed to 4170 ESH at solar factor of 10.6 suns.
263*	20	1963	~298	~0°		S-13; SP-500 ZnO in G. E. LTV-602 methyl silicone binder; PBR 2.40; PVC 30%; aluminum substrate; spray application; cured 16 hrs at ~298 K; computed from spectral reflectance data.
264*	20	1963	~298	~0°		Above specimen and conditions except exposed to 4170 ESH at solar factor of 10.6 suns.
265*	56	1969	298	~0°		S-13G over B-1056; ZnO in methyl silicone binder (0.0508 mm thick) over ZnO in RTV-602 silicone resin binder substrate (0.254 mm thick); property calculated from reflectance; lab data taken on sample to be tested on Lunar Orbiter IV.
266*	56	1969	298	~0°		Similar to above specimen and conditions except sample to be tested on Lunar Orbiter V.
267*	56	1969	298	~0°		S-13G; ZnO in methyl silicone binder (0.254 mm thick); property calculated from reflectance; lab data taken on sample to be tested on Lunar Orbiter IV.
268*	56	1969	298	~0°		B-1060; silicate treated SP-500 ZnO in silicone binder (0.264 mm thick); property calculated from reflectance; lab data taken on sample to be tested on Lunar Orbiter IV.
269*	56	1969	298	~0°		Z-93; SP-500 ZnO in PS-7 potassium silicate binder; property calculated from reflectance; lab data taken on sample to be tested on Lunar Orbiter V.
270*	39	1969	298	~0°		SP-500 ZnO in K_2SiO_3 binder; 53.4 ZnO; ball-milled for 2 hrs, then compacted; calculated from normal spectral reflectance; property measured in air.
271*	39	1969	298	~0°		Above specimen and conditions except exposed to 250 eV protons from Ortec rf ion source in vacuum (5×10^{-7} mm Hg); vacuum maintained by diffusion pump; flux 9×10^{9} p cm^{-2} sec^{-1}; fluence 1.1×10^{14} p cm^{-2}.

*Not shown on plot

SPECIFICATION TABLE NO. 129 NORMAL SOLAR ABSORPTANCE OF ZINC OXIDE PIGMENTED COATINGS (continued)

Curve No.	Ref. No.	Year	Temperature Range, K	Geometry θ	Reported Error, %	Composition (weight percent), Specifications and Remarks
272	39	1969	298	~0°		Similar to above specimen and conditions except fluence 1.9×10^{14} p cm^{-2}; flux 1.3×10^{10} p cm^{-2} sec^{-1}.
273	39	1969	298	~0°		Similar to curve 271 specimen and conditions except proton energy 500 eV; flux 5×10^{10} p cm^{-2} sec^{-1}; fluence 3.5×10^{15} p cm^{-2}.
274	39	1969	298	~0°		Similar to above specimen and conditions except flux 4.5×10^{10} p cm^{-2} sec^{-1}; fluence 5.5×10^{15} p cm^{-2}.
275	39	1969	298	~0°		Similar to curve 271 specimen and conditions except proton energy 1200 eV; flux 2.1×10^{11} p cm^{-2} sec^{-1}; fluence 3.2×10^{16} p cm^{-2}.
276	39	1969	298	~0°		Similar to above specimen and conditions except flux 1.9×10^{11} p cm^{-2} sec^{-1}; fluence 4.4×10^{16} p cm^{-2}.
277	39	1969	298	~0°		Similar to curve 271 specimen and conditions except proton energy 3200 eV; flux 2×10^{11} p cm^{-2} sec^{-1}; fluence 1×10^{16} p cm^{-2}.
278	39	1969	298	~0°		Similar to above specimen and conditions except flux 2.4×10^{11} p cm^{-2} sec^{-1}; fluence 2.6×10^{16} p cm^{-2}.
279	39	1969	298	~0°		Similar to above specimen and conditions except flux 2.1×10^{11} p cm^{-2} sec^{-1}; fluence 3.1×10^{16} p cm^{-2}.
280	39	1969	298	~0°		Similar to above specimen and conditions except flux 2.5×10^{11} p cm^{-2} sec^{-1}; fluence 3.3×10^{16} p cm^{-2}.
281	39	1969	298	~0°		Similar to above specimen and conditions except flux 2.6×10^{11} p cm^{-2} sec^{-1}; fluence 1×10^{17} p cm^{-2}.
282*	39	1969	298	~0°		Similar to curve 270 specimen and conditions.
283*	39	1969	298	~0°		Above specimen and conditions except exposed to 500 eV protons from Ortec rf ion source in vacuum (5×10^{-7} mm Hg); vacuum maintained by diffusion pump; flux 3.7×10^{10} p cm^{-2} sec^{-1}; fluence 4.2×10^{15} p cm^{-2}; property measured in situ.
284*	39	1969	298	~0°		Curve 283 specimen and conditions except final measurement in air.
285*	39	1969	298	~0°		Similar to curve 270 specimen and conditions.
286*	39	1969	298	~0°		Above specimen and conditions except exposed to 3200 eV protons from Ortec rf ion source in vacuum (5×10^{-7} mm Hg); vacuum maintained by diffusion pump; flux 7×10^{10} p cm^{-2} sec^{-1}; fluence 3.4×10^{15} p cm^{-2}; property measured in situ.
287*	39	1969	298	~0°		Above specimen and conditions except final measurement in air.
288*	39	1969	298	~0°		Similar to curve 270 specimen and conditions.
289*	39	1969	298	~0°		Above specimen and conditions except exposed to 1200 eV protons from Ortec rf ion source in vacuum (5×10^{-7} mm Hg); vacuum maintained by diffusion pump; flux 6.4×10^{10} p cm^{-2} sec^{-1}; fluence 2.8×10^{15} p cm^{-2}; property measured in situ.
290*	39	1969	298	~0°		Above specimen and conditions except final measurement in air.

*Not shown on plot

SPECIFICATION TABLE NO. 129 NORMAL SOLAR ABSORPTANCE OF ZINC OXIDE PIGMENTED COATINGS (continued)

Curve No.	Ref. No.	Year	Temperature Range, K	Geometry θ	Reported Error, %	Composition (weight percent), Specifications and Remarks
291*	39	1969	423	~0°		Similar to curve 289 specimen and conditions except flux 6.1 x 10^{10} p cm^{-2} sec^{-1}; fluence 2.1 x 10^{15} p cm^{-2}.
292*	39	1969	298	~0°		Above specimen and conditions except final measurement in air.
293*	39	1969	77	~0°		Similar to curve 291 specimen and conditions except flux 6.5 x 10^{10} p cm^{-2} sec^{-1}.
294*	39	1969	298	~0°		Above specimen and conditions except final measurement in air.
295*	39	1969	77	~0°		Similar to curve 270 specimen and conditions.
296*	39	1969	77	~0°		Similar to above specimen and conditions except exposed to 1200 eV protons (flux 2.0 x 10^{10} p cm^{-2} sec^{-1}; fluence 4.6 x 10^{15} p cm^{-2}), and UV radiation (HA/UOU/A 500 w, d.c. xenon arc lamp source) for 254 ESH in vacuum (5 x 10^{-7} mm Hg); vacuum maintained by diffusion pump; property measured in situ.
297*	39	1969	298	~0°		Above specimen and conditions except final measurement in air.
298*	39	1969	77	~0°		Similar to curve 295 specimen and conditions except exposed to UV source for 254 ESH; in vacuum (5 x 10^{-7} mm Hg); vacuum maintained by diffusion pump; property measured in situ.
299*	39	1969	298	~0°		Above specimen and conditions except final measurement in air.
300*	39	1969	298	~0°		Similar to curve 296 specimen and conditions except flux 1.6 x 10^{10} p cm^{-2} sec^{-1}; fluence 3.7 x 10^{15} p cm^{-2}; ESH 260.
301*	39	1969	298	~0°		Above specimen and conditions except final measurement in air.
302*	39	1969	298	~0°		Similar to curve 298 specimen and conditions except ESH 260.
303*	39	1969	298	~0°		Above specimen and conditions except final measurement in air.
304*	39	1969	298	~0°		Similar to curve 270 specimen and conditions.
305*	39	1969	423	~0°		Above specimen and conditions except exposed to 1200 eV protons from Ortec rf ion source (flux 2 x 10^{10} p cm^{-2} sec^{-1}; fluence 4.7 x 10^{15} p cm^{-2}) and UV for 260 ESH from Hanovia xenon arc lamp in vacuum (5 x 10^{-7} mm Hg); vacuum maintained by diffusion pump; property measured in situ.
306*	39	1969	423	~0°		Similar to curve 302 specimen and conditions.
307*	39	1969	77	~0°		Similar to curve 289 specimen and conditions except exposed to UV for 230 ESH for $\lambda < 0.4$ μm with Corning filter 7-54 in vacuum (5 x 10^{-7} mm Hg); flux 2.2 x 10^{10} p cm^{-2} sec^{-1}; fluence 4.5 x 10^{15} p cm^{-2}.
308*	39	1969	298	~0°		Above specimen and conditions except final measurement in air.
309*	39	1969	77	~0°		Similar to curve 288 specimen and conditions except exposed to UV for 230 ESH for $\lambda < 0.4$ μm with Corning filter 7-54 in vacuum (5 x 10^{-7} mm Hg).
310*	39	1969	298	~0°		Above specimen and conditions except final measurement in air.
311*	39	1969	298	~0°		Similar to curve 307 specimen and conditions except flux 1.8 x 10^{10} p cm^{-2} sec^{-1}; fluence 4.4 x 10^{15} p cm^{-2}; 270 ESH.

*Not shown on plot

SPECIFICATION TABLE NO. 129 NORMAL SOLAR ABSORPTANCE OF ZINC OXIDE PIGMENTED COATINGS (continued)

Curve No.	Ref. No.	Year	Temperature Range, K	Geometry θ	Reported Error, %	Composition (weight percent), Specifications and Remarks
312*	39	1969	298	~0°		Above specimen and conditions except final measurement in air.
313*	39	1969	298	~0°		Similar to curve 309 specimen and conditions except 270 ESH.
314*	39	1969	298	~0°		Above specimen and conditions except final measurement in air.
315*	39	1969	423	~0°		Similar to curve 307 specimen and conditions except flux 2.1 x 10^{10} p cm^{-2} sec^{-1}; fluence 4.7 x 10^{15} p cm^{-2}; 248 ESH.
316*	39	1969	298	~0°		Above specimen and conditions except final measurement in air.
317*	39	1969	298	~0°		Similar to curve 309 specimen and conditions except 248 ESH.
318*	39	1969	298	~0°		Above specimen and conditions except final measurement in air.
319*	39	1969	298	~0°		Similar to curve 270 specimen and conditions.
320*	39	1969	77	~0°		Similar to above specimen and conditions except exposed to 1200 eV protons (flux 2.0 x 10^{10} p cm^{-2} sec^{-1}; fluence 4.6 x 10^{15} p cm^{-2}), and UV (254 ESH) for $\lambda > 0.4$ μm with Corning filter 3-73 in vacuum (5 x 10^{-7} mm Hg); vacuum maintained by diffusion pump; property measured in situ.
321*	39	1969	298	~0°		Above specimen and conditions except final measurement in air.
322*	39	1969	77	~0°		Similar to curve 319 specimen and conditions except exposed to UV (254 ESH) for $\lambda > 0.4$ μm with Corning filter 3-73 in vacuum (5 x 10^{-7} mm Hg); vacuum maintained by diffusion pump; property measured in situ.
323*	39	1969	298	~0°		Above specimen and conditions except final measurement in air.
324*	39	1969	298	~0°		Similar to curve 320 specimen and conditions except flux 1.9 x 10^{10} p cm^{-2} sec^{-1}; fluence 4.4 x 10^{15} p cm^{-2}; 260 ESH.
325*	39	1969	298	~0°		Above specimen and conditions except final measurement in air.
326*	39	1969	298	~0°		Similar to curve 322 specimen and conditions except 260 ESH.
327*	39	1969	298	~0°		Above specimen and conditions except final measurement in air.
328*	39	1969	423	~0°		Similar to curve 320 specimen and conditions except flux 2.4 x 10^{10} p cm^{-2} sec^{-1}; fluence 4.9 x 10^{15} p cm^{-2}; 230 ESH.
329*	39	1969	298	~0°		Above specimen and conditions except final measurement in air.
330*	39	1969	423	~0°		Similar to curve 322 specimen and conditions except 230 ESH.
331*	39	1969	298	~0°		Above specimen and conditions except final measurement in air.
332*	45	1969	298	~0°		Silicated-treated ZnO in methyl silicone binder; aluminum substrate; absorptance calculated from normal spectral reflectance; property measured in air. [Authors' designation: 101-7]
333	45	1969	281	~0°		Above specimen and conditions except exposed to 20 keV electrons (flux 1 x 10^{10}-5 x 10^{11} e cm^{-2} sec^{-1}) in dark in vacuum (10^{-8} mm Hg); vacuum maintained by ion pump; substrate held at 281 ± 2 K; property measured in situ after 1 x 10^{13} e cm^{-2}.

*Not shown on plot

SPECIFICATION TABLE NO. 129 NORMAL SOLAR ABSORPTANCE OF ZINC OXIDE PIGMENTED COATINGS (continued)

Curve No.	Ref. No.	Year	Temperature Range, K	Geometry θ	Reported Error, %	Composition (weight percent), Specifications and Remarks
334	45	1969	281	~0°		Above specimen and conditions except property measured in situ after 5 x 10^{13} e cm^{-2}.
335	45	1969	281	~0°		Above specimen and conditions except property measured in situ after 1 x 10^{14} e cm^{-2}.
336	45	1969	281	~0°		Above specimen and conditions except property measured in situ after 3 x 10^{14} e cm^{-2}.
337	45	1969	281	~0°		Above specimen and conditions except property measured in situ after 1 x 10^{15} e cm^{-2}.
338	45	1969	281	~0°		Above specimen and conditions except property measured in situ after 1 x 10^{16} e cm^{-2}.
339	45	1969	281	~0°		Similar to curve 333 specimen and conditions except exposed to 80 keV electrons; property measured in situ after 1 x 10^{13} e cm^{-2}.
340	45	1969	281	~0°		Above specimen and conditions except property measured in situ after 5 x 10^{13} e cm^{-2}.
341	45	1969	281	~0°		Above specimen and conditions except property measured in situ after 1 x 10^{14} e cm^{-2}.
342	45	1969	281	~0°		Above specimen and conditions except property measured in situ after 3 x 10^{14} e cm^{-2}.
343	45	1969	281	~0°		Above specimen and conditions except property measured in situ after 1 x 10^{15} e cm^{-2}.
344	45	1969	281	~0°		Above specimen and conditions except property measured in situ after 1 x 10^{16} e cm^{-2}.
345*	35	1964	298	~0°		ZnO in potassium silicate binder; air dried; no heat-treatment; measured in vacuum (10^{-7} mm Hg). [Authors' designation: Sample 4]
346*	35	1964	298	~0°		Above specimen and conditions except exposed to 2560 ESH at solar factor of 11.0 suns.
347*	35	1964	298	~0°		Similar to curve 345 specimen and conditions except heat-treated (673 K for 2 hrs). [Authors' designation: Sample 3]
348*	35	1964	298	~0°		Above specimen and conditions except exposed to 2560 ESH at solar factor of 11.0 suns.

*Not shown on plot

DATA TABLE NO. 129 NORMAL SOLAR ABSORPTANCE OF ZINC OXIDE PIGMENTED COATINGS

[Temperature, T, K; Absorptance, α]

T	α
CURVE 1	
298	0.132
CURVE 2*	
298	0.138
CURVE 3*	
298	0.146
CURVE 4*	
298	0.150
CURVE 5*	
298	0.258
CURVE 6*	
298	0.269
CURVE 7*	
298	0.139
CURVE 8*	
298	0.142
CURVE 9	
289	0.170
CURVE 10	
289	0.178
CURVE 11	
289	0.207
CURVE 12	
394	0.187
CURVE 13	
394	0.216
CURVE 14	
450	0.227
CURVE 15*	
298	0.14
CURVE 16*	
298	0.14
CURVE 17*	
298	0.18
CURVE 18*	
298	0.14
CURVE 19*	
298	0.18
CURVE 20*	
298	0.15
CURVE 21*	
298	0.15
CURVE 22*	
298	0.17
CURVE 23*	
298	0.22
CURVE 24*	
298	0.21
CURVE 25*	
298	0.22
CURVE 26*	
298	0.21
CURVE 27*	
298	0.21
CURVE 28*	
298	0.24
CURVE 29*	
298	0.147
mm	α
CURVE 30*	
0.113	0.144
0.115	0.144
0.115	0.140
0.118	0.141
0.119	0.139
0.120	0.137
0.123	0.138
0.129	0.134
T	α
CURVE 31*	
298	0.14
CURVE 32*	
298	0.16
CURVE 33*	
298	0.16
CURVE 34*	
298	0.14
CURVE 35*	
298	0.16
CURVE 36*	
298	0.10
CURVE 37*	
298	0.18
CURVE 38*	
298	0.16
CURVE 39*	
298	0.13
CURVE 40*	
300	0.20
550	0.21
CURVE 41*	
298	0.164
CURVE 42*	
298	0.158
CURVE 43*	
298	0.149
CURVE 44*	
298	0.160
CURVE 45*	
298	0.165
CURVE 46*	
298	0.158
CURVE 47*	
298	0.158
CURVE 48*	
298	0.159
CURVE 49*	
298	0.161
CURVE 50*	
298	0.168
CURVE 51*	
298	0.159
CURVE 52*	
298	0.160
CURVE 53*	
298	0.16
CURVE 54*	
298	0.165
CURVE 55*	
298	0.153
CURVE 56*	
298	0.159
CURVE 57*	
279	0.153
CURVE 58*	
279	0.159
CURVE 59*	
279	0.168
CURVE 60*	
279	0.174
CURVE 61*	
279	0.161
CURVE 62*	
279	0.165
CURVE 63*	
279	0.196
CURVE 64*	
279	0.204
CURVE 65*	
279	0.197
CURVE 66*	
279	0.210
CURVE 67*	
279	0.180
CURVE 68*	
279	0.198
CURVE 69*	
278	0.114
CURVE 70*	
278	0.124
CURVE 71*	
278	0.112
CURVE 72*	
278	0.134
CURVE 73*	
278	0.151
CURVE 74*	
278	0.136
CURVE 75*	
278	0.159
CURVE 76*	
278	0.200
CURVE 77*	
278	0.215
CURVE 78*	
278	0.199
CURVE 79*	
278	0.206
CURVE 80*	
279	0.147
CURVE 81*	
279	0.150
CURVE 82*	
279	0.150
CURVE 83*	
279	0.174
CURVE 84*	
279	0.150
CURVE 85*	
279	0.199
CURVE 86*	
279	0.198
CURVE 87*	
279	0.199
CURVE 88*	
279	0.196
CURVE 89*	
279	0.211
CURVE 90*	
279	0.219
CURVE 91*	
279	0.232
CURVE 92*	
279	0.190

*Not shown on plot

DATA TABLE NO. 129 NORMAL SOLAR ABSORPTANCE OF ZINC OXIDE PIGMENTED COATINGS (continued)

T	α	T	α	T	α	T	α	T	α	T	α	T	α	T	α
CURVE 93*		CURVE 105*		CURVE 117*		CURVE 129		CURVE 141*		CURVE 153*		CURVE 165*		CURVE 177*	
279	0.186	279	0.152	279	0.189	~298	0.185	298	0.17	298	0.17	225	0.19	298	0.12
CURVE 94*		CURVE 106*		CURVE 118*		CURVE 130		CURVE 142*		CURVE 154*		CURVE 166*		CURVE 178*	
279	0.138	279	0.140	279	0.206	~298	0.196	298	0.33	298	0.18	225	0.19	298	0.22
CURVE 95*		CURVE 107*		CURVE 119		CURVE 131		CURVE 143*		CURVE 155*		CURVE 167*		CURVE 179*	
279	0.135	279	0.119	298	0.33	~298	0.210	323	0.17	298	0.17	225	0.20	298	0.26
CURVE 96*		CURVE 108*		CURVE 120		CURVE 132		CURVE 144*		CURVE 156*		CURVE 168*		CURVE 180*	
279	0.144	279	0.172	298	0.27	~298	0.263	323	0.27	298	0.18	225	0.20	298	0.23
CURVE 97*		CURVE 109*		CURVE 121		CURVE 133*		CURVE 145*		CURVE 157*		CURVE 169*		CURVE 181*	
279	0.163	279	0.120	298	0.25	~298	0.169	298	0.17	298	0.159	225	0.21	298	0.26
CURVE 98*		CURVE 110*		CURVE 122		CURVE 134		CURVE 146*		CURVE 158*		CURVE 170*		CURVE 182*	
279	0.139	279	0.125	298	0.23	~298	0.193	298	0.30	298	0.185	297	0.20	298	0.27
CURVE 99*		CURVE 111*		CURVE 123		CURVE 135		CURVE 147*		CURVE 159*		CURVE 171*		CURVE 183*	
279	0.187	279	0.133	298	0.22	~298	0.222	298	0.17	298	0.162	225	0.19	298	0.26
CURVE 100*		CURVE 112*		CURVE 124		CURVE 136		CURVE 148*		CURVE 160*		CURVE 172*		CURVE 184*	
279	0.138	279	0.157	298	0.21	~298	0.247	298	0.21	298	0.167	297	0.19	298	0.27
CURVE 101*		CURVE 113*		CURVE 125		CURVE 137		CURVE 149*		CURVE 161*		CURVE 173*		CURVE 185*	
279	0.119	279	0.124	298	0.20	~298	0.288	298	0.17	298	0.190	225	0.22	298	0.27
CURVE 102*		CURVE 114*		CURVE 126		CURVE 138*		CURVE 150*		CURVE 162*		CURVE 174*		CURVE 186*	
279	0.124	279	0.120	298	0.19	193 263	0.231 0.231	298	0.19	298	0.182	225	0.22	298	0.27
CURVE 103*		CURVE 115*		CURVE 127		CURVE 139*		CURVE 151*		CURVE 163*		CURVE 175*		CURVE 187*	
279	0.131	279	0.238	298	0.18	193 263	0.216 0.216	298	0.17	298	0.144	225	0.21	298	0.23
CURVE 104*		CURVE 116*		CURVE 128*		CURVE 140*		CURVE 152*		CURVE 164*		CURVE 176*		CURVE 188*	
279	0.126	279	0.245	298	0.175	193 263	0.189 0.189	298	0.20	298	0.136	297	0.21	298	0.25

*Not shown on plot

DATA TABLE NO. 129 NORMAL SOLAR ABSORPTANCE OF ZINC OXIDE PIGMENTED COATINGS (continued)

T	α	T	α	T	α	T	α	T	α	T	α	T	α	T	α
CURVE 189*		CURVE 201*		CURVE 213*		CURVE 225*		CURVE 237*		CURVE 249*		CURVE 261*		CURVE 273	
298	0.20	298	0.230	298	0.230	298	0.224	298	0.269	298	0.261	298	0.282	298	0.135
CURVE 190*		CURVE 202*		CURVE 214*		CURVE 226*		CURVE 238*		CURVE 250*		CURVE 262*		CURVE 274	
298	0.23	298	0.240	298	0.260	298	0.226	298	0.283	298	0.277	298	0.316	298	0.164
CURVE 191*		CURVE 203*		CURVE 215*		CURVE 227*		CURVE 239*		CURVE 251*		CURVE 263*		CURVE 275	
298	0.259	298	0.218	298	0.230	298	0.237	298	0.299	298	0.260	298	0.211	298	0.261
CURVE 192*		CURVE 204*		CURVE 216*		CURVE 228*		CURVE 240*		CURVE 252*		CURVE 264*		CURVE 276	
298	0.267	298	0.218	298	0.260	298	0.237	298	0.315	298	0.266	298	0.269	298	0.300
CURVE 193*		CURVE 205*		CURVE 217*		CURVE 229*		CURVE 241*		CURVE 253*		CURVE 265*		CURVE 277	
298	0.266	298	0.224	298	0.175	298	0.225	298	0.280	298	0.283	298	0.191	298	0.198
CURVE 194*		CURVE 206*		CURVE 218*		CURVE 230*		CURVE 242*		CURVE 254*		CURVE 266*		CURVE 278	
298	0.274	298	0.226	298	0.192	298	0.245	298	0.286	298	0.433	298	0.191	298	0.266
CURVE 195*		CURVE 207*		CURVE 219*		CURVE 231*		CURVE 243*		CURVE 255*		CURVE 267*		CURVE 279	
298	0.250	298	0.223	298	0.161	298	0.236	298	0.241	298	0.279	298	0.184	298	0.266
CURVE 196*		CURVE 208*		CURVE 220*		CURVE 232*		CURVE 244*		CURVE 256*		CURVE 268*		CURVE 280	
298	0.260	298	0.237	298	0.173	298	0.249	298	0.257	298	0.390	298	0.178	298	0.266
CURVE 197*		CURVE 209*		CURVE 221*		CURVE 233*		CURVE 245*		CURVE 257*		CURVE 269*		CURVE 281	
298	0.259	298	0.221	298	0.230	298	0.235	298	0.253	298	0.237	298	0.184	298	0.421
CURVE 198*		CURVE 210*		CURVE 222*		CURVE 234*		CURVE 246*		CURVE 258*		CURVE 270*		CURVE 282*	
298	0.271	298	0.226	298	0.250	298	0.243	298	0.266	298	0.248	298	0.133	298	0.130
CURVE 199*		CURVE 211*		CURVE 223*		CURVE 235*		CURVE 247*		CURVE 259*		CURVE 271*		CURVE 283*	
298	0.230	298	0.260	298	0.230	298	0.248	298	0.265	298	0.216	298	0.136	298	0.156
CURVE 200*		CURVE 212*		CURVE 224*		CURVE 236*		CURVE 248*		CURVE 260*		CURVE 272		CURVE 284*	
298	0.250	298	0.260	298	0.240	298	0.263	298	0.285	298	0.236	298	0.136	298	0.145

*Not shown on plot

DATA TABLE NO. 129 NORMAL SOLAR ABSORPTANCE OF ZINC OXIDE PIGMENTED COATINGS (continued)

T	α	T	α	T	α	T	α	T	α	T	α
CURVE 285*		CURVE 297*		CURVE 309*		CURVE 321*		CURVE 333		CURVE 345*	
298	0.133	298	0.161	77	0.160	298	0.143	281	0.22	298	0.182
CURVE 286*		CURVE 298*		CURVE 310*		CURVE 322*		CURVE 334		CURVE 346*	
298	0.147	77	0.145	298	0.141	77	0.134	281	0.23	298	0.194
CURVE 287*		CURVE 299*		CURVE 311*		CURVE 323*		CURVE 335		CURVE 347*	
298	0.140	298	0.129	298	0.181	298	0.129	281	0.23	298	0.176
CURVE 288*		CURVE 300*		CURVE 312*		CURVE 324*		CURVE 336		CURVE 348*	
298	0.130	298	0.166	298	0.156	298	0.150	281	0.24	298	0.178
CURVE 289*		CURVE 301*		CURVE 313*		CURVE 325*		CURVE 337			
298	0.143	298	0.150	298	0.175	298	0.142	281	0.26		
CURVE 290*		CURVE 302*		CURVE 314*		CURVE 326*		CURVE 338			
298	0.140	298	0.160	298	0.142	298	0.128	281	0.29		
CURVE 291*		CURVE 303*		CURVE 315*		CURVE 327*		CURVE 339			
423	0.140	298	0.132	423	0.178	298	0.125	281	0.23		
CURVE 292*		CURVE 304*		CURVE 316*		CURVE 328*		CURVE 340			
298	0.138	298	0.124	298	0.160	423	0.157	281	0.24		
CURVE 293*		CURVE 305*		CURVE 317*		CURVE 329*		CURVE 341			
77	0.178	423	0.159	423	0.151	298	0.148	281	0.26		
CURVE 294*		CURVE 306*		CURVE 318*		CURVE 330*		CURVE 342			
298	0.140	423	0.144	298	0.141	423	0.130	281	0.32		
CURVE 295*		CURVE 307*		CURVE 319*		CURVE 331*		CURVE 343			
77	0.122	77	0.194	298	0.120	298	0.121	281	0.37		
CURVE 296*		CURVE 308*		CURVE 320*		CURVE 332*		CURVE 344			
77	0.182	298	0.167	77	0.169	298	0.22	281	0.41		

*Not shown on plot

FIGURE 130A

CHANGE IN NORMAL SOLAR ABSORPTANCE OF ZINC OXIDE PIGMENTED COATINGS

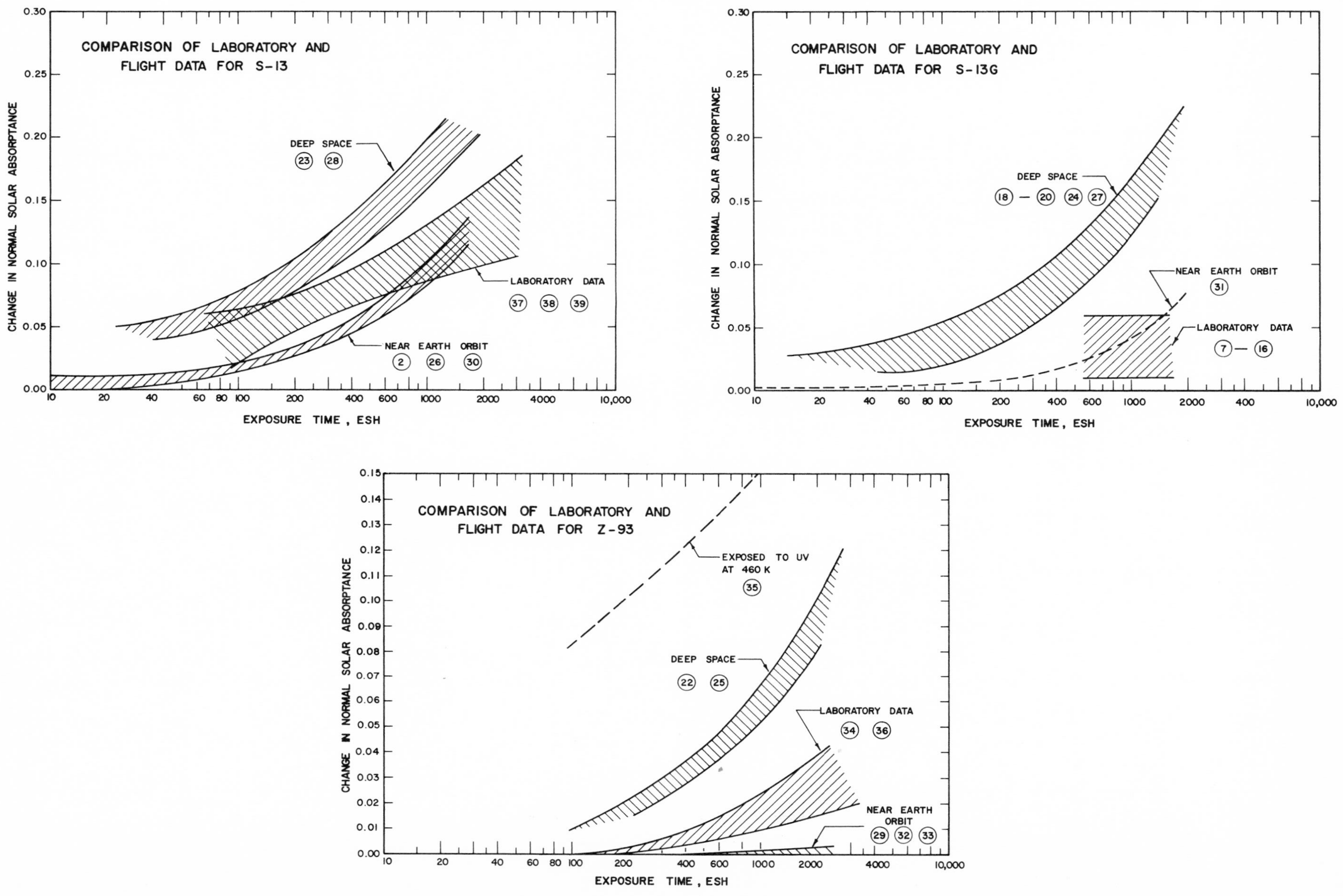

COMPARISON OF LABORATORY AND FLIGHT DATA FOR S-13

DEEP SPACE (23) (28)

LABORATORY DATA (37) (38) (39)

NEAR EARTH ORBIT (2) (26) (30)

CHANGE IN NORMAL SOLAR ABSORPTANCE

0.30 0.25 0.20 0.15 0.10 0.05 0.00

10 20 40 60 80 100 200 400 600 1000 2000 4000 10,000

EXPOSURE TIME, ESH

COMPARISON OF LABORATORY AND FLIGHT DATA FOR S-13G

DEEP SPACE (18) — (20) (24) (27)

NEAR EARTH ORBIT (31)

LABORATORY DATA (7) — (16)

CHANGE IN NORMAL SOLAR ABSORPTANCE

0.30 0.25 0.20 0.15 0.10 0.05 0.00

10 20 40 60 80 100 200 400 600 1000 2000 400 10,000

EXPOSURE TIME, ESH

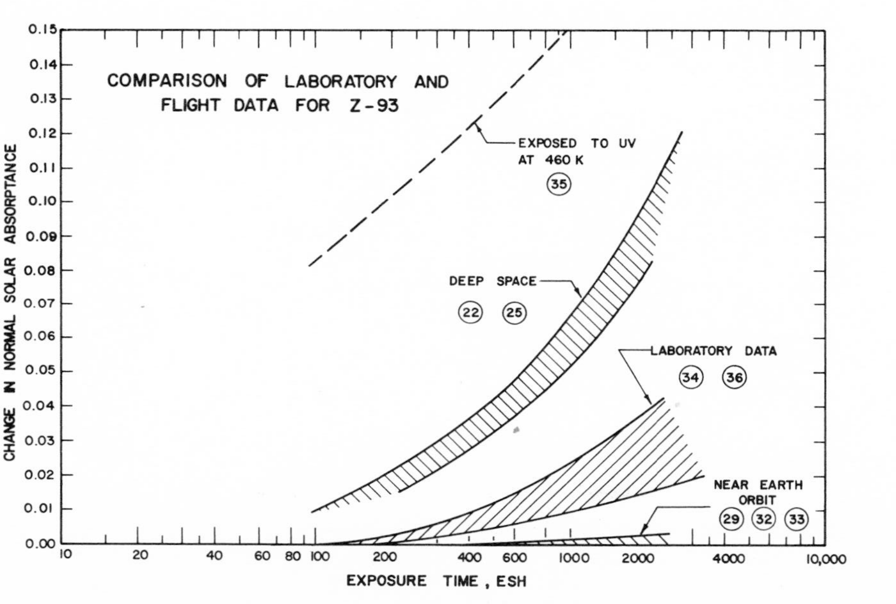

SPECIFICATION TABLE NO. 130 CHANGE IN NORMAL SOLAR ABSORPTANCE OF ZINC OXIDE PIGMENTED COATINGS

Curve No.	Ref. No.	Year	Temperature Range, K	Geometry θ	Reported Error, %	Composition (weight percent), Specifications and Remarks
1	52	1966	~298	~0°		ZnO in S-83-14 resin sol binder; data taken from OSO-II flight experiment; variable is equivalent length of time exposed to the sun.
2	52	1966	~298	~0°		ZnO in silicone binder; data taken from OSO-II flight experiment; variable is equivalent length of time exposed to the sun.
3	52	1966	~298	~0°		Z-52; ZnO in potassium silicate binder; data taken from OSO-II flight experiment; variable is equivalent length of time exposed to the sun.
4	63	1969	~298	~0°		ZnO in polydimethyl siloxane (RTV-602) binder; exposed to UV for 800 ESH from a G.E. AH-6 lamp in vacuum; absorptance calculated from reflectance; property measured in situ; property indicates change in absorptance from pre-irradiated state.
5	63	1969	~298	~0°		Z-93; SP-500 ZnO in potassium silicate binder; exposed to UV for 800 ESH from a G.E. AH-6 lamp in vacuum; absorptance calculated from reflectance; property measured in situ; property indicates change in absorptance from pre-irradiated state.
6	63	1969	~298	~0°		ZnO in methyl silicone binder; ZnO silicate treated; exposed to UV for 1000 ESH from a G.E. AH-6 lamp in vacuum; absorptance calculated from reflectance; property measured in situ; property indicates change in absorptance from pre-irradiated state.
7	63	1969	~298	~0°		S-13G; ZnO in methyl silicone binder; prepared from 16 hr sweated pigment; only toluene used as solvent; exposed to UV for 600 ESH from a G.E. AH-6 lamp in vacuum; absorptance calculated from reflectance; property measured in situ; property indicates change in absorptance from preirradiated state.
8	63	1969	~298	~0°		S-13G; ZnO in methyl silicone binder; prepared from sweated pigment; calcined for 16 hr at 923 K; exposed to UV for 600 ESH from a G.E. AH-6 lamp in vacuum; absorptance calculated from reflectance; property measured in situ; property indicates change in absorptance from preirradiated state.
9	63	1969	~298	~0°		S-13G; ZnO in methyl silicone binder; prepared from sweated pigment; toluene and petroleum ether solvent; exposed to UV for 800 ESH from a G.E. AH-6 lamp in vacuum; absorptance calculated from reflectance; property measured in situ; property indicates change in absorptance from preirradiated state.
10	63	1969	~298	~0°		S-13G; ZnO in methyl silicone binder; ZnO silicate treated; pigment was only sifted prior to wet-grinding; paint grind time 3 hrs; exposed to UV for 1400 ESH from a G.E. AH-6 lamp in vacuum; absorptance calculated from reflectance; property measured in situ; property indicates change in absorptance from pre-irradiated state.
11	63	1969	~298	~0°		S-13G; ZnO in methyl silicone binder; ZnO silicate treated; pigment was unsifted and unground prior to wet-grinding; paint grind time 4 hrs; exposed to UV for 1400 ESH from a G.E. AH-6 lamp in vacuum; absorptance calculated from reflectance; property measured in situ; property indicates change in absorptance from preirradiated state.

SPECIFICATION TABLE NO. 130 CHANGE IN NORMAL SOLAR ABSORPTANCE OF ZINC OXIDE PIGMENTED COATINGS (continued)

Curve No.	Ref. No.	Year	Temperature Range, K	Geometry θ	Reported Error, %	Composition (weight percent), Specifications and Remarks
12	63	1969	~298	~0°		S-13G; ZnO in methyl silicone binder; ZnO silicate treated; pigment dry-ground 30 min; paint grind time 3 hrs; exposed to UV for 1400 ESH from a G.E. AH-6 lamp in vacuum; absorptance calculated from reflectance; property measured in situ; property indicates change in absorptance from preirradiated state.
13	63	1969	~298	~0°		S-13G; ZnO in methyl silicone binder; ZnO silicate treated; pigment hand-mulled prior to wet-grinding; paint grind time 3 hrs; exposed to UV for 1400 ESH from a G.E. AH-6 lamp in vacuum; absorptance calculated from reflectance; property measured in situ; property indicates change in absorptance from pre-irradiated state.
14	63	1969	~298	~0°		S-13G; ZnO in methyl silicone binder; ZnO silicate treated; pigment remulled from first hand mulling; paint grind time 5 hrs; exposed to UV for 1400 ESH from a G.E. AH-6 lamp in vacuum; absorptance calculated from reflectance; property measured in situ; property indicates change in absorptance from pre-irradiated state.
15	63	1969	~298	~0°		S-13G; ZnO in methyl silicone binder; ZnO silicate treated; paint prepared from sweated pigment neutralized with formic acid and calcined for 16 hrs at 923 K; exposed to UV for 600 ESH from a G.E. AH-6 lamp in vacuum; absorptance calculated from relfectance; property measured in situ; property indicates change in absorptance from preirradiated state.
16	63	1969	~298	~0°		S-13G; ZnO in methyl silicone binder; ZnO silicate treated; paint prepared from sweated pigment neutralized with sodium acid phosphate and calcined for 16 hrs at 923 K; exposed to UV for 600 ESH from a G.E. AH-6 lamp in vacuum; absorptance calculated from reflectance; property measured in situ; property indicates change in absorptance from preirradiated state.
17	63	1969	~298	~0°		ZnO in Owens-Illinois 650 resin binder; ZnO silicate treated; PVC 32%; pigment slurried for 15 min with a 3% solution of NaH_2PO_4, filtered, redispersed in 3% solution of NaH_2PO_4, slurried 15 min, filtered, redispersed in distilled water, filtered, and dried for 18 hrs at 372 K; exposed to UV for 1200 ESH from a G.E. AH-6 lamp in vacuum; absorptance calculated from reflectance; property measured in situ; property indicates change in absorptance from pre-irradiated state.
18	56	1969	261-316	~0°		S-13G; ZnO in methyl silicone binder (0.254 mm thick) over B-1056 paint (0.0508 mm thick); property calculated from temperature of sample; flight data of Lunar Orbiter IV; ESH is variable; data extracted from smooth curve.
19	56	1969	261-314	~0°		Similar to above specimen and conditions except flight data of Lunar Orbiter V.
20	56	1969	261-303	~0°		S-13G; ZnO in methyl silicone binder (0.254 mm thick); property calculated from temperature of sample; flight data of Lunar Orbiter IV; ESH is variable; data extracted from smooth curve.
21	56	1969	261-294	~0°		B-1060; SP-500 ZnO in PS-7 potassium silicate binder; property calculated from temperature of sample; flight data of Lunar Orbiter IV; ESH is variable; data extracted from smooth curve.

SPECIFICATION TABLE NO. 130 CHANGE IN NORMAL SOLAR ABSORPTANCE OF ZINC OXIDE PIGMENTED COATINGS (continued)

Curve No.	Ref. No.	Year	Temperature Range, K	Geometry θ	Reported Error, %	Composition (weight percent), Specifications and Remarks
22	56	1969	258-280	$\sim 0°$		Z-93; SP-500 ZnO in PS-7 potassium silicate binder; property calculated from temperature of sample; flight data of Lunar Orbiter I; ESH is variable; data extracted from smooth curve.
23	56	1969	unknown	$\sim 0°$		B-1056; SP-500 ZnO in RTV-602 silicone resin binder; property calculated from temperature of sample; flight data of Lunar Orbiter I; ESH is variable; data extracted from smooth curve.
24	56	1969	unknown	$\sim 0°$		S-13G over B-1056; ZnO in methyl silicone binder (0.254 mm thick) over B-1056 paint (0.0508 mm thick); property calculated from temperature of sample; flight data of Lunar Orbiter II; ESH is variable; data extracted from smooth curve.
25	65	1969	unknown	$\sim 0°$		Z-93; SP-500 ZnO in PS-7 K_2SiO_4 binder; property calculated from flight data of Mariner IV; ESH is variable.
26	65	1969	unknown	$\sim 0°$		S-13; SP-500 ZnO in RTV-602 silicone, toluene, SRC-05 catalyst binder; ball milled 3 hrs; property calculated from temperature of substrate from flight data of Pegasus I; ESH is variable.
27	65	1969	unknown	$\sim 0°$		S-13G; SP-500 ZnO in RTV-602 silicone, K_2SiO_3, toluene, SRC-05 catalyst binder; property calculated from temperature of substrate from flight data of Mariner V; ESH is variable.
28	65	1969	unknown	$\sim 0°$		S-13; SP-500 ZnO in RTV-602 silicone, toluene, SRC-05 catalyst binder; ball milled 3 hrs; property calculated from temperature of substrate from flight data of Mariner V; ESH is variable.
29	65	1969	unknown	$\sim 0°$		Z-93; SP-500 ZnO in PS-7 potassium silicate binder (0.127 mm thick); PRB 4.30; ball milled 6 hrs; property calculated from temperature of substrate from flight data of OSO III; ESH is variable.
30	65	1969	unknown	$\sim 0°$		S-13; SP-500 ZnO in RTV-602 silicone, toluene, SRC-05 catalyst binder; ball milled 3 hrs; property calculated from temperature of substrate from flight data of OSO III; ESH is variable.
31	65	1969	unknown	$\sim 0°$		S-13G; SP-500 ZnO in RTV-602 silicone, K_2SiO_3, toluene, SRC-05 catalyst binder; property calculated from temperature of substrate from flight data of OSO III; ESH is variable.
32	65	1969	unknown	$\sim 0°$		Z-93; SP-500 ZnO in potassium silicate binder; PBR 4.30; ball milled 6 hrs; property calculated from temperature of substrate from flight data of Pegasus II; data extracted from smooth curve; ESH is variable.
33	65	1969	unknown	$\sim 0°$		Z-93; SP-500 ZnO in potassium silicate binder; PBR 4.30; ball milled 6 hrs; property calculated from temperature of substrate from flight data of OSO II; data extracted from smooth curve; ESH is variable.
34	65	1969	~ 298	$\sim 0°$		Z-93; SP-500 ZnO in PS-7 potassium silicate binder; PBR 4.30; ball milled 6 hrs; exposed to UV radiation from AH-6 lamp (6 x solar intensity) in vacuum; property measured in situ; data extracted from smooth curve; ESH is variable.

SPECIFICATION TABLE NO. 130 CHANGE IN NORMAL SOLAR ABSORPTANCE OF ZINC OXIDE PIGMENTED COATINGS (continued)

Curve No.	Ref. No.	Year	Temperature Range, K	Geometry θ	Reported Error, %	Composition (weight percent), Specifications and Remarks
35	65	1969	~298	~0°		Z-93; SP-500 ZnO in PS-7 potassium silicate binder; PBR 4.30; ball milled 6 hrs; exposed to UV radiation from xenon lamp (1 x solar intensity) in vacuum at 460 K; property measured in situ; data extracted from smooth curve; ESH is variable.
36	65	1969	~298	~0°		Z-93; SP-500 ZnO in PS-7 potassium silicate binder; PBR 4.30; ball milled 6 hrs; exposed to UV radiation from xenon lamp (1 x solar intensity) in vacuum; property measured in situ; data extracted from smooth curve; ESH is variable.
37	65	1969	~298	~0°		S-13; SP-500 ZnO in RTV-602 silicone, toluene, SRC-05 catalyst binder; ball milled 3 hrs; exposed to UV radiation from AH-6 lamp (1 x solar intensity) in vacuum; property measured in situ; data extracted from smooth curve; ESH is variable.
38	65	1969	~298	~0°		S-13; SP-500 ZnO in RTV-602 silicone, toluene, SRC-05 catalyst binder; ball milled 3 hrs; exposed to UV radiation from AH-6 lamp (6 x solar intensity) in vacuum; property measured in situ; data extracted from smooth curve; ESH is variable.
39	65	1969	~298	~0°		S-13; SP-500 ZnO in RTV-602 silicone, toluene, SRC-05 catalyst binder; ball milled 3 hrs; exposed to UV radiation from AH-6 lamp (10/20 x solar intensity) in vacuum; property measured in situ; data extracted from smooth curve; ESH is variable.

DATA TABLE NO. 130 CHANGE IN NORMAL SOLAR ABSORPTANCE OF ZINC OXIDE PIGMENTED COATINGS

[Temperature, T, K; Absorptance, α]

ESH	Δα
CURVE 1 T ~ 298	
1.00	0.000
3.98	0.000
10.0	0.002
39.8	0.010
100.0	0.028
200.0	0.048
398.0	0.072
508.0	0.081
CURVE 2 T ~ 298	
1.00	0.000
3.98	0.000
10.0	0.000
39.8	0.006
100.0	0.015
200.0	0.029
398.0	0.050
504.0	0.059
CURVE 3 T ~ 298	
1.00	0.000
3.98	0.000
10.0	0.000
39.8	0.000
100.0	0.000
200.0	0.000
398.0	0.000
538.0	0.000

T	Δα
CURVE 4	
298	0.08
CURVE 5	
298	0.01
CURVE 6	
298	0.01
CURVE 7	
298	0.01
CURVE 8	
298	0.03
CURVE 9	
298	0.03
CURVE 10	
298	0.01
CURVE 11	
298	0.02
CURVE 12	
298	0.03
CURVE 13	
298	0.05
CURVE 14	
298	0.06
CURVE 16	
298	0.06
CURVE 17	
298	0.02
CURVE 18	
298	0.0

ESH	Δα
CURVE 18	
14	0.000
17	0.028
51	0.041
126	0.060
264	0.083
395	0.101
679	0.134
1101	0.176
CURVE 19	
14	0.000
46	0.016
114	0.033
191	0.048
352	0.073
520	0.095
722	0.117
966	0.142
1375	0.180
1849	0.220
CURVE 20	
14	0.000
104	0.022
218	0.039
456	0.067
872	0.111
1086	0.132
CURVE 21	
14	0.000
74	0.010
197	0.026
378	0.044
590	0.062
1098	0.098
CURVE 22	
0	0.003
310	0.022
444	0.029
CURVE 22 (cont.)	
661	0.038
881	0.047
1194	0.058
1629	0.071
1995	0.081
CURVE 23	
0	0.020
37	0.039
104	0.058
181	0.076
285	0.092
386	0.104
518	0.118
741	0.138
944	0.153
1288	0.175
1793	0.202
CURVE 24	
50	0.020
118	0.042
209	0.062
360	0.087
506	0.108
691	0.129
872	0.148
944	0.153
1288	0.175
1793	0.202
CURVE 25	
100	0.010
500	0.046
650	0.050
1000	0.068
2650	0.118
CURVE 26	
151	0.021
500	0.058
CURVE 26 (cont.)	
1000	0.100
1660	0.137
CURVE 27	
100	0.025
204	0.041
500	0.079
1000	0.126
1200	0.141
CURVE 28	
22.9	0.050
56.3	0.064
100	0.078
500	0.150
1000	0.198
1230	0.213
CURVE 29	
0.46	0.002
1.1	-0.001
3.9	-0.004
4.0	-0.001
4.3	-0.001
4.9	-0.001
5.2	0.002
7.1	-0.003
7.4	0.001
7.7	-0.001
8.1	0.000
8.5	-0.004
8.7	0.001
721	-0.004
717	0.000
727	0.002
751	0.000
1336	-0.004
1336	-0.001
1336	0.000
1336	0.003
1513	0.005
1577	0.002
1577	0.003
1577	0.005
CURVE 30	
0.15	0.001
0.46	0.000
1.1	0.004
4.0	0.006
4.6	0.007
4.8	0.009
5.1	0.007
5.5	0.009
5.7	0.000
6.5	0.010
7.1	0.006
7.4	0.011
7.5	0.001
7.7	0.008
8.0	0.010
8.3	0.010
8.7	0.012
9.3	0.014
9.5	0.001
12.4	0.014
12.6	0.013
13.2	0.013
13.4	0.014
13.7	0.017
16.7	0.011
16.9	0.012
17.9	0.015
18.3	0.011
18.8	0.005
19.1	0.013
19.8	0.005
19.8	0.012
22.1	0.015
22.7	0.013
66.1	0.021
66.1	0.022
68.9	0.021
73.8	0.021
83.2	0.018
103	0.023
157	0.030
162	0.031
162	0.033
169	0.035
173	0.030
CURVE 30 (cont.)	
181	0.031
338	0.044
338	0.046
346	0.045
346	0.046
744	0.017
772	0.017
772	0.073
772	0.076
779	0.071
1032	0.081
1032	0.083
1032	0.085
1052	0.086
1052	0.087
1361	0.099
1374	0.101
1374	0.103
1393	0.098
1636	0.103
1636	0.104
1636	0.107
1636	0.109
1636	0.111
CURVE 31	
0.14	0.001
0.45	0.004
1.1	0.003
3.9	0.001
4.1	0.002
4.5	0.003
5.4	0.006
6.4	0.004
7.0	0.006
7.6	0.001
8.1	0.004
8.4	0.003
8.6	0.006
8.7	0.006
12.1	0.001
13.1	0.000
13.1	0.005
14.1	-0.001
CURVE 31 (cont.)	
16.5	0.001
16.9	0.003
17.4	0.004
18.0	0.002
18.0	0.006
18.6	0.014
19.6	0.001
19.6	0.012
21.5	0.001
21.5	0.016
22.9	0.003
23.3	-0.001
67.0	-0.004
67.0	0.005
67.0	0.007
69.2	-0.001
70.8	-0.004
160	0.007
165	0.010
174	0.007
182	0.007
182	0.010
313	0.011
322	0.012
322	0.015
741	0.032
765	0.028
776	0.031
1009	0.041
1009	0.042
1037	0.042
1037	0.044
1386	0.053
1386	0.057
1386	0.059
1386	0.061
1432	0.058
1577	0.060
1577	0.064
1599	0.060
1599	0.064

DATA TABLE NO. 130 CHANGE IN NORMAL SOLAR ABSORPTANCE OF ZINC OXIDE PIGMENTED COATINGS (continued)

ESH	$\Delta\alpha$
CURVE 32	
0.1	0.0
1.0	0.0
10	0.0
100	0.0
1000	0.0
2500	0.0
CURVE 33	
0.1	0.0
1.0	0.0
10	0.0
100	0.0
1000	0.0
2500	0.0
CURVE 34 T ~ 298	
100	0.0
500	0.005
1000	0.009
3000	0.019
CURVE 35 T ~ 460	
100	0.083
500	0.129
1000	0.152
1800	0.155
CURVE 36 T ~ 298	
100	0.0
500	0.012
1000	0.024
2100	0.041
CURVE 37 T ~ 298	
50	0.012
100	0.022
500	0.070
1000	0.087
1600	0.094

ESH	$\Delta\alpha$
CURVE 38 T ~ 298	
70	0.061
100	0.064
500	0.107
1000	0.132
3000	0.182
CURVE 39 T ~ 298	
100	0.059
500	0.086
1000	0.097
1300	0.103

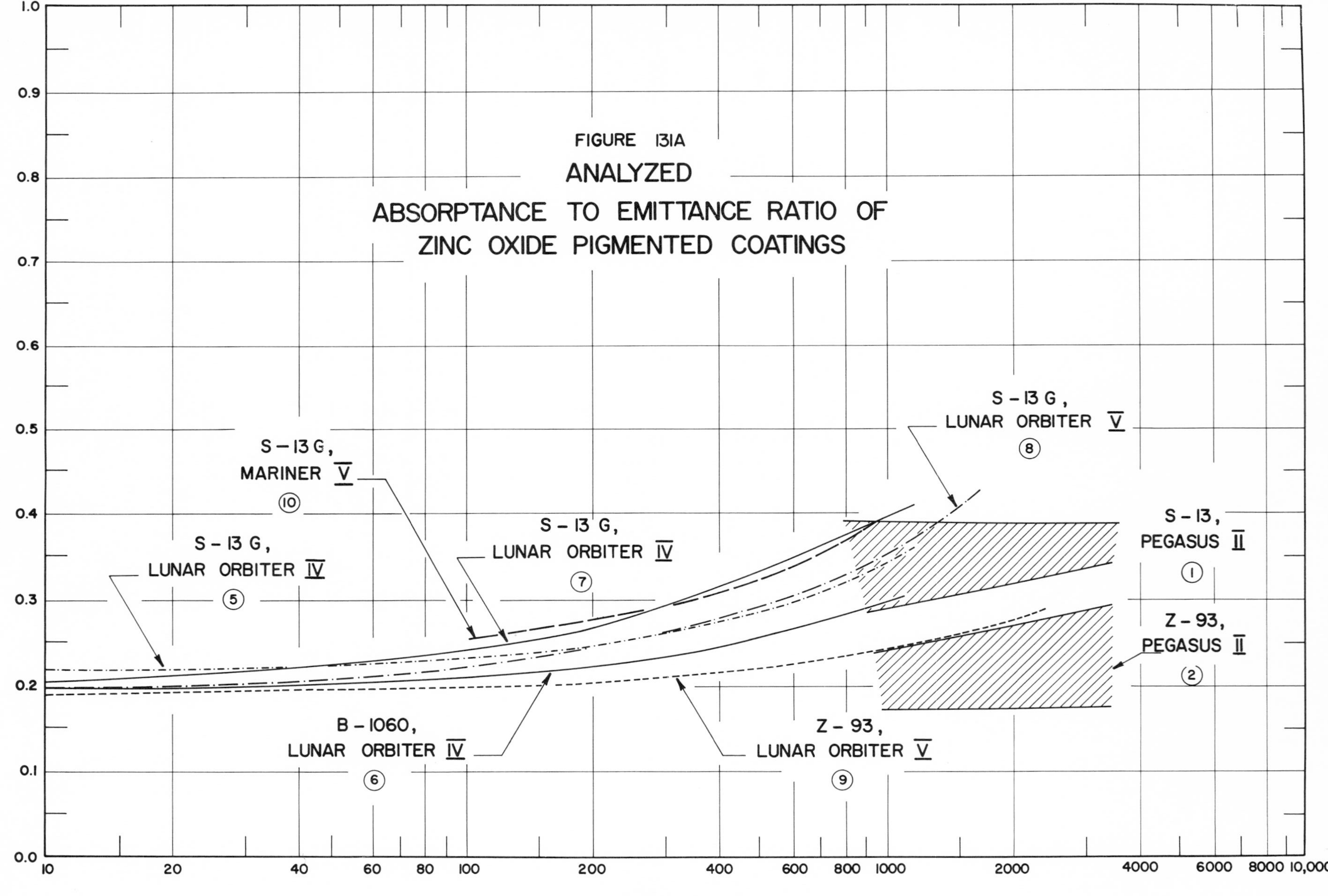

FIGURE 131A
ANALYZED
ABSORPTANCE TO EMITTANCE RATIO OF
ZINC OXIDE PIGMENTED COATINGS

SPECIFICATION TABLE NO. 131 ABSORPTANCE TO EMITTANCE RATIO OF ZINC OXIDE PIGMENTED COATINGS

Curve No.	Ref. No.	Year	Temperature Range, K	Reported Error, %	Composition (weight percent), Specifications and Remarks
1	57	1967	~298		S-13; ZnO in methyl silicone binder; data taken from flight of Pegasus II; property deduced from temp of substrate; ESH is variable.
2	57	1967	~298		Z-93; ZnO in potassium silicate binder; data taken from flight of Pegasus II; property deduced from temp of substrate; ESH is variable.
3*	58	1967	~298	9.5	S-13; ZnO in methyl silicone binder; aluminum (6061 Al alloy) substrate; aluminum substrate given a prime coat; property deduced from reflectance; lab data measured 6 mo before flight of ATS-1.
4*	58	1967	~298	6.2	Above specimen and conditions except property deduced from temp of substrate from "initial" point flight data (48 hrs after launch) of ATS-1.
5	56	1969	261-303		S-13G; ZnO in methyl silicone binder (0.254 mm thick); property calculated from temp of sample; in-flight data of Lunar Orbiter IV; ESH is variable; data extracted from smooth curve.
6	56	1969	261-294		B-1060; SP-500 ZnO in silicone binder (0.264 mm thick); ZnO silicate treated; calculated from temp of sample; in-flight data of Lunar Orbiter IV; ESH is variable; data extracted from smooth curve.
7	56	1969	261-316		S-13G over B-1056; ZnO in methyl silicone binder (0.254 mm thick) over B-1056 paint (0.0508 mm thick); property calculated from temp of sample; in-flight data of Lunar Orbiter IV; ESH is variable; data extracted from smooth curve.
8	56	1969	261-314		Similar to above specimen and conditions except in-flight data of Lunar Orbiter V.
9	56	1969	258-280		Z-93; SP-500 ZnO in PS-7 potassium silicate binder; property calculated from temp of sample; in-flight data of Lunar Orbiter V; ESH is variable; data extracted from smooth curve.
10	295	1968	unknown		S-13G; silicate treated ZnO in methyl silicone binder; flight data from Mariner V; variable is exposure time in ESH.

*Not shown on plot

DATA TABLE NO. 131 ABSORPTANCE TO EMITTANCE RATIO OF ZINC OXIDE PIGMENTED COATINGS

[Temperature, T, K; Absorptance, α; Emittance, ϵ]

ESH	α/ϵ
CURVE 1	
1191	0.304
1201	0.325
1261	0.391
1277	0.302
1287	0.337
1309	0.331
1537	0.331
2053	0.321
2061	0.352
2090	0.366
2100	0.355
2114	0.333
2133	0.371
2244	0.371
2244	0.391
2251	0.377
2263	0.383
2290	0.330
2300	0.338
2300	0.371
2421	0.338
2427	0.362
2432	0.384
2442	0.364
CURVE 2	
1160	0.183
1169	0.226
1178	0.199
1198	0.231
1258	0.212
1291	0.240
1308	0.208
1537	0.259
2048	0.188
2070	0.200
2089	0.221
2100	0.200
2110	0.173
2133	0.254
2141	0.282
2243	0.245
2246	0.229
2260	0.243
2291	0.173
CURVE 2 (cont.)	
2300	0.234
2300	0.279
2423	0.173
2431	0.240
2431	0.270
2442	0.216
T	α/ϵ
CURVE 3*	
298	0.21
CURVE 4*	
298	0.27
ESH	α/ϵ
CURVE 5	
0	0.215
184	0.245
397	0.275
719	0.312
1128	0.356
CURVE 6	
0	0.187
58	0.203
208	0.225
365	0.244
551	0.262
795	0.283
919	0.292
1098	0.305
CURVE 7	
0	0.200
195	0.265
331	0.301
500	0.334
629	0.355
771	0.374
CURVE 7 (cont.)	
951	0.392
1142	0.408
CURVE 8	
0	0.187
130	0.226
238	0.251
438	0.285
612	0.309
830	0.336
1564	0.416
CURVE 9	
0	0.185
54	0.193
120	0.200
214	0.207
346	0.214
783	0.232
2038	0.277
CURVE 10	
120	0.258
199	0.273
301	0.291
401	0.308
499	0.324
602	0.340
700	0.354
800	0.368
900	0.381

*Not shown on plot

SPECIFICATION TABLE NO. 132 NORMAL TOTAL EMITTANCE OF ZINC SULFIDE PIGMENTED COATINGS

Curve No.	Ref. No.	Year	Temperature Range, K	Geometry θ'	Reported Error, %	Composition (weight percent), Specifications and Remarks
1*	100	1962	300	~0°		Zinc sulfide in silicone (ZW 60) binder; these values represent the max degradation that test data indicate may occur during a 3-mo. Venus mission; integrated over the wavelength range 2 to 23 μm.
2*	100	1962	300	~0°		Zinc sulfide in silicone (ZW 40) binder; these values represent the max degradation that test data indicate may occur during a 3-mo. Venus mission; integrated over the wavelength range 2 to 23 μm.
3*	100	1962	300	~0°		Zinc sulfide in silicone (ZW 60) binder; integrated over the wavelength range 2 to 23 μm.
4*	100	1962	300	~0°		Zinc sulfide in silicone (ZW 40) binder; integrated over the wavelength range of 2 to 23 μm.
5*	8	1961	373	~0°		Zinc sulfide in silicone binder (0.0508 mm thick) on anodized aluminum substrate; 30% pigment volume ratio; measured in vacuum ($\leq 10^{-5}$ mm Hg).

DATA TABLE NO. 132 NORMAL TOTAL EMITTANCE OF ZINC SULFIDE PIGMENTED COATINGS

[Temperature, T, K; Emittance, ϵ]

T	ϵ
CURVE 1*	
300	0.91
CURVE 2*	
300	0.90
CURVE 3*	
300	0.91
CURVE 4*	
300	0.90
CURVE 5*	
373	0.77

* No plot given

FIGURE 133A

ANALYZED NORMAL SPECTRAL REFLECTANCE OF ZINC SULFIDE PIGMENTED COATINGS

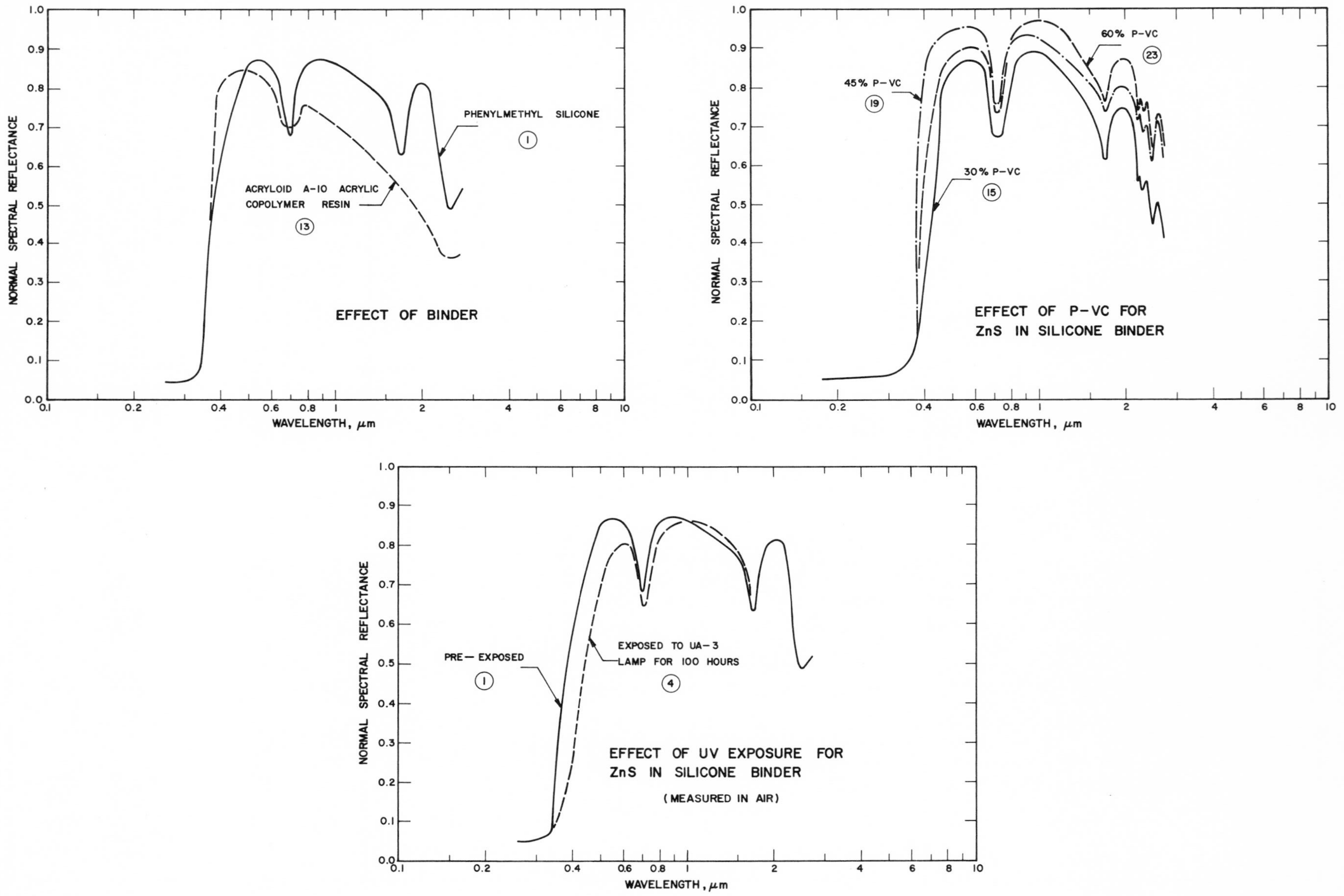

NORMAL SPECTRAL REFLECTANCE
WAVELENGTH, μm
PHENYLMETHYL SILICONE (1)
ACRYLOID A-10 ACRYLIC COPOLYMER RESIN (13)
EFFECT OF BINDER

NORMAL SPECTRAL REFLECTANCE
WAVELENGTH, μm
60% P-VC (23)
45% P-VC (19)
30% P-VC (15)
EFFECT OF P-VC FOR ZnS IN SILICONE BINDER

NORMAL SPECTRAL REFLECTANCE
WAVELENGTH, μm
PRE-EXPOSED (1)
EXPOSED TO UA-3 LAMP FOR 100 HOURS (4)
EFFECT OF UV EXPOSURE FOR ZnS IN SILICONE BINDER
(MEASURED IN AIR)

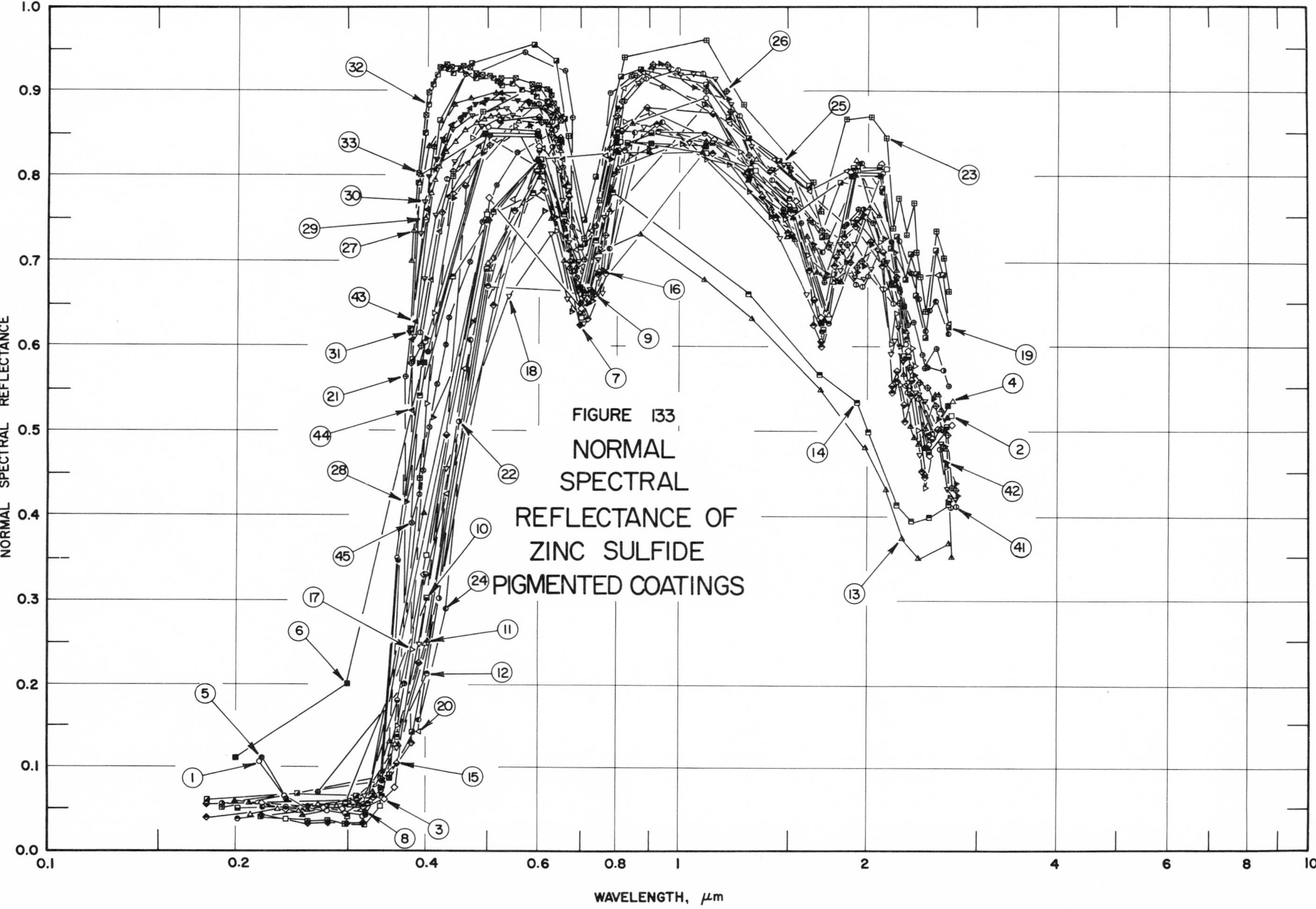
FIGURE 133
NORMAL SPECTRAL REFLECTANCE OF ZINC SULFIDE PIGMENTED COATINGS
NORMAL SPECTRAL REFLECTANCE
WAVELENGTH, μm

SPECIFICATION TABLE NO. 133 NORMAL SPECTRAL REFLECTANCE OF ZINC SULFIDE PIGMENTED COATINGS

Curve No.	Ref. No.	Year	Temperature, K	Wavelength Range, μm	Geometry θ	θ'	ω'	Reported Error, %	Composition (weight percent), Specifications, and Remarks
1	1	1960	298	0.218-2.701	~0°		2π		Zinc sulfide in phenylmethyl silicone (50 Dow No. 805 and 50 Dow No. 806A) binder (0.0508 mm thick) on 24S-T aluminum alloy substrate; 30% pigment-volume ratio; substrate anodized; measured relative to MgO.
2	1	1960	298	0.219-2.705	~0°		2π		Above specimen and conditions except exposed in vacuum (10^{-5} mm Hg) at ~366 K to UV radiation from G. E. type UA-3 lamp (lamp-sample distance 8.5 cm) for 20 hrs.
3	1	1960	298	0.220-2.708	~0°		2π		Above specimen and conditions except exposed for 60 hrs.
4	1	1960	298	0.212-2.725	~0°		2π		Above specimen and conditions except exposed for 100 hrs.
5	7	1961	298	0.220-0.800	~0°		2π		Zinc sulfide in silicone resin binder (~0.0508 mm thick) on anodized aluminum substrate; 30% pigment-volume ratio; draw-down applied; measured relative to MgO.
6	7	1961	298	0.200-2.695	~0°		2π		Above specimen and conditions.
7	7	1961	298	0.220-0.800	~0°		2π		Above specimen and conditions except exposed in vacuum (≤10^{-5} mm Hg) to UV radiation from UA-3 mercury vapor lamp for 20 hrs.
8	7	1961	298	0.199-2.693	~0°		2π		Above specimen and conditions.
9	7	1961	298	0.200-0.804	~0°		2π		Curve 7 specimen and conditions except exposed for 60 hrs.
10	7	1961	298	0.201-2.697	~0°		2π		Above specimen and conditions.
11	7	1961	298	0.212-0.800	~0°		2π		Curve 7 specimen and conditions except exposed for 100 hrs.
12	7	1961	298	0.202-2.713	~0°		2π		Above specimen and conditions.
13	2	1962	~298	0.21-2.70	~0°		2π		ZnS in Acryloid A-10 (Rohm and Haas) acrylic copolymer resin binder (0.0508 mm thick); 30% pigment-volume ratio; data extracted from smooth curve; measured relative to and corrected for reflectance of MgO.
14	2	1962	~298	0.20-2.69	~0°		2π		Similar to above specimen and conditions except exposed at 333 K in vacuum (10^{-5} mm Hg) to 100 hrs of UV radiation from a UA-3 lamp (lamp-sample distance 8.5 cm).
15	2	1962	~298	0.18-2.70	~0°		2π		ZnS in silicone binder (0.0508 mm thick); 30% pigment-volume ratio; data extracted from smooth curve; measured relative to and corrected for reflectance of MgO.
16	2	1962	~298	0.18-2.70	~0°		2π		Similar to above specimen and conditions except exposed at 332 K in vacuum (10^{-5} mm Hg) to 138 hrs of UV radiation from UA-3 lamp (lamp-sample distance 8.5 cm).
17	2	1962	~298	0.20-2.70	~0°		2π		ZnS in silicone binder (0.0508 mm thick); 40% pigment-volume ratio; data extracted from smooth curve; measured relative to and corrected for reflectance of MgO.
18	2	1962	~298	0.20-2.70	~0°		2π		Similar to above specimen and conditions except exposed at 334 K in vacuum (10^{-5} mm Hg) to 138 hrs of UV radiation from UA-3 lamp (lamp-sample distance 8.5 cm).
19	2	1962	~298	0.18-2.70	~0°		2π		ZnS in silicone binder (0.0508 mm thick); 45% pigment-volume ratio; data extracted from smooth curve; measured relative to and corrected for reflectance of MgO.
20	2	1962	~298	0.18-2.70	~0°		2π		Similar to above specimen and conditions except exposed at 339 K in vacuum (10^{-5} mm Hg) to 144 hrs of UV radiation from UA-3 lamp (lamp-sample distance 8.5 cm).

SPECIFICATION TABLE NO. 133 NORMAL SPECTRAL REFLECTANCE OF ZINC SULFIDE PIGMENTED COATINGS (continued)

Curve No.	Ref. No.	Year	Temperature, K	Wavelength Range, μm	Geometry θ	θ'	ω'	Reported Error, %	Composition (weight percent), Specifications, and Remarks
21	2	1962	~298	0.20-2.70	~0°		2π		ZnS in silicone binder (0.0508 mm thick); 50% pigment-volume ratio; data extracted from smooth curve; measured relative to and corrected for reflectance of MgO.
22	2	1962	~298	0.19-2.70	~0°		2π		Similar to above specimen and conditions except exposed at 333 K in vacuum (10^{-5} mm Hg) to 144 hrs of UV radiation from UA-3 lamp (lamp-sample distance 8.5 cm).
23	2	1962	~298	0.19-2.70	~0°		2π		ZnS in silicone binder (0.0508 mm thick); 60% pigment-volume ratio; data extracted from smooth curve; measured relative to and corrected for reflectance of MgO.
24	2	1962	~298	0.19-2.70	~0°		2π		Similar to above specimen and conditions except exposed at 331 K in vacuum (10^{-5} mm Hg) to 138 hrs of UV radiation from UA-3 lamp (lamp-sample distance 8.5 cm).
25	20	1963	~298	0.383-2.765	~0°		2π		Zinc sulfide in PS7 potassium silicate binder on aluminum substrate; PBR 3.19; solids content 59.0%; paint ball milled; substrate grit blasted with 40 mesh silicon carbide; airbrush application; sample cured at 413 K for 18 hrs; data extracted from smooth curve. [Authors' designation: Sample C18]
26	20	1963	~298	0.377-2.765	~0°		2π		Above specimen and conditions except exposed in vacuum (10^{-5} mm Hg) to 260 ESH of UV radiation (G. E. AH-6 lamp) at solar factor of 4 suns.
27	20	1963	~298	0.378-2.724	~0°		2π		Dow Corning Q-9-0108, zinc sulfide in methyl silicone RTV elastomer binder on aluminum substrate; air brush application; data extracted from smooth curve. [Authors' designation: Sample Q-9-0108]
28	20	1963	~298	0.371-2.724	~0°		2π		Above specimen and conditions except exposed in vacuum (10^{-5} mm Hg) to 1530 ESH of radiation from G.E. AH-6 lamp.
29	101	1964	~298	0.3988-0.6500	~0°		2π		ZnS (Superlith XXXN, C.J. Osborn Co.) in melamine modified polyvinyl butyral (90 Butvar B-98 polyvinyl butyral resin, Shawinigan Resins Corp., and 10 CYMEL 300 hexamethoxymethyl melamine, American Cyanamid Co.) binder (~0.09 mm thick) on aluminum (Alclad 2024-T3) substrate; PVC 30%; freshly prepared,then applied with Bird film applicator; data extracted from smooth curve.
30	101	1964	~298	0.3767-0.6500	~0°		2π		Similar to above specimen and conditions.
31	101	1964	~298	0.3767-0.6500	~0°		2π		Similar to curve 29 specimen and conditions.
32	101	1964	~298	0.3708-0.6500	~0°		2π		Similar to curve 29 specimen and conditions except paint aged several wks before application.
33	101	1964	~298	0.3708-0.6500	~0°		2π		Similar to curve 29 specimen and conditions except paint aged several wks before application.
34*	27	1966	~298	0.44-0.60	~0°		2π		51.8 ZnS (Superlith XXXN, C.J. Osborn Co.) in 48.2 methyl-phenyl silicone resin (Dow Corning 806A) binder on aluminum substrate; PVC 40%; paint ground in porcelain ball mill for 16 hrs. [Authors' designation: Sample S1]
35*	27	1964	~298	0.44-0.60	~0°		2π		ZnS in PS-7 potassium silicate binder on aluminum substrate; PBR 3.19; solids content 59.0%; substrate abraded with No. 60 Aloxite cloth; sample cured at 413 K for 18 hrs. [Authors' designation: Sample C18]
36*	27	1964	~298	0.44-0.60	~0°		2π		Above specimen and conditions except irradiated in vacuum (10^{-6} mm Hg) with 260 ESH of UV radiation (AH-6 lamp) at solar factor of 4 suns.

* Not shown on plot

SPECIFICATION TABLE NO. 133 NORMAL SPECTRAL REFLECTANCE OF ZINC SULFIDE PIGMENTED COATINGS (continued)

Curve No.	Ref. No.	Year	Temperature, K	Wavelength Range, μm	Geometry θ	θ'	ω'	Reported Error, %	Composition (weight percent), Specifications, and Remarks
37*	27	1964	~298	0.44-0.60	~0°		2π		ZnS in PS7 potassium silicate binder on aluminum substrate; PBR 4.30; solids content 56.9%; aluminum substrate abraded with No. 60 Aloxite cloth; sample cured at 413 K for 18 hrs. [Authors' designation: Sample C19]
38*	27	1964	~298	0.44-0.60	~0°		2π		Above specimen and conditions except irradiated in vacuum (10^{-6} mm Hg) with 250 ESH of UV radiation (AH-6 lamp) at solar factor of 2.5 suns.
39*	27	1964	~298	0.44-0.60	~0°		2π		ZnS in 3M Kel-F 800 copolymer resin binder on aluminum substrate; PBR 5.0; paint ground in porcelain ball mill for 16 hrs. [Authors' designation: Sample P12]
40*	27	1964	~298	0.44-0.60	~0°		2π		Above specimen and conditions except irradiated in vacuum (10^{-6} mm Hg) with 108 ESH of UV radiation at solar factor of 4 suns.
41	29	1962	298	0.39-2.78	0°		2π		ZnS in potassium silicate binder on 6061-T6 aluminum substrate; PBR 3.22; substrate grit blasted with 40 mesh SiC; air-brush application; air dried overnight, then heat cured at 395-415 K for 24 hrs; data extracted from smooth curve. [Author's designation: 2-10-0]
42	29	1962	298	0.39-2.78	0°		2π		Above specimen and conditions except exposed in vacuum (~10^{-6} mm Hg) to UV radiation (from three AH-6 lamps) at 4 suns for 67 hrs.
43	29	1962	298	0.380-0.700	0°		2π		ZnS in Leonite 201-S binder on 6061-T6 aluminum substrate; PBR 0.67; substrate grit blasted with 40 mesh SiC; sprayed; cured at 398 K for 24 hrs; data extracted from smooth curve. [Author's designation: Sample No. 406-1B]
44	29	1962	298	0.380-0.700	0°		2π		Above specimen and conditions except exposed in vacuum (~10^{-6} mm Hg) to UV radiation (from three AH-6 lamps) at 4 suns for 18.5 hrs.
45	29	1962	298	0.380-0.700	0°		2π		Above specimen and conditions except exposed in vacuum (~10^{-6} mm Hg) to UV radiation (from three AH-6 lamps) at 4 suns for 78.5 hrs.

* Not shown on plot

DATA TABLE NO. 133 NORMAL SPECTRAL REFLECTANCE OF ZINC SULFIDE PIGMENTED COATINGS

[Wavelength, λ, μm; Reflectance, ρ; Temperature, T, K]

CURVE 1
T = 358

λ	ρ
0.218	0.105
0.239	0.065
0.259	0.050
0.279	0.047
0.299	0.044
0.318	0.041
0.339	0.072
0.360	0.350
0.399	0.603
0.500	0.857
0.600	0.853
0.700	0.662
0.801	0.856
1.107	0.843
1.309	0.804
1.502	0.765
1.698	0.618
1.903	0.805
2.105	0.802
2.301	0.601
2.502	0.484
2.701	0.532

CURVE 2
T = 298

λ	ρ
0.219	0.040
0.240	0.037
0.259	0.034
0.279	0.034
0.298	0.030
0.319	0.030
0.338	0.052
0.359	0.137
0.400	0.353
0.499	0.754
0.602	0.821
0.702	0.633
0.802	0.841
1.105	0.828
1.328	0.806
1.500	0.750
1.928	0.806
2.137	0.808
2.298	0.590
2.536	0.497
2.705	0.517

CURVE 3
T = 298

λ	ρ
0.220	0.056
0.240	0.048
0.275	0.050
0.296	0.049
0.318	0.052
0.343	0.061
0.358	0.175
0.398	0.329
0.501	0.773
0.600	0.835
0.702	0.644
0.807	0.840*
1.136	0.843
1.306	0.820
1.533	0.769
1.694	0.634
1.896	0.805
2.103	0.816
2.328	0.595
2.523	0.471
2.708	0.507

CURVE 4
T = 298

λ	ρ
0.212	0.043
0.234	0.048
0.253	0.046
0.271	0.054
0.290	0.052
0.314	0.051
0.333	0.066
0.358	0.126
0.399	0.247
0.500	0.697
0.603	0.804
0.706	0.649
0.802	0.811
1.103	0.853
1.327	0.820
1.507	0.783
1.736	0.637
1.921	0.820
2.127	0.816*
2.302	0.617
2.512	0.500
2.725	0.534

CURVE 5
T = 298

λ	ρ
0.220	0.110
0.240	0.062
0.260	0.052
0.279	0.047*
0.299	0.043*
0.320	0.043
0.339	0.074
0.361	0.348
0.401	0.594
0.501	0.849
0.600	0.847
0.700	0.651
0.800	0.848

CURVE 6
T = 298

λ	ρ
0.200	0.110
0.300	0.200
0.396	0.581
0.495	0.850
0.597	0.850
0.699	0.669
0.796	0.839
0.902	0.839
1.101	0.836
1.298	0.799
1.501	0.761
1.693	0.628
1.893	0.800
2.099	0.800
2.291	0.598
2.493	0.482
2.695	0.529

CURVE 7
T = 298

λ	ρ
0.220	0.042
0.240	0.037*
0.260	0.032
0.280	0.032
0.300	0.032
0.319	0.032
0.339	0.052*
0.360	0.137*
0.400	0.352*
0.501	0.748
0.600	0.817
0.700	0.626
0.800	0.832

CURVE 8
T = 298

λ	ρ
0.199	0.059
0.322	0.046
0.398	0.402
0.498	0.747
0.598	0.820
0.699	0.650*
0.804	0.827
0.898	0.827
1.099	0.827
1.320	0.800
1.503	0.750
1.712	0.631
1.912	0.801
2.116	0.798
2.294	0.587
2.519	0.479
2.693	0.516

CURVE 9
T = 298

λ	ρ
0.220	0.050
0.239	0.050
0.275	0.050*
0.296	0.050*
0.318	0.054
0.343	0.062*
0.360	0.179
0.400	0.331
0.501	0.772*
0.600	0.830
0.700	0.639
0.804	0.832

CURVE 10
T = 298

λ	ρ
0.201	0.049
0.298	0.054
0.401	0.301
0.499	0.680
0.599	0.806
0.713	0.666
0.828	0.839
0.906	0.853
1.129	0.838
1.303	0.811
1.524	0.759
1.702	0.620
1.893	0.810
2.097	0.810
2.323	0.588
2.513	0.467
2.697	0.504

CURVE 11
T = 298

λ	ρ
0.212	0.042*
0.235	0.048*
0.254	0.042
0.272	0.053*
0.291	0.053*
0.314	0.053
0.334	0.066*
0.361	0.125
0.401	0.251
0.502	0.690
0.600	0.798
0.705	0.639
0.800	0.806

CURVE 12
T = 298

λ	ρ
0.202	0.037
0.323	0.055
0.401	0.212
0.500	0.671
0.601	0.798*
0.738	0.661
0.802	0.854
0.925	0.854
1.100	0.851
1.323	0.810*
1.508	0.773
1.740	0.628
1.923	0.814*
2.121	0.809
2.304	0.608
2.507	0.496
2.713	0.531*

CURVE 13
T ~ 298

λ	ρ
0.21	0.056
0.26	0.047
0.33	0.068
0.35	0.129
0.39	0.800
0.49	0.849
0.60	0.808
0.63	0.750
0.66	0.701
0.69	0.694
0.75	0.710
0.78	0.760
0.87	0.731
1.10	0.678
1.32	0.633
1.68	0.537
1.99	0.480
2.14	0.431
2.27	0.373
2.42	0.350
2.70	0.367

CURVE 14
T ~ 298

λ	ρ
0.20	0.049*
0.30	0.040
0.34	0.064
0.36	0.134
0.39	0.541
0.44	0.681
0.52	0.757
0.59	0.778
0.66	0.754
0.70	0.715
0.74	0.725
0.79	0.779
1.29	0.662
1.68	0.567
1.93	0.533
2.05	0.498
2.23	0.413
2.35	0.394
2.52	0.398
2.69	0.413

CURVE 15
T ~ 298

λ	ρ
0.18	0.054
0.31	0.059
0.36	0.103
0.39	0.225
0.43	0.494
0.46	0.788
0.52	0.865
0.60	0.869
0.65	0.847
0.67	0.775
0.68	0.697
0.73	0.664
0.75	0.687
0.81	0.853
0.89	0.882
1.13	0.864
1.20	0.841
1.31	0.808*
1.36	0.776
1.47	0.756
1.52	0.732
1.61	0.708
1.65	0.655
1.69	0.604
1.73	0.639*
1.77	0.704
1.92	0.749
2.12	0.711
2.18	0.553
2.23	0.572
2.29	0.530
2.36	0.560
2.47	0.447
2.54	0.491
2.60	0.502
2.70	0.416

CURVE 16
T ~ 298

λ	ρ
0.18	0.039
0.33	0.063
0.38	0.127
0.51	0.647
0.55	0.759
0.61	0.783
0.66	0.701*
0.72	0.632
0.77	0.686
0.82	0.831
0.92	0.861
1.13	0.827
1.28	0.806
1.36	0.776*
1.42	0.751
1.49	0.733
1.61	0.689
1.64	0.626
1.69	0.597
1.74	0.626*
1.77	0.694
1.83	0.729
1.91	0.740
2.12	0.696
2.19	0.545
2.23	0.559
2.30	0.511
2.35	0.541
2.44	0.452
2.47	0.446

* Not shown on plot

DATA TABLE NO. 133 NORMAL SPECTRAL REFLECTANCE OF ZINC SULFIDE PIGMENTED COATINGS (continued)

λ	ρ
CURVE 16 (cont.) T ~ 298	
2.54	0.494*
2.60	0.501*
2.70	0.419*
CURVE 17 T ~ 298	
0.20	0.053
0.30	0.057
0.34	0.086
0.36	0.148
0.38	0.240
0.40	0.531
0.41	0.639
0.44	0.777
0.47	0.845
0.55	0.859
0.61	0.794
0.64	0.734
0.70	0.671*
0.75	0.714
0.82	0.865
0.90	0.875
1.25	0.828
1.47	0.749
1.60	0.707
1.68	0.621*
1.76	0.680
1.89	0.746
2.09	0.722
2.13	0.668
2.18	0.601
2.22	0.639
2.26	0.555
2.28	0.532*
2.33	0.549
2.36	0.597
2.39	0.577
2.41	0.500
2.47	0.433
2.56	0.498
2.70	0.422

λ	ρ
CURVE 18 T ~ 298	
0.20	0.052*
0.32	0.065
0.36	0.124
0.39	0.247
0.43	0.454
0.54	0.658
0.63	0.732
0.67	0.656
0.71	0.627
0.76	0.663
0.81	0.830*
0.88	0.858
1.08	0.838
1.45	0.728
1.60	0.662
1.70	0.602*
1.81	0.689
1.91	0.719
2.12	0.668
2.18	0.592
2.21	0.607
2.30	0.510*
2.33	0.558
2.35	0.572
2.40	0.546
2.43	0.472
2.47	0.432*
2.56	0.499*
2.70	0.422*
CURVE 19 T ~ 298	
0.18	0.061
0.25	0.067
0.31	0.064
0.35	0.086
0.38	0.140
0.38	0.585
0.42	0.866
0.47	0.934
0.59	0.955
0.64	0.937
0.66	0.868
0.67	0.782

λ	ρ
CURVE 19 (cont.)	
0.71	0.748
0.74	0.799
0.81	0.918
0.87	0.927
1.11	0.910
1.20	0.870
1.29	0.845
1.45	0.819
1.62	0.787
1.69	0.727
1.81	0.791
1.89	0.810*
2.11	0.785
2.18	0.714
2.22	0.729
2.30	0.678
2.35	0.708
2.39	0.685
2.39	0.659
2.47	0.611
2.57	0.713
2.64	0.684
2.70	0.624
CURVE 20 T ~ 298	
0.18	0.039*
0.34	0.073*
0.39	0.141
0.43	0.424
0.46	0.573
0.55	0.771
0.61	0.833
0.65	0.805
0.69	0.713
0.73	0.736
0.77	0.824
0.82	0.903
0.88	0.922
1.22	0.862
1.43	0.819
1.63	0.781
1.69	0.728*
1.86	0.806
1.93	0.811

λ	ρ
CURVE 20 (cont.)	
2.11	0.785*
2.18	0.716
2.23	0.729*
2.30	0.680
2.36	0.707*
2.39	0.686*
2.39	0.660*
2.48	0.612*
2.57	0.713*
2.64	0.680
2.70	0.626
CURVE 21 T ~ 298	
0.20	0.057*
0.27	0.069
0.34	0.085*
0.37	0.200
0.37	0.564
0.41	0.833
0.46	0.920
0.57	0.943
0.66	0.924
0.68	0.869
0.68	0.738
0.71	0.701
0.74	0.740
0.78	0.898
0.84	0.918
0.94	0.905
1.11	0.887
1.47	0.759
1.57	0.744
1.71	0.675
1.83	0.739
1.94	0.761
2.04	0.744
2.17	0.675
2.24	0.668
2.32	0.625
2.37	0.628
2.45	0.591
2.47	0.576
2.58	0.597
2.70	0.554

λ	ρ
CURVE 22 T ~ 298	
0.19	0.055
0.27	0.070*
0.34	0.083
0.38	0.139*
0.42	0.301
0.45	0.509
0.51	0.702
0.63	0.835
0.71	0.719
0.77	0.825*
0.85	0.918*
1.10	0.886
1.46	0.760*
1.57	0.743*
1.71	0.675*
1.84	0.742
1.97	0.759
2.11	0.710*
2.17	0.673
2.25	0.665
2.31	0.625*
2.37	0.629*
2.49	0.576
2.57	0.598*
2.65	0.573
2.70	0.555*
CURVE 23 T ~ 298	
0.19	0.051
0.32	0.061*
0.35	0.089
0.39	0.443
0.44	0.799
0.49	0.876
0.60	0.907
0.67	0.848
0.71	0.726
0.75	0.771
0.82	0.941
1.11	0.961
1.27	0.886
1.64	0.793
1.69	0.758

λ	ρ
CURVE 23 (cont.)	
1.85	0.867
2.03	0.870
2.14	0.846
2.20	0.737
2.24	0.772
2.31	0.730
2.37	0.766
2.40	0.710
2.43	0.680
2.47	0.640
2.57	0.734
2.65	0.703
2.70	0.655
CURVE 24 T ~ 298	
0.19	0.051*
0.32	0.061*
0.39	0.156
0.43	0.290
0.47	0.606
0.61	0.810
0.66	0.745
0.70	0.644*
0.78	0.714
0.85	0.852
0.94	0.864
1.33	0.805*
1.70	0.711
1.88	0.804*
1.96	0.815
2.11	0.779
2.18	0.721
2.24	0.723
2.32	0.679*
2.35	0.686
2.42	0.656
2.48	0.617
2.52	0.642
2.58	0.653
2.70	0.616

λ	ρ
CURVE 25 T ~ 298	
0.383	0.611
0.397	0.677
0.413	0.753
0.442	0.816
0.488	0.857
0.526	0.870
0.562	0.878
0.603	0.881
0.625	0.869
0.656	0.769
0.671	0.720
0.696	0.699
0.709	0.657
0.725	0.666
0.747	0.701
0.764	0.750
0.791	0.846
0.797	0.867
0.860	0.910
0.907	0.928
0.963	0.928
1.003	0.920
1.055	0.920
1.139	0.914
1.222	0.886
1.312	0.836
1.399	0.803
1.470	0.812
1.496	0.812
1.683	0.727*
1.747	0.717
1.813	0.717
1.839	0.710
1.906	0.678
1.967	0.677
2.023	0.690
2.077	0.711
2.112	0.712*
2.147	0.698
2.238	0.622
2.283	0.582
2.332	0.556
2.411	0.542
2.481	0.534
2.586	0.500*

λ	ρ
CURVE 25 (cont.)	
2.626	0.484
2.684	0.430
2.720	0.418*
2.765	0.418
CURVE 26 T ~ 298	
0.377	0.578
0.391	0.599
0.423	0.756
0.440	0.803
0.502	0.835
0.608	0.864
0.627	0.848
0.663	0.727*
0.675	0.699
0.702	0.649*
0.730	0.759
0.746	0.684
0.767	0.739
0.800	0.869
0.811	0.887
0.878	0.922*
0.908	0.933
0.954	0.933
1.103	0.900
1.188	0.900
1.234	0.876
1.291	0.824
1.323	0.811*
1.402	0.805
1.500	0.810
1.647	0.747
1.687	0.735
1.726	0.730
1.824	0.730*
1.856	0.718
1.886	0.697
1.925	0.693
1.957	0.700
2.014	0.722
2.117	0.700
2.212	0.681
2.250	0.650
2.294	0.593*

* Not shown on plot

DATA TABLE NO. 133 NORMAL SPECTRAL REFLECTANCE OF ZINC SULFIDE PIGMENTED COATINGS (continued)

λ	ρ
CURVE 26 (cont.) T ~ 298	
2.340	0.575
2.435	0.557
2.500	0.551
2.580	0.528
2.633	0.500
2.697	0.445
2.729	0.433
2.765	0.430
CURVE 27 T ~ 298	
0.378	0.700
0.381	0.735
0.403	0.835
0.441	0.874
0.468	0.892
0.513	0.897
0.590	0.884
0.638	0.867
0.667	0.791
0.683	0.700
0.694	0.664*
0.714	0.664
0.779	0.833
0.796	0.856*
0.843	0.863
1.108	0.841
1.188	0.824
1.278	0.794
1.406	0.763
1.490	0.731
1.538	0.725
1.601	0.729
1.645	0.715
1.707	0.687
1.755	0.675
1.810	0.675
1.916	0.739*
1.956	0.753
2.019	0.763
2.082	0.753
2.215	0.690
2.230	0.671
2.261	0.598*
2.280	0.567

λ	ρ
CURVE 27 (cont.)	
2.344	0.505
2.370	0.492
2.416	0.485
2.474	0.501
2.579	0.543
2.600	0.542
2.635	0.525
2.651	0.499
2.670	0.479
2.724	0.351
CURVE 28 T ~ 298	
0.371	0.416
0.409	0.515
0.472	0.627
0.612	0.757
0.638	0.751
0.679	0.640
0.699	0.629*
0.730	0.654
0.792	0.801
0.837	0.824
0.896	0.832
1.011	0.837
1.108	0.820
1.201	0.814
1.289	0.781
1.400	0.752
1.488	0.735*
1.541	0.728*
1.637	0.717
1.719	0.685
1.756	0.676*
1.800	0.676
1.856	0.700
1.914	0.739*
1.960	0.755
2.022	0.763*
2.134	0.726
2.176	0.719*
2.202	0.710
2.259	0.600
2.280	0.567*
2.344	0.505*
2.370	0.492*

λ	ρ
CURVE 28 (cont.)	
2.416	0.485*
2.475	0.501*
2.579	0.543*
2.600	0.542*
2.636	0.524*
2.662	0.483
2.672	0.457
2.699	0.424*
2.724	0.351*
CURVE 29 T ~ 298	
0.3988	0.750
0.4053	0.778
0.4116	0.795
0.4195	0.811
0.4296	0.827
0.4417	0.840
0.4592	0.854
0.4795	0.863
0.4911	0.866
0.5033	0.871
0.5189	0.872
0.5582	0.865
0.6086	0.863*
0.6200	0.860
0.6280	0.856
0.6397	0.845
0.6449	0.837
0.6500	0.830
CURVE 30 T ~ 298	
0.3767	0.616*
0.3919	0.732
0.3986	0.767
0.4063	0.797*
0.4144	0.819*
0.4275	0.842*
0.4392	0.854
0.4525	0.865
0.4706	0.872
0.5150	0.878
0.5327	0.879
0.5446	0.873

λ	ρ
CURVE 30 (cont.)	
0.6129	0.864
0.6280	0.856*
0.6397	0.845*
0.6449	0.837*
0.6500	0.830*
CURVE 31 T ~ 298	
0.3767	0.616
0.3919	0.732*
0.3986	0.767*
0.4063	0.797
0.4144	0.819
0.4275	0.842
0.4392	0.854*
0.4597	0.873
0.4783	0.884
0.4928	0.890
0.5290	0.891
0.5596	0.889
0.5879	0.885
0.6192	0.882
0.6291	0.876*
0.6391	0.864*
0.6458	0.853*
0.6500	0.844*
CURVE 32 T ~ 298	
0.3708	0.621
0.3792	0.698*
0.3884	0.792
0.3960	0.852
0.4014	0.884
0.4082	0.907
0.4147	0.920
0.4242	0.928
0.4328	0.927
0.4398	0.922
0.4501	0.924
0.4636	0.922*
0.4787	0.914
0.5090	0.915
0.5154	0.913
0.5208	0.910

λ	ρ
CURVE 32 (cont.)	
0.5294	0.908
0.5415	0.908
0.5641	0.901
0.5875	0.898
0.5954	0.894
0.6174	0.888
0.6291	0.876*
0.6391	0.864*
0.6458	0.853*
0.6500	0.844*
CURVE 33 T ~ 298	
0.3708	0.621*
0.3880	0.803
0.3973	0.871
0.4025	0.897
0.4118	0.919*
0.4203	0.929
0.4290	0.932
0.4387	0.929
0.4527	0.931
0.4669	0.929
0.4788	0.922
0.4897	0.919
0.5172	0.917
0.5237	0.915
0.5517	0.915
0.5842	0.908
0.5972	0.904
0.6186	0.902
0.6278	0.896
0.6360	0.886
0.6500	0.865
CURVE 34* T ~ 298	
0.44	0.844
0.60	0.890
CURVE 35* T ~ 298	
0.44	0.850
0.60	0.905

λ	ρ
CURVE 36* T ~ 298	
0.44	0.810
0.60	0.885
CURVE 37* T ~ 298	
0.44	0.885
0.60	0.880
CURVE 38* T ~ 298	
0.44	0.865
0.60	0.875
CURVE 39* T ~ 298	
0.44	0.895
0.60	0.920
CURVE 40* T ~ 298	
0.44	0.495
0.60	0.780
CURVE 41 T = 298	
0.39	0.615
0.43	0.796
0.50	0.859*
0.56	0.879*
0,60	0.884
0.63	0.864
0.69	0.695*
0.72	0.662
0.75	0.682*
0.78	0.834*
0.82	0.889
0.88	0.916
0.98	0.926
1.09	0.922
1.18	0.900*
1.25	0.871

λ	ρ
CURVE 41 (cont.)	
1.37	0.806*
1.41	0.795
1.48	0.810*
1.63	0.746
1.73	0.716
1.87	0.700*
1.92	0.673
1.97	0.670
2.01	0.683
2.11	0.713*
2.15	0.700*
2.20	0.650
2.26	0.600*
2.29	0.568*
2.36	0.546
2.43	0.537
2.59	0.499*
2.63	0.478
2.71	0.416*
2.73	0.410
2.78	0.411
CURVE 42 T = 298	
0.39	0.579
0.40	0.607
0.44	0.773
0.49	0.826
0.58	0.860
0.62	0.858*
0.68	0.700*
0.72	0.650
0.76	0.688
0.78	0.820
0.80	0.880
0.87	0.925*
0.93	0.934
1.06	0.905
1.20	0.892
1.24	0.870*
1.31	0.816
1.38	0.808*
1.51	0.804
1.64	0.745*
1.71	0.731
1.83	0.728*

λ	ρ
CURVE 42 (cont.)	
1.92	0.687
1.97	0.694
2.03	0.719
2.22	0.677
2.28	0.619
2.30	0.587*
2.37	0.566
2.55	0.537
2.60	0.514
2.66	0.462
2.71	0.437
2.78	0.424
CURVE 43 T = 298	
0.380	0.607
0.385	0.628
0.392	0.677*
0.402	0.758
0.413	0.807
0.425	0.839
0.438	0.859
0.454	0.868
0.469	0.879
0.492	0.886
0.522	0.898
0.561	0.900*
0.594	0.896*
0.621	0.900
0.632	0.896*
0.642	0.877
0.654	0.839
0.662	0.788
0.668	0.727
0.672	0.711
0.677	0.701*
0.682	0.703*
0.686	0.699*
0.695	0.672*
0.700	0.648*
CURVE 44 T = 298	
0.380	0.523
0.386	0.543*

* Not shown on plot

DATA TABLE NO. 133 NORMAL SPECTRAL REFLECTANCE OF ZINC SULFIDE PIGMENTED COATINGS (continued)

λ	ρ
CURVE 44 (cont.) T = 298	
0.396	0.606*
0.407	0.677
0.421	0.734
0.433	0.765
0.465	0.816
0.483	0.835
0.498	0.848*
0.548	0.871*
0.574	0.878
0.623	0.882*
0.636	0.873
0.644	0.857
0.651	0.838*
0.656	0.820
0.664	0.760
0.667	0.730*
0.672	0.709*
0.675	0.701*
0.682	0.699*
0.691	0.683*
0.700	0.647*
CURVE 45 T = 298	
0.380	0.391
0.390	0.424
0.396	0.452
0.405	0.503
0.417	0.554
0.429	0.600
0.443	0.633
0.469	0.699
0.490	0.744
0.516	0.788
0.556	0.827
0.599	0.853*
0.628	0.860*
0.639	0.854
0.648	0.835
0.655	0.809
0.662	0.760*
0.668	0.720*
0.674	0.693
0.679	0.689
0.685	0.688

λ	ρ
CURVE 45 (cont.)	
0.690	0.679
0.700	0.643*

* Not shown on plot

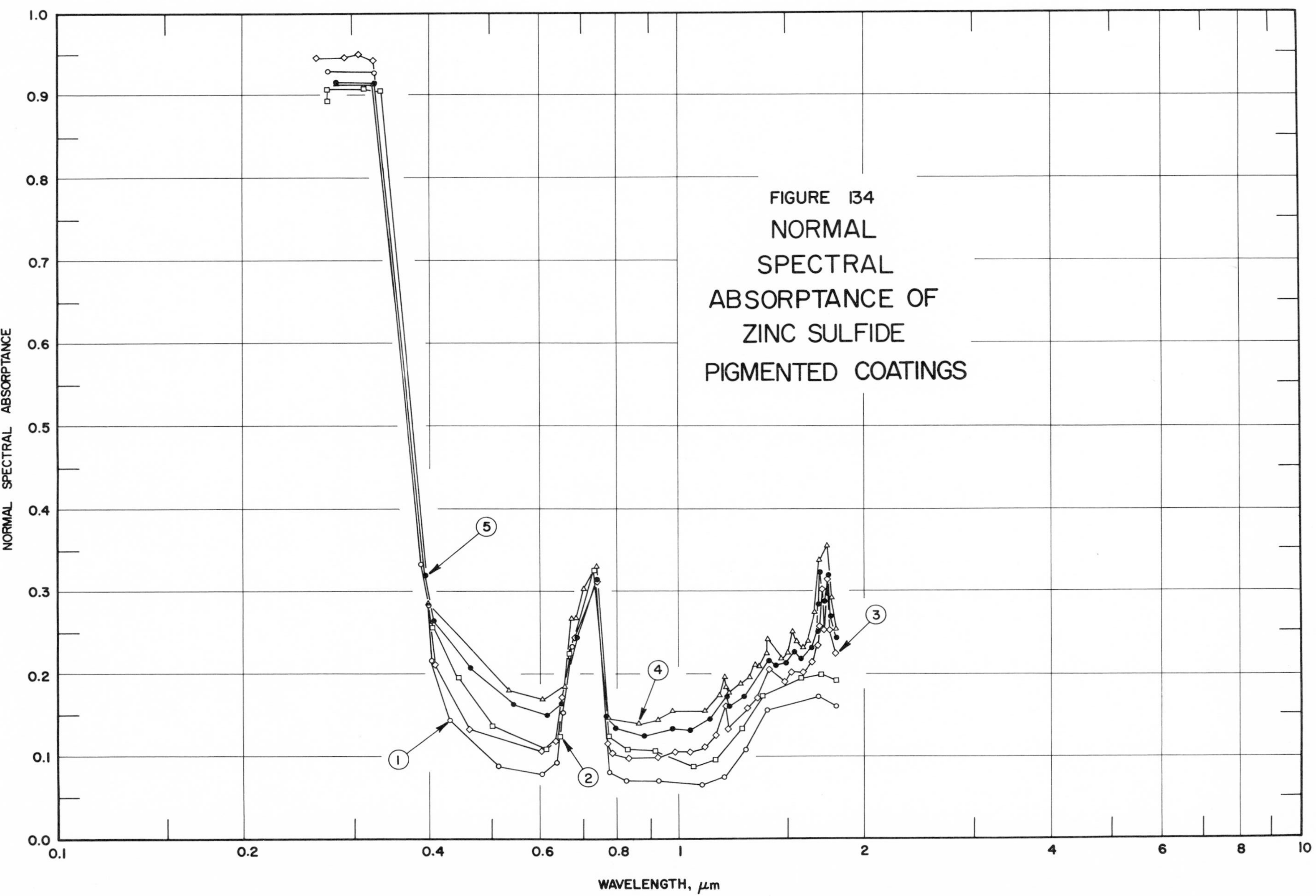

FIGURE 134 NORMAL SPECTRAL ABSORPTANCE OF ZINC SULFIDE PIGMENTED COATINGS

SPECIFICATION TABLE NO. 134 NORMAL SPECTRAL ABSORPTANCE OF ZINC SULFIDE PIGMENTED COATINGS

Curve No.	Ref. No.	Year	Temperature, K	Wavelength Range, μm	Geometry θ	Reported Error, %	Composition (weight percent), Specifications, and Remarks
1	24	1965	298	0.275-1.800	~0°		Zinc sulfide in sodium silicate binder; unknown substrate; data extracted from smooth curve.
2	24	1965	298	0.272-1.800	~0°		Similar to above specimen and conditions except exposed in vacuum to a nuclear radiation of approx 10^8 R of gamma and 5 x 10^{14} neutrons cm^{-2} (E ≥ 2.9 MeV).
3	24	1965	298	0.266-1.799	~0°		Zinc sulfide in silicone (Dow Corning 432) binder; unknown substrate; data extracted from smooth curve.
4	24	1965	298	0.284-1.800	~0°		Similar to above specimen and conditions except exposed in vacuum to a nuclear radiation of approx 10^8 R of gamma and 5 x 10^{14} neutrons cm^{-2} (E ≥ 2.9 MeV).
5	24	1965	298	0.282-1.800	~0°		Similar to curve 3 specimen and conditions except exposed in vacuum simultaneously to nuclear radiation of approx 10^8 R of gamma and 5 x 10^{14} neutrons cm^{-2} (E ≥ 2.9 MeV) and ~500 sun hrs of ultraviolet radiation.

DATA TABLE NO. 134 NORMAL SPECTRAL ABSORPTANCE OF ZINC SULFIDE PIGMENTED COATINGS

[Wavelength, λ, μm; Absorptance, α; Temperature, T, K]

λ	α
CURVE 1 T = 298	
0.275	0.930
0.326	0.929
0.388	0.334
0.397	0.284
0.403	0.216
0.430	0.143
0.514	0.088
0.603	0.078
0.639	0.092
0.653	0.152
0.675	0.234
0.735	0.326
0.779	0.082
0.826	0.071
0.929	0.070
1.089	0.065
1.184	0.075
1.281	0.106
1.394	0.156
1.680	0.173
1.800	0.161
CURVE 2 T = 298	
0.272	0.894
0.272	0.909
0.312	0.909
0.332	0.905
0.401	0.257
0.443	0.196
0.504	0.136
0.614	0.109
0.648	0.123
0.670	0.225
0.733	0.327
0.777	0.124
0.830	0.109
0.915	0.107
1.062	0.088
1.150	0.096
1.269	0.132
1.368	0.172
1.582	0.195
1.701	0.199
1.800	0.192
CURVE 3 T = 298	
0.266	0.948
0.293	0.948
0.309	0.951
0.325	0.944
0.409	0.209
0.462	0.133
0.602	0.107
0.638	0.119
0.650	0.172
0.684	0.244
0.742	0.312
0.773	0.116
0.788	0.103
0.832	0.096
0.924	0.096
0.989	0.105
1.049	0.105
1.109	0.111
1.153	0.126
1.198	0.162
1.212	0.132
1.293	0.159
1.344	0.173
1.403	0.205
1.475	0.191
1.534	0.202
1.590	0.202
1.646	0.215
1.677	0.234
1.688	0.259
1.709	0.302
1.725	0.252
1.744	0.314
1.757	0.253
1.799	0.226
CURVE 4 T = 298	
0.284	0.915
0.326	0.913
0.399	0.286
0.535	0.180
0.606	0.169
0.656	0.185
0.675	0.268
CURVE 4 (cont.)	
0.686	0.267
0.706	0.302
0.740	0.330
0.773	0.147
0.864	0.139
0.926	0.143
0.976	0.154
1.101	0.154
1.174	0.174
1.186	0.196
1.192	0.184
1.211	0.178
1.264	0.189
1.304	0.196
1.336	0.212
1.353	0.209
1.387	0.224
1.396	0.241
1.472	0.218
1.503	0.225
1.532	0.251
1.552	0.238
1.586	0.232
1.622	0.239
1.660	0.275
1.693	0.338
1.721	0.288*
1.748	0.355
1.770	0.293
1.800	0.253
CURVE 5 T = 298	
0.282	0.915
0.327	0.913
0.392	0.320
0.406	0.266
0.465	0.208
0.544	0.163
0.616	0.150
0.649	0.164
0.685	0.245
0.741	0.313
0.769	0.148
0.796	0.134
0.880	0.125
CURVE 5 (cont.)	
0.975	0.133
1.047	0.131
1.127	0.145
1.195	0.172
1.216	0.161
1.281	0.174
1.394	0.217
1.434	0.210
1.497	0.214
1.534	0.227
1.575	0.219
1.638	0.231
1.669	0.251
1.674	0.285
1.699	0.322
1.724	0.287
1.750	0.320
1.763	0.270
1.800	0.243

* Not shown on plot

FIGURE 135A

ANALYZED NORMAL SOLAR ABSORPTANCE OF ZINC SULFIDE PIGMENTED COATINGS

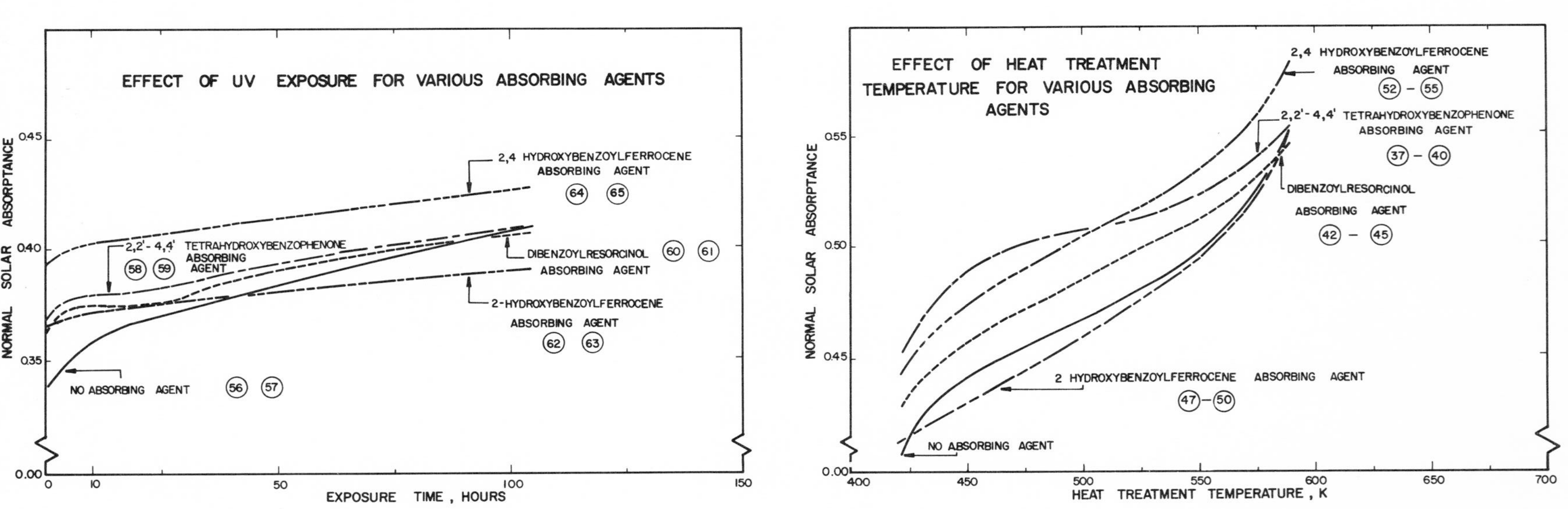

Curve No.	Ref. No.	Year	Temperature Range, K	Geometry θ	Reported Error, %	Composition (weight percent), Specifications and Remarks
1*	100	1962	300	~0°		Zinc sulfide in silicone (ZW 60) binder.
2*	100	1962	300	~0°		Zinc sulfide in silicone (ZW 40) binder.
3*	100	1962	300	~0°		Zinc sulfide in silicone (ZW 60) binder; values represent the max degradation that test data indicate may occur during a 3-month Venus mission.
4*	100	1962	300	~0°		Zinc sulfide in silicone (ZW 40) binder; values represent the max degradation that test data indicate may occur during a 3-month Venus mission.
5*	101	1964	~298	~0°		ZnS (Superlith XXXN, C. J. Osborn Co.) in melamine modified polyvinyl butyral (90 Butvar B-98 polyvinyl butyral resin, Shawinigan Resins Corp, and 10 CYMEL 300 hexamethoxymethyl melamine, American Cyanamid Co.) binder (0.072 mm thick) on quartz substrate; PVC 30%; applied with Bird film applicator; cured 30 min at 366 K; computed from spectral reflectance data. [Authors' designation: sample 245]
6*	101	1964	~298	~0°		Similar to above specimen and conditions except exposed in vacuum (10^{-6} mm Hg) to 100 ESH of UV radiation (from AH-6 lamp); measured in situ.
7*	101	1964	~298	~0°		Similar to curve 5 specimen and conditions except coating (~0.05 mm thick) on aluminum substrate. [Authors' designation: sample 300]
8*	101	1964	~298	~0°		Similar to above specimen and conditions except exposed in vacuum (10^{-6} mm Hg) to 100 ESH of UV radiation (from AH-6 lamp); measured in situ.
9*	101	1964	~298	~0°		Similar to curve 5 specimen and conditions except paint contains 0.25% hydroxybenzoylferrocene additive; thickness 0.05 to 0.1 mm. [Authors' designation: sample 269]
10*	101	1964	~298	~0°		Similar to above specimen and conditions except exposed in vacuum (10^{-6} mm Hg) to 100 ESH of UV radiation (from AH-6 lamp); measured in situ.
11*	101	1964	~298	~0°		Similar to curve 5 specimen and conditions except paint contains 1.0% osmocene stabilizing agent; thickness 0.05 to 0.10 mm. [Authors' designation: sample 277]
12*	101	1964	~298	~0°		Similar to above specimen and conditions except exposed in vacuum (10^{-6} mm Hg) to 100 ESH of UV radiation (from AH-6 lamp); measured in situ.
13*	101	1964	~298	~0°		Similar to curve 5 specimen and conditions except paint contains 1% benzoyl osmocene stabilizing agent. [Authors' designation: sample 278]
14*	101	1964	~298	~0°		Similar to above specimen and conditions except exposed in vacuum (10^{-6} mm Hg) to 100 ESH of UV radiation (from AH-6 lamp); measured in situ.
15*	101	1964	~298	~0°		Similar to curve 5 specimen and conditions except paint contains 25% dibenzoyl resorcinol screening agent; thickness 0.05 to 0.10 mm. [Authors' designation: sample 279]
16*	101	1964	~298	~0°		Similar to above specimen and conditions except exposed in vacuum (10^{-6} mm Hg) to 100 ESH of UV radiation (from AH-6 lamp); measured in situ.
17*	101	1964	~298	~0°		Similar to curve 5 specimen and conditions except paint contains 25% methyl salicylate screening agent. [Authors' designation: sample 280]
18*	101	1964	~298	~0°		Similar to above specimen and conditions except exposed in vacuum (10^{-6} mm Hg) to 100 ESH of UV radiation (from AH-6 lamp); measured in situ.

* No plot given

SPECIFICATION TABLE NO. 135 NORMAL SOLAR ABSORPTANCE OF ZINC SULFIDE PIGMENTED COATINGS (continued)

Curve No.	Ref. No.	Year	Temperature Range, K	Geometry θ	Reported Error, %	Composition (weight percent), Specifications and Remarks
19*	101	1964	~298	~0°		Similar to curve 5 specimen and conditions except paint contains 25% methyl salicylate screening agent and applied on aluminum substrate; thickness 0.05 to 0.10 mm. [Authors' designation: sample 304]
20*	101	1964	~298	~0°		Similar to above specimen and conditions except exposed in vacuum (10^{-6} mm Hg) to 100 ESH of UV radiation (from AH-6 lamp); measured in situ.
21*	101	1964	~298	~0°		ZnS (Superlith XXXN, C. J. Osborn Co.) in melamine modified polyvinyl butyral (90 Butvar B-98 polyvinyl butyral, Shawinigan Resins Corp, and 10 CYMEL 300 hexamethoxymethyl melamine, American Cyanamid Co.) binder (0.085 mm thick) on aluminum substrate; PVC 30%; applied with Bird film applicator; cured 45 min at 366 K; calculated from spectral reflectance data. [Authors' designation: sample 351]
22*	101	1964	~298	~0°		Similar to above specimen and conditions except exposed to 100 ESH of UV radiation (from G. E. AH-6 lamp) in vacuum (10^{-6} mm Hg); measured in situ.
23*	101	1964	~298	~0°		Similar to curve 21 specimen and conditions except paint prepared with 5% perylene additive. [Authors' designation: sample 356]
24*	101	1964	~298	~0°		Similar to above specimen and conditions except exposed to 100 ESH of UV radiation (from G. E. AH-6 lamp) in vacuum (10^{-6} mm Hg); measured in situ.
25*	101	1964	~298	~0°		Similar to curve 21 specimen and conditions except paint prepared with 5% europium trisdibenzoylmethide; film thickness 0.098 mm. [Authors' designation: sample 367]
26*	101	1964	~298	~0°		Similar to above specimen and conditions except exposed to 100 ESH of UV radiation (from G. E. AH-6 lamp) in vacuum (10^{-6} mm Hg); measured in situ.
27*	101	1964	~298	~0°		Similar to above specimen and conditions except paint prepared with 5% Lupersol DDM methyl ethyl ketone peroxide. [Authors' designation: sample 316]
28*	101	1964	~298	~0°		Similar to above specimen and conditions except exposed to 100 ESH of UV radiation (from G. E. AH-6 lamp) in vacuum (10^{-6} mm Hg); measured in situ.
29*	101	1964	~298	~0°		Similar to curve 21 specimen and conditions except paint prepared with 5% aniline additive. [Authors' designation: sample 384]
30*	101	1964	~298	~0°		Similar to above specimen and conditions except exposed to 100 ESH of UV radiation (from G. E. AH-6 lamp) in vacuum (10^{-6} mm Hg); measured in situ.
31*	77	1962	~298	~0°		26.1 ZnS, 30 calcined clay (southern clay Al SiL Ate W), and 21.6 organic solvents in 22.3 Plaskon ST-873 silicone-alkyd binder (~0.0635 mm thick) on zinc chromate primer (0.0127 mm thick) and 2024 aluminum alloy substrates; aluminum substrate roughened, then primed with zinc chromate primer (MIL-P-8585); Fischer-Payne Dipcoater application; air cured at ~298 K; computed from spectral reflectance data.
32*	77	1962	~298	~0°		Similar to above specimen and conditions except heat treated to 422 K in a 24-hr cycle in vacuum (10^{-5} mm Hg); vacuum maintained by diffusion pump.
33*	77	1962	~298	~0°		Similar to curve 31 specimen and conditions except heat treated to 478 K in a 24-hr cycle in vacuum (10^{-5} mm Hg); vacuum maintained by diffusion pump.
34*	77	1962	~298	~0°		Similar to curve 31 specimen and conditions except heat treated to 533 K in a 24-hr cycle in vacuum (10^{-5} mm Hg); vacuum maintained by diffusion pump.

* No plot given

SPECIFICATION TABLE NO. 135 NORMAL SOLAR ABSORPTANCE OF ZINC SULFIDE PIGMENTED COATINGS (continued)

Curve No.	Ref. No.	Year	Temperature Range, K	Geometry θ	Reported Error, %	Composition (weight percent), Specifications and Remarks
35*	77	1962	~298	~0°		Similar to curve 31 specimen and conditions except heat treated to 589 K in a 24-hr cycle in vacuum (10^{-5} mm Hg); vacuum maintained by diffusion pump.
36*	77	1962	~298	~0°		Similar to curve 31 specimen and conditions except 2, 2'-4, 4' tetrahydroxybenzophenone (D-50) absorbing agent added to the coating formulation at 1% by weight of resin content.
37*	77	1962	~298	~0°		Similar to above specimen and conditions except heat treated to 422 K in a 24-hr cycle in vacuum (10^{-5} mm Hg); vacuum maintained by diffusion pump.
38*	77	1962	~298	~0°		Similar to curve 36 specimen and conditions except heat treated to 478 K in a 24-hr cycle in vacuum (10^{-5} mm Hg); vacuum maintained by diffusion pump.
39*	77	1962	~298	~0°		Similar to curve 36 specimen and conditions except heat treated to 533 K in a 24-hr cycle in vacuum (10^{-5} mm Hg); vacuum maintained by diffusion pump.
40*	77	1962	~298	~0°		Similar to curve 36 specimen and conditions except heat treated to 589 K in a 24-hr cycle in vacuum (10^{-5} mm Hg); vacuum maintained by diffusion pump.
41*	77	1962	~298	~0°		Similar to curve 31 specimen and conditions except dibenzoylresorcinol (DBR) absorbing agent added to the coating formulation at 1% by weight of resin content.
42*	77	1962	~298	~0°		Similar to above specimen and conditions except heat treated to 422 K in a 24-hr cycle in vacuum (10^{-5} mm Hg); vacuum maintained by diffusion pump.
43*	77	1962	~298	~0°		Similar to curve 41 specimen and conditions except heat treated to 478 K in a 24-hr cycle in vacuum (10^{-5} mm Hg); vacuum maintained by diffusion pump.
44*	77	1962	~298	~0°		Similar to curve 41 specimen and conditions except heat treated to 533 K in a 24-hr cycle in vacuum (10^{-5} mm Hg); vacuum maintained by diffusion pump.
45*	77	1962	~298	~0°		Similar to curve 41 specimen and conditions except heat treated to 589 K in a 24-hr cycle in vacuum (10^{-5} mm Hg); vacuum maintained by diffusion pump.
46*	77	1962	~298	~0°		Similar to curve 31 specimen and conditions except 2-hydroxybenzoylferrocene (HBF) absorbing agent added to the coating formulation at 1% by weight of resin content.
47*	77	1962	~298	~0°		Similar to above specimen and conditions except heat treated to 422 K in a 24-hr cycle in vacuum (10^{-5} mm Hg); vacuum maintained by diffusion pump.
48*	77	1962	~298	~0°		Similar to curve 46 specimen and conditions except heat treated to 478 K in a 24-hr cycle in vacuum (10^{-5} mm Hg); vacuum maintained by diffusion pump.
49*	77	1962	~298	~0°		Similar to curve 46 specimen and conditions except heat treated to 533 K in a 24-hr cycle in vacuum (10^{-5} mm Hg); vacuum maintained by diffusion pump.
50*	77	1962	~298	~0°		Similar to curve 46 specimen and conditions except heat treated to 589 K in a 24-hr cycle in vacuum (10^{-5} mm Hg); vacuum maintained by diffusion pump.
51*	77	1962	~298	~0°		Similar to curve 31 specimen and conditions except 2, 4-hydroxybenzoylferrocene (DHBF) absorbing agent added to the coating formulation at 1% by weight of resin content.
52*	77	1962	~298	~0°		Similar to above specimen and conditions except heat treated to 422 K in a 24-hr cycle in vacuum (10^{-5} mm Hg); vacuum maintained by diffusion pump.

* No plot given

SPECIFICATION TABLE NO. 135 NORMAL SOLAR ABSORPTANCE OF ZINC SULFIDE PIGMENTED COATINGS (continued)

Curve No.	Ref. No.	Year	Temperature Range, K	Geometry θ	Reported Error, %	Composition (weight percent), Specifications and Remarks
53*	77	1962	~298	~0°		Similar to curve 51 specimen and conditions except heat treated to 478 K in a 24-hr cycle in vacuum (10^{-5} mm Hg); vacuum maintained by diffusion pump.
54*	77	1962	~298	~0°		Similar to curve 51 specimen and conditions except heat treated to 533 K in a 24-hr cycle in vacuum (10^{-5} mm Hg); vacuum maintained by diffusion pump.
55*	77	1962	~298	~0°		Similar to curve 51 specimen and conditions except heat treated to 589 K in a 24-hr cycle in vacuum (10^{-5} mm Hg); vacuum maintained by diffusion pump.
56*	77	1962	~298	~0°		Similar to curve 31 specimen and conditions.
57*	77	1962	~298	~0°		Above specimen and conditions except exposed to UV radiation from G. E. B-H6 lamp 15. 24 cm from sample; solar factor ~5 suns; exposure time (hrs) is variable.
58*	77	1962	~298	~0°		Similar to curve 36 specimen and conditions.
59*	77	1962	~298	~0°		Above specimen and conditions except exposed to UV radiation from G. E. B-H6 lamp 15. 24 cm from sample; solar factor ~5 suns; exposure time (hrs) is variable.
60*	77	1962	~298	~0°		Similar to curve 41 specimen and conditions.
61*	77	1962	~298	~0°		Above specimen and conditions except exposed to UV radiation from G. E. B-H6 lamp 15. 24 cm from sample; solar factor ~5 suns; exposure time (hrs) is variable.
62*	77	1962	~298	~0°		Similar to curve 46 specimen and conditions.
63*	77	1962	~298	~0°		Above specimen and conditions except exposed to UV radiation from G. E. B-H6 lamp 15. 24 cm from sample; solar factor ~5 suns; exposure time (hrs) is variable.
64*	77	1962	~298	~0°		Similar to curve 51 specimen and conditions.
65*	77	1962	~298	~0°		Above specimen and conditions except exposed to UV radiation from G. E. B-H6 lamp 15. 24 cm from sample; solar factor ~5 suns; exposure time (hrs) is variable.
66*	27	1964	~298	~0°		ZnS in PS-7 potassium silicate binder on aluminum substrate; PBR 3. 19; solids content 59. 0%; aluminum substrate abraded with No. 60 Aloxite cloth; sample cured at 413 K for 18 hrs. [Authors' designation: sample C18]
67*	27	1964	~298	~0°		Above specimen and conditions except irradiated in vacuum (10^{-6} mm Hg) with 260 ESH of UV radiation (AH-6 lamp) at solar factor of 4 suns.

* No plot given

DATA TABLE NO. 135 NORMAL SOLAR ABSORPTANCE OF ZINC SULFIDE PIGMENTED COATINGS

[Temperature, T, K; Absorptance, α]

T	α
CURVE 1*	
300	0.21
CURVE 2*	
300	0.26
CURVE 3*	
300	0.29
CURVE 4*	
300	0.34
CURVE 5*	
298	0.209
CURVE 6*	
298	0.359
CURVE 7*	
298	0.209
CURVE 8*	
298	0.443
CURVE 9*	
298	0.346
CURVE 10*	
298	0.458
CURVE 11*	
298	0.236
CURVE 12*	
298	0.431
CURVE 13*	
298	0.257
CURVE 14*	
298	0.426
CURVE 15*	
298	0.283
CURVE 16*	
298	0.374
CURVE 17*	
298	0.241
CURVE 18*	
298	0.374
CURVE 19*	
298	0.222
CURVE 20*	
298	0.436
CURVE 21*	
298	0.178
CURVE 22*	
298	0.338
CURVE 23*	
298	0.255
CURVE 24*	
298	0.355
CURVE 25*	
298	0.252
CURVE 26*	
298	0.362
CURVE 27*	
298	0.237
CURVE 28*	
298	0.297
CURVE 29*	
298	0.253
CURVE 30*	
298	0.383
CURVE 31*	
298	0.338
CURVE 32*	
298	0.408
CURVE 33*	
298	0.457
CURVE 34*	
298	0.486
CURVE 35*	
298	0.553
CURVE 36*	
298	0.368
CURVE 37*	
298	0.453
CURVE 38*	
298	0.502
CURVE 39*	
298	0.516
CURVE 40*	
298	0.555
CURVE 41*	
298	0.362
CURVE 42*	
298	0.429
CURVE 43*	
298	0.474
CURVE 44*	
298	0.503
CURVE 45*	
298	0.547
CURVE 46*	
298	0.366
CURVE 47*	
298	0.414
CURVE 48*	
298	0.446
CURVE 49*	
298	0.482
CURVE 50*	
298	0.553
CURVE 51*	
298	0.393
CURVE 52*	
298	0.444
CURVE 53*	
298	0.492
CURVE 54*	
298	0.523
CURVE 55*	
298	0.584
CURVE 56*	
298	0.338

Exp. Time	α
CURVE 57* T ~ 298	
6	0.352
24	0.370
54	0.386
104	0.410

T	α
CURVE 58*	
298	0.368

Exp. Time	α
CURVE 59* T ~ 298	
6	0.379
24	0.383
54	0.395
104	0.410

T	α
CURVE 60*	
298	0.362

Exp. Time	α
CURVE 61* T ~ 298	
6	0.374
24	0.376
54	0.392
104	0.407

T	α
CURVE 62*	
298	0.366

Exp. Time	α
CURVE 63* T ~ 298	
6	0.370
24	0.376
54	0.382
104	0.391

T	α
CURVE 64*	
298	0.393

Exp. Time	α
CURVE 65* T ~ 298	
6	0.401
24	0.407
54	0.415
104	0.428

T	α
CURVE 66*	
298	0.220
CURVE 67*	
298	0.231

*No plot given

SPECIFICATION TABLE NO. 136 NORMAL SOLAR ABSORPTANCE OF ZINC TITANATE PIGMENTED COATINGS

Curve No.	Ref. No.	Year	Temperature Range, K	Geometry θ	Reported Error, %	Composition (weight percent), Specifications and Remarks
1*	86	1964	~298	~0°		Zinc titanate (A-54-2) in potassium silicate binder; substrate unknown; pigment calcined at 973 K for 16 hrs; PBR 4.30; cured by air drying; calculated from spectral reflectance. [Authors' designation: Sample No. 7008]
2*	86	1964	~298	~0°		Above specimen and conditions except exposed to 690 ESH of UV radiation in vacuum.
3*	86	1964	~298	~0°		Similar to curve 1 specimen and conditions except heat treated for 2 hrs at 773 K. [Authors' designation: Sample No. 7009]
4*	86	1964	~298	~0°		Above specimen and conditions except exposed to 690 ESH of UV radiation in vacuum.
5*	86	1964	~298	~0°		Zinc titanate (602-26-1M) in potassium silicate binder; substrate unknown; pigment calcined at 973 K for 4 hrs; air dried; calculated from spectral reflectance. [Authors' designation: Sample No. H-19-53]
6*	86	1964	~298	~0°		Above specimen and conditions except heat treated at 773 K for 2 hrs.
7*	86	1964	~298	~0°		Above specimen and conditions except exposed in vacuum (< 10^{-5} mm Hg) to 170 ESH of UV radiation at 3.5 suns.

DATA TABLE NO. 136 NORMAL SOLAR ABSORPTANCE OF ZINC TITANATE PIGMENTED COATINGS

[Temperature, T, K; Absorptance, α]

T	α	T	α
CURVE 1*		CURVE 5*	
298	0.122	298	0.139
CURVE 2*		CURVE 6*	
298	0.141	298	0.128
CURVE 3*		CURVE 7*	
298	0.118	298	0.159
CURVE 4*			
298	0.150		

* No plot given.

SPECIFICATION TABLE NO. 137 HEMISPHERICAL TOTAL EMITTANCE OF ZIRCONIUM CARBIDE PIGMENTED COATINGS

Curve No.	Ref. No.	Year	Temperature Range, K	Reported Error, %	Composition (weight percent), Specifications and Remarks
1*	97	1961	873-1673	± 2.5	88.5 zirconium metal powder and 11.5 finely powdered graphite suspended in necoloidine binder molybdenum substrate; sprayed to a density of 7.5 mg cm^{-2}; sintered at 1773 K; measured in vacuum (<5 x 10^{-6} mm Hg); data extracted from smooth curve.

DATA TABLE NO. 137 HEMISPHERICAL TOTAL EMITTANCE OF ZIRCONIUM CARBIDE PIGMENTED COATINGS

[Temperature, T, K; Emittance, ϵ]

T	ϵ
CURVE 1*	
873	0.700
973	0.755
1073	0.800
1173	0.825
1273	0.840
1373	0.850
1473	0.860
1573	0.865
1673	0.865

* No plot given

SPECIFICATION TABLE NO. 138 HEMISPHERICAL TOTAL EMITTANCE OF ZIRCONIUM HYDRIDE PIGMENTED COATINGS

Curve No.	Ref. No.	Year	Temperature Range, K	Reported Error, %	Composition (weight percent), Specifications and Remarks
1*	97	1961	873-1473	± 2.5	75 zirconium hydride and 25 zirconia suspended in necoloidine binder on molybdenum substrate; sprayed to a density of 7.5 mg cm^{-2}; measured in vacuum (<5 x 10^{-6} mm Hg); data extracted from smooth curve.
2*	97	1961	873-1273	± 2.5	85 zirconium hydride and 15 ferric oxide suspended in necoloidine binder on iron substrate; sprayed to a density of 7.5 mg cm^{-2}; sintered at 1073 K; measured in vacuum (<5 x 10^{-6} mm Hg); data extracted from smooth curve.

DATA TABLE NO. 138 HEMISPHERICAL TOTAL EMITTANCE OF ZIRCONIUM HYDRIDE PIGMENTED COATINGS

[Temperature, T, K; Emittance, ϵ]

T	ϵ
CURVE 1*	
873	0.660
973	0.695
1073	0.720
1173	0.745
1273	0.755
1373	0.760
1473	0.765
CURVE 2*	
873	0.630
973	0.655
1073	0.675
1173	0.695
1273	0.705

* No plot given

SPECIFICATION TABLE NO. 139 HEMISPHERICAL TOTAL EMITTANCE OF ZIRCONIUM OXIDE PIGMENTED COATINGS

Curve No.	Ref. No.	Year	Temperature Range, K	Reported Error, %	Composition (weight percent), Specifications and Remarks
1*	43	1962	348-391	± 3	Zirconium oxide (~0.0763 mm thick) in nitrocellulose medium paint binder; measured in vacuum (10^{-3} mm Hg); data extracted from smooth curve.

DATA TABLE NO. 139 HEMISPHERICAL TOTAL EMITTANCE OF ZIRCONIUM OXIDE PIGMENTED COATINGS

[Temperature, T, K; Emittance, ϵ]

T	ϵ
CURVE 1*	
348	0.870
355	0.880
366	0.875
373	0.900
382	0.885
391	0.770

* No plot given

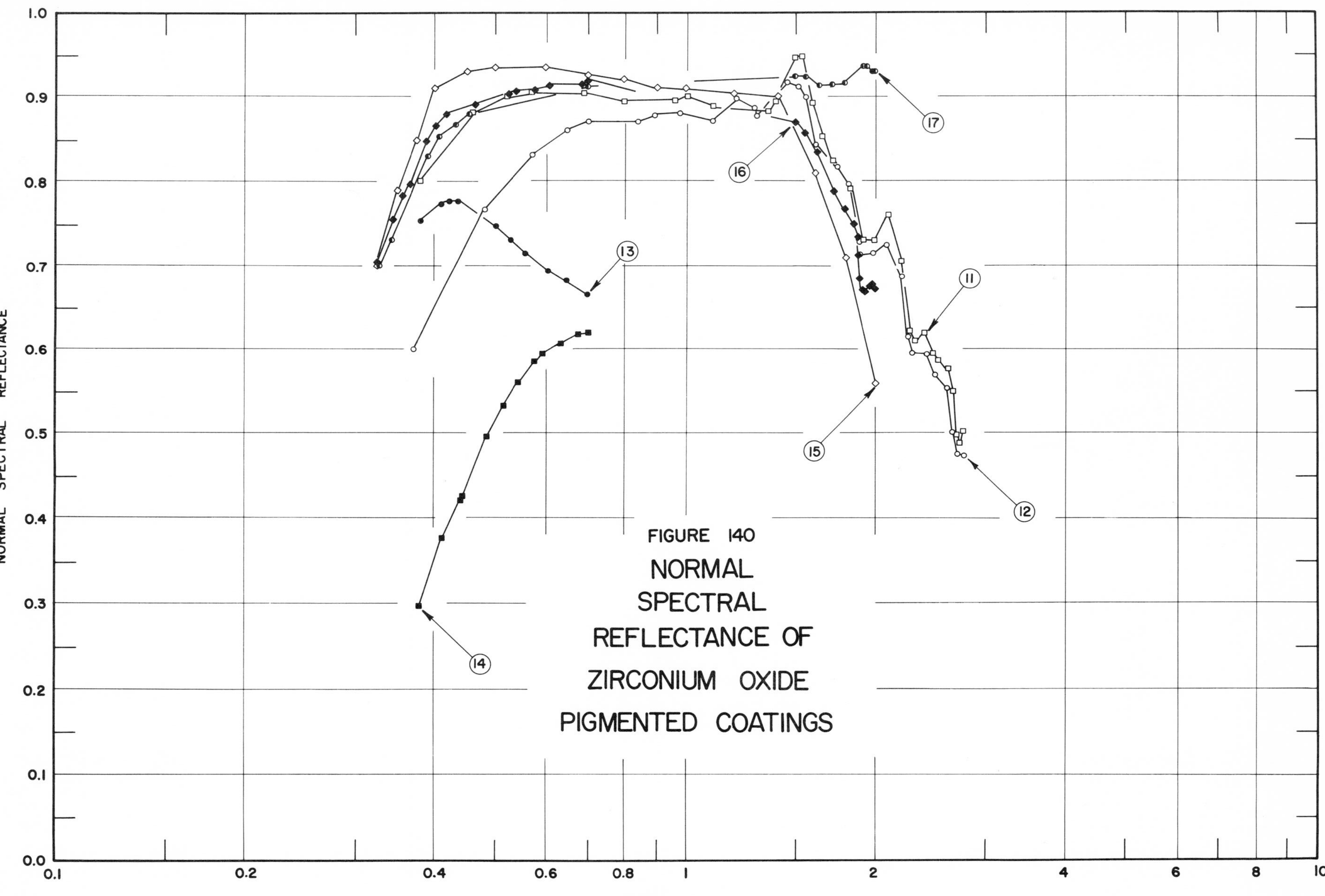

FIGURE 140 NORMAL SPECTRAL REFLECTANCE OF ZIRCONIUM OXIDE PIGMENTED COATINGS

SPECIFICATION TABLE NO. 140 NORMAL SPECTRAL REFLECTANCE OF ZIRCONIUM OXIDE PIGMENTED COATINGS

Curve No.	Ref. No.	Year	Temperature, K	Wavelength Range, μm	Geometry θ	θ'	ω'	Reported Error, %	Composition (weight percent), Specifications, and Remarks
1*	27	1964	~298	0.44-0.60	~0°		2π		ZrO_2 in PS-7 potassium silicate binder on aluminum substrate; PBR 4.30; solids content 73.0%; substrate abraded with No. 60 Aloxite cloth; heat treated at 413 K for 18 hrs. [Authors' designation: sample C45]
2*	27	1964	~298	0.44-0.60	~0°		2π		Above specimen and conditions except irradiated in vacuum (10^{-6} mm Hg) with 75 ESH of UV radiation (AH-6 lamp) at solar factor of 1.5 suns.
3*	27	1964	~298	0.44-0.60	~0°		2π		ZrO_2 in aluminum acid phosphate binder on aluminum substrate; PBR 2.80; solids content 72.0%; aluminum substrate abraded with No. 60 Aloxite cloth; sample cured at 413 K for 18 hrs. [Authors' designation: sample C46]
4*	27	1964	~298	0.44-0.60	~0°		2π		Above specimen and conditions except irradiated in vacuum (10^{-6} mm Hg) with 75 ESH of UV radiation (AH-6 lamp) at solar factor of 1.5 suns.
5*	27	1964	~298	0.44-0.60	~0°		2π		ZrO_2 in colloidal silica binder on aluminum substrate; PBR 5.33; solids content 74.8%; substrate abraded with No. 60 Aloxite cloth; sample cured at 413 K for 18 hrs. [Authors' designation: sample C47]
6*	27	1964	~298	0.44-0.60	~0°		2π		Above specimen and conditions except irradiated in vacuum (10^{-6} mm Hg) with 75 ESH of UV radiation (AH-6 lamp) at solar factor of 1.5 suns.
7*	27	1964	~298	0.44-0.60	~0°		2π		ZrO_2 in PS-7 potassium silicate binder on aluminum substrate; PBR 4.30; solids content 64.4%; substrate abraded with No. 60 Aloxite cloth; sample cured at 413 K for 18 hrs. [Authors' designation: sample C52]
8*	27	1964	~298	0.44-0.60	~0°		2π		Above specimen and conditions except irradiated in vacuum (10^{-6} mm Hg) with 200 ESH of UV radiation (AH-6 lamp) at solar factor of 3 suns.
9*	27	1966	~298	0.44-0.60	~0°		2π		ZrO_2 in DuPont Teflon-30 polytetrafluoroethylene resin binder on aluminum substrate; PBR 0.66; paint ground in porcelain jar mill for 16 hrs. [Authors' designation: sample P5]
10*	27	1964	~298	0.44-0.60	~0°		2π		Above specimen and conditions except irradiated in vacuum (10^{-6} mm Hg) with 74 ESH of UV radiation (AH-6 lamp) at solar factor of 4 suns.
11	29	1962	298	0.38-2.75	0°		2π		ZrO_2 in potassium silicate binder on 6061-T6 aluminum substrate; substrate grit blasted with 40 mesh SiC; sprayed with air-brush; air dried overnight, then heat cured at 395-415 K for 24 hrs; data extracted from smooth curve. [Author's designation: 2-11-3]
12	29	1962	298	0.37-2.76	0°		2π		Above specimen and conditions except exposed in vacuum (~10^{-6} mm Hg) to UV radiation (from three AH-6 lamps) at 4 suns for 67 hrs.
13	29	1962	298	0.380-0.700	0°		2π		ZrO_2 in TFE 30 binder on 6061-T6 aluminum substrate; PBR 0.66; substrate grit blasted with 40 mesh SiC; sprayed; dried at 298 K. [Author's designation: Sample No. 30]
14	29	1962	298	0.380-0.700	0°		2π		Above specimen and conditions except exposed in vacuum (~10^{-6} mm Hg) to UV radiation (from three AH-6 lamps) at 4 suns for 18.5 hrs.
15	34	1964	298	0.325-2.00	~0°		2π		Zirconium oxide in potassium silicate binder; PBR 4.30; measured relative to MgO.

*Not shown on plot

SPECIFICATION TABLE NO. 140 NORMAL SPECTRAL REFLECTANCE OF ZIRCONIUM OXIDE PIGMENTED COATINGS (continued)

Curve No.	Ref. No.	Year	Temperature, K	Wavelength Range, μm	Geometry θ	Geometry θ'	Geometry ω'	Reported Error, %	Composition (weight percent), Specifications, and Remarks
16	35	1964	298	0.325-2.000	$\sim 0^\circ$		2π		Zirconium oxide in potassium silicate binder (0.127 mm thick) on pyrex substrate; air dried; data extracted from smooth curve; λ for 0.325-0.7 and 1.5-2.0 measured on different systems.
17	35	1964	298	0.325-2.000	$\sim 0^\circ$		2π		Above specimen and conditions except heat treated at 773 K for 2 hrs.

DATA TABLE NO. 140 NORMAL SPECTRAL REFLECTANCE OF ZIRCONIUM OXIDE PIGMENTED COATINGS

[Wavelength, λ, μm; Reflectance, ρ; Temperature, T, K]

λ	ρ
CURVE 1*	
T ~ 298	
0.44	0.890
0.60	0.920
CURVE 2*	
T ~ 298	
0.44	0.835
0.60	0.905
CURVE 3*	
T ~ 298	
0.44	0.760
0.60	0.860
CURVE 4*	
T ~ 298	
0.44	0.715
0.60	0.850
CURVE 5*	
T ~ 298	
0.44	0.880
0.60	0.955
CURVE 6*	
T ~ 298	
0.44	0.615
0.60	0.880
CURVE 7*	
T ~ 298	
0.44	0.905
0.60	0.930
CURVE 8*	
T ~ 298	
0.44	0.735
0.60	0.870
CURVE 9*	
T ~ 298	
0.44	0.878
0.60	0.700
CURVE 10*	
T ~ 298	
0.44	0.420
0.60	0.596
CURVE 11	
T = 298	
0.38	0.800
0.46	0.881
0.52	0.901
0.57	0.906
0.69	0.905
0.80	0.895
0.96	0.896
1.01	0.900
1.11	0.890
1.35	0.884
1.39	0.895
1.48	0.948
1.52	0.949
1.59	0.894
1.65	0.854
1.72	0.826
1.83	0.792
1.92	0.733
1.99	0.730
2.10	0.761
2.19	0.708
2.26	0.624
2.31	0.610
2.39	0.621
2.47	0.597
2.51	0.587
2.61	0.578
2.65	0.550
2.68	0.498
2.71	0.489
2.75	0.501
CURVE 12	
T = 298	
0.37	0.600
0.48	0.767
0.57	0.832
0.65	0.862
0.70	0.871
0.84	0.871
0.89	0.879
0.98	0.882
1.10	0.871
1.21	0.898
1.28	0.887
1.30	0.878
1.45	0.917
1.52	0.913
1.55	0.900
1.61	0.843
1.75	0.818
1.81	0.797
1.88	0.728
1.90	0.714
1.99	0.716
2.08	0.724
2.20	0.687
2.26	0.615
2.29	0.596
2.42	0.595
2.49	0.570
2.59	0.554
2.65	0.500
2.70	0.476
2.76	0.473
CURVE 13	
T = 298	
0.380	0.755
0.412	0.773
0.423	0.777
0.438	0.777
0.501	0.748
0.527	0.731
0.558	0.716
0.607	0.695
0.647	0.681
0.700	0.666
CURVE 14	
T = 298	
0.380	0.297
0.413	0.375
0.440	0.421
0.446	0.429
0.485	0.495
0.515	0.533
0.543	0.560
0.577	0.585
0.595	0.595
0.636	0.608
0.679	0.618
0.700	0.620
CURVE 15	
T = 298	
0.325	0.700
0.350	0.790
0.375	0.850
0.400	0.910
0.450	0.930
0.500	0.935
0.600	0.935
0.700	0.925
0.800	0.920
0.900	0.910
1.00	0.910
1.20	0.905
1.40	0.900
1.60	0.810
1.80	0.710
2.00	0.560
CURVE 16	
T = 298	
0.325	0.707
0.343	0.755
0.356	0.784
0.365	0.798
0.389	0.849
0.402	0.867
0.421	0.880
0.466	0.892
CURVE 16 (cont.)	
0.523	0.904
0.537	0.908
0.578	0.909
0.608	0.914
0.685	0.915
0.700	0.918
1.497	0.871
1.550	0.859
1.628	0.835
1.734	0.787
1.793	0.768
1.853	0.750
1.876	0.736
1.888	0.712
1.897	0.685
1.909	0.671
1.929	0.669
1.957	0.675
1.979	0.677
2.000	0.672
CURVE 17	
T = 298	
0.329	0.700
0.341	0.733
0.392	0.830
0.408	0.854
0.431	0.869
0.455	0.880
0.529	0.903*
0.700	0.913
1.497	0.926
1.551	0.924
1.638	0.914
1.712	0.916
1.782	0.918
1.921	0.939
1.947	0.938
1.984	0.931
2.000	0.931

* Not shown on plot.

SPECIFICATION TABLE NO. 141 NORMAL SOLAR ABSORPTANCE OF ZIRCONIUM OXIDE PIGMENTED COATINGS

Curve No.	Ref. No.	Year	Temperature Range, K	Geometry θ	Reported Error, %	Composition (weight percent), Specifications and Remarks
1*	86	1964	~298	~0°		ZrO_2 in potassium silicate binder; PBR 4.30; cured by air drying; calculated from spectral reflectance. [Authors' designation: Sample No. 7010]
2*	86	1964	~298	~0°		Above specimen and conditions except exposed to 690 ESH of UV radiation in vacuum ($< 10^{-6}$ mm Hg); solar factor of 9.4 suns.
3*	27	1964	~298	~0°		ZrO_2 in PS-7 potassium silicate binder on aluminum substrate; PBR 4.30; solids content 64.4%; substrate abraded with No. 60 Aloxite cloth; sample cured at 413 K for 18 hrs. [Authors' designation: Sample C52]
4*	27	1964	~298	~0°		Above specimen and conditions except irradiated in vacuum (10^{-6} mm Hg) with 200 ESH of UV radiation at solar factor of 3 suns.
5*	51	1965	298	~0°		Zirconium oxide in potassium silicate binder. [Authors' designation: Sample 7070]
6*	51	1965	298	~0°		Above specimen and conditions except exposed in vacuum (10^{-6} mm Hg) at 279 K to 2160 ESH of UV radiation (AH-6 lamp) at solar factor of 10.0 suns; vacuum maintained by VacIon pump.
7*	51	1965	298	~0°		Zirconium oxide in potassium silicate binder. [Authors' designation: Sample 7090]
8*	51	1965	298	~0°		Above specimen and conditions except exposed in vacuum (3×10^{-6} mm Hg) at 278 K to 1020 ESH of UV radiation (AH-6 lamp) at solar factor of 3.9 suns; vacuum maintained by Ultek ion pump.
9*	51	1965	298	~0°		Similar to curve 7 specimen and conditions. [Authors' designation: Sample 7091]
10*	51	1965	298	~0°		Above specimen and conditions except exposed in vacuum (3×10^{-6} mm Hg) at 278 K to 1240 ESH of UV radiation (AH-6 lamp) at solar factor of 14.4 suns; vacuum maintained by Ultek ion pump.
11*	51	1965	298	~0°		Zirconium oxide in potassium silicate binder; PBR 6.45; preparation aged 4 mo.; sprayed. [Authors' designation: Sample 7139]
12*	51	1965	298	~0°		Above specimen and conditions except exposed in vacuum ($<10^{-6}$ mm Hg) at 279 K to 300 ESH of UV radiation (AH-6 lamp) at solar factor of 4.8 suns; vacuum maintained by Ultek ion pump.
13*	51	1965	298	~0°		Similar to curve 11 specimen and conditions. [Authors' designation: Sample 7144]
14*	51	1965	298	~0°		Above specimen and conditions except exposed in vacuum (5×10^{-6} mm Hg) at 279 K to 4500 ESH of UV radiation (AH-6 lamp) at solar factor of 6.1 suns; vacuum maintained by Ultek ion pump.
15*	51	1965	298	~0°		Similar to curve 11 specimen and conditions. [Authors' designation: Sample 7134]
16*	51	1965	298	~0°		Above specimen and conditions except exposed in vacuum (3.5×10^{-6} mm Hg) at 279 K to 7900 ESH of UV radiation (AH-6 lamp) at solar factor of 5.8 suns; vacuum maintained by Ultek ion pump.

* No plot given

DATA TABLE NO. 141 NORMAL SOLAR ABSORPTANCE OF ZIRCONIUM OXIDE PIGMENTED COATINGS

[Temperature, T, K; Absorptance, α]

T	α	T	α
CURVE 1*		CURVE 13*	
298	0.191	298	0.130
CURVE 2*		CURVE 14*	
298	0.222	298	0.247
CURVE 3*		CURVE 15*	
298	0.140	298	0.099
CURVE 4*		CURVE 16*	
298	0.205	298	0.333
CURVE 5*			
298	0.200		
CURVE 6*			
298	0.262		
CURVE 7*			
298	0.157		
CURVE 8*			
298	0.213		
CURVE 9*			
298	0.158		
CURVE 10*			
298	0.291		
CURVE 11*			
298	0.104		
CURVE 12*			
298	0.136		

* No plot given

SPECIFICATION TABLE NO. 142 HEMISPHERICAL TOTAL EMITTANCE OF ZIRCONIUM SILICATE PIGMENTED COATINGS

Curve No.	Ref. No.	Year	Temperature Range, K	Reported Error, %	Composition (weight percent), Specifications and Remarks
1*	49	1961	278		Zirconium silicate in sodium silicate K binder on Dow 17 treated Mg alloy substrate.
2*	49	1961	278		Above specimen and conditions except exposed in vacuum (10^{-6}-8 x 10^{-6} mm Hg) to UV radiation from an argon-filled A-H6 high pressure Hg arc lamp for 127 hrs.
3*	49	1961	278		Zirconium silicate in sodium silicate D binder on Dow 17 treated Mg alloy substrate; exposed in vacuum (10^{-6}-8 x 10^{-6} mm Hg) to UV radiation from an argon-filled A-H6 high pressure Hg arc lamp for 127 hrs.
4*	49	1961	278		Zirconium silicate in silicone resin binder on Dow 17 treated Mg alloy substrate; exposed in vacuum (10^{-6}-8 x 10^{-6} mm Hg) to UV radiation from an argon-filled A-H6 high pressure Hg arc lamp for 127 hrs.

DATA TABLE NO. 142 HEMISPHERICAL TOTAL EMITTANCE OF ZIRCONIUM SILICATE PIGMENTED COATINGS

[Temperature, T, K; Emittance, ϵ]

T	ϵ
CURVE 1*	
278	0.84
CURVE 2*	
278	0.83
CURVE 3*	
278	0.83
CURVE 4*	
278	0.85

* Not plot given

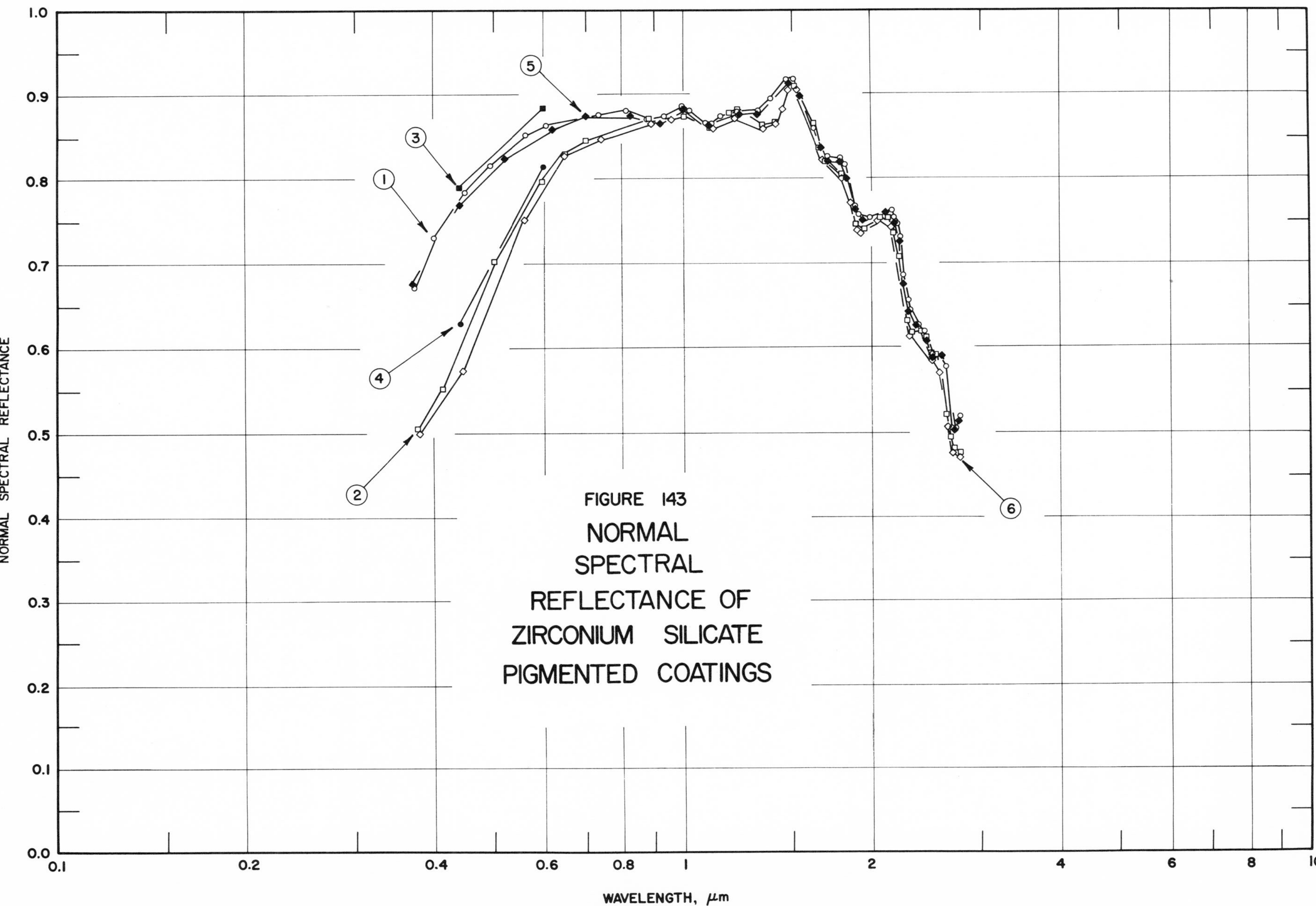

FIGURE 143 NORMAL SPECTRAL REFLECTANCE OF ZIRCONIUM SILICATE PIGMENTED COATINGS

SPECIFICATION TABLE NO. 143   NORMAL SPECTRAL REFLECTANCE OF ZIRCONIUM SILICATE PIGMENTED COATINGS

Curve No.	Ref. No.	Year	Temperature, K	Wavelength Range, μm	Geometry θ	Geometry θ'	Geometry ω'	Reported Error, %	Composition (weight percent), Specifications, and Remarks
1	20	1963	~298	0.372-2.760	~0°		2π		Superpax zirconium silicate in PS-7 potassium silicate binder on aluminum substrate; PBR 4.30; solids content 56.9%; substrate grit blasted with 40 mesh silicon carbide; paint ball milled; air brush application; sample cured at 413 K for 18 hrs; data extracted from smooth curve. [Authors' designation: Sample C57]
2	20	1963	~298	0.378-2.768	~0°		2π		Above specimen and conditions except exposed in vacuum (10^{-5} mm Hg) to 200 ESH of UV radiation at solar factor of 3 suns.
3	27	1964	~298	0.44-0.60	~0°		2π		Superpax $ZrSiO_4$ in PS-7 potassium silicate binder on aluminum substrate; PBR 4.30; solids content 56.9%; substrate abraded with No. 60 Aloxite cloth; sample cured at 413 K for 18 hrs. [Authors' designation: Sample C57]
4	27	1964	~298	0.44-0.60	~0°		2π		Above specimen and conditions except irradiated in vacuum (10^{-6} mm Hg) with 200 ESH of UV radiation (AH-6 lamp) at solar factor of 3 suns.
5	29	1962	298	0.37-2.75	0°		2π		$ZrSiO_4$ in potassium silicate binder on 6061-T6 aluminum substrate; substrate grit blasted with 40 mesh SiC; sprayed with air-brush; air dried overnight, then heat-cured at 395-415 K for 24 hrs; data extracted from smooth curve. [Author's designation: Sample 2-13-3]
6	29	1962	298	0.38-2.76	0°		2π		Above specimen and conditions except exposed in vacuum (~10^{-6} mm Hg) to UV radiation (from three AH-6 lamps) at 4 suns for 67 hrs.

DATA TABLE NO. 143 NORMAL SPECTRAL REFLECTANCE OF ZIRCONIUM SILICATE PIGMENTED COATINGS

[Wavelength, λ, μm; Reflectance, ρ; Temperature, T, K]

CURVE 1 $T \sim 298$

λ	ρ
0.372	0.673
0.403	0.732
0.451	0.785
0.495	0.818
0.563	0.854
0.608	0.866
0.737	0.878
0.815	0.882
0.881	0.872
0.936	0.876
0.996	0.887
1.032	0.882
1.081	0.869
1.129	0.868
1.151	0.875
1.230	0.880
1.332	0.881
1.394	0.895
1.467	0.919
1.507	0.919
1.633	0.860
1.712	0.828
1.786	0.825
1.822	0.819
1.897	0.767
1.925	0.759
1.994	0.754
2.151	0.763
2.184	0.754
2.200	0.747
2.212	0.732
2.245	0.686
2.280	0.657
2.304	0.646
2.372	0.629
2.435	0.620
2.516	0.591
2.599	0.591
2.628	0.579
2.697	0.507
2.708	0.506
2.728	0.506
2.760	0.519

CURVE 2 $T \sim 298$

λ	ρ
0.378	0.507
0.412	0.553
0.500	0.702
0.598	0.799
0.650	0.831
0.704	0.847
0.886	0.872
1.005	0.875
1.107	0.861
1.186	0.878
1.223	0.881
1.342	0.865
1.402	0.868
1.498	0.917
1.529	0.911
1.620	0.865
1.655	0.836
1.692	0.820
1.800	0.805
1.899	0.746
1.951	0.740
2.071	0.753
2.117	0.752
2.167	0.736
2.211	0.708
2.283	0.632
2.319	0.617
2.416	0.620
2.460	0.613
2.506	0.593
2.531	0.591
2.649	0.521
2.690	0.495
2.716	0.482
2.768	0.478

CURVE 3 $T \sim 298$

λ	ρ
0.44	0.790
0.60	0.885

CURVE 4 $T \sim 298$

λ	ρ
0.44	0.630
0.60	0.815

CURVE 5 $T = 298$

λ	ρ
0.37	0.673
0.44	0.770
0.52	0.825
0.62	0.860
0.70	0.873
0.83	0.876
0.92	0.868
1.00	0.881
1.11	0.863
1.24	0.879
1.33	0.877
1.49	0.916
1.55	0.899
1.67	0.839
1.72	0.823
1.79	0.821
1.84	0.800
1.90	0.764
1.95	0.751
2.14	0.760
2.18	0.748
2.21	0.725
2.25	0.675
2.30	0.643
2.36	0.629
2.45	0.609
2.51	0.589
2.60	0.590
2.71	0.503
2.75	0.514

CURVE 6 $T = 298$

λ	ρ
0.38	0.500
0.44	0.575
0.56	0.752

CURVE 6 (cont.)

λ	ρ
0.65	0.829
0.74	0.849
0.89	0.867
0.96	0.871
1.13	0.862
1.22	0.873
1.35	0.860
1.41	0.869
1.45	0.884
1.48	0.905
1.53	0.906
1.68	0.822
1.81	0.800
1.86	0.772
1.90	0.738
1.93	0.734
2.05	0.750
2.15	0.744
2.30	0.614
2.39	0.620
2.46	0.604
2.51	0.584
2.57	0.572
2.66	0.509
2.70	0.476
2.76	0.471

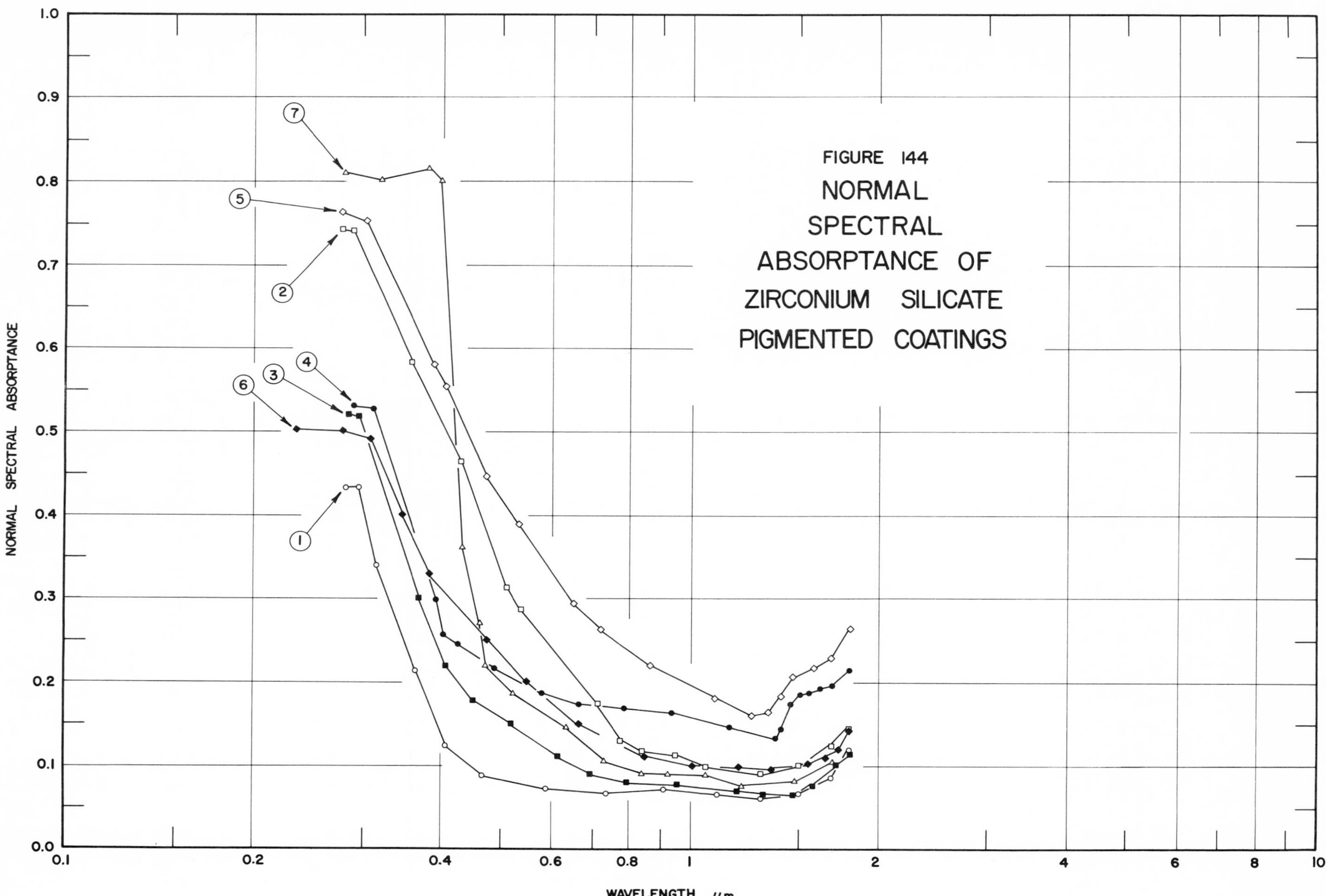
FIGURE 144
NORMAL
SPECTRAL
ABSORPTANCE OF
ZIRCONIUM SILICATE
PIGMENTED COATINGS
NORMAL SPECTRAL ABSORPTANCE
WAVELENGTH, μm
0.0
0.1
0.2
0.3
0.4
0.5
0.6
0.7
0.8
0.9
1.0
0.1
0.2
0.4
0.6
0.8
1
2
4
6
8
10
1
2
3
4
5
6
7

SPECIFICATION TABLE NO. 144 NORMAL SPECTRAL ABSORPTANCE OF ZIRCONIUM SILICATE PIGMENTED COATINGS

Curve No.	Ref. No.	Year	Temperature, K	Wavelength Range, μm	Geometry θ	Reported Error, %	Composition (weight percent), Specifications, and Remarks
1	24	1965	298	0.281-1.799	$\sim 0^\circ$		Ultrox, $ZrSiO_4$, in potassium silicate binder; data extracted from smooth curve; calculated from spectral reflectance.
2	24	1965	298	0.277-1.799	$\sim 0^\circ$		Similar to the above specimen and conditions except exposed in vacuum to ~200 sun hrs of UV radiation.
3	24	1965	298	0.285-1.800	$\sim 0^\circ$		Similar to curve 1 specimen and conditions except exposed in vacuum to a nuclear-radiation dose of approx 10^8R of gamma and 5 x 10^{14} neutrons cm^{-2} ($E \geq 2.9$ MeV).
4	24	1965	298	0.290-1.800	$\sim 0^\circ$		Ultrox, $ZrSiO_4$, in aluminum phosphate binder; data extracted from smooth curve.
5	24	1965	298	0.275-1.800	$\sim 0^\circ$		Similar to above specimen and conditions except exposed in vacuum to ~200 sun hrs of UV radiation.
6	24	1965	298	0.235-1.800	$\sim 0^\circ$		Similar to curve 4 specimen and conditions except exposed in vacuum to a nuclear-radiation dose of approx 10^8R of gamma and 5 x 10^{14} neutrons cm^{-2} ($E \geq 2.9$ MeV).
7	24	1965	298	0.276-1.800	$\sim 0^\circ$		Similar to curve 4 specimen and conditions except exposed in vacuum at 77 K to a nuclear-radiation dose of approx 10^8R of gamma and 5 x 10^{14} neutrons cm^{-2} ($E \geq 2.9$ MeV).

DATA TABLE NO. 144 NORMAL SPECTRAL ABSORPTANCE OF ZIRCONIUM SILICATE PIGMENTED COATINGS

[Wavelength, λ, μm; Absorptance, α; Temperature, T, K]

λ	α	λ	α	λ	α	λ	α	λ	α	λ	α	λ	α
CURVE 1 T = 298		CURVE 2 T = 298		CURVE 3 T = 298		CURVE 4 T = 298		CURVE 5 T = 298		CURVE 6 T = 298		CURVE 7 T = 298	
0.281	0.434	0.277	0.744	0.285	0.522	0.290	0.532	0.275	0.764	0.235	0.503	0.276	0.812
0.295	0.434	0.289	0.742	0.295	0.521	0.313	0.528	0.301	0.756	0.277	0.503	0.318	0.805
0.316	0.340	0.358	0.584	0.367	0.300	0.394	0.298	0.388	0.582	0.303	0.493	0.380	0.819
0.362	0.214	0.429	0.466	0.406	0.220	0.404	0.258	0.405	0.555	0.345	0.401	0.397	0.801
0.408	0.125	0.507	0.314	0.450	0.179	0.425	0.246	0.471	0.447	0.383	0.330	0.433	0.361
0.465	0.090	0.535	0.288	0.518	0.151	0.489	0.219	0.532	0.390	0.474	0.251	0.462	0.272
0.587	0.074	0.712	0.175	0.613	0.113	0.578	0.189	0.650	0.295	0.547	0.202	0.471	0.222
0.732	0.069	0.775	0.132	0.692	0.091	0.664	0.174	0.719	0.264	0.668	0.151	0.521	0.189
0.901	0.073	0.840	0.119	0.791	0.083	0.783	0.171	0.864	0.222	0.845	0.115	0.636	0.148
1.101	0.068	0.946	0.115	0.951	0.080	0.930	0.165	1.081	0.181	1.011	0.102	0.730	0.107
1.299	0.063	1.054	0.100	1.187	0.072	1.156	0.147	1.261	0.160	1.193	0.100	0.835	0.091
1.489	0.067	1.298	0.094	1.310	0.068	1.364	0.136	1.334	0.165	1.353	0.097	0.919	0.092
1.668	0.087	1.480	0.102	1.456	0.069	1.395	0.145	1.388	0.184	1.544	0.106	1.057	0.090
1.799	0.121	1.672	0.124	1.565	0.078	1.448	0.175	1.465	0.209	1.645	0.110	1.221	0.079
		1.799	0.146	1.707	0.101	1.500	0.185	1.577	0.218	1.733	0.122	1.467	0.084
				1.800	0.115	1.550	0.186	1.686	0.230	1.800	0.144	1.692	0.105
						1.616	0.193	1.800	0.264			1.800	0.142
						1.673	0.196						
						1.800	0.215						

FIGURE 145A

ANALYZED NORMAL SOLAR ABSORPTANCE OF ZIRCONIUM SILICATE PIGMENTED COATINGS

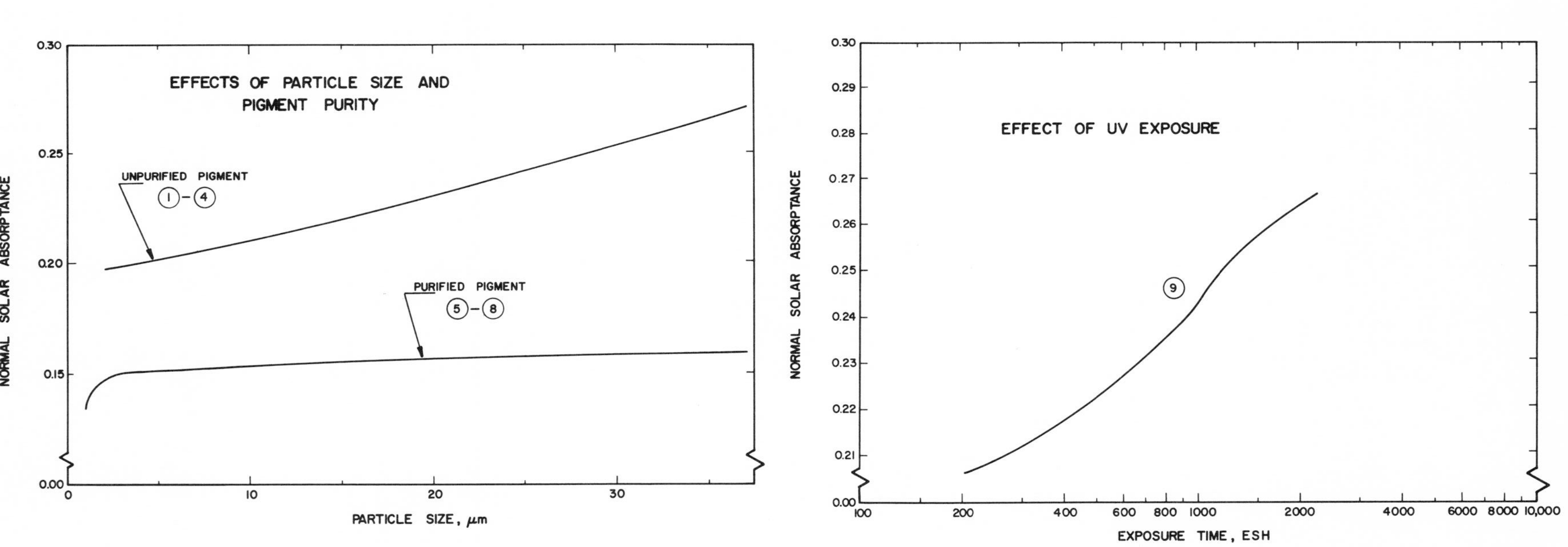

SPECIFICATION TABLE NO. 145 NORMAL SOLAR ABSORPTANCE OF ZIRCONIUM SILICATE PIGMENTED COATINGS

Curve No.	Ref. No.	Year	Temperature Range, K	Geometry θ	Reported Error, %	Composition (weight percent), Specifications and Remarks
1*	78	1961	298	~0°		$ZrSiO_4$ in K_2SiO_3 (K_2O to SiO_2 ratio 0.455) binder (~0.127 mm thick) on aluminum substrate; pigment nonpurified; particle size 2 μm; pigment to binder volume ratio 4; sprayed several coats ~0.0381 mm thick each, each coat cured at ~298 K for 1 hr, then at 373 K for 1 hr; final coat baked at 473 K for 2 hrs; data extracted from smooth curve; calculated from reflectance.
2*	78	1961	298	~0°		Similar to above specimen and conditions except particle size 10 μm.
3*	78	1961	298	~0°		Similar to above specimen and conditions except particle size 20 μm.
4*	78	1961	298	~0°		Similar to above specimen and conditions except particle size 37 μm.
5*	78	1961	298	~0°		Similar to curve 1 specimen and conditions except pigment purified; particle size 1 μm.
6*	78	1961	298	~0°		Similar to above specimen and conditions except particle size 3 μm.
7*	78	1961	298	~0°		Similar to above specimen and conditions except particle size 10 μm.
8*	78	1961	298	~0°		Similar to above specimen and conditions except particle size 37 μm.
9*	78	1961	298	~0°		$ZrSiO_4$ in K_2SiO_3 (K_2O to SiO_2 ratio 0.455) binder (~0.127 mm thick) on aluminum substrate; pigment to binder volume ratio 4; sprayed several coats ~0.0381 mm thick each, each coat cured at ~298 K for 1 hr then at 373 K for 1 hr; final coat baked at 473 K for 2 hrs; exposed to UV radiation in vacuum (10^{-7} mm Hg); data extracted from smooth curve; ESH of UV radiation is variable; property calculated from reflectance.
10*	49	1961	278	~0°		Zircon in sodium silicate D binder on Dow 17 treated Mg alloy substrate; substrate 3.18 mm thick.
11*	49	1961	278	~0°		Above specimen and conditions except exposed to UV radiation (from an argon-filled A-H6 high pressure Hg arc lamp) in vacuum (10^{-6}-8 x 10^{-6} mm Hg) for 127 hrs; vacuum maintained by VacIon pump (Varian model V-11404); 6 times intensity of solar UV radiation.
12*	86	1964	~298	~0°		Zircon in potassium silicate binder; pigment calcined at 1073 K for 16 hrs; PBR 4.30; cured by air-drying; calculated from spectral reflectance. [Authors' designation: Sample No. 7004]
13*	86	1964	~298	~0°		Above specimen and conditions except exposed to 690 ESH of UV radiation in vacuum.
14*	86	1964	~298	~0°		Similar to curve 12 specimen and conditions except heat treated at 773 K for 2 hrs. [Authors' designation: Sample No. 7005]
15*	86	1964	~298	~0°		Above specimen and conditions except exposed to 690 ESH of UV radiation in vacuum.
16*	49	1961	278	~0°		Zircon in silicone resin binder on Dow 17 treated Mg alloy substrate; exposed in vacuum (10^{-6}-8 x 10^{-6} mm Hg) to UV radiation from an argon-filled A-H6 high pressure Hg arc lamp for 127 hrs.
17*	49	1961	278	~0°		Zircon in sodium silicate K binder on Dow 17 treated Mg alloy substrate.
18*	49	1961	278	~0°		Similar to above specimen and conditions except exposed in vacuum (10^{-6}-8 x 10^{-6} mm Hg) to UV radiation from an argon-filled A-H6 high pressure Hg arc lamp for 127 hrs.
19*	27	1964	~298	~0°		Superpax $ZrSiO_4$ in PS-7 potassium silicate binder on aluminum substrate; PBR 4.30; solids content 56.9%; substrate abraded with No. 60 Aloxite cloth; sample cured at 413 K for 18 hrs. [Authors' designation: Sample C57]

* No plot given

SPECIFICATION TABLE NO. 145 NORMAL SOLAR ABSORPTANCE OF ZIRCONIUM SILICATE PIGMENTED COATINGS (continued)

Curve No.	Ref. No.	Year	Temperature Range, K	Geometry θ	Reported Error, %	Composition (weight percent), Specifications and Remarks
20*	27	1964	~298	~0°		Above specimen and conditions except irradiated in vacuum (10^{-6} mm Hg) with 200 ESH of UV radiation (AH-6 lamp) at solar factor of 3 suns.
21*	52	1965	193-263	~0°		Zircon in potassium silicate binder; measured in earth orbit on OSO-II; property is avg over temp range.
22*	52	1965	193-263	~0°		Zirconium silicate in potassium silicate binder; measured in earth orbit on OSO-II; property is avg over temp range.

* No plot given

DATA TABLE NO. 145 NORMAL SOLAR ABSORPTANCE OF ZIRCONIUM SILICATE PIGMENTED COATINGS

[Temperature, T, K; Absorptance, α]

T	α
CURVE 1*	
298	0.197
CURVE 2*	
298	0.210
CURVE 3*	
298	0.230
CURVE 4*	
298	0.272
CURVE 5*	
298	0.135
CURVE 6*	
298	0.150
CURVE 7*	
298	0.153
CURVE 8*	
298	0.160
ESH	α
CURVE 9* T = 298	
200	0.205
500	0.222
1000	0.243
1500	0.258
2250	0.266
T	α
CURVE 10*	
278	0.184

T	α
CURVE 11*	
278	0.220
CURVE 12*	
298	0.168
CURVE 13*	
298	0.240
CURVE 14*	
298	0.155
CURVE 15*	
298	0.225
CURVE 16*	
278	0.45
CURVE 17*	
278	0.20
CURVE 18*	
278	0.21
CURVE 19*	
298	0.180
CURVE 20*	
298	0.249
CURVE 21*	
193	0.152
263	0.152
CURVE 22*	
193	0.163
263	0.163

* No plot given

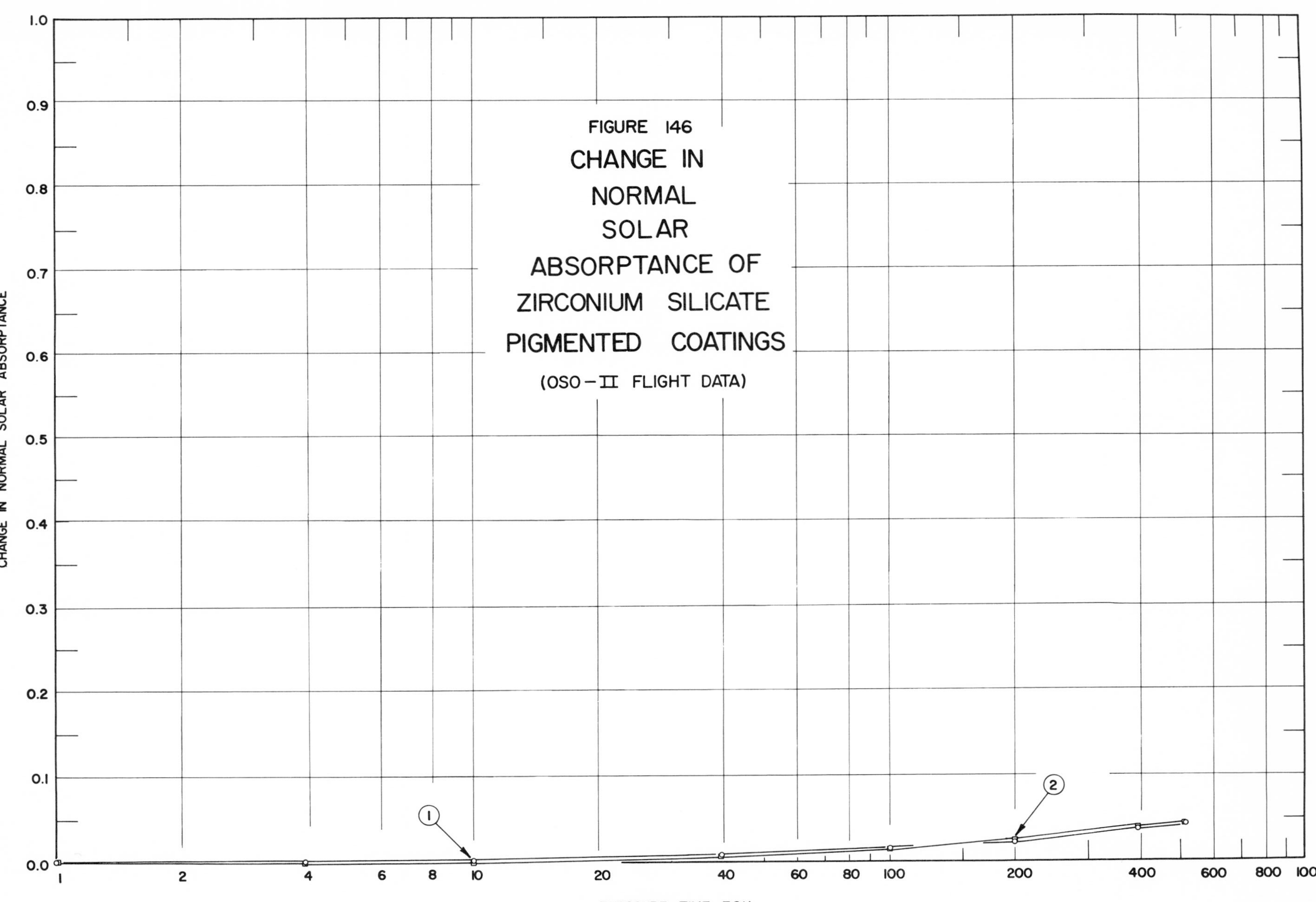

FIGURE 146
CHANGE IN NORMAL SOLAR ABSORPTANCE OF ZIRCONIUM SILICATE PIGMENTED COATINGS
(OSO-II FLIGHT DATA)

SPECIFICATION TABLE NO. 146 CHANGE IN NORMAL SOLAR ABSORPTANCE OF ZIRCONIUM SILICATE PIGMENTED COATINGS

Curve No.	Ref. No.	Year	Temperature Range, K	Geometry θ	Reported Error, %	Composition (weight percent), Specifications and Remarks
1	52	1966	~298	~0°		Zirconium silicate in potassium silicate binder; data taken from OSO-II flight experiment; variable is equivalent length of time exposed to the sun. [Author's designation: Coating No. 5]
2	52	1966	~298	~0°		Zircon in potassium silicate binder; data taken from OSO-II flight experiment; variable is equivalent length of time exposed to the sun.

DATA TABLE NO. 146 CHANGE IN NORMAL SOLAR ABSORPTANCE OF ZIRCONIUM SILICATE PIGMENTED COATINGS

[Temperature, T, K; Absorptance, α]

ESH	$\Delta\alpha$
CURVE 1 T ~ 298	
1.00	0.000
3.98	0.000
10.0	0.003
39.8	0.009
100	0.015
200	0.023
398	0.037
513	0.044
CURVE 2 T ~ 298	
1.00	0.000
3.98	0.000
10.0	0.002
39.8	0.007
100	0.014
CURVE 2 (cont.)	
200	0.024
398	0.038
506	0.045

1. PIGMENTED COATINGS (continued)

C. Enamels (Fused Vitreous Porcelain)

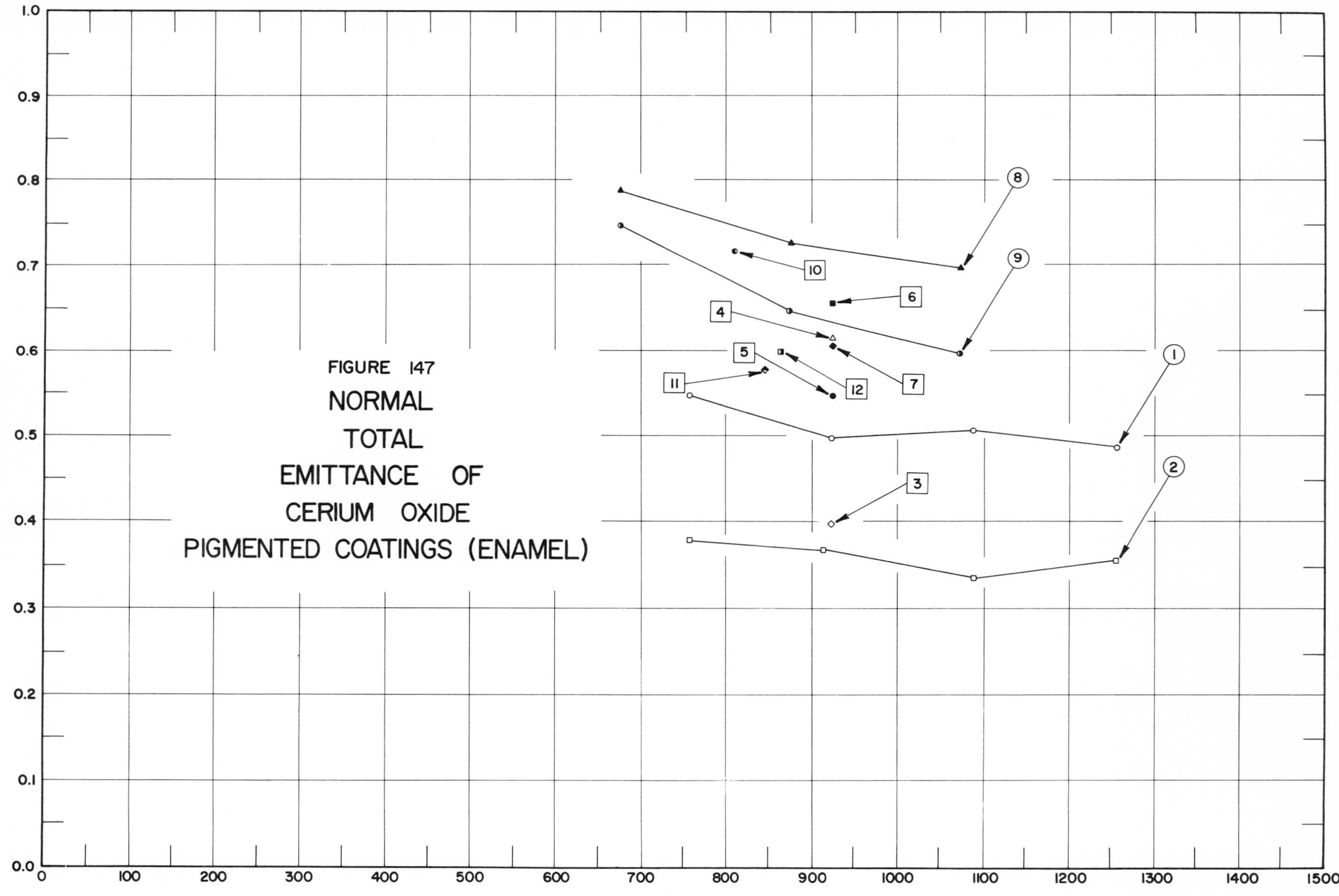
FIGURE 147
NORMAL
TOTAL
EMITTANCE OF
CERIUM OXIDE
PIGMENTED COATINGS (ENAMEL)
NORMAL TOTAL EMITTANCE
TEMPERATURE, K
0.0
0.1
0.2
0.3
0.4
0.5
0.6
0.7
0.8
0.9
1.0
0
100
200
300
400
500
600
700
800
900
1000
1100
1200
1300
1400
1500
1
2
3
4
5
6
7
8
9
10
11
12

SPECIFICATION TABLE NO. 147 NORMAL TOTAL EMITTANCE OF CERIUM OXIDE PIGMENTED COATINGS (ENAMEL)

Curve No.	Ref. No.	Year	Temperature Range, K	Geometry θ'	Reported Error, %	Composition (weight percent), Specifications and Remarks
1	40	1959	755-1255	~0°		NBS Coating N-143 (0.051 mm thick); Inconel substrate; coating consists of boron-free barium beryllium silicate frit with a refractory mill addition of CeO_2.
2	40	1959	755-1255	~0°		NBS Coating N-143 (0.051 mm thick) stainless steel 321 substrate; coating consists of boron-free barium beryllium silicate frit with a refractory mill addition of CeO_2; calculated from normal spectral emittance.
3	40	1959	922	~0°		NBS Coating N-143 (0.056 mm thick); stainless steel 321 substrate; coating consists of boron-free barium beryllium silicate frit with a refractory mill addition of CeO_2.
4	40	1959	922	~0°		Similar to above specimen and conditions except 0.130 mm thick.
5	40	1959	922	~0°		NBS Coating N-143 (0.066 mm thick); Inconel substrate; coating consists of boron-free barium beryllium silicate frit with a refractory mill addition of CeO_2.
6	40	1959	922	~0°		Similar to above specimen and conditions except 0.125 mm thick.
7	40	1959	922	~0°		Similar to curve 5 specimen and conditions except 0.218 mm thick.
8	41	1960	673-1073	~0°	±4	20 CeO_2 and 20 SnO_2 in 60 NBS Frit No. 332 binder (0.051 mm thick); Inconel substrate; NBS Frit No. 332 composed of 44.0 BaO, 37.5 SiO_2, 6.5 B_2O_3, 5.0 ZnO, 3.5 CaO, 2.5 ZrO_2, and 1.0 Al_2O_3; grit blasted Inconel substrate from Whitehead Metals, Inc.; sprayed; fired 3-10 min at 1298 K. [Authors' designation: Enamel W-3]
9	41	1960	673-1073	~0°	±4	Similar to above specimen and conditions except coating 0.152 mm thick.
10	69	1963	~810	0°		9.09 CeO_2 (No. 217) in 90.91 base glaze No. 1 binder (0.127-0.178 mm thick); alumina (6.35 mm thick) substrate; base glaze No. 1 composed of 34.5 Frit No. 71, 19.5 Feldspar, 23.0 SiO_2, 7.0 ZnO, 7.0 E.P. Kaolin, and 9.0 $CaCO_3$; Frit No. 71 composed of 71.0 PbO, 24.96 SiO_2, 2.35 Al_2O_3, and 1.69 NaKO; sprayed; fired at 1366 K; cream glossy finish; measured relative to graphite. [Author's designation: E-114]
11	69	1963	849	9°		66.6 CeO_2 (No. 210 B) in 33.3 base glaze No. 1 binder (0.0508-0.0762 mm thick); alumina (6.35 mm thick) substrate; base glaze No. 1 composed of 34.5 Frit No. 71, 19.5 Feldspar, 23.0 SiO_2, 7.0 ZnO, 7.0 E.P. Kaolin, and 9.0 $CaCO_3$; Frit No. 71 composed of 71.0 PbO, 24.96 SiO_2, 2.35 Al_2O_3, and 1.69 NaKO; sprayed; fired at 1561 K; yellow-orange matte finish; measured relative to graphite. [Author's designation: E-117]
12	69	1963	867	0°		8.7 CeO_2 (No. 210 B) and 4.35 Co_2O_3 in 86.95 base glaze No. 3 binder (0.0508-0.1016 mm thick); alumina (6.35 mm thick) substrate; base glaze No. 3 composed of 46.1 Frit No. 14, 13.1 Feldspar, 16.3 SiO_2, 3.7 ZnO, 3.7 $CaCO_3$, 4.2 E.P. Kaolin, and 12.9 $ZrSiO_4$; Frit No. 14 composed of 46.15 SiO_2, 23.25 B_2O_3, 20.2 CaO, and 10.4 NaKO; sprayed; fired at 1366 K; blue-black semi-matte finish; measured relative to graphite. [Author's designation: E-317]

DATA TABLE NO. 147 NORMAL TOTAL EMITTANCE OF CERIUM OXIDE PIGMENTED COATINGS (ENAMEL)

[Temperature, T, K; Emittance, ϵ]

T	ϵ	T	ϵ
CURVE 1		CURVE 10	
755	0.55	~810	0.72
922	0.50	CURVE 11	
1089	0.51	849	0.58
1255	0.49	CURVE 12	
CURVE 2		867	0.60
755	0.38		
922	0.37		
1089	0.34		
1255	0.36		
CURVE 3			
922	0.40		
CURVE 4			
922	0.62		
CURVE 5			
922	0.55		
CURVE 6			
922	0.66		
CURVE 7			
922	0.61		
CURVE 8			
673	0.79		
873	0.73		
1073	0.70		
CURVE 9			
673	0.75		
873	0.65		
1073	0.60		

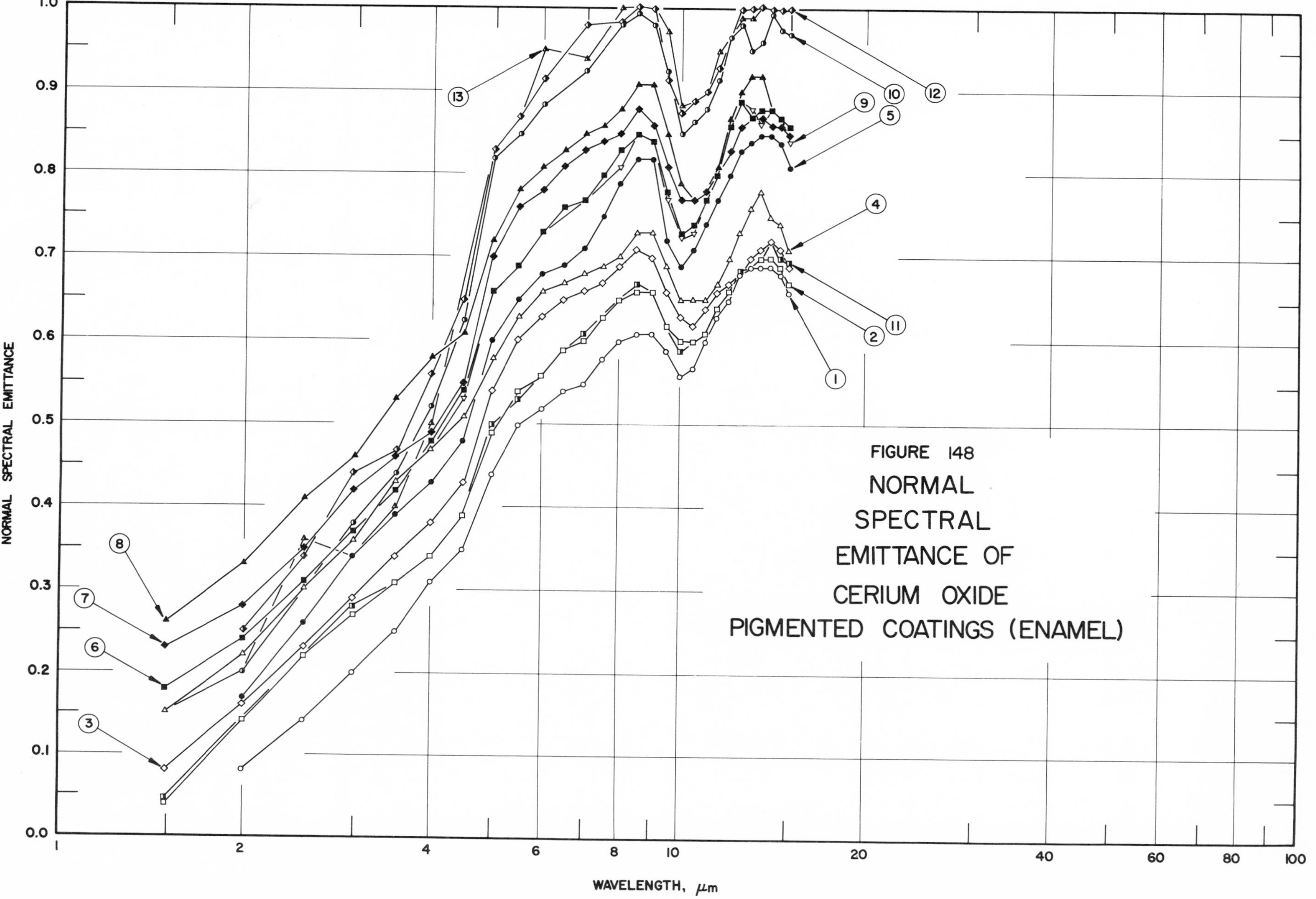

FIGURE 148 NORMAL SPECTRAL EMITTANCE OF CERIUM OXIDE PIGMENTED COATINGS (ENAMEL)

SPECIFICATION TABLE NO. 148 NORMAL SPECTRAL EMITTANCE OF CERIUM OXIDE PIGMENTED COATINGS (ENAMEL)

Curve No.	Ref. No.	Year	Temperature, K	Wavelength Range, μm	Geometry θ'	Reported Error, %	Composition (weight percent), Specifications, and Remarks
1	40	1959	755	2.0-15.0	~0°	<±4	NBS Coating N-143 (0.051 mm thick); stainless steel 321 substrate; coating consists of boron-free barium beryllium silicate frit with a refractory mill addition of CeO_2.
2	40	1959	922	1.5-15.0	~0°	<±4	Above specimen and conditions.
3	40	1959	1089	1.5-15.0	~0°	<±4	Above specimen and conditions.
4	40	1959	1255	1.5-15.0	~0°	<±4	Above specimen and conditions.
5	40	1959	755	2.0-15.0	~0°	<±4	NBS Coating N-143 (0.051 mm thick); Inconel substrate; coating consists of boron-free barium beryllium silicate frit with a refractory mill addition of CeO_2.
6	40	1959	922	1.5-15.0	~0°	<±4	Above specimen and conditions.
7	40	1959	1089	1.5-15.0	~0°	<±4	Above specimen and conditions.
8	40	1959	1255	1.5-15.0	~0°	<±4	Above specimen and conditions.
9	40	1959	922	2.0-15.0	~0°	<±4	NBS Coating N-143 (0.056 mm thick); stainless steel 321 substrate; coating consists of boron-free barium beryllium silicate frit with a refractory mill addition of CeO_2.
10	40	1959	922	1.5-15.0	~0°	<±4	Similar to above specimen and conditions except 0.129 mm thick.
11	40	1959	922	1.5-15.0	~0°	<±4	NBS Coating N-143 (0.066 mm thick); Inconel substrate; coating consists of boron-free barium beryllium silicate frit with a refractory mill addition of CeO_2.
12	40	1959	922	2.0-15.0	~0°	<±4	Similar to above specimen and conditions except 0.125 mm thick.
13	40	1959	922	2.0-15.0	~0°	<±4	Similar to curve 11 specimen and conditions except 0.218 mm thick.

DATA TABLE NO. 148 NORMAL SPECTRAL EMITTANCE OF CERIUM OXIDE PIGMENTED COATINGS (ENAMEL)

[Wavelength, λ, μm; Emittance, ∈; Temperature, T, K]

λ	∈
CURVE 1 T = 755	
2.0	0.08
2.5	0.14
3.0	0.20
3.5	0.25
4.0	0.31
4.5	0.35
5.0	0.44
5.5	0.50
6.0	0.52
6.5	0.54
7.0	0.55
7.5	0.58
8.0	0.60
8.5	0.61
9.0	0.61
9.5	0.59
10.0	0.56
10.5	0.57
11.0	0.60
11.5	0.63
12.0	0.65
12.5	0.68
13.0	0.69
13.5	0.69
14.0	0.69
14.5	0.68
15.0	0.66
CURVE 2 T = 922	
1.5	0.04
2.0	0.14
2.5	0.22
3.0	0.27
3.5	0.31
4.0	0.34
4.5	0.39
5.0	0.49
5.5	0.54
6.0	0.56
6.5	0.59
7.0	0.60
7.5	0.63
CURVE 2 (cont.)	
8.0	0.65
8.5	0.66
9.0	0.66
9.5	0.62
10.0	0.60
10.5	0.60
11.0	0.61
11.5	0.64
12.0	0.66
12.5	0.68*
13.0	0.69*
13.5	0.70
14.0	0.70
14.5	0.69
15.0	0.67
CURVE 3 T = 1089	
1.5	0.08
2.0	0.16
2.5	0.23
3.0	0.29
3.5	0.34
4.0	0.38
4.5	0.43
5.0	0.54
5.5	0.60
6.0	0.63
6.5	0.65
7.0	0.66
7.5	0.67
8.0	0.69
8.5	0.71
9.0	0.70
9.5	0.66
10.0	0.63
10.5	0.62
11.0	0.64
11.5	0.66
12.0	0.67
12.5	0.68*
13.0	0.70
13.5	0.71
14.0	0.72
CURVE 3 (cont.)	
14.5	0.71
15.0	0.69
CURVE 4 T = 1255	
1.5	0.15
2.0	0.22
2.5	0.30
3.0	0.36
3.5	0.43
4.0	0.47
4.5	0.51
5.0	0.58
5.5	0.63
6.0	0.66
6.5	0.67
7.0	0.68
7.5	0.69
8.0	0.70
8.5	0.73
9.0	0.73
9.5	0.69
10.0	0.65
10.5	0.65
11.0	0.65
11.5	0.67
12.0	0.70
12.5	0.73
13.0	0.76
13.5	0.78
14.0	0.75
14.5	0.74
15.0	0.71
CURVE 5 T = 755	
2.0	0.17
2.5	0.26
3.0	0.34
3.5	0.39
4.0	0.43
4.5	0.48
5.0	0.60
CURVE 5 (cont.)	
5.5	0.65
6.0	0.68
6.5	0.69
7.0	0.71
7.5	0.75
8.0	0.79
8.5	0.82
9.0	0.82
9.5	0.72
10.0	0.69
10.5	0.71
11.0	0.74
11.5	0.77
12.0	0.80
12.5	0.83
13.0	0.84
13.5	0.85
14.0	0.85
14.5	0.84
15.0	0.81
CURVE 6 T = 922	
1.5	0.18
2.0	0.24
2.5	0.31
3.0	0.37
3.5	0.42
4.0	0.48
4.5	0.54
5.0	0.66
5.5	0.69
6.0	0.73
6.5	0.76
7.0	0.77
7.5	0.80
8.0	0.83
8.5	0.85
9.0	0.84
9.5	0.78
10.0	0.73
10.5	0.74
11.0	0.77
11.5	0.80
CURVE 6 (cont.)	
12.0	0.86
12.5	0.89
13.0	0.87
13.5	0.88
14.0	0.88
14.5	0.87
15.0	0.86
CURVE 7 T = 1089	
1.5	0.23
2.0	0.28
2.5	0.35
3.0	0.42
3.5	0.46
4.0	0.49
4.5	0.55
5.0	0.70
5.5	0.76
6.0	0.78
6.5	0.81
7.0	0.83
7.5	0.84
8.0	0.85
8.5	0.88
9.0	0.86
9.5	0.81
10.0	0.77
10.5	0.77
11.0	0.78
11.5	0.80*
12.0	0.83
12.5	0.86
13.0	0.87*
13.5	0.87
14.0	0.86
14.5	0.86
15.0	0.85
CURVE 8 T = 1255	
1.5	0.26
2.0	0.33
CURVE 8 (cont.)	
2.5	0.41
3.0	0.46
3.5	0.53
4.0	0.58
4.5	0.61
5.0	0.72
5.5	0.78
6.0	0.81
6.5	0.83
7.0	0.85
7.5	0.86
8.0	0.88
8.5	0.91
9.0	0.91
9.5	0.85
10.0	0.79
10.5	0.77*
11.0	0.78*
11.5	0.81
12.0	0.87
12.5	0.90
13.0	0.92
13.5	0.92
14.0	0.88*
14.5	0.87*
15.0	0.86*
CURVE 9 T = 922	
2.0	0.24*
2.5	0.30*
3.0	0.37*
3.5	0.42*
4.0	0.47*
4.5	0.53
5.0	0.66*
5.5	0.69*
6.0	0.73*
7.0	0.78*
8.0	0.81
8.5	0.85*
9.0	0.84*
9.5	0.77
10.0	0.726
CURVE 9 (cont.)	
10.5	0.73
11.0	0.77*
11.5	0.80*
12.0	0.87*
12.5	0.89*
13.0	0.88
13.5	0.868
14.0	0.88*
14.5	0.86*
15.0	0.84
CURVE 10 T = 922	
1.5	0.15*
2.0	0.20
3.0	0.38
3.5	0.44*
4.0	0.52
4.5	0.625
5.0	0.82
5.5	0.85
6.0	0.885
7.0	0.925
8.0	0.98
8.5	0.995
9.0	0.98
9.5	0.925
10.0	0.85
10.5	0.865
11.0	0.880
11.5	0.915
12.0	0.965
12.5	0.98
13.0	0.95
13.5	0.96
14.0	0.995
14.5	0.975
15.0	0.97
CURVE 11 T = 922	
1.5	0.045
2.0	0.14*
2.5	0.22*
CURVE 11 (cont.)	
3.0	0.28
3.5	0.31*
4.0	0.34*
4.5	0.39*
5.0	0.50
5.5	0.53
6.0	0.56*
6.5	0.59*
7.0	0.61
8.0	0.65*
8.5	0.67
9.0	0.65*
9.5	0.62*
10.0	0.59
10.5	0.60*
11.0	0.61*
11.5	0.64*
12.0	0.66*
12.5	0.685
13.0	0.69*
13.5	0.70*
14.0	0.72*
14.5	0.70
15.0	0.695
CURVE 12 T = 922	
2.0	0.25
2.5	0.34
3.0	0.44
3.5	0.47
4.0	0.56
4.5	0.65
5.0	0.83
5.5	0.87
6.0	0.915
7.0	0.98
8.0	0.985
8.5	1.015
9.0	1.00
9.5	0.915
10.0	0.875
10.5	0.89
11.0	0.90

*Not shown on plot

DATA TABLE NO. 148 NORMAL SPECTRAL EMITTANCE OF CERIUM OXIDE PIGMENTED COATINGS (ENAMEL) (continued)

CURVE 12 (cont.)	
T = 922	
11.5	0.93
12.0	0.97*
12.5	1.00
13.0	1.015
13.5	1.045
13.9	1.05
14.5	1.04
15.0	1.025
CURVE 13	
T = 922	
2.0	0.20*
2.5	0.36
3.0	0.34*
3.5	0.40
4.0	0.50
5.0	0.83*
5.5	0.87*
6.0	0.95
7.0	0.94
8.0	1.015
8.5	1.05*
9.0	1.05*
9.5	0.975
10.0	0.885
10.5	0.890*
11.0	0.90*
11.5	0.95
12.0	0.97*
12.5	0.99
13.0	0.99
13.5	1.03*
14.0	1.05*
14.5	1.03*
15.0	1.02*

*Not shown on plot

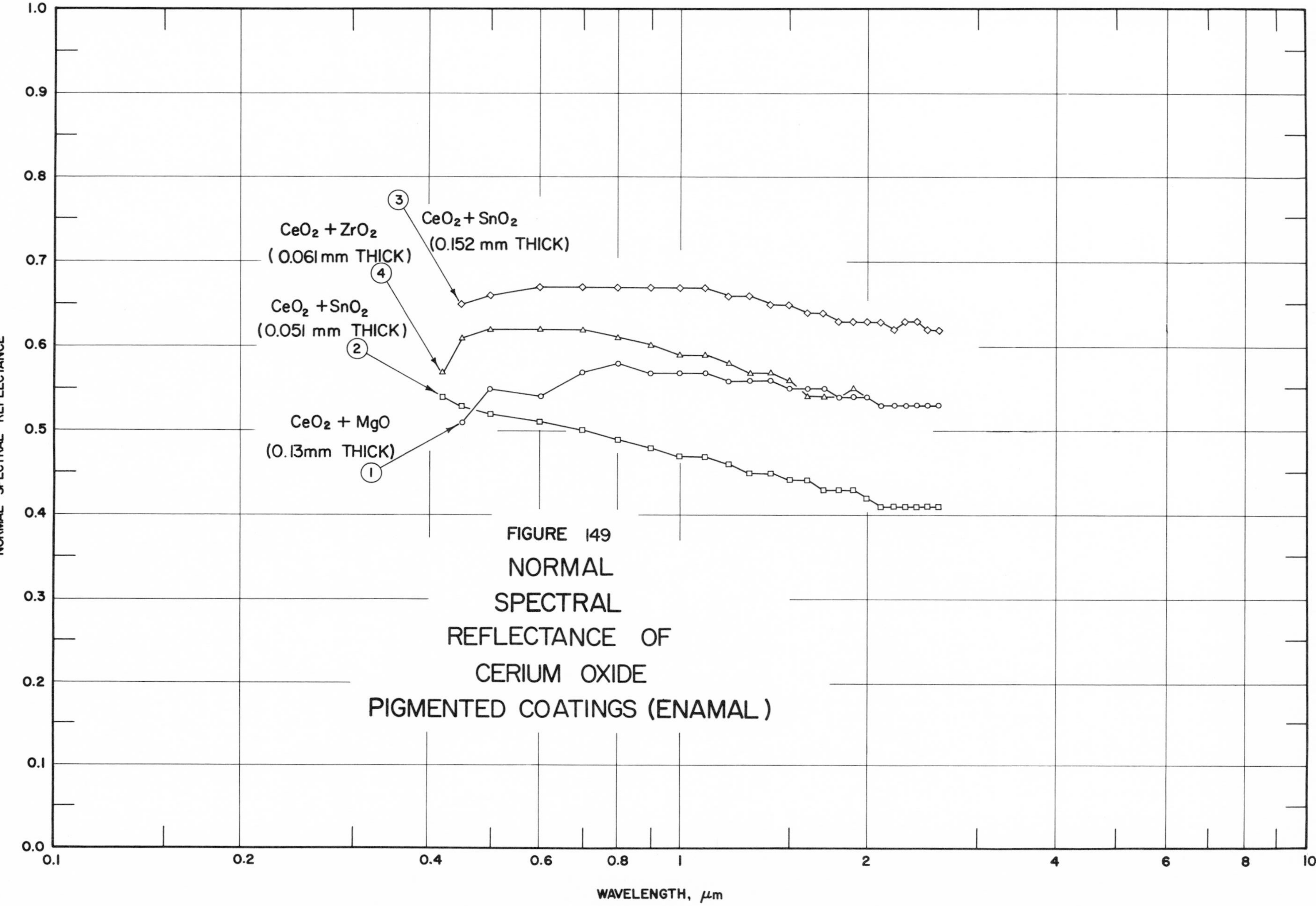

FIGURE 149 NORMAL SPECTRAL REFLECTANCE OF CERIUM OXIDE PIGMENTED COATINGS (ENAMAL)

SPECIFICATION TABLE NO. 149 NORMAL SPECTRAL REFLECTANCE OF CERIUM OXIDE PIGMENTED COATINGS (ENAMEL)

Curve No.	Ref. No.	Year	Temperature, K	Wavelength Range, μm	Geometry θ	θ'	ω'	Reported Error, %	Composition (weight percent), Specifications, and Remarks
1	41	1960	298	0.45-2.60	~0°		2π		CeO_2+ MgO in NBS Frit No. 332, 44.0 BaO, 37.5 SiO_2, 6.5 B_2O_3, 5.0 ZnO, 3.5 CaO, 2.5 ZrO_2, and 1.0 Al_2O_3 (0.130 mm thick); 60 frit, 30 CeO_2, and 10 MgO; grit blasted Inconel substrate from Whitehead Metals, Inc.; measured relative to MgO; [Author's designation: Enamel W-1].
2	41	1960	298	0.42-2.60	~0°		2π		CeO_2 + SnO_2 in NBS Frit No. 332, 44.0 BaO, 37.5 SiO_2, 6.5 B_2O_3, 5.0 ZnO, 3.5 CaO, 2.5 ZrO_2, and 1.0 Al_2O_3 (0.051 mm thick); 60 frit, 20 CeO_2, and 20 SnO_2; grit blasted Inconel substrate from Whitehead Metals, Inc.; measured relative to MgO; [Author's designation: Enamel W-3].
3	41	1960	298	0.45-2.60	~0°		2π		CeO_2 + SnO_2 in NBS Frit No. 332, 44.0 BaO, 37.5 SiO_2, 6.5 B_2O_3, 5.0 ZnO, 3.5 CaO, 2.5 ZrO_2, and 1.0 Al_2O_3 (0.152 mm thick); 60 frit, 20 CeO_2, and 20 SnO_2; grit blasted Inconel substrate from Whitehead Metals, Inc.; measured relative to MgO; [Author's designation: Enamel W-3].
4	41	1960	298	0.42-2.60	~0°		2π		CeO_2 + ZrO_2 in NBS Frit No. 332, 44.0 BaO, 37.5 SiO_2, 6.5 B_2O_3, 5.0 ZnO, 3.5 CaO, 2.5 ZrO_2, and 1.0 Al_2O_3 (0.061 mm thick); 60 frit, 20 CeO_2, and 20 ZrO_2; grit blasted Inconel substrate from Whitehead Metals, Inc.; measured relative to MgO; [Author's designation: Enamel W-4].

DATA TABLE NO. 149 NORMAL SPECTRAL REFLECTANCE OF CERIUM OXIDE PIGMENTED COATINGS (ENAMEL)

[Wavelength, λ, μm; Reflectance, ρ; Temperature, T, K]

λ	ρ
CURVE 1	
T = 298	
0.45	0.51
0.50	0.55
0.60	0.54
0.70	0.57
0.80	0.58
0.90	0.57
1.00	0.57
1.10	0.57
1.20	0.56
1.30	0.56
1.40	0.56
1.50	0.55
1.60	0.55
1.70	0.55
1.80	0.54
1.90	0.54
2.00	0.54
2.10	0.53
2.20	0.53
2.30	0.53
2.40	0.53
2.50	0.53
2.60	0.53
CURVE 2	
T = 298	
0.42	0.54
0.45	0.53
0.50	0.52
0.60	0.51
0.70	0.50
0.80	0.49
0.90	0.48
1.00	0.47
1.10	0.47
1.20	0.46
1.30	0.45
1.40	0.45
1.50	0.44
1.60	0.44
1.70	0.43
1.80	0.43
1.90	0.43
CURVE 2 (cont.)	
2.00	0.42
2.10	0.41
2.20	0.41
2.30	0.41
2.40	0.41
2.50	0.41
2.60	0.41
CURVE 3	
T = 298	
0.45	0.65
0.50	0.66
0.60	0.67
0.70	0.67
0.80	0.67
0.90	0.67
1.00	0.67
1.10	0.67
1.20	0.66
1.30	0.66
1.40	0.65
1.50	0.65
1.60	0.64
1.70	0.64
1.80	0.63
1.90	0.63
2.00	0.63
2.10	0.63
2.20	0.62
2.30	0.63
2.40	0.63
2.50	0.62
2.60	0.62
CURVE 4	
T = 298	
0.42	0.57
0.45	0.61
0.50	0.62
0.60	0.62
0.70	0.62
0.80	0.61
0.90	0.60
CURVE 4 (cont.)	
1.00	0.59
1.10	0.59
1.20	0.58
1.30	0.57
1.40	0.57
1.50	0.56
1.60	0.54
1.70	0.54
1.80	0.54*
1.90	0.55
2.00	0.54*
2.10	0.53*
2.20	0.53*
2.30	0.53*
2.40	0.53*
2.50	0.53*
2.60	0.53*

* Not shown on plot

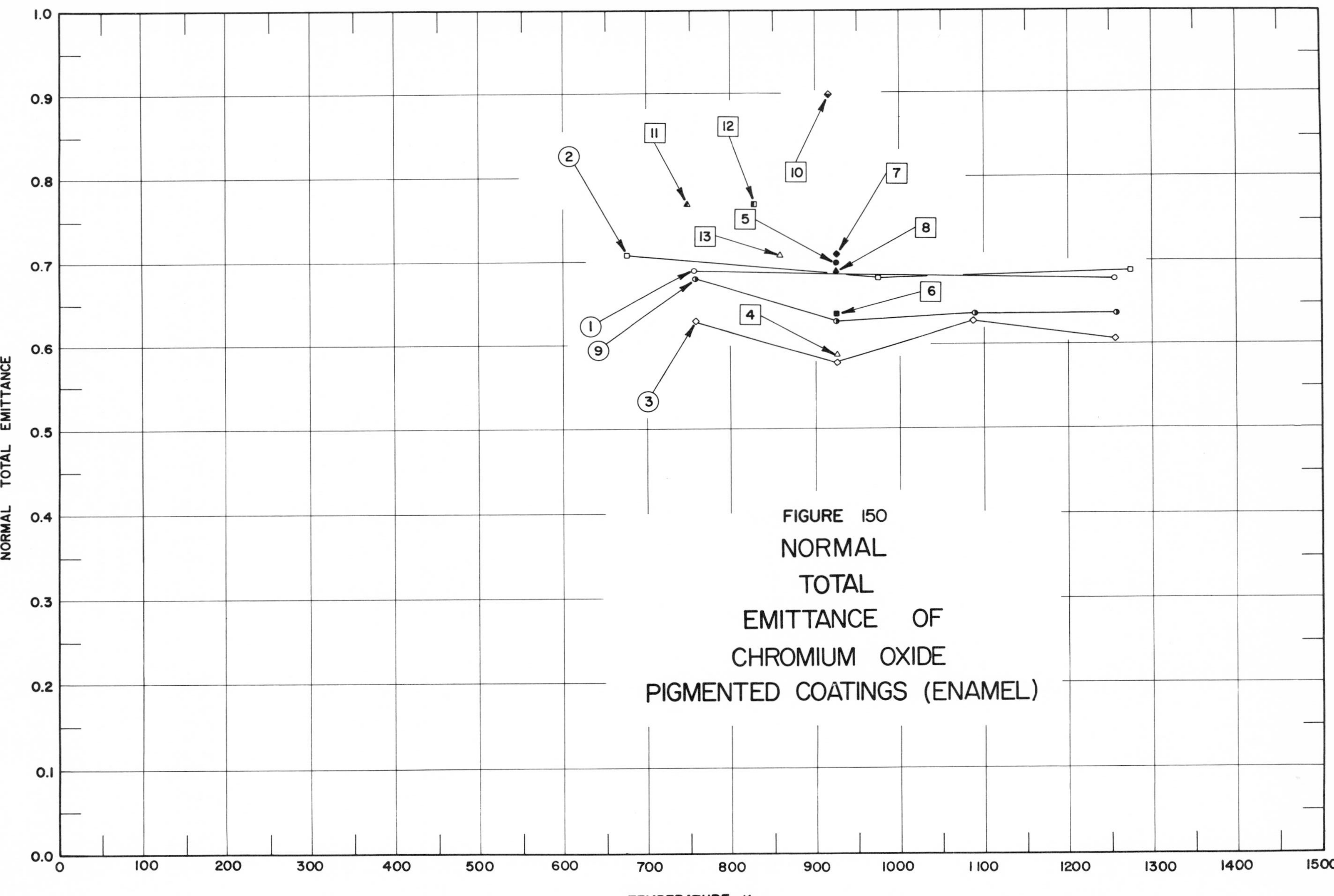

FIGURE 150
NORMAL
TOTAL
EMITTANCE OF
CHROMIUM OXIDE
PIGMENTED COATINGS (ENAMEL)
NORMAL TOTAL EMITTANCE
TEMPERATURE, K
0.0
0.1
0.2
0.3
0.4
0.5
0.6
0.7
0.8
0.9
1.0
0
100
200
300
400
500
600
700
800
900
1000
1100
1200
1300
1400
1500
1
2
3
4
5
6
7
8
9
10
11
12
13

SPECIFICATION TABLE NO. 150 NORMAL TOTAL EMITTANCE OF CHROMIUM OXIDE PIGMENTED COATINGS (ENAMEL)

Curve No.	Ref. No.	Year	Temperature Range, K	Geometry θ'	Reported Error, %	Composition (weight percent), Specifications and Remarks
1	41	1960	753-1253	~0°	±4	Cr_2O_3 in NBS Frit No. 332, 44.0 BaO, 37.5 SiO_2, 6.5 B_2O_3, 5.0 ZnO, 3.5 CaO, 2.5 ZrO_2, and 1.0 Al_2O_3 (0.137 mm thick); 70 frit and 30 Cr_2O_3; grit blasted Inconel substrate from Whitehead Metals, Inc. [Authors' designation: Enamel A418]
2	41	1960	673-1273	~0°	±4	Cr_2O_3 + black stain in NBS Frit No. 332, 44.0 BaO, 37.5 SiO_2, 6.5 B_2O_3, 5.0 ZnO, 3.5 CaO, 2.5 ZrO_2, and 1 Al_2O_3 (0.152 mm thick); 60 frit, 25 black stain, and 15 Cr_2O_3; black stain: 37 Fe_2O_3, 28 Co_2O_3, 14 NiO, 11 MnO_2, 10 Cr_2O_3; grit blasted Inconel substrate from Whitehead Metals, Inc. [Authors' designation: Enamel B-1]
3	40	1959	755-1255	~0°		NBS Coating A-418 (0.051 mm thick); stainless steel 321 substrate; coating consists of an alkali-free barium borosilicate frit with a refractory mill addition of Cr_2O_3; calculated from normal spectral emittance.
4	40	1959	922	~0°		NBS Coating A-418 (0.056 mm thick); stainless steel 321 substrate; coating consists of an alkali-free barium borosilicate frit with a refractory mill addition of Cr_2O_3.
5	40	1959	922	~0°		Similar to above specimen and conditions except 0.140 mm thick.
6	40	1959	922	~0°		NBS Coating A-418 (0.051 mm thick); Inconel substrate; consists of an alkali-free barium borosilicate frit with a refractory mill addition of Cr_2O_3; calculated from normal spectral emittance.
7	40	1959	922	~0°		Similar to above specimen and conditions except 0.122 mm thick.
8	40	1959	922	~0°		Similar to above specimen and conditions except 0.147 mm thick.
9	40	1959	755-1255	~0°		NBS Coating A-418 (0.051 mm thick); Inconel substrate; coating consists of an alkali-free barium borosilicate frit with a refractory mill addition of Cr_2O_3; calculated from normal spectral emittance.
10	69	1963	913	0°		4.55 Cr_2O_3 and 4.55 Co_2O_3 in 90.9 base glaze No. 1 binder (0.178-0.254 mm thick) on alumina (6.35 mm thick) substrate; base glaze No. 1 composed of 34.5 Frit No. 71, 19.5 Feldspar, 23.0 SiO_2, 7.0 ZnO, 7.0 E.P. Kaolin, and 9.0 $CaCO_3$; Frit No. 71 composed of 71.0 PbO, 24.96 SiO_2, 2.35 Al_2O_3, and 1.69 NaKO; sprayed; fired at 1422 K; blue-green glossy finish; measured relative to graphite. [Author's designation: E-111]
11	69	1963	748	0°		33.3 Cr_2O_3 and 33.3 Co_2O_3 in base glaze No. 1 binder (0.0508-0.0762 mm thick) on alumina (6.35 mm thick) substrate; base glaze No. 1 composed of 34.5 Frit No. 71, 19.5 Feldspar, 23.0 SiO_2, 7.0 ZnO, 7.0 E.P. Kaolin, and 9.0 $CaCO_3$; Frit No. 71 composed of 71.0 PbO, 24.96 SiO_2, 2.35 Al_2O_3, and 1.69 NaKC; sprayed; fired at 1561 K; blue-green matte finish; measured relative to graphite. [Author's designation: E-116]

SPECIFICATION TABLE NO. 150 NORMAL TOTAL EMITTANCE OF CHROMIUM OXIDE PIGMENTED COATINGS (ENAMEL) (continued)

Curve No.	Ref. No.	Year	Temperature Range, K	Geometry θ'	Reported Error, %	Composition (weight percent), Specifications and Remarks
12	69	1963	825	0°		4.55 Cr_2O_3 and 4.55 Co_2O_3 in 90.9 base glaze No. 2 binder (0.254-0.381 mm thick) on alumina (6.35 mm thick) substrate; base glaze No. 2 composed of 49.0 Frit No. 44, 19.0 Feldspar, 16.0 SiO_2, 5.0 ZnO, 6.0 E.P. Kaolin, and 5.0 $CaCO_3$; Frit No. 44 composed of 38.4 PbO, 32.76 SiO_2, 8.40 ZnO, 7.98 B_2O_3, 5.36 Al_2O_3, 5.00 CaO, and 2.10 NaKO; sprayed; fired at 1422 K; blue-green glossy finish; measured relative to graphite. [Author's designation: E-211]
13	69	1963	854	0°		33.3 Cr_2O_3 and 33.3 Fe_2O_3 in 33.3 base glaze No. 3 binder (0.0508-0.0763 mm thick) on alumina (6.35 mm thick) substrate; base glaze No. 3 composed of 46.1 Frit No. 14, 13.1 Feldspar, 16.3 SiO_2, 3.7 ZnO, 3.7 $CaCO_3$, 4.2 E.P. Kaolin, and 12.9 $ZrSiO_4$; Frit No. 14 composed of 46.15 SiO_2, 23.25 B_2O_3, 20.2 CaO, and 10.4 NaKO; sprayed; fired at 1561 K; black matte finish; measured relative to graphite. [Author's designation: E-322]

DATA TABLE NO. 150 NORMAL TOTAL EMITTANCE OF CHROMIUM OXIDE PIGMENTED COATINGS (ENAMEL)

[Temperature, T, K; Emittance, ∈]

T	∈	T	∈
CURVE 1		CURVE 10	
753	0.69	913	0.90
1253	0.68		
CURVE 2		CURVE 11	
673	0.71	748	0.77
973	0.68		
1273	0.69		
CURVE 3		CURVE 12	
755	0.63	825	0.77
922	0.58		
1089	0.63		
1255	0.61		
CURVE 4		CURVE 13	
922	0.59	854	0.71
CURVE 5			
922	0.70		
CURVE 6			
922	0.64		
CURVE 7			
922	0.71		
CURVE 8			
922	0.69		
CURVE 9			
755	0.68		
922	0.63		
1089	0.64		
1255	0.64		

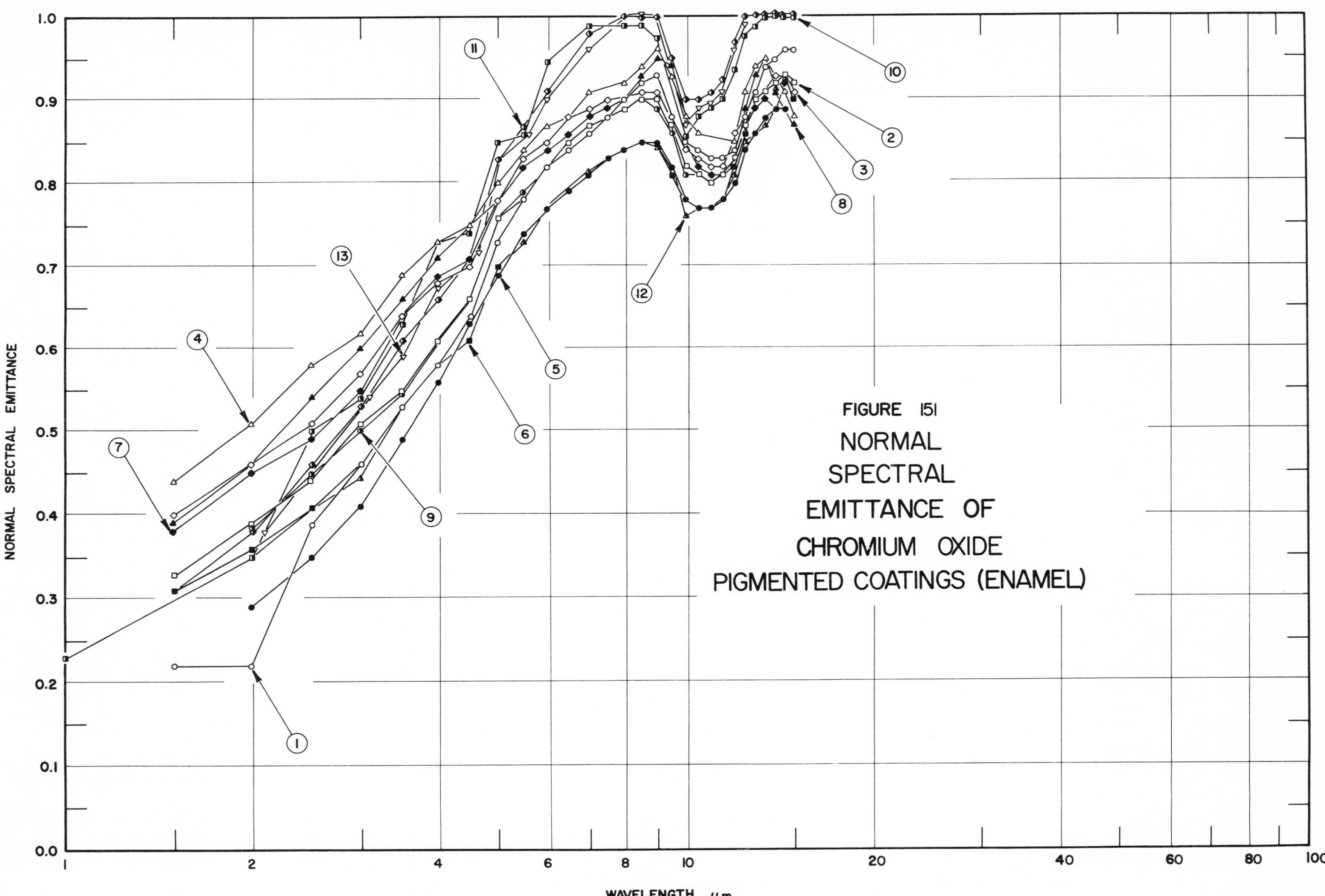

FIGURE 151 NORMAL SPECTRAL EMITTANCE OF CHROMIUM OXIDE PIGMENTED COATINGS (ENAMEL)

SPECIFICATION TABLE NO. 151 NORMAL SPECTRAL EMITTANCE OF CHROMIUM OXIDE PIGMENTED COATINGS (ENAMEL)

Curve No.	Ref. No.	Year	Temperature K	Wavelength Range, μm	Geometry θ'	Reported Error, %	Composition (weight percent), Specifications, and Remarks
1	40	1959	755	1.5-15.0	~0°	<±4	NBS coating A-418 (0.051 mm thick); Inconel substrate; coating consists of an alkali-free barium borosilicate frit with a refractory mill addition of Cr_2O_3.
2	40	1959	922	1.5-15.0	~0°	<±4	Above specimen and conditions.
3	40	1959	1089	1.5-15.0	~0°	<±4	Above specimen and conditions.
4	40	1959	1255	1.5-15.0	~0°	<±4	Above specimen and conditions.
5	40	1959	755	2.0-15.0	~0°	<±4	NBS coating A-418 (0.051 mm thick); stainless steel 321 substrate; coating consists of an alkali-free barium borosilicate frit with a refractory mill addition of Cr_2O_3.
6	40	1959	922	1.5-15.0	~0°	<±4	Above specimen and conditions.
7	40	1959	1089	1.5-15.0	~0°	<±4	Above specimen and conditions.
8	40	1959	1255	1.5-15.0	~0°	<±4	Above specimen and conditions.
9	40	1959	922	2.0-15.0	~0°		NBS coating A-418 (0.051 mm thick); Inconel substrate; coating consists of an alkali-free barium borosilicate frit with a refractory mill addition of Cr_2O_3.
10	40	1959	922	1.0-15.0	~0°		Similar to above specimen and conditions except 0.122 mm thick.
11	40	1959	922	1.5-15.0	~0°		Similar to curve 9 specimen and conditions except 0.147 mm thick.
12	40	1959	922	2.0-15.0	~0°		NBS coating A-418 (0.056 mm thick); stainless steel 321 substrate; coating consists of an alkali-free barium borosilicate frit with a refractory mill addition of Cr_2O_3.
13	40	1959	922	2.1-15.0	~0°		Similar to above specimen and conditions except 0.140 mm thick.

DATA TABLE NO. 151 NORMAL SPECTRAL EMITTANCE OF CHROMIUM OXIDE PIGMENTED COATINGS (ENAMEL)

[Wavelength, λ, μm; Emittance, ε; Temperature, T, K]

CURVE 1, T = 755

λ	ε
1.5	0.22
2.0	0.22
2.5	0.39
3.0	0.46
3.5	0.53
4.0	0.58
4.5	0.64
5.0	0.73
5.5	0.78
6.0	0.82
6.5	0.84
7.0	0.86
7.5	0.88
8.0	0.90
8.5	0.92
9.0	0.93
9.5	0.88
10.0	0.85
10.5	0.84
11.0	0.83
11.5	0.83
12.0	0.84
12.5	0.88
13.0	0.91
13.5	0.94
14.0	0.95
14.5	0.96
15.0	0.96

CURVE 2, T = 922

λ	ε
1.5	0.33
2.0	0.39
2.5	0.44
3.0	0.51
3.5	0.55
4.0	0.61
4.5	0.66
5.0	0.76
5.5	0.78*
6.0	0.82*
6.5	0.85
7.0	0.87
7.5	0.88*
8.0	0.89
8.5	0.90
9.0	0.90
9.5	0.87
10.0	0.82
10.5	0.81
11.0	0.80
11.5	0.81
12.0	0.83
12.5	0.87
13.0	0.90
13.5	0.91
14.0	0.92
14.5	0.93
15.0	0.92

CURVE 3, T = 1089

λ	ε
1.5	0.40
2.0	0.46
2.5	0.51
3.0	0.57
3.5	0.64
4.0	0.68
4.5	0.70
5.0	0.78
5.5	0.83
6.0	0.85
6.5	0.88
7.0	0.89
7.5	0.90
8.0	0.90*
8.5	0.91
9.0	0.91
9.5	0.88*
10.0	0.84
10.5	0.83
11.0	0.82
11.5	0.82
12.0	0.86
12.5	0.88
13.0	0.90*
13.5	0.91*
14.0	0.93
14.5	0.93*
15.0	0.91

CURVE 4, T = 1255

λ	ε
1.5	0.44
2.0	0.51
2.5	0.58
3.0	0.62
3.5	0.69
4.0	0.73
4.5	0.75
5.0	0.80
5.5	0.84
6.0	0.87
6.5	0.88*
7.0	0.91
7.5	0.90*
8.0	0.92
8.5	0.94
9.0	0.96
9.5	0.93
10.0	0.88
10.5	0.86
11.0	0.83*
11.5	0.83*
12.0	0.85
12.5	0.91
13.0	0.94
13.5	0.95
14.0	0.93*
14.5	0.91
15.0	0.88

CURVE 5, T = 755

λ	ε
2.0	0.29
2.5	0.35
3.0	0.41
3.5	0.49
4.0	0.56
4.5	0.63
5.0	0.69
5.5	0.74
6.0	0.77
6.5	0.79
7.0	0.81
7.5	0.83
8.0	0.84
8.5	0.85
9.0	0.85
9.5	0.82
10.0	0.78
10.5	0.77
11.0	0.77
11.5	0.78
12.0	0.80
12.5	0.84
13.0	0.86
13.5	0.88
14.0	0.89
14.5	0.89
15.0	0.88*

CURVE 6, T = 922

λ	ε
1.5	0.31
2.0	0.36
2.5	0.41
3.0	0.46*
3.5	0.53*
4.0	0.58*
4.5	0.61
5.0	0.70
5.5	0.74*
6.0	0.77*
6.5	0.79*
7.0	0.81*
7.5	0.83*
8.0	0.84*
8.5	0.85*
9.0	0.85*
9.5	0.81
10.0	0.78*
10.5	0.77*
11.0	0.77*
11.5	0.78*
12.0	0.82
12.5	0.86
13.0	0.86*
13.5	0.88*
14.0	0.93*
14.5	0.93*
15.0	0.90

CURVE 7, T = 1089

λ	ε
1.5	0.38
2.0	0.45
2.5	0.49
3.0	0.55
3.5	0.64*
4.0	0.69
4.5	0.71
5.0	0.78*
5.5	0.82
6.0	0.84
6.5	0.86
7.0	0.88
7.5	0.89
8.0	0.89*
8.5	0.90*
9.0	0.90*
9.5	0.88*
10.0	0.85*
10.5	0.83
11.0	0.82
11.5	0.82*
12.0	0.83*
12.5	0.87*
13.0	0.89
13.5	0.90
14.0	0.89*
14.5	0.92
15.0	0.92*

CURVE 8, T = 1255

λ	ε
1.5	0.39
2.0	0.46*
2.5	0.54
3.0	0.60
3.5	0.66
4.0	0.71
4.5	0.75*
5.0	0.78*
5.5	0.82*
6.0	0.85*
6.5	0.86*
7.0	0.89*
7.5	0.89*
8.0	0.90*
8.5	0.93
9.0	0.95
9.5	0.94
10.0	0.88*
10.5	0.86*
11.0	0.83*
11.5	0.82*
12.0	0.84*
12.5	0.89
13.0	0.93
13.5	0.95*
14.0	0.91
14.5	0.89*
15.0	0.87

CURVE 9, T = 922

λ	ε
2.0	0.38
2.5	0.45
3.0	0.50
3.5	0.54
4.0	0.61*
4.5	0.66*
5.0	0.76*
5.5	0.79
6.0	0.82*
7.0	0.87*
8.0	0.89*
8.5	0.90*
9.0	0.89
9.5	0.86
10.0	0.81
10.5	0.81*
11.0	0.80*
11.5	0.81*
12.0	0.83*
12.5	0.88*
13.0	0.89*
13.5	0.91*
14.0	0.93*
14.5	0.92
15.0	0.91*

CURVE 10, T = 922

λ	ε
1.0	0.23
2.0	0.35
2.5	0.50
3.0	0.54
3.5	0.63
4.0	0.73*
4.5	0.74
5.0	0.85
5.5	0.86
6.0	0.94
6.5	--
7.0	0.99
7.5	--
8.0	0.99
8.5	0.99
9.0	0.98
9.5	0.93*
10.0	0.86
10.5	0.88
11.0	0.89
11.5	0.90
12.0	0.94
12.5	0.98
13.0	0.98
13.5	1.00
14.0	1.02
14.5	1.01
15.0	1.00

CURVE 11, T = 922

λ	ε
1.5	0.31*
2.0	0.38
2.5	0.46
3.0	0.53
3.5	0.61
4.0	0.66
4.5	0.71*
5.0	0.83
5.5	0.87
6.0	0.91
6.5	--
7.0	0.98
7.5	--
8.0	1.01
8.5	1.00
9.0	1.01
9.5	0.95
10.0	0.90
10.5	0.90
11.0	0.91
11.5	0.92
12.0	0.97
12.5	1.00
13.0	1.02
13.5	1.04
14.0	1.07
14.5	1.05
15.0	1.04

CURVE 12, T = 922

λ	ε
2.0	0.35*
2.5	0.41*
3.0	0.44
3.5	0.53*
4.0	0.58*
4.5	0.61*
5.0	0.70*
5.5	0.73
6.0	0.77*
6.5	--
7.0	0.81
7.5	--
8.0	0.84*
8.5	0.85*
9.0	0.85

* Not shown on plot

DATA TABLE NO. 151 NORMAL SPECTRAL EMITTANCE OF CHROMIUM OXIDE PIGMENTED COATINGS (ENAMEL) (continued)

λ	ϵ
CURVE 12 (cont.)	
T = 922	
9.5	0.82*
10.0	0.76
10.5	0.77*
11.0	0.77*
11.5	0.78*
12.0	0.81
12.5	0.85
13.0	0.86*
13.5	0.87
14.0	0.91
14.5	0.91*
15.0	0.90*
CURVE 13	
T = 922	
2.1	0.38
2.5	0.46*
3.1	0.54
3.5	0.59
4.0	0.67
4.65	0.72
5.0	0.83*
5.6	0.86
6.0	0.90
7.0	0.96
8.0	1.00*
8.5	1.01
9.0	1.00*
9.5	0.94*
10.0	0.87
10.5	0.89
11.0	0.89
11.5	0.91
12.0	0.96
12.5	0.99
13.0	1.00*
13.5	0.02*
14.05	1.03*
14.5	1.04*
15.0	1.02*

* Not shown on plot

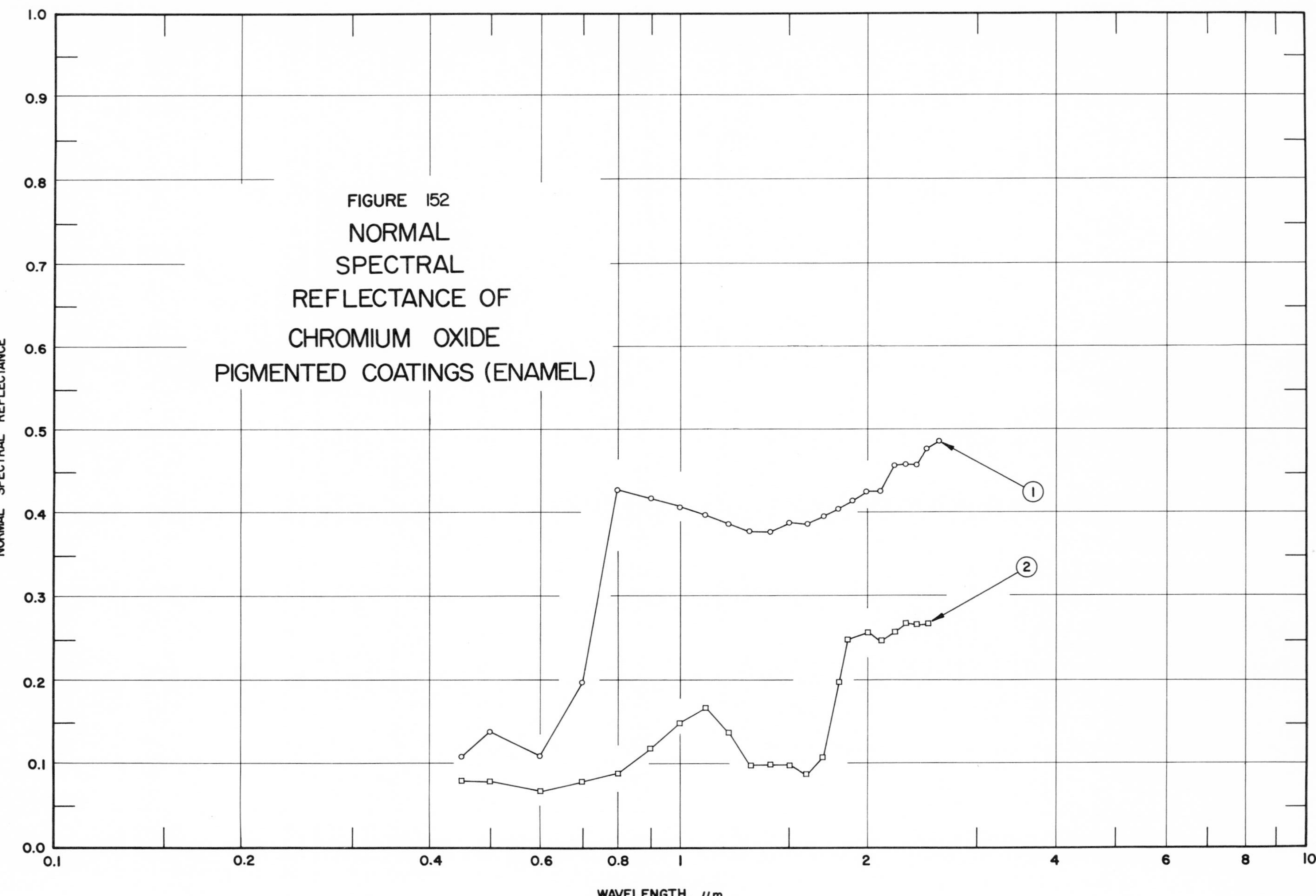

FIGURE 152 NORMAL SPECTRAL REFLECTANCE OF CHROMIUM OXIDE PIGMENTED COATINGS (ENAMEL)

SPECIFICATION TABLE NO. 152 NORMAL SPECTRAL REFLECTANCE OF CHROMIUM OXIDE PIGMENTED COATINGS (ENAMEL)

Curve No.	Ref. No.	Year	Temperature K	Wavelength Range, μm	Geometry θ	θ'	ω'	Reported Error, %	Composition (weight percent), Specifications, and Remarks
1	41	1960	298	0.45-2.60	~0°		2π		Cr_2O_3 in NBS Frit No. 332, 44.0 BaO, 37.5 SiO_2, 6.5 B_2O_3, 5.0 ZnO, 3.5 CaO, 2.5 ZrO_2, and 1.0 Al_2O_3 (0.107 mm thick); 70 frit and 30 Cr_2O_3; grit blasted Inconel substrate from Whitehead Metals, Inc.; measured relative to MgO. [Authors' designation: Enamel A-418]
2	41	1960	298	0.45-2.60	~0°		2π		Cr_2O_3 + black stain in NBS Frit No. 332, 44.0 BaO, 37.5 SiO_2, 6.5 B_2O_3, 5.0 ZnO, 3.5 CaO, 2.5 ZrO_2, and 1.0 Al_2O_3 (0.025 mm thick); 60 frit, 25 black stain, and 15 Cr_2O_3; black stain: 37 Fe_2O_3, 28 Co_2O_3, 14 NiO, 11 11 MnO_2, 10 Cr_2O_3; grit blasted Inconel substrate from Whitehead Metals, Inc.; measured relative to MgO. [Authors' designation: Enamel B-1]

DATA TABLE NO. 152 NORMAL SPECTRAL REFLECTANCE OF CHROMIUM OXIDE PIGMENTED COATINGS (ENAMEL)

[Wavelength, λ, μm; Reflectance, ρ; Temperature, T, K]

λ	ρ	λ	ρ	λ	ρ
CURVE 1 T = 298		CURVE 1 (cont.)		CURVE 2 (cont.)	
0.45	0.11	2.30	0.46	1.60	0.09
0.50	0.14	2.40	0.46	1.70	0.11
0.60	0.11	2.50	0.48	1.80	0.20
0.70	0.20	2.60	0.49	1.90	0.25
0.80	0.43			2.00	0.26
0.90	0.42	CURVE 2 T = 298		2.10	0.25
1.00	0.41			2.20	0.26
1.10	0.40	0.45	0.08	2.30	0.27
1.20	0.39	0.50	0.08	2.40	0.27
1.30	0.38	0.60	0.07	2.50	0.27
1.40	0.38	0.70	0.08	2.60	0.39*
1.50	0.39	0.80	0.09		
1.60	0.39	0.90	0.12		
1.70	0.40	1.00	0.15		
1.80	0.41	1.10	0.17		
1.90	0.42	1.20	0.14		
2.00	0.43	1.30	0.10		
2.10	0.43	1.40	0.10		
2.20	0.46	1.50	0.10		

* Not shown on plot

SPECIFICATION TABLE NO. 153 NORMAL TOTAL EMITTANCE OF COBALT OXIDE PIGMENTED COATINGS (ENAMEL)

Curve No.	Ref. No.	Year	Temperature Range, K	Geometry θ'	Reported Error, %	Composition (weight percent), Specifications and Remarks
1*	69	1963	913	0°		4.55 Co_2O_3 and 4.55 Cr_2O_3 in 90.9 base glaze No. 1 binder (0.1778-0.2540 mm thick) on alumina (6.35 mm thick) substrate; base glaze No. 1 composed of 34.5 Frit No. 71, 19.5 Feldspar, 23.0 SiO_2, 7.0 ZnO, 7.0 E.P. Kaolin, and 9.0 $CaCO_3$; Frit No. 71 composed of 71.0 PbO, 24.96 SiO_2, 2.35 Al_2O_3, and 1.69 NaKO; sprayed; fired at 1422 K; blue-green glossy finish; measured relative to graphite. [Author's designation: E-111]
2*	69	1963	748	0°		33.3 Co_2O_3 and 33.3 Cr_2O_3 in 33.3 base glaze No. 1 binder (0.0508-0.0762 mm thick) on alumina (6.35 mm thick) substrate; base glaze No. 1 composed of 34.5 Frit No. 71, 19.5 Feldspar, 23.0 SiO_2, 7.0 ZnO, 7.0 E.P. Kaolin, and 9.0 $CaCO_3$; Frit No. 71 composed of 71.0 PbO, 24.96 SiO_2, 2.35 Al_2O_3, and 1.69 NaKO; sprayed; fired at 1561 K; blue-green matte finish; measured relative to graphite. [Author's designation: E-116]
3*	69	1963	825	0°		4.55 Co_2O_3 and 4.55 Cr_2O_3 in 90.9 base glaze No. 2 binder (0.254-0.381 mm thick) on alumina (6.35 mm thick) substrate; base glaze No. 2 composed of 49.0 Frit No. 44, 19.0 Feldspar, 16.0 SiO_2, 5.0 ZnO, 6.0 E.P. Kaolin, and 5.0 $CaCO_3$; Frit No. 44 composed of 38.4 PbO, 32.76 SiO_2, 8.40 ZnO, 7.98 B_2O_3, 5.36 Al_2O_3, 5.00 CaO, and 2.10 NaKO; sprayed; fired at 1422 K; blue-green glossy finish; measured relative to graphite. [Author's designation: E-211]
4*	69	1963	887	0°		4.55 Co_2O_3 and 4.55 MnO_2 in 90.9 base glaze No. 2 binder (0.254-0.381 mm thick) on alumina (6.35 mm thick) substrate; base glaze No. 2 composed of 49.0 Frit No. 44, 19.0 Feldspar, 16.0 SiO_2, 5.0 ZnO, 6.0 E.P. Kaolin, and 5.0 $CaCO_3$; Frit No. 44 composed of 38.4 PbO, 32.76 SiO_2, 8.40 ZnO, 7.98 B_2O_3, 5.36 Al_2O_3, 5.00 CaO, and 2.10 NaKO; sprayed; fired at 1422 K; blue glossy finish; measured relative to graphite. [Author's designation: E-212]
5*	69	1963	839	0°		50 Co_2O_3 in 50 base glaze No. 3 binder (0.0508-0.0762 mm thick) on alumina (6.35 mm thick) substrate; base glaze No. 3 composed of 46.1 Frit No. 14, 13.1 Feldspar, 16.3 SiO_2, 3.7 ZnO, 3.7 $CaCO_3$, 4.2 E.P. Kaolin, and 12.9 $ZrSiO_4$; Frit No. 14 composed of 46.15 SiO_2, 23.25 B_2O_3, 20.2 CaO, and 10.4 NaKO; sprayed; fired at 1561 K; blue-black semi-matte finish; measured relative to graphite. [Author's designation: E-323]
6*	69	1963	783	0°		33.3 Co_2O_3 and 33.3 NiO in 33.3 base glaze No. 3 binder (0.0508-0.0762 mm thick) on alumina (6.35 mm thick) substrate; base glaze No. 3 composed of 46.1 Frit No. 14, 13.1 Feldspar, 16.3 SiO_2, 3.7 ZnO, 3.7 $CaCO_3$, 4.2 E.P. Kaolin, and 12.9 $ZrSiO_4$; Frit No. 14 composed of 46.15 SiO_2, 23.25 B_2O_3, 20.2 CaO, and 10.4 NaKO; sprayed; fired at 1561 K; blue-black matte finish; measured relative to graphite. [Author's designation: E-324]
7*	69	1963	761	0°		25 Co_2O_3 and 25 NiO in 50 base glaze No. 3 binder (0.0508-0.0762 mm thick) on alumina (6.35 mm thick) substrate; base glaze No. 3 composed of 46.1 Frit No. 14, 13.1 Feldspar, 16.3 SiO_2, 3.7 ZnO, 3.7 $CaCO_3$, 4.2 E.P. Kaolin, and 12.9 $ZrSiO_4$; Frit No. 14 composed of 46.15 SiO_2, 23.25 B_2O_3, 20.2 CaO, and 10.4 NaKO; sprayed; fired at 1561 K; blue-black matte finish; measured relative to graphite. [Author's designation: E-325]

* No plot given

DATA TABLE NO. 153 NORMAL TOTAL EMITTANCE OF COBALT OXIDE PIGMENTED COATINGS (ENAMEL)

[Temperature, T, K; Emittance, ∈]

T	∈
CURVE 1*	
913	0.90
CURVE 2*	
748	0.77
CURVE 3*	
825	0.77
CURVE 4*	
887	0.84
CURVE 5*	
839	0.69
CURVE 6*	
783	0.80
CURVE 7*	
761	0.86

* No plot given

SPECIFICATION TABLE NO. 154 NORMAL TOTAL EMITTANCE OF IRON OXIDE PIGMENTED COATINGS (ENAMEL)

Curve No.	Ref. No.	Year	Temperature Range, K	Geometry θ'	Reported Error, %	Composition (weight percent), Specifications and Remarks
1*	41	1960	673-1273	~0°		20 Fe_2O_3, 15 CoO, and 5 Cr_2O_3 in 60 NBS Frit No. 332 binder (0.096 mm thick) on Inconel substrate; NBS Frit No. 332 composed of 37.5 SiO_2, 6.5 B_2O_3, 44 BaO, 3.5 CaO, 5 ZnO, 1 Al_2O_3, and 2.5 ZrO_2; grit-blasted Inconel substrate from Whitehead Metals, Inc; sprayed; fired 3-10 min at 1298 K. [Authors' designation: Enamel B-4]
2*	69	1963	854	0°		33.3 Fe_2O_3 and 33.3 Cr_2O_3 in 33.3 base glaze No. 3 binder (0.0508-0.0763 mm thick) on alumina (6.35 mm thick) substrate; base glaze No. 3 composed of 46.1 Frit No. 14, 13.1 Feldspar, 16.3 SiO_2, 3.7 ZnO, 3.7 $CaCO_3$, 4.2 E. P. Kaolin, and 12.9 $ZrSiO_4$; Frit No. 14 composed of 46.15 SiO_2, 23.25 B_2O_3, 20.2 CaO, and 10.4 NaKO; sprayed; fired at 1561 K; black matte finish; measured relative to graphite. [Authors' designation: E-322]
3*	69	1963	785	0°		33.3 Fe_2O_3 and 33.3 NiO in 33.3 base glaze No. 3 binder (0.0508-0.0763 mm thick) on alumina (6.35 mm thick) substrate; base glaze No. 3 composed of 46.1 Frit No. 14, 13.1 Feldspar, 16.3 SiO_2, 3.7 ZnO, 3.7 $CaCO_3$, 4.2 E. P. Kaolin, and 12.9 $ZrSiO_4$; Frit No. 14 composed of 46.15 SiO_2, 23.25 B_2O_3, 20.2 CaO, and 10.4 NaKO; sprayed; fired at 1561 K; black matte finish; measured relative to graphite. [Authors' designation: E-326]
4*	69	1963	840	0°		33.3 Fe_2O_3 and 33.3 MnO_2 in 33.3 base glaze No. 3 binder (0.0508-0.0762 mm thick) on alumina (6.35 mm thick) substrate; base glaze No. 3 composed of 46.1 Frit No. 14, 13.1 Feldspar, 16.3 SiO_2, 3.7 ZnO, 3.7 $CaCO_3$, 4.2 E. P. Kaolin, and 12.9 $ZrSiO_4$; Frit No. 14 composed of 46.15 SiO_2, 23.25 B_2O_3, 20.2 CaO, and 10.4 NaKO; sprayed; fired at 1561 K; black matte finish; measured relative to graphite. [Author's designation: E-327]

DATA TABLE NO. 154 NORMAL TOTAL EMITTANCE OF IRON OXIDE PIGMENTED COATINGS (ENAMEL)

[Temperature, T, K; Emittance, ϵ]

T	ϵ	T	ϵ
CURVE 1*		CURVE 4*	
673	0.78	840	0.82
973	0.78		
1273	0.77		
CURVE 2*			
854	0.71		
CURVE 3*			
785	0.83		

* No plot given

SPECIFICATION TABLE NO. 155 NORMAL SPECTRAL REFLECTANCE OF IRON OXIDE PIGMENTED COATINGS (ENAMEL)

Curve No.	Ref. No.	Year	Temperature, K	Wavelength Range, μm	Geometry θ	θ'	ω'	Reported Error, %	Composition (weight percent), Specifications, and Remarks
1*	41	1960	298	0.45-2.60	~0°		2π		20 Fe_2O_3, 15 CoO, and 5 Cr_2O_3 in 60 NBS Frit No. 332 binder (0.028 mm thick) on Inconel substrate; NBS Frit No. 332 composed of 37.5 SiO_2, 6.5 B_2O_3, 44 BaO, 3.5 CaO, 5 ZnO, 1 Al_2O_3, and 2.5 ZrO_2; grit blasted Inconel substrate from Whitehead Metals, Inc; sprayed; fired 3-10 min at 1298 K. [Authors' designation: Enamel B-4]

DATA TABLE NO. 155 NORMAL SPECTRAL REFLECTANCE OF IRON OXIDE PIGMENTED COATINGS (ENAMEL)

[Wavelength, λ, μm; Reflectance, ρ; Temperature, T, K]

λ	ρ
CURVE 1*	
T = 298	
0.45	0.10
0.50	0.10
0.60	0.10
0.70	0.12
0.80	0.17
0.90	0.21
1.00	0.19
1.10	0.15
1.20	0.13
1.30	0.13
1.40	0.13
1.50	0.13
1.60	0.13
1.70	0.16
1.80	0.26
1.90	0.26
2.00	0.26
2.10	0.26
2.20	0.27
2.30	0.28
2.40	0.28
2.50	0.28
2.60	0.28

* No plot given

SPECIFICATION TABLE NO. 156 NORMAL TOTAL EMITTANCE OF MANGANESE OXIDE PIGMENTED COATINGS (ENAMEL)

Curve No.	Ref. No.	Year	Temperature Range, K	Geometry θ'	Reported Error, %	Composition (weight percent), Specifications and Remarks
1*	69	1963	887	0°		4.55 MnO_2 and 4.55 Co_2O_3 in 90.9 base glaze No. 2 binder (0.254-0.381 mm thick) on alumina (6.35 mm thick) substrate; base glaze No. 2 composed of 49.0 Frit No. 44, 19.0 Feldspar, 16.0 SiO_2, 5.0 ZnO, 6.0 E.P. Kaolin, and 5.0 $CaCO_3$; Frit No. 44 composed of 38.40 PbO, 32.76 SiO_2, 8.40 ZnO, 7.98 B_2O_3, 5.36 Al_2O_3, 5.00 CaO, and 2.10 NaKO; sprayed; fired at 1422 K; blue glossy finish; measured relative to graphite. [Author's designation: E-212]
2*	69	1963	861	0°		40 MnO_2 and 20 Co_2O_3 in 40 base glaze No. 3 binder (0.1270-0.1778 mm thick) on alumina (6.35 mm thick) substrate; base glaze No. 3 composed of 46.1 Frit No. 14, 13.1 Feldspar, 16.3 SiO_2, 3.7 ZnO, 3.7 $CaCO_3$, 4.2 E.P. Kaolin, and 12.9 $ZrSiO_4$; Frit No. 14 composed of 46.15 SiO_2, 23.25 B_2O_3, 20.2 CaO, and 10.4 NaKO; sprayed; fired at 1366 K; black matte finish; measured relative to graphite. [Author's designation: E-318]
3*	69	1963	840	0°		33.3 MnO_2 and 33.3 Fe_2O_3 in 33.3 base glaze No. 3 binder (0.0508-0.0762 mm thick) on alumina (6.35 mm thick) substrate; base glaze No. 3 composed of 46.1 Frit No. 14, 13.1 Feldspar, 16.3 SiO_2, 3.7 ZnO, 3.7 $CaCO_3$, 4.2 E.P. Kaolin, and 12.9 $ZrSiO_4$; Frit No. 14 composed of 46.15 SiO_2, 23.25 B_2O_3, 20.2 CaO, and 10.4 NaKO; sprayed; fired at 1561 K; black matte finish; measured relative to graphite. [Author's designation: E-327]

DATA TABLE NO. 156 NORMAL TOTAL EMITTANCE OF MANGANESE OXIDE PIGMENTED COATINGS (ENAMEL)

[Temperature, T, K; Emittance, ϵ]

T	ϵ
CURVE 1*	
887	0.84
CURVE 2*	
861	0.84
CURVE 3*	
840	0.82

* No plot given

SPECIFICATION TABLE NO. 157 NORMAL TOTAL EMITTANCE OF NICKEL OXIDE PIGMENTED COATINGS (ENAMEL)

Curve No.	Ref. No.	Year	Temperature Range, K	Geometry θ'	Reported Error, %	Composition (weight percent), Specifications and Remarks
1*	69	1963	785	0°		33.3 NiO and 33.3 Fe_2O_3 in 33.3 base glaze No. 3 binder (0.0508-0.0763 mm thick) on alumina (6.35 mm thick) substrate; base glaze No. 3 composed of 46.1 Frit No. 14, 13.1 Feldspar, 16.3 SiO_2, 3.7 ZnO, 3.7 $CaCO_3$, 4.2 E.P. Kaolin, and 12.9 $ZrSiO_4$; Frit No. 14 composed of 46.15 SiO_2, 23.25 B_2O_3, 20.2 CaO and 10.4 NaKO; sprayed; fired at 1561 K; black matte finish; measured relative to graphite. [Author's designation: E-326]
2*	69	1962	783	0°		33.3 NiO and 33.3 Co_2O_3 in 33.3 base glaze No. 3 binder (0.0508-0.0762 mm thick) on alumina (6.35 mm thick) substrate; base glaze No. 3 composed of 46.1 Frit No. 14, 13.1 Feldspar, 16.3 SiO_2, 3.7 ZnO, 3.7 $CaCO_3$, 4.2 E.P. Kaolin, and 12.9 $ZrSiO_4$; Frit No. 14 composed of 46.15 SiO_2, 23.25 B_2O_3, 20.2 CaO, and 10.4 NaKO; sprayed; fired at 1561 K; blue-black matte finish; measured relative to graphite. [Author's designation: E-324]
3*	69	1963	761	0°		25 NiO and 25 Co_2O_3 in 50 base glaze No. 3 binder (0.0508-0.0762 mm thick) on alumina (6.35 mm thick) substrate; base glaze No. 3 composed of 46.1 Frit No. 14, 13.1 Feldspar, 16.3 SiO_2, 3.7 ZnO, 3.7 $CaCO_3$, 4.2 E.P. Kaolin, and 12.9 $ZrSiO_4$; Frit No. 14 composed of 46.15 SiO_2, 23.25 B_2O_3, 20.2 CaO, and 10.4 NaKO; sprayed; fired at 1561 K; blue-black matte finish; measured relative to graphite. [Author's designation: E-325]

DATA TABLE NO. 157 NORMAL TOTAL EMITTANCE OF NICKEL OXIDE PIGMENTED COATINGS (ENAMEL)

[Temperature, T, K; Emittance, ϵ]

T	ϵ
CURVE 1*	
785	0.83
CURVE 2*	
783	0.80
CURVE 3*	
761	0.86

* No plot given

SPECIFICATION TABLE NO. 158 NORMAL TOTAL EMITTANCE OF POTASSIUM TITANATE PIGMENTED COATINGS (ENAMEL)

Curve No.	Ref. No.	Year	Temperature Range, K	Geometry θ'	Reported Error, %	Composition (weight percent), Specifications and Remarks
1*	83	1963	355	~0°		Potassium titanate porcelain enamel; measured in vacuum (1.6 x 10^{-4} mm Hg).

DATA TABLE NO. 158 NORMAL TOTAL EMITTANCE OF POTASSIUM TITANATE PIGMENTED COATINGS (ENAMEL)

[Temperature, T, K; Emittance, ϵ]

T	ϵ
CURVE 1*	
355	0.821

* No plot given

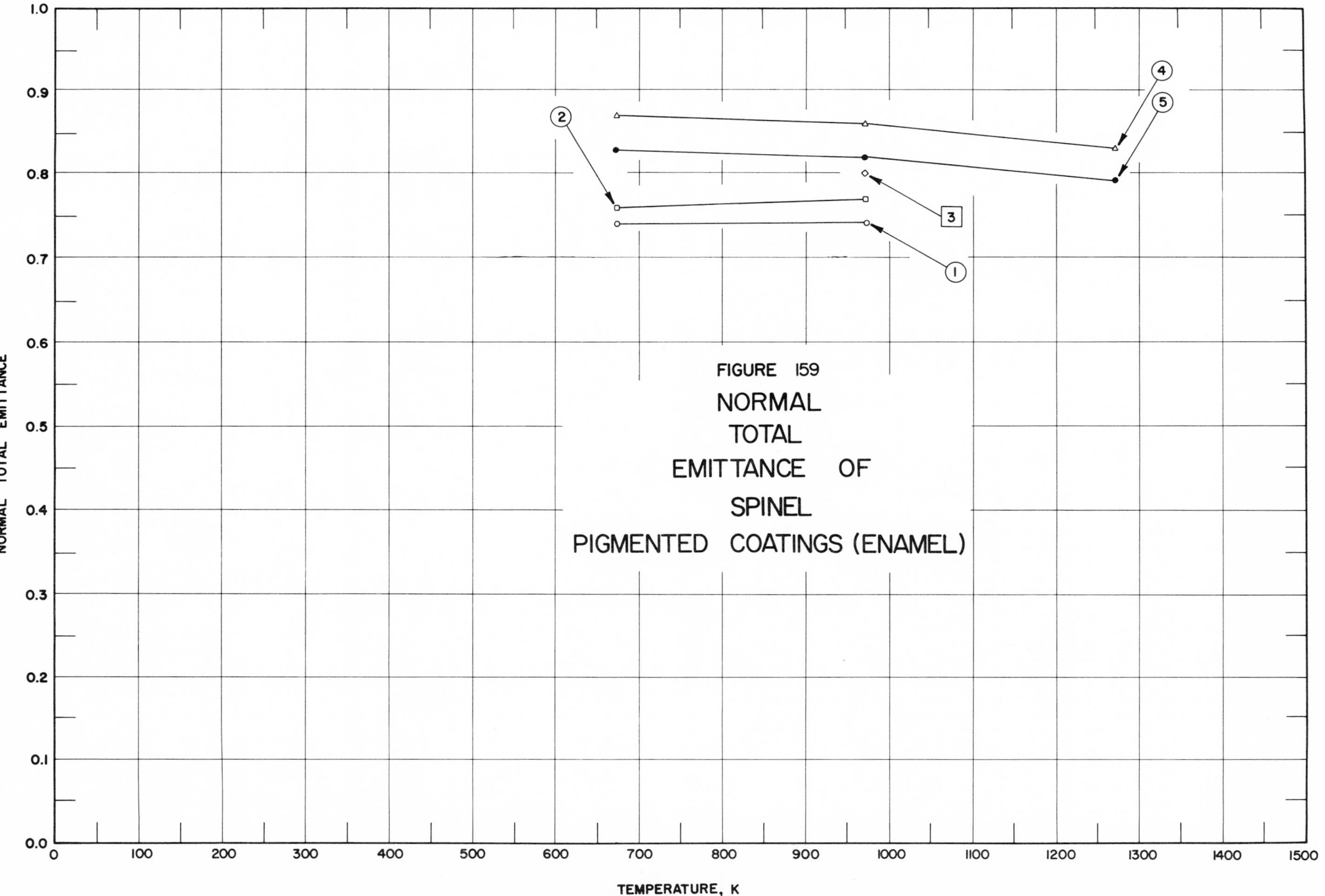
FIGURE 159
NORMAL
TOTAL
EMITTANCE OF
SPINEL
PIGMENTED COATINGS (ENAMEL)
NORMAL TOTAL EMITTANCE
TEMPERATURE, K
0.0
0.1
0.2
0.3
0.4
0.5
0.6
0.7
0.8
0.9
1.0
0
100
200
300
400
500
600
700
800
900
1000
1100
1200
1300
1400
1500
1
2
3
4
5

SPECIFICATION TABLE NO. 159 NORMAL TOTAL EMITTANCE OF SPINEL PIGMENTED COATINGS (ENAMEL)

Curve No.	Ref. No.	Year	Temperature Range, K	Geometry θ'	Reported Error, %	Composition (weight percent), Specifications and Remarks
1	41	1960	673-973	~0°	±4	40 CoO · Cr_2O_3 spinel in 60 NBS Frit No. 332 binder (0.187 mm thick) on Inconel substrate; NBS Frit No. 332 composed of 44.0 BaO, 37.5 SiO_2, 6.5 B_2O_3, 5.0 ZnO, 3.5 CaO, 2.5 ZrO_2, and 1.0 Al_2O_3; grit blasted Inconel substrate from Whitehead Metals, Inc.; sprayed; fired 3-10 min at 1298 K. [Authors' designation: Enamel B-7]
2	41	1960	673-973	~0°	±4	50 CoO · Cr_2O_3 spinel in 50 NBS Frit No. 332 binder (0.132 mm thick) on Inconel substrate; NBS Frit No. 332 composed of 44.0 BaO, 37.5 SiO_2, 6.5 B_2O_3, 5.0 ZnO, 3.5 CaO, 2.5 ZrO_2, and 1.0 Al_2O_3; grit blasted Inconel substrate from Whitehead Metals, Inc.; sprayed; fired 3-10 min at 1298 K. [Authors' designation: Enamel B-13]
3	41	1960	973	~0°	±4	40 CoO · Fe_2O_3 spinel in 60 NBS Frit No. 332 binder (0.175 mm thick) on Inconel substrate; NBS Frit No. 332 composed of 44.0 BaO, 37.5 SiO_2, 6.5 B_2O_3, 5.0 ZnO, 3.5 CaO, 2.5 ZrO_2, and 1.0 Al_2O_3; grit blasted Inconel substrate from Whitehead Metals, Inc.; sprayed; fired 3-10 min at 1298 K. [Authors' designation: Enamel B-11]
4	41	1960	673-1273	~0°	±4	40 CoO · Mn_2O_3 spinel in 60 NBS Frit No. 332 binder (0.051 mm thick) on Inconel substrate; NBS Frit No. 332 composed of 44.0 BaO, 37.5 SiO_2, 6.5 B_2O_3, 5.0 ZnO, 3.5 CaO, 2.5 ZrO_2, and 1.0 Al_2O_3; grit blasted Inconel substrate from Whitehead Metals, Inc.; sprayed; fired 3-10 min at 1298 K. [Authors' designation: Enamel B-12]
5	41	1960	673-1273	~0°	±4	Similar to above specimen and conditions except coating 0.175 mm thick.
6*	41	1960	973	~0°	±4	40 NiO · Cr_2O_3 spinel in 60 NBS Frit No. 332 binder (0.196 mm thick) on Inconel substrate; NBS Frit No. 332 composed of 44.0 BaO, 37.5 SiO_2, 6.5 B_2O_3, 5.0 ZnO, 3.5 CaO, 2.5 ZrO_2, and 1.0 Al_2O_3; grit blasted Inconel substrate from Whitehead Metals, Inc.; sprayed; fired 3-10 min at 1298 K. [Authors' designation: Enamel B-8]
7*	41	1960	973	~0°	±4	40 NiO · Fe_2O_3 spinel in 60 NBS Frit No. 332 binder (0.206 mm thick) on Inconel substrate; NBS Frit No. 332 composed of 44.0 BaO, 37.5 SiO_2, 6.5 B_2O_3, 5.0 ZnO, 3.5 CaO, 2.5 ZrO_2, and 1.0 Al_2O_3; grit blasted Inconel substrate from Whitehead Metals, Inc.; sprayed; fired 3-10 min at 1298 K. [Authors' designation: Enamel B-9]

* Not shown on plot

DATA TABLE NO. 159 NORMAL TOTAL EMITTANCE OF SPINEL PIGMENTED COATINGS (ENAMEL)

[Temperature, T, K; Emittance, ϵ]

T	ϵ
CURVE 1	
673	0.74
973	0.74
CURVE 2	
673	0.76
973	0.77
CURVE 3	
973	0.80
CURVE 4	
673	0.87
973	0.86
1273	0.83
CURVE 5	
673	0.83
973	0.82
1273	0.79
CURVE 6*	
973	0.81
CURVE 7*	
973	0.80

* Not shown on plot

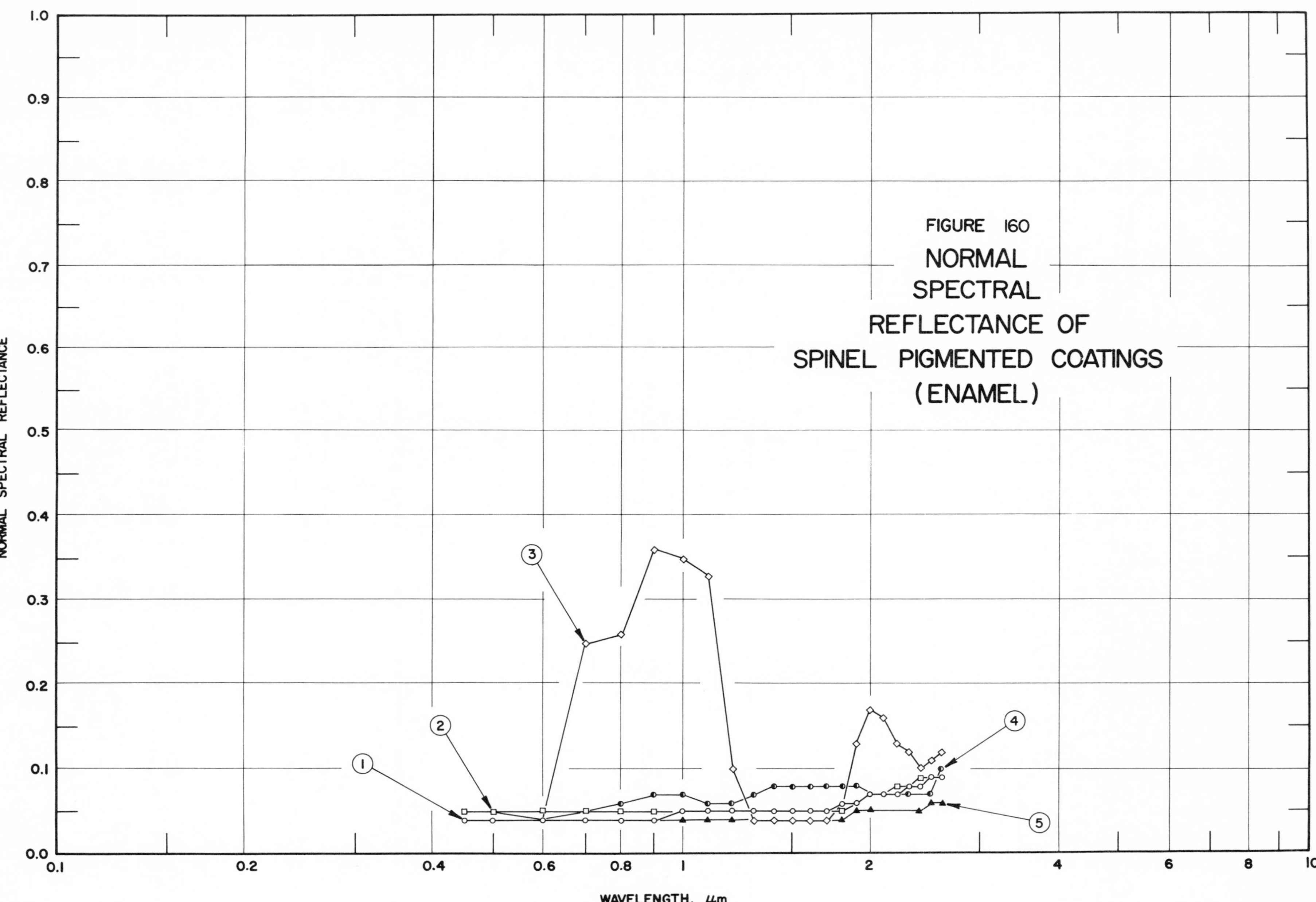
FIGURE 160
NORMAL
SPECTRAL
REFLECTANCE OF
SPINEL PIGMENTED COATINGS
(ENAMEL)
NORMAL SPECTRAL REFLECTANCE
WAVELENGTH, μm
0.0
0.1
0.2
0.3
0.4
0.5
0.6
0.7
0.8
0.9
1.0
0.1
0.2
0.4
0.6
0.8
1
2
4
6
8
10
1
2
3
4
5

SPECIFICATION TABLE NO. 160 NORMAL SPECTRAL REFLECTANCE OF SPINEL PIGMENTED COATINGS (ENAMEL)

Curve No.	Ref. No.	Year	Temperature, K	Wavelength Range, μm	Geometry θ	Geometry θ'	Geometry ω'	Reported Error, %	Composition (weight percent), Specifications, and Remarks
1	41	1960	298	0.45-2.60	~0°		2π		40 $CoO \cdot Fe_2O_3$ spinel in 60 NBS Frit No. 332 binder (0.081 mm thick) on Inconel substrate; NBS Frit No. 332 composed of 44.0 BaO, 37.5 SiO_2, 6.5 B_2O_3, 5.0 ZnO, 3.5 CaO, 2.5 ZrO_2, and 1.0 Al_2O_3; grit blasted Inconel substrate from Whitehead Metals, Inc.; sprayed; fired 3-10 min at 1298 K; measured relative to MgO. [Authors' designation: Enamel B-11]
2	41	1960	298	0.45-2.60	~0°		2π		40 $CoO \cdot Mn_2O_3$ spinel in 60 NBS Frit No. 332 binder (0.099 mm thick) on Inconel substrate; NBS Frit No. 332 composed of 44.0 BaO, 37.5 SiO_2, 6.5 B_2O_3, 5.0 ZnO, 3.5 CaO, 2.5 ZrO_2, and 1.0 Al_2O_3; grit blasted Inconel substrate from Whitehead Metals, Inc.; sprayed; fired 3-10 min at 1298 K; measured relative to MgO. [Authors' designation: Enamel B-12]
3	41	1960	298	0.45-2.60	~0°		2π		50 $CoO \cdot Cr_2O_3$ spinel in 50 NBS Frit No. 332 binder (0.124 mm thick) on Inconel substrate; NBS Frit No. 332 composed of 44.0 BaO, 37.5 SiO_2, 6.5 B_2O_3, 5.0 ZnO, 3.5 CaO, 2.5 ZrO_2, and 1.0 Al_2O_3; grit blasted Inconel substrate from Whitehead Metals, Inc.; sprayed; fired 3-10 min at 1298 K; measured relative to MgO. [Authors' designation: Enamel B-13]
4	41	1960	298	0.45-2.60	~0°		2π		40 $NiO \cdot Cr_2O_3$ spinel in 60 NBS Frit No. 332 binder (0.079 mm thick) on Inconel substrate; NBS Frit No. 332 composed of 44.0 BaO, 37.5 SiO_2, 6.5 B_2O_3, 5.0 ZnO, 3.5 CaO, 2.5 ZrO_2, and 1.0 Al_2O_3; grit blasted Inconel substrate from Whitehead Metals, Inc.; sprayed; fired 3-10 min at 1298 K; measured relative to MgO. [Authors' designation: Enamel B-8]
5	41	1960	298	0.45-2.60	~0°		2π		40 $NiO \cdot Fe_2O_3$ spinel in 60 NBS Frit No. 332 binder (0.104 mm thick) on Inconel substrate; NBS Frit No. 332 composed of 44.0 BaO, 37.5 SiO_2, 6.5 B_2O_3, 5.0 ZnO, 3.5 CaO, 2.5 ZrO_2, and 1.0 Al_2O_3; grit blasted Inconel substrate from Whitehead Metals, Inc.; sprayed; fired 3-10 min at 1298 K; measured relative to MgO. [Authors' designation: Enamel B-9]

DATA TABLE NO. 160 NORMAL SPECTRAL REFLECTANCE OF SPINEL PIGMENTED COATINGS (ENAMEL)

[Wavelength, λ, μm; Reflectance, ρ; Temperature, T, K]

λ	ρ
CURVE 1 T = 298	
0.45	0.04
0.50	0.04
0.60	0.04
0.70	0.04
0.80	0.04
0.90	0.04
1.00	0.05
1.10	0.05
1.20	0.05
1.30	0.05
1.40	0.05
1.50	0.05
1.60	0.05
1.70	0.05
1.80	0.06
1.90	0.06
2.00	0.07
2.10	0.07
2.20	0.07
2.30	0.08
2.40	0.08
2.50	0.09
2.60	0.09
CURVE 2 T = 298	
0.45	0.05
0.50	0.05
0.60	0.05
0.70	0.05
0.80	0.05
0.90	0.05
1.00	0.05*
1.10	0.05*
1.20	0.05*
1.30	0.05*
1.40	0.05*
1.50	0.05*
1.60	0.05*
1.70	0.05*
1.80	0.05
1.90	0.06*
2.00	0.07*
CURVE 2 (cont.)	
2.10	0.07*
2.20	0.08
2.30	0.08*
2.40	0.09
2.50	0.09*
2.60	0.09*
CURVE 3 T = 298	
0.45	0.04*
0.50	0.04*
0.60	0.04*
0.70	0.25
0.80	0.26
0.90	0.36
1.00	0.35
1.10	0.33
1.20	0.10
1.30	0.04
1.40	0.04
1.50	0.04
1.60	0.04
1.70	0.04
1.80	0.06*
1.90	0.13
2.00	0.17
2.10	0.16
2.20	0.13
2.30	0.12
2.40	0.10
2.50	0.11
2.60	0.12
CURVE 4 T = 298	
0.45	0.05*
0.50	0.05*
0.60	0.04*
0.70	0.05*
0.80	0.06
0.90	0.07
1.00	0.07
1.10	0.06
CURVE 4 (cont.)	
1.20	0.06
1.30	0.07
1.40	0.08
1.50	0.08
1.60	0.08
1.70	0.08
1.80	0.08
1.90	0.08
2.00	0.07*
2.10	0.07*
2.20	0.07*
2.30	0.07
2.40	0.07*
2.50	0.07
2.60	0.10
CURVE 5 T = 298	
0.45	0.04*
0.50	0.04*
0.60	0.04*
0.70	0.04*
0.80	0.04*
0.90	0.04*
1.00	0.04
1.10	0.04
1.20	0.04
1.30	0.04*
1.40	0.04*
1.50	0.04*
1.60	0.04*
1.70	0.04*
1.80	0.04
1.90	0.05
2.00	0.05
2.10	0.05*
2.20	0.05*
2.30	0.05*
2.40	0.05
2.50	0.06
2.60	0.06

* Not shown on plot

SPECIFICATION TABLE NO. 161 NORMAL TOTAL EMITTANCE OF TIN OXIDE PIGMENTED COATINGS (ENAMEL)

Curve No.	Ref. No.	Year	Temperature Range, K	Geometry θ'	Reported Error, %	Composition (weight percent), Specifications and Remarks
1*	41	1960	673-1073	~0°	± 4	20 SnO_2 and 20 CeO_2 in 60 NBS Frit No. 332 binder (0.051 mm thick) on Inconel substrate; NBS Frit No. 332 composed of 44.0 BaO, 37.5 SiO_2, 6.5 B_2O_3, 5.0 ZnO, 3.5 CaO, 2.5 ZrO_2, and 1.0 Al_2O_3; grit blasted Inconel substrate from Whitehead Metals, Inc.; sprayed; fired 3-10 min at 1298 K. [Authors' designation: Enamel W-3]
2*	41	1960	673-1073	~0°	± 4	Similar to above specimen and conditions except coating 0.152 mm thick.

DATA TABLE NO. 161 NORMAL TOTAL EMITTANCE OF TIN OXIDE PIGMENTED COATINGS (ENAMEL)

[Temperature, T, K; Emittance, ϵ]

T	ϵ
CURVE 1*	
673	0.79
873	0.73
1073	0.70
CURVE 2*	
673	0.75
873	0.65
1073	0.60

* No plot given

SPECIFICATION TABLE NO. 162 NORMAL SPECTRAL REFLECTANCE OF TIN OXIDE PIGMENTED COATINGS (ENAMEL)

Curve No.	Ref. No.	Year	Temperature, K	Wavelength Range, μm	Geometry θ	θ'	ω'	Reported Error, %	Composition (weight percent), Specifications, and Remarks
1*	41	1960	298	0.42-2.60	~0°		2π		20 SnO_2 and 20 CeO_2 in 60 NBS Frit No. 332 binder (0.051 mm thick) on Inconel substrate; NBS Frit No. 332 composed of 44.0 BaO, 37.5 SiO_2, 6.5 B_2O_3, 5.0 ZnO, 3.5 CaO, 2.5 ZrO_2, and 1.0 Al_2O_3; grit blasted Inconel substrate from Whitehead Metals, Inc.; sprayed; fired 3-10 min at 1298 K; measured relative to MgO. [Authors' designation: Enamel W-3]
2*	41	1960	298	0.45-2.60	~0°		2π		Similar to above specimen and conditions except coating 0.152 mm thick.

DATA TABLE NO. 162 NORMAL SPECTRAL REFLECTANCE OF TIN OXIDE PIGMENTED COATINGS (ENAMEL)

[Wavelength, λ, μm; Reflectance, ρ; Temperature, T, K]

λ	ρ	λ	ρ	λ	ρ
CURVE 1*		**CURVE 1 (cont.)***		**CURVE 2 (cont.)***	
T = 298		1.90	0.43	0.90	0.67
0.42	0.54	2.00	0.42	1.00	0.67
0.45	0.53	2.10	0.41	1.10	0.67
0.50	0.52	2.20	0.41	1.20	0.66
0.60	0.51	2.30	0.41	1.30	0.66
0.70	0.50	2.40	0.41	1.40	0.65
0.80	0.49	2.50	0.41	1.50	0.65
0.90	0.48	2.60	0.41	1.60	0.64
1.00	0.47			1.70	0.64
1.10	0.47	**CURVE 2***		1.80	0.63
1.20	0.46	T = 298		1.90	0.63
1.30	0.45			2.00	0.63
1.40	0.45	0.45	0.65	2.10	0.63
1.50	0.44	0.50	0.66	2.20	0.62
1.60	0.44	0.60	0.67	2.30	0.63
1.70	0.43	0.70	0.67	2.40	0.63
1.80	0.43	0.80	0.67	2.50	0.62
				2.60	0.62

* No plot given

SPECIFICATION TABLE NO. 163 NORMAL SPECTRAL REFLECTANCE OF ZIRCONIUM OXIDE PIGMENTED COATINGS (ENAMEL)

Curve No.	Ref. No.	Year	Temperature, K	Wavelength Range, μm	Geometry θ	θ'	ω'	Reported Error, %	Composition (weight percent), Specifications, and Remarks
1*	41	1960	298	0.42-2.60	~0°		2π		20 ZrO_2 and 20 CeO_2 in 60 NBS Frit No. 332 binder (0.061 mm thick) on Inconel substrate; NBS Frit No. 332 composed of 44.0 BaO, 37.5 SiO_2, 6.5 B_2O_3, 5.0 ZnO, 3.5 CaO, 2.5 ZrO_2, and 1.0 Al_2O_3; grit blasted Inconel substrate from Whitehead Metals, Inc.; sprayed; fired 3-10 min at 1298 K; measured relative to MgO. [Authors' designation: Enamel W-4]

DATA TABLE NO. 163 NORMAL SPECTRAL REFLECTANCE OF ZIRCONIUM OXIDE PIGMENTED COATINGS (ENAMEL)

[Wavelength, λ, μm; Reflectance, ρ; Temperature, T, K]

λ	ρ	λ	ρ
CURVE 1*		CURVE 1 (cont.)*	
0.42	0.57	2.10	0.53
0.45	0.61	2.20	0.53
0.50	0.62	2.30	0.53
0.60	0.62	2.40	0.53
0.70	0.62	2.50	0.53
0.80	0.61	2.60	0.53
0.90	0.60		
1.00	0.59		
1.10	0.59		
1.20	0.58		
1.30	0.57		
1.40	0.57		
1.50	0.56		
1.60	0.54		
1.70	0.54		
1.80	0.54		
1.90	0.55		
2.00	0.54		

* No plot given

1. PIGMENTED COATINGS (continued)

D. Trade Names

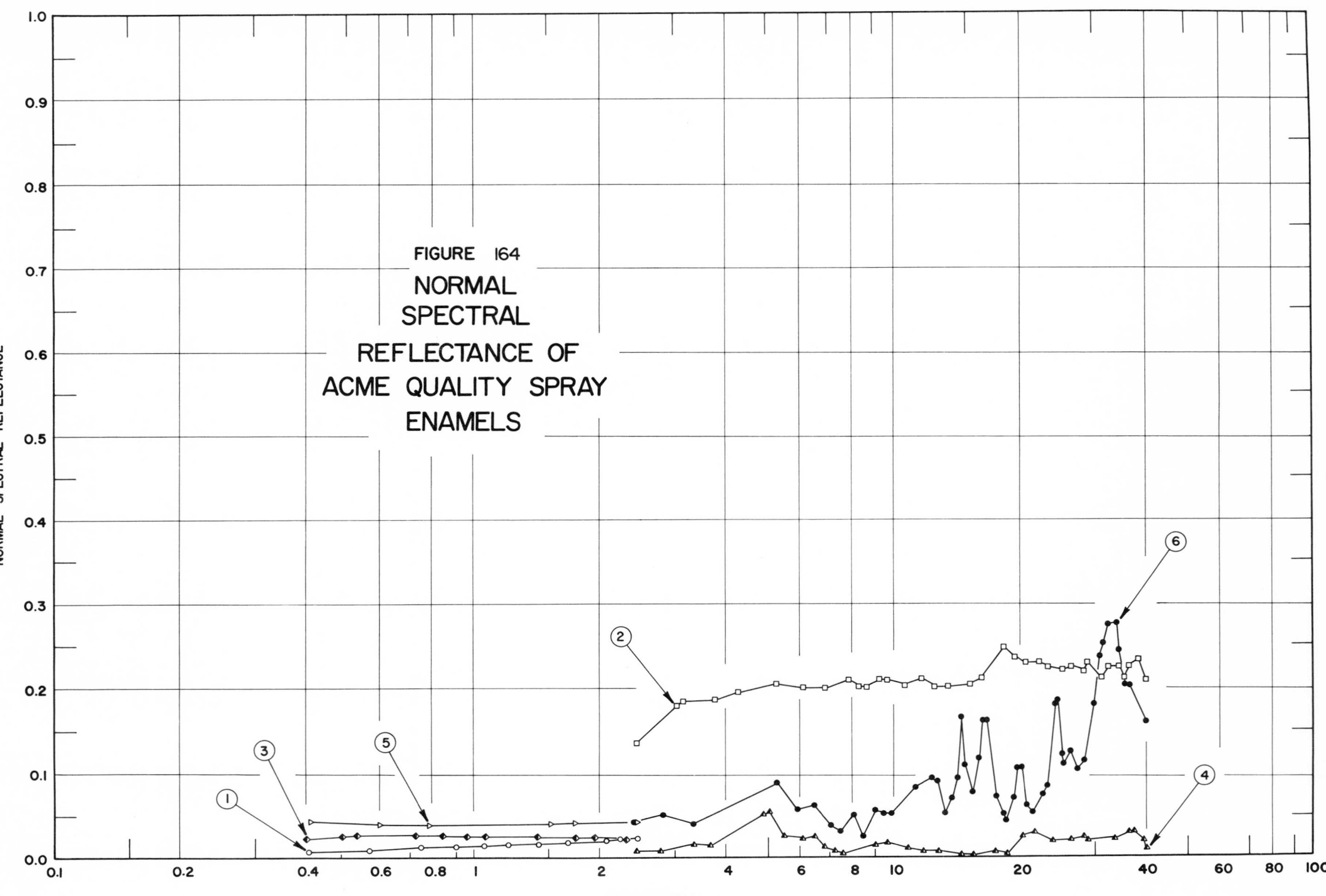

FIGURE 164 NORMAL SPECTRAL REFLECTANCE OF ACME QUALITY SPRAY ENAMELS

SPECIFICATION TABLE NO. 164 NORMAL SPECTRAL REFLECTANCE OF ACME QUALITY SPRAY ENAMELS

Curve No.	Ref. No.	Year	Temperature, K	Wavelength Range, μm	Geometry θ	θ'	ω'	Reported Error, %	Composition (weight percent), Specifications, and Remarks
1	102	1964	~296	0.407-2.484	8°	8°			Acme Quality Spray Enamel No. 803 aluminum (~1.06 x 10^{-4} mm thick) on aluminum substrate; data extracted from smooth curve; measured at least 24 hrs after final coating.
2	102	1964	~313	2.45-40.00	30°	30°			Above specimen and conditions.
3	102	1964	~296	0.400-2.493	8°	8°			Acme Quality Spray Enamel No. 800 Appliance White (~1.44 x 10^{-4} mm thick) on aluminum substrate; data extracted from smooth curve; measured at least 24 hrs after final coating.
4	102	1964	~313	2.45-40.00	30°	30°			Above specimen and conditions.
5	102	1964	~296	0.411-2.491	8°	8°			Acme Quality Spray Enamel No. 801 Brilliant Black (~1.11 x 10^{-4} mm thick) on aluminum substrate; data extracted from smooth curve; measured at least 24 hrs after final coating.
6	102	1964	~313	2.45-40.00	30°	30°			Above specimen and conditions.

DATA TABLE NO. 164 NORMAL SPECTRAL REFLECTANCE OF ACME QUALITY SPRAY ENAMELS

[Wavelength, λ, μm; Reflectance, ρ; Temperature, T, K]

λ	ρ
CURVE 1	
T ~ 296	
0.407	0.009
0.563	0.011
0.752	0.012
0.909	0.013
1.065	0.015
1.221	0.017
1.420	0.018
1.695	0.020
2.090	0.022
2.251	0.023
2.365	0.023*
2.484	0.024
CURVE 2	
T ~ 313	
2.45	0.138
3.05	0.180
3.18	0.186
3.78	0.188
4.26	0.199
5.22	0.209
6.11	0.201
6.94	0.201
7.81	0.211
8.38	0.202
8.69	0.202
9.39	0.212
9.76	0.212
10.75	0.206
11.70	0.213
12.67	0.202
13.54	0.202
15.16	0.209
16.31	0.215
17.41	0.230*
18.40	0.250
18.80	0.250*
19.56	0.238
20.81	0.232
22.16	0.232
22.82	0.234*
23.66	0.227
24.32	0.227*

λ	ρ
CURVE 2 (cont.)	
25.14	0.224
26.66	0.229
27.83	0.222*
28.56	0.222
29.08	0.231
29.81	0.223*
31.13	0.214
31.51	0.218*
31.99	0.215*
32.54	0.228
34.92	0.228
35.19	0.226*
35.63	0.214
36.60	0.229
37.57	0.225*
38.57	0.236
39.19	0.222*
40.00	0.211
CURVE 3	
T ~ 296	
0.400	0.024
0.489	0.027
0.525	0.028
0.736	0.028
0.848	0.028
0.970	0.027
1.072	0.027
1.413	0.027
1.598	0.027*
1.777	0.026
1.989	0.026
2.066	0.025*
2.244	0.023*
2.319	0.022
2.493	0.022*
CURVE 4	
T ~ 313	
2.45	0.010
2.80	0.010
3.35	0.018
3.68	0.017

λ	ρ
CURVE 4 (cont.)	
4.94	0.054
5.09	0.057
5.26	0.053*
5.51	0.029
5.69	0.024*
6.01	0.023
6.51	0.026
6.98	0.014
7.21	0.010
7.60	0.007
9.02	0.019
9.24	0.020*
9.65	0.020
10.89	0.013
11.74	0.010
12.70	0.010
14.59	0.004
15.57	0.004
17.51	0.010
18.77	0.009
20.19	0.029
20.94	0.031*
21.71	0.031
24.03	0.022
26.50	0.024
27.32	0.022
28.50	0.025
29.03	0.022
33.67	0.026
36.23	0.032
37.77	0.032
39.19	0.021
40.00	0.011
CURVE 5	
T ~ 296	
0.411	0.043
0.602	0.040
0.784	0.040
1.540	0.041
1.741	0.042
2.491	0.045

λ	ρ
CURVE 6	
T ~ 313	
2.45	0.043
2.83	0.052
3.37	0.041
5.37	0.090
5.98	0.060
6.53	0.064
7.15	0.040
7.53	0.034
8.06	0.052
8.52	0.029
9.09	0.058
9.54	0.055
9.99	0.055
11.38	0.086
12.36	0.098
12.63	0.091
13.23	0.055
13.51	0.055*
13.86	0.071
14.13	0.098
14.50	0.169
14.86	0.112
15.38	0.080
15.60	0.083*
16.02	0.120
16.38	0.162
16.70	0.162
16.91	0.153*
17.65	0.075
18.13	0.056
18.55	0.046
19.25	0.072
19.92	0.110
20.13	0.110
20.83	0.065
21.20	0.058
22.80	0.079
23.16	0.088
24.18	0.183
24.57	0.186
24.78	0.178*
25.24	0.123
25.49	0.115
26.39	0.129

λ	ρ
CURVE 6 (cont.)	
27.51	0.108
27.82	0.108*
28.59	0.118
30.01	0.184
31.09	0.240
31.46	0.252
32.33	0.276
34.00	0.279
34.52	0.246
34.98	0.246*
35.51	0.205
36.39	0.205
40.00	0.161

* Not shown on plot

SPECIFICATION TABLE NO. 165 NORMAL SPECTRAL ABSORPTANCE OF AISI 99 GRAY PAINT

Curve No.	Ref. No.	Year	Temperature, K	Wavelength Range, μm	Geometry θ	Reported Error, %	Composition (weight percent), Specifications, and Remarks
1*	103	1954	~298	0.400-2.600	~0°		AISI 99 gray paint on quartz substrate; data extracted from smooth curve.

DATA TABLE NO. 165 NORMAL SPECTRAL ABSORPTANCE OF AISI 99 GRAY PAINT

[Wavelength, λ, μm; Absorptance, α; Temperature, T, K]

λ	α
CURVE 1*	
T ~ 298	
0.400	0.600
0.417	0.545
0.477	0.517
0.597	0.531
0.719	0.545
1.024	0.555
1.163	0.528
1.307	0.535
1.478	0.504
1.612	0.492
1.800	0.525
1.999	0.579
2.200	0.646
2.400	0.656
2.600	0.687

* No plot given

SPECIFICATION TABLE NO. 166 NORMAL SPECTRAL TRANSMITTANCE OF AISI 99 GRAY PAINT

Curve No.	Ref. No.	Year	Temperature, K	Wavelength Range, μm	Geometry θ	θ'	ω'	Reported Error, %	Composition (weight percent), Specifications, and Remarks
1*	103	1954	~298	0.400-2.600	~0°		2π		AISI 99 gray paint on quartz substrate; data extracted from smooth curve; the accuracies of the measurements in the ultraviolet, visible, and infrared regions are approx. 2, 1, and 2 percent respectively.

DATA TABLE NO. 166 NORMAL SPECTRAL TRANSMITTANCE OF AISI 99 GRAY PAINT

[Wavelength, λ, μm; Transmittance, τ; Temperature, T, K]

λ	τ
CURVE 1*	T ~ 298
0.400	0.003
0.596	0.018
0.800	0.028
0.999	0.033
1.189	0.033
1.407	0.044
1.800	0.040
2.000	0.040
2.200	0.026
2.400	0.024
2.600	0.024

* No plot given

SPECIFICATION TABLE NO. 167 NORMAL SPECTRAL REFLECTANCE OF BOYSEN PAINTS

Curve No.	Ref. No.	Year	Temperature, K	Wavelength Range, μm	Geometry θ	θ'	ω'	Reported Error, %	Composition (weight percent), Specifications, and Remarks
1*	104	1953	298	1.00-15.0	5°		2π	2	Boysen No. 11 flat black paint (0.15 mm thick); substrate unknown; data extracted from smooth curve; computed from R(2π, 5°).

DATA TABLE NO. 167 NORMAL SPECTRAL REFLECTANCE OF BOYSEN PAINTS

[Wavelength, λ, μm; Reflectance, ρ; Temperature, T, K]

λ	ρ	λ	ρ	λ	ρ
CURVE 1* T = 298		CURVE 1 (cont.)*		CURVE 1 (cont.)*	
		6.20	0.100	12.2	0.110
1.00	0.050	6.50	0.105	12.8	0.100
1.80	0.060	7.00	0.095	13.0	0.105
2.00	0.060	7.20	0.105	13.2	0.100
2.20	0.065	7.50	0.095	13.5	0.110
2.60	0.060	7.60	0.095	14.0	0.100
2.80	0.060	8.00	0.090	14.2	0.110
3.00	0.070	9.00	0.110	14.5	0.110
3.50	0.070	9.20	0.110	14.7	0.105
3.70	0.060	9.50	0.110	15.0	0.125
4.00	0.070	9.80	0.130		
4.20	0.100	10.0	0.130		
4.40	0.070	10.5	0.115		
4.80	0.070	11.0	0.115		
5.00	0.075	11.4	0.110		
5.50	0.075	11.7	0.115		
5.80	0.070	12.0	0.110		
6.00	0.090				

* No plot given

SPECIFICATION TABLE NO. 168 NORMAL SPECTRAL REFLECTANCE OF BROMA ALKYD PAINTS

Curve No.	Ref. No.	Year	Temperature, K	Wavelength Range, μm	Geometry θ θ' ω'	Reported Error, %	Composition (weight percent), Specifications, and Remarks
1*	102	1964	~296	0.398-2.480	8° 8°		Broma Alkyd Enamel No. 113, light blue, (~0.80 x 10^{-4} mm thick) on aluminum substrate; data extracted from smooth curve; measured at least 24 hrs after final coating.
2*	102	1964	~313	2.45-40.00	30° 30°		Above specimen and conditions.

DATA TABLE NO. 168 NORMAL SPECTRAL REFLECTANCE OF BROMA ALKYD PAINTS

[Wavelength, λ, μm; Reflectance, ρ; Temperature, T, K]

λ	ρ	λ	ρ	λ	ρ	λ	ρ	λ	ρ
CURVE 1*		CURVE 2 (cont.)*		CURVE 2 (cont.)*		CURVE 2 (cont.)*		CURVE 2 (cont.)*	
T ~ 296									
		3.92	0.040	17.40	0.043	30.41	0.089	38.93	0.115
0.398	0.038	4.90	0.083	18.30	0.053	30.94	0.082	39.31	0.115
0.462	0.041	5.05	0.083	19.10	0.054	31.32	0.099	40.00	0.111
0.492	0.041	5.70	0.057	20.05	0.086	31.89	0.095		
0.561	0.041	6.23	0.042	20.30	0.086	32.51	0.099		
0.671	0.041	7.41	0.025	22.26	0.067	33.24	0.093		
0.889	0.044	7.98	0.036	22.90	0.063	33.47	0.095		
1.346	0.044	8.51	0.021	23.36	0.063	33.79	0.103		
1.983	0.047	8.91	0.041	25.35	0.083	33.99	0.105		
2.207	0.047	9.13	0.043	25.96	0.083	34.44	0.103		
2.480	0.047	9.66	0.040	26.43	0.078	34.93	0.111		
		10.00	0.040	27.12	0.062	35.51	0.093		
CURVE 2*		10.99	0.026	27.33	0.062	36.46	0.111		
T ~ 313		11.59	0.024	28.26	0.079	36.74	0.111		
		12.91	0.003	29.14	0.085	37.61	0.105		
2.45	0.027	14.16	0.025	29.65	0.090	37.88	0.115		
2.97	0.033	14.70	0.031	29.94	0.088	38.16	0.115		
3.54	0.049	15.68	0.039	30.09	0.088	38.57	0.114		

* No plot given

SPECIFICATION TABLE NO. 169 NORMAL SPECTRAL REFLECTANCE OF BROMA METALLIC PAINTS

Curve No.	Ref. No.	Year	Temperature, K	Wavelength Range, μm	Geometry θ θ' ω'	Reported Error, %	Composition (weight percent), Specifications, and Remarks
1*	102	1964	~296	0.406-2.494	8° 8°		Broma Metallic Enamel No. 102 gold leaf (~0.20 x 10^{-4} mm thick) on aluminum substrate; data extracted from smooth curve; measured at least 24 hrs after final coating.
2*	102	1964	~313	2.45-40.00	30° 30°		Above specimen and conditions.

DATA TABLE NO. 169 NORMAL SPECTRAL REFLECTANCE OF BROMA METALLIC PAINTS

[Wavelength, λ, μm; Reflectance, ρ; Temperature, T, K]

λ	ρ	λ	ρ	λ	ρ	λ	ρ
CURVE 1* T ~ 296		CURVE 2* T ~ 313		CURVE 2 (cont.)*		CURVE 2 (cont.)*	
0.406	0.009	2.45	0.015	15.35	0.029	37.45	0.081
0.562	0.017	3.29	0.022	16.39	0.035	38.04	0.101
0.629	0.020	4.03	0.019	17.97	0.035	38.36	0.104
0.685	0.021	4.38	0.035	19.48	0.046	38.97	0.104
0.792	0.022	4.97	0.022	22.62	0.062	39.36	0.100
1.029	0.023	5.57	0.018	23.70	0.062	40.00	0.089
1.361	0.024	6.49	0.022	26.75	0.076		
1.767	0.025	7.09	0.022	27.34	0.070		
2.086	0.026	7.89	0.014	27.87	0.076		
2.270	0.026	8.34	0.014	28.54	0.076		
2.394	0.026	9.36	0.022	28.89	0.081		
2.494	0.026	11.56	0.025	29.41	0.084		
		12.01	0.020	30.44	0.078		
		12.56	0.028	31.66	0.093		
		12.93	0.025	32.41	0.089		
		13.65	0.033	33.96	0.102		
		14.80	0.029	34.72	0.095		
				36.85	0.103		

* No plot given

SPECIFICATION TABLE NO. 170 NORMAL SPECTRAL REFLECTANCE OF BURCH PAINTS

Curve No.	Ref. No.	Year	Temperature, K	Wavelength Range, μm	Geometry θ	θ'	ω'	Reported Error, %	Composition (weight percent), Specifications, and Remarks
1*	102	1964	~296	0.401-2.490	8°	8°			Burch Photometric Sphere White No. 2210 flat white (~1.06 x 10^{-4} mm thick) on aluminum substrate; data extracted from smooth curve; measured at least 24 hrs after final coating.
2*	102	1964	~313	2.45-40.0	30°	30°			Above specimen and conditions.

DATA TABLE NO. 170 NORMAL SPECTRAL REFLECTANCE OF BURCH PAINTS

λ	ρ	λ	ρ
CURVE 1* T ~ 296		CURVE 2* T ~ 313	
0.401	0.008	2.45	~0
0.481	0.009	5.00	~0
0.541	0.010	10.0	~0
0.641	0.010	15.0	~0
0.722	0.010	20.0	~0
0.902	0.010	25.0	~0
1.002	0.010	30.0	~0
1.188	0.010	35.0	~0
1.363	0.010	40.0	~0
1.810	0.088		
2.088	0.008		
2.203	0.007		
2.311	0.006		
2.434	0.006		
2.490	0.005		

* No plot given

SPECIFICATION TABLE NO. 171 HEMISPHERICAL TOTAL EMITTANCE OF CAT-A-LAC PAINTS

Curve No.	Ref. No.	Year	Temperature Range, K	Reported Error, %	Composition (weight percent), Specifications and Remarks
1*	17	1968	302.7		Cat-A-Lac white paint.

DATA TABLE NO. 171 HEMISPHERICAL TOTAL EMITTANCE OF CAT-A-LAC PAINTS

[Temperature, T, K; Emittance, ϵ]

T	ϵ
CURVE 1*	
302.7	0.89

* No plot given

SPECIFICATION TABLE NO. 172 NORMAL INTEGRATED ABSORPTANCE OF CAT-A-LAC PAINTS

Curve No.	Ref. No.	Year	Temperature Range, K	Geometry θ	Reported Error, %	Composition (weight percent), Specifications and Remarks
1*	17	1968	298	~0°		Cat-A-Lac white paint; Finch Paint and Chemical Co.; white epoxy gloss paint; incident energy is Albedo, approximated by the solar spectrum truncated below 0.35 microns; supplied by ESRO I Project Group.
2*	17	1968	298	~0°		Similar to above specimen and conditions except exposed in vacuum (~10^{-7} mm Hg) for 2 hrs to UV radiation from an Osram XBO 900-watt xenon bulb; vacuum maintained by Galileo Type V3C oil diffusion pump.

DATA TABLE NO. 172 NORMAL INTEGRATED ABSORPTANCE OF CAT-A-LAC PAINTS

[Temperature, T, K; Absorptance, α]

T	α
CURVE 1*	
298	0.12
CURVE 2*	
298	0.206

* No plot given

SPECIFICATION TABLE NO. 173 NORMAL SOLAR ABSORPTANCE OF CAT-A-LAC PAINTS

Curve No.	Ref. No.	Year	Temperature Range, K	Geometry θ	Reported Error, %	Composition (weight percent), Specifications and Remarks
1*	17	1968	298	~0°		Cat-A-Lac white paint; Finch Paint and Chemical Co.; white epoxy gloss paint; supplied by ESRO I Project Group.
2*	17	1968	298	~0°		Similar to above specimen and conditions except exposed in vacuum (~10^{-7} mm Hg) for 2 hrs to UV radiation from an Osram XBO 900-watt xenon bulb; vacuum maintained by Galileo Type V3C oil diffusion pump.

DATA TABLE NO. 173 NORMAL SOLAR ABSORPTANCE OF CAT-A-LAC PAINTS

[Temperature, T, K; Absorptance, α]

T	α
CURVE 1*	
298	0.12
CURVE 2*	
298	0.21

* No plot given

SPECIFICATION TABLE NO. 174 NORMAL SPECTRAL REFLECTANCE OF CHROMATONE

Curve No.	Ref. No.	Year	Temperature, K	Wavelength Range, μm	Geometry θ	θ'	ω'	Reported Error, %	Composition (weight percent), Specifications, and Remarks
1*	62	1949	~298	0.98-15.00	~0°		2π	5	Chromatone, stabilized silver finish, (Alumatone Corp); data extracted from smooth curve; converted from R (2π,~0°).

DATA TABLE NO. 174 NORMAL SPECTRAL REFLECTANCE OF CHROMATONE

[Wavelength, λ, μm; Reflectance, ρ; Temperature, T, K]

λ	ρ	λ	ρ	λ	ρ
CURVE 1* T ~ 298		CURVE 1 (cont.)*		CURVE 1 (cont.)*	
		5.48	0.748	10.23	0.690
0.98	0.738	5.73	0.716	10.49	0.688
1.21	0.777	6.00	0.733	10.73	0.692
1.49	0.787	6.24	0.728	10.98	0.688
1.72	0.767	6.48	0.732	11.24	0.701
1.97	0.772	6.75	0.712	11.50	0.698
2.20	0.769	6.98	0.717	11.74	0.703
2.47	0.766	7.24	0.711	11.98	0.692
2.72	0.760	7.49	0.718	12.25	0.697
2.97	0.755	7.76	0.718	12.49	0.699
3.21	0.757	7.99	0.697	12.75	0.693
3.47	0.739	8.24	0.704	12.99	0.694
3.70	0.753	8.51	0.701	13.26	0.673
3.98	0.762	8.73	0.706	13.50	0.690
4.21	0.722	8.99	0.698	13.74	0.689
4.47	0.751	9.25	0.702	14.00	0.673
4.72	0.745	9.50	0.697	14.24	0.676
5.00	0.746	9.75	0.696	14.48	0.672
5.21	0.749	10.00	0.695	14.72	0.685
				15.00	0.700

* No plot given

SPECIFICATION TABLE NO. 175 NORMAL SPECTRAL REFLECTANCE OF DREEM PAINTS

Curve No.	Ref. No.	Year	Temperature, K	Wavelength Range, μm	Geometry θ θ' ω'	Reported Error, %	Composition (weight percent), Specifications, and Remarks
1*	102	1964	~296	0.400-2.500	8° 8°		Dreem Wall Enamel No. 13N27ES4 (~1.31 x 10^{-4} mm thick) on aluminum substrate; data extracted from smooth curve; measured at least 24 hrs after final coating.
2*	102	1964	~313	2.45-40.00	30° 30°		Above specimen and conditions.

DATA TABLE NO. 175 NORMAL SPECTRAL REFLECTANCE OF DREEM PAINTS

[Wavelength, λ, μm; Reflectance, ρ; Temperature, T, K]

λ	ρ	λ	ρ
CURVE 1* T ~ 296		CURVE 2 (cont.)*	
		18.79	0.005
0.400	0.000	20.38	0.008
0.930	0.000	22.35	0.008
1.382	0.001	23.78	0.004
1.573	0.001	24.78	0.004
2.168	0.001	25.86	0.009
2.299	0.001	26.96	0.018
2.398	0.001	28.23	0.045
2.500	0.001	31.88	0.129
		32.17	0.133
CURVE 2* T ~ 313		32.51	0.133
		33.95	0.112
		34.66	0.098
2.45	0.000	35.63	0.085
4.95	0.003	37.31	0.070
9.06	0.003	39.48	0.058
13.19	0.005	40.00	0.045

* No plot given

SPECIFICATION TABLE NO. 176 HEMISPHERICAL TOTAL EMITTANCE OF DULITE PAINTS

Curve No.	Ref. No.	Year	Temperature Range, K	Reported Error, %	Composition (weight percent), Specifications and Remarks
1*	105	1964	307		Dulite II black oxide on 1015 steel substrate; measured in vacuum (10^{-6} mm Hg) maintained by diffusion pump. [Authors' designation: Test No. 120]
2*	105	1964	307		Above specimen and conditions. [Authors' designation: Test No. 121]
3*	105	1964	307		1015 St. Dulite on Cat-A-Lac Flat Black 463-3-8; measured in vacuum (10^{-6} mm Hg) maintained by diffusion pump. [Authors' designation: Test No. 156, LDA Sample No. 1]
4*	105	1964	307		Similar to above specimen and conditions. [Authors' designation: Test No. 157, LPJ Sample No. 1]
5*	105	1964	307		Similar to above specimen and conditions. [Authors' designation: Test No. 158, LDA Sample No. 2]
6*	105	1964	307		Similar to above specimen and conditions. [Authors' designation: Test No. 159, LPJ Sample No. 2]
7*	105	1964	307		1015 St. Dulite on Cat-A-Lac Gloss White 443-1-500 substrate; measured in vacuum (10^{-6} mm Hg) maintained by diffusion pump. [Authors' designation: Test No. 160, LDA Sample No. 3]
8*	105	1964	307		Similar to above specimen and conditions. [Authors' designation: Test No. 161, LPJ Sample No. 3]
9*	105	1964	307		Similar to above specimen and conditions. [Authors' designation: Test No. 162, LDA Sample No. 4]
10*	105	1964	307		Similar to above specimen and conditions. [Authors' designation: Test No. 163, LPJ Sample No. 4]
11*	105	1964	307		1015 St. Dulite on Vit-A-Var, P. V.-100 White No. 15966 substrate; measured in vacuum (10^{-6} mm Hg) maintained by diffusion pump. [Authors' designation: Test No. 167, LDA Sample No. 5]
12*	105	1964	307		Similar to above specimen and conditions. [Authors' designation: Test No. 168, LPJ Sample No. 5]
13*	105	1964	307		Similar to above specimen and conditions. [Authors' designation: Test No. 170, LDA Sample No. 6]
14*	105	1964	307		Similar to above specimen and conditions. [Authors' designation: Test No. 171, LPJ Sample No. 6]
15*	105	1964	307		1015 St. Dulite on 4-B-2 Laminar X-500 satin black poly substrate; measured in vacuum (10^{-6} mm Hg) maintained by diffusion pump. [Authors' designation: Test No. 172, LDA Sample No. 7]
16*	105	1964	307		Similar to above specimen and conditions. [Authors' designation: Test No. 173, LPJ Sample No. 7]
17*	105	1964	307		Similar to above specimen and conditions. [Authors' designation: Test No. 174, LDA Sample No. 8]
18*	105	1964	307		Similar to above specimen and conditions. [Authors' designation: Test No. 175, LPJ Sample No. 8]
19*	105	1964	307		1015 St. Dulite on Laminar X-4-83 dark gray poly substrate; measured in vacuum (10^{-6} mm Hg) maintained by diffusion pump. [Authors' designation: Test No. 176, LDA Sample No. 9]
20*	105	1964	307		Similar to above specimen and conditions. [Authors' designation: Test No. 177, LDA Sample No. 10]
21*	105	1964	307		Similar to above specimen and conditions. [Authors' designation: Test No. 178, LPJ Sample No. 9]
22*	105	1964	307		Similar to above specimen and conditions. [Authors' designation: Test No. 179, LPJ Sample No. 10]

* No plot given

DATA TABLE NO. 176 HEMISPHERICAL TOTAL EMITTANCE OF DULITE PAINTS

[Temperature, T, K; Emittance, ∈]

T	∈	T	∈
CURVE 1*		CURVE 13*	
307	0.520	307	0.852
CURVE 2*		CURVE 14*	
307	0.712	307	0.857
CURVE 3*		CURVE 15*	
307	0.865	307	0.871
CURVE 4*		CURVE 16*	
307	0.865	307	0.857
CURVE 5*		CURVE 17*	
307	0.900	307	0.870
CURVE 6*		CURVE 18*	
307	0.861	307	0.865
CURVE 7*		CURVE 19*	
307	0.886	307	0.881
CURVE 8*		CURVE 20*	
307	0.886	307	0.875
CURVE 9*		CURVE 21*	
307	0.869	307	0.894
CURVE 10*		CURVE 22*	
307	0.882	307	0.894
CURVE 11*			
307	0.847		
CURVE 12*			
307	0.852		

* No plot given

SPECIFICATION TABLE NO. 177 NORMAL SPECTRAL REFLECTANCE OF DUPONT DUCO PAINTS

Curve No.	Ref. No.	Year	Temperature, K	Wavelength Range, μm	Geometry θ θ' ω'	Reported Error, %	Composition (weight percent), Specifications, and Remarks
1*	102	1964	~296	0.400-2.495	8° 8°		Dupont Duco Wrought Iron No. 71 Black (~0.20 x 10^{-4} mm thick) on aluminum substrate; data extracted from smooth curve; measured at least 24 hrs after final coating.
2*	102	1964	~313	2.45-40.00	30° 30°		Above specimen and conditions.

DATA TABLE NO. 177 NORMAL SPECTRAL REFLECTANCE OF DUPONT DUCO PAINTS

[Wavelength, λ, μm; Reflectance, ρ; Temperature, T, K]

λ	ρ	λ	ρ	λ	ρ	λ	ρ
CURVE 1* T ~ 296		CURVE 2 (cont.)*		CURVE 2 (cont.)*		CURVE 2 (cont.)*	
		4.15	0.034	16.96	0.219	29.27	0.430
0.400	0.004	5.45	0.077	17.39	0.197	31.16	0.465
0.596	0.004	6.03	0.057	18.02	0.187	32.30	0.455
0.759	0.004	6.50	0.083	18.91	0.176	33.01	0.424
0.951	0.005	7.54	0.029	19.70	0.159	33.76	0.449
1.417	0.007	7.97	0.023	20.57	0.096	35.46	0.476
1.727	0.009	9.56	0.061	20.83	0.089	36.12	0.510
1.947	0.012	10.04	0.103	21.15	0.085	36.54	0.492
2.133	0.014	10.41	0.114	21.52	0.093	36.96	0.500
2.241	0.015	11.16	0.126	22.55	0.149	37.70	0.500
2.432	0.019	11.95	0.132	23.30	0.169	38.60	0.516
2.495	0.020	12.45	0.127	23.48	0.187	39.08	0.513
		12.98	0.169	23.87	0.379	39.63	0.518
CURVE 2*		13.42	0.118	23.99	0.383	40.00	0.512
T ~ 313		13.60	0.125	24.59	0.349		
		13.90	0.217	25.88	0.390		
2.45	0.010	14.93	0.277	26.75	0.395		
3.50	0.018	15.79	0.252	28.84	0.430		

* No plot given

SPECIFICATION TABLE NO. 178 NORMAL TOTAL EMITTANCE OF DUTCH BOY PAINTS

Curve No.	Ref. No.	Year	Temperature Range, K	Geometry θ'	Reported Error, %	Composition (weight percent), Specifications and Remarks
1*	109	1962	422	~0°		Dutch Boy 46 H 47 (National Lead High Heat Black Paint); baked for 20 min at 450 K.
2*	109	1962	422	~0°		Similar to above specimen and conditions except calculated from spectral data.

DATA TABLE NO. 178 NORMAL TOTAL EMITTANCE OF DUTCH BOY PAINTS

[Temperature, T, K; Emittance, ϵ]

T	ϵ
CURVE 1*	
422	0.925
CURVE 2*	
422	0.919

* No plot given

SPECIFICATION TABLE NO. 179 NORMAL SPECTRAL REFLECTANCE OF DUTCH BOY PAINTS

Curve No.	Ref. No.	Year	Temperature, K	Wavelength Range, μm	Geometry θ	θ'	ω'	Reported Error, %	Composition (weight percent), Specifications, and Remarks
1*	109	1962	298	2.0-20.0	$\sim 0°$		2π		Dutch Boy 46 H 47 (National Lead High Heat Black); baked for 20 min at 450 K; calculated from R ($2\pi, \sim 0°$).

DATA TABLE NO. 179 NORMAL SPECTRAL REFLECTANCE OF DUTCH BOY PAINTS

[Wavelength, λ, μm; Reflectance, ρ; Temperature, T, K]

λ	ρ
CURVE 1*	
T = 298	
2	0.083
10	0.100
20	0.109

* No plot given

SPECIFICATION TABLE NO. 180 HEMISPHERICAL TOTAL EMITTANCE OF FULLER PAINTS

Curve No.	Ref. No.	Year	Temperature Range, K	Reported Error, %	Composition (weight percent), Specifications and Remarks
1*	15	1965	~298		Fuller arcylic lacquer white No. 171 W 560 on aluminum substrate; property measured in vacuum (10^{-5} mm Hg); vacuum maintained by vac-ion and diffusion pumps; value is avg of 8 specimens.
2*	15	1965	~298		Above specimen and conditions except weathered for 24 hrs; weathering consists of spraying with distilled water for 10 min at each hr of test interval and continuous exposure to a carbon arc which simulates earth sunlight.
3*	15	1965	~298		Above specimen and conditions except in addition maintained in vacuum (10^{-6} mm Hg) for 24 hrs at 450 K.

DATA TABLE NO. 180 HEMISPHERICAL TOTAL EMITTANCE OF FULLER PAINTS

[Temperature, T, K; Emittance, ϵ]

T	ϵ
CURVE 1*	
298	0.90
CURVE 2*	
298	0.86
CURVE 3*	
298	0.86

* No plot given

SPECIFICATION TABLE NO. 181 NORMAL TOTAL EMITTANCE OF FULLER PAINTS

Curve No.	Ref. No.	Year	Temperature Range, K	Geometry θ'	Reported Error, %	Composition (weight percent), Specifications and Remarks
1*	59	1948	344-437	~0°		Fuller D-70-6342 (external air drying).

DATA TABLE NO. 181 NORMAL TOTAL EMITTANCE OF FULLER PAINTS

[Temperature, T, K; Emittance, ϵ]

T	ϵ	T	ϵ
CURVE 1*		CURVE 1 (cont.)*	
344	0.850	436	0.855
344	0.840	437	0.850
345	0.810	437	0.845
347	0.825		
354	0.845		
354	0.840		
355	0.810		
369	0.855		
370	0.860		
370	0.855		
371	0.854		
387	0.900		
388	0.880		
389	0.881		
411	0.845		
412	0.840		
412	0.825		

* No plot given

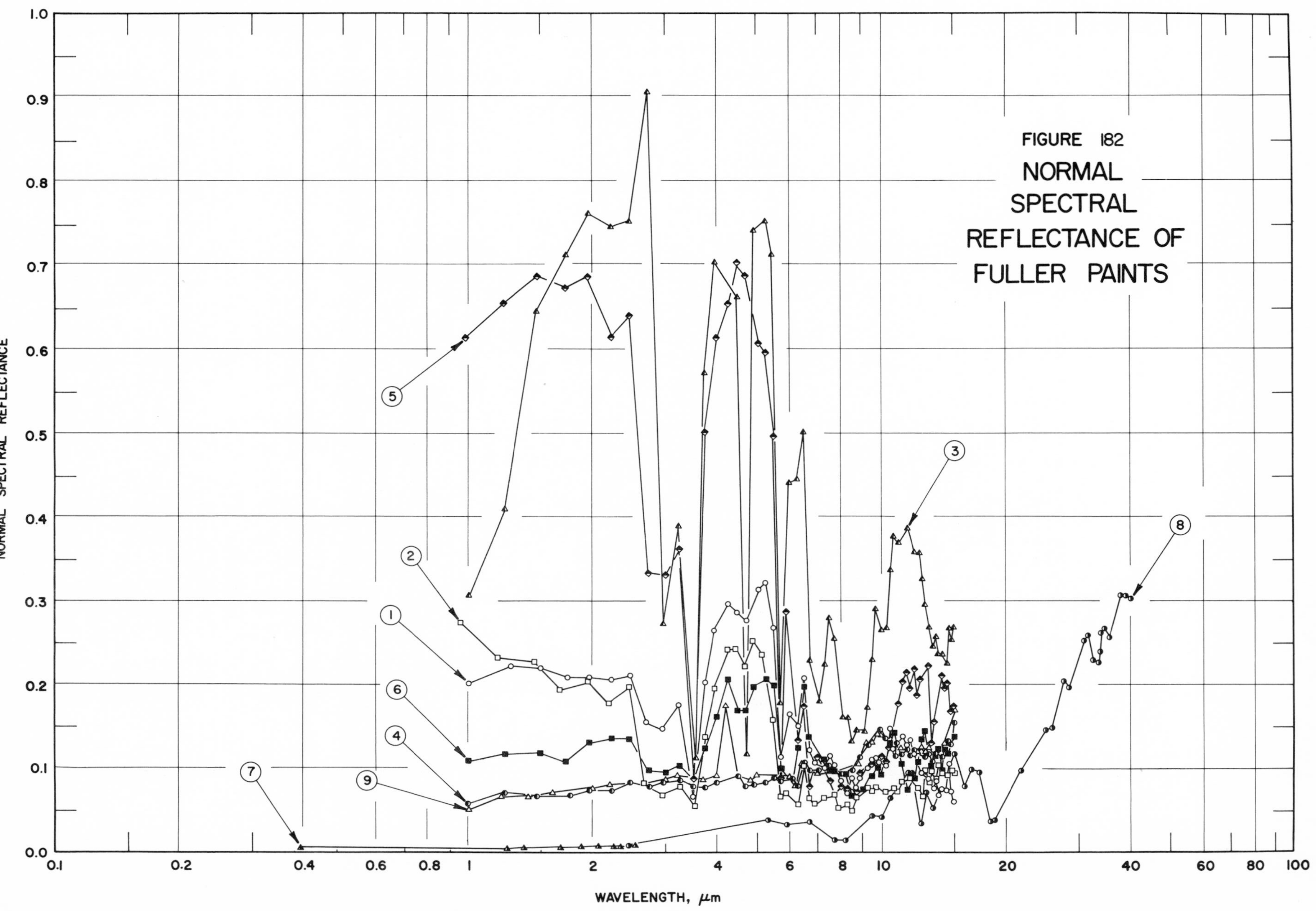
FIGURE 182
NORMAL SPECTRAL REFLECTANCE OF FULLER PAINTS
NORMAL SPECTRAL REFLECTANCE
WAVELENGTH, μm
0.0
0.1
0.2
0.3
0.4
0.5
0.6
0.7
0.8
0.9
1.0
0.1
0.2
0.4
0.6
0.8
1
2
4
6
8
10
20
40
60
80
100
1
2
3
4
5
6
7
8
9

SPECIFICATION TABLE NO. 182 NORMAL SPECTRAL REFLECTANCE OF FULLER PAINTS

Curve No.	Ref. No.	Year	Temperature, K	Wavelength Range, μm	Geometry θ	θ'	ω'	Reported Error, %	Composition (weight percent), Specifications, and Remarks
1	62	1949	~298	1.00-15.00	~0°		2π	5	Fuller Medal Mixed Paint No. 2946 Harvard Gray; data extracted from smooth curve; converted from R $(2\pi, \sim 0°)$.
2	62	1949	~298	0.96-15.00	~0°		2π	5	Fuller Medal Mixed Paint No. 2909 Light Brown; data extracted from smooth curve; converted from R $(2\pi, \sim 0°)$.
3	62	1949	~298	1.00-14.99	~0°		2π	5	Fuller Mariposa Blue Decoret Enamel No. 2889; data extracted from smooth curve; converted from R $(2\pi, \sim 0°)$.
4	62	1949	~298	1.00-15.00	~0°		2π	5	Fuller Flat Black Decoret; data extracted from smooth curve; converted from R $(2\pi, \sim 0°)$.
5	62	1949	~298	0.99-15.00	~0°		2π	5	Fuller TL-9465 No. 45, Insignia Aircraft Finish, Red Cam. En. Spec. 14109b; data extracted from smooth curve; converted from R $(2\pi, \sim 0°)$.
6	62	1949	~298	1.00-15.00	~0°		2π	5	Fuller TL-8606 No. 43 Neutral Aircraft Finish, Gray Cam. Dope Spec. 14160a; data extracted from smooth curve; converted from R $(2\pi, \sim 0°)$.
7	102	1964	~296	0.391-2.514	8°	8°			Fuller Heavy Duty Plastic No. 1518 Velvet Black (~0.60 x 10^{-4} mm thick) on aluminum substrate; data extracted from smooth curve; measured at least 24 hrs after final coating.
8	102	1964	~313	2.45-40.00	30°	30°			Above specimen and conditions.
9	104	1953	298	1.00-15.0	5°		2π	±2	Fuller Flat Black Decoret; data extracted from smooth curve; converted from R $(2\pi, 5°)$.

DATA TABLE NO. 182 NORMAL SPECTRAL REFLECTANCE OF FULLER PAINTS

[Wavelength, λ, μm; Reflectance, ρ; Temperature, T, K]

λ	ρ
CURVE 1	
T ~ 298	
1.00	0.201
1.26	0.222
1.50	0.220
1.74	0.210
1.98	0.210
2.21	0.207
2.48	0.211
2.70	0.154
2.99	0.146
3.22	0.175
3.50	0.065
3.76	0.202
3.97	0.264
4.22	0.298
4.51	0.287
4.75	0.276
5.01	0.314
5.24	0.322
5.49	0.269
5.75	0.114
6.00	0.164
6.27	0.150
6.51	0.210
6.77	0.121
6.99	0.106
7.25	0.100
7.51	0.115
7.75	0.103
8.00	0.084
8.24	0.070
8.49	0.088
8.74	0.077
8.99	0.089
9.26	0.100
9.51	0.110
9.74	0.111
9.98	0.108
10.23	0.101
10.50	0.148
10.77	0.123
10.99	0.125*
11.27	0.139
11.52	0.128
11.77	0.132
CURVE 1 (cont.)	
12.02	0.118
12.30	0.119*
12.53	0.094
12.79	0.090*
13.03	0.087
13.32	0.075
13.55	0.083
13.83	0.068
14.07	0.075
14.33	0.072
14.58	0.105
14.82	0.071
15.00	0.060
CURVE 2	
T ~ 298	
0.96	0.273
1.19	0.232
1.45	0.228
1.66	0.194
1.97	0.205
2.20	0.179
2.45	0.199
2.67	0.081
2.98	0.069
3.27	0.077
3.52	0.055
3.73	0.137
3.95	0.198
4.21	0.242
4.47	0.242
4.67	0.223
4.94	0.252
5.19	0.236
5.46	0.158
5.71	0.067
5.96	0.070
6.22	0.057
6.49	0.101
6.73	0.065
6.98	0.059
7.21	0.066
7.44	0.064*
7.74	0.069
CURVE 2 (cont.)	
7.96	0.052
8.22	0.059
8.46	0.050
8.71	0.068
8.97	0.066*
9.23	0.076
9.48	0.074
9.72	0.076
9.95	0.073*
10.22	0.073
10.50	0.073*
10.73	0.075
10.96	0.072
11.24	0.084
11.50	0.083*
11.75	0.088
11.98	0.085*
12.24	0.077
12.52	0.068
12.78	0.097
13.02	0.082
13.26	0.098
13.51	0.087
13.75	0.101
14.02	0.094
14.26	0.096
14.52	0.093
14.79	0.099
15.00	0.095
CURVE 3	
T ~ 298	
1.00	0.309
1.24	0.410
1.49	0.647
1.72	0.711
1.97	0.762
2.22	0.746
2.48	0.753
2.71	0.907
2.99	0.272
3.22	0.390
3.54	0.111
3.73	0.572
CURVE 3 (cont.)	
3.97	0.702
4.49	0.662
4.76	0.116
4.99	0.742
5.23	0.752
5.46	0.661
5.76	0.178
6.00	0.441
6.23	0.445
6.49	0.501
6.73	0.230
7.02	0.180
7.25	0.225
7.51	0.280
7.75	0.255
8.02	0.160
8.26	0.160
8.52	0.131
8.78	0.145
9.02	0.143
9.28	0.171
9.52	0.230
9.77	0.290
10.06	0.265
10.28	0.269
10.53	0.339
10.79	0.378
11.02	0.370
11.28	0.378*
11.55	0.389
11.78	0.389*
12.05	0.359
12.31	0.359
12.55	0.329
12.81	0.299
13.04	0.269
13.31	0.247
13.54	0.256
13.76	0.238
14.04	0.237
14.30	0.228
14.54	0.269
14.79	0.253
14.99	0.269
CURVE 4	
T ~ 298	
1.00	0.058
1.23	0.070
1.48	0.066
1.76	0.068
1.99	0.073
2.22	0.073
2.48	0.081
2.76	0.079
2.99	0.081
3.27	0.085
3.50	0.078
3.73	0.077
4.00	0.081
4.26	0.283
4.50	0.090
4.76	0.078
4.99	0.080
5.24	0.081
5.48	0.088
5.74	0.083
6.01	0.086
6.26	0.079
6.52	0.104
6.77	0.097
7.01	0.098
7.24	0.098*
7.51	0.101
7.74	0.097*
8.00	0.091
8.25	0.095*
8.48	0.098
8.76	0.101*
8.98	0.112
9.23	0.128
9.49	0.129*
9.74	0.141*
9.98	0.146
10.25	0.135
10.48	0.125*
10.72	0.115
10.98	0.130
11.25	0.116
11.51	0.121
11.74	0.115
CURVE 4 (cont.)	
12.02	0.121
12.26	0.114*
12.53	0.117*
12.77	0.113
13.02	0.118
13.27	0.109
13.53	0.119
13.76	0.117*
14.02	0.115
14.26	0.125*
14.51	0.133
14.76	0.129
15.00	0.153
CURVE 5	
T ~ 298	
0.99	0.613
1.22	0.654
1.47	0.688
1.71	0.671
1.96	0.686
2.23	0.615
2.47	0.640
2.73	0.334
3.00	0.331
3.25	0.362
3.50	0.087
3.75	0.501
4.00	0.614
4.25	0.655
4.50	0.701
4.76	0.689
5.01	0.609
5.25	0.599
5.51	0.499
5.75	0.087
5.99	0.287
6.26	0.132
6.51	0.173
6.74	0.079
7.00	0.113
7.23	0.112*
7.49	0.084
7.74	0.067*
CURVE 5 (cont.)	
8.00	0.077
8.23	0.075
8.48	0.079*
8.75	0.072
8.99	0.093
9.26	0.094*
9.51	0.105
9.77	0.112*
10.00	0.114
10.26	0.108
10.49	0.124
10.75	0.124*
11.01	0.178
11.24	0.203
11.51	0.216
11.75	0.196
12.00	0.220
12.25	0.188
12.53	0.208
12.76	0.208*
13.00	0.224
13.26	0.130
13.51	0.155
13.76	0.181*
14.02	0.211
14.25	0.197
14.50	0.203
14.75	0.168
15.00	0.175
CURVE 6	
T ~ 298	
1.00	0.108
1.24	0.116
1.50	0.118
1.71	0.109
1.99	0.130
2.22	0.135
2.47	0.134
2.71	0.097
3.00	0.094
3.25	0.101
3.51	0.085*
3.73	0.123
CURVE 6 (cont.)	
4.00	0.160
4.24	0.207
4.52	0.168
4.75	0.169
4.99	0.196
5.24	0.207
5.53	0.200
5.76	0.100
6.01	0.089*
6.23	0.125
6.49	0.199
6.76	0.138
7.01	0.118*
7.24	0.110
7.49	0.097
7.74	0.096
7.99	0.093*
8.26	0.091
8.51	0.067
8.78	0.068*
9.01	0.073
9.26	0.090*
9.56	0.090
9.81	0.100
10.02	0.091
10.27	0.111*
10.52	0.130
10.78	0.142
11.28	0.104
11.53	0.074
11.78	0.084*
12.02	0.088
12.28	0.107
12.53	0.134
12.74	0.143
13.02	0.113*
13.27	0.101
13.53	0.116
13.77	0.121
14.02	0.099
14.26	0.121
14.52	0.118
14.75	0.127*
15.00	0.137

* Not shown on plot

DATA TABLE NO. 182 NORMAL SPECTRAL REFLECTANCE OF FULLER PAINTS (continued)

λ	ρ
CURVE 7	
T ~ 296	
0.391	0.004
1.252	0.004
1.378	0.005
1.674	0.005
1.870	0.006
2.083	0.006
2.248	0.006
2.350	0.007
2.514	0.008
CURVE 8	
T ~ 313	
2.45	0.008
5.34	0.039
5.88	0.031
6.66	0.036
7.62	0.014
8.20	0.014
9.43	0.043
10.00	0.041
10.51	0.065
10.86	0.065*
11.60	0.093
11.93	0.093
12.08	0.085*
12.46	0.038
12.95	0.070
13.42	0.051
13.99	0.094*
15.02	0.118
15.93	0.079
16.55	0.098
16.97	0.098*
17.27	0.094
18.44	0.039
18.89	0.039
21.89	0.099
25.04	0.145
26.01	0.148
27.43	0.205
27.67	0.205*
28.41	0.198
28.90	0.198*
31.01	0.252
CURVE 8 (cont.)	
31.58	0.258
32.68	0.230
32.92	0.227*
33.43	0.227
33.74	0.240
34.01	0.261
34.49	0.267
35.30	0.256*
35.77	0.256
38.36	0.308
39.06	0.308
40.00	0.302
CURVE 9	
T = 298	
1.00	0.050
1.20	0.065
1.40	0.065
1.60	0.070
2.00	0.075
2.20	0.080
2.70	0.080*
3.00	0.085
3.20	0.090
3.70	0.085
4.00	0.090
4.20	0.275
4.50	0.090*
4.80	0.085
5.00	0.090
5.70	0.090*
6.00	0.090
6.20	0.080
6.40	0.105
6.80	0.099*
7.00	0.095
7.20	0.095*
7.50	0.100*
7.70	0.095*
8.00	0.090*
8.80	0.105
9.00	0.115*
9.20	0.130
9.50	0.130
9.80	0.140
CURVE 9 (cont.)	
10.0	0.140
10.2	0.135*
10.5	0.130*
10.8	0.130*
11.0	0.130*
11.2	0.125
11.7	0.125*
12.0	0.125*
12.2	0.120
12.5	0.125
12.8	0.120
13.0	0.125
13.2	0.115*
13.5	0.120*
13.8	0.115*
14.3	0.130
14.5	0.135
14.8	0.130
15.0	0.170

* Not shown on plot

SPECIFICATION TABLE NO. 183 NORMAL SOLAR ABSORPTANCE OF FULLER PAINTS

Curve No.	Ref. No.	Year	Temperature Range, K	Geometry θ	Reported Error, %	Composition (weight percent), Specifications and Remarks
1*	15	1965	~298	~0°		Fuller acrylic lacquer white No. 171 W 560 on aluminum substrate; calculated from $\rho(\sim 0°, 2\pi)$; value avg of 8 specimens.
2*	15	1965	298	~0°		Above specimen and conditions except weathered for 24 hrs; weathering consists of spraying with distilled water for 10 min at each hr of test interval and continuous exposure to a carbon arc which simulates earth sunlight.
3*	15	1965	298	~0°		Above specimen and conditions except in addition maintained in vacuum (10^{-6} mm Hg) for 24 hrs at 450 K.

DATA TABLE NO. 183 NORMAL SOLAR ABSORPTANCE OF FULLER PAINTS

[Temperature, T, K; Absorptance, α]

T	α
CURVE 1*	
298	0.26
CURVE 2*	
298	0.26
CURVE 3*	
298	0.28

* No plot given

SPECIFICATION TABLE NO. 184 NORMAL SOLAR ABSORPTANCE OF GLASURIT PAINTS

Curve No.	Ref. No.	Year	Temperature Range, K	Geometry θ	Reported Error, %	Composition (weight percent), Specifications and Remarks
1*	17	1968	298	~0°		Glasurit white epoxy paint; sample supplied by ESRO I Project Group.

DATA TABLE NO. 184 NORMAL SOLAR ABSORPTANCE OF GLASURIT PAINTS

[Temperature, T, K; Absorptance, α]

T	α
CURVE 1*	
298	0.23

* No plot given

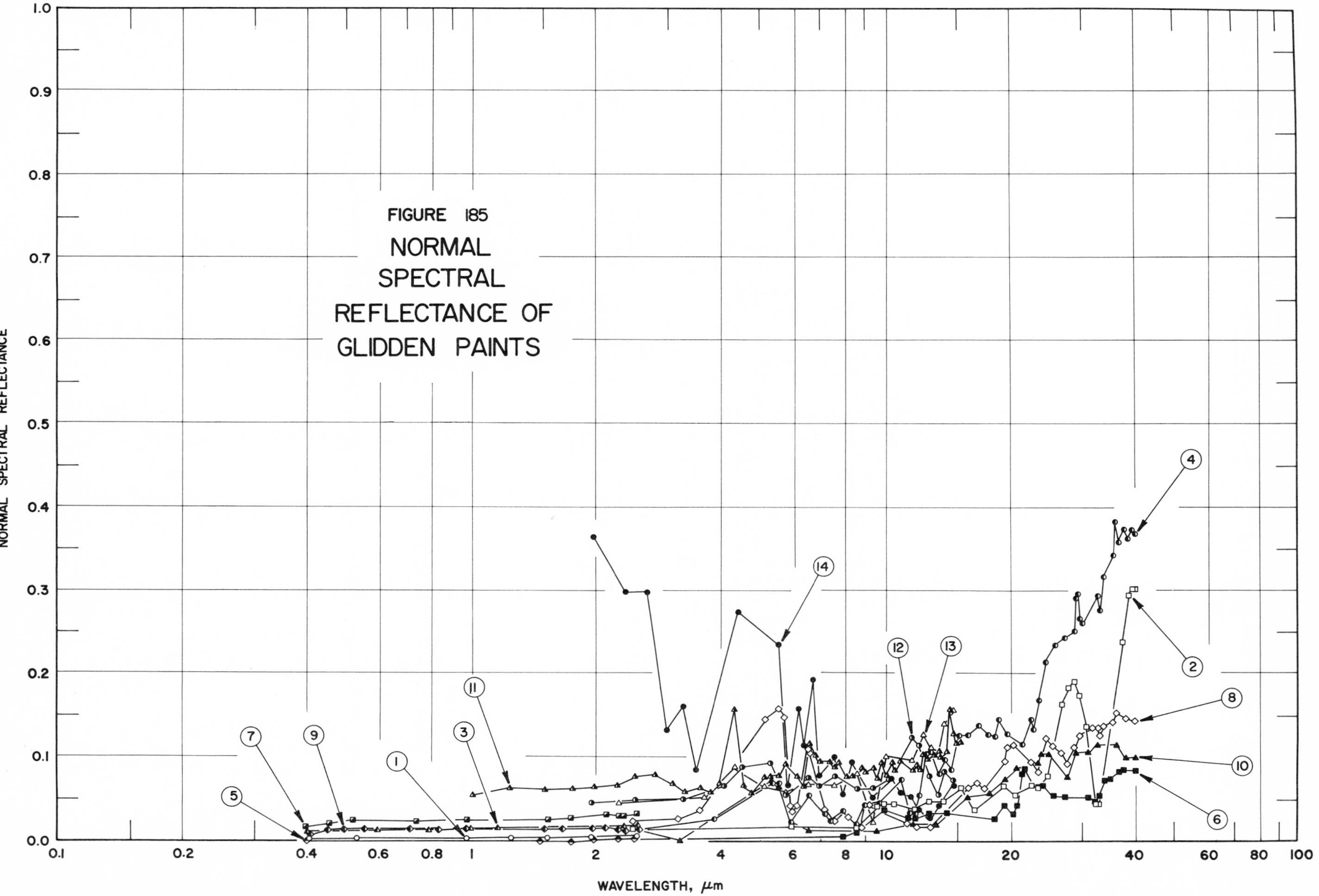

FIGURE 185
NORMAL SPECTRAL REFLECTANCE OF GLIDDEN PAINTS
NORMAL SPECTRAL REFLECTANCE
WAVELENGTH, μm
0.0
0.1
0.2
0.3
0.4
0.5
0.6
0.7
0.8
0.9
1.0
0.1
0.2
0.4
0.6
0.8
1
2
4
6
8
10
20
40
60
80
100
1
2
3
4
5
6
7
8
9
10
11
12
13
14

SPECIFICATION TABLE NO. 185 NORMAL SPECTRAL REFLECTANCE OF GLIDDEN PAINTS

Curve No.	Ref. No.	Year	Temperature, K	Wavelength Range, μm	Geometry θ	θ'	ω'	Reported Error, %	Composition (weight percent), Specifications, and Remarks
1	102	1964	~296	0.401-2.503	8°	8°			Glidair (Zapon) Lacquer No. 131-B-190B Black (~0.30 x 10^{-4} mm thick) on aluminum substrate; data extracted from smooth curve; measured at least 24 hrs after final coating.
2	102	1964	~313	2.45-40.00	30°	30°			Above specimen and conditions.
3	102	1964	~298	0.400-2.500	8°	8°			Glidden Flat Black Lacquer No. 131-B-216 (~0.20 x 10^{-4} mm thick) on aluminum substrate; data extracted from smooth curve; measured at least 24 hrs after final coating.
4	102	1964	~313	2.45-40.00	30°	30°			Above specimen and conditions.
5	102	1964	~296	0.400-2.499	8°	8°			Glidden Japalac Flat Black No. 1208 (~1.08 x 10^{-4} mm thick) on aluminum substrate; data extracted from smooth curve; measured at least 24 hrs after final coating.
6	102	1964	~313	2.45-40.00	30°	30°			Above specimen and conditions.
7	102	1964	~296	0.397-2.500	8°	8°			Glidden Flat Sheen White No. 2995 (~0.75 x 10^{-4} mm thick) on aluminum substrate; data extracted from smooth curve; measured at least 24 hrs after final coating.
8	102	1964	~313	2.45-40.00	30°	30°			Above specimen and conditions.
9	102	1964	~296	0.401-2.529	8°	8°			Glidden Flat Sheen White No. 5064 (~0.15 x 10^{-4} mm thick) on aluminum substrate; data extracted from smooth curve; measured at least 24 hrs after final coating.
10	102	1964	~313	2.45-40.00	30°	30°			Above specimen and conditions.
11	62	1949	~298	1.0-15.0	~0°		2π	5	Glidden Japalac quick drying enamel No. 1207 brilliant black; data extracted from smooth curve; converted from R (2π, ~0°).
12	99	1966	~300	1.96-14.51	~0°		2π	< ±2	Glidden Missile Black No. RGL-22818 (~0.0381 mm thick) on AISI 316 stainless steel substrate; sprayed; air dried at least 24 hrs; converted from R (2π, ~0°). [Authors' designation: Specimen No. 1]
13	99	1966	~300	2.27-14.42	~0°		2π	< ±2	Glidden black lacquer No. 9099 (~0.0381 mm thick) on AISI 316 stainless steel substrate; sprayed; air dried at least 24 hrs; converrnted from R (2π, ~0°). [Authors' designation: Specimen No. 3]

DATA TABLE NO. 185 NORMAL SPECTRAL REFLECTANCE OF GLIDDEN PAINTS

[Wavelength, λ, μm; Reflectance, ρ; Temperature, T, K]

λ	ρ
CURVE 1 T ~296	
0.401	0.001
0.522	0.002
0.977	0.002
1.248	0.003
1.525	0.003
1.959	0.004
2.298	0.004
2.503	0.005
CURVE 2 T ~313	
2.45	0.014
5.96	0.019
8.41	0.016
9.59	0.041
9.94	0.045
10.49	0.045
11.63	0.039
12.76	0.048
13.96	0.048
15.13	0.062
16.13	0.042*
16.45	0.038
16.82	0.038*
19.14	0.065
19.56	0.066*
20.54	0.055
20.93	0.055*
22.15	0.066
22.58	0.066*
23.25	0.063
23.87	0.063*
24.63	0.079
26.70	0.161
27.48	0.184
27.82	0.190*
28.43	0.190
29.14	0.174
30.27	0.137
32.05	0.045
32.66	0.045
37.16	0.237
38.90	0.295

λ	ρ
CURVE 2 (cont.)	
39.20	0.301
40.00	0.301
CURVE 3 T ~ 296	
0.400	0.010
0.589	0.011
0.789	0.011
0.991	0.013
1.150	0.014
2.313	0.019
2.386	0.019*
2.500	0.019
CURVE 4 T ~ 313	
2.45	0.013
3.83	0.027
5.30	0.069
5.49	0.069
5.96	0.021
6.53	0.053
7.05	0.031
7.39	0.023
7.95	0.036
8.41	0.020
8.85	0.042
9.68	0.045
10.90	0.071
11.45	0.032
11.68	0.032
12.00	0.054
12.52	0.102
12.83	0.106
13.40	0.080
13.95	0.097
14.42	0.083
15.02	0.125
15.89	0.126
16.71	0.137
16.93	0.137*
17.78	0.127
18.31	0.125

λ	ρ
CURVE 4 (cont.)	
18.95	0.143
19.99	0.128
21.25	0.116
22.41	0.145
22.84	0.131
23.36	0.167
24.34	0.212
25.81	0.232
27.09	0.241
28.62	0.250
28.81	0.258*
28.95	0.290
29.06	0.296
29.49	0.266
29.77	0.260
32.47	0.293
32.89	0.382
33.52	0.317
35.26	0.341
35.89	0.382
36.38	0.357
36.81	0.353*
37.91	0.372
38.95	0.361
39.53	0.373
40.00	0.367
CURVE 5 T ~ 296	
0.400	0.000
1.461	0.000
1.722	0.000
1.999	0.001
2.259	0.001
2.370	0.001*
2.499	0.001
CURVE 6 T ~ 313	
2.45	0.003*
7.99	0.005
8.52	0.010
9.92	0.038

λ	ρ
CURVE 6 (cont.)	
11.21	0.029
11.72	0.029
12.84	0.033
14.00	0.033
18.21	0.028
19.47	0.042
20.14	0.032
20.41	0.032*
20.67	0.042
21.41	0.080
21.73	0.088
22.04	0.085*
22.71	0.068*
24.01	0.065
25.28	0.053
27.07	0.052
30.93	0.052
31.80	0.049
32.42	0.049*
32.82	0.054
33.71	0.072
34.81	0.073
36.48	0.082
37.52	0.083
40.00	0.083
CURVE 7 T ~ 296	
0.397	0.017
0.454	0.020
0.519	0.023
0.735	0.023
0.977	0.025
1.526	0.027
1.723	0.029
2.103	0.031
2.260	0.030
2.313	0.030
2.384	0.031*
2.500	0.032

λ	ρ
CURVE 8 T ~313	
2.45	0.025
3.13	0.029
3.55	0.037
5.13	0.145
5.45	0.159
5.64	0.147
5.83	0.042
5.94	0.036
6.06	0.041
6.43	0.110
6.59	0.105
7.10	0.039
7.45	0.024
7.61	0.024*
7.90	0.032
8.10	0.030
8.52	0.016*
8.70	0.019
8.98	0.041
9.14	0.044
9.53	0.044
11.20	0.022
11.88	0.019
12.95	0.019
15.89	0.064
16.41	0.070
16.66	0.070*
17.27	0.064
17.48	0.064*
19.27	0.097
19.76	0.112
20.20	0.114
22.36	0.095
23.25	0.083
23.55	0.087*
23.99	0.113
24.47	0.123
25.11	0.115
25.67	0.117*
26.65	0.106
27.26	0.092
27.55	0.094*
28.25	0.113
29.40	0.126

λ	ρ
CURVE 8 (cont.)	
31.69	0.138
32.58	0.138
32.98	0.128
33.46	0.139
33.92	0.142*
35.10	0.142
36.06	0.154
36.55	0.154*
38.00	0.148
40.00	0.145
CURVE 9 T ~ 296	
0.401	0.008
0.455	0.011
0.496	0.012
0.548	0.013
0.704	0.013
0.838	0.012
0.969	0.013
1.507	0.014
1.655	0.015
1.765	0.015*
1.963	0.016
2.107	0.016
2.279	0.014
2.340	0.014
2.430	0.015
2.529	0.015
CURVE 10 T ~ 313	
2.45	0.012*
3.19	0.000
5.28	0.072
5.57	0.063
5.86	0.022*
6.45	0.012
9.56	0.012
11.66	0.020
13.15	0.020
15.79	0.052
17.81	0.059

λ	ρ
CURVE 10 (cont.)	
19.52	0.074
20.68	0.089
21.74	0.089*
22.34	0.093
23.14	0.092
23.93	0.103
24.79	0.103
27.33	0.078
28.93	0.106
30.87	0.109
32.13	0.116
36.09	0.116
38.05	0.100
40.00	0.100
CURVE 11 T ~ 298	
1.00	0.054
1.23	0.062
1.50	0.060
1.74	0.061
1.99	0.063
2.24	0.066
2.49	0.076
2.76	0.079
3.01	0.069
3.25	0.059
3.51	0.063
3.75	0.057
3.99	0.069
4.25	0.156
4.50	0.064
4.75	0.058
5.01	0.075
5.25	0.076
5.52	0.078
5.75	0.091
6.02	0.078
6.24	0.065
6.52	0.117
6.78	0.101
6.99	0.093
7.25	0.094
7.51	0.089

λ	ρ
CURVE 11 (cont.)	
7.76	0.093
8.00	0.076
8.26	0.077
8.51	0.077*
8.75	0.086
8.98	0.081
9.22	0.087
9.50	0.073
9.73	0.092
10.00	0.078
10.26	0.092
10.52	0.085
10.74	0.095
11.01	0.091*
11.25	0.090*
11.50	0.082
11.77	0.090
12.01	0.085
12.26	0.101
12.51	0.097
12.75	0.103*
13.00	0.101
13.27	0.109
13.50	0.100
13.77	0.103*
14.01	0.107
14.24	0.159
14.52	0.128
14.74	0.118
15.00	0.119
CURVE 12 T ~ 300	
1.96	0.045
2.49	0.050
3.23	0.050
4.01	0.066
4.54	0.089
5.23	0.091
5.76	0.055
6.42	0.076
6.99	0.065
7.58	0.076
8.50	0.061

* Not shown on plot

DATA TABLE NO. 185 NORMAL SPECTRAL REFLECTANCE OF GLIDDEN PAINTS (continued)

λ	ρ	λ	ρ
CURVE 12 (cont.) T ~ 300		CURVE 14 (cont.)	
9.36	0.063	6.93	0.068
10.17	0.075	7.53	0.100
10.98	0.085*	7.89	0.056
11.57	0.123	8.28	0.094
12.09	0.115	8.50	0.080*
12.70	0.077	9.37	0.053
13.31	0.056	10.18	0.076
13.89	0.082	10.96	0.059
14.51	0.072	11.57	0.053
CURVE 13 T ~ 300		12.07	0.039
2.27	0.046	12.65	0.029
3.24	0.051*	13.30	0.041
3.61	0.051	13.90	0.048*
4.32	0.086	14.52	0.066
5.01	0.065		
5.76	0.057		
6.42	0.068		
7.50	0.066		
8.49	0.080		
9.34	0.021		
10.08	0.100		
10.85	0.095*		
11.56	0.096		
12.20	0.128		
12.81	0.111		
13.38	0.104*		
13.88	0.140		
14.42	0.156		
CURVE 14 T ~ 300			
1.96	0.363		
2.35	0.299		
2.64	0.299		
2.98	0.131		
3.23	0.160		
3.48	0.085		
4.40	0.274		
5.50	0.233		
5.80	0.066		
6.18	0.158		
6.33	0.115		
6.62	0.193		

* Not shown on plot

SPECIFICATION TABLE NO. 186 NORMAL SPECTRAL EMITTANCE OF GLYPTAL PAINTS

Curve No.	Ref. No.	Year	Temperature, K	Wavelength Range, μm	Geometry θ'	Reported Error, %	Composition (weight percent), Specifications, and Remarks
1*	74	1959	325	0.46-19.20	~0°		Glyptal black, manufactured by General Electric; data extracted from smooth curve; property converted from $R(2\pi, \sim 0°)$.

DATA TABLE NO. 186 NORMAL SPECTRAL EMITTANCE OF GLYPTAL PAINTS

[Wavelength, λ, μm; Emittance, ϵ; Temperature, T, K]

λ	ϵ	λ	ϵ
CURVE 1* T = 325		CURVE 1 (cont.)*	
0.46	0.967	14.00	0.798
1.82	0.956	14.54	0.759
2.82	0.944	14.79	0.755
3.23	0.912	15.57	0.780
3.39	0.894	16.07	0.784
3.67	0.836	16.60	0.778
3.89	0.819	18.25	0.724
4.28	0.802	19.20	0.699
4.91	0.802		
5.57	0.817		
6.58	0.846		
7.22	0.855		
7.97	0.857		
8.79	0.847		
9.60	0.828		
10.19	0.821		
13.78	0.806		

* No plot given

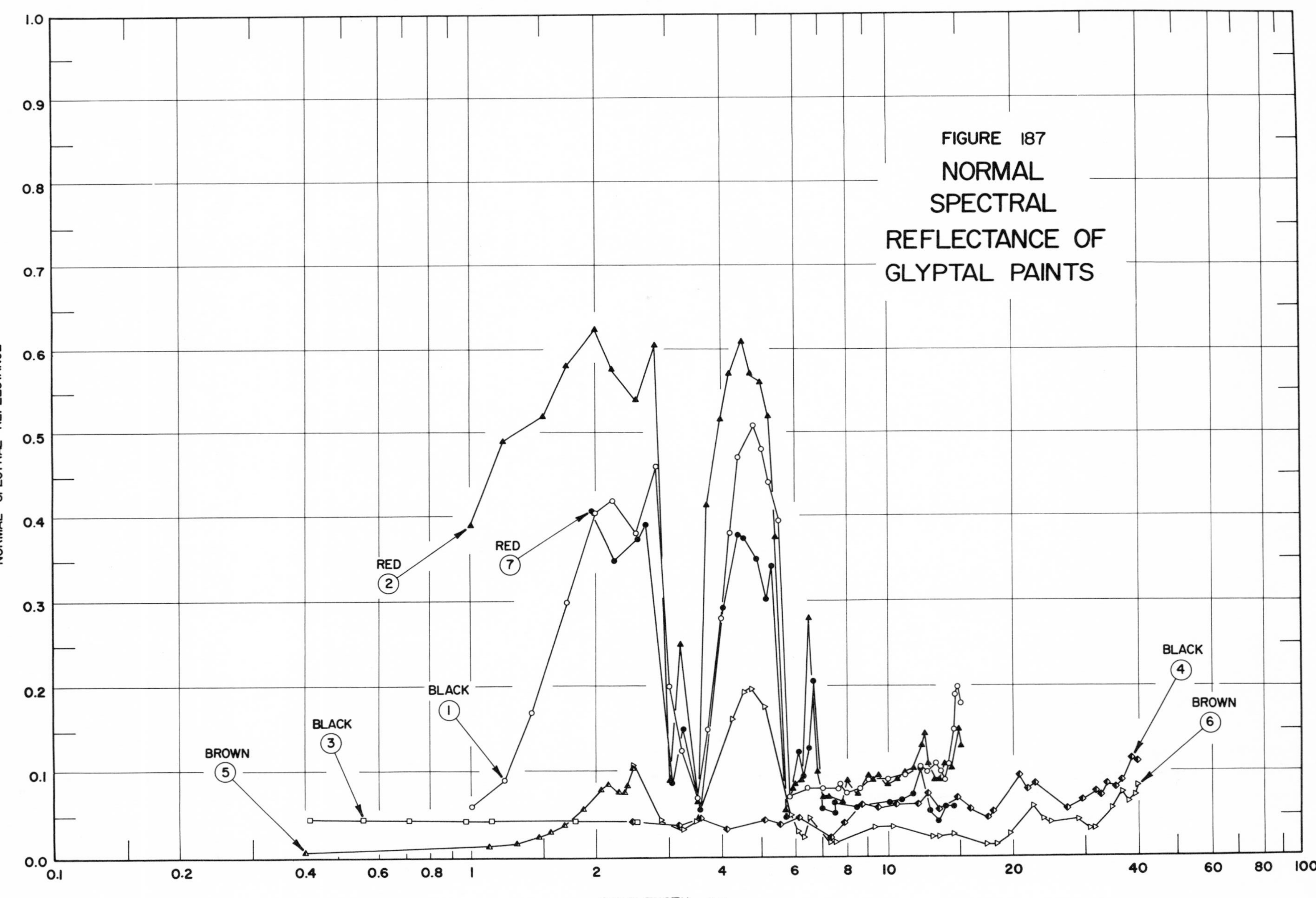

FIGURE 187 NORMAL SPECTRAL REFLECTANCE OF GLYPTAL PAINTS

SPECIFICATION TABLE NO. 187 NORMAL SPECTRAL REFLECTANCE OF GLYPTAL PAINTS

Curve No.	Ref. No.	Year	Temperature, K	Wavelength Range, μm	Geometry θ	θ'	ω'	Reported Error, %	Composition (weight percent), Specifications, and Remarks
1	104	1953	298	1.00-15.0	5°		2π	± 2	Glyptal black (0.2286 mm thick) on aluminum substrate; substrate polished; data extracted from smooth curve; converted from R (2π, 5°).
2	104	1953	298	1.00-15.0	5°		2π	± 2	Glyptal red (0.1016 mm thick) on aluminum substrate; substrate polished; data extracted from smooth curve; converted from R (2π, 5°).
3	102	1964	~296	0.409-2.503	8°	8°			Glyptal black No. 10021A (~0.15×10^{-4} mm thick) on aluminum substrate; data extracted from smooth curve; measured at least 24 hrs after final coating.
4	102	1964	~313	2.45-40.0	30°	30°			Above specimen and conditions.
5	102	1964	~296	0.400-2.430	8°	8°			Glyptal brown (~0.45×10^{-4} mm thick) on aluminum substrate; data extracted from smooth curve; measured at least 24 hrs after final coating.
6	102	1964	~313	2.45-40.0	30°	30°			Above specimen and conditions.
7	99	1966	~300	1.99-14.50	~0°		2π	$< \pm 2$	Glyptal red enamel No. 1201, manufactured by General Electric (~0.0381 mm thick) on AISI 316 stainless steel substrate; sprayed; air dried at least 24 hrs; converted from R (2π, ~0°). [Authors' designation: Specimen No. 10]

DATA TABLE NO. 187 NORMAL SPECTRAL REFLECTANCE OF GLYPTAL PAINTS

[Wavelength, λ, μm; Reflectance, ρ; Temperature, T, K]

CURVE 1
T = 298

λ	ρ
1.00	0.060
1.20	0.090
1.40	0.170
1.70	0.300
2.00	0.405
2.20	0.420
2.50	0.380
2.80	0.460
3.00	0.200
3.20	0.125
3.50	0.070
3.70	0.150
4.00	0.280
4.20	0.380
4.40	0.470
4.80	0.510
5.00	0.480
5.20	0.440
5.50	0.395
5.80	0.070
6.40	0.080
7.00	0.080
7.60	0.080
7.70	0.085
8.00	0.075
8.60	0.080
9.00	0.090
10.0	0.090
10.6	0.090*
11.0	0.095
11.4	0.095*
11.5	0.100*
12.0	0.115
12.2	0.100*
12.4	0.100
12.7	0.110*
13.0	0.110
13.2	0.100*
13.4	0.100
13.7	0.090
14.0	0.110
14.3	0.150
14.5	0.190
14.8	0.200
15.0	0.180

CURVE 2
T = 298

λ	ρ
1.00	0.390
1.20	0.490
1.50	0.520
1.70	0.580
2.00	0.625
2.20	0.575
2.50	0.540
2.80	0.605
3.00	0.090
3.20	0.250
3.50	0.065
3.70	0.415
4.00	0.515
4.20	0.570
4.50	0.610
4.70	0.570
5.00	0.560
5.20	0.520
5.40	0.375
5.70	0.055
5.90	0.080
6.00	0.085
6.20	0.090
6.50	0.280
6.80	0.100
7.00	0.070
7.20	0.070
7.40	0.065*
7.80	0.065
8.00	0.090
8.50	0.075
9.00	0.095
9.20	0.090
9.50	0.095
10.0	0.085
10.5	0.090
10.7	0.100*
11.0	0.100
11.2	0.105*
11.5	0.105
11.8	0.110*
12.0	0.130
12.2	0.145
12.5	0.110
12.8	0.100*

CURVE 2 (cont.)

λ	ρ
13.0	0.090
13.2	0.090
13.8	0.110
14.2	0.105
14.8	0.150
15.0	0.130

CURVE 3
T ~ 296

λ	ρ
0.409	0.045
0.551	0.043
0.716	0.043
0.972	0.043
1.116	0.042
1.793	0.042
2.503	0.041

CURVE 4
T ~ 313

λ	ρ
2.45	0.041
3.18	0.038
3.56	0.046
4.11	0.032
5.03	0.043
5.59	0.039
6.17	0.046
7.39	0.023
7.92	0.040
8.65	0.061
9.58	0.058
10.58	0.062
11.91	0.062
12.53	0.074
13.25	0.057
13.68	0.057*
14.88	0.070
15.85	0.056
17.44	0.049
18.07	0.054
18.84	0.075*
20.95	0.099
21.82	0.080
22.66	0.087
24.35	0.083*

CURVE 4 (cont.)

λ	ρ
27.07	0.057
27.42	0.057*
29.51	0.068
31.06	0.068*
31.95	0.079
32.50	0.073
33.05	0.081*
33.66	0.086
34.80	0.082*
35.37	0.082
36.56	0.090
37.93	0.103*
38.82	0.118
39.27	0.118*
40.00	0.113

CURVE 5
T ~ 296

λ	ρ
0.400	0.006
1.103	0.011
1.281	0.015
1.455	0.024
1.553	0.030
1.676	0.039
1.860	0.059
2.051	0.080
2.111	0.085
2.160	0.085*
2.287	0.078
2.315	0.078
2.348	0.083
2.430	0.104

CURVE 6
T ~ 313

λ	ρ
2.45	0.108
2.88	0.044
3.21	0.031
3.48	0.041
4.21	0.161
4.58	0.194
4.71	0.197
4.79	0.194*
5.03	0.176

CURVE 6 (cont.)

λ	ρ
5.83	0.050
6.04	0.030
6.28	0.023
6.53	0.047
7.21	0.019
7.58	0.019
8.15	0.031*
9.22	0.035
10.25	0.035
12.93	0.023
13.35	0.023
14.45	0.029
17.37	0.014
18.29	0.014
19.86	0.028
22.46	0.060
23.84	0.045
24.86	0.040
28.89	0.043
30.89	0.034
31.37	0.033
32.17	0.041*
34.95	0.058
36.34	0.076
36.76	0.076*
38.04	0.065
38.73	0.065*
39.35	0.072
40.00	0.082

CURVE 7
T ~ 300

λ	ρ
1.99	0.408
2.23	0.348
2.53	0.372
2.63	0.390
3.01	0.088
3.25	0.150
3.56	0.055
4.05	0.291
4.40	0.379
4.58	0.375
4.81	0.350
5.14	0.304
5.27	0.341

CURVE 7 (cont.)

λ	ρ
5.34	0.318*
5.76	0.048
6.19	0.123
6.32	0.094
6.42	0.126
6.62	0.208
7.00	0.057
7.56	0.051
7.56	0.064
8.49	0.059
9.38	0.048
10.17	0.064
10.97	0.068
11.55	0.073
12.09	0.104
12.66	0.052
13.28	0.041
13.88	0.060
14.50	0.060

* Not shown on plot

SPECIFICATION TABLE NO. 188 HEMISPHERICAL TOTAL EMITTANCE OF GRUMMAN AIRCRAFT ENGINEERING CORP. PAINTS

Curve No.	Ref. No.	Year	Temperature Range, K	Reported Error, %	Composition (weight percent), Specifications and Remarks
1*	107	1967	100-302	~8.5	Grumman Aircraft Engineering Corp. black epoxy paint No. 1019 on G. A. E. C. No. 1020 clear epoxy primer and 304 stainless steel substrates; two coats of No. 1019; coating and primer thickness 0.048 mm; measured in vacuum (~10^{-6} mm Hg).

DATA TABLE NO. 188 HEMISPHERICAL TOTAL EMITTANCE OF GRUMMAN AIRCRAFT ENGINEERING CORP. PAINTS

[Temperature, T, K; Emittance, ϵ]

T	ϵ
CURVE 1*	
100	0.858
163	0.858
220	0.865
268	0.853
302	0.846

* No plot given

SPECIFICATION TABLE NO. 189 NORMAL SPECTRAL REFLECTANCE OF ILLINOIS BRONZE POWDER CO. PAINTS

Curve No.	Ref. No.	Year	Temperature, K	Wavelength Range, μm	Geometry θ	θ'	ω'	Reported Error, %	Composition (weight percent), Specifications, and Remarks
1*	99	1966	~300	1.99-14.50	~0°		2π	<±2	Illinois Bronze Powder Co. white lacquer (~0.0381 mm thick) on AISI 316 stainless steel substrate; sprayed; air dried at least 24 hrs; converted from R(2π,~0°). [Authors' designation: Specimen No. 4]

DATA TABLE NO. 189 NORMAL SPECTRAL REFLECTANCE OF ILLINOIS BRONZE POWDER CO. PAINTS

[Wavelength, λ, μm; Reflectance, ρ; Temperature, T, K]

λ	ρ	λ	ρ	λ	ρ
CURVE 1*		CURVE 1 (cont.)*		CURVE 1 (cont.)*	
T ~ 300					
		6.10	0.062	13.88	0.013
1.99	0.400	6.38	0.231	14.50	0.025
2.09	0.415	6.56	0.295		
2.42	0.379	6.92	0.075		
2.53	0.367	7.00	0.090		
2.71	0.398	7.09	0.118		
3.04	0.115	7.27	0.074		
3.21	0.182	7.49	0.095		
3.34	0.197	7.54	0.089		
3.54	0.060	7.72	0.043		
4.01	0.323	8.47	0.028		
4.45	0.387	9.36	0.023		
4.57	0.368	10.17	0.053		
5.27	0.363	11.00	0.061		
5.48	0.289	11.58	0.048		
5.75	0.055	12.06	0.037		
5.80	0.044	12.68	0.018		
5.96	0.079	13.26	0.010		

* No plot given

SPECIFICATION TABLE NO. 190 NORMAL SPECTRAL REFLECTANCE OF JET DRY SATIN FINISH PAINTS

Curve No.	Ref. No.	Year	Temperature, K	Wavelength Range, μm	Geometry θ θ' ω'	Reported Error, %	Composition (weight percent), Specifications, and Remarks
1*	102	1964	~296	0.401-2.502	8° 8°		Jet Dry Satin Finish Black No. 78 (~0.96 x 10^{-4} mm thick) on aluminum substrate; data extracted from smooth curve; measured at least 24 hrs after final coating.
2*	102	1964	~313	2.45-40.00	30° 30°		Above specimen and conditions.

DATA TABLE NO. 190 NORMAL SPECTRAL REFLECTANCE OF JET DRY SATIN FINISH PAINTS

[Wavelength, λ, μm; Reflectance, ρ; Temperature, T, K]

λ	ρ	λ	ρ	λ	ρ	λ	ρ
CURVE 1*		CURVE 2*		CURVE 2 (cont.)*		CURVE 2 (cont.)*	
T ~ 296		T ~ 313					
				22.38	0.054	32.82	0.073
0.401	0.005	2.45	0.005	23.81	0.054	33.11	0.078
0.641	0.005	4.20	0.004	24.53	0.061	33.37	0.101
0.826	0.005	6.19	0.004	24.97	0.049	33.73	0.104
1.039	0.006	8.04	0.003	25.25	0.046	34.62	0.087
1.799	0.007	9.45	0.007	25.76	0.052	35.21	0.079
1.920	0.008	11.83	0.009	26.04	0.052	35.91	0.081
2.013	0.009	12.78	0.026	26.54	0.039	36.32	0.070
2.076	0.009	13.75	0.021	27.00	0.034	37.69	0.059
2.217	0.007	14.56	0.027	27.45	0.034	38.83	0.063
2.347	0.008	16.08	0.027	28.63	0.044	39.41	0.078
2.502	0.008	16.99	0.030	29.00	0.036	40.00	0.082
		17.60	0.027	29.87	0.068		
		17.93	0.027	30.31	0.082		
		19.31	0.041	30.82	0.099		
		20.10	0.032	31.14	0.104		
		20.77	0.032	32.36	0.089		

* No plot given

SPECIFICATION TABLE NO. 191 NORMAL SPECTRAL REFLECTANCE OF KERPO PAINTS

Curve No.	Ref. No.	Year	Temperature, K	Wavelength Range, μm	Geometry θ	θ'	ω'	Reported Error, %	Composition (weight percent), Specifications, and Remarks
1*	102	1964	~296	0.407-2.502	8°	8°			Kerpo Q. D. Spray Metallic Enamel WB-S-N 52-E-4 grey (~0.43 x 10^{-4} mm thick) on aluminum substrate; data extracted from smooth curve; measured at least 24 hrs after final coating.
2*	102	1964	~313	2.45-40.00	30°	30°			Above specimen and conditions.

DATA TABLE NO. 191 NORMAL SPECTRAL REFLECTANCE OF KERPO PAINTS

[Wavelength, λ, μm; Reflectance, ρ; Temperature, T, K]

λ	ρ	λ	ρ	λ	ρ
CURVE 1* T ~ 296		CURVE 2 (cont.)*		CURVE 2 (cont.)*	
		11.71	0.031	38.74	0.086
0.407	0.042	12.71	0.022	38.98	0.104
0.648	0.040	15.06	0.024	39.11	0.109
0.863	0.040	15.92	0.033	39.22	0.104
1.224	0.038	16.62	0.022	39.68	0.075
1.956	0.039	17.06	0.024	40.00	0.056
2.502	0.040	17.54	0.034		
		18.29	0.024		
CURVE 2* T ~ 313		18.85	0.024		
		19.48	0.036		
		20.32	0.061		
2.45	0.044	20.79	0.067		
2.89	0.031	21.36	0.070		
3.44	0.050	21.79	0.070		
4.04	0.031	23.11	0.058		
5.20	0.038	25.66	0.064		
6.74	0.024	35.78	0.064		
8.98	0.039	38.04	0.069		

* No plot given

SPECIFICATION TABLE NO. 192 NORMAL SPECTRAL EMITTANCE OF KODAK PAINTS

Curve No.	Ref. No.	Year	Temperature, K	Wavelength Range, μm	Geometry θ'	Reported Error, %	Composition (weight percent), Specifications, and Remarks
1*	74	1959	299	1.75-21.00	~0°		Kodak Brushing Lacquer Black on brass substrate; data extracted from smooth curve; property converted from R (2π, ~0°).

DATA TABLE NO. 192 NORMAL SPECTRAL EMITTANCE OF KODAK PAINTS

[Wavelength, λ, μm; Emittance, ϵ; Temperature, T, K]

λ	ϵ	λ	ϵ	λ	ϵ
CURVE 1*		CURVE 1 (cont.)*		CURVE 1 (cont.)*	
T = 299		11.30	0.802	20.10	0.470
1.75	0.937	11.51	0.842	20.48	0.470
3.33	0.892	11.76	0.872	20.73	0.447
3.85	0.867	11.95	0.881	20.87	0.452
4.33	0.826	12.22	0.860	21.00	0.465
4.86	0.804	12.55	0.788		
5.57	0.804	12.77	0.751		
6.76	0.827	13.01	0.733		
7.70	0.858	13.36	0.753		
8.26	0.825	13.56	0.753		
9.03	0.843	13.98	0.719		
9.34	0.872	14.60	0.730		
9.51	0.872	15.85	0.602		
10.31	0.801	16.64	0.542		
10.71	0.751	17.47	0.501		
10.87	0.751	19.14	0.493		
11.07	0.764	19.68	0.485		

* No plot given

SPECIFICATION TABLE NO. 193 NORMAL SPECTRAL REFLECTANCE OF KODAK PAINTS

Curve No.	Ref. No.	Year	Temperature, K	Wavelength Range, μm	Geometry θ θ' ω'	Reported Error, %	Composition (weight percent), Specifications, and Remarks
1*	102	1964	~296	0.398-2.498	8° 8°		Kodak Dull Black Brushing Lacquer No. 4 (~0.44 x 10^{-4} mm thick) on aluminum substrate; data extracted from smooth curve; measured at least 24 hrs after final coating.
2*	102	1964	~313	2.45-40.00	30° 30°		Above specimen and conditions.

DATA TABLE NO. 193 NORMAL SPECTRAL REFLECTANCE OF KODAK PAINTS

[Wavelength, λ, μm; Reflectance, ρ; Temperature, T, K]

λ	ρ	λ	ρ	λ	ρ	λ	ρ
CURVE 1*		CURVE 2 (cont.)*		CURVE 2 (cont.)*		CURVE 2 (cont.)*	
T ~ 296		5.41	0.078	15.47	0.159	33.90	0.236
0.398	0.005	6.01	0.055	16.07	0.111	34.37	0.279
0.840	0.013	6.46	0.055	16.72	0.111	35.58	0.346
0.980	0.017	7.13	0.038	17.52	0.079	36.90	0.404
1.112	0.019	7.99	0.055	18.30	0.105	38.01	0.437
1.253	0.023	8.87	0.058	19.55	0.201	38.39	0.445
1.634	0.029	9.97	0.085	21.22	0.327	38.63	0.445
2.054	0.034	10.56	0.111	21.72	0.351	39.14	0.441
2.498	0.039	10.96	0.091	22.11	0.358	40.00	0.456
		11.45	0.057	22.59	0.358		
CURVE 2*		12.29	0.140	23.37	0.334		
T ~ 313		12.49	0.148	26.96	0.182		
		12.92	0.077	28.64	0.105		
2.45	0.052	13.25	0.063	29.17	0.097		
3.11	0.048	13.63	0.075	30.52	0.101		
3.65	0.056	14.06	0.132	32.02	0.085		
4.38	0.049	14.61	0.129	32.77	0.093		
5.23	0.078	15.20	0.159	33.15	0.131		

* No plot given

SPECIFICATION TABLE NO. 194 NORMAL TOTAL EMITTANCE OF KRY-KOTE PAINTS

Curve No.	Ref. No.	Year	Temperature Range, K	Geometry θ'	Reported Error, %	Composition (weight percent), Specifications and Remarks
1*	83	1963	360	~0°		Kry-Kote white paint; measured in vacuum (1.6×10^{-4} mm Hg).

DATA TABLE NO. 194 NORMAL TOTAL EMITTANCE OF KRY-KOTE PAINTS

[Temperature, T, K; Emittance, ϵ]

T	ϵ
CURVE 1*	
360	0.880

* No plot given

SPECIFICATION TABLE NO. 195 NORMAL SPECTRAL EMITTANCE OF KRYLON PAINTS

Curve No.	Ref. No.	Year	Temperature, K	Wavelength Range, μm	Geometry θ'	Reported Error, %	Composition (weight percent), Specifications, and Remarks
1*	96	1963	323.2	2.97-20.02	0°		Krylon Flat Black paint on aluminum substrate. [Authors' designation: Sample BEC-3]

DATA TABLE NO. 195 NORMAL SPECTRAL EMITTANCE OF KRYLON PAINTS

[Wavelength, λ, μm; Emittance, ϵ; Temperature, T, K]

λ	ϵ
CURVE 1*	
T = 323.2	
2.97	0.963
4.35	0.949
4.78	0.947
5.24	0.939
5.79	0.950
7.07	0.954
7.54	0.959
9.21	0.957
9.92	0.956
11.55	0.966
12.60	0.981
14.56	0.972
16.39	0.972
16.68	0.968
16.95	0.973
18.01	0.969
18.63	0.975
19.56	0.973
20.02	0.976

* No plot given

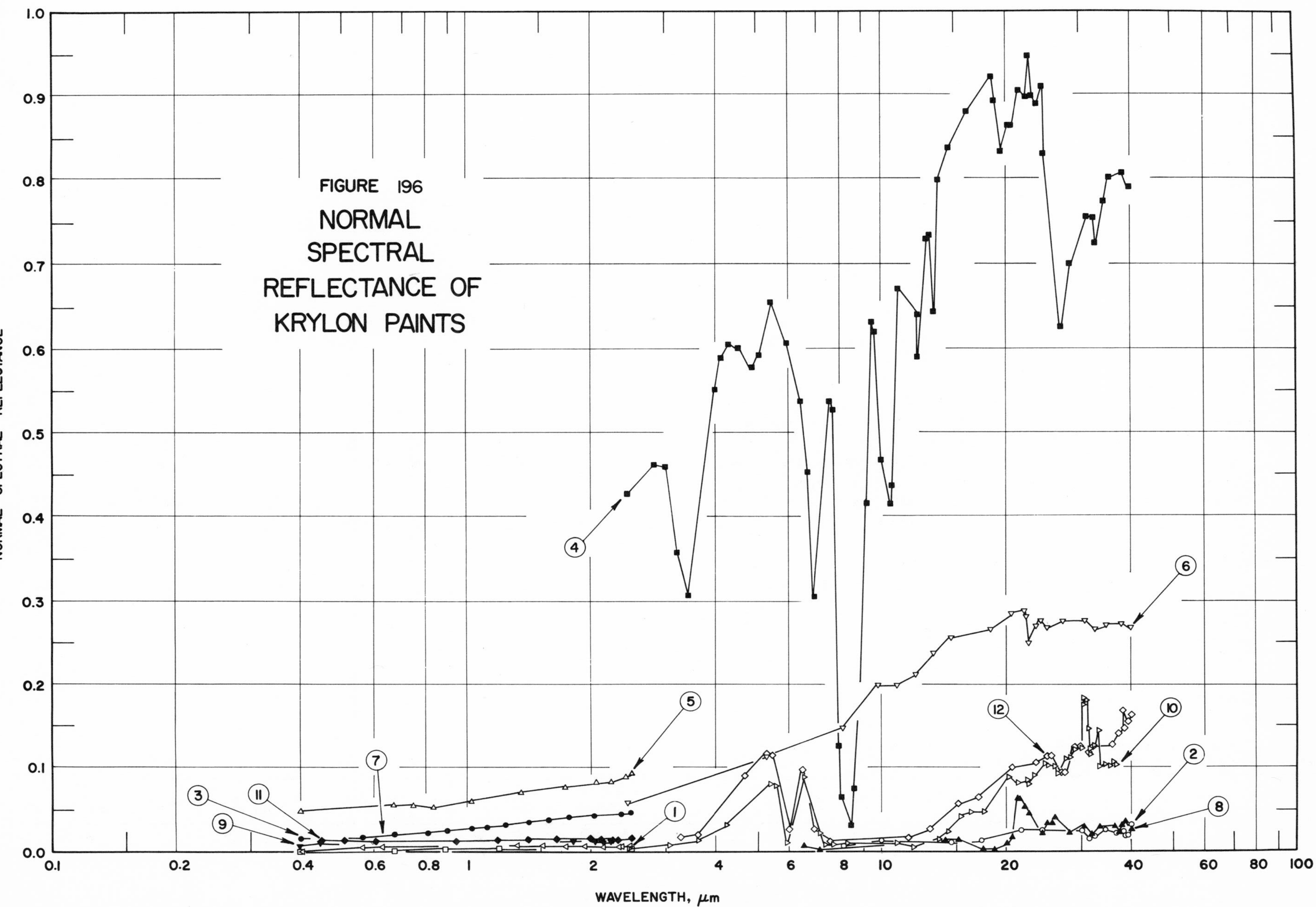

FIGURE 196 NORMAL SPECTRAL REFLECTANCE OF KRYLON PAINTS

SPECIFICATION TABLE NO. 196 NORMAL SPECTRAL REFLECTANCE OF KRYLON PAINTS

Curve No.	Ref. No.	Year	Temperature, K	Wavelength Range, μm	Geometry θ	θ'	ω'	Reported Error, %	Composition (weight percent), Specifications, and Remarks
1	102	1964	~296	0.400-2.500	8°	8°			Krylon Black No. 1602 (~0.50 x 10^{-4} mm thick) on aluminum substrate; data extracted from smooth curve; measured at least 24 hrs after final coating.
2	102	1964	~313	2.45-40.00	30°	30°			Above specimen and conditions.
3	102	1964	~296	0.400-2.494	8°	8°			Krylon Crystal Clear No. 1301 (~0.53 x 10^{-4} mm thick) on aluminum substrate; data extracted from smooth curve; measured at least 24 hrs after final coating.
4	102	1964	~313	2.45-40.00	30°	30°			Above specimen and conditions.
5	102	1964	~296	0.408-2.504	8°	8°			Krylon Bright Silver Aluminum No. 1401 (~0.53 x 10^{-4} mm thick) on aluminum substrate; data extracted from smooth curve; measured at least 24 hrs after final coating.
6	102	1964	~313	2.45-40.00	30°	30°			Above specimen and conditions.
7	102	1964	~296	0.409-2.500	8°	8°			Krylon Yellow Spray (~1.66 x 10^{-4} mm thick) on aluminum substrate; data extracted from smooth curve; measured at least 24 hrs after final coating.
8	102	1964	~313	2.45-40.00	30°	30°			Above specimen and conditions.
9	102	1964	~296	0.400-2.481	8°	8°			Krylon Flat White No. 1502 (~ 0.30 x 10^{-4} mm thick) on aluminum substrate; data extracted from smooth curve; measured at least 24 hrs after final coating.
10	102	1964	~313	2.45-40.00	30°	30°			Above specimen and conditions.
11	102	1964	~296	0.401-2.500	8°	8°			Krylon Flat White No. 1502A (~0.70 x 10^{-4} mm thick) on aluminum substrate; data extracted from smooth curve; measured at least 24 hrs after final coating.
12	102	1964	~313	2.45-40.00	30°	30°			Above specimen and conditions.

DATA TABLE NO. 196 NORMAL SPECTRAL REFLECTANCE OF KRYLON PAINTS

[Wavelength, λ, μm; Reflectance, ρ; Temperature, T, K]

λ	ρ
CURVE 1 T ~296	
0.400	0.000
0.674	0.000
0.891	0.001
1.221	0.001
2.500	0.001
CURVE 2 T ~313	
2.45	0.008
8.12	0.008
10.07	0.014
14.99	0.011
17.37	0.013
21.65	0.026
24.47	0.026
30.13	0.023
31.77	0.018
32.65	0.019
34.70	0.027
36.94	0.022
38.24	0.023
38.84	0.019
39.13	0.019
40.00	0.032
CURVE 3 T ~296	
0.400	0.013
0.562	0.018
0.679	0.020
0.811	0.021
0.904	0.023
1.041	0.027
1.127	0.029
1.251	0.031
1.412	0.035
1.576	0.037
1.787	0.040
2.040	0.042
2.385	0.044
2.494	0.046
CURVE 4 T ~313	
2.45	0.427
2.85	0.461
3.02	0.459
3.21	0.357
3.43	0.307
3.99	0.550
4.15	0.589
4.37	0.604
4.55	0.601
4.90	0.576
5.12	0.592
5.43	0.656
5.58	0.656
5.99	0.608
6.41	0.535
6.65	0.453
6.95	0.308
7.46	0.537
7.61	0.526
7.91	0.125
8.06	0.066
8.44	0.031
8.63	0.074
9.28	0.415
9.56	0.632
9.68	0.621
10.08	0.465
10.36	0.418
10.58	0.436
11.00	0.670
11.90	0.633
12.12	0.640
12.76	0.729
13.05	0.734
13.47	0.643
13.73	0.799
14.30	0.837
16.15	0.880
18.35	0.921
18.90	0.893
19.50	0.832
20.17	0.866
20.75	0.866
CURVE 4 (cont.)	
21.74	0.905
22.41	0.891
22.76	0.945
23.46	0.898
23.96	0.880
24.44	0.911
24.78	0.831
27.18	0.626
27.58	0.626
28.69	0.700
31.60	0.757
32.33	0.757
32.99	0.724
34.03	0.777
35.64	0.802
38.64	0.807
40.00	0.791
CURVE 5 T ~296	
0.408	0.048
0.673	0.053
0.748	0.053
0.839	0.052
1.025	0.060
1.361	0.070
1.730	0.076
2.077	0.082
2.236	0.082
2.405	0.089
2.504	0.091
CURVE 6 T ~313	
2.45	0.069
5.36	0.114
8.16	0.147
9.85	0.199
10.93	0.199
12.10	0.212
13.29	0.238
14.94	0.255
CURVE 6 (cont.)	
18.31	0.266
20.65	0.284
22.09	0.288
22.41	0.281
22.70	0.249
23.75	0.270
24.35	0.277
25.05	0.269
27.61	0.275
31.29	0.275
32.81	0.265
35.04	0.272
38.03	0.272
40.00	0.267
CURVE 7 T ~296	
0.409	0.001
0.567	0.005
0.631	0.006
1.337	0.008
1.539	0.008
1.656	0.008
1.750	0.008
1.930	0.008
2.169	0.007
2.309	0.005
2.391	0.006
2.500	0.006
CURVE 8 T ~313	
2.45	0.007
6.51	0.009
7.08	0.004
10.44	0.013
13.42	0.009
13.89	0.012
14.52	0.010
15.06	0.016
17.60	0.003
18.89	0.003
CURVE 8 (cont.)	
20.10	0.010
20.46	0.018
21.28	0.063
21.51	0.063*
22.22	0.053
22.97	0.048
24.48	0.023
25.35	0.035
25.90	0.035*
26.41	0.040
28.56	0.024
30.22	0.033
30.90	0.033
32.09	0.022
33.14	0.031
36.57	0.033
37.43	0.024
38.32	0.033
38.95	0.037
40.00	0.029
CURVE 9 T ~296	
0.400	0.007
0.449	0.009
0.513	0.011
1.815	0.011
2.062	0.011
2.151	0.011
2.241	0.011
2.313	0.011
2.481	0.013
CURVE 10 T ~313	
2.45	0.005
3.06	0.008
3.64	0.015
4.24	0.034
5.40	0.080
5.60	0.078
5.98	0.011
CURVE 10 (cont.)	
6.48	0.089
7.08	0.023
7.37	0.008
7.66	0.008
8.32	0.011
10.09	0.011
11.91	0.006
13.76	0.015
14.57	0.025
15.68	0.043
16.34	0.049
17.71	0.049
20.13	0.089
21.07	0.081
22.31	0.086
22.70	0.080
23.38	0.091
24.78	0.105
25.29	0.102
26.03	0.102
26.66	0.093
27.84	0.111
28.69	0.112
29.29	0.122
30.03	0.124
30.26	0.174
30.61	0.183
31.09	0.181
31.50	0.146
31.73	0.123
31.95	0.117
32.51	0.128
33.01	0.145
33.40	0.151
34.59	0.154
35.50	0.152
36.09	0.158
36.54	0.154
37.56	0.160
38.12	0.158
39.16	0.165
39.65	0.188
40.00	0.176
CURVE 11 T ~296	
0.401	0.008*
0.457	0.011
0.517	0.012
0.605	0.013
0.950	0.013
1.178	0.013
1.402	0.015
1.654	0.015
2.000	0.017
2.113	0.017
2.207	0.017*
2.259	0.016
2.325	0.016
2.420	0.017
2.500	0.017
CURVE 12 T ~313	
2.45	0.017*
3.28	0.017
3.64	0.022
4.67	0.090
5.30	0.118
5.48	0.115
6.00	0.028
6.43	0.099
6.94	0.027
7.55	0.013
11.74	0.018
13.07	0.028
15.26	0.059
17.20	0.066
20.79	0.100
23.56	0.105
25.11	0.112
25.77	0.112
27.33	0.093
27.62	0.093
29.33	0.124
30.24	0.124
31.40	0.119
32.19	0.124
CURVE 12 (cont.)	
36.07	0.128
37.32	0.140
38.11	0.169
38.66	0.145
39.40	0.155
40.00	0.161

* Not shown on plot

SPECIFICATION TABLE NO. 197 NORMAL SPECTRAL REFLECTANCE OF MAGIC IRON CEMENT CO. WHITE PORCELAIN ENAMEL

Curve No.	Ref. No.	Year	Temperature, K	Wavelength Range, μm	Geometry θ	θ'	ω'	Reported Error, %	Composition (weight percent), Specifications, and Remarks
1*	99	1966	~300	1.95-14.50	~0°		2π		Magic Iron Cement Co. white porcelain enamel (~0.0381 mm thick) on AISI 316 stainless steel substrate; sprayed; air dried at least 24 hrs; converted from R(2π,~0°). [Authors' designation: Specimen No. 16]

DATA TABLE NO. 197 NORMAL SPECTRAL REFLECTANCE OF MAGIC IRON CEMENT CO. WHITE PORCELAIN ENAMEL

[Wavelength, λ, μm; Reflectance, ρ; Temperature, T, K]

λ	ρ	λ	ρ
CURVE 1* T ~ 300		CURVE 1 (cont.)*	
		6.98	0.075
1.95	0.418	7.11	0.124
2.19	0.345	7.25	0.061
2.48	0.366	7.43	0.093
2.99	0.109	7.52	0.082
3.18	0.179	8.47	0.035
3.22	0.198	9.37	0.022
3.59	0.041	10.15	0.034
4.00	0.329	10.96	0.049
4.38	0.399	11.56	0.021
4.54	0.377	12.04	0.021
5.24	0.373	12.67	0.018
5.74	0.138	13.29	0.014
5.79	0.039	13.90	0.024
6.05	0.034	14.50	0.034
6.37	0.241		
6.58	0.293		
6.93	0.060		

* No plot given

SPECIFICATION TABLE NO. 198 NORMAL SPECTRAL REFLECTANCE OF METALTONE HAMMER FINISH SILVER PAINTS

Curve No.	Ref. No.	Year	Temperature K	Wavelength Range, μm	Geometry θ	θ'	ω'	Reported Error, %	Composition (weight percent), Specifications, and Remarks
1*	102	1964	~296	0.400-2.513	8°	8°			Metaltone Hammer Finish Silver (~0.48 x 10^{-4} mm thick) on aluminum substrate; data extracted from smooth curve; measured at least 24 hrs after final coating.
2*	102	1964	~296	2.45-40.00	30°	30°			Above specimen and conditions.

DATA TABLE NO. 198 NORMAL SPECTRAL REFLECTANCE OF METALTONE HAMMER FINISH SILVER PAINTS

[Wavelength, λ, μm; Reflectance, ρ; Temperature, T, K]

λ	ρ	λ	ρ	λ	ρ	λ	ρ
CURVE 1* T ~ 296		CURVE 2 (cont.)		CURVE 2 (cont.)		CURVE 2 (cont.)	
		5.61	0.043	15.10	0.051	34.46	0.118
0.400	0.054	6.09	0.045	16.11	0.049	34.92	0.120
0.561	0.053	7.42	0.028	17.13	0.057	35.52	0.114
0.829	0.052	7.74	0.033	17.78	0.055	36.83	0.131
2.184	0.052	8.07	0.048	18.34	0.055	37.06	0.129
2.319	0.052	8.60	0.023	18.97	0.046	37.52	0.104
2.429	0.052	9.01	0.045	19.20	0.046	37.77	0.122
2.513	0.052	9.24	0.045	19.97	0.091	38.01	0.126
		9.79	0.042	21.66	0.089	39.14	0.121
CURVE 2* T ~ 313		11.86	0.042	22.59	0.100	40.00	0.112
		12.19	0.041	22.98	0.100		
		12.59	0.047	23.83	0.091		
2.45	0.050	12.96	0.044	26.21	0.103		
3.52	0.052	13.44	0.053	27.31	0.102		
4.50	0.061	13.94	0.051	30.28	0.111		
4.76	0.056	14.46	0.061	31.03	0.106		
5.39	0.043	14.69	0.060	33.94	0.124		

* No plot given

SPECIFICATION TABLE NO. 199 HEMISPHERICAL TOTAL EMITTANCE OF MICABOND PAINTS

Curve No.	Ref. No.	Year	Temperature Range, K	Reported Error, %	Composition (weight percent), Specifications and Remarks
1*	49	1961	278	10	Dull black Micabond; vinyl (phenolic) binder; value avg of several determinations.

DATA TABLE NO. 199 HEMISPHERICAL TOTAL EMITTANCE OF MICABOND PAINTS

[Temperature, T, K; Emittance, ϵ]

T	ϵ
CURVE 1*	
278	0.84

* No plot given

SPECIFICATION TABLE NO. 200 NORMAL SOLAR ABSORPTANCE OF MICABOND PAINTS

Curve No.	Ref. No.	Year	Temperature Range, K	Geometry θ	Reported Error, %	Composition (weight percent), Specifications and Remarks
1*	49	1961	278	~0°		Dull black Micabond; vinyl (phenolic) binder; value avg of several determinations; measured for extra terrestrial conditions.

DATA TABLE NO. 200 NORMAL SOLAR ABSORPTANCE OF MICABOND PAINTS

[Temperature, T, K; Absorptance, α]

T	α
CURVE 1*	
278	0.93

* No plot given

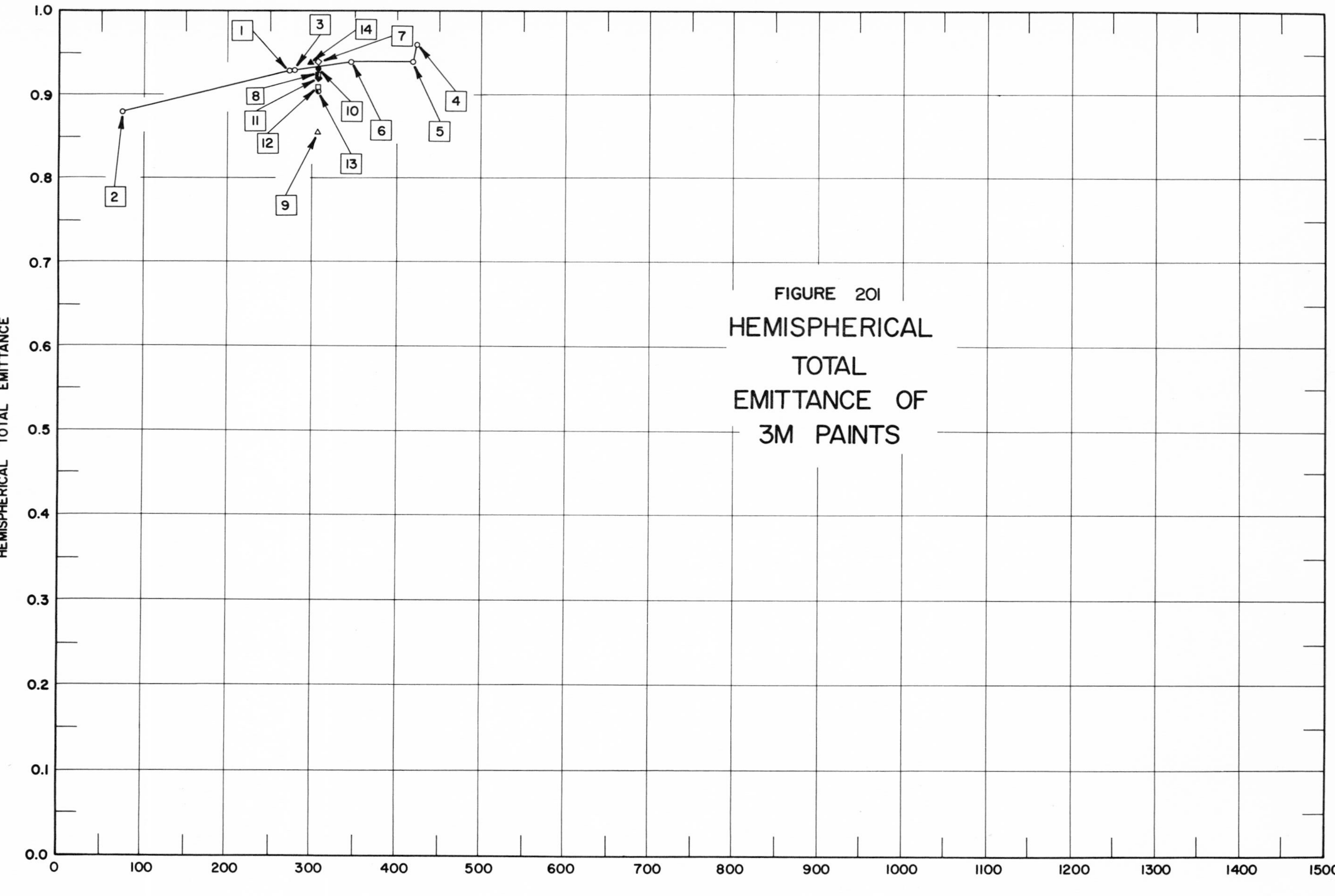

FIGURE 201
HEMISPHERICAL TOTAL EMITTANCE OF 3M PAINTS

SPECIFICATION TABLE NO. 201 HEMISPHERICAL TOTAL EMITTANCE OF 3M PAINTS

Curve No.	Ref. No.	Year	Temperature Range, K	Reported Error, %	Composition (weight percent), Specifications and Remarks
1	105	1964	277		3M Black Velvet No. 9564 (0.089 mm ± 0.038 mm thick) on stainless steel substrate. [Authors' designation: Test No. 1055a]
2	105	1964	78		Above specimen and conditions. [Authors' designation: Test No. 1055b]
3	105	1964	279		Above specimen and conditions. [Authors' designation: Test No. 1055c]
4	105	1964	422		Similar to above specimen and conditions except temp measurement ± 1.665 K. [Authors' designation: Test No. 1056a]
5	105	1964	420		Above specimen and conditions except temp measurement ± 2.78 K. [Authors' designation: Test No. 1056b]
6	105	1964	347		Curve 4 specimen and conditions. [Authors' designation: Test No. 1056c]
7	105	1964	307		3M Black Velvet No. 9564 on aluminum substrate; measured in vacuum (10^{-6} mm Hg) maintained by diffusion pump. [Authors' designation: Test No. 8]
8	105	1964	307		Similar to above specimen and conditions. [Authors' designation: Test No. 9]
9	105	1964	307		3M Black Velvet No. 9564 on copper substrate; measured in vacuum (10^{-6} mm Hg) maintained by diffusion pump. [Authors' designation: Test No. 124]
10	105	1964	307		3M Black Velvet No. 9564 on copper substrate; measured in vacuum (10^{-6} mm Hg) maintained by diffusion pump. [Authors' designation: Test No. 130]
11	105	1964	307		3M Black Velvet No. 9564 on copper (0.00318 mm thick) substrate; measured in vacuum (10^{-6} mm Hg) maintained by diffusion pump. [Authors' designation: Test No. 151]
12	105	1964	307		3M Black Velvet No. 9564 on copper (0.00318 mm thick) substrate; measured in vacuum (10^{-6} mm Hg) maintained by diffusion pump. [Author's designation: Test No. 154]
13	105	1964	307		3M Black Velvet No. 9564; measured in vacuum (10^{-6} mm Hg) maintained by diffusion pump. [Authors' designation: Test No. 164]
14	17	1968	298	±2.13	3M 401-C10 black paint.

DATA TABLE NO. 201 HEMISPHERICAL TOTAL EMITTANCE OF 3M PAINTS

[Temperature, T, K; Emittance, ∈]

T	∈	T	∈
CURVE 1		CURVE 13	
277	0.93	307	0.906
CURVE 2		CURVE 14	
78	0.88	298	0.94
CURVE 3*			
279	0.93		
CURVE 4			
422	0.96		
CURVE 5			
420	0.94		
CURVE 6			
347	0.94		
CURVE 7			
307	0.94		
CURVE 8			
307	0.926		
CURVE 9			
307	0.859		
CURVE 10			
307	0.933		
CURVE 11			
307	0.923		
CURVE 12			
307	0.910		

* Not shown on plot

SPECIFICATION TABLE NO. 202 NORMAL TOTAL EMITTANCE OF 3M PAINTS

Curve No.	Ref. No.	Year	Temperature Range, K	Geometry θ'	Reported Error, %	Composition (weight percent), Specifications and Remarks
1*	19	1965	298	0°		3M black on aluminum substrate; reported error ± 0.02 unit.
2*	19	1965	298	0°		Above specimen and conditions except computed from 1-R $(2\pi, 0°)$.

DATA TABLE NO. 202 NORMAL TOTAL EMITTANCE OF 3M PAINTS

[Temperature, T, K; Emittance, ϵ]

T	ϵ
CURVE 1*	
298	0.92
CURVE 2*	
298	0.92

* No plot given

SPECIFICATION TABLE NO. 203 NORMAL SPECTRAL EMITTANCE OF 3M PAINTS

Curve No.	Ref. No.	Year	Temperature, K	Wavelength Range, μm	Geometry θ'	Reported Error, %	Composition (weight percent), Specifications, and Remarks
1*	93	1965	77	3.49-39.81	~0°		3M Black paint (~0.089 mm thick); data extracted from smooth curve.
2*	93	1965	373	2.49-39.81	~0°		Similar to above specimen and conditions.

DATA TABLE NO. 203 NORMAL SPECTRAL EMITTANCE OF 3M PAINTS

[Wavelength, λ, μm; Emittance, ∈; Temperature, T, K]

λ	∈	λ	∈	λ	∈	λ	∈	λ	∈	λ	∈
CURVE 1* T = 77		CURVE 1 (cont.)*		CURVE 1 (cont.)*		CURVE 2 (cont.)*		CURVE 2 (cont.)*		CURVE 2 (cont.)*	
		21.55	0.919	32.99	0.963	5.41	0.967	20.64	0.938	35.99	0.944
3.49	0.952	21.98	0.925	33.66	0.943	6.25	0.949	21.07	0.921	36.60	0.940
5.77	0.962	22.43	0.940	34.15	0.943	6.66	0.954	21.31	0.921	37.86	0.920
7.67	0.948	22.73	0.940	34.96	0.953	6.87	0.958	21.99	0.940	39.81	0.908
8.47	0.933	23.27	0.929	36.01	0.953	7.34	0.948	22.40	0.954		
8.67	0.916	23.50	0.929	37.55	0.936	8.02	0.922	24.43	0.938		
8.91	0.866	23.85	0.941	38.21	0.936	8.81	0.818	25.03	0.938		
9.02	0.859	24.51	0.941	39.00	0.937	8.96	0.813	25.87	0.944		
9.22	0.868	25.24	0.970	39.81	0.925	9.74	0.894	26.57	0.937		
9.52	0.907	26.03	0.972			10.48	0.929	27.11	0.937		
10.25	0.938	26.86	0.966	CURVE 2*		11.17	0.946	27.98	0.949		
11.14	0.938	27.61	0.949	T = 373		12.19	0.946	29.30	0.939		
11.67	0.945	28.08	0.949			12.92	0.940	29.97	0.939		
15.60	0.945	29.04	0.978	2.49	0.921	13.35	0.945	30.90	0.952		
16.43	0.953	29.32	0.978	3.33	0.930	13.98	0.945	32.85	0.948		
19.33	0.953	29.94	0.963	3.94	0.959	15.14	0.963	33.52	0.940		
19.94	0.943	31.31	0.963	4.74	0.960	19.31	0.963	33.97	0.940		
20.75	0.905	31.64	0.959	5.15	0.967	20.06	0.957	35.00	0.952		
20.99	0.905										

* No plot given

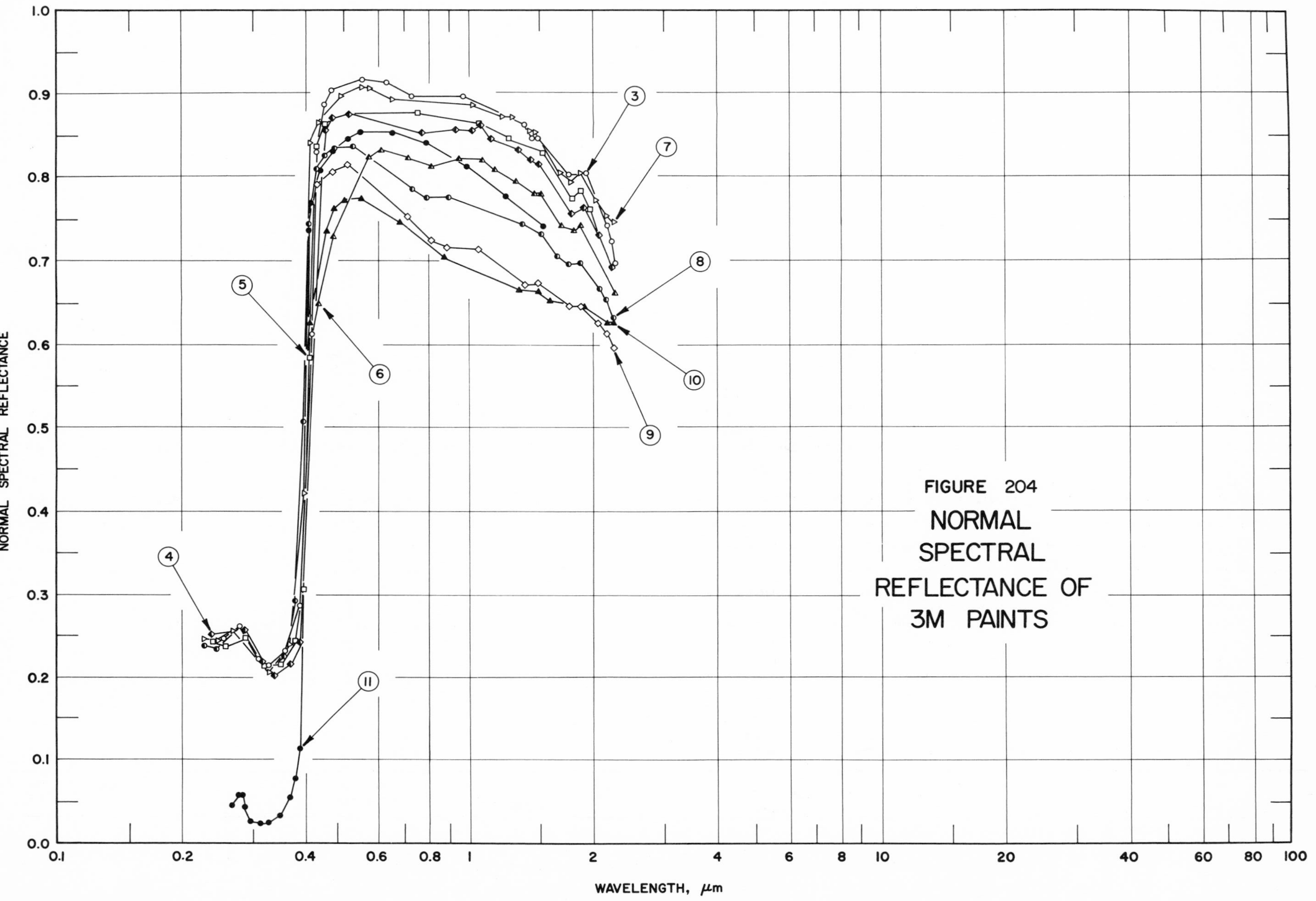

FIGURE 204 NORMAL SPECTRAL REFLECTANCE OF 3M PAINTS

SPECIFICATION TABLE NO. 204 NORMAL SPECTRAL REFLECTANCE OF 3M PAINTS

Curve No.	Ref. No.	Year	Temperature, K	Wavelength Range, μm	Geometry θ	θ'	ω'	Reported Error, %	Composition (weight percent), Specifications, and Remarks
1*	102	1964	~296	0.400-2.500	8°	8°			3M Brand Velvet Coating 9560 Series Optical Black (~0.41 x 10^{-4} mm thick) on aluminum substrate; data extracted from smooth curve; measured at least 24 hrs after final coating.
2*	102	1964	~313	2.45-40.00	30°	30°			Above specimen and conditions.
3	79	1967	~298	0.251-2.270	12.5°		2π		3M White Velvet 202-A-10 on stainless steel substrate; measured in vacuum (2 x 10^{-7}-5 x 10^{-7} mm Hg); data extracted from smooth curve; converted from reflectance factor. [Authors' designation: Sample No. 8-1]
4	79	1967	~298	0.237-2.240	12.5°		2π		Above specimen and conditions except irradiated in a vacuum (~4 x 10^{-6} mm Hg) with protons of nominal energy 3 keV, total dose 10^{14} p cm^{-2}; measured in situ after irradiation.
5	79	1967	~298	0.239-2.266	12.5°		2π		Above specimen and conditions except irradiated additionally to a total dose of 10^{15} p cm^{-2}.
6	79	1967	~298	0.239-2.262	12.5°		2π		Above specimen and conditions except irradiated additionally to a total dose of 10^{16} p cm^{-2}.
7	79	1967	~298	0.225-2.247	12.5°		2π		Similar to curve 3 specimen and conditions. [Authors' designation: Sample No. 7-1]
8	79	1967	~298	0.228-2.249	12.5°		2π		Above specimen and conditions except irradiated in a vacuum (~4 x 10^{-6} mm Hg) with electrons of energy 145 keV, total dose 1.3 x 10^{15} e cm^{-2}; measured in situ after irradiation.
9	79	1967	~298	0.228-2.258	12.5°		2π		Above specimen and conditions except irradiated additionally to a total dose of 1.3 x 10^{16} e cm^{-2}.
10	79	1967	~298	0.228-2.246	12.5°		2π		Above specimen and conditions except irradiated additionally to a total dose of 4 x 10^{16} e cm^{-2}.
11	108	1966	~298	0.2642-1.514	~0°		2π		3M White Velvet paint; substrate unknown; data extracted from smooth curve; property measured relative to freshly smoked MgO.

DATA TABLE NO. 204 NORMAL SPECTRAL REFLECTANCE OF 3M PAINTS

[Wavelength, λ, μm; Reflectance, ρ; Temperature, T, K]

CURVE 1* T ~ 296

λ	ρ
0.400	0.000
0.600	0.000
2.500	0.000

CURVE 2* T ~ 313

λ	ρ
2.45	~0
5	~0
10	~0
15	~0
20	~0
25	~0
30	~0
35	~0
40	~0

CURVE 3 T ~ 298

λ	ρ
0.251	0.246
0.278	0.261
0.308	0.224
0.326	0.215
0.355	0.231
0.388	0.286
0.417	0.612
0.431	0.830
0.443	0.888
0.467	0.904
0.552	0.918
0.630	0.914
0.725	0.899
0.961	0.898
1.373	0.862
1.414	0.849
1.482	0.849
1.740	0.801
1.912	0.802
2.188	0.743
2.230	0.724
2.270	0.699

CURVE 4 T ~ 298

λ	ρ
0.237	0.251
0.282	0.258
0.312	0.220
0.336	0.202
0.369	0.218
0.386	0.242
0.416	0.611*
0.445	0.858
0.461	0.872
0.515	0.877
0.767	0.853
0.928	0.858
1.016	0.858
1.072	0.863
1.129	0.849
1.310	0.834
1.419	0.821
1.496	0.817
1.792	0.759
1.901	0.763
2.090	0.731
2.240	0.694

CURVE 5 T ~ 298

λ	ρ
0.239	0.242
0.255	0.239
0.285	0.249
0.317	0.216
0.347	0.217
0.373	0.244
0.391	0.308
0.413	0.585
0.432	0.839
0.452	0.865
0.509	0.878*
0.756	0.879
1.076	0.866
1.250	0.849
1.517	0.830
1.795	0.776
1.868	0.785
1.976	0.761
2.266	0.691*

CURVE 6 T ~ 298

λ	ρ
0.239	0.242*
0.255	0.239*
0.285	0.249*
0.317	0.216*
0.347	0.217*
0.373	0.244*
0.391	0.308*
0.413	0.585*
0.435	0.650
0.471	0.730
0.572	0.825
0.613	0.834
0.715	0.825
0.809	0.812
0.943	0.821
1.084	0.821
1.154	0.810
1.303	0.795
1.432	0.781
1.505	0.781
1.696	0.744
1.802	0.739
1.876	0.744
2.262	0.661

CURVE 7 T ~ 298

λ	ρ
0.225	0.247
0.242	0.244
0.267	0.257
0.326	0.207
0.364	0.240
0.394	0.421
0.415	0.842
0.426	0.868
0.442	0.886
0.488	0.899
0.543	0.908
0.577	0.907
0.641	0.892
1.010	0.888
1.207	0.872
1.265	0.872
1.328	0.858*

CURVE 7 (cont.)

λ	ρ
1.400	0.856
1.457	0.856
1.667	0.805
1.749	0.801*
1.787	0.794
1.869	0.802
2.035	0.771
2.162	0.753
2.247	0.748

CURVE 8 T ~ 298

λ	ρ
0.228	0.237
0.241	0.233
0.269	0.258*
0.321	0.206
0.355	0.225
0.378	0.293
0.396	0.507
0.406	0.745
0.427	0.810
0.444	0.826
0.467	0.835
0.520	0.837
0.728	0.786
0.796	0.778
0.896	0.777
1.335	0.745
1.502	0.731
1.640	0.705
1.734	0.696
1.795	0.697*
1.893	0.699
2.077	0.669
2.189	0.653
2.249	0.634

CURVE 9 T ~ 298

λ	ρ
0.228	0.237
0.241	0.233
0.269	0.258
0.321	0.206
0.355	0.225

CURVE 9 (cont.)

λ	ρ
0.378	0.293
0.396	0.507*
0.406	0.745*
0.431	0.791
0.461	0.807
0.506	0.815
0.702	0.751
0.805	0.728
0.889	0.718
1.036	0.716
1.382	0.671
1.477	0.674
1.756	0.649
1.795	0.642*
1.863	0.647
2.060	0.627
2.187	0.613
2.258	0.598

CURVE 10 T ~ 298

λ	ρ
0.228	0.237
0.241	0.233
0.269	0.258
0.321	0.206
0.355	0.225
0.378	0.293
0.396	0.507
0.408	0.628
0.456	0.737
0.473	0.762
0.500	0.771
0.543	0.775
0.680	0.746
0.868	0.705
1.312	0.665
1.483	0.665
1.575	0.652
1.903	0.645
2.173	0.627
2.246	0.627

CURVE 11 T ~ 298

λ	ρ
0.2642	0.046
0.2735	0.059
0.2805	0.059
0.2844	0.045
0.2938	0.027
0.3105	0.025
0.3273	0.027
0.3467	0.035
0.3656	0.055
0.3793	0.077
0.3899	0.114
0.4085	0.739
0.4169	0.770
0.4385	0.808
0.4634	0.831
0.5024	0.847
0.5433	0.853
0.6501	0.853
0.7852	0.841
0.9931	0.814
1.247	0.779
1.514	0.741

* Not shown on plot

SPECIFICATION TABLE NO. 205 NORMAL SOLAR ABSORPTANCE OF 3M PAINTS

Curve No.	Ref. No.	Year	Temperature Range, K	Geometry θ	Reported Error, %	Composition (weight percent), Specifications and Remarks
1*	17	1968	298	~0°		3M Black Velvet 401-C10.

DATA TABLE NO. 205 NORMAL SOLAR ABSORPTANCE OF 3M PAINTS

[Temperature, T, K; Absorptance, α]

T	α
CURVE 1*	
298	0.98

* No plot given

SPECIFICATION TABLE NO. 206 NORMAL SPECTRAL REFLECTANCE OF MOREWEAR DURO-LAC PAINTS

Curve No.	Ref. No.	Year	Temperature, K	Wavelength Range, μm	Geometry θ θ'	ω'	Reported Error, %	Composition (weight percent), Specifications, and Remarks
1*	62	1949	~298	1.00-15.00	~0°	2π		Morewear Duro-Lac Black Brushing Lacquer No. 519; data extracted from smooth curve; converted from R(2π,~0°).

DATA TABLE NO. 206 NORMAL SPECTRAL REFLECTANCE OF MOREWEAR DURO-LAC PAINTS

[Wavelength, λ, μm; Reflectance, ρ; Temperature, T, K]

λ	ρ	λ	ρ	λ	ρ	λ	ρ
CURVE 1*		CURVE 1 (cont.)*		CURVE 1 (cont.)*		CURVE 1 (cont.)*	
T ~ 298		5.23	0.088	9.74	0.075	14.23	0.147
1.00	0.079	5.50	0.083	9.99	0.072	14.47	0.158
1.21	0.078	5.70	0.068	10.23	0.082	14.73	0.181
1.50	0.082	5.99	0.075	10.52	0.086	15.00	0.190
1.71	0.083	6.24	0.077	10.74	0.086		
1.99	0.076	6.49	0.101	11.01	0.085		
2.23	0.077	6.72	0.072	11.25	0.084		
2.49	0.080	6.99	0.069	11.49	0.082		
2.72	0.076	7.23	0.065	11.73	0.087		
2.98	0.073	7.48	0.058	12.01	0.100		
3.21	0.075	7.74	0.064	12.24	0.119		
3.46	0.075	8.00	0.063	12.49	0.132		
3.73	0.074	8.24	0.077	12.74	0.128		
3.98	0.078	8.48	0.057	12.98	0.149		
4.22	0.192	8.74	0.066	13.25	0.133		
4.48	0.084	9.00	0.063	13.50	0.149		
4.75	0.082	9.24	0.075	13.72	0.159		
4.99	0.086	9.49	0.069	13.97	0.140		

* No plot given

SPECIFICATION TABLE NO. 207 NORMAL SPECTRAL REFLECTANCE OF MURPHY DACOTE PAINTS

Curve No.	Ref. No.	Year	Temperature, K	Wavelength Range, μm	Geometry θ	θ'	ω'	Reported Error, %	Composition (weight percent), Specifications, and Remarks
1*	102	1964	~296	0.400-2.481	8°	8°			Murphy DaCote 4 Hour Enamel Black (~0.77 x 10^{-4} mm thick) on aluminum substrate; data extracted from smooth curve; measured at least 24 hrs after final coating.
2*	102	1964	~313	2.45-40.00	30°	30°			Above specimen and conditions.

DATA TABLE NO. 207 NORMAL SPECTRAL REFLECTANCE OF MURPHY DACOTE PAINTS

[Wavelength, λ, μm; Reflectance, ρ; Temperature, T, K]

λ	ρ	λ	ρ	λ	ρ	λ	ρ
CURVE 1*		CURVE 2 (cont.)*		CURVE 2 (cont.)*		CURVE 2 (cont.)*	
T ~ 296							
		8.77	0.070	20.09	0.216	37.39	0.478
0.400	0.001	9.45	0.082	20.20	0.197	38.04	0.456
0.856	0.001	11.21	0.179	20.39	0.130	38.53	0.482
1.395	0.002	11.88	0.188	20.64	0.119	38.67	0.482
1.666	0.003	12.58	0.165	21.49	0.154	39.25	0.444
1.960	0.005	13.01	0.179	22.80	0.243	39.47	0.431
2.216	0.007	13.18	0.170	24.22	0.348	40.00	0.416
2.481	0.011	13.33	0.134	25.16	0.386		
		13.50	0.129	26.31	0.410		
CURVE 2*		13.88	0.185	28.90	0.433		
T ~ 313		14.35	0.223	31.89	0.448		
		15.06	0.259	33.26	0.463		
2.45	0.007	15.40	0.257	34.45	0.481		
3.59	0.027	16.32	0.257	34.97	0.449		
5.29	0.071	17.51	0.236	35.53	0.476		
5.60	0.076	17.94	0.236	36.00	0.437		
6.12	0.076	19.30	0.241	36.54	0.460		
7.90	0.038	19.77	0.231	36.90	0.471		

* No plot given

SPECIFICATION TABLE NO. 208 NORMAL SPECTRAL REFLECTANCE OF NUCLEAR ENTERPRISE LTD. PAINTS

Curve No.	Ref. No.	Year	Temperature, K	Wavelength Range, μm	Geometry θ	θ'	ω'	Reported Error, %	Composition (weight percent), Specifications, and Remarks
1*	80	1966	~298	0.4260-0.5400	0°		2π		Nuclear Enterprise Ltd. 561 on aluminum substrate; measured relative to smoked MgO; converted from R(2π, 0°). [Author's designation: Reflector V]
2*	80	1966	~298	0.3237-0.5949	0°		2π		Above specimen and conditions except measured relative to pressed MgO powder; data extracted from smooth curve; converted from R(0°, 2π).
3*	80	1966	~298	0.4260-0.5400	0°		2π		Curve 1 specimen and conditions except exposed to deionized water at 373 K for 10 hrs.
4*	80	1966	~298	0.3164-0.4935	45°	0°			Nuclear Enterprise Ltd. 561 paint on aluminum substrate; data extracted from smooth curve; measured relative to pressed MgO powder. [Author's designation: Reflector No. V]

DATA TABLE NO. 208 NORMAL SPECTRAL REFLECTANCE OF NUCLEAR ENTERPRISE LTD. PAINTS

[Wavelength, λ, μm; Reflectance, ρ; Temperature, T, K]

λ	ρ	λ	ρ	λ	ρ
CURVE 1*		CURVE 2 (cont.)		CURVE 4*	
T ~ 298				T ~ 298	
		0.4001	0.691		
0.4260	0.773	0.4079	0.738	0.3164	0.019
0.4570	0.823	0.4230	0.778	0.3379	0.019
0.4950	0.85	0.4380	0.805	0.2972	0.031
0.5400	0.85	0.4609	0.824	0.3564	0.059
		0.5002	0.842	0.3604	0.089
CURVE 2*		0.5949	0.850	0.3662	0.142
T ~ 298				0.3841	0.465
		CURVE 3*		0.3910	0.555
		T ~ 298		0.3988	0.624
0.3237	0.053			0.4114	0.678
0.3317	0.053			0.4270	0.718
0.3439	0.066	0.4260	0.544	0.4573	0.757
0.3517	0.081	0.4570	0.665	0.4935	0.804
0.3569	0.102	0.4950	0.752		
0.3626	0.146	0.5400	0.804		
0.3699	0.261				
0.3821	0.487				
0.3887	0.587				

* No plot given

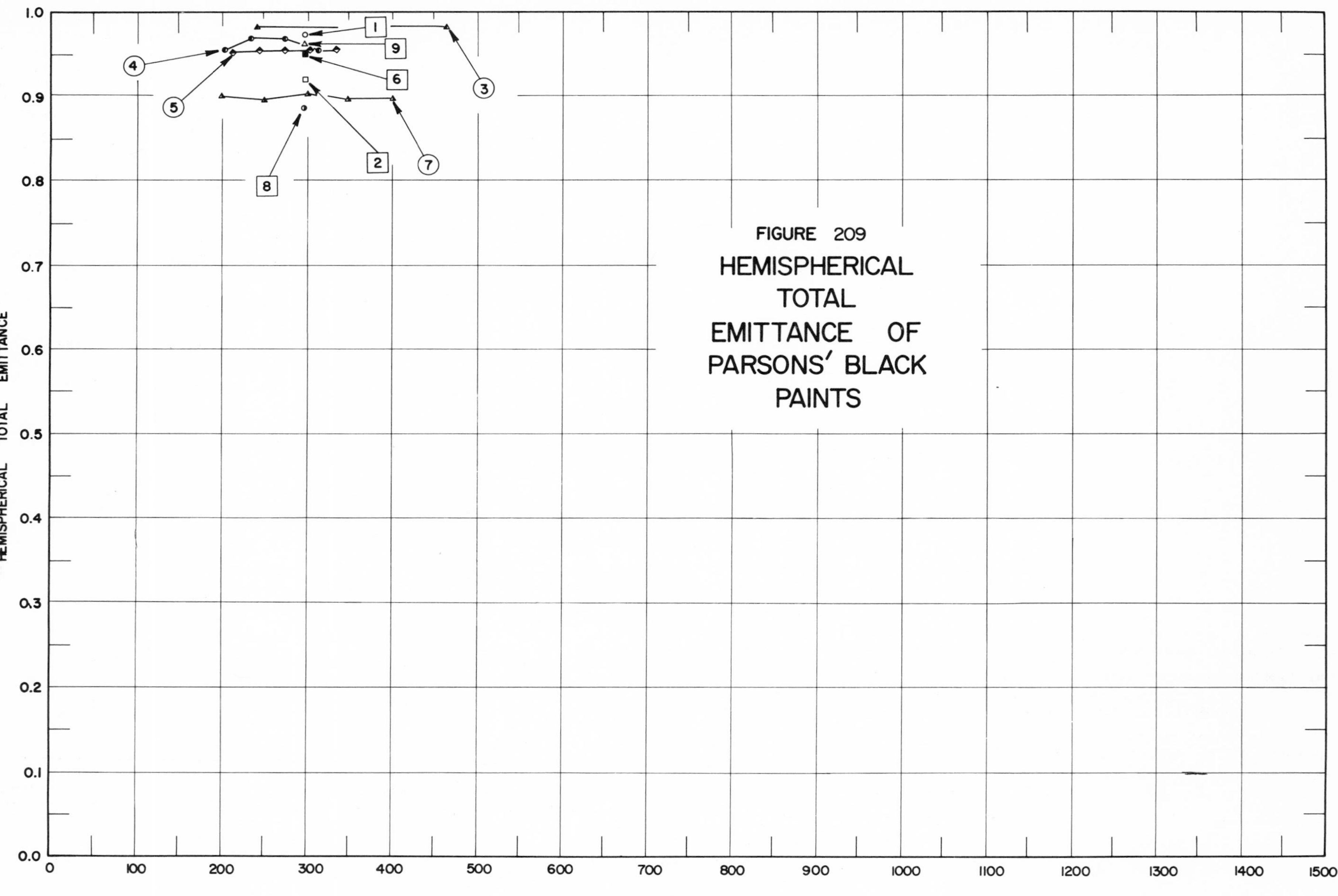
FIGURE 209
HEMISPHERICAL
TOTAL
EMITTANCE OF
PARSONS' BLACK
PAINTS
HEMISPHERICAL TOTAL EMITTANCE
TEMPERATURE, K
1.0
0.9
0.8
0.7
0.6
0.5
0.4
0.3
0.2
0.1
0.0
0
100
200
300
400
500
600
700
800
900
1000
1100
1200
1300
1400
1500
1
2
3
4
5
6
7
8
9

SPECIFICATION TABLE NO. 209 HEMISPHERICAL TOTAL EMITTANCE OF PARSONS' BLACK PAINTS

Curve No.	Ref. No.	Year	Temperature Range, K	Reported Error, %	Composition (weight percent), Specifications and Remarks
1	71	1962	298		Parsons' optical black lacquer; error in emittance ±0.003 units.
2	71	1962	298		Parsons' optical black lacquer; error in emittance ±0.06 units.
3	71	1962	240-462		Parsons' optical black lacquer.
4	111	1969	203-314		Parsons' black lacquer (~0.075 mm thick); substrate unknown; approx 9 coats applied; property measured by steady state calorimetric method.
5	111	1969	213-333		Curve 4 specimen and conditions except property measured by transient calorimetric method; property of second side of sample assumed.
6	111	1969	298		Curve 4 specimen and conditions except property converted from $\rho(2\pi, 7.5^\circ)$ measured by ellipsoid method.
7	111	1969	200-400		Curve 4 specimen and conditions except property converted from $\rho(2\pi, 15^\circ-75^\circ)$ measured by heated cavity method relative to a platinum surface.
8	111	1969	295		Curve 4 specimen and conditions except property measured by a portable Quick Emittance Device.
9	111	1969	295		Curve 4 specimen and conditions except property measured by a portable emissometer.

DATA TABLE NO. 209 HEMISPHERICAL TOTAL EMITTANCE OF PARSONS' BLACK PAINTS

[Temperature, T, K; Emittance, ∈]

T	∈	T	∈	T	∈
CURVE 1		CURVE 5		CURVE 8	
298	0.971	213	0.952	295	0.887
		244	0.954		
CURVE 2		273	0.954	CURVE 9	
		303	0.955		
298	0.92	333	0.956	295	0.961
CURVE 3		CURVE 6			
240	0.981	298	0.950		
462	0.981				
		CURVE 7			
CURVE 4					
		200	0.900		
203	0.955	250	0.896		
233	0.969	300	0.901		
273	0.969	350	0.898		
314	0.955	400	0.899		

SPECIFICATION TABLE NO. 210 NORMAL TOTAL EMITTANCE OF PARSONS' BLACK PAINTS

Curve No.	Ref. No.	Year	Temperature Range, K	Geometry θ	Reported Error, %	Composition (weight percent), Specifications and Remarks
1*	19	1965	298	0°		Parsons' black on aluminum substrate; reported error ± 0.02 unit.
2*	19	1965	298	0°		Above specimen and conditions except computed from 1 - R (2π, 0°).

DATA TABLE NO. 210 NORMAL TOTAL EMITTANCE OF PARSONS' BLACK PAINTS

[Temperature, T, K; Emittance, ϵ]

T	ϵ
CURVE 1*	
298	0.91
CURVE 2*	
298	0.92

* No plot given

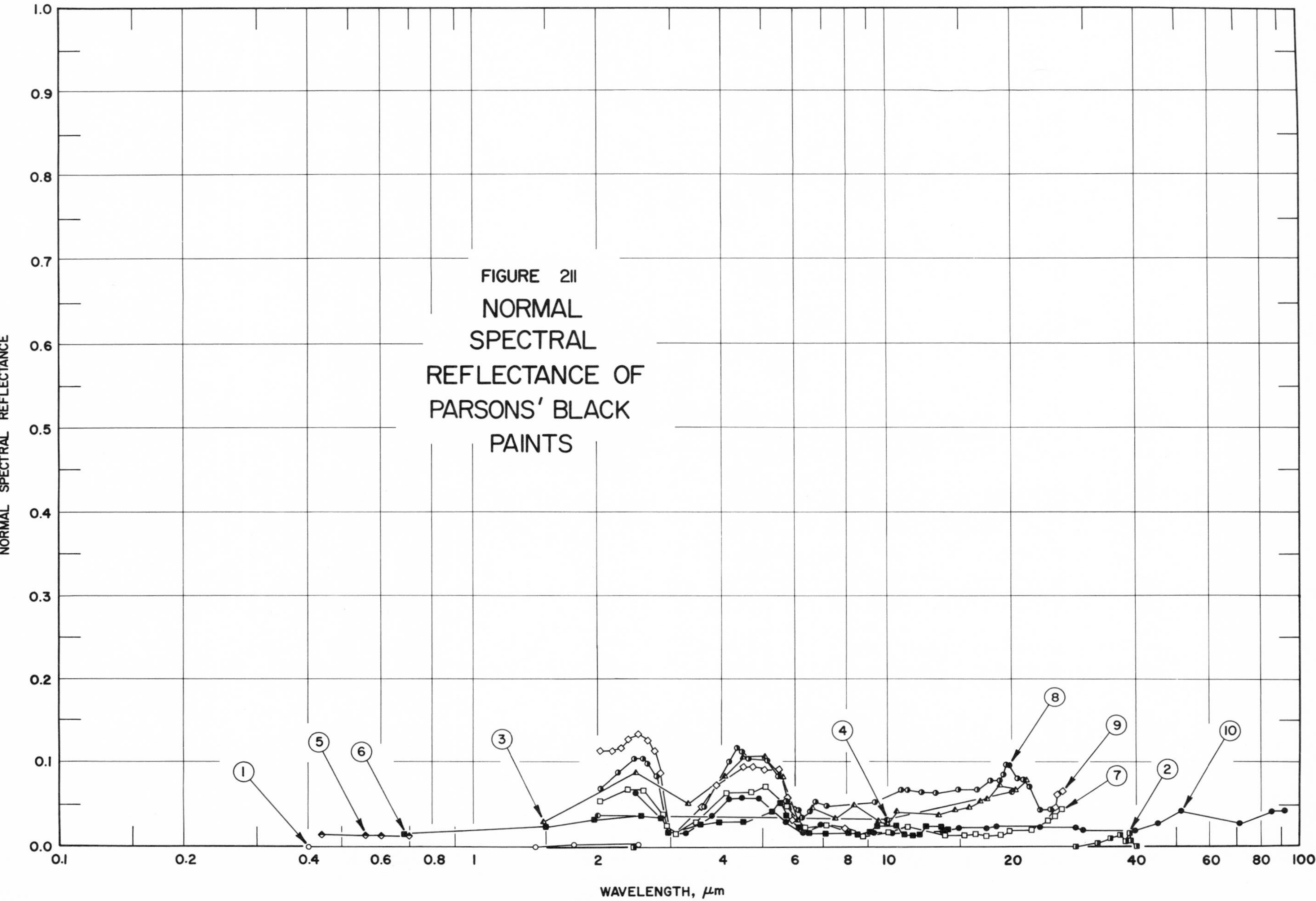

FIGURE 211
NORMAL SPECTRAL REFLECTANCE OF PARSONS' BLACK PAINTS

SPECIFICATION TABLE NO. 211 NORMAL SPECTRAL REFLECTANCE OF PARSONS' BLACK PAINTS

Curve No.	Ref. No.	Year	Temperature, K	Wavelength Range, μm	Geometry θ	θ'	ω'	Reported Error, %	Composition (weight percent), Specifications, and Remarks
1	102	1964	~296	0.400-2.500	8°	8°			Parsons' optical black (~0.82 x 10^{-4} mm thick) on aluminum substrate; data extracted from smooth curve; measured at least 24 hrs after final coating.
2	102	1964	~313	2.45-40.00	30°	30°			Above specimen and conditions.
3	112	1963	~298	1.48-21.99	~0°		2π	±2	Parsons' black paint (0.0508 mm thick) on copper substrate; sprayed with two coats; data extracted from smooth curve; converted from R (2π, ~0°).
4	109	1962	298	2-20	~0°		2π		Parsons' optical black (3 coats) on Parsons primer (1 coat) and cold-rolled steel substrates.
5	113	1967	~298	0.436-0.700	6°				Parsons' optical black lacquer on glass substrate; property measured using a Hardy spectrophotometer with an integrating sphere (with a port removed from that part of the sphere which might have collected the specular reflection).
6	114	1965	~298	0.68-14.0	10°		2π		Parsons' optical black lacquer (2 coats) on brass substrate.
7	111	1969	~298	2.01-26.2	17°		2π		Parsons' black lacquer (~0.075 mm thick); ~9 coats applied; data extracted from smooth curve; converted from R (2π, 17°) measured relative to gold.
8	111	1969	~298	2.01-24.7	17°		2π		Above specimen and conditions except measured relative to platinum; corrected for reflectance of platinum.
9	111	1969	~298	2.01-26.5	17°		2π		Above specimen and conditions except property measured by paraboloid method relative to gold after heated cavity measurement.
10	111	1969	~298	2.46-91.4	7.5°		2π		Similar to curve 1 specimen and conditions.

DATA TABLE NO. 211 NORMAL SPECTRAL REFLECTANCE OF PARSONS' BLACK PAINTS

[Wavelength, λ, μm; Reflectance, ρ; Temperature, T, K]

CURVE 1, T ~ 296

λ	ρ
0.400	0.000
1.414	0.000
1.752	0.001
2.500	0.001

CURVE 2, T ~ 313

λ	ρ
2.45	0.000
28.55	0.000
32.04	0.005
34.47	0.010
36.56	0.014
37.68	0.006
37.86	0.007*
38.40	0.018
38.54	0.018*
38.95	0.007
39.38	0.000*
40.00	0.000

CURVE 3, T ~ 298

λ	ρ
1.48	0.030
2.49	0.087
3.30	0.051
4.02	0.082
4.56	0.105
5.03	0.107
5.60	0.081
6.04	0.032
6.60	0.048
7.56	0.033
8.16	0.050
9.58	0.031
10.03	0.029
10.53	0.042
13.03	0.039
14.06	0.044
15.07	0.048
16.08	0.055
17.04	0.057
18.08	0.071
19.05	0.071*
20.14	0.067
21.99	0.078

CURVE 4, T = 298

λ	ρ
2	0.038
10	0.032
20	0.065

CURVE 5, T ~ 298

λ	ρ
0.436	0.014
0.546	0.013
0.600	0.012
0.700	0.012

CURVE 6, T ~ 298

λ	ρ
0.68	0.015
1.50	0.025
1.99	0.032
2.55	0.037
2.83	0.034
2.98	0.017
3.51	0.027
3.99	0.030
4.55	0.030
5.22	0.042
5.54	0.051
5.78	0.037
6.01	0.022
6.57	0.017
7.04	0.017
7.50	0.017*
8.01	0.017
8.56	0.014
9.04	0.019
9.54	0.025
10.0	0.031*
10.5	0.027
11.0	0.015
11.5	0.014
12.0	0.015
12.5	0.024
13.0	0.024*
13.5	0.024
14.0	0.021

CURVE 7, T ~ 298

λ	ρ
2.01	0.054
2.08	0.054*
2.38	0.066
2.58	0.066
2.87	0.037
2.93	0.024
3.08	0.015
3.46	0.029
4.08	0.061
4.65	0.064
5.08	0.070
5.66	0.047
6.15	0.021*
6.38	0.021
7.16	0.026
8.66	0.012
9.14	0.017*
10.0	0.017
11.2	0.021
13.7	0.013
15.4	0.013
16.3	0.015
17.3	0.013
18.6	0.014
19.9	0.020
22.1	0.020
24.3	0.031
25.4	0.038
26.2	0.046

CURVE 8, T ~ 298

λ	ρ
2.01	0.068
2.24	0.086
2.43	0.102
2.56	0.102
2.63	0.099
2.67	0.089*
2.79	0.081
2.87	0.037*
2.93	0.024*
3.03	0.018*
3.60	0.049
4.16	0.100
4.28	0.114*

CURVE 8 (cont.)

λ	ρ
4.34	0.117
4.44	0.111
4.60	0.105
5.16	0.104
5.45	0.085
5.71	0.054
6.08	0.043
6.20	0.035
6.51	0.041
6.76	0.052
7.14	0.049
9.35	0.053
10.8	0.067
11.1	0.067
12.0	0.064
13.0	0.064
14.8	0.068
16.4	0.068
17.9	0.078
18.7	0.078
19.0	0.085
19.3	0.096
19.9	0.096
20.6	0.080
21.5	0.080
22.0	0.070
22.3	0.070
23.2	0.044
23.9	0.044*
24.7	0.044

CURVE 9, T ~ 298

λ	ρ
2.01	0.112
2.17	0.112
2.28	0.118
2.38	0.127
2.50	0.131
2.64	0.126
2.74	0.112
2.82	0.085
2.93	0.025*
3.03	0.019*
3.59	0.047
3.85	0.071
4.48	0.094

CURVE 9 (cont.)

λ	ρ
4.76	0.094
5.02	0.090
5.43	0.090
5.78	0.059
5.91	0.031
6.25	0.021*
7.14	0.027*
7.96	0.022
8.66	0.012*
9.14	0.017*
10.0	0.017*
11.2	0.021*
13.7	0.013*
15.4	0.013*
16.3	0.015*
17.3	0.013*
18.6	0.014*
19.9	0.020*
22.1	0.020*
24.3	0.031*
25.5	0.049
25.8	0.061
26.5	0.066

CURVE 10, T ~ 298

λ	ρ
2.46	0.062
3.08	0.016*
3.27	0.016
3.77	0.036
4.18	0.056
4.43	0.058*
4.94	0.058
5.77	0.030
5.85	0.030*
6.24	0.018
6.52	0.018*
6.97	0.026
9.23	0.018
10.2	0.018
11.4	0.021*
13.6	0.019
14.9	0.022
17.3	0.022
18.2	0.025
22.2	0.022*

CURVE 10 (cont.)

λ	ρ
23.3	0.024
28.3	0.024
29.7	0.020
39.9	0.020
45.5	0.029
51.1	0.042
70.3	0.029
84.9	0.041
91.4	0.041

* Not shown on plot

SPECIFICATION TABLE NO. 212 NORMAL SOLAR ABSORPTANCE OF PARSONS' BLACK PAINTS

Curve No.	Ref. No.	Year	Temperature Range, K	Geometry θ	Reported Error, %	Composition (weight percent), Specifications and Remarks
1*	115	1960	298	15°		Parsons' optical black lacquer; two coats; computed from spectral reflectance data. [Authors' designation: Sample 25]

DATA TABLE NO. 212 NORMAL SOLAR ABSORPTANCE OF PARSONS' BLACK PAINTS

[Temperature, T, K; Absorptance, α]

T	α
CURVE 1*	
298	0.98

* No plot given

SPECIFICATION TABLE NO. 213 NORMAL SPECTRAL REFLECTANCE OF PEDIGREE PAINTS

Curve No.	Ref. No.	Year	Temperature, K	Wavelength Range, μm	Geometry θ θ'	ω'	Reported Error, %	Composition (weight percent), Specifications, and Remarks
1*	104	1953	298	1.00-15.0	5°	2π	± 2	Pedigree red (0.0762 mm thick) on aluminum substrate; substrate polished; data extracted from smooth curve; converted from R (2π, 5°).

DATA TABLE NO. 213 NORMAL SPECTRAL REFLECTANCE OF PEDIGREE PAINTS

[Wavelength, λ, μm; Reflectance, ρ; Temperature, T, K]

λ	ρ	λ	ρ	λ	ρ
CURVE 1*		CURVE 1 (cont.)*		CURVE 1 (cont.)*	
T = 298		6.00	0.110	12.4	0.095
1.00	0.420	6.20	0.110	13.0	0.095
1.20	0.480	6.40	0.140	13.0	0.095
1.50	0.450	7.00	0.105	13.6	0.100
2.00	0.410	7.60	0.100	14.0	0.095
2.20	0.370	8.00	0.100	14.2	0.100
2.50	0.340	8.20	0.100	14.4	0.100
2.80	0.260	8.70	0.090	14.7	0.090
3.00	0.120	8.80	0.090	15.0	0.100
3.20	0.140	9.50	0.115		
3.50	0.095	9.80	0.105		
3.80	0.190	10.0	0.120		
4.00	0.210	10.2	0.120		
4.20	0.275	10.7	0.100		
4.40	0.290	11.0	0.105		
5.00	0.270	11.3	0.105		
5.50	0.215	11.6	0.120		
5.70	0.090	12.0	0.120		

* No plot given

SPECIFICATION TABLE NO. 214 NORMAL SPECTRAL REFLECTANCE OF PITTSBURGH PAINTS

Curve No.	Ref. No.	Year	Temperature, K	Wavelength Range, μm	Geometry ϑ	θ'	ω'	Reported Error, %	Composition (weight percent), Specifications, and Remarks
1*	62	1949	~298	1.00-15.00	~0°		2π	5	Pittsburgh flat white enamel Undercoater LA 404; data extracted from smooth curve; converted from R (2π, ~0°).

DATA TABLE NO. 214 NORMAL SPECTRAL REFLECTANCE OF PITTSBURGH PAINTS

[Wavelength, λ, μm; Reflectance, ρ; Temperature, T, K]

λ	ρ	λ	ρ	λ	ρ	λ	ρ
CURVE 1* T ~ 298		CURVE 1 (cont.)*		CURVE 1 (cont.)*		CURVE 1 (cont.)*	
1.00	0.512	5.00	0.508	9.25	0.141	13.53	0.174
1.23	0.599	5.24	0.490	9.49	0.134	13.76	0.133
1.46	0.624	5.51	0.353	9.74	0.131	13.98	0.138
1.75	0.565	5.78	0.094	9.99	0.118	14.24	0.170
1.97	0.592	5.99	0.194	10.23	0.125	14.49	0.137
2.23	0.469	6.26	0.143	10.50	0.126	15.00	0.110
2.51	0.482	6.50	0.206	10.74	0.149		
2.75	0.250	6.77	0.114	10.98	0.155		
2.98	0.186	7.00	0.093	11.23	0.177		
3.23	0.248	7.24	0.100	11.51	0.198		
3.49	0.072	7.50	0.105	11.75	0.227		
3.72	0.317	7.74	0.099	11.94	0.212		
3.99	0.404	8.00	0.061	12.27	0.232		
4.23	0.490	8.26	0.053	12.51	0.220		
4.50	0.435	8.50	0.077	12.76	0.207		
4.78	0.389	8.76	0.062	13.00	0.199		
		8.98	0.098	13.26	0.197		

* No plot given

SPECIFICATION TABLE NO. 215 NORMAL SPECTRAL REFLECTANCE OF PROVEN PAINTS

Curve No.	Ref. No.	Year	Temperature, K	Wavelength Range, μm	Geometry θ θ' ω'	Reported Error, %	Composition (weight percent), Specifications, and Remarks
1*	102	1964	~296	0.396-2.496	8° 8°		Proven flat white SP-15 (~1.45 x 10^{-4} mm thick) on aluminum substrate; data extracted from smooth curve; measured at least 24 hrs after final coating.
2*	102	1964	~313	2.45-40.00	30° 30°		Above specimen and conditions.

DATA TABLE NO. 215 NORMAL SPECTRAL REFLECTANCE OF PROVEN PAINTS

[Wavelength, λ, μm; Reflectance, ρ; Temperature, T, K]

λ	ρ	λ	ρ	λ	ρ	λ	ρ
CURVE 1* T ~ 296		CURVE 2* T ~ 313		CURVE 2 (cont.)*		CURVE 2 (cont.)*	
				19.66	0.086	31.36	0.173
0.396	0.006	2.45	0.012	20.26	0.101	32.03	0.181
0.500	0.009	3.46	0.005	20.78	0.120	32.71	0.181
0.563	0.010	4.04	0.042	21.13	0.144	34.25	0.174
0.618	0.010	4.58	0.046	21.57	0.185	35.97	0.174
0.751	0.009	5.11	0.066	22.88	0.170	37.38	0.179
1.259	0.009	5.38	0.066	23.90	0.143	38.38	0.170
1.570	0.010	6.11	0.031	24.90	0.132	39.18	0.156
1.862	0.011	7.03	0.016	25.54	0.173	40.00	0.152
2.153	0.013	7.95	0.023	26.08	0.182		
2.277	0.013	9.43	0.005	27.24	0.179		
2.365	0.014	10.02	0.038	28.06	0.179		
2.496	0.017	11.62	0.023	28.54	0.168		
		12.85	0.017	28.96	0.182		
		14.61	0.054	29.56	0.167		
		15.37	0.063	30.05	0.180		
		18.75	0.077	30.81	0.173		

* No plot given

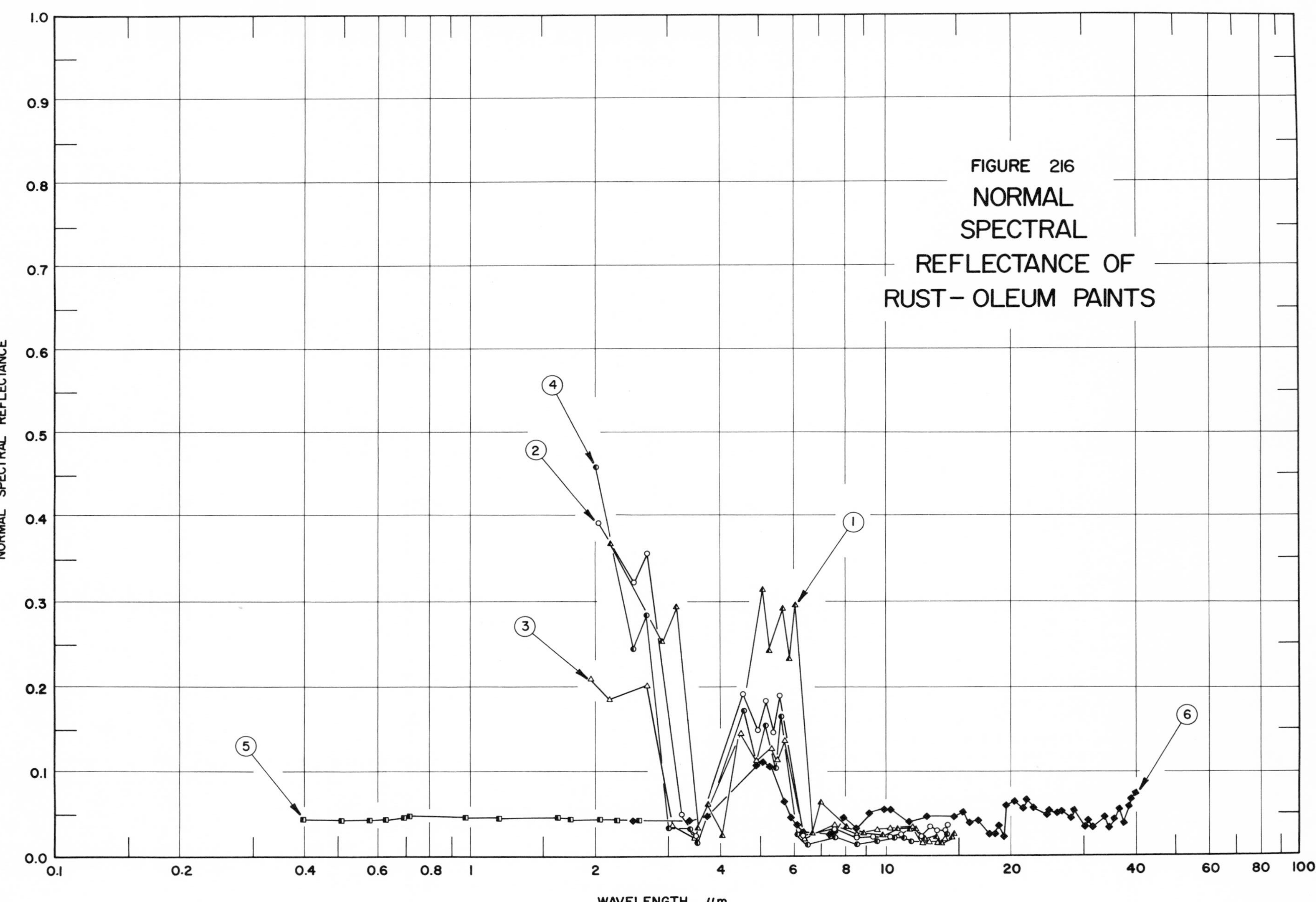

FIGURE 216 NORMAL SPECTRAL REFLECTANCE OF RUST-OLEUM PAINTS

SPECIFICATION TABLE NO. 216 NORMAL SPECTRAL REFLECTANCE OF RUST-OLEUM PAINTS

Curve No.	Ref. No.	Year	Temperature, K	Wavelength Range, μm	Geometry θ	θ'	ω'	Reported Error, %	Composition (weight percent), Specifications, and Remarks
1	99	1966	~300	2.19-14.44	~0°		2π	< ±2	Rust-Oleum Green No. 205 (~0.0381 mm thick) on AISI 316 stainless steel substrate; sprayed; air dried at least 24 hrs; converted from R (2π, ~0°). [Authors' designation: Specimen No. 9]
2	99	1966	~300	2.01-14.02	~0°		2π	< ±2	Rust-Oleum Red No. 215 (~0.0381 mm thick) on AISI 316 stainless steel substrate; sprayed; air dired at least 24 hrs; converted from R (2π, ~0°). [Authors' designation: Specimen No. 8]
3	99	1966	~300	1.96-14.00	~0°		2π	< ±2	Rust-Oleum Silver Gray No. 208 (~0.0381 mm thick) on ASIS 316 stainless steel substrate; sprayed; air dried at least 24 hrs; converted from R (2π, ~0°). [Authors' designation: Specimen No. 7]
4	99	1966	~300	2.00-14.00	~0°		2π	< ±2	Rust-Oleum White No. 225 (~0.0381 mm thick) on AISI 316 stainless steel substrate; sprayed; air dried at least 24 hrs; converted from R (2π, ~0°). [Authors' designation: Specimen No. 6]
5	102	1964	~296	0.397-2.505	8°	8°			Rust-Oleum Fire Hydrant Red No. 1210 (~ 0.77 x 10^{-4} mm thick) on aluminum substrate; data extracted from smooth curve; measured at least 24 hrs after final coating.
6	102	1964	~313	2.45-40.00	30°	30°			Above specimen and conditions.

DATA TABLE NO. 216 NORMAL SPECTRAL REFLECTANCE OF RUST-OLEUM PAINTS

[Wavelength, λ, μm; Reflectance, ρ; Temperature, T, K]

λ	ρ
CURVE 1	
T ~ 300	
2.19	0.369
2.90	0.253
3.11	0.295
3.52	0.034
3.74	0.061
4.04	0.025
5.04	0.316
5.32	0.243
5.63	0.292
5.86	0.233
6.03	0.296
6.75	0.028
7.00	0.064
8.03	0.035
8.98	0.028
9.86	0.025
10.65	0.031
11.46	0.031
11.98	0.031
12.53	0.020
13.15	0.021
13.78	0.015
14.36	0.022
14.44	0.028
CURVE 2	
T ~ 300	
2.01	0.392
2.47	0.324
2.65	0.357
3.20	0.050
3.48	0.025
4.52	0.191
4.86	0.150
5.14	0.185
5.37	0.148
5.58	0.190
6.28	0.025
7.56	0.032
8.48	0.021
9.38	0.024
10.17	0.024
10.98	0.028
11.48	0.033*
CURVE 2 (cont.)	
12.07	0.025
12.68	0.036
13.26	0.031
13.90	0.031
14.02	0.038
CURVE 3	
T ~ 300	
1.96	0.210
2.18	0.185
2.63	0.201
3.06	0.037
3.47	0.022
4.42	0.146
4.91	0.112
5.25	0.129
5.42	0.114
5.63	0.138
6.32	0.021
7.59	0.037
8.52	0.024
9.41	0.031
10.21	0.033
10.99	0.030*
11.50	0.033
12.08	0.018
12.69	0.018
13.31	0.018
13.88	0.020
14.00	0.032*
CURVE 4	
T ~ 300	
2.00	0.459
2.48	0.243
2.69	0.284
3.00	0.034
3.36	0.032
3.51	0.016
4.57	0.171
4.89	0.112*
5.16	0.155
5.40	0.105
5.60	0.165
CURVE 4 (cont.)	
6.18	0.027
6.44	0.014
7.58	0.024
8.54	0.015
9.42	0.019
10.23	0.021
11.00	0.021
11.51	0.018
12.11	0.016*
12.69	0.016*
13.30	0.020
13.91	0.021
14.00	0.028
CURVE 5	
T ~ 296	
0.397	0.045
0.488	0.043
0.565	0.043
0.622	0.045
0.690	0.048
0.712	0.049
0.968	0.048
1.174	0.047
1.615	0.048
1.740	0.046
2.056	0.046
2.265	0.043
2.340	0.043*
2.505	0.043
CURVE 6	
T ~ 313	
2.45	0.041
3.37	0.042
3.70	0.049
4.81	0.108
5.01	0.111
5.21	0.108
5.64	0.066
5.87	0.049
6.08	0.039
6.34	0.030
6.70	0.027*
CURVE 6 (cont.)	
7.33	0.027
7.50	0.030
7.98	0.047
8.47	0.035
9.07	0.051
9.99	0.057
10.20	0.057
11.34	0.041
11.55	0.041*
12.53	0.049
14.50	0.049
15.23	0.052
15.99	0.040
16.67	0.044
17.72	0.029
18.07	0.029
18.48	0.037
19.11	0.025
19.77	0.060
20.20	0.065
21.15	0.059
21.84	0.069
22.01	0.069*
22.86	0.058
24.03	0.050
24.56	0.055
25.35	0.055*
25.97	0.051
26.56	0.054
27.88	0.046
28.30	0.055
28.88	0.046*
30.00	0.036
30.88	0.043
31.35	0.036
32.18	0.045*
33.52	0.048
34.39	0.036
35.02	0.044
35.70	0.048*
36.26	0.058
36.44	0.058*
37.27	0.040
37.47	0.040*
38.31	0.060
38.80	0.069
40.00	0.074

* Not shown on plot

SPECIFICATION TABLE NO. 217 NORMAL SPECTRAL EMITTANCE OF SHERWIN-WILLIAMS PAINTS

Curve No.	Ref. No.	Year	Temperature, K	Wavelength Range, μm	Geometry θ'	Reported Error, %	Composition (weight percent), Specifications, and Remarks
1*	74	1959	~298	1.50-20.00	~0°		Sherwin-Williams Moroon enamel (color No. 10049 per F5595, M49 MC2) on titanium substrate; data extracted from smooth curve; property converted from R (2π, ~0°); property avg of 3 runs.

DATA TABLE NO. 217 NORMAL SPECTRAL EMITTANCE OF SHERWIN-WILLIAMS PAINTS

[Wavelength, λ, μm; Emittance, ϵ; Temperature, T, K]

λ	ϵ	λ	ϵ
CURVE 1*		CURVE 1 (cont.)*	
T ~ 298		10.50	0.831
1.50	0.856	11.39	0.749
2.26	0.848	12.03	0.675
3.00	0.817	12.20	0.664
5.00	0.683	12.32	0.664
6.01	0.580	12.95	0.725
6.50	0.606	13.40	0.741
7.07	0.644	14.15	0.753
7.50	0.770	15.10	0.753
7.80	0.824	15.71	0.785
8.10	0.860	16.23	0.821
8.45	0.873	16.59	0.837
8.58	0.875	16.93	0.840
8.93	0.875	17.35	0.840
9.33	0.865	18.49	0.809
10.11	0.822	18.93	0.809
10.30	0.831	19.23	0.814
		20.00	0.838

* No plot given

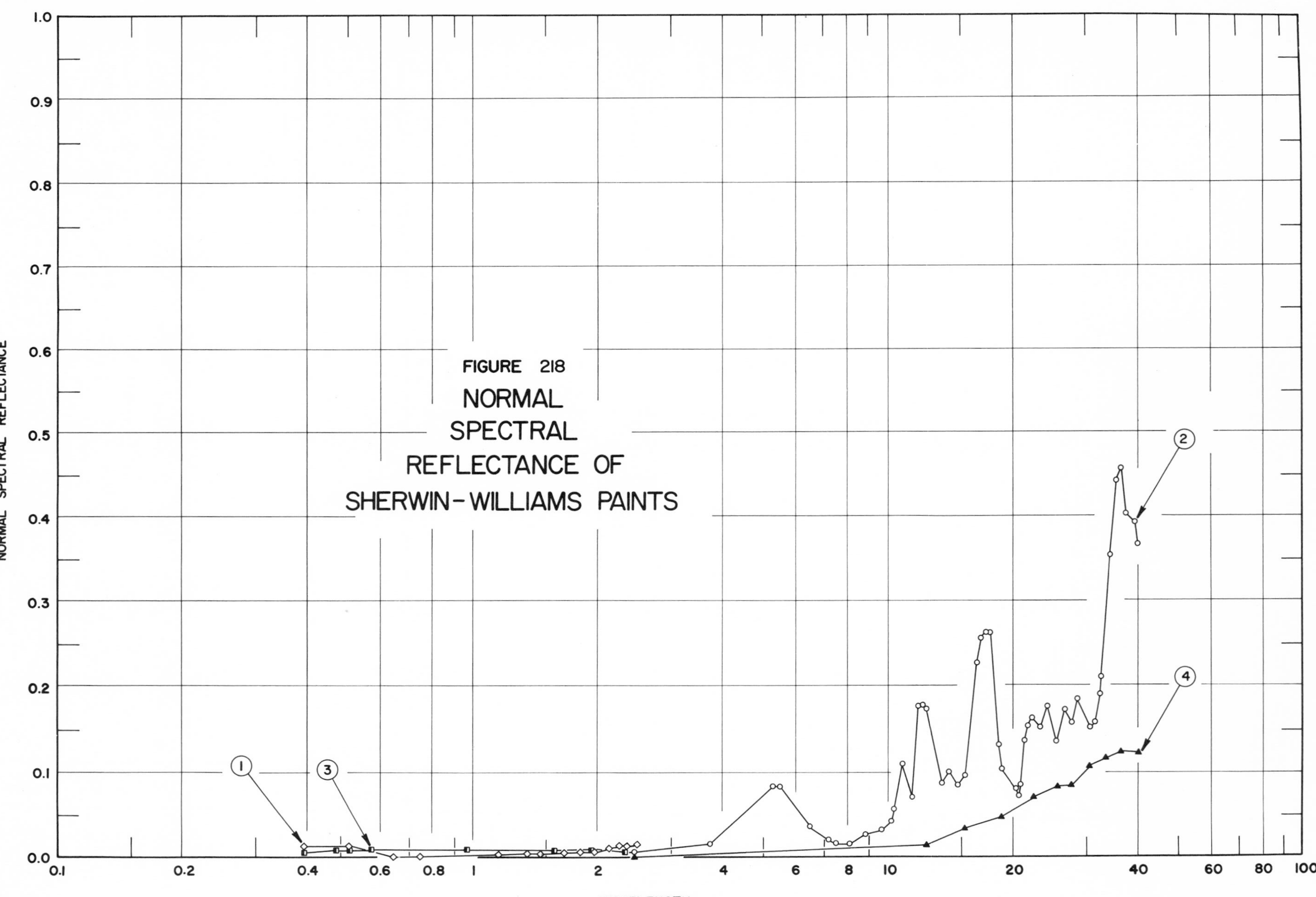

FIGURE 218 NORMAL SPECTRAL REFLECTANCE OF SHERWIN-WILLIAMS PAINTS

SPECIFICATION TABLE NO. 218 NORMAL SPECTRAL REFLECTANCE OF SHERWIN-WILLIAMS PAINTS

Curve No.	Ref. No.	Year	Temperature K	Wavelength Range, μm	Geometry θ θ' ω'	Reported Error, %	Composition (weight percent), Specifications, and Remarks
1	102	1964	~296	0.399-2.503	8° 8°		Sherwin-Williams Enameloid Flat Black (~0.83 x 10^{-4} mm thick) on aluminum substrate; data extracted from smooth curve; measured at least 24 hrs after final coating.
2	102	1964	~313	2.45-40.00	30° 30°		Above specimen and conditions.
3	102	1964	~296	0.396-2.488	8° 8°		Sherwin-Williams Super Kemtone Shasta White No. 793 (~0.25 x 10^{-4} mm thick) on aluminum substrate; data extracted from smooth curve; measured at least 24 hrs after final coating.
4	102	1964	~313	2.45-40.00	30° 30°		Above specimen and conditions.

DATA TABLE NO. 218 NORMAL SPECTRAL REFLECTANCE OF SHERWIN-WILLIAMS PAINTS

[Wavelength, λ, μm; Reflectance, ρ; Temperature, T, K]

λ	ρ	λ	ρ	λ	ρ	λ	ρ	λ	ρ	λ	ρ
CURVE 1 T ~296		CURVE 2 T ~313		CURVE 2 (cont.)		CURVE 2 (cont.)		CURVE 2 (cont.)		CURVE 4 T ~313	
0.399	0.001	2.45	0.006	12.40	0.173	22.30	0.161	37.82	0.404	2.45	0.000
0.502	0.001	3.71	0.015	13.50	0.087	23.47	0.151	39.19	0.394	12.47	0.012
0.646	0.000	5.30	0.082	14.00	0.100	24.16	0.176	40.00	0.366	15.43	0.032
0.750	0.000	5.51	0.082	14.82	0.084	24.47	0.176*	CURVE 3 T ~296		18.98	0.047
1.157	0.002	6.52	0.036	15.28	0.096	25.50	0.136	0.396	0.005	22.39	0.070
1.350	0.003	7.20	0.020	16.37	0.229	25.68	0.136*	0.465	0.008	25.77	0.081
1.473	0.003	7.49	0.016	16.77	0.256	26.93	0.161	0.504	0.008	27.85	0.085
1.674	0.005	8.06	0.016	17.07	0.262	27.76	0.159	0.571	0.009	30.59	0.107
1.812	0.006	8.86	0.026	17.43	0.262	28.89	0.186	0.970	0.009	33.10	0.116
1.971	0.007	9.65	0.031	18.56	0.131	30.94	0.151	1.594	0.008	36.96	0.123
2.152	0.010	10.17	0.041	18.99	0.103	31.44	0.159	1.690	0.008*	40.00	0.122
2.282	0.012	10.42	0.058	20.24	0.080	32.19	0.190	1.959	0.008		
2.352	0.012	10.98	0.110	20.49	0.071	32.60	0.211	2.138	0.008*		
2.503	0.015	11.49	0.070	20.71	0.071	34.16	0.353	2.337	0.006		
		11.98	0.177	20.99	0.085	35.72	0.443	2.488	0.006*		
		12.15	0.179	21.44	0.136	36.08	0.458				
				21.90	0.154	36.63	0.458*				

* Not shown on plot

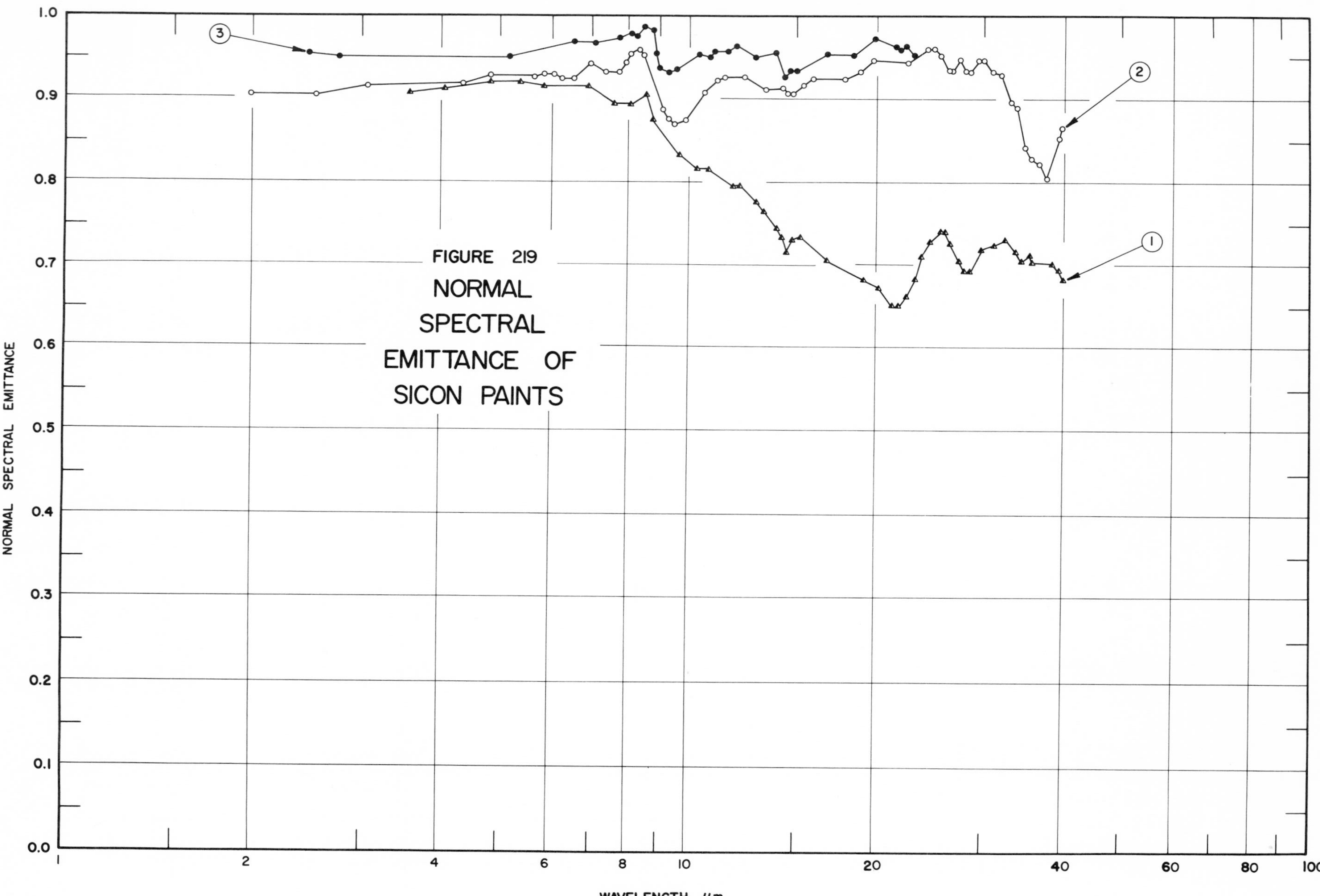

FIGURE 219
NORMAL SPECTRAL EMITTANCE OF SICON PAINTS

SPECIFICATION TABLE NO. 219 NORMAL SPECTRAL EMITTANCE OF SICON PAINTS

Curve No.	Ref. No.	Year	Temperature, K	Wavelength Range, μm	Geometry θ'	Reported Error, %	Composition (weight percent), Specifications, and Remarks
1	93	1965	77	3.58-40.00	$\sim 0°$		Sicon gloss black paint; data extracted from smooth curve.
2	93	1965	373	1.98-39.78	$\sim 0°$		Similar to above specimen and conditions.
3	96	1963	323.2	2.46-23.00	0°		Sicon black on aluminum substrate. [Authors' designation: Sample BEC-2]

DATA TABLE NO. 219 NORMAL SPECTRAL EMITTANCE OF SICON PAINTS

[Wavelength, λ, μm; Emittance, ϵ; Temperature, T, K]

CURVE 1
T = 77

λ	ϵ
3.58	0.908
4.06	0.911
4.80	0.920
5.37	0.920
5.85	0.915
6.90	0.915
7.60	0.895
8.03	0.895
8.52	0.907
8.74	0.876
9.66	0.835
10.33	0.819
10.75	0.819
11.76	0.799
12.11	0.799
12.79	0.779
13.05	0.766

CURVE 1 (cont.)

λ	ϵ
13.96	0.747
14.17	0.736
14.33	0.719
14.68	0.731
15.09	0.735
16.78	0.709
19.16	0.682
20.10	0.674
21.25	0.651
21.66	0.651
22.34	0.662
23.06	0.685
23.65	0.712
24.43	0.730
25.40	0.741
25.85	0.741
26.43	0.728
27.01	0.708

CURVE 1 (cont.)

λ	ϵ
27.63	0.695
28.10	0.695
29.89	0.720
31.00	0.725
32.07	0.731
32.51	0.730*
33.78	0.718
34.21	0.709
34.75	0.713*
35.22	0.713
35.60	0.705
38.56	0.703
39.08	0.699
40.00	0.685

CURVE 2
T = 373

λ	ϵ
1.98	0.902
2.51	0.902
3.05	0.913
4.31	0.919
4.80	0.928
5.65	0.928
5.85	0.930
6.09	0.930
6.29	0.925
6.55	0.925
6.91	0.941
7.32	0.934
7.71	0.934
7.97	0.943
8.09	0.953
8.30	0.959
8.49	0.952

CURVE 2 (cont.)

λ	ϵ
9.02	0.888
9.21	0.877
9.49	0.870
9.82	0.875
10.65	0.909
11.04	0.921
11.45	0.927
12.13	0.927
13.30	0.912
14.10	0.913
14.43	0.909
14.71	0.909
15.15	0.918
15.88	0.923
17.80	0.923
18.90	0.933
19.85	0.949
22.39	0.943

CURVE 2 (cont.)

λ	ϵ
24.19	0.960
24.77	0.960
25.44	0.952
26.09	0.935
26.38	0.935
26.94	0.949*
27.19	0.949
27.97	0.935
28.47	0.935
29.17	0.949
29.81	0.949
30.92	0.933
31.91	0.930
33.07	0.897
33.95	0.890
34.99	0.843
35.52	0.830
36.63	0.822

CURVE 2 (cont.)

λ	ϵ
37.50	0.806
37.99	0.806*
39.28	0.854
39.78	0.868

CURVE 3
T = 323.2

λ	ϵ
2.46	0.953
2.75	0.950
5.15	0.950
6.53	0.967
7.03	0.968
7.77	0.973
8.02	0.979
8.21	0.976
8.45	0.986
8.72	0.981
8.83	0.954
8.98	0.939

CURVE 3 (cont.)

λ	ϵ
9.26	0.933
9.58	0.936
10.34	0.952
10.73	0.950
11.00	0.957
11.58	0.957
11.94	0.961
12.79	0.950
13.70	0.955
14.31	0.928
14.54	0.934
14.90	0.934
16.60	0.952
18.40	0.952
19.95	0.972
21.58	0.961
21.82	0.959
22.29	0.961
22.61	0.960*
23.00	0.952

* Not shown on plot

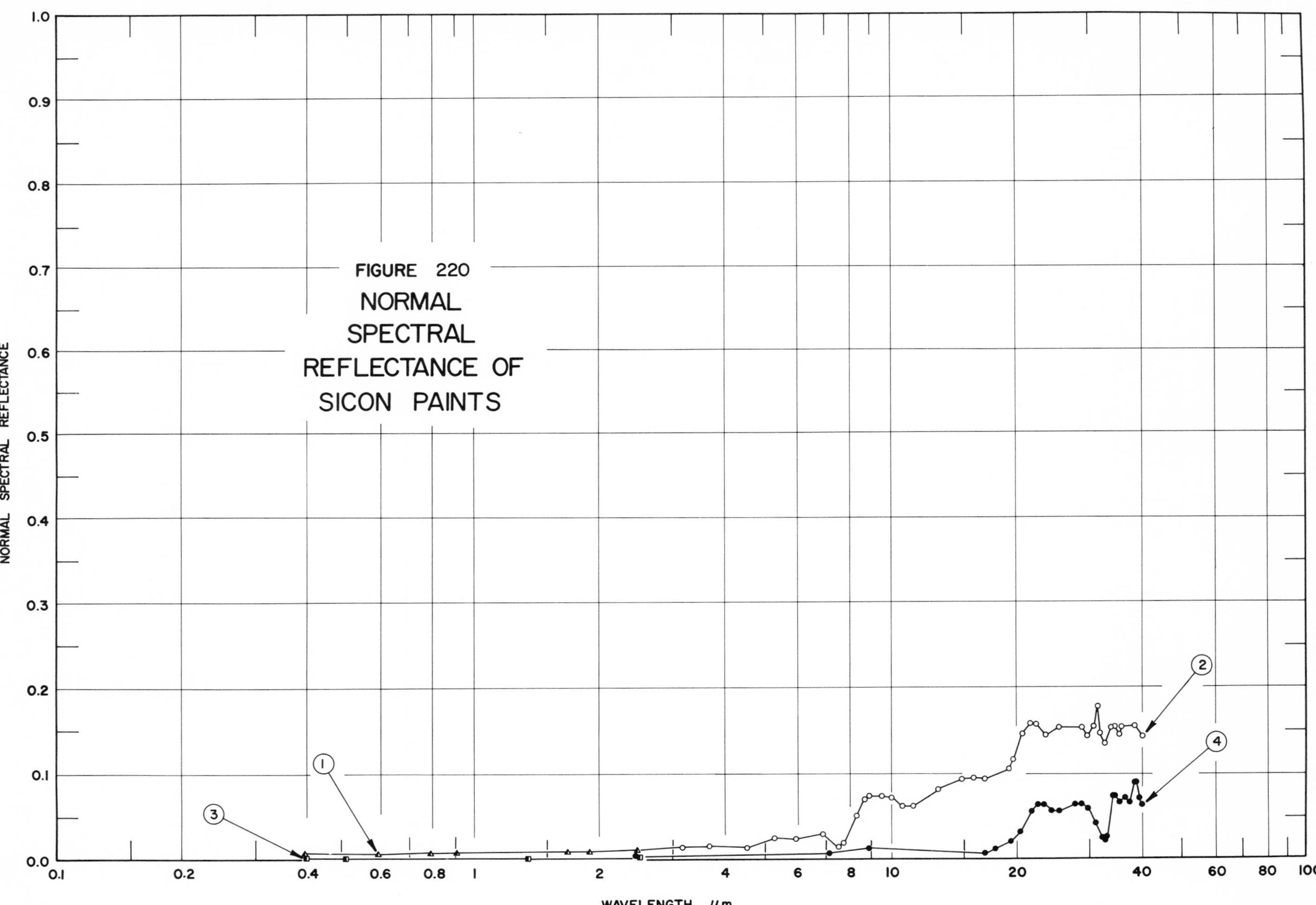

FIGURE 220 NORMAL SPECTRAL REFLECTANCE OF SICON PAINTS

SPECIFICATION TABLE NO. 220 NORMAL SPECTRAL REFLECTANCE OF SICON PAINTS

Curve No.	Ref. No.	Year	Temperature, K	Wavelength Range, μm	Geometry θ θ' ω'	Reported Error, %	Composition (weight percent), Specifications, and Remarks
1	102	1964	~296	0.396-2.492	8° 8°		Sicon Black (3X923) No. 1-5-378 (~0.53 x 10^{-4} mm thick) on aluminum substrate; data extracted from smooth curve; measured at least 24 hrs after final coating.
2	102	1964	~313	2.45-40.00	30° 30°		Above specimen and conditions.
3	102	1964	~296	0.399-2.500	8° 8°		Sicon Black (8X906) (~0.36 x 10^{-4} mm thick) on aluminum substrate; data extracted from smooth curve; measured at least 24 hrs after final coating.
4	102	1964	~313	2.45-40.00	30° 30°		Above specimen and conditions.

DATA TABLE NO. 220 NORMAL SPECTRAL REFLECTANCE OF SICON PAINTS

[Wavelength, λ, μm; Reflectance, ρ; Temperature, T, K]

λ	ρ	λ	ρ	λ	ρ	λ	ρ	λ	ρ	λ	ρ
CURVE 1		CURVE 2 (cont.)		CURVE 2 (cont.)		CURVE 3		CURVE 4 (cont.)		CURVE 4 (cont.)	
T ~ 296						T ~ 296					
		6.85	0.030	23.60	0.145			24.29	0.058	39.42	0.072
0.396	0.008	7.43	0.015	25.20	0.153	0.399	0.002	25.32	0.058	40.00	0.065
0.596	0.008	7.78	0.020	28.76	0.153	0.498	0.001	27.68	0.065		
0.796	0.008	8.32	0.051	29.79	0.144	1.356	0.001	28.64	0.065		
0.910	0.009	8.71	0.070	30.07	0.144*	2.500	0.002	29.89	0.060		
1.377	0.009*	8.97	0.074	30.60	0.154			31.04	0.043		
1.677	0.010	9.54	0.074	31.01	0.178	CURVE 4		32.01	0.027		
1.903	0.010	10.08	0.071	31.60	0.148	T ~ 313		32.36	0.025		
2.492	0.011	10.71	0.062	32.52	0.135			32.61	0.025*		
		11.28	0.062	33.67	0.153	2.45	0.005	32.91	0.028		
CURVE 2		13.08	0.082	34.36	0.153	7.18	0.008	34.01	0.075		
T ~ 313		14.95	0.092	35.11	0.145	8.83	0.013	34.42	0.075		
		15.94	0.095	35.84	0.154	16.77	0.009	35.32	0.067		
2.45	0.012*	16.91	0.095	38.48	0.154	17.97	0.012	36.55	0.071		
3.13	0.014	19.23	0.108	40.00	0.143	19.42	0.021	37.21	0.067		
3.67	0.017	19.65	0.117			20.44	0.034	37.48	0.067*		
4.51	0.013	20.84	0.148			21.80	0.057	37.78	0.071*		
5.24	0.025	21.58	0.157			22.48	0.063	38.46	0.090		
5.96	0.025	22.37	0.157			23.13	0.063	38.61	0.090		

* Not shown on plot

SPECIFICATION TABLE NO. 221 NORMAL SPECTRAL REFLECTANCE OF SPRAINT PAINTS

Curve No.	Ref. No.	Year	Temperature, K	Wavelength Range, μm	Geometry θ θ' ω'	Reported Error, %	Composition (weight percent), Specifications, and Remarks
1*	102	1964	~296	0.399-2.500	8° 8°		Spraint Grey No. 63 (~1.22 x 10^{-4} mm thick) on aluminum substrate; data extracted from smooth curve; measured at least 24 hrs after final coating.
2*	102	1964	~313	2.45-40.00	30° 30°		Above specimen and conditions.

DATA TABLE NO. 221 NORMAL SPECTRAL REFLECTANCE OF SPRAINT PAINTS

[Wavelength, λ, μm; Reflectance, ρ; Temperature, T, K]

λ	ρ	λ	ρ	λ	ρ	λ	ρ
CURVE 1* T ~ 296		CURVE 2* T ~ 313		CURVE 2 (cont.)*		CURVE 2 (cont.)*	
				11.40	0.040	31.35	0.109
0.399	0.035	2.45	0.045	12.20	0.026	31.57	0.109
0.877	0.035	3.14	0.050	12.81	0.021	32.10	0.101
0.982	0.035	3.40	0.050	15.22	0.048	32.33	0.101
1.029	0.035	4.29	0.041	17.39	0.064	33.81	0.120
1.183	0.034	5.61	0.041	18.14	0.061	34.26	0.118
1.518	0.034	6.98	0.035	18.48	0.073	34.94	0.118
1.880	0.035	7.38	0.039	18.93	0.071	35.55	0.112
2.135	0.036	7.89	0.053	19.97	0.099	36.77	0.130
2.500	0.037	8.02	0.053	23.74	0.071	37.09	0.127
		8.49	0.037	25.27	0.090	37.52	0.103
		8.94	0.060	25.98	0.090	37.85	0.120
		9.22	0.060	27.40	0.069	38.09	0.124
		9.64	0.036	28.45	0.085	38.87	0.122
		9.81	0.031	29.03	0.090	40.00	0.121
		10.39	0.040	30.07	0.090		
		10.87	0.031	30.48	0.094		

* No plot given

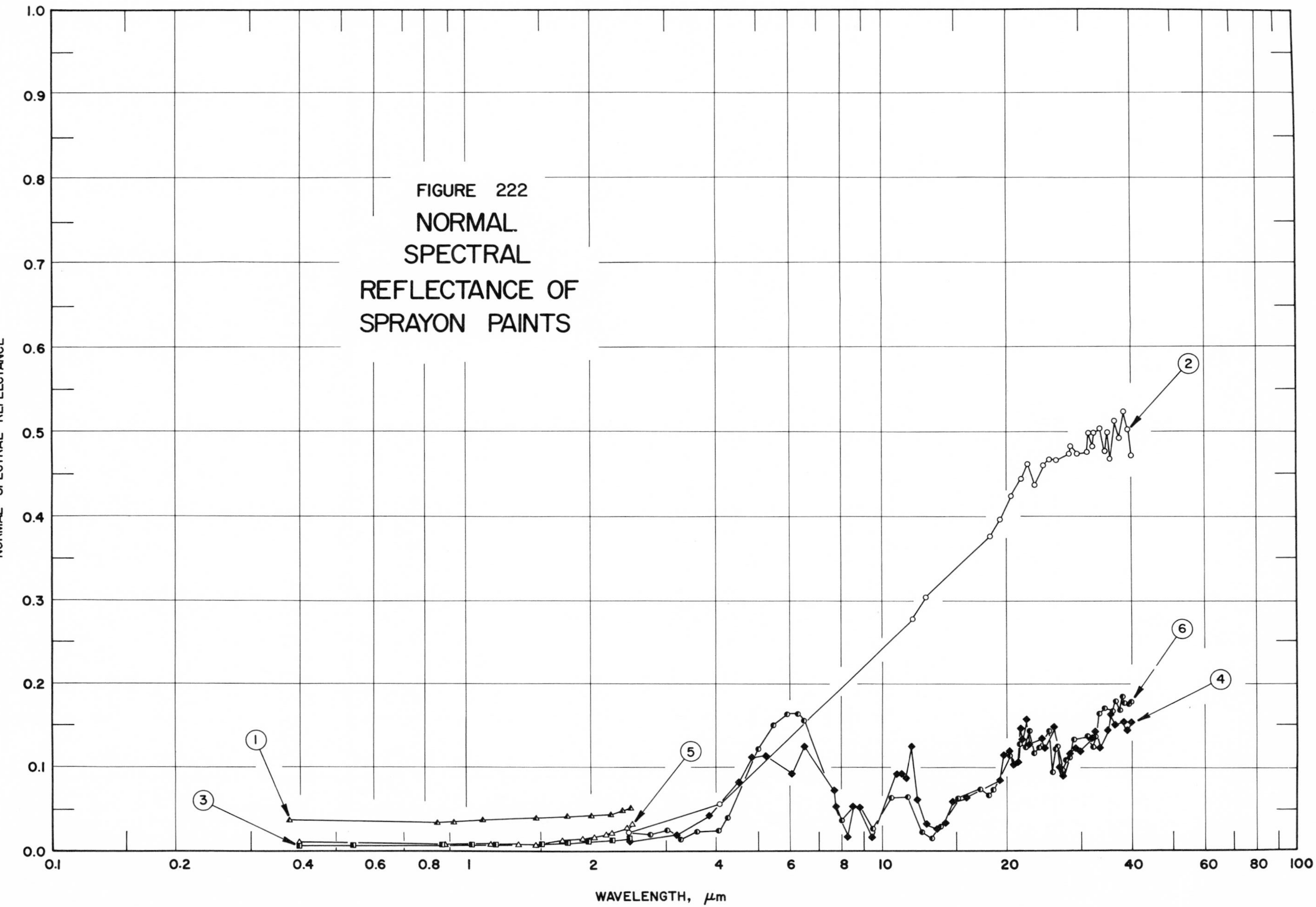

FIGURE 222 NORMAL SPECTRAL REFLECTANCE OF SPRAYON PAINTS

SPECIFICATION TABLE NO. 222 NORMAL SPECTRAL REFLECTANCE OF SPRAYON PAINTS

Curve No.	Ref. No.	Year	Temperature, K	Wavelength Range, μm	Geometry θ	θ'	ω'	Reported Error, %	Composition (weight percent), Specifications, and Remarks
1	102	1964	~296	0.374-2.493	8°	8°			High-heat (Sprayon) No. 324 aluminum (~0.68 x 10^{-4} mm thick) on aluminum substrate; data extracted from smooth curve; measured at least 24 hrs after final coating.
2	102	1964	~313	2.45-40.00	30°	30°			Above specimen and conditions.
3	102	1964	~296	0.398-2.494	8°	8°			Sprayon Machinery Dark Grey Enamel No. 325 (~0.15 x 10^{-4} mm thick) on aluminum substrate; data extracted from smooth curve; measured at least 24 hrs after final coating.
4	102	1964	~313	2.45-40.00	30°	30°			Above specimen and conditions.
5	102	1964	~296	0.397-2.506	8°	8°			Sprayon Machinery Light Grey Enamel No. 326 (~0.15 x 10^{-4} mm thick) on aluminum substrate; data extracted from smooth curve; measured at least 24 hrs after final coating.
6	102	1964	~313	2.45-40.00	30°	30°			Above specimen and conditions.

DATA TABLE NO. 222 NORMAL SPECTRAL REFLECTANCE OF SPRAYON PAINTS

[Wavelength, λ, μm; Reflectance, ρ; Temperature, T, K]

λ	ρ
CURVE 1	
T ~ 296	
0.374	0.037
0.849	0.035
0.933	0.036
1.099	0.039
1.474	0.040
1.743	0.042
2.007	0.044
2.214	0.045
2.366	0.050
2.493	0.051
CURVE 2	
T ~ 313	
2.45	0.022
4.04	0.058
11.90	0.278
12.96	0.304
18.27	0.376
19.21	0.399
20.05	0.423
20.78	0.436*
21.85	0.444
22.62	0.461
22.87	0.461*
23.56	0.436
24.78	0.460
25.31	0.466
26.54	0.466
28.06	0.474
28.86	0.484
29.79	0.472
31.12	0.476
31.61	0.500
32.07	0.482
32.51	0.500
33.86	0.504
34.30	0.497*
34.56	0.479
35.01	0.500
35.24	0.500*
35.70	0.468
36.79	0.511
37.65	0.493
CURVE 2 (cont.)	
38.20	0.517*
38.42	0.521
38.74	0.520*
39.30	0.502
39.67	0.477*
40.00	0.472
CURVE 3	
T ~ 296	
0.398	0.007
0.530	0.008
0.863	0.008
1.020	0.009
1.177	0.009
1.515	0.010
1.754	0.010
1.980	0.011
2.238	0.013
2.494	0.016
CURVE 4	
T ~ 313	
2.45	0.013
3.20	0.020
3.82	0.043
4.59	0.082
4.92	0.111
5.24	0.114
6.02	0.091
6.51	0.125
7.61	0.073
7.79	0.053
8.23	0.019
8.59	0.054
8.92	0.054
9.48	0.019
9.63	0.020*
10.93	0.091
11.11	0.091
11.47	0.088
11.81	0.125
11.93	0.120*
12.17	0.061
CURVE 4 (cont.)	
12.43	0.046*
12.80	0.033
13.54	0.028
14.14	0.035
14.95	0.060
15.20	0.063*
16.04	0.064
19.24	0.086
19.43	0.088*
19.86	0.116
20.06	0.120
20.94	0.102
21.14	0.106
21.50	0.146
21.76	0.142*
21.98	0.133
22.31	0.156
22.72	0.128
24.43	0.133
24.90	0.124
25.54	0.148*
26.01	0.148
26.56	0.141*
26.90	0.100
27.19	0.090
27.51	0.091*
28.40	0.117
29.19	0.122
30.01	0.120
32.03	0.135
32.66	0.143
33.05	0.125
33.60	0.143*
35.05	0.145
35.96	0.162
36.47	0.150
37.18	0.153*
38.55	0.153
39.09	0.144
39.50	0.150*
40.00	0.152
CURVE 5	
T ~ 296	
0.397	0.008
0.895	0.009
1.152	0.009
1.333	0.009
1.477	0.009
1.701	0.011
1.903	0.014
2.015	0.016
2.098	0.018*
2.199	0.020
2.246	0.021
2.442	0.029
2.506	0.031
CURVE 6	
T ~ 313	
2.45	0.023*
2.78	0.020
3.04	0.026
3.29	0.015
3.60	0.023
4.04	0.025
4.25	0.040
5.04	0.121
5.56	0.150
5.97	0.162
6.24	0.162
6.54	0.154
8.00	0.036
8.46	0.053*
8.63	0.053*
9.49	0.026
10.50	0.063
11.03	0.062*
11.51	0.064
11.71	0.062*
12.53	0.024
12.93	0.016*
13.15	0.016
13.91	0.030
15.15	0.063
17.11	0.073
18.09	0.067
CURVE 6 (cont.)	
18.54	0.072
19.95	0.113*
20.23	0.113
20.90	0.105*
21.15	0.107*
21.57	0.128
22.02	0.125
22.68	0.143
23.41	0.118
24.06	0.124
24.98	0.126*
25.49	0.144
25.94	0.095
26.33	0.121
26.51	0.124
27.29	0.096
27.47	0.096*
27.87	0.110
28.42	0.112
29.02	0.134
31.15	0.136
32.06	0.125
32.51	0.136
33.28	0.164
33.86	0.169*
34.67	0.170
35.72	0.164*
36.06	0.165
36.82	0.179
37.12	0.179*
37.54	0.167
38.03	0.184
38.43	0.176
39.06	0.179*
39.63	0.176
40.00	0.178

* Not shown on plot

SPECIFICATION TABLE NO. 223 NORMAL SPECTRAL EMITTANCE OF SYLVANIA PHOSPHOR PAINTS

Curve No.	Ref. No.	Year	Temperature, K	Wavelength Range, μm	Geometry θ'	Reported Error, %	Composition (weight percent), Specifications, and Remarks
1*	74	1959	306	1.51-21.00	~0°		Sylvania Phosphor; data extracted from smooth curve; property converted from R (2π, ~0°).

DATA TABLE NO. 223 NORMAL SPECTRAL EMITTANCE OF SYLVANIA PHOSPHOR PAINTS

[Wavelength, λ, μm; Emittance, ∈; Temperature, T, K]

λ	∈	λ	∈
CURVE 1* T = 306		CURVE 1 (cont.)*	
		11.19	0.828
1.51	0.786	11.95	0.847
2.11	0.813	12.69	0.876
2.83	0.832	12.94	0.880
3.94	0.832	13.46	0.880
4.43	0.840	14.53	0.858
5.04	0.855	15.37	0.848
6.74	0.935	16.42	0.848
7.19	0.943	17.64	0.852
7.36	0.943	18.43	0.830
7.87	0.922	18.86	0.821
8.20	0.922	19.31	0.821
8.68	0.939	19.98	0.837
8.83	0.939	20.26	0.837
9.30	0.918	20.59	0.821
9.89	0.862	21.00	0.821
10.25	0.839		
10.58	0.828		

* No plot given

SPECIFICATION TABLE NO. 224 NORMAL SPECTRAL REFLECTANCE OF U. S. ARMY OLIVE DRAB PAINT

Curve No.	Ref. No.	Year	Temperature, K	Wavelength Range, μm	Geometry θ	θ'	ω'	Reported Error, %	Composition (weight percent), Specifications, and Remarks
1*	110	1969	~298	0.400-1.80	~0°		2π	1	U. S. Army Olive Drab paint, specification MIL-E-46096 (MR), 0.04 mm thick on glossy black baking enamel and metal substrates; sprayed; data extracted from smooth curve; measured relative to smoked MgO; corrected for reflectance of MgO.

DATA TABLE NO. 224 NORMAL SPECTRAL REFLECTANCE OF U. S. ARMY OLIVE DRAB PAINT

[Wavelength, λ, μm; Reflectance, ρ; Temperature, T, K]

λ	ρ	λ	ρ
CURVE 1*		CURVE 1 (cont.)*	
T = 298			
		1.10	0.544
0.400	0.073	1.12	0.571
0.504	0.073	1.18	0.560
0.525	0.109	1.21	0.555
0.534	0.120	1.39	0.499
0.566	0.109	1.40	0.468
0.596	0.124	1.40	0.497
0.622	0.105	1.69	0.392
0.684	0.111	1.70	0.378
0.724	0.096	1.72	0.378
0.750	0.159	1.80	0.355
0.808	0.587		
0.867	0.628		
0.912	0.616		
0.966	0.624		
1.01	0.599		
1.06	0.601		

* No plot given

SPECIFICATION TABLE NO. 225 NORMAL SPECTRAL ABSORPTANCE OF VITA VAR PAINTS

Curve No.	Ref. No.	Year	Temperature, K	Wavelength Range, μm	Geometry θ	Reported Error, %	Composition (weight percent), Specifications, and Remarks
1*	103	1954	~298	0.401-2.600	~0°		Vita Var grey paint on quartz substrate; data extracted from smooth curve.

DATA TABLE NO. 225 NORMAL SPECTRAL ABSORPTANCE OF VITA VAR PAINTS

[Wavelength, λ, μm; Absorptance, α; Temperature, T, K]

λ	α
CURVE 1*	
T ~ 298	
0.401	0.676
0.490	0.582
0.543	0.570
0.693	0.575
0.844	0.588
1.022	0.588
1.139	0.583
1.208	0.569
1.335	0.596
1.507	0.560
1.656	0.556
1.799	0.580
2.000	0.580
2.200	0.648
2.400	0.718
2.600	0.717

* No plot given

SPECIFICATION TABLE NO. 226 NORMAL SPECTRAL TRANSMITTANCE OF VITA VAR PAINTS

Curve No.	Ref. No.	Year	Temperature, K	Wavelength Range, μm	Geometry θ	θ'	ω'	Reported Error, %	Composition (weight percent), Specifications, and Remarks
1*	103	1954	~298	0.900-2.600	~0°		2π		Vita Var grey paint on quartz substrate; data extracted from smooth curve; the accuracy of the measurement in the infrared region is 2%.

DATA TABLE NO. 226 NORMAL SPECTRAL TRANSMITTANCE OF VITA VAR PAINTS

[Wavelength, λ, μm; Transmittance, τ; Temperature, T, K]

λ	τ
CURVE 1*	
T ~ 298	
0.900	0.005
1.006	0.016
1.200	0.018
1.400	0.017
1.596	0.023
1.792	0.023
2.000	0.035
2.200	0.035
2.400	0.033
2.600	0.045

* No plot given

SPECIFICATION TABLE NO. 227 NORMAL SPECTRAL REFLECTANCE OF ZAPON BLACK PAINTS

Curve No.	Ref. No.	Year	Temperature, K	Wavelength Range, μm	Geometry θ θ' ω'	Reported Error, %	Composition (weight percent), Specifications, and Remarks
1*	102	1964	~296	0.400-2.490	8° 8°		Zapon black (~0.15 x 10^{-4} mm thick) on aluminum substrate; data extracted from smooth curve; measured at least 24 hrs after final coating.
2*	102	1964	~313	2.45-40.00	30° 30°		Above specimen and conditions.

DATA TABLE NO. 227 NORMAL SPECTRAL REFLECTANCE OF ZAPON BLACK PAINTS

[Wavelength, λ, μm; Reflectance, ρ; Temperature, T, K]

λ	ρ	λ	ρ	λ	ρ	λ	ρ	λ	ρ
CURVE 1* T ~ 296		CURVE 2* T ~ 313		CURVE 2 (cont.)*		CURVE 2 (cont.)*		CURVE 2 (cont.)*	
0.400	0.008	2.45	0.011	8.62	0.040	19.66	0.046	31.04	0.067
0.857	0.010	3.53	0.011	9.12	0.040	20.38	0.063	31.97	0.063
1.225	0.012	4.17	0.020	9.47	0.042	21.28	0.058	32.45	0.071
1.496	0.013	4.55	0.017	9.96	0.028	21.83	0.063	33.20	0.132
1.743	0.014	5.21	0.023	10.24	0.026	22.57	0.082	33.55	0.148
1.884	0.015	6.01	0.023	11.56	0.036	22.77	0.082	35.31	0.195
2.082	0.015	6.32	0.025	12.24	0.058	24.72	0.066	35.85	0.205
2.208	0.015	7.06	0.020	12.46	0.058	25.31	0.065	36.57	0.194
2.263	0.016	7.44	0.008	13.09	0.048	27.12	0.080	37.44	0.191
2.333	0.017	7.91	0.030	13.47	0.053	28.04	0.102	39.02	0.164
2.438	0.017	8.27	0.023	14.14	0.042	28.39	0.113	39.43	0.193
2.490	0.017	8.43	0.035	15.19	0.048	28.66	0.113	40.00	0.176
				15.76	0.042	30.21	0.075		
				17.14	0.049				
				17.64	0.049				
				18.83	0.040				

* No plot given

1. PIGMENTED COATINGS (continued)

E. Miscellaneous

SPECIFICATION TABLE NO. 228 HEMISPHERICAL TOTAL EMITTANCE OF AQUABLACK PAINTS

Curve No.	Ref. No.	Year	Temperature Range, K	Reported Error, %	Composition (weight percent), Specifications and Remarks
1*	49	1961	278		Aquablack B in sodium silicate binder on Dow 17 treated Mg alloy substrate; exposed in vacuum (10^{-6} - 8 x 10^{-6} mm Hg) to UV radiation from an argon filled A-H6 high pressure Hg arc lamp for 127 hrs.

DATA TABLE NO. 228 HEMISPHERICAL TOTAL EMITTANCE OF AQUABLACK PAINTS

[Temperature, T, K; Emittance, ϵ]

T	ϵ
CURVE 1*	
278	0.78

* No plot given

SPECIFICATION TABLE NO. 229 NORMAL SPECTRAL ABSORPTANCE OF AQUABLACK PAINTS

Curve No.	Ref. No.	Year	Temperature Range, K	Geometry θ	Reported Error, %	Composition (weight percent), Specifications and Remarks
1*	49	1961	278	$\sim 0°$		Aquablack B in sodium silicate binder on Dow 17 treated Mg alloy substrate; exposed in vacuum (10^{-6} - 8 x 10^{-6} mm Hg) to UV radiation from an argon filled A-H6 high pressure Hg arc lamp for 127 hrs.

DATA TABLE NO. 229 NORMAL SPECTRAL ABSORPTANCE OF AQUABLACK PAINTS

[Temperature, T, K; Absorptance, α]

T	α
CURVE 1*	
278	0.91

* No plot given

SPECIFICATION TABLE NO. 230 HEMISPHERICAL TOTAL EMITTANCE OF FERRO WHITE PAINTS

Curve No.	Ref. No.	Year	Temperature Range, K	Reported Error, %	Composition (weight percent), Specifications and Remarks
1*	106	1963	207-309	<2.5	Ferro white porcelain enamel (0.127 mm thick) on 2014 aluminum; measured in vacuum (5×10^{-6} mm Hg).

DATA TABLE NO. 230 HEMISPHERICAL TOTAL EMITTANCE OF FERRO WHITE PAINTS

[Temperature, T, K; Emittance, ϵ]

T	ϵ
CURVE 1*	
207	0.682
235	0.798
258	0.775
309	0.714

* No plot given

SPECIFICATION TABLE NO. 231 HEMISPHERICAL TOTAL EMITTANCE OF JERSEY STANDARD PAINTS

Curve No.	Ref. No.	Year	Temperature Range, K	Reported Error, %	Composition (weight percent), Specifications and Remarks
1*	105	1964	307		Optical black; measured in vacuum (10^{-6} mm Hg) maintained by diffusion pump. [Authors' designation: Test no. 219, Jersey Standard optical black]

DATA TABLE NO. 231 HEMISPHERICAL TOTAL EMITTANCE OF JERSEY STANDARD PAINTS

[Temperature, T, K; Emittance, ϵ]

T	ϵ
CURVE 1*	
307	0.900

* No plot given

SPECIFICATION TABLE NO. 232 NORMAL SPECTRAL REFLECTANCE OF WHITE PORCELAIN ENAMEL

Curve No.	Ref. No.	Year	Temperature, K	Wavelength Range, μm	Geometry θ	θ'	ω'	Reported Error, %	Composition (weight percent), Specifications, and Remarks
1*	116	1963	311	2.00-35.00	~0°		2π		White porcelain enamel P-110; converted from R (2π, ~0°).

DATA TABLE NO. 232 NORMAL SPECTRAL REFLECTANCE OF WHITE PORCELAIN ENAMEL

[Wavelength, λ, μm; Reflectance, ρ; Temperature, T, K]

λ	ρ	λ	ρ
CURVE 1* T = 311		CURVE 1 (cont.)*	
		13.66	0.131
2.00	0.575	16.18	0.195
2.56	0.534	17.88	0.250
2.77	0.499	19.81	0.278
2.90	0.445	20.99	0.274
3.00	0.340	23.00	0.289
3.25	0.265	25.00	0.332
4.24	0.233	26.24	0.324
5.00	0.243	26.99	0.308
5.35	0.244	28.88	0.349
5.78	0.211	30.56	0.365
6.27	0.148	32.59	0.358
6.80	0.129	33.92	0.361
7.55	0.107	35.00	0.371
9.00	0.050		
10.99	0.154		
12.35	0.092		
13.00	0.089		

* No plot given

2. CONTACT COATINGS

A. Metallic Contact Coatings

SPECIFICATION TABLE NO. 233 HEMISPHERICAL TOTAL EMITTANCE OF ALUMINUM CONTACT COATINGS

Curve No.	Ref. No.	Year	Temperature Range, K	Reported Error, %	Composition (weight percent), Specifications and Remarks
1*	194	1955	76	5	Aluminum; plastic mylar (0.0127 mm thick) and aluminum substrates; produced by vaporizing aluminum on both sides of mylar; measured in vacuum (10^{-6} to 10^{-7} mm Hg); authors assumed $\alpha = \epsilon$ for 300 K blackbody incident radiation.
2*	194	1955	76	5	Aluminum; stainless steel substrate; Al sprayed on substrate; measured in vacuum (10^{-6} to 10^{-7} mm Hg); authors assumed $\alpha = \epsilon$ for 300 K blackbody incident radiation.
3*	194	1955	76	5	Aluminum; stainless steel substrate; Al sprayed on substrate; wire brushed; measured in vacuum (10^{-6} to 10^{-7} mm Hg); authors assumed $\alpha = \epsilon$ for 300 K blackbody incident radiation.
4*	195	1953	76		Aluminum; plastic mylar (0.0127 mm thick) and aluminum substrates; Al vapor deposited on both sides of mylar; measured in vacuum ($< 10^{-6}$ mm Hg); authors assumed $\alpha = \epsilon$ for 294 K blackbody radiation.
5*	195	1953	76		Similar to above specimen and conditions.
6*	97	1961	773-1273	±2.5	Aluminum; iron substrate; substrate sandblasted and degreased; coated to a density of 7.5 g cm^{-2} with aluminum powder suspension; sintered at 1223 K in vacuum; measured in vacuum ($< 5 \times 10^{-6}$ mm Hg); data extracted from smooth curve.
7*	43	1962	378-486	±3	Aluminum (0.127 to 0.203 mm thick); Al substrate; sprayed; measured in vacuum (10^{-8} mm Hg).
8*	106	1963	333	<3.5	Aluminum; mylar substrate; measured in vacuum (5×10^{-6} mm Hg).
9*	16	1965	300-415	±5	Aluminum (1 μm thick); mylar(12 μm thick) and stainless steel(0.25 mm thick) substrates; mylar cemented to stainless steel; Al vapor deposited; measured in vacuum (10^{-7} mm Hg).
10*	111	1969	258-348		Aluminum (2×10^{-4} mm thick); stainless steel substrate; Al vapor deposited on hand-polished substrate; property measured by steady-state calorimetric method.
11*	111	1969	248-348		Curve 10 specimen and conditions except property measured by transient calorimetric method; property of second side of sample assumed.
12*	111	1969	303		Curve 10 specimen and conditions except property calculated from $\rho(10°, 10°)$ measured by specular method relative to a front surface aluminized mirror.
13*	111	1969	300		Curve 10 specimen and conditions except property calculated from $\rho(7.5°, 2\pi)$ measured by ellipsoid method.
14*	111	1969	200-400		Curve 10 specimen and conditions except property calculated from $\rho(15°, 2\pi\text{-}75°)$ measured by heated cavity method relative to a platinum surface.
15*	111	1969	295		Curve 10 specimen and conditions except property measured by a portable Quick Emittance Device.
16*	111	1969	295		Curve 10 specimen and conditions except property measured by a portable emissometer.
17*	105	1964	~298		Aluminum (0.0455 μm thick); polyester film (6.35 μm thick) and aluminum (0.0378 μm thick) substrates; aluminum vapor deposited on both sides of polyester film. [Authors' designation: Insulation System No. 5, Shield 1, Side A]
18*	105	1964	~298		Aluminum (0.0370 μm thick); polyester film (6.35 μm thick) and aluminum (0.0455 μm thick) substrates; aluminum vapor deposited on both sides of polyester film. [Authors' designation: Insulation System No. 5, Shield 1, Side B]

* No plot given

SPECIFICATION TABLE NO. 233 HEMISPHERICAL TOTAL EMITTANCE OF ALUMINUM CONTACT COATINGS (continued)

Curve No.	Ref. No.	Year	Temperature Range, K	Reported Error, %	Composition (weight percent), Specifications and Remarks
19*	105	1964	~298		Aluminum (0.0453 μm thick); polyester film (6.35 μm thick) and aluminum (0.0395 μm thick) substrates; aluminum vapor deposited on both sides of polyester film. [Authors' designation: Insulation System No. 5, Shield 2, Side A]
20*	105	1964	~298		Aluminum (0.0395 μm thick); polyester film (6.35 μm thick) and aluminum (0.0453 μm thick) substrates; aluminum vapor deposited on both sides of polyester film. [Authors' designation: Insulation System No. 5, Shield 2, Side B]
21*	105	1964	~298		Aluminum (0.0450 μm thick); polyester film (6.35 μm thick) and aluminum (0.0395 μm thick) substrates; aluminum vapor deposited on both sides of polyester film. [Authors' designation: Insulation System No. 5, Shield 3, Side A]
22*	105	1964	~298		Aluminum (0.0395 μm thick); polyester film (6.35 μm thick) and aluminum (0.0450 μm thick) substrates; aluminum vapor deposited on both sides of polyester film. [Authors' designation: Insulation System No. 5, Shield 3, Side B]
23*	105	1964	~298		Aluminum (0.0450 μm thick); polyester film (6.35 μm thick) and aluminum (0.0400 μm thick) substrates; aluminum vapor deposited on both sides of polyester film. [Authors' designation: Insulation System No. 5, Shield 4, Side A]
24*	105	1964	~298		Aluminum (0.0400 μm thick); polyester film (6.35 μm thick) and aluminum (0.0400 μm thick) substrates; aluminum vapor deposited on both sides of polyester film. [Authors' designation: Insulation System No. 5, Shield 4, Side B]
25*	105	1964	~298		Aluminum (0.0456 μm thick); polyester film (6.35 μm thick) and aluminum (0.0395 μm thick) substrates; aluminum vapor deposited on both sides of polyester film. [Authors' designation: Insulation System No. 5, Shield 5, Side A]
26*	105	1964	~298		Aluminum (0.0395 μm thick); polyester film (6.35 μm thick) and aluminum (0.0456 μm thick) substrates; aluminum vapor deposited on both sides of polyester film. [Authors' designation: Insulation System No. 5, Shield 5, Side B]
27*	105	1964	307		Aluminum; quartz substrate; measured in vacuum (10^{-6} mm Hg) maintained by diffusion pump.
28*	105	1964	307		Aluminum; polyester film (6.35 μm thick) substrate; vacuum deposited; measured in vacuum (10^{-6} mm Hg) maintained by diffusion pump.
29*	105	1964	307		Aluminum; polyester film (6.35 μm thick) and aluminum substrates; aluminum vacuum deposited on both sides of polyester film; measured in vacuum (10^{-6} mm Hg) maintained by diffusion pump. [Authors' designation: Side B]
30*	105	1964	307		Above specimen and conditions except opposite side measured. [Authors' designation: Side B]
31*	105	1964	307		Aluminum; polyester film (6.35 μm thick) substrate; vacuum deposited; measured in vacuum (10^{-6} mm Hg) maintained by diffusion pump.
32*	105	1964	307		Similar to above specimen and conditions.
33*	105	1964	307		Aluminum (0.0235 μm thick); polyester film (6.35 μm thick) substrate; vacuum deposited; measured in vacuum (10^{-6} mm Hg) maintained by diffusion pump.
34*	105	1964	307		Aluminum; polyester film (6.35 μm thick) and aluminum substrates; aluminum vapor deposited on both sides of polyester film; measured in vacuum (10^{-6} mm Hg) maintained by diffusion pump. [Authors' designation: Side B]

* No plot given

SPECIFICATION TABLE NO. 233 HEMISPHERICAL TOTAL EMITTANCE OF ALUMINUM CONTACT COATINGS (continued)

Curve No.	Ref. No.	Year	Temperature Range, K	Reported Error, %	Composition (weight percent), Specifications and Remarks
35*	105	1964	307		Aluminum (0.0334 μm thick); polyester film (6.35 μm thick) and aluminum (0.0248 μm thick) substrates; aluminum vapor deposited on both sides of polyester film; measured in vacuum (10^{-6} mm Hg) maintained by diffusion pump. [Authors' designation: Side A]
36*	105	1964	307		Aluminum (0.0248 μm thick); polyester film (6.35 μm thick) and aluminum (0.0334 μm thick) substrates; aluminum vapor deposited on both sides of polyester film; measured in vacuum (10^{-6} mm Hg) maintained by diffusion pump. [Authors' designation: Side B]
37*	105	1964	307		Aluminum; polyester film (6.35 μm thick) substrate; vacuum deposited; measured in vacuum (10^{-6} mm Hg) maintained by diffusion pump.
38*	105	1964	307		Aluminum; mylar and aluminum substrates; aluminum vacuum deposited on both sides of mylar; measured in vacuum (10^{-6} mm Hg) maintained by diffusion pump. [Authors' designation: Side A]
39*	105	1964	307		Aluminum; reinforced polyester film substrate (aluminized scrim); measured in vacuum (10^{-6} mm Hg) maintained by diffusion pump. [Authors' designation: Test No. 4]
40*	105	1964	307		Aluminum; quartz substrate; measured in vacuum (10^{-6} mm Hg) maintained by diffusion pump. [Authors' designation: Test No. 23]
41*	105	1964	307		Aluminum; polyester film (6.35 μm thick) and aluminum substrates; measured in vacuum (10^{-6} mm Hg) maintained by diffusion pump. [Authors' designation: Test No. 101, Sample 2E, Side 1]
42*	105	1964	307		Aluminum; polyester film (6.35 μm thick) and aluminum substrates; measured in vacuum (10^{-6} mm Hg) maintained by diffusion pump. [Authors' designation: Test No. 102, Sample 2E, Side 2]
43*	105	1964	307		Aluminum; polyester film (6.35 μm thick) and aluminum substrates; vapor deposited; measured in vacuum (10^{-6} mm Hg) maintained by diffusion pump. [Authors' designation: Test No. 103, Sample 2E, Side 1]
44*	105	1964	307		Similar to above specimen and conditions. [Authors' designation: Test No. 104, Sample 1E, Side 1]
45*	105	1964	307		Similar to above specimen and conditions. [Authors' designation: Test No. 105, Sample 1E, Side 2]
46*	105	1964	307		Aluminum (0.0455 μm thick); polyester film (6.35 μm thick) and aluminum (0.0378 μm thick) substrates; vapor deposited; measured in vacuum (10^{-6} mm Hg) maintained by diffusion pump. [Authors' designation: Test No. 109, Sample 1BE, Side A]
47*	105	1964	307		Aluminum (0.0453 μm thick); polyester film (6.35 μm thick) and aluminum (0.0395 μm thick) substrates; vapor deposited; measured in vacuum (10^{-6} mm Hg) maintained by diffusion pump. [Authors' designation: Test No. 110, Sample 2BE, Side A]
48*	105	1964	307		Aluminum (0.0450 μm thick); polyester film (6.35 μm thick) and aluminum (0.0380 μm thick) substrates; vacuum deposited; measured in vacuum (10^{-6} mm Hg) maintained by diffusion pump. [Authors' designation: Test No. 111, Sample 3BE, Side A]
49*	105	1964	307		Aluminum (0.0450 μm thick); polyester film (6.35 μm thick) and aluminum (0.0400 μm thick) substrates; vacuum deposited; measured in vacuum (10^{-6} mm Hg) maintained by diffusion pump. [Authors' designation: Test No. 112, Sample 4BE, Side A]
50*	105	1964	307		Aluminum (0.0456 μm thick); polyester film (6.35 μm thick) and aluminum (0.0395 μm thick) substrates; vapor deposited; measured in vacuum (10^{-6} mm Hg) maintained by diffusion pump. [Authors' designation: Test No. 113, Sample 5BE, Side A]

* No plot given

SPECIFICATION TABLE NO. 233 HEMISPHERICAL TOTAL EMITTANCE OF ALUMINUM CONTACT COATINGS (continued)

Curve No.	Ref. No.	Year	Temperature Range, K	Reported Error, %	Composition (weight percent), Specifications and Remarks
51*	105	1964	307		Aluminum (0.0378 μm thick); polyester film (6.35 μm thick) and aluminum (0.0455 μm thick) substrates; vapor deposited; measured in vacuum (10^{-6} mm Hg) maintained by diffusion pump. [Authors' designation: Test No. 114, Sample 1BE, Side B]
52*	105	1964	307		Aluminum (0.0395 μm thick); polyester film (6.35 μm thick) and aluminum (0.0453 μm thick) substrates; vapor deposited; measured in vacuum (10^{-6} mm Hg) maintained by diffusion pump. [Authors' designation: Test No. 115, Sample 2BE, Side B]
53*	105	1964	307		Aluminum (0.0380 μm thick); polyester film (6.35 μm thick) and aluminum (0.0450 μm thick) substrates; vacuum deposited; measured in vacuum (10^{-6} mm Hg) maintained by diffusion pump. [Authors' designation: Test No. 117, Sample 3BE, Side B]
54*	105	1964	307		Aluminum (0.0400 μm thick); polyester film (6.35 μm thick) and aluminum (0.0450 μm thick) substrates; vacuum deposited; measured in vacuum (10^{-6} mm Hg) maintained by diffusion pump. [Authors' designation: Test No. 118, Sample 4BE, Side B]
55*	105	1964	307		Aluminum (0.0395 μm thick); polyester film (6.35 μm thick) and aluminum (0.0456 μm thick) substrates; vapor deposited; measured in vacuum (10^{-6} mm Hg) maintained by diffusion pump. [Authors' designation: Test No. 119, Sample 5BE, Side B]
56*	105	1964	307		Aluminum; polyester (6.35 μm thick) and aluminum substrates; vapor deposited; measured in vacuum (10^{-6} mm Hg) maintained by diffusion pump. [Authors' designation: Test No. 181, Sample 1B2, Side B]
57*	105	1964	307		Similar to above specimen and conditions. [Authors' designation: Test No. 183, Sample 2B2, Side B]
58*	105	1964	307		Similar to above specimen and conditions. [Authors' designation: Test No. 185, Sample 3B2, Side B]
59*	105	1964	307		Similar to above specimen and conditions. [Authors' designation: Test No. 193, Sample 4B2, Side B]
60*	105	1964	307		Similar to above specimen and conditions. [Authors' designation: Test No. 194, Sample 5B2, Side B]
61*	105	1964	307		Similar to above specimen and conditions. [Authors' designation: Test No. 195, Sample 1B2, Side A]
62*	105	1964	307		Similar to above specimen and conditions. [Authors' designation: Test No. 196, Sample 2B2, Side A]
63*	105	1964	307		Similar to above specimen and conditions. [Authors' designation: Test No. 197, Sample 3B2, Side A]
64*	105	1964	307		Similar to above specimen and conditions. [Authors' designation: Test No. 198]
65*	105	1964	307		Similar to above specimen and conditions. [Authors' designation: Test No. 199]
66*	105	1964	307		Aluminum; polyester film (6.35 μm thick) substrate; measured in vacuum (10^{-6} mm Hg) maintained by diffusion pump. [Authors' designation: Test No. 204, Test No. 1]
67*	105	1964	307		Above specimen and conditions. [Authors' designation: Test No. 205, Test No. 2]
68*	105	1964	307		Above specimen and conditions. [Authors' designation: Test No. 206, Test No. 3]
69*	105	1964	307		Aluminum (0.0218 μm thick); polyester film substrate; measured in vacuum (10^{-6} mm Hg) maintained by diffusion pump. [Authors' designation: Test No. 209, Mfr. A]
70*	105	1964	307		Aluminum (0.003 μm thick); polyester film substrate; measured in vacuum (10^{-6} mm Hg) maintained by diffusion pump. [Authors' designation: Test No. 212, Sample 9-1E, Mfr. A]

* No plot given

SPECIFICATION TABLE NO. 233 HEMISPHERICAL TOTAL EMITTANCE OF ALUMINUM CONTACT COATINGS (continued)

Curve No.	Ref. No.	Year	Temperature Range, K	Reported Error, %	Composition (weight percent), Specifications and Remarks
71*	105	1964	307		Aluminum (0.0108 μm thick); polyester film substrate; measured in vacuum (10^{-6} mm Hg) maintained by diffusion pump. [Authors' designation: Test No. 213, Sample 8-1E, Mfr. A]
72*	106	1963	333		Aluminum; mylar substrate; measured in vacuum (5×10^{-6} mm Hg). [Authors' designation: NRC-2 Insulation]
73*	217	1960	77		Aluminum household type foil wrapped loosely around G.E. No. 7031 surface; cleaned with acetone; measured in vacuum ($\sim 1 \times 10^{-5}$ mm Hg).

* No plot given

DATA TABLE NO. 233 HEMISPHERICAL TOTAL EMITTANCE OF ALUMINUM CONTACT COATINGS

[Temperature, T, K; Emittance, ∈]

T	∈
CURVE 1*	
76	0.04
CURVE 2*	
76	0.07
CURVE 3*	
76	0.06
CURVE 4*	
76	0.040
CURVE 5*	
76	0.043
CURVE 6*	
773	0.690
873	0.700
973	0.715
1073	0.725
1173	0.750
1273	0.780
CURVE 7*	
378	0.330
396	0.305
406	0.320
420	0.325
435	0.300
443	0.320
469	0.300
486	0.270
CURVE 8*	
333	0.047
CURVE 9*	
300	0.06
415	0.06
CURVE 10*	
258	0.0428
288	0.0428
318	0.0442
348	0.0454
CURVE 11*	
248	0.0293
274	0.0293
298	0.0307
324	0.0311
348	0.0311
CURVE 12*	
303	0.0240
CURVE 13*	
300	0.0482
CURVE 14*	
200	0.0224
251	0.0224
300	0.0224
351	0.0233
400	0.0237
CURVE 15*	
295	0.0279
CURVE 16*	
295	0.0401
CURVE 17*	
298	0.0334
CURVE 18*	
298	0.0335
CURVE 19*	
298	0.0335
CURVE 20*	
298	0.0378
CURVE 21*	
298	0.0300
CURVE 22*	
298	0.0331
CURVE 23*	
298	0.0369
CURVE 24*	
298	0.0358
CURVE 25*	
298	0.0335
CURVE 26*	
298	0.0243
CURVE 27*	
307	0.0246
CURVE 28*	
307	0.0143
CURVE 29*	
307	0.0543
CURVE 30*	
307	0.0543
CURVE 31*	
307	0.0613
CURVE 32*	
307	0.0609
CURVE 33*	
307	0.0450
CURVE 34*	
307	0.0532
CURVE 35*	
307	0.0773
CURVE 36*	
307	0.0513
CURVE 37*	
307	0.0476
CURVE 38*	
307	0.0377
CURVE 39*	
307	0.472
CURVE 40*	
307	0.0248
CURVE 41*	
307	0.0176
CURVE 42*	
307	0.0410
CURVE 43*	
307	0.0416
CURVE 44*	
307	0.0422
CURVE 45*	
307	0.0398
CURVE 46*	
307	0.0333
CURVE 47*	
307	0.0334
CURVE 48*	
307	0.0299
CURVE 49*	
307	0.0369
CURVE 50*	
307	0.0334
CURVE 51*	
307	0.0335
CURVE 52*	
307	0.0347
CURVE 53*	
307	0.03810
CURVE 54*	
307	0.0358
CURVE 55*	
307	0.0244
CURVE 56*	
307	0.0349
CURVE 57*	
307	0.0337
CURVE 58*	
307	0.0350
CURVE 59*	
307	0.0835
CURVE 60*	
307	0.0306
CURVE 61*	
307	0.0284
CURVE 62*	
307	0.0369
CURVE 63*	
307	0.0369
CURVE 64*	
307	0.0270
CURVE 65*	
307	0.0289
CURVE 66*	
307	0.0326
CURVE 67*	
307	0.030
CURVE 68*	
307	0.0337
CURVE 69*	
307	0.0231
CURVE 70*	
307	0.0722
CURVE 71*	
307	0.0399
CURVE 72*	
333	0.038
CURVE 73*	
77	0.043

* No plot given

SPECIFICATION TABLE NO. 234 NORMAL TOTAL EMITTANCE OF ALUMINUM CONTACT COATINGS

Curve No.	Ref. No.	Year	Temperature Range, K	Geometry θ'	Reported Error, %	Composition (weight percent), Specifications and Remarks
1*	100	1962	300	~0°		Aluminum (0.127 mm thick); FEP Teflon substrate (type A); calculated from reflectance (2 to 23 μm).
2*	100	1962	300	~0°		Above specimen and conditions except exposed to 384 hr of simulated solar (10 x solar intensity) radiation; data indicates max degradation which may occur during a Mariner mission.
3*	196	1960	423-666	~0°		Aluminum; fabric substrate; substrate aluminized; data extracted from smooth curve.

DATA TABLE NO. 234 NORMAL TOTAL EMITTANCE OF ALUMINUM CONTACT COATINGS

[Temperature, T, K; Emittance, ϵ]

T	ϵ	T	ϵ
CURVE 1*		CURVE 3 (cont.)	
300	0.84	655	0.355
		659	0.382
CURVE 2*		662	0.413
		666	0.462
300	0.84		
CURVE 3*			
423	0.267		
448	0.246		
480	0.227		
517	0.216		
544	0.216		
574	0.225		
604	0.240		
618	0.254		
629	0.274		
641	0.298		
648	0.326		

* No plot given.

SPECIFICATION TABLE NO. 235 HEMISPHERICAL SPECTRAL REFLECTANCE OF ALUMINUM CONTACT COATINGS

Curve No.	Ref. No.	Year	Temperature, K	Wavelength Range, μm	Geometry ω θ' ω'	Reported Error, %	Composition (weight percent), Specifications, and Remarks
1*	197	1963	298	0.250-0.900	2π 45°	< 10	Aluminum; glass substrate.

DATA TABLE NO. 235 HEMISPHERICAL SPECTRAL REFLECTANCE OF ALUMINUM CONTACT COATINGS

[Wavelength, λ, μm; Reflectance, ρ; Temperature, T, K]

λ	ρ
CURVE 1*	
0.250	0.75
0.275	0.80
0.300	0.83
0.325	0.83
0.350	0.87
0.370	0.84
0.400	0.91
0.425	0.87
0.450	0.88
0.475	0.88
0.500	0.85
0.550	0.84
0.600	0.81
0.650	0.80
0.700	0.75
0.750	0.74
0.800	0.78
0.850	0.83
0.900	0.86

* No plot given

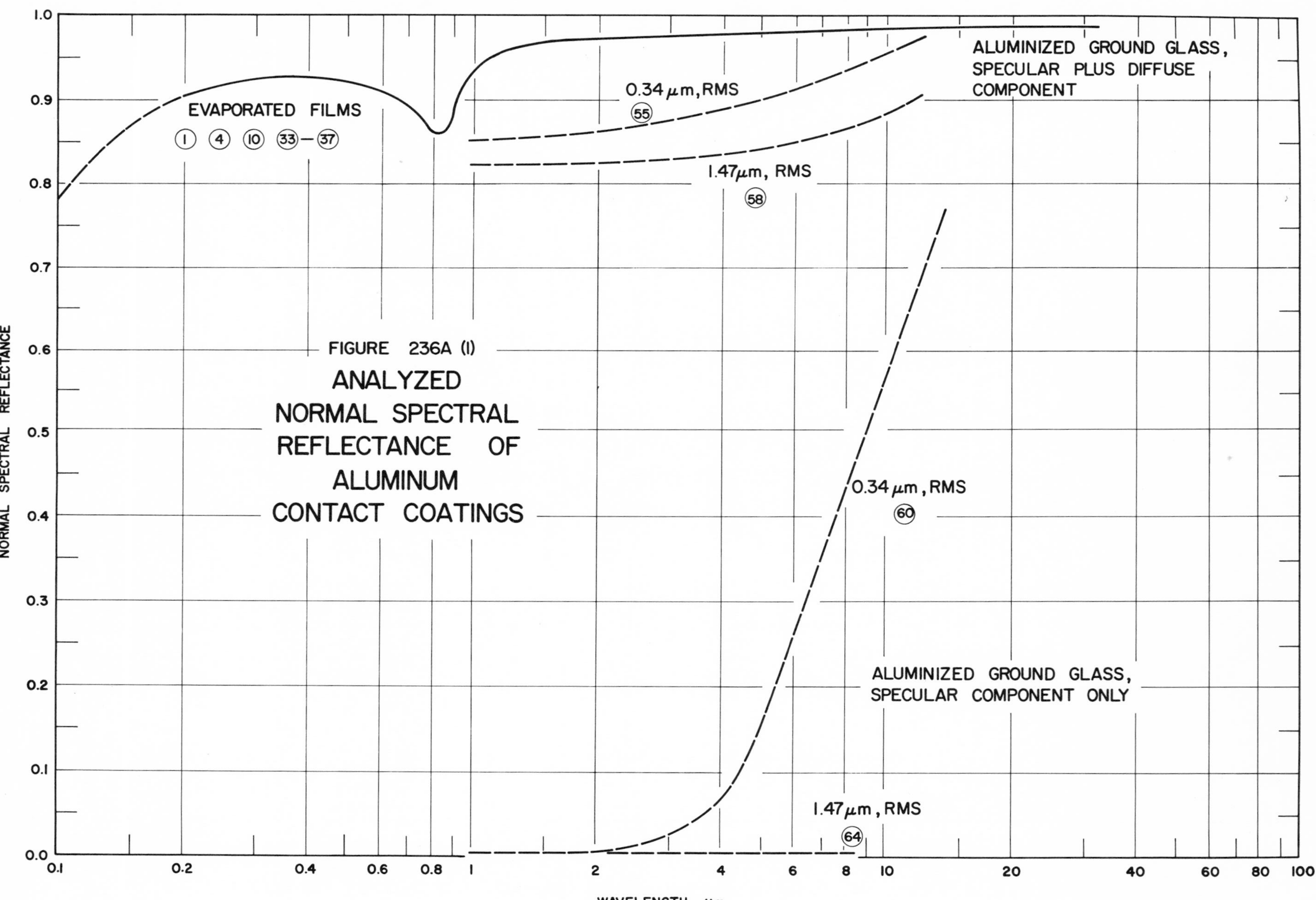

FIGURE 236A (I) ANALYZED NORMAL SPECTRAL REFLECTANCE OF ALUMINUM CONTACT COATINGS

FIGURE 236A (2)

ANALYZED NORMAL SPECTRAL REFLECTANCE OF ALUMINUM CONTACT COATINGS

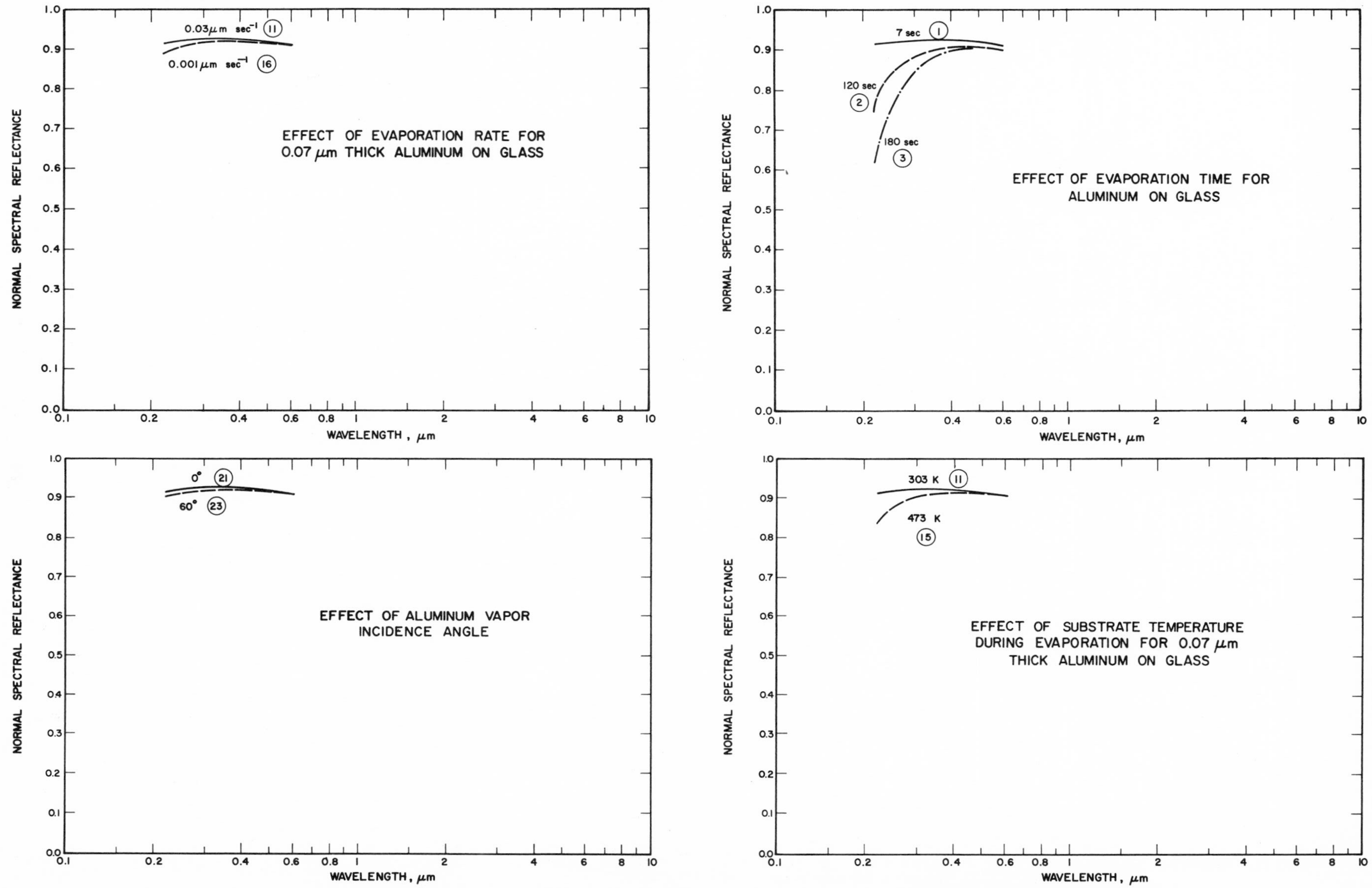
EFFECT OF EVAPORATION RATE FOR 0.07 μm THICK ALUMINUM ON GLASS
0.03 μm sec⁻¹ (11)
0.001 μm sec⁻¹ (16)
EFFECT OF EVAPORATION TIME FOR ALUMINUM ON GLASS
7 sec (1)
120 sec (2)
180 sec (3)
EFFECT OF ALUMINUM VAPOR INCIDENCE ANGLE
0° (21)
60° (23)
EFFECT OF SUBSTRATE TEMPERATURE DURING EVAPORATION FOR 0.07 μm THICK ALUMINUM ON GLASS
303 K (11)
473 K (15)
NORMAL SPECTRAL REFLECTANCE
WAVELENGTH, μm
0.0
0.1
0.2
0.3
0.4
0.5
0.6
0.7
0.8
0.9
1.0
1
2
4
6
8
10

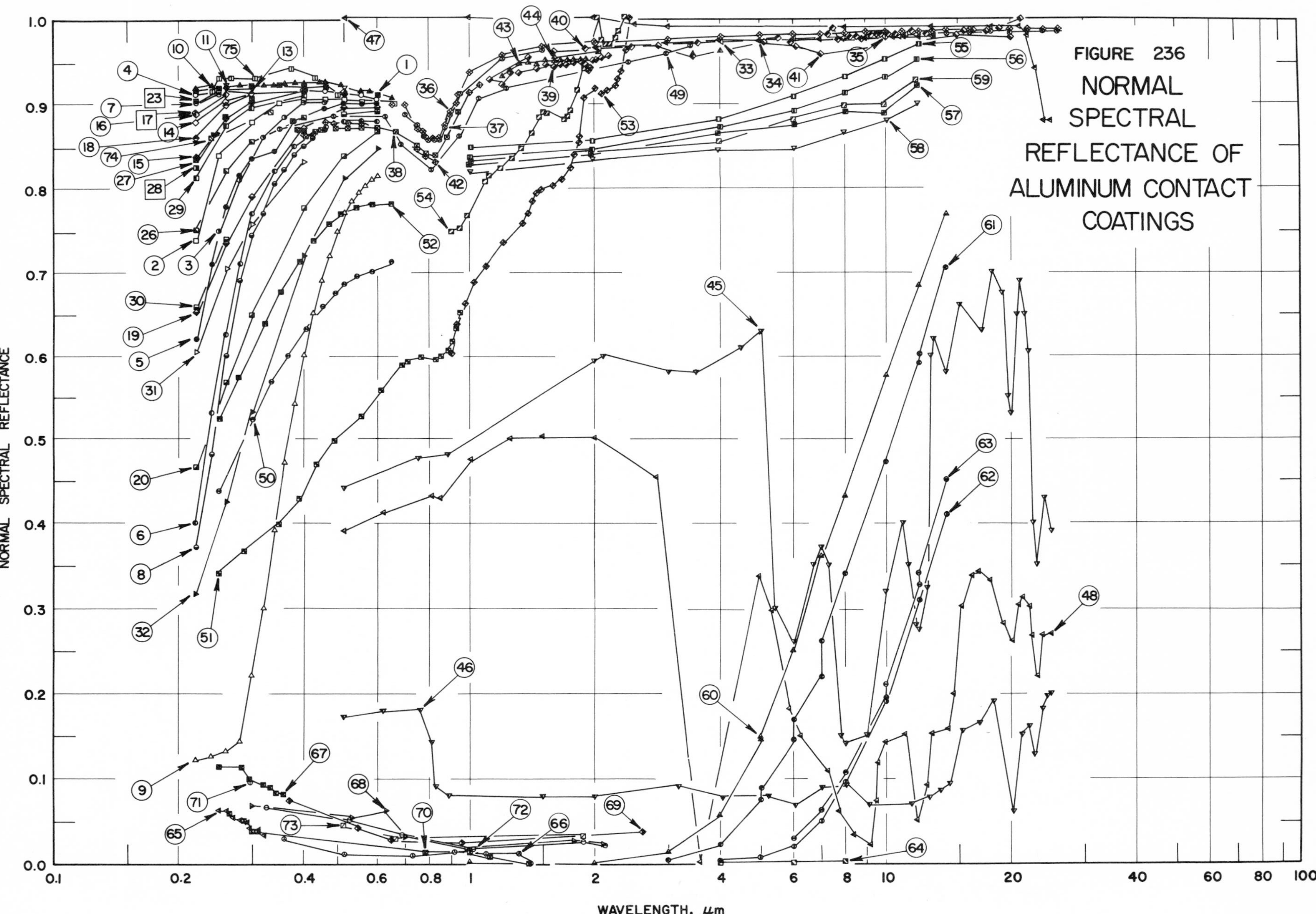
FIGURE 236
NORMAL SPECTRAL REFLECTANCE OF ALUMINUM CONTACT COATINGS
NORMAL SPECTRAL REFLECTANCE
WAVELENGTH, μm

SPECIFICATION TABLE NO. 236 NORMAL SPECTRAL REFLECTANCE OF ALUMINUM CONTACT COATINGS

Curve No.	Ref. No.	Year	Temperature, K	Wavelength Range, μm	Geometry θ θ' ω'	Reported Error, %	Composition (weight percent), Specifications, and Remarks
1	198	1961	298	0.22-0.60	~0° ~0°	0.3	99.99 Al (0.06 to 0.07 μm thick); glass substrate; deposited in vacuum (1 to 2 x 10^{-5} mm Hg) for 7 sec; data extracted from smooth curve.
2	198	1961	298	0.22-0.60	~0° ~0°	0.3	Similar to above specimen and conditions except deposited for 120 sec.
3	198	1961	298	0.22-0.60	~0° ~0°	0.3	Similar to above specimen and conditions except deposited for 180 sec.
4	198	1961	298	0.22-0.60	~0° ~0°	0.3	99.99 Al (0.06 to 0.07 μm thick); glass substrate; deposited in vacuum (1 to 2 x 10^{-4} mm Hg) for 6 sec; data extracted from smooth curve.
5	198	1961	298	0.22-0.60	~0° ~0°	0.3	Similar to above specimen and conditions except deposited for 135 sec.
6	198	1961	298	0.22-0.60	~0° ~0°	0.3	Similar to above specimen and conditions except deposited for 180 sec.
7	198	1961	298	0.22-0.60	~0° ~0°	0.3	99.99 Al (0.06 to 0.07 μm thick); glass substrate; deposited in vacuum (1 x 10^{-3} mm Hg) for 4 sec; data extracted from smooth curve.
8	198	1961	298	0.22-0.60	~0° ~0°	0.3	Similar to above specimen and conditions except deposited for 60 sec.
9	198	1961	298	0.22-0.60	~0° ~0°	0.3	Similar to above specimen and conditions except deposited for 145 sec.
10	198	1961	298	0.220-0.650	~0° ~0°	0.3	> 99.99 Al; glass substrate; deposited at rate of more than 0.03 μm sec^{-1} at 1 x 10^{-5} mm Hg.
11	198	1961	298	0.220-0.600	~0° ~0°	0.3	99.99 Al (~0.07 μm thick); glass substrate; deposited at ~0.03 μm sec^{-1} in vacuum (1 to 2 x 10^{-5} mm Hg)with substrate temp of 303 K.
12*	198	1961	298	0.220-0.600	~0° ~0°	0.3	Similar to above specimen and conditions except substrate temp was 323 K.
13	198	1961	298	0.220-0.600	~0° ~0°	0.3	Similar to above specimen and conditions except substrate temp was 373 K.
14	198	1961	298	0.220-0.600	~0° ~0°	0.3	Similar to above specimen and conditions except substrate temp was 423 K.
15	198	1961	298	0.220-0.600	~0° ~0°	0.3	Similar to above specimen and conditions except substrate temp was 473 K.
16	198	1961	298	0.220-0.600	~0° ~0°	0.3	99.99 Al (~0.07 μm thick); glass substrate; deposited at 0.0010 to 0.0015 μm sec^{-1} in vacuum (1 to 2 x 10^{-5} mm Hg) with substrate temp of 303 K.
17	198	1961	298	0.220-0.600	~0° ~0°	0.3	Similar to above specimen and conditions except substrate temp was 323 K.
18	198	1961	298	0.220-0.600	~0° ~0°	0.3	Similar to above specimen and conditions except substrate temp was 373 K.
19	198	1961	298	0.220-0.600	~0° ~0°	0.3	Similar to above specimen and conditions except substrate temp was 423 K.
20	198	1961	298	0.220-0.600	~0° ~0°	0.3	Similar to above specimen and conditions except substrate temp was 473 K.
21*	198	1961	298	0.220-0.600	~0° ~0°	0.3	99.99 Al (~0.06 μm thick); glass substrate; deposited at 0.03 μm sec^{-1} in vacuum (1 x 10^{-5} mm Hg) with 0° angle of vapor incidence.
22*	198	1961	298	0.220-0.600	~0° ~0°	0.3	Similar to above specimen and conditions except 30° angle of vapor incidence.
23	198	1961	298	0.220-0.600	~0° ~0°	0.3	Similar to above specimen and conditions except 60° angle of vapor incidence.
24*	198	1961	298	0.220-0.600	~0° ~0°	0.3	99.99 Al (~0.20 μm thick); glass substrate; deposited at 0.03 μm sec^{-1} in vacuum (1 x 10^{-5} mm Hg) with 0° angle of vapor incidence.
25*	198	1961	298	0.220-0.600	~0° ~0°	0.3	Similar to above specimen and conditions except 30° angle of vapor incidence.

* Not shown on plot

SPECIFICATION TABLE NO. 236 NORMAL SPECTRAL REFLECTANCE OF ALUMINUM CONTACT COATINGS (continued)

Curve No.	Ref. No.	Year	Temperature, K	Wavelength Range, μm	Geometry θ	θ'	ω'	Reported Error, %	Composition (weight percent), Specifications, and Remarks
26	198	1961	298	0.220-0.600	~0°	~0°		0.3	Similar to above specimen and conditions except 60° angle of vapor incidence.
27	198	1961	298	0.200-0.600	~0°	~0°		0.3	99.99 Al (~0.06 μm thick); glass substrate; deposited at 0.001 μm sec^{-1} in vacuum (1 x 10^{-4} mm Hg) with 0° angle of vapor incidence.
28	198	1961	298	0.220-0.600	~0°	~0°		0.3	Similar to above specimen and conditions except 30° angle of vapor incidence.
29	198	1961	298	0.220-0.600	~0°	~0°		0.3	Similar to above specimen and conditions except 60° angle of vapor incidence.
30	198	1961	298	0.220-0.600	~0°	~0°		0.3	99.99 Al (~2000 Å thick); glass substrate; deposited at 10 Å sec^{-1} in vacuum (1 x 10^{-4} mm Hg) with 0° angle of vapor incidence.
31	198	1961	298	0.220-0.600	~0°	~0°		0.3	Similar to above specimen and conditions except 30° angle of vapor incidence.
32	198	1961	298	0.220-0.600	~0°	~0°		0.3	Similar to above specimen and conditions except 60° angle of vapor incidence.
33	199	1960	298	2.06-13.07	~5°	~5°			Aluminum; substrate unknown; vapor deposited.
34	199	1960	298	2.01-13.09	~5°	~5°			Similar to above specimen and conditions.
35	109	1962	298	2.00-20.0	~0°		2π		Aluminum (0.20 μm thick); Mylar substrate; vapor deposited.
36	200	1962	298	0.550-32.0	~5°	~5°		±0.1	Aluminum (0.065 to 0.110 μm thick); clean and smooth fused quartz optical flat substrate; deposited by evaporating aluminum rod of 99.998 purity in vacuum (1 x 10^{-5} mm Hg); freshly prepared.
37	200	1962	298	0.550-32.0	~5°	~5°		±0.1	Similar to above specimen and conditions except aged for several wks.
38	201	1961	298	0.375-0.940	~0°	~0°			High purity aluminum; glass substrate; vacuum deposited; measured in vacuum (~5 x 10^{-6} mm Hg); tungsten lamp source (0.4 to 1.0 μm)and globar source(1.0 to 21.15 μm). [Authors' designation: Run No. 1]
39	201	1961	298	1.13-2.16	~0°	~0°			Above specimen and conditions.
40	201	1961	298	1.90-12.60	~0°	~0°			Above specimen and conditions.
41	201	1961	298	1.60-21.15	~0°	~0°			Above specimen and conditions.
42	201	1961	298	0.450-1.060	~0°	~0°			Curve 38 specimen and conditions. [Authors' designation: Run No. 2]
43	201	1961	298	1.20-2.16	~0°	~0°			Above specimen and conditions.
44	201	1961	298	1.60-4.00	~0°	~0°			Above specimen and conditions.
45	202	1962	~322	0.50-25.00	~0°		2π	<2.0	Aluminum; Mylar substrate; aluminized; grit blasted with 60 grit silicon carbide with air pressure of 110 to 120 psi for 30 to 45 sec; data extracted from smooth curve; hohlraum at 1273 K; converted from R (2π, 0°).
46	202	1962	~322	0.50-25.00	~0°		2π	<2.0	Above specimen and conditions; diffuse component only.
47	202	1962	~322	0.50-25.00	~0°		2π	<2.0	Similar to curve 45 specimen and conditions except exposed to vacuum (<4 x 10^{-8} mm Hg) for 24 hrs.
48	202	1962	~322	0.50-25.00	~0°		2π	<2.0	Above specimen and conditions; diffuse component only.
49	203	1962	298	0.351-3.400	~0°	~0°			Aluminum; substrate unknown; vacuum evaporated; data extracted from smooth curve.
50	101	1964	~298	0.2500-0.6500	0°		2π		Aluminum, commercially pure; Al alloy substrate; data extracted from smooth curve; sample supplied by North American Aviation. [Authors' designation: Disk No. 7075]

SPECIFICATION TABLE NO. 236 NORMAL SPECTRAL REFLECTANCE OF ALUMINUM CONTACT COATINGS (continued)

Curve No.	Ref. No.	Year	Temperature, K	Wavelength Range, μm	Geometry θ	θ'	ω'	Reported Error, %	Composition (weight percent), Specifications, and Remarks
51	101	1964	~298	0.2500-0.9500	~0°		2π		Similar to above specimen and conditions except polished with DuPont Duco-7 polish.
52	101	1964	~298	0.2500-0.6500	~0°		2π		Alclad 2024-T3; 1050 Al alloy (composition: 0.40 Fe, 0.25 Si, balance Al) cladding on 2024 Al alloy; nominal cladding thickness ~0.0254 mm; sample thickness 0.508 mm; data extracted from smooth curve; cladding composition may be altered due to diffusion from substrate.
53	101	1964	~298	0.900-2.445	~0°		2π		Similar to above specimen and conditions.
54	101	1964	~298	0.900-2.364	~0°		2π		Similar to above specimen and conditions.
55	204	1965	~298	1.01-12.01	9°		2π		Aluminum (0.203 μm thick); ground glass substrate; vapor deposited; surface roughness measured with profilometer 0.34 μm (RMS); avg grit size used for grinding glass was 9.5 μm.
56	204	1965	~298	1.00-11.99	9°		2π		Similar to above specimen and conditions except 0.38 μm (RMS) surface roughness; avg grit size 5.0 μm.
57	204	1965	~298	0.99-11.99	9°		2π		Similar to above specimen and conditions except surface roughness 0.61 μm (RMS); avg grit size 22.5 μm.
58	204	1965	~298	1.00-11.98	9°		2π		Similar to above specimen and conditions except surface roughness 1.47 μm (RMS); avg grit size 32.0 μm.
59	204	1965	~298	1.00-11.99	9°		2π		Similar to above specimen and conditions except surface roughness 0.61 μm (RMS); avg grit size 600 mesh.
60	204	1965	~298	1-14	10°	10°			Aluminum (0.203 μm thick); ground glass substrate; vapor deposited; surface roughness measured with profilometer 0.34 μm (RMS); avg grit size used for grinding glass was 9.5 μm; measured relative to polished aluminum surface.
61	204	1965	~298	1-14	10°	10°			Similar to above specimen and conditions except surface roughness 0.38 μm (RMS); avg grit size 5.0 μm.
62	204	1965	~298	2-14	10°	10°			Similar to above specimen and conditions except surface roughness 0.61 μm (RMS); avg grit size 22.5 μm.
63	204	1965	~298	2-14	10°	10°			Similar to above specimen and conditions except surface roughness 0.61 μm (RMS); avg grit size 600 mesh.
64	204	1965	~298	2-8	10°	10°			Similar to above specimen and conditions except surface roughness 1.47 μm; avg grit size 32.0 μm.
65	31	1965	~300	0.250-0.362	~0°		2π		Aluminum (~1 μm thick); Alcoa 1199 (H-18) substrate; vapor-deposited; measured in vacuum (10^{-6} mm Hg); data extracted from smooth curve. [Authors' designation: Control Sample No. 180]
66	31	1965	~300	0.36-1.40	~0°		2π		Above specimen and conditions.
67	31	1965	~300	0.250-0.359	~0°		2π		Similar to the above specimen and conditions except exposed to 1.0×10^{16} p cm^{-2}. [Authors' designation: Exposed Sample No. 178]
68	31	1965	~300	0.30-2.11	~0°		2π		Above specimen and conditions.

SPECIFICATION TABLE NO. 236 NORMAL SPECTRAL REFLECTANCE OF ALUMINUM CONTACT COATINGS (continued)

Curve No.	Ref. No.	Year	Temperature, K	Wavelength Range, μm	Geometry θ	θ'	ω'	Reported Error, %	Composition (weight percent), Specifications, and Remarks
69	31	1965	~300	0.37-2.60	~0°		2π		Above specimen and conditions.
70	33	1965	~298	0.250-1.392	5°		2π		Aluminum (1.0 μm thick); Alcoa 1199 (H-18) aluminum substrate; substrate chemically-brightened with Alcoa R5 solution; vapor deposited; data extracted from smooth curve. [Authors' designation: Sample No. 180]
71	33	1965	~298	0.252-0.359	5°		2π		Similar to above specimen and conditions except irradiated in vacuum at 300 K with 7.4 keV protons to a total dose of 1.0 x 10^{16} p cm^{-2}. [Authors' designation: Sample No. 178]
72	33	1965	~298	0.326-2.110	5°		2π		Above specimen and conditions except diffuse reflectance measured with Beckman DK-2A.
73	33	1965	~298	0.373-2.602	5°		2π		Above specimen and conditions except diffuse reflectance measured with Gier Dunkle reflectometer.
74	205	1950	298	0.240-0.490	~0°	~0°		±2.0	100 Al coating; glass substrate; vapor deposited.
75	205	1950	298	0.240-0.488	~0°	~0°		±2.0	98 Al + 2 Ag (δ phase); glass substrate; vapor deposited.
76*	112	1963	~298	1.43-22.01	~0°		2π	±2	Fasson foil (0.0508 mm thick); copper (3.18 mm thick) substrate; sample maintained at ~298 K; hohlraum at 1073 K; converted from R(2π, ~0°).

* Not shown on plot

DATA TABLE NO. 236 NORMAL SPECTRAL REFLECTANCE OF ALUMINUM CONTACT COATINGS

[Wavelength, λ, μm; Reflectance, ρ; Temperature, T, K]

λ	ρ
CURVE 1	
T = 298	
0.22	0.915
0.25	0.918
0.30	0.922
0.35	0.923
0.40	0.922
0.45	0.920
0.50	0.915
0.55	0.912
0.60	0.910
CURVE 2	
T = 298	
0.22	0.740
0.25	0.840
0.30	0.880
0.35	0.900
0.40	0.905
0.45	0.905
0.50	0.905
0.55	0.903
0.66	0.900
CURVE 3	
T = 298	
0.22	0.620*
0.25	0.750
0.28	0.810
0.32	0.860
0.36	0.885
0.40	0.902
0.45	0.903
0.50	0.902*
0.55	0.901*
0.60	0.900*
CURVE 4	
T = 298	
0.22	0.910
0.25	0.915
0.30	0.920*
0.35	0.920*
0.40	0.918
CURVE 4 (cont.)	
0.45	0.916
0.50	0.915*
0.55	0.910
0.60	0.906
CURVE 5	
T = 298	
0.22	0.620
0.24	0.710
0.26	0.780
0.28	0.815
0.30	0.835
0.34	0.845
0.38	0.880
0.45	0.895
0.50	0.900
0.55	0.900
0.60	0.900
CURVE 6	
T = 298	
0.22	0.400
0.24	0.530
0.26	0.625
0.28	0.710
0.30	0.770
0.34	0.820
0.38	0.850
0.42	0.866
0.45	0.874
0.50	0.880
0.55	0.880
0.60	0.880
CURVE 7	
T = 298	
0.22	0.900
0.25	0.910
0.30	0.920*
0.35	0.920*
0.40	0.920*
0.45	0.920*
0.50	0.915*
CURVE 7 (cont.)	
0.55	0.913*
0.60	0.910*
CURVE 8	
T = 298	
0.22	0.370
0.24	0.480
0.26	0.600
0.28	0.690
0.30	0.745
0.32	0.770
0.34	0.805
0.36	0.823
0.38	0.840
0.40	0.850
0.42	0.860
0.45	0.870
0.50	0.878*
0.55	0.882*
0.60	0.882*
CURVE 9	
T = 298	
0.22	0.120
0.24	0.126
0.26	0.132
0.28	0.143
0.30	0.220
0.32	0.300
0.34	0.390
0.36	0.470
0.38	0.540
0.40	0.600
0.42	0.650
0.44	0.690
0.46	0.720
0.48	0.750
0.50	0.770
0.52	0.785
0.54	0.795
0.56	0.804
0.58	0.810
0.60	0.815
CURVE 10	
T = 298	
0.220	0.918
0.240	0.921
0.260	0.920
0.280	0.922
0.300	0.921*
0.320	0.922
0.340	0.923
0.360	0.924
0.380	0.926
0.400	0.926
0.436	0.926
0.450	0.925
0.492	0.922
0.546	0.916
0.578	0.915
0.650	0.907
CURVE 11	
T = 298	
0.220	0.915*
0.260	0.922
0.300	0.923*
0.400	0.924*
0.500	0.917
0.600	0.910*
CURVE 12*	
T = 298	
0.220	0.913
0.260	0.921
0.300	0.923
0.400	0.923
0.500	0.917
0.600	0.909
CURVE 13	
T = 298	
0.220	0.900*
0.260	0.915
0.300	0.919
0.400	0.920*
0.500	0.915*
0.600	0.908*
CURVE 14	
T = 298	
0.220	0.879
0.260	0.904
0.300	0.912
0.400	0.917
0.500	0.914
0.600	0.907*
CURVE 15	
T = 298	
0.220	0.838
0.260	0.884
0.300	0.902
0.400	0.915
0.500	0.913
0.600	0.905*
CURVE 16	
T = 298	
0.220	0.890
0.260	0.912
0.300	0.918*
0.400	0.920*
0.500	0.916*
0.600	0.909*
CURVE 17	
T = 298	
0.220	0.889
0.260	0.910*
0.300	0.917*
0.400	0.919*
0.500	0.915*
0.600	0.909*
CURVE 18	
T = 298	
0.220	0.860
0.260	0.899
0.300	0.912
0.400	0.917*
0.500	0.915*
0.600	0.908*
CURVE 19	
T = 298	
0.220	0.652
0.260	0.735
0.300	0.792
0.400	0.874
0.500	0.895
0.600	0.896
CURVE 20	
T = 298	
0.220	0.466
0.260	0.567
0.300	0.649
0.400	0.779
0.500	0.839
0.600	0.867
CURVE 21*	
T = 298	
0.220	0.915
0.260	0.922
0.300	0.923
0.400	0.924
0.500	0.917
0.600	0.910
CURVE 22*	
T = 298	
0.220	0.914
0.260	0.920
0.300	0.921
0.400	0.921
0.500	0.916
0.600	0.909
CURVE 23	
T = 298	
0.220	0.904
0.260	0.915*
0.300	0.918*
0.400	0.918*
0.500	0.915*
0.600	0.908*
CURVE 24*	
T = 298	
0.220	0.912
0.260	0.919
0.300	0.922
0.400	0.922
0.500	0.917
0.600	0.909
CURVE 25*	
T = 298	
0.220	0.907
0.260	0.917
0.300	0.918
0.400	0.919
0.500	0.916
0.600	0.908
CURVE 26	
T = 298	
0.220	0.750
0.260	0.822
0.300	0.855
0.400	0.885
0.500	0.900*
0.600	0.896*
CURVE 27	
T = 298	
0.220	0.835
0.260	0.884
0.300	0.905
0.400	0.913*
0.500	0.910*
0.600	0.903*
CURVE 28	
T = 298	
0.220	0.826
0.260	0.882*
0.300	0.904*
0.400	0.913*
0.500	0.910*
0.600	0.902*
CURVE 29	
T = 298	
0.220	0.814
0.260	0.876
0.300	0.896
0.400	0.910
0.500	0.908*
0.600	0.900*
CURVE 30	
T = 298	
0.220	0.695
0.260	0.741
0.300	0.795*
0.400	0.868
0.500	0.888
0.600	0.893
CURVE 31	
T = 298	
0.220	0.605
0.260	0.705
0.300	0.759
0.400	0.832
0.500	0.875
0.600	0.890*
CURVE 32	
T = 298	
0.220	0.318
0.260	0.425
0.300	0.532
0.400	0.720
0.500	0.813
0.600	0.847
CURVE 33	
T = 298	
2.06	0.9676
3.00	0.9717
4.03	0.9753
5.06	0.9772
6.06	0.9780
7.05	0.9786

* Not shown on plot

DATA TABLE NO. 236 NORMAL SPECTRAL REFLECTANCE OF ALUMINUM CONTACT COATINGS (continued)

CURVE 33 (cont.) T = 298

λ	ρ
8.04	0.9788
9.06	0.9796
10.08	0.9798
11.08	0.9802
12.07	0.9804
13.07	0.9812

CURVE 34 T = 298

λ	ρ
2.01	0.9662
3.00	0.9706
4.01	0.9742
5.02	0.9759
6.02	0.9767
7.02	0.9770
8.05	0.9778
9.08	0.9786
10.06	0.9785
11.06	0.9799
12.07	0.9794
13.09	0.9797

CURVE 35 T = 298

λ	ρ
2.00	0.972
10.0	0.978
20.0	0.978

CURVE 36 T = 298

λ	ρ
0.550	0.9094*
0.600	0.9048*
0.650	0.8989
0.700	0.8900
0.750	0.8761
0.775	0.8678
0.800	0.8604
0.825	0.8569
0.850	0.8622
0.875	0.8759
0.900	0.8920
0.925	0.9072
0.950	0.9192

CURVE 36 (cont.)

λ	ρ
1.00	0.9360
1.20	0.9596
1.50	0.9676
2.00	0.9718
3.00	0.9765
4.00	0.9795
5.00	0.9812
6.00	0.9823
7.00	0.9831
8.00	0.9837
9.00	0.9841
10.0	0.9845
11.0	0.9849
12.0	0.9854
13.0	0.9857
14.0	0.9861
16.0	0.9868
18.0	0.9873
20.0	0.9878
22.0	0.9883
24.0	0.9887
26.0	0.9890
28.0	0.9893
30.0	0.9896
32.0	0.9898

CURVE 37 T = 298

λ	ρ
0.550	0.9049*
0.600	0.9021*
0.650	0.8976*
0.700	0.8886
0.750	0.8761
0.775	0.8678
0.800	0.8596
0.825	0.8556
0.850	0.8596
0.875	0.8730
0.900	0.8894
0.925	0.9030
0.950	0.9154
1.00	0.9360
1.20	0.9596
1.50	0.9676
2.00	0.9718*
3.00	0.9765*

CURVE 37 (cont.)

λ	ρ
4.00	0.9795*
5.00	0.9772*
6 00	0.9784*
7.00	0.9794*
8.00	0.9801*
9.00	0.9807
10.0	0.9812*
11.0	0.9816*
12.0	0.9821*
13.0	0.9826*
14.0	0.9830
16.0	0.9838
18.0	0.9845
20.0	0.9852
22.0	0.9856
24.0	0.9861
26.0	0.9864
28.0	0.9867
30.0	0.9870*
32.0	0.9872

CURVE 38 T = 298

λ	ρ
0.375	0.880
0.385	0.871
0.405	0.864
0.425	0.869
0.450	0.870
0.475	0.872
0.510	0.873
0.550	0.874
0.600	0.872
0.665	0.867
0.750	0.850
0.790	0.842
0.834	0.840
0.880	0.864
0.940	0.893

CURVE 39 T = 298

λ	ρ
1.13	0.928
1.20	0.924
1.26	0.936
1.32	0.935

CURVE 39 (cont.)

λ	ρ
1.39	0.955
1.4y	0.941
1.53	0.945
1.60	0.943
1.66	0.946
1.72	0.945
1.79	0.948
1.85	0.948
1.91	0.946
1.97	0.951
2.02	0.953
2.08	0.954
2.16	0.956

CURVE 40 T = 298

λ	ρ
1.90	0.966
2.45	0.967
3.00	0.974*
3.55	0.972
4.06	0.974*
4.52	0.969
4.95	0.973
5.35	0.976
7.04	0.982*
7.25	0.983
7.53	0.988
7.78	0.976
8.05	0.978*
8.26	0.979
8.55	0.978
8.75	0.977
9.05	0.978*
9.28	0.981
9.50	0.983
9.73	0.980
9.93	0.983*
10.15	0.983
10.35	0.981*
10.55	0.981
10.75	0.983*
11.13	0.986*
11.32	0.985*
11.66	0.985
12.00	0.985*
12.30	0.985*
12.60	0.981

CURVE 41 T = 298

λ	ρ
1.60	0.963
2.05	0.969*
2.85	0.968
3.45	0.955
4.00	0.981*
5.10	0.972
6.10	0.968
7.00	0.959
7.75	0.975*
8.50	0.980*
9.87	0.980*
11.05	0.981
12.20	0.983*
13.15	0.984*
14.03	0.986*
14.82	0.981
15.56	0.986
16.28	0.987
17.67	0.986
18.90	0.988
20.04	0.996*
21.15	0.998

CURVE 42

λ	ρ
0.450	0.878
0.475	0.875
0.510	0.877
0.550	0.876
0.600	0.872*
0.750	0.849
0.790	0.839
0.834	0.831
0.880	0.862*
0.940	0.892*
1.000	0.914
1.060	0.923

CURVE 43 T = 298

λ	ρ
1.20	0.934
1.26	0.939
1.32	0.947
1.39	0.947
1.46	0.947

CURVE 43 (cont.)

λ	ρ
1.53	0.952
1.60	0.950
1.60	0.951
1.72	0.951
1.78	0.951
1.85	0.950
1.91	0.953
1.97	0.951*
2.02	0.953*
2.08	0.955*
2.16	0.956*

CURVE 44 T = 298

λ	ρ
1.60	0.956
2.05	0.962
3.45	0.958
4.00	0.963

CURVE 45 T ~ 322

λ	ρ
0.50	0.440
0.75	0.475
0.88	0.480
2.00	0.595
2.10	0.600
3.00	0.580
3.50	0.580
4.50	0.610
5.05	0.630
5.40	0.300
6.00	0.260
6.70	0.350
7.00	0.370
7.30	0.350
7.80	0.150
8.00	0.140
9.00	0.150
10.00	0.320
11.00	0.400
11.40	0.350
11.70	0.280
12.00	0.275
12.50	0.325
12.80	0.600

CURVE 45 (cont.)

λ	ρ
13.00	0.620
14.00	0.580
15.00	0.660
17.00	0.630
18.00	0.700
19.00	0.675
19.70	0.550
20.00	0.530
20.60	0.650
21.00	0.690
21.60	0.650
22.00	0.605
22.50	0.400
23.00	0.350
24.00	0.430
25.00	0.390

CURVE 46 T ~ 322

λ	ρ
0.50	0.170
0.62	0.179
0.76	0.179
0.81	0.141
0.83	0.090
0.89	0.078
1.50	0.078
2.00	0.078
3.18	0.090
4.08	0.077
5.22	0.079
6.06	0.068
7.00	0.089
8.05	0.090
9.10	0.069
11.65	0.069
12.74	0.077
13.56	0.086
14.25	0.094
15.17	0.156
16.85	0.166
18.04	0.190
19.10	0.151
20.10	0.062
21.05	0.153
22.03	0.163
22.99	0.128

CURVE 46 (cont.)

λ	ρ
23.78	0.182
24.38	0.198
25.00	0.200

CURVE 47 T ~ 322

λ	ρ
0.50	1.000
1.00	1.000
2.00	1.000
2.50	0.994
3.00	0.990
8.00	0.990
13.00	0.990
18.00	0.990
21.00	0.990
22.00	0.980
23.00	0.940
24.00	0.880
25.00	0.880

CURVE 48 T ~ 322

λ	ρ
0.50	0.388
0.62	0.414
0.81	0.431
0.85	0.428
1.01	0.474
1.25	0.500
1.50	0.502
2.00	0.502
2.81	0.454
3.57	0.002
4.99	0.337
5.30	0.298
5.81	0.181
6.22	0.148
7.29	0.109
7.71	0.063
8.39	0.033
9.19	0.024
9.48	0.074
9.59	0.119
9.99	0.143
11.02	0.102
11.96	0.052

* Not shown on plot

DATA TABLE NO. 236 NORMAL SPECTRAL REFLECTANCE OF ALUMINUM CONTACT COATINGS (continued)

CURVE 48 (cont.) T ~ 322

λ	ρ
12.40	0.092
12.98	0.152
14.08	0.158
14.60	0.198
15.34	0.304
16.03	0.339
16.74	0.344
17.92	0.332
19.34	0.281
20.04	0.262
20.79	0.305
21.28	0.313
22.13	0.304
22.54	0.268
23.01	0.220
23.98	0.269
25.00	0.270

CURVE 49 T = 298

λ	ρ
0.351	0.870
0.468	0.886
0.629	0.885
0.688	0.852
0.808	0.824
0.943	0.863
1.062	0.905
1.234	0.918
1.725	0.947
2.404	0.958
2.999	0.967
3.400	0.969

CURVE 50 T ~ 298

λ	ρ
0.2500	0.438
0.3011	0.523
0.3349	0.569
0.3656	0.601
0.4027	0.633
0.4400	0.658
0.4775	0.675
0.5049	0.685
0.5338	0.695

CURVE 50 (cont.)

λ	ρ
0.5802	0.702
0.6500	0.714

CURVE 51 T ~ 298

λ	ρ
0.2500	0.340
0.2885	0.366
0.3492	0.399
0.3872	0.428
0.4289	0.466
0.4786	0.497
0.5460	0.526
0.6097	0.557
0.6842	0.589
0.7120	0.594
0.7631	0.598
0.8387	0.595
0.8580	0.598
0.8846	0.606
0.9090	0.617
0.9295	0.632
0.9500	0.650

CURVE 52 T ~ 298

λ	ρ
0.2500	0.524
0.2794	0.576
0.3212	0.638
0.3522	0.676
0.3920	0.714
0.4208	0.738
0.4614	0.758
0.4997	0.770
0.5388	0.778
0.5859	0.782
0.6500	0.782

CURVE 53 T ~ 298

λ	ρ
0.900	0.604
0.932	0.638
0.977	0.663
1.030	0.688
1.096	0.707

CURVE 53 (cont.)

λ	ρ
1.221	0.737
1.344	0.759
1.389	0.771
1.421	0.785
1.448	0.796
1.481	0.799
1.591	0.803
1.675	0.811
1.745	0.826
1.797	0.840
1.839	0.852
1.857	0.863
1.898	0.906
1.943	0.939
1.959	0.941
2.003	0.918
2.086	0.913
2.163	0.915
2.255	0.921
2.297	0.928
2.332	0.938
2.387	0.964
2.445	1.000

CURVE 54 T ~ 298

λ	ρ
0.900	0.749
0.945	0.753
0.992	0.767
1.088	0.808
1.112	0.815
1.188	0.826
1.271	0.837
1.333	0.848
1.409	0.866
1.475	0.886
1.511	0.890
1.557	0.889
1.681	0.882
1.715	0.885
1.759	0.893
1.820	0.915
1.883	0.942
1.990	1.000*
2.013	1.000
2.091	0.976

CURVE 54 (cont.)

λ	ρ
2.137	0.970
2.194	0.971
2.256	0.978
2.297	0.985
2.364	1.000

CURVE 55 T = 298

λ	ρ
1.01	0.848
1.99	0.855
3.99	0.881
6.02	0.908
8.00	0.930
10.00	0.953
12.01	0.972

CURVE 56 T = 298

λ	ρ
1.00	0.837
1.98	0.845
4.00	0.874
6.02	0.892
8.00	0.913
10.00	0.932
11.99	0.952

CURVE 57 T = 298

λ	ρ
0.99	0.829
1.97	0.842
3.98	0.865
6.01	0.876
8.01	0.890
9.99	0.889
11.99	0.920

CURVE 58 T = 298

λ	ρ
1.00	0.822
1.98	0.836
3.97	0.845
6.02	0.848
7.99	0.867

CURVE 58 (cont.)

λ	ρ
10.02	0.881
11.98	0.899

CURVE 59 T = 298

λ	ρ
1.00	0.834
1.98	0.847*
3.99	0.855
6.00	0.882
8.00	0.897
9.99	0.897
11.99	0.929

CURVE 60 T ~ 298

λ	ρ
1	0.003
2	0.006
3	0.013
4	0.057
5	0.145
5	0.149
6	0.250
7	0.360
8	0.434
10	0.576
12	0.685
14	0.770

CURVE 61 T ~ 298

λ	ρ
1	0.002*
2	0.004*
3	0.005
4	0.024
5	0.089
5	0.075
6	0.143
6	0.168
7	0.220
7	0.262
8	0.340
10	0.470
12	0.601
12	0.590
14	0.705

CURVE 62 T ~ 298

λ	ρ
2	0.003*
4	0.004
5	0.008
6	0.022
7	0.050
8	0.095
10	0.190
12	0.311
14	0.410

CURVE 63 T ~ 298

λ	ρ
2	0.003*
4	0.004*
5	0.009*
6	0.030
7	0.063
8	0.106
10	0.210
10	0.195
12	0.341
12	0.329
14	0.450

CURVE 64 T ~ 298

λ	ρ
2	0.002*
4	0.002
6	0.002
8	0.003

CURVE 65 T ~ 300

λ	ρ
0.250	0.062
0.264	0,062
0.271	0.054
0.282	0.052
0.295	0.049
0.299	0.041
0.311	0.039
0.321	0.034
0.362	0.029

CURVE 66 T ~ 300

λ	ρ
0.36	0.029
0.50	0.012
0.73	0.009
0.92	0.014
1.07	0.012
1.12	0.008
1.32	0.012
1.40	0.000

CURVE 67 T ~ 300

λ	ρ
0.250	0.114
0.284	0.114
0.297	0.098
0.320	0.091
0.332	0.089
0.343	0.083
0.359	0.080

CURVE 68 T ~ 300

λ	ρ
0.30	0.067
0.52	0.053
0.63	0.063
0.70	0.034
0.98	0.018
1.78	0.028
2.11	0.022

CURVE 69 T ~ 300

λ	ρ
0.37	0.074
0.54	0.042
0.65	0.029
0.96	0.026
2.60	0.038

CURVE 70 T ~ 298

λ	ρ
0.250	0.062*
0.267	0.058
0.289	0.050

CURVE 70 (cont.)

λ	ρ
0.300	0.039
0.310	0.039
0.360	0.030*
0.360	0.029*
0.500	0.013*
0.785	0.013
1.000	0.014
1.137	0.009
1.321	0.012*
1.392	0.000

CURVE 71 T ~ 298

λ	ρ
0.252	0.116*
0.282	0.116*
0.299	0.096
0.336	0.087*
0.359	0.080*

CURVE 72 T ~ 298

λ	ρ
0.326	0.066
0.517	0.053
0.631	0.065*
0.691	0.036
1.035	0.019
1.887	0.027
2.110	0.021

CURVE 73 T ~ 298

λ	ρ
0.373	0.075*
0.500	0.046
0.667	0.029
0.973	0.026*
1.891	0.034
2.602	0.038*

CURVE 74 T = 298

λ	ρ
0.240	0.860
0.333	0.890
0.490	0.920*

* Not shown on plot

DATA TABLE NO. 236 NORMAL SPECTRAL REFLECTANCE OF ALUMINUM CONTACT COATINGS (continued)

λ	ρ
CURVE 75 T = 298	
0.240	0.915
0.253	0.930
0.269	0.930
0.305	0.930
0.372	0.940
0.426	0.930
0.488	0.910
CURVE 76* T ~ 298	
1.43	0.936
1.98	0.959
2.43	0.956
3.36	0.986
4.55	0.983
5.50	0.980
6.06	0.985
7.06	0.986
7.73	0.986
7.98	0.986
9.03	0.983
10.04	0.981
11.04	0.973
12.02	0.955
13.05	0.948
14.05	0.964
15.07	0.978
16.06	0.977
18.09	0.966
19.07	0.954
22.01	0.965

* Not shown on plot

FIGURE 237

CHANGE IN NORMAL SPECTRAL REFLECTANCE OF ALUMINUM CONTACT COATINGS

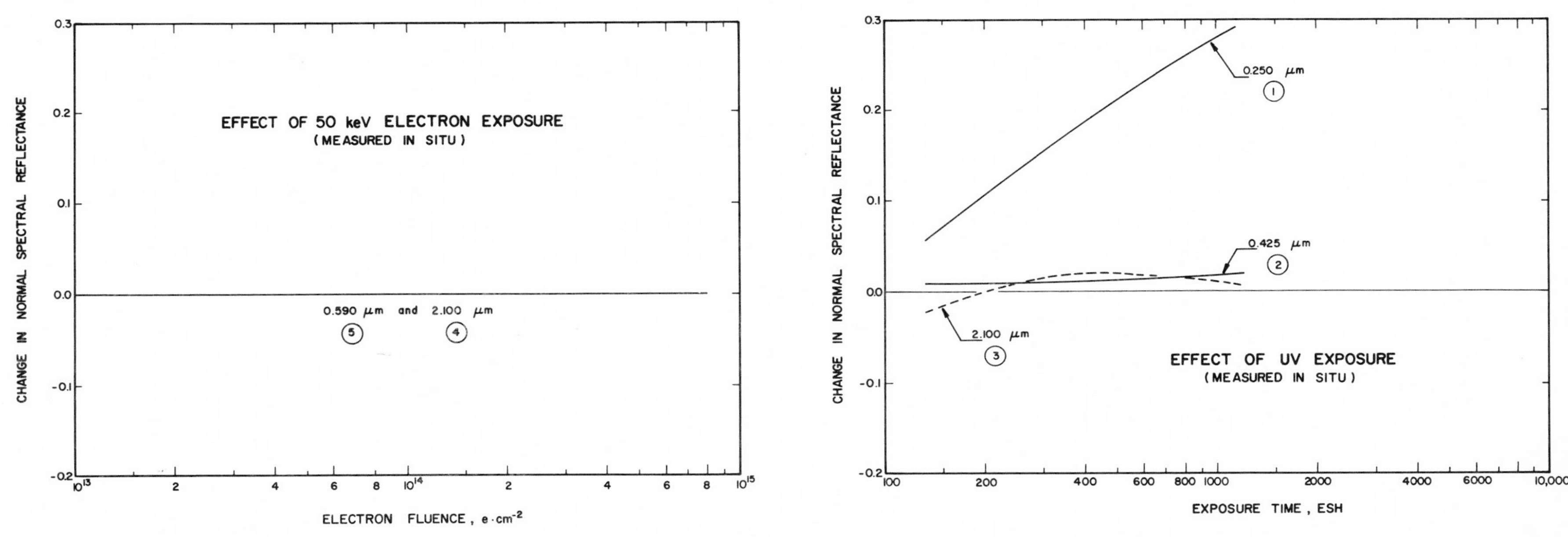

SPECIFICATION TABLE NO. 237 CHANGE IN NORMAL SPECTRAL REFLECTANCE OF ALUMINUM CONTACT COATINGS

Curve No.	Ref. No.	Year	Temperature, K	Wavelength Range, μm	Geometry θ θ'	ω'	Reported Error, %	Composition (weight percent), Specifications, and Remarks
1	46	1969	295	0.250	~0°	2π		Aluminum on thermosetting lacquer (0.203 mm thick) substrate; vapor deposited; exposed to UV radiation (4.7 UV sun rate) in vacuum (10^{-8} mm Hg); vacuum maintained by ion pump; ESH is variable; reflectance measured in situ; positive change indicates decrease in reflectance from preirradiation, in air reflectance. [Authors' designation: J]
2	46	1969	295	0.425	~0°	2π		Above specimen and conditions.
3	46	1969	295	2.100	~0°	2π		Above specimen and conditions.
4	46	1969	295	2.100	~0°	2π		Aluminum on thermosetting lacquer (0.203 mm thick) substrate; vapor deposited; exposed to 50 keV electrons (4×10^{8} - 1.7×10^{12} e cm^{-2} sec^{-1}) in vacuum (10^{-8} mm Hg); vacuum maintained by ion pump; electron fluence (e cm^{-2}) is variable; reflectance measured in situ; positive change indicates decrease in reflectance from preirradiation, in air reflectance. [Authors' designation: J]
5	46	1969	295	0.590	~0°	2π		Above specimen and conditions.

DATA TABLE NO. 237 CHANGE IN NORMAL SPECTRAL REFLECTANCE OF ALUMINUM CONTACT COATINGS

[Wavelength, λ, μm; Reflectance, ρ; Temperature, T, K]

ESH	$\Delta\rho$
CURVE 1 $\lambda = 0.250$ T = 295	
135	0.06
250	0.13
490	0.21
1130	0.29
CURVE 2 $\lambda = 0.425$ T = 295	
135	0.01
250	0.01
490	0.01
1130	0.02
CURVE 3 $\lambda = 2.100$ T = 295	
135	-0.02
250	0.01
490	0.02
1130	0.01

electron fluence	$\Delta\rho$
CURVE 4	
2×10^{14}	0
6×10^{14}	0
CURVE 5 $\lambda = 0.590$ T = 295	
1×10^{13}	0
CURVE 5 (cont.)	
2×10^{14}	0
6×10^{14}	0
8×10^{14}	0

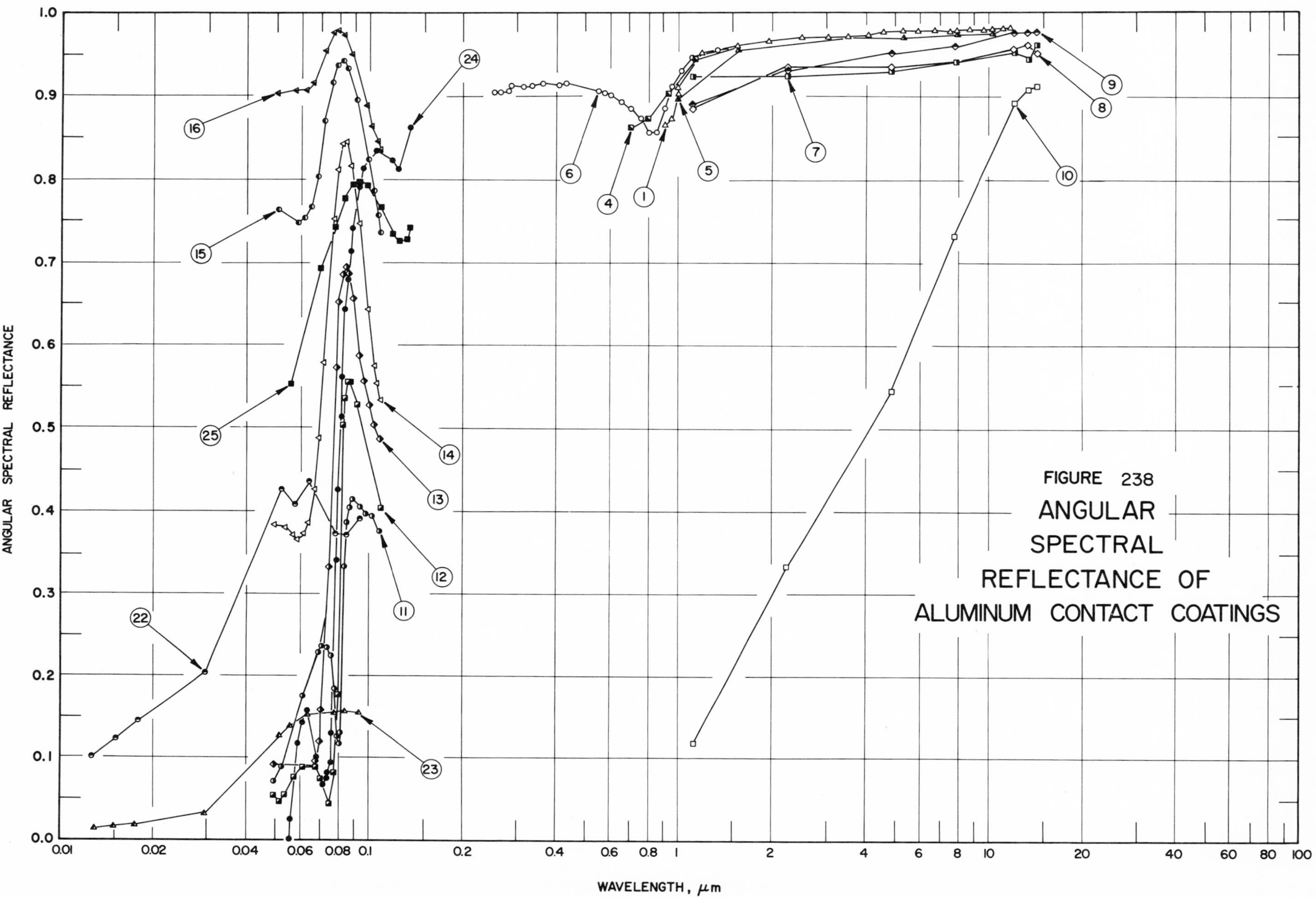
FIGURE 238
ANGULAR
SPECTRAL
REFLECTANCE OF
ALUMINUM CONTACT COATINGS
ANGULAR SPECTRAL REFLECTANCE
WAVELENGTH, μm
0.0
0.1
0.2
0.3
0.4
0.5
0.6
0.7
0.8
0.9
1.0
0.01
0.02
0.04
0.06
0.08
0.1
0.2
0.4
0.6
0.8
1
2
4
6
8
10
20
40
60
80
100
1
4
5
6
7
8
9
10
11
12
13
14
15
16
22
23
24
25

SPECIFICATION TABLE NO. 238 ANGULAR SPECTRAL REFLECTANCE OF ALUMINUM CONTACT COATINGS

Curve No.	Ref. No.	Year	Temperature, K	Wavelength Range, μm	Geometry θ	θ'	ω'	Reported Error, %	Composition (weight percent), Specifications, and Remarks
1	206	1958	~298	0.90-11.87	20°	20°			Aluminum; substrate unknown; vapor deposited.
2*	206	1958	~298	1.00-11.87	30°	30°			Above specimen and conditions.
3*	206	1958	~298	1.00-11.87	40°	40°			Above specimen and conditions.
4	206	1958	~298	0.700-11.815	50°	50°			Above specimen and conditions.
5	206	1958	~298	1.00-11.87	60°	60°			Above specimen and conditions.
6	207	1957	298	0.2537-1.3570	~20°	~20°		< 2	99.99 Aluminum (0.07 μm thick); fused quartz substrate; vapor deposited.
7	231	1962	~298	1.120-14.38	45°	45°			Aluminum (950 to 3000 Å thick); epoxy resin and stainless steel 316 substrates; stainless steel cleaned, etched, and vapor degreased, then coated with Maraset 617-C epoxy resin (Marblette Co.), applied by flowing the resin onto the steel at 408 K; plastic surface was cleaned by scrubbing in Alconox solution, then rinsed with water and methanol; Al vapor deposited on heated substrate in vacuum (~10^{-5} mm Hg); data shown is avg of two trials on the specimen. [Author's designation: Al 121]
8	231	1962	~298	1.120-14.38	45°	45°			Similar to above specimen and conditions. [Author's designation: Al 122]
9	231	1962	~298	1.120-14.38	45°	45°			Similar to curve 7 specimen and conditions except substrate consists of 316 stainless steel coated with SY627-119 polyurethane (Febert Shorndorfer Co.) by a dip process. [Author's designation: Al 123]
10	231	1962	~298	1.120-14.38	45°	45°			Similar to curve 7 specimen and conditions except substrate consists of uncoated, polished 316 stainless steel. [Author's designation: AlSS 1]
11	209	1968	~298	0.048-0.108	15°	15°			Aluminum(~150 Å thick); glass substrate; vapor deposited in vacuum (2 x 10^{-9} mm Hg); parallel-polarized incident light; data extracted from smooth curve.
12	209	1968	~298	0.049-0.108	30°	30°			Above specimen and conditions.
13	209	1968	~298	0.049-0.108	45°	45°			Above specimen and conditions.
14	209	1968	~298	0.049-0.108	60°	60°			Above specimen and conditions.
15	209	1968	~298	0.050-0.108	75°	75°			Above specimen and conditions.
16	209	1968	~298	0.050-0.108	82.5°	82.5°			Above specimen and conditions.
17*	209	1968	~298	0.051-0.109	30°	30°			Similar to curve 11 specimen and conditions.
18*	209	1968	~298	0.051-0.109	30°	30°			Above specimen and conditions except vented for 5 min at 10^{-7} mm Hg with O_2.
19*	209	1968	~298	0.051-0.109	30°	30°			Above specimen and conditions except vented for 5 min at 10^{-5} mm Hg with O_2.
20*	209	1968	~298	0.051-0.109	30°	30°			Above specimen and conditions except vented for 5 min at 10^{-2} mm Hg with O_2.
21*	209	1968	~298	0.051-0.109	30°	30°			Above specimen and conditions except vented for 5 min at 760 mm Hg with O_2.
22	210	1968	~298	0.0123-0.0934	75°	75°		±5	Aluminum (~2000 Å thick); glass substrate; rapid evaporation deposition in vacuum (10^{-6} mm Hg); exposed to air for several days.

* Not shown on plot

SPECIFICATION TABLE NO. 238 ANGULAR SPECTRAL REFLECTANCE OF ALUMINUM CONTACT COATINGS (continued)

Curve No.	Ref. No.	Year	Temperature, K	Wavelength Range, μm	Geometry θ	θ'	ω'	Reported Error, %	Composition (weight percent), Specifications, and Remarks
23	210	1968	~298	0.0128-0.0924	60°	60°		±5	Above specimen and conditions.
24	211	1966	~298	0.0552-0.135	10°	10°			Aluminum (~500 Å thick); glass substrate; vapor deposited.
25	211	1966	~298	0.0551-0.135	~45°	~45°			Aluminum (600 Å thick); Ag substrate.

DATA TABLE NO. 238 ANGULAR SPECTRAL REFLECTANCE OF ALUMINUM CONTACT COATINGS

[Wavelength, λ, μm; Reflectance, ρ; Temperature, T, K]

λ	ρ
CURVE 1 T ~ 298	
0.90	0.866
0.95	0.875
1.00	0.905
1.00	0.913
1.20	0.955
1.57	0.964
1.96	0.970
2.50	0.972
3.02	0.974
3.51	0.975
4.07	0.976
4.60	0.980
5.31	0.980
5.90	0.980
6.65	0.981
7.50	0.980
7.81	0.981
8.78	0.982
9.66	0.981
10.40	0.982
11.20	0.984
11.87	0.984
CURVE 2* T ~ 298	
1.00	0.906
1.57	0.966
3.51	0.977
5.31	0.980
7.81	0.981
10.40	0.982
11.87	0.984
CURVE 3* T ~ 298	
1.00	0.905
1.57	0.965
3.51	0.977
5.31	0.980
7.81	0.981
10.41	0.981
11.87	0.982
CURVE 4 T ~ 298	
0.700	0.864
0.800	0.875
0.930	0.905
1.156	0.949
1.570	0.965*
1.956	0.969*
2.494	0.971*
3.025	0.973*
4.068	0.977*
5.314	0.980*
6.646	0.981*
7.810	0.980*
8.785	0.982*
9.666	0.981*
11.815	0.982*
CURVE 5 T ~ 298	
1.00	0.900
1.57	0.960
3.51	0.972*
5.31	0.975
7.81	0.979
10.41	0.978
11.87	0.981*
CURVE 6 T = 298	
0.2537	0.906
0.2653	0.906
0.2804	0.909
0.2894	0.914
0.2968	0.913*
0.3022	0.912*
0.3132	0.912
0.3342	0.913
0.3650	0.916
0.4047	0.914
0.4358	0.916
0.5461	0.908
0.5780	0.906
0.6000	0.901
CURVE 6 (cont.)	
0.6500	0.894
0.7000	0.886
0.7500	0.874
0.8000	0.858
0.8500	0.859
0.9000	0.888
0.9500	0.912
1.014	0.931
1.129	0.949
1.357	0.958
CURVE 7 T ~ 298	
1.120	0.928
2.240	0.928
4.824	0.931
7.780	0.943
12.10	0.952
13.53	0.948
14.38	0.962
CURVE 8 T ~ 298	
1.120	0.886
2.240	0.939
4.824	0.936
7.780	0.942*
12.10	0.959
13.53	0.963
14.38	0.951
CURVE 9 T ~ 298	
1.120	0.891
2.240	0.938
4.824	0.953
7.780	0.962
12.10	0.978
13.53	0.978
14.38	0.978
CURVE 10 T ~ 298	
1.120	0.118
2.240	0.334
4.824	0.546
7.780	0.732
12.10	0.891
13.53	0.910
14.38	0.915
CURVE 11 T ~ 298	
0.049	0.069
0.052	0.088
0.061	0.175
0.068	0.226
0.070	0.235
0.073	0.233
0.075	0.223
0.077	0.184
0.079	0.124
0.080	0.115
0.081	0.129
0.083	0.333
0.084	0.387
0.086	0.405
0.088	0.413
0.093	0.406
0.097	0.398
0.102	0.396
0.108	0.378
CURVE 12 T ~ 298	
0.049	0.051
0.051	0.046
0.053	0.052
0.057	0.074
0.061	0.087
0.067	0.086
0.070	0.071
0.074	0.041
0.075	0.041*
0.077	0.080
CURVE 12 (cont.)	
0.079	0.177
0.082	0.504
0.083	0.536
0.085	0.556
0.086	0.556
0.091	0.528
0.108	0.403
CURVE 13 T ~ 298	
0.049	0.090
0.066	0.088*
0.067	0.095
0.069	0.119
0.070	0.159
0.072	0.234*
0.074	0.332
0.078	0.573
0.080	0.653
0.082	0.686
0.084	0.693
0.085	0.688
0.088	0.658
0.093	0.589
0.096	0.556
0.100	0.526
0.104	0.503
0.108	0.488
CURVE 14 T ~ 298	
0.049	0.385
0.053	0.381
0.056	0.371
0.058	0.367
0.061	0.373
0.063	0.388
0.066	0.426
0.068	0.488
0.071	0.580
0.076	0.751
0.079	0.811
0.082	0.842
CURVE 14 (cont.)	
0.084	0.845
0.088	0.818
0.093	0.748
0.099	0.644
0.104	0.578
0.106	0.555
0.108	0.531
CURVE 15 T ~ 298	
0.050	0.763
0.058	0.749
0.061	0.752
0.064	0.769
0.067	0.804
0.071	0.870
0.075	0.916
0.078	0.937
0.082	0.942
0.085	0.934
0.091	0.895
0.099	0.825
0.103	0.787
0.106	0.758
0.108	0.736
CURVE 16 T ~ 298	
0.050	0.903
0.057	0.906
0.062	0.907
0.065	0.916
0.072	0.955
0.076	0.976
0.079	0.980
0.083	0.975
0.088	0.951
0.098	0.890
0.102	0.864
0.106	0.847
0.108	0.836
CURVE 17* T ~ 298	
0.051	0.047
0.053	0.053
0.056	0.074
0.060	0.087
0.064	0.093
0.067	0.091
0.070	0.079
0.073	0.056
0.074	0.039
0.075	0.037
0.076	0.046
0.077	0.104
0.079	0.187
0.082	0.450
0.083	0.517
0.084	0.538
0.085	0.551
0.086	0.553
0.089	0.537
0.102	0.447
0.106	0.420
0.109	0.407
CURVE 18* T ~ 298	
0.051	0.047
0.053	0.053
0.056	0.074
0.060	0.087
0.064	0.093
0.067	0.091
0.070	0.079
0.073	0.056
0.074	0.039
0.075	0.037
0.076	0.046
0.077	0.104
0.079	0.187
0.082	0.450
0.083	0.490
0.084	0.520
0.085	0.526
0.086	0.523
CURVE 18 (cont.)*	
0.095	0.460
0.099	0.433
0.105	0.398
0.109	0.374
CURVE 19* T ~ 298	
0.051	0.032
0.055	0.046
0.058	0.063
0.061	0.073
0.065	0.075
0.069	0.064
0.072	0.043
0.075	0.026
0.076	0.025
0.077	0.045
0.078	0.072
0.079	0.144
0.082	0.283
0.084	0.373
0.085	0.389
0.087	0.393
0.096	0.357
0.102	0.337
0.107	0.325
0.109	0.320
CURVE 20* T ~ 298	
0.051	0.022
0.057	0.042
0.061	0.055
0.065	0.058
0.068	0.051
0.072	0.036
0.074	0.021
0.076	0.019
0.077	0.033
0.078	0.056
0.079	0.088
0.082	0.182
0.083	0.218

* Not shown on plot

DATA TABLE NO. 238 ANGULAR SPECTRAL REFLECTANCE OF ALUMINUM CONTACT COATINGS (continued)

λ	ρ
CURVE 20 (cont.)*	
T ~ 298	
0.085	0.240
0.086	0.245
0.091	0.239
0.096	0.229
0.103	0.223
0.109	0.218
CURVE 21*	
T ~ 298	
0.051	0.014
0.058	0.035
0.063	0.044
0.067	0.042
0.071	0.033
0.075	0.007
0.076	0.009
0.078	0.020
0.079	0.042
0.080	0.081
0.082	0.131
0.084	0.158
0.086	0.170
0.109	0.160
CURVE 22	
T ~ 298	
0.0123	0.100
0.0150	0.120
0.0179	0.144
0.0297	0.204
0.0514	0.429
0.0561	0.409
0.0630	0.438
0.0766	0.374
0.0840	0.371
0.0934	0.391
CURVE 23	
T ~ 298	
0.0128	0.011
0.0149	0.013
0.0173	0.016
0.0296	0.030

λ	ρ
CURVE 23 (cont.)	
0.0505	0.125
0.0558	0.138
0.0631	0.151
0.0761	0.155
0.0834	0.156
0.0924	0.155
CURVE 24	
T ~ 298	
0.0552	0.000
0.0556	0.024
0.0594	0.116
0.0607	0.140
0.0629	0.157
0.0666	0.100
0.0689	0.074*
0.0708	0.066
0.0726	0.072
0.0735	0.080
0.0746	0.092
0.0757	0.128
0.0780	0.342
0.0790	0.426
0.0802	0.513
0.0813	0.562
0.0833	0.644
0.0849	0.680
0.0862	0.713
0.0880	0.742
0.0924	0.791
0.0954	0.814
0.0994	0.825*
0.105	0.833
0.109	0.833*
0.118	0.824
0.124	0.814
0.135	0.861
CURVE 25	
T ~ 298	
0.0551	0.553
0.0690	0.692
0.0761	0.742
0.0834	0.778
0.0882	0.794

λ	ρ
CURVE 25 (cont.)	
0.0930	0.799
0.0994	0.793
0.109	0.768
0.118	0.735
0.124	0.726
0.131	0.729
0.135	0.742

* Not shown on plot

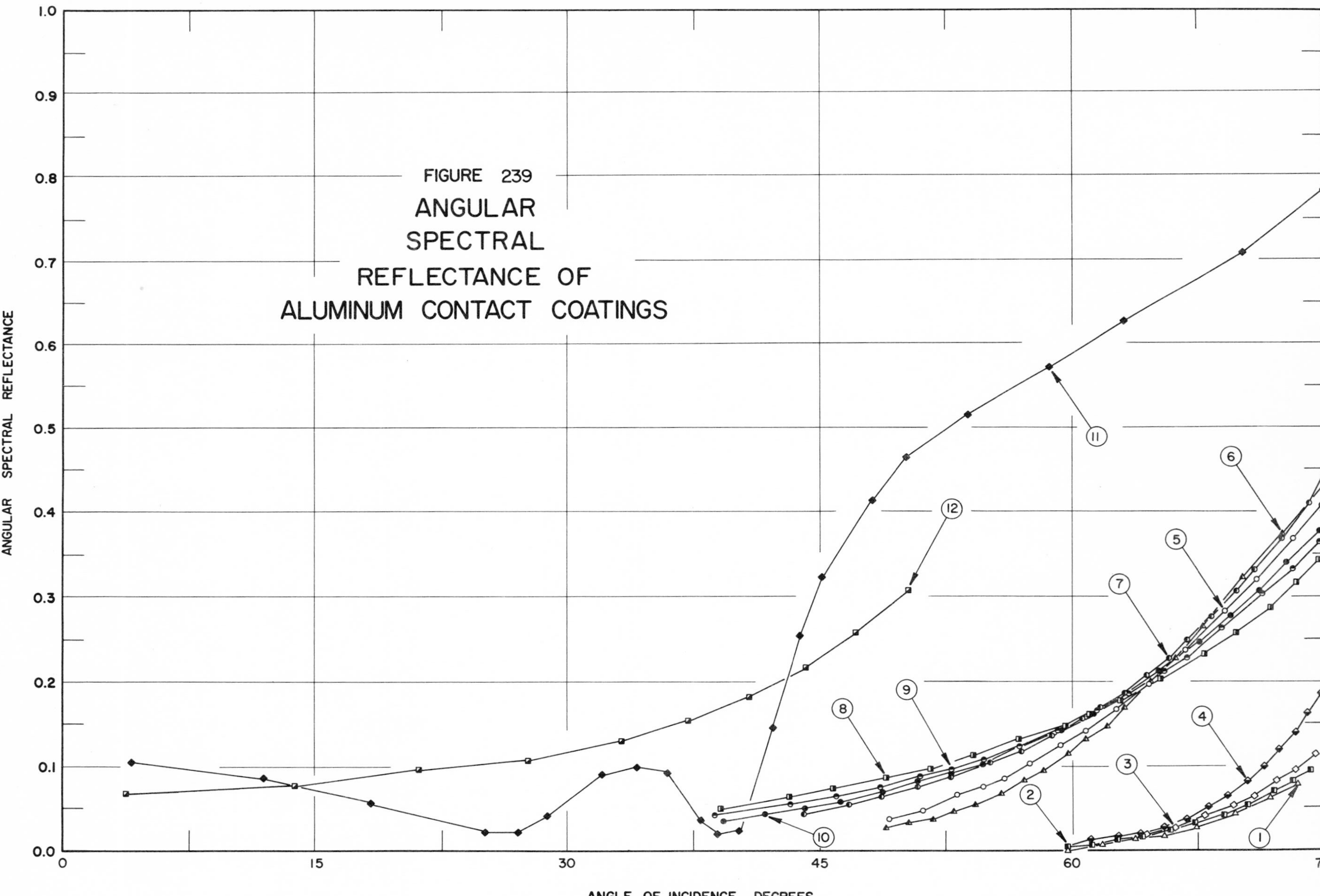

FIGURE 239 ANGULAR SPECTRAL REFLECTANCE OF ALUMINUM CONTACT COATINGS

SPECIFICATION TABLE NO. 239 ANGULAR SPECTRAL REFLECTANCE OF ALUMINUM CONTACT COATINGS

Curve No.	Ref. No.	Year	Temperature, K	Wavelength, µm	Angular Range,°	Geometry θ	θ'	ω'	Reported Error, %	Composition (weight percent), Specifications and Remarks
1	210	1968	~298	0.0123	59.8-76.8	θ	θ		±5	Aluminum (~2000 Å thick); glass substrate; rapid evaporation deposition in vacuum (10^{-6} mm Hg); $\theta = \theta'$; exposed to air for several days; data extracted from smooth curve.
2	210	1968	~298	0.0150	59.8-76.9	θ	θ		±5	Curve 1 specimen and conditions.
3	210	1968	~298	0.0173	59.8-76.8	θ	θ		±5	Curve 1 specimen and conditions.
4	210	1968	~298	0.0248	59.8-75.4	θ	θ		±5	Curve 1 specimen and conditions.
5	210	1968	~298	0.0555	49.1-76.3	θ	θ		±5	Curve 1 specimen and conditions.
6	210	1968	~298	0.0508	49.0-75.1	θ	θ		±5	Curve 1 specimen and conditions.
7	210	1968	~298	0.0630	44.1-76.5	θ	θ		±5	Curve 1 specimen and conditions.
8	210	1968	~298	0.0760	39.2-75.8	θ	θ		±5	Curve 1 specimen and conditions.
9	210	1968	~298	0.0835	38.9-76.2	θ	θ		±5	Curve 1 specimen and conditions.
10	210	1968	~298	0.0923	39.4-76.2	θ	θ		±5	Curve 1 specimen and conditions.
11	212	1961	~300	0.0584	4.2-87.2	θ	θ		1	Aluminum (~700 Å thick); glass substrate; fast-fired in vacuum; $\theta' = \theta$; measured in vacuum (~5 x 10^{-6} mm Hg); unexposed to air.
12	212	1961	~300	0.0584	3.8-87.6	θ	θ		1	Similar to above specimen and conditions except ~350 Å thick.

DATA TABLE NO. 239 ANGULAR SPECTRAL REFLECTANCE OF ALUMINUM CONTACT COATINGS

[Angle, θ, °; Reflectance, ρ; Temperature, T, K; Wavelength, λ, μm]

θ	ρ
CURVE 1	
T ~ 298	
λ = 0.0123	
59.8	0.000
61.8	0.007
63.7	0.012
65.5	0.019
67.5	0.029
69.7	0.044
71.9	0.062
73.5	0.080
75.1	0.102*
76.0	0.113*
76.8	0.126*
CURVE 2	
T ~ 298	
λ = 0.0150	
59.8	0.004
61.3	0.007
62.7	0.011
64.2	0.016
65.9	0.024
67.4	0.032
69.0	0.043
70.5	0.054
72.0	0.070
73.1	0.082
74.2	0.096
75.3	0.112*
76.0	0.122*
76.9	0.139*
CURVE 3	
T ~ 298	
λ = 0.0173	
59.8	0.007*
61.3	0.009*
62.9	0.012*
64.5	0.020
66.2	0.029
67.9	0.041
69.6	0.053
70.9	0.065
CURVE 3 (cont.)	
72.2	0.082
73.4	0.097
74.5	0.113
75.5	0.128*
76.1	0.141*
76.8	0.156*
CURVE 4	
T ~ 298	
λ = 0.0248	
59.8	0.008*
61.1	0.011
62.7	0.015
64.1	0.020
65.5	0.029
66.9	0.039
68.1	0.051
69.3	0.065
70.5	0.083
71.5	0.100
72.4	0.120
73.3	0.141
74.0	0.163
74.8	0.189
75.4	0.214*
CURVE 5	
T ~ 298	
λ = 0.0555	
49.1	0.039
51.1	0.049
53.2	0.065
54.7	0.075
56.0	0.086
57.5	0.102
59.4	0.124
60.8	0.142
62.6	0.169
64.6	0.198
66.8	0.238
69.2	0.282
71.0	0.320
73.2	0.368
CURVE 5 (cont.)	
74.9	0.409
76.3	0.446*
CURVE 6	
T ~ 298	
λ = 0.0508	
49.0	0.028
50.4	0.032
51.7	0.039
53.0	0.047
54.3	0.055
55.8	0.067
57.3	0.083
58.4	0.096
59.8	0.115
60.9	0.131
62.1	0.149
63.1	0.170
64.7	0.200
66.1	0.227
67.8	0.266
70.2	0.322
72.5	0.372
75.1	0.436
CURVE 7	
T ~ 298	
λ = 0.0630	
44.1	0.042
46.8	0.053
48.7	0.063
50.8	0.074
52.9	0.088
55.2	0.103
57.0	0.119
58.9	0.137
60.6	0.158
61.7	0.170
63.1	0.187
64.5	0.208
65.8	0.228
66.9	0.250
68.4	0.277
CURVE 7 (cont.)	
69.9	0.307
70.9	0.331
72.5	0.369
74.1	0.410
75.4	0.445
76.5	0.480*
CURVE 8	
T ~ 298	
λ = 0.0760	
39.2	0.050
43.3	0.063
45.8	0.073
49.0	0.086
51.6	0.098
54.1	0.112
56.8	0.131
59.6	0.149
61.1	0.162
62.9	0.178
65.3	0.202
67.9	0.232
69.8	0.257
71.8	0.288
73.4	0.316
74.6	0.342
75.8	0.369*
CURVE 9	
T ~ 298	
λ = 0.0835	
38.9	0.043
43.3	0.055
46.0	0.064
48.6	0.074
51.0	0.087
52.8	0.097
54.8	0.109
56.9	0.123
59.2	0.141
61.5	0.166
63.4	0.187
65.5	0.211
CURVE 9 (cont.)	
66.9	0.229
68.9	0.261
71.4	0.302
73.1	0.331
74.7	0.363
76.2	0.393*
CURVE 10	
T ~ 298	
λ = 0.0923	
39.4	0.035
41.8	0.042
44.2	0.050
46.3	0.058
48.7	0.069
50.8	0.081
52.9	0.091
54.8	0.104
57.0	0.122*
59.5	0.144
61.4	0.165
63.4	0.189*
65.2	0.211
67.6	0.246
69.5	0.277
71.2	0.306
72.8	0.340
74.7	0.378
76.2	0.410*
CURVE 11	
T ~ 300	
λ = 0.0584	
4.2	0.106
12.0	0.086
18.4	0.057
25.1	0.021
27.2	0.021
28.9	0.041
32.1	0.090
34.1	0.100
36.0	0.091
38.0	0.036
CURVE 11 (cont.)	
39.0	0.020
40.3	0.024
42.4	0.148
43.9	0.256
45.2	0.325
48.2	0.414
50.3	0.462
53.9	0.519
58.7	0.571
63.2	0.627
70.2	0.710
75.0	0.781
80.4	0.858*
87.2	0.962*
CURVE 12	
T ~ 300	
λ = 0.0584	
3.8	0.068
13.9	0.077
21.2	0.092
27.6	0.109
33.3	0.130
37.3	0.153
40.8	0.183
44.2	0.219
47.1	0.259
50.4	0.309
80.6	0.820*
87.6	0.959*

* Not shown on plot

SPECIFICATION TABLE NO. 240 HEMISPHERICAL INTEGRATED ABSORPTANCE OF ALUMINUM CONTACT COATINGS

Curve No.	Ref. No.	Year	Temperature Range, K	Reported Error, %	Composition (weight percent), Specifications and Remarks
1*	195	1953	76		Aluminum; mylar (0.0127 mm thick) and aluminum substrates; Al vapor deposited on both sides of mylar; measured in vacuum ($<10^{-6}$ mm Hg); absorptance for 294 K blackbody incident radiation.
2*	195	1953	76		Similar to above specimen and conditions.
3*	194	1955	76	5	Aluminum; mylar (0.127 mm thick) and aluminum substrates; Al vaporized on both sides of mylar; measured in vacuum (10^{-6} to 10^{-7} mm Hg); absorptance for 300 K blackbody incident radiation.
4*	194	1955	76	5	Aluminum; stainless steel substrate; sprayed; measured in vacuum (10^{-6} to 10^{-7} mm Hg); absorptance for 300 K blackbody incident radiation.
5*	194	1955	76	5	Aluminum; stainless steel substrate; sprayed; wire brushed; measured in vacuum (10^{-6} to 10^{-7} mm Hg); absorptance for 300 K blackbody incident radiation.

DATA TABLE NO. 240 HEMISPHERICAL INTEGRATED ABSORPTANCE OF ALUMINUM CONTACT COATINGS

[Temperature, T, K; Absorptance, α]

T	α
CURVE 1*	
76	0.040
CURVE 2*	
76	0.043
CURVE 3*	
76	0.04
CURVE 4*	
76	0.07
CURVE 5*	
76	0.06

* No plot given

SPECIFICATION TABLE NO. 241 NORMAL SOLAR ABSORPTANCE OF ALUMINUM CONTACT COATINGS

Curve No.	Ref. No.	Year	Temperature Range, K	Geometry θ	Reported Error, %	Composition (weight percent), Specifications and Remarks
1*	115	1960	298	15°		Aluminum (0.254 to 0.381 μm thick); thermo-setting epoxy resin (0.0254 to 0.0381 mm thick) and aluminum substrates; Al deposited by vacuum evaporation; computed from spectral reflectance data for above atm conditions. [Authors' designation: Sample 5]
2*	115	1960	298	15°		Similar to above specimen and conditions except exposed to vacuum (10^{-6} mm Hg) for 6 days and cleaned with isopropyl alcohol. [Authors' designation: Sample 6]
3*	115	1960	298	15°		Similar to curve 1 specimen and conditions except peel coat applied and removed; cleaned with isopropyl alcohol. [Authors' designation: Sample 7]
4*	100	1962	300	~0°		Aluminum (0.127 mm thick); FEP teflon (type A) substrate.
5*	100	1962	300	~0°		Above specimen and conditions except exposed to 384 hrs of simulated solar (10 x solar intensity) radiation; data indicates maximum degradation which may occur during a Mariner mission.

DATA TABLE NO. 241 NORMAL SOLAR ABSORPTANCE OF ALUMINUM CONTACT COATINGS

[Temperature, T, K; Absorptance, α]

T	α
CURVE 1*	
298	0.10
CURVE 2*	
298	0.10
CURVE 3*	
298	0.09
CURVE 4*	
300	0.26
CURVE 5*	
300	0.31

* No plot given

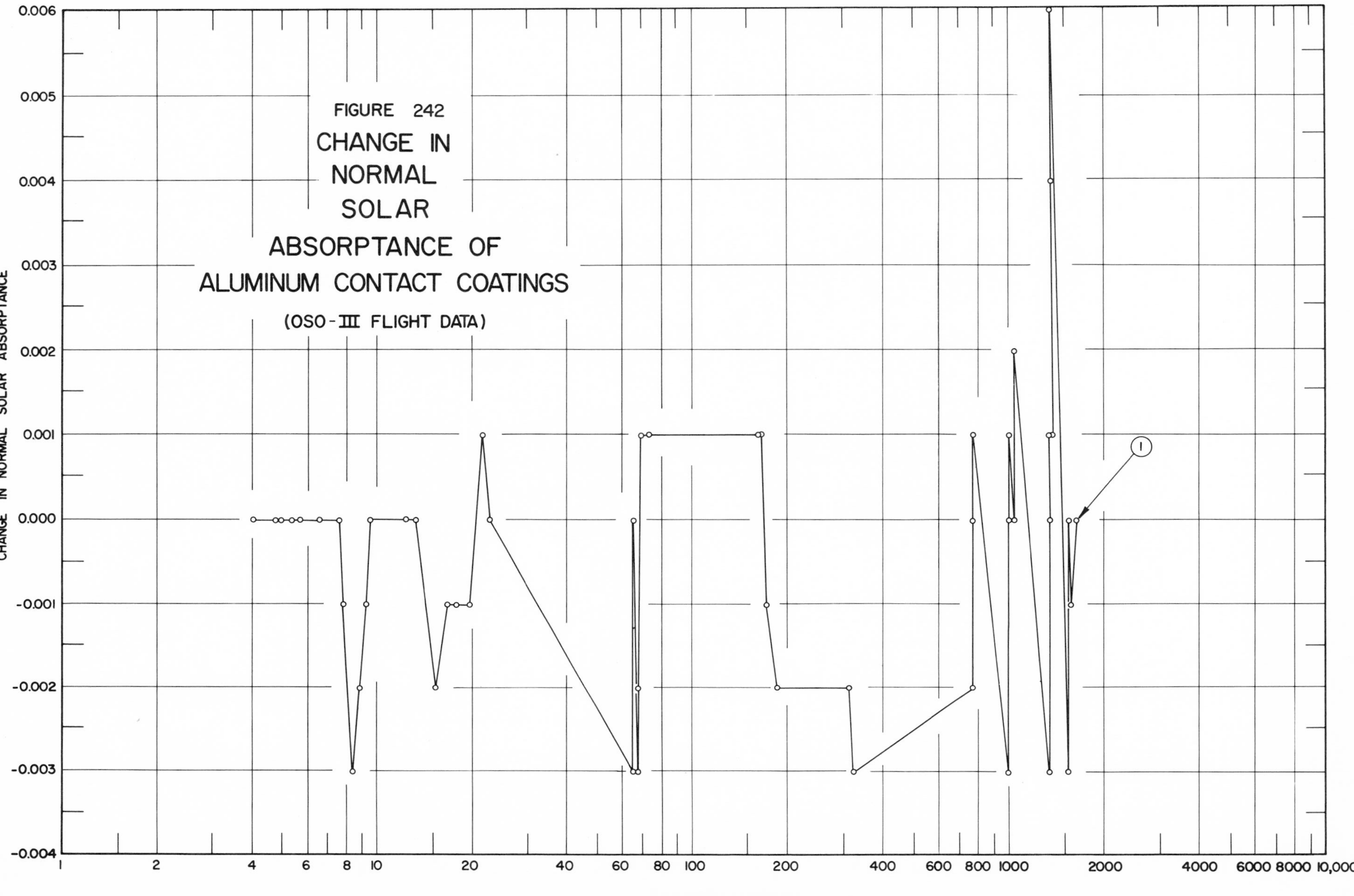

FIGURE 242
CHANGE IN NORMAL SOLAR ABSORPTANCE OF ALUMINUM CONTACT COATINGS
(OSO-III FLIGHT DATA)

SPECIFICATION TABLE NO. 242 CHANGE IN NORMAL SOLAR ABSORPTANCE OF ALUMINUM CONTACT COATINGS

Curve No.	Ref. No.	Year	Temperature Range, K	Geometry θ	Reported Error, %	Composition (weight percent), Specifications and Remarks
1	65	1969	~300	~0°		Aluminum (2×10^{-4} mm thick); 302 stainless steel substrate; Al vapor deposited; property calculated from temp of substrate from in-flight data of OSO III; ESH is variable. [Author's designation: 2000 Å Al]

DATA TABLE NO. 242 CHANGE IN NORMAL SOLAR ABSORPTANCE OF ALUMINUM CONTACT COATINGS

[Temperature, T, K; Absorptance, α]

ESH	$\Delta\alpha$	ESH	$\Delta\alpha$	ESH	$\Delta\alpha$
CURVE 1		CURVE 1 (cont.)		CURVE 1 (cont.)	
T ~ 300		19.9	-0.001	1000.	0.001
0.48	0.001*	21.8	0.001	1032.	0.000
4.0	0.000	22.7	0.000	1032.	0.002
4.7	0.000	65.8	-0.003	1355.	-0.003
4.9	0.000	65.8	0.000	1355.	0.000
5.3	0.000	67.3	-0.003	1355.	0.001
5.6	0.000	67.3	-0.002	1367.	0.001
6.5	0.000	68.9	0.001	1367.	0.004
7.5	0.000	73.8	0.001	1367.	0.006
7.8	-0.001	163.	0.000	1584.	-0.003
8.3	-0.003	168.	0.001	1592.	-0.001
8.8	-0.002	173.	-0.001	1584.	0.000
9.2	-0.001	186.	-0.002	1629.	0.000
9.5	0.000	313.	-0.002		
12.1	0.000	322.	-0.003		
13.1	0.000	765.	-0.002		
15.1	-0.002	765.	0.000		
16.8	-0.001	765.	0.001		
17.9	-0.001	1000.	-0.003		
		1000.	0.000		

* Not shown on plot

SPECIFICATION TABLE NO. 243 NORMAL SPECTRAL REFLECTANCE OF ALUMINUM + MAGNESIUM CONTACT COATINGS

Curve No.	Ref. No.	Year	Temperature, K	Wavelength Range, μm	Geometry θ θ' ω'	Reported Error, %	Composition (weight percent), Specifications, and Remarks
1*	205	1950	298	0.255-0.490	$\sim 0°$ $\sim 0°$	± 2	97 Al and 3 Mg (α phase); glass substrate; vapor deposited.
2*	205	1950	298	0.245-0.424	$\sim 0°$ $\sim 0°$	± 2	63 Al and 37 Mg (β phase); glass substrate; vapor deposited.

DATA TABLE NO. 243 NORMAL SPECTRAL REFLECTANCE OF ALUMINUM + MAGNESIUM CONTACT COATINGS

[Wavelength, λ, μm; Reflectance, ρ; Temperature, T, K]

λ	ρ
CURVE 1*	
T = 298	
0.255	0.660
0.268	0.670
0.280	0.720
0.306	0.755
0.375	0.780
0.490	0.785
CURVE 2*	
T = 298	
0.245	0.780
0.255	0.785
0.265	0.790
0.280	0.790
0.305	0.780
0.345	0.770
0.377	0.770
0.424	0.770

* No plot given

SPECIFICATION TABLE NO. 244 ANGULAR SPECTRAL REFLECTANCE OF ANTIMONY CONTACT COATINGS

Curve No.	Ref. No.	Year	Temperature, K	Wavelength Range, μm	Geometry θ	θ'	ω'	Reported Error, %	Composition (weight percent), Specifications, and Remarks
1*	213	1961	298	0.0453-0.1642	20°	20°			Sb (860 Å thick); glass substrate; vapor deposited in vacuum (8×10^{-5} mm Hg); measured in vacuum

DATA TABLE NO. 244 ANGULAR SPECTRAL REFLECTANCE OF ANTIMONY CONTACT COATINGS

[Wavelength, λ, μm; Reflectance, ρ; Temperature, T, K]

λ	ρ	λ	ρ	λ	ρ	λ	ρ
CURVE 1*		CURVE 1 (cont.)*		CURVE 1 (cont.)*		CURVE 1 (cont.)*	
T = 298							
		0.0708	0.040	0.0979	0.225	0.1450	0.415
0.0453	0.011	0.0721	0.034	0.0983	0.241	0.1483	0.420
0.0474	0.015	0.0726	0.040	0.1024	0.258	0.1501	0.418
0.0484	0.016	0.0750	0.059	0.1036	0.271	0.1544	0.419
0.0490	0.009	0.0765	0.070	0.1060	0.276	0.1544	0.429
0.0504	0.005	0.0771	0.103	0.1123	0.323	0.1606	0.429
0.0509	0.005	0.0791	0.101	0.1142	0.320	0.1634	0.427
0.0531	0.014	0.0832	0.128	0.1171	0.335	0.1642	0.429
0.0558	0.021	0.0834	0.151	0.1176	0.341		
0.0568	0.026	0.0852	0.154	0.1190	0.350		
0.0577	0.017	0.0859	0.156	0.1200	0.353		
0.0592	0.011	0.0880	0.161	0.1224	0.355		
0.0617	0.005	0.0881	0.176	0.1248	0.364		
0.0640	0.013	0.0902	0.178	0.1267	0.376		
0.0644	0.019	0.0926	0.181	0.1282	0.387		
0.0667	0.040	0.0927	0.203	0.1367	0.405		
0.0689	0.040	0.0932	0.218	0.1392	0.413		
0.0699	0.040	0.0942	0.209	0.1406	0.411		
0.0705	0.034	0.0963	0.229	0.1420	0.419		

* No plot given

SPECIFICATION TABLE NO. 245 NORMAL SPECTRAL TRANSMITTANCE OF ANTIMONY CONTACT COATINGS

Curve No.	Ref. No.	Year	Temperature, K	Wavelength Range, μm	Geometry θ θ' ω'	Reported Error, %	Composition (weight percent), Specifications, and Remarks
1*	213	1961	298	0.0454-0.0833	~0° ~0°		Sb (1330 Å thick); stilbene crystal substrate; vapor deposited in vacuum (8×10^{-5} mm Hg); measured in vacuum.
2*	214	1965	298	0.0288-0.0701	~0° ~0°		Sb, 99.9 pure (1200 Å thick): 99.999 pure Al (400 Å thick) substrate; Sb evaporated on Al; all evaporations were made at 2×10^{-5} mm Hg and completed within 10-15 sec; above 0.035 μm the error is approx 10%, while below this wavelength, error high as 25% has been observed.

DATA TABLE NO. 245 NORMAL SPECTRAL TRANSMITTANCE OF ANTIMONY CONTACT COATINGS

[Wavelength, λ, μm; Transmittance, τ; Temperature, T, K]

λ	τ	λ	τ	λ	τ	λ	τ
CURVE 1*		CURVE 1 (cont.)*		CURVE 2 (cont.)*		CURVE 2 (cont.)*	
T = 298							
		0.0763	0.006	0.0380	0.020	0.0579	0.062
0.0454	0.214	0.0790	0.005	0.0394	0.189	0.0602	0.039
0.0478	0.209	0.0812	0.003	0.0399	0.219	0.0608	0.032
0.0485	0.192	0.0829	0.006	0.0404	0.245	0.0614	0.026
0.0492	0.165	0.0833	0.004	0.0423	0.253	0.0626	0.025
0.0506	0.157			0.0434	0.231	0.0629	0.024
0.0512	0.144	CURVE 2*		0.0438	0.220	0.0645	0.018
0.0528	0.127	T = 298		0.0453	0.185	0.0674	0.013
0.0532	0.121			0.0464	0.178	0.0681	0.006
0.0557	0.106	0.0288	0.001	0.0473	0.184	0.0701	0.003
0.0568	0.093	0.0302	0.044	0.0481	0.172		
0.0578	0.078	0.0308	0.051	0.0484	0.154		
0.0592	0.064	0.0320	0.076	0.0492	0.121		
0.0639	0.047	0.0328	0.037	0.0508	0.118		
0.0646	0.042	0.0341	0.000	0.0525	0.099		
0.0690	0.023	0.0350	0.000	0.0532	0.099		
0.0719	0.014	0.0356	0.000	0.0539	0.090		
0.0725	0.012	0.0364	0.000	0.0556	0.070		
0.0748	0.008	0.0368	0.000	0.0574	0.076		

* No plot given

SPECIFICATION TABLE NO. 246 HEMISPHERICAL TOTAL EMITTANCE OF BARIUM + STRONTIUM CONTACT COATINGS

Curve No.	Ref. No.	Year	Temperature Range, K	Reported Error, %	Composition (weight percent), Specifications and Remarks
1*	97	1961	923-1273	±2.5	Barium + Strontium; nickel substrate; Marconi-Osram Valve Co. (G.E.C.) coating; specification S.J.C. 123; coating area density 6 mg cm^{-2}; substrate standard cathode nickel (>98.6 Ni, 0.87 Co, 0.20 Si, 0.18 Fe, 0.06 Mg, 0.05 Mn, 0.04 Cu, < 0.04 C, < 0.007 S); measured in vacuum (< 5 x 10^{-6} mm Hg); data extracted from smooth curve.
2*	97	1961	923-1273	±2.5	Similar to above specimen and conditions except area density 10 mg cm^{-2}.

DATA TABLE NO. 246 HEMISPHERICAL TOTAL EMITTANCE OF BARIUM + STRONTIUM CONTACT COATINGS

[Temperature, T, K; Emittance, ∈]

T	∈
CURVE 1*	
923	0.330
973	0.330
1023	0.330
1073	0.330
1123	0.330
1173	0.330
1223	0.330
1273	0.330
CURVE 2*	
923	0.420
973	0.420
1023	0.420
1073	0.420
1123	0.420
1173	0.420
1223	0.420
1273	0.420

* No plot given

SPECIFICATION TABLE NO. 247 NORMAL SPECTRAL TRANSMITTANCE OF CALCIUM CONTACT COATINGS

Curve No.	Ref. No.	Year	Temperature, K	Wavelength Range, μm	Geometry θ θ' ω'	Reported Error, %	Composition (weight percent), Specifications, and Remarks
1*	215	1962	~298	0.4-2.0	~0° ~0°		Calcium; glass (1.59 mm thick) substrate; prepared from solution of calcium resinate, diluted with aromatic solvents and essential oils; applied to a soda-lime glass substrate, then fired to 873 K in a continuous lehr on a 1.5 hr cycle; application of solution was by dropping an excess amount on the substrate, which was kept spinning at 1550 rpm, until no more solution was flung from the edges; τ is for coating plus substrate; substrate is 1.59 mm thick.

DATA TABLE NO. 247 NORMAL SPECTRAL TRANSMITTANCE OF CALCIUM CONTACT COATINGS

[Wavelength, λ, μm; Transmittance, τ; Temperature, T, K]

λ	τ
CURVE 1* T ~ 298	
0.4	0.160
0.5	0.300
0.6	0.416
0.7	0.516
0.8	0.586
0.9	0.640
1.0	0.680
1.1	0.706
1.2	0.716
1.3	0.724
1.4	0.597
1.5	0.434
1.6	0.344
1.8	0.268
2.0	0.250

* No plot given

SPECIFICATION TABLE NO. 248 HEMISPHERICAL TOTAL EMITTANCE OF CHROMIUM CONTACT COATINGS

Curve No.	Ref. No.	Year	Temperature Range, K	Reported Error, %	Composition (weight percent), Specifications and Remarks
1*	216	1948	90	< 20	Chromium (0.05 mm thick); copper substrate; electrolytic deposition; measured in vacuum (10^{-5} mm Hg).
2*	194	1955	76	5	Chromium; copper substrate; plated; measured in vacuum (10^{-6} to 10^{-7} mm Hg); emittance for 300 K blackbody incident radiation; authors assumed $\alpha = \epsilon$.
3*	217	1960	77		Chromium; Monel substrate; plated; bright; cleaned with acetone; measured in vacuum ($\sim 1 \times 10^{-5}$ mm Hg).

DATA TABLE NO. 248 HEMISPHERICAL TOTAL EMITTANCE OF CHROMIUM CONTACT COATINGS

[Temperature, T, K; Emittance, ϵ]

T	ϵ
CURVE 1*	
90	0.065
CURVE 2*	
76	0.08
CURVE 3*	
77	0.084

* No plot given

SPECIFICATION TABLE NO. 249 NORMAL TOTAL EMITTANCE OF CHROMIUM CONTACT COATINGS

Curve No.	Ref. No.	Year	Temperature Range, K	Geometry θ'	Reported Error, %	Composition (weight percent), Specifications and Remarks
1*	89	1958	366-672	~0°		Chromium from North American Aviation (0.0127 mm thick); 321 stainless steel substrate; plated. [Author's designation: Specimen 14]

DATA TABLE NO. 249 NORMAL TOTAL EMITTANCE OF CHROMIUM CONTACT COATINGS

[Temperature, T, K; Emittance, ϵ]

T	ϵ
CURVE 1*	
366	0.133
444	0.160
522	0.156
600	0.176
672	0.123

* No plot given

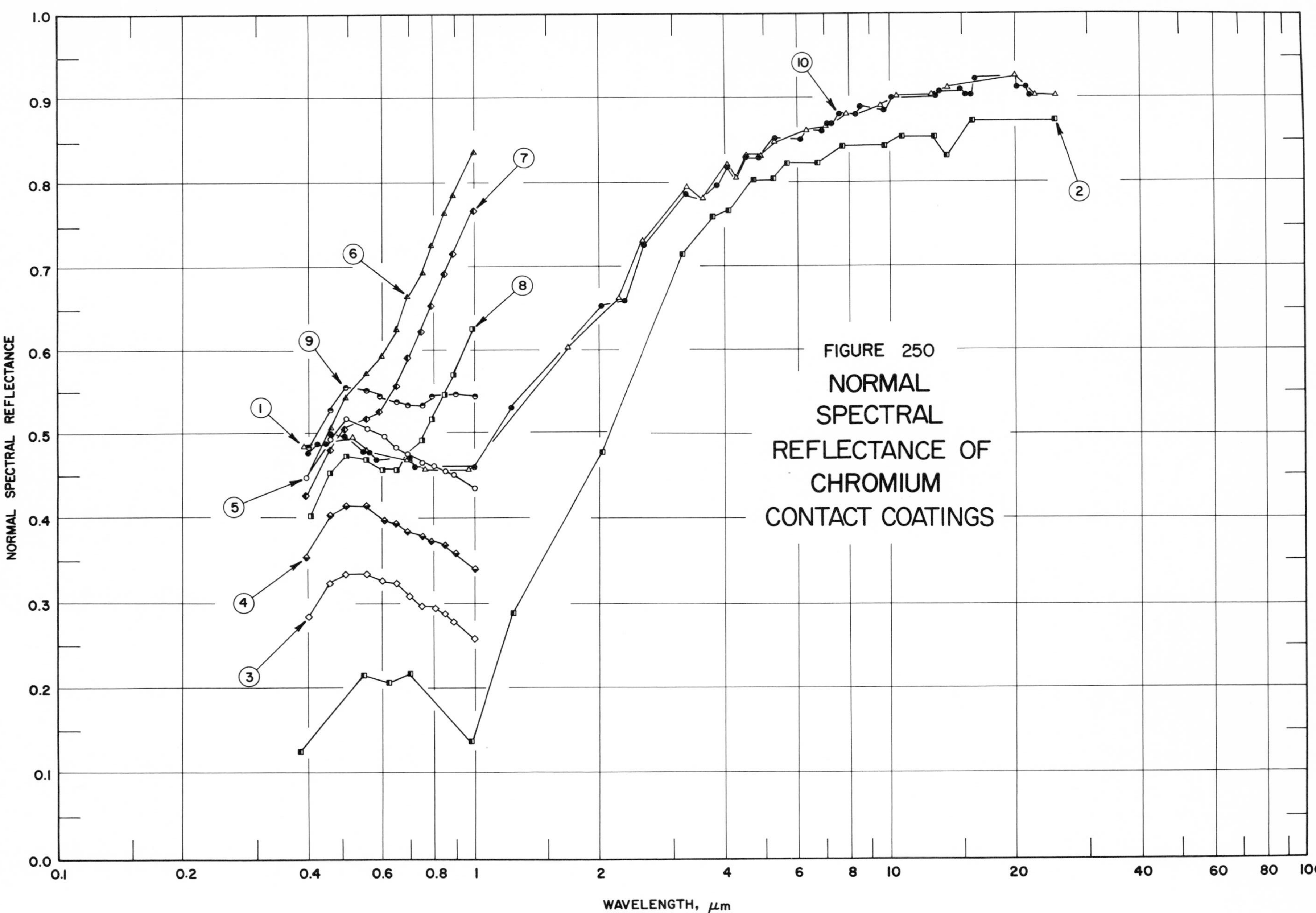
FIGURE 250
NORMAL
SPECTRAL
REFLECTANCE OF
CHROMIUM
CONTACT COATINGS
NORMAL SPECTRAL REFLECTANCE
WAVELENGTH, μm
1.0
0.9
0.8
0.7
0.6
0.5
0.4
0.3
0.2
0.1
0.0
0.1
0.2
0.4
0.6
0.8
1
2
4
6
8
10
20
40
60
80
100
1
2
3
4
5
6
7
8
9
10

SPECIFICATION TABLE NO. 250 NORMAL SPECTRAL REFLECTANCE OF CHROMIUM CONTACT COATINGS

Curve No.	Ref. No.	Year	Temperature, K	Wavelength Range, μm	Geometry θ	Geometry θ'	Geometry ω'	Reported Error, %	Composition (weight percent), Specifications, and Remarks
1	203	1962	298	0.390-25.0	~0°	~0°			Chromium; 321 stainless steel substrate; substrate as rolled; plated; data extracted from smooth curve.
2	203	1962	298	0.380-25.0	~0°	~0°			Chromium (0.00254 mm thick); nickel (0.0127 mm thick) and stainless steel 321 substrates; nickel plated on stainless steel, then chromium plated on nickel; exposed to JP-4 Combustion Products for 50 hrs at 865 K; data extracted from smooth curve.
3	218	1967	~298	0.400-1.000	~0°	~0°			Chromium (76 Å thick); wedgelike glass substrate; deposited by thermal sublimation in vacuum (10^{-4} mm Hg); deposition rate 10 Å sec^{-1}.
4	218	1967	~298	0.399-1.00	~0°	~0°			Above specimen and conditions except film thickness 105 Å.
5	218	1967	~298	0.399-1.00	~0°	~0°			Above specimen and conditions except film thickness 185 Å.
6	218	1967	~298	0.399-1.00	~0°	~0°			Chromium(76 Å thick); silver and glass substrates; deposited by thermal sublimation in vacuum (10^{-4} mm Hg); deposition rate 10 Å sec^{-1}; opaque silver film (~900 Å thick), deposited on wedgelike glass substrate; deposition rate of Ag film 20 Å sec^{-1}.
7	218	1967	~298	0.399-1.00	~0°	~0°			Above specimen and conditions except film thickness 105 Å.
8	218	1967	~298	0.402-0.998	~0°	~0°			Above specimen and conditions except film thickness 185 Å.
9	218	1967	~298	0.400-1.00	~0°	~0°			Above specimen and conditions except film thickness 760 Å.
10	89	1958	~298	0.400-25.0	~0°	~0°			Chromium from North American Aviation (0.0127 mm thick); 321 stainless steel substrate; plated; data extracted from smooth curve. [Author's designation: Sample 14]

DATA TABLE NO. 250 NORMAL SPECTRAL REFLECTANCE OF CHROMIUM CONTACT COATINGS

[Wavelength, λ, μm; Reflectance, ρ; Temperature, T, K]

CURVE 1
T = 298

λ	ρ
0.390	0.485
0.510	0.498
0.550	0.480
0.690	0.469
0.760	0.457
0.970	0.457
1.69	0.602
2.21	0.663
2.54	0.733
3.23	0.795
3.56	0.782
4.03	0.822
4.24	0.807
4.52	0.833
4.92	0.833
5.27	0.850
6.30	0.861
7.00	0.868
7.87	0.882
9.57	0.891
10.4	0.902
12.6	0.902
13.9	0.911
20.0	0.926
22.3	0.902
25.0	0.902

CURVE 2
T = 298

λ	ρ
0.380	0.125
0.540	0.215
0.620	0.206
0.700	0.218
0.980	0.136
1.24	0.289
2.01	0.478
3.19	0.716
3.75	0.760
4.10	0.769
4.70	0.802
5.21	0.805
5.61	0.822
6.72	0.822

CURVE 2 (cont.)

λ	ρ
7.73	0.844
9.77	0.844
10.6	0.855
12.7	0.855
13.7	0.830
15.9	0.874
25.0	0.874

CURVE 3
T ~ 298

λ	ρ
0.400	0.282
0.450	0.322
0.499	0.333
0.548	0.333
0.600	0.327
0.650	0.322
0.700	0.308
0.749	0.297
0.801	0.293
0.859	0.288
0.899	0.278
1.000	0.257

CURVE 4
T ~ 298

λ	ρ
0.399	0.353
0.450	0.403
0.498	0.415
0.548	0.415
0.601	0.399
0.650	0.393
0.698	0.383
0.750	0.379
0.799	0.372
0.858	0.369
0.900	0.358
1.00	0.340

CURVE 5
T ~ 298

λ	ρ
0.399	0.448
0.450	0.492

CURVE 5 (cont.)

λ	ρ
0.498	0.519
0.548	0.505
0.601	0.498
0.649	0.482
0.698	0.474
0.749	0.464
0.800	0.460
0.859	0.454
0.897	0.450
1.00	0.433

CURVE 6
T ~ 298

λ	ρ
0.399	0.447*
0.449	0.508
0.496	0.541
0.547	0.571
0.600	0.594
0.648	0.626
0.698	0.666
0.746	0.694
0.797	0.728
0.859	0.765
0.898	0.787
1.00	0.839

CURVE 7
T ~ 298

λ	ρ
0.399	0.426
0.451	0.480
0.497	0.506
0.548	0.517
0.598	0.526
0.649	0.557
0.698	0.591
0.749	0.621
0.799	0.653
0.859	0.691
0.898	0.717
1.00	0.768

CURVE 8
T ~ 298

λ	ρ
0.402	0.402
0.449	0.451
0.498	0.472
0.548	0.468
0.600	0.457
0.648	0.457
0.698	0.474*
0.748	0.491
0.798	0.517
0.858	0.546
0.899	0.570
0.998	0.628

CURVE 9
T ~ 298

λ	ρ
0.400	0.483
0.449	0.528
0.496	0.555
0.548	0.550
0.599	0.544
0.649	0.538
0.698	0.533
0.747	0.533
0.799	0.544
0.858	0.546*
0.900	0.546
1.00	0.543

CURVE 10
T ~ 298

λ	ρ
0.400	0.476
0.420	0.489
0.440	0.489
0.450	0.500
0.490	0.499
0.540	0.479
0.560	0.479
0.580	0.468
0.700	0.470
0.720	0.460
1.00	0.460
1.24	0.530

CURVE 10 (cont.)

λ	ρ
2.04	0.653
2.30	0.660
2.57	0.729
3.24	0.789
3.55	0.781*
3.85	0.799
4.01	0.820
4.24	0.806*
4.56	0.830
4.83	0.830
5.37	0.852
6.11	0.852
6.33	0.862*
6.84	0.861
7.09	0.870
7.32	0.870
7.60	0.881
8.36	0.881
8.57	0.890
9.79	0.889
10.1	0.900
12.9	0.902
13.1	0.909
14.9	0.910
15.2	0.902
15.6	0.902
16.0	0.922
20.0	0.922*
20.3	0.912
21.2	0.912
21.6	0.901
25.0	0.904*

* Not shown on plot

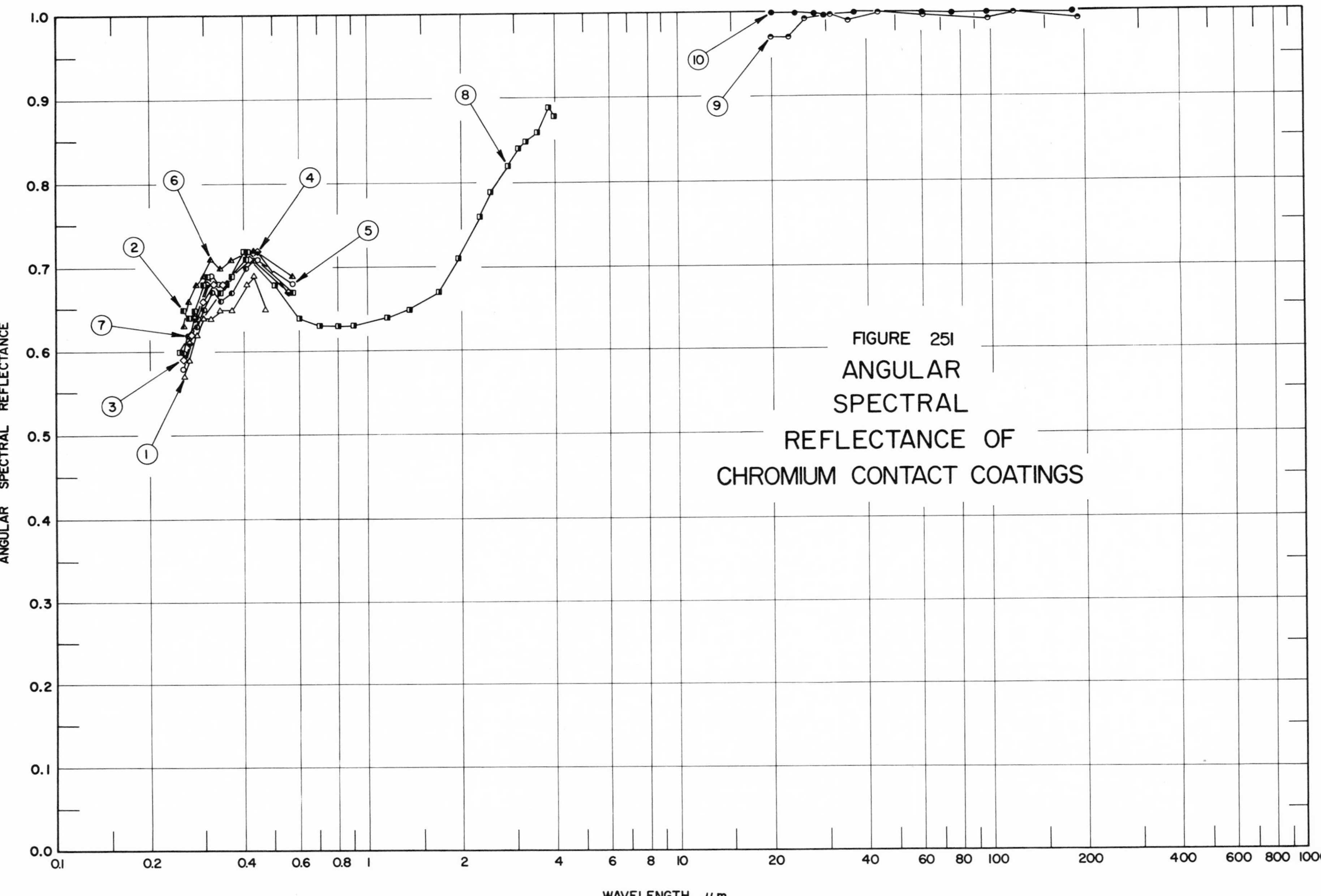

FIGURE 251 ANGULAR SPECTRAL REFLECTANCE OF CHROMIUM CONTACT COATINGS

SPECIFICATION TABLE NO. 251 ANGULAR SPECTRAL REFLECTANCE OF CHROMIUM CONTACT COATINGS

Curve No.	Ref. No.	Year	Temperature, K	Wavelength Range, μm	Geometry θ	θ'	ω'	Reported Error, %	Composition (weight percent), Specifications, and Remarks
1	219	1928	298	0.255-0.462	45°	45°			Chromium (0.127 mm thick); hardened, highly polished steel substrate; electroplated; highly polished and free from scratches.
2	219	1928	298	0.257-0.565	45°	45°			Similar to above specimen and conditions.
3	219	1928	298	0.257-0.417	45°	45°			Above specimen and conditions except exposed to UV for 30 hrs from Hg lamp.
4	219	1928	298	0.252-0.557	45°	45°			Similar to curve 1 specimen and conditions.
5	219	1928	298	0.255-0.562	45°	45°			Similar to above specimen and conditions except 0.1 mm thick.
6	219	1928	298	0.255-0.562	45°	45°			Similar to above specimen and conditions.
7	219	1928	298	0.255-0.562	45°	45°			Chromium; hardened, highly polished steel substrate; electroplated; highly polished and free from scratches.
8	219	1928	298	0.250-4.000	45°	45°			Similar to above specimen and conditions.
9	220	1964	298	20.0-192	45°	45°		±5	Chromium; stainless steel substrate, electrpolated; data extracted from smooth curve; measured relative to Al 2024.
10	220	1964	8.5	20.0-188	45°	45°		±5	Above specimen and conditions.

DATA TABLE NO. 251 ANGULAR SPECTRAL REFLECTANCE OF CHROMIUM CONTACT COATINGS

[Wavelength, λ, μm; Reflectance, ρ; Temperature, T, K]

λ	ρ
CURVE 1	
T = 298	
0.255	0.57
0.265	0.59
0.280	0.62
0.300	0.64
0.312	0.64
0.335	0.65
0.365	0.65
0.402	0.68
0.435	0.69
0.462	0.65
CURVE 2	
T = 298	
0.257	0.65
0.265	0.64
0.277	0.65
0.300	0.68
0.315	0.69
0.335	0.67
0.365	0.69
0.402	0.71
0.410	0.72
0.565	0.67
CURVE 3	
T = 298	
0.257	0.59
0.270	0.62
0.300	0.66
0.320	0.68
0.342	0.68
0.417	0.71
CURVE 4	
T = 298	
0.252	0.59*
0.265	0.61
0.277	0.64
0.302	0.68
0.315	0.68*
0.337	0.68*
CURVE 4 (cont.)	
0.370	0.69*
0.415	0.71*
0.440	0.72
0.557	0.67
CURVE 5	
T = 298	
0.255	0.58
0.280	0.64
0.300	0.68*
0.312	0.69
0.332	0.68
0.365	0.69*
0.407	0.71*
0.440	0.71
0.562	0.68
CURVE 6	
T = 298	
0.255	0.63
0.265	0.66
0.280	0.68
0.300	0.69
0.312	0.71
0.335	0.70
0.365	0.71
0.402	0.72*
0.435	0.72
0.562	0.69
CURVE 7	
0.255	0.60
0.265	0.62
0.280	0.63
0.300	0.65
0.315	0.67
0.335	0.66
0.365	0.67
0.402	0.70
0.435	0.71
0.562	0.67*
CURVE 8	
T = 298	
0.250	0.60
0.350	0.68
0.400	0.72
0.500	0.68
0.600	0.64
0.700	0.63
0.800	0.63
0.900	0.63
1.150	0.64
1.375	0.65
1.700	0.67
1.975	0.71
2.300	0.76
2.550	0.79
2.825	0.82
3.050	0.84
3.250	0.85
3.575	0.86
3.825	0.89
4.000	0.88
CURVE 9	
T = 298	
20.0	0.973
22.5	0.973
25.5	0.995
31.0	0.999
35.4	0.991
44.0	1.011
61.3	0.998
99.0	0.994
120	1.000
192	0.993
CURVE 10	
T = 8.5	
20.0	1.000
23.8	1.000
27.4	1.000
29.9	0.998
37.0	1.000
44.2	1.000*
CURVE 10 (cont.)	
60.9	1.000
75.1	1.000
98.0	1.000
188	1.000

* Not shown on plot

SPECIFICATION TABLE NO. 252 ANGULAR SPECTRAL ABSORPTANCE OF CHROMIUM CONTACT COATINGS

Curve No.	Ref. No.	Year	Temperature, K	Wavelength Range, μm	Geometry θ	Reported Error, %	Composition (weight percent), Specifications, and Remarks
1*	221	1965	306	0.340-22.9	25°		Chromium; polished nickel substrate; electroplated; polished with optical grade rouge with cotton moistened with ethyl alcohol; measured in dry nitrogen; heated cavity at 1056 K; measured relative to platinum; authors assumed $\alpha = 1 - R(2\pi, 25°)$.

DATA TABLE NO. 252 ANGULAR SPECTRAL ABOSPRTANCE OF CHROMIUM CONTACT COATINGS

[Wavelength, λ, μm; Absorptance, α; Temperature, T, K]

λ	α	λ	α
CURVE 1*		CURVE 1 (cont.)*	
T = 306			
		4.00	0.148
0.340	0.484	4.58	0.127
0.388	0.449	5.10	0.115
0.409	0.410	5.46	0.093
0.502	0.393	5.96	0.089
0.604	0.390	6.22	0.082
0.703	0.379	7.03	0.080
0.804	0.384	7.43	0.079
0.902	0.411	8.17	0.077
0.995	0.402	9.06	0.071
1.29	0.367	10.2	0.063
1.50	0.344	11.1	0.059
1.80	0.306	11.9	0.053
2.00	0.283	12.9	0.057
2.30	0.246	14.9	0.049
2.48	0.224	16.9	0.049
2.97	0.201	18.9	0.048
3.42	0.177	20.8	0.050
3.76	0.168	22.9	0.040

* No plot given

SPECIFICATION TABLE NO. 253 NORMAL SOLAR ABSORPTANCE OF CHROMIUM CONTACT COATINGS

Curve No.	Ref. No.	Year	Temperature Range, K	Geometry θ	Reported Error, %	Composition (weight percent), Specifications and Remarks
1*	203	1962	298	$\sim 0°$		Chromium (0.00254 mm thick); nickel (0.0127 mm thick) and 321 stainless steel substrates; nickel plated on stainless steel, then chromium plated on nickel; substrate exposed to JP4 combustion products for 50 hrs at 865 K.

DATA TABLE NO. 253 NORMAL SOLAR ABSORPTANCE OF CHROMIUM CONTACT COATINGS

[Temperature, T, K; Absorptance, α]

T	α
CURVE 1*	
298	0.78

* No plot given

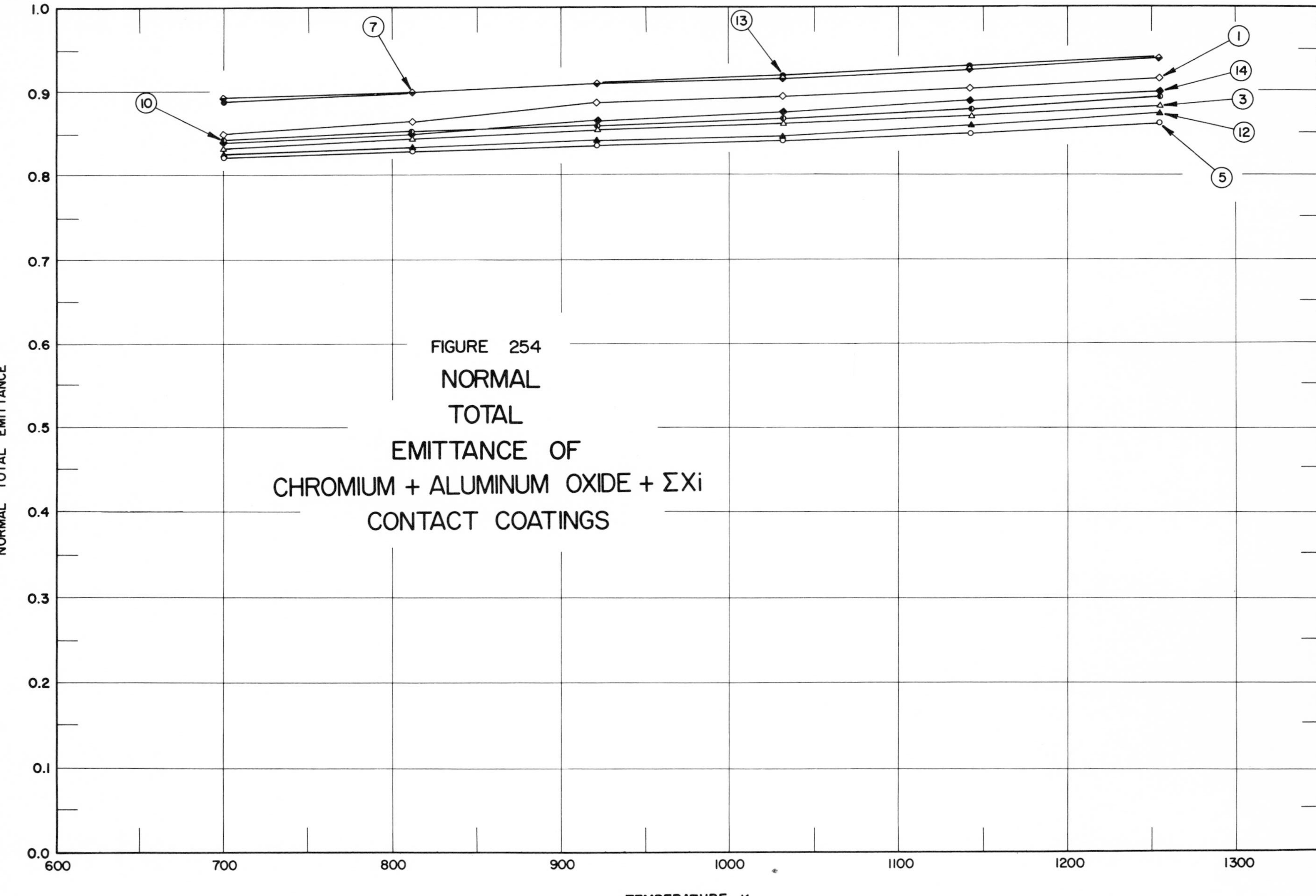
FIGURE 254
NORMAL
TOTAL
EMITTANCE OF
CHROMIUM + ALUMINUM OXIDE + ΣXi
CONTACT COATINGS
NORMAL TOTAL EMITTANCE
TEMPERATURE, K
1.0
0.9
0.8
0.7
0.6
0.5
0.4
0.3
0.2
0.1
0.0
600
700
800
900
1000
1100
1200
1300
7
13
1
14
3
12
5
10

SPECIFICATION TABLE NO. 254 NORMAL TOTAL EMITTANCE OF CHROMIUM + ALUMINUM OXIDE + ΣX_i CONTACT COATINGS

Curve No.	Ref. No.	Year	Temperature Range, K	Geometry θ'	Reported Error, %	Composition (weight percent), Specifications and Remarks
1	73	1962	700-1255	$\sim 0^\circ$		77 Cr and 23 Al_2O_3, Haynes LT-1 cermet (0.102 mm thick); Inconel substrate; substrate sandblasted and oxidized; flame sprayed.
2	73	1962	700-1255	$\sim 0^\circ$		Similar to above specimen and conditions except 0.155 mm thick.
3	73	1962	700-1255	$\sim 0^\circ$		Similar to above specimen and conditions except 0.203 mm thick.
4*	73	1962	700-1255	$\sim 0^\circ$		Similar to above specimen and conditions except 0.254 mm thick.
5	73	1962	700-1255	$\sim 0^\circ$		Similar to above specimen and conditions except 0.305 mm thick.
6*	73	1962	700-1255	$\sim 0^\circ$		Similar to above specimen and conditions except 0.102 mm thick.
7	73	1962	700-1255	$\sim 0^\circ$		60 Cr, 19 Al_2O_3, 19 Mo and 2 TiO_2, Haynes LT-1B cermet (0.102 mm thick); Inconel substrate; substrate sandblasted and oxidized; flame sprayed.
8*	73	1962	700-1255	$\sim 0^\circ$		Similar to above specimen and conditions except 0.152 mm thick.
9*	73	1962	700-1255	$\sim 0^\circ$		Similar to above specimen and conditions except 0.203 mm thick.
10	73	1962	700-1255	$\sim 0^\circ$		Similar to above specimen and conditions except 0.254 mm thick.
11*	73	1962	700-1255	$\sim 0^\circ$		Similar to above specimen and conditions except 0.305 mm thick.
12	73	1962	700-1255	$\sim 0^\circ$		Similar to above specimen and conditions except 0.356 mm thick.
13	73	1962	700-1255	$\sim 0^\circ$		60 Cr, 19 Al_2O_3, 19 Mo and 2 TiO_2, Haynes LT-1B cermet; Inconel substrate; substrate sandblasted and oxidized; flame sprayed.
14	73	1962	700-1255	$\sim 0^\circ$		77 Cr and 23 Al_2O_3, Haynes LT-1 cermet; Inconel substrate; substrate sandblasted and oxidized; flame sprayed.

* Not shown on plot

DATA TABLE NO. 254 NORMAL TOTAL EMITTANCE OF CHROMIUM + ALUMINUM OXIDE + ΣX_i CONTACT COATINGS

[Temperature, T, K; Emittance, ϵ]

T	ϵ	T	ϵ	T	ϵ
CURVE 1		CURVE 6*		CURVE 11*	
700	0.850	700	0.822	700	0.833
811	0.865	811	0.829	811	0.841
922	0.889	922	0.834	922	0.849
1033	0.899	1033	0.839	1033	0.857
1144	0.904	1144	0.845	1144	0.867
1255	0.915	1255	0.857	1255	0.883
CURVE 2*		CURVE 7		CURVE 12	
700	0.843	700	0.893	700	0.825
811	0.855	811	0.900	811	0.832
922	0.871	922	0.910	922	0.841
1033	0.878	1033	0.916	1033	0.848
1144	0.885	1144	0.926	1144	0.860
1255	0.898	1255	0.940	1255	0.875
CURVE 3		CURVE 8*		CURVE 13	
700	0.835	700	0.875	700	0.890
811	0.845	811	0.883	811	0.900*
922	0.855	922	0.891	922	0.910*
1033	0.862	1033	0.899	1033	0.920
1144	0.870	1144	0.909	1144	0.930
1255	0.882	1255	0.925	1255	0.940
CURVE 4*		CURVE 9*		CURVE 14	
700	0.827	700	0.859	700	0.840
811	0.835	811	0.866	811	0.850
922	0.843	922	0.876	922	0.865
1033	0.849	1033	0.883	1033	0.875
1144	0.858	1144	0.893	1144	0.890
1255	0.872	1255	0.908	1255	0.900
CURVE 5		CURVE 10			
700	0.823	700	0.845		
811	0.830	811	0.852		
922	0.836	922	0.860		
1033	0.842	1033	0.868		
1144	0.850	1144	0.879		
1255	0.862	1255	0.894		

* Not shown on plot

SPECIFICATION TABLE NO. 255 NORMAL TOTAL EMITTANCE OF COBALT CONTACT COATINGS

Curve No.	Ref. No.	Year	Temperature Range, K	Geometry θ'	Reported Error, %	Composition (weight percent), Specifications and Remarks
1*	89	1958	366-672	$\sim 0^\circ$		Cobalt (0.0127 mm thick); 321 stainless steel substrate; plated. [Author's designation: Specimen 10]
2*	222	1962	367	$\sim 0^\circ$		Cobalt (0.00508 mm thick); platinum clad carpenter No. 20 stainless steel substrate; electroplated.

DATA TABLE NO. 255 NORMAL TOTAL EMITTANCE OF COBALT CONTACT COATINGS

[Temperature, T, K; Emittance, ϵ]

T	ϵ
CURVE 1*	
366	0.129
444	0.128
522	0.134
600	0.163
672	0.305
CURVE 2*	
367	0.15

* No plot given

SPECIFICATION TABLE NO. 256 NORMAL SPECTRAL REFLECTANCE OF COBALT CONTACT COATINGS

Curve No.	Ref. No.	Year	Temperature, K	Wavelength Range, μm	Geometry θ θ' ω'	Reported Error, %	Composition (weight percent), Specifications, and Remarks
1*	89	1958	~298	0.400-25.0	~0° ~0°		Cobalt (0.0127 mm thick); 321 stainless steel substrate; plated; data extracted from smooth curve. [Author's designation: Specimen 10]
2*	215	1962	~298	0.40-4.0	~0° ~0°		Cobalt; glass substrate; solution of cobalt resinate, diluted with aromatic solvents and essential oils, applied to a soda-lime glass substrate, then fired to 873 K in a continuous lehr on a 1.5 hr cycle; application of solution was by dropping an excess amount on the substrate, which was kept spinning at 1550 rpm until no more solution was flung from the edges; measured relative to plane, polished aluminum; specular component only.

DATA TABLE NO. 256 NORMAL SPECTRAL REFLECTANCE OF COBALT CONTACT COATINGS

[Wavelength, λ, μm; Reflectance, ρ; Temperature, T, K]

λ	ρ	λ	ρ	λ	ρ	λ	ρ
CURVE 1*		CURVE 1 (cont.)*		CURVE 1 (cont.)*		CURVE 2 (cont.)*	
T ~ 298		1.24	0.331	10.0	0.907	1.0	0.06
0.400	0.200	1.50	0.362	11.6	0.911	1.2	0.09
0.450	0.200	1.79	0.425	11.8	0.919	1.4	0.16
0.460	0.210	2.27	0.500	16.0	0.920	1.6	0.22
0.480	0.210	2.65	0.598	16.3	0.929	1.8	0.22
0.500	0.219	3.01	0.600	19.5	0.929	2.0	0.21
0.580	0.221	3.27	0.660	19.7	0.920	2.2	0.20
0.600	0.232	3.50	0.680	20.4	0.920	2.4	0.19
0.720	0.232	3.81	0.743	20.8	0.911	2.6	0.19
0.740	0.242	4.06	0.772	22.9	0.911	2.8	0.17
0.750	0.242	4.93	0.810	23.2	0.900	3.0	0.12
0.780	0.252	5.02	0.819	25.0	0.900	3.2	0.14
0.810	0.252	5.76	0.850			3.4	0.14
0.830	0.261	6.56	0.850	CURVE 2*		3.6	0.07
0.880	0.262	6.81	0.868	T ~ 298		3.8	0.12
0.890	0.270	7.58	0.868			4.0	0.10
0.910	0.270	8.10	0.889	0.40	0.12		
0.930	0.282	8.85	0.889	0.50	0.09		
0.950	0.282	9.02	0.899	0.60	0.08		
0.980	0.300	9.78	0.899	0.80	0.12		

* No plot given

SPECIFICATION TABLE NO. 257 NORMAL SOLAR ABSORPTANCE OF COBALT CONTACT COATINGS

Curve No.	Ref. No.	Year	Temperature Range, K	Geometry θ	Reported Error, %	Composition (weight percent), Specifications and Remarks
1*	222	1962	367	$\sim 0°$		Cobalt (0.00508 mm thick); platinum clad carpenter No. 20 stainless steel substrate; electroplated; error in measurement ± 0.01.

DATA TABLE NO. 257 NORMAL SOLAR ABSORPTANCE OF COBALT CONTACT COATINGS

[Temperature, T, K; Absorptance, α]

T	α
CURVE 1*	
367	0.81

* No plot given

SPECIFICATION TABLE NO. 258 NORMAL SPECTRAL TRANSMITTANCE OF COBALT CONTACT COATINGS

Curve No.	Ref. No.	Year	Temperature, K	Wavelength Range, μm	Geometry θ	θ'	ω'	Reported Error, %	Composition (weight percent), Specifications, and Remarks
1*	215	1962	~298	0.40-2.0	~0°	~0°			Cobalt; glass (1.59 mm thick) substrate; solution of cobalt resinate, diluted with aromatic solvents and essential oils, applied to a soda-lime glass substrate, then fired to 873 K in a continuous lehr on a 1.5 hr cycle; application of solution was by dropping an excess amount on the substrate, which was kept spinning at 1550 rpm until no more solution was flung from the edges; τ is for coating plus substrate.

DATA TABLE NO. 258 NORMAL SPECTRAL TRANSMITTANCE OF COBALT CONTACT COATINGS

[Wavelength, λ, μm; Transmittance, τ; Temperature, T, K]

λ	τ
CURVE 1*	
T ~ 298	
0.40	0.019
0.50	0.056
0.60	0.210
0.70	0.280
0.80	0.333
0.90	0.720
1.0	0.800
1.1	0.766
1.2	0.655
1.3	0.580
1.4	0.487
1.5	0.363
1.6	0.270
1.8	0.164
2.0	0.105

* No plot given

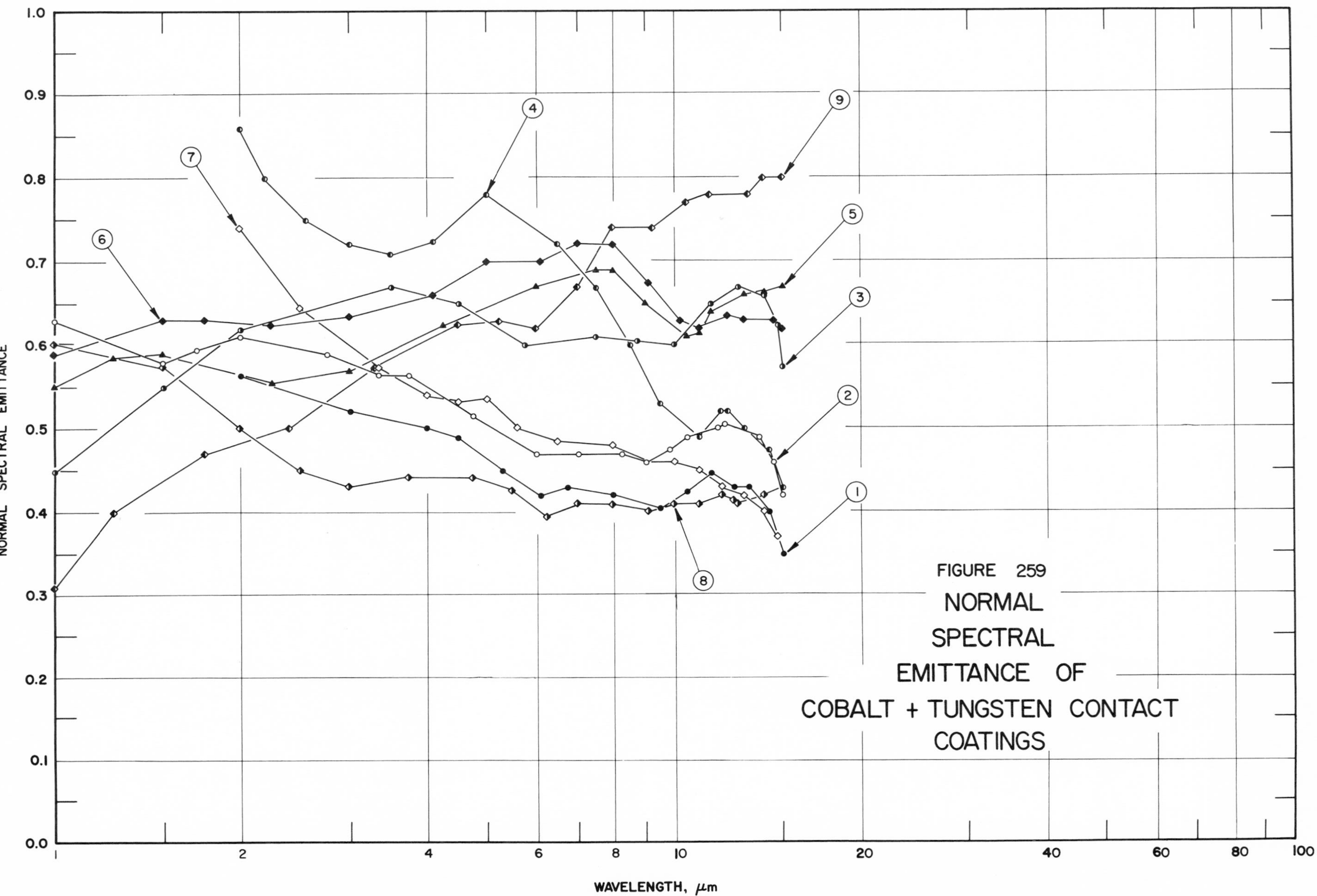

FIGURE 259 NORMAL SPECTRAL EMITTANCE OF COBALT + TUNGSTEN CONTACT COATINGS

SPECIFICATION TABLE NO. 259 NORMAL SPECTRAL EMITTANCE OF COBALT + TUNGSTEN CONTACT COATINGS

Curve No.	Ref. No.	Year	Temperature, K	Wavelength Range, μm	Geometry θ'	Reported Error, %	Composition (weight percent), Specifications, and Remarks
1	139	1961	523	2.00-15.00	$\sim 0^\circ$	±5	50 Co and 50 W; Inconel X substrate; substrate sandblasted; flame sprayed using a plasmatron; as received; data extracted from smooth curve.
2	139	1961	773	1.00-15.00	$\sim 0^\circ$	±5	Similar to above specimen and conditions.
3	139	1961	1023	1.00-15.00	$\sim 0^\circ$	±5	Similar to curve 1 specimen and conditions.
4	139	1961	523	2.00-15.00	$\sim 0^\circ$	±5	Similar to curve 1 specimen and conditions except heated in air at 1089 K for 30 min.
5	139	1961	773	1.00-15.00	$\sim 0^\circ$	±5	Similar to above specimen and conditions.
6	139	1961	1023	1.00-15.00	$\sim 0^\circ$	±5	Similar to curve 4 specimen and conditions.
7	139	1961	523	2.00-15.00	$\sim 0^\circ$	±5	Similar to curve 1 specimen and conditions except heated in vacuum (6.8×10^{-5} mm Hg) at 1089 K for 30 min.
8	139	1961	773	1.00-15.00	$\sim 0^\circ$	±5	Similar to above specimen and conditions.
9	139	1961	1023	1.00-15.00	$\sim 0^\circ$	±5	Similar to curve 7 specimen and conditions.

DATA TABLE NO. 259 NORMAL SPECTRAL EMITTANCE OF COBALT + TUNGSTEN CONTACT COATINGS

[Wavelength, λ, μm; Emittance, ∈; Temperature, T, K]

λ	∈
CURVE 1	**T = 523**
2.00	0.565
3.00	0.520
4.00	0.500
4.50	0.490
5.30	0.450
6.10	0.420
6.75	0.430
8.00	0.420
9.50	0.405
10.50	0.425
11.50	0.445
12.50	0.430
13.25	0.430
14.25	0.400
15.00	0.350
CURVE 2	**T = 773**
1.00	0.630
1.50	0.580
1.70	0.595
2.00	0.610
2.75	0.590
3.35	0.565
3.75	0.565
4.75	0.515
6.00	0.470
7.00	0.470
8.25	0.470
9.00	0.460
9.80	0.475
10.50	0.490
11.75	0.500
12.20	0.505
13.75	0.490
14.50	0.460
15.00	0.420
CURVE 3	**T = 1023**
1.00	0.450
1.50	0.550
2.00	0.620
CURVE 3 (cont.)	
3.50	0.670
4.50	0.650
5.75	0.600
7.50	0.610
8.75	0.605
10.00	0.600
11.50	0.650
12.75	0.670
14.00	0.660
14.75	0.625
15.00	0.575
CURVE 4	**T = 523**
2.00	0.860
2.20	0.800
2.55	0.750
3.00	0.720
3.50	0.710
4.10	0.725
5.00	0.780
6.50	0.720
7.50	0.670
8.50	0.600
9.50	0.530
11.00	0.490
11.80	0.520
12.25	0.520
13.00	0.500
14.25	0.475
15.00	0.430
CURVE 5	**T = 773**
1.00	0.550
1.25	0.585
1.50	0.590
2.25	0.555
3.00	0.570
4.25	0.625
6.00	0.670
7.50	0.690
8.00	0.690
9.00	0.650
CURVE 5 (cont.)	
10.50	0.610
11.00	0.615
11.50	0.640
13.00	0.660
14.00	0.665
15.00	0.670
CURVE 6	**T = 1023**
1.00	0.590
1.50	0.630
1.75	0.632
2.25	0.625
3.00	0.635
4.10	0.660
5.00	0.700
6.10	0.700
7.00	0.720
8.00	0.720
9.10	0.675
10.25	0.630
11.00	0.620
12.25	0.635
13.00	0.630
14.50	0.630
15.00	0.620
CURVE 7	**T = 523**
2.00	0.740
2.50	0.645
3.35	0.575
4.00	0.540
4.50	0.530
5.00	0.535
5.60	0.500
6.50	0.485
8.00	0.480
9.00	0.460
10.00	0.460
11.00	0.450
12.00	0.430
13.00	0.420
14.00	0.400
CURVE 7 (cont.)	
14.75	0.370
15.00	0.350
CURVE 8	**T = 773**
1.00	0.600
1.50	0.575
2.00	0.500
2.50	0.450
3.00	0.430
3.75	0.440
4.75	0.440
5.50	0.425
6.25	0.395
7.00	0.410
8.00	0.410
9.10	0.400
10.00	0.410
11.00	0.410
12.00	0.420
12.50	0.415
12.75	0.410
14.00	0.420
15.00	0.430
CURVE 9	**T = 1023**
1.00	0.310
1.25	0.400
1.75	0.470
2.40	0.500
3.30	0.575
4.50	0.625
5.25	0.630
6.00	0.620
7.00	0.670
8.00	0.740
9.25	0.740
10.50	0.770
11.50	0.780
13.25	0.780
14.00	0.800
15.00	0.800

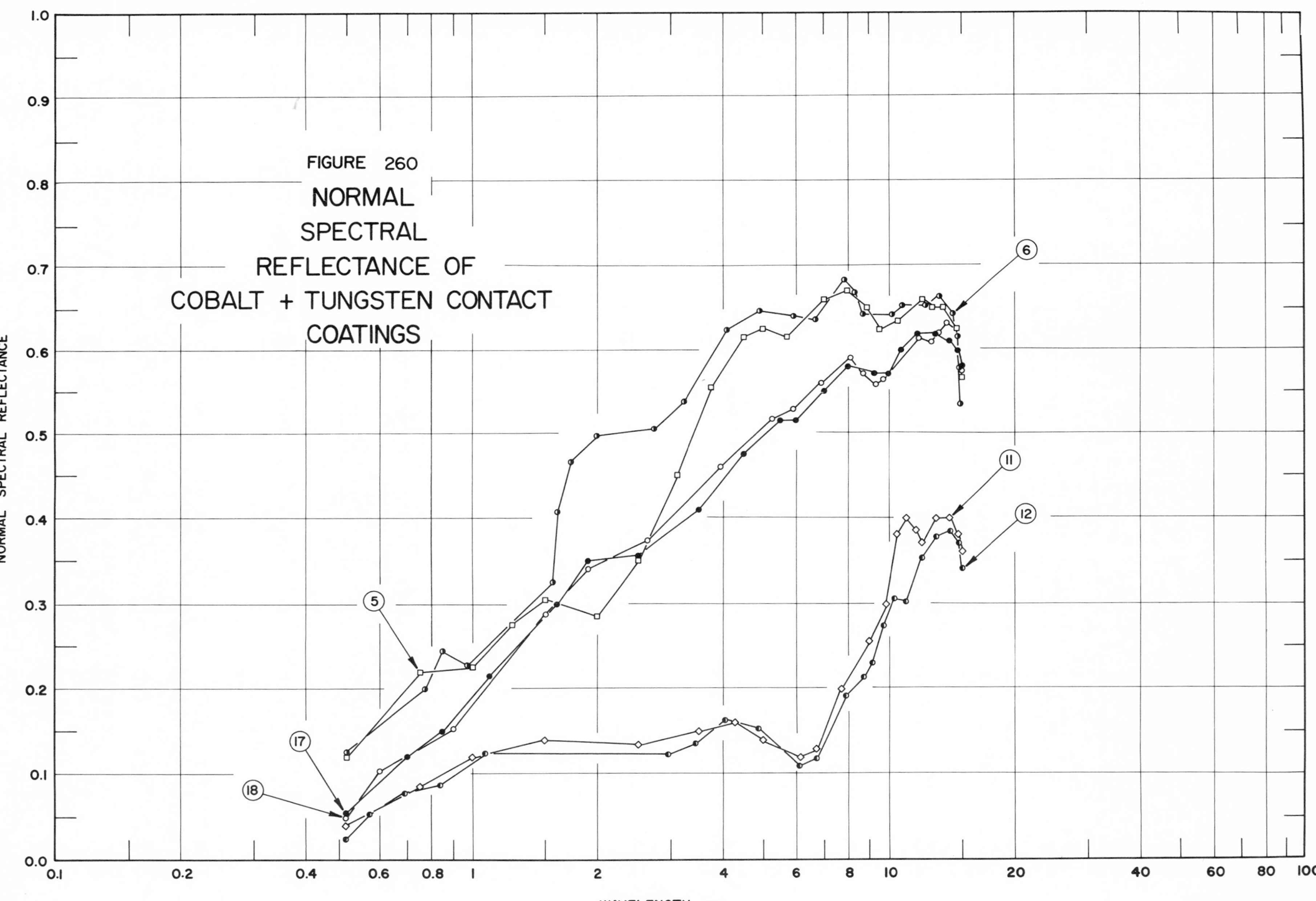

FIGURE 260 NORMAL SPECTRAL REFLECTANCE OF COBALT + TUNGSTEN CONTACT COATINGS

SPECIFICATION TABLE NO. 260 NORMAL SPECTRAL REFLECTANCE OF COBALT + TUNGSTEN CONTACT COATINGS

Curve No.	Ref. No.	Year	Temperature, K	Wavelength Range, μm	Geometry θ	θ'	ω'	Reported Error, %	Composition (weight percent), Specifications, and Remarks
1*	139	1961	~322	2.00-15.00	~0°		2π	< 2	50 Co and 50 W; Inconel X substrate; flame sprayed using a plasmatron; as received; data extracted from smooth curve; hohlraum at 523 K; converted from R(2π, 0°).
2*	139	1961	~322	2.00-15.00	~0°		2π	< 2	Above specimen and conditions; diffuse component only.
3*	139	1961	~322	1.00-15.00	~0°		2π	< 2	Similar to curve 1 specimen and conditions except hohlraum at 773 K.
4*	139	1961	~322	1.00-15.00	~0°		2π	< 2	Above specimen and conditions; diffuse component only.
5	139	1961	~322	0.50-15.00	~0°		2π	< 2	Similar to curve 1 specimen and conditions except hohlraum at 1273 K.
6	139	1961	~322	0.50-14.99	~0°		2π	< 2	Above specimen and conditions; diffuse component only.
7*	139	1961	~322	2.00-15.00	~0°		2π	< 2	Similar to curve 1 specimen and conditions except heated in air at 1089 K for 30 min.
8*	139	1961	~322	2.00-15.00	~0°		2π	< 2	Above specimen and conditions; diffuse component only.
9*	139	1961	~322	1.00-15.00	~0°		2π	< 2	Similar to curve 7 specimen and conditions except hohlraum at 773 K.
10*	139	1961	~322	1.00-15.00	~0°		2π	< 2	Above specimen and conditions; diffuse component only.
11	139	1961	~322	0.50-15.00	~0°		2π	< 2	Similar to curve 7 specimen and conditions except hohlraum at 1273 K.
12	139	1961	~322	0.50-15.00	~0°		2π	< 2	Above specimen and conditions; diffuse component only.
13*	139	1961	~322	2.00-15.00	~0°		2π	< 2	Similar to curve 1 specimen and conditions except heated in vacuum (6.8 x 10^{-5} mm Hg) at 1089 K for 30 min.
14*	139	1961	~322	2.00-15.00	~0°		2π	< 2	Above specimen and conditions; diffuse component only.
15*	139	1961	~322	1.00-15.00	~0°		2π	< 2	Similar to curve 13 specimen and conditions except hohlraum at 773 K.
16*	139	1961	~322	1.00-15.00	~0°		2π	< 2	Above specimen and conditions; diffuse component only.
17	139	1961	~322	0.50-15.00	~0°		2π	< 2	Similar to curve 13 specimen and conditions except hohlraum at 1273 K.
18	139	1961	~322	0.50-15.00	~0°		2π	< 2	Above specimen and conditions; diffuse component only.

* Not shown on plot

DATA TABLE NO. 260 NORMAL SPECTRAL REFLECTANCE OF COBALT + TUNGSTEN CONTACT COATINGS

[Wavelength, λ, μm; Reflectance, ρ; Temperature, T, K]

CURVE 1* T ~322

λ	ρ
2.00	0.380
2.75	0.410
3.50	0.475
4.50	0.550
5.10	0.565
6.00	0.545
6.50	0.555
7.25	0.605
8.00	0.615
9.00	0.600
10.00	0.600
11.25	0.585
13.00	0.600
13.50	0.600
14.50	0.565
15.00	0.530

CURVE 2* T ~ 322

λ	ρ
2.00	0.394
2.39	0.390
3.05	0.425
3.97	0.527
4.45	0.564
5.06	0.581
5.85	0.553
6.16	0.542
6.78	0.563
7.32	0.597
8.04	0.613
8.89	0.586
9.55	0.593
10.19	0.591
11.05	0.581
12.25	0.587
13.16	0.582
14.08	0.562
14.72	0.530
15.00	0.502

CURVE 3* T ~ 322

λ	ρ
1.00	0.300
1.50	0.285
1.75	0.300
1.80	0.350
2.00	0.440
2.70	0.465
3.50	0.560
4.00	0.610
4.80	0.635
6.00	0.620
7.25	0.650
8.00	0.685
9.35	0.660
10.50	0.665
11.50	0.670
13.00	0.670
14.00	0.650
14.80	0.600
15.00	0.560

CURVE 4* T ~ 322

λ	ρ
1.00	0.309
1.45	0.323
1.75	0.376
1.86	0.415
2.11	0.431
2.56	0.426
3.03	0.452
3.99	0.561
4.54	0.589
5.13	0.600
6.01	0.569
7.01	0.600
7.91	0.620
8.56	0.604
9.06	0.584
10.01	0.584
11.01	0.602
12.01	0.611
13.11	0.611
14.14	0.597
14.86	0.554
15.00	0.525

CURVE 5 T ~ 322

λ	ρ
0.50	0.120*
0.75	0.220*
1.00	0.225
1.25	0.275
1.50	0.305
2.00	0.285
2.50	0.350
3.10	0.450
3.75	0.555
4.50	0.615
5.00	0.625
5.75	0.615
7.00	0.660
8.00	0.670
8.85	0.650
9.50	0.625
10.50	0.635
12.00	0.660
12.75	0.650
13.50	0.650
14.50	0.625
15.00	0.565

CURVE 6 T ~ 322

λ	ρ
0.50	0.125*
0.77	0.200*
0.85	0.245*
0.97	0.229*
1.25	0.275
1.56	0.326
1.60	0.408
1.72	0.464
2.00	0.499
2.72	0.505
3.24	0.538
4.11	0.625
4.99	0.647
5.99	0.640
6.65	0.638
7.89	0.681
8.36	0.669
8.77	0.641
10.03	0.641
10.78	0.651
12.03	0.651
13.03	0.662
14.03	0.642
14.67	0.617
14.96	0.576
14.99	0.533

CURVE 7* T ~ 322

λ	ρ
2.00	0.125
2.75	0.110
4.00	0.125
5.00	0.080
6.00	0.060
7.00	0.100
8.25	0.175
8.75	0.200
9.75	0.300
10.50	0.365
11.00	0.390
12.00	0.360
13.00	0.335
13.60	0.375
14.00	0.400
14.60	0.375
15.00	0.340

CURVE 8* T ~ 322

λ	ρ
2.00	0.115
2.44	0.100
2.82	0.097
3.93	0.117
4.34	0.111
5.73	0.066
6.06	0.068
7.77	0.128
9.00	0.180
9.84	0.274
10.89	0.353
11.26	0.359
11.98	0.327
12.40	0.321
13.16	0.354
13.63	0.352
14.00	0.359
14.59	0.339
15.00	0.308

CURVE 9* T ~ 322

λ	ρ
1.00	0.080
1.50	0.110
2.15	0.085
2.50	0.080
4.00	0.100
4.50	0.105
5.25	0.125
6.00	0.090
6.55	0.100
7.75	0.140
9.00	0.165
10.00	0.240
10.50	0.275
11.00	0.290
11.80	0.270
13.00	0.290
14.00	0.320
14.50	0.325
15.00	0.300

CURVE 10* T ~ 322

λ	ρ
1.00	0.082
2.00	0.094
2.79	0.118
3.99	0.163
4.57	0.152
5.26	0.124
5.69	0.121
6.24	0.108
7.01	0.133
8.25	0.217
8.82	0.229
9.76	0.309
10.77	0.384
11.26	0.390
12.25	0.374
12.77	0.374
14.01	0.399
14.72	0.401
15.00	0.389

CURVE 11 T ~ 322

λ	ρ
0.50	0.040*
0.75	0.085*
1.00	0.120
1.50	0.140
2.50	0.135
3.50	0.150
4.25	0.160
5.00	0.140
6.15	0.120
6.75	0.130
7.75	0.200
9.00	0.255
9.85	0.300
10.50	0.380
11.00	0.400
11.60	0.385
12.00	0.370
13.00	0.400
14.00	0.400
14.75	0.380
15.00	0.360

CURVE 12 T ~ 322

λ	ρ
0.50	0.025*
0.57	0.053*
0.69	0.078*
0.84	0.087*
1.07	0.125
2.97	0.122
3.43	0.136
4.03	0.161
4.84	0.151
6.11	0.110
6.74	0.119
7.99	0.190
8.75	0.214
9.15	0.230
9.73	0.273
10.26	0.306
11.00	0.301
12.00	0.351
13.00	0.376
14.00	0.382
14.68	0.370
15.00	0.341

CURVE 13* T ~ 322

λ	ρ
2.00	0.330
2.50	0.260
3.00	0.240
3.50	0.270
4.00	0.340
5.00	0.375
6.00	0.340
6.70	0.375
7.00	0.400
7.50	0.410
8.50	0.410
9.50	0.410
11.00	0.440
12.00	0.450
13.25	0.490
14.50	0.515
15.00	0.500

CURVE 14* T ~ 322

λ	ρ
2.00	0.450
2.65	0.395
2.98	0.387
3.38	0.406
3.95	0.454
4.95	0.511
6.22	0.545
7.22	0.564
7.99	0.584
8.98	0.564
10.74	0.593
11.99	0.609
13.09	0.615
14.00	0.600
14.67	0.575
15.00	0.550

CURVE 15* T ~ 322

λ	ρ
1.00	0.270
1.50	0.280
2.10	0.325
2.80	0.350
3.50	0.410
5.00	0.490
5.75	0.500
6.50	0.520
8.00	0.575
9.25	0.565
11.25	0.620
12.00	0.625
13.00	0.650
14.50	0.670
15.00	0.660

CURVE 16* T ~ 322

λ	ρ
1.00	0.239
1.79	0.303
2.19	0.330
2.76	0.338
3.24	0.376
3.86	0.418
5.19	0.489
6.04	0.505
7.23	0.548
8.16	0.572
9.04	0.551
9.79	0.563
10.53	0.586
11.29	0.610
12.26	0.624
13.00	0.631
14.01	0.650
15.00	0.650

* Not shown on plot

DATA TABLE NO. 260 NORMAL SPECTRAL REFLECTANCE OF COBALT + TUNGSTEN CONTACT COATINGS (continued)

λ	ρ
CURVE 17	
$T \sim 322$	
0.50	0.055*
0.70	0.120*
0.85	0.150*
1.10	0.215
1.60	0.300
1.90	0.350
2.50	0.355
3.50	0.410
4.50	0.475
5.50	0.515
6.00	0.515
7.00	0.550
8.00	0.580
9.25	0.570
10.00	0.570
10.75	0.600
11.75	0.620
13.00	0.620
14.00	0.610
14.60	0.600
15.00	0.580
CURVE 18	
$T \sim 322$	
0.50	0.050*
0.60	0.103*
0.90	0.151*
1.50	0.289
1.90	0.340
2.62	0.372
3.99	0.460
5.27	0.518
5.99	0.529
6.99	0.560
8.11	0.590
8.76	0.570
9.25	0.558
9.77	0.562
11.19	0.615
12.17	0.610
13.01	0.620*
13.99	0.631
14.58	0.621
15.00	0.575

* Not shown on plot

SPECIFICATION TABLE NO. 261 NORMAL SPECTRAL REFLECTANCE OF COPPER CONTACT COATINGS

Curve No.	Ref. No.	Year	Temperature, K	Wavelength Range, μm	Geometry θ	θ'	ω'	Reported Error, %	Composition (weight percent), Specifications, and Remarks
1*	215	1962	~298	0.4-4.0	~0°	~0°			Copper; glass substrate; solution of copper resinate, diluted with aromatic solvents and essential oils, applied to a soda-lime glass substrate, then fired to 873 K in a continuous lehr on a 1.5 hr cycle; application of solution was by dropping an excess amount on the substrate, which was kept spinning at 1550 rpm until no more solution was flung from the edges; measured relative to plane, polished aluminum; specular component only.

DATA TABLE NO. 261 NORMAL SPECTRAL REFLECTANCE OF COPPER CONTACT COATINGS

[Wavelength, λ, μm: Reflectance, ρ; Temperature, T, K]

λ	ρ
CURVE 1*	
T ~ 298	
0.4	0.12
0.5	0.05
0.6	0.03
0.8	0.03
1.0	0.06
1.2	0.09
1.4	0.09
1.6	0.09
1.8	0.09
2.0	0.09
2.2	0.08
2.4	0.08
2.6	0.08
2.8	0.07
3.0	0.05
3.2	0.05
3.4	0.05
3.6	0.06
CURVE 1 (cont.)*	
3.8	0.05
4.0	0.06

* No plot given

SPECIFICATION TABLE NO. 262 ANGULAR SPECTRAL REFLECTANCE OF COPPER CONTACT COATINGS

Curve No.	Ref. No.	Year	Temperature, K	Wavelength Range, μm	Geometry θ	θ'	ω'	Reported Error, %	Composition (weight percent), Specifications, and Remarks
1*	208	1962	~298	1.120-14.38	45°	45°			Copper (950 to 3000 Å thick); epoxy and 316 stainless steel substrates; substrate cleaned, etched, and vapor degreased, then coated with Maraset 617-C epoxy resin (Marblette Co.), applied by flowing the resin onto the steel at 408 K; plastic surface was cleaned by scrubbing in Alconox solution, then rinsing with water and methanol; copper vapor deposited in vacuum (~10^{-5} mm Hg) on heated substrate; specular component only. [Authors' designation: Cu 65]
2*	208	1962	~298	1.120-14.38	45°	45°			Similar to above specimen and conditions except substrate consists of 316 stainless steel coated with SY627-119 polyurethane (Febert Shorndorfer Co.) by a dip process. [Authors' designation: Cu 66B]

DATA TABLE NO. 262 ANGULAR SPECTRAL REFLECTANCE OF COPPER CONTACT COATINGS

[Wavelength, λ, μm; Reflectance, ρ; Temperature, T, K]

λ	ρ
CURVE 1*	
T ~ 298	
1.120	0.815
2.240	0.759
4.824	0.794
7.780	0.806
12.10	0.832
13.53	0.843
14.38	0.867
CURVE 2*	
T ~ 298	
1.120	0.871
2.240	0.956
4.824	0.980
7.780	0.985
12.10	0.981
13.53	0.988
14.38	1.00

* No plot given

SPECIFICATION TABLE NO. 263 NORMAL SPECTRAL TRANSMITTANCE OF COPPER CONTACT COATINGS

Curve No.	Ref. No.	Year	Temperature, K	Wavelength Range, μm	Geometry θ θ' ω'	Reported Error, %	Composition (weight percent), Specifications, and Remarks
1*	215	1962	~298	0.4-2.0	~0° ~0°		Copper (1.59 mm thick); glass substrate; solution of copper resinate, diluted with aromatic solvents and essential oils, applied to a soda-lime glass substrate, then fired to 873 K in a continuous lehr on a 1.5 hr cycle; application of solution was by dropping an excess amount on the substrate, which was kept spinning at 1550 rpm until no more solution was flung from the edges; τ is for coating plus substrate.

DATA TABLE NO. 263 NORMAL SPECTRAL TRANSMITTANCE OF COPPER CONTACT COATINGS

[Wavelength, λ, μm; Transmittance, τ; Temperature, T, K]

λ	τ
CURVE 1*	
T ~ 298	
0.4	0.034
0.5	0.096
0.6	0.234
0.7	0.460
0.8	0.690
0.9	0.798
1.0	0.820
1.1	0.832
1.2	0.827
1.3	0.810
1.4	0.666
1.5	0.450
1.6	0.326
1.8	0.198
2.0	0.140

* No plot given

SPECIFICATION TABLE NO. 264 NORMAL SPECTRAL REFLECTANCE OF COPPER + TIN CONTACT COATINGS

Curve No.	Ref. No.	Year	Temperature, K	Wavelength Range, μm	Geometry θ	θ'	ω'	Reported Error, %	Composition (weight percent), Specifications, and Remarks
1*	223	1947	298	0.45-0.65	~0°	~0°			55 Cu and 45 Sn, speculum; glass substrate; vacuum evaporated.
2*	223	1947	298	0.45-0.65	~0°	~0°			55 Cu and 45 Sn, speculum (0.0254 mm thick); surface-ground steel substrate; electrodeposited; polished on a buffing wheel and then rubbed with cotton-wool.
3*	223	1947	298	0.45-0.65	~0°	~0°			Similar to above specimen and conditions.

DATA TABLE NO. 264 NORMAL SPECTRAL REFLECTANCE OF COPPER + TIN CONTACT COATINGS

[Wavelength, λ, μm; Reflectance, ρ; Temperature, T, K]

λ	ρ	λ	ρ
CURVE 1* T ~ 298		CURVE 3* T ~ 298	
0.45	0.685	0.45	0.636
0.50	0.724	0.50	0.675
0.55	0.750	0.55	0.700
0.60	0.770	0.60	0.730
0.65	0.781	0.65	0.750
CURVE 2* T ~ 298			
0.45	0.608		
0.50	0.625		
0.55	0.640		
0.60	0.655		
0.65	0.660		

* No plot given

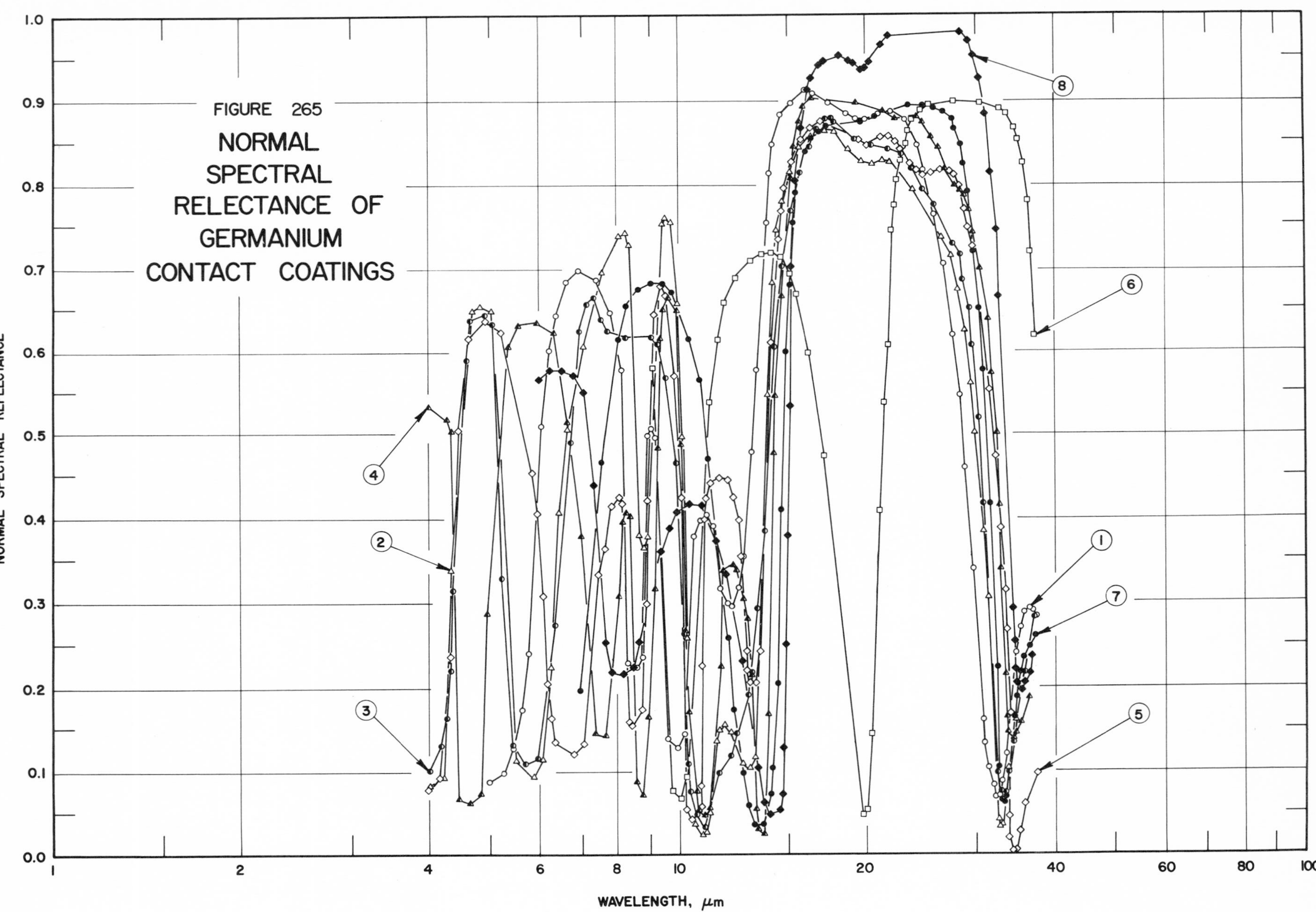

FIGURE 265 NORMAL SPECTRAL RELECTANCE OF GERMANIUM CONTACT COATINGS

SPECIFICATION TABLE NO. 265 NORMAL SPECTRAL REFLECTANCE OF GERMANIUM CONTACT COATINGS

Curve No.	Ref. No.	Year	Temperature, K	Wavelength Range, μm	Geometry θ	θ'	ω'	Reported Error, %	Composition (weight percent), Specifications, and Remarks
1	224	1963	298	5.00-37.2	~0°	~0°			Germanium (6.3 μm optical thickness); LiF (8 μm optical thickness) and glass substrates; LiF vacuum (5-10 x 10^{-5} mm Hg) evaporated at 473 K at 0.5 μm (optical) min^{-1} deposition rate; Ge vacuum (5-10 x 10^{-5} mm Hg) evaporated onto substrate at 374-413 K at 0.3 μm (optical) min^{-1} deposition rate; data extracted from smooth curve.
2	224	1963	298	4.00-35.2	~0°	~0°			Similar to above specimen and conditions except LiF deposition temp 423 K.
3	224	1963	298	4.00-37.0	~0°	~0°			Germanium (6.3 μm optical thickness); LiF (8 μm optical thickness) and BaF_2 substrates; LiF vacuum (5-10 x 10^{-5} mm Hg) evaporated at 423 K at 0.5 μm (optical) min^{-1} deposition rate; Ge vacuum (5-10 x 10^{-5} mm Hg) evaporated onto substrate at 373-413 K at 0.3 μm (optical) min^{-1} deposition rate; data extracted from smooth curve.
4	224	1963	298	4.00-36.0	~0°	~0°			Germanium (5 μm optical thickness); $PbCl_2$ (9.4 μm thick), LiF (8 μm optical thickness), and glass substrates; LiF vacuum (5-10 x 10^{-5} mm Hg) evaporated at 473 K at 0.5 μm (optical) min^{-1} deposition rate; $PbCl_2$ vacuum evaporated onto substrate at 373 K at 0.1 μm (optical) min^{-1} deposition rate; Ge vacuum evaporated onto substrate at 373-413 K at 0.3 μm (optical) min^{-1} deposition rate; data extracted from smooth curve.
5	224	1963	298	4.00-37.3	~0°	~0°			Similar to above specimen and conditions except LiF deposition temp 423 K.
6	164	1963	298	9.10-37.0	~0°	~0°			Germanium (9.825 μm optical thickness); CaF_2 substrate; vacuum deposited; data extracted from smooth curve.
7	164	1963	298	6.96-37.0	~0°	~0°			Similar to above specimen and conditions except Ge coating has an optical thickness of 6.3 μm.
8	164	1963	298	6.00-36.9	~0°	~0°			Germanium (5 μm optical thickness); $PbCl_2$ (9.4 μm thick) and LiF substrates; vacuum deposited; data extracted from smooth curve.

DATA TABLE NO. 265 NORMAL SPECTRAL REFLECTANCE OF GERMANIUM CONTACT COATINGS

[Wavelength, λ, μm; Reflectance, ρ; Temperature, T, K]

λ	ρ
CURVE 1	
T = 298	
5.00	0.086
5.27	0.099
5.45	0.126
5.63	0.173
5.77	0.240
6.01	0.509
6.23	0.600
6.40	0.641
6.68	0.682
6.98	0.697
7.42	0.685
7.84	0.646
8.14	0.578
8.30	0.228
8.54	0.224
8.75	0.234
8.93	0.497
9.06	0.505
9.19	0.496
9.60	0.137
9.94	0.126
10.2	0.141
10.6	0.377
10.8	0.397
11.1	0.401
11.4	0.390
11.7	0.315
12.0	0.299
12.2	0.297
12.5	0.317
12.7	0.353
13.1	0.477
13.4	0.578
13.8	0.752
14.0	0.812
14.2	0.848
14.7	0.881
15.3	0.899
16.0	0.911
16.5	0.909
17.5	0.899
18.7	0.884
19.4	0.879
20.0	0.877
21.1	0.884
CURVE 1 (cont.)	
22.0	0.886
23.1	0.876
24.1	0.846
24.8	0.816
25.7	0.761
26.6	0.703
27.5	0.619
28.2	0.546
28.9	0.459
29.6	0.338
30.7	0.160
30.9	0.133
31.3	0.102
31.6	0.084
32.0	0.069
32.5	0.070
32.9	0.089
33.3	0.120
34.6	0.240
35.1	0.270
35.7	0.287
36.1	0.292
36.6	0.290
37.2	0.284
CURVE 2	
T = 298	
4.00	0.083
4.21	0.092
4.33	0.337
4.70	0.649
4.82	0.653
5.01	0.649
5.50	0.112
5.88	0.093
6.08	0.113
6.27	0.223
6.45	0.405
6.69	0.504
7.09	0.605
7.58	0.696
8.03	0.738
8.22	0.740
8.36	0.726
8.64	0.379
CURVE 2 (cont.)	
8.80	0.364
8.92	0.376
9.45	0.752
9.55	0.760
9.71	0.755
9.92	0.657
10.1	0.496
10.3	0.257
10.6	0.038
10.9	0.024
11.1	0.027
11.3	0.050
11.5	0.135
11.6	0.149
11.8	0.155
12.1	0.145
12.7	0.109
13.0	0.104
13.2	0.116
13.8	0.547
14.2	0.680
14.4	0.741
14.7	0.779
15.2	0.814
15.7	0.842
16.4	0.858
17.2	0.863
17.9	0.861
18.8	0.840
19.7	0.828
20.5	0.823
21.4	0.827
21.9	0.825
23.9	0.791
26.4	0.735
27.2	0.711
28.0	0.672
28.7	0.622
29.4	0.560
29.9	0.500
30.8	0.381
31.2	0.304
32.2	0.070*
32.3	0.041
32.6	0.031
32.9	0.035
CURVE 2 (cont.)	
33.7	0.101
34.6	0.160
35.2	0.201
CURVE 3	
T = 298	
4.00	0.100
4.19	0.130
4.26	0.163
4.33	0.220
4.37	0.313
4.60	0.590
4.69	0.637
4.92	0.644
5.08	0.631
5.22	0.328
5.41	0.131
5.70	0.108
5.99	0.115
6.37	0.272
6.75	0.490
6.98	0.622
7.15	0.656
7.38	0.664
7.54	0.636
7.78	0.623
8.29	0.616
9.02	0.616
9.24	0.608
9.56	0.569
9.90	0.464
10.2	0.268
10.3	0.109
10.4	0.076
10.7	0.047
11.0	0.031
11.6	0.098
12.1	0.118
12.4	0.144
12.9	0.190
13.1	0.219
13.4	0.292
13.7	0.381
14.3	0.603
14.7	0.700
CURVE 3 (cont.)	
15.3	0.767
15.7	0.813
16.2	0.843
16.7	0.866
17.3	0.877
17.6	0.879
17.8	0.874
19.3	0.855
20.5	0.846
21.8	0.840
22.9	0.834
23.8	0.819
24.8	0.792
25.6	0.774
27.7	0.729
28.2	0.714
28.7	0.683
29.2	0.650
29.5	0.606
30.2	0.519
30.8	0.416
31.9	0.220
32.4	0.098
32.6	0.075
32.9	0.065
33.4	0.075
33.9	0.098
34.4	0.134
35.7	0.217
37.0	0.281
CURVE 4	
T = 298	
4.00	0.533
4.29	0.518
4.35	0.502
4.48	0.067
4.65	0.062
4.82	0.072
4.96	0.287
5.37	0.605
5.58	0.630
5.99	0.632
6.34	0.621
6.67	0.514
CURVE 4 (cont.)	
7.00	0.378
7.37	0.145
7.69	0.141
8.04	0.307
8.13	0.395
8.21	0.404
8.37	0.400
8.57	0.087
8.76	0.071
8.96	0.164
9.12	0.316
9.21	0.484
9.39	0.618
9.46	0.650
9.69	0.663
9.92	0.649
10.08	0.488
10.28	0.266
10.48	0.170
10.72	0.078
10.76	0.053
11.00	0.048
11.24	0.057
11.57	0.223
11.83	0.337
12.15	0.344
12.40	0.337
12.78	0.301
12.91	0.280
13.05	0.216
13.21	0.118*
13.28	0.055
13.38	0.030
13.56	0.024
13.95	0.167
14.29	0.479
14.41	0.546
14.69	0.667
15.22	0.808
15.44	0.846
15.63	0.873
15.92	0.892
16.42	0.901
19.44	0.898
21.25	0.887
22.32	0.879
CURVE 4 (cont.)	
24.02	0.880
24.79	0.874
25.55	0.856
26.10	0.841
26.96	0.818
27.87	0.799
28.48	0.790
28.74	0.785
29.28	0.766
29.88	0.740
30.50	0.699
31.27	0.637
31.82	0.571
32.26	0.500
32.67	0.415
32.98	0.339
33.32	0.216
33.59	0.164
33.79	0.144
34.02	0.138
34.36	0.142
35.02	0.156
36.03	0.187
CURVE 5	
T = 298	
4.00	0.079
4.19	0.092
4.31	0.236
4.49	0.505
4.61	0.616
4.92	0.636
5.24	0.621
5.83	0.453
5.97	0.405
6.09	0.309
6.17	0.203
6.27	0.161
6.34	0.133
6.80	0.120
7.03	0.132
7.47	0.331
7.61	0.362
7.89	0.414
8.08	0.424
CURVE 5 (cont.)	
8.18	0.418
8.31	0.159
8.44	0.153
8.73	0.172
8.90	0.297
8.97	0.420
9.12	0.643
9.37	0.677
9.54	0.666
9.87	0.570
10.1	0.423
10.2	0.251*
10.3	0.054
10.5	0.044
10.8	0.059
10.8	0.083
10.9	0.225
11.0	0.399
11.1	0.424
11.3	0.440
11.6	0.447
12.0	0.445
12.3	0.423
12.5	0.398
12.6	0.354
12.8	0.242
12.8	0.220
13.0	0.205
13.3	0.205
13.5	0.242
14.1	0.610
14.5	0.734
14.6	0.766
14.8	0.795
15.3	0.828
15.8	0.853
16.4	0.869
17.0	0.874
17.9	0.872*
19.5	0.852
20.1	0.846
20.5	0.850*
21.1	0.855
21.7	0.857
22.3	0.850
22.9	0.840

* Not shown on plot

DATA TABLE NO. 265 NORMAL SPECTRAL REFLECTANCE OF GERMANIUM CONTACT COATINGS (continued)

λ	ρ	λ	ρ	λ	ρ	λ	ρ	λ	ρ
CURVE 5 (cont.)		CURVE 6 (cont.)		CURVE 7 (cont.)		CURVE 8 (cont.)		CURVE 8 (cont.)	
T = 298									
		19.7	0.049	14.2	0.101	7.34	0.437	32.5	0.664
23.6	0.824	20.0	0.051	14.4	0.201	7.67	0.254	34.1	0.291
24.1	0.816	20.4	0.141	14.6	0.410	7.85	0.218	34.3	0.251
24.7	0.810	21.0	0.408	14.9	0.599	8.13	0.216	34.5	0.220
25.5	0.811	21.4	0.536	15.1	0.679	8.41	0.223	34.8	0.201
26.4	0.816	21.7	0.607	15.3	0.751	8.66	0.252	35.3	0.196
27.1	0.816	22.0	0.743	15.5	0.790	9.39	0.360	35.7	0.201
27.6	0.811	22.2	0.774	15.6	0.814*	9.66	0.387	36.3	0.216
28.2	0.799	22.4	0.803	16.0	0.839	9.96	0.407	36.9	0.236
28.5	0.786*	22.8	0.827	16.4	0.854	10.4	0.416		
28.9	0.770	23.1	0.849	16.8	0.863	10.8	0.415		
29.3	0.749	23.4	0.862	17.4	0.869	11.1	0.403*		
29.7	0.723	23.9	0.877	19.8	0.873	11.5	0.372		
31.5	0.551	24.6	0.889	20.8	0.880	11.9	0.332		
32.2	0.472	25.4	0.894	23.5	0.892	12.7	0.230		
32.9	0.387	27.8	0.899	24.7	0.893	13.4	0.101		
33.3	0.312	30.5	0.897	25.8	0.890	13.7	0.062		
33.5	0.268	32.9	0.890	26.7	0.886	14.0	0.048		
33.7	0.168	33.7	0.883	27.3	0.879	14.5	0.053		
33.7	0.100*	34.5	0.868	27.9	0.866	14.6	0.071		
33.8	0.046	35.0	0.851	28.4	0.849	14.7	0.128		
33.8	0.020	35.6	0.823	28.9	0.823	14.9	0.250		
34.1	0.004	36.1	0.779	29.3	0.790	15.0	0.377		
34.3	0.005	36.6	0.716	29.9	0.718	15.1	0.532		
35.0	0.029	37.0	0.617	30.4	0.650	15.3	0.700		
35.9	0.061			30.9	0.578	15.5	0.803		
37.3	0.096	CURVE 7		31.6	0.418	15.8	0.867		
		T = 298		32.2	0.222	16.2	0.911		
CURVE 6				32.5	0.102	16.4	0.926		
T = 298		6.96	0.196	32.7	0.067*	16.8	0.941		
		7.53	0.465	33.0	0.062	17.2	0.948		
9.10	0.580	8.02	0.613	33.2	0.068	18.3	0.952		
9.72	0.078	8.28	0.654	34.2	0.161	18.8	0.949		
10.1	0.069	8.64	0.673	34.6	0.188	19.2	0.944		
10.3	0.093	9.01	0.680	35.2	0.216	19.6	0.938		
11.3	0.536	9.43	0.680	35.7	0.234	20.0	0.939		
11.6	0.613	9.75	0.670	36.3	0.248	20.4	0.947		
11.9	0.658	10.4	0.614	37.0	0.260	21.2	0.967		
12.4	0.689	10.8	0.564			21.9	0.976		
13.1	0.709	11.3	0.470	CURVE 8		28.5	0.980		
13.6	0.715	12.0	0.258	T = 298		28.5	0.980*		
14.1	0.717	12.2	0.171			29.2	0.970		
14.7	0.711	12.6	0.098	6.00	0.566	29.9	0.952		
15.2	0.692	12.9	0.060	6.24	0.578	30.4	0.927		
15.5	0.668	13.2	0.037	6.51	0.577	31.1	0.881		
16.2	0.599	13.6	0.039	6.81	0.571	31.7	0.813		
17.1	0.473	14.0	0.071	7.02	0.550	32.2	0.744		

* Not shown on plot

SPECIFICATION TABLE NO. 266 NORMAL SPECTRAL TRANSMITTANCE OF GERMANIUM CONTACT COATINGS

Curve No.	Ref. No.	Year	Temperature, K	Wavelength Range, μm	Geometry θ θ' ω'	Reported Error, %	Composition (weight percent), Specifications, and Remarks
1*	224	1963	298	4.00-15.4	~0° ~0°		Germanium (6.3 μm optical thickness); LiF (8 μm optical thickness) and BaF_2 substrates; LiF vacuum (5-10 x 10^{-5} mm Hg) evaporated at 423 K at 0.5 μm (optical) min^{-1} deposition rate; Ge vacuum (5-10 x 10^{-5} mm Hg) evaporated onto substrate at 373-413 K at 0.3 μm (optical) min^{-1} deposition rate; data extracted from smooth curve.

DATA TABLE NO. 266 NORMAL SPECTRAL TRANSMITTANCE OF GERMANIUM CONTACT COATINGS

[Wavelength, λ, μm; Transmittance, τ; Temperature, T, K]

λ	τ	λ	τ
CURVE 1* T = 298		CURVE 1 (cont.)*	
		9.06	0.359
4.00	0.822	9.49	0.413
4.16	0.796	9.83	0.487
4.34	0.577	10.2	0.563
4.73	0.324	10.8	0.762
4.94	0.337	11.0	0.768
5.23	0.459	11.2	0.762
5.40	0.592	12.2	0.578
5.57	0.769	13.3	0.292
5.68	0.829	14.1	0.094
5.84	0.843	14.4	0.059
5.98	0.834	14.7	0.032
6.12	0.779	15.0	0.016
6.71	0.490	15.4	0.000
6.85	0.417		
6.96	0.341		
7.22	0.323		
7.59	0.346		
8.00	0.358		

* No plot given

SPECIFICATION TABLE NO. 267 HEMISPHERICAL TOTAL EMITTANCE OF GOLD CONTACT COATINGS

Curve No.	Ref. No.	Year	Temperature Range, K	Reported Error, %	Composition (weight percent), Specifications and Remarks
1*	15	1965	699-866		Gold, Hanovia liquid brite type 8146, Au + B_2O_3 + Cr_2O_3 + Rh, (~0.254 μm thick); molybdenum (0.0254 mm thick) substrate; two coats.
2*	15	1965	811		Similar to above specimen and conditions except measured during exposure to vacuum (10^{-6} mm Hg) at 811 K; measured in situ; exposure time, hrs, is variable.
3*	225	1968	480-536		Gold; 347 stainless steel substrate; plated; measured in vacuum (10^{-6} mm Hg).
4*	194	1955	76	5	24 K gold; stainless steel substrate; gold plated; measured in vacuum (10^{-6} to 10^{-7} mm Hg); authors assumed $\alpha = \epsilon$ for 300 K blackbody incident radiation.
5*	194	1955	76	5	Au + 1 Ag (5.1 μm thick); copper substrate; plated; measured in vacuum (10^{-6} to 10^{-7} mm Hg); emittance for 300 K blackbody incident radiation; authors assumed $\alpha = \epsilon$.
6*	194	1955	76	5	Au + 1 Ag (5.1 μm thick); stainless steel substrate; measured in vacuum (10^{-6} to 10^{-7} mm Hg); emittance for 300 K blackbody incident radiation; authors assumed $\alpha = \epsilon$.
7*	195	1953	76		Gold; mylar (0.0127 mm thick), gold and lucite substrates; gold vapor deposited on both sides of mylar; measured in vacuum ($<10^{-6}$ mm Hg); authors assumed $\alpha = \epsilon$ for 294 K blackbody radiation.
8*	195	1953	76		Similar to above specimen and conditions except final substrate is Pyrex.
9*	106	1963	322-366	<3.5	Gold; smooth fiberglass substrate; measured in vacuum (5 x 10^{-6} mm Hg); resistance 2 ohms.
10*	106	1963	333	<3.0	Gold; sandblasted fiberglass substrate; measured in vacuum (5 x 10^{-6} mm Hg); resistance 2 ohms.
11*	106	1963	335	<3.5	Gold; smooth, degreased fiberglass; measured in vacuum (5 x 10^{-6} mm Hg); resistance 2 ohms.
12*	106	1963	333	<3.5	Gold; smooth fiberglass; measured in vacuum (5 x 10^{-6} mm Hg); resistance 5 ohms.
13*	106	1963	332	<3.5	Gold; smooth fiberglass; measured in vacuum (5 x 10^{-6} mm Hg); resistance 5 ohms.
14*	106	1963	339	<3.0	Gold; sandblasted fiberglass; measured in vacuum (5 x 10^{-6} mm Hg); resistance 1 ohm.
15*	105	1964	307		Gold (0.090-0.15 μm thick); polyester film (6.35 μm thick) substrate; vacuum deposited; measured in vacuum (10^{-6} mm Hg) maintained by diffusion pump.
16*	105	1964	307		Gold; polyester film (6.35 μm thick) and gold substrates; gold vacuum deposited on both sides of polyester film; measured in vacuum (10^{-6} mm Hg) maintained by diffusion pump. [Authors' designation: Side A]
17*	105	1964	307		Above specimen and conditions except opposite side measured. [Authors' designation: Side B]
18*	105	1964	307		Gold (0.090-0.15 μm thick); polyester film (6.35 μm thick) substrate; vacuum deposited; measured in vacuum (10^{-6} mm Hg) maintained by diffusion pump.
19*	105	1964	307		Gold; polyester film (6.35 μm thick) and gold substrates; gold vacuum deposited on both sides of polyester film; measured in vacuum (10^{-6} mm Hg) maintained by diffusion pump. [Authors' designation: Test No. 7, Side B]
20*	105	1964	307		Gold (7.62 μm thick); 304 stainless steel substrate; gold plated; measured in vacuum (10^{-6} mm Hg) maintained by diffusion pump. [Authors' designation: Test No. 127, XHV-7]
21*	105	1964	307		Gold; 304 stainless steel substrate; gold plated; measured in vacuum (10^{-6} mm Hg) maintained by diffusion pump. [Authors' designation: Test No. 131, XHV-5]

* No plot given

SPECIFICATION TABLE NO. 267 HEMISPHERICAL TOTAL EMITTANCE OF GOLD CONTACT COATINGS (continued)

Curve No.	Ref. No.	Year	Temperature Range, K	Reported Error, %	Composition (weight percent), Specifications and Remarks
22*	105	1964	307		Gold; ETP copper substrate; gold plated; measured in vacuum (10^{-6} mm Hg) maintained by diffusion pump. [Authors' designation: Test No. 132, XHV-3]
23*	105	1964	307		Gold; 304 stainless steel substrate; gold plated; measured in vacuum (10^{-6} mm Hg) maintained by diffusion pump. [Authors' designation: Test No. 133, XHV-5]
24*	105	1964	307		Gold; 304 stainless steel substrate; gold plated; measured in vacuum (10^{-6} mm Hg) maintained by diffusion pump. [Authors' designation: Test No. 134, XHV-8]
25*	105	1964	307		Gold; aluminum foil (0.127 mm thick) and gold substrates; gold plated; measured in vacuum (10^{-6} mm Hg) maintained by diffusion pump. [Authors' designation: Test No. 135, Side B]
26*	105	1964	307		Similar to above specimen and conditions. [Authors' designation: Test No. 136, Side B]
27*	105	1964	307		Gold matte (7.62 μm thick); dull nickel substrate; measured in vacuum (10^{-6} mm Hg) maintained by diffusion pump. [Authors' designation: Test No. 140]
28*	105	1964	307		Gold; ETP copper substrate; gold plated; measured in vacuum (10^{-6} mm Hg) maintained by diffusion pump. [Authors' designation: Test No. 144, XHV-6]
29*	105	1964	307		Gold; 304 stainless steel substrate; gold plated; measured in vacuum (10^{-6} mm Hg) maintained by diffusion pump. [Authors' designation: Test No. 145, XHV-4]
30*	105	1964	307		Similar to above specimen and conditions. [Authors' designation: Test No. 146, XHV-5]
31*	105	1964	307		Similar to above specimen and conditions. [Authors' designation: Test No. 147, XHV-1]
32*	105	1964	307		Similar to above specimen and conditions. [Authors' designation: Test No. 148, XHV-2]
33*	105	1964	307		Gold (7.62 μm thick); ETP copper substrate; gold plated; measured in vacuum (10^{-6} mm Hg) maintained by diffusion pump. [Authors' designation: Test No. 152]
34*	105	1964	307		Gold; 100 ETP copper (3.18 μm thick) and gold substrates; gold plated; measured in vacuum (10^{-6} mm Hg) maintained by diffusion pump. [Authors' designation: Test No. 155, Sample No. XHV-6, Dull Side]
35*	105	1964	307		Gold; 100 ETP copper substrate; gold plated; measured in vacuum (10^{-6} mm Hg) maintained by diffusion pump. [Authors' designation: Test No. 165, Sample XHV-6]
36*	105	1964	307		Similar to above specimen and conditions. [Authors' designation: Test No. 166, Sample XHV-6]
37*	105	1964	307		Gold; ETP copper substrate; gold plated; measured in vacuum (10^{-6} mm Hg) maintained by diffusion pump. [Authors' designation: Test No. 169, Sample XHV-6]
38*	105	1964	307		Gold; ETP copper (3.18μm thick) substrate; gold plated; measured in vacuum (10^{-6} mm Hg) maintained by diffusion pump. [Authors' designation: Test No. 188, Sample C]
39*	105	1964	307		Similar to above specimen and conditions. [Authors' designation: Test No. 192, Sample D]
40*	105	1964	307		Gold; ETP copper substrate; gold plated; measured in vacuum (10^{-6} mm Hg) maintained by diffusion pump. [Authors' designation: Test No. 184, Sample XHV-6]
41*	105	1964	307		Similar to above specimen and conditions. [Authors' designation: Test No. 186, Sample XHV-6]

* No plot given

SPECIFICATION TABLE NO. 267 HEMISPHERICAL TOTAL EMITTANCE OF GOLD CONTACT COATINGS (continued)

Curve No.	Ref. No.	Year	Temperature Range, K	Reported Error, %	Composition (weight percent), Specifications and Remarks
42*	105	1964	307		Similar to above specimen and conditions. [Authors' designation: Test No. 190, Sample XHV-6]
43*	105	1964	307		Gold; ETP copper (3.18 μm thick) substrate; gold plated; polished substrate; measured in vacuum (10^{-6} mm Hg) maintained by diffusion pump. [Authors' designation: Test No. 187, Sample A]
44*	105	1964	307		Similar to above specimen and conditions. [Authors' designation: Test No. 189, Sample A]
45*	105	1964	307		Similar to above specimen and conditions. [Authors' designation: Test No. 191, Sample B]
46*	105	1964	307		Gold; ETP copper (7.62 μm thick); gold plated; measured in vacuum (10^{-6} mm Hg) maintained by diffusion pump. [Authors' designation: Test No. 201, Sample XHV-6]
47*	105	1964	307		Gold (1.45 Å thick); polyester film substrate; measured in vacuum (10^{-6} mm Hg) maintained by diffusion pump. [Authors' designation: Test No. 202, NRL No. 92264-1-E1, Mfr. A]
48*	105	1964	307		Gold (29.3 Å thick); polyester film substrate; measured in vacuum (10^{-6} mm Hg) maintained by diffusion pump. [Authors' designation: Test No. 203, NRL No. 92264-1-E1, Mfr. A]
49*	105	1964	307		Gold (1.0 Å thick); polyester film substrate; measured in vacuum (10^{-6} mm Hg) maintained by diffusion pump. [Authors' designation: Test No. 207, Mfr. A]
50*	105	1964	307		Gold (2.94 Å thick); polyester film substrate; measured in vacuum (10^{-6} mm Hg) maintained by diffusion pump. [Authors' designation: Test No. 208]
51*	105	1964	307		Gold (500 Å thick); aluminum and polyester film substrates; measured in vacuum (10^{-6} mm Hg) maintained by diffusion pump. [Authors' designation: Test No. 210]
52*	105	1964	307		Gold (412 Å thick); aluminum and polyester film substrates; measured in vacuum (10^{-6} mm Hg) maintained by diffusion pump. [Authors' designation: Test No. 211]
53*	105	1964	307		Gold (78 Å thick); polyester film substrate; measured in vacuum (10^{-6} mm Hg) maintained by diffusion pump. [Authors' designation: Test No. 214, Mfr. K, Sample 10664-1-1E]
54*	105	1964	307		Gold (147 Å thick); polyester film substrate; measured in vacuum (10^{-6} mm Hg) maintained by diffusion pump. [Authors' designation: Test No. 215, Mfr. K, Sample 10664-2-1E]
55*	105	1964	307		Gold; copper (3.18 μm thick) substrate; gold plated; measured in vacuum (10^{-6} mm Hg) maintained by diffusion pump. [Authors' designation: Test No. 216, Sample XHV-6]
56*	105	1964	307		Above specimen and conditions. [Authors' designation: Test No. 217, Sample XHV-6]
57*	105	1964	307		Gold (224 Å thick); polyester film substrate; measured in vacuum (10^{-6} mm Hg) maintained by diffusion pump. [Authors' designation: Test No. 220, Mfr. K, N0640, Sample 10664-3-1E]
58*	105	1964	307		Gold (0.235 μm thick); polyester film substrate; measured in vacuum (10^{-6} mm Hg) maintained by diffusion pump. [Authors' designation: Test No. 221, Mfr. K, 645, Sample 10664-4-1E]
59*	105	1964	307		Gold (135 Å thick); polyester film substrate; vacuum deposited; measured in vacuum (10^{-6} mm Hg) maintained by diffusion pump. [Authors' designation: Test No. 224, Mfr. A, Sample 101469-1-1E]
60*	105	1964	307		Gold (52 Å thick); polyester film substrate; vacuum deposited; measured in vacuum (10^{-6} mm Hg) maintained by diffusion pump. [Authors' designation: Test No. 226, Mfr. A, Sample 101464-2-1E]

* No plot given

DATA TABLE NO. 267 HEMISPHERICAL TOTAL EMITTANCE OF GOLD CONTACT COATINGS

[Temperature, T, K; Emittance, ε]

T	ε
CURVE 1*	
699	0.050
755	0.050
810	0.050
866	0.060
Exposure time	ε
CURVE 2* T = 811	
0	0.048
1000	0.055
1465	0.055
2400	0.055
3000	0.055
T	ε
CURVE 3*	
480	0.038
504	0.041
514	0.043
536	0.047
CURVE 4*	
76	0.017
CURVE 5*	
76	0.025
CURVE 6*	
76	0.025
CURVE 7*	
76	0.021
CURVE 8*	
76	0.025
CURVE 9*	
366	0.078
322	0.080
CURVE 10*	
333	0.288
CURVE 11*	
335	0.062
CURVE 12*	
333	0.055
CURVE 13*	
332	0.062
CURVE 14*	
339	0.222
339	0.224
CURVE 15*	
307	0.043
CURVE 16*	
307	0.047
CURVE 17*	
307	0.048
CURVE 18*	
307	0.048
CURVE 19*	
307	0.038
CURVE 20*	
307	0.040
CURVE 21*	
307	0.104
CURVE 22*	
307	0.070
CURVE 23*	
307	0.106
CURVE 24*	
307	0.041
CURVE 25*	
307	0.362
CURVE 26*	
307	0.364
CURVE 27*	
307	0.083
CURVE 28*	
307	0.061
CURVE 29*	
307	0.059
CURVE 30*	
307	0.105
CURVE 31*	
307	0.037
CURVE 32*	
307	0.063
CURVE 33*	
307	0.076
CURVE 34*	
307	0.063
CURVE 35*	
307	0.066
CURVE 36*	
307	0.056
CURVE 37*	
307	0.055
CURVE 38*	
307	0.035
CURVE 39*	
307	0.024
CURVE 40*	
307	0.064
CURVE 41*	
307	0.063
CURVE 42*	
307	0.053
CURVE 43*	
307	0.020
CURVE 44*	
307	0.020
CURVE 45*	
307	0.022
CURVE 46*	
307	0.051
CURVE 47*	
307	0.247
CURVE 48*	
307	0.077
CURVE 49*	
307	0.857
CURVE 50*	
307	0.356
CURVE 51*	
307	0.025
CURVE 52*	
307	0.024
CURVE 53*	
307	0.049
CURVE 54*	
307	0.029
CURVE 55*	
307	0.054
CURVE 56*	
307	0.054
CURVE 57*	
307	0.030
CURVE 58*	
307	0.020
CURVE 59*	
307	0.044
CURVE 60*	
307	0.063

* No plot given

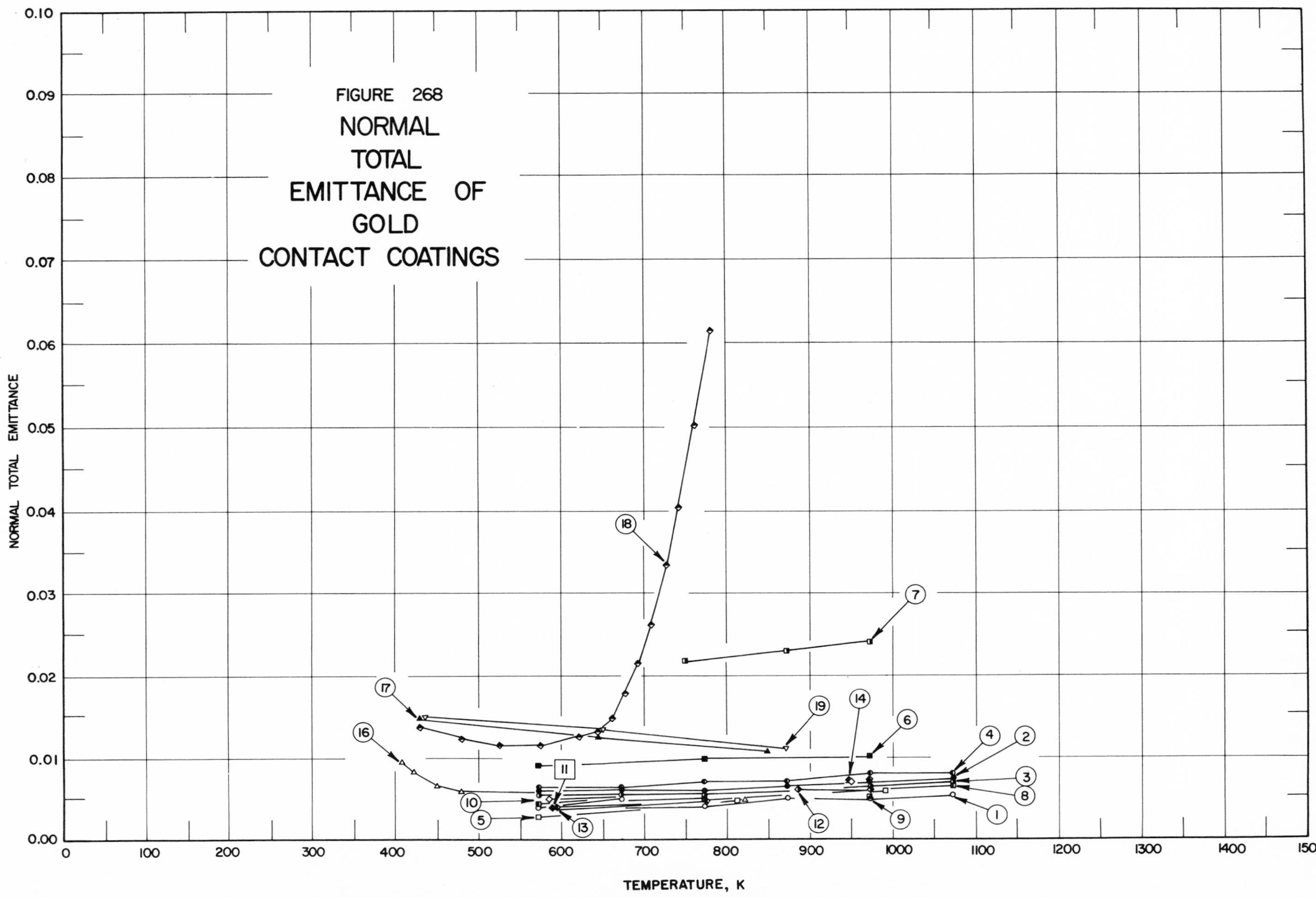
FIGURE 268
NORMAL
TOTAL
EMITTANCE OF
GOLD
CONTACT COATINGS
NORMAL TOTAL EMITTANCE
0.00
0.01
0.02
0.03
0.04
0.05
0.06
0.07
0.08
0.09
0.10
TEMPERATURE, K
0
100
200
300
400
500
600
700
800
900
1000
1100
1200
1300
1400
1500
1
2
3
4
5
6
7
8
9
10
11
12
13
14
16
17
18
19

SPECIFICATION TABLE NO. 268 NORMAL TOTAL EMITTANCE OF GOLD CONTACT COATINGS

Curve No.	Ref. No.	Year	Temperature Range, K	Geometry θ'	Reported Error, %	Composition (weight percent), Specifications and Remarks
1	226	1961	573-1073	~0°	~10	Gold (1-5 μm thick); Inconel X substrate; substrate surface washed, etched in HNO_3-HF solution for 45 min, rinsed, electropolished ~7 min in H_3PO_4-H_2SO_4-HF solution, rinsed, and degreased in trichloroethylene for 30 min; vapor deposited in vacuum ($< 10^{-4}$ mm Hg) on substrate heated to ~573 K.
2	226	1961	573-1073	~0°	~10	Similar to curve 1 specimen and conditions except substrate surface was not electropolished.
3	226	1961	573-1073	~0°	~10	Gold, Hanovia Liquid Bright No. 6854; Inconel X substrate; substrate surface washed, etched in HNO_3-HF solution for 45 min, rinsed, electropolished ~7 min in H_3PO_4-H_2SO_4-HF solution, rinsed and degreased in trichlorethylene for 30 min; brush application; air dried 20 min, baked 20 min at 473 K, fired 1 hr at 673 K.
4	226	1961	573-1073	~0°	~10	Similar to curve 3 specimen and conditions except substrate surface was not electropolished.
5	226	1961	573-994	~0°		Gold (6-8.4 mg cm^{-2} area density); NBS ceramic A418 (0.0508 mm thick) and Inconel X substrates; NBS ceramic A418 applied to Inconel X and fired 3 min at 1283 K; gold vapor deposited in vacuum ($< 10^{-4}$ mm Hg).
6	226	1961	573-973	~0°	~10	Similar to above specimen and conditions except heat treated 51 hrs at 1073 K.
7	226	1961	750-974	~0°	~10	Similar to curve 5 specimen and conditions except heat treated 260 hrs at 1073 K.
8	226	1961	573-1073	~0°	~10	Similar to curve 5 specimen and conditions except gold coating has 2-4 mg cm^{-2} area density.
9	226	1961	573-973	~0°		Gold (5 mg cm^{-2} area density); nickel oxide (0.00254 mm thick) and Inconel X substrates; prepared as follows: Inconel X surface pumice cleaned, alkaline cleaned; etched in acid solution (conc. HNO_3, 46 parts; 46% HF, 8 parts; H_2O, 46 parts), rinsed in demineralized H_2O, vapor degreased in trichloroethylene 30 min, electroplated with nickel oxide, oxidized 48 hrs at 1073 K; gold vapor deposited in vacuum (10^{-4} mm Hg).
10	226	1961	586-950	~0°	~10	Similar to above specimen and conditions except heat treated 24 hrs at 1073 K.
11	226	1961	589	~0°		Similar to curve 9 specimen and conditions except heat treated 73 hrs at 1073 K.
12	226	1961	885-973	~0°		Similar to curve 9 specimen and conditions except heat treated 110 hrs at 1073 K.
13	226	1961	595-973	~0°	~10	Gold (5.5 mg cm^{-2} area density); ceric oxide (0.80 mg cm^{-2} area density) and Inconel X substrates; prepared as follows: Inconel X surface pumice cleaned, alkaline cleaned, etched in acid solution (46 parts conc. HNO_3, 8 parts 46% HF, 46 parts H_2O), rinsed in demineralized H_2O, vapor degreased 30 min in trichloroethylene, ceric oxide vacuum deposited (10^{-4} mm Hg) from dimpled tungsten boat; gold vapor deposited in vacuum (10^{-4} mm Hg) from molybdenum boat.
14	226	1961	573-949	~0°	~10	Similar to above specimen and conditions except heat treated 73 hrs at 1073 K.
15	226	1961	873	~0°	~10	Similar to curve 13 specimen and conditions except heat treated 110 hrs at 1073 K.
16	196	1960	407-823	~0°		Gold; glass substrate; vapor plated; data extracted from smooth curve.
17	196	1960	429-848	~0°		Hanovia "liquid gold"; glass substrate; painted on substrate, then fired at 866 K; data extracted from smooth curve.

SPECIFICATION TABLE NO. 268 NORMAL TOTAL EMITTANCE OF GOLD CONTACT COATINGS (continued)

Curve No.	Ref. No.	Year	Temperature Range, K	Geometry θ'	Reported Error, %	Composition (weight percent), Specifications and Remarks
18	196	1960	428-781	$\sim 0^\circ$		Hanovia "liquid gold"; ceramic and steel substrates; painted on ceramic coated steel, then fired at 866 K; data extracted from smooth curve.
19	196	1960	435-871	$\sim 0^\circ$		Hanovia "liquid gold"; glass substrate; painted on substrate, then fired at 866 K; data extracted from smooth curve.

DATA TABLE NO. 268 NORMAL TOTAL EMITTANCE OF GOLD CONTACT COATINGS

[Temperature, T, K; Emittance, ϵ]

T	ϵ
CURVE 1	
573	0.040
673	0.050
773	0.041
873	0.050
973	0.050
1073	0.056
CURVE 2	
573	0.060
673	0.061
773	0.060
873	0.065
973	0.070
1073	0.075
CURVE 3	
573	0.055
673	0.055
773	0.055
873	0.060
973	0.065
1073	0.070
CURVE 4	
573	0.065
673	0.066
773	0.070
873	0.070
973	0.080
1073	0.080
CURVE 5	
573	0.029
814	0.048
994	0.060
CURVE 6	
573	0.090
773	0.099
973	0.101
CURVE 7	
750	0.219
873	0.230
974	0.240
CURVE 8	
573	0.045
672	0.050
773	0.050
873	0.050
873	0.057
1073	0.066
CURVE 9	
573	0.040
774	0.060
973	0.051
CURVE 10	
586	0.050
950	0.070
CURVE 11	
589	0.040
CURVE 12	
885	0.062
973	0.060
CURVE 13	
595	0.040
775	0.049
973	0.073
CURVE 14	
573	0.040
949	0.073
CURVE 15	
873	0.060
CURVE 16	
407	0.096
421	0.083
450	0.065
478	0.059
672	0.053
823	0.048
CURVE 17	
429	0.148
644	0.126
848	0.107
CURVE 18	
428	0.137
478	0.123
526	0.115
577	0.115
621	0.124
644	0.134
660	0.149
677	0.178
692	0.216
708	0.262
728	0.336
743	0.405
761	0.501
781	0.615
CURVE 19	
435	0.148
650	0.134
871	0.112

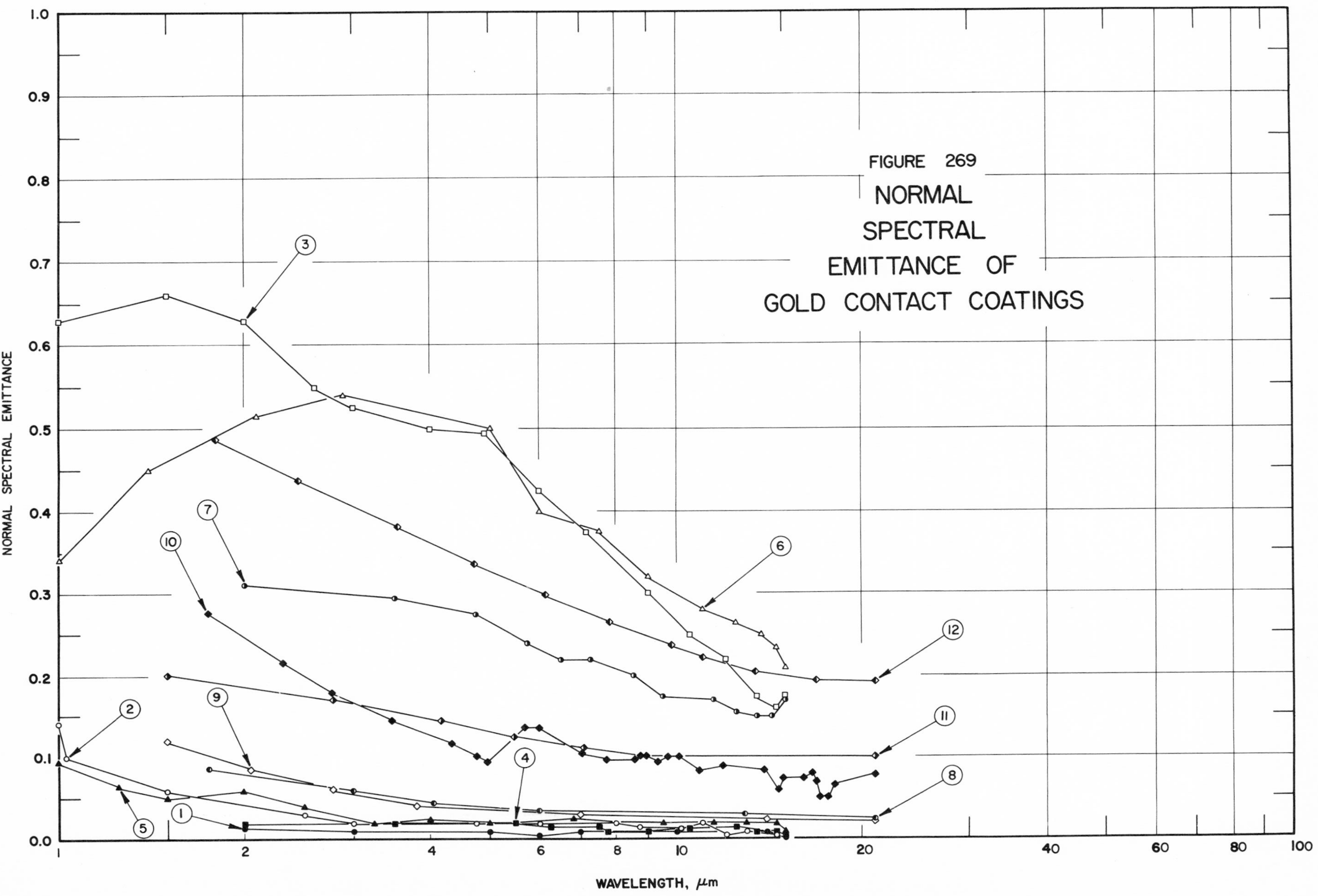

FIGURE 269 NORMAL SPECTRAL EMITTANCE OF GOLD CONTACT COATINGS

SPECIFICATION TABLE NO. 269 NORMAL SPECTRAL EMITTANCE OF GOLD CONTACT COATINGS

Curve No.	Ref. No.	Year	Temperature, K	Wavelength Range, μm	Geometry θ'	Reported Error, %	Composition (weight percent), Specifications, and Remarks
1	139	1961	523	2.00-15.00	$\sim 0°$	±5	Gold, Bright Gold No. 6854 from Engelhard Industries; titanium substrate; sprayed; fired at 873 K for 5 min; shiny finish; as received; data extracted from smooth curve.
2	139	1961	773	1.00-15.00	$\sim 0°$	±5	Similar to above specimen and conditions.
3	139	1961	1023	1.00-15.00	$\sim 0°$	±5	Similar to curve 1 specimen and conditions.
4	139	1961	523	2.00-15.00	$\sim 0°$	±5	Similar to curve 1 specimen and conditions except matte finish.
5	139	1961	773	1.00-15.00	$\sim 0°$	±5	Similar to above specimen and conditions.
6	139	1961	1023	1.00-15.00	$\sim 0°$	±5	Similar to curve 4 specimen and conditions.
7	139	1961	1023	2.00-15.00	$\sim 0°$	±5	Similar to curve 4 specimen and conditions except heated in vacuum (6.8×10^{-5} mm Hg) at 1089 K for 30 min.
8	74	1959	306	1.75-21.00	$\sim 0°$		Gold; magnesium substrate; data extracted from smooth curve; property converted from $R(2\pi, \sim 0°)$.
9	74	1959	301	1.50-21.00	$\sim 0°$		Gold; titanium substrate; data extracted from smooth curve; property converted from $R(2\pi, \sim 0°)$.
10	74	1959	298	1.75-21.00	$\sim 0°$		Gold; Dow 17 and magnesium thorium substrates; data extracted from smooth curve; property converted from $R(2\pi, \sim 0°)$.
11	74	1959	302	1.50-21.00	$\sim 0°$		Gold (0.02 mm thick); titanium substrate; gold plate; data extracted from smooth curve; property converted from $R(2\pi, \sim 0°)$.
12	74	1959	296	1.80-21.00	$\sim 0°$		White gold; fiberglass substrate (obtained from Englehard Industries); data extracted from smooth curve; property converted from $R(2\pi, \sim 0°)$.

DATA TABLE NO. 269 NORMAL SPECTRAL EMITTANCE OF GOLD CONTACT COATINGS

[Wavelength, λ, μm; Emittance, ∈; Temperature, T, K]

λ	∈
CURVE 1	
T = 523	
2.00	0.015
3.00	0.010
5.00	0.010
6.00	0.005
7.00	0.010
10.00	0.010
14.00	0.010
15.00	0.000
CURVE 2	
T = 773	
1.00	0.140
1.05	0.100
1.50	0.060
2.50	0.030
3.00	0.020
4.75	0.020
6.00	0.020
8.00	0.020
8.70	0.015
10.25	0.015
11.00	0.020
12.10	0.005
13.00	0.010
14.50	0.005
15.00	0.000
CURVE 3	
T = 1023	
1.00	0.630
1.50	0.660
2.00	0.630
2.60	0.550
3.00	0.525
4.00	0.500
4.90	0.495
6.00	0.425
7.15	0.375
9.00	0.300
10.50	0.250
12.00	0.220
13.50	0.175
CURVE 3 (cont.)	
14.50	0.160
15.00	0.175
CURVE 4	
T = 523	
2.00	0.020
3.50	0.020
5.50	0.020
6.25	0.015
7.50	0.015
7.75	0.010
9.00	0.010
10.50	0.015
12.50	0.015
13.50	0.010
14.50	0.010
15.00	0.005
CURVE 5	
T = 773	
1.00	0.095
1.25	0.065
1.50	0.050
2.00	0.060
2.50	0.040
3.25	0.020
4.00	0.025
5.00	0.020
6.00	0.020
6.80	0.025
8.00	0.020
9.50	0.020
11.50	0.020
13.00	0.020
14.50	0.020
15.00	0.010
CURVE 6	
T = 1023	
1.00	0.340
1.40	0.450
2.10	0.515
2.90	0.540
CURVE 6 (cont.)	
5.00	0.500
6.00	0.400
7.50	0.375
9.00	0.320
11.00	0.280
12.50	0.265
13.75	0.250
14.50	0.235
15.00	0.210
CURVE 7	
T = 1023	
2.00	0.310
3.50	0.295
4.75	0.275
5.75	0.240
6.50	0.220
7.25	0.220
8.50	0.200
9.50	0.175
11.50	0.170
12.50	0.155
13.50	0.150
14.25	0.150
15.00	0.170
CURVE 8	
T = 306	
1.75	0.087
3.00	0.060
4.05	0.046
6.00	0.036
12.97	0.032
21.00	0.026
CURVE 9	
T = 301	
1.50	0.120
2.05	0.085
2.79	0.063
3.81	0.043
7.00	0.031
CURVE 9 (cont.)	
14.00	0.025
21.00	0.023
CURVE 10	
T = 298	
1.75	0.278
2.31	0.217
2.78	0.180
3.49	0.145
4.35	0.119
4.79	0.102
4.99	0.095
5.70	0.137
6.04	0.137
7.07	0.106
7.73	0.098
8.56	0.098
8.74	0.103
8.95	0.103
9.30	0.096
9.69	0.101
10.17	0.101
10.82	0.085
11.90	0.090
13.94	0.086
14.68	0.061
14.95	0.077
16.12	0.077
16.60	0.081
16.98	0.071
17.33	0.052
17.69	0.052
18.12	0.069
21.00	0.080
CURVE 11	
T = 302	
1.50	0.202
2.79	0.172
4.17	0.145
5.49	0.126
7.07	0.112
8.75	0.101
CURVE 11 (cont.)	
21.00	0.101
CURVE 12	
T = 296	
1.80	0.488
2.46	0.438
3.54	0.384
4.71	0.337
6.12	0.299
7.82	0.265
9.80	0.238
11.21	0.223
13.49	0.205
16.96	0.193
21.00	0.193

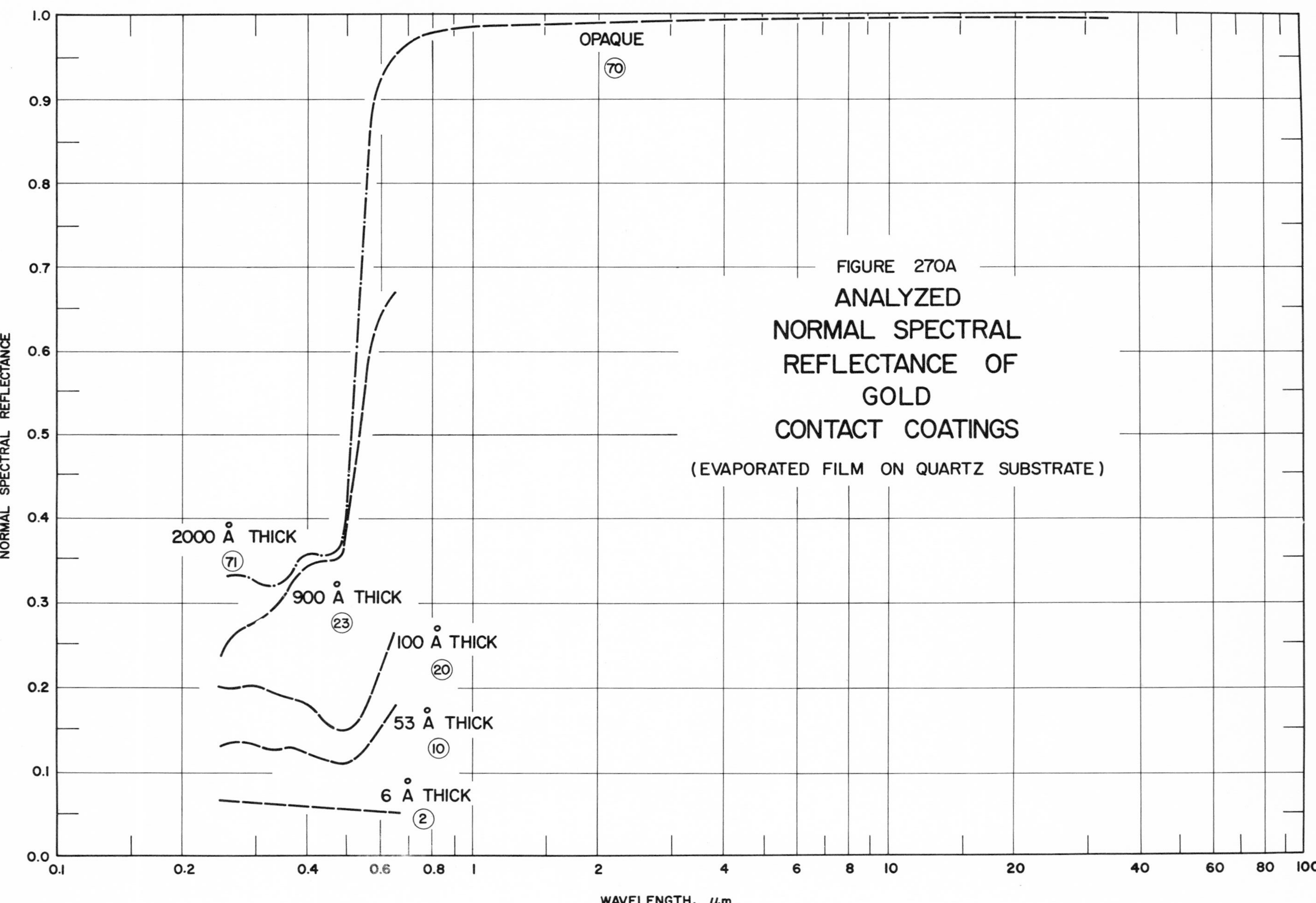
FIGURE 270A
ANALYZED
NORMAL SPECTRAL
REFLECTANCE OF
GOLD
CONTACT COATINGS
(EVAPORATED FILM ON QUARTZ SUBSTRATE)
OPAQUE
(70)
2000 Å THICK
(71)
900 Å THICK
(23)
100 Å THICK
(20)
53 Å THICK
(10)
6 Å THICK
(2)
NORMAL SPECTRAL REFLECTANCE
WAVELENGTH, μm
0.0
0.1
0.2
0.3
0.4
0.5
0.6
0.7
0.8
0.9
1.0
0.1
0.2
0.4
0.6
0.8
1
2
4
6
8
10
20
40
60
80
100

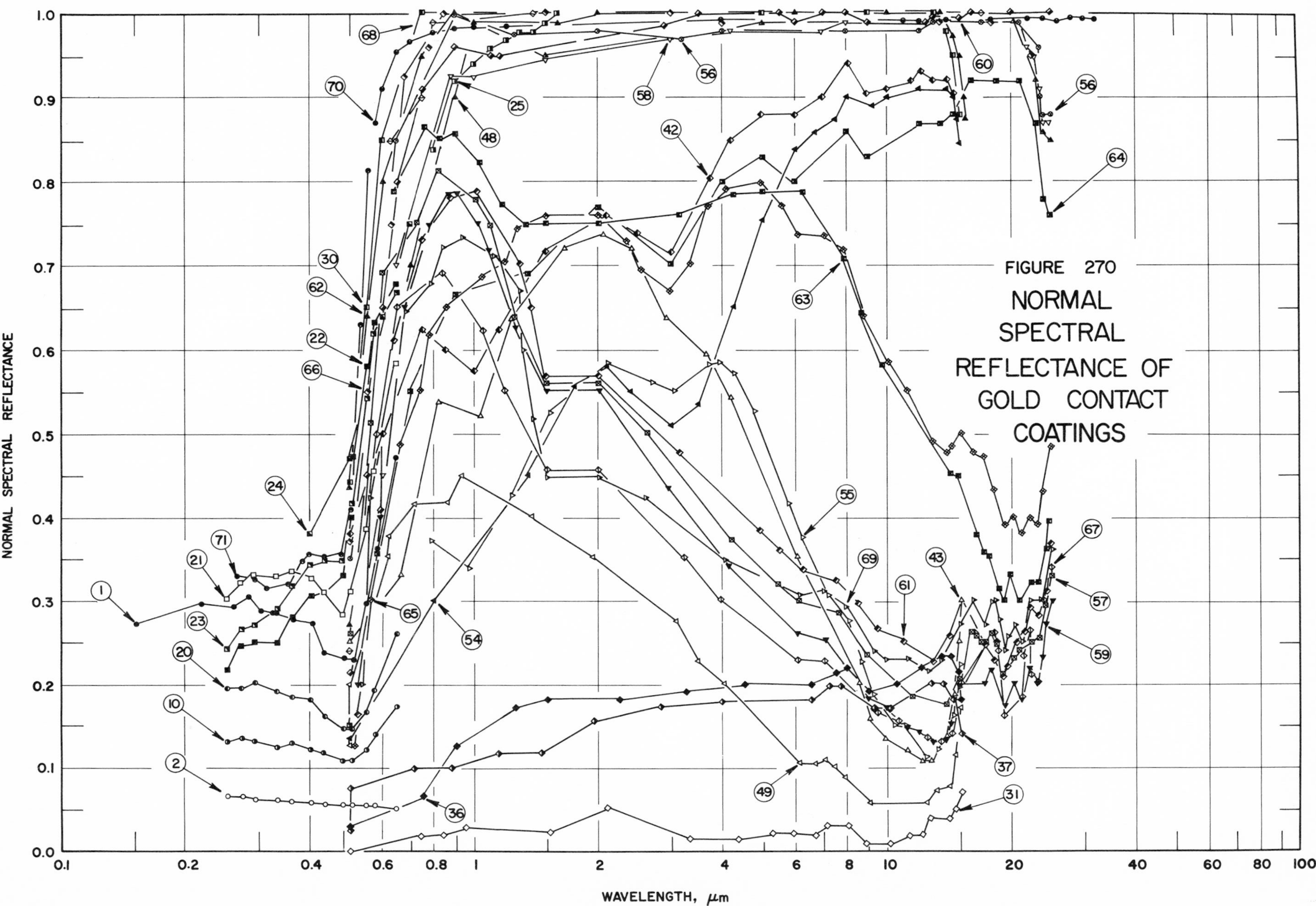
FIGURE 270
NORMAL SPECTRAL REFLECTANCE OF GOLD CONTACT COATINGS
NORMAL SPECTRAL REFLECTANCE
WAVELENGTH, μm

SPECIFICATION TABLE NO. 270 NORMAL SPECTRAL REFLECTANCE OF GOLD CONTACT COATINGS

Curve No.	Ref. No.	Year	Temperature, K	Wavelength Range, μm	Geometry θ	θ'	ω'	Reported Error, %	Composition (weight percent), Specifications, and Remarks
1	227	1958	298	0.153-0.643	~0°	~0°			Gold; quartz substrate; vacuum deposited.
2	228	1959	298	0.2536-0.6438	~0°	~0°			Gold (6 Å thick); quartz substrate; vacuum deposited.
3*	228	1959	298	0.2536-0.6438	~0°	~0°			Similar to above specimen and conditions except 12 Å thick.
4*	228	1959	298	0.2536-0.6438	~0°	~0°			Similar to above specimen and conditions except 18 Å thick.
5*	228	1959	298	0.2536-0.6438	~0°	~0°			Similar to above specimen and conditions except 23.5 Å thick.
6*	228	1959	298	0.2536-0.6438	~0°	~0°			Similar to above specimen and conditions except 29.5 Å thick.
7*	228	1959	298	0.2536-0.6438	~0°	~0°			Similar to above specimen and conditions except 35 Å thick.
8*	228	1959	298	0.2536-0.6438	~0°	~0°			Similar to above specimen and conditions except 41 Å thick.
9*	228	1959	298	0.2536-0.6438	~0°	~0°			Similar to above specimen and conditions except 47 Å thick.
10	228	1959	298	0.2536-0.6438	~0°	~0°			Similar to above specimen and conditions except 53 Å thick.
11*	228	1959	298	0.2536-0.6438	~0°	~0°			Similar to above specimen and conditions except 105.5 Å thick.
12*	228	1959	298	0.2536-0.6438	~0°	~0°			Similar to above specimen and conditions except 211 Å thick.
13*	228	1959	298	0.2536-0.6438	~0°	~0°			Similar to above specimen and conditions except 317 Å thick.
14*	228	1959	298	0.2536-0.6438	~0°	~0°			Similar to above specimen and conditions except 422 Å thick.
15*	228	1959	298	0.2536-0.6438	~0°	~0°			Similar to above specimen and conditions except 528 Å thick.
16*	228	1959	298	0.2536-0.6438	~0°	~0°			Similar to above specimen and conditions except 623 Å thick.
17*	228	1959	298	0.2536-0.6438	~0°	~0°			Similar to above specimen and conditions except 739 Å thick.
18*	228	1959	298	0.2536-0.6438	~0°	~0°			Similar to above specimen and conditions except 844 Å thick.
19*	228	1959	298	0.2536-0.6438	~0°	~0°			Similar to above specimen and conditions except 950 Å thick.
20	228	1959	298	0.2536-0.6438	~0°	~0°			Similar to above specimen and conditions except 100 Å thick.
21	228	1959	298	0.2536-0.6438	~0°	~0°			Similar to above specimen and conditions except 300 Å thick.
22	228	1959	298	0.2536-0.6438	~0°	~0°			Similar to above specimen and conditions except 600 Å thick.
23	228	1959	298	0.2536-0.6438	~0°	~0°			Similar to above specimen and conditions except 900 Å thick.
24	215	1962	~298	0.4-0.7	~0°		2π		Gold (~1250 Å thick); glass substrate; Hanovia Liquid Bright White Gold No. 10, manufactured by Englehard Industries; exact composition proprietary, contains gold, rhodium, bismuth, chromium, and vanadium; prepared by thermal decomposition in air of a solution of gold sulforesinate, rhodium sulforesinate, and bismuth, chromium, and vanadium resinates; application consisted of dropping solution onto substrate spinning at 1550 rpm; measured relative to $MgCO_3$.
25	215	1962	~298	0.8-2.0	~0°	~0°			Above specimen and conditions except measured relative to plane, polished aluminum; specular component only.

* Not shown on plot

SPECIFICATION TABLE NO. 270 NORMAL SPECTRAL REFLECTANCE OF GOLD CONTACT COATINGS (continued)

Curve No.	Ref. No.	Year	Temperature, K	Wavelength Range, μm	Geometry θ	θ'	ω'	Reported Error, %	Composition (weight percent), Specifications, and Remarks
26*	139	1961	322	1.50-15.0	~0°		2π	<2	Gold, Bright Gold No. 6854 from Engelhard Industries; titanium substrate; sprayed; fired at 873 K for 5 min; shiny finish; as received; data extracted from smooth curve; hohlraum at 523 K; sample maintained at ~322 K; converted from R (2π, ~0°).
27*	139	1961	322	2.0-15.0	~0°		2π	<2	Above specimen and conditions; diffuse component only.
28*	139	1961	322	1.0-15.0	~0°		2π	<2	Similar to curve 26 specimen and conditions except hohlraum at 773 K.
29*	139	1961	322	1.0-14.99	~0°		2π	<2	Above specimen and conditions; diffuse component only.
30	139	1961	322	0.50-15.0	~0°		2π	<2	Similar to curve 26 specimen and conditions except hohlraum at 1273 K.
31	139	1961	322	0.50-15.0	~0°		2π	<2	Above specimen and conditions; diffuse component only.
32*	139	1961	322	2.0-15.0	~0°		2π	<2	Similar to curve 26 specimen and conditions except heated in air at 1089 K for 30 min.
33*	139	1961	322	2.0-15.0	~0°		2π	<2	Above specimen and conditions; diffuse component only.
34*	139	1961	322	1.0-15.0	~0°		2π	<2	Similar to curve 32 specimen and conditions except hohlraum at 773 K.
35*	139	1961	322	1.0-15.0	~0°		2π	<2	Above specimen and conditions; diffuse component only.
36	139	1961	322	0.50-15.0	~0°		2π	<2	Similar to curve 32 specimen and conditions except hohlraum at 1273 K.
37	139	1961	322	0.50-15.0	~0°		2π	<2	Above specimen and conditions; diffuse component only.
38*	139	1961	322	2.0-15.5	~0°		2π	<2	Similar to curve 26 specimen and conditions except heated in vacuum (6.8 x 10^{-5} mm Hg) at 1089 K for 30 min.
39*	139	1961	322	2.0-15.0	~0°		2π	<2	Above specimen and conditions; diffuse component only.
40*	139	1961	322	1.0-15.0	~0°		2π	<2	Similar to curve 38 specimen and conditions except hohlraum at 773 K.
41*	139	1961	322	1.0-15.0	~0°		2π	<2	Above specimen and conditions; diffuse component only.
42	139	1961	322	0.50-15.0	~0°		2π	<2	Similar to curve 38 specimen and conditions except hohlraum at 1273 K.
43	139	1961	322	0.50-15.0	~0°		2π	<2	Above specimen and conditions; diffuse component only.
44*	139	1961	322	2.0-15.0	~0°		2π	<2	Similar to curve 26 specimen and conditions except matte finish.
45*	139	1961	322	2.0-15.0	~0°		2π	<2	Above specimen and conditions; diffuse component only.
46*	139	1961	322	1.0-14.0	~0°		2π	<2	Similar to curve 44 specimen and conditions except hohlraum at 773 K.
47*	139	1961	322	1.0-15.0	~0°		2π	<2	Above specimen and conditions; diffuse component only.
48	139	1961	322	0.50-15.5	~0°		2π	<2	Similar to curve 44 specimen and conditions except hohlraum at 1273 K.
49	139	1961	322	0.50-15.0	~0°		2π	<2	Above specimen and conditions; diffuse component only.
50*	139	1961	322	2.0-15.0	~0°		2π	<2	Similar to curve 26 specimen and conditions except heated in vacuum (6.8 x 10^{-5} mm Hg) at 1089 K for 30 min; matte finish.
51*	139	1961	322	2.0-15.0	~0°		2π	<2	Above specimen and conditions; diffuse component only.
52*	139	1961	322	1.0-15.0	~0°		2π	<2	Similar to curve 50 specimen and conditions except hohlraum at 773 K.

* Not shown on plot

SPECIFICATION TABLE NO. 270 NORMAL SPECTRAL REFLECTANCE OF GOLD CONTACT COATINGS (continued)

Curve No.	Ref. No.	Year	Temperature, K	Wavelength Range, μm	Geometry θ	θ'	ω'	Reported Error, %	Composition (weight percent), Specifications, and Remarks
53*	139	1961	322	1.0-15.0	~0°		2π	<2	Above specimen and conditions; diffuse component only.
54	139	1961	322	0.50-15.0	~0°		2π	<2	Similar to curve 50 specimen and conditions except hohlraum at 1273 K.
55	139	1961	322	0.50-15.0	~0°		2π	<2	Above specimen and conditions; diffuse component only.
56	202	1962	322	0.50-25.0	~0°		2π	<2	Gold, Selrex 24 K (5.08 μm thick); as received, cleaned aluminum 6061-T6 alloy substrate; plated; sample maintained at ~322 K; hohlraum at 1273 K; data extracted from smooth curve; converted from R (2π,~0°).
57	202	1962	322	0.50-25.0	~0°		2π	<2	Above specimen and conditions; diffuse component only.
58	202	1962	322	0.50-25.0	~0°		2π	<2	Similar to curve 56 specimen and conditions except exposed to vacuum (<4 x 10^{-8} mm Hg) for 24 hrs.
59	202	1962	322	0.52-25.0	~0°		2π	<2	Above specimen and conditions; diffuse component only.
60	202	1962	322	0.50-25.0	~0°		2π	<2	Similar to curve 56 specimen and conditions except x-ray exposed in vacuum (4 x 10^{-8} mm Hg) for 24 hrs.
61	202	1962	322	0.52-24.99	~0°		2π	<2	Above specimen and conditions; diffuse component only.
62	202	1962	322	0.50-25.0	~0°		2π	<2	Gold, Selrex 24 K (5.08 μm thick); chemically polished aluminum 6061-T6 alloy substrate; plated; sample maintained at ~322 K; hohlraum at 1273 K; data extracted from smooth curve; converted from R (2π,~0°).
63	202	1962	322	0.51-24.99	~0°		2π	<2	Above specimen and conditions; diffuse component only.
64	202	1962	322	0.50-25.0	~0°		2π	<2	Similar to curve 62 specimen and conditions except grit blasted using 60 grit silicon carbide with air pressure of 110 to 120 psi for 30 to 45 sec.
65	202	1962	322	0.51-25.0	~0°		2π	<2	Above specimen and conditions; diffuse component only.
66	202	1962	322	0.50-25.0	~0°		2π	<2	Similar to curve 62 specimen and conditions except exposed to vacuum (<4 x 10^{-8} mm Hg) for 24 hrs.
67	202	1962	322	0.51-25.0	~0°		2π	<2	Above specimen and conditions; diffuse component only.
68	202	1962	322	0.50-25.0	~0°		2π	<2	Similar to curve 62 specimen and conditions except x-ray exposed in vacuum (4 x 10^{-8} mm Hg) for 24 hrs.
69	202	1962	322	0.53-25.0	~0°		2π	<2	Above specimen and conditions; diffuse component only.
70	229	1965	298	0.575-32.0	~5°	~5°			Gold; fused quartz substrate; evaporated in ultrahigh vacuum (5 x 10^{-9} mm Hg); measured in dry nitrogen; 1623 K globar source.
71	230	1965	298	0.268-0.559	~0°	~0°			Gold (~2000 Å thick); quartz substrate; evaporated in vacuum (<3 x 10^{-6} mm Hg) on substrate at 423 K; data extracted from smooth curve.

* Not shown on plot

DATA TABLE NO. 270 NORMAL SPECTRAL REFLECTANCE OF GOLD CONTACT COATINGS

[Wavelength, λ, μm; Reflectance, ρ; Temperature, T, K]

λ	ρ
CURVE 1 T = 298	
0.153	0.271
0.217	0.297
0.262	0.294
0.285	0.305
0.304	0.287
0.327	0.286
0.334	0.285
0.364	0.278
0.407	0.274
0.435	0.238
0.481	0.234
0.510	0.231
0.546	0.298
0.580	0.364
0.643	0.472
CURVE 2 T = 298	
0.2536	0.065
0.2753	0.065
0.2967	0.063
0.3341	0.061
0.3610	0.059
0.4046	0.057
0.4358	0.056
0.4800	0.056
0.5085	0.056
0.5461	0.056
0.5780	0.056
0.6438	0.052
CURVE 3* T = 298	
0.2536	0.083
0.2753	0.081
0.2967	0.077
0.3341	0.074
0.3610	0.071
0.4046	0.069
0.4358	0.068
0.4800	0.066
0.5085	0.068
0.5461	0.071
CURVE 3 (cont.)*	
0.5780	0.074
0.6438	0.075
CURVE 4* T = 298	
0.2536	0.097
0.2753	0.095
0.2967	0.092
0.3341	0.086
0.3610	0.084
0.4046	0.081
0.4358	0.079
0.4800	0.077
0.5085	0.078
0.5461	0.085
0.5780	0.092
0.6438	0.103
CURVE 5* T = 298	
0.2536	0.108
0.2753	0.107
0.2967	0.104
0.3341	0.098
0.3610	0.096
0.4046	0.091
0.4358	0.086
0.4800	0.085
0.5085	0.085
0.5461	0.094
0.5780	0.104
0.6438	0.125
CURVE 6* T = 298	
0.2536	0.118
0.2753	0.117
0.2967	0.115
0.3341	0.108
0.3610	0.105
0.4046	0.101
0.4358	0.097
0.4800	0.092
CURVE 6 (cont.)*	
0.5085	0.093
0.5461	0.102
0.5780	0.115
0.6438	0.142
CURVE 7* T = 298	
0.2536	0.127
0.2753	0.128
0.2967	0.124
0.3341	0.117
0.3610	0.113
0.4046	0.109
0.4358	0.105
0.4800	0.099
0.5085	0.099
0.5461	0.109
0.5780	0.123
0.6438	0.154
CURVE 8* T = 298	
0.2536	0.131
0.2753	0.134
0.2967	0.131
0.3341	0.124
0.3610	0.119
0.4046	0.114
0.4358	0.110
0.4800	0.104
0.5085	0.104
0.5461	0.114
0.5780	0.130
0.6438	0.161
CURVE 9* T = 298	
0.2536	0.133
0.2753	0.137
0.2967	0.134
0.3341	0.125
0.3610	0.123
0.4046	0.119
CURVE 9 (cont.)*	
0.4358	0.115
0.4800	0.108
0.5085	0.108
0.5461	0.119
0.5780	0.134
0.6438	0.165
CURVE 10 T = 298	
0.2536	0.130
0.2753	0.135
0.2967	0.133
0.3341	0.126
0.3610	0.129
0.4046	0.122
0.4358	0.117
0.4800	0.107
0.5085	0.108
0.5461	0.120
0.5780	0.139
0.6438	0.172
CURVE 11* T = 298	
0.2536	0.207
0.2753	0.210
0.2967	0.207
0.3341	0.195
0.3610	0.190
0.4046	0.181
0.4358	0.167
0.4800	0.145
0.5085	0.145
0.5461	0.172
0.5780	0.205
0.6438	0.273
CURVE 12* T = 298	
0.2536	0.272
0.2753	0.303
0.2967	0.301
0.3341	0.289
CURVE 12 (cont.)*	
0.3610	0.285
0.4046	0.278
0.4358	0.259
0.4800	0.223
0.5085	0.231
0.5461	0.300
0.5780	0.366
0.6438	0.478
CURVE 13* T = 298	
0.2536	0.300
0.2753	0.323
0.2967	0.330
0.3341	0.329
0.3610	0.335
0.4046	0.333
0.4358	0.321
0.4800	0.286
0.5085	0.315
0.5461	0.418
0.5780	0.493
0.6438	0.593
CURVE 14* T = 298	
0.2536	0.259
0.2753	0.301
0.2967	0.316
0.3341	0.322
0.3610	0.342
0.4046	0.352
0.4358	0.345
0.4800	0.324
0.5085	0.369
0.5461	0.489
0.5780	0.567
0.6438	0.660
CURVE 15* T = 298	
0.2536	0.256
0.2753	0.278
CURVE 15 (cont.)*	
0.2967	0.274
0.3341	0.294
0.3610	0.331
0.4046	0.349
0.4358	0.346
0.4800	0.338
0.5085	0.402
0.5461	0.530
0.5780	0.605
0.6438	0.683
CURVE 16* T = 298	
0.2536	0.232
0.2753	0.259
0.2967	0.265
0.3341	0.276
0.3610	0.326
0.4046	0.339
0.4358	0.339
0.4800	0.338
0.5085	0.414
0.5461	0.550
0.5780	0.620
0.6438	0.674
CURVE 17* T = 298	
0.2536	0.233
0.2753	0.255
0.2967	0.261
0.3341	0.264
0.3610	0.306
0.4046	0.320
0.4358	0.328
0.4800	0.333
0.5085	0.422
0.5461	0.557
0.5780	0.623
0.6438	0.675
CURVE 18* T = 298	
0.2536	0.218
0.2753	0.248
0.2967	0.250
0.3341	0.253
0.3610	0.294
0.4046	0.315
0.4358	0.322
0.4800	**0.332**
0.5085	0.426
0.5461	0.569
0.5780	0.627
0.6438	0.668
CURVE 19* T = 298	
0.2536	0.212
0.2753	0.228
0.2967	0.249
0.3341	0.243
0.3610	0.256
0.4046	0.302
0.4358	0.308
0.4800	0.327
0.5085	0.431
0.5461	0.580
0.5780	0.629
0.6438	0.679
CURVE 20 T = 298	
0.2536	0.195
0.2753	0.196
0.2967	0.202
0.3341	0.191
0.3610	0.186
0.4046	0.180
0.4358	0.160
0.4800	0.145
0.5085	0.146
0.5461	0.166
0.5780	0.194
0.6438	0.260
CURVE 21 T = 298	
0.2536	0.304
0.2753	0.322
0.2967	0.331
0.3341	0.330
0.3610	0.335
0.4046	0.327
0.4358	0.310
0.4800	0.283
0.5085	0.311
0.5461	0.386
0.5780	0.456
0.6438	0.585
CURVE 22 T = 298	
0.2536	0.218
0.2753	0.245
0.2967	0.250
0.3341	0.250
0.3610	0.279
0.4046	0.307
0.4358	0.310*
0.4800	0.330
0.5085	0.473
0.5461	0.582
0.5780	0.634
0.6438	0.677
CURVE 23 T = 298	
0.2536	0.242
0.2753	0.266
0.2967	0.270
0.3341	0.290
0.3610	0.318
0.4046	0.344
0.4358	0.348
0.4800	0.347
0.5085	0.417
0.5461	0.542
0.5780	0.619
0.6438	0.668

* Not shown on plot

DATA TABLE NO. 270 NORMAL SPECTRAL REFLECTANCE OF GOLD CONTACT COATINGS (continued)

λ	ρ
CURVE 24 T ~ 298	
0.4	0.38
0.5	0.47
0.6	0.64
0.7	0.75
CURVE 25 T ~ 298	
0.8	0.84
0.9	0.92
1.0	0.94
1.1	0.96
1.2	0.97
1.3	0.98
1.4	0.98
1.5	0.99
1.6	1.00
1.7	1.01*
1.8	1.01*
1.9	1.02*
2.0	1.02*
CURVE 26* T = 322	
1.50	0.915
1.85	0.975
2.50	1.000
6.00	1.000
10.00	1.000
14.00	1.000
14.75	0.975
15.00	0.910
CURVE 27* T = 322	
2.00	0.007
5.99	0.007
7.01	0.013
7.93	0.000
10.53	0.000
15.00	0.000
CURVE 28* T = 322	
1.00	1.000
2.00	0.990
3.50	0.985
4.00	0.980
5.00	0.990
6.25	0.970
7.00	0.970
8.00	1.000
10.00	1.000
11.25	0.990
12.30	0.995
14.00	0.970
14.75	0.930
15.05	0.875
15.00	0.830
CURVE 29* T = 322	
1.00	0.000
1.50	0.030
1.94	0.001
6.00	0.001
6.89	0.013
7.25	0.013
7.75	0.000
11.00	0.000
14.99	0.000
CURVE 30 T = 322	
0.50	0.440
0.55	0.650
0.60	0.850
0.75	1.000
5.00	1.000
9.00	1.000
13.00	1.000*
14.00	0.980*
14.50	0.950*
15.00	0.880*
CURVE 31 T = 322	
0.50	0.000
0.74	0.019
0.84	0.019
0.95	0.029
1.54	0.024
2.08	0.052
3.30	0.015
4.37	0.015
5.25	0.021
5.94	0.023
6.69	0.019
7.13	0.031
8.00	0.031
8.87	0.009
10.17	0.009
11.38	0.019
12.01	0.020
12.78	0.040
14.02	0.040
14.65	0.051
15.00	0.072
CURVE 32* T = 322	
2.00	0.190
2.25	0.150
2.90	0.130
4.00	0.140
4.75	0.135
5.25	0.130
6.25	0.145
7.10	0.160
7.75	0.150
9.00	0.150
10.00	0.155
11.00	0.130
12.00	0.130
13.00	0.150
14.00	0.180
15.00	0.135
CURVE 33* T = 322	
2.00	0.168
2.49	0.134
2.90	0.121
6.21	0.120
6.97	0.140
7.33	0.138
8.14	0.131
8.52	0.122
9.49	0.122
9.77	0.129
10.15	0.129
12.09	0.099
13.18	0.119
14.49	0.139
15.00	0.139
CURVE 34* T = 322	
1.00	0.120
2.00	0.155
2.50	0.140
3.00	0.130
4.05	0.150
5.00	0.170
6.40	0.175
7.75	0.195
9.00	0.170
9.50	0.180
10.00	0.190
10.80	0.170
12.00	0.170
13.00	0.170
14.00	0.200
14.50	0.200
15.00	0.190
CURVE 35* T = 322	
1.00	0.150
1.56	0.166
3.00	0.166
4.70	0.157
5.03	0.164
5.68	0.164
CURVE 35 (cont.)*	
6.01	0.170
7.54	0.184
7.96	0.182
9.08	0.155
10.02	0.167
12.23	0.168
13.90	0.189
14.58	0.188
15.00	0.166
CURVE 36 T = 322	
0.50	0.030
0.75	0.065
0.90	0.125
1.25	0.170
1.50	0.180
2.25	0.180
3.25	0.190
4.50	0.200
6.25	0.200
7.50	0.215
8.00	0.220
9.00	0.190
10.50	0.200
12.00	0.220
13.50	0.235
14.25	0.235
14.75	0.215
15.00	0.180
CURVE 37 T = 322	
0.50	0.025
0.50	0.075
0.71	0.099
0.88	0.099
1.16	0.117
1.45	0.119
1.95	0.155
2.83	0.171
3.99	0.179
6.50	0.180
7.23	0.197
7.70	0.197
CURVE 37 (cont.)	
9.23	0.170
10.11	0.170
12.87	0.201
13.51	0.201
14.46	0.180
15.00	0.140
CURVE 38* T = 322	
2.00	0.920
2.30	0.850
2.75	0.800
3.10	0.790
4.00	0.835
4.75	0.885
6.00	0.910
7.50	0.930
8.25	0.935
9.25	0.925
10.50	0.925
11.00	0.930
12.75	0.930
13.50	0.900
14.50	0.840
15.50	0.790
CURVE 39* T = 322	
2.00	0.799
2.37	0.700
3.08	0.601
3.99	0.507
4.65	0.420
6.00	0.292
6.85	0.249
7.99	0.173
9.29	0.118
10.77	0.084
12.20	0.066
13.23	0.064
14.14	0.074
15.00	0.100
CURVE 40* T = 322	
1.00	0.730
1.25	0.755
1.50	0.760
2.20	0.740
3.30	0.800
4.25	0.865
4.75	0.885
6.00	0.890
7.00	0.910
7.60	0.930
9.35	0.915
10.50	0.925
11.75	0.940
12.35	0.940
13.50	0.965
14.50	0.985
15.00	0.980
CURVE 41* T = 322	
1.00	0.616
1.38	0.606
1.67	0.614
2.33	0.652
2.65	0.647
3.19	0.620
3.90	0.552
4.65	0.471
5.34	0.402
7.00	0.276
8.01	0.223
9.00	0.154
9.91	0.124
11.00	0.105
11.77	0.082
13.02	0.084
14.02	0.084
15.00	0.114
CURVE 42 T = 322	
0.50	0.380
0.60	0.500
0.75	0.625
CURVE 42 (cont.)	
0.85	0.600
1.00	0.575
1.15	0.625
1.50	0.715
2.10	0.760
2.50	0.740
3.00	0.715
3.75	0.805
4.20	0.850
5.00	0.880
6.00	0.880
7.00	0.900
8.10	0.940
9.00	0.905
10.00	0.910
11.50	0.920
12.15	0.930
13.00	0.920
14.00	0.920
14.65	0.905
15.00	0.880*
CURVE 43 T = 322	
0.50	0.250
0.66	0.331
0.82	0.537
1.04	0.523
1.24	0.637
1.66	0.721
2.05	0.739
2.40	0.720
2.94	0.639
3.64	0.596
4.19	0.543
6.01	0.351
7.59	0.277
8.57	0.203
9.09	0.158
9.81	0.134
11.29	0.120
12.01	0.109
12.77	0.109
13.81	0.139
14.52	0.187
14.92	0.252
15.00	0.300

* Not shown on plot

DATA TABLE NO. 270 NORMAL SPECTRAL REFLECTANCE OF GOLD CONTACT COATINGS (continued)

CURVE 44*
T = 322

λ	ρ
2.00	0.935
2.25	0.985
2.50	1.000
3.00	1.000
14.00	1.000
14.65	0.950
15.00	0.900

CURVE 45*
T = 322

λ	ρ
2.00	0.212
2.73	0.210
4.22	0.165
6.09	0.121
6.98	0.100
8.51	0.065
9.51	0.050
10.11	0.050
11.51	0.035
12.11	0.035
13.01	0.021
15.00	0.021

CURVE 46*
T = 322

λ	ρ
1.00	1.000
1.50	1.000
2.00	0.965
3.00	0.950
3.75	0.990
4.00	0.990
5.00	0.965
5.75	0.955
6.50	0.955
7.50	0.980
8.20	0.995
9.00	1.000
10.00	1.000
11.00	0.985
11.75	0.980
13.00	0.980
14.00	0.955

CURVE 47*
T = 322

λ	ρ
1.00	0.599
1.46	0.526
1.67	0.451
1.81	0.350
2.02	0.308
3.19	0.237
4.01	0.181
4.99	0.137
5.71	0.109
7.08	0.084
7.71	0.073
8.26	0.051
9.91	0.032
10.87	0.023
11.71	0.015
13.48	0.015
14.33	0.009
15.00	0.019

CURVE 48
T = 322

λ	ρ
0.50	0.270
0.60	0.500*
0.70	0.700
0.90	0.900
1.00	0.990
1.60	0.990
2.00	1.000
6.00	1.000
10.00	1.000
13.50	1.000
14.50	0.975
15.00	0.950
15.45	0.900
15.50	0.875

CURVE 49
T = 322

λ	ρ
0.50	0.125
0.50	0.200
0.62	0.352
0.62	0.377
0.72	0.418
0.86	0.418

CURVE 49 (cont.)

λ	ρ
0.93	0.450
1.38	0.402
1.94	0.352
3.09	0.275
3.49	0.229
4.01	0.201
6.14	0.106
6.65	0.105
7.07	0.109
7.45	0.100
7.99	0.088
9.07	0.058
12.59	0.058
13.23	0.073
14.03	0.078
14.58	0.115
15.00	0.171

CURVE 50*
T = 322

λ	ρ
2.00	0.750
3.00	0.620
3.15	0.500
3.35	0.400
3.70	0.300
4.00	0.250
4.50	0.210
5.00	0.190
5.50	0.190
7.00	0.150
8.00	0.150
8.75	0.160
10.00	0.160
10.75	0.130
11.50	0.085
12.00	0.080
13.50	0.080
14.25	0.090
15.00	0.110

CURVE 51*
T = 322

λ	ρ
2.00	0.669
2.30	0.616
2.70	0.565

CURVE 51 (cont.)*

λ	ρ
3.06	0.546
3.32	0.557
3.85	0.593
4.44	0.582
4.98	0.552
5.28	0.528
5.76	0.477
6.10	0.436
6.75	0.401
7.26	0.357
8.01	0.281
8.59	0.226
9.30	0.186
10.52	0.147
12.43	0.091
13.51	0.091
14.26	0.104
15.00	0.111

CURVE 52*
T = 322

λ	ρ
1.00	0.590
2.00	0.620
3.00	0.610
3.50	0.650
3.75	0.720
5.00	0.800
6.00	0.825
7.00	0.865
8.00	0.900
8.75	0.890
10.25	0.915
10.75	0.915
12.50	0.940
14.00	0.970
14.50	0.970
15.00	0.950

CURVE 53*
T = 322

λ	ρ
1.00	0.550
1.27	0.604
1.58	0.612
2.02	0.603
2.68	0.554

CURVE 53 (cont.)*

λ	ρ
3.14	0.527
3.84	0.567
4.44	0.548
4.99	0.501
5.87	0.417
6.60	0.352
7.03	0.327
7.64	0.250
7.98	0.225
9.07	0.188
9.80	0.156
11.07	0.131
12.00	0.098
12.51	0.089
12.99	0.089
13.50	0.080
14.25	0.080
15.00	0.125

CURVE 54
T = 322

λ	ρ
0.50	0.135
0.80	0.300
1.35	0.450
1.75	0.555
2.10	0.580
2.40	0.550
3.00	0.510
3.50	0.535
4.25	0.650
5.00	0.755
6.00	0.840
6.75	0.860
7.50	0.875
8.00	0.900
9.25	0.890
10.00	0.900
12.00	0.910
14.00	0.910
14.50	0.900
14.85	0.875
15.00	0.845

CURVE 55
T = 322

λ	ρ
0.50	0.125*
0.79	0.371
0.97	0.339
1.22	0.427
1.55	0.526
2.10	0.586
2.67	0.562
3.01	0.551
3.70	0.583
3.91	0.585
4.25	0.571
4.79	0.527
5.79	0.417
6.23	0.376
7.20	0.309
8.14	0.276
8.63	0.229
9.25	0.187
10.32	0.150
11.02	0.150
12.44	0.114
13.22	0.122
14.38	0.164
14.87	0.198
15.00	0.225

CURVE 56
T = 322

λ	ρ
0.50	0.350
0.55	0.650*
0.65	0.850
0.90	1.000
1.25	0.975
2.00	0.980
3.20	0.970
4.00	0.980
8.00	0.980
12.00	0.980
13.00	0.990
17.00	0.990
21.00	0.990
23.00	0.960
23.60	0.900
24.00	0.880
25.00	0.880

CURVE 57
T = 322

λ	ρ
0.50	0.261
0.56	0.513
0.60	0.694
0.73	0.752
0.82	0.814
1.01	0.780
1.09	0.749
1.25	0.640
1.50	0.560
2.00	0.560
2.62	0.501
4.20	0.373
5.43	0.320
6.16	0.301
7.60	0.286
8.95	0.236
11.48	0.185
13.94	0.175
14.90	0.209
15.88	0.261
16.54	0.258
16.92	0.252
17.33	0.252
17.91	0.262
18.31	0.249
18.99	0.213
20.01	0.233
20.93	0.240
22.06	0.252
23.22	0.254
24.03	0.296
25.00	0.330

CURVE 58
T = 322

λ	ρ
0.50	0.200*
0.60	0.450
0.65	0.700
0.88	0.925
1.00	0.925
1.50	0.945
2.00	0.980*
3.00	0.970
4.20	0.980
7.00	0.980

CURVE 58 (cont.)

λ	ρ
8.00	0.990
13.00	0.990*
17.00	0.990*
21.00	0.990*
22.00	0.960
23.00	0.950
23.50	0.910
24.00	0.870
25.00	0.870

CURVE 59
T = 322

λ	ρ
0.52	0.200
0.59	0.400
0.68	0.650
0.78	0.749
0.87	0.786
0.91	0.788
1.03	0.750
1.09	0.719
1.26	0.627
1.50	0.552
2.01	0.552
2.94	0.436
4.12	0.341
6.02	0.261
7.02	0.254
8.97	0.181
9.96	0.170
11.07	0.149
11.97	0.143
12.95	0.130
13.70	0.134
14.23	0.154
14.85	0.199
17.04	0.201
17.97	0.219
19.03	0.173
20.03	0.203
21.00	0.182
22.02	0.221
23.02	0.201
23.61	0.235
24.01	0.273
25.00	0.300

* Not shown on plot

DATA TABLE NO. 270 NORMAL SPECTRAL REFLECTANCE OF GOLD CONTACT COATINGS (continued)

CURVE 60 T = 322

λ	ρ
0.50	0.215
0.55	0.450
0.60	0.650
0.65	0.800
0.75	0.910
0.90	0.960
1.10	0.950
1.50	0.950
2.90	0.985
3.50	1.000
5.50	1.000
6.00	0.990
7.00	1.000
8.00	1.000
9.00	0.990
12.50	0.990
13.00	0.995
15.00	0.995
16.00	1.000
20.00	1.000
25.00	1.000

CURVE 61 T = 322

λ	ρ
0.52	0.162
0.59	0.409
0.64	0.612
0.75	0.733
0.88	0.781
1.03	0.789
1.29	0.701
1.38	0.650
1.50	0.568
2.00	0.568
3.13	0.478
4.97	0.383
5.46	0.360
6.22	0.337
7.48	0.326
8.50	0.297
9.40	0.265
10.99	0.252
12.98	0.229
14.09	0.257
15.02	0.300*

CURVE 61 (cont.)

λ	ρ
17.08	0.249
18.04	0.280
18.49	0.252
19.00	0.211
19.46	0.224
20.05	0.250
21.02	0.230
21.60	0.264
22.01	0.292
23.08	0.281
24.09	0.314
24.99	0.369

CURVE 62 T = 322

λ	ρ
0.50	0.435
0.55	0.640
0.60	0.800
0.75	0.950
0.90	1.000
1.50	0.950
2.00	0.980*
4.00	0.980*
5.00	0.990
6.00	0.990*
7.00	1.000*
8.00	1.000*
9.00	0.990*
13.00	0.990*
17.00	0.990*
20.50	0.990
23.00	0.920
24.00	0.860
25.00	0.850

CURVE 63 T = 322

λ	ρ
0.51	0.400
0.57	0.649
0.64	0.789
0.76	0.866
0.83	0.851
0.90	0.858
1.04	0.825
1.18	0.774

CURVE 63 (cont.)

λ	ρ
1.34	0.749
1.50	0.750
2.00	0.750
3.16	0.762
4.28	0.786
5.00	0.787
6.26	0.739
7.93	0.708
8.69	0.644
9.79	0.582
14.34	0.451
14.99	0.449
16.45	0.378
17.23	0.357
17.83	0.354
18.60	0.315
19.04	0.301
19.97	0.332
20.95	0.301
22.17	0.323
23.01	0.323
24.21	0.361
24.99	0.396

CURVE 64 T = 322

λ	ρ
0.50	0.150
0.58	0.355
0.70	0.550
0.90	0.665
1.35	0.690
2.00	0.770
3.00	0.700
4.00	0.800
5.00	0.830
6.00	0.800
8.00	0.860
9.00	0.830
12.00	0.870
13.40	0.870
14.50	0.880
15.10	0.880*
16.00	0.920
18.50	0.920
21.00	0.920
23.00	0.870

CURVE 64 (cont.)

λ	ρ
24.00	0.780
25.00	0.760

CURVE 65 T = 322

λ	ρ
0.51	0.125
0.56	0.301
0.66	0.487
0.74	0.551
0.78	0.619
0.86	0.650
1.03	0.661
1.16	0.687
1.28	0.705
1.37	0.745
1.50	0.760
2.00	0.760
2.36	0.730
2.54	0.695
2.93	0.670
3.36	0.702
3.70	0.771
4.10	0.794
4.99	0.799
5.60	0.774
6.10	0.739
7.12	0.738
7.90	0.718
8.79	0.641
10.11	0.586
11.28	0.551
12.99	0.490
13.95	0.478
14.44	0.486
15.04	0.500
16.19	0.477
17.03	0.472
18.26	0.432
19.36	0.390
20.03	0.400
21.11	0.381
22.05	0.400
23.07	0.391
23.99	0.430
25.00	0.485

CURVE 66 T = 322

λ	ρ
0.50	0.370
0.55	0.550
0.63	0.750
0.68	0.925
0.78	0.960
0.85	1.000
1.50	1.000
3.00	1.000
4.00	0.990
8.00	0.990*
13.50	0.990
18.00	0.990
21.10	0.990*
22.50	0.950
23.00	0.950*
24.00	0.880*
25.00	0.880*

CURVE 67 T = 322

λ	ρ
0.51	0.259
0.58	0.500
0.65	0.651
0.84	0.691
1.05	0.623
1.19	0.551
1.51	0.457
2.00	0.457
3.24	0.351
3.96	0.302
6.06	0.230
7.03	0.229
9.48	0.165
10.65	0.155
12.59	0.135
13.48	0.130
14.25	0.141
15.10	0.202
16.06	0.262
17.10	0.252*
18.07	0.261
18.52	0.243
19.07	0.212
20.05	0.253*
21.05	0.234

CURVE 67 (cont.)

λ	ρ
22.11	0.265
23.01	0.254*
23.92	0.303
25.00	0.340

CURVE 68 T = 322

λ	ρ
0.50	0.240
0.55	0.550*
0.63	0.850
0.75	0.900
0.80	0.990
1.00	0.990*
1.40	1.000
9.00	1.000*
17.00	1.000
25.00	1.000*

CURVE 69 T = 322

λ	ρ
0.53	0.200
0.56	0.424
0.64	0.613*
0.68	0.647
0.78	0.678
0.84	0.722
0.93	0.735
1.12	0.711
1.27	0.670
1.32	0.600
1.36	0.519
1.50	0.447
2.00	0.447
2.57	0.424
4.05	0.349
6.16	0.308
7.00	0.312
7.98	0.293
9.33	0.242
9.87	0.232
11.19	0.232
11.92	0.223*
12.46	0.218
13.45	0.230
14.95	0.271

CURVE 69 (cont.)

λ	ρ
16.04	0.303
17.02	0.272
17.84	0.299
18.25	0.302
18.60	0.277
19.10	0.243
19.58	0.257
20.13	0.272
21.14	0.251
22.10	0.302
23.41	0.303
25.00	0.362

CURVE 70 T = 298

λ	ρ
0.575	0.8708
0.600	0.9116
0.650	0.9566
0.700	0.9695
0.800	0.9795
0.900	0.9839
1.00	0.9860
1.20	0.9878
1.50	0.9896*
2.00	0.9914
3.00	0.9930*
4.00	0.9938
5.00	0.9938*
6.00	0.9939*
7.00	0.9939*
8.00	0.9939*
9.00	0.9939*
10.0	0.9939*
11.0	0.9940
12.0	0.9940
13.0	0.9940*
14.0	0.9940
16.0	0.9940*
18.0	0.9940
20.0	0.9940*
22.0	0.9941
24.0	0.9941
26.0	0.9941
28.0	0.9941
30.0	0.9942
32.0	0.9942

CURVE 71 T = 298

λ	ρ
0.268	0.333
0.296	0.325
0.314	0.316
0.355	0.320
0.384	0.348
0.397	0.355
0.432	0.352
0.475	0.356
0.502	0.410
0.515	0.473*
0.532	0.632
0.559	0.816

* Not shown on plot

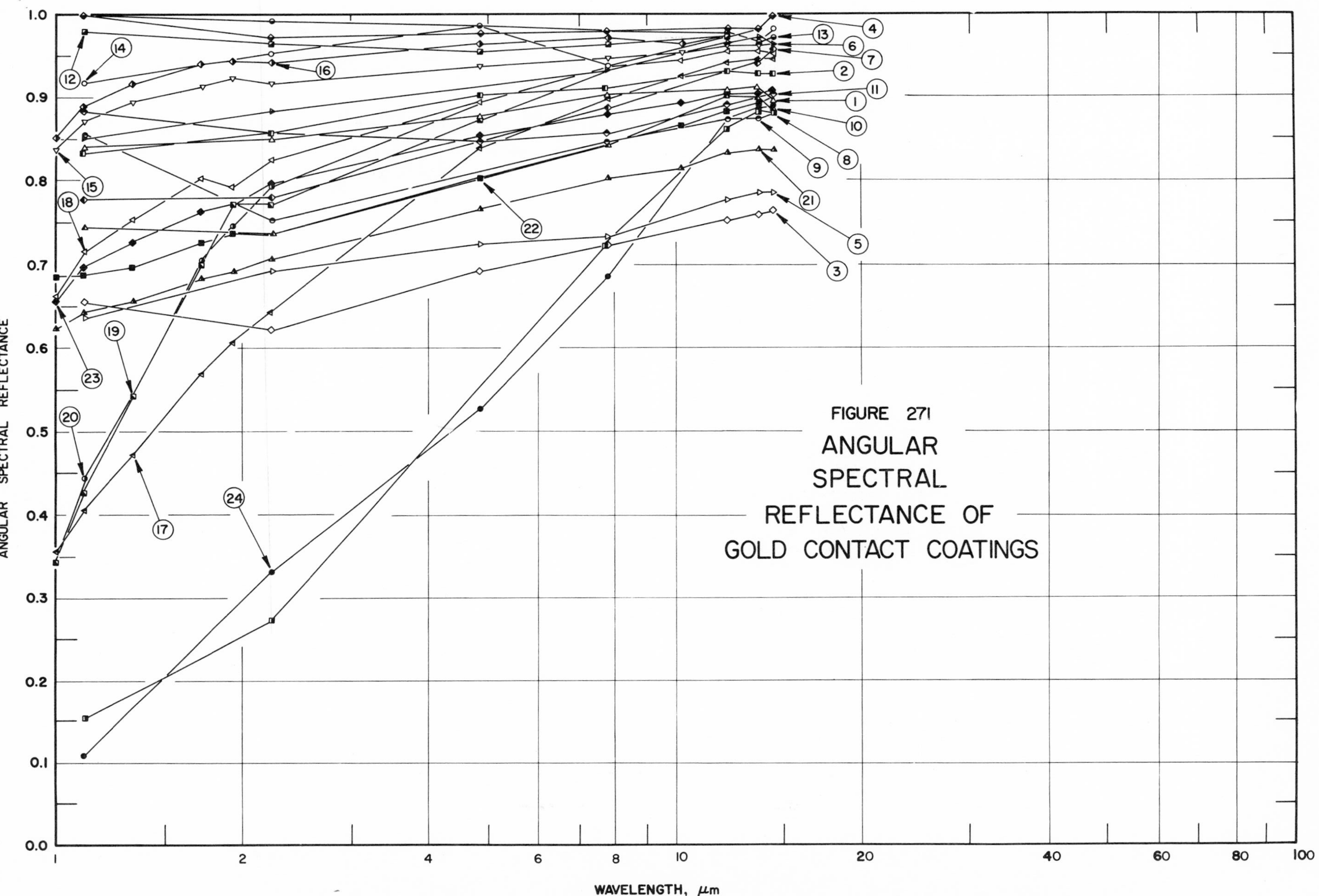
FIGURE 271
ANGULAR
SPECTRAL
REFLECTANCE OF
GOLD CONTACT COATINGS
ANGULAR SPECTRAL REFLECTANCE
WAVELENGTH, μm
1.0
0.9
0.8
0.7
0.6
0.5
0.4
0.3
0.2
0.1
0.0
1
2
4
6
8
10
20
40
60
80
100
1
2
3
4
5
6
7
8
9
10
11
12
13
14
15
16
17
18
19
20
21
22
23
24

SPECIFICATION TABLE NO. 271 ANGULAR SPECTRAL REFLECTANCE OF GOLD CONTACT COATINGS

Curve No.	Ref. No.	Year	Temperature, K	Wavelength Range, μm	Geometry θ	θ'	ω'	Reported Error, %	Composition (weight percent), Specifications, and Remarks
1	231	1962	~298	1.120-14.38	45°	45°			Gold: nickel, epoxy, and 316 stainless steel substrates; stainless steel coated with Maraset 617-C epoxy resin (Marblette Co.), then coated with nickel by immersion in an electroless, alkaline, nickel hypophosphite solution for 30 seconds; gold deposited by immersion in Lustralloy G (gold hypophosphite type) for 2 min at 370 K, then rinsed with water and dried; specular component only. [Authors' designation: NiAu 1C]
2	231	1962	~298	1.120-14.38	45°	45°			Similar to above specimen and conditions. [Authors' designation: NiAu 2C]
3	231	1962	~298	1.120-14.38	45°	45°			Similar to above specimen and conditions. [Authors' designation: NiAu 3C]
4	231	1962	~298	1.120-14.38	45°	45°			Similar to above specimen and conditions. [Authors' designation: NiAu 4C]
5	231	1962	~298	1.120-14.38	45°	45°			Similar to above specimen and conditions. [Authors' designation: NiAu 5C]
6	231	1962	~298	1.120-14.38	45°	45°			Gold, Hanovia Liquid Bright (Englehard Industries): ceramic wall tile (Tiffany Tile Corp.) substrate: substrate cleaned and glazed; applied to substrate spinning at several hundred rpm; fired in air at ~1000 K; specular component only. [Authors' designation: TH Au 2]
7	231	1962	~298	1.120-14.38	45°	45°			Similar to above specimen and conditions. [Authors' designation: TH Au 3]
8	231	1962	~298	1.120-14.38	45°	45°			Similar to above specimen and conditions except glazed ceramic wall tile provided by Sears, Roebuck, and Co. [Authors' designation: SH Au 1]
9	231	1962	~298	1.120-14.38	45°	45°			Similar to above specimen and conditions. [Authors' designation: SH Au 2]
10	231	1962	~298	1.120-14.38	45°	45°			Gold, DuPont Liquid Bright Gold No. 4942; ceramic wall tile (Tiffany Tile Corp.) substrate; substrate cleaned and glazed; applied to substrate spinning at several hundred rpm; fired in air at ~1000 K; specular component only. [Authors' designation: TD Au 1]
11	231	1962	~298	1.120-14.38	45°	45°			Gold (950-3000 Å thick): epoxy and 316 stainless steel substrates; stainless steel cleaned with Maraset 617-C epoxy resin (Marblette Co.), applied by flowing the resin onto the steel at 408 K; plastic surface was cleaned by scrubbing in Alconox solution, then rinsing with water and methanol; gold vapor deposited on heated substrate at ~10^{-5} mm Hg; specular component only. [Authors' designation: Au 187]
12	231	1962	~298	1.120-14.38	45°	45°			Similar to above specimen and conditions. [Authors' designation: Au 188]
13	231	1962	~298	1.120-14.38	45°	45°			Similar to curve 11 specimen and conditions except substrate consists of 316 stainless steel coated with SY627-119 polyurethane (Febert Shorndorfer Co.) by a dip process. [Authors' designation: Au 189B]
14	231	1962	~298	1.009-14.38	45°	45°			Similar to curve 11 specimen and conditions except substrate consists of 316 stainless steel coated with General Electric SR-111 silicone resin by a dip process. [Authors' designation: Au 190]
15	231	1962	~298	1.009-14.38	45°	45°			Similar to curve 11 specimen and conditions except substrate consists of 316 stainless steel coated with SY627-119 polyurethane by a dip process, then coated with vapor-deposited SiO (~5000 Å thick). [Authors' designation: SiOAu 25B]
16	231	1962	~298	1.009-14.38	45°	45°			Similar to above specimen and conditions. [Authors' designation: SiOAu 26B]
17	231	1962	~298	1.009-14.38	45°	45°			Similar to above specimen and conditions. [Authors' designation: SiOAu 28]

SPECIFICATION TABLE NO. 271 ANGULAR SPECTRAL REFLECTANCE OF GOLD CONTACT COATINGS (continued)

Curve No.	Ref. No.	Year	Temperature K	Wavelength Range, μm	Geometry θ	θ'	ω'	Reported Error, %	Composition (weight percent), Specifications, and Remarks
18	231	1962	~298	1.009-14.38	45°	45°			Similar to above specimen and conditions. [Authors' designation: SiOAu 29]
19	231	1962	~298	1.009-14.38	45°	45°			Similar to above specimen and conditions. [Authors' designation: SiOAu 31]
20	231	1962	~298	1.009-14.38	45°	45°			Similar to above specimen and conditions. [Authors' designation: SiOAu 32]
21	231	1962	~298	1.009-14.38	45°	45°			Similar to above specimen and conditions except relatively rough surface. [Authors' designation: SiOAu 27B]
22	231	1962	~298	1.009-14.38	45°	45°			Similar to above specimen and conditions. [Authors' designation: SiOAu 30]
23	231	1962	~298	1.009-14.38	45°	45°			Similar to above specimen and conditions. [Authors' designation: SiOAu 33]
24	231	1962	~298	1.120-14.38	45°	45°			Similar to curve 11 specimen and conditions except substrate consists of uncoated 316 stainless steel. [Authors' designation: AuSS 2]

DATA TABLE NO. 271 ANGULAR SPECTRAL REFLECTANCE OF GOLD CONTACT COATINGS

[Wavelength, λ, μm; Reflectance, ρ; Temperature, T, K]

λ	ρ
CURVE 1	
T ~ 298	
1.120	0.840
2.240	0.850
4.824	0.877
7.780	0.901
12.10	0.910
13.53	0.912
14.38	0.897
CURVE 2	
T ~ 298	
1.120	0.833
2.240	0.859
4.824	0.901
7.780	0.911
12.10	0.932
13.53	0.930
14.38	0.930
CURVE 3	
T ~ 298	
1.120	0.658
2.240	0.624
4.824	0.694
7.780	0.722
12.10	0.752
13.53	0.760
14.38	0.766
CURVE 4	
T ~ 298	
1.120	1.000
2.240	0.972
4.824	0.977
7.780	0.980
12.10	0.983
13.53	0.982
14.38	1.000
CURVE 5	
T ~ 298	
1.120	0.638
CURVE 5 (cont.)	
2.240	0.692
4.824	0.723
7.780	0.734
12.10	0.777
13.53	0.786
14.38	0.787
CURVE 6	
T ~ 298	
1.120	0.851
2.240	0.885
7.780	0.933
12.10	0.966
13.53	0.974
14.38	0.965
CURVE 7	
T ~ 298	
1.120	0.779
2.240	0.780
7.780	0.887
12.10	0.932*
13.53	0.941
14.38	0.959
CURVE 8	
T ~ 298	
1.120	0.152
2.240	0.271
7.780	0.722
12.10	0.864
13.53	0.884
14.38	0.882
CURVE 9	
T ~ 298	
1.120	0.855
2.240	0.751
7.780	0.849
12.10	0.874
13.53	0.875
14.38	0.882*
CURVE 10	
T ~ 298	
1.120	0.745
2.240	0.737
7.780	0.845
12.10	0.900
13.53	0.900
14.38	0.888
CURVE 11	
T ~ 298	
1.120	0.883
2.240	0.856*
4.824	0.849
7.780	0.859
12.10	0.891
13.53	0.900*
14.38	0.903
CURVE 12	
T ~ 298	
1.120	0.980
2.240	0.965
4.824	0.956
7.780	0.964
12.10	0.974
13.53	0.985*
14.38	0.966*
CURVE 13	
T ~ 298	
1.120	1.000*
2.240	0.994
4.824	0.987
7.780	0.980*
12.10	0.979
13.53	0.969
14.38	0.972
CURVE 14	
T ~ 298	
1.120	0.918
2.240	0.952
CURVE 14 (cont.)	
4.824	0.987*
7.780	0.939
12.10	0.975*
13.53	0.971*
14.38	0.982
CURVE 15	
T ~ 298	
1.009	0.837
1.120	0.870
1.345	0.893
1.720	0.912
1.945	0.922
2.240	0.917
4.824	0.939
7.780	0.948
10.20	0.954
12.10	0.963
13.53	0.961
14.38	0.965*
CURVE 16	
T ~ 298	
1.009	0.851
1.120	0.889
1.345	0.917
1.720	0.940
1.945	0.943
2.240	0.941
4.824	0.965
7.780	0.972
10.20	0.965
12.10	0.970*
13.53	0.971*
14.38	0.975*
CURVE 17	
T ~ 298	
1.009	0.357
1.120	0.407
1.345	0.472
1.720	0.570
1.945	0.607
CURVE 17 (cont.)	
2.240	0.643
4.824	0.840
7.780	0.900
10.20	0.929
12.10	0.942
13.53	0.944
14.38	0.949
CURVE 18	
T ~ 298	
1.009	0.662
1.120	0.716
1.345	0.753
1.720	0.801
1.945	0.794
2.240	0.826
4.824	0.897
7.780	0.935*
10.20	0.946
12.10	0.956
13.53	0.958
14.38	0.953
CURVE 19	
T ~ 298	
1.009	0.342
1.120	0.429
1.345	0.541
1.720	0.700
1.945	0.772
2.240	0.772
4.824	0.875
7.780	0.932*
10.20	0.948*
12.10	0.961*
13.53	0.963*
14.38	0.971*
CURVE 20	
T ~ 298	
1.009	0.342*
1.120	0.442
1.345	0.549*
CURVE 20 (cont.)	
1.720	0.702
1.945	0.748
2.240	0.794
4.824	0.896*
7.780	0.947*
10.20	0.963*
12.10	0.972*
13.53	0.976*
14.38	0.977*
CURVE 21	
T ~ 298	
1.009	0.624
1.120	0.643
1.345	0.658
1.720	0.683
1.945	0.692
2.240	0.708
4.824	0.768
7.780	0.803
10.20	0.816
12.10	0.834
13.53	0.837
14.38	0.838
CURVE 22	
T ~ 298	
1.009	0.686
1.120	0.689
1.345	0.699
1.720	0.725
1.945	0.737
2.240	0.739*
4.824	0.803
7.780	0.843*
10.20	0.868
12.10	0.883
13.53	0.893
14.38	0.903*
CURVE 23	
T ~ 298	
1.009	0.657
CURVE 23 (cont.)	
1.120	0.697
1.345	0.726
1.720	0.762
1.945	0.774*
2.240	0.796
4.824	0.855
7.780	0.880
10.20	0.893
12.10	0.902
13.53	0.902
14.38	0.907
CURVE 24	
T ~ 298	
1.120	0.108
2.240	0.331
4.824	0.527
7.780	0.688
12.10	0.871*
13.53	0.889
14.38	0.892

* Not shown on plot

SPECIFICATION TABLE NO. 272 HEMISPHERICAL INTEGRATED ABSORPTANCE OF GOLD CONTACT COATINGS

Curve No.	Ref. No.	Year	Temperature Range, K	Reported Error, %	Composition (weight percent), Specifications and Remarks
1*	194	1955	76	5	24 K gold; stainless steel substrate; gold plated; measured in vacuum (10^{-6} to 10^{-7} mm Hg); absorptance for 300 K blackbody incident radiation.
2*	194	1955	76	5	Gold; plastic mylar (0.0127 mm thick) and gold substrates; gold vaporized on both sides of mylar; measured in vacuum (10^{-6} to 10^{-7} mm Hg); absorptance for 300 K blackbody incident radiation.
3*	194	1955	76	5	Au + 1 Ag (5.08 μm thick); stainless steel substrate; plated; measured in vacuum (10^{-6} to 10^{-7} mm Hg); absorptance for 300 K blackbody incident radiation.
4*	194	1955	76	5	Similar to above specimen and conditions except 2.54 μm thick.
5*	194	1955	76	5	Similar to above specimen and conditions except 0.127 μm thick.
6*	194	1955	76	5	Au + 1 Ag (5.08 μm thick); copper substrate; plated; measured in vacuum (10^{-6} to 10^{-7} mm Hg); absorptance for 300 K blackbody incident radiation.
7*	195	1953	76		Gold; plastic mylar (0.0127 mm thick), gold and lucite substrates; vaporized on both sides of mylar, then wrapped around lucite cylinder; measured in vacuum ($<10^{-6}$ mm Hg); absorptance for 294 K blackbody incident radiation.
8*	195	1953	76		Gold; plastic mylar (0.0127 mm thick), gold and pyrex substrates; gold vaporized on both sides of mylar, then wrapped around pyrex cylinder; measured in vacuum ($<10^{-6}$ mm Hg); absorptance for 294 K blackbody incident radiation.

* No plot given

DATA TABLE NO. 272 HEMISPHERICAL INTEGRATED ABSORPTANCE OF GOLD CONTACT COATINGS

[Temperature, T, K; Absorptance, α]

T	α
CURVE 1*	
76	0.017
CURVE 2*	
76	0.02
CURVE 3*	
76	0.025
CURVE 4*	
76	0.027
CURVE 5*	
76	0.028
CURVE 6*	
76	0.025
CURVE 7*	
76	0.021
CURVE 8*	
76	0.025

* No plot given

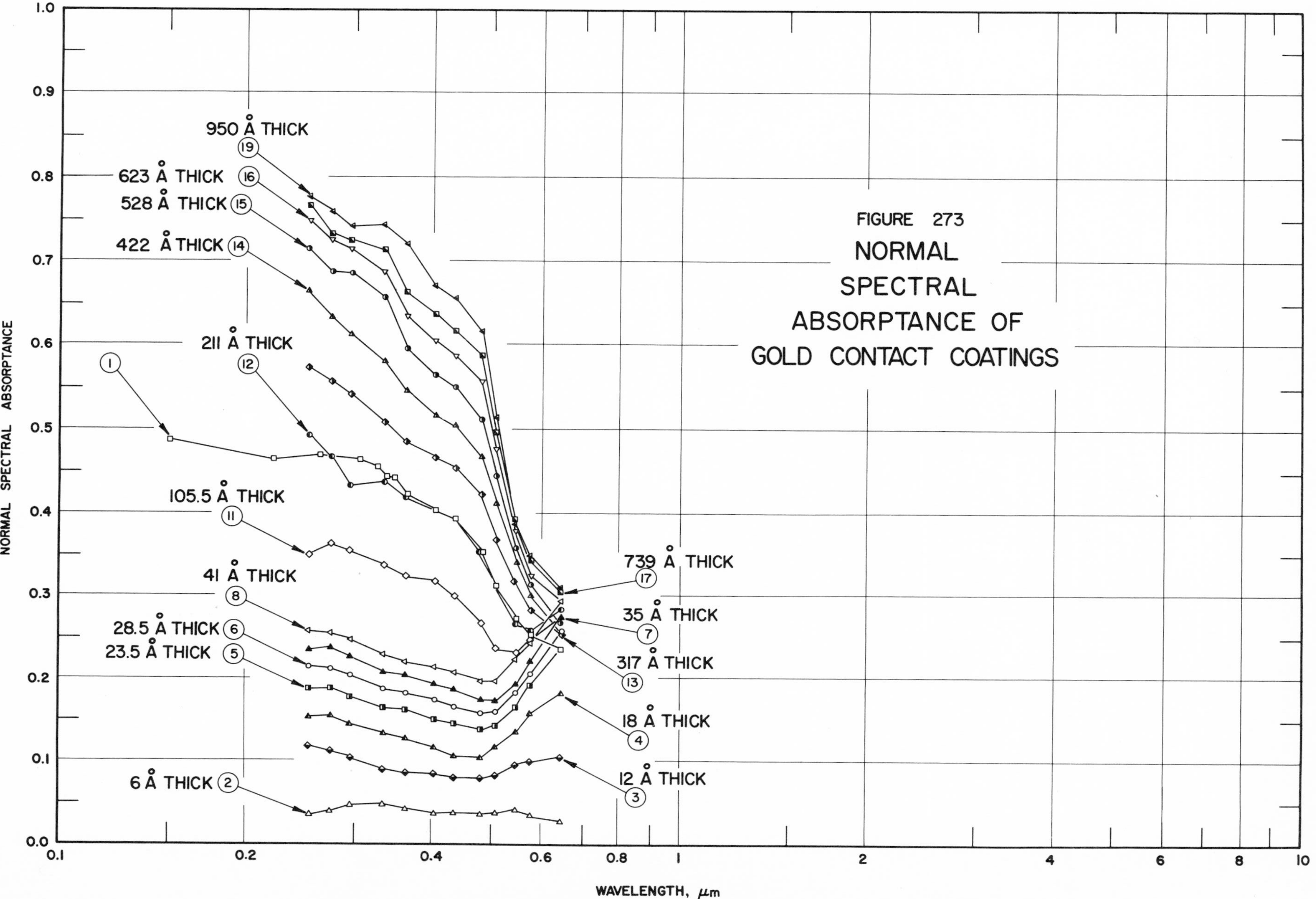
FIGURE 273
NORMAL
SPECTRAL
ABSORPTANCE OF
GOLD CONTACT COATINGS
950 Å THICK (19)
623 Å THICK (16)
528 Å THICK (15)
422 Å THICK (14)
211 Å THICK (12)
(1)
105.5 Å THICK (11)
41 Å THICK (8)
28.5 Å THICK (6)
23.5 Å THICK (5)
6 Å THICK (2)
739 Å THICK (17)
35 Å THICK (7)
317 Å THICK (13)
18 Å THICK (4)
12 Å THICK (3)
NORMAL SPECTRAL ABSORPTANCE
1.0
0.9
0.8
0.7
0.6
0.5
0.4
0.3
0.2
0.1
0.0
0.1
0.2
0.4
0.6
0.8
1
2
4
6
8
10
WAVELENGTH, μm

SPECIFICATION TABLE NO. 273 NORMAL SPECTRAL ABSORPTANCE OF GOLD CONTACT COATINGS

Curve No.	Ref. No.	Year	Temperature, K	Wavelength Range, μm	Geometry θ	Reported Error, %	Composition (weight percent), Specifications, and Remarks
1	227	1958	298	0.2447-0.6438	~0°		Gold; quartz substrate; vacuum deposited.
2	232	1959	298	0.2536-0.6438	~0°		Gold (6 Å thick); quartz substrate; vacuum deposited.
3	232	1959	298	0.2536-0.6438	~0°		Similar to above specimen and conditions except 12 Å thick.
4	232	1959	298	0.2536-0.6438	~0°		Similar to above specimen and conditions except 18 Å thick.
5	232	1959	298	0.2536-0.6438	~0°		Similar to above specimen and conditions except 23.5 Å thick.
6	232	1959	298	0.2536-0.6438	~0°		Similar to above specimen and conditions except 28.5 Å thick.
7	232	1959	298	0.2536-0.6438	~0°		Similar to above specimen and conditions except 35.0 Å thick.
8	232	1959	298	0.2536-0.6438	~0°		Similar to above specimen and conditions except 41.0 Å thick.
9*	232	1959	298	0.2536-0.6438	~0°		Similar to above specimen and conditions except 47.0 Å thick.
10*	232	1959	298	0.2536-0.6438	~0°		Similar to above specimen and conditions except 53.0 Å thick.
11	232	1959	298	0.2536-0.6438	~0°		Similar to above specimen and conditions except 105.5 Å thick.
12	232	1959	298	0.2536-0.6438	~0°		Similar to above specimen and conditions except 211.0 Å thick.
13	232	1959	298	0.2536-0.6438	~0°		Similar to above specimen and conditions except 317.0 Å thick.
14	232	1959	298	0.2536-0.6438	~0°		Similar to above specimen and conditions except 422.0 Å thick.
15	232	1959	298	0.2536-0.6438	~0°		Similar to above specimen and conditions except 528.0 Å thick.
16	232	1959	298	0.2536-0.6438	~0°		Similar to above specimen and conditions except 623 Å thick.
17	232	1959	298	0.2536-0.6438	~0°		Similar to above specimen and conditions except 739.0 Å thick.
18*	232	1959	298	0.2536-0.6438	~0°		Similar to above specimen and conditions except 844.0 Å thick.
19	232	1959	298	0.2536-0.6438	~0°		Similar to above specimen and conditions except 950.0 Å thick.
20*	232	1959	298	0.2536-0.6438	~0°		Similar to above specimen and conditions except 100.0 Å thick.
21*	232	1959	298	0.2536-0.6438	~0°		Similar to above specimen and conditions except 300.0 Å thick.
22*	232	1959	298	0.2536-0.6438	~0°		Similar to above specimen and conditions except 600.0 Å thick.
23*	232	1959	298	0.2536-0.6438	~0°		Similar to above specimen and conditions except 900.0 Å thick.

* Not shown on plot

DATA TABLE NO. 273 NORMAL SPECTRAL ABSORPTANCE OF GOLD CONTACT COATINGS

[Wavelength, λ, μm; Absorptance, α; Temperature, T, K]

CURVE 1
T = 298

λ	α
0.1524	0.488
0.2223	0.464
0.2622	0.470
0.3067	0.463
0.3287	0.455
0.3369	0.441
0.3484	0.441
0.3637	0.423
0.4050	0.403
0.4363	0.393
0.4815	0.352
0.5093	0.312
0.5480	0.272
0.5794	0.253
0.6434	0.236

CURVE 2
T = 298

λ	α
0.2536	0.036
0.2753	0.040
0.2967	0.048
0.3341	0.049
0.3610	0.043
0.4046	0.038
0.4358	0.039
0.4800	0.038
0.5085	0.039
0.5461	0.041
0.5780	0.036
0.6438	0.029

CURVE 3
T = 298

λ	α
0.2536	0.119
0.2753	0.113
0.2967	0.102
0.3341	0.090
0.3610	0.088
0.4046	0.085
0.4358	0.080
0.4800	0.080
0.5085	0.084
0.5461	0.096

CURVE 3 (cont.)

λ	α
0.5780	0.100
0.6438	0.105

CURVE 4
T = 298

λ	α
0.2536	0.151
0.2753	0.153
0.2967	0.145
0.3341	0.134
0.3610	0.129
0.4046	0.117
0.4358	0.106
0.4800	0.104
0.5085	0.117
0.5461	0.136
0.5780	0.159
0.6438	0.181

CURVE 5
T = 298

λ	α
0.2536	0.189
0.2753	0.189
0.2967	0.177
0.3341	0.164
0.3610	0.161
0.4046	0.150
0.4358	0.146
0.4800	0.139
0.5085	0.142
0.5461	0.166
0.5780	0.191
0.6438	0.235*

CURVE 6
T = 298

λ	α
0.2536	0.215
0.2753	0.211
0.2967	0.204
0.3341	0.187
0.3610	0.182
0.4046	0.174
0.4358	0.165
0.4800	0.158

CURVE 6 (cont.)

λ	α
0.5085	0.160
0.5461	0.181
0.5780	0.207
0.6438	0.259

CURVE 7
T = 298

λ	α
0.2536	0.236
0.2753	0.238
0.2967	0.228
0.3341	0.209
0.3610	0.205
0.4046	0.193
0.4358	0.187
0.4800	0.174
0.5085	0.173
0.5461	0.194
0.5780	0.221
0.6438	0.274

CURVE 8
T = 298

λ	α
0.2536	0.258
0.2753	0.255
0.2967	0.249
0.3341	0.230
0.3610	0.221
0.4046	0.214
0.4358	0.209
0.4800	0.199
0.5085	0.198
0.5461	0.223
0.5780	0.244
0.6438	0.294

CURVE 9*
T = 298

λ	α
0.2536	0.241
0.2753	0.270
0.2967	0.264
0.3341	0.244
0.3610	0.231
0.4046	0.221

CURVE 9 (cont.)*

λ	α
0.4358	0.211
0.4800	0.197
0.5085	0.197
0.5461	0.215
0.5780	0.238
0.6438	0.285

CURVE 10*
T = 298

λ	α
0.2536	0.280
0.2753	0.282
0.2967	0.271
0.3341	0.255
0.3610	0.236
0.4046	0.227
0.4358	0.216
0.4800	0.201
0.5085	0.203
0.5461	0.220
0.5780	0.243
0.6438	0.286

CURVE 11
T = 298

λ	α
0.2536	0.350
0.2753	0.362
0.2967	0.355
0.3341	0.337
0.3610	0.325
0.4046	0.318
0.4358	0.300
0.4800	0.266
0.5085	0.239
0.5461	0.232
0.5780	0.247
0.6438	0.273*

CURVE 12
T = 298

λ	α
0.2536	0.491
0.2753	0.467
0.2967	0.433
0.3341	0.437

CURVE 12 (cont.)

λ	α
0.3610	0.420
0.4046	0.402*
0.4358	0.392*
0.4800	0.354
0.5085	0.306*
0.5461	0.267
0.5780	0.256
0.6438	0.282

CURVE 13
T = 298

λ	α
0.2536	0.572
0.2753	0.557
0.2967	0.540
0.3341	0.509
0.3610	0.485
0.4046	0.466
0.4358	0.453
0.4800	0.421
0.5085	0.369
0.5461	0.317
0.5780	0.281
0.6438	0.254

CURVE 14
T = 298

λ	α
0.2536	0.665
0.2753	0.633
0.2967	0.612
0.3341	0.581
0.3610	0.546
0.4046	0.518
0.4358	0.503
0.4800	0.469
0.5085	0.411
0.5461	0.340
0.5780	0.300
0.6438	0.256*

CURVE 15
T = 298

λ	α
0.2536	0.714
0.2753	0.689

CURVE 15 (cont.)

λ	α
0.2967	0.688
0.3341	0.659
0.3610	0.596
0.4046	0.565
0.4358	0.550
0.4800	0.511
0.5085	0.443
0.5461	0.359
0.5780	0.313
0.6438	0.268

CURVE 16
T = 298

λ	α
0.2536	0.749
0.2753	0.723
0.2967	0.713
0.3341	0.689
0.3610	0.634
0.4046	0.604
0.4358	0.589
0.4800	0.556
0.5085	0.477
0.5461	0.378
0.5780	0.325
0.6438	0.295*

CURVE 17
T = 298

λ	α
0.2536	0.767
0.2753	0.732
0.2967	0.725
0.3341	0.713
0.3610	0.662
0.4046	0.638
0.4358	0.619
0.4800	0.588
0.5085	0.497
0.5461	0.392
0.5780	0.342
0.6438	0.304

CURVE 18*
T = 298

λ	α
0.2536	0.771
0.2753	0.741
0.2967	0.739
0.3341	0.730
0.3610	0.681
0.4046	0.653
0.4358	0.637
0.4800	0.603
0.5085	0.510
0.5461	0.394
0.5780	0.347
0.6438	0.317

CURVE 19
T = 298

λ	α
0.2536	0.778
0.2753	0.760
0.2967	0.741
0.3341	0.742
0.3610	0.720
0.4046	0.671
0.4358	0.658
0.4800	0.619
0.5085	0.516
0.5461	0.389
0.5780	0.350
0.6438	0.309

CURVE 20*
T = 298

λ	α
0.2536	0.342
0.2753	0.361
0.2967	0.341
0.3341	0.319
0.3610	0.311
0.4046	0.306
0.4358	0.297
0.4800	0.250
0.5085	0.245
0.5461	0.234
0.5780	0.242
0.6438	0.274

CURVE 21*
T = 298

λ	α
0.2536	0.563
0.2753	0.549
0.2967	0.539
0.3341	0.514
0.3610	0.493
0.4046	0.464
0.4358	0.450
0.4800	0.415
0.5085	0.366
0.5461	0.312
0.5780	0.282
0.6438	0.249

CURVE 22*
T = 298

λ	α
0.2536	0.741
0.2753	0.718
0.2967	0.713
0.3341	0.683
0.3610	0.658
0.4046	0.601
0.4358	0.589
0.4800	0.547
0.5085	0.474
0.5461	0.377
0.5780	0.329
0.6438	0.291

CURVE 23*
T = 298

λ	α
0.2536	0.780
0.2753	0.765
0.2967	0.751
0.3341	0.745
0.3610	0.712
0.4046	0.673
0.4358	0.659
0.4800	0.624
0.5085	0.520
0.5461	0.399
0.5780	0.359
0.6438	0.321

* Not shown on plot

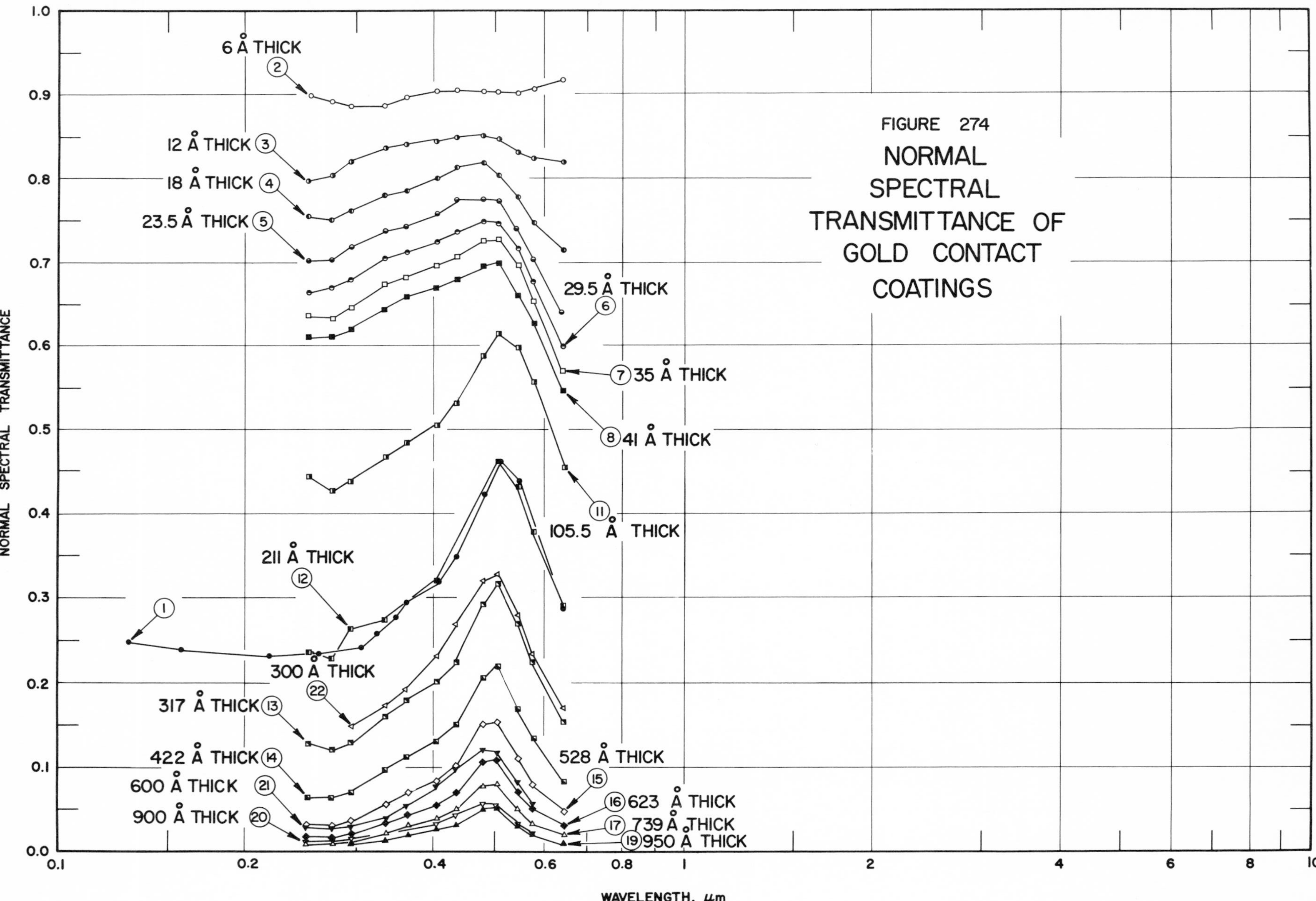

FIGURE 274 NORMAL SPECTRAL TRANSMITTANCE OF GOLD CONTACT COATINGS

SPECIFICATION TABLE NO. 274 NORMAL SPECTRAL TRANSMITTANCE OF GOLD CONTACT COATINGS

Curve No.	Ref. No.	Year	Temperature, K	Wavelength Range, μm	Geometry θ θ' ω'	Reported Error, %	Composition (weight percent), Specifications, and Remarks
1	227	1958	298	0.1308-0.6425	$\sim 0° \sim 0°$		Gold; quartz substrate; vacuum deposited.
2	232	1959	298	0.2536-0.6438	$\sim 0° \sim 0°$		Gold (6 Å thick); quartz substrate; vacuum deposited.
3	232	1959	298	0.2536-0.6438	$\sim 0° \sim 0°$		Similar to above specimen and conditions except 12 Å thick.
4	232	1959	298	0.2536-0.6438	$\sim 0° \sim 0°$		Similar to above specimen and conditions except 18 Å thick.
5	232	1959	298	0.2536-0.6438	$\sim 0° \sim 0°$		Similar to above specimen and conditions except 23.5 Å thick.
6	232	1959	298	0.2536-0.6438	$\sim 0° \sim 0°$		Similar to above specimen and conditions except 29.5 Å thick.
7	232	1959	298	0.2536-0.6438	$\sim 0° \sim 0°$		Similar to above specimen and conditions except 35.0 Å thick.
8	232	1959	298	0.2536-0.6438	$\sim 0° \sim 0°$		Similar to above specimen and conditions except 41.0 Å thick.
9*	232	1959	298	0.2536-0.6438	$\sim 0° \sim 0°$		Similar to above specimen and conditions except 47 Å thick.
10*	232	1959	298	0.2536-0.6438	$\sim 0° \sim 0°$		Similar to above specimen and conditions except 53 Å thick.
11	232	1959	298	0.2536-0.6438	$\sim 0° \sim 0°$		Similar to above specimen and conditions except 105.5 Å thick.
12	232	1959	298	0.2536-0.6438	$\sim 0° \sim 0°$		Similar to above specimen and conditions except 211.0 Å thick.
13	232	1959	298	0.2536-0.6438	$\sim 0° \sim 0°$		Similar to above specimen and conditions except 317 Å thick.
14	232	1959	298	0.2536-0.6438	$\sim 0° \sim 0°$		Similar to above specimen and conditions except 422 Å thick.
15	232	1959	298	0.2536-0.6438	$\sim 0° \sim 0°$		Similar to above specimen and conditions except 528 Å thick.
16	232	1959	298	0.2536-0.6438	$\sim 0° \sim 0°$		Similar to above specimen and conditions except 623 Å thick.
17	232	1959	298	0.2536-0.6438	$\sim 0° \sim 0°$		Similar to above specimen and conditions except 739 Å thick.
18*	232	1959	298	0.2536-0.6438	$\sim 0° \sim 0°$		Similar to above specimen and conditions except 844 Å thick.
19	232	1959	298	0.2536-0.6438	$\sim 0° \sim 0°$		Similar to above specimen and conditions except 950 Å thick.
20	232	1959	298	0.2536-0.6438	$\sim 0° \sim 0°$		Similar to above specimen and conditions except 900 Å thick.
21	232	1959	298	0.2536-0.6438	$\sim 0° \sim 0°$		Similar to above specimen and conditions except 600 Å thick.
22	232	1959	298	0.2967-0.6438	$\sim 0° \sim 0°$		Similar to above specimen and conditions except 300 Å thick.
23*	232	1959	298	0.2967-0.6438	$\sim 0° \sim 0°$		Similar to above specimen and conditions except 100 Å thick.

* Not shown on plot

DATA TABLE NO. 274 NORMAL SPECTRAL TRANSMITTANCE OF GOLD CONTACT COATINGS

[Wavelength, λ, μm; Transmittance, τ; Temperature, T, K]

λ	τ
CURVE 1 T = 298	
0.1308	0.248
0.1582	0.238
0.2192	0.232
0.2637	0.236
0.3035	0.243
0.3278	0.259
0.3493	0.277
0.3615	0.295
0.4086	0.319
0.4359	0.349
0.4824	0.423
0.5109	0.461
0.5475	0.437
0.6425	0.288
CURVE 2 T = 298	
0.2536	0.899
0.2753	0.895
0.2967	0.889
0.3341	0.890
0.3610	0.898
0.4046	0.906
0.4358	0.906
0.4800	0.906
0.5085	0.906
0.5461	0.903
0.5780	0.909
0.6438	0.919
CURVE 3 T = 298	
0.2536	0.798
0.2753	0.806
0.2967	0.821
0.3341	0.837
0.3610	0.841
0.4046	0.846
0.4358	0.852
0.4800	0.853
0.5085	0.849
0.5461	0.833
CURVE 3 (cont.)	
0.5780	0.826
0.6438	0.821
CURVE 4 T = 298	
0.2536	0.757
0.2753	0.752
0.2967	0.764
0.3341	0.780
0.3610	0.787
0.4046	0.802
0.4358	0.815
0.4800	0.819
0.5085	0.805
0.5461	0.779
0.5780	0.749
0.6438	0.716
CURVE 5 T = 298	
0.2536	0.703
0.2753	0.704
0.2967	0.719
0.3341	0.738
0.3610	0.743
0.4046	0.759
0.4358	0.766
0.4800	0.776
0.5085	0.773
0.5461	0.740
0.5780	0.704
0.6438	0.640
CURVE 6 T = 298	
0.2536	0.667
0.2753	0.672
0.2967	0.681
0.3341	0.705
0.3610	0.713
0.4046	0.726
0.4358	0.738
CURVE 6 (cont.)	
0.4800	0.750
0.5085	0.747
0.5461	0.717
0.5780	0.678
0.6438	0.599
CURVE 7 T = 298	
0.2536	0.637
0.2753	0.634
0.2967	0.648
0.3341	0.675
0.3610	0.682
0.4046	0.698
0.4358	0.708
0.4800	0.727
0.5085	0.728
0.5461	0.697
0.5780	0.656
0.6438	0.572
CURVE 8 T = 298	
0.2536	0.611
0.2753	0.611
0.2967	0.620
0.3341	0.646
0.3610	0.660
0.4046	0.672
0.4358	0.681
0.4800	0.697
0.5085	0.699
0.5461	0.663
0.5780	0.626
0.6438	0.548
CURVE 9* T = 298	
0.2536	0.626
0.2753	0.593
0.2967	0.602
0.3341	0.631
CURVE 9 (cont.)	
0.3610	0.646
0.4046	0.660
0.4358	0.674
0.4800	0.696
0.5085	0.695
0.5461	0.667
0.5780	0.628
0.6438	0.550
CURVE 10* T = 298	
0.2536	0.590
0.2753	0.583
0.2967	0.595
0.3341	0.619
0.3610	0.635
0.4046	0.652
0.4358	0.667
0.4800	0.690
0.5085	0.689
0.5461	0.660
0.5780	0.618
0.6438	0.542
CURVE 11 T = 298	
0.2536	0.444
0.2753	0.427
0.2967	0.438
0.3341	0.468
0.3610	0.485
0.4046	0.506
0.4358	0.533
0.4800	0.589
0.5085	0.616
0.5461	0.597
0.5780	0.558
0.6438	0.454
CURVE 12 T = 298	
0.2536	0.238
CURVE 12 (cont.)	
0.2753	0.230
0.2967	0.265
0.3341	0.274
0.3610	0.295*
0.4046	0.320
0.4358	0.349*
0.4800	0.424*
0.5085	0.464
0.5461	0.433
0.5780	0.378
0.6438	0.290
CURVE 13 T = 298	
0.2536	0.128
0.2753	0.121
0.2967	0.130
0.3341	0.163
0.3610	0.180
0.4046	0.201
0.4358	0.226
0.4800	0.293
0.5085	0.317
0.5461	0.270
0.5780	0.226
0.6438	0.153
CURVE 14 T = 298	
0.2536	0.066
0.2753	0.066
0.2967	0.072
0.3341	0.097
0.3610	0.112
0.4046	0.130
0.4358	0.152
0.4800	0.207
0.5085	0.220
0.5461	0.170
0.5780	0.133
0.6438	0.084
CURVE 15 T = 298	
0.2536	0.033
0.2753	0.033
0.2967	0.039
0.3341	0.057
0.3610	0.072
0.4046	0.086
0.4358	0.104
0.4800	0.152
0.5085	0.155
0.5461	0.111
0.5780	0.082
0.6438	0.049
CURVE 16 T = 298	
0.2536	0.019
0.2753	0.019
0.2967	0.023
0.3341	0.036
0.3610	0.046
0.4046	0.057
0.4358	0.072
0.4800	0.107
0.5085	0.109
0.5461	0.072
0.5780	0.053
0.6438	0.031
CURVE 17 T = 298	
0.2536	0.010
0.2753	0.012
0.2967	0.014
0.3341	0.024
0.3610	0.033
0.4046	0.042
0.4358	0.053
0.4800	0.079
0.5085	0.081
0.5461	0.051
0.5780	0.035
0.6438	0.022
CURVE 18* T = 298	
0.2536	0.011
0.2753	0.012
0.2967	0.011
0.3341	0.018
0.3610	0.025
0.4046	0.033
0.4358	0.041
0.4800	0.063
0.5085	0.064
0.5461	0.037
0.5780	0.026
0.6438	0.015
CURVE 19 T = 298	
0.2536	0.010*
0.2753	0.010*
0.2967	0.010
0.3341	0.015
0.3610	0.021
0.4046	0.027
0.4358	0.034
0.4800	0.051
0.5085	0.053
0.5461	0.031
0.5780	0.021
0.6438	0.012
CURVE 20 T = 298	
0.2536	0.013
0.2753	0.015
0.2967	0.015
0.3341	0.015*
0.3610	0.023*
0.4046	0.033
0.4358	0.045
0.4800	0.058
0.5085	0.055
0.5461	0.033
0.5780	0.023
0.6438	0.012*
CURVE 21 T = 298	
0.2536	0.027
0.2753	0.027
0.2967	0.030
0.3341	0.042
0.3610	0.055
0.4046	0.079
0.4358	0.099
0.4800	0.123
0.5085	0.119
0.5461	0.084
0.5780	0.058
0.6438	0.036*
CURVE 22 T = 298	
0.2967	0.149
0.3341	0.176
0.3610	0.196
0.4046	0.233
0.4358	0.270
0.4800	0.321
0.5085	0.329
0.5461	0.282
0.5780	0.236
0.6438	0.173
CURVE 23* T = 298	
0.2967	0.447
0.3341	0.468
0.3610	0.488
0.4046	0.521
0.4358	0.564
0.4800	0.609
0.5085	0.617
0.5461	0.599
0.5780	0.558
0.6438	0.459

* Not shown on plot

SPECIFICATION TABLE NO. 275 NORMAL SPECTRAL REFLECTANCE OF GOLD + PALLADIUM + ΣX_i CONTACT COATINGS

Curve No.	Ref. No.	Year	Temperature, K	Wavelength Range, μm	Geometry θ θ' ω'	Reported Error, %	Composition (weight percent), Specifications, and Remarks
1	215	1962	~ 298	0.4-0.7	~0° 2π		Hanovia Liquid Bright Palladium No. 62, manufactured by Englehard Industries; glass substrate; exact composition proprietary, but contains 4.5 parts gold to 1.5 parts palladium, plus a small amount of rhodium and other metals; prepared by thermal decomposition of a solution of gold sulforesinate, rhodium sulforesinate, and other metallic resinates after application to the substrate; measured relative to $MgCO_3$.
2	215	1962	~298	0.8-2.0	~0° ~0°		Above specimen and conditions except measured relative to plane, polished aluminum; specular component only.

DATA TABLE NO. 275 NORMAL SPECTRAL REFLECTANCE OF GOLD + PALLADIUM + ΣX_i CONTACT COATINGS

[Wavelength, λ, μm; Reflectance, ρ; Temperature, T, K]

λ	ρ	λ	ρ
CURVE 1 T ~ 298		CURVE 2 (cont.)	
0.4	0.26	1.8	0.51
0.5	0.27	1.9	0.53
0.6	0.24	2.0	0.56
0.7	0.26		
CURVE 2 T ~ 298			
0.8	0.29		
0.9	0.30		
1.0	0.30		
1.1	0.31		
1.2	0.33		
1.3	0.36		
1.4	0.39		
1.5	0.43		
1.6	0.46		
1.7	0.49		

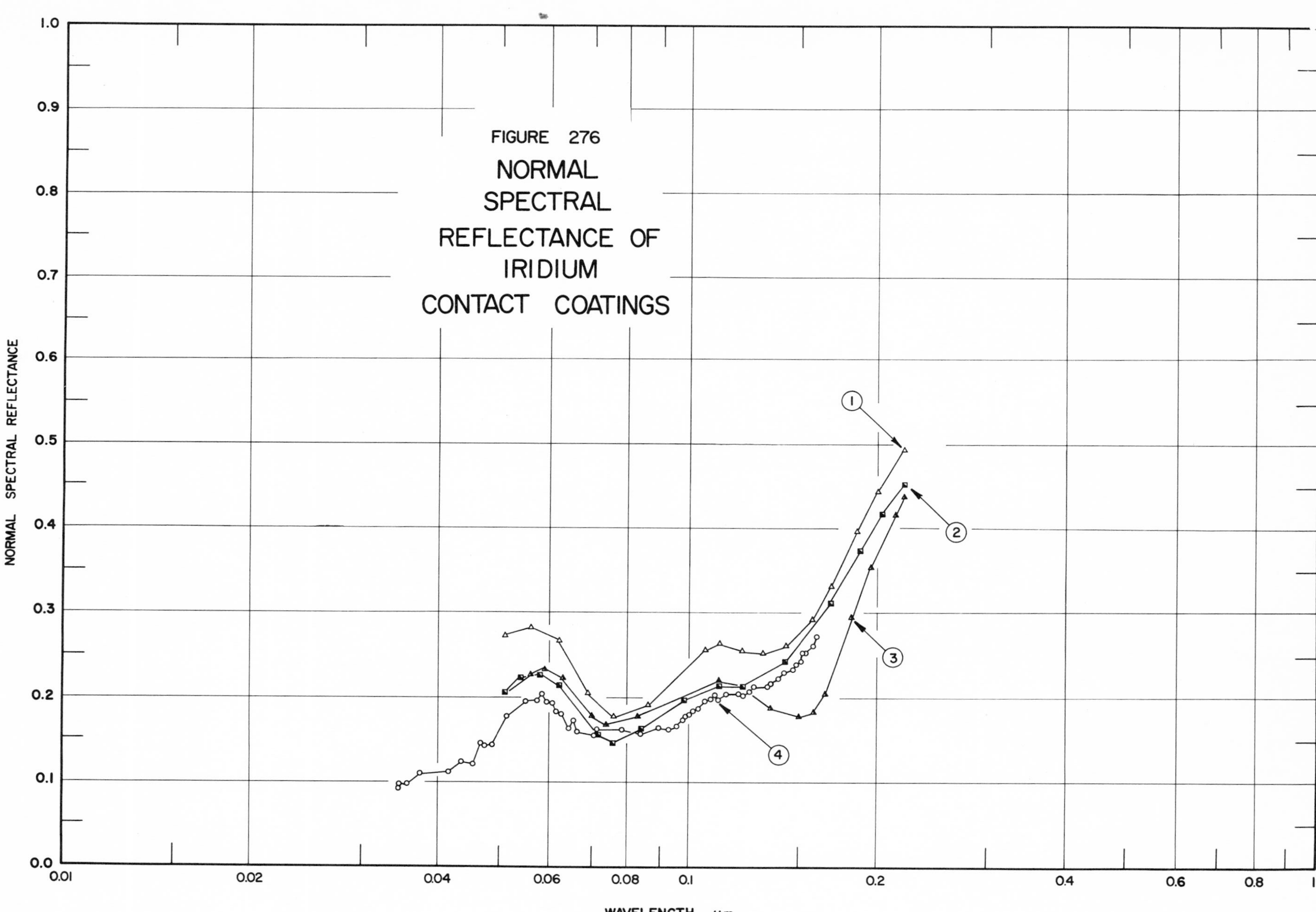
FIGURE 276
NORMAL
SPECTRAL
REFLECTANCE OF
IRIDIUM
CONTACT COATINGS
NORMAL SPECTRAL REFLECTANCE
1.0
0.9
0.8
0.7
0.6
0.5
0.4
0.3
0.2
0.1
0.0
0.01
0.02
0.04
0.06
0.08
0.1
0.2
0.4
0.6
0.8
1
WAVELENGTH, μm
1
2
3
4

SPECIFICATION TABLE NO. 276 NORMAL SPECTRAL REFLECTANCE OF IRIDIUM CONTACT COATINGS

Curve No.	Ref. No.	Year	Temperature, K	Wavelength Range, μm	Geometry θ θ' ω'	Reported Error, %	Composition (weight percent), Specifications, and Remarks
1	233	1967	298	0.051-0.220	~0° ~0°		Iridium, 99.98 pure (150 Å thick); glass substrate; vacuum (1 x 10^{-5} mm Hg) evaporated at 30 Å sec^{-1} on substrate at 313 K; coating semi-transparent.
2	233	1967	298	0.051-0.220	~0° ~0°		Similar to above specimen and conditions except 135 Å thick.
3	233	1967	298	0.051-0.220	~0° ~0°		Similar to above specimen and conditions except substrate is aged aluminum.
4	234	1967	298	0.0346-0.1607	~10°~10°		Iridium (several hundred angstroms thick); substrate unknown; laser deposited by single pulse of ruby laser with laser action lasting ~500 μ sec and having an energy of 100 J per pulse; additional information received from author.

DATA TABLE NO. 276 NORMAL SPECTRAL REFLECTANCE OF IRIDIUM CONTACT COATINGS

[Wavelength, λ, μm; Reflectance, ρ; Temperature, T, K]

λ	ρ	λ	ρ	λ	ρ	λ	ρ	λ	ρ	λ	ρ
CURVE 1 T = 298		CURVE 2 T = 298		CURVE 3 T = 298		CURVE 4 T = 298		CURVE 4 (cont.)		CURVE 4 (cont.)	
								0.0658	0.172	0.1158	0.203
0.051	0.273	0.051	0.205	0.051	0.206*	0.0346	0.092	0.0666	0.160	0.1172	0.203*
0.056	0.283	0.054	0.223	0.056	0.228	0.0346	0.098	0.0702	0.155	0.1188	0.201*
0.062	0.269	0.058	0.227	0.059	0.232	0.0359	0.098	0.0713	0.162	0.1204	0.202*
0.069	0.205	0.062	0.215	0.063	0.223	0.0373	0.110	0.0786	0.162	0.1209	0.200*
0.076	0.179	0.072	0.153	0.070	0.179	0.0417	0.111	0.0836	0.158	0.1222	0.201
0.086	0.192	0.076	0.146	0.074	0.167	0.0432	0.123	0.0896	0.165	0.1256	0.208
0.106	0.258	0.084	0.163	0.083	0.178	0.0451	0.121	0.0923	0.162	0.1272	0.211
0.112	0.266	0.098	0.197	0.112	0.220	0.0466	0.144	0.0953	0.167	0.1347	0.212
0.122	0.257	0.112	0.213	0.122	0.213*	0.0471	0.141	0.0974	0.173	0.1356	0.217
0.131	0.254	0.123	0.213	0.135	0.189	0.0488	0.143	0.0988	0.178	0.1375	0.218*
0.143	0.262	0.142	0.243	0.150	0.179	0.0511	0.177	0.1002	0.180	0.1393	0.222
0.157	0.293	0.169	0.311	0.158	0.183	0.0550	0.195	0.1013	0.180*	0.1433	0.230
0.169	0.331	0.187	0.374	0.165	0.205	0.0573	0.196	0.1023	0.184	0.1468	0.233
0.185	0.396	0.204	0.419	0.182	0.296	0.0582	0.203	0.1045	0.188	0.1487	0.240
0.200	0.443	0.220	0.451	0.195	0.355	0.0595	0.195	0.1063	0.197	0.1514	0.243
0.220	0.495			0.212	0.418	0.0606	0.194	0.1088	0.199	0.1527	0.253
				0.220	0.438	0.0613	0.184	0.1100	0.202	0.1546	0.253
						0.0629	0.180	0.1113	0.197	0.1589	0.264
						0.0641	0.164	0.1144	0.201*	0.1607	0.273

* Not shown on plot

SPECIFICATION TABLE NO. 277 NORMAL SPECTRAL TRANSMITTANCE OF IRIDIUM CONTACT COATINGS

Curve No.	Ref. No.	Year	Temperature, K	Wavelength Range, μm	Geometry θ	θ'	ω'	Reported Error, %	Composition (weight percent), Specifications, and Remarks
1*	234	1967	298	0.0233-0.1609	0°	0°			Iridium (~400 Å thick); NE 102 plastic scintillator; laser deposited by single pulse of ruby laser with laser action lasting ~500 μsec and having an energy of 100 J per pulse; additional information received from author.

DATA TABLE NO. 277 NORMAL SPECTRAL TRANSMITTANCE OF IRIDIUM CONTACT COATINGS

[Wavelength, λ, μm; Transmittance, τ; Temperature, T, K]

λ	τ	λ	τ	λ	τ	λ	τ
CURVE 1* T = 298		CURVE 1 (cont.)*		CURVE 1 (cont.)*		CURVE 1 (cont.)*	
0.0233	0.219	0.0608	0.084	0.1027	0.132	0.1364	0.225
0.0246	0.252	0.0628	0.093	0.1044	0.137	0.1374	0.235
0.0262	0.286	0.0644	0.090	0.1077	0.146	0.1398	0.227
0.0282	0.199	0.0680	0.076	0.1094	0.138	0.1429	0.240
0.0307	0.199	0.0753	0.082	0.1094	0.151	0.1462	0.239
0.0320	0.182	0.0774	0.084	0.1117	0.154	0.1488	0.250
0.0341	0.180	0.0781	0.076	0.1142	0.160	0.1520	0.272
0.0376	0.194	0.0834	0.087	0.1160	0.157	0.1535	0.274
0.0386	0.165	0.0925	0.096	0.1177	0.160	0.1544	0.269
0.0415	0.164	0.0957	0.099	0.1191	0.173	0.1585	0.270
0.0432	0.160	0.0959	0.094	0.1202	0.182	0.1609	0.259
0.0448	0.156	0.0968	0.108	0.1218	0.185		
0.0469	0.150	0.0968	0.115	0.1237	0.184		
0.0506	0.124	0.0974	0.120	0.1241	0.193		
0.0527	0.115	0.0985	0.123	0.1260	0.197		
0.0553	0.082	0.0997	0.116	0.1283	0.210		
0.0586	0.111	0.1001	0.128	0.1339	0.208		
0.0595	0.080	0.1005	0.132	0.1346	0.215		
		0.1014	0.135	0.1360	0.220		

* No plot given

SPECIFICATION TABLE NO. 278 NORMAL SPECTRAL REFLECTANCE OF IRON CONTACT COATINGS

Curve No.	Ref. No.	Year	Temperature, K	Wavelength Range, μm	Geometry θ	θ'	ω'	Reported Error, %	Composition (weight percent), Specifications, and Remarks
1*	215	1962	~298	0.4-4.0	~0°	~0°			Iron; glass substrate; prepared by applying a solution of iron resinate, diluted with aromatic solvents and essential oils, to a soda-lime glass substrate, then firing to 873 K in a continuous lehr on a 1.5 hr cycle; application of solution was by dropping an excess amount on the substrate, which was kept spinning at 1550 rpm until no more solution was flung from the edges; measured relative to plane, polished aluminum; specular component only.

DATA TABLE NO. 278 NORMAL SPECTRAL REFLECTANCE OF IRON CONTACT COATINGS

[Wavelength, λ, μm; Reflectance, ρ; Temperature, T, K]

λ	ρ	λ	ρ
CURVE 1*		CURVE 1 (cont.)*	
T ~ 298			
0.4	0.14	3.8	0.09
0.5	0.13	4.0	0.07
0.6	0.16		
0.8	0.06		
1.0	0.11		
1.2	0.14		
1.4	0.17		
1.6	0.17		
1.8	0.17		
2.0	0.16		
2.2	0.15		
2.4	0.15		
2.6	0.15		
2.8	0.12		
3.0	0.10		
3.2	0.10		
3.4	0.11		
3.6	0.09		

* No plot given

SPECIFICATION TABLE NO. 279 NORMAL SPECTRAL TRANSMITTANCE OF IRON CONTACT COATINGS

Curve No.	Ref. No.	Year	Temperature, K	Wavelength Range, μm	Geometry θ	θ'	ω'	Reported Error, %	Composition (weight percent), Specifications, and Remarks
1*	215	1962	~298	0.4-2.0	~0°	~0°			Iron; glass (1.59 mm thick) substrate; prepared by applying a solution of iron resinate, diluted with aromatic solvents and essential oils, to a soda-lime glass substrate, then firing to 873 K in a continuous lehr on a 1.5 hr cycle: application of solution was by dropping an excess amount on the substrate, which was kept spinning at 1550 rpm until no more solution was flung from the edges; τ is for coating plus substrate.

DATA TABLE NO. 279 NORMAL SPECTRAL TRANSMITTANCE OF IRON CONTACT COATINGS

[Wavelength, λ, μm; Transmittance, τ; Temperature, T, K]

λ	τ
CURVE 1*	
T ~ 298	
0.4	0.011
0.5	0.120
0.6	0.520
0.7	0.776
0.8	0.840
0.9	0.812
1.0	0.783
1.1	0.763
1.2	0.743
1.3	0.724
1.4	0.613
1.5	0.438
1.6	0.324
1.8	0.195
2.0	0.124

* No plot given

SPECIFICATION TABLE NO. 280 NORMAL SPECTRAL REFLECTANCE OF LANTHANUM ANTIMONIDE CONTACT COATINGS

Curve No.	Ref. No.	Year	Temperature, K	Wavelength Range, μm	Geometry θ	θ'	ω'	Reported Error, %	Composition (weight percent), Specifications, and Remarks
1*	235	1967	~298	1.0-17.6	~0°	~0°			LaSb; glass substrate; vacuum deposited on substrate at 773 K. [Authors' designation: Sample No. 2]

DATA TABLE NO. 280 NORMAL SPECTRAL REFLECTANCE OF LANTHANUM ANTIMONIDE CONTACT COATINGS

[Wavelength, λ, μm; Reflectance, ρ; Temperature, T, K]

λ	ρ
CURVE 1* T ~ 298	
1.0	0.071
1.0	0.085
1.0	0.112
1.3	0.200
1.5	0.254
1.7	0.243
1.9	0.195
2.2	0.100
2.2	0.074
2.3	0.055
2.6	0.021
2.8	0.040
2.9	0.074
3.3	0.094
3.7	0.068
4.0	0.094
4.6	0.258
5.1	0.409
CURVE 1 (cont.)*	
5.7	0.559
6.2	0.630
8.4	0.748
10.1	0.766
12.7	0.780
15.5	0.777
17.6	0.777

* No plot given

SPECIFICATION TABLE NO. 281 HEMISPHERICAL TOTAL EMITTANCE OF LEAD + TIN CONTACT COATINGS

Curve No.	Ref. No.	Year	Temperature Range, K	Reported Error, %	Composition (weight percent), Specifications and Remarks
1*	194	1955	76	5	50 Pb and 50 Sn (0.0508 mm thick); copper (0.127 mm thick) substrate; measured in vacuum (10^{-6} to 10^{-7} mm Hg); emittance for 300 K blackbody incident radiation; authors assumed $\alpha = \epsilon$.

DATA TABLE NO. 281 HEMISPHERICAL TOTAL EMITTANCE OF LEAD + TIN CONTACT COATINGS

[Temperature, T, K; Emittance, ϵ]

T	ϵ
CURVE 1*	
76	0.03

* No plot given

SPECIFICATION TABLE NO. 282 HEMISPHERICAL INTEGRATED ABSORPTANCE OF LEAD + TIN CONTACT COATINGS

Curve No.	Ref. No.	Year	Temperature Range, K	Reported Error, %	Composition (weight percent), Specifications and Remarks
1*	194	1955	76	5	50 Pb and 50 Sn (0.0508 mm thick); copper (0.127 mm thick) substrate; measured in vacuum (10^{-6} to 10^{-7} mm Hg); absorptance for 300 K blackbody incident radiation.

DATA TABLE NO. 282 HEMISPHERICAL INTEGRATED ABSORPTANCE OF LEAD + TIN CONTACT COATINGS

[Temperature, T, K; Absorptance, α]

T	α
CURVE 1*	
76	0.03

* No plot given

SPECIFICATION TABLE NO. 283 NORMAL SPECTRAL REFLECTANCE OF MAGNESIUM CONTACT COATINGS

Curve No.	Ref. No.	Year	Temperature, K	Wavelength Range, μm	Geometry θ	θ'	ω'	Reported Error, %	Composition (weight percent), Specifications, and Remarks
1*	205	1950	298	0.268-0.490	$\sim 0°$	$\sim 0°$		±2	Magnesium; glass substrate; vaporized.

DATA TABLE NO. 283 NORMAL SPECTRAL REFLECTANCE OF MAGNESIUM CONTACT COATINGS

[Wavelength, λ, μm; Reflectance, ρ; Temperature, T, K]

λ	ρ
CURVE 1* T = 298	
0.268	0.710
0.280	0.730
0.306	0.740
0.345	0.740
0.377	0.740
0.425	0.730
0.490	0.735

* No plot given

SPECIFICATION TABLE NO. 284 NORMAL SPECTRAL REFLECTANCE OF MAGNESIUM + ALUMINUM CONTACT COATINGS

Curve No.	Ref. No.	Year	Temperature, K	Wavelength Range, μm	Geometry θ	θ'	ω'	Reported Error, %	Composition (weight percent), Specifications and Remarks
1*	205	1950	298	0.245-0.491	~0°	~0°			58 Mg (δ phase) and 42 Al; glass substrate; vaporized.

DATA TABLE NO. 284 NORMAL SPECTRAL REFLECTANCE OF MAGNESIUM + ALUMINUM CONTACT COATINGS

[Wavelength, λ, μm; Reflectance, ρ; Temperature, T, K]

λ	ρ
CURVE 1* T = 298	
0.245	0.720
0.254	0.740
0.265	0.760
0.280	0.775
0.305	0.770
0.350	0.750
0.375	0.700
0.425	0.660
0.491	0.680

* No plot given

SPECIFICATION TABLE NO. 285 NORMAL SPECTRAL REFLECTANCE OF MANGANESE CONTACT COATINGS

Curve No.	Ref. No.	Year	Temperature, K	Wavelength Range, μm	Geometry θ	θ'	ω'	Reported Error, %	Composition (weight percent), Specifications, and Remarks
1*	215	1962	~298	0.4-4.0	~0°	~0°			Manganese; glass substrate; prepared by applying a solution of manganese resinate, diluted with aromatic solvents and essential oils, to a soda-lime glass substrate, then firing to 873 K in a continuous lehr on a 1.5 hr cycle; application of solution was by dropping an excess amount on the substrate, which was kept spinning at 1550 rpm until no more solution was flung from the edges; measured relative to plane, polished aluminum; specular component only.

DATA TABLE NO. 285 NORMAL SPECTRAL REFLECTANCE OF MANGANESE CONTACT COATINGS

[Wavelength, λ, μm; Reflectance, ρ; Temperature, T, K]

λ	ρ
CURVE 1* T ~ 298	
0.4	0.12
0.5	0.10
0.6	0.11
0.8	0.09
1.0	0.11
1.2	0.14
1.4	0.16
1.6	0.17
1.8	0.15
2.0	0.14
2.2	0.13
2.4	0.12
2.6	0.11
2.8	0.09
3.0	0.08
3.2	0.07
3.4	0.07
3.6	0.07
CURVE 1 (cont.)*	
3.8	0.06
4.0	0.05

* No plot given

SPECIFICATION TABLE NO. 286 NORMAL SPECTRAL TRANSMITTANCE OF MANGANESE CONTACT COATINGS

Curve No.	Ref. No.	Year	Temperature, K	Wavelength Range, μm	Geometry θ θ' ω'	Reported Error, %	Composition (weight percent), Specifications, and Remarks
1*	215	1962	~298	0.4-2.0	~0° ~0°		Manganese; glass (1.59 mm thick) substrate; prepared by applying a solution of manganese resinate, diluted with aromatic solvents and essential oils, to a soda-lime glass substrate, then firing to 873 K in a continuous lehr on a 1.5 hr cycle; application of solution was by dropping an excess amount on the substrate, which was kept spinning at 1550 rpm until no more solution was flung from the edges; τ is for coating plus substrate.

DATA TABLE NO. 286 NORMAL SPECTRAL TRANSMITTANCE OF MANGANESE CONTACT COATINGS

[Wavelength, λ, μm; Transmittance, τ; Temperature, T, K]

λ	τ
CURVE 1*	
T ~ 298	
0.4	0.090
0.5	0.187
0.6	0.328
0.7	0.480
0.8	0.575
0.9	0.654
1.0	0.710
1.1	0.736
1.2	0.746
1.3	0.730
1.4	0.585
1.5	0.400
1.6	0.294
1.8	0.210
2.0	0.193

* No plot given

SPECIFICATION TABLE NO. 287 HEMISPHERICAL TOTAL EMITTANCE OF MOLYBDENUM CONTACT COATINGS

Curve No.	Ref. No.	Year	Temperature Range, K	Reported Error, %	Composition (weight percent), Specifications and Remarks
1*	7	1961	468-1093	< 10	Molybdenum, granular, Metco XP-1103 (-140 mesh ± 325 mesh); Armco iron substrate; plasma flame sprayed; measured in vacuum (10^{-5} mm Hg).

DATA TABLE NO. 287 HEMISPHERICAL TOTAL EMITTANCE OF MOLYBDENUM CONTACT COATINGS

[Temperature, T, K; Emittance, ϵ]

T	ϵ
CURVE 1*	
468	0.29
668	0.32
873	0.40
1093	0.40

* No plot given

SPECIFICATION TABLE NO. 288 NORMAL SOLAR ABSORPTANCE OF MOLYBDENUM CONTACT COATINGS

Curve No.	Ref. No.	Year	Temperature Range, K	Geometry θ	Reported Error, %	Composition (weight percent), Specifications and Remarks
1*	7	1961	468-1093	$\sim 0^\circ$		Molybdenum, granular, Metco XP-1103 (-140 mesh ± 325 mesh); Armco iron substrate; flame sprayed; measured in vacuum (10^{-5} mm Hg).

DATA TABLE NO. 288 NORMAL SOLAR ABSORPTANCE OF MOLYBDENUM CONTACT COATINGS

[Temperature, T, K; Absorptance, α]

T	α
CURVE 1*	
468	0.76
668	0.77
873	0.79
1093	0.65

* No plot given

SPECIFICATION TABLE NO. 289 NORMAL SPECTRAL TRANSMITTANCE OF NEODYMIUM CONTACT COATINGS

Curve No.	Ref. No.	Year	Temperature, K	Wavelength Range, μm	Geometry θ	θ'	ω'	Reported Error, %	Composition (weight percent), Specifications, and Remarks
1*	236	1967	~298	0.69-5.17	~0°	~0°			Neodymium; fused quartz substrate; vacuum (5×10^{-5} mm Hg) deposited; very thick coating.
2*	236	1967	~298	0.81-6.55	~0°	~0°			Similar to above specimen and conditions except thin coating.

DATA TABLE NO. 289 NORMAL SPECTRAL TRANSMITTANCE OF NEODYMIUM CONTACT COATINGS

[Wavelength, λ, μm; Transmittance, τ; Temperature, T, K]

λ	τ	λ	τ
CURVE 1* T ~ 298		CURVE 2* T ~ 298	
0.69	0.118	0.81	0.882
0.73	0.108	0.91	0.821
0.80	0.075	1.01	0.768
0.91	0.060	1.22	0.769
0.96	0.064	1.34	0.778
1.00	0.086	1.65	0.810
1.07	0.128	2.00	0.855
1.23	0.154	2.93	0.880
1.37	0.161	3.60	0.860
2.09	0.150	4.23	0.860
3.01	0.139	4.63	0.860
3.74	0.128	5.03	0.879
4.36	0.080	5.33	0.878
4.41	0.061	5.64	0.858
4.78	0.020	5.92	0.868
5.17	0.002	6.34	0.879
		6.55	0.886

* No plot given

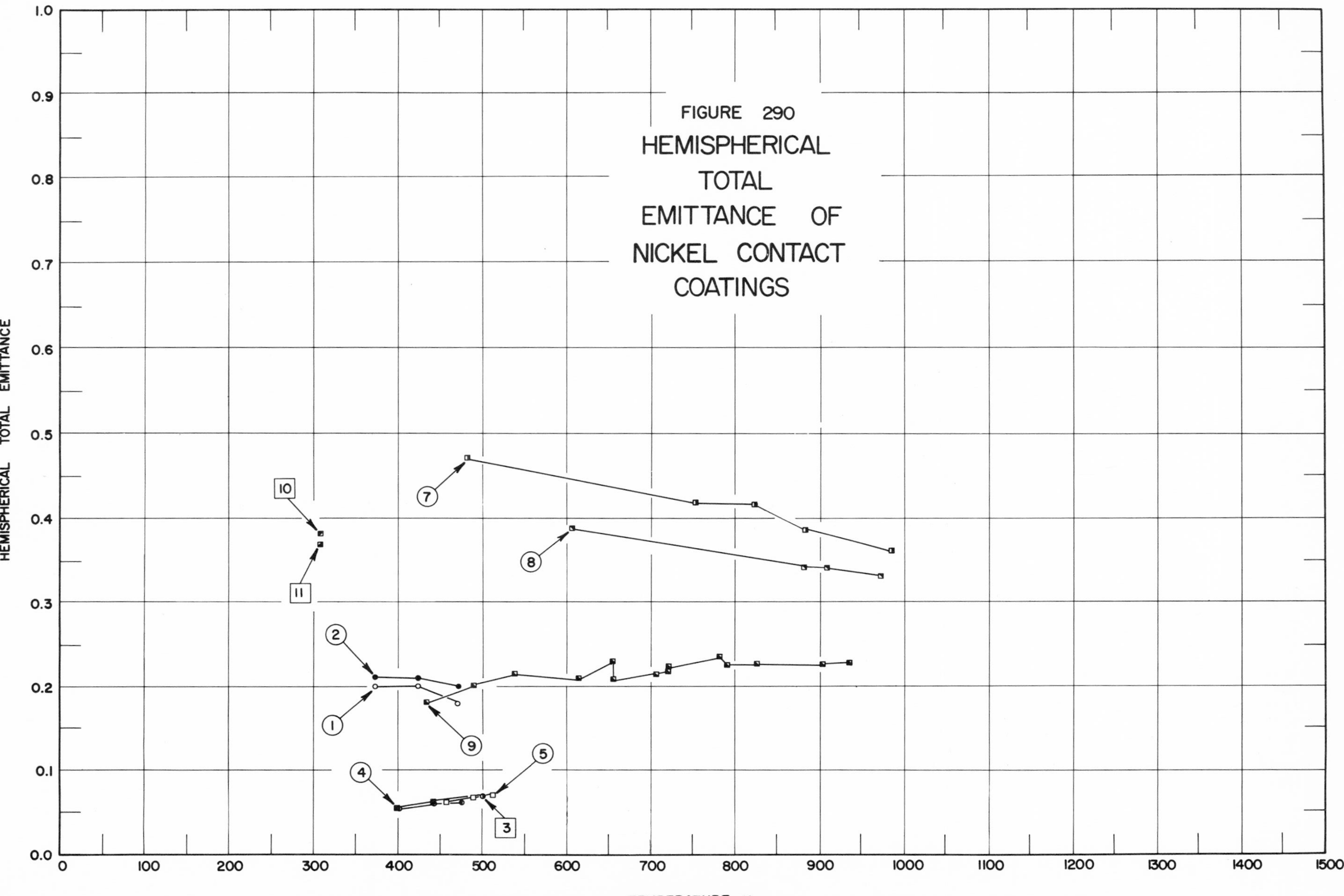

FIGURE 290 HEMISPHERICAL TOTAL EMITTANCE OF NICKEL CONTACT COATINGS

SPECIFICATION TABLE NO. 290 HEMISPHERICAL TOTAL EMITTANCE OF NICKEL CONTACT COATINGS

Curve No.	Ref. No.	Year	Temperature Range, K	Reported Error, %	Composition (weight percent), Specifications and Remarks
1	231	1962	373-473	10-20	Nickel; epoxy and 316 stainless steel substrates; stainless steel cleaned, etched, and degreased, then coated with Maraset 617-C epoxy resin (Marblette Co.) by flowing the resin onto the stainless steel at 408 K; resin prepared for plating by roughening with chromic acid, sensitizing with $SnCl_2$, and activating with $PdCl_2$; nickel deposited by immersing for less than 1 min in an aqueous solution of nickelous chloride, sodium hypophosphite, sodium citrate, ammonium hydroxide, and ammonium chloride; measured in vacuum ($\sim 10^{-5}$ mm Hg). [Authors' designation: Ni 5C]
2	231	1962	373-473	10-20	Above specimen and conditions.
3	237	1963	503	< 26	Nickel; copper substrate; plated; polished and cleaned with trichlorethylene; measured in vacuum (10^{-3} mm Hg).
4	237	1963	503-404	< 26	Above specimen and conditions; decreasing temp.
5	237	1963	404-513	< 26	Above specimen and conditions; increasing temp.
6	237	1963	513-399	< 26	Above specimen and conditions; decreasing temp.
7	121	1967	480-986	±10.1	Nickel; Kh18NIOT steel substrate; plasma deposited; measured in vacuum (10^{-4} mm Hg).
8	121	1967	605-973	±10.1	Above specimen and conditions except heated in vacuum (10^{-4} mm Hg) at 973 K for 2 hrs; cooling cycle.
9	121	1967	433-936	±10.1	Nickel; Kh18NIOT steel substrate; plasma deposited; machine-finished to class 10 surface finish (Russian standard); sample heated in vacuum (10^{-4} mm Hg) at 973 K for 3 hrs; measured in situ.
10	105	1964	307		Black nickel; copper (0.00318 mm thick) substrate; measured in vacuum (10^{-6} mm Hg) maintained by diffusion pump. [Authors' designation: Test No. 22]
11	105	1964	307		Similar to above specimen and conditions. [Authors' designation: Test No. 25]

DATA TABLE NO. 290 HEMISPHERICAL TOTAL EMITTANCE OF NICKEL CONTACT COATINGS

[Temperature, T, K; Emittance, ∈]

T	∈
CURVE 1	
373	0.20
423	0.20
473	0.18
CURVE 2	
373	0.21
423	0.21
473	0.20
CURVE 3	
503	0.069
CURVE 4	
503	0.069*
476	0.063
444	0.061
404	0.054
CURVE 5	
404	0.054*
457	0.063
489	0.068
513	0.069
CURVE 6	
513	0.069
442	0.062
399*	0.054
CURVE 7	
480	0.471
754	0.419
824	0.417
881	0.386
986	0.363
CURVE 8	
605	0.387
880	0.342
908	0.342
973	0.331
CURVE 9	
433	0.183
488	0.202
538	0.217
613	0.210
653	0.210
652	0.231
706	0.213
719	0.218
719	0.224
779	0.237
789	0.225
825	0.227
901	0.227
936	0.227
CURVE 10	
307	0.384
CURVE 11	
307	0.369

* Not shown on plot

SPECIFICATION TABLE NO. 291 NORMAL SPECTRAL REFLECTANCE OF NICKEL CONTACT COATINGS

Curve No.	Ref. No.	Year	Temperature, K	Wavelength Range, μm	Geometry θ	θ'	ω'	Reported Error, %	Composition (weight percent), Specifications, and Remarks
1*	203	1962	298	0.50-25.00	~0°	~0°			Nickel; copper substrate; plated; data extracted from smooth curve.

DATA TABLE NO. 291 NORMAL SPECTRAL REFLECTANCE OF NICKEL CONTACT COATINGS

[Wavelength, λ, μm; Reflectance, ρ; Temperature, T, K]

λ	ρ
CURVE 1*	
0.50	0.535
0.62	0.604
0.85	0.656
0.96	0.664
1.69	0.776
2.51	0.836
3.48	0.873
4.09	0.895
4.97	0.908
6.27	0.926
7.00	0.928
9.81	0.928
11.17	0.939
14.57	0.939
15.19	0.949
18.63	0.944
22.80	0.922
25.00	0.915

* No plot given

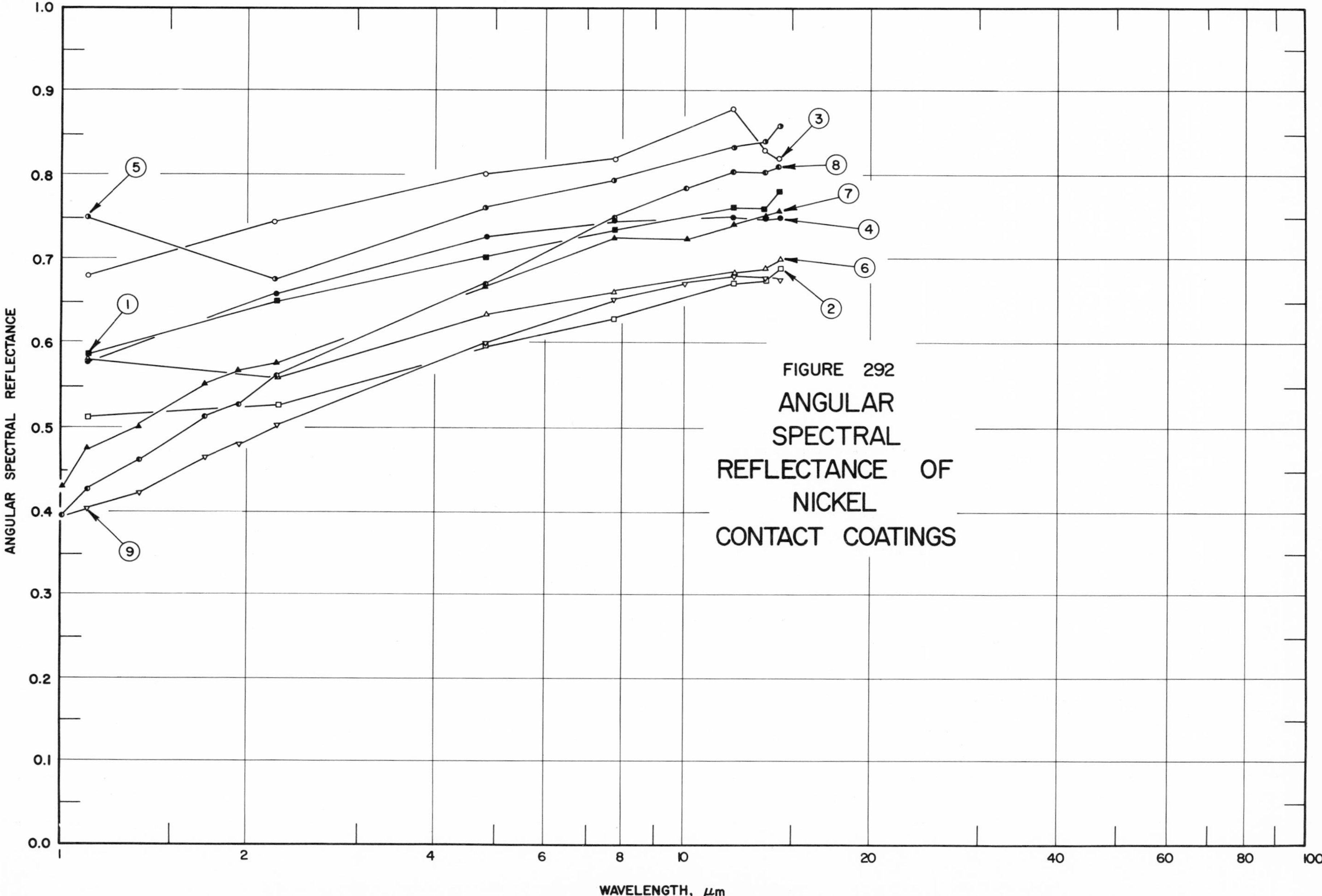

FIGURE 292 ANGULAR SPECTRAL REFLECTANCE OF NICKEL CONTACT COATINGS

SPECIFICATION TABLE NO. 292 ANGULAR SPECTRAL REFLECTANCE OF NICKEL CONTACT COATINGS

Curve No.	Ref. No.	Year	Temperature, K	Wavelength Range, μm	Geometry θ	θ'	ω'	Reported Error, %	Composition (weight percent), Specifications, and Remarks
1	231	1962	~298	1.120-14.38	45°	45°			Nickel; epoxy and 316 stainless steel substrates; stainless steel cleaned, etched, and degreased, then coated with Maraset 617-C epoxy resin (Marblette Co.) by flowing the resin onto the stainless steel at 408 K; resin prepared for plating by roughening with chromic acid, sensitizing with $SnCl_2$, and activating with $PdCl_2$; nickel deposited by immersing for less than 1 min in an aqueous solution of nickelous chloride, sodium hypophosphite, sodium citrate, ammonium hydroxide, and ammonium chloride; specular component only. [Authors' designation: Ni 1 C]
2	231	1962	~298	1.120-14.38	45°	45°			Similar to above specimen and conditions. [Authors' designation: Ni 2 C]
3	231	1962	~298	1.120-14.38	45°	45°			Similar to above specimen and conditions. [Authors' designation: Ni 3 C]
4	231	1962	~298	1.120-14.38	45°	45°			Similar to above specimen and conditions. [Authors' designation: Ni 4 C]
5	231	1962	~298	1.120-14.38	45°	45°			Similar to above specimen and conditions. [Authors' designation: Ni 5 C]
6	231	1962	~298	1.120-14.38	45°	45°			Similar to above specimen and conditions. [Authors' designation: Ni 6 C]
7	231	1962	~298	1.009-14.38	45°	45°			Similar to curve 1 specimen and conditions except substrate consists of 316 stainless steel coated with SY627-119 polyurethane (Febert Shorndorfer Co.) by a dip process. [Authors' designation: Ni 7 C]
8	231	1962	~298	1.009-14.38	45°	45°			Similar to above specimen and conditions. [Authors' designation: Ni 8 C]
9	231	1962	~298	1.009-14.38	45°	45°			Similar to above specimen and conditions. [Authors' designation: Ni 9 C]

DATA TABLE NO. 292 ANGULAR SPECTRAL REFLECTANCE OF NICKEL CONTACT COATINGS

[Wavelength, λ, μm; Reflectance, ρ; Temperature, T, K]

λ	ρ
CURVE 1	
T ~ 298	
1.120	0.588
2.240	0.650
4.824	0.702
7.780	0.735
12.10	0.764
13.53	0.763
14.38	0.782
CURVE 2	
T ~ 298	
1.120	0.513
2.240	0.527
4.824	0.599
7.780	0.630
12.10	0.672
13.53	0.676
14.38	0.690
CURVE 3	
T ~ 298	
1.120	0.680
2.240	0.745
4.824	0.801
7.780	0.820
12.10	0.881
13.53	0.831
14.38	0.821
CURVE 4	
T ~ 298	
1.120	0.578
2.240	0.659
4.824	0.728
7.780	0.745
12.10	0.751
13.53	0.748
14.38	0.751
CURVE 5	
T ~ 298	
1.120	0.750
CURVE 5 (cont.)	
2.240	0.676
4.824	0.762
7.780	0.796
12.10	0.835
13.53	0.841
14.38	0.860
CURVE 6	
T ~ 298	
1.120	0.582
2.240	0.560
4.824	0.635
7.780	0.661
12.10	0.686
13.53	0.691
14.38	0.700
CURVE 7	
T ~ 298	
1.009	0.431
1.120	0.476
1.345	0.501
1.720	0.552
1.945	0.569
2.240	0.577
4.824	0.669
7.780	0.725
10.20	0.725
12.10	0.740
13.53	0.753
14.38	0.757
CURVE 8	
T ~ 298	
1.009	0.395
1.120	0.429
1.345	0.461
1.720	0.514
1.945	0.529
2.240	0.564
4.824	0.671
7.780	0.750
10.20	0.785
CURVE 8 (cont.)	
12.10	0.806
13.53	0.806
14.38	0.811
CURVE 9	
T ~ 298	
1.009	0.397*
1.120	0.404
1.345	0.424
1.720	0.466
1.945	0.480
2.240	0.504
4.824	0.599
7.780	0.653
10.20	0.671
12.10	0.679
13.53	0.678
14.38	0.675

* Not shown on plot

SPECIFICATION TABLE NO. 293 NORMAL TOTAL EMITTANCE OF NICKEL ALUMINIDE CONTACT COATINGS

Curve No.	Ref. No.	Year	Temperature Range, K	Geometry θ'	Reported Error, %	Composition (weight percent), Specifications and Remarks
1*	238	1963	1223	~0°		NiAl (0.0127 mm thick); Inconel substrate; integrated over the range 1-15 μm. [Authors' designation: Sample No. 137]
2*	238	1963	1223	~0°		NiAl and Ni_2Al_3 (0.0508 mm thick); Inconel substrate; integrated over the range 1-15 μm. [Authors' designation: Sample No. 135]
3*	238	1963	1223	~0°		NiAl and Ni_3Al (0.0381 mm thick); Inconel substrate; integrated over the range 1-15 μm. [Authors' designation: Sample No. 136]

DATA TABLE NO. 293 NORMAL TOTAL EMITTANCE OF NICKEL ALUMINIDE CONTACT COATINGS

[Temperature, T, K; Emittance, ε]

T	ε
CURVE 1*	
1223	0.69
CURVE 2*	
1223	0.68
CURVE 3*	
1223	0.73

* No plot given

SPECIFICATION TABLE NO. 294 HEMISPHERICAL TOTAL EMITTANCE OF NICKEL + CHROMIUM + ΣX_i CONTACT COATINGS

Curve No.	Ref. No.	Year	Temperature Range, K	Reported Error, %	Composition (weight percent), Specifications and Remarks
1*	81	1963	434-1217		48.85 Ni, 22 Cr, 20 Fe, 9 Mo and 0.15 C, oxidized Hastelloy X (0.203 mm thick); AISI-310 stainless steel substrate; plasma arc sprayed; particles of size 44-105 μm; measured in vacuum (< 9.7 x 10^{-6} mm Hg); temp measured with thermocouple; heating cycle. [Author's designation: Run No. 1]
2*	81	1963	851-1217		Above specimen and conditions; cooling cycle.
3*	81	1963	1095-1214		Curve 1 specimen and conditions except temp measured with optical pyrometer.

DATA TABLE NO. 294 HEMISPHERICAL TOTAL EMITTANCE OF NICKEL + CHROMIUM + ΣX_i CONTACT COATINGS

[Temperature, T, K; Emittance, ∈]

T	∈	T	∈
CURVE 1*		CURVE 3*	
434	0.522	1095	0.642
546	0.538	1152	0.650
658	0.551	1214	0.647
770	0.575		
826	0.581		
882	0.589		
965	0.585		
993	0.602		
1049	0.611		
1105	0.619		
1162	0.627		
1217	0.636		
CURVE 2*			
1217	0.636		
964	0.606		
851	0.587		

* No plot given

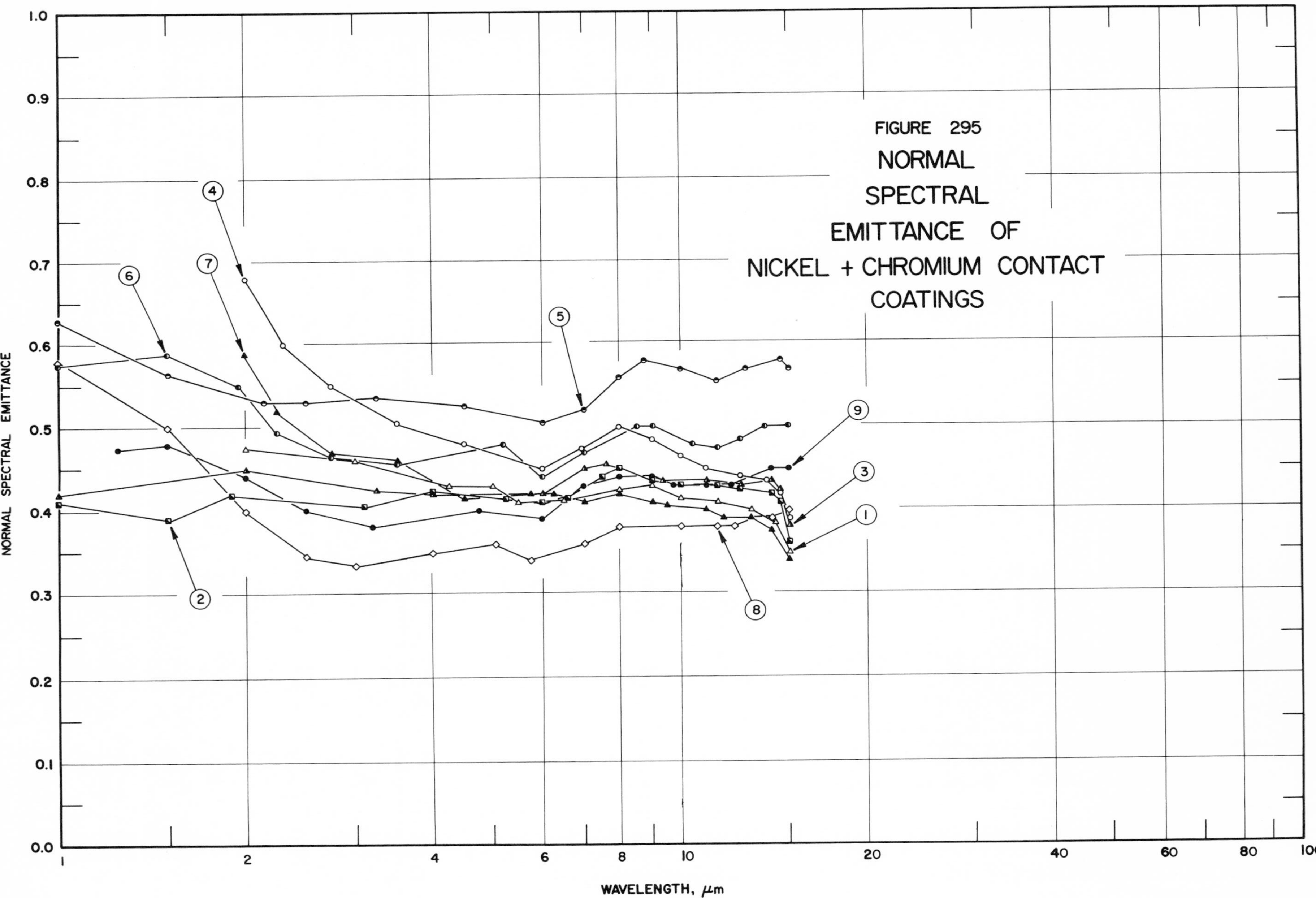
FIGURE 295
NORMAL
SPECTRAL
EMITTANCE OF
NICKEL + CHROMIUM CONTACT
COATINGS
NORMAL SPECTRAL EMITTANCE
WAVELENGTH, μm
0.0
0.1
0.2
0.3
0.4
0.5
0.6
0.7
0.8
0.9
1.0
1
2
4
6
8
10
20
40
60
80
100
1
2
3
4
5
6
7
8
9

SPECIFICATION TABLE NO. 295 NORMAL SPECTRAL EMITTANCE OF NICKEL + CHROMIUM CONTACT COATINGS

Curve No.	Ref. No.	Year	Temperature, K	Wavelength Range, μm	Geometry θ'	Reported Error, %	Composition (weight percent), Specifications, and Remarks
1	139	1961	523	2.00-15.00	$\sim 0^{\circ}$	±5	80 Ni and 20 Cr; sandblasted Inconel X substrate; flame sprayed using a Plasmatron; as received; data extracted from smooth curve.
2	139	1961	773	1.00-15.00	$\sim 0^{\circ}$	±5	Similar to curve 1 specimen and conditions.
3	139	1961	1023	1.00-15.00	$\sim 0^{\circ}$	±5	Similar to curve 1 specimen and conditions.
4	139	1961	523	2.00-15.00	$\sim 0^{\circ}$	±5	Similar to curve 1 specimen and conditions except heated in air at 1089 K for 30 min.
5	139	1961	773	1.00-15.00	$\sim 0^{\circ}$	±5	Similar to curve 4 specimen and conditions.
6	139	1961	1023	1.00-15.00	$\sim 0^{\circ}$	±5	Similar to curve 4 specimen and conditions.
7	139	1961	523	2.00-15.00	$\sim 0^{\circ}$	±5	Similar to curve 1 specimen and conditions except heated in vacuum (6.8×10^{-5} mm Hg) at 1089 K for 30 min.
8	139	1961	773	1.00-15.00	$\sim 0^{\circ}$	±5	Similar to curve 7 specimen and conditions.
9	139	1961	1023	1.25-15.00	$\sim 0^{\circ}$	±5	Similar to curve 7 specimen and conditions.

DATA TABLE NO. 295 NORMAL SPECTRAL EMITTANCE OF NICKEL + CHROMIUM CONTACT COATINGS

[Wavelength, λ, μm; Emittance, ϵ; Temperature, T, K]

λ	ϵ
CURVE 1	**T = 523**
2.00	0.475
3.00	0.460
4.25	0.430
5.00	0.430
5.50	0.410
6.50	0.412
8.00	0.425
9.00	0.430
10.00	0.415
11.50	0.410
13.00	0.400
14.25	0.385
15.00	0.350
CURVE 2	**T = 773**
1.00	0.410
1.50	0.390
1.90	0.420
3.10	0.405
4.00	0.425
5.25	0.415
6.00	0.410
6.60	0.415
7.50	0.440
8.00	0.450
9.00	0.435
10.00	0.430
11.50	0.430
12.50	0.425
14.00	0.420
14.50	0.410
15.00	0.360
CURVE 3	**T = 1023**
1.00	0.420
2.00	0.450
3.25	0.425
4.00	0.420
6.00	0.420
7.00	0.450
CURVE 3 (cont.)	
7.60	0.455
9.35	0.435
11.00	0.435
12.50	0.430
14.00	0.435
14.50	0.425
15.00	0.380
CURVE 4	**T = 523**
2.00	0.680
2.30	0.600
2.75	0.550
3.50	0.505
4.50	0.480
6.00	0.450
6.95	0.475
8.00	0.500
9.00	0.485
10.00	0.465
11.00	0.450
12.50	0.440
13.75	0.435
14.50	0.420
15.00	0.390
CURVE 5	**T = 773**
1.00	0.630
1.50	0.565
2.15	0.530
2.50	0.530
3.25	0.535
4.50	0.525
6.00	0.505
7.00	0.520
8.00	0.560
8.75	0.580
10.00	0.570
11.50	0.555
12.75	0.570
14.50	0.580
15.00	0.570
CURVE 6	**T = 1023**
1.00	0.575
1.50	0.590
1.95	0.550
2.25	0.495
2.75	0.465
3.50	0.455
5.20	0.480
6.00	0.440
7.00	0.470
8.50	0.500
9.00	0.500
10.50	0.480
11.50	0.475
12.50	0.485
13.75	0.500
15.00	0.500
CURVE 7	**T = 523**
2.00	0.590
2.25	0.520
2.75	0.470
3.50	0.460
4.50	0.415
5.75	0.420
6.25	0.420
7.00	0.410
8.00	0.420
9.00	0.410
9.50	0.405
11.00	0.400
11.75	0.390
13.00	0.390
14.00	0.375
15.00	0.340
CURVE 8	**T = 773**
1.00	0.580
1.50	0.500
2.00	0.400
2.50	0.345
CURVE 8 (cont.)	
3.00	0.335
4.00	0.350
5.05	0.360
5.75	0.340
7.00	0.360
8.00	0.380
10.00	0.380
11.50	0.380
12.25	0.380
13.00	0.390*
14.00	0.390
15.00	0.400
CURVE 9	**T = 1023**
1.25	0.475
1.50	0.480
2.00	0.440
2.50	0.400
3.20	0.380
4.75	0.400
6.00	0.390
7.00	0.430
8.00	0.440
9.00	0.440
9.75	0.430
11.00	0.430
12.10	0.430
14.00	0.450
15.00	0.450

* Not shown on plot

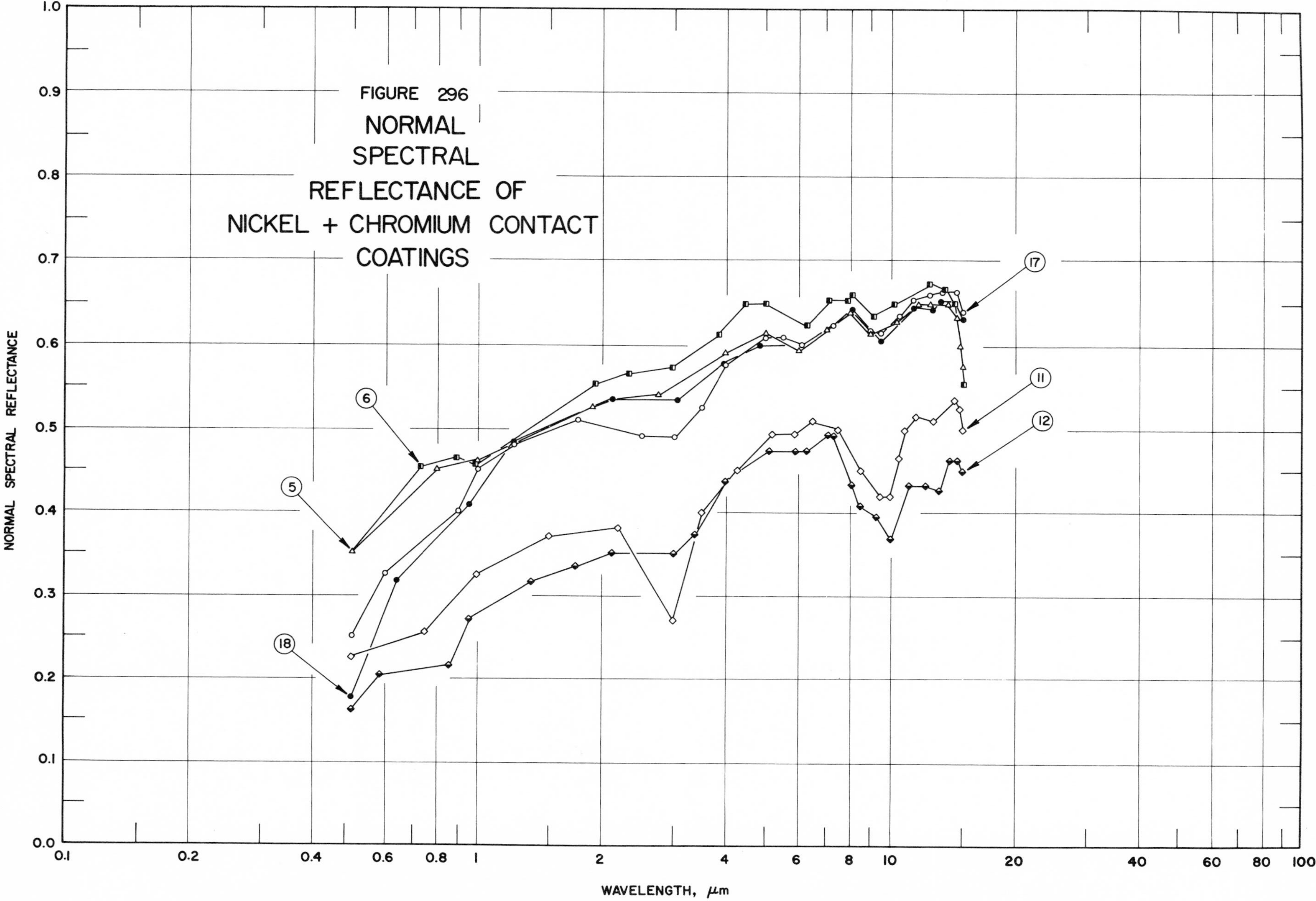

FIGURE 296 NORMAL SPECTRAL REFLECTANCE OF NICKEL + CHROMIUM CONTACT COATINGS

SPECIFICATION TABLE NO. 296 NORMAL SPECTRAL REFLECTANCE OF NICKEL + CHROMIUM CONTACT COATINGS

Curve No.	Ref. No.	Year	Temperature, K	Wavelength Range, μm	Geometry θ	θ'	ω'	Reported Error, %	Composition (weight percent), Specifications, and Remarks
1*	139	1961	~322	2.00-15.00	~0°		2π	< 2	80 Ni and 20 Cr; sandblasted Inconel X substrate; flame sprayed using a Plasmatron; as received; data extracted from smooth curve; sample maintained at ~322 K; converted from R (2π, ~0°); hohlraum at 523 K.
2*	139	1961	~322	2.00-14.99	~0°		2π	< 2	Above specimen and conditions; diffuse component only.
3*	139	1961	~322	1.00-15.00	~0°		2π	< 2	Similar to curve 1 specimen and conditions except hohlraum at 773 K.
4*	139	1961	~322	1.00-15.00	~0°		2π	< 2	Above specimen and conditions; diffuse component only.
5	139	1961	~322	0.50-15.00	~0°		2π	< 2	Similar to curve 1 specimen and conditions except hohlraum at 1273 K.
6	139	1961	~322	0.50-15.00	~0°		2π	< 2	Above specimen and conditions; diffuse component only.
7*	139	1961	~322	2.00-15.00	~0°		2π	< 2	Similar to curve 1 specimen and conditions except heated in air at 1089 K for 30 min.
8*	139	1961	~322	2.00-15.00	~0°		2π	< 2	Above specimen and conditions; diffuse component only.
9*	139	1961	~322	1.00-15.00	~0°		2π	< 2	Similar to curve 7 specimen and conditions except hohlraum at 773 K.
10*	139	1961	~322	1.01-15.00	~0°		2π	< 2	Above specimen and conditions; diffuse component only.
11	139	1961	~322	0.50-15.00	~0°		2π	< 2	Similar to curve 7 specimen and conditions except hohlraum at 1273 K.
12	139	1961	~322	0.50-15.00	~0°		2π	< 2	Above specimen and conditions; diffuse component only.
13*	139	1961	~322	2.00-15.00	~0°		2π	< 2	Similar to curve 1 specimen and conditions except heated in vacuum (6.8 x 10^{-5} mm Hg) at 1089 K for 30 min.
14*	139	1961	~322	2.00-15.00	~0°		2π	< 2	Above specimen and conditions; diffuse component only.
15*	139	1961	~322	1.00-15.00	~0°		2π	< 2	Similar to curve 13 specimen and conditions except hohlraum at 773 K.
16*	139	1961	~322	1.00-15.00	~0°		2π	< 2	Above specimen and conditions; diffuse component only.
17	139	1961	~322	0.50-15.00	~0°		2π	< 2	Similar to curve 13 specimen and conditions except hohlraum at 1273 K.
18	139	1961	~322	0.50-15.00	~0°		2π	< 2	Above specimen and conditions; diffuse component only.

* Not shown on plot

DATA TABLE NO. 296 NORMAL SPECTRAL REFLECTANCE OF NICKEL + CHROMIUM CONTACT COATINGS

[Wavelength, λ, μm; Reflectance, ρ; Temperature, T, K]

CURVE 1* T ~ 322

λ	ρ
2.00	0.475
3.00	0.520
4.00	0.570
5.00	0.605
5.50	0.605
6.10	0.590
7.00	0.605
7.85	0.620
8.90	0.600
10.50	0.620
12.00	0.640
13.00	0.640
14.00	0.620
15.00	0.560

CURVE 2* T ~ 322

λ	ρ
2.00	0.446
2.60	0.475
3.65	0.542
4.17	0.567
4.95	0.584
5.62	0.581
6.09	0.569
6.65	0.578
7.23	0.601
7.62	0.610
8.22	0.601
8.86	0.589
10.00	0.598
11.37	0.612
12.46	0.620
13.01	0.618
13.94	0.597
14.48	0.564
14.84	0.516
14.98	0.475
14.99	0.420

CURVE 3* T ~ 322

λ	ρ
1.00	0.520

CURVE 3 (cont.)*

λ	ρ
1.50	0.510
1.75	0.575
2.50	0.600
3.00	0.600
4.00	0.650
4.50	0.660
5.75	0.660
7.20	0.660
8.00	0.695
9.00	0.690
11.00	0.700
12.10	0.710
13.50	0.710
14.00	0.700
14.60	0.675
14.85	0.650
15.00	0.600

CURVE 4* T ~ 322

λ	ρ
1.00	0.521
2.31	0.521
3.00	0.532
4.35	0.600
4.80	0.611
5.47	0.607
6.40	0.607
6.69	0.616
7.43	0.644
7.94	0.642
8.97	0.612
11.50	0.647
12.17	0.662
12.98	0.653
13.60	0.654
14.03	0.661
14.59	0.644
14.91	0.621
15.00	0.601
15.00	0.551

CURVE 5 T ~ 322

λ	ρ
0.50	0.350
0.80	0.450
1.00	0.460
1.90	0.525
2.10	0.535*
2.75	0.540
4.00	0.590
5.00	0.615
6.00	0.595
7.00	0.620
8.00	0.640
9.00	0.615
10.25	0.630
11.75	0.650
12.50	0.650
13.95	0.650
14.50	0.635
14.85	0.600
15.00	0.575

CURVE 6 T ~ 322

λ	ρ
0.50	0.350*
0.73	0.451
0.89	0.463
0.96	0.458
1.93	0.551
2.31	0.565
2.96	0.572
3.85	0.612
4.46	0.650
5.00	0.650
6.35	0.625
7.19	0.654
7.86	0.654
8.09	0.660
9.08	0.635
10.02	0.650
12.52	0.674
13.67	0.667
14.31	0.650
14.91	0.601*
15.00	0.574*
15.00	0.552

CURVE 7* T ~ 322

λ	ρ
2.00	0.445
2.50	0.395
3.00	0.370
3.65	0.400
4.20	0.450
5.50	0.490
6.25	0.500
7.00	0.505
8.00	0.475
8.50	0.430
9.00	0.400
10.00	0.435
11.00	0.500
11.50	0.510
12.50	0.500
13.00	0.495
13.50	0.500
14.00	0.500
14.50	0.490
14.80	0.475
15.00	0.450

CURVE 8* T ~ 322

λ	ρ
2.00	0.413
3.05	0.350
3.43	0.364
4.17	0.426
4.73	0.447
5.40	0.474
6.42	0.491
6.87	0.494
7.47	0.484
8.02	0.462
9.01	0.400
9.60	0.416
11.03	0.495
11.34	0.498
12.52	0.492
14.01	0.496
14.55	0.482
14.84	0.468
15.00	0.444

CURVE 9* T ~ 322

λ	ρ
1.00	0.300
1.75	0.325
2.15	0.345
2.50	0.340
3.50	0.370
5.25	0.430
6.00	0.420
6.75	0.435
8.00	0.340
9.00	0.310
9.75	0.295
10.20	0.310
11.10	0.400
12.00	0.400
13.00	0.400
13.50	0.435
14.50	0.450
15.00	0.450

CURVE 10* T ~ 322

λ	ρ
1.01	0.312
1.50	0.326
2.63	0.326
2.99	0.323
3.32	0.335
3.86	0.398
4.98	0.453
5.59	0.463
6.12	0.463
6.99	0.483
8.83	0.403
9.98	0.403
10.30	0.416
11.00	0.492
11.33	0.490
12.01	0.472
12.53	0.474
13.78	0.525
14.50	0.538
15.00	0.534

CURVE 11 T ~ 322

λ	ρ
0.50	0.225
0.75	0.255
1.00	0.325
1.50	0.370
2.20	0.380
3.00	0.270
3.50	0.400
4.25	0.450
5.20	0.495
5.90	0.495
6.50	0.510
7.50	0.500
8.50	0.450
9.50	0.420
10.00	0.420
10.50	0.465
10.90	0.500
11.50	0.515
12.75	0.510
14.25	0.535
14.75	0.525
15.00	0.500

CURVE 12 T ~ 322

λ	ρ
0.50	0.160
0.58	0.202
0.86	0.216
0.96	0.270
1.35	0.316
1.71	0.335
2.11	0.350
3.00	0.350
3.37	0.371
4.00	0.436
5.15	0.473
5.99	0.472
6.26	0.473
7.01	0.493
7.28	0.492
8.01	0.432
8.53	0.409
9.28	0.395

CURVE 12 (cont.)

λ	ρ
10.00	0.369
11.01	0.431
12.20	0.431
13.01	0.427
14.00	0.461
14.52	0.461
15.00	0.450

CURVE 13* T ~ 322

λ	ρ
2.00	0.605
2.50	0.620
3.00	0.490
3.50	0.500
4.20	0.550
5.00	0.575
5.75	0.570
7.00	0.600
7.75	0.610
8.75	0.590
9.50	0.585
10.30	0.600
11.00	0.620
12.50	0.620
13.00	0.625
13.50	0.625
14.85	0.600
15.00	0.550

CURVE 14* T ~ 322

λ	ρ
2.00	0.635
2.44	0.569
2.99	0.531
3.64	0.544
4.96	0.601
5.82	0.602
6.75	0.620
7.50	0.632
8.04	0.629
9.35	0.606
9.71	0.609
11.23	0.644

CURVE 14 (cont.)*

λ	ρ
11.70	0.644
12.12	0.650
12.51	0.650
13.52	0.635
14.33	0.621
14.79	0.599
15.00	0.550

CURVE 15* T ~ 322

λ	ρ
1.00	0.480
1.50	0.460
2.00	0.490
2.25	0.500
2.75	0.495
3.50	0.535
4.25	0.580
5.25	0.600
6.15	0.590
7.00	0.620
7.85	0.630
8.80	0.610
10.00	0.625
11.00	0.650
12.50	0.660
14.00	0.690
14.65	0.695
15.00	0.690

CURVE 16* T ~ 322

λ	ρ
1.00	0.472
1.36	0.458
1.70	0.465
2.32	0.490
2.99	0.494
3.21	0.502
3.99	0.561
5.81	0.595
5.78	0.596
6.30	0.589
7.30	0.615
8.05	0.621

* Not shown on plot

DATA TABLE NO. 296 NORMAL SPECTRAL REFLECTANCE OF NICKEL + CHROMIUM CONTACT COATINGS (continued)

λ	ρ
CURVE 16 (cont.)*	
T ~ 322	
8.70	0.605
9.75	0.605
10.22	0.614
11.42	0.641
12.02	0.652
12.80	0.655
14.39	0.683
15.00	0.683
CURVE 17	
T ~ 322	
0.50	0.250
0.60	0.325
0.90	0.400
1.00	0.450
1.25	0.480
1.75	0.510
2.50	0.490
3.00	0.490
3.50	0.525
4.00	0.575
5.00	0.610
5.50	0.610
6.10	0.600
7.25	0.625
8.15	0.640*
9.00	0.615
9.50	0.615
10.50	0.635
11.25	0.655
12.50	0.660
13.50	0.665
14.50	0.665
15.00	0.640
CURVE 18	
T ~ 322	
0.50	0.175
0.64	0.318
0.96	0.408
1.25	0.482
2.11	0.535
3.03	0.532
3.98	0.579
CURVE 18 (cont.)	
4.81	0.600
6.15	0.600*
8.12	0.643
9.00	0.613*
9.53	0.606
11.15	0.643
12.64	0.643
13.04	0.652
14.35	0.652*
15.00	0.632

* Not shown on plot

SPECIFICATION TABLE NO. 297 NORMAL TOTAL EMITTANCE OF NICKEL + COBALT CONTACT COATINGS

Curve No.	Ref. No.	Year	Temperature Range, K	Geometry θ'	Reported Error, %	Composition (weight percent), Specifications and Remarks
1*	89	1958	366-672	$\sim 0°$		Ni-Co alloy, North American Aviation (0.0127 mm thick); 321 stainless steel substrate; plated. [Author's designation: Specimen 11]

DATA TABLE NO. 297 NORMAL TOTAL EMITTANCE OF NICKEL + COBALT CONTACT COATINGS

[Temperature, T, K; Emittance, ϵ]

T	ϵ
CURVE 1*	
366	0.111
444	0.147
522	0.092
600	0.109
672	0.116

* No plot given

SPECIFICATION TABLE NO. 298 NORMAL SPECTRAL REFLECTANCE OF NICKEL + COBALT CONTACT COATINGS

Curve No.	Ref. No.	Year	Temperature, K	Wavelength Range, μm	Geometry θ θ' ω'	Reported Error, %	Composition (weight percent), Specifications, and Remarks
1*	89	1958	~298	0.400-24.9	~0° ~0°		Ni-Co, North American Aviation, (0.0127 mm thick); 321 stainless steel substrate; plated; data extracted from smooth curve. [Author's designation: Specimen 11]

DATA TABLE NO. 298 NORMAL SPECTRAL REFLECTANCE OF NICKEL + COBALT CONTACT COATINGS

[Wavelength, λ, μm; Reflectance, ρ; Temperature, T, K]

λ	ρ	λ	ρ	λ	ρ
CURVE 1* T ~ 298		CURVE 1 (cont.)*		CURVE 1 (cont.)*	
		1.28	0.633	11.9	0.921
0.400	0.330	1.71	0.687	24.9	0.922
0.450	0.350	1.80	0.712		
0.460	0.369	2.09	0.741		
0.560	0.419	2.25	0.748		
0.590	0.419	2.53	0.789		
0.600	0.441	2.80	0.810		
0.620	0.441	4.03	0.861		
0.660	0.459	4.29	0.861		
0.740	0.459	4.52	0.872		
0.760	0.470	4.81	0.872		
0.780	0.470	4.98	0.881		
0.800	0.480	5.27	0.881		
0.840	0.480	5.58	0.892		
0.860	0.491	7.61	0.892		
0.900	0.491	7.82	0.901		
0.920	0.501	10.1	0.901		
0.960	0.501	10.5	0.911		
1.00	0.518	11.6	0.911		

* No plot given

SPECIFICATION TABLE NO. 299 HEMISPHERICAL TOTAL EMITTANCE OF NICKEL + MOLYBDENUM + ΣX_i CONTACT COATINGS

Curve No.	Ref. No.	Year	Temperature Range, K	Reported Error, %	Composition (weight percent), Specifications and Remarks
1*	81	1963	519-1219		54 Ni, 17 Mo, 15 Cr, 5 Fe and 4 W, oxidized Hastelloy C (0.178 mm thick); AISI-310 stainless steel substrate; plasma arc sprayed; measured in vacuum (< 5.5×10^{-6} mm Hg); temp measured with thermocouple; heating cycle. [Author's designation: Run No. 1]
2*	81	1963	1219-794		Above specimen and conditions; cooling cycle.
3*	81	1963	1091-1205		Curve 1 specimen and conditions except temp measured with optical pyrometer.

DATA TABLE NO. 299 HEMISPHERICAL TOTAL EMITTANCE OF NICKEL + MOLYBDENUM + ΣX_i CONTACT COATINGS

[Temperature, T, K; Emittance, ϵ]

T	ϵ	T	ϵ
CURVE 1*		CURVE 3*	
519	0.476	1091	0.592
547	0.493	1143	0.557
658	0.502	1205	0.602
767	0.525		
824	0.530		
879	0.535		
936	0.541		
993	0.549		
1048	0.548		
1105	0.563		
1162	0.577		
1219	0.575		
CURVE 2*			
1219	0.575		
964	0.523		
794	0.524		

* No plot given

SPECIFICATION TABLE NO. 300 NORMAL TOTAL EMITTANCE OF PALLADIUM CONTACT COATINGS

Curve No.	Ref. No.	Year	Temperature Range, K	Geometry θ'	Reported Error, %	Composition (weight percent), Specifications and Remarks
1*	226	1961	570-965	~0°	~10	Palladium (0.12 mg cm^{-2} area density); Inconel X substrate; substrate prepared as follows: surface pumice cleaned, alkaline cleaned, etched in acid solution (46 parts conc HNO_3, 8 parts 46% HF, 46 parts H_2O), rinsed in H_2O, electropolished (60 parts 85% H_3PO_4, 30 parts 98% H_2SO_4, 10 parts 46% HF) for 5-10 min, rinsed in demineralized H_2O, vapor degreased 30 min in trichloroethylene; vapor deposited in vacuum (< 10^{-4} mm Hg) on substrate at 523-573 K.
2*	196	1960	446-771	~0°		"Liquid palladium"; glass substrate; painted on substrate, then fired at 866 K; data extracted from smooth curve.

DATA TABLE NO. 300 NORMAL TOTAL EMITTANCE OF PALLADIUM CONTACT COATINGS

[Temperature, T, K; Emittance, ϵ]

T	ϵ
CURVE 1*	
570	0.172
773	0.200
965	0.223
CURVE 2*	
446	0.194
603	0.189
771	0.181

* No plot given

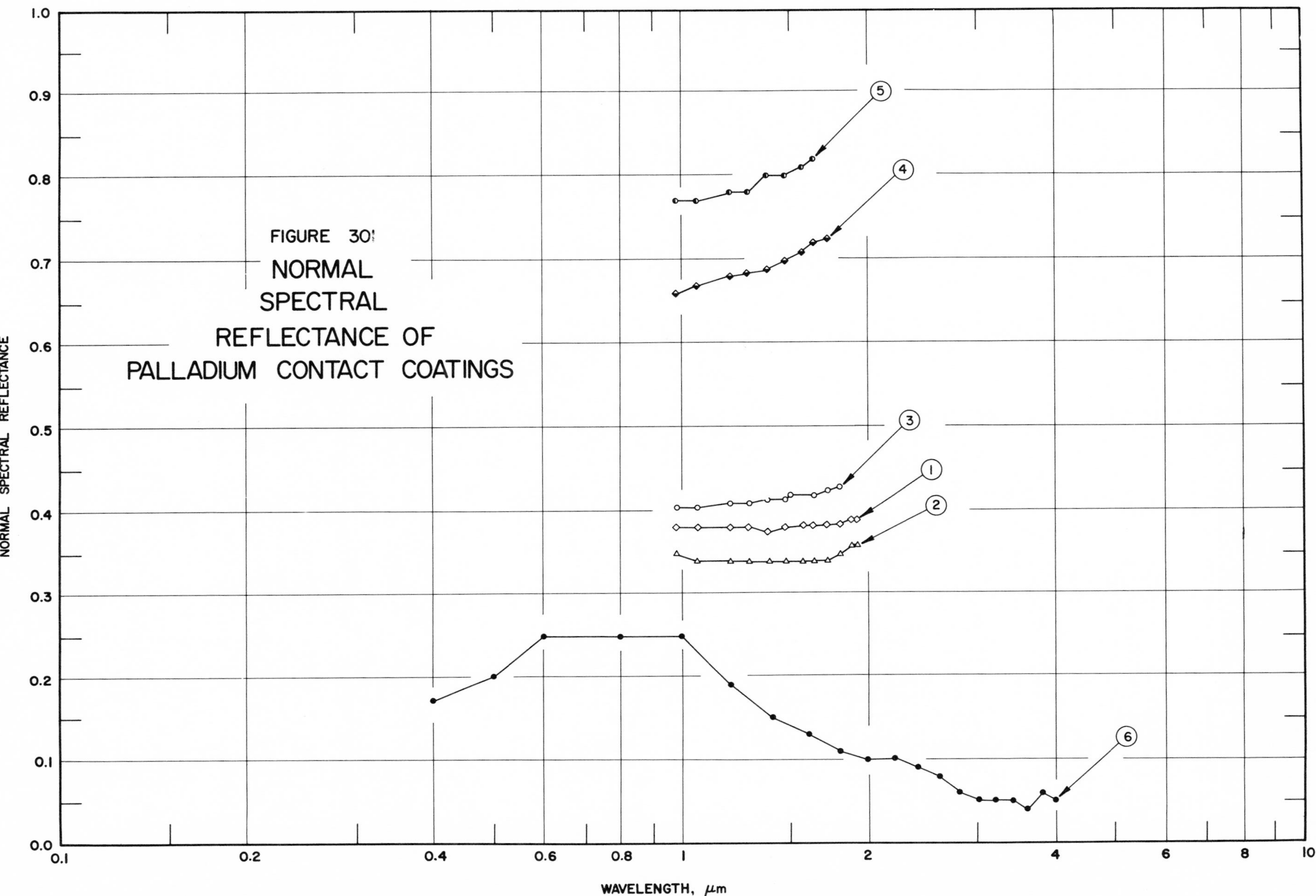

FIGURE 30[illegible] NORMAL SPECTRAL REFLECTANCE OF PALLADIUM CONTACT COATINGS

SPECIFICATION TABLE NO. 301 NORMAL SPECTRAL REFLECTANCE OF PALLADIUM CONTACT COATINGS

Curve No.	Ref. No.	Year	Temperature, K	Wavelength Range, μm	Geometry θ	θ'	ω'	Reported Error, %	Composition (weight percent), Specifications, and Remarks
1	239	1962	298	0.98-1.92	0°		2π		Palladium, 99.9 pure (125 Å thick); cleaved silicon substrate; vacuum evaporated; measured relative to silver mirror.
2	239	1962	298	0.98-1.92	0°		2π		Similar to above specimen and conditions except 130 Å thick.
3	239	1962	298	0.98-1.80	0°		2π		Similar to above specimen and conditions except 146 Å thick.
4	239	1962	298	0.98-1.72	0°		2π		Similar to above specimen and conditions except 250 Å thick.
5	239	1962	298	0.98-1.64	0°		2π		Similar to above specimen and conditions except 410 Å thick.
6	215	1962	~298	0.4-4.0	~0°	~0°			Palladium; glass substrate; prepared by applying a solution of palladium resinate, diluted with aromatic solvents and essential oils, to a soda-lime glass substrate, then firing to 873 K in a continuous lehr on a 1.5 hr cycle; application of solution was by dropping an excess amount on the substrate, which was kept spinning at 1550 rpm until no more solution was flung from the edges; measured relative to plane, polished aluminum; specular component only.

DATA TABLE NO. 301 NORMAL SPECTRAL REFLECTANCE OF PALLADIUM CONTACT COATINGS

[Wavelength, λ, μm; Reflectance, ρ; Temperature, T, K]

CURVE 1
T = 298

λ	ρ
0.98	0.380
1.06	0.380
1.20	0.380
1.28	0.380
1.37	0.375
1.47	0.380
1.56	0.382
1.64	0.382
1.72	0.382
1.80	0.385
1.88	0.390
1.92	0.390

CURVE 2
T = 298

λ	ρ
0.98	0.35
1.06	0.34

CURVE 2 (cont.)

λ	ρ
1.20	0.34
1.28	0.34
1.37	0.34
1.47	0.34
1.56	0.34
1.64	0.34
1.72	0.34
1.80	0.35
1.88	0.36
1.92	0.36

CURVE 3
T = 298

λ	ρ
0.98	0.405
1.06	0.405
1.20	0.41
1.28	0.41
1.37	0.415

CURVE 3 (cont.)

λ	ρ
1.47	0.415
1.50	0.420
1.64	0.42
1.72	0.425
1.80	0.430

CURVE 4
T = 298

λ	ρ
0.98	0.66
1.06	0.67
1.20	0.68
1.28	0.685
1.37	0.69
1.47	0.70
1.56	0.71
1.64	0.72
1.72	0.725

CURVE 5
T = 298

λ	ρ
0.98	0.77
1.06	0.77
1.20	0.78
1.28	0.78
1.37	0.80
1.47	0.80
1.56	0.81
1.64	0.82

CURVE 6
T ~ 298

λ	ρ
0.4	0.17
0.5	0.20
0.6	0.25
0.8	0.25
1.0	0.25
1.2	0.19

CURVE 6 (cont.)

λ	ρ
1.4	0.15
1.6	0.13
1.8	0.11
2.0	0.10
2.2	0.10
2.4	0.09
2.6	0.08
2.8	0.06
3.0	0.05
3.2	0.05
3.4	0.05
3.6	0.04
3.8	0.06
4.0	0.05

SPECIFICATION TABLE NO. 302 NORMAL SPECTRAL TRANSMITTANCE OF PALLADIUM CONTACT COATINGS

Curve No.	Ref. No.	Year	Temperature, K	Wavelength Range, μm	Geometry θ	θ'	ω'	Reported Error, %	Composition (weight percent), Specifications, and Remarks
1*	215	1962	~298	0.4-2.0	~0°	~0°			Palladium; glass substrate (1.59 mm thick); prepared by applying a solution of palladium resinate, diluted with aromatic solvents and essential oils, to a soda-lime glass substrate, then firing to 873 K in a continuous lehr on a 1.5 hr cycle; application of solution was by dropping an excess amount on the substrate, which was kept spinning at 1550 rpm until no more solution was flung from the edges; τ is for coating plus substrate.

DATA TABLE NO. 302 NORMAL SPECTRAL TRANSMITTANCE OF PALLADIUM CONTACT COATINGS

[Wavelength, λ, μm; Transmittance, τ; Temperature, T, K]

λ	τ
CURVE 1*	
T ~ 298	
0.4	0.150
0.5	0.196
0.6	0.400
0.7	0.510
0.8	0.576
0.9	0.623
1.0	0.663
1.1	0.693
1.2	0.776
1.3	0.716
1.4	0.600
1.5	0.416
1.6	0.326
1.8	0.245
2.0	0.210

* No plot given

SPECIFICATION TABLE NO. 303 NORMAL TOTAL EMITTANCE OF PLATINUM CONTACT COATINGS

Curve No.	Ref. No.	Year	Temperature Range, K	Geometry θ'	Reported Error, %	Composition (weight percent), Specifications and Remarks
1*	196	1960	409-618	$\sim 0^\circ$		"Liquid platinum"; glass substrate; painted on substrate, then fired at 866 K; data extracted from smooth curve.

DATA TABLE NO. 303 NORMAL TOTAL EMITTANCE OF PLATINUM CONTACT COATINGS

[Temperature, T, K; Emittance, ϵ]

T	ϵ
CURVE 1*	
409	0.463
513	0.477
618	0.491

* No plot given

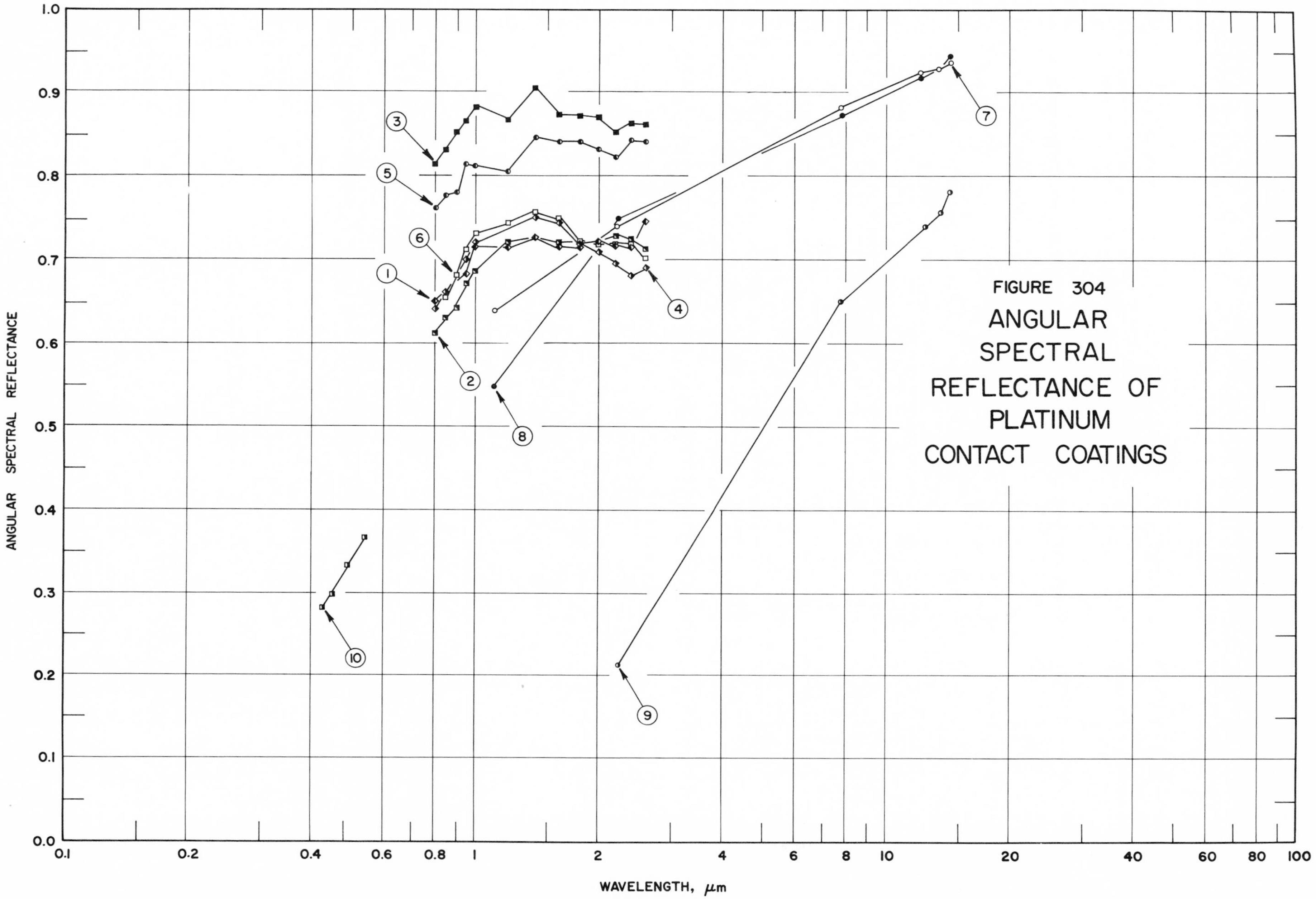

FIGURE 304 ANGULAR SPECTRAL REFLECTANCE OF PLATINUM CONTACT COATINGS

SPECIFICATION TABLE NO. 304 ANGULAR SPECTRAL REFLECTANCE OF PLATINUM CONTACT COATINGS

Curve No.	Ref. No.	Year	Temperature, K	Wavelength Range, μm	Geometry θ	θ'	ω'	Reported Error, %	Composition (weight percent), Specifications, and Remarks
1	240	1964	298	0.80-2.60	45°	45°			Platinum (~0.102 μm thick); fused quartz substrate; applied by brushing on a platinum solution and firing; tungsten filament source; measured relative to aluminum mirror; data beyond 2.2 μm subject to sizable errors. [Author's designation: Specimen No. 3A]
2	240	1964	298	0.80-2.60	15°	15°			Above specimen and conditions.
3	240	1964	298	0.80-2.60	45°	45°			Similar to curve 1 specimen and conditions. [Author's designation: Specimen No. 4A]
4	240	1964	298	0.80-2.60	15°	15°			Above specimen and conditions.
5	240	1964	298	0.80-2.60	45°	45°			Similar to curve 1 specimen and conditions. [Author's designation: Specimen No. 5A]
6	240	1964	298	0.85-2.60	15°	15°			Above specimen and conditions.
7	231	1962	~298	1.120-14.38	45°	45°			Platinum, Hanovia Liquid Bright Platinum No. 1 (Englehard Industries); glazed, cleaned ceramic wall tile (Tiffany Tile Corp.) substrate; metal suspension deposited on substrate rotating at several hundred rpm, then fired in air at ~1000 K; specular component only. [Authors' designation: TH Pt 2]
8	231	1962	~298	1.120-14.38	45°	45°			Similar to above specimen and conditions. [Author's designation: TH Pt 3]
9	231	1962	~298	2.240-14.38	45°	45°			Similar to above specimen and conditions except glazed ceramic wall tile provided by Sears, Roebuck, and Co. [Author's designation: SH Pt 1]
10	80	1966	~298	0.4260-0.5400	0°		2π		Liquid Bright Platinum Ga, supplied by Johnson, Matthey and Co., Ltd.; quartz substrate; measured relative to MgO; converted from R (2π, 0°). [Author's designation: Reflector No. XI]

DATA TABLE NO. 304 ANGULAR SPECTRAL REFLECTANCE OF PLATINUM CONTACT COATINGS

[Wavelength, λ, μm; Reflectance, ρ; Temperature, T, K]

λ	ρ
CURVE 1	
T = 298	
0.80	0.650
0.85	0.659*
0.90	0.681*
0.95	0.684
1.00	0.715
1.20	0.715
1.40	0.728
1.60	0.716
1.80	0.716
2.00	0.721
2.20	0.718
2.40	0.715
2.60	0.748
CURVE 2	
T = 298	
0.80	0.611
0.85	0.631
0.90	0.643
0.95	0.673
1.00	0.687
1.20	0.723
1.40	0.728*
1.60	0.721
1.80	0.721*
2.00	0.721*
2.20	0.730
2.40	0.726
2.60	0.713
CURVE 3	
T = 298	
0.80	0.815
0.85	0.830
0.90	0.853
0.95	0.865
1.00	0.881
1.20	0.867
1.40	0.904
1.60	0.873
1.80	0.871
2.00	0.871
2.20	0.853
CURVE 3 (cont.)	
2.40	0.863
2.60	0.863
CURVE 4	
T = 298	
0.80	0.643
0.85	0.662
0.90	0.681*
0.95	0.700
1.00	0.721
1.20	0.745*
1.40	0.752
1.60	0.746
1.80	0.719
2.00	0.710
2.20	0.697
2.40	0.682
2.60	0.690
CURVE 5	
T = 298	
0.80	0.762
0.85	0.778
0.90	0.780
0.95	0.815
1.00	0.813
1.20	0.807
1.40	0.847
1.60	0.841
1.80	0.841
2.00	0.832
2.20	0.824
2.40	0.843
2.60	0.843
CURVE 6	
T = 298	
0.85	0.655
0.90	0.681
0.95	0.713
1.00	0.733
1.20	0.745
1.40	0.757
CURVE 6 (cont.)	
1.60	0.752
1.80	0.723
2.00	0.721
2.20	0.721
2.40	0.721
2.60	0.703
CURVE 7	
T ~ 298	
1.120	0.639
2.240	0.741
7.780	0.883
12.10	0.924
13.53	0.928
14.38	0.935
CURVE 8	
T ~ 298	
1.120	0.547
2.240	0.750
7.780	0.873
12.10	0.917
13.53	0.928*
14.38	0.942
CURVE 9	
T ~ 298	
2.240	0.213
7.780	0.650
12.10	0.741
13.53	0.757
14.38	0.783
CURVE 10	
T ~ 298	
0.4260	0.281
0.4570	0.299
0.4950	0.334
0.5400	0.365

* Not shown on plot

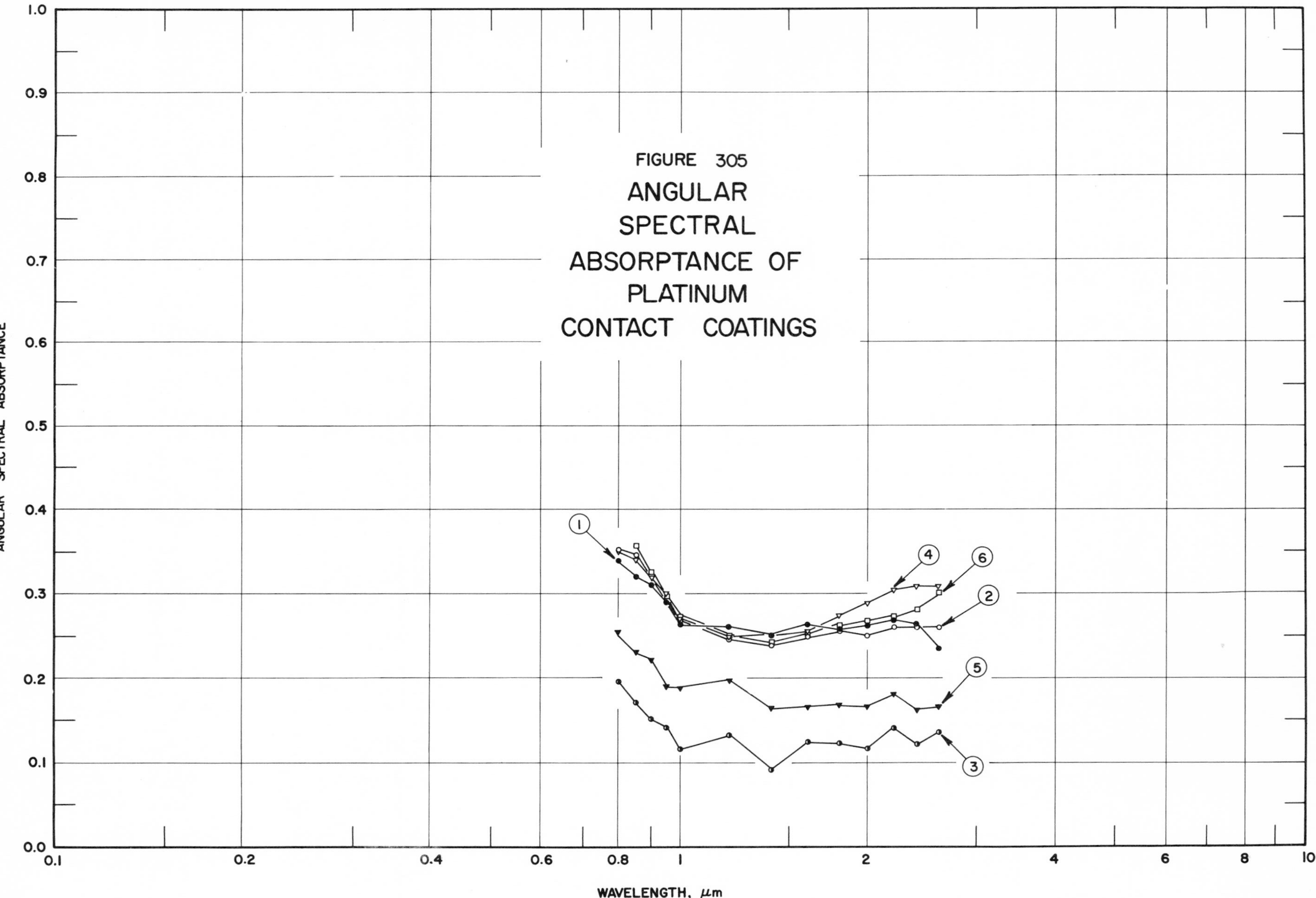
FIGURE 305
ANGULAR
SPECTRAL
ABSORPTANCE OF
PLATINUM
CONTACT COATINGS
ANGULAR SPECTRAL ABSORPTANCE
WAVELENGTH, μm
1.0
0.9
0.8
0.7
0.6
0.5
0.4
0.3
0.2
0.1
0.0
0.1
0.2
0.4
0.6
0.8
1
2
4
6
8
10
1
4
6
2
5
3

SPECIFICATION TABLE NO. 305 ANGULAR SPECTRAL ABSORPTANCE OF PLATINUM CONTACT COATINGS

Curve No.	Ref. No.	Year	Temperature, K	Wavelength Range, μm	Geometry θ	Reported Error, %	Composition (weight percent), Specifications, and Remarks
1	240	1964	298	0.80-2.60	45°		Platinum (~0.102 μm thick); fused quartz substrate; applied by brushing on a platinum solution and firing; author assumed $\alpha(45°) = 1 - \rho(45°, 45°) - \tau(0°, 0°)$; data beyond 2.2 μm subject to sizable errors. [Author's designation: Specimen No. 3A]
2	240	1964	298	0.80-2.60	15°		Above specimen and conditions except author assumed $\alpha(15°) = 1 - \rho(15°, 15°) - \tau(0°, 0°)$.
3	240	1964	298	0.80-2.60	45°		Similar to curve 1 specimen and conditions. [Author's designation: Specimen No. 4A]
4	240	1964	298	0.80-2.60	15°		Above specimen and conditions except author assumed $\alpha(15°) = 1 - \rho(15°, 15°) - \tau(0°,0°)$.
5	240	1964	298	0.80-2.60	45°		Similar to curve 1 specimen and conditions. [Author's designation: Specimen No. 5A]
6	240	1964	298	0.85-2.60	15°		Above specimen and conditions except author assumed $\alpha(15°) = 1 - \rho(15°, 15°) - \tau(0°, 0°)$.

DATA TABLE NO. 305 ANGULAR SPECTRAL ABSORPTANCE OF PLATINUM CONTACT COATINGS

[Wavelength, λ, μm; Absorptance, α; Temperature, T, K]

λ	α
CURVE 1 T = 298	
0.80	0.340
0.85	0.321
0.90	0.310
0.95	0.291
1.00	0.266
1.20	0.264
1.40	0.252
1.60	0.265
1.80	0.259
2.00	0.264
2.20	0.270
2.40	0.265
2.60	0.236
CURVE 2 T = 298	
0.80	0.353
0.85	0.347
0.90	0.321
0.95	0.292*
1.00	0.269
1.20	0.247
1.40	0.239
1.60	0.250
1.80	0.257
2.00	0.251
2.20	0.261
2.40	0.261
2.60	0.261
CURVE 3 T = 298	
0.80	0.198
0.85	0.172
0.90	0.151
0.95	0.142
1.00	0.117
1.20	0.134
1.40	0.094
1.60	0.125
1.80	0.124
2.00	0.118
2.20	0.141
2.40	0.122
2.60	0.136
CURVE 4 T = 298	
0.80	0.351
0.85	0.340
0.90	0.320
0.95	0.301
1.00	0.273
1.20	0.250
1.40	0.251*
1.60	0.255
1.80	0.275
2.00	0.290
2.20	0.306
2.40	0.310
2.60	0.310
CURVE 5 T = 298	
0.80	0.256
0.85	0.231
0.90	0.224
0.95	0.191
1.00	0.189
1.20	0.197
1.40	0.166
1.60	0.167
1.80	0.169
2.00	0.167
2.20	0.182
2.40	0.163
2.60	0.167
CURVE 6 T = 298	
0.85	0.357
0.90	0.327
0.95	0.297
1.00	0.276
1.20	0.251
1.40	0.246
1.60	0.252
1.80	0.266
2.00	0.269
2.20	0.275
2.40	0.283
2.60	0.303

* Not shown on plot

SPECIFICATION TABLE NO. 306 NORMAL SPECTRAL TRANSMITTANCE OF PLATINUM CONTACT COATINGS

Curve No.	Ref. No.	Year	Temperature, K	Wavelength Range, μm	Geometry θ	θ'	ω'	Reported Error, %	Composition (weight percent), Specifications, and Remarks
1*	240	1964	298	0.80-2.60	0°	0°			Platinum (~0.102 μm thick); fused quartz substrate; applied by brushing on a platinum solution and firing; data beyond 2.2 μm subject to sizable errors. [Author's designation: Specimen No. 3A]
2*	240	1964	298	0.80-2.60	0°	0°			Similar to above specimen and conditions. [Author's designation: Specimen No. 4A]
3*	240	1964	298	0.80-2.60	0°	0°			Similar to above specimen and conditions. [Author's designation: Specimen No. 5A]

DATA TABLE NO. 306 NORMAL SPECTRAL TRANSMITTANCE OF PLATINUM CONTACT COATINGS

[Wavelength, λ, μm; Transmittance, τ; Temperature, T, K]

λ	τ	λ	τ	λ	τ
CURVE 1* T = 298		CURVE 2 (cont.)		CURVE 3 (cont.)	
		0.90	0.020	1.20	0.006
0.80	0.006	1.00	0.021	1.40	0.007
0.85	0.009	1.20	0.022	1.60	0.008
0.90	0.008	1.40	0.021	1.80	0.010
0.95	0.011	1.60	0.023	2.00	0.013
1.00	0.012	1.80	0.028	2.20	0.013
1.20	0.012	2.00	0.032	2.40	0.010
1.40	0.011	2.20	0.034	2.60	0.007
1.60	0.010	2.40	0.031		
1.80	0.013	2.60	0.024		
2.00	0.015				
2.40	0.016	CURVE 3* T = 298			
2.60	0.019				
		0.80	0.004		
CURVE 2* T = 298		0.85	0.004		
		0.90	0.005		
0.80	0.020	0.95	0.005		
0.85	0.020	1.00	0.006		

* No plot given

SPECIFICATION TABLE NO. 307 HEMISPHERICAL TOTAL EMITTANCE OF RHODIUM CONTACT COATINGS

Curve No.	Ref. No.	Year	Temperature Range, K	Reported Error, %	Composition (weight percent), Specifications and Remarks
1*	194	1955	76	5	Rhodium; stainless steel substrate; plated; measured in vacuum (10^{-6} to 10^{-7} mm Hg); emittance calculated from absorptance for 300 K blackbody incident radiation; authors assumed $\alpha = \epsilon$.

DATA TABLE NO. 307 HEMISPHERICAL TOTAL EMITTANCE OF RHODIUM CONTACT COATINGS

[Temperature, T, K; Emittance, ϵ]

T	ϵ
CURVE 1*	
76	0.078

* No plot given

SPECIFICATION TABLE NO. 308 NORMAL TOTAL EMITTANCE OF RHODIUM CONTACT COATINGS

Curve No.	Ref. No.	Year	Temperature Range, K	Geometry θ'	Reported Error, %	Composition (weight percent), Specifications and Remarks
1*	226	1961	760-964	~0°	~10	Rhodium (1.0 mg cm^{-2} area density); Inconel X substrate; substrate prepared as follows: surface pumice cleaned, alkaline cleaned, etched in acid solution (46 parts conc. HNO_3, 8 parts 46% HF, 46 parts H_2O), rinsed in H_2O, electropolished (60 parts 85% H_3PO_4, 30 parts 98% H_2SO_4, 10 parts 46% HF) for 5-10 min, rinsed in demineralized H_2O, vapor degreased 30 min in trichloroethylene; vapor deposited in vacuum ($<10^{-4}$ mm Hg) on substrate at 523-573 K.

DATA TABLE NO. 308 NORMAL TOTAL EMITTANCE OF RHODIUM CONTACT COATINGS

[Temperature, T, K; Emittance, ϵ]

T	ϵ
CURVE 1*	
760	0.180
964	0.209

* No plot given

SPECIFICATION TABLE NO. 309 HEMISPHERICAL INTEGRATED ABSORPTANCE OF RHODIUM CONTACT COATINGS

Curve No.	Ref. No.	Year	Temperature Range, K	Reported Error, %	Composition (weight percent), Specifications and Remarks
1*	194	1955	76	5	Rhodium; stainless steel substrate; plated; measured in vacuum (10^{-6} to 10^{-7} mm Hg); absorptance for 300 K blackbody incident radiation.

DATA TABLE NO. 309 HEMISPHERICAL INTEGRATED ABSORPTANCE OF RHODIUM CONTACT COATINGS

[Temperature, T, K; Absorptance, α]

T	α
CURVE 1*	
76	0.078

* No plot given

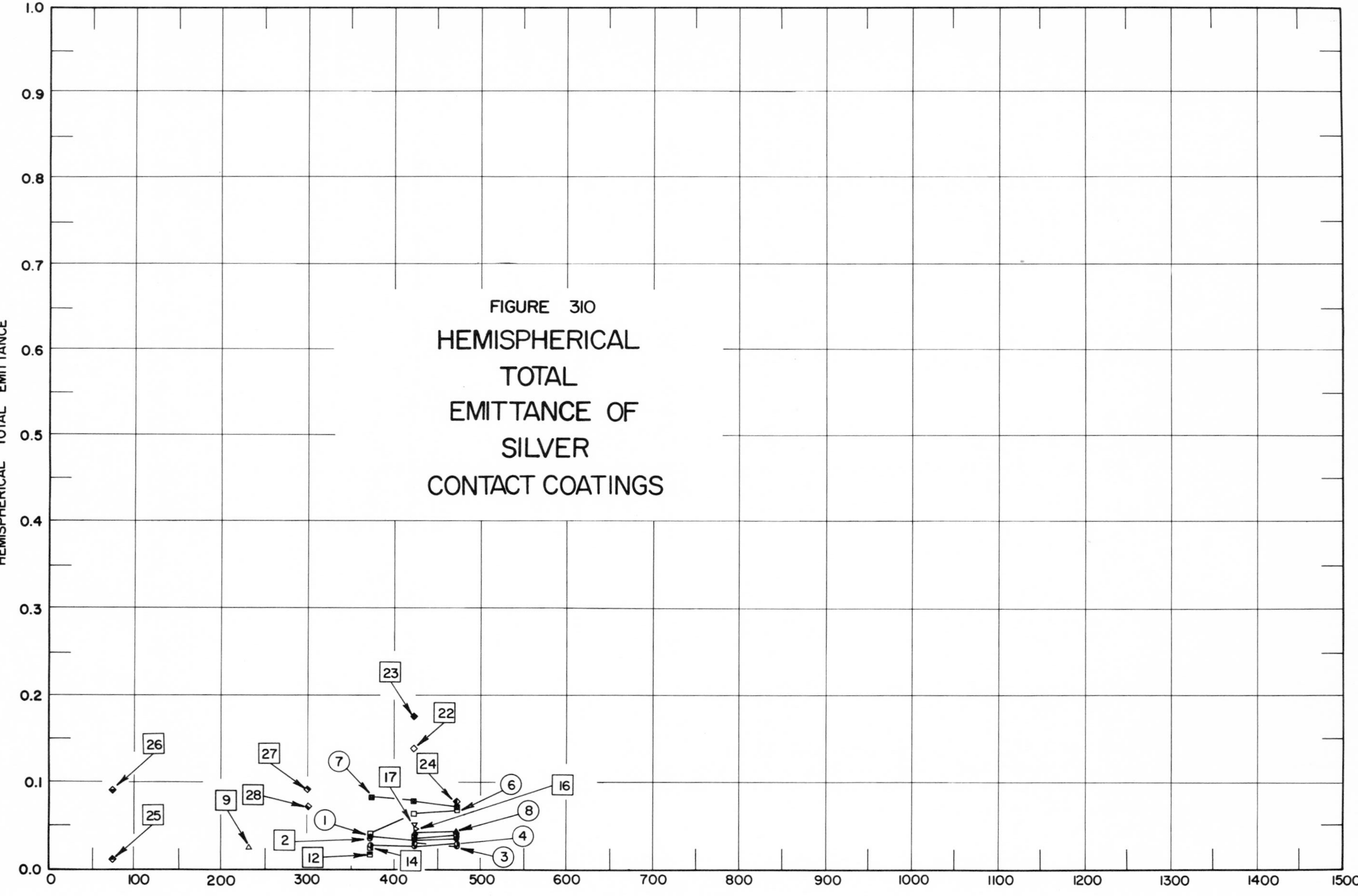
FIGURE 310
HEMISPHERICAL TOTAL EMITTANCE OF SILVER CONTACT COATINGS
HEMISPHERICAL TOTAL EMITTANCE
TEMPERATURE, K
0.0
0.1
0.2
0.3
0.4
0.5
0.6
0.7
0.8
0.9
1.0
0
100
200
300
400
500
600
700
800
900
1000
1100
1200
1300
1400
1500
1
2
3
4
6
7
8
9
12
14
16
17
22
23
24
25
26
27
28

SPECIFICATION TABLE NO. 310 HEMISPHERICAL TOTAL EMITTANCE OF SILVER CONTACT COATINGS

Curve No.	Ref. No.	Year	Temperature Range, K	Reported Error, %	Composition (weight percent), Specifications and Remarks
1	231	1962	373-473	10-20	Silver; General Electric SR-111 silicone resin and 316 stainless steel substrates; stainless steel cleaned, etched, and vapor degreased, then coated with silicone by a dip process; plastic surface was cleaned by scrubbing in Alconox solution, rinsing in tap water, dipped in diluted hot chromic acid, then rinsed in distilled water; silver deposited by immersing in distilled water at 296 K, then adding an equal volume of caustic ammoniacal silver nitrate precooled to 283 K, to which dextrose had been added immediately before use; after plating for about 3 min the specimen was rinsed in distilled water and dried; measured in vacuum ($\sim 10^{-5}$ mm Hg). [Authors' designation: Ag 79 C]
2	231	1962	373-473	10-20	Above specimen and conditions.
3	231	1962	373-473	10-20	Similar to curve 1 specimen and conditions. [Authors' designation: Ag 82 C]
4	231	1962	373-473	10-20	Above specimen and conditions.
5*	231	1962	373-473	10-20	Above specimen and conditions.
6	231	1962	373-473	10-20	Similar to curve 1 specimen and conditions except washed in detergent after plating. [Authors' designation: Ag 56 C]
7	231	1962	373-473	10-20	Above specimen and conditions.
8	231	1962	373-473	10-20	Similar to above specimen and conditions. [Authors' designation: Ag 57 C]
9	231	1962	373	10-20	Similar to curve 1 specimen and conditions except substrate consists of 316 stainless steel coated with Maraset 617-C epoxy resin (Marblette Co.); applied by flowing the resin onto the steel at 408 K. [Authors' designation: Ag 63 C]
10*	231	1962	373	10-20	Above specimen and conditions.
11*	231	1962	373	10-20	Above specimen and conditions.
12	231	1962	373	10-20	Similar to above specimen and conditions. [Authors' designation: Ag 73 C]
13*	231	1962	373-473	10-20	Similar to above specimen and conditions except turned slightly yellow during test. [Authors' designation: Ag 74 C]
14	231	1962	373-473	10-20	Above specimen and conditions.
15*	231	1962	373-473	10-20	Above specimen and conditions.
16	231	1962	423	10-20	Similar to curve 9 specimen and conditions except cleaned with detergent after plating. [Authors' designation: Ag 58 C]
17	231	1962	423	10-20	Similar to above specimen and conditions. [Authors' designation: Ag 59 C]
18*	231	1962	373-473	10-20	Silver (950-3000 Å thick); SY 627-119 polyurethane, Febert Shorndorfer Co., and 316 stainless steel substrates; stainless steel cleaned, etched, and vapor degreased, then coated with polyurethane by a dip process; plastic surface cleaned by scrubbing in Alconox solution, then rinsing with water and methanol; silver vapor deposited in vacuum ($\sim 10^{-5}$ mm Hg) onto heated substrate; measured in vacuum ($\sim 10^{-5}$ mm Hg). [Authors' designation: Ag 77]
19*	231	1962	373-473	10-20	Above specimen and conditions.
20*	231	1962	373-473	10-20	Above specimen and conditions.

* Not shown on plot

SPECIFICATION TABLE NO. 310 HEMISPHERICAL TOTAL EMITTANCE OF SILVER CONTACT COATINGS (continued)

Curve No.	Ref. No.	Year	Temperature Range, K	Reported Error, %	Composition (weight percent), Specifications and Remarks
21*	231	1962	373-473	10-20	Above specimen and conditions.
22	231	1962	423	10-20	Similar to curve 18 specimen and conditions except film appeared relatively dull. [Authors' designation: Ag 38]
23	231	1962	423	10-20	Similar to above specimen and conditions. [Authors' designation: Ag 39]
24	231	1962	473	10-20	Similar to above specimen and conditions. [Authors' designation: Ag 40]
25	194	1955	76	5	Silver; copper substrate; sprayed; measured in vacuum (10^{-6} to 10^{-7} mm Hg); emittance calculated from absorptance for 300 K blackbody incident radiation; authors assumed $\alpha = \epsilon$.
26	194	1955	76	5	Silver; stainless steel substrate; Allegheny Silver Spray Process; measured in vacuum (10^{-6} to 10^{-7} mm Hg); emittance calculated from absorptance for 300 K blackbody incident radiation; authors assumed $\alpha = \epsilon$.
27	194	1955	300	5	Silver; nickel strike and stainless steel substrates; silver plated; measured in vacuum (10^{-6} to 10^{-7} mm Hg); emittance calculated from absorptance for 300 K blackbody radiation; authors assumed $\alpha = \epsilon$.
28	194	1955	300	5	Silver; nickel strike, copper strike, and stainless steel substrates; silver plated; measured in vacuum (10^{-6} to 10^{-7} mm Hg); emittance calculated from absorptance for 300 K blackbody incident radiation; authors assumed $\alpha = \epsilon$.

* Not shown on plot

DATA TABLE NO. 310 HEMISPHERICAL TOTAL EMITTANCE OF SILVER CONTACT COATINGS

[Temperature, T, K; Emittance, ∈]

T	∈	T	∈	T	∈
CURVE 1		CURVE 9		CURVE 19*	
373	0.035	231	0.024	373	0.037
423	0.033			423	0.040
473	0.034	CURVE 10*		473	0.037
CURVE 2		373	0.027	CURVE 20*	
373	0.036	CURVE 11*		373	0.040
423	0.034			423	0.037
473	0.035	373	0.035	473	0.038
CURVE 3		CURVE 12		CURVE 21*	
373	0.027	373	0.15	373	0.040
423	0.025			423	0.038
473	0.028	CURVE 13*		473	0.037
CURVE 4		373	0.022	CURVE 22	
		423	0.026		
373	0.027*	473	0.024	423	0.137
423	0.027				
473	0.026	CURVE 14		CURVE 23	
CURVE 5*		373	0.020	423	0.175
		423	0.027*		
373	0.026	473	0.026*	CURVE 24	
423	0.026				
473	0.025	CURVE 15*		473	0.078
CURVE 6		373	0.021	CURVE 25	
		423	0.027		
373	0.029	473	0.026	76	0.01
423	0.064				
473	0.066	CURVE 16		CURVE 26	
CURVE 7		423	0.041	76	0.009
373	0.082	CURVE 17		CURVE 27	
423	0.078				
473	0.070	423	0.050	300	0.009
CURVE 8		CURVE 18*		CURVE 28	
373	0.038*	373	0.038	300	0.007
423	0.036	423	0.039		
473	0.040	473	0.035		

* Not shown on plot

SPECIFICATION TABLE NO. 311 NORMAL TOTAL EMITTANCE OF SILVER CONTACT COATINGS

Curve No.	Ref. No.	Year	Temperature Range, K	Geometry θ'	Reported Error, %	Composition (weight percent), Specifications and Remarks
1*	89	1958	366-672	~0°		Silver, Southwest Plating Co. (7.87 μm thick); 321 stainless steel substrate; plated. [Author's designation: Sample 12]

DATA TABLE NO. 311 NORMAL TOTAL EMITTANCE OF SILVER CONTACT COATINGS

[Temperature, T, K; Emittance, ϵ]

T	ϵ
CURVE 1*	
366	0.064
444	0.037
522	0.029
600	0.026
672	0.028

* No plot given

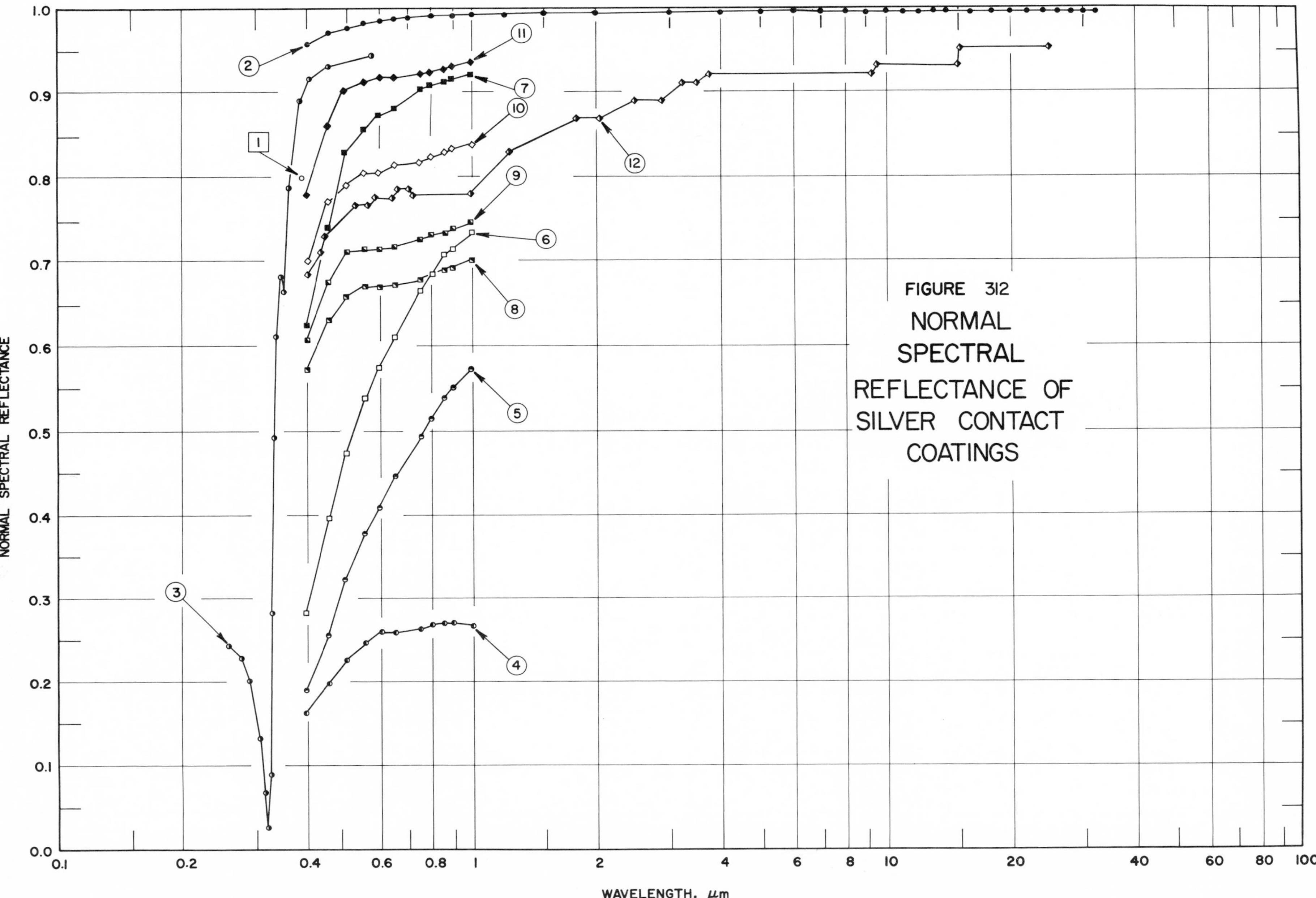

FIGURE 312 NORMAL SPECTRAL REFLECTANCE OF SILVER CONTACT COATINGS

SPECIFICATION TABLE NO. 312 NORMAL SPECTRAL REFLECTANCE OF SILVER CONTACT COATINGS

Curve No.	Ref. No.	Year	Temperature, K	Wavelength Range, μm	Geometry θ	θ'	ω'	Reported Error, %	Composition (weight percent), Specifications, and Remarks
1	205	1950	298	0.388	~0°	~0°		±2	100 Ag; glass substrate; vaporized.
2	229	1965	298	0.400-32.0	~5°	~5°			Silver on fused quartz substrate; evaporated in ultrahigh vacuum (10^{-9} mm Hg); measured in dry nitrogen; 1623 K globar source.
3	230	1965	298	0.257-0.571	~0°	~0°			Silver (~2000 Å thick); quartz substrate; evaporated on substrate at 423 K in vacuum (< 3 x 10^{-6} mm Hg); data extracted from smooth curve.
4	218	1967	~298	0.3980-0.9970	~0°	~0°			Silver (90 Å thick); wedgelike glass substrate; deposited by thermal sublimation in vacuum (10^{-4} mm Hg) at the rate of 20 Å sec^{-1}.
5	218	1967	~298	0.3990-0.9970	~0°	~0°			Similar to above specimen and conditions except film thickness 150 Å.
6	218	1967	~298	0.3970-0.9980	~0°	~0°			Similar to above specimen and conditions except film thickness 205 Å.
7	218	1967	~298	0.3990-0.9970	~0°	~0°			Similar to above specimen and conditions except film thickness 366 Å.
8	218	1967	~298	0.3980-0.9980	~0°	~0°			Silver (90 Å thick); chromium (~900 Å thick) and wedgelike glass substrates; chromium opaque, then Ag deposited by thermal sublimation in vacuum (10^{-4} mm Hg); Cr deposited at 10 Å sec^{-1} and Ag at 20 Å sec^{-1}.
9	218	1967	~298	0.3990-0.9970	~0°	~0°			Similar to above specimen and conditions except Ag film thickness 150 Å.
10	218	1967	~298	0.3970-0.9970	~0°	~0°			Similar to above specimen and conditions except Ag film thickness 205 Å.
11	218	1967	~298	0.3990-0.9950	~0°	~0°			Similar to above specimen and conditions except Ag film thickness 366 Å.
12	89	1958	~298	0.40-24.9	~0°	~0°			Silver, Southwest Plating Co. (7.62 μm thick); 321 stainless steel substrate; plated; data extracted from smooth curve. [Author's designation: Specimen 12]

DATA TABLE NO. 312 NORMAL SPECTRAL REFLECTANCE OF SILVER CONTACT COATINGS

[Wavelength, λ, μm; Reflectance, ρ; Temperature, T, K]

CURVE 1
T = 298

λ	ρ
0.388	0.800

CURVE 2
T = 298

λ	ρ
0.400	0.9564
0.450	0.9706
0.500	0.9786
0.550	0.9831
0.600	0.9860
0.650	0.9880
0.700	0.9894
0.800	0.9916
0.900	0.9929
1.000	0.9936
1.200	0.9938
1.500	0.9939
2.000	0.9940
3.000	0.9942
4.000	0.9944
5.000	0.9946
6.000	0.9948
7.000	0.9950
8.000	0.9951
9.000	0.9952
10.00	0.9953
11.00	0.9954
12.00	0.9954
13.00	0.9955
14.00	0.9955
16.00	0.9956
18.00	0.9956
20.00	0.9956
22.00	0.9956
24.00	0.9957
26.00	0.9957
28.00	0.9958
30.00	0.9958
32.00	0.9958

CURVE 3
T = 298

λ	ρ
0.257	0.243
0.276	0.229

CURVE 3 (cont.)

λ	ρ
0.289	0.200
0.306	0.131
0.315	0.067
0.320	0.026
0.326	0.089
0.328	0.281
0.332	0.492
0.336	0.612
0.345	0.683
0.350	0.664
0.363	0.789
0.383	0.890
0.404	0.917
0.451	0.931
0.571	0.943

CURVE 4
T ~ 298

λ	ρ
0.3980	0.161
0.4460	0.198
0.4990	0.225
0.5480	0.246
0.6000	0.259
0.6500	0.257
0.7490	0.262
0.8000	0.267
0.8590	0.269
0.8990	0.269
0.9970	0.226

CURVE 5
T ~ 298

λ	ρ
0.3990	0.188
0.4450	0.256
0.4960	0.323
0.5470	0.377
0.5990	0.409
0.6480	0.446
0.7470	0.494
0.7980	0.514
0.8560	0.537
0.8980	0.549
0.9970	0.572

CURVE 6
T ~ 298

λ	ρ
0.3970	0.281
0.4470	0.397
0.4960	0.473
0.5480	0.539
0.5970	0.574
0.6490	0.610
0.7490	0.666
0.8000	0.685
0.8580	0.709
0.8990	0.716
0.9980	0.735

CURVE 7
T ~ 298

λ	ρ
0.3990	0.626
0.4460	0.744
0.4960	0.830
0.5480	0.859
0.5960	0.874
0.6460	0.882
0.7500	0.904
0.7960	0.907
0.8580	0.911
0.8980	0.915
0.9970	0.921

CURVE 8
T ~ 298

λ	ρ
0.3980	0.570
0.4490	0.631
0.4960	0.659
0.5500	0.670
0.5980	0.670
0.6470	0.673
0.7490	0.679
0.8000	0.683*
0.8590	0.690
0.8960	0.693
0.9980	0.701

CURVE 9
T ~ 298

λ	ρ
0.3990	0.608
0.4460	0.677
0.4960	0.711
0.5480	0.715
0.5970	0.715
0.6460	0.718
0.7470	0.728
0.7980	0.734
0.8600	0.736
0.8990	0.741
0.9970	0.748

CURVE 10
T ~ 298

λ	ρ
0.3970	0.701
0.4470	0.771
0.4960	0.791
0.5470	0.807
0.5970	0.807
0.6470	0.816
0.7460	0.819
0.7970	0.826
0.8580	0.830
0.8990	0.835
0.9970	0.839

CURVE 11
T ~ 298

λ	ρ
0.3990	0.781
0.4450	0.863
0.4940	0.901
0.5480	0.914
0.5980	0.917
0.6480	0.917
0.7500	0.922
0.7960	0.924
0.8560	0.927
0.8960	0.930
0.9950	0.936

CURVE 12
T ~ 298

λ	ρ
0.40	0.685
0.43	0.711
0.44	0.731
0.52	0.768
0.56	0.768
0.58	0.779
0.64	0.778
0.66	0.788
0.70	0.788
0.72	0.780
0.99	0.782
1.23	0.830
1.78	0.870
2.04	0.870
2.48	0.891
2.87	0.891
3.23	0.911
3.49	0.911
3.73	0.921
9.25	0.922
9.57	0.932
14.9	0.931
15.1	0.952
24.9	0.952

* Not shown on plot

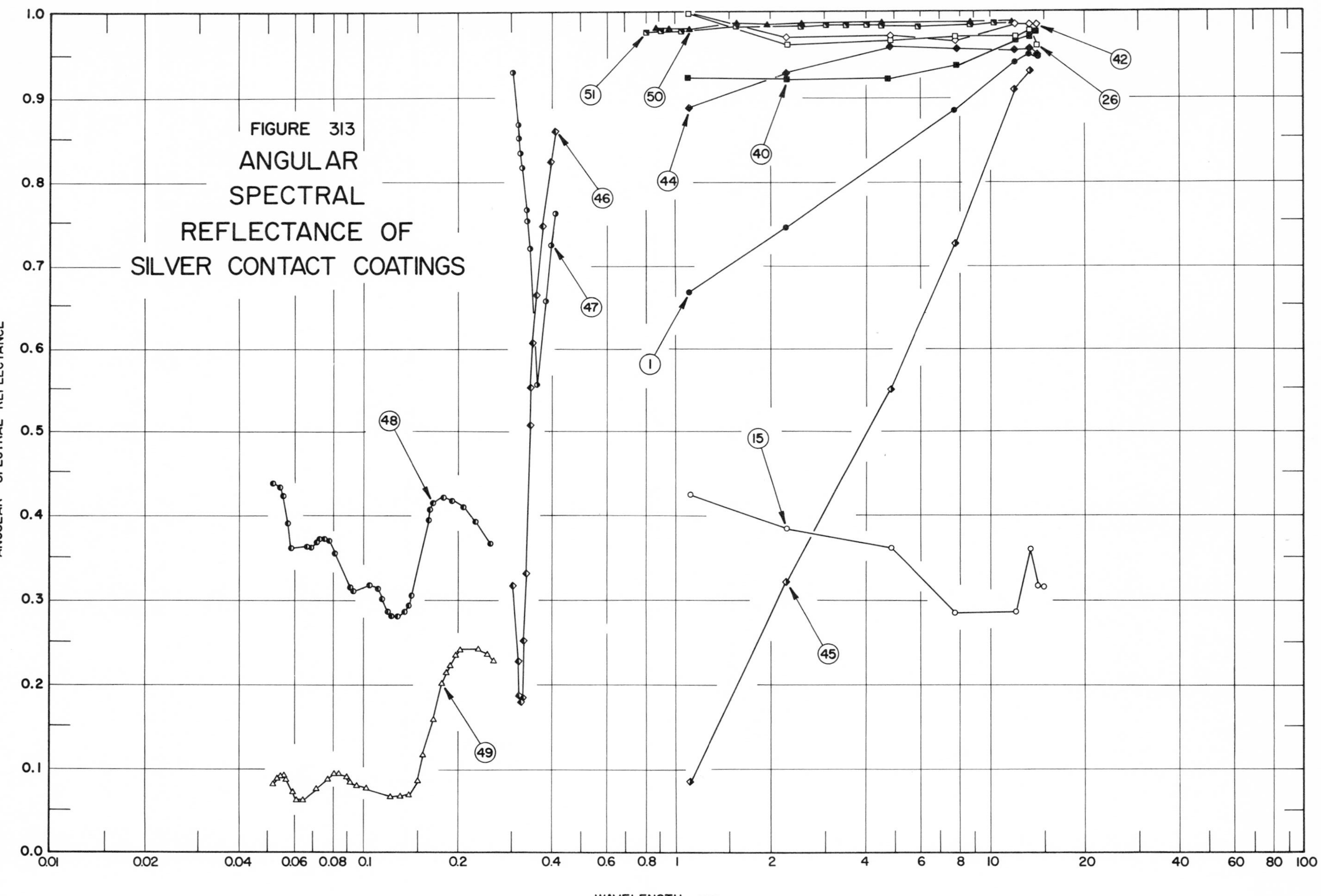

FIGURE 313 ANGULAR SPECTRAL REFLECTANCE OF SILVER CONTACT COATINGS

SPECIFICATION TABLE NO. 313 ANGULAR SPECTRAL REFLECTANCE OF SILVER CONTACT COATINGS

Curve No.	Ref. No.	Year	Temperature, K	Wavelength Range, μm	Geometry θ	θ'	ω'	Reported Error, %	Composition (weight percent), Specifications, and Remarks
1	231	1962	~298	1.120-14.375	45°	45°			Silver; SY627-119 polyurethane, Febert Shorndorder Co., and 316 stainless steel substrates; stainless steel cleaned, etched, and vapor degreased, then coated with polyurethane by a dip process; plastic surface was cleaned by scrubbing in Alconox solution, rinsing in tap water, dipping in diluted hot chromic acid, then rinsing in distilled water; silver deposited by immersing in distilled water at 296 K, then adding an equal volume of caustic ammoniacal silver nitrate precooled to 283 K, to which dextrose had been added immediately before use; after plating for about 3 min the specimen was rinsed in distilled water and dried; specular component only. [Authors' designation: Ag 42 CP]
2*	231	1962	~298	1.120-14.375	45°	45°			Similar to above specimen and conditions. [Authors' designation: Ag 43 C]
3*	231	1962	~298	1.120-14.375	45°	45°			Similar to above specimen and conditions. [Authors' designation: Ag 44 C]
4*	231	1962	~298	1.120-14.375	45°	45°			Similar to above specimen and conditions. [Authors' designation: Ag 45 CP]
5*	231	1962	~298	1.120-14.375	45°	45°			Similar to above specimen and conditions. [Authors' designation: Ag 46 CP]
6*	231	1962	~298	1.120-14.375	45°	45°			Similar to above specimen and conditions. [Authors' designation: Ag 47 C]
7*	231	1962	~298	1.120-14.375	45°	45°			Similar to above specimen and conditions. [Authors' designation: Ag 49 C]
8*	231	1962	~298	1.120-14.375	45°	45°			Similar to above specimen and conditions. [Authors' designation: Ag 51 C]
9*	231	1962	~298	1.009-14.375	45°	45°			Similar to above specimen and conditions. [Authors' designation: Ag 80 C]
10*	231	1962	~298	1.009-14.375	45°	45°			Similar to above specimen and conditions. [Authors' designation: Ag 84 C]
11*	231	1962	~298	1.009-14.375	45°	45°			Similar to above specimen and conditions. [Authors' designation: Ag 85 C]
12*	231	1962	~298	1.009-14.375	45°	45°			Similar to above specimen and conditions. [Authors' designation: Ag 86 C]
13*	231	1962	~298	1.120-14.375	45°	45°			Similar to above specimen and conditions except data shown is avg of six trials on the specimen. [Authors' designation: Ag 48 CP]
14*	231	1962	~298	1.120-14.375	45°	45°			Similar to above specimen and conditions. [Authors' designation: Ag 50 C]
15	231	1962	~298	1.120-14.375	45°	45°			Similar to curve 1 specimen and conditions except substrate consists of 316 stainless steel coated with General Electric SR-111 silicone resin by a dip process. [Authors' designation: Ag 52 C]
16*	231	1962	~298	1.120-14.375	45°	45°			Similar to above specimen and conditions. [Authors' designation: Ag 53 C]
17*	231	1962	~298	1.120-14.375	45°	45°			Similar to above specimen and conditions. [Authors' designation: Ag 54 C]
18*	231	1962	~298	1.120-14.375	45°	45°			Similar to above specimen and conditions. [Authors' designation: Ag 55 C]
19*	231	1962	~298	1.120-14.375	45°	45°			Similar to above specimen and conditions. [Authors' designation: Ag 56 C]
20*	231	1962	~298	1.120-14.375	45°	45°			Similar to above specimen and conditions. [Authors' designation: Ag 57 C]
21*	231	1962	~298	1.009-14.375	45°	45°			Similar to above specimen and conditions. [Authors' designation: Ag 78 C]
22*	231	1962	~298	1.009-14.375	45°	45°			Similar to above specimen and conditions. [Authors' designation: Ag 79 C]

* Not shown on plot

SPECIFICATION TABLE NO. 313 ANGULAR SPECTRAL REFLECTANCE OF SILVER CONTACT COATINGS (continued)

Curve No.	Ref. No.	Year	Temperature, K	Wavelength Range, μm	Geometry θ	θ'	ω'	Reported Error, %	Composition (weight percent), Specifications, and Remarks
23*	231	1962	~298	1.009-14.375	45°	45°			Similar to above specimen and conditions. [Authors' designation: Ag 81 C]
24*	231	1962	~298	1.009-14.375	45°	45°			Similar to above specimen and conditions. [Authors' designation: Ag 82 C]
25*	231	1962	~298	1.009-14.375	45°	45°			Similar to above specimen and conditions. [Authors' designation: Ag 83 C]
26	231	1962	~298	1.120-14.375	45°	45°			Similar to curve 1 specimen and conditions except substrate consists of 316 stainless steel coated with Maraset 617-C epoxy resin (Marblette Co.), applied by flowing the resin onto the steel at 408 K. [Authors' designation: Ag 58 C]
27*	231	1962	~298	1.120-14.375	45°	45°			Similar to above specimen and conditions. [Authors' designation: Ag 59 C]
28*	231	1962	~298	1.120-14.375	45°	45°			Similar to above specimen and conditions. [Authors' designation: Ag 60 C]
29*	231	1962	~298	1.120-14.375	45°	45°			Similar to above specimen and conditions. [Authors' designation: Ag 62 C]
30*	231	1962	~298	1.120-14.375	45°	45°			Similar to above specimen and conditions. [Authors' designation: Ag 63 C]
31*	231	1962	~298	1.120-14.375	45°	45°			Similar to above specimen and conditions. [Authors' designation: Ag 69 C]
32*	231	1962	~298	1.120-14.375	45°	45°			Similar to above specimen and conditions. [Authors' designation: Ag 70 C]
33*	231	1962	~298	1.120-14.375	45°	45°			Similar to above specimen and conditions. [Authors' designation: Ag 71 C]
34*	231	1962	~298	1.120-14.375	45°	45°			Similar to above specimen and conditions. [Authors' designation: Ag 72 C]
35*	231	1962	~298	1.009-14.375	45°	45°			Similar to above specimen and conditions. [Authors' designation: Ag 73 C]
36*	231	1962	~298	1.009-14.375	45°	45°			Similar to above specimen and conditions. [Authors' designation: Ag 74 C]
37*	231	1962	~298	1.009-14.375	45°	45°			Similar to above specimen and conditions. [Authors' designation: Ag 75 C]
38*	231	1962	~298	1.009-14.375	45°	45°			Similar to above specimen and conditions. [Authors' designation: Ag 76 C]
39*	231	1962	~298	1.009-14.375	45°	45°			Similar to above specimen and conditions except data shown is avg of three trials on the specimen. [Authors' designation: Ag 61 C]
40	231	1962	~298	1.120-14.375	45°	45°			Silver (950-3000 Å thick); Maraset 617-C epoxy resin, Marblette Co., and 316 stainless steel substrates; stainless steel cleaned, etched, and vapor degreased, then epoxy resin flowed onto stainless steel at 408 K; plastic surface was cleaned by scrubbing in Alconox solution, then rinsing with water and methanol; silver vapor deposited in vacuum (~10^{-5} mm Hg) onto heated substrate; specular component only. [Authors' designation: Ag 64]
41*	231	1962	~298	1.120-14.375	45°	45°			Similar to above specimen and conditions. [Authors' designation: Ag 65]
42	231	1962	~298	1.120-14.375	45°	45°			Similar to curve 40 specimen and conditions except substrate consists of 316 stainless steel coated with SY627-119 polyurethane (Febert Shorndorder Co.) by a dip process. [Authors' designation: Ag 66]
43*	231	1962	~298	1.009-14.375	45°	45°			Similar to above specimen and conditions. [Authors' designation: Ag 77]
44	231	1962	~298	1.120-14.375	45°	45°			Similar to curve 40 specimen and conditions except substrate consists of 316 stainless steel coated with General Electric SR-111 silicone resin by a dip process. [Authors' designation: Ag 68]
45	231	1962	~298	1.120-14.375	45°	45°			Similar to curve 40 specimen and conditions except substrate consists of uncoated 316 stainless steel. [Authors' designation: AgSS 3]

* Not shown on plot

SPECIFICATION TABLE NO. 313 ANGULAR SPECTRAL REFLECTANCE OF SILVER CONTACT COATINGS (continued)

Curve No.	Ref. No.	Year	Temperature, K	Wavelength Range, μm	Geometry θ	θ'	ω'	Reported Error, %	Composition (weight percent), Specifications, and Remarks
46	239	1962	~298	0.3000-0.4111	45°	45°			Silver (424 Å thick); fused quartz substrate; evaporated on substrate at room temp at a deposition rate of ~30 Å sec^{-1}; s-polarized incident light.
47	239	1962	~298	0.2994-0.4111	45°	45°			Above specimen and conditions except p-polarized incident light.
48	211	1966	~298	0.0509-0.255	70°	70°			Silver (1000 Å thick); substrate unknown; vapor deposited; data extracted from smooth curve.
49	211	1966	~298	0.0508-0.261	10°	10°			Similar to above specimen and conditions.
50	206	1958	~298	0.87-11.87	20°	20°			Silver; glass substrate; vapor deposited.
51	206	1958	~298	0.80-11.87	50°	50°			Above specimen and conditions.

DATA TABLE NO. 313 ANGULAR SPECTRAL REFLECTANCE OF SILVER CONTACT COATINGS

[Wavelength, λ, μm; Reflectance, ρ; Temperature, T, K]

λ	ρ
CURVE 1 T ~ 298	
1.120	0.667
2.240	0.746
7.780	0.885
12.099	0.944
13.530	0.954
14.375	0.951
CURVE 2* T ~ 298	
1.120	0.736
2.240	0.807
7.780	0.935
12.099	0.963
13.530	0.973
14.375	0.972
CURVE 3* T ~ 298	
1.120	0.627
2.240	0.637
7.780	0.852
12.099	0.908
13.530	0.909
14.375	0.893
CURVE 4* T ~ 298	
1.120	0.656
2.240	0.718
7.780	0.840
12.099	0.906
13.530	0.904
14.375	0.909
CURVE 5* T ~ 298	
1.120	0.793
2.240	0.885
7.780	0.952
12.099	0.974
CURVE 5 (cont.)*	
13.530	0.979
14.375	0.967
CURVE 6* T ~ 298	
1.120	0.645
2.240	0.675
7.780	0.845
12.099	0.908
13.530	0.915
14.375	0.918
CURVE 7* T ~ 298	
1.120	0.600
2.240	0.670
7.780	0.874
12.099	0.920
13.530	0.934
14.375	0.920
CURVE 8* T ~ 298	
1.120	0.740
2.240	0.789
7.780	0.888
12.099	0.936
13.530	0.945
14.375	0.930
CURVE 9* T ~ 298	
1.009	0.668
1.120	0.685
1.345	0.699
1.720	0.739
1.945	0.749
2.240	0.748
4.824	0.816
7.780	0.844
10.198	0.868
CURVE 9 (cont.)*	
12.099	0.881
13.530	0.880
14.375	0.883
CURVE 10* T ~ 298	
1.009	0.652
1.120	0.662
1.345	0.683
1.720	0.725
1.945	0.740
2.240	0.750
4.824	0.830
7.780	0.866
10.198	0.879
12.099	0.887
13.530	0.898
14.375	0.930
CURVE 11* T ~ 298	
1.009	0.515
1.120	0.530
1.345	0.558
1.720	0.609
1.945	0.627
2.240	0.665
4.824	0.785
7.780	0.824
10.198	0.852
12.099	0.862
13.530	0.859
14.375	0.864
CURVE 12* T ~ 298	
1.009	0.580
1.120	0.595
1.345	0.615
1.720	0.658
1.945	0.673
2.240	0.711
CURVE 12 (cont.)*	
4.824	0.806
7.780	0.842
10.198	0.854
12.099	0.866
13.530	0.875
14.375	0.863
CURVE 13* T ~ 298	
1.120	0.783
2.240	0.792
7.780	0.884
12.099	0.921
13.530	0.928
14.375	0.937
CURVE 14* T ~ 298	
1.120	0.761
2.240	0.825
7.780	0.961
12.099	0.951
13.530	0.962
14.375	0.952
CURVE 15 T ~ 298	
1.120	0.425
2.240	0.386
4.824	0.363
7.780	0.286
12.099	0.285
13.530	0.360
14.375	0.319
CURVE 16* T ~ 298	
1.120	0.797
2.240	0.860
4.824	0.877
7.780	0.826
CURVE 16 (cont.)*	
12.099	0.920
13.530	0.927
14.375	0.916
CURVE 17* T ~ 298	
1.120	0.760
2.240	0.840
4.824	0.882
7.780	0.905
12.099	0.925
13.530	0.920
14.375	0.921
CURVE 18* T ~ 298	
1.120	0.250
2.240	0.264
4.824	0.232
7.780	0.0464
12.099	0.0587
13.530	0.9693
14.375	0.104
CURVE 19* T ~ 298	
1.120	0.800
2.240	0.782
4.824	0.815
7.780	0.844
12.099	0.890
13.530	0.895
14.375	0.888
CURVE 20* T ~ 298	
1.120	0.500
2.240	0.757
4.824	0.838
7.780	0.879
12.099	0.913
CURVE 20 (cont.)*	
13.530	0.926
14.375	0.916
CURVE 21* T ~ 298	
1.009	0.326
1.120	0.378
1.345	0.457
1.720	0.584
1.945	0.623
2.240	0.670
4.824	0.802
7.780	0.839
10.198	0.870
12.099	0.883
13.530	0.887
14.375	0.876
CURVE 22* T ~ 298	
1.009	0.277
1.120	0.319
1.345	0.375
1.720	0.484
1.945	0.526
2.240	0.597
4.824	0.763
7.780	0.817
10.198	0.838
12.099	0.850
13.530	0.856
14.375	0.860
CURVE 23* T ~ 298	
1.009	0.665
1.120	0.697
1.345	0.723
1.720	0.754
1.945	0.761
2.240	0.806
4.824	0.853
CURVE 23 (cont.)*	
7.780	0.874
10.198	0.889
12.099	0.908
13.530	0.914
14.375	0.922
CURVE 24* T ~ 298	
1.009	0.673
1.120	0.682
1.345	0.701
1.720	0.737
1.945	0.750
2.240	0.775
4.824	0.838
7.780	0.860
10.198	0.875
12.099	0.884
13.530	0.894
14.375	0.893
CURVE 25* T ~ 298	
1.009	0.582
1.120	0.625
1.345	0.669
1.720	0.726
1.945	0.745
2.240	0.763
4.824	0.850
7.780	0.881
10.198	0.928
12.099	0.913
13.530	0.918
14.375	0.932
CURVE 26 T ~ 298	
1.120	1.000
2.240	0.967
4.824	0.971
7.780	0.975
CURVE 26 (cont.)	
12.099	0.976
13.530	0.981
14.375	0.965
CURVE 27* T ~ 298	
1.120	1.000
2.240	0.948
4.824	0.950
7.780	0.972
12.099	0.986
13.530	0.989
14.375	1.000
CURVE 28* T ~ 298	
1.120	0.935
2.240	0.920
4.824	0.925
7.780	0.944
12.099	0.979
13.530	0.988
14.375	0.982
CURVE 29* T ~ 298	
1.120	0.948
2.240	0.970
4.824	0.964
7.780	0.975
12.099	0.981
13.530	0.989
14.375	0.993
CURVE 30* T ~ 298	
1.120	0.863
2.240	0.895
4.824	0.911
7.780	0.952
12.099	0.985

* Not shown on plot

DATA TABLE NO. 313 ANGULAR SPECTRAL REFLECTANCE OF SILVER CONTACT COATINGS (continued)

λ	ρ
CURVE 30 (cont.)* T ~ 298	
13.530	0.968
14.375	0.935
CURVE 31* T ~ 298	
1.120	1.000
2.240	0.955
4.824	0.958
7.780	0.960
12.099	0.948
13.530	0.955
14.375	0.965
CURVE 32* T ~ 298	
1.120	0.898
2.240	0.933
4.824	0.939
7.780	0.945
12.099	0.957
13.530	0.957
14.375	0.965
CURVE 33* T ~ 298	
1.120	0.884
2.240	0.887
4.824	0.900
7.780	0.912
12.099	0.917
13.530	0.913
14.375	0.901
CURVE 34* T ~ 298	
1.120	0.865
2.240	0.912
4.824	0.942
7.780	0.946
12.099	0.960
13.530	0.953
14.375	0.953

λ	ρ
CURVE 35* T ~ 298	
1.009	0.945
1.120	0.945
1.345	0.946
1.720	0.952
1.945	0.950
2.240	0.955
4.824	0.967
7.780	0.970
10.198	0.966
12.099	0.966
13.530	0.966
14.375	0.958
CURVE 36* T ~ 298	
1.009	0.928
1.120	0.935
1.345	0.933
1.720	0.931
1.945	0.929
2.240	0.937
4.824	0.935
7.780	0.936
10.198	0.957
12.099	0.963
13.530	0.955
14.375	0.956
CURVE 37* T ~ 298	
1.009	0.891
1.120	0.916
1.345	0.918
1.720	0.918
1.945	0.911
2.240	0.929
4.824	0.930
7.780	0.931
10.198	0.934
12.099	0.936
13.530	0.929
14.375	0.929

λ	ρ
CURVE 38* T ~ 298	
1.009	0.834
1.120	0.852
1.345	0.863
1.720	0.866
1.945	0.865
2.240	0.869
4.824	0.883
7.780	0.896
10.198	0.906
12.099	0.916
13.530	0.913
14.375	0.905
CURVE 39* T ~ 298	
1.009	0.923
1.120	0.933
1.345	0.938
1.720	0.944
1.945	0.946
2.240	0.954
4.824	0.962
7.780	0.969
10.198	0.976
12.099	0.978
13.530	0.978
14.375	0.977
CURVE 40 T ~ 298	
1.120	0.927
2.240	0.925
4.824	0.926
7.780	0.942
12.099	0.970
13.530	0.975
14.375	0.980
CURVE 41* T ~ 298	
1.120	0.999
2.240	0.977
4.824	0.975
7.780	0.979

λ	ρ
CURVE 41 (cont.)*	
7.780	0.979
12.099	0.989
13.530	0.996
14.375	0.989
CURVE 42 T ~ 298	
1.120	1.000*
2.240	0.975
4.824	0.975
7.780	0.973
12.099	0.988
13.530	0.988
14.375	0.989
CURVE 43* T ~ 298	
1.009	0.895
1.120	0.900
1.345	0.909
1.945	0.909
2.240	0.918
4.824	0.930
7.780	0.942
10.198	0.951
12.099	0.960
13.530	0.958
14.375	0.945
CURVE 44 T ~ 298	
1.120	0.891
2.240	0.934
4.824	0.965
7.780	0.960
12.099	0.959
13.530	0.961
14.375	0.952
CURVE 45 T ~ 298	
1.120	0.086
2.240	0.321

λ	ρ
CURVE 45 (cont.)	
4.824	0.553
7.780	0.727
12.099	0.911
13.530	0.935
14.375	0.952*
CURVE 46 T ~ 298	
0.3000	0.319
0.3123	0.227
0.3166	0.187
0.3196	0.179
0.3223	0.186
0.3268	0.254
0.3318	0.331
0.3385	0.468
0.3410	0.509
0.3443	0.554
0.3500	0.608
0.3599	0.666
0.3792	0.748
0.3996	0.826
0.4111	0.863
CURVE 47 T ~ 298	
0.2994	0.935
0.3124	0.868
0.3153	0.854
0.3191	0.834
0.3222	0.818
0.3317	0.769
0.3350	0.753
0.3409	0.722
0.3603	0.557
0.3817	0.659
0.3999	0.726
0.4111	0.764
CURVE 48 T ~ 298	
0.0509	0.438
0.0538	0.433
0.0552	0.424
0.0570	0.391

λ	ρ
CURVE 48 (cont.)	
0.0570	0.391
0.0583	0.361
0.0591	0.354
0.0657	0.364
0.0675	0.363
0.0704	0.368
0.0720	0.372
0.0748	0.373
0.0775	0.370
0.0813	0.356
0.0916	0.315
0.0935	0.311
0.105	0.317
0.111	0.313
0.114	0.301
0.118	0.287
0.122	0.281
0.129	0.281
0.135	0.286
0.138	0.294
0.142	0.305
0.161	0.396
0.164	0.408
0.169	0.416
0.180	0.421
0.194	0.419
0.210	0.410
0.229	0.394
0.255	0.368
CURVE 49 T ~ 298	
0.0508	0.081
0.0523	0.087
0.0539	0.090
0.0554	0.090
0.0566	0.087
0.0596	0.071
0.0611	0.064
0.0639	0.063
0.0707	0.075
0.0768	0.087
0.0807	0.094
0.0836	0.094
0.0884	0.089
0.0909	0.084
0.0955	0.079

λ	ρ
CURVE 49 (cont.)	
0.103	0.077
0.120	0.066
0.131	0.065
0.139	0.068
0.147	0.085
0.155	0.115
0.167	0.159
0.177	0.201
0.183	0.215
0.189	0.224
0.196	0.235
0.205	0.241
0.232	0.242
0.248	0.236
0.261	0.227
CURVE 50 T ~ 298	
0.87	0.985
0.95	0.985
1.12	0.985
1.57	0.990
1.96	0.989
2.50	0.989
4.57	0.990
8.70	0.990
11.87	0.990
CURVE 51 T ~ 298	
0.80	0.979
0.90	0.981
1.05	0.981
1.57	0.989
2.50	0.988
3.01	0.987
3.52	0.987
4.03	0.988
4.57	0.988
5.94	0.986
8.70	0.988
10.41	0.989
11.87	0.989

* Not shown on plot

SPECIFICATION TABLE NO. 314 NORMAL SOLAR REFLECTANCE OF SILVER CONTACT COATINGS

Curve No.	Ref. No.	Year	Temperature Range, K	Geometry θ	θ'	ω'	Reported Error, %	Composition (weight percent), Specifications and Remarks
1*	222	1962	298	~0°		2π		Silver; Mylar (6.35 μm thick) substrate; measured relative to MgO using an integrating hemisphere; corrected for reflectance of MgO.
2*	222	1962	298	~0°		2π		Silver; Mylar (6.35 μm thick) substrate; measured relative to aluminum using an integrating sphere.

DATA TABLE NO. 314 NORMAL SOLAR REFLECTANCE OF SILVER CONTACT COATINGS

[Temperature, T, K; Reflectance, ρ]

T	ρ
CURVE 1*	
298	0.94
CURVE 2*	
298	0.94

* No plot given

SPECIFICATION TABLE NO. 315 HEMISPHERICAL INTEGRATED ABSORPTANCE OF SILVER CONTACT COATINGS

Curve No.	Ref. No.	Year	Temperature Range, K	Reported Error, %	Composition (weight percent), Specifications and Remarks
1*	194	1955	76	5	Silver; copper substrate; sprayed; measured in vacuum (10^{-6} to 10^{-7} mm Hg); absorptance for 300 K blackbody incident radiation.
2*	194	1955	300	5	Silver; stainless steel substrate; deposited by Allegheny Silver Spray process; measured in vacuum (10^{-6} to 10^{-7} mm Hg); absorptance for 300 K blackbody incident radiation.
3*	194	1955	300	5	Silver; nickel strike and stainless steel substrate; silver plated; measured in vacuum (10^{-6} to 10^{-7} mm Hg); absorptance for 300 K blackbody incident radiation.
4*	194	1955	300	5	Silver; nickel strike, copper strike, and stainless steel substrates; silver plated; measured in vacuum (10^{-6} to 10^{-7} mm Hg); absorptance for 300 K blackbody incident radiation.
5*	240	1962	75.8	<2	Silver (~50 μm thick); copper substrate; electroplated; put in a helium environment immediately following plating to halt oxidation, surface rinsed with distilled water and ethyl alcohol, calorimeter assembled in helium atm; sample accidentally oxidized by heating to 367 K for 16 hrs in air (10^{-3} mm Hg); measured in vacuum (3×10^{-6} mm Hg); radiation source temp 302 K.
6*	240	1962	75.8	<2	Above specimen and conditions except radiation source temp 321 K.
7*	240	1962	75.8	<2	Above specimen and conditions except radiation source temp 366 K.
8*	240	1962	75.8	<2	Silver (~25 μm thick); copper substrate; electroplated; put in a helium environment immediately following plating to halt oxidation, surface rinsed with distilled water and ethyl alcohol, calorimeter assembled in helium atm; measured in vacuum (3×10^{-6} mm Hg); radiation source temp 268 K.
9*	240	1962	75.8	<2	Above specimen and conditions.
10*	240	1962	75.8	<2	Above specimen and conditions except radiation source temp 299 K.
11*	240	1962	75.8	<2	Above specimen and conditions except radiation source temp 305 K.
12*	240	1962	75.8	<2	Above specimen and conditions except radiation source temp 315 K.
13*	240	1962	75.8	<2	Above specimen and conditions except radiation source temp 323 K.

* No plot given

DATA TABLE NO. 315 HEMISPHERICAL INTEGRATED ABSORPTANCE OF SILVER CONTACT COATINGS

[Temperature, T, K; Absorptance, α]

T	α	T	α
CURVE 1*		CURVE 13*	
76	0.01	75.8	0.782
CURVE 2*			
300	0.009		
CURVE 3*			
300	0.009		
CURVE 4*			
300	0.007		
CURVE 5*			
75.8	0.822		
CURVE 6*			
75.8	0.795		
CURVE 7*			
75.8	0.757		
CURVE 8*			
75.8	0.843		
CURVE 9*			
75.8	0.840		
CURVE 10*			
75.8	0.815		
CURVE 11*			
75.8	0.808		
CURVE 12*			
75.8	0.791		

* No plot given

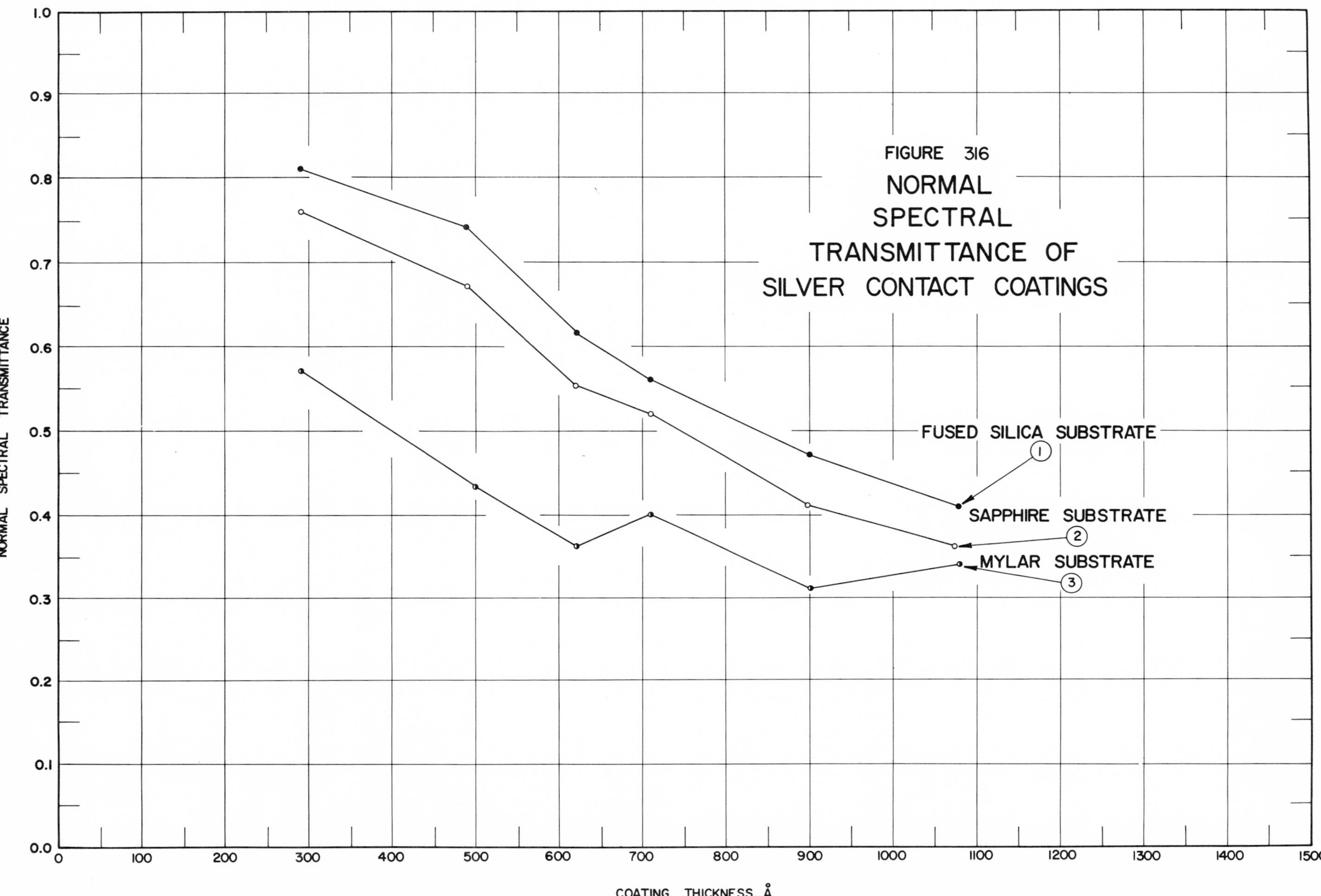

FIGURE 316 NORMAL SPECTRAL TRANSMITTANCE OF SILVER CONTACT COATINGS

SPECIFICATION TABLE NO. 316 NORMAL SPECTRAL TRANSMITTANCE OF SILVER CONTACT COATINGS

Curve No.	Ref. No.	Year	Temperature, K	Wavelength Range, μm	Geometry θ	θ'	ω'	Reported Error, %	Composition (weight percent), Specifications, and Remarks
1	122	1968	~298	0.322	~0°	~0°			Silver; fused silica (0.787 mm thick) substrate; coating thickness, Å, is variable; deposition rate 75 Å sec^{-1}.
2	122	1968	~298	0.322	~0°	~0°			Similar to above specimen and conditions except deposited on sapphire (0.787 mm thick) substrate.
3	122	1968	~298	0.322	~0°	~0°			Similar to curve 1 specimen and conditions except substrate Dupont Mylar D polyester film (0.1905 mm thick).

DATA TABLE NO. 316 NORMAL SPECTRAL TRANSMITTANCE OF SILVER CONTACT COATINGS

[Wavelength, λ, μm; Transmittance, τ; Temperature, T, K]

Thickness	τ
CURVE 1 T ~ 298, λ = 0.322	
290	0.812
489	0.743
621	0.618
711	0.562
900	0.472
1079	0.411
CURVE 2 T ~ 298, λ = 0.322	
290	0.760
489	0.671
620	0.553
710	0.521
898	0.413
1077	0.363
CURVE 3 T ~ 298, λ = 0.322	
291	0.571
500	0.435
621	0.363
711	0.401
901	0.313
1080	0.342

FIGURE 317

ANGULAR SPECTRAL TRANSMITTANCE OF SILVER CONTACT COATINGS

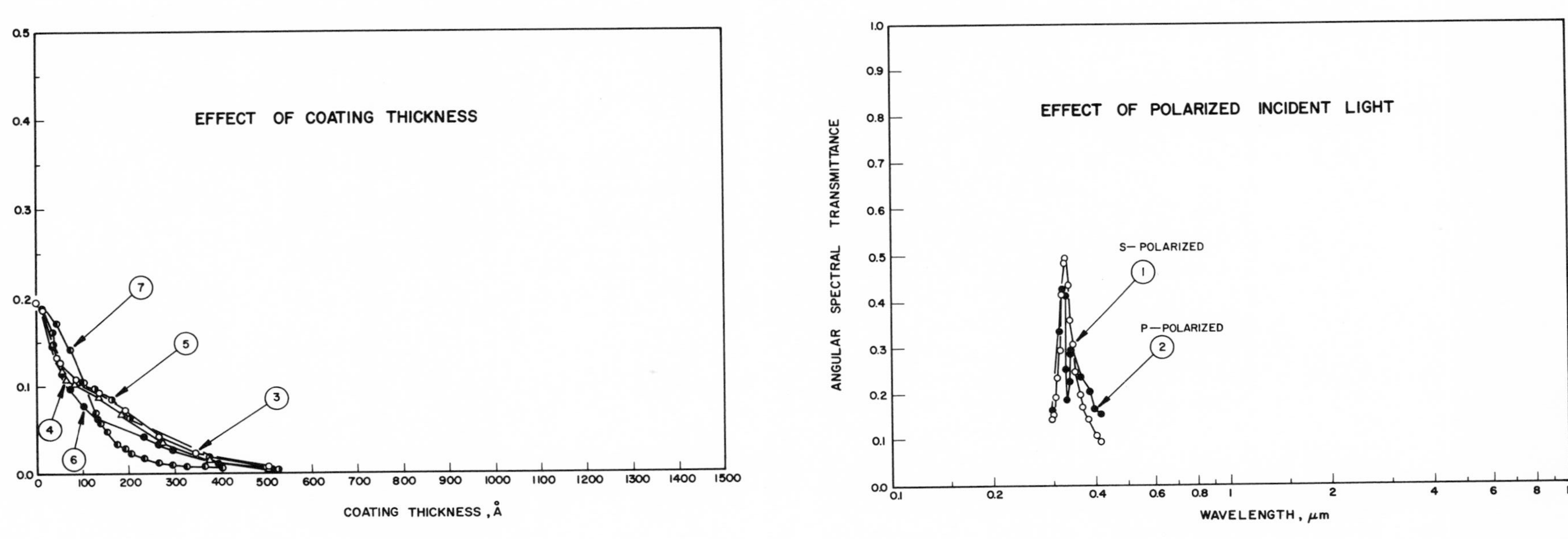

SPECIFICATION TABLE NO. 317 ANGULAR SPECTRAL TRANSMITTANCE OF SILVER CONTACT COATINGS

Curve No.	Ref. No.	Year	Temperature, K	Wavelength Range, μm	Geometry θ θ' ω'	Reported Error, %	Composition (weight percent), Specifications, and Remarks
1	239	1962	~298	0.2986-0.4111	45° 45°		Silver (424 Å thick); fused quartz substrate; evaporated on substrate at room temp at a deposition rate of ~30 Å per sec; s-polarized incident light.
2	239	1962	~298	0.2994-0.4111	45° 45°		Above specimen and conditions except p-polarized incident light.
3	241	1969	293	0.5461	78°~45°		Ag; BK 7 substrate; right angle prism of BK 7 glass of refractive index 1.519 at λ = 5461 Å; silver evaporated in vacuum (~10^{-5} mm Hg) from a tungsten wire conical basket ~30 cm away at the rate of 2.7 Å sec^{-1}; data extracted from smooth curve; coating thickness, Å, is variable.
4	241	1969	293	0.5461	78°~45°		Above specimen and conditions except continuous evaporation at the rate of 0.2 Å sec^{-1}.
5	241	1969	293	0.5461	78°~45°		Above specimen and conditions except continuous evaporation at the rate of 0.8 Å sec^{-1}.
6	241	1969	353	0.5461	78°~45°		Above specimen and conditions.
7	241	1969	368	0.5461	78°~45°		Above specimen and conditions.

DATA TABLE NO. 317 ANGULAR SPECTRAL TRANSMITTANCE OF SILVER CONTACT COATINGS

[Wavelength, λ, μm; Transmittance, τ; Temperature, T, K]

CURVE 1 T ~ 298

λ	τ
0.2986	0.141
0.3002	0.153
0.3049	0.192
0.3088	0.236
0.3118	0.295
0.3180	0.417
0.3220	0.483
0.3240	0.493
0.3271	0.491*
0.3320	0.435
0.3368	0.360
0.3412	0.308
0.3488	0.249
0.3601	0.196
0.3682	0.170
0.3802	0.144
0.4000	0.107
0.4111	0.094

CURVE 2 T ~ 298

λ	τ
0.2994	0.167
0.3117	0.332
0.3186	0.414*
0.3205	0.428
0.3219	0.411
0.3267	0.253
0.3293	0.187
0.3302	0.208*
0.3317	0.229
0.3342	0.265*
0.3375	0.288
0.3397	0.293
0.3609	0.239
0.3817	0.202
0.3999	0.169
0.4111	0.155

CURVE 3 T = 293, λ = 0.5461

Thickness	τ
0.00	0.197
17.0	0.187
43.0	0.134
50.0	0.127
89.0	0.107
138	0.094
191	0.073
267	0.041
344	0.023
502	0.007

CURVE 4 T = 293, λ = 0.5461

Thickness	τ
0.00	0.197*
17.0	0.187*
43.0	0.134*
52.0	0.119
61.0	0.108
83.0	0.101
137	0.086
183	0.069
274	0.035
378	0.015
502	0.006

CURVE 5 T = 293, λ = 0.5461

Thickness	τ
0.00	0.197*
12.0	0.190*
33.0	0.162
38.0	0.147
58.0	0.119*
94.0	0.105
127	0.099
163	0.085
204	0.064
378	0.017
529	0.005

CURVE 6 T = 353, λ = 0.5461

Thickness	τ
0.00	0.197*
12.0	0.187*
32.0	0.147
55.0	0.115
74.0	0.097
101	0.078
135	0.064
231	0.041
263	0.034
299	0.026
399	0.012
512	0.005

CURVE 7 T = 368, λ = 0.5461

Thickness	τ
0.00	0.197*
19.0	0.191*
46.0	0.173
73.0	0.142
102	0.105
130	0.070
140	0.058
155	0.047
176	0.034
192	0.028
208	0.024
232	0.018
266	0.013
299	0.010
328	0.008
367	0.007
405	0.006

* Not shown on plot

SPECIFICATION TABLE NO. 318 ABSORPTANCE TO EMITTANCE RATIO OF SILVER CONTACT COATINGS

Curve No.	Ref. No.	Year	Temperature Range, K	Reported Error, %	Composition (weight percent), Specifications and Remarks
1*	58	1967	~298	27	Silver; BeCu (0.254 mm thick) and aluminum (6061 Al alloy) substrates; Ag electroplated on BeCu; Ag-BeCu bonded to Al substrate with conductive epoxy; property deduced from reflectance; laboratory data measured 20 days before flight of ATS-1.
2*	58	1967	~300	6.9	Above specimen and conditions except property deduced from temp of substrate from "initial" point flight data (48 hrs after launch) of ATS-1.

DATA TABLE NO. 318 ABSORPTANCE TO EMITTANCE RATIO OF SILVER CONTACT COATINGS

[Temperature, T, K; Absorptance, α; Emittance, ϵ]

T	α/ϵ
CURVE 1*	
~298	2.92
CURVE 2*	
~300	3.20

* No plot given

SPECIFICATION TABLE NO. 319 NORMAL SPECTRAL REFLECTANCE OF SILVER + ALUMINUM CONTACT COATINGS

Curve No.	Ref. No.	Year	Temperature, K	Wavelength Range, μm	Geometry θ θ' ω'	Reported Error, %	Composition (weight percent), Specifications, and Remarks
1*	205	1950	298	0.255-0.425	~0° ~0°	± 2	99 Ag (α-phase) and 1 Al; glass substrate; vaporized.
2*	205	1950	298	0.255-0.490	~0° ~0°	± 2	96.5 Ag (α-phase) and 3.5 Al; glass substrate; vaporized.
3*	205	1950	298	0.255-0.490	~0° ~0°	± 2	92.3 Ag (β-phase) and 7.7 Al; glass substrate; vaporized.
4*	205	1950	298	0.255-0.377	~0° ~0°	± 2	89 Ag (γ-phase) and 11 Al; glass substrate; vaporized.

DATA TABLE NO. 319 NORMAL SPECTRAL REFLECTANCE OF SILVER + ALUMINUM CONTACT COATINGS

[Wavelength, λ, μm; Reflectance, ρ; Temperature, T, K]

λ	ρ	λ	ρ
CURVE 1* T = 298		CURVE 3* T = 298	
0.255	0.280	0.255	0.419
0.268	0.250	0.267	0.520
0.280	0.210	0.282	0.580
0.305	0.172	0.309	0.620
0.339	0.200	0.332	0.610
0.377	0.650	0.379	0.630
0.425	0.820	0.425	0.620
		0.490	0.610
CURVE 2* T = 298			
		CURVE 4* T = 298	
0.255	0.340		
0.268	0.340	0.255	0.430
0.281	0.335	0.267	0.580
0.305	0.320	0.306	0.610
0.336	0.440	0.382	0.630
0.376	0.680	0.377	0.660
0.490	0.920		

* No plot given

SPECIFICATION TABLE NO. 320 HEMISPHERICAL TOTAL EMITTANCE OF TIN CONTACT COATINGS

Curve No.	Ref. No.	Year	Temperature Range, K	Reported Error, %	Composition (weight percent), Specifications and Remarks
1*	194	1955	76	5	Tin; copper substrate; measured in vacuum (10^{-6} to 10^{-7} mm Hg); emittance calculated from absorptance for 300 K blackbody incident radiation; authors assumed $\alpha = \epsilon$.

DATA TABLE NO. 320 HEMISPHERICAL TOTAL EMITTANCE OF TIN CONTACT COATINGS

[Temperature, T, K; Emittance, ϵ]

T	ϵ
CURVE 1*	
76	0.02

* No plot given

SPECIFICATION TABLE NO. 321 HEMISPHERICAL INTEGRATED ABSOPRTANCE OF TIN CONTACT COATINGS

Curve No.	Ref. No.	Year	Temperature Range, K	Reported Error, %	Composition (weight percent), Specifications and Remarks
1*	194	1955	76	5	Tin; copper substrate; measured in vacuum (10^{-6} to 10^{-7} mm Hg); absorptance for 300 K blackbody incident radiation.

DATA TABLE NO. 321 HEMISPHERICAL INTEGRATED ABSORPTANCE OF TIN CONTACT COATINGS

[Temperature, T, K; Absorptance, α]

T	α
CURVE 1*	
76	0.02

* No plot given

SPECIFICATION TABLE NO. 322 NORMAL SPECTRAL REFLECTANCE OF TIN + COPPER CONTACT COATINGS

Curve No.	Ref. No.	Year	Temperature, K	Wavelength Range, μm	Geometry θ	θ'	ω'	Reported Error, %	Composition (weight percent), Specifications, and Remarks
1*	223	1947	298	0.45-0.65	~0°	~0°			55 Sn and 45 Cu, speculum (0.0254 mm thick); surface-ground steel substrate; electrodeposited; polished on buffing wheel and then rubbed with cotton-wool.

DATA TABLE NO. 322 NORMAL SPECTRAL REFLECTANCE OF TIN + COPPER CONTACT COATINGS

[Wavelength, λ, μm; Reflectance, ρ; Temperature, T, K]

λ	ρ
CURVE 1*	
T = 298	
0.45	0.610
0.50	0.642
0.55	0.670
0.60	0.695
0.65	0.715

* No plot given

SPECIFICATION TABLE NO. 323 HEMISPHERICAL TOTAL EMITTANCE OF TIN + LEAD CONTACT COATINGS

Curve No.	Ref. No.	Year	Temperature Range, K	Reported Error, %	Composition (weight percent), Specifications and Remarks
1*	194	1955	76	5	50 Sn and 50 Pb, 50-50 solder (0.0508 mm thick); copper (0.127 mm thick) substrate; measured in vacuum (10^{-6} to 10^{-7} mm Hg); emittance calculated from absorptance for 300 K blackbody incident radiation; authors assumed $\alpha = \epsilon$.

DATA TABLE NO. 323 HEMISPHERICAL TOTAL EMITTANCE OF TIN + LEAD CONTACT COATINGS

[Temperature, T, K; Emittance, ϵ]

T	ϵ
CURVE 1*	
76	0.03

* No plot given

SPECIFICATION TABLE NO. 324 HEMISPHERICAL INTEGRATED ABSORPTANCE OF TIN + LEAD CONTACT COATINGS

Curve No.	Ref. No.	Year	Temperature Range, K	Reported Error, %	Composition (weight percent), Specifications and Remarks
1*	194	1955	76	5	50 Sn and 50 Pb, 50-50 solder (0.0508 mm thick); copper (0.127 mm thick) substrate; measured in vacuum (10^{-6} to 10^{-7} mm Hg); absorptance for 300 K blackbody incident radiation.

DATA TABLE NO. 324 HEMISPHERICAL INTEGRATED ABSORPTANCE OF TIN + LEAD CONTACT COATINGS

[Temperature, T, K; Absorptance, α]

T	α
CURVE 1*	
76	0.03

* No plot given

SPECIFICATION TABLE NO. 325 NORMAL SPECTRAL REFLECTANCE OF TITANIUM CONTACT COATINGS

Curve No.	Ref. No.	Year	Temperature, K	Wavelength Range, μm	Geometry θ θ' ω'	Reported Error, %	Composition (weight percent), Specifications, and Remarks
1*	203	1962	298	0.25-24.99	~0° ~0°		Titanium; Reynold's Wrap aluminum foil substrate (800-1000 Å thick); vapor deposited on shiny side of foil; data extracted from smooth curve.
2*	203	1962	298	0.25-25.01	~0° ~0°		Titanium; Reynold's Wrap aluminum foil substrate; vapor deposited on shiny side of substrate; heated in air for 3 hrs at 698 K; data extracted from smooth curve.

DATA TABLE NO. 325 NORMAL SPECTRAL REFLECTANCE OF TITANIUM CONTACT COATINGS

[Wavelength, λ, μm; Reflectance, ρ; Temperature, T, K]

λ	ρ	λ	ρ	λ	ρ
CURVE 1*		CURVE 1 (cont.)*		CURVE 2 (cont.)*	
T = 298					
		21.70	0.955	3.90	0.665
0.25	0.183	22.23	0.961	5.55	0.761
0.34	0.315	24.99	0.960	6.29	0.777
0.63	0.484			6.50	0.778
1.01	0.514	CURVE 2*		10.10	0.843
1.80	0.544	T = 298		12.96	0.873
2.23	0.580			14.97	0.872
2.41	0.580	0.25	0.257	18.16	0.903
3.35	0.728	0.34	0.420	20.81	0.914
3.95	0.799	0.39	0.201	22.25	0.904
5.01	0.859	0.44	0.053	23.47	0.916
6.06	0.878	0.51	0.000	24.21	0.896
7.00	0.884	0.61	0.079	25.01	0.895
9.90	0.922	0.69	0.214		
14.35	0.939	0.83	0.321		
15.18	0.952	1.00	0.400		
20.00	0.959	2.04	0.526		
20.64	0.954	2.98	0.579		

* No plot given

SPECIFICATION TABLE NO. 326 ANGULAR SPECTRAL REFLECTANCE OF TITANIUM CONTACT COATINGS

Curve No.	Ref. No.	Year	Temperature, K	Wavelength Range, μm	Geometry θ	θ'	ω'	Reported Error, %	Composition (weight percent), Specifications, and Remarks
1*	213	1961	298	0.0452-0.1586	20°	20°			Titanium (324 Å thick); glass substrate; vapor deposited in vacuum (8×10^{-5} mm Hg); exposed to air; measured in vacuum.

DATA TABLE NO. 326 ANGULAR SPECTRAL REFLECTANCE OF TITANIUM CONTACT COATINGS

[Wavelength, λ, μm; Reflectance, ρ; Temperature, T, K]

λ	ρ	λ	ρ	λ	ρ	λ	ρ
CURVE 1* T = 298		CURVE 1 (cont.)*		CURVE 1 (cont.)*		CURVE 1 (cont.)*	
		0.0697	0.033	0.0923	0.079	0.1254	0.145
0.0452	0.012	0.0701	0.034	0.0950	0.074	0.1330	0.152
0.0474	0.012	0.0714	0.035	0.0962	0.085	0.1330	0.157
0.0481	0.009	0.0718	0.038	0.0976	0.073	0.1364	0.158
0.0489	0.006	0.0742	0.044	0.1011	0.080	0.1411	0.158
0.0501	0.005	0.0758	0.048	0.1020	0.086	0.1429	0.163
0.0509	0.003	0.0768	0.050	0.1044	0.093	0.1474	0.163
0.0523	0.002	0.0786	0.053	0.1096	0.108	0.1505	0.166
0.0529	0.002	0.0823	0.058	0.1110	0.111	0.1552	0.163
0.0554	0.003	0.0830	0.066	0.1144	0.116	0.1570	0.160
0.0566	0.000	0.0841	0.064	0.1162	0.119	0.1586	0.162
0.0575	0.006	0.0853	0.064	0.1170	0.122		
0.0590	0.006	0.0870	0.070	0.1175	0.126		
0.0614	0.012	0.0874	0.068	0.1191	0.128		
0.0635	0.016	0.0879	0.079	0.1199	0.134		
0.0641	0.017	0.0894	0.070	0.1206	0.135		
0.0664	0.022	0.0914	0.070	0.1214	0.138		
0.0684	0.030	0.0921	0.082	0.1228	0.147		

* No plot given

SPECIFICATION TABLE NO. 327 ANGULAR SPECTRAL ABSORPTANCE OF TITANIUM CONTACT COATINGS

Curve No.	Ref. No.	Year	Temperature, K	Wavelength Range, μm	Geometry θ	Reported Error, %	Composition (weight percent), Specifications, and Remarks
1*	221	1965	306	0.340-22.9	25°		Titanium; polished brass substrate; electropolished; polished with optical grade rouge with cotton moistened in ethyl alcohol; measured in dry nitrogen; heated cavity at 1056 K; measured relative to platinum; authors assumed $\alpha = 1 - R(2\pi, 25°)$.

DATA TABLE NO. 327 ANGULAR SPECTRAL ABSORPTANCE OF TITANIUM CONTACT COATINGS

[Wavelength, λ, μm; Absorptance, α; Temperature, T, K]

λ	α	λ	α
CURVE 1*		CURVE 1 (cont.)*	
T = 306		4.98	0.210
0.340	0.651	6.00	0.195
0.390	0.590	7.00	0.174
0.492	0.518	8.11	0.168
0.602	0.476	9.04	0.155
0.698	0.460	10.2	0.148
0.798	0.421	12.0	0.128
0.902	0.427	15.0	0.123
0.986	0.436	17.2	0.117
1.14	0.423	18.9	0.108
1.30	0.408	20.6	0.101
1.50	0.362	22.9	0.099
1.78	0.348		
1.98	0.330		
2.48	0.310		
3.05	0.283		
3.47	0.258		
4.01	0.240		

* No plot given

SPECIFICATION TABLE NO. 328 NORMAL SOLAR ABSORPTANCE OF TITANIUM CONTACT COATINGS

Curve No.	Ref. No.	Year	Temperature Range, K	Geometry θ	Reported Error, %	Composition (weight percent), Specifications and Remarks
1*	203	1962	298	~0°		Titanium; Reynold's Wrap aluminum foil substrate (80 to 100 μm thick); vapor deposited on shiny side of foil; heated 3 hrs at 670 K.

DATA TABLE NO. 328 NORMAL SOLAR ABSORPTANCE OF TITANIUM CONTACT COATINGS

[Temperature, T, K; Absorptance, α]

T	α
CURVE 1*	
298	0.75

* No plot given

SPECIFICATION TABLE NO. 329 HEMISPHERICAL TOTAL EMITTANCE OF TUNGSTEN CONTACT COATINGS

Curve No.	Ref. No.	Year	Temperature Range, K	Reported Error, %	Composition (weight percent), Specifications and Remarks
1*	146	1961	468-1093	<10	Tungsten, Metco XP-1106, crystalline, (-200 mesh, +30 μm); Armco iron substrate; plasma flame sprayed; measured in vacuum (10^{-5} mm Hg).

DATA TABLE NO. 329 HEMISPHERICAL TOTAL EMITTANCE OF TUNGSTEN CONTACT COATINGS

[Temperature, T, K; Emittance, ϵ]

T	ϵ
CURVE 1*	
468	0.34
668	0.41
873	0.50
1093	0.49

* No plot given

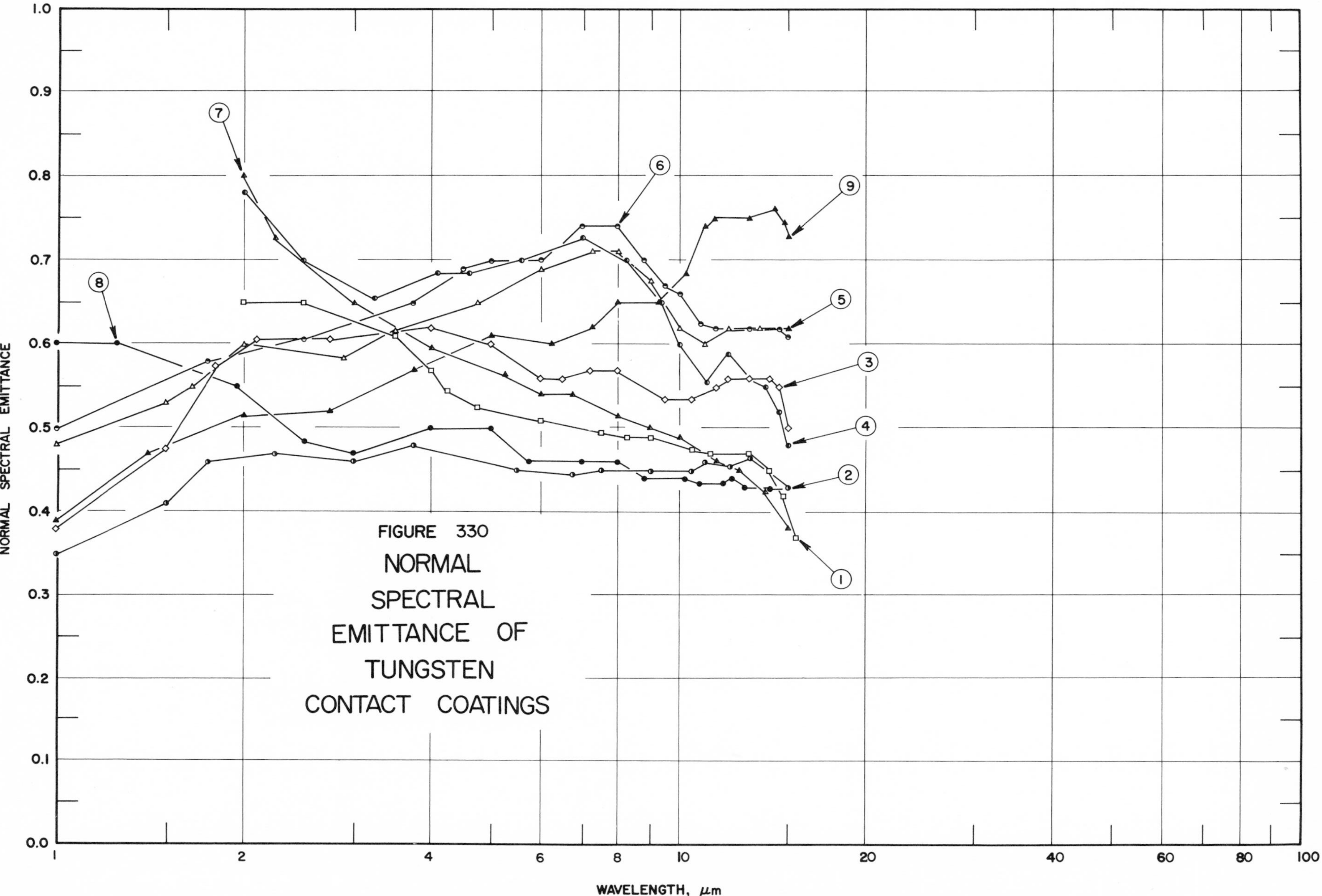

FIGURE 330 NORMAL SPECTRAL EMITTANCE OF TUNGSTEN CONTACT COATINGS

SPECIFICATION TABLE NO. 330 NORMAL SPECTRAL EMITTANCE OF TUNGSTEN CONTACT COATINGS

Curve No.	Ref. No.	Year	Temperature, K	Wavelength Range, μm	Geometry θ'	Reported Error, %	Composition (weight percent), Specifications, and Remarks
1	139	1961	523	2.00-15.50	~0°	± 5	Tungsten; sandblasted Inconel X substrate; flame sprayed using a Plasmatron; as received; data extracted from smooth curve.
2	139	1961	773	1.0-15.0	~0°	± 5	Similar to above specimen and conditions.
3	139	1961	1023	1.0-15.0	~0°	± 5	Similar to curve 1 specimen and conditions.
4	139	1961	523	2.0-15.0	~0°	± 5	Similar to curve 1 specimen and conditions except heated in air at 1089 K for 30 min.
5	139	1961	773	1.0-15.0	~0°	± 5	Similar to above specimen and conditions.
6	139	1961	1023	1.0-15.0	~0°	± 5	Similar to curve 4 specimen and conditions.
7	139	1961	523	2.0-15.0	~0°	± 5	Similar to curve 1 specimen and conditions except heated in vacuum (6.8×10^{-5} mm Hg) at 1089 K for 30 min.
8	139	1961	773	1.0-15.0	~0°	± 5	Similar to above specimen and conditions.
9	139	1961	1023	1.0-15.0	~0°	± 5	Similar to curve 7 specimen and conditions.

DATA TABLE NO. 330 NORMAL SPECTRAL EMITTANCE OF TUNGSTEN CONTACT COATINGS

[Wavelength, λ, μm; Emittance, ∈; Temperature, T, K]

λ	∈
CURVE 1	**T = 523**
2.00	0.650
2.50	0.650
3.50	0.610
4.00	0.570
4.25	0.545
4.75	0.525
6.00	0.510
7.50	0.495
8.25	0.490
9.00	0.490
10.50	0.475
11.25	0.470
13.00	0.470
14.00	0.450
14.75	0.420
15.50	0.370
CURVE 2	**T = 773**
1.00	0.350
1.50	0.410
1.75	0.460
2.25	0.470
3.00	0.460
3.75	0.480
5.50	0.450
6.75	0.445
7.50	0.450
9.00	0.450
10.50	0.450
11.00	0.460
12.10	0.455
13.00	0.465
14.00	0.450*
15.00	0.430
CURVE 3	**T = 1023**
1.00	0.380
1.50	0.475
1.80	0.575
2.10	0.605
2.75	0.605
CURVE 3 (cont.)	
4.00	0.620
5.00	0.600
6.00	0.560
6.50	0.560
7.20	0.570
8.00	0.570
9.50	0.535
10.50	0.535
11.50	0.550
12.00	0.560
13.00	0.560
14.00	0.560
14.50	0.550
15.00	0.500
CURVE 4	**T = 523**
2.00	0.780
2.50	0.700
3.25	0.655
4.10	0.685
4.60	0.685
5.60	0.700
7.00	0.725
8.25	0.700
9.35	0.650
10.00	0.600
11.10	0.555
12.00	0.590
12.90	0.560*
13.75	0.550
14.50	0.520
15.00	0.480
CURVE 5	**T = 773**
1.00	0.480
1.50	0.530
1.65	0.550
2.00	0.600
2.90	0.585
3.50	0.615
4.75	0.650
6.00	0.690
CURVE 5 (cont.)	
7.25	0.710
8.00	0.710
9.00	0.675
10.00	0.620
11.00	0.600
12.00	0.620
13.50	0.620
15.00	0.620
CURVE 6	**T = 1023**
1.00	0.500
1.75	0.580
2.50	0.605
3.75	0.650
4.50	0.690
5.00	0.700
6.00	0.700
7.00	0.740
8.00	0.740
8.80	0.700
9.50	0.670
10.00	0.660
10.75	0.625
11.50	0.620
13.00	0.620
14.50	0.620
15.00	0.610
CURVE 7	**T = 523**
2.00	0.800
2.25	0.725
3.00	0.650
4.00	0.595
5.25	0.565
6.00	0.540
6.75	0.540
8.00	0.515
9.00	0.500
10.00	0.490
11.50	0.460
12.50	0.450
13.75	0.425
15.00	0.380
CURVE 8	**T = 773**
1.00	0.600
1.25	0.600
1.95	0.550
2.50	0.485
3.00	0.470
4.00	0.500
5.00	0.500
5.75	0.460
7.00	0.460
8.00	0.460
8.80	0.440
10.25	0.440
10.75	0.435
11.75	0.435
12.25	0.440
12.75	0.430
14.00	0.430
15.00	0.430*
CURVE 9	**T = 1023**
1.00	0.390
1.40	0.470
2.00	0.515
2.75	0.520
3.75	0.570
5.00	0.610
6.25	0.600
7.25	0.620
8.00	0.650
9.25	0.650
10.25	0.685
11.00	0.740
11.40	0.750
13.00	0.750
14.25	0.760
14.75	0.745
15.00	0.730

* Not shown on plot

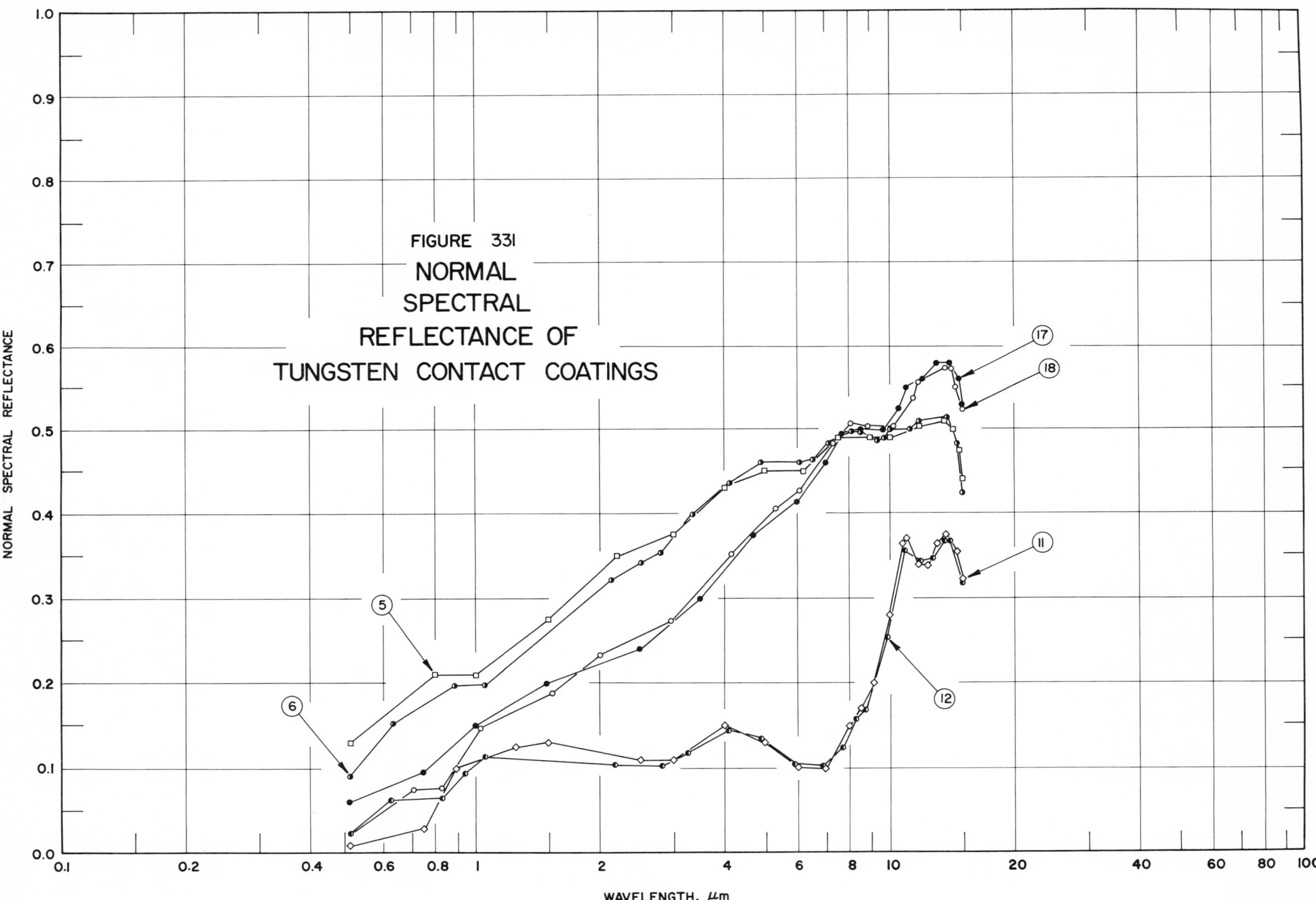

FIGURE 331
NORMAL
SPECTRAL
REFLECTANCE OF
TUNGSTEN CONTACT COATINGS
NORMAL SPECTRAL REFLECTANCE
WAVELENGTH, μm
1.0
0.9
0.8
0.7
0.6
0.5
0.4
0.3
0.2
0.1
0.0
0.1
0.2
0.4
0.6
0.8
1
2
4
6
8
10
20
40
60
80
100
5
6
11
12
17
18

SPECIFICATION TABLE NO. 331 NORMAL SPECTRAL REFLECTANCE OF TUNGSTEN CONTACT COATINGS

Curve No.	Ref. No.	Year	Temperature, K	Wavelength Range, μm	Geometry θ	θ'	ω'	Reported Error, %	Composition (weight percent), Specifications, and Remarks
1*	139	1961	~322	2.00-15.50	~0°		2π	<2	Tungsten; sandblasted Inconel X substrate; flame sprayed using a Plasmatron; as received; data extracted from smooth curve; hohlraum at 523 K; converted from R (2π, ~0°).
2*	139	1961	~322	2.0-15.50	~0°		2π	<2	Above specimen and conditions; diffuse component only.
3*	139	1961	~322	1.0-15.00	~0°		2π	<2	Similar to curve 1 specimen and conditions except hohlraum at 773 K.
4*	139	1961	~322	1.0-15.00	~0°		2π	<2	Above specimen and conditions; diffuse component only.
5	139	1961	~322	0.5-15.00	~0°		2π	<2	Similar to curve 1 specimen and conditions except hohlraum at 1273 K.
6	139	1961	~322	0.5-15.00	~0°		2π	<2	Above specimen and conditions; diffuse component only.
7*	139	1961	~322	2.0-15.00	~0°		2π	<2	Similar to curve 1 specimen and conditions except heated in air at 1089 K for 30 min.
8*	139	1961	~322	2.0-14.99	~0°		2π	<2	Above specimen and conditions; diffuse component only.
9*	139	1961	~322	1.0-15.00	~0°		2π	<2	Similar to curve 7 specimen and conditions except hohlraum at 773 K.
10*	139	1961	~322	1.0-15.00	~0°		2π	<2	Above specimen and conditions; diffuse component only.
11	139	1961	~322	0.5-15.00	~0°		2π	<2	Similar to curve 7 specimen and conditions except hohlraum at 1273 K.
12	139	1961	~322	0.50-15.00	~0°		2π	<2	Above specimen and conditions; diffuse component only.
13*	139	1961	~322	2.0-15.00	~0°		2π	<2	Similar to curve 1 specimen and conditions except heated in vacuum (6.8 x 10^{-5} mm Hg) at 1089 K for 30 min.
14*	139	1961	~322	2.0-15.00	~0°		2π	<2	Above specimen and conditions; diffuse component only.
15*	139	1961	~322	1.0-15.00	~0°		2π	<2	Similar to curve 13 specimen and conditions except hohlraum at 773 K.
16*	139	1961	~322	1.0-15.00	~0°		2π	<2	Above specimen and conditions; diffuse component only.
17	139	1961	~322	0.5-15.00	~0°		2π	<2	Similar to curve 13 specimen and conditions except hohlraum at 1273 K.
18	139	1961	~322	0.5-15.00	~0°		2π	<2	Above specimen and conditions; diffuse component only.

* Not shown on plot

DATA TABLE NO. 331 NORMAL SPECTRAL REFLECTANCE OF TUNGSTEN CONTACT COATINGS

[Wavelength, λ, μm; Reflectance, ρ; Temperature, T, K]

CURVE 1*
T ~ 322

λ	ρ
2.00	0.300
2.50	0.310
3.50	0.385
4.50	0.450
5.25	0.470
5.75	0.465
6.75	0.495
7.90	0.540
9.00	0.540
9.75	0.550
11.25	0.550
12.50	0.560
14.00	0.550
15.00	0.525
15.50	0.500

CURVE 2*
T ~ 322

λ	ρ
2.00	0.289
2.40	0.293
3.35	0.350
4.27	0.431
5.00	0.460
5.46	0.459
5.88	0.448
6.38	0.458
6.76	0.475
7.06	0.504
7.54	0.525
8.03	0.529
9.30	0.529
9.88	0.541
11.24	0.540
11.96	0.551
14.04	0.523
14.84	0.496
15.50	0.461

CURVE 3*
T ~ 322

λ	ρ
1.00	0.250
1.50	0.260
1.90	0.290

CURVE 3 (cont.)*

λ	ρ
2.50	0.290
3.25	0.300
4.30	0.350
5.75	0.400
6.50	0.425
7.50	0.475
8.00	0.500
9.00	0.475
9.50	0.470
10.75	0.500
12.50	0.500
14.00	0.500
14.50	0.485
15.00	0.435

CURVE 4*
T ~ 322

λ	ρ
1.00	0.265
2.84	0.351
4.46	0.474
5.08	0.502
5.27	0.500
5.78	0.490
6.53	0.512
7.22	0.552
7.95	0.572
8.81	0.551
9.48	0.524
11.00	0.577
13.01	0.578
13.44	0.583
14.00	0.583
14.48	0.553
15.00	0.510

CURVE 5
T ~ 322

λ	ρ
0.50	0.130
0.80	0.210
1.00	0.210
1.50	0.275
2.20	0.350
3.00	0.375
4.00	0.430

CURVE 5 (cont.)

λ	ρ
5.00	0.450
6.20	0.450
7.50	0.490
9.00	0.490
10.00	0.490
11.75	0.505
13.50	0.510
14.25	0.500
14.75	0.475
15.00	0.440

CURVE 6
T ~ 322

λ	ρ
0.50	0.090
0.63	0.151
0.89	0.199
1.06	0.199
2.13	0.324
2.50	0.342
2.80	0.354
3.35	0.400
4.14	0.435
4.89	0.460
6.01	0.460
6.34	0.463
7.11	0.483
8.01	0.499
8.59	0.499
9.27	0.486
9.67	0.490
10.02	0.500
11.20	0.500
11.77	0.510
13.77	0.512
14.58	0.482
14.90	0.442*
15.00	0.425

CURVE 7*
T ~ 322

λ	ρ
2.00	0.210
2.50	0.115
2.75	0.090
3.25	0.080

CURVE 7 (cont.)*

λ	ρ
4.00	0.115
5.00	0.090
6.45	0.045
7.00	0.060
7.50	0.050
8.40	0.100
9.50	0.200
10.15	0.265
10.50	0.315
11.00	0.350
11.50	0.330
12.00	0.310
13.00	0.340
14.00	0.355
14.50	0.350
15.00	0.300

CURVE 8*
T ~ 322

λ	ρ
2.00	0.100
2.46	0.080
2.90	0.080
3.94	0.110
4.64	0.092
5.88	0.047
6.34	0.043
7.72	0.081
8.69	0.125
9.75	0.207
11.02	0.319
11.97	0.301
13.00	0.321
13.99	0.333
14.83	0.308
14.99	0.276

CURVE 9*
T ~ 322

λ	ρ
1.00	0.060
1.50	0.095
2.00	0.100
2.50	0.100
3.50	0.125
4.25	0.140

CURVE 9 (cont.)*

λ	ρ
5.25	0.125
6.50	0.090
7.50	0.110
8.75	0.150
9.50	0.200
10.25	0.275
11.00	0.350
11.75	0.330
12.50	0.340
13.75	0.370
14.35	0.380
15.00	0.360

CURVE 10*
T ~ 322

λ	ρ
1.00	0.075
1.47	0.111
1.99	0.121
2.67	0.110
3.10	0.103
3.66	0.125
3.99	0.143
4.49	0.141
5.74	0.103
6.42	0.089
6.82	0.092
8.53	0.145
9.50	0.201
10.34	0.276
10.80	0.328
11.16	0.343
12.06	0.317
12.85	0.333
14.30	0.370
14.76	0.366
15.00	0.350

CURVE 11
T ~ 322

λ	ρ
0.50	0.010
0.75	0.030
0.90	0.100
1.25	0.125
1.50	0.130

CURVE 11 (cont.)

λ	ρ
2.50	0.110
3.00	0.110
4.00	0.150
5.00	0.130
6.00	0.100
7.00	0.100
8.00	0.150
8.50	0.170
9.10	0.200
10.00	0.280
10.75	0.365
11.00	0.370
11.75	0.340
12.25	0.340
13.00	0.365
13.75	0.375
14.50	0.355
15.00	0.325

CURVE 12
T ~ 322

λ	ρ
0.50	0.023
0.62	0.061
0.83	0.064
0.94	0.093
1.46	0.113
2.19	0.102
2.81	0.102
3.25	0.119
4.01	0.143
4.98	0.133
5.82	0.103
6.93	0.101
7.67	0.125
8.30	0.158
8.75	0.168
9.85	0.253
10.97	0.356
11.72	0.342
12.84	0.347
13.50	0.368
14.06	0.368
15.00	0.320

CURVE 13*
T ~ 322

λ	ρ
2.00	0.230
2.50	0.185
3.00	0.170
3.50	0.200
4.00	0.260
5.00	0.295
5.80	0.275
6.60	0.310
7.70	0.370
8.50	0.370
9.50	0.400
10.80	0.450
12.00	0.510
13.00	0.535
13.50	0.545
14.50	0.535
15.00	0.480

CURVE 14*
T ~ 322

λ	ρ
2.00	0.275
2.42	0.237
2.98	0.224
3.43	0.251
4.15	0.331
4.98	0.360
5.71	0.360
6.18	0.374
7.19	0.428
8.01	0.450
8.74	0.445
9.24	0.435
9.99	0.449
11.32	0.496
12.17	0.499
13.20	0.515
13.99	0.526
14.79	0.499
15.00	0.465

CURVE 15*
T ~ 322

λ	ρ
1.00	0.170
2.00	0.220
2.75	0.240
3.75	0.300
5.00	0.380
5.75	0.400
7.00	0.460
8.00	0.500
8.75	0.490
9.25	0.485
10.00	0.500
11.00	0.550
12.50	0.575
14.00	0.625
14.50	0.630
15.00	0.620

CURVE 16*
T ~ 322

λ	ρ
1.00	0.160
1.47	0.164
2.00	0.199
2.69	0.242
4.67	0.370
5.19	0.386
6.04	0.419
6.71	0.438
8.13	0.491
9.06	0.490
9.61	0.490
10.28	0.505
10.78	0.524
11.27	0.548
12.04	0.559
12.72	0.567
13.35	0.590
14.02	0.608
14.52	0.613
15.00	0.600

* Not shown on plot

DATA TABLE NO. 331 NORMAL SPECTRAL REFLECTANCE OF TUNGSTEN CONTACT COATINGS (continued)

λ	ρ
CURVE 17	
$T \sim 322$	
0.50	0.060
0.75	0.095
1.00	0.150
1.50	0.200
2.50	0.240
3.50	0.300
4.75	0.375
6.00	0.415
7.00	0.460
7.75	0.495
8.50	0.500
9.75	0.500
10.50	0.525
11.00	0.550
12.00	0.560
13.00	0.580
14.00	0.580
14.75	0.560
15.00	0.530
CURVE 18	
$T \sim 322$	
0.50	0.022*
0.71	0.075
0.83	0.077
1.03	0.149
1.53	0.187
2.00	0.233
2.99	0.274
4.18	0.351
5.34	0.407
6.01	0.428
7.27	0.483
8.01	0.507
8.89	0.502
10.28	0.502
11.06	0.539
11.78	0.556
13.06	0.572
14.01	0.572
14.59	0.550
15.00	0.523

* Not shown on plot

SPECIFICATION TABLE NO. 332 NORMAL SOLAR ABSORPTANCE OF TUNGSTEN CONTACT COATINGS

Curve No.	Ref. No.	Year	Temperature Range, K	Geometry θ	Reported Error, %	Composition (weight percent), Specifications and Remarks
1*	146	1961	468-1093	$\sim 0°$		Tungsten, Metco XP-1106, crystalline (-200 mesh, +30 μm); Armco iron substrate; plasma flame sprayed; measured in vacuum (10^{-5} mm Hg).

DATA TABLE NO. 332 NORMAL SOLAR ABSORPTANCE OF TUNGSTEN CONTACT COATINGS

[Temperature, T, K; Absorptance, α]

T	α
CURVE 1*	
468	0.90
668	0.86
873	0.81
873	0.73
1093	0.70

* No plot given

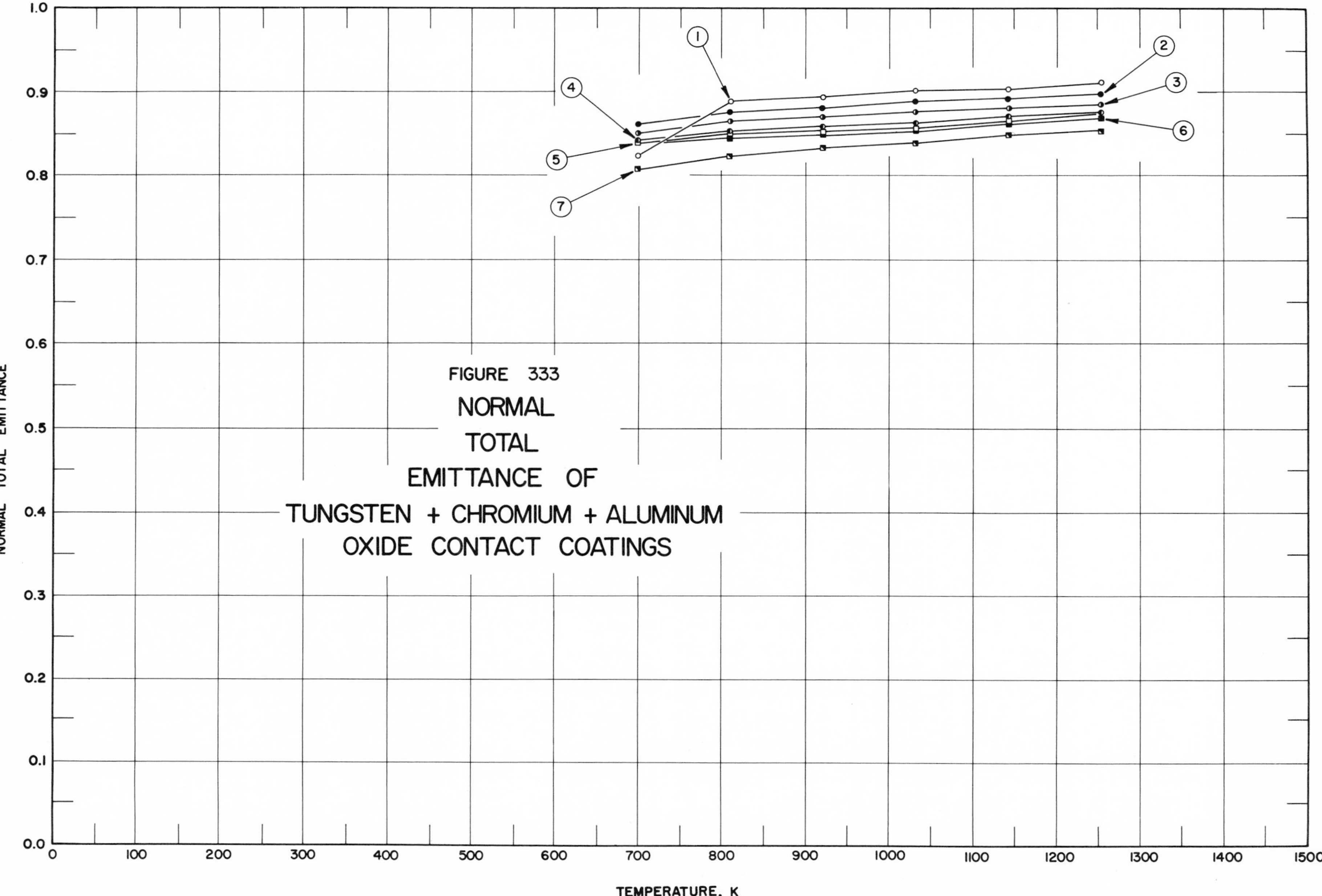

FIGURE 333 NORMAL TOTAL EMITTANCE OF TUNGSTEN + CHROMIUM + ALUMINUM OXIDE CONTACT COATINGS

SPECIFICATION TABLE NO. 333 NORMAL TOTAL EMITTANCE OF TUNGSTEN + CHROMIUM + ALUMINUM OXIDE CONTACT COATINGS

Curve No.	Ref. No.	Year	Temperature Range, K	Geometry θ'	Reported Error, %	Composition (weight percent), Specifications and Remarks
1	73	1962	700-1255	~0°		60 W, 25 Cr and 15 Al_2O_3, Haynes LT-2 cermet (0.102 mm thick); sandblasted oxidized Inconel substrate; flame sprayed.
2	73	1962	700-1255	~0°		Similar to above specimen and conditions except 0.152 mm thick.
3	73	1962	700-1255	~0°		Similar to above specimen and conditions except 0.203 mm thick.
4	73	1962	700-1255	~0°		Similar to above specimen and conditions except 0.254 mm thick.
5	73	1962	700-1255	~0°		Similar to above specimen and conditions except 0.305 mm thick.
6	73	1962	700-1255	~0°		Similar to above specimen and conditions except 0.356 mm thick.
7	73	1962	700-1255	~0°		60 W, 25 Cr and 15 Al_2O_3, Haynes LT-2 cermet; sandblasted oxidized Inconel substrate; flame sprayed.

DATA TABLE NO. 333 NORMAL TOTAL EMITTANCE OF TUNGSTEN + CHROMIUM + ALUMINUM OXIDE CONTACT COATINGS

[Temperature, T, K; Emittance, ϵ]

T	ϵ	T	ϵ	T	ϵ	T	ϵ
CURVE 1		CURVE 3		CURVE 5		CURVE 7	
700	0.825	700	0.853	700	0.841	700	0.81
811	0.89	811	0.866	811	0.85	811	0.825
922	0.897	922	0.872	922	0.854	922	0.835
1033	0.903	1033	0.878	1033	0.859	1033	0.84
1144	0.908	1144	0.883	1144	0.868	1144	0.85
1255	0.915	1255	0.887	1255	0.873*	1255	0.855
CURVE 2		CURVE 4		CURVE 6			
700	0.864	700	0.845	700	0.84*		
811	0.878	811	0.856	811	0.846		
922	0.882	922	0.86	922	0.85		
1033	0.89	1033	0.865	1033	0.855		
1144	0.895	1144	0.873	1144	0.865		
1255	0.90	1255	0.879	1255	0.87		

* Not shown on plot

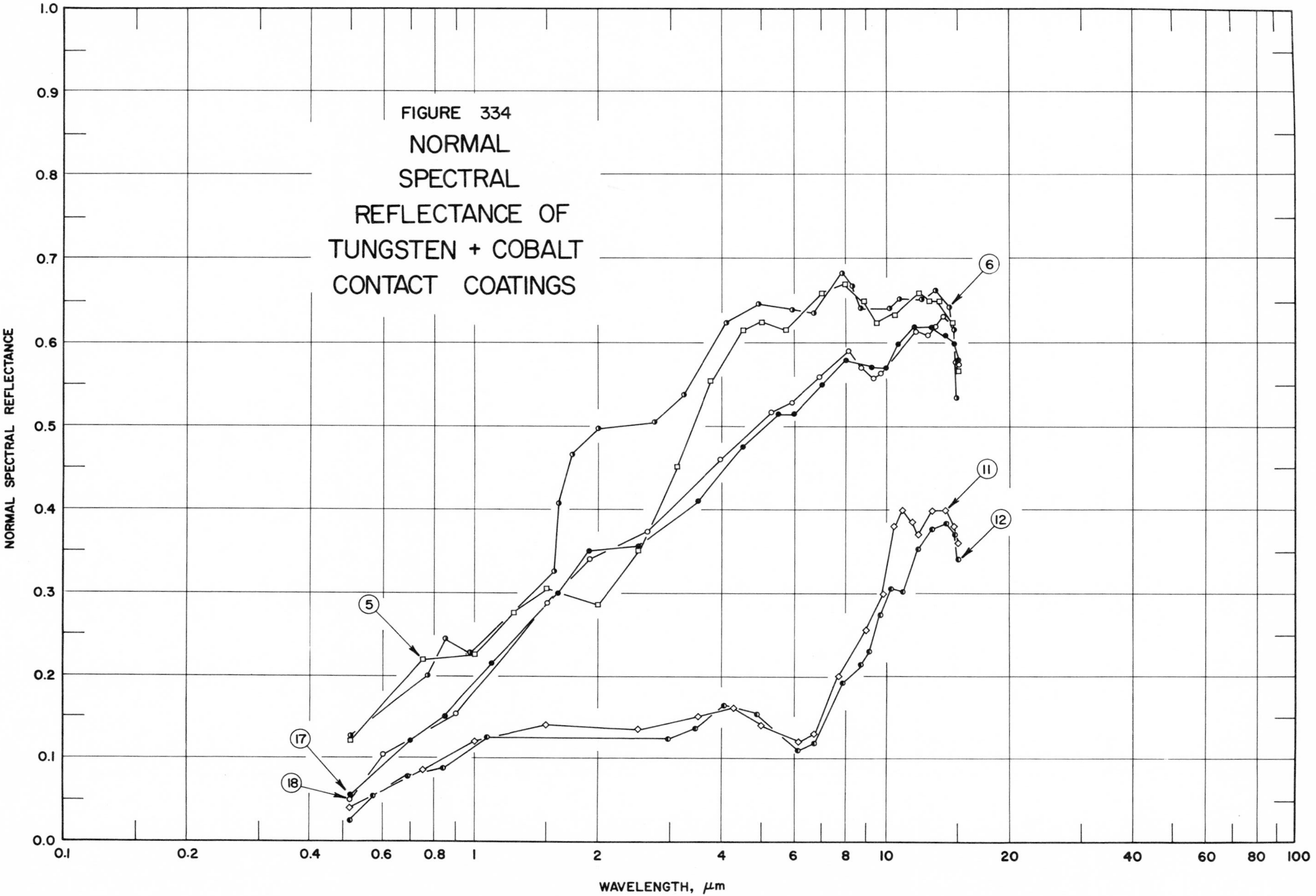
FIGURE 334
NORMAL
SPECTRAL
REFLECTANCE OF
TUNGSTEN + COBALT
CONTACT COATINGS
NORMAL SPECTRAL REFLECTANCE
WAVELENGTH, μm
0.0
0.1
0.2
0.3
0.4
0.5
0.6
0.7
0.8
0.9
1.0
0.1
0.2
0.4
0.6
0.8
1
2
4
6
8
10
20
40
60
80
100
5
6
11
12
17
18

SPECIFICATION TABLE NO. 334 NORMAL SPECTRAL REFLECTANCE OF TUNGSTEN + COBALT CONTACT COATINGS

Curve No.	Ref. No.	Year	Temperature, K	Wavelength Range, μm	Geometry θ	θ'	ω'	Reported Error, %	Composition (weight percent), Specifications, and Remarks
1*	139	1961	~322	2.00-15.00	~0°		2π	<2	50 W and 50 Co on Inconel X substrate; flame sprayed using a Plasmatron; as received; data extracted from smooth curve; sample maintained at ~322 K; hohlraum at 523 K; converted from R (2π, ~0°).
2*	139	1961	~322	2.00-15.00	~0°		2π	<2	Similar to above specimen and conditions; diffuse component only.
3*	139	1961	~322	1.00-15.00	~0°		2π	<2	Similar to curve 1 specimen and conditions except hohlraum at 773 K.
4*	139	1961	~322	1.00-15.00	~0°		2π	<2	Above specimen and conditions; diffuse component only.
5	139	1961	~322	0.50-15.00	~0°		2π	<2	Similar to curve 1 specimen and conditions except hohlraum at 1273 K.
6	139	1961	~322	0.50-14.99	~0°		2π	<2	Above specimen and conditions; diffuse component only.
7*	139	1961	~322	2.00-15.00	~0°		2π	<2	Similar to curve 1 specimen and conditions except heated in air at 1089 K for 30 min.
8*	139	1961	~322	2.00-15.00	~0°		2π	<2	Above specimen and conditions; diffuse component only.
9*	139	1961	~322	1.00-15.00	~0°		2π	<2	Similar to curve 7 specimen and conditions except hohlraum at 773 K.
10*	139	1961	~322	1.00-15.00	~0°		2π	<2	Above specimen and conditions; diffuse component only.
11	139	1961	~322	0.50-15.00	~0°		2π	<2	Similar to curve 7 specimen and conditions except hohlraum at 1273 K.
12	139	1961	~322	0.50-15.00	~0°		2π	<2	Above specimen and conditions; diffuse component only.
13*	139	1961	~322	2.00-15.00	~0°		2π	<2	Similar to curve 1 specimen and conditions except heated in vacuum (6.8×10^{-5} mm Hg) at 1089 K for 30 min.
14*	139	1961	~322	2.00-15.00	~0°		2π	<2	Above specimen and conditions; diffuse component only.
15*	139	1961	~322	1.00-15.00	~0°		2π	<2	Similar to curve 13 specimen and conditions except hohlraum at 773 K.
16*	139	1961	~322	1.00-15.00	~0°		2π	<2	Above specimen and conditions; diffuse component only.
17	139	1961	~322	0.50-15.00	~0°		2π	<2	Similar to curve 13 specimen and conditions except hohlraum at 1273 K.
18	139	1961	~322	0.50-15.00	~0°		2π	<2	Above specimen and conditions; diffuse component only.

* Not shown on plot

DATA TABLE NO. 334 NORMAL SPECTRAL REFLECTANCE OF TUNGSTEN + COBALT CONTACT COATINGS

[Wavelength, λ, μm; Reflectance, ρ; Temperature, T, K]

λ	ρ
CURVE 1*	
T ~ 322	
2.00	0.380
2.75	0.410
3.50	0.475
4.50	0.550
5.10	0.565
6.00	0.545
6.50	0.555
7.25	0.605
8.00	0.615
9.00	0.600
10.00	0.600
11.25	0.585
13.00	0.600
13.50	0.600
14.50	0.565
15.00	0.530
CURVE 2*	
T ~ 322	
2.00	0.394
2.39	0.390
3.05	0.425
3.97	0.527
4.45	0.564
5.06	0.581
5.85	0.553
6.16	0.542
6.78	0.563
7.32	0.597
8.04	0.613
8.89	0.586
9.55	0.593
10.19	0.591
11.05	0.581
12.25	0.587
13.16	0.582
14.08	0.562
14.72	0.530
15.00	0.502
CURVE 3*	
T ~ 322	
1.00	0.300
1.50	0.285
1.75	0.300
1.80	0.350
2.00	0.440
2.70	0.465
3.50	0.560
4.00	0.610
4.80	0.635
6.00	0.620
7.25	0.650
8.00	0.685
9.35	0.660
10.50	0.665
11.50	0.670
13.00	0.670
14.00	0.650
14.80	0.600
15.00	0.560
CURVE 4*	
T ~ 322	
1.00	0.309
1.45	0.323
1.75	0.376
1.86	0.415
2.11	0.431
2.56	0.426
3.03	0.452
3.99	0.561
4.54	0.589
5.13	0.600
6.01	0.569
7.01	0.600
7.91	0.620
8.56	0.604
9.06	0.584
10.01	0.584
11.01	0.602
12.01	0.611
13.11	0.611
14.14	0.597
14.86	0.554
15.00	0.525
CURVE 5	
T ~ 322	
0.50	0.120
0.75	0.220
1.00	0.225
1.25	0.275
1.50	0.305
2.00	0.285
2.50	0.350
3.10	0.450
3.75	0.555
4.50	0.615
5.00	0.625
5.75	0.615
7.00	0.660
8.00	0.670
8.85	0.650
9.50	0.625
10.50	0.635
12.00	0.660
12.75	0.650
13.50	0.650
14.50	0.625
15.00	0.565
CURVE 6	
T ~ 322	
0.50	0.125
0.77	0.200
0.85	0.245
0.97	0.229
1.25	0.275*
1.56	0.326
1.60	0.408
1.72	0.464
2.00	0.499
2.72	0.505
3.24	0.538
4.11	0.625
4.99	0.647
5.99	0.640
6.65	0.638
7.89	0.681
8.36	0.669
8.77	0.641
10.03	0.641
CURVE 6 (cont.)	
10.78	0.651
12.03	0.651
13.03	0.662
14.03	0.642
14.67	0.617
14.96	0.576
14.99	0.533
CURVE 7*	
T ~ 322	
2.00	0.125
2.75	0.110
4.00	0.125
5.00	0.080
6.00	0.060
7.00	0.100
8.25	0.175
8.75	0.200
9.75	0.300
10.50	0.365
11.00	0.390
12.00	0.360
13.00	0.335
13.60	0.375
14.00	0.400
14.60	0.375
15.00	0.340
CURVE 8*	
T ~ 322	
2.00	0.115
2.44	0.100
2.82	0.097
3.93	0.117
4.34	0.111
5.73	0.066
6.06	0.068
7.77	0.128
9.00	0.180
9.84	0.274
10.89	0.353
11.26	0.359
11.98	0.327
12.40	0.321
CURVE 8 (cont.)*	
13.16	0.354
13.63	0.352
14.00	0.359
14.59	0.339
15.00	0.308
CURVE 9*	
T ~ 322	
1.00	0.080
1.50	0.110
2.15	0.085
2.50	0.080
4.00	0.100
4.50	0.105
5.25	0.125
6.00	0.090
6.55	0.100
7.75	0.140
9.00	0.165
10.00	0.240
10.50	0.275
11.00	0.290
11.80	0.270
13.00	0.290
14.00	0.320
14.50	0.325
15.00	0.300
CURVE 10*	
T ~ 322	
1.00	0.082
2.00	0.094
2.79	0.118
3.99	0.163
4.57	0.152
5.26	0.124
5.69	0.121
6.24	0.108
7.01	0.133
8.25	0.217
8.82	0.229
9.76	0.309
10.77	0.384
11.26	0.390
CURVE 10 (cont.)*	
12.25	0.374
12.77	0.374
14.01	0.399
14.72	0.401
15.00	0.389
CURVE 11	
T ~ 322	
0.50	0.040
0.75	0.085
1.00	0.120
1.50	0.140
2.50	0.135
3.50	0.150
4.25	0.160
5.00	0.140
6.15	0.120
6.75	0.130
7.75	0.200
9.00	0.255
9.85	0.300
10.50	0.380
11.00	0.400
11.60	0.385
12.00	0.370
13.00	0.400
14.00	0.400
14.75	0.380
15.00	0.360
CURVE 12	
T ~ 322	
0.50	0.025
0.57	0.053
0.69	0.078
0.84	0.087
1.07	0.125
2.97	0.122
3.43	0.136
4.03	0.161
4.84	0.151
6.11	0.110
6.74	0.119
7.99	0.190
CURVE 12 (cont.)	
8.75	0.214
9.15	0.230
9.73	0.273
10.26	0.306
11.00	0.301
12.00	0.351
13.00	0.376
14.00	0.382
14.68	0.370
15.00	0.341
CURVE 13*	
T ~ 322	
2.00	0.330
2.50	0.260
3.00	0.240
3.50	0.270
4.00	0.340
5.00	0.375
6.00	0.340
6.70	0.375
7.00	0.400
7.50	0.410
8.50	0.410
9.50	0.410
11.00	0.440
12.00	0.450
13.25	0.490
14.50	0.515
15.00	0.500
CURVE 14*	
T ~ 322	
2.00	0.450
2.65	0.395
2.98	0.387
3.38	0.406
3.95	0.454
4.95	0.511
6.22	0.545
7.22	0.564
7.99	0.584
8.98	0.564
10.74	0.593
CURVE 14 (cont.)*	
11.99	0.609
13.09	0.615
14.00	0.600
14.67	0.575
15.00	0.550
CURVE 15*	
T ~ 322	
1.00	0.270
1.50	0.280
2.10	0.325
2.80	0.350
3.50	0.410
5.00	0.490
5.75	0.500
6.50	0.520
8.00	0.575
9.25	0.565
11.25	0.620
12.00	0.625
13.00	0.650
14.50	0.670
15.00	0.660
CURVE 16*	
T ~ 322	
1.00	0.239
1.79	0.303
2.19	0.330
2.76	0.338
3.24	0.376
3.86	0.418
5.19	0.489
6.04	0.505
7.23	0.548
8.16	0.572
9.04	0.551
9.79	0.563
10.53	0.586
11.29	0.610
12.26	0.624
13.00	0.631
14.01	0.650
15.00	0.650

* Not shown on plot

DATA TABLE NO. 334 NORMAL SPECTRAL REFLECTANCE OF TUNGSTEN + COBALT CONTACT COATINGS (continued)

λ	ρ
CURVE 17	
T ~ 322	
0. 50	0. 055
0. 70	0. 120
0. 85	0. 150
1. 10	0. 215
1. 60	0. 300
1. 90	0. 350
2. 50	0. 355
3. 50	0. 410
4. 50	0. 475
5. 50	0. 515
6. 00	0. 515
7. 00	0. 550
8. 00	0. 580
9. 25	0. 570
10. 00	0. 570
10. 75	0. 600
11. 75	0. 620
13. 00	0. 620
14. 00	0. 610
14. 60	0. 600
15. 00	0. 580
CURVE 18	
T ~ 322	
0. 50	0. 050
0. 60	0. 103
0. 90	0. 151
1. 50	0. 289
1. 90	0. 340
2. 62	0. 372
3. 99	0. 460
5. 27	0. 518
5. 99	0. 529
6. 99	0. 560
8. 11	0. 590
8. 76	0. 570
9. 25	0. 558
9. 77	0. 562
11. 19	0. 615
12. 17	0. 610
13. 01	0. 620
13. 99	0. 631
14. 58	0. 621*
15. 00	0. 575

* Not shown on plot

SPECIFICATION TABLE NO. 335 NORMAL SPECTRAL REFLECTANCE OF URANIUM CONTACT COATINGS

Curve No.	Ref. No.	Year	Temperature, K	Wavelength Range, μm	Geometry θ θ' ω'	Reported Error, %	Composition (weight percent), Specifications, and Remarks
1*	215	1962	~298	0.4-4.0	~0° ~0°		Uranium; glass substrate; prepared by applying a solution of uranium resinate, diluted with aromatic solvents and essential oils, to a soda-lime glass substrate, then firing to 873 K in a continuous lehr on a 1.5 hr cycle; application of solution was by dropping an excess amount on the substrate, which was kept spinning at 1550 rpm until no more solution was flung from the edges; measured relative to plane, polished aluminum; specular component only.

DATA TABLE NO. 335 NORMAL SPECTRAL REFLECTANCE OF URANIUM CONTACT COATINGS

[Wavelength, λ, μm; Reflectance, ρ; Temperature, T, K]

T	ρ
CURVE 1*	
T ~298	
0.4	0.14
0.5	0.11
0.6	0.06
0.8	0.06
1.0	0.08
1.2	0.10
1.4	0.10
1.6	0.10
1.8	0.10
2.0	0.10
2.2	0.09
2.4	0.07
2.6	0.07
2.8	0.06
3.0	0.04
3.2	0.04
3.4	0.03
3.6	0.03
3.8	0.03
4.0	0.01

* No plot given

SPECIFICATION TABLE NO. 336 NORMAL SPECTRAL TRANSMITTANCE OF URANIUM CONTACT COATINGS

Curve No.	Ref. No.	Year	Temperature, K	Wavelength Range, μm	Geometry θ θ' ω'	Reported Error, %	Composition (weight percent), Specifications, and Remarks
1*	215	1962	~298	0.4-2.0	~0° ~0°		Uranium; glass (1.59 mm thick) substrate; prepared by applying a solution of uranium resinate, diluted with aromatic solvents and essential oils, to a soda-lime glass substrate, then firing to 873 K in a continuous lehr on a 1.5 hr cycle; application of solution was by dropping an excess amount on the substrate, which was kept spinning at 1550 rpm until no more solution was flung from the edges; τ is for coating plus substrate.

DATA TABLE NO. 336 NORMAL SPECTRAL TRANSMITTANCE OF URANIUM CONTACT COATINGS

[Wavelength, λ, μm; Transmittance, τ; Temperature, T, K]

λ	τ
CURVE 1*	
T ~ 298	
0.4	0.136
0.5	0.352
0.6	0.606
0.7	0.694
0.8	0.653
0.9	0.613
1.0	0.587
1.1	0.586
1.2	0.593
1.3	0.597
1.4	0.514
1.5	0.374
1.6	0.296
1.8	0.236
2.0	0.243

* No plot given

SPECIFICATION TABLE NO. 337 NORMAL SPECTRAL REFLECTANCE OF ZINC CONTACT COATINGS

Curve No.	Ref. No.	Year	Temperature, K	Wavelength Range, μm	Geometry θ θ' ω'	Reported Error, %	Composition (weight percent), Specifications, and Remarks
1*	203	1962	298	0.50-14.99	~0° ~0°		Galvanized iron; zinc on iron substrate; commercial finish; data extracted from smooth curve.

DATA TABLE NO. 337 NORMAL SPECTRAL REFLECTANCE OF ZINC CONTACT COATINGS

[Wavelength, λ, μm; Reflectance, ρ; Temperature, T, K]

λ	ρ
CURVE 1* T = 203	
0.50	0.432
0.74	0.361
0.88	0.301
0.99	0.234
1.18	0.386
1.26	0.495
1.49	0.616
2.71	0.749
2.95	0.737
3.86	0.807
5.56	0.850
6.64	0.863
7.00	0.858
10.98	0.870
12.91	0.885
14.32	0.881
14.99	0.866

* No plot given

SPECIFICATION TABLE NO. 338 HEMISPHERICAL TOTAL EMITTANCE OF ZIRCONIUM CONTACT COATINGS

Curve No.	Ref. No.	Year	Temperature Range, K	Reported Error, %	Composition (weight percent), Specifications and Remarks
1*	97	1961	873-1473	±2.5	Zirconium, 200-300 mesh powder, area density 5.0 mg cm^{-2}; sandblasted molybdenum substrate; sprayed; measured in vacuum (< 5 x 10^{-6} mm Hg); data extracted from smooth curve.

DATA TABLE NO. 338 HEMISPHERICAL TOTAL EMITTANCE OF ZIRCONIUM CONTACT COATINGS

[Temperature, T, K; Emittance, ∈]

T	∈
CURVE 1*	
873	0.410
973	0.470
1073	0.510
1173	0.555
1273	0.590
1373	0.620
1473	0.645

* No plot given

2. CONTACT COATINGS (continued)

B. Nonmetallic Inorganic Contact Coatings

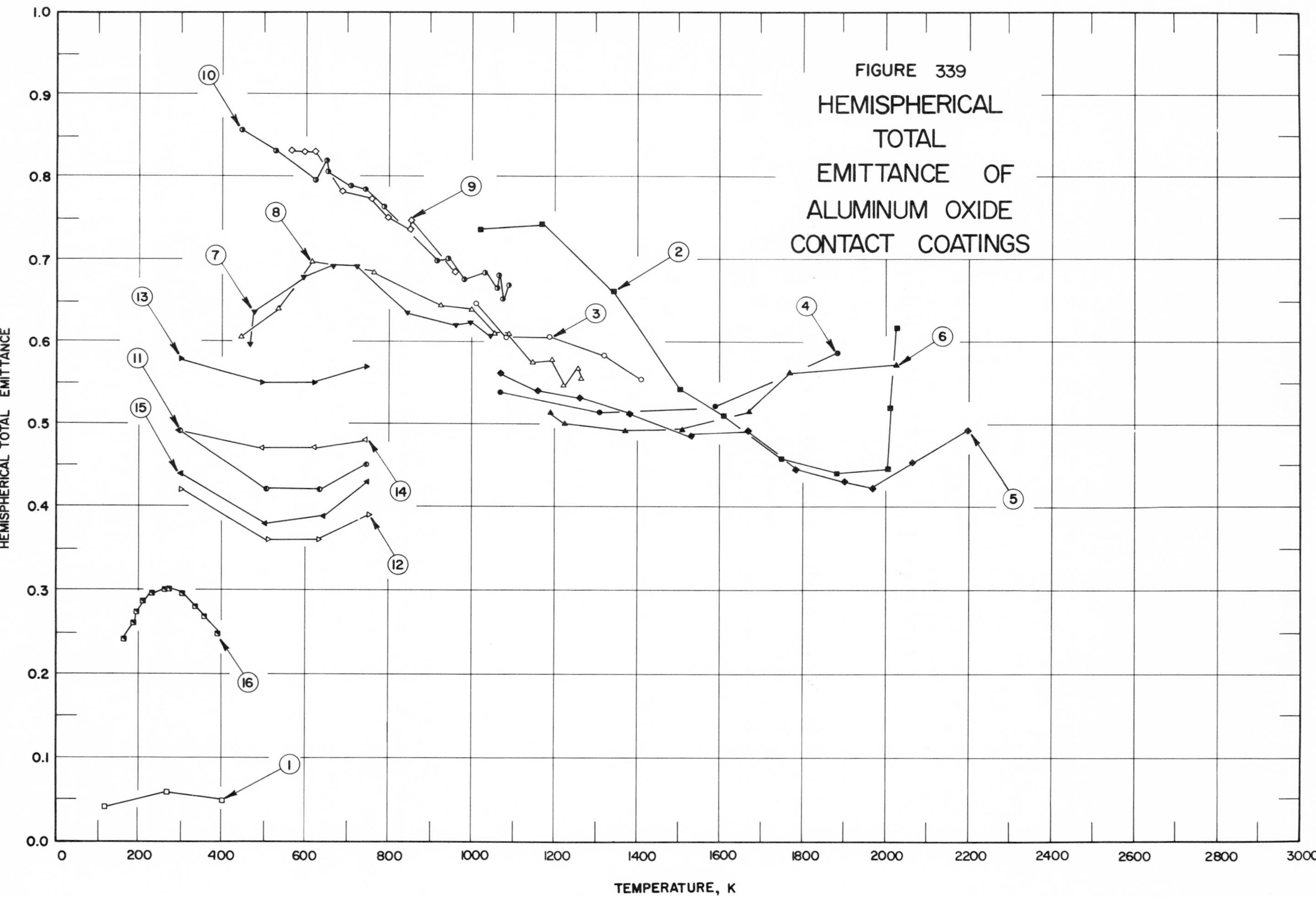

FIGURE 339 HEMISPHERICAL TOTAL EMITTANCE OF ALUMINUM OXIDE CONTACT COATINGS

SPECIFICATION TABLE NO. 339 HEMISPHERICAL TOTAL EMITTANCE OF ALUMINUM OXIDE CONTACT COATINGS

Curve No.	Ref. No.	Year	Temperature Range, K	Reported Error, %	Composition (weight percent), Specifications and Remarks
1	119	1967	120-402		Sapphire (alumina) (0.100 μm thick); silver (0.148 μm thick) and mill-finish 6061-T6 aluminum alloy substrates; high-purity silver vapor-deposited on aluminum, sapphire vapor-deposited on silver.
2	120	1966	1023-2243		98.5 Al_2O_3, 0.05 Fe, 0.1 alkalis, 0.1 silic acid, 0.2 sulfides, 0.05 chlorides, (130-400 μm thick); baked Mo substrate; particle size: 50% with dia 30-40 μm and 50% with dia 60-70 μm; plasma sprayed; sample kept at each temp 15 min; measured in vacuum (10^{-4}-10^{-5} mm Hg).
3	120	1966	1015-1407		Similar to above specimen and conditions except measured after heating to 1500 K.
4	120	1966	1072-1882		Similar to curve 2 specimen and conditions except measured after heating to 2245 K.
5	120	1966	1070-2199		Similar to curve 2 specimen and conditions except polished; measured after heating to 1530 K.
6	120	1966	1092-2027		Similar to curve 2 specimen and conditions except polished; measured after heating to 220 K. [Authors' designation: Norton LA-603]
7	121	1967	463-1044	±10.1	Alundum (0.3-1.5 mm thick); Nb substrate; substrate polished to class 9 surface finish (Russian standard); heated in vacuum (10^{-4} mm Hg) at 1373 for 50 hrs; measured in situ.
8	121	1967	446-1262	±10.1	Similar to above specimen and conditions.
9	121	1967	573-959	±10.1	Similar to curve 7 specimen and conditions except heated in vacuum (10^{-4} mm Hg) at 1073 K for 2 hrs.
10	121	1967	447-1088	±10.1	Similar to above specimen and conditions.
11	122	1968	300-745		Al_2O_3 (0.0540 μm thick); silver and stainless steel substrates; silver vacuum (2-5 x 10^{-7} mm Hg) deposited on stainless steel, Al_2O_3 vacuum deposited on silver.
12	122	1968	300-756		Similar to above specimen and conditions except coating 0.0245 μm thick.
13	122	1968	300-747		Al_2O_3 (0.1045 μm thick); silver (~0.120 μm thick) and fused silica substrates; silver vacuum (2-5 x 10^{-7} mm Hg) deposited on fused silica; Al_2O_3 vacuum deposited on silver.
14	122	1968	300-747		Similar to above specimen and conditions except coating 0.540 μm thick.
15	122	1968	300-754		Similar to curve 13 specimen and conditions except coating 0.245 μm thick.
16	85	1969	161-390		Sapphire (Al_2O_3) (~0.002 mm thick); anodized 1199 aluminum substrate; vapor deposited; aluminum substrate alkaline electropolished (sodium phosphate and sodium carbonate) for 15 min at 353 K and 12 VDC, than anodized for 15 min at 18 VDC in 10% sulfuric acid; property measured in vacuum by calorimetric method; data extracted from smooth curve. [Author's designation: A-3 Vapor Deposited Sapphire (Al_2O_3), NRDL-RTD-81-3]

DATA TABLE NO. 339 HEMISPHERICAL TOTAL EMITTANCE OF ALUMINUM OXIDE CONTACT COATINGS

[Temperature, T, K; Emittance, ε]

T	ε
CURVE 1	
120	0.041
264	0.057
402	0.049
CURVE 2	
1023	0.736
1170	0.742
1341	0.664
1501	0.541
1612	0.510
1745	0.457
1882	0.440
2015	0.446
2174	0.520
2243	0.617
CURVE 3	
1015	0.647
1081	0.605
1193	0.606
1319	0.583
1407	0.555
CURVE 4	
1072	0.538
1307	0.516
1590	0.523
1882	0.587
CURVE 5	
1070	0.564
1161	0.540
1262	0.532
1383	0.512
1530	0.485
1666	0.491
1786	0.446
1902	0.430
1972	0.423
2068	0.455
2199	0.540
CURVE 6	
1092	0.516
1221	0.499
1371	0.493
1509	0.493
1665	0.516
1768	0.563
2027	0.573
CURVE 7	
463	0.598
476	0.635
599	0.678
665	0.692
723	0.692
844	0.633
960	0.622
998	0.623
1044	0.606
CURVE 8	
446	0.604
536	0.641
616	0.697
761	0.683
923	0.645
999	0.640
1059	0.609
1087	0.609
1144	0.575
1194	0.578
1222	0.546
1258	0.568
1262	0.556
CURVE 9	
573	0.832
599	0.832
626	0.832
689	0.783
756	0.774
799	0.750
CURVE 9 (cont.)	
850	0.736
852	0.747
959	0.684
CURVE 10	
447	0.858
527	0.833
624	0.795
653	0.820
655	0.805
703	0.789
742	0.784
783	0.765
917	0.698
943	0.700
980	0.676
1035	0.684
1061	0.667
1065	0.680
1076	0.653
1088	0.669
CURVE 11	
300	0.049
506	0.042
635	0.042
745	0.045
CURVE 12	
300	0.042
510	0.036
634	0.036
756	0.039
CURVE 13	
300	0.058
496	0.055
623	0.055
747	0.057
CURVE 14	
300	0.049
496	0.047
622	0.047
747	0.048
CURVE 15	
300	0.044
503	0.038
642	0.039
754	0.043
CURVE 16	
161	0.242
188	0.261
194	0.275
212	0.288
233	0.296
261	0.300
277	0.300
305	0.295
339	0.281
359	0.269
390	0.249

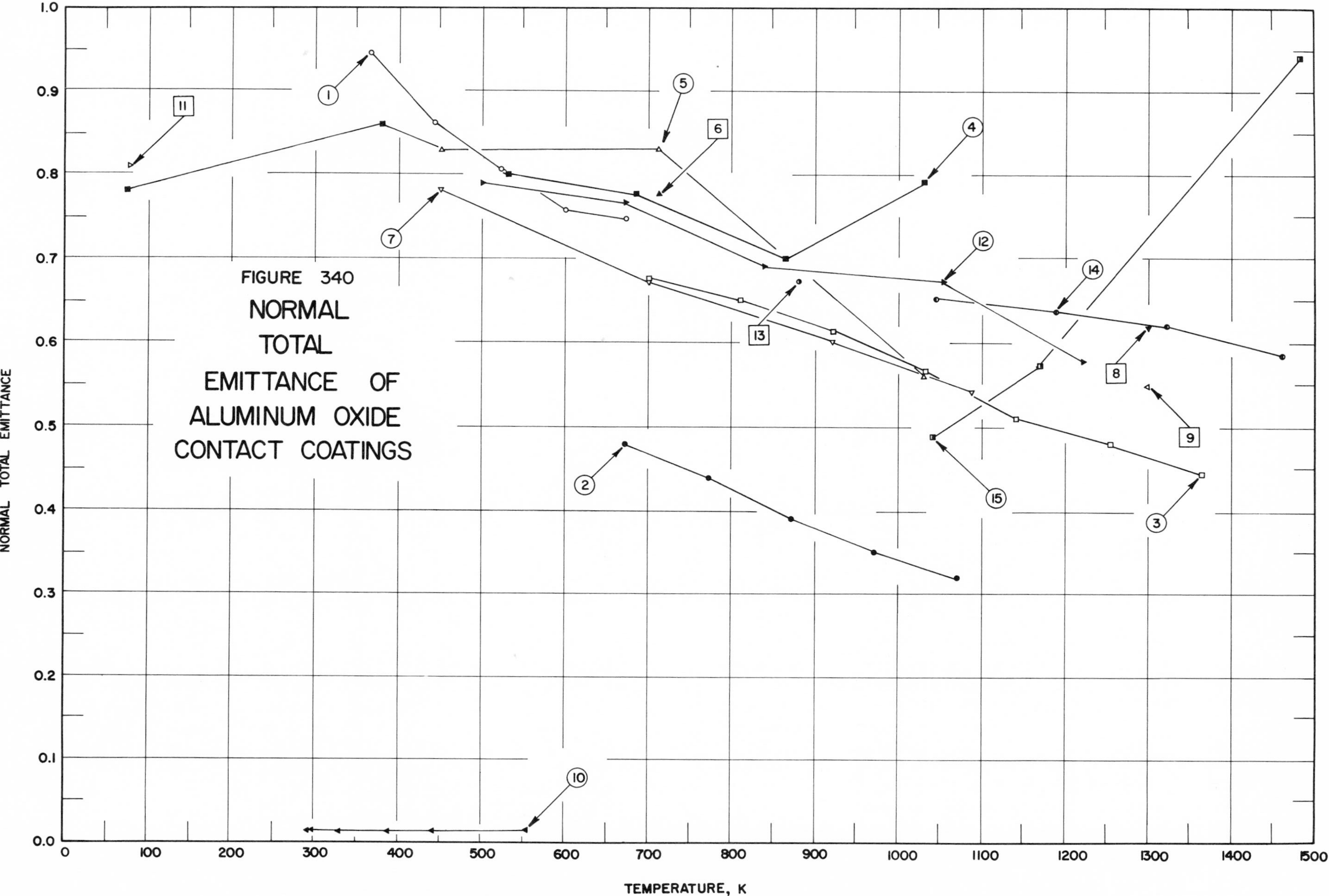

FIGURE 340 NORMAL TOTAL EMITTANCE OF ALUMINUM OXIDE CONTACT COATINGS

SPECIFICATION TABLE NO. 340 NORMAL TOTAL EMITTANCE OF ALUMINUM OXIDE CONTACT COATINGS

Curve No.	Ref. No.	Year	Temperature Range, K	Geometry θ'	Reported Error, %	Composition (weight percent), Specifications and Remarks
1	89	1958	366-672	$\sim 0^\circ$		Rokide A (0.254 mm thick); stainless steel substrate; flame-sprayed. [Author's designation: specimen 18]
2	123	1952	673-1073	$\sim 0^\circ$		Alumina powder; Nimonic 75 substrate.
3	124	1959	700-1366	$\sim 0^\circ$		Aluminum oxide; Inconel substrate; substrate sand blasted and oxidized; flame sprayed.
4	125	1958	75-1033	$\sim 0^\circ$		Rokide A, aluminum oxide; molybdenum substrate; substrate oxidized and volatilized at ~1030 K; increasing temp, cycle 1.
5	125	1958	450-1033	$\sim 0^\circ$		Above specimen and conditions except increasing temp, cycle 2.
6	125	1958	714	$\sim 0^\circ$		Above specimen and conditions except decreasing temp, cycle 2.
7	125	1958	450-1089	$\sim 0^\circ$		Curve 4 specimen and conditions except increasing temp, cycle 3.
8	126	1965	1300	$\sim 0^\circ$		98.55 Al_2O_3 (gamma), 0.58 SiO_2, 0.31 Na_2O, 0.23 MgO, 0.19 CaO, 0.10 Fe_2O_3, and 0.04 TiO_2 (0.305 mm thick); mild steel substrate; flame sprayed; surface roughness 5.72 μm measured with profilometer and optical comparator; density 3.3 g cm^{-3}; porosity 8 to 12%; measured in vacuum (3.5 to 5.0 x 10^{-2} mm Hg); computed from spectral data (0 to 10 μm).
9	126	1965	1300	$\sim 0^\circ$		98.55 Al_2O_3, 0.58 SiO_2, 0.31 Na_2O, 0.23 MgO, 0.19 CaO, 0.10 Fe_2O_3, and 0.04 TiO_2 (0.381 mm thick); mild steel substrate; flame sprayed; polished with polishing papers; density 3.3 g cm^{-3}; porosity 8 to 12%; measured in vacuum (3.5 to 5.0 x 10^{-2} mm Hg); computed from spectral data (0 to 10 μm).
10	119	1967	294-555	$\sim 0^\circ$		Sapphire (0.100 μm thick); silver (0.148 μm thick) and mill-finish 6061-T6 aluminum alloy substrates; high-purity silver vapor-deposited on aluminum, sapphire vapor-deposited on silver; computed from spectral reflectance (2-25 μm).
11	127	1959	77.6	$\sim 0^\circ$		Rokide A, aluminum oxide coating on stainless steel 446 substrate; cycle 1.
12	127	1959	499-1222	$\sim 0^\circ$		Above specimen and conditions.
13	127	1959	883-1570	$\sim 0^\circ$		Above specimen and conditions except cycle 2.
14	127	1959	1046-1460	$\sim 0^\circ$		Above specimen and conditions except cycle 3.
15	128	1961	1044-1483	$\sim 0^\circ$		Al_2O_3, polycrystalline; unknown substrate; cleaned; measured in vacuum (10^{-5} mm Hg); data extracted from smooth curve.

DATA TABLE NO. 340 NORMAL TOTAL EMITTANCE OF ALUMINUM OXIDE CONTACT COATINGS

[Temperature, T, K; Emittance, ϵ]

T	ϵ
CURVE 1	
366	0.945
444	0.863
522	0.805
600	0.758
672	0.698
CURVE 2	
673	0.48
773	0.44
873	0.39
973	0.35
1073	0.32
CURVE 3	
700	0.676
811	0.650
922	0.615
1033	0.565
1144	0.510
1255	0.480
1366	0.445
CURVE 4	
75	0.78
380	0.86
533	0.80
686	0.77
866	0.70
1033	0.79
CURVE 5	
450	0.83
714	0.83
1033	0.56
CURVE 6	
714	0.77
CURVE 7	
450	0.78
700	0.67
922	0.60
1089	0.54
CURVE 8	
1300	0.619
CURVE 9	
1300	0.547
CURVE 10	
294	0.014
300	0.014
333	0.014
389	0.014
444	0.014
555	0.015
CURVE 11	
77.6	0.809
CURVE 12	
499	0.788
671	0.766
838	0.690
1054	0.621
1222	0.577
CURVE 13	
883	0.672
1504	0.477*
1570	0.379*
CURVE 14	
1046	0.651
1190	0.638
CURVE 14 (cont.)	
1324	0.621
1460	0.585
CURVE 15	
1044	0.489
1170	0.571
1483	0.940

* Not shown on plot

SPECIFICATION TABLE NO. 341 NORMAL SPECTRAL EMITTANCE OF ALUMINUM OXIDE CONTACT COATINGS

Curve No.	Ref. No.	Year	Wavelength, μm	Temperature Range, K	Geometry θ'	Reported Error, %	Composition (weight percent), Specifications, and Remarks
1*	127	1959	0.665	1193-1460	$\sim 0°$		Rokide A, aluminum oxide; on stainless steel 446 substrate.

DATA TABLE NO. 341 NORMAL SPECTRAL EMITTANCE OF ALUMINUM OXIDE CONTACT COATINGS

[Wavelength, λ, μm; Emittance, ϵ; Temperature, T, K]

T	ϵ
CURVE 1* $\lambda = 0.665$	
1193	0.444
1328	0.469
1460	0.517

* No plot given

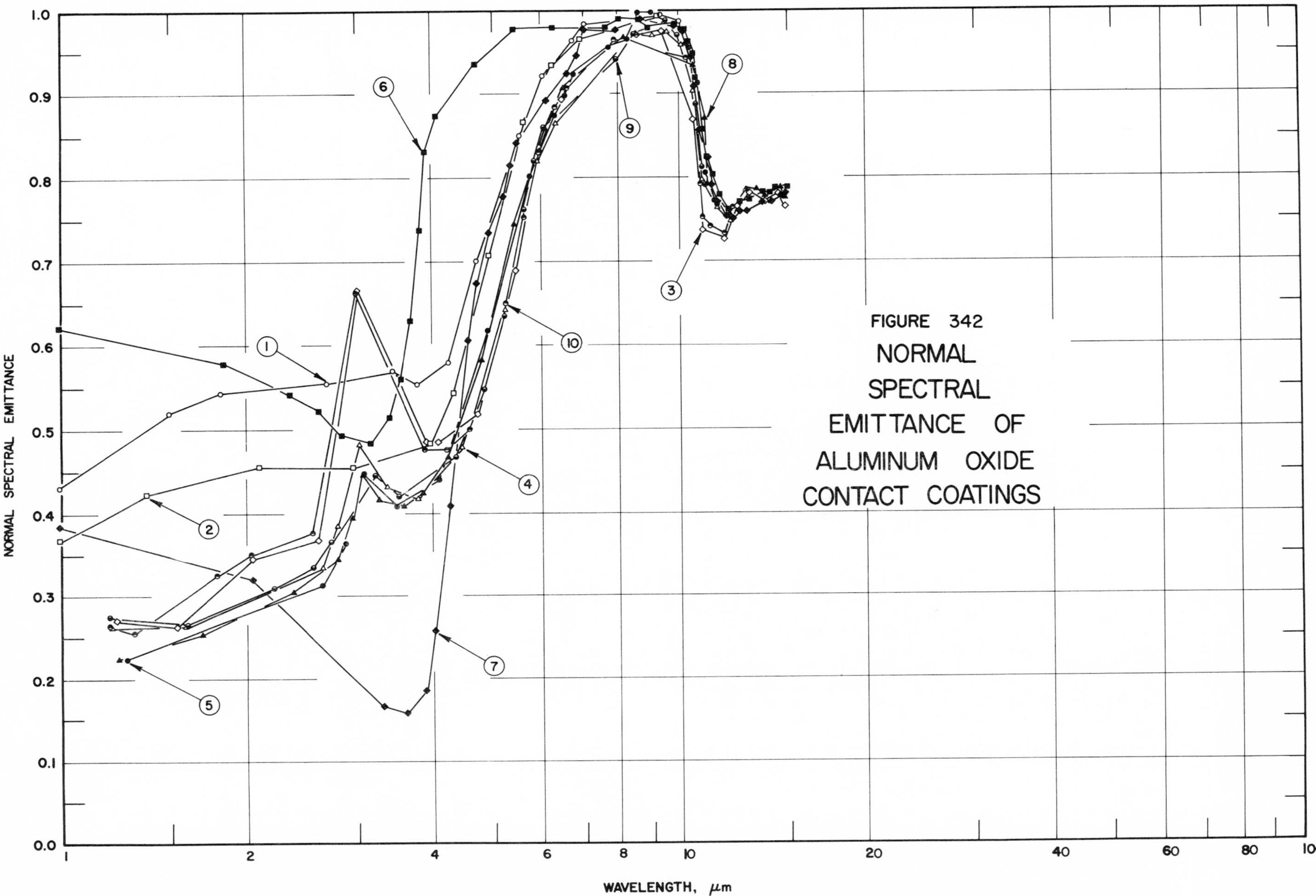

FIGURE 342 NORMAL SPECTRAL EMITTANCE OF ALUMINUM OXIDE CONTACT COATINGS

SPECIFICATION TABLE NO. 342 NORMAL SPECTRAL EMITTANCE OF ALUMINUM OXIDE CONTACT COATINGS

Curve No.	Ref. No.	Year	Temperature, K	Wavelength Range, μm	Geometry θ'	Reported Error, %	Composition (weight percent), Specifications, and Remarks
1	126	1965	1300	1.00-10.01	~0°		98.55 Al_2O_3 (gamma), 0.58 SiO_2, 0.31 Na_2O, 0.23 MgO, 0.19 CaO, 0.10 Fe_2O_3, and 0.04 TiO_2 (0.305 mm thick); mild steel substrate; flame sprayed; surface roughness 5.72 μm measured with a profilometer and optical comparator; density 3.3 g cm^{-3}; porosity 8 to 12%; measured in vacuum (3.0 to 5.0 x 10^{-2} mm Hg); data extracted from smooth curve.
2	126	1965	1300	1.00-10.00	~0°		98.55 Al_2O_3 (gamma), 0.58 SiO_2, 0.31 Na_2O, 0.23 MgO, 0.19 CaO, 0.10 Fe_2O_3, and 0.04 TiO_2 (0.381 mm thick); mild steel substrate; flame sprayed; polished with polishing papers; density 3.3 g cm^{-3}; porosity 8 to 12%; measured in vacuum (3.5 to 5.0 x 10^{-2} mm Hg); data extracted from smooth curve.
3	129	1967	300	1.23-14.88	~0°		Alumina (0.4 mm thick); stainless steel substrate; substrate roughened; flame-sprayed; data extracted from smooth curve.
4	129	1967	882	1.21-14.88	~0°		Above specimen and conditions.
5	129	1967	1074	1.28-14.88	~0°		Above specimen and conditions.
6	130	1966	1255	1.00-15.0	~0°		Rokide A (0.127 mm thick); oxidized Inconel substrate; flame sprayed; data extracted from smooth curve.
7	130	1966	1255	1.00-14.9	~0°		Similar to curve 6 specimen and conditions except 1.067 mm thick.
8	131	1968	1074	1.24-14.8	~0°		Al_2O_3 (0.4 mm thick); stainless steel substrate; substrate roughened; flame-sprayed; data extracted from smooth curve.
9	131	1968	882	1.20-14.8	~0°		Above specimen and conditions.
10	131	1968	300	1.20-14.8	~6°		Above specimen and conditions except calculated from reflectance.

DATA TABLE NO. 342 NORMAL SPECTRAL EMITTANCE OF ALUMINUM OXIDE CONTACT COATINGS

[Wavelength, λ, μm; Emittance, ∈; Temperature, T, K]

λ	∈
CURVE 1	**T = 1300**
1.00	0.431
1.50	0.520
1.82	0.543
2.69	0.556
3.43	0.570
3.78	0.554
4.13	0.579
4.81	0.700
5.56	0.851
6.03	0.921
6.77	0.963
8.06	0.984
9.33	0.995
10.01	0.988
CURVE 2	**T = 1300**
1.00	0.369
1.38	0.423
2.10	0.456
2.97	0.454
3.96	0.483
4.31	0.543
4.93	0.707
5.67	0.863
6.26	0.934
6.96	0.967
9.28	0.994
10.00	0.985
CURVE 3	**T = 300**
1.23	0.270
1.54	0.262
2.04	0.345
2.61	0.367
3.03	0.663
3.91	0.476
4.18	0.476
4.72	0.517
5.46	0.689
CURVE 3 (cont.)	
5.97	0.832
6.51	0.893
8.56	0.972
9.39	0.976
10.09	0.958*
10.55	0.869
10.95	0.737
11.86	0.727
12.48	0.765
13.02	0.781
13.74	0.771
14.50	0.779
14.88	0.765
CURVE 4	**T = 882**
1.21	0.262
1.59	0.262
2.64	0.334
2.80	0.384
3.03	0.482
3.36	0.432
3.77	0.417
4.45	0.479
5.23	0.642
5.91	0.820
6.35	0.866
8.55	0.973
9.09	0.971
9.60	0.975
10.09	0.958
10.58	0.902
10.92	0.797
11.56	0.764
12.13	0.748
12.48	0.765*
13.02	0.781*
13.74	0.771*
14.50	0.779*
14.88	0.765*
CURVE 5	**T = 1074**
1.28	0.224
2.64	0.313
2.89	0.364
3.09	0.448
3.49	0.409
4.09	0.439
4.92	0.617
5.76	0.802
6.32	0.876
6.59	0.898
6.80	0.922
7.76	0.954
8.30	0.967
8.61	0.998
9.05	0.998
10.09	0.978
10.40	0.954
10.70	0.913
11.08	0.806
11.46	0.771
12.13	0.748*
12.48	0.765*
13.02	0.781*
13.74	0.771*
14.50	0.779*
14.88	0.765*
CURVE 6	**T = 1255**
1.00	0.623
1.84	0.579
2.35	0.542
2.62	0.522
2.84	0.493
3.07	0.484
3.40	0.515
3.54	0.560
3.68	0.630
3.82	0.737
3.89	0.832
4.06	0.874
4.71	0.937
CURVE 6 (cont.)	
5.42	0.978
6.27	0.980
7.66	0.979
8.02	0.992
8.64	0.989
8.93	0.979
9.40	0.976*
9.87	0.982
10.14	0.977
10.36	0.961
10.59	0.947
10.68	0.918
10.92	0.857
11.16	0.824
11.38	0.802
11.66	0.779
12.01	0.762
12.54	0.771
13.01	0.773
13.69	0.782
14.01	0.781
14.33	0.788
15.00	0.789
CURVE 7	**T = 1255**
1.00	0.385
2.04	0.320
3.31	0.166
3.61	0.157
3.88	0.185
4.01	0.257
4.26	0.410
4.57	0.616
4.71	0.674
4.93	0.734
5.21	0.777
5.36	0.815
5.49	0.842
6.11	0.893
6.63	0.925
6.88	0.946
7.08	0.977
CURVE 7 (cont.)	
7.98	0.977
8.67	0.990
9.83	0.981
10.12	0.975
10.30	0.960
10.45	0.943
10.60	0.917
10.87	0.856
11.19	0.824
11.32	0.791
11.62	0.769
11.96	0.753
12.29	0.750
12.55	0.758
12.94	0.758
13.67	0.770
14.14	0.770
14.69	0.778
14.93	0.781
CURVE 8	**T = 1074**
1.24	0.225
1.69	0.253
2.37	0.304
2.80	0.344
2.96	0.395
3.07	0.446
3.27	0.417
3.57	0.409
3.84	0.425
4.22	0.467
4.80	0.582
5.42	0.745
6.10	0.860
6.52	0.907
8.19	0.969
8.66	1.002*
9.05	1.005*
9.96	0.985*
10.42	0.963*
10.61	0.934
11.04	0.821*
CURVE 8 (cont.)	
11.25	0.792
11.62	0.771*
12.12	0.756
12.84	0.786
13.34	0.786
13.74	0.777
14.39	0.788*
14.56	0.788
14.86	0.774
CURVE 9	**T = 882**
1.20	0.274
1.60	0.266
2.21	0.310
2.56	0.335
2.73	0.365
3.02	0.481*
3.22	0.445
3.51	0.421
3.84	0.424*
4.35	0.468
4.85	0.547
5.23	0.635
5.63	0.763
5.85	0.820
6.17	0.861
7.98	0.941
8.47	0.972
9.09	0.972*
9.54	0.988
9.96	0.971
10.34	0.941*
10.64	0.887
10.91	0.813
11.05	0.791
11.52	0.769
12.09	0.755
12.84	0.786*
13.34	0.786*
13.74	0.777*
14.39	0.788*
14.56	0.788*
CURVE 9 (cont.)	
14.86	0.774*
CURVE 10	**T = 300**
1.20	0.264
1.33	0.255
1.78	0.326
2.03	0.350
2.55	0.376
3.01	0.663
3.87	0.476
4.21	0.476
4.57	0.500
5.25	0.650
5.62	0.754
5.96	0.831
6.32	0.885
6.62	0.906
7.90	0.964
8.50	0.973*
9.01	0.973*
9.52	0.988*
9.98	0.970*
10.28	0.942
10.81	0.792
10.91	0.752
11.21	0.740
11.81	0.733
12.28	0.764
12.84	0.786*
13.34	0.786*
13.74	0.777*
14.39	0.788*
14.56	0.788*
14.86	0.774*

* Not shown on plot

SPECIFICATION TABLE NO. 343 ANGULAR INTEGRATED REFLECTANCE OF ALUMINUM OXIDE CONTACT COATINGS

Curve No.	Ref. No.	Year	Temperature, K	Angular Range, °	Geometry θ	Geometry θ'	Geometry ω'	Reported Error, %	Composition (weight percent), Specifications, and Remarks
1*	132	1965	298	-40-+80	65°	θ'			Sapphire; aluminum substrate; substrate sandblasted; sapphire pure; flame sprayed; measured in plane of incidence; tungsten source.
2*	132	1965	298	-25-+80	45°	θ'			Aluminum oxide; aluminum substrate; substrate sandblasted; aluminum oxide commercially pure; flame sprayed; measured in plane of incidence; tungsten source.
3*	132	1965	298	-40-+80	65°	θ'			Aluminum oxide; aluminum substrate; substrate sandblasted; aluminum oxide commercially pure; flame sprayed; measured in plane of incidence; tungsten source.
4*	132	1965	298	-40-+80	45°	θ'			Sapphire (~0.38 mm thick); aluminum substrate; substrate sandblasted; sapphire pure; flame sprayed; measured in plane of incidence; tungsten source.

DATA TABLE NO. 343 ANGULAR INTEGRATED REFLECTANCE OF ALUMINUM OXIDE CONTACT COATINGS

[Angle, *°; Reflectance, ρ; Temperature, T, K]

θ'	ρ
CURVE 1*	
T = 298	
-40	0.29
-30	0.31
-20	0.32
-10	0.35
0	0.36
20	0.36
30	0.36
40	0.35
50	0.33
60	0.31
70	0.25
80	0.16
CURVE 2*	
T = 298	
-25	0.45
-20	0.47
-15	0.49
-10	0.50
- 5	0.50
0	0.50
5	0.50
10	0.50
15	0.49
20	0.49
25	0.48
30	0.47
35	0.45
40	0.42
50	0.38
55	0.34
CURVE 2 (cont.)*	
60	0.30
65	0.30
70	0.20
75	0.16
80	0.10
CURVE 3*	
T = 298	
-40	0.31
-30	0.34
-20	0.37
-10	0.39
0	0.40
10	0.40
20	0.41
30	0.41
CURVE 3 (cont.)*	
40	0.41
50	0.41
60	0.40
70	0.35
80	0.22
CURVE 4*	
T = 298	
-20	0.41
-10	0.44
0	0.45
10	0.46
20	0.45
30	0.44
40	0.40
50	0.37
CURVE 4 (cont.)*	
60	0.30
70	0.22
80	0.11

* No plot given

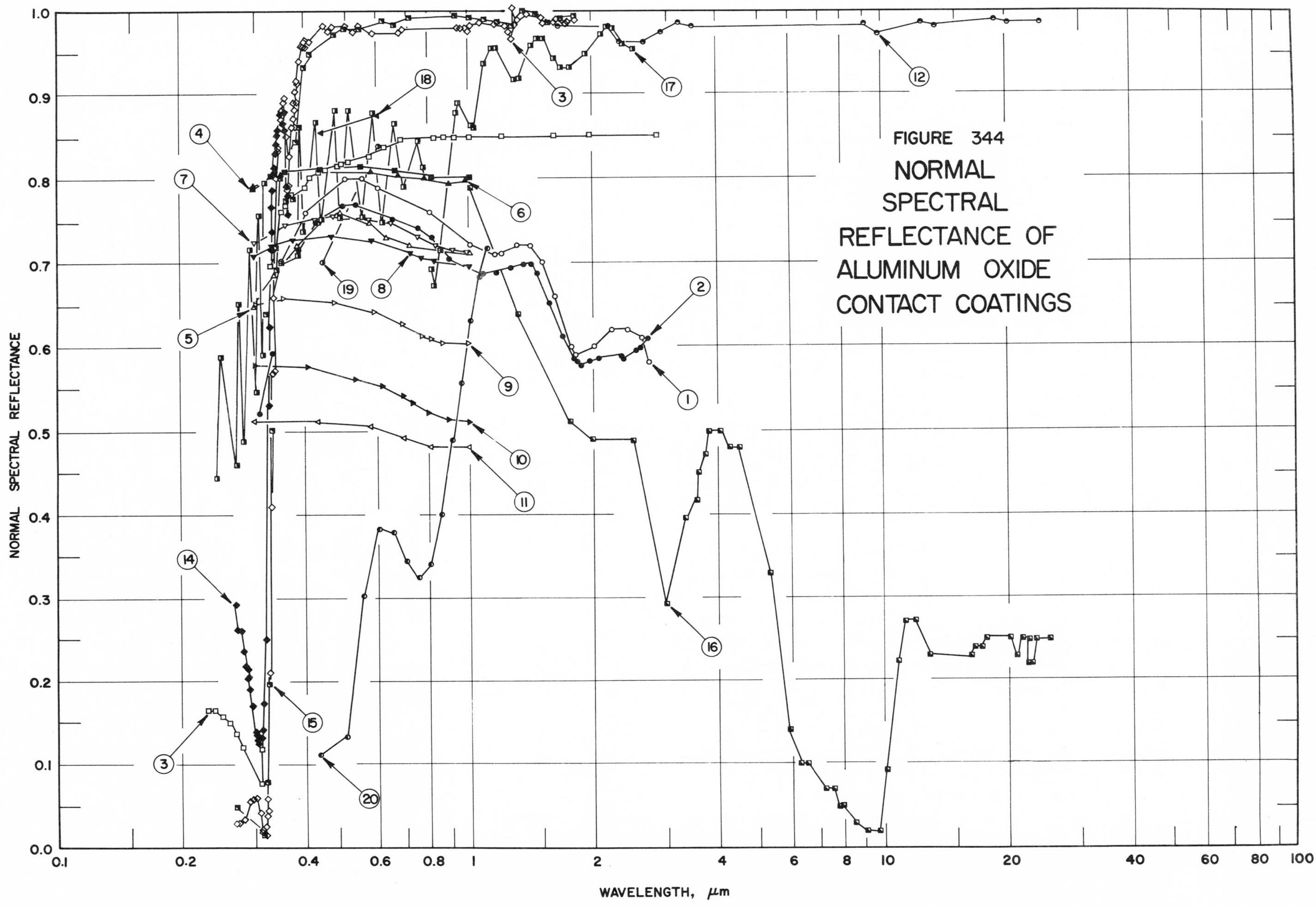
FIGURE 344
NORMAL SPECTRAL REFLECTANCE OF ALUMINUM OXIDE CONTACT COATINGS
NORMAL SPECTRAL REFLECTANCE
WAVELENGTH, μm
0.0
0.1
0.2
0.3
0.4
0.5
0.6
0.7
0.8
0.9
1.0
0.1
0.2
0.4
0.6
0.8
1
2
4
6
8
10
20
40
60
80
100

SPECIFICATION TABLE NO. 344 NORMAL SPECTRAL REFLECTANCE OF ALUMINUM OXIDE CONTACT COATINGS

Curve No.	Ref. No.	Year	Temperature, K	Wavelength Range, μm	Geometry θ	θ'	ω'	Reported Error, %	Composition (weight percent), Specifications, and Remarks
1	125	1958	298	0.30-2.70	~0°		2π	4	Aluminum oxide, Rokide A; molybdenum substrate.
2	125	1958	298	0.309-2.699	9°		2π	4	Aluminum oxide, Rokide A; stainless steel 446 substrate; measured relative to magnesium carbonate.
3	133	1966	298	0.23-2.88	~5°	~5°			Sapphire; silver substrate; silver vapor deposited.
4	132	1965	298	0.300-1.000	~0°		2π		Sapphire (0.25 mm thick); substrate unknown; flame-sprayed.
5	132	1965	298	0.300-1.000	~0°		2π		Aluminum oxide; substrate unknown; aluminum oxide commercially pure; flame sprayed.
6	132	1965	298	0.300-0.979	~0°		2π		Sapphire (0.25 mm thick); substrate unknown; flame sprayed.
7	132	1965	298	0.300-0.994	~0°		2π		Similar to above specimen and conditions except 0.18 mm thick.
8	132	1965	298	0.300-0.995	~0°		2π		Similar to curve 6 specimen and conditions except 0.13 mm thick.
9	132	1965	298	0.300-0.993	~0°		2π		Similar to curve 6 specimen and conditions except 0.08 mm thick.
10	132	1965	298	0.300-0.994	~0°		2π		Similar to curve 6 specimen and conditions except 0.05 mm thick.
11	132	1965	298	0.300-0.994	~0°		2π		Similar to curve 6 specimen and conditions except 0.025 mm thick
12	119	1967	~298	1.75-24.05	15°		2π		Sapphire (0.100 μm thick); silver (0.148μm thick) and mill-finish 6061-T6 aluminum alloy substrates; high-purity silver vapor-deposited on aluminum, sapphire vapor-deposited on silver.
13	122	1968	~298	0.270-1.80	~0°	~0°			Al_2O_3 (0.0955 μm thick); Ag (0.100 μm thick) and fused silica substrates; fused silica cleaned with ultrasonic cleaner and freon emulsion; Ag vacuum deposited on fused silica, Al_2O_3 vacuum deposited on Ag; data extracted from smooth curve.
14	122	1968	~298	0.270-1.80	~0°	~0°			Similar to above specimen and conditions except Al_2O_3 0.0550 μm thick.
15	122	1968	~298	0.271-1.80	~0°	~0°			Similar to curve 13 specimen and conditions except Al_2O_3 0.0210 μm thick.
16	89	1958	~298	0.40-25.0	~0°	~0°			Rokide A (0.254 mm thick); 321 stainless steel substrate; flame-sprayed; data extracted from smooth curve. [Author's designation: Specimen 18]
17	85	1969	~298	0.254-2.500	~0°		2π		Sapphire (~0.002 mm thick); anodized 1199 aluminum substrate; vapor deposited; aluminum substrate alkaline electropolished (sodium phosphate and sodium carbonate) for 15 min at 353 K and 12 VDC, then anodized for 15 min at 18 VDC in 10% sulfuric acid; data extracted from smooth curve. [Author's designation: A-3 Vapor Deposited Sapphire (Al_2O_3)]
18	20	1963	298	0.440-0.600	~0°		2π		Rokide A; substrate unknown; flame sprayed.
19	20	1963	298	0.440-0.600	~0°		2π		Above specimen and conditions except exposed to UV irradiation; 615 ESH with solar factor 8.9.
20	153	1963	~298	0.430-1.100	~0°		2π		Aluminum oxide on Englehard Industries Film No. 94, aluminum oxide, and honed Inconel (surface finish 0.051-0.102 μm) substrates; aluminum oxide diffusion barrier applied to Inconel by spraying 4 coats of mixture of aluminum resinate, aromatic solvents, and essential oils, then firing to 873 K in air through a continuous lehr on a 1.5 hr cycle; film No. 94 (89.5 Au + 0.4 Rh + 4.5 Bi_2O_3 + 0.2 Cr_2O_3 + 1.7 SiO_2 + 3.7 BaO) applied by spraying solution of gold sulforesinate, rhodium sulforesinate, and bismuth, chromium, and barium resinates, diluted with aromatic solvents and essential oils, then firing to 873 K in air through a continuous lehr on a 1.5 hr cycle; top aluminum oxide coating applied in 2 coats in the same manner as the diffusion barrier; measured relative to MgO; converted from R (2π, ~0°).

DATA TABLE NO. 344 NORMAL SPECTRAL REFLECTANCE OF ALUMINUM OXIDE CONTACT COATINGS

[Wavelength, λ, μm; Reflectance, ρ; Temperature, T, K]

CURVE 1
T = 298

λ	ρ
0.30	0.58
0.35	0.70
0.40	0.76
0.50	0.80
0.55	0.80
0.60	0.79
0.80	0.76
1.00	0.72
1.15	0.71
1.20	0.71
1.30	0.72
1.40	0.72
1.50	0.70
1.60	0.66
1.75	0.60
1.80	0.59
2.00	0.60
2.20	0.62
2.40	0.62
2.60	0.61
2.70	0.58

CURVE 2
T = 298

λ	ρ
0.309	0.520
0.333	0.594
0.386	0.715
0.424	0.749
0.492	0.769
0.527	0.770
0.651	0.754
0.749	0.744
0.812	0.732
0.892	0.705
0.963	0.693*
1.079	0.687
1.155	0.689
1.261	0.696
1.357	0.698
1.401	0.698
1.457	0.687
1.550	0.653
1.676	0.613
1.789	0.585

CURVE 2 (cont.)

λ	ρ
1.806	0.583
1.836	0.577
1.956	0.582
2.051	0.585
2.332	0.587
2.361	0.585
2.518	0.595
2.599	0.599
2.699	0.610

CURVE 3
T = 298

λ	ρ
0.23	0.163
0.24	0.163
0.25	0.156
0.26	0.147
0.27	0.136
0.28	0.119
0.31	0.076
0.31	0.117
0.33	0.696
0.34	0.718
0.35	0.762
0.36	0.769
0.37	0.782
0.40	0.791
0.41	0.801
0.43	0.809
0.45	0.812
0.48	0.815
0.49	0.818
0.51	0.820
0.57	0.828
0.62	0.838
0.68	0.848
0.82	0.849
0.87	0.849
0.92	0.849
1.00	0.849
1.20	0.850
1.61	0.850
1.96	0.852
2.88	0.852

CURVE 4
T = 298

λ	ρ
0.300	0.790
0.331	0.803
0.359	0.809
0.437	0.813
0.545	0.815
0.658	0.810
0.813	0.803
1.000	0.803

CURVE 5
T = 298

λ	ρ
0.300	0.648
0.336	0.685
0.381	0.722
0.433	0.749
0.485	0.761
0.573	0.747
0.629	0.732
0.717	0.722
1.000	0.711

CURVE 6
T = 298

λ	ρ
0.300	0.793
0.354	0.805*
0.431	0.810
0.579	0.810
0.674	0.805
0.784	0.800
0.894	0.795
0.979	0.797

CURVE 7
T = 298

λ	ρ
0.300	0.726
0.358	0.746
0.421	0.753
0.467	0.755
0.571	0.752
0.645	0.747
0.748	0.732
0.832	0.720

CURVE 7 (cont.)

λ	ρ
0.903	0.715
0.994	0.713

CURVE 8
T = 298

λ	ρ
0.300	0.708
0.331	0.720
0.373	0.728
0.464	0.731
0.578	0.727
0.719	0.713
0.761	0.706
0.824	0.700
0.995	0.695

CURVE 9
T = 298

λ	ρ
0.300	0.651
0.353	0.657
0.467	0.654
0.581	0.642
0.686	0.628
0.760	0.614
0.803	0.608
0.860	0.605
0.993	0.605

CURVE 10
T = 298

λ	ρ
0.300	0.578
0.404	0.575
0.523	0.561
0.612	0.554
0.682	0.543
0.724	0.533
0.791	0.522
0.881	0.515
0.994	0.513

CURVE 11
T = 298

λ	ρ
0.300	0.512
0.428	0.512
0.570	0.505
0.693	0.494
0.802	0.481
0.994	0.481

CURVE 12
T ~ 298

λ	ρ
1.75	0.981
2.18	0.981
2.32	0.964
2.65	0.964
2.94	0.975
3.24	0.986
3.49	0.981
9.14	0.984
9.86	0.972
12.59	0.986
13.49	0.981
18.71	0.988
20.04	0.985
24.05	0.985

CURVE 13
T ~ 298

λ	ρ
0.270	0.028
0.275	0.028*
0.282	0.034*
0.294	0.056
0.297	0.058
0.302	0.058
0.308	0.042
0.313	0.022
0.314	0.018
0.318	0.015
0.320	0.026
0.320	0.038
0.322	0.057
0.323	0.045
0.326	0.210
0.329	0.411
0.333	0.567

CURVE 13 (cont.)

λ	ρ
0.333	0.570
0.335	0.657
0.337	0.717
0.341	0.801
0.344	0.838
0.345	0.835
0.348	0.862
0.351	0.873
0.352	0.882
0.353	0.860
0.354	0.893
0.355	0.878
0.356	0.895
0.357	0.887*
0.359	0.878*
0.362	0.852
0.364	0.787
0.366	0.793
0.366	0.827
0.371	0.862*
0.373	0.872*
0.374	0.862*
0.375	0.877
0.376	0.873*
0.377	0.890
0.378	0.884
0.380	0.902
0.382	0.891
0.383	0.914
0.383	0.911
0.389	0.937
0.395	0.956
0.398	0.956
0.398	0.952
0.399	0.962
0.400	0.955
0.400	0.966
0.408	0.962
0.443	0.981
0.456	0.975
0.467	0.980
0.502	0.981
0.523	0.974
0.542	0.980
0.581	0.973
0.678	0.974

CURVE 13 (cont.)

λ	ρ
0.690	0.977
0.944	0.978
0.958	0.974
0.980	0.978
0.955	0.976
1.00	0.981
1.06	0.985
1.15	0.984
1.23	0.983
1.25	0.975
1.26	0.965
1.27	1.000
1.28	0.983
1.31	0.987
1.32	0.983
1.37	0.995
1.38	0.988
1.40	0.992
1.50	0.992
1.52	0.985
1.62	0.985
1.63	0.988
1.67	0.984
1.69	0.989
1.72	0.986
1.75	0.988
1.78	0.987
1.79	0.989
1.80	0.987

CURVE 14
T ~ 298

λ	ρ
0.270	0.294
0.274	0.262
0.277	0.257
0.280	0.236
0.285	0.217
0.287	0.215
0.288	0.203
0.290	0.206
0.293	0.189
0.296	0.168
0.301	0.137
0.304	0.134
0.305	0.127

CURVE 14 (cont.)

λ	ρ
0.306	0.124
0.311	0.130
0.313	0.140
0.316	0.174
0.319	0.250
0.325	0.530
0.328	0.626
0.331	0.716
0.332	0.737*
0.333	0.713
0.333	0.767
0.335	0.787
0.337	0.808
0.338	0.814
0.339	0.830
0.341	0.843
0.343	0.852
0.344	0.859
0.350	0.877
0.351	0.874
0.353	0.854
0.354	0.879
0.355	0.866
0.357	0.863*
0.359	0.855
0.363	0.793
0.363	0.758
0.364	0.782
0.364	0.744
0.367	0.797
0.367	0.783
0.369	0.803
0.376	0.813
0.376	0.825
0.378	0.804
0.379	0.843
0.380	0.837
0.385	0.879
0.396	0.920
0.416	0.931
0.447	0.947
0.495	0.962
0.508	0.953
0.513	0.962
0.604	0.970
0.695	0.980

* Not shown on plot

DATA TABLE NO. 344 NORMAL SPECTRAL REFLECTANCE OF ALUMINUM OXIDE CONTACT COATINGS (continued)

λ	ρ
CURVE 14 (cont.) T ~ 298	
0.723	0.986
0.828	0.989
0.942	0.986
0.947	0.985
0.961	0.985
0.971	0.987
0.983	0.982
0.994	0.988
1.04	0.992
1.09	0.995
1.11	0.991
1.14	0.993
1.19	0.990
1.22	0.990
1.25	0.986
1.26	0.967
1.27	1.000
1.28	0.990
1.29	0.988
1.35	0.997
1.35	0.990
1.37	1.000
1.38	0.987
1.39	0.998
1.48	0.997
1.52	0.988
1.54	0.995
1.56	0.986
1.61	0.992
1.63	0.989
1.76	0.989
1.77	0.986
1.80	0.990
CURVE 15 T ~ 298	
0.271	0.046
0.285	0.042*
0.292	0.036*
0.299	0.032*
0.306	0.018*
0.311	0.013*
0.314	0.016*
0.318	0.033*
0.321	0.078

λ	ρ
CURVE 15 (cont.)	
0.325	0.196
0.332	0.502
0.333	0.526*
0.333	0.521*
0.335	0.610*
0.340	0.693
0.343	0.743*
0.347	0.775*
0.351	0.800
0.353	0.792*
0.354	0.814*
0.355	0.802*
0.356	0.808*
0.357	0.798*
0.358	0.802*
0.361	0.777
0.362	0.752*
0.364	0.718*
0.365	0.722*
0.367	0.747*
0.368	0.741*
0.373	0.784*
0.374	0.777
0.377	0.811*
0.378	0.807*
0.380	0.845
0.381	0.841*
0.383	0.870*
0.388	0.897*
0.393	0.925*
0.397	0.930
0.398	0.946*
0.400	0.937*
0.416	0.945
0.424	0.957*
0.471	0.970
0.481	0.976*
0.500	0.978
0.510	0.973*
0.528	0.979*
0.543	0.977
0.561	0.984*
0.576	0.978*
0.618	0.986
0.660	0.984
0.717	0.993
0.931	0.996

λ	ρ
CURVE 15 (cont.)	
0.943	0.990*
1.00	0.990
1.06	0.995*
1.09	0.989
1.11	0.992*
1.18	0.986
1.22	0.990*
1.24	0.980*
1.25	0.956*
1.26	1.003*
1.28	0.982
1.32	0.995*
1.34	0.991*
1.35	0.998
1.37	0.962*
1.37	0.990*
1.39	0.997*
1.46	0.995
1.53	0.990*
1.56	0.985
1.67	0.990
1.71	0.988*
1.76	0.989*
1.80	0.990
CURVE 16	
0.40	0.813*
0.47	0.850*
0.68	0.850*
0.69	0.831*
0.72	0.828*
0.78	0.830*
0.80	0.819*
0.84	0.819*
0.86	0.809*
0.92	0.809*
0.94	0.800*
0.98	0.800*
1.00	0.791
1.32	0.640
1.76	0.512
1.98	0.491
2.48	0.489
2.98	0.294
3.30	0.397
3.52	0.419

λ	ρ
CURVE 16 (cont.)	
3.55	0.451
3.69	0.473
3.77	0.500
4.03	0.500
4.27	0.481
4.49	0.481
5.32	0.334
5.93	0.143
6.28	0.101
6.49	0.101
7.20	0.070
7.56	0.070
7.78	0.050
7.97	0.050
8.54	0.030
8.81	0.030
9.02	0.021
9.75	0.021
10.2	0.092
10.7	0.225
11.3	0.273
11.8	0.273
12.8	0.232
16.2	0.231
16.5	0.241
17.2	0.241*
17.5	0.252*
20.3	0.253*
20.7	0.231
20.9	0.231*
21.2	0.250*
21.6	0.250
22.0	0.223
22.2	0.250*
22.5	0.250
22.8	0.222*
23.0	0.250*
25.0	0.250
CURVE 17 T ~ 298	
0.254	0.444
0.257	0.587
0.272	0.459
0.276	0.653
0.282	0.487

λ	ρ
CURVE 17 (cont.)	
0.293	0.716
0.302	0.545
0.308	0.756
0.316	0.590
0.320	0.796
0.325	0.640
0.341	0.835*
0.350	0.700
0.357	0.847*
0.381	0.709
0.388	0.862
0.398	0.738
0.426	0.868
0.438	0.754
0.471	0.882
0.484	0.754
0.514	0.881
0.548	0.755
0.580	0.879
0.614	0.749
0.659	0.867
0.694	0.744
0.750	0.845
0.773	0.816
0.805	0.694
0.819	0.673
0.849	0.716
0.921	0.879
0.939	0.890
1.002	0.864
1.028	0.860
1.093	0.936
1.131	0.954
1.167	0.954
1.288	0.918
1.314	0.918
1.418	0.956
1.467	0.966
1.513	0.966
1.601	0.941
1.667	0.930
1.768	0.930
1.912	0.948
2.088	0.971
2.143	0.978
2.243	0.978
2.367	0.959
2.500	0.953

λ	ρ
CURVE 18 T = 298	
0.440	0.855
0.600	0.875
CURVE 19 T = 298	
0.440	0.700
0.600	0.840
CURVE 20 T ~ 298	
0.430	0.112
0.500	0.133
0.550	0.304
0.600	0.382
0.650	0.378
0.700	0.346
0.750	0.326
0.800	0.341
0.850	0.401
0.900	0.489
0.950	0.555
1.000	0.633
1.050	0.684
1.100	0.718

* Not shown on plot

SPECIFICATION TABLE NO. 345 CHANGE IN NORMAL SPECTRAL REFLECTANCE OF ALUMINUM OXIDE CONTACT COATINGS

Curve No.	Ref. No.	Year	Temperature, K	Wavelength Range, μm	Geometry θ	θ'	ω'	Reported Error, %	Composition (weight percent), Specifications, and Remarks
1*	46	1969	295	2.100	~0°		2π		Al_2O_3 (1.1 μm thick); aluminum (0.1 μm thick) substrate; Al then Al_2O_3 vapor-deposited; exposed to 50 keV electrons (4×10^{8}–1.7×10^{12} e cm^{-2} sec^{-1}) in vacuum (10^{-6} mm Hg); vacuum maintained by ion pump; electron fluence (e cm^{-2}) is variable; reflectance measured in situ; positive change indicates decrease in reflectance from preirradiation, in air reflectance. [Authors' designation: G]
2*	46	1969	295	0.590	~0°		2π		Above specimen and conditions.

DATA TABLE NO. 345 CHANGE IN NORMAL SPECTRAL REFLECTANCE OF ALUMINUM OXIDE CONTACT COATINGS

[Wavelength, λ, μm; Reflectance, ρ; Temperature, T, K]

Electron fluence	Δρ
CURVE 1*	
λ = 2.100	
T = 295	
1×10^{13}	0
2×10^{14}	0
6×10^{14}	0
8×10^{14}	0
CURVE 2*	
λ = 0.590	
T = 295	
1×10^{13}	0
2×10^{14}	0
6×10^{14}	0
8×10^{14}	0

* No plot given

SPECIFICATION TABLE NO. 346 ANGULAR SPECTRAL REFLECTANCE OF ALUMINUM OXIDE CONTACT COATINGS

Curve No.	Ref. No.	Year	Temperature K	Wavelength, μm	Angular Range, °	Geometry θ	θ'	ω'	Reported Error, %	Composition (weight percent), Specifications and Remarks
1*	132	1965	298	~0.56	-20-+80	45°	θ'			Sapphire (~0.38 mm thick); aluminum substrate; substrate sand blasted; sapphire pure; flame sprayed; measured in plane of incidence; tungsten filament source used with green filter (λ ~0.56 μm).
2*	132	1965	298	~0.46	-20-+80	45°	θ'			Curve 1 specimen and conditions except used blue filter (λ ~0.46 μm).
3*	132	1965	298	~0.59	-20-+80	45°	θ'			Curve 1 specimen and conditions except used amber filter (λ ~0.59 μm).

DATA TABLE NO. 346 ANGULAR SPECTRAL REFLECTANCE OF ALUMINUM OXIDE CONTACT COATINGS

[Angle, *,°; Reflectance, ρ; Temperature, T, K; Wavelength, λ, μm]

θ'	ρ	θ'	ρ	θ'	ρ
CURVE 1* T = 298 λ ~ 0.56		CURVE 2* T = 298 λ ~ 0.46		CURVE 3* T = 298 λ ~ 0.59	
-20	0.039	-20	0.013	-20	0.035
-10	0.041	-10	0.013	-10	0.037
0	0.044	0	0.014	0	0.038
10	0.044	10	0.014	10	0.038
20	0.044	20	0.014	20	0.037
30	0.044	30	0.014	30	0.036
40	0.039	50	0.012	40	0.034
50	0.035	60	0.009	50	0.030
60	0.029	70	0.007	60	0.024
70	0.020	80	0.004	70	0.018
80	0.010			80	0.008

* No plot given

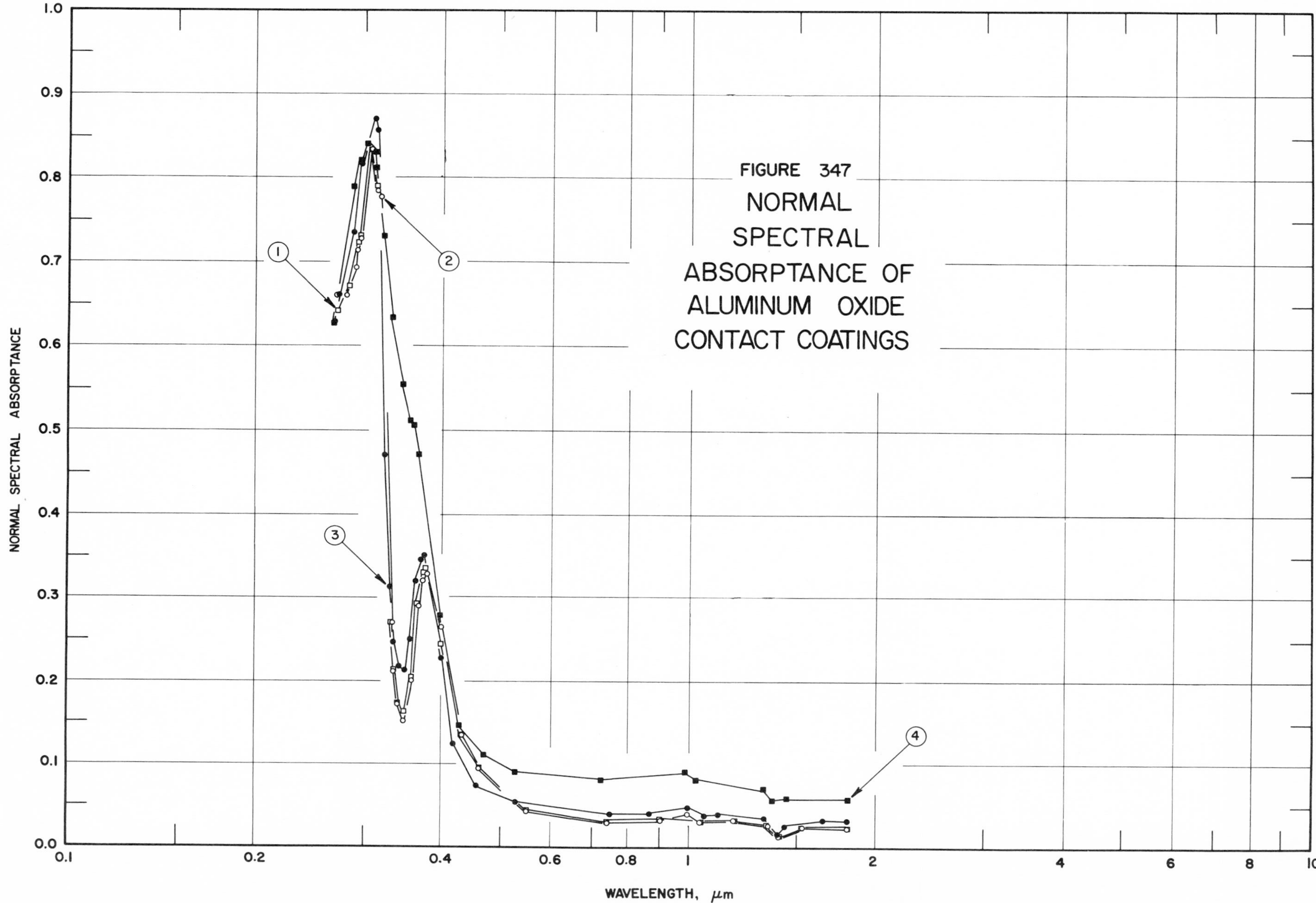
FIGURE 347
NORMAL
SPECTRAL
ABSORPTANCE OF
ALUMINUM OXIDE
CONTACT COATINGS
NORMAL SPECTRAL ABSORPTANCE
WAVELENGTH, μm
0.0
0.1
0.2
0.3
0.4
0.5
0.6
0.7
0.8
0.9
1.0
0.1
0.2
0.4
0.6
0.8
1
2
4
6
8
10
1
2
3
4

SPECIFICATION TABLE NO. 347 NORMAL SPECTRAL ABSORPTANCE OF ALUMINUM OXIDE CONTACT COATINGS

Curve No.	Ref. No.	Year	Temperature, K	Wavelength Range, μm	Geometry θ	Reported Error, %	Composition (weight percent), Specifications, and Remarks
1	119	1967	~298	0.271-1.800	~0°		Sapphire (0.100 μm thick); silver (0.148 μm thick) and mill-finish 6061-T6 aluminum alloy substrates; high-purity silver vapor-deposited on aluminum, sapphire vapor-deposited on silver; data extracted from smooth curve.
2	119	1967	~298	0.270-1.800	~0°		Above specimen and conditions except ultraviolet irradiated in vacuum at 294 K with 10-sun intensity for 200 hrs.
3	119	1967	~298	0.267-1.800	~0°		Similar to curve 1 specimen and conditions.
4	119	1967	~298	0.267-1.800	~0°		Above specimen and conditions except ultraviolet irradiated in vacuum at 533 K with 10-sun intensity for 200 hrs.

DATA TABLE NO. 347 NORMAL SPECTRAL ABSORPTANCE OF ALUMINUM OXIDE CONTACT COATINGS

[Wavelength, λ, μm; Absorptance, α; Temperature, T, K]

CURVE 1
T ~298

λ	α
0.271	0.643
0.283	0.671
0.293	0.724
0.296	0.734
0.308	0.835
0.313	0.793
0.318	0.781*
0.331	0.270
0.335	0.214
0.340	0.174
0.347	0.162
0.357	0.204
0.366	0.294
0.374	0.331
0.378	0.335
0.400	0.245
0.431	0.135
0.460	0.097
0.550	0.046
0.740	0.031
0.900	0.033
0.992	0.043*
1.047	0.032
1.175	0.035
1.344	0.028
1.370	0.014
1.525	0.025
1.800	0.024

CURVE 2
T ~298

λ	α
0.270	0.660
0.280	0.660
0.290	0.696
0.292	0.715
0.295	0.727
0.307	0.836
0.314	0.785
0.319	0.779
0.331	0.270
0.334	0.213
0.339	0.173
0.347	0.153
0.357	0.201
0.368	0.290
0.374	0.322
0.380	0.329
0.400	0.265
0.432	0.135
0.461	0.097
0.551	0.046
0.740	0.031
0.900	0.033
0.992	0.043
1.047	0.032
1.175	0.035
1.344	0.028
1.370	0.014
0.525	0.025
1.800	0.024

CURVE 3
T ~298

λ	α
0.267	0.629
0.273	0.662
0.287	0.736
0.296	0.818
0.311	0.872
0.313	0.857
0.324	0.473
0.330	0.313
0.336	0.248
0.341	0.219
0.348	0.216
0.355	0.251
0.364	0.322
0.370	0.347
0.376	0.352
0.400	0.229
0.420	0.126
0.457	0.076
0.526	0.056
0.749	0.043
0.867	0.043
0.995	0.048
1.055	0.039
1.125	0.042
1.334	0.037
1.388	0.017
1.431	0.027
1.641	0.034
1.800	0.034

CURVE 4
T ~298

λ	α
0.267	0.629
0.289	0.793*
0.296	0.823
0.303	0.843
0.311	0.833
0.314	0.813
0.323	0.734
0.333	0.636
0.345	0.556*
0.356	0.511
0.360	0.508*
0.367	0.473*
0.399	0.279
0.428	0.147
0.468	0.112
0.529	0.092
0.724	0.084
0.984	0.093
1.055	0.084
1.327	0.071
1.369	0.056
1.441	0.061
1.800	0.057

* Not shown on plot

SPECIFICATION TABLE NO. 348 NORMAL SOLAR ABSORPTANCE OF ALUMINUM OXIDE CONTACT COATINGS

Curve No.	Ref. No.	Year	Temperature Range, K	Geometry θ	Reported Error, %	Composition (weight percent), Specifications and Remarks
1*	125	1958	298	9°		Rokide A, aluminum oxide; molybdenum substrate; substrate oxidized and volatilized at ~1030 K; computed from spectral reflectance for above atm conditions.
2*	125	1958	298	9°		Above specimen and conditions except calculated for sea level conditions.
3*	119	1967	~298	~0°	<10	Sapphire (0.100 μm thick); gold (0.200 μm thick) and Corning 7058 glass (rms roughness 0.006 μm) substrates; high-purity gold vapor-deposited on glass, sapphire vapor-deposited on gold; computed from spectral reflectance data.
4*	119	1967	~298	~0°	<10	Above specimen and conditions except ultraviolet irradiated in vacuum at 533 K for 250 ESH and thermal cycled 10 times between 294 and 533 K.
5*	119	1967	~298	~0°	<10	Sapphire (0.100 μm thick); silver (0.200 μm thick) and Corning 7058 glass (rms roughness 0.006 μm) substrates; high-purity silver vapor-deposited on glass, sapphire vapor-deposited on silver; computed from spectral reflectance data.
6*	119	1967	~298	~0°	<10	Above specimen and conditions except ultraviolet irradiated in vacuum at 533 K for 250 ESH and thermal cycled 10 times between 294 and 533 K.
7*	119	1967	~298	~0°	<10	Sapphire (0.100 μm thick); aluminum (0.200 μm thick) and Corning 7058 glass (rms roughness 0.006 μm) substrates; high-purity aluminum vapor-deposited on glass, sapphire vapor-deposited on aluminum; computed from spectral reflectance data.
8*	119	1967	~298	~0°	<10	Above specimen and conditions except ultraviolet irradiated in vacuum at 533 K for 250 ESH and thermal cycled 10 times between 294 and 533 K.
9*	119	1967	~298	~0°	<10	Sapphire (0.100 μm thick); gold (0.200 μm thick) and 6061-T6 aluminum alloy substrates; high-purity gold vapor-deposited on highly polished aluminum, sapphire vapor-deposited on gold; computed from spectral reflectance data.
10*	119	1967	~298	~0°	<10	Above specimen and conditions except ultraviolet irradiated in vacuum at 533 K for 250 ESH and thermal cycled 10 times between 294 and 533 K.
11*	119	1967	~298	~0°	<10	Sapphire (0.100 μm thick); silver (0.200 μm thick) and 6061-T6 aluminum alloy substrates; high-purity silver vapor-deposited on highly polished aluminum, sapphire vapor-deposited on silver; computed from spectral reflectance data.
12*	119	1967	~298	~0°	<10	Above specimen and conditions except ultraviolet irradiated in vacuum at 533 K for 250 ESH and thermal cycled 10 times between 294 and 533 K.
13*	119	1967	~298	~0°	<10	Sapphire (0.100 μm thick); aluminum (0.200 μm thick) and 6061-T6 aluminum alloy substrates; high-purity aluminum vapor-deposited on highly polished aluminum alloy, sapphire vapor-deposited on aluminum; computed from spectral reflectance data.
14*	119	1967	~298	~0°	<10	Above specimen and conditions except ultraviolet irradiated in vacuum at 533 K for 250 ESH and thermal cycled 10 times between 294 and 533 K.
15*	119	1967	~298	~0°	<10	Sapphire (0.100 μm thick); gold (0.200 μm thick) and mill-finish 6061-T6 aluminum alloy substrates; high-purity gold vapor-deposited on aluminum, sapphire vapor-deposited on gold; computed from spectral reflectance data.

* No plot given

SPECIFICATION TABLE NO. 348 NORMAL SOLAR ABSORPTANCE OF ALUMINUM OXIDE CONTACT COATINGS (continued)

Curve No.	Ref. No.	Year	Temperature Range, K	Geometry θ	Reported Error, %	Composition (weight percent), Specifications and Remarks
16*	119	1967	~298	~0°	<10	Above specimen and conditions except ultraviolet irradiated in vacuum at 533 K for 250 ESH and thermal cycled 10 times between 294 and 533 K.
17*	119	1967	~298	~0°	<10	Sapphire (0. 100 μm thick); silver (0. 200 μm thick) and mill-finish 6061-T6 aluminum alloy substrates; high-purity silver vapor-deposited on aluminum, sapphire vapor-deposited on silver; computed from spectral reflectance.
18*	119	1967	~298	~0°	<10	Above specimen and conditions except ultraviolet irradiated in vacuum at 533 K for 250 ESH and thermal cycled 10 times between 294 and 533 K.
19*	119	1967	~298	~0°	<10	Sapphire (0. 100 μm thick); aluminum (0. 200 μm thick) and mill-finish 6061-T6 aluminum alloy substrates; high-purity aluminum vapor-deposited on aluminum alloy, sapphire vapor-deposited on aluminum; computed from spectral reflectance data.
20*	119	1967	~298	~0°	<10	Above specimen and conditions except ultraviolet irradiated in vacuum at 533 K for 250 ESH and thermal cycled 10 times between 294 and 533 K.
21*	119	1967	~298	~0°	<10	Sapphire (0. 100 μm thick); gold (0. 200 μm thick) and commercial bright stainless steel substrates; high-purity gold vapor-deposited on stainless steel, sapphire vapor-deposited on gold; computed from spectral reflectance data.
22*	119	1967	~298	~0°	<10	Above specimen and conditions except ultraviolet irradiated in vacuum at 533 K for 250 ESH and thermal cycled 10 times between 294 and 533 K.
23*	119	1967	~298	~0°	<10	Sapphire (0. 100 μm thick); silver (0. 200 μm thick) and commercial bright stainless steel substrates; high-purity silver vapor-deposited on stainless steel, sapphire vapor-deposited on silver; computed from spectral reflectance data.
24*	119	1967	~298	~0°	<10	Above specimen and conditions except ultraviolet irradiated in vacuum at 533 K for 250 ESH and thermal cycled 10 times between 294 and 533 K.
25*	119	1967	~298	~0°	<10	Sapphire (0. 100 μm thick); alumina (0. 200 μm thick) and commercial bright stainless steel substrates; high-purity aluminum vapor-deposited on stainless steel, sapphire vapor-deposited on aluminum; computed from spectral reflectance data.
26*	119	1967	~298	~0°	<10	Above specimen and conditions except ultraviolet irradiated in vacuum at 533 K for 250 ESH and thermal cycled 10 times between 294 and 533 K.
27*	119	1967	~298	~0°	<10	Sapphire (0. 100 μm thick); silver (0. 200 μm thick) and mill-finish 6061-T6 aluminum alloy substrates; high-purity silver vapor-deposited on aluminum, sapphire vapor-deposited on silver; computed from spectral reflectance data.
28*	119	1967	~298	~0°	<10	Above specimen and conditions except ultraviolet irradiated in vacuum at 294 K for 2000 ESH.
29*	119	1967	~298	~0°	<10	Similar to curve 27 specimen and conditions.
30*	119	1967	~298	~0°	<10	Above specimen and conditions except ultraviolet irradiated in vacuum at 533 K for 2000 ESH.

* No plot given

SPECIFICATION TABLE NO. 348 NORMAL SOLAR ABSORPTANCE OF ALUMINUM OXIDE CONTACT COATINGS (continued)

Curve No.	Ref. No.	Year	Temperature Range, K	Geometry θ	Reported Error, %	Composition (weight percent), Specifications and Remarks
31*	119	1967	~298	~0°	<10	Similar to curve 27 specimen and conditions except silver 0.1480 μm thick.
32*	127	1959	298	9°		Rokide A; aluminum oxide, stainless steel 446 substrate; computed from spectral reflectance data (0.3 to 3 μ) for above atm conditions.
33*	127	1959	298	9°		Above specimen and conditions except computed for sea level conditions.
34*	85	1969	156-399	~0°		Sapphire (~0.002 mm thick); anodized 1199 aluminum substrate; vapor deposited; aluminum substrate alkaline electropolished (sodium phosphate and sodium carbonate) for 15 min at 353 K and 12 VDC, then anodized for 15 min at 18 VDC in 10% sulfuric acid; property measured in vacuum by calorimetric method; data extracted from smooth curve. [Author's designation: A-3 Vapor Deposited Sapphire (Al_2O_3), NRDL-RTD-81-3]

* No plot given

DATA TABLE NO. 348 NORMAL SOLAR ABSORPTANCE OF ALUMINUM OXIDE CONTACT COATINGS

[Temperature, T, K; Absorptance, α]

T	α	T	α	T	α
CURVE 1*		CURVE 13*		CURVE 25*	
298	0.259	298	0.129	298	0.141
CURVE 2*		CURVE 14*		CURVE 26*	
298	0.245	298	0.134	298	0.135
CURVE 3*		CURVE 15*		CURVE 27*	
298	0.181	298	0.215	298	0.074
CURVE 4*		CURVE 16*		CURVE 28*	
298	0.180	298	0.217	298	0.076
CURVE 5*		CURVE 17*		CURVE 29*	
298	0.056	298	0.105	298	0.074
CURVE 6*		CURVE 18*		CURVE 30*	
298	0.060	298	0.101	298	0.125
CURVE 7*		CURVE 19*		CURVE 31*	
298	0.139	298	0.163	298	0.072
CURVE 8*		CURVE 20*		CURVE 32*	
298	0.128	298	0.149	298	0.289
CURVE 9*		CURVE 21*		CURVE 33*	
298	0.183	298	0.199	298	0.275
CURVE 10*		CURVE 22*		CURVE 34*	
298	0.186	298	0.204	156	0.148
				278	0.151
CURVE 11*		CURVE 23*		399	0.153
298	0.070	298	0.076		
CURVE 12*		CURVE 24*			
298	0.074	298	0.077		

* No plot given

SPECIFICATION TABLE NO. 349 NORMAL SPECTRAL TRANSMITTANCE OF ALUMINUM OXIDE CONTACT COATINGS

Curve No.	Ref. No.	Year	Temperature K	Wavelength Range, μm	Geometry θ θ' ω'	Reported Error, %	Composition (weight percent), Specifications, and Remarks
1*	134	1967	~298	0.1435-0.3000	~0° ~0°		Synthetic sapphire (3.18 mm thick) from Linde Division of Union Carbide Corp.; substrate unknown; optically polished.
2*	134	1967	~298	0.1435-0.3000	~0° ~0°		Above specimen and conditions except measured after irradiation by a total flux of 8.25×10^{10} protons cm^{-2} evenly distributed over 3, 3.8 and 4.6 MeV.

DATA TABLE NO. 349 NORMAL SPECTRAL TRANSMITTANCE OF ALUMINUM OXIDE CONTACT COATINGS

[Wavelength, λ, μm; Transmittance, τ; Temperature, T, K]

λ	τ	λ	τ	λ	τ	λ	τ	λ	τ
CURVE 1*		CURVE 1 (cont.)*		CURVE 1 (cont.)*		CURVE 2 (cont.)*		CURVE 2 (cont.)*	
T ~ 298		0.1894	0.823	0.2751	0.854	0.1543	0.682	0.2296	0.836
0.1435	0.029	0.1943	0.805	0.2801	0.853	0.1570	0.714	0.2346	0.837
0.1439	0.094	0.1994	0.820	0.2853	0.858	0.1600	0.729	0.2398	0.846
0.1455	0.425	0.2045	0.808	0.2903	0.838	0.1628	0.719	0.2450	0.845
0.1459	0.526	0.2094	0.820	0.2952	0.854	0.1663	0.731	0.2499	0.846
0.1482	0.599	0.2144	0.841	0.3000	0.868	0.1695	0.758	0.2549	0.831
0.1492	0.606	0.2196	0.832			0.1744	0.769	0.2599	0.843
0.1518	0.647	0.2245	0.851	CURVE 2*		0.1795	0.785	0.2650	0.847
0.1542	0.670	0.2296	0.845	T ~ 298		0.1843	0.787	0.2700	0.841
0.1572	0.703	0.2347	0.842			0.1894	0.793	0.2751	0.844
0.1601	0.703	0.2397	0.847	0.1435	0.029	0.1943	0.817	0.2802	0.844
0.1628	0.702	0.2448	0.846	0.1439	0.095	0.1994	0.829	0.2850	0.853
0.1663	0.714	0.2500	0.835	0.1453	0.445	0.2045	0.825	0.2905	0.861
0.1694	0.767	0.2548	0.840	0.1458	0.552	0.2095	0.821	0.2952	0.854
0.1742	0.791	0.2599	0.853	0.1483	0.610	0.2145	0.828	0.3000	0.854
0.1794	0.792	0.2649	0.848	0.1493	0.625	0.2198	0.831		
0.1843	0.788	0.2701	0.843	0.1518	0.662	0.2245	0.843		

* No plot given

SPECIFICATION TABLE NO. 350 HEMISPHERICAL TOTAL EMITTANCE OF ALUMINUM OXIDE + ALUMINUM TITANATE CONTACT COATINGS

Curve No.	Ref. No.	Year	Temperature Range, K	Reported Error, %	Composition (weight percent), Specifications and Remarks
1*	135	1969	1200		88 aluminum oxide and 12 titanium dioxide (powder from Zirconium Corp. of America) (0.102 mm thick) on Nb-1 Zr substrate; substrate cleaned in a solution of sulfuric acid and potassium dichromate and grit blasted with 60 mesh silicon carbide to roughness height of 2.79 μm; plasma sprayed; aged 4 hrs at 1200 K; measured in vacuum (10^{-7} mm Hg).
2*	135	1969	298-1200		Similar to above specimen and conditions except exposed to thermal cycling between 298 and 1200 K at the rate of 63 cycles per 10 000 hrs (starting at 1200 K); thermal cycling time (hrs) is variable.

DATA TABLE NO. 350 HEMISPHERICAL TOTAL EMITTANCE OF ALUMINUM OXIDE + ALUMINUM TITANATE CONTACT COATINGS

[Temperature, T, K; Emittance, ϵ]

T	ϵ	Thermal Cycling Time	ϵ	Thermal Cycling Time	ϵ	Thermal Cycling Time	ϵ	Thermal Cycling Time	ϵ	Thermal Cycling Time	ϵ
CURVE 1*		CURVE 2 (cont.)		CURVE 2 (cont.)		CURVE 2 (cont.)		CURVE 2 (cont.)		CURVE 2 (cont.)	
1200	0.870	1200	0.835	2910	0.837	4570	0.828	6900	0.821	9800	0.813
Thermal Cycling Time	ϵ	1300	0.831	3060	0.834	4720	0.826	7070	0.817	9960	0.808
		1390	0.831	3160	0.830	4820	0.826	7250	0.817	10030	0.810
CURVE 2*		1540	0.839	3220	0.826	4920	0.826	7420	0.820		
T = 298-1200		1640	0.837	3320	0.832	4990	0.827	7590	0.816		
		1710	0.837	3420	0.832	5100	0.822	7760	0.813		
130	0.863	1850	0.835	3560	0.833	5250	0.822	7930	0.818		
190	0.855	1980	0.840	3680	0.833	5410	0.824	8030	0.817		
300	0.852	2060	0.842	3730	0.835	5590	0.821	8290	0.813		
370	0.848	2210	0.843	3900	0.839	5740	0.820	8400	0.813		
490	0.843	2330	0.843	3980	0.825	5910	0.817	8650	0.816		
620	0.842	2400	0.846	4080	0.831	6080	0.818	8840	0.810		
720	0.837	2530	0.842	4160	0.831	6250	0.821	9040	0.811		
850	0.836	2640	0.838	4250	0.833	6380	0.820	9250	0.811		
970	0.835	2710	0.835	4370	0.830	6540	0.818	9430	0.810		
1050	0.839	2830	0.837	4500	0.825	6740	0.821	9650	0.815		

* No plot given

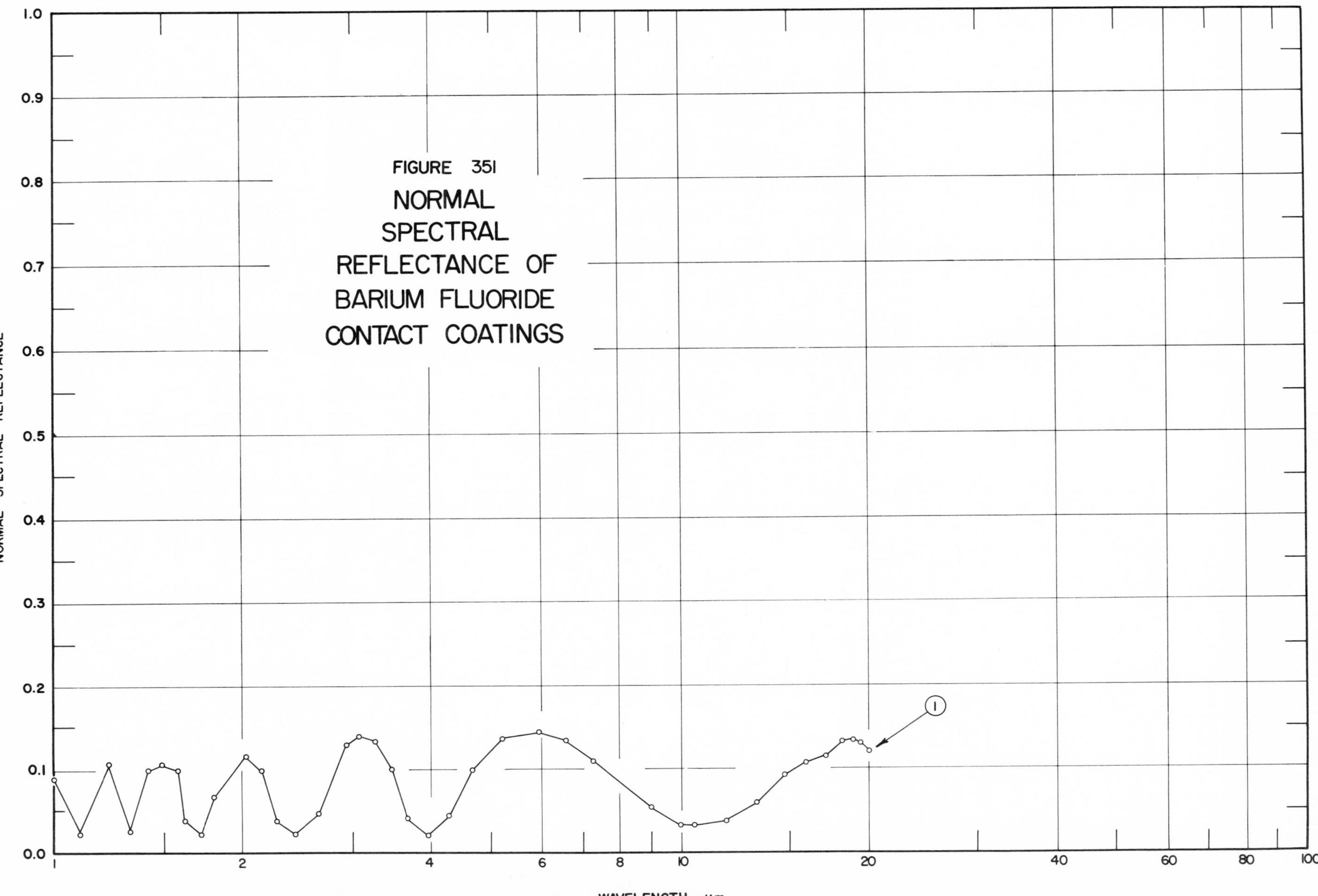

FIGURE 351 NORMAL SPECTRAL REFLECTANCE OF BARIUM FLUORIDE CONTACT COATINGS

SPECIFICATION TABLE NO. 351 NORMAL SPECTRAL REFLECTANCE OF BARIUM FLUORIDE CONTACT COATINGS

Curve No.	Ref. No.	Year	Temperature, K	Wavelength Range, μm	Geometry θ θ' ω'	Reported Error, %	Composition (weight percent), Specifications, and Remarks
1	136	1964	298	1.00-20.00	~0° ~0°		BaF_2; ZnSe substrate; vacuum deposited. [Author's designation: sample no. 6522-8]
2*	136	1964	523	1.00-20.00	~0° ~0°		Above specimen and conditions.
3*	136	1964	623	1.00-20.00	~0° ~0°		Above specimen and conditions.
4*	136	1964	673	1.00-20.00	~0° ~0°		Above specimen and conditions.

DATA TABLE NO. 351 NORMAL SPECTRAL REFLECTANCE OF BARIUM FLUORIDE CONTACT COATINGS

[Wavelength, λ, μm; Reflectance, ρ; Temperature, T, K]

λ	ρ	λ	ρ	λ	ρ	λ	ρ	λ	ρ	λ	ρ	λ	ρ	λ	ρ
CURVE 1		CURVE 1 (cont.)		CURVE 2*		CURVE 2 (cont.)		CURVE 3*		CURVE 3 (cont.)		CURVE 4*		CURVE 4 (cont.)	
T = 298		3.67	0.042	T = 523		3.67	0.042	T = 623		3.67	0.042	T = 673		3.67	0.042
1.00	0.090	3.98	0.022	1.00	0.090	3.98	0.022	1.00	0.090	3.98	0.022	1.00	0.090	3.98	0.022
1.10	0.024	4.29	0.046	1.10	0.024	4.29	0.046	1.10	0.024	4.29	0.046	1.10	0.024	4.29	0.046
1.23	0.109	4.66	0.100	1.23	0.109	4.66	0.100	1.23	0.109	4.66	0.100	1.23	0.109	4.66	0.100
1.33	0.026	5.20	0.136	1.33	0.026	5.20	0.136	1.33	0.026	5.20	0.136	1.33	0.026	5.20	0.136
1.42	0.100	5.98	0.145	1.42	0.100	5.98	0.145	1.42	0.100	5.98	0.145	1.42	0.100	5.98	0.145
1.49	0.109	6.57	0.135	1.49	0.109	6.57	0.135	1.49	0.109	6.57	0.135	1.49	0.109	6.57	0.135
1.57	0.100	7.24	0.110	1.57	0.100	7.24	0.110	1.57	0.100	7.24	0.110	1.57	0.100	7.24	0.110
1.62	0.040	8.98	0.053	1.62	0.040	8.98	0.053	1.62	0.040	8.98	0.053	1.62	0.040	8.98	0.053
1.72	0.024	9.97	0.035	1.72	0.024	9.97	0.035	1.72	0.024	9.97	0.035	1.72	0.024	9.97	0.035
1.80	0.069	10.58	0.031	1.80	0.069	10.58	0.031	1.80	0.069	10.58	0.031	1.80	0.069	10.58	0.031
2.02	0.117	11.76	0.039	2.02	0.117	11.76	0.039	2.02	0.117	11.76	0.039	2.02	0.117	11.76	0.039
2.15	0.100	13.22	0.060	2.15	0.100	13.22	0.060	2.15	0.100	13.22	0.060	2.15	0.100	13.22	0.060
2.27	0.040	14.61	0.093	2.27	0.040	14.61	0.093	2.27	0.040	14.61	0.093	2.27	0.040	14.61	0.093
2.41	0.024	15.97	0.109	2.41	0.024	15.97	0.109	2.41	0.024	15.97	0.109	2.41	0.024	15.97	0.109
2.63	0.049	17.00	0.115	2.63	0.049	17.00	0.115	2.63	0.049	17.00	0.115	2.63	0.049	17.00	0.115
2.93	0.130	18.19	0.131	2.93	0.130	18.19	0.131	2.93	0.130	18.19	0.131	2.93	0.130	18.19	0.131
3.07	0.140	18.73	0.133	3.07	0.140	18.73	0.133	3.07	0.140	18.73	0.133	3.07	0.140	18.73	0.133
3.27	0.134	19.43	0.130	3.27	0.134	19.43	0.130	3.27	0.134	19.43	0.130	3.27	0.134	19.43	0.130
3.45	0.100	20.00	0.120	3.45	0.100	20.00	0.120	3.45	0.100	20.00	0.120	3.45	0.100	20.00	0.120

* Not shown on plot.

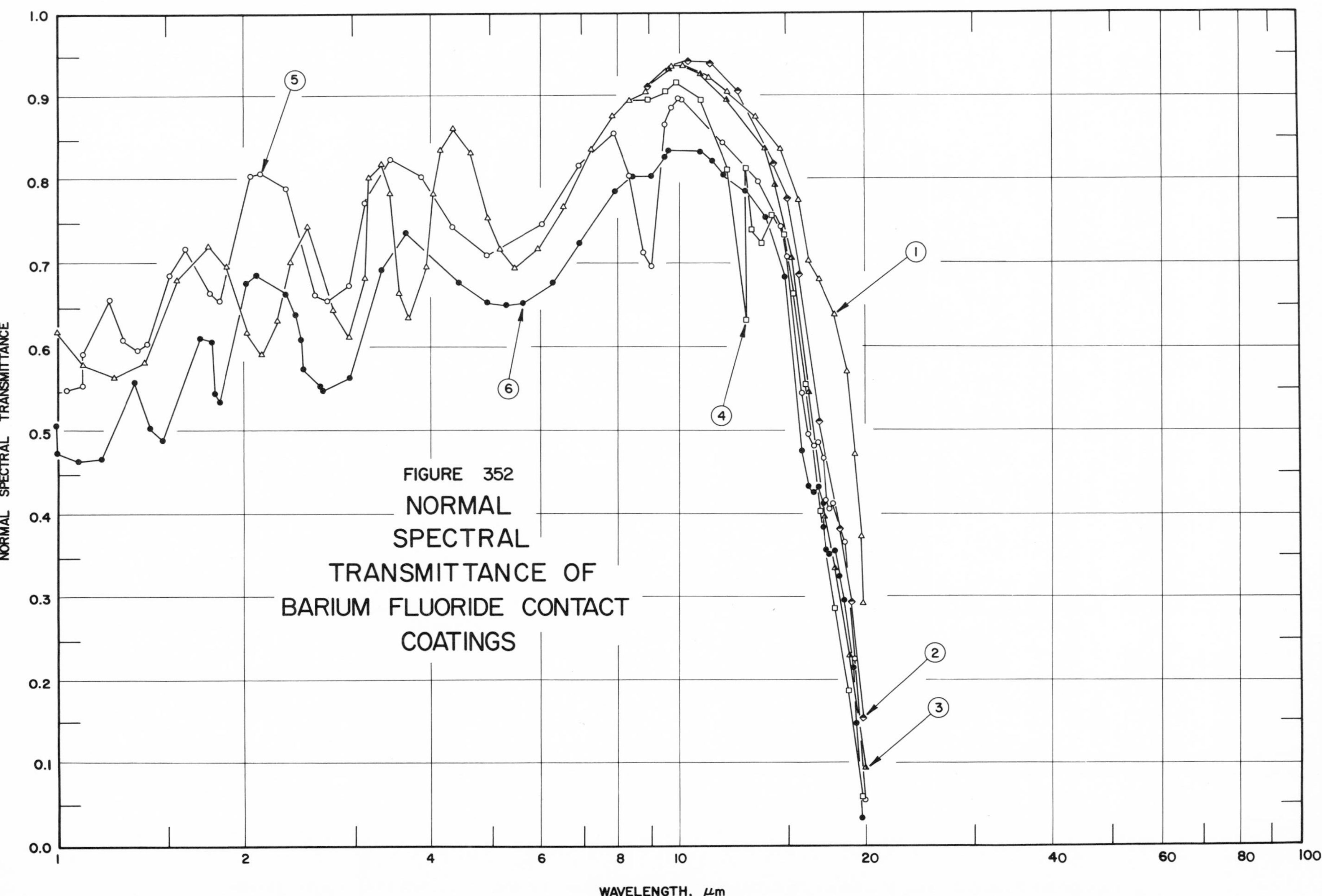

FIGURE 352 NORMAL SPECTRAL TRANSMITTANCE OF BARIUM FLUORIDE CONTACT COATINGS

SPECIFICATION TABLE NO. 352 NORMAL SPECTRAL TRANSMITTANCE OF BARIUM FLUORIDE CONTACT COATINGS

Curve No.	Ref. No.	Year	Temperature, K	Wavelength Range, μm	Geometry θ	θ'	ω'	Reported Error, %	Composition (weight percent), Specifications, and Remarks
1	136	1964	298	1.00-19.99	~0°	~0°			BaF_2; ZnSe (4.72 mm thick) substrate; vacuum deposited. [Author's designation: sample no. 6518-2]
2	136	1964	523	1.00-19.99	~0°	~0°			Above specimen and conditions.
3	136	1964	623	1.00-20.00	~0°	~0°			Above specimen and conditions.
4	136	1964	673	1.00-19.99	~0°	~0°			Above specimen and conditions.
5	136	1964	298	0.93-20.00	~0°	~0°			Above specimen and conditions except substrate 10.0 mm thick. [Author's designation: sample no. 6519]
6	136	1964	298	1.00-19.98	~0°	~0°			Above specimen and conditions except substrate 12.9 mm thick. [Author's designation: sample no. 6527]

DATA TABLE NO. 352 NORMAL SPECTRAL TRANSMITTANCE OF BARIUM FLUORIDE CONTACT COATINGS

[Wavelength, λ, μm; Transmittance, τ; Temperature, T, K]

λ	τ
CURVE 1	
T = 298	
1.00	0.620
1.10	0.580
1.24	0.566
1.37	0.583
1.56	0.680
1.75	0.720
1.87	0.699
2.01	0.620
2.14	0.593
2.27	0.633
2.38	0.701
2.53	0.743
2.79	0.648
2.95	0.614
3.11	0.684
3.19	0.801
3.34	0.820
3.43	0.782
3.56	0.666
3.69	0.637
3.92	0.699
4.04	0.784
4.15	0.836
4.35	0.861
4.62	0.834
4.97	0.755
5.19	0.718
5.46	0.699
5.95	0.718
6.53	0.769
7.29	0.838
7.82	0.878
8.35	0.898
8.89	0.909
9.76	0.938
10.29	0.939
11.25	0.923
12.06	0.906
13.47	0.877
14.60	0.839
15.53	0.775
16.36	0.702
16.98	0.681

λ	τ
CURVE 1 (cont.)	
17.98	0.640
18.77	0.571
19.35	0.474
19.70	0.373
19.99	0.294
CURVE 2	
T = 523	
1.00	0.620*
1.10	0.580*
1.24	0.566*
1.37	0.583*
1.56	0.680*
1.75	0.720*
1.87	0.699*
2.01	0.620*
2.14	0.593*
2.27	0.633*
2.38	0.701*
2.53	0.743*
2.79	0.648*
2.95	0.614*
3.11	0.684*
3.19	0.801*
3.34	0.820*
3.43	0.782*
3.56	0.666*
3.69	0.637*
3.92	0.699*
4.04	0.784*
4.15	0.836*
4.35	0.861*
4.62	0.834*
4.97	0.755*
5.19	0.718*
5.46	0.699*
5.95	0.718*
6.53	0.769*
7.29	0.838*
7.82	0.878*
8.35	0.898*
8.94	0.911
9.79	0.937

λ	τ
CURVE 2 (cont.)	
10.49	0.941
11.34	0.940
12.56	0.908
14.37	0.820
15.00	0.776
15.67	0.687
16.96	0.511
18.17	0.382
19.03	0.295
19.99	0.154
CURVE 3	
T = 623	
1.00	0.620*
1.10	0.580*
1.24	0.566*
1.37	0.583*
1.56	0.680*
1.75	0.720*
1.87	0.699*
2.01	0.620*
2.14	0.593*
2.27	0.633*
2.38	0.701*
2.53	0.743*
2.79	0.648*
2.95	0.614*
3.11	0.684*
3.19	0.801*
3.34	0.820*
3.43	0.782*
3.56	0.666*
3.69	0.637*
3.92	0.699*
4.04	0.784*
4.15	0.836*
4.35	0.861*
4.62	0.834*
4.97	0.755*
5.19	0.718*
5.46	0.699*
5.95	0.718*
6.53	0.769*

λ	τ
CURVE 3 (cont.)	
7.29	0.838*
7.82	0.878*
8.35	0.898*
8.96	0.910*
9.66	0.934
10.19	0.938*
10.98	0.929
12.07	0.896
13.72	0.837
14.48	0.793
15.29	0.707
16.29	0.548
17.33	0.397
17.89	0.335
18.93	0.230
20.00	0.094
CURVE 4	
T = 673	
1.00	0.620*
1.10	0.580*
1.24	0.566*
1.37	0.583*
1.56	0.680*
1.75	0.720*
1.87	0.699*
2.01	0.620*
2.14	0.593*
2.27	0.633*
2.38	0.701*
2.53	0.743*
2.79	0.648*
2.95	0.614*
3.11	0.684*
3.19	0.801*
3.34	0.820*
3.43	0.782*
3.56	0.666*
3.69	0.637*
3.92	0.699*
4.04	0.784*
4.15	0.836*
4.35	0.861*

λ	τ
CURVE 4 (cont.)	
4.62	0.834*
4.97	0.755*
5.19	0.718*
5.46	0.699*
5.95	0.718*
6.53	0.769*
7.29	0.838*
7.82	0.878*
8.35	0.898*
8.93	0.899
9.54	0.908
9.99	0.918
10.96	0.898
12.08	0.811
12.96	0.634
12.96	0.812
13.28	0.740
13.61	0.723
14.26	0.757
14.80	0.733
15.38	0.661
16.07	0.556
16.92	0.401
17.72	0.289
18.84	0.186
19.99	0.060
CURVE 5	
T ~298	
0.93	0.560*
1.04	0.550
1.11	0.554
1.11	0.592
1.23	0.659
1.27	0.610
1.35	0.597
1.40	0.603
1.52	0.686
1.61	0.718
1.76	0.664
1.83	0.656
2.05	0.804
2.12	0.806

λ	τ
CURVE 5 (cont.)	
2.32	0.790
2.60	0.661
2.71	0.658
2.95	0.673
3.16	0.771
3.47	0.824
3.86	0.802
4.32	0.743
4.91	0.710
6.04	0.747
6.94	0.817
7.90	0.857
8.39	0.805
8.80	0.711
9.07	0.699
9.58	0.868
9.72	0.888
10.04	0.899
10.23	0.898
11.90	0.847
13.56	0.799
14.62	0.742
15.01	0.709
15.95	0.547
16.16	0.499
16.52	0.484
16.70	0.489
16.99	0.483*
17.16	0.469
17.24	0.437*
17.39	0.417
17.59	0.409
17.82	0.413
18.54	0.367
19.38	0.229
20.00	0.057
CURVE 6	
T ~298	
1.00	0.507
1.00	0.472
1.08	0.463
1.17	0.469

λ	τ
CURVE 6 (cont.)	
1.34	0.560
1.41	0.502
1.48	0.490
1.70	0.612
1.78	0.608
1.80	0.545
1.84	0.535
2.03	0.677
2.10	0.688
2.33	0.664
2.41	0.640
2.48	0.610
2.50	0.575
2.63	0.554
2.69	0.550
2.96	0.565
3.34	0.693
3.63	0.737
4.44	0.676
4.91	0.655
5.30	0.651
5.61	0.653
6.30	0.678
6.97	0.724
7.98	0.786
8.43	0.804
9.01	0.804
9.51	0.827
9.68	0.833
10.81	0.833
11.40	0.821
11.89	0.807
12.95	0.787
13.98	0.754
14.98	0.686
15.96	0.479
16.21	0.434
16.51	0.429
16.64	0.433*
16.88	0.433
17.12	0.416
17.19	0.385
17.37	0.359
17.53	0.353

λ	τ
CURVE 6 (cont.)	
17.66	0.353*
17.80	0.357
18.33	0.329
18.59	0.299
19.10	0.217
19.48	0.149
19.98	0.031

* Not shown on plot

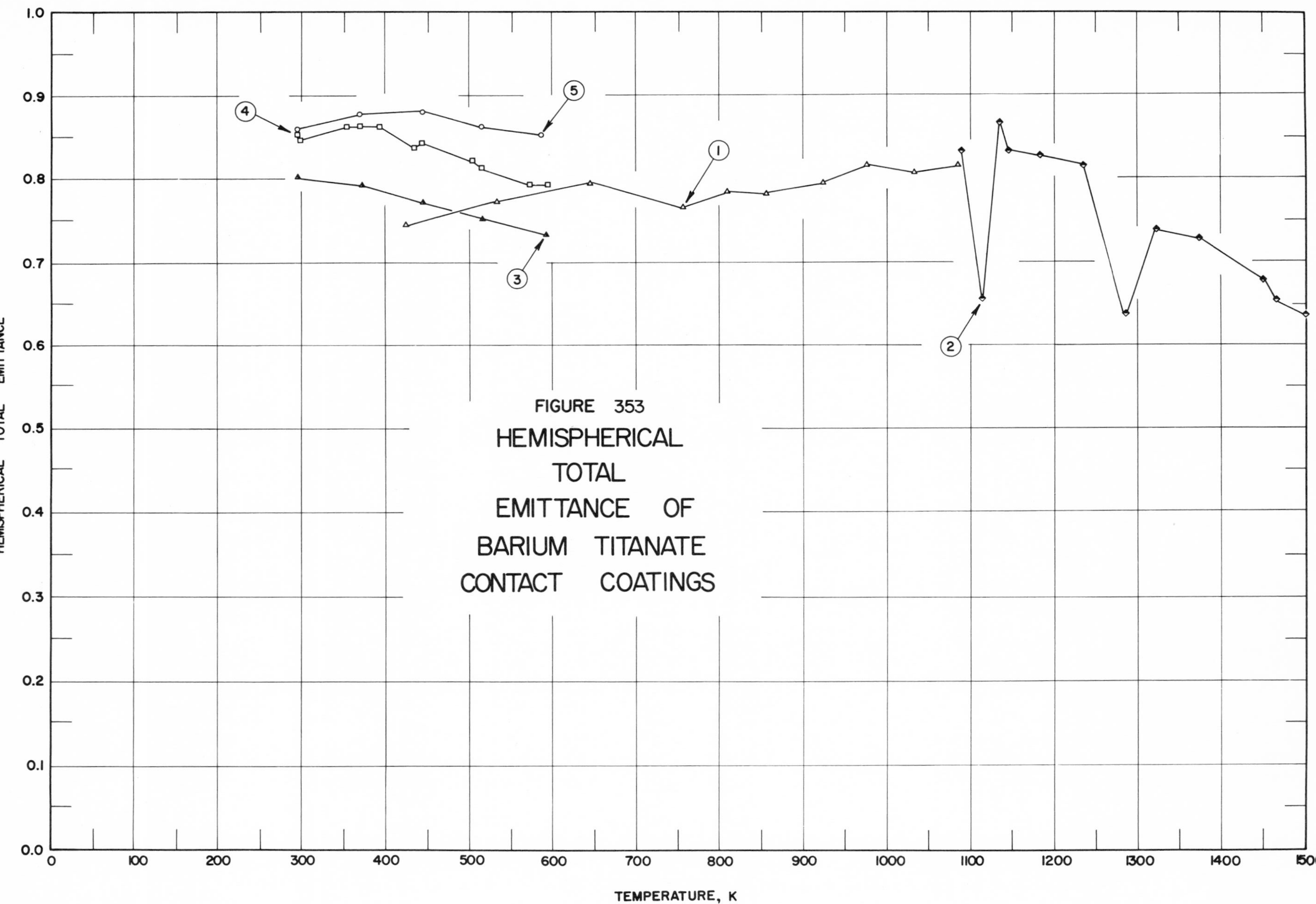

FIGURE 353
HEMISPHERICAL TOTAL EMITTANCE OF BARIUM TITANATE CONTACT COATINGS

SPECIFICATION TABLE NO. 353 HEMISPHERICAL TOTAL EMITTANCE OF BARIUM TITANATE CONTACT COATINGS

Curve No.	Ref. No.	Year	Temperature Range, K	Reported Error, %	Composition (weight percent), Specifications and Remarks
1	118	1962	424-1089		$BaO \cdot TiO_2$, powder from Continental Coatings Corp. (FCE-11), (0.127 mm thick); Nb-1 Zr alloy substrate; plasma-arc-sprayed; coating grey-black with fine grit texture; measured in vacuum ($<1.7 \times 10^{-6}$ mm Hg). [Author's designation: Run 1]
2	118	1962	1090-1502		Above specimen and conditions except vacuum $< 9.0 \times 10^{-7}$ mm Hg. [Author's designation: Run 2]
3	137	1965	298-591	±5	Barium titanate (~0.03 mm thick, 5.9 mg cm^{-2}); Al substrate; measured in vacuum (10^{-7} mm Hg).
4	137	1965	294-594	±5	Similar to above specimen and conditions except coating ~0.08 mm thick and 17 mg cm^{-2}.
5	137	1965	295-589	±5	Similar to curve 3 specimen and conditions except coating ~0.13 mm thick and 49 mg cm^{-2}.

DATA TABLE NO. 353 HEMISPHERICAL TOTAL EMITTANCE OF BARIUM TITANATE CONTACT COATINGS

[Temperature, T, K; Emittance, ϵ]

T	ϵ	T	ϵ	T	ϵ
CURVE 1		CURVE 2 (cont.)		CURVE 4 (cont.)	
424	0.745	1375	0.727	353	0.861
534	0.772	1450	0.676	370	0.864
645	0.794	1502	0.636	393	0.862
756	0.765	1469	0.651	435	0.836
810	0.783	1289	0.635	443	0.842
858	0.781	1114	0.653	505	0.821
925	0.794			508	0.821*
978	0.815	CURVE 3		515	0.812
1033	0.806			571	0.792
1089	0.813	298	0.802	594	0.792
		371	0.792		
CURVE 2		444	0.771	CURVE 5	
		517	0.751		
1090	0.831	591	0.732	295	0.860
1148	0.832			370	0.879
1136	0.865	CURVE 4		443	0.880
1182	0.827			515	0.861
1238	0.811	294	0.851	589	0.851
1322	0.738	298	0.849		

* Not shown on plot

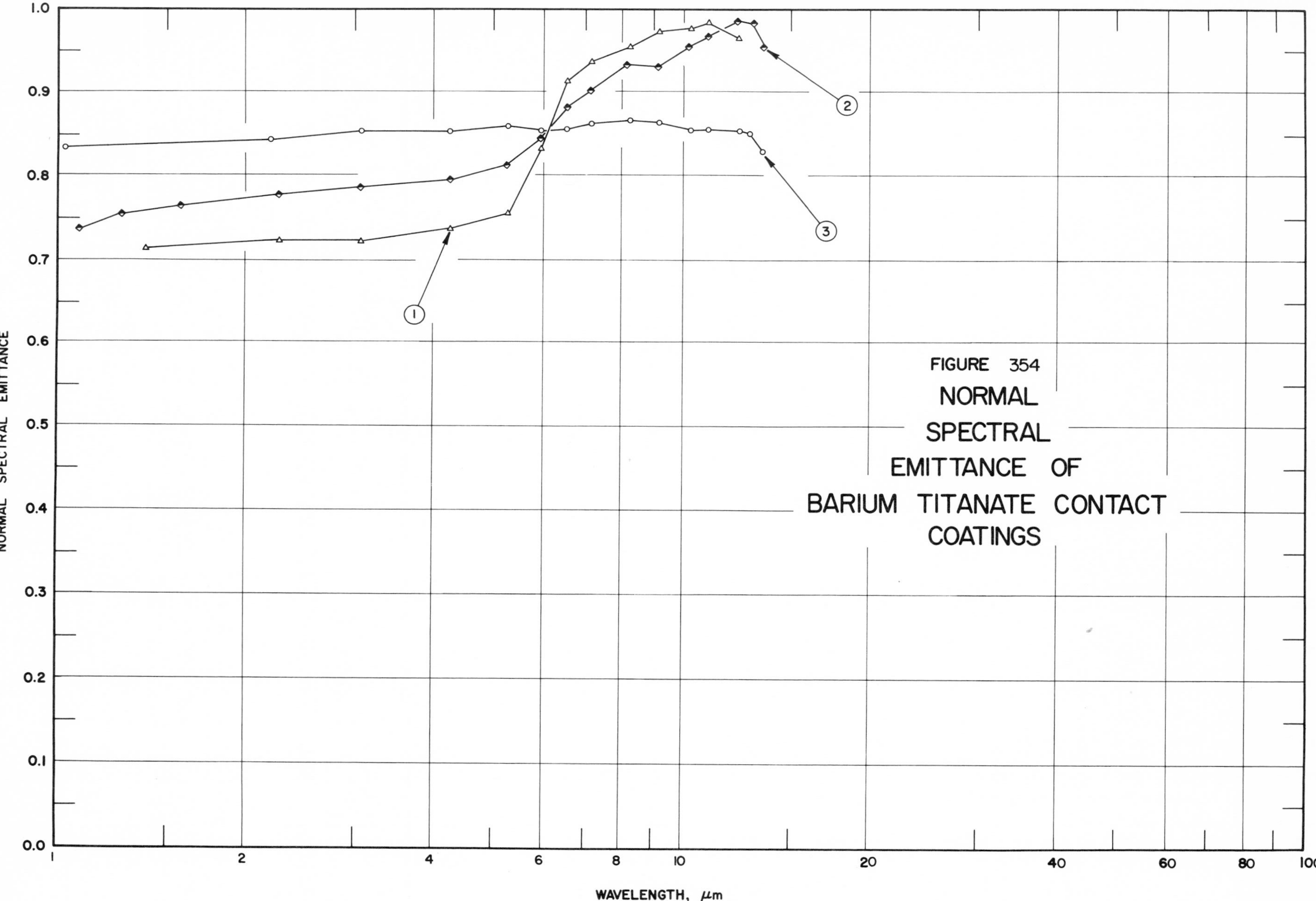

FIGURE 354
NORMAL
SPECTRAL
EMITTANCE OF
BARIUM TITANATE CONTACT
COATINGS
NORMAL SPECTRAL EMITTANCE
WAVELENGTH, μm
1.0
0.9
0.8
0.7
0.6
0.5
0.4
0.3
0.2
0.1
0.0
1
2
4
6
8
10
20
40
60
80
100
1
2
3

SPECIFICATION TABLE NO. 354 NORMAL SPECTRAL EMITTANCE OF BARIUM TITANATE CONTACT COATINGS

Curve No.	Ref. No.	Year	Temperature, K	Wavelength Range, μm	Geometry θ'	Reported Error, %	Composition (weight percent), Specifications, and Remarks
1	118	1962	755	1.39-12.39	~0°		$BaO \cdot TiO_2$, powder from Continental Coating Corp. (FCE-11), (0.127 mm thick); Nb-1 Zr alloy substrate; plasma-arc-sprayed; grey-black, hard, fine grit texture.
2	118	1962	1061	1.09-13.56	~0°		Above specimen and conditions.
3	118	1962	1366	0.59-13.55	~0°		Curve 1 specimen and conditions.

DATA TABLE NO. 354 NORMAL SPECTRAL EMITTANCE OF BARIUM TITANATE CONTACT COATINGS

[Wavelength, λ, μm; Emittance, ϵ; Temperature, T, K]

λ	ϵ	λ	ϵ	λ	ϵ
CURVE 1 T = 755		CURVE 2 (cont.)		CURVE 3 T = 1366	
		1.27	0.754		
1.39	0.713	1.59	0.764	0.59	0.800*
2.27	0.721	2.27	0.778	0.87	0.834*
3.08	0.722	3.08	0.788	1.04	0.835
4.29	0.738	4.27	0.797	2.20	0.843
5.26	0.756	5.28	0.815	3.08	0.853
5.96	0.833	5.96	0.847	4.25	0.853
6.56	0.915	6.58	0.884	5.29	0.860
7.19	0.939	7.18	0.903	5.97	0.855
8.22	0.955	8.19	0.934	6.59	0.857
9.16	0.974	9.13	0.933	7.17	0.862
10.34	0.978	10.34	0.956	8.25	0.868
11.07	0.986	11.08	0.969	9.16	0.865
12.39	0.966	12.39	0.988	10.35	0.857
		13.01	0.983	11.07	0.859
CURVE 2 T = 1061		13.56	0.955	12.41	0.855
				12.99	0.851
1.09	0.737			13.55	0.830

* Not shown on plot

SPECIFICATION TABLE NO. 355 NORMAL SOLAR ABSORPTANCE OF BARIUM TITANATE CONTACT COATINGS

Curve No.	Ref. No.	Year	Temperature Range, K	Geometry θ	Reported Error, %	Composition (weight percent), Specifications and Remarks
1*	137	1965	400-600	0°	±5	Barium titanate (~0.03 mm thick, 5.9 mg cm^{-2}); Al substrate; plasma-arc sprayed; measured in vacuum (10^{-7} mm Hg) after heating once to 600 K.
2*	137	1965	400-600	0°	±5	Similar to above specimen and conditions except coating ~0.08 mm thick and 17 mg cm^{-2}.
3*	137	1965	400-600	0°	±5	Similar to curve 2 specimen and conditions except coating ~0.13 mm thick and 49 mg cm^{-2}.

DATA TABLE NO. 355 NORMAL SOLAR ABSORPTANCE OF BARIUM TITANATE CONTACT COATINGS

[Temperature, T, K; Absorptance, α]

T	α
CURVE 1*	
400	0.65
600	0.65
CURVE 2*	
400	0.61
600	0.61
CURVE 3*	
400	0.74
600	0.74

* No plot given

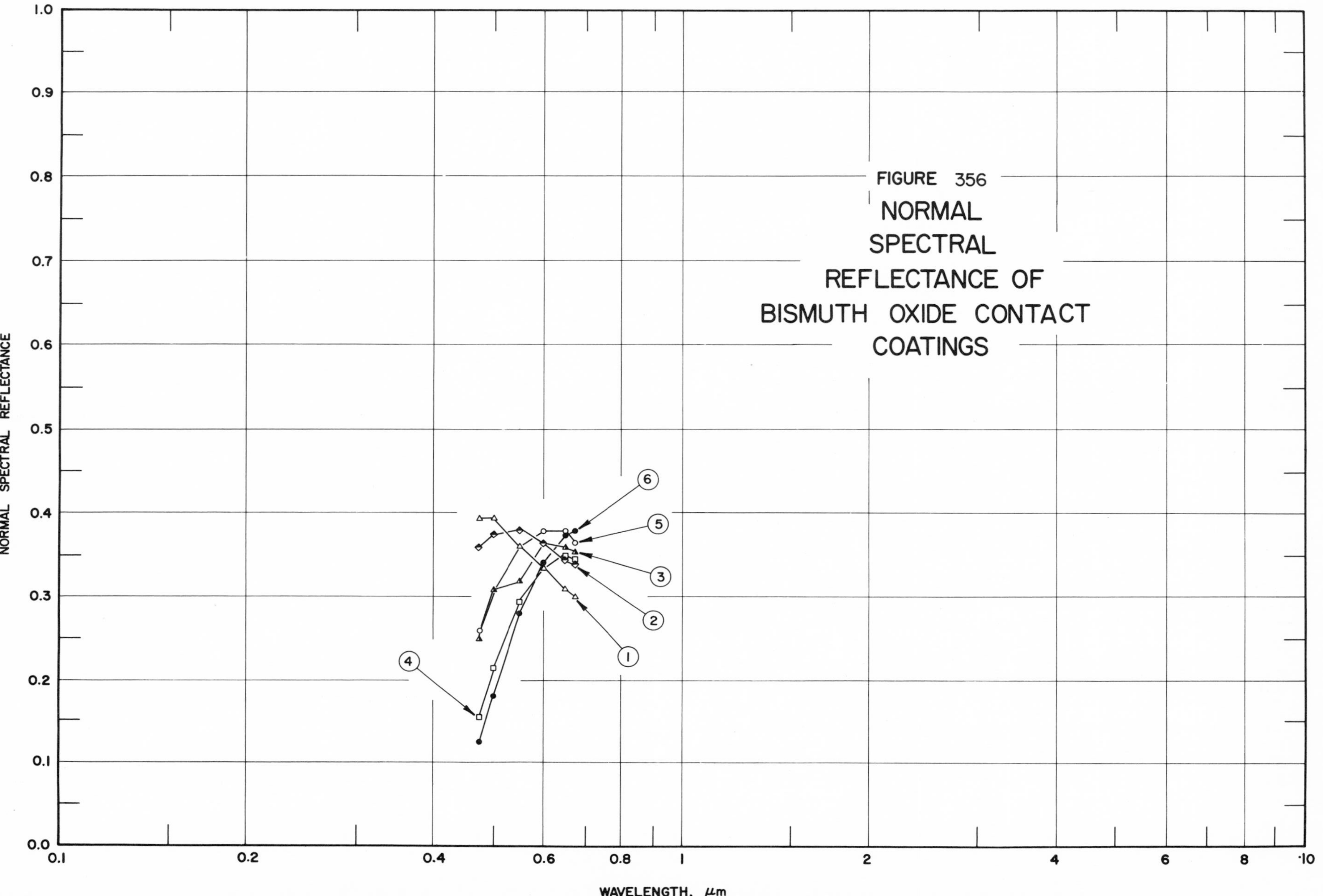
FIGURE 356
NORMAL
SPECTRAL
REFLECTANCE OF
BISMUTH OXIDE CONTACT
COATINGS
NORMAL SPECTRAL REFLECTANCE
WAVELENGTH, μm
1.0
0.9
0.8
0.7
0.6
0.5
0.4
0.3
0.2
0.1
0.0
0.1
0.2
0.4
0.6
0.8
1
2
4
6
8
10
1
2
3
4
5
6

SPECIFICATION TABLE NO. 356 NORMAL SPECTRAL REFLECTANCE OF BISMUTH OXIDE CONTACT COATINGS

Curve No.	Ref. No.	Year	Temperature, K	Wavelength Range, μm	Geometry θ	θ'	ω'	Reported Error, %	Composition (weight percent), Specifications, and Remarks
1	138	1967	298	0.4750-0.6750	$\sim 0°$	$\sim 0°$			Bismuth oxide coating ($\sim$0.0440 μm thick); glass substrate; produced by sputtering bismuth in argon + oxygen atm onto glass substrate; argon:oxygen flow rate 100:1; sputtering time 2.5 min; substrate to sputtering source distance 9 mm; data includes effect of back-surface reflections.
2	138	1967	298	0.4750-0.6750	$\sim 0°$	$\sim 0°$			Bismuth oxide coating ($\sim$0.0525 μm thick); glass substrate; produced by sputtering bismuth in argon + oxygen atm onto glass substrate; argon:oxygen flow rate 100:1; sputtering time 3 min; substrate to sputtering source distance 9 mm; data includes effect of back-surface reflections.
3	138	1967	298	0.4750-0.6750	$\sim 0°$	$\sim 0°$			Bismuth oxide coating ($\sim$0.0610 μm thick); glass substrate; produced by sputtering bismuth in argon + oxygen atm onto glass substrate; argon:oxygen flow rate 100:1; sputtering time 3.5 min; substrate to sputtering source distance 9 mm; data includes effect of back-surface reflections.
4	138	1967	298	0.4750-0.6750	$\sim 0°$	$\sim 0°$			Bismuth oxide coating ($\sim$0.0700 μm thick); glass substrate; produced by sputtering bismuth in argon + oxygen atm onto glass substrate; argon:oxygen flow rate 100:1; sputtering time 4 min; substrate to sputtering source distance 9 mm; data includes effect of back-surface reflections.
5	138	1967	298	0.4750-0.6750	$\sim 0°$	$\sim 0°$			Bismuth oxide coating ($\sim$0.0525 μm thick); glass substrate; produced by sputtering bismuth in argon + oxygen atm onto glass substrate; argon:oxygen flow rate 100:2; sputtering time 3 min; substrate to sputtering source distance 9 mm; data includes effect of back-surface reflections.
6	138	1967	298	0.4750-0.6750	$\sim 0°$	$\sim 0°$			Bismuth oxide coating ($\sim$0.0525 μm thick); glass substrate; produced by sputtering bismuth in argon + oxygen atm onto glass substrate; argon:oxygen flow rate 100:7; sputtering time 3 min; substrate to sputtering source distance 9 mm; data includes effect of back-surface reflections.

DATA TABLE NO. 356 NORMAL SPECTRAL REFLECTANCE OF BISMUTH OXIDE CONTACT COATINGS

[Wavelength, λ, μm; Reflectance, ρ; Temperature, T, K]

λ	ρ
CURVE 1	
T = 298	
0.4750	0.395
0.5000	0.395
0.5500	0.36
0.6000	0.335
0.6500	0.31
0.6750	0.30
CURVE 2	
T = 298	
0.4750	0.36
0.5000	0.375
0.5500	0.380
0.6000	0.365
0.6500	0.345
0.6750	0.34
CURVE 3	
T = 298	
0.4750	0.25
0.5000	0.31
0.5500	0.32
0.6000	0.365*
0.6500	0.36
0.6750	0.355
CURVE 4	
T = 298	
0.4750	0.155
0.5000	0.215
0.5500	0.295
0.6000	0.335*
0.6500	0.35
0.6750	0.345
CURVE 5	
T = 298	
0.4750	0.26
0.5000	0.31*
0.5500	0.36*
0.6000	0.38
CURVE 5 (cont.)	
0.6500	0.38
0.6750	0.365
CURVE 6	
T = 298	
0.4750	0.125
0.5000	0.18
0.5500	0.28
0.6000	0.34
0.6500	0.375
0.6750	0.38

* Not shown on plot

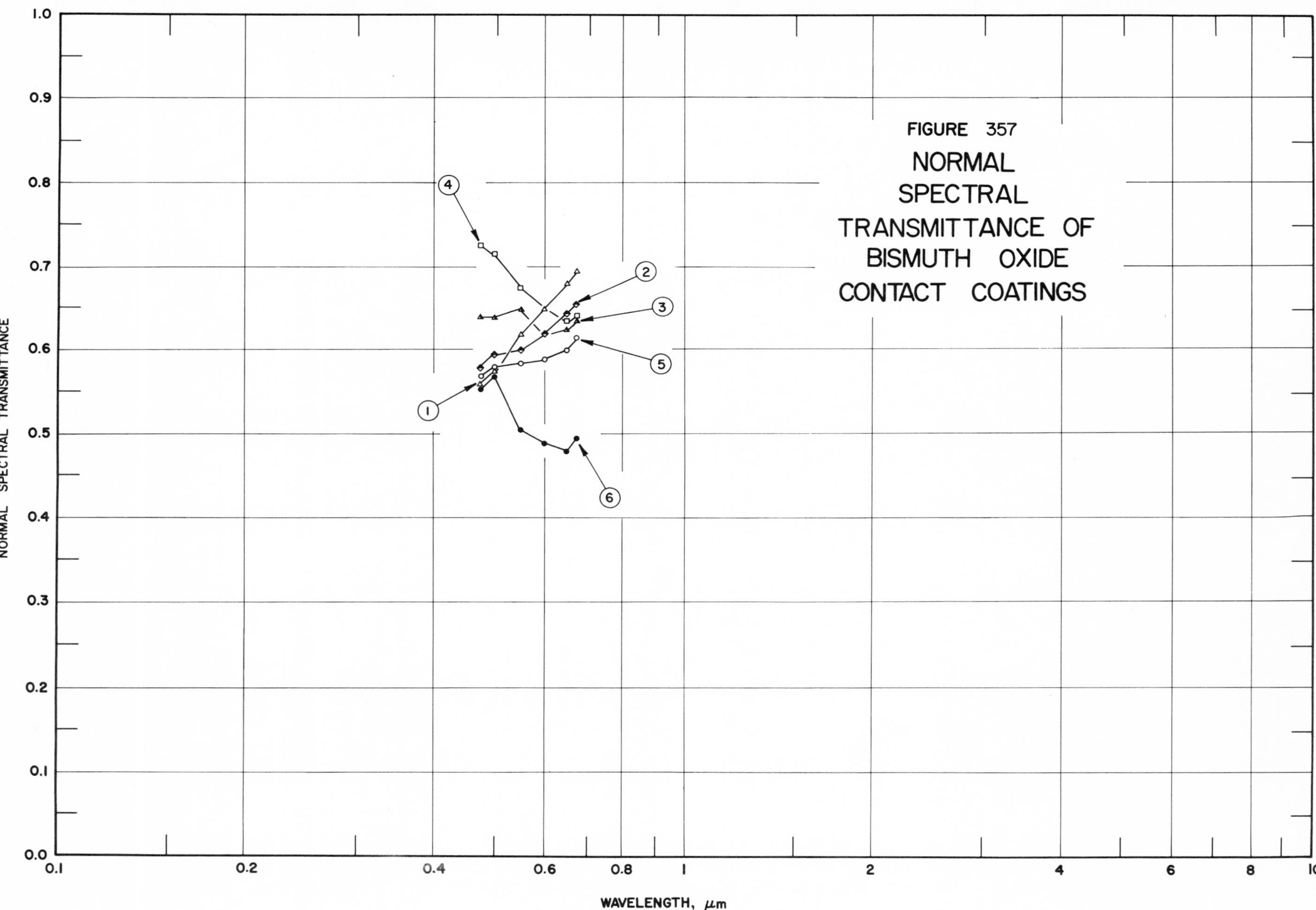
FIGURE 357
NORMAL
SPECTRAL
TRANSMITTANCE OF
BISMUTH OXIDE
CONTACT COATINGS
NORMAL SPECTRAL TRANSMITTANCE
WAVELENGTH, μm
1.0
0.9
0.8
0.7
0.6
0.5
0.4
0.3
0.2
0.1
0.0
0.1
0.2
0.4
0.6
0.8
1
2
4
6
8
10
4
2
3
5
1
6

SPECIFICATION TABLE NO. 357 NORMAL SPECTRAL TRANSMITTANCE OF BISMUTH OXIDE CONTACT COATINGS

Curve No.	Ref. No.	Year	Temperature, K	Wavelength Range, μm	Geometry θ	θ'	ω'	Reported Error, %	Composition (weight percent), Specifications, and Remarks
1	138	1967	298	0.4750-0.6750	$\sim 0^\circ$	$\sim 0^\circ$			Bismuth oxide coating ($\sim$0.0440 μm thick); glass substrate; produced by sputtering bismuth in argon + oxygen atm onto glass substrate; argon:oxygen flow rate 100:1; sputtering time 2.5 min; substrate to sputtering source distance 9 mm; data includes effect of back-surface reflections.
2	138	1967	298	0.4750-0.6750	$\sim 0^\circ$	$\sim 0^\circ$			Bismuth oxide coating ($\sim$0.0525 μm thick); glass substrate; produced by sputtering bismuth in argon + oxygen atm onto glass substrate; argon:oxygen flow rate 100:1; sputtering time 3 min; substrate to sputtering source distance 9 mm; data includes effect of back-surface reflections.
3	138	1967	298	0.4750-0.6750	$\sim 0^\circ$	$\sim 0^\circ$			Bismuth oxide coating ($\sim$0.0610 μm thick); glass substrate; produced by sputtering bismuth in argon + oxygen atm onto glass substrate; argon:oxygen flow rate 100:1; sputtering time 3.5 min; substrate to sputtering source distance 9 mm; data includes effect of back-surface reflections.
4	138	1967	298	0.4750-0.6750	$\sim 0^\circ$	$\sim 0^\circ$			Bismuth oxide coating ($\sim$0.0700 μm thick); glass substrate; produced by sputtering bismuth in argon + oxygen atm onto glass substrate; argon:oxygen flow rate 100:1; sputtering time 4 min; substrate to sputtering source distance 9 mm; data includes effect of back-surface reflections.
5	138	1967	298	0.4750-0.6750	$\sim 0^\circ$	$\sim 0^\circ$			Bismuth oxide coating ($\sim$0.0525 μm thick); glass substrate; produced by sputtering bismuth in argon + oxygen atm onto glass substrate; argon:oxygen flow rate 100:2; sputtering time 3 min; substrate to sputtering source distance 9 mm; data includes effect of back-surface reflections.
6	138	1967	298	0.4750-0.6750	$\sim 0^\circ$	$\sim 0^\circ$			Bismuth oxide coating ($\sim$0.0525 μm thick); glass substrate; produced by sputtering bismuth in argon + oxygen atm onto glass substrate; argon:oxygen flow rate 100:7; sputtering time 3 min; substrate to sputtering source distance 9 mm; data includes effect of back-surface reflections.

DATA TABLE NO. 357 NORMAL SPECTRAL TRANSMITTANCE OF BISMUTH OXIDE CONTACT COATINGS

[Wavelength, λ, μm; Transmittance, τ; Temperature, T, K]

λ	τ
CURVE 1	
T = 298	
0.4750	0.560
0.5000	0.575
0.5500	0.62
0.6000	0.650
0.6500	0.68
0.6750	0.695
CURVE 2	
T = 298	
0.4750	0.58
0.5000	0.595
0.5500	0.60
0.6000	0.620
0.6500	0.645
0.6750	0.655
CURVE 3	
T = 298	
0.4750	0.64
0.5000	0.64
0.5500	0.65
0.6000	0.620*
0.6500	0.625
0.6750	0.635
CURVE 4	
T = 298	
0.4750	0.725
0.5000	0.715
0.5500	0.675
0.6000	0.650*
0.6500	0.635
0.6750	0.640
CURVE 5	
T = 298	
0.4750	0.57
0.5000	0.58
0.5500	0.585
0.6000	0.59
CURVE 5 (cont.)	
0.6500	0.60
0.6750	0.615
CURVE 6	
T = 298	
0.4750	0.555
0.5000	0.57
0.5500	0.505
0.6000	0.49
0.6500	0.480
0.6750	0.495

* Not shown on plot

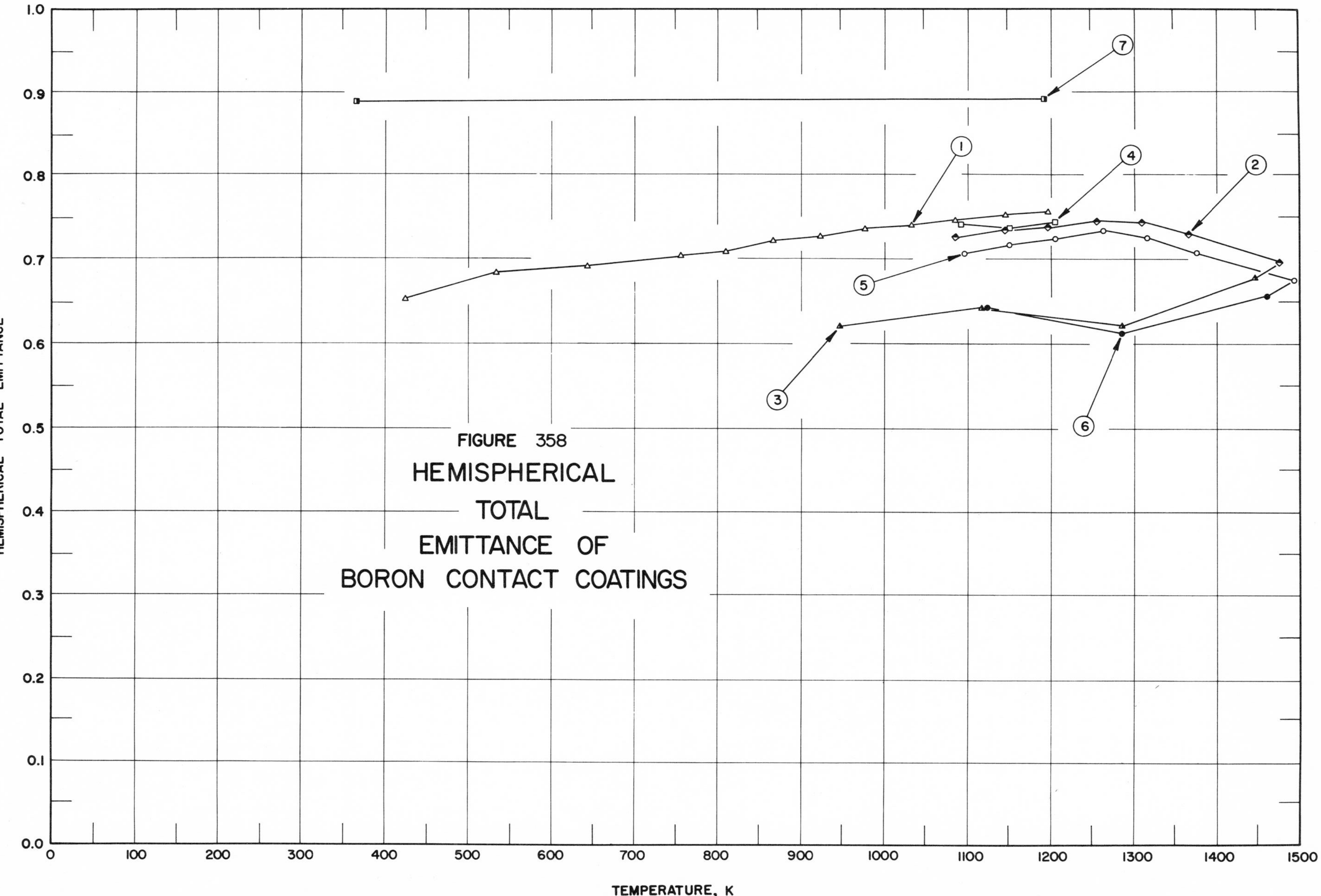

FIGURE 358
HEMISPHERICAL TOTAL EMITTANCE OF BORON CONTACT COATINGS

SPECIFICATION TABLE NO. 358 HEMISPHERICAL TOTAL EMITTANCE OF BORON CONTACT COATINGS

Curve No.	Ref. No.	Year	Temperature Range, K	Reported Error, %	Composition (weight percent), Specifications and Remarks
1	81	1963	422-1199		Crystalline boron (< 0.0254 mm thick); Nb-1 Zr substrate; plasma arc sprayed with particles of size 62 to 74 μm; measured in vacuum (4.0 x 10^{-7} to 1.8 x 10^{-6} mm Hg); temp measured with thermocouple; heating cycle. [Author's designation: Run No. 1]
2	81	1963	1088-1477		Above specimen and conditions. [Author's designation: Run No. 2]
3	81	1963	1477-949		Above specimen and conditions; cooling cycle.
4	81	1963	1091-1205		Curve 1 specimen and conditions except temp measured with optical pyrometer.
5	81	1963	1096-1491		Above specimen and conditions. [Author's designation: Run No. 2]
6	81	1963	1491-1123		Above specimen and conditions; cooling cycle.
7	82	1962	367-1191	< ±2.5	Crystalline boron (~0.0762 mm thick); niobium substrate; sprayed by Linde Plasmarc process; measured in vacuum; data extracted from smooth curve.

DATA TABLE NO. 358 HEMISPHERICAL TOTAL EMITTANCE OF BORON CONTACT COATINGS

[Temperature, T, K; Emittance, ϵ]

T	ϵ	T	ϵ	T	ϵ
CURVE 1		CURVE 2 (cont.)		CURVE 5	
422	0.651	1310	0.742	1096	0.707
533	0.685	1366	0.730	1150	0.718
644	0.691	1477	0.699	1205	0.724
755	0.705			1261	0.731
810	0.710	CURVE 3		1318	0.725
866	0.721			1377	0.707
922	0.728	1477	0.699*	1491	0.673
977	0.736	1449	0.677		
1033	0.740	1283	0.622	CURVE 6	
1088	0.747	1116	0.643		
1144	0.752	949	0.621	1491	0.673*
1199	0.757			1460	0.657
		CURVE 4		1288	0.612
CURVE 2				1123	0.643
		1091	0.740		
1088	0.727	1150	0.736	CURVE 7	
1144	0.734	1205	0.742		
1199	0.738			367	0.890
1255	0.745			1191	0.891

* Not shown on plot

SPECIFICATION TABLE NO. 359 HEMISPHERICAL TOTAL EMITTANCE OF BORON CARBIDE CONTACT COATINGS

Curve No.	Ref. No.	Year	Temperature Range, K	Reported Error, %	Composition (weight percent), Specifications and Remarks
1*	82	1962	461-915	<±2.5	Boron carbide (0.0762 mm thick); molybdenum substrate; applied by Linde Plasmarc process; measured in vacuum; data extracted from smooth curve.

DATA TABLE NO. 359 HEMISPHERICAL TOTAL EMITTANCE OF BORON CARBIDE CONTACT COATINGS

[Temperature, T, K; Emittance, ϵ]

T	ϵ
CURVE 1*	
461	0.745
585	0.762
715	0.775
792	0.780
915	0.787

* No plot given

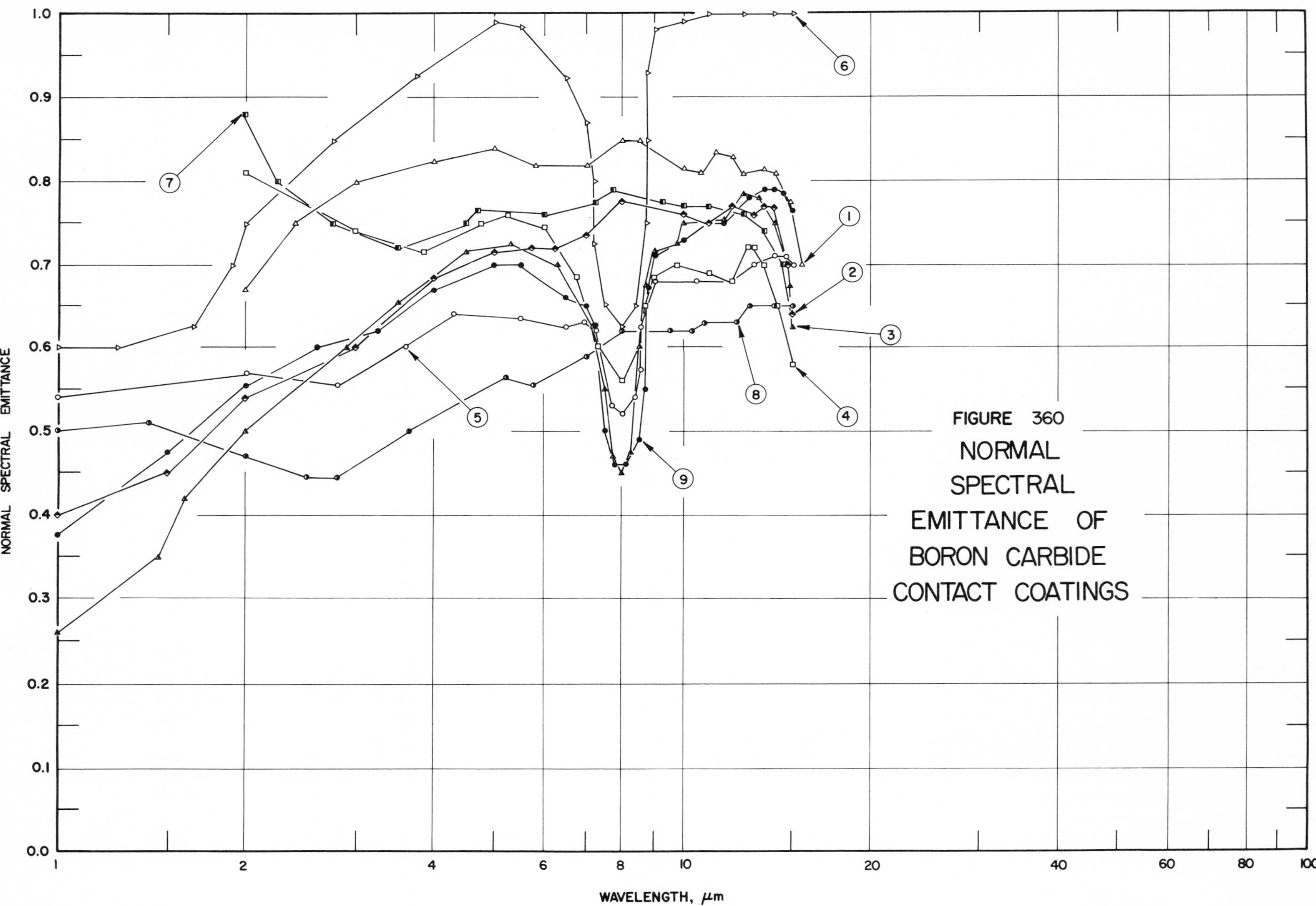
FIGURE 360
NORMAL
SPECTRAL
EMITTANCE OF
BORON CARBIDE
CONTACT COATINGS
NORMAL SPECTRAL EMITTANCE
WAVELENGTH, μm
1.0
0.9
0.8
0.7
0.6
0.5
0.4
0.3
0.2
0.1
0.0
1
2
4
6
8
10
20
40
60
80
100
1
2
3
4
5
6
7
8
9

SPECIFICATION TABLE NO. 360 NORMAL SPECTRAL EMITTANCE OF BORON CARBIDE CONTACT COATINGS

Curve No.	Ref. No.	Year	Temperature, K	Wavelength Range, μm	Geometry θ'	Reported Error, %	Composition (weight percent), Specifications, and Remarks
1	139	1961	523	2.00-15.50	~0°	±5	B_4C on Inconel X substrate; substrate sand blasted; flame sprayed using a plasmatron; data extracted from smooth curve.
2	139	1961	773	1.00-15.00	~0°	±5	Similar to above specimen and conditions.
3	139	1961	1023	1.00-15.00	~0°	±5	Similar to curve 1 specimen and conditions.
4	139	1961	523	2.00-15.00	~0°	±5	Similar to curve 1 specimen and conditions except heated in air at 1089 K for 30 min.
5	139	1961	773	1.00-15.00	~0°	±5	Similar to above specimen and conditions.
6	139	1961	1023	1.00-15.00	~0°	±5	Similar to curve 4 specimen and conditions.
7	139	1961	523	2.00-15.00	~0°	±5	Similar to curve 1 specimen and conditions except heated in vacuum (6.8×10^{-5} mm Hg) at 1089 K for 30 min.
8	139	1961	773	1.00-15.00	~0°	±5	Similar to above specimen and conditions.
9	139	1961	1023	1.00-15.00	~0°	±5	Similar to curve 7 specimen and conditions.

DATA TABLE NO. 360 NORMAL SPECTRAL EMITTANCE OF BORON CARBIDE CONTACT COATINGS

[Wavelength, λ, μm; Emittance, ∈; Temperature, T, K]

λ	∈
CURVE 1	
T = 523	
2.00	0.670
2.40	0.750
3.00	0.800
4.00	0.825
5.00	0.840
5.80	0.820
7.00	0.820
8.00	0.850
8.50	0.850
10.00	0.815
10.60	0.810
11.35	0.835
12.00	0.830
12.50	0.810
13.50	0.815
14.00	0.810
14.90	0.775
15.50	0.700
CURVE 2	
T = 773	
1.00	0.400
1.50	0.450
2.00	0.540
3.00	0.600
4.00	0.685
5.00	0.715
5.75	0.720
6.25	0.720
7.00	0.735
8.00	0.775
8.75	0.750*
10.00	0.760
11.00	0.750
12.00	0.770
13.00	0.760
13.50	0.770
14.00	0.770
14.75	0.700
15.00	0.640
CURVE 3	
T = 1023	
1.00	0.260
1.45	0.350
1.60	0.420
2.00	0.500
2.90	0.600
3.50	0.655
4.50	0.715
5.30	0.725
6.30	0.700
7.15	0.625
7.50	0.550
7.75	0.470
8.00	0.450
8.25	0.475
8.50	0.600
8.70	0.675
9.00	0.715
9.75	0.725
10.00	0.750
11.65	0.755
12.50	0.785
13.25	0.780
14.00	0.750
14.70	0.700*
14.85	0.675
15.00	0.625
CURVE 4	
T = 523	
2.00	0.810
3.00	0.740
3.85	0.715
4.75	0.750
5.25	0.760
6.00	0.745
6.75	0.685
7.30	0.600
8.00	0.560
8.50	0.600*
8.70	0.650
9.00	0.685
9.75	0.700
11.00	0.690
CURVE 4 (cont.)	
12.00	0.680
12.75	0.720
13.00	0.720
13.50	0.700
14.05	0.650
15.00	0.580
CURVE 5	
T = 773	
1.00	0.540
2.00	0.570
2.80	0.555
3.60	0.600
4.30	0.640
5.50	0.635
6.50	0.625
6.95	0.630
7.25	0.620
7.70	0.530
8.00	0.520
8.35	0.540
8.55	0.575
8.55	0.625
9.00	0.680
10.50	0.680
12.00	0.680*
13.00	0.700
14.00	0.710
14.60	0.710
15.00	0.700
CURVE 6	
T = 1023	
1.00	0.600
1.25	0.600
1.65	0.625
1.90	0.700
2.00	0.750
2.75	0.850
3.75	0.925
5.00	0.990
5.50	0.985
6.50	0.925
CURVE 6 (cont.)	
7.00	0.870
7.20	0.800
7.20	0.725
7.50	0.650
8.00	0.625
8.40	0.650
8.75	0.750
8.75	0.850
8.75	0.930
9.00	0.980
10.00	0.990
11.00	1.000
12.50	1.000
14.00	1.000
15.00	1.000
CURVE 7	
2.00	0.880
2.25	0.800
2.75	0.750
3.50	0.720
4.50	0.750
4.70	0.765
6.00	0.760
7.25	0.775
7.75	0.790
9.25	0.775
10.00	0.770
11.00	0.770
12.50	0.760
13.50	0.740
14.50	0.700
15.00	0.640*
CURVE 8	
1.00	0.500
1.40	0.510
2.00	0.470
2.50	0.445
2.80	0.445
3.65	0.500
5.20	0.565
5.75	0.555
7.00	0.590
CURVE 8 (cont.)	
8.00	0.620
9.50	0.620
10.35	0.620
10.75	0.630
12.20	0.630
12.80	0.650
14.00	0.650
15.00	0.650
CURVE 9	
T = 1023	
1.00	0.375
1.50	0.475
2.00	0.555
2.60	0.600
3.25	0.620
4.00	0.670
5.00	0.700
5.50	0.700
6.50	0.660
7.00	0.650
7.25	0.625
7.50	0.500
7.80	0.460
8.10	0.460
8.50	0.490
8.70	0.550
8.70	0.650*
8.75	0.675
9.00	0.710
10.00	0.730
11.00	0.750*
11.60	0.750
12.75	0.780
13.50	0.790
14.00	0.790
14.50	0.785
15.00	0.765

* Not shown on plot

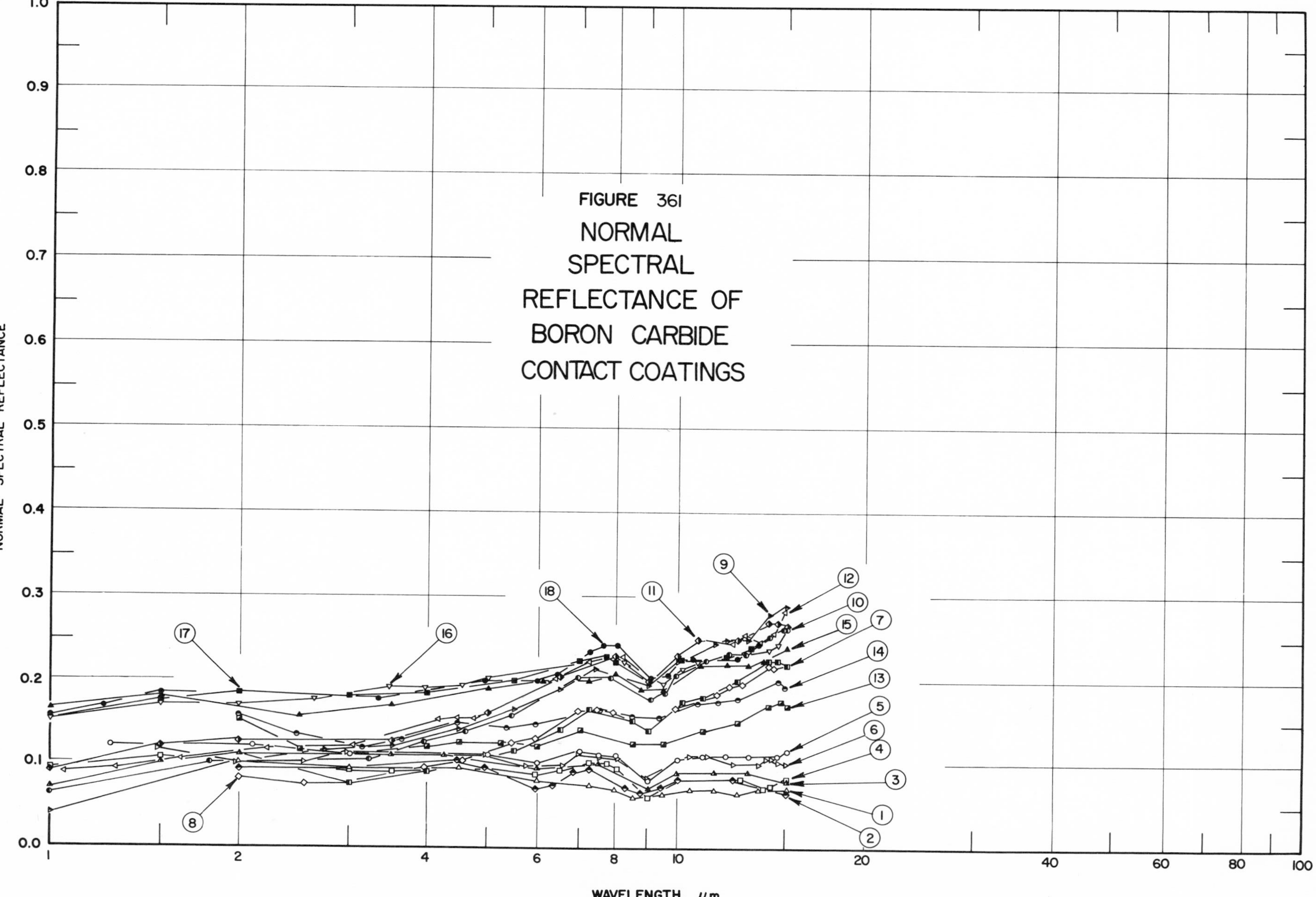

FIGURE 361 NORMAL SPECTRAL REFLECTANCE OF BORON CARBIDE CONTACT COATINGS

SPECIFICATION TABLE NO. 361 NORMAL SPECTRAL REFLECTANCE OF BORON CARBIDE CONTACT COATINGS

Curve No.	Ref. No.	Year	Temperature, K	Wavelength Range, μm	Geometry θ	Geometry θ'	Geometry ω'	Reported Error, %	Composition (weight percent), Specifications, and Remarks
1	139	1961	322	2.00-15.00	0°		2π	< 2	B_4C on Inconel X substrate; substrate sand blasted; flame sprayed using a plasmatron; sample maintained at 322 K; data extracted from smooth curve; converted from R $(2\pi, 0^\circ)$; hohlraum at 523 K used as radiation source.
2	139	1961	322	2.00-15.00	0°		2π	< 2	Above specimen and conditions; diffuse component only.
3	139	1961	322	1.00-15.00	0°		2π	< 2	Similar to curve 1 specimen and conditions except hohlraum at 773 K.
4	139	1961	322	1.00-15.00	0°		2π	< 2	Above specimen and conditions; diffuse component only.
5	139	1961	322	0.50-15.00	0°		2π	< 2	Similar to curve 1 specimen and conditions except hohlraum at 1273 K.
6	139	1961	322	0.50-14.99	0°		2π	< 2	Above specimen and conditions; diffuse component only.
7	139	1961	322	2.00-15.00	0°		2π	< 2	Similar to curve 1 specimen and conditions except heated in air at 1089 K for 30 min.
8	139	1961	322	2.00-15.00	0°		2π	< 2	Above specimen and conditions; diffuse component only.
9	139	1961	322	1.00-15.00	0°		2π	< 2	Similar to curve 7 specimen and conditions except hohlraum at 773 K.
10	139	1961	322	1.00-15.00	0°		2π	< 2	Above specimen and conditions; diffuse component only.
11	139	1961	322	0.50-15.00	0°		2π	< 2	Similar to curve 7 specimen and conditions except hohlraum at 1273 K.
12	139	1961	322	0.50-14.99	0°		2π	< 2	Above specimen and conditions; diffuse component only.
13	139	1961	322	2.00-15.00	0°		2π	< 2	Similar to curve 1 specimen and conditions except heated in vacuum (6.8×10^{-5} mm Hg) at 1089 K for 30 min.
14	139	1961	322	2.00-14.99	0°		2π	< 2	Above specimen and conditions; diffuse component only.
15	139	1961	322	1.00-15.00	0°		2π	< 2	Similar to curve 13 specimen and conditions except hohlraum at 773 K.
16	139	1961	322	1.00-15.00	0°		2π	< 2	Above specimen and conditions; diffuse component only.
17	139	1961	322	0.50-15.00	0°		2π	< 2	Similar to curve 13 specimen and conditions except hohlraum at 1273 K.
18	139	1961	322	0.50-15.00	0°		2π	< 2	Above specimen and conditions; diffuse component only.

DATA TABLE NO. 361 NORMAL SPECTRAL REFLECTANCE OF BORON CARBIDE CONTACT COATINGS

[Wavelength, λ, μm; Reflectance, ρ; Temperature, T, K]

λ	ρ
CURVE 1	**T = 322**
2.00	0.100
3.00	0.095
4.50	0.095
6.00	0.080
7.25	0.075
8.00	0.070
8.50	0.060
9.50	0.065
10.50	0.070
11.50	0.070
12.50	0.065
13.50	0.070
15.00	0.070
CURVE 2	**T = 322**
2.00	0.092
3.00	0.092*
4.49	0.101
4.98	0.096
5.98	0.070
6.31	0.075
6.88	0.090
7.27	0.093
8.26	0.071
8.75	0.066
9.49	0.072
10.01	0.081
12.36	0.081
15.00	0.067
CURVE 3	**T = 322**
1.00	0.070
1.50	0.100
2.00	0.110
3.50	0.110
4.25	0.110
6.00	0.095
7.50	0.100
9.00	0.070
10.00	0.090
CURVE 3 (cont.)	
11.50	0.090
13.00	0.090
15.00	0.080
CURVE 4	**T = 322**
1.00	0.093
1.51	0.104
3.00	0.091
3.51	0.090
4.97	0.095*
5.98	0.088
6.50	0.092
7.21	0.100
7.76	0.100
8.03	0.095
9.00	0.060
10.00	0.081*
12.67	0.081
13.71	0.070
14.20	0.074
15.00	0.082
CURVE 5	**T = 322**
0.50	0.060*
0.70	0.100*
1.25	0.120
2.10	0.120
3.00	0.110
4.25	0.110*
5.00	0.110
6.00	0.100
7.00	0.115
7.50	0.110
8.00	0.110
9.00	0.080
10.00	0.105
11.00	0.110
12.00	0.110
13.25	0.110
14.50	0.110
15.00	0.115
CURVE 6	**T = 322**
0.50	0.059*
0.64	0.089*
0.82	0.101*
0.91	0.101*
0.99	0.114*
1.49	0.116
1.98	0.100
2.54	0.100
2.98	0.111
4.99	0.111
5.71	0.099
6.55	0.099
6.99	0.111
7.51	0.107
8.07	0.107
8.99	0.081
10.00	0.106*
10.51	0.110
11.20	0.110
12.36	0.100
13.57	0.100
13.91	0.107
14.27	0.107
14.53	0.101
14.99	0.100
CURVE 7	**T = 322**
2.00	0.110*
3.00	0.075
4.00	0.090
5.50	0.115
6.50	0.140
7.25	0.165
8.50	0.150
9.00	0.140
10.25	0.175
11.00	0.180
12.50	0.200
14.00	0.225
14.50	0.225
15.00	0.220
CURVE 8	**T = 322**
2.00	0.082
2.54	0.075
3.00	0.075*
3.98	0.093
4.56	0.103
5.46	0.125
5.98	0.130
6.99	0.162
7.47	0.164
7.98	0.161
8.49	0.151*
9.00	0.142*
9.91	0.168
10.23	0.176*
11.65	0.182
12.23	0.195
12.74	0.198
14.00	0.220
14.33	0.218
15.00	0.200*
CURVE 9	**T = 322**
1.00	0.040
2.00	0.100*
2.75	0.115
3.50	0.120
4.50	0.140
5.50	0.165
6.50	0.190
7.40	0.215
9.00	0.195
10.10	0.225
11.50	0.245
12.00	0.250
13.00	0.250
14.00	0.280
15.00	0.290
CURVE 10	**T = 322**
1.00	0.062
CURVE 10 (cont.)	
1.80	0.100
3.23	0.104
3.97	0.125
4.60	0.138
5.48	0.158
6.99	0.203
7.81	0.203
9.01	0.179
9.59	0.185
9.99	0.208
11.15	0.224
12.18	0.232
12.99	0.232
14.01	0.252
14.76	0.261
15.00	0.261
CURVE 11	**T = 322**
0.50	0.000*
0.75	0.040*
1.00	0.090
1.50	0.120
2.00	0.125
3.50	0.125
5.00	0.160
6.50	0.205
8.00	0.230
9.00	0.200
10.00	0.230
10.75	0.250
12.50	0.250
14.00	0.270
14.50	0.270
15.00	0.265
CURVE 12	**T = 322**
0.50	0.144*
0.57	0.144*
1.05	0.089
1.27	0.091
1.61	0.103
CURVE 12 (cont.)	
2.22	0.116
3.01	0.120
4.17	0.151
4.48	0.154
4.74	0.154
6.41	0.201
7.25	0.224
8.20	0.230
9.00	0.200*
10.02	0.231*
10.82	0.250*
12.34	0.246
12.91	0.252
13.59	0.249
14.44	0.258
14.99	0.282
CURVE 13	**T = 322**
2.00	0.150
2.50	0.115
3.00	0.110*
4.00	0.120
4.50	0.125
5.25	0.125
6.00	0.120
7.00	0.140
8.50	0.125
9.50	0.125
11.00	0.140
12.50	0.150
14.00	0.170
14.65	0.175
15.00	0.170
CURVE 14	**T = 322**
2.00	0.156
2.48	0.131
3.13	0.119
3.64	0.127
4.48	0.149
5.33	0.142
CURVE 14 (cont.)	
5.99	0.146
7.50	0.165
8.50	0.154
9.37	0.153
10.75	0.172
11.65	0.173
12.51	0.178
14.53	0.200
14.99	0.192
CURVE 15	**T = 322**
1.00	0.165
1.50	0.180
2.50	0.155
3.50	0.170
5.00	0.190
6.10	0.200
7.25	0.200
8.00	0.210
8.75	0.190
9.50	0.190
10.90	0.220
12.00	0.220
13.00	0.220
13.75	0.225
15.00	0.240
CURVE 16	**T = 322**
1.00	0.150
1.50	0.170
2.00	0.170
2.63	0.176
3.49	0.191
3.99	0.191
4.53	0.194
5.00	0.201
7.96	0.230*
8.27	0.224
9.01	0.194*
9.50	0.196
10.21	0.215
CURVE 16 (cont.)	
10.87	0.221
12.02	0.221*
14.01	0.237
14.53	0.245
14.99	0.260*
CURVE 17	**T = 322**
0.50	0.030*
0.60	0.075*
0.80	0.100*
1.00	0.150*
1.50	0.175
2.00	0.185
3.00	0.180
4.00	0.185
5.50	0.200
7.00	0.225
7.75	0.230
8.00	0.225
9.00	0.200*
10.20	0.225
12.00	0.230
13.25	0.240
15.00	0.260*
CURVE 18	**T = 322**
0.50	0.038*
0.65	0.075*
0.79	0.101*
1.00	0.153
1.23	0.169
1.50	0.182
2.00	0.186*
3.32	0.179
4.01	0.188*
4.99	0.200
6.00	0.200
6.47	0.209
7.26	0.236
7.63	0.241
8.01	0.241

* Not shown on plot

DATA TABLE NO. 361 NORMAL SPECTRAL REFLECTANCE OF BORON CARBIDE CONTACT COATINGS (continued)

λ	ρ
CURVE 18 (cont.)	
T = 322	
9.02	0.201
9.62	0.207
10.24	0.226*
10.63	0.229
12.25	0.229
13.53	0.245
15.00	0.261*

* Not shown on plot

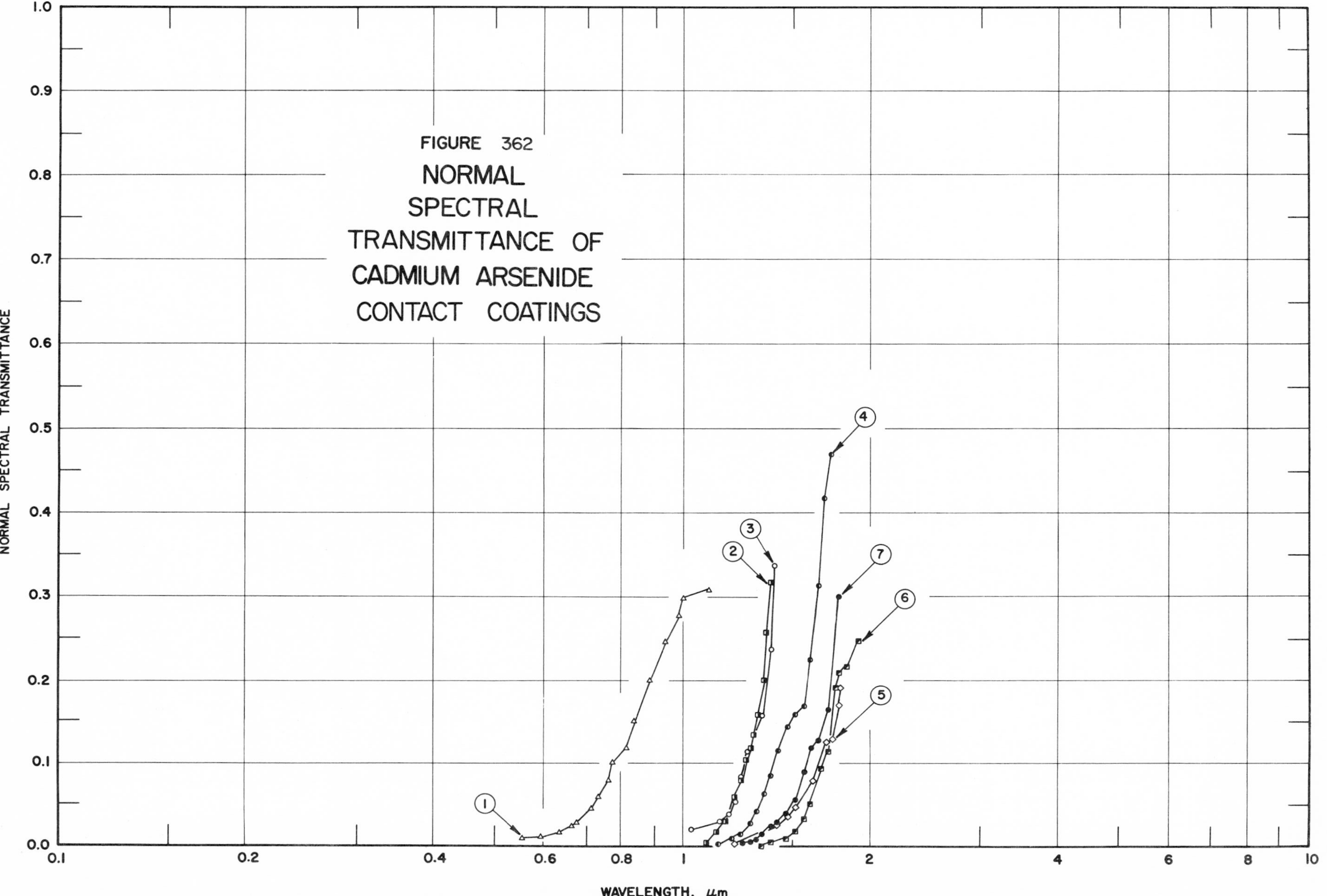
FIGURE 362
NORMAL
SPECTRAL
TRANSMITTANCE OF
CADMIUM ARSENIDE
CONTACT COATINGS
NORMAL SPECTRAL TRANSMITTANCE
WAVELENGTH, μm
1.0
0.9
0.8
0.7
0.6
0.5
0.4
0.3
0.2
0.1
0.0
0.1
0.2
0.4
0.6
0.8
1
2
4
6
8
10
1
2
3
4
5
6
7

SPECIFICATION TABLE NO. 362 NORMAL SPECTRAL TRANSMITTANCE OF CADMIUM ARSENIDE CONTACT COATINGS

Curve No.	Ref. No.	Year	Temperature, K	Wavelength Range, μm	Geometry θ θ' ω'	Reported Error, %	Composition (weight percent), Specifications, and Remarks
1	140	1967	298	0.555-1.113	~0° ~0°		Cd_3As_2 (0.13 μm thick); BK-7 glass (~2 mm thick) substrate; substrate optically polished and held at 433 K during vacuum (10^{-4} to 10^{-5} mm Hg) depositing of coating. [Author's designation: No. 49]
2	140	1967	298	1.093-1.389	~0° ~0°		Similar to above specimen and conditions except 1.25 μm thick. [Author's designation: No. 48b]
3	140	1967	298	1.104-1.409	~0° ~0°		Above specimen and conditions.
4	140	1967	298	1.148-1.740	~0° ~0°		Similar to above specimen and conditions except 1.3 μm thick. [Author's designation: No. 39]
5	140	1967	298	1.218-1.796	~0° ~0°		Similar to above specimen and conditions except 1.4 μm thick. [Author's designation: No. 36]
6	140	1967	298	1.349-1.920	~0° ~0°		Cd_3As_2; glass substrate. [Author's designation: No. 51]
7	140	1967	298	1.250-1.782	~0° ~0°		Cd_3As_2; glass substrate. [Author's designation: No. 52]

DATA TABLE NO. 362 NORMAL SPECTRAL TRANSMITTANCE OF CADMIUM ARSENIDE CONTACT COATINGS

[Wavelength, λ, μm; Transmittance, τ; Temperature, T, K]

λ	τ
CURVE 1	**T = 298**
0.555	0.010
0.595	0.011
0.639	0.019
0.664	0.025
0.679	0.029
0.711	0.046
0.737	0.060
0.764	0.080
0.784	0.100
0.812	0.119
0.837	0.150
0.888	0.200
0.933	0.248
0.983	0.279
1.003	0.299
1.113	0.309
CURVE 2	**T = 298**
1.092	0.005
1.135	0.017
1.173	0.030
1.218	0.060
1.247	0.080
1.267	0.103
1.294	0.119
1.305	0.134
1.329	0.159
1.351	0.200
1.369	0.257
1.389	0.318
CURVE 3	**T = 298**
1.104	0.020
1.156	0.030
1.185	0.039
1.220	0.053
1.242	0.086
1.278	0.114
1.345	0.158
1.387	0.239
1.409	0.337
CURVE 4	**T = 298**
1.148	0.002
1.205	0.010
1.249	0.015
1.279	0.028
1.317	0.041
1.351	0.063
1.387	0.084
1.433	0.115
1.475	0.142
1.525	0.159
1.566	0.168
1.608	0.226
1.653	0.312
1.698	0.419
1.740	0.470
CURVE 5	**T = 298**
1.218	0.004
1.429	0.026
1.477	0.036
1.527	0.048
1.613	0.080
1.702	0.125
1.741	0.129
1.783	0.170
1.796	0.191
CURVE 6	**T = 298**
1.349	0.000
1.384	0.005
1.468	0.010
1.513	0.019
1.560	0.033
1.605	0.050
1.662	0.094
1.717	0.114
1.764	0.191
1.783	0.210
1.845	0.218
1.920	0.249
CURVE 7	**T = 298**
1.250	0.002
1.281	0.005
1.313	0.009
1.346	0.015
1.386	0.023
1.423	0.030
1.466	0.040
1.515	0.055
1.555	0.090
1.600	0.119
1.656	0.126
1.720	0.164
1.782	0.300

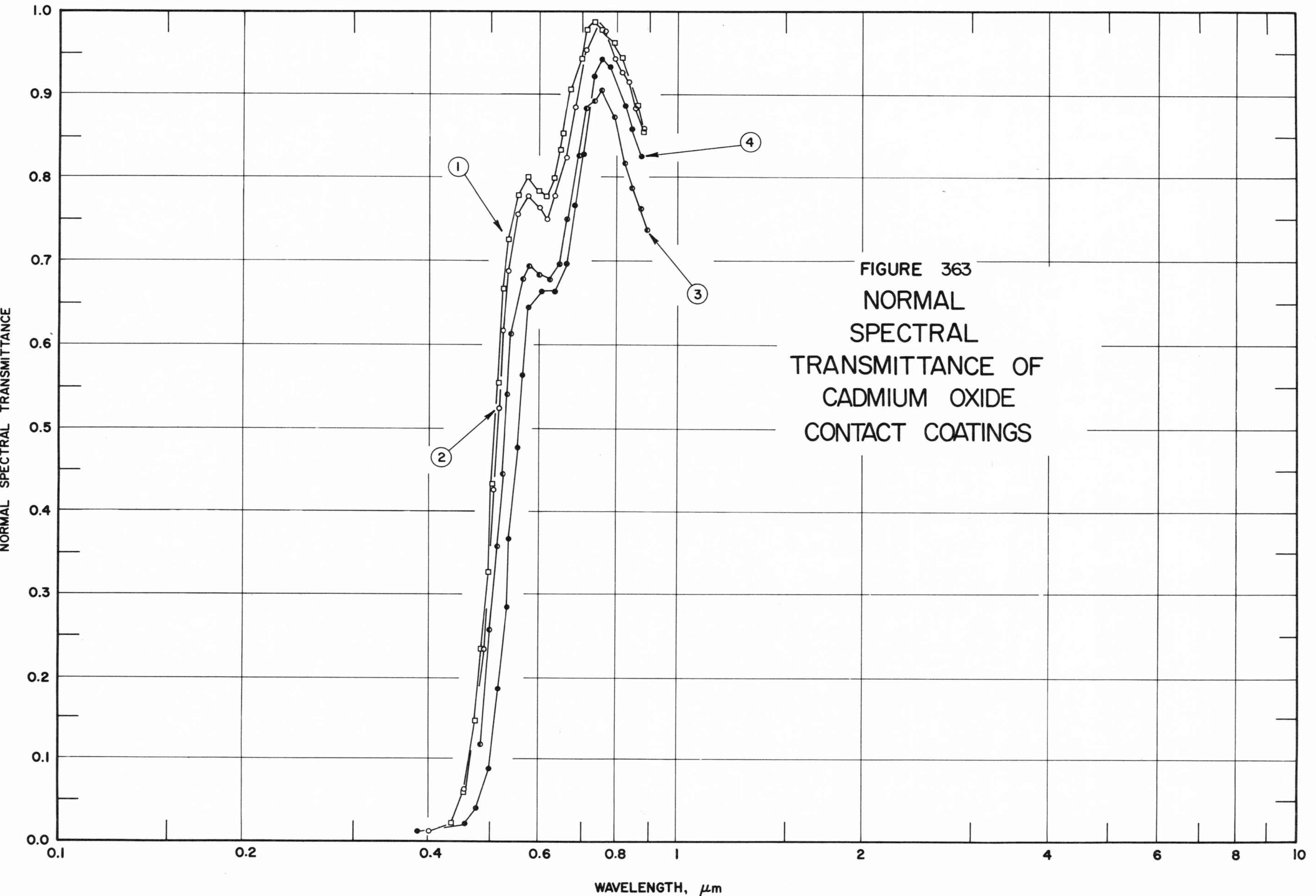
FIGURE 363
NORMAL
SPECTRAL
TRANSMITTANCE OF
CADMIUM OXIDE
CONTACT COATINGS
NORMAL SPECTRAL TRANSMITTANCE
WAVELENGTH, μm
1.0
0.9
0.8
0.7
0.6
0.5
0.4
0.3
0.2
0.1
0.0
0.1
0.2
0.4
0.6
0.8
1
2
4
6
8
10
1
2
3
4

SPECIFICATION TABLE NO. 363 NORMAL SPECTRAL TRANSMITTANCE OF CADMIUM OXIDE CONTACT COATINGS

Curve No.	Ref. No.	Year	Temperature, K	Wavelength Range, μm	Geometry θ	θ'	ω'	Reported Error, %	Composition (weight percent), Specifications, and Remarks
1	141	1969	~298	0.437-0.887	~0°	~0°			CdO; Pyrex glass substrate; sputtered at 5 x 10^{-1} mm Hg with dry air or argon-oxygen mixture present, 700 volts, 0.42 mA cm^{-2}; no annealing.
2	141	1969	~298	0.400-0.887	~0°	~0°			Above specimen and conditions except annealed at 473 K.
3	141	1969	~298	0.481-0.893	~0°	~0°			Above specimen and conditions except annealed at 573 K.
4	141	1969	~298	0.384-0.873	~0°	~0°			Above specimen and conditions except annealed at 773 K.

DATA TABLE NO. 363 NORMAL SPECTRAL TRANSMITTANCE OF CADMIUM OXIDE CONTACT COATINGS

[Wavelength, λ, μm; Transmittance, τ; Temperature, T, K]

λ	τ	λ	τ	λ	τ	λ	τ	λ	τ	λ	τ
CURVE 1 T ~298		CURVE 1 (cont.)		CURVE 2 (cont.)		CURVE 3 T ~298		CURVE 3 (cont.)		CURVE 4 (cont.)	
		0.733	0.988	0.538	0.687			0.875	0.761	0.709	0.830
0.437	0.023	0.751	0.979	0.556	0.755	0.481	0.117	0.893	0.736	0.737	0.921
0.456	0.606	0.769	0.976*	0.578	0.776	0.500	0.258			0.758	0.942
0.476	0.147	0.798	0.962	0.600	0.763	0.512	0.359	CURVE 4 T ~298		0.780	0.933
0.488	0.235	0.815	0.945	0.619	0.750	0.522	0.445			0.822	0.887
0.500	0.327	0.861	0.889	0.639	0.778	0.532	0.540			0.845	0.860
0.506	0.431	0.887	0.858	0.661	0.825	0.540	0.613	0.384	0.011	0.873	0.827
0.516	0.554			0.684	0.886	0.561	0.679	0.457	0.021		
0.527	0.669	CURVE 2 T ~298		0.713	0.952	0.579	0.694	0.479	0.040		
0.539	0.723			0.735	0.987	0.600	0.682	0.500	0.087		
0.556	0.779			0.752	0.981*	0.621	0.677	0.518	0.186		
0.578	0.800	0.400	0.012	0.769	0.978	0.642	0.698	0.531	0.284		
0.600	0.784	0.439	0.022*	0.793	0.944	0.668	0.750	0.539	0.366		
0.617	0.778	0.458	0.062	0.811	0.929	0.693	0.828	0.551	0.477		
0.635	0.800	0.477	0.148*	0.832	0.915	0.715	0.883	0.562	0.563		
0.650	0.833	0.490	0.235	0.859	0.883	0.736	0.892	0.578	0.644		
0.658	0.852	0.500	0.327*	0.887	0.860	0.753	0.903	0.606	0.662		
0.679	0.907	0.508	0.428			0.791	0.871	0.635	0.663		
0.700	0.942	0.516	0.523			0.824	0.819	0.661	0.696		
0.719	0.979	0.527	0.618			0.849	0.788	0.687	0.766		

* Not shown on plot

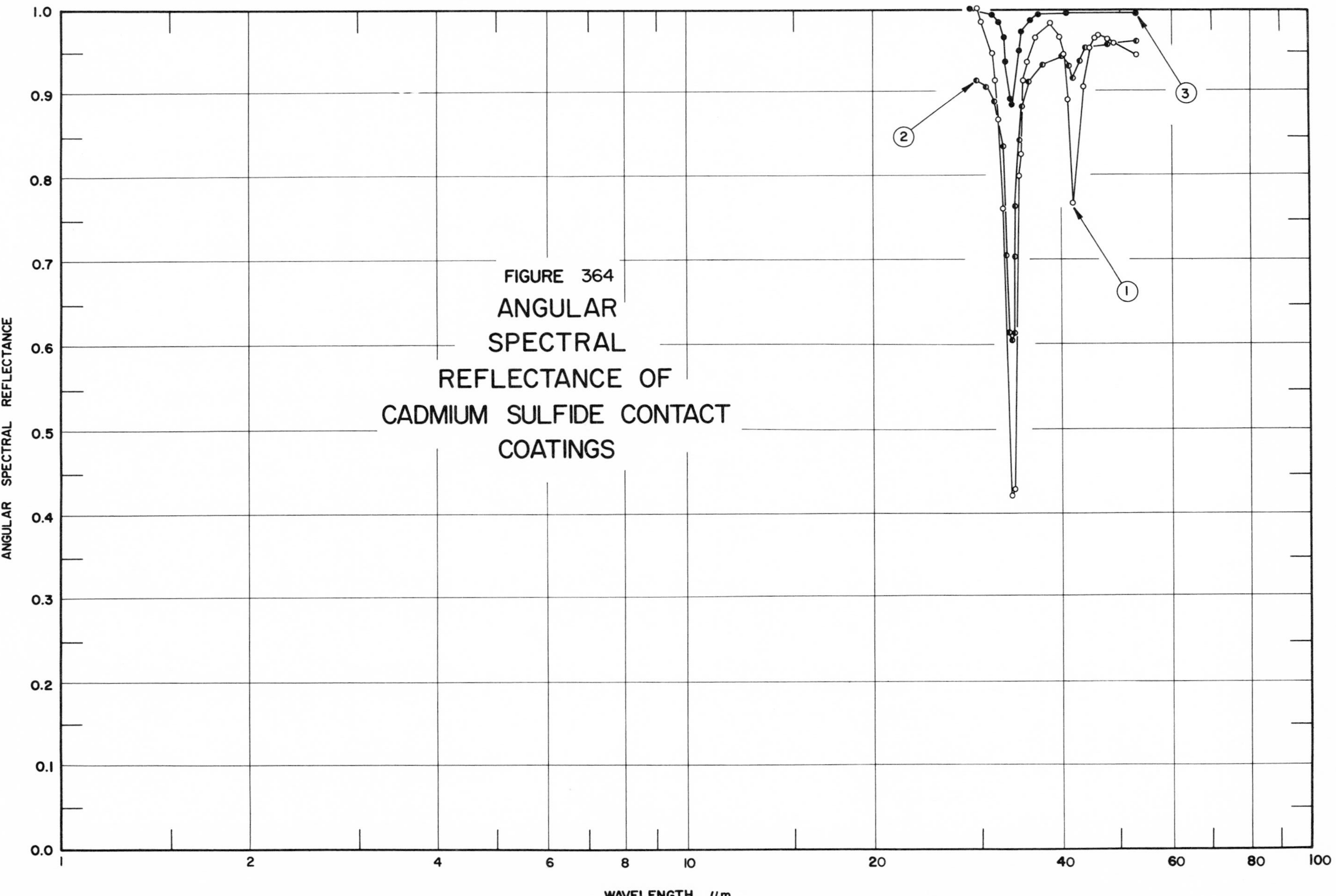
FIGURE 364
ANGULAR
SPECTRAL
REFLECTANCE OF
CADMIUM SULFIDE CONTACT
COATINGS
ANGULAR SPECTRAL REFLECTANCE
WAVELENGTH, μm
1.0
0.9
0.8
0.7
0.6
0.5
0.4
0.3
0.2
0.1
0.0
1
2
4
6
8
10
20
40
60
80
100
1
2
3

SPECIFICATION TABLE NO. 364 ANGULAR SPECTRAL REFLECTANCE OF CADMIUM SULFIDE CONTACT COATINGS

Curve No.	Ref. No.	Year	Temperature, K	Wavelength Range, μm	Geometry θ	θ'	ω'	Reported Error, %	Composition (weight percent), Specifications, and Remarks
1	142	1969	~298	29.2-52.9	45°	45°			CdS (0.66 μm thick); aluminum and glass substrates; aluminum vapor deposited on glass, CdS vapor deposited on aluminum; annealed in air at ~573 K for 1 hr; radiation polarized parallel to plane of incidence; measured relative to aluminum vapor deposited on glass.
2	142	1969	~298	29.2-52.9	45°	45°			Above specimen and conditions except 0.33 μm thick.
3	142	1969	~298	28.6-52.9	45°	45°			Above specimen and conditions except 0.086 μm thick.

DATA TABLE NO. 364 ANGULAR SPECTRAL REFLECTANCE OF CADMIUM SULFIDE CONTACT COATINGS

[Wavelength, λ, μm; Reflectance, ρ; Temperature, T, K]

λ	ρ
CURVE 1 T ~298	
29.2	1.000
29.9	0.983
31.0	0.947
31.4	0.915
31.7	0.869
32.3	0.761
33.2	0.424
33.3	0.430
34.1	0.800
34.2	0.827
34.4	0.846*
34.5	0.881*
34.8	0.914
35.2	0.937
36.2	0.964
37.0	0.975*
38.2	0.981
38.9	0.975*
CURVE 1 (cont.)	
39.8	0.964
40.5	0.946
40.8	0.925*
41.0	0.892
41.8	0.768
43.1	0.906
43.3	0.927*
44.1	0.951
45.0	0.964
45.9	0.967
47.4	0.961
48.3	0.957
52.9	0.944
CURVE 2 T ~298	
29.2	0.915
30.3	0.907
CURVE 2 (cont.)	
31.2	0.890
31.6	0.870*
32.1	0.836
32.5	0.759*
32.6	0.707
33.0	0.615
33.1	0.605
33.3	0.615
33.6	0.704
33.8	0.764
34.1	0.829*
34.4	0.843
34.7	0.884
35.7	0.914
37.2	0.932
40.0	0.942
41.0	0.932
41.7	0.918
42.9	0.939
CURVE 2 (cont.)	
43.7	0.952
47.2	0.957
52.9	0.960
CURVE 3 T ~298	
28.6	1.000
31.0	0.993
31.6	0.982
32.2	0.965
32.5	0.939
33.0	0.892
33.2	0.886
34.1	0.950
34.6	0.971
35.7	0.987
36.6	0.992
40.7	0.997
52.9	0.996

* Not shown on plot

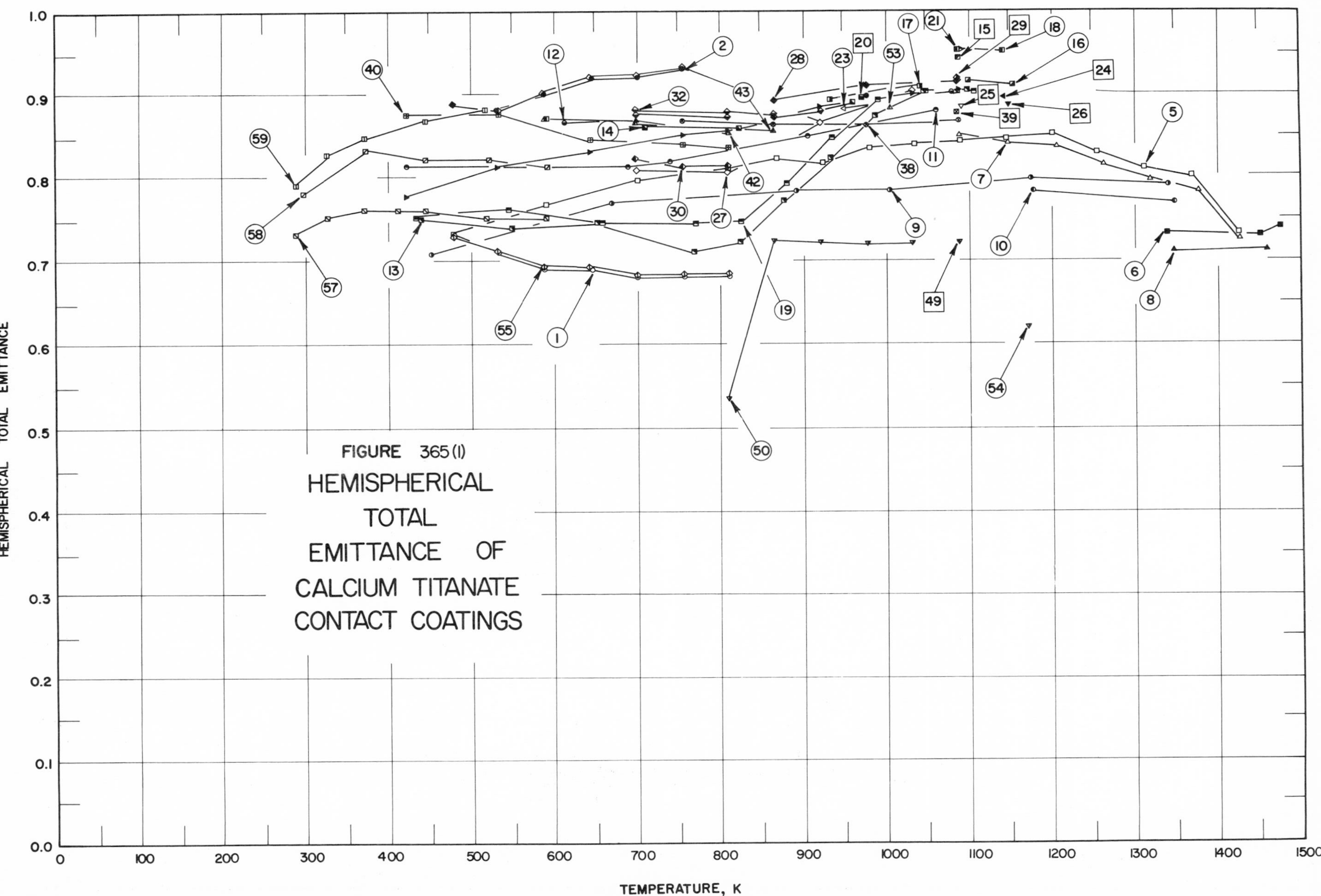

FIGURE 365 (1) HEMISPHERICAL TOTAL EMITTANCE OF CALCIUM TITANATE CONTACT COATINGS

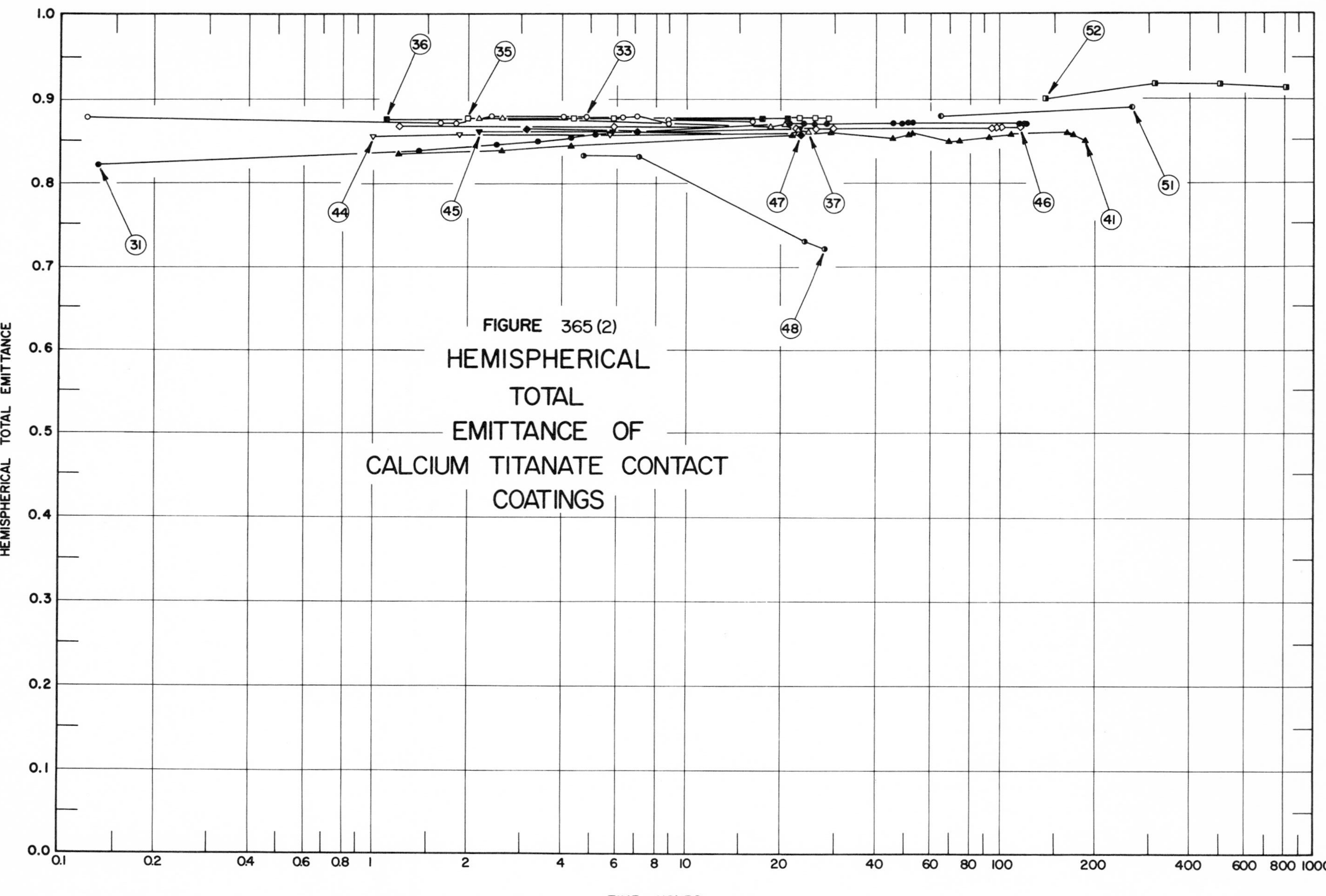
FIGURE 365 (2)
HEMISPHERICAL TOTAL EMITTANCE OF CALCIUM TITANATE CONTACT COATINGS
HEMISPHERICAL TOTAL EMITTANCE
TIME, HOURS
1.0
0.9
0.8
0.7
0.6
0.5
0.4
0.3
0.2
0.1
0.0
0.1
0.2
0.4
0.6
0.8
1
2
4
6
8
10
20
40
60
80
100
200
400
600
800
1000
36
35
33
52
31
44
45
47
37
46
41
51
48

SPECIFICATION TABLE NO. 365 HEMISPHERICAL TOTAL EMITTANCE OF CALCIUM TITANATE CONTACT COATINGS

Curve No.	Ref. No.	Year	Temperature Range, K	Reported Error, %	Composition (weight percent), Specifications and Remarks
1	135	1969	478-812		$CaTiO_3$, National Lead Co.; high purity IS-2 beryllium ingot substrate; plasma arc sprayed; measured in vacuum ($<10^{-5}$ mm Hg).
2	135	1969	478-756		Similar to above specimen and conditions except aged 1000 hrs at 922 K.
3	135	1969	1005		$CaTiO_3$, powder from National Lead Co.; AISI-310 stainless steel substrate; substrate cleaned in saturated soln. of sulfuric acid and potassium dichromate and grit blasted with 60 mesh silicon carbide to a roughness height of 2.79 μm; plasma arc sprayed; aged 12 hrs at 1005 K; measured in vacuum ($\sim 10^{-7}$ mm Hg).
4*	135	1969	298-1005		Similar to above specimen and conditions except exposed to thermal cycling between 298 and 1005 K at the rate of 112 cycles per 20 000 hrs (starting at 1005 K); thermal cycling time (hrs) is variable.
5	118	1962	479-1476		Calcium titanate, powder from Metco, Inc., (0.127 mm thick); niobium substrate; plasma arc sprayed; extremely hard; rough texture; measured in vacuum ($<3.6 \times 10^{-6}$ mm Hg); temp measured with thermocouple; rising temp. [Authors' designation: Run 1]
6	118	1962	1476-1339		Above specimen and conditions except decending temp.
7	118	1962	1089-1476		Curve 5 specimen and conditions except temp measured with optical pyrometer.
8	118	1962	1476-1348		Curve 6 specimen and conditions except temp measured with optical pyrometer.
9	118	1962	1952-352		Curve 5 specimen and conditions except vacuum ($<2.4 \times 10^{-6}$ mm Hg); descending temp. [Authors' designation: Run 2]
10	118	1962	1348-1178		Above specimen and conditions except temp measured with optical pyrometer.
11	82	1962	422-1062	<±2.5	Calcium titanate ($\sim$0.0762 mm thick); stainless steel 310 substrate; plasma arc sprayed; measured in vacuum (10^{-7} mm Hg); data extracted from smooth curve.
12	82	1962	613-1080	<±2.5	Above specimen and conditions except heated at 1160 K for 17 hrs.
13	81	1963	439-1099		Calcium titanate (National Lead Co.), $CaO \cdot TiO_2$ (0.127 mm thick); Nb-1Zr substrate; plasma arc sprayed; measured in vacuum ($<9.5 \times 10^{-6}$ mm Hg); temp measured with thermocouple; Run No. 1, heating cycle.
14	81	1963	1099-710		Above specimen and conditions; cooling cycle.
15	81	1963	1087		Curve 13 specimen and conditions except temp measured with optical pyrometer.
16	81	1963	1099-1154		Curve 13 specimen and conditions; Run No. 2.
17	81	1963	1154-934		Above specimen and conditions; cooling cycle.
18	81	1963	1087-1141		Curve 16 specimen and conditions except temp measured with optical pyrometer.
19	81	1963	434-1107		Calcium titanate (National Lead Co.), $CaO \cdot TiO_2$ (0.102 mm thick); Nb-1Zr substrate; plasma arc sprayed; measured in vacuum ($<2.2 \times 10^{-6}$ mm Hg); temp measured with thermocouple; Run No. 1, heating cycle.
20	81	1963	1107-970		Above specimen and conditions; cooling cycle.

* Not shown on plot

SPECIFICATION TABLE NO. 365 HEMISPHERICAL TOTAL EMITTANCE OF CALCIUM TITANATE CONTACT COATINGS (continued)

Curve No.	Ref. No.	Year	Temperature Range, K	Reported Error, %	Composition (weight percent), Specifications and Remarks
21	81	1963	1092		Curve 19 specimen and conditions except temp measured with optical pyrometer.
22	81	1963	422-1088		Calcium titanate (0. 127 mm thick); Nb-1Zr substrate; plasma arc sprayed; measured in vacuum (3. 1 x 10^{-7} to 1. 6 x 10^{-6} mm Hg); temp measured with thermocouple; Run No. 1, heating cycle.
23	81	1963	1088-950		Above specimen and conditions; cooling cycle.
24	81	1963	1144		Curve 22 specimen and conditions; Run No. 2.
25	81	1963	1094		Curve 22 specimen and conditions except temp measured with optical pyrometer.
26	81	1963	1148		Above specimen and conditions; cooling cycle.
27	81	1963	699-1088		Calcium titanate (0. 0508 mm thick); Nb-1Zr substrate; plasma arc sprayed; measured in vacuum (1. 2 x 10^{-6} mm Hg); temp measured with thermocouple; Run No. 1, heating cycle.
28	81	1963	1088-866		Above specimen and conditions; cooling cycle.
29	81	1963	1088		Curve 27 specimen and conditions except temp measured with optical pyrometer.
30	81	1963	699-810		Calcium titanate (0. 127 mm thick); Nb-1Zr substrate; plasma arc sprayed; measured in vacuum (<1. 1 x 10^{-7} mm Hg); temp measured with thermocouple; Run No. 1, heating cycle.
31	81	1963	810		Above specimen and conditions except in addition sample maintained at 810 K for 124. 1 hrs endurance test; endurance time, hrs, (time from first measurement at 810 K) is variable.
32	81	1963	699-866		Curve 30 specimen and conditions; after curve 31; Run No. 2.
33	81	1963	866		Above specimen and conditions except in addition sample maintained at 866 K for 160. 5 hrs endurance test; endurance time, hrs (time from first measurement at 866 K) is variable.
34	81	1963	699-922		Curve 30 specimen and conditions; after curve 33; Run No. 3.
35	81	1963	922		Above specimen and conditions except in addition sample maintained at 922 K during 28. 6 hrs endurance test; endurance time, hrs (time from first measurement at 922 K) is variable.
36	81	1963	977		Above specimen and conditions except in addition sample maintained at 977 K during 21. 2 hrs endurance test; endurance time, hrs (time from first measurement at 977 K) is variable.
37	81	1963	1033		Above specimen and conditions except in addition sample maintained at 1033 K during 24. 5 hrs endurance test; endurance time, hrs (time from first measurement at 1033 K) is variable.
38	81	1963	1088-755		Curve 34 specimen and conditions; after curve 37; cooling cycle.
39	81	1963	1085		Above specimen and conditions except temp measured with optical pyrometer.
40	81	1963	422-810		Calcium titanate (0. 127 mm thick); Nb-1Zr substrate; plasma arc sprayed; measured in vacuum (<1. 7 x 10^{-6} mm Hg); temp measured with thermocouple; Run No. 1, heating cycle.
41	81	1963	810		Above specimen and conditions except in addition maintained at 810 K during 189. 7 endurance test; endurance time, hrs (time from first measurement at 810 K) is variable.
42	81	1963	811-589		Curve 40 specimen and conditions; after curve 41; cooling cycle.
43	81	1963	699-866		Curve 40 specimen and conditions; after curve 42; Run No. 2.

SPECIFICATION TABLE NO. 365 HEMISPHERICAL TOTAL EMITTANCE OF CALCIUM TITANATE CONTACT COATINGS (continued)

Curve No.	Ref. No.	Year	Temperature Range, K	Reported Error, %	Composition (weight percent), Specifications and Remarks
44	81	1963	866		Above specimen and conditions except in addition sample maintained at 866 K during 22.3 hrs endurance test; endurance time, hrs (time from first measurement at 866 K) is variable.
45	81	1963	922		Above specimen and conditions except in addition sample maintained at 922 K during 22.8 hrs endurance test; endurance time, hrs (time from first measurement at 922 K) is variable.
46	81	1963	977		Above specimen and conditions except in addition sample maintained at 977 K during 118.3 hrs endurance test; endurance time, hrs (time from first measurement at 977 K) is variable.
47	81	1963	1033		Above specimen and conditions except in addition sample maintained at 1033 K during 23.4 hrs endurance test; endurance time, hrs (time from first measurement at 1033 K), is variable.
48	81	1963	1088		Above specimen and conditions except in addition sample maintained at 1088 K during 27.7 hrs endurance test; endurance time, hrs (time from first measurement at 1088 K), is variable.
49	81	1963	1088		Above specimen and conditions except temp measured with optical pyrometer; endurance time 27.7 hrs; temp is variable.
50	81	1963	1088-810		Curve 43 specimen and conditions; after curve 49; cooling cycle.
51	143	1964	1006	±2.3-2.7	$CaTiO_3$ (0.102 mm thick); AISI-310 stainless steel substrate; plasma arc sprayed; measured in vacuum ($<10^{-8}$ mm Hg); time (hrs) at 1006 K is variable; thermal shock consisting of cooling to ~300 K in 6.5 min then reheating slowly to 1006 K occurred after initial heating, 100 hrs, and 200 hrs of test.
52	143	1964	1006	±2.7	Similar to above specimen and conditions except sample maintained at 1006 K during ~2000 hrs endurance test; endurance test time, hrs, is variable; no thermal shock.
53	143	1964	1006	±2.3-2.7	$CaTiO_3$ (0.127 mm thick); Nb-1Zr alloy substrate; plasma arc sprayed; sample maintained 100 hrs at 1006 K.
54	143	1964	1172	±2.3-2.7	Above specimen and conditions except in addition sample maintained at 1172 K for 68 hrs.
55	144	1966	478-810		Calcium titanate (0.102 mm thick); high purity Be ingot substrate; measured in vacuum (10^{-5} mm Hg).
56	144	1966	478-755		Above specimen and conditions except aged 1000 hrs in vacuum at 922 K.
57	137	1965	287-592	±5	Calcium titanate (~0.05 mm thick, 6.2 mg cm^{-2}); Al substrate; measured in vacuum (10^{-7} mm Hg).
58	137	1965	298-593	±5	Similar to above specimen and conditions except coating 0.088 mm thick, 11.3 mg cm^{-2}.
59	137	1965	288-591	±5	Similar to curve 57 specimen and conditions except coating 0.114 mm thick, 23 mg cm^{-2}.

DATA TABLE NO. 365 HEMISPHERICAL TOTAL EMITTANCE OF CALCIUM TITANATE CONTACT COATINGS

[Temperature, T, K; Emittance, ∈]

T	∈
CURVE 1	
478	0.730
533	0.710
589	0.690
645	0.689
700	0.679
756	0.679
812	0.680
CURVE 2	
478	0.889
533	0.881
589	0.900
645	0.920
700	0.921
756	0.930
CURVE 3*	
1005	0.908

Thermal cycling time	∈
CURVE 4* T = 298-1005	
310	0.920
400	0.914
480	0.925
710	0.915
880	0.911
1130	0.907
1380	0.913
1550	0.913
1680	0.912
1870	0.914
2110	0.910
2320	0.912
2540	0.907
2790	0.907
2960	0.901
3190	0.896
3380	0.897
3530	0.900
CURVE 4 (cont.)*	
3780	0.901
3960	0.899
4170	0.893
4430	0.890
4620	0.894
4860	0.897
5060	0.895
5280	0.895
5490	0.897
5640	0.897
5830	0.894
6070	0.900
6320	0.902
6520	0.893
6640	0.889
6820	0.892
7030	0.900
7280	0.889
7480	0.889
7670	0.896
7820	0.898
8060	0.893
8300	0.888
8470	0.892
8690	0.892
8970	0.898
9170	0.898
9280	0.890
9490	0.893
9690	0.894
9860	0.894
10120	0.894
10420	0.889
10640	0.890
11070	0.894
11390	0.895
11830	0.898
12170	0.895
12450	0.892
12740	0.892
12960	0.892
13410	0.883
13870	0.883
14360	0.883
CURVE 4 (cont.)*	
14720	0.879
15020	0.880
15360	0.887
15680	0.882
15950	0.883
16290	0.884
16690	0.881
17010	0.881
17430	0.885
17780	0.888
18080	0.882
18400	0.871
18730	0.874
19290	0.881
19710	0 878
20000	0.896

T	∈
CURVE 5	
479	0.732
590	0.768
700	0.796
810	0.810
869	0.823
924	0.817
981	0.834
1035	0.839
1090	0.844
1145	0.846
1200	0.850
1256	0.828
1311	0.810
1369	0.800
1425	0.731
1476	0.738
CURVE 6	
1476	0.738
1450	0.729
1339	0.732
CURVE 7	
1089	0.848
1147	0.840
1205	0.836
1261	0.813
1318	0.795
1376	0.782
1428	0.724
1476	0.738*
CURVE 8	
1476	0.738*
1459	0.712
1348	0.711
CURVE 9	
1340	0.788
1173	0.797
1006	0.784
894	0.784
669	0.769
451	0.708
CURVE 10	
1348	0.784
1178	0.769
CURVE 11	
422	0.814
689	0.812
740	0.818
905	0.849
1062	0.877
CURVE 12	
613	0.867
977	0.897
1080	0.901
CURVE 13	
439	0.750
547	0.738
651	0.745
768	0.710
824	0.724
879	0.774
934	0.823
988	0.871
1046	0.901
1099	0.904
CURVE 14	
1099	0.904*
962	0.889
823	0.857
710	0.859
CURVE 15	
1087	0.944
CURVE 16	
1099	0.915
1154	0.910
CURVE 17	
1154	0.910*
1043	0.907
934	0.893
CURVE 18	
1087	0.953
1141	0.951
CURVE 19	
434	0.752
546	0.762
659	0.745
770	0.744
825	0.746
CURVE 19 (cont.)	
881	0.791
936	0.846
991	0.891
1049	0.901
1107	0.902
CURVE 20	
1107	0.902*
970	0.896
CURVE 21	
1092	0.953
CURVE 22	
422	0.778
533	0.814
644	0.830
755	0.850
810	0.856
866	0.861
920	0.884
977	0.897
1033	0.900
1088	0.904
CURVE 23	
1088	0.904
950	0.880
CURVE 24	
1144	0.895
CURVE 25	
1094	0.884
CURVE 26	
1148	0.883
CURVE 27	
699	0.807
810	0.805
922	0.866
1033	0.904
1088	0.915
CURVE 28	
1088	0.915
977	0.909
866	0.894
CURVE 29	
1088	0.917
CURVE 30	
699	0.822
755	0.812
810	0.812

Endurance time	∈
CURVE 31 T = 810	
0.0	0.812*
0.3	0.823
1.4	0.837
2.5	0.847
3.4	0.850
4.3	0.856
5.1	0.858
21.7	0.870
23.9	0.871
26.0	0.871
28.3	0.871
46.4	0.872
49.3	0.872
51.3	0.872
53.2	0.872
118.1	0.872
120.3	0.872
CURVE 31 (cont.)	
122.2	0.872
124.1	0.872*

T	∈
CURVE 32	
699	0.876
810	0.872
866	0.872

Endurance time	∈
CURVE 33 T = 866	
0.0	0.872*
0.2	0.881
16.6	0.874
18.6	0.874
24.0	0.882
40.6	0.883
48.3	0.883
65.9	0.883
70.9	0.883
88.5	0.872
89.0	0.878
93.4	0.878*
160.5	0.875

T	∈
CURVE 34	
699	0.882
810	0.878
866	0.876
922	0.878

* Not shown on plot

DATA TABLE NO. 365 HEMISPHERICAL TOTAL EMITTANCE OF CALCIUM TITANATE CONTACT COATINGS (continued)

Endurance time	ϵ
CURVE 35	
T = 922	
0.0	0.878
2.0	0.877
4.4	0.877
5.9	0.877
23.0	0.878
26.2	0.877
28.6	0.878
CURVE 36	
T = 977	
0.0	0.877
1.1	0.878
17.7	0.877
21.2	0.877
CURVE 37	
T = 1033	
0.0	0.881*
2.2	0.877
2.6	0.877
18.9	0.868
20.8	0.870
22.7	0.864
24.5	0.862

T	ϵ
CURVE 38	
1088	0.865
977	0.862
866	0.863
755	0.867
CURVE 39	
1085	0.876
CURVE 40	
422	0.875
533	0.876
644	0.843
CURVE 40 (cont.)	
755	0.838
810	0.834

Endurance time	ϵ
CURVE 41	
T = 810	
0.0	0.834
1.2	0.835
2.6	0.839
4.3	0.846
21.8	0.858
29.0	0.860
45.9	0.855
51.1	0.857
52.9	0.859
69.6	0.853
74.4	0.853
93.6	0.856
100.9	0.857
166.3	0.861
173.3	0.857
189.7	0.853

T	ϵ
CURVE 42	
811	0.853
699	0.867
589	0.870
CURVE 43	
699	0.864
810	0.855*
866	0.856

Endurance time	ϵ
CURVE 44	
T = 866	
0.0	0.856*
1.0	0.855
1.9	0.858
5.7	0.858
22.3	0.858
CURVE 45	
T = 922	
0.0	0.863*
2.2	0.862
5.8	0.862
22.8	0.866
CURVE 46	
T = 977	
0.0	0.866*
1.2	0.867
5.9	0.867
22.6	0.865
26.2	0.865
29.8	0.866
95.0	0.865
98.9	0.866
101.7	0.865
118.3	0.866
CURVE 47	
T = 1033	
0.0	0.865
3.1	0.865
7.0	0.863
23.4	0.858
CURVE 48	
T = 1088	
0.0	0.858
4.7	0.833
7.1	0.833
23.9	0.730
27.7	0.722

T	ϵ
CURVE 49	
1088	0.723
CURVE 50	
1088	0.722*
1033	0.720
922	0.719
866	0.724
810	0.555

Endurance time	ϵ
CURVE 51	
T = 1006	
65	0.88
265	0.89
CURVE 52	
T = 1006	
0.0	0.900*
140	0.900
310	0.920
500	0.920
810	0.915
1010	0.910*
1220	0.915*
1480	0.915*
1620	0.915*
1720	0.915*
2000	0.915*

T	ϵ
CURVE 53	
1006	0.88
CURVE 54	
1172	0.62
CURVE 55	
478	0.728
533	0.711
589	0.693
644	0.693
700	0.681
755	0.681
810	0.681
CURVE 56	
478	0.889
533	0.880
589	0.900
644	0.921
700	0.923
755	0.934
CURVE 57	
287	0.733
327	0.752
372	0.763
412	0.761
445	0.760
517	0.750
592	0.750
CURVE 58	
298	0.781
374	0.832
446	0.821
521	0.821
593	0.811
CURVE 59	
288	0.790
326	0.829
372	0.849
445	0.869
517	0.880
591	0.870

* Not shown on plot

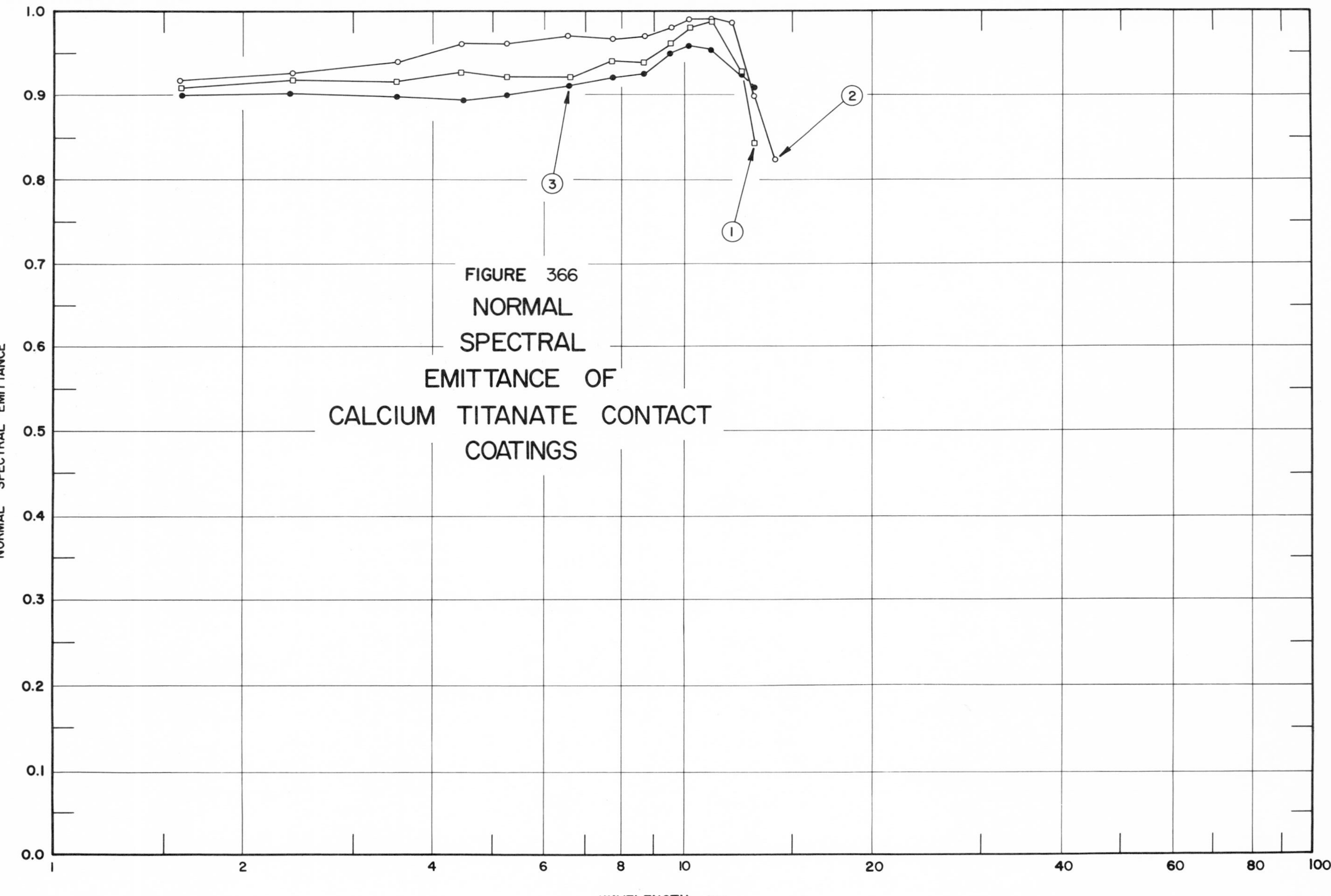

FIGURE 366 NORMAL SPECTRAL EMITTANCE OF CALCIUM TITANATE CONTACT COATINGS

SPECIFICATION TABLE NO. 366 NORMAL SPECTRAL EMITTANCE OF CALCIUM TITANATE CONTACT COATINGS

Curve No.	Ref. No.	Year	Temperature, K	Wavelength Range, μm	Geometry θ'	Reported Error, %	Composition (weight percent), Specifications, and Remarks
1	118	1962	755	1.59-13.00	~0°		Calcium titanate, powder from Metco, Inc., (0.0889 mm thick); niobium substrate; extremely hard, rough texture.
2	118	1962	1061	1.59-14.00	~0°		Above specimen and conditions.
3	118	1962	1366	1.62-13.00	~0°		Curve 1 specimen and conditions.

DATA TABLE NO. 366 NORMAL SPECTRAL EMITTANCE OF CALCIUM TITANATE CONTACT COATINGS

[Wavelength, λ, μm; Emittance, ϵ; Temperature, T, K]

λ	ϵ	λ	ϵ	λ	ϵ
CURVE 1 T = 755		CURVE 2 T = 1061		CURVE 3 T = 1366	
1.59	0.910	1.59	0.918	1.62	0.900
2.41	0.918	2.41	0.928	2.38	0.904
3.52	0.917	3.53	0.941	3.52	0.899
4.45	0.929	4.45	0.962	4.47	0.895
5.25	0.923	5.25	0.962	5.25	0.900
6.62	0.924	6.59	0.974	6.61	0.914
7.74	0.943	7.74	0.969	7.76	0.924
8.68	0.943	8.69	0.972	8.68	0.927
9.55	0.964	9.58	0.983	9.54	0.950
10.36	0.983	10.35	0.992	10.36	0.959
11.07	0.989	11.08	0.994	11.08	0.956
12.43	0.931	11.99	0.987	12.42	0.935
13.00	0.844	13.00	0.900	13.00	0.910
		14.00	0.826		

SPECIFICATION TABLE NO. 367 NORMAL SOLAR ABSORPTANCE OF CALCIUM TITANATE CONTACT COATINGS

Curve No.	Ref. No.	Year	Temperature Range, K	Geometry θ	Reported Error, %	Composition (weight percent), Specifications and Remarks
1*	137	1965	400-600	0°	±5	Calcium titanate (~0.050 mm thick, 6.2 mg cm^{-2}); Al substrate; measured in vacuum (10^{-7} mm Hg) after heating once to 600 K.
2*	137	1965	400-600	0°	±5	Similar to above specimen and conditions except coating 0.088 mm thick, 11.3 mg cm^{-2}.
3*	137	1965	400-600	0°	±5	Similar to curve 1 specimen and conditions except coating 0.144 mm thick, 23 mg cm^{-2}.

DATA TABLE NO. 367 NORMAL SOLAR ABSORPTANCE OF CALCIUM TITANATE CONTACT COATINGS

[Temperature, T, K; Absorptance, α]

T	α
CURVE 1*	
400	0.72
600	0.72
CURVE 2*	
400	0.70
600	0.70
CURVE 3*	
400	0.70
600	0.70

* No plot given

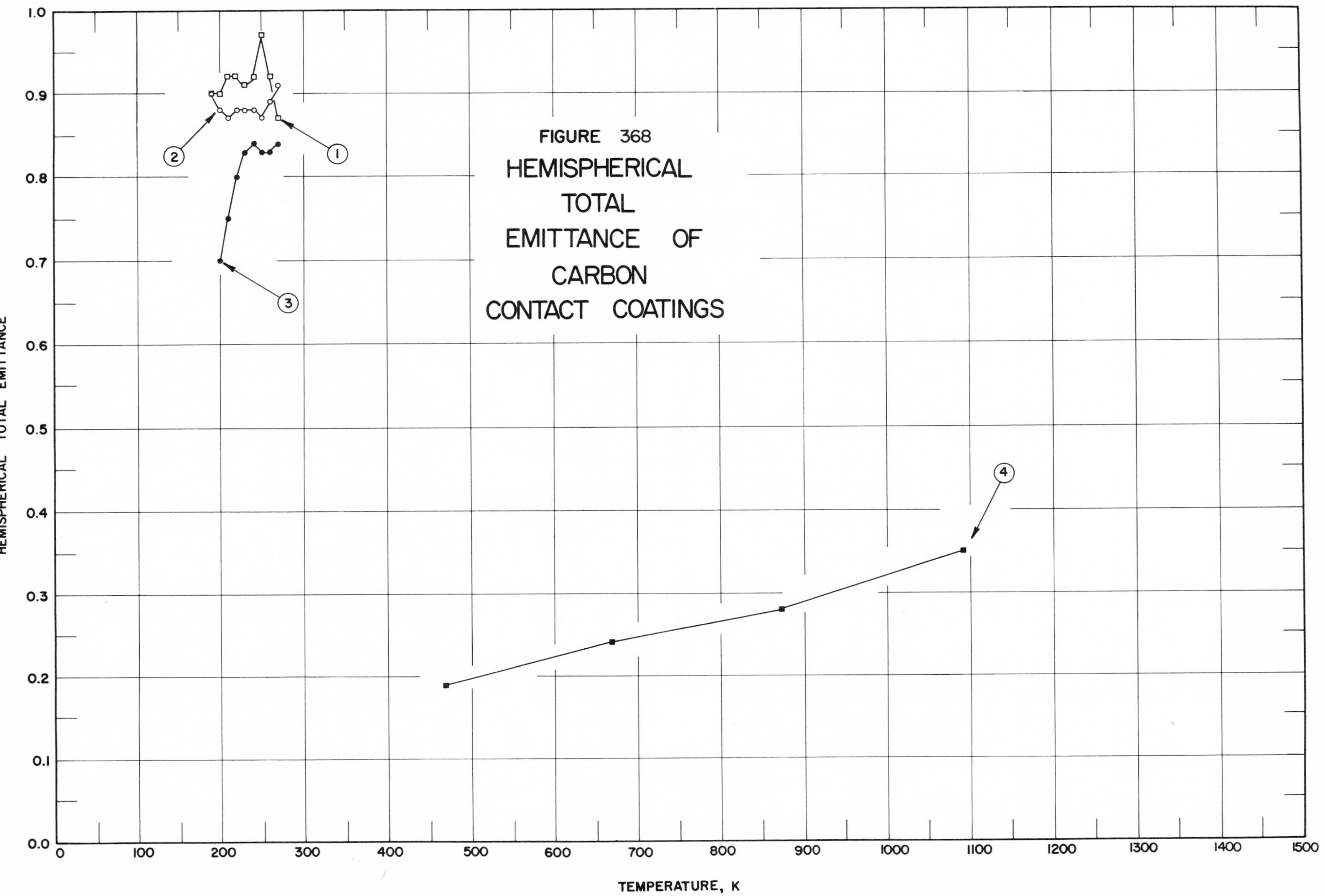
FIGURE 368
HEMISPHERICAL
TOTAL
EMITTANCE OF
CARBON
CONTACT COATINGS
HEMISPHERICAL TOTAL EMITTANCE
TEMPERATURE, K
0.0
0.1
0.2
0.3
0.4
0.5
0.6
0.7
0.8
0.9
1.0
0
100
200
300
400
500
600
700
800
900
1000
1100
1200
1300
1400
1500
1
2
3
4

SPECIFICATION TABLE NO. 368 HEMISPHERICAL TOTAL EMITTANCE OF CARBON CONTACT COATINGS

Curve No.	Ref. No.	Year	Temperature Range, K	Reported Error, %	Composition (weight percent), Specifications and Remarks
1	145	1957	190-270	4	Acetylene soot; copper substrate; measured in vacuum ($<2 \times 10^{-3}$ mm Hg); run 5.
2	145	1957	190-270	4	Above specimen and conditions; run 6.
3	145	1957	200-270	4	Above specimen and conditions; run 7.
4	146	1961	468-1093	<10	Carbon black; molybdenum substrate; substrate polished; applied with flame of burning kerosene; measured in vacuum (10^{-5} mm Hg).

DATA TABLE NO. 368 HEMISPHERICAL TOTAL EMITTANCE OF CARBON CONTACT COATINGS

[Temperature, T, K; Emittance, ϵ]

T	ϵ	T	ϵ	T	ϵ
CURVE 1		CURVE 2 (cont.)		CURVE 4 (cont.)	
190	0.90	250	0.87	668	0.24
200	0.90	260	0.89	873	0.28
210	0.92	270	0.91	1093	0.35
220	0.92				
230	0.91	CURVE 3			
240	0.92				
250	0.97	200	0.70		
260	0.92	210	0.75		
270	0.87	220	0.80		
		230	0.83		
CURVE 2		240	0.84		
		250	0.83		
190	0.90	260	0.83		
200	0.88	270	0.84		
210	0.87				
220	0.88	CURVE 4			
230	0.88				
240	0.88	468	0.19		

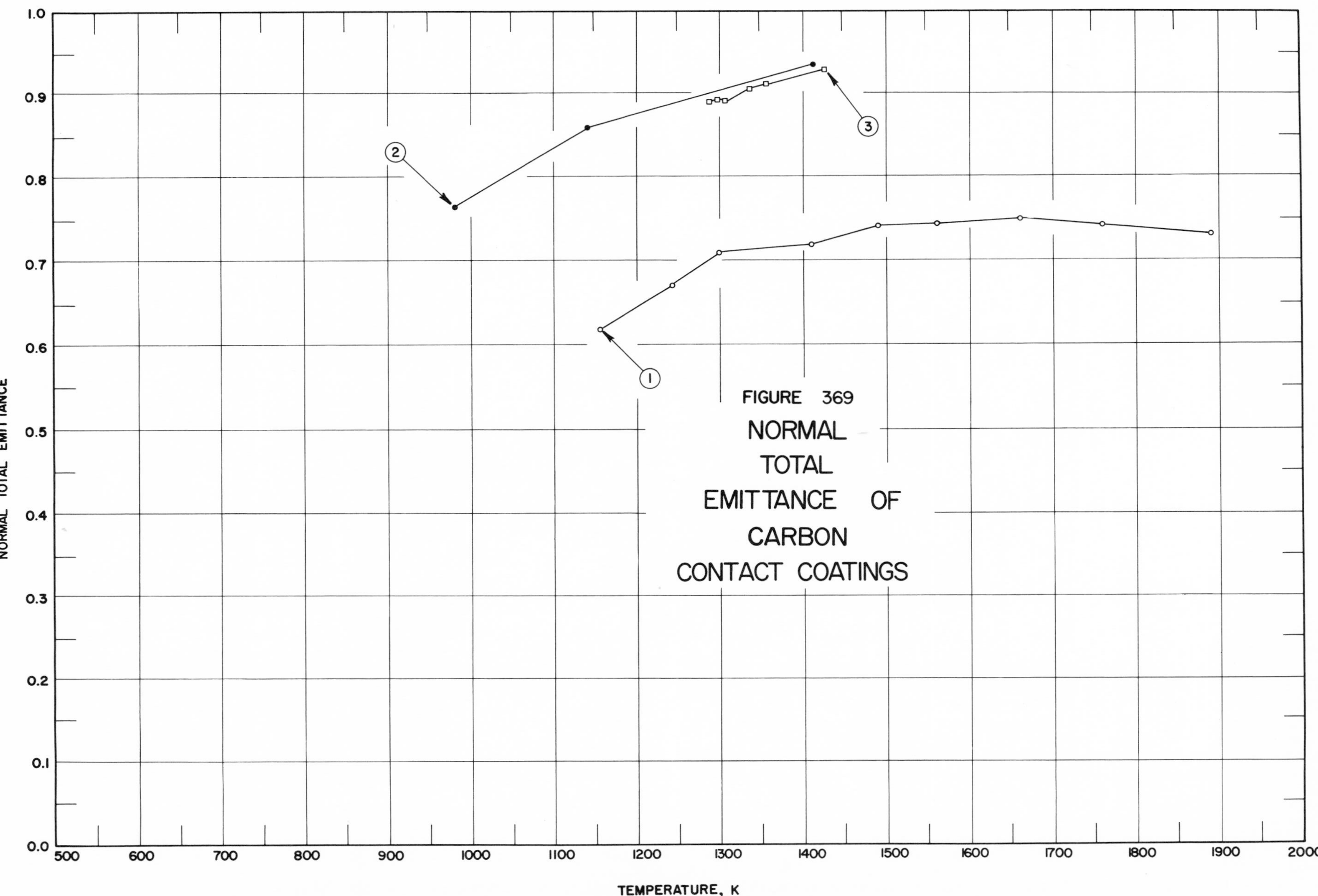

FIGURE 369 NORMAL TOTAL EMITTANCE OF CARBON CONTACT COATINGS

SPECIFICATION TABLE NO. 369 NORMAL TOTAL EMITTANCE OF CARBON CONTACT COATINGS

Curve No.	Ref. No.	Year	Temperature Range, K	Geometry θ'	Reported Error, %	Composition (weight percent), Specifications and Remarks
1	147	1963	1155-1894	~0°		Pyrolytic graphite (0.0203 mm thick); tantalum substrate; substrate grit blasted; vapor deposited; surface roughness 5 μm (RMS) measured with a profilometer; measured in vacuum (3 to 4 x 10^{-4} mm Hg).
2	128	1961	984-1416	~0°		Graphite; Al_2O_3 (polycrystalline) substrate; measured in vacuum (10^{-5} mm Hg); each value is the avg of several measurements.
3	128	1961	1289-1429	~0°		Similar to above specimen and conditions.

DATA TABLE NO. 369 NORMAL TOTAL EMITTANCE OF CARBON CONTACT COATINGS

[Temperature, T, K; Emittance, ε]

T	ε	T	ε
CURVE 1		CURVE 3	
1155	0.620	1289	0.890
1244	0.672	1300	0.894
1300	0.710	1308	0.892
1411	0.720	1337	0.906
1491	0.740	1357	0.912
1561	0.743	1429	0.928
1664	0.750	1429	0.928*
1761	0.742		
1894	0.730		
CURVE 2			
984	0.768		
1143	0.860		
1416	0.935		

* Not shown on plot

SPECIFICATION TABLE NO. 370 NORMAL SPECTRAL EMITTANCE OF CARBON CONTACT COATINGS

Curve No.	Ref. No.	Year	Wavelength, μm	Temperature Range, K	Geometry θ'	Reported Error, %	Composition (weight percent), Specifications, and Remarks
1*	147	1963	0.65	1150-1894	$\sim 0°$		Pyrolytic graphite (0.0203 mm thick); tantalum substrate; substrate grit blasted; vapor deposited; surface roughness 5 μm (RMS) measured with a profilometer; measured in vacuum (3 to 4 x 10^{-4} mm Hg); authors assumed specimen was a grey body.

DATA TABLE NO. 370 NORMAL SPECTRAL EMITTANCE OF CARBON CONTACT COATINGS

[Temperature, T, K; Emittance, ϵ; Wavelength, λ, μm]

T	ϵ
CURVE 1*	
$\lambda = 0.65$	
1150	0.700
1244	0.750
1300	0.785
1408	0.840
1489	0.837
1555	0.860
1664	0.767
1761	0.767
1894	0.670

* No plot given

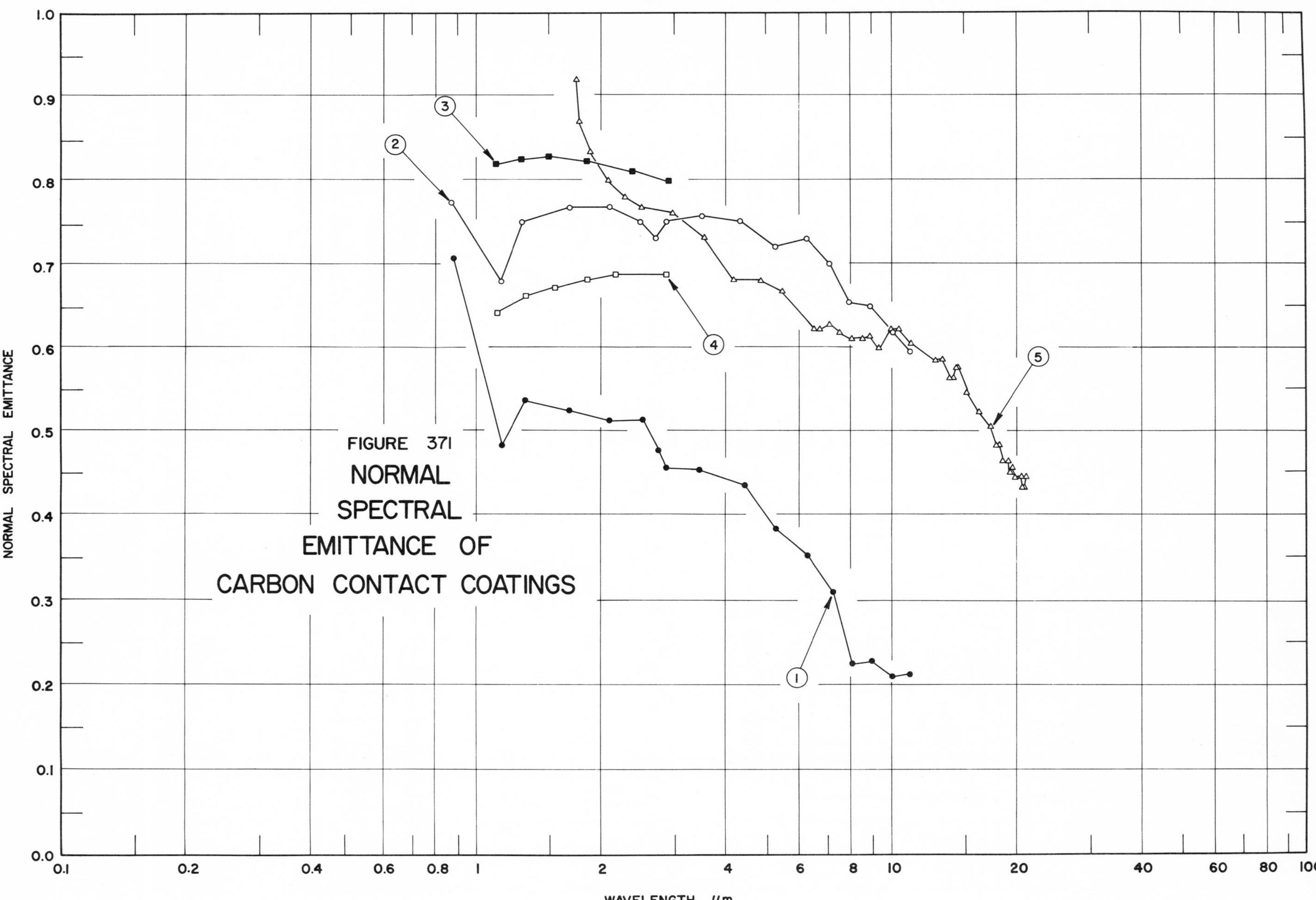

FIGURE 371 NORMAL SPECTRAL EMITTANCE OF CARBON CONTACT COATINGS

SPECIFICATION TABLE NO. 371 NORMAL SPECTRAL EMITTANCE OF CARBON CONTACT COATINGS

Curve No.	Ref. No.	Year	Temperature, K	Wavelength Range, μm	Geometry θ'	Reported Error, %	Composition (weight percent), Specifications, and Remarks
1	148	1968	1133	0.883-11.4	$\sim 0^\circ$		Pyrolytic graphite; substrate unknown; mechanically polished; vapor deposited at $\sim$2473 K; density $\sim$2.213 g cm^{-3}; measured from c-face in purified H_2 atm.
2	148	1968	1133	0.873-11.2	$\sim 0^\circ$		Similar to above specimen and conditions except measured from a-face.
3	148	1968	1133	1.14-2.90	$\sim 0^\circ$		Above specimen and conditions except measured with polarizer axis parallel to c-axis; data extracted from smooth curve.
4	148	1968	1133	1.14-2.88	$\sim 0^\circ$		Above specimen and conditions except measured with polarizer axis perpendicular to c-axis; data extracted from smooth curve.
5	74	1959	299	1.75-21.00	$\sim 0^\circ$		Aqua Dag (graphite); copper substrate; data extracted from smooth curve; property converted from R (2π, $\sim 0^\circ$).

DATA TABLE NO. 371 NORMAL SPECTRAL EMITTANCE OF CARBON CONTACT COATINGS

[Wavelength, λ, μm; Emittance, ϵ; Temperature, T, K]

λ	ϵ
CURVE 1 T = 1133	
0.883	0.709
1.16	0.481
1.32	0.536
1.69	0.524
2.10	0.511
2.51	0.513
2.76	0.476
2.88	0.454
3.44	0.453
4.42	0.435
5.27	0.383
6.28	0.353
7.28	0.310
8.07	0.225
9.04	0.229
10.2	0.210
11.4	0.214

λ	ϵ
CURVE 2 T = 1133	
0.873	0.776
1.15	0.680
1.29	0.750
1.69	0.767
2.10	0.769
2.48	0.751
2.70	0.733
2.87	0.751
3.50	0.757
4.36	0.752
5.22	0.722
6.21	0.733
7.15	0.701
7.93	0.654
8.89	0.649
10.2	0.619
11.2	0.596

λ	ϵ
CURVE 3 T = 1133	
1.14	0.820
1.29	0.826
1.50	0.828
1.85	0.824
2.37	0.813
2.90	0.799
CURVE 4 T = 1133	
1.14	0.642
1.33	0.661
1.55	0.673
1.84	0.682
2.19	0.687
2.88	0.687

λ	ϵ
CURVE 5 T = 299	
1.75	0.920
1.78	0.873
1.89	0.835
2.08	0.800
2.29	0.780
2.51	0.768
2.97	0.760
3.56	0.733
4.18	0.682
4.87	0.682
5.48	0.668
6.56	0.624
6.79	0.624
7.10	0.627
7.44	0.618
8.04	0.611
8.57	0.611

λ	ϵ
CURVE 5 (cont.)	
8.90	0.614
9.39	0.599
9.97	0.624
10.35	0.624
11.06	0.605
12.78	0.585
13.31	0.585
13.81	0.562
14.04	0.562
14.26	0.576
14.51	0.576
15.22	0.544
16.06	0.522
17.35	0.504
17.72	0.482
18.04	0.482
18.48	0.463
19.03	0.463

λ	ϵ
CURVE 5 (cont.)	
19.28	0.449
19.47	0.449
19.63	0.454
19.94	0.442
20.51	0.444
20.65	0.430
20.80	0.430
21.00	0.446

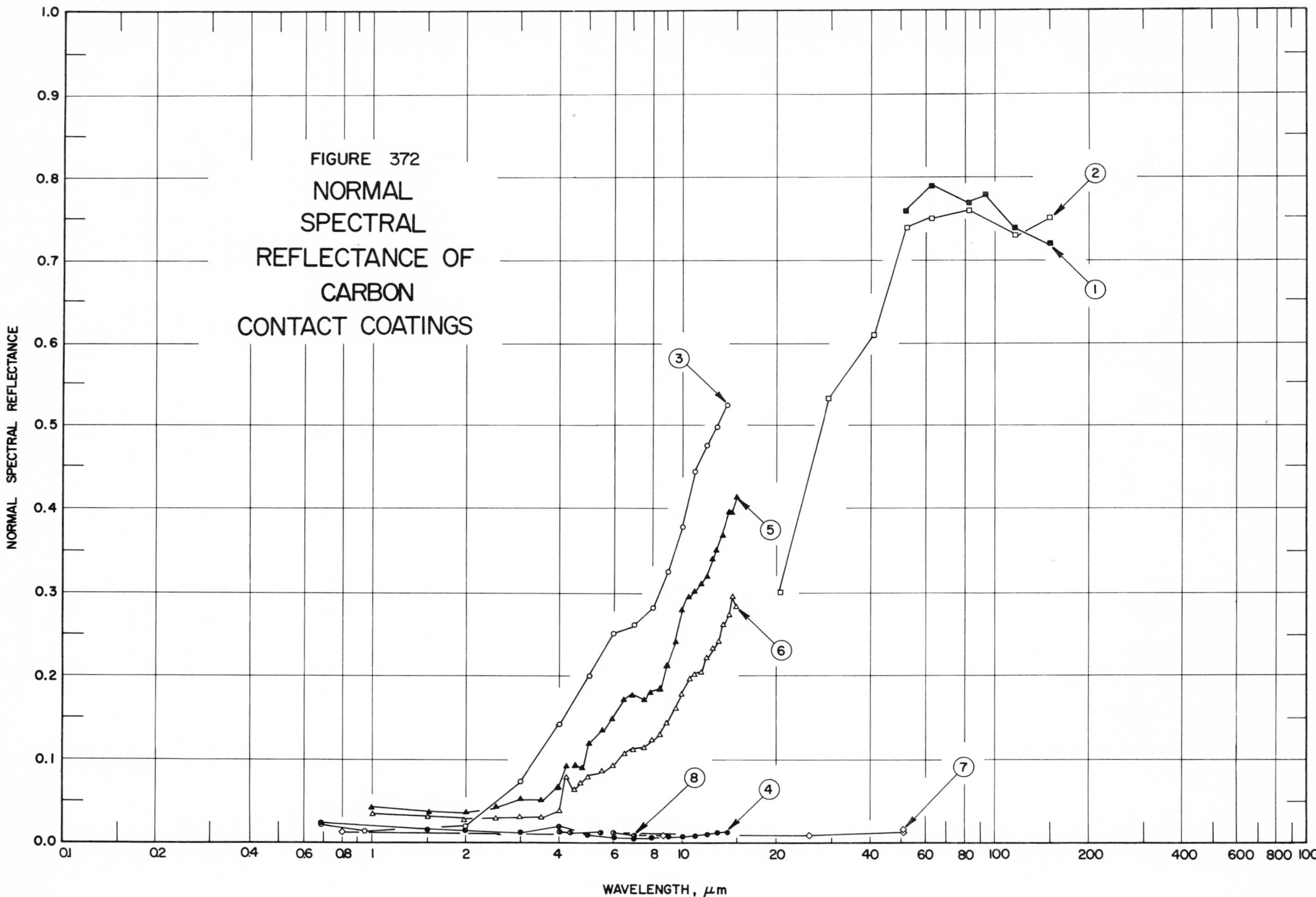

FIGURE 372
NORMAL SPECTRAL REFLECTANCE OF CARBON CONTACT COATINGS

SPECIFICATION TABLE NO. 372 NORMAL SPECTRAL REFLECTANCE OF CARBON CONTACT COATINGS

Curve No.	Ref. No.	Year	Temperature, K	Wavelength Range, μm	Geometry θ	θ'	ω'	Reported Error, %	Composition (weight percent), Specifications, and Remarks
1	149	1940	298	52-152	$\sim 0^\circ$	$\sim 0^\circ$			Aquadag graphite; SiO_2 (0.25 mm thick) substrate. [Authors' designation: Sample No. 10A]
2	149	1940	298	20.7-152	$\sim 0^\circ$	$\sim 0^\circ$			Aquadag graphite; brass substrate. [Authors' designation: Sample No. 33]
3	114	1965	~298	0.68-14.0	10°		2π		Smoke from burning camphor (0.01 mm thick); brass substrate; measured relative to front-surface Al mirror.
4	114	1965	~298	0.68-14.0	10°		2π		Similar to above specimen and conditions except 0.23 mm thick.
5	62	1949	~298	1.00-15.00	0°		2π		Lampblack; copper substrate; light deposit; data extracted from smooth curve; converted from R (2π, 0°).
6	62	1949	~298	1.00-15.00	0°		2π		Similar to above specimen and conditions except heavy deposit.
7	150	1911	~298	0.8-51.0	$\sim 0^\circ$		2π		Lampblack (0.205 mm thick); silver (0.1 mm thick) substrate; deposited from flame of small petroleum lamp.
8	113	1967	~298	0.400-0.700	$\sim 10^\circ$		2π		Acetylene black; unknown substrate; property measured with a Hardy spectrophotometer with an integrating sphere. [Author's designation: Sample 1]
9*	113	1967	~298	0.400-0.700	$\sim 10^\circ$		2π		Similar to above specimen and conditions. [Author's designation: Sample H]
10*	113	1967	~298	0.400-0.700	$\sim 10^\circ$		2π		Similar to above specimen and conditions. [Author's designation: Sample B]
11*	113	1967	~298	0.400-0.700	$\sim 10^\circ$		2π		Similar to above specimen and conditions. [Author's designation: Sample C]
12*	113	1967	~298	0.400-0.700	$\sim 10^\circ$		2π		Similar to above specimen and conditions. [Author's designation: Sample D]
13*	113	1967	~298	0.400-0.700	$\sim 10^\circ$		2π		Similar to above specimen and conditions. [Author's designation: Sample 2]

* Not shown on plot

DATA TABLE NO. 372 NORMAL SPECTRAL REFLECTANCE OF CARBON CONTACT COATINGS

[Wavelength, λ, μm; Reflectance, ρ; Temperature, T, K]

CURVE 1
T = 298

λ	ρ
52	0.76
63	0.79
83	0.77
94	0.78
117	0.74
152	0.72

CURVE 2
T = 298

λ	ρ
20.7	0.30
29.4	0.53
41	0.61
52	0.74
63	0.75
83	0.76
94	0.78*
117	0.73
152	0.75

CURVE 3
T ~298

λ	ρ
0.68	0.0202
0.95	0.0139
2.00	0.0188
3.00	0.0734
4.01	0.142
5.00	0.200
6.00	0.251
7.02	0.263
8.01	0.281
9.01	0.326
10.0	0.378
11.0	0.446
12.0	0.476
13.0	0.498
14.0	0.524

CURVE 4
T ~298

λ	ρ
0.68	0.0239
1.50	0.0167
2.00	0.0136

CURVE 4 (cont.)

λ	ρ
3.00	0.0113
4.00	0.0119
4.99	0.00731
6.01	0.00407
7.03	0.00310
7.99	0.00418
9.02	0.00631
10.0	0.00631
11.0	0.00631
12.0	0.00968
13.0	0.0106
14.0	0.0130

CURVE 5
T ~298

λ	ρ
1.00	0.042
1.52	0.036
2.00	0.036
2.50	0.042
3.00	0.052
3.49	0.051
3.99	0.065
4.21	0.093
4.50	0.093
4.74	0.089
4.99	0.117
5.46	0.136
5.98	0.149
6.48	0.171
6.97	0.177
7.45	0.171
7.96	0.181
8.48	0.186
8.97	0.213
9.49	0.241
9.99	0.279
10.50	0.296
11.00	0.303
11.50	0.311
12.00	0.319
12.50	0.342
12.99	0.353
13.49	0.368
14.01	0.396
14.48	0.395

CURVE 5 (cont.)

λ	ρ
15.00	0.415

CURVE 6
T ~298

λ	ρ
1.00	0.036
1.51	0.030
1.98	0.025
2.51	0.027
2.99	0.029
3.49	0.029
4.00	0.037
4.23	0.078
4.48	0.063
4.73	0.070
4.97	0.077
5.47	0.085
5.98	0.093
6.49	0.107
6.97	0.112
7.48	0.114
7.99	0.123
8.54	0.129
8.95	0.144
9.49	0.162
9.99	0.178
10.50	0.197
10.98	0.201
11.49	0.205
11.98	0.222
12.51	0.234
13.01	0.241
13.51	0.264
14.01	0.274
14.50	0.296
15.00	0.283

CURVE 7
T ~ 298

λ	ρ
0.8	0.011
8.7	0.007
25.5	0.008
51.0	0.014
51.0	0.016

CURVE 8
T ~298

λ	ρ
0.400	0.0130
0.436	0.0127
0.546	0.0112
0.600	0.0110
0.700	0.0095

CURVE 9*
T ~298

λ	ρ
0.400	0.0115
0.436	0.0112
0.546	0.0110
0.600	0.0095
0.700	0.0086

CURVE 10*
T ~298

λ	ρ
0.400	0.0125
0.436	0.0126
0.546	0.0100
0.600	0.0100
0.700	0.0085

CURVE 11*
T ~298

λ	ρ
0.400	0.0120
0.436	0.0119
0.546	0.0110
0.600	0.0092
0.700	0.0080

CURVE 12*
T ~298

λ	ρ
0.400	0.0130
0.436	0.0126
0.546	0.0110
0.600	0.0100
0.700	0.0085

CURVE 13*
T ~298

λ	ρ
0.400	0.0140
0.436	0.0137
0.546	0.0217
0.600	0.0122
0.700	0.0109

* Not shown on plot

SPECIFICATION TABLE NO. 373 NORMAL SPECTRAL TRANSMITTANCE OF CARBON CONTACT COATINGS

Curve No.	Ref. No.	Year	Temperature, K	Wavelength Range, μm	Geometry θ θ' ω'	Reported Error, %	Composition (weight percent), Specifications, and Remarks
1*	149	1940	298	20.7-152	$\sim 0^\circ$ $\sim 0^\circ$		Aquadag, graphite; pyroxylin substrate. [Authors' designation: Sample No. 11]

DATA TABLE NO. 373 NORMAL SPECTRAL TRANSMITTANCE OF CARBON CONTACT COATINGS

[Wavelength, λ, μm; Transmittance, τ; Temperature, T, K]

λ	τ
CURVE 1* T = 298	
20.7	0.00
29.4	0.00
52	0.00
63	0.00
83	0.00
94	0.00
117	0.00
152	0.00

* No plot given

SPECIFICATION TABLE NO. 374 NORMAL SOLAR ABSORPTANCE OF CARBON CONTACT COATINGS

Curve No.	Ref. No.	Year	Temperature Range, K	Geometry θ	Reported Error, %	Composition (weight percent), Specifications and Remarks
1*	7	1968	468-1093	$\sim 0^{\circ}$		Carbon black; polish d molybdenum substrate; applied by flame of burning kerosene; measured in vacuum (10^{-5} mm Hg).

DATA TABLE NO. 374 NORMAL SOLAR ABSORPTANCE OF CARBON CONTACT COATINGS

[Temperature, T, K; Absorptance, α]

T	α
CURVE 1*	
468	0.68
668	0.65
873	0.62
1093	0.60

* No plot given

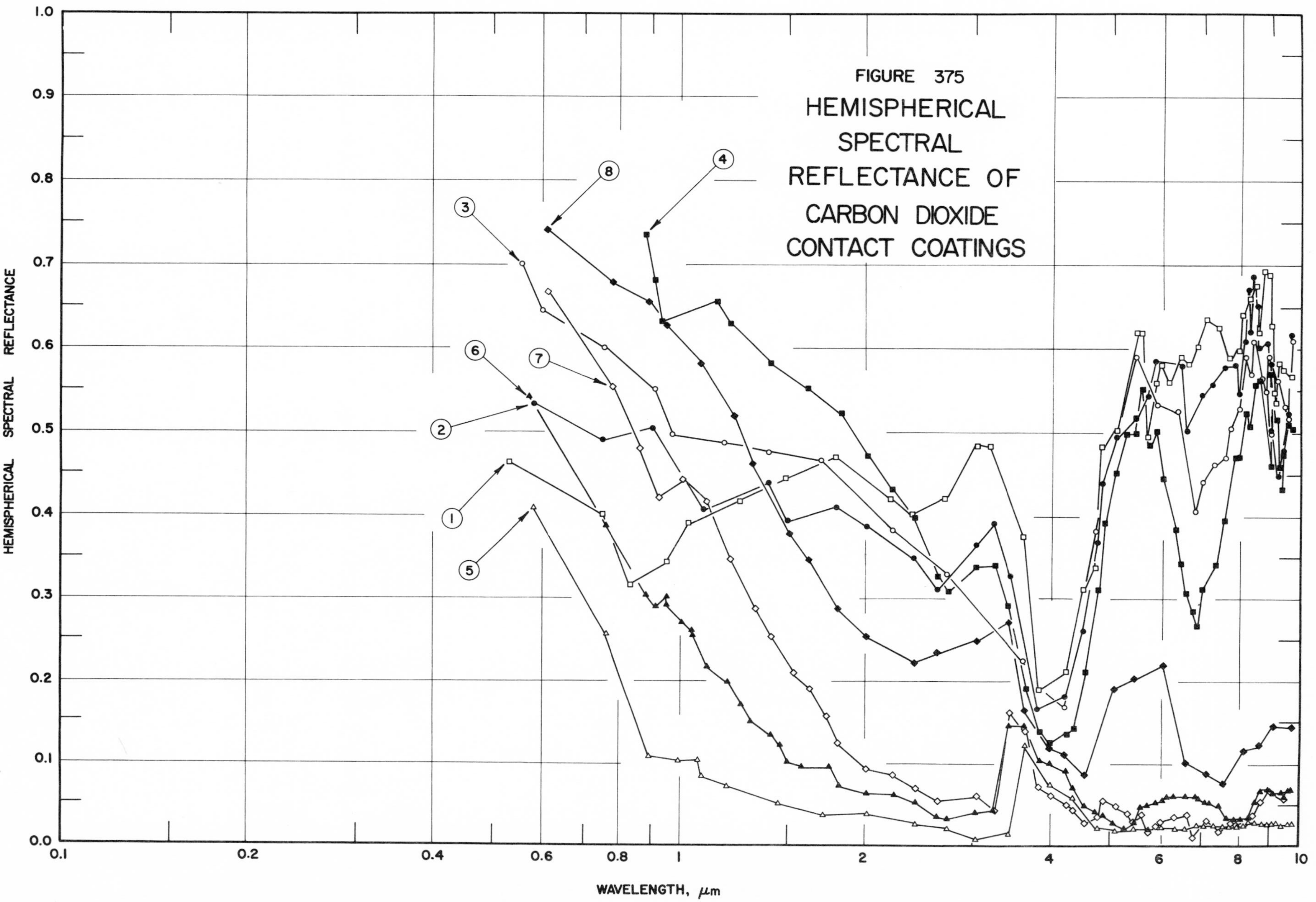

FIGURE 375
HEMISPHERICAL SPECTRAL REFLECTANCE OF CARBON DIOXIDE CONTACT COATINGS

SPECIFICATION TABLE NO. 375 HEMISPHERICAL SPECTRAL REFLECTANCE OF CARBON DIOXIDE CONTACT COATINGS

Curve No.	Ref. No.	Year	Temperature, K	Wavelength Range, μm	Geometry ω	θ'	ω'	Reported Error, %	Composition (weight percent), Specifications, and Remarks
1	151	1967	77	0.53-9.70	2π	20°	.02	6	CO_2 (0.318 mm thick, 1.5 gm cm^{-3}); stainless steel substrate; substrate highly polished; cryodeposited; measured in vacuum ($< 10^{-6}$ mm Hg).
2	151	1967	77	0.58-9.69	2π	20°	.02	6	Above specimen and conditions except CO_2 coating 0.635 mm thick.
3	151	1967	77	0.56-9.70	2π	20°	.02	6	Above specimen and conditions except CO_2 coating 1.27 mm thick.
4	151	1967	77	0.88-9.72	2π	20°	.02	6	Above specimen and conditions except CO_2 coating 2.5 mm thick.
5	151	1967	77	0.58-9.69	2π	20°	.02	6	CO_2 (0.318 mm thick); Cat-A-Lac Black (~200 μm thick) and stainless steel substrates; stainless steel highly polished; cryodeposited; measured in vacuum ($< 10^{-6}$ mm Hg).
6	151	1967	77	0.57-9.67	2π	20°	.02	6	Above specimen and conditions except CO_2 coating 0.635 mm thick.
7	151	1967	77	0.61-9.69	2π	20°	.02	6	Above specimen and conditions except CO_2 coating 1.27 mm thick.
8	151	1967	77	0.61-9.70	2π	20°	.02	6	Above specimen and conditions except CO_2 coating 2.5 mm thick.

DATA TABLE NO. 375 HEMISPHERICAL SPECTRAL REFLECTANCE OF CARBON DIOXIDE CONTACT COATINGS

[Wavelength, λ, μm; Reflectance, ρ; Temperature, T, K]

CURVE 1
T = 77

λ	ρ
0.53	0.466
0.75	0.402
0.83	0.317
0.95	0.346
1.04	0.392
1.25	0.418
1.48	0.445
1.78	0.471
2.18	0.420
2.39	0.403
2.66	0.423
3.02	0.486
3.19	0.486
3.58	0.376
3.80	0.190
4.20	0.212
4.48	0.312
4.69	0.388
4.79	0.484
5.06	0.503
5.41	0.519
5.55	0.620
5.68	0.495
5.85	0.560
5.99	0.581
6.14	0.560
6.40	0.591
6.57	0.583
6.80	0.603
7.05	0.636
7.39	0.627
7.68	0.590
7.94	0.598
8.05	0.641
8.28	0.661
8.48	0.676
8.58	0.620
8.71	0.696
8.89	0.689
8.95	0.629
9.04	0.547
9.13	0.535
9.23	0.584
9.38	0.576
9.65	0.568
9.70	0.520

CURVE 2
T = 77

λ	ρ
0.58	0.533
0.75	0.490
0.90	0.502
1.08	0.408
1.37	0.438
1.49	0.395
1.78	0.411
2.01	0.388
2.39	0.350
2.60	0.311
3.00	0.366
3.21	0.392
3.41	0.329
3.78	0.166
4.19	0.181
4.49	0.260
4.72	0.369
4.80	0.439
5.05	0.495
5.41	0.518
5.69	0.544
5.84	0.585
6.43	0.580
6.55	0.502
6.93	0.546
7.20	0.558
7.54	0.578
7.87	0.582
7.94	0.546
8.18	0.610
8.26	0.672
8.30	0.623
8.39	0.687
8.51	0.652
8.58	0.600
8.80	0.609
8.92	0.583
8.95	0.503
9.14	0.448
9.39	0.480
9.54	0.524
9.69	0.619

CURVE 3
T = 77

λ	ρ
0.56	0.700
0.60	0.645
0.75	0.600
0.91	0.552
0.97	0.497
1.17	0.487
1.38	0.475
1.68	0.467
2.20	0.383
2.70	0.330
3.58	0.225
4.19	0.169
4.70	0.382
5.42	0.591
5.88	0.534
6.34	0.525
6.78	0.408
6.95	0.440
7.29	0.461
7.56	0.470
7.70	0.506
7.97	0.529
8.14	0.593
8.30	0.570
8.39	0.608
8.69	0.565
8.77	0.549
8.89	0.592
8.96	0.499
9.17	0.564
9.41	0.531
9.54	0.517
9.70	0.610

CURVE 4
T = 77

λ	ρ
0.88	0.735
0.91	0.682
0.93	0.634
1.15	0.655
1.22	0.630
1.40	0.584
1.61	0.552
1.83	0.522
2.01	0.473
2.20	0.432
2.41	0.399
2.62	0.329
2.70	0.309
3.01	0.337
3.24	0.340
3.39	0.293
3.63	0.193
3.83	0.138
3.97	0.124
4.21	0.134
4.31	0.142
4.50	0.210
4.71	0.310
4.84	0.393
5.06	0.453
5.26	0.499
5.42	0.499
5.57	0.552
5.71	0.486
5.89	0.501
6.03	0.445
6.32	0.385
6.40	0.343
6.56	0.307
6.69	0.283
6.81	0.267
6.96	0.313
7.32	0.342
7.55	0.397
7.87	0.471
7.99	0.471
8.17	0.523
8.26	0.508
8.43	0.557
8.60	0.563
8.82	0.563
8.92	0.573
8.96	0.460
9.17	0.515
9.23	0.457
9.31	0.433
9.38	0.475
9.56	0.510
9.72	0.504

CURVE 5
T = 77

λ	ρ
0.58	0.407
0.76	0.257
0.89	0.109
0.99	0.104
1.07	0.104
1.08	0.085
1.19	0.072
1.44	0.050
1.71	0.038
2.01	0.038
2.42	0.026
2.72	0.020
3.01	0.009
3.42	0.016
3.60	0.121
3.98	0.075
4.31	0.060
4.73	0.022
5.06	0.018
5.40	0.020
5.70	0.020
6.00	0.022
6.30	0.020
6.55	0.020
6.81	0.025
7.03	0.025
7.29	0.025
7.53	0.024
7.70	0.024
7.94	0.024
8.12	0.024
8.29	0.029
8.47	0.029
8.68	0.029
8.88	0.027
9.02	0.027
9.17	0.027
9.37	0.025
9.63	0.027
9.69	0.027

CURVE 6
T = 77

λ	ρ
0.57	0.543
0.76	0.389
0.88	0.302
0.91	0.290
0.95	0.302
0.95	0.293
1.01	0.270
1.04	0.262
1.04	0.255
1.11	0.218
1.19	0.199
1.26	0.173
1.30	0.149
1.40	0.134
1.46	0.122
1.48	0.100
1.57	0.095
1.74	0.095
1.81	0.074
2.01	0.064
2.23	0.062
2.40	0.051
2.60	0.036
2.71	0.033
3.01	0.040
3.21	0.043
3.41	0.145
3.60	0.145
3.82	0.102
3.97	0.099
4.20	0.091
4.32	0.073
4.50	0.049
4.68	0.044
4.81	0.037
5.02	0.027
5.24	0.021
5.41	0.028
5.55	0.047
5.68	0.048
5.85	0.053
6.02	0.056
6.14	0.062
6.29	0.062
6.55	0.062
6.79	0.062
6.88	0.057
7.04	0.054
7.19	0.054
7.39	0.048
7.63	0.034
7.81	0.034
8.04	0.034
8.30	0.036
8.48	0.054
8.65	0.068
8.87	0.068
9.13	0.065
9.30	0.065
9.40	0.060
9.64	0.068
9.67	0.068

CURVE 7
T = 77

λ	ρ
0.61	0.667
0.65	0.635*
0.78	0.551
0.86	0.480
0.92	0.422
0.93	0.439*
0.98	0.473*
1.02	0.441
1.06	0.401*
1.10	0.418
1.10	0.377*
1.22	0.349
1.33	0.289
1.40	0.253
1.42	0.210*
1.53	0.210
1.62	0.191
1.73	0.155
1.80	0.125
2.01	0.095
2.22	0.087
2.43	0.070
2.62	0.053
2.72	0.053*
3.02	0.061
3.23	0.044
3.41	0.161
3.61	0.139
3.79	0.071
3.97	0.062
4.22	0.050
4.31	0.045
4.50	0.028
4.71	0.036
4.81	0.054
5.06	0.049
5.25	0.040
5.38	0.028
5.58	0.039
5.70	0.018
5.85	0.024
5.99	0.027
6.18	0.034*
6.29	0.034
6.38	0.021*
6.56	0.036
6.70	0.010
6.78	0.028*
6.93	0.038*
7.04	0.030
7.15	0.037*
7.40	0.019
7.53	0.031*
7.63	0.028*
7.66	0.027*
7.76	0.027
7.84	0.021*
7.92	0.026
8.04	0.027*
8.13	0.027*
8.29	0.033*
8.36	0.039
8.47	0.043*
8.59	0.054*
8.67	0.054
8.79	0.061*

* Not shown on plot

DATA TABLE NO. 375 HEMISPHERICAL SPECTRAL REFLECTANCE OF CARBON DIOXIDE CONTACT COATINGS (continued)

λ	ρ	λ	ρ
CURVE 7 (cont.)		CURVE 8 (cont.)	
T = 77			
		4.77	0.169*
8.89	0.069*	5.03	0.190
8.94	0.069	5.22	0.182*
9.01	0.069*	5.40	0.204
9.14	0.069*	5.56	0.096*
9.20	0.069*	5.68	0.100*
9.29	0.060*	5.83	0.120*
9.36	0.060*	6.02	0.120
9.46	0.057	6.16	0.111*
9.53	0.060*	6.33	0.101*
9.58	0.066*	6.38	0.101*
9.69	0.058*	6.56	0.101
		6.67	0.101*
CURVE 8		6.79	0.093*
T = 77		6.94	0.088*
		7.05	0.088
0.61	0.741	7.19	0.089*
0.78	0.678	7.26	0.079*
0.89	0.656	7.40	0.079*
0.95	0.627	7.53	0.078
1.07	0.582	7.61	0.086*
1.11	0.567*	7.70	0.110*
1.11	0.535*	7.80	0.103*
1.23	0.519	7.92	0.120*
1.29	0.494*	7.98	0.129*
1.32	0.463	8.11	0.115
1.40	0.441*	8.22	0.080*
1.43	0.411*	8.28	0.080*
1.50	0.378	8.36	0.089*
1.62	0.349	8.47	0.109*
1.70	0.316*	8.56	0.120
1.80	0.289	8.77	0.124*
2.01	0.256	8.85	0.124*
2.19	0.246*	8.91	0.154*
2.40	0.224	9.01	0.144
2.61	0.235	9.10	0.135*
2.70	0.235*	9.15	0.138*
3.02	0.248	9.26	0.135*
3.21	0.248*	9.36	0.141*
3.39	0.272*	9.46	0.138*
3.60	0.164	9.52	0.138*
3.81	0.135*	9.65	0.145*
3.95	0.119	9.70	0.144
4.19	0.111		
4.27	0.087*		
4.49	0.087		
4.69	0.132*		

* Not shown on plot

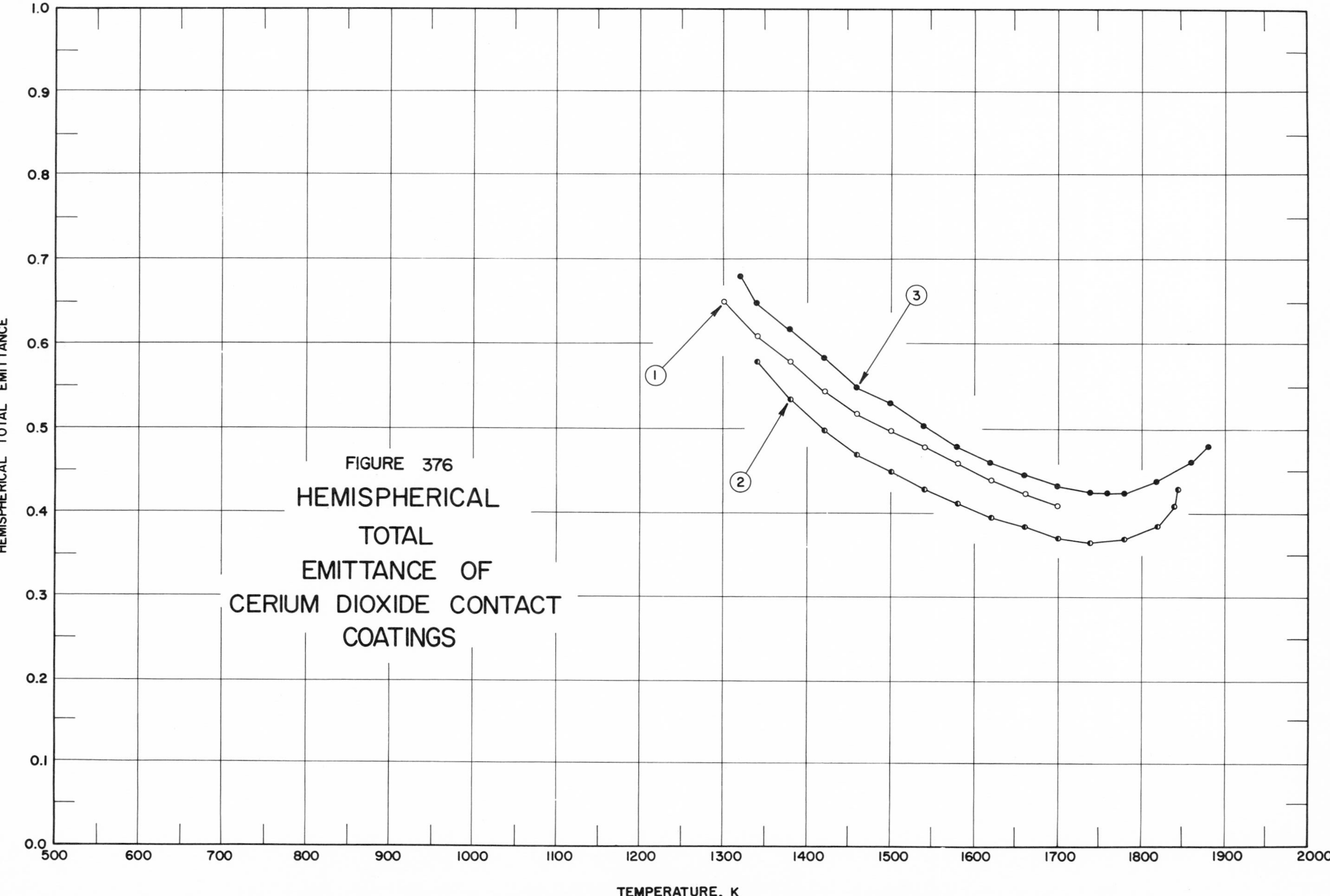

FIGURE 376
HEMISPHERICAL TOTAL EMITTANCE OF CERIUM DIOXIDE CONTACT COATINGS

SPECIFICATION TABLE NO. 376 HEMISPHERICAL TOTAL EMITTANCE OF CERIUM DIOXIDE CONTACT COATINGS

Curve No.	Ref. No.	Year	Temperature Range, K	Reported Error, %	Composition (weight percent), Specifications and Remarks
1	152	1952	1300-1700		CeO_2 (50 μm thick); tungsten substrate; commercial grade CeO_2; heat treated at less than 1750 K; measured in vacuum; coating may not have been opaque.
2	152	1952	1340-1845		Above specimen and conditions.
3	152	1952	1320-1880		Above specimen and conditions except heated above 1750 K.

DATA TABLE NO. 376 HEMISPHERICAL TOTAL EMITTANCE OF CERIUM DIOXIDE CONTACT COATINGS

[Temperature, T, K; Emittance, ϵ]

T	ϵ
CURVE 1	
1300	0.650
1340	0.610
1380	0.580
1420	0.545
1460	0.520
1500	0.500
1540	0.480
1580	0.460
1620	0.440
1660	0.425
1700	0.410
CURVE 2	
1340	0.580
1380	0.535
1420	0.500
1460	0.470
1500	0.450
CURVE 2 (cont.)	
1540	0.430
1580	0.412
1620	0.395
1660	0.385
1700	0.370
1740	0.365
1780	0.370
1820	0.385
1840	0.410
1845	0.430
CURVE 3	
1320	0.680
1340	0.650
1380	0.620
1420	0.585
1460	0.550
1500	0.530
CURVE 3 (cont.)	
1540	0.505
1580	0.480
1620	0.460
1660	0.446
1700	0.434
1740	0.426
1760	0.425
1780	0.426
1820	0.440
1860	0.460
1880	0.480

FIGURE 377

NORMAL TOTAL EMITTANCE OF CERIUM DIOXIDE CONTACT COATINGS

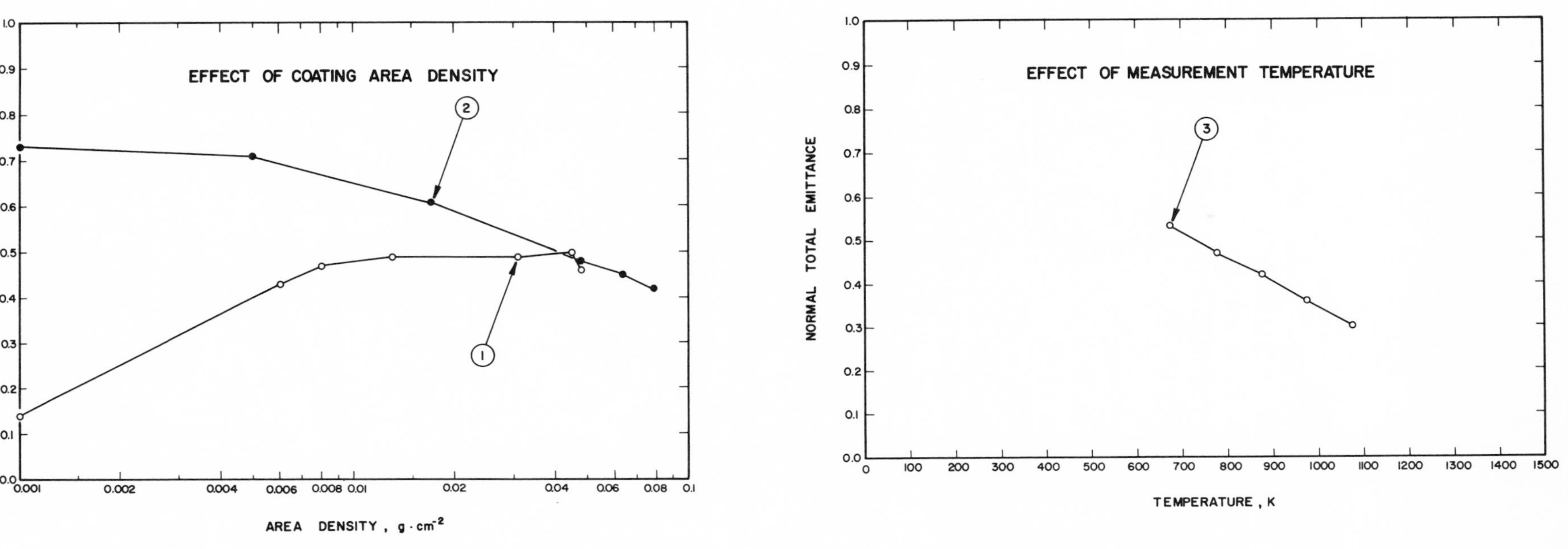

SPECIFICATION TABLE NO. 377 NORMAL TOTAL EMITTANCE OF CERIUM DIOXIDE CONTACT COATINGS

Curve No.	Ref. No.	Year	Temperature Range, K	Geometry θ'	Reported Error, %	Composition (weight percent), Specifications and Remarks
1	123	1952	673	~0°		Cerium dioxide; Nimonic 75 substrate; substrate buffed; area density (g cm^{-2}) is variable.
2	123	1952	673	~0°		Cerium dioxide; Nimonic 75 substrate; substrate oxidized; area density (g cm^{-2}) is variable.
3	123	1952	673-1073	~0°		CeO_2, powder; Nimonic 75 substrate.

DATA TABLE NO. 377 NORMAL TOTAL EMITTANCE OF CERIUM DIOXIDE CONTACT COATINGS

[Temperature, T, K; Emittance, ∈]

Area Density	∈
CURVE 1 T = 673	
0.001	0.14
0.006	0.43
0.008	0.47
0.013	0.49
0.031	0.49
0.045	0.50
0.048	0.46
CURVE 2 T = 673	
0.001	0.73
0.005	0.71
0.017	0.61
0.048	0.48
0.064	0.45
0.079	0.42

T	∈
CURVE 3	
673	0.53
773	0.47
873	0.42
973	0.36
1073	0.30

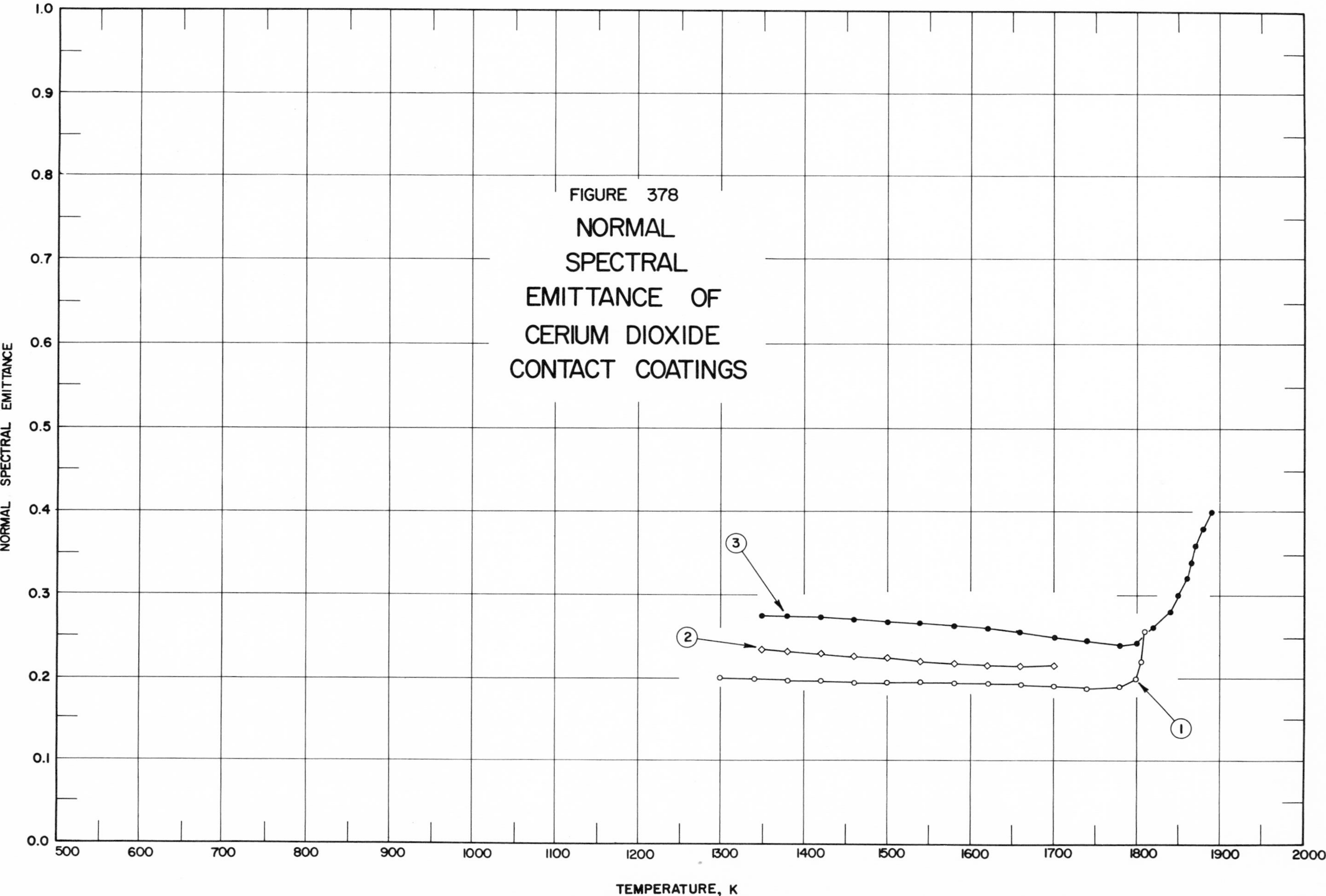

FIGURE 378 NORMAL SPECTRAL EMITTANCE OF CERIUM DIOXIDE CONTACT COATINGS

SPECIFICATION TABLE NO. 378 NORMAL SPECTRAL EMITTANCE OF CERIUM DIOXIDE CONTACT COATINGS

Curve No.	Ref. No.	Year	Wavelength, μm	Temperature Range, K	Geometry θ'	Reported Error, %	Composition (weight percent), Specifications, and Remarks
1	152	1952	0.665	1300-1810	~0°		CeO_2, commercial grade (50 μm thick); tungsten substrate; heat treated at less than 1750 K; measured in vacuum; coating may not have been opaque.
2	152	1952	0.665	1350-1700	~0°		Above specimen and conditions.
3	152	1952	0.665	1350-1890	~0°		Above specimen and conditions except heat treated above 1750 K.

DATA TABLE NO. 378 NORMAL SPECTRAL EMITTANCE OF CERIUM DIOXIDE CONTACT COATINGS

[Temperature, T, K; Emittance, ∈; Wavelength, λ, μm]

T	∈	T	∈	T	∈
CURVE 1		CURVE 2		CURVE 3 (cont.)	
λ = 0.665		λ = 0.665			
				1460	0.270
1300	0.200	1350	0.235	1500	0.268
1340	0.200	1380	0.232	1540	0.266
1380	0.199	1420	0.230	1580	0.263
1420	0.198	1460	0.226	1620	0.260
1460	0.197	1500	0.225	1660	0.254
1500	0.196	1540	0.220	1700	0.250
1540	0.195	1580	0.218	1740	0.245
1580	0.194	1620	0.216	1780	0.240
1620	0.193	1660	0.215	1800	0.242
1660	0.192	1700	0.214	1820	0.260
1700	0.190			1840	0.280
1740	0.188	CURVE 3		1850	0.300
1780	0.190	λ = 0.665		1860	0.320
1800	0.200			1865	0.340
1805	0.220	1350	0.274	1870	0.360
1808	0.240*	1380	0.273	1880	0.380
1810	0.255	1420	0.272	1890	0.400

* Not shown on plot

SPECIFICATION TABLE NO. 379 HEMISPHERICAL TOTAL EMITTANCE OF CHROMIUM CARBIDE + COBALT CONTACT COATINGS

Curve No.	Ref. No.	Year	Temperature Range, K	Reported Error, %	Composition (weight percent), Specifications and Remarks
1*	7	1961	468-1093	<10	Metco XP-1109, 60 chromium carbide + cobalt blend; Armco iron substrate; plasma flame sprayed; measured in vacuum (10^{-5} mm Hg).

DATA TABLE NO. 379 HEMISPHERICAL TOTAL EMITTANCE OF CHROMIUM CARBIDE + COBALT CONTACT COATINGS

[Temperature, T, K; Emittance, ϵ]

T	ϵ
CURVE 1*	
468	0.33
668	0.39
873	0.48
1093	0.50

* No plot given

SPECIFICATION TABLE NO. 380 NORMAL SOLAR ABSORPTANCE OF CHROMIUM CARBIDE + COBALT CONTACT COATINGS

Curve No.	Ref. No.	Year	Temperature Range, K	Geometry θ	Reported Error, %	Composition (weight percent), Specifications and Remarks
1*	154	1961	468-1093	~0°		Metco XP-1109, 60 chromium carbide + 40 cobalt; Armco ingot iron substrate; plasma flame sprayed; measured in vacuum (10^{-5} mm Hg).

DATA TABLE NO. 380 NORMAL SOLAR ABSORPTANCE OF CHROMIUM CARBIDE + COBALT CONTACT COATINGS

[Temperature, T, K; Absorptance, α]

T	α
CURVE 1*	
468	0.81
668	0.84
873	0.90
1093	0.89

* No plot given

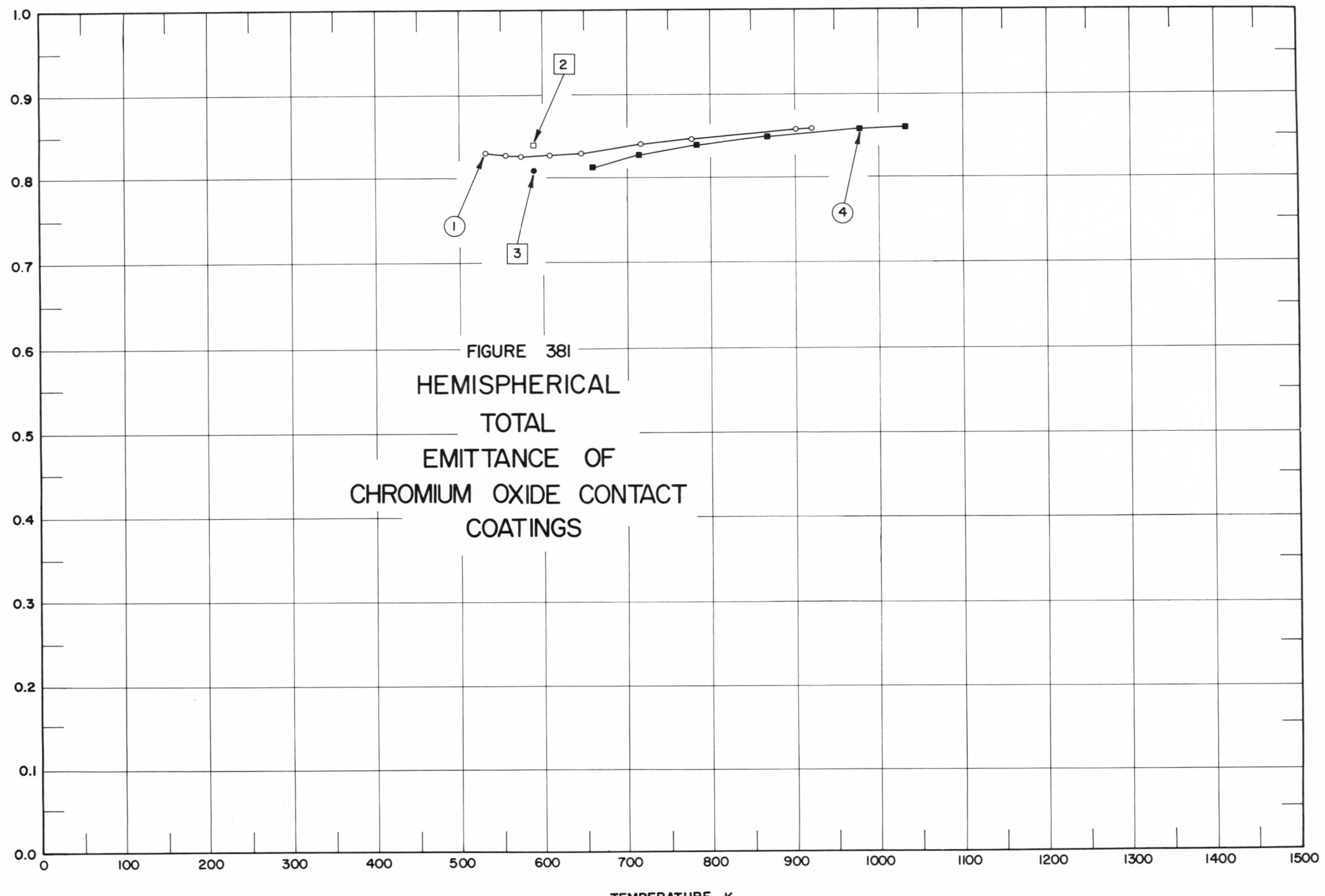

FIGURE 381
HEMISPHERICAL TOTAL EMITTANCE OF CHROMIUM OXIDE CONTACT COATINGS

SPECIFICATION TABLE NO. 381 HEMISPHERICAL TOTAL EMITTANCE OF CHROMIUM OXIDE CONTACT COATINGS

Curve No.	Ref. No.	Year	Temperature Range, K	Reported Error, %	Composition (weight percent), Specifications and Remarks
1	15	1965	533-921		Cr_2O_3 (0.0762 mm thick); aluminum 1100 substrate; plasma sprayed; measured in vacuum (10^{-5} mm Hg); data extracted from smooth curve.
2	53	1966	588		Cr_2O_3 (0.051 to 0.076 mm thick); substrate unknown; plasma sprayed; measured in vacuum (10^{-5} mm Hg).
3	53	1966	588		Above specimen and conditions except exposed to fast neutron irradiation for 3.0 x 10^8 neutrons cm^{-2} and 1.9 x 10^{10} R of gamma irradiation.
4	82	1962	657-1033	<±2.5	Chromia from Linde Flameplating Co. (~0.0762 mm thick); niobium substrate; plasma arc sprayed; measured in vacuum; data extracted from smooth curve.

DATA TABLE NO. 381 HEMISPHERICAL TOTAL EMITTANCE OF CHROMIUM OXIDE CONTACT COATINGS

[Temperature, T, K; Emittance, ϵ]

T	ϵ	T	ϵ
CURVE 1		CURVE 4	
533	0.831	657	0.814
553	0.828	714	0.828
575	0.827	783	0.840
609	0.828	869	0.850
646	0.831	979	0.857
718	0.842	1033	0.860
777	0.847		
901	0.857		
921	0.859		
CURVE 2			
588	0.84		
CURVE 3			
588	0.81		

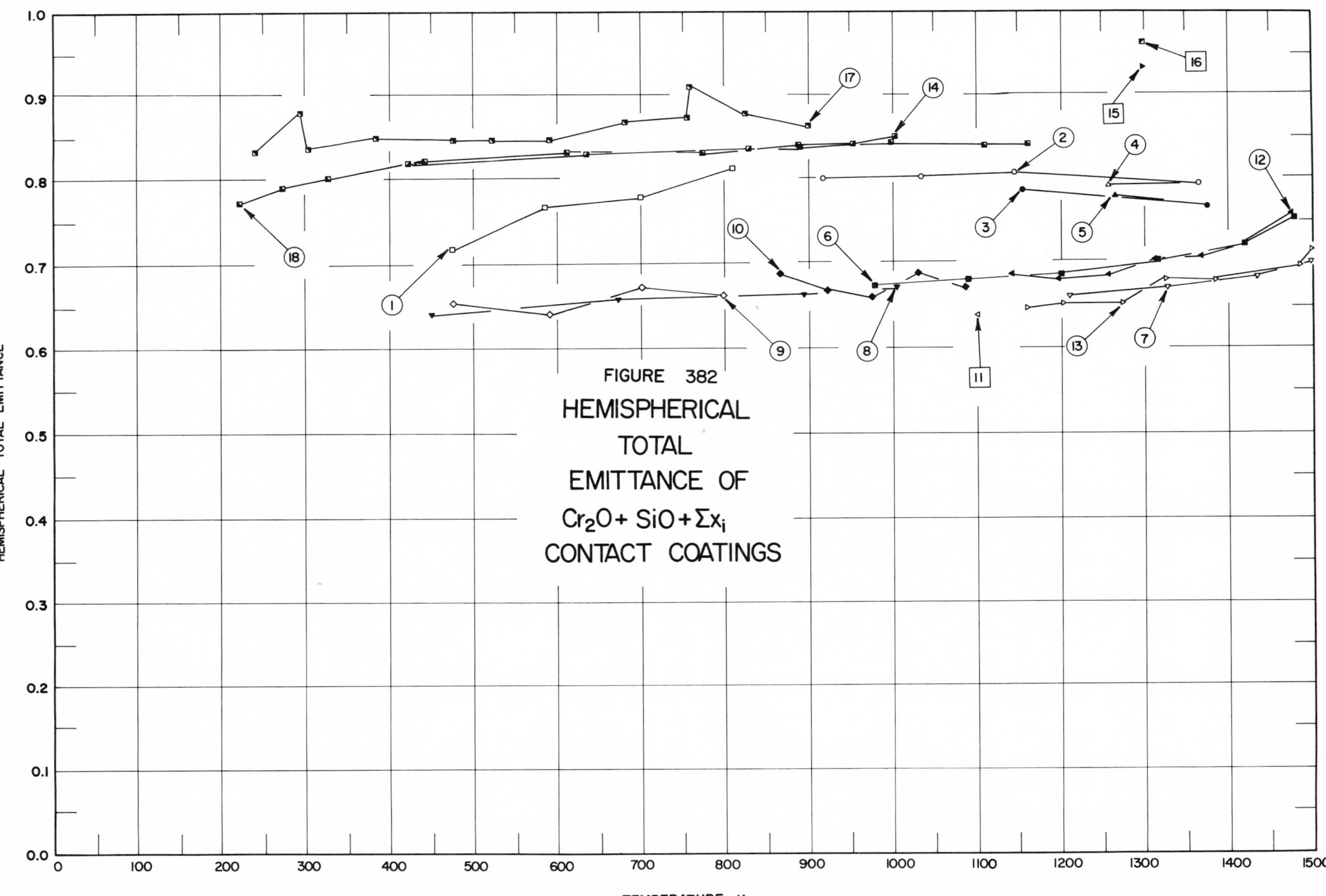

FIGURE 382
HEMISPHERICAL
TOTAL
EMITTANCE OF
$Cr_2O + SiO + \Sigma x_i$
CONTACT COATINGS

SPECIFICATION TABLE NO. 382 HEMISPHERICAL TOTAL EMITTANCE OF CHROMIUM OXIDE + SILICON DIOXIDE + ΣX_i CONTACT COATINGS

Curve No.	Ref. No.	Year	Temperature Range, K	Reported Error, %	Composition (weight percent), Specifications and Remarks
1	118	1962	475-810		Rokide C, Norton Co., (0.0762 mm thick); niobium substrate; flame sprayed (Rokide process); extremely hard, fine grit texture; measured in vacuum (3.3×10^{-6} mm Hg); temp measured with thermocouple; rising temp. [Authors' designation: Run 1]
2	118	1962	918-1366		Above specimen and conditions. [Authors' designation: Run 2]
3	118	1962	1153-1497		Above specimen and conditions except temp measured with optical pyrometer.
4	118	1962	1366-1258		Above specimen and conditions except temp descending.
5	118	1962	1376-1264		Above specimen and conditions except temp measured with optical pyrometer.
6	118	1962	1478-979		Curve 2 specimen and conditions except temp descending.
7	118	1962	1497-1212		Above specimen and conditions except temp measured with optical pyrometer.
8	118	1962	450-1006		Curve 6 specimen and conditions. [Authors' designation: Run 3]
9	118	1962	476-799		Above specimen and conditions except rising temp. [Authors' designation: Run 4]
10	118	1962	866-1101		Above specimen and conditions. [Authors' designation: Run 5]
11	118	1962	1101		Above specimen and conditions except temp measured with optical pyrometer.
12	118	1962	1142-1498		Curve 10 specimen and conditions. [Authors' designation: Run 6]
13	118	1962	1159-1498		Above specimen and conditions except temp measured with optical pyrometer.
14	82	1962	422-1002	$< \pm 2.5$	Rokide C (0.0762 mm thick); stainless steel 310 substrate; plasma-arc-sprayed; measured in vacuum (10^{-8} mm Hg); data extracted from smooth curve.
15	155	1961	1299-1899		Rokide C, 82.94 Cr_2O_3, 8.39 SiO_2, 3.16 Al_2O_3, 2.96 MgO, 1.28 CaO, 0.78 Fe_2O_3, 0.28 Na_2O, and 0.16 TiO_2 (0.0508 mm thick); molybdenum substrate.
16	155	1961	1299-1899		Similar to above specimen and conditions except coating 0.0762 mm thick.
17	156	1962	240-900		Rokide C; substrate unknown.
18	156	1962	222-1163		Similar to above specimen and conditions except computed from spectral reflectance data.

DATA TABLE NO. 382 HEMISPHERICAL TOTAL EMITTANCE OF CHROMIUM OXIDE + SILICON DIOXIDE + ΣX_i CONTACT COATINGS

[Temperature, T, K; Emittance, ϵ]

T	ϵ
CURVE 1	
475	0.717
586	0.767
699	0.778
810	0.811
CURVE 2	
918	0.798
1036	0.802
1145	0.807
1366	0.793
CURVE 3	
1153	0.786
1376	0.767
CURVE 4	
1366	0.793
1258	0.790
CURVE 5	
1376	0.767
1264	0.777
CURVE 6	
1478	0.751
1418	0.722
1313	0.701
1200	0.686
1091	0.679
979	0.673
CURVE 7	
1497	0.702
1433	0.682
1328	0.669
1212	0.660
CURVE 8	
1006	0.670
895	0.664
673	0.657
450	0.642
CURVE 9	
476	0.656
591	0.642
700	0.620
799	0.664
CURVE 10	
866	0.688
922	0.667
976	0.661
1031	0.687
1089	0.670
CURVE 11	
1101	0.640
CURVE 12	
1142	0.686
1198	0.680
1257	0.686
1313	0.702
1368	0.709
1420	0.724
1475	0.759
CURVE 13	
1159	0.647
1211	0.651
1273	0.652
1324	0.679
1384	0.678
1484	0.697
1498	0.714
CURVE 14	
422	0.819
636	0.829
829	0.833
890	0.836
952	0.839
1002	0.848
CURVE 15	
1299	0.93
1899	0.93*
CURVE 16	
1299	0.96
1899	0.96*
CURVE 17	
240	0.833
295	0.879
303	0.836
385	0.849
478	0.845
522	0.845
591	0.845
680	0.866
755	0.871
758	0.909
825	0.874
900	0.861
CURVE 18	
222	0.771
273	0.790
329	0.800
443	0.821
612	0.830
776	0.830
889	0.839
999	0.840
1110	0.838
1163	0.839

* Not shown on plot

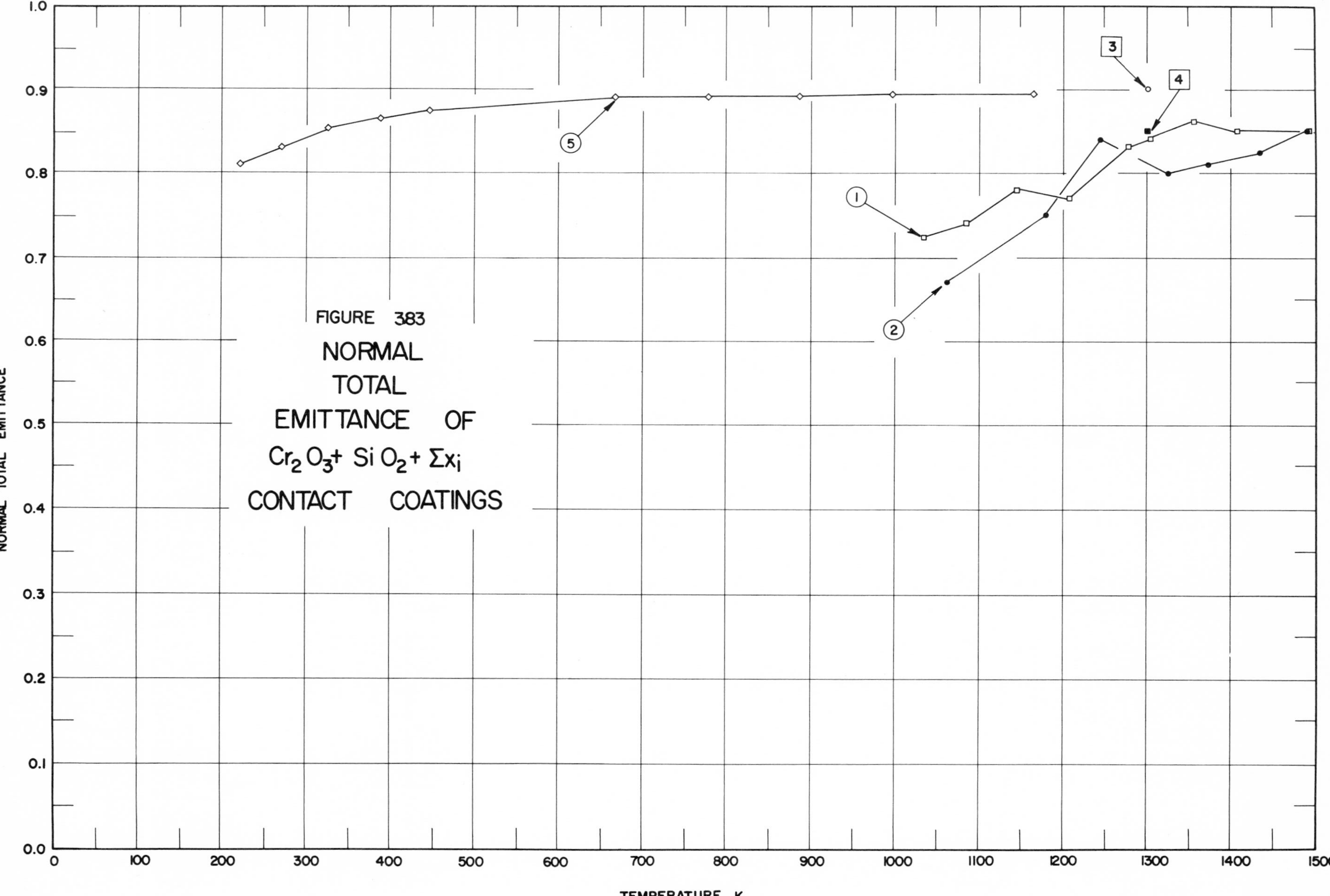

FIGURE 383 NORMAL TOTAL EMITTANCE OF $Cr_2O_3 + SiO_2 + \Sigma x_i$ CONTACT COATINGS

SPECIFICATION TABLE NO. 383 NORMAL TOTAL EMITTANCE OF CHROMIUM OXIDE + SILICON DIOXIDE + ΣX_i CONTACT COATINGS

Curve No.	Ref. No.	Year	Temperature Range, K	Geometry θ'	Reported Error, %	Composition (weight percent), Specifications and Remarks
1	147	1963	1035-1491	~0°		Rokide C, 82.94 Cr_2O_3, 8.39 SiO_2, 3.16 Al_2O_3, 2.96 MgO, 1.28 CaO, 0.78 Fe_2O_3, 0.28 Na_2O, and 0.16 TiO_2 (0.102 mm thick); titanium alloy 6Al-4V substrate; substrate grit blasted; flame sprayed; surface roughness 30 to 45 μm (RMS) measured with a profilometer; measured in vacuum (3 to 4 x 10^{-4} mm Hg); heating cycle.
2	147	1963	1491-1064	~0°		Above specimen and conditions except cooling cycle.
3	126	1965	1300	~0°		82.94 Cr_2O_3 (hexagonal), 8.39 SiO_2, 2.96 MgO, 1.28 CaO, 0.78 Fe_2O_3, 0.28 Na_2O, and 0.16 TiO_2 (0.305 mm thick); mild steel substrate; flame sprayed; surface roughness 9.14 μm measured with profilometer and optical comparator; density 4.6 g cm^{-3}; porosity 4%; measured in vacuum (3.5 to 5.0 x 10^{-2} mm Hg); computed from spectral data (0 to 10 μm).
4	126	1965	1300	~0°		82.94 Cr_2O_3, 8.39 SiO_2, 2.96 MgO, 1.28 CaO, 0.78 Fe_2O_3, 0.28 Na_2O, and 0.16 TiO_2 (0.381 mm thick) on mild steel substrate; flame sprayed; polished with polishing papers; density 4.6 g cm^{-3}; porosity 4%; measured in vacuum (3.5 to 5.0 x 10^{-2} mm Hg); computed from spectral data (0 to 10 μm).
5	156	1962	222-1166	~0°	5	Rokide C; substrate unknown.

DATA TABLE NO. 383 NORMAL TOTAL EMITTANCE OF CHROMIUM OXIDE + SILICON DIOXIDE + ΣX_i CONTACT COATINGS

[Temperature, T, K; Emittance, ϵ]

T	ϵ	T	ϵ
CURVE 1		CURVE 3	
1035	0.725	1300	0.902
1086	0.740		
1146	0.780	CURVE 4	
1209	0.770		
1278	0.832	1300	0.851
1304	0.840		
1355	0.860	CURVE 5	
1409	0.850		
1491	0.850	222	0.8116
		272	0.8339
CURVE 2		328	0.8531
		389	0.8652
1491	0.850	446	0.8763
1435	0.825	667	0.8903
1373	0.810	778	0.8928
1326	0.800	887	0.8926
1244	0.840	998	0.8949
1180	0.750	1166	0.8940
1064	0.720		

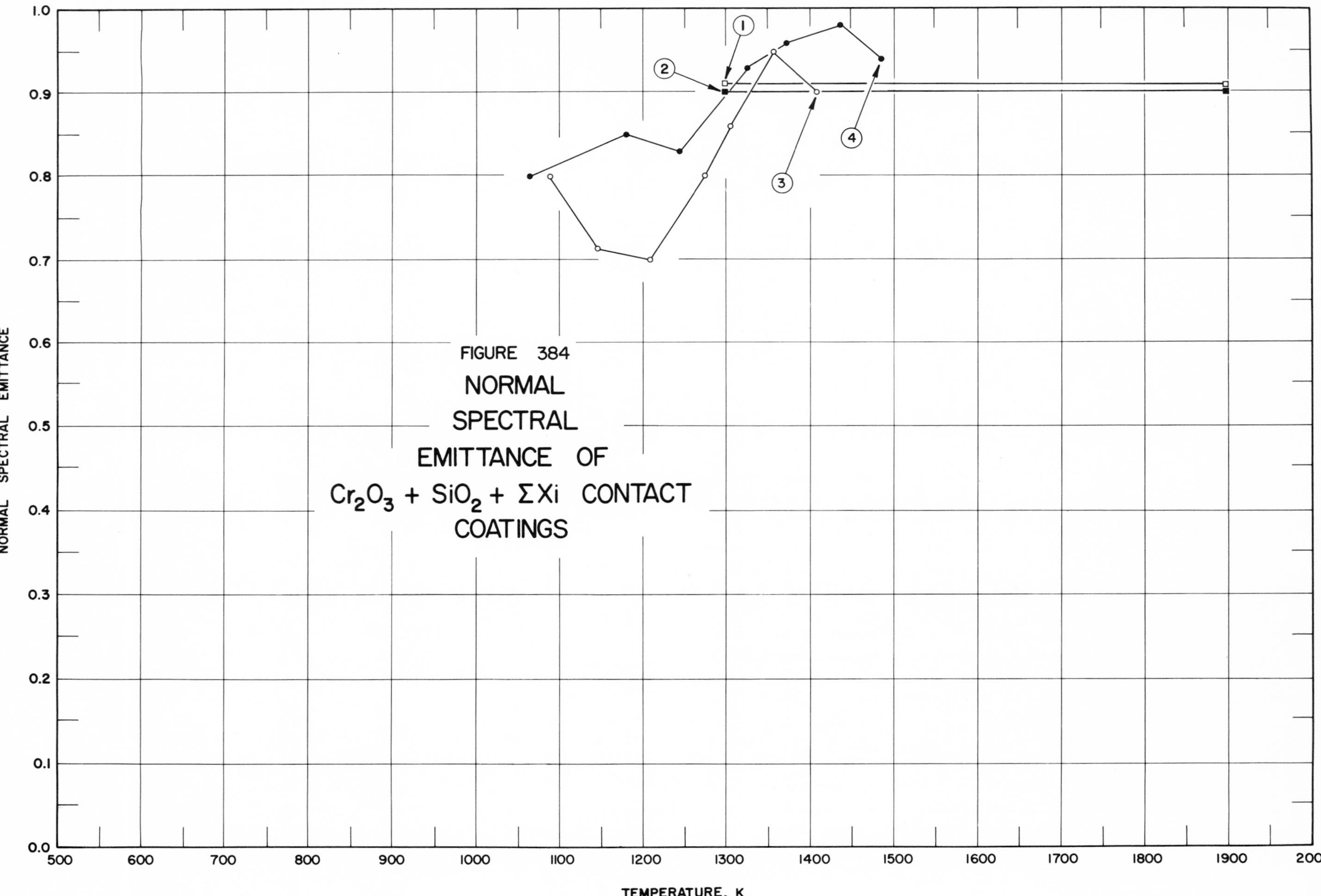

FIGURE 384
NORMAL SPECTRAL EMITTANCE OF $Cr_2O_3 + SiO_2 + \Sigma Xi$ CONTACT COATINGS

SPECIFICATION TABLE NO. 384 NORMAL SPECTRAL EMITTANCE OF CHROMIUM OXIDE + SILICON DIOXIDE + ΣX_i CONTACT COATINGS

Curve No.	Ref. No.	Year	Wavelength, μm	Temperature Range, K	Geometry θ'	Reported Error, %	Composition (weight percent), Specifications, and Remarks
1	155	1961	0.65	1299-1899	~0°		Rokide C, 82.94 Cr_2O_3, 8.39 SiO_2, 3.16 Al_2O_3, 2.96 MgO, 1.28 CaO, 0.78 Fe_2O_3, 0.28 Na_2O, and 0.16 TiO_2 (0.0508 mm thick); molybdenum substrate.
2	155	1961	0.65	1299-1899	~0°		Rokide C (0.0762 mm thick); molybdenum substrate.
3	147	1963	0.65	1089-1489	~0°		Rokide C (0.102 mm thick); titanium alloy 6A1-4V substrate; substrate grit blasted; flame sprayed; surface roughness 30 to 45 μm (RMS) measured with a profilometer; measured in vacuum (3 to 4 x 10^{-4} mm Hg); authors assumed specimen was a grey body; ascending temp.
4	147	1963	0.65	1489-1066	~0°		Above specimen and conditions; descending temp.

DATA TABLE NO. 384 NORMAL SPECTRAL EMITTANCE OF CHROMIUM OXIDE + SILICON DIOXIDE + ΣX_i CONTACT COATINGS

[Temperature, T, K; Emittance, ϵ; Wavelength, λ, μm]

T	ϵ	T	ϵ
CURVE 1 λ = 0.65		CURVE 3 (cont.)	
1299	0.91	1358	0.950
1899*	0.91	1409	0.900
		1489	0.940*
CURVE 2 λ = 0.65		CURVE 4 λ = 0.65	
1299	0.90	1489	0.940
1899*	0.90	1438	0.980
		1373	0.960
CURVE 3 λ = 0.65		1326	0.930
		1244	0.830
		1180	0.850
1089	0.800	1066	0.800
1146	0.715		
1209	0.700		
1275	0.800		
1304	0.860		

* Not shown on plot

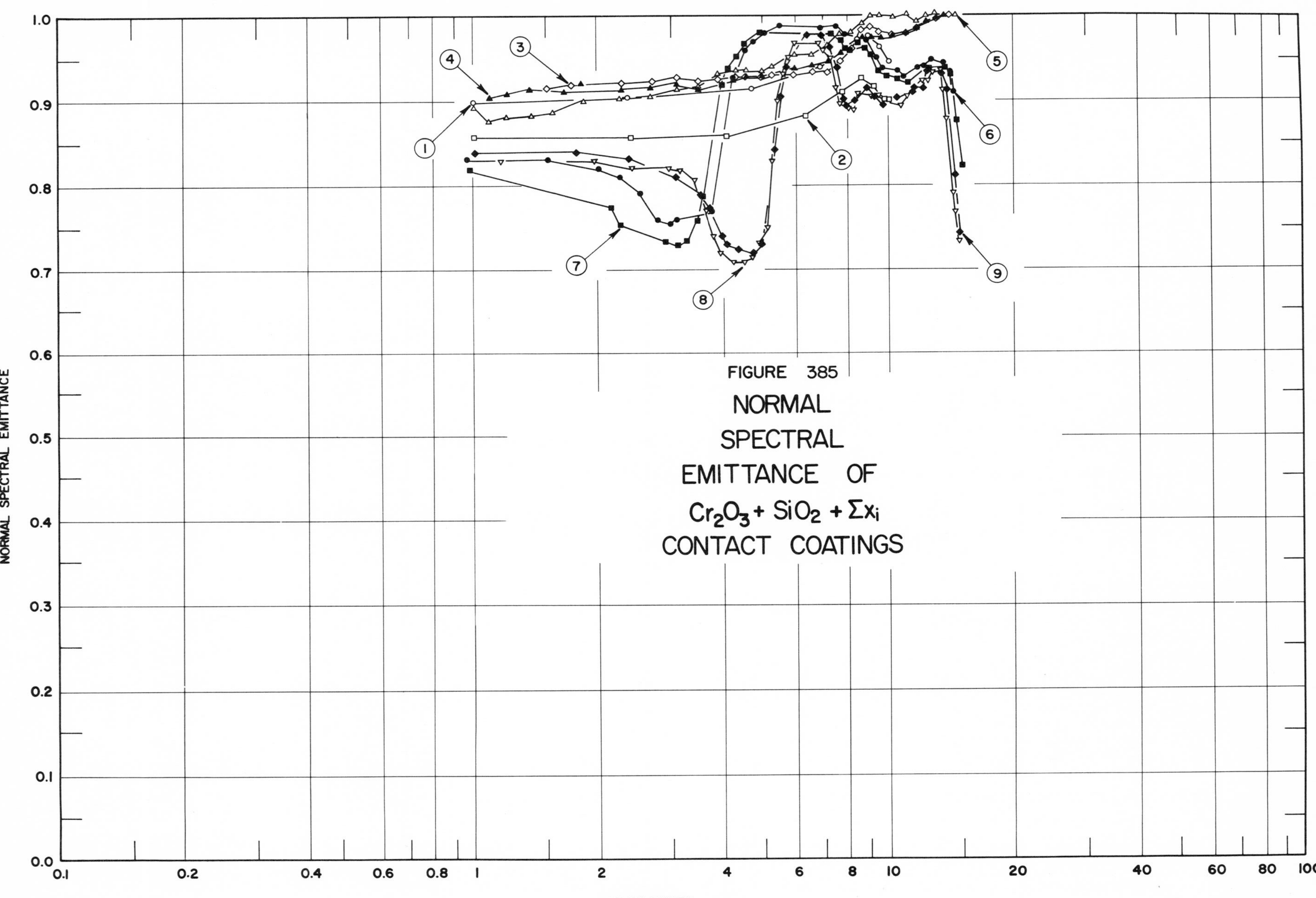

FIGURE 385 NORMAL SPECTRAL EMITTANCE OF $Cr_2O_3 + SiO_2 + \Sigma x_i$ CONTACT COATINGS

SPECIFICATION TABLE NO. 385 NORMAL SPECTRAL EMITTANCE OF CHROMIUM OXIDE + SILICON DIOXIDE + ΣX_i CONTACT COATINGS

Curve No.	Ref. No.	Year	Temperature, K	Wavelength Range, μm	Geometry θ'	Reported Error, %	Composition (weight percent), Specifications, and Remarks
1	126	1965	1300	1.00-10.01	$\sim 0°$		82.94 Cr_2O_3 (hexagonal), 8.39 SiO_2, 2.96 MgO, 1.28 CaO, 0.78 Fe_2O_3, 0.28 Na_2O, and 0.16 TiO_2 (0.305 mm thick); mild steel substrate; flame sprayed; surface roughness 9.14 μm measured with profilometer and optical comparator; density 4.6 g cm^{-3}; porosity 4%; measured in vacuum (3.5 to 5.0 x 10^{-2} mm Hg); data extracted from smooth curve.
2	126	1965	1300	1.01-10.01	$\sim 0°$		82.94 Cr_2O_3 (hexagonal), 8.39 SiO_2, 2.96 MgO, 1.28 CaO, 0.78 Fe_2O_3, 0.28 NaO_2, and 0.16 TiO_2 (0.381 mm thick); mild steel substrate; flame sprayed; polished with polishing papers; density 4.6 g cm^{-3}; porosity 4%; measured in vacuum (3.5 to 5.0 x 10^{-2} mm Hg); data extracted from smooth curve.
3	118	1962	755	1.51-14.07	$\sim 0°$		Rokide C, Norton Co., (0.0762 mm thick); niobium substrate; flame sprayed (Rokide process); extremely hard, fine grit texture; measured in vacuum.
4	118	1962	1061	1.09-14.06	$\sim 0°$		Above specimen and conditions.
5	118	1962	1366	1.00-14.05	$\sim 0°$		Curve 3 specimen and conditions.
6	130	1966	1255	0.96-15.0	$\sim 0°$		Rokide C (0.102 mm thick); Inconel substrate; flame sprayed; data extracted from smooth curve.
7	130	1966	1255	0.98-15.0	$\sim 0°$		Similar to above specimen and conditions except 1.04 mm thick.
8	130	1966	755	0.95-15.0	$\sim 0°$		Similar to curve 6 specimen and conditions.
9	130	1966	755	1.01-14.9	$\sim 0°$		Similar to above specimen and conditions except 1.04 mm thick.

DATA TABLE NO. 385 NORMAL SPECTRAL EMITTANCE OF CHROMIUM OXIDE + SILICON DIOXIDE + ΣX_i CONTACT COATINGS

[Wavelength, λ, μm; Emittance, ∈; Temperature, T, K]

λ	∈
CURVE 1	
T = 1300	
1.00	0.899
2.37	0.902
4.68	0.915
6.86	0.940
8.45	0.973
8.96	0.978
9.56	0.965
10.01	0.947
CURVE 2	
T = 1300	
1.01	0.858
2.40	0.857
4.10	0.859
6.28	0.882
7.80	0.910
8.72	0.928
9.26	0.918
9.66	0.901
10.01	0.901
CURVE 3	
T = 755	
1.51	0.915
1.83	0.919
2.28	0.921
2.70	0.922
3.10	0.927
3.48	0.923
3.88	0.925
4.26	0.928
4.55	0.928
4.95	0.927
5.25	0.930
5.97	0.931
6.60	0.934
7.18	0.934
7.72	0.947
8.23	0.961
8.67	0.985
9.13	0.988
9.56	0.983

λ	∈
CURVE 3 (cont.)	
10.35	0.979
11.10	0.981
11.75	0.986
13.07	0.998
14.07	1.000
CURVE 4	
T = 1061	
1.09	0.903
1.24	0.908
1.37	0.913
1.65	0.912
1.82	0.921
2.28	0.913
2.68	0.915
3.09	0.920
3.49	0.913
3.87	0.924*
4.26	0.927
4.58	0.927
4.95	0.930
5.24	0.931*
5.97	0.938
6.57	0.942
7.19	0.948
7.68	0.957
8.22	0.961*
8.65	0.977
9.17	0.974
9.60	0.977
10.34	0.979
11.09	0.979
11.77	0.988
12.38	0.996
13.06	0.998
13.55	1.000
14.06	1.000*
CURVE 5	
T = 1366	
1.00	0.893
1.08	0.878
1.20	0.882

λ	∈
CURVE 5 (cont.)	
1.38	0.883
1.56	0.889
1.85	0.900
2.25	0.903
2.67	0.906
3.09	0.914
3.48	0.916
3.89	0.933
4.28	0.937
4.55	0.937
4.96	0.935
5.26	0.941
5.96	0.955
6.55	0.955
7.16	0.966
7.68	0.980
8.19	0.982
8.66	0.992
9.05	1.000
9.55	1.000
10.33	0.999
11.08	1.001
11.73	0.994
12.39	1.000
12.99	1.001
13.58	0.999*
14.05	1.000
CURVE 6	
T = 1255	
0.96	0.830
1.52	0.830
2.00	0.820
2.26	0.810
2.52	0.790
2.78	0.758
2.99	0.755
3.08	0.760
3.74	0.868
4.22	0.928
4.55	0.961
4.77	0.973
5.04	0.982
5.44	0.989

λ	∈
CURVE 6 (cont.)	
6.97	0.989
7.48	0.990
7.90	0.981
8.65	0.978*
8.94	0.973
9.40	0.952
9.72	0.938
10.5	0.936
10.9	0.929
11.7	0.939
12.1	0.941
12.5	0.948
13.6	0.944
14.0	0.936
14.4	0.912
15.0	0.837*
CURVE 7	
T = 1255	
0.98	0.820
2.14	0.775
2.63	0.753
2.90	0.733
3.10	0.731
3.24	0.736
3.46	0.758
3.58	0.787
3.99	0.918
4.12	0.937
4.29	0.953
4.50	0.968
4.89	0.982
5.43	0.991*
6.95	0.987
7.35	0.981
7.67	0.973
7.84	0.963
8.09	0.961
8.47	0.969
8.80	0.964
9.11	0.954
9.55	0.936
9.93	0.929
10.4	0.926

λ	∈
CURVE 7 (cont.)	
11.1	0.922
12.5	0.937
13.7	0.937
14.1	0.932
14.5	0.877
15.0	0.825
CURVE 8	
T = 755	
0.95	0.829*
1.16	0.827
1.96	0.829
2.41	0.821
2.95	0.821
3.15	0.817
3.40	0.805
3.54	0.788*
3.64	0.770
3.79	0.740
3.95	0.720
4.22	0.709
4.50	0.709
4.71	0.715
4.96	0.732
5.07	0.751
5.25	0.828
5.42	0.900
5.60	0.932
5.78	0.953
5.99	0.968
6.82	0.967
7.00	0.959
7.14	0.950
7.54	0.914
7.77	0.896
8.05	0.890
8.25	0.898
8.50	0.907
9.19	0.904
9.58	0.906
9.80	0.897
10.70	0.894
11.0	0.906
11.5	0.916

λ	∈
CURVE 8 (cont.)	
12.1	0.923
12.5	0.923
12.9	0.934
13.4	0.936
13.6	0.912
13.9	0.879
14.3	0.790
14.4	0.769
14.8	0.734
15.0	0.721*
CURVE 9	
T = 755	
1.01	0.840
1.77	0.840
2.37	0.833
2.93	0.821*
3.08	0.811
3.52	0.789
3.70	0.773
3.99	0.741
4.14	0.731
4.37	0.725
4.71	0.720
4.97	0.731
5.07	0.752*
5.33	0.843
5.54	0.904
5.71	0.938
6.00	0.968*
6.38	0.977
6.93	0.977
7.23	0.964
7.58	0.938
7.82	0.902
7.96	0.894
8.26	0.899
8.51	0.908*
8.98	0.914
9.30	0.904
9.78	0.896
10.4	0.903
11.3	0.909
11.5	0.915

λ	∈
CURVE 9 (cont.)	
12.0	0.915
12.4	0.934
12.9	0.934*
13.3	0.932
13.7	0.912
14.4	0.812
14.9	0.744

* Not shown on plot

SPECIFICATION TABLE NO. 386 NORMAL TOTAL EMITTANCE OF COBALT OXIDE CONTACT COATINGS

Curve No.	Ref. No.	Year	Temperature Range, K	Geometry θ'	Reported Error, %	Composition (weight percent), Specifications and Remarks
1*	157	1959	413	$\sim 0°$		Co_3O_4 (mixture of CoO and Co_2O_3) (0.68 μm thick); Ag (1 mm thick) substrate; substrate treated with acid, highly polished with rouge and cleaned with carbon tetrachloride; electroplated from hot solution of 50 $CoSO_4 \cdot 7\ H_2O$, 1.5 NaCl, 4.5 H_3BO_3 at 313 K at current density 0.05 amps cm^{-2}.
2*	157	1959	413	$\sim 0°$		Similar to above specimen and conditions except 1.2 μm thick.
3*	157	1959	413	$\sim 0°$		Similar to curve 1 specimen and conditions except 1.9 μm thick.
4*	147	1963	1033-2189	$\sim 0°$		CoO ($\sim$0.0127 mm thick); tantalum substrate; substrate grit blasted; plasma sprayed; surface roughness 30 to 35 μm (RMS) measured with profilometer; measured in vacuum (3 to 4 x 10^{-4} mm Hg).

DATA TABLE NO. 386 NORMAL TOTAL EMITTANCE OF COBALT OXIDE CONTACT COATINGS

[Temperature, T, K; Emittance, ϵ]

T	ϵ
CURVE 1*	
413	0.27
CURVE 2*	
413	0.29
CURVE 3*	
413	0.30
CURVE 4*	
1033	0.700
1116	0.700
1172	0.750
1311	0.777
1436	0.737
1550	0.710
1589	0.757
CURVE 4 (cont.)	
1650	0.830
1700	0.780
2189	0.420

* No plot given

SPECIFICATION TABLE NO. 387 NORMAL SPECTRAL EMITTANCE OF COBALT OXIDE CONTACT COATINGS

Curve No.	Ref. No.	Year	Wavelength, μm	Temperature Range, K	Geometry θ'	Reported Error, %	Composition (weight percent), Specifications, and Remarks
1*	147	1963	0.65	1108-2194	~0°		CoO (~0.0127 mm thick); tantalum substrate; substrate grit blasted; plasma sprayed; surface roughness 30 to 35 μm (RMS) measured with profilometer; measured in vacuum (3 to 4 x 10^{-4} mm Hg); authors assumed specimen was a grey body.

DATA TABLE NO. 387 NORMAL SPECTRAL EMITTANCE OF COBALT OXIDE CONTACT COATINGS

[Temperature, T, K; Emittance, ϵ; Wavelength, λ, μm]

T	ϵ
CURVE 1*	
$\lambda = 0.65$	
1108	0.897
1141	0.850
1172	0.800
1308	0.885
1428	0.900
1547	0.850
1586	0.790
1647	0.770
1694	0.730
2194	0.390

* No plot given

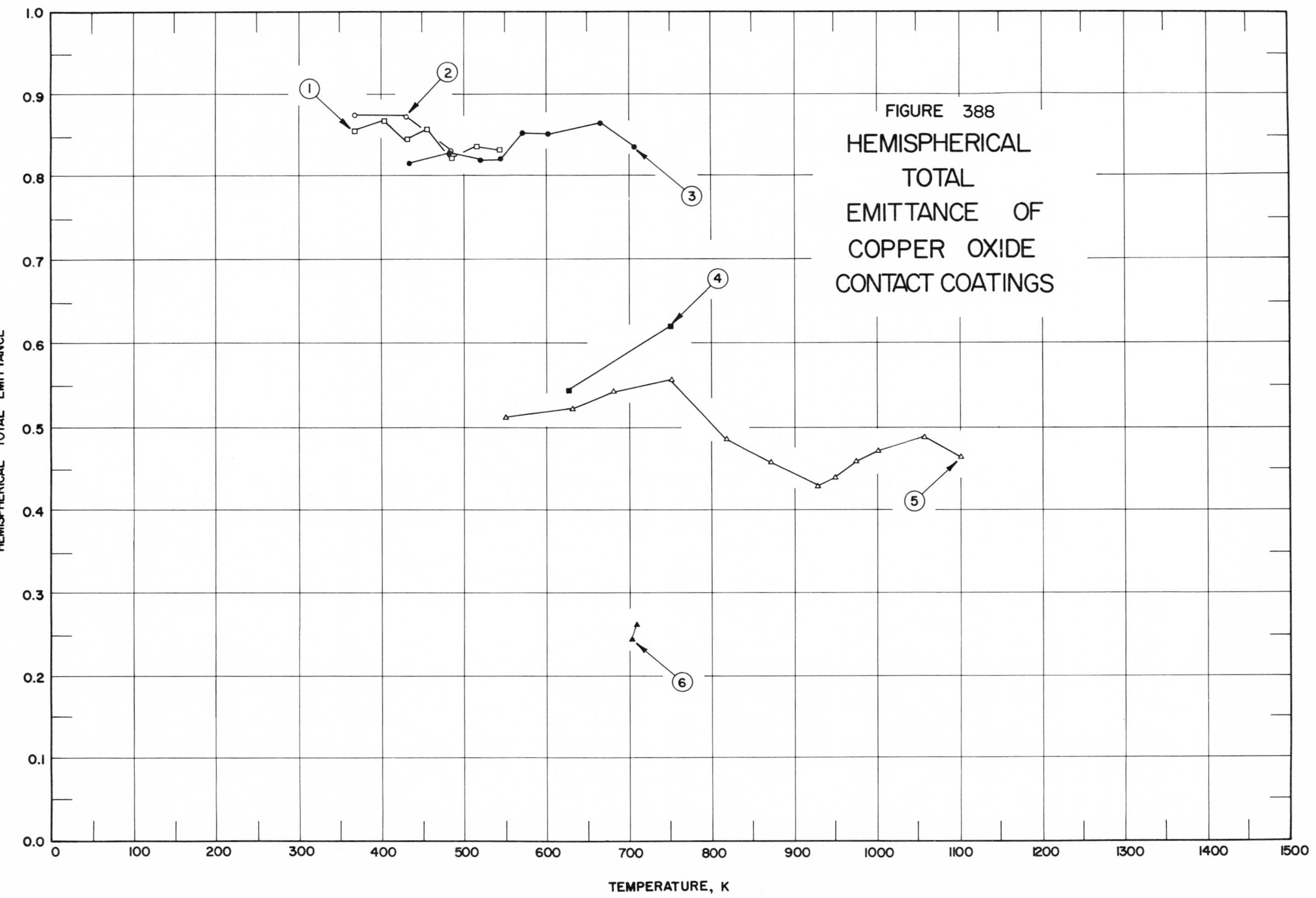

FIGURE 388 HEMISPHERICAL TOTAL EMITTANCE OF COPPER OXIDE CONTACT COATINGS

SPECIFICATION TABLE NO. 388 HEMISPHERICAL TOTAL EMITTANCE OF COPPER OXIDE CONTACT COATINGS

Curve No.	Ref. No.	Year	Temperature Range, K	Reported Error, %	Composition (weight percent), Specifications and Remarks
1	158	1962	369-544	±2.3	Copper oxide, CuO, on AISI 310 stainless steel substrate; applied by Ebonol C process; measured in vacuum ($<10^{-5}$ mm Hg); rising temp. [Authors' designation: Run No. 1]
2	158	1962	484-369	±2.3	Above specimen and conditions except descending temp. [Authors' designation: Run No. 2A]
3	158	1962	432-707	±2.3	Above specimen and conditions except rising temp. [Authors' designation: Run No. 2B]
4	158	1962	750-627	±2.3	Above specimen and conditions except descending temp. [Authors' designation: Run No. 2C]
5	158	1962	550-1101	±2.3	Above specimen and conditions except rising temp. [Authors' designation: Run No. 3]
6	158	1962	707-702	±2.3	Above specimen and conditions except descending temp. [Authors' designation: Run No. 4]

DATA TABLE NO. 388 HEMISPHERICAL TOTAL EMITTANCE OF COPPER OXIDE CONTACT COATINGS

[Temperature, T, K; Emittance, ϵ]

T	ϵ	T	ϵ	T	ϵ
CURVE 1		CURVE 3 (cont.)		CURVE 5 (cont.)	
369	0.856	545	0.822	929	0.429
403	0.869	573	0.853	950	0.440
431	0.846	601	0.850	975	0.458
456	0.857	666	0.865	1000	0.471
484	0.822	707	0.836	1058	0.488
518	0.835			1101	0.464
544	0.832	CURVE 4			
				CURVE 6	
CURVE 2		750	0.622		
		627	0.544	707	0.262
484	0.831			702	0.246
428	0.872	CURVE 5			
369	0.875				
		550	0.513		
CURVE 3		633	0.524		
		681	0.543		
432	0.817	752	0.557		
484	0.829	819	0.485		
520	0.820	872	0.458		

SPECIFICATION TABLE NO. 389 NORMAL TOTAL EMITTANCE OF COPPER OXIDE CONTACT COATINGS

Curve No.	Ref. No.	Year	Temperature Range, K	Geometry θ'	Reported Error, %	Composition (weight percent), Specifications and Remarks
1*	157	1959	413	~0°		CuO (0.24 μm thick); Ag (1 mm thick) substrate; substrate treated with acid, highly polished with rouge and cleaned with carbon tetrachloride; electroplated from a solution of 2.25 CuCN, 3.4 NaCN, and 1.5 Na_2CO_3 at current density 0.003-0.015 amps cm^{-2}.
2*	157	1959	413	~0°		Similar to above specimen and conditions except 0.28 μm thick.
3*	157	1959	413	~0°		Similar to curve 1 specimen and conditions except 0.49 μm thick.
4*	157	1959	413	~0°		CuO (0.24 μm); nickel substrate; substrate treated with acid, highly polished with rouge and cleaned with carbon tetrachloride; electroplated from a solution of 2.25 CuCN, 3.4 NaCN, and 1.5 Na_2CO_3 at current density 0.003-0.015 amps cm^{-2}.
5*	157	1959	413	~0°		Similar to above specimen and conditions except 0.28 μm thick.
6*	157	1959	413	~0°		Similar to curve 4 specimen and conditions except 0.49 μm thick.

DATA TABLE NO. 389 NORMAL TOTAL EMITTANCE OF COPPER OXIDE CONTACT COATINGS

[Temperature, T, K; Emittance, ε]

T	ε	T	ε
CURVE 1*		CURVE 6*	
413	0.10	413	0.25
CURVE 2*			
413	0.11		
CURVE 3*			
413	0.14		
CURVE 4*			
413	0.17		
CURVE 5*			
413	0.19		

* No plot given

SPECIFICATION TABLE NO. 390 NORMAL SPECTRAL REFLECTANCE OF COPPER PHOSPHOROUS SELENIDE CONTACT COATINGS

Curve No.	Ref. No.	Year	Temperature, K	Wavelength Range, μm	Geometry θ θ' ω'	Reported Error, %	Composition (weight percent), Specifications, and Remarks
1*	159	1968	~298	0.37-1.00	~0° ~0°		Cu_3PSe (synthesized by fusion of high purity copper, red phosphorous, and selenium at 1173 K) (0.1 μm thick); fluorite substrate; vacuum (10^{-4}-10^{-5} mm Hg) deposited; data extracted from smooth curve.
2*	159	1968	~298	0.40-15.00	~0° ~0°		Similar to above specimen and conditions except oxidized in air at 473 K.

DATA TABLE NO. 390 NORMAL SPECTRAL REFLECTANCE OF COPPER PHOSPHOROUS SELENIDE CONTACT COATINGS

[Wavelength, λ, μm; Reflectance, ρ; Temperature, T, K]

λ	ρ	λ	ρ	λ	ρ
CURVE 1* T = 298		CURVE 2 (cont.)*		CURVE 2 (cont.)*	
		0.75	0.241	8.66	0.692
0.37	0.412	0.81	0.175	9.05	0.711
0.42	0.479	0.86	0.141	9.75	0.680
0.48	0.508	0.91	0.126	10.34	0.651
0.54	0.529	0.99	0.108	11.03	0.641
0.63	0.512	1.07	0.115	11.85	0.632
0.81	0.431	1.29	0.199	12.60	0.632
0.92	0.369	1.68	0.288	13.63	0.645
1.00	0.314	2.19	0.366	14.38	0.659
		2.62	0.411	15.00	0.646
CURVE 2* T ~ 298		3.11	0.466		
		3.66	0.494		
		4.22	0.521		
0.40	0.228	5.24	0.561		
0.46	0.362	5.88	0.613		
0.48	0.388	6.31	0.613		
0.54	0.412	7.03	0.599		
0.63	0.359	7.51	0.611		
0.68	0.310	8.04	0.638		

* No plot given

SPECIFICATION TABLE NO. 391 NORMAL SPECTRAL TRANSMITTANCE OF COPPER PHOSPHOROUS SELENIDE CONTACT COATINGS

Curve No.	Ref. No.	Year	Temperature, K	Wavelength Range, μm	Geometry θ θ' ω'	Reported Error, %	Composition (weight percent), Specifications, and Remarks
1*	159	1968	~298	0.37-1.00	~0° ~0°		Cu_3PSe (synthesized by fusion of high purity copper, red phosphorous, and selenium at 1173 K) (0.1 μm thick); fluorite substrate; vacuum (10^{-4}-10^{-5} mm Hg) deposited; data extracted from smooth curve.
2*	159	1968	~298	0.37-3.74	~0° ~0°		Similar to above specimen and conditions except oxidized in air at 473 K.

DATA TABLE NO. 391 NORMAL SPECTRAL TRANSMITTANCE OF COPPER PHOSPHOROUS SELENIDE CONTACT COATINGS

[Wavelength, λ, μm; Transmittance, τ; Temperature, T, K]

λ	τ	λ	τ
CURVE 1* T ~ 298		CURVE 2 (cont.)*	
0.37	0.109	1.22	0.527
0.71	0.417	1.37	0.447
1.00	0.657	1.64	0.354
		2.09	0.276
CURVE 2* T ~ 298		2.43	0.200
0.37	0.128	2.76	0.137
0.45	0.226	3.16	0.073
0.61	0.462	3.74	0.000
0.68	0.533		
0.74	0.579		
0.81	0.607		
0.87	0.619		
0.94	0.610		
1.00	0.592		
1.13	0.581		

* No plot given

SPECIFICATION TABLE NO. 392 HEMISPHERICAL TOTAL EMITTANCE OF COPPER SULFIDE CONTACT COATINGS

Curve No.	Ref. No.	Year	Temperature Range, K	Reported Error, %	Composition (weight percent), Specifications and Remarks
1*	105	1964	307		Copper sulfide black; OFHC copper substrate; measured in vacuum (10^{-6} mm Hg) maintained by diffusion pump. [Authors' designation: Test No. 122]
2*	105	1964	307		Above specimen and conditions. [Authors' designation: Test No. 123]

DATA TABLE NO. 392 HEMISPHERICAL TOTAL EMITTANCE OF COPPER SULFIDE CONTACT COATINGS

[Temperature, T, K; Emittance, ϵ]

T	ϵ
CURVE 1*	
307	0.676
CURVE 2*	
307	0.684

* No plot given

SPECIFICATION TABLE NO. 393 HEMISPHERICAL TOTAL EMITTANCE OF HAFNIUM OXIDE CONTACT COATINGS

Curve No.	Ref. No.	Year	Temperature Range, K	Reported Error, %	Composition (weight percent), Specifications and Remarks
1*	160	1963	1199-2698		HfO_2 (Y_2O_3 stabilized); tungsten substrate; plasma arc sprayed; coating was opaque; measured in vacuum.
2*	160	1963	1802-2623		Similar to above specimen and conditions.

DATA TABLE NO. 393 HEMISPHERICAL TOTAL EMITTANCE OF HAFNIUM OXIDE CONTACT COATINGS

[Temperature, T, K; Emittance, ϵ]

T	ϵ
CURVE 1*	
1199	0.621
1809	0.610
1921	0.625
2033	0.662
2143	0.708
2255	0.759
2367	0.805
2477	0.836
2589	0.849
2698	0.840
CURVE 2*	
1802	0.502
2118	0.657
2382	0.843
2623	0.851

* No plot given

SPECIFICATION TABLE NO. 394 NORMAL SPECTRAL EMITTANCE OF HAFNIUM OXIDE CONTACT COATINGS

Curve No.	Ref. No.	Year	Wavelength, μm	Temperature Range, K	Geometry θ'	Reported Error, %	Composition (weight percent), Specifications, and Remarks
1*	160	1963	0.65	1922-2700	~0°		HfO_2 (Y_2O_3 stabilized); tungsten substrate; plasma arc sprayed to opacity; measured in vacuum.

DATA TABLE NO. 394 NORMAL SPECTRAL EMITTANCE OF HAFNIUM OXIDE CONTACT COATINGS

[Temperature, T, K; Emittance, ∈; Wavelength, λ, μm]

T	∈
CURVE 1*	
λ = 0.65	
1922	0.690
2033	0.701
2144	0.722
2256	0.730
2367	0.769
2478	0.773
2588	0.800
2700	0.804

* No plot given

SPECIFICATION TABLE NO. 395 NORMAL SPECTRAL TRANSMITTANCE OF HAFNIUM OXIDE CONTACT COATINGS

Curve No.	Ref. No.	Year	Temperature, K	Wavelength Range, μm	Geometry θ	θ'	ω'	Reported Error, %	Composition (weight percent), Specifications, and Remarks
1*	161	1966	298	0.189-0.418	0°	0°			HfO_2 (0.7 μm optical thickness) on plane-parallel fused quartz (2 mm thick) substrate; deposited from solution of hafnium oxychloride and stored for 24 hrs; amorphous film structure; 86.4 dry residue and 10-12 chlorine content; data extracted from smooth curve.
2*	161	1966	473	0.189-0.418	0°	0°			Similar to above specimen and conditions except coating 0.5 μm optical thickness; heated to 473 K; 63.7 dry residue and 5-7 chlorine content.
3*	161	1966	673	0.193-0.421	0°	0°			Similar to above specimen and conditions except coating <0.5 μm optical thickness; fine crystalline film structure; heated to 673 K; 54.67 dry residue and 0.5 chlorine content.

DATA TABLE NO. 395 NORMAL SPECTRAL TRANSMITTANCE OF HAFNIUM OXIDE CONTACT COATINGS

[Wavelength, λ, μm; Transmittance, τ; Temperature, T, K]

CURVE 1* T = 298

λ	τ
0.189	0.649
0.199	0.812
0.210	0.814
0.221	0.895
0.224	0.905
0.235	0.883
0.251	0.921
0.266	0.901
0.290	0.942
0.319	0.898
0.361	0.947
0.371	0.923
0.391	0.912
0.418	0.908

CURVE 2* T = 473

λ	τ
0.189	0.213
0.194	0.394
0.207	0.718
0.210	0.806
0.215	0.823
0.220	0.806
0.225	0.783
0.227	0.798
0.230	0.765
0.235	0.801
0.240	0.859
0.249	0.906
0.259	0.883
0.279	0.808
0.288	0.763
0.296	0.782
0.302	0.818

CURVE 2 (cont.)*

λ	τ
0.313	0.902
0.330	0.944
0.342	0.906
0.364	0.853
0.380	0.829
0.398	0.814
0.418	0.873

CURVE 3* T = 673

λ	τ
0.193	0.052
0.201	0.151
0.210	0.406
0.215	0.578
0.220	0.733
0.220	0.810
0.227	0.792

CURVE 3 (cont.)*

λ	τ
0.233	0.741
0.240	0.698
0.243	0.744
0.248	0.816
0.254	0.885
0.260	0.881
0.281	0.799
0.291	0.756
0.302	0.823
0.315	0.915
0.330	0.942
0.342	0.918
0.365	0.849
0.380	0.821
0.398	0.801
0.421	0.857

* No plot given

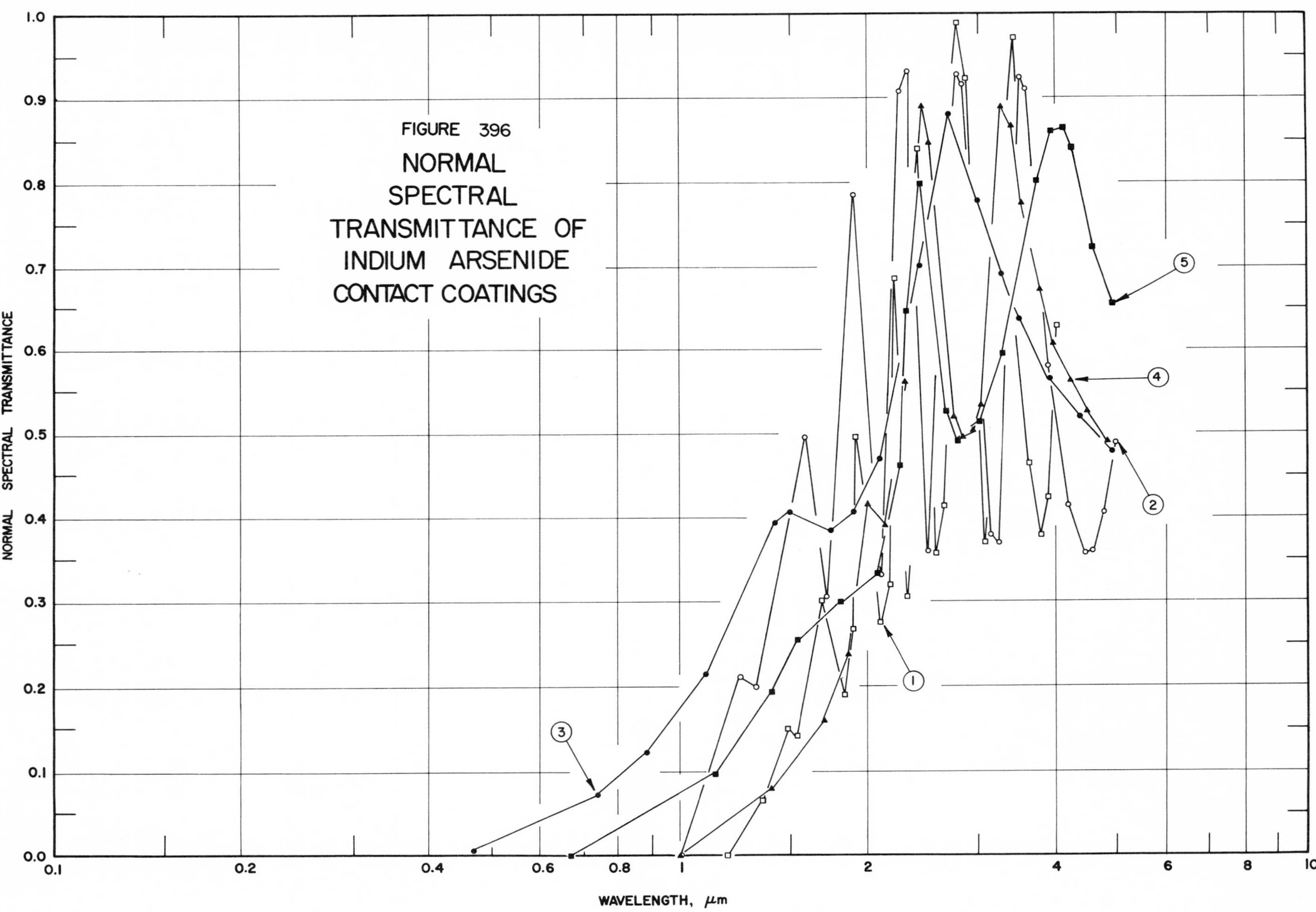

FIGURE 396 NORMAL SPECTRAL TRANSMITTANCE OF INDIUM ARSENIDE CONTACT COATINGS

SPECIFICATION TABLE NO. 396 NORMAL SPECTRAL TRANSMITTANCE OF INDIUM ARSENIDE CONTACT COATINGS

Curve No.	Ref. No.	Year	Temperature, K	Wavelength Range, μm	Geometry θ	θ'	ω'	Reported Error, %	Composition (weight percent), Specifications, and Remarks
1	162	1968	~298	1.19-4.01	0°	0°			InAs (<2.5 μm thick); glass (0.254 mm thick) substrate; vapor deposited in vacuum (10^{-7} mm Hg) at 0.001 $\mu m\ s^{-1}$ on substrate maintained at 303 K; data extracted from smooth curve. [Author's designation: Sample 38]
2	162	1968	~298	1.00-4.99	0°	0°			Similar to above specimen and conditions except substrate heated to 383 K during deposition. [Author's designation: Sample 21]
3	162	1968	~298	0.47-4.93	0°	0°			Similar to above specimen and conditions except substrate heated to 478 K during deposition. [Author's designation: Sample 62]
4	162	1968	~298	1.00-4.83	0°	0°			Similar to above specimen and conditions except substrate heated to 583 K during deposition. [Author's designation: Sample 59]
5	162	1968	~298	0.67-4.92	0°	0°			Similar to above specimen and conditions except substrate heated to 678 K during deposition. [Author's designation: Sample 14]

DATA TABLE NO. 396 NORMAL SPECTRAL TRANSMITTANCE OF INDIUM ARSENIDE CONTACT COATINGS

[Wavelength, λ, μm; Transmittance, τ; Temperature, T, K]

CURVE 1
T ~298

λ	τ
1.19	0.000
1.36	0.068
1.48	0.150
1.54	0.141
1.68	0.302
1.83	0.190
1.89	0.267
1.92	0.495
2.10	0.275
2.17	0.321
2.21	0.686
2.33	0.305
2.43	0.840
2.57	0.359
2.65	0.416
2.79	0.991
2.87	0.924
3.09	0.370
3.44	0.973
3.65	0.466
3.77	0.376
3.86	0.425
4.01	0.629

CURVE 2
T ~298

λ	τ
1.00	0.000
1.25	0.212
1.33	0.199
1.58	0.495
1.72	0.305
1.90	0.785
2.10	0.333
2.25	0.909
2.32	0.933
2.52	0.360
2.78	0.928
2.84	0.917
3.15	0.379
3.24	0.370
3.52	0.925
3.57	0.912
3.89	0.581
4.19	0.415
4.45	0.357
4.55	0.359
4.76	0.407
4.99	0.488

CURVE 3
T ~298

λ	τ
0.47	0.007
0.74	0.074
0.88	0.123
1.10	0.215
1.42	0.394
1.50	0.408
1.75	0.386
1.91	0.408
2.09	0.470
2.43	0.702
2.70	0.884
3.00	0.779
3.28	0.692
3.53	0.636
3.94	0.566
4.36	0.519
4.93	0.479

CURVE 4
T ~298

λ	τ
1.00	0.000
1.42	0.081
1.72	0.160
1.86	0.239
2.00	0.417
2.14	0.391
2.30	0.562
2.46	0.892
2.52	0.848
2.75	0.519
2.84	0.495
2.96	0.504
3.04	0.533
3.29	0.890
3.40	0.867
3.54	0.775
3.77	0.673
3.97	0.607
4.21	0.564
4.48	0.526
4.83	0.490

CURVE 5
T ~298

λ	τ
0.67	0.000
1.14	0.097
1.40	0.196
1.54	0.256
1.81	0.300
2.07	0.334
2.25	0.462
2.31	0.647
2.43	0.798
2.67	0.526
2.79	0.491
3.03	0.514
3.30	0.597
3.74	0.803
3.96	0.860
4.12	0.867
4.25	0.840
4.60	0.721
4.92	0.653

SPECIFICATION TABLE NO. 397 NORMAL TOTAL EMITTANCE OF IRON OXIDE CONTACT COATINGS

Curve No.	Ref. No.	Year	Temperature Range, K	Geometry θ'	Reported Error, %	Composition (weight percent), Specifications and Remarks
1*	147	1963	1039-1397	$\sim 0°$		Fe_2O_3 (0.0254 mm thick); Haynes Alloy 25 (L-605) substrate; substrate grit blasted; plasma sprayed; surface roughness 30 to 45 μm (RMS) measured with profilometer; measured in vacuum (3 to 4 x 10^{-4} mm Hg).

DATA TABLE NO. 397 NORMAL TOTAL EMITTANCE OF IRON OXIDE CONTACT COATINGS

[Temperature, T, K; Emittance, ϵ]

T	ϵ
CURVE 1*	
1039	0.725
1133	0.800
1191	0.830
1278	0.835
1358	0.850
1397	0.875

* No plot given

SPECIFICATION TABLE NO. 398 NORMAL SPECTRAL EMITTANCE OF IRON OXIDE CONTACT COATINGS

Curve No.	Ref. No.	Year	Wavelength, μm	Temperature Range, K	Geometry θ'	Reported Error, %	Composition (weight percent), Specifications, and Remarks
1*	147	1963	0.65	1133-1397	$\sim 0^{\circ}$		Fe_2O_3 (0.0254 mm thick); Haynes Alloy 25 (L-605) substrate; substrate grit blasted; plasma sprayed; surface roughness 30 to 45 μm (RMS) measured with profilometer; measured in vacuum (3 to 4 x 10^{-4} mm Hg); authors assumed specimen was a grey body.

DATA TABLE NO. 398 NORMAL SPECTRAL EMITTANCE OF IRON OXIDE CONTACT COATINGS

[Temperature, T, K; Emittance, ϵ; Wavelength, λ, μm]

T	ϵ
CURVE 1*	
1133	0.760
1189	0.885
1279	0.905
1355	0.955
1397	0.965

* No plot given

SPECIFICATION TABLE NO. 399 NORMAL SPECTRAL EMITTANCE OF IRON OXIDE CONTACT COATINGS

Curve No.	Ref. No.	Year	Temperature, K	Wavelength Range, μm	Geometry θ'	Reported Error, %	Composition (weight percent), Specifications, and Remarks
1*	96	1963	323.2	3.53-21.00	0°		Iron oxide; substrate unknown; opaque.

DATA TABLE NO. 399 NORMAL SPECTRAL EMITTANCE OF IRON OXIDE CONTACT COATINGS

[Wavelength, λ, μm; Emittance, ϵ; Temperature, T, K]

λ	ϵ	λ	ϵ
CURVE 1*		CURVE 1 (cont.)	
T = 323.2			
		12.01	0.882
3.53	0.907	12.28	0.880
3.92	0.906	12.92	0.866
4.48	0.899	13.44	0.850
5.08	0.912	13.61	0.837
5.29	0.909	14.15	0.810
5.80	0.947	14.41	0.804
6.07	0.929	15.53	0.807
6.38	0.928	16.43	0.807
6.78	0.949	16.84	0.820
7.10	0.937	17.44	0.825
8.42	0.966	18.11	0.838
8.79	0.959	18.87	0.838
9.08	0.952	19.63	0.846
9.30	0.951	20.01	0.857
10.05	0.924	20.46	0.879
11.06	0.873	21.00	0.906
11.42	0.866		
11.73	0.879		

* No plot given

SPECIFICATION TABLE NO. 400 NORMAL SPECTRAL REFLECTANCE OF IRON OXIDE CONTACT COATINGS

Curve No.	Ref. No.	Year	Temperature, K	Wavelength Range, μm	Geometry θ θ' ω'	Reported Error, %	Composition (weight percent), Specifications, and Remarks
1*	163	1964	~298	0.500-0.663	~0° ~0°		Iron oxide; aluminum substrate; substrate washed and then cleaned in vacuum by ion bombardment; vapor deposited in vacuum until first reflection minima achieved (~0.25 λ); measured in vacuum (10^{-6} mm Hg); data extracted from smooth curve.
2*	163	1964	~298	0.399-0.662	~0° ~0°		Similar to above specimen and conditions except film deposited until second reflectance maxima achieved (~0.75 λ).
3*	163	1964	~298	0.400-0.701	~0° ~0°		Similar to above specimen and conditions.
4*	163	1964	~298	0.401-0.702	~0° ~0°		Iron oxide; gold and aluminum substrates; gold fired on aluminum; iron oxide vapor deposited in vacuum until second reflectance maxima achieved (~0.75 λ); measured in vacuum (10^{-6} mm Hg); data extracted from smooth curve.
5*	163	1964	~298	0.400-0.701	~0° ~0°		Similar to above specimen and conditions.
6*	163	1964	~298	0.400-0.700	~0° ~0°		Similar to above specimen and conditions except film deposited until second reflection minima achieved (~0.75 λ).

DATA TABLE NO. 400 NORMAL SPECTRAL REFLECTANCE OF IRON OXIDE CONTACT COATINGS

[Wavelength, λ, μm; Reflectance, ρ; Temperature, T, K]

λ	ρ	λ	ρ	λ	ρ
CURVE 1* T ~ 298		CURVE 3* T ~ 298		CURVE 5* T ~ 298	
0.500	0.091	0.400	0.170	0.400	0.164
0.517	0.089	0.500	0.168	0.500	0.117
0.563	0.070	0.601	0.101	0.601	0.054
0.600	0.073	0.701	0.136	0.701	0.190
0.663	0.091				
		CURVE 4* T ~ 298		CURVE 6* T ~ 298	
CURVE 2* T ~ 298		0.401	0.201	0.400	0.148
0.399	0.176	0.500	0.205	0.500	0.146
0.500	0.199	0.601	0.130	0.600	0.084
0.562	0.191	0.702	0.089	0.700	0.100
0.601	0.171				
0.662	0.149				

* No plot given

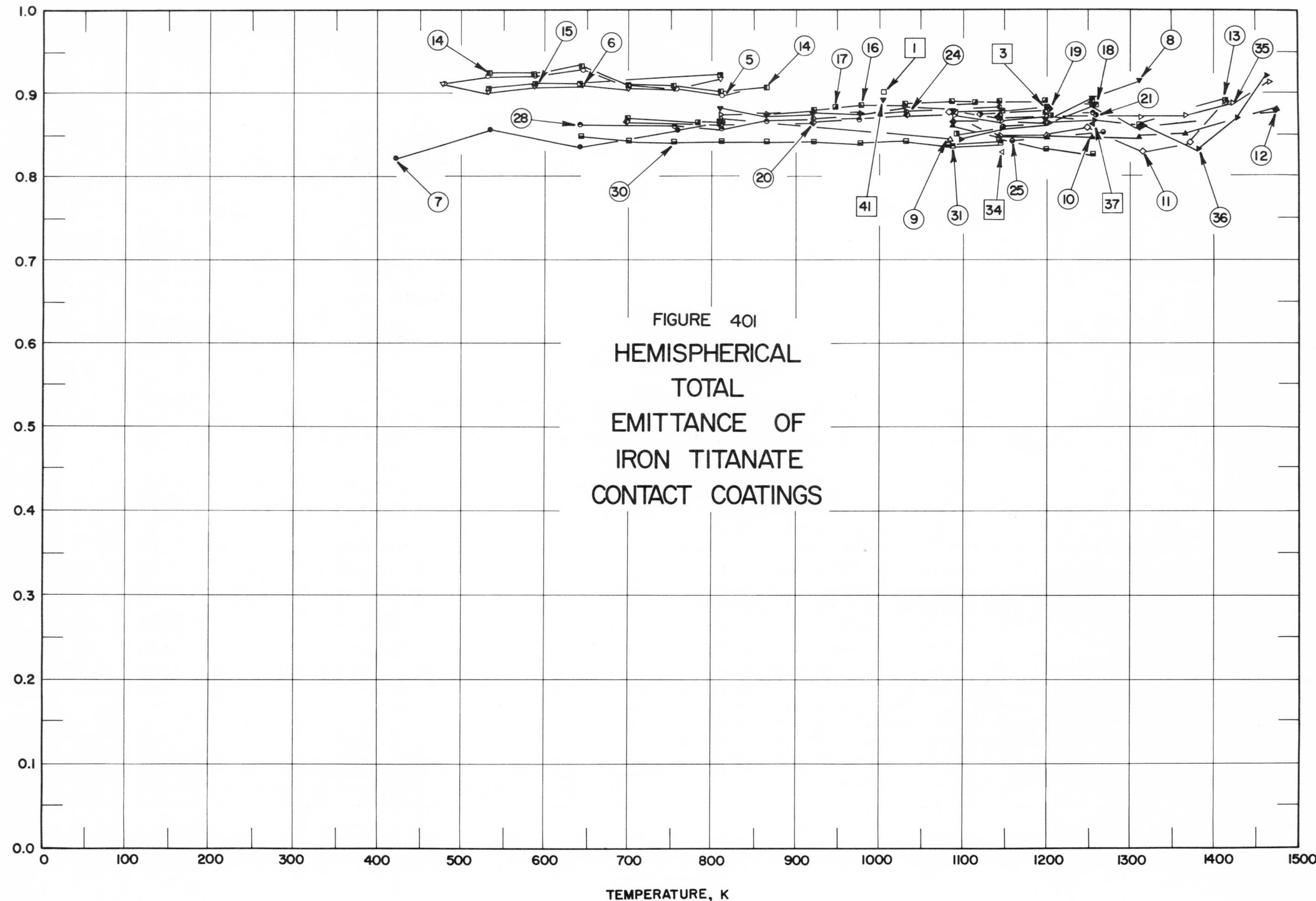
FIGURE 401
HEMISPHERICAL TOTAL EMITTANCE OF IRON TITANATE CONTACT COATINGS
HEMISPHERICAL TOTAL EMITTANCE
TEMPERATURE, K
0.0
0.1
0.2
0.3
0.4
0.5
0.6
0.7
0.8
0.9
1.0
0
100
200
300
400
500
600
700
800
900
1000
1100
1200
1300
1400
1500

SPECIFICATION TABLE NO. 401 HEMISPHERICAL TOTAL EMITTANCE OF IRON TITANATE CONTACT COATINGS

Curve No.	Ref. No.	Year	Temperature Range, K	Reported Error, %	Composition (weight percent), Specifications and Remarks
1	135	1969	1005		Fe_2TiO_5, powder from Continental Coating Corp, FCT-11; AISI-310 stainless steel substrate; cleaned in saturated solution of sulfuric acid and potassium dichromate and grit blasted with 60 mesh silicon carbide to roughness height of 2.79 μm; plasma arc sprayed; aged 10 hrs at 1005 K; measured in vacuum ($\sim 2 \times 10^{-8}$ mm Hg).
2*	135	1969	298-1005		Similar to above specimen and conditions except exposed to thermal cycling between 1005 K and 298 K (starting at 1005 K) at the rate of 114 cycles per 20 000 hrs.
3	135	1969	1200		Fe_2TiO_5, powder from Continental Coatings Corp, FCT-11 (0.102 mm thick); Nb-1 Zr substrate; substrate cleaned in solution of sulfuric acid and potassium dichromate and grit blasted with 60 mesh silicon carbide to roughness height of 2.79 μm; plasma arc sprayed; aged 92 hrs at 1200 K; measured in vacuum ($\sim 10^{-7}$ mm Hg).
4*	135	1969	298-1200		Similar to curve 3 specimen and conditions except exposed to thermal cycling between 298 and 1200 K (starting at 1200 K) at the rate of 51 cycles per 10 000 hrs.
5	135	1969	478-866		Fe_2TiO_5, Continental Coatings Corp, FCT-11; high purity IS-2 beryllium ingot (0.254 cm thick) substrate; plasma arc sprayed; measured in vacuum ($<10^{-6}$ mm Hg).
6	135	1969	478-812		Similar to above specimen and conditions except aged 1000 hrs at 922 K.
7	118	1962	422-813		Fe_2TiO_5, powder from Continental Coatings Corp (FCT-11) (0.102 mm thick); Nb-1Zr substrate; plasma arc sprayed; hard, medium grit texture; measured in vacuum ($<3.2 \times 10^{-7}$ mm Hg); temp measured with thermocouple. [Authors' designation: Run 1]
8	118	1962	811-1312		Above specimen and conditions except vacuum $<3.4 \times 10^{-6}$ mm Hg. [Authors' designation: Run 2]
9	118	1962	811-1085		Above specimen and conditions except vacuum $<2.6 \times 10^{-6}$ mm Hg. [Authors' designation: Run 3]
10	118	1962	1084-1475		Above specimen and conditions except vacuum $<2.0 \times 10^{-6}$ mm Hg. [Authors' designation: Run 4]
11	118	1962	1084-1415		Above specimen and conditions except temp measured with optical pyrometer.
12	118	1962	1475-1089		Curve 10 specimen and conditions; cooling cycle.
13	118	1962	1415-1093		Above specimen and conditions except temp measured with optical pyrometer.
14	144	1966	478-866		Fe_2TiO_5; high purity Be substrate; measured with optical pyrometer.
15	144	1966	478-810		Above specimen and conditions except aged in vacuum for 1000 hrs at 922 K.
16	81	1963	699-1255		Fe_2TiO_5, Continental Coatings Corp (0.0508 mm thick); Nb-1Zr substrate; plasma arc sprayed; measured in vacuum ($<1.6 \times 10^{-6}$ mm Hg); temp measured with thermocouple; heating cycle. [Author's designation: Run No. 1]
17	81	1963	1255-783		Above specimen and conditions; cooling cycle.
18	81	1963	1092-1257		Curve 16 specimen and conditions except temp measured with optical pyrometer.
19	81	1963	1257-1120		Curve 17 specimen and conditions except temp measured with optical pyrometer.
20	81	1963	699-1255		Fe_2TiO_5 (0.0762 mm thick); Nb-1Zr substrate; plasma arc sprayed; measured in vacuum ($<8.2 \times 10^{-6}$ mm Hg); temp measured with thermocouple; heating cycle. [Author's designation: Run No. 1]
21	81	1963	1148-1259		Above specimen and conditions except temp measured with optical pyrometer.

* Not shown on plot

SPECIFICATION TABLE NO. 401 HEMISPHERICAL TOTAL EMITTANCE OF IRON TITANATE CONTACT COATINGS (continued)

Curve No.	Ref. No.	Year	Temperature Range, K	Reported Error, %	Composition (weight percent), Specifications and Remarks
22*	81	1963	1255		Curve 20 specimen and conditions; measurements taken during 4.4 hr endurance test.
23*	81	1963	1259		Above specimen and conditions except temp measured with optical pyrometer.
24	81	1963	699-1255		Curve 20 specimen and conditions. [Author's designation: Run No. 2]
25	81	1963	1157-1268		Above specimen and conditions except temp measured with optical pyrometer.
26*	81	1963	1232-1255		Curve 24 specimen and conditions; measurements taken during 173.4 hr endurance test.
27*	81	1963	1259-1276		Above specimen and conditions except temp measured with optical pyrometer.
28	81	1963	644-1255		Fe_2TiO_5 (0.102 mm thick); Nb-1Zr substrate; plasma arc sprayed; measured in vacuum ($<7.8 \times 10^{-6}$ mm Hg); temp measured with thermocouple; heating cycle. [Author's designation: Run No. 1]
29	81	1963	1089-1255		Above specimen and conditions except temp measured with optical pyrometer.
30	81	1963	644-1255		Curve 28 specimen and conditions.
31	81	1963	1088-1255		Above specimen and conditions except temp measured with optical pyrometer.
32*	81	1963	1254.3-1254.8		Curve 30 specimen and conditions; measurements taken during 1.5 hr endurance test.
33*	81	1963	1254-1256		Above specimen and conditions except temp measured with optical pyrometer during 187.1 hrs endurance test.
34	81	1963	1254-1147		Curve 31 specimen and conditions; cooling cycle.
35	81	1963	810-1466		Fe_2TiO_5, Continental Coatings Corp (0.127 mm thick); Nb-1Zr substrate; plasma arc sprayed; measured in vacuum ($<8.9 \times 10^{-6}$ mm Hg); temp measured with thermocouple; heating cycle. [Author's designation: Run No. 1]
36	81	1963	1097-1462		Above specimen and conditions except temp measured with optical pyrometer.
37	143	1964	1256	±2.3-2.7	Fe_2TiO_5 (0.102 mm thick); Nb-1Zr substrate; plasma arc sprayed; measured in vacuum ($<10^{-8}$ mm Hg); data obtained throughout a 29.8 hr endurance test at 1256 K.
38	143	1964	1200	±2.3-2.7	Similar to above specimen and conditions except maintained at 1200 K for 200 hrs.
39	143	1964	1256	±2.3-2.7	Above specimen and conditions except maintained at 1256 K for an additional 118 hrs.
40*	143	1964	1200	±2.3	Similar to above specimen and conditions except sample maintained at 1200 K for over 1900 hr endurance test; endurance test time, hrs, is variable; thermal cycling consisting of cooling to ~300 K in 6.6 min then slowly reheating to 1200 K occurred after initial heating, 100 hrs, 200 hrs, and 4 cycles between 1870 and 1950 hrs.
41	143	1964	1006	±2.3-2.7	Fe_2TiO_5 (0.102 mm thick); AISI-310 stainless steel substrate; plasma arc sprayed; measured in vacuum ($<10^{-8}$ mm Hg); data obtained during a 300 hr endurance test; thermal cycling consisting of cooling to ~300 K in 6.5 min then reheating slowly to 1006 K occurred after initial heating, 100 hrs, and 200 hrs.
42*	143	1964	1006	±2.7	Similar to above specimen and conditions except endurance test over 900 hrs; thermal cycling occurred after 200, 400, and 500 hrs.

* Not shown on plot

DATA TABLE NO. 401 HEMISPHERICAL TOTAL EMITTANCE OF IRON TITANATE CONTACT COATINGS

[Temperature, T, K; Emittance, ∈]

CURVE 1

T	∈
1005	0.900

CURVE 2* (T = 298-1005)

Thermal Cycling time	∈
290	0.897
490	0.899
620	0.894
840	0.896
1100	0.894
1290	0.894
1500	0.891
1700	0.898
1870	0.898
2100	0.897
2280	0.894
2440	0.895
2610	0.889
2840	0.889
3030	0.880
3290	0.887
3450	0.887
3680	0.883
3850	0.885
4060	0.885
4330	0.879
4510	0.881
4650	0.884
4850	0.884
5070	0.884
5200	0.884
5310	0.884
5480	0.884
5650	0.875
5870	0.874
6180	0.874
6390	0.874
6570	0.873
6770	0.876
7010	0.884
7240	0.885
7410	0.878

CURVE 2 (cont.)*

Thermal Cycling time	∈
7570	0.884
7720	0.884
7890	0.889
8040	0.878
8190	0.878
8370	0.882
8500	0.889
8710	0.886
8930	0.885
9230	0.884
9460	0.886
9680	0.884
9910	0.883
10240	0.885
10600	0.886
10940	0.883
11240	0.882
11520	0.886
11760	0.880
12080	0.880
12400	0.874
12770	0.880
13030	0.882
13420	0.877
13730	0.880
14070	0.880
14410	0.884
14700	0.877
14970	0.873
15210	0.870
15450	0.872
15700	0.880
15980	0.871
16280	0.881
16600	0.876
16870	0.875
17050	0.872
17360	0.869
17590	0.877
17990	0.877
18310	0.879
18630	0.866
18950	0.863
19340	0.872
19640	0.872
19980	0.866

CURVE 3

T	∈
1200	0.880

CURVE 4* (T = 298-1200)

Thermal Cycling time	∈
182	0.879
335	0.879
504	0.879
659	0.874
806	0.875
925	0.862
1134	0.860
1281	0.866
1467	0.862
1680	0.856
1783	0.856
1939	0.852
2076	0.851
2259	0.856
2442	0.854
2619	0.848
2776	0.847
3000	0.850
3175	0.857
3336	0.857
3506	0.850
3650	0.851
3880	0.849
4000	0.851
4129	0.849
4293	0.848
4473	0.849
4731	0.846
4938	0.845
5115	0.846
5400	0.851
5602	0.844
5766	0.840
6000	0.845
6146	0.848
6387	0.849
6642	0.846
6875	0.854
7084	0.847

CURVE 4 (cont.)*

Thermal Cycling time	∈
7232	0.850
7394	0.851
7644	0.851
7887	0.853
8168	0.846
8333	0.842
8627	0.840
8814	0.845
9001	0.837
9185	0.841
9460	0.840
9591	0.840
9771	0.840
9981	0.840

CURVE 5

T	∈
478	0.911
533	0.921
589	0.921
645	0.928
700	0.907
756	0.905
812	0.898
866	0.906

CURVE 6

T	∈
478	0.911
533	0.901
589	0.908
645	0.909
700	0.906
756	0.904
812	0.917

CURVE 7

T	∈
422	0.822
536	0.856
641	0.836
758	0.854
813	0.863

CURVE 8

T	∈
811	0.881
866	0.871
922	0.873
978	0.874
1033	0.882
1089	0.875
1142	0.867
1199	0.865
1255	0.893
1312	0.914

CURVE 9

T	∈
811	0.868
1085	0.843

CURVE 10

T	∈
1089	0.860
1144	0.848
1200	0.846
1255	0.845
1310	0.845
1367	0.849
1475	0.879

CURVE 11

T	∈
1084	0.874
1144	0.846
1200	0.847
1250	0.857
1316	0.829
1371	0.839
1415	0.888

CURVE 12

T	∈
1475	0.879
1311	0.858
1200	0.866
1089	0.866

CURVE 13

T	∈
1415	0.888
1311	0.859

CURVE 13 (cont.)

T	∈
1205	0.869
1093	0.849

CURVE 14

T	∈
478	0.912*
533	0.923
589	0.923
644	0.932
700	0.908
755	0.908
810	0.901
866	0.906

CURVE 15

T	∈
478	0.912*
533	0.905
589	0.910
644	0.910
700	0.910*
755	0.910*
810	0.922

CURVE 16

T	∈
699	0.869
810	0.866
922	0.878
977	0.882
1033	0.884
1088	0.887
1144	0.887
1199	0.873
1255	0.889

CURVE 17

T	∈
1255	0.889*
1199	0.888
1116	0.887
949	0.880
783	0.864

CURVE 18

T	∈
1092	0.876
1147	0.876
1203	0.863
1257	0.882

CURVE 19

T	∈
1257	0.882*
1203	0.878
1120	0.873

CURVE 20

T	∈
699	0.866
810	0.860
922	0.864
1033	0.871
1144	0.871
1255	0.883

CURVE 21

T	∈
1148	0.858
1259	0.871

CURVE 22*

T	∈
1255	0.883
1255	0.884
1255	0.883
1255	0.885
1255	0.884

CURVE 23*

T	∈
1259	0.871
1259	0.871
1258	0.874
1259	0.874
1259	0.873

CURVE 24

T	∈
699	0.866*
810	0.864*
922	0.867
1033	0.875

CURVE 24 (cont.)

T	∈
1144	0.882
1255	0.887*

CURVE 25

T	∈
1157	0.842
1268	0.850

CURVE 26*

T	∈
1255	0.887
1255	0.888
1255	0.888
1255	0.891
1255	0.891
1255	0.924
1237	0.922
1237	0.922
1237	0.926
1238	0.933
1237	0.932
1237	0.937
1237	0.943
1237	0.944
1237	0.943
1232	0.949
1233	0.947
1234	0.946
1232	0.950
1232	0.951
1232	0.953

CURVE 27*

T	∈
1268	0.850
1268	0.853
1268	0.853
1267	0.857
1267	0.894
1276	0.864
1259	0.860
1259	0.858
1260	0.864
1260	0.868
1259	0.868
1260	0.871
1260	0.871

* Not shown on plot

DATA TABLE NO. 401 HEMISPHERICAL TOTAL EMITTANCE OF IRON TITANATE CONTACT COATINGS (continued)

T	ε
CURVE 27 (cont.)*	
1259	0.868
1260	0.871
1260	0.877
1260	0.877
1260	0.877
1259	0.871
1259	0.870
1260	0.870
1260	0.867
1260	0.868
1260	0.868
CURVE 28	
644	0.863
699	0.865*
755	0.859
811	0.857
865	0.866
921	0.864*
977	0.867*
1033	0.870*
1088	0.869*
1144	0.869
1199	0.872*
1255	0.873
1255	0.871
1255	0.872
CURVE 29*	
1089	0.866
1144	0.868
1199	0.872
1255	0.873
1255	0.871
1255	0.872
CURVE 30	
644	0.847
700	0.844
755	0.842
811	0.842
865	0.840
922	0.840
978	0.837

T	ε
CURVE 30 (cont.)	
1033	0.840
1088	0.836
1144	0.844
1199	0.831
1255	0.825
CURVE 31	
1088	0.836
1145	0.839
1200	0.828
1255	0.823*
CURVE 32*	
1254.8	0.825
1254.8	0.824
1254.3	0.825
CURVE 33*	
1255	0.823
1256	0.821
1256	0.821
1256	0.819
1255	0.815
1256	0.814
1254	0.816
1254	0.817
1255	0.818
1254	0.817
1254	0.818
1254	0.816
1254	0.818
1255	0.822
1255	0.821
1254	0.826
CURVE 34	
1254	0.826*
1147	0.828

T	ε
CURVE 35	
810	0.872
866	0.872
921	0.874
977	0.972
1033	0.871*
1089	0.871
1144	0.866
1199	0.869
1255	0.872*
1312	0.869
1367	0.870
1421	0.886
1466	0.911
CURVE 36	
1097	0.843
1147	0.856
1202	0.863
1258	0.865
1316	0.859
1382	0.831
1429	0.867
1462	0.919
CURVE 37	
1256	0.86
CURVE 38*	
1200	0.86
CURVE 39*	
1256	0.88

Endurance time	ε
CURVE 40* T = 1200	
0	0.875
180	0.880
580	0.875
920	0.860
CURVE 40 (cont.)*	
1180	0.860
1420	0.860
1660	0.855
1940	0.855

T	ε
CURVE 41	
1006	0.89

Endurance time	ε
CURVE 42* T = 1006	
0	0.895
165	0.895
400	0.890
575	0.895
740	0.895
935	0.895

* Not shown on plot

SPECIFICATION TABLE NO. 402 HEMISPHERICAL TOTAL EMITTANCE OF IRON TITANIUM ALUMINUM OXIDE CONTACT COATINGS

Curve No.	Ref. No.	Year	Temperature Range, K	Reported Error, %	Composition (weight percent), Specifications and Remarks
1*	118	1962	421-1449		Iron titanium aluminum oxide (Continental Coatings Corp. FCT-12) (0.127 mm thick); Nb-1Zr alloy substrate; plasma arc sprayed; fairly hard, medium grit texture; poor adhesion to substrate; measured in vacuum (< 4.4 x 10^{-5} mm Hg); heating cycle.
2*	118	1962	1449-771		Above specimen and conditions except cooling cycle.

DATA TABLE NO. 402 HEMISPHERICAL TOTAL EMITTANCE OF IRON TITANIUM ALUMINUM OXIDE CONTACT COATINGS

[Temperature, T, K; Emittance, ϵ]

T	ϵ	T	ϵ
CURVE 1*		CURVE 2*	
421	0.831	1449	0.891
531	0.855	1283	0.901
644	0.864	1117	0.893
757	0.879	950	0.881
810	0.876	771	0.867
866	0.874		
921	0.880		
978	0.878		
1033	0.872		
1088	0.879		
1144	0.875		
1199	0.872		
1255	0.865		
1309	0.871		
1366	0.882		
1420	0.888		
1449	0.891		

* No plot given

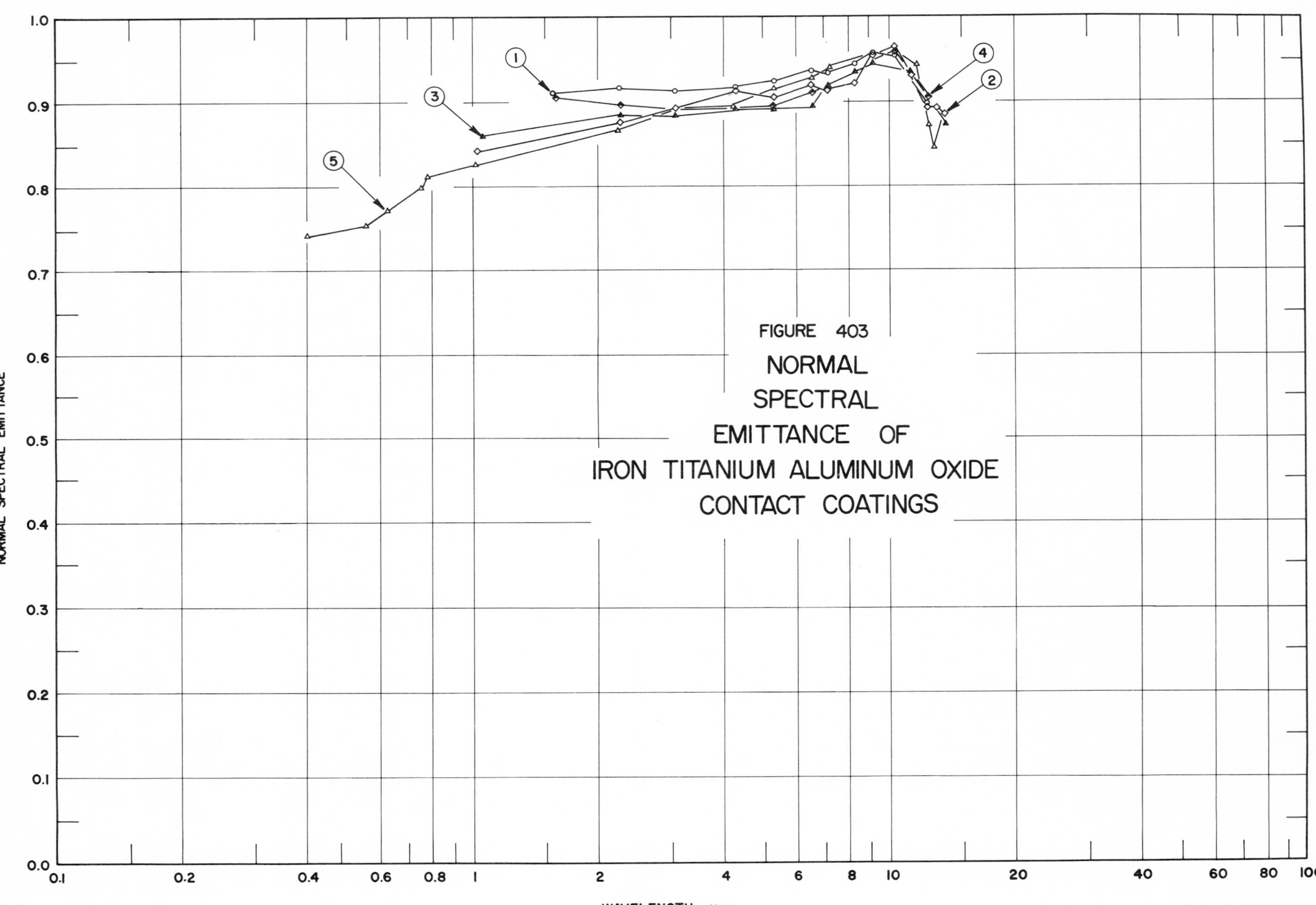
FIGURE 403
NORMAL
SPECTRAL
EMITTANCE OF
IRON TITANIUM ALUMINUM OXIDE
CONTACT COATINGS
NORMAL SPECTRAL EMITTANCE
WAVELENGTH, μm
1
2
3
4
5

SPECIFICATION TABLE NO. 403 NORMAL SPECTRAL EMITTANCE OF IRON TITANIUM ALUMINUM OXIDE CONTACT COATINGS

Curve No.	Ref. No.	Year	Temperature, K	Wavelength Range, μm	Geometry θ'	Reported Error, %	Composition (weight percent), Specifications, and Remarks
1	118	1962	755	1. 56-12. 40	~0°		Iron titanium aluminum oxide (Continental Coatings Corp, FCT-12) (0. 102 mm thick); Nb-1 Zr alloy substrate; plasma arc sprayed; fairly hard, medium grit texture, poor adhesion to substrate.
2	118	1962	1450	1. 03-13. 56	~0°		Above specimen and conditions.
3	118	1962	1061	1. 06-13. 56	~0°		Curve 2 specimen and conditions except maintained at 1061 K overnight.
4	118	1962	755	1. 59-12. 42	~0°		Curve 3 specimen and conditions.
5	118	1962	1366	0. 40-13. 52	~0°		Curve 3 specimen and conditions.

DATA TABLE NO. 403 NORMAL SPECTRAL EMITTANCE OF IRON TITANIUM ALUMINUM OXIDE CONTACT COATINGS

[Wavelength, λ, μm; Emittance, ϵ; Temperature, T, K]

λ	ϵ
CURVE 1	**T = 755**
1. 56	0. 910
2. 24	0. 916
3. 05	0. 912
4. 24	0. 917
5. 25	0. 923
6. 58	0. 936
7. 16	0. 933
8. 23	0. 945
9. 16	0. 958
10. 35	0. 954
12. 40	0. 899
CURVE 2	**T = 1061**
1. 03	0. 843
2. 27	0. 879
3. 08	0. 893
4. 24	0. 913
5. 28	0. 905
CURVE 2 (cont.)	
6. 57	0. 920
7. 18	0. 912
8. 22	0. 921
9. 14	0. 955
10. 35	0. 965
11. 04	0. 930
12. 44	0. 895
13. 00	0. 894
13. 56	0. 886
CURVE 3	**T = 1061**
1. 06	0. 860
2. 28	0. 884
3. 03	0. 885
4. 23	0. 891
5. 25	0. 891
6. 58	0. 896
7. 14	0. 919
8. 21	0. 933
CURVE 3 (cont.)	
9. 11	0. 944
11. 05	0. 934
12. 99	0. 890*
13. 56	0. 874
CURVE 4	**T = 755**
1. 59	0. 905
2. 28	0. 899
3. 03	0. 891*
5. 23	0. 893
6. 59	0. 910
7. 18	0. 917*
8. 20	0. 933*
10. 32	0. 959
12. 42	0. 903
CURVE 5	**T = 1366**
0. 40	0. 744
0. 53	0. 756
0. 62	0. 774
0. 75	0. 800
0. 78	0. 812
1. 01	0. 828
2. 24	0. 869
3. 09	0. 887*
4. 22	0. 893
5. 27	0. 919
6. 58	0. 929
7. 20	0. 940
8. 23	0. 945*
9. 17	0. 958*
11. 08	0. 942
12. 40	0. 874
12. 99	0. 847
13. 52	0. 884*

* Not shown on plot

SPECIFICATION TABLE NO. 404 NORMAL SPECTRAL REFLECTANCE OF LEAD CHLORIDE CONTACT COATINGS

Curve No.	Ref. No.	Year	Temperature, K	Wavelength Range, μm	Geometry θ θ' ω'	Reported Error, %	Composition (weight percent), Specifications, and Remarks
1*	164	1963	298	6.00-36.98	~0° ~0°		$PbCl_2$ (11 μm thick); Ge (5 μm optical thickness), $PbCl_2$ (9.4 μm thick), and LiF substrates; data extracted from smooth curve.

DATA TABLE NO. 404 NORMAL SPECTRAL REFLECTANCE OF LEAD CHLORIDE CONTACT COATINGS

[Wavelength, λ, μm; Reflectance, ρ; Temperature, T, K]

λ	ρ	λ	ρ	λ	ρ	λ	ρ
CURVE 1* T = 298		CURVE 1 (cont.)		CURVE 1 (cont.)		CURVE 1 (cont.)	
6.00	0.592	13.23	0.110	21.57	0.923	33.84	0.377
6.35	0.604	13.81	0.149	23.23	0.921	34.12	0.360
6.52	0.596	14.20	0.183	24.35	0.946	34.52	0.349
6.90	0.521	14.35	0.207	25.67	0.966	35.32	0.346
7.20	0.400	14.75	0.362	26.43	0.970	35.94	0.336
7.33	0.320	15.23	0.565	27.33	0.966	36.39	0.324
7.47	0.179	15.37	0.605	28.19	0.956	36.98	0.304
7.65	0.122	15.51	0.643	28.91	0.942		
7.83	0.074	15.66	0.668	29.61	0.921		
8.10	0.070	15.82	0.693	30.35	0.891		
8.82	0.113	16.17	0.737	31.14	0.845		
9.14	0.120	16.73	0.786	31.70	0.802		
9.45	0.108	17.14	0.805	32.21	0.748		
9.95	0.071	17.65	0.833	32.69	0.686		
10.28	0.038	18.06	0.858	33.05	0.611		
10.79	0.017	18.61	0.876	33.25	0.537		
11.29	0.016	19.09	0.889	33.33	0.486		
11.76	0.029	19.89	0.907	33.48	0.430		
		20.52	0.916	33.62	0.403		

* No plot given

SPECIFICATION TABLE NO. 405 NORMAL SPECTRAL REFLECTANCE OF LEAD MOLYBDENUM TETRAOXIDE CONTACT COATINGS

Curve No.	Ref. No.	Year	Temperature, K	Wavelength Range, μm	Geometry θ θ' ω'	Reported Error, %	Composition (weight percent), Specifications, and Remarks
1*	165	1965	~298	2.0-38.0	~0° ~0°		$PbMoO_4$, Wulfenite (Chihauhua, Mexico); KBr substrate; evaporated in vacuum (3×10^{-5} mm Hg); deposition rate 0.25 μm per min.
2*	166	1967	~298	10.00-16.93	~0° ~0°		$PbMoO_4$ (~2.0 μm thick); glass substrate; vapor deposited in vacuum (1.2×10^{-5} mm Hg) onto substrate at 573 K at the rate of 0.25 wavelength of optical thickness per min using monitor wavelength of 0.549 μm; data extracted from smooth curve.

DATA TABLE NO. 405 NORMAL SPECTRAL REFLECTANCE OF LEAD MOLYBDENUM TETRAOXIDE CONTACT COATINGS

[Wavelength, λ, μm; Reflectance, ρ; Temperature, T, K]

λ	ρ	λ	ρ	λ	ρ	λ	ρ	λ	ρ	λ	ρ
CURVE 1*		CURVE 1 (cont.)		CURVE 1 (cont.)		CURVE 2 (cont.)		CURVE 2 (cont.)		CURVE 2 (cont.)	
T ~ 298											
2.0	0.157	12.6	0.886	28.9	0.103	10.18	0.145	12.08	0.590	14.36	0.265
3.0	0.155	12.8	0.883	29.9	0.070	10.24	0.122	12.22	0.608	14.45	0.231
4.0	0.151	13.1	0.794	30.9	0.034	10.38	0.085	12.35	0.631	14.67	0.155
5.0	0.150	13.4	0.617	31.3	0.005	10.46	0.071	12.46	0.650	14.78	0.129
6.0	0.149	13.6	0.523	31.5	0.042	10.53	0.061	12.56	0.672	14.88	0.119
7.0	0.141	14.0	0.490	31.9	0.307	10.62	0.063	12.66	0.693	15.00	0.114
8.1	0.134	14.9	0.395	31.9	0.360	10.69	0.066	12.78	0.721	15.11	0.117
9.0	0.082	15.9	0.326	32.4	0.524	10.81	0.079	13.00	0.759	15.47	0.152
10.0	0.050	16.9	0.286	32.9	0.568	11.03	0.123	13.06	0.763	16.14	0.229
10.5	0.020	17.9	0.274	33.4	0.548	11.23	0.168	13.11	0.755	16.93	0.319
11.0	0.005	18.9	0.252	33.9	0.443	11.38	0.214	13.18	0.720		
11.2	0.146	19.9	0.238	34.9	0.324	11.46	0.240	13.28	0.659		
11.4	0.725	20.9	0.230	36.0	0.250	11.51	0.260	13.42	0.606		
11.5	0.792	21.9	0.219	37.0	0.221	11.58	0.313	13.85	0.491		
11.6	0.831	22.8	0.206	38.0	0.217	11.73	0.489	13.96	0.459		
11.9	0.824	23.8	0.193			11.78	0.522	14.03	0.440		
12.1	0.815	25.9	0.183	CURVE 2*		11.81	0.542	14.08	0.421		
12.4	0.856	26.8	0.159			11.86	0.560	14.18	0.372		
		27.9	0.145	10.00	0.224	11.90	0.569	14.26	0.317		

* No plot given

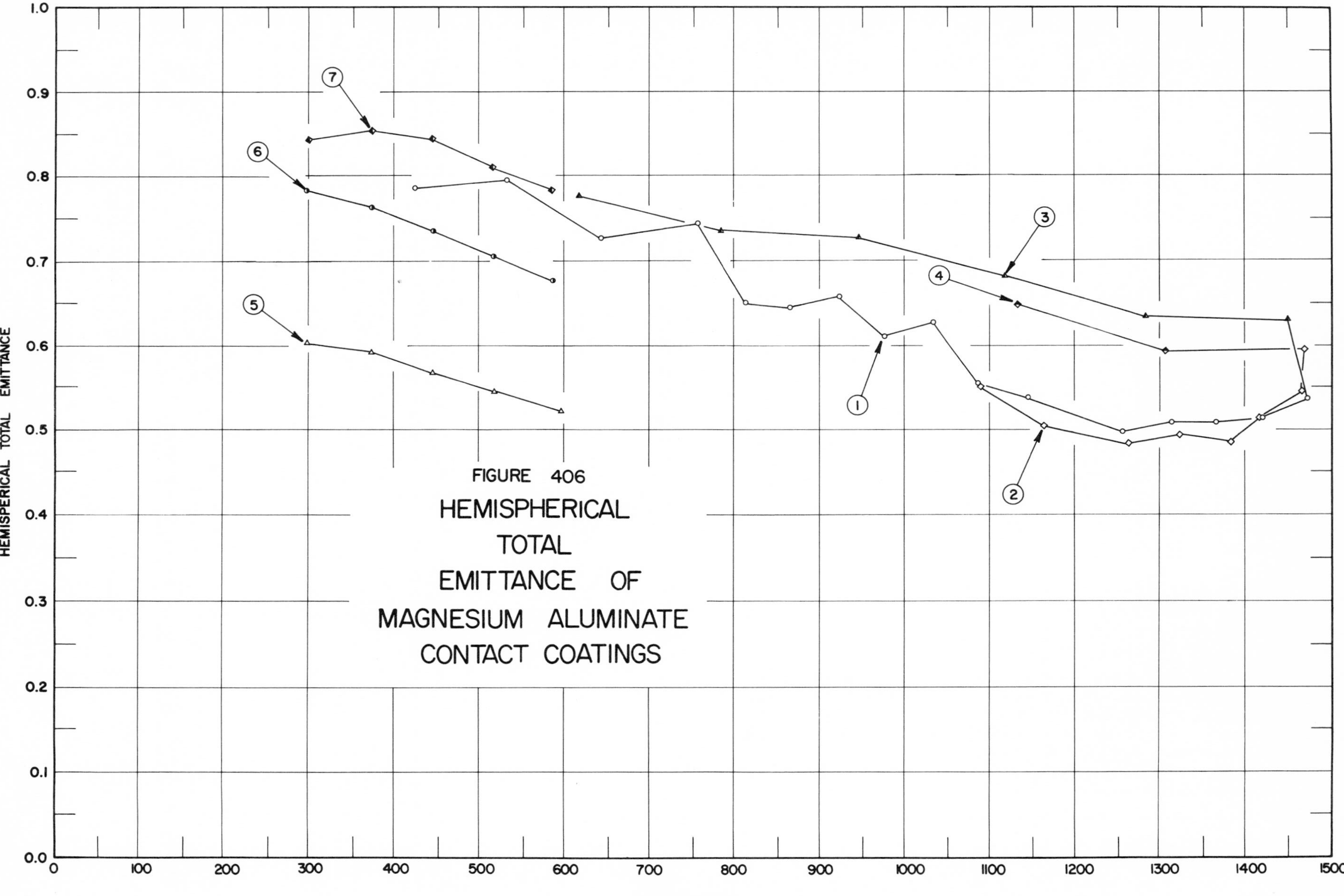
FIGURE 406
HEMISPHERICAL
TOTAL
EMITTANCE OF
MAGNESIUM ALUMINATE
CONTACT COATINGS
HEMISPERICAL TOTAL EMITTANCE
TEMPERATURE, K
1.0
0.9
0.8
0.7
0.6
0.5
0.4
0.3
0.2
0.1
0.0
0
100
200
300
400
500
600
700
800
900
1000
1100
1200
1300
1400
1500
1
2
3
4
5
6
7

SPECIFICATION TABLE NO. 406 HEMISPHERICAL TOTAL EMITTANCE OF MAGNESIUM ALUMINATE CONTACT COATINGS

Curve No.	Ref. No.	Year	Temperature Range, K	Reported Error, %	Composition (weight percent), Specifications and Remarks
1	118	1962	424-1473		$MgO\cdot Al_2O_3$, Rokide MA from Norton Co; composition: 66.8 Al_2O_3, 29.5 MgO, 2.9 SiO_2, 0.69 Cr_2O_3, and 0.08 Fe_2O_3 + TiO_2 + NaO_2 (0.127 mm thick); Nb-1 Zr alloy substrate; flame sprayed by Rokide process; hard, rough glossy appearance; measured in vacuum (~3.8 x 10^{-7} mm Hg); temp measured with thermocouple; heating cycle.
2	118	1962	1090-1470		Above specimen and conditions except temp measured with optical pyrometer.
3	118	1962	1473-618		Curve 1 specimen and conditions; cooling cycle.
4	118	1962	1470-1131		Above specimen and conditions except temp measured with optical pyrometer.
5	137	1965	299-599	±5	Rokide MA (~0.025 mm thick, 2.3 mg cm^{-2}); Al substrate; applied by Rokide process; measured in vacuum (10^{-7} mm Hg).
6	137	1965	297-589	±5	Similar to above specimen and conditions except 0.063 mm thick, 6.0 mg cm^{-2}.
7	137	1965	300-589	±5	Similar to curve 5 specimen and conditions except 0.140 mm thick, 31 mg cm^{-2}.

DATA TABLE NO. 406 HEMISPHERICAL TOTAL EMITTANCE OF MAGNESIUM ALUMINATE CONTACT COATINGS

[Temperature, T, K; Emittance, ∈]

T	∈	T	∈	T	∈	T	∈
CURVE 1		CURVE 2		CURVE 4		CURVE 7	
424	0.787	1090	0.550	1470	0.596*	300	0.842
534	0.795	1163	0.503	1306	0.591	374	0.853
644	0.727	1264	0.483	1131	0.649	445	0.844
756	0.742	1322	0.491			517	0.810
811	0.650	1382	0.485	CURVE 5		589	0.782
867	0.643	1419	0.513				
923	0.656	1468	0.542	299	0.604		
978	0.610	1470	0.596	373	0.591		
1033	0.626			445	0.569		
1089	0.552	CURVE 3		518	0.544		
1144	0.537			599	0.521		
1256	0.497	1473	0.534*				
1311	0.509	1450	0.630	CURVE 6			
1366	0.509	1283	0.633				
1420	0.512	1118	0.680	297	0.782		
1473	0.534	949	0.728	371	0.762		
		784	0.734	445	0.735		
		618	0.777	516	0.705		
				589	0.675		

* Not shown on plot

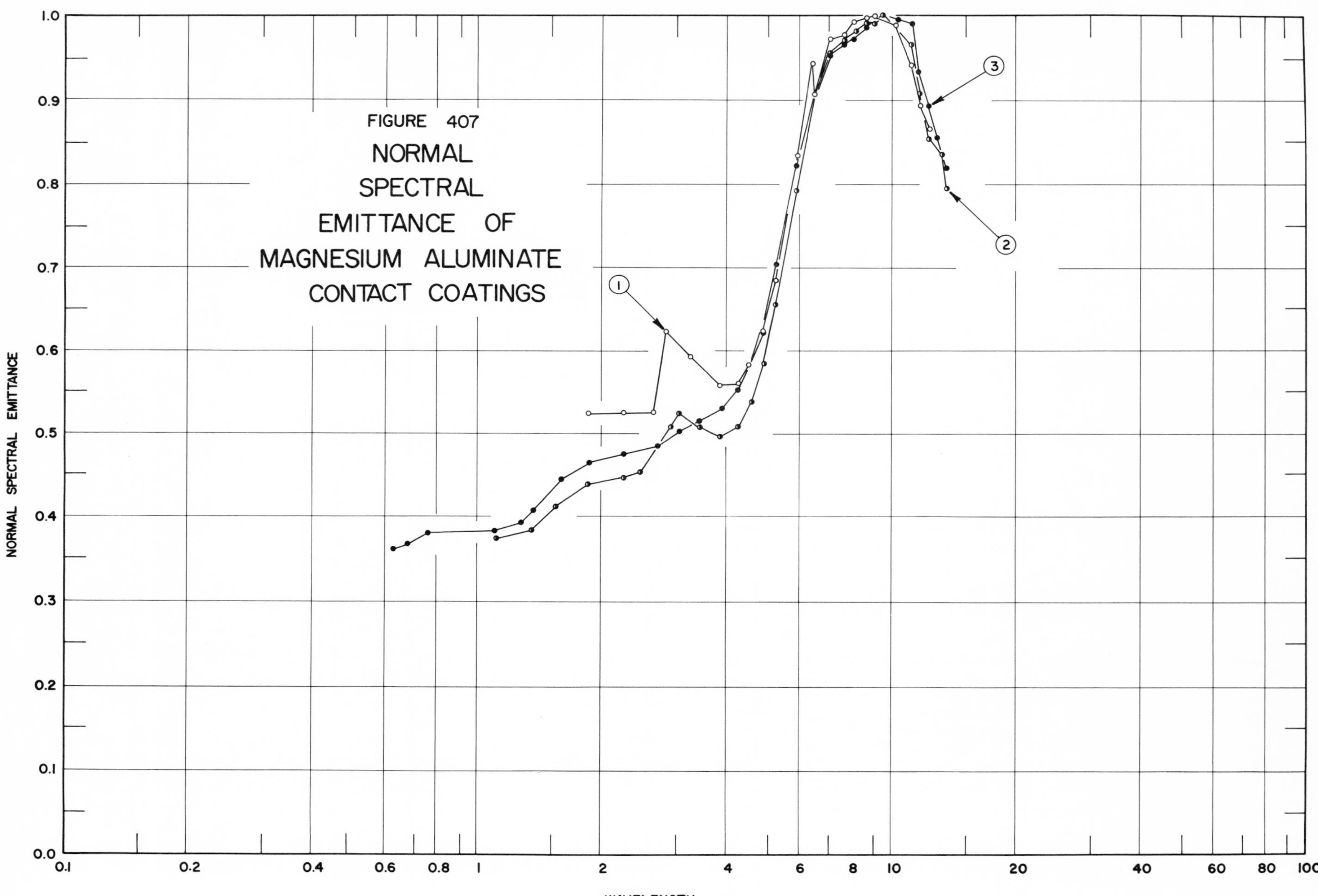

FIGURE 407 NORMAL SPECTRAL EMITTANCE OF MAGNESIUM ALUMINATE CONTACT COATINGS

SPECIFICATION TABLE NO. 407 NORMAL SPECTRAL EMITTANCE OF MAGNESIUM ALUMINATE CONTACT COATINGS

Curve No.	Ref. No.	Year	Temperature, K	Wavelength Range, μm	Geometry θ'	Reported Error, %	Composition (weight percent), Specifications, and Remarks
1	118	1962	755	1.86-12.43	~0°		$MgO \cdot Al_2O_3$, Rokide MA from Norton Co.; composition: 66.8 Al_2O_3, 29.5 MgO, 2.9 SiO_2, 0.69 Cr_2O_3, and 0.08 Fe_2O_3 + TiO_2 + NaO_2 (0.127 mm thick); Nb-1 Zr alloy substrate; flame sprayed by Rokide process; very hard, rough, glossy appearance.
2	118	1962	1061	1.12-13.59	~0°		Above specimen and conditions.
3	118	1962	1366	0.63-13.59	~0°		Curve 1 specimen and conditions.

DATA TABLE NO. 407 NORMAL SPECTRAL EMITTANCE OF MAGNESIUM ALUMINATE CONTACT COATINGS

[Wavelength, λ, μm; Emittance, ∈; Temperature, T, K]

λ	∈	λ	∈	λ	∈	λ	∈	λ	∈
CURVE 1		CURVE 1 (cont.)		CURVE 2 (cont.)		CURVE 3		CURVE 3 (cont.)	
T = 755						T = 1366			
		10.36	0.990	4.95	0.585			6.56	0.909*
1.86	0.523	11.19	0.942	5.22	0.655	0.63	0.360	7.14	0.955
2.27	0.524	11.76	0.895	5.95	0.795	0.68	0.366	7.73	0.968
2.66	0.525	12.43	0.867	6.55	0.909*	0.76	0.380	8.19	0.973
2.86	0.622			7.19	0.958	1.10	0.381	8.68	0.988
3.27	0.594	CURVE 2		7.69	0.972	1.12	0.366*	9.14	1.000*
3.87	0.559	T = 1061		8.20	0.984	1.28	0.391	9.50	1.001*
4.26	0.560			8.67	0.995	1.38	0.409	10.37	0.998
4.58	0.584	1.12	0.372	9.17	0.992	1.60	0.444	11.06	0.991
4.93	0.623	1.35	0.384	9.50	1.000	1.89	0.463	11.77	0.935
5.23	0.685	1.55	0.411	10.41	1.001*	2.28	0.474	12.40	0.893
5.96	0.836	1.85	0.436	11.11	0.968	2.72	0.484	12.99	0.858
6.57	0.909	2.26	0.446	11.77	0.909	3.09	0.501	13.59	0.820
6.47	0.943	2.49	0.451	12.42	0.854	3.44	0.515		
7.16	0.975	2.92	0.509	13.03	0.838	3.90	0.530		
7.68	0.979	3.05	0.524	13.59	0.796	4.24	0.551		
8.19	0.994	3.45	0.508			4.58	0.583*		
8.67	0.999	3.86	0.499			4.95	0.620		
9.13	1.000	4.22	0.508			5.26	0.704		
9.54	1.001*	4.60	0.539			5.97	0.824		

* Not shown on plot

SPECIFICATION TABLE NO. 408 NORMAL SOLAR ABSORPTANCE OF MAGNESIUM ALUMINATE CONTACT COATINGS

Curve No.	Ref. No.	Year	Temperature Range, K	Geometry θ	Reported Error, %	Composition (weight percent), Specifications and Remarks
1*	137	1965	400-600	0°	±5	Rokide MA (0.025 mm thick, 2.3 mg cm^{-2}); Al substrate; measured in vacuum (10^{-7} mm Hg).
2*	137	1965	400-600	0°	±5	Similar to above specimen and conditions except 0.063 mm thick, 6.0 mg cm^{-2}.
3*	137	1965	400-600	0°	±5	Similar to curve 1 specimen and conditions except 0.140 mm thick, 31 mg cm^{-2}.

DATA TABLE NO. 408 NORMAL SOLAR ABSORPTANCE OF MAGNESIUM ALUMINATE CONTACT COATINGS

[Temperature, T, K; Absorptance, α]

T	α
CURVE 1*	
400	0.55
600	0.55
CURVE 2*	
400	0.58
600	0.58
CURVE 3*	
400	0.41
600	0.41

* No plot given

SPECIFICATION TABLE NO. 409 NORMAL TOTAL EMITTANCE OF MAGNESIUM OXIDE CONTACT COATINGS

Curve No.	Ref. No.	Year	Temperature Range, K	Geometry θ'	Reported Error, %	Composition (weight percent), Specifications and Remarks
1*	123	1952	673-1073	$\sim 0^\circ$		MgO powder; Nimonic 75 substrate; prepared from super pure magnesium.

DATA TABLE NO. 409 NORMAL TOTAL EMITTANCE OF MAGNESIUM OXIDE CONTACT COATINGS

[Temperature, T, K; Emittance, ϵ]

T	ϵ
CURVE 1*	
673	0.60
773	0.55
873	0.51
973	0.46
1073	0.42

* No plot given

SPECIFICATION TABLE NO. 410 NORMAL SPECTRAL EMITTANCE OF MAGNESIUM OXIDE CONTACT COATINGS

Curve No.	Ref. No.	Year	Temperature, K	Wavelength Range, μm	Geometry θ'	Reported Error, %	Composition (weight percent), Specifications, and Remarks
1*	96	1963	323.2	2.49-21.12	0°		MgO powder; substrate unknown; opaque. [Authors' designation: Sample BEC-4]

DATA TABLE NO. 410 NORMAL SPECTRAL EMITTANCE OF MAGNESIUM OXIDE CONTACT COATINGS

[Wavelength, λ, μm; Emittance, ϵ; Temperature, T, K]

λ	ϵ	λ	ϵ
CURVE 1*		CURVE 1 (cont.)	
T = 323.2			
2.49	0.729	11.67	0.963
2.71	0.739	12.05	0.970
3.02	0.743	12.50	0.973
4.25	0.671	13.03	0.978
4.63	0.674	13.60	0.972
4.91	0.705	14.41	0.965
5.15	0.734	15.25	0.954
5.51	0.856	15.76	0.952
5.74	0.929	16.16	0.952
5.90	0.944	17.00	0.947
6.26	0.946	17.46	0.945
6.70	0.961	18.55	0.948
6.97	0.951	18.84	0.945
7.52	0.950	19.13	0.948
7.93	0.955	19.98	0.945
9.13	0.955	21.12	0.939
9.65	0.963		
10.14	0.958		

* No plot given

SPECIFICATION TABLE NO. 411 NORMAL INTEGRATED REFLECTANCE OF MAGNESIUM OXIDE CONTACT COATINGS

Curve No.	Ref. No.	Year	Temperature, K	Angular Range, °	Geometry θ	Geometry θ'	Reported Error, %	Composition (weight percent), Specifications, and Remarks
1*	167	1950	~298	30.1-88.6	0°	θ'		MgO; substrate unknown; θ' is variable.

DATA TABLE NO. 411 NORMAL INTEGRATED REFLECTANCE OF MAGNESIUM OXIDE CONTACT COATINGS

[Angle, θ', °; Reflectance, ρ; Temperature, T, K]

θ'	ρ
CURVE 1*	
T ~ 298	
30.1	1.050
35.2	1.040
40.2	1.022
45.3	1.000
50.0	0.980
55.1	0.961
59.9	0.921
64.9	0.889
69.9	0.852
74.8	0.782
79.7	0.740
84.3	0.639
88.6	0.487

* No plot given

SPECIFICATION TABLE NO. 412 ANGULAR INTEGRATED REFLECTANCE OF MAGNESIUM OXIDE CONTACT COATINGS

Curve No.	Ref. No.	Year	Temperature, K	Angular Range, °	Geometry θ	θ'	ω'	Reported Error, %	Composition (weight percent), Specifications, and Remarks
1*	132	1965	298	-25-+80	45°	θ'			MgO (~0.76 mm thick); substrate unknown; $\theta = 45°$, $\theta' = \theta'$; tungsten filament lamp source; measured in plane of incidence.

DATA TABLE NO. 412 ANGULAR INTEGRATED REFLECTANCE OF MAGNESIUM OXIDE CONTACT COATINGS

[Angle, *°; Reflectance, ρ; Temperature, T, K]

θ'	ρ	θ'	ρ
CURVE 1*		CURVE 1 (cont.)	
T = 298			
		65	0.28
-25	0.52	70	0.23
-20	0.54	75	0.18
-15	0.56	80	0.13
-10	0.57		
-5	0.58		
0	0.58		
5	0.57		
10	0.57		
15	0.57		
20	0.55		
25	0.53		
30	0.51		
35	0.49		
40	0.46		
45	0.44		
50	0.40		
55	0.38		
60	0.32		

* No plot given

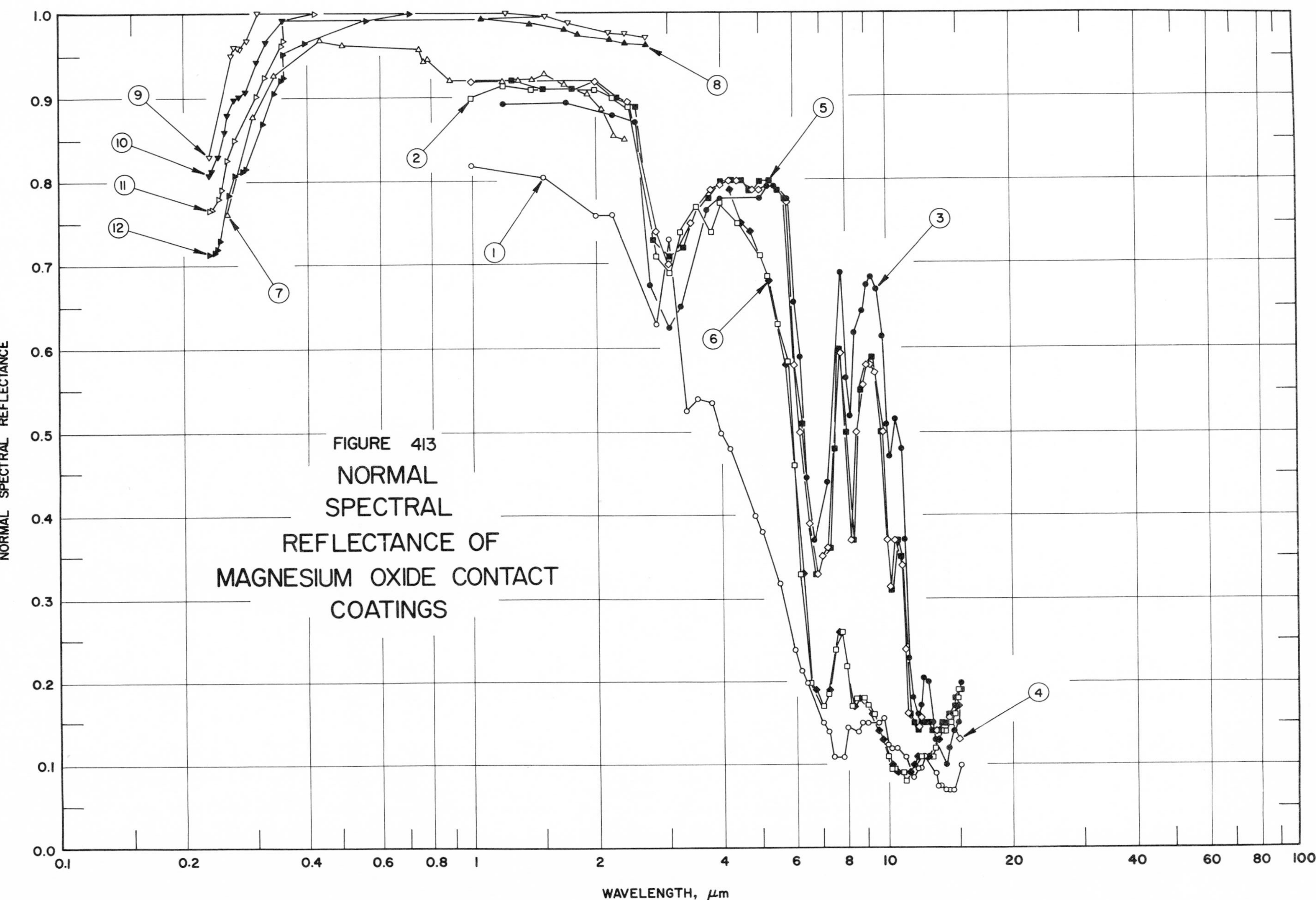

FIGURE 413 NORMAL SPECTRAL REFLECTANCE OF MAGNESIUM OXIDE CONTACT COATINGS

SPECIFICATION TABLE NO. 413 NORMAL SPECTRAL REFLECTANCE OF MAGNESIUM OXIDE CONTACT COATINGS

Curve No.	Ref. No.	Year	Temperature, K	Wavelength Range, μm	Geometry θ	θ'	ω'	Reported Error, %	Composition (weight percent), Specifications, and Remarks
1	104	1953	298	1.0-15.0	5°		2π	±2	Magnesium oxide (0.152 mm thick) on flat black paint (0.0762 mm thick) and aluminum substrates; aluminum substrate polished; data extracted from smooth curve; converted from R(2π, 5°).
2	104	1953	298	1.0-14.8	5°		2π	±2	Similar to above specimen and conditions except magnesium oxide 1.02 mm thick.
3	104	1953	298	1.2-15.0	5°		2π		Magnesium oxide (0.254 mm thick) on aluminum substrate; substrate polished; data extracted from smooth curve; converted from R(2π, 5°).
4	104	1953	298	1.0-15.0	5°		2π	±2	Similar to above specimen and conditions except magnesium oxide 0.914 mm thick.
5	168	1954	1089	1.0-15.0	5°		2π		MgO (0.914 mm thick) on Al substrate; polished substrate.
6	168	1954	1089	4.25-15.0	5°		2π		MgO (1.02 mm thick) on black paint and aluminum substrates.
7	169	1966	~298	0.256-2.36	~7°		2π		MgO (2 mm thick) on aluminum substrate; freshly smoked; data extracted from smooth curve; measured relative to MgO.
8	170	1966	~298	0.230-2.650	~0°		2π		MgO; unknown substrate; smoked; data extracted from smooth curve; measured relative to MgO.
9	170	1966	~298	0.230-2.650	~0°		2π		Above specimen and conditions except after 5 min in vacuum (10^{-3} mm Hg).
10	170	1966	~298	0.230-2.650	~0°		2π		Above specimen and conditions except after 35 min in vacuum.
11	170	1966	~298	0.230-2.650	~0°		2π		Above specimen and conditions except after 17 hrs in vacuum.
12	170	1966	~298	0.230-2.650	~0°		2π		Above specimen and conditions except after 134 hrs in vacuum.

DATA TABLE NO. 413 NORMAL SPECTRAL REFLECTANCE OF MAGNESIUM OXIDE CONTACT COATINGS

[Wavelength, λ, μm; Reflectance, ρ; Temperature, T, K]

λ	ρ
CURVE 1	
T = 298	
1.0	0.820
1.5	0.805
2.0	0.760
2.2	0.760
2.8	0.630
3.0	0.730
3.3	0.525
3.5	0.540
3.8	0.535
4.0	0.500
4.2	0.480
4.8	0.400
5.0	0.380
5.5	0.320
6.0	0.240
6.2	0.215
6.4	0.200
7.0	0.150
7.2	0.140
7.4	0.110
7.8	0.110
8.0	0.145
8.5	0.140
8.7	0.150
9.0	0.150
9.2	0.150*
9.5	0.150
9.8	0.155
10.0	0.125
10.2	0.120
10.5	0.120
11.0	0.110
11.5	0.085
11.7	0.095
12.0	0.095
12.2	0.110
13.0	0.090
13.2	0.075
13.4	0.075
13.8	0.070
14.0	0.070
14.4	0.070
15.0	0.100
CURVE 2	
T = 298	
1.0	0.900
1.2	0.915
1.4	0.910
2.0	0.910
2.2	0.900
2.4	0.890
2.8	0.710
3.0	0.690
3.2	0.740
3.5	0.770
3.8	0.740
4.0	0.775
4.4	0.750
5.0	0.710
5.2	0.685
5.5	0.630
5.8	0.585
6.0	0.460
6.2	0.330
6.5	0.200
7.0	0.170
7.2	0.185
7.5	0.240
7.8	0.260
8.0	0.220
8.2	0.170
8.4	0.180
8.8	0.180
9.0	0.170
9.3	0.160
10.0	0.110
10.2	0.095
10.4	0.095
10.8	0.090
11.0	0.080
11.8	0.110*
12.0	0.110
12.8	0.110
13.0	0.120
13.4	0.140*
13.8	0.140
14.0	0.150
14.2	0.150*
14.5	0.160
CURVE 2 (cont.)	
14.7	0.180
14.8	0.190
CURVE 3	
T = 298	
1.2	0.895
1.7	0.895
2.2	0.880
2.5	0.870
2.7	0.675
3.0	0.625
3.2	0.650
3.7	0.765
4.0	0.780
5.0	0.780
5.2	0.795
5.4	0.795
5.8	0.780
6.0	0.655
6.2	0.590
6.4	0.445
6.7	0.370
7.2	0.440
7.8	0.690
8.0	0.565
8.2	0.520
8.4	0.620
8.8	0.645
9.0	0.675
9.2	0.685
9.5	0.670
9.0	0.615
10.0	0.510
10.2	0.470
10.5	0.515
10.8	0.480
11.0	0.370
11.2	0.230
11.5	0.180
11.8	0.160
12.0	0.170
12.2	0.205
12.5	0.200
12.8	0.150
CURVE 3 (cont.)	
13.0	0.130
13.8	0.100
14.0	0.120
14.4	0.140
14.7	0.150
15.0	0.200
CURVE 4	
T = 298	
1.0	0.920
2.0	0.920
2.2	0.900
2.4	0.895
2.8	0.740
3.0	0.700
3.4	0.750
3.5	0.770
3.8	0.790
4.0	0.795
4.2	0.800
4.4	0.800
4.8	0.790
5.0	0.790
5.2	0.795
5.8	0.775
6.0	0.580
6.2	0.500
6.5	0.390
6.8	0.330
7.0	0.350
7.2	0.360
7.8	0.595
8.2	0.370
8.5	0.500
8.8	0.555
9.0	0.580
9.2	0.580
9.4	0.570
9.8	0.500
10.0	0.370
10.2	0.315
10.4	0.370
10.8	0.340
11.0	0.240
CURVE 4 (cont.)	
11.2	0.150
11.8	0.145
12.0	0.155
12.8	0.140
13.0	0.140
14.0	0.155
14.8	0.180
15.0	0.200
CURVE 5	
T = 1089	
1.00	0.92
1.25	0.92
1.50	0.91
1.75	0.91
2.00	0.92
2.25	0.90
2.50	0.89
2.75	0.73
3.00	0.71
3.25	0.72
3.50	0.77
3.75	0.78
4.00	0.80
4.25	0.80
4.50	0.80
4.75	0.79
5.00	0.80
5.25	0.80
5.50	0.79
5.75	0.78
6.00	0.58
6.25	0.51
6.50	0.39
6.75	0.33
7.00	0.35
7.25	0.36
7.50	0.48
7.75	0.60
8.00	0.50
8.25	0.37
8.50	0.50
8.75	0.55
9.00	0.58
CURVE 5 (cont.)	
9.25	0.59
9.50	0.57
9.75	0.50
10.00	0.37
10.25	0.31
10.50	0.37
10.75	0.35
11 00	0.24
11.25	0.16
11.50	0.15
11.75	0.14
12.00	0.15
12.25	0.15
12.50	0.15
12.75	0.14
13.00	0.14
13.25	0.14
13.50	0.15
13.75	0.15
14.00	0.16
14.75	0.16
14.50	0.17
14.75	0.18
15.00	0.19
CURVE 6	
T = 1089	
4.25	0.79
4.50	0.75
4.75	0.74
5.00	0.71
5.25	0.68
5.50	0.63
5.75	0.58
6.00	0.46
6.25	0.33
6.50	0.20
6.75	0.19
7.00	0.17
7.25	0.19
7.50	0.24
7.75	0.26
8.00	0.22
8.25	0.17
CURVE 6 (cont.)	
8.50	0.18
8.75	0.18
9.00	0.17
9.25	0.16
9.50	0.14
9.75	0.13
10.00	0.11
10.25	0.10
10.50	0.09
10.75	0.09
11.00	0.08
11.25	0.09
11.50	0.10
11.75	0.11
12.00	0.11
12.25	0.11
12.50	0.11
12.75	0.11
13.00	0.12
13.75	0.13
13.50	0.14
13.75	0.14
14.00	0.15
14.25	0.15
14.50	0.16
14.75	0.17
15.00	0.19
CURVE 7	
T ~ 298	
0.256	0.762
0.296	0.878
0.333	0.927
0.431	0.970
0.488	0.964
0.743	0.958
0.771	0.944
0.783	0.947
0.890	0.922
1.19	0.918
1.30	0.922
1.42	0.922
1.52	0.929
1.68	0.915
CURVE 7 (cont.)	
1.92	0.904
2.07	0.887
2.22	0.855
2.36	0.851
CURVE 8	
T ~ 298	
0.230	1.047*
0.254	1.045*
0.290	1.043*
0.303	1.037*
0.309	1.033*
0.326	1.032*
0.336	1.024*
0.447	1.019*
0.923	1.002*
1.052	0.994
1.387	0.987
1.693	0.981
1.803	0.974
2.117	0.969
2.355	0.963
2.650	0.961
CURVE 9	
T ~ 298	
0.230	0.830
0.261	0.950
0.266	0.959
0.272	0.959
0.274	0.957
0.285	0.968
0.302	1.000
0.313	1.014*
0.326	1.016*
0.366	1.010*
0.445	1.013*
1.151	1.008*
1.218	1.005
1.506	0.996
1.722	0.987
2.150	0.976
2.376	0.975
2.650	0.971

* Not shown on plot

DATA TABLE NO. 413 NORMAL SPECTRAL REFLECTANCE OF MAGNESIUM OXIDE CONTACT COATINGS (continued)

λ	ρ
CURVE 10	
T ~ 298	
0.230	0.807
0.233	0.812
0.242	0.829
0.251	0.859
0.256	0.880
0.266	0.897
0.272	0.901
0.277	0.901
0.282	0.906
0.301	0.942
0.318	0.966
0.350	0.993
0.351	1.008*
0.401	1.012*
0.445	1.013*
1.151	1.008*
1.218	1.005*
1.506	0.996*
1.722	0.987*
2.150	0.976*
2.376	0.975*
2.650	0.971*
CURVE 11	
T ~ 298	
0.230	0.765
0.233	0.766
0.243	0.780
0.247	0.791
0.255	0.825
0.266	0.850
0.300	0.902
0.314	0.924
0.346	0.962
0.350	0.967
0.351	0.993*
0.407	0.999
0.538	1.001*
0.650	1.011*
0.922	1.003*
1.052	0.994*
1.387	0.987*
1.693	0.981*
1.803	0.974*
2.117	0.969*

λ	ρ
CURVE 11 (cont.)	
2.355	0.963*
2.650	0.961*
CURVE 12	
T ~ 298	
0.230	0.713
0.236	0.714
0.240	0.718
0.244	0.729
0.258	0.785
0.266	0.807
0.277	0.813
0.281	0.816
0.311	0.870
0.332	0.904
0.345	0.920
0.350	0.924
0.350	0.952
0.393	0.965
0.558	0.992
0.701	1.012*
0.922	1.003*
1.052	0.994*
1.387	0.987*
1.693	0.981*
1.803	0.974*
2.117	0.969*
2.355	0.963*
2.650	0.961*

* Not shown on plot

SPECIFICATION TABLE NO. 414 HEMISPHERICAL TOTAL EMITTANCE OF MOLYBDENUM DISILICIDE CONTACT COATINGS

Curve No.	Ref. No.	Year	Temperature Range, K	Reported Error, %	Composition (weight percent), Specifications and Remarks
1*	171	1966	1036-2026	4.4	$MoSi_2$, 75.2 Mo and 24.8 Si (85 μm thick); VM-1 molybdenum substrate; plasma sprayed; measured in vacuum (10^{-5} mm Hg).
2*	171	1966	1070-1909	4.4	Similar to above specimen and conditions except 95 μm thick; heat treated in air at 1473 K for 1 hr.

DATA TABLE NO. 414 HEMISPHERICAL TOTAL EMITTANCE OF MOLYBDENUM DISILICIDE CONTACT COATINGS

[Temperature, T, K; Emittance, ε]

T	ε
CURVE 1*	
1036	0.675
1159	0.756
1226	0.725
1377	0.648
1543	0.609
1716	0.603
1885	0.603
2026	0.601
CURVE 2*	
1070	0.780
1203	0.790
1322	0.796
1474	0.767
1708	0.633
1766	0.708
1822	0.763
1909	0.806

* No plot given

SPECIFICATION TABLE NO. 415 NORMAL TOTAL EMITTANCE OF MOLYBDENUM DISILICIDE CONTACT COATINGS

Curve No.	Ref. No.	Year	Temperature Range, K	Geometry θ'	Reported Error, %	Composition (weight percent), Specifications and Remarks
1*	171	1966	331-1017	0°	5	$MoSi_2$ (60 μm thick), 75.2 Mo and 24.8 Si; molybdenum substrate; plasma flame sprayed.
2*	171	1966	334-1136	0°	5	$MoSi_2$ (50 μm thick), 75.2 Mo and 24.8 Si; bronze substrate; plasma flame sprayed.

DATA TABLE NO. 415 NORMAL TOTAL EMITTANCE OF MOLYBDENUM DISILICIDE CONTACT COATINGS

[Temperature, T, K; Emittance, ∈]

T	∈	T	∈
CURVE 1*		CURVE 2 (cont.)	
331	0.694	1001	0.748
372	0.705	1136	0.754
436	0.686		
525	0.674		
621	0.709		
714	0.889		
800	0.871		
888	0.780		
1017	0.752		
CURVE 2*			
334	0.656		
372	0.666		
439	0.648		
565	0.644		
690	0.772		
747	0.839		
915	0.745		

* No plot given

SPECIFICATION TABLE NO. 416 NORMAL SPECTRAL EMITTANCE OF MOLYBDENUM DISILICIDE CONTACT COATINGS

Curve No.	Ref. No.	Year	Wavelength, μm	Temperature Range, K	Geometry θ'	Reported Error, %	Composition (weight percent), Specifications, and Remarks
1*	174	1962	0.65	1060-1700	$\sim 0°$		$MoSi_2$, Pfaudler Co., PRF-6; molybdenum substrate; vacuum deposited; measured in vacuum.

DATA TABLE NO. 416 NORMAL SPECTRAL EMITTANCE OF MOLYBDENUM DISILICIDE CONTACT COATINGS

[Temperature, T, K; Emittance, ϵ; Wavelength, λ, μm]

T	ϵ
CURVE 1*	
$\lambda = 0.65$	
1060	0.98
1185	0.95
1188	0.92
1312	0.93
1312	0.92
1403	0.91
1406	0.88
1501	0.88
1504	0.89
1691	0.82
1700	0.79

* No plot given

SPECIFICATION TABLE NO. 417 NORMAL SPECTRAL REFLECTANCE OF MOLYBDENUM DISULFIDE CONTACT COATINGS

Curve No.	Ref. No.	Year	Temperature, K	Wavelength Range, μm	Geometry θ	θ'	ω'	Reported Error, %	Composition (weight percent), Specifications, and Remarks
1*	139	1961	~322	0.50-15.00	~0°		2π	<2	MoS_2 on sand blasted Inconel X substrate; flame sprayed using a plasmatron; as received; data extracted from smooth curve; converted from $R(2\pi, 0°)$; hohlraum at 523 K; sample maintained at ~322 K.
2*	139	1961	~322	0.50-15.00	~0°		2π	<2	Above specimen and conditions; diffuse component only.

DATA TABLE NO. 417 NORMAL SPECTRAL REFLECTANCE OF MOLYBDENUM DISULFIDE CONTACT COATINGS

[Wavelength, λ, μm; Reflectance, ρ; Temperature, T, K]

λ	ρ	λ	ρ
CURVE 1* T ~ 322		CURVE 2* T ~ 322	
0.50	0.090	0.50	0.100
0.75	0.100	0.75	0.110
1.00	0.150	1.00	0.150
1.50	0.225	1.50	0.225
2.00	0.300	2.50	0.275
3.00	0.340	3.50	0.325
4.50	0.385	4.50	0.350
5.00	0.400	5.50	0.365
6.50	0.420	7.50	0.410
7.25	0.435	8.00	0.420
8.00	0.435	8.50	0.420
9.00	0.440	9.50	0.390
10.00	0.420	10.00	0.380
10.50	0.425	10.50	0.385
11.50	0.450	12.00	0.420
12.50	0.470	13.50	0.435
13.00	0.480	14.50	0.440
14.25	0.470	15.00	0.420
15.00	0.460		

* No plot given

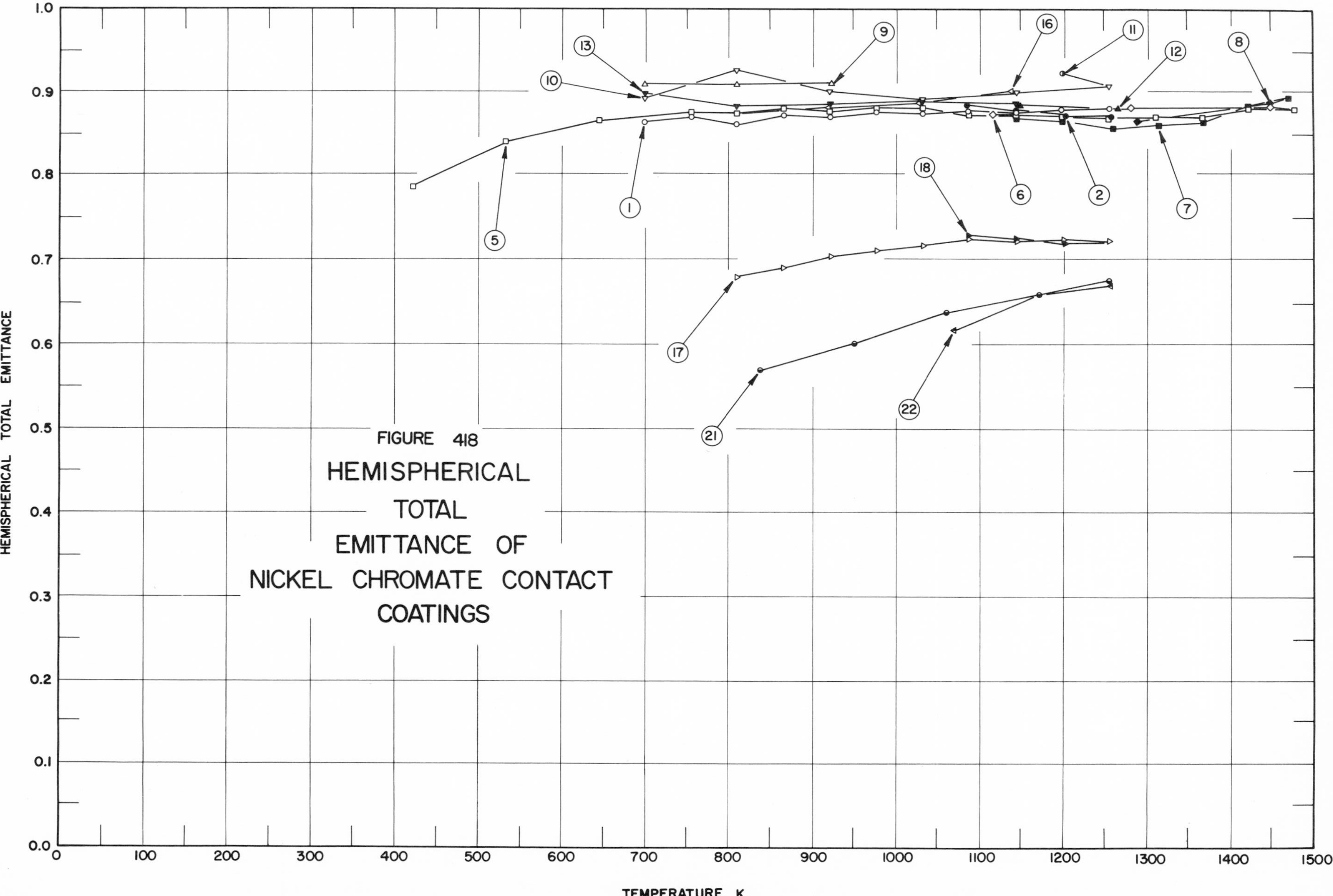

FIGURE 418 HEMISPHERICAL TOTAL EMITTANCE OF NICKEL CHROMATE CONTACT COATINGS

SPECIFICATION TABLE NO. 418 HEMISPHERICAL TOTAL EMITTANCE OF NICKEL CHROMATE CONTACT COATINGS

Curve No.	Ref. No.	Year	Temperature Range, K	Reported Error, %	Composition (weight percent), Specifications and Remarks
1	81	1963	699-1255		Nickel chrome spinel, $NiO \cdot Cr_2O_3$; Nb-1 Zr substrate; plasma arc sprayed; measured in vacuum ($< 8.0 \times 10^{-6}$ mm Hg); temp measured with thermocouple; heating cycle. [Author's designation: Run No. 1]
2	81	1963	1086-1258		Above specimen and conditions except temp measured with optical pyrometer.
3*	81	1963	1255		Curve 1 specimen and conditions; measurements taken during 67.1 hr endurance test.
4*	81	1963	1114-1273		Above specimen and conditions except temp measured with optical pyrometer and endurance test lasted 94.8 hrs.
5	81	1963	422-1477		Nickel chrome spinel, $NiO \cdot Cr_2O_3$ (0.0508 mm thick); Nb-1 Zr substrate; plasma arc sprayed; measured in vacuum ($< 8.4 \times 10^{-6}$ mm Hg); temp measured with thermocouple; heating cycle. [Author's designation: Run No. 1]
6	81	1963	1477-1116		Above specimen and conditions; cooling cycle.
7	81	1963	1088-1471		Curve 5 specimen and conditions except temp measured with optical pyrometer.
8	81	1963	1471-1289		Curve 6 specimen and conditions except temp measured with optical pyrometer.
9	81	1963	699-922		Nickel chrome spinel, $NiO \cdot Cr_2O_3$ (0.102 mm thick); Nb-1 Zr substrate; plasma arc sprayed; measured in vacuum ($< 2.0 \times 10^{-6}$ mm Hg); temp measured with thermocouple; heating cycle. [Author's designation: Run No. 1]
10	81	1963	699-1255		Above specimen and conditions. [Author's designation: Run No. 2]
11	81	1963	1255-1199		Above specimen and conditions; cooling cycle.
12	81	1963	1148-1264		Curve 9 specimen and conditions except temp measured with optical pyrometer.
13	81	1963	699-1144		Nickel chrome spinel, $NiO \cdot Cr_2O_3$ (0.102 mm thick); Nb-1 Zr substrate; plasma arc sprayed; measured in vacuum ($< 2.6 \times 10^{-6}$ mm Hg); temp measured with thermocouple; heating cycle. [Author's designation: Run No. 1]
14*	81	1963	~1144		Above specimen and conditions except measurements taken during 262.4 hr endurance test at 1144 K; endurance test time, hrs, is variable.
15*	81	1963	~1146		Above specimen and conditions except temp measured with optical pyrometer.
16	81	1963	1138-809		Curve 13 specimen and conditions; cooling cycle.
17	81	1963	810-1255		Nickel chrome spinel, $NiO \cdot Cr_2O_3$ (0.102 mm thick); Nb-1 Zr substrate; plasma arc sprayed; measured in vacuum ($< 2.0 \times 10^{-6}$ mm Hg); temp measured with thermocouple; heating cycle. [Author's designation: Run No. 1]
18	81	1963	1087-1255		Above specimen and conditions except temp measured with optical pyrometer.
19*	81	1963	1255		Curve 17 specimen and conditions except measurements taken during 17.5 hr endurance test at 1255 K; endurance test time, hrs, is variable.
20*	81	1963	1255		Above specimen and conditions except temp measured with optical pyrometer.
21	81	1963	1255-838		Curve 17 specimen and conditions; cooling cycle.
22	81	1963	1258-1070		Above specimen and conditions except temp measured with optical pyrometer.

* Not shown on plot

DATA TABLE NO. 418 HEMISPHERICAL TOTAL EMITTANCE OF NICKEL CHROMATE CONTACT COATINGS

[Temperature, T, K; Emittance, ∈]

T	∈
CURVE 1	
699	0.863
755	0.868
810	0.861
866	0.870
922	0.868
978	0.874
1032	0.872
1088	0.875
1144	0.874
1199	0.877
1255	0.878
CURVE 2	
1086	0.883
1143	0.876
1202	0.870
1258	0.869
CURVE 3*	
1255	0.878
1255	0.832
1255	0.875
CURVE 4*	
1259	0.869
1259	0.820
1272	0.830
1272	0.829
1272	0.841
1273	0.839
1114	0.835
CURVE 5	
422	0.786
533	0.839
644	0.865
755	0.874
810	0.873
866	0.878
922	0.876
977	0.878
CURVE 5 (cont.)	
1033	0.879
1088	0.871
1144	0.870
1199	0.867
1855	0.866
1310	0.868
1366	0.867
1422	0.878
1477	0.878
CURVE 6	
1477	0.878
1449	0.883
1283	0.880
1116	0.872
CURVE 7	
1088	0.871
1144	0.869
1200	0.865
1259	0.855
1314	0.859
1368	0.863
1420	0.881
1471	0.893
CURVE 8	
1471	0.893
1448	0.885
1289	0.864
CURVE 9	
699	0.909
810	0.908
922	0.910
CURVE 10	
699	0.891
810	0.925
922	0.899
CURVE 10 (cont.)	
1033	0.889
1144	0.898
1255	0.906
CURVE 11	
1255	0.906
1199	0.923
CURVE 12	
1148	0.884
1264	0.879
CURVE 13	
699	0.896
810	0.882
922	0.885
1033	0.887
1144	0.885

Endurance Time	∈
CURVE 14 T ~ 1144	
1.0	0.885
1.9	0.885
5.8	0.884
22.0	0.885
27.0	0.885
29.0	0.885
94.6	0.883
98.0	0.883
101.7	0.883
118.7	0.883
121.9	0.883
125.0	0.883
142.1	0.883
146.1	0.883
149.8	0.883
166.5	0.888
170.1	0.887
CURVE 14 (cont.)*	
173.8	0.887
191.4	0.895
194.5	0.893
197.4	0.893
262.4	0.901
CURVE 15* T ~ 1146	
1.0	0.886
1.9	0.886
5.8	0.886
22.0	0.880
27.0	0.880
29.0	0.880
94.6	0.876
98.0	0.876
101.7	0.877
118.7	0.876
121.9	0.876
125.0	0.876
142.1	0.876
146.1	0.877
149.8	0.876
166.5	0.881
170.1	0.880
173.8	0.880
191.4	0.876
194.5	0.878
197.4	0.876
262.4	0.875

T	∈
CURVE 16	
1138	0.901
1030	0.888
920	0.879
809	0.874
CURVE 17	
810	0.678
CURVE 17 (cont.)	
866	0.689
921	0.703
977	0.711
1033	0.717
1088	0.724
1144	0.722
1200	0.724
1255	0.722
CURVE 18	
1087	0.728
1143	0.724
1202	0.720
1255	0.722

Endurance Time	∈
CURVE 19* T = 1255	
0.0	0.722
0.9	0.704
1.4	0.704
3.4	0.681
5.4	0.670
6.4	0.665
15.7	0.690
17.5	0.675
CURVE 20* T = 1255	
0.0	0.722
0.9	0.705
1.4	0.704
3.4	0.679
5.4	0.669
6.4	0.664
15.7	0.675
17.5	0.669

T	∈
CURVE 21	
1255	0.675
1172	0.660
1061	0.639
950	0.603
838	0.570
CURVE 22	
1258	0.669
1172	0.660
1070	0.618

* Not shown on plot

SPECIFICATION TABLE NO. 419 NORMAL SPECTRAL EMITTANCE OF POTASSIUM BROMIDE CONTACT COATINGS

Curve No.	Ref. No.	Year	Temperature, K	Wavelength Range, μm	Geometry θ'	Reported Error, %	Composition (weight percent), Specifications, and Remarks
1*	175	1969	620	63.8 -143.	$\sim 0°$		KBr (7.46 μm thick); platinum substrate.
2*	175	1969	620	67.9 -123.	$\sim 0°$		Similar to above specimen and conditions except 4.76 μm thick.
3*	175	1969	620	66.6 -117.	$\sim 0°$		Similar to curve 1 specimen and conditions except 2.86 μm thick.

DATA TABLE NO. 419 NORMAL SPECTRAL EMITTANCE OF POTASSIUM BROMIDE CONTACT COATINGS

[Wavelength, λ, μm; Emittance, ϵ; Temperature, T, K]

λ	ϵ	λ	ϵ	λ	ϵ	λ	ϵ
CURVE 1* T = 620		CURVE 1 (cont.)		CURVE 2 (cont.)		CURVE 3* T = 620	
		119.	0.509	87.8	0.219		
63.8	0.439	121.	0.448	89.6	0.241	66.6	0.063
66.0	0.430	124.	0.384	91.1	0.274	84.3	0.061
69.3	0.408	128.	0.317	92.7	0.313	86.4	0.068
76.7	0.346	132.	0.259	93.9	0.362	88.0	0.075
82.5	0.346	138.	0.212	97.9	0.590	89.6	0.087
86.2	0.351	143.	0.183	99.3	0.644	91.4	0.114
87.9	0.365			100.	0.644	94.7	0.188
89.6	0.390	CURVE 2* T = 620		102.	0.607	96.5	0.212
91.9	0.428			103.	0.532	98.0	0.225
93.6	0.476			105.	0.399	99.9	0.217
95.7	0.550	67.9	0.179	107.	0.318	102.	0.150
97.0	0.596	71.5	0.172	110.	0.239	107.	0.089
98.1	0.662	73.9	0.178	113.	0.179	109.	0.064
104.	0.980	75.8	0.183	116.	0.133	113.	0.051
106.	0.996	79.1	0.179	123.	0.073	117.	0.042
108.	0.980	82.2	0.176				
111.	0.876	84.3	0.183				
116.	0.617	86.3	0.201				

* No plot given

SPECIFICATION TABLE NO. 420 ANGULAR SPECTRAL TRANSMITTANCE OF POTASSIUM CHLORIDE CONTACT COATINGS

Curve No.	Ref. No.	Year	Temperature, K	Wavelength Range, μm	Geometry θ θ' ω'	Reported Error, %	Composition (weight percent), Specifications, and Remarks
1*	176	1968	~298	0.118-0.180	0° 0°		KCl (250 Å thick); LiF substrate; vapor deposited; measured in vacuum; p-polarization; data extracted from smooth curve.
2*	176	1968	~298	0.118-0.181	45° 45°		Above specimen and conditions.
3*	176	1968	~298	0.118-0.180	60° 60°		Above specimen and conditions.

DATA TABLE NO. 420 ANGULAR SPECTRAL TRANSMITTANCE OF POTASSIUM CHLORIDE CONTACT COATINGS

[Wavelength, λ, μm; Transmittance, τ; Temperature, T, K]

λ	τ	λ	τ	λ	τ	λ	τ	λ	τ	λ	τ
CURVE 1*		CURVE 1 (cont.)		CURVE 2*		CURVE 2 (cont.)		CURVE 3*		CURVE 3 (cont.)	
T ~ 298		0.147	0.487	T ~ 298		0.147	0.448	T ~ 298		0.149	0.482
0.118	0.428	0.151	0.570	0.118	0.348	0.149	0.476	0.118	0.335	0.150	0.495
0.119	0.420	0.152	0.573	0.120	0.339	0.150	0.497	0.119	0.314	0.151	0.487
0.121	0.422	0.153	0.556	0.120	0.349	0.151	0.532	0.121	0.310	0.153	0.441
0.122	0.431	0.159	0.203	0.122	0.353	0.152	0.520	0.123	0.294	0.155	0.356
0.123	0.435	0.160	0.148	0.124	0.330	0.153	0.471	0.124	0.258	0.156	0.233
0.124	0.418	0.161	0.117	0.125	0.316	0.156	0.283	0.126	0.158	0.158	0.163
0.125	0.385	0.163	0.123	0.126	0.237	0.158	0.175	0.128	0.109	0.161	0.107
0.127	0.318	0.163	0.142	0.127	0.171	0.161	0.108	0.129	0.081	0.162	0.111
0.128	0.185	0.164	0.201	0.128	0.124	0.162	0.110	0.130	0.072	0.164	0.145
0.129	0.138	0.167	0.465	0.129	0.093	0.164	0.142	0.132	0.085	0.165	0.201
0.130	0.114	0.168	0.574	0.131	0.083	0.165	0.201	0.134	0.113	0.167	0.592
0.131	0.102	0.170	0.655	0.132	0.098	0.167	0.465	0.135	0.154	0.169	0.733
0.133	0.133	0.173	0.723	0.134	0.140	0.168	0.587	0.136	0.198	0.171	0.801
0.135	0.189	0.176	0.773	0.136	0.200	0.170	0.701	0.138	0.292	0.174	0.855
0.137	0.303	0.180	0.800	0.136	0.242	0.172	0.755	0.140	0.348	0.176	0.883
0.140	0.396			0.139	0.332	0.176	0.810	0.142	0.387	0.180	0.900
0.143	0.445			0.142	0.386	0.181	0.845	0.146	0.408		
0.146	0.469			0.143	0.418			0.147	0.422		

* No plot given

SPECIFICATION TABLE NO. 421 NORMAL SPECTRAL REFLECTANCE OF POTASSIUM IODIDE CONTACT COATINGS

Curve No.	Ref. No.	Year	Temperature, K	Wavelength Range, μm	Geometry θ	θ'	ω'	Reported Error, %	Composition (weight percent), Specifications, and Remarks
1*	172	1968	~298	0.164-0.248	14°	14°			Evaporated film; LiF substrate; optical density 0.102 at 0.203 μm; measured in vacuum (< 5 x 10^{-5} mm Hg) without exposure to air after film application; data extracted from smooth curve.
2*	172	1968	~298	0.166-0.248	14°	14°			Similar to above specimen and conditions except optical density 0.300.
3*	172	1968	~298	0.166-0.247	14°	14°			Similar to above specimen and conditions except optical density 0.700.
4*	172	1968	~298	0.166-0.248	14°	14°			Similar to above specimen and conditions except optical density > 2.

DATA TABLE NO. 421 NORMAL SPECTRAL REFLECTANCE OF POTASSIUM IODIDE CONTACT COATINGS

[Wavelength, λ, μm; Reflectance, ρ; Temperature, T, K]

λ	ρ	λ	ρ	λ	ρ	λ	ρ	λ	ρ
CURVE 1*		CURVE 1 (cont.)		CURVE 2 (cont.)		CURVE 3 (cont.)		CURVE 4*	
T ~ 298								T ~ 298	
		0.241	0.099	0.213	0.143	0.185	0.215		
0.165	0.130	0.248	0.092	0.215	0.177	0.186	0.215	0.166	0.027
0.168	0.142			0.217	0.257	0.188	0.201	0.169	0.044
0.171	0.161	CURVE 2*		0.220	0.289	0.191	0.148	0.171	0.054
0.173	0.166	T ~ 298		0.223	0.303	0.194	0.131	0.175	0.049
0.177	0.161			0.225	0.303	0.198	0.120	0.179	0.043
0.181	0.167	0.166	0.168	0.235	0.258	0.204	0.112	0.188	0.067
0.183	0.180	0.171	0.228	0.241	0.238	0.207	0.101	0.194	0.039
0.187	0.213	0.173	0.238	0.248	0.222	0.211	0.064	0.197	0.031
0.188	0.214	0.174	0.238			0.212	0.068	0.209	0.027
0.191	0.196	0.178	0.230	CURVE 3*		0.215	0.116	0.212	0.025
0.195	0.152	0.179	0.234	T ~ 298		0.218	0.206	0.215	0.036
0.202	0.125	0.184	0.271			0.219	0.221	0.220	0.076
0.211	0.101	0.186	0.275	0.166	0.118	0.221	0.187	0.222	0.070
0.215	0.110	0.186	0.271	0.169	0.161	0.223	0.161	0.225	0.051
0.220	0.154	0.188	0.243	0.171	0.176	0.228	0.203	0.229	0.048
0.221	0.157	0.190	0.226	0.172	0.177	0.229	0.207	0.233	0.052
0.230	0.127	0.204	0.192	0.177	0.171	0.234	0.198	0.236	0.049
0.234	0.110	0.208	0.170	0.180	0.183	0.247	0.149	0.248	0.057

* No plot given

SPECIFICATION TABLE NO. 422 NORMAL SPECTRAL REFLECTANCE OF POTASSIUM SILICATE CONTACT COATINGS

Curve No.	Ref. No.	Year	Temperature, K	Wavelength Range, μm	Geometry θ	θ'	ω'	Reported Error, %	Composition (weight percent), Specifications, and Remarks
1*	31	1965	~300	0.250-0.362	~0°		2π		Potassium silicate; aluminum substrate; vapor deposited in vacuum; measured in vacuum (10^{-6} mm Hg); data extracted from smooth curve; measured relative to MgO. [Authors' designation: Control Sample No. 233]
2*	31	1965	~300	0.33-2.60	~0°		2π		Above specimen and conditions.
3*	31	1965	~300	0.33-2.60	~0°		2π		Similar to above specimen and conditions except exposed to 1 x 10^{16} p cm^{-2}. [Authors' designation: Exposed Sample No. 232]

DATA TABLE NO. 422 NORMAL SPECTRAL REFLECTANCE OF POTASSIUM SILICATE CONTACT COATINGS

[Wavelength, λ, μm; Reflectance, ρ; Temperature, T, K]

λ	ρ	λ	ρ	λ	ρ
CURVE 1* T ~ 300		CURVE 2 (cont.)		CURVE 3* T ~ 300	
		0.61	0.989		
0.250	0.005	0.76	0.981	0.33	0.032
0.283	0.014	0.81	0.985	0.36	0.032
0.330	0.011	0.92	0.977	0.37	0.148
0.343	0.013	0.97	0.982	0.54	0.834
0.362	0.021	1.02	0.976	0.59	0.927
		1.38	0.974	0.67	0.954
CURVE 2* T ~ 300		1.42	0.965	0.81	0.952
		1.50	0.966	1.23	0.923
		1.63	0.948	1.38	0.909
0.33	0.032	1.76	0.909	1.41	0.893
0.36	0.032	1.85	0.874	1.50	0.887
0.37	0.148	1.91	0.835	1.62	0.859
0.40	0.965	1.94	0.844	1.77	0.810
0.42	1.019	2.14	0.801	1.85	0.771
0.43	1.031	2.30	0.762	1.92	0.723
0.45	1.035	2.39	0.749	1.97	0.727
0.50	1.008	2.45	0.713	2.06	0.706
0.58	1.001	2.60	0.683	2.40	0.608
				2.60	0.540

* No plot given

SPECIFICATION TABLE NO. 423 NORMAL SOLAR ABSORPTANCE OF POTASSIUM SILICATE CONTACT COATINGS

Curve No.	Ref. No.	Year	Temperature Range, K	Geometry θ	Reported Error, %	Composition (weight percent), Specifications and Remarks
1*	86	1964	~298	~0°		Potassium silicate (0.0254 mm thick); quartz substrate; air-dried; calculated from normal solar reflectance. [Authors' designation: Sample No. 7022]
2*	86	1964	~298	~0°		Above specimen and conditions except exposed to 710 ESH of ultraviolet radiation (avg solar factor 9.2 suns) in vacuum ($<10^{-6}$ mm Hg).
3*	86	1964	~298	~0°		Similar to curve 1 specimen and conditions except 0.127 mm thick. [Authors' designation: Sample No. 7023]
4*	86	1964	~298	~0°		Above specimen and conditions except exposed to 710 ESH of ultraviolet radiation (avg solar factor 9.2 suns) in vacuum ($<10^{-6}$ mm Hg).
5*	86	1964	~298	~0°		Similar to curve 1 specimen and conditions except cured at 383 K. [Authors' designation: Sample No. 7024]
6*	86	1964	~298	~0°		Above specimen and conditions except exposed to 710 ESH of ultraviolet radiation (avg solar factor 9.2 suns) in vacuum ($<10^{-6}$ mm Hg).
7*	86	1964	~298	~0°		Similar to curve 1 specimen and conditions except 0.127 mm thick; cured at 383 K. [Authors' designation: Sample No. 7025]
8*	86	1964	~298	~0°		Above specimen and conditions except exposed to 710 ESH of ultraviolet radiation (avg solar factor 9.2 suns) in vacuum ($<10^{-6}$ mm Hg).
9*	86	1964	~298	~0°		Similar to curve 3 specimen and conditions. [Authors' designation: Sample No. 7027]
10*	86	1964	~298	~0°		Above specimen and conditions except wrapped in aluminum and then exposed to 710 ESH of ultraviolet radiation (avg solar factor 9.2 suns) in vacuum ($<10^{-6}$ mm Hg); unwrapped before measured.
11*	86	1964	~298	~0°		Similar to curve 5 specimen and conditions. [Authors' designation: Sample No. 7028]
12*	86	1964	~298	~0°		Above specimen and conditions except wrapped in aluminum, then exposed to 710 ESH of ultraviolet radiation (avg solar factor 9.2 suns) in vacuum ($<10^{-6}$ mm Hg); unwrapped before measured.
13*	86	1964	~298	~0°		Similar to curve 7 specimen and conditions. [Authors' designation: Sample No. 7029]
14*	86	1964	~298	~0°		Above specimen and conditions except wrapped in aluminum, then exposed to 710 ESH of ultraviolet radiation (avg solar factor 9.2 suns) in vacuum ($<10^{-6}$ mm Hg); unwrapped before measured.

* No plot given

DATA TABLE NO. 423 NORMAL SOLAR ABSORPTANCE OF POTASSIUM SILICATE CONTACT COATINGS

[Temperature, T, K; Absorptance, α]

T	α	T	α
CURVE 1*		CURVE 13*	
298	0.567	298	0.292
CURVE 2*		CURVE 14*	
298	0.628	298	0.268
CURVE 3*			
298	0.613		
CURVE 4*			
298	0.558		
CURVE 5*			
298	0.560		
CURVE 6*			
298	0.617		
CURVE 7*			
298	0.275		
CURVE 8*			
298	0.326		
CURVE 9*			
209	0.60		
CURVE 10*			
298	0.427		
CURVE 11*			
298	0.562		
CURVE 12*			
298	0.568		

* No plot given

SPECIFICATION TABLE NO. 424 NORMAL SPECTRAL REFLECTANCE OF RUBIDIUM IODIDE CONTACT COATINGS

Curve No.	Ref. No.	Year	Temperature, K	Wavelength Range, μm	Geometry θ	θ'	ω'	Reported Error, %	Composition (weight percent), Specifications, and Remarks
1*	172	1968	~298	0.169-0.248	14°	14°			Rubidium iodide; LiF substrate; vapor deposited; optical density 0.065 at 0.207 μm; measured in vacuum ($< 5 \times 10^{-5}$ mm Hg) without exposure to air after film application; data extracted from smooth curve.
2*	172	1968	~298	0.169-0.247	14°	14°			Similar to above specimen and conditions except optical density 0.162.
3*	172	1968	~298	0.169-0.248	14°	14°			Similar to above specimen and conditions except optical density 0.728.
4*	172	1968	~298	0.170-0.248	14°	14°			Similar to above specimen and conditions except optical density > 2.
5*	172	1968	~298	0.169-0.248	14°	14°			Similar to above specimen and conditions except optical density 0.605.
6*	172	1968	~298	0.169-0.248	14°	14°			Above specimen and conditions except annealed for 15 min at 373 K; measured in situ after annealing.
7*	172	1968	~298	0.169-0.248	14°	14°			Above specimen and conditions except annealed for 30 additional min.
8*	172	1968	~298	0.169-0.248	14°	14°			Similar to curve 4 specimen and conditions.
9*	172	1968	~298	0.169-0.248	14°	14°			Above specimen and conditions except annealed for 2 hrs at 393 K; measured in situ after annealing.

* No plot given

DATA TABLE NO. 424 NORMAL SPECTRAL REFLECTANCE OF RUBIDIUM IODIDE CONTACT COATINGS

[Wavelength, λ, μm; Reflectance, ρ; Temperature, T, K]

CURVE 1*
T ~ 298

λ	ρ
0.169	0.114
0.173	0.114
0.176	0.119
0.179	0.137
0.180	0.137
0.185	0.130
0.188	0.146
0.191	0.148
0.193	0.158
0.195	0.152
0.199	0.119
0.205	0.096
0.212	0.083
0.216	0.082
0.219	0.090
0.222	0.111
0.224	0.136
0.226	0.139
0.231	0.111
0.237	0.097
0.248	0.082

CURVE 2*
T ~ 298

λ	ρ
0.169	0.153
0.172	0.151
0.175	0.164
0.177	0.187
0.180	0.223
0.184	0.215
0.186	0.235
0.190	0.269
0.191	0.274
0.194	0.293
0.195	0.296
0.197	0.274
0.201	0.216
0.205	0.176
0.217	0.120
0.219	0.130
0.224	0.237
0.226	0.261
0.228	0.262

CURVE 2 (cont.)

λ	ρ
0.234	0.199
0.239	0.167
0.247	0.141

CURVE 3*
T ~ 298

λ	ρ
0.169	0.076
0.172	0.076
0.174	0.092
0.176	0.117
0.177	0.141
0.178	0.159
0.182	0.145
0.184	0.147
0.185	0.160
0.187	0.184
0.189	0.194
0.191	0.191
0.193	0.200
0.195	0.193
0.198	0.136
0.200	0.119
0.202	0.110
0.207	0.101
0.211	0.101
0.216	0.081
0.218	0.088
0.221	0.125
0.223	0.191
0.225	0.198
0.226	0.196
0.230	0.124
0.231	0.132
0.236	0.183
0.238	0.199
0.242	0.205
0.244	0.205
0.248	0.196

CURVE 4*
T ~ 298

λ	ρ
0.170	0.030
0.172	0.034

CURVE 4 (cont.)

λ	ρ
0.175	0.048
0.177	0.068
0.179	0.068
0.182	0.062
0.183	0.065
0.188	0.085
0.189	0.088
0.191	0.086
0.193	0.090
0.194	0.086
0.198	0.060
0.202	0.050
0.210	0.044
0.215	0.037
0.219	0.051
0.220	0.063
0.222	0.094
0.225	0.105
0.227	0.098
0.230	0.076
0.231	0.076
0.235	0.093
0.240	0.093
0.248	0.104

CURVE 5*
T ~ 298

λ	ρ
0.169	0.096
0.172	0.108
0.174	0.131
0.177	0.191
0.178	0.195
0.182	0.181
0.183	0.184
0.185	0.193
0.187	0.234
0.189	0.240
0.191	0.238
0.192	0.243
0.194	0.227
0.197	0.177
0.198	0.167
0.207	0.151
0.212	0.131

CURVE 5 (cont.)

λ	ρ
0.216	0.105
0.216	0.102
0.218	0.106
0.222	0.222
0.225	0.248
0.226	0.243
0.228	0.220
0.231	0.276
0.232	0.282
0.237	0.256
0.242	0.200
0.248	0.149

CURVE 6*
T ~ 298

λ	ρ
0.169	0.103
0.171	0.110
0.173	0.132
0.175	0.162
0.176	0.205
0.177	0.216
0.181	0.204
0.182	0.203
0.184	0.214
0.188	0.264
0.188	0.270
0.190	0.267
0.192	0.275
0.193	0.274
0.196	0.218
0.197	0.211
0.198	0.206
0.200	0.199
0.207	0.166
0.212	0.135
0.215	0.107
0.216	0.101
0.218	0.109
0.222	0.225
0.225	0.278
0.228	0.261
0.228	0.269
0.230	0.320
0.232	0.327

CURVE 6 (cont.)

λ	ρ
0.239	0.223
0.245	0.159
0.248	0.128

CURVE 7*
T ~ 298

λ	ρ
0.169	0.118
0.172	0.124
0.173	0.139
0.175	0.187
0.177	0.236
0.178	0.245
0.181	0.217
0.182	0.213
0.184	0.230
0.187	0.284
0.188	0.290
0.189	0.294
0.190	0.288
0.192	0.299
0.193	0.293
0.196	0.244
0.197	0.219
0.200	0.209
0.207	0.166
0.212	0.135
0.215	0.107
0.216	0.101
0.218	0.109
0.223	0.289
0.225	0.303
0.228	0.272
0.230	0.332
0.231	0.359
0.232	0.352
0.239	0.227
0.245	0.140
0.248	0.123

CURVE 8*
T ~ 298

λ	ρ
0.169	0.019
0.173	0.023

CURVE 8 (cont.)

λ	ρ
0.175	0.032
0.177	0.040
0.178	0.043
0.180	0.033
0.182	0.033
0.187	0.051
0.190	0.047
0.193	0.050
0.197	0.035
0.208	0.029
0.214	0.023
0.218	0.030
0.221	0.055
0.224	0.062
0.226	0.059
0.229	0.045
0.232	0.063
0.233	0.045
0.235	0.034
0.239	0.070
0.241	0.097
0.243	0.104
0.248	0.078

CURVE 9*
T ~ 298

λ	ρ
0.169	0.106
0.171	0.113
0.173	0.135
0.176	0.225
0.177	0.238
0.182	0.205
0.185	0.228
0.187	0.270
0.188	0.277
0.190	0.272
0.193	0.282
0.193	0.278
0.195	0.246
0.197	0.193
0.200	0.161
0.207	0.125
0.209	0.125
0.211	0.121

CURVE 9 (cont.)

λ	ρ
0.216	0.091
0.217	0.088
0.219	0.128
0.223	0.281
0.225	0.304
0.228	0.267
0.229	0.239
0.230	0.287
0.233	0.114
0.234	0.122
0.237	0.270
0.238	0.283
0.239	0.273
0.245	0.125
0.248	0.093

* No plot given

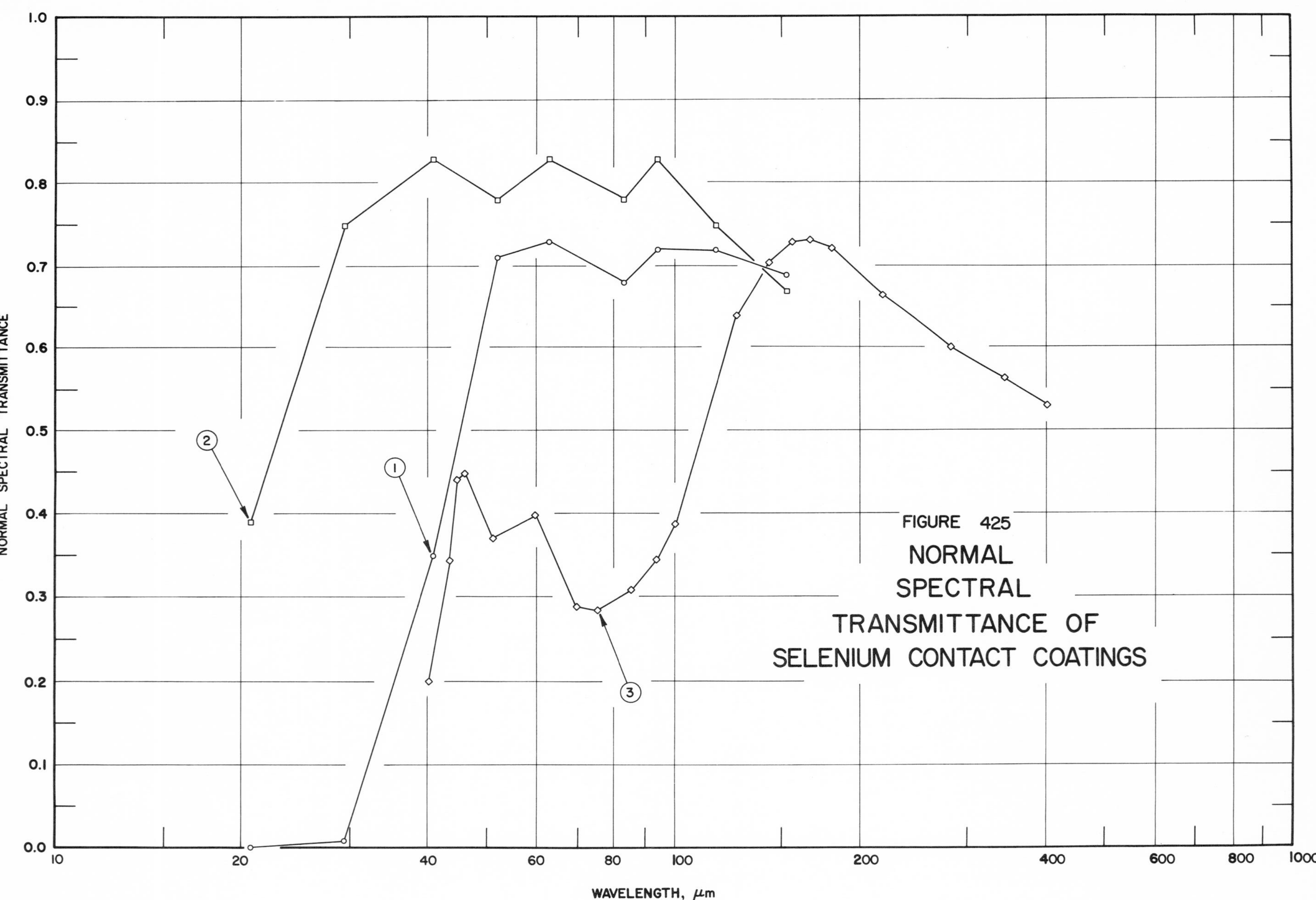

FIGURE 425
NORMAL
SPECTRAL
TRANSMITTANCE OF
SELENIUM CONTACT COATINGS
NORMAL SPECTRAL TRANSMITTANCE
WAVELENGTH, μm
0.0
0.1
0.2
0.3
0.4
0.5
0.6
0.7
0.8
0.9
1.0
10
20
40
60
80
100
200
400
600
800
1000
1
2
3

SPECIFICATION TABLE NO. 425 NORMAL SPECTRAL TRANSMITTANCE OF SELENIUM CONTACT COATINGS

Curve No.	Ref. No.	Year	Temperature, K	Wavelength Range, μm	Geometry θ θ' ω'	Reported Error, %	Composition (weight percent), Specifications, and Remarks
1	149	1940	298	20.7-152	~0° ~0°		Selenium; SiO_2 substrate. [Authors' designation: Sample No. 27]
2	149	1940	298	20.7-152	~0° ~0°		Selenium; pliofilm substrate. [Authors' designation: Sample No. 29]
3	177	1965	300	40.2-401.	~0° ~0°		Selenium (35.8 μm thick); germanium (2 mm thick) and selenium (34.6 μm thick) substrates; prepared by vacuum evaporating selenium on germanium; data extracted from smooth curve.

DATA TABLE NO. 425 NORMAL SPECTRAL TRANSMITTANCE OF SELENIUM CONTACT COATINGS

[Wavelength, λ, μm; Transmittance, τ; Temperature, T, K]

λ	τ	λ	τ	λ	τ
CURVE 1 T = 298		CURVE 2 (cont.)		CURVE 3 (cont.)	
		83	0.78	126.	0.638
20.7	0.00	94	0.83	143.	0.704
29.4	0.09	117	0.75	156.	0.730
41	0.35	152	0.67	166.	0.733
52	0.71			180.	0.723
63	0.73	CURVE 3 T = 300		217.	0.665
83	0.68			280.	0.600
94	0.72			344.	0.564
117	0.72	40.2	0.200	401.	0.531
152	0.69	43.3	0.344		
		44.6	0.441		
CURVE 2 T = 298		46.0	0.448		
		51.0	0.370		
		59.7	0.348		
20.7	0.39	69.6	0.289		
29.4	0.75	75.1	0.285		
41	0.83	85.1	0.309		
52	0.78	93.7	0.346		
63	0.83	100.	0.387		

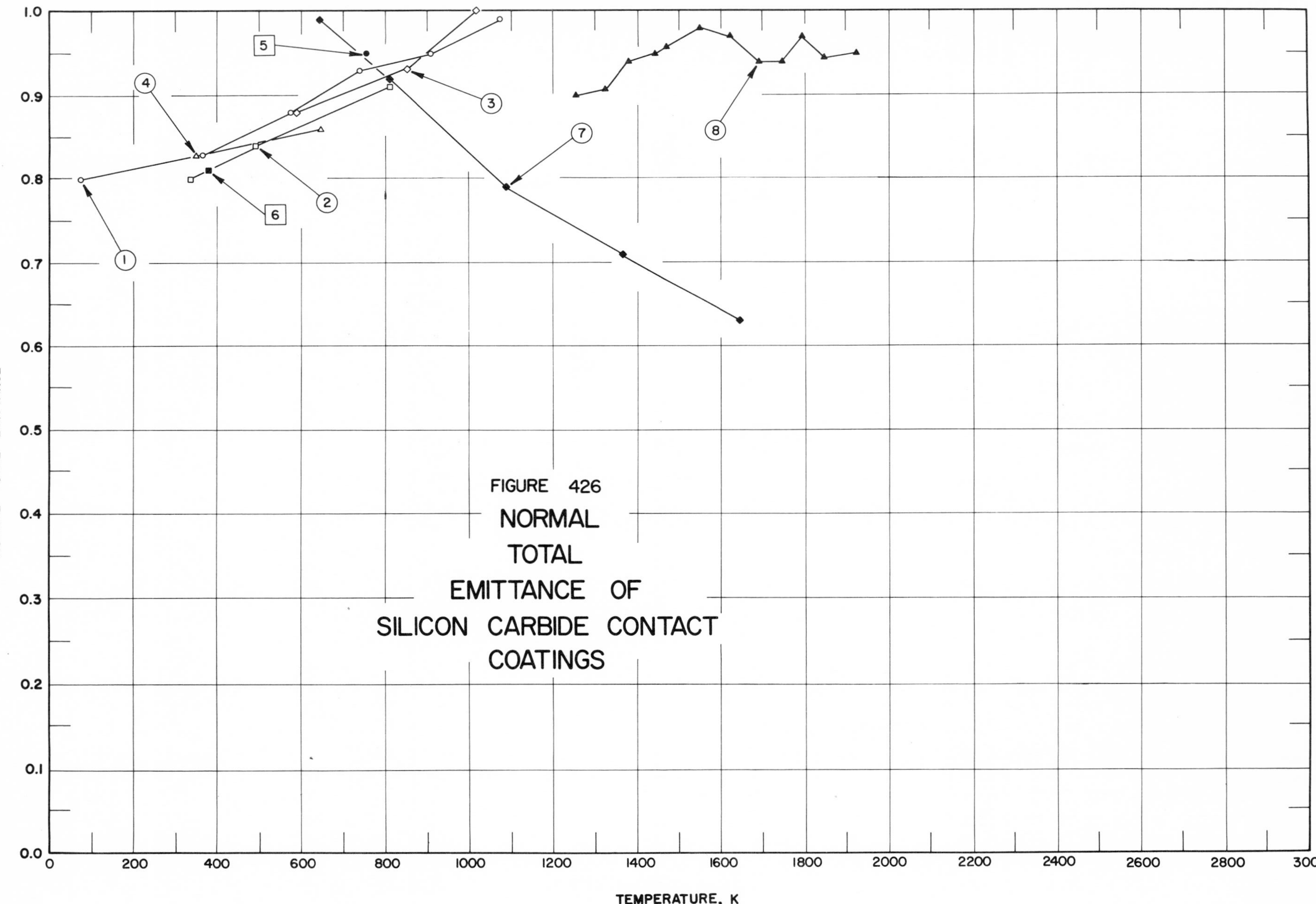

FIGURE 426 NORMAL TOTAL EMITTANCE OF SILICON CARBIDE CONTACT COATINGS

SPECIFICATION TABLE NO. 426 NORMAL TOTAL EMITTANCE OF SILICON CARBIDE CONTACT COATINGS

Curve No.	Ref. No.	Year	Temperature Range, K	Geometry θ'	Reported Error, %	Composition (weight percent), Specifications and Remarks
1	125	1958	75-1075	~0°		Silicon carbide; graphite substrate; heating cycle. [Authors' designation: Cycle 1]
2	125	1958	811-339	~0°		Above specimen and conditions; cooling cycle.
3	125	1958	589-1019	~0°		Curve 1 specimen and conditions. [Authors' designation: Cycle 2]
4	125	1958	644-352	~0°		Above specimen and conditions; cooling cycle.
5	125	1958	755	~0°		Curve 1 specimen and conditions. [Authors' designation: Cycle 3]
6	125	1958	380	~0°		Above specimen and conditions; cooling cycle.
7	178	1960	644-1644	~0°	±20	Silicon carbide; graphite substrate; measured in demoisturized helium gas.
8	147	1963	1255-1925	~0°		SiC (~0.0762 mm thick); tantalum substrate; substrate grit blasted; vapor deposited; surface roughness 80 to 100 μm (RMS) measured with profilometer; measured in vacuum (3 to 4 x 10^{-4} mm Hg).

DATA TABLE NO. 426 NORMAL TOTAL EMITTANCE OF SILICON CARBIDE CONTACT COATINGS

[Temperature, T, K; Emittance, ∈]

T	∈	T	∈	T	∈
CURVE 1		CURVE 4		CURVE 8	
75	0.80	644	0.86	1255	0.900
366	0.83	352	0.83	1325	0.907
575	0.88			1380	0.940
741	0.93	CURVE 5		1444	0.950
908	0.95			1469	0.957
1075	0.99	755	0.95	1553	0.980
				1622	0.970
CURVE 2		CURVE 6		1694	0.940
				1747	0.940
811	0.91	380	0.81	1794	0.970
491	0.84			1847	0.945
339	0.80	CURVE 7		1925	0.950
CURVE 3		644	0.99		
		811	0.92		
589	0.88	1089	0.79		
852	0.93	1367	0.71		
1019	1.00	1644	0.63		

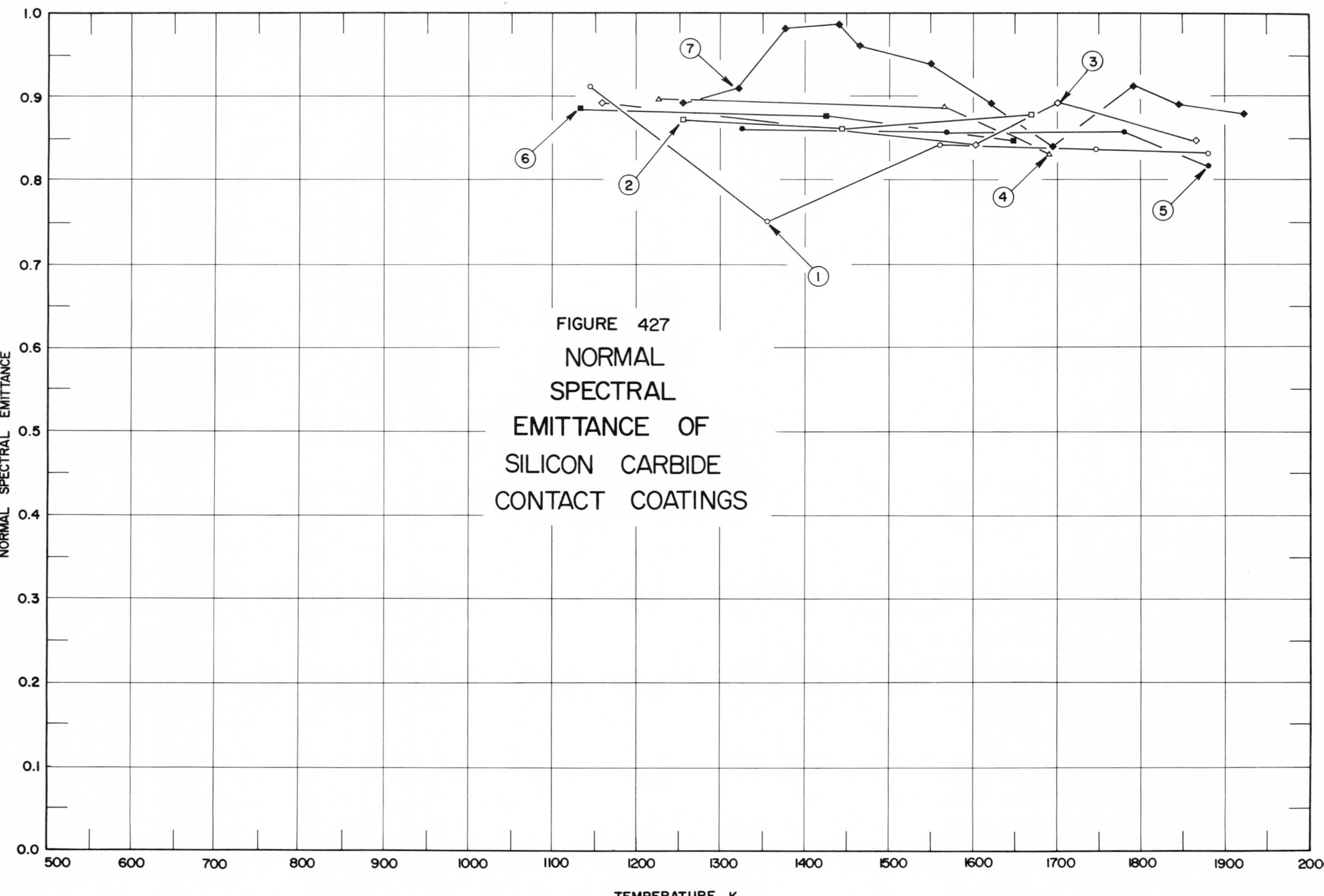

FIGURE 427 NORMAL SPECTRAL EMITTANCE OF SILICON CARBIDE CONTACT COATINGS

SPECIFICATION TABLE NO. 427 NORMAL SPECTRAL EMITTANCE OF SILICON CARBIDE CONTACT COATINGS

Curve No.	Ref. No.	Year	Wavelength, μm	Temperature Range, K	Geometry θ'	Reported Error, %	Composition (weight percent), Specifications, and Remarks
1	179	1957	0.665	1144-1880	~0°		Silicon carbide (Norton Co.); graphite substrate; as received; measured in air (5×10^{-4} mm Hg); heating cycle. [Authors' designation: Cycle 1]
2	179	1957	0.665	1669-1255	~0°		Above specimen and conditions; cooling cycle.
3	179	1957	0.665	1158-1866	~0°		Curve 1 specimen and conditions. [Authors' designation: Cycle 2]
4	179	1957	0.665	1691-1225	~0°		Above specimen and conditions; cooling cycle.
5	179	1957	0.665	1325-1880	~0°		Curve 1 specimen and conditions. [Authors' designation: Cycle 3]
6	179	1957	0.665	1647-1133	~0°		Above specimen and conditions; cooling cycle.
7	147	1963	0.65	1255-1922	~0°		SiC (~0.0762 mm thick); tantalum substrate; substrate grit blasted; vapor deposited; surface roughness 80 to 100 μm (RMS) measured with profilometer; measured in vacuum (3 to 4×10^{-4} mm Hg); authors assumed specimen was a grey body.

DATA TABLE NO. 427 NORMAL SPECTRAL EMITTANCE OF SILICON CARBIDE CONTACT COATINGS

[Temperature, T, K; Emittance, ϵ; Wavelength, λ, μm]

T	ϵ	T	ϵ	T	ϵ
CURVE 1 λ = 0.665		CURVE 3 (cont.)		CURVE 6 λ = 0.665	
		1602	0.840		
1144	0.910	1866	0.845	1647	0.845
1355	0.750			1425	0.875
1561	0.840	CURVE 4 λ = 0.665		1133	0.885
1747	0.835				
1880	0.830			CURVE 7 λ = 0.65	
		1691	0.830		
CURVE 2 λ = 0.665		1566	0.885		
		1225	0.895	1255	0.890
				1322	0.907
1669	0.875	CURVE 5 λ = 0.665		1378	0.980
1444	0.860			1441	0.985
1255	0.870			1466	0.960
		1325	0.860	1550	0.937
CURVE 3 λ = 0.665		1569	0.855	1622	0.890
		1780	0.855	1694	0.837
		1880	0.815	1791	0.910
1158	0.890			1844	0.887
1700	0.890			1922	0.877

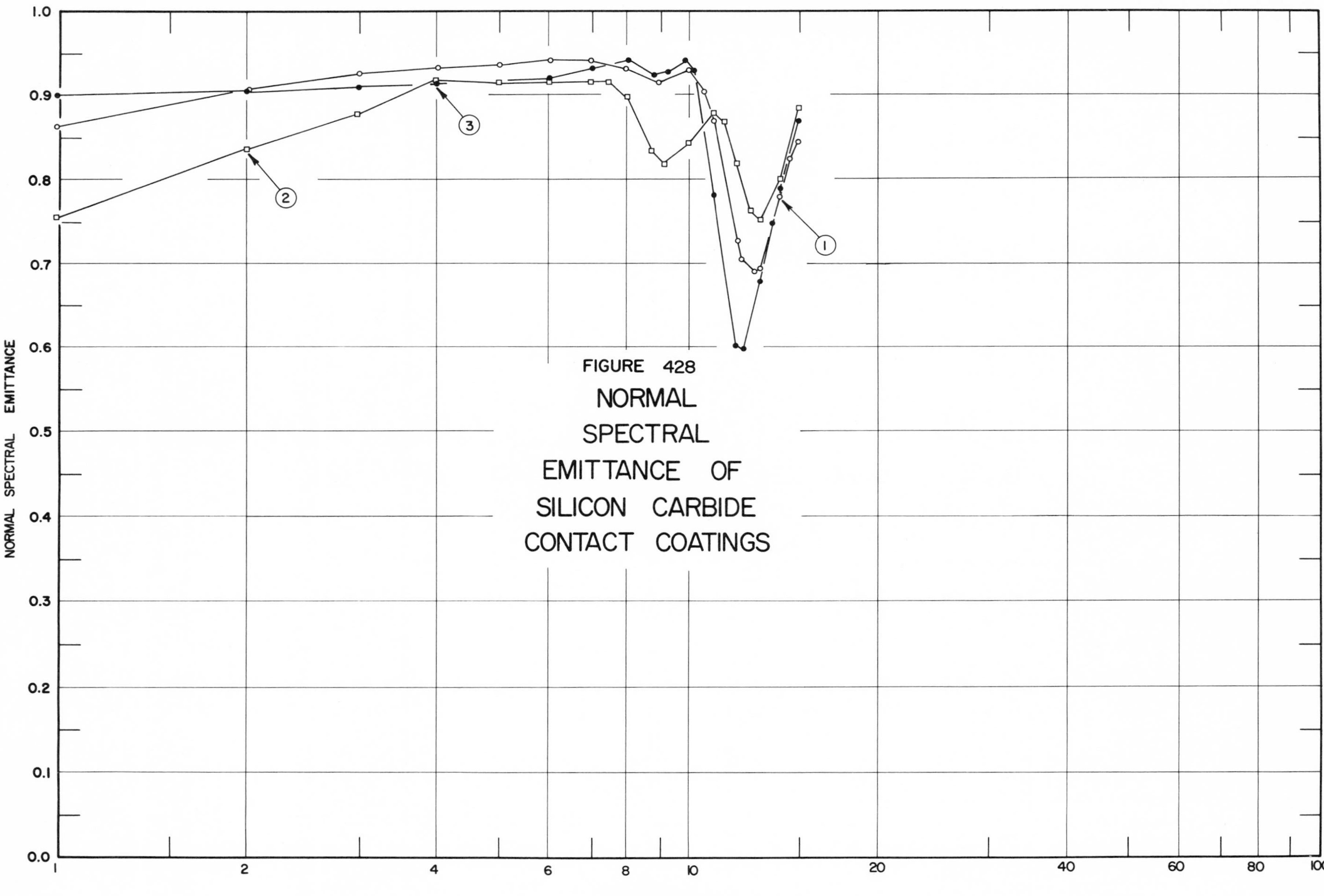

FIGURE 428 NORMAL SPECTRAL EMITTANCE OF SILICON CARBIDE CONTACT COATINGS

SPECIFICATION TABLE NO. 428 NORMAL SPECTRAL EMITTANCE OF SILICON CARBIDE CONTACT COATINGS

Curve No.	Ref. No.	Year	Temperature, K	Wavelength Range, μm	Geometry θ'	Reported Error, %	Composition (weight percent), Specifications, and Remarks
1	180	1964	1203	1.00-14.99	~0°		Silicon carbide; aluminum oxide substrate; data extracted from smooth curve.
2	180	1964	1513	1.00-15.00	~0°		Similar to above specimen and conditions.
3	180	1964	813	1.00-15.00	~0°		Similar to above specimen and conditions.

DATA TABLE NO. 428 NORMAL SPECTRAL EMITTANCE OF SILICON CARBIDE CONTACT COATINGS

[Wavelength, λ, μm; Emittance, ϵ; Temperature, T, K]

λ	ϵ	λ	ϵ	λ	ϵ
CURVE 1 T = 1203		CURVE 2 T = 1513		CURVE 3 T = 813	
1.00	0.864	1.00	0.755	1.00	0.900
2.03	0.907	2.00	0.836	2.00	0.904
3.02	0.926	2.99	0.877	3.01	0.910
4.01	0.933	3.99	0.908	3.99	0.914
5.02	0.937	5.01	0.915	5.01	0.915*
6.05	0.942	6.01	0.915	6.02	0.919
7.00	0.942	7.00	0.915	7.02	0.933
7.99	0.932	7.48	0.915	8.03	0.942
8.99	0.915	8.01	0.897	8.81	0.924
10.00	0.931	8.73	0.834	9.28	0.929
10.65	0.904	9.15	0.817	9.86	0.942
10.99	0.870	10.00	0.843	10.18	0.929
11.99	0.727	10.98	0.878	10.99	0.771
12.25	0.706	11.41	0.868	11.89	0.601
12.75	0.691	11.98	0.818	12.21	0.588
13.01	0.696	12.61	0.763	13.00	0.678
13.99	0.779	13.01	0.753	13.58	0.749
14.50	0.824	14.01	0.800	14.00	0.790
14.99	0.845	15.00	0.885	15.00	0.870

* Not shown on plot

SPECIFICATION TABLE NO. 429 NORMAL SPECTRAL REFLECTANCE OF SILICON CARBIDE CONTACT COATINGS

Curve No.	Ref. No.	Year	Temperature, K	Wavelength Range, μm	Geometry θ	Geometry θ'	Geometry ω'	Reported Error, %	Composition (weight percent), Specifications, and Remarks
1*	181	1956	298	0.38-0.70	9°		2π		Silicon carbide; graphite substrate; as received; data extracted from smooth curve; measured relative to MgO.
2*	179	1957	298	0.30-2.60	9°		2π	±4	Silicon carbide (Norton Co.); graphite substrate; as received; data extracted from smooth curve; measured relative to magnesium carbonate.

DATA TABLE NO. 429 NORMAL SPECTRAL REFLECTANCE OF SILICON CARBIDE CONTACT COATINGS

[Wavelength, λ, μm; Reflectance, ρ; Temperature, T, K]

λ	ρ	λ	ρ
CURVE 1* T = 298		CURVE 2 (cont.)	
		0.78	0.950
0.38	0.104	0.90	0.100
0.40	0.109	0.96	0.100
0.42	0.113	1.10	0.090
0.44	0.115	1.30	0.090
0.46	0.117	1.36	0.080
0.50	0.118	1.50	0.080
0.54	0.118	1.62	0.090
0.58	0.116	1.70	0.090
0.62	0.114	1.90	0.090
0.66	0.113	2.10	0.090
0.70	0.112	2.30	0.085
		2.50	0.085
CURVE 2* T = 298		2.60	0.090
0.30	0.090		
0.50	0.115		
0.62	0.110		
0.70	0.105		

* No plot given

SPECIFICATION TABLE NO. 430 NORMAL SOLAR ABSORPTANCE OF SILICON CARBIDE CONTACT COATINGS

Curve No.	Ref. No.	Year	Temperature Range, K	Geometry θ	Reported Error, %	Composition (weight percent), Specifications and Remarks
1*	179	1957	298	9°		Silicon carbide (Norton Co.); graphite substrate; as received; computed from spectral reflectance for sea level conditions.
2*	179	1957	298	9°		Above specimen and conditions except above atm conditions.

DATA TABLE NO. 430 NORMAL SOLAR ABSORPTANCE OF SILICON CARBIDE CONTACT COATINGS

[Temperature, T, K; Absorptance, α]

T	α
CURVE 1*	
298	0.897
CURVE 2*	
298	0.899

* No plot given

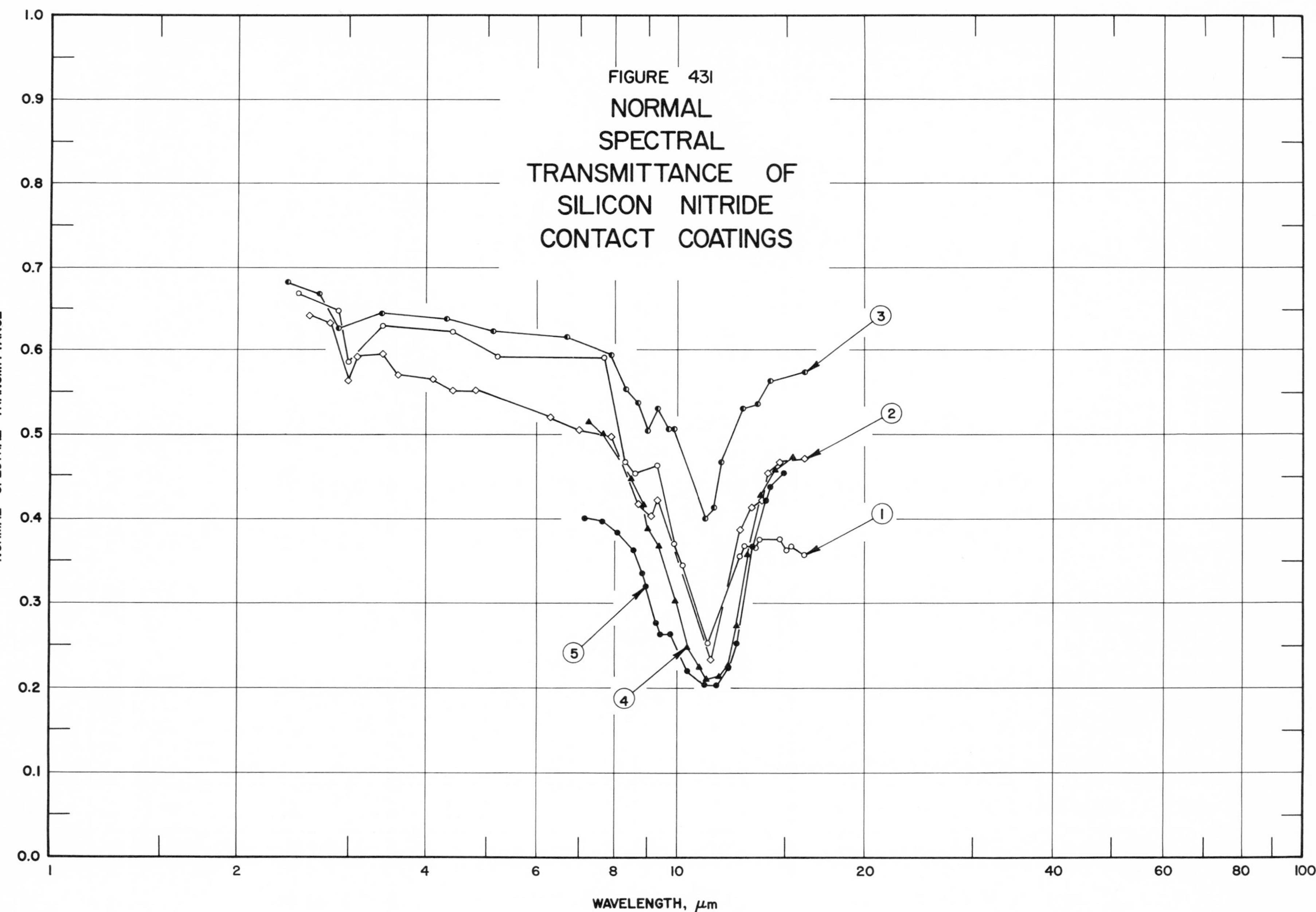
FIGURE 431
NORMAL
SPECTRAL
TRANSMITTANCE OF
SILICON NITRIDE
CONTACT COATINGS
1
2
3
4
5
NORMAL SPECTRAL TRANSMITTANCE
1.0
0.9
0.8
0.7
0.6
0.5
0.4
0.3
0.2
0.1
0.0
WAVELENGTH, μm
1
2
4
6
8
10
20
40
60
80
100

SPECIFICATION TABLE NO. 431 NORMAL SPECTRAL TRANSMITTANCE OF SILICON NITRIDE CONTACT COATINGS

Curve No.	Ref. No.	Year	Temperature, K	Wavelength Range, μm	Geometry θ θ' ω'	Reported Error, %	Composition (weight percent), Specifications, and Remarks
1	182	1967	~298	2.5-16.0	~0° ~0°		Silicon nitride; GaAs substrate; vapor deposited at 823 K; deposition rate 150 Å min^{-1}; data extracted from smooth curve.
2	182	1967	~298	2.6-16.0	~0° ~0°		Silicon nitride; Si substrate; vapor deposited at 823 K; data extracted from smooth curve.
3	182	1967	~298	2.4-16.0	~0° ~0°		Silicon nitride; Si substrate; vapor deposited at 723 K; deposition rate ~30 Å min^{-1}; data extracted from smooth curve.
4	184	1968	~298	7.21-15.4	~0° ~0°		Si_3N_4; phosphorus-doped n-type silicon wafer (of ~5 Ω cm^{-1} resistivity, with (111) surface) substrate; substrate etched away ~30 μm by nitric acid and hydrofluoric acid etchant; deposited on substrate at 1123 K by gas phase reaction between silane, SiH_4, and ammonia, NH_3, using N_2 as the carrier gas; flow rate of nitrogen through ammonia 100 cm^3 min^{-1}; data extracted from smooth curve.
5	184	1968	~298	7.12-14.9	~0° ~0°		Similar to the above specimen and conditions except flow rate of nitrogen through ammonia 1000 cm^3 min^{-1}.

DATA TABLE NO. 431 NORMAL SPECTRAL TRANSMITTANCE OF SILICON NITRIDE CONTACT COATINGS

[Wavelength, λ, μm; Transmittance, τ; Temperature, T, K]

λ	τ	λ	τ	λ	τ	λ	τ	λ	τ	λ	τ
CURVE 1 T ~ 298		CURVE 1 (cont.)		CURVE 2 (cont.)		CURVE 3 (cont.)		CURVE 4 T ~ 298		CURVE 5 T ~ 298	
		15.0	0.364	8.7	0.418	4.3	0.639				
2.5	0.670	15.3	0.367	9.1	0.402	5.1	0.623	7.21	0.514	7.12	0.400
2.9	0.649	16.0	0.357	9.3	0.421	6.7	0.617	7.63	0.500	7.62	0.397
3.0	0.587			10.2	0.349*	7.9	0.596	8.46	0.446	8.09	0.383
3.4	0.630	CURVE 2		11.4	0.234	8.3	0.552	8.81	0.418	8.55	0.361
4.4	0.622	T ~ 298		12.7	0.389	8.7	0.537	9.00	0.389	8.82	0.335
5.2	0.593			13.2	0.415	9.0	0.503	9.35	0.368	8.96	0.320
7.7	0.593	2.6	0.642	13.7	0.422	9.3	0.530	9.99	0.301	9.22	0.276
8.3	0.466	2.8	0.633	14.0	0.455	9.7	0.506	10.4	0.249	9.40	0.264
8.6	0.453	3.0	0.565	14.6	0.469	9.9	0.506	10.8	0.224	9.79	0.264
9.3	0.461	3.1	0.594	16.0	0.471	11.1	0.400	11.2	0.210	10.4	0.220
9.9	0.370	3.4	0.598			11.5	0.413	11.7	0.213	11.1	0.201
10.2	0.346	3.6	0.571	CURVE 3		11.8	0.467	12.1	0.226	11.6	0.202
11.3	0.251	4.1	0.566	T ~ 298		12.8	0.530	12.5	0.273	12.1	0.223
12.7	0.356	4.4	0.552			13.5	0.536	13.0	0.358	12.5	0.254
12.9	0.369	4.8	0.552	2.4	0.682	14.1	0.565	13.7	0.426	13.3	0.368
13.4	0.366	6.3	0.520	2.7	0.669	16.0	0.575	14.4	0.457	13.9	0.422
13.6	0.375	7.0	0.504	2.9	0.626			15.4	0.473	14.2	0.437
14.6	0.375	7.9	0.498	3.4	0.644					14.9	0.453

* Not shown on plot

SPECIFICATION TABLE NO. 432 HEMISPHERICAL TOTAL EMITTANCE OF SILVER SULFIDE CONTACT COATINGS

Curve No.	Ref. No.	Year	Temperature Range, K	Reported Error, %	Composition (weight percent), Specifications and Remarks
1*	146	1961	468-1093	<10	Ag_2S; silver substrate; electrolytically deposited; dried and buffed; measured in vacuum (10^{-5} mm Hg). [Authors' designation: second cycle]

DATA TABLE NO. 432 HEMISPHERICAL TOTAL EMITTANCE OF SILVER SULFIDE CONTACT COATINGS

[Temperature, T, K; Emittance, ϵ]

T	ϵ
CURVE 1*	
468	0.10
668	0.13
873	0.14
1093	0.15

* No plot given

SPECIFICATION TABLE NO. 433 NORMAL SOLAR ABSORPTANCE OF SILVER SULFIDE CONTACT COATINGS

Curve No.	Ref. No.	Year	Temperature Range, K	Geometry θ	Reported Error, %	Composition (weight percent), Specifications and Remarks
1*	146	1961	468-1093	$\sim 0^{\circ}$		Ag_2S; silver substrate; electrolytically deposited; dried and buffed; measured in vacuum (10^{-5} mm Hg); carbon arc source at $\sim$5800 K.

DATA TABLE NO. 433 NORMAL SOLAR ABSORPTANCE OF SILVER SULFIDE CONTACT COATINGS

[Temperature, T, K; Absorptance, α]

T	α
CURVE 1*	
468	0.66
668	0.65
873	0.59
1093	0.43

* No plot given

SPECIFICATION TABLE NO. 434 NORMAL SPECTRAL REFLECTANCE OF SODIUM SALICYLATE CONTACT COATINGS

Curve No.	Ref. No.	Year	Temperature, K	Wavelength Range, μm	Geometry θ	θ'	ω'	Reported Error, %	Composition (weight percent), Specifications, and Remarks
1*	30	1965	298	0.3852-0.5456	<15°		2π		Sodium salicylate (1.5 mg cm^{-2}); MgO pigmented paint substrate; measured in vacuum; data extracted from smooth curve.

DATA TABLE NO. 434 NORMAL SPECTRAL REFLECTANCE OF SODIUM SALICYLATE CONTACT COATINGS

[Wavelength, λ, μm; Reflectance, ρ; Temperature, T, K]

λ	ρ
CURVE 1*	
T = 298	
0.3852	0.782
0.3893	0.803
0.3944	0.822
0.4024	0.843
0.4143	0.861
0.4266	0.872
0.4444	0.880
0.4741	0.892
0.5276	0.936
0.5456	0.947

* No plot given

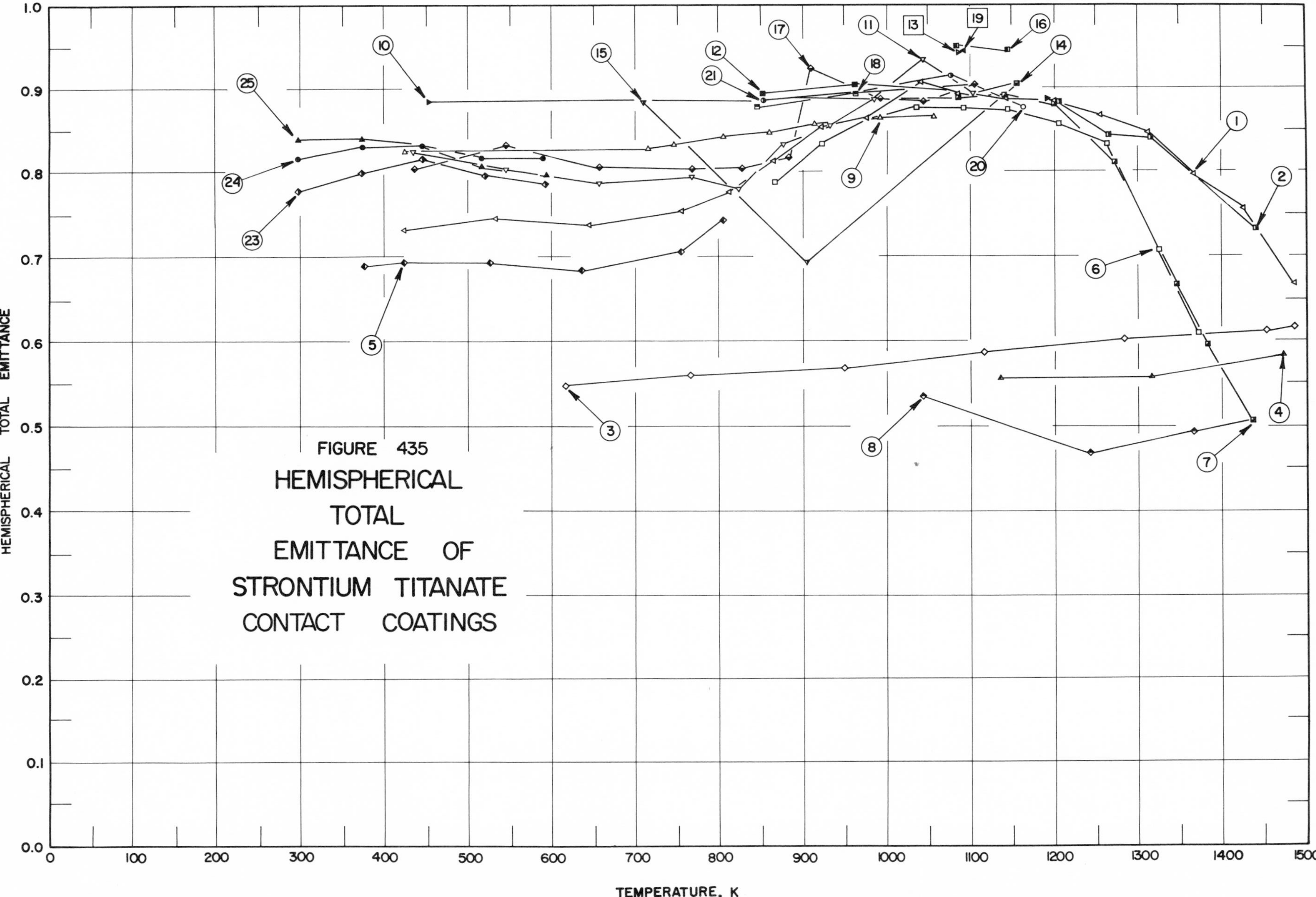

FIGURE 435
HEMISPHERICAL TOTAL EMITTANCE OF STRONTIUM TITANATE CONTACT COATINGS

SPECIFICATION TABLE NO. 435 HEMISPHERICAL TOTAL EMITTANCE OF STRONTIUM TITANATE CONTACT COATINGS

Curve No.	Ref. No.	Year	Temperature Range, K	Reported Error, %	Composition (weight percent), Specifications and Remarks
1	118	1962	422-1489		Strontium titanate, powder from Plasmadyne Corp. (0.102 mm thick); Nb-1 Zr alloy substrate; plasma arc sprayed; hard, medium grit texture; measured in vacuum (1.7×10^{-6} mm Hg); temp measured with thermocouple; heating cycle.
2	118	1962	1144-1503		Above specimen and conditions except temp measured with optical pyrometer.
3	118	1962	1489-616		Curve 1 specimen and conditions; cooling cycle.
4	118	1962	1503-1133		Above specimen and conditions except temp measured with optical pyrometer.
5	118	1962	376-806		Strontium titanate, powder from Plasmadyne Corp. (0.0508 mm thick); Nb-1 Zr alloy substrate; plasma arc sprayed; moderately hard with smooth texture; measured in vacuum ($< 2.5 \times 10^{-6}$ mm Hg); temp measured with thermocouple. [Authors' designation: Run 1]
6	118	1962	866-1438		Curve 5 specimen and conditions except vacuum $< 2.9 \times 10^{-6}$ mm Hg . [Authors' designation: Run 2]
7	118	1962	1088-1438		Above specimen and conditions except temp measured with optical pyrometer.
8	118	1962	1438-1043		Above specimen and conditions; cooling cycle.
9	82	1962	422-1058	$< \pm 2.5$	Strontium titanate (~ 0.0762 mm thick); stainless steel 310 substrate; plasma arc sprayed; measured in vacuum (10^{-7} mm Hg); data extracted from smooth curve.
10	82	1962	452-1191	$< \pm 2.5$	Above specimen and conditions except after heated at 1060 K for 17 hrs.
11	81	1963	434-1101		Strontium titanate (0.0762 mm thick), from National Lead Co; Nb-1 Zr substrate; plasma arc sprayed; measured in vacuum ($< 8.6 \times 10^{-6}$ mm Hg); temp measured with thermocouple; heating cycle. [Author's designation: Run No. 1]
12	81	1963	1101-851		Above specimen and conditions; cooling cycle.
13	81	1963	1086		Curve 11 specimen and conditions except temp measured with optical pyrometer.
14	81	1963	1102-1156		Curve 11 specimen and conditions. [Author's designation: Run No. 2]
15	81	1963	1156-710		Above specimen and conditions; cooling cycle.
16	81	1963	1085-1143		Curve 15 specimen and conditions except temp measured with optical pyrometer.
17	81	1963	435-1105		Strontium titanate (0.127 mm thick), from National Lead Co; Nb-1 Zr substrate; plasma arc sprayed; measured in vacuum ($< 9.8 \times 10^{-6}$ mm Hg); temp measured with thermocouple; heating cycle. [Author's designation: Run 1]
18	81	1963	1105-849		Above specimen and conditions; cooling cycle.
19	81	1963	1093		Curve 17 specimen and conditions except temp measured with optical pyrometer.
20	81	1963	1105-1162		Curve 17 specimen and conditions. [Author's designation: Run No. 2]
21	81	1963	1162-852		Above specimen and conditions; cooling cycle.
22*	81	1963	1093-1153		Curve 20 specimen and conditions except temp measured with optical pyrometer.
23	137	1965	299-591	± 5	Strontium titanate (~ 0.038 mm thick, 12 mg cm^{-2}); Al substrate; plasma arc sprayed; measured in vacuum (10^{-7} mm Hg).

* Not shown on plot

SPECIFICATION TABLE NO. 435 HEMISPHERICAL TOTAL EMITTANCE OF STRONTIUM TITANATE CONTACT COATINGS (continued)

Curve No.	Ref. No.	Year	Temperature Range, K	Reported Error, %	Composition (weight percent), Specifications and Remarks
24	137	1965	299-590	±5	Similar to above specimen and conditions except 0.089 mm thick, 28 mg cm^{-2}.
25	137	1965	299-591	±5	Similar to curve 23 specimen and conditions except 0.128 mm thick, 40 mg cm^{-2}.

DATA TABLE NO. 435 HEMISPHERICAL TOTAL EMITTANCE OF STRONTIUM TITANATE CONTACT COATINGS

[Temperature, T, K; Emittance, ϵ]

T	ϵ
CURVE 1	
422	0.732
533	0.747
645	0.738
756	0.755
811	0.778
866	0.813
921	0.853
977	0.864
1041	0.907
1087	0.894
1144	0.887
1200	0.884
1256	0.866
1313	0.847
1369	0.798
1427	0.758
1489	0.669
CURVE 2	
1144	0.887*
1205	0.884
1265	0.842
1315	0.840
1370	0.796*
1440	0.731
1503	0.644*
CURVE 3	
1489	0.669
1452	0.613
1283	0.602
1118	0.588
950	0.569
765	0.560
616	0.549
CURVE 4	
1503	0.644*
1472	0.581
1316	0.557
1133	0.556
CURVE 5	
376	0.690
424	0.694
529	0.693
636	0.683
754	0.708
806	0.743
CURVE 6	
866	0.789
921	0.832
978	0.862*
1036	0.875
1092	0.875
1146	0.873
1207	0.859
1263	0.834
1329	0.708
1372	0.610
CURVE 7	
1088	0.889
1140	0.890
1200	0.880
1272	0.810
1349	0.667
1383	0.597
1438	0.507
CURVE 8	
1438	0.507*
1366	0.493
1241	0.468
1043	0.535
CURVE 9	
422	0.827
717	0.829
747	0.832
805	0.841
860	0.849
CURVE 9 (cont.)	
913	0.857
994	0.864
1058	0.865
CURVE 10	
452	0.885
1191	0.886
CURVE 11	
434	0.826
546	0.802
657	0.788
769	0.794
824	0.780
879	0.831
934	0.854
989	0.885
1046	0.933
1101	0.891
CURVE 12	
1101	0.891*
962	0.901
851	0.894
CURVE 13	
1086	0.941
CURVE 14	
1102	0.893*
1156	0.905
CURVE 15	
1156	0.905*
905	0.693
710	0.883
CURVE 16	
1085	0.950
1143	0.946
CURVE 17	
435	0.803
547	0.831
658	0.805
769	0.802
827	0.804
882	0.818
910	0.924
993	0.887
1048	0.882
1105	0.904
CURVE 18	
1105	0.904*
965	0.893
849	0.878
CURVE 19	
1093	0.945
CURVE 20	
1105	0.902*
1162	0.875
CURVE 21	
1162	0.875*
1077	0.916
964	0.893*
852	0.885
CURVE 22*	
1093	0.943
1153	0.904
CURVE 23	
299	0.779
372	0.800
446	0.818
520	0.799
591	0.787
CURVE 24	
299	0.818
371	0.830
445	0.831
517	0.817
590	0.817
CURVE 25	
299	0.840
371	0.840
445	0.831*
516	0.808
591	0.796

* Not shown on plot

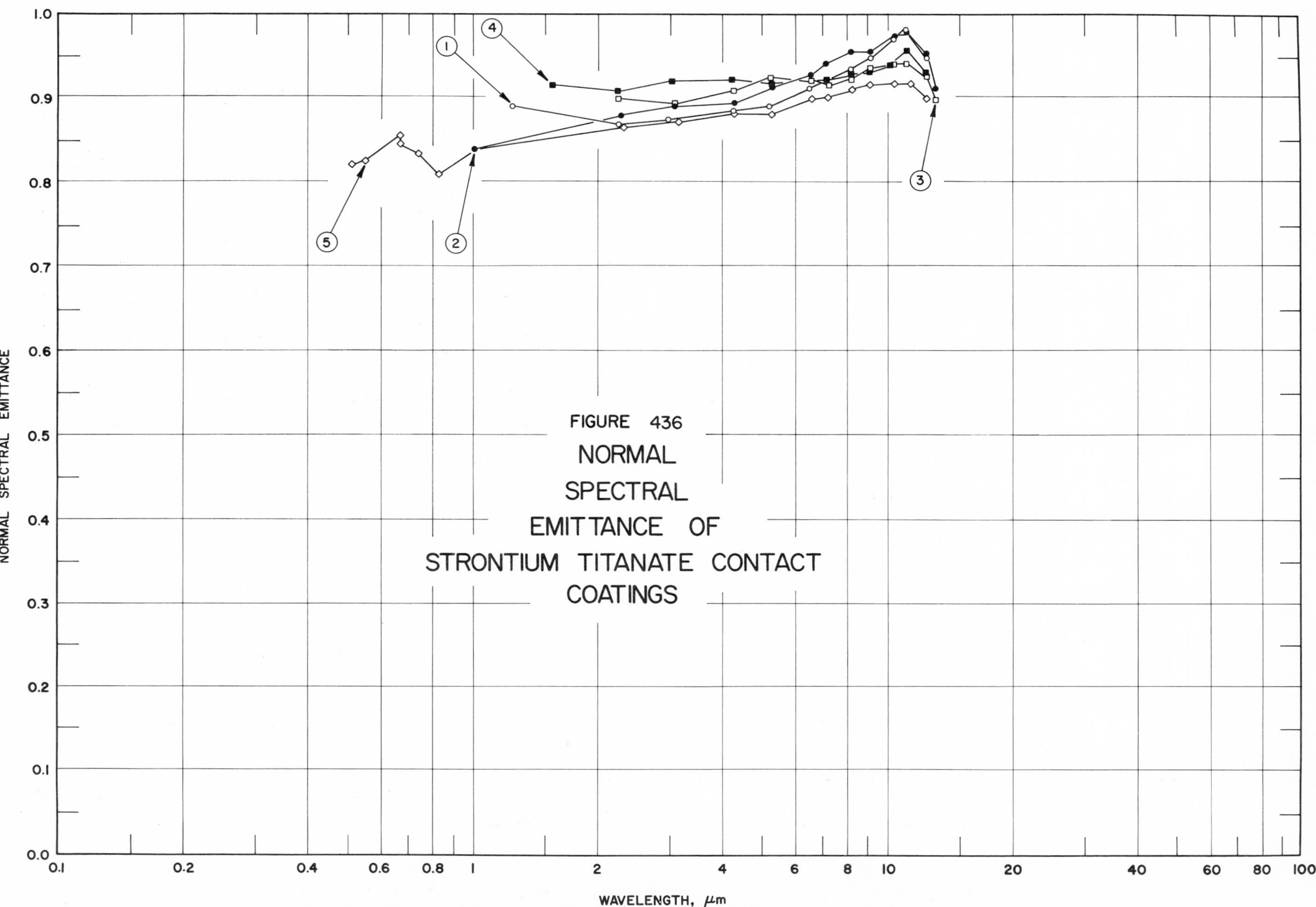

FIGURE 436 NORMAL SPECTRAL EMITTANCE OF STRONTIUM TITANATE CONTACT COATINGS

SPECIFICATION TABLE NO. 436 NORMAL SPECTRAL EMITTANCE OF STRONTIUM TITANATE CONTACT COATINGS

Curve No.	Ref. No.	Year	Temperature, K	Wavelength Range, μm	Geometry θ'	Reported Error, %	Composition (weight percent), Specifications, and Remarks
1	118	1962	755	1.24-12.41	$\sim 0^{\circ}$		Strontium titanate, powder from Plasmadyne Corp., (0.102 mm thick); Nb-1 Zr alloy substrate; plasma arc sprayed; hard with smooth texture.
2	118	1962	1061	1.07-13.02	$\sim 0^{\circ}$		Above specimen and conditions.
3	118	1962	1061	2.25-13.03	$\sim 0^{\circ}$		Above specimen and conditions except maintained at 1061 K overnight.
4	118	1962	755	1.56-13.03	$\sim 0^{\circ}$		Above specimen and conditions.
5	118	1962	1366	0.51-12.43	$\sim 0^{\circ}$		Above specimen and conditions.

DATA TABLE NO. 436 NORMAL SPECTRAL EMITTANCE OF STRONTIUM TITANATE CONTACT COATINGS

[Wavelength, λ, μm; Emittance, ϵ; Temperature, T, K]

λ	ϵ	λ	ϵ	λ	ϵ	λ	ϵ	λ	ϵ
CURVE 1		CURVE 2 (cont.)		CURVE 3 (cont.)		CURVE 4 (cont.)		CURVE 5 (cont.)	
T = 755									
		5.29	0.913	9.15	0.936	13.03	0.898	10.39	0.917
1.24	0.890	6.59	0.927	10.35	0.940			11.13	0.918
2.26	0.869	7.16	0.940	11.09	0.941	CURVE 5		12.43	0.899
2.98	0.874	8.21	0.955	12.41	0.926	T = 1366			
4.24	0.885	9.12	0.956	13.03	0.897				
5.21	0.890	10.38	0.974			0.51	0.822		
6.56	0.911	11.08	0.978	CURVE 4		0.55	0.826		
8.20	0.934	12.41	0.953	T = 755		0.67	0.854		
9.15	0.948	13.02	0.911			0.67	0.846		
10.36	0.971			1.56	0.916	0.74	0.834		
11.06	0.983	CURVE 3		2.24	0.907	0.83	0.810		
12.41	0.947	T = 1061		3.05	0.919	1.04	0.840		
				4.21	0.922	2.32	0.864		
CURVE 2		2.25	0.897	5.27	0.917	3.13	0.872		
T = 1061		3.07	0.894	7.18	0.922	4.27	0.882		
		4.25	0.908	8.20	0.928	5.26	0.881		
1.07	0.840	5.22	0.924	9.14	0.930	6.60	0.899		
2.29	0.879	6.59	0.920	10.32	0.939	7.21	0.900		
3.08	0.890	7.22	0.915	11.08	0.956	8.25	0.909		
4.26	0.894	8.21	0.923	12.39	0.930	9.13	0.916		

SPECIFICATION TABLE NO. 437 NORMAL SOLAR ABSORPTANCE OF STRONTIUM TITANATE CONTACT COATINGS

Curve No.	Ref. No.	Year	Temperature Range, K	Geometry θ	Reported Error, %	Composition (weight percent), Specifications and Remarks
1*	137	1965	400-600	0°	±5	Strontium titanate (~0.038 mm thick, 12 mg cm^{-2}); Al substrate; plasma arc sprayed; measured in vacuum (10^{-7} mm Hg).
2*	137	1965	400-600	0°	±5	Similar to above specimen and conditions except 0.089 mm thick, 28 mg cm^{-2}.
3*	137	1965	400-600	0°	±5	Similar to above specimen and conditions except 0.128 mm thick, 40 mg cm^{-2}.

DATA TABLE NO. 437 NORMAL SOLAR ABSORPTANCE OF STRONTIUM TITANATE CONTACT COATINGS

[Temperature, T, K; Absorptance, α]

T	α
CURVE 1*	
400	0.73
600	0.73
CURVE 2*	
400	0.76
600	0.76
CURVE 3*	
400	0.64
600	0.64

* No plot given

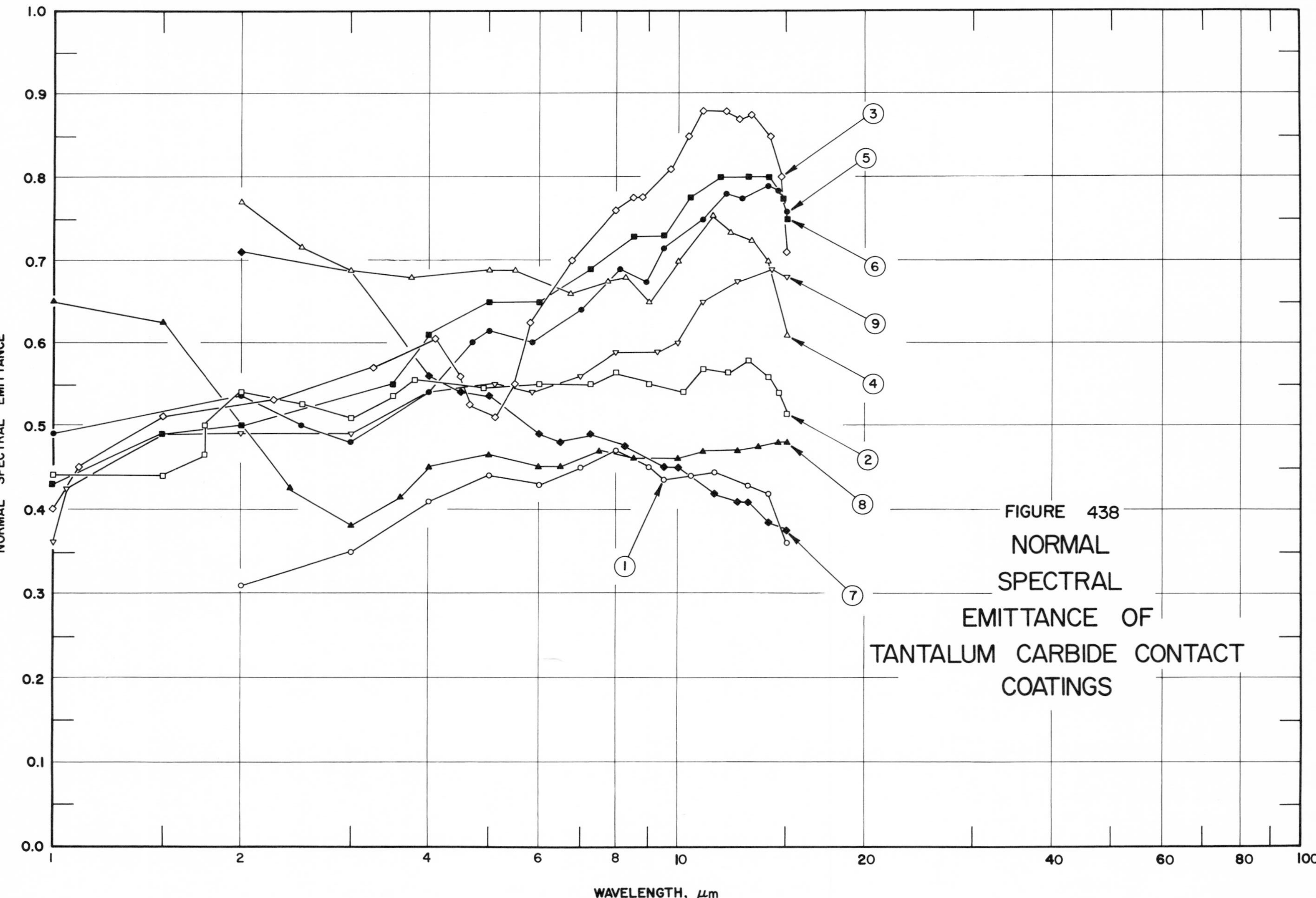

FIGURE 438 NORMAL SPECTRAL EMITTANCE OF TANTALUM CARBIDE CONTACT COATINGS

SPECIFICATION TABLE NO. 438 NORMAL SPECTRAL EMITTANCE OF TANTALUM CARBIDE CONTACT COATINGS

Curve No.	Ref. No.	Year	Temperature, K	Wavelength Range, μm	Geometry θ'	Reported Error, %	Composition (weight percent), Specifications, and Remarks
1	139	1961	523	2.00-15.00	$\sim 0^{\circ}$	± 5	TaC; Inconel X substrate; substrate sandblasted; flame sprayed using a plasmatron; as received; data extracted from smooth curve.
2	139	1961	773	1.00-15.00	$\sim 0^{\circ}$	± 5	Similar to above specimen and conditions.
3	139	1961	1023	1.00-15.00	$\sim 0^{\circ}$	± 5	Similar to curve 1 specimen and conditions.
4	139	1961	523	2.00-15.00	$\sim 0^{\circ}$	± 5	Similar to curve 1 specimen and conditions except heated in air at 1089 K for 30 min.
5	139	1961	773	1.00-15.00	$\sim 0^{\circ}$	± 5	Similar to curve 4 specimen and conditions.
6	139	1961	1023	1.00-15.00	$\sim 0^{\circ}$	± 5	Similar to curve 4 specimen and conditions.
7	139	1961	523	2.00-15.00	$\sim 0^{\circ}$	± 5	Similar to curve 1 specimen and conditions except heated in vacuum (6.8×10^{-5} mm Hg) at 1089 K for 30 min.
8	139	1961	773	1.00-15.00	$\sim 0^{\circ}$	± 5	Similar to curve 7 specimen and conditions.
9	139	1961	1023	1.00-15.00	$\sim 0^{\circ}$	± 5	Similar to curve 7 specimen and conditions.

DATA TABLE NO. 438 NORMAL SPECTRAL EMITTANCE OF TANTALUM CARBIDE CONTACT COATINGS

[Wavelength, λ, μm; Emittance, ∈; Temperature, T, K]

λ	∈
CURVE 1 T = 523	
2.00	0.310
3.00	0.350
4.00	0.410
5.00	0.440
6.00	0.430
7.00	0.450
8.00	0.470
9.00	0.450
9.50	0.435
10.50	0.440
11.50	0.445
13.00	0.430
14.00	0.420
15.00	0.360
CURVE 2 T = 773	
1.00	0.440
1.50	0.440
1.75	0.465
1.75	0.500
2.00	0.540
2.50	0.525
3.00	0.510
3.50	0.535
3.80	0.555
4.90	0.545
6.00	0.550
7.25	0.550
8.00	0.565
9.00	0.550
10.25	0.540
11.00	0.570
12.10	0.565
13.00	0.580
14.00	0.560
14.60	0.540
15.00	0.515
CURVE 3 T = 1023	
1.00	0.400
CURVE 3 (cont.)	
1.10	0.450
1.50	0.510
2.25	0.530
3.25	0.570
4.10	0.605
4.50	0.560
4.65	0.525
5.10	0.510
5.50	0.550
5.80	0.625
6.80	0.700
8.00	0.760
8.50	0.775
8.80	0.775
9.75	0.810
10.40	0.850
11.00	0.880
12.00	0.880
12.60	0.870
13.25	0.875
14.25	0.850
14.75	0.800
15.00	0.710
CURVE 4 T = 523	
2.00	0.770
2.50	0.715
3.00	0.690
3.75	0.680
5.00	0.690
5.50	0.690
6.75	0.660
7.75	0.675
8.25	0.680
9.00	0.650
10.00	0.700
11.40	0.755
12.25	0.735
13.25	0.725
14.00	0.700
15.00	0.610
CURVE 5 T = 773	
1.00	0.490
2.00	0.535
2.50	0.500
3.00	0.480
4.00	0.540
4.70	0.600
5.00	0.615
5.85	0.600
7.00	0.640
8.10	0.690
8.90	0.675
9.50	0.715
11.00	0.750
12.00	0.780
12.75	0.775
14.00	0.790
14.50	0.785
15.00	0.760
CURVE 6 T = 1023	
1.00	0.430
1.50	0.490
2.00	0.500
3.50	0.550
4.00	0.610
5.00	0.650
6.00	0.650
7.25	0.690
8.50	0.730
9.50	0.730
10.50	0.775
11.75	0.800
13.00	0.800
14.00	0.800
14.75	0.775
15.00	0.750
CURVE 7 T = 523	
2.00	0.710
3.00	0.690
CURVE 7 (cont.)	
4.00	0.560
4.50	0.540
5.00	0.535
6.00	0.490
6.50	0.480
7.25	0.490
8.25	0.475
9.50	0.450
10.00	0.450
11.50	0.420
12.50	0.410
13.00	0.410
14.00	0.385
15.00	0.325
CURVE 8 T = 773	
1.00	0.650
1.50	0.625
2.00	0.500
2.40	0.425
3.00	0.380
3.60	0.415
4.00	0.450
5.00	0.465
6.00	0.450
6.50	0.450
7.50	0.470
8.50	0.460
10.00	0.460
11.00	0.470
12.50	0.470
13.50	0.475
14.50	0.480
15.00	0.480
CURVE 9 T = 1023	
1.00	0.360
1.05	0.425
1.50	0.490
2.00	0.490
3.00	0.490
CURVE 9 (cont.)	
4.00	0.540
5.10	0.550
5.85	0.540
7.00	0.560
8.00	0.590
9.25	0.590
10.00	0.600
11.00	0.650
12.50	0.675
14.25	0.690
15.00	0.680

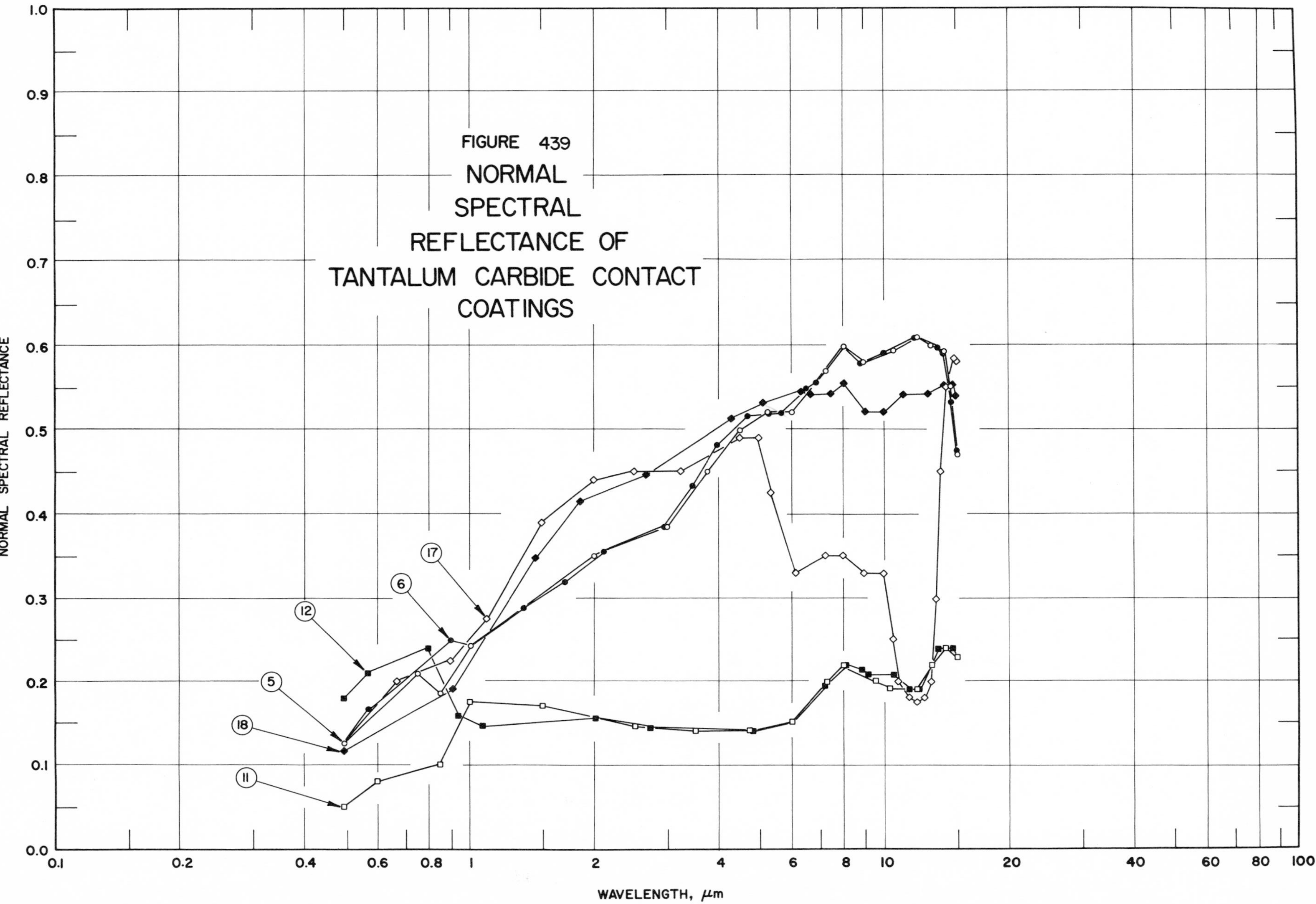
FIGURE 439
NORMAL
SPECTRAL
REFLECTANCE OF
TANTALUM CARBIDE CONTACT
COATINGS
NORMAL SPECTRAL REFLECTANCE
1.0
0.9
0.8
0.7
0.6
0.5
0.4
0.3
0.2
0.1
0.0
0.1
0.2
0.4
0.6
0.8
1
2
4
6
8
10
20
40
60
80
100
WAVELENGTH, μm
17
6
12
5
18
11

SPECIFICATION TABLE NO. 439 NORMAL SPECTRAL REFLECTANCE OF TANTALUM CARBIDE CONTACT COATINGS

Curve No.	Ref. No.	Year	Temperature, K	Wavelength Range, μm	Geometry θ θ'	ω'	Reported Error, %	Composition (weight percent), Specifications, and Remarks
1*	139	1961	322	2.00-15.00	$\sim 0^{\circ}$	2π	< 2	TaC; Inconel X substrate; substrate sandblasted; flame sprayed using a plasmatron; sample maintained at $\sim$322 K; data extracted from smooth curve; converted from R(2π, $\sim 0^{\circ}$); hohlraum at 523 K.
2*	139	1961	322	2.00-15.00	$\sim 0^{\circ}$	2π	< 2	Above specimen and conditions; diffuse component only.
3*	139	1961	322	1.00-15.00	$\sim 0^{\circ}$	2π	< 2	Similar to curve 1 specimen and conditions except hohlraum at 773 K.
4*	139	1961	322	1.00-15.00	$\sim 0^{\circ}$	2π	< 2	Above specimen and conditions; diffuse component only.
5	139	1961	322	0.50-15.00	$\sim 0^{\circ}$	2π	< 2	Similar to curve 1 specimen and conditions except hohlraum at 1273 K.
6	139	1961	322	0.50-15.00	$\sim 0^{\circ}$	2π	< 2	Above specimen and conditions; diffuse component only.
7*	139	1961	322	2.00-15.00	$\sim 0^{\circ}$	2π	< 2	Similar to curve 1 specimen and conditions except heated in air at 1089 K for 30 min.
8*	139	1961	322	2.01-14.99	$\sim 0^{\circ}$	2π	< 2	Above specimen and conditions; diffuse component only.
9*	139	1961	322	1.00-15.00	$\sim 0^{\circ}$	2π	< 2	Similar to curve 7 specimen and conditions except hohlraum at 773 K.
10*	139	1961	322	1.00-15.00	$\sim 0^{\circ}$	2π	< 2	Above specimen and conditions; diffuse component only.
11	139	1961	322	0.50-15.00	$\sim 0^{\circ}$	2π	< 2	Similar to curve 7 specimen and conditions except hohlraum at 1273 K.
12	139	1961	322	0.50-15.00	$\sim 0^{\circ}$	2π	< 2	Above specimen and conditions; diffuse component only.
13*	139	1961	322	2.00-15.00	$\sim 0^{\circ}$	2π	< 2	Similar to curve 1 specimen and conditions except heated in vacuum (6.8×10^{-5} mm Hg) at 1089 K for 30 min.
14*	139	1961	322	2.00-15.00	$\sim 0^{\circ}$	2π	< 2	Above specimen and conditions; diffuse component only.
15*	139	1961	322	1.00-15.00	$\sim 0^{\circ}$	2π	< 2	Similar to curve 13 specimen and conditions except hohlraum at 773 K.
16*	139	1961	322	1.00-15.00	$\sim 0^{\circ}$	2π	< 2	Above specimen and conditions; diffuse component only.
17	139	1961	322	0.50-15.00	$\sim 0^{\circ}$	2π	< 2	Similar to curve 13 specimen and conditions except hohlraum at 1273 K.
18	139	1961	322	0.50-14.99	$\sim 0^{\circ}$	2π	< 2	Above specimen and conditions; diffuse component only.

* Not shown on plot

DATA TABLE NO. 439 NORMAL SPECTRAL REFLECTANCE OF TANTALUM CARBIDE CONTACT COATINGS

[Wavelength, λ, μm; Reflectance, ρ; Temperature, T, K]

λ	ρ
CURVE 1*	
T = 322	
2.00	0.300
2.75	0.325
3.50	0.375
4.50	0.420
5.36	0.430
6.50	0.415
7.00	0.435
7.80	0.455
9.00	0.450
10.50	0.460
12.00	0.470
13.75	0.480
14.50	0.465
15.00	0.440
CURVE 2*	
T = 322	
2.00	0.594
2.48	0.600
3.18	0.585
3.87	0.551
4.17	0.535
5.00	0.511
6.00	0.480
7.50	0.474
8.49	0.467
8.78	0.464
9.30	0.473
9.95	0.481
13.31	0.481
14.00	0.474
14.49	0.456
15.00	0.420
CURVE 3*	
T = 322	
1.00	0.290
1.20	0.225
1.60	0.200
1.75	0.225
1.75	0.325
2.00	0.350
CURVE 3 (cont.)*	
2.60	0.350
3.20	0.400
4.00	0.460
5.00	0.495
6.50	0.530
6.80	0.535
8.20	0.600
9.00	0.590
10.50	0.600
11.75	0.610
13.00	0.600
14.00	0.600
14.75	0.570
15.00	0.525
CURVE 4*	
T = 322	
1.00	0.275
2.64	0.332
4.18	0.424
4.99	0.447
5.46	0.445
6.08	0.432
6.77	0.463
7.22	0.475
9.63	0.472
11.24	0.489
13.98	0.503
14.73	0.483
15.00	0.443
CURVE 5	
T = 322	
0.50	0.125
0.75	0.210
0.85	0.185
1.00	0.245
2.00	0.350
3.00	0.385
3.75	0.450
4.50	0.500
5.25	0.520
6.00	0.520
CURVE 5 (cont.)	
7.25	0.570
8.00	0.600
9.00	0.580
10.50	0.595
12.00	0.610
13.00	0.600
14.00	0.595
14.50	0.550
15.00	0.470
CURVE 6	
T = 322	
0.50	0.124
0.57	0.165
0.90	0.249
1.01	0.243*
1.35	0.288
1.70	0.320
2.11	0.355
2.96	0.384
3.44	0.433
3.97	0.482
4.67	0.516
5.26	0.518
5.63	0.518
6.48	0.548
6.91	0.554
8.00	0.599
8.86	0.580
10.00	0.592
11.80	0.610
13.40	0.598
13.97	0.591
14.66	0.532
15.00	0.474
CURVE 7*	
T = 322	
2.00	0.150
2.50	0.165
3.00	0.100
4.25	0.100
5.25	0.100
CURVE 7 (cont.)*	
6.00	0.110
7.00	0.150
8.50	0.170
9.50	0.165
10.00	0.155
12.00	0.150
12.75	0.175
13.35	0.200
14.00	0.210
15.00	0.210
CURVE 8*	
T = 322	
2.01	0.175
2.54	0.139
3.02	0.119
3.99	0.119
5.18	0.119
5.75	0.126
6.32	0.158
6.92	0.171
7.74	0.174
8.31	0.185
8.88	0.180
9.57	0.165
10.47	0.151
11.76	0.143
12.16	0.141
12.70	0.157
13.23	0.186
13.99	0.200
14.99	0.200
CURVE 9*	
T = 322	
1.00	0.120
1.50	0.140
2.00	0.125
2.50	0.105
4.00	0.130
5.00	0.130
6.00	0.135
7.00	0.175
CURVE 9 (cont.)*	
8.25	0.200
9.00	0.200
10.00	0.185
11.25	0.180
12.00	0.180
13.25	0.215
14.50	0.250
15.00	0.250
CURVE 10*	
T = 322	
1.00	0.142
1.40	0.156
1.82	0.148
2.76	0.120
3.14	0.124
4.19	0.145
4.58	0.146
4.97	0.141
5.33	0.142
6.30	0.167
7.39	0.200
8.31	0.216
9.30	0.209
10.66	0.199
11.21	0.188
11.81	0.188
12.53	0.200
13.60	0.239
14.12	0.242
14.59	0.250
15.00	0.270
CURVE 11	
T = 322	
0.50	0.050
0.60	0.080
0.85	0.100
1.00	0.175
1.50	0.170
2.50	0.145
3.50	0.140
4.75	0.140
CURVE 11 (cont.)	
6.00	0.150
7.25	0.200
8.00	0.220
9.50	0.200
10.25	0.190
12.00	0.190
13.00	0.220
14.00	0.240
15.00	0.230
CURVE 12	
T = 322	
0.50	0.180
0.57	0.211
0.80	0.239
0.94	0.157
1.08	0.146
2.02	0.156
2.72	0.144
4.84	0.140
6.01	0.151
7.21	0.195
8.04	0.220
8.83	0.214
9.18	0.209
10.61	0.209
11.41	0.190
12.03	0.190
13.44	0.239
14.59	0.239
15.00	0.230
CURVE 13*	
T = 322	
2.00	0.360
2.40	0.300
3.00	0.270
3.50	0.300
4.00	0.375
4.60	0.405
5.40	0.375
6.10	0.330
6.50	0.350
CURVE 13 (cont.)*	
7.00	0.400
7.75	0.410
9.00	0.410
10.50	0.425
11.75	0.450
12.75	0.515
13.75	0.550
14.25	0.550
14.75	0.525
15.00	0.465
CURVE 14*	
T = 322	
2.00	0.509
2.73	0.437
3.11	0.427
3.71	0.453
4.11	0.478
5.01	0.502
5.99	0.507
7.00	0.523
8.01	0.523
8.94	0.491
9.91	0.490
11.24	0.518
11.68	0.518
12.06	0.511
13.80	0.522
14.71	0.504
15.00	0.476
CURVE 15*	
T = 322	
1.00	0.350
1.50	0.385
2.25	0.395
4.00	0.380
5.50	0.355
6.00	0.360
7.00	0.330
7.75	0.340
9.25	0.320
10.25	0.320
CURVE 15 (cont.)*	
11.75	0.290
13.00	0.290
14.00	0.300
15.00	0.310
CURVE 16*	
T = 322	
1.00	0.340
2.35	0.393
3.05	0.418
4.23	0.479
5.05	0.499
6.05	0.500
7.52	0.525
8.03	0.531
8.63	0.520
9.51	0.503
10.51	0.518
11.40	0.536
12.26	0.541
12.91	0.546
14.50	0.577
15.00	0.575
CURVE 17	
T = 322	
0.50	0.125
0.65	0.200
0.90	0.225
1.10	0.275
1.50	0.390
2.00	0.440
2.50	0.450
3.25	0.450
4.50	0.490
5.00	0.490
5.35	0.425
6.15	0.330
7.25	0.350
8.00	0.350
9.00	0.330
10.00	0.330
10.50	0.250

* Not shown on plot.

DATA TABLE NO. 439 NORMAL SPECTRAL REFLECTANCE OF TANTALUM CARBIDE CONTACT COATINGS (continued)

λ	ρ
CURVE 17 (cont.) T = 322	
11.50	0.180
12.00	0.175
12.50	0.180
13.00	0.200
13.35	0.300
13.75	0.450
14.20	0.550
14.75	0.585
15.00	0.580
CURVE 18 T = 322	
0.50	0.114
0.91	0.190
1.42	0.347
1.85	0.414
2.68	0.445
4.34	0.512
5.17	0.531
6.37	0.543
6.76	0.541
7.44	0.541
8.02	0.553
9.02	0.520
10.00	0.520
11.11	0.541
12.97	0.541
13.99	0.552
14.65	0.552
14.99	0.538

SPECIFICATION TABLE NO. 440 ANGULAR SPECTRAL REFLECTANCE OF TELLURIUM CONTACT COATINGS

Curve No.	Ref. No.	Year	Temperature, K	Wavelength Range, μm	Geometry θ	θ'	ω'	Reported Error, %	Composition (weight percent), Specifications, and Remarks
1*	213	1961	298	0.0525-0.1608	20°	20°			Te (400 Å thick); glass substrate; vapor deposited in vacuum (8 x 10^{-6} mm Hg); measured in vacuum.

DATA TABLE NO. 440 ANGULAR SPECTRAL REFLECTANCE OF TELLURIUM CONTACT COATINGS

[Wavelength, λ, μm; Reflectance, ρ; Temperature, T, K]

λ	ρ	λ	ρ	λ	ρ
CURVE 1* T = 298		CURVE 1 (cont.)*		CURVE 1 (cont.)*	
		0.0859	0.104	0.1176	0.202
0.0525	0.007	0.0876	0.120	0.1191	0.209
0.0556	0.005	0.0883	0.113	0.1202	0.213
0.0564	0.012	0.0900	0.113	0.1224	0.222
0.0601	0.017	0.0921	0.130	0.1242	0.226
0.0638	0.026	0.0934	0.134	0.1271	0.231
0.0641	0.029	0.0970	0.134	0.1349	0.233
0.0687	0.051	0.0981	0.138	0.1395	0.220
0.0700	0.056	0.0983	0.150	0.1409	0.223
0.0705	0.064	0.1017	0.152	0.1449	0.215
0.0713	0.066	0.1032	0.162	0.1449	0.225
0.0719	0.074	0.1050	0.147	0.1471	0.232
0.0746	0.082	0.1050	0.163	0.1512	0.238
0.0759	0.089	0.1050	0.178	0.1540	0.252
0.0788	0.088	0.1112	0.187	0.1576	0.263
0.0806	0.100	0.1130	0.190	0.1608	0.257
0.0830	0.097	0.1158	0.194	0.1608	0.251
0.0845	0.107	0.1169	0.198		

* No plot given

SPECIFICATION TABLE NO. 441 NORMAL SPECTRAL TRANSMITTANCE OF TELLURIUM CONTACT COATINGS

Curve No.	Ref. No.	Year	Temperature, K	Wavelength Range, μm	Geometry θ θ' ω'	Reported Error, %	Composition (weight percent), Specifications, and Remarks
1*	213	1961	298	0.0452-0.0832	~0° ~0°		Te (930 Å thick); stilbene crystal substrate; vapor deposited in vacuum (8×10^{-5} mm Hg); exposed to air; measured in vacuum.

DATA TABLE NO. 441 NORMAL SPECTRAL TRANSMITTANCE OF TELLURIUM CONTACT COATINGS

[Wavelength, λ, μm; Transmittance, τ; Temperature, T, K]

λ	τ	λ	τ
CURVE 1*		CURVE 1 (cont.)*	
T = 298		0.0706	0.027
0.0452	0.202	0.0713	0.023
0.0476	0.188	0.0718	0.027
0.0482	0.186	0.0745	0.023
0.0489	0.164	0.0759	0.021
0.0503	0.156	0.0787	0.019
0.0510	0.143	0.0827	0.024
0.0523	0.116	0.0832	0.018
0.0527	0.117		
0.0557	0.081		
0.0568	0.096		
0.0576	0.067		
0.0600	0.071		
0.0617	0.064		
0.0638	0.052		
0.0644	0.060		
0.0685	0.031		
0.0699	0.024		

* No plot given

SPECIFICATION TABLE NO. 442 NORMAL TOTAL EMITTANCE OF THORIUM DIOXIDE CONTACT COATINGS

Curve No.	Ref. No.	Year	Temperature Range, K	Geometry θ'	Reported Error, %	Composition (weight percent), Specifications and Remarks
1*	123	1952	673-1073	$\sim 0°$		ThO_2; Nimonic 75 substrate.

DATA TABLE NO. 442 NORMAL TOTAL EMITTANCE OF THORIUM DIOXIDE CONTACT COATINGS

[Temperature, T, K; Emittance, ϵ]

T	ϵ
CURVE 1*	
673	0.60
773	0.56
873	0.52
973	0.48
1073	0.44

* No plot given

SPECIFICATION TABLE NO. 445 NORMAL SPECTRAL TRANSMITTANCE OF THORIUM DIOXIDE CONTACT COATINGS

Curve No.	Ref. No.	Year	Temperature, K	Wavelength Range, μm	Geometry θ	θ'	ω'	Reported Error, %	Composition (weight percent), Specifications, and Remarks
1*	161	1966	473	0.190-0.428	0°	0°			ThO_2 (0.6 μm optical thickness); plane-parallel fused quartz (2 mm thick) substrate; deposited from solution of thorium tetrachloride; amorphous film structure; 64.3 dry residue and 14.4 chlorine content.
2*	161	1966	573	0.195-0.404	0°	0°			Similar to above specimen and conditions except < 0.6 μm optical thickness; heated to 573 K; 54.3 dry residue and 11.7 chlorine content.
3*	161	1966	773	0.190-0.404	0°	0°			Similar to above specimen and conditions except heated to 773 K; crystalline film structure; 51.01 dry residue and 1.45 chlorine content.
4*	161	1966	473	0.212-0.406	0°	0°			ThO_2 (0.6 μm optical thickness); plane-parallel fused quartz plate (2 mm thick) substrate; deposited from solution of thorium nitrate; amorphous film structure; 74.7 dry residue and 20-25 NO_3 content.
5*	161	1966	573	0.190-0.406	0°	0°			Similar to the above specimen and conditions except < 0.6 μm optical thickness; heated to 573 K; 51.9 dry residue and 6-8 NO_3 content.
6*	161	1966	773	0.189-0.395	0°	0°			Similar to the above specimen and conditions except heated to 773 K; crystalline film structure; 48.3 dry residue and 1-2 NO_3 content.

* No plot given

DATA TABLE NO. 445 NORMAL SPECTRAL TRANSMITTANCE OF THORIUM DIOXIDE CONTACT COATINGS

[Wavelength, λ, μm; Transmittance, τ; Temperature, T, K]

λ	τ
CURVE 1*	
T = 473	
0.190	0.305
0.194	0.522
0.197	0.630
0.203	0.728
0.208	0.786
0.217	0.875
0.252	0.898
0.285	0.913
0.428	0.916
CURVE 2*	
T = 573	
0.195	0.322
0.207	0.689
0.211	0.768
0.221	0.840
0.230	0.785
0.239	0.713
0.249	0.706
0.260	0.749
0.268	0.851
0.279	0.889
0.306	0.790
0.316	0.765
0.325	0.772
0.339	0.811
0.353	0.843
0.360	0.863
0.375	0.891
0.388	0.908
0.404	0.905
CURVE 3*	
T = 773	
0.190	0.106
0.197	0.209
0.201	0.317
0.207	0.592
0.213	0.701
0.217	0.753
0.222	0.787
0.231	0.744
CURVE 3 (cont.)*	
0.240	0.707
0.251	0.682
0.258	0.729
0.263	0.799
0.268	0.856
0.279	0.885
0.292	0.852
0.301	0.812
0.310	0.765
0.321	0.739
0.334	0.771
0.341	0.801
0.351	0.829
0.361	0.856
0.374	0.894
0.383	0.901
0.392	0.906
0.404	0.894
CURVE 4*	
T = 473	
0.212	0.002
0.221	0.018
0.226	0.062
0.232	0.164
0.238	0.454
0.243	0.651
0.246	0.809
0.256	0.774
0.259	0.746
0.268	0.782
0.275	0.846
0.287	0.901
0.299	0.861
0.305	0.830
0.313	0.805
0.321	0.828
0.329	0.851
0.346	0.901
0.361	0.920
0.371	0.916
0.378	0.901
0.391	0.873
0.406	0.840
CURVE 5*	
T = 573	
0.190	0.033
0.200	0.050
0.204	0.087
0.213	0.176
0.228	0.480
0.235	0.595
0.242	0.746
0.249	0.849
0.257	0.751
0.268	0.784
0.276	0.856
0.289	0.897
0.297	0.889
0.312	0.802
0.345	0.894
0.356	0.913
0.373	0.913
0.392	0.875
0.406	0.840
CURVE 6*	
T = 773	
0.189	0.089
0.193	0.212
0.197	0.295
0.204	0.571
0.207	0.676
0.214	0.643
0.217	0.733
0.220	0.804
0.226	0.838
0.234	0.796
0.243	0.716
0.253	0.724
0.261	0.814
0.269	0.865
0.274	0.887
0.282	0.861
0.291	0.809
0.298	0.749
0.309	0.747
0.313	0.766
0.323	0.820
CURVE 6 (cont.)*	
0.343	0.890
0.354	0.912
0.362	0.903
0.368	0.890
0.380	0.854
0.390	0.825
0.395	0.788

* No plot given

FIGURE 444

NORMAL SPECTRAL REFLECTANCE OF TIN OXIDE CONTACT COATINGS

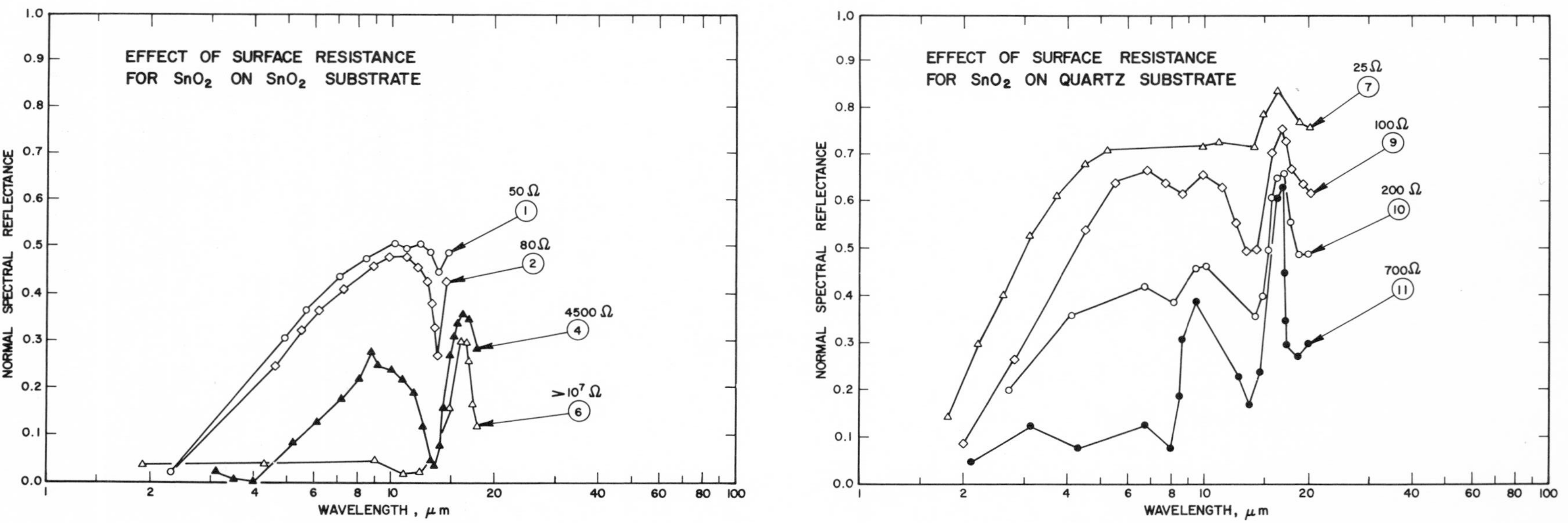

SPECIFICATION TABLE NO. 444 NORMAL SPECTRAL REFLECTANCE OF TIN OXIDE CONTACT COATINGS

Curve No.	Ref. No.	Year	Temperature, K	Wavelength Range, μm	Geometry θ	θ'	ω'	Reported Error, %	Composition (weight percent), Specifications, and Remarks
1	183	1969	~298	2.3-14.7	~0°	~0°			SnO_2; SnO_2 substrate; substrate sintered at 1673 K; deposited by pyrolysis of $SnCl_2$; surface resistance 50 Ω.
2	183	1969	~298	2.3-14.7	~0°	~0°			Similar to above specimen and conditions except heated in air at 1173 K to increase surface resistance to 80 Ω.
3*	183	1969	~298	3.7-14.5	~0°	~0°			Similar to curve 1 specimen and conditions except heated at 1173 K in air to increase surface resistance to 500 Ω.
4	183	1969	~298	3.1-17.7	~0°	~0°			Similar to curve 1 specimen and conditions except heated at 1173 K in air to increase surface resistance to 4500 Ω.
5*	183	1969	~298	3.1-17.8	~0°	~0°			Similar to curve 1 specimen and conditions except heated at 1173 K in air to increase surface resistance to 1.5×10^4 Ω.
6	183	1969	~298	1.9-17.8	~0°	~0°			Similar to curve 1 specimen and conditions except heated at 1173 K in air to increase surface resistance to $>10^7$ Ω.
7	183	1969	~298	1.8-20.2	~0°	~0°			SnO_2; fused quartz substrate; deposited by pyrolysis of $SnCl_2$; surface resistance 25 Ω.
8*	183	1969	~298	1.8-20.2	~0°	~0°			Similar to above specimen and conditions except heated at 1073-1173 K in air to increase surface resistance to 40 Ω.
9	183	1969	~298	2.0-20.2	~0°	~0°			Similar to curve 7 specimen and conditions except heated at 1073-1173 K in air to increase surface resistance to 100 Ω.
10	183	1969	~298	2.7-20.0	~0°	~0°			Similar to curve 7 specimen and conditions except heated at 1073-1173 K in air to increase surface resistance to 200 Ω
11	183	1969	~298	2.1-20.0	~0°	~0°			Similar to curve 7 specimen and conditions except heated at 1073-1173 K in air to increase surface resistance to 7000 Ω.

* Not shown on plot

DATA TABLE NO. 444 NORMAL SPECTRAL REFLECTANCE OF TIN OXIDE CONTACT COATINGS

[Wavelength, λ, μm; Reflectance, ρ; Temperature, T, K]

CURVE 1
T ~ 298

λ	ρ
2.3	0.020*
4.9	0.302
5.7	0.364
7.1	0.431
8.5	0.473
10.3	0.505
11.1	0.498
12.1	0.502
13.0	0.488
13.8	0.445
14.7	0.485

CURVE 2
T ~ 298

λ	ρ
2.3	0.020*
4.6	0.244
5.5	0.320
6.2	0.362
7.3	0.409
8.9	0.458
9.9	0.476
11.1	0.475
12.0	0.458
12.7	0.421
13.2	0.378
13.4	0.328
13.7	0.266
14.5	0.423

CURVE 3*
T ~ 298

λ	ρ
3.7	0.061
6.6	0.275
7.5	0.316
8.4	0.347
9.9	0.368
11.0	0.375
11.9	0.352
12.5	0.310
12.9	0.214
13.4	0.089
13.9	0.154
14.5	0.350

CURVE 4
T ~ 298

λ	ρ
3.1	0.020
3.5	0.005
4.0	0.000
5.2	0.084
6.1	0.128
7.2	0.174
8.1	0.218
8.8	0.275
9.2	0.248
10.0	0.236
10.8	0.219
11.6	0.188
12.4	0.119
13.0	0.041
13.4	0.032
13.8	0.077
14.1	0.154
14.9	0.267
15.2	0.308
15.5	0.337
16.1	0.355
16.7	0.346
17.7	0.280

CURVE 5*
T ~ 298

λ	ρ
3.1	0.009
5.6	0.054
6.8	0.091
7.9	0.131
8.7	0.193
9.2	0.146
9.8	0.110
10.6	0.077
11.3	0.056
12.2	0.046
13.2	0.052
15.1	0.248
15.5	0.311
15.9	0.337
16.5	0.331
17.2	0.264
17.8	0.208

CURVE 6
T ~ 298

λ	ρ
1.9	0.033
4.3	0.039
9.0	0.043
10.8	0.014
12.1	0.020
13.9	0.077*
14.7	0.156
16.0	0.299
16.5	0.293
16.9	0.256
17.3	0.162
17.8	0.111

CURVE 7
T ~ 298

λ	ρ
1.8	0.140
2.2	0.295
2.6	0.400
3.1	0.523
3.7	0.610
4.5	0.676
5.2	0.705
9.9	0.711
11.0	0.721
13.9	0.711
14.8	0.783
16.3	0.831
18.7	0.769
20.2	0.759

CURVE 8*
T ~ 298

λ	ρ
1.8	0.140
2.2	0.295
2.6	0.400
3.7	0.511
5.1	0.661
6.9	0.680
8.6	0.651
10.1	0.669
11.9	0.633
13.0	0.581
13.4	0.544

CURVE 8 (cont.)

λ	ρ
14.2	0.594
14.9	0.739
16.4	0.813
17.9	0.763
20.2	0.759

CURVE 9
T ~ 298

λ	ρ
2.0	0.088
2.8	0.269
4.5	0.537
5.5	0.639
6.8	0.664
7.7	0.637
8.6	0.614
9.9	0.653
11.3	0.629
12.4	0.556
13.3	0.492
14.2	0.499
15.7	0.704
16.8	0.753
17.2	0.728
17.8	0.664
19.2	0.632
20.2	0.618

CURVE 10
T ~ 298

λ	ρ
2.7	0.196
4.1	0.357
6.7	0.419
8.1	0.383
9.4	0.455
10.1	0.460
14.0	0.352
14.7	0.399
15.4	0.494
15.7	0.606
16.3	0.647
17.0	0.656
17.7	0.551
18.8	0.482
20.0	0.486

CURVE 11
T ~ 298

λ	ρ
2.1	0.046
3.1	0.120
4.3	0.073
6.7	0.121
7.9	0.076
8.5	0.188
8.6	0.306
9.5	0.382
12.6	0.223
13.5	0.166
14.5	0.238
16.3	0.606
16.9	0.625
17.0	0.446
17.2	0.343
17.4	0.293
18.7	0.270
20.0	0.297

* Not shown on plot

SPECIFICATION TABLE NO. 445 NORMAL SPECTRAL TRANSMITTANCE OF TIN OXIDE CONTACT COATINGS

Curve No.	Ref. No.	Year	Temperature, K	Wavelength Range, μm	Geometry θ	θ'	ω'	Reported Error, %	Composition (weight percent), Specifications, and Remarks
1*	173	1966	298	0.479-0.620	0°	0°			Tin oxide; glass substrate.

DATA TABLE NO. 445 NORMAL SPECTRAL TRANSMITTANCE OF TIN OXIDE CONTACT COATINGS

[Wavelength, λ, μm; Transmittance, τ; Temperature, T, K]

λ	τ
CURVE 1*	
T = 298	
0.479	0.811
0.490	0.817
0.496	0.815
0.501	0.835
0.505	0.839
0.511	0.851
0.521	0.875
0.530	0.896
0.541	0.917
0.551	0.918
0.561	0.914
0.581	0.886
0.601	0.858
0.620	0.821

* No plot given

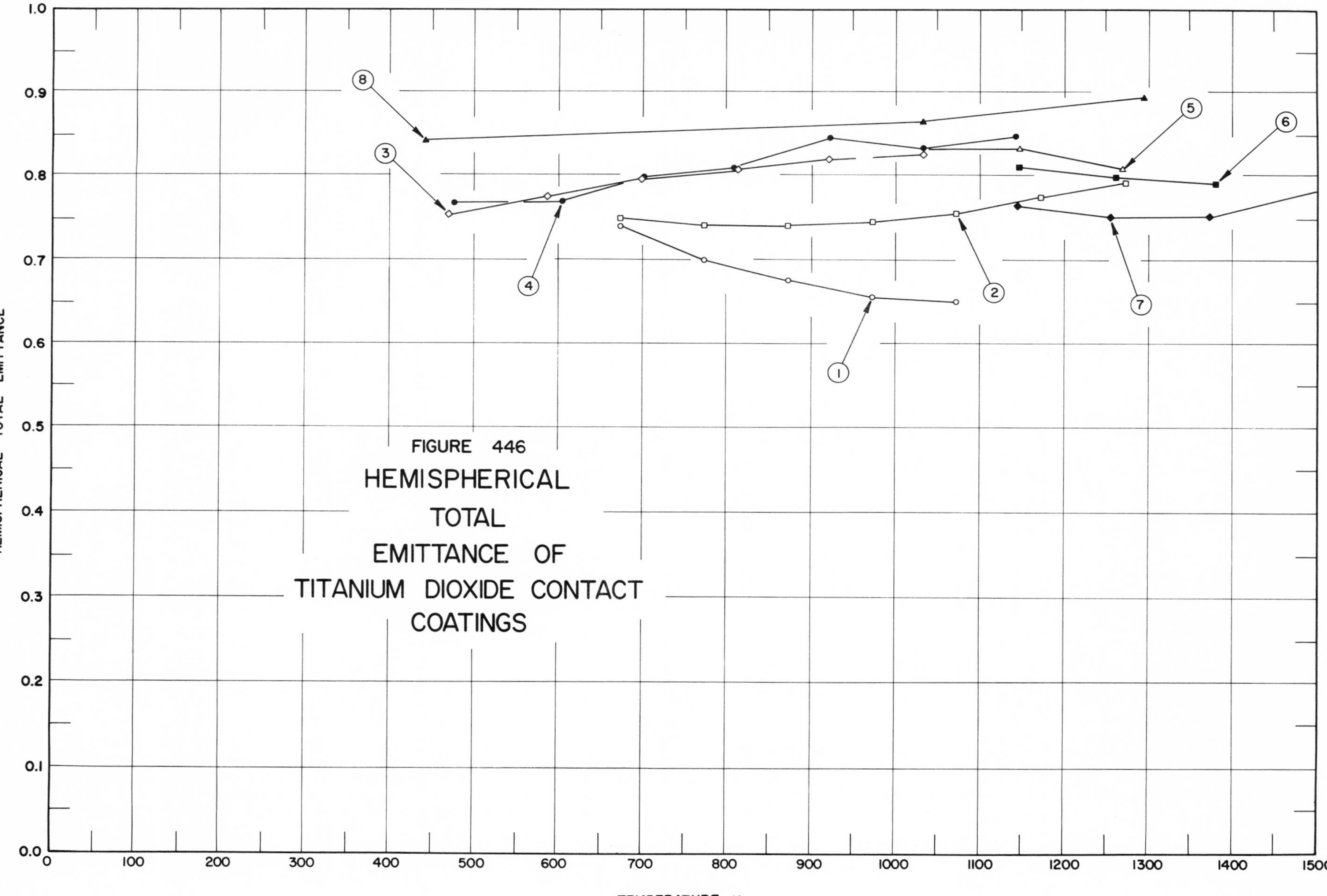

FIGURE 446 HEMISPHERICAL TOTAL EMITTANCE OF TITANIUM DIOXIDE CONTACT COATINGS

SPECIFICATION TABLE NO. 446 HEMISPHERICAL TOTAL EMITTANCE OF TITANIUM DIOXIDE CONTACT COATINGS

Curve No.	Ref. No.	Year	Temperature Range, K	Reported Error, %	Composition (weight percent), Specifications and Remarks
1	97	1961	673-1073	±2.5	TiO_2 (8 mg cm^{-2} area density); nickel substrate; 200-300 mesh titanium dioxide; sprayed; sintered at 1273 K in vacuum; measured in vacuum (< 5 x 10^{-6} mm Hg); data extracted from smooth curve.
2	97	1961	673-1273	±2.5	TiO_2 (8 mg cm^{-2} area density); soft iron substrate; sprayed; sintered at 1373 K in vacuum; measured in vacuum (< 5 x 10^{-6} mm Hg); data extracted from smooth curve.
3	118	1962	470-1033		TiO_2, Metco, Inc. (XP-1114) (0.0762 mm thick); niobium substrate; plasma arc sprayed; coating blue-black in color; hard; coating-substrate bond strength fair; measured in vacuum (< 2.2 x 10^{-6} mm Hg); temp measured with thermocouple. [Authors' designation: Run 1]
4	118	1962	476-1269		Above specimen and conditions except vacuum < 3.0 x 10^{-6} mm Hg. [Authors' designation: Run 2]
5	118	1962	1033-1269		Above specimen and conditions except temp measured with optical pyrometer.
6	118	1962	1147-1380		Curve 3 specimen and conditions except vacuum < 8.0 x 10^{-6} mm Hg. [Authors' designation: Run 3]
7	118	1962	1146-1505		Curve 3 specimen and conditions except vacuum < 2.3 x 10^{-6} mm Hg. [Authors' designation: Run 4]
8	82	1962	443-1295	±2.5	TiO_2, powder from Plasmadyne (0.0762 mm thick); aluminum substrate; plasma arc sprayed; measured in vacuum (10^{-7} mm Hg); data extracted from smooth curve.
9*	82	1962	443-1295	< ±2.5	TiO_2 (Metco XP 1114 titania powder) (0.0762 mm thick); stainless steel 310 substrate; plasma arc sprayed; measured in vacuum (10^{-7} mm Hg); data extracted from smooth curve.

* Not shown on plot.

DATA TABLE NO. 446 HEMISPHERICAL TOTAL EMITTANCE OF TITANIUM DIOXIDE CONTACT COATINGS

[Temperature, T, K; Emittance, ∈]

T	∈
CURVE 1	
673	0.740
773	0.700
873	0.675
973	0.655
1073	0.650
CURVE 2	
673	0.750
773	0.740
873	0.740
973	0.745
1073	0.755
1173	0.775
1273	0.790
CURVE 3	
470	0.754
589	0.776
699	0.796
812	0.809
922	0.820
1033	0.825
CURVE 4	
476	0.767
604	0.770
701	0.799
810	0.809
923	0.845
1033	0.833
1142	0.847
CURVE 5	
1033	0.832*
1147	0.832
1269	0.807
CURVE 6	
1147	0.810
1261	0.799
1380	0.791
CURVE 7	
1146	0.764
1255	0.751
1373	0.751
1505	0.680*
CURVE 8	
443	0.842
1033	0.865
1295	0.892
CURVE 9*	
443	0.842
1033	0.865
1295	0.892

* Not shown on plot

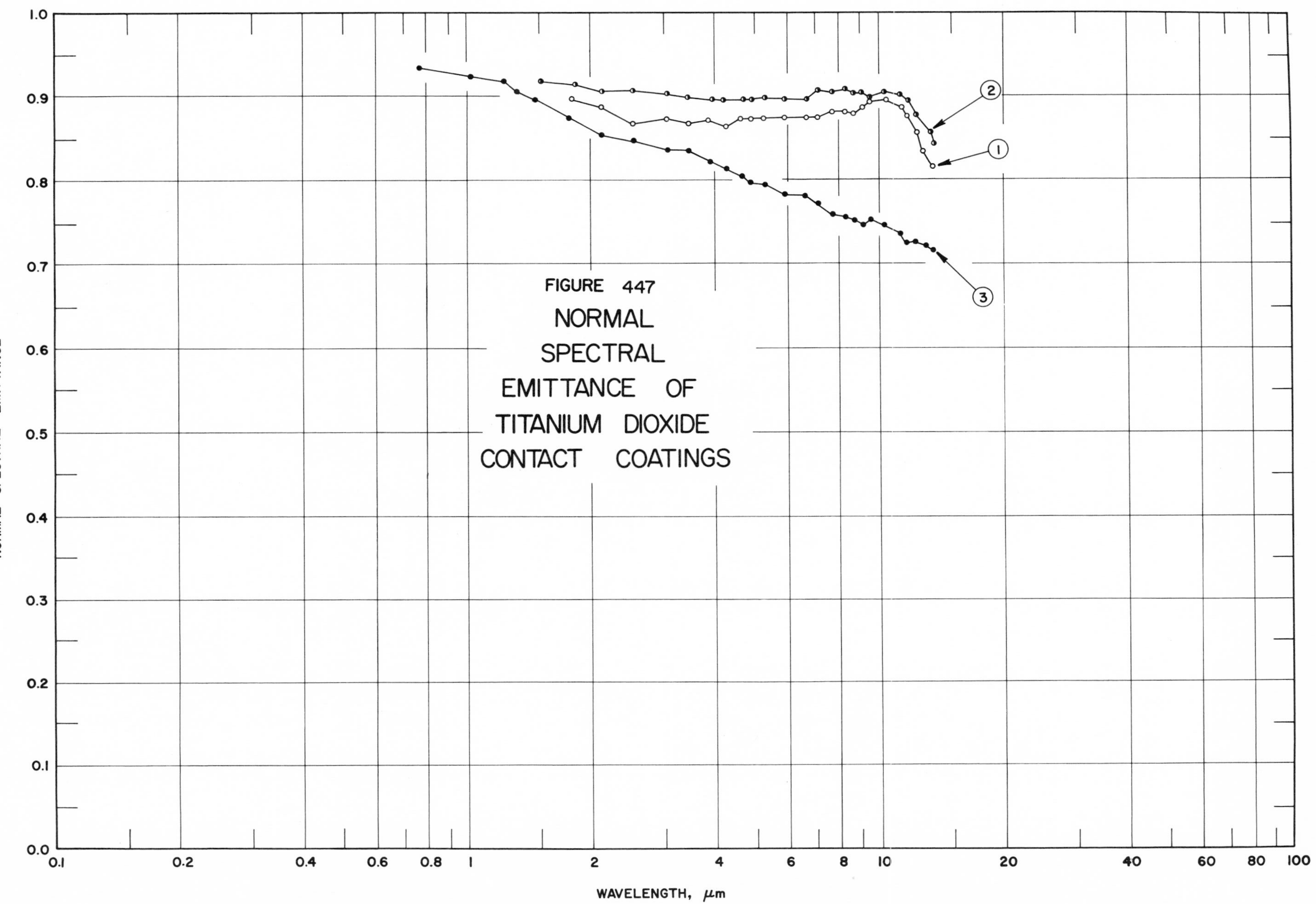

FIGURE 447 NORMAL SPECTRAL EMITTANCE OF TITANIUM DIOXIDE CONTACT COATINGS

SPECIFICATION TABLE NO. 447 NORMAL SPECTRAL EMITTANCE OF TITANIUM DIOXIDE CONTACT COATINGS

Curve No.	Ref. No.	Year	Temperature, K	Wavelength Range, μm	Geometry θ'	Reported Error, %	Composition (weight percent), Specifications, and Remarks
1	118	1962	755	1.80-13.53	$\sim 0°$		TiO_2, Metco, Inc. (XP-1114) (0.0635 mm thick); niobium substrate; plasma arc sprayed; coating blue-black in color; hard; coating-substrate bond strength fair.
2	118	1962	1061	1.51-13.56	$\sim 0°$		Above specimen and conditions.
3	118	1962	1340	0.77-13.53	$\sim 0°$		Curve 1 specimen and conditions.

DATA TABLE NO. 447 NORMAL SPECTRAL EMITTANCE OF TITANIUM DIOXIDE CONTACT COATINGS

[Wavelength, λ, μm; Emittance, ϵ; Temperature, T, K]

λ	ϵ
CURVE 1	**T = 755**
1.80	0.897
2.11	0.887
2.53	0.868
3.08	0.871
3.47	0.869
3.88	0.871
4.28	0.861
4.60	0.871
4.93	0.872
5.22	0.873
5.96	0.875
6.65	0.875
7.19	0.875
7.77	0.881
8.27	0.881
8.70	0.880
9.14	0.887
9.55	0.893
10.34	0.895
11.09	0.888
11.77	0.877
12.42	0.857
12.98	0.834
13.53	0.817
CURVE 2	**T = 1061**
1.51	0.917
1.81	0.913
2.11	0.905
2.52	0.906
3.04	0.901
3.46	0.899
3.91	0.898
4.20	0.896
4.66	0.896
4.96	0.897
5.27	0.898
5.96	0.897
6.61	0.897
7.19	0.907
7.77	0.902
8.25	0.909
8.68	0.901
9.12	0.902
9.59	0.899
10.35	0.903
11.04	0.900
11.74	0.893
12.41	0.877
13.03	0.858
13.56	0.843
CURVE 3	**T = 1340**
0.77	0.934
1.04	0.922
1.24	0.919
1.32	0.905
1.49	0.896
1.77	0.875
2.11	0.853
2.54	0.847
3.07	0.839
3.45	0.836
3.90	0.822
4.23	0.814
4.61	0.806
4.92	0.798
5.25	0.796
5.98	0.784
6.60	0.783
7.19	0.773
7.76	0.760
8.26	0.757
8.66	0.751
9.12	0.747
9.59	0.752
10.34	0.747
11.08	0.739
11.74	0.725
12.40	0.727
13.00	0.722
13.53	0.717

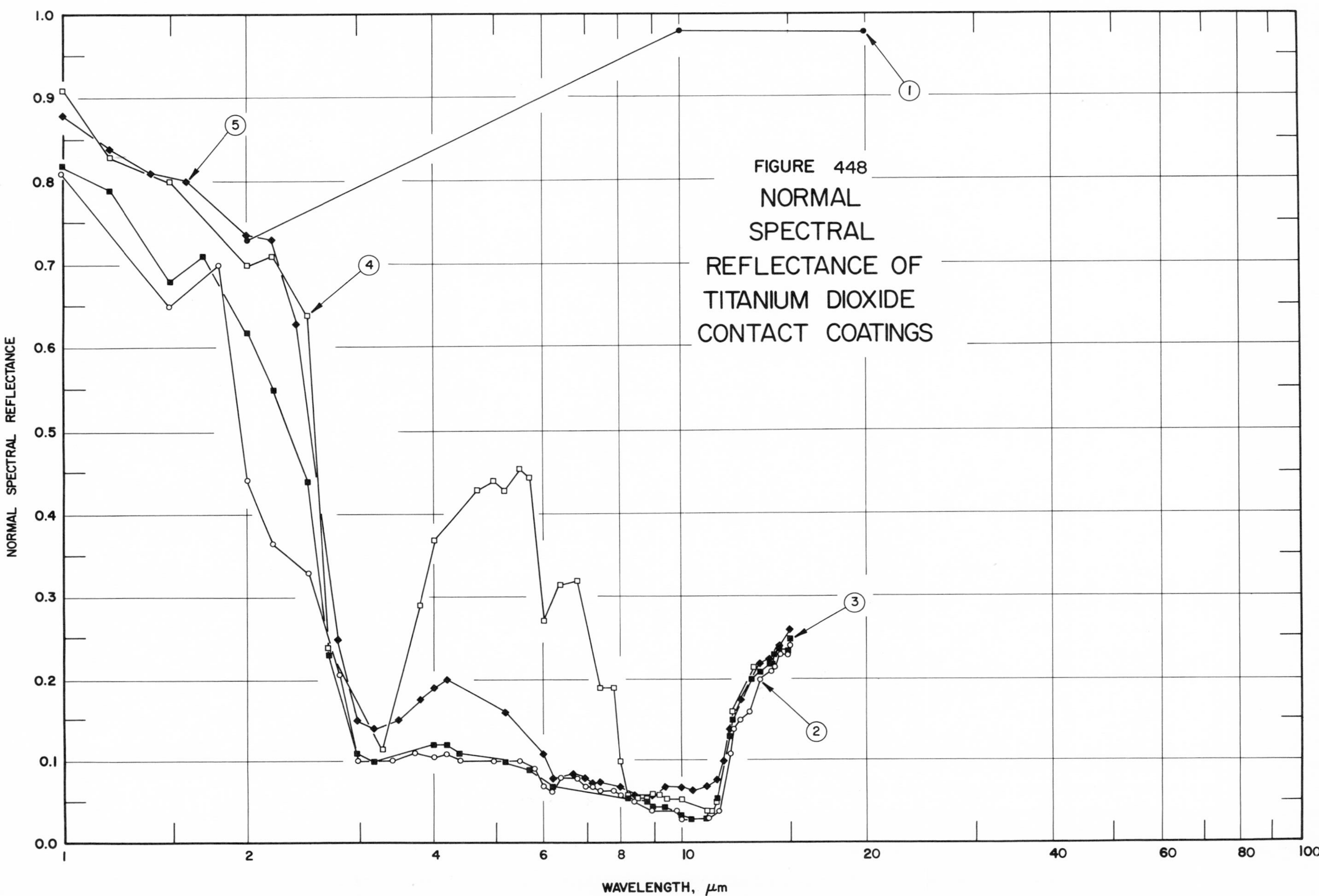

FIGURE 448 NORMAL SPECTRAL REFLECTANCE OF TITANIUM DIOXIDE CONTACT COATINGS

SPECIFICATION TABLE NO. 448 NORMAL SPECTRAL REFLECTANCE OF TITANIUM DIOXIDE CONTACT COATINGS

Curve No.	Ref. No.	Year	Temperature, K	Wavelength Range, μm	Geometry θ	θ'	ω'	Reported Error, %	Composition (weight percent), Specifications, and Remarks
1	109	1962	298	2.00-20.0	$\sim 0°$		2π		Titanium dioxide (0.08 μm thick); aluminum and scotchcal substrates; vacuum evaporated; converted from R(2π, 0°).
2	104	1953	298	1.0-15.0	5°		2π	±2	TiO_2 (0.0762 mm thick); aluminum substrate; substrate polished; data extracted from smooth curve; converted from R(2π, 5°).
3	104	1953	298	1.0-15.0	5°		2π	±2	Similar to above specimen and conditions except 0.102 mm thick.
4	104	1953	298	1.0-15.0	5°		2π	±2	Similar to above specimen and conditions except 0.203 mm thick.
5	104	1953	298	1.0-15.0	5°		2π	±2	Similar to above specimen and conditions except 1.02 mm thick.
6*	104	1953	298	1.0-15.0	5°		2π	±2	TiO_2 (0.178 mm thick); flat black paint and aluminum substrates; data extracted from smooth curve; converted from R(2π, 5°).
7*	104	1953	298	1.0-15.0	5°		2π	±2	Similar to above specimen and conditions except 0.813 mm thick.

* Not shown on plot

DATA TABLE NO. 448 NORMAL SPECTRAL REFLECTANCE OF TITANIUM DIOXIDE CONTACT COATINGS

[Wavelength, λ, μm; Reflectance, ρ; Temperature, T, K]

CURVE 1
T = 298

λ	ρ
2.00	0.730
10.0	0.980
20.0	0.980

CURVE 2
T = 298

λ	ρ
1.0	0.810
1.5	0.650
1.8	0.700
2.0	0.440
2.2	0.365
2.5	0.330
2.8	0.205
3.0	0.100
3.2	0.100
3.4	0.100
3.7	0.110
4.0	0.105
4.2	0.110
4.4	0.100
5.0	0.100
5.5	0.100
5.8	0.090
6.0	0.070
6.2	0.065
6.4	0.080
6.8	0.080
7.0	0.070
7.2	0.070
7.4	0.065
7.8	0.065
8.0	0.060
8.4	0.050
9.0	0.040
9.8	0.040
10.0	0.030
11.0	0.030
11.5	0.040
12.0	0.110
12.2	0.140
12.4	0.150
12.8	0.160
13.0	0.160
13.2	0.200
13.4	0.200
14.0	0.210
14.2	0.215
14.4	0.230
14.8	0.230
15.0	0.240

CURVE 3
T = 298

λ	ρ
1.0	0.820
1.2	0.790
1.5	0.680
1.7	0.710
2.0	0.620
2.2	0.550
2.5	0.440
2.7	0.230
3.0	0.110
3.2	0.100
3.4	0.100*
3.7	0.110*
4.0	0.120
4.2	0.120
4.4	0.110
5.0	0.100*
5.2	0.100
5.7	0.090
6.0	0.070*
6.2	0.070
6.4	0.080*
6.8	0.080*
7.0	0.070*
8.0	0.060*
8.2	0.055
8.8	0.050
9.0	0.045
9.4	0.045
10.0	0.035
10.4	0.030
11.0	0.030
11.4	0.055
12.0	0.130
12.2	0.150
13.0	0.200
13.4	0.210
13.8	0.220
14.0	0.220
14.2	0.230
14.4	0.235
14.8	0.235
15.0	0.250

CURVE 4
T = 298

λ	ρ
1.0	0.910
1.2	0.830
1.5	0.800
2.0	0.700
2.2	0.710
2.5	0.640
2.7	0.240
3.0	0.110*
3.3	0.115
3.8	0.290
4.0	0.370
4.7	0.430
5.0	0.440
5.2	0.430
5.5	0.455
5.7	0.445
6.0	0.270
6.4	0.315
6.8	0.320
7.4	0.190
7.8	0.190
8.0	0.100
8.2	0.060
8.5	0.055
8.8	0.055
9.0	0.060
9.2	0.060
9.5	0.055
10.0	0.055
11.0	0.040
11.2	0.040
11.4	0.050
12.0	0.130*
12.2	0.160
13.0	0.200*
13.2	0.215
14.0	0.225
14.2	0.230*
14.8	0.240*
15.0	0.250*

CURVE 5
T = 298

λ	ρ
1.0	0.880
1.2	0.840
1.4	0.810
1.6	0.800
2.0	0.735
2.2	0.730
2.4	0.630
2.8	0.250
3.0	0.150
3.2	0.140
3.5	0.150
3.8	0.175
4.0	0.190
4.2	0.200
5.2	0.160
6.0	0.110
6.2	0.080
6.7	0.085
7.0	0.080
7.2	0.075
7.4	0.075
8.0	0.070
8.4	0.060
9.0	0.060
9.4	0.070
10.0	0.070
10.4	0.065
10.8	0.070
11.0	0.070
11.4	0.080
11.6	0.100
12.0	0.140
12.5	0.175
13.0	0.200*
13.4	0.220
14.0	0.225
14.4	0.240
15.0	0.260

CURVE 6*
T = 298

λ	ρ
1.0	0.810
1.5	0.650
1.8	0.700
2.0	0.440
2.2	0.365
2.5	0.330
2.8	0.205
3.0	0.100
3.2	0.100
3.4	0.100
3.7	0.110
4.0	0.105
4.2	0.110
4.4	0.100
5.0	0.100
5.5	0.100
5.8	0.090
6.0	0.070
6.2	0.065
6.4	0.080
6.8	0.080
7.0	0.070
7.2	0.070
7.4	0.065
7.8	0.065
8.0	0.060
8.4	0.050
9.0	0.040
9.8	0.040
10.0	0.030
11.0	0.030
11.5	0.040
12.0	0.110
12.2	0.140
12.4	0.150
12.8	0.160
13.0	0.160
13.2	0.200
13.4	0.200
14.0	0.210
14.2	0.215
14.4	0.230
14.8	0.230
15.0	0.240

CURVE 7*
T = 298

λ	ρ
1.0	0.820
1.2	0.790
1.5	0.680
1.7	0.710
2.0	0.620
2.2	0.550
2.5	0.440
2.7	0.230
3.0	0.110
3.2	0.100
3.4	0.100
3.7	0.110
4.0	0.120
4.2	0.120
4.4	0.110
5.0	0.100
5.2	0.100
5.7	0.090
6.0	0.070
6.2	0.070
6.4	0.080
6.8	0.080
7.0	0.070
8.0	0.060
8.2	0.055
8.8	0.050
9.0	0.045
9.4	0.045
10.0	0.035
10.4	0.030
11.0	0.030
11.4	0.055
12.0	0.130
12.2	0.150
13.0	0.200
13.4	0.210
13.8	0.220
14.0	0.220
14.2	0.230
14.4	0.235
14.8	0.235
15.0	0.250

* Not shown on plot

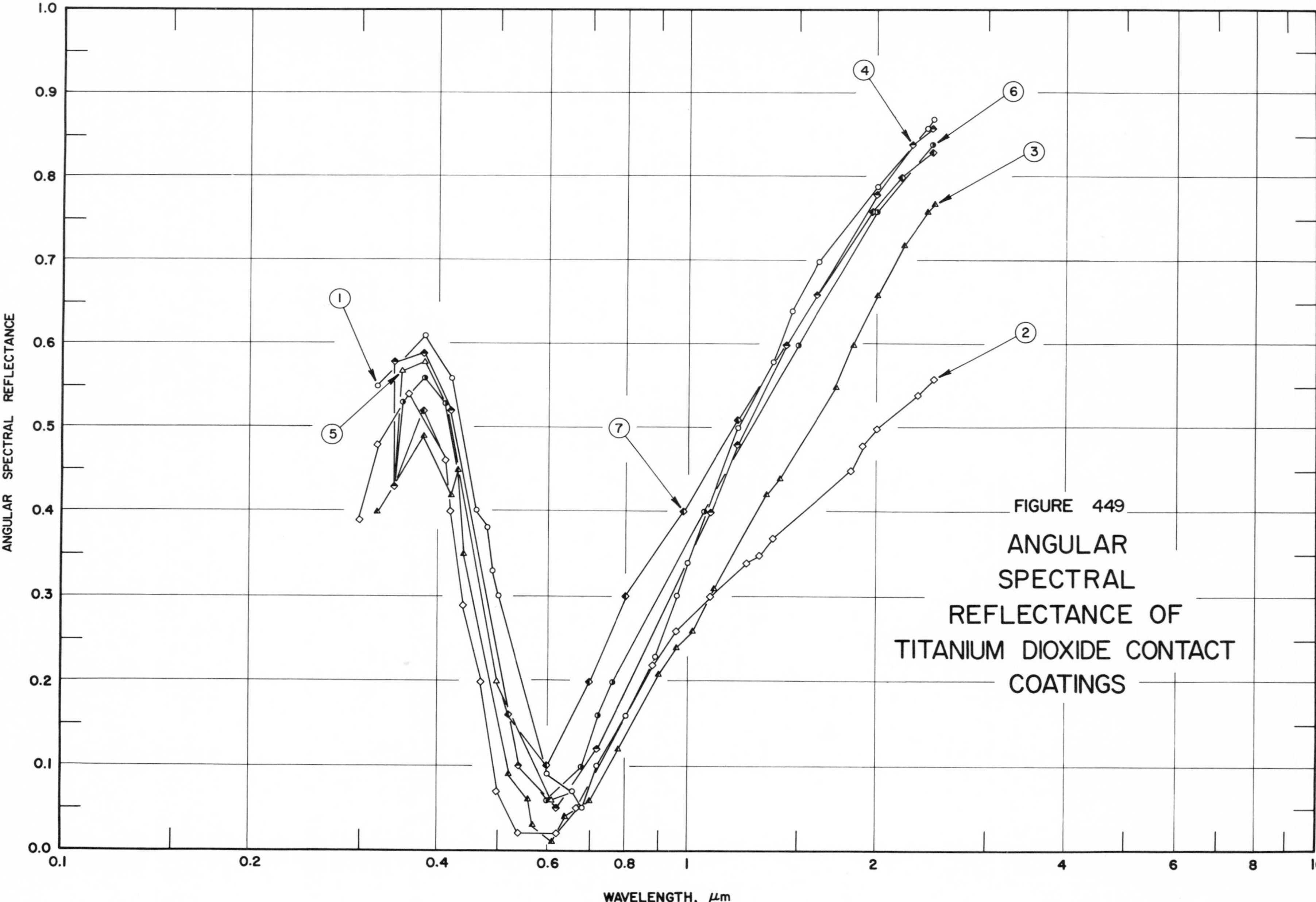
FIGURE 449
ANGULAR
SPECTRAL
REFLECTANCE OF
TITANIUM DIOXIDE CONTACT
COATINGS
ANGULAR SPECTRAL REFLECTANCE
WAVELENGTH, μm
1.0
0.9
0.8
0.7
0.6
0.5
0.4
0.3
0.2
0.1
0.0
0.1
0.2
0.4
0.6
0.8
1
2
4
6
8
10
1
2
3
4
5
6
7

SPECIFICATION TABLE NO. 449 ANGULAR SPECTRAL REFLECTANCE OF TITANIUM DIOXIDE CONTACT COATINGS

Curve No.	Ref. No.	Year	Temperature, K	Wavelength Range, μm	Geometry θ	Geometry θ'	Geometry ω'	Reported Error, %	Composition (weight percent), Specifications, and Remarks
1	115	1960	298	0.32-2.45	15°		2π		Titanium dioxide (0.8 to 1.0 μm thick) on aluminum foil substrate (Reynolds Wrap); vacuum deposited; oxidized 3 hrs at 673 K; color of surface, blue. [Authors' designation: Sample 10]
2	115	1960	298	0.30-2.46	15°		2π		Similar to above specimen and conditions.
3	115	1960	298	0.32-2.46	15°		2π		Similar to above specimen and conditions except exposed to 700 MeV proton irradiation, 1.9 x 10^{13} protons cm^{-2}. [Authors' designation: Sample 12]
4	115	1960	298	0.34-2.45	15°		2π		Titanium dioxide (0.8 to 1.0 μm thick); aluminum foil substrate (Reynolds Wrap); vacuum deposited; oxidized 3 hrs at 673 K; color of surface, blue.
5	115	1960	298	0.34-2.45	30°		2π		Above specimen and conditions.
6	115	1960	298	0.34-2.45	45°		2π		Above specimen and conditions.
7	115	1960	298	0.34-2.45	60°		2π		Above specimen and conditions.

DATA TABLE NO. 449 ANGULAR SPECTRAL REFLECTANCE OF TITANIUM DIOXIDE CONTACT COATINGS

[Wavelength, λ, μm; Reflectance, ρ; Temperature, T, K]

λ	ρ
CURVE 1	**T = 298**
0.32	0.55
0.38	0.61
0.42	0.56
0.46	0.40
0.48	0.38
0.49	0.33
0.50	0.30
0.60	0.09
0.66	0.07
0.68	0.05
0.72	0.10
0.80	0.16
0.89	0.23
0.96	0.30
1.00	0.34
1.20	0.50
1.36	0.58
1.46	0.64
1.61	0.70
2.00	0.79
CURVE 1 (cont.)	
2.40	0.86
2.45	0.87
CURVE 2	**T = 298**
0.30	0.39
0.32	0.48
0.36	0.54
0.41	0.46
0.42	0.40
0.44	0.29
0.47	0.20
0.50	0.07
0.54	0.02
0.62	0.02
0.67	0.05
0.88	0.22
0.96	0.26
1.08	0.30
1.24	0.34
CURVE 2 (cont.)	
1.30	0.35
1.36	0.37
1.82	0.45
1.90	0.48
2.00	0.50
2.32	0.54
2.46	0.56
CURVE 3	**T = 298**
0.32	0.40
0.38	0.49
0.42	0.42
0.43	0.45
0.44	0.35
0.52	0.09
0.56	0.06
0.57	0.03
0.61	0.01
0.64	0.04
CURVE 3 (cont.)	
0.70	0.06
0.78	0.12
0.90	0.21
0.96	0.24
1.02	0.26
1.10	0.31
1.34	0.42
1.40	0.44
1.72	0.55
1.84	0.60
2.00	0.66
2.20	0.72
2.40	0.76
2.46	0.77
CURVE 4	**T = 298**
0.34	0.43
0.34	0.58
0.38	0.59
CURVE 4 (cont.)	
0.42	0.52
0.54	0.10
0.62	0.05
0.72	0.12
1.08	0.40
1.20	0.48
1.44	0.60
1.60	0.66
2.00	0.78
2.28	0.84
2.45	0.86
CURVE 5	**T = 298**
0.34	0.43*
0.35	0.57
0.38	0.58
0.42	0.52*
0.50	0.20
0.61	0.06
CURVE 5 (cont.)	
0.66	0.07*
1.08	0.40*
1.20	0.48*
1.44	0.60*
1.60	0.66*
2.00	0.78*
2.28	0.84*
2.45	0.86*
CURVE 6	**T = 298**
0.34	0.43*
0.35	0.53
0.38	0.56
0.41	0.53
0.50	0.20*
0.60	0.06
0.68	0.10
0.72	0.16
0.76	0.20
CURVE 6 (cont.)	
1.06	0.40
1.50	0.60
2.00	0.76
2.45	0.84
CURVE 7	**T = 298**
0.34	0.43*
0.38	0.52
0.42	0.46*
0.52	0.16
0.60	0.10
0.70	0.20
0.80	0.30
0.98	0.40
1.20	0.51
1.44	0.60*
1.60	0.66*
1.96	0.76
2.18	0.80
2.45	0.83

* Not shown on plot

SPECIFICATION TABLE NO. 450 ANGULAR SPECTRAL REFLECTANCE OF TITANIUM DIOXIDE CONTACT COATINGS

Curve No.	Ref. No.	Year	Temperature K	Wavelength, μm	Angular Range, °	Geometry θ	Geometry θ'	Geometry ω'	Reported Error, %	Composition (weight percent), Specifications and Remarks
1*	115	1960	298	0.37	15-60	θ		2π		Titanium dioxide (0.8 to 1.0 μm thick) on aluminum foil (Reynolds wrap) substrate; vacuum deposited; oxidized 3 hrs at 673 K; color of surface, blue.
2*	115	1960	298	0.60	15-60	θ		2π		Above specimen and conditions.
3*	115	1960	298	0.70	15-60	θ		2π		Above specimen and conditions.
4*	115	1960	298	1.00	15-60	θ		2π		Above specimen and conditions.
5*	115	1960	298	1.54	15-60	θ		2π		Above specimen and conditions.

DATA TABLE NO. 450 ANGULAR SPECTRAL REFLECTANCE OF TITANIUM DIOXIDE CONTACT COATINGS

[Angle, *,°; Reflectance, ρ; Temperature, T, K; Wavelength, λ, μm]

θ	ρ
CURVE 1*	
T = 298	
λ = 0.37	
15	0.59
20	0.59
30	0.58
40	0.56
50	0.54
60	0.52
CURVE 2*	
T = 298	
λ = 0.60	
15	0.06
30	0.05
40	0.05
50	0.06
60	0.10
CURVE 3*	
T = 298	
λ = 0.70	
15	0.10
30	0.10
40	0.12
50	0.15
60	0.20
CURVE 4*	
T = 298	
λ = 1.00	
15	0.34
30	0.34
40	0.35
50	0.36
60	0.40
CURVE 5*	
T = 298	
λ = 1.54	
15	0.63
30	0.63
40	0.64
50	0.63
60	0.64

* No plot given

SPECIFICATION TABLE NO. 451 ANGULAR SOLAR ABSORPTANCE OF TITANIUM DIOXIDE CONTACT COATINGS

Curve No.	Ref. No.	Year	Temperature, K	Angular Range, °	Geometry θ	Reported Error, %	Composition (weight percent), Specifications, and Remarks
1*	115	1960	298	15-60	θ		Titanium dioxide (0.8 to 1.0 μm thick); aluminum foil (Reynolds wrap) substrate; vacuum deposited; oxidized 3 hrs at 673 K; color of surface, blue; computed from spectral reflectance data for above atm conditions.

DATA TABLE NO. 451 ANGULAR SOLAR ABSORPTANCE OF TITANIUM DIOXIDE CONTACT COATINGS

[Angle, θ, °; Absorptance, α; Temperature, T, K]

θ	α
CURVE 1*	
T = 298	
15	0.665
30	0.667
45	0.666
60	0.633

* No plot given

SPECIFICATION TABLE NO. 452 ANGULAR SOLAR TRANSMITTANCE OF TITANIUM DIOXIDE CONTACT COATINGS

Curve No.	Ref. No.	Year	Temperature Range, K	Geometry θ	Reported Error, %	Composition (weight percent), Specifications and Remarks
1*	115	1960	298	15°		Titanium dioxide (0.8 to 1.0 μm thick) on aluminum foil (Reynolds wrap) substrate; vacuum deposited; oxidized 3 hrs at 673 K; color of surface, blue; computed from spectral reflectance data for above atm conditions. [Authors' designation: Sample 10]
2*	115	1960	298	15°		Similar to above specimen and conditions except exposed to vacuum (10^{-6} mm Hg) for 6 days and cleaned with isopropyl alcohol. [Authors' designation: Sample 11]
3*	115	1960	298	15°		Similar to curve 1 specimen and conditions except exposed to 700 MeV proton irradiation, 1.9 x 10^{13} p cm^{-2}. [Authors' designation: Sample 12]

DATA TABLE NO. 452 ANGULAR SOLAR TRANSMITTANCE OF TITANIUM DIOXIDE CONTACT COATINGS

[Temperature, T, K; Absorptance, α]

T	α
CURVE 1*	
298	0.66
CURVE 2*	
298	0.68
CURVE 3*	
298	0.73

* No plot given

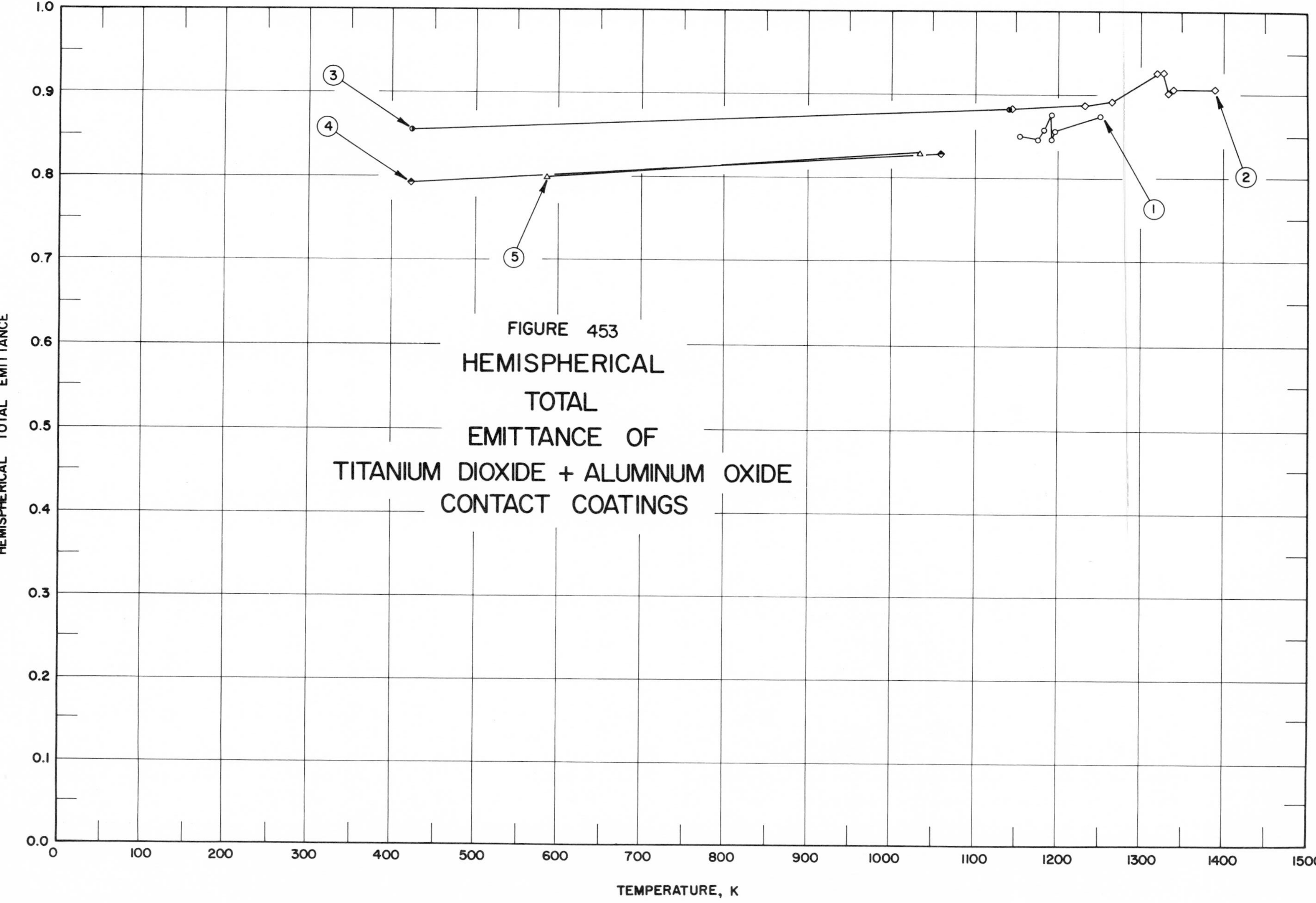

FIGURE 453 HEMISPHERICAL TOTAL EMITTANCE OF TITANIUM DIOXIDE + ALUMINUM OXIDE CONTACT COATINGS

SPECIFICATION TABLE NO. 453 HEMISPHERICAL TOTAL EMITTANCE OF TITANIUM DIOXIDE + ALUMINUM OXIDE CONTACT COATINGS

Curve No.	Ref. No.	Year	Temperature Range, K	Reported Error, %	Composition (weight percent), Specifications and Remarks
1	185	1962	1155-1251	~5	50 TiO_2 and 50 Al_2O_3, Metco XP-1121 (0.0381 mm thick); Nb-1 Zr alloy substrate; substrate grit blasted; flame sprayed; coating surface degreased with acetone; measured in vacuum (10^{-4} mm Hg).
2	185	1962	1148-1391	~5	Similar to above specimen and conditions except heat treated 312 hrs at 1478 K in helium.
3	82	1962	421,1144	<±2.5	50 TiO_2 and 50 Al_2O_3, Metco XP-1121 (0.0762 mm thick); stainless steel 310 substrate; plasma arc sprayed; measured in vacuum ($<10^{-8}$ mm Hg); data extracted from smooth curve. [Authors' designation: Sample No. 1]
4	82	1962	422,1060	<±2.5	Similar to above specimen and conditions. [Authors' designation: Sample No. 2]
5	186	1965	589-1033		50 TiO_2 and 50 Al_2O_3; substrate unknown; flame-sprayed.

DATA TABLE NO. 453 HEMISPHERICAL TOTAL EMITTANCE OF TITANIUM DIOXIDE + ALUMINUM OXIDE CONTACT COATINGS

[Temperature, T, K; Emittance, ∈]

T	∈	T	∈
CURVE 1		CURVE 3	
1155	0.850	421	0.856
1176	0.846	1144	0.882
1181	0.857		
1192	0.874	CURVE 4	
1192	0.845		
1196	0.856	422	0.792
1251	0.872	1060	0.830
CURVE 2		CURVE 5	
1148	0.881	589	0.80
1231	0.889	1033	0.83
1267	0.891		
1320	0.927		
1329	0.928		
1333	0.901		
1341	0.908		
1391	0.908		

SPECIFICATION TABLE NO. 454 HEMISPHERICAL TOTAL EMITTANCE OF TUNGSTEN CARBIDE + COBALT CONTACT COATINGS

Curve No.	Ref. No.	Year	Temperature Range, K	Reported Error, %	Composition (weight percent), Specifications and Remarks
1*	146	1961	468-1093	<10	Tungsten carbide + 12 cobalt aggregate (-270 mesh, +15 μm), Metco XP-1110; Armco iron substrate; plasma flame sprayed; measured in vacuum (10^{-5} mm Hg).

DATA TABLE NO. 454 HEMISPHERICAL TOTAL EMITTANCE OF TUNGSTEN CARBIDE + COBALT CONTACT COATINGS

[Temperature, T, K; Emittance, ϵ]

T	ϵ
CURVE 1*	
468	0.31
668	0.36
873	0.41
1093	0.45

* No plot given

SPECIFICATION TABLE NO. 455 NORMAL SOLAR ABSORPTANCE OF TUNGSTEN CARBIDE + COBALT CONTACT COATINGS

Curve No.	Ref. No.	Year	Temperature Range, K	Geometry θ	Reported Error, %	Composition (weight percent), Specifications and Remarks
1*	146	1961	468-1093	$\sim 0^\circ$		Tungsten carbide + 12 cobalt aggregate (-270 mesh, +15 μm), Metco XP-1110; Armco iron substrate; plasma flame sprayed; measured in vacuum (10^{-5} mm Hg).

DATA TABLE NO. 455 NORMAL SOLAR ABSORPTANCE OF TUNGSTEN CARBIDE + COBALT CONTACT COATINGS

[Temperature, T, K; Absorptance, α]

T	α
CURVE 1*	
468	0.62
668	0.70
873	0.79
1093	0.69

* No plot given

SPECIFICATION TABLE NO. 456 NORMAL SPECTRAL EMITTANCE OF URANIUM DIOXIDE CONTACT COATINGS

Curve No.	Ref. No.	Year	Wavelength, μm	Temperature Range, K	Geometry θ'	Reported Error, %	Composition (weight percent), Specifications, and Remarks
1*	187	1958	0.65	2860	$\sim 0^\circ$		UO_2; tungsten substrate; melted in vacuum; emittance uncertainty: ± 0.026; temp within 373 K of melting point of uranium dioxide.

DATA TABLE NO. 456 NORMAL SPECTRAL EMITTANCE OF URANIUM DIOXIDE CONTACT COATINGS

[Temperature, T, K; Emittance, ϵ; Wavelength, λ, μm]

T	ϵ
CURVE 1* $\lambda = 0.65$	
2860	0.416

* No plot given

SPECIFICATION TABLE NO. 457 NORMAL SPECTRAL EMITTANCE OF VANADIUM OXIDE CONTACT COATINGS

Curve No.	Ref. No.	Year	Temperature, K	Wavelength Range, μm	Geometry θ'	Reported Error, %	Composition (weight percent), Specifications, and Remarks
1*	188	1914	1433	0.65	$\sim 0°$	1	V_2O_3; tungsten substrate; melted in hydrogen, then oxidized in air by heating; measured in air relative to Pt; corrected for emittance of Pt.

DATA TABLE NO. 457 NORMAL SPECTRAL EMITTANCE OF VANADIUM OXIDE CONTACT COATINGS

[Wavelength, λ, μm; Emittance, ϵ; Temperature, T, K]

λ	ϵ
CURVE 1* T = 1433	
0.65	0.69

* No plot given

SPECIFICATION TABLE NO. 458 NORMAL SPECTRAL REFLECTANCE OF VANADIUM OXIDE CONTACT COATINGS

Curve No.	Ref. No.	Year	Temperature, K	Wavelength Range, μm	Geometry θ θ' ω'	Reported Error, %	Composition (weight percent), Specifications, and Remarks
1*	189	1968	~298	0.249-4.960	~0° ~0°		VO_2 (0.1 μm thick); thick sapphire substrate; vapor deposited.

DATA TABLE NO. 458 NORMAL SPECTRAL REFLECTANCE OF VANADIUM OXIDE CONTACT COATINGS

[Wavelength, λ, μm; Reflectance, ρ; Temperature, T, K]

λ	ρ	λ	ρ	λ	ρ	λ	ρ
CURVE 1*		CURVE 1 (cont.)*		CURVE 1 (cont.)*		CURVE 1 (cont.)*	
T ~ 298		0.574	0.191	0.992	0.359	3.875	0.244
0.249	0.203	0.596	0.168	1.087	0.395	4.275	0.218
0.262	0.205	0.623	0.165	1.180	0.415	4.960	0.187
0.275	0.216	0.649	0.169	1.291	0.436		
0.299	0.246	0.670	0.184	1.377	0.448		
0.325	0.281	0.692	0.202	1.476	0.455		
0.349	0.288	0.716	0.219	1.569	0.463		
0.362	0.292	0.746	0.242	1.653	0.458		
0.374	0.292	0.765	0.256	1.771	0.449		
0.400	0.296	0.794	0.270	1.937	0.437		
0.426	0.301	0.821	0.282	2.175	0.420		
0.449	0.301	0.843	0.294	2.339	0.395		
0.460	0.315	0.867	0.307	2.431	0.389		
0.473	0.324	0.885	0.319	2.695	0.353		
0.488	0.311	0.918	0.326	2.952	0.326		
0.510	0.285	0.932	0.330	3.179	0.302		
0.521	0.257	0.953	0.340	3.351	0.288		
0.548	0.223	0.961	0.351	3.647	0.267		

* No plot given

SPECIFICATION TABLE NO. 459 NORMAL SPECTRAL TRANSMITTANCE OF VANADIUM OXIDE CONTACT COATINGS

Curve No.	Ref. No.	Year	Temperature, K	Wavelength Range, μm	Geometry θ θ' ω'	Reported Error, %	Composition (weight percent), Specifications, and Remarks
1*	189	1968	~298	0.250-4.960	~0° ~0°		VO_2 (0.1 μm thick); thick sapphire substrate; vapor deposited.

DATA TABLE NO. 459 NORMAL SPECTRAL TRANSMITTANCE OF VANADIUM OXIDE CONTACT COATINGS

[Wavelength, λ, μm; Transmittance, τ; Temperature, T, K]

λ	τ	λ	τ	λ	τ	λ	τ
CURVE 1*		CURVE 1 (cont.)*		CURVE 1 (cont.)*		CURVE 1 (cont.)*	
T ~ 298		0.525	0.238	0.892	0.372	3.444	0.756
0.250	0.006	0.536	0.268	0.953	0.370	3.542	0.780
0.263	0.006	0.543	0.287	1.033	0.370	3.757	0.800
0.274	0.006	0.553	0.308	1.097	0.370	4.000	0.821
0.302	0.003	0.563	0.330	1.180	0.377	4.275	0.841
0.327	0.009	0.571	0.345	1.291	0.387	4.428	0.864
0.348	0.009	0.576	0.360	1.393	0.407	4.960	0.883
0.375	0.012	0.582	0.377	1.493	0.430		
0.402	0.018	0.590	0.395	1.589	0.448		
0.418	0.026	0.601	0.409	1.653	0.475		
0.430	0.037	0.613	0.423	1.722	0.493		
0.439	0.049	0.629	0.436	1.823	0.513		
0.449	0.068	0.663	0.436	1.937	0.541		
0.457	0.081	0.688	0.434	2.137	0.572		
0.467	0.101	0.712	0.425	2.296	0.602		
0.476	0.122	0.738	0.416	2.530	0.633		
0.488	0.147	0.765	0.404	2.755	0.677		
0.497	0.172	0.800	0.392	3.024	0.713		
0.512	0.201	0.849	0.380	3.179	0.729		

* No plot given

SPECIFICATION TABLE NO. 460 NORMAL SPECTRAL EMITTANCE OF YTTRIUM OXIDE CONTACT COATINGS

Curve No.	Ref. No.	Year	Temperature, K	Wavelength Range, μm	Geometry θ'	Reported Error, %	Composition (weight percent), Specifications, and Remarks
1*	188	1914	1673	0.65	$\sim 0°$	1	Y_2O_3; tungsten substrate; melted in hydrogen, then oxidized in air by heating; measured in air relative to Pt; corrected for emittance of Pt.

DATA TABLE NO. 460 NORMAL SPECTRAL EMITTANCE OF YTTRIUM OXIDE CONTACT COATINGS

[Wavelength, λ, μm; Emittance, ϵ; Temperature, T, K]

λ	ϵ
CURVE 1*	
T = 1673	
0.65	0.61

* No plot given

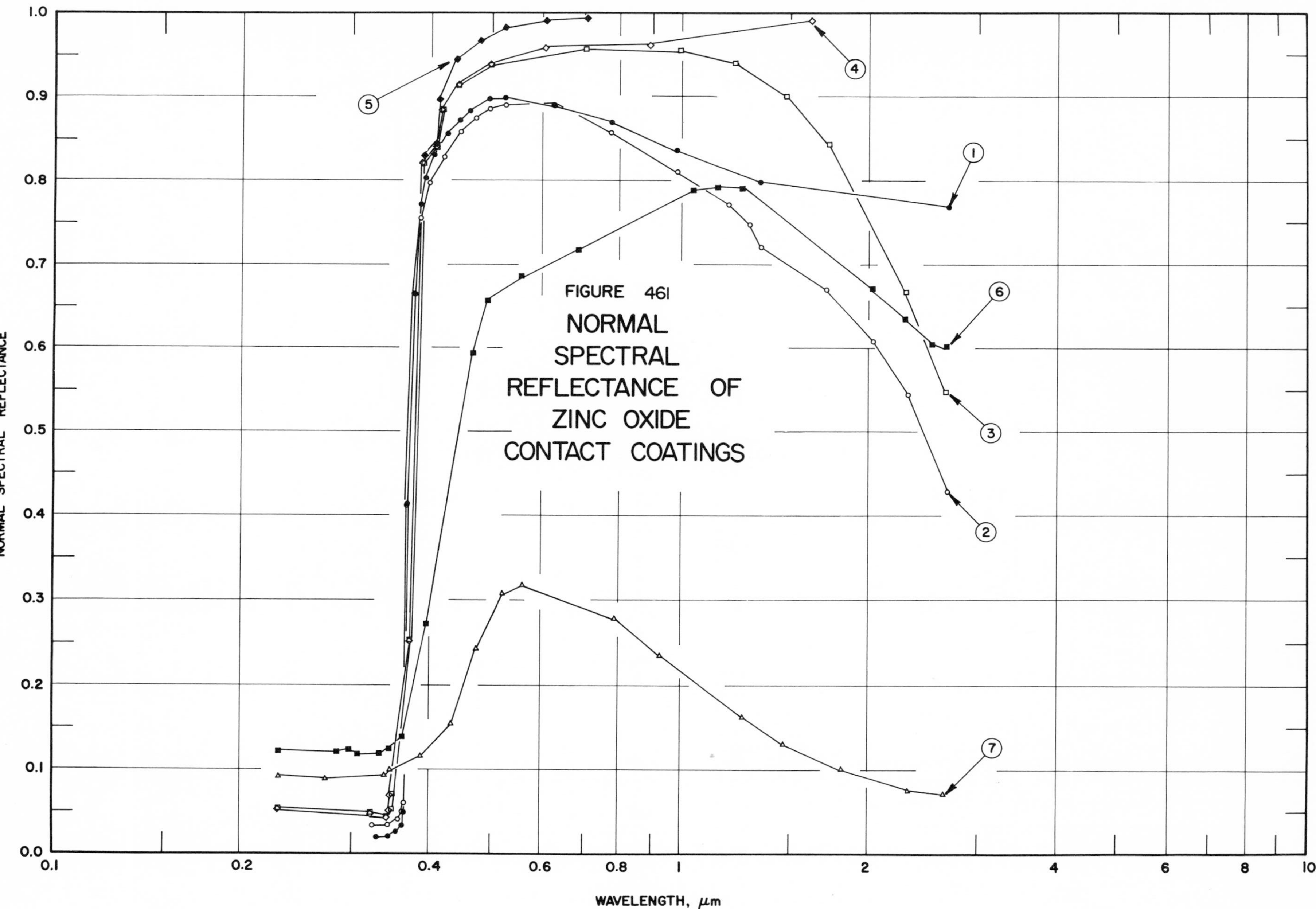

FIGURE 461
NORMAL SPECTRAL REFLECTANCE OF ZINC OXIDE CONTACT COATINGS

SPECIFICATION TABLE NO. 461 NORMAL SPECTRAL REFLECTANCE OF ZINC OXIDE CONTACT COATINGS

Curve No.	Ref. No.	Year	Temperature, K	Wavelength Range, μm	Geometry θ	θ'	ω'	Reported Error, %	Composition (weight percent), Specifications, and Remarks
1	63	1969	~298	0.330-2.670	~0°		2π		SP-500 ZnO; substrate unknown; water-sprayed; exposed to vacuum; reflectance measured in situ.
2	63	1969	~298	0.325-2.697	~0°		2π		Above specimen and conditions except exposed to UV radiation (from a G.E. AH-6 lamp) in vacuum; ESH 1000.
3	170	1966	~298	0.230-2.650	~0°		2π		ZnO; titanium substrate; compacted at 18 900 psi; data extracted from smooth curve; measured relative to MgO.
4	170	1966	~298	0.230-2.650	~0°		2π		Similar to above specimen and conditions except sintered 1 hr at 373 K and 1 hr at 973 K.
5	170	1966	~298	0.230-2.650	~0°		2π		Similar to above specimen and conditions except sintered 1 hr at 1273 K.
6	170	1966	~298	0.230-2.650	~0°		2π		Similar to above specimen and conditions except sintered 1 hr at 1723 K.
7	170	1966	~298	0.230-2.650	~0°		2π		Similar to above specimen and conditions except sintered 1 hr at 1823 K.

DATA TABLE NO. 461 NORMAL SPECTRAL REFLECTANCE OF ZINC OXIDE CONTACT COATINGS

[Wavelength, λ, μm; Reflectance, ρ; Temperature, T, K]

CURVE 1 (T ~ 298)

λ	ρ
0.330	0.019
0.345	0.019
0.356	0.025
0.362	0.033
0.367	0.048
0.369	0.414
0.378	0.665
0.385	0.771
0.392	0.802
0.406	0.830
0.425	0.857
0.442	0.873
0.460	0.883
0.495	0.897
0.525	0.899
0.630	0.890
0.776	0.868
0.985	0.835
1.340	0.798
2.670	0.769

CURVE 2 (T ~ 298)

λ	ρ
0.325	0.033
0.343	0.033
0.357	0.040
0.362	0.048
0.367	0.060
0.369	0.414
0.378	0.665
0.384	0.753
0.399	0.797
0.420	0.829
0.445	0.859
0.471	0.875
0.497	0.885
0.525	0.890
0.630	0.890
0.776	0.857
0.985	0.810
1.192	0.772
1.284	0.747
1.340	0.720
1.715	0.671
2.039	0.609
2.322	0.544
2.697	0.429

CURVE 3 (T ~ 298)

λ	ρ
0.230	0.054
0.324	0.047
0.342	0.044
0.347	0.052
0.349	0.070
0.373	0.254
0.389	0.822
0.408	0.840
0.418	0.885
0.442	0.916
0.499	0.937
0.605	0.957
0.995	0.956
1.220	0.941
1.478	0.901
1.732	0.844
2.282	0.669
2.650	0.548

CURVE 4 (T ~ 298)

λ	ρ
0.230	0.054
0.324	0.047
0.342	0.044
0.347	0.052
0.349	0.070
0.373	0.254
0.389	0.822
0.408	0.840
0.418	0.885
0.442	0.916
0.499	0.937
0.605	0.957
0.890	0.961
1.607	0.994
1.686	1.002
1.806	1.017
1.932	1.008
2.184	1.027
2.290	1.024
2.421	1.010
2.528	1.005
2.650	1.005

CURVE 5 (T ~ 298)

λ	ρ
0.230	0.054
0.324	0.047
0.342	0.044
0.347	0.052
0.349	0.070
0.373	0.254
0.388	0.828
0.407	0.845
0.413	0.897
0.438	0.945
0.477	0.968
0.525	0.983
0.608	0.992
0.709	0.996
1.085	1.038
1.216	1.045
1.337	1.065
1.426	1.065
1.583	1.071
1.846	1.104
1.895	1.101
2.086	1.111
2.166	1.111
2.385	1.113
2.510	1.110
2.650	1.109

CURVE 6 (T ~ 298)

λ	ρ
0.230	0.121
0.286	0.121
0.298	0.123
0.308	0.118
0.333	0.119
0.346	0.124
0.363	0.138
0.395	0.272
0.469	0.596
0.492	0.657
0.559	0.685
0.688	0.717
1.043	0.789
1.146	0.794
1.252	0.791
2.023	0.673
2.282	0.634
2.531	0.605
2.650	0.603

CURVE 7 (T ~ 298)

λ	ρ
0.230	0.091
0.274	0.088
0.338	0.092
0.345	0.098
0.389	0.116
0.434	0.154
0.476	0.244
0.522	0.308
0.564	0.317
0.788	0.277
0.928	0.234
1.258	0.163
1.483	0.130
1.814	0.100
2.305	0.076
2.650	0.071

SPECIFICATION TABLE NO. 462 ANGULAR SPECTRAL REFLECTANCE OF ZINC SELENIDE CONTACT COATINGS

Curve No.	Ref. No.	Year	Temperature, K	Wavelength Range, μm	Geometry θ	θ'	ω'	Reported Error, %	Composition (weight percent), Specifications, and Remarks
1*	190	1968	~298	40.0-54.1	<30°	<30°			ZnSe (1.7 μm thick); quartz substrate; substrate crystalline and cut parallel to the z axis; high purity grade ZnSe vapor deposited on substrate at 548 K.

DATA TABLE NO. 462 ANGULAR SPECTRAL REFLECTANCE OF ZINC SELENIDE CONTACT COATINGS

[Wavelength, λ, μm; Reflectance, ρ; Temperature, T, K]

λ	ρ
CURVE 1* T ~ 298	
40.0	0.433
40.9	0.684
41.6	0.765
42.7	0.834
43.5	0.857
44.5	0.883
45.4	0.891
46.6	0.877
47.8	0.830
48.7	0.758
49.7	0.645
50.2	0.594
50.7	0.563
51.3	0.529
52.4	0.482
54.1	0.451

* No plot given

SPECIFICATION TABLE NO. 463 NORMAL SPECTRAL TRANSMITTANCE OF ZINC SELENIDE CONTACT COATINGS

Curve No.	Ref. No.	Year	Temperature, K	Wavelength Range, μm	Geometry θ	θ'	ω'	Reported Error, %	Composition (weight percent), Specifications, and Remarks
1*	190	1968	~298	42.8-51.6	~0°	~0°			ZnSe (1.7 μm thick); quartz substrate; substrate crystalline and cut parallel to the z axis; high purity grade ZnSe vapor deposited on substrate at 548 K.
2*	190	1968	295	39.0-50.4	45°	45°			Similar to above specimen and conditions.

DATA TABLE NO. 463 NORMAL SPECTRAL TRANSMITTANCE OF ZINC SELENIDE CONTACT COATINGS

[Wavelength, λ, μm; Transmittance, τ; Temperature, T, K]

λ	τ	λ	τ	λ	τ	λ	τ	λ	τ
CURVE 1*		CURVE 1 (cont.)*		CURVE 2 (cont.)*		CURVE 2 (cont.)*		CURVE 2 (cont.)*	
T ~ 298		49.2	0.072	40.4	0.670	45.3	0.190	50.0	0.161
42.8	0.581	49.4	0.087	40.7	0.675	45.5	0.163	50.4	0.195
43.5	0.459	49.7	0.119	40.9	0.675	45.9	0.116		
44.5	0.306	50.0	0.151	41.2	0.666	46.0	0.100		
44.7	0.277	50.5	0.202	41.5	0.647	46.3	0.076		
45.3	0.186	50.7	0.221	41.6	0.632	46.5	0.063		
45.6	0.157	51.1	0.258	42.0	0.601	46.7	0.047		
45.9	0.128	51.6	0.302	42.1	0.590	46.9	0.031		
46.1	0.105			42.3	0.572	47.1	0.023		
46.3	0.089	CURVE 2*		42.5	0.553	47.4	0.019		
46.5	0.074	T = 295		42.7	0.531	47.6	0.011		
46.8	0.054			43.0	0.492	47.9	0.013		
47.0	0.042	39.0	0.891	43.2	0.461	48.0	0.017		
47.2	0.030	39.1	0.793	43.6	0.427	48.2	0.022		
47.4	0.024	39.1	0.773	43.9	0.370	48.6	0.035		
47.7	0.012	39.3	0.764	44.2	0.341	48.8	0.046		
48,1	0.015	39.7	0.735	44.4	0.301	49.1	0.065		
48.3	0.022	39.8	0.715	44.7	0.278	49.4	0.087		
48.6	0.034	40.0	0.697	45.0	0.235	49.7	0.113		
48.8	0.045	40.2	0.686	45.1	0.214	50.0	0.142		

* No plot given

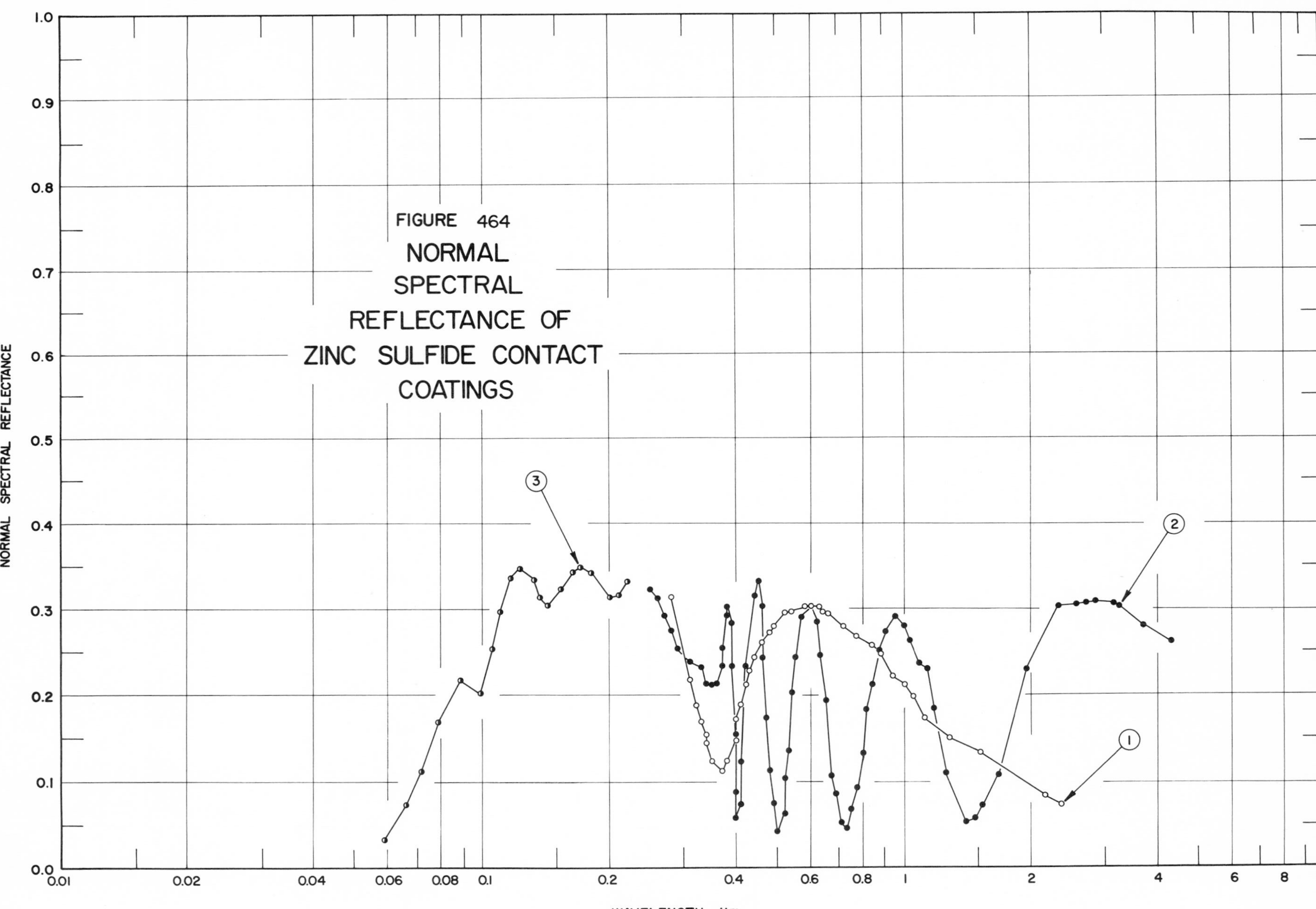

FIGURE 464 NORMAL SPECTRAL REFLECTANCE OF ZINC SULFIDE CONTACT COATINGS

SPECIFICATION TABLE NO. 464 NORMAL SPECTRAL REFLECTANCE OF ZINC SULFIDE CONTACT COATINGS

Curve No.	Ref. No.	Year	Temperature, K	Wavelength Range, μm	Geometry θ	θ'	ω'	Reported Error, %	Composition (weight percent), Specifications, and Remarks
1	191	1968	~298	0.28-2.38	~0°	~0°			Zinc sulfide (0.065 μm thick); glass substrate; vacuum evaporated.
2	191	1968	~298	0.25-4.34	~0°	~0°			Similar to the above specimen and conditions except 0.326 μm thick.
3	192	1959	298	0.0588-0.2208	~6°	~6°			Zinc sulfide (0.12 μm thick); glass substrate; vacuum deposited.

DATA TABLE NO. 464 NORMAL SPECTRAL REFLECTANCE OF ZINC SULFIDE CONTACT COATINGS

[Wavelength, λ, μm; Reflectance, ρ; Temperature, T, K]

λ	ρ	λ	ρ	λ	ρ	λ	ρ	λ	ρ	λ	ρ	λ	ρ	λ	ρ
CURVE 1		CURVE 1 (cont.)		CURVE 1 (cont.)		CURVE 2 (cont.)		CURVE 2 (cont.)		CURVE 2 (cont.)		CURVE 2 (cont.)		CURVE 3 (cont.)	
T ~ 298		0.49	0.280	2.17	0.082	0.36	0.213	0.49	0.073	0.84	0.211	3.22	0.305	0.1536	0.323
0.28	0.314	0.52	0.295	2.38	0.072	0.37	0.233	0.50	0.042	0.87	0.251	3.70	0.280	0.1631	0.344
0.31	0.217	0.54	0.298			0.37	0.253	0.52	0.061	0.90	0.272	4.34	0.261	0.1708	0.349
0.32	0.186	0.55	0.301*	CURVE 2		0.38	0.292	0.52	0.102	0.95	0.290			0.1800	0.342
0.33	0.168	0.58	0.302	T ~ 298		0.38	0.302	0.53	0.135	1.00	0.280	CURVE 3		0.2001	0.314
0.34	0.153	0.59	0.302*			0.39	0.283	0.54	0.201	1.04	0.263	T = 298		0.2105	0.319
0.34	0.142	0.60	0.302	0.25	0.322	0.39	0.232	0.55	0.242	1.08	0.239			0.2208	0.332
0.35	0.121	0.63	0.301	0.25	0.321*	0.40	0.152	0.57	0.290	1.12	0.230	0.0588	0.033		
0.37	0.111	0.64	0.298	0.26	0.312	0.40	0.086	0.59	0.304*	1.17	0.183	0.0662	0.073		
0.38	0.122	0.66	0.296	0.27	0.291	0.40	0.057	0.62	0.285	1.26	0.110	0.0724	0.111		
0.40	0.146	0.72	0.280	0.27	0.282*	0.41	0.072	0.63	0.246	1.40	0.054	0.0791	0.169		
0.40	0.170	0.77	0.268	0.28	0.273	0.41	0.122	0.65	0.191	1.47	0.058	0.0891	0.218		
0.41	0.188	0.84	0.257	0.28	0.267*	0.42	0.232	0.67	0.106	1.53	0.072	0.0993	0.203		
0.42	0.211	0.88	0.247	0.29	0.253	0.44	0.317	0.69	0.085	1.69	0.109	0.1061	0.253		
0.43	0.229	0.94	0.221	0.31	0.239	0.45	0.332	0.71	0.053	1.96	0.230	0.1101	0.299		
0.44	0.244	1.00	0.211	0.31	0.233*	0.45	0.327*	0.73	0.047	2.32	0.302	0.1171	0.337		
0.45	0.252*	1.05	0.199	0.33	0.231	0.46	0.302	0.75	0.067	2.56	0.306	0.1221	0.348		
0.46	0.260	1.11	0.171	0.34	0.212	0.46	0.242	0.77	0.092	2.70	0.308	0.1324	0.334		
0.48	0.271	1.28	0.150	0.35	0.211	0.47	0.172	0.80	0.132	2.85	0.310	0.1360	0.313		
0.49	0.271*	1.51	0.132			0.48	0.112	0.81	0.182	3.12	0.307	0.1427	0.303		

* Not shown on plot

SPECIFICATION TABLE NO. 465 ANGULAR SPECTRAL REFLECTANCE OF ZINC SULFIDE CONTACT COATINGS

Curve No.	Ref. No.	Year	Temperature K	Wavelength, μm	Angular Range°	Geometry θ	θ'	ω'	Reported Error, %	Composition (weight percent), Specifications and Remarks
1*	192	1959	298	0.0584	10-86	θ	θ'			ZnS (0.12 μm thick); glass substrate; vacuum deposited using commercial pure powder; data extracted from smooth curve; angle of incidence, θ, is equal to angle of viewing, θ'; θ is variable.
2*	192	1959	298	0.0735	10-86	θ	θ'			Similar to above specimen and conditions.
3*	192	1959	298	0.0900	10-86	θ	θ'			Similar to above specimen and conditions.
4*	192	1959	298	0.1105	10-86	θ	θ'			Similar to above specimen and conditions.
5*	192	1959	298	0.1216	10-86	θ	θ'			Similar to above specimen and conditions.
6*	192	1959	298	0.1606	10-86	θ	θ'			Similar to above specimen and conditions.
7*	192	1959	298	0.2200	10-86	θ	θ'			Similar to above specimen and conditions.

DATA TABLE NO. 465 ANGULAR SPECTRAL REFLECTANCE OF ZINC SULFIDE CONTACT COATINGS

[Angle, *,°; Reflectance, ρ; Temperature, T, K; Wavelength, λ, μm]

θ	ρ
CURVE 1*	
T = 298	
λ = 0.0584	
10	0.040
20	0.042
30	0.051
40	0.093
50	0.172
60	0.296
70	0.473
80	0.675
86	0.871
CURVE 2*	
T = 298	
λ = 0.0735	
10	0.118
20	0.127
30	0.144
CURVE 2 (cont.)*	
40	0.190
50	0.266
60	0.380
70	0.512
80	0.718
86	0.868
CURVE 3*	
T = 298	
λ = 0.0900	
10	0.223
20	0.227
30	0.238
40	0.264
50	0.317
60	0.394
70	0.501
80	0.698
CURVE 3 (cont.)*	
86	0.852
CURVE 4*	
T = 298	
λ = 0.1105	
10	0.296
20	0.301
30	0.309
40	0.325
50	0.358
60	0.417
70	0.514
80	0.688
86	0.851
CURVE 5*	
T = 298	
λ = 0.1216	
10	0.351
20	0.351
30	0.355
40	0.367
50	0.387
60	0.431
70	0.519
80	0.690
86	0.841
CURVE 6*	
T = 298	
λ = 0.1606	
10	0.338
20	0.340
30	0.344
CURVE 6 (cont.)*	
40	0.352
50	0.364
60	0.387
70	0.442
80	0.579
86	0.800
CURVE 7*	
T = 298	
λ = 0.2200	
10	0.336
20	0.336
30	0.337
40	0.339
50	0.343
60	0.346
70	0.366
80	0.459
CURVE 7 (cont.)*	
86	0.776

* No plot given

SPECIFICATION TABLE NO. 466 NORMAL SPECTRAL TRANSMITTANCE OF ZINC TELLURIDE CONTACT COATINGS

Curve No.	Ref. No.	Year	Temperature, K	Wavelength Range, μm	Geometry θ	θ'	ω'	Reported Error, %	Composition (weight percent), Specifications, and Remarks
1*	190	1968	~298	52.69-58.87	~0°	~0°			ZnTe (1.5 μm thick) on quartz substrate; substrate crystalline and cut parallel to z axis; high purity grade ZnTe evaporated on substrate at 348 K.
2*	190	1968	~298	50.99-58.87	~0°	~0°			Similar to the above specimen and conditions except 3.0 μm thick.
3*	190	1968	~298	48.70-58.87	~0°	~0°			Similar to the above specimen and conditions except 9.2 μm thick.
4*	190	1968	295	46.7-59.1	45°	45°			Similar to curve 1 specimen and conditions.

DATA TABLE NO. 466 NORMAL SPECTRAL TRANSMITTANCE OF ZINC TELLURIDE CONTACT COATINGS

[Wavelength, λ, μm; Transmittance, τ; Temperature, T, K]

CURVE 1* (T ~ 298)

λ	τ
52.69	0.388
52.97	0.319
53.19	0.277
53.48	0.220
53.78	0.177
54.07	0.136
54.38	0.094
54.77	0.054
54.97	0.041
55.60	0.016
56.01	0.032
56.54	0.066
56.88	0.091
57.19	0.113
57.46	0.138
57.76	0.166
58.06	0.193
58.47	0.232
58.87	0.276

CURVE 2* (T ~ 298)

λ	τ
50.99	0.405
51.30	0.359
51.55	0.311
51.88	0.254
52.07	0.222
52.37	0.178
52.70	0.142
52.99	0.105
53.17	0.089
53.48	0.067
53.81	0.046
54.07	0.033
54.39	0.022
54.79	0.010
54.98	0.008
55.60	0.002
56.01	0.008
56.50	0.030
56.86	0.051
57.17	0.075
57.45	0.094
57.78	0.120
58.07	0.145
58.47	0.183
58.87	0.233

CURVE 3* (T ~ 298)

λ	τ
48.70	0.365
48.88	0.300
49.18	0.207
49.41	0.164
49.69	0.115
50.00	0.079
50.11	0.071
50.40	0.047
50.67	0.032
51.00	0.020
51.31	0.013
52.00	0.003
56.99	0.003
57.51	0.016
57.79	0.029
58.09	0.046
58.49	0.076
58.87	0.116

CURVE 4* (T = 295)

λ	τ
46.7	0.866
46.9	0.858
47.1	0.842
47.3	0.821
47.6	0.792
47.8	0.759
48.0	0.714
48.2	0.688
48.4	0.680
48.6	0.687
48.8	0.705
49.1	0.725
49.4	0.725
49.6	0.712
49.9	0.698
50.3	0.667
50.7	0.642
50.9	0.614
51.2	0.578
51.5	0.546
51.8	0.500
52.1	0.473
52.3	0.432
52.6	0.380
52.9	0.326
53.1	0.293
53.4	0.237
53.7	0.192
54.0	0.149
54.3	0.115
54.7	0.074
55.0	0.058
55.3	0.042
55.6	0.026
55.9	0.038
56.5	0.067
56.8	0.091
57.1	0.113
57.5	0.146
57.7	0.178
58.1	0.207
58.4	0.252
58.8	0.293
59.1	0.325

* No plot given

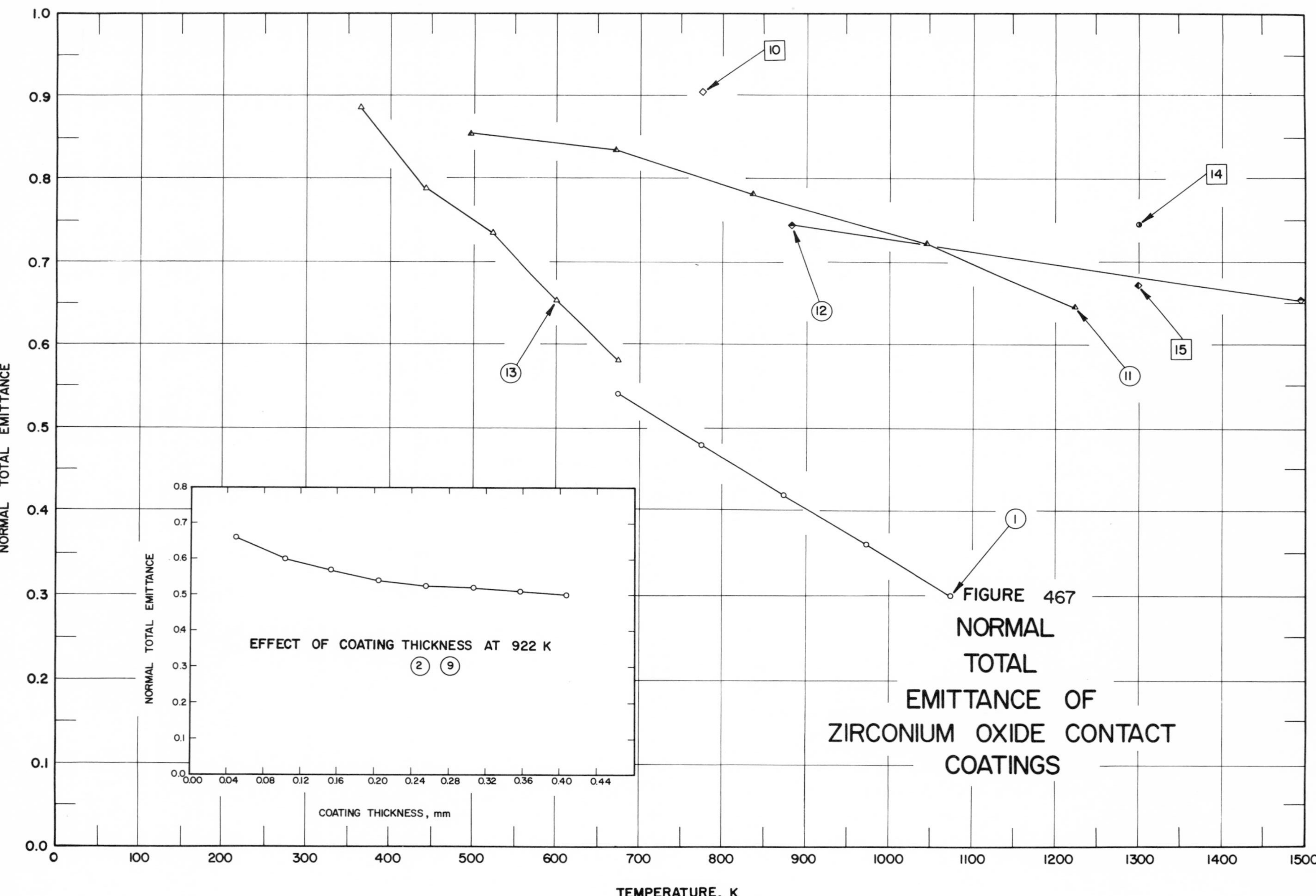
NORMAL TOTAL EMITTANCE
1.0
0.9
0.8
0.7
0.6
0.5
0.4
0.3
0.2
0.1
0.0
0
100
200
300
400
500
600
700
800
900
1000
1100
1200
1300
1400
1500
TEMPERATURE, K
10
14
12
13
15
11
1
FIGURE 467
NORMAL
TOTAL
EMITTANCE OF
ZIRCONIUM OXIDE CONTACT
COATINGS
NORMAL TOTAL EMITTANCE
0.8
0.7
0.6
0.5
0.4
0.3
0.2
0.1
0.0
EFFECT OF COATING THICKNESS AT 922 K
2 9
0.00
0.04
0.08
0.12
0.16
0.20
0.24
0.28
0.32
0.36
0.40
0.44
COATING THICKNESS, mm

SPECIFICATION TABLE NO. 467 NORMAL TOTAL EMITTANCE OF ZIRCONIUM OXIDE CONTACT COATINGS

Curve No.	Ref. No.	Year	Temperature Range, K	Geometry θ'	Reported Error, %	Composition (weight percent), Specifications and Remarks
1	123	1952	673-1073	$\sim 0°$		Zirconium oxide; Nimonic 75 substrate.
2	124	1959	922	$\sim 0°$		Zirconium oxide (0.0508 mm thick); Inconel substrate; substrate sandblasted and oxidized; flame sprayed.
3	124	1959	922	$\sim 0°$		Similar to above specimen and conditions except 0.102 mm thick.
4	124	1959	922	$\sim 0°$		Similar to above specimen and conditions except 0.152 mm thick.
5	124	1959	922	$\sim 0°$		Similar to above specimen and conditions except 0.203 mm thick.
6	124	1959	922	$\sim 0°$		Zirconium oxide (0.254 mm thick); Inconel substrate; substrate sandblasted and oxidized; flame sprayed; opaque coating.
7	124	1959	922	$\sim 0°$		Similar to above specimen and conditions except 0.305 mm thick.
8	124	1959	922	$\sim 0°$		Similar to above specimen and conditions except 0.356 mm thick.
9	124	1959	922	$\sim 0°$		Similar to above specimen and conditions except 0.406 mm thick.
10	127	1959	77.6	$\sim 0°$		ZrO_2; Inconel substrate. [Authors' designation: Cycle one]
11	127	1959	499-1222	$\sim 0°$		Above specimen and conditions.
12	127	1959	882-1496	$\sim 0°$		Above specimen and conditions. [Authors' designation: Cycle two]
13	89	1958	366-672	$\sim 0°$		Zirconium oxide, Rokide Z (0.254 mm thick); 321 stainless steel substrate; flame sprayed. [Author's designation: Sample 19]
14	126	1965	1300	$\sim 0°$		94.57 ZrO_2 (cubic), 3.73 CaO, 0.39 TiO_2, 0.33 Fe_2O_3, 0.33 SiO_2, and 0.02 NaO_2 (0.305 mm thick, density 5.2 g cm^{-3}, porosity 8-12%); mild steel substrate; flame sprayed; surface roughness 5.72 μm measured with profilometer and optical comparator; measured in vacuum (3.5 to 5.0 x 10^{-2} mm Hg); computed from spectral data (0 to 10 μm).
15	126	1965	1300	$\sim 0°$		Similar to above specimen and conditions.

DATA TABLE NO. 467 NORMAL TOTAL EMITTANCE OF ZIRCONIUM OXIDE CONTACT COATINGS

[Temperature, T, K; Emittance, ϵ]

T	ϵ
CURVE 1	
673	0.54
773	0.48
873	0.42
973	0.36
1073	0.30
CURVE 2	
922	0.66
CURVE 3	
922	0.60
CURVE 4	
922	0.57
CURVE 5	
922	0.54
CURVE 6	
922	0.525
CURVE 7	
922	0.520
CURVE 8	
922	0.51
CURVE 9	
922	0.50
CURVE 10	
77.6	0.907
CURVE 11	
499	0.857
670	0.837
CURVE 11 (cont.)	
838	0.783
1048	0.723
1222	0.648
CURVE 12	
882	0.745
1496	0.656
CURVE 13	
366	0.889
444	0.790
522	0.738
600	0.653
672	0.582
CURVE 14	
1300	0.747
CURVE 15	
1300	0.671

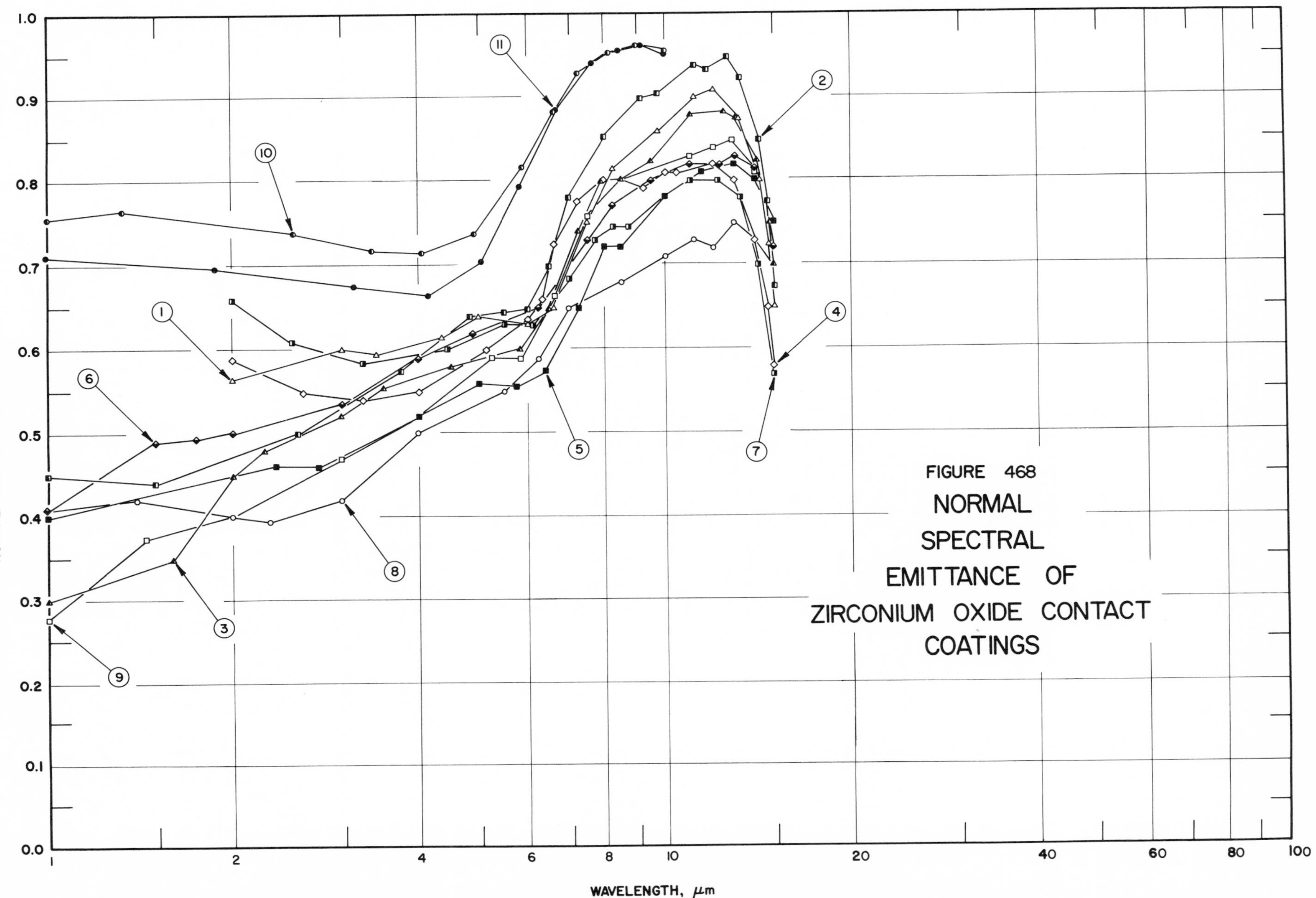

FIGURE 468 NORMAL SPECTRAL EMITTANCE OF ZIRCONIUM OXIDE CONTACT COATINGS

SPECIFICATION TABLE NO. 468 NORMAL SPECTRAL EMITTANCE OF ZIRCONIUM OXIDE CONTACT COATINGS

Curve No.	Ref. No.	Year	Temperature, K	Wavelength Range, μm	Geometry θ'	Reported Error, %	Composition (weight percent), Specifications, and Remarks
1	139	1961	523	2.00-15.00	~0°	±5	Zirconium oxide; Inconel X substrate; substrate sandblasted; flame sprayed using a plasmatron; as received; data extracted from smooth curve.
2	139	1961	773	1.00-15.00	~0°	±5	Similar to above specimen and conditions.
3	139	1961	1023	1.00-15.00	~0°	±5	Similar to above specimen and conditions.
4	139	1961	523	2.00-15.00	~0°	±5	Similar to curve 1 specimen and conditions except heated in air at 1089 K for 30 min.
5	139	1961	773	1.00-15.00	~0°	±5	Similar to above specimen and conditions.
6	139	1961	1023	1.00-15.00	~0°	±5	Similar to above specimen and conditions.
7	139	1961	523	2.00-15.00	~0°	±5	Similar to curve 1 specimen and conditions except heated in vacuum (6.8×10^{-5} mm Hg) at 1089 K for 30 min.
8	139	1961	773	1.00-15.00	~0°	±5	Similar to above specimen and conditions.
9	139	1961	1023	1.00-15.00	~0°	±5	Similar to above specimen and conditions.
10	126	1965	1300	1.00-10.01	~0°		94.57 ZrO_2 (cubic), 3.73 CaO, 0.39 TiO_2, 0.33 Fe_2O_3, 0.33 SiO_2, and 0.02 NaO_2 (0.305 mm thick, density 5.2 g cm^{-3}, porosity 8-12%); mild steel substrate; flame sprayed; surface roughness 5.72 μm measured with profilometer and optical comparator; measured in vacuum (3.5 to 5.0×10^{-2} mm Hg); computed from spectral data (0 to 10 μm).
11	126	1965	1300	1.00-10.00	~0°		94.57 ZrO_2 (cubic), 3.73 CaO, 0.39 TiO_2, 0.33 Fe_2O_3, 0.33 SiO_2, and 0.02 NaO_2 (0.432 mm thick, density 5.2 g cm^{-3}, porosity 8 to 12%); mild steel substrate; substrate polished with polishing papers; flame sprayed; measured in vacuum (3.5 to 5.0×10^{-2} mm Hg); data extracted from smooth curve.

DATA TABLE NO. 468 NORMAL SPECTRAL EMITTANCE OF ZIRCONIUM OXIDE CONTACT COATINGS

[Wavelength, λ, μm; Emittance, ρ; Temperature, T, K]

λ	∈
CURVE 1	
T = 523	
2.00	0.565
3.00	0.600
3.40	0.595
4.35	0.615
5.00	0.640
6.00	0.630
6.60	0.650
7.50	0.750
8.25	0.815
9.75	0.860
11.25	0.900
12.00	0.910
13.25	0.875
14.25	0.800
14.80	0.725
15.00	0.650
CURVE 2	
T = 773	
1.00	0.450
1.50	0.440
2.55	0.500
3.75	0.575
4.85	0.640
5.50	0.645
6.00	0.650
6.50	0.700
7.00	0.780
8.00	0.855
9.15	0.900
9.75	0.905
11.25	0.940
11.75	0.935
12.60	0.950
13.35	0.925
14.20	0.850
14.60	0.775
15.00	0.675
CURVE 3	
T = 1023	
1.00	0.300
1.60	0.350
CURVE 3 (cont.)	
2.00	0.450
2.25	0.480
3.00	0.520
3.50	0.555
4.50	0.580
5.80	0.600
6.50	0.650
7.25	0.740
7.90	0.800
8.50	0.800
9.50	0.825
11.00	0.880
12.50	0.885
13.10	0.875
14.10	0.825
14.80	0.750
15.00	0.700
15.00	0.650*
CURVE 4	
T = 523	
2.00	0.590
2.60	0.550
3.25	0.540
4.00	0.550
5.15	0.600
6.00	0.635
6.35	0.660
6.65	0.725
7.25	0.775
8.00	0.800
8.50	0.800*
9.25	0.790
10.00	0.810
10.50	0.810
12.00	0.820
13.00	0.800
14.00	0.730
14.75	0.650
15.00	0.580
CURVE 5	
T = 773	
1.00	0.400
CURVE 5 (cont.)	
2.35	0.460
2.75	0.460
4.00	0.520
5.00	0.560
5.75	0.555
6.40	0.575
7.25	0.650
8.00	0.720
8.50	0.720
10.00	0.780
11.50	0.810
13.00	0.820
14.00	0.800
15.00	0.750
CURVE 6	
T = 1023	
1.00	0.410
1.50	0.490
1.75	0.495
2.00	0.500
3.00	0.535
4.00	0.590
4.90	0.620
5.55	0.620*
6.25	0.650
7.50	0.730
8.25	0.770
9.50	0.800
11.00	0.820
12.25	0.820
13.00	0.830
14.00	0.815
14.60	0.775*
15.00	0.720
CURVE 7	
T = 523	
2.00	0.660
2.50	0.610
3.25	0.585
4.45	0.600
5.50	0.630
6.10	0.630
CURVE 7 (cont.)	
7.00	0.685
7.75	0.730
8.25	0.745
8.75	0.745
10.00	0.780*
11.00	0.800
12.25	0.800
13.25	0.780
14.05	0.700
15.00	0.570
CURVE 8	
T = 773	
1.00	0.410*
1.40	0.420
2.00	0.400
2.30	0.395
3.00	0.420
4.00	0.500
5.50	0.550
6.25	0.590
7.00	0.650
8.50	0.680
10.00	0.710
11.25	0.730
12.00	0.720
13.00	0.750
14.00	0.730*
15.00	0.700*
CURVE 9	
T = 1023	
1.00	0.280
1.45	0.375
2.00	0.400*
3.00	0.470
4.00	0.520*
5.25	0.590
5.85	0.590
6.65	0.665
7.50	0.755
8.50	0.800*
9.50	0.800*
11.00	0.830
CURVE 9 (cont.)	
12.00	0.840
12.90	0.850
14.00	0.810
15.00	0.750*
CURVE 10	
T = 1300	
1.00	0.755
1.34	0.763
2.51	0.739
3.38	0.718
4.03	0.713
4.91	0.736
5.90	0.818
6.61	0.884
7.26	0.930
8.15	0.953
9.00	0.961
10.01	0.956
CURVE 11	
T = 1300	
1.00	0.710
1.89	0.699
3.15	0.675
4.17	0.665
5.07	0.704
5.82	0.792
6.70	0.886
7.67	0.942
8.48	0.957
9.15	0.961
10.00	0.955

* Not shown on plot

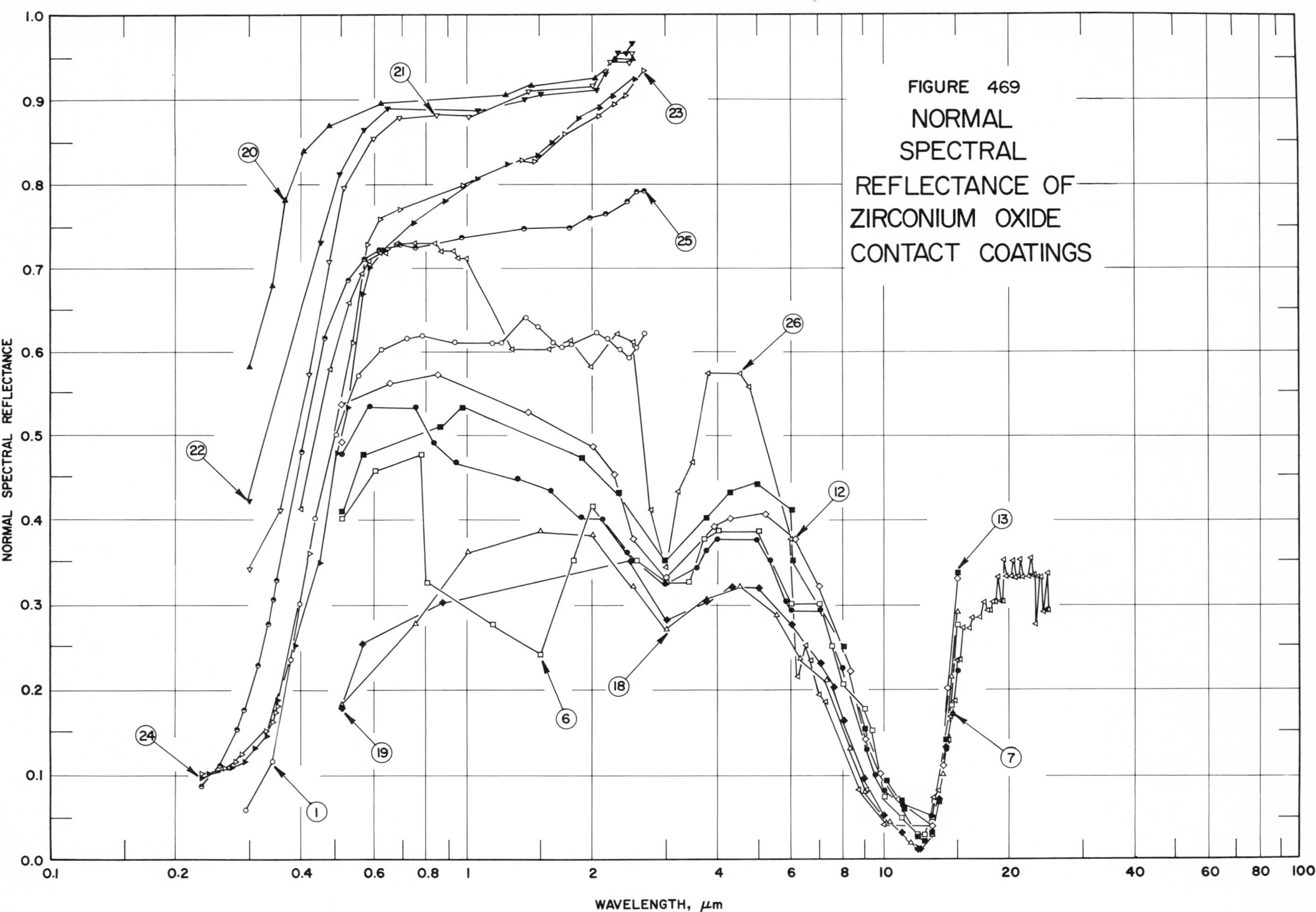
FIGURE 469
NORMAL
SPECTRAL
REFLECTANCE OF
ZIRCONIUM OXIDE
CONTACT COATINGS
NORMAL SPECTRAL REFLECTANCE
WAVELENGTH, μm
1.0
0.9
0.8
0.7
0.6
0.5
0.4
0.3
0.2
0.1
0.0
0.1
0.2
0.4
0.6
0.8
1
2
4
6
8
10
20
40
60
80
100
1
6
7
12
13
18
19
20
21
22
23
24
25
26

SPECIFICATION TABLE NO. 469 NORMAL SPECTRAL REFLECTANCE OF ZIRCONIUM OXIDE CONTACT COATINGS

Curve No.	Ref. No.	Year	Temperature, K	Wavelength Range, μm	Geometry θ	Geometry θ'	Geometry ω'	Reported Error, %	Composition (weight percent), Specifications, and Remarks
1	127	1959	298	0.295-2.698	9°		2π	4	ZrO_2; Inconel substrate; measured relative to magnesium carbonate; data extracted from smooth curve.
2*	139	1961	~322	1.00-15.00	$\sim 0^\circ$		2π	< 2	ZrO_2; Inconel X substrate; substrate sandblasted; flame sprayed using a plasmatron; as received; converted from R $(2\pi, 0^\circ)$; hohlraum source at 523 K.
3*	139	1961	~322	2.00-15.00	$\sim 0^\circ$		2π	< 2	Above specimen and conditions; diffuse component only.
4*	139	1961	~322	1.00-15.00	$\sim 0^\circ$		2π	< 2	Similar to curve 2 specimen and conditions except hohlraum at 773 K.
5*	139	1961	~322	1.00-15.00	$\sim 0^\circ$		2π	< 2	Above specimen and conditions; diffuse component only.
6	139	1961	~322	0.50-15.00	$\sim 0^\circ$		2π	< 2	Similar to curve 2 specimen and conditions except hohlraum at 1273 K.
7	139	1961	~322	0.50-15.00	$\sim 0^\circ$		2π	< 2	Above specimen and conditions; diffuse component only.
8*	139	1961	~322	2.00-15.00	$\sim 0^\circ$		2π	< 2	Similar to curve 2 specimen and conditions except heated in air at 1089 K for 30 min.
9*	139	1961	~322	2.00-15.00	$\sim 0^\circ$		2π	< 2	Above specimen and conditions; diffuse component only.
10*	139	1961	~322	1.00-15.00	$\sim 0^\circ$		2π	< 2	Similar to curve 8 specimen and conditions except hohlraum at 773 K.
11*	139	1961	~322	1.00-15.00	$\sim 0^\circ$		2π	< 2	Above specimen and conditions; diffuse component only.
12	139	1961	~322	0.50-15.00	$\sim 0^\circ$		2π	< 2	Similar to curve 8 specimen and conditions except hohlraum at 1273 K.
13	139	1961	~322	0.50-15.00	$\sim 0^\circ$		2π	< 2	Above specimen and conditions; diffuse component only.
14*	139	1961	~322	2.00-15.00	$\sim 0^\circ$		2π	< 2	Similar to curve 2 specimen and conditions except heated in vacuum (6.8×10^{-5} mm Hg) at 1089 K for 30 min.
15*	139	1961	~322	1.00-15.00	$\sim 0^\circ$		2π	< 2	Above specimen and conditions; diffuse component only.
16*	139	1961	~322	1.00-15.00	$\sim 0^\circ$		2π	< 2	Similar to curve 14 specimen and conditions except hohlraum at 773 K.
17*	139	1961	~322	2.00-15.00	$\sim 0^\circ$		2π	< 2	Above specimen and conditions; diffuse component only.
18	139	1961	~322	0.50-15.00	$\sim 0^\circ$		2π	< 2	Similar to curve 14 specimen and conditions except hohlraum at 1273 K.
19	139	1961	~322	0.50-15.00	$\sim 0^\circ$		2π	< 2	Above specimen and conditions; diffuse component only.
20	32	1967	~298	0.300-2.500	7°		2π		ZrO_2, TAM CP powder; aluminum substrate; water mull of ZrO_2 sprayed on warmed substrate; measured in situ after 17 hrs in vacuum ($< 10^{-6}$ mm Hg); data extracted from smooth curve.
21	32	1967	~298	0.300-2.500	7°		2π		Above specimen and conditions except ultraviolet irradiated in vacuum with 6-8 sun intensity for 800 ESH; measured in situ.
22	32	1967	~298	0.300-2.500	7°		2π		Above specimen and conditions except exposed to air for about 1 hr; measured in air.
23	193	1966	~298	0.230-2.650	$\sim 0^\circ$		2π		ZrO_2; stainless steel substrate; powder compacted at 23 500 psi; data extracted from smooth curve; measured relative to MgO.
24	193	1966	~298	0.230-2.650	$\sim 0^\circ$		2π		ZrO_2 (-170 mesh); stainless steel substrate; powder compacted at 23 500 psi; data extracted from smooth curve; measured relative to MgO.

* Not shown on plot

SPECIFICATION TABLE NO. 469 NORMAL SPECTRAL REFLECTANCE OF ZIRCONIUM OXIDE CONTACT COATINGS (continued)

Curve No.	Ref. No.	Year	Temperature, K	Wavelength Range, μm	Geometry θ	θ'	ω'	Reported Error, %	Composition (weight percent), Specifications, and Remarks
25	193	1966	~298	0.230-2.650	~0°		2π		ZrO_2 (-400 mesh); stainless steel substrate; powder compacted at 23 500 psi; data extracted from smooth curve; measured relative to MgO.
26	89	1958	~298	0.40-24.9	~0°	~0°			Zirconium oxide, Rokide Z, (0.254 mm thick); 321 stainless steel substrate; flame sprayed; data extracted from smooth curve. [Author's designation: Sample 19]

DATA TABLE NO. 469 NORMAL SPECTRAL REFLECTANCE OF ZIRCONIUM OXIDE CONTACT COATINGS

[Wavelength, λ, μm; Reflectance, ρ; Temperature, T, K]

CURVE 1, T = 298

λ	ρ
0.295	0.059
0.341	0.116
0.378	0.234
0.399	0.301
0.431	0.401
0.484	0.500
0.554	0.570
0.621	0.603
0.718	0.616
0.781	0.618
0.935	0.611
1.144	0.609
1.206	0.610
1.393	0.640
1.489	0.629
1.604	0.610
1.687	0.604
1.794	0.608
2.036	0.622
2.193	0.615
2.326	0.602
2.439	0.593
2.547	0.604
2.698	0.621

CURVE 2*, T ~ 322

λ	ρ
1.00	0.320
2.60	0.290
3.00	0.300
4.00	0.345
4.50	0.355
5.50	0.315
6.00	0.280
6.50	0.275
7.00	0.280
7.50	0.250
8.50	0.160
9.50	0.085
10.50	0.040
11.50	0.020
12.00	0.000
13.00	0.000
13.50	0.025
14.00	0.085
14.50	0.135
15.00	0.200

CURVE 3, T ~ 322

λ	ρ
2.00	0.290
2.42	0.271
2.66	0.275
3.01	0.283
4.02	0.334
4.51	0.349
5.00	0.340
5.51	0.316
6.14	0.277
6.52	0.275
7.00	0.283
7.51	0.248
8.01	0.200
9.01	0.126
10.02	0.074
11.00	0.035
12.00	0.000
13.02	0.000
13.61	0.026
14.12	0.076
14.53	0.118
15.00	0.180

CURVE 4*, T ~ 322

λ	ρ
1.00	0.490
2.00	0.410
2.50	0.360
3.00	0.315
3.50	0.340
4.00	0.370
5.00	0.360
6.00	0.285
6.75	0.285
8.00	0.200
9.00	0.120
10.00	0.060
10.75	0.025
11.75	0.000
13.00	0.000
13.60	0.060
14.50	0.125
15.00	0.200

CURVE 5*, T ~ 322

λ	ρ
1.00	0.470
1.49	0.433
2.24	0.349
2.72	0.306
2.98	0.301
3.48	0.327
3.98	0.360
4.74	0.355
5.56	0.326
6.03	0.301
6.98	0.295
8.00	0.200
9.00	0.117
10.02	0.059
11.02	0.021
11.86	0.003
12.00	0.000
13.00	0.000
13.50	0.024
14.00	0.074
14.51	0.122
14.91	0.161
15.00	0.200

CURVE 6, T ~ 322

λ	ρ
0.50	0.400
0.60	0.455
0.75	0.475
0.75	0.475*
0.80	0.325
1.15	0.275
1.50	0.240
1.80	0.350
2.00	0.415
2.55	0.350
3.00	0.325
3.40	0.325
3.70	0.375
4.00	0.385
5.00	0.385
6.00	0.300
7.00	0.300
7.50	0.250
8.00	0.205
9.00	0.175
9.35	0.150
10.00	0.075
11.00	0.050
12.00	0.030
12.50	0.030
13.00	0.050
14.00	0.135
14.50	0.180
15.00	0.275

CURVE 7, T ~ 322

λ	ρ
0.50	0.475
0.58	0.532
0.75	0.532
0.83	0.489
0.94	0.466
1.33	0.445
1.58	0.433
1.87	0.402
2.11	0.400
2.42	0.360
2.98	0.324
3.53	0.343
3.74	0.363
3.98	0.375
4.98	0.375
5.34	0.351
5.79	0.303
5.99	0.293
7.00	0.293
7.98	0.224
9.01	0.128
9.49	0.099
9.99	0.081
11.01	0.064
12.00	0.030*
12.51	0.030*
12.98	0.053
14.01	0.129
14.51	0.169
15.00	0.221

CURVE 8*, T ~ 322

λ	ρ
2.00	0.410
2.50	0.290
3.00	0.250
3.50	0.280
4.00	0.325
4.50	0.335
5.10	0.325
6.00	0.265
6.25	0.260
7.00	0.260
7.55	0.200
8.50	0.110
9.40	0.050
10.50	0.000
13.00	0.000
14.00	0.085
15.00	0.180

CURVE 9*, T ~ 322

λ	ρ
2.00	0.350
2.46	0.286
3.01	0.238
3.50	0.252
4.31	0.302
4.84	0.302
5.51	0.284
6.25	0.256
7.02	0.241
7.74	0.174
8.03	0.148
8.80	0.081
9.52	0.037
10.00	0.018
10.50	0.000
13.00	0.000
14.00	0.066
15.00	0.125

CURVE 10*, T ~ 322

λ	ρ
1.00	0.320
1.50	0.330
2.00	0.370
2.50	0.375
3.25	0.365
4.25	0.370
5.00	0.380
6.00	0.365
7.00	0.300
8.15	0.200
8.75	0.140
9.75	0.080
10.75	0.040
11.25	0.035
12.15	0.005
13.00	0.005
14.00	0.095
14.50	0.150
15.00	0.240

CURVE 11*, T ~ 322

λ	ρ
1.00	0.475
2.00	0.397
4.49	0.401
5.25	0.413
5.83	0.400
6.65	0.350
7.21	0.306
7.53	0.291
8.02	0.274
8.67	0.175
9.42	0.104
10.00	0.078
11.00	0.051
11.61	0.030
12.09	0.004
13.00	0.004
14.02	0.100
14.51	0.162
15.00	0.250

CURVE 12, T ~ 322

λ	ρ
0.50	0.490
0.50	0.535
0.65	0.560
0.85	0.570
1.40	0.525
2.00	0.485
2.25	0.450
2.50	0.375
3.00	0.330
3.90	0.390
4.25	0.400
5.20	0.405
6.10	0.375
7.00	0.320
8.25	0.220
9.00	0.140
9.75	0.100
10.75	0.070
12.00	0.030*
12.50	0.030*
13.00	0.040
13.75	0.110
14.25	0.200
15.00	0.330

CURVE 13, T ~ 322

λ	ρ
0.50	0.407
0.56	0.474
0.86	0.507
0.97	0.531
1.88	0.470
2.31	0.430
2.98	0.350
3.74	0.401
4.25	0.430
4.98	0.440
6.00	0.412
7.01	0.350
8.00	0.249
9.00	0.153
10.02	0.093
11.00	0.068
11.27	0.059
12.01	0.027
12.51	0.023
13.00	0.034
13.52	0.067
14.01	0.140
15.00	0.326

CURVE 14*, T ~322

λ	ρ
2.00	0.210
2.75	0.145
3.00	0.140
3.60	0.175
4.00	0.210
4.75	0.200
5.50	0.150
6.15	0.065
6.60	0.085
7.00	0.095
7.75	0.050
9.00	0.000
13.00	0.000
14.00	0.050
14.75	0.125
15.00	0.160

CURVE 15*, T ~ 322

λ	ρ
1.00	0.380
1.99	0.323
2.97	0.323
3.98	0.303
5.01	0.303
6.04	0.251
7.00	0.221
8.00	0.151
9.02	0.077

* Not shown on plot

DATA TABLE NO. 469 NORMAL SPECTRAL REFLECTANCE OF ZIRCONIUM OXIDE CONTACT COATINGS (continued)

λ	ρ
CURVE 15 (cont.)*	
T ~ 322	
10.00	0.038
10.99	0.008
11.99	0.000
12.99	0.000
13.99	0.099
14.67	0.177
15.00	0.250
CURVE 16*	
T ~ 322	
1.00	0.350
1.50	0.325
2.50	0.335
3.00	0.330
4.00	0.300
5.00	0.300
6.00	0.240
6.90	0.225
7.90	0.150
8.90	0.075
10.00	0.030
11.50	0.000
13.00	0.000
14.00	0.095
14.70	0.175
15.00	0.240
CURVE 17*	
T ~ 322	
2.00	0.225
2.33	0.174
3.00	0.125
3.24	0.126
3.49	0.160
3.74	0.224
4.00	0.245
4.74	0.243
5.38	0.208
5.72	0.151
6.25	0.100
7.02	0.148
7.52	0.126
8.01	0.067
9.01	0.022
CURVE 17 (cont.)	
10.00	0.000
13.00	0.000
13.51	0.033
14.00	0.066
14.51	0.103
15.00	0.168
CURVE 18	
T ~ 322	
0.50	0.180
0.75	0.275
1.00	0.360
1.50	0.385
2.00	0.380
2.50	0.320
3.00	0.270
3.75	0.305
4.50	0.320
5.50	0.285
6.25	0.235
7.25	0.210
8.25	0.130
9.00	0.080
10.25	0.045
11.50	0.020
12.25	0.015
13.00	0.030
13.75	0.100
14.50	0.215
15.00	0.290
CURVE 19	
T ~ 322	
0.50	0.175
0.66	0.252
0.87	0.301
1.01	0.360*
1.50	0.385*
2.00	0.380
2.45	0.348
3.00	0.280
3.75	0.303
4.29	0.319
5.00	0.308
6.00	0.275
CURVE 19 (cont.)	
7.02	0.229
7.58	0.201
8.00	0.162
8.99	0.094
9.99	0.053
11.00	0.033
12.00	0.013
12.24	0.013
13.00	0.029*
13.52	0.071
14.00	0.132
14.52	0.214*
15.00	0.290*
CURVE 20	
T ~ 298	
0.300	0.581
0.341	0.680
0.366	0.783
0.404	0.840
0.467	0.869
0.620	0.896
1.238	0.906
1.431	0.916
2.023	0.925
2.279	0.948
2.500	0.957
CURVE 21	
T ~ 298	
0.300	0.342
0.356	0.410
0.416	0.571
0.463	0.707
0.524	0.795
0.602	0.855
0.687	0.878
0.850	0.882
1.005	0.880
1.407	0.908
2.001	0.915
2.217	0.944
2.455	0.944
2.500	0.955
CURVE 22	
T ~ 298	
0.300	0.422
0.443	0.732
0.498	0.812
0.563	0.864
0.645	0.890
1.065	0.887
1.377	0.899
1.515	0.906
2.047	0.911
2.182	0.930
2.302	0.955
2.430	0.956
2.500	0.966
CURVE 23	
T ~ 298	
0.230	0.101
0.262	0.108
0.278	0.115
0.288	0.124
0.330	0.150
0.342	0.161
0.347	0.172
0.350	0.179
0.414	0.357
0.529	0.609
0.577	0.728
0.615	0.757
0.685	0.769
0.970	0.798
1.348	0.827
1.461	0.826
1.715	0.858
2.066	0.881
2.259	0.895
2.401	0.904
2.650	0.934
CURVE 24	
T ~ 298	
0.230	0.097
0.239	0.101
0.272	0.108
0.292	0.116
CURVE 24 (cont.)	
0.310	0.131
0.330	0.146
0.342	0.162*
0.350	0.186
0.385	0.251
0.442	0.347
0.488	0.477
0.518	0.532
0.561	0.668
0.581	0.701
0.630	0.722
0.742	0.754
0.881	0.782
1.062	0.805
1.252	0.823
1.477	0.836
1.601	0.848
1.854	0.876
2.085	0.893
2.257	0.904
2.531	0.924
2.650	0.935*
CURVE 25	
T ~ 298	
0.230	0.085
0.255	0.112
0.280	0.152
0.292	0.175
0.315	0.227
0.332	0.276
0.343	0.305
0.349	0.328
0.400	0.479
0.457	0.615
0.518	0.686
0.564	0.711
0.614	0.723
0.691	0.728
0.748	0.725
0.976	0.737
1.368	0.747
1.763	0.748
1.975	0.759
2.153	0.764
2.409	0.779
CURVE 25 (cont.)	
2.559	0.791
2.650	0.792
CURVE 26	
T ~ 298	
0.40	0.413
0.47	0.577
0.52	0.657
0.56	0.694
0.58	0.708
0.61	0.720
0.62	0.719
0.64	0.729
0.73	0.731
0.84	0.731
0.87	0.721
0.93	0.721
0.95	0.712
1.00	0.712
1.28	0.602
1.57	0.602
1.77	0.612
1.99	0.580
2.31	0.620
2.52	0.611
2.76	0.410
3.00	0.343
3.21	0.432
3.47	0.466
3.80	0.572
4.54	0.572
4.76	0.556
5.99	0.375
6.23	0.214
6.52	0.250
6.76	0.233
7.00	0.192
7.23	0.183
8.73	0.082
9.02	0.082
9.99	0.042
12.8	0.042
13.3	0.073
13.5	0.081
14.2	0.139
14.4	0.167
CURVE 26 (cont.)	
14.7	0.184
15.0	0.233
15.2	0.233
15.4	0.271
16.0	0.271
16.3	0.283
17.0	0.283
17.4	0.302
17.7	0.291
18.0	0.291
18.2	0.302
18.5	0.302
18.7	0.332
19.0	0.303
19.2	0.302
19.4	0.350
19.7	0.332
20.2	0.332
20.4	0.348
20.7	0.331
21.0	0.332
21.2	0.351
21.5	0.331
22.2	0.332
22.5	0.353
22.7	0.333
23.0	0.333
23.2	0.275
23.5	0.332
23.9	0.332
24.2	0.291
24.5	0.292
24.7	0.336
24.9	0.292

* Not shown on plot

SPECIFICATION TABLE NO. 470 NORMAL SOLAR ABSORPTANCE OF ZIRCONIUM OXIDE CONTACT COATINGS

Curve No.	Ref. No.	Year	Temperature Range, K	Geometry θ	Reported Error, %	Composition (weight percent), Specifications and Remarks
1*	127	1959	298	9°		ZrO_2; Inconel substrate; computed from spectral reflectance data (0.3 to 3 μm) for above atm conditions.
2*	127	1959	298	9°		Above specimen and conditions except computed for sea level conditions.

DATA TABLE NO. 470 NORMAL SOLAR ABSORPTANCE OF ZIRCONIUM OXIDE CONTACT COATINGS

[Temperature, T, K; Absorptance, α]

T	α
CURVE 1*	
298	0.459
CURVE 2*	
298	0.431

* No plot given

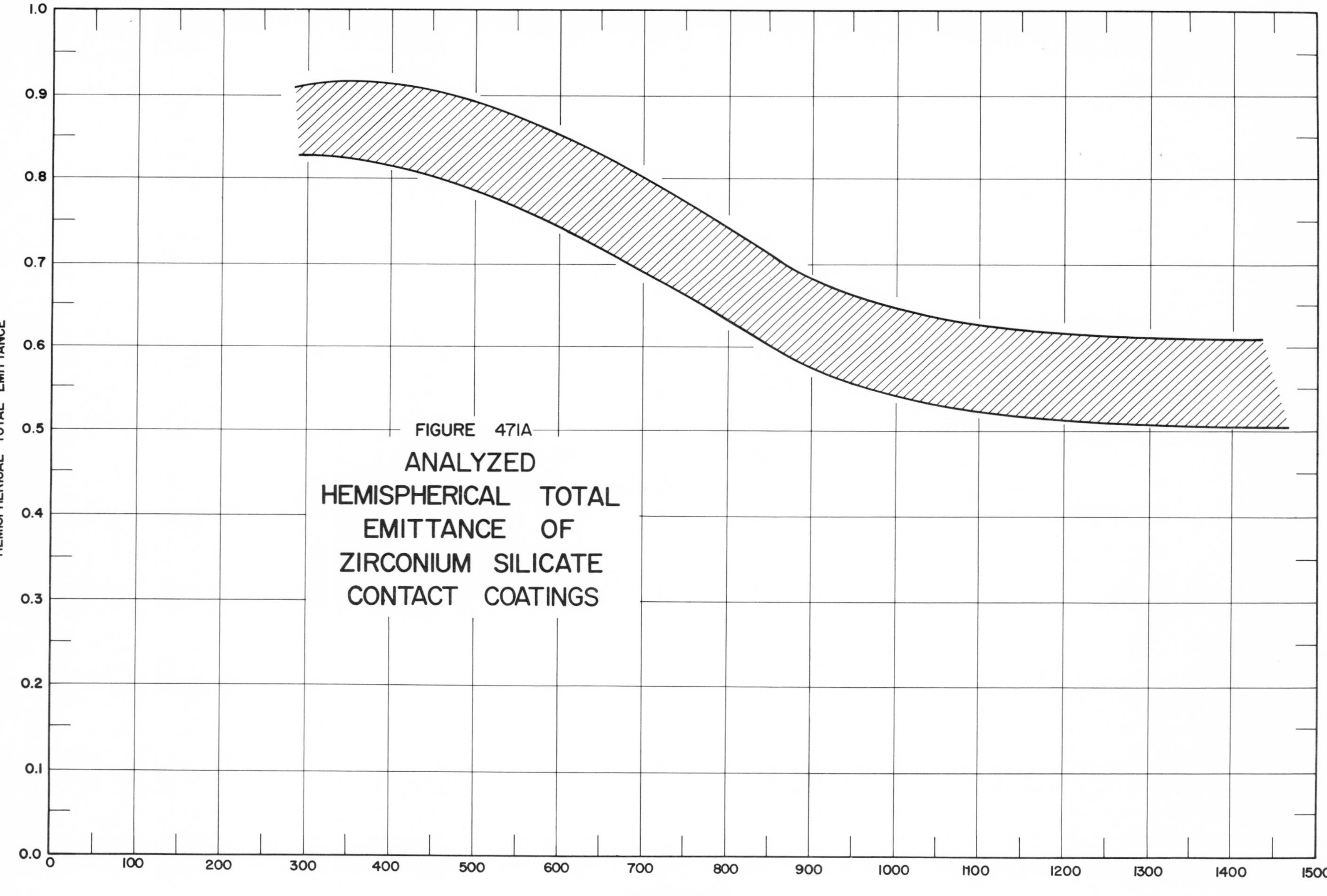

FIGURE 471A ANALYZED HEMISPHERICAL TOTAL EMITTANCE OF ZIRCONIUM SILICATE CONTACT COATINGS

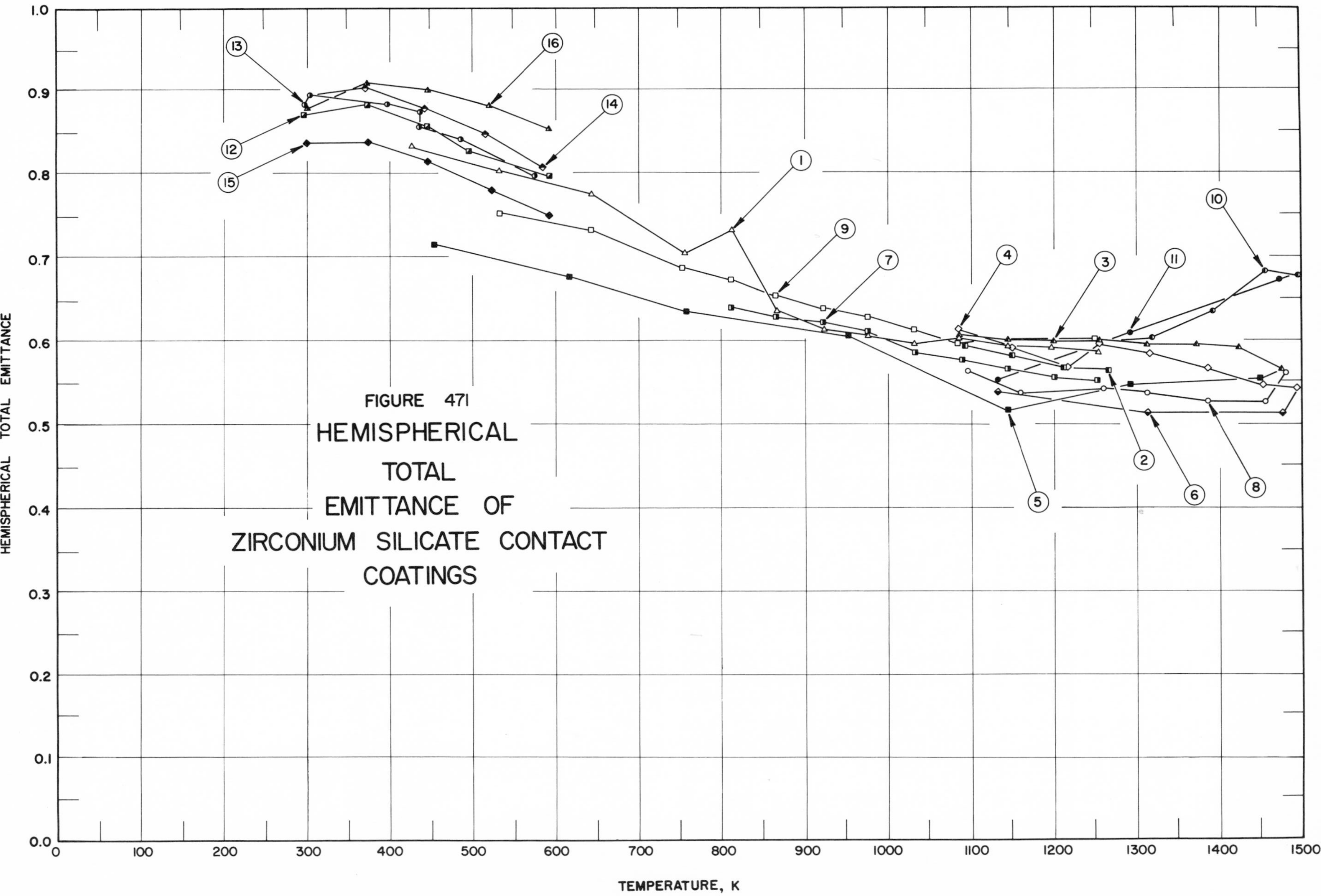

FIGURE 471 HEMISPHERICAL TOTAL EMITTANCE OF ZIRCONIUM SILICATE CONTACT COATINGS

SPECIFICATION TABLE NO. 471 HEMISPHERICAL TOTAL EMITTANCE OF ZIRCONIUM SILICATE CONTACT COATINGS

Curve No.	Ref. No.	Year	Temperature Range, K	Reported Error, %	Composition (weight percent), Specifications and Remarks
1	118	1962	426-1255		$ZrO_2 \cdot SiO_2$, FCZ-11 powder from Continental Coatings Corp (0.102 mm thick); Nb-1Zr substrate; plasma arc sprayed; coating white, rough texture, well bonded to substrate; measured in vacuum (2.0-2.6 x 10^{-6} mm Hg); temp measured by thermocouple; heating cycle.
2	118	1962	1039-1269		Above specimen and conditions except temp measured by optical pyrometer.
3	118	1962	1089-1475		Curve 1 specimen and conditions; measured immediately after curve 1 run.
4	118	1962	1086-1491		Above specimen and conditions except temp measured with optical pyrometer.
5	118	1962	1475-451		Curve 3 specimen and conditions; cooling cycle.
6	118	1962	1491-1133		Above specimen and conditions except temp measured with optical pyrometer.
7	118	1962	811-1253		Rokide ZS, 64.12 ZrO_2, 33.22 SiO_2, 1.42 Al_2O_3, and 0.97 CaO + TiO_2 + Fe_2O_3 + NaO_2 (0.102 mm thick); 310 stainless steel substrate; flame sprayed; coating white, rough but glossy appearane, very hard, fair coating-substrate bond; measured in vacuum (<10^{-5} mm Hg); temp measured with thermocouple; heating cycle.
8	118	1962	1097-1480		Above specimen and conditions except temp measured with optical pyrometer.
9	118	1962	531-1251		Rokide ZS, 64.12 ZrO_2, 33.22 SiO_2, 1.42 Al_2O_3, and 0.97 CaO + TiO_2 + Fe_2O_3 + NaO_2 (0.127 mm thick); Nb-1Zr alloy substrate; flame sprayed; coating white, rough, glossy in appearance, hard, well bonded to substrate; measured in vacuum (< 8.8 x 10^{-7} mm Hg); temp measured with thermocouple; heating cycle.
10	118	1962	1148-1495		Above specimen and conditions except temp measured with optical pyrometer.
11	118	1962	1495-1132		Above specimen and conditions; cooling cycle.
12	137	1965	299-591	±5	Zirconium silicate (~0.063 mm thick, 8.3 mg cm^{-2} area density); Al substrate; plasma arc sprayed; measured in vacuum (10^{-7} mm Hg); surface light grey before and after test.
13	137	1965	298-575	±5	Similar to curve 12 specimen and conditions except 0.076 mm thick, 9.5 mg cm^{-2} area density.
14	137	1965	296-587	±5	Similar to curve 12 specimen and conditions except 0.127 mm thick, 29 mg cm^{-2} area density.
15	137	1965	300-593	±5	Rokide ZS (~0.050 mm thick, 6.5 mg cm^{-2} area density); Al substrate; applied by Rokide process; measured in vacuum (10^{-7} mm Hg); surface appeared light grey with blue tint before and after test.
16	137	1965	300-593	±5	Similar to above specimen and conditions except 0.188 mm thick, 32 mg cm^{-2} area density.

DATA TABLE NO. 471 HEMISPHERICAL TOTAL EMITTANCE OF ZIRCONIUM SILICATE CONTACT COATINGS

[Temperature, T, K; Emittance, ϵ]

CURVE 1

T	ϵ
426	0.832
531	0.801
643	0.775
756	0.705
811	0.731
866	0.639
922	0.615
977	0.608
1031	0.599
1089	0.602
1144	0.594
1199	0.591
1255	0.589

CURVE 2

T	ϵ
1093	0.593
1150	0.583
1211	0.569
1269	0.564

CURVE 3

T	ϵ
1089	0.607
1145	0.601
1200	0.600
1255	0.600
1311	0.597
1371	0.597
1424	0.592
1475	0.566

CURVE 4

T	ϵ
1086	0.614
1150	0.591
1215	0.569
1256	0.598
1316	0.585
1389	0.567
1451	0.549
1491	0.541

CURVE 5

T	ϵ
1475	0.566*
1450	0.554
1292	0.548
1145	0.518
951	0.607
759	0.639
618	0.677
451	0.717

CURVE 6

T	ϵ
1491	0.541*
1479	0.512
1313	0.514
1133	0.539

CURVE 7

T	ϵ
811	0.641
868	0.630
923	0.623
979	0.611
1034	0.586
1090	0.577
1145	0.568
1200	0.566
1253	0.552

CURVE 8

T	ϵ
1097	0.562
1160	0.539
1260	0.541
1316	0.537
1385	0.528
1456	0.527
1480	0.561

CURVE 9

T	ϵ
531	0.752
644	0.731
752	0.689
813	0.672
867	0.655

CURVE 9 (cont.)

T	ϵ
921	0.640
978	0.630
1033	0.613
1086	0.599
1145	0.603*
1251	0.602

CURVE 10

T	ϵ
1148	0.598*
1253	0.599*
1319	0.604
1391	0.638
1455	0.681
1495	0.679

CURVE 11

T	ϵ
1495	0.679*
1474	0.672
1293	0.610
1132	0.551

CURVE 12

T	ϵ
299	0.870
373	0.881
445	0.859
495	0.828
591	0.797

CURVE 13

T	ϵ
298	0.881
302	0.894
399	0.882
435	0.873
435	0.855
485	0.840
575	0.798

CURVE 14

T	ϵ
296	0.891*
370	0.901
444	0.879
516	0.847
587	0.807

CURVE 15

T	ϵ
300	0.838
374	0.838
447	0.812
521	0.780
593	0.750

CURVE 16

T	ϵ
300	0.879
374	0.908
447	0.900
520	0.881
593	0.851

* Not shown on plot

SPECIFICATION TABLE NO. 472 NORMAL TOTAL EMITTANCE OF ZIRCONIUM SILICATE CONTACT COATINGS

Curve No.	Ref. No.	Year	Temperature Range, K	Geometry θ'	Reported Error, %	Composition (weight percent), Specifications and Remarks
1*	89	1958	366-672	$\sim 0^{\circ}$		Rokide ZS (0.254 mm thick); 321 stainless steel substrate; flame sprayed. [Author's designation: Specimen 20]

DATA TABLE NO. 472 NORMAL TOTAL EMITTANCE OF ZIRCONIUM SILICATE CONTACT COATINGS

[Temperature, T, K; Emittance, ϵ]

T	ϵ
CURVE 1*	
366	0.937
444	0.929
522	0.867
600	0.815
672	0.760

* No plot given

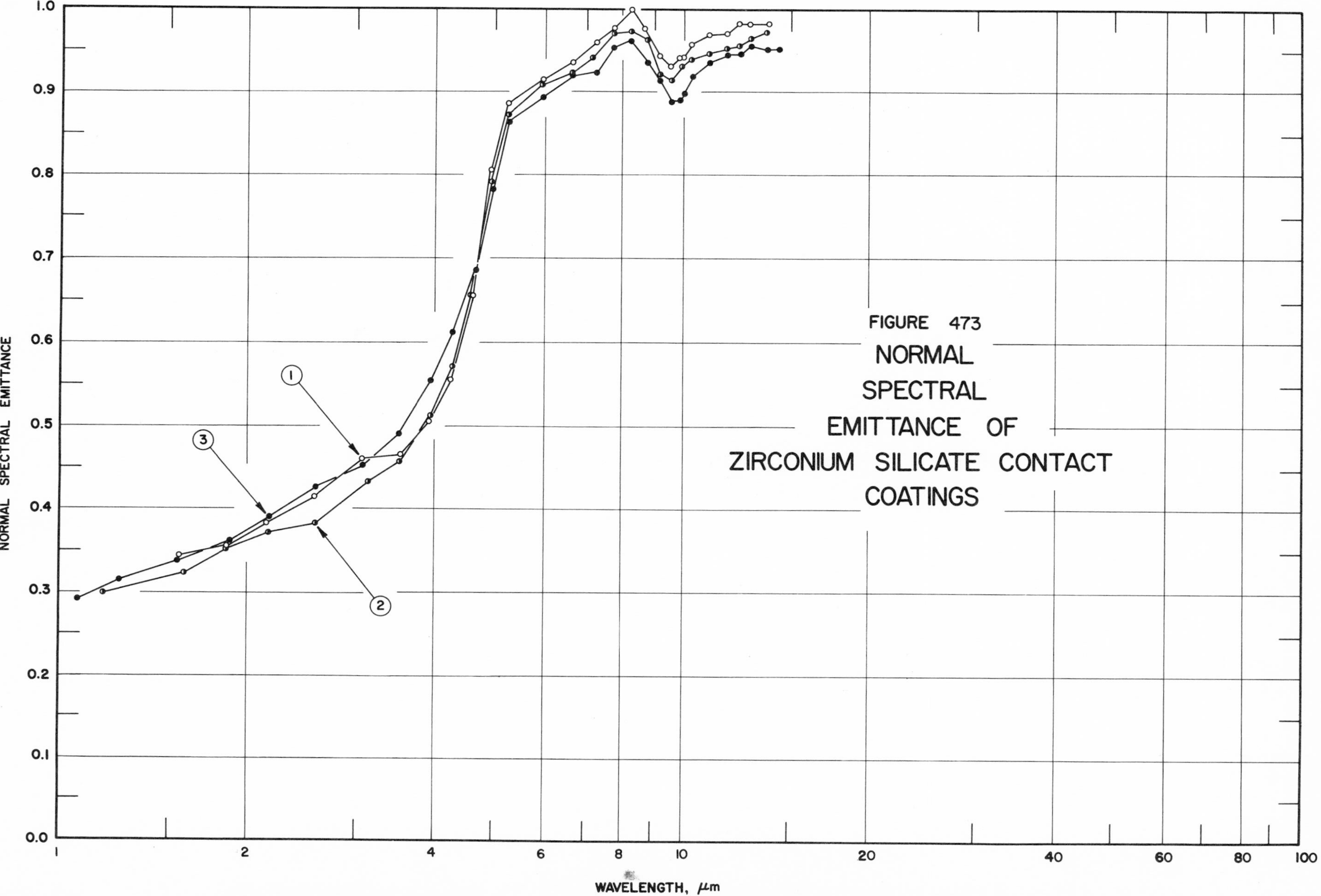

FIGURE 473 NORMAL SPECTRAL EMITTANCE OF ZIRCONIUM SILICATE CONTACT COATINGS

SPECIFICATION TABLE NO. 473 NORMAL SPECTRAL EMITTANCE OF ZIRCONIUM SILICATE CONTACT COATINGS

Curve No.	Ref. No.	Year	Temperature, K	Wavelength Range, μm	Geometry θ'	Reported Error, %	Composition (weight percent), Specifications, and Remarks
1	118	1962	755	1.56-13.57	$\sim 0°$		$ZrO_2 \cdot SiO_2$; FCZ-11 powder from Continental Coatings Corp. (0.102 mm thick); Nb-1 Zr alloy substrate; plasma arc sprayed; coating white, rough in texture, well bonded to substrate; measured in vacuum (2.0-2.6 x 10^{-6} mm Hg).
2	118	1962	1061	1.17-13.56	$\sim 0°$		Curve 1 specimen and conditions.
3	118	1962	1366	1.07-14.13	$\sim 0°$		Curve 1 specimen and conditions.

DATA TABLE NO. 473 NORMAL SPECTRAL EMITTANCE OF ZIRCONIUM SILICATE CONTACT COATINGS

[Wavelength, λ, μm; Emittance, ∈; Temperature, T, K]

λ	∈
CURVE 1 T = 755	
1.56	0.344
1.86	0.355
2.17	0.382
2.59	0.415
3.08	0.460
3.51	0.466
3.92	0.506
4.25	0.557
4.60	0.659
4.91	0.806
5.24	0.889
5.96	0.915
6.61	0.937
7.22	0.960
7.76	0.979
8.24	1.000
8.69	0.977
9.15	0.944
CURVE 1 (cont.)	
9.53	0.931
9.89	0.941
10.03	0.943
10.32	0.959
11.03	0.970
11.72	0.971
12.37	0.984
12.98	0.984
13.57	0.984
CURVE 2 T = 1061	
1.17	0.300
1.58	0.323
1.85	0.351
2.17	0.372
2.57	0.383
3.11	0.433
CURVE 2 (cont.)	
3.50	0.457
3.94	0.513
4.26	0.572
4.59	0.659
4.92	0.791
5.27	0.874
5.97	0.910
6.61	0.924
7.18	0.943
7.78	0.971
8.26	0.974
8.71	0.963
9.15	0.923
9.57	0.916
9.91	0.931
10.05	0.934*
10.35	0.940
11.05	0.949
11.74	0.954
CURVE 2 (cont.)	
12.39	0.957
12.99	0.966
13.56	0.974
CURVE 3 T = 1366	
1.07	0.293
1.25	0.316
1.55	0.339
1.87	0.362
2.18	0.391
2.59	0.429
3.06	0.452
3.50	0.491
3.94	0.556
4.26	0.612
4.65	0.686
4.96	0.783
CURVE 3 (cont.)	
5.25	0.867
5.95	0.895
6.61	0.920
7.21	0.924
7.77	0.955
8.23	0.962
8.73	0.939
9.19	0.916
9.55	0.890
9.85	0.892
10.04	0.900
10.35	0.920
11.08	0.937
11.74	0.946
12.41	0.949
12.99	0.959
13.57	0.952
14.13	0.953

* Not shown on plot

SPECIFICATION TABLE NO. 474 NORMAL SPECTRAL REFLECTANCE OF ZIRCONIUM SILICATE CONTACT COATINGS

Curve No.	Ref. No.	Year	Temperature, K	Wavelength Range, μm	Geometry θ	Geometry θ'	Geometry ω	Reported Error, %	Composition (weight percent), Specifications, and Remarks
1*	89	1958	~298	0.40-25.0	~0°	~0°			Rokide ZS (0.254 mm thick); 321 stainless steel substrate; flame sprayed; data extracted from smooth curve. [Author's designation: Sample 20]

DATA TABLE NO. 474 NORMAL SPECTRAL REFLECTANCE OF ZIRCONIUM SILICATE CONTACT COATINGS

[Wavelength, λ, μm; Reflectance, ρ; Temperature, T, K]

λ	ρ	λ	ρ	λ	ρ	λ	ρ	λ	ρ
CURVE 1* T ~298		CURVE 1 (cont.)		CURVE 1 (cont.)		CURVE 1 (cont.)		CURVE 1 (cont.)	
		0.96	0.753	5.73	0.102	11.6	0.081	19.5	0.301
0.40	0.488	0.97	0.752	6.25	0.082	12.3	0.080	19.7	0.294
0.41	0.532	1.00	0.750	6.52	0.092	12.6	0.074	20.0	0.294
0.42	0.563	1.21	0.621	6.74	0.092	14.0	0.074	20.2	0.334
0.44	0.591	1.52	0.604	7.27	0.073	14.8	0.105	20.7	0.334
0.45	0.648	1.77	0.621	7.73	0.032	15.0	0.123	21.2	0.354
0.49	0.699	1.94	0.621	8.04	0.032	15.2	0.115	21.7	0.354
0.50	0.718	2.24	0.612	8.28	0.052	15.7	0.133	22.0	0.325
0.52	0.741	2.45	0.559	8.51	0.087	16.2	0.133	22.2	0.325
0.55	0.762	2.76	0.276	8.78	0.099	16.7	0.162	22.4	0.333
0.58	0.762	3.01	0.292	8.94	0.137	16.9	0.163	22.7	0.295
0.60	0.770	3.25	0.362	8.98	0.152	17.3	0.183	23.0	0.330
0.69	0.771	3.75	0.453	9.23	0.161	17.7	0.183	23.2	0.294
0.72	0.780	4.07	0.448	9.44	0.135	17.9	0.192	25.0	0.295
0.82	0.780	4.62	0.283	9.75	0.124	18.1	0.213		
0.83	0.771	4.74	0.253	9.99	0.102	18.4	0.213		
0.87	0.771	5.03	0.111	10.3	0.102	18.7	0.249		
0.89	0.761	5.20	0.093	10.5	0.091	18.9	0.221		
0.94	0.761	5.52	0.093	11.3	0.091	19.2	0.248		

* No plot given

SPECIFICATION TABLE NO. 475 NORMAL SOLAR ABSORPTANCE OF ZIRCONIUM SILICATE CONTACT COATINGS

Curve No.	Ref. No.	Year	Temperature Range, K	Geometry θ	Reported Error, %	Composition (weight percent), Specifications and Remarks
1*	137	1965	400-600	0°	±5	Zirconium silicate (~0.063 mm thick, 8.3 mg cm^{-2} area density); Al substrate; plasma arc sprayed; measured in vacuum (10^{-7} mm Hg); surface appeared light grey before and after test.
2*	137	1965	400-600	0°	±5	Similar to above specimen and conditions except 0.127 mm thick, 29 mg cm^{-2} area density.
3*	137	1965	400-600	0°	±5	Similar to above specimen and conditions except 0.076 mm thick, 9.5 mg cm^{-2}.
4*	137	1965	400-600	0°	±5	Rokide ZS (~0.050 mm thick, 6.5 mg cm^{-2} area density); Al substrate; applied by Rokide process; measured in vacuum (10^{-7} mm Hg); surface appeared light grey with blue tint before and after test.
5*	137	1965	400-600	0°	±5	Similar to above specimen and conditions except 0.188 mm thick, 32 mg cm^{-2} area density.

DATA TABLE NO. 475 NORMAL SOLAR ABSORPTANCE OF ZIRCONIUM SILICATE CONTACT COATINGS

[Temperature, T, K; Absorptance, α]

T	α	T	α
CURVE 1*		CURVE 5*	
400	0.46	400	0.45
600	0.46	600	0.45
CURVE 2*			
400	0.37		
600	0.37		
CURVE 3*			
400	0.38		
600	0.38		
CURVE 4*			
400	0.54		
600	0.54		

* No plot given

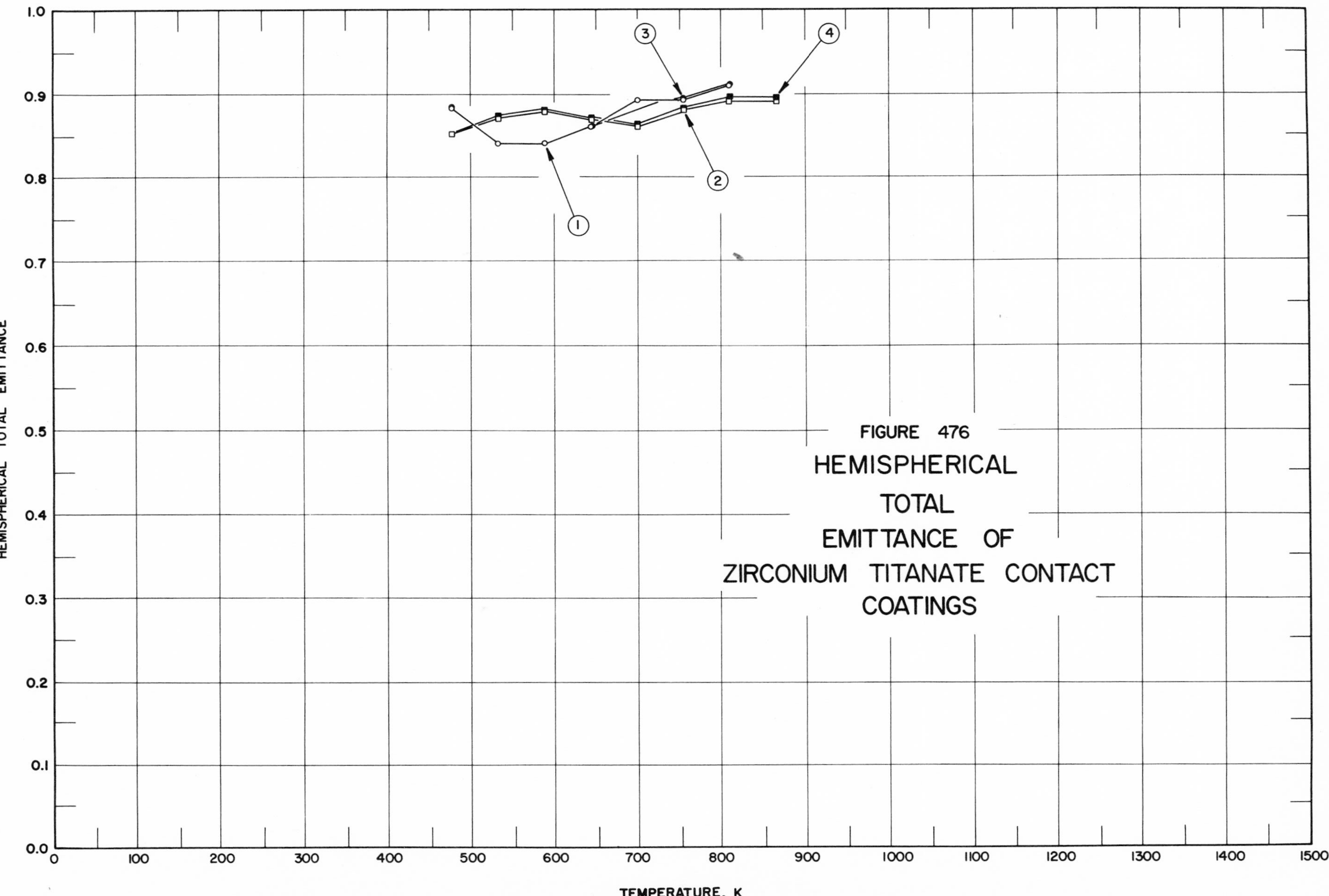

FIGURE 476 HEMISPHERICAL TOTAL EMITTANCE OF ZIRCONIUM TITANATE CONTACT COATINGS

SPECIFICATION TABLE NO. 476 HEMISPHERICAL TOTAL EMITTANCE OF ZIRCONIUM TITANATE CONTACT COATINGS

Curve No.	Ref. No.	Year	Temperature Range, K	Reported Error, %	Composition (weight percent), Specifications and Remarks
1	144	1966	478-810		Zirconium titanate; high purity Be substrate; substrate grit blasted to surface roughness of 2.16 μm AA; measured in vacuum (10^{-5} mm Hg).
2	144	1966	478-866		Above specimen and conditions except aged in vacuum for 1000 hrs at 922 K.
3	135	1969	478-812		$ZrTiO_4$, from Zirconium Corp. of America; high purity IS-2 beryllium substrate; plasma arc sprayed; measured in vacuum (10^{-5} mm Hg).
4	135	1969	478-866		Similar to above specimen and conditions except aged 1000 hrs at 922 K.

DATA TABLE NO. 476 HEMISPHERICAL TOTAL EMITTANCE OF ZIRCONIUM TITANATE CONTACT COATINGS

[Temperature, T, K; Emittance, ϵ]

T	ϵ	T	ϵ
CURVE 1		CURVE 3	
478	0.883	478	0.881
533	0.841	533	0.840*
589	0.841	589	0.841*
644	0.861	645	0.862
700	0.893	700	0.893*
755	0.893	756	0.894
810	0.910	812	0.911
CURVE 2		CURVE 4	
478	0.852	478	0.852*
533	0.870	533	0.874
589	0.878	589	0.881
644	0.870	645	0.872
700	0.862	700	0.864
755	0.880	756	0.884
810	0.892	812	0.897
866	0.892	866	0.895

* Not shown on plot

2. CONTACT COATINGS (continued)

C. Second Surface Mirrors

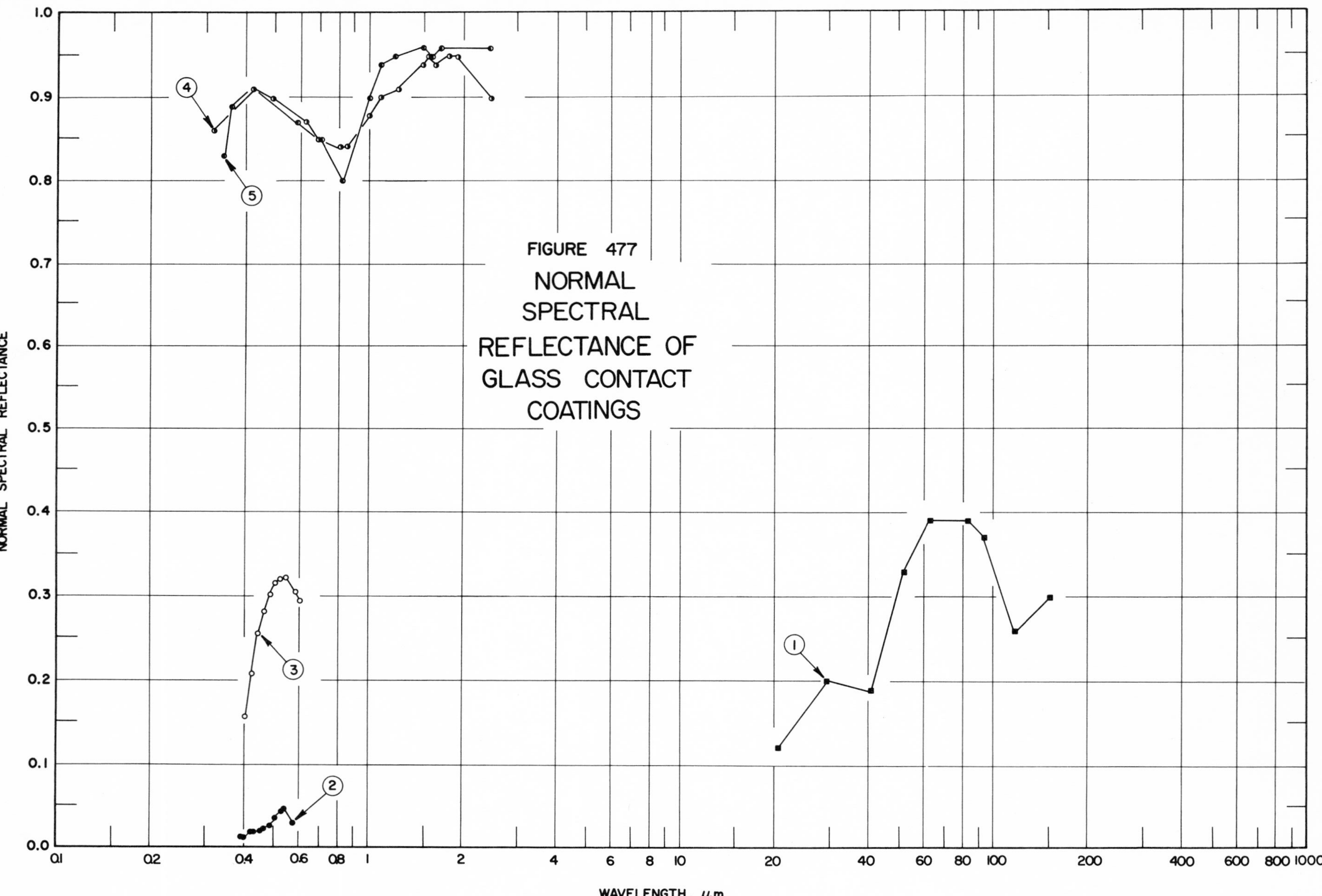

FIGURE 477 NORMAL SPECTRAL REFLECTANCE OF GLASS CONTACT COATINGS

SPECIFICATION TABLE NO. 477 NORMAL SPECTRAL REFLECTANCE OF GLASS CONTACT COATINGS

Curve No.	Ref. No.	Year	Temperature, K	Wavelength Range, μm	Geometry θ	θ'	ω'	Reported Error, %	Composition (weight percent), Specifications, and Remarks
1	149	1940	298	20.7-152	~0°	~0°			Kimball glass; aluminum substrate. [Authors' designation: Sample No. 45]
2	243	1968	298	0.390-0.570	~0°	~0°			Glass; silver (80 Å thick) substrate; silver vapor deposited in vacuum (10^{-5} mm Hg) on glass at 573 K; corrected for effects of glass.
3	243	1968	298	0.401-0.600	~0°	~0°			Similar to the above specimen and conditions except silver 160 Å thick.
4	115	1960	298	0.330-2.45	~15°		2π		Cover glass (0.158 mm thick); aluminum substrate; Al vacuum deposited on glass. [Authors' designation: Sample 23]
5	115	1960	298	0.340-2.45	~15°		2π		Microscope slide glass (1.0 mm thick); aluminum substrate; Al vacuum deposited on glass. [Authors' designation: Sample 24]

DATA TABLE NO. 477 NORMAL SPECTRAL REFLECTANCE OF GLASS CONTACT COATINGS

[Wavelength, λ, μm; Reflectance, ρ; Temperature, T, K]

λ	ρ	λ	ρ	λ	ρ	λ	ρ
CURVE 1	T = 298	CURVE 2 (cont.)		CURVE 3 (cont.)		CURVE 4 (cont.)	
		0.460	0.022	0.600	0.296	1.92	0.95
20.7	0.12	0.480	0.026			2.45	0.90
29.4	0.20	0.500	0.035	CURVE 4	T = 298		
41.0	0.19	0.521	0.044			CURVE 5	T = 298
52.0	0.33	0.538	0.047				
63.0	0.39	0.570	0.029	0.330	0.86		
83.0	0.39			0.420	0.91	0.340	0.83
94.0	0.37	CURVE 3	T = 298	0.480	0.90	0.360	0.89
117	0.26			0.620	0.87	0.420	0.91*
152	0.30			0.680	0.85	0.580	0.87
		0.401	0.157	0.800	0.79	0.700	0.85
CURVE 2	T = 298	0.421	0.209	0.850	0.79	0.820	0.80
		0.440	0.257	1.00	0.88	1.00	0.90
		0.460	0.281	1.08	0.90	1.08	0.94
0.390	0.014	0.480	0.302	1.24	0.91	1.20	0.95
0.400	0.014	0.500	0.315	1.48	0.94	1.48	0.96
0.420	0.018	0.519	0.320	1.56	0.95	1.60	0.95
0.430	0.018	0.540	0.323	1.64	0.94	1.72	0.96
0.449	0.018	0.580	0.306	1.80	0.95	2.45	0.96

* Not shown on plot

SPECIFICATION TABLE NO. 478 NORMAL SOLAR ABSORPTANCE OF GLASS CONTACT COATINGS

Curve No.	Ref. No.	Year	Temperature Range, K	Geometry θ	Reported Error, %	Composition (weight percent), Specifications and Remarks
1*	115	1960	298	15°		Microscope slide glass (1.0 mm thick); aluminum substrate; Al vacuum deposited on glass; computed from spectral reflectance data for above atm conditions. [Authors' designation: Sample 24]
2*	115	1960	298	15°		Cover glass (0.158 mm thick); aluminum substrate; Al vacuum deposited on glass; computed from spectral reflectance data for above atm conditions. [Authors' designation: Sample 23]

DATA TABLE NO. 478 NORMAL SOLAR ABSORPTANCE OF GLASS CONTACT COATINGS

[Temperature, T, K; Absorptance, α]

T	α
CURVE 1*	
298	0.12
CURVE 2*	
298	0.11

* No plot given

SPECIFICATION TABLE NO. 479 NORMAL SPECTRAL REFLECTANCE OF KAPTON CONTACT COATINGS

Curve No.	Ref. No.	Year	Temperature, K	Wavelength Range, μm	Geometry θ θ' ω'	Reported Error, %	Composition (weight percent), Specifications, and Remarks
1*	244	1966	300	0.27-14.00	~0° ~0°		"H" film (0.0254 mm thick); aluminum substrate; aluminized; data extracted from smooth curve.

DATA TABLE NO. 479 NORMAL SPECTRAL REFLECTANCE OF KAPTON CONTACT COATINGS

[Wavelength, λ, μm; Reflectance, ρ; Temperature, T, K]

λ	ρ	λ	ρ	λ	ρ	λ	ρ
CURVE 1* T = 300		CURVE 1 (cont.)*		CURVE 1 (cont.)*		CURVE 1 (cont.)*	
		3.98	0.830	7.38	0.181	10.73	0.174
0.27	0.111	4.27	0.646	7.51	0.181	11.09	0.398
0.45	0.086	4.40	0.698	7.67	0.134	11.26	0.110
0.64	0.777	4.48	0.688	7.84	0.109	11.51	0.206
0.94	0.871	4.55	0.793	7.91	0.114	12.00	0.125
1.14	0.917	4.75	0.761	7.99	0.174	12.10	0.125
1.42	0.948	4.94	0.814	8.28	0.131	12.23	0.144
2.00	0.950	5.13	0.649	8.42	0.140	12.50	0.576
2.33	0.938	5.27	0.691	8.64	0.210	12.71	0.363
2.47	0.920	5.53	0.548	8.76	0.131	12.80	0.345
2.55	0.896	5.57	0.611	8.90	0.112	12.90	0.438
2.63	0.908	5.82	0.118	9.18	0.132	13.08	0.344
2.78	0.598	6.01	0.164	9.37	0.186	13.28	0.575
3.03	0.772	6.15	0.222	9.52	0.186	13.47	0.135
3.48	0.436	6.26	0.135	9.78	0.558	13.56	0.151
3.55	0.749	6.53	0.275	9.83	0.527	13.80	0.553
3.63	0.742	6.76	0.148	9.86	0.245	14.00	0.674
3.78	0.797	6.91	0.158	9.92	0.268		
3.84	0.790	7.17	0.113	10.35	0.743		

* No plot given

FIGURE 480

CHANGE IN NORMAL SPECTRAL REFLECTANCE OF KAPTON CONTACT COATINGS

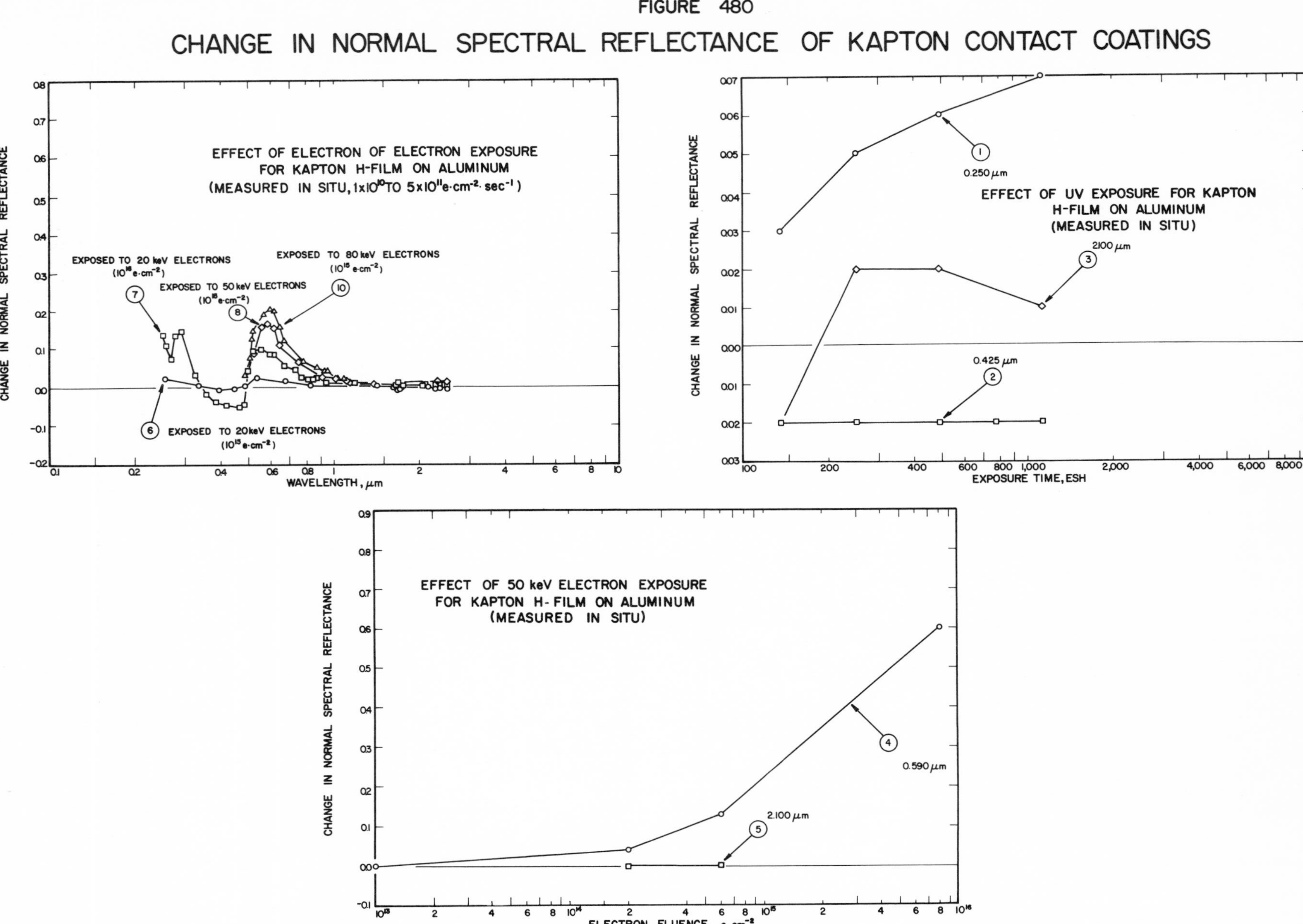

SPECIFICATION TABLE NO. 480 CHANGE IN NORMAL SPECTRAL REFLECTANCE OF KAPTON CONTACT COATINGS

Curve No.	Ref. No.	Year	Temperature, K	Wavelength Range, μm	Geometry θ	θ'	ω'	Reported Error, %	Composition (weight percent), Specifications, and Remarks
1	46	1969	295	0.250	~0°		2π		H-film (0.0508 mm thick); aluminum substrate; exposed to UV radiation (4.7 UV sun rate) in vacuum (10^{-8} mm Hg); vacuum maintained by ion pump; ESH is variable; reflectance measured in situ; positive change indicates decrease in reflectance from preirradiation, in air reflectance. [Authors' designation: N]
2	46	1969	295	0.425	~0°		2π		Above specimen and conditions.
3	46	1969	295	2.100	~0°		2π		Above specimen and conditions.
4	46	1969	295	0.590	~0°		2π		H-film (0.0508 mm thick); aluminum substrate; exposed to 50 keV electrons (4×10^{9}-1.7×10^{12} e cm^{-2} sec^{-1}) in vacuum (10^{-9} mm Hg); vacuum maintained by ion pump; electron fluence (e cm^{-2}) is variable; reflectance measured in situ; positive change indicates decrease in reflectance from preirradiation, in air reflectance. [Authors' designation: N]
5	46	1969	295	2.100	~0°		2π		Above specimen and conditions.
6	45	1970	281	0.255-2.500	~0°		2π		H-film (0.0508 mm thick); aluminum substrate; substrate held at 281 ± 2 K; exposed to 20 keV electrons in dark in vacuum (10^{-8} mm Hg); vacuum maintained by ion pump; flux 1×10^{10}-5×10^{11} e cm^{-2} sec^{-1}; reflectance measured in situ after 10^{15} e cm^{-2}; data extracted from smooth curve.
7	45	1970	281	0.250-2.490	~0°		2π		Above specimen and conditions except reflectance measured after 10^{16} e cm^{-2}.
8	45	1970	281	0.517-2.494	~0°		2π		Similar to curve 6 specimen and conditions except exposed to 50 keV electrons.
9	45	1970	281	0.491-2.490	~0°		2π		Similar to above specimen and conditions except exposed to 80 keV electrons in dark in vacuum (10^{-8} mm Hg); reflectance measured in situ after 3×10^{14} e cm^{-2}.
10	45	1970	281	0.489-2.494	~0°		2π		Above specimen and conditions except reflectance measured in situ after 10^{15} e cm^{-2}.
11	45	1970	281	0.250-2.497	~0°		2π		Above specimen and conditions except reflectance measured in situ after 1×10^{16} e cm^{-2}.

DATA TABLE NO. 480 CHANGE IN NORMAL SPECTRAL REFLECTANCE OF KAPTON CONTACT COATINGS

[Wavelength, λ, μm; Reflectance, ρ; Temperature, T, K]

CURVE 1 (λ = 0.250, T = 295)

ESH	Δρ
135	0.03
250	0.05
490	0.06
1130	0.07

CURVE 2 (λ = 0.425, T = 295)

ESH	Δρ
135	-0.02
250	-0.02
490	-0.02
770	-0.02
1130	-0.02

CURVE 3 (λ = 2.100, T = 295)

ESH	Δρ
135	-0.02
250	0.02
490	0.02
1130	0.01

CURVE 4 (λ = 0.590, T = 295)

Electron fluence	Δρ
1×10^{13}	0.00
2×10^{14}	0.04
6×10^{14}	0.13
8×10^{15}	0.60

CURVE 5 (λ = 2.100, T = 295)

Electron fluence	Δρ
2×10^{14}	0.0
6×10^{14}	0.0

CURVE 6 (T = 281)

λ	Δρ
0.255	0.024
0.334	0.006
0.395	-0.007
0.449	-0.001
0.487	0.005
0.532	0.026
0.673	0.018
0.824	0.008
1.412	0.004
1.634	0.000
1.657	-0.006
1.673	-0.006
1.702	0.000
2.144	0.000
2.276	-0.005
2.359	-0.002
2.500	-0.005

CURVE 7 (T = 281)

λ	Δρ
0.250	0.134
0.259	0.108
0.266	0.075
0.279	0.133
0.290	0.144
0.325	0.031
0.358	0.019
0.384	0.039
0.416	0.047
0.464	0.051
0.480	0.045
0.498	0.045
0.518	0.093
0.551	0.099
0.595	0.088
0.615	0.088
0.670	0.057
0.732	0.045
0.768	0.029
0.802	0.022
0.849	0.022
0.863	0.024

CURVE 7 (cont.)

λ	Δρ
0.940	0.011
1.178	0.011
1.648	0.006
1.673	0.011
1.745	0.005
2.029	0.003
2.049	0.007
2.141	0.003
2.267	0.007
2.399	0.004
2.434	0.007
2.490	0.007

CURVE 8 (T = 281)

λ	Δρ
0.517	0.094
0.551	0.158
0.581	0.167
0.615	0.155
0.641	0.110
0.758	0.065
0.903	0.029
1.005	0.024
1.110	0.018
1.134	0.014
1.398	0.010
2.038	0.010
2.355	0.013
2.494	0.013

CURVE 9 (T = 281)

λ	Δρ
0.491	0.015
0.511	0.030
0.526	0.052
0.565	0.068
0.600	0.074
0.659	0.046
0.802	0.022
0.849	0.022
0.863	0.024
0.940	0.011

CURVE 9 (cont.)

λ	Δρ
1.178	0.011
1.648	0.006
1.673	0.011
1.745	0.005
2.029	0.003
2.049	0.007
2.141	0.003
2.267	0.007
2.399	0.004
2.434	0.007
2.490	0.007

CURVE 10 (T = 281)

λ	Δρ
0.489	0.033
0.506	0.078
0.511	0.125
0.518	0.148
0.562	0.192
0.598	0.207
0.614	0.200
0.641	0.158
0.668	0.121
0.780	0.066
0.870	0.052
0.922	0.041
0.945	0.041
1.035	0.022
1.091	0.022
1.134	0.014
1.393	0.010
2.038	0.010
2.355	0.013
2.494	0.013

CURVE 11 (T = 281)

λ	Δρ
0.250	0.000
0.250	0.008
0.267	0.015
0.301	0.015
0.365	0.000

CURVE 11 (cont.)

λ	Δρ
0.415	0.000
0.474	0.014
0.483	0.024
0.524	0.543
0.553	0.636
0.572	0.660
0.600	0.666
0.616	0.657
0.672	0.484
0.702	0.428
0.718	0.424
0.755	0.342
0.800	0.285
0.893	0.205
1.032	0.109
1.137	0.057
1.197	0.040
1.237	0.037
1.262	0.029
1.612	0.019
1.628	0.014
1.873	0.019
2.250	0.020
2.497	0.022

SPECIFICATION TABLE NO. 481 NORMAL SOLAR ABSORPTANCE OF KAPTON CONTACT COATINGS

Curve No.	Ref. No.	Year	Temperature Range, K	Geometry θ	Reported Error, %	Composition (weight percent), Specifications and Remarks
1*	45	1969	298	$\sim 0°$		Type H (0.0508 mm thick); aluminum substrate; calculated from normal spectral reflectance.
2*	45	1969	281	$\sim 0°$		Above specimen and conditions except exposed to 20 keV electrons (flux 1 x 10^{10} – 5 x 10^{11} e cm^{-2} sec^{-1}) in dark in vacuum (10^{-8} mm Hg); vacuum maintained by ion pump; substrate held at 281 ± 2 K; property measured in situ after 10^{15} e cm^{-2}.
3*	45	1969	281	$\sim 0°$		Above specimen and conditions except property measured in situ after 10^{16} e cm^{-2}.
4*	45	1969	281	$\sim 0°$		Similar to curve 2 specimen and conditions except exposed to 80 keV electrons; property measured in situ after 3 x 10^{14} e cm^{-2}.
5*	45	1969	281	$\sim 0°$		Above specimen and conditions except property measured in situ after 10^{15} e cm^{-2}.
6*	45	1969	281	$\sim 0°$		Above specimen and conditions except property measured in situ after 10^{16} e cm^{-2}.

DATA TABLE NO. 481 NORMAL SOLAR ABSORPTANCE OF KAPTON CONTACT COATINGS

[Temperature, T, K; Absorptance, α]

T	α
CURVE 1*	
298	0.37
CURVE 2*	
281	0.38
CURVE 3*	
281	0.39
CURVE 4*	
281	0.39
CURVE 5*	
281	0.41
CURVE 6*	
281	0.52

* No plot given

SPECIFICATION TABLE NO. 482 HEMISPHERICAL TOTAL EMITTANCE OF MYLAR CONTACT COATINGS

Curve No.	Ref. No.	Year	Temperature Range, K	Reported Error, %	Composition (weight percent), Specifications and Remarks
1*	106	1963	311		Mylar; aluminum substrate; measured in vacuum (5 x 10^{-6} mm Hg). [Authors' designation: NRC-2]
2*	49	1961	278		Fascal chrome aluminized mylar film.

DATA TABLE NO. 482 HEMISPHERICAL TOTAL EMITTANCE OF MYLAR CONTACT COATINGS

[Temperature, T, K; Emittance, ϵ]

T	ϵ
CURVE 1*	
311	0.300
CURVE 2*	
278	0.09

* No plot given

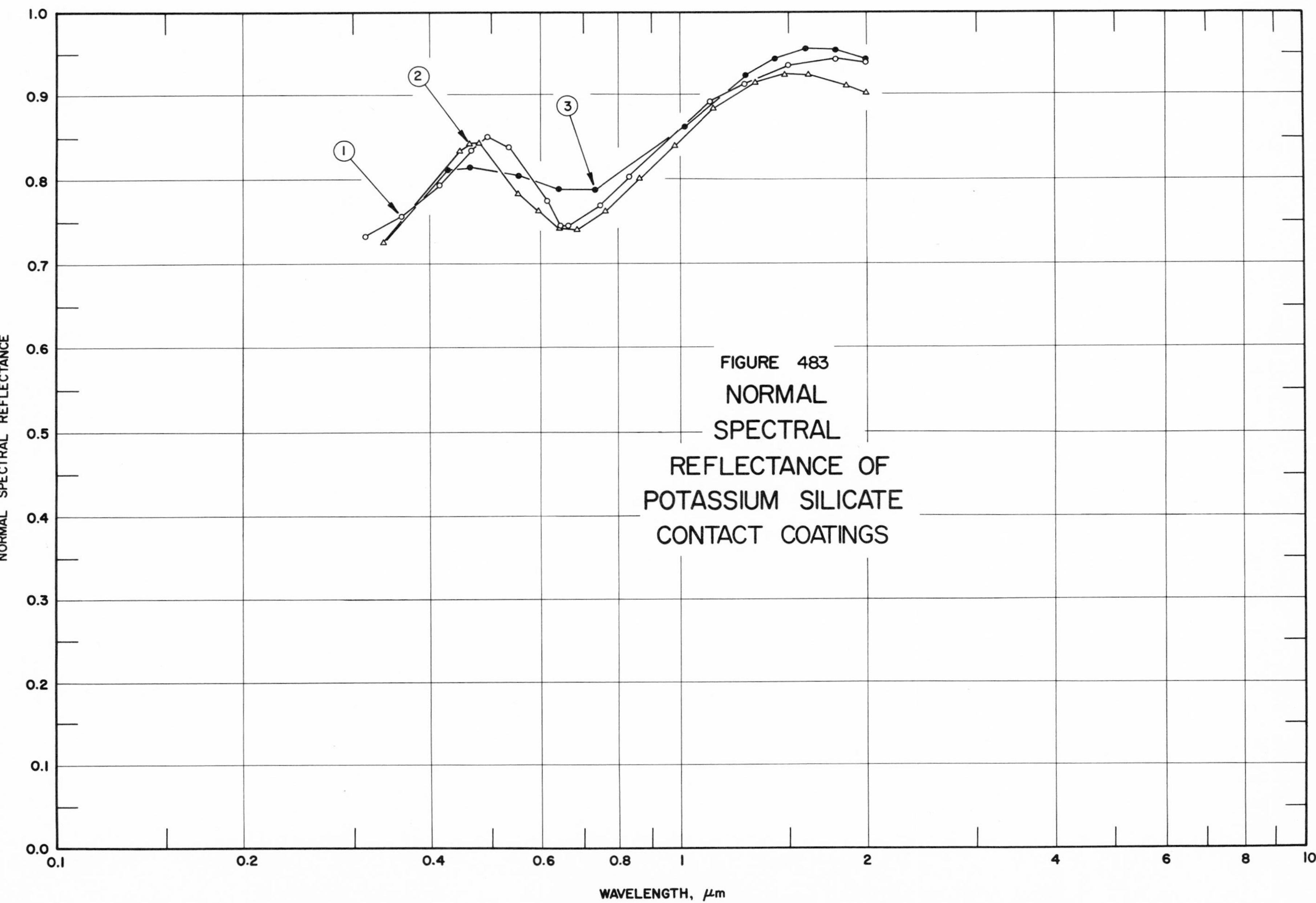

FIGURE 483
NORMAL SPECTRAL REFLECTANCE OF POTASSIUM SILICATE CONTACT COATINGS

SPECIFICATION TABLE NO. 483 NORMAL SPECTRAL REFLECTANCE OF POTASSIUM SILICATE CONTACT COATINGS

Curve No.	Ref. No.	Year	Temperature, K	Wavelength Range, μm	Geometry θ θ'	ω'	Reported Error, %	Composition (weight percent), Specifications, and Remarks
1	39	1970	298	0.316-2.000	~0°	2π		PS-7 K_2SiO_3; aluminum substrate; coating cast on substrate; exposed to 1200 eV protons from Ortec rf ion source in vacuum (5×10^{-7} mm Hg); fluence 2×10^{15} p cm^{-2}; measured in situ.
2	39	1970	298	0.338-2.000	~0°	2π		Similar to above specimen and conditions except also exposed to UV source for 265 ESH.
3	39	1970	298	0.338-2.000	~0°	2π		PS-7 K_2SiO_3; aluminum substrate; coating cast on substrate; exposed to UV from a Hanovia 500 w, d.c. xenon arc lamp for 265 ESH.

DATA TABLE NO. 483 NORMAL SPECTRAL REFLECTANCE OF POTASSIUM SILICATE CONTACT COATINGS

[Wavelength, λ, μm; Reflectance, ρ; Temperature, T, K]

λ	ρ	λ	ρ	λ	ρ
CURVE 1	T = 298	CURVE 2	T = 298	CURVE 3	T = 298
0.316	0.733	0.338	0.729	0.338	0.729*
0.360	0.758	0.447	0.836	0.429	0.811
0.414	0.795	0.461	0.845	0.463	0.817
0.465	0.837	0.480	0.845	0.558	0.806
0.493	0.851	0.553	0.785	0.642	0.790
0.535	0.840	0.598	0.763	0.733	0.790
0.616	0.775	0.645	0.744	1.027	0.864
0.647	0.748	0.688	0.741	1.287	0.926
0.689	0.748	0.766	0.762	1.427	0.946
0.750	0.770	0.866	0.803	1.604	0.956
0.832	0.804	0.983	0.841	1.790	0.956
1.139	0.893	1.140	0.887	2.000	0.943
1.271	0.915	1.330	0.916		
1.503	0.938	1.489	0.926		
1.790	0.944	1.618	0.926		
2.000	0.940	1.852	0.913		
		2.000	0.901		

* Not shown on plot

SPECIFICATION TABLE NO. 484 NORMAL SPECTRAL REFLECTANCE OF POTASSIUM TANTALATE CONTACT COATINGS

Curve No.	Ref. No.	Year	Temperature, K	Wavelength Range, μm	Geometry θ	θ'	ω'	Reported Error, %	Composition (weight percent), Specifications, and Remarks
1*	245	1966	298	0.463-0.750	15°	15°			$KTaO_3$; gold (3000 Å thick) substrate; gold vapor deposited in vacuum (2 x 10^{-6} mm Hg) for 30 sec; $KTaO_3$ flat and parallel single crystal.
2*	245	1966	298	0.480-0.750	15°	15°			$KTaO_3$; platinum (3000 Å thick) substrate; platinum evaporated in vacuum (4-6 x 10^{-6} mm Hg) for 6-7 min on flat and parallel single crystal $KTaO_3$.

DATA TABLE NO. 484 NORMAL SPECTRAL REFLECTANCE OF POTASSIUM TANTALATE CONTACT COATINGS

[Wavelength, λ, μm; Reflectance, ρ; Temperature, T, K]

λ	ρ	λ	ρ
CURVE 1* T = 298		CURVE 2* T = 298	
0.463	0.250	0.480	0.431
0.482	0.268	0.500	0.440
0.501	0.399	0.520	0.431
0.522	0.580	0.540	0.450
0.541	0.684	0.551	0.450
0.551	0.722	0.560	0.460
0.561	0.771	0.580	0.460
0.581	0.798	0.600	0.491
0.601	0.826	0.619	0.491
0.620	0.852	0.639	0.481
0.641	0.864	0.650	0.481
0.651	0.883	0.660	0.491
0.660	0.883	0.680	0.491
0.680	0.883	0.700	0.521
0.700	0.892	0.750	0.521
0.750	0.909		

* No plot given

SPECIFICATION TABLE NO. 485 NORMAL SPECTRAL REFLECTANCE OF SILICON CONTACT COATINGS

Curve No.	Ref. No.	Year	Temperature, K	Wavelength Range, μm	Geometry θ	θ'	ω'	Reported Error, %	Composition (weight percent), Specifications, and Remarks
1*	203	1962	298	0.339-3.399	~0°	~0°			Silicon, non-diffused (0.5 mm thick); aluminum substrate; aluminum evaporated on silicon; data extracted from smooth curve.

DATA TABLE NO. 485 NORMAL SPECTRAL REFLECTANCE OF SILICON CONTACT COATINGS

[Wavelength, λ, μm; Reflectance, ρ; Temperature, T, K]

λ	ρ
CURVE 1*	
T = 298	
0.339	0.518
0.404	0.427
0.553	0.358
0.732	0.341
0.840	0.327
0.958	0.330
0.998	0.384
1.055	0.603
1.115	0.800
1.164	0.855
1.332	0.879
1.445	0.874
1.872	0.885
2.052	0.870
2.502	0.872
2.645	0.865
3.194	0.860
3.399	0.858

* No plot given

SPECIFICATION TABLE NO. 486 NORMAL SOLAR ABSORPTANCE OF SILICON CONTACT COATINGS

Curve No.	Ref. No.	Year	Temperature Range, K	Geometry θ	Reported Error, %	Composition (weight percent), Specifications and Remarks
1*	203	1962	298	$\sim 0^\circ$		Silicon, crystalline and undoped (0.5 mm thick); aluminum substrate; aluminum evaporated on silicon.

DATA TABLE NO. 486 NORMAL SOLAR ABSORPTANCE OF SILICON CONTACT COATINGS

[Temperature, T, K; Absorptance, α]

T	α
CURVE 1*	
298	0.50

* No plot given

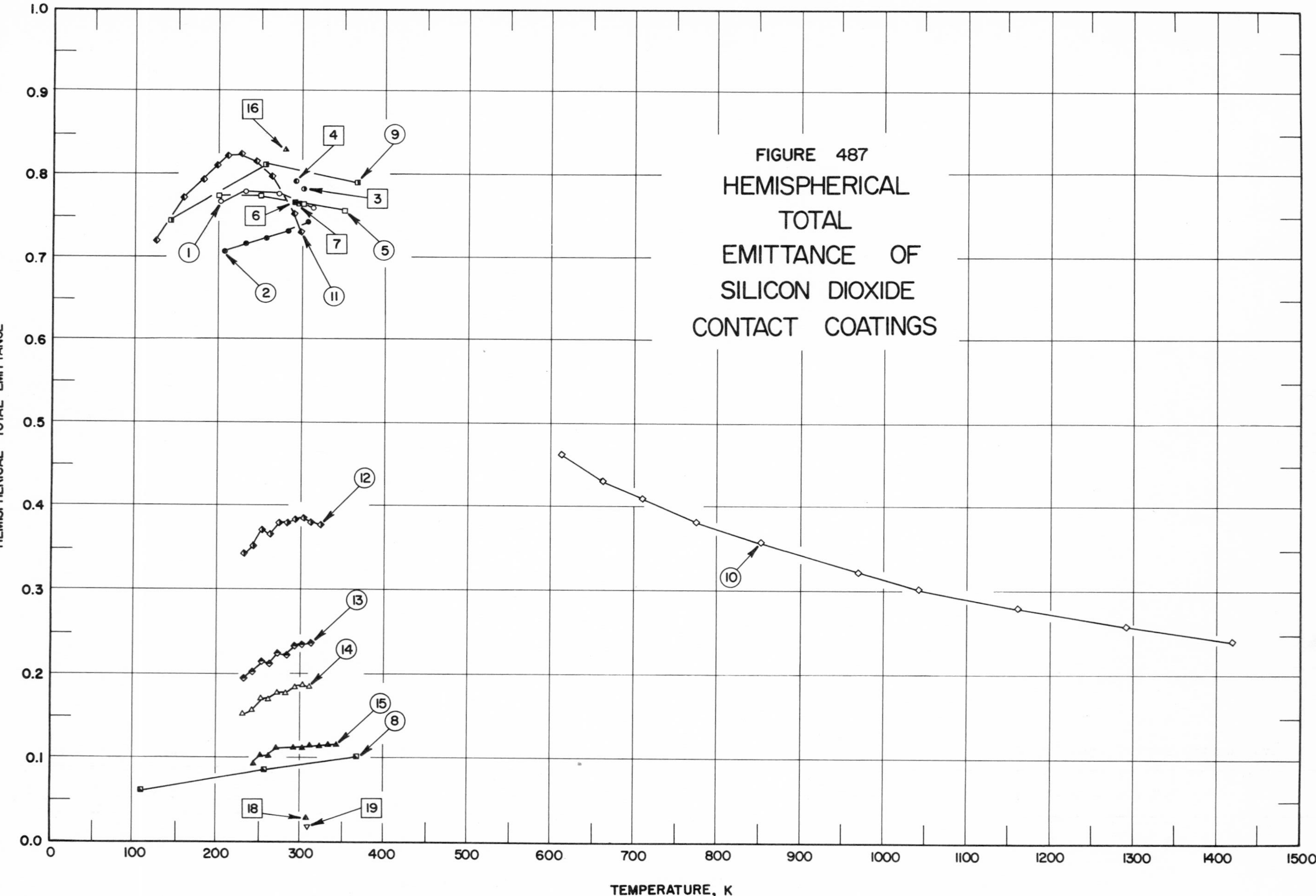

FIGURE 487
HEMISPHERICAL
TOTAL
EMITTANCE OF
SILICON DIOXIDE
CONTACT COATINGS
HEMISPHERICAL TOTAL EMITTANCE
TEMPERATURE, K
0.0
0.1
0.2
0.3
0.4
0.5
0.6
0.7
0.8
0.9
1.0
0
100
200
300
400
500
600
700
800
900
1000
1100
1200
1300
1400
1500
1
2
3
4
5
6
7
8
9
10
11
12
13
14
15
16
18
19

SPECIFICATION TABLE NO. 487 HEMISPHERICAL TOTAL EMITTANCE OF SILICON DIOXIDE CONTACT COATINGS

Curve No.	Ref. No.	Year	Temperature Range, K	Reported Error, %	Composition (weight percent), Specifications and Remarks
1	111	1969	203-313		Corning 7940 fused silica (0.2 mm thick); silver (~1 x 10^{-4} mm thick) and Inconel (~0.5 x 10^{-4} mm thick) substrates; silver then Inconel vapor deposited on the fused silica; property measured by steady state calorimetric method. [Authors' designation: OSR]
2	111	1969	208-308		Curve 1 specimen and conditions except property measured by transient calorimetric method; property of second side of sample assumed.
3	111	1969	301		Curve 1 specimen and conditions except property calculated from $\rho(10^\circ, 10^\circ)$ measured by specular method relative to a front surface aluminized mirror.
4	111	1969	293		Curve 1 specimen and conditions except property calculated from $R(2\pi, 7.5^\circ)$ measured by ellipsoid method.
5	111	1969	200-350		Curve 1 specimen and conditions except property calculated from $R(2\pi, 15^\circ\text{-}75^\circ)$ measured by heated cavity method relative to a platinum surface.
6	111	1969	290		Curve 1 specimen and conditions except property measured by a portable Quick Emittance Device.
7	111	1969	295		Curve 1 specimen and conditions except property measured by a portable emissometer.
8	119	1967	110-367		Quartz (silica), optical quality (3000 Å thick); silver, high purity (1810 Å thick) and highly polished 6061-T6 aluminum alloy substrates; silver then quartz vapor deposited.
9	119	1967	144-366		Quartz (fused silica); TP-060 silver (~2000 Å thick) substrate. [Authors' designation: LMSC Optical Solar Reflector]
10	246	1966	612-1573		Fused quartz; Taylor wire with a platinum-10% rhodium core substrate; measured in vacuum.
11	85	1969	126-299		Fused quartz; silver and Inconel substrates (~0.178 ± 0.025 mm thick); silver and Inconel vapor deposited; property measured in vacuum by calorimetric method; data extracted from smooth curve. [Author's designation: B-2 Optical Solar Reflector (OSR)]
12	247	1969	234-324		SiO_2 (~1.91 μm thick); Al and highly polished 304 stainless steel substrates; Al then SiO_2 vapor deposited.
13	247	1969	234-313		Similar to above specimen and conditions except SiO_2 ~1.18 μm thick.
14	247	1969	233-313		Similar to above specimen and conditions except SiO_2 ~0.974 μm thick.
15	247	1969	246-343		Similar to above specimen and conditions except SiO_2 ~0.606 μm thick.
16	49	1969	278	10	Silica (0.127 mm thick); magnesium substrate.
17*	67	1967	1750		Synar, silica mortar (Pennsalt Chemicals Corp.); D-36 niobium substrate.
18	105	1964	307		SiO_2; silver and polyester film substrates; measured in vacuum (10^{-6} mm Hg) maintained by diffusion pump. [Authors' designation: Test No. 222, Mfr. K, 647, Sample 10864-1-1E]
19	105	1964	307		Similar to above specimen and conditions. [Authors' designation: Test No. 223, Mfr. K, 648, Sample 10864-2-1E]

* Not shown on plot

DATA TABLE NO. 487 HEMISPHERICAL TOTAL EMITTANCE OF SILICON DIOXIDE CONTACT COATINGS

[Temperature, T, K; Emittance, ε]

T	ε
CURVE 1	
203	0.768
233	0.779
273	0.776
313	0.759
CURVE 2	
208	0.707
233	0.717
258	0.724
283	0.731
308	0.741
CURVE 3	
301	0.780
CURVE 4	
293	0.791
CURVE 5	
200	0.774
251	0.773
301	0.764
350	0.756
CURVE 6	
290	0.765
CURVE 7	
295	0.764
CURVE 8	
110	0.061
257	0.085
367	0.101
CURVE 9	
144	0.744
255	0.81
366	0.790
CURVE 10	
612	0.463
662	0.431
710	0.410
775	0.384
851	0.358
970	0.324
1044	0.304
1162	0.279
1292	0.258
1417	0.241
1502	0.230*
1573	0.222*
CURVE 11	
126	0.720
159	0.770
181	0.794
198	0.810
212	0.822
228	0.825
246	0.817
263	0.798
288	0.752
299	0.730
CURVE 12	
234	0.344
244	0.354
254	0.370
263	0.366
274	0.380
283	0.379
293	0.384
303	0.385
313	0.380
324	0.378
CURVE 13	
234	0.195
244	0.203
253	0.216
263	0.214
273	0.226
283	0.224
293	0.234
303	0.236
313	0.237
CURVE 14	
233	0.154
244	0.157
252	0.170
263	0.169
273	0.179
283	0.178
293	0.185
304	0.186
313	0.185
CURVE 15	
246	0.094
253	0.102
263	0.102
273	0.110
293	0.112
303	0.112
313	0.114
324	0.114
333	0.115
343	0.115
CURVE 16	
278	0.83
CURVE 17*	
1750	0.61
CURVE 18	
307	0.0227
CURVE 19	
307	0.0175

* Not shown on plot

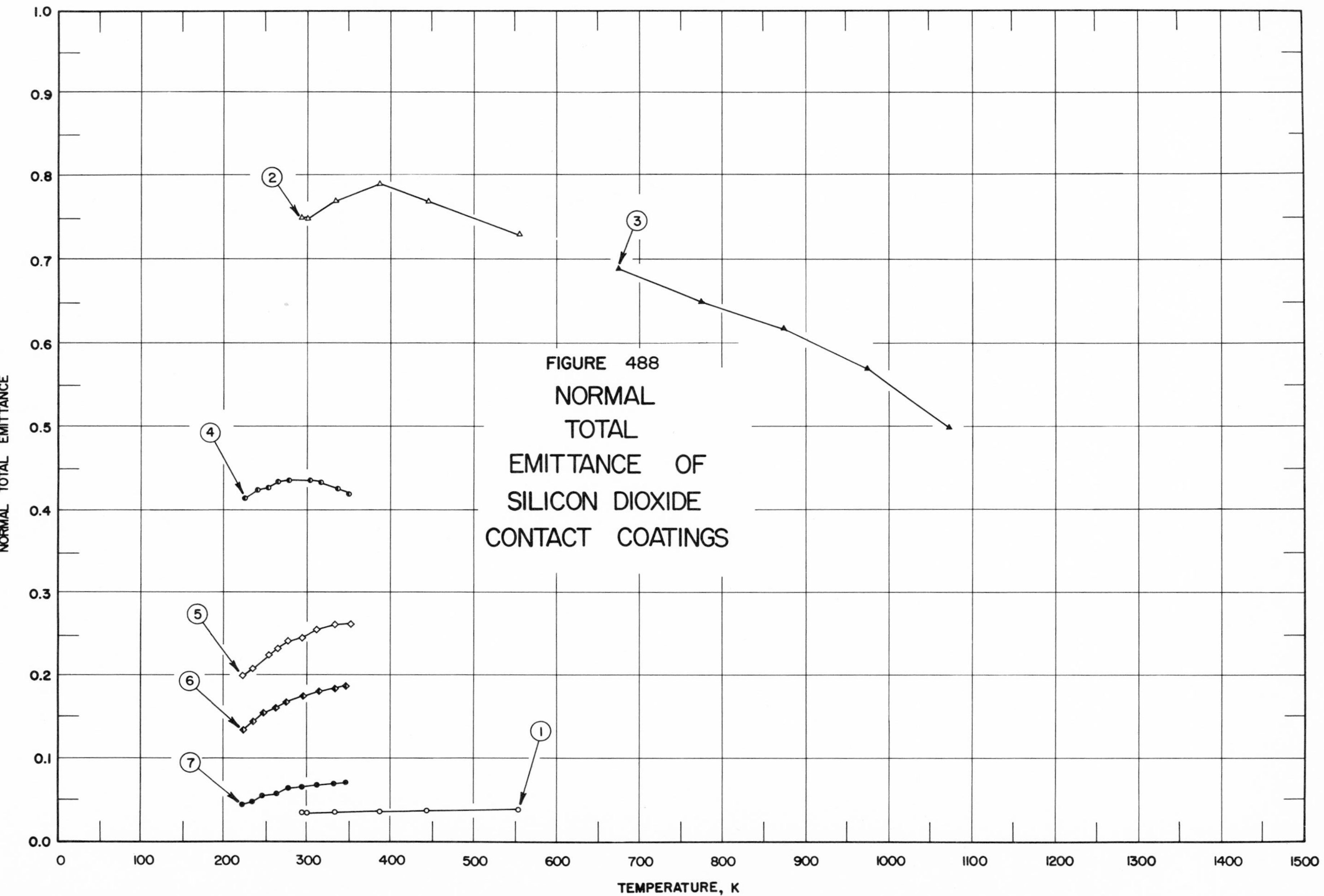

FIGURE 488 NORMAL TOTAL EMITTANCE OF SILICON DIOXIDE CONTACT COATINGS

SPECIFICATION TABLE NO. 488 NORMAL TOTAL EMITTANCE OF SILICON DIOXIDE CONTACT COATINGS

Curve No.	Ref. No.	Year	Temperature Range, K	Geometry θ'	Reported Error, %	Composition (weight percent), Specifications and Remarks
1	119	1967	294-555	$\sim 0^\circ$		Quartz (silica), optical quality (3000 Å thick); silver, high purity (1810 Å thick) and highly polished 6061-T6 aluminum alloy substrates; silver then quartz vapor deposited.
2	119	1967	294-555	$\sim 0^\circ$		Quartz (fused silica); TP-060 silver ($\sim$2000 Å thick) substrate; computed from spectral reflectance data. [Authors' designation: LMSC Optical Solar Reflector]
3	123	1952	673-1073	$\sim 0^\circ$		SiO_2; Nimonic 75 substrate; powder coating produced by crushed silica treated with 50 percent HCl.
4	247	1969	225-350	$\sim 0^\circ$		SiO_2 ($\sim$1.91 μm thick); Al and 304 stainless steel substrates; Al then SiO_2 vapor deposited; data extracted from smooth curve.
5	247	1969	224-351	$\sim 0^\circ$		Similar to above specimen and conditions except SiO_2 ($\sim$1.21 μm thick).
6	247	1969	224-349	$\sim 0^\circ$		Similar to above specimen and conditions except SiO_2 ($\sim$0.974 μm thick).
7	247	1969	221-349	$\sim 0^\circ$		Similar to above specimen and conditions except SiO_2 ($\sim$0.606 μm thick).

DATA TABLE NO. 488 NORMAL TOTAL EMITTANCE OF SILICON DIOXIDE CONTACT COATINGS

[Temperature, T, K; Emittance, ϵ]

T	ϵ
CURVE 1	
294	0.034
300	0.034
333	0.035
389	0.035
444	0.036
555	0.038
CURVE 2	
294	0.75
300	0.75
333	0.77
389	0.79
444	0.77
555	0.73
CURVE 3	
673	0.69
773	0.65
873	0.62
973	0.57
1073	0.50
CURVE 4	
225	0.413
240	0.423
252	0.429
265	0.435
279	0.437
302	0.437
318	0.434
338	0.427
350	0.420
CURVE 5	
224	0.200
234	0.209
252	0.225
265	0.233
276	0.241
292	0.248
312	0.257
333	0.261
351	0.262
CURVE 6	
224	0.132
235	0.143
249	0.153
262	0.160
275	0.166
295	0.174
314	0.180
332	0.184
349	0.187
CURVE 7	
221	0.043
234	0.048
247	0.053
262	0.058
279	0.062
294	0.065
312	0.067
333	0.069
349	0.070

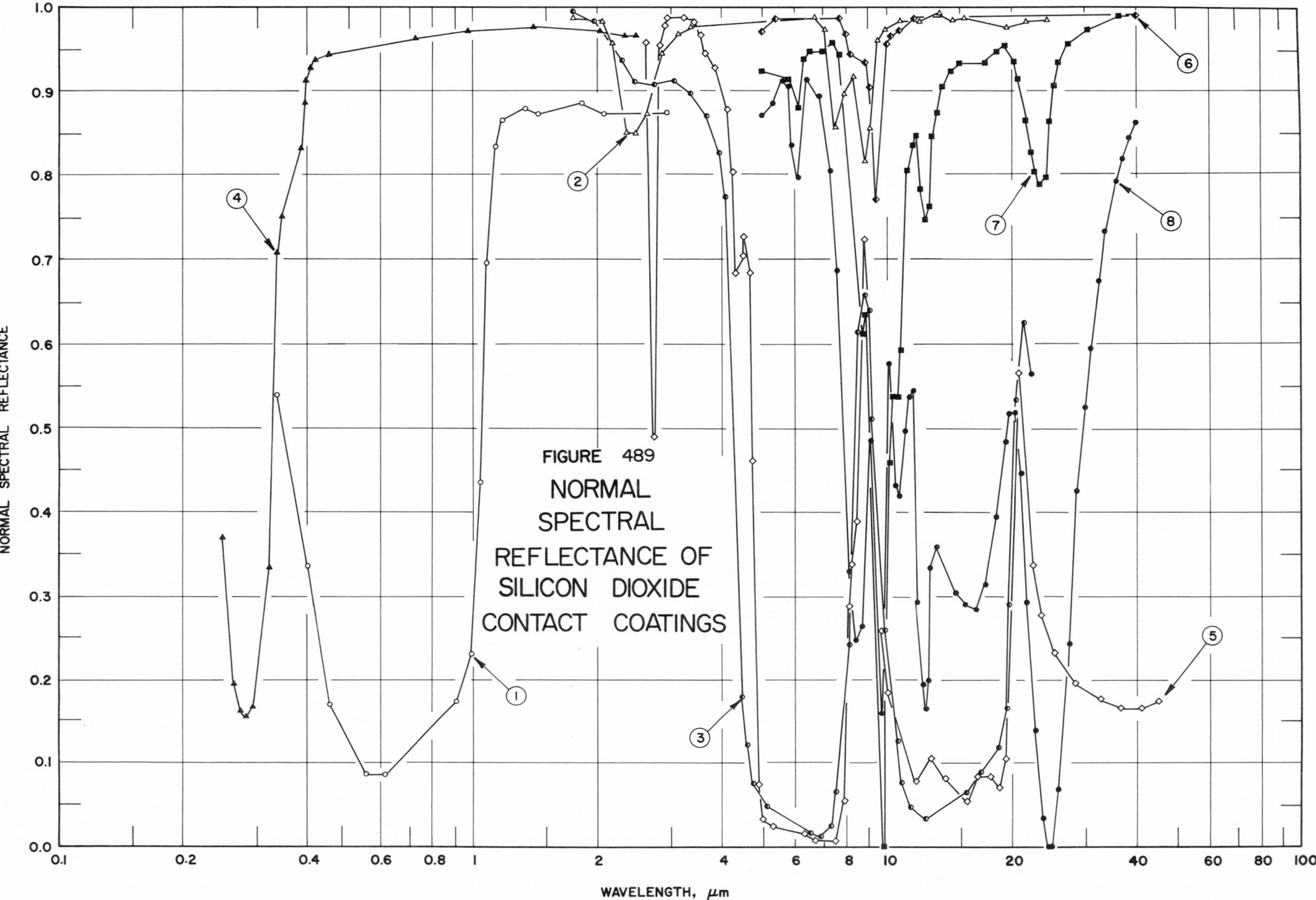

FIGURE 489 NORMAL SPECTRAL REFLECTANCE OF SILICON DIOXIDE CONTACT COATINGS

SPECIFICATION TABLE NO. 489 NORMAL SPECTRAL REFLECTANCE OF SILICON DIOXIDE CONTACT COATINGS

Curve No.	Ref. No.	Year	Temperature, K	Wavelength Range, μm	Geometry θ	θ'	ω'	Reported Error, %	Composition (weight percent), Specifications, and Remarks
1	203	1962	298	0.338-3.39	~0°	~0°			Silicon dioxide (0.0833 μm thick); aluminum substrate; aluminum vacuum evaporated on SiO_2; data extracted from smooth curve.
2	119	1967	~298	1.75 -24.0	15°		2π		Quartz (silica), optical quality (3000 Å thick); silver, high purity (1810 Å thick) and highly polished 6061-T6 aluminum alloy substrates; silver then quartz vapor deposited.
3	119	1967	~298	1.75 -22.2	15°		2π		Quartz, fused silica; TP-060 silver (~2000 Å thick) substrate. [Authors' designation: LMSC Optical Solar Reflector]
4	85	1969	~298	0.25 -2.4	~0°		2π		Fused quartz; silver and Inconel substrates (~0.178 ± 0.025 mm thick); silver and Inconel vapor deposited; data extracted from smooth curve. [Author's designation: B-2 Optical Solar Reflector (OSR)]
5	111	1969	~298	2.62-44.8	10°	10°			Corning 7940 fused silica (0.2 mm thick); silver (~1 x 10^{-4} mm thick) and Inconel (~0.5 x 10^{-4} mm thick) substrates; silver then Inconel vapor deposited on the fused silica; data extracted from smooth curve; property measured by specular method relative to a front surface aluminized mirror. [Authors' designation: OSR]
6	247	1969	~298	5.0-40.0	~0°	~0°			SiO_2 (0.40 μm thick); Al and highly polished 304 stainless steel substrates; Al then SiO_2 evaporation deposited; data extracted from smooth curve; measured relative to aluminum.
7	247	1969	~298	5.0-40.0	~0°	~0°			Similar to above specimen and conditions except SiO_2 0.97 μm thick.
8	247	1969	~298	5.0-40.0	~0°	~0°			Similar to above specimen and conditions except SiO_2 2.59 μm thick.

DATA TABLE NO. 489 NORMAL SPECTRAL REFLECTANCE OF SILICON DIOXIDE CONTACT COATINGS

[Wavelength, λ, μm; Reflectance, ρ; Temperature, T, K]

CURVE 1
T = 298

λ	ρ
0.338	0.538
0.400	0.335
0.449	0.168
0.556	0.085
0.611	0.084
0.908	0.171
0.997	0.230
1.04	0.432
1.09	0.695
1.14	0.835
1.19	0.865
1.33	0.879
1.44	0.873
1.84	0.886
2.09	0.871
2.93	0.875
3.39	0.878

CURVE 2
T ~ 298

λ	ρ
1.75	0.988
2.04	0.984
2.19	0.959
2.35	0.850
2.61	0.872
2.87	0.943
3.11	0.969
3.36	0.979
6.74	0.989
7.14	0.975
7.55	0.858
7.94	0.898
8.33	0.919
8.85	0.818
9.14	0.856
9.44	0.960
9.81	0.973
10.8	0.983
12.0	0.983
13.3	0.992
14.3	0.985
15.3	0.987
19.5	0.978
21.9	0.985
24.0	0.985

CURVE 3
T ~ 298

λ	ρ
1.75	0.995
1.99	0.984
2.30	0.936
2.46	0.910
2.73	0.907
3.08	0.911
3.37	0.899
3.69	0.870
3.95	0.828
4.06	0.775
4.44	0.179
4.60	0.120
4.79	0.074
5.12	0.047
6.47	0.017
6.98	0.012
7.29	0.025
7.56	0.064
8.01	0.240
8.47	0.614
8.85	0.658
9.03	0.640
9.20	0.510
9.84	0.259
10.5	0.125
10.8	0.075
11.3	0.048
12.2	0.034
15.4	0.064
16.9	0.088
18.5	0.119
19.3	0.163
19.7	0.290
20.2	0.531
21.4	0.626
22.2	0.561

CURVE 4
T ~ 298

λ	ρ
0.250	0.369
0.268	0.193
0.277	0.160
0.285	0.152
0.295	0.165
0.321	0.334
0.339	0.707
0.349	0.750
0.381	0.834
0.391	0.889
0.393	0.913
0.403	0.927
0.418	0.936
0.452	0.942
0.722	0.962
0.972	0.971
1.40	0.977
2.03	0.972
2.33	0.968
2.49	0.968

CURVE 5
T ~ 298

λ	ρ
2.62	0.959
2.74	0.489
2.84	0.954
2.91	0.978
2.99	0.988
3.22	0.988
3.42	0.981
3.56	0.969
3.64	0.945
3.88	0.929
4.12	0.878
4.21	0.802
4.35	0.684
4.45	0.703
4.50	0.727
4.63	0.683
4.75	0.460
4.84	0.073
5.00	0.033
5.34	0.025
6.29	0.017
6.72	0.008
7.43	0.008
7.81	0.054
8.03	0.289
8.20	0.338
8.41	0.389
8.81	0.724
9.66	0.257
10.0	0.183
11.7	0.077
12.8	0.105
13.8	0.080
15.5	0.054
16.4	0.081
17.8	0.081
18.6	0.070
19.1	0.102
20.8	0.563
22.5	0.338
23.4	0.277
25.4	0.231
28.4	0.196
32.5	0.177
36.8	0.163
41.6	0.163
44.8	0.172

CURVE 6
T ~ 298

λ	ρ
5.0	0.971
5.4	0.986
7.7	0.986
8.0	0.969
8.2	0.942
8.9	0.933
9.1	0.904
9.4	0.770
10.0	0.955
10.2	0.968
10.7	0.971
11.7	0.988
40.0	0.990

CURVE 7
T ~ 298

λ	ρ
5.0	0.924
5.8	0.912
6.1	0.880
6.3	0.938
6.5	0.947
7.0	0.947
7.4	0.957
7.7	0.942
8.7	0.611
8.9	0.637
9.8	0.000
10.1	0.458
10.4	0.538
10.6	0.538
10.8	0.593
11.3	0.804
11.5	0.837
11.7	0.847
12.0	0.783
12.4	0.749
12.6	0.762
12.9	0.848
13.2	0.875
13.7	0.903
14.3	0.923
15.3	0.933
17.2	0.933
18.4	0.949
19.1	0.951
20.2	0.932
20.8	0.913
21.6	0.864
22.1	0.828
22.8	0.801
23.5	0.790
24.1	0.798
24.6	0.862
25.3	0.909
26.0	0.933
27.3	0.957
30.6	0.973
36.4	0.990
40.0	0.990*

CURVE 8
T ~ 298

λ	ρ
5.0	0.871
5.3	0.886
5.6	0.911
5.8	0.906
5.9	0.838
6.1	0.799
6.4	0.913
6.9	0.893
7.3	0.803
7.6	0.687
8.1	0.330
8.4	0.248
8.7	0.263
9.1	0.485
9.7	0.158
10.1	0.575
10.5	0.430
10.6	0.419
11.0	0.497
11.2	0.536
11.3	0.545*
11.4	0.545
11.9	0.294
12.1	0.193
12.4	0.163
12.5	0.200
12.8	0.334
13.1	0.359
14.7	0.303
15.4	0.290
16.2	0.285
17.1	0.315
18.3	0.392
19.2	0.484
19.8	0.518
20.1	0.518
21.0	0.445
21.8	0.292
22.9	0.139
23.9	0.033
24.6	0.000
24.9	0.000
25.9	0.069
27.4	0.241
28.9	0.424
30.0	0.523
31.0	0.595
32.5	0.675
33.8	0.733
35.6	0.792
37.0	0.820
38.4	0.845
40.0	0.862

* Not shown on plot

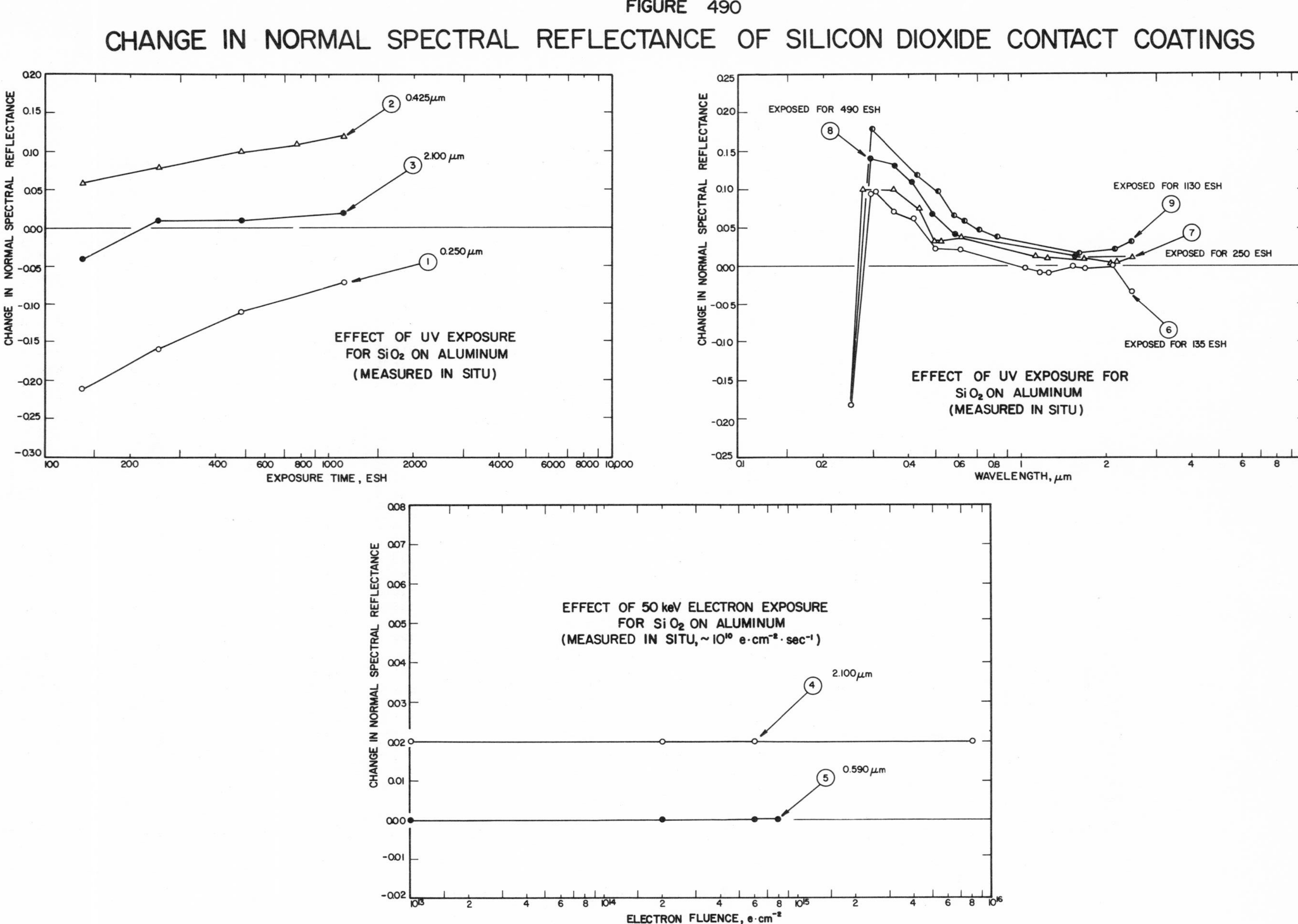
FIGURE 490
CHANGE IN NORMAL SPECTRAL REFLECTANCE OF SILICON DIOXIDE CONTACT COATINGS
CHANGE IN NORMAL SPECTRAL REFLECTANCE
0.425 μm
2.100 μm
0.250 μm
EFFECT OF UV EXPOSURE FOR SiO2 ON ALUMINUM (MEASURED IN SITU)
EXPOSURE TIME, ESH
EXPOSED FOR 490 ESH
EXPOSED FOR 1130 ESH
EXPOSED FOR 250 ESH
EXPOSED FOR 135 ESH
EFFECT OF UV EXPOSURE FOR SiO2 ON ALUMINUM (MEASURED IN SITU)
WAVELENGTH, μm
EFFECT OF 50 keV ELECTRON EXPOSURE FOR SiO2 ON ALUMINUM (MEASURED IN SITU, ~10^10 e·cm^-2·sec^-1)
2.100 μm
0.590 μm
ELECTRON FLUENCE, e·cm^-2

SPECIFICATION TABLE NO. 490 CHANGE IN NORMAL SPECTRAL REFLECTANCE OF SILICON DIOXIDE CONTACT COATINGS

Curve No.	Ref. No.	Year	Temperature, K	Wavelength Range, μm	Geometry θ	θ'	ω'	Reported Error, %	Composition (weight percent), Specifications, and Remarks
1	46	1969	295	0.250	~0°		2π		Silicon dioxide (2.5 x 10^{-3} mm thick); aluminum substrate; SiO_2 vacuum deposited; exposed to UV radiation (4.7 UV sun rate) in vacuum (10^{-8} mm Hg); vacuum maintained by ion pump; ESH is variable; reflectance measured in situ; positive change indicates decrease in reflectance from preirradiation, in air reflectance. [Authors' designation: H]
2	46	1969	295	0.425	~0°		2π		Above specimen and conditions.
3	46	1969	295	2.100	~0°		2π		Above specimen and conditions.
4	46	1969	295	2.100	~0°		2π		Silicon dioxide (2.5 x 10^{-3} mm thick); aluminum substrate; SiO_2 vacuum deposited; exposed to 50 keV electrons (4 x 10^{8} - 1.7 x 10^{12} e cm^{-2} sec^{-1}) in vacuum (10^{-8} mm Hg); vacuum maintained by ion pump; electron fluence (e cm^{-2}) is variable; reflectance measured in situ; positive change indicates decrease in reflectance from preirradiation, in air reflectance. [Authors' designation: H]
5	46	1969	295	0.590	~0°		2π		Above specimen and conditions.
6	46	1969	295	0.250-2.450	~0°		2π		Silicon dioxide (2.5 x 10^{-3} mm thick); aluminum substrate; SiO_2 vacuum deposited; exposed to UV radiation (with rates equivalent to 5 suns in UV) in vacuum (10^{-7} mm Hg); ESH is 135; vacuum maintained by ion pump; reflectance measured in situ; positive change indicates decrease in reflectance from preirradiation, in air reflectance; data extracted from smooth curve. [Authors' designation: H]
7	46	1969	295	0.250-2.450	~0°		2π		Above specimen and conditions except ESH is 250.
8	46	1969	295	2.50-2.450	~0°		2π		Above specimen and conditions except ESH is 490.
9	46	1969	295	0.250-2.450	~0°		2π		Above specimen and conditions except ESH is 1130.

DATA TABLE NO. 490 CHANGE IN NORMAL SPECTRAL REFLECTANCE OF SILICON DIOXIDE CONTACT COATINGS

[Wavelength, λ, μm; Reflectance, ρ; Temperature, T, K]

CURVE 1 ($\lambda = 0.250$, T = 295)

ESH	$\Delta\rho$
135	-0.21
250	-0.16
490	-0.11
1130	-0.07

CURVE 2 ($\lambda = 0.425$, T = 295)

ESH	$\Delta\rho$
135	0.06
250	0.08
490	0.10
770	0.11
1130	0.12

CURVE 3 ($\lambda = 2.100$, T = 295)

ESH	$\Delta\rho$
135	-0.04
250	0.01
490	0.01
1130	0.02

CURVE 4 ($\lambda = 2.100$, T = 295)

electron fluence	$\Delta\rho$
1×10^{13}	0.02
2×10^{14}	0.02
6×10^{14}	0.02
8×10^{15}	0.02

CURVE 5 ($\lambda = 0.590$, T = 295)

electron fluence	$\Delta\rho$
1×10^{13}	0.0
2×10^{14}	0.0
6×10^{14}	0.0
8×10^{14}	0.0

CURVE 6 (T = 295)

λ	$\Delta\rho$
0.250	-0.182*
0.294	0.095
0.307	0.096
0.358	0.072
0.415	0.062
0.494	0.024
0.617	0.022
1.040	-0.003
1.172	-0.008
1.255	-0.008
1.549	0.000
1.669	-0.003
2.101	0.000
2.450	-0.039

CURVE 7 (T = 295)

λ	$\Delta\rho$
0.250	-0.182*
0.275	0.100
0.356	0.100
0.437	0.075
0.493	0.033
0.514	0.033
0.606	0.039
1.136	0.013
1.240	0.010
1.655	0.010
2.074	0.002
2.198	0.005
2.450	0.012*

CURVE 8 (T = 295)

λ	$\Delta\rho$
0.250	-0.182*
0.292	0.140*
0.357	0.131*
0.410	0.110*
0.487	0.067
0.588	0.041
1.552	0.014
2.450	0.012*

CURVE 9 (T = 295)

λ	$\Delta\rho$
0.250	-0.182*
0.299	0.178*
0.430	0.118*
0.507	0.097
0.577	0.065
0.623	0.057
0.707	0.046
0.820	0.037
1.582	0.016
2.132	0.022
2.450	0.031

* Not shown on plot

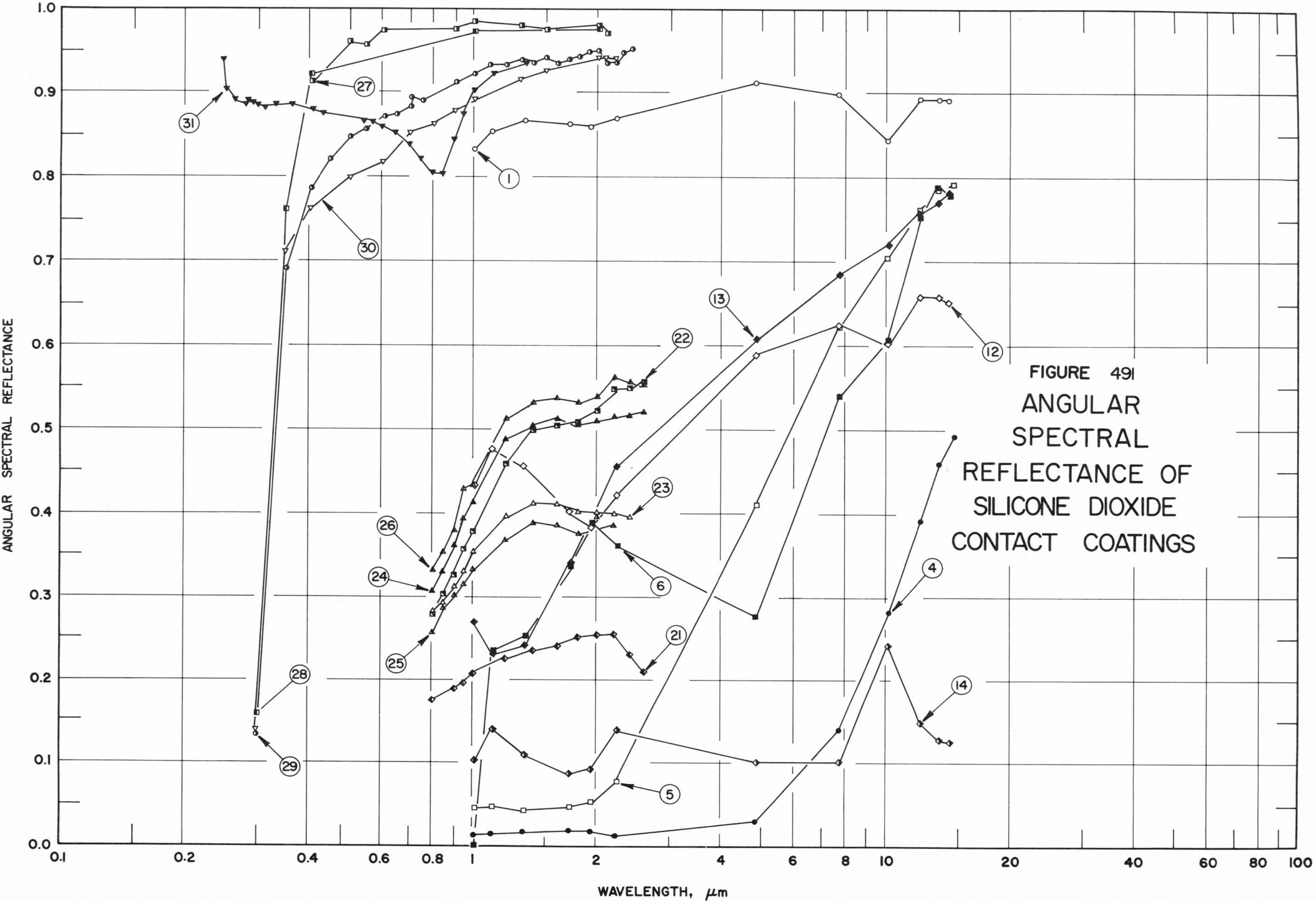
FIGURE 491
ANGULAR SPECTRAL REFLECTANCE OF SILICONE DIOXIDE CONTACT COATINGS
ANGULAR SPECTRAL REFLECTANCE
WAVELENGTH, μm
0.0
0.1
0.2
0.3
0.4
0.5
0.6
0.7
0.8
0.9
1.0
0.1
0.2
0.4
0.6
0.8
1
2
4
6
8
10
20
40
60
80
100
1
4
5
6
12
13
14
21
22
23
24
25
26
27
28
29
30
31

SPECIFICATION TABLE NO. 491 ANGULAR SPECTRAL REFLECTANCE OF SILICON DIOXIDE CONTACT COATINGS

Curve No.	Ref. No.	Year	Temperature, K	Wavelength Range, μm	Geometry θ	θ'	ω'	Reported Error, %	Composition (weight percent), Specifications, and Remarks
1	231	1962	~298	1.009-14.375	45°	45°			SiO_2 (~5000 Å thick); silver, epoxy, and 316 stainless steel substrates; stainless steel coated with Maraset 617-C epoxy resin (Marblette Co.), then with silver by immersion in a silver nitrate and dextrose solution for ~3 min (Brashear method), then SiO_2 evaporated in a residual atm of oxygen at a pressure of ~4 x 10^{-4} mm Hg; specular component only. [Authors' designation: AgSiO 71 C]
2*	231	1962	~298	1.009-14.375	45°	45°			Similar to above specimen and conditions. [Authors' designation: AgSiO 75 C]
3*	231	1962	~298	1.009-14.375	45°	45°			Similar to above specimen and conditions. [Authors' designation: AgSiO 76 C]
4	231	1962	~298	1.009-14.375	45°	45°			Similar to curve 1 specimen and conditions except stainless steel is coated with General Electric SR-111 silicone resin, then coated with silver and SiO_2. [Authors' designation: AgSiO 78 C]
5	231	1962	~298	1.009-14.375	45°	45°			Similar to curve 1 specimen and conditions except stainless steel is coated with SY627-119 polyurethane (Febert Shorndorfer Co.), then coated with silver and SiO_2. [Authors' designation: AgSiO 86 C]
6	231	1962	~298	1.009-14.375	45°	45°			Similar to curve 1 specimen and conditions except stainless steel is coated with SY627-119 polyurethane, then coated with gold (vapor deposited at ~10^{-5} mm Hg), then SiO_2. [Authors' designation: AuSiO 1 B]
7*	231	1962	~298	1.009-14.375	45°	45°			Similar to above specimen and conditions. [Authors' designation: AuSiO 2 B]
8*	231	1962	~298	1.009-14.375	45°	45°			Similar to above specimen and conditions. [Authors' designation: AuSiO 4 B]
9*	231	1962	~298	1.009-14.375	45°	45°			Similar to above specimen and conditions. [Authors' designation: AuSiO 5 B]
10*	231	1962	~298	1.009-14.375	45°	45°			Similar to above specimen and conditions except relatively rougher surface. [Authors' designation: AuSiO 3 B]
11*	231	1962	~298	1.009-14.375	45°	45°			Similar to above specimen and conditions. [Authors' designation: AuSiO 6 B]
12	231	1962	~298	1.009-14.375	45°	45°			Similar to curve 1 specimen and conditions except stainless steel is coated with Maraset 617-C epoxy resin, then coated with nickel by immersion in an electroless, alkaline, nickel hypophosphite solution, then SiO_2. [Authors' designation: NiSiO 2 C]
13	231	1962	~298	1.009-14.375	45°	45°			Similar to curve 1 specimen and conditions except stainless steel is coated with SY627-119 polyurethane, then coated with nickel by immersion in an electroless, alkaline, nickel hypophosphite solution, then SiO_2. [Authors' designation: NiSiO C]
14	231	1962	~298	1.009-14.375	45°	45°			Similar to curve 1 specimen and conditions except substrate is polished glass. [Authors' designation: SiO 7]
15*	231	1962	~298	1.009-14.375	45°	45°			Similar to above specimen and conditions. [Authors' designation: SiO 8]
16*	231	1962	~298	1.009-14.375	45°	45°			Similar to above specimen and conditions. [Authors' designation: SiO 9]
17*	231	1962	~298	1.009-14.375	45°	45°			Similar to above specimen and conditions. [Authors' designation: SiO 10]
18*	231	1962	~298	1.009-14.375	45°	45°			Similar to above specimen and conditions. [Authors' designation: SiO 11]

* Not shown on plot

SPECIFICATION TABLE NO. 491 ANGULAR SPECTRAL REFLECTANCE OF SILICON DIOXIDE CONTACT COATINGS (continued)

Curve No.	Ref. No.	Year	Temperature, K	Wavelength Range, μm	Geometry θ	θ'	ω'	Reported Error, %	Composition (weight percent), Specifications, and Remarks
19*	231	1962	~298	1.009-14.375	45°	45°			Similar to above specimen and conditions. [Authors' designation: SiO 12]
20*	231	1962	~298	1.009-14.375	45°	45°			Similar to above specimen and conditions. [Authors' designation: SiO 14]
21	208	1964	298	0.80-2.60	45°	45°			Fused quartz; platinum (~0.102 μm thick) substrate; platinum applied to rear surface of quartz by brushing on a platinum solution and firing; measured relative to aluminum mirror; data beyond 2.2 μm subject to sizable errors. [Author's designation: Specimen No. 3B]
22	208	1964	298	0.80-2.60	15°	15°			Above specimen and conditions.
23	208	1964	298	0.80-2.40	45°	45°			Similar to curve 21 specimen and conditions. [Author's designation: Specimen No. 4B]
24	208	1964	298	0.80-2.60	15°	15°			Above specimen and conditions.
25	208	1964	298	0.80-2.40	45°	45°			Similar to curve 21 specimen and conditions. [Author's designation: Specimen No. 5B]
26	208	1964	298	0.80-2.60	15°	15°			Above specimen and conditions.
27	248	1967	~298	0.407-2.114	25°		2π		Quartz (1.02 mm thick); opaque vacuum-deposited silver film substrate. [Author's designation: Clear No. 1]
28	248	1967	~298	0.303-2.018	25°		2π		Similar to above specimen and conditions. [Author's designation: Clear No. 2]
29	248	1967	~298	0.300-2.438	25°		2π		Quartz (1.02 mm thick); quartz roughened by conventional glass grinding techniques (grit size 3 and 9 μm dia.); roughness height 5 μm rms; opaque vacuum-deposited silver film substrate. [Authors' designation: Sample A]
30	248	1967	~298	0.300-2.218	25°		2π		Similar to above specimen and conditions except roughness height 9 μm rms. [Authors' designation: Sample B]
31	207	1957	298	0.2482-1.3570	~20°	~20°		<2	Fused quartz; Al, 99.99 pure, substrate; Al vapor deposited; data is mean of five specimens.

* Not shown on plot

DATA TABLE NO. 491 ANGULAR SPECTRAL REFLECTANCE OF SILICON DIOXIDE CONTACT COATINGS

[Wavelength, λ, μm; Reflectance, ρ; Temperature, T, K]

CURVE 1 T ~ 298

λ	ρ
1.009	0.835
1.120	0.854
1.345	0.867
1.720	0.863
1.945	0.860
2.240	0.870
4.824	0.913
7.780	0.898
10.198	0.844
12.099	0.894
13.530	0.893
14.375	0.892

CURVE 2* T ~ 298

λ	ρ
1.009	0.905
1.120	0.898
1.345	0.894
1.720	0.888
1.945	0.880
2.240	0.860
4.824	0.857
7.780	0.853
10.198	0.867
12.099	0.881
13.530	0.880
14.375	0.897

CURVE 3* T ~ 298

λ	ρ
1.009	0.890
1.120	0.884
1.345	0.888
1.720	0.892
1.945	0.895
2.240	0.895
4.824	0.900
7.780	0.899
10.198	0.905
12.099	0.925
13.530	0.920
14.375	0.913

CURVE 4 T ~ 298

λ	ρ
1.009	0.014
1.120	0.016
1.345	0.017
1.720	0.018
1.945	0.018
2.240	0.013
4.824	0.030
7.780	0.138
10.198	0.281
12.099	0.391
13.530	0.458
14.375	0.492

CURVE 5 T ~ 298

λ	ρ
1.009	0.046
1.120	0.047
1.345	0.044
1.720	0.047
1.945	0.054
2.240	0.078
4.824	0.410
7.780	0.624
10.198	0.705
12.099	0.764
13.530	0.788
14.375	0.792

CURVE 6 T ~ 298

λ	ρ
1.009	0.008
1.120	0.0236
1.345	0.0251
1.720	0.0337
1.945	0.0388
2.240	0.0360
4.824	0.275
7.780	0.537
10.198	0.607
12.099	0.753
13.530	0.789
14.375	0.780

CURVE 7* T ~ 298

λ	ρ
1.009	0.0231
1.120	0.0236
1.345	0.0235
1.720	0.0275
1.945	0.0273
2.240	0.0371
4.824	0.0591
7.780	0.320
10.198	0.472
12.099	0.625
13.530	0.685
14.375	0.713

CURVE 8* T ~ 298

λ	ρ
1.009	0.0486
1.120	0.0344
1.345	0.0324
1.720	0.0306
1.945	0.0366
2.240	0.0575
4.824	0.0792
7.780	0.361
10.198	0.522
12.099	0.658
13.530	0.702
14.375	0.730

CURVE 9* T ~ 298

λ	ρ
1.009	0.0278
1.120	0.0316
1.345	0.0273
1.720	0.0307
1.945	0.0305
2.240	0.0410
4.824	0.0443
7.780	0.0265
10.198	0.425
12.099	0.595
13.530	0.658
14.375	0.688

CURVE 10* T ~ 298

λ	ρ
1.009	0.236
1.120	0.219
1.345	0.229
1.720	0.294
1.945	0.343
2.240	0.403
4.824	0.706
7.780	0.816
10.198	0.783
12.099	0.860
13.530	0.868
14.375	0.868

CURVE 11* T ~ 298

λ	ρ
1.009	0.598
1.120	0.597
1.345	0.607
1.720	0.633
1.945	0.627
2.240	0.670
4.824	0.771
7.780	0.788
10.198	0.766
12.099	0.818
13.530	0.832
14.375	0.856

CURVE 12 T ~ 298

λ	ρ
1.009	0.432
1.120	0.476
1.345	0.456
1.720	0.402
1.945	0.384
2.240	0.422
4.824	0.588
7.780	0.625
10.198	0.604
12.099	0.658
13.530	0.658
14.375	0.654

CURVE 13 T ~ 298

λ	ρ
1.009	0.268
1.120	0.231
1.345	0.242
1.720	0.340
1.945	0.383*
2.240	0.456
4.824	0.607
7.780	0.685
10.198	0.720
12.099	0.757
13.530	0.771
14.375	0.786

CURVE 14 T ~ 298

λ	ρ
1.009	0.103
1.120	0.139
1.345	0.109
1.720	0.085
1.945	0.091
2.240	0.139
4.824	0.100
7.780	0.100
10.198	0.242
12.099	0.147
13.530	0.127
14.375	0.125

CURVE 15* T ~ 298

λ	ρ
1.009	0.086
1.120	0.078
1.345	0.076
1.720	0.073
1.945	0.071
2.240	0.063
4.824	0.030
7.780	0.000
10.198	0.181
12.099	0.067
13.530	0.071
14.375	0.060

CURVE 16* T ~ 298

λ	ρ
1.009	0.090
1.120	0.087
1.345	0.080
1.720	0.072
1.945	0.068
2.240	0.062
4.824	0.025
7.780	0.000
10.198	0.165
12.099	0.062
13.530	0.067
14.375	0.054

CURVE 17* T ~ 298

λ	ρ
1.009	0.124
1.120	0.146
1.345	0.149
1.720	0.143
1.945	0.135
2.240	0.123
4.824	0.048
7.780	0.005
10.198	0.187
12.099	0.071
13.530	0.080
14.375	0.067

CURVE 18* T ~ 298

λ	ρ
1.009	0.098
1.120	0.076
1.345	0.100
1.720	0.140
1.945	0.144
2.240	0.149
4.824	0.068
7.780	0.016
10.198	0.200
12.099	0.078
13.530	0.079
14.375	0.074

CURVE 19* T ~ 298

λ	ρ
1.009	0.133
1.120	0.092
1.345	0.077
1.720	0.116
1.945	0.126
2.240	0.135
4.824	0.072
7.780	0.002
10.198	0.188
12.099	0.083
13.530	0.082
14.375	0.073

CURVE 20* T ~ 298

λ	ρ
1.009	0.098
1.120	0.100
1.345	0.087
1.720	0.077
1.945	0.075
2.240	0.072
4.824	0.029
7.780	0.003
10.198	0.174
12.099	0.070
13.530	0.073
14.375	0.060

CURVE 21 T = 298

λ	ρ
0.80	0.173
0.90	0.187
0.95	0.195
1.00	0.207
1.20	0.226
1.40	0.235
1.60	0.239
1.80	0.250
2.00	0.253
2.20	0.254
2.40	0.230
2.60	0.210

CURVE 22 T = 298

λ	ρ
0.80	0.277
0.85	0.303
0.90	0.327
0.95	0.356
1.00	0.378
1.20	0.458
1.40	0.499
1.60	0.504
1.80	0.507
2.00	0.520
2.20	0.547
2.40	0.547
2.60	0.555

CURVE 23 T = 298

λ	ρ
0.80	0.281
0.85	0.292
0.90	0.312
0.95	0.330
1.00	0.354
1.20	0.395
1.40	0.411
1.60	0.411
1.80	0.401
2.00	0.401
2.20	0.399
2.40	0.396

CURVE 24 T = 298

λ	ρ
0.80	0.308
0.85	0.330
0.90	0.361
0.95	0.394
1.00	0.414
1.20	0.488
1.40	0.505
1.60	0.511
1.80	0.506
2.00	0.509
2.20	0.515
2.40	0.515
2.60	0.519

* Not shown on plot

DATA TABLE NO. 491 ANGULAR SPECTRAL REFLECTANCE OF SILICON DIOXIDE CONTACT COATINGS (continued)

CURVE 25
T = 298

λ	ρ
0.80	0.256
0.85	0.285
0.90	0.301
0.95	0.315
1.00	0.333
1.20	0.368
1.40	0.387
1.60	0.386
1.80	0.376
2.00	0.397
2.20	0.386
2.40	0.396*

CURVE 26
T = 298

λ	ρ
0.80	0.333
0.85	0.352
0.90	0.379
0.95	0.428
1.00	0.431
1.20	0.512
1.40	0.530
1.60	0.535
1.80	0.531
2.00	0.538
2.20	0.561
2.40	0.556
2.60	0.552

CURVE 27
T ~ 298

λ	ρ
0.407	0.914
0.502	0.962
0.556	0.957
0.605	0.975
0.903	0.976
1.009	0.986
1.318	0.982
1.514	0.977
2.014	0.984
2.114	0.974

CURVE 28
T ~ 298

λ	ρ
0.303	0.156
0.353	0.763
0.404	0.922
1.009	0.975
2.018	0.977

CURVE 29
T ~ 298

λ	ρ
0.300	0.134
0.354	0.693
0.404	0.787*
0.454	0.821*
0.502	0.847
0.553	0.856
0.609	0.871
0.653	0.876
0.708	0.884
0.709	0.895
0.758	0.890
0.906	0.913
1.011	0.922
1.112	0.934
1.213	0.933
1.315	0.939
1.409	0.935
1.510	0.941
1.614	0.935
1.710	0.939
1.811	0.942
1.910	0.949
2.014	0.950
2.104	0.935
2.229	0.935
2.328	0.948
2.438	0.954

CURVE 30
T ~ 298

λ	ρ
0.300	0.138
0.352	0.711
0.402	0.763

CURVE 30 (cont.)

λ	ρ
0.504	0.798
0.605	0.818
0.703	0.853
0.805	0.864
0.903	0.879
1.007	0.892
1.312	0.915
1.514	0.926
2.018	0.940
2.119	0.941
2.218	0.941

CURVE 31
T = 298

λ	ρ
0.2482	0.937
0.2537	0.901
0.2653	0.891
0.2804	0.886
0.2894	0.890
0.2968	0.887
0.3022	0.886
0.3132	0.881
0.3342	0.886
0.3650	0.885
0.4047	0.880
0.4358	0.876
0.5461	0.867
0.5780	0.865
0.6000	0.860
0.6500	0.852
0.7000	0.840
0.7500	0.822
0.8000	0.805
0.8500	0.804
0.9000	0.846
0.9500	0.876
1.0140	0.901
1.1287	0.923
1.3570	0.935

* Not shown on plot

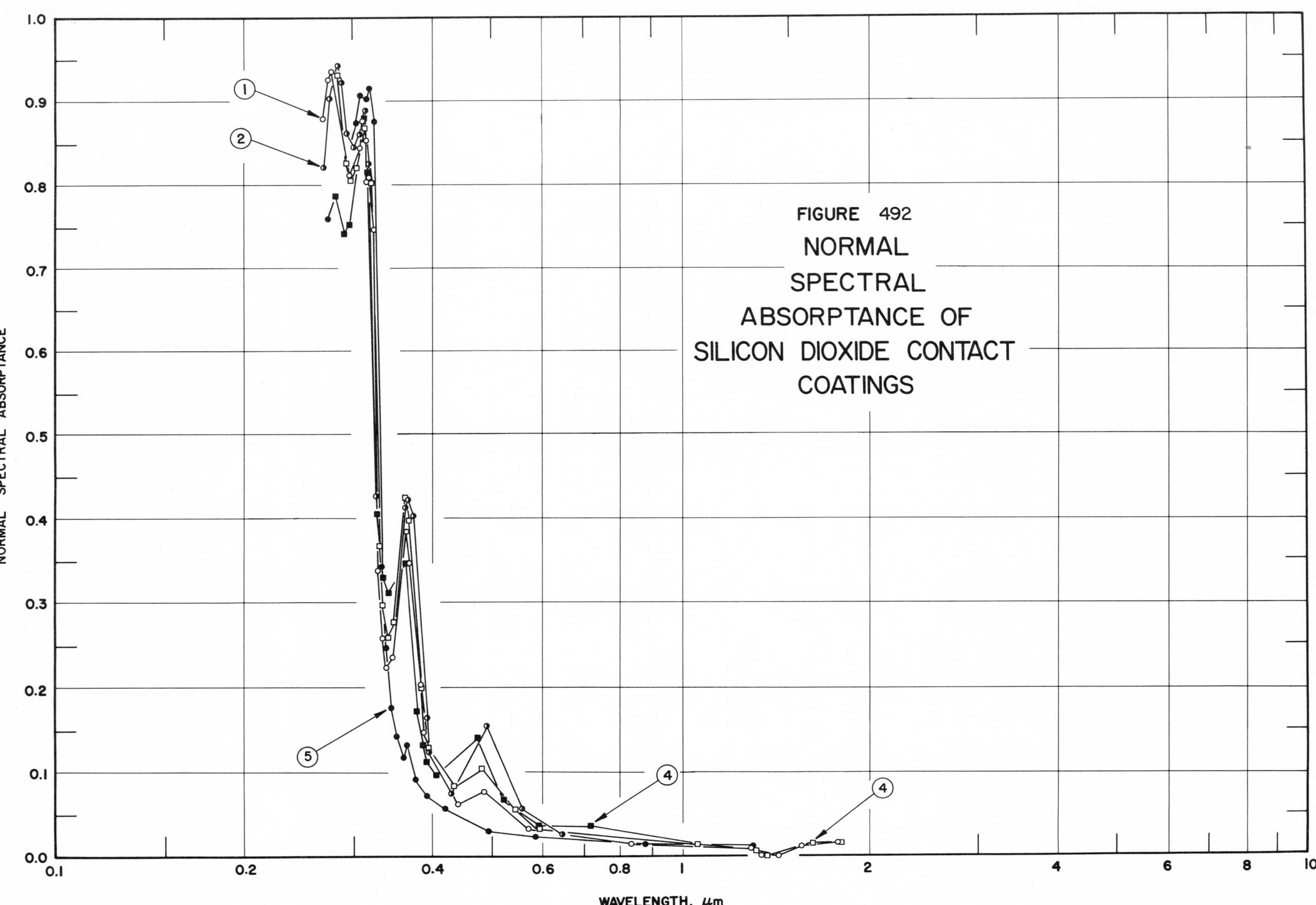

FIGURE 492 NORMAL SPECTRAL ABSORPTANCE OF SILICON DIOXIDE CONTACT COATINGS

SPECIFICATION TABLE NO. 492 NORMAL SPECTRAL ABSORPTANCE OF SILICON DIOXIDE CONTACT COATINGS

Curve No.	Ref. No.	Year	Temperature, K	Wavelength Range, μm	Geometry θ	Reported Error, %	Composition (weight percent), Specifications, and Remarks
1	119	1967	~298	0.269-1.770	~0°		Quartz (silica), optical quality (3000 Å thick); silver, high-purity (1810 Å thick) and highly polished 6061-T6 aluminum alloy substrates; silver, then quartz vapor deposited; data extracted from smooth curve.
2	119	1967	~298	0.269-1.770	~0°		Above specimen and conditions except UV irradiated in vacuum at 294 K with 10-sun intensity for 200 hrs.
3	119	1967	~298	0.281-1.800	~0°		Similar to curve 1 specimen and conditions.
4	119	1967	~298	0.280-1.800	~0°		Above specimen and conditions except UV irradiated in vacuum at 533 K with 10-sun intensity for 200 hrs.
5	119	1967	~298	0.271-1.800	~0°		Quartz, fused silica; TP-060 silver (~2000 Å thick) substrate; data extracted from smooth curve. [Authors' designation: LMSC Optical Solar Reflector]

DATA TABLE NO. 492 NORMAL SPECTRAL ABSORPTANCE OF SILICON DIOXIDE CONTACT COATINGS

[Wavelength, λ, μm; Absorptance, α; Temperature, T, K]

λ	α	λ	α	λ	α	λ	α	λ	α	λ	α	λ	α	λ	α
CURVE 1		CURVE 1 (cont.)		CURVE 2 (cont.)		CURVE 2 (cont.)		CURVE 3 (cont.)		CURVE 4		CURVE 4 (cont.)		CURVE 5 (cont.)	
T ~ 298		0.387	0.149	0.294	0.861	0.647	0.026	0.330	0.368	T ~ 298		1.057	0.013*	0.337	0.247
0.269	0.880	0.440	0.062	0.300	0.846	0.830	0.015*	0.333	0.299	0.280	0.787	1.327	0.008*	0.343	0.176
0.272	0.927	0.483	0.077	0.306	0.861	1.294	0.010*	0.340	0.260	0.290	0.742	1.370	0.000*	0.350	0.142
0.276	0.937	0.570	0.031	0.312	0.890	1.341	0.000*	0.346	0.279	0.296	0.755	1.613	0.016*	0.360	0.118
0.297	0.812	0.830	0.015	0.317	0.829	1.431	0.000*	0.361	0.426	0.312	0.880	1.800	0.018*	0.364	0.132
0.306	0.847	1.294	0.010	0.320	0.824*	1.554	0.011*	0.369	0.399	0.316	0.818			0.376	0.091
0.310	0.876	1.341	0.000	0.327	0.428	1.770	0.017*	0.385	0.200	0.320	0.818	CURVE 5		0.391	0.072
0.314	0.854	1.431	0.000	0.334	0.258*			0.395	0.130	0.329	0.406	T ~ 298		0.420	0.057
0.316	0.804	1.554	0.011	0.339	0.223*	CURVE 3		0.399	0.118*	0.334	0.330			0.491	0.030
0.318	0.810	1.770	0.017	0.345	0.235*	T ~ 298		0.432	0.084	0.340	0.312	0.271	0.760	0.585	0.024
0.322	0.747			0.361	0.414			0.480	0.105	0.363	0.349	0.281	0.789	0.878	0.015
0.329	0.339	CURVE 2		0.366	0.422	0.281	0.933	0.541	0.057	0.379	0.172	0.293	0.828*	1.303	0.011
0.334	0.258	T ~ 298		0.371	0.403	0.293	0.827	0.594	0.031	0.387	0.132	0.302	0.875	1.377	0.000*
0.339	0.223			0.394	0.163	0.298	0.807	1.057	0.013	0.392	0.111	0.309	0.908	1.539	0.013*
0.345	0.235	0.269	0.821	0.399	0.124	0.303	0.821	1.327	0.008	0.406	0.098	0.315	0.902	1.800	0.017*
0.363	0.383	0.274	0.904	0.430	0.075	0.312	0.870	1.370	0.000	0.472	0.141	0.319	0.917		
0.369	0.348	0.281	0.944	0.490	0.155	0.318	0.804*	1.613	0.016	0.520	0.069	0.322	0.879		
0.381	0.202	0.288	0.921	0.557	0.056	0.320	0.804	1.800	0.018	0.592	0.038	0.332	0.344		
										0.719	0.038				

* Not shown on plot

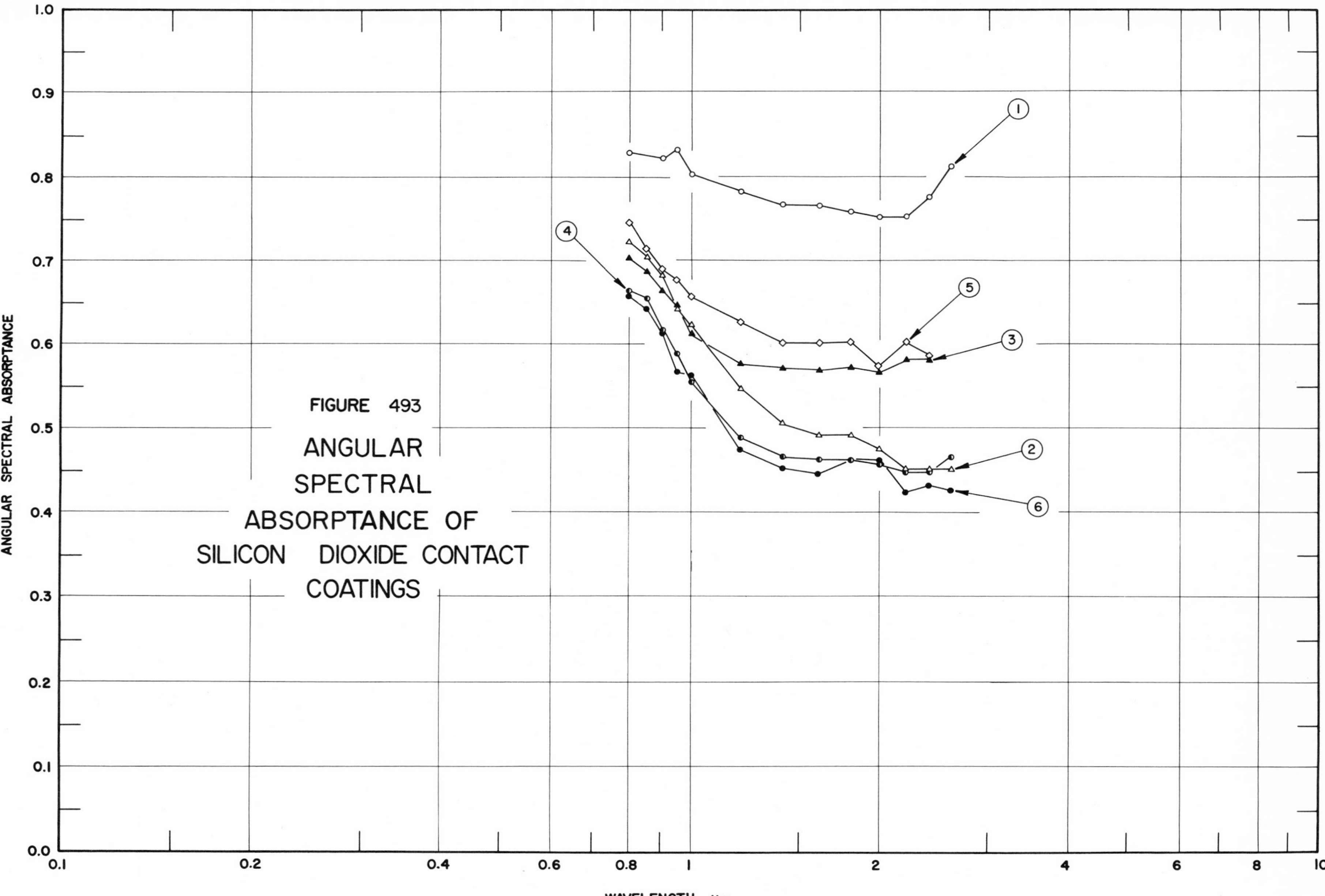

FIGURE 493 ANGULAR SPECTRAL ABSORPTANCE OF SILICON DIOXIDE CONTACT COATINGS

SPECIFICATION TABLE NO. 493 ANGULAR SPECTRAL ABSORPTANCE OF SILICON DIOXIDE CONTACT COATINGS

Curve No.	Ref. No.	Year	Temperature, K	Wavelength Range, μm	Geometry θ	Reported Error, %	Composition (weight percent), Specifications, and Remarks
1	208	1964	298	0.80-2.60	45°		Fused quartz; platinum (~0.102 μm thick) substrate; platinum applied to rear surface of quartz by brushing on a platinum solution and firing; author assumed $\alpha(45°) = 1 - \rho(45°, 45°) - \tau(0°, 0°)$; data beyond 2.2 μm subject to sizable errors. [Author's designation: Specimen No. 3B]
2	208	1964	298	0.80-2.60	15°		Above specimen and conditions except author assumed $\alpha(15°) = 1 - \rho(15°, 15°) - \tau(0°, 0°)$.
3	208	1964	298	0.80-2.60	45°		Similar to curve 1 specimen and conditions. [Author's designation: Specimen No. 4B]
4	208	1964	298	0.80-2.60	15°		Above specimen and conditions except author assumed $\alpha(15°) = 1 - \rho(15°, 15°) - \tau(0°, 0°)$.
5	208	1964	298	0.80-2.40	45°		Similar to curve 1 specimen and conditions. [Author's designation: Specimen No. 5B]
6	208	1964	298	0.80-2.60	15°		Above specimen and conditions except author assumed $\alpha(15°) = 1 - \rho(15°, 15°) - \tau(0°, 0°)$.

DATA TABLE NO. 493 ANGULAR SPECTRAL ABSORPTANCE OF SILICON DIOXIDE CONTACT COATINGS

[Wavelength, λ, μm; Absorptance, α; Temperature, T, K]

λ	α	λ	α	λ	α	λ	α	λ	α	λ	α
CURVE 1 T = 298		CURVE 2 T = 298		CURVE 3 T = 298		CURVE 4 T = 298		CURVE 5 T = 298		CURVE 6 T = 298	
0.80	0.830	0.80	0.723	0.80	0.702	0.80	0.664	0.80	0.748	0.80	0.659
0.90	0.822	0.85	0.703	0.85	0.689	0.85	0.655	0.85	0.715	0.85	0.641
0.95	0.832	0.90	0.682	0.90	0.665	0.90	0.617	0.90	0.690	0.90	0.614
1.00	0.802	0.95	0.644	0.95	0.647	0.95	0.590	0.95	0.678	0.95	0.568
1.20	0.782	1.00	0.625	1.00	0.614	1.00	0.555	1.00	0.658	1.00	0.564
1.40	0.769	1.20	0.546	1.20	0.577	1.20	0.490	1.20	0.627	1.20	0.475
1.60	0.767	1.40	0.506	1.40	0.572	1.40	0.468	1.40	0.601	1.40	0.454
1.80	0.760	1.60	0.493	1.60	0.570	1.60	0.463	1.60	0.601	1.60	0.449
2.00	0.752	1.80	0.493	1.80	0.574	1.80	0.463	1.80	0.603	1.80	0.462*
2.20	0.753	2.00	0.476	2.00	0.568	2.00	0.459	2.00	0.575	2.00	0.451
2.40	0.776	2.20	0.453	2.20	0.581	2.20	0.450	2.20	0.603	2.20	0.426
2.60	0.811	2.40	0.453	2.40	0.582	2.40	0.450	2.40	0.589	2.40	0.433
		2.60	0.452			2.60	0.467			2.60	0.429

* Not shown on plot

SPECIFICATION TABLE NO. 494 NORMAL SOLAR ABSORPTANCE OF SILICON DIOXIDE CONTACT COATINGS

Curve No.	Ref. No.	Year	Temperature Range, K	Geometry θ	Reported Error, %	Composition (weight percent), Specifications and Remarks
1*	119	1967	~298	~0°	<10	Quartz (silica), optical quality (1000 Å thick); gold, high purity (2000 Å thick) and Corning 7058 glass (rms roughness 60 Å) substrates; gold then quartz vapor deposited; computed from spectral reflectance data.
2*	119	1967	~298	~0°	<10	Above specimen and conditions except ultraviolet irradiated in vacuum at 533 K for 250 equivalent sun hrs and thermal cycled 10 times between 294 and 533 K.
3*	119	1967	~298	~0°	<10	Similar to curve 1 specimen and conditions except silver, high purity (2000 Å thick) and Corning 7058 glass (rms roughness 60 Å) substrates; silver then quartz vapor deposited.
4*	119	1967	~298	~0°	<10	Above specimen and conditions except ultraviolet irradiated in vacuum at 533 K for 250 equivalent sun hrs and thermal cycled 10 times between 294 and 533 K.
5*	119	1967	~298	~0°	<10	Similar to curve 1 specimen and conditions except aluminum high purity (2000 Å thick) and Corning 7058 glass (rms roughness 60 Å) substrates; aluminum then quartz vapor deposited.
6*	119	1967	~298	~0°	<10	Above specimen and conditions except ultraviolet irradiated in vacuum at 533 K for 250 equivalent sun hrs and thermal cycled 10 times between 294 and 533 K.
7*	119	1967	~298	~0°	<10	Similar to curve 1 specimen and conditions except substrate is highly polished 6061-T6 aluminum alloy.
8*	119	1967	~298	~0°	<10	Similar to above specimen and conditions except high purity silver (2000 Å thick) and highly polished 6061-T6 aluminum alloy substrates; silver then quartz vapor deposited.
9*	119	1967	~298	~0°	<10	Above specimen and conditions except ultraviolet irradiated in vacuum at 533 K for 250 equivalent sun hrs and thermal cycled 10 times between 294 and 533 K.
10*	119	1967	~298	~0°	<10	Similar to curve 7 specimen and conditions except high purity aluminum (2000 Å thick) and 6061-T6 aluminum alloy substrates; aluminum then quartz vapor deposited.
11*	119	1967	~298	~0°	<10	Above specimen and conditions except ultraviolet irradiated in vacuum at 533 K for 250 equivalent sun hrs and thermal cycled 10 times between 294 and 533 K.
12*	119	1967	~298	~0°	<10	Similar to curve 1 specimen and conditions except substrate is mill finish 6061-T6 aluminum alloy.
13*	119	1967	~298	~0°	<10	Similar to above specimen and conditions except high purity silver (2000 Å thick) and mill finish 6061-T6 aluminum alloy substrates; silver then quartz vapor deposited.
14*	119	1967	~298	~0°	<10	Above specimen and conditions except ultraviolet irradiated in vacuum at 533 K for 250 equivalent sun hrs and thermal cycled 10 times between 294 and 533 K.
15*	119	1967	~298	~0°	<10	Similar to curve 12 specimen and conditions except high purity aluminum (2000 Å thick) and mill finish 6061-T6 aluminum alloy substrates; aluminum then quartz vapor deposited.
16*	119	1967	~298	~0°	<10	Above specimen and conditions except ultraviolet irradiated in vacuum at 533 K for 250 equivalent sun hrs and thermal cycled 10 times between 294 and 533 K.
17*	119	1967	~298	~0°	<10	Similar to curve 1 specimen and conditions except substrate is commercial bright stainless steel.

* No plot given

SPECIFICATION TABLE NO. 494 NORMAL SOLAR ABSORPTANCE OF SILICON DIOXIDE CONTACT COATINGS (continued)

Curve No.	Ref. No.	Year	Temperature Range, K	Geometry θ	Reported Error, %	Composition (weight percent), Specifications and Remarks
18*	119	1967	~298	~0°	< 10	Above specimen and conditions except ultraviolet irradiated in vacuum at 533 K for 250 equivalent sun hrs and thermal cycled 10 times between 294 and 533 K.
19*	119	1967	~298	~0°	< 10	Similar to curve 17 specimen and conditions except high purity silver (2000 Å thick), vapor deposited, and commercial bright stainless steel substrates.
20*	119	1967	~298	~0°	< 10	Above specimen and conditions except ultraviolet irradiated in vacuum at 533 K for 250 equivalent sun hrs and thermal cycled 10 times between 294 and 533 K.
21*	119	1967	~298	~0°	< 10	Similar to curve 17 specimen and conditions except high purity aluminum (2000 Å thick), vapor deposited, and commercial bright stainless steel substrates.
22*	119	1967	~298	~0°	< 10	Above specimen and conditions except ultraviolet irradiated in vacuum at 533 K for 250 equivalent sun hrs and thermal cycled 10 times between 294 and 533 K.
23*	119	1967	~298	~0°	< 10	Quartz (silica), optical quality (1000 Å thick); silver, high purity (2000 Å thick) and highly polished 6061-T6 aluminum alloy substrates; silver then quartz vapor deposited; computed from spectral reflectance data.
24*	119	1967	~298	~0°	< 10	Above specimen and conditions except ultraviolet irradiated in vacuum at 294 K for 2000 equivalent sun hrs.
25*	119	1967	~298	~0°	< 10	Similar to curve 23 specimen and conditions.
26*	119	1967	~298	~0°	< 10	Above specimen and conditions except ultraviolet irradiated in vacuum at 533 K for 2000 equivalent sun hrs.
27*	119	1967	~298	~0°	< 10	Similar to curve 23 specimen and conditions except quartz 2000 Å thick.
28*	119	1967	~298	~0°	< 10	Above specimen and conditions except ultraviolet irradiated in vacuum at 294 K for 2000 equivalent sun hrs.
29*	119	1967	~298	~0°	< 10	Similar to curve 27 specimen and conditions.
30*	119	1967	~298	~0°	< 10	Above specimen and conditions except ultraviolet irradiated in vacuum at 533 K for 2000 equivalent sun hrs.
31*	119	1967	~298	~0°	< 10	Similar to curve 23 specimen and conditions except quartz 3000 Å thick.
32*	119	1967	~298	~0°	< 10	Above specimen and conditions except ultraviolet irradiated in vacuum at 294 K for 2000 equivalent sun hrs.
33*	119	1967	~298	~0°	< 10	Similar to curve 31 specimen and conditions.
34*	119	1967	~298	~0°	< 10	Above specimen and conditions except ultraviolet irradiated in vacuum at 533 K for 2000 equivalent sun hrs.
35*	119	1967	~298	~0°	< 10	Similar to curve 31 specimen and conditions except silver 1810 Å thick.
36*	119	1967	~298	~0°	< 10	Quartz, fused silica; TP-060 silver (~2000 Å thick) substrate; computed from spectral reflectance data. [Authors' designation: LMSC Optical Solar Reflector]
37*	119	1967	~298	~0°	< 10	Above specimen and conditions except ultraviolet irradiated in vacuum at 533 K for 250 equivalent sun hrs and thermal cycled 10 times between 294 and 533 K.
38*	119	1967	~298	~0°	< 10	Similar to curve 36 specimen and conditions.
39*	119	1967	~298	~0°	< 10	Above specimen and conditions except ultraviolet irradiated in vacuum at 294 K for 2000 equivalent sun hrs.

* No plot given

SPECIFICATION TABLE NO. 494 NORMAL SOLAR ABSORPTANCE OF SILICON DIOXIDE CONTACT COATINGS (continued)

Curve No.	Ref. No.	Year	Temperature Range, K	Geometry θ	Reported Error, %	Composition (weight percent), Specifications and Remarks
40*	119	1967	~298	~0°	< 10	Similar to curve 36 specimen and conditions.
41*	119	1967	~298	~0°	< 10	Above specimen and conditions except ultraviolet irradiated in vacuum at 533 K for 2000 equivalent sun hrs.
42*	119	1967	~298	~0°	< 10	Similar to curve 36 specimen and conditions.
43*	85	1969	~300	~0°		Fused quartz; silver and Inconel substrates (~0.178 ± 0.025 mm thick); silver and Inconel vapor deposited; exposed to UV in vacuum; data extracted from smooth curve; property converted from spectral reflectance measured in situ; simulated flight time (hrs) is variable; actual UV ESH is one-third of flight time. [Author's designation: B-2 Optical Solar Reflector (OSR)]
44*	85	1969	~300	~0°		Similar to above specimen and conditions.
45*	49	1969	278	~0°	10	Silica (0.127 mm thick); magnesium substrate; extraterrestrial; avg of several runs.
46*	247	1969	~298	~0°		SiO_2; Al and highly polished 304 stainless steel substrates; Al evaporated; SiO_2 evaporated; SiO_2 thickness, μm, is variable.

* No plot given

DATA TABLE NO. 494 NORMAL SOLAR ABSORPTANCE OF SILICON DIOXIDE CONTACT COATINGS

[Temperature, T, K; Absorptance, α]

T	α
CURVE 1*	
298	0.187
CURVE 2*	
298	0.178
CURVE 3*	
298	0.051
CURVE 4*	
298	0.051
CURVE 5*	
298	0.126
CURVE 6*	
298	0.122
CURVE 7*	
298	0.199
CURVE 8*	
298	0.061
CURVE 9*	
298	0.061
CURVE 10*	
298	0.181
CURVE 11*	
298	0.185
CURVE 12*	
298	0.206

T	α
CURVE 13*	
298	0.085
CURVE 14*	
298	0.080
CURVE 15*	
298	0.184
CURVE 16*	
298	0.177
CURVE 17*	
298	0.202
CURVE 18*	
298	0.202
CURVE 19*	
298	0.071
CURVE 20*	
298	0.061
CURVE 21*	
298	0.164
CURVE 22*	
298	0.169
CURVE 23*	
298	0.084
CURVE 24*	
298	0.108

T	α
CURVE 25*	
298	0.087
CURVE 26*	
298	0.109
CURVE 27*	
298	0.067
CURVE 28*	
298	0.077
CURVE 29*	
298	0.063
CURVE 30*	
298	0.068
CURVE 31*	
298	0.063
CURVE 32*	
298	0.078
CURVE 33*	
298	0.072
CURVE 34*	
298	0.076
CURVE 35*	
298	0.063
CURVE 36*	
298	0.045

T	α
CURVE 37*	
298	0.042
CURVE 38*	
298	0.048
CURVE 39*	
298	0.047
CURVE 40*	
298	0.052
CURVE 41*	
298	0.049
CURVE 42*	
298	0.048
Flight Time	α
CURVE 43*	
$T \sim 300$	
11.8	0.0600
84.9	0.0604
647	0.0607
CURVE 44*	
$T \sim 300$	
11.8	0.0500
84.9	0.0504
647	0.0506
T	α
CURVE 45*	
278	0.21

Thickness	α
CURVE 46*	
$T \sim 298$	
0.000	0.070
0.087	0.131
0.212	0.110
0.235	0.109
0.401	0.115
0.609	0.114
0.791	0.114
0.978	0.114
1.186	0.114
1.212	0.114
1.516	0.114
1.870	0.114

* No plot given

SPECIFICATION TABLE NO. 495 CHANGE IN NORMAL SOLAR ABSORPTANCE OF SILICON DIOXIDE CONTACT COATINGS

Curve No.	Ref. No.	Year	Temperature Range, K	Geometry θ	Reported Error, %	Composition (weight percent), Specifications and Remarks
1*	65	1969	~300	~0°		Corning 7940 fused silica (0.203 mm thick); silver (1000 Å thick) and Inconel (500 Å thick); silver then Inconel vapor deposited on back of fused silica; property calculated from temp of substrate from in flight data of OSOIII; data extracted from smooth curve; ESH is variable. [Author's designation: OSR]

DATA TABLE NO. 495 CHANGE IN NORMAL SOLAR ABSORPTANCE OF SILICON DIOXIDE CONTACT COATINGS

[Temperature, T, K; Absorptance, α]

ESH	$\Delta\alpha$
CURVE 1*	
T ~ 300	
0.1	0.0
1.0	0.0
10	0.0
100	0.0
1000	0.0
1580	0.0

* No plot given

SPECIFICATION TABLE NO. 496 NORMAL SPECTRAL TRANSMITTANCE OF SILICON DIOXIDE CONTACT COATINGS

Curve No.	Ref. No.	Year	Temperature, K	Wavelength Range, μm	Geometry θ θ' ω'	Reported Error, %	Composition (weight percent), Specifications, and Remarks
1*	249	1968	~298	5.63-15.31	~0° ~0°		Silicon dioxide (~9000 Å thick); Si substrate; deposited on etched Si substrate by oxidation of SiH_4 in atm of 15 ($SiH_4 + O_2$) and 85 N_2, by volume; heat treated 5 hrs at 573 K; data extracted from smooth curve.
2*	208	1964	298	0.80-2.60	~0° ~0°		Fused quartz; platinum (0.102 μm thick) substrate; platinum applied to rear surface of quartz by brushing on a platinum solution and firing; data beyond 2.2 μm subject to sizable errors. [Author's designation: Specimen No. 3B]
3*	208	1964	298	0.80-2.60	~0° ~0°		Similar to above specimen and conditions. [Author's designation: Specimen No. 4B]
4*	208	1964	298	0.80-2.60	~0° ~0		Similar to above specimen and conditions. [Author's designation: Specimen No. 5B]

DATA TABLE NO. 496 NORMAL SPECTRAL TRANSMITTANCE OF SILICON DIOXIDE CONTACT COATINGS

[Wavelength, λ, μm; Transmittance, τ; Temperature, T, K]

λ	τ
CURVE 1* T ~ 298	
5.63	0.949
6.06	0.944
6.29	0.957
6.79	0.932
7.52	0.827
7.69	0.762
7.79	0.703
7.89	0.600
8.00	0.443
8.14	0.346
8.33	0.285
8.92	0.149
9.31	0.060
9.43	0.079
9.58	0.139
9.75	0.227
10.00	0.448
10.05	0.600
CURVE 1 (cont.)*	
10.17	0.693
10.38	0.749
10.52	0.766
10.81	0.772
11.26	0.745
11.43	0.757
11.69	0.755
12.06	0.673
12.50	0.626
12.89	0.653
13.42	0.670
14.53	0.642
15.31	0.600
CURVE 2* T = 298	
0.80	0.008
0.85	0.006
CURVE 2 (cont.)*	
0.90	0.006
0.95	0.006
1.00	0.007
1.20	0.008
1.40	0.008
1.60	0.009
1.80	0.012
2.00	0.013
2.20	0.014
2.40	0.014
2.60	0.009
CURVE 3* T = 298	
0.80	0.021
0.85	0.022
0.90	0.024
0.95	0.023
CURVE 3 (cont.)*	
1.00	0.029
1.20	0.026
1.40	0.021
1.60	0.023
1.80	0.032
2.00	0.037
2.20	0.037
2.40	0.035
2.60	0.025
CURVE 4* T = 298	
0.80	0.003
0.85	0.004
0.90	0.006
0.95	0.006
1.00	0.006
1.20	0.007
CURVE 4 (cont.)*	
1.40	0.007
1.60	0.008
1.80	0.010
2.00	0.012
2.20	0.012
2.40	0.011
2.60	0.018

* No plot given

SPECIFICATION TABLE NO. 497 HEMISPHERICAL TOTAL EMITTANCE OF SILICON MONOXIDE CONTACT COATINGS

Curve No.	Ref. No.	Year	Temperature Range, K	Reported Error, %	Composition (weight percent), Specifications and Remarks
1*	16	1965	300–415	±5	SiO (0.6 μm thick); Al (1.0 μm thick), SiO(1.0 μm thick), and stainless steel (250 μm thick) substrates; SiO then Al then SiO vapor deposited; measured in vacuum (10^{-7} mm Hg).
2*	16	1965	300–415	±5	SiO (1.2 μm thick) on Al (1.0 μm thick), mylar (12 μm thick), and stainless steel (250 μm thick) substrates; Al then SiO vapor deposited; measured in vacuum (10^{-7} mm Hg).
3*	16	1965	300–415	±5	Similar to above specimen and conditions except SiO 1.1 μm thick, Al 0.1 μm thick, and mylar 6 μm thick.
4*	16	1965	300–415	±5	Similar to above specimen and conditions except SiO 0.7 μm thick, Al 0.1 μm thick, and mylar 6 μm thick.
5*	16	1965	300–415	±5	Similar to curve 2 specimen and conditions except SiO 0.6 μm thick.

DATA TABLE NO. 497 HEMISPHERICAL TOTAL EMITTANCE OF SILICON MONOXIDE CONTACT COATINGS

[Temperature, T, K; Emittance, ∈]

T	∈	T	∈
CURVE 1*		CURVE 5*	
300	0.15	300	0.16
415	0.15	415	0.16
CURVE 2*			
300	0.46		
415	0.46		
CURVE 3*			
300	0.26		
415	0.26		
CURVE 4*			
300	0.15		
415	0.15		

* No plot given

SPECIFICATION TABLE NO. 498 NORMAL SPECTRAL EMITTANCE OF SILICON MONOXIDE CONTACT COATINGS

Curve No.	Ref. No.	Year	Temperature, K	Wavelength Range, μm	Geometry θ'	Reported Error, %	Composition (weight percent), Specifications, and Remarks
1*	250	1960	873	2.0-14.0	~0°	±4	SiO (1000 Å thick); polished platinum substrate; vacuum evaporated.
2*	250	1960	1273	1.0-14.0	~0°	±4	SiO (1000 Å thick); polished Inconel substrate; some oxide growth noted at the coating-substrate interface.

DATA TABLE NO. 498 NORMAL SPECTRAL EMITTANCE OF SILICON MONOXIDE CONTACT COATINGS

[Wavelength, λ, μm; Emittance, ϵ; Temperature, T, K]

λ	ϵ	λ	ϵ
CURVE 1*		CURVE 2 (cont.)*	
T = 873			
		2.0	0.40
2.0	0.32	3.0	0.52
3.0	0.38	4.0	0.60
4.0	0.33	5.0	0.46
5.0	0.26	6.0	0.36
6.0	0.24	7.0	0.26
7.0	0.22	8.0	0.21
8.0	0.22	9.0	0.26
9.0	0.29	10.0	0.52
10.0	0.38	11.0	0.30
11.0	0.25	12.0	0.24
12.0	0.22	13.0	0.22
13.0	0.22	14.0	0.21
14.0	0.21		
CURVE 2*			
T = 1273			
1.0	0.66		

* No plot given

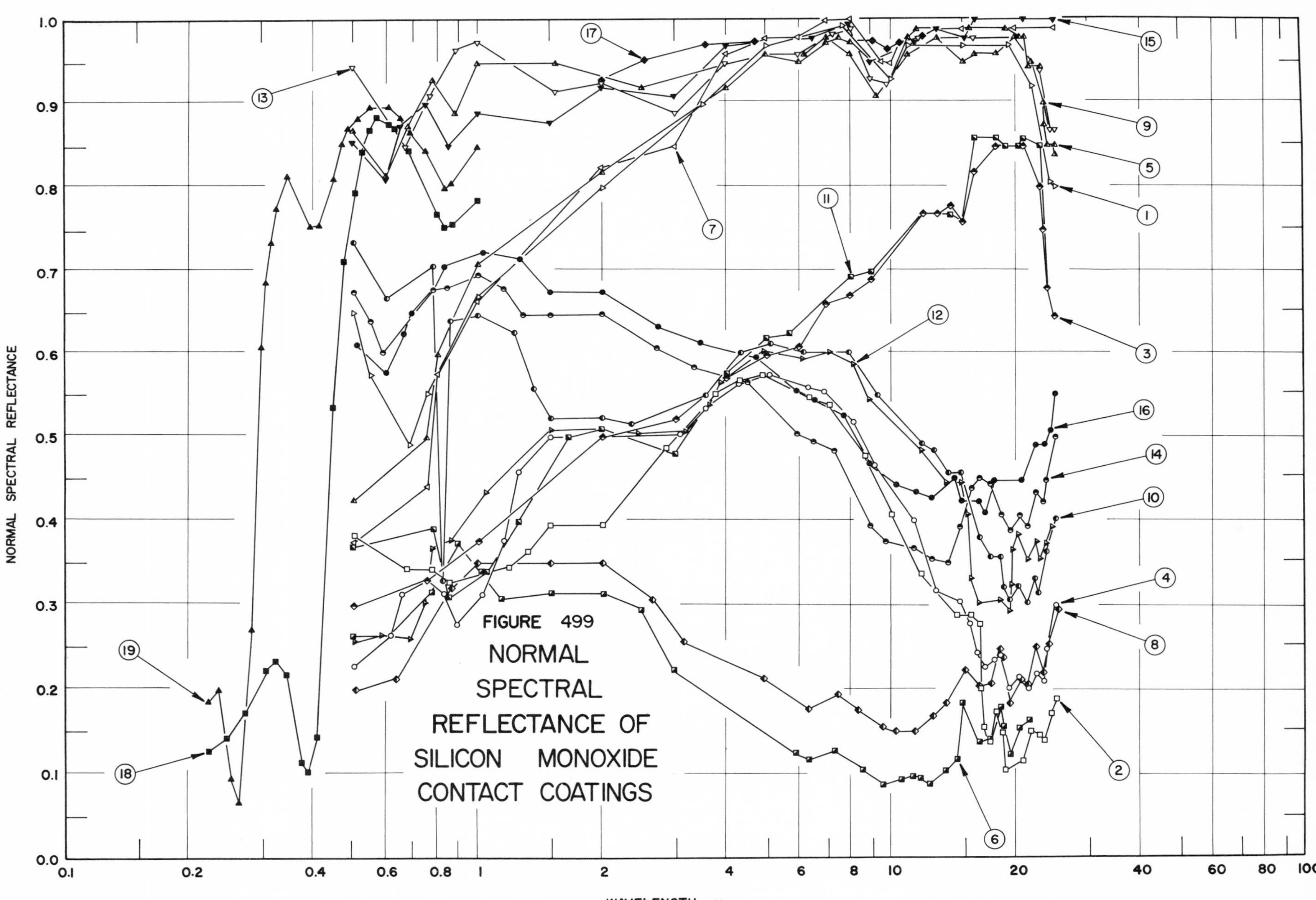

FIGURE 499 NORMAL SPECTRAL REFLECTANCE OF SILICON MONOXIDE CONTACT COATINGS

SPECIFICATION TABLE NO. 499 NORMAL SPECTRAL REFLECTANCE OF SILICON MONOXIDE CONTACT COATINGS

Curve No.	Ref. No.	Year	Temperature, K	Wavelength Range, μm	Geometry θ	θ'	ω'	Reported Error, %	Composition (weight percent), Specifications, and Remarks
1	202	1962	~322	0.50-25.00	~0°		2π	<2	SiO (1200 to 1400 Å thick); as received, cleaned, 6061-T6 alloy substrate; SiO vacuum deposited; data extracted from smooth curve; hohlraum at 1273 K; converted from R (2π, 0°).
2	202	1962	~322	0.50-25.00	~0°		2π	<2	Above specimen and conditions; diffuse component only.
3	202	1962	~322	0.50-25.00	~0°		2π	<2	Similar to curve 1 specimen and conditions except grit blasted using 60 grit silicon carbide with air pressure of 110 to 120 psi for 30 to 45 sec.
4	202	1962	~322	0.50-25.00	~0°		2π	<2	Above specimen and conditions; diffuse component only.
5	202	1962	~322	0.50-25.00	~0°		2π	<2	Similar to curve 1 specimen and conditions except exposed to vacuum (<4 x 10^{-8} mm Hg) for 24 hrs.
6	202	1962	~322	0.50-25.00	~0°		2π	<2	Above specimen and conditions; diffuse component only.
7	202	1962	~322	0.50-25.00	~0°		2π	<2	Similar to curve 1 specimen and conditions except x-ray exposed in vacuum (4 x 10^{-8} mm Hg) for 24 hrs.
8	202	1962	~322	0.51-25.01	~0°		2π	<2	Above specimen and conditions; diffuse component only.
9	202	1962	~322	0.50-25.00	~0°		2π	<2	SiO (1200 to 1400 Å thick); chemically polished 6061-T6 aluminum alloy substrate; SiO vacuum deposited; data extracted from smooth curve; hohlraum at 1273 K; converted from R (2π, 0°).
10	202	1962	~322	0.50-25.00	~0°		2π	<2	Above specimen and conditions; diffuse component only.
11	202	1962	~322	0.50-25.00	~0°		2π	<2	Similar to curve 9 specimen and conditions except grit blasted using 60 grit silicon carbide with air pressure of 110 to 120 psi for 30 to 45 sec.
12	202	1962	~322	0.50-24.99	~0°		2π	<2	Above specimen and conditions; diffuse component only.
13	202	1962	~322	0.50-25.00	~0°		2π	<2	Similar to curve 1 specimen and conditions except exposed to vacuum (<4 x 10^{-8} mm Hg) for 24 hrs.
14	202	1962	~322	0.50-25.00	~0°		2π	<2	Above specimen and conditions; diffuse component only.
15	202	1962	~322	0.50-25.00	~0°		2π	<2	Similar to curve 1 specimen and conditions except x-ray exposed in vacuum (4 x 10^{-8} mm Hg) for 24 hrs.
16	202	1962	~322	0.51-25.00	~0°		2π	<2	Above specimen and conditions; diffuse component only.
17	251	1955	298	2.00-12.00	~0°	~0°			SiO (125 μm thick); aluminum substrate; SiO evaporated in vacuum (1 x 10^{-5} mm Hg); data extracted from smooth curve.
18	251	1955	298	0.224-1.000	~0°	~0°			SiO (125 μm thick); strongly oxidized aluminum substrate; SiO evaporated in vacuum (1 x 10^{-5} mm Hg); evaporation time 2 min, 9 sec; deposition rate 10 to 15 Å sec^{-1}; data extracted from smooth curve.
19	251	1955	298	0.221-1.000	~0°	~0°			SiO (150 μm thick); strongly oxidized aluminum substrate; SiO evaporated in vacuum (1 x 10^{-4} mm Hg); evaporation time 10 min at 10 to 15 Å sec^{-1}; data extracted from smooth curve.

DATA TABLE NO. 499 NORMAL SPECTRAL REFLECTANCE OF SILICON MONOXIDE CONTACT COATINGS

[Wavelength, λ, μm; Reflectance, ρ; Temperature, T, K]

CURVE 1, T ~ 322

λ	ρ
0.50	0.650
0.55	0.575
0.68	0.490
0.75	0.550
1.00	0.670
2.00	0.800
3.50	0.900
5.00	0.970
7.70	0.995
10.00	0.930
11.00	0.970
15.00	0.970
19.20	0.970
22.00	0.920
24.20	0.805
25.00	0.800

CURVE 2, T ~ 322

λ	ρ
0.50	0.382
0.67	0.342
0.77	0.342
0.85	0.329
1.19	0.345
1.32	0.361
1.50	0.396
2.00	0.396
2.89	0.486
3.78	0.550
4.30	0.567
4.87	0.572
6.38	0.547
7.10	0.538
8.63	0.479
10.00	0.409
11.97	0.338
14.21	0.289
15.58	0.289
16.05	0.277
16.44	0.200
16.65	0.154
17.11	0.139
17.99	0.171
18.52	0.117
18.98	0.105
20.82	0.117
21.99	0.150
22.71	0.145
23.03	0.140
24.05	0.170
25.00	0.188

CURVE 3, T ~ 322

λ	ρ
0.50	0.300
0.75	0.330
1.00	0.375
2.00	0.500
3.00	0.520
4.00	0.570
5.00	0.600
6.00	0.610
7.00	0.660
8.00	0.680
9.00	0.690
12.00	0.770
13.00	0.770
14.00	0.780
15.00	0.760
16.00	0.820
18.00	0.850
21.00	0.850
23.00	0.800
23.60	0.750
24.00	0.680
25.00	0.645

CURVE 4, T ~ 322

λ	ρ
0.50	0.228
0.61	0.263
0.65	0.314
0.75	0.330*
0.82	0.315
0.88	0.277
1.01	0.312
1.16	0.376
1.25	0.458
1.50	0.500
2.00	0.500*
3.09	0.502
3.53	0.532
4.26	0.564
5.05	0.572
6.25	0.558
6.96	0.552
8.15	0.517
9.13	0.465
11.08	0.400
12.99	0.319
14.69	0.305
15.42	0.278
16.00	0.241
16.99	0.226
17.65	0.234
18.04	0.240
19.05	0.200
20.05	0.213
21.08	0.200
22.03	0.218
23.05	0.210
23.95	0.248
25.00	0.300

CURVE 5, T ~ 322

λ	ρ
0.50	0.425
0.75	0.500
0.80	0.600
1.00	0.710
2.00	0.820
4.00	0.920
5.00	0.960
6.00	0.950
7.00	0.975
7.50	0.980
8.00	0.975
10.00	0.930*
11.00	0.980
11.60	0.990
15.50	0.990
19.00	0.990
20.50	0.980
21.00	0.980
22.45	0.945
23.00	0.945
23.50	0.875
24.00	0.850
25.00	0.850

CURVE 6, T ~ 322

λ	ρ
0.50	0.262
0.61	0.263*
0.77	0.317
0.89	0.372
1.01	0.340
1.25	0.309
1.50	0.315
2.00	0.314
2.49	0.292
2.98	0.224
5.82	0.125
6.27	0.118
7.24	0.129
8.55	0.105
9.59	0.087
10.51	0.093
11.11	0.099
11.78	0.095
12.21	0.089
13.58	0.104
14.22	0.118
14.96	0.182
16.20	0.139
17.15	0.140
18.12	0.179
18.62	0.154
19.04	0.122
20.02	0.151
21.15	0.111
22.11	0.140
23.43	0.089
24.21	0.098
25.00	0.098

CURVE 7, T ~ 322

λ	ρ
0.50	0.375
0.75	0.440
0.80	0.575
1.00	0.665
2.00	0.825
3.00	0.850
4.00	0.960
5.00	0.980
6.00	0.980
7.00	1.000
8.00	1.000
9.50	0.950
10.00	0.950
11.50	0.990*
15.00	0.990
20.00	0.990
25.00	0.990

CURVE 8, T ~ 322

λ	ρ
0.51	0.200
0.63	0.212
0.86	0.321
0.99	0.350
1.50	0.350
2.00	0.350
2.63	0.307
3.13	0.256
4.96	0.213
6.26	0.176
7.40	0.192
8.33	0.175
9.54	0.154
10.26	0.150
11.37	0.150
12.51	0.169
13.54	0.181
15.05	0.221
16.12	0.201
17.38	0.208
18.02	0.249
18.39	0.238
19.08	0.181
20.17	0.212
21.20	0.202
22.07	0.250
23.03	0.220
24.00	0.252
25.01	0.296

CURVE 9, T ~ 322

λ	ρ
0.50	0.870
0.60	0.815
0.68	0.875
0.78	0.930
0.88	0.890
1.00	0.950
1.55	0.950
2.50	0.920
5.00	0.960*
6.20	0.960
7.00	0.980
8.00	0.960
9.20	0.910
11.00	0.960
13.00	0.980
15.00	0.950
16.00	0.960
18.00	0.960
20.00	0.980
22.00	0.950
23.40	0.900
24.00	0.850*
25.00	0.840

CURVE 10, T ~ 322

λ	ρ
0.50	0.737
0.60	0.669
0.78	0.709
0.82	0.330
0.86	0.640
1.00	0.649
1.22	0.626
1.37	0.557
1.50	0.521
2.00	0.522
2.35	0.516
3.54	0.550
4.38	0.601
5.11	0.611
6.18	0.601
7.99	0.601
9.22	0.550
11.99	0.491
12.81	0.483
13.93	0.457
14.98	0.457
16.41	0.380
17.06	0.358
18.07	0.358
18.74	0.321
19.28	0.307
20.02	0.322
21.01	0.301
22.02	0.331
22.87	0.315
23.99	0.361
25.00	0.401

CURVE 11, T ~ 322

λ	ρ
0.50	0.370
0.78	0.390
0.85	0.310
1.05	0.340
1.25	0.400
1.65	0.500
2.00	0.510
3.00	0.480
4.00	0.575
5.00	0.620
5.70	0.625
8.00	0.695
9.00	0.700
12.00	0.770*
14.00	0.770
15.00	0.760*
16.00	0.860
18.00	0.860
19.00	0.850
20.50	0.850
21.00	0.860
23.00	0.800
24.00	0.680*
25.00	0.645*

CURVE 12, T ~ 322

λ	ρ
0.50	0.259
0.58	0.264
0.68	0.260
0.74	0.304
0.77	0.369
0.86	0.377
1.13	0.433
1.50	0.509
2.00	0.509*
2.42	0.504
3.17	0.507
3.60	0.538
3.85	0.565
4.96	0.604
6.06	0.593
7.03	0.603
8.04	0.589
8.86	0.545
11.92	0.483
13.51	0.446
14.95	0.446
15.28	0.409
15.55	0.331
16.15	0.301
18.25	0.309
19.07	0.292
19.53	0.326
19.68	0.366
20.03	0.382
21.11	0.353
22.10	0.375
22.98	0.354
23.75	0.373
24.40	0.393
24.99	0.402*

* Not shown on plot

DATA TABLE NO. 499 NORMAL SPECTRAL REFLECTANCE OF SILICON MONOXIDE CONTACT COATINGS (continued)

λ	ρ
CURVE 13 T ~ 322	
0.50	0.945
0.65	0.850
0.75	0.910
0.88	0.965
1.00	0.975
1.55	0.915
2.00	0.925
3.00	0.890
4.00	0.950
6.00	0.960
7.30	0.985
8.10	0.990
9.00	0.930
9.90	0.925
11.80	0.980
16.00	0.980
21.00	0.980*
23.00	0.940
23.50	0.900*
24.20	0.870
25.00	0.870
CURVE 14 T ~ 322	
0.50	0.675
0.55	0.640
0.59	0.602
0.78	0.679
0.84	0.681
1.00	0.697
1.15	0.680
1.29	0.649
1.50	0.649
2.00	0.649
2.70	0.608
3.33	0.582
4.57	0.565
5.86	0.504
6.42	0.493
7.28	0.481
8.94	0.394
9.62	0.373
11.23	0.368
12.52	0.354
13.88	0.350
CURVE 14 (cont.)	
14.69	0.393
15.60	0.439
16.16	0.450
17.38	0.443
18.64	0.407
19.17	0.389
20.09	0.409
21.02	0.392
22.08	0.433
23.04	0.421
23.94	0.448
25.00	0.500
CURVE 15 T ~ 322	
0.50	0.855
0.60	0.810
0.65	0.875
0.75	0.900
0.85	0.850
1.00	0.890
1.50	0.880
2.00	0.920
3.00	0.910
4.00	0.970
6.50	0.980
8.00	0.995
9.00	0.950
10.00	0.950*
11.50	0.975
13.00	0.990
15.10	0.980
16.00	1.000
21.00	1.000
25.00	1.000
CURVE 16 T ~ 322	
0.51	0.612
0.60	0.577
0.66	0.625
0.69	0.650
0.78	0.678*
0.83	0.708
1.07	0.723
CURVE 16 (cont.)	
1.27	0.717
1.50	0.676
2.00	0.676
2.71	0.633
3.46	0.614
4.78	0.597
5.90	0.556
6.58	0.544
7.66	0.526
8.83	0.469
10.23	0.441
11.58	0.435
12.50	0.428
14.01	0.450
14.96	0.423
16.31	0.422
16.99	0.410
17.97	0.449
20.88	0.449
22.14	0.490
23.42	0.490
24.18	0.508
25.00	0.550
CURVE 17 T = 298	
2.00	0.929
2.53	0.952
3.58	0.971
4.86	0.977
9.17	0.977
9.97	0.969
10.65	0.975
11.20	0.979*
12.00	0.982
CURVE 18 T = 298	
0.224	0.128
0.249	0.142
0.274	0.171
0.307	0.222
0.324	0.233
0.342	0.219
0.373	0.113
CURVE 18 (cont.)	
0.389	0.101
0.405	0.143
0.443	0.532
0.479	0.714
0.502	0.796
0.524	0.843
0.547	0.870
0.569	0.885
0.607	0.879
0.632	0.871
0.680	0.847
0.800	0.770
0.838	0.755
0.869	0.758
1.000	0.788
CURVE 19 T = 298	
0.221	0.185
0.236	0.200
0.251	0.095
0.261	0.068
0.273	0.171*
0.281	0.270
0.300	0.608
0.309	0.687
0.316	0.739
0.326	0.777
0.348	0.818
0.396	0.756
0.415	0.757
0.447	0.811
0.475	0.852
0.492	0.871
0.519	0.885
0.554	0.897
0.606	0.899
0.650	0.886
0.692	0.869
0.747	0.845
0.837	0.800
0.863	0.806
1.000	0.850

* Not shown on plot

SPECIFICATION TABLE NO. 500 NORMAL SOLAR ABSORPTANCE OF SILICON MONOXIDE CONTACT COATINGS

Curve No.	Ref. No.	Year	Temperature Range, K	Geometry θ	Reported Error, %	Composition (weight percent), Specifications and Remarks
1*	137	1965	300-415	0°	± 5	SiO (0.6 μm thick); Al (1.0 μm thick), SiO (1.0 μm thick), and stainless steel (250 μm thick) substrates; SiO then Al then SiO vapor deposited; measured in vacuum (10^{-7} mm Hg).
2*	137	1965	300-415	0°	± 5	SiO (0.6 μm thick); Al (1.0 μm thick), Mylar (12 μm thick), and stainless steel (250 μm thick) substrates; Al then SiO vapor deposited; measured in vacuum (10^{-7} mm Hg).
3*	137	1965	300-415	0°	± 5	Similar to above specimen and conditions except SiO 0.7 μm thick, Al 0.1 μm thick and Mylar 6.0 μm thick.
4*	137	1965	300-415	0°	± 5	Similar to above specimen and conditions except SiO 1.1 μm thick.
5*	137	1965	300-415	0°	± 5	Similar to curve 2 specimen and conditions except SiO 1.2 μm thick.

DATA TABLE NO. 500 NORMAL SOLAR ABSORPTANCE OF SILICON MONOXIDE CONTACT COATINGS

[Temperature, T, K; Absorptance, α]

T	α	T	α
CURVE 1*		CURVE 5*	
300	0.18	300	0.15
415	0.18	415	0.15
CURVE 2*			
300	0.19		
415	0.19		
CURVE 3*			
300	0.13		
415	0.13		
CURVE 4*			
300	0.13		
415	0.13		

* No plot given

SPECIFICATION TABLE NO. 501 ANGULAR SPECTRAL REFLECTANCE OF SODIUM CHLORIDE CONTACT COATINGS

Curve No.	Ref. No.	Year	Temperature, K	Wavelength Range, μm	Geometry θ	θ'	ω'	Reported Error, %	Composition (weight percent), Specifications, and Remarks
1*	252	1968	~298	34.8-40.8	30°	30°			NaCl (1.35 μm thick); aluminum mirror substrate; NaCl evaporated on substrate; measured relative to aluminum mirror.
2*	252	1968	~298	34.0-41.8	30°	30°			Similar to the above specimen and conditions except NaCl 2.34 μm thick.
3*	252	1968	~298	25.9-41.8	30°	30°			Similar to the above specimen and conditions except NaCl 5.51 μm thick.

DATA TABLE NO. 501 ANGULAR SPECTRAL REFLECTANCE OF SODIUM CHLORIDE CONTACT COATINGS

[Wavelength, λ, μm; Reflectance, ρ; Temperature, T, K]

λ	ρ	λ	ρ	λ	ρ	λ	ρ
CURVE 1* T ~ 298		CURVE 2 (cont.)*		CURVE 3 (cont.)*		CURVE 3 (cont.)*	
		35.8	0.818	30.4	0.817	40.3	0.840
34.8	0.996	36.3	0.813	30.9	0.804	40.8	0.871
35.3	0.928	36.7	0.730	31.4	0.791	41.8	0.919
35.8	0.848	37.3	0.712	31.9	0.778		
36.3	0.823	37.8	0.752	32.4	0.776		
36.9	0.811	38.3	0.752	32.8	0.743		
37.3	0.794	38.9	0.743	33.4	0.714		
37.8	0.834	39.3	0.768	33.8	0.690		
38.4	0.847	39.8	0.821	34.3	0.668		
38.9	0.880	40.9	0.956	34.9	0.656		
39.5	0.890	41.8	0.959	35.4	0.656		
40.0	0.916			35.8	0.637		
40.8	0.973	CURVE 3* T ~ 298		36.3	0.613		
				36.9	0.584		
CURVE 2* T ~ 298				37.4	0.595		
		25.9	0.950	37.8	0.617		
		27.9	0.921	38.4	0.621		
34.0	0.965	28.9	0.849	38.7	0.614		
34.9	0.959	29.4	0.834	39.3	0.650		
35.3	0.946	29.9	0.841	40.0	0.699		

* No plot given

SPECIFICATION TABLE NO. 502 NORMAL SOLAR ABSORPTANCE OF TESSLAR CONTACT COATINGS

Curve No.	Ref. No.	Year	Temperature Range, K	Geometry θ	Reported Error, %	Composition (weight percent), Specifications and Remarks
1*	115	1960	298	15°		DuPont type 20 Tesslar (0.102 mm thick); aluminum substrate; aluminum vapor deposited; computed from spectral reflectance data for above atm conditions. [Authors' designation: Sample 19]

DATA TABLE NO. 502 NORMAL SOLAR ABSORPTANCE OF TESSLAR CONTACT COATINGS

[Temperature, T, K; Absorptance, α]

T	α
CURVE 1*	
298	0.18

* No plot given

FIGURE 503A

ANALYZED NORMAL SPECTRAL REFLECTANCE OF TEFLON CONTACT COATINGS

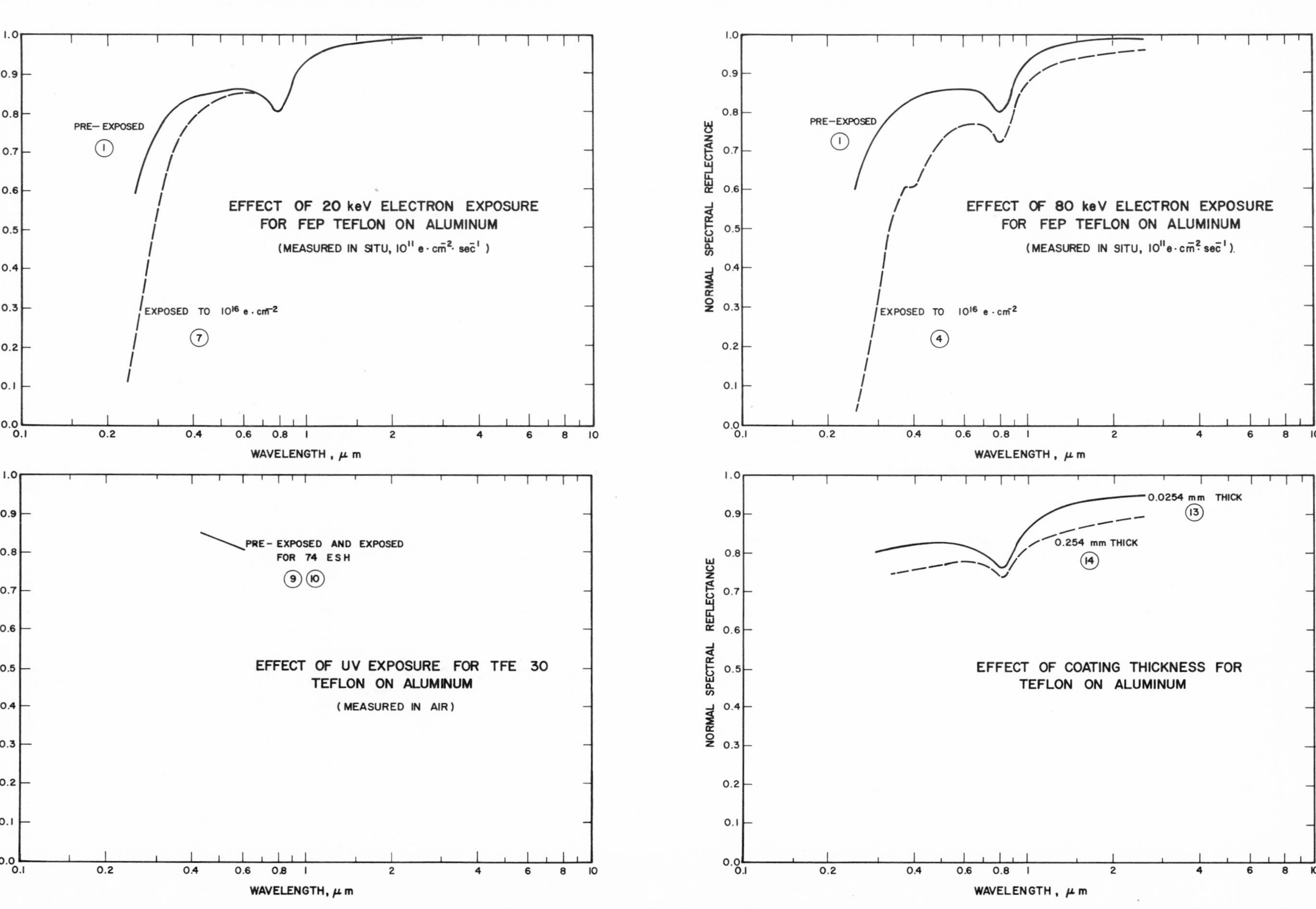

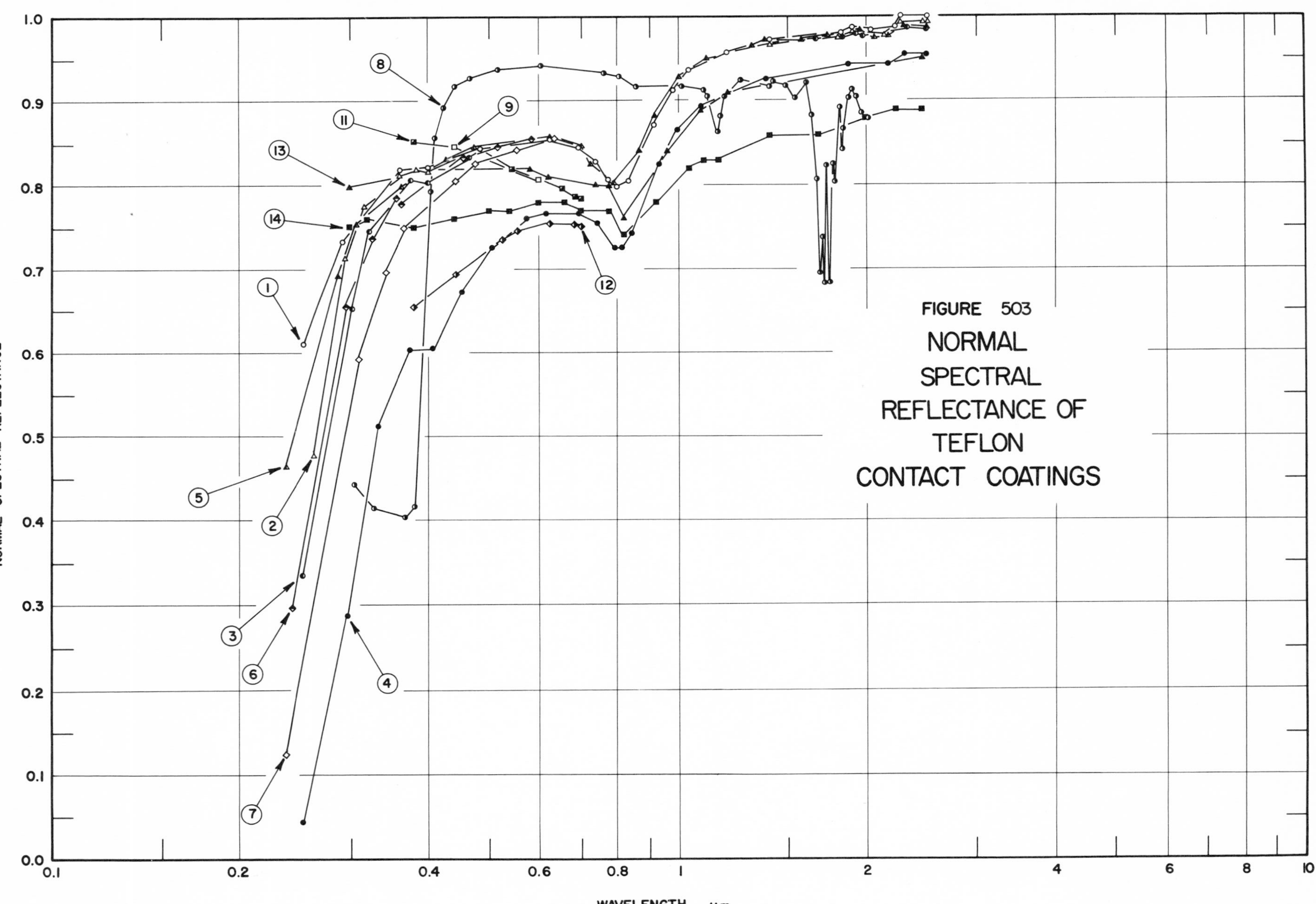
FIGURE 503
NORMAL
SPECTRAL
REFLECTANCE OF
TEFLON
CONTACT COATINGS
NORMAL SPECTRAL REFLECTANCE
WAVELENGTH, μm
1.0
0.9
0.8
0.7
0.6
0.5
0.4
0.3
0.2
0.1
0.0
0.1
0.2
0.4
0.6
0.8
1
2
4
6
8
10
8
9
11
13
14
1
12
5
2
3
6
4
7

SPECIFICATION TABLE NO. 503 NORMAL SPECTRAL REFLECTANCE OF TEFLON CONTACT COATINGS

Curve No.	Ref. No.	Year	Temperature, K	Wavelength Range, μm	Geometry θ	Geometry θ'	Geometry ω'	Reported Error, %	Composition (weight percent), Specifications, and Remarks
1	45	1970	298	0.254-2.503	~0°		2π		FEP teflon (0.0508 mm thick); aluminum substrate. [Authors' designation: TA-2]
2	45	1970	281	0.261-2.505	~0°		2π		Above specimen and conditions except exposed to 80 keV electrons in dark in vacuum (10^{-8} mm Hg) vacuum maintained by ion pump; flux 1 x 10^{10} - 5 x 10^{11} e cm^{-2} sec^{-1}; substrate held at 281 ± 2 K; reflectance measured in situ after 3 x 10^{14} e cm^{-2}.
3	45	1970	281	0.251-2.505	~0°		2π		Above specimen and conditions except reflectance measured after 1 x 10^{15} e cm^{-2}.
4	45	1970	281	0.252-2.505	~0°		2π		Above specimen and conditions except reflectance measured after 1 x 10^{16} e cm^{-2}.
5	45	1970	281	0.238-2.500	~0°		2π		Similar to curve 2 specimen and conditions except 20 keV electrons; reflectance measured after 3 x 10^{14} e cm^{-2}.
6	45	1970	281	0.243-2.500	~0°		2π		Above specimen and conditions except reflectance measured after 1 x 10^{15} e cm^{-2}.
7	45	1970	281	0.239-2.500	~0°		2π		Above specimen and conditions except reflectance measured after 1 x 10^{16} e cm^{-2}.
8	34	1964	298	0.305-2.018	~0°		2π		Teflon (0.0762 mm thick); S-13 paint substrate; data extracted from smooth curve.
9	27	1964	~298	0.440-0.600	~0°		2π		Teflon TFE 30, E.I. duPont de Nemours and Co.; aluminum substrate.
10	27	1964	~298	0.440-0.600	~0°		2π		Above specimen and conditions except irradiated in vacuum (10^{-6} mm Hg) with 74 ESH of UV at solar factor of 4 suns.
11	29	1962	298	0.380-0.700	0°		2π		Teflon (TFE 30 paint vehicle); 6061-T6 Al (grit blasted with 40 mesh SiC) substrate; sprayed; cured at 433 K for 16 hrs. [Author's designation: Sample No. 68]
12	29	1962	298	0.380-0.700	0°		2π		Above specimen and conditions except being exposed to simulated solar radiation at 4 suns for 18.5 hrs.
13	115	1960	298	0.30-2.45	15°		2π		DuPont teflon (0.0254 mm thick); aluminum substrate; Al vacuum deposited on back of teflon. [Authors' designation: Sample 16]
14	115	1960	298	0.30-2.45	15°		2π		Similar to above specimen and conditions except teflon (0.254 mm thick). [Authors' designation: Sample 18]

DATA TABLE NO. 503 NORMAL SPECTRAL REFLECTANCE OF TEFLON CONTACT COATINGS

[Wavelength, λ, μm; Reflectance, ρ; Temperature, T, K]

CURVE 1 T = 298

λ	ρ
0.254	0.610
0.293	0.734
0.360	0.820
0.400	0.821
0.485	0.845
0.623	0.853
0.698	0.847
0.740	0.829
0.776	0.806
0.800	0.798
0.834	0.805
0.913	0.871
0.980	0.912
1.048	0.936
1.200	0.957
1.401	0.973
1.823	0.980
1.901	0.987
2.146	0.982
2.218	0.988
2.263	1.000
2.503	1.000

CURVE 2 T = 281

λ	ρ
0.261	0.478
0.295	0.713
0.316	0.775
0.360	0.812
0.383	0.820
0.400	0.819
0.485	0.845
0.623	0.853
0.698	0.847
0.740	0.829
0.776	0.806
0.800	0.798
0.834	0.805
0.913	0.871
0.980	0.912
1.048	0.936
1.200	0.957
1.408	0.969
1.800	0.976

CURVE 2 (cont.)

λ	ρ
1.909	0.982
2.026	0.982
2.140	0.978
2.267	0.993
2.465	0.993
2.505	0.995

CURVE 3 T = 281

λ	ρ
0.251	0.335
0.301	0.651
0.322	0.746
0.377	0.808
0.400	0.802
0.466	0.835
0.485	0.845
0.623	0.853
0.698	0.847
0.740	0.829
0.776	0.806
0.800	0.798
0.834	0.805
0.913	0.871
0.980	0.912
1.048	0.936
1.200	0.957
1.401	0.967
1.665	0.973
1.845	0.974
1.912	0.979
1.972	0.977
2.128	0.974
2.313	0.987
2.354	0.984
2.505	0.984

CURVE 4 T = 281

λ	ρ
0.252	0.043
0.299	0.289
0.333	0.512
0.373	0.602
0.407	0.604
0.454	0.671

CURVE 4 (cont.)

λ	ρ
0.506	0.728
0.572	0.760
0.617	0.768
0.695	0.768
0.747	0.753
0.791	0.728
0.813	0.728
0.841	0.741
0.930	0.825
0.998	0.868
1.091	0.894
1.380	0.926
1.889	0.942
2.179	0.942
2.309	0.955
2.505	0.955

CURVE 5 T = 281

λ	ρ
0.238	0.464
0.286	0.691
0.308	0.754
0.361	0.800
0.429	0.831
0.474	0.848
0.621	0.856
0.700	0.845
0.726	0.827
0.797	0.801
0.869	0.842
0.917	0.884
1.008	0.930
1.122	0.950
1.318	0.964
1.378	0.971
1.588	0.971
1.741	0.978
1.952	0.982
2.077	0.976
2.172	0.976
2.302	0.989
2.500	0.985

CURVE 6 T = 281

λ	ρ
0.243	0.298
0.296	0.653
0.326	0.738
0.356	0.784
0.361	0.779
0.459	0.833
0.518	0.849
0.585	0.856
0.700	0.845
0.726	0.827
0.797	0.801
0.869	0.842
0.917	0.884
1.008	0.930
1.122	0.950
1.318	0.964
1.378	0.971
1.588	0.971
1.741	0.978
1.952	0.982
2.077	0.976
2.172	0.976
2.302	0.989
2.500	0.985

CURVE 7 T = 281

λ	ρ
0.239	0.123
0.310	0.593
0.343	0.699
0.368	0.750
0.441	0.806
0.479	0.826
0.559	0.841
0.635	0.856
0.700	0.845
0.726	0.827
0.797	0.801
0.869	0.842
0.917	0.884
1.008	0.930
1.122	0.950
1.318	0.964
1.378	0.971

CURVE 7 (cont.)

λ	ρ
1.588	0.971
1.741	0.978
1.952	0.982
2.077	0.976
2.172	0.976
2.302	0.989
2.500	0.985

CURVE 8 T = 298

λ	ρ
0.305	0.441
0.329	0.414
0.366	0.404
0.380	0.419
0.401	0.794
0.410	0.858
0.422	0.892
0.440	0.919
0.467	0.929
0.519	0.938
0.606	0.941
0.769	0.934
0.808	0.930
0.858	0.917
1.027	0.918
1.103	0.913
1.123	0.905
1.142	0.884
1.149	0.863
1.162	0.863
1.170	0.881
1.195	0.906
1.225	0.919
1.270	0.923
1.400	0.918
1.428	0.921
1.468	0.917
1.499	0.917
1.540	0.902
1.563	0.909
1.569	0.921
1.604	0.921
1.617	0.917
1.647	0.884
1.677	0.806

CURVE 8 (cont.)

λ	ρ
1.691	0.697
1.704	0.739
1.717	0.682
1.735	0.822
1.755	0.682
1.773	0.826
1.779	0.804
1.784	0.828
1.812	0.891
1.832	0.843
1.835	0.867
1.848	0.843
1.861	0.885
1.880	0.904
1.903	0.913
1.923	0.912
1.946	0.902
1.964	0.886
2.018	0.880

CURVE 9 T ~ 298

λ	ρ
0.440	0.847
0.600	0.806

CURVE 10 T ~ 298

λ	ρ
0.440	0.847
0.600	0.806

CURVE 11 T = 298

λ	ρ
0.380	0.852
0.445	0.849
0.482	0.840
0.543	0.820
0.655	0.798
0.689	0.787
0.700	0.784

CURVE 12 T = 298

λ	ρ
0.380	0.653
0.441	0.695
0.528	0.736
0.555	0.746
0.621	0.753
0.682	0.753
0.700	0.751

CURVE 13 T = 298

λ	ρ
0.30	0.80
0.36	0.81
0.40	0.82
0.58	0.82
0.62	0.81
0.74	0.80
0.78	0.80
0.82	0.76
0.96	0.84
1.08	0.89
1.20	0.91
2.45	0.95

CURVE 14 T = 298

λ	ρ
0.30	0.75
0.32	0.76
0.38	0.75
0.44	0.76
0.50	0.77
0.54	0.77
0.60	0.78
0.66	0.78
0.70	0.77
0.78	0.77
0.82	0.74
0.92	0.78
1.04	0.82
1.10	0.83
1.16	0.83
1.40	0.86
1.68	0.86
2.00	0.88

CURVE 14 (cont.)

λ	ρ
2.22	0.89
2.45	0.89

SPECIFICATION TABLE NO. 504 NORMAL SOLAR ABSORPTANCE OF TEFLON CONTACT COATINGS

Curve No.	Ref. No.	Year	Temperature Range, K	Geometry θ	Reported Error, %	Composition (weight percent), Specifications and Remarks
1*	45	1969	298	$\sim 0^{\circ}$		FEP Teflon, Type A (5.08×10^{-2} mm thick); aluminum substrate; absorptance calculated from normal spectral reflectance; property measured in air. [Authors' designation: TA-2]
2*	45	1969	281	$\sim 0^{\circ}$		Above specimen and conditions except exposed to 80 keV electrons (flux 1×10^{10} - 5×10^{11} e cm^{-2} sec^{-1}) in dark in vacuum (10^{-8} mm Hg); vacuum maintained by ion pump; substrate held at 281 ± 2 K; property measured in situ after 3×10^{14} e cm^{-2}.
3*	45	1969	281	$\sim 0^{\circ}$		Above specimen and conditions except property measured in situ after 10^{15} e cm^{-2}.
4*	45	1969	281	$\sim 0^{\circ}$		Above specimen and conditions except property measured in situ after 10^{16} e cm^{-2}.
5*	45	1969	281	$\sim 0^{\circ}$		Similar to curve 2 specimen and conditions except exposed to 20 keV electrons; property measured in situ after 3×10^{14} e cm^{-2}.
6*	45	1969	281	$\sim 0^{\circ}$		Above specimen and conditions except property measured in situ after 1×10^{15} e cm^{-2}.
7*	45	1969	281	$\sim 0^{\circ}$		Above specimen and conditions except property measured in situ after 1×10^{16} e cm^{-2}.
8*	115	1960	298	15°		DuPont Teflon (0.0254 mm thick); aluminum substrate; Al vacuum deposited on back of teflon; computed from spectral reflectance data for above atm conditions. [Authors' designation: Sample 16]
9*	115	1960	298	15°		Similar to above specimen and conditions except teflon (0.127 mm thick). [Authors' designation: Sample 17]
10*	115	1960	298	15°		Similar to above specimen and conditions except teflon (0.254 mm thick). [Authors' designation: Sample 18]

* No plot given

DATA TABLE NO. 504 NORMAL SOLAR ABSORPTANCE OF TEFLON CONTACT COATINGS

[Temperature, T, K; Absorptance, α]

T	α
CURVE 1*	
298	0.12
CURVE 2*	
281	0.12
CURVE 3*	
281	0.13
CURVE 4*	
281	0.23
CURVE 5*	
281	0.12
CURVE 6*	
281	0.13
CURVE 7*	
281	0.14
CURVE 8*	
298	0.16
CURVE 9*	
298	0.15
CURVE 10*	
298	0.22

* No plot given

2. CONTACT COATINGS (continued)

D. Anti-Reflection Coatings

SPECIFICATION TABLE NO. 505 NORMAL SPECTRAL REFLECTANCE OF LITHIUM FLUORIDE CONTACT COATINGS

Curve No.	Ref. No.	Year	Temperature, K	Wavelength Range, μm	Geometry θ	θ'	ω'	Reported Error, %	Composition (weight percent), Specifications, and Remarks
1*	253	1968	~298	0.0911-0.1900	~0°	~0°			LiF (140 Å thick); Al 99.99 pure, and unknown substrate; Al vacuum deposited at ~400 Å sec^{-1}; LiF evaporation deposited at 20 Å sec^{-1} on substrate at ~298 K; measured in vacuum 10 min after preparation; data extracted from smooth curve.
2*	224	1963	298	13.98-36.02	~0°	~0°			LiF (8 μm optical thickness); glass substrate; vacuum evaporated at 5-10 x 10^{-5} mm Hg and 0.5 μm (optical) min^{-1} on substrate at 473 K; data extracted from smooth curve.
3*	164	1963	298	14.00-36.07	~0°	~0°			LiF (6 μm thick); glass substrate; vacuum deposited on substrate at 308 K.

DATA TABLE NO. 505 NORMAL SPECTRAL REFLECTANCE OF LITHIUM FLUORIDE CONTACT COATINGS

[Wavelength, λ, μm; Reflectance, ρ; Temperature, T, K]

CURVE 1* T ~ 298

λ	ρ
0.0911	0.038
0.0917	0.034
0.0952	0.245
0.0969	0.300
0.0985	0.294
0.1018	0.690
0.1023	0.725
0.1038	0.744
0.1180	0.716
0.1287	0.708
0.1430	0.676
0.1593	0.695
0.1826	0.768
0.1900	0.778

CURVE 2* T = 298

λ	ρ
13.98	0.307
15.00	0.454
15.67	0.532
16.41	0.583
17.01	0.616
17.48	0.628
18.12	0.634
19.42	0.619
20.20	0.617
20.65	0.624
22.79	0.679
23.61	0.690
24.90	0.686
26.54	0.664
28.75	0.617
30.28	0.578
30.99	0.575
31.97	0.576

CURVE 2 (cont.)

λ	ρ
32.86	0.559
33.65	0.529
34.67	0.473
36.02	0.364

CURVE 3* T = 298

λ	ρ
14.00	0.191
16.03	0.400
17.96	0.384
19.88	0.385
21.96	0.468
23.96	0.557
26.05	0.614
28.03	0.629
30.00	0.625
32.07	0.586
33.98	0.519

CURVE 3 (cont.)

λ	ρ
36.07	0.433

* No plot given

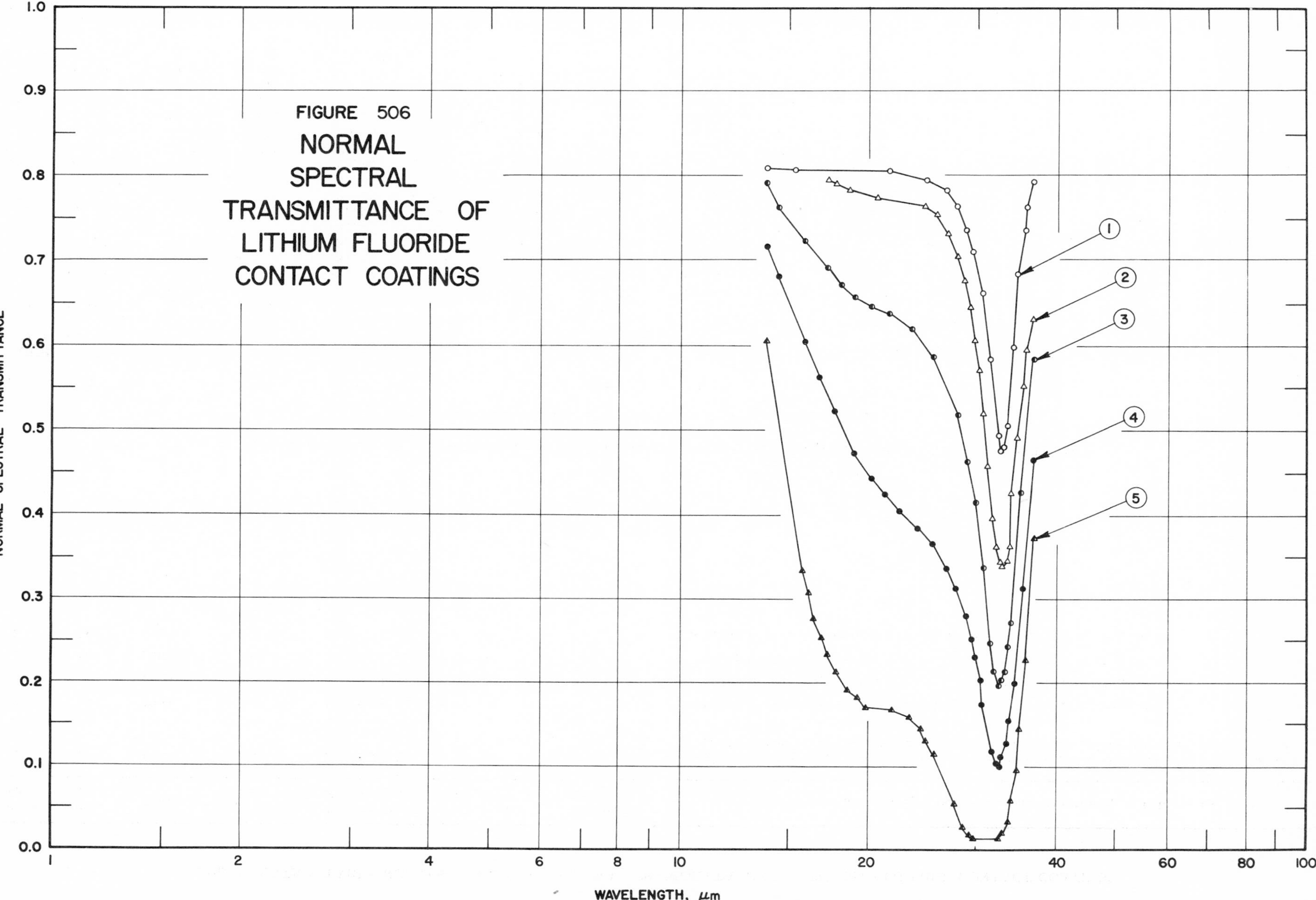

FIGURE 506 NORMAL SPECTRAL TRANSMITTANCE OF LITHIUM FLUORIDE CONTACT COATINGS

SPECIFICATION TABLE NO. 506 NORMAL SPECTRAL TRANSMITTANCE OF LITHIUM FLUORIDE CONTACT COATINGS

Curve No.	Ref. No.	Year	Temperature, K	Wavelength Range, μm	Geometry θ θ' ω'	Reported Error, %	Composition (weight percent), Specifications, and Remarks
1	164	1963	298	13.79-36.50	~0° ~0°		LiF (0.22 μm thick); KRS-5 substrate; vacuum evaporated on substrate at ~298 K; data corrected for second surface reflection of the substrate; data extracted from smooth curve.
2	164	1963	298	17.28-36.64	~0° ~0°		Similar to above specimen and conditions except LiF 0.32 μm thick.
3	164	1963	298	13.79-36.66	~0° ~0°		Similar to above specimen and conditions except LiF 0.72 μm thick.
4	164	1963	298	13.79-36.65	~0° ~0°		Similar to above specimen and conditions except LiF 1.44 μm thick.
5	164	1963	298	13.78-36.85	~0° ~0°		Similar to above specimen and conditions except LiF 3.00 μm thick.

DATA TABLE NO. 506 NORMAL SPECTRAL TRANSMITTANCE OF LITHIUM FLUORIDE CONTACT COATINGS

[Wavelength, λ, μm; Transmittance, τ; Temperature, T, K]

CURVE 1, T = 298

λ	τ
13.79	0.810
15.22	0.809
21.45	0.806
24.94	0.796
26.60	0.783
27.68	0.765
28.63	0.739
29.38	0.711
30.12	0.661
31.15	0.584
32.13	0.494
32.46	0.478
32.92	0.481*
33.20	0.505
34.06	0.598
34.88	0.685
35.51	0.737
35.92	0.764
36.50	0.793

CURVE 2, T = 298

λ	τ
17.28	0.798
17.82	0.792
18.52	0.784
20.64	0.777
24.88	0.767
25.88	0.757
26.90	0.734
27.80	0.708
28.47	0.679
29.09	0.648
29.66	0.608
30.06	0.572
30.55	0.520
31.04	0.458
31.56	0.398
32.03	0.363
32.41	0.344
32.79	0.340
33.16	0.347
33.42	0.362
33.89	0.426
34.52	0.491
35.23	0.552
35.95	0.598
36.64	0.633

CURVE 3, T = 298

λ	τ
13.79	0.792
14.47	0.764
15.86	0.724
17.15	0.692
18.08	0.672
19.03	0.659
20.16	0.649
21.69	0.639
23.46	0.620
25.41	0.587
27.77	0.518
28.96	0.461
29.72	0.416
30.54	0.336
31.24	0.249
31.73	0.212
32.13	0.197
32.62	0.201
33.09	0.213
33.48	0.241
33.80	0.271
35.09	0.429
36.66	0.584

CURVE 4, T = 298

λ	τ
13.79	0.719
14.41	0.681
15.83	0.604
16.61	0.562
17.68	0.521
19.00	0.473
20.21	0.442
21.10	0.424
22.37	0.406
24.01	0.385
25.33	0.366
26.58	0.338
27.65	0.311
28.62	0.280
29.33	0.254
29.86	0.230
30.39	0.202
30.82	0.173
31.56	0.117
32.02	0.102
32.36	0.100
32.78	0.111
33.17	0.129
33.58	0.152
34.13	0.200
35.23	0.311
36.65	0.468

CURVE 5, T = 298

λ	τ
13.78	0.606
15.67	0.335
16.00	0.308
16.39	0.279
16.78	0.255
17.23	0.234
17.85	0.211
18.56	0.191
19.11	0.182
19.94	0.170
21.93	0.169
23.13	0.159
24.16	0.145
24.96	0.130
25.59	0.114
27.29	0.055
28.25	0.028
29.04	0.017
29.59	0.013
32.23	0.013
32.84	0.020
33.34	0.033
33.97	0.060
34.50	0.096
34.97	0.143
35.67	0.227
36.85	0.372

* Not shown on plot

SPECIFICATION TABLE NO. 507 HEMISPHERICAL TOTAL EMITTANCE OF MAGNESIUM FLUORIDE CONTACT COATINGS

Curve No.	Ref. No.	Year	Temperature Range, K	Reported Error, %	Composition (weight percent), Specifications and Remarks
1*	85	1969	162-395		MgF_2 (1.4 x 10^{-4} mm thick); Fe_2O_3 (2.8 x 10^{-4} mm thick) and 2024 aluminum substrates; MgF_2 and Fe_2O_3 vacuum deposited; property measured in vacuum by calorimetric method; data extracted from smooth curve. [Author's designation: A-4 Solar Absorber NRDL-RTD-81-5]

DATA TABLE NO. 507 HEMISPHERICAL TOTAL EMITTANCE OF MAGNESIUM FLUORIDE CONTACT COATINGS

[Temperature, T, K; Emittance, ϵ]

T	ϵ
CURVE 1*	
162	0.082
193	0.101
235	0.121
270	0.132
319	0.141
395	0.149

* No plot given

SPECIFICATION TABLE NO. 508 NORMAL SPECTRAL EMITTANCE OF MAGNESIUM FLUORIDE CONTACT COATINGS

Curve No.	Ref. No.	Year	Temperature, K	Wavelength Range, μm	Geometry θ'	Reported Error, %	Composition (weight percent), Specifications, and Remarks
1*	254	1961	298	1.07-15.96	$\sim 0°$		MgF_2; SiO and silicon substrates; SiO then MgF_2 vacuum deposited; data extracted from smooth curve.

DATA TABLE NO. 508 NORMAL SPECTRAL EMITTANCE OF MAGNESIUM FLUORIDE CONTACT COATINGS

[Wavelength, λ, μm; Emittance, ϵ; Temperature, T, K]

λ	ϵ	λ	ϵ
CURVE 1*		CURVE 1 (cont.)*	
T = 298			
		9.39	0.869
1.07	0.950	10.42	0.854
1.52	0.957	11.67	0.872
2.10	0.947	12.38	0.915
2.84	0.895	13.26	0.933
3.12	0.905	14.55	0.914
3.45	0.948	15.96	0.876
3.75	0.955		
4.09	0.948		
4.61	0.864		
5.18	0.717		
5.45	0.633		
6.10	0.615		
7.10	0.645		
7.88	0.709		
8.32	0.765		
8.48	0.815		
8.58	0.862		
8.94	0.875		

* No plot given

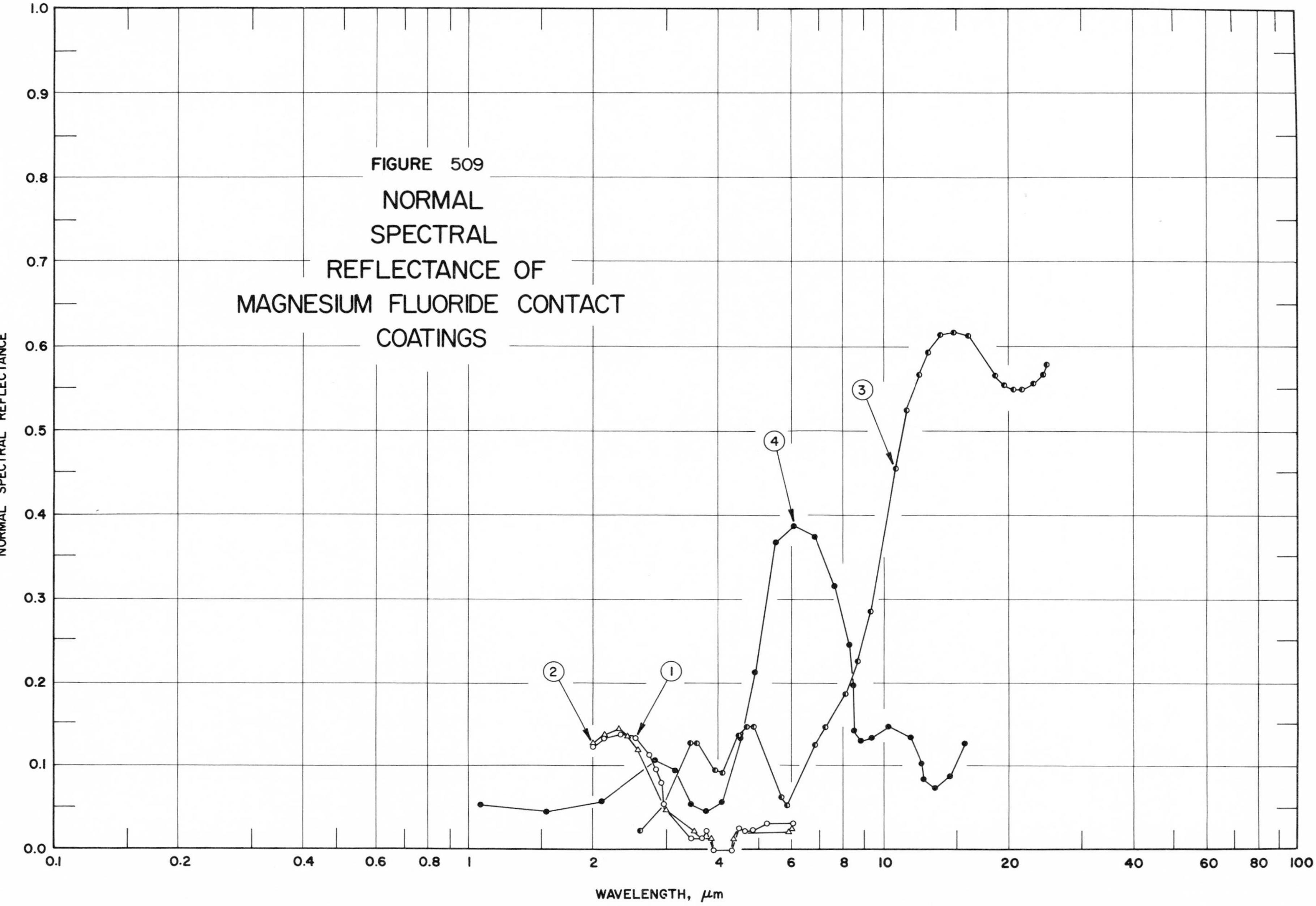

FIGURE 509 NORMAL SPECTRAL REFLECTANCE OF MAGNESIUM FLUORIDE CONTACT COATINGS

SPECIFICATION TABLE NO. 509 NORMAL SPECTRAL REFLECTANCE OF MAGNESIUM FLUORIDE CONTACT COATINGS

Curve No.	Ref. No.	Year	Temperature, K	Wavelength Range, μm	Geometry θ θ' ω'	Reported Error, %	Composition (weight percent), Specifications, and Remarks
1	255	1957	~298	2.00-6.05	~0° ~0°		MgF_2 (560 μm thick); calcium aluminate glass (Melt No. 1A 5561, 4 mm thick) substrate; data extracted from smooth curve.
2	255	1957	~298	2.00-6.00	~0° ~0°		MgF_2 (560 μm thick); SiO_2 (560 μm thick) and calcium aluminate glass (Melt No. 1A 5561, 4.0 mm thick) substrate; data extracted from smooth curve.
3	85	1969	~298	0.260-2.491	~0° 2π		MgF_2 (1.4 x 10^{-4} mm thick); Fe_2O_3 (2.8 x 10^{-4} mm thick) and 2024 aluminum substrates; MgF_2 and Fe_2O_3 vacuum vapor deposited; data extracted from smooth curve. [Author's designation: A-4 Solar Absorber]
4	254	1961	298	1.07-15.94	~0° ~0°		MgF_2; SiO and silicon substrates; SiO then MgF_2 vacuum deposited; data extracted from smooth curve.

DATA TABLE NO. 509 NORMAL SPECTRAL REFLECTANCE OF MAGNESIUM FLUORIDE CONTACT COATINGS

[Wavelength, λ, μm; Reflectance, ρ; Temperature, T, K]

λ	ρ	λ	ρ	λ	ρ	λ	ρ	λ	ρ	λ	ρ
CURVE 1		CURVE 1 (cont.)		CURVE 2 (cont.)		CURVE 3 (cont.)		CURVE 3 (cont.)		CURVE 4 (cont.)	
T ~ 298											
		5.21	0.032	4.39	0.014	0.571	0.064	2.168	0.550	4.91	0.212
2.00	0.123	6.05	0.032	4.47	0.023	0.594	0.053	2.300	0.557	5.47	0.369
2.11	0.132			5.91	0.023	0.688	0.126	2.413	0.569	6.01	0.389
2.32	0.139	CURVE 2		6.00	0.026	0.728	0.147	2.491	0.580	6.84	0.374
2.51	0.132	T ~ 298				0.818	0.187			7.65	0.317
2.72	0.113			CURVE 3		0.873	0.228	CURVE 4		8.30	0.246
2.84	0.097	2.00	0.128	T ~ 298		0.928	0.285	T ~ 298		8.50	0.196
2.91	0.080	2.13	0.138			1.066	0.457			8.59	0.142
2.98	0.055	2.30	0.145	0.260	0.025	1.143	0.525	1.07	0.054	8.91	0.130
3.43	0.015	2.44	0.136	0.298	0.055	1.212	0.568	1.55	0.047	9.40	0.134
3.67	0.015	2.58	0.120	0.344	0.129	1.281	0.593	2.10	0.058	10.40	0.149
3.75	0.023	3.00	0.049	0.359	0.129	1.397	0.614	2.84	0.109	11.60	0.133
3.90	0.000	3.50	0.022	0.391	0.095	1.498	0.618	3.16	0.095	12.27	0.102
4.38	0.000	3.62	0.015	0.410	0.091	1.600	0.612	3.44	0.055	12.56	0.083
4.51	0.028	3.88	0.015	0.454	0.137	1.855	0.566	3.75	0.049	13.24	0.072
4.61	0.022	3.93	0.000	0.469	0.147	1.970	0.554	4.08	0.059	14.53	0.089
4.83	0.025	4.36	0.000	0.483	0.147	2.060	0.550	4.59	0.134	15.94	0.129

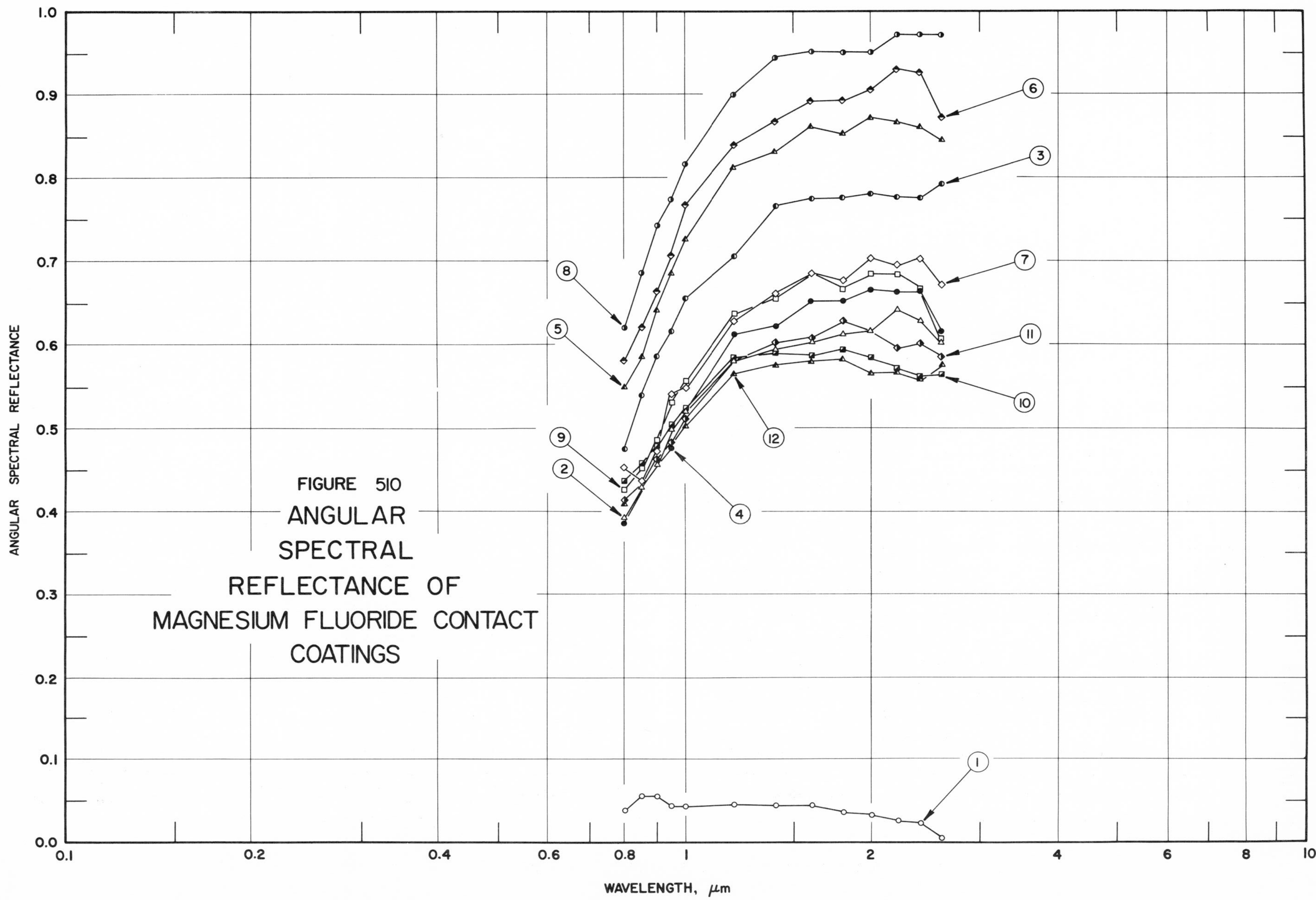

FIGURE 510 ANGULAR SPECTRAL REFLECTANCE OF MAGNESIUM FLUORIDE CONTACT COATINGS

SPECIFICATION TABLE NO. 510 ANGULAR SPECTRAL REFLECTANCE OF MAGNESIUM FLUORIDE CONTACT COATINGS

Curve No.	Ref. No.	Year	Temperature, K	Wavelength Range, μm	Geometry θ	θ'	ω'	Reported Error, %	Composition (weight percent), Specifications, and Remarks
1	208	1964	298	0.80-2.60	45°	45°			MgF_2; fused quartz substrate; measured relative to aluminum mirror; data beyond 2.2 μm subject to sizable errors.
2	208	1964	298	0.80-2.60	45°	45°			MgF_2; platinum (~0.102 μm thick), fused quartz, and MgF_2 substrate; platinum applied by brushing on a platinum solution and firing; measured relative to aluminum mirror; data beyond 2.2 μm subject to sizable errors. [Author's designation: Specimen No. 6]
3	208	1964	298	0.80-2.60	45°	45°			Similar to above specimen and conditions. [Author's designation: Specimen No. 7]
4	208	1964	298	0.80-2.60	15°	15°			Above specimen and conditions.
5	208	1964	298	0.80-2.60	45°	45°			Similar to above specimen and conditions. [Author's designation: Specimen No. 8]
6	208	1964	298	0.80-2.60	45°	45°			Similar to above specimen and conditions. [Author's designation: Specimen No. 9]
7	208	1964	298	0.80-2.60	15°	15°			Above specimen and conditions.
8	208	1964	298	0.80-2.60	45°	45°			Similar to above specimen and conditions. [Author's designation: Specimen No. 10]
9	208	1964	298	0.80-2.60	15°	15°			Above specimen and conditions.
10	208	1964	298	0.80-2.60	15°	15°			MgF_2; fused quartz, platinum (~0.102 μm thick), and MgF_2 substrates; platinum applied to quartz by brushing on a platinum solution (Hanovia 05-X) and firing; measured relative to aluminum mirror; tungsten filament source; data beyond 2.2 μm subject to sizable errors. [Author's designation: Specimen No. 7]
11	208	1964	298	0.80-2.60	15°	15°			Similar to above specimen and conditions. [Author's designation: Specimen No. 9]
12	208	1964	298	0.80-2.60	45°	45°			Similar to above specimen and conditions. [Author's designation: Specimen No. 10]

DATA TABLE NO. 510 ANGULAR SPECTRAL REFLECTANCE OF MAGNESIUM FLUORIDE CONTACT COATINGS

[Wavelength, λ, μm; Reflectance, ρ; Temperature, T, K]

CURVE 1
T = 298

λ	ρ
0.80	0.040
0.85	0.057
0.90	0.057
0.95	0.045
1.00	0.045
1.20	0.047
1.40	0.046
1.60	0.047
1.80	0.038
2.00	0.034
2.20	0.027
2.40	0.024
2.60	0.007

CURVE 2
T = 298

λ	ρ
0.80	0.394
0.85	0.430
0.90	0.458
0.95	0.500
1.00	0.521
1.20	0.583
1.40	0.596
1.60	0.603
1.80	0.614
2.00	0.619
2.20	0.643
2.40	0.630
2.60	0.604

CURVE 3
T = 298

λ	ρ
0.80	0.476
0.85	0.540
0.90	0.589
0.95	0.619
1.00	0.656
1.20	0.706
1.40	0.766
1.60	0.775
1.80	0.776
2.00	0.780
2.20	0.776
2.40	0.776
2.60	0.793

CURVE 4
T = 298

λ	ρ
0.80	0.388
0.85	0.429*
0.90	0.457*
0.95	0.478
1.00	0.522*
1.20	0.611
1.40	0.622
1.60	0.652
1.80	0.653
2.00	0.668
2.20	0.664
2.40	0.664
2.60	0.619

CURVE 5
T = 298

λ	ρ
0.80	0.550
0.85	0.589
0.90	0.643
0.95	0.687
1.00	0.726
1.20	0.813
1.40	0.832
1.60	0.863
1.80	0.855
2.00	0.873
2.20	0.869
2.40	0.863
2.60	0.849

CURVE 6
T = 298

λ	ρ
0.80	0.582
0.85	0.622
0.90	0.664
0.95	0.706
1.00	0.767
1.20	0.840
1.40	0.869
1.60	0.893
1.80	0.893
2.00	0.906
2.20	0.931
2.40	0.929
2.60	0.873

CURVE 7
T = 298

λ	ρ
0.80	0.454
0.85	0.439
0.90	0.475
0.95	0.542
1.00	0.550
1.20	0.630
1.40	0.661
1.60	0.686
1.80	0.679
2.00	0.702
2.20	0.697
2.40	0.702
2.60	0.671

CURVE 8
T = 298

λ	ρ
0.80	0.621
0.85	0.687
0.90	0.741
0.95	0.775
1.00	0.817
1.20	0.900
1.40	0.946
1.60	0.951
1.80	0.951
2.00	0.951
2.20	0.973
2.40	0.973
2.60	0.973

CURVE 9
T = 298

λ	ρ
0.80	0.429
0.85	0.452
0.90	0.488
0.95	0.532
1.00	0.557
1.20	0.638
1.40	0.656
1.60	0.686*
1.80	0.668
2.00	0.684
2.20	0.684
2.40	0.667
2.60	0.608

CURVE 10
T = 298

λ	ρ
0.80	0.438
0.85	0.459
0.90	0.480
0.95	0.505
1.00	0.524
1.20	0.585
1.40	0.592
1.60	0.590
1.80	0.596
2.00	0.588
2.20	0.573
2.40	0.565
2.60	0.566

CURVE 11
T = 298

λ	ρ
0.80	0.415
0.85	0.435*
0.90	0.463
0.95	0.483
1.00	0.511
1.20	0.581*
1.40	0.601
1.60	0.608
1.80	0.630
2.00	0.619*
2.20	0.598
2.40	0.601
2.60	0.588

CURVE 12
T = 298

λ	ρ
0.80	0.411
0.85	0.430*
0.90	0.458*
0.95	0.486*
1.00	0.506
1.20	0.569
1.40	0.577
1.60	0.582
1.80	0.585
2.00	0.569
2.20	0.570
2.40	0.562
2.60	0.577

* Not shown on plot

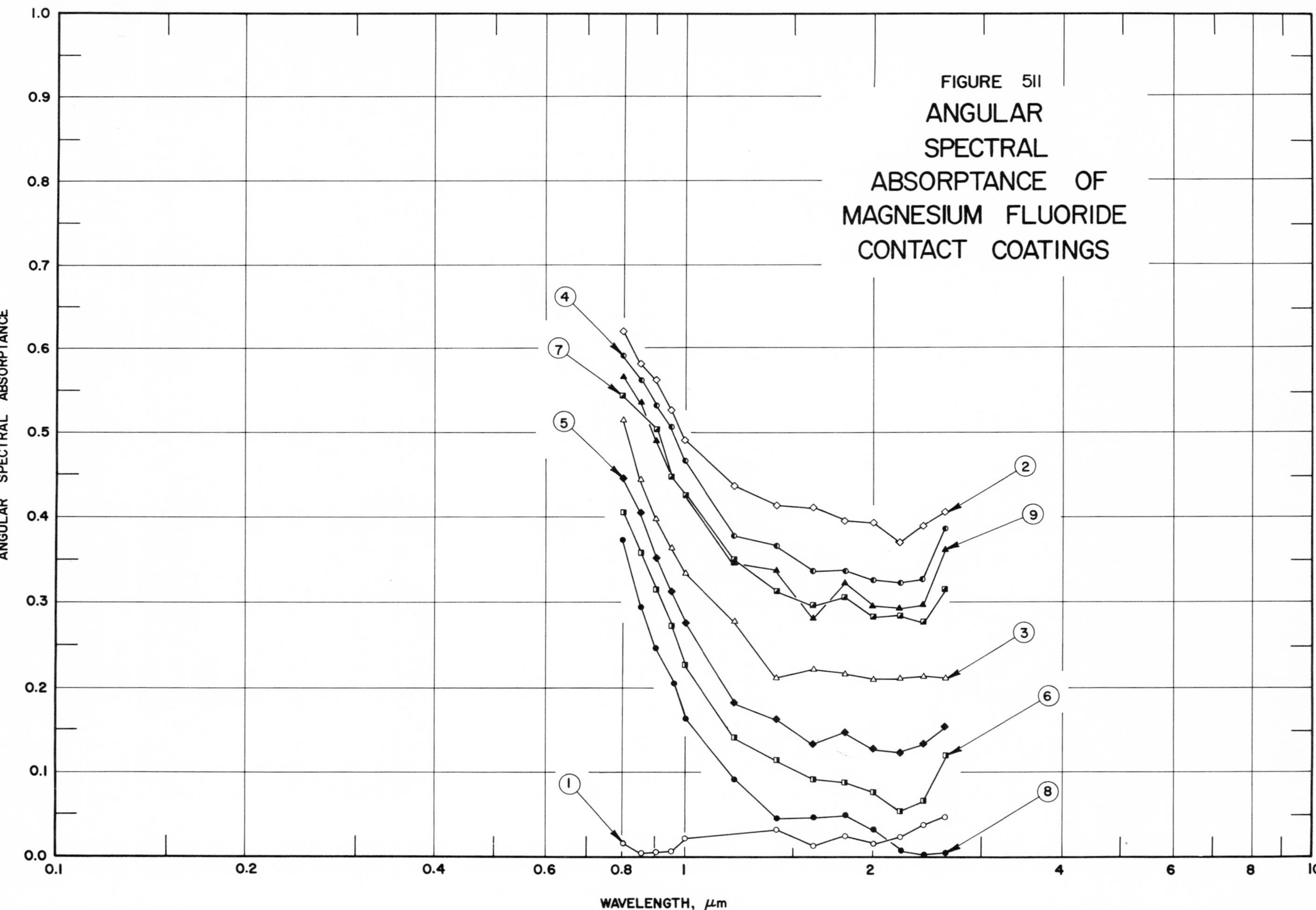

FIGURE 511
ANGULAR SPECTRAL ABSORPTANCE OF MAGNESIUM FLUORIDE CONTACT COATINGS
ANGULAR SPECTRAL ABSORPTANCE
WAVELENGTH, μm
1.0
0.9
0.8
0.7
0.6
0.5
0.4
0.3
0.2
0.1
0.0
0.1
0.2
0.4
0.6
0.8
1
2
4
6
8
10
4
7
5
1
2
9
3
6
8

SPECIFICATION TABLE NO. 511 ANGULAR SPECTRAL ABSORPTANCE OF MAGNESIUM FLUORIDE CONTACT COATINGS

Curve No.	Ref. No.	Year	Temperature, K	Wavelength Range, μm	Geometry θ	Reported Error, %	Composition (weight percent), Specifications, and Remarks
1	208	1964	298	0.80-2.60	45°		MgF_2; fused quartz substrate; author assumed $\alpha(45°) = 1 - \rho(45°, 45°) - \tau(0°, 0°)$; data beyond 2.2 μm subject to sizable errors.
2	208	1964	298	0.80-2.60	45°		MgF_2; platinum (~0.102 μm thick), fused quartz, and MgF_2 substrates; platinum applied to quartz by brushing on a platinum solution (Hanovia 05-X) and firing; author assumed $\alpha(45°) = 1 - \rho(45°, 45°) - \tau(0°, 0°)$; data beyond 2.2 μm subject to sizable errors. [Author's designation: Specimen No. 6]
3	208	1964	298	0.80-2.60	45°		Similar to above specimen and conditions. [Author's designation: Specimen No. 7]
4	208	1964	298	0.80-2.60	15°		Above specimen and conditions except author assumed $\alpha(15°) = 1 - \rho(15°, 15°) - \tau(0°, 0°)$.
5	208	1964	298	0.80-2.60	45°		Similar to curve 2 specimen and conditions. [Author's designation: Specimen No. 8]
6	208	1964	298	0.80-2.60	45°		Similar to curve 2 specimen and conditions. [Author's designation: Specimen No. 9]
7	208	1964	298	0.85-2.60	15°		Above specimen and conditions except author assumed $\alpha(15°) = 1 - \rho(15°, 15°) - \tau(0°, 0°)$.
8	208	1964	298	0.80-2.60	45°		Similar to curve 2 specimen and conditions. [Author's designation: Specimen No. 10]
9	208	1964	298	0.80-2.60	15°		Above specimen and conditions except author assumed $\alpha(15°) = 1 - \rho(15°, 15°) - \tau(0°, 0°)$.

DATA TABLE NO. 511 ANGULAR SPECTRAL ABSORPTANCE OF MAGNESIUM FLUORIDE CONTACT COATINGS

[Wavelength, λ, μm; Absorptance, α; Temperature, T, K]

CURVE 1
T = 298

λ	α
0.80	0.018
0.85	0.004
0.90	0.006
0.95	0.009
1.00	0.021
1.40	0.031
1.60	0.013
1.80	0.026
2.00	0.017
2.20	0.023
2.40	0.039
2.60	0.048

CURVE 2
T = 298

λ	α
0.80	0.621
0.85	0.585
0.90	0.564
0.95	0.528
1.00	0.492
1.20	0.438
1.40	0.415
1.60	0.412
1.80	0.396
2.00	0.394
2.20	0.370
2.40	0.390
2.60	0.406

CURVE 3
T = 298

λ	α
0.80	0.516
0.85	0.445
0.90	0.398
0.95	0.363
1.00	0.334
1.20	0.279
1.40	0.211
1.60	0.221
1.80	0.218
2.00	0.210
2.20	0.210

CURVE 3 (cont.)

λ	α
2.40	0.215
2.60	0.211

CURVE 4
T = 298

λ	α
0.80	0.593
0.85	0.564
0.90	0.532
0.95	0.509
1.00	0.466
1.20	0.377
1.40	0.366
1.60	0.336
1.80	0.337
2.00	0.327
2.20	0.324
2.40	0.329
2.60	0.386

CURVE 5
T = 298

λ	α
0.80	0.446
0.85	0.406
0.90	0.352
0.95	0.313
1.00	0.277
1.20	0.183
1.40	0.164
1.60	0.133
1.80	0.148
2.00	0.129
2.20	0.124
2.40	0.134
2.60	0.154

CURVE 6
T = 298

λ	α
0.80	0.407
0.85	0.359
0.90	0.315
0.95	0.273
1.00	0.227

CURVE 6 (cont.)

λ	α
1.20	0.140
1.40	0.114
1.60	0.091
1.80	0.087
2.00	0.076
2.20	0.054
2.40	0.067
2.60	0.120

CURVE 7
T = 298

λ	α
0.85	0.545
0.90	0.502
0.95	0.449
1.00	0.429
1.20	0.350
1.40	0.313
1.60	0.296
1.80	0.305
2.00	0.283
2.20	0.284
2.40	0.278
2.60	0.315

CURVE 8
T = 298

λ	α
0.80	0.372
0.85	0.295
0.90	0.247
0.95	0.205
1.00	0.165
1.20	0.091
1.40	0.046
1.60	0.047
1.80	0.050
2.00	0.031
2.20	0.007
2.40	0.002
2.60	0.005

CURVE 9
T = 298

λ	α
0.80	0.570
0.85	0.537
0.90	0.491
0.95	0.447*
1.00	0.429*
1.20	0.348
1.40	0.337
1.60	0.282
1.80	0.323
2.00	0.297
2.20	0.294
2.40	0.299
2.60	0.361

* Not shown on plot

SPECIFICATION TABLE NO. 512 NORMAL SOLAR ABSORPTANCE OF MAGNESIUM FLUORIDE CONTACT COATINGS

Curve No.	Ref. No.	Year	Temperature Range, K	Geometry θ	Reported Error, %	Composition (weight percent), Specifications and Remarks
1*	85	1969	159-398	$\sim 0^\circ$		MgF_2 (1.4×10^{-4} mm thick); Fe_2O_3 (2.8×10^{-4} mm thick) and 2024 aluminum substrates; MgF_2 and Fe_2O_3 vacuum vapor deposited; property measured in vacuum by calorimetric method; data extracted from smooth curve. [Author's designation: A-4 Solar Absorber NRDL-RTD-81-5]

DATA TABLE NO. 512 NORMAL SOLAR ABSORPTANCE OF MAGNESIUM FLUORIDE CONTACT COATINGS

[Temperature, T, K; Absorptance, α]

T	α
CURVE 1*	
159	0.831
211	0.860
264	0.882
327	0.905
398	0.918

* No plot given

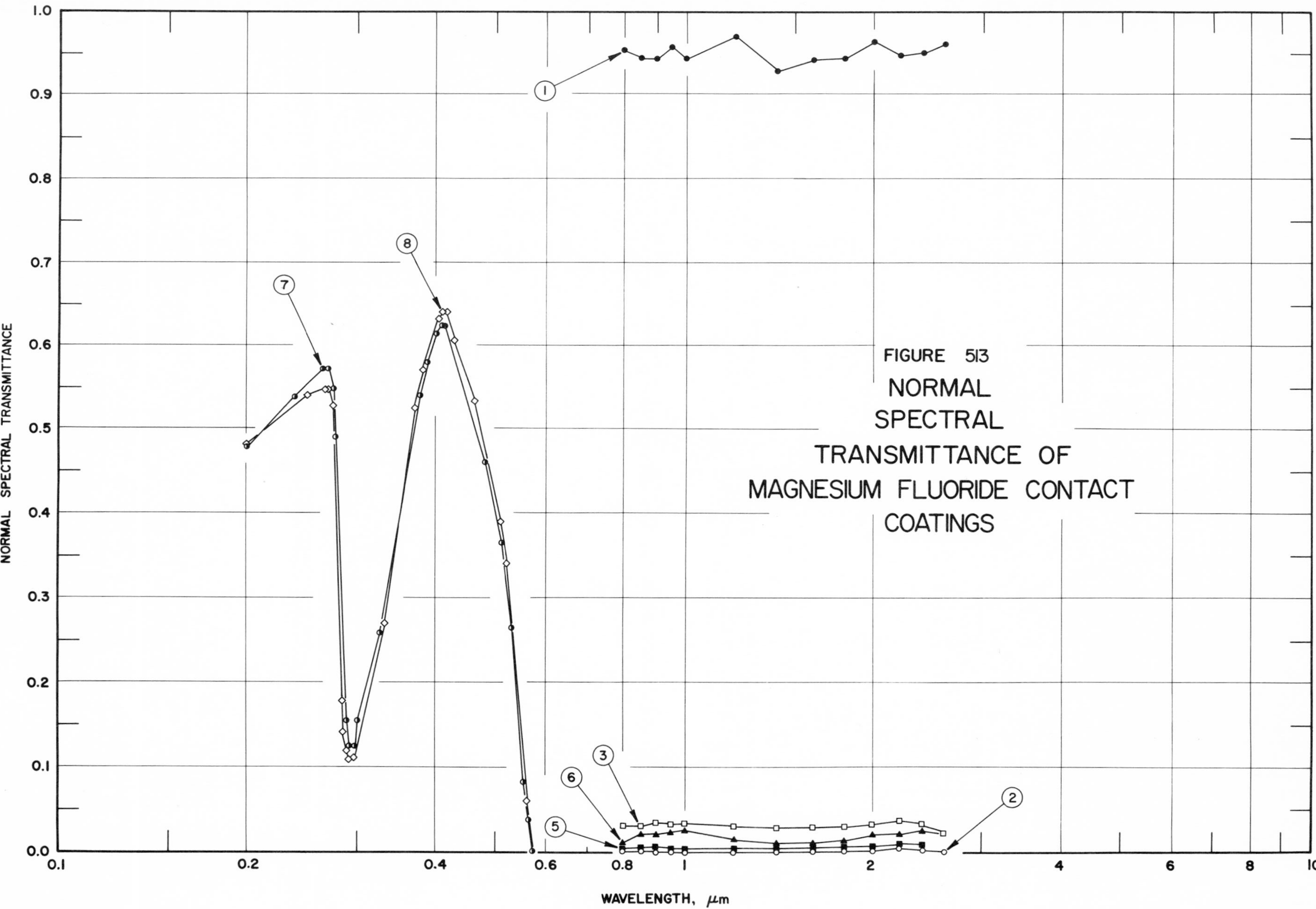

FIGURE 513
NORMAL
SPECTRAL
TRANSMITTANCE OF
MAGNESIUM FLUORIDE CONTACT
COATINGS
NORMAL SPECTRAL TRANSMITTANCE
WAVELENGTH, μm
1.0
0.9
0.8
0.7
0.6
0.5
0.4
0.3
0.2
0.1
0.0
0.1
0.2
0.4
0.6
0.8
1
2
4
6
8
10
1
7
8
3
6
5
2

SPECIFICATION TABLE NO. 513 NORMAL SPECTRAL TRANSMITTANCE OF MAGNESIUM FLUORIDE CONTACT COATINGS

Curve No.	Ref. No.	Year	Temperature, K	Wavelength Range, μm	Geometry θ	θ'	ω'	Reported Error, %	Composition (weight percent), Specifications, and Remarks
1	208	1964	298	0.80-2.60	0°	0°			MgF_2; fused quartz substrate; data beyond 2.2 μm subject to sizable errors.
2	208	1964	298	0.80-2.60	0°	0°			MgF_2; platinum (~0.102 μm thick), fused quartz, and MgF_2 substrates; platinum applied to quartz by brushing on a platinum solution (Hanovia 05-X) and firing; data beyond 2.2 μm subject to sizable errors. [Author's designation: Specimen No. 6]
3	208	1964	298	0.80-2.60	0°	0°			Similar to above specimen and conditions. [Author's designation: Specimen No. 7]
4	208	1964	298	0.80-2.40	0°	0°			Similar to above specimen and conditions. [Author's designation: Specimen No. 8]
5	208	1964	298	0.80-2.40	0°	0°			Similar to above specimen and conditions. [Author's designation: Specimen No. 9]
6	208	1964	298	0.80-2.60	0°	0°			Similar to above specimen and conditions. [Author's designation: Specimen No. 10]
7	255	1957	~298	2.00-5.72	~0°	~0°			MgF_2 (560 μm thick); calcium aluminate glass (Melt No. 1A 5561, 4 mm thick) substrate; data extracted from smooth curve.
8	255	1957	~298	2.00-5.73	~0°	~0°			MgF_2 (560 μm thick); SiO_2 (560 μm thick) and calcium aluminate glass (Melt No. 1A 5561, 4.0 mm thick) substrate; data extracted from smooth curve.

DATA TABLE NO. 513 NORMAL SPECTRAL TRANSMITTANCE OF MAGNESIUM FLUORIDE CONTACT COATINGS

[Wavelength, λ, μm; Transmittance, τ; Temperature, T, K]

λ	τ
CURVE 1	**T = 298**
0.80	0.953
0.85	0.946
0.90	0.946
0.95	0.957
1.00	0.946
1.20	0.971
1.40	0.931
1.60	0.944
1.80	0.946
2.00	0.966
2.20	0.948
2.40	0.953
2.60	0.962
CURVE 2	**T = 298**
0.80	0.002
0.85	0.002
0.90	0.002
0.95	0.002
1.00	0.002
1.20	0.002
1.40	0.002
1.60	0.003
1.80	0.004
2.00	0.004
2.20	0.006
2.40	0.004
2.60	0.002
CURVE 3	**T = 298**
0.80	0.033
0.85	0.033
0.90	0.035
0.95	0.034
1.00	0.034
1.20	0.032
1.40	0.028
1.60	0.029
1.80	0.029
2.00	0.034
CURVE 3 (cont.)	
2.20	0.037
2.40	0.034
2.60	0.024
CURVE 4	**T = 298**
0.80	0.002*
0.85	0.002*
0.90	0.002*
0.95	0.003
1.00	0.004
1.20	0.003
1.40	0.003
1.60	0.003*
1.80	0.004*
2.00	0.006*
2.20	0.007
2.40	0.006
CURVE 5	**T = 298**
0.80	0.004
0.85	0.007
0.90	0.008
0.95	0.006
1.00	0.005
1.20	0.005
1.40	0.005
1.60	0.005
1.80	0.007
2.00	0.008
2.20	0.009
2.40	0.011
CURVE 6	**T = 298**
0.80	0.014
0.85	0.022
0.90	0.022
0.95	0.024
1.00	0.026
1.20	0.015
CURVE 6 (cont.)	
1.40	0.013
1.60	0.012
1.80	0.015
2.00	0.021
2.20	0.021
2.40	0.026
2.60	0.022*
CURVE 7	**T ~ 298**
2.00	0.480*
2.37	0.538
2.65	0.574
2.70	0.574
2.76	0.549
2.78	0.491
2.87	0.154
2.91	0.124
2.97	0.124
3.07	0.156
3.27	0.259
3.79	0.540
3.88	0.581
4.02	0.616
4.10	0.625
4.15	0.625
4.83	0.463
5.11	0.365
5.32	0.265
5.55	0.083
5.64	0.039
5.72	0.000
CURVE 8	**T ~ 298**
2.00	0.482
2.53	0.540
2.67	0.548
2.71	0.548
2.74	0.529
2.85	0.179
2.86	0.142
2.90	0.120
CURVE 8 (cont.)	
2.94	0.110
2.97	0.112
3.32	0.270
3.73	0.524
3.83	0.572
4.06	0.634
4.11	0.640
4.19	0.640
4.27	0.607
4.62	0.534
5.07	0.391
5.20	0.340
5.61	0.063
5.73	0.000*

* Not shown on plot

2. CONTACT COATINGS (continued)

E. Resin Coatings

SPECIFICATION TABLE NO. 514 HEMISPHERICAL TOTAL EMITTANCE OF ACRYLIC CONTACT COATINGS

Curve No.	Ref. No.	Year	Temperature Range, K	Reported Error, %	Composition (weight percent), Specifications and Remarks
1*	43	1962	362-513	± 3	Acrylic (~0.0762 mm thick); aluminum substrate; measured in vacuum (10^{-3} mm Hg); data extracted from smooth curve.

DATA TABLE NO. 514 HEMISPHERICAL TOTAL EMITTANCE OF ACRYLIC CONTACT COATINGS

[Temperature, T, K; Emittance, ϵ]

T	ϵ
CURVE 1*	
362	0.650
372	0.640
388	0.610
408	0.630
432	0.600
466	0.610
491	0.585
513	0.565

* No plot given

SPECIFICATION TABLE NO. 515 NORMAL TOTAL EMITTANCE OF ACRYLIC CONTACT COATINGS

Curve No.	Ref. No.	Year	Temperature Range, K	Geometry θ'	Reported Error, %	Composition (weight percent), Specifications and Remarks
1*	7	1961	373	~0°		Krylon; acrylic resin (0.051 mm thick); anodized aluminum substrate; measured in vacuum ($\leq 10^{-5}$ mm Hg).

DATA TABLE NO. 515 NORMAL TOTAL EMITTANCE OF ACRYLIC CONTACT COATINGS

[Temperature, T, K; Emittance, ϵ]

T	ϵ
CURVE 1*	
373	0.90

* No plot given

SPECIFICATION TABLE NO. 516 NORMAL SPECTRAL REFLECTANCE OF ACRYLIC CONTACT COATINGS

Curve No.	Ref. No.	Year	Temperature, K	Wavelength Range, μm	Geometry θ	θ'	ω'	Reported Error, %	Composition (weight percent), Specifications, and Remarks
1*	100	1962	298	0.371-0.747	~0°		2π		Acrylic, T-40-C-C-9 (binder used in Sherwin Williams High Heat Aluminum Paint); polished aluminum substrate; data extracted from smooth curve.
2*	100	1962	298	0.371-0.747	~0°		2π		Above specimen and conditions except exposed to UV at ~10 suns for 22.75 hrs.

DATA TABLE NO. 516 NORMAL SPECTRAL REFLECTANCE OF ACRYLIC CONTACT COATINGS

[Wavelength, λ, μm; Reflectance, ρ; Temperature, T, K]

λ	ρ	λ	ρ
CURVE 1* T = 298		CURVE 2 (cont.)*	
		0.499	0.210
0.371	0.668	0.548	0.312
0.399	0.683	0.598	0.412
0.448	0.716	0.649	0.478
0.498	0.736	0.697	0.533
0.548	0.745	0.747	0.578
0.599	0.748		
0.649	0.756		
0.661	0.766		
0.696	0.766		
0.747	0.746		
CURVE 2* T = 298			
0.371	0.091		
0.401	0.091		
0.450	0.129		

* No plot given

SPECIFICATION TABLE NO. 517 NORMAL SPECTRAL REFLECTANCE OF ALKYD RESIN CONTACT COATINGS

Curve No.	Ref. No.	Year	Temperature, K	Wavelength Range, μm	Geometry θ θ'	ω'	Reported Error, %	Composition (weight percent), Specifications, and Remarks
1*	104	1953	298	1.00-15.0	5°	2π	± 2	Clear glyptal (0.203 mm thick), alkyd resin; polished aluminum substrate; data extracted from smooth curve; converted from R (2π, 5°).

DATA TABLE NO. 517 NORMAL SPECTRAL REFLECTANCE OF ALKYD RESIN CONTACT COATINGS

[Wavelength, λ, μm; Reflectance, ρ; Temperature, T, K]

λ	ρ	λ	ρ	λ	ρ
CURVE 1* T = 298		CURVE 1 (cont.)*		CURVE 1 (cont.)*	
		5.50	0.160	11.0	0.085
1.00	0.590	5.80	0.060	11.2	0.085
1.20	0.650	6.00	0.075	11.4	0.090
1.40	0.680	6.20	0.070	11.7	0.085
1.80	0.690	6.40	0.140	12.0	0.090
2.00	0.660	6.80	0.125	12.8	0.090
2.20	0.565	7.00	0.070	13.0	0.100
2.50	0.415	7.20	0.070	14.0	0.100
2.70	0.510	7.50	0.065	14.2	0.100
3.00	0.070	7.80	0.070	14.4	0.105
3.60	0.070	8.00	0.095	14.8	0.105
3.80	0.140	8.50	0.080	15.0	0.120
4.00	0.220	9.00	0.100		
4.20	0.310	9.20	0.085		
4.50	0.400	9.40	0.090		
4.70	0.320	10.0	0.085		
5.00	0.330	10.2	0.090		
5.20	0.310	10.5	0.085		

* No plot given

SPECIFICATION TABLE NO. 518 HEMISPHERICAL TOTAL EMITTANCE OF BAKELITE LACQUER CONTACT COATINGS

Curve No.	Ref. No.	Year	Temperature Range, K	Reported Error, %	Composition (weight percent), Specifications and Remarks
1*	217	1960	77		Bakelite lacquer; unknown substrate; baked on at 408 K; cleaned with acetone; measured in vacuum (~1 x 10^{-5} mm Hg).

DATA TABLE NO. 518 HEMISPHERICAL TOTAL EMITTANCE OF BAKELITE LACQUER CONTACT COATINGS

[Temperature, T, K; Emittance, ϵ]

T	ϵ
CURVE 1*	
77	0.87

* No plot given

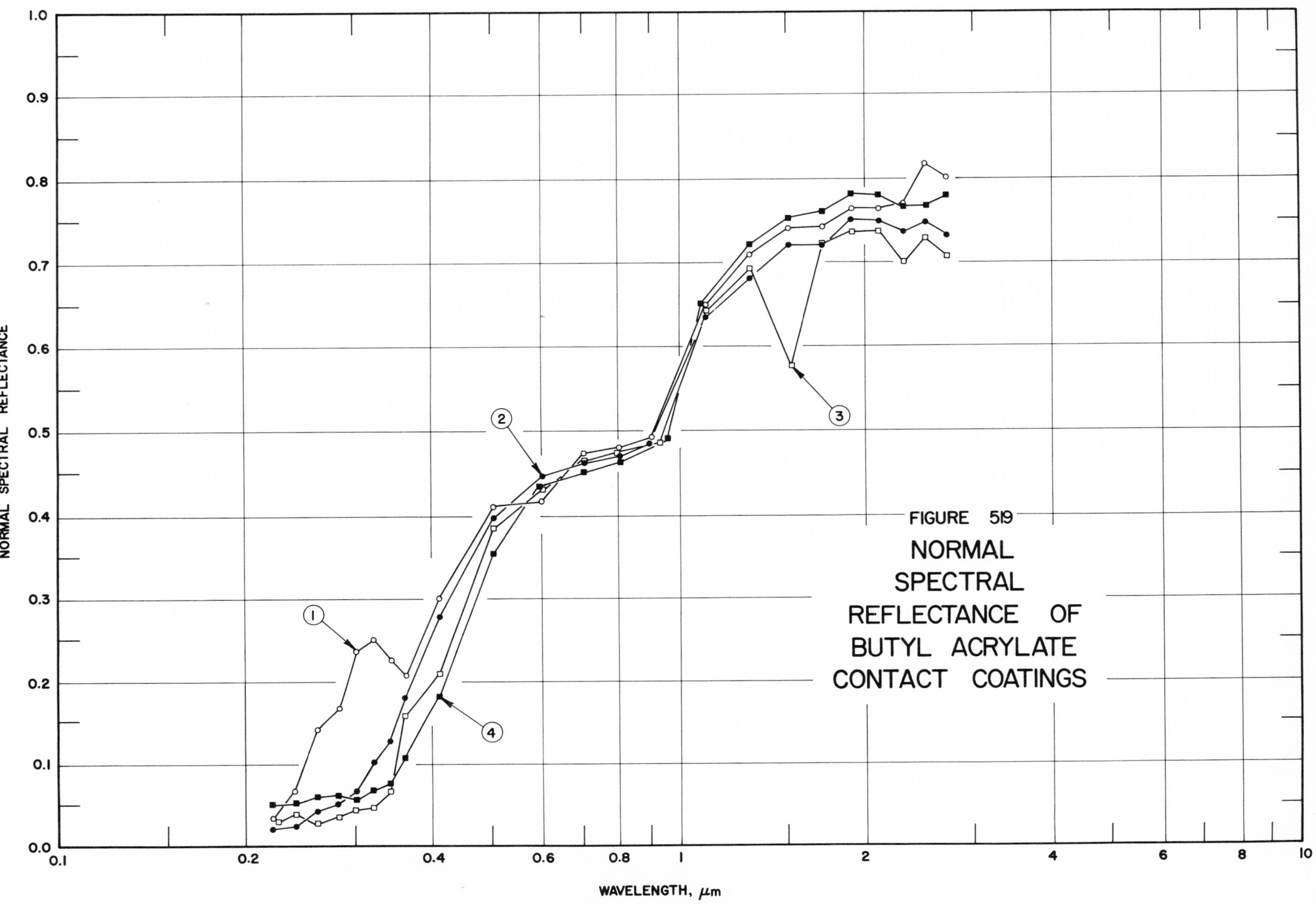

FIGURE 519 NORMAL SPECTRAL REFLECTANCE OF BUTYL ACRYLATE CONTACT COATINGS

SPECIFICATION TABLE NO. 519 NORMAL SPECTRAL REFLECTANCE OF BUTYL ACRYLATE CONTACT COATINGS

Curve No.	Ref. No.	Year	Temperature, K	Wavelength Range, μm	Geometry θ	θ'	ω'	Reported Error, %	Composition (weight percent), Specifications, and Remarks
1	1	1960	358	0.221-2.710	~0°		2π		Butyl acrylate copolymer (0.0508 mm thick); anodized aluminum alloy 24S-T substrate; measured in vacuum (10^{-5} mm Hg); measured relative to MgO.
2	1	1960	358	0.221-2.709	~0°		2π		Above specimen and conditions except exposed to UV radiation for 20 hrs from G. E. UA-3 lamp.
3	1	1960	358	0.225-2.710	~0°		2π		Above specimen and conditions except exposed to UV radiation for 42 hrs.
4	1	1960	358	0.222-2.707	~0°		2π		Above specimen and conditions except exposed to UV radiation for 105 hrs.

DATA TABLE NO. 519 NORMAL SPECTRAL REFLECTANCE OF BUTYL ACRYLATE CONTACT COATINGS

[Wavelength, λ, μm; Reflectance, ρ; Temperature, T, K]

CURVE 1, T = 358

λ	ρ
0.221	0.035
0.240	0.068
0.261	0.140
0.283	0.168
0.301	0.238
0.322	0.252
0.342	0.228
0.362	0.209
0.410	0.300
0.503	0.411
0.601	0.418
0.702	0.475
0.803	0.481
0.902	0.494
1.115	0.652
1.312	0.712
1.511	0.742
1.715	0.743
1.911	0.766
2.111	0.766
2.302	0.720
2.504	0.769
2.710	0.751

CURVE 2, T = 358

λ	ρ
0.221	0.022
0.241	0.026
0.261	0.045
0.282	0.052
0.301	0.068
0.322	0.102
0.342	0.129
0.361	0.182
0.412	0.278
0.503	0.397
0.602	0.449
0.701	0.463
0.802	0.472
0.899	0.485
1.115	0.636
1.313	0.681
1.509	0.722
1.705	0.722
1.908	0.751
2.109	0.751
2.302	0.738
2.505	0.749
2.709	0.734

CURVE 3, T = 358

λ	ρ
0.225	0.031
0.242	0.040
0.261	0.029
0.283	0.038
0.301	0.045
0.321	0.047
0.341	0.069
0.361	0.158
0.412	0.210
0.502	0.386
0.604	0.431
0.708	0.465
0.798	0.473
0.931	0.486
1.139	0.646
1.308	0.693
1.533	0.578
1.722	0.723
1.907	0.737
2.107	0.738
2.302	0.702
2.504	0.730
2.710	0.709

CURVE 4, T = 358

λ	ρ
0.222	0.051
0.241	0.054
0.261	0.061
0.282	0.063
0.301	0.057
0.321	0.069
0.341	0.076
0.361	0.107
0.412	0.182
0.502	0.354
0.597	0.434
0.701	0.452
0.807	0.464
0.959	0.492
1.086	0.653
1.310	0.723
1.512	0.754
1.713	0.762
1.906	0.781
2.110	0.780
2.304	0.768
2.526	0.769
2.707	0.780

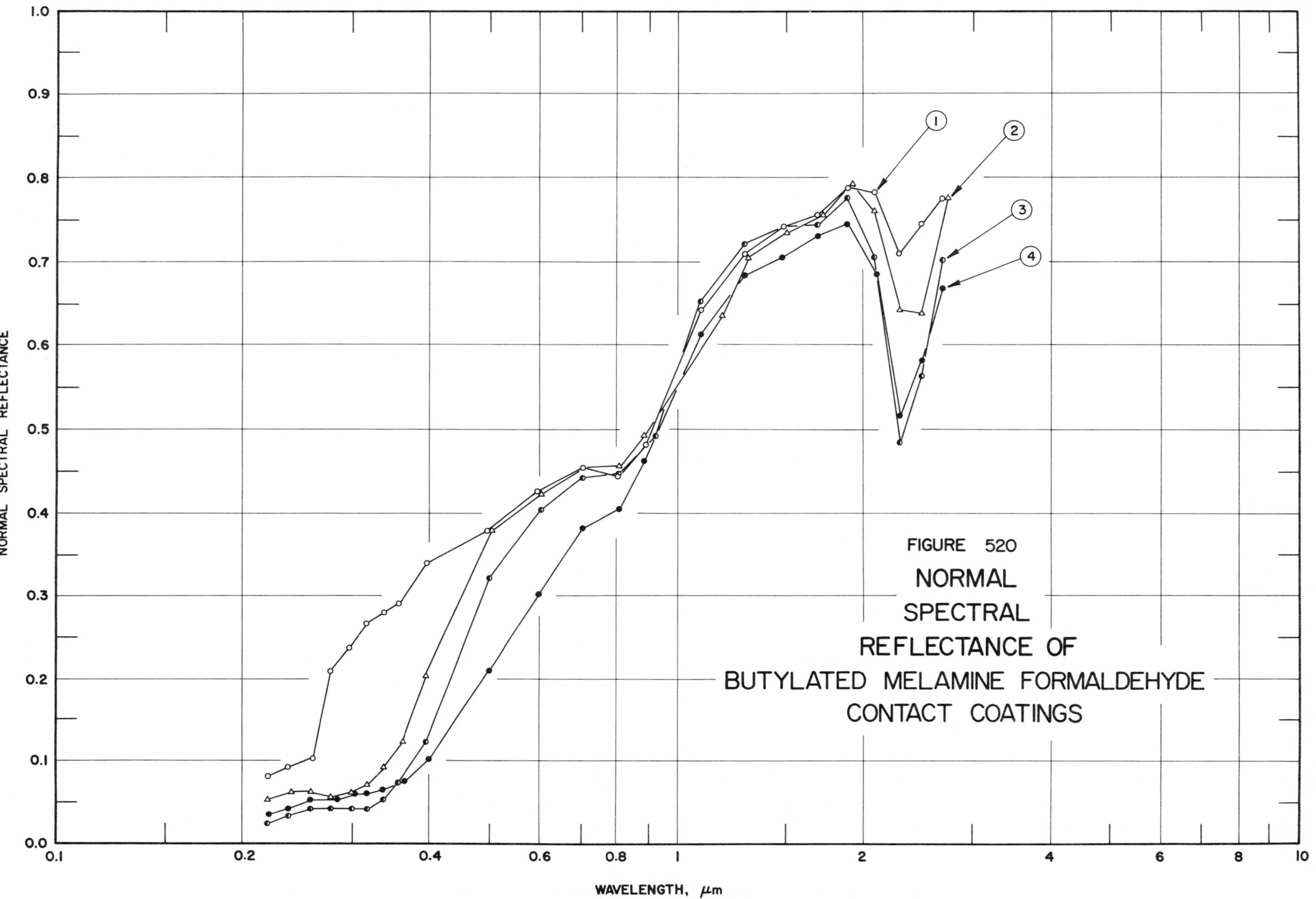

FIGURE 520 NORMAL SPECTRAL REFLECTANCE OF BUTYLATED MELAMINE FORMALDEHYDE CONTACT COATINGS

SPECIFICATION TABLE NO. 520 NORMAL SPECTRAL REFLECTANCE OF BUTYLATED MELAMINE FORMALDEHYDE CONTACT COATINGS

Curve No.	Ref. No.	Year	Temperature, K	Wavelength Range, μm	Geometry θ	θ'	ω'	Reported Error, %	Composition (weight percent), Specifications, and Remarks
1	1	1960	358	0.220-2.694	~0°		2π		Butylated melamine formaldehyde (0.0508 mm thick); anodized aluminum alloy 24S-T substrate; measured in vacuum (10^{-5} mm Hg); measured relative to MgO.
2	1	1960	358	0.220-2.722	~0°		2π		Above specimen and conditions except exposed to UV radiation for 8 hrs from G. E. UA-3 lamp.
3	1	1960	358	0.220-2.695	~0°		2π		Above specimen and conditions except exposed to UV radiation for 40 hrs.
4	1	1960	358	0.221-2.694	~0°		2π		Above specimen and conditions except exposed to UV radiation for 104 hrs.

DATA TABLE NO. 520 NORMAL SPECTRAL REFLECTANCE OF BUTYLATED MELAMINE FORMALDEHYDE CONTACT COATINGS

[Wavelength, λ, μm; Reflectance, ρ; Temperature, T, K]

λ	ρ	λ	ρ	λ	ρ	λ	ρ	λ	ρ	λ	ρ
CURVE 1		CURVE 1 (cont.)		CURVE 2 (cont.)		CURVE 3 (cont.)		CURVE 4		CURVE 4 (cont.)	
T = 358								T = 358			
		2.093	0.783	0.803	0.457	0.319	0.043			2.102	0.689
0.220	0.081	2.292	0.710	0.886	0.495	0.338	0.053	0.221	0.036	2.297	0.519
0.238	0.092	2.492	0.749	1.117	0.639	0.358	0.075	0.239	0.043	2.494	0.583
0.260	0.102	2.694	0.778	1.309	0.707	0.399	0.122	0.259	0.053	2.694	0.670
0.278	0.210			1.508	0.738	0.500	0.322	0.285	0.053		
0.299	0.239	CURVE 2		1.714	0.759	0.601	0.404	0.303	0.061		
0.319	0.269	T = 358		1.926	0.794	0.701	0.441	0.319	0.061		
0.339	0.280			2.094	0.761	0.080	0.447	0.338	0.066		
0.358	0.291	0.220	0.054	2.296	0.645	0.920	0.494	0.365	0.076		
0.399	0.340	0.240	0.063	2.495	0.640	1.094	0.654	0.400	0.101		
0.499	0.380	0.259	0.063	2.722	0.778	1.292	0.721	0.500	0.210		
0.599	0.426	0.279	0.058			1.493	0.740	0.600	0.303		
0.701	0.455	0.300	0.062	CURVE 3		1.699	0.744	0.701	0.382		
0.802	0.443	0.319	0.071	T = 358		1.897	0.778	0.803	0.408		
0.890	0.482	0.339	0.094			2.093	0.707	0.884	0.463		
1.091	0.642	0.361	0.124	0.220	0.025	2.292	0.486	1.096	0.614		
1.287	0.710	0.399	0.204	0.239	0.033	2.492	0.562	1.293	0.687		
1.489	0.741	0.504	0.380	0.259	0.043	2.695	0.702	1.489	0.708		
1.685	0.758	0.605	0.424	0.279	0.043			1.694	0.731		
1.896	0.790	0.705	0.455	0.300	0.043			1.893	0.749		

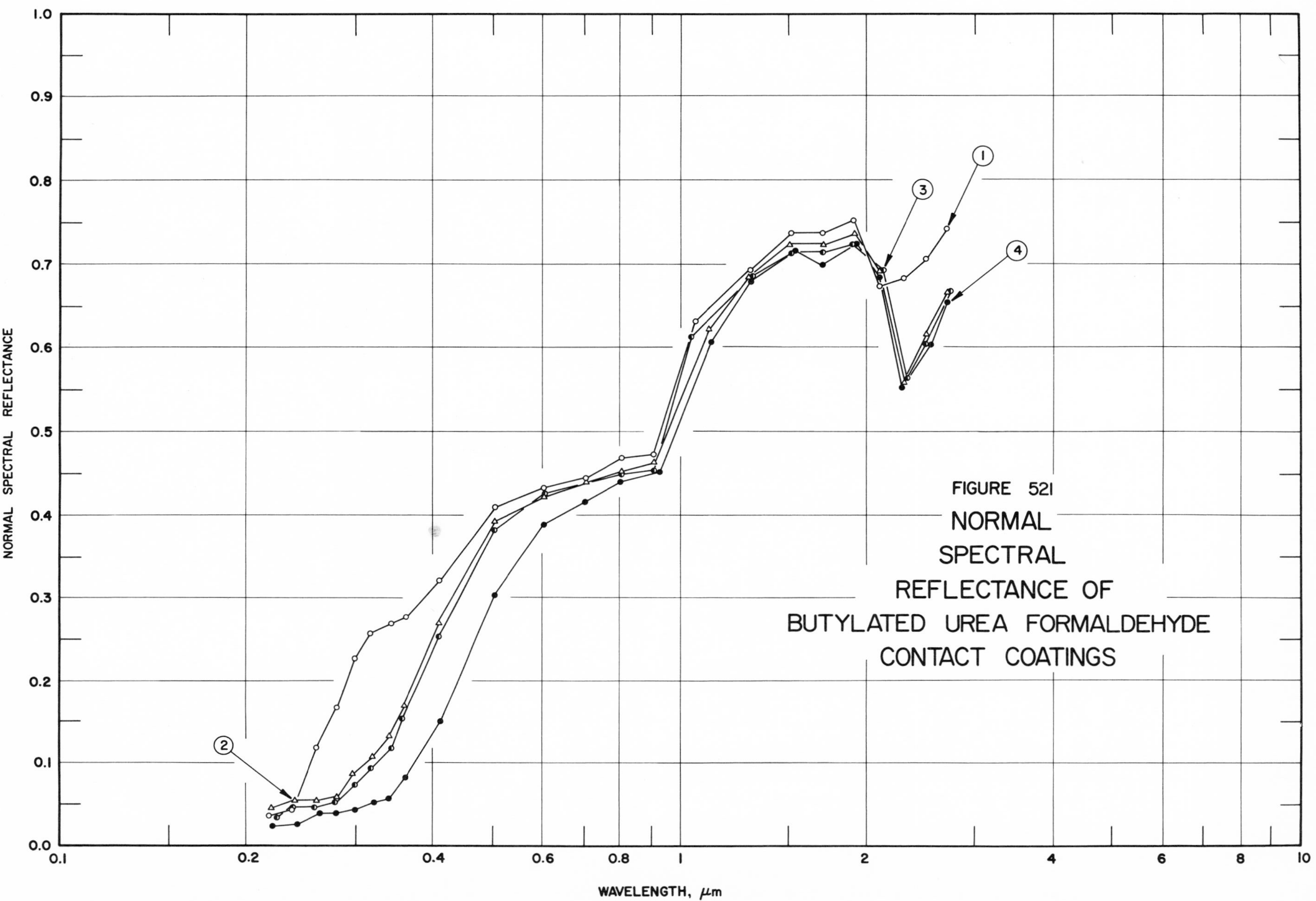

FIGURE 521 NORMAL SPECTRAL REFLECTANCE OF BUTYLATED UREA FORMALDEHYDE CONTACT COATINGS

SPECIFICATION TABLE NO. 521 NORMAL SPECTRAL REFLECTANCE OF BUTYLATED UREA FORMALDEHYDE CONTACT COATINGS

Curve No.	Ref. No.	Year	Temperature, K	Wavelength Range, μm	Geometry θ	Geometry θ'	Geometry ω'	Reported Error, %	Composition (weight percent), Specifications, and Remarks
1	1	1960	358	0.219-2.708	~0°		2π		Butylated urea formaldehyde (0.0508 mm thick); anodized aluminum alloy 24 S-T substrate; measured in vacuum (10^{-5} mm Hg); measured relative to MgO.
2	1	1960	358	0.220-2.706	~0°		2π		Above specimen and conditions except exposed to UV radiation for 20 hrs from G. E. UA-3 lamp.
3	1	1960	358	0.225-2.730	~0°		2π		Above specimen and conditions except exposed to UV radiation for 42 hrs.
4	1	1960	358	0.221-2.709	~0°		2π		Above specimen and conditions except exposed to UV radiation for 105 hrs.

DATA TABLE NO. 521 NORMAL SPECTRAL REFLECTANCE OF BUTYLATED UREA FORMALDEHYDE CONTACT COATINGS

[Wavelength, λ, μm; Reflectance, ρ; Temperature, T, K]

λ	ρ	λ	ρ	λ	ρ	λ	ρ	λ	ρ	λ	ρ
CURVE 1		CURVE 1 (cont.)		CURVE 2 (cont.)		CURVE 3		CURVE 3 (cont.)		CURVE 4 (cont.)	
T = 358						T = 358					
		1.512	0.737	0.340	0.133			1.705	0.715	0.361	0.081
0.219	0.038	1.702	0.739	0.360	0.170	0.225	0.033	1.901	0.724	0.411	0.150
0.239	0.043	1.910	0.751	0.410	0.270	0.239	0.045	2.134	0.694	0.501	0.303
0.260	0.119	2.108	0.674	0.501	0.395	0.259	0.046	2.339	0.564	0.602	0.390
0.280	0.166	2.306	0.682	0.602	0.422	0.279	0.051	2.504	0.605	0.702	0.417
0.300	0.228	2.504	0.708	0.707	0.440	0.300	0.072	2.730	0.667	0.801	0.440
0.319	0.257	2.708	0.742	0.803	0.453	0.319	0.092			0.928	0.451
0.341	0.270			0.905	0.465	0.341	0.118	CURVE 4		1.131	0.608
0.361	0.279	CURVE 2		1.121	0.624	0.359	0.152	T = 358		1.302	0.680
0.410	0.321	T = 358		1.293	0.686	0.410	0.254			1.540	0.717
0.501	0.410			1.507	0.724	0.501	0.381	0.221	0.024	1.708	0.700
0.601	0.434	0.220	0.046	1.709	0.724	0.607	0.426	0.244	0.028	1.929	0.725
0.702	0.445	0.239	0.055	1.912	0.739	0.803	0.450	0.264	0.040	2.103	0.685
0.801	0.470	0.260	0.055	2.105	0.693	0.901	0.453	0.280	0.040	2.295	0.551
0.906	0.473	0.280	0.060	2.301	0.560	1.104	0.614	0.300	0.043	2.530	0.604
1.106	0.631	0.299	0.089	2.505	0.618	1.326	0.688	0.321	0.051	2.709	0.655
1.309	0.695	0.320	0.108	2.706	0.669	1.511	0.712	0.340	0.058		

SPECIFICATION TABLE NO. 522 NORMAL SPECTRAL TRANSMITTANCE OF DIACETYL CELLULOSE CONTACT COATINGS

Curve No.	Ref. No.	Year	Temperature, K	Wavelength Range, μm	Geometry θ	θ'	ω'	Reported Error, %	Composition (weight percent), Specifications, and Remarks
1*	177	1965	300	100-698	~0°	~0°			Diacetyl cellulose (43 μm thick); varnish, germanium (1 mm thick) varnish and diacetyl cellulse (43 μm thick) substrates; prepared by cementing with General Electric No. 7031 varnish; data extracted from smooth curve.

DATA TABLE NO. 522 NORMAL SPECTRAL TRANSMITTANCE OF DIACETYL CELLULOSE CONTACT COATINGS

[Wavelength, λ, μm; Transmittance, τ; Temperature, T, K]

λ	τ
CURVE 1*	T = 300
100	0.217
107	0.227
115	0.227
127	0.214
134	0.242
147	0.284
171	0.304
190	0.369
206	0.446
248	0.649
261	0.700
281	0.733
319	0.756
358	0.750
425	0.719
517	0.670
590	0.635
698	0.603

* No plot given

SPECIFICATION TABLE NO. 523 HEMISPHERICAL TOTAL EMITTANCE OF EPOXY CONTACT COATINGS

Curve No.	Ref. No.	Year	Temperature Range, K	Reported Error, %	Composition (weight percent), Specifications and Remarks
1*	48	1963	178-270	± 3	Epoxy; aluminum alloy substrate; measured in vacuum (10^{-3} mm Hg).

DATA TABLE NO. 523 HEMISPHERICAL TOTAL EMITTANCE OF EPOXY CONTACT COATINGS

[Temperature, T, K; Emittance, ϵ]

T	ϵ
CURVE 1*	
178	0.69
191	0.71
201	0.71
212	0.72
222	0.72
231	0.73
241	0.72
253	0.73
263	0.71
270	0.73

* No plot given

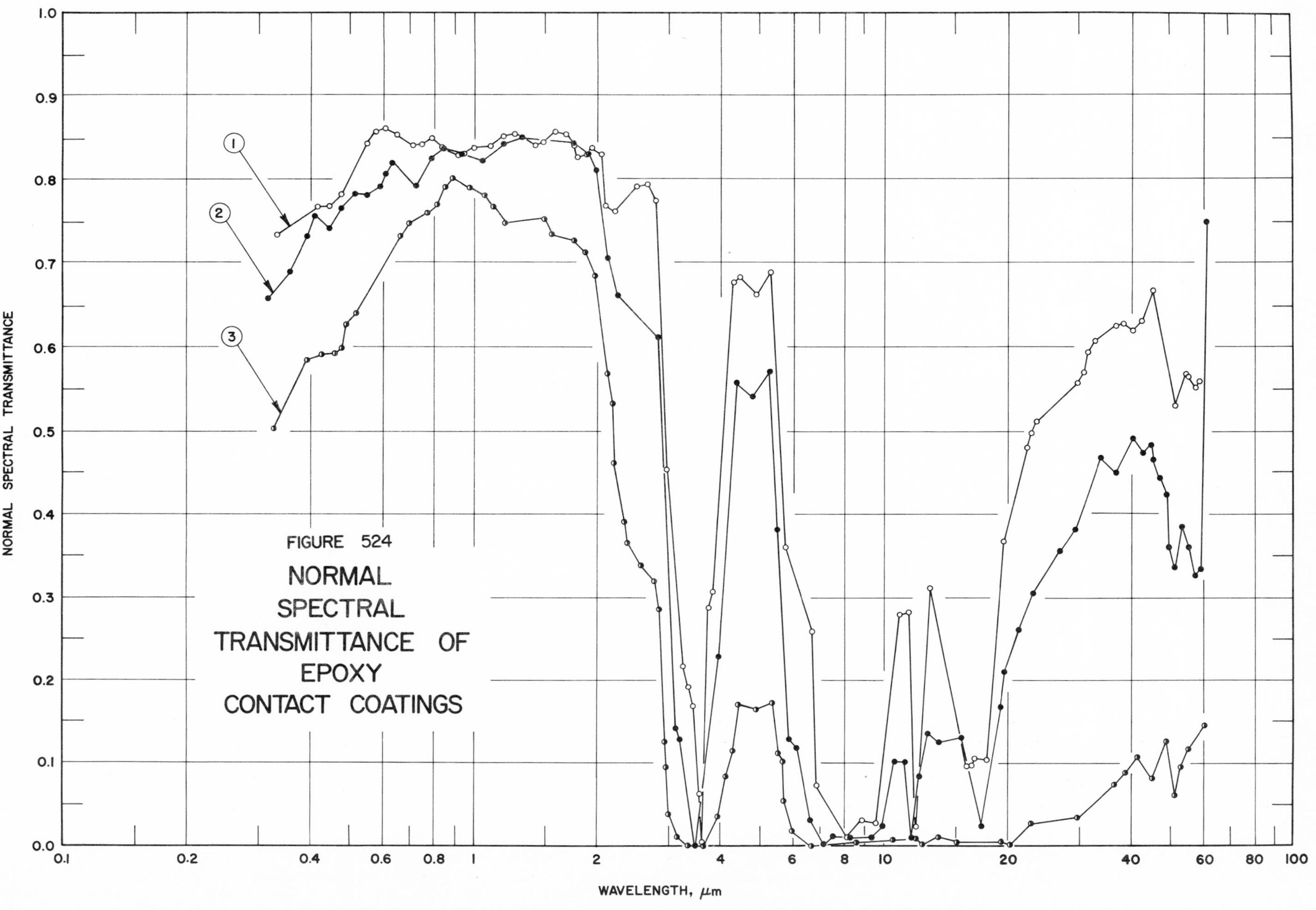

FIGURE 524
NORMAL SPECTRAL TRANSMITTANCE OF EPOXY CONTACT COATINGS

SPECIFICATION TABLE NO. 524 NORMAL SPECTRAL TRANSMITTANCE OF EPOXY CONTACT COATINGS

Curve No.	Ref. No.	Year	Temperature, K	Wavelength Range, μm	Geometry θ	θ'	ω'	Reported Error, %	Composition (weight percent), Specifications, and Remarks
1	91	1966	~298	0.331-58.8	~0°	~0°			Cat-A-Lac clear, epoxy resin (43 μm thick); polyethylene (20 μm thick) substrate.
2	91	1966	~298	0.316-61.3	~0°	~0°			Similar to above specimen and conditions except 75 μm thick.
3	91	1966	~298	0.325-60.3	~0°	~0°			Similar to above specimen and conditions except 250 μm thick.

DATA TABLE NO. 524 NORMAL SPECTRAL TRANSMITTANCE OF EPOXY CONTACT COATINGS

[Wavelength, λ, μm; Transmittance, τ; Temperature, T, K]

λ	τ
CURVE 1	**T ~ 298**
0.331	0.734
0.416	0.767
0.443	0.768
0.474	0.783
0.547	0.845
0.574	0.858
0.605	0.862
0.656	0.856
0.711	0.841
0.743	0.844
0.792	0.850
0.839	0.839
0.907	0.829
0.948	0.832
1.00	0.838
1.09	0.842
1.17	0.853
1.26	0.856
1.42	0.842
1.47	0.846
1.57	0.858
1.67	0.856
1.74	0.841
1.79	0.827
1.88	0.830
1.94	0.839
2.05	0.832
2.10	0.768
2.21	0.760
2.49	0.793
2.65	0.796
2.79	0.777
2.95	0.453
3.24	0.217
3.31	0.190
3.41	0.168
3.54	0.065
3.60	0.005
3.73	0.288
3.82	0.308
4.33	0.678
4.45	0.684
4.89	0.665
5.32	0.692
5.74	0.360
CURVE 1 (cont.)	
6.62	0.259
6.83	0.073
8.10	0.012
8.81	0.033
9.59	0.028
10.8	0.278
11.4	0.282
11.9	0.026
12.9	0.313
15.7	0.097
16.2	0.097
16.6	0.106
17.7	0.104
19.5	0.368
22.2	0.480
22.7	0.499
23.5	0.511
29.5	0.558
30.6	0.570
31.4	0.597
32.5	0.609
36.8	0.627
38.1	0.629
40.5	0.623
42.8	0.634
45.1	0.669
51.4	0.530
54.5	0.569
55.4	0.567
57.5	0.552
58.8	0.561
CURVE 2	**T ~ 298**
0.316	0.658
0.359	0.689
0.391	0.733
0.408	0.756
0.443	0.742
0.478	0.766
0.509	0.783
0.549	0.783
0.587	0.793
0.606	0.807
0.633	0.820
CURVE 2 (cont.)	
0.721	0.792
0.788	0.825
0.849	0.837
0.939	0.831
1.04	0.823
1.18	0.844
1.31	0.851
1.75	0.845
1.90	0.831
1.99	0.813
2.12	0.705
2.25	0.662
2.81	0.613
3.10	0.141
3.17	0.127
3.45	0.000
3.98	0.229
4.37	0.558
4.79	0.542
5.24	0.570
5.50	0.380
5.83	0.128
6.12	0.117
6.60	0.032
7.12	0.001
7.49	0.014
8.29	0.010
9.39	0.012
9.93	0.026
10.6	0.101
11.3	0.101
11.6	0.011
12.3	0.084
12.7	0.136
13.6	0.126
15.4	0.131
17.3	0.027
19.3	0.167
19.6	0.212
21.3	0.263
23.0	0.306
26.8	0.355
29.1	0.380
33.6	0.468
36.9	0.448
40.2	0.491
CURVE 2 (cont.)	
42.7	0.473
44.4	0.484
45.6	0.466
46.9	0.443
48.6	0.424
49.7	0.362
51.1	0.337
53.0	0.386
55.8	0.360
57.1	0.328
59.0	0.336
61.3	0.750
CURVE 3	**T ~ 298**
0.325	0.503
0.393	0.585
0.422	0.592
0.459	0.593
0.477	0.599
0.492	0.628
0.518	0.640
0.660	0.734
0.698	0.747
0.765	0.760
0.809	0.770
0.853	0.793
0.883	0.802
0.977	0.791
1.05	0.781
1.12	0.767
1.18	0.748
1.48	0.752
1.55	0.734
1.75	0.728
1.87	0.714
1.97	0.686
2.11	0.568
2.18	0.533
2.20	0.460
2.34	0.390
2.36	0.364
2.56	0.339
2.75	0.320
2.83	0.286
CURVE 3 (cont.)	
2.92	0.126
2.95	0.096
2.99	0.038
3.14	0.013
3.34	0.000
3.64	0.000
3.96	0.037
4.16	0.084
4.27	0.115
4.40	0.169
4.86	0.164
5.35	0.171
5.50	0.113
5.63	0.101
5.67	0.056
5.99	0.019
6.65	0.000
8.57	0.005
10.5	0.007
11.9	0.008
12.4	0.004
13.5	0.012
14.9	0.006
19.2	0.007
20.1	0.003
22.7	0.029
29.5	0.036
36.3	0.076
38.5	0.089
41.5	0.108
44.8	0.081
48.3	0.127
50.8	0.062
52.6	0.095
55.2	0.117
60.3	0.146

SPECIFICATION TABLE NO. 525 NORMAL TOTAL EMITTANCE OF LACQUER CONTACT COATINGS

Curve No.	Ref. No.	Year	Temperature Range, K	Geometry θ'	Reported Error, %	Composition (weight percent), Specifications and Remarks
1*	60	1954	97-467	~0°		Clear lacquer AN-TT-V-116; Dow metal substrate; applied with a brush and allowed to dry; measured in argon (10^{-3} mm Hg). [Author's designation: Sample 1]
2*	60	1954	97-449	~0°		Clear lacquer AN-L-29; Dow metal substrate; applied with a camel's hair brush, dried and reapplied and dried; measured in argon (10^{-3} mm Hg). [Author's designation: Sample 1]
3*	60	1954	97-445	~0°		Similar to above specimen and conditions except aluminum alloy 24ST substrate. [Author's designation: Sample 2]
4*	60	1954	90-477	~0°		Clear lacquer AN-TT-V-116; aluminum alloy 75ST substrate; applied with a brush and allowed to dry; measured in argon (10^{-3} mm Hg). [Author's designation: Sample 2]
5*	60	1954	90-452	~0°		Clear lacquer AN-L-29; aluminum alloy 75ST substrate; applied with a camel's hair brush, dried and another coat applied and dried; measured in argon (10^{-3} mm Hg). [Author's designation: Sample 3]
6*	60	1954	94-470	~0°		Clear lacquer AN-TT-V-116; aluminum alloy 24ST substrate; applied with a brush and allowed to dry; measured in argon (10^{-3} mm Hg). [Author's designation: Sample 3]
7*	59	1948	378-542	~0°		Clear lacquer 1234; 75ST Alclad substrate; thermal decomposition of lacquer occurs at ~519 K.

DATA TABLE NO. 525 NORMAL TOTAL EMITTANCE OF LACQUER CONTACT COATINGS

[Temperature, T, K; Emittance, ∈]

T	∈
CURVE 1*	
203	0.712
97	0.754
369	0.854
467	0.805
205	0.778
102	0.777
402	0.816
475	0.846
108	0.777
473	0.837
CURVE 2*	
200	0.988
97	0.922
385	0.962
CURVE 2 (cont.)*	
449	0.834
209	0.900
102	0.892
442	0.903
102	0.892
434	0.914
CURVE 3*	
208	0.983
402	0.946
445	0.855
204	0.942
97	0.913
412	0.949
445	0.920
CURVE 3 (cont.)*	
97	0.912
428	0.915
CURVE 4*	
203	0.900
90	0.899
389	0.887
459	0.856
204	0.855
90	0.920
367	0.891
474	0.851
204	0.871
90	0.871
477	0.857
CURVE 5*	
90	0.970
202	0.992
397	0.918
440	0.835
202	0.957
95	0.936
395	0.965
452	0.872
96	0.885
446	0.912
202	0.965
96	0.893
CURVE 6*	
204	0.967
97	0.962
397	0.878
463	0.878
203	0.898
95	0.912
390	0.878
470	0.885
94	0.870
422	0.885
100	0.871
CURVE 7*	
378	0.525
444	0.500
495	0.425
542	0.375

* No plot given

SPECIFICATION TABLE NO. 526 NORMAL SPECTRAL REFLECTANCE OF LACQUER CONTACT COATINGS

Curve No.	Ref. No.	Year	Temperature, K	Wavelength Range, μm	Geometry θ	θ'	ω'	Reported Error, %	Composition (weight percent), Specifications, and Remarks
1*	62	1949	~298	1.00-15.00	0°		2π	5	Clear lacquer; copper substrate; data extracted from smooth curve; specimen maintained at ~298 K; converted from R (2π, 0°).

DATA TABLE NO. 526 NORMAL SPECTRAL REFLECTANCE OF LACQUER CONTACT COATINGS

[Wavelength, λ, μm; Reflectance, ρ; Temperature, T, K]

λ	ρ	λ	ρ	λ	ρ	λ	ρ
CURVE 1* T ~ 298		CURVE 1 (cont.)*		CURVE 1 (cont.)*		CURVE 1 (cont.)*	
		5.00	0.860	9.55	0.118	14.03	0.497
1.00	0.820	5.24	0.829	9.79	0.121	14.30	0.453
1.26	0.875	5.51	0.724	10.01	0.147	14.58	0.421
1.50	0.895	5.79	0.084	10.27	0.161	14.77	0.471
1.72	0.889	6.04	0.099	10.55	0.274	15.00	0.523
2.00	0.894	6.28	0.512	10.76	0.343		
2.25	0.859	6.50	0.593	11.01	0.317		
2.50	0.855	6.79	0.220	11.26	0.303		
2.75	0.594	7.04	0.260	11.52	0.169		
3.01	0.402	7.29	0.171	11.77	0.126		
3.28	0.593	7.54	0.250	12.01	0.151		
3.53	0.106	7.80	0.088	12.29	0.388		
3.70	0.552	8.04	0.102	12.52	0.548		
3.76	0.685	8.29	0.137	12.79	0.504		
4.00	0.774	8.55	0.103	13.03	0.531		
4.23	0.804	8.76	0.098	13.28	0.426		
4.48	0.863	9.02	0.111	13.55	0.530		
4.74	0.860	9.33	0.099	13.79	0.513		

* No plot given

SPECIFICATION TABLE NO. 527 NORMAL SPECTRAL ABSORPTANCE OF LACQUER CONTACT COATINGS

Curve No.	Ref. No.	Year	Temperature, K	Wavelength Range, μm	Geometry θ	Reported Error, %	Composition (weight percent), Specifications, and Remarks
1*	103	1954	~298	0.221-2.600	~0°		Clear lacquer (~0.0762 mm thick); steel substrate; data extracted from smooth curve.
2*	103	1954	~298	0.400-2.600	~0°		Lacquer; quartz substrate; data extracted from smooth curve.

DATA TABLE NO. 527 NORMAL SPECTRAL ABSORPTANCE OF LACQUER CONTACT COATINGS

[Wavelength, λ, μm; Absorptance, α; Temperature, T, K]

λ	α	λ	α	λ	α
CURVE 1* T ~ 298		CURVE 1 (cont.)*		CURVE 2 (cont.)*	
		2.000	0.437	2.000	0.023
0.221	0.843	2.200	0.515	2.200	0.053
0.236	0.895	2.400	0.456	2.400	0.065
0.278	0.915	2.600	0.456	2.600	0.028
0.308	0.886				
0.333	0.802	CURVE 2* T ~ 298			
0.369	0.700				
0.440	0.619				
0.635	0.596	0.400	0.234		
0.990	0.526	0.456	0.157		
1.107	0.498	0.530	0.094		
1.187	0.502	0.679	0.059		
1.317	0.452	0.813	0.055		
1.385	0.454	0.958	0.057		
1.513	0.412	1.191	0.046		
1.598	0.418	1.398	0.028		
1.735	0.384	1.597	0.023		
1.803	0.412	1.800	0.023		

* No plot given

SPECIFICATION TABLE NO. 528 NORMAL SOLAR ABSORPTANCE OF LACQUER CONTACT COATINGS

Curve No.	Ref. No.	Year	Temperature Range, K	Geometry θ	Reported Error, %	Composition (weight percent), Specifications and Remarks
1*	60	1954	311	~0°		Clear lacquer AN-TT-V-116; Dow metal substrate; heated to 308 K; measured for sea level conditions.
2*	60	1954	311	~0°		Above specimen and conditions except reheated to 434 K.
3*	60	1954	311	~0°		Above specimen and conditions except reheated to 478 K.
4*	60	1954	311	~0°		Clear lacquer AN-L-29; Dow metal substrate; heated to 314 K; measured for sea level conditions.
5*	60	1954	311	~0°		Above specimen and conditions except reheated to 423 K.
6*	60	1954	311	~0°		Above specimen and conditions except reheated to 450 K.
7*	60	1954	311	~0°		Clear lacquer AN-TT-V-116; aluminum alloy 24 ST substrate; applied with a brush; heated to 310 K; measured for sea level conditions.
8*	60	1954	311	~0°		Above specimen and conditions except reheated to 425 K.
9*	60	1954	311	~0°		Above specimen and conditions except reheated to 469 K.
10*	60	1954	311	~0°		Clear lacquer AN-L-29; aluminum alloy 24 ST substrate; applied with a brush; heated to 310 K; measured for sea level conditions.
11*	60	1954	311	~0°		Above specimen and conditions except reheated to 423 K.
12*	60	1954	311	~0°		Above specimen and conditions except reheated to 450 K.
13*	60	1954	311	~0°		Clear lacquer AN-TT-V-116; aluminum alloy 75 ST substrate; heated to 307 K; measured for sea level conditions.
14*	60	1954	311	~0°		Above specimen and conditions except reheated to 474 K.
15*	60	1954	311	~0°		Above specimen and conditions except reheated to 478 K.
16*	60	1954	311	~0°		Clear lacquer AN-L-29; aluminum alloy 75 ST substrate; heated to 306 K; measured for sea level conditions.
17*	60	1954	311	~0°		Above specimen and conditions except reheated to 419 K.
18*	60	1954	311	~0°		Above specimen and conditions except reheated to 453 K.

* No plot given

DATA TABLE NO. 528 NORMAL SOLAR ABSORPTANCE OF LACQUER CONTACT COATINGS

[Temperature, T, K; Absorptance, α]

T	α	T	α
CURVE 1*		CURVE 13*	
311	0.74	311	0.74
CURVE 2*		CURVE 14*	
311	0.75	311	0.74
CURVE 3*		CURVE 15*	
311	0.84	311	0.80
CURVE 4*		CURVE 16*	
311	0.80	311	0.67
CURVE 5*		CURVE 17*	
311	0.76	311	0.74
CURVE 6*		CURVE 18*	
311	0.84	311	0.84
CURVE 7*			
311	0.78		
CURVE 8*			
311	0.78		
CURVE 9*			
311	0.83		
CURVE 10*			
311	0.72		
CURVE 11*			
311	0.77		
CURVE 12*			
311	0.88		

* No plot given

SPECIFICATION TABLE NO. 529 NORMAL SPECTRAL TRANSMITTANCE OF LACQUER CONTACT COATINGS

Curve No.	Ref. No.	Year	Temperature, K	Wavelength Range, μm	Geometry θ	θ'	ω'	Reported Error, %	Composition (weight percent), Specifications, and Remarks
1*	103	1954	~298	0.401-2.600	~0°		2π		Lacquer; quartz substrate; data extracted from smooth curve; the accuracies of the measurements in the UV, visible, and infrared regions are approx 2, 1, and 2 percent, respectively.

DATA TABLE NO. 529 NORMAL SPECTRAL TRANSMITTANCE OF LACQUER CONTACT COATINGS

[Wavelength, λ, μm; Transmittance, τ; Temperature, T, K]

λ	τ
CURVE 1*	
T ~ 298	
0.401	0.701
0.479	0.801
0.599	0.861
0.802	0.878
1.205	0.882
1.406	0.887
1.637	0.863
1.800	0.851
2.000	0.843
2.200	0.801
2.400	0.767
2.600	0.804

* No plot given

SPECIFICATION TABLE NO. 530 NORMAL SOLAR ABSORPTANCE OF MELAMINE FORMALDEHYDE CONTACT COATINGS

Curve No.	Ref. No.	Year	Temperature Range, K	Geometry θ	Reported Error, %	Composition (weight percent), Specifications and Remarks
1*	101	1964	~298	~0°		Melamine formaldehyde, Cymel 405 resin from Amer. Cyanamid Co. (0.015 mm thick); quartz substrate; prepared as 20% solution in spectro grade methyl cellosolve and filtered through flannel; applied with Bird film applicator; cured 20 min at 414 K; calculated from specular reflectance and transmittance data. [Authors' designation: Sample 337]
2*	101	1964	~298	~0°		Similar to above specimen and conditions except exposed to 100 ESH.
3*	101	1964	~298	~0°		Similar to curve 1 specimen and conditions except 0.024 mm thick. [Authors' designation: Sample 338]
4*	101	1964	~298	~0°		Similar to above specimen and conditions except exposed to 100 ESH.
5*	101	1964	~298	~0°		Similar to curve 1 specimen and conditions except 0.055 mm thick. [Authors' designation: Sample 523]
6*	101	1964	~298	~0°		Similar to above specimen and conditions except exposed to 100 ESH.

DATA TABLE NO. 530 NORMAL SOLAR ABSORPTANCE OF MELAMINE FORMALDEHYDE CONTACT COATINGS

[Temperature, T, K; Absorptance, α]

T	α	T	α
CURVE 1*		CURVE 6*	
298	0.006	298	0.087
CURVE 2*			
298	0.056		
CURVE 3*			
298	0.008		
CURVE 4*			
298	0.058		
CURVE 5*			
298	0.027		

* No plot given

SPECIFICATION TABLE NO. 531 HEMISPHERICAL TOTAL EMITTANCE OF NITROCELLULOSE CONTACT COATINGS

Curve No.	Ref. No.	Year	Temperature Range, K	Reported Error, %	Composition (weight percent), Specifications and Remarks
1*	43	1962	343-409	±3	Nitrocellulose (~0.0762 mm thick); copper substrate; measured in vacuum (10^{-3} mm Hg); data extracted from smooth curve.

DATA TABLE NO. 531 HEMISPHERICAL TOTAL EMITTANCE OF NITROCELLULOSE CONTACT COATINGS

[Temperature, T, K; Emittance, ϵ]

T	ϵ
CURVE 1*	
343	0.800
352	0.850
361	0.850
368	0.840
378	0.860
386	0.820
396	0.700
409	0.670

* No plot given

SPECIFICATION TABLE NO. 532 NORMAL SPECTRAL REFLECTANCE OF NYLON CONTACT COATINGS

Curve No.	Ref. No.	Year	Temperature, K	Wavelength Range, μm	Geometry θ	θ'	ω'	Reported Error, %	Composition (weight percent), Specifications, and Remarks
1*	99	1966	~300	2.00-14.50	~0°		2π	< ± 2	Nylon (~0.0508 mm thick); vapor blasted AISI 316 stainless steel substrate; 15% solution in ethyl alcohol, manufactured by DuPont Co.; sprayed; air dried for a minimum of 24 hrs; converted from R (2π, ~0°). [Authors' designation: Specimen 15]
2*	99	1966	~300	1.98-14.50	~0°		2π	< ± 2	Similar to above specimen and conditions except white lacquer and AISI 316 stainless steel substrates. [Authors' designation: Specimen 13]
3*	99	1966	~300	1.98-14.49	~0°		2π	< ± 2	Similar to curve 1 specimen and conditions except zinc chromate and AISI 316 stainless steel substrates. [Authors' designation: Specimen 14]

DATA TABLE NO. 532 NORMAL SPECTRAL REFLECTANCE OF NYLON CONTACT COATINGS

[Wavelength, λ, μm; Reflectance, ρ; Temperature, T, K]

λ	ρ	λ	ρ	λ	ρ	λ	ρ	λ	ρ
CURVE 1*		CURVE 1 (cont.)*		CURVE 2 (cont.)*		CURVE 2 (cont.)*		CURVE 3 (cont.)*	
T ~ 300									
		7.89	0.181	3.11	0.023	14.50	0.023	10.18	0.019
2.00	0.350	8.20	0.255	3.30	0.049			10.98	0.017
2.76	0.373	9.08	0.365	3.56	0.015	CURVE 3*		11.57	0.021
3.08	0.111	9.35	0.361	4.41	0.251	T ~ 300		12.06	0.023
3.19	0.235	10.13	0.382	5.31	0.242			12.66	0.023
3.27	0.218	10.96	0.402	5.87	0.020	1.98	0.391	13.25	0.023
3.39	0.266	11.55	0.432	6.34	0.033	2.23	0.344	13.86	0.020
3.49	0.127	12.04	0.419	6.49	0.014	2.67	0.362	14.49	0.028
4.38	0.431	12.62	0.369	6.73	0.058	3.11	0.025		
5.05	0.426	13.24	0.323	7.58	0.024	3.31	0.070		
5.54	0.441	13.84	0.273	8.50	0.013	3.54	0.029		
6.06	0.065	14.50	0.272	9.36	0.011	3.86	0.137		
6.28	0.189			10.17	0.019	4.45	0.223		
6.48	0.066	CURVE 2*		11.01	0.026	5.86	0.020		
6.78	0.225	T ~ 300		11.58	0.026	6.71	0.056		
6.86	0.157			12.09	0.023	6.98	0.024		
7.19	0.250	1.98	0.296	12.68	0.021	7.54	0.026		
7.35	0.187	2.38	0.228	13.29	0.020	8.49	0.018		
7.61	0.245	2.63	0.243	13.89	0.020	9.39	0.013		

* No plot given

SPECIFICATION TABLE NO. 533 HEMISPHERICAL TOTAL EMITTANCE OF POLYMIDE FLUORINATED ETHYLENE PROPYLENE CONTACT COATINGS

Curve No.	Ref. No.	Year	Temperature Range, K	Reported Error, %	Composition (weight percent), Specifications and Remarks
1*	107	1967	273-418	~8.5	Laminated polymide fluorinated ethylene propylene; silver on copper stranded conductor, No. 12 guage, natural (super temp) substrate; measured in vacuum (~10^{-6} mm Hg).
2*	107	1967	284-434	~8.5	Similar to above specimen and conditions except silver on copper stranded conductor, No. 18 guage, natural (super temp) substrate.

DATA TABLE NO. 533 HEMISPHERICAL TOTAL EMITTANCE OF POLYMIDE FLUORINATED ETHYLENE PROPYLENE CONTACT COATINGS

[Temperature, T, K; Emittance, ϵ]

T	ϵ
CURVE 1*	
273	0.881
322	0.851
350	0.831
418	0.820
CURVE 2*	
284	0.890
327	0.872
378	0.877
434	0.862

* No plot given

SPECIFICATION TABLE NO. 534 NORMAL SPECTRAL TRANSMITTANCE OF POLYBUTADIENE CONTACT COATINGS

Curve No.	Ref. No.	Year	Temperature, K	Wavelength Range, μm	Geometry θ	θ'	ω'	Reported Error, %	Composition (weight percent), Specifications, and Remarks
1*	173	1966	298	0.4797-0.6214	0°	0°			Polybutadiene (>1000 Å thick); tin oxide substrate; deposited by glow discharge.

DATA TABLE NO. 534 NORMAL SPECTRAL TRANSMITTANCE OF POLYBUTADIENE CONTACT COATINGS

[Wavelength, λ, μm; Transmittance, τ; Temperature, T, K]

λ	τ
CURVE 1* T = 298	
0.4797	0.621
0.4895	0.661
0.4946	0.681
0.4997	0.710
0.5052	0.731
0.5103	0.750
0.5202	0.790
0.5302	0.813
0.5408	0.822
0.5499	0.815
0.5603	0.800
0.5808	0.749
0.6003	0.697
0.6214	0.686

* No plot given

SPECIFICATION TABLE NO. 535 HEMISPHERICAL TOTAL EMITTANCE OF POLYESTER CONTACT COATINGS

Curve No.	Ref. No.	Year	Temperature Range, K	Reported Error, %	Composition (weight percent), Specifications and Remarks
1*	105	1964	307		Polyester film (0.00635 mm thick); aluminum substrate; aluminum vacuum deposited; measured in vacuum (10^{-6} mm Hg) maintained by diffusion pump.
2*	105	1964	307		Reinforced polyester film; aluminum substrate (aluminum scrim); measured in vacuum (10^{-6} mm Hg) maintained by diffusion pump. [Authors' designation: Test No. 3, Scrim Side]
3*	105	1964	307		Polyester film (0.00635 mm thick); gold substrate; gold vapor deposited on polyester film; measured in vacuum (10^{-6} mm Hg) maintained by diffusion pump. [Authors' designation: Test No. 225, Mfr. A, Sample 101464-1-2E, uncoated side]

DATA TABLE NO. 535 HEMISPHERICAL TOTAL EMITTANCE OF POLYESTER CONTACT COATINGS

[Temperature, T, K; Emittance, ϵ]

T	ϵ
CURVE 1*	
307	0.364
CURVE 2*	
307	0.669
CURVE 3*	
307	0.365

* No plot given

SPECIFICATION TABLE NO. 536 ANGULAR SPECTRAL REFLECTANCE OF POLYSTYRENE CONTACT COATINGS

Curve No.	Ref. No.	Year	Temperature, K	Wavelength Range, μm	Geometry θ	θ'	ω'	Reported Error, %	Composition (weight percent), Specifications, and Remarks
1*	256	1968	~298	0.058-0.199	30°	30°		< 5	Polystyrene; polished glass substrate; crystalline polystyrene dissolved in benzene and flowed over substrate.

DATA TABLE NO. 536 ANGULAR SPECTRAL REFLECTANCE OF POLYSTYRENE CONTACT COATINGS

[Wavelength, λ, μm; Reflectance, ρ; Temperature, T, K]

λ	ρ	λ	ρ
CURVE 1* T ~ 298		CURVE 1 (cont.)*	
0.580	0.069	1.79	0.058
0.580	0.066	1.79	0.079
0.730	0.106	1.88	0.134
0.730	0.113	1.88	0.155
1.02	0.104	1.99	0.166
1.02	0.097	1.99	0.159
1.11	0.098		
1.11	0.091		
1.21	0.085		
1.21	0.077		
1.34	0.075		
1.34	0.070		
1.45	0.056		
1.45	0.052		
1.60	0.024		
1.60	0.021		
1.69	0.026		
1.69	0.019		

* No plot given

SPECIFICATION TABLE NO. 537 ANGULAR SPECTRAL REFLECTANCE OF POLYSTYRENE CONTACT COATINGS

Curve No.	Ref. No.	Year	Temperature K	Wavelength, μm	Angular Range, °	Geometry θ	Geometry θ'	Geometry ω'	Reported Error, %	Composition (weight percent), Specifications and Remarks
1*	256	1968	~298	0.1216	30-75	θ	θ'		< 5	Polystyrene; polished glass substrate; crystalline polystyrene dissolved in benzene and flowed over substrate.

DATA TABLE NO. 537 ANGULAR SPECTRAL REFLECTANCE OF POLYSTYRENE CONTACT COATINGS

[Angle, θ, °; Reflectance, ρ; Temperature, T, K; Wavelength, λ, μm]

θ	ρ
CURVE 1*	
T ~ 298	
λ = 0.1216	
30	0.084
45	0.093
60	0.135
75	0.301

* No plot given

SPECIFICATION TABLE NO. 538 NORMAL SPECTRAL REFLECTANCE OF POLYURETHANE CONTACT COATINGS

Curve No.	Ref. No.	Year	Temperature, K	Wavelength Range, μm	Geometry θ	θ'	ω'	Reported Error, %	Composition (weight percent), Specifications, and Remarks
1*	100	1962	298	0.369-0.750	~0°		2π		Polyurethane (Magna Paint Co. Laminar X-500 clear 4-C-8); polished aluminum substrate; data extracted from smooth curve.
2*	100	1962	298	0.370-0.749	~0°		2π		Above specimen and conditions except exposed to UV at 10 suns for 22.75 hrs.

DATA TABLE NO. 538 NORMAL SPECTRAL REFLECTANCE OF POLYURETHANE CONTACT COATINGS

[Wavelength, λ, μm; Reflectance, ρ; Temperature, T, K]

λ	ρ	λ	ρ
CURVE 1* T = 298		CURVE 2 (cont.)*	
		0.450	0.252
0.369	0.294	0.499	0.383
0.374	0.406	0.528	0.441
0.400	0.556	0.549	0.483
0.421	0.608	0.581	0.532
0.448	0.629	0.600	0.557
0.549	0.675	0.627	0.587
0.599	0.679	0.650	0.607
0.649	0.689	0.673	0.625
0.699	0.693	0.701	0.634
0.750	0.693	0.749	0.645
CURVE 2* T = 298			
0.370	0.119		
0.392	0.136		
0.401	0.146		
0.429	0.204		

* No plot given

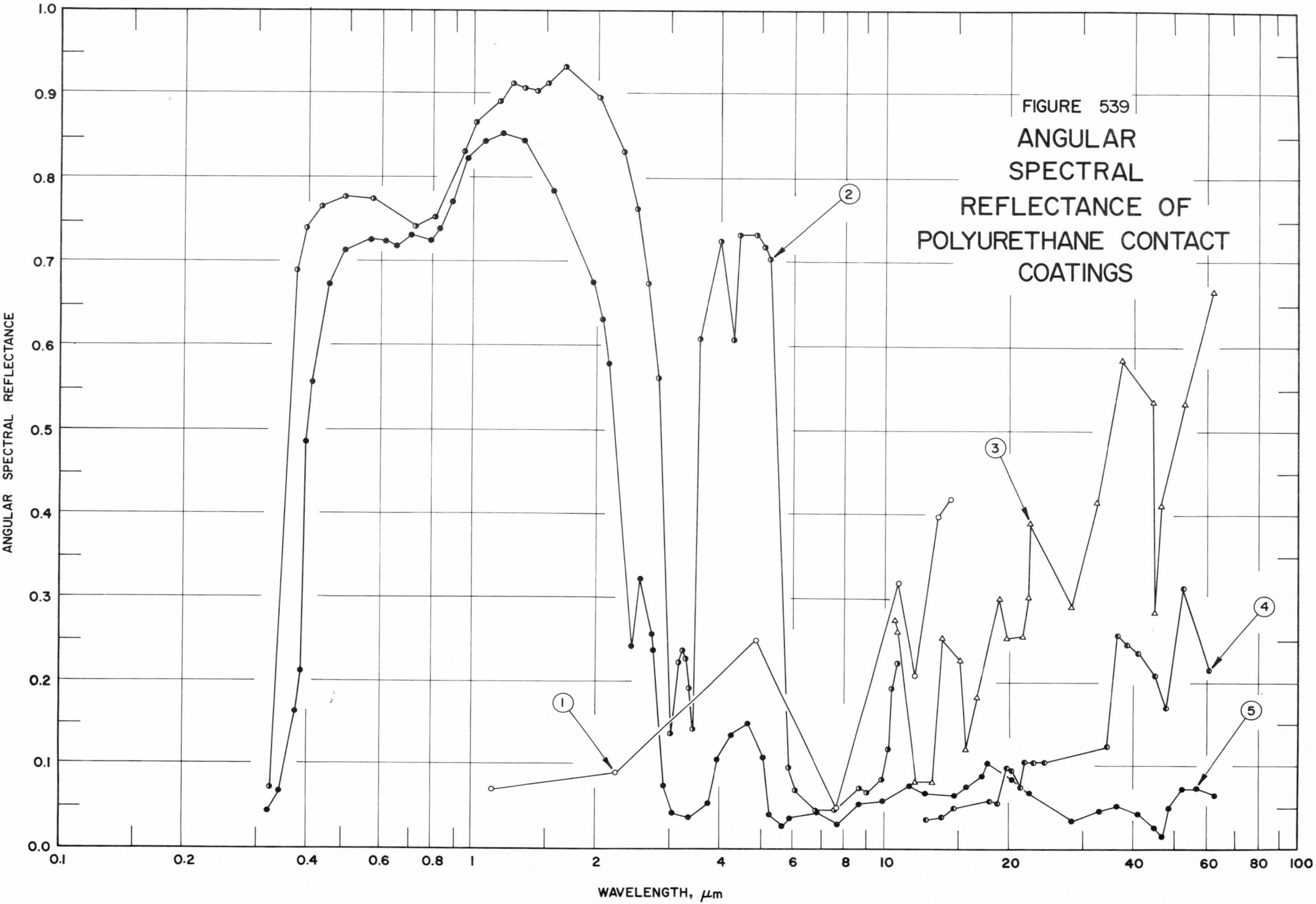

FIGURE 539 ANGULAR SPECTRAL REFLECTANCE OF POLYURETHANE CONTACT COATINGS

SPECIFICATION TABLE NO. 539 ANGULAR SPECTRAL REFLECTANCE OF POLYURETHANE CONTACT COATINGS

Curve No.	Ref. No.	Year	Temperature, K	Wavelength Range, μm	Geometry θ	θ'	ω'	Reported Error, %	Composition (weight percent), Specifications, and Remarks
1	231	1962	~298	1.120-14.375	45°	45°			SY627-119 polyurethane (Febert Shorndorfer Co.); consists mainly of DuPont Hylene isocyanates; 316 stainless steel substrate; two coats applied by dipping substrate in SY627-119 mixed with a polyester leveling agent, ethyl acetate and xylene; substrate cleaned, etched, and vapor degreased; specular component only. [Authors' designation: SS + P]
2	91	1966	~298	0.329-10.8	~20°		2π		Polyurethane, Laminar X-500, Magna Coatings and Chemical Corp (25 μm thick); aluminum substrate; 3 apparatus used in ranges 0.33-2.5 μm (θ=20°), 1.5-23 μm (θ=25°), and 20-61 μm (θ=17°).
3	91	1966	~298	10.6-62.3	~20°		2π		Similar to above specimen and conditions.
4	91	1966	~298	12.7-61.3	~20°		2π		Similar to above specimen and conditions except thickness 71 μm.
5	91	1966	~298	0.322-63.0	~20°		2π		Similar to above specimen and conditions except thickness 260 μm.

DATA TABLE NO. 539 ANGULAR SPECTRAL REFLECTANCE OF POLYURETHANE CONTACT COATINGS

[Wavelength, λ, μm; Reflectance, ρ; Temperature, T, K]

λ	ρ
CURVE 1	
T ~ 298	
1.120	0.07
2.240	0.09
4.824	0.25
7.780	0.05
10.198	0.32
12.099	0.21
13.530	0.40
14.375	0.42
CURVE 2	
T ~ 298	
0.329	0.071
0.375	0.690
0.397	0.741
0.430	0.768
0.497	0.780
0.572	0.777
0.721	0.744
0.812	0.755
0.955	0.833
1.01	0.867
1.16	0.892
1.25	0.911
1.32	0.909
1.43	0.905
1.51	0.912
1.67	0.932
2.01	0.899
2.31	0.833
2.50	0.765
2.68	0.676
2.81	0.562
3.05	0.137
3.16	0.225
3.23	0.239
3.30	0.229
3.38	0.192
3.41	0.141
3.56	0.610
4.00	0.728
4.30	0.610
4.41	0.734
4.90	0.734
5.10	0.720
CURVE 2 (cont.)	
5.21	0.705
5.84	0.099
6.06	0.070
6.80	0.047
7.60	0.049
8.79	0.072
9.03	0.069
9.90	0.082
10.2	0.120
10.3	0.194
10.8	0.225
CURVE 3	
T ~ 298	
10.6	0.274
10.7	0.210
11.9	0.080
13.0	0.080
13.9	0.252
15.1	0.228
15.8	0.120
16.9	0.181
19.0	0.301
19.8	0.252
21.6	0.255
22.1	0.301
22.4	0.391
28.5	0.292
32.5	0.418
37.5	0.588
44.1	0.537
44.7	0.285
46.7	0.411
53.0	0.534
62.3	0.668
CURVE 4	
T ~ 298	
12.7	0.037
13.8	0.039
14.8	0.050
18.0	0.059
18.9	0.056
19.9	0.099
CURVE 4 (cont.)	
20.3	0.094
21.2	0.073
21.9	0.103
23.0	0.103
24.3	0.103
34.2	0.123
36.7	0.257
38.8	0.246
41.2	0.235
45.9	0.210
48.0	0.170
52.2	0.315
61.3	0.218
CURVE 5	
T ~ 298	
0.322	0.043
0.343	0.067
0.377	0.163
0.386	0.211
0.397	0.488
0.411	0.559
0.452	0.674
0.498	0.714
0.561	0.727
0.616	0.726
0.645	0.720
0.701	0.733
0.785	0.727
0.829	0.743
0.885	0.773
0.975	0.826
1.06	0.846
1.18	0.855
1.32	0.847
1.57	0.787
1.97	0.676
2.07	0.632
2.14	0.580
2.44	0.241
2.57	0.323
2.71	0.257
2.73	0.236
2.91	0.075
3.09	0.042
CURVE 5 (cont.)	
3.38	0.039
3.74	0.054
3.97	0.109
4.22	0.135
4.65	0.150
5.10	0.110
5.34	0.040
5.61	0.028
5.98	0.037
6.85	0.045
7.70	0.030
8.70	0.053
9.84	0.057
11.5	0.076
12.5	0.066
14.9	0.065
15.9	0.074
17.1	0.089
17.7	0.102
20.2	0.083
22.5	0.069
28.3	0.035
33.0	0.047
36.8	0.051
41.6	0.042
45.0	0.029
47.4	0.017
48.9	0.050
52.9	0.072
56.7	0.072
63.0	0.066

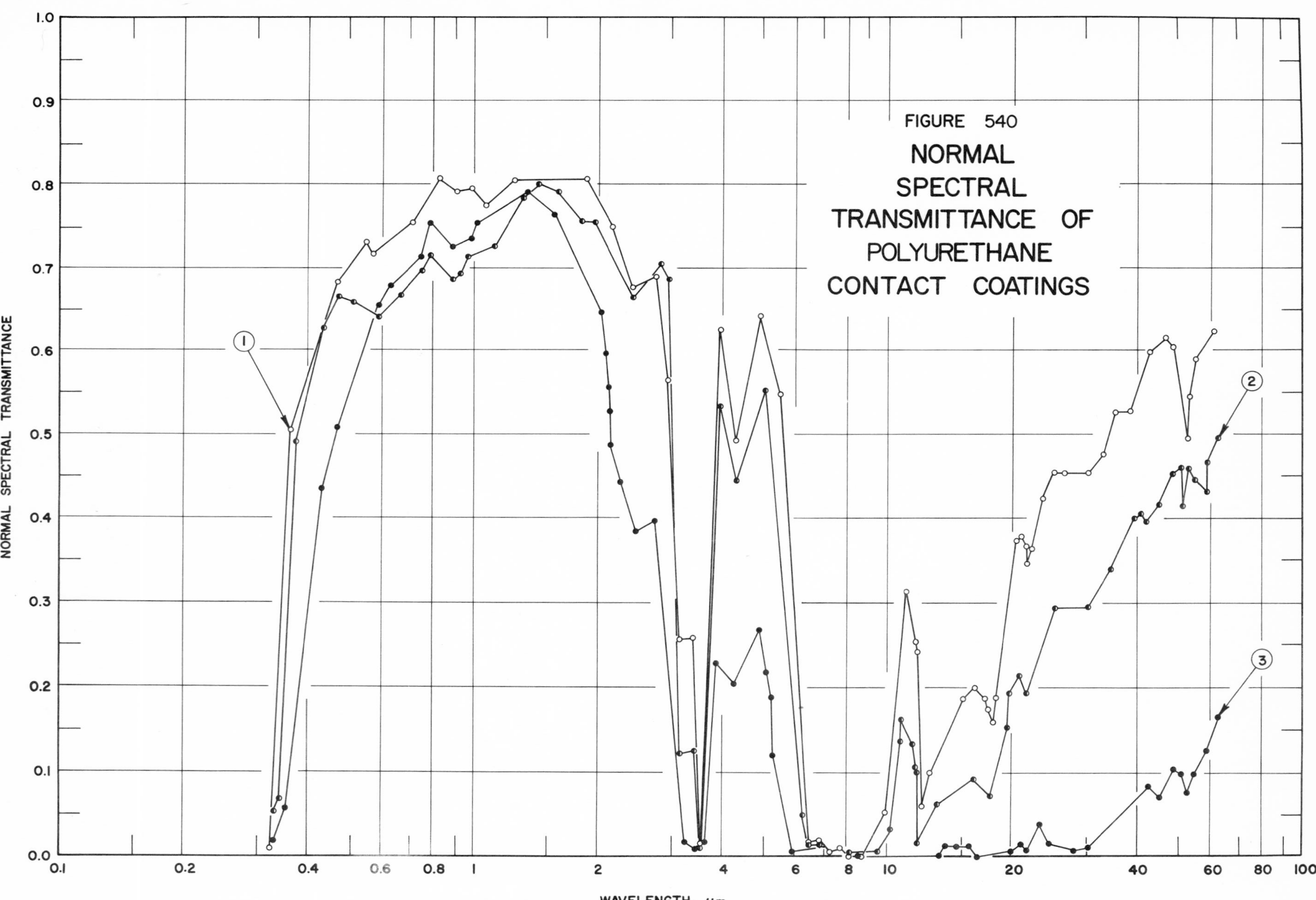

FIGURE 540 NORMAL SPECTRAL TRANSMITTANCE OF POLYURETHANE CONTACT COATINGS

SPECIFICATION TABLE NO. 540 NORMAL SPECTRAL TRANSMITTANCE OF POLYURETHANE CONTACT COATINGS

Curve No.	Ref. No.	Year	Temperature, K	Wavelength Range, μm	Geometry θ θ' ω'	Reported Error, %	Composition (weight percent), Specifications, and Remarks
1	91	1966	~298	0.329-61.8	~0° ~0°		Polyurethane, Laminar X-500, Magna Coatings and Chemical Corp (46 μm thick); polyethylene (20 μm thick) substrate; 3 apparatae used in ranges 0.33-2.5 μm, 1.5-23 μm, and 20-61 μm.
2	91	1966	~298	0.333-62.5	~0° ~0°		Similar to above specimen and conditions except coating thickness 71 μm.
3	91	1966	~298	0.331-62.2	~0° ~0°		Similar to above specimen and conditions except coating thickness 260 μm.

DATA TABLE NO. 540 NORMAL SPECTRAL TRANSMITTANCE OF POLYURETHANE CONTACT COATINGS

[Wavelength, λ, μm; Transmittance, τ; Temperature, T, K]

CURVE 1
T ~ 298

λ	τ
0.329	0.009
0.363	0.503
0.475	0.682
0.547	0.730
0.572	0.718
0.717	0.754
0.837	0.809
0.910	0.792
0.990	0.797
1.08	0.778
1.25	0.805
1.87	0.809
2.16	0.750
2.42	0.678
2.76	0.690
2.95	0.563
3.15	0.256
3.40	0.258
3.58	0.016
3.96	0.625
4.37	0.492
4.94	0.641
5.50	0.549
6.51	0.019
6.85	0.020
7.22	0.006
7.74	0.010
8.61	0.003
9.97	0.053
11.1	0.312
11.6	0.251
11.8	0.241
12.1	0.060
12.6	0.100
15.1	0.187
16.2	0.200
17.3	0.188
17.5	0.175
18.0	0.160
18.2	0.187
20.5	0.373
21.0	0.378
21.8	0.367
21.8	0.346
22.2	0.363

CURVE 1 (cont.)

λ	τ
23.6	0.425
25.2	0.454
26.7	0.454
30.3	0.454
33.0	0.477
35.4	0.527
38.5	0.528
43.4	0.600
46.7	0.616
48.8	0.606
52.1	0.499
53.0	0.546
55.5	0.591
61.8	0.623

CURVE 2
T ~ 298

λ	τ
0.333	0.053
0.344	0.067
0.375	0.490
0.439	0.628
0.479	0.664
0.517	0.656
0.588	0.640
0.665	0.666
0.748	0.698
0.796	0.714
0.897	0.686
0.937	0.691
0.970	0.712
1.12	0.726
1.31	0.785
1.44	0.800
1.60	0.791
1.83	0.756
1.98	0.756
2.44	0.662
2.84	0.705
2.97	0.687
3.19	0.121
3.42	0.125
3.53	0.010
3.98	0.531
4.37	0.444
5.01	0.551

CURVE 2 (cont.)

λ	τ
6.26	0.050
6.50	0.015
6.98	0.015
8.09	0.005
9.59	0.008
10.2	0.033
10.8	0.137
10.8	0.162
11.5	0.135
11.7	0.106
11.8	0.100
11.8	0.017
13.2	0.061
16.1	0.091
17.9	0.073
19.5	0.153
19.8	0.192
20.8	0.215
21.7	0.191
25.2	0.293
30.1	0.296
34.6	0.340
39.3	0.400
41.4	0.406
42.0	0.397
45.7	0.419
48.1	0.453
50.8	0.460
51.6	0.416
53.0	0.460
55.2	0.448
58.7	0.431
59.1	0.467
62.5	0.499

CURVE 3
T ~ 298

λ	τ
0.331	0.017
0.354	0.056
0.433	0.433
0.469	0.506
0.591	0.654
0.638	0.679
0.744	0.713
0.798	0.754

CURVE 3 (cont.)

λ	τ
0.893	0.726
0.986	0.735
1.02	0.751
1.35	0.791
1.58	0.765
2.02	0.646
2.09	0.598
2.11	0.556
2.13	0.527
2.13	0.487
2.28	0.441
2.48	0.382
2.74	0.397
3.28	0.018
3.47	0.010
3.62	0.017
3.87	0.228
4.23	0.202
4.98	0.268
5.11	0.217
5.21	0.188
5.26	0.120
5.97	0.007
6.85	0.012*
8.57	0.003
13.4	0.003
13.9	0.013
14.6	0.011
15.7	0.011
16.5	0.002
20.0	0.008
21.0	0.016
21.7	0.009
23.1	0.040
24.7	0.017
28.1	0.009
30.5	0.011
42.9	0.085
44.6	0.072
48.9	0.105
51.1	0.100
53.7	0.078
54.7	0.100
59.0	0.129
62.2	0.165

* Not shown on plot

SPECIFICATION TABLE NO. 541 NORMAL SPECTRAL REFLECTANCE OF POLYVINYL ALCOHOL CONTACT COATINGS

Curve No.	Ref. No.	Year	Temperature, K	Wavelength Range, μm	Geometry θ	θ'	ω'	Reported Error, %	Composition (weight percent), Specifications, and Remarks
1*	257	1964	~298	0.229-1.800	~0°	~0°			DuPont elvanol 71-30 polyvinyl alcohol; plain woven ECD 450 3/2 fiberglass fabric (0.019 g cm^{-2}) substrate; coating applied in 8 layers by dipping sample in 10% solution of above, drying overnight and heating 10 min at 323 K before next coating; final coating baked 1 hr at 383 K; data extracted from smooth curve.

DATA TABLE NO. 541 NORMAL SPECTRAL REFLECTANCE OF POLYVINYL ALCOHOL CONTACT COATINGS

[Wavelength, λ, μm; Reflectance, ρ; Temperature, T, K]

λ	ρ	λ	ρ
CURVE 1* T ~ 298		CURVE 1 (cont.)*	
		1.208	0.638
0.229	0.087	1.287	0.676
0.270	0.104	1.329	0.683
0.290	0.113	1.358	0.679
0.299	0.122	1.411	0.678
0.351	0.258	1.432	0.629
0.395	0.440	1.450	0.558
0.458	0.547	1.458	0.514
0.542	0.598	1.506	0.472
0.652	0.602	1.569	0.479
0.705	0.513	1.602	0.478
0.715	0.529	1.622	0.487
0.753	0.625	1.644	0.523
0.812	0.635	1.658	0.533
0.849	0.633	1.709	0.537
0.914	0.609	1.731	0.510
1.017	0.626	1.745	0.487
1.071	0.647	1.772	0.491
1.165	0.663	1.800	0.513

* No plot given

SPECIFICATION TABLE NO. 542 NORMAL SOLAR ABSORPTANCE OF POLYVINYL ALCOHOL CONTACT COATINGS

Curve No.	Ref. No.	Year	Temperature Range, K	Geometry θ	Reported Error, %	Composition (weight percent), Specifications and Remarks
1*	257	1964	~298	~0°		DuPont elvanol 71-30 polyvinyl alcohol; plain woven ECD 450 3/2 fiberglass fabric (0.019 g cm^{-2}) substrate; coating applied in 8 layers by dipping sample in 10% solution of above, drying overnight and heating 10 min at 323 K before next coating; final coating baked 1 hr at 383 K; data calculated from spectral reflectance data.

DATA TABLE NO. 542 NORMAL SOLAR ABSORPTANCE OF POLYVINYL ALCOHOL CONTACT COATINGS

[Temperature, T, K; Absorptance, α]

T	α
CURVE 1*	
298	0.44

* No plot given

SPECIFICATION TABLE NO. 543 NORMAL SOLAR ABSORPTANCE OF POLYVINYL BUTYRAL CONTACT COATINGS

Curve No.	Ref. No.	Year	Temperature Range, K	Geometry θ	Reported Error, %	Composition (weight percent), Specifications and Remarks
1*	101	1964	~298	~0°		Clear melamine modified polyvinyl butyral (0.062 mm thick): 90 Butvar B-98 polyvinyl butyral resin (Shawinigan Resin Corp.) +10 Cymel 300 hexamethoxymethyl melamine; quartz substrate; coated as solution in organic solvents and a catalyst with Bird film applicator; cured 45 min at 366 K; calculated from specular reflectance data. [Authors' designation: Sample 294]
2*	101	1964	~298	~0°		Similar to curve 1 specimen and conditions except film thickness ~0.025 mm. [Authors' designation: Sample 325]
3*	101	1964	~298	~0°		Similar to curve 1 specimen and conditions except film thickness 0.043 mm. [Authors' designation: Sample 327]
4*	101	1964	~298	~0°		Similar to curve 1 specimen and conditions except film thickness 0.060 mm. [Authors' designation: Sample 326]
5*	101	1964	~298	~0°		Similar to curve 1 specimen and conditions except coating prepared with 5% Lupersol DDM methyl ethyl ketone peroxide; film thickness 0.050 mm. [Authors' designation: Sample 336]
6*	101	1964	~298	~0°		Polyvinyl butyral, Butvar B-98 resin, Shawinigan Resins Corp. (0.073 mm thick); solvent: diacetone alcohol, cellosolve, xylene in 40:40:20 ratio by weight; quartz substrate; applied with Bird film applicator; cured 20 min at 339 K; calculated from specular reflectance and transmittance data. [Authors' designation: Sample 521]
7*	101	1964	~298	~0°		Similar to above specimen and conditions except exposed to 100 ESH.

* No plot given

DATA TABLE NO. 543 NORMAL SOLAR ABSORPTANCE OF POLYVINYL BUTYRAL CONTACT COATINGS

[Temperature, T, K; Absorptance, α]

T	α
CURVE 1*	
298	0.014
CURVE 2*	
298	0.008
CURVE 3*	
298	0.013
CURVE 4*	
298	0.014
CURVE 5*	
298	0.012
CURVE 6*	
298	0.015
CURVE 7*	
298	0.045

* No plot given

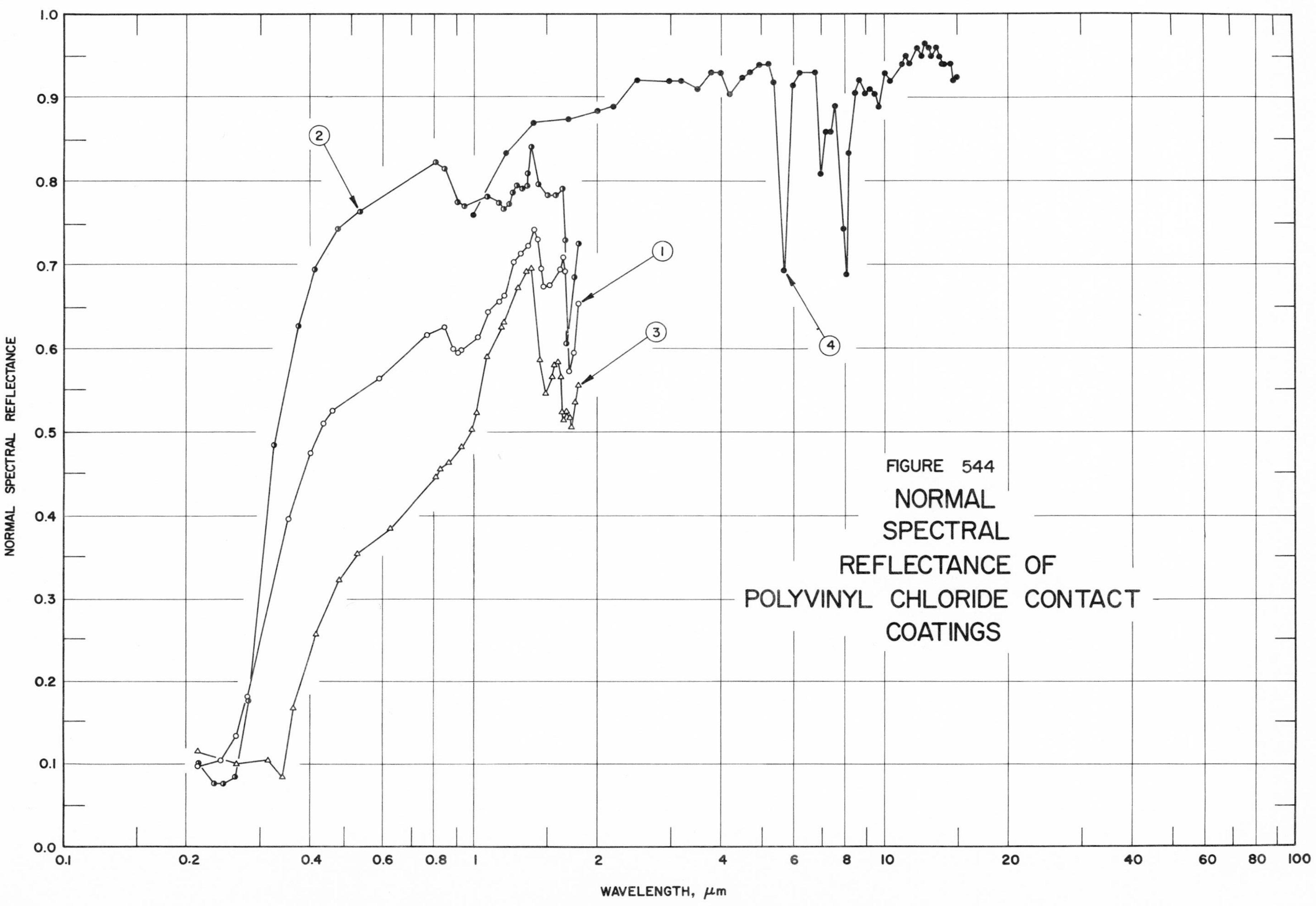

FIGURE 544 NORMAL SPECTRAL REFLECTANCE OF POLYVINYL CHLORIDE CONTACT COATINGS

SPECIFICATION TABLE NO. 544 NORMAL SPECTRAL REFLECTANCE OF POLYVINYL CHLORIDE CONTACT COATINGS

Curve No.	Ref. No.	Year	Temperature, K	Wavelength Range, μm	Geometry θ	θ'	ω'	Reported Error, %	Composition (weight percent), Specifications, and Remarks
1	257	1964	~298	0.214-1.800	~0°	~0°			Vinyl plastisol based on polyvinyl chloride resin; composition, parts by weight: 100 polyvinyl chloride, 79.5 dioctyl phthalate, 2.3 Thermolite, 1 Cyasorb UV-24; plain woven ECD 450 3/2 fiberglass fabric (0.019 g cm^{-2}) substrate; brush application; cured 20 min at 478 K; data extracted from smooth curve. [Authors' designation: Sample W-1, V-19]
2	257	1964	~298	0.215-1.800	~0°	~0°			Similar to curve 1 specimen and conditions except coating was a proprietary formula of the Duracote Corp containing Union Carbide polyvinyl chloride and polyvinyl chloride (86), trioctyl phosphate, dioctyl phthalate, clorafin 40, nonferral filler, diisobutyl ketone, tylol and apco solvent. [Authors' designation: Sample W-1, V-1]
3	257	1964	~298	0.214-1.800	~0°	~0°			Similar to curve 1 specimen and conditions except substrate plain woven (Nomex 200/100, 8 tpi Z), high temp resistant nylon fabric. [Authors' designation: Sample W-46, V-19]
4	104	1953	298	1.00-15.0	5°		2π	±2	Polyvinyl chloride, Bakelite Corp (0.102 mm thick); aluminum foil (0.0254 mm thick) substrate; data extracted from smooth curve; converted from R (2π, 5°).

DATA TABLE NO. 544 NORMAL SPECTRAL REFLECTANCE OF POLYVINYL CHLORIDE CONTACT COATINGS

[Wavelength, λ, μm; Reflectance, ρ; Temperature, T, K]

CURVE 1
T ~ 298

λ	ρ
0.214	0.098
0.243	0.103
0.265	0.132
0.281	0.180
0.355	0.399
0.400	0.473
0.434	0.510
0.458	0.525
0.588	0.563
0.771	0.618
0.842	0.626
0.850	0.620*
0.894	0.600
0.916	0.596
0.937	0.598
1.011	0.613
1.095	0.645
1.155	0.658
1.194	0.664
1.258	0.703
1.300	0.714
1.347	0.724
1.402	0.743
1.422	0.731
1.447	0.696
1.492	0.677
1.546	0.678
1.619	0.698
1.651	0.710
1.666	0.695
1.707	0.571
1.744	0.597
1.800	0.655

CURVE 2
T ~ 298

λ	ρ
0.215	0.100
0.231	0.076
0.247	0.076
0.263	0.083
0.283	0.176
0.327	0.485
0.372	0.626
0.409	0.695
0.462	0.742
0.522	0.763
0.806	0.822
0.846	0.816
0.910	0.776
0.948	0.771
1.096	0.783
1.158	0.776
1.193	0.769
1.213	0.773
1.248	0.787
1.286	0.796
1.318	0.792
1.331	0.795
1.357	0.810
1.392	0.842
1.447	0.798
1.514	0.784
1.587	0.784
1.641	0.791
1.666	0.730
1.698	0.606
1.754	0.687
1.800	0.727

CURVE 3
T ~ 298

λ	ρ
0.214	0.116
0.266	0.100
0.316	0.103
0.344	0.084
0.366	0.166
0.416	0.258
0.472	0.321
0.520	0.355
0.624	0.385
0.808	0.448
0.838	0.455
0.875	0.463
0.939	0.481
0.984	0.503
1.018	0.521
1.086	0.590
1.172	0.626
1.197	0.632
1.280	0.674
1.347	0.693
1.391	0.696
1.452	0.588
1.508	0.545
1.560	0.566
1.577	0.580
1.600	0.583
1.621	0.566
1.647	0.523
1.658	0.514
1.677	0.524
1.712	0.518*
1.745	0.506
1.781	0.535
1.800	0.556

CURVE 4
T = 298

λ	ρ
1.00	0.760
1.20	0.835
1.40	0.870
1.70	0.875
2.00	0.885
2.20	0.890
2.50	0.920
3.00	0.920
3.20	0.920
3.50	0.910
3.80	0.930
4.00	0.930
4.20	0.905
4.50	0.925
4.70	0.930
5.00	0.940
5.20	0.940
5.40	0.920
5.70	0.695
6.00	0.915
6.20	0.930
6.80	0.930
7.00	0.810
7.20	0.860
7.40	0.860
7.60	0.890
8.00	0.745
8.10	0.690
8.20	0.835
8.50	0.905
8.70	0.920
9.00	0.905
9.20	0.910
9.50	0.905
9.70	0.890
10.0	0.930
10.2	0.920
11.0	0.940
11.2	0.950
11.4	0.940
12.0	0.960
12.2	0.950
12.5	0.965
12.8	0.960
13.0	0.950
13.3	0.960
13.6	0.950
13.8	0.940
14.0	0.940
14.4	0.940
14.8	0.920
15.0	0.925

* Not shown on plot

SPECIFICATION TABLE NO. 545 NORMAL SOLAR ABSORPTANCE OF POLYVINYL CHLORIDE CONTACT COATINGS

Curve No.	Ref. No.	Year	Temperature Range, K	Geometry θ	Reported Error, %	Composition (weight percent), Specifications and Remarks
1*	257	1964	~298	~0°		Vinyl plastisol based on polyvinyl chloride resin; composition, parts by weight: 100 polyvinyl chloride, 79.5 dioctyl phthalate, 23 Thermolite, 1 Cyaserb UV-24; plain woven ECD 450 3/2 fiberglass fabric (0.019 g cm^{-2}) substrate; brush application; cured 20 min at 478 K; data calculated from spectral reflectance data. [Authors' designation: Sample W-1, V-19]
2*	257	1964	~298	~0°		Similar to above specimen and conditions except coating was a proprietary formula of the Duracote Corp. containing Union Carbide polyvinyl chloride and polyvinyl chloride (86), trioctyl phosphate, dioctyl phlhalate, clorafin 40, nonferral filler, diisobutyl ketone, tylol and apco solvent. [Authors' designation: Sample W-1, V-1]
3*	257	1964	~298	~0°		Similar to curve 1 specimen and conditions except substrate plain woven (Nomex 200/100, 8 tpi Z) high temp resistant nylon fabric. [Authors' designation: Sample W-46, V-19]

DATA TABLE NO. 545 NORMAL SOLAR ABSORPTANCE OF POLYVINYL CHLORIDE CONTACT COATINGS

[Temperature, T, K; Absorptance, α]

T	α
CURVE 1*	
298	0.43
CURVE 2*	
298	0.25
CURVE 3*	
298	0.44

* No plot given

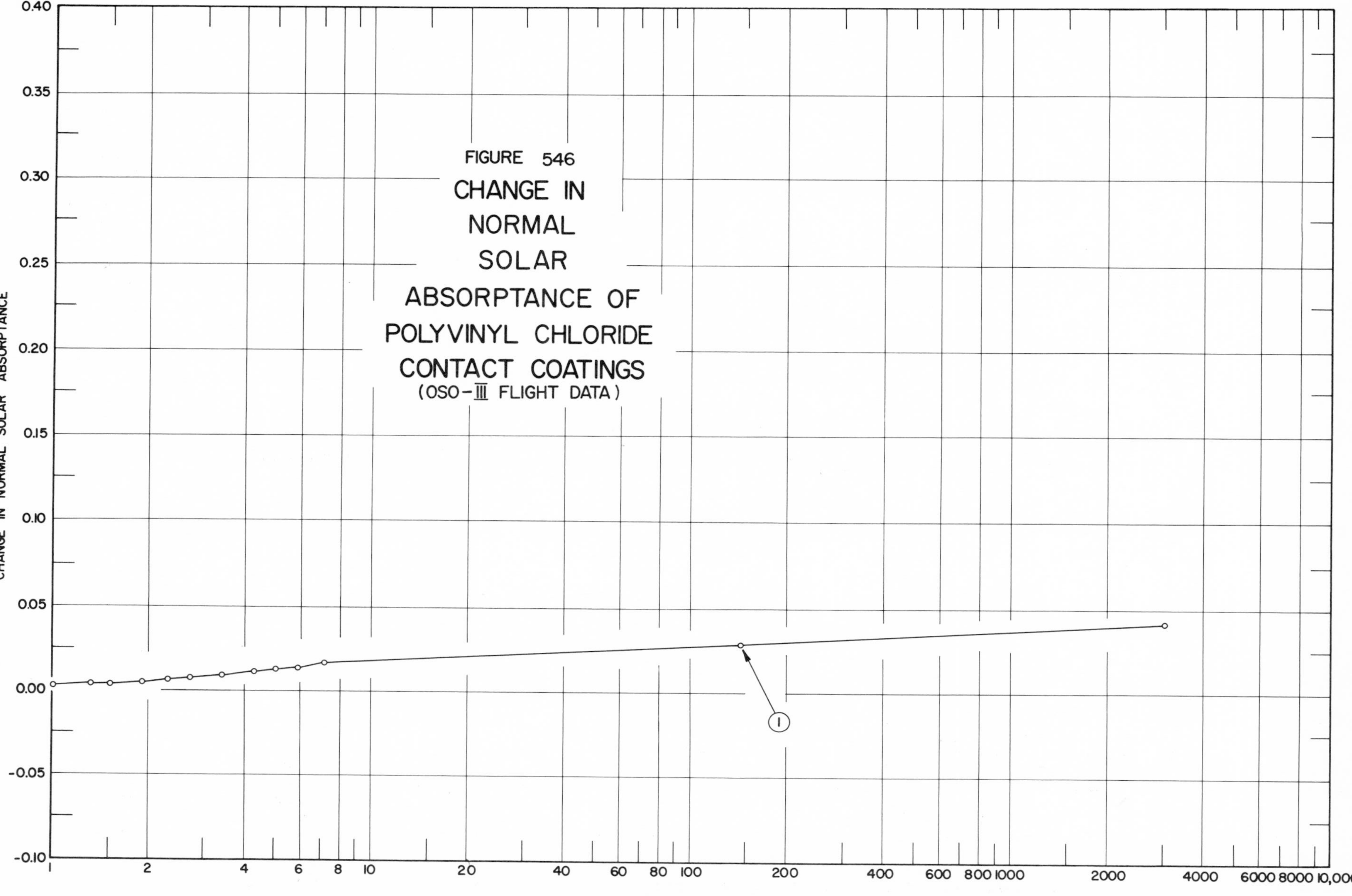
FIGURE 546
CHANGE IN NORMAL SOLAR ABSORPTANCE OF POLYVINYL CHLORIDE CONTACT COATINGS
(OSO-III FLIGHT DATA)
CHANGE IN NORMAL SOLAR ABSORPTANCE
0.40
0.35
0.30
0.25
0.20
0.15
0.10
0.05
0.00
-0.05
-0.10
1
2
4
6
8
10
20
40
60
80
100
200
400
600
800
1000
2000
4000
6000
8000
10,000
EXPOSURE TIME, ESH
1

SPECIFICATION TABLE NO. 546 CHANGE IN NORMAL SOLAR ABSORPTANCE OF POLYVINYL CHLORIDE CONTACT COATINGS

Curve No.	Ref. No.	Year	Temperature Range, K	Geometry θ	Reported Error, %	Composition (weight percent), Specifications and Remarks
1	65	1969	~300	~0°		PVC terpolymer (0.0178 mm thick); aluminum (5052 H19 Al alloy) substrate; property calculated from temp of substrate from in-flight data of OSO III; ESH is variable; data extracted from smooth curve. [Author's designation: VMCH]

DATA TABLE NO. 546 CHANGE IN NORMAL SOLAR ABSORPTANCE OF POLYVINYL CHLORIDE CONTACT COATINGS

[Temperature, T, K; Absorptance, α]

ESH	$\Delta\alpha$
CURVE 1	
T ~300	
0.80	0.002*
1.0	0.003
1.3	0.004
1.5	0.004
1.9	0.005
2.3	0.006
2.7	0.007
3.4	0.009
4.3	0.011
5.0	0.012
5.9	0.013
7.1	0.016
145.9	0.029
3019.9	0.041

* Not shown on plot

SPECIFICATION TABLE NO. 547 NORMAL SPECTRAL TRANSMITTANCE OF POLYVINYLIDENE CHLORIDE CONTACT COATINGS

Curve No.	Ref. No.	Year	Temperature, K	Wavelength Range, μm	Geometry θ θ' ω'	Reported Error, %	Composition (weight percent), Specifications, and Remarks
1*	177	1965	300	18.6-220	~0° ~0°		Polyvinylidene chloride; silicon (2 mm thick) and polyvinylidene chloride substrates; data extracted from smooth curve.

DATA TABLE NO. 547 NORMAL SPECTRAL TRANSMITTANCE OF POLYVINYLIDENE CHLORIDE CONTACT COATINGS

[Wavelength, λ, μm; Transmittance, τ; Temperature, T, K]

λ	τ	λ	τ
CURVE 1*		CURVE 1 (cont.)*	
T = 300			
		35.4	0.692
18.6	0.440	38.2	0.801
18.9	0.169	40.5	0.817
19.6	0.786	47.8	0.834
20.1	0.829	55.8	0.859
20.8	0.813	60.3	0.922
21.5	0.743	65.9	0.936
22.1	0.463	71.1	0.906
22.5	0.642	82.0	0.823
23.1	0.629	90.1	0.809
23.8	0.719	102	0.797
24.8	0.753	125	0.820
27.0	0.755	153	0.820
28.1	0.716	188	0.852
29.2	0.832	220	0.883
30.9	0.850		
33.0	0.818		
33.7	0.760		
34.3	0.692		

* No plot given

SPECIFICATION TABLE NO. 548 NORMAL TOTAL EMITTANCE OF SILICONE CONTACT COATINGS

Curve No.	Ref. No.	Year	Temperature Range, K	Geometry θ'	Reported Error, %	Composition (weight percent), Specifications and Remarks
1*	60	1954	127-584	~0°		Clear silicone (0.0254-0.0381 mm thick) aluminum alloy 24-ST substrate; applied with a camel's hair brush, dried and baked at 478 K for 16 hrs; measured in argon (10^{-3} mm Hg). [Author's designation: Sample 7]
2*	60	1954	128-552	~0°		Similar to above specimen and conditions except aluminum alloy 75-ST substrate. [Author's designation: Sample 8]
3*	60	1954	120-600	~0°		Similar to above specimen and conditions except stainless steel 374 substrate. [Author's designation: Sample 4]
4*	60	1954	140-610	~0°		Similar to above specimen and conditions except stainless steel 316 substrate. [Author's designation: Sample 2]
5*	60	1954	117-518	~0°		Similar to above specimen and conditions except Dow metal substrate. [Author's designation: Sample 10]
6*	60	1954	129-612	~0°		Similar to above specimen and conditions except stainless steel 301 substrate. [Author's designation: Sample 3]
7*	60	1954	110-580	~0°		Similar to above specimen and conditions except mild steel substrate. [Author's designation: Sample 5]
8*	7	1961	373	~0°		Clear silicone varnish (0.0508 mm thick); anodized aluminum substrate; measured in vacuum.
9*	56	1969	311	~0°		RTV-602 silicone (0.9065 mm thick); aluminum (1145-0 aluminum alloy foil) substrate (6.35 μm thick); property calculated from reflectance; lab data taken on sample to be tested on Lunar Orbiter V.

* No plot given

DATA TABLE NO. 548 NORMAL TOTAL EMITTANCE OF SILICONE CONTACT COATINGS

[Temperature, T, K; Emittance, ∈]

T	∈
CURVE 1*	
218	0.978
127	0.873
411	0.911
488	0.868
584	0.772
220	0.884
134	0.810
554	0.760
137	0.801
CURVE 2*	
218	0.913
128	0.831
404	0.929
542	0.830
219	0.941
128	0.849
402	0.899
552	0.808
129	0.821
CURVE 3*	
419	0.889
217	0.890
120	0.921
502	0.869
585	0.801
125	0.852
600	0.712
137	0.819
585	0.706
225	0.809
CURVE 4*	
292	0.800
140	0.785
393	0.751
505	0.727
610	0.701
217	0.786
597	0.663
148	0.786
CURVE 5*	
211	0.846
117	0.838
410	0.801
508	0.775
136	0.830
217	0.381
518	0.743
230	0.811
CURVE 6*	
218	0.891
129	0.803
354	0.906
452	0.848
567	0.759
612	0.740
264	0.841
604	0.713
218	0.880
404	0.832
CURVE 7*	
205	0.835
112	0.839
381	0.825
507	0.747
580	0.672
110	0.809
564	0.600
212	0.839
559	0.641
116	0.808
CURVE 8*	
373	0.74
CURVE 9*	
311	0.800

* No plot given

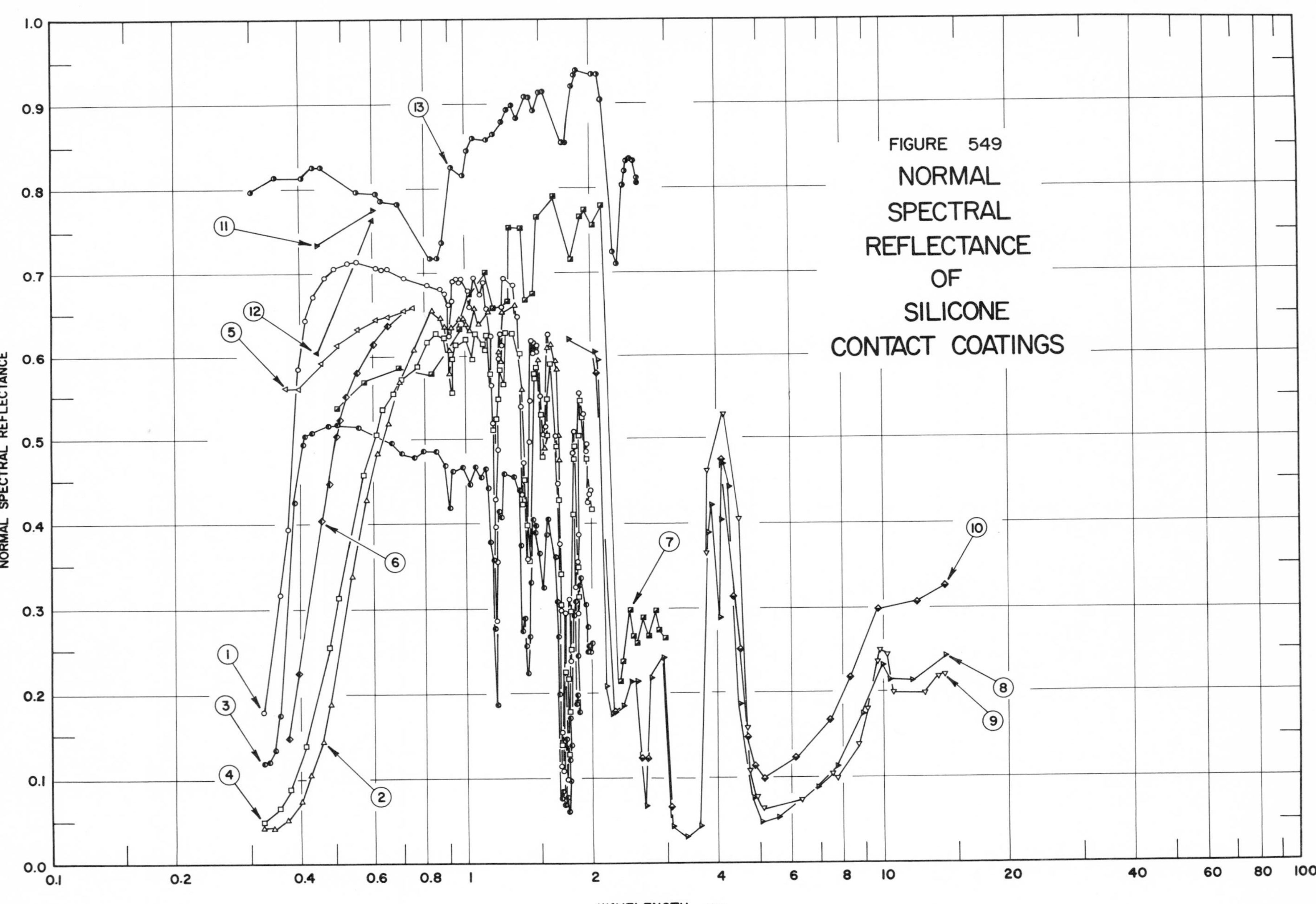
FIGURE 549
NORMAL
SPECTRAL
REFLECTANCE
OF
SILICONE
CONTACT COATINGS
NORMAL SPECTRAL REFLECTANCE
WAVELENGTH, μm
0.0
0.1
0.2
0.3
0.4
0.5
0.6
0.7
0.8
0.9
1.0
0.1
0.2
0.4
0.6
0.8
1
2
4
6
8
10
20
40
60
80
100
1
2
3
4
5
6
7
8
9
10
11
12
13

SPECIFICATION TABLE NO. 549 NORMAL SPECTRAL REFLECTANCE OF SILICONE CONTACT COATINGS

Curve No.	Ref. No.	Year	Temperature, K	Wavelength Range, μm	Geometry θ	θ'	ω'	Reported Error, %	Composition (weight percent), Specifications, and Remarks
1	86	1964	~298	0.325-2.000	~0°	~0°			LTV 602 silicone polymer (linear polydimethylsiloxane with a methyl-to-silicon atom ratio of 2.0); Al substrate; foamed; contains 10% porofor N; data extracted from smooth curve.
2	86	1964	~298	0.325-2.000	~0°	~0°			Above specimen and conditions except after UV exposure.
3	86	1964	~298	0.325-2.000	~0°	~0°			Similar to curve 1 specimen and conditions except sample is unfoamed.
4	86	1964	~298	0.325-2.000	~0°	~0°			Similar to above specimen and conditions except after UV exposure.
5	100	1962	298	0.021-0.751	~0°		2π		Silicone (DC 808 silicone 3 parts by wt TBT 2 parts by wt); polished aluminum substrate; data extracted from smooth curve.
6	100	1962	298	0.372-0.751	~0°		2π		Above specimen and conditions except exposed to UV at ~10 suns for 22.75 hrs.
7	258	1963	300	0.49-3.00	5°		2π		Silicone, Dow Corning 6510 (0.432 mm thick); aluminum substrate. [Author's designation: Sample 29R]
8	258	1963	389	1.76-14.00	5°		2π		Similar to above specimen and conditions. [Author's designation: Sample 31Ra]
9	258	1963	422	3.72-14.00	5°		2π		Similar to above specimen and conditions. [Author's designation: Sample 30Ra]
10	258	1963	450	1.76-14.00	5°		2π		Similar to above specimen and conditions. [Author's designation: Sample 31Rb]
11	27	1964	~298	0.440-0.600	~0°		2π		RTV-11 silicone paste, General Electric; aluminum substrate.
12	27	1964	~298	0.440-0.600	~0°		2π		Above specimen and conditions except irradiated in vacuum (10^{-6} mm Hg) with 108 ESH of UV at solar factor of 4 suns.
13	56	1969	298	0.307-2.600	~0°		2π		RTV-602 silicone (0.0965 mm thick); aluminum (1145-0 aluminum alloy foil) substrate (0.00635 mm thick); data extracted from smooth curve.

DATA TABLE NO. 549 NORMAL SPECTRAL REFLECTANCE OF SILICONE CONTACT COATINGS

[Wavelength, λ, μm; Reflectance, ρ; Temperature, T, K]

λ	ρ
CURVE 1	
T ~ 298	
0.325	0.177
0.359	0.318
0.372	0.391
0.399	0.584
0.415	0.642
0.432	0.670
0.460	0.694
0.481	0.706
0.520	0.711
0.556	0.713
0.609	0.709
0.633	0.705
0.652	0.706
0.718	0.692
0.811	0.684
0.877	0.680
0.895	0.674
0.906	0.660
0.915	0.625
0.921	0.665
0.933	0.690
0.950	0.691
0.963	0.687
0.979	0.691
1.011	0.678
1.027	0.659
1.034	0.681*
1.047	0.693
1.069	0.692*
1.097	0.671
1.106	0.689
1.129	0.682*
1.136	0.663*
1.136	0.654
1.145	0.635*
1.142	0.622
1.152	0.576*
1.150	0.561
1.166	0.545*
1.166	0.518
1.173	0.487*
1.174	0.466*
1.180	0.427
1.177	0.413*
1.183	0.395

λ	ρ
CURVE 1 (cont.)	
1.189	0.284
1.196	0.352
1.198	0.486
1.205	0.597
1.212	0.626
1.221	0.611
1.229	0.656
1.241	0.691
1.251	0.689*
1.256	0.689*
1.267	0.685*
1.283	0.687*
1.301	0.685
1.319	0.687*
1.329	0.679*
1.341	0.676*
1.349	0.646
1.363	0.559*
1.366	0.538
1.373	0.431
1.378	0.456*
1.384	0.470
1.393	0.423*
1.403	0.356
1.412	0.408*
1.418	0.497
1.435	0.543
1.440	0.577*
1.455	0.617
1.468	0.601
1.476	0.616
1.490	0.604
1.495	0.611
1.511	0.550
1.520	0.517*
1.531	0.504
1.546	0.506*
1.556	0.514
1.579	0.618*
1.596	0.625
1.609	0.623*
1.633	0.581*
1.648	0.502
1.649	0.473*
1.666	0.446
1.678	0.375

λ	ρ
CURVE 1 (cont.)	
1.681	0.299
1.680	0.282*
1.689	0.113
1.696	0.152
1.704	0.107
1.716	0.294
1.744	0.098
1.774	0.310
1.786	0.237
1.797	0.396*
1.808	0.481
1.816	0.508
1.826	0.427*
1.837	0.324
1.843	0.353
1.850	0.292
1.856	0.386
1.872	0.549*
1.881	0.551
1.891	0.547*
1.898	0.534*
1.914	0.528
1.926	0.507*
1.939	0.483
1.947	0.491
1.959	0.449*
1.965	0.425
1.975	0.433*
1.980	0.433
1.990	0.423*
2.000	0.438
CURVE 2	
T ~ 298	
0.325	0.043
0.341	0.042
0.370	0.052
0.400	0.074
0.420	0.104
0.449	0.142
0.469	0.186
0.534	0.339
0.579	0.427
0.608	0.481
0.645	0.519

λ	ρ
CURVE 2 (cont.)	
0.693	0.568
0.757	0.608
0.833	0.653
0.877	0.645
0.893	0.633
0.901	0.607
0.910	0.578
0.919	0.612*
0.923	0.634
0.936	0.640*
0.965	0.644
0.993	0.644
1.008	0.638
1.016	0.630*
1.027	0.630
1.036	0.650*
1.052	0.657
1.068	0.655*
1.094	0.638
1.104	0.646*
1.117	0.653
1.129	0.647*
1.142	0.622*
0.052	0.576*
1.150	0.561*
1.166	0.545*
1.166	0.518*
1.173	0.487*
1.174	0.466*
1.180	0.427*
1.177	0.413*
1.183	0.395*
1.189	0.284*
1.196	0.352*
1.198	0.486*
1.205	0.597*
1.211	0.605
1.218	0.592
1.222	0.606*
1.229	0.651
1.237	0.658*
1.314	0.660
1.336	0.657*
1.349	0.646*
1.363	0.559
1.366	0.538*

λ	ρ
CURVE 2 (cont.)	
1.373	0.431*
1.378	0.456*
1.384	0.470*
1.393	0.423*
1.403	0.356*
1.412	0.408*
1.418	0.497*
1.435	0.543*
1.440	0.577*
1.456	0.602*
1.465	0.583
1.480	0.604*
1.491	0.586*
1.500	0.595
1.511	0.550*
1.527	0.492*
1.545	0.489
1.589	0.610
1.602	0.612
1.611	0.603*
1.622	0.594
1.633	0.581
1.648	0.502
1.649	0.473
1.666	0.446*
1.678	0.375*
1.681	0.299*
1.680	0.282*
1.689	0.113*
1.696	0.152*
1.704	0.107*
1.716	0.294*
1.744	0.098*
1.774	0.310*
1.786	0.237*
1.797	0.396*
1.808	0.481*
1.816	0.508*
1.826	0.427*
1.837	0.324*
1.843	0.353*
1.850	0.292*
1.856	0.386*
1.872	0.549*
1.881	0.551*
1.891	0.547*

λ	ρ
CURVE 2 (cont.)	
1.898	0.534*
1.914	0.528*
1.926	0.507*
1.939	0.483*
1.947	0.491*
1.959	0.449*
1.965	0.421*
1.975	0.423*
1.987	0.419*
2.000	0.438*
CURVE 3	
T ~ 298	
0.325	0.118
0.335	0.119
0.346	0.132
0.356	0.172
0.387	0.425
0.404	0.492
0.413	0.501
0.424	0.508
0.463	0.517
0.496	0.518
0.559	0.514
0.660	0.495
0.700	0.461
0.747	0.477
0.794	0.484
0.841	0.484
0.890	0.467
0.913	0.419
0.919	0.447*
0.933	0.460
0.973	0.462*
0.980	0.464
0.995	0.461*
1.024	0.443
1.039	0.459*
1.053	0.465
1.095	0.451
1.113	0.461
1.131	0.440
1.144	0.396*
1.147	0.377
1.157	0.356*

λ	ρ
CURVE 3 (cont.)	
1.161	0.355
1.177	0.273
1.187	0.183
1.193	0.288*
1.206	0.414
1.216	0.407
1.230	0.453*
1.244	0.457
1.258	0.455*
1.305	0.452
1.316	0.454*
1.338	0.437
1.358	0.371
1.371	0.271
1.380	0.287
1.390	0.254
1.400	0.221
1.405	0.257*
1.410	0.265
1.415	0.326*
1.422	0.330
1.453	0.403
1.462	0.389
1.474	0.396
1.482	0.392*
1.489	0.393*
1.494	0.385*
1.504	0.361
1.518	0.324*
1.545	0.323
1.561	0.361*
1.573	0.385
1.577	0.403
1.612	0.392*
1.632	0.359
1.642	0.307
1.661	0.265
1.669	0.198
1.675	0.193*
1.680	0.116*
1.690	0.074
1.694	0.083
1.700	0.069
1.707	0.112*
1.714	0.144
1.726	0.107*

λ	ρ
CURVE 3 (cont.)	
1.732	0.076
1.742	0.060
1.751	0.096
1.759	0.120
1.759	0.136*
1.769	0.168
1.782	0.136
1.791	0.225*
1.801	0.290
1.814	0.309
1.822	0.281*
1.833	0.185
1.842	0.197
1.848	0.175
1.855	0.243
1.863	0.278*
1.868	0.329
1.889	0.335
1.911	0.315*
1.931	0.302
1.936	0.296*
1.944	0.296*
1.956	0.276
1.964	0.249
1.974	0.254
1.988	0.244*
1.993	0.248
1.991	0.255*
2.000	0.257
CURVE 4	
T ~ 298	
0.325	0.050
0.355	0.065
0.378	0.087
0.412	0.138
0.468	0.254
0.494	0.311
0.563	0.456
0.603	0.505
0.635	0.534
0.664	0.551
0.700	0.570
0.761	0.586
0.809	0.616

* Not shown on plot

DATA TABLE NO. 549 NORMAL SPECTRAL REFLECTANCE OF SILICONE CONTACT COATINGS (continued)

CURVE 4 (cont.)

λ	ρ
0.854	0.626
0.885	0.621
0.911	0.606*
0.911	0.574*
0.921	0.555
0.931	0.597
0.943	0.612
0.967	0.614*
1.000	0.620
1.008	0.616*
1.028	0.596
1.052	0.627
1.081	0.624*
1.101	0.615
1.111	0.608
1.119	0.626*
1.132	0.626
1.147	0.577
1.161	0.510
1.172	0.521
1.173	0.437*
1.183	0.397*
1.205	0.547
1.215	0.581
1.221	0.563
1.230	0.598*
1.245	0.627
1.297	0.625*
1.302	0.627
1.331	0.627*
1.354	0.602
1.376	0.421
1.387	0.450
1.401	0.399
1.410	0.354
1.430	0.489*
1.462	0.578
1.472	0.570
1.486	0.583
1.499	0.565*
1.517	0.529
1.529	0.476
1.551	0.480*
1.569	0.504
1.577	0.547
1.590	0.582*
1.597	0.589*
1.607	0.590
1.619	0.583*
1.651	0.490
1.656	0.434*
1.673	0.427
1.684	0.340
1.683	0.303
1.699	0.137
1.710	0.223
1.720	0.293*
1.730	0.216
1.747	0.126
1.758	0.178
1.777	0.297
1.784	0.250
1.798	0.315*
1.801	0.410
1.813	0.474
1.818	0.490
1.838	0.308*
1.849	0.349
1.853	0.313
1.857	0.355*
1.857	0.391*
1.869	0.503
1.882	0.544
1.898	0.538*
1.905	0.522
1.947	0.475
1.971	0.420*
2.000	0.416

CURVE 5
T = 298

λ	ρ
0.021	0.560*
0.370	0.560
0.399	0.560
0.449	0.591
0.499	0.613
0.550	0.631
0.601	0.644
0.649	0.648
0.701	0.655
0.751	0.658

CURVE 6
T = 298

λ	ρ
0.372	0.149
0.399	0.224
0.449	0.403
0.466	0.447
0.487	0.502
0.500	0.522
0.518	0.550
0.549	0.580
0.600	0.613
0.649	0.637
0.702	0.654*
0.751	0.658*

CURVE 7
T = 300

λ	ρ
0.49	0.536
0.57	0.567
0.69	0.585
0.83	0.578
0.97	0.631
1.02	0.672
1.11	0.700
1.17	0.656
1.26	0.664
1.27	0.753
1.36	0.753
1.40	0.667
1.45	0.673
1.50	0.766
1.63	0.790
1.80	0.715
1.90	0.769
1.94	0.773
2.02	0.757
2.13	0.780
2.33	0.211
2.39	0.237
2.47	0.299
2.50	0.265
2.57	0.259
2.63	0.289
2.71	0.266
2.85	0.296
2.90	0.273
3.00	0.263

CURVE 8
T = 389

λ	ρ
1.76	0.620
2.03	0.604
2.09	0.592
2.16	0.209
2.22	0.172
2.38	0.183
2.48	0.211
2.56	0.211
2.69	0.067
2.79	0.217
2.98	0.240
3.10	0.041
3.32	0.030
3.60	0.044
3.80	0.386
3.84	0.420
4.01	0.288
4.06	0.293*
4.05	0.441
4.15	0.468
4.25	0.440
4.51	0.184
4.81	0.073
5.02	0.048
5.59	0.052
6.95	0.088
7.66	0.111
8.85	0.171
9.83	0.230
10.34	0.213
11.63	0.212
14.00	0.240

CURVE 9
T = 422

λ	ρ
3.72	0.363
3.79	0.460
4.18	0.529
4.45	0.402
4.69	0.158
4.78	0.107
4.94	0.075
5.13	0.062
6.32	0.071
7.50	0.102
7.72	0.099
8.71	0.139
9.14	0.179
9.61	0.235
9.80	0.249
10.09	0.241
10.49	0.199
12.47	0.199
12.61	0.209*
13.56	0.218
14.00	0.220

CURVE 10
T = 450

λ	ρ
1.76	0.620*
2.03	0.604*
2.07	0.579
2.25	0.175
2.39	0.183*
2.49	0.211*
2.56	0.211*
2.61	0.121
2.70	0.121
2.80	0.218*
2.97	0.239*
3.08	0.065
3.31	0.031*
3.58	0.045
3.78	0.362*
3.87	0.420*
4.02	0.289*
4.13	0.471
4.38	0.311
4.57	0.250
4.69	0.146
4.88	0.112
5.19	0.099
6.13	0.121
7.40	0.165
8.38	0.218
9.73	0.298
12.00	0.305
14.00	0.323

CURVE 11
T ~ 298

λ	ρ
0.440	0.735
0.600	0.775

CURVE 12
T ~ 298

λ	ρ
0.440	0.605
0.600	0.765

CURVE 13
T = 298

λ	ρ
0.307	0.798
0.350	0.814
0.403	0.814
0.432	0.828
0.453	0.828
0.548	0.797
0.611	0.795
0.627	0.786
0.690	0.782
0.828	0.719
0.860	0.719
0.880	0.738
0.930	0.825
0.993	0.816
1.017	0.845
1.058	0.860
1.137	0.859
1.181	0.864
1.232	0.880
1.260	0.893
1.301	0.899
1.340	0.884
1.360	0.886*
1.403	0.907
1.427	0.907
1.466	0.898*
1.485	0.891
1.499	0.902*
1.512	0.911
1.535	0.911
1.614	0.893*
1.725	0.853
1.747	0.853
1.810	0.920
1.846	0.933
1.873	0.937
2.010	0.934
2.108	0.934
2.128	0.902
2.203	0.881
2.295	0.725
2.314	0.710
2.332	0.714*
2.404	0.802
2.437	0.820
2.468	0.831
2.508	0.834
2.553	0.832
2.600	0.811

* Not shown on plot

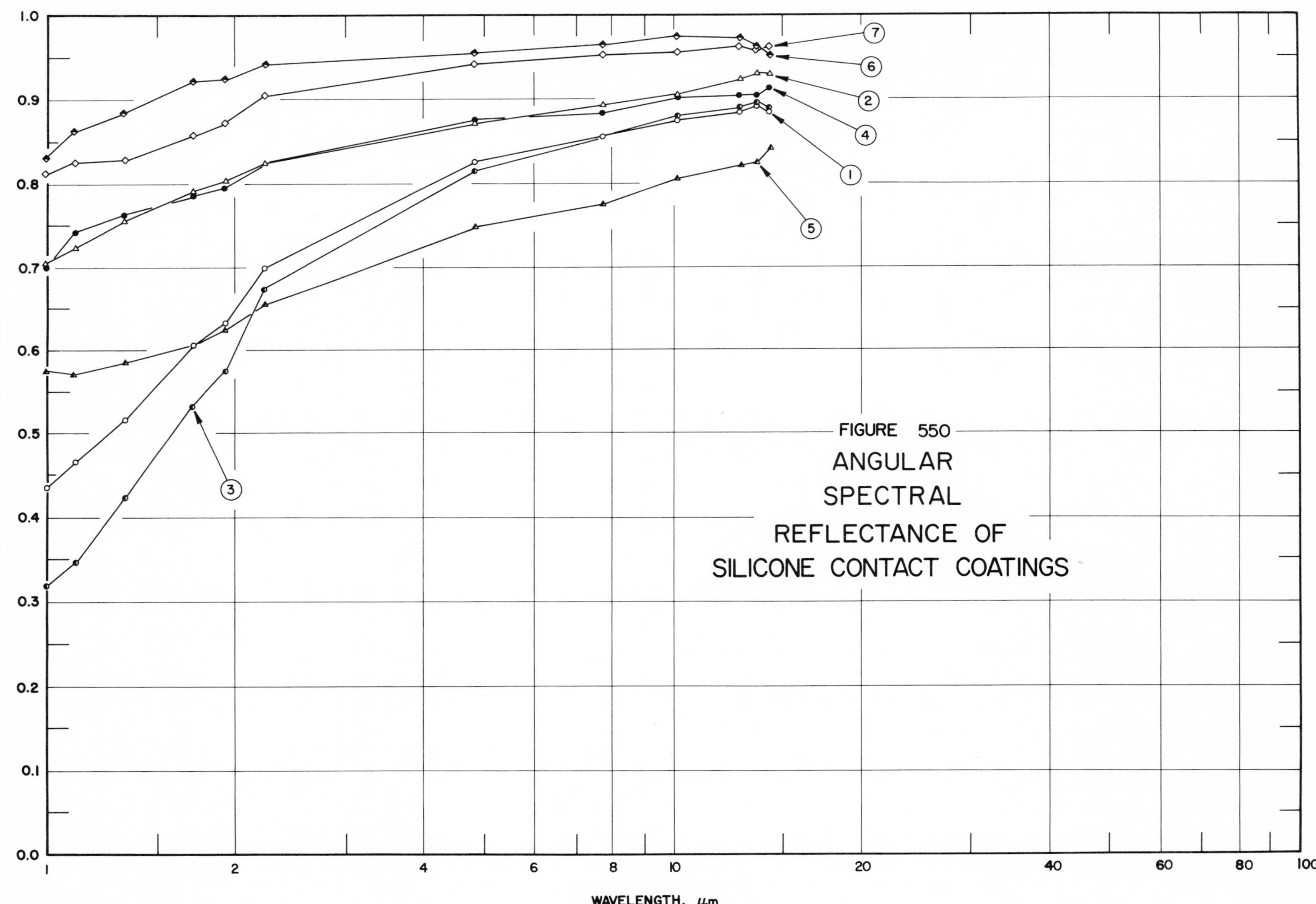

FIGURE 550
ANGULAR SPECTRAL REFLECTANCE OF SILICONE CONTACT COATINGS

SPECIFICATION TABLE NO. 550 ANGULAR SPECTRAL REFLECTANCE OF SILICONE CONTACT COATINGS

Curve No.	Ref. No.	Year	Temperature, K	Wavelength Range, μm	Geometry θ	θ'	ω'	Reported Error, %	Composition (weight percent), Specifications, and Remarks
1	231	1962	~298	1.009-14.375	45°	45°			Siliclad, silicone resin; silver, General Electric SR-111 silicone resin, and 316 stainless steel substrates; silver applied by immersion in a silver nitrate and dextrose solution for about 3 min (Brashear method); specular component only. [Authors' designation: Ag 87 CS]
2	231	1962	~298	1.009-14.375	45°	45°			Similar to above specimen and conditions. [Authors' designation: Ag 88 CS]
3	231	1962	~298	1.009-14.375	45°	45°			Similar to above specimen and conditions. [Authors' designation: Ag 89 CS]
4	231	1962	~298	1.009-14.375	45°	45°			Similar to curve 1 specimen and conditions except silver SY627-119 polyurethane (Febert Shorndorfer Co.), and 316 stainless steel substrates. [Authors' designation: Ag 90 CS]
5	231	1962	~298	1.009-14.375	45°	45°			Similar to above specimen and conditions. [Authors' designation: Ag 91 CS]
6	231	1962	~298	1.009-14.375	45°	45°			Similar to curve 1 specimen and conditions except silver, Maraset 617-C epoxy resin (Marblette Co.), and 316 stainless steel substrates. [Authors' designation: Ag 92 CS]
7	231	1962	~298	1.009-14.375	45°	45°			Similar to above specimen and conditions. [Authors' designation: Ag 93 CS]

DATA TABLE NO. 550 ANGULAR SPECTRAL REFLECTANCE OF SILICONE CONTACT COATINGS

[Wavelength, λ, μm; Reflectance, ρ; Temperature, T, K]

λ	ρ
CURVE 1	**T ~ 298**
1.009	0.435
1.120	0.463
1.345	0.519
1.720	0.606
1.945	0.633
2.240	0.700
4.824	0.828
7.780	0.859
10.198	0.877
12.099	0.888
13.530	0.892
14.375	0.889
CURVE 2	**T ~ 298**
1.009	0.706
1.120	0.725
1.345	0.757
1.720	0.791
1.945	0.805
2.240	0.827
4.824	0.875
7.780	0.896
10.198	0.909
12.099	0.928
13.530	0.934
14.375	0.932
CURVE 3	**T ~ 298**
1.009	0.320
1.120	0.349
1.345	0.422
1.720	0.531
1.945	0.577
2.240	0.675
4.824	0.818
7.780	0.859*
10.198	0.881
12.099	0.893
13.530	0.896
14.375	0.892
CURVE 4	**T ~ 298**
1.009	0.700
1.120	0.741
1.345	0.767
1.720	0.786
1.945	0.795
2.240	0.827*
4.824	0.878
7.780	0.884
10.198	0.902
12.099	0.908
13.530	0.908
14.375	0.917
CURVE 5	**T ~ 298**
1.009	0.577
1.120	0.572
1.345	0.586
1.720	0.608*
1.945	0.626
2.240	0.656
4.824	0.750
7.780	0.778
10.198	0.809
12.099	0.821
13.530	0.827
14.375	0.842
CURVE 6	**T ~ 298**
1.009	0.831
1.120	0.864
1.345	0.886
1.720	0.923
1.945	0.929
2.240	0.945
4.824	0.958
7.780	0.967
10.198	0.977
12.099	0.974
13.530	0.965
14.375	0.953
CURVE 7	**T ~ 298**
1.009	0.814
1.120	0.826
1.345	0.830
1.720	0.860
1.945	0.874
2.240	0.907
4.824	0.944
7.780	0.953
10.198	0.957
12.099	0.965
13.530	0.960
14.375	0.962

* Not shown on plot

SPECIFICATION TABLE NO. 551 NORMAL SOLAR ABSORPTANCE OF SILICONE CONTACT COATINGS

Curve No.	Ref. No.	Year	Temperature Range, K	Geometry θ	Reported Error, %	Composition (weight percent), Specifications and Remarks
1*	60	1954	311	$\sim 0^\circ$		Silicone (0.0254-0.0381 mm thick); mild steel substrate; applied with a brush; heated to 478 K; measured for sea level conditions.
2*	60	1954	311	$\sim 0^\circ$		Above specimen and conditions except reheated to 578 K.
3*	60	1954	311	$\sim 0^\circ$		Similar to curve 1 specimen and conditions except stainless steel 316 substrate.
4*	60	1954	311	$\sim 0^\circ$		Above specimen and conditions except reheated to 611 K.
5*	60	1954	311	$\sim 0^\circ$		Similar to curve 1 specimen and conditions except stainless steel 347 substrate.
6*	60	1954	311	$\sim 0^\circ$		Above specimen and conditions except reheated to 600 K.
7*	60	1954	311	$\sim 0^\circ$		Similar to curve 1 specimen and conditions except Dow Metal substrate.
8*	60	1954	311	$\sim 0^\circ$		Above specimen and conditions except reheated to 519 K.
9*	60	1954	311	$\sim 0^\circ$		Similar to curve 1 specimen and conditions except aluminum alloy 75 ST substrate.
10*	60	1954	311	$\sim 0^\circ$		Above specimen and conditions except reheated to 553 K.
11*	60	1954	311	$\sim 0^\circ$		Similar to curve 1 specimen and conditions except aluminum alloy 24 ST substrate.
12*	60	1954	311	$\sim 0^\circ$		Above specimen and conditions except reheated to 583 K.
13*	60	1954	311	$\sim 0^\circ$		Similar to curve 1 specimen and conditions except stainless steel 301 substrate.
14*	60	1954	311	$\sim 0^\circ$		Above specimen and conditions except reheated to 617 K.
15*	56	1969	298	$\sim 0^\circ$		RTV-602 silicone (0.0965 mm thick); aluminum (1145-0 aluminum alloy foil) substrate (0.00635 mm thick); property calculated from reflectance; lab data taken on sample to be tested on Lunar Orbiter V.

* No plot given

DATA TABLE NO. 551 NORMAL SOLAR ABSORPTANCE OF SILICONE CONTACT COATINGS

[Temperature, T, K; Absorptance, α]

T	α
CURVE 1*	
311	0.91
CURVE 2*	
311	0.91
CURVE 3*	
311	0.85
CURVE 4*	
311	0.77
CURVE 5*	
311	0.83
CURVE 6*	
311	0.83
CURVE 7*	
311	0.71
CURVE 8*	
311	0.76
CURVE 9*	
311	0.71
CURVE 10*	
311	0.75
CURVE 11*	
311	0.75
CURVE 12*	
311	0.68
CURVE 13*	
311	0.91
CURVE 14*	
311	0.83
CURVE 15*	
298	0.197

* No plot given

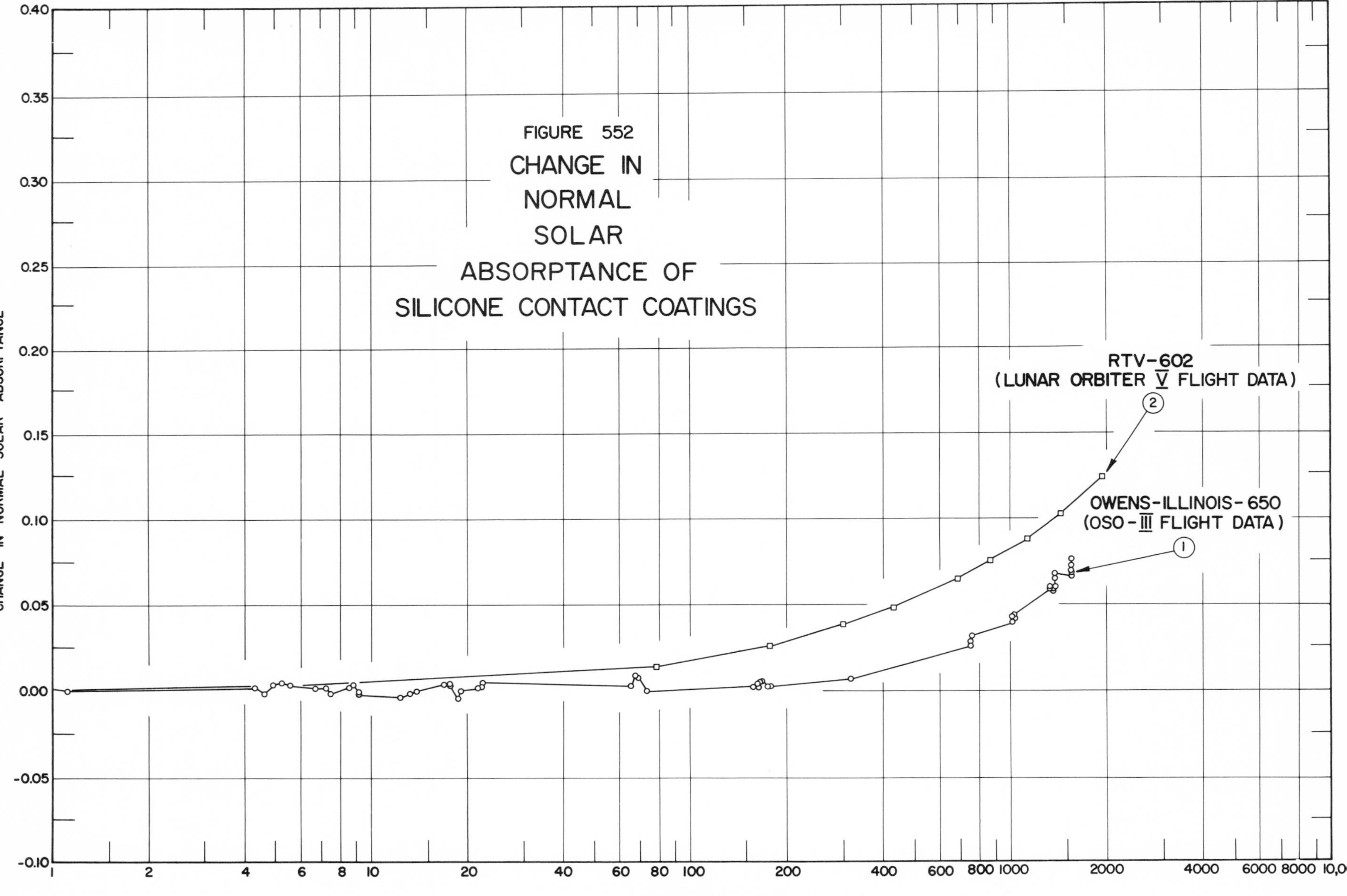

FIGURE 552 CHANGE IN NORMAL SOLAR ABSORPTANCE OF SILICONE CONTACT COATINGS

SPECIFICATION TABLE NO. 552 CHANGE IN NORMAL SOLAR ABSORPTANCE OF SILICONE CONTACT COATINGS

Curve No.	Ref. No.	Year	Temperature Range, K	Geometry θ	Reported Error, %	Composition (weight percent), Specifications and Remarks
1	65	1969	unknown	~0°		Owens-Illinois glass resin type 650 methyl silicone (0.0405 mm thick); aluminum (5052 H19 Al alloy) substrate; property calculated from temp of substrate from in-flight data of OSO III; ESH is variable.
2	65	1969	278-305	~0°		RTV-602 silicone (0.0965 mm thick); aluminum (1145-0 aluminum alloy foil) substrate (0.00635 mm thick); property calculated from temp of sample; in-flight data of Lunar Orbiter V; ESH is variable; data extracted from smooth curve.

DATA TABLE NO. 552 CHANGE IN NORMAL SOLAR ABSORPTANCE OF SILICONE CONTACT COATINGS

[Temperature, T, K; Absorptance, α]

ESH	Δα	ESH	Δα	ESH	Δα	ESH	Δα
CURVE 1		CURVE 1 (cont.)		CURVE 1 (cont.)		CURVE 1 (cont.)	
T = unknown							
0.13	0.000*	17.0	0.004	180.3	0.003	1599.5	0.073
0.47	0.004*	17.6	0.005	320.6	0.007	1599.5	0.076
1.1	0.000	18.0	0.004	748.2	0.026		
4.3	0.001	18.6	-0.004	748.2	0.029	CURVE 2	
4.6	-0.001	19.0	0.000	758.6	0.032	T = 278-305	
4.9	0.004	21.8	0.002	1009.2	0.039		
5.2	0.005	21.8	0.004	1009.2	0.042	0	0.003*
5.5	0.004	22.3	0.002	1023.3	0.041	78	0.014
6.6	0.002	22.3	0.005	1023.3	0.043	179	0.026
7.2	0.002	65.8	0.004	1342.8	0.058	300	0.038
7.4	-0.001	67.0	0.009	1342.8	0.060	438	0.048
8.5	0.002	68.2	0.007	1361.4	0.057	686	0.065
8.8	0.004	73.4	0.000	1361.4	0.058	873	0.076
9.2	-0.002	158.5	0.002	1367.7	0.060	1118	0.088
9.2	-0.001	162.2	0.002	1367.7	0.065	1441	0.102
12.2	-0.003	161.4	0.005	1367.7	0.068	1995	0.124
13.1	-0.001	165.9	0.006	1599.5	0.066		
13.8	0.000	169.8	0.006	1599.5	0.068		
		177.0	0.003	1599.5	0.070		

* Not shown on plot

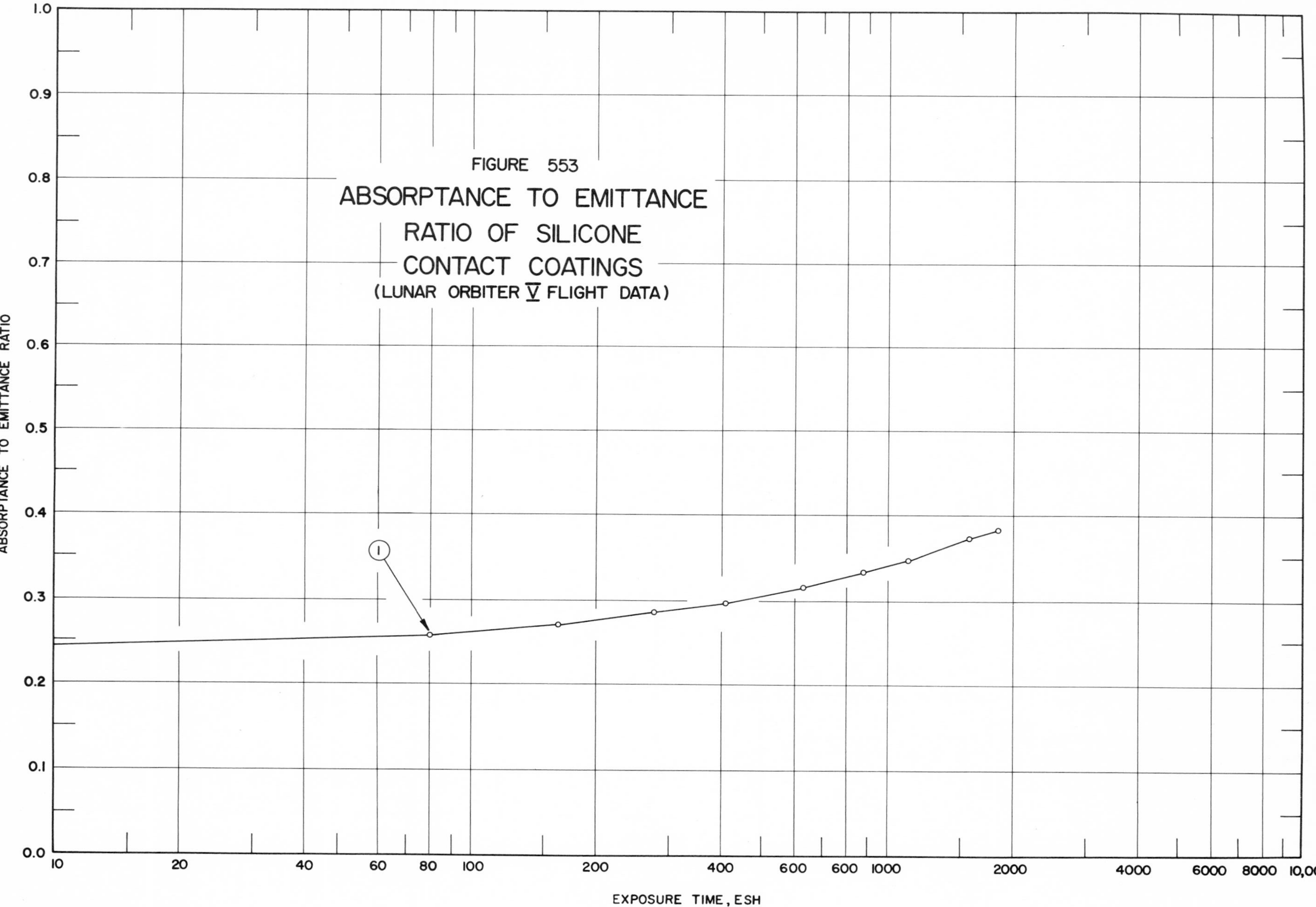

FIGURE 553
ABSORPTANCE TO EMITTANCE RATIO OF SILICONE CONTACT COATINGS
(LUNAR ORBITER V FLIGHT DATA)

SPECIFICATION TABLE NO. 553 ABSORPTANCE TO EMITTANCE RATIO OF SILICONE CONTACT COATINGS

Curve No.	Ref. No.	Year	Temperature Range, K	Reported Error, %	Composition (weight percent), Specifications and Remarks
1	56	1969	278-305		RTV-602 silicone (0.0965 mm thick); aluminum (1145-0 aluminum alloy foil) substrate (0.00635 mm thick); property calculated from temp of sample; in-flight data of Lunar Orbiter V; ESH is variable; data extracted from smooth curve.

DATA TABLE NO. 553 ABSORPTANCE TO EMITTANCE RATIO OF SILICONE CONTACT COATINGS

[Temperature, T, K; Absorptance, α]

ESH	α/ϵ
CURVE 1 T = 278-305	
0	0.242*
80	0.257
161	0.270
276	0.285
415	0.299
634	0.316
888	0.334
1122	0.348
1577	0.371
1851	0.383

* Not shown on plot

2. CONTACT COATINGS (continued)

F. Metallic Black Coatings

SPECIFICATION TABLE NO. 554 NORMAL SPECTRAL TRANSMITTANCE OF ANTIMONY BLACK CONTACT COATINGS

Curve No.	Ref. No.	Year	Temperature, K	Wavelength Range, μm	Geometry θ	θ'	ω'	Reported Error, %	Composition (weight percent), Specifications, and Remarks
1*	113	1967	~298	0.50-15.03	0°	0°			Antimony black (122 x 10^{-6} g cm^{-2}); cellulose nitrate substrate; evaporation deposited (slow rate) in "prepurified" nitrogen (< 0.002 oxygen, < 0.002 hydrogen) at a pressure of 1 mm Hg; data extracted from smooth curve; property corrected for transmittance of substrate.
2*	113	1967	~298	2.38-15.11	0°	0°			Antimony black; KRS-5 substrate; evaporation deposited (slow rate) in "prepurified" nitrogen (< 0.002 oxygen, < 0.002 hydrogen) at a pressure of 1 mm Hg; data extracted from smooth curve; property corrected for transmittance of substrate.
3*	113	1967	~298	4.27-15.00	0°	0°			Antimony black; KRS-5 substrate; evaporation deposited (slow rate) in "prepurified" nitrogen (< 0.002 oxygen, < 0.002 hydrogen) at a pressure of 1 mm Hg; data extracted from smooth curve; property corrected for transmittance of substrate.

DATA TABLE NO. 554 NORMAL SPECTRAL TRANSMITTANCE OF ANTIMONY BLACK CONTACT COATINGS

[Wavelength, λ, μm; Transmittance, τ; Temperature, T, K]

λ	τ	λ	τ	λ	τ
CURVE 1* T ~ 298		CURVE 1 (cont.)*		CURVE 2 (cont.)*	
		9.70	0.508	7.41	0.402
0.50	0.008	15.03	0.492	8.13	0.393
0.78	0.014			12.57	0.312
1.13	0.033	CURVE 2* T ~ 298		13.73	0.289
1.58	0.065			15.11	0.254
2.06	0.107				
2.50	0.152	2.38	0.000	CURVE 3* T ~ 298	
2.79	0.188	2.57	0.013		
3.08	0.231	2.80	0.033		
3.52	0.315	3.00	0.059	4.27	0.000
3.81	0.371	3.30	0.105	5.15	0.016
4.18	0.421	3.78	0.195	6.18	0.029
4.57	0.459	4.25	0.264	7.65	0.035
4.97	0.484	4.67	0.311	13.73	0.035
5.78	0.520	5.13	0.348	15.00	0.032
6.48	0.538	5.52	0.371		
6.93	0.542	5.96	0.388		
7.29	0.542	6.50	0.398		
9.02	0.513	7.08	0.402		

* No plot given

SPECIFICATION TABLE NO. 555 HEMISPHERICAL TOTAL EMITTANCE OF CHROMIUM BLACK CONTACT COATINGS

Curve No.	Ref. No.	Year	Temperature Range, K	Reported Error, %	Composition (weight percent), Specifications and Remarks
1*	82	1962	645-1062	< ±2.5	Chromium black, Pratt&Whitney Aircraft (~0.0762 mm thick); stainless steel 310 substrate; electroplated; measured in vacuum (5 x 10^{-7} mm Hg); data extracted from smooth curve. [Authors' designation: Sample No. 1]
2*	82	1962	840-1065	< ±2.5	Similar to above specimen and conditions. [Authors' designation: Sample No. 2]
3*	105	1964	307		Black chromium; aluminum, polyester film (0.00635 mm thick), aluminum, and black chromium substrates; measured in vacuum (10^{-6} mm Hg) maintained by diffusion pump. [Authors' designation: Test No. 30, Sample 1E, Side 1]
4*	105	1964	307		Black chromium; copper (0.00318 mm thick) substrate; measured in vacuum (10^{-6} mm Hg) maintained by diffusion pump. [Authors' designation: Test No. 116]

DATA TABLE NO. 555 HEMISPHERICAL TOTAL EMITTANCE OF CHROMIUM BLACK CONTACT COATINGS

[Temperature, T, K; Emittance, ∈]

T	∈
CURVE 1*	
645	0.881
955	0.887
1062	0.891
CURVE 2*	
840	0.908
992	0.885
1020	0.887
1065	0.908
CURVE 3*	
307	0.0528
CURVE 4*	
307	0.603

* No plot given

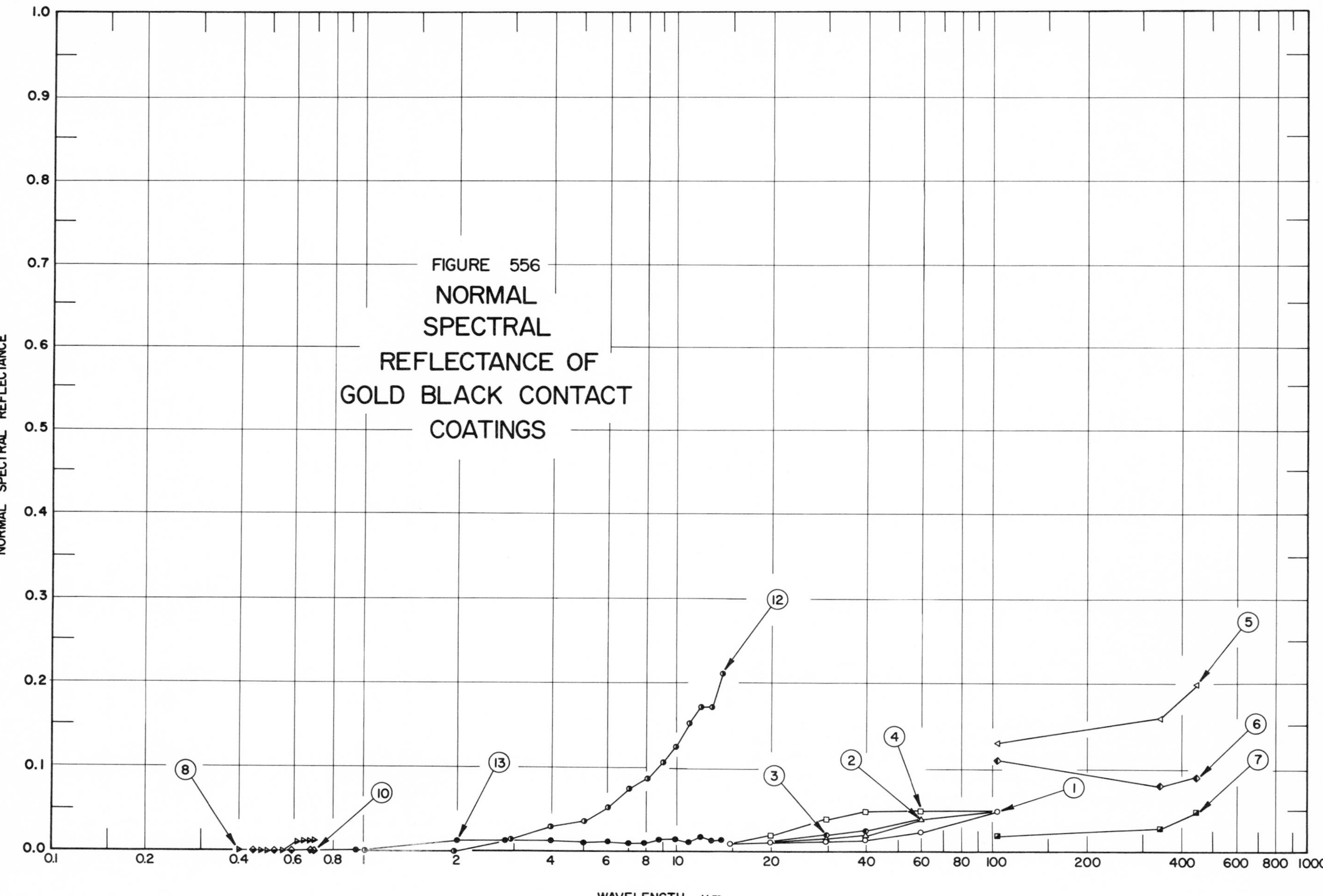

FIGURE 556 NORMAL SPECTRAL REFLECTANCE OF GOLD BLACK CONTACT COATINGS

SPECIFICATION TABLE NO. 556 NORMAL SPECTRAL REFLECTANCE OF GOLD BLACK CONTACT COATINGS

Curve No.	Ref. No.	Year	Temperature, K	Wavelength Range, μm	Geometry θ	θ'	ω'	Reported Error, %	Composition (weight percent), Specifications, and Remarks
1	113	1967	~298	15-105	7.5°	7.5°			Gold black (103 x 10^{-6} g cm^{-2}); cellulose nitrate film substrate (~5 x 10^{-5} mm thick); evaporation deposited with a tungsten filament in Airco "prepurified" nitrogen (<0.002 oxygen, <0.002 hydrogen) at a pressure of 3 mm Hg; property measured using a Mass. Inst. Tech. grating spectrometer. [Author's designation: Sample 43]
2	113	1967	~298	15-105	7.5°	7.5°			Similar to above specimen and conditions except gold black 71 x 10^{-6} g cm^{-2}. [Author's designation: Sample 95]
3	113	1967	~298	15-105	7.5°	7.5°			Similar to above specimen and conditions except gold black 60 x 10^{-6} g cm^{-2}. [Author's designation: Sample 85]
4	113	1967	~298	15-105	7.5°	7.5°			Similar to above specimen and conditions except gold black 41 x 10^{-6} g cm^{-2}. [Author's designation: Sample 89]
5	113	1967	~298	105-455	40°	40°			Similar to curve 1 specimen and conditions except gold black 121 x 10^{-6} g cm^{-2}; property measured using a Johns Hopkins grating spectrometer. [Author's designation: Sample 53]
6	113	1967	~298	105-455	40°	40°			Similar to above specimen and conditions except gold black 66 x 10^{-6} g cm^{-2}. [Author's designation: Sample 57]
7	113	1967	~298	105-455	40°	40°			Similar to above specimen and conditions except gold black 26 x 10^{-6} g cm^{-2}. [Author's designation: Sample 52]
8	113	1967	~298	0.400-0.700	~10°		2π		Gold black (150 x 10^{-6} g cm^{-2}); glass substrate; evaporation deposited with a tungsten filament in Airco "prepurified" nitrogen (<0.002 oxygen, <0.002 nitrogen) at a pressure of 3 mm Hg; data extracted from smooth curve; property measured using a Hardy spectrometer with an integrating sphere.
9*	113	1967	~298	0.400-0.700	~10°		~2π		Above specimen and conditions except property includes only radiation reflected at angles greater than 8.6° from the specular angle and some radiation reflected from 1.8°-8.6°.
10	113	1967	~298	0.400-0.700	~10°		2π		Similar to curve 8 specimen and conditions except gold black 205 x 10^{-6} g cm^{-2}. [Author's designation: Sample 426]
11*	113	1967	~298	0.400-0.700	~10°		~2π		Above specimen and conditions except property includes only radiation reflected at angles greater than 8.6° from the specular angle and some radiation reflected from 1.8°-8.6°.
12	259	1965	~298	1.01-14.0	10°		2π		Gold black; brass substrate; evaporated in a N_2 atm at 1 mm Hg; measured relative to front-surface Al mirror.
13	259	1965	~298	0.68-14.0	10°		2π		Similar to above specimen and conditions except evaporated at 2 mm Hg.

* Not shown on plot

DATA TABLE NO. 556 NORMAL SPECTRAL REFLECTANCE OF GOLD BLACK CONTACT COATINGS

[Wavelength, λ, μm; Reflectance, ρ; Temperature, T, K]

λ	ρ
CURVE 1	
T ~ 298	
15	0.010
20	0.011
30	0.012
40	0.015
60	0.025
105	0.050
CURVE 2	
T ~ 298	
15	0.010*
20	0.010
30	0.015
40	0.020
60	0.040
105	0.050*
CURVE 3	
T ~ 298	
15	0.010*
20	0.010*
30	0.020
40	0.025
60	0.040*
105	0.050*
CURVE 4	
T ~ 298	
15	0.01 *
20	0.02
30	0.04
40	0.05
60	0.05
105	0.05 *
CURVE 5	
T ~ 298	
105	0.13
345	0.16
455	0.20
CURVE 6	
T ~ 298	
105	0.11
345	0.08
455	0.09
CURVE 7	
T ~ 298	
105	0.02
345	0.03
455	0.05
CURVE 8	
T ~ 298	
0.400	0.009
0.468	0.007
0.498	0.007
0.550	0.008
0.610	0.010
0.640	0.011
0.675	0.011
0.700	0.011
CURVE 9*	
T ~ 298	
0.400	0.009
0.474	0.007
0.498	0.007
0.550	0.008
0.638	0.010
0.676	0.011
0.700	0.011
CURVE 10	
T ~ 298	
0.400	0.006*
0.443	0.006
0.492	0.005*
0.520	0.005
0.544	0.005*
0.590	0.006
CURVE 10 (cont.)	
0.629	0.006*
0.680	0.006
0.700	0.006
CURVE 11*	
T ~ 298	
0.400	0.006
0.443	0.006
0.492	0.005
0.520	0.005
0.544	0.005
0.590	0.006
0.629	0.006
0.677	0.006
0.700	0.006
CURVE 12	
T ~ 298	
1.01	0.005
1.98	0.007
2.98	0.014
3.97	0.027
5.02	0.038
6.04	0.051
7.04	0.073
8.01	0.087
9.02	0.108
10.0	0.125
11.0	0.152
12.0	0.171
13.0	0.171
14.0	0.213
CURVE 13	
T ~ 298	
0.68	0.005*
0.95	0.008
2.00	0.011
2.84	0.011
4.00	0.011
5.04	0.010
CURVE 13 (cont.)	
6.02	0.011
7.02	0.010
7.98	0.010
8.99	0.014
10.0	0.014
11.0	0.011
12.0	0.017
13.0	0.014
14.0	0.015

* Not shown on plot

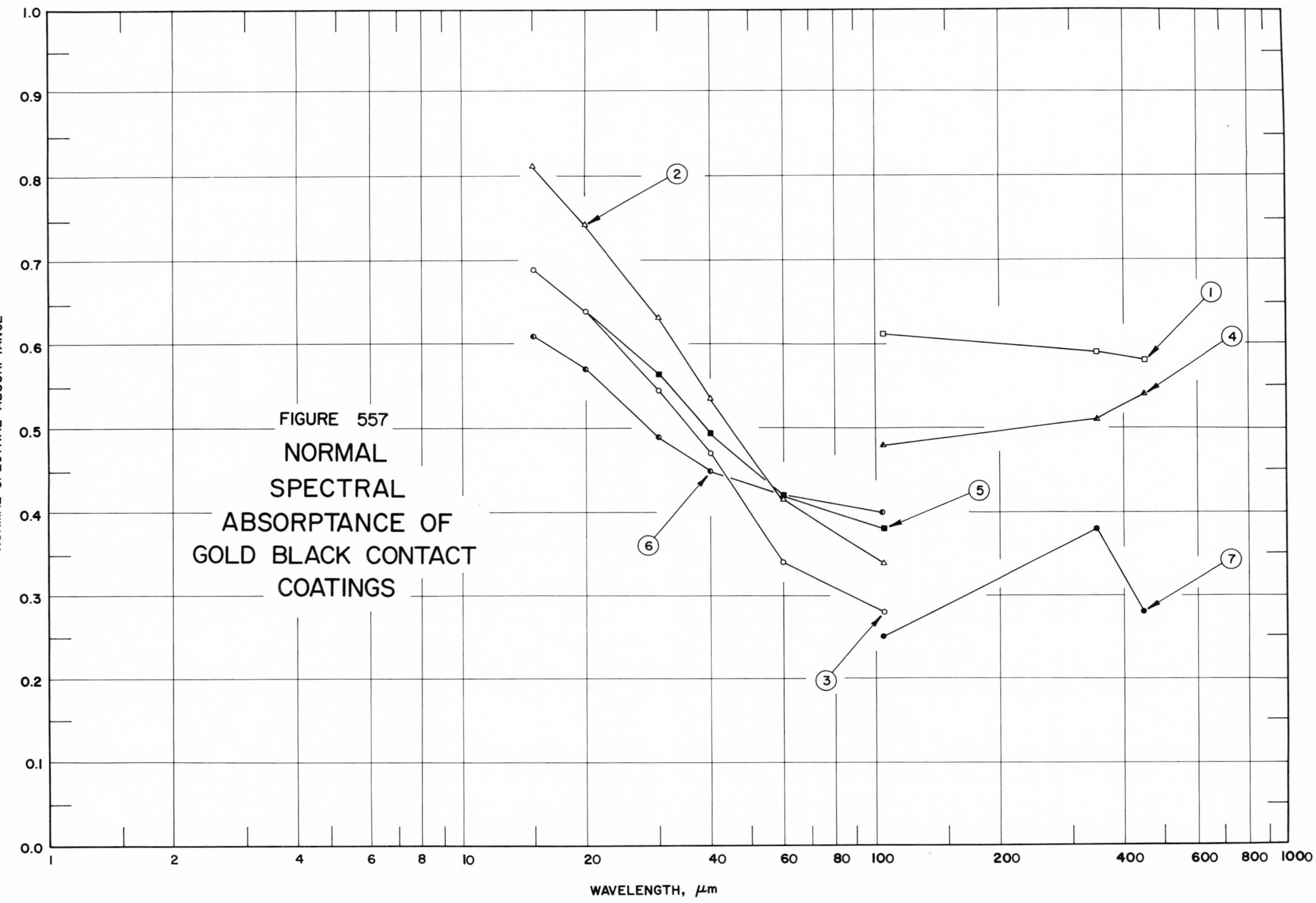

FIGURE 557 NORMAL SPECTRAL ABSORPTANCE OF GOLD BLACK CONTACT COATINGS

SPECIFICATION TABLE NO. 557 NORMAL SPECTRAL ABSORPTANCE OF GOLD BLACK CONTACT COATINGS

Curve No.	Ref. No.	Year	Temperature, K	Wavelength Range, μm	Geometry θ	Reported Error, %	Composition (weight percent), Specifications, and Remarks
1	113	1967	~298	105-455	40°		Gold Black (121 x 10^{-6} g cm^{-2} area density); ellulose nitrate (~500 Å thick) substrate; evaporated in N_2 at 2.5-3 mm Hg; 24 hrs after deposition sample annealed at 342 K for 24 hrs; calculated from transmittance and reflectance. [Author's designation: Sample 53]
2	113	1967	~298	15-105	7.5°		Similar to above specimen and conditions except 103 x 10^{-6} g cm^{-2} area density. [Author's designation: Sample 43]
3	113	1967	~298	15-105	7.5°		Similar to above specimen and conditions except 71 x 10^{-6} g cm^{-2} area density. [Author's designation: Sample 95]
4	113	1967	~298	105-455	40°		Similar to above specimen and conditions except 66 x 10^{-6} g cm^{-2} area density. [Author's designation: Sample 57]
5	113	1967	~298	15-105	7.5°		Similar to above specimen and conditions except 60 x 10^{-6} g cm^{-2} area density. [Author's designation: Sample 85]
6	113	1967	~298	15-105	7.5°		Similar to above specimen and conditions except 41 x 10^{-6} g cm^{-2} area density. [Author's designation: Sample 89]
7	113	1967	~298	105-455	40°		Similar to above specimen and conditions except 26 x 10^{-6} g cm^{-2} area density. [Author's designation: Sample 52]

DATA TABLE NO. 557 NORMAL SPECTRAL ABSORPTANCE OF GOLD BLACK CONTACT COATINGS

[Wavelength, λ, μm; Absorptance, α; Temperature, T, K]

λ	α
CURVE 1	
T ~ 298	
105	0.61
345	0.59
455	0.58
CURVE 2	
T ~ 298	
15	0.815
20	0.744
30	0.633
40	0.535
60	0.415
105	0.34
CURVE 3	
T ~ 298	
15	0.69
20	0.64
30	0.545
40	0.47
60	0.34
105	0.28
CURVE 4	
T ~ 298	
105	0.48
345	0.51
455	0.54
CURVE 5	
T ~ 298	
15	0.69*
20	0.64*
30	0.563
40	0.495
60	0.42
105	0.38
CURVE 6	
T ~ 298	
15	0.61
20	0.57
30	0.49
40	0.45
60	0.42*
105	0.40
CURVE 7	
T ~ 298	
105	0.25
345	0.38
455	0.28

* Not shown on plot

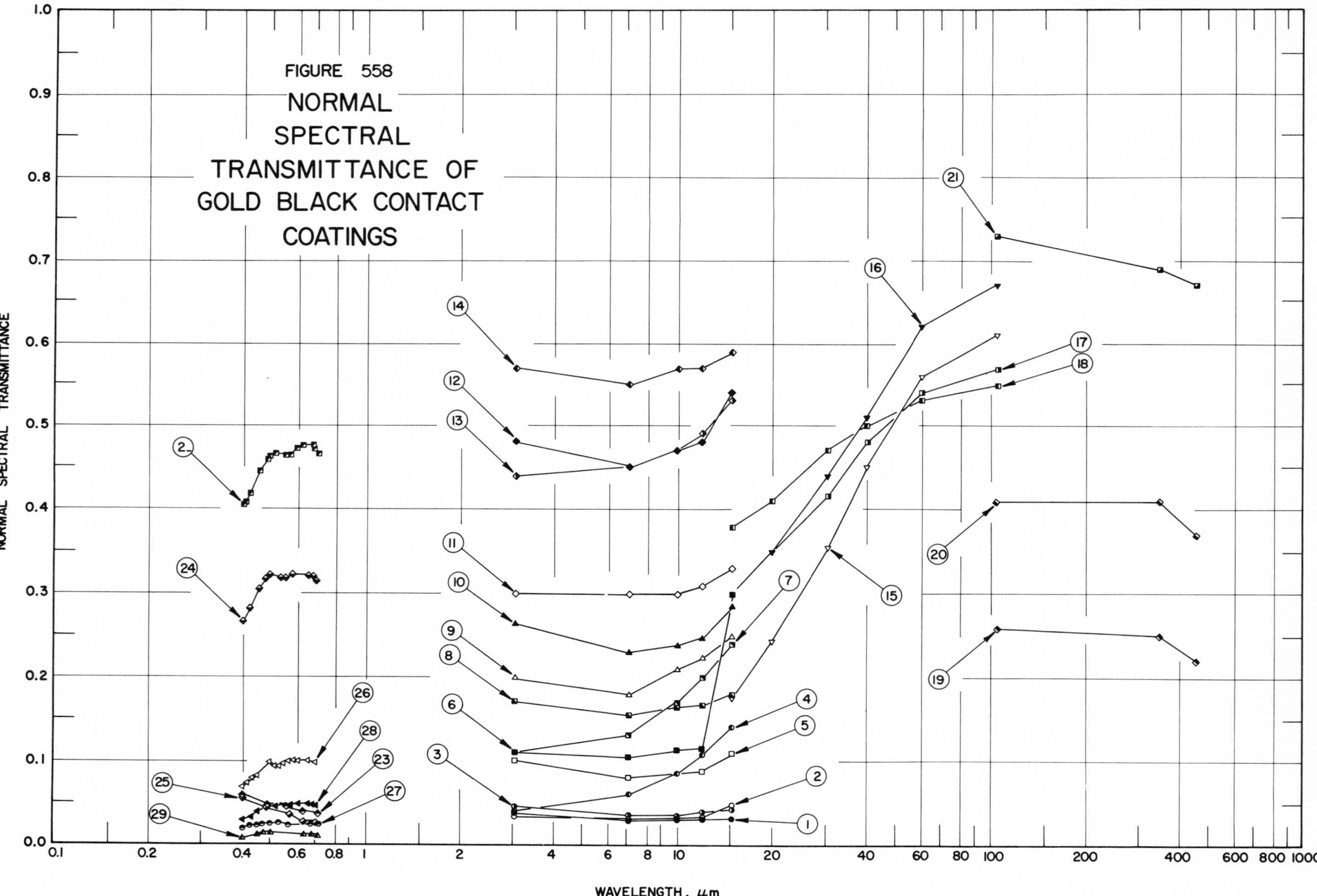
FIGURE 558
NORMAL
SPECTRAL
TRANSMITTANCE OF
GOLD BLACK CONTACT
COATINGS
NORMAL SPECTRAL TRANSMITTANCE
WAVELENGTH, μm
1.0
0.9
0.8
0.7
0.6
0.5
0.4
0.3
0.2
0.1
0.0
0.1
0.2
0.4
0.6
0.8
1
2
4
6
8
10
20
40
60
80
100
200
400
600
800
1000

SPECIFICATION TABLE NO. 558 NORMAL SPECTRAL TRANSMITTANCE OF GOLD BLACK CONTACT COATINGS

Curve No.	Ref. No.	Year	Temperature, K	Wavelength Range, μm	Geometry θ	θ'	ω'	Reported Error, %	Composition (weight percent), Specifications, and Remarks
1	113	1967	~298	3-15	0°	0°			Gold black (130×10^{-6} g cm^{-2}); cellulose nitrate film substrate (~5×10^{-5} mm thick); evaporation deposited with a tungsten filament in Airco "prepurified" nitrogen (<0.002% oxygen, <0.002% hydrogen) at a pressure of 3 mm Hg; property measured using a prism spectrometer. [Author's designation: Sample 61]
2	113	1967	~298	3-15	0°	0°			Similar to above specimen and conditions except gold black 124×10^{-6} g cm^{-2}. [Author's designation: Sample 96]
3	113	1967	~298	3-15	0°	0°			Similar to above specimen and conditions except gold black 121×10^{-6} g cm^{-2}. [Author's designation: Sample 53]
4	113	1967	~298	3-15	0°	0°			Similar to above specimen and conditions except gold black 103×10^{-6} g cm^{-2}. [Author's designation: Sample 43]
5	113	1967	~298	3-15	0°	0°			Similar to above specimen and conditions except gold black 91×10^{-6} g cm^{-2}. [Author's designation: Sample 33]
6	113	1967	~298	3-15	0°	0°			Similar to above specimen and conditions except gold black 81×10^{-6} g cm^{-2}. [Author's designation: Sample 58]
7	113	1967	~298	3-15	0°	0°			Similar to above specimen and conditions except gold black 71×10^{-6} g cm^{-2}. [Author's designation: Sample 95]
8	113	1967	~298	3-15	0°	0°			Similar to above specimen and conditions except gold black 66×10^{-6} g cm^{-2}. [Author's designation: Sample 57]
9	113	1967	~298	3-15	0°	0°			Similar to above specimen and conditions except gold black 60×10^{-6} g cm^{-2}. [Author's designation: Sample 85]
10	113	1967	~298	3-15	0°	0°			Similar to above specimen and conditions except gold black 48×10^{-6} g cm^{-2}. [Author's designation: Sample 38]
11	113	1967	~298	3-15	0°	0°			Similar to above specimen and conditions except gold black 41×10^{-6} g cm^{-2}. [Author's designation: Sample 89]
12	113	1967	~298	3-15	0°	0°			Similar to above specimen and conditions except gold black 26×10^{-6} g cm^{-2}. [Author's designation: Sample 34]
13	113	1967	~298	3-15	0°	0°			Similar to above specimen and conditions. [Author's designation: Sample 52]
14	113	1967	~298	3-15	0°	0°			Similar to above specimen and conditions except gold black 17×10^{-6} g cm^{-2}. [Author's designation: Sample 31]
15	113	1967	~298	15-105	0°	0°			Curve 4 specimen and conditions except property measured using a Mass. Inst. Tech. grating spectrometer.
16	113	1967	~298	15-105	0°	0°			Curve 7 specimen and conditions except property measured using a Mass. Inst. Tech. grating spectrometer.
17	113	1967	~298	15-105	0°	0°			Curve 9 specimen and conditions except property measured using a Mass. Inst. Tech. grating spectrometer.
18	113	1967	~298	15-105	0°	0°			Curve 11 specimen and conditions except property measured using a Mass. Inst. Tech. grating spectrometer.

SPECIFICATION TABLE NO. 558 NORMAL SPECTRAL TRANSMITTANCE OF GOLD BLACK CONTACT COATINGS (continued)

Curve No.	Ref. No.	Year	Temperature, K	Wavelength Range, μm	Geometry θ	θ'	ω'	Reported Error, %	Composition (weight percent), Specifications, and Remarks
19	113	1967	~298	105-455	0°	0°			Curve 3 specimen and conditions except property measured using a Johns Hopkins grating spectrometer.
20	113	1967	~298	105-455	0°	0°			Curve 8 specimen and conditions except property measured using a Johns Hopkins grating spectrometer.
21	113	1967	~298	105-455	0°	0°			Curve 13 specimen and conditions except property measured using a Johns Hopkins grating spectrometer.
22	113	1967	~298	0.4000-0.7004	0°		2π		Curve 11 specimen and conditions except data extracted from a smooth curve; property measured using a Hardy spectrophotometer with an integrating sphere.
23	113	1967	~298	0.4000-0.7005	0°		ω'		Above specimen and conditions except property includes only transmitted radiation that was scattered through angles greater than 7.5° from the incident axis and some radiation scattered through 2.5°-7.5°.
24	113	1967	~298	0.4000-0.6995	0°		2π		Curve 9 specimen and conditions except data extracted from a smooth curve; property measured using a Hardy spectrophotometer with an integrating sphere.
25	113	1967	~298	0.4000-0.6995	0°		ω'		Above specimen and conditions except property includes only transmitted radiation that was scattered through angles greater than 7.5° from the incident axis and some radiation scattered through 2.5°-7.5°.
26	113	1967	~298	0.4018-0.6998	0°		2π		Gold black (150×10^{-6} g cm^{-2}); glass substrate; evaporation deposited with a tungsten filament in Airco "prepurified" nitrogen (<0.002% oxygen, <0.002% hydrogen) at a pressure of 3 mm Hg; data extracted from smooth curve; property measured using a Hardy spectrometer with an integrating sphere.
27	113	1967	~298	0.4018-0.7013	0°		ω'		Above specimen and conditions except property includes only transmitted radiation that was scattered through angles greater than 7.5° from the incident axis and some radiation scattered through 2.5°-7.5°.
28	113	1967	~298	0.4000-0.6995	0°		ω'		Similar to curve 26 specimen and conditions except gold black 205×10^{-6} g cm^{-2}. [Author's designation: Sample 426]
29	113	1967	~298	0.4000-0.7000	0°		ω'		Above specimen and conditions except property includes only transmitted radiation that was scattered through angles greater than 7.5° from the incident axis and some radiation scattered through 2.5°-7.5°.

DATA TABLE NO. 558 NORMAL SPECTRAL TRANSMITTANCE OF GOLD BLACK CONTACT COATINGS

[Wavelength, λ, μm; Transmittance, τ; Temperature, T, K]

λ	τ
CURVE 1	
T ~ 298	
3	0.038
7	0.028
10	0.029
12	0.030
15	0.032
CURVE 2	
T ~ 298	
3	0.034
7	0.032
10	0.032
12	0.034
15	0.048
CURVE 3	
T ~ 298	
3	0.046
7	0.036
10	0.036
12	0.039
15	0.043
CURVE 4	
T ~ 298	
3	0.04
7	0.06
10	0.085
12	0.109
15	0.140
CURVE 5	
T ~ 298	
3	0.102
7	0.080
10	0.085*
12	0.088
15	0.110
CURVE 6	
T ~ 298	
3	0.112
7	0.105
10	0.114
12	0.116
15	0.300
CURVE 7	
T ~ 298	
3	0.11 *
7	0.13
10	0.17
12	0.20
15	0.24
CURVE 8	
T ~ 298	
3	0.172
7	0.156
10	0.166
12	0.167
15	0.180
CURVE 9	
T ~ 298	
3	0.20
7	0.18
10	0.21
12	0.225
15	0.25
CURVE 10	
T ~ 298	
3	0.265
7	0.230
10	0.239
12	0.249
15	0.285
CURVE 11	
T ~ 298	
3	0.30
7	0.30
10	0.30
12	0.31
15	0.33
CURVE 12	
T ~ 298	
3	0.48
7	0.45
10	0.47
12	0.48
15	0.54
CURVE 13	
T ~ 298	
3	0.44
7	0.45*
10	0.47*
12	0.49
15	0.53
CURVE 14	
T ~ 298	
3	0.57
7	0.55
10	0.57
12	0.57
15	0.59
CURVE 15	
T ~ 298	
15	0.175
20	0.245
30	0.355
40	0.450
60	0.560
105	0.610
CURVE 16	
T ~ 298	
15	0.30*
20	0.35
30	0.44
40	0.51
60	0.62
105	0.67
CURVE 17	
T ~ 298	
15	0.300*
20	0.350*
30	0.417
40	0.480
60	0.540
105	0.570
CURVE 18	
T ~ 298	
15	0.38
20	0.41
30	0.47
40	0.50
60	0.53
105	0.55
CURVE 19	
T ~ 298	
105	0.26
345	0.25
455	0.22
CURVE 20	
T ~ 298	
105	0.41
345	0.41
455	0.37
CURVE 21	
T ~ 298	
105	0.73
345	0.69
455	0.67
CURVE 22	
T ~ 298	
0.4000	0.405
0.4086	0.408
0.4263	0.419
0.4560	0.445
0.4719	0.457
0.4850	0.463
0.5069	0.465
0.5438	0.462
0.5634	0.464
0.5953	0.472
0.6230	0.476
0.6606	0.476
0.6791	0.472
0.7004	0.465
CURVE 23	
T ~ 298	
0.4000	0.057
0.4833	0.047
0.5576	0.045
0.6236	0.039
0.7005	0.037
CURVE 24	
T ~ 298	
0.4000	0.267
0.4203	0.282
0.4501	0.306
0.4708	0.318
0.4866	0.322
0.5255	0.318
0.5425	0.318
0.5797	0.323
0.6473	0.323
CURVE 24 (cont.)	
0.6745	0.320
0.6995	0.314
CURVE 25	
T ~ 298	
0.4000	0.056
0.4715	0.044
0.5768	0.035
0.6395	0.027
0.6995	0.025
CURVE 26	
T ~ 298	
0.4018	0.068
0.4199	0.073
0.4320	0.077
0.4465	0.082
0.4583	0.087*
0.4828	0.097*
0.4856	0.099*
0.4886	0.099*
0.4934	0.099
0.4982	0.098*
0.5091	0.096
0.5176	0.094*
0.5248	0.094
0.5417	0.097
0.5548	0.098*
0.5677	0.099
0.5811	0.100
0.6008	0.101
0.6512	0.101
0.6794	0.100*
0.6998	0.099
CURVE 27	
T ~ 298	
0.4018	0.019
0.4226	0.020
0.4426	0.022
0.4668	0.023
CURVE 27 (cont.)	
0.4921	0.024
0.5205	0.024
0.5676	0.023
0.6325	0.024*
0.6668	0.024
0.7013	0.023
CURVE 28	
T ~ 298	
0.4000	0.029
0.4131	0.030*
0.4257	0.032
0.4382	0.035*
0.4481	0.038*
0.4650	0.043
0.4836	0.047*
0.4930	0.049*
0.4990	0.049*
0.5145	0.046
0.5655	0.047
0.6070	0.048
0.6419	0.048
0.6708	0.047
0.6995	0.046
CURVE 29	
T ~ 298	
0.4000	0.008
0.4436	0.011
0.4672	0.013
0.4919	0.013
0.6331	0.012
0.6669	0.010
0.6839	0.010*
0.7000	0.009

* Not shown on plot

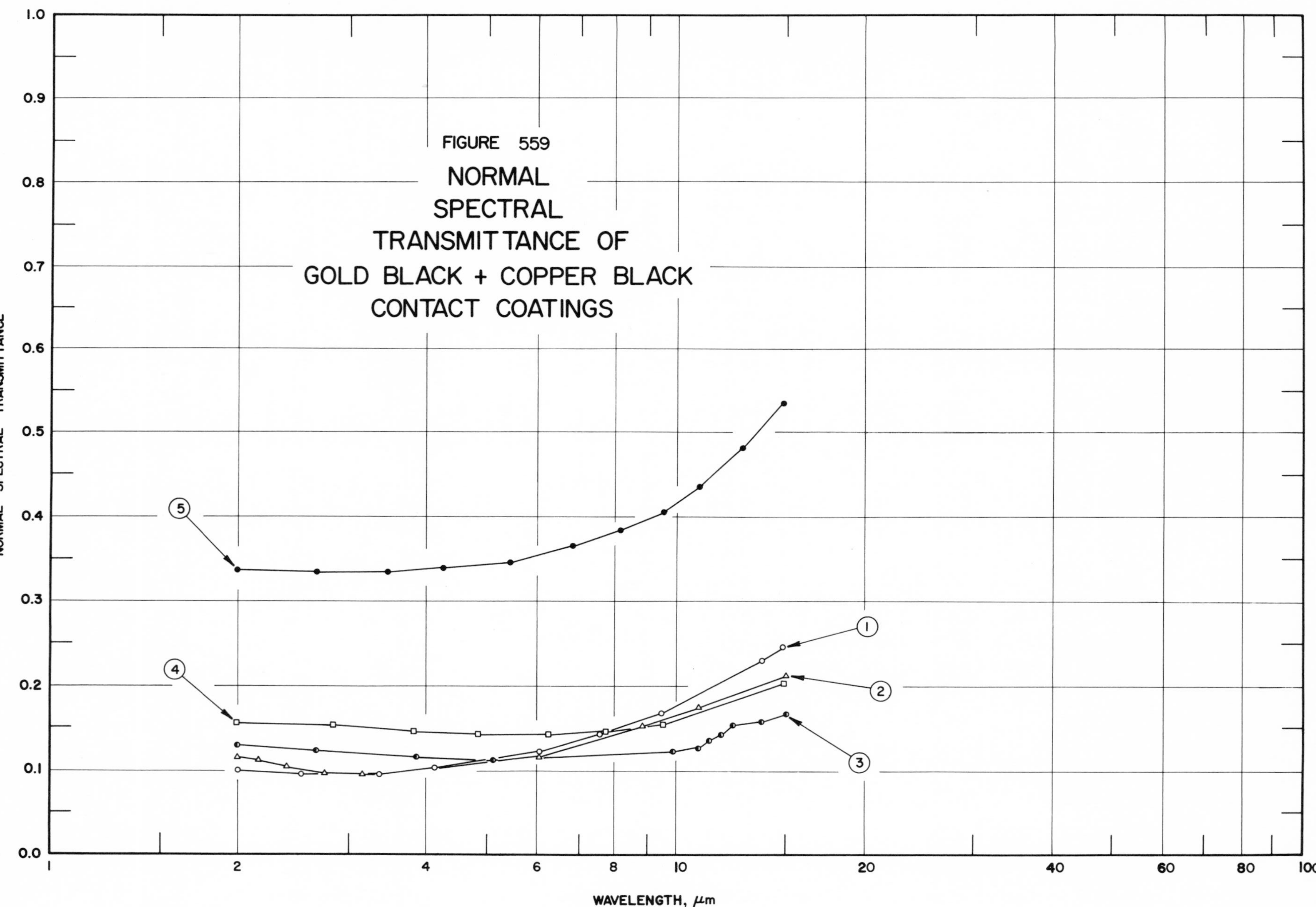
FIGURE 559
NORMAL
SPECTRAL
TRANSMITTANCE OF
GOLD BLACK + COPPER BLACK
CONTACT COATINGS
NORMAL SPECTRAL TRANSMITTANCE
WAVELENGTH, μm
1.0
0.9
0.8
0.7
0.6
0.5
0.4
0.3
0.2
0.1
0.0
1
2
4
6
8
10
20
40
60
80
100
1
2
3
4
5

SPECIFICATION TABLE NO. 559 NORMAL SPECTRAL TRANSMITTANCE OF GOLD BLACK + COPPER BLACK CONTACT COATINGS

Curve No.	Ref. No.	Year	Temperature, K	Wavelength Range, μm	Geometry θ	θ'	ω'	Reported Error, %	Composition (weight percent), Specifications, and Remarks
1	113	1967	~298	2.00-14.99	0°	0°			97.5 gold black and 2.5 copper black; NaCl substrate; evaporation deposited; data extracted from smooth curve; property measured immediately after preparation; property measured with a prism spectrometer; property corrected for NaCl substrate. [Author's designation: Sample 310 B]
2	113	1967	~298	2.00-15.00	0°	0°			Above specimen and conditions except property measured after 24 hrs at 343 K.
3	113	1967	~298	2.00-15.00	0°	0°			Above specimen and conditions except property measured after 3 hrs at 472 K.
4	113	1967	~298	2.00-14.99	0°	0°			Above specimen and conditions except property measured after 1 wk at 423 K.
5	113	1967	~298	2.00-14.99	0°	0°			Above specimen and conditions except property measured after 1 wk at 453 K.

DATA TABLE NO. 559 NORMAL SPECTRAL TRANSMITTANCE OF GOLD BLACK + COPPER BLACK CONTACT COATINGS

[Wavelength, λ, μm; Transmittance, τ; Temperature, T, K]

λ	τ	λ	τ	λ	τ	λ	τ	λ	τ
CURVE 1 T ~ 298		CURVE 2 T ~ 298		CURVE 3 T ~ 298		CURVE 4 T ~ 298		CURVE 5 T ~ 298	
2.00	0.100	2.00	0.117	2.00	0.130	2.00	0.157	2.00	0.338
2.53	0.096	2.16	0.112	2.68	0.123	2.84	0.152	2.69	0.336
3.37	0.096	2.40	0.105	3.83	0.117	3.81	0.146	3.48	0.336
4.14	0.101	2.75	0.099	5.10	0.114	4.84	0.143	4.27	0.340
6.02	0.121	3.17	0.096	9.82	0.122	6.25	0.143	5.47	0.349
7.59	0.142	3.37	0.096*	10.80	0.128	7.78	0.148	6.88	0.366
9.42	0.169	4.14	0.101*	11.32	0.135	9.59	0.155	8.17	0.385
13.57	0.230	6.06	0.118	11.72	0.143	14.99	0.204	9.56	0.409
14.99	0.248	8.87	0.151	12.36	0.151			10.95	0.438
		10.88	0.175	13.62	0.159			12.79	0.483
		15.00	0.212	15.00	0.169			14.99	0.536

* Not shown on plot

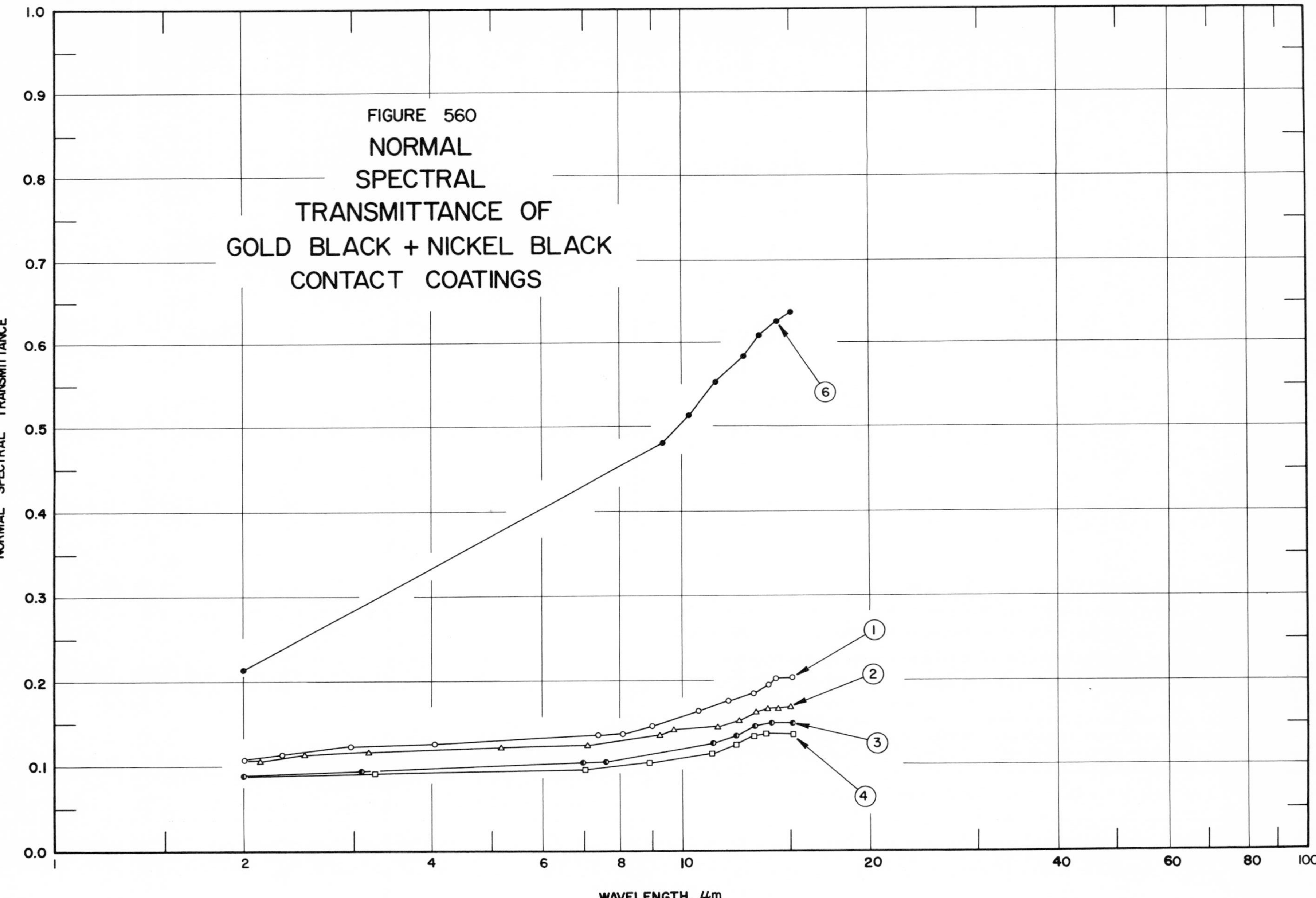

FIGURE 560
NORMAL SPECTRAL TRANSMITTANCE OF GOLD BLACK + NICKEL BLACK CONTACT COATINGS

SPECIFICATION TABLE NO. 560 NORMAL SPECTRAL TRANSMITTANCE OF GOLD BLACK + NICKEL BLACK CONTACT COATINGS

Curve No.	Ref. No.	Year	Temperature, K	Wavelength Range, μm	Geometry θ	θ'	ω'	Reported Error, %	Composition (weight percent), Specifications, and Remarks
1	113	1967	~298	2.00-15.00	0°	0°			95 gold black and 5 nickel black; NaCl substrate; evaporation deposited; data extracted from smooth curve; property measured immediately after preparation; property measured with a prism spectrometer and corrected for NaCl substrate. [Author's designation: Sample 376 B]
2	113	1967	~298	2.00-14.99	0°	0°			Above specimen and conditions except property measured after 24 hrs at 342 K.
3	113	1967	~298	2.00-15.00	0°	0°			Above specimen and conditions except property measured after 3 hrs at 462 K.
4	113	1967	~298	2.00-15.00	0°	0°			Above specimen and conditions except property measured after 1 wk at 423 K.
5*	113	1967	~298	2.00-14.99	0°	0°			Above specimen and conditions except property measured after 1 wk at 453 K.
6	113	1967	~298	2.00-15.00	0°	0°			Above specimen and conditions except property measured after 4 wks at 453 K.

DATA TABLE NO. 560 NORMAL SPECTRAL TRANSMITTANCE OF GOLD BLACK + NICKEL BLACK CONTACT COATINGS

[Wavelength, λ, μm; Transmittance, τ; Temperature, T, K]

λ	τ
CURVE 1 T ~ 298	
2.00	0.107
2.30	0.114
2.97	0.122
4.01	0.128
7.31	0.137
8.05	0.139
8.97	0.147
10.69	0.165
11.94	0.178
13.01	0.185
13.79	0.196
14.10	0.201
15.00	0.204
CURVE 2 T ~ 298	
2.00	0.106*
2.12	0.108
2.50	0.113
3.16	0.117
5.14	0.121
7.05	0.123
9.11	0.136
9.69	0.141
11.49	0.145
12.34	0.152
13.11	0.162
13.74	0.167
14.23	0.169
14.99	0.170
CURVE 3 T ~ 298	
2.00	0.090
3.09	0.095
6.95	0.103
7.52	0.104
11.20	0.126
12.06	0.135
13.01	0.146
13.82	0.150
15.00	0.150
CURVE 4 T ~ 298	
2.00	0.089*
3.23	0.092
7.00	0.096
CURVE 4 (cont.)	
8.89	0.104
11.15	0.116
12.26	0.125
13.00	0.134
13.58	0.137
15.00	0.137
CURVE 5* T ~ 298	
2.00	0.106
2.12	0.108
2.50	0.113
3.16	0.117
5.14	0.121
7.05	0.123
9.11	0.136
CURVE 5 (cont.)*	
9.69	0.141
11.49	0.145
12.34	0.152
13.11	0.162
13.74	0.167
14.23	0.169
14.99	0.170
CURVE 6 T ~ 298	
2.00	0.216
9.33	0.482
10.30	0.515
11.44	0.553
12.56	0.586
13.48	0.610
CURVE 6 (cont.)	
14.22	0.626
15.00	0.637

* Not shown on plot

SPECIFICATION TABLE NO. 561 NORMAL SPECTRAL REFLECTANCE OF PLATINUM BLACK CONTACT COATINGS

Curve No.	Ref. No.	Year	Temperature, K	Wavelength Range, μm	Geometry θ	Geometry θ'	Geometry ω'	Reported Error, %	Composition (weight percent), Specifications, and Remarks
1*	150	1911	~298	0.8-51	~0°		2π		Platinum black; substrate unknown; deposited for 3 min.
2*	150	1911	~298	0.8-51	~0°		2π		Platinum black; substrate unknown; deposited for 15 min.

DATA TABLE NO. 561 NORMAL SPECTRAL REFLECTANCE OF PLATINUM BLACK CONTACT COATINGS

[Wavelength, λ, μm; Reflectance, ρ; Temperature, T, K]

λ	ρ
CURVE 1*	
T = 298	
0.8	0.013
8.7	0.057
25.5	0.0708
51	0.068
51	0.074
CURVE 2*	
T = 298	
0.8	0.0017
8.7	0.0059
25.5	0.0093
51	0.0079
51	0.011

* No plot given

SPECIFICATION TABLE NO. 562 NORMAL SPECTRAL TRANSMITTANCE OF SILVER BLACK CONTACT COATINGS

Curve No.	Ref. No.	Year	Temperature, K	Wavelength Range, μm	Geometry θ	θ'	ω'	Reported Error, %	Composition (weight percent), Specifications, and Remarks
1*	113	1967	~298	2.20-11.82	0°	0°			Silver black; evaporation deposited (fast rate) in nitrogen at 2 mm Hg; data extracted from smooth curve; property measured with a prism spectrometer.

DATA TABLE NO. 562 NORMAL SPECTRAL TRANSMITTANCE OF SILVER BLACK CONTACT COATINGS

[Wavelength, λ, μm; Transmittance, τ; Temperature, T, K]

λ	τ
CURVE 1*	
T ~ 298	
2.20	0.267
2.75	0.260
3.31	0.256
3.81	0.256
5.60	0.281
8.25	0.309
8.93	0.318
11.82	0.371

* No plot given

SPECIFICATION TABLE NO. 563 NORMAL SPECTRAL REFLECTANCE OF ZINC BLACK CONTACT COATINGS

Curve No.	Ref. No.	Year	Temperature, K	Wavelength Range, μm	Geometry θ	θ'	ω'	Reported Error, %	Composition (weight percent), Specifications, and Remarks
1*	149	1940	298	52.0-152	~0°	~0°			Zinc black; metastyrene substrate. [Authors' designation: Sample No. 15]
2*	149	1940	298	20.7-152	~0°	~0°			Zinc black; brass substrate. [Authors' designation: Sample No. 32]

DATA TABLE NO. 563 NORMAL SPECTRAL REFLECTANCE OF ZINC BLACK CONTACT COATINGS

[Wavelength, λ, μm; Reflectance, ρ; Temperature, T, K]

λ	ρ
CURVE 1* T = 298	
52.0	0.06
63.0	0.07
83.0	0.07
94.0	0.10
117	0.10
152	0.14
CURVE 2* T = 298	
20.7	0.15
29.4	0.51
41.0	0.68
52.0	0.72
63.0	0.75
83.0	0.83
94.0	0.88
117	0.80
152	0.83

* No plot given

SPECIFICATION TABLE NO. 564 NORMAL SPECTRAL TRANSMITTANCE OF ZINC BLACK CONTACT COATINGS

Curve No.	Ref. No.	Year	Temperature, K	Wavelength Range, μm	Geometry θ	θ'	ω'	Reported Error, %	Composition (weight percent), Specifications, and Remarks
1*	149	1940	298	20.7-152	$\sim 0°$	$\sim 0°$			Zinc black; pliofilm substrate. [Authors' designation: Sample No. 31]
2*	149	1940	298	20.7-152	$\sim 0°$	$\sim 0°$			Zinc black; pyroxylin substrate. [Authors' designation: Sample No. 14]

DATA TABLE NO. 564 NORMAL SPECTRAL TRANSMITTANCE OF ZINC BLACK CONTACT COATINGS

[Wavelength, λ, μm; Transmittance, τ; Temperature, T, K]

λ	τ
CURVE 1* T = 298	
20.7	0.11
29.4	0.34
41.0	0.49
52.0	0.52
63.0	0.57
83.0	0.61
94.0	0.60
117	0.63
152	0.61
CURVE 2* T = 298	
20.7	0.08
29.4	0.22
52.0	0.46
63.0	0.51
83.0	0.59
94.0	0.58
CURVE 2 (cont.)*	
117	0.68
152	0.61

* No plot given

3. CONVERSION COATINGS

A. Anodized Coatings

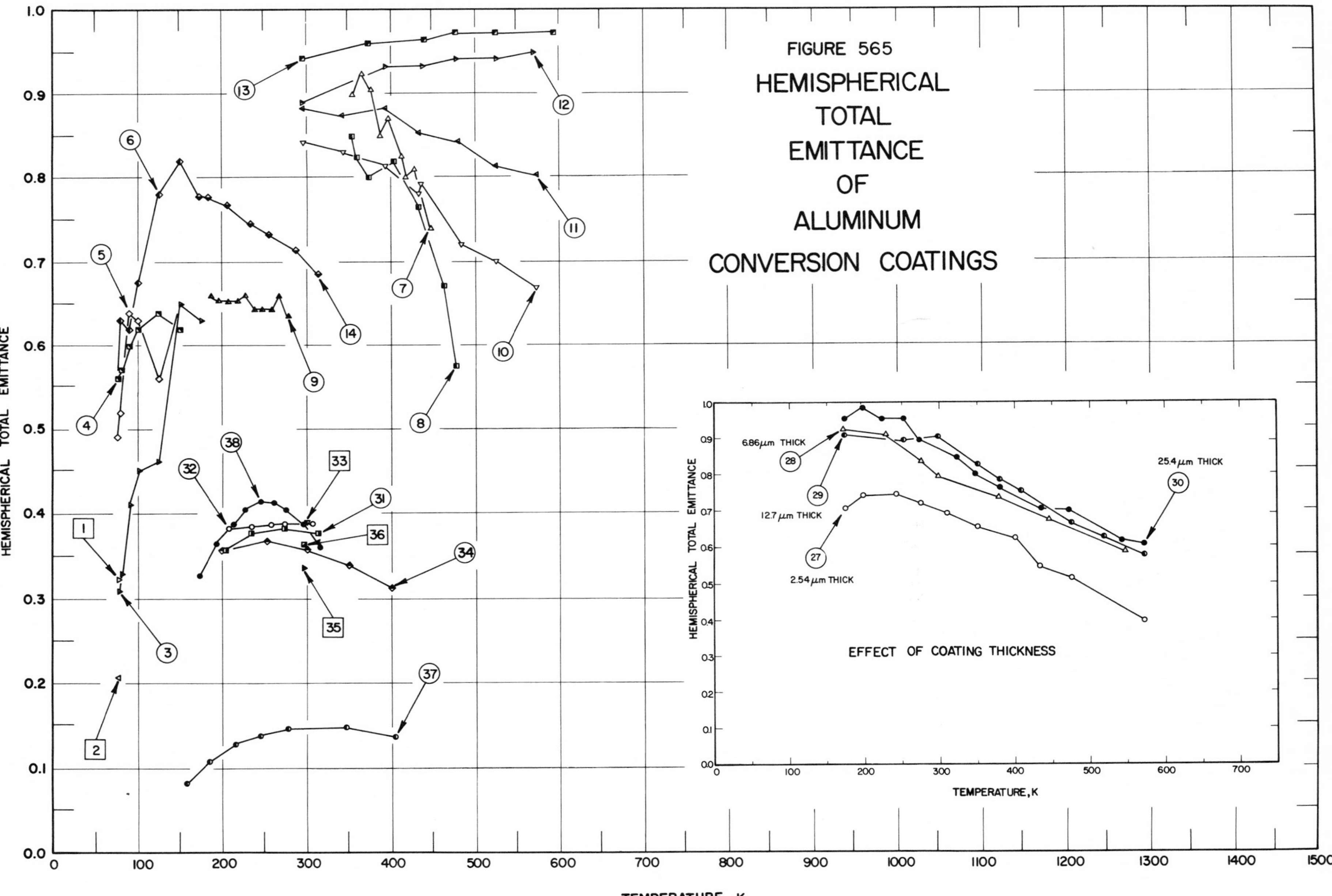
FIGURE 565
HEMISPHERICAL
TOTAL
EMITTANCE
OF
ALUMINUM
CONVERSION COATINGS
HEMISPHERICAL TOTAL EMITTANCE
TEMPERATURE, K
1.0
0.9
0.8
0.7
0.6
0.5
0.4
0.3
0.2
0.1
0.0
0
100
200
300
400
500
600
700
800
900
1000
1100
1200
1300
1400
1500
13
12
6
11
5
7
14
10
9
4
8
38
33
32
31
1
36
34
35
3
37
2
EFFECT OF COATING THICKNESS
HEMISPHERICAL TOTAL EMITTANCE
TEMPERATURE, K
6.86 μm THICK
28
29
12.7 μm THICK
27
2.54 μm THICK
25.4 μm THICK
30

SPECIFICATION TABLE NO. 565 HEMISPHERICAL TOTAL EMITTANCE OF ALUMINUM CONVERSION COATINGS

Curve No.	Ref. No.	Year	Temperature Range, K	Reported Error, %	Composition (weight percent), Specifications and Remarks
1	195	1966	76		Aluminum (0.51 mm thick); electrolytically polished and heavily anodized; measured in vacuum ($< 10^{-6}$ mm Hg); authors assumed $\alpha = \epsilon$.
2	195	1966	76		Aluminum (0.51 mm thick); electrolytically polished and lightly anodized; measured in vacuum ($< 10^{-6}$ mm Hg); authors assumed $\alpha = \epsilon$.
3	264	1963	77-175		Anodized Al (2 μm thick); sulfuric acid anodized; measured in vacuum (5×10^{-5} mm Hg).
4	264	1963	77-150	±10	Similar to above specimen and conditions except anodized layer 9 μm thick.
5	264	1963	77-150	±10	Similar to above specimen and conditions except anodized layer 18 μm thick.
6	264	1963	77-150	±10	Similar to above specimen and conditions except anodized layer 28 μm thick.
7	43	1962	353-446	±3	Aluminum; sulfuric acid anodized; measured in vacuum (10^{-3} mm Hg); data extracted from smooth curve.
8	43	1962	353-475	±3	Above specimen and conditions except vapor blasted after anodization.
9	48	1963	188-278	±3	Aluminum (5.1 mm thick); hand polished; sulfuric acid anodized; measured in vacuum (10^{-3} mm Hg).
10	137	1965	296-574	±5	Aluminum; anodized; measured in vacuum (10^{-7} mm Hg).
11	137	1965	295-574	±5	Similar to above specimen and conditions except surface appeared duller.
12	137	1965	294-570	±5	Aluminum; anodized; layer of carbon electrophoretically deposited into the pores of the surface; measured in vacuum (10^{-7} mm Hg).
13	137	1965	295-595	±5	Similar to above specimen and conditions.
14	107	1967	172-314	~8.5	High purity Alclad on Al substrate (Alzak SI grade lighting sheet from Alcoa); thin anodized to 4.8 μm thick; measured in vacuum ($\sim 10^{-6}$ mm Hg).
15*	265	1963	338		1199 aluminum anodized for 10 min in sulfuric acid; coating thickness 8.4 μm; polished in phosphoric/nitric acid bath for 2 min at 364 K.
16*	265	1963	338		Similar to above specimen and conditions except anodized 15 min; thickness 9.5 μm.
17*	265	1963	338		Similar to above specimen and conditions except anodized 20 min; thickness 11.0 μm.
18*	265	1963	338		Similar to above specimen and conditions except anodized 25 min; thickness 13.2 μm.
19*	265	1963	339		1199 aluminum anodized for 15 min in sulfuric acid; coating thickness 9.5 μm; polished by Alzak process (electrolytic fluoboric acid bath).
20*	265	1963	339		Similar to above specimen and conditions except heat treated 24 hrs at 589 K in vacuum (5×10^{-5} mm Hg).
21*	265	1963	339		Similar to above specimen and conditions except heat treated 48 hrs at 589 K in vacuum (5×10^{-5} mm Hg).
22*	265	1963	339		Similar to above specimen and conditions except heat treated 96 hrs at 589 K in vacuum (5×10^{-5} mm Hg).
23*	265	1963	339		Similar to above specimen and conditions except anodized 25 min; coating thickness 3.2 μm.

* Not shown on plot

SPECIFICATION TABLE NO. 565 HEMISPHERICAL TOTAL EMITTANCE OF ALUMINUM CONVERSION COATINGS (continued)

Curve No.	Ref. No.	Year	Temperature Range, K	Reported Error, %	Composition (weight percent), Specifications and Remarks
24*	265	1963	339		Similar to above specimen and conditions except heat treated 24 hrs at 589 K in vacuum (5×10^{-5} mm Hg).
25*	265	1963	339		Similar to above specimen and conditions except heat treated 48 hrs at 589 K in vacuum (5×10^{-5} mm Hg).
26*	265	1963	339		Similar to above specimen and conditions except heat treated 96 hrs at 589 K in vacuum (5×10^{-5} mm Hg).
27	265	1963	175-573		1199 aluminum anodized in sulfuric acid; polished substrate; coating thickness 2.54 μm.
28	265	1963	173-546		Similar to above specimen and conditions except coating thickness 6.86 μm.
29	265	1963	174-572		Similar to above specimen and conditions except coating thickness 12.7 μm.
30	265	1963	174-573		Similar to above specimen and conditions except coating thickness 25.4 μm.
31	111	1969	204-312		Anodized 1199 aluminum alloy; electropolished in fluoboric acid solution, then anodized in ammonium tartrate solution; property measured using steady state calorimetric method. [Authors' designation: Anodic]
32	111	1969	208-308		Above specimen and conditions except property measured by transient calorimetric method; property of second side of sample assumed.
33	111	1969	300		Above specimen and conditions except property converted from $\rho(10^\circ, 10^\circ)$ measured by specular method relative to a front surface aluminized mirror.
34	111	1969	200-400		Above specimen and conditions except property converted from $\rho(2\pi, 15^\circ - 75^\circ)$ measured by heated cavity method relative to a platinum surface.
35	111	1969	295		Above specimen and conditions except property measured by a portable Quick Emittance Device.
36	111	1969	295		Above specimen and conditions except property measured by a portable emissometer.
37	85	1969	157-401		Anodized 1199 aluminum; substrate alkaline electropolished (sodium phosphate and sodium carbonate) for 15 min at 353 K and 12 VDC, then anodized for 15 min at 18 VDC in 10% sulfuric acid; property measured in vacuum by calorimetric method; data extracted from smooth curve. [Author's designation: B-3 Sulfuric Acid Anodize, NRDL-RTD-81-7]
38	85	1969	171-315		Anodized 1199 aluminum; substrate ~0.254 mm thick; barrier anodized; property measured in vacuum by calorimetric method; data extracted from smooth curve. [Author's designation: A-2 Barrier Anodize]

* Not shown on plot

DATA TABLE NO. 565 HEMISPHERICAL TOTAL EMITTANCE OF ALUMINUM CONVERSION COATINGS

[Temperature, T, K; Emittance, ϵ]

T	ϵ
CURVE 1	
76	0.324
CURVE 2	
76	0.2082
CURVE 3	
77	0.31
80	0.33
90	0.41
100	0.45
125	0.46
150	0.65
175	0.63
CURVE 4	
77	0.56
80	0.57
90	0.60
100	0.62
125	0.64
150	0.62
CURVE 5	
77	0.49
80	0.52
90	0.64
100	0.63
125	0.56
150	0.65*
CURVE 6	
77	0.56*
80	0.63
90	0.62
100	0.675
125	0.780
150	0.820
CURVE 7	
353	0.900
CURVE 7 (cont.)	
367	0.925
377	0.905
389	0.850
397	0.870
413	0.825
418	0.800
429	0.810
446	0.740
CURVE 8	
353	0.85
360	0.825
373	0.800
403	0.820
431	0.765
461	0.670
476	0.575
CURVE 9	
188	0.660
198	0.652
208	0.652
218	0.652
228	0.660
238	0.644
248	0.644
258	0.644
268	0.660
278	0.636
CURVE 10	
296	0.842
343	0.830
394	0.812
432	0.780
435	0.791
484	0.720
526	0.700
574	0.669
CURVE 11	
295	0.882
CURVE 11 (cont.)	
341	0.873
394	0.882
433	0.852
480	0.842
525	0.812
574	0.802
CURVE 12	
294	0.890
393	0.932
439	0.932
478	0.941
525	0.941
570	0.950
CURVE 13	
295	0.941
372	0.960
440	0.962
478	0.971
525	0.971
595	0.971
CURVE 14	
172	0.778
183	0.778
206	0.769
234	0.745
255	0.731
286	0.714
314	0.686
CURVE 15*	
338	0.69
CURVE 16*	
338	0.71
CURVE 17*	
338	0.73
CURVE 18*	
338	0.75
CURVE 19*	
339	0.69
CURVE 20*	
339	0.70
CURVE 21*	
339	0.69
CURVE 22*	
339	0.69
CURVE 23*	
339	0.72
CURVE 24*	
339	0.74
CURVE 25*	
339	0.74
CURVE 26*	
339	0.74
CURVE 27	
175	0.709
198	0.744
241	0.744
275	0.720
310	0.694
350	0.655
400	0.622
434	0.547
475	0.514
573	0.395
CURVE 28	
173	0.928
226	0.910
275	0.835
299	0.794
379	0.736
445	0.678
546	0.586
CURVE 29	
174	0.910
251	0.892
299	0.903
350	0.826
380	0.785
409	0.753
475	0.661
519	0.625
572	0.576
CURVE 30	
174	0.953
197	0.982
224	0.953
251	0.953
274	0.895
324	0.847
348	0.800
380	0.761
435	0.703
474	0.700
544	0.615
573	0.602
CURVE 31	
204	0.359
233	0.379
274	0.381
312	0.377
CURVE 32	
208	0.382
233	0.384
CURVE 32 (cont.)	
258	0.387
283	0.387
308	0.389
CURVE 33	
300	0.390
CURVE 34	
200	0.356
251	0.368
300	0.358
350	0.340
400	0.312
CURVE 35	
295	0.338
CURVE 36	
295	0.362
CURVE 37	
157	0.082
183	0.107
214	0.126
241	0.137
277	0.147
347	0.147
401	0.137
CURVE 38	
171	0.327
193	0.363
212	0.388
229	0.404
244	0.412
260	0.412
275	0.404
295	0.387
315	0.360

* Not shown on plot

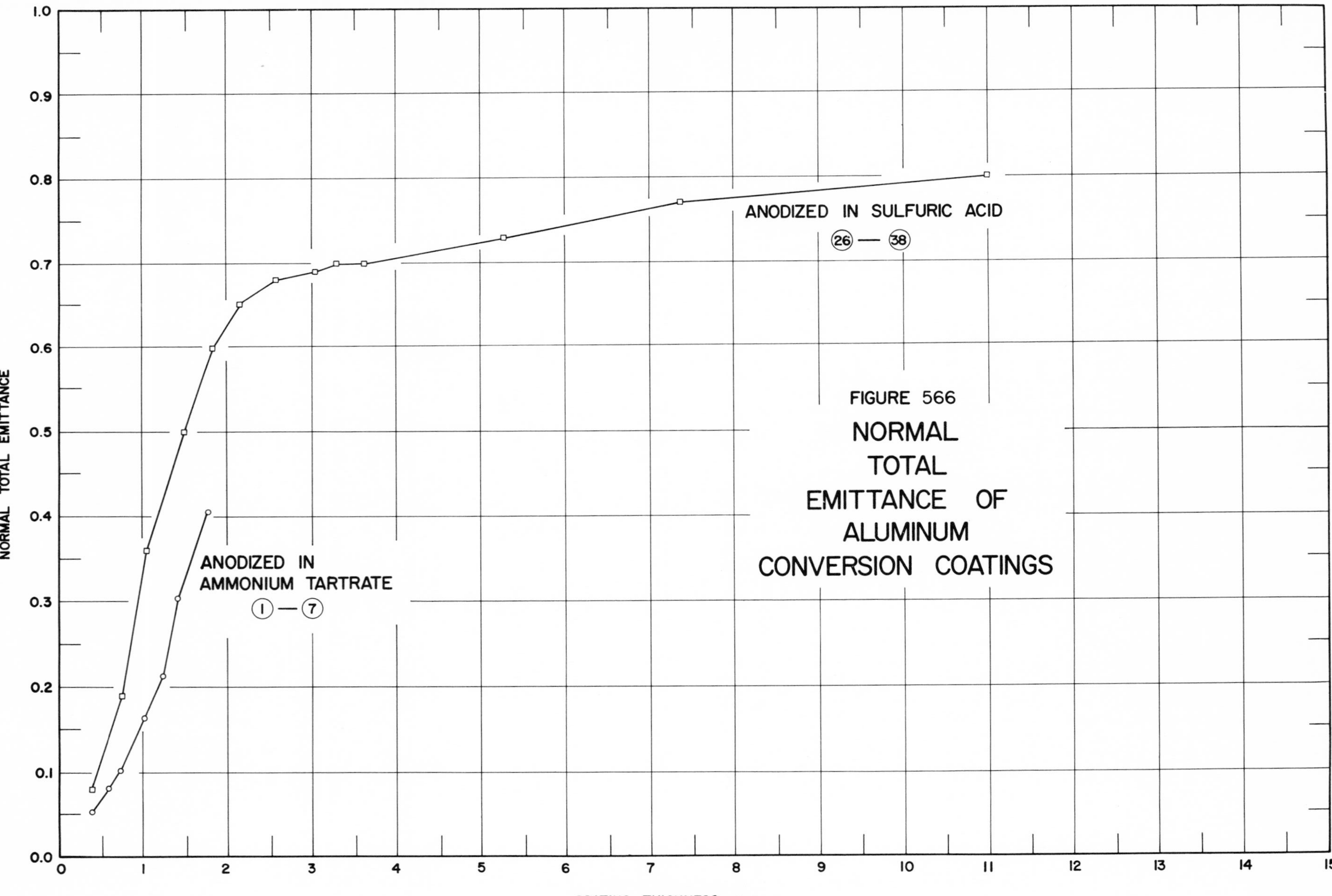

FIGURE 566 NORMAL TOTAL EMITTANCE OF ALUMINUM CONVERSION COATINGS

SPECIFICATION TABLE NO. 566 NORMAL TOTAL EMITTANCE OF ALUMINUM CONVERSION COATINGS

Curve No.	Ref. No.	Year	Temperature Range, K	Geometry θ'	Reported Error, %	Composition (weight percent), Specifications and Remarks
1	260	1967	~298	~0°		Barrier-layer anodic coating on a high-purity aluminum substrate; coating thickness 0.40 μm; coating applied as follows: substrate thoroughly cleaned, electropolished in a fluoboric acid solution, anodized in aqueous ammonium tartrate solution at 300 volts, then immersed in a phosphoric acid solution; measured in vacuum (~10^{-6} mm Hg).
2	260	1967	~298	~0°		Similar to above specimen and conditions except coating thickness 0.60 μm; anodized in aqueous ammonium tartrate solution diluted with ethyl alcohol; anodization voltage 458 volts.
3	260	1967	~298	~0°		Similar to above specimen and conditions except coating thickness 0.74 μm; anodization voltage 563 volts.
4	260	1967	~298	~0°		Similar to above specimen and conditions except coating thickness 7.07 μm; anodization voltage 774 volts.
5	260	1967	~298	~0°		Similar to above specimen and conditions except coating thickness 1.24 μm; anodization voltage 948 volts.
6	260	1967	~298	~0°		Similar to above specimen and conditions except coating thickness 7.47 μm; anodization voltage 1076 volts.
7	260	1967	~298	~0°		Similar to above specimen and conditions except coating thickness 1.77 μm; anodization voltage 1357 volts.
8*	260	1967	~298	~0°		Similar to curve 1 specimen and conditions except coating thickness not specified; ultraviolet irradiated in vacuum for 1300 equivalent space sun hrs; (for all specimens, the relationship between coating thickness and anodization voltage is, on the avg, 13.4 Å per volt).
9*	260	1967	~298	~0°		Similar to above specimen and conditions.
10*	260	1967	~298	~0°		Similar to above specimen and conditions.
11*	260	1967	~298	~0°		Similar to above specimen and conditions.
12*	260	1967	~298	~0°		Similar to above specimen and conditions.
13*	260	1967	~298	~0°		Similar to above specimen and conditions except ultraviolet irradiated in vacuum for 1600 equivalent space sun hrs.
14*	260	1967	~298	~0°		Similar to above specimen and conditions except ultraviolet irradiated in vacuum for 1100 equivalent space sun hrs.
15*	260	1967	~298	~0°		Similar to above specimen and conditions.
16*	260	1967	~298	~0°		Similar to above specimen and conditions except ultraviolet irradiated in vacuum for 21 000 equivalent space sun hrs.
17*	260	1967	~298	~0°		Similar to above specimen and conditions.
18*	260	1967	~298	~0°		Similar to above specimen and conditions.
19*	260	1967	~298	~0°		Similar to above specimen and conditions.
20*	260	1967	~298	~0°		Similar to above specimen and conditions.
21*	260	1967	~298	~0°		Similar to above specimen and conditions.

* Not shown on plot

SPECIFICATION TABLE NO. 566 NORMAL TOTAL EMITTANCE OF ALUMINUM CONVERSION COATINGS (continued)

Curve No.	Ref. No.	Year	Temperature Range, K	Geometry θ'	Reported Error, %	Composition (weight percent), Specifications and Remarks
22*	260	1967	~298	~0°		Similar to above specimen and conditions except anodized at 1250 volts in aqueous ammonium tartrate solution highly diluted with ethyl alcohol; ultraviolet irradiated in vacuum for 1300 equivalent sun hrs.
23*	260	1967	~298	~0°		Similar to above specimen and conditions.
24*	260	1967	~298	~0°		Similar to above specimen and conditions.
25*	260	1967	~298	~0°		Similar to above specimen and conditions.
26	261	1939	311	~0°		Foil; oxidized electrolytically in 15% sulfuric acid at 294 K; oxide film 0.406 μm thick; measured relative to aluminum.
27	261	1939	311	~0°		Similar to above specimen and conditions except oxide film 0.762 μm thick.
28	261	1939	311	~0°		Similar to above specimen and conditions except oxide film 1.04 μm thick.
29	261	1939	311	~0°		Similar to above specimen and conditions except oxide film 1.50 μm thick.
30	261	1939	311	~0°		Similar to above specimen and conditions except oxide film 1.83 μm thick.
31	261	1939	311	~0°		Similar to above specimen and conditions except oxide film 2.16 μm thick.
32	261	1939	311	~0°		Similar to above specimen and conditions except oxide film 2.59 μm thick.
33	261	1939	311	~0°		Similar to above specimen and conditions except oxide film 3.05 μm thick.
34	261	1939	311	~0°		Similar to above specimen and conditions except oxide film 3.30 μm thick.
35*	261	1939	311	~0°		Similar to above specimen and conditions except oxide film 3.61 μm thick.
36	261	1939	311	~0°		Similar to above specimen and conditions except oxide film 5.28 μm thick.
37	261	1939	311	~0°		Similar to above specimen and conditions except oxide film 7.37 μm thick.
38	261	1939	311	~0°		Similar to above specimen and conditions except oxide film 11.0 μm thick.
39*	262	1967	394-413	0°	2	Pure; anodized. [Author's designation: Specimen 384]
40*	262	1967	411-750	0°	2	Above specimen and conditions.
41*	260	1967	~298	~0°		Phosphoric/nitric porous anodized aluminum; measured in vacuum (~10^{-6} mm Hg); ultraviolet irradiated in vacuum for 1000 equivalent sun hrs.
42*	260	1967	~298	~0°		Similar to above specimen and conditions.
43*	263	1966	400	~0°		Aluminum; electrolytically oxidized to a thickness of 0.25 μm.
44*	263	1966	400	~0°		Similar to above specimen and conditions except 0.35 μm thick.
45*	263	1966	400	~0°		Similar to above specimen and conditions except 0.5 μm thick.
46*	263	1966	400	~0°		Similar to above specimen and conditions except 0.65 μm thick.
47*	263	1966	400	~0°		Similar to above specimen and conditions except 1.0 μm thick.
48*	263	1966	400	~0°		Similar to above specimen and conditions except 3.0 μm thick.
49*	263	1966	400	~0°		Similar to above specimen and conditions except 5.0 μm thick.
50*	263	1966	400	~0°		Similar to above specimen and conditions except 7.0 μm thick.

* Not shown on plot

DATA TABLE NO. 566 NORMAL TOTAL EMITTANCE OF ALUMINUM CONVERSION COATINGS

[Temperature, T, K; Emittance, ϵ]

Curve	T	ϵ
CURVE 1	298	0.054
CURVE 2	298	0.082
CURVE 3	298	0.102
CURVE 4	298	0.164
CURVE 5	298	0.214
CURVE 6	298	0.302
CURVE 7	298	0.405
CURVE 8*	298	0.05
CURVE 9*	298	0.05
CURVE 10*	298	0.05
CURVE 11*	298	0.05
CURVE 12*	298	0.05
CURVE 13*	298	0.05
CURVE 14*	298	0.05
CURVE 15*	298	0.05
CURVE 16*	298	0.05
CURVE 17*	298	0.05
CURVE 18*	298	0.05
CURVE 19*	298	0.05
CURVE 20*	298	0.05
CURVE 21*	298	0.05
CURVE 22*	298	0.33
CURVE 23*	298	0.32
CURVE 24*	298	0.31
CURVE 25*	298	0.31
CURVE 26	311	0.08
CURVE 27	311	0.19
CURVE 28	311	0.36
CURVE 29	311	0.50
CURVE 30	311	0.60
CURVE 31	311	0.65
CURVE 32	311	0.68
CURVE 33	311	0.69
CURVE 34	311	0.70
CURVE 35*	311	0.70
CURVE 36	311	0.73
CURVE 37	311	0.77
CURVE 38	311	0.80
CURVE 39*	394	0.826
	413	0.805
CURVE 40*	411	0.806
	503.3	0.700
	653	0.571
	750	0.506
CURVE 41*	298	0.15
CURVE 42*	298	0.23
CURVE 43*	400	0.11
CURVE 44*	400	0.12
CURVE 45*	400	0.20
CURVE 46*	400	0.27
CURVE 47*	400	0.41
CURVE 48*	400	0.63
CURVE 49*	400	0.60
CURVE 50*	400	0.62

* Not shown on plot

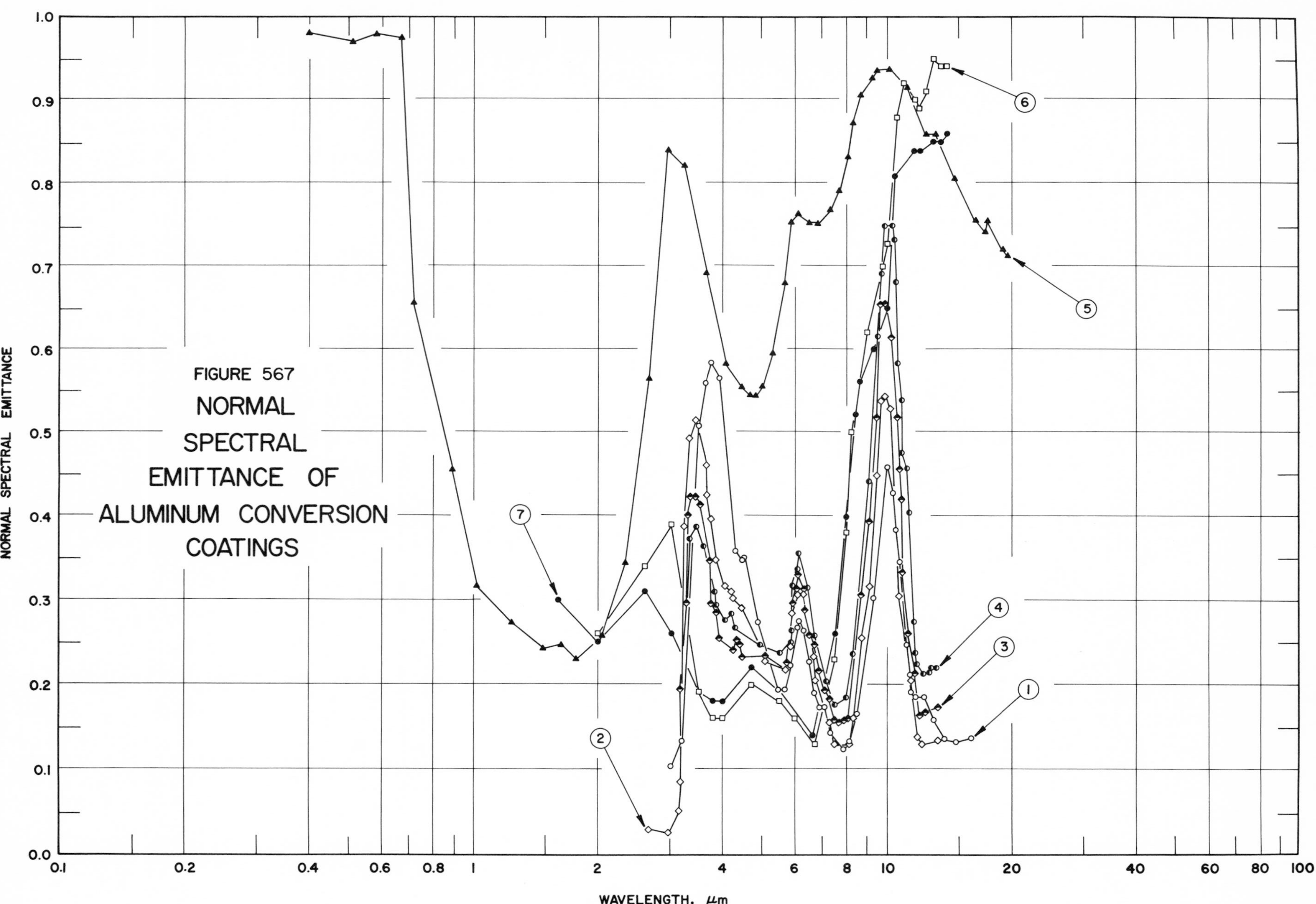

FIGURE 567 NORMAL SPECTRAL EMITTANCE OF ALUMINUM CONVERSION COATINGS

SPECIFICATION TABLE NO. 567 NORMAL SPECTRAL EMITTANCE OF ALUMINUM CONVERSION COATINGS

Curve No.	Ref. No.	Year	Temperature, K	Wavelength Range, μm	Geometry θ'	Reported Error, %	Composition (weight percent), Specifications, and Remarks
1	273	1964	295	3.0-16.0	~0°		Alodine 401-45; produced by reacting an aluminum surface with an aqueous solution of chromic, phosphoric, and hydrofluoric acid; substrate: Echo II laminate consisting of 8.89 μm Mylar film sandwiched between two layers of 4.57 μm 1080 aluminum alloy foil; data extracted from smooth curve; converted from R (2π, 0°).
2	273	1964	295	2.65-13.15	~0°		Similar to above specimen and conditions except reacted for 20 sec in a 1 normal sodium hydroxide solution at 298 K, then washed in water and dried in air.
3	273	1964	295	2.65-13.23	~0°		Similar to above specimen and conditions except reacted for 40 sec.
4	273	1964	295	2.65-13.14	~0°		Similar to above specimen and conditions except reacted for 60 sec.
5	74	1959	323	0.40-19.63	~0°		Black anodized aluminum; data extracted from smooth curve; property converted from R (2π, 0°). [Authors' designation: Sample No. 6]
6	274	1961	461	2.0-14.0	~0°	± 10	99.7 Al, 0.11 Fe, 0.11 Si, 0.01 Cu, 0.01 Mg, <0.01 Mn, Ni, and Zn; anodized for 30 min at 1 amp dm^{-2} in 4N analar sulphuric acid at 293 K and sealed for 30 min in boiling distilled water; thickness of anodic film 2.54 μm; heated at 456 K for 15 hrs; data extracted from smooth curve.
7	274	1961	575	1.6-14.0	~0°	± 10	Above specimen and conditions except heated at 575 K for 8 hrs before measurement.

DATA TABLE NO. 567 NORMAL SPECTRAL EMITTANCE OF ALUMINUM CONVERSION COATINGS

[Wavelength, λ, μm; Emittance, ∈; Temperature, T, K]

CURVE 1
T = 295

λ	∈
3.00	0.103
3.20	0.133
3.50	0.508
3.64	0.559
3.79	0.583
3.92	0.563
4.37	0.359
4.43	0.348
4.52	0.350
4.88	0.273
5.43	0.194
5.63	0.194
5.89	0.221
6.05	0.269
6.16	0.274
6.31	0.264
6.54	0.229
6.74	0.190
6.98	0.171
7.08	0.171
7.33	0.142
7.89	0.124
8.17	0.131
8.52	0.165
9.32	0.302
10.00	0.457
10.30	0.427
10.54	0.382
10.78	0.345
11.11	0.249
11.32	0.212
11.57	0.190
11.75	0.185
12.15	0.185
13.04	0.159
13.98	0.136
14.65	0.131
16.00	0.136

CURVE 2
T = 295

λ	∈
2.65	0.030
2.98	0.027
3.12	0.051

CURVE 2 (cont.)

λ	∈
3.18	0.086
3.22	0.388
3.36	0.492
3.47	0.513
3.68	0.460
3.68	0.425
3.78	0.399
3.87	0.348
4.05	0.317
4.20	0.310
4.21	0.301
4.47	0.290
5.10	0.229
5.72	0.219
5.89	0.245
5.97	0.283
6.08	0.308
6.23	0.308
6.61	0.232
6.70	0.205
7.28	0.154
7.48	0.130
8.17	0.130
8.35	0.160
8.74	0.255
9.04	0.319
9.41	0.449
9.66	0.536
9.94	0.542
10.16	0.529
10.77	0.305
11.02	0.244*
11.42	0.205
11.94	0.139
12.10	0.130
13.15	0.134

CURVE 3
T = 295

λ	∈
2.65	0.030*
2.98	0.027*
3.12	0.051*
3.18	0.086*
3.19	0.195
3.29	0.298

CURVE 3 (cont.)

λ	∈
3.30	0.401
3.36	0.422
3.46	0.422
3.53	0.415
3.72	0.347
3.78	0.296
3.87	0.285
3.94	0.255
4.25	0.240
4.36	0.251
4.40	0.249
4.42	0.232
5.08	0.232
5.63	0.219*
5.78	0.226
5.90	0.249*
5.98	0.298
6.02	0.315
6.13	0.330
6.31	0.312
6.36	0.289
6.45	0.284*
6.59	0.258
6.71	0.247
6.86	0.217
7.08	0.191
7.26	0.181
7.42	0.158
7.73	0.154
7.82	0.159
8.08	0.160
8.72	0.307
9.12	0.394
9.43	0.518
9.77	0.655
9.95	0.655
10.32	0.612
10.64	0.516
10.74	0.453
10.82	0.420
10.82	0.393*
10.93	0.332
11.28	0.260
11.61	0.214
12.06	0.163
12.44	0.167
13.23	0.171

CURVE 4
T = 295

λ	∈
2.65	0.030*
2.98	0.027*
3.12	0.051*
3.18	0.086*
3.19	0.195*
3.29	0.298*
3.32	0.371
3.39	0.375*
3.47	0.388
3.60	0.362
3.85	0.310
3.88	0.293
4.06	0.277
4.20	0.282
4.30	0.266
4.99	0.249
5.58	0.239
5.84	0.250
5.90	0.264
5.98	0.319
6.04	0.336
6.11	0.355
6.40	0.315
6.68	0.258
7.19	0.203
7.47	0.175
8.00	0.183
8.23	0.236
9.03	0.440
9.50	0.615
9.74	0.691
9.91	0.750
10.06	0.750
10.24	0.734
10.41	0.681
10.64	0.581
10.81	0.539
10.91	0.473
11.02	0.456
11.04	0.404
11.59	0.273
11.78	0.238
11.94	0.223
12.17	0.212
12.61	0.213

CURVE 4 (cont.)

λ	∈
12.76	0.220
13.14	0.220

CURVE 5
T = 323

λ	∈
0.40	0.980
0.51	0.970
0.58	0.980
0.67	0.975
0.72	0.659
0.89	0.456
1.05	0.318
1.25	0.271
1.49	0.242
1.62	0.246
1.78	0.230
2.06	0.257
2.33	0.343
2.68	0.563
2.96	0.841
3.26	0.821
3.66	0.694
4.12	0.582
4.47	0.552
4.66	0.544
4.80	0.544
5.00	0.553
5.34	0.598
5.78	0.681
5.98	0.756
6.07	0.767
6.50	0.753
6,81	0.753
7.39	0.770
7.78	0.792
8.08	0.832
8.34	0.874
8.75	0.906
9.22	0.928
9.58	0.935
10.11	0.936
11.14	0.913
12.54	0.860
13.16	0.860
14.61	0.809

CURVE 5 (cont.)

λ	∈
16.36	0.759
17.25	0.743
17.50	0.758
19.01	0.724
19.63	0.714

CURVE 6
T = 461

λ	∈
2.0	0.26
2.6	0.34
3.0	0.39
3.5	0.19
3.8	0.16
4.0	0.16
4.7	0.20
5.5	0.18
6.0	0.16
6.7	0.13
7.5	0.23
8.0	0.38
8.2	0.50
9.0	0.62
9.8	0.70
10.0	0.73
10.6	0.88
11.0	0.92
11.7	0.90
12.0	0.89
12.5	0.91
13.0	0.95
13.5	0.94
14.0	0.94

CURVE 7
T = 575

λ	∈
1.6	0.30
2.0	0.25
2.6	0.31
3.0	0.26
3.5	0.19*
3.8	0.18
4.0	0.18
4.7	0.22
5.5	0.19*

CURVE 7 (cont.)

λ	∈
6.0	0.16*
6.6	0.14
7.5	0.26
8.0	0.40
8.4	0.52
8.6	0.56
9.3	0.60
10.0	0.65
10.5	0.81
11.6	0.84
12.0	0.84
13.0	0.85
13.5	0.85
14.0	0.86

* Not shown on plot

FIGURE 568A

ANALYZED NORMAL SPECTRAL REFLECTANCE OF ALUMINUM CONVERSION COATINGS

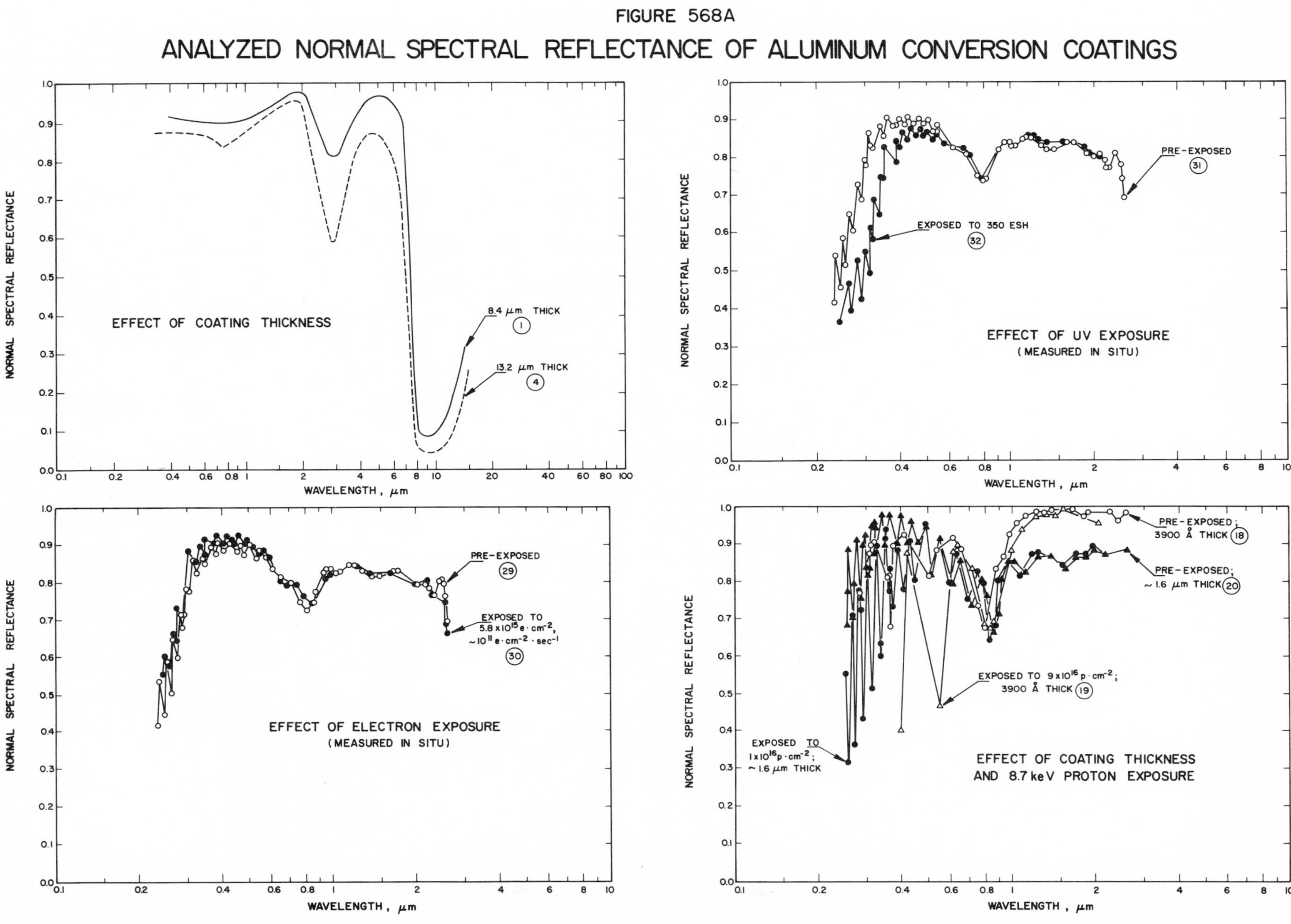

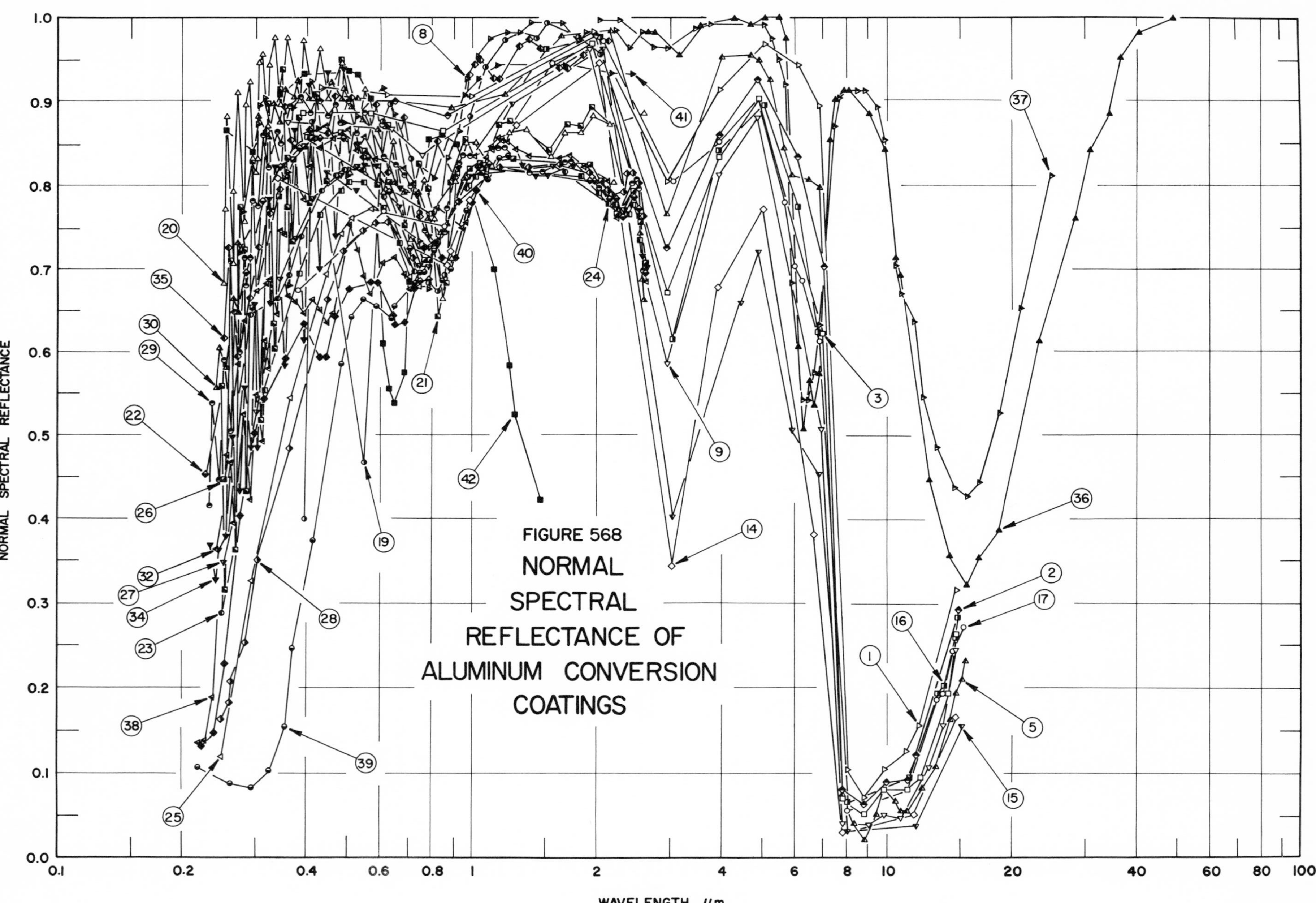

FIGURE 568
NORMAL SPECTRAL REFLECTANCE OF ALUMINUM CONVERSION COATINGS
NORMAL SPECTRAL REFLECTANCE
WAVELENGTH, μm
0.0
0.1
0.2
0.3
0.4
0.5
0.6
0.7
0.8
0.9
1.0
0.1
0.2
0.4
0.6
0.8
1
2
4
6
8
10
20
40
60
80
100

SPECIFICATION TABLE NO. 568 NORMAL SPECTRAL REFLECTANCE OF ALUMINUM CONVERSION COATINGS

Curve No.	Ref. No.	Year	Temperature, K	Wavelength Range, μm	Geometry θ	θ'	ω'	Reported Error, %	Composition (weight percent), Specifications, and Remarks
1	265	1963	~298	0.43-14.98	~0°	~0°			1199 aluminum anodized in sulfuric acid; polished substrate; coating thickness 8.4 μm.
2	265	1963	~298	0.43-15.00	~0°	~0°			Similar to above specimen and conditions except coating thickness 9.4 μm.
3	265	1963	~298	0.39-14.98	~0°	~0°			Similar to above specimen and conditions except coating thickness 11.4 μm.
4	265	1963	~298	0.36-14.96	~0°	~0°			Similar to above specimen and conditions except coating thickness 13.2 μm.
5	265	1963	~298	0.53-15.34	~0°	~0°			1199 aluminum anodized 10 min in 10% sulfuric acid; substrate polished in phosphoric/nitric acid bath.
6*	265	1963	~298	0.380-0.700	~0°	~0°			1199 aluminum anodized 15 min in sulfuric acid; polished substrate.
7*	265	1963	~298	0.380-0.700	~0°	~0°			Similar to above specimen and conditions except heated in vacuum (5×10^{-5} mm Hg) at 589 K for 24 hrs.
8*	265	1963	~298	0.380-0.700	~0°	~0°			Similar to above specimen and conditions except heated in vacuum (5×10^{-5} mm Hg) at 589 K for 48 hrs.
9*	265	1963	~298	0.380-0.700	~0°	~0°			Similar to above specimen and conditions except heated in vacuum (5×10^{-5} mm Hg) at 589 K for 96 hrs.
10*	265	1963	~298	0.380-0.700	~0°	~0°			1199 aluminum anodized 25 min in sufluric acid; polished substrate.
11*	265	1963	~298	0.380-0.700	~0°	~0°			Similar to above specimen and conditions except heated in vacuum (5×10^{-5} mm Hg) at 589 K for 24 hrs.
12*	265	1963	~298	0.380-0.700	~0°	~0°			Similar to above specimen and conditions except heated in vacuum (5×10^{-5} mm Hg) at 589 K for 48 hrs.
13*	265	1963	~298	0.380-0.700	~0°	~0°			Similar to above specimen and conditions except heated in vacuum (5×10^{-5} mm Hg) at 589 K for 96 hrs.
14	265	1963	~298	0.34-14.97	~0°	~0°			1199 aluminum anodized in sulfuric acid; polished substrate; coating thickness 22.9 μm.
15	265	1963	~298	0.44-15.02	~0°	~0°			Similar to above specimen and conditions except exposed to 65 hrs of vacuum (10^{-5} mm Hg).
16	265	1963	~298	0.40-14.99	~0°	~0°			1199 aluminum anodized in sulfuric acid; polished substrate; coating thickness 7.6 μm.
17	265	1963	~298	0.38-15.07	~0°	~0°			Similar to above specimen and conditions except heat treated (48 hrs at 589 K) in air.
18	33	1965	~298	0.280-2.590	5°		2π		Barrier-layer anodic-coated Alcoa 1199 (H-18) aluminum; coating thickness 3900 ±15 Å; measured in vacuum (10^{-6} mm Hg); data extracted from smooth curve. [Authors' designation: Sample No. 25]
19	33	1965	~298	0.395-2.028	5°		2π		Similar to above specimen and conditions except irradiated in vacuum at 300 K with 8.7 keV protons to a total dose of 9.25×10^{16} p cm^{-2}. [Authors' designation: Sample No. 57]
20	33	1965	~298	0.251-2.602	5°		2π		Similar to curve 18 specimen and conditions except coating thickness 1.6 to 1.7 μm; measured relative to MgO. [Authors' designation: Sample No. 162A]
21	33	1965	~298	0.250-2.602	5°		2π		Similar to above specimen and conditions except irradiated in vacuum at 300 K with 8.7 keV protons to a total dose of 1×10^{16} p cm^{-2}. [Authors' designation: Sample No. 162B]

* Not shown on plot

SPECIFICATION TABLE NO. 568 NORMAL SPECTRAL REFLECTANCE OF ALUMINUM CONVERSION COATINGS (continued)

Curve No.	Ref. No.	Year	Temperature K	Wavelength Range, μm	Geometry θ	θ'	ω'	Reported Error, %	Composition (weight percent), Specifications, and Remarks
22	75	1969	77	0.229-2.614	~0°		2π		1199 aluminum anodized in sulfuric acid; exposed to vacuum (5×10^{-9} - 2×10^{-7} mm Hg); vacuum maintained with diffusion pump; reflectance calculated from R(2π, 0°); property measured in situ; data extracted from smooth curve.
23	75	1969	77	0.250-2.614	~0°		2π		Above specimen and conditions except simultaneous exposure to electron (3.5×10^{10} e cm^{-2} sec^{-1}) and UV (8 sun intensity) radiation in vacuum; 1.2×10^{15} e cm^{-2}; 66 ESH.
24	75	1969	77	0.250-2.609	~0°		2π		Above specimen and conditions except exposed to 2.3×10^{15} e cm^{-2} and 130 ESH.
25	75	1969	77	0.250-2.614	~0°		2π		Above specimen and conditions except exposed to 5.8×10^{15} e cm^{-2} and 310 ESH.
26	75	1969	77	0.251-2.611	~0°		2π		Similar to curve 22 specimen and conditions except exposed to electron radiation (8.6×10^{10} - 1.6×10^{12} e cm^{-2} sec^{-1}) in vacuum; 5.8×10^{15} e cm^{-2}.
27	75	1969	77	0.251-2.611	~0°		2π		Above specimen and conditions except exposed to 1.2×10^{16} e cm^{-2}.
28	75	1969	77	0.250-2.600	~0°		2π		Similar to curve 22 specimen and conditions except exposed to 6 sun intensity UV radiation in vacuum for 350 ESH.
29	75	1969	77	0.233-2.605	~0°		2π		1199 aluminum (0.254 mm thick); barrier anodized; exposed to vacuum (5×10^{-9} - 2×10^{-7} mm Hg); vacuum maintained with diffusion pump; reflectance calculated from R (2π, 0°); property measured in situ; data extracted from smooth curve.
30	75	1969	77	0.242-2.600	~0°		2π		Above specimen and conditions except exposed to electron radiation (8.6×10^{10} - 1.6×10^{12} e cm^{-2} sec^{-1}) in vacuum; 5.8×10^{15} e cm^{-2}.
31*	75	1969	77	0.232-2.615	~0°		2π		Similar to curve 29 specimen and conditions.
32	75	1969	77	0.362-2.614	~0°		2π		Above specimen and conditions except exposed to 6 sun intensity UV radiation in vacuum for 350 ESH.
33*	75	1969	77	0.237-2.612	~0°		2π		Similar to curve 29 specimen and conditions.
34	75	1969	77	0.235-2.622	~0°		2π		Above specimen and conditions except simultaneous exposure to electron (3.5×10^{10} e cm^{-2} sec^{-1}) and UV (8 sun intensity) radiation in vacuum; 5.8×10^{15} e cm^{-2}; 290 ESH.
35	85	1969	~298	0.254-2.162	~0°		2π		Anodized 1199 aluminum; substrate ~0.254 mm thick; barrier anodized; data extracted from smooth curve. [Author's designation: A-2 Barrier Anodize]
36	111	1969	~298	2.67-49.5	10°	10°			Anodized 1199 aluminum alloy; electropolished in fluoboric acid solution, then anodized in ammonium tartrate solution; data extracted from smooth curve; property measured by specular method relative to a front surface aluminized mirror. [Authors' designation: Anodic]
37	111	1969	~298	2.01-25.4	~0°		2π		Anodized 1199 aluminum alloy; electropolished in fluoboric acid solution, then anodized in ammonium tartrate solution; data extracted from smooth curve; property corrected for value measured by heated cavity method relative to a platinum surface. [Authors' designation: Anodic]
38	29	1963	300	0.220-1.220	~0°		2π		Alodine 401-45; coating produced by the reaction of Al surface with aqueous solution of chromic, phosphoric, and hydrofluoric acids; substrate, Echo A-12 thermal control surface (4.57 μm Al foil glued to both sides of 8.89 μm mylar film); data extracted from smooth curve.

* Not shown on plot

SPECIFICATION TABLE NO. 568 NORMAL SPECTRAL REFLECTANCE OF ALUMINUM CONVERSION COATINGS (continued)

Curve No.	Ref. No.	Year	Temperature, K	Wavelength Range, μm	Geometry θ	θ'	ω'	Reported Error, %	Composition (weight percent), Specifications, and Remarks
39	29	1963	300	0.220-1.221	$\sim 0°$		2π		Similar to above specimen and conditions except exposed to 353 hrs ultraviolet radiation at solar factor of 3-4 suns; substrate temp 278 K during exposure.
40	29	1963	300	0.220-1.220	$\sim 0°$		2π		Similar to above specimen and conditions except exposed to 358 hrs ultraviolet radiation at solar factor of 3-4 suns; substrate temp 248 K during exposure.
41	260	1967	~298	0.277-2.437	$\sim 0°$		2π		Barrier-layer anodic coating on a high-purity aluminum substrate; coating applied as follows: substrate thoroughly cleaned, electropolished in a fluoboric acid solution, anodized in an aqueous ammonium tartrate solution, then immersed in a phosphoric acid solution; measured in vacuum ($\sim 10^{-6}$ mm Hg); data extracted from smooth curve.
42	260	1967	~298	2.56-14.76	$\sim 0°$		2π		Similar to above specimen and conditions.
43	37	1967	293	0.242-2.50	$\sim 0°$		2π		Alzak anodized aluminum; coating thickness 2.54-7.63 μm; measured in situ under vacuum ($< 4 \times 10^{-7}$ mm Hg); data extracted from smooth curve; measured relative to MgO. [Authors' designation: Sample 50 Alzak (No. 21)]
44*	37	1967	293	0.239-2.50	$\sim 0°$		2π		Above specimen and conditions except exposed in situ to 1130 ESH at solar factor of 4-5 suns.

* Not shown on plot

DATA TABLE NO. 568 NORMAL SPECTRAL REFLECTANCE OF ALUMINUM CONVERSION COATINGS

[Wavelength, λ, μm; Reflectance, ρ; Temperature, T, K]

λ	ρ
CURVE 1	
T ~ 298	
0.43	0.918
0.62	0.907
0.99	0.907
1.97	0.987
2.99	0.807
3.98	0.915
5.01	0.970
6.03	0.944
6.99	0.896
8.02	0.106
8.98	0.071
9.97	0.108
11.02	0.128
12.00	0.157
14.98	0.317
CURVE 2	
T ~ 298	
0.43	0.900
0.65	0.900
0.87	0.884
1.98	0.982
2.95	0.729
3.95	0.861
4.96	0.928
6.05	0.836
7.02	0.702
7.98	0.083
8.99	0.063
10.07	0.091
11.09	0.095
11.98	0.123
15.00	0.293
CURVE 3	
T ~ 298	
0.39	0.888
0.57	0.888
0.85	0.866
1.95	0.970
2.99	0.672
3.98	0.838
4.98	0.904
CURVE 3 (cont.)	
7.00	0.623
7.97	0.072
8.98	0.053
9.97	0.081
11.23	0.081
12.01	0.097
14.01	0.194
14.98	0.265
CURVE 4	
T ~ 298	
0.36	0.875
0.64	0.869
0.76	0.837
1.91	0.961
2.98	0.587
3.95	0.816
4.96	0.882
7.00	0.509
7.97	0.042
9.11	0.042
9.94	0.052
10.98	0.050
12.99	0.110
13.97	0.159
14.96	0.248
CURVE 5	
T ~ 298	
0.53	0.902
0.89	0.892
1.20	0.909
1.64	0.980
2.16	0.986
2.95	0.768
4.00	0.954
4.71	0.958
4.97	0.950
5.21	0.927
5.65	0.846
5.89	0.815
6.46	0.810
6.95	0.800
7.96	0.079
CURVE 5 (cont.)	
8.40	0.044
8.92	0.023
9.47	0.053
9.90	0.083*
10.53	0.069
10.98	0.058
11.29	0.058
12.04	0.084
13.01	0.110
14.12	0.163
14.85	0.196
15.10	0.211
15.34	0.231
CURVE 6*	
T ~ 298	
0.380	0.894
0.399	0.892
0.497	0.873
0.599	0.846
0.700	0.803
CURVE 7*	
T ~ 298	
0.380	0.725
0.399	0.745
0.498	0.813
0.597	0.807
0.700	0.787
CURVE 8*	
T ~ 298	
0.380	0.653
0.399	0.697
0.497	0.803
0.598	0.800
0.700	0.783
CURVE 9*	
T ~ 298	
0.380	0.643
0.400	0.689
CURVE 9 (cont.)	
0.498	0.796
0.598	0.792
0.700	0.778
CURVE 10*	
T ~ 298	
0.380	0.848
0.404	0.843
0.508	0.828
0.602	0.815
0.700	0.795
CURVE 11*	
T ~ 298	
0.380	0.639
0.509	0.804
0.601	0.807
0.700	0.801
CURVE 12*	
T ~ 298	
0.380	0.616
0.505	0.785
0.601	0.802
0.700	0.791
CURVE 13*	
T ~ 298	
0.380	0.552
0.404	0.596
0.509	0.753
0.600	0.796
0.700	0.781
CURVE 14	
T ~ 298	
0.34	0.810
0.87	0.705
0.89	0.721
0.99	0.784
1.29	0.871
CURVE 14 (cont.)	
2.01	0.949
3.01	0.343
3.93	0.680
5.02	0.773
6.75	0.381
7.99	0.031
11.87	0.054
14.97	0.167
CURVE 15	
T ~ 298	
0.44	0.818
0.79	0.728
1.25	0.897
2.07	0.962
3.02	0.403
4.44	0.660
4.99	0.721
5.96	0.508
6.90	0.455
8.02	0.034
11.89	0.040
15.02	0.156
CURVE 16	
T ~ 298	
0.40	0.859
0.80	0.767
1.51	0.961
2.03	0.977
3.04	0.615
3.95	0.843
5.05	0.897
6.05	0.779
6.96	0.625
8.04	0.069
11.32	0.096
13.12	0.194
13.91	0.203
14.67	0.260
14.99	0.285
CURVE 17	
T ~ 298	
0.38	0.675
0.61	0.770
0.86	0.751
1.56	0.947
2.06	0.972
3.06	0.806
3.97	0.853
4.99	0.903*
5.66	0.781
6.00	0.705
6.26	0.687
6.94	0.611
8.01	0.059
11.34	0.083*
13.22	0.187
13.91	0.194
14.68	0.246
15.07	0.271
CURVE 18	
T ~ 298	
0.280	0.770
0.301	0.872
0.309	0.897
0.316	0.902
0.351	0.810
0.360	0.818
0.361	0.680
0.370	0.894
0.409	0.921
0.500	0.812
0.530	0.883
0.608	0.916
0.656	0.886
0.753	0.735
0.788	0.676
0.878	0.833
0.929	0.864
0.968	0.926
1.027	0.955
1.105	0.975
1.213	0.982
1.308	0.982
1.391	0.994
CURVE 18 (cont.)	
1.650	0.994
1.807	0.972
1.881	0.982
2.227	0.985
2.404	0.962
2.590	0.982
CURVE 19	
T ~ 298	
0.395	0.400
0.411	0.876
0.548	0.466
0.606	0.879
0.639	0.899
0.719	0.832
0.784	0.683
0.821	0.673
0.860	0.694
0.999	0.883
1.087	0.940
1.202	0.973
1.306	0.980*
1.440	0.978
1.516	0.994
1.800	0.973
2.028	0.958
CURVE 20	
T ~ 298	
0.251	0.683
0.254	0.772
0.257	0.881
0.265	0.706
0.268	0.794
0.272	0.910
0.281	0.757
0.287	0.897
0.291	0.922
0.297	0.834
0.300	0.818
0.303	0.831
0.309	0.946
0.312	0.956
0.314	0.943*
CURVE 20 (cont.)	
0.322	0.871
0.325	0.885*
0.329	0.945*
0.335	0.977
0.349	0.896
0.360	0.977
0.360	0.973*
0.384	0.900
0.397	0.973
0.417	0.901
0.430	0.962
0.455	0.902
0.484	0.949
0.506	0.815
0.546	0.913
0.605	0.792
0.643	0.854
0.710	0.739
0.775	0.808
0.813	0.761
0.851	0.664
0.891	0.713
0.919	0.808
0.963	0.851
1.014	0.851
1.114	0.825
1.244	0.863
1.359	0.867
1.554	0.836
1.696	0.862
1.824	0.865
1.976	0.884
2.169	0.874
2.602	0.886
CURVE 21	
T ~ 298	
0.250	0.559
0.255	0.316
0.263	0.710
0.270	0.362
0.279	0.776
0.282	0.724
0.288	0.431
0.299	0.831

* Not shown on plot

DATA TABLE NO. 568 NORMAL SPECTRAL REFLECTANCE OF ALUMINUM CONVERSION COATINGS (continued)

λ	ρ
CURVE 21 (cont.)	
T ~ 298	
0.310	0.517
0.322	0.897
0.325	0.887*
0.332	0.634
0.335	0.601
0.338	0.634*
0.347	0.914
0.350	0.938
0.354	0.914*
0.360	0.775
0.360	0.835
0.370	0.733
0.383	0.881
0.401	0.780
0.421	0.909
0.446	0.806
0.486	0.950
0.510	0.816*
0.545	0.913*
0.590	0.800
0.626	0.874
0.682	0.752
0.742	0.828
0.784	0.799
0.828	0.642
0.865	0.685
0.887	0.804
0.967	0.854
1.062	0.817
1.180	0.872
1.248	0.879
1.520	0.843
1.706	0.874
1.840	0.872
1.952	0.894
2.169	0.874*
2.602	0.886*
CURVE 22	
T = 77	
0.229	0.451
0.289	0.698
0.315	0.782
0.347	0.822
0.382	0.846
CURVE 22 (cont.)	
0.430	0.859
0.485	0.859
0.560	0.835
0.633	0.794
0.713	0.728
0.756	0.689
0.783	0.689
0.967	0.797*
1.049	0.822
1.103	0.825
1.920	0.809
1.993	0.809
2.188	0.786
2.221	0.786
2.318	0.764
2.484	0.789
2.528	0.773
2.614	0.696
CURVE 23	
T = 77	
0.250	0.288
0.294	0.644
0.340	0.663
0.363	0.692
0.388	0.740
0.437	0.793
0.470	0.813
0.499	0.820
0.542	0.820
0.616	0.789
0.737	0.698
0.757	0.685*
0.787	0.685*
0.968	0.799
1.049	0.822*
1.103	0.825*
1.698	0.818
1.993	0.809*
2.188	0.786*
2.221	0.786*
2.318	0.764*
2.484	0.789*
2.528	0.773*
2.614	0.696*
CURVE 24	
T = 77	
0.250	0.288*
0.317	0.552
0.382	0.695
0.436	0.766
0.481	0.795
0.504	0.804
0.545	0.804
0.613	0.773
0.676	0.732
0.736	0.695*
0.755	0.686*
0.795	0.686*
0.988	0.800
1.020	0.811
1.065	0.820
1.847	0.812
2.018	0.791
2.123	0.780
2.223	0.780
2.318	0.764*
2.401	0.769
2.456	0.780
2.523	0.765
2.568	0.737
2.609	0.696*
CURVE 25	
T = 77	
0.250	0.120
0.297	0.326
0.369	0.545
0.442	0.695
0.484	0.744
0.518	0.761
0.577	0.772
0.618	0.772*
0.718	0.692
0.751	0.680
0.792	0.680*
0.988	0.800*
1.020	0.813*
1.049	0.822*
1.103	0.825*
1.815	0.815*
1.944	0.815
CURVE 25 (cont.)	
2.188	0.786*
2.221	0.786*
2.318	0.764*
2.484	0.789*
2.528	0.773*
2.614	0.696*
CURVE 26	
T = 77	
0.251	0.446
0.297	0.652
0.342	0.788
0.371	0.832
0.395	0.849
0.416	0.859
0.484	0.864
0.559	0.842
0.638	0.805
0.772	0.711
0.787	0.711
0.986	0.821
1.035	0.829
1.277	0.834
1.916	0.826
2.059	0.809
2.251	0.768
2.337	0.768*
2.466	0.794
2.489	0.794*
2.553	0.758
2.611	0.707
CURVE 27	
T = 77	
0.251	0.347
0.301	0.526
0.346	0.689
0.377	0.738
0.419	0.785
0.482	0.814
0.530	0.823
0.587	0.823
0.672	0.777
0.759	0.703
0.792	0.703
CURVE 27 (cont.)	
0.982	0.813
1.023	0.824*
1.589	0.824
1.952	0.811*
2.137	0.789*
2.258	0.771
2.323	0.771
2.441	0.795*
2.479	0.795*
2.553	0.758*
2.611	0.707*
CURVE 28	
T = 77	
0.250	0.163
0.261	0.181
0.262	0.208
0.286	0.253
0.306	0.350
0.361	0.483
0.445	0.662
0.498	0.723
0.541	0.748
0.584	0.757
0.625	0.757
0.721	0.713
0.759	0.689*
0.780	0.689*
0.933	0.784
0.996	0.811*
1.052	0.820*
1.185	0.824
1.369	0.824
1.631	0.827
1.932	0.815
2.115	0.796
2.190	0.791
2.319	0.794
2.392	0.818
2.454	0.818
2.519	0.801
2.600	0.764
CURVE 29	
T = 77	
0.233	0.416
0.236	0.538
0.249	0.446
0.253	0.590
0.261	0.504
0.268	0.649
0.275	0.600
0.282	0.713
0.285	0.680
0.287	0.714*
0.292	0.714
0.296	0.782
0.301	0.776
0.314	0.860
0.321	0.821
0.340	0.864
0.349	0.850
0.364	0.895
0.380	0.878*
0.387	0.903
0.407	0.886
0.427	0.903*
0.454	0.884
0.468	0.898
0.480	0.875
0.501	0.900
0.531	0.863
0.544	0.881*
0.556	0.887
0.579	0.867*
0.610	0.836
0.649	0.816
0.689	0.811
0.709	0.800
0.763	0.747
0.812	0.729
0.860	0.747
0.876	0.775
0.939	0.828
0.957	0.837
0.997	0.837
1.041	0.823*
1.076	0.828*
1.157	0.846
1.209	0.846
1.294	0.830*
CURVE 29 (cont.)	
1.397	0.819
1.498	0.819
1.680	0.830
1.734	0.830
2.013	0.797
2.190	0.797*
2.240	0.784*
2.285	0.764*
2.339	0.764*
2.449	0.803
2.484	0.809*
2.503	0.809
2.519	0.797
2.566	0.761*
2.605	0.699
CURVE 30	
T = 77	
0.242	0.556
0.249	0.604
0.256	0.580
0.266	0.664
0.272	0.649
0.272	0.731
0.284	0.712*
0.294	0.715*
0.305	0.883
0.322	0.859
0.333	0.898
0.349	0.877
0.349	0.916*
0.358	0.916
0.372	0.905
0.380	0.925
0.405	0.902
0.416	0.925*
0.438	0.915*
0.444	0.901*
0.454	0.901*
0.460	0.925
0.482	0.902
0.497	0.916
0.514	0.893
0.541	0.879
0.549	0.886*
0.565	0.886*
CURVE 30 (cont.)	
0.598	0.869
0.654	0.805
0.686	0.794
0.742	0.794
0.781	0.761
0.844	0.743
0.878	0.754
0.957	0.810
0.981	0.820*
1.076	0.823*
1.158	0.845*
1.233	0.845
1.376	0.825*
1.642	0.825*
2.007	0.796*
2.080	0.796*
2.206	0.804
2.298	0.766*
2.338	0.766*
2.479	0.800*
2.511	0.795*
2.564	0.747
2.589	0.689
2.600	0.661
CURVE 31*	
T = 77	
0.232	0.419
0.236	0.539
0.247	0.452
0.250	0.585
0.255	0.511
0.267	0.649
0.273	0.601
0.285	0.724
0.292	0.683
0.301	0.792
0.305	0.777
0.312	0.863
0.317	0.830
0.326	0.821
0.343	0.880
0.355	0.855
0.366	0.902
0.385	0.880
0.397	0.885

* Not shown on plot

DATA TABLE NO. 568 NORMAL SPECTRAL REFLECTANCE OF ALUMINUM CONVERSION COATINGS (continued)

CURVE 31 (cont.)* T = 77

λ	ρ
0.408	0.900
0.422	0.886
0.436	0.905
0.459	0.887
0.478	0.900
0.495	0.887
0.511	0.899
0.537	0.869
0.552	0.869
0.559	0.882
0.638	0.821
0.704	0.806
0.769	0.748
0.812	0.732
0.838	0.737
0.922	0.817
0.968	0.837
1.011	0.837
1.041	0.828
1.069	0.828
1.138	0.844
1.171	0.849
1.213	0.849
1.325	0.826
1.391	0.818
1.468	0.818
1.648	0.833
1.745	0.833
1.943	0.802
2.045	0.796
2.129	0.801
2.167	0.801
2.232	0.785
2.278	0.763
2.319	0.763
2.448	0.802
2.489	0.807
2.554	0.772
2.589	0.738
2.615	0.686

CURVE 32 T = 77

λ	ρ
0.243	0.362
0.261	0.466
0.266	0.395
0.282	0.524
0.294	0.421
0.301	0.548
0.313	0.491
0.319	0.611
0.323	0.580
0.326	0.687
0.340	0.645
0.345	0.748
0.352	0.742
0.359	0.827
0.393	0.785
0.393	0.846*
0.404	0.821
0.415	0.864
0.433	0.847
0.444	0.871
0.461	0.858
0.482	0.871*
0.497	0.857*
0.497	0.869*
0.505	0.869
0.537	0.844
0.552	0.859
0.588	0.832
0.685	0.821
0.729	0.803
0.800	0.737
0.838	0.737
0.922	0.817
0.968	0.837*
1.011	0.837
1.041	0.828*
1.069	0.828*
1.138	0.844
1.192	0.853
1.241	0.853
1.285	0.843
1.387	0.837
1.589	0.837*
1.629	0.833
1.727	0.836
1.881	0.822
1.971	0.808*
2.127	0.808
2.189	0.799*
2.232	0.785*
2.278	0.763
2.319	0.763*
2.448	0.802*
2.489	0.807*
2.534	0.783*
2.580	0.738*
2.614	0.685

CURVE 33* T = 77

λ	ρ
0.237	0.416
0.245	0.536
0.245	0.440
0.258	0.581
0.258	0.502
0.270	0.648
0.283	0.558
0.286	0.718
0.297	0.666
0.305	0.785
0.312	0.774
0.325	0.858
0.325	0.817
0.339	0.805
0.346	0.875
0.355	0.852
0.366	0.871
0.377	0.900
0.392	0.876
0.406	0.920
0.412	0.892
0.425	0.886
0.446	0.906
0.455	0.888
0.479	0.900
0.507	0.883
0.530	0.890
0.549	0.884
0.551	0.868
0.569	0.868
0.593	0.876
0.604	0.843
0.638	0.822
0.687	0.813
0.730	0.790
0.759	0.753
0.804	0.730
0.831	0.730
0.881	0.770
0.932	0.822
0.970	0.835
1.001	0.835
1.038	0.826
1.073	0.826
1.152	0.845
1.226	0.845
1.324	0.824
1.402	0.815
1.495	0.815
1.661	0.830
1.741	0.830
1.901	0.804
1.980	0.798
2.037	0.798
2.082	0.802
2.150	0.802
2.208	0.794
2.291	0.762
2.310	0.762
2.401	0.782
2.472	0.806
2.500	0.806
2.543	0.783
2.576	0.751
2.612	0.687

CURVE 34 T = 77

λ	ρ
0.235	0.369
0.241	0.326
0.247	0.362
0.258	0.379
0.265	0.498
0.279	0.433
0.282	0.556
0.295	0.485
0.300	0.655
0.307	0.485
0.325	0.785
0.329	0.659
0.337	0.822
0.358	0.584
0.372	0.867
0.393	0.613
0.406	0.920
0.432	0.700
0.451	0.933
0.472	0.740
0.498	0.938
0.522	0.795
0.536	0.850
0.550	0.887*
0.558	0.822
0.679	0.799
0.740	0.780
0.794	0.740
0.804	0.730*
0.831	0.730
0.869	0.756*
0.936	0.820
0.975	0.839
1.005	0.839*
1.046	0.827*
1.079	0.828*
1.166	0.844*
1.224	0.844*
1.319	0.827
1.425	0.812
1.547	0.812
1.658	0.826*
1.764	0.826
1.948	0.801*
2.048	0.793*
2.133	0.793*
2.198	0.799*
2.316	0.764*
2.428	0.800*
2.532	0.740*
2.579	0.719
2.622	0.702

CURVE 35 T ~ 298

λ	ρ
0.254	0.617
0.260	0.726
0.272	0.596
0.279	0.777*
0.292	0.665
0.299	0.832*
0.307	0.729
0.320	0.867
0.327	0.767
0.337	0.899*
0.346	0.797
0.355	0.893
0.369	0.855
0.385	0.904*
0.406	0.866
0.426	0.906*
0.449	0.865
0.469	0.907
0.499	0.852
0.469	0.907
0.499	0.852*
0.517	0.904
0.552	0.835
0.586	0.897
0.629	0.818
0.687	0.881
0.759	0.763
0.830	0.854
0.905	0.810
0.995	0.931
1.014	0.945
1.044	0.950
1.067	0.950*
1.143	0.927
1.174	0.927
1.305	0.967
1.362	0.973
1.424	0.973*
1.492	0.963
1.583	0.945*
1.640	0.940
1.703	0.940
1.880	0.959
2.162	0.974

CURVE 36 T ~ 298

λ	ρ
2.67	0.984
2.79	0.984
3.16	0.955
3.52	0.990
4.37	1.004
4.75	0.992
5.15	1.000
5.43	1.000
5.66	0.978
6.12	0.608
6.32	0.509
6.56	0.565
6.77	0.536
6.95	0.573
7.34	0.856
7.58	0.903
7.88	0.915
8.16	0.915
9.16	0.888
9.93	0.845
10.5	0.715
10.8	0.694
12.7	0.449
14.2	0.357
15.7	0.321
16.9	0.353
18.8	0.389
23.4	0.614
28.7	0.761
31.0	0.846
34.2	0.889
36.5	0.952
40.3	0.984
49.5	1.003

CURVE 37 T ~ 298

λ	ρ
2.01	0.997
2.28	0.997
2.72	0.964
2.95	0.963
3.45	0.989
3.75	0.992
5.02	0.992
5.33	0.977
5.57	0.950
5.75	0.920
5.97	0.685
6.28	0.541
6.50	0.541
6.77	0.577
6.99	0.632
7.46	0.872
7.63	0.905
7.85	0.917*
8.18	0.917*
8.49	0.913
8.81	0.913
9.54	0.894
9.86	0.854
10.5	0.703
10.9	0.670
11.6	0.639
12.3	0.544
13.2	0.486
14.6	0.438
15.7	0.429
16.9	0.443
18.9	0.529
21.0	0.654
25.4	0.812

CURVE 38 T = 300

λ	ρ
0.220	0.135
0.227	0.139
0.238	0.189
0.259	0.475
0.273	0.585
0.284	0.636
0.301	0.672
0.320	0.683
0.345	0.687*
0.362	0.669
0.377	0.660
0.392	0.647
0.416	0.661
0.431	0.650
0.446	0.636
0.460	0.646
0.493	0.744*
0.509	0.754
0.532	0.722
0.548	0.691
0.575	0.670

* Not shown on plot

DATA TABLE NO. 568 NORMAL SPECTRAL REFLECTANCE OF ALUMINUM CONVERSION COATINGS (continued)

CURVE 38 (cont.) T = 300

λ	ρ
0.615	0.708
0.651	0.714
0.682	0.693
0.718	0.678
0.799	0.765*
0.822	0.775
0.941	0.750
0.976	0.758
1.039	0.819*
1.090	0.834
1.112	0.849
1.165	0.859
1.220	0.862*

CURVE 39 T = 300

λ	ρ
0.220	0.107
0.261	0.088
0.298	0.082
0.327	0.101
0.357	0.153
0.380	0.247
0.416	0.374
0.487	0.586
0.516	0.642
0.542	0.662
0.589	0.654
0.640	0.640
0.654	0.654
0.728	0.683
0.751	0.696
0.777	0.696
0.794	0.705*
0.859	0.689
0.892	0.700
0.971	0.771
0.993	0.784*
1.081	0.812
1.092	0.808
1.172	0.834
1.221	0.837

CURVE 40 T = 300

λ	ρ
0.223	0.132
0.240	0.148
0.255	0.229
0.277	0.403
0.300	0.501
0.316	0.543
0.358	0.591
0.394	0.633
0.437	0.594
0.442	0.594
0.472	0.642
0.504	0.678
0.567	0.685
0.586	0.685
0.647	0.643
0.658	0.634
0.698	0.636
0.733	0.678
0.777	0.727
0.801	0.733
0.856	0.715
0.918	0.715
1.001	0.783
1.030	0.796
1.043	0.820*
1.089	0.832*
1.114	0.845*
1.181	0.851*
1.220	0.855*

CURVE 41 T ~ 298

λ	ρ
0.277	0.720
0.372	0.847
0.426	0.869
0.490	0.875
0.558	0.870
0.812	0.810
0.960	0.908
1.034	0.927
1.249	0.943
1.696	0.943
2.183	0.933
2.437	0.933

CURVE 42 T ~ 298

λ	ρ
2.56	0.867
2.99	0.840
3.45	0.909
3.96	0.932
5.05	0.937
5.34	0.931
5.71	0.903
6.15	0.610
6.31	0.555
6.48	0.538
6.82	0.575
7.08	0.686
7.38	0.804*
7.84	0.855
8.47	0.861
9.20	0.850
9.87	0.819*
11.14	0.700
12.12	0.584
12.68	0.522
14.76	0.424

CURVE 43 T = 293

λ	ρ
0.242	0.509
0.245	0.541
0.251	0.589
0.264	0.649
0.275	0.673
0.282	0.696*
0.285	0.710
0.289	0.742
0.295	0.771
0.309	0.794
0.329	0.810
0.360	0.822
0.360	0.813*
0.363	0.821*
0.366	0.811
0.369	0.824
0.372	0.813
0.375	0.832*
0.378	0.815*
0.382	0.830
0.385	0.816
0.388	0.832
0.391	0.813
0.395	0.836
0.398	0.820
0.402	0.839
0.405	0.819
0.409	0.843
0.412	0.821*
0.416	0.845*
0.420	0.824
0.425	0.845*
0.430	0.821
0.433	0.849*
0.437	0.823*
0.441	0.853
0.446	0.825
0.450	0.853
0.455	0.823
0.459	0.857*
0.464	0.828*
0.468	0.859*
0.473	0.827
0.478	0.861*
0.482	0.825
0.488	0.864*
0.493	0.828
0.498	0.865*
0.504	0.829
0.509	0.868*
0.514	0.826
0.519	0.864
0.522	0.864*
0.525	0.825*
0.527	0.835
0.531	0.870
0.533	0.870*
0.540	0.827
0.544	0.861
0.546	0.869*
0.552	0.826*
0.558	0.871
0.562	0.855
0.565	0.822
0.569	0.854
0.572	0.869
0.580	0.820
0.587	0.872
0.595	0.820
0.602	0.873
0.611	0.819
0.613	0.830*
0.617	0.865*
0.620	0.868
0.627	0.815*
0.636	0.866*
0.643	0.819
0.646	0.814*
0.650	0.852*
0.654	0.863
0.658	0.851*
0.663	0.815
0.666	0.815*
0.673	0.855*
0.677	0.857
0.684	0.815*
0.696	0.845
0.705	0.822
0.710	0.826*
0.710	0.805
0.748	0.769
0.762	0.748*
0.812	0.724
0.845	0.735
0.895	0.781
0.955	0.826
0.974	0.830
0.990	0.848
1.00	0.845
1.03	0.865
1.04	0.865*
1.05	0.860
1.06	0.862*
1.08	0.878
1.10	0.868
1.12	0.869*
1.13	0.886
1.14	0.886*
1.17	0.871*
1.20	0.895
1.23	0.878
1.25	0.878*
1.27	0.898*
1.28	0.898
1.30	0.880
1.32	0.880*
1.34	0.901
1.36	0.901*
1.39	0.884
1.43	0.900
1.45	0.900*
1.48	0.878
1.50	0.878*
1.53	0.903
1.60	0.877
1.63	0.889
1.65	0.897
1.72	0.869
1.74	0.866*
1.79	0.885
1.82	0.882*
1.88	0.851
1.96	0.872
2.00	0.859
2.04	0.835
2.10	0.845
2.14	0.854
2.21	0.847
2.25	0.831
2.30	0.823
2.38	0.831
2.44	0.823
2.50	0.803*

CURVE 44* T = 293

λ	ρ
0.239	0.183
0.265	0.219
0.277	0.236
0.308	0.296
0.337	0.348
0.360	0.387
0.360	0.401
0.362	0.384
0.365	0.418
0.368	0.394
0.371	0.430
0.374	0.404
0.377	0.446
0.381	0.414
0.384	0.459
0.388	0.427
0.391	0.471
0.394	0.436
0.398	0.489
0.401	0.446
0.405	0.503
0.408	0.459
0.412	0.517
0.416	0.472
0.420	0.535
0.425	0.483
0.429	0.549
0.433	0.493
0.437	0.563
0.442	0.507
0.446	0.583
0.451	0.519
0.455	0.596
0.460	0.536
0.464	0.614
0.469	0.548
0.474	0.629
0.478	0.561
0.483	0.644
0.489	0.574
0.494	0.663
0.499	0.584
0.505	0.680
0.510	0.602
0.516	0.693
0.522	0.614
0.527	0.707
0.535	0.627
0.541	0.721
0.548	0.640
0.555	0.733
0.561	0.650
0.568	0.746
0.575	0.660
0.582	0.758
0.590	0.672
0.598	0.768
0.606	0.683
0.614	0.776
0.623	0.692
0.631	0.783
0.640	0.705
0.647	0.779
0.651	0.787
0.658	0.717
0.660	0.716
0.668	0.784
0.672	0.784
0.679	0.731
0.682	0.735
0.689	0.777
0.693	0.777
0.699	0.749
0.703	0.746
0.710	0.764
0.710	0.733
0.735	0.719
0.747	0.702
0.778	0.690
0.809	0.686
0.839	0.700
0.860	0.720
0.935	0.789
0.962	0.803
0.981	0.822
1.00	0.821
1.02	0.844
1.04	0.836
1.07	0.862
1.10	0.847
1.13	0.873
1.14	0.870
1.16	0.859
1.19	0.881
1.20	0.881
1.22	0.863
1.26	0.891
1.30	0.868
1.33	0.894
1.36	0.890
1.37	0.876
1.39	0.876
1.42	0.896
1.44	0.890
1.46	0.871
1.48	0.871
1.52	0.896
1.53	0.896
1.58	0.868
1.62	0.891
1.65	0.892
1.71	0.863
1.77	0.883
1.81	0.878
1.85	0.849

* Not shown on plot

DATA TABLE NO. 568 NORMAL SPECTRAL REFLECTANCE OF ALUMINUM CONVERSION COATINGS (continued)

λ	ρ
CURVE 44 (cont.)*	
T = 293	
1.89	0.853
1.90	0.861
1.94	0.870
1.98	0.861
2.03	0.835
2.05	0.835
2.10	0.852
2.16	0.852
2.27	0.825
2.33	0.832
2.37	0.833
2.43	0.820
2.47	0.805
2.50	0.799

* Not shown on plot

FIGURE 569A

ANALYZED CHANGE IN NORMAL SPECTRAL REFLECTANCE OF ALUMINUM CONVERSION COATINGS

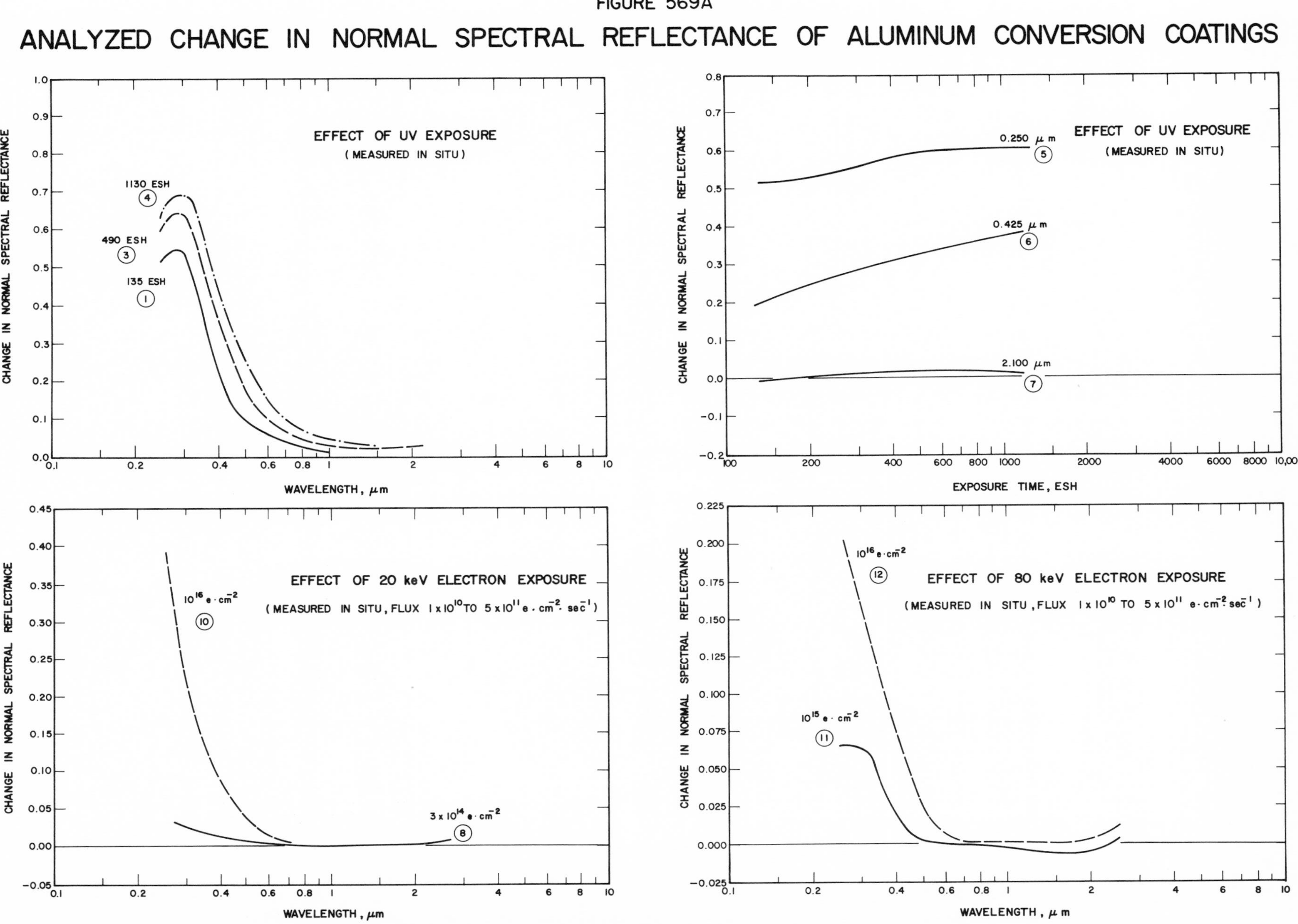

SPECIFICATION TABLE NO. 569 CHANGE IN NORMAL SPECTRAL REFLECTANCE OF ALUMINUM CONVERSION COATINGS

Curve No.	Ref. No.	Year	Temperature, K	Wavelength Range, μm	Geometry θ	Geometry θ'	Geometry ω'	Reported Error, %	Composition (weight percent), Specifications, and Remarks
1	46	1969	295	0.250-0.920	~0°		2π		Alzak No. 21 anodized aluminum (7.36 x 10^{-3} mm thick); exposed to UV radiation (with rates equivalent to 5 suns in UV) in vacuum (10^{-7} mm Hg); 135 ESH; vacuum maintained by ion pump; reflectance measured in situ; positive change indicates decrease in reflectance from preirradiation, in air reflectance; data extracted from smooth curve. [Authors' designation: Z]
2	46	1969	295	0.250-0.938	~0°		2π		Above specimen and conditions except 250 ESH.
3	46	1969	295	0.250-1.564	~0°		2π		Above specimen and conditions except 490 ESH.
4	46	1969	295	0.250-2.450	~0°		2π		Above specimen and conditions except 1130 ESH.
5	46	1969	295	0.250	~0°		2π		Alzak anodized aluminum (7.36 x 10^{-3} mm thick); exposed to UV radiation (4.7 UV sun rate) in vacuum (10^{-8} mm Hg); vacuum maintained by ion pump; ESH is variable; reflectance measured in situ; positive change indicates decrease in reflectance from preirradiation, in air reflectance. [Authors' designation: Z]
6	46	1969	295	0.425	~0°		2π		Above specimen and conditions.
7	46	1969	295	2.100	~0°		2π		Above specimen and conditions.
8	45	1970	281	0.274-2.492	~0°		2π		Alzak anodized aluminum (3.81 x 10^{-3} mm thick); substrate held at 281 ± 2 K; exposed to 20 keV electrons in dark in vacuum (10^{-8} mm Hg); vacuum maintained by ion pump; flux 1 x 10^{10} - 5 x 10^{11} e cm^{-2} sec^{-1}; reflectance measured in situ after 3 x 10^{14} e cm^{-2}; data extracted from smooth curve. [Authors' designation: Z3]
9	45	1970	281	0.245-2.492	~0°		2π		Above specimen and conditions except reflectance measured in situ after 10^{15} e cm^{-2}.
10	45	1970	281	0.250-2.500	~0°		2π		Above specimen and conditions except reflectance measured in situ after 10^{16} e cm^{-2}.
11	45	1970	281	0.257-2.500	~0°		2π		Similar to curve 8 specimen and conditions except exposed to 80 keV electrons; reflectance measured in situ after 10^{15} e cm^{-2}.
12	45	1970	281	0.263-2.500	~0°		2π		Above specimen and conditions except reflectance measured in situ after 10^{16} e cm^{-2}.

DATA TABLE NO. 569 CHANGE IN NORMAL SPECTRAL REFLECTANCE OF ALUMINUM CONVERSION COATINGS

[Wavelength, λ, μm; Reflectance, ρ; Temperature, T, K]

λ	Δρ
CURVE 1 T = 295	
0.250	0.515
0.293	0.549
0.420	0.188
0.447	0.143
0.477	0.111
0.516	0.087
0.566	0.073
0.920	0.008
CURVE 2 T = 295	
0.250	0.561
0.301	0.605
0.476	0.167
0.503	0.135
0.534	0.111
0.569	0.093
0.623	0.072
0.938	0.016
CURVE 3 T = 295	
0.250	0.600
0.295	0.651
0.458	0.247
0.498	0.182
0.534	0.143
0.570	0.117
0.612	0.093
0.679	0.070
0.797	0.043
0.936	0.028
1.133	0.021
1.564	0.021
CURVE 4 T = 295	
0.250	0.626
0.301	0.700
0.340	0.590
0.402	0.426
0.456	0.329
CURVE 4 (cont.)	
0.494	0.264
0.536	0.211
0.581	0.161
0.628	0.130
0.691	0.098
0.761	0.076
0.847	0.055
0.952	0.037
1.040	0.028
1.206	0.023
1.560	0.023*
1.707	0.014
2.077	0.014
2.196	0.018
2.274	0.022
2.450	0.012

ESH	Δρ
CURVE 5 λ = 0.250 T = 295	
135	0.51
250	0.54
490	0.59
1130	0.60
CURVE 6 λ = 0.425 T = 295	
135	0.20
250	0.27
490	0.32
770	0.35
1130	0.38
CURVE 7 λ = 2.100 T = 295	
135	-0.01
250	0.01
490	0.01
1130	0.01

λ	Δρ
CURVE 8 T = 281	
0.274	0.030
0.323	0.030
0.404	0.014
0.465	0.011
0.548	0.008
0.731	0.004
0.779	0.000
1.407	-0.002
1.846	-0.002
2.492	0.009
CURVE 9 T = 281	
0.245	0.131
0.272	0.093
0.301	0.090
0.316	0.079
0.351	0.041
0.423	0.024
0.441	0.024
0.514	0.008
0.731	0.005
0.779	0.000
1.407	-0.002
1.846	-0.002
2.492	0.009
CURVE 10 T = 281	
0.250	0.390
0.265	0.369
0.277	0.334
0.315	0.259
0.344	0.139
0.400	0.098
0.513	0.035
0.572	0.009
0.631	0.009
0.681	0.013
0.778	0.000*
1.011	0.000
1.096	0.007
1.140	0.000
CURVE 10 (cont.)	
1.167	0.004
1.561	0.004
1.710	0.010
2.048	0.010
2.079	0.015
2.092	0.009*
2.180	0.009
2.232	0.015
2.275	0.010
2.340	0.016*
2.500	0.016
CURVE 11 T = 281	
0.257	0.065
0.285	0.065
0.314	0.061
0.337	0.047
0.370	0.031
0.471	0.003
0.779	0.000
0.897	0.000
0.934	-0.004
1.065	0.000
1.310	-0.002
1.446	-0.008
1.549	-0.008
1.568	-0.011
1.611	-0.011
1.639	-0.004
1.678	-0.004*
1.722	-0.010
1.816	-0.003
1.945	-0.003
2.022	0.000
2.213	0.000
2.233	-0.006*
2.254	-0.006
2.292	0.000
2.317	-0.006
2.402	0.004
2.500	0.004
CURVE 12 T = 281	
0.263	0.200
0.357	0.103
0.399	0.076
0.455	0.041
0.512	0.020
0.585	0.009
0.717	0.009
0.778	0.000*
1.616	0.002
1.707	0.007
2.208	0.007
2.310	0.010
2.500	0.012

* Not shown on plot

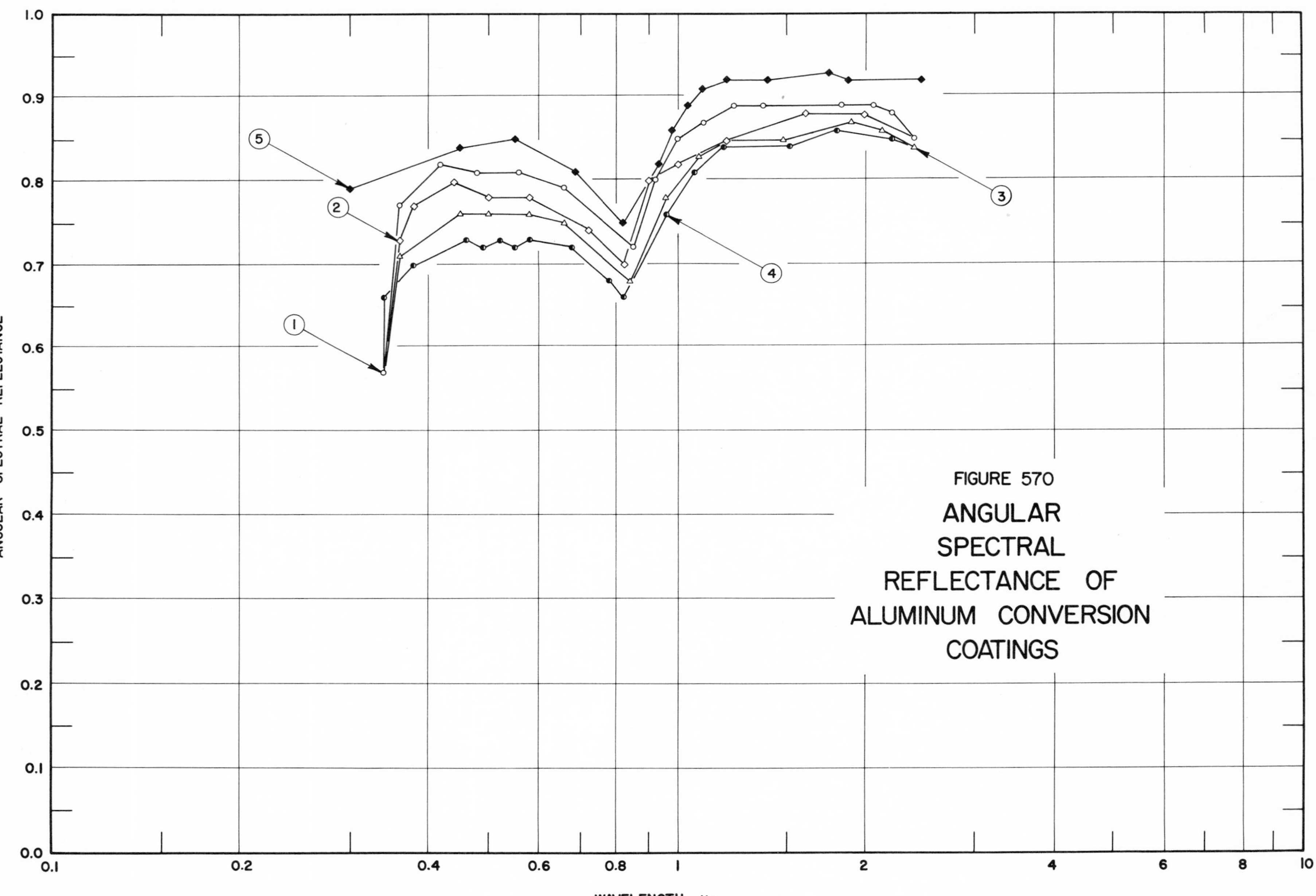

FIGURE 570
ANGULAR SPECTRAL REFLECTANCE OF ALUMINUM CONVERSION COATINGS

SPECIFICATION TABLE NO. 570 ANGULAR SPECTRAL REFLECTANCE OF ALUMINUM CONVERSION COATINGS

Curve No.	Ref. No.	Year	Temperature, K	Wavelength Range, μm	Geometry θ	θ'	ω'	Reported Error, %	Composition (weight percent), Specifications, and Remarks
1	115	1960	298	0.34-2.40	15°		2π		1199 aluminum foil (0.051 mm thick); bright dipped and anodized on both sides (0.0076 mm thick) in dilute sulfuric acid.
2	115	1960	298	0.34-2.40	30°		2π		Above specimen and conditions.
3	115	1960	298	0.34-2.40	45°		2π		Above specimen and conditions.
4	115	1960	298	0.34-2.40	60°		2π		Above specimen and conditions.
5	115	1960	298	0.30-2.48	15°		2π		Similar to curve 1 specimen and conditions. [Authors' designation: Sample 1]

DATA TABLE NO. 570 ANGULAR SPECTRAL REFLECTANCE OF ALUMINUM CONVERSION COATINGS

[Wavelength, λ, μm; Reflectance, ρ; Temperature, T, K]

λ	ρ
CURVE 1 T = 298	
0.34	0.57
0.36	0.77
0.42	0.82
0.48	0.81
0.56	0.81
0.66	0.79
0.84	0.72
0.92	0.80
1.00	0.85
1.10	0.87
1.24	0.89
1.37	0.89
1.84	0.89
2.06	0.89
2.20	0.88
2.40	0.85
CURVE 2 T = 298	
0.34	0.57*
0.36	0.73
0.38	0.77
0.44	0.80
0.50	0.78
0.58	0.78
0.72	0.74
0.82	0.70
0.90	0.80
1.00	0.82
1.20	0.85
1.60	0.88
2.00	0.88
2.40	0.85*
CURVE 3 T = 298	
0.34	0.57*
CURVE 3 (cont.)	
0.36	0.71
0.45	0.76
0.50	0.76
0.58	0.76
0.66	0.75
0.84	0.68
0.96	0.78
1.08	0.83
1.20	0.85*
1.48	0.85
1.90	0.87
2.12	0.86
2.40	0.84
CURVE 4 T = 298	
0.34	0.57*
0.34	0.66
0.38	0.70
CURVE 4 (cont.)	
0.46	0.73
0.49	0.72
0.52	0.73
0.55	0.72
0.58	0.73
0.68	0.72
0.78	0.68
0.82	0.66
0.96	0.76
1.06	0.81
1.28	0.84
1.52	0.84
1.80	0.86
2.20	0.85
2.40	0.84*
CURVE 5 T = 298	
0.30	0.79
CURVE 5 (cont.)	
0.45	0.84
0.55	0.85
0.69	0.81
0.82	0.75
0.93	0.82
0.98	0.86
1.04	0.89
1.10	0.91
1.20	0.92
1.40	0.92
1.76	0.93
1.88	0.92
2.48	0.92

* Not shown on plot

SPECIFICATION TABLE NO. 571 ANGULAR SPECTRAL REFLECTANCE OF ALUMINUM CONVERSION COATINGS

Curve No.	Ref. No.	Year	Temperature, K	Wavelength, μm	Angular Range, °	Geometry θ	θ'	ω'	Reported Error, %	Composition (weight percent), Specifications and Remarks
1*	115	1960	298	0.37	15-60	θ		2π		1199 aluminum foil (0.051 mm thick); bright dipped; anodized on both sides (0.0076 mm thick) in dilute sulfuric acid.
2*	115	1960	298	0.60	15-60	θ		2π		Above specimen and conditions.
3*	115	1960	298	0.70	15-60	θ		2π		Above specimen and conditions.
4*	115	1960	298	1.00	15-60	θ		2π		Above specimen and conditions.
5*	115	1960	298	1.54	15-60	θ		2π		Above specimen and conditions.
6*	267	1965	298	0.0507	30.0-69.6	θ	θ			Al evaporated on Pyrex glass substrate at about 400 Å sec^{-1} at pressures between 10^{-5} to 2 x 10^{-5} mm Hg (3000 Å thick); anodized in a bath of freshly prepared 3 percent ammonium tartrate solution with a pure Al cathode; rinsed in distilled water and dried; angle of viewing the same as angle of incidence for each point.
7*	267	1965	298	0.0762	30.0-69.9	θ	θ			Above specimen and conditions.
8*	267	1965	298	0.1032	30.0-69.5	θ	θ			Above specimen and conditions.

DATA TABLE NO. 571 ANGULAR SPECTRAL REFLECTANCE OF ALUMINUM CONVERSION COATINGS

[Angle, θ, °; Reflectance, ρ; Temperature, T, K; Wavelength, λ, μm]

θ	ρ
CURVE 1*	
λ = 0.37	
T = 298	
15	0.79
30	0.78
45	0.73
55	0.72
60	0.71
CURVE 2*	
λ = 0.60	
T = 298	
15	0.79
30	0.78
45	0.76
60	0.73
CURVE 3*	
λ = 0.70	
T = 298	
15	0.77
30	0.75
40	0.74
50	0.71
60	0.70
CURVE 4*	
λ = 1.00	
T = 298	
15	0.84
30	0.82
45	0.80
60	0.78
CURVE 5*	
λ = 1.54	
T = 298	
15	0.89
20	0.89
30	0.88
40	0.87
45	0.86
50	0.85
60	0.84
CURVE 6*	
λ = 0.0507	
T = 298	
30.0	0.084
34.8	0.123
CURVE 6 (cont.)*	
39.9	0.135
44.6	0.177
49.8	0.226
54.8	0.291
57.0	0.322
59.9	0.355
62.1	0.383
64.7	0.423
67.1	0.461
69.6	0.530
CURVE 7*	
λ = 0.0762	
T = 298	
30.0	0.191
35.1	0.208
39.9	0.224
44.8	0.223
49.9	0.243
54.9	0.271
60.0	0.318
64.8	0.343
69.9	0.423
CURVE 8*	
λ = 0.1032	
T = 298	
30.0	0.188
35.0	0.208
40.0	0.205
44.9	0.224
50.0	0.224
55.0	0.239
59.8	0.259
65.1	0.299
67.2	0.335
69.5	0.374

* No plot given

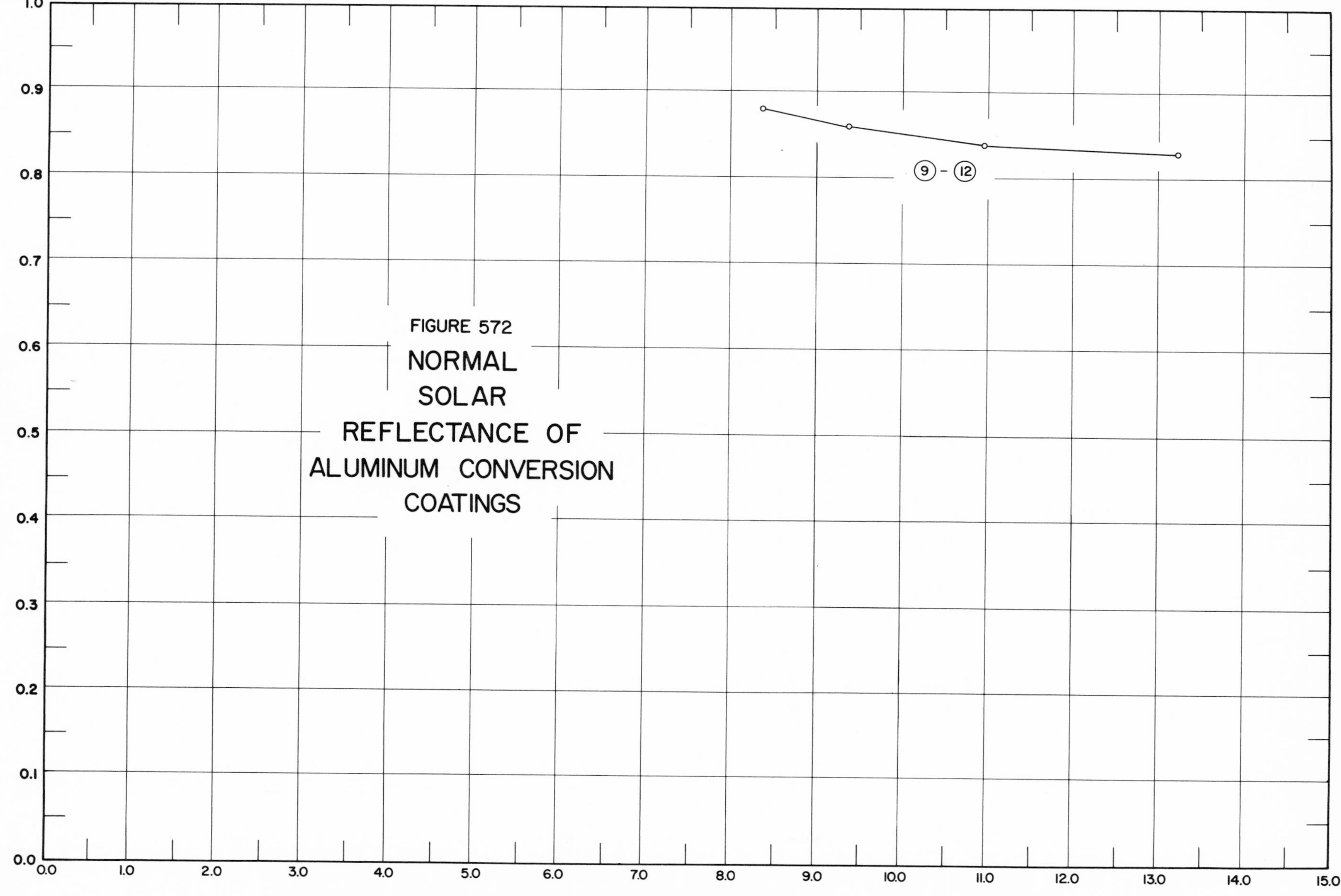

FIGURE 572 NORMAL SOLAR REFLECTANCE OF ALUMINUM CONVERSION COATINGS

SPECIFICATION TABLE NO. 572 NORMAL SOLAR REFLECTANCE OF ALUMINUM CONVERSION COATINGS

Curve No.	Ref. No.	Year	Temperature Range, K	Geometry θ	θ'	ω'	Reported Error, %	Composition (weight percent), Specifications and Remarks
1*	265	1963	~293	~0°	~0°			1199 aluminum anodized 15 min in sulfuric acid; coating thickness 9.38 μm; substrate polished by Alzak process (electrolytic fluoboric acid bath); calculated from spectral reflectance data.
2*	265	1963	~298	~0°	~0°			Similar to above specimen and conditions except heat treated (24 hrs at 589 K) in vacuum (5 x 10^{-5} mm Hg).
3*	265	1963	~298	~0°	~0°			Similar to above specimen and conditions except heat treated (48 hrs at 589 K) in vacuum (5 x 10^{5} mm Hg).
4*	265	1963	~298	~0°	~0°			Similar to above specimen and conditions except heat treated (96 hrs at 589 K) in vacuum (5 x 10^{-5} mm Hg).
5*	265	1963	~298	~0°	~0°			Similar to above specimen and conditions except anodized 25 min; coating thickness 13.21 μm.
6*	265	1963	~298	~0°	~0°			Similar to curve 5 specimen and conditions except heat treated (24 hrs at 589 K) in vacuum (5 x 10^{-5} mm Hg).
7*	265	1963	~298	~0°	~0°			Similar to curve 5 specimen and conditions except heat treated (48 hrs at 589 K) in vacuum (5 x 10^{-5} mm Hg).
8*	265	1963	~298	~0°	~0°			Similar to curve 5 specimen and conditions except heat treated (96 hrs at 589 K) in vacuum (5 x 10^{-5} mm Hg).
9	265	1963	~298	~0°	~0°			1199 aluminum anodized 10 min in sulfuric acid; coating thickness 8.36 μm; substrate polished in phosphoric/nitric acid bath for 2 min at 364 K; calculated from spectral reflectance data.
10	265	1963	~298	~0°	~0°			Similar to above specimen and conditions except anodizing time 15 min; coating thickness 9.38 μm.
11	265	1963	~298	~0°	~0°			Similar to above specimen and conditions except anodized 20 min; coating thickness 10.96 μm.
12	265	1963	~298	~0°	~0°			Similar to above specimen and conditions except anodized 25 min; coating thickness 13.21 μm.

* Not shown on plot

DATA TABLE NO. 572 NORMAL SOLAR REFLECTANCE OF ALUMINUM CONVERSION COATINGS

[Temperature, T, K; Reflectance, ρ]

T	ρ
CURVE 1*	
298	0.84
CURVE 2*	
298	0.82
CURVE 3*	
298	0.81
CURVE 4*	
298	0.82
CURVE 5*	
298	0.82
CURVE 6*	
298	0.78
CURVE 7*	
298	0.77
CURVE 8*	
298	0.78
CURVE 9	
298	0.88
CURVE 10	
298	0.86
CURVE 11	
298	0.84
CURVE 12	
298	0.83

* Not shown on plot

SPECIFICATION TABLE NO. 573 HEMISPHERICAL INTEGRATED ABSORPTANCE OF ALUMINUM CONVERSION COATINGS

Curve No.	Ref. No.	Year	Temperature Range, K	Reported Error, %	Composition (weight percent), Specifications and Remarks
1*	195	1953	76		Sheet (0.508 mm thick); electrolytically polished and heavily anodized; measured in vacuum ($<10^{-6}$ mm Hg); absorptance for 294 K blackbody incident radiation.
2*	195	1953	76		Sheet (0.508 mm thick); electrolytically polished and lightly anodized; measured in vacuum ($<10^{-6}$ mm Hg); absorptance for 294 K blackbody incident radiation.

DATA TABLE NO. 573 HEMISPHERICAL INTEGRATED ABSORPTANCE OF ALUMINUM CONVERSION COATINGS

[Temperature, T, K; Absorptance, α]

T	α
CURVE 1*	
76	0.324
CURVE 2*	
76	0.2082

* No plot given

FIGURE 574

NORMAL SOLAR ABSORPTANCE OF ALUMINUM CONVERSION COATINGS

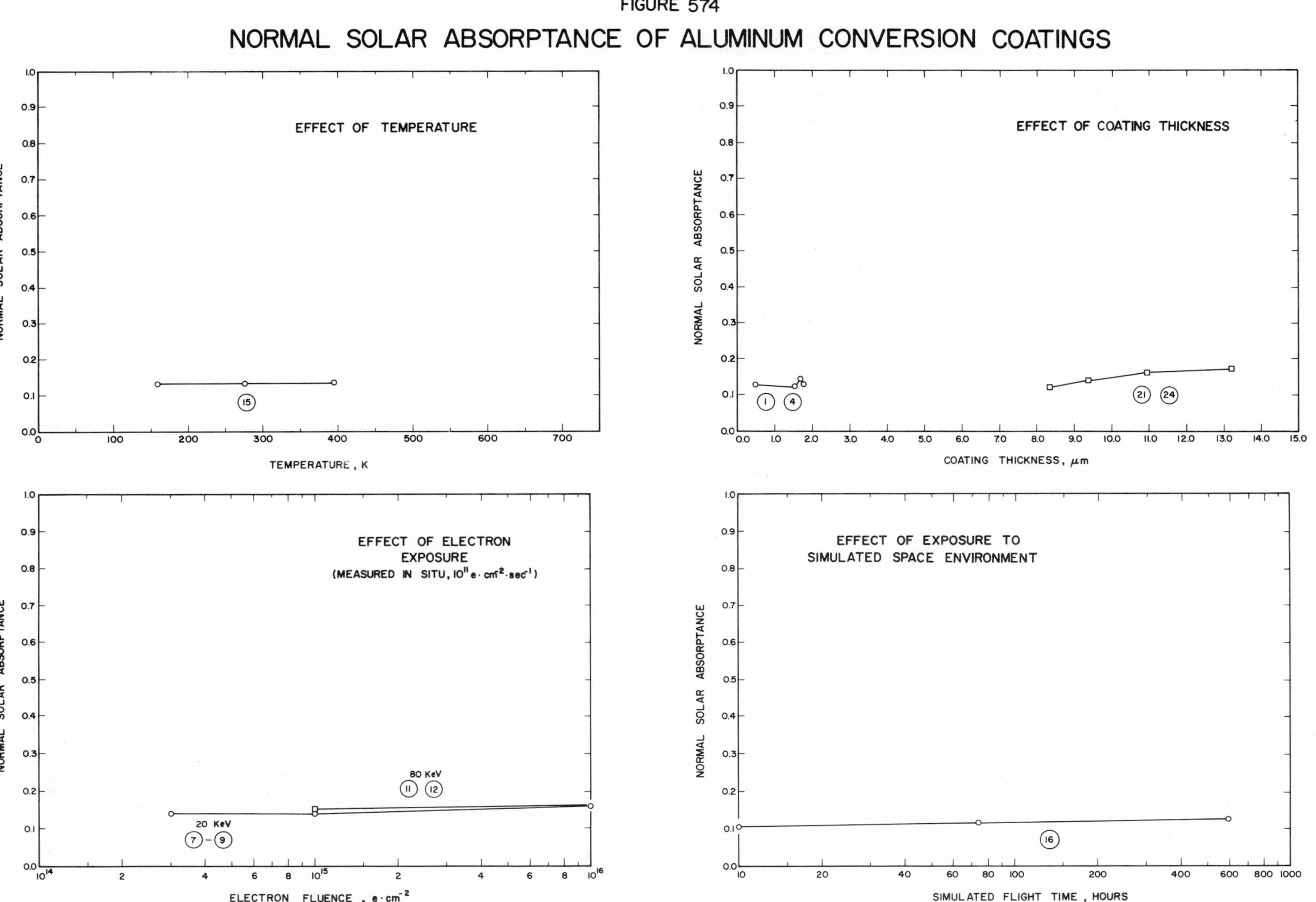

SPECIFICATION TABLE NO. 574 NORMAL SOLAR ABSORPTANCE OF ALUMINUM CONVERSION COATINGS

Curve No.	Ref. No.	Year	Temperature Range, K	Geometry θ	Reported Error, %	Composition (weight percent), Specifications and Remarks
1	260	1967	~298	~0°		Barrier-layer anodic coating on a high-purity aluminum substrate; coating thickness 0.47 μm; coating applied as follows: substrate thoroughly cleaned, electropolished in a fluoboric acid solution, anodized in aqueous ammonium tartrate solution diluted with ethyl alcohol, then immersed in a phosphoric acid solution; anodization voltage 361 V; measured in vacuum (~10^{-6} mm Hg); computed from spectral reflectance data.
2	260	1967	~298	~0°		Similar to above specimen and conditions except coating thickness 1.43 μm; anodization voltage 1089 V.
3	260	1967	~298	~0°		Similar to above specimen and conditions except coating thickness 1.58 μm; anodization voltage 1208 V.
4	260	1967	~298	~0°		Similar to above specimen and conditions except coating thickness 1.78 μm; anodization voltage 1360 V.
5*	260	1967	~298	~0°		Similar to above specimen and conditions.
6*	45	1969	298	~0°		Alzak anodized aluminum (3.81 μm thick); absorptance calculated from normal spectral reflectance; property measured in air. [Authors' designation: Z_3]
7	45	1969	281	~0°		Above specimen and conditions except exposed to 20 keV electrons (flux 1 x 10^{10}-5 x 10^{11} e cm^{-2} sec^{-1}) in dark in vacuum (10^{-8} mm Hg); vacuum maintained by ion pump; substrate held at 281 ± 2 K; property measured in situ after 3 x 10^{14} e cm^{-2}.
8	45	1969	281	~0°		Above specimen and conditions except property measured in situ after 10^{15} e cm^{-2}.
9	45	1969	281	~0°		Above specimen and conditions except property measured in situ after 10^{16} e cm^{-2}.
10*	45	1969	298	~0°		Similar to curve 6 specimen and conditions.
11	45	1969	281	~0°		Above specimen and conditions except exposed to 80 keV electrons (flux 1 x 10^{10}-5 x 10^{11} e cm^{-2} sec^{-1}) in dark in vacuum (10^{-8} mm Hg); vacuum maintained by ion pump; substrate held at 281 ± 2 K; property measured in situ after 10^{15} e cm^{-2}.
12	45	1969	281	~0°		Above specimen and conditions except property measured in situ after 10^{16} e cm^{-2}.
13*	266	1965	400-600	0°	± 5	Anodized; layer of carbon electrophoretically deposited into the pores of the surface; measured in vacuum (10^{-7} mm Hg).
14*	266	1965	400-600	0°	± 5	Similar to above specimen and conditions.
15	85	1969	158-395	~0°		Anodized 1199 aluminum; substrate alkaline electropolished (sodium phosphate and sodium carbonate) for 15 min at 353 K and 12 VDC, then anodized for 15 min at 18 VDC in 10% sulfuric acid; property measured in vacuum by calorimetric method; data extracted from smooth curve. [Author's designation: B-3 Sulfuric Acid Anodize, NRDL-RTD-81-7]
16	85	1969	unknown	~0°		Barrier anodized 1199 aluminum (~0.254 mm thick); exposed to UV in vacuum; data extracted from smooth curve; property converted from spectral reflectance measured in situ; simulated flight time (hrs) is variable; actual UV ESH is one-third of flight time. [Author's designation: A-2 Barrier Anodize]
17*	265	1963	~293	~0°		1199 aluminum anodized 15 min in sulfuric acid; coating thickness 9.38 μm; substrate polished by Alzak process; calculated from spectral reflectance data.

* Not shown on plot

SPECIFICATION TABLE NO. 574 NORMAL SOLAR ABSORPTANCE OF ALUMINUM CONVERSION COATINGS (continued)

Curve No.	Ref. No.	Year	Temperature Range, K	Geometry θ	Reported Error, %	Composition (weight percent), Specifications and Remarks
18*	265	1963	~298	~0°		Similar to above specimen and conditions except heat treated 96 hrs at 589 K in vacuum (5×10^{-5} mm Hg).
19*	265	1963	~298	~0°		Similar to above specimen and conditions except anodized for 25 min; coating thickness 9.38 μm.
20*	265	1963	~298	~0°		Similar to above specimen and conditions except heat treated 96 hrs at 589 K in vacuum (5×10^{-5} mm Hg).
21	265	1963	~298	~0°		1199 aluminum anodized 10 min in sulfuric acid; coating thickness 8.36 μm; substrate polished in phosphoric/nitric acid bath for 2 min at 364 K; calculated from spectral reflectance data.
22	265	1963	~298	~0°		Similar to above specimen and conditions except anodized 15 min; coating thickness 9.38 μm.
23	265	1963	~298	~0°		Similar to above specimen and conditions except anodized 20 min; coating thickness 10.96 μm.
24	265	1963	~298	~0°		Similar to above specimen and conditions except anodized 25 min; coating thickness 13.21 μm.
25*	75	1969	77	~0°		Barrier anodized 1199 aluminum (0.254 mm thick); exposed to vacuum (5×10^{-9}-2×10^{-7} mm Hg); vacuum maintained with diffusion pump; absorptance calculated from R (2π, 0°); property measured in situ.
26*	75	1969	77	~0°		Similar to above specimen and conditions except exposed to 6 sun intensity UV radiation in vacuum; 350 ESH.
27*	75	1969	77	~0°		Similar to curve 25 specimen and conditions except exposed to electron radiation (8.6×10^{10}-1.6×10^{12} e cm^{-2} sec^{-1}) in vacuum; 5.8×10^{15} e cm^{-2}.
28*	75	1969	77	~0°		Similar to curve 25 specimen and conditions except simultaneous exposure to electron (3.5×10^{10} e cm^{-2} sec^{-1}) and UV (8 sun intensity) radiation in vacuum; 5.8×10^{15} e cm^{-2}; 350 ESH.
29*	75	1969	77	~0°		1199 aluminum; sulfuric acid-anodized; exposed to vacuum (5×10^{-9}-2×10^{-7} mm Hg); vacuum maintained with diffusion pump; absorptance calculated from R (2π, 0°); property measured in situ.
30*	75	1969	77	~0°		Similar to above specimen and conditions except exposed to 6 sun intensity UV radiation in vacuum; 350 ESH.
31*	75	1969	77	~0°		Similar to curve 29 specimen and conditions except exposed to electron radiation (8.6×10^{10}-1.6×10^{12} e cm^{-2} sec^{-1}) in vacuum; 5.8×10^{15} e cm^{-2}.
32*	75	1969	77	~0°		Similar to curve 29 specimen and conditions except simultaneous exposure to electron (3.5×10^{10} e cm^{-2} sec^{-1}) and UV (8 sun intensity) radiation in vacuum; 5.8×10^{15} e cm^{-2}; 350 ESH.
33*	115	1960	298	15°		1199 aluminum foil (0.0508 mm thick); bright dipped; anodized on both sides (0.0076 mm thick) in dilute sulfuric acid; computed from spectral reflectance data for above atm conditions. [Authors' designation: Sample 1]

* Not shown on plot

DATA TABLE NO. 574 NORMAL SOLAR ABSORPTANCE OF ALUMINUM CONVERSION COATINGS

[Temperature, T, K; Absorptance, α]

T	α
CURVE 1	
298	0.129
CURVE 2	
298	0.122
CURVE 3	
298	0.142
CURVE 4	
298	0.129
CURVE 5*	
298	0.136
CURVE 6*	
298	0.13
CURVE 7	
281	0.14
CURVE 8	
281	0.14
CURVE 9	
281	0.16
CURVE 10*	
298	0.14
CURVE 11	
281	0.15
CURVE 12	
281	0.16
CURVE 13*	
400	0.97
600	0.97
CURVE 14*	
400	0.97
600	0.97
CURVE 15	
158	0.131
276	0.133
395	0.136
Flight time	**α**
CURVE 16 T = unknown	
10.0	0.109
74.6	0.116
596	0.126
T	**α**
CURVE 17*	
298	0.16
CURVE 18*	
298	0.18
CURVE 19*	
298	0.18
CURVE 20*	
298	0.22
CURVE 21	
298	0.12
CURVE 22	
298	0.14
CURVE 23	
298	0.16
CURVE 24	
298	0.17
CURVE 25*	
77	0.17
CURVE 26*	
77	0.19
CURVE 27*	
77	0.16
CURVE 28*	
77	0.20
CURVE 29*	
77	0.20
CURVE 30*	
77	0.28
CURVE 31*	
77	0.20
CURVE 32*	
77	0.27
CURVE 33*	
298	0.16

* Not shown on plot

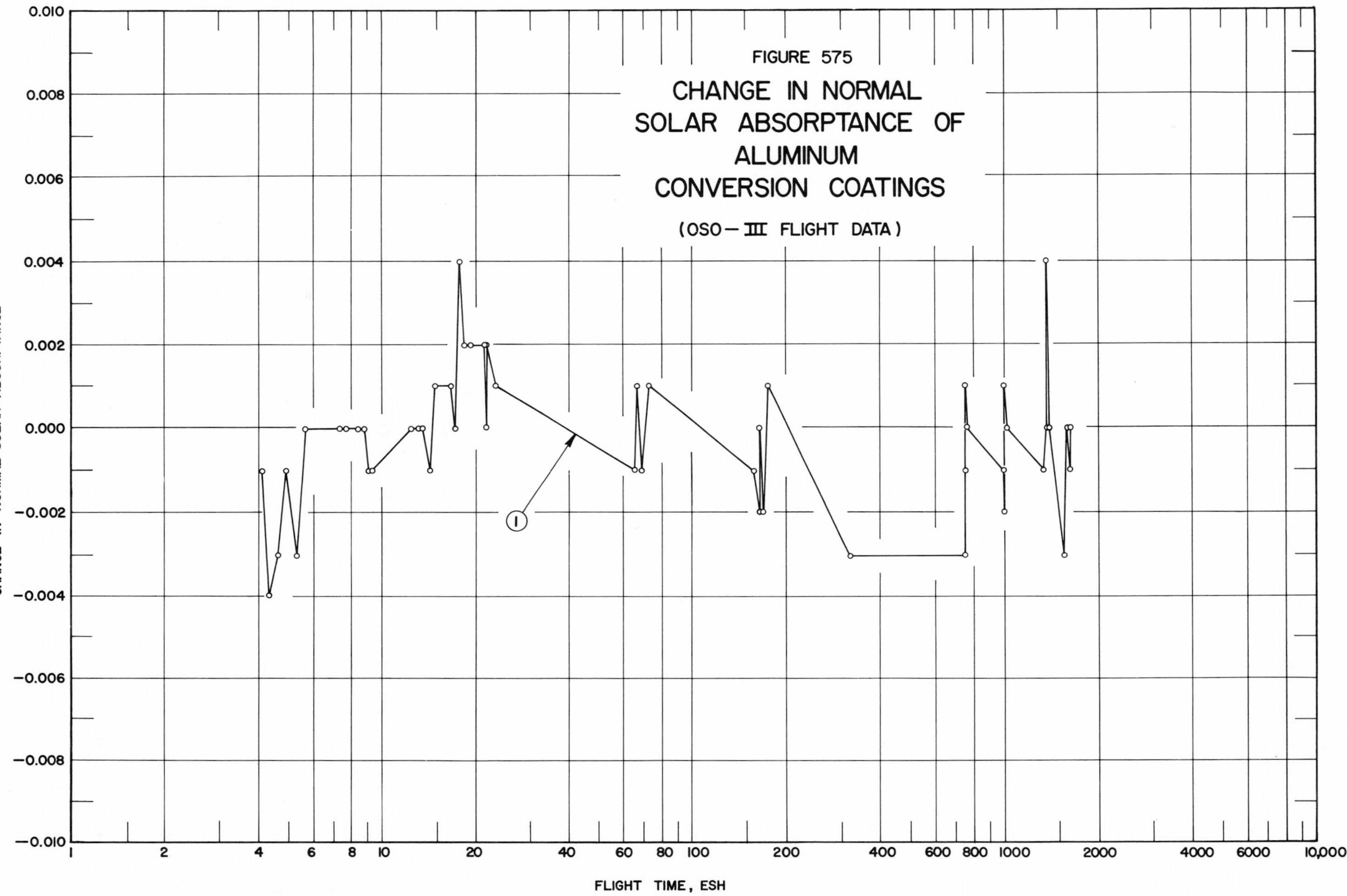

FIGURE 575
CHANGE IN NORMAL SOLAR ABSORPTANCE OF ALUMINUM CONVERSION COATINGS
(OSO-III FLIGHT DATA)

SPECIFICATION TABLE NO. 575 CHANGE IN NORMAL SOLAR ABSORPTANCE OF ALUMINUM CONVERSION COATINGS

Curve No.	Ref. No.	Year	Temperature Range, K	Geometry θ	Reported Error, %	Composition (weight percent), Specifications and Remarks
1	65	1969	~300	~0°		Anodized 1199 Al alloy; electropolished in a solution of fluoboric acid; anodized in a solution of ammonium tartrate; property calculated from temp of substrate from in flight data of OSO III; ESH is variable. [Author's designation: Anodic]

DATA TABLE NO. 575 CHANGE IN NORMAL SOLAR ABSORPTANCE OF ALUMINUM CONVERSION COATINGS

[Temperature, T, K; Absorptance, α]

ESH	$\Delta\alpha$	ESH	$\Delta\alpha$	ESH	$\Delta\alpha$
CURVE 1 T~300		CURVE 1 (cont.)		CURVE 1 (cont.)	
		16.9	0.001	748.2	-0.001
0.15	0.003*	17.4	0.000	748.2	0.001
0.47	0.000*	17.9	0.004	758.6	0.000
4.1	-0.001	18.4	0.002	1000	-0.002
4.3	-0.004	19.3	0.002	995.4	-0.001
4.6	-0.003	21.6	0.002	1000	0.001
4.9	-0.001	21.9	0.000	1018.6	0.000
5.3	-0.003	21.9	0.002	1349.0	-0.001
5.6	0.000	23.4	0.001	1355.2	0.000
7.3	0.000	65.2	-0.001	1367.7	0.004
7.6	0.000	66.4	0.001	1386.7	0.000
8.4	0.000	68.2	-0.001	1577.6	-0.003
8.8	0.000	72.4	0.001	1577.6	0.000
9.1	-0.001	159.2	-0.001	1629.3	-0.001
9.4	-0.001	165.9	-0.002	1629.3	0.000
12.4	0.000	165.2	0.000		
13.1	0.000	170.6	-0.002		
13.5	0.000	176.2	0.001		
14.2	-0.001	323.6	-0.003		
14.7	0.001	748.2	-0.003		

SPECIFICATION TABLE NO. 576 ANGULAR SOLAR ABSORPTANCE OF ALUMINUM CONVERSION COATINGS

Curve No.	Ref. No.	Year	Temperature, K	Angular Range, °	Geometry θ	Reported Error, %	Composition (weight percent), Specifications, and Remarks
1*	115	1960	298	15-60	θ		Type 1199 aluminum foil (0.051 mm thick) from Alcoa Research Lab; bright dipped; anodized on both sides (0.0076 mm thick) in dilute sulfuric acid; computed from spectral reflectance data for above atm conditions.

DATA TABLE NO. 576 ANGULAR SOLAR ABSORPTANCE OF ALUMINUM CONVERSION COATINGS

[Angle, θ, °; Absorptance, α; Temperature, T, K]

θ	α
CURVE 1* T = 298	
15	0.190
30	0.208
45	0.243
60	0.254

* No plot given

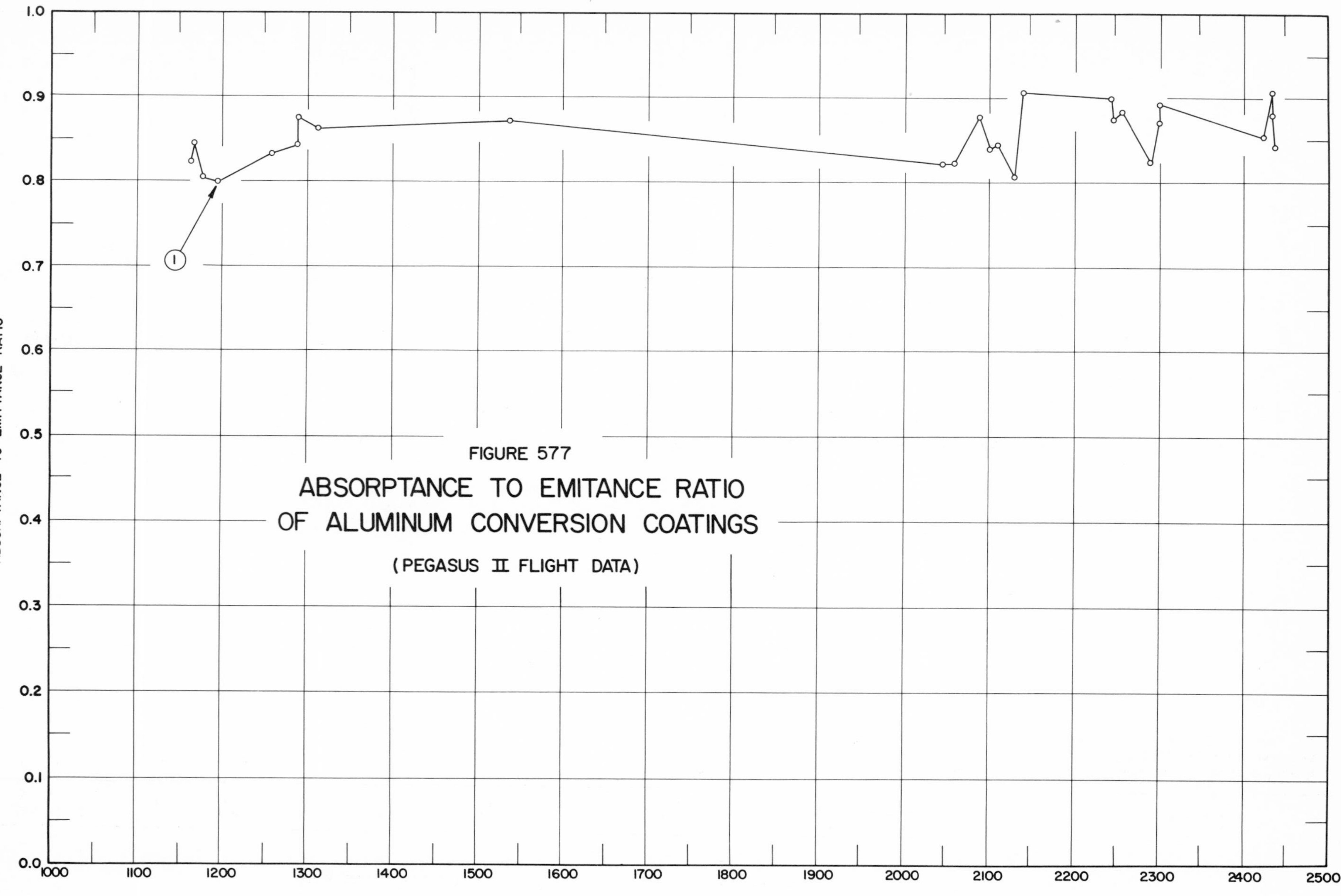

FIGURE 577
ABSORPTANCE TO EMITANCE RATIO
OF ALUMINUM CONVERSION COATINGS
(PEGASUS II FLIGHT DATA)

SPECIFICATION TABLE NO. 577 ABSORPTANCE TO EMITTANCE RATIO OF ALUMINUM CONVERSION COATINGS

Curve No.	Ref. No.	Year	Temperature Range, K	Reported Error, %	Composition (weight percent), Specifications and Remarks
1	57	1967	~300		Alodine aluminum; data taken from in-flight data of Pegasus II; property deduced from temp of substrate; equivalent sun hrs is variable.

DATA TABLE NO. 577 ABSORPTANCE TO EMITTANCE RATIO OF ALUMINUM CONVERSION COATINGS

[Temperature, T, K; Absorptance, α; Emittance, ϵ]

ESH	α/ϵ	ESH	α/ϵ
CURVE 1 T ~ 300		CURVE 1 (cont.)	
		2140	0.908
1162	0.824	2243	0.900
1168	0.845	2246	0.873
1178	0.804	2259	0.882
1196	0.800	2290	0.825
1259	0.835	2300	0.870
1289	0.843	2300	0.893
1289	0.876	2423	0.854
1312	0.862	2432	0.880
1539	0.871	2431	0.907
2049	0.821	2439	0.841
2060	0.823		
2090	0.878		
2100	0.840		
2110	0.846		
2130	0.809		

SPECIFICATION TABLE NO. 578 HEMISPHERICAL TOTAL EMITTANCE OF ALUMINUM + COPPER + ΣX_i CONVERSION COATINGS

Curve No.	Ref. No.	Year	Temperature Range, K	Reported Error, %	Composition (weight percent), Specifications and Remarks
1*	106	1963	196-319	<2.5	Anodized 2014 T6 aluminum; nominal composition: 3.9-5.0 Cu, 0.50-1.2 Si, 0.40-1.2 Mn, 1.0 Fe, 0.20-0.8 Mg, 0.25 Zn, 0.15 Ti, 0.10 Cr and 0.15 others; measured in vacuum (5×10^{-6} mm Hg).
2*	107	1967	175-359	~8.5	Alodine 1200; substrate 2024-T3 aluminum; nominal composition: 3.8-4.9 Cu, 1.2-1.8 Mg, 0.30-0.9 Mn, 0.50 Fe, 0.50 Si, 0.25 Zn, 0.10 Cr and 0.15 others; measured in vacuum ($\sim 10^{-6}$ mm Hg).

DATA TABLE NO. 578 HEMISPHERICAL TOTAL EMITTANCE OF ALUMINUM + COPPER + ΣX_i CONVERSION COATINGS

[Temperature, T, K; Emittance, ϵ]

T	ϵ
CURVE 1*	
196	0.626
227	0.582
260	0.675
285	0.657
319	0.595
CURVE 2*	
175	0.175
196	0.165
216	0.140
249	0.126
293	0.106
353	0.099
359	0.099

* No plot given

SPECIFICATION TABLE NO. 579 NORMAL TOTAL EMITTANCE OF ALUMINUM + COPPER + ΣX_i CONVERSION COATINGS

Curve No.	Ref. No.	Year	Temperature Range, K	Geometry θ'	Reported Error, %	Composition (weight percent), Specifications and Remarks
1*	7	1961	373	~0°		Aluminum 24-ST (2024-T) anodized in chromic acid; nominal composition: 4.5 Cu, 1.5 Mg, 0.6 Mn, Al balance; measured in vacuum (<1 x 10^{-5} mm Hg).
2*	68	1969	300	~0°		Aluminum 2024 (~1.59 mm thick) anodized in sulfuric acid; (coating thickness ~0.0127 mm); emittance calculated from reflectance; property measured in air.
3*	68	1969	300	~0°		Similar to above specimen and conditions except exposed to vacuum (10^{-8} mm Hg) for 300 hrs; vacuum maintained by ion pump; property measured in air after vacuum exposure.
4*	68	1969	300	~0°		Similar to curve 2 specimen and conditions except exposed to UV radiation in vacuum (10^{-5} mm Hg) for 300 hrs; vacuum maintained by oil-diffusion pump; H_2 gas UV source; property measured in air after exposure.
5*	68	1969	300	~0°		Similar to above specimen and conditions except He gas UV source.

DATA TABLE NO. 579 NORMAL TOTAL EMITTANCE OF ALUMINUM + COPPER + ΣX_i CONVERSION COATINGS

[Temperature, T, K; Emittance, ϵ]

T	ϵ	T	ϵ
CURVE 1*		CURVE 5*	
373	0.55	300	0.78
CURVE 2*			
300	0.78		
CURVE 3*			
300	0.78		
CURVE 4*			
300	0.78		

* No plot given

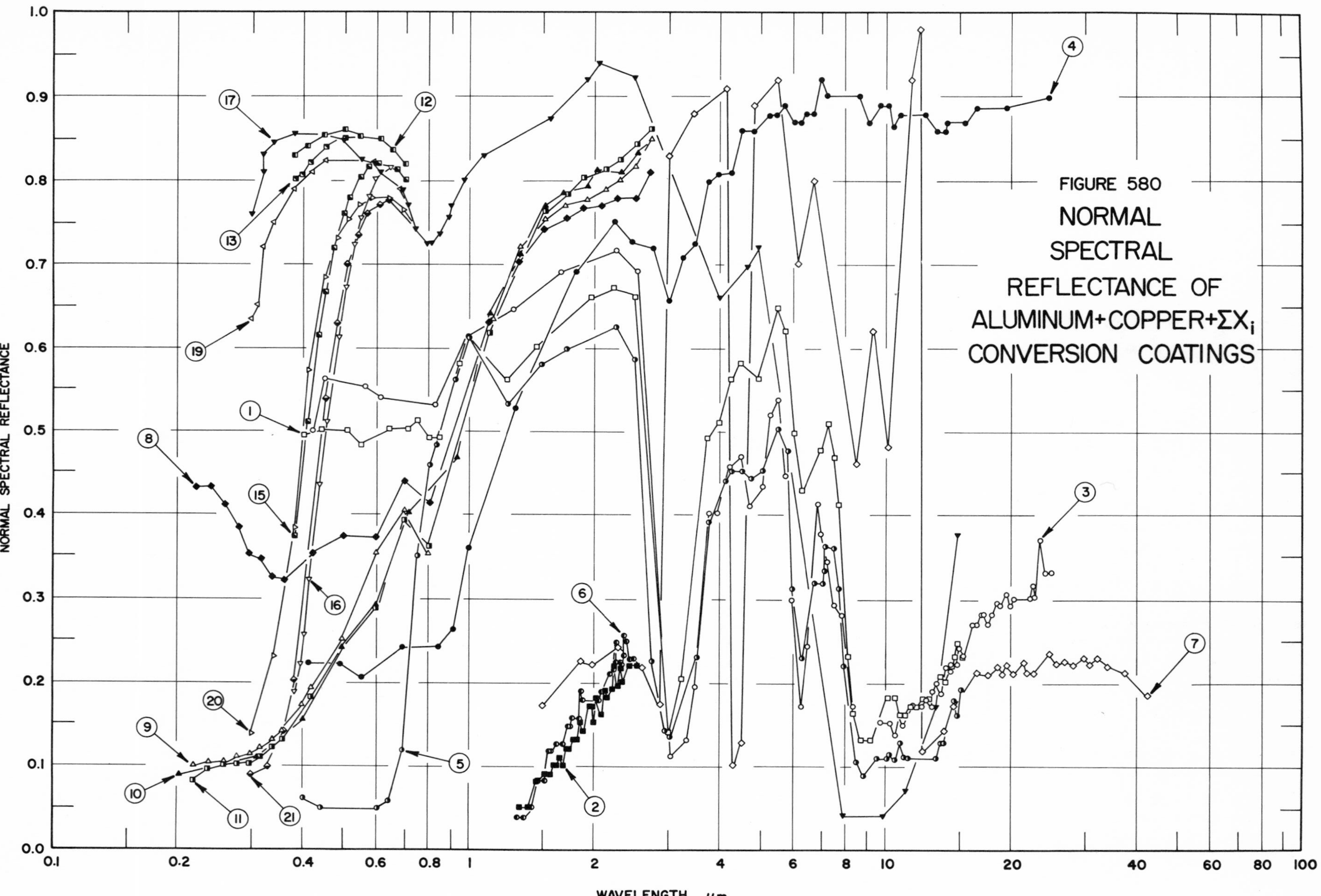
FIGURE 580
NORMAL
SPECTRAL
REFLECTANCE OF
ALUMINUM+COPPER+ΣX_i
CONVERSION COATINGS
NORMAL SPECTRAL REFLECTANCE
WAVELENGTH, μm
0.0
0.1
0.2
0.3
0.4
0.5
0.6
0.7
0.8
0.9
1.0
0.1
0.2
0.4
0.6
0.8
1
2
4
6
8
10
20
40
60
80
100
1
2
3
4
5
6
7
8
9
10
11
12
13
15
16
17
19
20
21

SPECIFICATION TABLE NO. 580 NORMAL SPECTRAL REFLECTANCE OF ALUMINUM + COPPER + ΣX_i CONVERSION COATINGS

Curve No.	Ref. No.	Year	Temperature, K	Wavelength Range, μm	Geometry θ	θ'	ω'	Reported Error, %	Composition (weight percent), Specifications, and Remarks
1	268	1960	298	0.40-15.22	$\sim 0^\circ$	$\sim 0^\circ$			Aluminum 24S-T81 (2024-T81) anodized in sulfuric acid; nominal composition: 4.5 Cu, 1.5 Mg, 0.6 Mn, Al balance; data extracted from smooth curve.
2	268	1960	298	13.26-25.04	$\sim 0^\circ$	$\sim 0^\circ$			Similar to above specimen and conditions.
3	268	1960	298	0.41-24.93	$\sim 0^\circ$	$\sim 0^\circ$			Similar to above specimen and conditions except substrate was chem-milled.
4	268	1960	298	0.41-24.28	$\sim 0^\circ$	$\sim 0^\circ$			Similar to curve 1 specimen and conditions except anodized in chromic acid.
5	268	1960	298	0.40-15.02	$\sim 0^\circ$	$\sim 0^\circ$			Similar to above specimen and conditions except anodized in black chromic acid.
6	268	1960	298	13.00-24.93	$\sim 0^\circ$	$\sim 0^\circ$			Similar to above specimen and conditions.
7	116	1963	311	1.50-42.3	$\sim 0^\circ$		2π		Aluminum 2014-T6 hard-coat anodized in sulfuric acid; coating thickness 0.0508 mm; nominal composition: 4.4 Cu, 0.8 Mn, 0.8 Si and 0.40 Mg; converted from R $(2\pi, 0^\circ)$.
8	1	1960	378	0.221-2.710	$\sim 0^\circ$		2π		Anodized aluminum alloy 24S-T (2024-T); nominal composition: 4.5 Cu, 1.5 Mg, 0.6 Mn, Al balance; cleaned; measured in vacuum (10^{-5} mm Hg) relative to MgO.
9	1	1960	378	0.219-2.705	$\sim 0^\circ$		2π		Above specimen and conditions except exposed to UV radiation for 20 hrs from GE UA-3 lamp.
10	1	1960	378	0.204-2.739	$\sim 0^\circ$		2π		Above specimen and conditions except exposed to UV radiation for 60 hrs.
11	1	1960	378	0.219-2.710	$\sim 0^\circ$		2π		Above specimen and conditions except exposed to UV radiation for 100 hrs.
12	68	1969	300	0.3800-0.7000	$\sim 0^\circ$		2π		Aluminum 2024 ($\sim$1.59 mm thick) anodized in sulfuric acid; coating thickness $\sim$0.0127 mm; nominal composition: 4.5 Cu, 1.5 Mg and 0.6 Mn; data extracted from smooth curve.
13	68	1969	300	0.3800-0.7000	$\sim 0^\circ$		2π		Similar to above specimen and conditions except exposed to vacuum (10^{-8} mm Hg) for 300 hrs; vacuum maintained by ion pump; reflectance measured in air after vacuum exposure.
14	68	1969	300	0.3800-0.7000	$\sim 0^\circ$		2π		Similar to curve 12 specimen and conditions except exposed to vacuum (10^{-5} mm Hg) for 300 hrs; vacuum maintained by oil-diffusion pump; reflectance measured in air after vacuum exposure.
15	68	1969	300	0.3800-0.7000	$\sim 0^\circ$		2π		Similar to above specimen and conditions except exposed to UV radiation in vacuum for 300 hrs; H_2 gas UV source.
16	68	1969	300	0.3800-0.7000	$\sim 0^\circ$		2π		Similar to above specimen and conditions except He gas UV source.
17	68	1969	300	0.30-14.97	$\sim 0^\circ$		2π		Similar to curve 12 specimen and conditions.
18	68	1969	300	0.30-14.97	$\sim 0^\circ$		2π		Similar to curve 13 specimen and conditions.
19	68	1969	300	0.30-14.97	$\sim 0^\circ$		2π		Similar to curve 14 specimen and conditions.
20	68	1969	300	0.30-14.97	$\sim 0^\circ$		2π		Similar to curve 15 specimen and conditions.
21	68	1969	300	0.30-14.97	$\sim 0^\circ$		2π		Similar to curve 16 specimen and conditions.

DATA TABLE NO. 580 NORMAL SPECTRAL REFLECTANCE OF ALUMINUM + COPPER + ΣX_i CONVERSION COATINGS

[Wavelength, λ, μm; Reflectance, ρ; Temperature, T, K]

CURVE 1, T = 298

λ	ρ
0.40	0.496
0.44	0.501
0.51	0.501
0.55	0.484
0.64	0.503
0.71	0.503
0.75	0.514
0.80	0.493
0.85	0.493
0.94	0.580
0.99	0.615
1.24	0.560
1.46	0.602
1.97	0.661
2.22	0.671
2.49	0.662
3.00	0.139
3.24	0.204
3.72	0.492
3.98	0.509
4.23	0.562
4.47	0.582
4.99	0.564
5.46	0.648
5.74	0.620
6.00	0.497
6.25	0.428
6.99	0.475
7.28	0.508
7.54	0.467
7.70	0.412
8.15	0.232
8.38	0.164
8.71	0.131
9.20	0.131
10.20	0.182
10.47	0.182
10.93	0.161
11.20	0.161
11.45	0.172
11.90	0.172
12.25	0.180
12.77	0.180
13.48	0.208
13.92	0.201

CURVE 1 (cont.)

λ	ρ
14.30	0.220
14.64	0.231
14.91	0.247
15.22	0.231

CURVE 2, T = 298

λ	ρ
13.26	0.051
13.92	0.051
14.48	0.080
14.69	0.080
15.21	0.089
15.74	0.089
16.03	0.100
16.33	0.100
16.51	0.110
16.74	0.101
17.26	0.120
17.55	0.120
17.80	0.132
18.27	0.132
18.53	0.151
18.72	0.140
18.98	0.140*
19.49	0.160
19.83	0.160
19.99	0.151
20.27	0.182
20.79	0.161
21.29	0.188
21.50	0.173
22.01	0.191
22.25	0.220
22.50	0.220*
22.77	0.196
23.06	0.217
23.23	0.200
23.58	0.238*
23.82	0.238*
24.06	0.219
25.04	0.219

CURVE 3, T = 298

λ	ρ
0.41	0.500
0.45	0.561
0.56	0.551
0.61	0.538
0.82	0.528
0.99	0.616
1.27	0.645
1.66	0.690
2.26	0.717
2.51	0.692
3.01	0.112
3.30	0.130
3.49	0.194
3.76	0.400
3.92	0.400
4.20	0.456
4.48	0.468
4.73	0.409
5.02	0.433
5.24	0.519
5.48	0.536
5.76	0.446
5.94	0.297
6.24	0.170
6.42	0.244
6.81	0.412
6.99	0.375
7.23	0.344
7.52	0.293
7.81	0.278
8.38	0.170
8.69	0.131*
9.20	0.131*
9.69	0.151
10.29	0.151
10.48	0.137
10.70	0.157*
10.97	0.149
11.44	0.172
12.23	0.170
12.47	0.179
12.75	0.170
12.84	0.188
13.20	0.198
13.58	0.185

CURVE 3 (cont.)

λ	ρ
13.91	0.219
14.17	0.214
14.29	0.221
14.75	0.221
14.96	0.242
15.24	0.233
16.16	0.268
16.69	0.268
16.93	0.280
17.20	0.280
17.41	0.269
17.79	0.281
17.95	0.281
18.44	0.295
18.62	0.291
19.50	0.305
19.95	0.290
20.23	0.299
22.18	0.299
22.43	0.316
22.78	0.301
23.59	0.367
24.02	0.330
24.93	0.331

CURVE 4, T = 298

λ	ρ
0.41	0.223
0.49	0.222
0.55	0.205
0.69	0.240
0.84	0.241
0.91	0.263
0.99	0.357
1.28	0.525
1.80	0.690
2.23	0.752
2.45	0.727
2.76	0.719
3.00	0.655
3.25	0.708
3.48	0.726
3.77	0.798
3.98	0.808
4.23	0.809

CURVE 4 (cont.)

λ	ρ
4.50	0.858
4.80	0.858
5.26	0.879
5.49	0.879
5.76	0.890
6.01	0.870
6.26	0.870
6.41	0.881
6.73	0.881
7.00	0.920
7.24	0.900
8.66	0.900
9.18	0.870
9.76	0.890
10.23	0.890
10.48	0.865
10.80	0.879
12.45	0.879
13.39	0.859
13.97	0.859
14.02	0.870
15.59	0.870
16.57	0.887
19.59	0.887
19.59	0.898*
24.78	0.899

CURVE 5, T = 298

λ	ρ
0.40	0.061
0.44	0.049
0.60	0.049
0.64	0.058
0.69	0.118
0.75	0.348
0.80	0.457
0.83	0.483
0.92	0.560
0.99	0.610
1.24	0.531
1.49	0.579
1.70	0.598
2.23	0.626
2.47	0.586
2.73	0.226

CURVE 5 (cont.)

λ	ρ
2.94	0.141
3.04	0.135
3.50	0.229
3.76	0.389
4.13	0.437
4.23	0.450
4.52	0.450
4.74	0.443
5.02	0.453
5.49	0.503
5.80	0.476
5.97	0.314
6.24	0.229
6.75	0.318
7.02	0.318
7.09	0.333
7.17	0.360
7.49	0.358
7.76	0.313
7.99	0.220
8.50	0.105
8.82	0.088
9.46	0.109
10.02	0.109
10.17	0.115
10.33	0.115*
10.47	0.109
10.75	0.128
11.00	0.112
11.22	0.110
13.26	0.110
13.49	0.127
13.74	0.127
14.49	0.179
14.77	0.160
15.02	0.191

CURVE 6, T = 298

λ	ρ
13.00	0.038
13.53	0.038
13.82	0.052*
14.15	0.052
14.62	0.080*
15.02	0.080

CURVE 6 (cont.)

λ	ρ
15.55	0.117
15.74	0.117
16.04	0.126
16.73	0.126
17.25	0.147
17.62	0.147
17.79	0.156
18.32	0.156
18.51	0.189
18.91	0.177
20.60	0.177
20.89	0.187
21.45	0.187*
21.81	0.210
22.10	0.217
22.27	0.247
22.58	0.226
23.05	0.226
23.21	0.221
23.44	0.233
23.50	0.255
23.79	0.255*
23.92	0.248
24.14	0.227
24.93	0.227

CURVE 7, T = 311

λ	ρ
1.50	0.172
1.86	0.225
1.98	0.222
2.27	0.242
2.61	0.219
2.87	0.174
3.02	0.830
3.45	0.880
4.16	0.910
4.34	0.100
4.50	0.128
4.83	0.890
5.54	0.920
6.16	0.700
6.74	0.800
8.51	0.460
9.31	0.620

CURVE 7 (cont.)

λ	ρ
10.3	0.480
11.6	0.920
12.1	0.980
12.3	0.118
13.7	0.144
14.5	0.172
16.5	0.213
17.5	0.210
18.5	0.219
19.0	0.211
19.5	0.222
20.3	0.211
21.5	0.225
21.8	0.214
22.7	0.214
24.9	0.236
25.8	0.224
26.9	0.226
28.2	0.223
29.9	0.230
30.9	0.223
32.2	0.229
34.0	0.219
37.5	0.214
42.3	0.185

CURVE 8, T = 378

λ	ρ
0.221	0.430
0.240	0.434
0.260	0.411
0.280	0.384
0.298	0.351
0.319	0.345
0.339	0.325
0.360	0.323
0.420	0.353
0.500	0.374
0.599	0.371
0.699	0.438
0.801	0.414
1.106	0.632
1.303	0.703
1.506	0.744
1.704	0.755

* Not shown on plot.

DATA TABLE NO. 580 NORMAL SPECTRAL REFLECTANCE OF ALUMINUM + COPPER + ΣX_i CONVERSION COATINGS (continued)

λ	ρ
CURVE 8 (cont.) T = 378	
1.897	0.769
2.098	0.771
2.298	0.781
2.501	0.780
2.710	0.810
CURVE 9 T = 378	
0.219	0.100
0.239	0.105
0.259	0.106
0.279	0.112
0.299	0.115
0.317	0.122
0.339	0.131
0.358	0.142
0.399	0.173
0.419	0.193
0.498	0.250
0.598	0.354
0.699	0.405
0.799	0.353
1.131	0.635
1.309	0.724
1.505	0.755
1.694	0.771
1.900	0.779
2.102	0.791
2.300	0.804
2.501	0.818
2.705	0.852
CURVE 10 T = 378	
0.204	0.089
0.310	0.109
0.400	0.156
0.495	0.241
0.599	0.290
0.719	0.403
0.934	0.467
1.106	0.643
1.305	0.714
1.506	0.771
1.696	0.787
CURVE 10 (cont.)	
1.904	0.795
2.099	0.814
2.301	0.814
2.507	0.835
2.739	0.852*
CURVE 11 T = 378	
0.219	0.083
0.238	0.095
0.259	0.100
0.278	0.103
0.297	0.103
0.318	0.112
0.338	0.122
0.358	0.130
0.419	0.183
0.499	0.243*
0.599	0.337
0.699	0.394
0.800	0.362
1.108	0.619
1.330	0.712*
1.528	0.766
1.722	0.786
1.893	0.805
2.120	0.813
2.300	0.826
2.508	0.846
2.710	0.864
CURVE 12 T = 300	
0.3800	0.830
0.4078	0.843
0.4492	0.856
0.5006	0.861
0.5495	0.854
0.6139	0.850
0.6563	0.838
0.7000	0.821
CURVE 13 T = 300	
0.3800	0.803
CURVE 13 (cont.)	
0.3995	0.808
0.4192	0.824
0.4523	0.840
0.5056	0.852
0.5495	0.854*
0.6139	0.850*
0.6563	0.838*
0.7000	0.821*
CURVE 14 T = 300	
0.3800	0.768
0.3988	0.781
0.4163	0.803
0.4494	0.822
0.4861	0.834
0.5276	0.842
0.5983	0.842
0.6506	0.830
0.7000	0.819
CURVE 15 T = 300	
0.3800	0.374
0.4097	0.510
0.4373	0.616
0.4532	0.668
0.4749	0.720
0.5000	0.761
0.5176	0.781
0.5446	0.805
0.5727	0.817
0.6046	0.821
0.6681	0.816
0.7000	0.803
CURVE 16 T = 300	
0.3800	0.187
0.3913	0.221
0.4014	0.256
0.4158	0.323
0.4391	0.436
0.4557	0.510
0.4851	0.613
CURVE 16 (cont.)	
0.5049	0.673
0.5268	0.724
0.5478	0.755
0.5736	0.783
0.5980	0.802
0.6415	0.817
0.6681	0.816*
0.7000	0.803*
CURVE 17 T = 300	
0.30	0.760
0.32	0.811
0.32	0.833
0.34	0.845
0.38	0.856
0.45	0.856*
0.50	0.849
0.55	0.826
0.59	0.823
0.61	0.812
0.68	0.790
0.71	0.772
0.74	0.744
0.79	0.727
0.81	0.727
0.85	0.738
0.89	0.757
0.90	0.771
0.97	0.801
1.08	0.832
1.57	0.876
1.91	0.920
2.05	0.939
2.50	0.924
4.00	0.659
4.65	0.697
4.98	0.723
7.93	0.048
9.84	0.048
11.03	0.071
13.07	0.173
14.97	0.375
CURVE 18 T = 300	
0.30	0.717
0.33	0.770
0.37	0.807
0.39	0.822
0.43	0.835
0.53	0.835
0.55	0.826
0.59	0.823
0.61	0.812
0.68	0.790
0.71	0.772
0.74	0.744
0.79	0.727
0.81	0.727
0.85	0.738
0.89	0.757
0.90	0.771
0.97	0.801
1.08	0.832
1.57	0.876
1.91	0.920
2.05	0.939
2.50	0.924
4.00	0.659
4.65	0.697
4.98	0.723
7.93	0.048
9.84	0.048
11.03	0.071
13.07	0.173
14.97	0.375
CURVE 19 T = 300	
0.30	0.636
0.31	0.651
0.32	0.720
0.34	0.751
0.38	0.792
0.42	0.811
0.45	0.826
0.55	0.826*
0.59	0.823
0.61	0.812*
0.68	0.790
0.71	0.772*
CURVE 19 (cont.)	
0.74	0.744*
0.79	0.727*
0.81	0.727*
0.85	0.738*
0.89	0.757*
0.90	0.771*
0.97	0.801*
1.08	0.832*
1.57	0.876*
1.91	0.920*
2.05	0.939*
2.50	0.924*
4.00	0.659*
4.65	0.697*
4.98	0.723*
7.93	0.048*
9.84	0.048*
11.03	0.071*
13.07	0.173*
14.97	0.375*
CURVE 20 T = 300	
0.30	0.137
0.34	0.229
0.38	0.382
0.41	0.571
0.45	0.685
0.48	0.732
0.51	0.755
0.54	0.772
0.58	0.780
0.64	0.780
0.69	0.767
0.74	0.744*
0.79	0.727*
0.81	0.727*
0.85	0.738*
0.89	0.757*
0.90	0.771*
0.97	0.801*
1.08	0.832*
1.57	0.876*
1.91	0.920*
2.05	0.939*
2.50	0.924*
4.00	0.659*
CURVE 20 (cont.)	
4.65	0.697*
4.98	0.723*
7.93	0.048*
9.84	0.048*
11.03	0.071*
13.07	0.173*
14.97	0.375*
CURVE 21 T = 300	
0.30	0.088
0.33	0.097
0.36	0.140
0.38	0.203
0.45	0.537
0.48	0.628
0.51	0.699
0.54	0.735
0.57	0.759
0.61	0.773
0.64	0.777
0.74	0.744*
0.79	0.727*
0.81	0.727*
0.85	0.738*
0.89	0.757*
0.90	0.771*
0.97	0.801*
1.08	0.832*
1.57	0.876*
1.91	0.920*
2.05	0.939*
2.50	0.924*
4.00	0.659*
4.65	0.697*
4.98	0.723*
7.93	0.048*
9.84	0.048*
11.03	0.071*
13.07	0.173*
14.97	0.375*

* Not shown on plot

SPECIFICATION TABLE NO. 581 NORMAL SOLAR REFLECTANCE OF ALUMINUM + COPPER + ΣX_i CONVERSION COATINGS

Curve No.	Ref. No.	Year	Temperature Range, K	Geometry θ	θ'	ω'	Reported Error, %	Composition (weight percent), Specifications and Remarks
1*	268	1960	298	$\sim 0°$		2π		Aluminum 24S-T81 (2024-T) anodized in chromic acid; nominal composition: 4.5 Cu, 1.5 Mg and 0.6 Mn.
2*	268	1960	298	$\sim 0°$		2π		Similar to above specimen and conditions except calculated from spectral data.
3*	268	1960	298	$\sim 0°$		2π		Similar to curve 1 specimen and conditions except anodized in black chromic acid.
4*	268	1960	298	$\sim 0°$		2π		Similar to above specimen and conditions except calculated from spectral data.
5*	268	1960	298	$\sim 0°$		2π		Similar to curve 1 specimen and conditions except anodized in sulfuric acid.
6*	268	1960	298	$\sim 0°$		2π		Similar to above specimen and conditions except calculated from spectral data.
7*	268	1960	298	$\sim 0°$		2π		Similar to curve 5 specimen and conditions except substrate was chem-milled.
8*	268	1960	298	$\sim 0°$		2π		Similar to above specimen and conditions except calculated from spectral data.

DATA TABLE NO. 581 NORMAL SOLAR REFLECTANCE OF ALUMINUM + COPPER + ΣX_i CONVERSION COATINGS

[Temperature, T, K; Reflectance, ρ]

T	ρ	T	ρ
CURVE 1*		CURVE 6*	
298	0.28	298	0.53
CURVE 2*		CURVE 7*	
298	0.30	298	0.64
CURVE 3*		CURVE 8*	
298	0.31	298	0.57
CURVE 4*			
298	0.29		
CURVE 5*			
298	0.56		

* No plot given

SPECIFICATION TABLE NO. 582 NORMAL SOLAR ABSORPTANCE OF ALUMINUM + COPPER + ΣX_i CONVERSION COATINGS

Curve No.	Ref. No.	Year	Temperature Range, K	Geometry θ	Reported Error, %	Composition (weight percent), Specifications and Remarks
1*	68	1969	300	~0°		Aluminum 2024 (~1.59 mm thick) anodized in sulfuric acid; coating thickness ~0.0127 mm; nominal composition: 4.5 Cu, 1.5 Mg and 0.6 Mn; absorptance calculated from reflectance measured in air.
2*	68	1969	300	~0°		Similar to above specimen and conditions except exposed to vacuum (10^{-8} mm Hg) for 300 hrs; vacuum maintained by ion pump; property measured in air after vacuum exposure.
3*	68	1969	300	~0°		Similar to curve 1 specimen and conditions except exposed to vacuum (10^{-5} mm Hg) for 300 hrs; vacuum maintained by oil-diffusion pump; property measured in air after vacuum exposure.
4*	68	1969	300	~0°		Similar to above specimen and conditions except exposed to UV radiation in vacuum for 300 hrs; H_2 gas UV source.
5*	68	1969	300	~0°		Similar to above specimen and conditions except He gas UV source.

DATA TABLE NO. 582 NORMAL SOLAR ABSORPTANCE OF ALUMINUM + COPPER + ΣX_i CONVERSION COATINGS

[Temperature, T, K; Absorptance, α]

T	α
CURVE 1*	
300	0.15
CURVE 2*	
300	0.16
CURVE 3*	
300	0.18
CURVE 4*	
300	0.26
CURVE 5*	
300	0.31

* No plot given

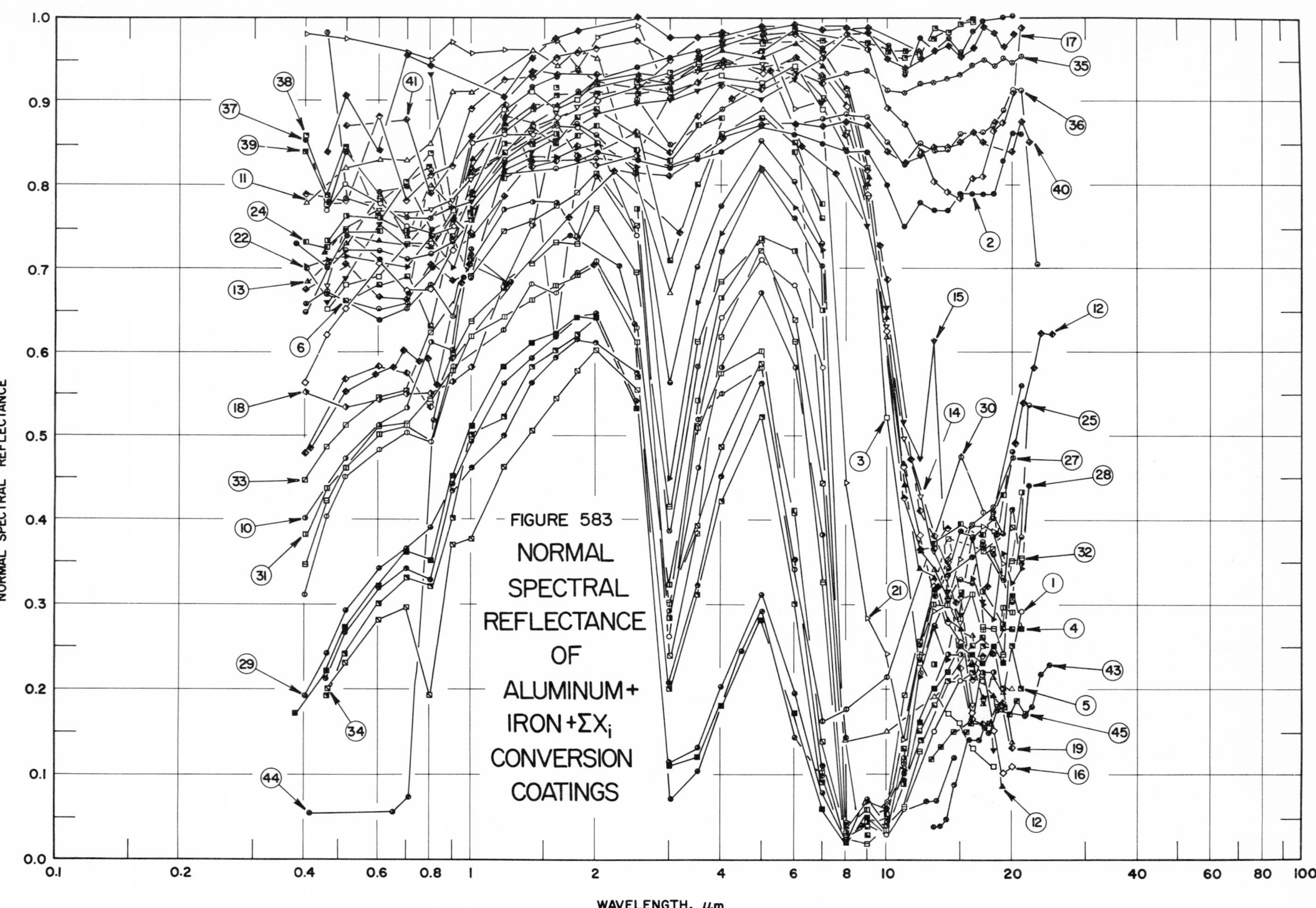

FIGURE 583 NORMAL SPECTRAL REFLECTANCE OF ALUMINUM + IRON + ΣX_i CONVERSION COATINGS

SPECIFICATION TABLE NO. 583 NORMAL SPECTRAL REFLECTANCE OF ALUMINUM + IRON + ΣX_i CONVERSION COATINGS

Curve No.	Ref. No.	Year	Temperature, K	Wavelength Range, μm	Geometry θ	θ'	ω'	Reported Error, %	Composition (weight percent), Specifications, and Remarks
1	270	1962	298	0.45-21.0	$\sim 0^\circ$		2π		1075 aluminum anodized in sulfuric acid for 30 min and sealed 15 min in boiling water; nominal composition: 0.20 Fe, 0.20 Si, 0.04 Cu, 0.04 Zn, 0.03 Mn, 0.03 Mg, 0.03 Ti and 0.15 others; substrate mechanically polished, lapped to roughness < 0.005 μm RMS and electropolished; measured in vacuum (10^{-5} mm Hg); converted from R (2π, 0°).
2	270	1962	298	0.38-21.0	$\sim 0^\circ$		2π		Similar to above specimen and conditions except boric acid anodized and not sealed.
3	270	1962	298	0.45-18.0	$\sim 0^\circ$		2π		Similar to above specimen and conditions except chromic acid anodized and not sealed.
4	270	1962	298	0.38-21.0	$\sim 0^\circ$		2π		Similar to curve 1 specimen and conditions except hard coat anodized for 80 min and sealed.
5	270	1962	298	0.45-21.0	$\sim 0^\circ$		2π		Similar to curve 1 specimen and conditions except not sealed.
6	270	1962	298	0.45-20.0	$\sim 0^\circ$		2π		Similar to curve 1 specimen and conditions except sulfuric acid anodized for 10 min and sealed.
7*	270	1962	298	0.5-20.0	$\sim 0^\circ$		2π		Similar to curve 1 specimen and conditions.
8*	270	1962	298	0.5-20.0	$\sim 0^\circ$		2π		Above specimen and conditions.
9*	270	1962	298	0.40-21.0	$\sim 0^\circ$		2π		Similar to curve 1 specimen and conditions.
10	270	1962	298	0.4-18.0	$\sim 0^\circ$		2π		Similar to curve 1 specimen and conditions except hard coat anodized for 1/3 of standard duration and sealed.
11	270	1962	298	0.4-20.0	$\sim 0^\circ$		2π		Similar to curve 1 specimen and conditions except anodized for 1/3 of standard duration.
12	201	1961	298	0.45-19	$\sim 0^\circ$		2π		1075 aluminum anodized in chromic acid for 30 min; nominal composition: 0.20 Fe, 0.20 Si, 0.04 Cu, 0.04 Zn, 0.03 Mn, 0.03 Mg, 0.03 Ti and 0.15 others; substrate mechanically polished and electropolished; measured in vacuum ($\sim 10^{-6}$ mm Hg); converted from R (2π, 0°). [Authors' designation: Specimen No. 55]
13	201	1961	298	0.4-20	$\sim 0^\circ$		2π		Above specimen and conditions.
14	201	1961	298	0.45-16	$\sim 0^\circ$		2π		Above specimen and conditions.
15	201	1961	298	0.45-18	$\sim 0^\circ$		2π		Above specimen and conditions.
16	201	1961	298	0.4-20	$\sim 0^\circ$		2π		Above specimen and conditions after heating in vacuum.
17	201	1961	298	0.45-21	$\sim 0^\circ$		2π		Similar to curve 12 specimen and conditions except chromic acid anodized for 10 min. [Authors' designation: Specimen No. 57]
18	201	1961	298	0.4-21	$\sim 0^\circ$		2π		Similar to above specimen and conditions except only mechanically polished. [Authors' designation: Specimen No. 83]
19	201	1961	298	0.4-20	$\sim 0^\circ$		2π		Similar to above specimen and conditions except chromic acid anodized for 30 min. [Authors' designation: Specimen No. 84]
20	201	1961	298	0.4-18	$\sim 0^\circ$		2π		Similar to curve 12 specimen and conditions except mill finished and electropolished. [Authors' designation: Specimen No. 92]

* Not shown on plot

SPECIFICATION TABLE NO. 583 NORMAL SPECTRAL REFLECTANCE OF ALUMINUM + IRON + ΣX_i CONVERSION COATINGS (continued)

Curve No.	Ref. No.	Year	Temperature, K	Wavelength Range, μm	Geometry θ	θ'	ω'	Reported Error, %	Composition (weight percent), Specifications, and Remarks
21	201	1961	298	0.4-19	~0°		2π		Similar to curve 12 specimen and conditions except sulfuric acid anodized for 10 min. [Authors' designation: Specimen No. 67]
22	201	1961	298	0.4-21	~0°		2π		Similar to above specimen and conditions except anodized for 30 min. [Authors' designation: Specimen No. 68]
23	201	1961	298	0.45-21	~0°		2π		Similar to above specimen and conditions. [Authors' designation: Specimen No. 69]
24	201	1961	298	0.4-19	~0°		2π		Above specimen and conditions.
25	201	1961	298	0.4-22	~0°		2π		Above specimen and conditions after heating in vacuum.
26*	201	1961	298	0.45-20	~0°		2π		Similar to above specimen and conditions except sulfuric acid anodized for 30 min. [Authors' designation: Specimen No. 80]
27	201	1961	298	0.45-20	~0°		2π		Similar to above specimen and conditions. [Authors' designation: Specimen No. 81]
28	201	1961	298	0.45-22	~0°		2π		Similar to curve 12 specimen and conditions except hard coat anodized for 80 min. [Authors' designation: Specimen No. 13]
29	201	1961	298	0.4-21	~0°		2π		Similar to above specimen and conditions. [Authors' designation: Specimen No. 15]
30	201	1961	589	0.4-17	~0°		2π		Above specimen and conditions.
31	201	1961	298	0.4-20	~0°		2π		Above specimen and conditions after heating in vacuum.
32	201	1961	298	0.4-21	~0°		2π		Similar to curve 12 specimen and conditions except hard coat anodized for 26.7 min. [Authors' designation: Specimen No. 20]
33	201	1961	298	0.4-18	~0°		2π		Similar to above specimen and conditions except mechanically polished only. [Authors' designation: Specimen No. 23]
34	201	1961	298	0.45-20	~0°		2π		Similar to above specimen and conditions except hard coat anodized for 80 min. [Authors' designation: Specimen No. 82]
35	201	1961	298	0.4-21	~0°		2π		Similar to curve 12 specimen and conditions except boric acid anodized for 15 min. [Authors' designation: Specimen No. 58]
36	201	1961	298	0.4-22	~0°		2π		Similar to above specimen and conditions. [Authors' designation: Specimen No. 59]
37	201	1961	422	0.4-22	~0°		2π		Above specimen and conditions.
38	201	1961	589	0.4-22	~0°		2π		Above specimen and conditions.
39	201	1961	714	0.4-22	~0°		2π		Above specimen and conditions.
40	201	1961	298	0.4-22	~0°		2π		Above specimen and conditions after heating in vacuum.
41	201	1961	298	0.5-21	~0°		2π		Similar to curve 12 specimen and conditions except boric acid anodized for 5 min. [Authors' designation: Specimen No. 60]
42	268	1960	298	0.41-25.01	~0°	~0°			Alclad 24-ST 81 (Alclad 2024-T, aluminum 1050 on aluminum 2024-T) anodized in chromic acid; nominal composition of 1050 aluminum: 0.40 Fe, 0.25 Si, 0.05 Cu, 0.05 Mn, 0.05 Mg, 0.05 Zn, 0.03 Ti and 0.15 others; data extracted from smooth curve.
43	268	1960	298	12.99-24.89	~0°	~0°			Similar to above specimen and conditions except anodized in black chromic acid.

* Not shown on plot

SPECIFICATION TABLE NO. 583 NORMAL SPECTRAL REFLECTANCE OF ALUMINUM + IRON + ΣX_i CONVERSION COATINGS (continued)

Curve No.	Ref. No.	Year	Temperature, K	Wavelength Range, μm	Geometry θ	Geometry θ'	Geometry ω'	Reported Error, %	Composition (weight percent), Specifications, and Remarks
44	268	1960	298	0.41-14.91	$\sim 0^\circ$	$\sim 0^\circ$			Similar to above specimen and conditions.
45	268	1960	298	12.92-24.95	$\sim 0^\circ$	$\sim 0^\circ$			Similar to curve 42 specimen and conditions except anodized in sulfuric acid.
46	268	1960	298	0.41-14.96	$\sim 0^\circ$	$\sim 0^\circ$			Similar to above specimen and conditions.

DATA TABLE NO. 583 NORMAL SPECTRAL REFLECTANCE OF ALUMINUM + IRON + ΣX_i CONVERSION COATINGS

[Wavelength, λ, μm; Reflectance, ρ; Temperature, T, K]

CURVE 1, T = 298

λ	ρ
0.45	0.77
0.50	0.80
0.60	0.78
0.70	0.75
0.80	0.74
0.90	0.75
1.0	0.85
1.2	0.86
1.4	0.86
1.6	0.87
1.8	0.87
2.0	0.85
2.5	0.74
3.0	0.26
3.5	0.51
4.0	0.64
5.0	0.71
6.0	0.68
7.0	0.58
8.0	0.03
9.0	0.04
10.0	0.03
11.0	0.06
13.0	0.15
15.0	0.21
17.0	0.21
19.0	0.18
21.0	0.29

CURVE 2, T = 298

λ	ρ
0.38	0.73
0.45	0.70
0.50	0.74
0.60	0.71
0.70	0.67
0.80	0.70
0.90	0.76
1.0	0.74
1.2	0.83*
1.4	0.83
1.6	0.83
1.8	0.85
2.0	0.84
2.5	0.83
3.0	0.82
3.5	0.83
4.0	0.84
5.0	0.87
6.0	0.86
7.0	0.85
8.0	0.84
9.0	0.84
10.0	0.80
11.0	0.75
12.0	0.78
13.0	0.77
14.0	0.77
15.0	0.79
16.0	0.79
17.0	0.79
18.0	0.79
19.0	0.83
20.0	0.86
21.0	0.86

CURVE 3, T = 298

λ	ρ
0.45	0.65
0.50	0.68
0.60	0.69
0.70	0.73
0.80	0.73
0.90	0.77
1.0	0.85*
1.2	0.89
1.4	0.91
1.6	0.89
1.8	0.90
2.0	0.92
3.0	0.91
4.0	0.93
5.0	0.92
6.0	0.94
7.0	0.89
8.0	0.84*
9.0	0.82
10.0	0.52
12.0	0.23
14.0	0.17
15.0	0.16
16.0	0.13
18.0	0.11

CURVE 4, T = 298

λ	ρ
0.38	0.17
0.45	0.22
0.5	0.27
0.6	0.32
0.7	0.36
0.8	0.35
0.9	0.45
1.0	0.51
1.2	0.58
1.4	0.61
1.6	0.62
1.8	0.64
2.0	0.64
2.5	0.53
3.0	0.11
3.5	0.12
4.0	0.18
5.0	0.28
6.0	0.17
7.0	0.06
8.0	0.02
9.0	0.05
10.0	0.03*
11.0	0.06*
12.0	0.16
13.0	0.20
14.0	0.22
15.0	0.25
16.0	0.24
17.0	0.23
18.0	0.25
19.0	0.23
20.0	0.31
21.0	0.27

CURVE 5, T = 298

λ	ρ
0.45	0.19
0.5	0.24
0.6	0.30
0.7	0.33
0.8	0.32
0.9	0.40
1.0	0.50
1.2	0.52
1.4	0.58
1.6	0.60
1.8	0.62
2.0	0.64*
2.5	0.57
3.0	0.20
3.5	0.31
4.0	0.42
5.0	0.52
6.0	0.30
7.0	0.09
8.0	0.02*
9.0	0.04*
10.0	0.04
11.0	0.06*
12.0	0.14
13.0	0.20*
14.0	0.21
15.0	0.21*
16.0	0.22
17.0	0.25
18.0	0.25*
19.0	0.20
20.0	0.25
21.0	0.20

CURVE 6, T = 298

λ	ρ
0.45	0.65*
0.50	0.66
0.60	0.68
0.70	0.69
0.80	0.63
0.90	0.74
1.0	0.79
1.2	0.84
1.4	0.87
1.6	0.85
1.8	0.88
2.0	0.89
2.5	0.86
3.0	0.71
3.5	0.80
4.0	0.86
5.0	0.88
6.0	0.85
7.0	0.76
8.0	0.02*
9.0	0.03
10.0	0.03*
11.0	0.13
12.0	0.15
13.0	0.18
14.0	0.22*
15.0	0.25*
16.0	0.25
17.0	0.26
18.0	0.25*
19.0	0.27
20.0	0.27

CURVE 7*, T = 298

λ	ρ
0.5	0.72
0.6	0.73
0.7	0.74
0.8	0.72
0.9	0.77
1.0	0.83
1.2	0.88
1.4	0.90
1.6	0.87
1.8	0.90
2.0	0.90
3.0	0.37
4.0	0.68
5.0	0.78
6.0	0.71
7.0	0.63
8.0	0.04
9.0	0.02
10.0	0.03
11.0	0.07
12.0	0.14
14.0	0.22
16.0	0.25
18.0	0.26
20.0	0.31

CURVE 8*, T = 298

λ	ρ
0.5	0.72
0.6	0.73
0.7	0.74
0.8	0.75
0.9	0.77
1.0	0.83
1.2	0.88
1.4	0.89
1.6	0.87
1.8	0.90
2.0	0.90
2.5	0.82
3.0	0.37
3.5	0.55
4.0	0.68
5.0	0.77
6.0	0.71
7.0	0.63
8.0	0.04
9.0	0.02
10.0	0.04
11.0	0.07
12.0	0.14
14.0	0.22
16.0	0.25
18.0	0.26
20.0	0.30

CURVE 9*, T = 298

λ	ρ
0.40	0.21
0.45	0.26
0.5	0.30
0.6	0.35
0.7	0.39
0.8	0.43
0.9	0.48
1.0	0.55
1.2	0.60
1.4	0.62
1.6	0.63
1.8	0.64
2.0	0.64
2.5	0.52
3.0	0.11
3.5	0.10
4.0	0.16
5.0	0.24
6.0	0.15
7.0	0.06
8.0	0.02
9.0	0.04
10.0	0.02
11.0	0.04
12.0	0.14
13.0	0.19
14.0	0.21
15.0	0.21
16.0	0.23
17.0	0.25
18.0	0.25
19.0	0.25
20.0	0.29
21.0	0.30

CURVE 10, T = 298

λ	ρ
0.4	0.40
0.5	0.47
0.6	0.51
0.7	0.53
0.8	0.61
0.9	0.60
1.0	0.71
1.2	0.77
1.4	0.78
1.6	0.78
1.8	0.74
2.0	0.81
2.5	0.74*
3.0	0.29
3.5	0.46
4.0	0.58
5.0	0.67
6.0	0.58
7.0	0.38
8.0	0.04
9.0	0.03*
10.0	0.04*
11.0	0.12
12.0	0.16*
13.0	0.20*
14.0	0.24
15.0	0.24
16.0	0.22*
17.0	0.22
18.0	0.22

CURVE 11, T = 298

λ	ρ
0.4	0.78
0.5	0.82
0.6	0.83
0.7	0.83
0.8	0.85
0.9	0.91
1.0	0.91
1.4	0.96
1.6	0.94
1.8	0.96
2.0	0.95
3.0	0.67
4.0	0.86*
5.0	0.89
6.0	0.86*
8.0	0.14
10.0	0.15
11.0	0.06*
12.0	0.16*
13.0	0.19
14.0	0.22*
15.0	0.25*
16.0	0.23
17.0	0.21*
18.0	0.24
19.0	0.27*
20.0	0.20

* Not shown on plot

DATA TABLE NO. 583 NORMAL SPECTRAL REFLECTANCE OF ALUMINUM + IRON + ΣX_i CONVERSION COATINGS (continued)

λ	ρ
CURVE 12	
T = 298	
0.45	0.718
0.5	0.740*
0.6	0.734
0.7	0.730*
0.8	0.725
0.9	0.764
1.0	0.832
1.2	0.874
1.4	0.895
1.6	0.895
1.8	0.909
2.0	0.926
2.5	0.930
3.0	0.928
4	0.959
5	0.953
6	0.967
7	0.941
8	0.906
9	0.810
10	0.641
11	0.439
12	0.343
13	0.332
14	0.280
15	0.270
16	0.228
17	0.189
18	0.191
19	0.086
CURVE 13	
T = 422	
0.4	0.685
0.45	0.707
0.5	0.731
0.6	0.788
0.7	0.740
0.8	0.799
0.9	0.770*
1.0	0.823
1.2	0.884
1.4	0.873
1.6	0.896*
1.8	0.908*
2.0	0.918
CURVE 13 (cont.)	
2.5	0.927
3.0	0.923
3.5	0.937
4	0.947
5	0.938
6	0.950
7	0.931
8	0.895
9	0.803
10	0.618
11	0.465
12	0.362
13	0.340
14	0.307
15	0.270*
16	0.264
17	0.182
18	0.215
19	0.195
20	0.137
CURVE 14	
T = 589	
0.45	0.680
0.5	0.749
0.6	0.770
0.7	0.766
0.8	0.770
0.9	0.784
1.0	0.806
1.2	0.890*
1.4	0.861
1.6	0.886
1.8	0.894
2.0	0.906
2.5	0.918
3.0	0.928*
3.5	0.936
4	0.940
5	0.944
6	0.934
7	0.904
8	0.865
9	0.787
10	0.629
11	0.495
12	0.427
CURVE 14 (cont.)	
13	0.362
14	0.332
15	0.278
16	0.182
CURVE 15	
T = 714	
0.45	0.659
0.5	0.688
0.6	0.752
0.7	0.728
0.8	0.931
0.9	0.774
1.0	0.780
1.2	0.826
1.4	0.842
1.6	0.856
1.8	0.870*
2.0	0.885
2.5	0.897
3.0	0.899
3.5	0.911
4	0.920
5	0.902
6	0.926
7	0.897
8	0.843
9	0.752
10	0.604
11	0.515
12	0.471
13	0.612
14	0.316
15	0.287
16	0.376
17	0.304
18	0.129
CURVE 16	
T = 298	
0.4	0.561
0.45	0.619
0.5	0.650
0.6	0.706
0.7	0.674
0.8	0.676
CURVE 16 (cont.)	
0.9	0.723
1.0	0.819
1.2	0.838*
1.4	0.871*
1.6	0.883
1.8	0.893*
2.0	0.900
2.5	0.918
3.0	0.922
3.5	0.941
4	0.948
5	0.933
6	0.950
7	0.932
8	0.894
9	0.790
10	0.625
11	0.461
12	0.383
13	0.321
14	0.278
15	0.256
16	0.173
17	0.217
18	0.153
19	0.103
20	0.110
CURVE 17	
T = 298	
0.45	0.840
0.5	0.906
0.6	0.841
0.7	0.956
0.8	0.942
0.9	1.025*
1.0	1.031*
1.2	0.904
1.4	0.952
1.6	0.976
1.8	0.984
2.0	1.069*
2.5	0.997
3.0	0.976
3.5	0.975
4	0.978
5	0.989
CURVE 17 (cont.)	
6	1.025*
7	1.001*
8	0.985
9	0.983
10	0.950
11	0.938
12	0.950
13	0.959
14	0.965
15	0.954
16	0.964
17	0.987
18	0.982
19	0.965
20	0.980
21	1.004*
CURVE 18	
T = 298	
0.4	0.550
0.5	0.533
0.6	0.541
0.7	0.547
0.8	0.548
0.9	0.563
1.0	0.582
1.2	0.678
1.4	0.752
1.6	0.882*
1.8	0.836
2.0	0.826
2.5	0.836
3.0	0.903
3.5	0.922
4	0.941
5	0.938*
6	0.933
7	0.953
8	0.980
9	0.960
10	0.893
11	0.872
12	0.840
13	0.806
14	0.793
15	0.785
16	0.808
CURVE 18 (cont.)	
17	0.812
18	0.865
19	0.874
20	0.911
21	0.987
CURVE 19	
0.4	0.477
0.5	0.566
0.6	0.582
0.7	0.574
0.8	0.533
0.9	0.595
1.0	0.744
1.2	0.896
1.4	0.823
1.6	0.828
1.8	0.847
2.0	0.840*
2.5	0.864
3.0	0.839
3.5	0.883
4	0.917
5	0.913
6	0.932*
7	0.921
8	0.861
10	0.687
12	0.411
13	0.380
14	0.348
15	0.226
16	0.165
17	0.158
18	0.242
19	0.184
20	0.133
CURVE 20	
T = 298	
0.4	0.788
0.5	0.784
0.6	0.885
0.7	0.784
0.8	0.813
0.9	0.824
CURVE 20 (cont.)	
1.0	0.894
1.2	0.925
1.4	0.927
1.6	0.952
1.8	0.957
2.0	0.963
2.5	0.972
3.0	0.947
3.5	0.976*
4	0.982
5	0.971
6	0.982
7	0.958
8	0.915
9	0.818
10	0.643*
11	0.426
12	0.364
13	0.366*
14	0.389
15	0.308
16	0.212
17	0.238
18	0.189*
CURVE 21	
T = 298	
0.4	0.980
0.5	0.977
0.6	0.751*
0.7	0.957
0.8	0.949
0.9	0.970
1.0	0.957
1.2	0.962
1.4	0.960*
1.6	0.973
1.8	0.938
2.0	0.976
2.5	0.988
3.0	0.910*
3.5	0.951
4	0.966
5	0.981
6	0.891
7	0.900
8	0.444
CURVE 21 (cont.)	
9	0.283
10	0.242
11	0.141
12	0.240
13	0.294
14	0.299
15	0.354
16	0.394
17	0.393
18	0.386
19	0.347
CURVE 22	
T = 298	
0.4	0.704*
0.45	0.705
0.5	0.715
0.6	0.709*
0.7	0.701
0.8	0.738
0.9	0.700
1.0	0.773
1.2	0.818
1.4	0.834
1.6	0.833
1.8	0.845*
2.0	0.849
2.5	0.823
3.0	0.448
3.5	0.658
4	0.745
5	0.822
6	0.776
7	0.724
8	0.038
9	0.068
10	0.063
11	0.106
12	0.215
13	0.272
14	0.235
15	0.313*
16	0.330
17	0.298
18	0.282
19	0.359
20	0.325
21	0.342

* Not shown on plot

DATA TABLE NO. 583 NORMAL SPECTRAL REFLECTANCE OF ALUMINUM + IRON + ΣX_i CONVERSION COATINGS (continued)

λ	ρ
CURVE 23 T = 298	
0.45	0.736
0.5	0.764
0.6	0.761
0.7	0.739*
0.8	0.727*
0.9	0.732
1.0	0.833
1.2	0.849
1.4	0.848
1.6	0.862
1.8	0.863
2.0	0.843
2.5	0.773
3.0	0.284
3.5	0.540
4	0.666
5	0.737
6	0.724
7	0.649
8	0.032
9	0.046
10	0.039*
11	0.089
13	0.229
15	0.315
17	0.316
19	0.276
21	0.432
CURVE 24 T = 298	
0.4	0.732
0.45	0.725
0.5	0.745
0.6	0.746
0.7	0.747*
0.8	0.741
0.9	0.753
1.0	0.764
1.2	0.807
1.4	0.837
1.6	0.847*
1.8	0.857
2.0	0.861
2.5	0.843
3.0	0.828
CURVE 24 (cont.)	
3.5	0.872
4	0.878
5	0.884
6	0.840
7	0.778
8	0.041*
9	0.037
10	0.055
11	0.143*
13	0.370
15	0.396
17	0.382
19	0.429
CURVE 25 T = 298	
0.4	0.646
0.45	0.701*
0.5	0.721
0.6	0.723
0.7	0.711
0.8	0.717
0.9	0.734
1.0	0.769*
1.2	0.830
1.4	0.852
1.6	0.861*
1.8	0.874
2.0	0.859
2.5	0.819
3.0	0.561
3.5	0.702
4	0.776
5	0.854
6	0.806
7	0.732
8	0.046
9	0.054*
10	0.060
11	0.129*
12	0.234
13	0.276
14	0.306
15	0.328
16	0.326
17	0.374
18	0.357
CURVE 25 (cont.)	
19	0.332
20	0.304
21	0.379
22	0.536
CURVE 26* T = 298	
0.45	0.712
0.5	0.720
0.6	0.722
0.7	0.706
0.8	0.706
0.9	0.724
1.0	0.775
1.2	0.817
1.4	0.849
1.6	0.856
1.8	0.881
2.0	0.902
2.5	0.881
3.0	0.732
3.5	0.827
4	0.886
5	0.914
6	0.888
7	0.829
8	0.020
9	0.045
10	0.036
11	0.203
12	0.244
13	0.280
14	0.347
15	0.386
16	0.393
17	0.413
18	0.395
19	0.434
20	0.432
CURVE 27 T = 298	
0.45	0.981
0.5	0.786
0.6	0.779*
0.7	0.760
CURVE 27 (cont.)	
0.8	0.760
0.9	0.766*
1.0	0.815
1.2	0.860*
1.4	0.874*
1.6	0.873
1.8	0.910
2.0	0.903*
2.5	0.830*
3.0	0.386
3.5	0.582
4	0.720
5	0.819
6	0.762
7	0.704
8	0.143
9	0.031*
10	0.067
11	0.114
12	0.219
14	0.352
16	0.394
18	0.416
20	0.473
CURVE 28 T = 298	
0.45	0.214
0.5	0.265
0.6	0.319
0.7	0.341
0.8	0.328
0.9	0.433
1.0	0.460
1.2	0.499
1.4	0.562
1.6	0.594
1.8	0.615
2.0	0.610
2.5	0.576
3.0	0.208
3.5	0.323
4	0.450
5	0.561
6	0.351
7	0.112
8	0.022*
CURVE 28 (cont.)	
9	0.052*
10	0.046
11	0.093
12	0.219*
13	0.317
14	0.336
15	0.330*
16	0.356
17	0.366
18	0.409
19	0.328
20	0.412
21	0.347
22	0.438
CURVE 29 T = 298	
0.4	0.190
0.45	0.242
0.5	0.292
0.6	0.342
0.7	0.366
0.8	0.389
0.9	0.442
1.0	0.494
1.2	0.562
1.4	0.593
1.6	0.618
1.8	0.639
2.0	0.645
2.5	0.539
3.0	0.115
3.5	0.131
4	0.203
5	0.310
6	0.196
7	0.077
8	0.026
9	0.071
10	0.049
11	0.100
12	0.251
13	0.307
14	0.343
15	0.386
16	0.379
17	0.366*
CURVE 29 (cont.)	
18	0.402
19	0.384
20	0.479
21	0.434*
CURVE 30 T = 589	
0.4	0.310
0.45	0.403
0.5	0.448
0.6	0.483
0.7	0.503
0.8	0.492
0.9	0.584
1.0	0.584*
1.2	0.625
1.4	0.682
1.6	0.670
1.8	0.696
2.0	0.709
2.5	0.628
3.0	0.298
3.5	0.571
4	0.547
5	0.580
6	0.342
7	0.162
8	0.175
10	0.215
12	0.232*
13	0.276*
15	0.474
17	0.408
CURVE 31 T = 298	
0.4	0.381
0.45	0.436
0.5	0.462
0.6	0.500
0.7	0.511
0.8	0.492*
0.9	0.575
1.0	0.617
1.2	0.642
1.4	0.661
CURVE 31 (cont.)	
1.6	0.679
1.8	0.690
2.0	0.707
2.5	0.611
3.0	0.324
3.5	0.508
4	0.572
5	0.600
6	0.412
7	0.104
8	0.033*
9	0.058
10	0.035
11	0.116
12	0.238
13	0.299
14	0.298
15	0.283
16	0.314
17	0.269
18	0.362
19	0.296
20	0.289
CURVE 32 T = 298	
0.4	0.345
0.45	0.421
0.5	0.459*
0.6	0.507
0.7	0.514
0.8	0.541
0.9	0.577*
1.0	0.635
1.2	0.677
1.4	0.706
1.6	0.732
1.8	0.730
2.0	0.773
2.5	0.696
3.0	0.415
3.5	0.611
4	0.686
5	0.735
6	0.612
7	0.327
8	0.026*
CURVE 32 (cont.)	
9	0.018
10	0.032*
11	0.063
12	0.128
13	0.180*
14	0.220*
15	0.253*
16	0.261*
17	0.274
18	0.270
19	0.240
20	0.349
21	0.353
CURVE 33 T = 298	
0.4	0.446
0.45	0.485
0.5	0.511
0.6	0.542
0.7	0.551
0.8	0.624
0.9	0.598
1.0	0.691
1.2	0.745
1.4	0.757
1.6	0.775
1.8	0.793
2.0	0.816
2.5	0.752
3.0	0.301
3.5	0.483
4	0.617
5	0.724
6	0.639
7	0.442
8	0.046*
9	0.039*
10	0.052*
11	0.193
12	0.255
13	0.316
14	0.377
15	0.385*
16	0.376*
17	0.361
18	0.365

* Not shown on plot

DATA TABLE NO. 583 NORMAL SPECTRAL REFLECTANCE OF ALUMINUM + IRON + ΣX_i CONVERSION COATINGS (continued)

λ	ρ
CURVE 34 T = 298	
0.45	0.199
0.5	0.229
0.6	0.276
0.7	0.295
0.8	0.194
0.9	0.368
1.0	0.375
1.2	0.462
1.4	0.506
1.6	0.544
1.8	0.576
2.0	0.603
2.5	0.554
3.0	0.237
3.5	0.394
4	0.486
5	0.585
6	0.409
7	0.137
8	0.029*
9	0.018*
10	0.058*
11	0.126*
12	0.254*
14	0.354*
16	0.356*
18	0.399*
20	0.389*
CURVE 35 T = 298	
0.4	0.655
0.45	0.673
0.5	0.687*
0.6	0.637
0.7	0.650
0.8	0.680
0.9	0.643
1.0	0.724
1.2	0.806*
1.4	0.831*
1.6	0.835*
1.8	0.837*
2.0	0.840*
2.5	0.899
3.0	0.848
CURVE 35 (cont.)	
3.5	0.870*
4	0.889
5	0.917
6	0.925*
7	0.926
8	0.932
9	0.936
10	0.914
11	0.909
12	0.921
13	0.921
14	0.926
15	0.932
16	0.943
17	0.948
18	0.940
19	0.950
20	0.945
21	0.953
CURVE 36 T = 298	
0.4	0.700
0.45	0.668
0.5	0.718*
0.6	0.650
0.7	0.657
0.8	0.817
0.9	0.741*
1.0	0.688
1.2	0.817
1.4	0.823*
1.6	0.820
1.8	0.837*
2.0	0.834
2.5	0.827
3.0	0.823
3.5	0.836
4	0.858*
5	0.880*
6	0.874
7	0.878
8	0.886
9	0.882
10	0.853
11	0.826
12	0.853
CURVE 36 (cont.)	
13	0.842
14	0.844
15	0.862
16	0.863
17	0.864
18	0.870
19	0.893
20	0.912
21	0.912
22	0.706
CURVE 37 T = 422	
0.4	0.853
0.45	0.783
0.5	0.837
0.6	0.792
0.7	0.797
0.8	0.748
0.9	0.765*
1.0	0.792
1.2	0.878
1.4	0.915
1.6	0.928
1.8	0.934
2.0	0.927*
2.5	0.939
3.0	0.950
3.5	0.957
4	0.967
5	0.979
6	0.982*
7	0.962
8	0.988*
9	0.988*
10	0.966
11	0.931
12	0.976
13	0.962*
14	0.976
15	0.958
16	0.984
17	0.995
18	1.005*
19	0.999
20	0.999
21	0.558
22	1.054*
CURVE 38 T = 589	
0.4	0.857
0.45	0.779
0.5	0.839*
0.6	0.780
0.7	0.804
0.8	0.837
0.9	0.756*
1.0	0.767
1.2	0.884*
1.4	0.910*
1.6	0.915
1.8	0.907*
2.0	0.924*
2.5	0.922
3.0	0.933
3.5	0.947
4	0.958*
5	0.968
6	0.980*
7	0.973
8	0.980*
9	0.982*
10	0.963
11	0.951
12	0.954
13	0.985
14	0.983
15	0.993
16	0.995
17	1.006*
18	1.026*
19	1.017*
20	1.027*
21	1.026*
22	1.025*
CURVE 39 T = 714	
0.4	0.839
0.45	0.787
0.5	0.845
0.6	0.764
0.7	0.804
0.8	0.821
0.9	0.763*
1.0	0.759
CURVE 39 (cont.)	
1.2	0.865
1.4	0.892
1.6	0.906
1.8	0.899*
2.0	0.908
2.5	0.907
3.0	0.914
3.5	0.940*
4	0.944*
5	0.958
6	0.984
7	0.970*
8	0.969
9	0.973
10	0.958
11	0.957
12	0.951*
13	0.975
14	0.984*
15	0.993*
16	0.997
17	1.009*
18	1.020*
19	1.025*
20	1.025*
21	1.016*
22	1.038*
CURVE 40 T = 298	
0.4	0.676
0.45	0.657*
0.5	0.706
0.6	0.665
0.7	0.662
0.8	0.706
0.9	0.685
1.0	0.691*
1.2	0.787
1.4	0.827*
1.6	0.834*
1.8	0.831*
2.0	0.815*
2.5	0.814
3.0	0.813
3.5	0.835*
4	0.855
CURVE 40 (cont.)	
5	0.872
6	0.874
7	0.870
8	0.876
9	0.873
10	0.843
11	0.828
12	0.838
13	0.846
14	0.846
15	0.853
16	0.864
17	0.853
18	0.876
19	0.876*
20	0.840
21	0.875
22	0.853
CURVE 41 T = 298	
0.5	0.873
0.6	0.842*
0.7	0.878
0.8	0.794
0.9	0.759*
1.0	0.857
1.2	0.891*
1.4	0.935
1.6	0.932
1.8	0.935*
2.0	0.933
2.5	0.931*
3.0	0.933*
3.5	0.954
4	0.972
5	0.985
6	0.993
7	0.981
8	0.986*
9	0.988
10	0.961*
11	0.948*
12	0.959
13	0.977*
14	0.986*
15	0.970*
CURVE 41 (cont.)	
16	0.997*
17	1.007*
18	1.017*
19	1.027*
20	1.026*
21	1.001*
CURVE 42 T = 298	
0.41	0.485
0.50	0.550
0.59	0.573
0.65	0.573
0.69	0.601
0.75	0.587
0.79	0.593*
0.83	0.559*
0.94	0.686
0.99	0.705
1.24	0.683
1.72	0.763
1.98	0.706
2.21	0.818
2.48	0.834*
2.78	0.818*
2.97	0.673*
3.19	0.744
3.47	0.744*
3.68	0.851*
4.21	0.904
4.50	0.904*
4.72	0.922*
4.96	0.916*
5.23	0.935
5.49	0.922*
5.61	0.907*
5.75	0.892*
5.99	0.877*
6.23	0.847
6.49	0.878*
6.73	0.891*
7.00	0.845*
7.58	0.816
7.93	0.780*
8.68	0.679*
9.06	0.707*
9.71	0.728
CURVE 42 (cont.)	
10.06	0.702*
10.45	0.605*
11.48	0.470
11.72	0.470*
12.94	0.380*
13.33	0.322
13.77	0.322*
14.09	0.301*
14.76	0.301
14.98	0.291*
16.78	0.291*
17.30	0.321*
17.59	0.321
18.09	0.349*
18.34	0.349*
18.55	0.383
18.75	0.383*
19.18	0.408*
19.61	0.442*
19.90	0.442*
20.32	0.490
20.83	0.490*
21.06	0.508*
21.33	0.537
21.76	0.520*
22.03	0.561*
22.64	0.582
23.05	0.582*
23.34	0.591*
23.66	0.621
24.08	0.612*
24.36	0.621*
25.01	0.621
CURVE 43 T = 298	
12.99	0.039
13.31	0.039
13.93	0.048
14.47	0.088
14.76	0.088
15.95	0.140
16.67	0.140
17.14	0.156
17.42	0.148
18.23	0.148
18.45	0.176

* Not shown on plot

DATA TABLE NO. 583 NORMAL SPECTRAL REFLECTANCE OF ALUMINUM + IRON + ΣX_i CONVERSION COATINGS (continued)

CURVE 43 (cont.) T = 298

λ	ρ
18.71	0.169*
21.93	0.168
22.21	0.178
22.89	0.178*
23.21	0.161*
23.50	0.218
23.73	0.218*
23.99	0.178*
24.89	0.178

CURVE 44 T = 298

λ	ρ
0.41	0.056
0.65	0.056
0.71	0.074
0.81	0.517
0.96	0.689
1.00	0.717
1.26	0.685
1.74	0.740
2.00	0.704*
2.27	0.704
2.49	0.633
2.73	0.164*
3.02	0.071
3.50	0.104
3.96	0.251*
4.27	0.281*
4.48	0.245
4.76	0.224*
5.02	0.291
5.26	0.265*
5.52	0.328*
5.78	0.246*
6.00	0.144
6.51	0.015*
6.78	0.022*
7.01	0.099
7.28	0.168*
7.57	0.149*
7.80	0.078*
8.75	0.040*
9.27	0.069*
12.45	0.069
12.72	0.079*
12.89	0.070

CURVE 44 (cont.)

λ	ρ
13.15	0.079
13.68	0.078*
14.26	0.098*
14.51	0.120
14.69	0.118*
14.91	0.138*

CURVE 45 T = 298

λ	ρ
12.92	0.120
13.24	0.120*
13.53	0.133
13.99	0.133*
14.45	0.150
15.52	0.150
15.88	0.162
17.74	0.161
18.40	0.179
19.27	0.179
19.44	0.188*
19.93	0.170
20.27	0.187*
20.52	0.187
20.86	0.170*
21.22	0.185*
21.40	0.170
21.97	0.171*
22.21	0.179*
24.47	0.179
24.64	0.211*
24.94	0.221*

CURVE 46 T = 298

λ	ρ
0.41	0.584
0.44	0.631
0.50	0.652
0.60	0.653
0.64	0.644
0.70	0.644
0.80	0.606
0.86	0.606
0.91	0.636
0.94	0.689
1.00	0.717
1.26	0.637

CURVE 46 (cont.)

λ	ρ
1.51	0.465
1.75	0.668
2.01	0.452
2.25	0.633
2.52	0.528
3.08	0.085
3.27	0.142
3.76	0.516
3.97	0.540
4.23	0.604
4.48	0.593
5.00	0.681
5.25	0.663
5.50	0.663
5.75	0.680
6.23	0.480
6.48	0.454
6.79	0.511
6.98	0.530
7.25	0.462
8.01	0.038
8.21	0.031
8.47	0.040
8.94	0.040
9.29	0.052
9.48	0.080
9.77	0.080
10.23	0.036
11.22	0.075
11.46	0.145
11.74	0.171
11.96	0.153
12.15	0.171
12.69	0.171
13.43	0.222
14.22	0.222
14.50	0.250
14.69	0.240
14.96	0.254

* Not shown on plot

SPECIFICATION TABLE NO. 584 ANGULAR SPECTRAL REFLECTANCE OF ALUMINUM + IRON + ΣX_i CONVERSION COATINGS

Curve No.	Ref. No.	Year	Temperature, K	Wavelength Range, μm	Geometry θ	θ'	ω'	Reported Error, %	Composition (weight percent), Specifications, and Remarks
1*	115	1960	298	0.38-2.46	15°		2π		1145 aluminum foil (0.038 mm thick); nominal composition: 99.45 min. Al, 0.55 Fe+Si, 0.05 Cu, 0.05 Mn, 0.03 other; bright dipped and anodized on both sides (0.0076 mm thick) in dilute sulfuric acid. [Authors' designation: Sample 2]

DATA TABLE NO. 584 ANGULAR SPECTRAL REFLECTANCE OF ALUMINUM + IRON + ΣX_i CONVERSION COATINGS

[Wavelength, λ, μm; Reflectance, ρ; Temperature, T, K]

λ	ρ	λ	ρ
CURVE 1*		CURVE 1 (cont.)*	
T = 298		1.48	0.88
0.38	0.77	1.56	0.87
0.46	0.78	1.68	0.87
0.54	0.79	1.88	0.88
0.64	0.78	2.18	0.85
0.68	0.77	2.40	0.85
0.72	0.76	2.46	0.86
0.84	0.71		
0.88	0.72		
0.96	0.80		
1.01	0.83		
1.06	0.85		
1.12	0.86		
1.22	0.86		
1.32	0.87		
1.37	0.87		
1.44	0.88		

* No plot given

SPECIFICATION TABLE NO. 585 NORMAL SOLAR REFLECTANCE OF ALUMINUM + IRON + ΣX_i CONVERSION COATINGS

Curve No.	Ref. No.	Year	Temperature Range, K	Geometry θ	θ'	ω'	Reported Error, %	Composition (weight percent), Specifications and Remarks
1*	269	1964	298	~0°		2π		L-34 (British Aircraft Material Specification) anodized to DTD 911; nominal composition: 99.0 min. Al, 0.7 Fe, 0.5 Si, 0.15 Ti, 0.10 Cu, 0.1 Mg, 0.1 Mn, 0.1 Ni, 0.10 Zn, 0.05 Sn, and 0.05 Pb; substrate machine polished; freshly prepared; measured relative to MgO; computed from spectral data.
2*	269	1964	298	~0°		2π		Above specimen and conditions; diffuse component only.
3*	269	1964	298	~0°		2π		Curve 1 specimen and conditions except exposed within clean lab area for 1 month.
4*	269	1964	298	~0°		2π		Above specimen and conditions; diffuse component only.
5*	269	1964	298	~0°		2π		Similar to curve 1 specimen and conditions.
6*	269	1964	298	~0°		2π		Above specimen and conditions; diffuse component only.
7*	269	1964	298	~0°		2π		Curve 5 specimen and conditions except exposed within clean lab area for 1 month.
8*	269	1964	298	~0°		2π		Above specimen and conditions; diffuse component only.
9*	268	1960	298	~0°		2π		Alclad 24-ST81 (Alclad 2024-T,aluminum 1050 on aluminum 2024-T), anodized in chromic acid; nominal composition of 1050 aluminum: 0.40 Fe, 0.25 Si, 0.05 Cu, 0.05 Mn, 0.05 Mg, 0.05 Zn, 0.03 Ti, and 0.15 others; calculated from spectral data.
10*	268	1960	298	~0°		2π		Similar to above specimen and conditions.
11*	268	1960	298	~0°		2π		Similar to curve 9 specimen and conditions except anodized in black chromic acid.
12*	268	1960	298	~0°		2π		Similar to above specimen and conditions.

DATA TABLE NO. 585 NORMAL SOLAR REFLECTANCE OF ALUMINUM + IRON + ΣX_i CONVERSION COATINGS

[Temperature, T, K; Reflectance, ρ]

T	ρ	T	ρ	T	ρ	T	ρ
CURVE 1*		CURVE 4*		CURVE 7*		CURVE 10*	
298	0.50	298	0.49	298	0.47	298	0.56
CURVE 2*		CURVE 5*		CURVE 8*		CURVE 11*	
298	0.49	298	0.47	298	0.46	298	0.33
CURVE 3*		CURVE 6*		CURVE 9*		CURVE 12*	
298	0.50	298	0.47	298	0.46	298	0.31

* No plot given

SPECIFICATION TABLE NO. 586 NORMAL SOLAR ABSORPTANCE OF ALUMINUM + IRON + ΣX_i CONVERSION COATINGS

Curve No.	Ref. No.	Year	Temperature Range, K	Geometry θ	Reported Error, %	Composition (weight percent), Specifications and Remarks
1*	269	1964	298	~0°		L-34 (British Aircraft Material Specification) anodized to DTD 911; nominal composition: 99.0 min. Al, 0.7 Fe, 0.5 Si, 0.15 Ti, 0.10 Cu, 0.1 Mg, 0.1 Mn, 0.1 Ni, 0.10 Zn, 0.005 Sn, and 0.05 Pb; substrate machine polished; freshly prepared; measured relative to MgO; computed from spectral data.
2*	269	1964	298	~0°		Above specimen and conditions except exposed within clean laboratory area for 1 month.
3*	269	1964	298	~0°		Similar to curve 1 specimen and conditions.
4*	269	1964	298	~0°		Above specimen and conditions except exposed within clean laboratory area for 1 month.
5*	115	1960	298	15°		1145 aluminum foil (0.038 mm thick); nominal composition: 99.45 min. Al, 0.55 Fe+Si, 0.05 Cu, 0.05 Mn, and 0.03 other; bright dipped and anodized on both sides (0.0076 mm thick) in dilute sulfuric acid; computed from spectral reflectance data for above atm conditions. [Authors' designation: Sample 2]
6*	115	1960	298	15°		Similar to above specimen and conditions except exposed to vacuum (10^{-6} mm Hg) for 6 days and cleaned with isopropyl alcohol. [Authors' designation: Sample 3]
7*	115	1960	298	15°		Similar to curve 5 specimen and conditions except exposed to 700 MeV proton irradiation, 1.9×10^{13} protons cm^{-2}. [Authors' designation: Sample 4]

DATA TABLE NO. 586 NORMAL SOLAR ABSORPTANCE OF ALUMINUM + IRON + ΣX_i CONVERSION COATINGS

[Temperature, T, K; Absorptance, α]

T	α	T	α
CURVE 1*		CURVE 5*	
298	0.50	298	0.20
CURVE 2*		CURVE 6*	
298	0.50	298	0.20
CURVE 3*		CURVE 7*	
298	0.53	298	0.20
CURVE 4*			
298	0.53		

* No plot given

SPECIFICATION TABLE NO. 587 NORMAL SPECTRAL REFLECTANCE OF ALUMINUM + MAGNESIUM CONVERSION COATINGS

Curve No.	Ref. No.	Year	Temperature, K	Wavelength Range, μm	Geometry θ	Geometry θ'	Geometry ω'	Reported Error, %	Composition (weight percent), Specifications, and Remarks
1*	80	1966	~298	0.4260-0.5400	0°		2π		Al + 0.5 Mg; anodized; alloyed from aluminum of quality >99.99 Al; substrate finely turned; measured relative to MgO; converted from R (2π, 0°). [Author's designation: Reflector No. III]
2*	80	1966	~298	0.3079-0.5968	0°		2π		Above specimen and conditions except converted from R (0°, 2π).

DATA TABLE NO. 587 NORMAL SPECTRAL REFLECTANCE OF ALUMINUM + MAGNESIUM CONVERSION COATINGS

[Wavelength, λ, μm; Reflectance, ρ; Temperature, T, K]

λ	ρ
CURVE 1*	
T ~ 298	
0.4260	0.766
0.4570	0.771
0.4950	0.768
0.5400	0.767
CURVE 2*	
T ~ 298	
0.3079	0.644
0.3275	0.665
0.3423	0.677
0.3646	0.691
0.3905	0.699
0.4158	0.705
0.5428	0.705
0.5968	0.693

* No plot given

SPECIFICATION TABLE NO. 588 ANGULAR SPECTRAL REFLECTANCE OF ALUMINUM + MAGNESIUM CONVERSION COATINGS

Curve No.	Ref. No.	Year	Temperature, K	Wavelength Range, μm	Geometry θ	θ'	ω'	Reported Error, %	Composition (weight percent), Specifications, and Remarks
1*	80	1966	~298	0.2568-0.4957	45°	0°			Al + 0.5 Mg; anodized; alloyed from aluminum of quality >99.99 Al; substrate finely turned; measured relative to MgO powder; data extracted from smooth curve. [Author's designation: Reflector No. III]

DATA TABLE NO. 588 ANGULAR SPECTRAL REFLECTANCE OF ALUMINUM + MAGNESIUM CONVERSION COATINGS

[Wavelength, λ, μm; Reflectance, ρ; Temperature, T, K]

λ	ρ
CURVE 1*	
T ~ 298	
0.2568	0.275
0.2653	0.275
0.2812	0.291
0.2931	0.305
0.3070	0.319
0.3333	0.332
0.3569	0.338
0.4957	0.340

* No plot given

SPECIFICATION TABLE NO. 589 HEMISPHERICAL TOTAL EMITTANCE OF ALUMINUM + ZINC + ΣX_i CONVERSION COATINGS

Curve No.	Ref. No.	Year	Temperature Range, K	Reported Error, %	Composition (weight percent), Specifications and Remarks
1*	106	1963	293		Alodined aluminum 7075-T6; nominal composition: 5.6 Zn, 2.5 Mg, 1.6 Cu, 0.30 Cr, Al balance; substrate sandblasted; measured in vacuum (5×10^{-6} mm Hg).
2*	106	1963	259		Similar to above specimen and conditions except substrate alclad.
3*	106	1963	273-313		Similar to above specimen and conditions except substrate smooth and unclad.
4*	106	1963	217-313		Aluminum 7075-T6; Martin Hardcoate Anodize; nominal composition: 5.6 Zn, 2.5 Mg, 1.6 Cu, 0.30 Cr, Al balance; measured in vacuum (5×10^{-6} mm Hg).

DATA TABLE NO. 589 HEMISPHERICAL TOTAL EMITTANCE OF ALUMINUM + ZINC + ΣX_i CONVERSION COATINGS

[Temperature, T, K; Emittance, ϵ]

T	ϵ	T	ϵ
CURVE 1*		CURVE 4*	
293	0.234	217	0.790
		242	0.849
CURVE 2*		285	0.869
		313	0.856
259	0.117		
CURVE 3*			
273	0.100		
307	0.091		
313	0.091		

* No plot given

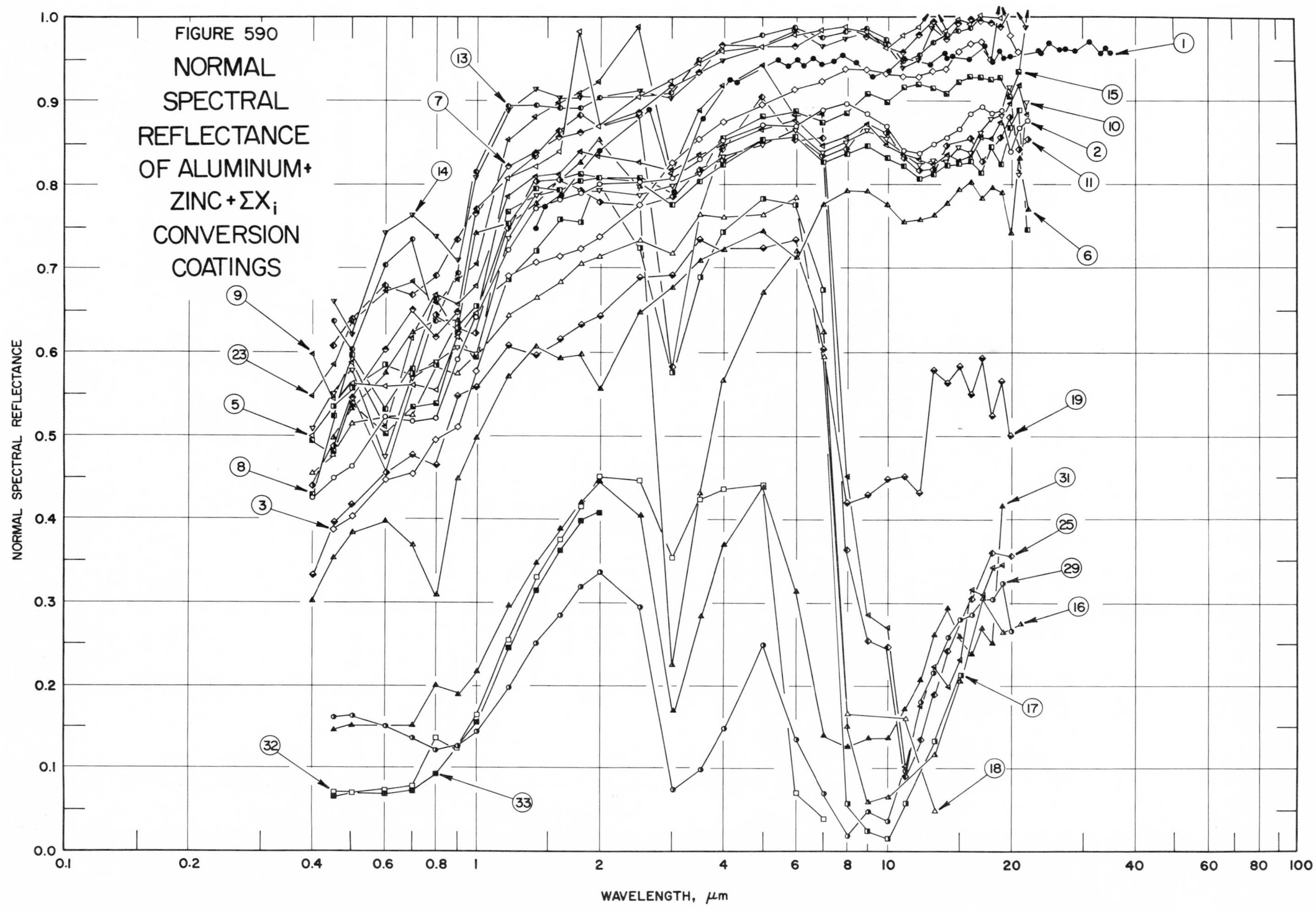

FIGURE 590 NORMAL SPECTRAL REFLECTANCE OF ALUMINUM + ZINC + ΣX_i CONVERSION COATINGS

SPECIFICATION TABLE NO. 590 NORMAL SPECTRAL REFLECTANCE OF ALUMINUM + ZINC + ΣX_i CONVERSION COATINGS

Curve No.	Ref. No.	Year	Temperature, K	Wavelength Range, μm	Geometry θ	θ'	ω'	Reported Error, %	Composition (weight percent), Specifications, and Remarks
1	116	1963	311	1.40-35.1	~0°		2π		Alodined 7075-T6 Aluminum; nominal composition: 5.6 Zn, 2.5 Mg, 1.6 Cu, 0.30 Cr, Al balance; alclad removed by compounding; converted from R (2π, 0°).
2	201	1961	298	0.4-22	~0°		2π		7075 Aluminum anodized in chromic acid for 30 min; nominal composition: 5.6 Zn, 2.5 Mg, 1.6 Cu, 0.30 Cr, and Al balance; mechanically polished and electropolished; measured in vacuum (~10^{-6} mm Hg); converted from R (2π, 0°). [Authors' designation: Specimen No. 50]
3	201	1961	589	0.45-21	~0°		2π		Above specimen and conditions.
4*	201	1961	298	0.4-21	~0°		2π		Above specimen and conditions after heating in vacuum.
5	201	1961	298	0.4-19	~0°		2π		Similar to curve 2 specimen and conditions except chromic acid anodized for 10 min. [Authors' designation: Specimen No. 51]
6	201	1961	295	0.4-22	~0°		2π		Similar to above specimen and conditions except only mechanically polished. [Authors' designation: Specimen No. 75]
7*	201	1961	298	0.5-21	~0°		2π		Similar to curve 2 specimen and conditions. [Authors' designation: Specimen No. 110]
8	201	1961	298	0.4-22	~0°		2π		7075 Aluminum anodized in boric acid for 15 min; nominal composition: 5.6 Zn, 2.5 Mg, 1.6 Cu, 0.30 Cr, Al balance; substrate mechanically polished and electropolished; measured in vacuum (~10^{-6} mm Hg); converted from R (2π, 0°). [Authors' designation: Specimen No. 52]
9	201	1961	422	0.4-22	~0°		2π		Above specimen and conditions.
10	201	1961	589	0.4-22	~0°		2π		Above specimen and conditions.
11	201	1961	714	0.4-22	~0°		2π		Above specimen and conditions.
12*	201	1961	298	0.4-22	~0°		2π		Above specimen and conditions after heating in vacuum.
13	201	1961	298	0.45-22	~0°		2π		Similar to curve 8 specimen and conditions except mechanically polished only. [Authors' designation: Specimen No. 54]
14	201	1961	298	0.45-22	~0°		2π		Similar to curve 8 specimen and conditions except boric acid anodized for 5 min. [Authors' designation: Specimen No. 62]
15	201	1961	298	0.4-22	~0°		2π		Similar to curve 8 specimen and conditions except mill finished and electropolished. [Authors' designation: Specimen No. 115]
16	201	1961	298	0.45-21	~0°		2π		7075 Aluminum anodized in sulfuric acid for 30 min; nominal composition: 5.6 Zn, 2.5 Mg, 1.6 Cu, 0.30 Cr, Al balance; substrate mechanically polished and electropolished; measured in vacuum (~10^{-6} mm Hg); converted from R (2π, 0°). [Authors' designation: Specimen No. 63]
17	201	1961	422	0.45-15	~0°		2π		Above specimen and conditions.
18	201	1961	589	0.4-13	~0°		2π		Above specimen and conditions.
19	201	1961	714	0.4-20	~0°		2π		Above specimen and conditions.
20*	201	1961	298	0.4-20	~0°		2π		Above specimen and conditions after heating in vacuum.

* Not shown on plot

SPECIFICATION TABLE NO. 590 NORMAL SPECTRAL REFLECTANCE OF ALUMINUM + ZINC + ΣX_i CONVERSION COATINGS

Curve No.	Ref. No.	Year	Temperature, K	Wavelength Range, μm	Geometry θ	θ'	ω'	Reported Error, %	Composition (weight percent), Specifications, and Remarks
21*	201	1961	298	0.45-21	~0°		2π		Similar to curve 16 specimen and conditions. [Authors' designation: Specimen No. 64]
22*	201	1961	298	0.4-19	~0°		2π		Similar to above specimen and conditions. [Authors' designation: Specimen No. 65]
23	201	1961	298	0.4-19	~0°		2π		Similar to above specimen and conditions except sulfuric acid anodized for 10 min. [Authors' designation: Specimen No. 66]
24*	201	1961	298	0.4-22	~0°		2π		Similar to above specimen and conditions. [Authors' designation: Specimen No. 73]
25	201	1961	298	0.45-20	~0°		2π		Similar to above specimen and conditions except mechanically polished only. [Authors' designation: Specimen No. 74]
26*	201	1961	298	0.5-20	~0°		2π		Similar to above specimen and conditions except sulfuric acid anodized for 30 min. [Authors' designation: Specimen No. 77]
27*	201	1961	298	0.45-20	~0°		2π		Similar to above specimen and conditions. [Authors' designation: Specimen No. 78]
28*	201	1961	298	0.45-13	~0°		2π		Similar to curve 16 specimen and conditions. [Authors' designation: Specimen No. 120]
29	201	1961	298	0.45-20	~0°		2π		7075 Aluminum hard coat anodized for 80 min; nominal composition: 5.6 Zn, 2.5 Mg, 1.6 Cu, 0.30 Cr, Al balance; substrate mechanically polished and electropolished; measured in vacuum (~10^{-6} mm Hg); converted from R (2π, 0°). [Authors' designation: Specimen No. 86]
30*	201	1961	298	0.45-20	~0°		2π		Similar to above specimen and conditions. [Authors' designation: Specimen No. 87]
31	201	1961	422	0.45-19	~0°		2π		Above specimen and conditions.
32	201	1961	589	0.45-7	~0°		2π		Above specimen and conditions.
33	201	1961	714	0.45-2.0	~0°		2π		Above specimen and conditions.
34*	201	1961	298	0.45-20	~0°		2π		Above specimen and conditions after heating in vacuum.
35*	201	1961	298	0.4-17	~0°		2π		Similar to curve 29 specimen and conditions. [Authors' designation: Specimen No. 88]
36*	201	1961	298	0.45-20	~0°		2π		Similar to above specimen and conditions except hard coat anodized for 26.7 min. [Authors' designation: Specimen No. 89]
37*	201	1961	298	0.4-20	~0°		2π		Similar to curve 29 specimen and conditions except mill finished and electropolished. [Authors' designation: Specimen No. 129]
38*	201	1961	298	0.45-20	~0°		2π		Similar to curve 29 specimen and conditions except mechanically polished only. [Authors' designation: Specimen No. 85]
39*	201	1961	298	0.4-20	~0°		2π		Similar to above specimen and conditions except hard coat anodized for 26.7 min. [Authors' designation: Specimen No. 11]

* Not shown on plot

DATA TABLE NO. 590 NORMAL SPECTRAL REFLECTANCE OF ALUMINUM + ZINC + ΣX_i CONVERSION COATINGS

[Wavelength, λ, μm; Reflectance, ρ; Temperature, T, K]

CURVE 1, T = 311

λ	ρ
1.40	0.747
1.49	0.776
1.62	0.787
2.00	0.842
2.66	0.890
3.02	0.793
3.59	0.879
4.18	0.925
4.34	0.923
5.44	0.949
5.78	0.942
6.06	0.948
6.31	0.943
6.63	0.949
6.99	0.944
7.44	0.947
7.81	0.955
8.43	0.945
9.24	0.929
10.2	0.936
11.0	0.950
12.7	0.944
13.9	0.956
14.1	0.949
14.5	0.952
15.9	0.947
17.4	0.965
18.3	0.945
18.9	0.957
19.4	0.948
20.0	0.951
23.6	0.958
23.9	0.955
25.0	0.967
26.6	0.959
27.4	0.962
28.9	0.958
31.1	0.969
33.1	0.956
34.2	0.962
35.1	0.955

CURVE 2, T = 298

λ	ρ
0.4	0.426
0.45	0.448
0.5	0.464
0.6	0.523
0.7	0.517
0.8	0.519
0.9	0.592
1.0	0.644
1.2	0.723
1.4	0.772
1.6	0.784
1.8	0.789
2.0	0.798
2.5	0.804
3.0	0.807
3.5	0.832
4	0.854
5	0.872
6	0.870
7	0.888
8	0.895
9	0.885
10	0.870
11	0.844
12	0.838
13	0.847
14	0.856
15	0.867
16	0.885
17	0.892
18	0.885
19	0.888
20	0.840
21	0.867
22	0.877

CURVE 3, T = 589

λ	ρ
0.45	0.387
0.5	0.401
0.6	0.446
0.7	0.454
0.8	0.495
0.9	0.510
1.0	0.576
1.2	0.690
1.4	0.705
1.6	0.713
1.8	0.724
2.0	0.738
2.5	0.777
3.0	0.827
3.5	0.855
4	0.875
5	0.895
6	0.913
7	0.922
8	0.938
9	0.936
10	0.931
11	0.929
12	0.928
13	0.935
14	0.936
15	0.959
16	0.966
17	0.970
18	0.948
19	1.021*
20	0.978
21	0.957

CURVE 4*, T = 298

λ	ρ
0.4	0.392
0.45	0.424
0.5	0.431
0.6	0.471
0.7	0.487
0.8	0.452
0.9	0.529
1.0	0.589
1.2	0.730
1.4	0.756
1.6	0.760
1.8	0.771
2.0	0.771
2.5	0.807
3.0	0.873
3.5	0.891
4	0.902
5	0.931
6	0.951
7	0.950
8	0.946
9	0.942
10	0.927
11	0.909
12	0.918
13	0.920
14	0.926
15	0.941
16	0.936
17	0.940
18	0.931
19	0.951
20	0.956
21	0.933

CURVE 5, T = 298

λ	ρ
0.4	0.500
0.5	0.561
0.6	0.559
0.7	0.560
0.8	0.554
0.9	0.617
1.0	0.645
1.2	0.809
1.4	0.822
1.6	0.840
1.8	0.983
2.0	0.870
2.5	0.903
3.0	0.922
3.5	0.943
4	0.959
5	0.963
6	0.979
7	0.984
8	0.986
9	0.981
10	0.962
11	0.977
12	0.989
13	1.015*
14	0.983
15	0.996
16	0.991
17	1.003
18	1.017*
19	0.998

CURVE 6, T = 295

λ	ρ
0.4	0.303
0.45	0.352
0.5	0.383
0.6	0.397
0.7	0.368
0.8	0.309
0.9	0.449
1.0	0.497
1.2	0.570
1.4	0.606
1.6	0.593
1.8	0.597
2.0	0.555
2.5	0.647
3.0	0.676
3.5	0.708
4	0.721
5	0.745
6	0.712
7	0.776
8	0.792
9	0.791
10	0.778
11	0.755
12	0.758
13	0.763
14	0.778
15	0.794
16	0.801
17	0.785
18	0.796
19	0.790
20	0.743
21	0.832
22	0.770

CURVE 7, T = 298

λ	ρ
0.5	0.545
0.6	0.601
0.7	0.650
0.8	0.617
0.9	0.646
1.0	0.764
1.2	0.822
1.4	0.839
1.6	0.857
1.8	0.861
2.0	0.870*
2.5	0.886
3.0	0.918
3.5	0.933
4	0.966
5	0.962*
6	0.965
7	0.981*
8	0.987*
9	0.979
10	0.963*
11	0.958
12	0.979
13	0.993
14	0.972
15	0.991
16	0.995
17	0.996
18	0.991
19	0.989
20	1.011*
21	1.016*

CURVE 8, T = 298

λ	ρ
0.4	0.492
0.45	0.521
0.5	0.596
0.6	0.530
0.7	0.579
0.8	0.661
0.9	0.621
1.0	0.645*
1.2	0.753
1.4	0.796
1.6	0.793
1.8	0.803
2.0	0.807
2.5	0.801*
3.0	0.778
3.5	0.802
4	0.825
5	0.852
6	0.857
7	0.828
8	0.836
9	0.846
10	0.831
11	0.821
12	0.808
13	0.811
14	0.821
15	0.825
16	0.827
17	0.814
18	0.845
19	0.825
20	0.867
21	0.888
22	0.746

CURVE 9, T = 422

λ	ρ
0.4	0.599
0.45	0.542
0.5	0.587
0.6	0.510
0.7	0.617
0.8	0.665
0.9	0.656
1.0	0.679
1.2	0.788
1.4	0.810
1.6	0.812
1.8	0.840
2.0	0.837
2.5	0.829
3.0	0.819
3.5	0.836
4	0.848
5	0.867
6	0.874
7	0.846
8	0.856
9	0.871
10	0.847
11	0.836
12	0.830
13	0.823
14	0.837
15	0.828
16	0.833
17	0.858
18	0.855
19	0.874
20	0.896
21	0.919
22	0.882

CURVE 10, T = 589

λ	ρ
0.4	0.509
0.45	0.550
0.5	0.577
0.6	0.475
0.7	0.567
0.8	0.587
0.9	0.605
1.0	0.597
1.2	0.735
1.4	0.788
1.6	0.793*
1.8	0.793*
2.0	0.794
2.5	0.788
3.0	0.799
3.5	0.813
4	0.832
5	0.848
6	0.867

* Not shown on plot

DATA TABLE NO. 590 NORMAL SPECTRAL REFLECTANCE OF ALUMINUM + ZINC + ΣX_i CONVERSION COATINGS (continued)

CURVE 10 (cont.) T = 589

λ	ρ
7	0.837
8	0.850
9	0.864
10	0.852
11	0.838*
12	0.824
13	0.829
14	0.826
15	0.845
16	0.837
17	0.860
18	0.878
19	0.881
20	0.915
21	0.812
22	0.898

CURVE 11 T = 714

λ	ρ
0.4	0.439
0.45	0.487
0.5	0.543*
0.6	0.455
0.7	0.562*
0.8	0.643
0.9	0.627
1.0	0.622
1.2	0.748
1.4	0.803
1.6	0.805
1.8	0.792
2.0	0.780
2.5	0.776*
3.0	0.785
3.5	0.819
4	0.830
5	0.855*
6	0.863*
7	0.836
8	0.841
9	0.867*
10	0.861
11	0.832
12	0.818
13	0.819
14	0.846

CURVE 11 (cont.)

λ	ρ
15	0.843*
16	0.855
17	0.827
18	0.847*
19	0.855
20	0.880
21	0.841
22	0.853

CURVE 12* T = 298

λ	ρ
0.4	0.428
0.45	0.450
0.5	0.513
0.6	0.421
0.7	0.510
0.8	0.574
0.9	0.539
1.0	0.550
1.2	0.710
1.4	0.737
1.6	0.748
1.8	0.771
2.0	0.759
2.5	0.760
3.0	0.784
3.5	0.806
4	0.821
5	0.859
6	0.871
7	0.851
8	0.858
9	0.874
10	0.864
11	0.856
12	0.849
13	0.842
14	0.839
15	0.847
16	0.855
17	0.856
18	0.857
19	0.888
20	0.862
21	0.874
22	0.791

CURVE 13 T = 298

λ	ρ
0.45	0.638
0.5	0.601
0.6	0.701
0.7	0.735
0.8	0.636
0.9	0.692
1.0	0.815
1.2	0.893
1.4	0.895
1.6	0.891
1.8	0.891
2.0	0.901
2.5	0.905*
3.0	0.909
3.5	0.949
4	0.961*
5	0.979
6	0.986
7	0.975
8	0.982
9	0.985
10	0.971
11	0.947
12	0.954
13	0.970
14	0.979
15	0.984
16	0.988
17	1.000*
18	1.015*
19	1.020*
20	1.012*
21	1.041*
22	1.054*

CURVE 14 T = 298

λ	ρ
0.45	0.660
0.5	0.620
0.6	0.744
0.7	0.763
0.8	0.739
0.9	0.709
1.0	0.809
1.2	0.888
1.4	0.912

CURVE 14 (cont.)

λ	ρ
1.6	0.901
1.8	0.902
2.0	0.904*
2.5	0.910
3.0	0.903
3.5	0.932*
4	0.948
5	0.964*
6	0.981
7	0.963
8	0.973
9	0.976*
10	0.968
11	0.938
12	0.949
13	0.987
14	0.976*
15	0.982*
16	0.996*
17	0.992*
18	0.996*
19	1.007*
20	1.024*
21	1.025*
22	0.986

CURVE 15 T = 298

λ	ρ
0.4	0.495
0.45	0.481
0.5	0.538
0.6	0.501
0.7	0.533
0.8	0.538
0.9	0.621*
1.0	0.595
1.2	0.767
1.4	0.788*
1.6	0.805*
1.8	0.812
2.0	0.810*
2.5	0.809
3.0	0.790*
3.5	0.813*
4	0.852
5	0.880
6	0.888

CURVE 15 (cont.)

λ	ρ
7	0.875
8	0.886
9	0.907
10	0.897
11	0.914
12	0.919
13	0.913
14	0.909
15	0.922
16	0.928
17	0.927
18	0.925
19	0.927
20	0.904
21	0.933
22	1.017

CURVE 16 T = 298

λ	ρ
0.45	0.499
0.5	0.532
0.6	0.575
0.7	0.623
0.8	0.637*
0.9	0.637
1.0	0.741
1.2	0.751*
1.4	0.778
1.6	0.804*
1.8	0.828
2.0	0.853
2.5	0.799
3.0	0.225
3.5	0.430
4	0.565
5	0.670
6	0.720
7	0.623
8	0.149
9	0.059
10	0.064
13	0.116
15	0.205
17	0.304
19	0.261
21	0.273

CURVE 17 T = 422

λ	ρ
0.45	0.533
0.5	0.556
0.6	0.584
0.7	0.573
0.8	0.583
0.9	0.632
1.0	0.652
1.2	0.689
1.4	0.720
1.6	0.758
1.8	0.755
2.0	0.801*
2.5	0.725
3.0	0.574
3.5	0.689
4	0.743
5	0.784
6	0.777
7	0.673
8	0.056
9	0.022
10	0.014
11	0.056
13	0.131
15	0.211

CURVE 18 T = 589

λ	ρ
0.4	0.453
0.45	0.477
0.5	0.514
0.6	0.521*
0.7	0.523
0.8	0.581*
0.9	0.573
1.0	0.595*
1.2	0.643
1.4	0.663
1.6	0.684
1.8	0.703
2.0	0.714
2.5	0.732
3.0	0.718
3.5	0.764
4	0.760
5	0.764

CURVE 18 (cont.)

λ	ρ
6	0.785
7	0.594
8	0.164
11	0.159
13	0.049

CURVE 19 T = 714

λ	ρ
0.4	0.332
0.45	0.395
0.5	0.417
0.6	0.454*
0.7	0.476
0.8	0.464
0.9	0.548
1.0	0.557
1.2	0.607
1.4	0.596
1.6	0.615
1.8	0.631
2.0	0.641
2.5	0.689
3.0	0.690
3.5	0.734
4	0.723*
5	0.724
6	0.735
7	0.601
8	0.419
9	0.427
10	0.447
11	0.450
12	0.430
13	0.579
14	0.561
15	0.581
16	0.548
17	0.593
18	0.522
19	0.565
20	0.500

CURVE 20* T = 298

λ	ρ
0.4	0.309
0.45	0.311
0.5	0.334
0.6	0.367
0.7	0.367
0.8	0.418
0.9	0.458
1.0	0.512
1.2	0.582
1.4	0.607
1.6	0.627
1.8	0.638
2.0	0.657
2.5	0.641
3.0	0.554
3.5	0.620
4	0.677
5	0.743
6	0.730
7	0.699
8	0.419
9	0.385
10	0.434
11	0.406
12	0.427
13	0.470
14	0.461
15	0.552
16	0.511
17	0.532
18	0.530
19	0.489
20	0.523

CURVE 21* T = 298

λ	ρ
0.45	0.553
0.5	0.611
0.6	0.656
0.7	0.658
0.8	0.639
0.9	0.675
1.0	0.804
1.2	0.813
1.4	0.827
1.6	0.850

* Not shown on plot

DATA TABLE NO. 590 NORMAL SPECTRAL REFLECTANCE OF ALUMINUM + ZINC + ΣX_i CONVERSION COATINGS (continued)

CURVE 21 (cont.)* T = 298

λ	ρ
1.8	0.851
2.0	0.851
2.5	0.786
3.0	0.217
3.5	0.426
4	0.576
5	0.676
6	0.644
7	0.622
8	0.141
9	0.015
10	0.070
11	0.094
12	0.132
13	0.160
14	0.220
15	0.292
16	0.329
17	0.348
18	0.397
19	0.357
20	0.495
21	0.340

CURVE 22* T = 298

λ	ρ
0.4	0.495
0.45	0.512
0.5	0.567
0.6	0.617
0.7	0.613
0.8	0.585
0.9	0.632
1.0	0.778
1.2	0.800
1.4	0.820
1.6	0.847
1.8	0.851
2.0	0.867
2.5	0.857
3.0	0.585
3.5	0.705
4	0.788
5	0.842
6	0.738
7	0.765

CURVE 22 (cont.)

λ	ρ
8	0.118
9	0.047
10	0.107
11	0.097
12	0.193
13	0.173
14	0.308
15	0.248
16	0.294
17	0.269
18	0.276
19	0.224

CURVE 23 T = 298

λ	ρ
0.4	0.548
0.45	0.585
0.5	0.635
0.6	0.671
0.7	0.683
0.8	0.658
0.9	0.687
1.0	0.705
1.2	0.853
1.4	0.881
1.6	0.898
1.8	0.908
2.0	0.922
2.5	0.987
3.0	0.814
3.5	0.889
4	0.916
5	0.940
6	0.881
7	0.886
8	0.450
9	0.283
10	0.269
11	0.100
12	0.173
13	0.221
14	0.199
15	0.230
16	0.314
17	0.306
18	0.340
19	0.342

CURVE 24* T = 298

λ	ρ
0.4	0.618
0.45	0.638
0.5	0.677
0.6	0.703
0.7	0.692
0.8	0.694
0.9	0.754
1.0	0.788
1.2	0.816
1.4	0.848
1.6	0.863
1.8	0.877
2.0	0.884
2.5	0.881
3.0	0.610
3.5	0.736
4	0.832
5	0.893
6	0.844
7	0.862
8	0.306
9	0.226
10	0.228
11	0.077
12	0.122
13	0.143
14	0.217
15	0.233
16	0.275
17	0.318
18	0.280
19	0.239
20	0.249
21	0.265
22	0.250

CURVE 25 T = 298

λ	ρ
0.45	0.607
0.5	0.639
0.6	0.679
0.7	0.668
0.8	0.690
0.9	0.733
1.0	0.770
1.2	0.808*

CURVE 25 (cont.)

λ	ρ
1.4	0.836
1.6	0.863
1.8	0.883
2.0	0.871*
2.5	0.884
3.0	0.580
3.5	0.764*
4	0.844
5	0.904
6	0.854
7	0.855
8	0.361
9	0.251
10	0.245
11	0.089
12	0.132
13	0.188
14	0.240
16	0.304
18	0.357
20	0.353

CURVE 26* T = 298

λ	ρ
0.5	0.486
0.6	0.547
0.7	0.557
0.8	0.543
0.9	0.633
1.0	0.691
1.2	0.743
1.4	0.778
1.6	0.800
1.8	0.814
2.0	0.827
2.5	0.821
3.0	0.543
3.5	0.616
4	0.790
5	0.839
6	0.799
7	0.753
8	0.033
9	0.041
10	0.075
12	0.139
14	0.164

CURVE 26 (cont.)

λ	ρ
16	0.258
18	0.199
20	0.249

CURVE 27* T = 298

λ	ρ
0.45	0.484
0.5	0.519
0.6	0.565
0.7	0.626
0.8	0.578
0.9	0.653
1.0	0.702
1.2	0.767
1.4	0.805
1.6	0.819
1.8	0.832
2.0	0.844
2.5	0.739
3.0	0.301
3.5	0.502
4	0.634
5	0.746
6	0.729
7	0.692
8	0.018
9	0.016
10	0.072
12	0.154
14	0.233
16	0.303
18	0.294
20	0.309

CURVE 28* T = 298

λ	ρ
0.45	0.591
0.5	0.653
0.6	0.676
0.7	0.694
0.8	0.717
0.9	0.769
1.0	0.826
1.2	0.837
1.4	0.953
1.6	0.968

CURVE 28 (cont.)

λ	ρ
1.8	0.893
2.0	0.872
2.5	0.796
3.0	0.265
3.5	0.431
4	0.586
5	0.693
6	0.671
7	0.640
8	0.039
9	0.000
10	0.000
11	0.000
12	0.170
13	0.000

CURVE 29 T = 298

λ	ρ
0.45	0.160
0.5	0.161
0.6	0.149
0.7	0.135
0.8	0.120
0.9	0.125
1.0	0.142
1.2	0.198
1.4	0.250
1.6	0.284
1.8	0.319
2.0	0.335
2.5	0.293
3.0	0.074
3.5	0.096
4	0.146
5	0.247
6	0.134
7	0.067
8	0.017
9	0.048
10	0.035
11	0.095
12	0.179
13	0.216
14	0.256
15	0.279
16	0.284
17	0.302*

CURVE 29 (cont.)

λ	ρ
18	0.302
19	0.321
20	0.263

CURVE 30* T = 298

λ	ρ
0.45	0.163
0.5	0.155
0.6	0.159
0.7	0.153
0.8	0.162
0.9	0.190
1.0	0.229
1.2	0.300
1.4	0.364
1.6	0.400
1.8	0.426
2.0	0.549
2.5	0.479
3.0	0.138
3.5	0.127
4	0.182
5	0.301
6	0.208
7	0.124
8	0.040
9	0.084
10	0.041
11	0.107
12	0.174
13	0.220
14	0.258
15	0.292
16	0.309
17	0.349
18	0.328
19	0.262
20	0.369

CURVE 31 T = 422

λ	ρ
0.45	0.144
0.5	0.150
0.6	0.149*
0.7	0.150
0.8	0.200

CURVE 31 (cont.)

λ	ρ
0.9	0.188
1.0	0.217
1.2	0.296
1.4	0.347
1.6	0.389
1.8	0.420
2.0	0.445
2.5	0.402
3.0	0.169
3.5	0.282
4	0.368
5	0.437
6	0.313
7	0.137
8	0.125
9	0.135
10	0.135
11	0.170
12	0.207
13	0.260
14	0.292
15	0.258
16	0.236
17	0.267
18	0.250
19	0.415

CURVE 32 T = 589

λ	ρ
0.45	0.070
0.5	0.070
0.6	0.071
0.7	0.076
0.8	0.135
0.9	0.123
1.0	0.163
1.2	0.253
1.4	0.330
1.6	0.373
1.8	0.415
2.0	0.450
2.5	0.445
3.0	0.351
3.5	0.422
4	0.435
5	0.439
6	0.068
7	0.037

* Not shown on plot.

DATA TABLE NO. 590 NORMAL SPECTRAL REFLECTANCE OF ALUMINUM + ZINC + ΣX_i CONVERSION COATINGS (continued)

CURVE 33, T = 714

λ	ρ
0.45	0.065
0.5	0.068*
0.6	0.068
0.7	0.071
0.8	0.091
0.9	0.122*
1.0	0.153
1.2	0.245
1.4	0.314
1.6	0.361
1.8	0.399
2.0	0.408

CURVE 34*, T = 298

λ	ρ
0.45	0.079
0.5	0.075
0.6	0.074
0.7	0.077
0.8	0.135
0.9	0.141
1.0	0.159
1.2	0.252
1.4	0.318
1.6	0.369
1.8	0.411
2.0	0.447
2.5	0.473
3.0	0.426
3.5	0.497
4	0.535
5	0.542
6	0.399
7	0.114
8	0.015
9	0.028
10	0.022
11	0.092
12	0.192
13	0.216
14	0.245
15	0.273
16	0.290
17	0.305
18	0.286
19	0.379
20	0.270

CURVE 35*, T = 298

λ	ρ
0.4	0.203
0.45	0.210
0.5	0.214
0.6	0.214
0.7	0.202
0.8	0.243
0.9	0.240
1.0	0.286
1.2	0.356
1.4	0.407
1.6	0.423
1.8	0.454
2.0	0.480
2.5	0.411
3.0	0.104
3.5	0.135
4	0.206
5	0.344
6	0.205
7	0.114
8	0.018
9	0.027
10	0.035
11	0.087
13	0.285
15	0.270
17	0.300

CURVE 36*, T = 298

λ	ρ
0.45	0.211
0.5	0.245
0.6	0.303
0.7	0.332
0.8	0.395
0.9	0.452
1.0	0.516
1.2	0.625
1.4	0.668
1.6	0.699
1.8	0.728
2.0	0.752
2.5	0.707
3.0	0.450
3.5	0.588
4	0.682
5	0.755

CURVE 36 (cont.)

λ	ρ
6	0.662
7	0.500
8	0.026
9	0.050
10	0.043
11	0.126
12	0.223
13	0.215
14	0.299
15	0.299
16	0.345
17	0.352
18	0.358
19	0.394
20	0.334

CURVE 37*, T = 298

λ	ρ
0.4	0.193
0.45	0.184
0.5	0.183
0.6	0.175
0.7	0.170
0.9	0.202
1.0	0.235
1.2	0.285
1.4	0.335
1.6	0.362
1.8	0.379
2.0	0.387
2.5	0.309
3.0	0.078
3.5	0.064
4	0.081
5	0.140
6	0.090
7	0.070
8	0.023
9	0.058
10	0.033
11	0.116
12	0.183
13	0.234
14	0.289
15	0.288
16	0.309
17	0.335
18	0.295

CURVE 37 (cont.)

λ	ρ
19	0.333
20	0.280

CURVE 38*, T = 298

λ	ρ
0.45	0.172
0.5	0.160
0.6	0.147
0.7	0.141
0.8	0.164
0.9	0.141
1.0	0.156
1.2	0.203
1.4	0.253
1.6	0.290
1.8	0.326
2.0	0.347
2.5	0.311
3.0	0.082
3.5	0.108
4	0.162
5	0.183
6	0.145
7	0.085
8	0.022
9	0.048
10	0.033
11	0.098
12	0.195
14	0.268
16	0.317
18	0.302
20	0.285

CURVE 39*, T = 298

λ	ρ
0.4	0.099
0.45	0.111
0.5	0.120
0.6	0.137
0.7	0.159
0.8	0.218
0.9	0.245
1.0	0.280
1.2	0.362
1.4	0.441
1.6	0.465

CURVE 39 (cont.)

λ	ρ
1.8	0.498
2.0	0.536
2.5	0.465
3.0	0.119
3.5	0.133
4	0.207
5	0.317
6	0.229
7	0.178
8	0.035
9	0.059
10	0.055
11	0.081
12	0.205
13	0.290
14	0.289
15	0.290
16	0.352
17	0.347
18	0.355
19	0.399
20	0.435

* Not shown on plot

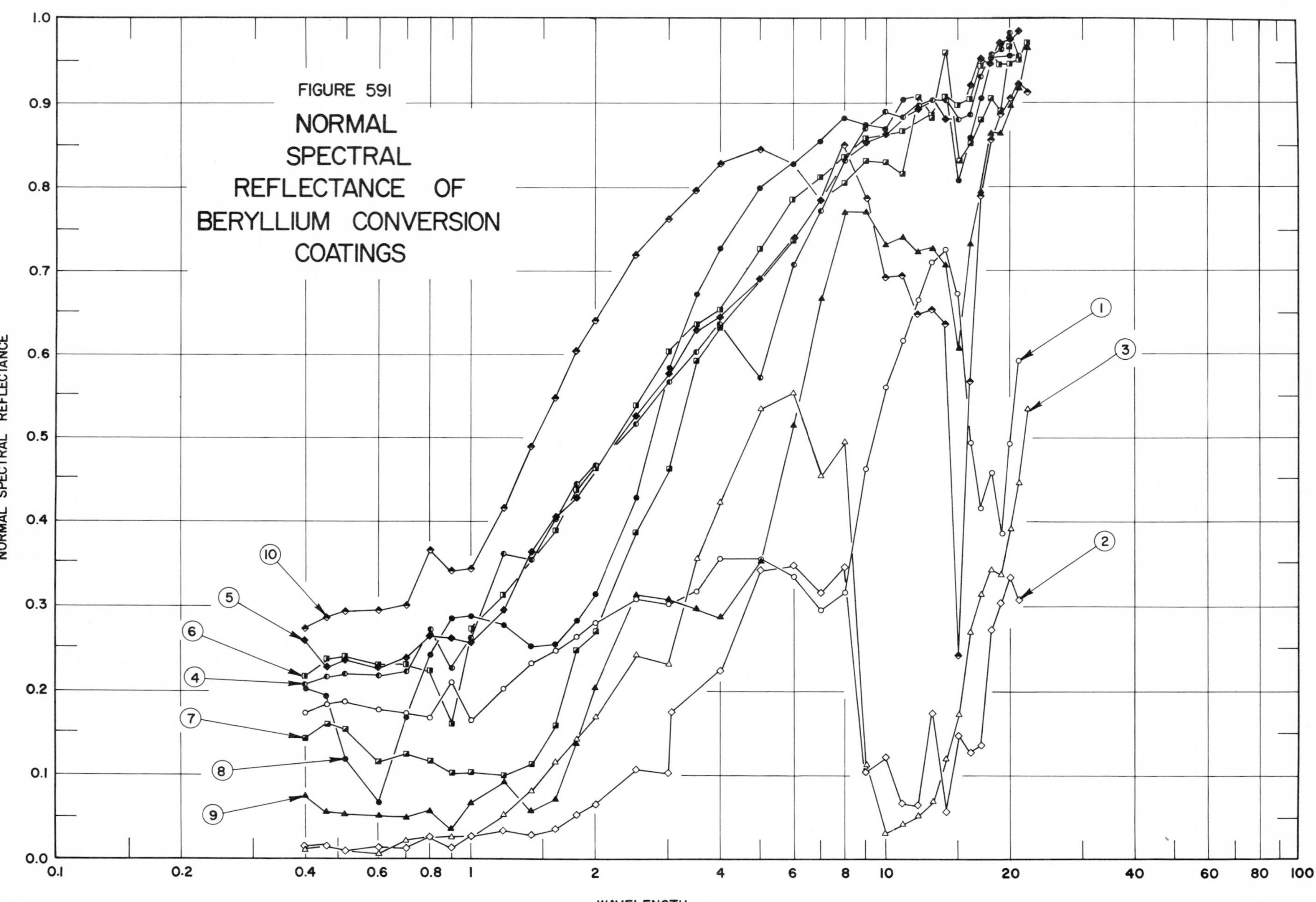

FIGURE 591
NORMAL SPECTRAL REFLECTANCE OF BERYLLIUM CONVERSION COATINGS
NORMAL SPECTRAL REFLECTANCE
WAVELENGTH, μm
0.0
0.1
0.2
0.3
0.4
0.5
0.6
0.7
0.8
0.9
1.0
0.1
0.2
0.4
0.6
0.8
1
2
4
6
8
10
20
40
60
80
100
1
2
3
4
5
6
7
8
9
10

SPECIFICATION TABLE NO. 591 NORMAL SPECTRAL REFLECTANCE OF BERYLLIUM CONVERSION COATINGS

Curve No.	Ref. No.	Year	Temperature, K	Wavelength Range, μm	Geometry θ	θ'	ω'	Reported Error, %	Composition (weight percent), Specifications, and Remarks
1	201	1961	298	0.4-21	$\sim 0^\circ$		2π		QMV beryllium ($\sim$99.7 beryllium) anodized in chromic acid for 30 min; mechanically polished; measured in vacuum ($\sim 10^{-6}$ mm Hg); converted from R (2π, 0°). [Authors' designation: Specimen No. 172]
2	201	1961	298	0.4-21	$\sim 0^\circ$		2π		QMV beryllium ($\sim$99.7 beryllium) anodized in nitric acid for 30 min; mechanically polished; measured in vacuum ($\sim 10^{-6}$ mm Hg); converted from R (2π, 0°). [Authors' designation: Specimen No. 174]
3	201	1961	298	0.4-22	$\sim 0^\circ$		2π		Similar to above specimen and conditions except nitric acid anodized for 10 min. [Authors' designation: Specimen No. 176]
4	201	1961	298	0.4-21	$\sim 0^\circ$		2π		QMV beryllium ($\sim$99.7 beryllium) anodized in sodium hydroxide for 30 min; mechanically polished; measured in vacuum ($\sim 10^{-6}$ mm Hg); converted from R (2π, 0°). [Authors' designation: Specimen No. 173]
5	201	1961	298	0.4-21	$\sim 0^\circ$		2π		Above specimen and conditions.
6	201	1961	298	0.4-22	$\sim 0^\circ$		2π		Above specimen and conditions.
7	201	1961	298	0.4-21	$\sim 0^\circ$		2π		Above specimen and conditions.
8	201	1961	298	0.4-21	$\sim 0^\circ$		2π		Above specimen and conditions after heating in vacuum.
9	201	1961	298	0.4-22	$\sim 0^\circ$		2π		Similar to curve 1 specimen and conditions except electropolished. [Authors' designation: Specimen No. 177]
10	201	1961	298	0.4-22	$\sim 0^\circ$		2π		Above specimen and conditions.

DATA TABLE NO. 591 NORMAL SPECTRAL REFLECTANCE OF BERYLLIUM CONVERSION COATINGS

[Wavelength, λ, μm; Reflectance, ρ; Temperature, T, K]

λ	ρ
CURVE 1 T = 298	
0.4	0.172
0.45	0.184
0.5	0.187
0.6	0.179
0.7	0.173
0.8	0.168
0.9	0.210
1.0	0.165
1.2	0.202
1.4	0.232
1.6	0.249
1.8	0.264
2.0	0.280
2.5	0.308
3.0	0.303
3.5	0.318
4	0.355
5	0.354
6	0.335
7	0.296
8	0.317
9	0.461
10	0.560
11	0.618
12	0.666
13	0.712
14	0.729
15	0.673
16	0.494
17	0.416
18	0.459
19	0.385
20	0.494
21	0.594
CURVE 2 T = 298	
0.4	0.016
0.45	0.017
0.5	0.011
0.6	0.015
0.7	0.015
0.8	0.029
0.9	0.015
CURVE 2 (cont.)	
1.0	0.029
1.2	0.036
1.4	0.030
1.6	0.039
1.8	0.052
2.0	0.065
2.5	0.107
3.0	0.104
3.5	0.175
4	0.225
5	0.342
6	0.347
7	0.317
8	0.345
9	0.105
10	0.121
11	0.069
12	0.065
13	0.175
14	0.059
15	0.149
16	0.127
17	0.136
18	0.274
19	0.303
20	0.334
21	0.309
CURVE 3 T = 298	
0.4	0.013
0.45	0.016*
0.5	0.012*
0.6	0.008
0.7	0.022
0.8	0.028*
0.9	0.026
1.0	0.027*
1.2	0.053
1.4	0.080
1.6	0.116
1.8	0.142
2.0	0.169
2.5	0.243
3.0	0.231
CURVE 3 (cont.)	
3.5	0.355
4	0.423
5	0.534
6	0.554
7	0.452
8	0.496
9	0.113
10	0.033
11	0.044
12	0.053
13	0.070
14	0.120
15	0.171
16	0.269
17	0.314
18	0.341
19	0.436
20	0.390
21	0.444
22	0.534
CURVE 4 T = 298	
0.4	0.208
0.45	0.217
0.5	0.220
0.6	0.219
0.7	0.223
0.8	0.271
0.9	0.226
1.0	0.262
1.2	0.360
1.4	0.353
1.6	0.402
1.8	0.444
2.0	0.465
2.5	0.517
3.0	0.566
3.5	0.602
4	0.637
5	0.572
6	0.708
7	0.772
8	0.831
9	0.870
CURVE 4 (cont.)	
10	0.890
11	0.884
12	0.899
13	0.904
14	0.903
15	0.881
16	0.887
17	0.931
18	0.959
19	0.964
20	0.985
21	0.958
CURVE 5 T = 298	
0.4	0.259
0.45	0.229
0.5	0.235
0.6	0.229
0.7	0.240
0.8	0.265
0.9	0.261
1.0	0.258
1.2	0.296
1.4	0.362
1.6	0.405
1.8	0.428
2.0	0.462*
2.5	0.525
3.0	0.577
3.5	0.630
4	0.645
5	0.691
6	0.741
7	0.787
8	0.833*
9	0.854
10	0.864
11	0.885*
12	0.894
13	0.904*
14	0.883
15	0.881*
16	0.921
17	0.951
CURVE 5 (cont.)	
18	0.950
19	0.971
20	0.975
21	0.988
CURVE 6 T = 298	
0.4	0.217
0.45	0.237
0.5	0.240
0.6	0.230
0.7	0.231
0.8	0.225
0.9	0.160
1.0	0.273
1.2	0.312
1.4	0.355*
1.6	0.389
1.8	0.437
2.0	0.461
2.5	0.539
3.0	0.603
3.5	0.634
4	0.654
5	0.729
6	0.787
7	0.812
8	0.839
9	0.857
10	0.862*
11	0.869
12	0.886
13	0.889
14	0.907
15	0.898
16	0.903
17	0.945
18	0.953*
19	0.945
20	0.947
21	0.952
22	0.972
CURVE 7 T = 298	
0.4	0.144
0.45	0.160
0.5	0.154
0.6	0.116
0.7	0.124
0.8	0.116
0.9	0.101
1.0	0.102
1.2	0.100
1.4	0.114
1.6	0.159
1.8	0.249
2.0	0.260
2.5	0.385
3.0	0.461
3.5	0.593
4	0.633
5	0.691*
6	0.739
7	0.789*
8	0.805
9	0.832
10	0.831
11	0.818
12	0.909
13	0.882
14	0.960
15	0.831
16	0.855*
17	0.881
18	0.906
19	0.892
20	0.968
21	1.034*
CURVE 8 T = 298	
0.4	0.201
0.45	0.192
0.5	0.119
0.6	0.068
0.7	0.167
0.8	0.241
0.9	0.284
CURVE 8 (cont.)	
1.0	0.287
1.2	0.277
1.4	0.252
1.6	0.254
1.8	0.282
2.0	0.314
2.5	0.427
3.0	0.584
3.5	0.671
4	0.729
5	0.800
6	0.829
7	0.855
8	0.882
9	0.874
10	0.870
11	0.903
12	0.909*
13	0.895*
14	0.903*
15	0.810
16	0.859
17	0.905
18	0.953
19	0.967*
20	0.959
21	0.956*
CURVE 9 T = 298	
0.4	0.075
0.45	0.057
0.5	0.054
0.6	0.052
0.7	0.050
0.8	0.058
0.9	0.037
1.0	0.068
1.2	0.092
1.4	0.059
1.6	0.071
1.8	0.139
2.0	0.203
2.5	0.312
3.0	0.309
CURVE 9 (cont.)	
3.5	0.298
4	0.286
5	0.351
6	0.514
7	0.667
8	0.772
9	0.771
10	0.733
11	0.742
12	0.725
13	0.729
14	0.708
15	0.606
16	0.735
17	0.796
18	0.865
19	0.865
20	0.899
21	0.919
22	0.968
CURVE 10 T = 298	
0.4	0.271
0.45	0.285
0.5	0.294
0.6	0.295
0.7	0.300
0.8	0.365
0.9	0.340
1.0	0.342
1.2	0.415
1.4	0.488
1.6	0.547
1.8	0.603
2.0	0.640
2.5	0.720
3.0	0.762
3.5	0.796
4	0.827
5	0.846
6	0.827*
7	0.784*
8	0.850
9	0.789

* Not shown on plot

DATA TABLE NO. 591 NORMAL SPECTRAL REFLECTANCE OF BERYLLIUM CONVERSION COATINGS (continued)

λ	ρ
CURVE 10 (cont.)	
10	0.693
11	0.696
12	0.647
13	0.654
14	0.638
15	0.242
16	0.568
17	0.793
18	0.859
19	0.890*
20	0.904
21	0.922
22	0.911

* Not shown on plot

SPECIFICATION TABLE NO. 592 HEMISPHERICAL TOTAL EMITTANCE OF BERYLLIUM + COPPER CONVERSION COATINGS

Curve No.	Ref. No.	Year	Temperature Range, K	Reported Error, %	Composition (weight percent), Specifications and Remarks
1*	271	1964	305-955	± 3	1 Cu; anodized in 7.5% NaOH at ~298 K; substrate hand polished with 1200 Crystolon lapping compound prior to anodizing; measured in vacuum (10^{-5} mm Hg). [Authors' designation: Sample C_7]

DATA TABLE NO. 592 HEMISPHERICAL TOTAL EMITTANCE OF BERYLLIUM + COPPER CONVERSION COATINGS

[Temperature, T, K; Emittance, ε]

T	ε
CURVE 1*	
305	0.37
537	0.31
673	0.28
816	0.27
955	0.28

* No plot given

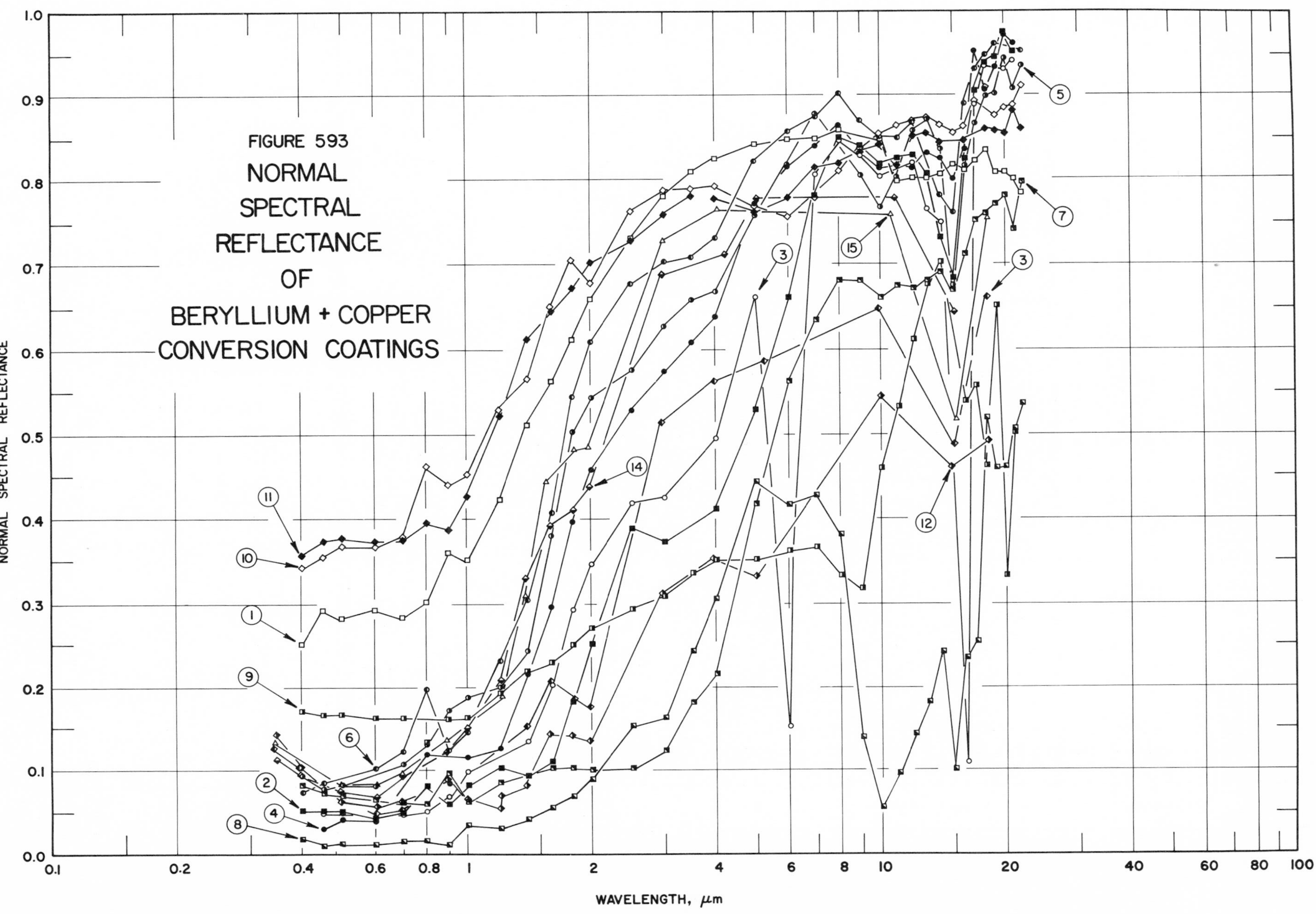
FIGURE 593
NORMAL
SPECTRAL
REFLECTANCE
OF
BERYLLIUM + COPPER
CONVERSION COATINGS
NORMAL SPECTRAL REFLECTANCE
WAVELENGTH, μm
1.0
0.9
0.8
0.7
0.6
0.5
0.4
0.3
0.2
0.1
0.0
0.1
0.2
0.4
0.6
0.8
1
2
4
6
8
10
20
40
60
80
100
1
2
3
4
5
6
7
8
9
10
11
12
14
15

SPECIFICATION TABLE NO. 593 NORMAL SPECTRAL REFLECTANCE OF BERYLLIUM + COPPER CONVERSION COATINGS

Curve No.	Ref. No.	Year	Temperature, K	Wavelength Range, μm	Geometry θ	θ'	ω'	Reported Error, %	Composition (weight percent), Specifications, and Remarks
1	201	1961	298	0.4-22	~0°		2π		1 Cu; anodized in sulfuric acid; mechanically polished; measured in vacuum (~10^{-6} mm Hg); converted from R (2π, 0°). [Authors' designation: Specimen No. 162]
2	201	1961	298	0.4-21	~0°		2π		Similar to above specimen and conditions except sodium hydroxide anodized for 30 min. [Authors' designation: Specimen No. 166]
3	201	1961	298	0.4-21	~0°		2π		Above specimen and conditions.
4	201	1961	298	0.4-21	~0°		2π		Above specimen and conditions.
5	201	1961	298	0.4-22	~0°		2π		Above specimen and conditions.
6	201	1961	298	0.4-22	~0°		2π		Above specimen and conditions after heating in vacuum.
7	201	1961	298	0.4-22	~0°		2π		Similar to curve 1 specimen and conditions except also electropolished. [Authors' designation: Specimen No. 164]
8	201	1961	298	0.4-22	~0°		2π		Similar to curve 1 specimen and conditions except nitric acid anodized for 30 min. [Authors' designation: Specimen No. 167]
9	201	1961	298	0.4-21	~0°		2π		Similar to curve 1 specimen and conditions except chromic acid anodized for 30 min. [Authors' designation: Specimen No. 168]
10	201	1961	298	0.4-22	~0°		2π		Similar to above specimen and conditions except chromic acid anodized for 10 min. [Authors' designation: Specimen No. 170]
11	201	1961	298	0.4-22	~0°		2π		Similar to curve 1 specimen and conditions except only mechanically polished. [Authors' designation: Specimen No. 171]
12	271	1964	~298	0.348-18.2	~0°		2π		1 Cu; anodized in 7.5% NaOH at ~298 K; substrate hand polished with 1200 crystolon lapping compound prior to anodizing; measured in vacuum (10^{-5} mm Hg); converted from R (2π, 0°).
13	271	1964	533	0.344-18.3	~0°		2π		Similar to above specimen and conditions.
14	271	1964	811	0.344-18.3	~0°		2π		Similar to above specimen and conditions.
15	271	1964	~298	0.346-18.4	~0°		2π		Similar to above specimen and conditions except after 811 K measurement.

DATA TABLE NO. 593 NORMAL SPECTRAL REFLECTANCE OF BERYLLIUM + COPPER CONVERSION COATINGS

[Temperature, T, K; Reflectance, ρ; Temperature, T, K]

λ	ρ
CURVE 1	**T = 298**
0.4	0.250
0.45	0.292
0.5	0.282
0.6	0.292
0.7	0.284
0.8	0.302
0.9	0.357
1.0	0.350
1.2	0.422
1.4	0.507
1.6	0.563
1.8	0.614
2.0	0.658
2.5	0.732
3.0	0.781
3.5	0.809
4	0.825
5	0.843
6	0.847
7	0.848
8	0.858
9	0.848
10	0.817
11	0.797
12	0.800
13	0.802
14	0.806
15	0.817
16	0.812
17	0.823
18	0.835
19	0.808
20	0.809
21	0.800
22	0.786
CURVE 2	**T = 298**
0.4	0.051
0.45	0.051
0.5	0.050
0.6	0.044
0.7	0.049
0.8	0.080
0.9	0.057
1.0	0.083
1.2	0.104
1.4	0.094
1.6	0.108
1.8	0.183
2.0	0.249
2.5	0.388
3.0	0.372
3.5	0.394
4	0.410
5	0.528
6	0.662
7	0.782
8	0.850
9	0.842
10	0.818
11	0.826
12	0.828
13	0.807
14	0.734
15	0.686
16	0.825
17	0.905
18	0.938
19	0.945
20	0.975
21	0.952
CURVE 3	**T = 298**
0.4	0.054*
0.45	0.047
0.5	0.047
0.6	0.040*
0.7	0.047
0.8	0.051
0.9	0.067
1.0	0.098
1.2	0.100*
1.4	0.134
1.6	0.201
1.8	0.291
2.0	0.345
2.5	0.416
3.0	0.424
3.5	0.478
4	0.495
5	0.661
6	0.150
7	0.805
8	0.847
9	0.829
10	0.803
11	0.803
12	0.820
13	0.807*
14	0.765
15	0.749
16	1.108
17	0.889
18	0.935
19	0.945*
20	0.931
21	0.940
CURVE 4	**T = 298**
0.4	0.050*
0.45	0.030
0.5	0.041
0.6	0.037
0.7	0.050*
0.8	0.117
0.9	0.084
1.0	0.116
1.2	0.125
1.4	0.215
1.6	0.296
1.8	0.396
2.0	0.458
2.5	0.527
3.0	0.573
3.5	0.609
4	0.639
5	0.770
6	0.815
7	0.842
8	0.866
9	0.831
10	0.814
11	0.815
12	0.814
13	0.832
14	0.826
15	0.799
16	0.837
17	0.954
18	0.906
19	0.934
20	0.974
21	0.962
CURVE 5	**T = 298**
0.4	0.093
0.45	0.077
0.5	0.080
0.6	0.078
0.7	0.105
0.8	0.128
0.9	0.169
1.0	0.186
1.2	0.198
1.4	0.241
1.6	0.378
1.8	0.501
2.0	0.543
2.5	0.576
3.0	0.628
3.5	0.658
4	0.668
5	0.757
6	0.817
7	0.878
8	0.849*
9	0.804
10	0.768
11	0.803
12	0.858
13	0.871
14	0.835
15	0.675
16	0.817
17	0.867
18	0.898
19	0.901
20	0.946
21	0.857
22	0.886
CURVE 6	**T = 298**
0.4	0.074
0.45	0.084
0.5	0.083*
0.6	0.100
0.7	0.121
0.8	0.197
0.9	0.123
1.0	0.144
1.2	0.230
1.4	0.304
1.6	0.406
1.8	0.542
2.0	0.609
2.5	0.678
3.0	0.703
3.5	0.707
4	0.731
5	0.822
6	0.857
7	0.875
8	0.901
9	0.870
10	0.851
11	0.849
12	0.867
13	0.871*
14	0.783
15	0.762
16	0.891
17	0.931
18	0.948
19	0.960
20	1.029*
21	0.953*
22	0.953
CURVE 7	**T = 298**
0.4	0.081
0.45	0.071
0.5	0.068
0.6	0.066
0.7	0.062
0.8	0.059
0.9	0.095
1.0	0.062
1.2	0.085
1.4	0.095*
1.6	0.101
1.8	0.102
2.0	0.099
2.5	0.100
3.0	0.123
3.5	0.180
4	0.215
5	0.415
6	0.562
7	0.636
8	0.682
9	0.680
10	0.660
11	0.675
12	0.672
13	0.681
14	0.691
15	0.670
16	0.713
17	0.752
18	0.759
19	0.773
20	0.783
21	0.743
22	0.797
CURVE 8	**T = 298**
0.4	0.018
0.45	0.009
0.5	0.013
0.6	0.013
0.7	0.015
0.8	0.016
0.9	0.011
1.0	0.035
1.2	0.032
1.4	0.043
1.6	0.055
1.8	0.068
2.0	0.088
2.5	0.151
3.0	0.161
3.5	0.242
4	0.305
5	0.441
6	0.415
7	0.426
8	0.379
9	0.137
10	0.055
11	0.095
12	0.142
13	0.180
14	0.242
15	0.099
16	0.234
17	0.254
18	0.518
19	0.459
20	0.460
21	0.506
22	0.535
CURVE 9	**T = 298**
0.4	0.169
0.45	0.164
0.5	0.166
0.6	0.162
0.7	0.162
0.8	0.133
0.9	0.160
1.0	0.161
1.2	0.192
1.4	0.218
1.6	0.228
1.8	0.249
2.0	0.269
2.5	0.293
3.0	0.307
3.5	0.335
4	0.349
5	0.350
6	0.361
7	0.365
8	0.333
9	0.317
10	0.459
11	0.531
12	0.613
13	0.677
14	0.701
15	0.678*
16	0.538
17	0.556
18	0.462
19	0.653
20	0.334
21	0.504
CURVE 10	**T = 298**
0.4	0.341
0.45	0.354
0.5	0.367
0.6	0.367
0.7	0.377
0.8	0.462
0.9	0.438
1.0	0.451
1.2	0.528
1.4	0.580
1.6	0.650
1.8	0.704
2.0	0.679
2.5	0.764
3.0	0.787
3.5	0.789
4	0.791
5	0.768
6	0.756
7	0.778
8	0.810

* Not shown on plot

DATA TABLE NO. 593 NORMAL SPECTRAL REFLECTANCE OF BERYLLIUM + COPPER CONVERSION COATINGS (continued)

λ	ρ
CURVE 10 (cont.)	
9	0.839
10	0.855
11	0.866
12	0.870
13	0.874
14	0.866
15	0.856
16	0.864
17	0.896
18	0.909
19	0.877
20	0.886
21	0.889
22	0.911
CURVE 11 T = 298	
0.4	0.356*
0.45	0.372
0.5	0.377
0.6	0.371
0.7	0.374
0.8	0.396
0.9	0.386
1.0	0.425
1.2	0.521
1.4	0.613
1.6	0.646
1.8	0.673
2.0	0.703
2.5	0.728
3.0	0.758
3.5	0.781
4	0.777
5	0.764
6	0.779
7	0.816
8	0.818
9	0.834
10	0.843
11	0.817
12	0.853
13	0.855
14	0.845
15	0.854
16	0.847
17	0.867*

λ	ρ
CURVE 11 (cont.)	
18	0.859
19	0.858
20	0.856
21	0.884
22	0.861
CURVE 12 T ~ 298	
0.348	0.141
0.399	0.102
0.495	0.063
0.601	0.056
0.696	0.064
0.889	0.087
1.00	0.065
1.21	0.055
1.38	0.082
1.58	0.144
1.79	0.141
1.97	0.135
2.97	0.314
3.96	0.354
5.00	0.334
10.0	0.546
14.9	0.460
18.2	0.491
CURVE 13 T = 533	
0.344	0.126
0.394	0.103
0.495	0.060*
0.601	0.049
0.696	0.054
0.889	0.064
1.00	0.065
1.21	0.069
1.38	0.151
1.59	0.207
1.80	0.185
1.97	0.176
2.97	0.513
3.99	0.562
5.02	0.585
9.95	0.648
15.1	0.489

λ	ρ
CURVE 13 (cont.)	
18.3	0.662
CURVE 14 T = 811	
0.348	0.112
0.396	0.095
0.495	0.026
0.601	0.069
0.698	0.092
0.885	0.120
1.00	0.149
1.20	0.208
1.37	0.330
1.57	0.390
1.82	0.409
1.99	0.436
2.99	0.687
4.02	0.711
5.00	0.779
10.0	0.779
15.1	0.645
CURVE 15 T ~ 298	
0.345	0.133
0.399	0.103*
0.498	0.081
0.601	0.082
0.698	0.095
0.889	0.135
1.00	0.164*
1.21	0.187
1.37	0.307
1.55	0.443
1.81	0.480
1.98	0.484
2.99	0.728
4.03	0.765
5.00	0.773*
10.0	0.759
15.1	0.517
18.4	0.755

* Not shown on plot

SPECIFICATION TABLE NO. 594 NORMAL SOLAR ABSORPTANCE OF BERYLLIUM + COPPER CONVERSION COATINGS

Curve No.	Ref. No.	Year	Temperature Range, K	Geometry θ	Reported Error, %	Composition (weight percent), Specifications and Remarks
1*	271	1964	298-811	~0°		1 Cu; anodized in 7.5% NaOH at ~298 K; substrate hand polished with 1200 Crystolon lapping compound prior to anodizing; measured in vacuum (10^{-5} mm Hg); computed from normal spectral reflectance. [Authors' designation: Sample C_7]

DATA TABLE NO. 594 NORMAL SOLAR ABSORPTANCE OF BERYLLIUM + COPPER CONVERSION COATINGS

[Temperature, T, K; Absorptance, α]

T	α
CURVE 1*	
298	0.91
533	0.91
811	0.84

* No plot given

SPECIFICATION TABLE NO. 595 HEMISPHERICAL TOTAL EMITTANCE OF MAGNESIUM CONVERSION COATINGS

Curve No.	Ref. No.	Year	Temperature Range, K	Reported Error, %	Composition (weight percent), Specifications and Remarks
1*	272	1964	304-305	3.3-7.7	Dow 7 coating on magnesium.
2*	272	1964	302	3.3-7.7	Above specimen and conditions.
3*	49	1961	273		Dow 15 finish on magnesium.
4*	49	1961	278		Above specimen and conditions except exposed to UV radiation in vacuum (8×10^{-6} to 10^{-6} mm Hg) from argon filled A-H6 high pressure Hg arc lamp for 12 hrs.
5*	49	1961	278		Dow 15 finish on magnesium.
6*	49	1961	278		Above specimen and conditions except exposed to UV radiation in vacuum (8×10^{-6} to 10^{-6} mm Hg) from an argon filled A-H6 high pressure Hg arc for 12 hrs.
7*	49	1961	278		Similar to curve 5 specimen and conditions.
8*	49	1961	278		Similar to curve 6 specimen and conditions.
9*	49	1961	278		Dow 17 finish on magnesium.

DATA TABLE NO. 595 HEMISPHERICAL TOTAL EMITTANCE OF MAGNESIUM CONVERSION COATINGS

[Temperature, T, K; Emittance, ∈]

T	∈	T	∈	T	∈
CURVE 1*		CURVE 5*		CURVE 9*	
304	0.252	278	0.07	278	0.50-0.82
305	0.254				
		CURVE 6*			
CURVE 2*		278	0.08		
302	0.248				
		CURVE 7*			
CURVE 3*		278	0.06		
273	0.09				
		CURVE 8*			
CURVE 4*		278	0.07		
278	0.14				

* No plot given

SPECIFICATION TABLE NO. 596 NORMAL SOLAR ABSORPTANCE OF MAGNESIUM CONVERSION COATINGS

Curve No.	Ref. No.	Year	Temperature Range, K	Geometry θ	Reported Error, %	Composition (weight percent), Specifications and Remarks
1*	49	1961	273	~0°		Dow 15 on magnesium.
2*	49	1961	278	~0°		Above specimen and conditions except exposed to UV radiation in vacuum (8×10^{-6} to 10^{-6} mm Hg) from an argon-filled A-H6 high pressure Hg arc lamp for 12 hrs.
3*	49	1961	278	~0°		Similar to curve 1 specimen and conditions.
4*	49	1961	278	~0°		Above specimen and conditions except exposed to UV radiation under same conditions as curve 2.
5*	49	1961	278	~0°		Dow 15 on magnesium.
6*	49	1961	278	~0°	10	Dow 17 on magnesium; measured for extraterrestrial conditions.

DATA TABLE NO. 596 NORMAL SOLAR ABSORPTANCE OF MAGNESIUM CONVERSION COATINGS

[Temperature, T, K; Absorptance, α]

T	α	T	α
CURVE 1*		CURVE 5*	
273	0.18	278	0.23
CURVE 2*		CURVE 6*	
278	0.38	278	0.53-0.72
CURVE 3*			
278	0.17		
CURVE 4*			
278	0.28		

* No plot given

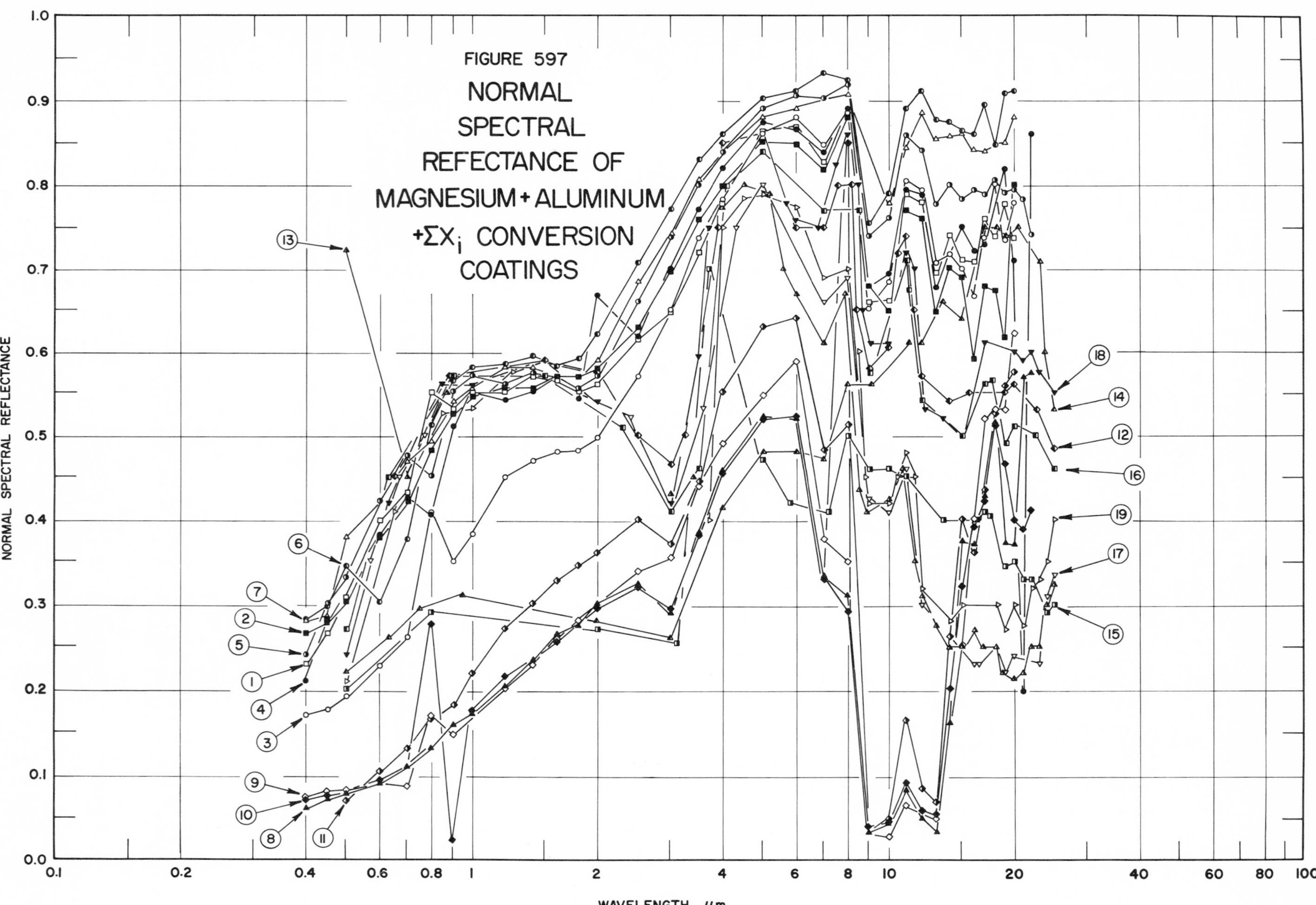

FIGURE 597 NORMAL SPECTRAL REFECTANCE OF MAGNESIUM + ALUMINUM $+\Sigma X_i$ CONVERSION COATINGS

SPECIFICATION TABLE NO. 597 NORMAL SPECTRAL REFLECTANCE OF MAGNESIUM + ALUMINUM + ΣX_i CONVERSION COATINGS

Curve No.	Ref. No.	Year	Temperature, K	Wavelength Range, μm	Geometry θ	θ'	ω'	Reported Error, %	Composition (weight percent), Specifications, and Remarks
1	201	1961	298	0.4-20	~0°		2π		AZ-31 B magnesium, HAE anodized (light coating); nominal composition: 3 Al, 1 Zn, 0.2 Mn, Mg balance; substrate mechanically polished and electropolished; measured in vacuum (~10^{-6} mm Hg); converted from R (2π, 0°). [Authors' designation: Specimen No. 25]
2	201	1961	450	0.4-20	~0°		2π		Above specimen and conditions.
3*	201	1961	298	0.4-20	~0°		2π		Above specimen and conditions.
4	201	1961	298	0.4-22	~0°		2π		Above specimen and conditions after heating in vacuum.
5*	201	1961	298	0.4-22	~0°		2π		Similar to curve 1 specimen and conditions. [Authors' designation: Specimen No. 36]
6	201	1961	298	0.4-20	~0°		2π		Similar to curve 1 specimen and conditions except substrate mechanically polished only. [Authors' designation: Specimen No. 138]
7	201	1961	298	0.4-20	~0°		2π		Above specimen and conditions except mill finish.
8	201	1961	298	0.4-22	~0°		2π		Similar to curve 6 specimen and conditions except dark coating. [Authors' designation: Specimen No. 140]
9*	201	1961	298	0.4-20	~0°		2π		Above specimen and conditions.
10	201	1961	298	0.4-22	~0°		2π		Above specimen and conditions except mill finish.
11	201	1961	298	0.4-20	~0°		2π		Similar to curve 1 specimen and conditions except dark coating. [Authors' designation: Specimen No. 155]
12	202	1962	~322	0.50-25.00	~0°		2π	< 2	AZ-31 magnesium HAE anodized as per Mil-C-13335-1; nominal composition: 3 Al, 1 Zn, 0.2 Mn, Mg balance; data extracted from smooth curve; hohlraum at 1273 K; converted from R (2π,0°).
13	202	1962	~322	0.50-25.00	~0°		2π	< 2	Above specimen and conditions; diffuse component only.
14	202	1962	~322	0.50-25.00	~0°		2π	< 2	Similar to curve 12 specimen and conditions except grit blasted using 60 grit silicon carbide with air pressure of 110 to 120 psi for 30 to 45 sec.
15	202	1962	~322	0.50-25.00	~0°		2π	< 2	Above specimen and conditions; diffuse component only.
16	202	1962	~322	0.50-25.00	~0°		2π	< 2	Similar to curve 12 specimen and conditions except exposed to vacuum (< 4 x 10^{-8} mm Hg) for 24 hrs.
17	202	1962	~322	0.50-25.00	~0°		2π	< 2	Above specimen and conditions; diffuse component only.
18	202	1962	~322	0.50-25.00	~0°		2π	< 2	Similar to curve 12 specimen and conditions except X-ray exposed in vacuum (4 x 10^{-8} mm Hg) for 24 hrs.
19	202	1962	~322	0.50-25.00	~0°		2π	< 2	Above specimen and conditions; diffuse component only.

* Not shown on plot

DATA TABLE NO. 597 NORMAL SPECTRAL REFLECTANCE OF MAGNESIUM + ALUMINUM + ΣX_i CONVERSION COATINGS

[Wavelength, λ, μm; Reflectance, ρ; Temperature, T, K]

λ	ρ
CURVE 1 T = 298	
0.4	0.229
0.45	0.266
0.5	0.308
0.6	0.398
0.7	0.433
0.8	0.549
0.9	0.530
1.0	0.554
1.2	0.553
1.4	0.569
1.6	0.565
1.8	0.553
2.0	0.559
2.5	0.616
3.0	0.647
3.5	0.721
4	0.781
5	0.865
6	0.858
7	0.827
8	0.886
9	0.660
10	0.663
11	0.790
12	0.781
13	0.695
14	0.743
15	0.713
16	0.710
17	0.760
18	0.740
19	0.778
20	0.737
CURVE 2 T = 450	
0.4	0.265
0.45	0.277
0.5	0.304
0.6	0.377
0.7	0.424
0.8	0.484
0.9	0.525
1.0	0.546
CURVE 2 (cont.)	
1.2	0.556
1.4	0.555
1.6	0.570
1.8	0.568
2.0	0.580
2.5	0.628
3.0	0.697
3.5	0.761
4	0.798
5	0.854
6	0.848
7	0.818
8	0.871
9	0.681
10	0.647
11	0.771
12	0.763
13	0.647
14	0.702
15	0.691
16	0.594
17	0.679
18	0.676
19	0.617
20	0.801
CURVE 3* T = 298	
0.4	0.168
0.45	0.175
0.5	0.190
0.6	0.227
0.7	0.260
0.8	0.408
0.9	0.350
1.0	0.383
1.2	0.449
1.4	0.468
1.6	0.480
1.8	0.484
2.0	0.497
2.5	0.569
3.0	0.651
3.5	0.737
4	0.786
CURVE 3 (cont.)	
5	0.861
6	0.880
7	0.848
8	0.885*
9	0.654
10	0.686
11	0.806
12	0.796
13	0.707
14	0.718
15	0.700
16	0.667
17	0.738
18	0.802
19	0.736
20	0.782
CURVE 4 T = 298	
0.4	0.209
0.45	0.282
0.5	0.309*
0.6	0.381
0.7	0.425
0.8	0.406
0.9	0.510
1.0	0.554*
1.2	0.541
1.4	0.553
1.6	0.568*
1.8	0.544
2.0	0.668
2.5	0.618
3.0	0.701
3.5	0.773
4	0.820
5	0.875
6	0.867
7	0.839
8	0.893
9	0.679*
10	0.695
11	0.795
12	0.788
13	0.678
CURVE 4 (cont.)	
14	0.719*
15	0.751
16	0.723
17	0.730
18	0.800*
19	0.819
20	0.710
21	0.198
22	0.860
CURVE 5* T = 298	
0.4	0.241
0.45	0.302
0.5	0.334
0.6	0.424
0.7	0.475
0.8	0.453
0.9	0.552
1.0	0.569
1.2	0.560
1.4	0.576
1.6	0.568*
1.8	0.556
2.0	0.570
2.5	0.659
3.0	0.739
3.5	0.800
4	0.840
5	0.893
6	0.905
7	0.904
8	0.917
9	0.743
10	0.764
11	0.858
12	0.843
13	0.778
14	0.801
15	0.786
16	0.796
17	0.790
18	0.807
19	0.794
20	0.795
CURVE 5 (cont.)	
21	0.785
22	0.743
CURVE 6 T = 298	
0.4	0.282
0.45	0.298
0.5	0.346
0.6	0.304
0.7	0.264
0.8	0.514
0.9	0.566
1.0	0.582
1.2	0.585
1.4	0.596
1.6	0.584
1.8	0.593
2.0	0.621
2.5	0.707
3.0	0.774
3.5	0.831
4	0.861
5	0.903
6	0.911
7	0.934
8	0.925
9	0.756
10	0.793
11	0.891
12	0.911
13	0.878
14	0.876
15	0.866
16	0.861
17	0.896
18	0.848
19	0.907
20	0.910
CURVE 7 T = 298	
0.4	0.279
0.45	0.286
0.5	0.378
CURVE 7 (cont.)	
0.6	0.420*
0.7	0.467
0.8	0.494
0.9	0.539
1.0	0.556
1.2	0.582
1.4	0.581
1.6	0.573*
1.8	0.571*
2.0	0.590
2.5	0.687
3.0	0.746
3.5	0.808
4	0.842*
5	0.880
6	0.894
7	0.900*
8	0.907
9	0.742*
10	0.779
11	0.846
12	0.885
13	0.856
14	0.858
15	0.861
16	0.844
17	0.841
18	0.849*
19	0.852
20	0.881
CURVE 8 T = 298	
0.4	0.061
0.45	0.073
0.5	0.077
0.6	0.090
0.7	0.111
0.8	0.134
0.9	0.157
1.0	0.171
1.2	0.205
1.4	0.235
1.6	0.265
1.8	0.275
CURVE 8 (cont.)	
2.0	0.303
2.5	0.325
3.0	0.289
3.5	0.386
4	0.457
5	0.522
6	0.521
7	0.336
8	0.313
9	0.035
10	0.046
11	0.084
12	0.052
13	0.035
14	0.161
15	0.376
16	0.372
17	0.427
18	0.515
19	0.374
20	0.371
21	0.569
22	0.576
CURVE 9* T = 298	
0.4	0.076
0.45	0.082
0.5	0.083
0.6	0.092*
0.7	0.087
0.8	0.169
0.9	0.147
1.0	0.158*
1.2	0.202
1.4	0.228
1.6	0.259
1.8	0.281
2.0	0.299
2.5	0.339
3.0	0.356
3.5	0.437
4	0.491
5	0.547
6	0.587
CURVE 9 (cont.)	
7	0.377
8	0.351
9	0.036*
10	0.027
11	0.066
13	0.051
15	0.254
16	0.401
17	0.519
18	0.530
19	0.529
20	0.623
CURVE 10 T = 298	
0.4	0.072
0.45	0.076
0.5	0.078*
0.6	0.096
0.7	0.109*
0.8	0.275
0.9	1.025
1.0	0.176
1.2	0.216
1.4	0.235*
1.6	0.256
1.8	0.276*
2.0	0.296
2.5	0.322
3.0	0.296
3.5	0.382
4	0.455
5	0.519
6	0.523
7	0.330
8	0.294
9	0.044
10	0.047
11	0.092
12	0.060
13	0.067
14	0.203
15	0.323
16	0.394
17	0.424

* Not shown on plot

DATA TABLE NO. 597 NORMAL SPECTRAL REFLECTANCE OF MAGNESIUM + ALUMINUM + ΣX_i CONVERSION COATINGS (continued)

λ	ρ
CURVE 10 (cont.)	
18	0.513
19	0.466
20	0.400
21	0.387
22	0.414
CURVE 11	
T = 298	
0.4	0.072*
0.45	0.072*
0.5	0.071
0.6	0.105
0.7	0.132
0.8	0.166
0.9	0.181
1.0	0.219
1.2	0.271
1.4	0.301
1.6	0.327
1.8	0.346
2.0	0.360
2.5	0.401
3.0	0.371
3.5	0.445
4	0.553
5	0.630
6	0.642
7	0.484
8	0.514
9	0.036
10	0.052
11	0.165
12	0.085
13	0.068
14	0.264
15	0.403
16	0.363
17	0.436
18	0.526
19	0.557
20	0.576
CURVE 12	
T ~ 322	
0.50	0.345*
0.65	0.450
0.88	0.570
CURVE 12 (cont.)	
1.50	0.590
2.00	0.575
2.50	0.500
3.00	0.465
3.25	0.500
3.90	0.750
4.00	0.850
5.00	0.860*
6.00	0.750
7.00	0.750
7.60	0.800
8.00	0.850
8.20	0.800
8.40	0.650
9.00	0.580
10.00	0.605
10.50	0.720
11.00	0.740
11.50	0.650
12.00	0.570
14.00	0.540
15.60	0.550
19.00	0.550
20.00	0.560
22.80	0.530
25.00	0.485
CURVE 13	
T ~ 322	
0.50	0.725
0.70	0.450
0.87	0.550
2.00	0.560*
2.50	0.500*
3.00	0.430
3.40	0.450
4.00	0.775
4.50	0.800
5.20	0.790
5.60	0.700
6.00	0.670
7.00	0.610
7.90	0.670
8.50	0.435
8.90	0.410
10.00	0.425
10.80	0.460
CURVE 13 (cont.)	
11.50	0.350
12.00	0.310
13.00	0.275
14.00	0.250
15.00	0.250
16.00	0.270
16.80	0.250
18.10	0.250
18.80	0.220
20.00	0.215
21.00	0.220
22.00	0.250
23.00	0.250
24.00	0.300
25.00	0.325
CURVE 14	
T ~ 322	
0.50	0.220
0.63	0.260
0.75	0.295
0.95	0.310
2.00	0.280
3.00	0.260
4.00	0.415
5.00	0.480
6.00	0.480
7.00	0.470
8.00	0.560
9.10	0.560
11.20	0.610
12.00	0.610
13.50	0.660
15.00	0.640
17.00	0.750
18.10	0.750
19.10	0.740
20.50	0.750
23.00	0.710
23.90	0.600
25.00	0.530
CURVE 15	
T ~ 322	
0.50	0.200
0.80	0.290
2.00	0.270
CURVE 15 (cont.)	
3.10	0.255
4.00	0.415*
5.00	0.470
5.80	0.420
7.20	0.410
8.00	0.500
9.00	0.460
10.00	0.460
11.00	0.450
13.50	0.400
16.30	0.400
17.00	0.410
17.50	0.405
19.00	0.345
20.00	0.350
21.00	0.330
22.00	0.330
24.00	0.290
25.00	0.300
CURVE 16	
T ~ 322	
0.50	0.270
0.63	0.450
0.90	0.570
1.50	0.570
2.30	0.510
3.00	0.410
3.50	0.460
3.70	0.700
4.10	0.800
5.00	0.840
7.00	0.770
8.00	0.850*
8.50	0.770
8.55	0.650*
9.00	0.575
10.00	0.560
11.10	0.710*
11.25	0.675
12.00	0.540
15.00	0.500
17.00	0.560
17.90	0.565
19.20	0.540
20.00	0.560
22.50	0.550
25.00	0.510
CURVE 17	
T ~ 322	
0.50	0.270*
0.57	0.350
0.67	0.450
0.77	0.500
0.90	0.570*
1.50	0.570*
2.40	0.520
3.00	0.410*
3.60	0.530
4.30	0.750
5.00	0.800
7.00	0.660
8.00	0.690
9.00	0.425
10.00	0.410
11.00	0.460
12.00	0.300
14.00	0.250*
15.00	0.250*
16.00	0.230
16.50	0.230
18.00	0.250*
19.20	0.220
20.00	0.240
21.00	0.220*
22.20	0.250*
23.00	0.230
24.00	0.310
25.00	0.335
CURVE 18	
T ~ 322	
0.50	0.240
0.63	0.420
0.85	0.560
1.00	0.560
1.50	0.590*
2.00	0.540
3.00	0.420
3.50	0.595
3.70	0.750
5.00	0.860*
5.70	0.780
6.00	0.760
6.80	0.750
7.50	0.825
8.00	0.860
CURVE 18 (cont.)	
8.50	0.800
8.70	0.650
9.10	0.610
10.00	0.610
11.00	0.720
11.50	0.700
12.20	0.530
13.50	0.520
15.00	0.500*
17.00	0.610
20.00	0.600
21.00	0.590
22.00	0.600
23.00	0.575
25.00	0.550
CURVE 19	
T ~ 322	
0.50	0.220
0.65	0.410
0.85	0.525
1.00	0.530
1.25	0.575
1.50	0.590*
2.00	0.540*
3.00	0.410*
3.70	0.400
4.00	0.750
4.50	0.785
5.00	0.790
6.00	0.775
7.00	0.690
8.00	0.700
8.50	0.600
8.80	0.450
9.00	0.420
10.00	0.420
10.50	0.450
11.00	0.480
11.50	0.450
12.00	0.320
14.00	0.280
15.00	0.300
18.20	0.300
19.00	0.270
20.00	0.300
21.00	0.275
22.00	0.320
CURVE 19 (cont.)	
23.00	0.330
24.00	0.350
25.00	0.400

* Not shown on plot

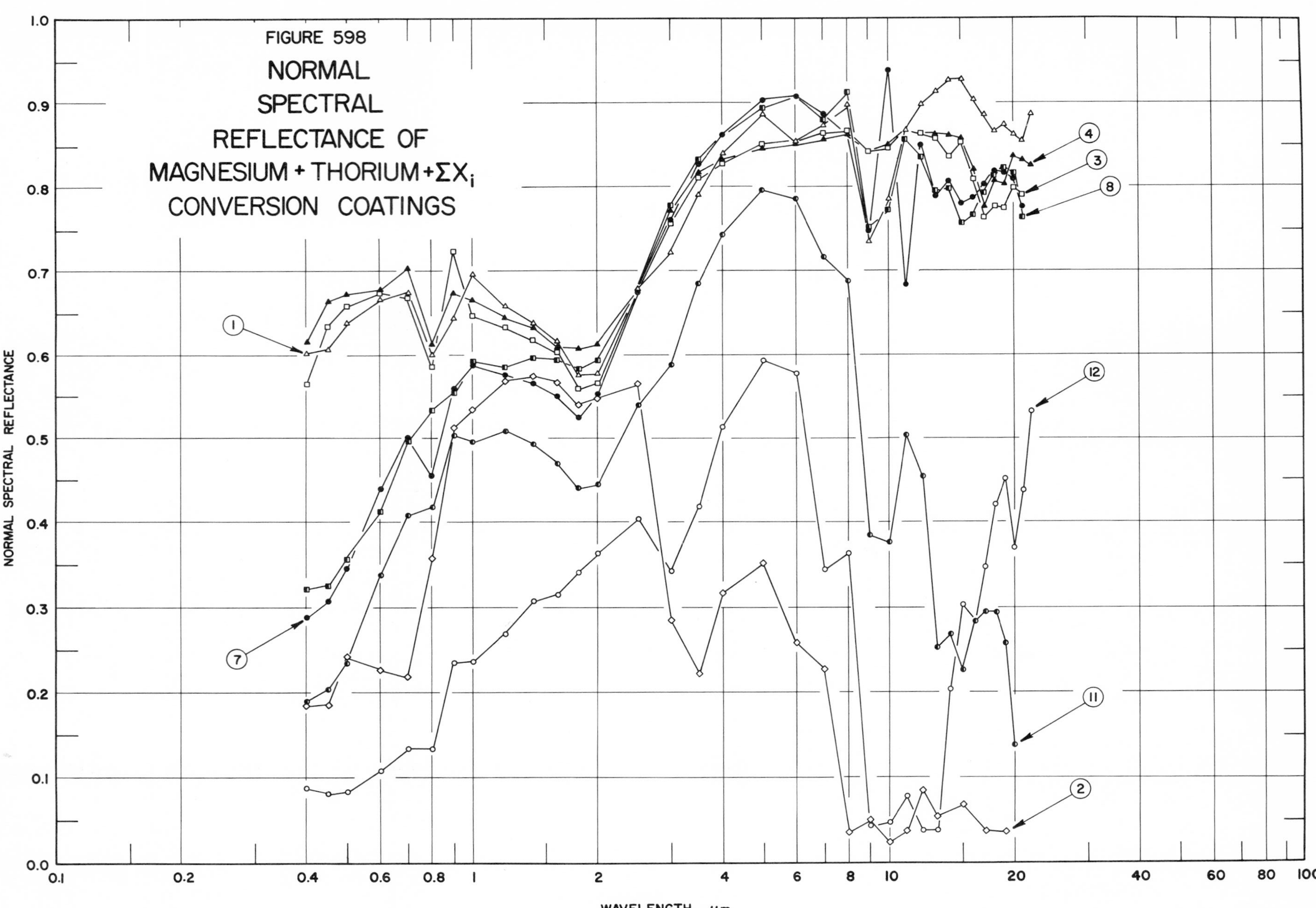

FIGURE 598 NORMAL SPECTRAL REFLECTANCE OF MAGNESIUM + THORIUM + ΣX_i CONVERSION COATINGS

SPECIFICATION TABLE NO. 598 NORMAL SPECTRAL REFLECTANCE OF MAGNESIUM + THORIUM + ΣX_i CONVERSION COATINGS

Curve No.	Ref. No.	Year	Temperature, K	Wavelength Range, μm	Geometry θ	Geometry θ'	Geometry ω'	Reported Error, %	Composition (weight percent), Specifications, and Remarks
1	201	1961	298	0.4-22	$\sim 0^\circ$		2π		HK-31A magnesium Dow 17 anodized (light coating); nominal composition: 3.25 Th, 0.7 Zr, Mg balance; substrate mechanically polished; measured in vacuum ($\sim 10^{-6}$ mm Hg); converted from R (2π, 0°). [Authors' designation: Specimen No. 212]
2	201	1961	298	0.4-19	$\sim 0^\circ$		2π		Similar to above specimen and conditions except dark coating. [Authors' designation: Specimen No. 213]
3	201	1961	298	0.4-21	$\sim 0^\circ$		2π		Similar to curve 1 specimen and conditions except also electropolished. [Authors' designation: Specimen No. 214]
4	201	1961	450	0.4-22	$\sim 0^\circ$		2π		Above specimen and conditions.
5*	201	1961	298	0.4-21	$\sim 0^\circ$		2π		Above specimen and conditions except heated in vacuum.
6*	201	1961	298	0.4-21	$\sim 0^\circ$		2π		Curve 3 specimen and conditions.
7	201	1961	298	0.4-21	$\sim 0^\circ$		2π		HK-31A magnesium HAE anodized (light coating); nominal composition: 3.25 Th, 0.7 Zr, Mg balance; substrate mechanically polished; measured in vacuum ($\sim 10^{-6}$ mm Hg); converted from R (2π, 0°). [Authors' designation: Specimen No. 211]
8	201	1961	450	0.4-21	$\sim 0^\circ$		2π		Above specimen and conditions.
9*	201	1961	298	0.4-21	$\sim 0^\circ$		2π		Above specimen and conditions after heating in vacuum.
10*	201	1961	298	0.4-22	$\sim 0^\circ$		2π		Curve 7 specimen and conditions.
11	201	1961	298	0.4-20	$\sim 0^\circ$		2π		Similar to above specimen and conditions except electropolished. [Authors' designation: Specimen No. 215]
12	201	1961	298	0.4-22	$\sim 0^\circ$		2π		Similar to curve 7 specimen and conditions except dark coating. [Authors' designation: Sample No. 210]

* Not shown on plot

DATA TABLE NO. 598 NORMAL SPECTRAL REFLECTANCE OF MAGNESIUM + THORIUM + ΣX$_i$ CONVERSION COATINGS

[Wavelength, λ, μm; Reflectance, ρ; Temperature, T, K]

λ	ρ	λ	ρ	λ	ρ	λ	ρ	λ	ρ	λ	ρ	λ	ρ	λ	ρ
CURVE 1		CURVE 2 (cont.)		CURVE 3 (cont.)		CURVE 4 (cont.)		CURVE 6*		CURVE 7 (cont.)		CURVE 8 (cont.)		CURVE 9 (cont.)	
T = 298								T = 298							
		0.9	0.512	7	0.865	15	0.858			1.0	0.588	3.5	0.832	11	0.865
0.4	0.601	1.0	0.532	8	0.867	16	0.821	0.4	0.274	1.2	0.573	4	0.860*	12	0.860
0.45	0.607	1.2	0.568	9	0.843	17	0.779	0.45	0.310	1.4	0.565	5	0.893	13	0.784
0.5	0.639	1.4	0.572	10	0.847	18	0.808	0.5	0.350	1.6	0.550	6	0.905*	14	0.817
0.6	0.665	1.6	0.565	11	0.869*	19	0.804	0.6	0.382	1.8	0.524	7	0.880	15	0.794
0.7	0.673	1.8	0.539	12	0.863	20	0.839	0.7	0.403	2.0	0.550	8	0.912	16	0.791
0.8	0.600	2.0	0.546	13	0.856	21	0.832	0.8	0.444	2.5	0.673	9	0.750	17	0.808
0.9	0.644	2.5	0.563	14	0.839	22	0.827	0.9	0.285	3.0	0.760	10	0.774	18	0.811
1.0	0.697	3.0	0.285	15	0.854			1.0	0.446	3.5	0.826	11	0.857	19	0.799
1.2	0.658	3.5	0.222	16	0.810	CURVE 5*		1.2	0.470	4	0.861	12	0.836	20	0.804
1.4	0.638	4	0.317	17	0.764	T = 298		1.4	0.476	5	0.902	13	0.793	21	0.779
1.6	0.616	5	0.350	18	0.789			1.6	0.481	6	0.906	14	0.799		
1.8	0.575	6	0.259	19	0.787	0.4	0.683	1.8	0.475	7	0.888	15	0.757	CURVE 10*	
2.0	0.577	7	0.227	20	0.800	0.45	0.880	2.0	0.484	8	0.867*	16	0.766	T = 298	
2.5	0.679	8	0.035	21	0.791	0.5	0.674	2.5	0.616	9	0.749	17	0.794		
3.0	0.723	9	0.050			0.6	0.713	3.0	0.738	10	0.939	18	0.816	0.4	0.178
3.5	0.791	10	0.024	CURVE 4		0.7	0.688	3.5	0.792	11	0.683	19	0.821	0.45	0.178
4	0.841	11	0.037	T = 450		0.8	0.630	4	0.820	12	0.850	20	0.815	0.5	0.209
5	0.888	12	0.084			0.9	0.646	5	0.848	13	0.790	21	0.764	0.6	0.258
6	0.854	13	0.053	0.4	0.616	1.0	0.659	6	0.858	14	0.807			0.7	0.294
7	0.874	15	0.069	0.45	0.663	1.2	0.628	7	0.864	15	0.781	CURVE 9*		0.8	0.381
8	0.898	17	0.037	0.5	0.671	1.4	0.619	8	0.873	16	0.789	T = 298		0.9	0.382
9	0.736	19	0.036	0.6	0.677	1.6	0.596	9	0.852	17	0.802			1.0	0.400
10	0.787			0.7	0.703	1.8	0.566	10	0.863	18	0.820	0.4	0.306	1.2	0.407
11	0.869	CURVE 3		0.8	0.613	2.0	0.569	11	0.887	19	0.818	0.45	0.327	1.4	0.423
12	0.899	T = 298		0.9	0.673	2.5	0.676	12	0.882	20	0.810	0.5	0.357	1.6	0.429
13	0.912			1.0	0.665	3.0	0.785	13	0.867	21	0.776	0.6	0.448	1.8	0.438
14	0.928	0.4	0.564	1.2	0.645	3.5	0.824	14	0.884			0.7	0.500	2.0	0.459
15	0.928	0.45	0.634	1.4	0.632	4	0.846	15	0.893	CURVE 8		0.8	0.571	2.5	0.572
16	0.901	0.5	0.657	1.6	0.610	5	0.860	16	0.846	T = 450		0.9	0.605	3.0	0.694
17	0.886	0.6	0.673	1.8	0.608	6	0.859	17	0.786			1.0	0.585	3.5	0.767
18	0.868	0.7	0.668	2.0	0.613	7	0.867	18	0.827	0.4	0.321	1.2	0.572	4	0.804
19	0.875	0.8	0.585	2.5	0.680*	8	0.875	19	0.840	0.45	0.325	1.4	0.572	5	0.866
20	0.862	0.9	0.724	3.0	0.774	9	0.851	20	0.847	0.5	0.355	1.6	0.560	6	0.886
21	0.855	1.0	0.646	3.5	0.817	10	0.858	21	0.895	0.6	0.413	1.8	0.535	7	0.886
22	0.887	1.2	0.632	4	0.834	11	0.873			0.7	0.497	2.0	0.547	8	0.903
		1.4	0.618	5	0.849	12	0.868	CURVE 7		0.8	0.533	2.5	0.617	9	0.749
CURVE 2		1.6	0.603	6	0.852	13	0.869	T = 298		0.9	0.553	3.0	0.786	10	0.795
T = 298		1.8	0.557	7	0.858	14	0.871			1.0	0.592	3.5	0.838	11	0.857
		2.0	0.565	8	0.865	15	0.865	0.4	0.287	1.2	0.585	4	0.872	12	0.846
0.4	0.183	2.5	0.677*	9	0.840*	16	0.819	0.45	0.308	1.4	0.596	5	0.904	13	0.786
0.45	0.184	3.0	0.758	10	0.850	17	0.780	0.5	0.345	1.6	0.595	6	0.909	14	0.831
0.5	0.241	3.5	0.811	11	0.868*	18	0.808	0.6	0.438	1.8	0.581	7	0.893	15	0.803
0.6	0.227	4	0.829	12	0.864*	19	0.842	0.7	0.500	2.0	0.592	8	0.921	16	0.816
0.7	0.219	5	0.851	13	0.862	20	0.796	0.8	0.454	2.5	0.675*	9	0.752	17	0.790
0.8	0.356	6	0.855*	14	0.861	21	0.844	0.9	0.558	3.0	0.779	10	0.785	18	0.805

* Not shown on plot

DATA TABLE NO. 598 NORMAL SPECTRAL REFLECTANCE OF MAGNESIUM + THORIUM + ΣX_i CONVERSION COATINGS (continued)

λ	ρ	λ	ρ
CURVE 10 (cont.)*		CURVE 12 (cont.)	
T = 298			
		0.5	0.083
19	0.819	0.6	0.108
20	0.847	0.7	0.134
21	0.806	0.8	0.134
22	0.925	0.9	0.235
		1.0	0.236
CURVE 11		1.2	0.269
T = 298		1.4	0.307
		1.6	0.315
0.4	0.188	1.8	0.340
0.45	0.204	2.0	0.361
0.5	0.235	2.5	0.403
0.6	0.338	3.0	0.341
0.7	0.407	3.5	0.418
0.8	0.418	4	0.511
0.9	0.501	5	0.593
1.0	0.495	6	0.576
1.2	0.509	7	0.344
1.4	0.491	8	0.361
1.6	0.468	9	0.044
1.8	0.439	10	0.046
2.0	0.443	11	0.078
2.5	0.539	12	0.038
3.0	0.588	13	0.039
3.5	0.685	14	0.203
4	0.745	15	0.302
5	0.798	16	0.281*
6	0.787	17	0.347
7	0.717	18	0.420
8	0.687	19	0.450
9	0.382	20	0.369
10	0.375	21	0.436
11	0.503	22	0.531
12	0.453		
13	0.252		
14	0.269		
15	0.226		
16	0.282		
17	0.295		
18	0.294		
19	0.259		
20	0.137		
CURVE 12			
T = 298			
0.4	0.087		
0.45	0.080		

* Not shown on plot

SPECIFICATION TABLE NO. 599 HEMISPHERICAL TOTAL EMITTANCE OF TANTALUM CONVERSION COATINGS

Curve No.	Ref. No.	Year	Temperature Range, K	Reported Error, %	Composition (weight percent), Specifications and Remarks
1*	271	1964	301-956	± 3	Evaporated tantalum film anodized in oxalic acid bath at 342 K; substrate: molybdenum film evaporated onto glass microscope slide with work-excited electron beam gun; measured in vacuum (10^{-5} mm Hg). [Authors' designation: Sample 280]

DATA TABLE NO. 599 HEMISPHERICAL TOTAL EMITTANCE OF TANTALUM CONVERSION COATINGS

[Temperature, T, K; Emittance, ϵ]

T	ϵ
CURVE 1*	
301	0.051
533	0.058
674	0.067
819	0.078
956	0.093

* No plot given

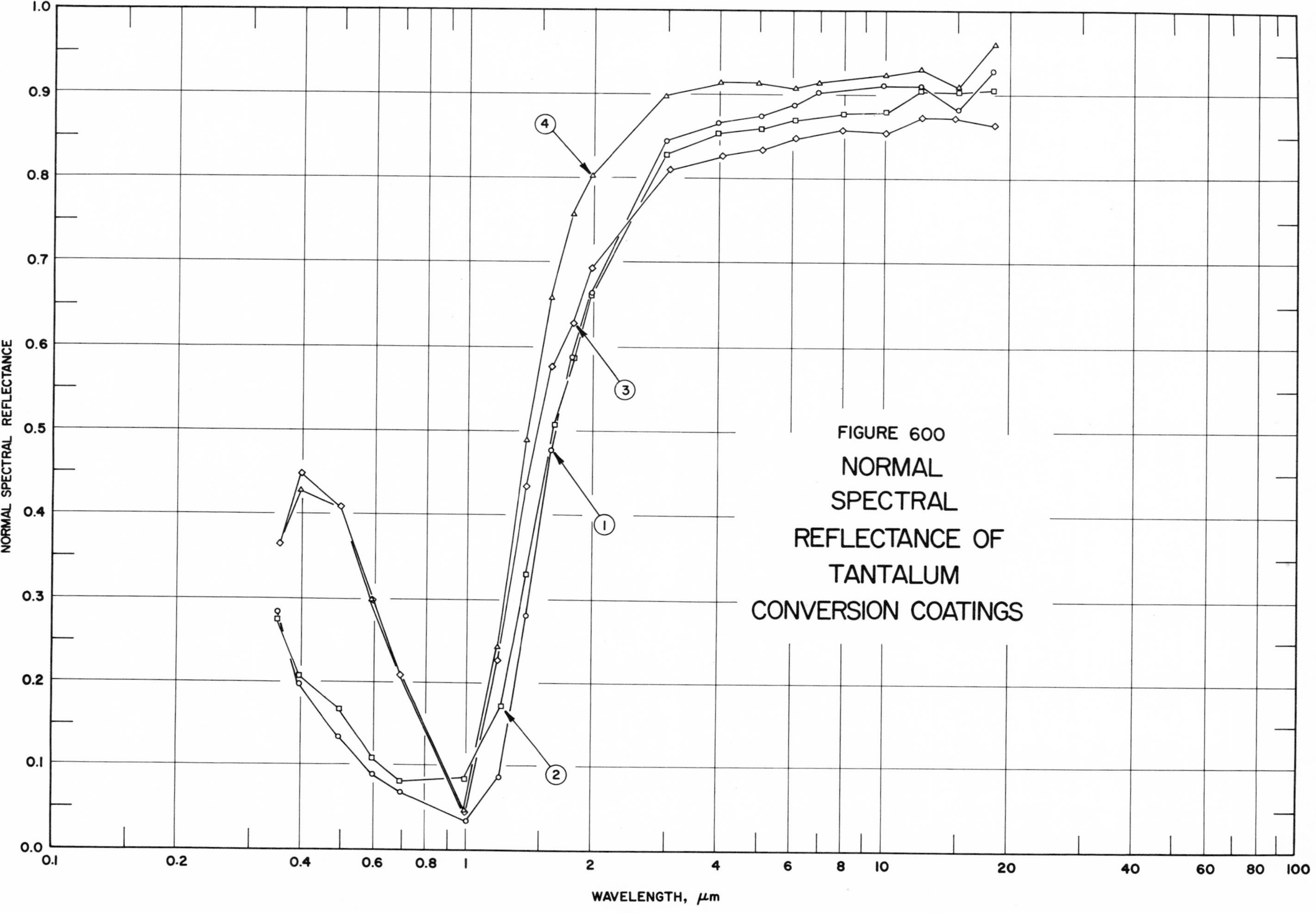

FIGURE 600 NORMAL SPECTRAL REFLECTANCE OF TANTALUM CONVERSION COATINGS

SPECIFICATION TABLE NO. 600 NORMAL SPECTRAL REFLECTANCE OF TANTALUM CONVERSION COATINGS

Curve No.	Ref. No.	Year	Temperature, K	Wavelength Range, μm	Geometry θ	θ'	ω'	Reported Error, %	Composition (weight percent), Specifications, and Remarks
1	271	1964	~298	0.351-18.2	~0°		2π		Evaporated tantalum film anodized in oxalicacid bath at 342 K; substrate - molybdenum film evaporated onto glass microscope slide with work-excited electron beam gun; measured in vacuum (10^{-5} mm Hg). [Authors' designation: Sample 280]
2	271	1964	533	0.351-18.2	~0°		2π		Similar to curve 1 specimen and conditions.
3	271	1964	811	0.351-18.2	~0°		2π		Similar to curve 1 specimen and conditions.
4	271	1964	~298	0.351-18.3	~0°		2π		Similar to curve 1 specimen and conditions except after 811 K measurement.

DATA TABLE NO. 600 NORMAL SPECTRAL REFLECTANCE OF TANTALUM CONVERSION COATINGS

[Wavelength, λ, μm; Reflectance, ρ; Temperature, T, K]

λ	ρ	λ	ρ	λ	ρ	λ	ρ
CURVE 1 T ~ 298		CURVE 2 T = 533		CURVE 3 T = 811		CURVE 4 T ~ 298	
0.350	0.282	0.350	0.275	0.351	0.363	0.351	0.363*
0.395	0.197	0.395	0.208	0.398	0.447	0.398	0.427
0.493	0.131	0.497	0.167	0.498	0.408	0.499	0.408*
0.598	0.088	0.596	0.108	0.593	0.299	0.592	0.299
0.693	0.067	0.695	0.080	0.693	0.209	0.693	0.209*
1.00	0.034	0.998	0.083	0.991	0.045	0.989	0.046
1.20	0.086	1.21	0.171	1.19	0.227	1.19	0.242
1.39	0.280	1.39	0.330	1.39	0.432	1.38	0.489
1.59	0.477	1.61	0.508	1.59	0.576	1.58	0.659
1.77	0.587	1.79	0.587	1.79	0.629	1.79	0.757
1.97	0.664	1.98	0.661	1.98	0.691	1.98	0.801
2.99	0.845	2.99	0.829	3.02	0.810	2.99	0.897
3.98	0.867	3.98	0.853	4.01	0.828	4.00	0.913
5.01	0.875	5.02	0.859	5.03	0.834	4.97	0.913
6.01	0.889	6.03	0.870	6.03	0.848	6.02	0.906
7.97	0.902	7.96	0.877	7.96	0.858	7.97	0.913
9.95	0.910	10.0	0.880	10.0	0.854	10.0	0.924
12.1	0.910	12.1	0.905	12.2	0.873	12.1	0.930
15.0	0.883	15.0	0.902	14.9	0.872	15.0	0.910
18.2	0.928	18.2	0.905	18.2	0.863	18.3	0.960

* Not shown on plot

SPECIFICATION TABLE NO. 601 NORMAL SOLAR ABSORPTANCE OF TANTALUM CONVERSION COATINGS

Curve No.	Ref. No.	Year	Temperature Range, K	Geometry θ	Reported Error, %	Composition (weight percent), Specifications and Remarks
1*	271	1964	298-811	~0°		Evaporated tantalum film anodized in oxalic acid bath at 342 K; substrate: molybdenum film evaporated onto glass microscope slide with work-excited electron beam gun; measured in vacuum (10^{-5} mm Hg); computed from normal spectral reflectance. [Authors' designation: Sample 280]

DATA TABLE NO. 601 NORMAL SOLAR ABSORPTANCE OF TANTALUM CONVERSION COATINGS

[Temperature, T, K; Absorptance, α]

T	α
CURVE 1*	
298	0.83
533	0.79
811	0.68

* No plot given

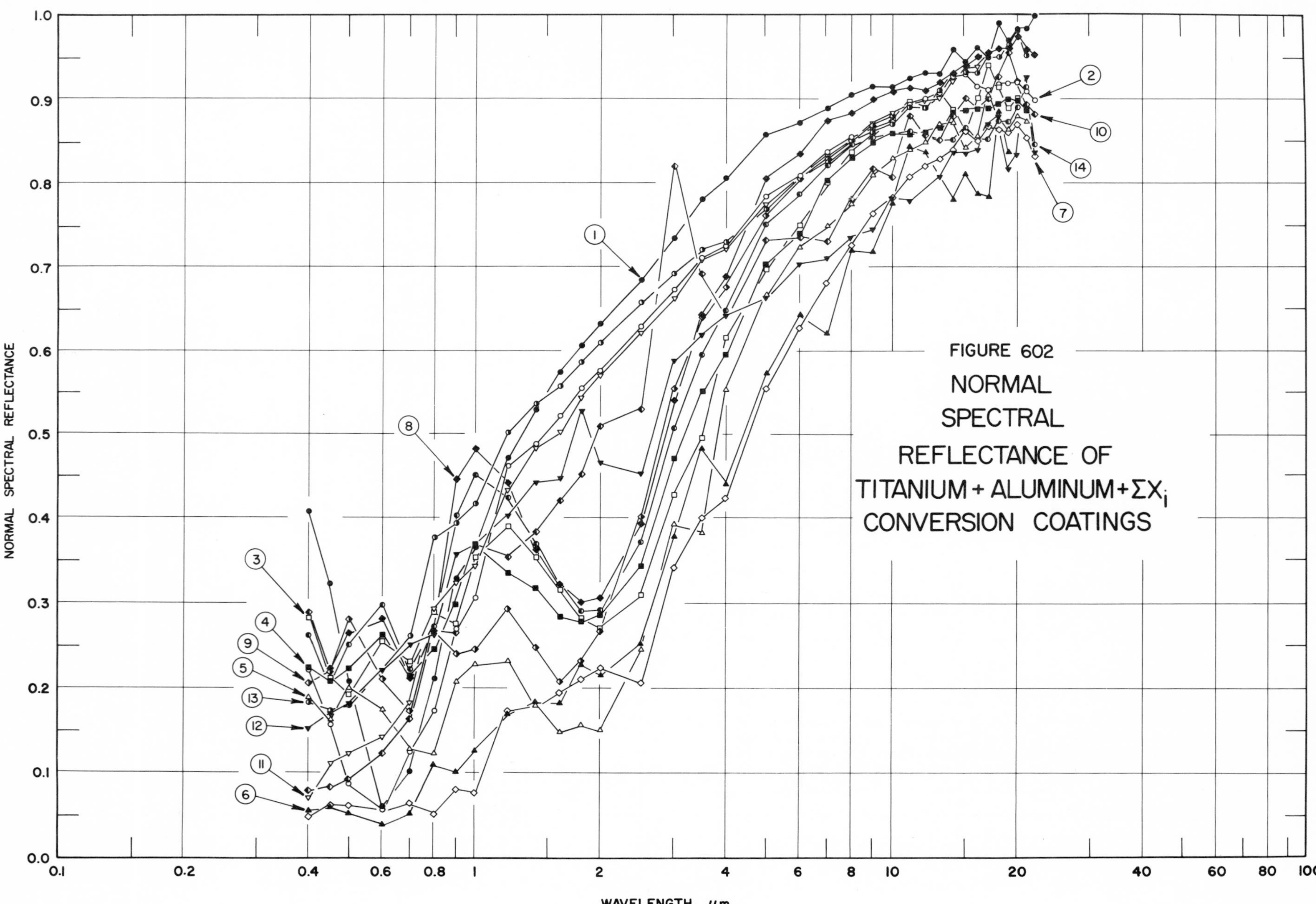

FIGURE 602 NORMAL SPECTRAL REFLECTANCE OF TITANIUM + ALUMINUM + ΣX_i CONVERSION COATINGS

SPECIFICATION TABLE NO. 602 NORMAL SPECTRAL REFLECTANCE OF TITANIUM + ALUMINUM + ΣX_i CONVERSION COATINGS

Curve No.	Ref. No.	Year	Temperature, K	Wavelength Range, μm	Geometry θ	θ'	ω'	Reported Error, %	Composition (weight percent), Specifications, and Remarks
1	201	1961	298	0.4-22	$\sim 0^\circ$		2π		A110-AT Titanium anodized in sulfuric acid for 20 min; nominal composition: 5 Al, 2.5 Sn and balance Ti; substrate mechanically polished and electropolished; measured in vacuum ($\sim 10^{-6}$ mm Hg); converted from R (2π, 0°). [Authors' designation: Specimen No. 38]
2	201	1961	298	0.4-22	$\sim 0^\circ$		2π		Similar to above specimen and conditions except pickled instead of electropolished. [Authors' designation: Specimen No. 43]
3	201	1961	298	0.4-21	$\sim 0^\circ$		2π		A110-AT Titanium anodized in sodium hydroxide for 20 min; nominal composition: 5 Al, 2.5 Sn and balance Ti; substrate mechanically polished and pickled; measured in vacuum ($\sim 10^{-6}$ mm Hg); converted from R (2π, 0°). [Authors' designation: Specimen No. 45]
4	201	1961	422	0.4-21	$\sim 0^\circ$		2π		Above specimen and conditions.
5	201	1961	589	0.4-21	$\sim 0^\circ$		2π		Above specimen and conditions.
6	201	1961	755	0.4-19	$\sim 0^\circ$		2π		Above specimen and conditions.
7	201	1961	298	0.4-22	$\sim 0^\circ$		2π		Above specimen and conditions after heating in vacuum.
8	201	1961	298	0.4-22	$\sim 0^\circ$		2π		Similar to curve 3 specimen and conditions except electropolished instead of pickled. [Authors' designation: Specimen No. 48]
9	201	1961	533	0.4-22	$\sim 0^\circ$		2π		Above specimen and conditions.
10	201	1961	811	0.4-22	$\sim 0^\circ$		2π		Above specimen and conditions.
11	201	1961	298	0.4-22	$\sim 0^\circ$		2π		Above specimen and conditions after heating in vacuum.
12	201	1961	978	0.4-22	$\sim 0^\circ$		2π		Above specimen and conditions.
13	201	1961	298	0.4-22	$\sim 0^\circ$		2π		Above specimen and conditions after heating in vacuum.
14	201	1961	298	0.4-22	$\sim 0^\circ$		2π		Similar to curve 8 specimen and conditions except sodium hydroxide anodized for 6.7 min. [Authors' designation: Specimen No. 178]

DATA TABLE NO. 602 NORMAL SPECTRAL REFLECTANCE OF TITANIUM + ALUMINUM + ΣX_i CONVERSION COATINGS

[Wavelength, λ, μm; Reflectance, ρ; Temperature, T, K]

λ	ρ
CURVE 1	
T = 298	
0.4	0.406
0.45	0.324
0.5	0.207
0.6	0.059
0.7	0.100
0.8	0.211
0.9	0.328
1.0	0.363
1.2	0.469
1.4	0.527
1.6	0.574
1.8	0.605
2.0	0.633
2.5	0.684
3.0	0.736
3.5	0.782
4	0.805
5	0.858
6	0.874
7	0.890
8	0.905
9	0.915
10	0.915
11	0.926
12	0.932
13	0.931
14	0.958
15	0.946
16	0.964
17	0.952
18	0.993
19	0.971
20	0.986
21	0.985
22	0.997
CURVE 2	
T = 298	
0.4	0.210
0.45	0.155
0.5	0.085
0.6	0.055
0.7	0.124
0.8	0.173
CURVE 2 (cont.)	
0.9	0.268
1.0	0.306
1.2	0.460
1.4	0.487
1.6	0.520
1.8	0.554
2.0	0.575
2.5	0.628
3.0	0.673
3.5	0.710
4	0.726
5	0.786
6	0.811
7	0.839
8	0.856
9	0.868
10	0.873
11	0.895
12	0.902
13	0.907
14	0.928
15	0.931
16	0.916
17	0.913
18	0.918
19	0.918
20	0.920
21	0.910
22	0.898
CURVE 3	
T = 298	
0.4	0.283
0.45	0.211
0.5	0.190
0.6	0.254
0.7	0.230
0.8	0.288
0.9	0.275
1.0	0.351
1.2	0.387
1.4	0.351
1.6	0.315
1.8	0.281
2.0	0.269
CURVE 3 (cont.)	
2.5	0.308
3.0	0.425
3.5	0.496
4	0.616
5	0.698
6	0.752
7	0.803
8	0.838
9	0.858
10	0.878
11	0.897
12	0.893
13	0.904
14	0.887
15	0.865
16	0.902
17	0.943
18	0.914
19	0.892
20	0.901
21	0.889
CURVE 4	
T = 422	
0.4	0.223
0.45	0.208
0.5	0.224
0.6	0.262
0.7	0.216
0.8	0.245
0.9	0.298
1.0	0.367
1.2	0.335
1.4	0.318
1.6	0.283
1.8	0.277
2.0	0.285
2.5	0.344
3.0	0.469
3.5	0.549
4	0.595
5	0.703
6	0.743
7	0.804
8	0.831
CURVE 4 (cont.)	
9	0.849
10	0.861
11	0.859
12	0.862
13	0.866
14	0.885
15	0.887
16	0.890
17	0.891
18	0.895
19	0.900
20	0.898
21	0.888
CURVE 5	
0.4	0.187
0.45	0.162
0.5	0.199
0.6	0.174
0.7	0.127
0.8	0.124
0.9	0.207
1.0	0.227
1.2	0.232
1.4	0.179
1.6	0.148
1.8	0.155
2.0	0.149
2.5	0.245
3.0	0.390
3.5	0.382
4	0.551
5	0.667
6	0.726
7	0.751
8	0.777
9	0.811
10	0.831
11	0.842
12	0.849
13	0.871
14	0.873
15	0.845
16	0.857
17	0.905
CURVE 5 (cont.)	
18	0.877
19	0.866
20	0.881
21	0.876
CURVE 6	
0.4	0.056
0.45	0.058
0.5	0.053
0.6	0.038
0.7	0.052
0.8	0.109
0.9	0.100
1.0	0.125
1.2	0.168
1.4	0.172
1.6	0.171
1.8	0.227
2.0	0.216
2.5	0.251
3.0	0.375
3.5	0.482
4	0.438
5	0.572
6	0.643
7	0.619
8	0.720
9	0.719
10	0.777
11	0.846
12	0.838
14	0.783
15	0.813
16	0.789
17	0.786
18	0.887
19	0.839
CURVE 7	
T = 298	
0.4	0.048
0.45	0.060
0.5	0.060
0.6	0.057*
CURVE 7 (cont.)	
0.7	0.064
0.8	0.051
0.9	0.079
1.0	0.075
1.2	0.171
1.4	0.179*
1.6	0.194
1.8	0.210
2.0	0.224
2.5	0.206
3.0	0.340
3.5	0.399
4	0.423
5	0.554
6	0.627
7	0.680
8	0.727
9	0.765
10	0.785
11	0.808
12	0.822
13	0.830
14	0.842
15	0.864
16	0.853
17	0.869
18	0.866
19	0.862
20	0.871
21	0.855
22	0.834
CURVE 8	
T = 298	
0.4	0.288
0.45	0.222
0.5	0.263
0.6	0.279
0.7	0.211
0.8	0.265
0.9	0.444
1.0	0.480
1.2	0.439
1.4	0.360
1.6	0.323
CURVE 8 (cont.)	
1.8	0.300
2.0	0.305
2.5	0.391
3.0	0.538
3.5	0.642
4	0.687
5	0.806
6	0.837
7	0.876
8	0.886
9	0.900
10	0.908
11	0.913
12	0.910
13	0.921
14	0.931
15	0.940
16	0.950
17	0.956
18	0.961
19	0.962
20	0.975
21	0.959
22	0.953
CURVE 9	
T = 533	
0.4	0.206
0.45	0.219
0.5	0.279
0.6	0.210
0.7	0.172
0.8	0.265*
0.9	0.239
1.0	0.246
1.2	0.294
1.4	0.247
1.6	0.208
1.8	0.231
2.0	0.266
2.5	0.400
3.0	0.554
3.5	0.639
4	0.675
5	0.761
CURVE 9 (cont.)	
6	0.806
7	0.831
8	0.852
9	0.870
10	0.881
11	0.913*
12	0.910*
13	0.921*
14	0.931*
15	0.940*
16	0.950*
17	0.956*
18	0.961*
19	0.962*
20	0.975*
21	0.959*
22	0.953*
CURVE 10	
T = 811	
0.4	0.077
0.45	0.083
0.5	0.092
0.6	0.122
0.7	0.162
0.8	0.211*
0.9	0.264
1.0	0.305*
1.2	0.351
1.4	0.381
1.6	0.419
1.8	0.449
2.0	0.508
2.5	0.527
3.0	0.822
3.5	0.692
4	0.644
5	0.733
6	0.737
7	0.733
8	0.783
9	0.818
10	0.808
11	0.882
12	0.858

* Not shown on plot

DATA TABLE NO. 602 NORMAL SPECTRAL REFLECTANCE OF TITANIUM + ALUMINUM + ΣX_i CONVERSION COATINGS (continued)

CURVE 10 (cont.), T = 811

λ	ρ
13	0.853
14	0.880
15	0.901
16	0.892*
17	0.899
18	0.927
19	0.955
20	0.921
21	0.896
22	0.834

CURVE 11, T = 298

λ	ρ
0.4	0.069
0.45	0.110
0.5	0.121
0.6	0.142
0.7	0.183
0.8	0.293
0.9	0.324
1.0	0.343
1.2	0.433
1.4	0.481
1.6	0.502
1.8	0.542
2.0	0.568
2.5	0.620
3.0	0.662
3.5	0.707
4	0.724
5	0.776
6	0.807*
7	0.834
8	0.853*
9	0.874
10	0.885
11	0.895*
12	0.899
13	0.903
14	0.922
15	0.937
16	0.937
17	0.951*
18	0.962
19	0.964
20	0.955

CURVE 11 (cont.)

λ	ρ
21	0.957
22	0.911

CURVE 12, T = 978

λ	ρ
0.4	0.152
0.45	0.169
0.5	0.179
0.6	0.220
0.7	0.250
0.8	0.260
0.9	0.356
1.0	0.364*
1.2	0.401
1.4	0.440
1.6	0.446
1.8	0.525
2.0	0.464
2.5	0.452
3.0	0.588
3.5	0.618
4	0.641
5	0.663
6	0.704
7	0.710
8	0.736
9	0.746
10	0.783
11	0.781
12	0.808
13	0.813
14	0.838
15	0.837
16	0.842
17	0.873
18	0.880
19	0.818
20	0.835
21	0.927
22	0.838

CURVE 13, T = 298

λ	ρ
0.4	0.182
0.45	0.171
0.5	0.178

CURVE 13 (cont.)

λ	ρ
0.6	0.219*
0.7	0.260
0.8	0.375
0.9	0.394
1.0	0.417
1.2	0.500
1.4	0.535
1.6	0.556
1.8	0.585
2.0	0.609
2.5	0.657
3.0	0.692
3.5	0.721
4	0.730
5	0.771
6	0.806
7	0.826
8	0.850*
9	0.862
10	0.873
11	0.891
12	0.891
13	0.911
14	0.931*
15	0.932
16	0.932
17	0.953*
18	0.952
19	0.966
20	0.982
21	0.954
22	1.016*

CURVE 14, T = 298

λ	ρ
0.4	0.262
0.45	0.209*
0.5	0.250
0.6	0.297
0.7	0.222
0.8	0.271
0.9	0.403
1.0	0.449
1.2	0.424
1.4	0.367
1.6	0.322*
1.8	0.289

CURVE 14 (cont.)

λ	ρ
2.0	0.290
2.5	0.369
3.0	0.505
3.5	0.595
4	0.647
5	0.751
6	0.787
7	0.823
8	0.847
9	0.856
10	0.860*
11	0.863
12	0.848*
13	0.853*
14	0.854
15	0.867
16	0.853*
17	0.854
18	0.876
19	0.876
20	0.894
21	0.915
22	0.847

* Not shown on plot

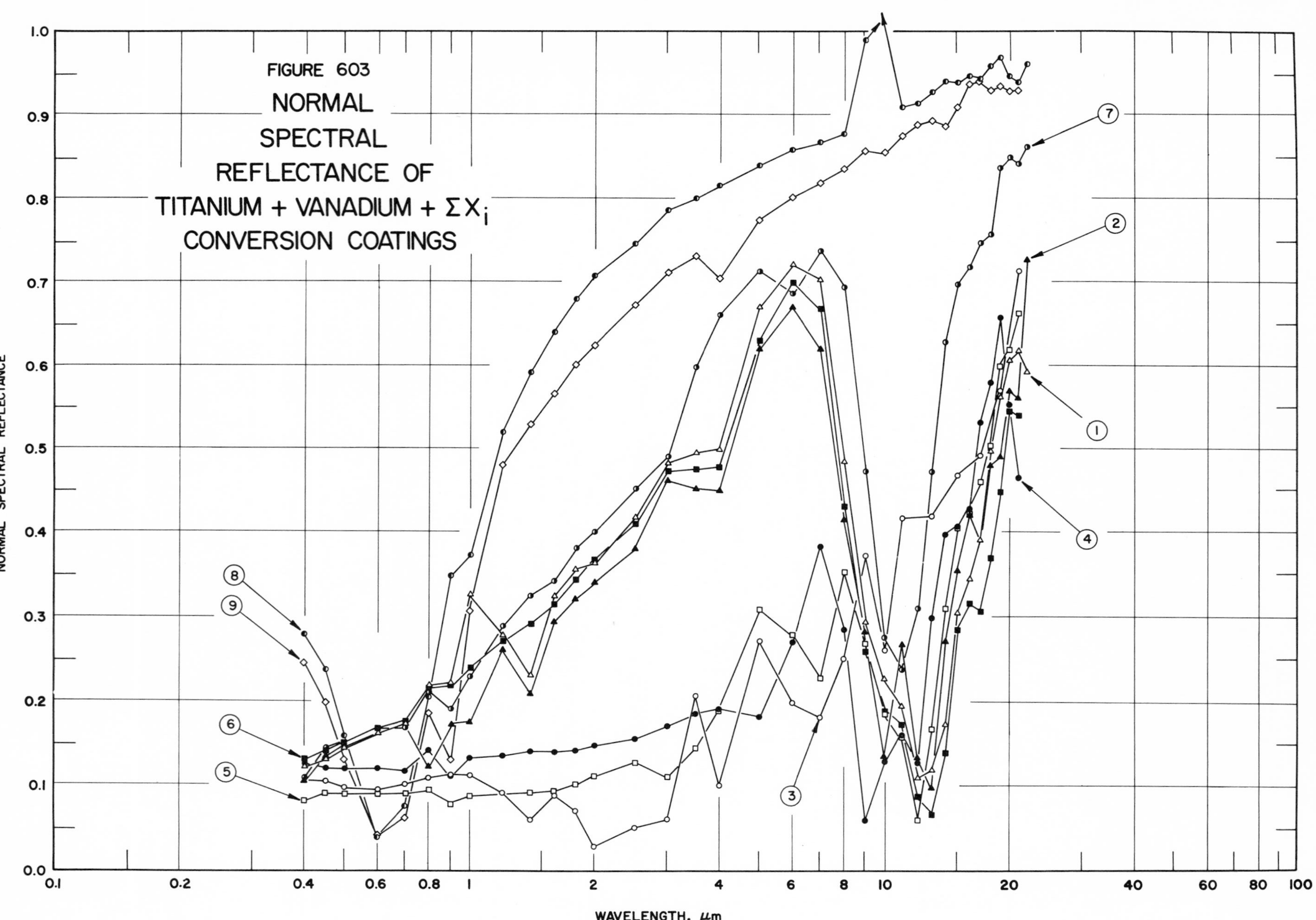
FIGURE 603
NORMAL
SPECTRAL
REFLECTANCE OF
TITANIUM + VANADIUM + ΣX$_i$
CONVERSION COATINGS
NORMAL SPECTRAL REFLECTANCE
WAVELENGTH, μm
1.0
0.9
0.8
0.7
0.6
0.5
0.4
0.3
0.2
0.1
0.0
0.1
0.2
0.4
0.6
0.8
1
2
4
6
8
10
20
40
60
80
100
1
2
3
4
5
6
7
8
9

SPECIFICATION TABLE NO. 603 NORMAL SPECTRAL REFLECTANCE OF TITANIUM + VANADIUM + ΣX_i CONVERSION COATINGS

Curve No.	Ref. No.	Year	Temperature, K	Wavelength Range, μm	Geometry θ	θ'	ω'	Reported Error, %	Composition (weight percent), Specifications, and Remarks
1	201	1961	298	0.4-22	~0°		2π		B120 VAC Titanium; anodized; nominal composition: 73 Ti, 13 V, 11 Cr and 3 Al; substrate mechanically polished and electropolished; measured in vacuum (~10^{-6} mm Hg); converted from R (2π, 0°). [Authors' designation: Specimen No. 202]
2	201	1961	533	0.4-22	~0°		2π		Above specimen and conditions.
3	201	1961	811	0.4-21	~0°		2π		Above specimen and conditions.
4	201	1961	298	0.4-21	~0°		2π		Above specimen and conditions.
5	201	1961	298	0.4-21	~0°		2π		Above specimen and conditions except heated to 1089 K in vacuum.
6	201	1961	298	0.4-21	~0°		2π		Similar to curve 1 specimen and conditions; thin coating. [Authors' designation: Specimen No. 197]
7	201	1961	298	0.4-22	~0°		2π		Similar to curve 1 specimen and conditions except pickled instead of electropolished. [Authors' designation: Specimen No. 199]
8	201	1961	298	0.4-22	~0°		2π		Similar to curve 1 specimen and conditions except sulfuric acid anodized for 20 min. [Authors' designation: Specimen No. 203]
9	201	1961	298	0.4-21	~0°		2π		Similar to above specimen and conditions except pickled instead of electropolished. [Authors' designation: Specimen No. 207]

DATA TABLE NO. 603 NORMAL SPECTRAL REFLECTANCE OF TITANIUM + VANADIUM + ΣX_i CONVERSION COATINGS

[Wavelength, λ, μm; Reflectance, ρ; Temperature, T, K]

CURVE 1 (T = 298)

λ	ρ
0.4	0.123
0.45	0.130
0.5	0.146
0.6	0.161
0.7	0.172
0.8	0.219
0.9	0.223
1.0	0.327
1.2	0.278
1.4	0.231
1.6	0.326
1.8	0.355
2.0	0.363
2.5	0.418
3.0	0.484
3.5	0.495
4	0.498
5	0.671
6	0.724
7	0.705
8	0.485
9	0.296
10	0.228
11	0.196
12	0.110
13	0.119
14	0.174
15	0.306
16	0.345
17	0.390
18	0.498
19	0.563
20	0.608
21	0.619
22	0.596

CURVE 2 (T = 533)

λ	ρ
0.4	0.106
0.45	0.136
0.5	0.147
0.6	0.160*
0.7	0.171*
0.8	0.123
0.9	0.173
1.0	0.176
1.2	0.262
1.4	0.210
1.6	0.296
1.8	0.322
2.0	0.342
2.5	0.379
3.0	0.464
3.5	0.453
4	0.449
5	0.622
6	0.672
7	0.623
8	0.416
9	0.284
10	0.136
11	0.269
12	0.136
13	0.098
14	0.273
15	0.355
16	0.420
17	0.393*
18	0.480
19	0.490
20	0.569
21	0.562
22	0.729

CURVE 3 (T = 811)

λ	ρ
0.4	0.105*
0.45	0.105
0.5	0.097
0.6	0.096
0.7	0.102
0.8	0.108
0.9	0.114
1.0	0.113
1.2	0.093
1.4	0.060
1.6	0.087
1.8	0.070
2.0	0.028
2.5	0.053
3.0	0.062
3.5	0.209
4	0.100
5	0.274
6	0.199
7	0.183
8	0.252
9	0.374
10	0.262
11	0.418
13	0.420
15	0.467
17	0.494
19	0.572
21	0.716

CURVE 4 (T = 298)

λ	ρ
0.4	0.127
0.45	0.121
0.5	0.120
0.6	0.121
0.7	0.117
0.8	0.144
0.9	0.111
1.0	0.134
1.2	0.135
1.4	0.141
1.6	0.139
1.8	0.144
2.0	0.147
2.5	0.156
3.0	0.170
3.5	0.186
4	0.192
5	0.184
6	0.270
7	0.383
8	0.286
9	0.062
10	0.128
11	0.161
12	0.127
13	0.299
14	0.398
15	0.408
16	0.428
17	0.533
18	0.580
19	0.658
20	0.554
21	0.466

CURVE 5 (T = 298)

λ	ρ
0.4	0.084
0.45	0.090
0.5	0.090
0.6	0.088
0.7	0.089
0.8	0.096
0.9	0.077
1.0	0.087
1.2	0.091*
1.4	0.092
1.6	0.096
1.8	0.104
2.0	0.114
2.5	0.127
3.0	0.113
3.5	0.145
4	0.189
5	0.308
6	0.278
7	0.228
8	0.354
9	0.269
10	0.185
11	0.159
12	0.062
13	0.167
14	0.312
15	0.407
16	0.424
17	0.459
18	0.505
19	0.601
20	0.621
21	0.665

CURVE 6 (T = 298)

λ	ρ
0.4	0.133
0.45	0.141
0.5	0.152
0.6	0.168
0.7	0.174
0.8	0.216
0.9	0.219
1.0	0.242
1.2	0.273
1.4	0.294
1.6	0.316
1.8	0.344
2.0	0.367
2.5	0.409
3.0	0.474
3.5	0.476
4	0.477
5	0.631
6	0.699
7	0.668
8	0.432
9	0.259
10	0.189
11	0.172
12	0.087
13	0.067
14	0.142
15	0.286
16	0.317
17	0.307
18	0.370
19	0.448
20	0.546
21	0.540

CURVE 7 (T = 298)

λ	ρ
0.4	0.113
0.45	0.144
0.5	0.149*
0.6	0.165
0.7	0.167
0.8	0.210
0.9	0.191
1.0	0.229
1.2	0.287
1.4	0.325
1.6	0.344
1.8	0.380
2.0	0.400
2.5	0.452
3.0	0.490
3.5	0.599
4	0.661
5	0.716
6	0.687
7	0.740
8	0.695
9	0.471
10	0.275
11	0.238
12	0.311
13	0.473
14	0.629
15	0.698
16	0.721
17	0.748
18	0.759
19	0.840
20	0.852
21	0.845
22	0.865

CURVE 8 (T = 298)

λ	ρ
0.4	0.279
0.45	0.237
0.5	0.158
0.6	0.038
0.7	0.076*
0.8	0.206
0.9	0.347
1.0	0.373
1.2	0.519
1.4	0.594
1.6	0.642
1.8	0.681
2.0	0.708
2.5	0.747
3.0	0.787
3.5	0.802
4	0.817
5	0.844
6	0.861
7	0.871
8	0.880
9	0.991
10	1.068*
11	0.910
12	0.916
13	0.928
14	0.942
15	0.941
16	0.947
17	0.946
18	0.960
19	0.970
20	0.947
21	0.941
22	0.964

CURVE 9 (T = 298)

λ	ρ
0.4	0.245
0.45	0.198
0.5	0.132
0.6	0.043
0.7	0.063
0.8	0.185
0.9	0.131
1.0	0.308
1.2	0.480
1.4	0.528
1.6	0.566
1.8	0.602
2.0	0.626
2.5	0.674
3.0	0.714
3.5	0.736
4	0.706
5	0.777
6	0.804
7	0.821
8	0.839
9	0.858
10	0.857
11	0.877
12	0.891
13	0.895
14	0.889
15	0.913
16	0.938
17	0.944
18	0.932
19	0.935
20	0.931
21	0.932

* Not shown on plot

3. CONVERSION COATINGS (continued)

B. Oxidized Coatings

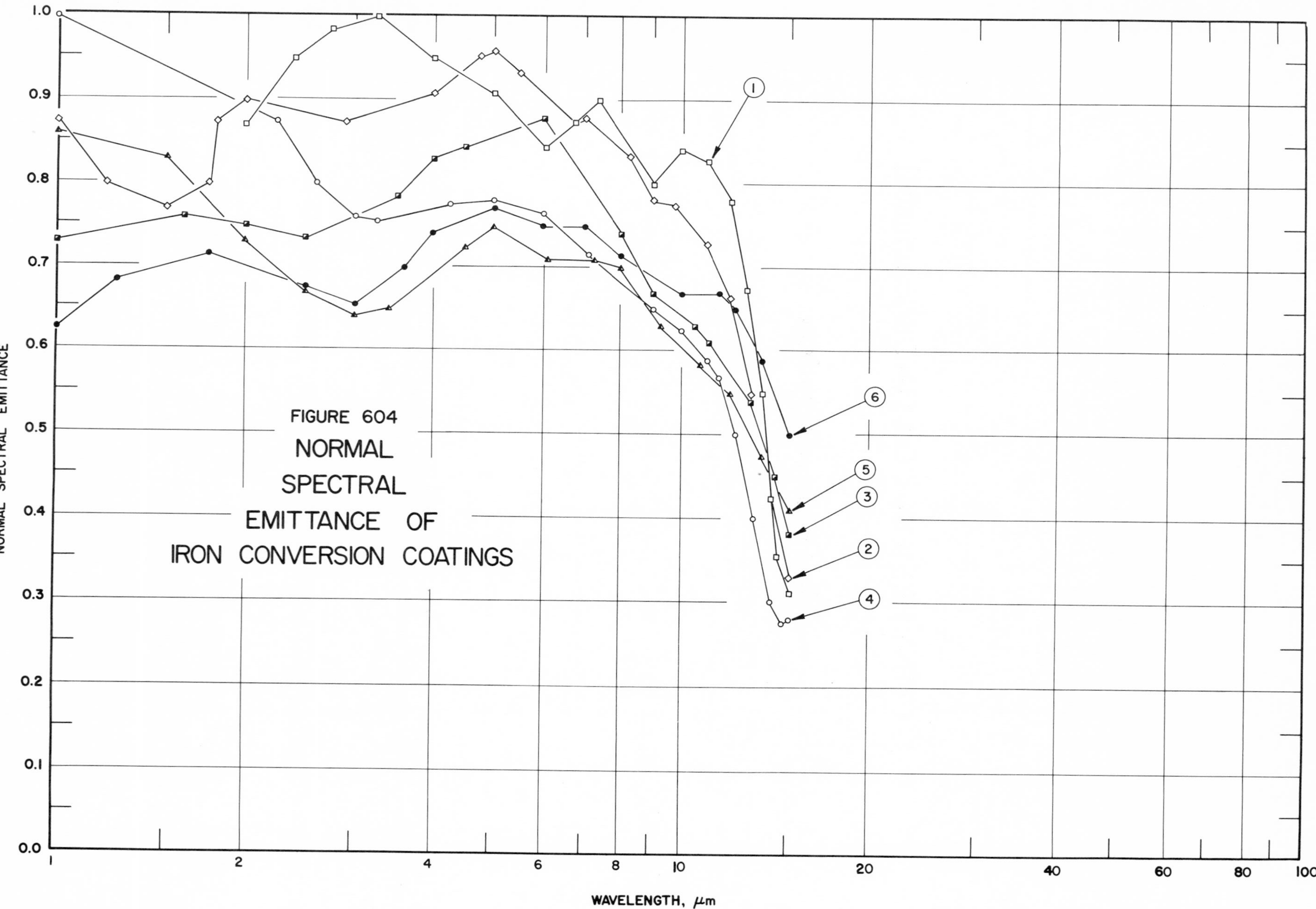

FIGURE 604 NORMAL SPECTRAL EMITTANCE OF IRON CONVERSION COATINGS

SPECIFICATION TABLE NO. 604 NORMAL SPECTRAL EMITTANCE OF IRON CONVERSION COATINGS

Curve No.	Ref. No.	Year	Temperature, K	Wavelength Range, μm	Geometry θ'	Reported Error, %	Composition (weight percent), Specifications, and Remarks
1	139	1961	523	2.00-15.00	$\sim 0^\circ$	±5	Armco blackened steel; as received; data extracted from smooth curve.
2	139	1961	773	1.00-15.00	$\sim 0^\circ$	±5	Similar to curve 1 specimen and conditions.
3	139	1961	1023	1.00-15.00	$\sim 0^\circ$	±5	Similar to curve 1 specimen and conditions.
4	139	1961	523	1.00-15.00	$\sim 0^\circ$	±5	Similar to curve 1 specimen and conditions except heated in vacuum (4.4×10^{-5} mm Hg) at 755 K for 30 min.
5	139	1961	773	1.00-15.00	$\sim 0^\circ$	±5	Similar to curve 4 specimen and conditions.
6	139	1961	1023	1.00-15.00	$\sim 0^\circ$	±5	Similar to curve 4 specimen and conditions.

DATA TABLE NO. 604 NORMAL SPECTRAL EMITTANCE OF IRON CONVERSION COATINGS

[Wavelength, λ, μm; Emittance, ϵ; Temperature, T, K]

λ	ϵ
CURVE 1	**T = 523**
2.00	0.870
2.40	0.950
2.75	0.985
3.25	1.000
4.00	0.950
5.00	0.910
6.05	0.845
6.75	0.875
7.35	0.900
9.00	0.800
10.00	0.840
11.00	0.830
12.00	0.780
12.85	0.675
13.55	0.550
14.00	0.425
14.40	0.355
15.00	0.310
CURVE 2	**T = 773**
1.00	0.875
1.20	0.800
1.50	0.770
1.75	0.800
1.80	0.875
2.00	0.900
2.90	0.875
4.00	0.910
4.75	0.955
5.00	0.960
5.50	0.935
6.75	0.875*
7.00	0.880
8.25	0.835
9.00	0.785
9.75	0.775
11.00	0.730
12.00	0.665
CURVE 2 (cont.)	
13.00	0.550
14.00	0.425*
15.00	0.330
CURVE 3	**T = 1023**
1.00	0.730
1.60	0.760
2.00	0.750
2.50	0.735
3.50	0.785
4.00	0.830
4.50	0.845
6.00	0.880
8.00	0.740
9.00	0.670
10.50	0.630
11.10	0.610
CURVE 3 (cont.)	
13.00	0.540
14.25	0.450
15.00	0.380
CURVE 4	**T = 523**
1.00	1.000
2.25	0.875
2.60	0.800
3.00	0.760
3.25	0.755
4.25	0.775
5.00	0.780
6.00	0.765
7.10	0.715
9.00	0.650
10.00	0.625
11.00	0.590
CURVE 4 (cont.)	
11.50	0.570
12.25	0.500
13.10	0.400
14.00	0.300
14.60	0.275
15.00	0.280
CURVE 5	**T = 773**
1.00	0.860
1.50	0.830
2.00	0.730
2.50	0.670
3.00	0.640
3.40	0.650
4.50	0.725
5.00	0.750
6.10	0.710
CURVE 5 (cont.)	
7.25	0.710
8.00	0.700
9.25	0.630
10.75	0.585
12.00	0.550
13.50	0.475
15.00	0.410
CURVE 6	**T = 1023**
1.00	0.625
1.25	0.685
1.75	0.715
2.50	0.675
3.00	0.655
3.60	0.700
4.00	0.740
5.00	0.770
CURVE 6 (cont.)	
6.00	0.750
7.00	0.750
8.00	0.715
10.00	0.670
11.50	0.670
12.30	0.650
13.50	0.590
15.00	0.500

* Not shown on plot

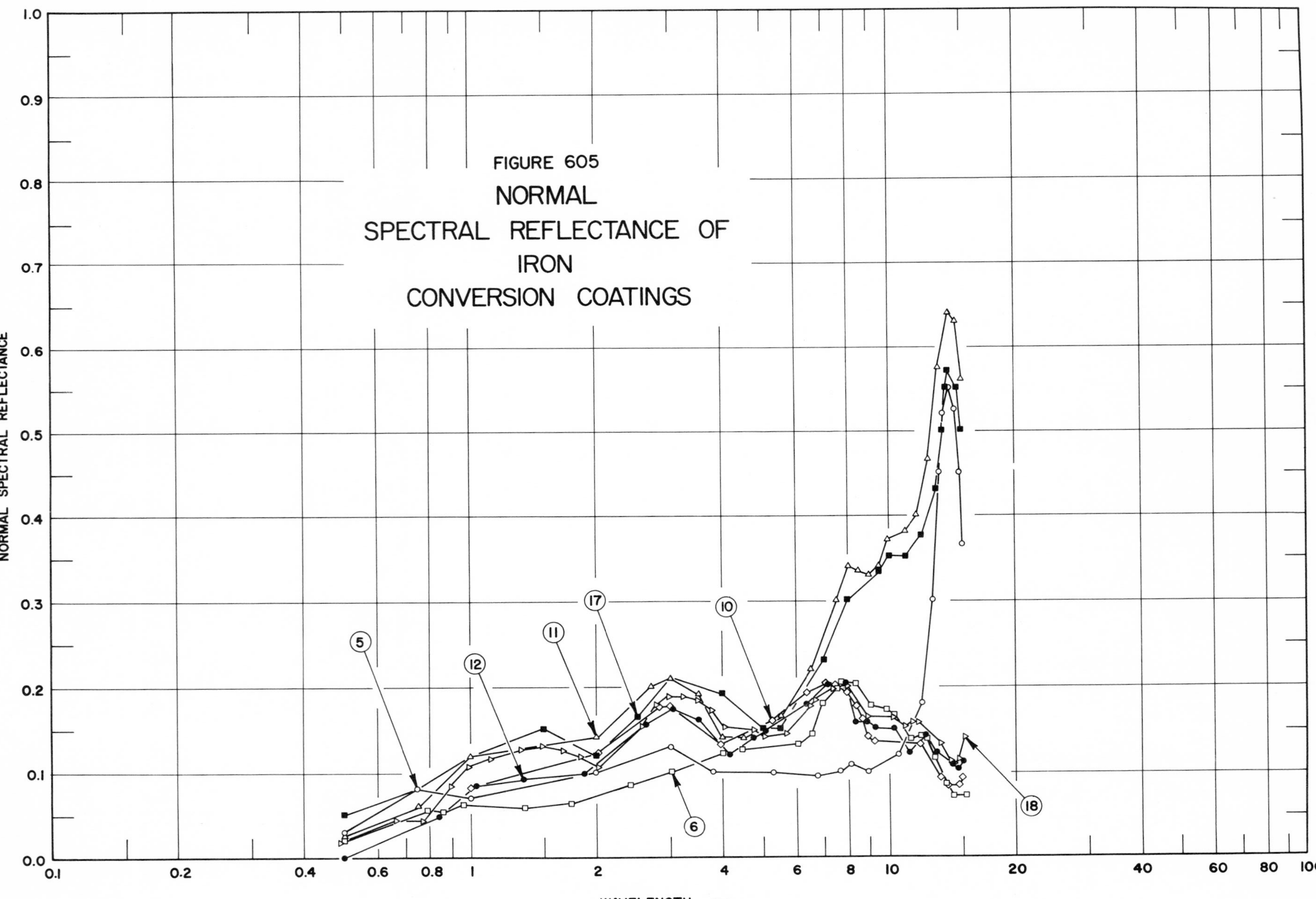
FIGURE 605
NORMAL
SPECTRAL REFLECTANCE OF
IRON
CONVERSION COATINGS
NORMAL SPECTRAL REFLECTANCE
WAVELENGTH, μm
0.0
0.1
0.2
0.3
0.4
0.5
0.6
0.7
0.8
0.9
1.0
0.1
0.2
0.4
0.6
0.8
1
2
4
6
8
10
20
40
60
80
100
5
12
11
17
10
6
18

SPECIFICATION TABLE NO. 605 NORMAL SPECTRAL REFLECTANCE OF IRON CONVERSION COATINGS

Curve No.	Ref. No.	Year	Temperature, K	Wavelength Range, μm	Geometry θ θ' ω'		Reported Error, %	Composition (weight percent), Specifications, and Remarks
1*	139	1961	~322	2.00-15.00	~0°	2π	< 2	Armco blackened steel; as received; data extracted from smooth curve; hohlraum at 523 K; converted from R (2π, 0°).
2*	139	1961	~322	2.00-15.01	~0°	2π	< 2	Above specimen and conditions; diffuse component only.
3*	139	1961	~322	1.00-15.00	~0°	2π	< 2	Similar to curve 1 specimen and conditions except hohlraum at 773 K.
4*	139	1961	~322	1.00-15.00	~0°	2π	< 2	Above specimen and conditions; diffuse component only.
5	139	1961	~322	0.50-15.00	~0°	2π	< 2	Similar to curve 1 specimen and conditions except hohlraum at 1273 K.
6	139	1961	~322	0.50-15.01	~0°	2π	< 2	Above specimen and conditions; diffuse component only.
7*	139	1961	~322	2.00-15.00	~0°	2π	< 2	Similar to curve 1 specimen and conditions except heated in air at 755 K for 30 min.
8*	139	1961	~322	2.01-15.01	~0°	2π	< 2	Above specimen and conditions; diffuse component only.
9*	139	1961	~322	1.00-15.00	~0°	2π	< 2	Similar to curve 7 specimen and conditions except hohlraum at 773 K.
10*	139	1961	~322	1.00-15.00	~0°	2π	< 2	Above specimen and conditions; diffuse component only.
11	139	1961	~322	0.50-15.00	~0°	2π	< 2	Similar to curve 7 specimen and conditions except hohlraum at 1273 K.
12	139	1961	~322	0.50-15.00	~0°	2π	< 2	Above specimen and conditions; diffuse component only.
13*	139	1961	~322	2.00-15.00	~0°	2π	< 2	Similar to curve 1 specimen and conditions except heated in vacuum (4.4×10^{-5} mm Hg) at 755 K for 30 min.
14*	139	1961	~322	2.00-14.99	~0°	2π	< 2	Above specimen and conditions; diffuse component only.
15*	139	1961	~322	1.00-15.00	~0°	2π	< 2	Similar to curve 13 specimen and conditions except hohlraum at 773 K.
16*	139	1961	~322	1.00-14.99	~0°	2π	< 2	Above specimen and conditions; diffuse component only.
17	139	1961	~322	0.50-15.00	~0°	2π	< 2	Similar to curve 13 specimen and conditions except hohlraum at 1273 K.
18	139	1961	~322	0.49-15.01	~0°	2π	< 2	Above specimen and conditions; diffuse component only.

* Not shown on plot

DATA TABLE NO. 605 NORMAL SPECTRAL REFLECTANCE OF IRON CONVERSION COATINGS

[Wavelength, λ, μm; Reflectance, ρ; Temperature, T, K]

CURVE 1* T ~ 322

λ	ρ
2.00	0.100
2.50	0.105
3.50	0.170
4.25	0.205
5.50	0.190
6.50	0.175
8.00	0.190
8.75	0.175
9.10	0.170
10.00	0.190
10.50	0.190
11.00	0.200
12.00	0.270
12.85	0.375
13.25	0.500
13.75	0.585
14.25	0.600
14.75	0.575
15.00	0.540

CURVE 2* T ~ 322

λ	ρ
2.00	0.067
2.64	0.075
3.45	0.099
4.22	0.123
5.11	0.125
5.51	0.128
6.41	0.149
7.67	0.188
7.97	0.196
8.31	0.189
8.59	0.182
9.21	0.162
10.22	0.159
10.50	0.153
11.00	0.140
11.99	0.140
12.26	0.131
13.01	0.100
14.03	0.057
14.52	0.038
15.01	0.029

CURVE 3* T ~ 322

λ	ρ
1.00	0.065
2.00	0.090
2.50	0.090
3.25	0.130
4.00	0.190
5.25	0.175
6.00	0.170
7.90	0.200
9.00	0.180
10.00	0.195
11.00	0.200
12.00	0.270
13.00	0.400
13.60	0.500
14.00	0.560
14.50	0.530
15.00	0.465

CURVE 4* T ~ 322

λ	ρ
1.00	0.025
1.30	0.041
1.67	0.050
2.01	0.050
2.63	0.065
3.32	0.090
4.01	0.110
4.59	0.110
4.95	0.109
5.44	0.116
6.13	0.135
7.02	0.164
7.78	0.179
8.38	0.174
9.00	0.161
9.99	0.148
11.00	0.131
11.50	0.126
11.71	0.126
12.06	0.134
12.32	0.132
12.88	0.119
13.93	0.119
14.40	0.124
14.75	0.135
15.00	0.148

CURVE 5 T ~ 322

λ	ρ
0.50	0.030
0.75	0.080
1.00	0.070
2.00	0.100
3.00	0.130
3.80	0.100
5.30	0.100
6.75	0.095
7.70	0.100
8.10	0.110
8.90	0.100
10.50	0.120
12.00	0.180
12.70	0.300
13.15	0.450
13.50	0.520
14.00	0.550
14.50	0.525
14.85	0.450
15.00	0.365

CURVE 6 T ~ 322

λ	ρ
0.50	0.020
0.79	0.055
0.86	0.053
0.96	0.061
1.35	0.059
1.74	0.064
2.40	0.085
3.01	0.100
4.00	0.121
4.47	0.127
6.05	0.132
6.53	0.145
6.99	0.180
7.51	0.197
7.86	0.205
8.34	0.202
9.01	0.177
9.86	0.171
10.26	0.167
11.16	0.138
11.90	0.141
12.25	0.139
12.90	0.117
13.63	0.085
14.11	0.071
15.01	0.071

CURVE 7* T ~ 322

λ	ρ
2.00	0.210
3.25	0.155
4.50	0.105
5.00	0.110
6.25	0.175
7.00	0.250
7.50	0.290
8.00	0.310
9.50	0.330
10.00	0.345
11.00	0.360
12.15	0.400
12.75	0.450
13.25	0.500
13.60	0.600
14.00	0.620
14.55	0.600
15.00	0.550

CURVE 8* T ~ 322

λ	ρ
2.01	0.170
2.38	0.143
3.01	0.119
4.14	0.093
4.51	0.098
5.07	0.110
5.77	0.123
6.25	0.144
6.77	0.153
7.68	0.147
8.25	0.131
9.00	0.100
10.65	0.101
10.93	0.095
11.52	0.092
12.01	0.100
12.33	0.097
13.20	0.078
14.31	0.068
14.64	0.068
15.01	0.073

CURVE 9* T ~ 322

λ	ρ
1.00	0.070
2.05	0.110
2.50	0.165
3.00	0.190
4.00	0.155
5.00	0.150
6.00	0.175
7.00	0.240
7.50	0.290
8.50	0.315
9.50	0.325
11.00	0.375
12.00	0.430
13.00	0.540
13.75	0.650
14.00	0.700
14.50	0.710
15.00	0.690

CURVE 10 T ~ 322

λ	ρ
1.00	0.081
2.01	0.122
2.83	0.176
3.00	0.177
3.99	0.132
5.27	0.160
6.34	0.191
7.01	0.202
7.50	0.202
7.97	0.192
8.37	0.177
8.66	0.162
8.99	0.142
9.25	0.137
11.98	0.134
13.37	0.094
13.87	0.085
14.58	0.085
15.00	0.094

CURVE 11 T ~ 322

λ	ρ
0.50	0.025
0.75	0.060
1.00	0.120
2.00	0.140
2.70	0.200
3.00	0.210
3.50	0.190
4.00	0.140
4.50	0.140
5.50	0.165
6.50	0.220
7.50	0.300
8.00	0.340
8.50	0.335
9.00	0.330
9.50	0.340
10.00	0.370
11.00	0.380
11.65	0.400
12.50	0.465
13.25	0.575
14.00	0.640
14.50	0.630
15.00	0.560

CURVE 12 T ~ 322

λ	ρ
0.50	0.000
0.84	0.049
1.03	0.084
1.34	0.091
1.87	0.099
2.61	0.156
3.05	0.173
3.50	0.160
4.17	0.120
4.71	0.139
5.02	0.149
6.33	0.179
7.02	0.201
7.90	0.202
8.37	0.159
8.82	0.159
9.23	0.151
10.25	0.150
11.07	0.122
12.08	0.142
13.00	0.123
14.01	0.109
14.54	0.103
15.00	0.113

CURVE 13* T ~ 322

λ	ρ
2.00	0.200
3.50	0.150
4.85	0.100
6.25	0.155
7.00	0.220
7.50	0.260
8.00	0.280
9.50	0.325
11.00	0.330
11.50	0.340
12.30	0.400
13.00	0.470
13.50	0.570
13.75	0.605
14.10	0.620
14.55	0.600
15.00	0.540

CURVE 14* T ~ 322

λ	ρ
2.00	0.175
3.00	0.154
4.45	0.118
4.76	0.113
4.99	0.113
5.99	0.142
6.72	0.160
6.99	0.163
7.72	0.162
8.10	0.160
8.72	0.137
10.23	0.136
10.98	0.124
11.98	0.135
14.00	0.094
14.65	0.084
14.99	0.084

CURVE 15* T ~ 322

λ	ρ
1.00	0.135
1.50	0.150
2.00	0.120
2.50	0.185
3.00	0.215
4.00	0.190
5.25	0.145
6.00	0.200
8.00	0.300
9.00	0.330
9.75	0.360
11.00	0.370
12.00	0.400
12.80	0.450
13.20	0.500
13.55	0.600
13.75	0.635
14.10	0.650
15.00	0.615

* Not shown on plot

DATA TABLE NO. 605 NORMAL SPECTRAL REFLECTANCE OF IRON CONVERSION COATINGS (continued)

λ	ρ
CURVE 16*	
T ~ 322	
1.00	0.104
1.34	0.141
1.50	0.141
1.85	0.120
2.00	0.116
2.16	0.122
2.49	0.161
2.99	0.188
3.43	0.180
3.70	0.168
3.98	0.153
4.42	0.142
4.99	0.142
5.99	0.171
6.58	0.187
7.00	0.191
7.97	0.191
9.10	0.155
9.24	0.154
11.97	0.154
12.48	0.149
13.49	0.112
14.01	0.093
14.99	0.092
CURVE 17	
T ~ 322	
0.50	0.050
0.75	0.080*
1.00	0.120*
1.50	0.150
2.00	0.120
2.50	0.165
3.00	0.210*
4.00	0.190
5.00	0.150
5.50	0.150
7.00	0.230
8.00	0.300
9.50	0.335
10.00	0.350
11.00	0.350
12.00	0.375
13.00	0.430
13.45	0.500
13.70	0.550
14.00	0.570
CURVE 17 (cont.)	
14.50	0.550
15.00	0.500
CURVE 18	
T ~ 322	
0.49	0.018
0.66	0.044
0.77	0.044
0.90	0.084
0.99	0.106
1.11	0.116
1.31	0.127
1.49	0.130
1.67	0.127
1.84	0.119
2.02	0.105
2.56	0.153
2.78	0.178
2.99	0.189
3.21	0.189
3.48	0.184
3.77	0.171
4.02	0.151
4.76	0.149
5.02	0.141
5.68	0.145
6.44	0.177
6.67	0.183
7.39	0.198
7.56	0.200
8.00	0.200
8.98	0.164
10.23	0.164
10.99	0.154
11.47	0.158
11.75	0.158
13.25	0.131
14.00	0.112
14.25	0.110
14.59	0.116
15.01	0.140

* Not shown on plot

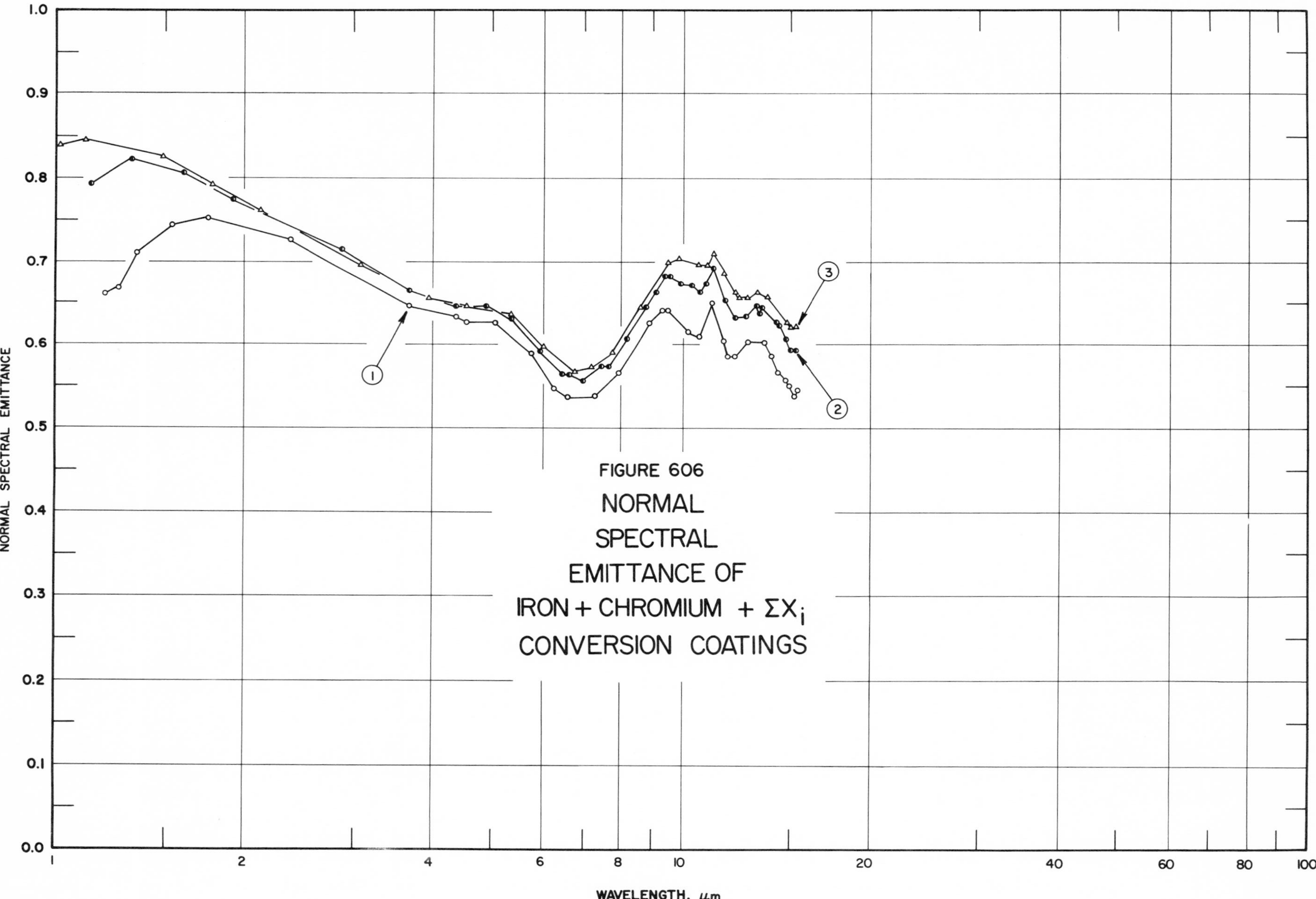

FIGURE 606
NORMAL
SPECTRAL
EMITTANCE OF
IRON + CHROMIUM + ΣX_i
CONVERSION COATINGS

SPECIFICATION TABLE NO. 606 NORMAL SPECTRAL EMITTANCE OF IRON + CHROMIUM + ΣX_i CONVERSION COATINGS

Curve No.	Ref. No.	Year	Temperature, K	Wavelength Range, μm	Geometry θ'	Reported Error, %	Composition (weight percent), Specifications, and Remarks
1	293	1963	800	1.20-15.17	~0°		Kanthal, oxidized at 1340 K for 400 hrs; nominal composition: 25 Cr, 5 Al, 3 Co and bal Fe; machined from 1.09 mm Kanthal sheet; cleaned with acetone, sand-blasted with 60-mesh fused alumina grit at 70 psi, cleaned ultrasonically in acetone, passivated for 1 min in 10% nitric acid at 316 K, rinsed in distilled water and distilled acetone; mean of 3 determinations on each of 6 specimens.
2	293	1963	1100	1.14-15.17	~0°		Above specimen and conditions.
3	293	1963	1300	1.05-15.10	~0°		Above specimen and conditions.

DATA TABLE NO. 606 NORMAL SPECTRAL EMITTANCE OF IRON + CHROMIUM + ΣX_i CONVERSION COATINGS

[Wavelength, λ, μm; Emittance, ϵ; Temperature, T, K]

λ	ϵ	λ	ϵ	λ	ϵ	λ	ϵ	λ	ϵ	λ	ϵ
CURVE 1 T = 800		CURVE 1 (cont.)		CURVE 2 T = 1100		CURVE 2 (cont.)		CURVE 3 T = 1300		CURVE 3 (cont.)	
1.20	0.660	10.67	0.610	1.14	0.793	9.56	0.681	1.05	0.840	11.27	0.710
1.26	0.668	11.10	0.650	1.33	0.822	9.90	0.674	1.22	0.848	11.65	0.688
1.35	0.710	11.23	0.650*	1.60	0.806	10.38	0.671	1.49	0.826	12.03	0.662
1.54	0.743	11.64	0.603	1.92	0.774	10.68	0.664	1.79	0.792	12.39	0.658
1.75	0.751	11.89	0.589	2.86	0.715	10.97	0.674	2.12	0.761	12.67	0.658
2.38	0.728	12.02	0.589	3.68	0.665	11.26	0.692	3.07	0.699	13.15	0.663
3.66	0.649	12.78	0.601	4.34	0.645	11.78	0.652	3.91	0.656	13.64	0.659
4.34	0.633	13.50	0.601	4.86	0.646	12.13	0.632	4.53	0.646	13.88	0.659*
4.52	0.628	13.90	0.589	5.31	0.630	12.52	0.634	4.87	0.646*	14.74	0.627
5.02	0.627	14.30	0.569	5.98	0.592	13.17	0.646	5.32	0.634	14.88	0.621
5.73	0.590	14.51	0.557	6.46	0.564	13.38	0.639	6.00	0.597	15.10	0.621
6.23	0.548	14.62	0.557*	6.60	0.564	13.48	0.645	6.71	0.569		
6.58	0.537	14.85	0.550	6.95	0.556	14.11	0.627	7.17	0.573		
7.25	0.537	14.93	0.539*	7.47	0.574	14.31	0.624	7.73	0.591		
7.96	0.568	15.06	0.539	7.63	0.574	14.62	0.607	8.54	0.645		
8.88	0.626	15.17	0.544	8.16	0.607	14.76	0.608*	9.41	0.700		
9.26	0.640			8.72	0.645	14.96	0.594	9.89	0.705		
9.41	0.640			9.09	0.662	15.17	0.594	10.54	0.699		
10.21	0.616			9.37	0.681			10.98	0.699		

* Not shown on plot

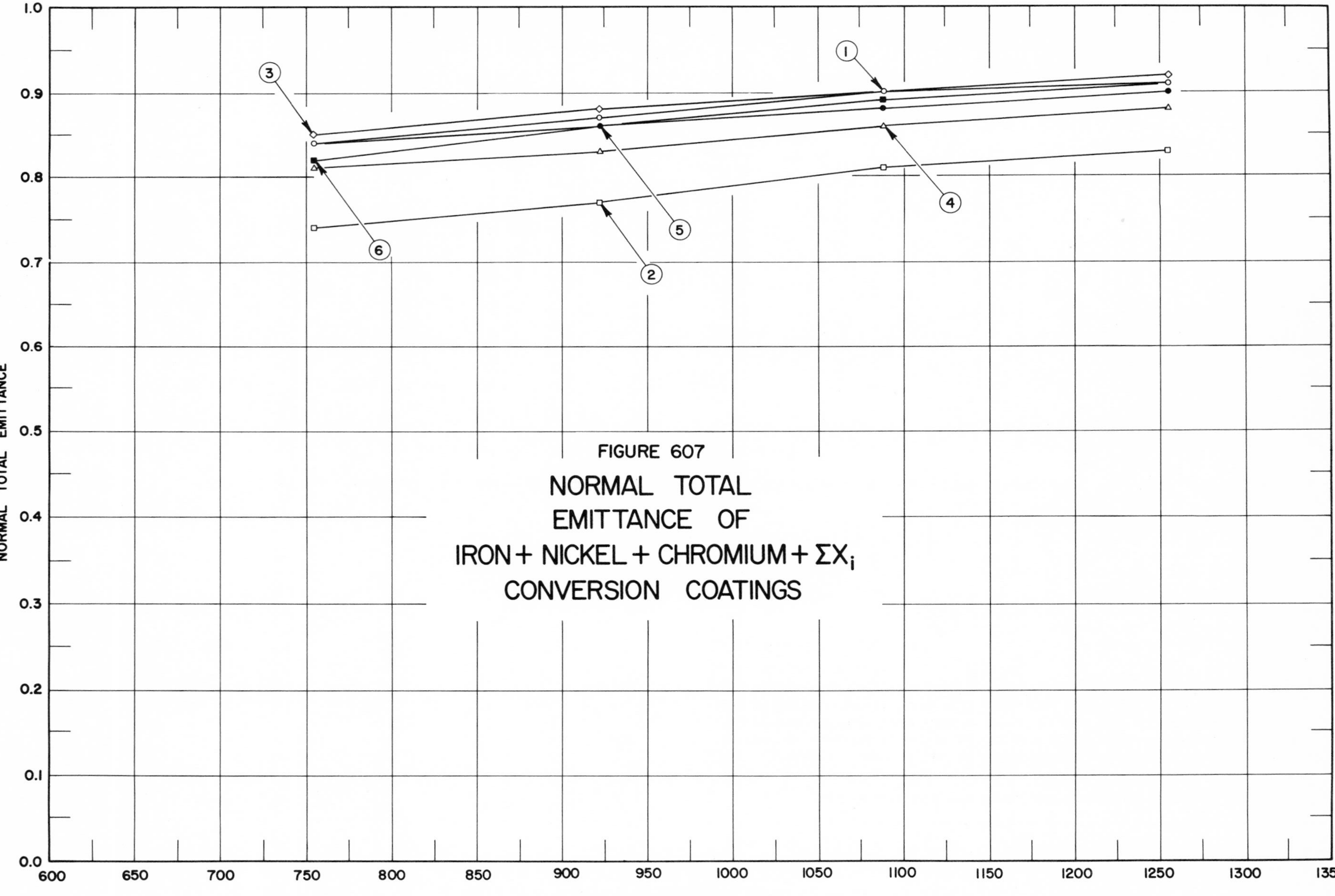

FIGURE 607
NORMAL TOTAL EMITTANCE OF IRON + NICKEL + CHROMIUM + ΣX_i CONVERSION COATINGS

SPECIFICATION TABLE NO. 607 NORMAL TOTAL EMITTANCE OF IRON + NICKEL + CHROMIUM + ΣX_i CONVERSION COATINGS

Curve No.	Ref. No.	Year	Temperature Range, K	Geometry θ'	Reported Error, %	Composition (weight percent), Specifications and Remarks
1	292	1964	755-1255	$\sim 0°$		347 stainless steel, oxidized 20 min at 1255 K; substrate surface degreased, etched in acid solution (100 ml 38% HCl, 25 g powdered CrO_3, 50 ml distilled H_2O) for 3-5 min, degreased, and air dried prior to oxidation.
2	292	1964	755-1255	$\sim 0°$		347 stainless steel, oxidized 20 min at 1255 K; substrate surface degreased, hand polished through Nos. 240, 320, 480, 600 grit papers, degreased, and air dried prior to oxidation.
3	292	1964	755-1255	$\sim 0°$		347 stainless steel, oxidized 20 min at 1255 K; substrate surface degreased with trichloroethylene vapor, blasted with No. 40-60 silicon carbide grit under 90 psi at distance of 6 in., degreased, and air dried prior to oxidation; surface blotched, nonuniform, and not reproducible.
4	292	1964	755-1255	$\sim 0°$		347 stainless steel, oxidized 20 min at 1255 K; substrate surface degreased with trichloroethylene vapor, blasted with No. 100 silicon carbide grit under 90 psi at distance of 6 in., degreased, etched in acid solution (100 ml 38% HCl, 25 g CrO_3, 50 ml distilled H_2O) for 3-5 min, degreased, and air dried prior to oxidation.
5	292	1964	755-1255	$\sim 0°$		Similar to above specimen and conditions except grit blasted with No. 40-60 silicon carbide grit.
6	292	1964	755-1255	$\sim 0°$		347 stainless steel, oxidized 20 min at 1255 K; surface of substrate as received from supplier except thoroughly degreased prior to oxidation.

DATA TABLE NO. 607 NORMAL TOTAL EMITTANCE OF IRON + NICKEL + CHROMIUM + ΣX_i CONVERSION COATINGS

[Temperature, T, K; Emittance, ϵ]

T	ϵ
CURVE 1	
755	0.84
922	0.87
1089	0.90
1255	0.91
CURVE 2	
755	0.74
922	0.77
1089	0.81
1255	0.83
CURVE 3	
755	0.85
922	0.88
1089	0.90*
1255	0.92
CURVE 4	
755	0.81
922	0.83
1089	0.86
1255	0.88
CURVE 5	
755	0.84*
922	0.86
1089	0.88
1255	0.90
CURVE 6	
755	0.82
922	0.86*
1089	0.89
1255	0.91*

* Not shown on plot

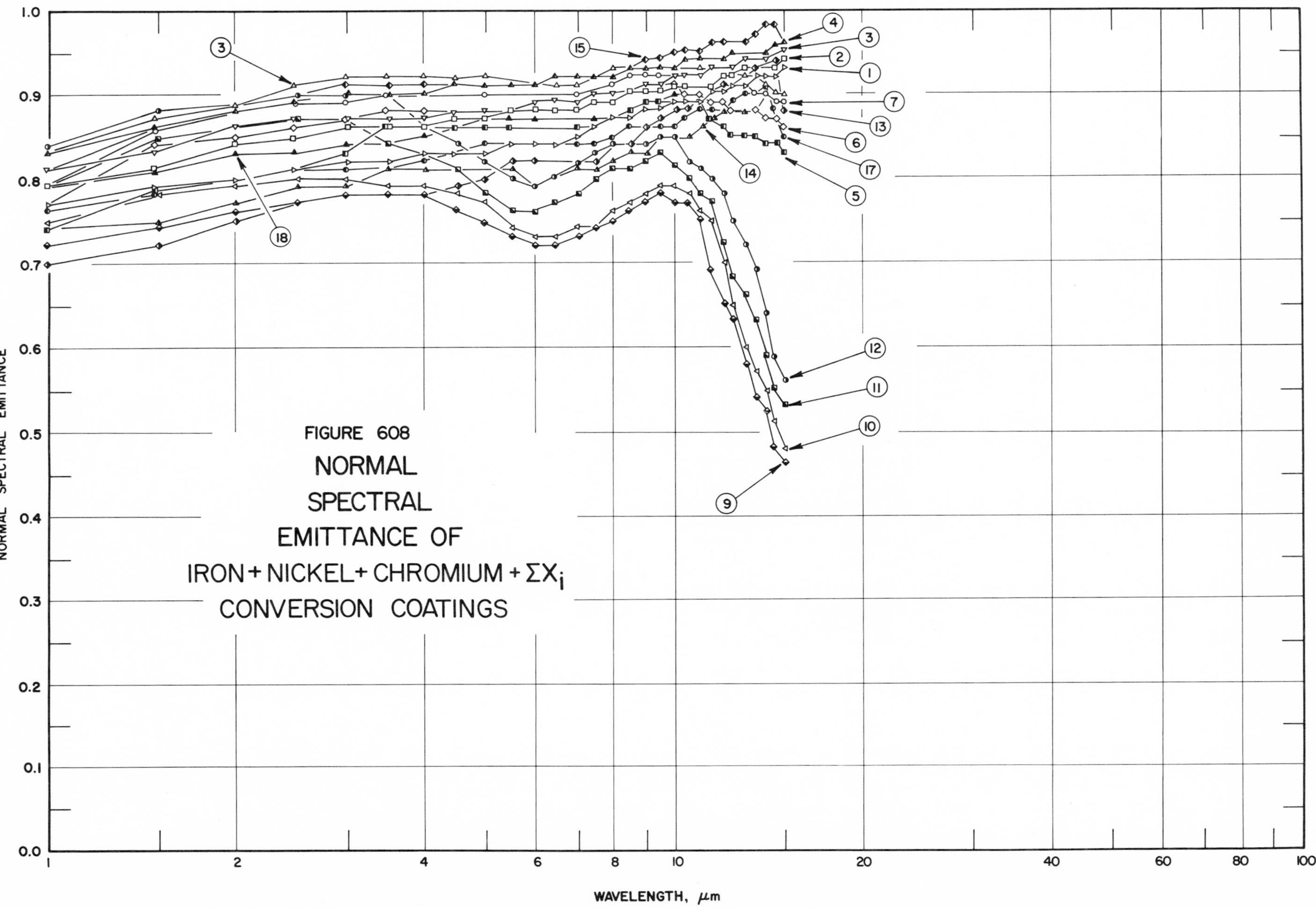

FIGURE 608
NORMAL
SPECTRAL
EMITTANCE OF
IRON + NICKEL + CHROMIUM + ΣX_i
CONVERSION COATINGS

SPECIFICATION TABLE NO. 608 NORMAL SPECTRAL EMITTANCE OF IRON + NICKEL + CHROMIUM + ΣX_i CONVERSION COATINGS

Curve No.	Ref. No.	Year	Temperature, K	Wavelength Range, μm	Geometry θ'	Reported Error, %	Composition (weight percent), Specifications, and Remarks
1	292	1964	755	1.00-15.00	$\sim 0°$		347 stainless steel, oxidized 20 min at 1255 K; substrate surface degreased, etched in acid solution (100 ml 38% HCl, 25 gm powdered CrO_3, 50 ml distilled H_2O), degreased, and air dried prior to oxidation.
2	292	1964	922	1.00-15.00	$\sim 0°$		Similar to curve 1 specimen and conditions.
3	292	1964	1089	1.00-15.00	$\sim 0°$		Similar to curve 1 specimen and conditions.
4	292	1964	1255	1.00-15.00	$\sim 0°$		Similar to curve 1 specimen and conditions.
5	292	1964	755	1.00-15.00	$\sim 0°$		347 stainless steel, oxidized 20 min at 1255 K; substrate surface degreased with trichloroethylene vapor, blasted with No. 40-60 silicon carbide grit under 90 psi at distance of 6 in., degreased, and air dried prior to oxidation.
6	292	1964	922	1.00-15.00	$\sim 0°$		Similar to curve 5 specimen and conditions.
7	292	1964	1089	1.00-15.00	$\sim 0°$		Similar to curve 5 specimen and conditions.
8	292	1964	1255	1.00-15.00	$\sim 0°$		Similar to curve 5 specimen and conditions.
9	292	1964	755	1.00-15.00	$\sim 0°$		347 stainless steel, oxidized 20 min at 1255 K; substrate surface thoroughly degreased, hand polished through Nos. 240, 320, 400, and 600 grit papers, degreased, and air dried prior to oxidation.
10	292	1964	922	1.00-15.00	$\sim 0°$		Similar to curve 9 specimen and conditions.
11	292	1964	1089	1.00-15.00	$\sim 0°$		Similar to curve 9 specimen and conditions.
12	292	1964	1255	1.00-15.00	$\sim 0°$		Similar to curve 9 specimen and conditions.
13	292	1964	755	1.00-15.00	$\sim 0°$		347 stainless steel, oxidized 20 min at 1255 K; substrate surface degreased with trichloroethylene vapor, blasted with No. 40-60 silicon carbide grit under 90 psi at distance of 6 in., degreased, etched in acid solution (100 ml 38% HCl, 25 g powdered CrO_3, 50 ml distilled H_2O) for 3-5 min, degreased, and air dried prior to oxidation.
14	292	1964	755	1.00-15.00	$\sim 0°$		Similar to curve 13 specimen and conditions except grit blasted with No. 100 silicon carbide prior to oxidation.
15	292	1964	1255	1.00-15.00	$\sim 0°$		Similar to curve 13 specimen and conditions.
16*	292	1964	1255	1.00-15.00	$\sim 0°$		Similar to curve 14 specimen and conditions.
17	292	1964	755	1.00-15.00	$\sim 0°$		347 stainless steel, oxidized 20 min at 1255 K; substrate surface as received from supplier and thoroughly degreased before oxidation.
18	292	1964	922	1.00-15.00	$\sim 0°$		Similar to curve 17 specimen and conditions.
19*	292	1964	1089	1.00-15.00	$\sim 0°$		Similar to curve 17 specimen and conditions.
20*	292	1964	1255	1.00-15.00	$\sim 0°$		Similar to curve 17 specimen and conditions.

* Not shown on plot

DATA TABLE NO. 608 NORMAL SPECTRAL EMITTANCE OF IRON + NICKEL + CHROMIUM + ΣX_i CONVERSION COATINGS

[Wavelength, λ, μm; Emittance, ∈; Temperature, T, K]

CURVE 1 T = 755

λ	∈
1.00	0.771
1.49	0.792
2.00	0.800
2.48	0.812
3.01	0.821
3.51	0.821
4.00	0.831
4.48	0.831
5.00	0.831
5.50	0.841
6.00	0.841
6.49	0.841
7.00	0.850
7.49	0.862
8.00	0.874
8.50	0.873
9.00	0.883
9.50	0.884
10.00	0.891
10.48	0.891
11.00	0.891
11.50	0.902
12.00	0.902
12.49	0.911
13.00	0.911
13.51	0.921
14.00	0.921
14.50	0.921
15.00	0.931

CURVE 2 T = 922

λ	∈
1.00	0.792
1.49	0.813
2.00	0.842
2.48	0.850
3.01	0.861
3.51	0.861
4.00	0.861
4.48	0.871
5.00	0.871
5.50	0.882
6.00	0.882
6.48	0.882
7.00	0.882
7.49	0.891
8.00	0.891
8.50	0.904
9.00	0.903
9.50	0.903
10.00	0.910
10.48	0.910
11.00	0.910*
11.50	0.910
12.00	0.922
12.49	0.922
13.00	0.931
13.51	0.931*
14.00	0.931*
14.50	0.931
15.00	0.941

CURVE 3 T = 1089

λ	∈
1.00	0.811
1.49	0.831
2.00	0.862
2.48	0.871
3.01	0.871
3.51	0.871
4.00	0.871
4.48	0.880
5.00	0.880
5.50	0.880*
6.00	0.891
6.48	0.894
7.00	0.891
7.49	0.900
8.00	0.900
8.50	0.904*
9.00	0.911
9.50	0.911
10.00	0.921
10.48	0.921
11.00	0.921
11.50	0.931
12.00	0.931*
12.49	0.931
13.00	0.941
13.51	0.941*
14.00	0.941
14.50	0.950*
15.00	0.951

CURVE 4 T = 1255

λ	∈
1.00	0.831
1.49	0.862
2.00	0.881
2.48	0.891
3.01	0.902
3.51	0.900
4.00	0.902
4.49	0.914
5.00	0.911
5.50	0.911
6.00	0.912
6.48	0.922
7.00	0.921
7.49	0.922
8.00	0.921
8.50	0.931
9.00	0.931
9.50	0.933
10.00	0.932
10.48	0.941
11.00	0.942
11.50	0.941*
12.00	0.942
12.49	0.950
13.00	0.950*
13.51	0.950*
14.00	0.950
14.50	0.960
15.00	0.962

CURVE 5 T = 755

λ	∈
1.00	0.740
1.49	0.790
2.00	0.800*
2.50	0.811*
3.00	0.831
3.49	0.863
4.00	0.862*
4.49	0.862
5.00	0.862
5.49	0.862
6.00	0.862*
6.50	0.862*
7.00	0.862
7.49	0.862*
8.00	0.872*
8.50	0.882
9.00	0.891
9.49	0.892
10.00	0.891*
10.49	0.891*
11.00	0.891*
11.49	0.872
12.00	0.863
12.47	0.851
13.00	0.851
13.50	0.851
14.00	0.842
14.51	0.842
15.00	0.831

CURVE 6 T = 922

λ	∈
1.00	0.771*
1.49	0.842
2.00	0.850
2.49	0.861
3.00	0.872
3.49	0.881
4.00	0.881
4.49	0.881*
5.00	0.881*
5.49	0.881*
6.00	0.881*
6.50	0.881*
7.00	0.881*
7.49	0.890*
8.00	0.900*
8.50	0.900*
9.00	0.912*
9.49	0.912*
10.00	0.912
10.49	0.900
11.00	0.900
11.50	0.891
12.00	0.891
12.47	0.881
13.00	0.881*
13.50	0.881
14.00	0.871
14.51	0.871
15.00	0.861

CURVE 7 T = 1089

λ	∈
1.00	0.790*
1.49	0.860
2.00	0.881*
2.49	0.892
3.00	0.892
3.49	0.900
4.00	0.900*
4.49	0.900*
5.00	0.900
5.49	0.900*
6.00	0.900*
6.50	0.900*
7.00	0.900
7.49	0.900*
8.00	0.911
8.50	0.922
9.00	0.922
9.49	0.922
10.00	0.922*
10.49	0.922*
11.00	0.921*
11.49	0.922*
12.00	0.922*
12.47	0.911*
13.00	0.911*
13.50	0.900
14.00	0.900
14.51	0.891
15.00	0.891

CURVE 8 T = 1255

λ	∈
1.00	0.831*
1.49	0.872
2.00	0.890
2.49	0.911
3.00	0.921
3.49	0.921
4.00	0.921
4.49	0.921*
5.00	0.921
5.49	0.920
6.00	0.911*
6.50	0.911
7.00	0.911
7.49	0.923*
8.00	0.931
8.50	0.931*
9.00	0.931*
9.49	0.932*
10.00	0.931*
10.49	0.931*
11.00	0.931
11.49	0.932*
12.00	0.921*
12.47	0.921*
13.00	0.922
13.50	0.921*
14.00	0.912
14.51	0.900
15.00	0.900

CURVE 9 T = 755

λ	∈
1.00	0.721
1.51	0.742
2.00	0.761
2.52	0.772
3.00	0.781
3.50	0.781
4.00	0.781
4.50	0.763
5.00	0.750
5.51	0.732
6.00	0.721
6.49	0.721
7.01	0.733
7.50	0.742
8.00	0.750
8.49	0.762
9.00	0.772
9.50	0.782
10.00	0.771
10.51	0.771
11.00	0.751
11.48	0.692
12.00	0.651
12.48	0.633
13.00	0.581
13.50	0.541
14.00	0.525
14.47	0.483
15.00	0.463

CURVE 10 T = 922

λ	∈
1.00	0.750
1.51	0.781
2.00	0.792
2.52	0.800
3.00	0.800
3.50	0.791
4.00	0.791
4.51	0.781
5.00	0.772
5.51	0.741
6.00	0.731
6.49	0.731
7.01	0.741
7.50	0.741*
8.00	0.763
8.50	0.771
9.00	0.782
9.50	0.792
10.00	0.792
10.51	0.781
11.00	0.762
11.50	0.750
12.00	0.700
12.49	0.650
13.00	0.600
13.50	0.573
14.00	0.550
14.49	0.511
15.00	0.481

CURVE 11 T = 1089

λ	∈
1.00	0.792*
1.51	0.850
2.00	0.861*
2.52	0.872
3.00	0.872*
3.50	0.842
4.00	0.833*
4.51	0.813
5.00	0.783
5.51	0.762
6.00	0.762
6.49	0.772
7.01	0.782
7.50	0.800
8.00	0.813
8.50	0.812
9.00	0.822
9.50	0.832
10.00	0.817
10.51	0.800
11.00	0.781
11.50	0.772
12.00	0.722
12.49	0.681
13.00	0.661
13.50	0.631
14.00	0.591
14.49	0.551
15.00	0.532

CURVE 12 T = 1255

λ	∈
1.00	0.840
1.51	0.881
2.00	0.890*
2.52	0.900
3.00	0.900

* Not shown on plot

DATA TABLE NO. 608 NORMAL SPECTRAL EMITTANCE OF IRON + NICKEL + CHROMIUM + ΣX_i CONVERSION COATINGS (continued)

λ	∈
CURVE 12 (cont.)	
3.50	0.872*
4.00	0.863*
4.51	0.843
5.00	0.821
5.51	0.800
6.00	0.791
6.49	0.801
7.01	0.820
7.50	0.831
8.00	0.841
8.50	0.841*
9.00	0.841
9.50	0.850
10.00	0.850
10.51	0.820
11.00	0.813
11.50	0.800
12.00	0.781
12.49	0.750
13.00	0.720
13.50	0.691
14.00	0.640
14.49	0.590
15.00	0.561
CURVE 13	
T = 755	
1.00	0.700
1.51	0.721
2.00	0.750
2.51	0.771*
3.00	0.781*
3.50	0.781*
4.01	0.781*
4.51	0.792
5.00	0.800
5.52	0.822
6.00	0.822
6.50	0.822*
7.00	0.822*
7.51	0.822
8.00	0.842*
8.51	0.842
9.02	0.861
9.51	0.872
10.01	0.882
10.51	0.882
CURVE 13 (cont.)	
11.01	0.900*
11.51	0.900*
12.00	0.911
12.51	0.910*
13.00	0.931*
13.52	0.931
14.01	0.940*
14.51	0.940
15.00	0.880
CURVE 14	
T = 755	
1.00	0.742*
1.51	0.750
2.00	0.771
2.51	0.792
3.00	0.792
3.50	0.812
4.01	0.812
4.52	0.812*
5.00	0.812*
5.52	0.812
6.00	0.792*
6.50	0.800
7.00	0.812
7.51	0.812
8.00	0.822
8.51	0.831
9.02	0.831
9.51	0.851*
10.01	0.850*
10.51	0.850
11.01	0.862
11.51	0.871
12.00	0.880
12.51	0.880*
13.00	0.880
13.52	0.880*
14.01	0.900*
14.51	0.892*
15.00	0.861*
CURVE 15	
T = 1255	
1.00	0.840*
1.52	0.850*
CURVE 15 (cont.)	
2.00	0.881*
2.50	0.900*
3.00	0.911
3.51	0.911
4.00	0.911
4.51	0.911*
5.00	0.911*
5.51	0.911*
6.00	0.911*
6.53	0.911*
7.00	0.921*
7.50	0.921*
8.00	0.932*
8.51	0.932*
9.00	0.942
9.50	0.944
10.00	0.950
10.48	0.952
11.00	0.952
11.50	0.962
12.00	0.962
12.50	0.962*
13.00	0.962
13.51	0.971
14.00	0.982
14.50	0.982
15.00	0.961*
CURVE 16*	
T = 1255	
1.00	0.812
1.52	0.861
2.00	0.890
2.50	0.890
3.00	0.900
3.51	0.881
4.00	0.861
4.51	0.863
5.00	0.862
5.51	0.854
6.02	0.852
6.53	0.852
7.00	0.852
7.50	0.852
8.00	0.872
8.51	0.883
9.00	0.882
CURVE 16 (cont.)	
9.50	0.892
9.99	0.892
10.51	0.901
11.00	0.901
11.50	0.912
12.00	0.912
12.52	0.912
13.00	0.922
13.51	0.922
14.02	0.932
14.50	0.932
15.00	0.911
CURVE 17	
T = 755	
1.00	0.762
1.49	0.781
2.00	0.791*
2.49	0.811*
3.00	0.811
3.50	0.813*
4.00	0.822
4.50	0.832*
5.00	0.841
5.48	0.841*
6.00	0.841*
6.48	0.841*
7.00	0.841
7.48	0.841
8.00	0.850
8.48	0.862
9.00	0.862*
9.50	0.862
10.00	0.862
10.49	0.873
11.00	0.881
11.50	0.881
12.00	0.881*
12.48	0.892
13.00	0.900
13.50	0.900
14.00	0.910
14.49	0.882
15.00	0.850
CURVE 18	
T = 922	
1.00	0.791*
1.49	0.810
2.00	0.831
2.49	0.831
3.00	0.841
3.48	0.841*
4.00	0.851
4.46	0.862*
5.00	0.871*
5.48	0.871
6.00	0.871
6.48	0.871*
7.00	0.871*
7.48	0.871
8.00	0.871*
8.48	0.882*
9.00	0.892*
9.50	0.892*
10.00	0.900
10.49	0.900*
11.00	0.900*
11.50	0.900*
12.00	0.911*
12.48	0.911*
13.00	0.911*
13.50	0.922*
14.00	0.922*
14.49	0.900*
15.00	0.881*
CURVE 19*	
T = 1089	
1.00	0.862
1.49	0.871
2.00	0.890
2.49	0.890
3.00	0.890
3.48	0.890
4.00	0.900
4.48	0.900
5.00	0.900
5.48	0.900
6.00	0.900
6.48	0.900
7.00	0.900
7.48	0.900
CURVE 19 (cont.)	
8.00	0.911
8.48	0.911
9.00	0.911
9.50	0.911
10.00	0.922
10.49	0.922
11.00	0.922
11.50	0.922
12.00	0.921
12.48	0.931
13.00	0.931
13.50	0.931
14.00	0.940
14.49	0.911
15.00	0.900
CURVE 20*	
T = 1255	
1.00	0.875
1.49	0.881
2.00	0.901
2.49	0.911
3.00	0.911
3.48	0.912
4.00	0.911
4.48	0.911
5.00	0.911
5.48	0.911
6.00	0.911
6.48	0.912
7.00	0.911
7.48	0.922
8.00	0.922
8.48	0.922
9.00	0.932
9.50	0.932
10.00	0.931
10.49	0.931
11.00	0.931
11.50	0.932
12.00	0.941
12.48	0.941
13.00	0.941
13.50	0.941
14.00	0.950
14.49	0.932
15.00	0.922

* Not shown on plot

SPECIFICATION TABLE NO. 609 NORMAL SPECTRAL TRANSMITTANCE OF MOLYBDENUM SILICIDE CONVERSION COATINGS

Curve No.	Ref. No.	Year	Temperature, K	Wavelength Range, μm	Geometry θ	θ'	ω'	Reported Error, %	Composition (weight percent), Specifications, and Remarks
1*	285	1964	298	2.0-15.2	~0°	~0°			$MoSi_2$ oxidized at 763 K.
2*	285	1964	298	2.0-15.2	~0°	~0°			Above specimen and conditions except vacuum heat treated at 923 K to evaporate MoO_3.
3*	285	1964	298	2.0-15.2	~0°	~0°			Mo_5Si_3 oxidized at 1433 K.

DATA TABLE NO. 609 NORMAL SPECTRAL TRANSMITTANCE OF MOLYBDENUM SILICIDE CONVERSION COATINGS

[Wavelength, λ, μm; Transmittance, τ, Temperature, T, K]

λ	τ
CURVE 1*	
T = 298	
2.0	0.181
2.6	0.275
2.8	0.173
3.0	0.234
4.2	0.349
5.7	0.343
5.9	0.331
6.9	0.319
7.2	0.284
7.8	0.180
8.0	0.124
8.2	0.099
8.3	0.074
8.7	0.074
9.3	0.049
9.6	0.028
9.8	0.068
10.0	0.143
CURVE 1 (cont.)*	
10.2	0.114
10.7	0.175
10.9	0.145
11.2	0.135
11.7	0.145
12.6	0.147
12.8	0.158
13.1	0.203
14.7	0.175
14.9	0.152
15.2	0.147
CURVE 2*	
T = 298	
2.0	0.235
2.6	0.235
2.9	0.207
3.1	0.238
CURVE 2 (cont.)*	
3.7	0.249
4.3	0.242
7.2	0.261
7.7	0.255
8.5	0.167
8.8	0.167
9.1	0.117
9.4	0.138
9.6	0.198
10.0	0.312
10.3	0.330
11.8	0.338
12.4	0.316
12.7	0.346
13.0	0.384
14.0	0.392
14.2	0.379
14.8	0.370
15.0	0.341
CURVE 2 (cont.)*	
15.1	0.359
15.2	0.353
CURVE 3*	
T = 298	
2.0	0.285
2.6	0.328
2.9	0.303
3.1	0.343
3.5	0.353
5.5	0.330
5.6	0.319
5.8	0.322
6.2	0.291
6.4	0.280
6.7	0.285
7.1	0.257
7.8	0.176
CURVE 3 (cont.)*	
8.1	0.128
8.4	0.077
8.6	0.077
8.9	0.042
9.1	0.018
9.5	0.018
10.1	0.114
11.9	0.464
12.1	0.470
12.2	0.307
12.6	0.147
13.0	0.324
13.2	0.404
13.6	0.469
13.9	0.512
14.3	0.535
14.7	0.535
14.9	0.499
15.0	0.454
15.2	0.496

* No plot given

SPECIFICATION TABLE NO. 610 HEMISPHERICAL TOTAL EMITTANCE OF NICKEL CONVERSION COATINGS

Curve No.	Ref. No.	Year	Temperature Range, K	Reported Error, %	Composition (weight percent), Specifications and Remarks
1*	286	1965	711-1168		Cleaned and annealed in hydrogen; oxidized (6.6 micrograms cm^{-2} oxygen); measured in vacuum (1×10^{-7} mm Hg). [Author's designation: Sample 35]
2*	286	1965	631-1197		Similar to above specimen and conditions except oxidized with 15.9 micrograms cm^{-2} oxygen; measured in vacuum (7×10^{-8} mm Hg). [Author's designation: Sample 36]
3*	286	1965	423-1209		Above specimen and conditions except measured at 1×10^{-7} mm Hg .
4*	286	1965	646-1180		Similar to above specimen and conditions except oxidized with 36.3 micrograms cm^{-2} oxygen; measured in vacuum (2×10^{-7} mm Hg). [Author's designation: Sample 33]
5*	286	1965	658-1171		Similar to above specimen and conditions except oxidized with 43.5 micrograms cm^{-2} oxygen; measured in vacuum (1×10^{-7} mm Hg). [Author's designation: Sample 34]
6*	286	1965	653-1089		Similar to above specimen and conditions except oxidized with 135 micrograms cm^{-2} oxygen; measured in vacuum (7×10^{-8} mm Hg). [Author's designation: Sample 18]
7*	286	1965	690-1133		Above specimen and conditions except measured at 2×10^{-7} mm Hg.
8*	286	1965	596-1201		Similar to above specimen and conditions except oxidized with 258 micrograms cm^{-2} oxygen; measured in vacuum (4.2×10^{-8} mm Hg). [Author's designation: Sample 32]
9*	286	1965	672-1198		Above specimen and conditions except measured at 5.2×10^{-8} mm Hg.
10*	286	1965	658-1203		Similar to above specimen and conditions except oxidized with 456 micrograms cm^{-2}; measured in vacuum (9×10^{-8} mm Hg). [Author's designation: Sample 38]

* No plot given

DATA TABLE NO. 610 HEMISPHERICAL TOTAL EMITTANCE OF NICKEL CONVERSION COATINGS

[Temperature, T, K; Emittance, ∈]

T	∈
CURVE 1*	
1102	0.172
1168	0.182
1039	0.155
956	0.143
875	0.130
773	0.121
711	0.110
CURVE 2*	
1003	0.176
1099	0.227
1133	0.254
1173	0.274
1197	0.285
846	0.158
897	0.174
783	0.142
631	0.118
692	0.120
CURVE 3*	
1209	0.285
917	0.273
828	0.233
714	0.186
633	0.169
561	0.153
510	0.143
423	0.119
CURVE 4*	
953	0.305
1023	0.331
1076	0.357
1129	0.377
1180	0.402
862	0.271
755	0.239
646	0.198
CURVE 5*	
941	0.324
1002	0.342
1075	0.377
1139	0.404
1171	0.417
846	0.290
744	0.252
658	0.211
682	0.220
CURVE 6*	
992	0.398
1089	0.434
873	0.368
780	0.352
739	0.336
653	0.296
CURVE 7*	
1133	0.458
953	0.376
826	0.350
690	0.303
CURVE 8*	
939	0.388
993	0.404
1064	0.441
1104	0.456
1153	0.477
1171	0.481
1201	0.495
886	0.357
756	0.322
818	0.336
701	0.296
596	0.305
664	0.278
CURVE 9*	
1198	0.504
CURVE 9 (cont.)*	
1155	0.476
1075	0.432
976	0.392
885	0.361
806	0.343
699	0.297
672	0.280
CURVE 10*	
1015	0.448
1203	0.538
1174	0.518
1108	0.487
920	0.418
804	0.400
695	0.358
658	0.341

* No plot given

SPECIFICATION TABLE NO. 611 NORMAL TOTAL EMITTANCE OF NICKEL + CHROMIUM + ΣX_i CONVERSION COATINGS

Curve No.	Ref. No.	Year	Temperature Range, K	Geometry θ'	Reported Error, %	Composition (weight percent), Specifications and Remarks
1*	95	1962	589-1255	$\sim 0^\circ$		Inconel oxidized by treatment with saturated aqueous solution of sodium dichromate; nominal composition: 80 Ni, 14 Cr and 6 Fe; data extracted from smooth curve.
2*	95	1962	589-1144	$\sim 0^\circ$		Similar to above specimen and conditions except held at 1144 K for 30 min.
3*	95	1962	589-1255	$\sim 0^\circ$		Similar to above specimen and conditions except held at 1255 K for 30 min.

DATA TABLE NO. 611 NORMAL TOTAL EMITTANCE OF NICKEL + CHROMIUM + ΣX_i CONVERSION COATINGS

[Temperature, T, K; Emittance, ϵ]

T	ϵ	T	ϵ
CURVE 1*		CURVE 3*	
589	0.905	589	0.889
740	0.924	617	0.858
901	0.947	642	0.834
1033	0.959	669	0.812
1108	0.965	700	0.793
1200	0.966	725	0.782
1255	0.965	749	0.774
		789	0.769
CURVE 2*		816	0.767
		865	0.771
589	0.900	975	0.789
681	0.915	1063	0.802
808	0.928	1143	0.811
921	0.942	1255	0.818
1023	0.948		
1080	0.948		
1144	0.942		

* No plot given

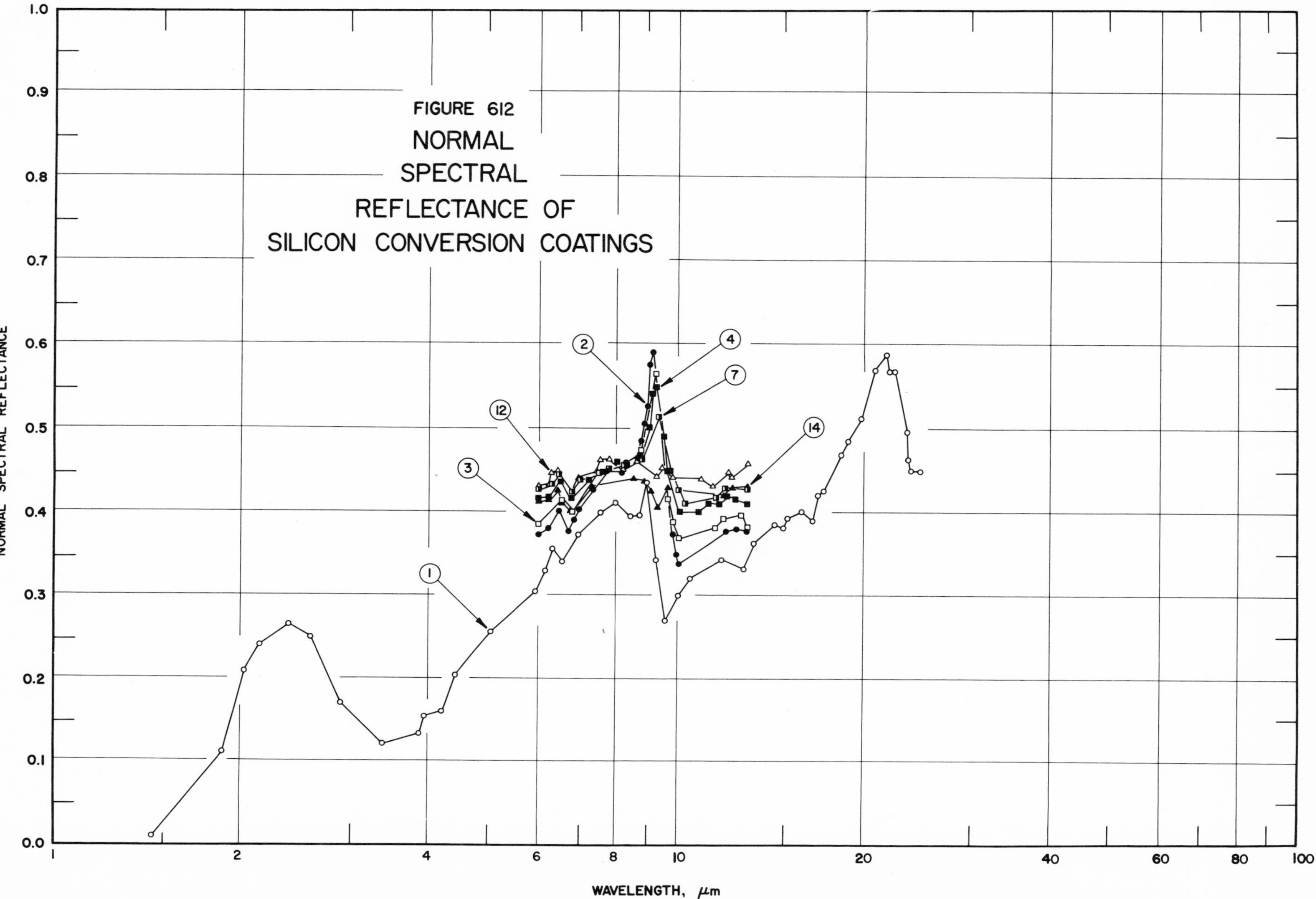

FIGURE 612 NORMAL SPECTRAL REFLECTANCE OF SILICON CONVERSION COATINGS

SPECIFICATION TABLE NO. 612 NORMAL SPECTRAL REFLECTANCE OF SILICON CONVERSION COATINGS

Curve No.	Ref. No.	Year	Temperature, K	Wavelength Range, μm	Geometry θ	θ'	ω'	Reported Error, %	Composition (weight percent), Specifications, and Remarks
1	287	1966	~298	1.54-24.86	10°	10°			Phosphorous doped (10^{17} cm^{-3}) silicon oxidized by heating to 1473 K in wet oxygen; formed an amorphous silica film (5500 Å thick); data extracted from smooth curve.
2	287	1966	~298	6.00-13.00	10°	10°			Phosphorous doped (10^{17} cm^{-3}) silicon oxidized by heating to 1473 K in oxygen reaction tube and then exposing to air; formed an amorphous silica film (5200 Å thick); coating etched in CP_4; data extracted from smooth curve.
3	287	1966	~298	6.00-13.00	10°	10°			Above specimen and conditions except film thickness reduced to 4650 Å by etching in CP_4.
4	287	1966	~298	6.00-13.00	10°	10°			Above specimen and conditions except film thickness reduced to 3650 Å by etching in CP_4.
5*	287	1966	~298	6.00-13.00	10°	10°			Above specimen and conditions except film thickness reduced to 3250 Å by etching in CP_4.
6*	287	1966	~298	6.00-13.00	10°	10°			Above specimen and conditions except film thickness reduced to 3000 Å by etching in CP_4.
7	287	1966	~298	6.00-13.00	10°	10°			Above specimen and conditions except film thickness reduced to 2750 Å by etching in CP_4.
8*	287	1966	~298	6.00-13.00	10°	10°			Above specimen and conditions except film thickness reduced to 2500 Å by etching in CP_4.
9*	287	1966	~298	6.00-13.00	10°	10°			Above specimen and conditions except film thickness reduced to 2250 Å by etching in CP_4.
10*	287	1966	~298	6.00-13.00	10°	10°			Above specimen and conditions except film thickness reduced to 2000 Å by etching in CP_4.
11*	287	1966	~298	6.00-13.00	10°	10°			Above specimen and conditions except film thickness reduced to 1250 Å by etching in CP_4.
12	287	1966	~298	6.00-13.00	10°	10°			Above specimen and conditions except film thickness reduced to 1000 Å by etching in CP_4.
13*	287	1966	~298	6.00-13.00	10°	10°			Above specimen and conditions except film thickness reduced to 750 Å by etching in CP_4.
14	287	1966	~298	6.00-13.00	10°	10°			Above specimen and conditions except film thickness reduced to $<$ 750 Å by etching in CP_4.
15*	287	1966	~298	2.01-14.00	10°	10°			Arsenic doped (10^{19} cm^{-3}) silicon oxidized by heating to 1473 K in dry oxygen; formed an amorphous silica film; data extracted from smooth curve.
16*	287	1966	~298	1.62-14.00	10°	10°			Similar to above specimen and conditions except surface film grown in wet oxygen.

* Not shown on plot

DATA TABLE NO. 612 NORMAL SPECTRAL REFLECTANCE OF SILICON CONVERSION COATINGS

[Wavelength, λ, μm; Reflectance, ρ; Temperature, T, K]

λ	ρ
CURVE 1 T ~ 298	
1.54	0.011
1.87	0.110
2.03	0.210
2.14	0.242
2.38	0.267
2.58	0.251
2.90	0.173
3.37	0.121
3.87	0.134
3.95	0.156
4.21	0.161
4.42	0.205
5.04	0.258
5.95	0.306
6.15	0.333
6.31	0.358
6.58	0.341
6.99	0.376
7.56	0.401
8.00	0.414
8.42	0.397
8.73	0.397
8.98	0.438
9.25	0.345
9.61	0.272
10.10	0.301
10.52	0.323
11.89	0.346
12.93	0.334
13.39	0.366
14.48	0.387
14.92	0.384
15.12	0.395
15.89	0.402
16.53	0.392
16.98	0.423
17.28	0.427
18.45	0.472
18.94	0.488
19.93	0.514
20.98	0.571
21.80	0.592
22.00	0.570
22.33	0.570
23.52	0.498
CURVE 1 (cont.)	
23.79	0.465
23.96	0.452
24.86	0.452
CURVE 2 T ~ 298	
6.00	0.375
6 21	0.382
6.49	0.404
6.72	0.378
6.85	0.393
6.91	0.407
7.35	0.429
7.72	0.451
8.18	0.449
8.29	0.461
8.61	0.467
8.79	0.488
8.83	0.506
8.96	0.528
9.02	0.577
9.16	0.594
9.83	0.374
9.98	0.352
10.16	0.340
12.00	0.379
12.50	0.382
13.00	0.378
CURVE 3 T ~ 298	
6.00	0.387
6.53	0.417
6.80	0.403
7.51	0.449
8.22	0.458
8.74	0.477
9.22	0.567
9.68	0.418
9.86	0.390
10.08	0.371
11.50	0.384
11.77	0.396
12.70	0.398
CURVE 3 (cont.)	
13.00	0.385
CURVE 4 T ~ 298	
6.00	0.419
6.21	0.420
6.51	0.438
6.78	0.419
7.23	0.440
7.62	0.450
8.04	0.464
8.31	0.459
8.72	0.471
9.02	0.503
9.13	0.544
9.26	0.551
9.54	0.494
9.67	0.453
10.02	0.402
10.76	0.402
11.31	0.413
11.69	0.413
12.13	0.421
12.49	0.417
13.00	0.414
CURVE 5* T ~ 298	
6.00	0.396
6.36	0.400
6.57	0.412
6.83	0.396
6.93	0.410
7.47	0.428
8.89	0.448
9.15	0.511
9.30	0.511
9.50	0.491
9.70	0.427
9.97	0.397
10.39	0.384
10.92	0.394
11.50	0.394
11.81	0.404
CURVE 5 (cont.)	
13.00	0.404
CURVE 6* T ~ 298	
6.00	0.420
6.32	0.420
6.48	0.439
6.69	0.422
7.91	0.459
8.16	0.451
8.36	0.456
8.49	0.456
8.88	0.474
9.06	0.497
9.31	0.523
9.85	0.422
10.04	0.410
10.63	0.407
11.01	0.415
11.62	0.410
12.00	0.417
12.53	0.417
13.00	0.422
CURVE 7 T ~ 298	
6.00	0.428
6.31	0.435
6.48	0.447
6.80	0.427
7.00	0.439
7.80	0.456
8.29	0.456*
8.80	0.465
9.34	0.515
9.79	0.451
10.04	0.428
10.39	0.414
11.57	0.419
11.99	0.431
13.00	0.429
CURVE 8* T ~ 298	
6.00	0.414
6.38	0.417
6.55	0.439
6.73	0.410
7.20	0.433
8.11	0.443
8.74	0.443
9.02	0.458
9.19	0.480
9.37	0.480
9.79	0.431
10.24	0.411
11.89	0.411
12.09	0.418
12.49	0.415
13.00	0.423
CURVE 9* T ~ 298	
6.00	0.436
6.51	0.442
6.79	0.426
7.05	0.441
7.58	0.455
8.72	0.455
9.09	0.469
9.24	0.485
9.43	0.485
9.93	0.437
10.54	0.426
12.01	0.431
12.40	0.433
13.00	0.427
CURVE 10* T ~ 298	
6.00	0.423
6.31	0.423
6.47	0.441
6.75	0.416
7.69	0.436
9.12	0.440
9.28	0.453
CURVE 10 (cont.)	
9.65	0.446
9.92	0.427
10.23	0.420
10.99	0.420
11.76	0.412
12.01	0.422
12.28	0.415
12.72	0.421
13.00	0.414
CURVE 11* T ~ 298	
6.00	0.425
6.47	0.443
6.88	0.426
7.56	0.445
9.20	0.447
9.41	0.453
10.50	0.428
12.52	0.428
13.00	0.431
CURVE 12 T ~ 298	
6.00	0.433
6.23	0.436
6.29	0.449
6.45	0.452
6.80	0.421
6.95	0.443
7.42	0.449
7.54	0.465
7.80	0.466
8.20	0.454
8.63	0.464
9.26	0.445
9.45	0.456
9.82	0.444
10.98	0.442
11.48	0.435
12.17	0.450
12.36	0.445
13.00	0.460
CURVE 13* T ~ 298	
6.00	0.429
6.41	0.442
6.59	0.420
6.75	0.415
6.85	0.429
7.06	0.440
7.67	0.451
8.08	0.448
8.67	0.453
8.85	0.448
9.30	0.428
9.77	0.441
11.23	0.435
13.00	0.445
CURVE 14 T ~ 298	
6.00	0.416
6.22	0.417
6.45	0.428
6.56	0.414
6.78	0.404
7.32	0.433
8.50	0.442
8.87	0.438
9.07	0.427
9.26	0.409
9.64	0.433
11.82	0.422
12.33	0.433
13.00	0.431
CURVE 15* T ~ 298	
2.01	0.000
2.83	0.101
3.28	0.147
3.99	0.164
4.13	0.147
4.36	0.199
4.95	0.206
5.46	0.206
6.01	0.198
CURVE 15 (cont.)	
6.22	0.212
6.45	0.201
6.81	0.215
6.99	0.229
7.87	0.243
8.13	0.267
8.61	0.296
8.89	0.343
9.00	0.408
9.17	0.437
9.48	0.404
9.65	0.321
9.86	0.268
10.24	0.212
11.91	0.206
12.35	0.219
14.00	0.211
CURVE 16* T ~ 298	
1.62	0.000
1.86	0.040
2.04	0.123
2.38	0.192
3.48	0.080
3.83	0.044
3.96	0.044
4.26	0.032
5.30	0.089
6.70	0.150
7.01	0.190
7.75	0.268
8.53	0.365
8.74	0.413
8.90	0.470
9.05	0.564
10.12	0.133
10.50	0.088
11.24	0.097
11.79	0.137
12.22	0.151
12.69	0.151
13.02	0.122
13.48	0.126
14.00	0.140

* Not shown on plot

SPECIFICATION TABLE NO. 613 ANGULAR SPECTRAL REFLECTANCE OF SILICON CONVERSION COATINGS

Curve No.	Ref. No.	Year	Temperature, K	Wavelength Range, μm	Geometry θ θ' ω'	Reported Error, %	Composition (weight percent), Specifications, and Remarks
1*	288	1966	~298	0.220-0.320	20° 20°		p-type (boron doped) silicon oxidized at 1398 K in dry oxygen; coating thickness ~8531 Å; substrate cut on <111> plane, lapped and chemically polished before oxidation; data extracted from smooth curve. [Authors' designation: Sample 16]
2*	288	1966	~298	0.220-0.320	20° 20°		Similar to above specimen and conditions except film thickness ~11,397 Å. [Authors' designation: Sample 18]

DATA TABLE NO. 613 ANGULAR SPECTRAL REFLECTANCE OF SILICON CONVERSION COATINGS

[Wavelength, λ, μm; Reflectance, ρ; Temperature, T, K]

λ	ρ	λ	ρ	λ	ρ	λ	ρ
CURVE 1*		CURVE 1 (cont.)*		CURVE 2 (cont.)*		CURVE 2 (cont.)*	
T ~ 298		0.276	0.654	0.228	0.646	0.272	0.566
0.220	0.546	0.279	0.662	0.230	0.637	0.276	0.665
0.223	0.435	0.281	0.652	0.233	0.541	0.279	0.692
0.225	0.412	0.286	0.574	0.234	0.485	0.282	0.669
0.226	0.430	0.290	0.475	0.235	0.465	0.288	0.508
0.230	0.576	0.293	0.437	0.236	0.465	0.289	0.503
0.233	0.599	0.296	0.423	0.238	0.511	0.293	0.520
0.237	0.554	0.299	0.450	0.240	0.598	0.299	0.606
0.240	0.459	0.307	0.548	0.242	0.634	0.301	0.614
0.241	0.423	0.310	0.565	0.244	0.637	0.305	0.585
0.243	0.406	0.314	0.553	0.246	0.612	0.314	0.416
0.245	0.434	0.320	0.477	0.248	0.538	0.315	0.409
0.249	0.582			0.251	0.492	0.318	0.420
0.253	0.634	CURVE 2*		0.253	0.529	0.320	0.457
0.255	0.638	T ~ 298		0.257	0.638		
0.258	0.615			0.259	0.676		
0.264	0.486	0.220	0.508	0.261	0.678		
0.265	0.479	0.222	0.487	0.264	0.628		
0.268	0.496	0.224	0.539	0.266	0.565		
0.273	0.617	0.227	0.637	0.268	0.540		

* No plot given

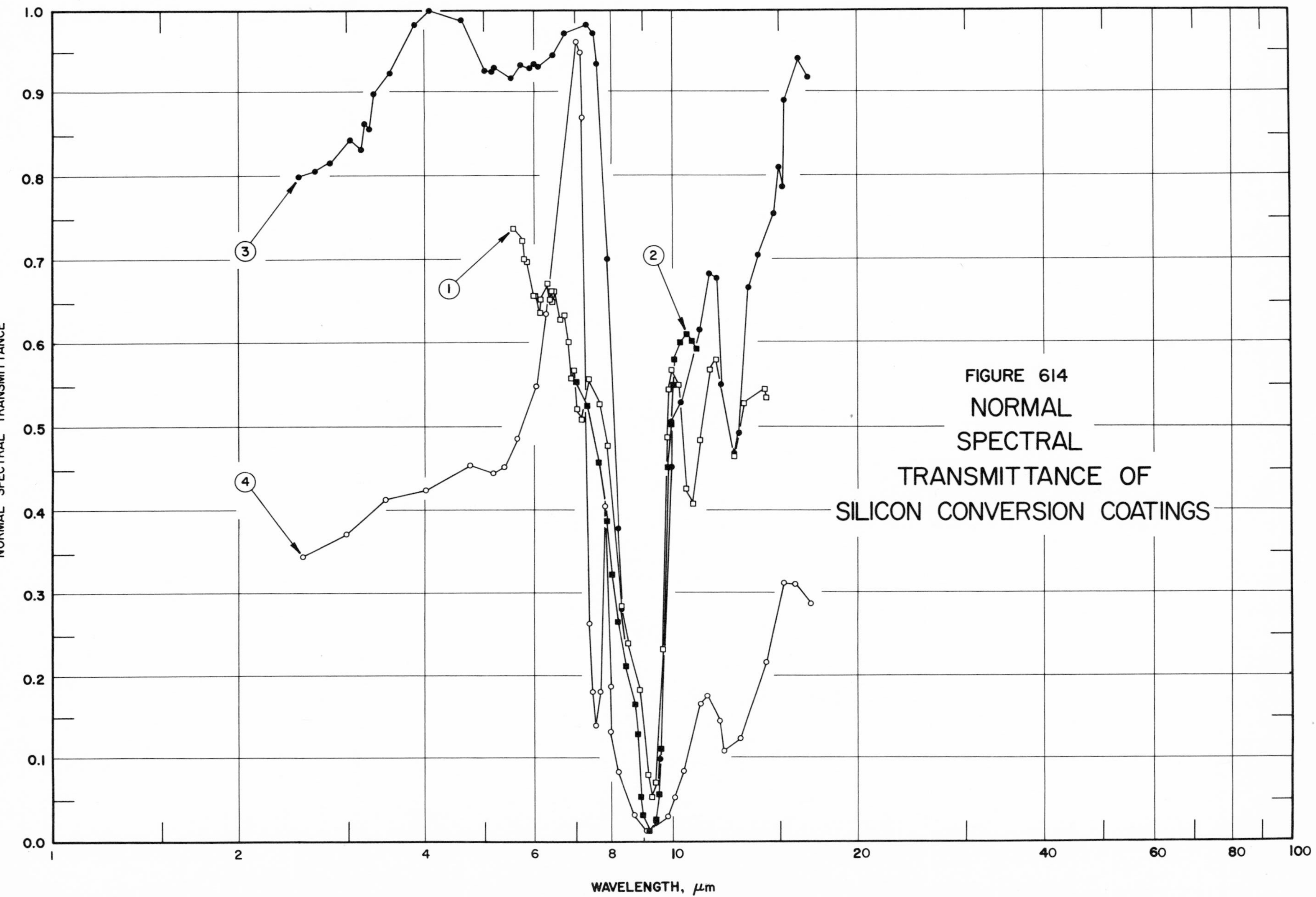

FIGURE 614
NORMAL
SPECTRAL
TRANSMITTANCE OF
SILICON CONVERSION COATINGS
NORMAL SPECTRAL TRANSMITTANCE
WAVELENGTH, μm
1.0
0.9
0.8
0.7
0.6
0.5
0.4
0.3
0.2
0.1
0.0
1
2
4
6
8
10
20
40
60
80
100
1
2
3
4

SPECIFICATION TABLE NO. 614 NORMAL SPECTRAL TRANSMITTANCE OF SILICON CONVERSION COATINGS

Curve No.	Ref. No.	Year	Temperature, K	Wavelength Range, μm	Geometry θ θ' ω'	Reported Error, %	Composition (weight percent), Specifications, and Remarks
1	290	1968	~298	5.55-14.26	~0° ~0°		n-type (boron doped) silicon oxidized at 1423 K in dry oxygen; resistivity of silicon 20 Ω cm; data extracted from smooth curve.
2	291	1968	~298	7.00-10.90	~0° ~0°		p-type silicon oxidized at 1423 K in 1 atm moist oxygen; coating thickness 1.17 μm; substrate thickness 250 μm; crystal substrate cut along (111) plane and polished; data extracted from smooth curve.
3	289	1964	298	2.50-16.66	~0° ~0°		SiO_2 film prepared by thermal oxidation; data extracted from smooth curve.
4	289	1964	298	2.53-16.66	~0° ~0°		Similar to the above specimen and conditions except doped with a phosphorous concentration of ~4.5×10^{21} cm^{-3}.

DATA TABLE NO. 614 NORMAL SPECTRAL TRANSMITTANCE OF SILICON CONVERSION COATINGS

[Wavelength, λ, μm; Transmittance, τ; Temperature, T, K]

CURVE 1
T ~ 298

λ	τ
5.55	0.738
5.75	0.723
5.78	0.702
5.85	0.698
5.98	0.657
6.05	0.659
6.12	0.638
6.15	0.653
6.32	0.672
6.36	0.653
6.39	0.664
6.41	0.650
6.47	0.665
6.62	0.629
6.71	0.634
6.84	0.601
6.89	0.558
6.94	0.568
7.04	0.521
7.16	0.509
7.36	0.557
7.66	0.527
7.88	0.479
8.27	0.285
8.46	0.239
8.84	0.184
9.08	0.080
9.20	0.054
9.35	0.071
9.61	0.234
9.78	0.489
9.86	0.545
9.99	0.568
10.20	0.550
10.51	0.427
10.72	0.409
11.06	0.485
11.50	0.569
11.79	0.580
12.53	0.465
13.05	0.527
14.12	0.545
14.26	0.535

CURVE 2
T ~ 298

λ	τ
7.00	0.555
7.28	0.526
7.60	0.457
7.83	0.387
7.99	0.323
8.15	0.265
8.39	0.213
8.66	0.165
8.76	0.129
8.89	0.054
8.95	0.031
9.13	0.013
9.33	0.026
9.50	0.057
9.56	0.111
9.88	0.453
9.97	0.504
10.05	0.550
10.19	0.583
10.37	0.600
10.55	0.611
10.78	0.604
10.90	0.595

CURVE 3
T = 298

λ	τ
2.50	0.800
2.66	0.807
2.81	0.818
3.04	0.844
3.17	0.833
3.20	0.864
3.26	0.857
3.31	0.898
3.52	0.923
3.85	0.984
4.07	1.000
4.59	0.989
5.00	0.927
5.13	0.926
5.19	0.931
5.52	0.917
5.73	0.934
5.91	0.930

CURVE 3 (cont.)

λ	τ
6.00	0.936
6.10	0.931
6.47	0.946
6.72	0.972
7.30	0.982
7.47	0.973
7.58	0.935
7.89	0.701
8.15	0.378
8.27	0.280
9.09	0.012
9.31	0.024
9.52	0.099
9.96	0.453
10.35	0.529
11.09	0.617
11.52	0.684
11.84	0.678
12.00	0.550
12.56	0.471
12.78	0.494
13.38	0.667
13.77	0.706
14.59	0.756
14.83	0.811
15.01	0.788
15.22	0.891
16.00	0.941
16.66	0.919

CURVE 4
T = 298

λ	τ
2.53	0.345
2.97	0.372
3.44	0.414
4.00	0.426
4.71	0.454
5.15	0.446
5.36	0.453
5.62	0.487
6.04	0.548
6.28	0.635
7.01	0.960
7.13	0.949
7.19	0.870

CURVE 4 (cont.)

λ	τ
7.35	0.264
7.40	0.180
7.50	0.138
7.62	0.181
7.80	0.406
7.95	0.188
7.98	0.131
8.16	0.084
8.65	0.032
9.04	0.012
9.77	0.030
10.11	0.052
10.42	0.086
11.01	0.165
11.35	0.176
11.77	0.144
12.12	0.108
12.73	0.124
14.18	0.218
15.08	0.314
15.72	0.312
16.66	0.289

SPECIFICATION TABLE NO. 615 NORMAL SPECTRAL REFLECTANCE OF TITANIUM CONVERSION COATINGS

Curve No.	Ref. No.	Year	Temperature, K	Wavelength Range, μm	Geometry θ	θ'	ω'	Reported Error, %	Composition (weight percent), Specifications, and Remarks
1*	109	1962	298	2.0-20.0	~0°		2π		Titanium oxide coating (0.15 μm thick); aluminum foil substrate; produced by evaporating titanium on the substrate and oxidizing at 672 K for 3 hrs in air.

DATA TABLE NO. 615 NORMAL SPECTRAL REFLECTANCE OF TITANIUM CONVERSION COATINGS

[Wavelength, λ, μm; Reflectance, ρ; Temperature, T, K]

λ	ρ
CURVE 1* T = 298	
2.0	0.522
10.0	0.875
20.0	0.950

* No plot given

3. CONVERSION COATINGS (continued)

C. Miscellaneous

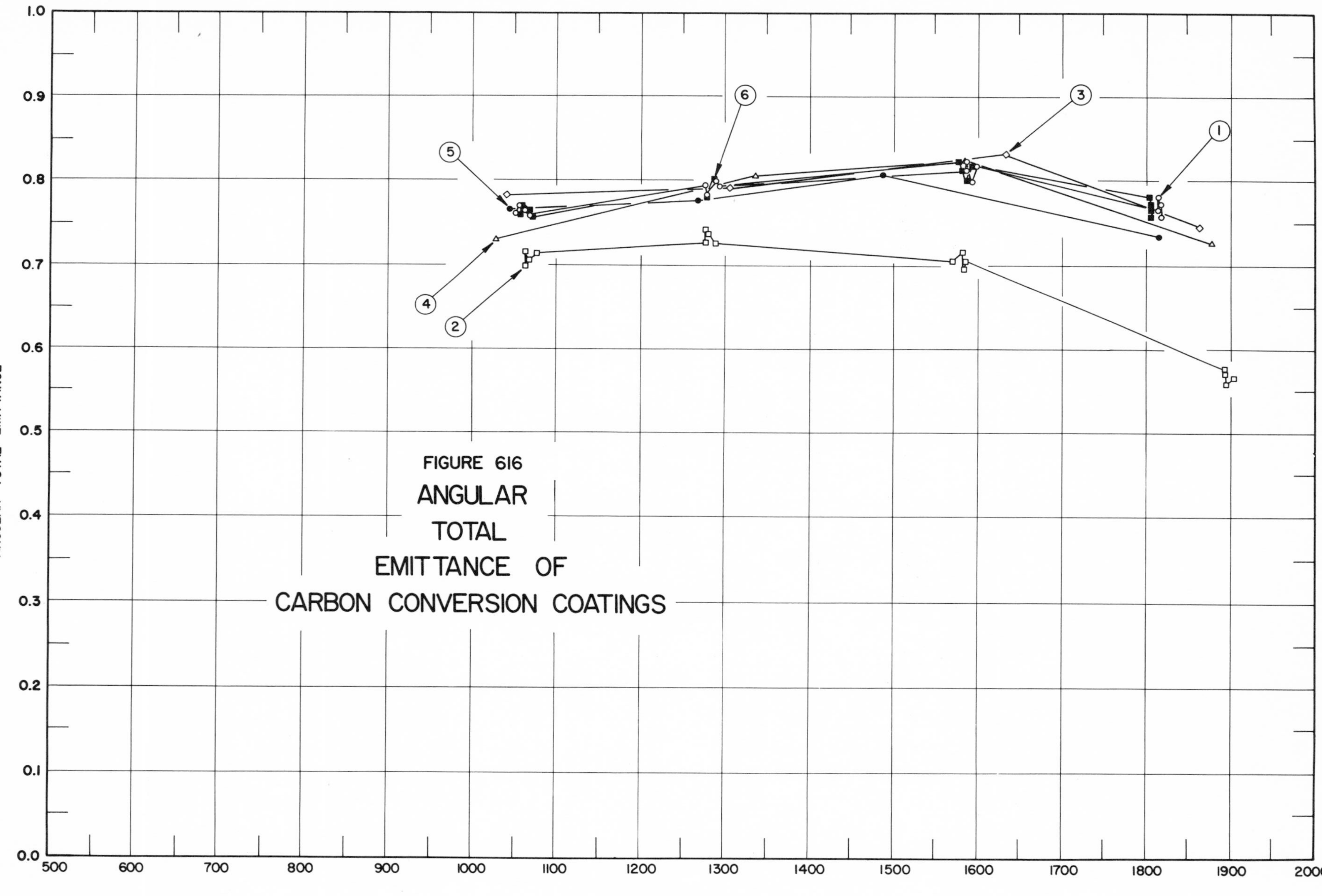

FIGURE 616 ANGULAR TOTAL EMITTANCE OF CARBON CONVERSION COATINGS

SPECIFICATION TABLE NO. 616 ANGULAR TOTAL EMITTANCE OF CARBON CONVERSION COATINGS

Curve No.	Ref. No.	Year	Temperature Range, K	Geometry θ'	Reported Error, %	Composition (weight percent), Specifications and Remarks
1	276	1960	1050-1816	~0°	±5	Siliconized ATJ graphite; measured in 90 Ar + 10 H_2 atm at 78 cm Hg.
2	276	1960	1061-1904	0°	±5	Siliconized ATJ graphite; oxidized 1 hr in air at 1366 K; measured in 90 Ar + 10 H_2 atm at 78 cm Hg.
3	276	1960	1040-1861	30°	±5	Siliconized ATJ graphite; measured in 90 Ar + 10 H_2 atm at 78 cm Hg.
4	276	1960	1029-1879	45°	±5	Above specimen and conditions.
5	276	1960	1042-1814	60°	±5	Above specimen and conditions.
6	283	1960	1055-1803	~0°		Siliconized graphite; rough surface; measured in 90 Ar + 10 H_2 atm.

DATA TABLE NO. 616 ANGULAR TOTAL EMITTANCE OF CARBON CONVERSION COATINGS

[Temperature, T, K; Emittance, ∈]

T	∈	T	∈	T	∈	T	∈
CURVE 1		CURVE 2 (cont.)		CURVE 3 (cont.)		CURVE 6 (cont.)	
1050	0.762	1061	0.717	1631	0.832	1059	0.772
1055	0.772	1066	0.708	1861	0.746	1067	0.767
1061	0.766	1076	0.716			1070	0.759
1068	0.760	1276	0.728	CURVE 4		1276	0.796*
1276	0.796	1276	0.743			1277	0.781
1279	0.783	1280	0.736	1029	0.731	1285	0.802
1289	0.800	1289	0.728	1335	0.806	1291	0.793*
1294	0.793	1570	0.706	1585	0.825	1578	0.824
1588	0.812	1581	0.716	1879	0.727	1580	0.813
1588	0.824	1582	0.697			1585	0.801
1593	0.800	1585	0.707	CURVE 5		1593	0.819
1599	0.819	1892	0.578			1801	0.782
1813	0.781	1892	0.570	1042	0.767	1803	0.774
1813	0.768	1893	0.559	1267	0.778	1803	0.768
1816	0.774	1904	0.567	1487	0.808	1803	0.759
1816	0.759			1814	0.733		
		CURVE 3					
CURVE 2				CURVE 6			
		1040	0.784				
1061	0.700	1305	0.793	1055	0.761		

* Not shown on plot

SPECIFICATION TABLE NO. 617 ANGULAR TOTAL EMITTANCE OF CARBON CONVERSION COATINGS

Curve No.	Ref. No.	Year	Temperature, K	Angular Range, °	Geometry θ'	Reported Error, %	Composition (weight percent), Specifications, and Remarks
1*	283	1960	1050	0-60	θ'		Siliconized graphite; rough surface; measured in 90 argon-10 hydrogen atm.
2*	283	1960	1600	0-60	θ'		Above specimen and conditions.

DATA TABLE NO. 617 ANGULAR TOTAL EMITTANCE OF CARBON CONVERSION COATINGS

[Angle, θ', °; Emittance, $\in$; Temperature, T, K]

θ'	$\in$
CURVE 1*	
T = 1050	
0	0.769
30	0.787
45	0.735
60	0.769
CURVE 2*	
T = 1600	
0	0.811
30	0.833
45	0.827
60	0.810

* No plot given

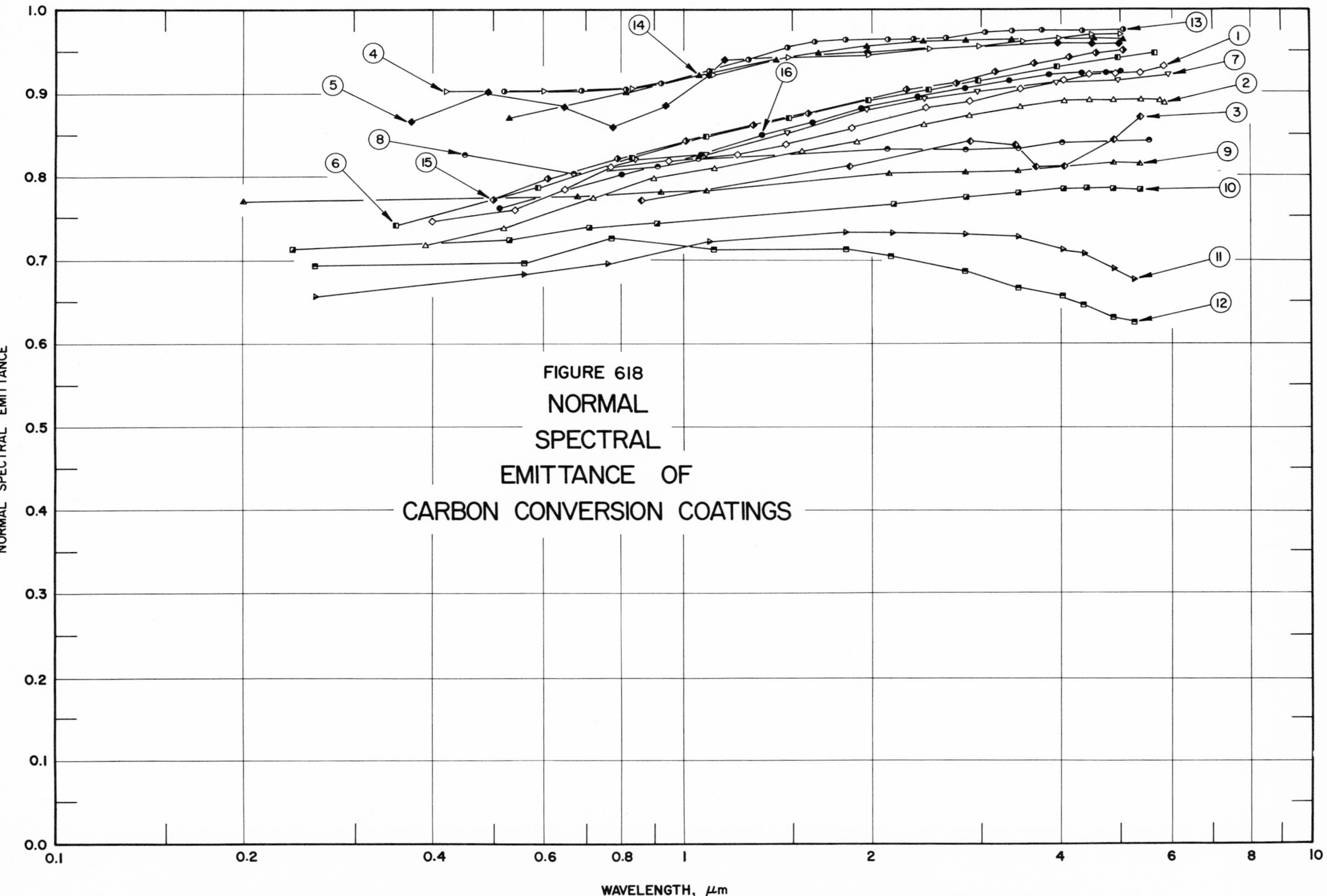

FIGURE 618 NORMAL SPECTRAL EMITTANCE OF CARBON CONVERSION COATINGS

SPECIFICATION TABLE NO. 618 NORMAL SPECTRAL EMITTANCE OF CARBON CONVERSION COATINGS

Curve No.	Ref. No.	Year	Temperature, K	Wavelength Range, μm	Geometry θ'	Reported Error, %	Composition (weight percent), Specifications, and Remarks
1	278	1963	2077	0.40-5.82	$\sim 0°$		Siliconized ATJ graphite from National Carbon Co.; polished with 240, 400, and 600 grit carbide paper, silk cloth, and felt cloth; washed in acetone, then alcohol, and dried with dry nitrogen; measured in argon (77.6 mm Hg).
2	278	1963	2241	0.39-5.82	$\sim 0°$		Above specimen and conditions.
3	278	1963	2360	0.86-5.38	$\sim 0°$		Above specimen and conditions.
4	278	1963	1573	0.42-4.97	$\sim 0°$		Siliconized ATJ graphite from National Carbon Co.; polished with 240, 400, and 600 grit carbide paper, silk cloth, and felt cloth; washed in acetone, then alcohol, and dried with dry nitrogen; measured in argon (1.33 atm); data extracted from smooth curve.
5	278	1963	1781	0.37-4.99	$\sim 0°$		Above specimen and conditions.
6	278	1963	2000	0.35-5.67	$\sim 0°$		Above specimen and conditions.
7	278	1963	2130	0.35-5.94	$\sim 0°$		Above specimen and conditions.
8	278	1963	1435	0.45-5.57	$\sim 0°$		Siliconized ATJ graphite from National Carbon Co.; polished with 240, 400, and 600 grit carbide paper, silk cloth, and felt cloth; washed in acetone, then alcohol, and dried with dry nitrogen; measured in argon (77.6 mm Hg).
9	278	1963	1830	0.20-5.38	$\sim 0°$		Above specimen and conditions.
10	278	1963	2040	0.24-5.36	$\sim 0°$		Above specimen and conditions.
11	278	1963	2236	0.26-5.21	$\sim 0°$		Above specimen and conditions.
12	278	1963	2270	0.26-5.22	$\sim 0°$		Above specimen and conditions.
13	284	1961	1600	0.52-5.01	$\sim 0°$		Siliconized ATJ graphite; measured in argon (250-760 mm Hg); data extracted from smooth curve.
14	284	1961	1800	0.53-5.01	$\sim 0°$		Above specimen and conditions.
15	284	1961	2000	0.50-5.01	$\sim 0°$		Above specimen and conditions.
16	284	1961	2150	0.51-5.00	$\sim 0°$		Above specimen and conditions.

DATA TABLE NO. 618 NORMAL SPECTRAL EMITTANCE OF CARBON CONVERSION COATINGS

[Wavelength, λ, μm; Emittance, ∈; Temperature, T, K]

λ	∈
CURVE 1	
T = 2077	
0.40	0.748
0.54	0.760
0.65	0.785
0.77	0.811
0.95	0.820
1.23	0.827
1.46	0.840
1.86	0.860
2.43	0.883
2.88	0.891
3.47	0.906
4.06	0.918
4.45	0.923
4.90	0.924
5.38	0.927
5.82	0.935
CURVE 2	
T = 2241	
0.39	0.719
0.52	0.739
0.72	0.775
0.90	0.799
1.13	0.810
1.55	0.831
1.90	0.848
2.44	0.864
2.87	0.874
3.46	0.884
4.05	0.891
4.43	0.893
4.89	0.893
5.37	0.893
5.75	0.893
5.82	0.890
CURVE 3	
T = 2360	
0.86	0.771
1.85	0.811
2.88	0.843
3.40	0.840
3.67	0.811
CURVE 3 (cont.)	
4.07	0.813
4.88	0.849
5.38	0.874
CURVE 4	
T = 1573	
0.42	0.903
0.60	0.903
0.83	0.907
1.12	0.923
1.47	0.945
1.98	0.949
2.47	0.955
2.97	0.958
3.48	0.962
3.99	0.968
4.48	0.971
4.97	0.971
CURVE 5	
T = 1781	
0.37	0.869
0.49	0.901
0.65	0.885
0.77	0.860
0.94	0.887
1.11	0.923
1.17	0.941
1.47	0.946*
1.98	0.950
2.47	0.955*
2.96	0.958*
3.48	0.960*
3.98	0.960
4.51	0.960
4.99	0.960
CURVE 6	
T = 2000	
0.35	0.741
0.59	0.788
0.83	0.825
1.09	0.850
CURVE 6 (cont.)	
1.48	0.871
1.99	0.893
2.48	0.906
2.97	0.917
3.97	0.933
4.98	0.945
5.67	0.950
CURVE 7	
T = 2130	
0.35	0.741*
0.58	0.789*
0.84	0.821
1.09	0.829
1.48	0.854
1.98	0.882
2.47	0.895
2.98	0.903
3.99	0.915
4.98	0.918
5.94	0.922
CURVE 8	
T = 1435	
0.45	0.828
0.67	0.804
0.91	0.812
1.06	0.821
2.14	0.832
2.82	0.832
3.42	0.836
4.01	0.841
5.57	0.844
CURVE 9	
T = 1830	
0.20	0.770
0.68	0.776
0.92	0.781
1.09	0.782
2.14	0.804
2.81	0.806
3.42	0.808
CURVE 9 (cont.)	
4.01	0.813*
4.88	0.819
5.36	0.818
CURVE 10	
T = 2040	
0.24	0.714
0.53	0.725
0.71	0.738
0.91	0.743
2.17	0.768
2.81	0.775
3.43	0.780
4.02	0.785
4.40	0.786
4.88	0.786
5.36	0.783
CURVE 11	
T = 2236	
0.26	0.658
0.56	0.683
0.76	0.697
1.10	0.721
1.82	0.732
2.16	0.732
2.81	0.730
3.42	0.729
4.03	0.712
4.38	0.709
4.89	0.690
5.21	0.678
CURVE 12	
T = 2270	
0.26	0.695
0.56	0.698
0.77	0.726
1.13	0.711
1.83	0.712
2.15	0.705
2.81	0.689
3.43	0.669
CURVE 12 (cont.)	
4.02	0.659
4.39	0.649
4.88	0.631
5.22	0.627
CURVE 13	
T = 1600	
0.52	0.902
0.69	0.903
0.81	0.906
0.92	0.914
1.10	0.929
1.28	0.942
1.47	0.956
1.63	0.961
1.81	0.963
2.12	0.964
2.33	0.965
2.63	0.968
3.03	0.974
3.35	0.975
3.71	0.976
4.31	0.978
5.01	0.979
CURVE 14	
T = 1800	
0.53	0.872
0.81	0.901
1.07	0.922
1.41	0.941
1.65	0.950
1.97	0.957
2.41	0.962
2.81	0.964
3.37	0.966
3.99	0.967*
4.52	0.968
5.01	0.968
CURVE 15	
T = 2000	
0.50	0.771
CURVE 15 (cont.)	
0.61	0.799
0.79	0.823
1.01	0.843
1.30	0.863
1.59	0.878
1.99	0.894*
2.29	0.903
2.72	0.914
3.18	0.928
3.64	0.938
4.11	0.945
4.58	0.950
5.01	0.952
CURVE 16	
T = 2150	
0.51	0.761
0.80	0.801
1.07	0.829
1.34	0.851
1.61	0.868
1.94	0.884
2.38	0.898
2.81	0.908
3.32	0.917
3.81	0.922
4.32	0.926
4.72	0.927
5.00	0.928

* Not shown on plot

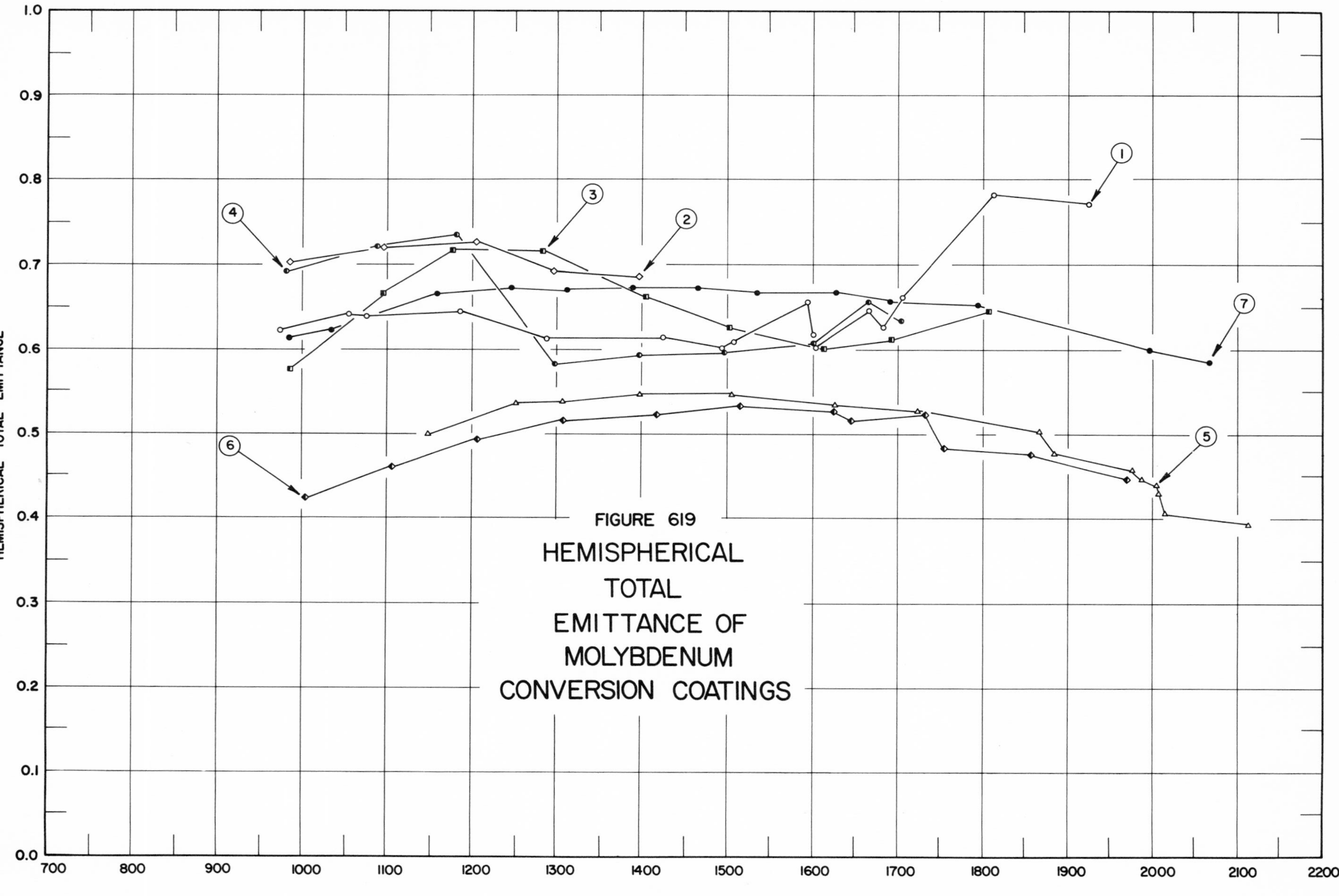

FIGURE 619
HEMISPHERICAL TOTAL EMITTANCE OF MOLYBDENUM CONVERSION COATINGS

SPECIFICATION TABLE NO. 619 HEMISPHERICAL TOTAL EMITTANCE OF MOLYBDENUM CONVERSION COATINGS

Curve No.	Ref. No.	Year	Temperature Range, K	Reported Error, %	Composition (weight percent), Specifications and Remarks
1	275	1963	974-1925		CV-IV (Chance Vought Corp.); silicon diffusion coating on molybdenum; measured in air. [Author's designation: Sample No. 1]
2	275	1963	984-1395		Similar to above specimen and conditions. [Author's designation: Sample No. 2]
3	275	1963	987-1808		Similar to above specimen and conditions. [Author's designation: Sample No. 3]
4	275	1963	980-1702		Similar to above specimen and conditions. [Author's designation: Sample No. 4]
5	275	1963	1147-2112		Durak-B (Chromizing Corp.); silicon diffusion coating on molybdenum; measured in air. [Author's designation: Sample No. 1]
6	275	1963	1002-1970		Similar to above specimen and conditions. [Author's designation: Sample No. 2]
7	275	1963	982-2068		PRF-6 (Pfaudler Corp.); silicon diffusion coating on molybdenum; measured in air.

DATA TABLE NO. 619 HEMISPHERICAL TOTAL EMITTANCE OF MOLYBDENUM CONVERSION COATINGS

[Temperature, T, K; Emittance, ∈]

T	∈	T	∈	T	∈	T	∈	T	∈	T	∈
CURVE 1		CURVE 2		CURVE 4		CURVE 5 (cont.)		CURVE 6 (cont.)		CURVE 7 (cont.)	
974	0.622	984	0.704	980	0.692	1722	0.527	1516	0.531	1627	0.668
1054	0.644	1095	0.721	1089	0.722	1867	0.501	1627	0.527	1690	0.657
1075	0.640	1201	0.728	1180	0.735	1883	0.476	1649	0.516	1793	0.651
1185	0.646	1296	0.692	1298	0.583	1885	0.470	1732	0.522	1998	0.600
1289	0.613	1395	0.686	1398	0.594	1978	0.457	1755	0.483	2068	0.587
1423	0.614			1497	0.599	1987	0.446	1858	0.476		
1491	0.601	CURVE 3		1600	0.608	2004	0.440	1970	0.446		
1509	0.610			1666	0.657	2006	0.430				
1593	0.657	987	0.578	1702	0.633	2104	0.405	CURVE 7			
1600	0.620	1096	0.666			2112	0.391				
1601	0.601	1177	0.719	CURVE 5				982	0.615		
1667	0.649	1282	0.719			CURVE 6		1032	0.624		
1681	0.628	1404	0.663	1147	0.500			1159	0.668		
1705	0.661	1503	0.626	1250	0.536	1002	0.425	1245	0.672		
1812	0.782	1612	0.602	1305	0.539	1105	0.460	1310	0.670		
1925	0.772	1692	0.612	1399	0.547	1203	0.494	1389	0.674		
		1808	0.646	1505	0.547	1307	0.518	1463	0.673		
				1629	0.533	1417	0.523	1532	0.669		

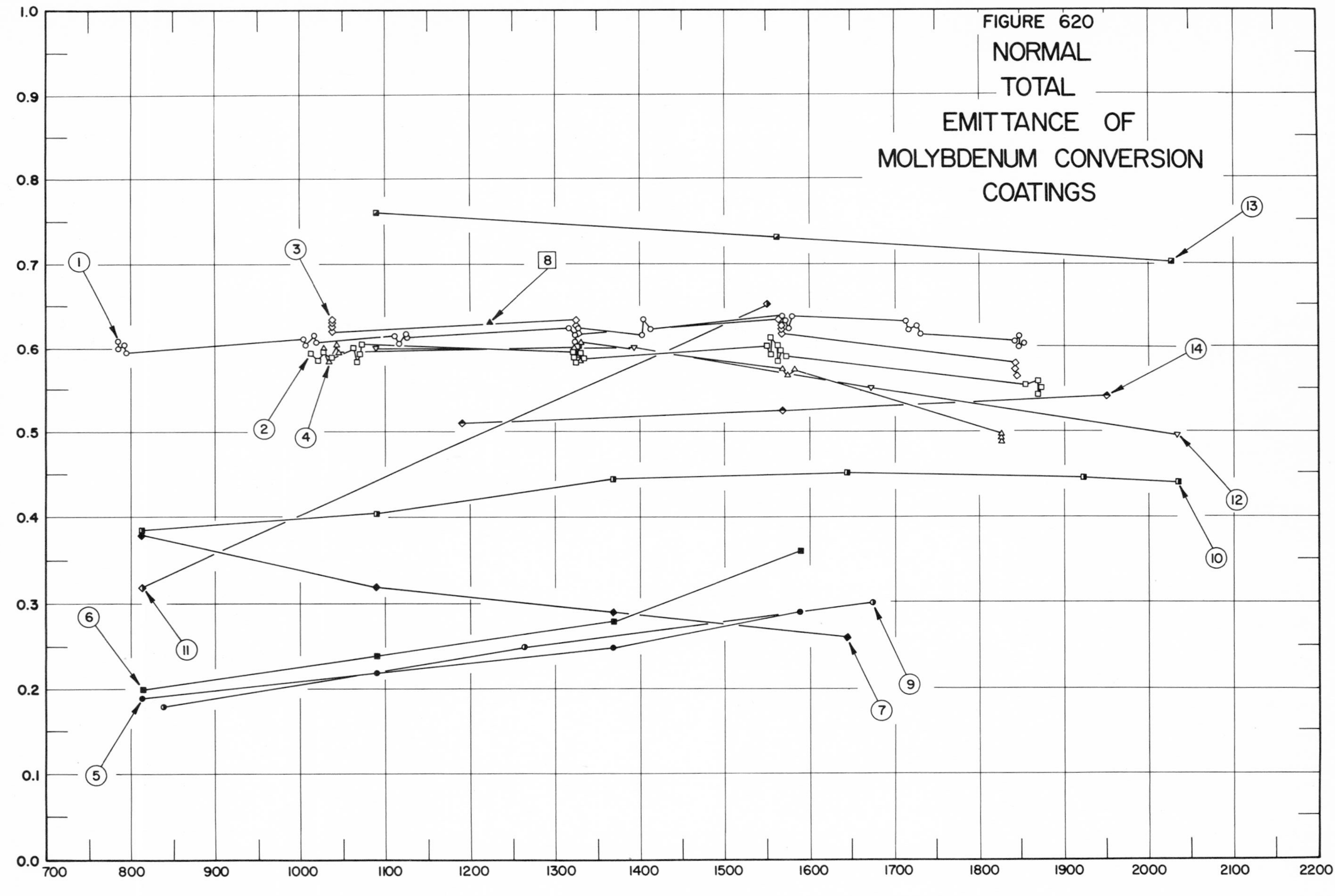
FIGURE 620
NORMAL
TOTAL
EMITTANCE OF
MOLYBDENUM CONVERSION
COATINGS
NORMAL TOTAL EMITTANCE
TEMPERATURE, K
0.0
0.1
0.2
0.3
0.4
0.5
0.6
0.7
0.8
0.9
1.0
700
800
900
1000
1100
1200
1300
1400
1500
1600
1700
1800
1900
2000
2100
2200
1
2
3
4
5
6
7
8
9
10
11
12
13
14

SPECIFICATION TABLE NO. 620 NORMAL TOTAL EMITTANCE OF MOLYBDENUM CONVERSION COATINGS

Curve No.	Ref. No.	Year	Temperature Range, K	Geometry θ'	Reported Error, %	Composition (weight percent), Specifications and Remarks
1	276	1960	785-1853	~0°	±5	W-2 (Chromalloy Corp.); silicon diffusion coating (0.102 mm thick) on Mo + 0.5 Ti; measured in 90 Ar + 10 H_2 atm at 78 cm Hg.
2	276	1960	1012-1873	~0°	±5	Similar to above specimen and conditions except substrate pre-oxidized.
3	276	1960	1036-1843	~0°	±5	Durak MG (Chromizing Corp.); silicon diffusion coating (0.051 mm thick) on Mo + 0.5 Ti; measured in 90 Ar + 10 H_2 atm at 78 cm Hg.
4	276	1960	1028-1827	~0°	±5	Similar to above specimen and conditions except substrate pre-oxidized 1 hr in air at 1366 K.
5	178	1960	811-1589	~0°	±20	W-2 (Chromalloy Corp.); silicon diffusion coating (0.038 mm thick) on molybdenum alloy; measured in demoisturized helium atm.
6	178	1960	811-1589	~0°	±20	Similar to above specimen and conditions except 0.064 mm thick.
7	178	1960	811-1644	~0°	±20	Similar to above specimen and conditions except coating oxidized.
8	238	1963	1223	~0°		Pack cementation coating (0.051 mm thick) on molybdenum substrate; pack composition: 73 Al_2O_3, 23 Si and 4 NaI; prepared by heating the pack in an inert atm at 1367 K for 16 hrs; integrated from spectral data over the range 1-15 μm. [Authors' designation: Sample No. 129]
9	277	1962	839-1672	~0°		W-2, with graphite (Chromalloy Corp.); silicon diffusion coating on molybdenum; measured in inert atm; data extracted from smooth curve. [Author's designation: Specimen 1, Run 1]
10	277	1962	811-2033	~0°		Above specimen and conditions. [Author's designation: Specimen 1, Run 8]
11	277	1962	811-1550	~0°		Similar to above specimen and conditions. [Author's designation: Specimen 2, Run 1]
12	277	1962	1089-2033	~0°		Above specimen and conditions. [Author's designation: Specimen 2, Run 2]
13	277	1962	1089-2028	~0°		Similar to above specimen and conditions. [Author's designation: Specimen 3, Run 2]
14	277	1962	1189-1950	~0°		Similar to above specimen and conditions. [Author's designation: Specimen 4, Run 2]

DATA TABLE NO. 620 NORMAL TOTAL EMITTANCE OF MOLYBDENUM CONVERSION COATINGS

[Temperature, T, K; Emittance, ∈]

T	∈
CURVE 1	
785	0.600
785	0.610
793	0.604
795	0.595
1004	0.611
1005	0.604
1015	0.616
1018	0.608
1110	0.615
1115	0.607
1123	0.619
1123	0.612
1318	0.624
1323	0.617
1323	0.609
1328	0.624
1401	0.617
1404	0.634
1411	0.624
1568	0.639
1571	0.631
1577	0.623
1580	0.638
1714	0.631
1718	0.622
1729	0.627
1731	0.619
1841	0.607
1849	0.614
1849	0.600
1853	0.605
CURVE 2	
1012	0.595
1020	0.586
1026	0.597
1036	0.590
1061	0.601
1066	0.585
1070	0.594
1071	0.607
1321	0.597
1321	0.590
1324	0.583
1325	0.604
CURVE 2 (cont.)	
1330	0.595
1334	0.589
1550	0.601
1555	0.593
1555	0.611
1561	0.604
1561	0.584
1566	0.598
1571	0.590
1855	0.555
1870	0.560
1870	0.543
1873	0.551
CURVE 3	
1036	0.635
1036	0.631
1036	0.626
1036	0.620
1325	0.626
1325	0.634
1326	0.619
1564	0.634
1568	0.626
1568	0.622
1568	0.617
1841	0.581
1841	0.575
1843	0.566
CURVE 4	
1028	0.600
1033	0.586
1041	0.604
1044	0.595
1324	0.601
1328	0.592
1328	0.585
1330	0.606
1569	0.575
1574	0.567
1581	0.573
1827	0.498
1827	0.493
CURVE 4 (cont.)	
1827	0.488
CURVE 5	
811	0.19
1089	0.22
1367	0.25
1589	0.29
CURVE 6	
811	0.20
1089	0.24
1367	0.28
1589	0.36
CURVE 7	
811	0.38
1089	0.32
1367	0.29
1644	0.26
CURVE 8	
1223	0.63
CURVE 9	
839	0.180
1261	0.250
1672	0.300
CURVE 10	
811	0.385
1089	0.403
1366	0.445
1644	0.450
1922	0.445
2033	0.440
CURVE 11	
811	0.320
1550	0.650
CURVE 12	
1089	0.600
1394	0.600
1672	0.550
2033	0.495
CURVE 13	
1089	0.76
1561	0.73
2028	0.70
CURVE 14	
1189	0.510
1569	0.525
1950	0.540

SPECIFICATION TABLE NO. 621 ANGULAR TOTAL EMITTANCE OF MOLYBDENUM CONVERSION COATINGS

Curve No.	Ref. No.	Year	Temperature Range, K	Geometry θ'	Reported Error, %	Composition (weight percent), Specifications and Remarks
1*	276	1960	1031-1599	30°	±5	W-2 (Chromalloy Corp.); silicon diffusion coating (0.102 mm thick) on Mo + 0.5 Ti; measured in 90 Ar + 10 H_2 atm at 78 cm Hg.
2*	276	1960	1046-1588	45°	±5	Above specimen and conditions.
3*	276	1960	1079-1597	60°	±5	Above specimen and conditions.
4*	276	1960	1034-1578	45°	±5	Similar to curve 1 specimen and conditions except substrate was pre-oxidized.
5*	276	1960	1063-1596	60°	±5	Above specimen and conditions.
6*	276	1960	1033-1591	30°	±5	Durak MG (Chromizing Corp.); silicon diffusion coating (0.051 mm thick) on Mo + 0.5 Ti; measured in 90 Ar + 10 H_2 atm at 78 cm Hg.
7*	276	1960	1066-1569	45°	±5	Above specimen and conditions.
8*	276	1960	1086-1584	60°	±5	Above specimen and conditions.

DATA TABLE NO. 621 ANGULAR TOTAL EMITTANCE OF MOLYBDENUM CONVERSION COATINGS

[Temperature, T, K; Emittance, ϵ]

T	ϵ	T	ϵ
CURVE 1*		CURVE 5*	
1031	0.616	1063	0.608
1599	0.640	1596	0.586
CURVE 2*		CURVE 6*	
1046	0.579	1033	0.620
1588	0.647	1591	0.621
CURVE 3*		CURVE 7*	
1079	0.632	1066	0.629
1597	0.632	1569	0.639
CURVE 4*		CURVE 8*	
1034	0.602	1086	0.628
1578	0.593	1584	0.651

* No plot given

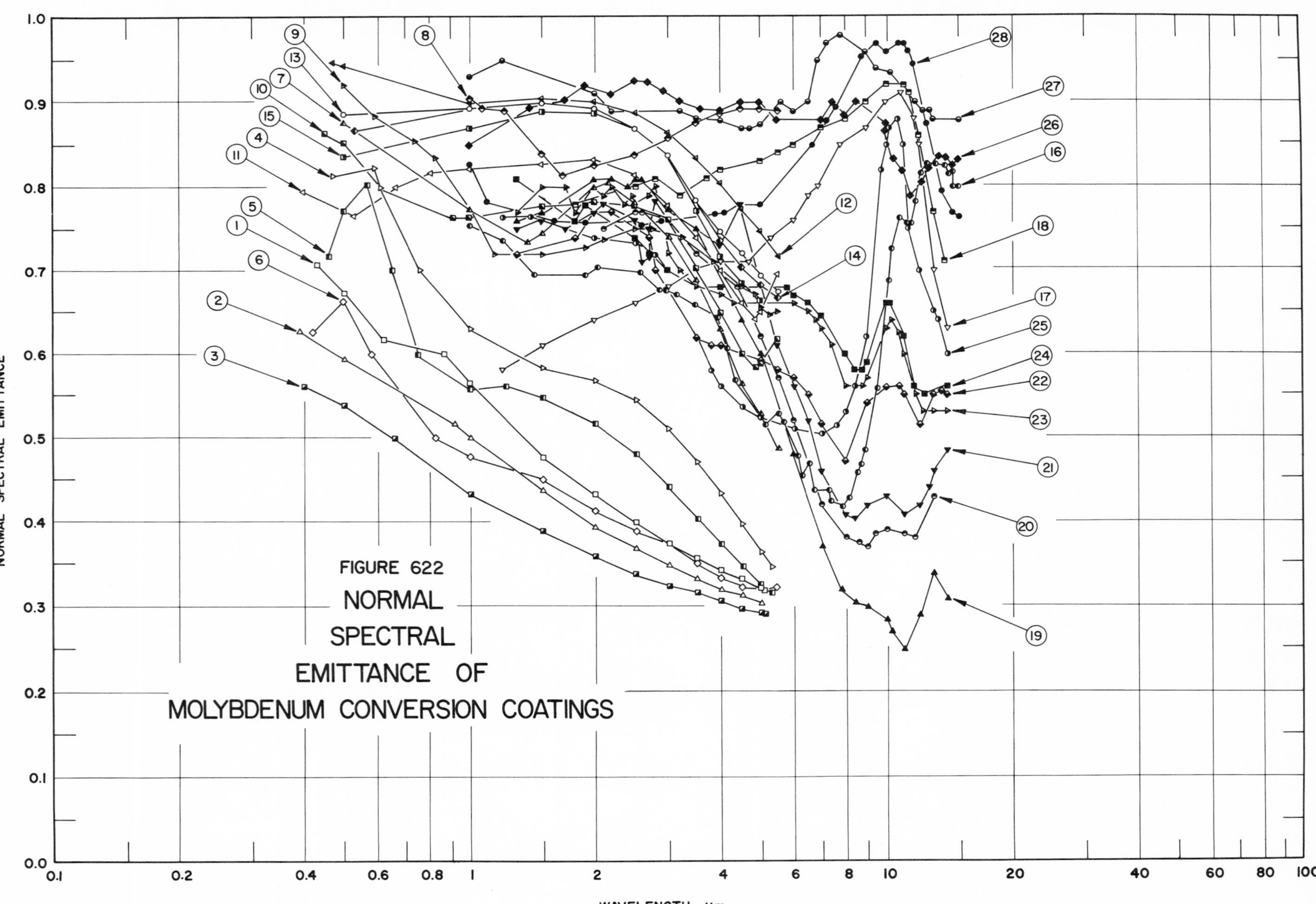

FIGURE 622 NORMAL SPECTRAL EMITTANCE OF MOLYBDENUM CONVERSION COATINGS

SPECIFICATION TABLE NO. 622 NORMAL SPECTRAL EMITTANCE OF MOLYBDENUM CONVERSION COATINGS

Curve No.	Ref. No.	Year	Temperature, K	Wavelength Range, μm	Geometry θ'	Reported Error, %	Composition (weight percent), Specifications, and Remarks
1	278	1963	1603	0.43-5.05	~0°		W-2 (Chromalloy Corp.); silicon diffusion coating on molybdenum substrate; washed in acetone, then alcohol, and dried with dry nitrogen; measured in argon (0.33 atm); data extracted from smooth curve; authors observed transformation of $MoSi_2$ to Mo_5Si_3. [Authors' designation: Specimen No. 4]
2	278	1963	1775	0.39-5.05	~0°		Above specimen and conditions.
3	278	1963	1905	0.40-5.05	~0°		Above specimen and conditions.
4	278	1963	1400	0.47-5.29	~0°		Similar to curve 1 specimen and conditions. [Authors' designation: Specimen No. 5]
5	278	1963	1600	0.46-5.30	~0°		Above specimen and conditions.
6	278	1963	1800	0.42-5.42	~0°		Above specimen and conditions.
7	278	1963	1400	0.50-5.50	~0°		Similar to above specimen and conditions except measured in air (1 atm). [Authors' designation: Specimen No. 2]
8	278	1963	1400	1.00-5.50	~0°		Above specimen and conditions except heated at 1400 K for 6 hrs.
9	278	1963	1400	0.50-5.50	~0°		Similar to curve 7 specimen and conditions. [Authors' designation: Specimen No. 6, Run 1]
10	278	1963	1600	0.45-5.50	~0°		Above specimen and conditions except held at 1400 K for 2 hrs and rerun at 1400 K before 1600 K run; first run.
11	278	1963	1800	0.40-5.50	~0°		Above specimen and conditions except held at 1600 K for 2 hrs and rerun at 1600 K before 1800 K run.
12	278	1963	1600	0.47-5.50	~0°		Similar to curve 9 specimen and conditions. [Authors' designation: Specimen No. 7]
13	278	1963	1600	0.50-5.50	~0°		Above specimen and conditions except heated at 1600 K for 2 hrs.
14	278	1963	1600	0.53-5.50	~0°		Above specimen and conditions except heated at 1600 K for 4.5 hrs.
15	278	1963	1600	0.50-5.50	~0°		Above specimen and conditions except heated at 1600 K for 7 hrs.
16	238	1963	1223	1.00-14.99	~0°		Pack cementation coating (0.051 mm thick) on molybdenum substrate; pack composition: 73 Al_2O_3, 23 Si, 4 NaI; prepared by heating the pack in an inert atm at 1367 K for 16 hrs; data extracted from smooth curve. [Authors' designation: Sample No. 129]
17	279	1963	1030	1.2-14.0	~0°		CV II + IX + TiO_2 (Chance Vought Corp.); heated at 1030 K for 4 hrs; measured in argon; data extracted from smooth curve.
18	279	1963	1295	1.8-13.8	~0°		Above specimen and conditions except heated at 778 K for 2 hrs.
19	279	1963	1044	1.3-14.0	~0°		W-2 (Chromalloy Corp.); silicon diffusion coating on molybdenum substrate; heated at 1000 K for 2 hrs; measured in argon; data extracted from smooth curve. [Author's designation: Sample 1]
20	279	1963	1172	2.1-13.0	~0°		Above specimen and conditions except heated at 1168 K for 27 hrs.
21	279	1963	1336	1.3-14.0	~0°		Above specimen and conditions except heated at 1333 K for 20 hrs.
22	279	1963	1386	1.3-14.0	~0°		Above specimen and conditions except heated at 1386 K for 3 hrs.
23	279	1963	1117	1.3-14.0	~0°		Above specimen and conditions except heated at 1111 K for 5 hrs.

SPECIFICATION TABLE NO. 622 NORMAL SPECTRAL EMITTANCE OF MOLYBDENUM CONVERSION COATINGS (continued)

Curve No.	Ref. No.	Year	Temperature, K	Wavelength Range, μm	Geometry θ'	Reported Error, %	Composition (weight percent), Specifications, and Remarks
24	279	1963	1283	1.3-14.0	$\sim 0°$		Above specimen and conditions except heated at 1277 K for 3 hrs.
25	279	1963	1447	1.2-14.0	$\sim 0°$		Similar to curve 19 specimen and conditions except heated at 1389 K for 2 hrs. [Author's designation: Sample 2]
26	280	1962	922	1.00-15.00	$\sim 0°$		CV II + IX (Chance Vought Corp.); silicon diffusion coating on molybdenum; measured in air; data extracted from smooth curve.
27	280	1962	922	1.00-15.00	$\sim 0°$		CV II + FeB (Chance Vought Corp.); silicon diffusion coating on molybdenum; measured in air; data extracted from smooth curve.
28	280	1962	922	1.00-15.00	$\sim 0°$		CV II + IX + TiO_2 (Chance Vought Corp.); silicon diffusion coating on molybdenum; measured in air; data extracted from smooth curve.

DATA TABLE NO. 622 NORMAL SPECTRAL EMITTANCE OF MOLYBDENUM CONVERSION COATINGS

[Wavelength, λ, μm; Emittance, ϵ; Temperature, T, K]

λ	ϵ
CURVE 1	
T = 1603	
0.43	0.708
0.50	0.672
0.62	0.616
0.87	0.600
1.00	0.564
1.50	0.476
2.00	0.432
2.50	0.400
3.00	0.372
3.50	0.356
4.00	0.342
4.50	0.332
5.00	0.322
5.05	0.320
CURVE 2	
T = 1775	
0.39	0.628
0.50	0.594
0.92	0.516
1.00	0.500
1.50	0.436
2.00	0.394
2.50	0.368
3.00	0.348
3.50	0.332
4.00	0.320
4.50	0.312
5.00	0.304
5.05	0.302
CURVE 3	
T = 1905	
0.40	0.560
0.50	0.538
0.66	0.500
1.00	0.432
1.50	0.388
2.00	0.358
2.50	0.338
3.00	0.324
3.50	0.316
4.00	0.306
CURVE 3 (cont.)	
4.50	0.298
5.00	0.292
5.05	0.292
CURVE 4	
T = 1400	
0.47	0.814
0.59	0.824
0.61	0.800
0.76	0.700
1.00	0.630
1.50	0.582
2.00	0.567
2.50	0.544
3.00	0.510
3.50	0.470
4.00	0.432
4.50	0.396
5.00	0.364
5.29	0.346
CURVE 5	
T = 1600	
0.46	0.718
0.50	0.772
0.57	0.804
0.65	0.700
0.75	0.600
1.00	0.558
1.22	0.560
1.50	0.548
2.00	0.516
2.50	0.480
3.00	0.440
3.50	0.402
4.00	0.372
4.50	0.346
5.00	0.326
5.30	0.316
CURVE 6	
T = 1800	
0.42	0.626
CURVE 6 (cont.)	
0.50	0.662
0.68	0.600
0.83	0.500
1.00	0.476
1.50	0.450
2.00	0.414
2.50	0.388
3.00	0.374*
3.50	0.350
4.00	0.334
4.50	0.322
5.00	0.320*
5.42	0.324
CURVE 7	
T = 1400	
0.50	0.878
1.00	0.772
1.38	0.736
1.50	0.746
2.00	0.800
2.16	0.804
2.50	0.782
3.00	0.740
3.50	0.688
4.00	0.630
4.50	0.562
5.00	0.528
5.50	0.486
CURVE 8	
T = 1400	
1.00	0.904
1.07	0.894
1.21	0.890
1.50	0.840
1.68	0.814
2.00	0.826
2.50	0.836
3.00	0.858
3.50	0.876
4.00	0.886
4.50	0.892
5.00	0.892
CURVE 8 (cont.)	
5.50	0.890
CURVE 9	
T = 1400	
0.50	0.920
0.59	0.885
0.74	0.855
0.83	0.835
1.00	0.775*
1.15	0.720
1.50	0.720
1.90	0.728
2.10	0.737
2.50	0.750
2.76	0.750
3.25	0.740
3.85	0.710
4.50	0.680
4.84	0.670
5.00	0.655
5.24	0.646
5.50	0.650
CURVE 10	
T = 1600	
0.45	0.864
0.50	0.852
0.61	0.800*
0.91	0.762
1.00	0.764
1.50	0.778
2.00	0.782
2.50	0.780
3.00	0.760
3.50	0.702
4.00	0.648
4.50	0.600
4.83	0.584
5.00	0.588
5.50	0.618
CURVE 11	
T = 1800	
0.40	0.796
0.53	0.768
0.67	0.800
0.81	0.818
1.00	0.822
1.50	0.828
2.00	0.832
2.50	0.814
3.00	0.778
3.50	0.740
4.00	0.700
4.50	0.660
4.82	0.640
5.00	0.650
5.50	0.696
CURVE 12	
T = 1600	
0.47	0.948
0.50	0.942
1.00	0.900
1.50	0.902
2.00	0.900
2.50	0.886
3.00	0.864
3.50	0.834
4.00	0.802
4.50	0.774
5.00	0.746
5.50	0.718
CURVE 13	
T = 1600	
0.50	0.886
1.00	0.892
1.50	0.900
2.00	0.894
2.50	0.870
3.00	0.836
3.50	0.784
4.00	0.748
4.50	0.720
5.00	0.694
CURVE 13 (cont.)	
5.50	0.674
CURVE 14	
T = 1600	
0.53	0.868
1.00	0.892*
1.50	0.900*
2.00	0.894*
2.50	0.870*
3.00	0.836*
3.50	0.784*
4.00	0.736
4.50	0.704
5.00	0.682
5.50	0.666
CURVE 15	
T = 1600	
0.50	0.838
1.00	0.870
1.50	0.890
2.00	0.888
2.50	0.870*
3.00	0.836*
3.50	0.772
4.00	0.717
4.50	0.684
5.00	0.664
5.50	0.650*
CURVE 16	
T = 1223	
1.00	0.754
1.20	0.736
1.42	0.695
1.90	0.695
2.01	0.701
2.57	0.697
2.86	0.676
2.99	0.676
3.13	0.670
3.49	0.658
3.95	0.642
CURVE 16 (cont.)	
4.18	0.608
4.36	0.569
4.50	0.562*
5.00	0.528*
5.16	0.515
5.50	0.527
5.63	0.519
5.72	0.516*
6.12	0.478
6.22	0.452
6.51	0.468
6.66	0.465*
6.79	0.436
7.26	0.436
7.39	0.423
7.89	0.419
8.19	0.428
8.51	0.457
8.73	0.467
8.83	0.485
8.91	0.479*
9.50	0.559
10.10	0.689
10.26	0.696*
10.41	0.728
10.87	0.762
11.05	0.757
11.52	0.757
11.80	0.783
11.93	0.805*
12.13	0.818
12.50	0.828
13.01	0.828
13.50	0.824*
13.90	0.824
14.02	0.813
14.40	0.818
14.59	0.800
14.99	0.800
CURVE 17	
T = 1030	
1.2	0.58
1.5	0.61
2.0	0.64
CURVE 17 (cont.)	
2.5	0.66
3.0	0.68
3.5	0.70*
4.0	0.71
4.7	0.71
5.3	0.74
6.0	0.76
6.5	0.79
6.9	0.80
7.8	0.85
9.0	0.87
10.0	0.90
10.8	0.91
11.6	0.88
12.0	0.85
13.0	0.70
14.0	0.63
CURVE 18	
T = 1295	
1.8	0.78
2.5	0.80
2.8	0.81
3.2	0.79
3.7	0.81
4.0	0.82
5.0	0.83
5.5	0.84
6.0	0.85
7.0	0.87
8.0	0.88
9.0	0.90
10.0	0.92
11.0	0.92
11.4	0.91
12.0	0.86
13.0	0.77
13.8	0.71
CURVE 19	
T = 1044	
1.3	0.760
1.5	0.770
2.0	0.810

* Not shown on plot

DATA TABLE NO. 622 NORMAL SPECTRAL EMITTANCE OF MOLYBDENUM CONVERSION COATINGS (continued)

λ	ε
CURVE 19 (cont.) T = 1044	
2.2	0.810
2.4	0.800
2.5	0.810
2.6	0.810
3.0	0.775
3.5	0.750
4.0	0.700*
4.5	0.640
5.0	0.600
6.0	0.480
7.0	0.370
7.8	0.320
8.4	0.305
9.0	0.300
10.0	0.285
10.4	0.270
11.0	0.250
12.0	0.290
13.0	0.340
14.0	0.310
CURVE 20 T = 1172	
2.1	0.750
2.5	0.770
2.6	0.770
3.0	0.760*
3.5	0.720
4.0	0.700*
4.4	0.680
5.0	0.620
5.5	0.570
6.0	0.520
7.0	0.420
8.0	0.380
8.6	0.375
9.0	0.370
9.4	0.385
10.0	0.390
11.0	0.385
11.6	0.380
13.0	0.430
CURVE 21 T = 1336	
1.3	0.750
1.5	0.760
1.7	0.750
2.0	0.770
2.1	0.780
2.5	0.760
2.6	0.710
2.7	0.715
2.7	0.750
2.8	0.785
3.0	0.760*
3.5	0.750*
4.0	0.730
4.5	0.780
5.0	0.650*
5.5	0.610
6.0	0.560
6.5	0.520
7.0	0.460
8.0	0.410
8.4	0.405
9.0	0.420
10.0	0.430
11.0	0.410
12.0	0.420
12.6	0.440
13.0	0.460
14.0	0.485
CURVE 22 T = 1386	
1.3	0.720
1.8	0.740
2.0	0.770*
2.2	0.770
2.7	0.740
2.8	0.700
3.5	0.620
3.8	0.610
4.0	0.610
5.0	0.590
5.5	0.580
6.0	0.570
6.5	0.550
7.0	0.515
8.0	0.470
CURVE 22 (cont.)	
9.0	0.540
10.0	0.560
10.6	0.560
11.0	0.550
12.0	0.515
12.1	0.515*
13.0	0.550
13.6	0.555
14.0	0.550
CURVE 23 T = 1117	
1.3	0.770
1.5	0.800
1.7	0.800
1.8	0.770
2.0	0.800*
2.1	0.790
2.2	0.800
2.35	0.780
2.45	0.790
2.55	0.770*
2.70	0.790
2.80	0.800
3.00	0.780*
3.00	0.720
3.2	0.700
3.5	0.680
4.0	0.670
4.3	0.660
5.0	0.660*
6.0	0.660
6.5	0.650
6.8	0.640
7.0	0.630
7.4	0.610
8.0	0.560
8.8	0.560
9.0	0.570
10.0	0.630
10.2	0.640
10.6	0.625
11.0	0.600
11.6	0.550
12.2	0.530
13.0	0.530
14.0	0.530
CURVE 24 T = 1283	
1.3	0.810
1.8	0.760
1.9	0.780
2.5	0.740
2.7	0.720
3.0	0.700
3.5	0.680*
4.0	0.680
4.5	0.680*
5.0	0.680*
5.8	0.680
6.0	0.670
6.5	0.660
7.0	0.645
8.0	0.600
8.4	0.580
8.8	0.580
9.0	0.590
10.0	0.660
10.2	0.660
11.0	0.620
11.8	0.560
12.4	0.550
14.0	0.560
CURVE 25 T = 1447	
1.2	0.765
1.4	0.765
1.5	0.760*
2.0	0.740
2.5	0.735
2.8	0.720
3.0	0.700*
3.5	0.620*
3.8	0.580
4.0	0.560
4.5	0.535
5.0	0.525
6.0	0.510
7.0	0.505
7.6	0.515
8.0	0.530
8.4	0.560
9.0	0.620
9.8	0.820
CURVE 25 (cont.)	
10.0	0.850
10.2	0.870
10.6	0.880
11.0	0.850
11.4	0.750
12.0	0.700
13.0	0.650
13.2	0.640
14.0	0.600
CURVE 26 T = 922	
1.00	0.850
1.40	0.895
1.70	0.902
1.90	0.920
2.20	0.910
2.50	0.925
2.70	0.925
2.95	0.913
3.20	0.900
3.60	0.892
4.00	0.890
4.50	0.900
5.00	0.900
5.50	0.880
7.00	0.880
7.50	0.900
8.00	0.885
8.50	0.900
9.00	0.900*
10.00	0.875
10.00	0.868
10.50	0.835
11.00	0.820
11.50	0.790
12.20	0.805
12.80	0.825
13.50	0.836
14.00	0.835
14.50	0.828
15.00	0.835
CURVE 27 T = 922	
1.00	0.930
CURVE 27 (cont.)	
1.20	0.950
2.00	0.910
2.20	0.890
3.20	0.890
3.50	0.880
4.00	0.880
4.50	0.870
4.70	0.870
5.00	0.875
5.60	0.900
6.00	0.890
6.50	0.900
6.90	0.950
7.20	0.970
7.80	0.980
9.00	0.960
9.50	0.940
10.20	0.935
11.80	0.900
12.20	0.890
12.80	0.890
13.00	0.880
15.00	0.880
CURVE 28 T = 922	
1.00	0.825
1.10	0.782
1.30	0.766*
1.60	0.760
1.90	0.758
2.50	0.760*
2.60	0.755
2.90	0.760
3.90	0.768
4.10	0.770
4.50	0.780*
5.00	0.780
6.65	0.850
7.20	0.880
7.60	0.895
8.80	0.955
9.50	0.970
10.00	0.960
10.80	0.970
11.00	0.970
11.40	0.960
CURVE 28 (cont.)	
11.70	0.945
12.50	0.875
13.00	0.825*
13.60	0.795
14.50	0.770
15.00	0.765

* Not shown on plot

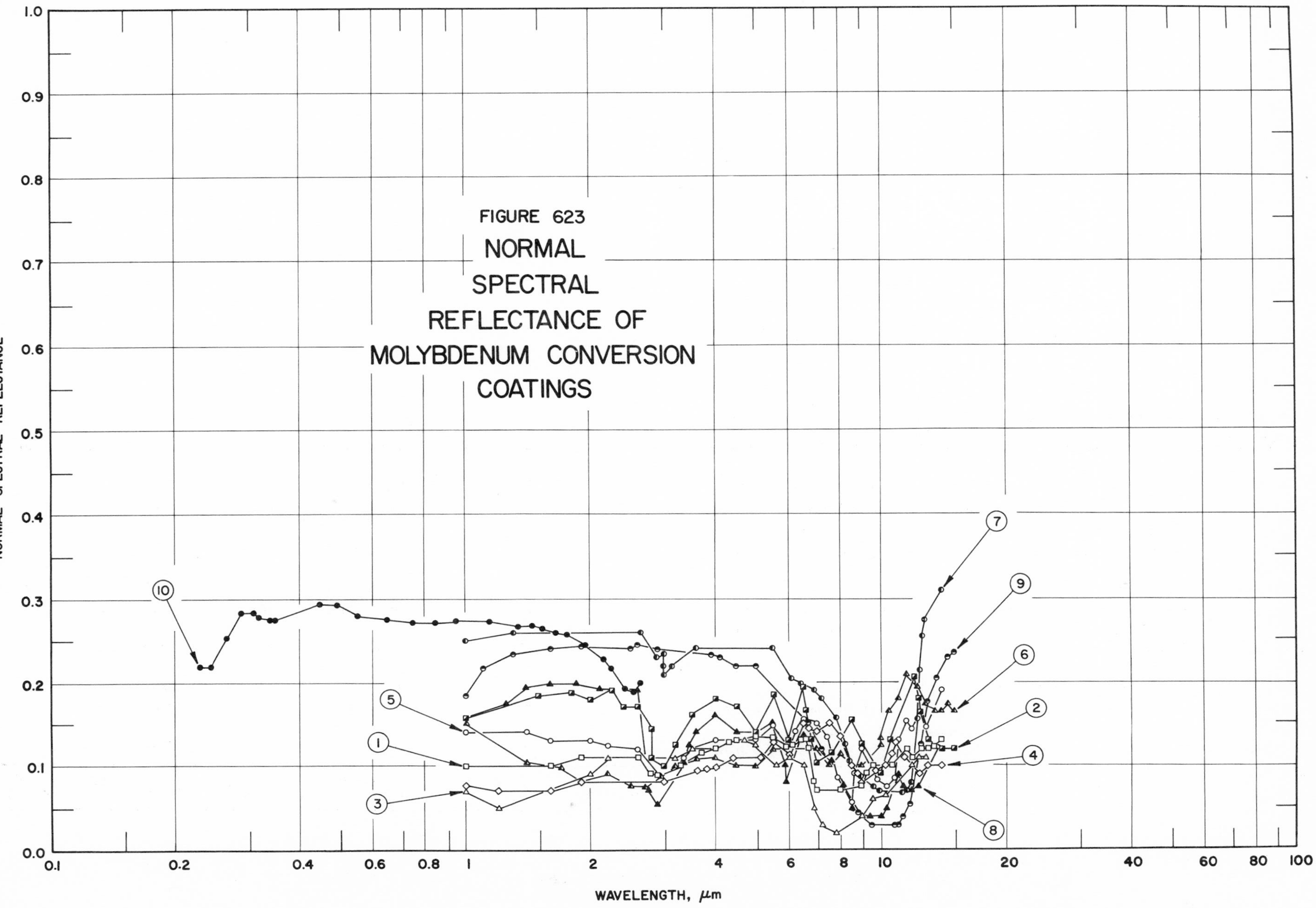

FIGURE 623 NORMAL SPECTRAL REFLECTANCE OF MOLYBDENUM CONVERSION COATINGS

SPECIFICATION TABLE NO. 623 NORMAL SPECTRAL REFLECTANCE OF MOLYBDENUM CONVERSION COATINGS

Curve No.	Ref. No.	Year	Temperature, K	Wavelength Range, μm	Geometry θ	θ'	ω'	Reported Error, %	Composition (weight percent), Specifications, and Remarks
1	280	1962	294	1.00-14.00	7°		2π		CV II + FeB (Chance Vought Corp.); silicon diffusion coating on molybdenum; measured in air; data extracted from smooth curve; converted from R (2π, 7°).
2	280	1962	294	1.00-15.00	7°		2π		Similar to above specimen and conditions except heated at 1366 K for 1 hr.
3	280	1962	922	1.00-15.00	7°		2π		Similar to curve 1 specimen and conditions except computed from normal spectral emittance.
4	280	1962	294	1.00-14.00	7°		2π		CV II + IX (Chance Vought Corp.); silicon diffusion coating on molybdenum; measured in air; data extracted from smooth curve; converted from R (2π, 7°).
5	280	1962	294	1.00-14.00	7°		2π		Similar to above specimen and conditions except heated at 1366 K for 1 hr.
6	280	1962	922	1.00-15.00	7°		2π		Similar to curve 4 specimen and conditions except computed from normal spectral emittance.
7	280	1962	294	1.00-14.00	7°		2π		CV II + IX + TiO_2 (Chance Vought Corp.); silicon diffusion coating on molybdenum; measured in air; data extracted from smooth curve; converted from R (2π, 7°).
8	280	1962	294	1.00-14.00	7°		2π		Similar to above specimen and conditions except heated at 1366 K for 1 hr.
9	280	1962	922	1.00-15.00	7°		2π		Similar to above specimen and conditions except computed from normal spectral emittance.
10	238	1963	298	0.230-2.650	~0°		2π		Pack cementation coating (0.051 mm thick) on molybdenum substrate; pack composition: 73 Al_2O_3, 23 Si, 4 NaI; prepared by heating the pack in an inert atm at 1367 K for 16 hrs; data extracted from smooth curve. [Authors' designation: Sample No. 129]

DATA TABLE NO. 623 NORMAL SPECTRAL REFLECTANCE OF MOLYBDENUM CONVERSION COATINGS

[Wavelength, λ, μm; Reflectance, ρ; Temperature, T, K]

λ	ρ
CURVE 1 T = 294	
1.00	0.100
1.60	0.100
1.90	0.110
2.60	0.110
2.80	0.090
2.90	0.087
3.30	0.100
3.70	0.115
4.00	0.120
4.30	0.128
4.50	0.130
5.00	0.134
5.50	0.132
5.90	0.121
6.10	0.123
6.40	0.130
6.60	0.130
6.73	0.120
6.90	0.080
7.00	0.070
8.00	0.070
9.00	0.075
9.20	0.090
9.60	0.100
10.60	0.100
11.50	0.120
11.90	0.110
12.50	0.120
13.00	0.120
13.40	0.122
14.00	0.130
CURVE 2 T = 294	
1.00	0.158
1.50	0.183
1.80	0.187
2.00	0.180
2.25	0.190
2.40	0.170
2.60	0.170
2.80	0.144
2.80	0.110
3.00	0.100
CURVE 2 (cont.)	
3.20	0.125
3.50	0.160
4.00	0.180
4.50	0.170
5.00	0.140
5.50	0.185
6.00	0.130
6.50	0.195
6.60	0.165
6.80	0.130
7.00	0.101
7.60	0.115
8.50	0.154
9.00	0.125
10.00	0.090
10.50	0.130
12.00	0.208
12.20	0.180
12.40	0.164
12.50	0.150*
13.00	0.130
14.00	0.120
15.00	0.120
CURVE 3 T = 922	
1.00	0.070
1.20	0.050
2.00	0.090
2.20	0.110
3.20	0.110
3.50	0.120
4.00	0.120*
4.50	0.130*
4.70	0.130
5.00	0.125
5.60	0.100
6.00	0.110
6.50	0.100
6.90	0.050
7.20	0.030
7.80	0.020
9.00	0.040
9.50	0.060
10.20	0.065
CURVE 3 (cont.)	
11.80	0.100
12.20	0.110
12.80	0.110
13.00	0.120*
15.00	0.120*
CURVE 4 T = 294	
1.00	0.076
1.20	0.070
1.60	0.070
1.90	0.080
3.00	0.080
3.60	0.094
3.80	0.096
4.00	0.098
4.40	0.110
5.10	0.110
5.50	0.128
6.00	0.112
6.50	0.150
6.70	0.145
7.00	0.140
7.50	0.150
8.00	0.135
8.50	0.100
8.80	0.090
9.00	0.080
9.70	0.094
10.20	0.100
10.60	0.115
10.90	0.129
11.40	0.110
12.40	0.090
13.00	0.100
14.00	0.100
CURVE 5 T = 294	
1.00	0.140
1.40	0.140
1.60	0.130
2.00	0.130
2.20	0.124
CURVE 5 (cont.)	
2.60	0.120
3.00	0.080*
3.35	0.110
3.50	0.120*
4.00	0.130
5.00	0.130
5.50	0.148
6.00	0.124*
6.20	0.140
6.50	0.155
7.00	0.150
7.40	0.135
7.60	0.115*
7.90	0.085
8.50	0.057
8.60	0.090
9.00	0.120
9.80	0.083
10.20	0.075
10.60	0.085
11.00	0.130
11.50	0.152
11.90	0.143
12.50	0.165*
12.90	0.145
14.00	0.190
CURVE 6 T = 922	
1.00	0.150
1.40	0.105
1.70	0.098
1.90	0.080*
2.20	0.090
2.50	0.075
2.70	0.075
2.95	0.087
3.20	0.100
3.60	0.108
4.00	0.110
4.50	0.100
5.00	0.100
5.50	0.120
7.00	0.120
7.50	0.100
CURVE 6 (cont.)	
8.00	0.115
8.50	0.100*
9.00	0.100
10.00	0.125
10.00	0.132
10.50	0.165
11.00	0.180
11.50	0.210
12.20	0.195
12.80	0.175
13.50	0.164
14.00	0.165
14.50	0.172
15.00	0.165
CURVE 7 T = 294	
1.00	0.250
1.30	0.260
2.65	0.260*
2.90	0.230
3.00	0.220
3.00	0.210
3.15	0.220
3.00	0.235
3.60	0.240
5.50	0.240
6.10	0.205
6.40	0.200
6.90	0.190
7.20	0.180
7.80	0.155
8.20	0.125
8.40	0.105
9.00	0.085
9.60	0.075
9.90	0.070
11.20	0.068
11.50	0.070
11.70	0.080
12.05	0.155
12.20	0.165*
12.40	0.215
12.55	0.225*
12.60	0.255
CURVE 7 (cont.)	
12.80	0.275
14.00	0.310
CURVE 8 T = 294	
1.00	0.155*
1.25	0.175
1.40	0.195
1.60	0.200
1.85	0.200
2.10	0.192
2.60	0.190
2.75	0.070
2.90	0.055
3.35	0.105
3.45	0.125
3.60	0.140
4.00	0.160
4.50	0.140
5.00	0.140*
5.45	0.150
5.85	0.100
5.95	0.080
6.50	0.135
7.60	0.105
8.10	0.075
8.50	0.050
9.40	0.040
10.00	0.040
10.40	0.050
11.00	0.090
11.30	0.075
11.80	0.070
12.40	0.075
14.00	0.130*
CURVE 9 T = 922	
1.00	0.185
1.10	0.218
1.30	0.234
1.60	0.240
1.90	0.242
2.50	0.240
CURVE 9 (cont.)	
2.60	0.245
2.90	0.240
3.90	0.232
4.10	0.230
4.50	0.220
5.00	0.220
6.65	0.150
7.20	0.120
7.60	0.105*
8.80	0.045
9.50	0.030
10.00	0.040*
10.80	0.030
11.00	0.030
11.40	0.040
11.70	0.055
12.50	0.125
13.00	0.175
13.60	0.205
14.50	0.230
15.00	0.235
CURVE 10 T = 298	
0.230	0.220
0.245	0.220
0.268	0.253
0.290	0.284
0.310	0.284
0.320	0.279
0.340	0.277
0.350	0.275
0.444	0.294
0.496	0.294
0.547	0.280
0.648	0.275
0.749	0.271
0.848	0.271
0.948	0.274
1.151	0.272
1.348	0.268
1.450	0.268
1.549	0.265
1.653	0.260
1.750	0.257
CURVE 10 (cont.)	
1.952	0.245
2.150	0.229
2.251	0.219
2.428	0.192
2.551	0.189
2.650	0.200

* Not shown on plot

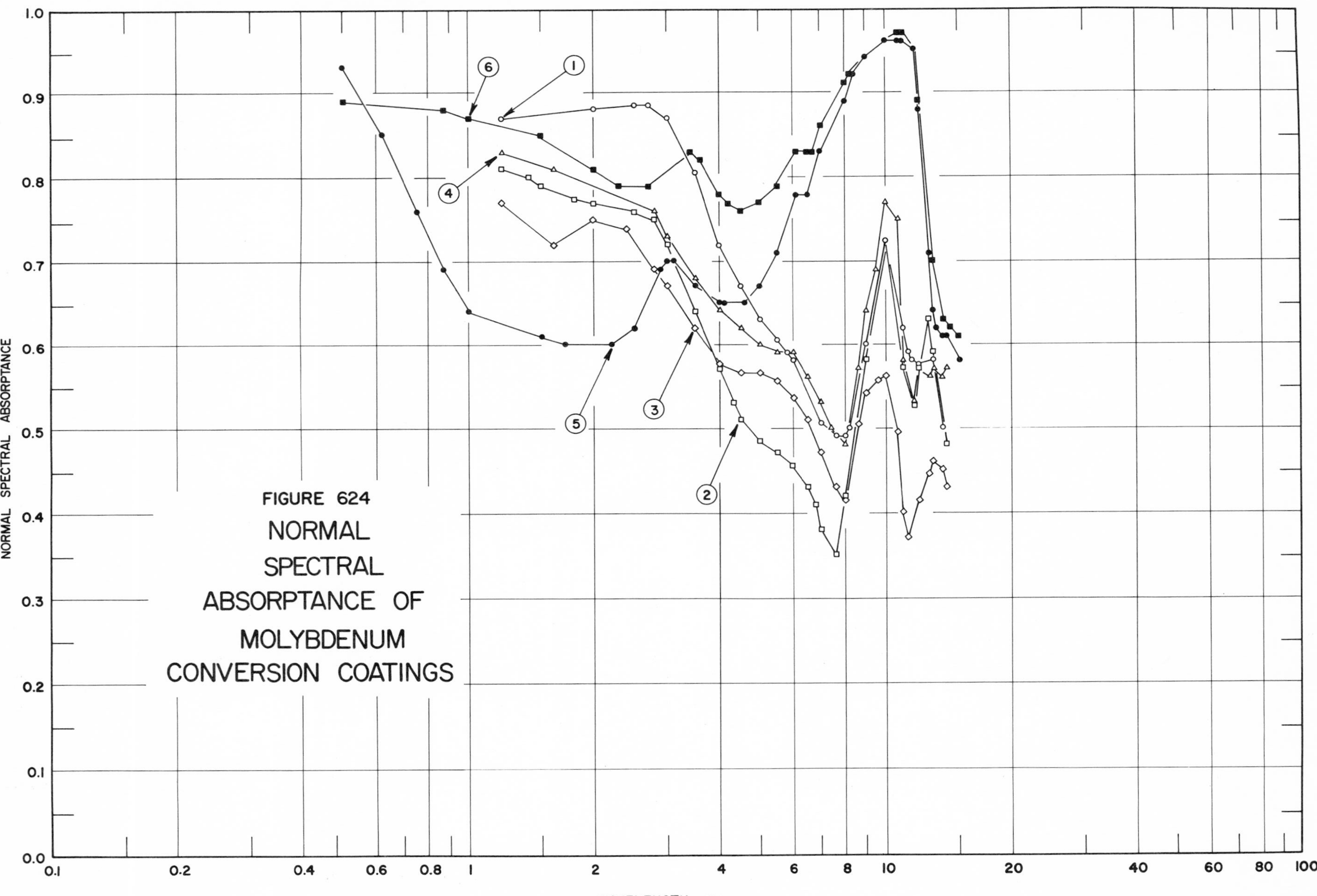

FIGURE 624 NORMAL SPECTRAL ABSORPTANCE OF MOLYBDENUM CONVERSION COATINGS

SPECIFICATION TABLE NO. 624 NORMAL SPECTRAL ABSORPTANCE OF MOLYBDENUM CONVERSION COATINGS

Curve No.	Ref. No.	Year	Temperature, K	Wavelength Range, μm	Geometry θ	Reported Error, %	Composition (weight percent), Specifications, and Remarks
1	279	1963	294	1.2-13.8	$\sim 0^{\circ}$		W-2 (Chromalloy Corp.); silicon diffusion coating on molybdenum substrate; measured in air; data extracted from smooth curve. [Author's designation: Sample 2]
2	279	1963	294	1.2-14.0	$\sim 0^{\circ}$		Above specimen and conditions except heated at 1389 K for 2 hrs.
3	279	1963	294	1.2-14.0	$\sim 0^{\circ}$		Similar to curve 1 specimen and conditions except heated at 1000 K for 2 hrs, at 1168 K for 27 hrs, at 1333 K for 20 hrs and at 1386 K for 3 hrs. [Author's designation: Sample 1]
4	279	1963	294	1.2-14.0	$\sim 0^{\circ}$		Above specimen and conditions except heated at 1111 K for 5 hrs and at 1277 K for 3 hrs.
5	279	1963	294	0.50-15.0	$\sim 0^{\circ}$		CV II + IX + TiO_2 (Chance Vought Corp.); silicon diffusion coating on molybdenum substrate; heated at 1700 K; measured in air; data extracted from smooth curve.
6	279	1963	294	0.50-15.0	$\sim 0^{\circ}$		Similar to above specimen and conditions except heated at 778 K for 30 hrs.

DATA TABLE NO. 624 NORMAL SPECTRAL ABSORPTANCE OF MOLYBDENUM CONVERSION COATINGS

[Wavelength, λ, μm; Absorptance, α; Temperature, T, K]

λ	α
CURVE 1	**T = 294**
1.2	0.870
2.0	0.880
2.5	0.885
2.7	0.885
3.0	0.870
3.5	0.805
4.0	0.720
4.5	0.670
5.0	0.630
5.5	0.605
5.8	0.590
6.0	0.580
7.0	0.505
7.6	0.490
8.0	0.490
8.2	0.500
9.0	0.600
10.0	0.725
11.0	0.620
11.4	0.590
11.6	0.580
12.0	0.575
13.0	0.580
13.8	0.500
CURVE 2	**T = 294**
1.2	0.810
1.4	0.800
1.5	0.790
1.8	0.775
2.0	0.770
2.5	0.760
2.8	0.750
3.0	0.720
3.5	0.640
4.0	0.570
4.3	0.530
4.5	0.510
5.0	0.485
5.5	0.470
6.0	0.455
6.5	0.430
6.8	0.410
CURVE 2 (cont.)	
7.0	0.380
7.6	0.350
8.0	0.420
9.0	0.580
10.0	0.725*
11.0	0.570
11.6	0.525
12.0	0.570
12.6	0.630
13.0	0.590
14.0	0.480
CURVE 3	**T = 294**
1.2	0.770
1.6	0.720
2.0	0.750
2.4	0.740
2.8	0.690
3.0	0.670
3.5	0.620
4.0	0.575
4.5	0.565
5.0	0.565
5.5	0.555
6.0	0.535
6.5	0.510
7.0	0.470
7.6	0.430
8.0	0.415
8.6	0.505
9.0	0.540
9.6	0.555
10.0	0.560
10.6	0.495
11.0	0.400
11.2	0.370
12.0	0.415
12.6	0.445
13.0	0.460
13.6	0.450
14.0	0.430
CURVE 4	**T = 294**
1.2	0.83
1.6	0.81
2.8	0.76
3.0	0.73
3.5	0.68
4.0	0.64
4.5	0.62
5.0	0.60
5.5	0.59
6.0	0.59
6.5	0.56
7.0	0.53
7.4	0.50
8.0	0.48
8.6	0.57
9.0	0.64
9.5	0.69
10.0	0.77
10.6	0.75
11.0	0.58
11.6	0.53
12.0	0.57*
12.6	0.56
13.0	0.57
13.6	0.56
14.0	0.57
CURVE 5	**T = 294**
0.50	0.93
0.62	0.85
0.75	0.76
0.87	0.69
1.0	0.64
1.5	0.61
1.7	0.60
2.2	0.60
2.5	0.62
2.9	0.69
3.0	0.70
3.1	0.70
3.5	0.67
4.0	0.65
4.1	0.65
CURVE 5 (cont.)	
4.6	0.65
5.0	0.67
5.5	0.71
6.1	0.78
6.5	0.78
7.0	0.83
8.0	0.89
8.4	0.92
9.0	0.94
10.0	0.96
10.6	0.96
11.0	0.96
11.6	0.95
12.0	0.88
12.6	0.71
13.0	0.64
13.1	0.62
13.6	0.61
14.0	0.61
15.0	0.58
CURVE 6	**T = 294**
0.50	0.89
0.87	0.88
1.0	0.87
1.5	0.85
2.0	0.81
2.3	0.79
2.7	0.79
3.4	0.83
3.6	0.82
4.0	0.78
4.2	0.77
4.5	0.76
5.0	0.77
5.5	0.79
6.1	0.83
6.5	0.83
6.7	0.83
7.0	0.86
8.0	0.91
8.2	0.92
9.0	0.94*
10.0	0.96*
CURVE 6 (cont.)	
10.6	0.97
11.0	0.97
11.6	0.95*
12.0	0.89
13.0	0.70
13.8	0.63
14.4	0.62
15.0	0.61

* Not shown on plot

SPECIFICATION TABLE NO. 625 NORMAL TOTAL EMITTANCE OF NIOBIUM CONVERSION COATINGS

Curve No.	Ref. No.	Year	Temperature Range, K	Geometry θ'	Reported Error, %	Composition (weight percent), Specifications and Remarks
1*	238	1963	1223	$\sim 0°$		Pack cementation coating (0.025 mm thick) on niobium substrate; pack composition: 59 Al_2O_3, 40 Al, 1 NH_4Cl; prepared by heating the pack in an inert atm at 1339 K for 12 hrs; integrated over the range 1-15 μm. [Author's designation: Sample No. 132]

DATA TABLE NO. 625 NORMAL TOTAL EMITTANCE OF NIOBIUM CONVERSION COATINGS

[Temperature, T, K; Emittance, ∈]

T	∈
CURVE 1*	
1223	0.90

* No plot given

SPECIFICATION TABLE NO. 626 NORMAL SPECTRAL EMITTANCE OF NIOBIUM CONVERSION COATINGS

Curve No.	Ref. No.	Year	Temperature, K	Wavelength Range, μm	Geometry θ'	Reported Error, %	Composition (weight percent), Specifications, and Remarks
1*	238	1963	1223	1.00-15.00	~0°		Pack cementation coating (0.025 mm thick) on niobium substrate; pack composition: 59 Al_2O_3, 40 Al, 1 NH_4Cl; prepared by heating the pack in an inert atm at 1339 K for 12 hrs; data extracted from smooth curve. [Authors' designation: Sample No. 132]

DATA TABLE NO. 626 NORMAL SPECTRAL EMITTANCE OF NIOBIUM CONVERSION COATINGS

[Wavelength, λ, μm; Emittance, ∈; Temperature, T, K]

λ	∈	λ	∈	λ	∈
CURVE 1* T = 1223		CURVE 1 (cont.)		CURVE 1 (cont.)	
		6.01	0.911	11.27	0.826
1.00	0.900	6.51	0.909	11.72	0.797
1.35	0.846	7.00	0.903	12.29	0.793
1.46	0.838	7.51	0.900	12.38	0.800
1.98	0.910	7.95	0.892	12.51	0.790
2.32	0.921	8.16	0.853	12.93	0.793
2.58	0.921	8.23	0.858	13.02	0.786
2.72	0.915	8.30	0.847	13.31	0.793
2.99	0.922	8.42	0.861	13.47	0.779
3.30	0.942	8.84	0.856	13.75	0.787
3.65	0.947	8.93	0.844	14.01	0.775
3.95	0.953	9.09	0.854	14.51	0.774
4.17	0.953	9.51	0.841	14.90	0.780
4.30	0.924	9.61	0.830	15.00	0.795
4.78	0.932	9.66	0.842		
4.87	0.916	9.87	0.831		
4.93	0.932	10.00	0.844		
4.99	0.916	10.75	0.845		
5.30	0.918	11.00	0.832		
5.44	0.908	11.19	0.816		

* No plot given

SPECIFICATION TABLE NO. 627 NORMAL SPECTRAL REFLECTANCE OF NIOBIUM CONVERSION COATINGS

Curve No.	Ref. No.	Year	Temperature, K	Wavelength Range, μm	Geometry θ	Geometry θ'	Geometry ω'	Reported Error, %	Composition (weight percent), Specifications, and Remarks
1*	238	1963	298	0.230-2.650	~0°		2π		Pack cementation coating (0.025 mm thick) on niobium substrate; pack composition: 59 Al_2O_3, 40 Al, 1 NH_4Cl; prepared by heating the pack in an inert atm at 1339 K for 12 hrs; data extracted from smooth curve. [Authors' designation: Sample No. 132]
2*	238	1963	298	0.230-2.650	~0°		2π		Pack cementation coating (0.025 mm thick) on niobium substrate; pack composition: 59 Al_2O_3, 40 $NbAl_3$, 1 NaI; prepared by heating the pack in an inert atm at 1394 K for 14 hrs; data extracted from smooth curve. [Authors' designation: Sample No. 131]

DATA TABLE NO. 627 NORMAL SPECTRAL REFLECTANCE OF NIOBIUM CONVERSION COATINGS

[Wavelength, λ, μm; Reflectance, ρ; Temperature, T, K]

CURVE 1* (T = 298)

λ	ρ
0.230	0.220
0.240	0.214
0.250	0.215
0.270	0.215
0.280	0.215
0.320	0.215
0.331	0.220
0.335	0.223
0.350	0.223
0.396	0.233
0.465	0.233
0.548	0.235
0.650	0.242
0.749	0.242
0.849	0.238
0.946	0.239
1.052	0.242
1.150	0.246
1.351	0.252
1.551	0.259
1.751	0.266
1.850	0.272
2.051	0.279
2.251	0.286
2.452	0.291
2.650	0.295

CURVE 2* (T = 298)

λ	ρ
0.230	0.343
0.240	0.337
0.250	0.337
0.260	0.343
0.280	0.343
0.300	0.340
0.310	0.338
0.320	0.337
0.333	0.330
0.350	0.325
0.365	0.315
0.455	0.321
0.550	0.340
0.590	0.349
0.650	0.344
0.749	0.356
0.849	0.368
0.949	0.386
1.049	0.403
1.117	0.411
1.206	0.412
1.251	0.418
1.349	0.422
1.450	0.435
1.550	0.443
1.750	0.455
1.948	0.464
2.051	0.468
2.252	0.480
2.449	0.504
2.551	0.520
2.650	0.545

* No plot given

SPECIFICATION TABLE NO. 628 NORMAL TOTAL EMITTANCE OF NIOBIUM + TITANIUM CONVERSION COATINGS

Curve No.	Ref. No.	Year	Temperature Range, K	Geometry θ'	Reported Error, %	Composition (weight percent), Specifications and Remarks
1*	178	1960	811-1589	~0°	±20	W-2 (Chromalloy Corp.); silicon diffusion coating (0.038 mm thick) on 10 Ti, 10 Mo and Nb balance substrate; measured in demoisturized helium atm.
2*	178	1960	811-1589	~0°	±20	Similar to above specimen and conditions except 0.064 mm thick.

DATA TABLE NO. 628 NORMAL TOTAL EMITTANCE OF NIOBIUM + TITANIUM CONVERSION COATINGS

[Temperature, T, K; Emittance, ϵ]

T	ϵ
CURVE 1*	
811	0.12
1089	0.21
1367	0.26
1589	0.35
CURVE 2*	
811	0.24
1089	0.26
1367	0.29
1589	0.32

* No plot given.

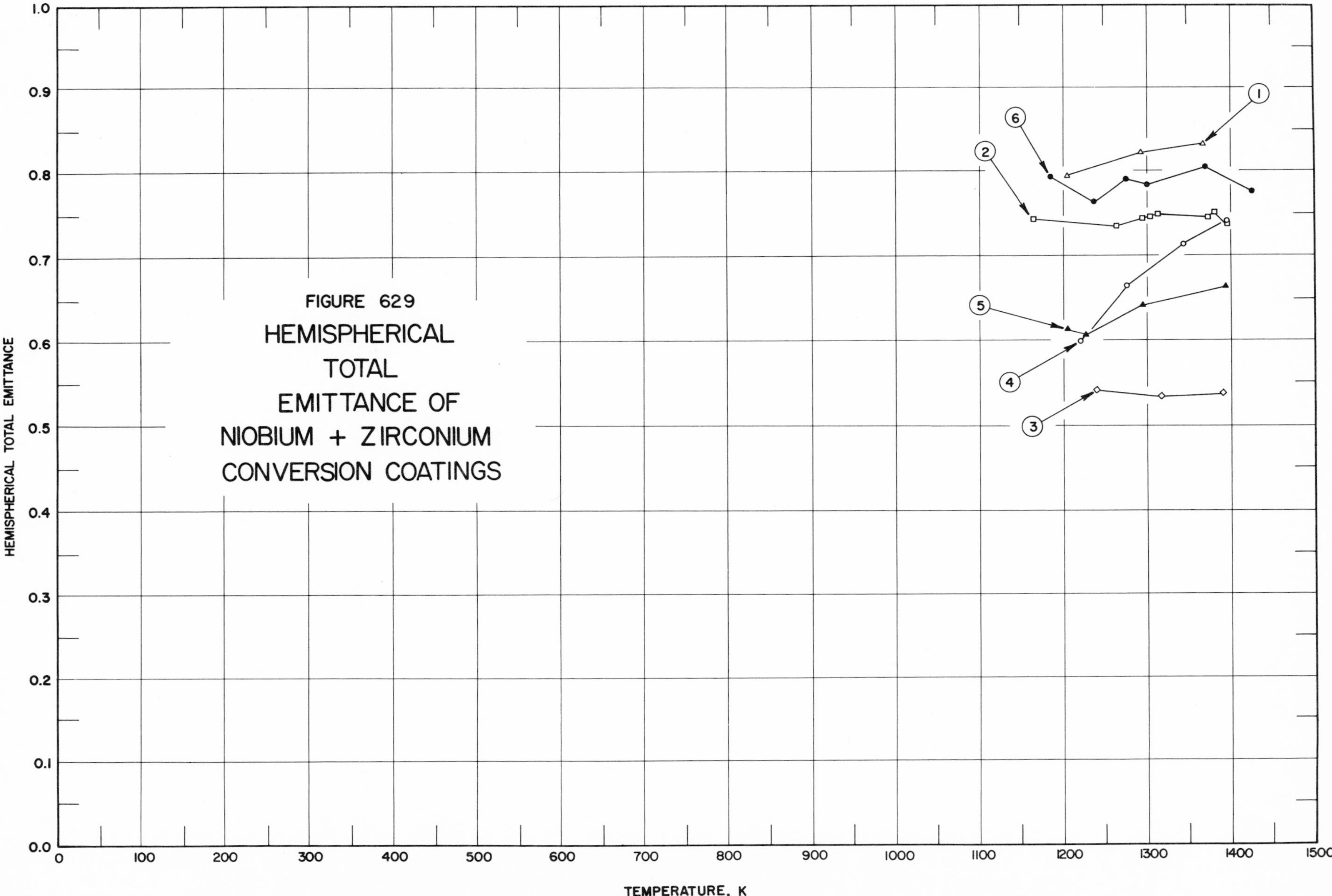

FIGURE 629
HEMISPHERICAL TOTAL EMITTANCE OF NIOBIUM + ZIRCONIUM CONVERSION COATINGS

SPECIFICATION TABLE NO. 629 HEMISPHERICAL TOTAL EMITTANCE OF NIOBIUM + ZIRCONIUM CONVERSION COATINGS

Curve No.	Ref. No.	Year	Temperature Range, K	Reported Error, %	Composition (weight percent), Specifications and Remarks
1	185	1962	1205-1369	~5	10 Cr-Al diffusion coating (0.025 to 0.038 mm thick); formed by diffusing 10 Cr-2 Si-Al alloy into Nb-1Zr substrate; coating surface degreased with acetone; measured in vacuum (10^{-4} mm Hg).
2	185	1962	1166-1398	~5	Similar to above specimen and conditions except sample heat treated 312 hrs at 1478 K in helium.
3	185	1962	1240-1390	~5	LB-2 aluminide diffusion coating (0.051 mm thick); formed by diffusing a 10 Cr-2 Si-Al alloy plus Al into a Nb-1Zr substrate; coating surface degreased with acetone; measured in vacuum (10^{-4} mm Hg).
4	185	1962	1220-1396	~5	ED84-2 ceramic coating; composition: barium-aluminum-silicate glass + 13% Cr_2O_3; thickness 1.5 mils; applied to Nb-1Zr alloy substrate in slurry form and fired in argon at 1783 K; coating surface degreased with acetone; measured in vacuum (10^{-4} mm Hg).
5	185	1962	1204-1393	~5	ED84-1 ceramic coating (0.038 mm thick); barium-aluminum-silicate glass, applied to Nb-1Zr alloy substrate in slurry form and fired in argon at 1783 K; coating surface degreased with acetone; measured in vacuum (10^{-4} mm Hg).
6	185	1962	1188-1429	~5	Similar to above specimen and conditions except heat treated 312 hrs at 1478 K in helium.

DATA TABLE NO. 629 HEMISPHERICAL TOTAL EMITTANCE OF NIOBIUM + ZIRCONIUM CONVERSION COATINGS

[Temperature, T, K; Emittance, ∈]

T	∈	T	∈	T	∈
CURVE 1		CURVE 3		CURVE 6	
1205	0.795	1240	0.540	1188	0.794
1291	0.822	1319	0.533	1239	0.765
1369	0.833	1390	0.537	1275	0.790
				1300	0.785
CURVE 2		CURVE 4		1370	0.805
				1429	0.777
1166	0.742	1220	0.600		
1264	0.734	1278	0.664		
1297	0.744	1344	0.715		
1303	0.745	1396	0.740		
1314	0.750				
1373	0.746	CURVE 5			
1380	0.751				
1398	0.737	1204	0.614		
		1229	0.608		
		1298	0.643		
		1393	0.663		

SPECIFICATION TABLE NO. 630 NORMAL TOTAL EMITTANCE OF TANTALUM CONVERSION COATINGS

Curve No.	Ref. No.	Year	Temperature Range, K	Geometry θ'	Reported Error, %	Composition (weight percent), Specifications and Remarks
1*	238	1963	1223	~0°		Pack cementation coating (0.025 mm thick) on tantalum substrate; pack composition: 59 Al_2O_3, 40 Al, 1 NH_4Cl; prepared by heating the pack in an inert atm at 1339 K for 12 hrs; integrated over the range 1-15 μm. [Authors' designation: Sample No. 134]
2*	238	1963	1223	~0°		Pack cementation coating (0.025 mm thick) on tantalum substrate; pack composition: 59 Al_2O_3, 40 $TaAl_3$, 1 NH_4Cl; prepared by heating the pack in an inert atm at 1394 K for 14 hrs; integrated over the range 1-15 μm. [Authors' designation: Sample 133]
3*	238	1963	1223	~0°		Pack cementation coating (0.051 mm thick) on tantalum substrate; pack composition: 72 Al_2O_3, 22 Si, 6 NaI; prepared by heating the pack in an inert atm at 1367 K for 16 hrs; integrated over the range 1-15 μm. [Authors' designation: Sample No. 127]
4*	238	1963	1223	~0°		Pack cementation coating (0.051 mm thick) on tantalum substrate; pack composition: 72 Al_2O_3, 22 Si, 6 NaI; prepared by heating the pack in an inert atm at 1367 K for 16 hrs; integrated over the range 1-15 μm. [Authors' designation: Sample No. 126]

DATA TABLE NO. 630 NORMAL TOTAL EMITTANCE OF TANTALUM CONVERSION COATINGS

[Temperature, T, K; Emittance, ϵ]

T	ϵ
CURVE 1*	
1223	0.93
CURVE 2*	
1223	0.86
CURVE 3*	
1223	0.76
CURVE 4*	
1223	0.85

* No plot given

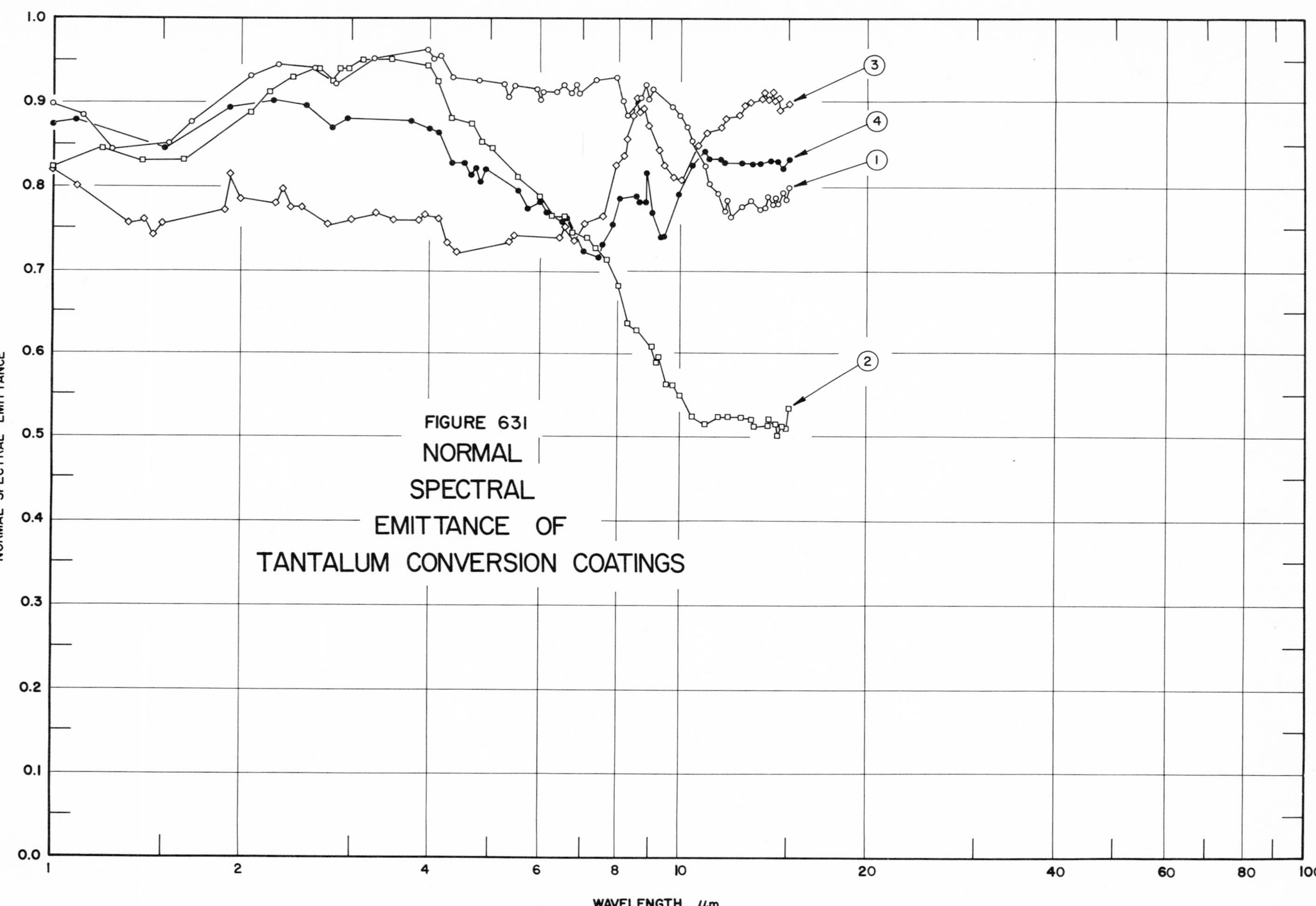

FIGURE 631 NORMAL SPECTRAL EMITTANCE OF TANTALUM CONVERSION COATINGS

SPECIFICATION TABLE NO. 631 NORMAL SPECTRAL EMITTANCE OF TANTALUM CONVERSION COATINGS

Curve No.	Ref. No.	Year	Temperature, K	Wavelength Range, μm	Geometry θ'	Reported Error, %	Composition (weight percent), Specifications, and Remarks
1	238	1963	1223	1.00-15.00	$\sim 0^\circ$		Pack cementation coating (0.025 mm thick) on tantalum substrate; pack composition: 59 Al_2O_3, 40 Al, 1 NH_4Cl; prepared by heating the pack in an inert atm at 1339 K for 12 hrs; data extracted from smooth curve. [Authors' designation: Sample No. 134]
2	238	1963	1223	1.00-15.00	$\sim 0^\circ$		Pack cementation coating (0.025 mm thick) on tantalum substrate; pack composition: 59 Al_2O_3, 40 $TaAl_3$, 1 NH_4Cl; prepared by heating the pack in an inert atm at 1394 K for 14 hrs; data extracted from smooth curve. [Authors' designation: Sample No. 133]
3	238	1963	1223	1.00-15.00	$\sim 0^\circ$		Pack cementation coating (0.051 mm thick) on tantalum substrate; pack composition: 72 Al_2O_3, 22 Si, 6 NaI; prepared by heating the pack in an inert atm at 1367 K for 16 hrs; data extracted from smooth curve. [Authors' designation: Sample No. 126]
4	238	1963	1223	1.00-15.00	$\sim 0^\circ$		Pack cementation coating (0.051 mm thick) on tantalum substrate; pack composition: 72 Al_2O_3, 22 Si, 6 NaI; prepared by heating the pack in an inert atm at 1367 K for 16 hrs; data extracted from smooth curve. [Authors' designation: Sample No. 127]

DATA TABLE NO. 631 NORMAL SPECTRAL EMITTANCE OF TANTALUM CONVERSION COATINGS

[Wavelength, λ, μm; Emittance, ∈; Temperature, T, K]

CURVE 1
T = 1223

λ	∈
1.00	0.900
1.12	0.887
1.25	0.847
1.54	0.854
1.66	0.879
2.07	0.933
2.30	0.946
2.62	0.941
2.83	0.924
3.26	0.954
3.99	0.963
4.07	0.953
4.18	0.957
4.35	0.931
4.80	0.929
5.26	0.924
5.33	0.909
5.48	0.922
5.93	0.918
6.00	0.903
6.08	0.915
6.39	0.915
6.56	0.922
6.73	0.912
6.88	0.924
6.93	0.912
7.36	0.929
7.98	0.932
8.12	0.902
8.26	9.887
8.70	0.908
8.86	0.922
8.95	0.907
9.07	0.918
9.78	0.898
10.00	0.887
10.28	0.873
10.50	0.857
11.00	0.826
11.17	0.804
11.51	0.793
11.85	0.771
11.95	0.785
12.10	0.767
12.52	0.776
13.07	0.785
13.50	0.774
13.75	0.776
13.93	0.790
14.18	0.780
14.28	0.789
14.45	0.780
14.65	0.795
14.90	0.786
15.00	0.800

CURVE 2
T = 1223

λ	∈
1.00	0.825
1.20	0.848
1.39	0.831
1.62	0.834
2.06	0.890
2.22	0.915
2.41	0.933
2.66	0.943
2.80	0.930
2.87	0.942
2.98	0.942
3.12	0.953
3.48	0.953
3.98	0.946
4.11	0.928
4.33	0.884
4.67	0.878
4.84	0.855
5.01	0.849
5.51	0.813
6.00	0.789
6.26	0.766
6.59	0.766
6.77	0.747
7.14	0.740
7.31	0.729
7.65	0.713
8.00	0.683
8.26	0.638
8.51	0.630
9.01	0.610
9.19	0.591
9.26	0.598
9.51	0.564
9.74	0.564
10.00	0.550
10.50	0.525
11.00	0.517
11.51	0.525
12.00	0.525
12.51	0.523
13.01	0.521
13.17	0.514
13.80	0.515
13.94	0.522
14.36	0.516
14.44	0.502
14.55	0.514
14.86	0.512
15.00	0.535

CURVE 3
T = 1223

λ	∈
1.00	0.820
1.09	0.801
1.33	0.759
1.40	0.761
1.45	0.744
1.50	0.757
1.87	0.773
1.93	0.817
2.00	0.786
2.27	0.781
2.33	0.799
2.40	0.777
2.50	0.777
2.75	0.757
3.00	0.761
3.29	0.770
3.50	0.761
3.85	0.761
3.94	0.768
4.14	0.763
4.28	0.735
4.41	0.722
5.35	0.736
5.45	0.742
6.47	0.740
6.59	0.753
6.80	0.736
7.06	0.758
7.51	0.767
7.99	0.828
8.18	0.839
8.27	0.859
8.42	0.886
8.55	0.907
8.65	0.891
8.76	0.895
8.92	0.875
9.25	0.845
9.48	0.829
9.77	0.812
10.01	0.810
10.72	0.851
11.01	0.866
11.69	0.872
11.83	0.883
12.50	0.888
12.73	0.899
13.00	0.901
13.51	0.906
13.73	0.915
13.93	0.906
14.07	0.915
14.21	0.904
14.45	0.907
14.59	0.893
15.00	0.900

CURVE 4
T = 1223

λ	∈
1.00	0.875
1.09	0.881
1.51	0.847
1.93	0.896
2.26	0.903
2.54	0.898
2.80	0.872
2.99	0.883
3.73	0.880
4.00	0.870
4.15	0.866
4.34	0.830
4.56	0.830
4.67	0.816
4.74	0.824
4.83	0.807
4.91	0.821
5.53	0.797
5.72	0.775
6.00	0.782
6.12	0.770
6.51	0.758
6.63	0.764
6.82	0.742
7.01	0.723
7.43	0.716
7.51	0.731
7.65	0.731
7.83	0.756
8.02	0.789
8.56	0.790
8.69	0.783
8.86	0.781
8.90	0.818
9.02	0.770
9.34	0.740
9.48	0.741
9.99	0.791
10.50	0.826
11.00	0.844
11.15	0.835
11.66	0.835
11.86	0.830
12.52	0.830
13.01	0.830
13.50	0.830
14.00	0.834
14.49	0.832
14.83	0.825
15.00	0.835

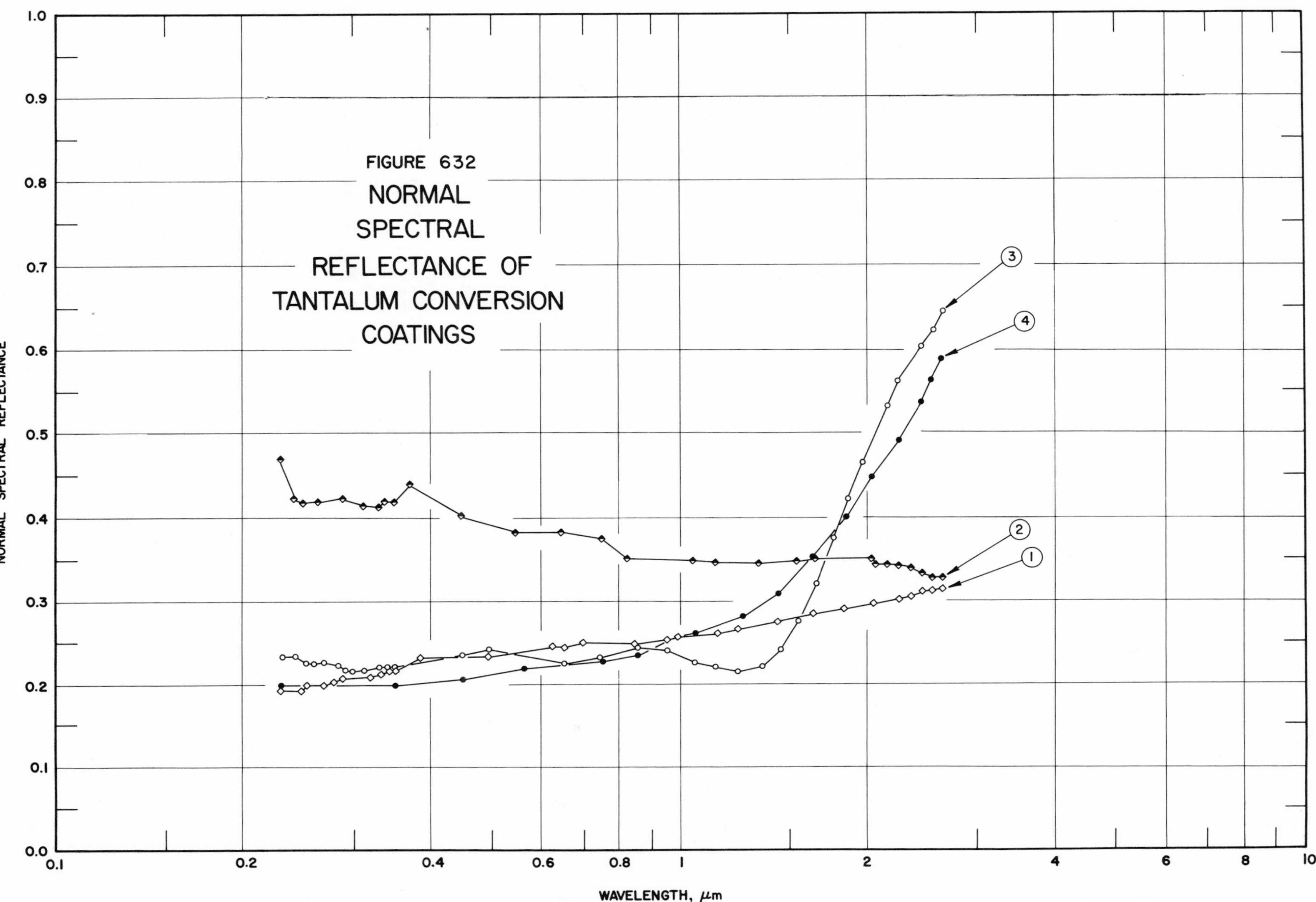

FIGURE 632 NORMAL SPECTRAL REFLECTANCE OF TANTALUM CONVERSION COATINGS

SPECIFICATION TABLE NO. 632 NORMAL SPECTRAL REFLECTANCE OF TANTALUM CONVERSION COATINGS

Curve No.	Ref. No.	Year	Temperature, K	Wavelength Range, μm	Geometry θ θ' ω'	Reported Error, %	Composition (weight percent), Specifications, and Remarks
1	238	1963	298	0.230-2.649	~0° 2π		Pack cementation coating (0.025 mm thick) on tantalum substrate; pack composition: 59 Al_2O_3, 40 Al, 1 NH_4Cl; prepared by heating the pack in an inert atm at 1339 K for 12 hrs; data extracted from smooth curve. [Authors' designation: Sample No. 134]
2	238	1963	298	0.230-2.650	~0° 2π		Pack cementation coating (0.025 mm thick) on tantalum substrate; pack composition: 59 Al_2O_3, 40 $TaAl_3$, 1 NH_4Cl; prepared by heating the pack in an inert atm at 1394 K for 14 hrs; data extracted from smooth curve. [Authors' designation: Sample No. 133]
3	238	1963	298	0.231-2.649	~0° 2π		Pack cementation coating (0.051 mm thick) on tantalum substrate; pack composition: 72 Al_2O_3, 22 Si, 6 NaI; prepared by heating the pack in an inert atm at 1367 K for 16 hrs; data extracted from smooth curve. [Authors' designation: Sample No. 126]
4	238	1963	298	0.230-2.649	~0° 2π		Pack cementation coating (0.076 mm thick) on tantalum substrate; pack composition: 85 Al_2O_3, 12 Al, 3 NaI; prepared by heating the pack in an inert atm at 1367 K for 14 hrs; data extracted from smooth curve. [Authors' designation: Sample No. 138]

DATA TABLE NO. 632 NORMAL SPECTRAL REFLECTANCE OF TANTALUM CONVERSION COATINGS

[Wavelength, λ, μm; Reflectance, ρ; Temperature, T, K]

λ	ρ
CURVE 1 T = 298	
0.230	0.194
0.248	0.194
0.252	0.200
0.270	0.200
0.280	0.205
0.290	0.209
0.320	0.210
0.331	0.212
0.341	0.217
0.350	0.218
0.385	0.234
0.495	0.235
0.623	0.248
0.651	0.246
0.700	0.251
0.848	0.250
0.951	0.253
0.990	0.259
CURVE 1 (cont.)	
1.150	0.263
1.249	0.268
1.449	0.276
1.647	0.286
1.849	0.293
2.049	0.299
2.252	0.304
2.350	0.307
2.453	0.312
2.552	0.313
2.649	0.317
CURVE 2 T = 298	
0.230	0.470
0.241	0.423
0.250	0.419
0.264	0.420
CURVE 2 (cont.)	
0.290	0.423
0.311	0.414
0.330	0.414
0.339	0.420
0.350	0.420
0.370	0.440
0.448	0.401
0.549	0.384
0.648	0.383
0.750	0.376
0.826	0.353
1.050	0.350
1.149	0.348
1.349	0.347
1.549	0.350
1.650	0.351
2.012	0.351
2.054	0.344
2.152	0.344
CURVE 2 (cont.)	
2.249	0.343
2.351	0.340
2.450	0.335
2.548	0.330
2.650	0.330
CURVE 3 T = 298	
0.231	0.235
0.243	0.235
0.252	0.227
0.260	0.227
0.270	0.228
0.285	0.223
0.291	0.219
0.300	0.218
0.311	0.219
0.330	0.222
CURVE 3 (cont.)	
0.340	0.222
0.350	0.221
0.450	0.236
0.499	0.243
0.651	0.227
0.749	0.234
0.853	0.246
0.951	0.241
1.051	0.229
1.149	0.221
1.249	0.217
1.352	0.221
1.450	0.242
1.550	0.277
1.650	0.323
1.752	0.376
1.851	0.422
1.954	0.466
2.154	0.532
CURVE 3 (cont.)	
2.248	0.561
2.449	0.603
2.552	0.623
2.649	0.647
CURVE 4 T = 298	
0.230	0.200
0.350	0.200
0.450	0.206
0.564	0.220
0.750	0.229
0.851	0.237
0.950	0.252
1.052	0.261
1.251	0.282
1.447	0.310
1.648	0.353
CURVE 4 (cont.)	
1.850	0.401
2.046	0.450
2.248	0.492
2.448	0.539
2.553	0.565
2.649	0.590

SPECIFICATION TABLE NO. 633 NORMAL TOTAL EMITTANCE OF TITANIUM CONVERSION COATINGS

Curve No.	Ref. No.	Year	Temperature Range, K	Geometry θ'	Reported Error, %	Composition (weight percent), Specifications and Remarks
1*	238	1963	1223	$\sim 0^{\circ}$		Pack cementation coating (0.051 mm thick) on titanium substrate; pack composition: 74 Al_2O_3, 24 Si, 2 NaF; prepared by heating the pack in an inert atm at 1367 K for 14 hrs; integrated over the range 1-15 μm. [Authors' designation: Sample No. 128]
2*	238	1963	1223	$\sim 0^{\circ}$		Pack cementation coating (0.076 mm thick) on titanium substrate; pack composition: 85 Al_2O_3, 12 Al, 3 NaI; prepared by heating the pack in an inert atm at 1367 K for 14 hrs; integrated over the range 1-15 μm. [Authors' designation: Sample No. 138]
3*	238	1963	1223	$\sim 0^{\circ}$		Pack cementation coating (0.102 mm thick) on titanium substrate; pack composition: 85 Al_2O_3, 12 TiAl, 3 NH_4Cl; prepared by heating the pack in an inert atm at 1367 K for 14 hrs; integrated over the range 1-15 μm. [Authors' designation: Sample No. 138]

DATA TABLE NO. 633 NORMAL TOTAL EMITTANCE OF TITANIUM CONVERSION COATINGS

[Temperature, T, K; Emittance, ϵ]

T	ϵ
CURVE 1*	
1223	0.79
CURVE 2*	
1223	0.77
CURVE 3*	
1223	0.82

* No plot given

SPECIFICATION TABLE NO. 634 NORMAL SPECTRAL EMITTANCE OF TITANIUM CONVERSION COATINGS

Curve No.	Ref. No.	Year	Temperature, K	Wavelength Range, μm	Geometry θ'	Reported Error, %	Composition (weight percent), Specifications, and Remarks
1*	238	1963	1223	1.00-15.00	~0°		Pack cementation coating (0.051 mm thick) on titanium substrate; pack composition: 74 Al_2O_3, 24 Si, 2 NaF; prepared by heating the pack in an inert atm at 1367 K for 16 hrs; data extracted from smooth curve. [Authors' designation: Sample No. 128]
2*	238	1963	1223	1.00-14.99	~0°		Pack cementation coating (0.076 mm thick) on titanium substrate; pack composition: 85 Al_2O_3, 12 Al, 3 NaI; prepared by heating the pack in an inert atm at 1367 K for 14 hrs; data extracted from smooth curve. [Authors' designation: Sample No. 138]
3*	238	1963	1223	1.00-15.00	~0°		Pack cementation coating (0.102 mm thick) on titanium substrate; pack composition: 85 Al_2O_3, 12 TiAl, 3 NH_4Cl; prepared by heating the pack in an inert atm at 1367 K for 14 hrs; data extracted from smooth curve. [Authors' designation: Sample No. 138]

* No plot given

DATA TABLE NO. 634 NORMAL SPECTRAL EMITTANCE OF TITANIUM CONVERSION COATINGS

[Wavelength, λ, μm; Emittance, ∈; Temperature, T, K]

CURVE 1*
T = 1223

λ	∈
1.00	0.915
1.13	0.898
1.22	0.898
1.47	0.844
1.99	0.844
2.31	0.833
2.45	0.808
2.84	0.761
2.99	0.769
3.55	0.751
3.88	0.755
4.15	0.737
4.28	0.709
4.35	0.695
4.65	0.699
4.81	0.711
4.95	0.732
5.03	0.727
5.21	0.732
5.33	0.750
5.55	0.749
5.70	0.743
6.52	0.743
6.61	0.743
6.74	0.746
6.99	0.768
7.29	0.773
7.49	0.801
7.65	0.814
7.75	0.844
8.05	0.901
8.51	0.949
8.64	0.944
8.78	0.932
8.93	0.913
9.07	0.913
9.22	0.896
9.41	0.868
9.71	0.855
9.99	0.863
10.50	0.882
11.00	0.903
11.51	0.908
11.63	0.906
11.99	0.914

CURVE 1 (cont.)

λ	∈
12.09	0.925
12.51	0.939
13.01	0.946
13.51	0.955
14.00	0.955
14.51	0.955
15.00	0.960

CURVE 2*
T = 1223

λ	∈
1.00	0.880
1.19	0.902
1.50	0.885
1.60	0.863
1.71	0.878
1.83	0.899
1.95	0.865
2.08	0.878
2.28	0.850
2.49	0.822
2.82	0.784
3.01	0.793
3.52	0.777
4.01	0.785
4.14	0.769
4.22	0.777
4.43	0.752
4.51	0.763
4.67	0.760
4.86	0.749
5.51	0.736
5.80	0.715
6.02	0.718
6.51	0.703
7.01	0.696
7.15	0.678
7.24	0.685
7.37	0.705
7.44	0.675
7.60	0.681
8.04	0.656
8.19	0.621
8.42	0.597
8.51	0.618
8.67	0.602

CURVE 2 (cont.)

λ	∈
8.81	0.602
8.91	0.571
8.99	0.595
9.09	0.579
9.21	0.595
9.53	0.573
10.00	0.562
10.50	0.567
10.71	0.584
10.88	0.606
11.01	0.618
11.50	0.628
11.96	0.634
12.07	0.626
12.76	0.662
13.37	0.662
13.50	0.670
13.72	0.665
13.90	0.657
14.01	0.667
14.79	0.668
14.90	0.661
14.99	0.670

CURVE 3*
T = 1223

λ	∈
1.00	0.870
1.19	0.808
1.25	0.806
1.31	0.776
1.37	0.790
1.59	0.797
2.06	0.845
2.49	0.832
2.57	0.817
2.64	0.832
2.88	0.814
3.18	0.826
4.02	0.824
4.18	0.839
4.29	0.788
4.35	0.798
5.28	0.788
5.57	0.778
5.73	0.787

CURVE 3 (cont.)

λ	∈
6.08	0.787
6.30	0.781
6.99	0.800
7.18	0.800
7.33	0.816
7.50	0.817
8.00	0.831
8.13	0.811
8.83	0.849
8.89	0.800
9.01	0.858
9.50	0.867
9.96	0.896
10.50	0.939
11.01	0.964
11.51	0.972
11.68	0.971
12.00	0.947
12.32	0.918
13.01	0.867
13.29	0.853
13.50	0.838
14.00	0.832
14.10	0.829
14.39	0.814
14.58	0.812
14.84	0.801
15.00	0.815

* No plot given

SPECIFICATION TABLE NO. 635 NORMAL SPECTRAL REFLECTANCE OF TITANIUM CONVERSION COATINGS

Curve No.	Ref. No.	Year	Temperature, K	Wavelength Range, μm	Geometry θ	θ'	ω'	Reported Error, %	Composition (weight percent), Specifications, and Remarks
1*	238	1963	298	0.230-2.650	~0°		2π		Pack cementation coating (0.051 mm thick) on titanium substrate; pack composition: 74 Al_2O_3, 24 Si, 2 NaF; prepared by heating the pack in an inert atm at 1367 K for 16 hrs; data extracted from smooth curve. [Authors' designation: Sample No. 128]
2*	238	1963	298	0.230-2.650	~0°		2π		Pack cementation coating (0.076 mm thick) on titanium substrate; pack composition: 77 Al_2O_3, 7 Al + 13 Ti, 3 NaI; prepared by heating the pack in an inert atm at 1367 K for 14 hrs; data extracted from smooth curve. [Authors' designation: Sample No. 140]
3*	238	1963	298	0.230-2.650	~0°		2π		Pack cementation coating (0.102 mm thick) on titanium substrate; pack composition: 85 Al_2O_3, 12 TiAl, 3 NH_4Cl; prepared by heating the pack in an inert atm at 1367 K for 14 hrs; data extracted from smooth curve. [Authors' designation: Sample No. 138]

DATA TABLE NO. 635 NORMAL SPECTRAL REFLECTANCE OF TITANIUM CONVERSION COATINGS

[Wavelength, λ, μm; Reflectance, ρ; Temperature, T, K]

λ	ρ
CURVE 1*	
T = 298	
0.230	0.210
0.239	0.201
0.247	0.199
0.253	0.191
0.271	0.191
0.281	0.185
0.291	0.185
0.301	0.185
0.311	0.183
0.321	0.189
0.351	0.189
0.360	0.185
0.427	0.185
0.462	0.189
0.517	0.189
0.575	0.200
0.683	0.200
0.851	0.223
0.952	0.235
CURVE 1 (cont.)	
1.151	0.245
1.248	0.260
1.348	0.281
1.453	0.313
1.647	0.369
1.748	0.393
1.850	0.424
2.050	0.495
2.249	0.568
2.452	0.641
2.552	0.673
2.650	0.715
CURVE 2*	
T = 298	
0.230	0.200
0.240	0.196
0.244	0.189
0.267	0.186
CURVE 2 (cont.)	
0.280	0.181
0.300	0.180
0.320	0.184
0.927	0.189
0.332	0.191
0.339	0.200
0.350	0.204
0.450	0.328
0.550	0.381
0.626	0.406
0.664	0.406
0.752	0.433
0.951	0.472
1.051	0.486
1.251	0.509
1.375	0.532
1.452	0.535
1.649	0.562
1.850	0.594
2.151	0.638
CURVE 2 (cont.)	
2.309	0.661
2.355	0.663
2.453	0.679
2.593	0.700
2.650	0.720
CURVE 3*	
T = 298	
0.230	0.089
0.235	0.104
0.242	0.111
0.260	0.114
0.270	0.118
0.290	0.119
0.310	0.122
0.320	0.128
0.330	0.133
0.340	0.139
0.350	0.145
CURVE 3 (cont.)	
0.448	0.274
0.548	0.363
0.631	0.391
0.658	0.391
0.752	0.426
0.849	0.459
0.949	0.486
1.149	0.527
1.347	0.559
1.550	0.588
1.747	0.619
1.949	0.648
2.150	0.677
2.347	0.713
2.542	0.748
2.582	0.751
2.650	0.770

* No plot given

SPECIFICATION TABLE NO. 636 NORMAL TOTAL EMITTANCE OF TUNGSTEN CONVERSION COATINGS

Curve No.	Ref. No.	Year	Temperature Range, K	Geometry θ'	Reported Error, %	Composition (weight percent), Specifications and Remarks
1*	238	1963	1223	$\sim 0^\circ$		Pack cementation coating (0.025 mm thick) on tungsten substrate; pack composition: 73 Al_2O_3, 23 Si, 4 NaI; prepared by heating the pack in an inert atm at 1367 K for 14 hrs; integrated over the range 1-15 μm. [Authors' designation: Sample No. 130]

DATA TABLE NO. 636 NORMAL TOTAL EMITTANCE OF TUNGSTEN CONVERSION COATINGS

[Temperature, T, K; Emittance, ϵ]

T	ϵ
CURVE 1*	
1223	0.66

* No plot given

SPECIFICATION TABLE NO. 637 NORMAL SPECTRAL EMITTANCE OF TUNGSTEN CONVERSION COATINGS

Curve No.	Ref. No.	Year	Temperature, K	Wavelength Range, μm	Geometry θ'	Reported Error, %	Composition (weight percent), Specifications, and Remarks
1*	238	1963	1223	1.00-15.00	~0°		Pack cementation coating (0.025 mm thick) on tungsten substrate; pack composition: 73 Al_2O_3, 23 Si, 4 NaI; prepared by heating the pack in an inert atm at 1367 K for 14 hrs; data extracted from smooth curve. [Authors' designation: Sample No. 130]

DATA TABLE NO. 637 NORMAL SPECTRAL EMITTANCE OF TUNGSTEN CONVERSION COATINGS

[Wavelength, λ, μm; Emittance, ϵ; Temperature, T, K]

λ	ϵ	λ	ϵ	λ	ϵ	λ	ϵ
CURVE 1* T = 1223		CURVE 1 (cont.)		CURVE 1 (cont.)		CURVE 1 (cont.)	
		3.50	0.628	8.91	0.507	14.85	0.522
1.00	0.825	3.60	0.620	9.10	0.522	14.93	0.443
1.05	0.835	3.76	0.594	9.27	0.538	15.00	0.495
1.12	0.828	3.90	0.588	9.67	0.613		
1.18	0.837	4.07	0.551	9.83	0.622		
1.27	0.811	4.16	0.560	10.24	0.619		
1.35	0.828	4.27	0.494	10.34	0.611		
1.44	0.812	4.59	0.453	10.43	0.622		
1.56	0.836	5.00	0.430	10.58	0.625		
1.63	0.823	5.20	0.415	10.89	0.649		
1.71	0.863	5.43	0.407	11.07	0.649		
1.78	0.853	5.76	0.388	11.54	0.586		
1.86	0.889	6.07	0.388	12.01	0.524		
1.92	0.875	6.27	0.372	12.53	0.480		
2.00	0.893	6.61	0.372	12.66	0.465		
2.29	0.870	6.86	0.346	12.75	0.469		
2.40	0.847	7.24	0.346	12.96	0.453		
2.69	0.757	7.42	0.338	13.44	0.454		
3.01	0.706	8.01	0.338	14.01	0.484		
3.25	0.674	8.52	0.415	14.51	0.516		

* No plot given

SPECIFICATION TABLE NO. 638 NORMAL SPECTRAL REFLECTANCE OF TUNGSTEN CONVERSION COATINGS

Curve No.	Ref. No.	Year	Temperature, K	Wavelength Range, μm	Geometry θ	θ'	ω'	Reported Error, %	Composition (weight percent), Specifications, and Remarks
1*	238	1963	298	0.230-2.649	~0°		2π		Pack cementation coating (0.025 mm thick) on tungsten substrate; pack composition: 73 Al_2O_3, 23 Si, 4 NaI; prepared by heating the pack in an inert atm at 1367 K for 14 hrs; data extracted from smooth curve. [Authors' designation: Sample No. 130]

DATA TABLE NO. 638 NORMAL SPECTRAL REFLECTANCE OF TUNGSTEN CONVERSION COATINGS

[Wavelength, λ, μm; Reflectance, ρ; Temperature, T, K]

λ	ρ	λ	ρ
CURVE 1*		CURVE 1 (cont.)	
T = 298		1.146	0.226
0.230	0.295	1.248	0.216
0.239	0.292	1.350	0.209
0.249	0.284	1.450	0.202
0.270	0.283	1.546	0.193
0.280	0.282	1.747	0.178
0.290	0.282	1.948	0.160
0.300	0.278	2.051	0.155
0.310	0.275	2.150	0.159
0.330	0.275	2.249	0.173
0.350	0.275	2.348	0.201
0.450	0.278	2.449	0.254
0.491	0.278	2.549	0.312
0.548	0.260	2.649	0.370
0.648	0.250		
0.779	0.250		
0.818	0.243		
0.888	0.243		
0.977	0.233		
1.053	0.233		

* No plot given

References to Data Sources

Ref. No.	TPRC No.	
1	16009	Noonan, F.M., Alexander, A.L., and Cowling, J.E., "The Properties of Paints as Affected by Ultraviolet Radiation in a Vacuum. Part I," NRL Rept. 5503, 39 pp., 1960. [AD 240 141]
2	28965	Field, D.E., Cowling, J.E., and Noonan, F.M., "The Properties of Paints as Affected by Ultraviolet Radiation in a Vacuum, Part 2," NRL Rept. 5737, 28 pp., 1962. [AD 273 716]
3	32408	Tompkins, E.H. and Ivanuski, V.R., "Coatings for the Reflection of Intense Thermal Radiation. Final Rept., 20 Nov. 1958 - 20 Feb. 1960," ARF-3141-15, 74 pp., 1960. [PB 160 781]
4	27534	Strain, R.N.C., "Solar Reflectivity of Paints," J. Oil Col. Chem. Assoc., 44(10), 689-712, 1961.
5	22520	Coblentz, W.W., "The Diffuse Reflecting Power of Various Substances," Natl. Bur. Standards Bull., 9, 283-325, 1913.
6	23559	Gardner, H.A., "Reflection Factors of Industrial Paints and Pigments," Trans. Illum. Eng. Soc. (N.Y.), 17, 318-22, 1922.
7	24808	Cowling, J.E., Alexander, A.L., and Noonan, F.M., "The Design of Organic Coatings for Use in the Space Environment," in Coatings for the Aerospace Environment, WADD-TR-60-773, 17-37, 1961. [AD 267 310]
8	30290	de la Perrelle, E.T., Moss, T.S., and Herbert, H., "Measurements of Absorption and Reflectance," Infrared Phys., 3, 35-43, 1963.
9	33142	Cartwright, C.H., "Black Bodies in the Extreme Infrared," Phys. Rev., 35, 415-20, 1930.
10	34682	Drisko, R.W., "Reflectivity of Airfield Marking Paints," NCEL TR-R-323, 26 pp., 1964. [AD 446 327]
11	35546	Zerlaut, G.A., "Utilization of Pigmented Coatings for the Control of Equilibrium Skin Temperatures of Space Vehicles," ABMA-MISC-32, 46 pp., 1960. [AD 463 738]
12	45667	de la Perrelle, E.T. and Herbert, H., "The Measurement of Absorptivity and Reflectivity," RAE-TN-RAD-ARC-20879, 22 pp., 1962.
13	45700	Wilcock, D.F. and Soller, W., "Paints to Reflect Ultraviolet Light," Ind. Eng. Chem., 32, 1446-51, 1940.
14	28999	Kronstein, M., "Research, Studies, and Investigations on Spectral Reflectance and Absorption Characteristics of Camouflage Paint Materials and Natural Objects," ASTIA, 68 pp., 1956. [AD 100 058]
15	40136	Crosby, J.R., "Development and Qualification of Thermal Control Coatings for SNAP Systems," NAA-SR-9908, 69 pp., 1965.
16	46810	Curtis, H.B., "Measurement of Hemispherical Total Emittance and Normal Solar Absorptance of Selected Materials in the Temperature Range 280° to 600°K," in Proc. Conf. on Spacecraft Coatings Develop., NASA-TM-X-56167, 18 pp., 1964; NASA-TM-X-56273, 20 pp., 1965; J. Spacecraft Rockets, 3(3), 383-7, 1966.
17	51637	Westcott, M., "The Measurement of Solar Absorptance and Thermal Emittance," ESRO-TN-23, 43 pp., 1968.
18	36536	Streed, E.R. and Beveridge, C.M., "The Study of Low Solar Absorptance Coatings for a Solar Probe Mission," in Symp. on Thermal Radiation of Solids, San Francisco, Calif., March 4,5,6, 1964, NASA-SP-55, 535-48, 1965. [AD 629 980]
19	34832	Heaney, J.B., "A Comparison of Two Emittance Measurement Techniques," NASA-TM-X-55294, 20 pp., 1965.
20	28755	Zerlaut, G.A. and Harada, Y., "Stable White Coatings. Summary Report, Sept. 21, 1961 - July 15, 1963," NASA CR-52134, 194 pp., 1963.
21	34455	Brandenberg, W.M. and Neu, J.T., "Unidirectional Reflectance of Imperfectly Diffuse Surfaces," J. Opt. Soc. Am., 56(1), 97-103, 1966.
22	48374	Miller, R.A. and Campbell, F.J., "Effects of Low Energy Protons on Thermal Control Coatings," Progr. Astronaut. Aeronaut., 18, 399-412, 1966.

Ref. No.	TPRC No.	
23	40417	Arvesen, J.C., Neel, C.B., and Shaw, C.C., "Preliminary Results from a Round-Robin Study of Ultraviolet Degradation of Spacecraft Thermal-Control Coatings," in Proc. of Conf. on Spacecraft Coatings Develop., NASA-TM-X-56167, 21 pp., 1964.
24	36522	Gilligan, J.E. and Caren, R.P., "Some Fundamental Aspects of Nuclear Radiation Effects in Spacecraft Thermal Control Materials," in Symp. on Thermal Radiation of Solids, San Francisco, Calif., March 4,5,6, 1964, NASA-SP-55, 351-64, 1965. [AD 629 980]
25	36526	Olson, R.L., McKellar, L.A., and Stewart, J.V., "The Effects of Ultraviolet Radiation on Low α/ϵ Surfaces," in Symp. on Thermal Radiation of Solids, San Francisco, Calif., March 4,5,6, 1964, NASA-SP-55, 421-32, 1965. [AD 629 980]
26	36527	Pezdirtz, G.F. and Jewell, R.A., "A Study of the Photodegradation of Selected Thermal-Control Surfaces," in Symp. on Thermal Radiation of Solids, San Francisco, Calif., March 4,5,6, 1964, NASA-SP-55, 433-441, 1965. [AD 629 980]
27	36525	Zerlaut, G.A., Harada, Y., and Tompkins, E.H., "Ultraviolet Irradiation of White Spacecraft Coatings in Vacuum," in Symp. on Thermal Radiation of Solids, San Francisco, Calif., March 4,5,6, 1964, NASA-SP-55, 391-420, 1965. [AD 629 980]
28	40416	Jewell, R.A., Pezdirtz, G.F., and Burks, H.D., "The Effects of Ultraviolet and Gamma Rays on Thermal Control Coatings," in Proc. of Conf. on Spacecraft Coatings Develop., NASA-TM-X-56167, 16 pp., 1964.
29	42292	Tompkins, E.H., "Stable White Coatings," NASA-CR-50814, 98 pp., 1962.
30	44271	Heaney, J.B., "An Integrating Sphere Coating for the Vacuum Ultraviolet Spectral Region," NASA-TM-X-55645, 16 pp., 1966.
31	44096	Gillette, R.B., Brown, R.R., Seiler, R.F., and Sheldon, W.R., "Effects of Protons and Alpha Particles on Thermal Properties of Spacecraft and Solar Concentrator Coatings," Progr. Astronaut. Aeronaut., 18, 413-40, 1966.
32	47278	Zerlaut, G.A. and Courtney, W.J., "Space-Simulation Facility for In Situ Reflectance Measurements," Progr. Astronaut. Aeronaut., 20, 349-68, 1967.
33	48375	Gillette, R.B., Brown, R.R., Seiler, R.F., and Sheldon, W.R., "Effects of Protons and Alpha Particles on Thermal Properties of Spacecraft and Solar Concentrator Coatings," Progr. Astronaut. Aeronaut., 18, 413-40, 1966.
34	49031	Zerlaut, G.A., Tompkins, E.H., and Harada, Y., "Development of Space-Stable Thermal-Control Coatings, Triannual Rept. Sept. 20, 1963 to Jan. 20, 1964," NASA-CR-60355, 55 pp., 1964.
35	50030	Zerlaut, G.A., Harada, Y., and Berman, L.U., "Development of Space-Stable Thermal-Control Coatings, Triannual Rept. Jan. 1964 to June 1964," IITRI-C6014-13, 49 pp., 1964.
36	48492	Zerlaut, G.A. and Rubin, G.A., "Development of Space-Stable Thermal-Control Coatings, Triannual Rept.," IITRI-U6002-36, 46 pp., 1966.
37	47756	Cannaday, S.S. and Fogdall, L.B., "In Situ Ultraviolet Radiation Effects on Thermal Control Coatings, 8th Quarterly Progress Report," NASA-CR-91351, 111 pp., 1967.
38	48366	MacMillan, H.F., Sklensky, A.F., and McKellar, L.A., "Apparatus for Spectral Bidirectional Reflectance Measurements during Ultraviolet Irradiation in Vacuum," Progr. Astronaut. Aeronaut., 18, 129-49, 1966.
39	59647	McCargo, M., Greenberg, S.A., and Douglas, N.J., "Radiation-Induced Absorption Bands in Spacecraft Thermal Control Coating Pigments," Progr. Astronaut. Aeronaut., 23, 189-218, 1970.
40	8277	Richmond, J.C. and Stewart, J.E., "Spectral Emittance of Uncoated and Ceramic-Coated Inconel and Type 321 Stainless Steel," NASA MEMO 4-9-59W, 30 pp., 1959. [AD 214 232]
41	16628	Burgess, D.G., Jasperse, J.R., Marcus, L., Martin, W.S., and Flint, E.P., "Research on Ceramic Coatings with Controlled Reflective and Emissive Properties," WADD-TR-60-317, 46 pp., 1960. [AD 245 233L]
42	36219	Zerlaut, G.A., "Investigation of Light Scattering in Highly Reflecting Pigmented Coatings," NASA-CR-55196, 11 pp., 1963.
43	25811	Carpenter, W.G.D. and Sewell, J.H., "Thermal Emissivities of Some Metallic and Nonmetallic Surfaces over the Range of Temperature 70 C to 250 C," RAE Rept. CHEM-538, 6 pp., 1962. [AD 295 648]
44	31781	Zerlaut, G.A., Thompkins, E.H., and Harada, Y., "Development of Space-Stable Thermal-Control Coatings, Triannual Rept., 20 May-20 Sept. 1963," NASA-CR-55826, 43 pp., 1963.
45	59648	Fogdall, L.B., Cannaday, S.S., and Brown, R.R., "Electron Energy Dependence for In-Vacuum Degradation and Recovery in Thermal Control Surfaces," Progr. Astronaut. Aeronaut., 23, 219-48, 1970.
46	41589	Brown, R.R., Fogdall, L.B., and Cannaday, S.S., "Electron-Ultraviolet Radiation Effects on Thermal Control Coatings," Progr. Astronaut. Aeronaut., 21, 697-724, 1969.

Ref. No.	TPRC No.	
47	47272	Streed, E. R., "An Experimental Study of the Combined Space Environmental Effects on a Zinc-Oxide/Potassium-Silicate Coating," Progr. Astronaut. Aeronaut., 20, 237-64, 1967.
48	27325	Carpenter, W.G.D. and Sewell, J.H., "An Apparatus for Measuring the Thermal Emissivity of Metal Surfaces and Surface Finishes at Low Temperatures," RAE TN CPM-20, 9 pp., 1963. [AD 418 612]
49	31646	Gaumer, R.E., Clauss, F.J., Sibert, M.E., and Shaw, C.C., "Materials Effects in Spacecraft Thermal Control," in Coatings for the Aerospace Environment, LMSD-704019, 20 pp., 1960; WADD-TR-60-773, 117-136, 1961. [AD 267 310] [AD 269 918]
50	47145	Breuch, R.A., Douglas, N.J., and Vance, D., "Effects of Electron Bombardment on the Optical Properties of Spacecraft Temperature Control Coatings," AIAA Journal, 3(12), 2318-27, 1965.
51	43840	Zerlaut, G.A., Gilligan, J.E., and Harada, Y., "Stable White Coatings," IITRI-C-6027, 110 pp., 1965.
52	48377	Pearson, B.D. Jr., "Preliminary Results from the Ames Emissivity Experiment on OSO-II," Progr. Astronaut. Aeronaut., 18, 459-72, 1966.
53	48378	Crosby, J.R. and Perlow, M.A., "SNAP 10A Thermal Control Coatings," Progr. Astronaut. Aeronaut., 18, 473-91, 1966.
54	49030	Zerlaut, G.A. and Kaye, B.H., "Development of Space-Stable Thermal-Control Coatings. Triannual Rept. Feb. 1965," NASA-CR-67559, 34 pp., 1965.
55	49665	Scannapieco, J.F., "Irradiation of Thermal Control Coatings. Final Report, Feb. 1967-Feb. 1968," NASA-CR-94684, 220 pp., 1968.
56	41945	Caldwell, C. R. and Nelson, P. A., "Thermal Control Experiments on the Lunar Orbiter Spacecraft," Progr. Astronaut. Aeronaut., 21, 819-52, 1969.
57	47280	Schafer, C. F. and Bannister, T. C., "Thermal Control Coating Degradation Data from the Pegasus Experiment Packages," Progr. Astronaut. Aeronaut., 20, 457-73, 1967.
58	47281	Reichard, P.J. and Triolo, J.J., "Preflight Testing of the ATS-1 Thermal Coatings Experiment," Progr. Astronaut. Aeronaut., 20, 491-513, 1967.
59	16503	Snyder, N.W., Gier, J.T., Dunkle, R.V., and Possner, L., "Low Temperature Emissivity Measurements," ATI 90 576, 16 pp., 1948. [PB 142 256]
60	6596	Wilkes, G.B., "Total Normal Emissivities and Solar Absorptivities of Materials," WADC-TR-54-42, 94 pp., 1954. [AD 88 066]
61	18836	Yoshimoto, H. and Rall, R.M., "Successful Use and Control of Emissivity Finishes on HK-31A Magnesium for Thermal Design," Light Metal Age, 18(9/10), 6-11, 1960.
62	29648	Gier, J.T., Possner, L., Test, A.J., Dunkle, R.V., and Bevans, J.T., "The Absolute Spectral Reflectivity of Certain Pigments and Metals in the Wavelength Range Between 2 and 16 Microns," ATI-59635, 4 pp., 1949.
63	41610	Zerlaut, G. A., Rogers, F. O., and Noble, G., "The Development of S-13G-Type Thermal-Control Coatings," Progr. Astronaut. Aeronaut., 21, 741-66, 1969.
64	48376	Lewis, D.W. and Thostesen, T.O., "Mariner-Mars Absorptance Experiment," Progr. Astronaut. Aeronaut., 18, 441-57, 1966.
65	41919	Millard, J.P., "Results from the Thermal Control Coatings Experiment on OSO-III," Progr. Astronaut. Aeronaut., 21, 769-95, 1969.
66	33482	McKellar, L.A., "Effects of the Spacecraft Environment on Thermal Control Materials Characteristics," Lockheed Missiles and Space Co., 24 pp., 1962. [AD 274 162]
67	37194	Kerlee, C., "High Emittance for Refractory Alloys," Symp. Mater., Space Vehicle Use, 3(11), 19 pp., 1963.
68	41573	Swofford, D.D., Mangold, V.L., and Johnson, S.W., "Effects of Extreme Ultraviolet on the Optical Properties of Thermal Control Coatings," Progr. Astronaut. Aeronaut., 21, 667-95, 1969.
69	52788	Filippi, F.J., "Composite Ceramic Radome Manufacture by Mosaic Techniques," Interim Engineering Progress Report May 1963-July 1963, IR-7-984-(III), 107 pp., 1963. [AD 415 269]
70	51599	Rethwisch, F.B., Babcock, G.M., and Riggs, E.C., "Reflectivity of Aluminum Pigments and Paint-II," Paint Oil Chem. Rev., 116(24), 22-24, 1953.
71	24863	Gaumer, R.E., McKellar, L.A., Streed, E.R., Frame, K.L., and Grammer, J.R., "Calorimetric Determinations of Thermal Radiation Characteristics," in Progress in International Research on Thermodynamic and Transport Properties, ASME Second Symp. on Thermophysical Properties, Princeton, N.J., Jan. 24-6, 575-87, 1962.

Ref. No.	TPRC No.	
72	26749	Kerlin, E.E., "Investigation of Combined Effects of Radiation and Vacuum and of Radiation and Cryotemperatures on Engineering Materials. Volume 1: Radiation-Vacuum Tests. Annual Report, 9 Nov. 1961-8 Nov. 1962," NASA-CR-50948, 322 pp., 1963.
73	26089	Wade, W.R. and Slemp, W.S., "Measurements of Total Emittance of Several Refractory Oxides, Cermets, and Ceramics for Temperatures from 600°F to 2000°F," NASA-TN-D-998, 35 pp., 1962. [AD 272 614]
74	21553	Shaw, C., Berry, J., and Lee, T., "Spectral and Total Emissivity Apparatus and Measurements of Opaque Solids," LMSD-48488, 35 pp., 1959. [AD 282 600]
75	41590	Miles, J.K., Cheever, P.R., and Romanko, J., "Effects of Combined Electron-Ultraviolet Irradiation on Thermal Control Coatings in Vacuo at 77°K," Progr. Astronaut. Aeronaut., 21, 667-95, 1969.
76	40583	Dubs, C.W., "High Reflectance Coatings," IEEE Trans. and Nuclear Sci., 13(1), 729-34, 1966; AFCRL-IP-102, AFCRL-66-414, 10 pp., 1966. [AD 637 893]
77	33255	Hormann, H.H., Weaver, J.H., and Mattice, J.J., "Improved Coatings for Temperature Control in a Space Environment," in Third Symp. on Effects of Space Environment on Materials, St. Louis, May 7, 8 and 9, 57 pp., 1962.
78	28970	Tsukamoto, J., "Inorganic Passive Thermal Surfaces," ASTIA Rept. 3-71-61-15, 20 pp., 1961. [AD 267 538]
79	47274	Cheever, P.R., Miles, J.K., and Romanko, J., "In Situ Measurements of Spectral Reflectance of Thermal Control Coatings Irradiated in Vacuo," Progr. Astronaut. Aeronaut., 20, 281-96, 1967.
80	34814	Strindehag, O.M., "Optical Reflectors for Use in Internal Sample Aqueous Cerenkov Counters," Rev. Sci. Instr., 37(3), 344-9, 1966.
81	31743	Pratt and Whitney Aircraft Co., "Determination of the Emissivity of Materials. Annual Progress Report, January 1 through December 31, 1963," Pratt and Whitney Aircraft Co. Rept. PWA-2309, NASA-CR-58054, 83 pp., 1964.
82	30648	Askwyth, W.H., Hayes, R.J., and Mikk, G., "The Emittance of Materials Suitable for Use as Space-Craft Radiator Coatings," Am. Rocket Soc. Paper 2538-62, 26 pp., 1962.
83	30622	Zerlaut, G.A., "An Apparatus for the Measurement of the Total Normal Emittance of Surfaces at Satellite Temperatures," in Proc. Symp. on Measurement of Thermal Radiation Properties of Solids, Dayton, Ohio, Sept. 5, 6, and 7, 1962, NASA-SP-31, 275-285, 1963.
84	34333	Clausen, W., Meyer, K.H., and Percy, J.L., "Ultraviolet Stability Tests for Two White Paints," GDA-ERR-AN-64-487, 43 pp., 1964.
85	35545	Boebel, C.P., "OVI-10 Thermal Control Coating Orbital Experiment," AFML-TR-68-392, Part 1, 109 pp., 1969.
86	34819	Zerlaut, G.A., Gilligan, J.E., and Harada, Y., "Development of Space-Stable Thermal-Control Coatings, Triannual Report, 20 June-20 Oct. 1964," NASA-CR-67295, 72 pp., 1964.
87	51403	Grum, F. and Luckey, G.W., "Optical Sphere Paint and a Working Standard of Reflectance," Appl. Opt., 7(11), 2289-94, 1968.
88	41159	Nonaka, M., Kashima, T., and Kondo, Y., "A 5-m Integrating Sphere," Appl. Opt., 6(4), 757-64, 1967.
89	50539	Klemm, R.E., "Emissivity and Reflectance of Selected Surface Coatings," NA-57-707-1, 30 pp., 1958.
90	39479	Davidson, H.R. and Hemmendinger, H., "Color Prediction Using the Two-Constant Turbid-Media Theory," J. Opt. Soc. Am., 56(8), 1102-9, 1966.
91	34928	Edwards, D.K. and Hall, W.M., "Far-Infrared Reflectance of Spacecraft Coatings," NASA-CR-71187, 11 pp., 1966.
92	43801	McCullough, B.A., Wood, B.E., and Smith, A.M., "A Vacuum Integrating Sphere for In Situ Reflectance Measurements at 77 K from 0.5 to 10 Microns," AEDC-TR-67-10, 27 pp., 1967. [AD 650 072]
93	36154	Hall, W.M., "Effect of Low Temperature on the Thermal Emittance of Three Black Paints; Comparison of Normal and Hemispherical Emittances," JPL-RS-37-31, Vol. IV, 108-9, 1965.
94	29580	Hormann, H.H., "Improved Organic Coatings for Temperature Control in a Space Environment," ASD-TDR-62-917, 54 pp., 1963. [AD 400 488]
95	41768	Wade, W.R. and Slemp, W.S., "Total Emittance Measurements of Refractory Materials," Space/Aeronautics, 38(9), 133-9, 1962.
96	33450	Stierwalt, D.L., Kirk, D.D., and Bernstein, J.B., "Spectral Emittance of Solids. Spectral Emittance Measurements on Some Commercial Opaque and Transparent Solids," NOLC-TM-43-14, 6 pp., 1963. [AD 442 866]
97	25564	Cade, C.M., "The Thermal Emissivity of Some Materials Used in Thermionic Valve Manufacture," IRE Transactions on Electron Devices, ED-8, 56-69, 1961.

Ref. No.	TPRC No.	
98	12280	Boelter, L.M.K., Bromberg, R., and Gier, J.T., "An Investigation of Aircraft Heaters. XV. The Emissivity of Several Materials," ARR-4A21, NACA Rept. W-19, 13 pp., 1944.
99	37595	Mimura, T., Anagnostou, E., and Colarusso, P.E., "Thermal Radiation Absorptance and Vacuum Outgassing Characteristics of Several Metallic and Coated Surfaces," NASA-TN-D-3234, 56 pp., 1966.
100	29599	Carroll, W.F., "Development of Stable Temperature Control Surfaces for Spacecraft," JPL-TR-32-340, 20 pp., 1962.
101	32760	Searle, N.Z., Hirt, R.C., and Daniel, J.H. Jr., "Pigmented Surface Coatings for Use in the Space Environment," ASD-TDR-62-840 (Pt. 2), 97 pp., 1964. [AD 430 529]
102	28896	Ramsey, W.Y., "Specular Spectral Reflectance of Paints from 0.4 to 40.0 Microns. Report No. 31," Meterological Satellite Lab., Weather Bureau, 1964. [AD 446 291]
103	32388	Byrne, R.F. and Mancinelli, L.N., "Report of Investigation of Optical Transmittance, Reflectance, and Absorptance of Materials. Final Report," PB 159 155, 39 pp., 1954.
104	28940	Dunkle, R.V. and Gier, J.T., "Progress Report for the Year Ending June 27, 1953," University of Calif., Berkeley, 73 pp., 1953. [AD 16 830]
105	35796	Little, Arthur D., Inc., "Basic Investigation of Multi-Layer Insulation Systems," NASA-CR-54191, 298 pp., 1964.
106	30615	Sadler, R., Hemmerdinger, L., and Rando, I., "Emissometer, a Device for Measuring Total Hemispherical Emittance," in <u>Proc. Symp. on Measurement of Thermal Radiation Properties of Solids</u>, Dayton, Ohio, Sept. 5, 6 and 7, 1962, NASA-SP-31, 217-223, 1963.
107	47268	Androulakis, J.G., "The Development and Test of a Low to Moderately High Temperature Emissometer," Progr. Astronaut. Aeronaut., <u>20</u>, 151-76, 1967.
108	47098	Purcell, G.V., "Thermophysical Properties of Solids," in <u>Summary of the Eleventh Refractory Composites Working Group Meeting</u>, AFML-TR-66-179, 651-76, 1966. [AD 804 083]
109	24861	Dunkle, R.V., Edwards, D.K., Gier, J.T., Nelson, K.E., and Roddick, R.D., "Heated Cavity Reflectometer for Angular Reflectance Measurements," in <u>Progress in International Research on Thermodynamic and Transport Properties</u>, ASME Second Symposium on Thermophysical Properties, Princeton, N.J., Jan. 24-6, 541-62, 1962.
110	54714	Krewinghaus, A.B., "Infrared Reflectance of Paints," Appl. Opt., <u>8</u>(4), 807-12, 1969.
111	55741	Millard, J.P. and Streed, E.R., "A Comparison of Infrared-Emittance Measurements and Measurement Techniques," Appl. Opt., <u>8</u>(7), 1485-92, 1969.
112	30618	Streed, E.R., McKellar, L.A., Rolling, R. Jr., and Smith, C.A., "Errors Associated with Hohlraum Radiation Characteristics Determinations," in <u>Proc. Symp. on Measurement of Thermal Radiation Properties of Solids</u>, Dayton, Ohio, Sept. 5, 6, and 7, 1962, NASA-SP-31, 237- 52, 1963.
113	52871	Harris, L., "The Optical Properties of Metal Blacks and Carbon Blacks," MIT Mono, Ser. 1, 116 pp., 1967. [AD 831 788]
114	35723	Blevin, W.R. and Brown, W.J., "An Infrared Reflectometer with a Spheroidal Mirror," J. Sci. Instr., <u>42</u>(6), 385-9, 1965.
115	23571	Bevans, J.T., LeVantine, A.D., and Luedke, E.E., "Final Report on Satellite Materials and Environmental Control Systems Investigations," AFBMD-TR-61-12, 32 pp., 1960. [AD 255 968]
116	30611	Hembach, R.J., Hemmerdinger, L., and Katz, A.J., "Heated Cavity Reflectometer Modifications," in <u>Proc. Symp. on Measurement of Thermal Radiation Properties of Solids</u>, Dayton, Ohio, Sept. 5, 6, and 7, 1962, NASA-SP-31, 153- 67, 1963.
117	48372	Breuch, R.A., "Exploratory Trapped-Particle and Trapped-Particle-Plus-Ultraviolet Effects on the Optical Properties of Spacecraft Thermal Control Coatings," Progr. Astronaut. Aeronaut., <u>18</u>, 365-88, 1966.
118	26273	Askwyth, W.H. and Hayes, R.J., "Determination of the Emissivity of Materials. Quarterly Progress Report July 1-September 30, 1962," Pratt & Whitney Aircraft Rept. PWA-2128, 96 pp., 1962.
119	47275	Greenberg, S.A., Vance, D.A., and Streed, E.R., "Low Solar Absorptance Surfaces with Controlled Emittance: A Second Generation of Thermal Control Coatings," Progr. Astronaut. Aeronaut., <u>20</u>, 297-314, 1967.
120	43996	Zhorov, G.A., Kovalev, A.I., and Sivakova, E.V., "Thermal Conductivity and Emissivity of Aluminum Oxide Coatings at High Temperatures," Teplofiz. Vysokikh Temperatur, <u>4</u>(5), 643-8, 1966; English Translation: High Temperature USSR, <u>4</u>(5), 603-8, 1966.
121	45327	Smirnov, E.V., and Kondrashov, Yu.A., "Spectral Emissivity of Materials in the Temperature Range 100 to 1100 C," Teplofiz. Vysokikh Temperatur, <u>5</u>(1), 44-7, 1967; English Translation: High Temp. USSR, <u>5</u>(1), 37-40, 1967.
122	49666	Greenberg, S.A. and Vance, D.A., "Low Solar Absorptance and Emittance Surfaces Utilizing Vacuum Deposited Techniques. Final Report, June 1967-March 1968," NASA-CR-73228, 83 pp., 1968.

Ref. No.	TPRC No.	
123	6523	Sully, A.H., Brandes, E.A., and Waterhouse, R.B., "Some Measurements of the Total Emissivity of Metals and Pure Refractory Oxides and the Variation of Emissivity with Temperature," Brit. J. Appl. Phys., 3, 97-101, 1952.
124	8135	Wade, W.R., "Measurements of Total Hemispherical Emissivity of Several Stably Oxidized Metals and Some Refractory Oxide Coatings," NASA MEMO 1-20-59L, 30 pp., 1959. [AD 209 192]
125	10017	Olson, O.H. and Morris, J.C., "Determination of Emissivity and Reflectivity Data on Aircraft Structural Materials. Part 2, Supplement 1," WADC-TR-56-222, 31 pp., 1958. [AD 202 494]
126	35902	Grenis, A.F. and Levitt, A.P., "Infrared Radiation of Solids. Refractory Materials," Am. Ceram. Soc. Bull., 44(11), 901-6, 1965.
127	10060	Olson, O.H. and Morris, J.C., "Determination of Emissivity and Reflectivity Data on Aircraft Structural Materials. Part 3. Techniques for Measurement," WADC-TR-56-222, 96 pp., 1959. [AD 239 302]
128	25074	Childers, H.M. and Cerceo, J.M., "Electron Beam Techniques for Measuring the Thermophysical Properties of Materials. Report for 1 May 59-31 March 60 on Materials Analysis and Evaluation Techniques," WADD-TR-60-190, 66 pp., 1961. [AD 272 691]
129	47262	Liebert, C.H., "Spectral Emissivity of Highly Doped Silicon," Progr. Astronaut. Aeronaut., 20, 17-40, 1967.
130	50298	Lewis, B.W., Wade, W.R., Slemp, W.S., and Progar, D.J., "The Thermal Radiation Characteristics of Some High-Emittance Coatings for Space Applications," NASA-TM-X-59389, 16 pp., 1966.
131	46986	Liebert, C.H. and Thomas, R.D., "Spectral Emissivity of Highly Doped Silicon," NASA-TN-D-4303, 25 pp., 1968.
132	36320	Davies, J.M. and Zagieboylo, W., "An Integrating Sphere System for Measuring Average Reflectance and Transmittance," Appl. Opt., 4(2), 167-74, 1965.
133	33882	Mueller, W.E., "Note on Reflectance Measurements on Metals," Appl. Opt., 5(5), 876-7, 1966.
134	33849	Sacher, P.A., "The Effects of a Simulated Proton Space Environment on the Ultraviolet Transmittance of Optical Materials Between 3000 A and 1050 A," NASA-TM-X-55911, 17 pp., 1967.
135	52160	Cleary, R.E., Emanuelson, R., Luoma, W., and Ammann, C., "Properties of High Emittance Materials," NASA-CR-1278, 120 pp., 1969.
136	20142	Eastman Kodak Co., "Long Wavelength Infrared Domes. Final Tech. Rept., July 62-Oct. 63," AL-TDR-64-9, 50 pp., 1964. [AD 433 950]
137	38109	Curtis, H.B. and Nyland, T.W., "Apparatus for Measuring Emittance and Absorptance and Results for Selected Materials," NASA-TN-D-2583, 22 pp., 1965.
138	41161	Clapham, P.B., "Preparation and Properties of Sputtered Bismuth Oxide Films," Brit. J. Appl. Phys., 18(3), 363-6, 1967.
139	29405	Adams, J.G., "The Determination of Spectral Emissivities, Reflectivities, and Absorptivities of Materials and Coatings," NOR-61-189, 259 pp., 1962. [AD 274 558]
140	44231	Zdanowicz, L., "Some Optical Properties of Thin Evaporated Cd_3As_2 Films," Phys. Stat. Sol., 20(2), 473-80, 1967.
141	54041	Tanaka, K., Kunioka, A., and Sakai, Y., "Electrical and Optical Properties of Sputtered CdO Films," Japan. J. Appl. Phys., 8(6), 681-91, 1969.
142	54014	Proix, F. and Balkanski, M., "Infrared Measurements on Cadmium Sulfide Thin Films Deposited on Aluminum," Phys. Status Solidi, 32(1), 119-26, 1969.
143	38618	Emanuelson, R.C., "Determination of the Emissivity of Materials," NASA-CR-54268, 52 pp., 1964.
144	40400	Walek, W.J., Luoma, W.L., and Emanuelson, R.C., "Determination of the Emissivity of Materials," NASA-CR-72058, 88 pp., 1966.
145	10566	Campbell, D.A. and Schulte, H.A., "Measurement of Emissivity at Low Temperatures," ASTIA Tech. Rept. MT-R2J, 49 pp., 1957. [AD 151 931]
146	31650	Butler, C.P., Jenkins, R.J., Rudkin, R.L., and Laughridge, F.I., "High Temperature Surface Parameters for Solar Power," in Coatings for the Aerospace Environment, WADD-TR-60-773, 229- 52, 1961. [AD 267 310]
147	23145	Sklarew, S. and Rabensteine, A.S., "Emittance Studies of Various High Temperature Materials and Coatings," PR-281-3Q-1, 37 pp., 1963. [AD 299 417]
148	52946	Autio, G.W. and Scala, E., "Anisotropy in Emissivity of Single-Crystal Refractory Materials," Anisotropy Single-Cryst. Refract. Compounds, Proc. Int. Symp., Dayton, Ohio, 1967, 1, 357-81, 1968.
149	12026	Seifert, H.S. and Randall, H.M., "Transmission and Reflection of Plastics and Metal Blacks in the Far Infrared," Rev. Sci. Instruments, 11, 365-8, 1940.

Ref. No.	TPRC No.	
150	54642	Royds, T., "The Reflective Power of Lamp- and Platinum-Black," Phil. Magazine, 21, 167-72, 1911.
151	47267	McCullough, B.A., Wood, B.E., Smith, A.M., and Birkebak, R.C., "A Vacuum Integrating Sphere for In Situ Reflectance Measurements at 77°K from 0.5 to 10 Mu," Progr. Astronaut. Aeronaut., 20, 137-50, 1967.
152	4886	Uzan, R., "Some Results of Spectral and Thermionic Emissivity Measurements on Ceric Oxide," Le Vide, 7, 1139-40, 1952.
153	33499	Langley, R.C., Hill, J.S., and Albert, H.J., "Inorganic Films for Solar Energy Absorption," ASD-TDR-62-92(Pt. 2), 46 pp., 1963. [AD 424 099]
154	31650	See Reference No. 146.
155	18323	Allen, R.D., "Spectral and Total Emissivities of Rokide C on Molybdenum Above 1800°F," J. Am. Ceram. Soc., 44(7), 374, 1961.
156	24711	Streed, E.R., "Experimental Determination of the Thermal Radiation Properties of Temperature Control Surfaces for Spacecraft," Lockheed Missiles and Space Co., 14 pp., 1962. [AD 274 052]
157	51573	Kokoropoulos, P., Salam, E., and Daniels, F., "Selective Radiation Coatings. Preparation and High Temperature Stability," Solar Energy, 3, 19-23, 1959.
158	20641	Askwyth, W.H., Yahes, R.J., House, R.D., and Mikk, G., "Determination of the Emissivity of Materials. Vols. I, II, and III. Report Period: July 1, 1959 through December 31, 1962," PWA-2206, NASA-CR-56496, NASA-CR-56497, NASA-CR-56498, 277 pp., 1964.
159	52644	Kryzhanovskii, B.P., "Some Optical and Electrical Properties of Thin Films of Cu_3PSe, Cu_3PSeS_x and Cu_3PSeI_x," Optika i Spektroskopiya, 25(4), 613-5, 1968; English Translation: Opt. Spectrosc., 25(4), 343-4, 1968.
160	30644	Moore, V.S., Stetson, A.R., and Metcalfe, A.G., "Emittance Measurements of Refractory Oxide Coatings up to 2900°K," in Proc. Symp. on Measurement of Thermal Radiation Properties of Solids, Dayton, Ohio, Sept. 5, 6, and 7, 1962, NASA-SP-31, 527- 33, 1963.
161	45017	Sviridova, A.A. and Suikovskaya, N.V., "Transparence Limits of Interference Films of Hafnium and Thorium Oxides in the Ultraviolet Region of the Spectrum," Optika i Spektroskopiya, 22(6), 940-5, 1967; English Translation: Opt. Spectr'y, 22(6), 509-12, 1967.
162	47827	Howson, R.P., "Optical Properties of Thin Films of Indium Arsenide," Brit. J. Appl. Phys. 2, 1(1), 15-23, 1968.
163	43423	Long, R.L., "New Selectively Absorbing Coatings for Solar Powered Thermoelectric Devices," ML-TDR-64-302, 12 pp., 1964. [AD 455 304]
164	26693	Bausch and Lomb Inc., "Infrared Coating Studies. First Quarter Report 1963," 13 pp., 1963. [AD 408 635]
165	42979	Chang, L., "Infrared Coating Studies. Ninth Quarterly Report," Bausch and Lomb, Inc., 23 pp., 1965. [AD 471 653 L]
166	49233	Maier, R.L., "Infrared Coating Studies. Sixth Quarterly Report," Bausch and Lomb, Inc., 16 pp., 1967. [AD 664 786]
167	46403	Hammond, H.K. III and Nimeroff, I., "Measurement of Sixty-Degree Specular Gloss," J. Res. Nat. Bur. Stand., 44, 585-589, 1950.
168	25436	Gier, J.T., Dunkle, R.V., and Bevans, J.T., "Measurement of Absolute Spectral Reflectivity from 1.0 to 15 Microns," J. Opt. Soc. Am., 44(7), 558- 62, 1954.
169	41146	Zerlaut, G.A. and Krupnick, A.C., "Integrating-Sphere Reflectometer for the Determination of Absolute Hemispherical Spectral Reflectance," AIAA J., 4(7), 1227-32, 1966.
170	40351	Schatz, E.A. and Counts, C.R. III, "Improved Radiator Coatings," AFML-TDR-64-146, Pt. III, 95 pp., 1966. [AD 486 446L]
171	41745	Zhorov, G.A. and Sivakova, E.V., "Emissivity of Molybdenum Disilicide Coatings," Teplofiz. Vysokikh Temperatur, 4(2), 182-8, 1966; English Translation: High Temperature, 4(2), 180-6, 1966.
172	47309	Watanabe, M. and Kato, R., "Optical Study on Alkali Halide Films in the Ultraviolet," Japan. J. Appl. Phys., 7(1), 21-6, 1968.
173	39638	Bashara, N.M. and Peterson, D.W., "Ellipsometer Study of Anomalous Absorption in Very Thin Dielectric Films on Evaporated Metals," J. Opt. Soc. Am., 56(10), 1320-31, 1966.
174	47952	Alvares, W.J., "Emittance Measurements at the U. S. Naval Radiological Defense Laboratory," in Summary of the Sixth Refractory Composites Working Group Meeting, Volume II, 1962, ASD-TDR-63-610, 682-3, 1962. [AD 427 181]
175	52592	Mooij, J.E., Van de Bunt, W.B., and Schrijvers, J.E., "Thin Film Emission of KBr, RbCl, RbBr and NaCl in the Far Infra-Red," Phys. Lett., 28A(8), 573-4, 1969.
176	48978	Ejiri, A., "Optical Absorption of Longitudinal and Transverse Excitons in KCl and KBr," J. Phys. Soc. Japan, 24(5), 1181, 1968.

Ref. No.	TPRC No.	
177	42590	Mitsuishi, A., Yoshinaga, H., Yata, K., and Manabe, A., "Optical Measurement of Several Materials in the Far Infrared Region," Japan. J. Appl. Phys., Suppl., 4(1), 581-7, 1964.
178	16727	Anthony, F.M. and Pearl, H.A., Eds., "Investigation of Feasibility of Utilizing Available Heat Resistant Materials for Hypersonic Leading Edge Applications. Vol. 3. Screening Test Results and Selection of Materials," WADC-TR-59-744, 347 pp., 1960. [AD 247 110L]
179	6979	Betz, H.T., Olson, O.H., Schurin, B.D., and Morris, J.C., "Determination of Emissivity and Reflectivity Data on Aircraft Structural Materials. Part II. Techniques for Measurement of Total Normal Emissivity, Normal Spectral Emissivity, Solar Absorptivity and Presentation of Results," WADC-TR-56-222, 184 pp., 1957. [AD 202 493]
180	29570	Folweiler, R.C., "Thermal Radiation Characteristics of Transparent, Semi-Transparent, and Translucent Materials Under Non-Isothermal Conditions," ASD-TDR-62-719, 115 pp., 1964. [AD 600 370]
181	8144	Betz, H.T., Olson, O.H., Schurin, B.D., and Morris, J.C., "Determination of Emissivity and Reflectivity Data on Aircraft Structural Materials. Pt. 1: Techniques for Measurement of Total Normal Emissivity and Reflectivity with Some Data on Copper and Nickel," WADC-TR-56-222, 43 pp., 1956. [AD 110 458]
182	45954	Seki, H. and Moriyama, K., "Vapor Deposition of Silicon Nitride on GaAs by $SiCl_4$-NH_3-N_2 System," Japan. J. Appl. Phys., 6(11), 1345-6, 1967.
183	52081	Kryzhanovskii, B.P., "Infrared Reflection from Semiconducting Stannic Oxide Layers," Optika i Spektrosk., 25(3), 442-4, 1968; English Translation: Opt. Spectr'y, 25(3), 240-1, 1968.
184	48136	Sugano, T., Hirai, K., Kuroiwa, K., and Hoh, K., "Vapor Deposition of Silicon Nitride Film on Silicon and Properties of MNS Diodes," Japan. J. Appl. Phys., 7(2), 122-7, 1968.
185	31247	Dotson, L.E., "Emittance Coating Studies on Cb-Zr Alloy," General Electric Co., R61FPD571, 32 pp., 1962.
186	34835	Grenis, A.F., "Thermal Radiation of Complex Ceramic Solids," AMRA-TR-65-14, 19 pp., 1965. [AD 620 004]
187	8271	Ehlert, T.C. and Margrave, J.L., "Melting Point and Spectral Emissivity of Uranium Dioxide," J. Am. Ceram. Soc., 41(8), 330, 1958.
188	22522	Burgess, G.K. and Waltenberg, R.G., "The Emissivity of Metals and Oxides. II. Measurements with the Micropyrometer," Natl. Bur. Standards Bull., 11, 591-605, 1915.
189	49482	Verleur, H.W., "Determination of Optical Constants from Reflectance or Transmittance Measurements on Bulk Crystals or Thin Films," J. Opt. Soc. Am., 58(10), 1356-64, 1968.
190	50379	Riccius, H.D., "Infrared Lattice Vibrations of Zinc Selenide and Zinc Telluride," J. Appl. Phys., 39(9), 4381-2, 1968.
191	49362	Lupashko, E.A., Miloslavskii, V.K., and Shklyarevskii, I.N., "Use of the Kramers-Kronig Dispersion Relations in Determining the Phase Shift Occurring upon Reflection of Light from Thin Dielectric Layers," Optika i Spektroskopiya, 24(2), 257-62, 1968; English Translation: Opt. Spectr'y, 24(2), 132-4, 1968.
192	25672	Cox, J.T., Waylonis, J.E., and Hunter, W.R., "Optical Properties of Zinc Sulfide in the Vacuum Ultraviolet," J. Opt. Soc. Am., 49, 807-10, 1959.
193	48364	Schatz, E.A., "Thermal Radiation Properties of Binary Mixtures," Progr. Astronaut. Aeronaut., 18, 75-100, 1966.
194	10185	Fulk, M.M., Reynolds, M.M., and Park, O.E., "Thermal Radiation Absorption by Metals," in Proceedings of the 1954 Cryogenic Engineering Conference, NBS Rept. 3517, 151-7, 1955. [AD 125 047]
195	19293	Reynolds, M.M., Fulk, M.M., Weitzel, D.H., and Park, O.E., "A Preliminary Report on the Infrared Absorption of Metals at Low Temperature," NBS Tech. Memo 16, 10 pp., 1953.
196	50317	Kelly, D.P., "An Emissivity Study of Various Materials," University of Cincinnati, M.S. Thesis, 32 pp., 1960.
197	30613	Martin, W.E., "Hemispheric Spectral Reflectance of Solids," in Proc. Symp. on Measurement of Thermal Radiation Properties of Solids, Dayton, Ohio, Sept. 5, 6, and 7, 1962, NASA-SP-31, 183- 92, 1963.
198	22507	Hass, G. and Waylonis, J.E., "Optical Constants and Reflectance and Transmittance of Evaporated Aluminum in the Visible and Ultraviolet," J. Opt. Soc. Am., 51(7), 719-23, 1961.
199	24473	Bennett, H.E. and Koehler, W.F., "Precision Measurement of Absolute Specular Reflectance with Minimized Systematic Errors," J. Opt. Soc. Am., 50(1), 1-6, 1960.
200	27424	Bennett, H.E., Bennett, J.M., and Ashley, E.J., "Infrared Reflectance of Evaporated Aluminum Films," J. Opt. Soc. Am., 52, 1245-50, 1962.
201	29316	Janssen, J.E., Torborg, R.H., Luck, J.R., and Schmidt, R.N., "Normal Spectral Reflectance of Anodized Coatings on Aluminum, Magnesium, Titanium and Beryllium," ASD-TR-61-147, 269 pp., 1961. [AD 270 453]

Ref. No.	TPRC No.	
202	29572	Adams, J.G., "The Measurement of Spectral Reflectivities of Materials for Thermal Control Surfaces," Northrop Space Labs., NSL-62-198, 101 pp., 1962.
203	29594	Edwards, D.K., Gier, J.T., Nelson, K.E., and Roddick, R.D., "Spectral and Directional Thermal Radiation Characteristics of Selective Surfaces for Solar Collectors," Solar Energy, 6(1), 1-8, 1962.
204	36229	Birkebak, R.C. and Eckert, E.R.G., "Effects of Roughness of Metal Surfaces on Angular Distribution of Monochromatic Reflected Radiation," J. Heat Transfer, 87(1), 85-94, 1965.
205	31732	Boettcher, A., "The High-Reflecting Power of Vaporized Al-Ag and Al-Mg Alloys," Z. Angew Phys., 2, 340-3, 1950.
206	9111	Gates, D.M., Shaw, C.C., and Beaumont, D., "Infrared Reflectance of Evaporated Metal Films," J. Opt. Soc. Amer., 48(2), 88-9, 1958.
207	9438	Twidle, G.G., "The Spectral Reflectivity of Back-Surface and Front-Surface Aluminized Mirrors," Brit. J. Appl. Phys., 8(8), 337-9, 1957.
208	27141	Bogdan, L., "Measurement of Radiative Heat Transfer with Thin-Film Resistance Thermometers," NASA-CR-27, 39 pp., 1964.
209	48567	Feuerbacher, B., Godwin, R.P., and Skibowski, M., "Plasma Resonance in the Reflection Spectrum of Thin Aluminium Films," Phys. Lett., 26A(12), 595-6, 1968.
210	49206	Morse, A.L., "Optical and Photoelectirc Properties Including Polarization Effects of Gold and Aluminum in the Extreme Ultraviolet. Tech. Rept. Jan. 1965-Jan. 1968," NASA-CR-93971, 94 pp., 1968.
211	50297	Stanford, J.L., "Photoelectric and Optical Properties of Commercial Platinum, Gold and Palladium Foils and Evaporated Aluminum and Silver Films in the Extreme Ultraviolet," Tennessee Univ., Ph.D. Thesis, 106 pp., 1966.
212	44317	Madden, R.P. and Canfield, L.R., "Apparatus for the Measurement of Vacuum Ultraviolet Optical Properties of Freshly Evaporated Films Before Exposure to Air," J. Opt. Soc. Am., 51(8), 838- 45, 1961.
213	40793	Rustgi, O.P., Walker, W.C., and Weissler, G.L., "Optical Properties of Sb, Te, and Ti Films in the Vacuum Ultraviolet," J. Opt. Soc. Am., 51(12), 1357-9, 1961.
214	36346	Rustgi, O.P., "Transmittance of Thin Metallic Films in the Vacuum-Ultraviolet Region below 1000 Angstroms," J. Opt. Soc. Am., 55(6), 630-4, 1965.
215	23258	Langley, R.C., "Inorganic Films for Solar Energy Absorption. Part I. Final Summary Technical Rept., 1 Feb.-31 Dec. 1961," ASD-TDR-62-92, 33 pp., 1962. [AD 282 008]
216	7487	Blackman, M., Egerton, A., and Truter, E.V., "Heat Transfer by Radiation to Surfaces at Low Temperatures," Proc. Roy. Soc., Ser. A, 194, 147- 69, 1948.
217	27333	Ziegler, W.T. and Cheung, H., "Total Emissivity of Some Surfaces at 77°K," Advan. Cryog. Eng., 2, 100-3, 1960.
218	45015	Idchak, E.F., "Optical Properties of Double Metallic Films," Optika i Spektroskopiya, 22(6), 935-9, 1967; English Translation: Opt. Spectr'y, 22(6), 507-9, 1967.
219	22516	Coblentz, W.W. and Stair, R., "Reflecting Power of Beryllium, Chromium, and Several Other Metals," J. Res., Natl. Bur. Stand., 2, 343- 54, 1929.
220	36486	Aronson, J.R. and McLinden, H.G., "Far-Infrared Spectra of Solids," in Symp. on Thermal Radiation of Solids, San Francisco, Calif., March 4, 5, 6, 1964, NASA-SP-55, 29-38, 1965. [AD 629 980]
221	38390	Edwards, D.K. and de Volo, B. N., "Useful Approximations for the Spectral and Total Emissivity of Smooth Bare Metals," in Advances in Thermophysical Properties at Extreme Temperatures and Pressures, Third ASME Symp. on Thermophysical Properties, Purdue Univ., Lafayette, Ind., March 22-25, 1965, 174-88, 1965.
222	24862	Birkebak, R.C., Hartnett, J.P., and Eckert, E.R.G., "Measurement of Radiation Properties of Solid Materials," in Progress in International Research on Thermodynamic and Transport Properties, ASME Second Symposium on Thermophysical Properties, Princeton, N.J., Jan. 24-6, 1962, 563-74, 1962.
223	7173	Tolansky, S. and Donaldson, W.K., "The Reflectivity of Speculum Metal," J. Sci. Instruments, 24(9), 248-9, 1947.
224	25235	Turner, A.F., "Infrared Coating Studies. Quarterly Rept. No. 3, 15 Sept-15 Dec. 1963," Bausch and Lomb, Inc., 12 pp., 1963. [AD 429 240]
225	50023	Deverall, J.E., "Total Hemispherical Emissivity Measurements by the Heat Pipe Method," LA-3834-MS, 10 pp., 1967.
226	25073	Macklin, B.A., Withers, J.C., and Schatz, E.A., "Heat Barrier Coatings," ASD-TR-61-5, 72 pp., 1961. [AD 266 335]

Ref. No.	TPRC No.	
227	15016	Philip, R., "Change of Optical Properties of a Thin Gold Film as a Function of Wavelength," Compt. Rend., 247(25), 2322-4, 1958.
228	21573	Crowell, C.R., Spitzer, W.G., Howarth, L.E., and LaBate, E.E., "Attenuation Length Measurements of Hot Electrons in Metal Films," Phys. Rev., 127, 2006-15, 1962.
229	36321	Bennett, J.M. and Ashley, E.J., "Infrared Reflectance and Emittance of Silver and Gold Evaporated in Ultrahigh Vacuum," Appl. Opt., 4(2), 221-4, 1965.
230	44314	Fukutani, H. and Sueoka, O., "Optical Properties of Ag-Au Alloys," in Optical Properties and Electronic Structure of Metals and Alloys, North Holland Publishing Co., Amsterdam, 565-73, 1966.
231	33388	Belser, R.B. and Carithers, M.D., "Reflective Coatings on Polymeric Substrates," ASD-TR-61-151 (Pt. 2), 101 pp., 1962. [AD 286 836]
232	18520	Philip, R., "Coefficients of Reflection, Transmission, and Absorption of Thin Layers of Gold for Radiation of Wavelengths between 2536 and 6438 A," J. Phys. Radium, 20, 535-40, 1959.
233	40221	Hass, G., Jacobus, G.F., and Hunter, W.R., "Optical Properties of Evaporated Iridium in the Vacuum Ultraviolet from 5000 Angstroms to 2000 Angstroms," J. Opt. Soc. Am., 57(6), 758-62, 1967.
234	40233	Samson, J.A.R., Padur, J.P., and Sharma, A., "Reflectance and Relative Transmittance of Laser-Deposited Iridium in the Vacuum Ultraviolet," J. Opt. Soc. Am., 57(7), 966-7, 1967.
235	44860	Goncharova, E.V. and Kukharskii, A.A., "Determination of Effective Mass in LaSb from the Infrared Reflection Spectrum," Fiz. Tverd. Tela, 9(5), 1543-5, 1967; English Translation: Soviet Physics-Solid State, 9(5), 1214-5, 1967.
236	49181	Okorie, O.A., "Optical and Electrical Properties of Rare Earth Thin Films," Howard University, Ph.D. Thesis, 116 pp., 1967.
237	33673	Duhig, J.J. Jr., "A Calorimetric Determination of Thermal Emissivity," Georgia Inst. of Tech., M.S. Thesis, 32 pp., 1963.
238	22272	Schatz, E.A., Goldberg, D.M., Pearson, E.G., and Burks, T.L., "High Temperature, High Emittance Intermetallic Coatings. Part 1. Emittance and Reflectance of Intermetallic Compounds," ASD-TDR-63-657, 181 pp., 1963. [AD 423 743]
239	45735	Yamaguchi, S., "Optical Properties of Silver Films - Direct Observation of the Im/e-Spectrum," J. Phys. Soc. Japan., 18(2), 266- 72, 1963.
240	27269	Cline, D., "Infrared Wavelength Dependence of the Total Absorptivity of Electroplated Silver," J. Appl. Phys., 33, 2310-11, 1962.
241	53852	Yamaguchi, T., Yoshida, S., and Kinbara, A., "Continuous Ellipsometric Determination of the Optical Constants and Thickness of a Silver Film During Deposition," Japan. J. Appl. Phys., 8(5), 559-67, 1969.
242	30646	Pears, C.D., "Some Problems in Emittance Measurements at the Higher Temperatures and Surface Characterization," in Proc. Symp. on Measurement of Thermal Radiation Properties of Solids, Dayton, Ohio, Sept. 5, 6, and 7, 1962, NASA-SP-31, 541- 51, 1963.
243	50483	Shklyarevskii, I.N. and Korneeva, T.I., "Absorption of Light in Thin Granular Silver Films," Optika i Spektroskopiya, 24(5), 744-50, 1968; English Translation: Opt. Spectr'y, 24(5), 398-401, 1968.
244	48383	Streed, E.R., Cunnington, G.R., and Zierman, C.A., "Performance of Multilayer Insulation Systems for the 300 to 800°K Temperature Range," Progr. Astronaut. Aeronaut., 18, 735-71, 1966.
245	39215	Rideout, V.L. and Wemple, S.H., "Optical Constants of Evaporated Gold and Platinum Films on Potassium Tantalate," J. Opt. Soc. Am., 56(6), 749-51, 1966.
246	39622	Bradley, D. and Entwistle, A.G., "The Total Hemispherical Emittance of Coated Wires," Brit. J. Appl. Phys., 17(9), 1155-64, 1966.
247	52920	Hass, G., Ramsey, J.B., Heaney, J.B., and Triolo, J.J., "Reflectance, Solar Absorptivity, and Thermal Emissivity of SiO_2-Coated Aluminum," Appl. Opt., 8(2), 275-81, 1969.
248	44933	Starner, K.E. and Stark, R.L., "Effects of Front Surface Roughening on Solar Absorptivity of Quartz Rear Surface Mirror Satellite Coatings," SSD-TR-67-69, 31 pp., 1967. [AD 815 036]
249	51932	Ikeda, Y., "Variation of Infrared Absorption of Silicon Oxide Films from Oxidation of Silane with H_2 Treatment," Jap. J. Appl. Phys., 7(12), 1543-4, 1968.
250	16606	Blau, H.H. Jr., March, J.B., Martin, W.S., Jasperse, J.R., and Chaffee, E., "Infrared Spectral Emittance Properties of Solid Materials," AFCRL-TR-60-416, 78 pp., 1960. [AD 248 276]
251	40651	Hass, G., "Filmed Surfaces for Reflecting Optics," J. Opt. Soc. Am., 45(11), 945-52, 1955.
252	50111	Kubota, T., Hisano, K., and Matumura, O., "Far-Infrared Absorption Due to the Virtual Modes in NaCl," J. Phys. Soc. Japan, 25(2), 642, 1968.

Ref. No.	TPRC No.	
253	42270	Cox, J.T., Hass, G., and Waylonis, J.E., "Further Studies on LiF-Overcoated Aluminum Mirrors with Highest Reflectance in the Vacuum Ultraviolet," Appl. Opt., 7(8), 1535-9, 1968.
254	29641	Bausch and Lomb Co., "Emissivity Enhancement of Solar Cells for Temperature Control. Technical Report No. 1," B/L-TR-1, 8 pp., 1961.
255	48109	Kreidl, N.J., Hafner, H.C., Hensler, J.R., Weidel, R.A., and Letter, E.C., "An Investigation of Infrared Transmitting Materials. Part 1," WADC-TR-55-500, 150 pp., 1957. [AD 130 904]
256	48720	Shapiro, J.T. and Madden, R.P., "Optical Constants of Polystyrene in the Vacuum Ultraviolet," J. Opt. Soc. Am., 58(6), 771-5, 1968.
257	33188	Hawthorn, A.T., Simon, A.B., Brock, C.L., and Sigel, L.A., "Development of Fabric Base Materials for Space Applications. Rept. for July 62-Nov. 63," ML-TDR-64-20, 140 pp., 1964. [AD 437 821]
258	40420	Wetmore, R.A., "Monochromatic Reflectance Tests," BSD-TR-66-140, 116 pp., 1963. [AD 483 037]
259	35723	Blevin, W.R. and Brown, W.J., "An Infrared Reflectometer with a Spheroidal Mirror," J. Sci. Instr., 42(6), 385-9, 1965.
260	47276	Clarke, D.R., Gillette, R.B., and Beck, T.R., "Development of a Barrier-Layer Anodic Coating for Reflective Aluminum in Space," Progr. Astronaut. Aeronaut., 20, 315-28, 1967.
261	6459	Taylor, C.S. and Edwards, J.D., "Some Reflection and Radiation Characteristics of Aluminium," Heating, Piping, and Air-Conditioning, 11(1), 59-63, 1939.
262	33206	Schmidt, E., "Report on the Measurement of the Total Radiation of Aluminum Surfaces," Hauszeitschr. V. A. W. Erftwerk A. G. Aluminium, 3, 91-6, 1930.
263	42368	Rogers, R.N. and Morris, E.D. Jr., "Determination of Emissivities with a Differential Scanning Calorimeter," Anal. Chem., 38(3), 410-12, 1966.
264	25038	Jenkins, R.J., Butler, C.P., and Parker, W.J., "Total Hemispherical Emittance Measurements over the Temperature Range 77 K to 300 K," SSD-TDR-62-189, 57 pp., 1963. [AD 419 067]
265	29493	Weaver, J.H., "Anodized Aluminum Coatings for Temperature Control of Space Vehicles," ASD-TDR-62-918, 29 pp., 1963. [AD 402 381]
266	46810	Curtis, H.B., "Measurement of Hemispherical Total Emittance and Normal Solar Absorptance of Selected Materials in the Temperature Range 280 to 600 K," J. Spacecraft Rockets, 3(3), 383-7, 1966.
267	36700	Freeman, G.H.C., "The Optical Constants of Aluminium Oxide from 12·0 to 24·4 eV," Brit. J. Appl. Phys., 16(7), 927- 31, 1965.
268	29596	Dunkle, R.V., Edwards, D.K., Gier, J.T., and Bevans, J.T., "Solar Reflectance Integrating Sphere," Solar Energy, 4(2), 27-39, 1960.
269	25561	Porter, J. and Butler, E.A.W., "Spectral Absorptance of Metal Surfaces within the Solar Spectral Bandwidth," RAE-TN-S-64, 24 pp., 1964. [AD 447 164]
270	32619	Janssen, J., Luck, J., and Torborg, R., "Thermal Reflectance of Anodized Metals for Spacecraft," Tech. Proc. Amer. Electroplaters' Soc., 49, 126- 32, 1962.
271	36533	Schmidt, R.N. and Janssen, J.E., "Selective Coatings for Vacuum-Stable High-Temperature Solar Absorbers," in Symp. on Thermal Radiation of Solids, San Francisco, Calif., March 4, 5, 6, 1964, NASA-SP-55, 509-24, 1965. [AD 629 980]
272	31790	Hall, W.M., "The Application of Temperature Rate Measurements to the Determination of Thermal Emittance," NASA-CR-56037, 8 pp., 1964.
273	40414	Wakelyn, N.T. and Pezdirtz, G.F., "Some Poly-Basic Phosphate Conversion Coatings for Thermal Control," in Proc. of Conf. on Spacecraft Coatings Development, NASA-TM-X-56167, 17 pp., 1964.
274	11723	Reynolds, P.M., "Spectral Emissivity of 99.7% Aluminium Between 200 and 540 C," Brit. J. Appl. Physics, 12(3), 111-114, 1961.
275	37477	Alvares, N.J., "Emittance Measurements of Disilicide-Type Coatings at the U. S. Naval Radiological Defense Laboratory," in Summary of the Seventh Refractory Composites Working Group Meeting, Volume 2, RTD-TDR-63-4131, 341-51, 1963. [AD 601 265]
276	16590	Fieldhouse, I.B., Lang, J.I., and Blau, H.H. Jr., "Investigation of Feasibility of Utilizing Available Heat Resistant Materials for Hypersonic Leading Edge Applications. Vol. IV. Thermal Properties of Molybdenum Alloy and Graphite," WADC-TR-59-744, 78 pp., 1960. [AD 249 166]
277	24864	Pears, C.D., "The Determination of the Emittance of Refractory Materials to 5000 F," in Progress in International Research on Thermodynamic and Transport Properties, ASME Second Symposium on Thermophysical Properties, Princeton, N.J., Jan. 24-6, 1962, 588-98, 1962.
278	23126	Coffman, J.A., Kibler, G.M., Lyon, T.F., and Acchione, B.D., "Carbonization of Plastics and Refractory Materials Research. Final Rept. for 1 July 60-30 June 61 on Material Thermal Properties," WADD-TR-60-646 (Pt. 2), 183 pp., 1963. [AD 297 946]

Ref. No.	TPRC No.	
279	27592	Seban, R.A., "Thermal Radiation Properties of Materials, Part III," WADD-TR-60-370, 68 pp., 1963. [AD 419 028]
280	33603	Seban, R.A., "Thermal Radiation Properties of Materials, Part II," WADD-TR-60-370, 72 pp., 1962. [AD 286 863]
281	32639	Lapina, E.A. and Chudnovskii, F.A., "Spectral Emissivity of an Oxide Cathode," Teplofiz. Vysokikh Temperatur, 3(5), 686-90, 1965; English Translation: High Temperature (USSR), 3(5), 639-42, 1965.
282	43458	Kerr, J.R. and Cox, J.D., "Effect of Environmental Exposure on Mechanical Properties of Several Foil Gage Refractory Alloys and Superalloys," AFML-TR-65-92, 146 pp., 1965. [AD 464 905]
283	10461	Blau, H.H. Jr., Chaffee, E., Jasperse, J.R., and Martin, W.S., "High Temperature Thermal Radiation Properties of Solid Materials," AFCRC-TN-60-165, 71 pp., 1960. [AD 236 394]
284	16963	Riethof, T.R., "High Temperature Spectral Emissivity Studies," R6ISD004, 34 pp., 1961. [AD 250 274]
285	43741	Bartlett, R.W. and Gage, P.R., "Investigation of Mechanisms for Oxidation Protection and Failure of Intermetallic Coatings for Refractory Metals," ASD-TDR-63-753 (Pt. II), 127 pp., 1964. [AD 609 167]
286	33749	Shelton, J.L., "The Total Hemispherical Thermal Emittance of Nickel as a Function of Oxide Thickness in the Temperature Range 400-900°C," Rice Univ., Ph.D. Thesis, 146 pp., 1965.
287	39494	Sato, K. and Shibata, M., "Reststrahl Reflection Characteristics of Amorphous Silica," J. Phys. Soc. Japan, 21(6), 1088-96, 1966.
288	40269	Wesson, R.A., Phillips, R.P., and Pliskin, W.A., "Phase-Shift-Corrected Thickness Determination of Silicon Dioxide on Silicon by Ultraviolet Interference," J. Appl. Phys., 38(6), 2455-60, 1967.
289	43768	Fraenz, I., Langheinrich, W., and Loecherer, K.H., "Investigation of the Diffusion in Silicon and in the Diffusion Inhibiting Films of Silicon Dioxide with Radioactive Methods," Telefunken-Zeitung, 37(3/4), 194-209, 1964; English Translation: TT-66-11027, 45 pp., 1966.
290	49766	Sugano, T., Hoh, K., Kudo, K., and Hishinuma, N., "Ordered Structure and Ion Migration in Silicon Dioxide Films," Japan. J. Appl. Phys., 7(7), 715-30, 1968.
291	51945	Rakov, A.V., Potapov, E.V., and Mizgireva, L.P., "Investigation of the Optical Properties of SiO_2 Films Obtained by Thermal Oxidation of Silicon in the 7.0-11.0 Micron Wavelength Range," Opt. Spektrosk., 25(1), 117-21, 1968; English Translation: Opt. Spectrosc., 25(1), 59-61, 1968.
292	32713	Slemp, W.S., "Effects of Preoxidation Treatments on Spectral Normal and Total Normal Emittance of Inconel, Inconel-X, and Type 347 Stainless Steel," NASA-TN-D-2300, 56 pp., 1964.
293	21284	Harrison, W.N., Richmond, J.C., Shorten, F.J., and Joseph, H.M., "Standardization of Thermal Emittance Measurements. Part 4. Normal Spectral Emittance, 800-1400°K," WADC-TR-59-510, 90 pp., 1963. [AD 426 846]
294	53923	Zhorov, G.A. and Sivakova, E.V., "Effect of Oxidation on the Emissivity of Molybdenum Disilicide Coatings," Teplofiz. Vys. Temp., 6(6), 1040-3, 1968; English Translation: High Temperature, 6(6), 995-8, 1968.
295	46990	Carroll, W.F., "Design, Test, and Performance of the Mariner V Temperature Control Reference," JPL-TR-32-1250, 25 pp., 1968.

Material Index

Material Index to Thermal Radiative Properties
Companion Volumes 7, 8, and 9

MATERIAL NAME	VOLUME	EMITTANCE						REFLECTANCE									ABSORPTANCE									TRANSMITTANCE									Absorptance to Emittance Ratio
		Total			Spectral			Integrated			Spectral			Solar			Integrated			Spectral			Solar			Integrated			Spectral			Solar			
		Hemispherical	Normal	Angular	Hemispherical	Normal	Angular	Hemispherical	Normal	Angular	Hemispherical	Normal	Angular	Hemispherical	Normal	Angular	Hemispherical	Normal	Angular	Hemispherical	Normal	Angular	Hemispherical	Normal	Angular	Hemispherical	Normal	Angular	Hemispherical	Normal	Angular	Hemispherical	Normal	Angular	
A110-AT, Anodized	9	-	-	-	-	-	-	-	-	-	-	1289	-	-	-	-	-	-	-	-	-	-	-	-	-	-	-	-	-	-	-	-	-	-	-
Acetylene Black Pigment in Xylol Binder	9	81	-	-	-	-	-	-	-	-	-	-	-	-	-	-	-	-	-	-	-	-	-	-	-	-	-	-	-	-	-	-	-	-	-
Acme Quality Enamel No. 800	9	-	-	-	-	-	-	-	-	-	-	483	-	-	-	-	-	-	-	-	-	-	-	-	-	-	-	-	-	-	-	-	-	-	-
Acme Quality Enamel No. 801	9	-	-	-	-	-	-	-	-	-	-	483	-	-	-	-	-	-	-	-	-	-	-	-	-	-	-	-	-	-	-	-	-	-	-
Acme Quality Enamel No. 803	9	-	-	-	-	-	-	-	-	-	-	483	-	-	-	-	-	-	-	-	-	-	-	-	-	-	-	-	-	-	-	-	-	-	-
Acrylic Binder Pigmented with:																																			
Aluminum	9	3	-	-	-	-	-	-	-	-	-	-	-	-	-	-	-	-	-	-	20	-	-	-	-	-	-	-	-	-	-	-	-	-	-
Carbon Black	9	81	-	-	-	-	-	-	-	-	-	86	-	-	-	-	-	-	-	-	-	-	-	89	-	-	-	-	-	-	-	-	-	-	-
Lead Carbonate	9	-	-	-	-	-	-	-	-	-	-	139	-	-	-	-	-	-	-	-	-	-	-	-	-	-	-	-	-	-	-	-	-	-	-
Magnesium Oxide	9	-	-	-	-	-	-	-	-	-	-	161	-	-	-	-	-	-	-	-	-	-	-	-	-	-	-	-	-	-	-	-	-	-	-
Strontium Molybdate	9	-	189	-	-	-	-	-	-	-	-	191	-	-	-	-	-	-	-	-	-	-	-	194	-	-	-	-	-	-	-	-	-	-	-
Titanium Dioxide	9	211	-	-	-	-	-	-	-	-	-	223	-	-	-	-	-	-	-	-	-	-	-	-	-	-	-	-	-	-	-	-	-	-	-
Titanium Dioxide + Talc	9	290	-	-	-	-	-	-	-	-	-	-	-	-	-	-	-	-	-	-	293	-	-	295	-	-	-	-	-	-	-	-	-	-	-
Zinc Sulfide	9	-	-	-	-	-	-	-	-	-	-	404	-	-	-	-	-	-	-	-	-	-	-	-	-	-	-	-	-	-	-	-	-	-	-
Acrylic Coating on:																																			
Aluminum Substrate	9	1108	-	-	-	-	-	-	-	-	-	1110	-	-	-	-	-	-	-	-	-	-	-	-	-	-	-	-	-	-	-	-	-	-	-
Aluminum Oxide Substrate	9	-	1109	-	-	-	-	-	-	-	-	-	-	-	-	-	-	-	-	-	-	-	-	-	-	-	-	-	-	-	-	-	-	-	-
Acryloid-A10 Binder Pigmented with:																																			
Lead Carbonate	9	-	-	-	-	-	-	-	-	-	-	138	-	-	-	-	-	-	-	-	-	-	-	-	-	-	-	-	-	-	-	-	-	-	-
Zinc Sulfide	9	-	-	-	-	-	-	-	-	-	-	404	-	-	-	-	-	-	-	-	-	-	-	-	-	-	-	-	-	-	-	-	-	-	-

MATERIAL NAME	VOLUME	EMITTANCE						REFLECTANCE									ABSORPTANCE									TRANSMITTANCE									Absorptance to Emittance Ratio
		Total			Spectral			Integrated			Spectral			Solar			Integrated			Spectral			Solar			Integrated			Spectral			Solar			
		Hemispherical	Normal	Angular	Hemispherical	Normal	Angular	Hemispherical	Normal	Angular	Hemispherical	Normal	Angular	Hemispherical	Normal	Angular	Hemispherical	Normal	Angular	Hemispherical	Normal	Angular	Hemispherical	Normal	Angular	Hemispherical	Normal	Angular	Hemispherical	Normal	Angular	Hemispherical	Normal	Angular	
ADP, See Ammonium Dihydrogen Phosphate	8																																		
Ag_2Al	8	-	-	-	-	-	-	-	-	-	-	1352	-	-	-	-	-	-	-	-	-	-	-	-	-	-	-	-	-	-	-	-	-	-	-
AgCd	8	-	-	-	-	-	-	-	-	-	-	-	1326	-	-	-	-	-	-	-	-	-	-	-	-	-	-	-	-	-	-	-	-	-	-
AgZn	8	-	-	-	-	-	-	-	-	-	-	-	1328	-	-	-	-	-	-	-	-	-	-	-	-	-	-	-	-	-	-	-	-	-	-
AI-93	9	201	-	-	-	-	-	-	-	-	-	-	-	-	-	-	-	-	-	-	-	-	-	-	-	-	-	-	-	-	-	-	-	-	-
AISI 99 Grey Paint	9	-	-	-	-	-	-	-	-	-	-	-	-	-	-	-	-	-	-	-	485	-	-	-	-	-	-	-	-	486	-	-	-	-	-
Al + SiO_2	8	-	-	-	-	-	-	-	-	-	-	1428	-	-	-	-	-	-	-	-	-	-	-	-	-	-	-	-	-	-	-	-	-	-	-
Al + $TaAl_3$	8	-	-	-	-	-	-	-	-	-	-	1431	-	-	-	-	-	-	-	-	-	-	-	-	-	-	-	-	-	-	-	-	-	-	-
Al + ZrO_2	8	-	-	-	-	-	-	-	-	-	-	1442	-	-	-	-	-	-	-	-	-	-	-	-	-	-	-	-	-	-	-	-	-	-	-
Alcoa No. 2 Reflector Plate	7	4 5	-	-	-	-	-	-	-	-	-	-	-	-	-	-	42 43	-	-	-	-	-	-	-	-	-	-	-	-	-	-	-	-	-	-
Alkyd Binder Pigmented with:																																			
Aluminum Oxide	9	-	-	-	-	-	-	-	-	-	-	33	-	-	-	-	-	-	-	-	-	-	-	-	-	-	-	-	-	-	-	-	-	-	-
Lead Carbonate	9	-	-	-	-	-	-	-	-	-	-	139	-	-	-	-	-	-	-	-	-	-	-	-	-	-	-	-	-	-	-	-	-	-	-
Magnesium Oxide	9	-	-	-	-	-	-	-	-	-	-	161	-	-	-	-	-	-	-	-	-	-	-	-	-	-	-	-	-	-	-	-	-	-	-
Titanium Dioxide	9	211	214	-	-	-	-	-	-	-	-	221	250	-	-	-	-	-	-	-	256	-	-	-	-	-	-	-	-	-	-	-	-	-	-
Titanium Dioxide + Calcium Sulfate	9	-	-	-	-	-	-	-	-	-	-	289	-	-	-	-	-	-	-	-	-	-	-	-	-	-	-	-	-	-	-	-	-	-	-
Zinc Sulfide + Clay	9	-	-	-	-	-	-	-	-	-	-	-	-	-	-	-	-	-	-	-	-	-	-	416	-	-	-	-	-	-	-	-	-	-	-
Alkyd Coating on Aluminum Substrate	9	-	-	-	-	-	-	-	-	-	-	1111	-	-	-	-	-	-	-	-	-	-	-	-	-	-	-	-	-	-	-	-	-	-	-
Alleghany Alloy No. 66	7	-	1180	-	-	-	-	-	-	-	-	-	-	-	-	-	-	-	-	-	-	-	-	-	-	-	-	-	-	-	-	-	-	-	-
Alleghany Metal	7	-	1225	-	-	-	-	-	-	-	-	-	-	-	-	-	-	-	-	-	-	-	-	-	-	-	-	-	-	-	-	-	-	-	-
Al_2O_3 + Cr + ΣX_i Cermet	8	-	1355	-	-	-	-	-	-	-	-	-	-	-	-	-	-	-	-	-	-	-	-	-	-	-	-	-	-	-	-	-	-	-	-
Al_2O_3 + Cr_2O_3 Powders	8	-	-	-	-	554	-	-	-	-	-	-	-	-	-	-	-	-	-	-	-	-	-	-	-	-	-	-	-	-	-	-	-	-	-
Al_2O_3 + NiAl Cermet	8	-	1358	-	-	1359	-	-	-	-	-	1363	-	-	-	-	-	-	-	-	-	-	-	-	-	-	-	-	-	-	-	-	-	-	-
Al_2O_3 + NiO Powders	8	-	-	-	-	556	-	-	-	-	-	-	-	-	-	-	-	-	-	-	-	-	-	-	-	-	-	-	-	-	-	-	-	-	-
Al_2O_3 + SiO_2 Powders	8	-	-	-	-	558	-	-	-	-	-	560	-	-	-	-	-	-	-	-	-	-	-	-	-	-	-	-	-	-	-	-	-	-	-
Al_2O_3 + W + ΣX_i Cermet	8	-	1375	-	-	-	-	-	-	-	-	-	-	-	-	-	-	-	-	-	-	-	-	-	-	-	-	-	-	-	-	-	-	-	-
Alodine 401-45	9	-	-	-	-	1203	-	-	-	-	-	1208	-	-	-	-	-	-	-	-	-	-	-	-	-	-	-	-	-	-	-	-	-	-	1233
Alodine 1200	9	1234	-	-	-	-	-	-	-	-	-	-	-	-	-	-	-	-	-	-	-	-	-	-	-	-	-	-	-	-	-	-	-	-	-

MATERIAL NAME	VOLUME	EMITTANCE						REFLECTANCE									ABSORPTANCE									TRANSMITTANCE									Absorptance to Emittance Ratio
		Total			Spectral			Integrated			Spectral			Solar			Integrated			Spectral			Solar			Integrated			Spectral			Solar			
		Hemispherical	Normal	Angular	Hemispherical	Normal	Angular	Hemispherical	Normal	Angular	Hemispherical	Normal	Angular	Hemispherical	Normal	Angular	Hemispherical	Normal	Angular	Hemispherical	Normal	Angular	Hemispherical	Normal	Angular	Hemispherical	Normal	Angular	Hemispherical	Normal	Angular	Hemispherical	Normal	Angular	
AlSb	8	–	–	–	–	–	–	–	–	–	–	1352	–	–	–	–	–	–	–	–	–	–	–	–	–	–	–	–	–	–	–	–	–	–	–
Alucer MC (Al_2O_3) Pigment in Potassium Silicate Binder	9	–	–	–	–	–	–	–	–	–	–	–	–	–	–	–	–	–	–	–	41	–	–	–	–	–	–	–	–	–	–	–	–	–	–
Aluminates:																																			
Calcium Aluminate	8	–	–	–	–	–	–	–	–	–	–	–	–	–	–	–	–	–	–	–	–	–	–	–	–	–	–	–	–	573	–	–	–	–	–
Magnesium Aluminate	8	–	–	–	–	–	–	–	–	–	–	576	–	–	–	–	–	–	–	–	–	–	–	–	–	–	–	–	–	577	–	–	–	–	–
Yttrium Aluminate	8	–	–	–	–	–	–	–	–	–	–	579	–	–	–	–	–	–	–	–	–	–	–	–	–	–	–	–	–	–	–	–	–	–	–
Aluminum	7	2	8	–	–	12	15	–	–	18	20	24	34 38	–	40	–	42	45	–	–	47	50	–	52	55	–	–	–	–	57	60	–	–	–	–
Aluminum, Anodized	9	1195	1199	–	–	1203	–	–	–	–	–	1207 1216	1219 1220	–	1222	–	1224	–	–	–	–	–	–	1226 1230	1231	–	–	–	–	–	–	–	–	–	1233
Aluminum Alloys, Anodized (See Anodized Aluminum Alloys)																																			
Aluminum Alloys:																																			
24-ST	7	1063	1068 1069	–	–	–	–	–	–	–	–	1078	–	–	1084	–	–	–	–	–	–	–	–	1087	–	–	–	–	–	–	–	–	–	–	–
53-SO	7	–	1101	–	–	–	–	–	–	–	–	–	–	–	–	–	–	–	–	–	–	–	–	–	–	–	–	–	–	–	–	–	–	–	–
75-ST	7	–	1119	–	–	–	–	–	–	–	–	–	–	–	1129	–	–	–	–	–	–	–	–	1132	–	–	–	–	–	–	–	–	–	–	–
1075	7	–	–	–	–	–	–	–	–	–	–	28	–	–	–	–	–	–	–	–	–	–	–	–	–	–	–	–	–	–	–	–	–	–	–
1100	7	–	10	–	–	–	16	–	–	–	–	–	–	–	–	–	–	–	–	–	–	–	–	–	–	–	–	–	–	–	–	–	–	–	–
2024	7	1063	–	–	–	1074	–	–	–	–	–	–	–	–	–	–	–	–	–	–	–	–	–	1087	–	–	–	–	–	–	–	–	–	–	–
2024-T (see 24-ST)																																			
2219	7	–	–	–	–	–	–	–	–	–	–	1078	–	–	–	–	–	–	–	–	–	–	–	–	–	–	–	–	–	–	–	–	–	–	–
3003	7	1112	1114	–	–	–	–	–	–	–	–	–	–	–	–	–	–	–	–	–	–	–	–	–	–	–	–	–	–	–	–	–	–	–	–
5053-O (see 53-SO)																																			
6061	7	1098	–	–	–	–	–	–	–	–	–	1106	–	–	–	–	–	–	–	–	–	–	–	1110	–	–	–	–	–	–	–	–	–	–	–
7075	7	–	–	–	–	1122	–	–	–	–	–	–	–	–	–	–	–	–	–	–	–	–	–	–	–	–	–	–	–	–	–	–	–	–	–
7075-T (see 75-ST)																																			
7075-T6	7	–	–	–	–	–	1126	–	–	–	–	–	–	–	–	–	–	–	–	–	–	–	–	–	–	–	–	–	–	–	–	–	–	–	–
Alclad 24-SO	7	–	–	–	–	–	–	–	–	–	–	–	–	–	1084	–	–	–	–	–	–	–	–	–	–	–	–	–	–	–	–	–	–	–	–
Alclad 24-ST	7	–	1068 1069	–	–	–	–	–	–	–	–	–	–	–	–	–	–	–	–	–	–	–	–	–	–	–	–	–	–	–	–	–	–	–	–
Alclad 75-ST	7	1116	1119	–	–	–	–	–	–	–	–	–	–	–	1129	–	–	–	–	–	–	–	–	–	–	–	–	–	–	–	–	–	–	–	–
Alclad 2024-O (see Alclad 24-SO)																																			
Alclad 2024-T (see Alclad 24-ST)																																			
Alclad 7075-T (see Alclad 75-ST)																																			

MATERIAL NAME	VOLUME	EMITTANCE						REFLECTANCE									ABSORPTANCE									TRANSMITTANCE									Absorptance to Emittance Ratio
		Total			Spectral			Integrated			Spectral			Solar			Integrated			Spectral			Solar			Integrated			Spectral			Solar			
		Hemispherical	Normal	Angular	Hemispherical	Normal	Angular	Hemispherical	Normal	Angular	Hemispherical	Normal	Angular	Hemispherical	Normal	Angular	Hemispherical	Normal	Angular	Hemispherical	Normal	Angular	Hemispherical	Normal	Angular	Hemispherical	Normal	Angular	Hemispherical	Normal	Angular	Hemispherical	Normal	Angular	
Aluminum Alloys: (continued)																																			
L34	7	-	-	-	-	-	-	-	-	-	-	1092	-	-	1095	-	-	-	-	-	-	-	-	-	-	-	-	-	-	-	-	-	-	-	-
L72	7	-	-	-	-	-	-	-	-	-	-	-	-	-	1084	-	-	-	-	-	-	-	-	1087	-	-	-	-	-	-	-	-	-	-	-
Aluminum Coating on:																																			
Aluminum Substrate	9	580	-	-	-	-	-	-	-	-	-	592	-	-	-	-	-	-	-	-	-	-	-	610	-	-	-	-	-	-	-	-	-	-	-
Fabric Substrate	9	-	586	-	-	-	-	-	-	-	-	-	-	-	-	-	-	-	-	-	-	-	-	-	-	-	-	-	-	-	-	-	-	-	-
Glass Substrate	9	-	-	-	-	-	-	-	-	-	-	591	602 607	-	-	-	-	-	-	-	-	-	-	-	-	-	-	-	-	-	-	-	-	-	-
Iron Substrate	9	580	-	-	-	-	-	-	-	-	-	-	-	-	-	-	-	-	-	-	-	-	-	-	-	-	-	-	-	-	-	-	-	-	-
Lacquer Substrate	9	-	-	-	-	-	-	-	-	-	-	600	-	-	-	-	-	-	-	-	-	-	-	-	-	-	-	-	-	-	-	-	-	-	-
Mylar Substrate	9	580	-	-	-	-	-	-	-	-	-	592	-	-	-	-	609	-	-	-	-	-	-	-	-	-	-	-	-	-	-	-	-	-	-
Polyester Substrate	9	580	-	-	-	-	-	-	-	-	-	-	-	-	-	-	-	-	-	-	-	-	-	-	-	-	-	-	-	-	-	-	-	-	-
Polyurethane Substrate	9	-	-	-	-	-	-	-	-	-	-	-	602	-	-	-	-	-	-	-	-	-	-	-	-	-	-	-	-	-	-	-	-	-	-
Quartz Substrate	9	581	-	-	-	-	-	-	-	-	-	592	602	-	-	-	-	-	-	-	-	-	-	-	-	-	-	-	-	-	-	-	-	-	-
Silver Substrate	9	-	-	-	-	-	-	-	-	-	-	-	603	-	-	-	-	-	-	-	-	-	-	-	-	-	-	-	-	-	-	-	-	-	-
Stainless Steel Substrate	9	580	-	-	-	-	-	-	-	-	-	-	602	-	-	-	609	-	-	-	-	-	-	612	-	-	-	-	-	-	-	-	-	-	-
Teflon Substrate	9	-	586	-	-	-	-	-	-	-	-	-	-	-	-	-	-	-	-	-	-	-	-	610	-	-	-	-	-	-	-	-	-	-	-
Aluminum Bronze	7	-	1160	-	-	-	-	-	-	-	-	-	-	-	-	-	-	-	-	-	-	-	-	-	-	-	-	-	-	-	-	-	-	-	-
Aluminum Foil	7	5 6	9	-	-	-	-	-	-	-	-	26 27	-	-	40	-	43	-	-	-	-	50	-	-	55	-	-	-	-	-	60	-	-	-	-
Aluminum Pigment in:																																			
Acrylic Binder	9	3	-	-	-	-	-	-	-	-	-	-	-	-	-	-	-	-	-	-	20	-	-	-	-	-	-	-	-	-	-	-	-	-	-
Lacquer Binder	9	-	6	-	-	-	-	-	-	-	-	-	-	-	-	-	-	-	-	-	-	-	-	-	-	-	-	-	-	-	-	-	-	-	-
Nitrocellulose Binder	9	3	-	-	-	-	-	-	-	-	-	-	-	-	-	-	-	-	-	-	-	-	-	-	-	-	-	-	-	-	-	-	-	-	-
Oil Binder	9	-	-	-	-	-	-	-	-	9	-	-	-	-	-	-	-	-	-	-	-	-	-	-	-	-	-	-	-	-	-	-	-	-	-
Silicone Binder	9	3	6	-	-	-	-	-	-	-	-	13 17	-	-	-	-	-	-	-	-	-	-	-	21	-	-	-	-	-	-	-	-	-	-	-
Aluminum Substrate Coated with:																																			
Acrylic	9	1108	-	-	-	-	-	-	-	-	-	1110	-	-	-	-	-	-	-	-	-	-	-	-	-	-	-	-	-	-	-	-	-	-	-
Alkyd	9	-	-	-	-	-	-	-	-	-	-	1111	-	-	-	-	-	-	-	-	-	-	-	-	-	-	-	-	-	-	-	-	-	-	-
Aluminum	9	580	-	-	-	-	-	-	-	-	-	592	-	-	-	-	-	-	-	-	-	-	-	610	-	-	-	-	-	-	-	-	-	-	-
Aluminum Oxide	9	785	-	-	-	-	-	-	-	794	-	796 799	800	-	-	-	-	-	-	-	-	-	-	803	-	-	-	-	-	-	-	-	-	-	-
Antimony	9	-	-	-	-	-	-	-	-	-	-	-	-	-	-	-	-	-	-	-	-	-	-	-	-	-	-	-	-	615	-	-	-	-	-
Barium Titanate	9	815	-	-	-	-	-	-	-	-	-	-	-	-	-	-	-	-	-	-	-	-	-	818	-	-	-	-	-	-	-	-	-	-	-
Butyl Acrylate	9	-	-	-	-	-	-	-	-	-	-	1114	-	-	-	-	-	-	-	-	-	-	-	-	-	-	-	-	-	-	-	-	-	-	-
Butylated Melamine Formaldehyde	9	-	-	-	-	-	-	-	-	-	-	1116	-	-	-	-	-	-	-	-	-	-	-	-	-	-	-	-	-	-	-	-	-	-	-

MATERIAL NAME	VOLUME	EMITTANCE						REFLECTANCE									ABSORPTANCE									TRANSMITTANCE									Absorptance to Emittance Ratio
		Total			Spectral			Integrated			Spectral			Solar			Integrated			Spectral			Solar			Integrated			Spectral			Solar			
		Hemispherical	Normal	Angular	Hemispherical	Normal	Angular	Hemispherical	Normal	Angular	Hemispherical	Normal	Angular	Hemispherical	Normal	Angular	Hemispherical	Normal	Angular	Hemispherical	Normal	Angular	Hemispherical	Normal	Angular	Hemispherical	Normal	Angular	Hemispherical	Normal	Angular	Hemispherical	Normal	Angular	
Aluminum Substrate Coated with: (continued)																																			
Butylated Urea Formaldehyde	9	–	–	–	–	–	–	–	–	–	–	1118	–	–	–	–	–	–	–	–	–	–	–	–	–	–	–	–	–	–	–	–	–	–	–
Cadmium Sulfide	9	–	–	–	–	–	–	–	–	–	–	–	–	–	–	–	–	–	–	–	–	–	–	–	–	–	–	–	–	840	–	–	–	–	–
Calcium Titanate	9	845	–	–	–	–	–	–	–	–	–	–	–	–	–	–	–	–	–	–	–	–	–	850	–	–	–	–	–	–	–	–	–	–	–
Chromium Black	9	1173	–	–	–	–	–	–	–	–	–	–	–	–	–	–	–	–	–	–	–	–	–	–	–	–	–	–	–	–	–	–	–	–	–
Chromium Oxide	9	876	–	–	–	–	–	–	–	–	–	–	–	–	–	–	–	–	–	–	–	–	–	–	–	–	–	–	–	–	–	–	–	–	–
Epoxy	9	1120	–	–	–	–	–	–	–	–	–	–	–	–	–	–	–	–	–	–	–	–	–	–	–	–	–	–	–	–	–	–	–	–	–
Glass	9	–	–	–	–	–	–	–	–	–	–	1035	–	–	–	–	–	–	–	–	–	–	–	1036	–	–	–	–	–	–	–	–	–	–	–
Gold	9	652	–	–	–	–	–	–	–	–	–	666	–	–	–	–	–	–	–	–	–	–	–	–	–	–	–	–	–	–	–	–	–	–	–
Iron Oxide	9	–	–	–	–	–	–	–	–	–	–	903	–	–	–	–	–	–	–	–	–	–	–	–	–	–	–	–	–	–	–	–	–	–	–
Kapton	9	–	–	–	–	–	–	–	–	–	–	1037 1039	–	–	–	–	–	–	–	–	–	–	–	1041	–	–	–	–	–	–	–	–	–	–	–
Lacquer	9	–	1124	–	–	–	–	–	–	–	–	–	–	–	–	–	–	–	–	–	–	–	–	1127	–	–	–	–	–	–	–	–	–	–	–
Lithium Fluoride	9	–	–	–	–	–	–	–	–	–	–	1090	–	–	–	–	–	–	–	–	–	–	–	–	–	–	–	–	–	–	–	–	–	–	–
Magnesium Aluminate	9	915	–	–	–	–	–	–	–	–	–	–	–	–	–	–	–	–	–	–	–	–	–	918	–	–	–	–	–	–	–	–	–	–	–
Magnesium Oxide	9	–	–	–	–	–	–	–	–	–	–	924	–	–	–	–	–	–	–	–	–	–	–	–	–	–	–	–	–	–	–	–	–	–	–
Mylar	9	1042	–	–	–	–	–	–	–	–	–	–	–	–	–	–	–	–	–	–	–	–	–	–	–	–	–	–	–	–	–	–	–	–	–
Polyester	9	1135	–	–	–	–	–	–	–	–	–	–	–	–	–	–	–	–	–	–	–	–	–	–	–	–	–	–	–	–	–	–	–	–	–
Polyurethane	9	–	–	–	–	–	–	–	–	–	–	1138	1140	–	–	–	–	–	–	–	–	–	–	–	–	–	–	–	–	–	–	–	–	–	–
Polyvinyl Chloride	9	–	–	–	–	–	–	–	–	–	–	1150	–	–	–	–	–	–	–	–	–	–	–	1154	–	–	–	–	–	–	–	–	–	–	–
Potassium Silicate	9	–	–	–	–	–	–	–	–	–	–	937 1044	–	–	–	–	–	–	–	–	–	–	–	–	–	–	–	–	–	–	–	–	–	–	–
Sapphire	9	785	–	–	–	–	–	–	–	–	–	–	–	–	–	–	–	–	–	–	–	–	–	–	–	–	–	–	–	–	–	–	–	–	–
Silicon	9	–	–	–	–	–	–	–	–	–	–	1046	–	–	–	–	–	–	–	–	–	–	–	1047	–	–	–	–	–	–	–	–	–	–	–
Silicon Dioxide	9	1049	1052	–	–	–	–	–	–	–	–	1054 1057	1061	–	–	–	–	–	–	–	–	–	–	1068	–	–	–	–	–	–	–	–	–	–	–
Silicon Monoxide	9	1074	–	–	–	–	–	–	–	–	–	1077	–	–	–	–	–	–	–	–	–	–	–	1080	–	–	–	–	–	–	–	–	–	–	–
Silicone	9	–	1156	–	–	–	–	–	–	–	–	1159	–	–	–	–	–	–	–	–	–	–	–	1164 1167	–	–	–	–	–	–	–	–	–	–	–
Sodium Chloride	9	–	–	–	–	–	–	–	–	–	–	–	1081	–	–	–	–	–	–	–	–	–	–	–	–	–	–	–	–	–	–	–	–	–	–
Strontium Titanate	9	958	–	–	–	–	–	–	–	–	–	–	–	–	–	–	–	–	–	–	–	–	–	963	–	–	–	–	–	–	–	–	–	–	–
Teflon	9	–	–	–	–	–	–	–	–	–	–	1085	–	–	–	–	–	–	–	–	–	–	–	1087	–	–	–	–	–	–	–	–	–	–	–
Tesslar	9	–	–	–	–	–	–	–	–	–	–	–	–	–	–	–	–	–	–	–	–	–	–	1082	–	–	–	–	–	–	–	–	–	–	–
Titanium	9	–	–	–	–	–	–	–	–	–	–	760	–	–	–	–	–	–	–	–	–	–	–	763	–	–	–	–	–	–	–	–	–	–	–
Titanium Dioxide	9	981	–	–	–	–	–	–	–	–	–	986	989 990	–	–	–	–	–	–	–	–	–	–	–	991	–	–	–	–	–	–	–	–	992	–
Zirconium Oxide	9	–	–	–	–	–	–	–	–	–	–	1017	–	–	–	–	–	–	–	–	–	–	–	–	–	–	–	–	–	–	–	–	–	–	–
Zirconium Silicate	9	1024	–	–	–	–	–	–	–	–	–	–	–	–	–	–	–	–	–	–	–	–	–	1030	–	–	–	–	–	–	–	–	–	–	–

MATERIAL NAME	VOLUME	EMITTANCE						REFLECTANCE									ABSORPTANCE									TRANSMITTANCE									Absorptance to Emittance Ratio
		Total			Spectral			Integrated			Spectral			Solar			Integrated			Spectral			Solar			Integrated			Spectral			Solar			
		Hemispherical	Normal	Angular	Hemispherical	Normal	Angular	Hemispherical	Normal	Angular	Hemispherical	Normal	Angular	Hemispherical	Normal	Angular	Hemispherical	Normal	Angular	Hemispherical	Normal	Angular	Hemispherical	**Normal**	Angular	Hemispherical	Normal	Angular	Hemispherical	Normal	Angular	Hemispherical	Normal	Angular	
Aluminum + Carbon Pigment in Silicone Binder	9	-	-	-	-	-	-	-	-	-	-	-	-	-	-	-	-	-	-	-	-	-	-	24	-	-	-	-	-	-	-	-	-	-	-
Aluminum + Cobalt	7	-	-	-	-	-	-	-	-	-	-	887	-	-	-	-	-	-	-	-	-	-	-	-	-	-	-	-	-	-	-	-	-	-	-
Aluminum + Copper + ΣX_i	7	1062	1066	-	-	1072	-	-	-	-	-	1076	-	-	1083	-	-	-	-	-	-	-	-	1086	-	-	-	-	-	-	-	-	-	-	-
Aluminum + Copper + ΣX_i, Anodized	9	1234	1235	-	-	-	-	-	-	-	-	1237	-	-	1240	-	-	-	-	-	-	-	-	1241	-	-	-	-	-	-	-	-	-	-	-
Aluminum + Iron + ΣX_i	7	-	-	-	-	-	-	-	-	-	-	1090	-	-	1094	-	-	-	-	-	-	-	-	-	-	-	-	-	-	-	-	-	-	-	-
Aluminum + Iron + ΣX_i, Anodized	9	-	-	-	-	-	-	-	-	-	-	1243	1251	-	1252	-	-	-	-	-	-	-	-	1253	-	-	-	-	-	-	-	-	-	-	-
Aluminum + Magnesium	7	-	-	-	-	-	-	-	-	-	-	890	-	-	-	-	-	-	-	-	-	-	-	-	-	-	-	-	-	-	-	-	-	-	-
Aluminum + Magnesium + ΣX_i	7	1098	1100	-	-	-	-	-	-	-	-	1105	-	-	-	-	-	-	-	-	-	-	-	1110	-	-	-	-	-	-	-	-	-	-	-
Aluminum + Magnesium + ΣX_i, Anodized	9	-	-	-	-	-	-	-	-	-	-	1254	1255	-	-	-	-	-	-	-	-	-	-	-	-	-	-	-	-	-	-	-	-	-	-
Aluminum + Magnesium Coating on Glass Substrate	9	-	-	-	-	-	-	-	-	-	-	613	-	-	-	-	-	-	-	-	-	-	-	-	-	-	-	-	-	-	-	-	-	-	-
Aluminum + Manganese + ΣX_i	7	1112	1114	-	-	-	-	-	-	-	-	-	-	-	-	-	-	-	-	-	-	-	-	-	-	-	-	-	-	-	-	-	-	-	-
Aluminum + Silicon	7	-	-	-	-	-	-	-	-	-	-	893	-	-	-	-	-	-	-	-	-	-	-	-	-	-	-	-	-	-	-	-	-	-	-
Aluminum + Silver	7	-	-	-	-	-	-	-	-	-	-	896	-	-	-	-	-	-	-	-	-	-	-	-	-	-	-	-	-	-	-	-	-	-	-
Aluminum + Zinc + ΣX_i	7	1116	1118	-	-	1121	1124	-	-	-	-	-	-	-	1128	-	-	-	-	-	-	-	-	1131	-	-	-	-	-	-	-	-	-	-	-
Aluminum + Zinc + ΣX_i, Anodized	9	1256	-	-	-	-	-	-	-	-	-	1258	-	-	-	-	-	-	-	-	-	-	-	-	-	-	-	-	-	-	-	-	-	-	-
Aluminum Dodecaboride	8	-	-	-	-	732	-	-	-	-	-	-	-	-	-	-	-	-	-	-	-	-	-	-	-	-	-	-	-	-	-	-	-	-	-
Aluminum Nitride	8	-	1030	-	-	1031 1033	-	-	-	-	-	1035	-	-	-	-	-	-	-	-	-	-	-	-	-	-	-	-	-	-	-	-	-	-	-
Aluminum Oxide	8	141	142	-	-	146 148	-	-	-	-	-	157	163	-	-	-	-	-	-	-	166	-	-	168	-	-	-	-	-	169	-	-	-	-	-
Aluminum Oxide (Ruby)	8	-	-	-	-	-	-	-	-	-	-	-	174	-	-	-	-	-	-	-	-	-	-	-	-	-	-	-	-	176	-	-	-	-	-
Aluminum Oxide (Sapphire)	8	179	181	-	-	183	-	-	-	-	-	-	187	-	-	-	-	-	-	-	-	-	-	-	-	-	-	-	-	190	-	-	-	-	-
Aluminum Oxide Coating on:																																			
Aluminum Substrate	9	785	-	-	-	-	-	-	-	794	-	796 799	800	-	-	-	-	-	-	-	-	-	-	803	-	-	-	-	-	-	-	-	-	-	-
Gold Substrate	9	-	-	-	-	-	-	-	-	-	-	-	-	-	-	-	-	-	-	-	-	-	-	803	-	-	-	-	-	-	-	-	-	-	-
Inconel Substrate	9	-	788	-	-	792	-	-	-	-	-	-	-	-	-	-	-	-	-	-	-	-	-	-	-	-	-	-	-	-	-	-	-	-	-
Mild Steel Substrate	9	-	788	-	-	792	-	-	-	-	-	-	-	-	-	-	-	-	-	-	-	-	-	-	-	-	-	-	-	-	-	-	-	-	-
Molybdenum Substrate	9	785	788	-	-	-	-	-	-	-	-	796	-	-	-	-	-	-	-	-	-	-	-	803	-	-	-	-	-	-	-	-	-	-	-
Nimonic 75 Substrate	9	-	788	-	-	-	-	-	-	-	-	-	-	-	-	-	-	-	-	-	-	-	-	-	-	-	-	-	-	-	-	-	-	-	-

MATERIAL NAME	VOLUME	EMITTANCE						REFLECTANCE									ABSORPTANCE									TRANSMITTANCE									Absorptance to Emittance Ratio
		Total			Spectral			Integrated			Spectral			Solar			Integrated			Spectral			Solar			Integrated			Spectral			Solar			
		Hemispherical	Normal	Angular	Hemispherical	Normal	Angular	Hemispherical	Normal	Angular	Hemispherical	Normal	Angular	Hemispherical	Normal	Angular	Hemispherical	Normal	Angular	Hemispherical	Normal	Angular	Hemispherical	Normal	Angular	Hemispherical	Normal	Angular	Hemispherical	Normal	Angular	Hemispherical	Normal	Angular	
Aluminum Oxide Coating on: (continued)																																			
Niobium Substrate	9	785	-	-	-	-	-	-	-	-	-	-	-	-	-	-	-	-	-	-	-	-	-	-	-	-	-	-	-	-	-	-	-	-	-
Silicon Monoxide Substrate	9	-	-	-	-	-	-	-	-	-	-	1077	-	-	-	-	-	-	-	-	-	-	-	-	-	-	-	-	-	-	-	-	-	-	-
Silver Substrate	9	785	788	-	-	-	-	-	-	-	-	796	-	-	-	-	-	-	-	-	802	-	-	803	-	-	-	-	-	-	-	-	-	-	-
Stainless Steel Substrate	9	-	788	-	-	790 792	-	-	-	-	-	796	-	-	-	-	-	-	-	-	-	-	-	804	-	-	-	-	-	-	-	-	-	-	-
Unknown Substrate	9	-	-	-	-	-	-	-	-	-	-	-	-	-	-	-	-	-	-	-	-	-	-	-	-	-	-	-	-	807	-	-	-	-	-
Aluminum Oxide Pigment in:																																			
Alkyd-Melamine Binder	9	-	-	-	-	-	-	-	-	-	-	33	-	-	-	-	-	-	-	-	-	-	-	-	-	-	-	-	-	-	-	-	-	-	-
DuPont RC 7007 Binder	9	-	-	-	-	-	-	-	-	-	-	33	-	-	-	-	-	-	-	-	-	-	-	-	-	-	-	-	-	-	-	-	-	-	-
Potassium Silicate Binder	9	-	28	-	-	-	-	-	-	-	-	33 39	-	-	-	-	-	-	-	-	41	-	-	42	-	-	-	-	-	-	-	-	-	-	45
Phosphoric Acid Cement Binder	9	-	-	-	-	30	-	-	-	-	-	-	-	-	-	-	-	-	-	-	-	-	-	-	-	-	-	-	-	-	-	-	-	-	-
PS-7 Binder	9	-	-	-	-	-	-	-	-	-	-	34	-	-	-	-	-	-	-	-	41	-	-	42	-	-	-	-	-	-	-	-	-	-	-
Aluminum Oxide Substrate Coated with:																																			
Acrylic	9	-	1109	-	-	-	-	-	-	-	-	-	-	-	-	-	-	-	-	-	-	-	-	-	-	-	-	-	-	-	-	-	-	-	-
Graphite	9	-	854	-	-	-	-	-	-	-	-	-	-	-	-	-	-	-	-	-	-	-	-	-	-	-	-	-	-	-	-	-	-	-	-
Silicon Carbide	9	-	-	-	-	949	-	-	-	-	-	-	-	-	-	-	-	-	-	-	-	-	-	-	-	-	-	-	-	-	-	-	-	-	-
Aluminum Oxide + Aluminum Titanate Coating on Nb-1Zr Substrate	9	808	-	-	-	-	-	-	-	-	-	-	-	-	-	-	-	-	-	-	-	-	-	-	-	-	-	-	-	-	-	-	-	-	-
Al_2O_3 + ZnO + TiO_2 Pigment in Potassium Silicate Binder	9	-	28	-	-	-	-	-	-	-	-	33	-	-	-	-	-	-	-	-	-	-	-	42	-	-	-	-	-	-	-	-	-	-	-
Aluminum Phosphate	8	-	-	-	-	-	-	-	-	-	-	602	-	-	-	-	-	-	-	-	-	-	-	-	-	-	-	-	-	-	-	-	-	-	-
Aluminum Phosphate Binder Pigmented with:																																			
Barium Titanate	9	63	-	-	-	-	-	-	-	-	-	-	-	-	-	-	-	-	-	-	-	-	-	-	-	-	-	-	-	-	-	-	-	-	-
Calcium Titanate	9	79	-	-	-	80	-	-	-	-	-	-	-	-	-	-	-	-	-	-	-	-	-	-	-	-	-	-	-	-	-	-	-	-	-
Cr-Co-Ni Spinel	9	186	-	-	-	-	-	-	-	-	-	-	-	-	-	-	-	-	-	-	-	-	-	-	-	-	-	-	-	-	-	-	-	-	-
FCE-11	9	63	-	-	-	-	-	-	-	-	-	-	-	-	-	-	-	-	-	-	-	-	-	-	-	-	-	-	-	-	-	-	-	-	-
Iron Titanate	9	123	-	-	-	-	-	-	-	-	-	-	-	-	-	-	-	-	-	-	-	-	-	-	-	-	-	-	-	-	-	-	-	-	-
Iron Titanate + Alumina	9	123	-	-	-	-	-	-	-	-	-	-	-	-	-	-	-	-	-	-	-	-	-	-	-	-	-	-	-	-	-	-	-	-	-
NiO · Cr_2O_3 Spinel + SiO_2	9	186	-	-	-	-	-	-	-	-	-	-	-	-	-	-	-	-	-	-	-	-	-	-	-	-	-	-	-	-	-	-	-	-	-
SiC + SiO_2	9	174	-	-	-	-	-	-	-	-	-	-	-	-	-	-	-	-	-	-	-	-	-	-	-	-	-	-	-	-	-	-	-	-	-
Silicon Carbide	9	174	-	-	-	-	-	-	-	-	-	-	-	-	-	-	-	-	-	-	-	-	-	-	-	-	-	-	-	-	-	-	-	-	-
Strontium Titanate	9	197	-	-	-	-	-	-	-	-	-	-	-	-	-	-	-	-	-	-	-	-	-	-	-	-	-	-	-	-	-	-	-	-	-

MATERIAL NAME	VOLUME	EMITTANCE						REFLECTANCE									ABSORPTANCE									TRANSMITTANCE									Absorptance to Emittance Ratio
		Total			Spectral			Integrated			Spectral			Solar			Integrated			Spectral			Solar			Integrated			Spectral			Solar			
		Hemispherical	Normal	Angular	Hemispherical	Normal	Angular	Hemispherical	Normal	Angular	Hemispherical	Normal	Angular	Hemispherical	Normal	Angular	Hemispherical	Normal	Angular	Hemispherical	Normal	Angular	Hemispherical	Normal	Angular	Hemispherical	Normal	Angular	Hemispherical	Normal	Angular	Hemispherical	Normal	Angular	
Aluminum Phosphate Binder Pigmented with: (continued)																																			
Tin Oxide	9	201	-	-	-	-	-	-	-	-	-	-	-	-	-	-	-	-	-	-	205	-	-	-	-	-	-	-	-	-	-	-	-	-	-
Zirconium Oxide	9	-	-	-	-	-	-	-	-	-	-	425	-	-	-	-	-	-	-	-	-	-	-	-	-	-	-	-	-	-	-	-	-	-	-
Zirconium Silicate	9	-	-	-	-	-	-	-	-	-	-	-	-	-	-	-	-	-	-	-	435	-	-	-	-	-	-	-	-	-	-	-	-	-	-
Aluminum Silicate	8	-	-	-	-	-	-	-	-	-	-	618	-	-	-	-	-	-	-	-	-	-	-	-	-	-	-	-	-	-	-	-	-	-	-
Aluminum Silicate Glass	8	-	1523	-	-	[illegible]	-	-	-	-	-	1525	-	-	-	-	-	-	-	-	-	-	-	1527	-	-	1528	-	-	1530	-	-	-	-	-
Aluminum Silicate Pigment in Potassium Silicate Binder	9	-	-	-	-	-	-	-	-	-	-	47	-	-	-	-	-	-	-	-	-	-	-	49	-	-	-	-	-	-	-	-	-	-	-
Alundum Coating on Niobium Substrate	9	785	-	-	-	-	-	-	-	-	-	-	-	-	-	-	-	-	-	-	-	-	-	-	-	-	-	-	-	-	-	-	-	-	-
Ammonium Dihydrogen Phosphate	8	-	-	-	-	-	-	-	-	-	-	-	-	-	-	-	-	-	-	-	-	-	-	-	-	-	-	-	-	604	-	-	-	-	-
Anatase	8	-	-	-	-	-	-	-	-	-	-	464	-	-	-	-	-	-	-	-	-	-	-	-	-	-	-	-	-	-	-	-	-	-	-
Anglesite	8	-	-	-	-	-	-	-	-	-	-	-	-	-	-	-	-	-	-	-	-	-	-	-	-	-	-	-	-	1702	-	-	-	-	-
Anhydrite	8	-	-	-	-	-	-	-	-	-	-	-	-	-	-	-	-	-	-	-	-	-	-	-	-	-	-	-	-	630	-	-	-	-	-
AN-L-29 Coating on:																																			
Aluminum Substrate	9	-	1124	-	-	-	-	-	-	-	-	-	-	-	-	-	-	-	-	-	-	-	-	1127	-	-	-	-	-	-	-	-	-	-	-
Dow Metal Substrate	9	-	1124	-	-	-	-	-	-	-	-	-	-	-	-	-	-	-	-	-	-	-	-	1127	-	-	-	-	-	-	-	-	-	-	-
Anodized Aluminum	9	1195	1199	-	-	1203	-	-	-	-	-	1207 1216	1219 1220	-	1222	-	1224	-	-	-	-	-	-	1226 1230	1231	-	-	-	-	-	-	-	-	-	1233
Anodized Aluminum Alloys (also see individual alloys):																																			
Alclad 2024	9	-	-	-	-	-	-	-	-	-	-	1244	-	-	1252	-	-	-	-	-	-	-	-	-	-	-	-	-	-	-	-	-	-	-	-
Aluminum 1075	9	-	-	-	-	-	-	-	-	-	-	1243	-	-	-	-	-	-	-	-	-	-	-	-	-	-	-	-	-	-	-	-	-	-	-
Aluminum 1145	9	-	-	-	-	-	-	-	-	-	-	-	1251	-	-	-	-	-	-	-	-	-	-	1253	-	-	-	-	-	-	-	-	-	-	-
Aluminum 1199	9	1195	-	-	-	-	-	-	-	-	-	1207	1219 1220	-	1222	-	-	-	-	-	-	-	-	1226 1230	1231	-	-	-	-	-	-	-	-	-	-
Aluminum 2014	9	1234	-	-	-	-	-	-	-	-	-	-	-	-	-	-	-	-	-	-	-	-	-	-	-	-	-	-	-	-	-	-	-	-	-
Aluminum 2024	9	-	1235	-	-	-	-	-	-	-	-	1237	-	-	1240	-	-	-	-	-	-	-	-	1241	-	-	-	-	-	-	-	-	-	-	-
Aluminum 7075	9	1256	-	-	-	-	-	-	-	-	-	1258	-	-	-	-	-	-	-	-	-	-	-	-	-	-	-	-	-	-	-	-	-	-	-
Aluminum L-34	9	-	-	-	-	-	-	-	-	-	-	-	-	-	1252	-	-	-	-	-	-	-	-	1253	-	-	-	-	-	-	-	-	-	-	-
Aluminum + Copper + ΣX_i	9	1234	1235	-	-	-	-	-	-	-	-	1237	-	-	1240	-	-	-	-	-	-	-	-	1241	-	-	-	-	-	-	-	-	-	-	-
Aluminum + Iron + ΣX_i	9	-	-	-	-	-	-	-	-	-	-	1243	1251	-	1252	-	-	-	-	-	-	-	-	1253	-	-	-	-	-	-	-	-	-	-	-
Aluminum + Magnesium	9	-	-	-	-	-	-	-	-	-	-	1254	1255	-	-	-	-	-	-	-	-	-	-	-	-	-	-	-	-	-	-	-	-	-	-
Aluminum + Zinc + ΣX_i	9	1256	-	-	-	-	-	-	-	-	-	1258	-	-	-	-	-	-	-	-	-	-	-	-	-	-	-	-	-	-	-	-	-	-	-
Anodized QMV Beryllium	9	-	-	-	-	-	-	-	-	-	-	1265	-	-	-	-	-	-	-	-	-	-	-	-	-	-	-	-	-	-	-	-	-	-	-

MATERIAL NAME	VOLUME	EMITTANCE						REFLECTANCE									ABSORPTANCE									TRANSMITTANCE									Absorptance to Emittance Ratio
		Total			Spectral			Integrated			Spectral			Solar			Integrated			Spectral			Solar			Integrated			Spectral			Solar			
		Hemispherical	Normal	Angular	Hemispherical	Normal	Angular	Hemispherical	Normal	Angular	Hemispherical	Normal	Angular	Hemispherical	Normal	Angular	Hemispherical	Normal	Angular	Hemispherical	Normal	Angular	Hemispherical	Normal	Angular	Hemispherical	Normal	Angular	Hemispherical	Normal	Angular	Hemispherical	Normal	Angular	
Anodized Beryllium + Copper	9	1268	–	–	–	–	–	–	–	–	–	1270	–	–	–	–	–	–	–	–	–	–	–	1273	–	–	–	–	–	–	–	–	–	–	–
Anodized Magnesium	9	1274	–	–	–	–	–	–	–	–	–	–	–	–	–	–	–	–	–	–	–	–	–	1275	–	–	–	–	–	–	–	–	–	–	–
Anodized Magnesium + Aluminum + ΣX_i	9	–	–	–	–	–	–	–	–	–	–	1277	–	–	–	–	–	–	–	–	–	–	–	–	–	–	–	–	–	–	–	–	–	–	–
Anodized Magnesium + Thorium + ΣX_i	9	–	–	–	–	–	–	–	–	–	–	1281	–	–	–	–	–	–	–	–	–	–	–	–	–	–	–	–	–	–	–	–	–	–	–
Anodized Tantalum	9	1284	–	–	–	–	–	–	–	–	–	1286	–	–	–	–	–	–	–	–	–	–	–	1287	–	–	–	–	–	–	–	–	–	–	–
Anodized Titanium + Aluminum + ΣX_i	9	–	–	–	–	–	–	–	–	–	–	1289	–	–	–	–	–	–	–	–	–	–	–	–	–	–	–	–	–	–	–	–	–	–	–
Anodized Titanium + Vanadium + ΣX_i	9	–	–	–	–	–	–	–	–	–	–	1293	–	–	–	–	–	–	–	–	–	–	–	–	–	–	–	–	–	–	–	–	–	–	–
Antimony	7	–	–	–	–	–	–	–	–	–	–	–	–	–	–	63	–	–	–	–	–	–	–	–	–	–	–	–	–	66	–	–	–	–	–
Antimony Coating on:																																			
Aluminum Substrate	9	–	–	–	–	–	–	–	–	–	–	–	–	–	–	–	–	–	–	–	–	–	–	–	–	–	–	–	–	615	–	–	–	–	–
Glass Substrate	9	–	–	–	–	–	–	–	–	–	–	–	614	–	–	–	–	–	–	–	–	–	–	–	–	–	–	–	–	–	–	–	–	–	–
Stilbene Substrate	9	–	–	–	–	–	–	–	–	–	–	–	–	–	–	–	–	–	–	–	–	–	–	–	–	–	–	–	–	615	–	–	–	–	–
Antimony Black Coating on:																																			
Cellulose Nitrate Substrate	9	–	–	–	–	–	–	–	–	–	–	–	–	–	–	–	–	–	–	–	–	–	–	–	–	–	–	–	–	1172	–	–	–	–	–
KRS-5 Substrate	9	–	–	–	–	–	–	–	–	–	–	–	–	–	–	–	–	–	–	–	–	–	–	–	–	–	–	–	–	1172	–	–	–	–	–
Antimony Oxide Pigment in:																																			
Potassium Silicate Binder	9	–	–	–	–	–	–	–	–	–	–	51	–	–	–	–	54	–	–	–	56	–	–	57 60	–	–	–	–	–	–	–	–	–	–	–
Silicone Binder	9	–	–	–	–	–	–	–	–	–	–	51	–	–	–	–	–	–	–	–	–	–	–	–	–	–	–	–	–	–	–	–	–	–	–
Diantimony Trioxide	8	–	–	–	–	–	–	–	–	–	–	198	–	–	–	–	–	–	–	–	–	–	–	200	–	–	–	–	–	–	–	–	–	–	–
AN-TT-V-116 Coating on:																																			
Aluminum Substrate	9	–	1124	–	–	–	–	–	–	–	–	–	–	–	–	–	–	–	–	–	–	–	–	1127	–	–	–	–	–	–	–	–	–	–	–
Dow Metal Substrate	9	–	1124	–	–	–	–	–	–	–	–	–	–	–	–	–	–	–	–	–	–	–	–	1127	–	–	–	–	–	–	–	–	–	–	–
AO-1053 Glass	8	–	–	–	–	–	–	–	–	–	–	–	–	–	–	–	–	–	–	–	–	–	–	–	–	–	–	–	–	1533	–	–	–	–	–
Aquablack B Pigment in Sodium Silicate Binder	9	574	–	–	–	–	–	–	–	–	–	–	–	–	–	–	–	–	–	–	–	–	–	575	–	–	–	–	–	–	–	–	–	–	–
Armco Iron	7	303	308	–	–	–	–	–	–	–	–	322	–	–	–	–	–	–	–	–	–	–	–	332	–	–	–	–	–	–	–	–	–	–	–
Armco Iron, Oxidized	9	–	–	–	–	1297	–	–	–	–	–	1299	–	–	–	–	–	–	–	–	–	–	–	–	–	–	–	–	–	–	–	–	–	–	–
Arsenic	8	–	–	–	–	–	–	–	–	–	–	3	–	–	–	–	–	–	–	–	–	–	–	–	–	–	–	–	–	–	–	–	–	–	–
Arsenic Glasses	8	–	–	–	–	–	–	–	–	–	–	–	–	–	–	–	–	–	–	–	–	–	–	–	–	–	–	–	–	1535	–	–	–	–	–

MATERIAL NAME	VOLUME	EMITTANCE						REFLECTANCE									ABSORPTANCE									TRANSMITTANCE									Absorptance to Emittance Ratio
		Total			Spectral			Integrated			Spectral			Solar			Integrated			Spectral			Solar			Integrated			Spectral			Solar			
		Hemispherical	Normal	Angular	Hemispherical	Normal	Angular	Hemispherical	Normal	Angular	Hemispherical	Normal	Angular	Hemispherical	Normal	Angular	Hemispherical	Normal	Angular	Hemispherical	Normal	Angular	Hemispherical	Normal	Angular	Hemispherical	Normal	Angular	Hemispherical	Normal	Angular	Hemispherical	Normal	Angular	
Arsenides:																																			
Gallium Arsenide	8	-	-	-	-	-	-	-	-	-	-	679	-	-	-	-	-	-	-	-	-	-	-	-	-	-	-	-	-	683	-	-	-	-	-
Indium Arsenide	8	-	-	-	-	685	-	-	-	-	-	687	-	-	-	-	-	-	-	-	-	-	-	-	-	-	-	-	-	689	-	-	-	-	-
Diarsenic Triselenide	8	-	-	-	-	-	-	-	-	-	-	1130	-	-	-	-	-	-	-	-	-	-	-	-	-	-	-	-	-	1133	-	-	-	-	-
Diarsenic Trisulfide	8	-	-	-	-	-	-	-	-	-	-	1177	-	-	-	-	-	-	-	-	-	-	-	-	-	-	-	-	-	1179	-	-	-	-	-
Astrolloy	7	-	-	-	-	1363 1412	-	-	-	-	-	1377 1418	-	-	-	-	-	-	-	-	-	-	-	-	-	-	-	-	-	-	-	-	-	-	-
ATJ Graphite, Siliconized	9	-	-	1325	-	1328	-	-	-	-	-	-	-	-	-	-	-	-	-	-	-	-	-	-	-	-	-	-	-	-	-	-	-	-	-
$AuAl_2$	8	-	-	-	-	-	-	-	-	-	-	1295	-	-	-	-	-	-	-	-	-	-	-	-	-	-	-	-	-	-	-	-	-	-	-
$AuGa_2$	8	-	-	-	-	-	-	-	-	-	-	1295	-	-	-	-	-	-	-	-	-	-	-	-	-	-	-	-	-	-	-	-	-	-	-
$AuIn_2$	8	-	-	-	-	-	-	-	-	-	-	1295	-	-	-	-	-	-	-	-	-	-	-	-	-	-	-	-	-	-	-	-	-	-	-
AuZn	8	-	-	-	-	-	-	-	-	-	-	1292	-	-	-	-	-	-	-	-	-	-	-	-	-	-	-	-	-	-	-	-	-	-	-
AZ-31	7	-	1328	-	-	-	-	-	-	-	-	1332	-	-	-	-	-	-	-	-	-	-	-	1334	-	-	-	-	-	-	-	-	-	-	-
AZ-31B	7	-	-	-	-	-	-	-	-	-	-	1332	-	-	-	-	-	-	-	-	-	-	-	-	-	-	-	-	-	-	-	-	-	-	-
AZ-31B, Anodized	9	-	-	-	-	-	-	-	-	-	-	1277	-	-	-	-	-	-	-	-	-	-	-	-	-	-	-	-	-	-	-	-	-	-	-
AZO-33	8	-	-	-	-	-	-	-	-	-	-	510-518	-	-	-	-	-	-	-	-	-	-	-	-	-	-	-	-	-	-	-	-	-	-	-
AZO-55LO	8	-	-	-	-	-	-	-	-	-	-	510-518	-	-	-	-	-	-	-	-	-	-	-	-	-	-	-	-	-	-	-	-	-	-	-
AZO-66	8	-	-	-	-	-	-	-	-	-	-	510-518	-	-	-	-	-	-	-	-	-	-	-	-	-	-	-	-	-	-	-	-	-	-	-
Azurite	8	-	-	-	-	-	-	-	-	-	-	-	-	-	-	-	-	-	-	-	-	-	-	-	-	-	-	-	-	1658	-	-	-	-	-
B + TiB_2 Powders	8	-	-	-	-	-	-	-	-	-	-	1447	-	-	-	-	-	-	-	-	-	-	-	-	-	-	-	-	-	-	-	-	-	-	-
B-44 Acrylic Binder Pigmented with Strontium Molybdate	9	-	-	-	-	-	-	-	-	-	-	191	-	-	-	-	-	-	-	-	-	-	-	194	-	-	-	-	-	-	-	-	-	-	-
B-1060	9	-	304	-	-	-	-	-	-	-	-	-	-	-	-	-	-	-	-	-	-	-	-	382 393	-	-	-	-	-	-	-	-	-	-	399
B120-VAC, Anodized	9	-	-	-	-	-	-	-	-	-	-	1293	-	-	-	-	-	-	-	-	-	-	-	-	-	-	-	-	-	-	-	-	-	-	-
Bakelite	8	-	-	-	-	-	-	-	-	-	-	1742	-	-	-	-	-	-	-	-	-	-	-	1744	-	-	-	-	-	1746	-	-	-	-	-
Bakelite Lacquer Coating on Unknown Substrate	9	1112	-	-	-	-	-	-	-	-	-	-	-	-	-	-	-	-	-	-	-	-	-	-	-	-	-	-	-	-	-	-	-	-	-
Barite	8	-	-	-	-	-	-	-	-	-	-	-	-	-	-	-	-	-	-	-	-	-	-	-	-	-	-	-	-	1702	-	-	-	-	-
Barium	7	-	-	-	-	-	-	-	-	-	-	-	68	-	-	-	-	-	-	-	-	-	-	-	-	-	-	-	-	-	-	-	-	-	-

MATERIAL NAME	VOLUME	EMITTANCE						REFLECTANCE									ABSORPTANCE									TRANSMITTANCE									Absorptance to Emittance Ratio
		Total			Spectral			Integrated			Spectral			Solar			Integrated			Spectral			Solar			Integrated			Spectral			Solar			
		Hemispherical	Normal	Angular	Hemispherical	Normal	Angular	Hemispherical	Normal	Angular	Hemispherical	Normal	Angular	Hemispherical	Normal	Angular	Hemispherical	Normal	Angular	Hemispherical	Normal	Angular	Hemispherical	Normal	Angular	Hemispherical	Normal	Angular	Hemispherical	Normal	Angular	Hemispherical	Normal	Angular	
Barium Beryllium Silicate Binder Pigmented with Cerium Dioxide (Enamel)	9	-	445	-	-	448	-	-	-	-	-	-	-	-	-	-	-	-	-	-	-	-	-	-	-	-	-	-	-	-	-	-	-	-	-
Barium Borosilicate Frit Binder Pigmented with Chromium Oxide (Enamel)	9	-	455	-	-	459	-	-	-	-	-	-	-	-	-	-	-	-	-	-	-	-	-	-	-	-	-	-	-	-	-	-	-	-	-
Barium Fluoride Coating on Zinc Selenide Substrate	9	-	-	-	-	-	-	-	-	-	-	810	-	-	-	-	-	-	-	-	-	-	-	-	-	-	-	-	-	812	-	-	-	-	-
Barium Hexaboride	8	-	-	-	-	732	-	-	-	-	-	-	-	-	-	-	-	-	-	-	-	-	-	-	-	-	-	-	-	-	-	-	-	-	-
Barium Carbonate	8	-	-	-	-	-	-	-	-	-	-	592	-	-	-	-	-	-	-	-	-	-	-	-	-	-	-	-	-	-	-	-	-	-	-
Barium Fluoride	8	-	-	-	-	-	-	-	-	-	-	-	909	-	-	-	-	-	-	-	-	-	-	-	-	-	-	-	-	912	-	-	-	-	-
Barium Oxide	8	-	-	-	-	-	-	-	-	-	-	546	-	-	-	-	-	-	-	-	-	-	-	-	-	-	-	-	-	-	-	-	-	-	-
Barium Phosphate	8	-	-	-	-	-	-	-	-	-	-	608	-	-	-	-	-	-	-	-	-	-	-	-	-	-	-	-	-	-	-	-	-	-	-
$Ba_xSr_{1-x}F_2$	8	-	-	-	-	-	-	-	-	-	-	992	-	-	-	-	-	-	-	-	-	-	-	-	-	-	-	-	-	-	-	-	-	-	-
Barium + Strontium Coating on Nickel Substrate	9	616	-	-	-	-	-	-	-	-	-	-	-	-	-	-	-	-	-	-	-	-	-	-	-	-	-	-	-	-	-	-	-	-	-
Barium Sulfate	8	-	-	-	-	-	-	-	-	-	-	623	-	-	-	-	-	-	-	-	-	-	-	627	-	-	-	-	-	-	-	-	-	-	-
Barium Sulfate Pigment in:																																			
Carboxy-Methyl-Cellulose Binder	9	-	-	-	-	-	-	-	-	-	-	62	-	-	-	-	-	-	-	-	-	-	-	-	-	-	-	-	-	-	-	-	-	-	-
Polyvinyl Alcohol Binder	9	-	-	-	-	-	-	-	-	-	-	62	-	-	-	-	-	-	-	-	-	-	-	-	-	-	-	-	-	-	-	-	-	-	-
Barium Titanate	8	-	-	-	-	-	-	-	-	-	-	635 637	-	-	-	-	-	-	-	-	-	-	-	-	-	-	-	-	-	642	-	-	-	-	-
Barium Titanate Coating on:																																			
Aluminum Substrate	9	815	-	-	-	-	-	-	-	-	-	-	-	-	-	-	-	-	-	-	-	-	-	818	-	-	-	-	-	-	-	-	-	-	-
Nb-1Zr Substrate	9	815	-	-	-	817	-	-	-	-	-	-	-	-	-	-	-	-	-	-	-	-	-	-	-	-	-	-	-	-	-	-	-	-	-
Barium Titanate Pigment in:																																			
Aluminum Phosphate Binder	9	63	-	-	-	-	-	-	-	-	-	-	-	-	-	-	-	-	-	-	-	-	-	-	-	-	-	-	-	-	-	-	-	-	-
Silicone Binder	9	63	-	-	-	-	-	-	-	-	-	-	-	-	-	-	-	-	-	-	-	-	-	64	-	-	-	-	-	-	-	-	-	-	-
Barium Tungstate	8	-	-	-	-	-	-	-	-	-	-	666	-	-	-	-	-	-	-	-	-	-	-	-	-	-	-	-	-	-	-	-	-	-	-
Barium Zirconate	8	-	-	-	-	-	-	-	-	-	-	676	-	-	-	-	-	-	-	-	-	-	-	-	-	-	-	-	-	-	-	-	-	-	-
Barium Zirconium Silicate	8	-	-	-	-	-	-	-	-	-	-	-	-	-	-	-	-	-	-	-	-	-	-	616	-	-	-	-	-	-	-	-	-	-	-
Barytes	8	-	-	-	-	-	-	-	-	-	-	-	-	-	-	-	-	-	-	-	-	-	-	704	-	-	-	-	-	-	-	-	-	-	-
Basalt	8	-	-	-	-	-	-	-	-	-	-	688	-	-	-	-	-	-	-	-	-	-	-	-	-	-	-	-	-	-	-	-	-	-	-

MATERIAL NAME	VOLUME	EMITTANCE						REFLECTANCE									ABSORPTANCE									TRANSMITTANCE									Absorptance to Emittance Ratio
		Total			Spectral			Integrated			Spectral			Solar			Integrated			Spectral			Solar			Integrated			Spectral			Solar			
		Hemispherical	Normal	Angular	Hemispherical	Normal	Angular	Hemispherical	Normal	Angular	Hemispherical	Normal	Angular	Hemispherical	Normal	Angular	Hemispherical	Normal	Angular	Hemispherical	Normal	Angular	Hemispherical	Normal	Angular	Hemispherical	Normal	Angular	Hemispherical	Normal	Angular	Hemispherical	Normal	Angular	
Base Glaze No. 1 Binder Pigmented with:																																			
Cerium Dioxide	9	-	445	-	-	-	-	-	-	-	-	-	-	-	-	-	-	-	-	-	-	-	-	-	-	-	-	-	-	-	-	-	-	-	-
Chromium Oxide + Cobalt Oxide	9	-	455	-	-	-	-	-	-	-	-	-	-	-	-	-	-	-	-	-	-	-	-	-	-	-	-	-	-	-	-	-	-	-	-
Cobalt Oxide + Chromium Oxide	9	-	464	-	-	-	-	-	-	-	-	-	-	-	-	-	-	-	-	-	-	-	-	-	-	-	-	-	-	-	-	-	-	-	-
Base Glaze No. 2 Binder Pigmented with:																																			
Chromium Oxide + Cobalt Oxide	9	-	456	-	-	-	-	-	-	-	-	-	-	-	-	-	-	-	-	-	-	-	-	-	-	-	-	-	-	-	-	-	-	-	-
Cobalt Oxide + Chromium Oxide	9	-	464	-	-	-	-	-	-	-	-	-	-	-	-	-	-	-	-	-	-	-	-	-	-	-	-	-	-	-	-	-	-	-	-
Cobalt Oxide + Manganese Oxide	9	-	464	-	-	-	-	-	-	-	-	-	-	-	-	-	-	-	-	-	-	-	-	-	-	-	-	-	-	-	-	-	-	-	-
Manganese Oxide + Cobalt Oxide	9	-	468	-	-	-	-	-	-	-	-	-	-	-	-	-	-	-	-	-	-	-	-	-	-	-	-	-	-	-	-	-	-	-	-
Base Glaze No. 3 Binder Pigmented with:																																			
Cerium Dioxide + Cobalt Oxide	9	-	445	-	-	-	-	-	-	-	-	-	-	-	-	-	-	-	-	-	-	-	-	-	-	-	-	-	-	-	-	-	-	-	-
Chromium Oxide + Iron Oxide	9	-	456	-	-	-	-	-	-	-	-	-	-	-	-	-	-	-	-	-	-	-	-	-	-	-	-	-	-	-	-	-	-	-	-
Cobalt Oxide	9	-	464	-	-	-	-	-	-	-	-	-	-	-	-	-	-	-	-	-	-	-	-	-	-	-	-	-	-	-	-	-	-	-	-
Cobalt Oxide + Nickel Oxide	9	-	464	-	-	-	-	-	-	-	-	-	-	-	-	-	-	-	-	-	-	-	-	-	-	-	-	-	-	-	-	-	-	-	-
Iron Oxide + Chromium Oxide	9	-	466	-	-	-	-	-	-	-	-	-	-	-	-	-	-	-	-	-	-	-	-	-	-	-	-	-	-	-	-	-	-	-	-
Iron Oxide + Manganese Oxide	9	-	466	-	-	-	-	-	-	-	-	-	-	-	-	-	-	-	-	-	-	-	-	-	-	-	-	-	-	-	-	-	-	-	-
Iron Oxide + Nickel Oxide	9	-	466	-	-	-	-	-	-	-	-	-	-	-	-	-	-	-	-	-	-	-	-	-	-	-	-	-	-	-	-	-	-	-	-
Manganese Oxide + Cobalt Oxide	9	-	468	-	-	-	-	-	-	-	-	-	-	-	-	-	-	-	-	-	-	-	-	-	-	-	-	-	-	-	-	-	-	-	-
Manganese Oxide + Iron Oxide	9	-	468	-	-	-	-	-	-	-	-	-	-	-	-	-	-	-	-	-	-	-	-	-	-	-	-	-	-	-	-	-	-	-	-
Nickel Oxide + Cobalt Oxide	9	-	469	-	-	-	-	-	-	-	-	-	-	-	-	-	-	-	-	-	-	-	-	-	-	-	-	-	-	-	-	-	-	-	-
Nickel Oxide + Iron Oxide	9	-	469	-	-	-	-	-	-	-	-	-	-	-	-	-	-	-	-	-	-	-	-	-	-	-	-	-	-	-	-	-	-	-	-
Bauxite	8	-	-	-	-	-	-	-	-	-	-	-	-	-	-	-	-	-	-	-	-	-	-	-	-	-	-	-	-	670	-	-	-	-	-
$B_4C + MoO_3$ Powders	8	-	-	-	-	-	-	-	-	-	-	1455	-	-	-	-	-	-	-	-	-	-	-	-	-	-	-	-	-	-	-	-	-	-	-
Be_2Cr	8	-	-	-	-	1275	-	-	-	-	-	-	-	-	-	-	-	-	-	-	-	-	-	-	-	-	-	-	-	-	-	-	-	-	-
$Be_{12}Nb$	8	-	1273	-	-	-	-	-	-	-	-	1280	-	-	-	-	-	-	-	-	-	-	-	-	-	-	-	-	-	-	-	-	-	-	-
$Be_{17}Nb_2$	8	-	1273	-	-	1277	-	-	-	-	-	1280	-	-	-	-	-	-	-	-	-	-	-	-	-	-	-	-	-	-	-	-	-	-	-
$BeO + Be_{12}Ta$ Cermet	8	-	1377	-	-	1378	-	-	-	-	-	1382	-	-	-	-	-	-	-	-	-	-	-	-	-	-	-	-	-	-	-	-	-	-	-
Be_2Re	8	-	-	-	-	1275	-	-	-	-	-	-	-	-	-	-	-	-	-	-	-	-	-	-	-	-	-	-	-	-	-	-	-	-	-
Beryllia (see Beryllium Oxide)																																			
Beryllium	7	-	71	-	-	74	-	-	-	-	-	78	-	-	-	-	-	-	-	-	-	-	-	-	-	-	-	-	-	82	-	-	-	-	-
Beryllium, Anodized	9	-	-	-	-	-	-	-	-	-	-	1265	-	-	-	-	-	-	-	-	-	-	-	-	-	-	-	-	-	-	-	-	-	-	-
Beryllium Extrusion No. 30	7	-	-	-	-	-	-	-	-	-	-	79	-	-	-	-	-	-	-	-	-	-	-	-	-	-	-	-	-	-	-	-	-	-	-

MATERIAL NAME	VOLUME	EMITTANCE						REFLECTANCE									ABSORPTANCE									TRANSMITTANCE									Absorptance to Emittance Ratio
		Total			Spectral			Integrated			Spectral			Solar			Integrated			Spectral			Solar			Integrated			Spectral			Solar			
		Hemispherical	Normal	Angular	Hemispherical	Normal	Angular	Hemispherical	Normal	Angular	Hemispherical	Normal	Angular	Hemispherical	Normal	Angular	Hemispherical	Normal	Angular	Hemispherical	Normal	Angular	Hemispherical	Normal	Angular	Hemispherical	Normal	Angular	Hemispherical	Normal	Angular	Hemispherical	Normal	Angular	
Beryllium + Copper, Anodized	9	1268	-	-	-	-	-	-	-	-	-	1270	-	-	-	-	-	-	-	-	-	-	-	1273	-	-	-	-	-	-	-	-	-	-	-
Beryllium + Iron + ΣX_i	7	-	-	-	-	-	-	-	-	-	-	-	1134	-	-	-	-	-	-	-	-	-	-	-	-	-	-	-	-	-	-	-	-	-	-
Pentaberyllium Boride	8	-	-	-	-	732	-	-	-	-	-	-	-	-	-	-	-	-	-	-	-	-	-	-	-	-	-	-	-	-	-	-	-	-	-
Beryllium Intermetallic Compounds:	8	-	1273	-	-	1274	-	-	-	-	-	1279	-	-	-	-	-	-	-	-	-	-	-	-	-	-	-	-	-	-	-	-	-	-	-
Be_2Cr	8	-	-	-	-	1275	-	-	-	-	-	-	-	-	-	-	-	-	-	-	-	-	-	-	-	-	-	-	-	-	-	-	-	-	-
Be_2Re	8	-	-	-	-	1275	-	-	-	-	-	-	-	-	-	-	-	-	-	-	-	-	-	-	-	-	-	-	-	-	-	-	-	-	-
Be_2Ti	8	-	-	-	-	1275	-	-	-	-	-	-	-	-	-	-	-	-	-	-	-	-	-	-	-	-	-	-	-	-	-	-	-	-	-
$Be_{12}Nb$	8	-	1273	-	-	-	-	-	-	-	-	1280	-	-	-	-	-	-	-	-	-	-	-	-	-	-	-	-	-	-	-	-	-	-	-
$Be_{12}Ta$	8	-	1273	-	-	1277	-	-	-	-	-	1280	-	-	-	-	-	-	-	-	-	-	-	-	-	-	-	-	-	-	-	-	-	-	-
$Be_{13}Sc$	8	-	-	-	-	1275	-	-	-	-	-	-	-	-	-	-	-	-	-	-	-	-	-	-	-	-	-	-	-	-	-	-	-	-	-
$Be_{13}Zr$	8	-	-	-	-	-	-	-	-	-	-	1280	-	-	-	-	-	-	-	-	-	-	-	-	-	-	-	-	-	-	-	-	-	-	-
$Be_{17}Nb_2$	8	-	1273	-	-	1277	-	-	-	-	-	1280	-	-	-	-	-	-	-	-	-	-	-	-	-	-	-	-	-	-	-	-	-	-	-
$Be_{17}Ta_2$	8	-	1273	-	-	1277	-	-	-	-	-	1280	-	-	-	-	-	-	-	-	-	-	-	-	-	-	-	-	-	-	-	-	-	-	-
$Be_{17}Zr_2$	8	-	-	-	-	-	-	-	-	-	-	1280	-	-	-	-	-	-	-	-	-	-	-	-	-	-	-	-	-	-	-	-	-	-	-
Beryllium Oxide	8	-	201	-	-	203 205	-	-	-	-	-	-	-	-	-	208	-	-	-	-	210	-	-	212	-	-	-	-	-	213	-	-	-	-	-
$Be_{13}Sc$	8	-	-	-	-	1275	-	-	-	-	-	-	-	-	-	-	-	-	-	-	-	-	-	-	-	-	-	-	-	-	-	-	-	-	-
Beryllium Substrate Coated with:																																			
Calcium Titanate	9	843	-	-	-	-	-	-	-	-	-	-	-	-	-	-	-	-	-	-	-	-	-	-	-	-	-	-	-	-	-	-	-	-	-
Iron Titanate	9	905	-	-	-	-	-	-	-	-	-	-	-	-	-	-	-	-	-	-	-	-	-	-	-	-	-	-	-	-	-	-	-	-	-
Zirconium Titanate	9	1032	-	-	-	-	-	-	-	-	-	-	-	-	-	-	-	-	-	-	-	-	-	-	-	-	-	-	-	-	-	-	-	-	-
BeCu Substrate Coated with Silver	9	-	-	-	-	-	-	-	-	-	-	-	-	-	-	-	-	-	-	-	-	-	-	-	-	-	-	-	-	-	-	-	-	-	753
$Be_{12}Ta$	8	-	1273	-	-	1277	-	-	-	-	-	1280	-	-	-	-	-	-	-	-	-	-	-	-	-	-	-	-	-	-	-	-	-	-	-
$Be_{17}Ta_2$	8	-	1273	-	-	1277	-	-	-	-	-	1280	-	-	-	-	-	-	-	-	-	-	-	-	-	-	-	-	-	-	-	-	-	-	-
Be_2Ti	8	-	-	-	-	1275	-	-	-	-	-	-	-	-	-	-	-	-	-	-	-	-	-	-	-	-	-	-	-	-	-	-	-	-	-
$Be_{13}Zr$	8	-	-	-	-	-	-	-	-	-	-	1280	-	-	-	-	-	-	-	-	-	-	-	-	-	-	-	-	-	-	-	-	-	-	-
$Be_{17}Zr_2$	8	-	-	-	-	-	-	-	-	-	-	1280	-	-	-	-	-	-	-	-	-	-	-	-	-	-	-	-	-	-	-	-	-	-	-
Bismuth	7	-	-	-	-	-	-	-	-	-	-	-	-	-	-	-	-	-	-	-	-	-	-	-	-	-	-	-	-	85	88	-	-	-	-
Bismuth + Tin	7	-	-	-	-	-	-	-	-	-	-	-	-	-	-	-	-	899	-	-	-	-	-	-	-	-	-	-	-	-	-	-	-	-	-
Bismuth Oxide Coating on Glass Substrate	9	-	-	-	-	-	-	-	-	-	-	820	-	-	-	-	-	-	-	-	-	-	-	-	-	-	-	-	-	823	-	-	-	-	-
Bismuth Triiodide	8	-	-	-	-	-	-	-	-	-	-	1027	-	-	-	-	-	-	-	-	-	-	-	-	-	-	-	-	-	-	-	-	-	-	-
Dibismuth Triselenide	8	-	-	-	-	-	-	-	-	-	-	1130	-	-	-	-	-	-	-	-	-	-	-	-	-	-	-	-	-	-	-	-	-	-	-

MATERIAL NAME	VOLUME	EMITTANCE						REFLECTANCE									ABSORPTANCE									TRANSMITTANCE									Absorptance to Emittance Ratio
		Total			Spectral			Integrated			Spectral			Solar			Integrated			Spectral			Solar			Integrated			Spectral			Solar			
		Hemispherical	Normal	Angular	Hemispherical	Normal	Angular	Hemispherical	Normal	Angular	Hemispherical	Normal	Angular	Hemispherical	Normal	Angular	Hemispherical	Normal	Angular	Hemispherical	Normal	Angular	Hemispherical	Normal	Angular	Hemispherical	Normal	Angular	Hemispherical	Normal	Angular	Hemispherical	Normal	Angular	
Bismuth Tellurium Selenide	8	–	–	–	–	–	–	–	–	–	–	1130	–	–	–	–	–	–	–	–	–	–	–	–	–	–	–	–	–	–	–	–	–	–	–
Dibismuth Tritelluride	8	–	–	–	–	–	–	–	–	–	–	1238	–	–	–	–	–	–	–	–	–	–	–	–	–	–	–	–	–	–	–	–	–	–	–
Bismuth Titanate	8	–	–	–	–	–	–	–	–	–	–	–	–	–	–	–	–	–	–	–	–	–	–	–	–	–	–	–	–	644	–	–	–	–	–
Black Nickel Coating on Copper Substrate	9	700	–	–	–	–	–	–	–	–	–	–	–	–	–	–	–	–	–	–	–	–	–	–	–	–	–	–	–	–	–	–	–	–	–
Borates	8	–	–	–	–	–	–	–	–	–	–	582	–	–	–	–	–	–	–	–	–	–	–	–	–	–	–	–	–	–	–	–	–	–	–
Borides:																																			
Molybdenum Borides	8	–	–	–	–	691	–	–	–	–	–	693	–	–	–	–	–	–	–	–	–	–	–	–	–	–	–	–	–	–	–	–	–	–	–
Niobium Diboride	8	–	697	–	–	701	–	–	–	–	–	–	–	–	–	–	–	–	–	–	–	–	–	–	–	–	–	–	–	–	–	–	–	–	–
Titanium Diboride	8	–	703	–	–	705 709	–	–	–	–	–	710	–	–	–	–	–	–	–	–	–	–	–	–	–	–	–	–	–	–	–	–	–	–	–
Zirconium Diboride	8	–	712	–	–	715 717	–	–	–	–	–	720	–	–	–	–	–	–	–	–	–	–	–	–	–	–	–	–	–	–	–	–	–	–	–
Rare Earth Borides	8	–	–	–	–	722	–	–	–	–	–	–	–	–	–	–	–	–	–	–	–	–	–	–	–	–	–	–	–	726	–	–	–	–	–
Miscellaneous Borides	8	–	729	–	–	721	–	–	–	–	–	734	–	–	–	–	–	–	–	–	–	–	–	–	–	–	–	–	–	–	–	–	–	–	–
Tetraboron Carbide	8	–	–	–	–	852	–	–	–	–	–	855	–	–	–	–	–	–	–	–	–	–	–	–	–	–	–	–	–	–	–	–	–	–	–
Boron Coating on Nb-1Zr Substrate	9	826	–	–	–	–	–	–	–	–	–	–	–	–	–	–	–	–	–	–	–	–	–	–	–	–	–	–	–	–	–	–	–	–	–
Boron Carbide Coating on:																																			
Inconel X Substrate	9	–	–	–	–	829	–	–	–	–	–	832	–	–	–	–	–	–	–	–	–	–	–	–	–	–	–	–	–	–	–	–	–	–	–
Molybdenum Substrate	9	827	–	–	–	–	–	–	–	–	–	–	–	–	–	–	–	–	–	–	–	–	–	–	–	–	–	–	–	–	–	–	–	–	–
Boron Carbide Pigment in Synar Binder	9	65	–	–	–	–	–	–	–	–	–	–	–	–	–	–	–	–	–	–	–	–	–	–	–	–	–	–	–	–	–	–	–	–	–
Boron Nitride	8	–	1037	–	–	1040 1042	–	–	–	–	–	1047	–	–	–	–	–	–	–	–	–	–	–	–	–	–	–	–	–	1054	–	–	–	–	–
Boron Nitride Pigment in:																																			
Potassium Silicate Binder	9	66	–	–	–	–	–	–	–	–	–	–	–	–	–	–	–	–	–	–	–	–	–	–	–	–	–	–	–	–	–	–	–	–	–
Silicone Binder	9	–	–	–	–	–	–	–	–	–	–	68	–	–	–	–	–	–	–	–	–	–	–	–	–	–	–	–	–	–	–	–	–	–	–
Boron Nitride + Diatomaceous Earth Pigment in Silicone Binder	9	–	–	–	–	–	–	–	–	–	–	70	–	–	–	–	–	–	–	–	–	–	–	–	–	–	–	–	–	–	–	–	–	–	–
Boron Phosphate	8	–	–	–	–	–	–	–	–	–	–	608	–	–	–	–	–	–	–	–	–	–	–	–	–	–	–	–	–	–	–	–	–	–	–
Boron Phosphide	8	–	–	–	–	1105	–	–	–	–	–	1107	–	–	–	–	–	–	–	–	–	–	–	–	–	–	–	–	–	–	–	–	–	–	–
Tetraboron Silicide	8	–	1134	–	–	1136	–	–	–	–	–	1138	–	–	–	–	–	–	–	–	–	–	–	–	–	–	–	–	–	–	–	–	–	–	–
Hexaboron Silicide	8	–	1134	–	–	1136	–	–	–	–	–	1138	–	–	–	–	–	–	–	–	–	–	–	–	–	–	–	–	–	–	–	–	–	–	–
Borosilicate Glass	8	–	1539	–	–	1541	–	–	–	–	–	1543	–	–	–	–	–	–	–	–	–	–	–	1545	–	–	1546	–	1547	1549	–	–	–	–	–

MATERIAL NAME	VOLUME	EMITTANCE Total Hemispherical	EMITTANCE Total Normal	EMITTANCE Total Angular	EMITTANCE Spectral Hemispherical	EMITTANCE Spectral Normal	EMITTANCE Spectral Angular	REFLECTANCE Integrated Hemispherical	REFLECTANCE Integrated Normal	REFLECTANCE Integrated Angular	REFLECTANCE Spectral Hemispherical	REFLECTANCE Spectral Normal	REFLECTANCE Spectral Angular	REFLECTANCE Solar Hemispherical	REFLECTANCE Solar Normal	REFLECTANCE Solar Angular	ABSORPTANCE Integrated Hemispherical	ABSORPTANCE Integrated Normal	ABSORPTANCE Integrated Angular	ABSORPTANCE Spectral Hemispherical	ABSORPTANCE Spectral Normal	ABSORPTANCE Spectral Angular	ABSORPTANCE Solar Hemispherical	ABSORPTANCE Solar Normal	ABSORPTANCE Solar Angular	TRANSMITTANCE Integrated Hemispherical	TRANSMITTANCE Integrated Normal	TRANSMITTANCE Integrated Angular	TRANSMITTANCE Spectral Hemispherical	TRANSMITTANCE Spectral Normal	TRANSMITTANCE Spectral Angular	TRANSMITTANCE Solar Hemispherical	TRANSMITTANCE Solar Normal	TRANSMITTANCE Solar Angular	Absorptance to Emittance Ratio
Boysen No. 11 Black Paint	9	-	-	-	-	-	-	-	-	-	-	487	-	-	-	-	-	-	-	-	-	-	-	-	-	-	-	-	-	-	-	-	-	-	-
Brass	7	912	915	-	-	-	-	-	-	-	-	-	-	-	-	-	-	925	-	-	-	-	-	-	-	-	-	-	-	-	-	-	-	-	-
Brass, α	7	-	-	-	-	-	-	-	-	-	-	-	-	921	-	-	-	-	-	-	-	-	-	-	-	-	-	-	-	-	-	-	-	-	-
Brass, β	7	-	-	-	-	-	-	-	-	-	-	-	-	921	-	-	-	-	-	-	-	-	-	-	-	-	-	-	-	-	-	-	-	-	-
Brass, yellow	7	912	-	-	-	-	-	-	-	-	-	-	-	-	-	-	923	-	-	-	-	-	-	-	-	-	-	-	-	-	-	-	-	-	-
Brass Substrate Coated with:																																			
Carbon	9	-	-	-	-	-	-	-	-	-	-	859	-	-	-	-	-	-	-	-	-	-	-	-	-	-	-	-	-	-	-	-	-	-	-
Gold Black	9	-	-	-	-	-	-	-	-	-	-	1175	-	-	-	-	-	-	-	-	-	-	-	-	-	-	-	-	-	-	-	-	-	-	-
Titanium	9	-	-	-	-	-	-	-	-	-	-	-	-	-	-	-	-	-	-	-	-	-	-	-	762	-	-	-	-	-	-	-	-	-	-
Broma Alkyd Enamel No. 113	9	-	-	-	-	-	-	-	-	-	-	488	-	-	-	-	-	-	-	-	-	-	-	-	-	-	-	-	-	-	-	-	-	-	-
Broma Metallic Enamel No. 102	9	-	-	-	-	-	-	-	-	-	-	489	-	-	-	-	-	-	-	-	-	-	-	-	-	-	-	-	-	-	-	-	-	-	-
Bromides:																																			
Cesium Bromide	8	-	-	-	-	-	-	-	-	-	-	-	-	-	-	737	-	-	-	-	-	-	-	-	-	-	-	-	-	739	-	-	-	-	-
Copper Bromide	8	-	-	-	-	-	-	-	-	-	-	741	-	-	-	-	-	-	-	-	-	-	-	-	-	-	-	-	-	743	-	-	-	-	-
Lead Bromide	8	-	-	-	-	-	-	-	-	-	-	-	-	-	-	-	-	-	-	-	-	-	-	-	-	-	-	-	-	745	-	-	-	-	-
Dimercury Dibromide	8	-	-	-	-	-	-	-	-	-	-	747	-	-	-	-	-	-	-	-	-	-	-	-	-	-	-	-	-	-	-	-	-	-	-
Potassium Bromide	8	-	-	-	-	749	-	-	-	-	-	751 753	-	-	-	-	-	-	-	-	-	-	-	-	-	-	-	-	-	759	-	-	-	-	-
Rubidium Bromide	8	-	-	-	-	764	-	-	-	-	-	-	766	-	-	-	-	-	-	-	-	-	-	-	-	-	-	-	-	768	-	-	-	-	-
Silver Bromide	8	-	-	-	-	-	-	-	-	-	-	770	773	-	-	-	-	-	-	-	-	-	-	-	-	-	-	-	-	775	-	-	-	-	-
Thallium Bromide	8	-	-	-	-	-	-	-	-	-	-	778	780 782	-	-	-	-	-	-	-	-	-	-	-	-	-	-	-	-	783	-	-	-	-	-
Bronze	7	-	-	-	-	1163	-	-	-	-	-	918 1167	-	-	-	-	-	-	-	-	-	-	-	1170	-	-	-	-	-	-	-	-	-	-	-
Bronze Substrate Coated with Molybdenum Disilicide	9	-	928	-	-	-	-	-	-	-	-	-	-	-	-	-	-	-	-	-	-	-	-	-	-	-	-	-	-	-	-	-	-	-	-
Brucite	8	-	-	-	-	-	-	-	-	-	-	-	-	-	-	-	-	-	-	-	-	-	-	-	-	-	-	-	-	1662	-	-	-	-	-
Brush S. P.	8	-	202	-	-	204	-	-	-	-	-	-	-	-	-	-	-	-	-	-	-	-	-	-	-	-	-	-	-	-	-	-	-	-	-
B. S. 1433 Copper	7	-	-	-	-	-	-	-	-	-	-	-	-	-	173	-	-	-	-	-	-	-	-	194	-	-	-	-	-	-	-	-	-	-	-
Burch Photometric Sphere White No. 2210	9	-	-	-	-	-	-	-	-	-	-	490	-	-	-	-	-	-	-	-	-	-	-	-	-	-	-	-	-	-	-	-	-	-	-
Butyl Acrylate Coating on Anodized Aluminum Substrate	9	-	-	-	-	-	-	-	-	-	-	1114	-	-	-	-	-	-	-	-	-	-	-	-	-	-	-	-	-	-	-	-	-	-	-
Butylated Melamine Formaldehyde Coating on Anodized Aluminum Substrate	9	-	-	-	-	-	-	-	-	-	-	1116	-	-	-	-	-	-	-	-	-	-	-	-	-	-	-	-	-	-	-	-	-	-	-

MATERIAL NAME	VOLUME	EMITTANCE						REFLECTANCE									ABSORPTANCE									TRANSMITTANCE									Absorptance to Emittance Ratio
		Total			Spectral			Integrated			Spectral			Solar			Integrated			Spectral			Solar			Integrated			Spectral			Solar			
		Hemispherical	Normal	Angular	Hemispherical	Normal	Angular	Hemispherical	Normal	Angular	Hemispherical	Normal	Angular	Hemispherical	Normal	Angular	Hemispherical	Normal	Angular	Hemispherical	Normal	Angular	Hemispherical	Normal	Angular	Hemispherical	Normal	Angular	Hemispherical	Normal	Angular	Hemispherical	Normal	Angular	
Butylated Urea Formaldehyde Coating on Anodized Aluminum Substrate	9	-	-	-	-	-	-	-	-	-	-	1118	-	-	-	-	-	-	-	-	-	-	-	-	-	-	-	-	-	-	-	-	-	-	-
Cabot RF-1 TiO_2 Pigment in Potassium Silicate Binder	9	214	-	-	-	-	-	-	-	-	-	-	-	-	-	-	-	-	-	-	-	-	-	-	-	-	-	-	-	-	-	-	-	-	-
Cadmium	7	91	-	-	-	-	-	-	-	-	-	-	93	-	-	-	96	-	-	-	-	98	-	-	-	-	-	-	-	-	-	-	-	-	-
Cadmium Antimonide	8	-	-	-	-	-	-	-	-	-	-	1282	-	-	-	-	-	-	-	-	-	-	-	-	-	-	-	-	-	-	-	-	-	-	-
Cadmium Arsenide Coating on Glass Substrate	9	-	-	-	-	-	-	-	-	-	-	-	-	-	-	-	-	-	-	-	-	-	-	-	-	-	-	-	-	836	-	-	-	-	-
Cadmium Fluoride	8	-	-	-	-	-	-	-	-	-	-	-	-	-	-	-	-	-	-	-	-	-	-	-	-	-	-	-	-	919	-	-	-	-	-
Cadmium Oxide	8	-	-	-	-	216	-	-	-	-	-	-	-	-	-	-	-	-	-	-	-	-	-	-	-	-	-	-	-	-	-	-	-	-	-
Cadmium Oxide Coating on Glass Substrate	9	-	-	-	-	-	-	-	-	-	-	-	-	-	-	-	-	-	-	-	-	-	-	-	-	-	-	-	-	838	-	-	-	-	-
Cadmium Selenide	8	-	-	-	-	-	-	-	1108	-	-	-	-	-	-	-	-	-	-	-	-	-	-	-	-	-	-	-	-	1110	-	-	-	-	-
Cadmium Sulfide	8	-	1181	-	-	1183	-	-	-	-	-	1188	-	-	-	-	-	-	-	-	-	-	-	-	-	-	-	-	-	1194	-	-	-	-	-
Cadmium Sulfide Coating on Aluminum Substrate	9	-	-	-	-	-	-	-	-	-	-	-	-	-	-	-	-	-	-	-	-	-	-	-	-	-	-	-	-	840	-	-	-	-	-
Cadmium Telluride	8	1239	-	-	-	-	-	-	-	-	-	1241	-	-	-	-	-	-	-	-	-	-	-	-	-	-	-	-	-	1244	-	-	-	-	-
Calcite	8	-	-	-	-	-	-	-	-	-	-	584 1653	-	-	-	-	-	-	-	-	-	-	-	-	-	-	-	-	-	586	-	-	-	-	-
Calcium Aluminate	8	-	-	-	-	-	-	-	-	-	-	-	-	-	-	-	-	-	-	-	-	-	-	-	-	-	-	-	-	573	-	-	-	-	-
Calcium Aluminate Glass	8	-	-	-	-	-	-	-	-	-	-	1551	-	-	-	-	-	-	-	-	-	-	-	-	-	-	-	-	-	1553	-	-	-	-	-
Calcium Hexaboride	8	-	-	-	-	732	-	-	-	-	-	-	-	-	-	-	-	-	-	-	-	-	-	-	-	-	-	-	-	-	-	-	-	-	-
Calcium Carbonate	8	-	-	-	-	-	-	-	-	-	-	583	-	-	-	-	-	-	-	-	-	-	-	-	-	-	-	-	-	585	-	-	-	-	-
Calcium Carbonate Pigment in Silicone Binder	9	-	-	-	-	-	-	-	-	-	-	73	-	-	-	-	-	-	-	-	-	-	-	-	-	-	-	-	-	-	-	-	-	-	-
Calcium Coating on Glass Substrate	9	-	-	-	-	-	-	-	-	-	-	-	-	-	-	-	-	-	-	-	-	-	-	-	-	-	-	-	-	617	-	-	-	-	-
Calcium Fluoride	8	-	-	-	-	921	-	-	-	-	-	-	924	-	-	-	-	-	-	-	929	-	-	-	-	-	-	-	-	931	-	-	-	-	-
Calcium Fluoride Pigment in Sodium Silicate Binder	9	-	-	-	-	-	-	-	-	-	-	-	-	-	-	-	-	-	-	-	75	-	-	-	-	-	-	-	-	-	-	-	-	-	-
Calcium Fluoride Substrate Coated with Germanium	9	-	-	-	-	-	-	-	-	-	-	647	-	-	-	-	-	-	-	-	-	-	-	-	-	-	-	-	-	-	-	-	-	-	-
Calcium Oxide	8	-	-	-	-	-	-	-	-	-	-	218	-	-	-	-	-	-	-	-	-	-	-	-	-	-	-	-	-	220	-	-	-	-	-
$CaSO_4$ + TiO Pigment in Alkyd-Melamine Binder	9	-	-	-	-	-	-	-	-	-	-	78	-	-	-	-	-	-	-	-	-	-	-	-	-	-	-	-	-	-	-	-	-	-	-

MATERIAL NAME	VOLUME	EMITTANCE						REFLECTANCE									ABSORPTANCE									TRANSMITTANCE									Absorptance to Emittance Ratio
		Total			Spectral			Integrated			Spectral			Solar			Integrated			Spectral			Solar			Integrated			Spectral			Solar			
		Hemispherical	Normal	Angular	Hemispherical	Normal	Angular	Hemispherical	Normal	Angular	Hemispherical	Normal	Angular	Hemispherical	Normal	Angular	Hemispherical	Normal	Angular	Hemispherical	Normal	Angular	Hemispherical	Normal	Angular	Hemispherical	Normal	Angular	Hemispherical	Normal	Angular	Hemispherical	Normal	Angular	
Calcium Metasilicate Pigment in Potassium Silicate Binder	9	-	-	-	-	-	-	-	-	-	-	76	-	-	-	-	-	-	-	-	-	-	-	-	-	-	-	-	-	-	-	-	-	-	-
Calcium Silicate	8	-	-	-	-	-	-	-	-	-	-	618	-	-	-	-	-	-	-	-	-	-	-	-	-	-	-	-	-	-	-	-	-	-	-
Calcium Sulfate	8	-	-	-	-	-	-	-	-	-	-	627	-	-	-	-	-	-	-	-	-	-	-	-	-	-	-	-	-	629	-	-	-	-	-
Calcium Titanate Coating on:																																			
Aluminum Substrate	9	845	-	-	-	-	-	-	-	-	-	-	-	-	-	-	-	-	-	-	-	-	-	850	-	-	-	-	-	-	-	-	-	-	-
Beryllium Substrate	9	843	-	-	-	-	-	-	-	-	-	-	-	-	-	-	-	-	-	-	-	-	-	-	-	-	-	-	-	-	-	-	-	-	-
Niobium Substrate	9	843	-	-	-	-	849	-	-	-	-	-	-	-	-	-	-	-	-	-	-	-	-	-	-	-	-	-	-	-	-	-	-	-	-
Nb-1Zr Substrate	9	843	-	-	-	-	-	-	-	-	-	-	-	-	-	-	-	-	-	-	-	-	-	-	-	-	-	-	-	-	-	-	-	-	-
Stainless Steel Substrate	9	843	-	-	-	-	-	-	-	-	-	-	-	-	-	-	-	-	-	-	-	-	-	-	-	-	-	-	-	-	-	-	-	-	-
Calcium Titanate Pigment in Aluminum Phosphate Binder	9	79	-	-	-	80	-	-	-	-	-	-	-	-	-	-	-	-	-	-	-	-	-	-	-	-	-	-	-	-	-	-	-	-	-
Calcium Tungstates	8	-	-	-	-	-	-	-	-	-	-	-	-	-	-	-	-	-	-	-	-	-	-	-	-	-	-	-	-	663	-	-	-	-	-
Calcium Zirconium Silicate	8	-	-	-	-	-	-	-	-	-	-	-	-	-	-	-	-	-	-	-	616	-	-	-	-	-	-	-	-	-	-	-	-	-	-
Carbides:		-	-	-	-	-	-	-	-	-	-	-	-	-	-	-	-	-	-	-	-	-	-	-	-	-	-	-	-	-	-	-	-	-	-
Niobium Monocarbide	8	785	787	-	-	789	-	-	-	-	-	-	-	-	-	-	-	-	-	-	-	-	-	-	-	-	-	-	-	-	-	-	-	-	-
Silicon Monocarbide	8	791	792	-	-	796 798	-	-	-	-	-	802	-	-	-	-	-	-	-	-	808	-	-	810	-	-	-	-	-	-	-	-	-	-	-
Tantalum Carbide	8	811	813	-	-	815 817	-	-	-	-	-	-	-	-	-	-	-	-	-	-	-	-	-	-	-	-	-	-	-	-	-	-	-	-	-
Titanium Carbide	8	819	821	-	-	823	825	-	-	-	-	-	-	-	-	-	-	-	-	-	-	-	-	-	-	-	-	-	-	-	-	-	-	-	-
Tungsten Carbides	8	-	826	-	-	828 830	-	-	-	-	-	-	-	-	-	-	-	-	-	-	-	-	-	-	-	-	-	-	-	-	-	-	-	-	-
Zirconium Carbide	8	833	835	-	-	837 841	-	-	-	-	-	843	-	-	-	-	-	-	-	-	-	-	-	-	-	-	-	-	-	-	-	-	-	-	-
Miscellaneous Carbides	8	847	849	-	-	851	-	-	-	-	-	854	-	-	-	-	-	-	-	-	-	-	-	-	-	-	-	-	-	-	-	-	-	-	-
Carbon	8	-	5	-	8 10	12 14	16	-	-	-	18 20	22	24 25	-	-	-	-	-	-	-	-	-	-	-	-	-	-	-	-	27	-	-	-	-	-
Carbon (Graphite)	8	30	31	38	40 42	44 51	57	-	-	-	59 61	63 65	70 71	-	-	-	74	-	-	-	-	-	-	76	-	-	-	-	-	-	-	-	-	-	-
Carbonates:																																			
Calcium Carbonate	8	-	-	-	-	-	-	-	-	-	-	583	-	-	-	-	-	-	-	-	-	-	-	-	-	-	-	-	-	585	-	-	-	-	-
Lead Carbonate	8	-	-	-	-	-	-	-	-	-	-	587	-	-	-	-	-	-	-	-	-	-	-	589	-	-	-	-	-	-	-	-	-	-	-
Magnesium Carbonate	8	-	-	-	-	-	-	-	-	-	-	590	-	-	-	-	-	-	-	-	-	-	-	-	-	-	-	-	-	-	-	-	-	-	-
Miscellaneous Carbonates	8	-	-	-	-	-	-	-	-	-	-	592	-	-	-	-	-	-	-	-	-	-	-	-	-	-	-	-	-	-	-	-	-	-	-
Carbon Coating on (also see Graphite):																																			
Aluminum Oxide Substrate	9	-	854	-	-	-	-	-	-	-	-	-	-	-	-	-	-	-	-	-	-	-	-	-	-	-	-	-	-	-	-	-	-	-	-

MATERIAL NAME	VOLUME	EMITTANCE						REFLECTANCE									ABSORPTANCE									TRANSMITTANCE									Absorptance to Emittance Ratio
		Total			Spectral			Integrated			Spectral			Solar			Integrated			Spectral			Solar			Integrated			Spectral			Solar			
		Hemispherical	Normal	Angular	Hemispherical	Normal	Angular	Hemispherical	Normal	Angular	Hemispherical	Normal	Angular	Hemispherical	Normal	Angular	Hemispherical	Normal	Angular	Hemispherical	Normal	Angular	Hemispherical	Normal	Angular	Hemispherical	Normal	Angular	Hemispherical	Normal	Angular	Hemispherical	Normal	Angular	
Carbon Coating on (also see Graphite): (continued)																																			
Brass Substrate	9	–	–	–	–	–	–	–	–	–	–	859	–	–	–	–	–	–	–	–	–	–	–	–	–	–	–	–	–	–	–	–	–	–	–
Copper Substrate	9	852	–	–	–	857	–	–	–	–	–	859	–	–	–	–	–	–	–	–	–	–	–	–	–	–	–	–	–	–	–	–	–	–	–
Molybdenum Substrate	9	852	–	–	–	–	–	–	–	–	–	–	–	–	–	–	–	–	–	–	–	–	–	862	–	–	–	–	–	–	–	–	–	–	–
Pyroxylin Substrate	9	–	–	–	–	–	–	–	–	–	–	–	–	–	–	–	–	–	–	–	–	–	–	–	–	–	–	–	–	861	–	–	–	–	–
Silicon Dioxide Substrate	9	–	–	–	–	–	–	–	–	–	–	859	–	–	–	–	–	–	–	–	–	–	–	–	–	–	–	–	–	–	–	–	–	–	–
Silver Substrate	9	–	–	–	–	–	–	–	–	–	–	859	–	–	–	–	–	–	–	–	–	–	–	–	–	–	–	–	–	–	–	–	–	–	–
Tantalum Substrate	9	–	854	–	–	855	–	–	–	–	–	–	–	–	–	–	–	–	–	–	–	–	–	–	–	–	–	–	–	–	–	–	–	–	–
Carbon Black Pigment in (also see Lampblack):																																			
Acrylic Binder	9	81	–	–	–	–	–	–	–	–	–	86	–	–	–	–	–	–	–	–	–	–	–	89	–	–	–	–	–	–	–	–	–	–	–
Decoret Binder	9	–	–	–	–	–	–	–	–	–	–	86	–	–	–	–	–	–	–	–	–	–	–	–	–	–	–	–	–	–	–	–	–	–	–
Epoxy Binder	9	–	82	–	–	84	–	–	–	–	–	86	–	–	–	–	–	–	–	–	–	–	–	89	–	–	–	–	–	–	–	–	–	–	–
Nitrocellulose Binder	9	–	–	–	–	–	–	–	–	–	–	86	–	–	–	–	–	–	–	–	–	–	–	–	–	–	–	–	–	–	–	–	–	–	–
Phthalic Alkyd Binder	9	–	–	–	–	–	–	–	–	–	–	–	–	–	–	–	–	–	–	–	–	–	–	89	–	–	–	–	–	–	–	–	–	–	–
Silicone Binder	9	81	82	–	–	–	–	–	–	–	–	86	–	–	–	–	–	–	–	–	–	–	–	89	–	–	–	–	–	–	–	–	–	–	–
Turpentine Binder	9	–	–	–	–	–	–	–	–	–	–	86	–	–	–	–	–	–	–	–	–	–	–	–	–	–	–	–	–	–	–	–	–	–	–
Xylol Binder	9	81	–	–	–	–	–	–	–	–	–	–	–	–	–	–	–	–	–	–	–	–	–	–	–	–	–	–	–	–	–	–	–	–	–
Carbon Dioxide Coating on:																																			
Cat-A-Lac Black Substrate	9	–	–	–	–	–	–	–	–	–	864	–	–	–	–	–	–	–	–	–	–	–	–	–	–	–	–	–	–	–	–	–	–	–	–
Stainless Steel Substrate	9	–	–	–	–	–	–	–	–	–	864	–	–	–	–	–	–	–	–	–	–	–	–	–	–	–	–	–	–	–	–	–	–	–	–
Carboxy-Methyl-Cellulose Binder Pigmented with Barium Sulfate	9	–	–	–	–	–	–	–	–	–	–	62	–	–	–	–	–	–	–	–	–	–	–	–	–	–	–	–	–	–	–	–	–	–	–
Cassiterite	8	–	–	–	–	–	–	–	–	–	–	451	–	–	–	–	–	–	–	–	–	–	–	–	–	–	–	–	–	–	–	–	–	–	–
Cat-A-Lac Black	9	–	82	–	–	84	–	–	–	–	–	86	–	–	–	–	–	–	–	–	–	–	–	89	–	–	–	–	–	–	–	–	–	–	–
Cat-A-Lac Black Substrate Coated with Carbon Dioxide	9	–	–	–	–	–	–	–	–	–	864	–	–	–	–	–	–	–	–	–	–	–	–	–	–	–	–	–	–	–	–	–	–	–	–
Cat-A-Lac Clear Coating on Polyethylene Substrate	9	–	–	–	–	–	–	–	–	–	–	–	–	–	–	–	–	–	–	–	–	–	–	–	–	–	–	–	–	1122	–	–	–	–	–
Cat-A-Lac White	9	491	–	–	–	–	–	–	–	–	–	–	–	–	–	–	–	492	–	–	–	–	–	493	–	–	–	–	–	–	–	–	–	–	–
CdSb	8	–	–	–	–	–	–	–	–	–	–	1282	–	–	–	–	–	–	–	–	–	–	–	–	–	–	–	–	–	–	–	–	–	–	–
Celestite	8	–	–	–	–	–	–	–	–	–	–	–	–	–	–	–	–	–	–	–	–	–	–	–	–	–	–	–	–	1701	–	–	–	–	–
Cellulose Nitrate Binder Pigmented with TiO_2	9	–	–	–	–	–	–	–	–	–	–	–	–	–	–	–	–	–	–	–	256	–	–	–	–	–	–	–	–	–	–	–	–	–	–
Cellulose Nitrate Substrate Coated with:																																			
Antimony Black	9	–	–	–	–	–	–	–	–	–	–	–	–	–	–	–	–	–	–	–	–	–	–	–	–	–	–	–	–	1172	–	–	–	–	–

MATERIAL NAME	VOLUME	EMITTANCE						REFLECTANCE									ABSORPTANCE									TRANSMITTANCE									Absorptance to Emittance Ratio
		Total			Spectral			Integrated			Spectral			Solar			Integrated			Spectral			Solar			Integrated			Spectral			Solar			
		Hemispherical	Normal	Angular	Hemispherical	Normal	Angular	Hemispherical	Normal	Angular	Hemispherical	Normal	Angular	Hemispherical	Normal	Angular	Hemispherical	Normal	Angular	Hemispherical	Normal	Angular	Hemispherical	Normal	Angular	Hemispherical	Normal	Angular	Hemispherical	Normal	Angular	Hemispherical	Normal	Angular	
Cellulose Nitrate Substrate Coated with: (continued)																																			
Gold Black	9	-	-	-	-	-	-	-	-	-	-	1175	-	-	-	-	-	-	-	-	1178	-	-	-	-	-	-	-	-	1181	-	-	-	-	-
Cerium Hexaboride	8	-	-	-	-	722	-	-	-	-	-	-	-	-	-	-	-	-	-	-	-	-	-	-	-	-	-	-	-	-	-	-	-	-	-
Cerium Dioxide	8	-	225	-	-	-	-	-	-	-	-	227	-	-	-	-	-	-	-	-	-	-	-	-	-	-	-	-	-	-	-	-	-	-	-
Cerium Dioxide Coating on:																																			
Nimonic 75 Substrate	9	-	870	-	-	-	-	-	-	-	-	-	-	-	-	-	-	-	-	-	-	-	-	-	-	-	-	-	-	-	-	-	-	-	-
Tungsten Substrate	9	868	-	-	-	872	-	-	-	-	-	-	-	-	-	-	-	-	-	-	-	-	-	-	-	-	-	-	-	-	-	-	-	-	-
Cerium Dioxide Pigment in:																																			
Barium Beryllium Silicate Binder (Enamel)	9	-	445	-	-	448	-	-	-	-	-	-	-	-	-	-	-	-	-	-	-	-	-	-	-	-	-	-	-	-	-	-	-	-	-
Base Glaze No. 1 Binder (Enamel)	9	-	445	-	-	-	-	-	-	-	-	-	-	-	-	-	-	-	-	-	-	-	-	-	-	-	-	-	-	-	-	-	-	-	-
Base Glaze No. 3 Binder (Enamel)	9	-	445	-	-	-	-	-	-	-	-	-	-	-	-	-	-	-	-	-	-	-	-	-	-	-	-	-	-	-	-	-	-	-	-
Cerium Dioxide Substrate Coated with Gold	9	-	656	-	-	-	-	-	-	-	-	-	-	-	-	-	-	-	-	-	-	-	-	-	-	-	-	-	-	-	-	-	-	-	-
Cerium Dioxide + Cobalt Oxide Pigment in Base Glaze No. 3 Binder (Enamel)	9	-	445	-	-	-	-	-	-	-	-	-	-	-	-	-	-	-	-	-	-	-	-	-	-	-	-	-	-	-	-	-	-	-	-
Cerium Dioxide + Magnesium Oxide Pigment in NBS Frit No. 332 Binder (Enamel)	9	-	-	-	-	-	-	-	-	-	-	452	-	-	-	-	-	-	-	-	-	-	-	-	-	-	-	-	-	-	-	-	-	-	-
Cerium Dioxide + Tin Oxide Pigment in NBS Frit No. 332 Binder (Enamel)	9	-	445	-	-	-	-	-	-	-	-	452	-	-	-	-	-	-	-	-	-	-	-	-	-	-	-	-	-	-	-	-	-	-	-
Cerium Dioxide + Zirconium Dioxide Pigment in NBS Frit No. 332 Binder (Enamel)	9	-	-	-	-	-	-	-	-	-	-	452	-	-	-	-	-	-	-	-	-	-	-	-	-	-	-	-	-	-	-	-	-	-	-
Dicerium Trisulfide	8	-	-	-	-	1231	-	-	-	-	-	-	-	-	-	-	-	-	-	-	-	-	-	-	-	-	-	-	-	-	-	-	-	-	-
Cesium Bromide	8	-	-	-	-	-	-	-	-	-	-	-	737	-	-	-	-	-	-	-	-	-	-	-	-	-	-	-	-	739	-	-	-	-	-
Cesium Iodide	8	-	-	-	-	-	-	-	-	-	-	995	-	-	-	-	-	-	-	-	-	-	-	-	-	-	-	-	-	997	-	-	-	-	-
China Clay Pigment in Silicone Binder	9	-	95	-	-	-	-	-	-	-	-	97	-	-	-	-	-	-	-	-	-	-	-	98 100	-	-	-	-	-	-	-	-	-	-	102
Chlorates:																																			
Sodium Chlorate	8	-	-	-	-	-	-	-	-	-	-	594 596	-	-	-	-	-	-	-	-	-	-	-	-	-	-	-	-	-	-	-	-	-	-	-

MATERIAL NAME	VOLUME	EMITTANCE						REFLECTANCE									ABSORPTANCE									TRANSMITTANCE									Absorptance to Emittance Ratio
		Total			Spectral			Integrated			Spectral			Solar			Integrated			Spectral			Solar			Integrated			Spectral			Solar			
		Hemispherical	Normal	Angular	Hemispherical	Normal	Angular	Hemispherical	Normal	Angular	Hemispherical	Normal	Angular	Hemispherical	Normal	Angular	Hemispherical	Normal	Angular	Hemispherical	Normal	Angular	Hemispherical	Normal	Angular	Hemispherical	Normal	Angular	Hemispherical	Normal	Angular	Hemispherical	Normal	Angular	
Chlorides:																																			
Copper Chloride	8	–	–	–	–	–	–	–	–	–	–	856	–	–	–	–	–	–	–	–	–	–	–	–	–	–	–	–	–	858	–	–	–	–	–
Potassium Chloride	8	–	–	–	–	862	–	–	–	–	–	–	864	–	–	–	–	–	–	–	–	–	–	–	–	–	–	–	–	873	–	–	–	–	–
Silver Chloride	8	–	–	–	–	–	–	–	–	–	–	–	876	–	–	–	–	–	–	–	–	–	–	–	–	–	–	–	–	879	–	–	–	–	–
Sodium Chloride	8	881	–	–	–	883	–	–	–	–	885 886	888	893	–	–	–	–	–	–	–	–	–	–	–	–	–	–	–	–	895	–	–	–	–	–
Thallium Chloride	8	–	–	–	–	–	–	–	–	–	–	899	901	–	–	–	–	–	–	–	–	–	–	–	–	–	–	–	–	903	–	–	–	–	–
Miscellaneous Chlorides	8	–	–	–	–	–	–	–	–	–	–	905	–	–	–	–	–	–	–	–	–	–	–	–	–	–	–	–	–	907	–	–	–	–	–
Chromalloy W-2	9	–	1333 1350	1335	–	1337	–	–	–	–	–	–	–	–	–	–	–	–	–	–	1345	–	–	–	–	–	–	–	–	–	–	–	–	–	–
Chromatone	9	–	–	–	–	–	–	–	–	–	–	494	–	–	–	–	–	–	–	–	–	–	–	–	–	–	–	–	–	–	–	–	–	–	–
Chromium	7	101	103	–	–	106	–	–	–	–	–	110	113	–	–	–	–	–	–	–	115	–	–	118	–	–	–	–	–	120	–	–	–	–	–
Chromium Coating on:																																			
Copper Substrate	9	618	–	–	–	–	–	–	–	–	–	–	–	–	–	–	–	–	–	–	–	–	–	–	–	–	–	–	–	–	–	–	–	–	–
Glass Substrate	9	–	–	–	–	–	–	–	–	–	–	621	–	–	–	–	–	–	–	–	–	–	–	–	–	–	–	–	–	–	–	–	–	–	–
Monel Substrate	9	618	–	–	–	–	–	–	–	–	–	–	–	–	–	–	–	–	–	–	–	–	–	–	–	–	–	–	–	–	–	–	–	–	–
Nickel Substrate	9	–	–	–	–	–	–	–	–	–	–	621	–	–	–	–	–	–	–	–	–	626	–	627	–	–	–	–	–	–	–	–	–	–	–
Silver Substrate	9	–	–	–	–	–	–	–	–	–	–	621	–	–	–	–	–	–	–	–	–	–	–	–	–	–	–	–	–	–	–	–	–	–	–
Stainless Steel Substrate	9	–	619	–	–	–	–	–	–	–	–	621	624	–	–	–	–	–	–	–	–	–	–	–	–	–	–	–	–	–	–	–	–	–	–
Chromium Substrate Coated with Silver	9	–	–	–	–	–	–	–	–	–	–	738	–	–	–	–	–	–	–	–	–	–	–	–	–	–	–	–	–	–	–	–	–	–	–
Chromium Black Coating on:																																			
Aluminum Substrate	9	1173	–	–	–	–	–	–	–	–	–	–	–	–	–	–	–	–	–	–	–	–	–	–	–	–	–	–	–	–	–	–	–	–	–
Copper Substrate	9	1173	–	–	–	–	–	–	–	–	–	–	–	–	–	–	–	–	–	–	–	–	–	–	–	–	–	–	–	–	–	–	–	–	–
Stainless Steel Substrate	9	1173	–	–	–	–	–	–	–	–	–	–	–	–	–	–	–	–	–	–	–	–	–	–	–	–	–	–	–	–	–	–	–	–	–
Chromium + Aluminum Oxide + ΣX_i Coating on Inconel Substrate	9	–	629	–	–	–	–	–	–	–	–	–	–	–	–	–	–	–	–	–	–	–	–	–	–	–	–	–	–	–	–	–	–	–	–
Chromium Boride	8	–	–	–	–	–	–	–	–	–	–	734	–	–	–	–	–	–	–	–	–	–	–	–	–	–	–	–	–	–	–	–	–	–	–
Chromium Diboride	8	–	–	–	–	731	–	–	–	–	–	–	–	–	–	–	–	–	–	–	–	–	–	–	–	–	–	–	–	–	–	–	–	–	–
Dichromium Boride	8	–	–	–	–	–	–	–	–	–	–	734	–	–	–	–	–	–	–	–	–	–	–	–	–	–	–	–	–	–	–	–	–	–	–
Trichromium Dicarbide	8	–	–	–	–	852	–	–	–	–	–	–	–	–	–	–	–	–	–	–	–	–	–	–	–	–	–	–	–	–	–	–	–	–	–
7-Chromium Tricarbide	8	–	–	–	–	852	–	–	–	–	–	–	–	–	–	–	–	–	–	–	–	–	–	–	–	–	–	–	–	–	–	–	–	–	–
Chromium Carbide + Cobalt Coating on Armco Iron Substrate	9	873	–	–	–	–	–	–	–	–	–	–	–	–	–	–	–	–	–	–	–	–	–	874	–	–	–	–	–	–	–	–	–	–	–
Chromium Nitride	8	–	–	–	–	1087	–	–	–	–	–	–	–	–	–	–	–	–	–	–	–	–	–	–	–	–	–	–	–	–	–	–	–	–	–
Dichromium Nitride	8	–	–	–	–	1087	–	–	–	–	–	–	–	–	–	–	–	–	–	–	–	–	–	–	–	–	–	–	–	–	–	–	–	–	–

MATERIAL NAME	VOLUME	EMITTANCE						REFLECTANCE									ABSORPTANCE									TRANSMITTANCE									Absorptance to Emittance Ratio
		Total			Spectral			Integrated			Spectral			Solar			Integrated			Spectral			Solar			Integrated			Spectral			Solar			
		Hemispherical	Normal	Angular	Hemispherical	Normal	Angular	Hemispherical	Normal	Angular	Hemispherical	Normal	Angular	Hemispherical	Normal	Angular	Hemispherical	Normal	Angular	Hemispherical	Normal	Angular	Hemispherical	Normal	Angular	Hemispherical	Normal	Angular	Hemispherical	Normal	Angular	Hemispherical	Normal	Angular	
Chromium Oxide Coating on:																																			
Aluminum Substrate	9	876	-	-	-	-	-	-	-	-	-	-	-	-	-	-	-	-	-	-	-	-	-	-	-	-	-	-	-	-	-	-	-	-	-
Niobium Substrate	9	876	-	-	-	-	-	-	-	-	-	-	-	-	-	-	-	-	-	-	-	-	-	-	-	-	-	-	-	-	-	-	-	-	-
Chromium Oxide Pigment in:																																			
Barium Borosilicate Frit Binder (Enamel)	9	-	455	-	-	459	-	-	-	-	-	-	-	-	-	-	-	-	-	-	-	-	-	-	-	-	-	-	-	-	-	-	-	-	-
Base Glaze No. 1 Binder (Enamel)	9	-	455	-	-	-	-	-	-	-	-	-	-	-	-	-	-	-	-	-	-	-	-	-	-	-	-	-	-	-	-	-	-	-	-
Base Glaze No. 2 Binder (Enamel)	9	-	455	-	-	-	-	-	-	-	-	-	-	-	-	-	-	-	-	-	-	-	-	-	-	-	-	-	-	-	-	-	-	-	-
Base Glaze No. 3 Binder (Enamel)	9	-	455	-	-	-	-	-	-	-	-	-	-	-	-	-	-	-	-	-	-	-	-	-	-	-	-	-	-	-	-	-	-	-	-
NBS Frit No. 332 Binder (Enamel)	9	-	455	-	-	-	-	-	-	-	-	463	-	-	-	-	-	-	-	-	-	-	-	-	-	-	-	-	-	-	-	-	-	-	-
Chromium Trioxide	8	-	-	-	-	-	-	-	-	-	-	236	-	-	-	-	-	-	-	-	-	-	-	-	-	-	-	-	-	-	-	-	-	-	-
Dichromium Trioxide	8	-	231	-	-	233	-	-	-	-	-	236	-	-	-	-	-	-	-	-	-	-	-	-	-	-	-	-	-	-	-	-	-	-	-
Chromium Oxide + Black Stain Pigment in NBS Frit No. 332 Binder (Enamel)	9	-	455	-	-	-	-	-	-	-	-	-	-	-	-	-	-	-	-	-	-	-	-	-	-	-	-	-	-	-	-	-	-	-	-
Chromium Oxide + Cobalt Oxide Pigment in Base Glaze No. 1 Binder (Enamel)	9	-	455	-	-	-	-	-	-	-	-	-	-	-	-	-	-	-	-	-	-	-	-	-	-	-	-	-	-	-	-	-	-	-	-
Chromium Oxide + Iron Oxide Pigment in Base Glaze No. 3 Binder (Enamel)	9	-	456	-	-	-	-	-	-	-	-	-	-	-	-	-	-	-	-	-	-	-	-	-	-	-	-	-	-	-	-	-	-	-	-
Chromium Oxide + Silicon Dioxide + ΣX_i Coating on:																																			
Inconel Substrate	9	-	-	-	-	885	-	-	-	-	-	-	-	-	-	-	-	-	-	-	-	-	-	-	-	-	-	-	-	-	-	-	-	-	-
Molybdenum Substrate	9	878	-	-	-	883	-	-	-	-	-	-	-	-	-	-	-	-	-	-	-	-	-	-	-	-	-	-	-	-	-	-	-	-	-
Niobium Substrate	9	878	-	-	-	885	-	-	-	-	-	-	-	-	-	-	-	-	-	-	-	-	-	-	-	-	-	-	-	-	-	-	-	-	-
Stainless Steel Substrate	9	878	-	-	-	-	-	-	-	-	-	-	-	-	-	-	-	-	-	-	-	-	-	-	-	-	-	-	-	-	-	-	-	-	-
Steel Substrate	9	-	881	-	-	885	-	-	-	-	-	-	-	-	-	-	-	-	-	-	-	-	-	-	-	-	-	-	-	-	-	-	-	-	-
Titanium 6Al-4V Substrate	9	-	881	-	-	883	-	-	-	-	-	-	-	-	-	-	-	-	-	-	-	-	-	-	-	-	-	-	-	-	-	-	-	-	-
Chromium Monosilicide	8	-	-	-	-	1140 1142	-	-	-	-	-	-	-	-	-	-	-	-	-	-	-	-	-	-	-	-	-	-	-	-	-	-	-	-	-
Chromium Disilicide	8	-	-	-	-	1140 1142	-	-	-	-	-	-	-	-	-	-	-	-	-	-	-	-	-	-	-	-	-	-	-	-	-	-	-	-	-
Trichromium Silicide	8	-	1139	-	-	1140 1142	-	-	-	-	-	1144	-	-	-	-	-	-	-	-	-	-	-	-	-	-	-	-	-	-	-	-	-	-	-

MATERIAL NAME	VOLUME	EMITTANCE Total Hemispherical	EMITTANCE Total Normal	EMITTANCE Total Angular	EMITTANCE Spectral Hemispherical	EMITTANCE Spectral Normal	EMITTANCE Spectral Angular	REFLECTANCE Integrated Hemispherical	REFLECTANCE Integrated Normal	REFLECTANCE Integrated Angular	REFLECTANCE Spectral Hemispherical	REFLECTANCE Spectral Normal	REFLECTANCE Spectral Angular	REFLECTANCE Solar Hemispherical	REFLECTANCE Solar Normal	REFLECTANCE Solar Angular	ABSORPTANCE Integrated Hemispherical	ABSORPTANCE Integrated Normal	ABSORPTANCE Integrated Angular	ABSORPTANCE Spectral Hemispherical	ABSORPTANCE Spectral Normal	ABSORPTANCE Spectral Angular	ABSORPTANCE Solar Hemispherical	ABSORPTANCE Solar Normal	ABSORPTANCE Solar Angular	TRANSMITTANCE Integrated Hemispherical	TRANSMITTANCE Integrated Normal	TRANSMITTANCE Integrated Angular	TRANSMITTANCE Spectral Hemispherical	TRANSMITTANCE Spectral Normal	TRANSMITTANCE Spectral Angular	TRANSMITTANCE Solar Hemispherical	TRANSMITTANCE Solar Normal	TRANSMITTANCE Solar Angular	Absorptance to Emittance Ratio
Trichromium Disilicide	8	-	-	-	-	1140 1142	-	-	-	-	-	-	-	-	-	-	-	-	-	-	-	-	-	-	-	-	-	-	-	-	-	-	-	-	-
Hexachromium Silicide	8	-	1139	-	-	1140 1142	-	-	-	-	-	-	-	-	-	-	-	-	-	-	-	-	-	-	-	-	-	-	-	-	-	-	-	-	-
Clay + TiO_2 Pigment in Silicone Binder	9	105	-	-	-	-	-	-	-	-	-	-	-	-	-	-	-	-	-	-	-	-	-	107	-	-	-	-	-	-	-	-	-	-	-
CM-145 Al_2O_3 Paint	9	-	-	-	-	-	-	-	-	-	-	33	-	-	-	-	-	-	-	-	-	-	-	42	-	-	-	-	-	-	-	-	-	-	-
CM-146 TiO_2 Paint	9	-	-	-	-	-	-	-	-	-	-	-	-	-	-	-	-	-	-	-	-	-	-	274	-	-	-	-	-	-	-	-	-	-	-
CM-147 ZnO Paint	9	-	-	-	-	-	-	-	-	-	-	-	-	-	-	-	-	-	-	-	-	-	-	378	-	-	-	-	-	-	-	-	-	-	-
Cobalt	7	-	123	-	-	126	-	-	-	-	-	-	132	-	-	-	-	-	-	-	-	-	-	-	-	-	-	-	-	-	-	-	-	-	-
Cobalt Alloy N-155	7	1180 1181 1222 1226	-	-	-	1186 1236	-	-	-	-	-	1198 1266	-	-	-	-	-	-	-	-	-	-	-	1207 1301 1303	-	-	-	-	-	-	-	-	-	-	-
Cobalt Coating on:																																			
Glass Substrate	9	-	-	-	-	-	-	-	-	-	-	632	-	-	-	-	-	-	-	-	-	-	-	-	-	-	-	-	-	634	-	-	-	-	-
Platinum Substrate	9	-	631	-	-	-	-	-	-	-	-	-	-	-	-	-	-	-	-	-	-	-	-	633	-	-	-	-	-	-	-	-	-	-	-
Stainless Steel Substrate	9	-	631	-	-	-	-	-	-	-	-	632	-	-	-	-	-	-	-	-	-	-	-	-	-	-	-	-	-	-	-	-	-	-	-
Cobalt + Aluminum	7	-	-	-	-	-	-	-	-	-	-	901	-	-	-	-	-	-	-	-	-	-	-	-	-	-	-	-	-	-	-	-	-	-	-
Cobalt + Chromium + ΣX_i	7	1138	1142	-	-	1145 1148	-	-	-	-	-	1152	-	-	-	-	-	-	-	-	-	-	-	-	-	-	-	-	-	-	-	-	-	-	-
Cobalt + Iron	7	-	-	-	-	904	-	-	-	-	-	-	-	-	-	-	-	-	-	-	-	-	-	-	-	-	-	-	-	-	-	-	-	-	-
Cobalt + Nickel	7	-	-	-	-	906	-	-	-	-	-	-	-	-	-	-	-	-	-	-	-	-	-	-	-	-	-	-	-	-	-	-	-	-	-
Cobalt + Tungsten Coating on Inconel X Substrate	9	-	-	-	-	636	-	-	-	-	-	639	-	-	-	-	-	-	-	-	-	-	-	-	-	-	-	-	-	-	-	-	-	-	-
CoAl	8	-	-	-	-	-	-	-	-	-	-	1352	-	-	-	-	-	-	-	-	-	-	-	-	-	-	-	-	-	-	-	-	-	-	-
Dicobalt Boride	8	-	-	-	-	731	-	-	-	-	-	-	-	-	-	-	-	-	-	-	-	-	-	-	-	-	-	-	-	-	-	-	-	-	-
Cobalt Monoxide	8	-	238	-	-	240	-	-	-	-	-	242	-	-	-	-	-	-	-	-	-	-	-	-	-	-	-	-	-	-	-	-	-	-	-
Cobalt Oxide Coating on:																																			
Silver Substrate	9	-	887	-	-	-	-	-	-	-	-	-	-	-	-	-	-	-	-	-	-	-	-	-	-	-	-	-	-	-	-	-	-	-	-
Tantalum Substrate	9	-	887	-	-	888	-	-	-	-	-	-	-	-	-	-	-	-	-	-	-	-	-	-	-	-	-	-	-	-	-	-	-	-	-
Cobalt Oxide Pigment in Base Glaze No. 3 Binder (Enamel)	9	-	464	-	-	-	-	-	-	-	-	-	-	-	-	-	-	-	-	-	-	-	-	-	-	-	-	-	-	-	-	-	-	-	-
Cobalt Oxide + Chromium Oxide Pigment in Base Glaze No. 1 Binder (Enamel)	9	-	464	-	-	-	-	-	-	-	-	-	-	-	-	-	-	-	-	-	-	-	-	-	-	-	-	-	-	-	-	-	-	-	-

MATERIAL NAME	VOLUME	EMITTANCE						REFLECTANCE									ABSORPTANCE									TRANSMITTANCE									Absorptance to Emittance Ratio
		Total			Spectral			Integrated			Spectral			Solar			Integrated			Spectral			Solar			Integrated			Spectral			Solar			
		Hemispherical	Normal	Angular	Hemispherical	Normal	Angular	Hemispherical	Normal	Angular	Hemispherical	Normal	Angular	Hemispherical	Normal	Angular	Hemispherical	Normal	Angular	Hemispherical	Normal	Angular	Hemispherical	Normal	Angular	Hemispherical	Normal	Angular	Hemispherical	Normal	Angular	Hemispherical	Normal	Angular	
Cobalt Oxide + Chromium Oxide Pigment in Base Glaze No. 2 Binder (Enamel)	9	-	464	-	-	-	-	-	-	-	-	-	-	-	-	-	-	-	-	-	-	-	-	-	-	-	-	-	-	-	-	-	-	-	-
Cobalt Oxide + Manganese Oxide Pigment in Base Glaze No. 2 Binder (Enamel)	9	-	464	-	-	-	-	-	-	-	-	-	-	-	-	-	-	-	-	-	-	-	-	-	-	-	-	-	-	-	-	-	-	-	-
Cobalt Oxide + Nickel Oxide Pigment in Base Glaze No. 3 Binder (Enamel)	9	-	464	-	-	-	-	-	-	-	-	-	-	-	-	-	-	-	-	-	-	-	-	-	-	-	-	-	-	-	-	-	-	-	-
CoO · Cr_2O_3 Spinel Pigment in NBS Frit No. 332 Binder (Enamel)	9	-	472	-	-	-	-	-	-	-	-	475	-	-	-	-	-	-	-	-	-	-	-	-	-	-	-	-	-	-	-	-	-	-	-
CoO · Fe_2O_3 Spinel Pigment in NBS Frit No. 332 Binder (Enamel)	9	-	472	-	-	-	-	-	-	-	-	475	-	-	-	-	-	-	-	-	-	-	-	-	-	-	-	-	-	-	-	-	-	-	-
CoO · Mn_2O_3 Spinel Pigment in NBS Frit No. 332 Binder (Enamel)	9	-	472	-	-	-	-	-	-	-	-	475	-	-	-	-	-	-	-	-	-	-	-	-	-	-	-	-	-	-	-	-	-	-	-
Cobalt Silicide	8	-	-	-	-	1172	-	-	-	-	-	-	-	-	-	-	-	-	-	-	-	-	-	-	-	-	-	-	-	-	-	-	-	-	-
Cockron Home Foil	7	4 5	-	-	-	-	-	-	-	-	-	-	-	-	-	-	-	-	-	42 43	-	-	-	-	-	-	-	-	-	-	-	-	-	-	-
Coors AD85	8	-	-	-	-	150	-	-	-	-	-	-	-	-	-	-	-	-	-	-	-	-	-	-	-	-	-	-	-	172	-	-	-	-	-
Coors AD94	8	-	-	-	-	150	-	-	-	-	-	-	-	-	-	-	-	-	-	-	-	-	-	-	-	-	-	-	-	172	-	-	-	-	-
Coors AD96	8	-	-	-	-	150	-	-	-	-	-	-	-	-	-	-	-	-	-	-	-	-	-	-	-	-	-	-	-	172	-	-	-	-	-
Coors AD99	8	-	144	-	-	147 150	-	-	-	-	-	-	-	-	-	-	-	-	-	-	-	-	-	-	-	-	-	-	-	172	-	-	-	-	-
Coors AD 995	8	-	-	-	-	150	-	-	-	-	-	-	-	-	-	-	-	-	-	-	-	-	-	-	-	-	-	-	-	-	-	-	-	-	-
Coors BD 96	8	-	-	-	-	206	-	-	-	-	-	-	-	-	-	-	-	-	-	-	-	-	-	-	-	-	-	-	-	214	-	-	-	-	-
Coors BD 98	8	-	-	-	-	206	-	-	-	-	-	-	-	-	-	-	-	-	-	-	-	-	-	-	-	-	-	-	-	214	-	-	-	-	-
Copper	7	136	142	-	-	149 152	-	-	-	-	-	158	165 169	-	172	-	177	179	-	-	181 184	188	191	193	-	-	-	-	-	199	-	-	-	-	-
Copper, B. S. 1433	7	-	-	-	-	-	-	-	-	-	-	-	-	-	173 174	-	-	-	-	-	-	-	-	194	-	-	-	-	-	-	-	-	-	-	-
Copper, OFHC	7	138	-	-	-	-	-	-	-	-	-	-	-	-	-	-	-	-	-	-	-	189	-	-	-	-	-	-	-	-	-	-	-	-	-
Copper Coating on:																																			
Epoxy Substrate	9	-	-	-	-	-	-	-	-	-	-	-	643	-	-	-	-	-	-	-	-	-	-	-	-	-	-	-	-	-	-	-	-	-	-
Glass Substrate	9	-	-	-	-	-	-	-	-	-	-	642	-	-	-	-	-	-	-	-	-	-	-	-	-	-	-	-	-	644	-	-	-	-	-
Polyurethane Substrate	9	-	-	-	-	-	-	-	-	-	-	-	643	-	-	-	-	-	-	-	-	-	-	-	-	-	-	-	-	-	-	-	-	-	-

MATERIAL NAME	VOLUME	EMITTANCE						REFLECTANCE									ABSORPTANCE									TRANSMITTANCE									Absorptance to Emittance Ratio
		Total			Spectral			Integrated			Spectral			Solar			Integrated			Spectral			Solar			Integrated			Spectral			Solar			
		Hemispherical	Normal	Angular	Hemispherical	Normal	Angular	Hemispherical	Normal	Angular	Hemispherical	Normal	Angular	Hemispherical	Normal	Angular	Hemispherical	Normal	Angular	Hemispherical	Normal	Angular	Hemispherical	Normal	Angular	Hemispherical	Normal	Angular	Hemispherical	Normal	Angular	Hemispherical	Normal	Angular	
Copper Substrate Coated with:																																			
Carbon	9	852	-	-	-	857	-	-	-	-	-	859	-	-	-	-	-	-	-	-	-	-	-	-	-	-	-	-	-	-	-	-	-	-	-
Chromium	9	618	-	-	-	-	-	-	-	-	-	-	-	-	-	-	-	-	-	-	-	-	-	-	-	-	-	-	-	-	-	-	-	-	-
Chromium Black	9	1173	-	-	-	-	-	-	-	-	-	-	-	-	-	-	-	-	-	-	-	-	-	-	-	-	-	-	-	-	-	-	-	-	-
Copper Sulfide	9	894	-	-	-	-	-	-	-	-	-	-	-	-	-	-	-	-	-	-	-	-	-	-	-	-	-	-	-	-	-	-	-	-	-
Fasson Foil	9	-	-	-	-	-	-	-	-	-	-	594	-	-	-	-	-	-	-	-	-	-	-	-	-	-	-	-	-	-	-	-	-	-	-
Gold	9	652	-	-	-	-	-	-	-	-	-	-	-	-	-	-	-	-	-	-	-	-	-	-	-	-	-	-	-	-	-	-	-	-	-
Gold + Silver	9	651	-	-	-	-	-	-	-	-	-	-	-	-	-	-	-	-	-	-	-	-	-	-	-	-	-	-	-	-	-	-	-	-	-
Lacquer	9	-	-	-	-	-	-	-	-	-	-	1125	-	-	-	-	-	-	-	-	-	-	-	-	-	-	-	-	-	-	-	-	-	-	-
Lead + Tin	9	690	-	-	-	-	-	-	-	-	-	-	-	-	-	-	691	-	-	-	-	-	-	-	-	-	-	-	-	-	-	-	-	-	-
Nickel	9	700	-	-	-	-	-	-	-	-	-	702	-	-	-	-	-	-	-	-	-	-	-	-	-	-	-	-	-	-	-	-	-	-	-
Nitrocellulose	9	1131	-	-	-	-	-	-	-	-	-	-	-	-	-	-	-	-	-	-	-	-	-	-	-	-	-	-	-	-	-	-	-	-	-
Silver	9	734	-	-	-	-	-	-	-	-	-	-	-	-	-	-	747	-	-	-	-	-	-	-	-	-	-	-	-	-	-	-	-	-	-
Tin	9	755	-	-	-	-	-	-	-	-	-	-	-	-	-	-	756	-	-	-	-	-	-	-	-	-	-	-	-	-	-	-	-	-	-
Tin + Lead	9	758	-	-	-	-	-	-	-	-	-	-	-	-	-	-	759	-	-	-	-	-	-	-	-	-	-	-	-	-	-	-	-	-	-
Copper + Aluminum + ΣX_i	7	-	1159	-	-	1162	-	-	-	-	-	1166	-	-	-	-	-	-	-	-	-	-	-	1169	-	-	-	-	-	-	-	-	-	-	-
Copper + Nickel	7	-	908	-	-	-	-	-	-	-	-	-	-	-	-	-	-	-	-	-	-	-	-	-	-	-	-	-	-	-	-	-	-	-	-
Copper + Nickel + ΣX_i	7	-	-	-	-	-	-	-	-	-	-	1172	-	-	-	-	-	-	-	-	-	-	-	-	-	-	-	-	-	-	-	-	-	-	-
Copper + Tin	7	-	-	-	-	-	-	-	-	-	-	910	-	-	-	-	-	-	-	-	-	-	-	-	-	-	-	-	-	-	-	-	-	-	-
Copper + Tin + ΣX_i	7	-	-	-	-	-	-	-	-	-	-	1174	-	-	-	-	-	-	-	-	-	-	-	-	-	-	-	-	-	-	-	-	-	-	-
Copper + Zinc	7	912	914	-	-	-	-	-	-	-	-	917	920	-	-	-	923	925	-	-	-	928	-	-	-	-	-	-	-	-	-	-	-	-	-
Copper Bromide	8	-	-	-	-	-	-	-	-	-	-	741	-	-	-	-	-	-	-	-	-	-	-	-	-	-	-	-	-	743	-	-	-	-	-
Copper Chloride	8	-	-	-	-	-	-	-	-	-	-	856	-	-	-	-	-	-	-	-	-	-	-	-	-	-	-	-	-	858	-	-	-	-	-
Copper Iodide	8	-	-	-	-	-	-	-	-	-	-	999	-	-	-	-	-	-	-	-	-	-	-	-	-	-	-	-	-	1001	-	-	-	-	-
Copper Monoxide	8	-	-	-	-	-	-	-	-	-	-	247	-	-	-	-	-	-	-	-	-	-	-	-	-	-	-	-	-	249	-	-	-	-	-
Dicopper Oxide	8	-	-	243	-	-	245	-	-	-	-	-	-	-	-	-	-	-	-	-	-	-	-	-	-	-	-	-	-	249	-	-	-	-	-
Copper Oxide Coating on:																																			
Nickel Substrate	9	-	891	-	-	-	-	-	-	-	-	-	-	-	-	-	-	-	-	-	-	-	-	-	-	-	-	-	-	-	-	-	-	-	-
Silver Substrate	9	-	891	-	-	-	-	-	-	-	-	-	-	-	-	-	-	-	-	-	-	-	-	-	-	-	-	-	-	-	-	-	-	-	-
Stainless Steel Substrate	9	890	-	-	-	-	-	-	-	-	-	-	-	-	-	-	-	-	-	-	-	-	-	-	-	-	-	-	-	-	-	-	-	-	-
Copper Phosphorous Selenide Coating on Fluorite Substrate	9	-	-	-	-	-	-	-	-	-	-	892	-	-	-	-	-	-	-	-	-	-	-	-	-	-	-	-	-	893	-	-	-	-	-
Copper Sulfide Coating on Copper Substrate	9	894	-	-	-	-	-	-	-	-	-	-	-	-	-	-	-	-	-	-	-	-	-	-	-	-	-	-	-	-	-	-	-	-	-

MATERIAL NAME	VOLUME	EMITTANCE						REFLECTANCE									ABSORPTANCE									TRANSMITTANCE									Absorptance to Emittance Ratio
		Total			Spectral			Integrated			Spectral			Solar			Integrated			Spectral			Solar			Integrated			Spectral			Solar			
		Hemispherical	Normal	Angular	Hemispherical	Normal	Angular	Hemispherical	Normal	Angular	Hemispherical	Normal	Angular	Hemispherical	Normal	Angular	Hemispherical	Normal	Angular	Hemispherical	Normal	Angular	Hemispherical	Normal	Angular	Hemispherical	Normal	Angular	Hemispherical	Normal	Angular	Hemispherical	Normal	Angular	
Copper + Tin Coating on:																																			
Glass Substrate	9	-	-	-	-	-	-	-	-	-	-	645	-	-	-	-	-	-	-	-	-	-	-	-	-	-	-	-	-	-	-	-	-	-	-
Steel Substrate	9	-	-	-	-	-	-	-	-	-	-	645	-	-	-	-	-	-	-	-	-	-	-	-	-	-	-	-	-	-	-	-	-	-	-
Cordierite	8	-	-	-	-	-	-	-	-	-	-	-	-	-	-	-	-	-	-	-	-	-	-	-	-	-	-	-	-	1650	-	-	-	-	-
Corning 0160 Glass	8	-	-	-	-	-	-	-	-	-	-	-	-	-	-	-	-	-	-	-	-	-	-	-	-	-	-	-	-	1642	-	-	-	-	-
Corning 1723 Glass	8	-	1524	-	-	-	-	-	-	-	-	1526	-	-	-	-	-	-	-	-	-	-	-	1527	-	-	1529	-	-	1531	-	-	-	-	-
Corning 7740 Glass	8	-	1580	-	-	-	-	-	-	-	-	-	1588	-	-	-	-	-	-	-	-	-	-	1589	-	-	1590	-	-	1594	-	-	-	-	-
Corning 7900 Glass	8	-	1622	-	-	-	-	-	-	-	-	1626	-	-	-	-	-	-	-	-	-	-	-	1627	-	-	1629	-	-	1633	-	-	-	-	-
Corning 7905 Glass	8	-	-	-	-	-	-	-	-	-	-	-	-	-	-	-	-	-	-	-	-	-	-	-	-	-	-	-	-	1633	-	-	-	-	-
Corning 7910 Glass	8	-	-	-	-	-	-	-	-	-	-	-	-	-	-	-	-	-	-	-	-	-	-	-	-	-	-	-	-	1633	-	-	-	-	-
Corning 7940 Glass	8	1566	1570	-	-	-	-	-	-	-	-	1572	-	-	-	-	-	-	-	-	-	-	-	1573	-	-	1575	-	-	1577	-	-	1578	-	-
Corning 7940 Coating on Silver Substrate	9	1049	-	-	-	-	-	-	-	-	-	1054	-	-	-	-	-	-	-	-	-	-	-	1072	-	-	-	-	-	-	-	-	-	-	-
Corning 8363 Glass	8	-	-	-	-	-	-	-	-	-	-	-	-	-	-	-	-	-	-	-	-	-	-	-	-	-	-	-	-	1642	-	-	-	-	-
Corning 9690 Glass	8	-	1622	-	-	-	-	-	-	-	-	-	-	-	-	-	-	-	-	-	-	-	-	-	-	-	-	-	-	-	-	-	-	-	-
Corning 9752 Glass	8	-	-	-	-	1637	-	-	-	-	-	1640	-	-	-	-	-	-	-	-	-	-	-	-	-	-	-	-	-	-	-	-	-	-	-
Corning 9863 Glass	8	-	-	-	-	-	-	-	-	-	-	-	-	-	-	-	-	-	-	-	-	-	-	-	-	-	-	-	-	1642	-	-	-	-	-
Cr–Co–Ni Spinel Pigment in Aluminum Phosphate Binder	9	186	-	-	-	-	-	-	-	-	-	-	-	-	-	-	-	-	-	-	-	-	-	-	-	-	-	-	-	-	-	-	-	-	-
Cr_2O_3 + Al_2O_3 Powders	8	-	-	-	-	553	-	-	-	-	-	-	-	-	-	-	-	-	-	-	-	-	-	-	-	-	-	-	-	-	-	-	-	-	-
Cr_2O_3 + Fe_3O_4 + NiO Pigment in Synar Binder	9	103	-	-	-	-	-	-	-	-	-	-	-	-	-	-	-	-	-	-	-	-	-	-	-	-	-	-	-	-	-	-	-	-	-
Cr_2O_3 + $TiCr_2$ + ΣX_i Cermet	8	-	1385	-	-	1386	-	-	-	-	-	1390	-	-	-	-	-	-	-	-	-	-	-	-	-	-	-	-	-	-	-	-	-	-	-
Cr_2O_3 + Y_2O_3 Powders	8	-	-	-	-	570	-	-	-	-	-	-	-	-	-	-	-	-	-	-	-	-	-	-	-	-	-	-	-	-	-	-	-	-	-
Cryolite	8	-	-	-	-	-	-	-	-	-	-	1660	-	-	-	-	-	-	-	-	-	-	-	-	-	-	-	-	-	-	-	-	-	-	-
Crystolon	8	-	-	-	-	800	-	-	-	-	-	805	-	-	-	-	-	-	-	-	-	-	-	-	-	-	-	-	-	-	-	-	-	-	-
Cupric Oxide (see Copper Monoxide)																																			
Cuprous Oxide (see Dicopper Oxide)																																			
CuZn	8	-	-	-	-	-	-	-	-	-	-	-	1285	-	-	-	-	-	-	-	-	-	-	-	-	-	-	-	-	-	-	-	-	-	-
Cu_4Sn	8	-	-	-	-	-	-	-	-	-	-	1352	-	-	-	-	-	-	-	-	-	-	-	-	-	-	-	-	-	-	-	-	-	-	-

MATERIAL NAME	VOLUME	EMITTANCE						REFLECTANCE									ABSORPTANCE									TRANSMITTANCE									Absorptance to Emittance Ratio
		Total			Spectral			Integrated			Spectral			Solar			Integrated			Spectral			Solar			Integrated			Spectral			Solar			
		Hemispherical	Normal	Angular	Hemispherical	Normal	Angular	Hemispherical	Normal	Angular	Hemispherical	Normal	Angular	Hemispherical	Normal	Angular	Hemispherical	Normal	Angular	Hemispherical	Normal	Angular	Hemispherical	Normal	Angular	Hemispherical	Normal	Angular	Hemispherical	Normal	Angular	Hemispherical	Normal	Angular	
Crystex	8	–	–	–	–	–	–	–	–	–	–	117	–	–	–	–	–	–	–	–	–	–	–	–	–	–	–	–	–	–	–	–	–	–	–
CV-II + FeB	9	–	–	–	–	1338	–	–	–	–	–	1342	–	–	–	–	–	–	–	–	–	–	–	–	–	–	–	–	–	–	–	–	–	–	–
CV-II-IX	9	–	–	–	–	1338	–	–	–	–	–	1342	–	–	–	–	–	–	–	–	–	–	–	–	–	–	–	–	–	–	–	–	–	–	–
CV-II + IX + TiO_2	9	–	–	–	–	1337	–	–	–	–	–	1342	–	–	–	–	–	–	–	–	1345	–	–	–	–	–	–	–	–	–	–	–	–	–	–
CV-IV	9	1331	–	–	–	–	–	–	–	–	–	–	–	–	–	–	–	–	–	–	–	–	–	–	–	–	–	–	–	–	–	–	–	–	–
Cymel 405 Coating on Quartz Substrate	9	–	–	–	–	–	–	–	–	–	–	–	–	–	–	–	–	–	–	–	–	–	–	1130	–	–	–	–	–	–	–	–	–	–	–
D-43	7	1470	–	–	–	–	–	–	–	–	–	–	–	–	–	–	–	–	–	–	–	–	–	–	–	–	–	–	–	–	–	–	–	–	–
DaCote Black Enamel	9	–	–	–	–	–	–	–	–	–	–	542	–	–	–	–	–	–	–	–	–	–	–	–	–	–	–	–	–	–	–	–	–	–	–
Decoret Binder Pigmented with Lampblack	9	–	–	–	–	–	–	–	–	–	–	86	–	–	–	–	–	–	–	–	–	–	–	–	–	–	–	–	–	–	–	–	–	–	–
Diacetyl Cellulose Coating on Varnish Substrate	9	–	–	–	–	–	–	–	–	–	–	–	–	–	–	–	–	–	–	–	–	–	–	–	–	–	–	–	–	1119	–	–	–	–	–
Diaspore	8	–	–	–	–	–	–	–	–	–	–	–	–	–	–	–	–	–	–	–	–	–	–	–	–	–	–	–	–	1664	–	–	–	–	–
Diatomaceous Earth Pigment in:																																			
Potassium Silicate Binder	9	–	–	–	–	–	–	–	–	–	–	113	–	–	–	–	–	–	–	–	–	–	–	116	–	–	–	–	–	–	–	–	–	–	–
Silicone Binder	9	–	–	–	–	–	–	–	–	–	–	113	–	–	–	–	–	–	–	–	–	–	–	–	–	–	–	–	–	–	–	–	–	–	–
Dicalite WB-5 Pigment in Potassium Silicate Binder	9	–	–	–	–	–	–	–	–	–	–	113	–	–	–	–	–	–	–	–	–	–	–	179	–	–	–	–	–	–	–	–	–	–	–
Diffusion Coating on:																																			
Molybdenum Substrate	9	1331	1333	1335	–	1337	–	–	–	–	–	1342	–	–	–	–	–	–	–	–	1345	–	–	–	–	–	–	–	–	–	–	–	–	–	–
Niobium + Titanium Substrate	9	–	1350	–	–	–	–	–	–	–	–	–	–	–	–	–	–	–	–	–	–	–	–	–	–	–	–	–	–	–	–	–	–	–	–
Niobium + Zirconium Substrate	9	1352	–	–	–	–	–	–	–	–	–	–	–	–	–	–	–	–	–	–	–	–	–	–	–	–	–	–	–	–	–	–	–	–	–
Dow 7	9	1274	–	–	–	–	–	–	–	–	–	–	–	–	–	–	–	–	–	–	–	–	–	–	–	–	–	–	–	–	–	–	–	–	–
Dow 15	9	1274	–	–	–	–	–	–	–	–	–	–	–	–	–	–	–	–	–	–	–	–	–	1275	–	–	–	–	–	–	–	–	–	–	–
Dow 17	9	1274	–	–	–	–	–	–	–	–	–	–	–	–	–	–	–	–	–	–	–	–	–	1275	–	–	–	–	–	–	–	–	–	–	–
Dow 17 Substrate Coated with Gold	9	–	–	–	–	660	–	–	–	–	–	–	–	–	–	–	–	–	–	–	–	–	–	–	–	–	–	–	–	–	–	–	–	–	–
Dow Corning 20 Binder Pigmented with Silicon Carbide	9	174	–	–	–	–	–	–	–	–	–	–	–	–	–	–	–	–	–	–	–	–	–	–	–	–	–	–	–	–	–	–	–	–	–
Dow Corning 806A Binder Pigmented with:																																			
Boron Nitride	9	–	–	–	–	–	–	–	–	–	–	68	–	–	–	–	–	–	–	–	–	–	–	–	–	–	–	–	–	–	–	–	–	–	–
Boron Nitride + Diatomaceous Earth	9	–	–	–	–	–	–	–	–	–	–	70	–	–	–	–	–	–	–	–	–	–	–	–	–	–	–	–	–	–	–	–	–	–	–

MATERIAL NAME	VOLUME	EMITTANCE						REFLECTANCE									ABSORPTANCE									TRANSMITTANCE									Absorptance to Emittance Ratio
		Total			Spectral			Integrated			Spectral			Solar			Integrated			Spectral			Solar			Integrated			Spectral			Solar			
		Hemispherical	Normal	Angular	Hemispherical	Normal	Angular	Hemispherical	Normal	Angular	Hemispherical	Normal	Angular	Hemispherical	Normal	Angular	Hemispherical	Normal	Angular	Hemispherical	Normal	Angular	Hemispherical	Normal	Angular	Hemispherical	Normal	Angular	Hemispherical	Normal	Angular	Hemispherical	Normal	Angular	
Dow Corning 806A Binder Pigmented with: (continued)																																			
Diatomaceous Earth	9	-	-	-	-	-	-	-	-	-	-	113	-	-	-	-	-	-	-	-	-	-	-	-	-	-	-	-	-	-	-	-	-	-	-
Magnesium Oxide	9	-	-	-	-	-	-	-	-	-	-	161	-	-	-	-	-	-	-	-	-	-	-	-	-	-	-	-	-	-	-	-	-	-	-
Magnesium Oxide + Diatomaceous Earth	9	-	-	-	-	-	-	-	-	-	-	165	-	-	-	-	-	-	-	-	-	-	-	-	-	-	-	-	-	-	-	-	-	-	-
SiC + TiC	9	174	-	-	-	-	-	-	-	-	-	-	-	-	-	-	-	-	-	-	-	-	-	-	-	-	-	-	-	-	-	-	-	-	-
Titanium Dioxide	9	211	-	-	-	-	-	-	-	-	-	222	-	-	-	-	-	-	-	-	-	-	-	-	-	-	-	-	-	-	-	-	-	-	-
Zinc Oxide	9	-	-	-	-	-	-	-	-	-	-	317	-	-	-	-	-	-	-	-	-	-	-	-	-	-	-	-	-	-	-	-	-	-	-
Dow Corning 807 Binder Pigmented with Lampblack	9	-	82	-	-	-	-	-	-	-	-	86	-	-	-	-	-	-	-	-	-	-	-	-	-	-	-	-	-	-	-	-	-	-	-
Dow Corning 6510 Coating on Aluminum Substrate	9	-	-	-	-	-	-	-	-	-	-	1159	-	-	-	-	-	-	-	-	-	-	-	-	-	-	-	-	-	-	-	-	-	-	-
Dow Corning Q90016 Binder Pigmented with Zinc Oxide	9	-	-	-	-	-	-	-	308	-	-	321	-	-	-	-	-	-	-	-	-	-	-	375	-	-	-	-	-	-	-	-	-	-	-
Dow Corning Q90090 Binder Pigmented with Zinc Oxide	9	-	-	-	-	-	-	-	-	-	-	-	-	-	-	-	-	-	-	-	362	-	-	-	-	-	-	-	-	-	-	-	-	-	-
Dow Metal Substrate Coated with:																																			
Lacquer	9	-	1124	-	-	-	-	-	-	-	-	-	-	-	-	-	-	-	-	-	-	-	-	1127	-	-	-	-	-	-	-	-	-	-	-
Silicone	9	-	1156	-	-	-	-	-	-	-	-	-	-	-	-	-	-	-	-	-	-	-	-	1164	-	-	-	-	-	-	-	-	-	-	-
Dreem Enamel No. 13N27ES4	9	-	-	-	-	-	-	-	-	-	-	495	-	-	-	-	-	-	-	-	-	-	-	-	-	-	-	-	-	-	-	-	-	-	-
Duco Wrought Iron No. 71 Black	9	-	-	-	-	-	-	-	-	-	-	498	-	-	-	-	-	-	-	-	-	-	-	-	-	-	-	-	-	-	-	-	-	-	-
Dulite II	9	496	-	-	-	-	-	-	-	-	-	-	-	-	-	-	-	-	-	-	-	-	-	-	-	-	-	-	-	-	-	-	-	-	-
Dulite 1015	9	496	-	-	-	-	-	-	-	-	-	-	-	-	-	-	-	-	-	-	-	-	-	-	-	-	-	-	-	-	-	-	-	-	-
DuPont R-100 Pigment in DuPont RC-7007 Binder	9	-	-	-	-	-	-	-	-	-	-	223	-	-	-	-	-	-	-	-	-	-	-	-	-	-	-	-	-	-	-	-	-	-	-
DuPont R-510 Pigment in DuPont RC-7007 Binder	9	-	-	-	-	-	-	-	-	-	-	221	-	-	-	-	-	-	-	-	-	-	-	-	-	-	-	-	-	-	-	-	-	-	-
DuPont R-960 Pigment in RTV-602 Binder	9	-	-	-	-	-	-	-	-	-	-	228	-	-	-	-	-	-	-	-	-	-	-	273	-	-	-	-	-	-	-	-	-	-	-
DuPont RC-7007 Binder Pigmented with:																																			
Aluminum Oxide	9	-	-	-	-	-	-	-	-	-	-	33	-	-	-	-	-	-	-	-	-	-	-	-	-	-	-	-	-	-	-	-	-	-	-
$CaSO_4 + TiO_2$	9	-	-	-	-	-	-	-	-	-	-	78	-	-	-	-	-	-	-	-	-	-	-	-	-	-	-	-	-	-	-	-	-	-	-
Magnesium Oxide	9	-	-	-	-	-	-	-	-	-	-	161	-	-	-	-	-	-	-	-	-	-	-	-	-	-	-	-	-	-	-	-	-	-	-
Titanium Dioxide	9	-	-	-	-	-	-	-	-	-	-	221	-	-	-	-	-	-	-	-	-	-	-	-	-	-	-	-	-	-	-	-	-	-	-
Titanox C-50	9	-	-	-	-	-	-	-	-	-	-	78 289	-	-	-	-	-	-	-	-	-	-	-	-	-	-	-	-	-	-	-	-	-	-	-

MATERIAL NAME	VOLUME	EMITTANCE						REFLECTANCE									ABSORPTANCE									TRANSMITTANCE									Absorptance to Emittance Ratio
		Total			Spectral			Integrated			Spectral			Solar			Integrated			Spectral			Solar			Integrated			Spectral			Solar			
		Hemispherical	Normal	Angular	Hemispherical	Normal	Angular	Hemispherical	Normal	Angular	Hemispherical	Normal	Angular	Hemispherical	Normal	Angular	Hemispherical	Normal	Angular	Hemispherical	Normal	Angular	Hemispherical	Normal	Angular	Hemispherical	Normal	Angular	Hemispherical	Normal	Angular	Hemispherical	Normal	Angular	
DuPont RC-7007 Binder Pigmented with: (continued)																																			
Titanox RC	9	-	-	-	-	-	-	-	-	-	-	78 289	-	-	-	-	-	-	-	-	-	-	-	-	-	-	-	-	-	-	-	-	-	-	-
DuPont Viton B Binder Pigmented with Zinc Oxide	9	-	-	-	-	-	-	-	-	-	-	314	-	-	-	-	-	-	-	-	-	-	-	-	-	-	-	-	-	-	-	-	-	-	-
Durak-B	9	1331	-	-	-	-	-	-	-	-	-	-	-	-	-	-	-	-	-	-	-	-	-	-	-	-	-	-	-	-	-	-	-	-	-
Durak-MG	9	-	1333	1335	-	-	-	-	-	-	-	-	-	-	-	-	-	-	-	-	-	-	-	-	-	-	-	-	-	-	-	-	-	-	-
Duro-Lac Black Brushing Lacquer No. 519	9	-	-	-	-	-	-	-	-	-	-	541	-	-	-	-	-	-	-	-	-	-	-	-	-	-	-	-	-	-	-	-	-	-	-
Dutch Boy Quick Drying Enamel (aluminum)	9	-	-	-	-	-	-	-	-	-	-	13	-	-	-	-	-	-	-	-	-	-	-	-	-	-	-	-	-	-	-	-	-	-	-
Dutch Boy 47 H 47	9	499	-	-	-	-	-	-	-	-	-	500	-	-	-	-	-	-	-	-	-	-	-	-	-	-	-	-	-	-	-	-	-	-	-
Dysprosium Hexaboride	8	-	-	-	-	-	-	-	-	-	-	-	-	-	-	-	-	-	-	-	-	-	-	-	-	-	-	-	-	727	-	-	-	-	-
Dysprosium Nitride	8	-	-	-	-	-	-	-	-	-	-	-	-	-	-	-	-	-	-	-	-	-	-	-	-	-	-	-	-	1090	-	-	-	-	-
Dysprosium Oxide	8	-	-	-	-	-	-	-	-	-	-	252	-	-	-	-	-	-	-	-	-	-	-	254	-	-	-	-	-	-	-	-	-	-	-
Eastman White Reflectance Paint	9	-	-	-	-	-	-	-	-	-	-	62	-	-	-	-	-	-	-	-	-	-	-	-	-	-	-	-	-	-	-	-	-	-	-
Electroconducting Glass	8	-	1559	-	-	-	-	-	-	-	-	1561	-	-	-	-	-	-	-	-	-	-	-	1563	-	-	1564	-	-	1566	-	-	-	-	-
Elvanol 71-30 Coating on Fiberglass Substrate	9	-	-	-	-	-	-	-	-	-	-	1145	-	-	-	-	-	-	-	-	-	-	-	1146	-	-	-	-	-	-	-	-	-	-	-
Enstatite	8	-	-	-	-	-	-	-	-	-	-		1689	-	-	-	-	-	-	-	-	-	-	1692	-	-	-	-	-	-	-	-	-	-	-
E-P730 Pigment in Potassium Silicate Binder	9	-	-	-	-	-	-	-	-	-	-	316	-	-	-	-	-	-	-	-	-	-	-	-	-	-	-	-	-	-	-	-	-	-	-
Epoxide Binder Pigmented with Titanium Dioxide	9	211	-	-	-	-	-	-	-	-	-	-	-	-	-	-	-	-	-	-	-	-	-	-	-	-	-	-	-	-	-	-	-	-	-
Epoxy Binder Pigmented with:																																			
Lampblack	9	-	82	-	-	84	-	-	-	-	-	86	-	-	-	-	-	-	-	-	-	-	-	89	-	-	-	-	-	-	-	-	-	-	-
Titanium Dioxide	9	211	-	-	-	-	-	-	-	-	-	-	-	-	-	-	-	252	-	-	255	-	-	263 281	-	-	-	-	-	-	-	-	-	-	-
Epoxy Coating on:																																			
Aluminum Substrate	9	1120	-	-	-	-	-	-	-	-	-	-	-	-	-	-	-	-	-	-	-	-	-	-	-	-	-	-	-	-	-	-	-	-	-
Polyethylene Substrate	9	-	-	-	-	-	-	-	-	-	-	-	-	-	-	-	-	-	-	-	-	-	-	-	-	-	-	-	-	1122	-	-	-	-	-
Epoxy Substrate Coated with:																																			
Aluminum	9	-	-	-	-	-	-	-	-	-	-	-	602	-	-	-	-	-	-	-	-	-	-	610	-	-	-	-	-	-	-	-	-	-	-
Copper	9	-	-	-	-	-	-	-	-	-	-	-	643	-	-	-	-	-	-	-	-	-	-	-	-	-	-	-	-	-	-	-	-	-	-

MATERIAL NAME	VOLUME	EMITTANCE						REFLECTANCE									ABSORPTANCE									TRANSMITTANCE									Absorptance to Emittance Ratio
		Total			Spectral			Integrated			Spectral			Solar			Integrated			Spectral			Solar			Integrated			Spectral			Solar			
		Hemispherical	Normal	Angular	Hemispherical	Normal	Angular	Hemispherical	Normal	Angular	Hemispherical	Normal	Angular	Hemispherical	Normal	Angular	Hemispherical	Normal	Angular	Hemispherical	Normal	Angular	Hemispherical	Normal	Angular	Hemispherical	Normal	Angular	Hemispherical	Normal	Angular	Hemispherical	Normal	Angular	
Epoxy Substrate Coated with: (continued)																																			
Gold	9	-	-	-	-	-	-	-	-	-	-	-	-	-	-	672	-	-	-	-	-	-	-	-	-	-	-	-	-	-	-	-	-	-	-
Nickel	9	700	-	-	-	-	-	-	-	-	-	-	-	-	-	704	-	-	-	-	-	-	-	-	-	-	-	-	-	-	-	-	-	-	-
Silver	9	733	-	-	-	-	-	-	-	-	-	-	-	-	-	742	-	-	-	-	-	-	-	-	-	-	-	-	-	-	-	-	-	-	-
Erbium	7	-	-	-	-	202	-	-	-	-	-	-	-	-	-	-	-	-	-	-	-	-	-	-	-	-	-	-	-	-	-	-	-	-	-
Erbium Hexaboride	8	-	-	-	-	-	-	-	-	-	-	-	-	-	-	-	-	-	-	-	-	-	-	-	-	-	-	-	-	727	-	-	-	-	-
Erbium Nitride	8	-	-	-	-	-	-	-	-	-	-	-	-	-	-	-	-	-	-	-	-	-	-	-	-	-	-	-	-	1090	-	-	-	-	-
Erbium Oxide	8	-	255	-	257	259 261	-	-	-	-	-	-	-	-	-	-	-	-	-	-	-	-	-	-	-	-	-	-	-	-	-	-	-	-	-
Ethylcellulose + Dow 7 Binder Pigmented with:																																			
Magnesium Carbonate	9	-	-	-	-	-	-	-	156	-	-	-	-	-	-	-	-	-	-	-	-	-	-	-	-	-	-	-	-	-	-	-	-	-	-
Magnesium Oxide	9	-	-	-	-	-	-	-	158	-	-	-	-	-	-	-	-	-	-	-	-	-	-	-	-	-	-	-	-	-	-	-	-	-	-
ETP Copper Substrate Coated with Gold	9	652	-	-	-	-	-	-	-	-	-	-	-	-	-	-	-	-	-	-	-	-	-	-	-	-	-	-	-	-	-	-	-	-	-
Europium Hexaboride	8	-	-	-	-	723	-	-	-	-	-	-	-	-	-	-	-	-	-	-	-	-	-	-	-	-	-	-	-	-	-	-	-	-	-
Dieuropium Trioxide	8	-	-	-	-	-	-	-	-	-	-	546	-	-	-	-	-	-	-	-	-	-	-	-	-	-	-	-	-	-	-	-	-	-	-
Europium Silicate	8	-	-	-	-	-	-	-	-	-	-	-	-	-	-	-	-	-	-	-	-	-	-	-	-	-	-	-	-	622	-	-	-	-	-
Europium Sulfide	8	-	-	-	-	-	-	-	-	-	-	1234	-	-	-	-	-	-	-	-	-	-	-	-	-	-	-	-	-	-	-	-	-	-	-
Fabric Substrate Coated with Aluminum	9	-	586	-	-	-	-	-	-	-	-	-	-	-	-	-	-	-	-	-	-	-	-	-	-	-	-	-	-	-	-	-	-	-	-
Fassion Foil on Copper Substrate	9	-	-	-	-	-	-	-	-	-	-	594	-	-	-	-	-	-	-	-	-	-	-	-	-	-	-	-	-	-	-	-	-	-	-
Fayalite	8	-	-	-	-	-	-	-	-	-	-	1689	-	-	-	-	-	-	-	-	-	-	-	-	-	-	-	-	-	-	-	-	-	-	-
FCZ-11 Coating on Nb-1Zr Substrate	9	1024	-	-	-	1028	-	-	-	-	-	-	-	-	-	-	-	-	-	-	-	-	-	-	-	-	-	-	-	-	-	-	-	-	-
Ferric Oxide (see Diiron Trioxide)																																			
Ferroferric Oxide (see Triiron Tetraoxide)																																			
Ferro White Porcelain Enamel	9	576	-	-	-	-	-	-	-	-	-	-	-	-	-	-	-	-	-	-	-	-	-	-	-	-	-	-	-	-	-	-	-	-	-
Fiberglass Substrate Coated with:																																			
Gold	9	651	-	-	-	660	-	-	-	-	-	-	-	-	-	-	-	-	-	-	-	-	-	-	-	-	-	-	-	-	-	-	-	-	-
Polyvinyl Alcohol	9	-	-	-	-	-	-	-	-	-	-	1145	-	-	-	-	-	-	-	-	-	-	-	1146	-	-	-	-	-	-	-	-	-	-	-
Polyvinyl Chloride	9	-	-	-	-	-	-	-	-	-	-	1150	-	-	-	-	-	-	-	-	-	-	-	1152	-	-	-	-	-	-	-	-	-	-	-

MATERIAL NAME	VOLUME	EMITTANCE						REFLECTANCE									ABSORPTANCE									TRANSMITTANCE									Absorptance to Emittance Ratio
		Total			Spectral			Integrated			Spectral			Solar			Integrated			Spectral			Solar			Integrated			Spectral			Solar			
		Hemispherical	Normal	Angular	Hemispherical	Normal	Angular	Hemispherical	Normal	Angular	Hemispherical	Normal	Angular	Hemispherical	Normal	Angular	Hemispherical	Normal	Angular	Hemispherical	Normal	Angular	Hemispherical	Normal	Angular	Hemispherical	Normal	Angular	Hemispherical	Normal	Angular	Hemispherical	Normal	Angular	
Flight Data:																																			
Aluminum Coating on Stainless Steel Substrate	9	–	–	–	–	–	–	–	–	–	–	–	–	–	–	–	–	–	–	–	–	–	–	612	–	–	–	–	–	–	–	–	–	–	–
Anodized Aluminum	9	–	–	–	–	–	–	–	–	–	–	–	–	–	–	–	–	–	–	–	–	–	–	1230	–	–	–	–	–	–	–	–	–	–	1233
Al_2O_3 Pigment in Potassium Silicate Binder	9	–	–	–	–	–	–	–	–	–	–	–	–	–	–	–	–	–	–	–	–	–	–	–	–	–	–	–	–	–	–	–	–	–	45
Antimony Oxide Pigment in Potassium Silicate Binder	9	–	–	–	–	–	–	–	–	–	–	–	–	–	–	–	–	–	–	–	–	–	–	60	–	–	–	–	–	–	–	–	–	–	–
ATS-1	9	–	–	–	–	–	–	–	–	–	–	–	–	–	–	–	–	–	–	–	–	–	–	–	–	–	–	–	–	–	–	–	–	–	45 283 399 753
China Clay Pigment in Silicone Binder	9	–	–	–	–	–	–	–	–	–	–	–	–	–	–	–	–	–	–	–	–	–	–	100	–	–	–	–	–	–	–	–	–	–	102
Lunar Orbiter I	9	–	–	–	–	–	–	–	–	–	–	–	–	–	–	–	–	–	–	–	–	–	–	394	–	–	–	–	–	–	–	–	–	–	–
Lunar Orbiter II	9	–	–	–	–	–	–	–	–	–	–	–	–	–	–	–	–	–	–	–	–	–	–	394	–	–	–	–	–	–	–	–	–	–	–
Lunar Orbiter IV	9	–	–	–	–	–	–	–	–	–	–	–	–	–	–	–	–	–	–	–	–	–	–	281 393	–	–	–	–	–	–	–	–	–	–	283 399
Lunar Orbiter V	9	–	–	–	–	–	–	–	–	–	–	–	–	–	–	–	–	–	–	–	–	–	–	100 393 1167	–	–	–	–	–	–	–	–	–	–	102 399 1169
Mariner IV	9	–	–	–	–	–	–	–	–	–	–	–	–	–	–	–	–	–	–	–	–	–	–	394	–	–	–	–	–	–	–	–	–	–	–
Mariner V	9	–	–	–	–	–	–	–	–	–	–	–	–	–	–	–	–	–	–	–	–	–	–	394	–	–	–	–	–	–	–	–	–	–	399
OSO-II	9	–	–	–	–	–	–	–	–	–	–	–	–	–	–	–	–	–	–	–	–	–	–	60 153 205 207 272 281 376 392 441	–	–	–	–	–	–	–	–	–	–	–
OSO-III	9	–	–	–	–	–	–	–	–	–	–	–	–	–	–	–	–	–	–	–	–	–	–	394 612 1072 1154 1167 1230	–	–	–	–	–	–	–	–	–	–	–
OVI-10	9	–	–	–	–	–	–	–	–	–	–	–	–	–	–	–	–	–	–	–	–	–	–	275	–	–	–	–	–	–	–	–	–	–	–
Pegasus I	9	–	–	–	–	–	–	–	–	–	–	–	–	–	–	–	–	–	–	–	–	–	–	394	–	–	–	–	–	–	–	–	–	–	–
Pegasus II	9	–	–	–	–	–	–	–	–	–	–	–	–	–	–	–	–	–	–	–	–	–	–	394	–	–	–	–	–	–	–	–	–	–	399 1233
Polyvinyl Chloride Coating on Aluminum Substrate	9	–	–	–	–	–	–	–	–	–	–	–	–	–	–	–	–	–	–	–	–	–	–	1154	–	–	–	–	–	–	–	–	–	–	–
Silicon Dioxide Coating on Silver Substrate	9	–	–	–	–	–	–	–	–	–	–	–	–	–	–	–	–	–	–	–	–	–	–	1072	–	–	–	–	–	–	–	–	–	–	–

MATERIAL NAME	VOLUME	EMITTANCE						REFLECTANCE									ABSORPTANCE									TRANSMITTANCE									Absorptance to Emittance Ratio
		Total			Spectral			Integrated			Spectral			Solar			Integrated			Spectral			Solar			Integrated			Spectral			Solar			
		Hemispherical	Normal	Angular	Hemispherical	Normal	Angular	Hemispherical	Normal	Angular	Hemispherical	Normal	Angular	Hemispherical	Normal	Angular	Hemispherical	Normal	Angular	Hemispherical	Normal	Angular	Hemispherical	Normal	Angular	Hemispherical	Normal	Angular	Hemispherical	Normal	Angular	Hemispherical	Normal	Angular	
Flight Data: (continued)																																			
Silicone Coating on Aluminum Substrate	9	-	-	-	-	-	-	-	-	-	-	-	-	-	-	-	-	-	-	-	-	-	-	1167	-	-	-	-	-	-	-	-	-	-	1169
Silver Coating on BeCu Substrate	9	-	-	-	-	-	-	-	-	-	-	-	-	-	-	-	-	-	-	-	-	-	-	-	-	-	-	-	-	-	-	-	-	-	753
Spodumene Pigment in Sodium Silicate Binder	9	-	-	-	-	-	-	-	-	-	-	-	-	-	-	-	-	-	-	-	-	-	-	153	-	-	-	-	-	-	-	-	-	-	-
Tin Oxide Pigment in Silicate Binder	9	-	-	-	-	-	-	-	-	-	-	-	-	-	-	-	-	-	-	-	-	-	-	205 207	-	-	-	-	-	-	-	-	-	-	-
Titanium Dioxide Pigment in Epoxy Binder	9	-	-	-	-	-	-	-	-	-	-	-	-	-	-	-	-	-	-	-	-	-	-	272	-	-	-	-	-	-	-	-	-	-	-
Titanium Dioxide Pigment in Potassium Silicate Binder	9	-	-	-	-	-	-	-	-	-	-	-	-	-	-	-	-	-	-	-	-	-	-	281	-	-	-	-	-	-	-	-	-	-	283
Titanium Dioxide Pigment in Silicone Binder	9	-	-	-	-	-	-	-	-	-	-	-	-	-	-	-	-	-	-	-	-	-	-	272	-	-	-	-	-	-	-	-	-	-	-
Zinc Oxide Pigment in Potassium Silicate Binder	9	-	-	-	-	-	-	-	-	-	-	-	-	-	-	-	-	-	-	-	-	-	-	376 392	-	-	-	-	-	-	-	-	-	-	-
Zinc Oxide Pigment in Silicone Binder	9	-	-	-	-	-	-	-	-	-	-	-	-	-	-	-	-	-	-	-	-	-	-	376 392	-	-	-	-	-	-	-	-	-	-	-
Zirconium Silicate Pigment in Potassium Silicate Binder	9	-	-	-	-	-	-	-	-	-	-	-	-	-	-	-	-	-	-	-	-	-	-	441	-	-	-	-	-	-	-	-	-	-	-
Flint Glass	8	-	1644	-	-	-	-	-	-	-	-	1646	-	-	-	-	-	-	-	-	-	-	-	-	-	-	-	-	-	1648	-	-	-	-	-
Fluorides:																																			
Barium Fluoride	8	-	-	-	-	-	-	-	-	-	-	-	909	-	-	-	-	-	-	-	-	-	-	-	-	-	-	-	-	912	-	-	-	-	-
Cadmium Fluoride	8	-	-	-	-	-	-	-	-	-	-	-	-	-	-	-	-	-	-	-	-	-	-	-	-	-	-	-	-	919	-	-	-	-	-
Calcium Fluoride	8	-	-	-	-	921	-	-	-	-	-	924	-	-	-	-	-	-	-	-	929	-	-	-	-	-	-	-	-	931	-	-	-	-	-
Lithium Fluoride	8	-	-	-	-	-	-	-	-	-	-	937	942	-	-	-	-	-	-	-	-	-	-	-	-	-	-	-	-	944	-	-	-	-	-
Magnesium Difluoride	8	-	-	-	-	950	-	-	-	-	-	952	-	-	-	-	-	-	-	-	955	-	-	-	-	-	-	-	-	957	-	-	-	-	-
Sodium Fluoride	8	-	-	-	-	-	-	-	-	-	-	963	-	-	-	-	-	-	-	-	-	-	-	-	-	-	-	-	-	966	-	-	-	-	-
Strontium Fluoride	8	-	-	-	-	-	-	-	-	-	-	968	-	-	-	-	-	-	-	-	-	-	-	-	-	-	-	-	-	971	-	-	-	-	-
Potassium Magnesium Trifluoride	8	-	-	-	-	-	-	-	-	-	-	975	-	-	-	-	-	-	-	-	-	-	-	-	-	-	-	-	-	977	-	-	-	-	-
Potassium Nickel Trifluoride	8	-	-	-	-	-	-	-	-	-	-	979	-	-	-	-	-	-	-	-	-	-	-	-	-	-	-	-	-	981	-	-	-	-	-
Rubidium Manganese Trifluoride	8	-	-	-	-	-	-	-	-	-	-	983	-	-	-	-	-	-	-	-	-	-	-	-	-	-	-	-	-	985	-	-	-	-	-
Miscellaneous Cubic Perovskite Fluorides	8	-	-	-	-	-	-	-	-	-	-	987	-	-	-	-	-	-	-	-	-	-	-	-	-	-	-	-	-	989	-	-	-	-	-
Miscellaneous Fluorides	8	-	-	-	-	-	-	-	-	-	-	991	-	-	-	-	-	-	-	-	-	-	-	-	-	-	-	-	-	993	-	-	-	-	-
Fluorite	8	-	-	-	-	-	-	-	-	-	-	-	-	-	-	-	-	-	-	-	-	-	-	-	-	-	-	-	-	933	-	-	-	-	-

MATERIAL NAME	VOLUME	EMITTANCE						REFLECTANCE									ABSORPTANCE									TRANSMITTANCE									Absorptance to Emittance Ratio
		Total			Spectral			Integrated			Spectral			Solar			Integrated			Spectral			Solar			Integrated			Spectral			Solar			
		Hemispherical	Normal	Angular	Hemispherical	Normal	Angular	Hemispherical	Normal	Angular	Hemispherical	Normal	Angular	Hemispherical	Normal	Angular	Hemispherical	Normal	Angular	Hemispherical	Normal	Angular	Hemispherical	Normal	Angular	Hemispherical	Normal	Angular	Hemispherical	Normal	Angular	Hemispherical	Normal	Angular	
Fluorite Substrate Coated with Copper Phosphorous Selenide	9	-	-	-	-	-	-	-	-	-	-	892	-	-	-	-	-	-	-	-	-	-	-	-	-	-	-	-	-	893	-	-	-	-	-
Flowers of Sulfur	8	-	-	-	-	-	-	-	-	-	-	117	-	-	-	-	-	-	-	-	-	-	-	-	-	-	-	-	-	-	-	-	-	-	-
Formaldehyde Binder Pigmented with Lead Carbonate	9	-	-	-	-	-	-	-	-	-	-	138	-	-	-	-	-	-	-	-	-	-	-	-	-	-	-	-	-	-	-	-	-	-	-
Forsterite	8	-	-	-	-	-	-	-	-	-	-	-	-	1689	-	-	-	-	-	-	-	-	-	1692	-	-	-	-	-	-	-	-	-	-	-
Frenchtown 4402	8	-	144	-	-	147	-	-	-	-	-	-	-	-	-	-	-	-	-	-	-	-	-	-	-	-	-	-	-	-	-	-	-	-	-
Fuller Paints:																																			
D-70-6342	9	-	502	-	-	-	-	-	-	-	-	-	-	-	-	-	-	-	-	-	-	-	-	-	-	-	-	-	-	-	-	-	-	-	-
Flat Black Decoret	9	-	-	-	-	-	-	-	-	-	-	504	-	-	-	-	-	-	-	-	-	-	-	-	-	-	-	-	-	-	-	-	-	-	-
Flat Black Silicone	9	81	-	-	-	-	-	-	-	-	-	-	-	-	-	-	-	-	-	-	-	-	-	89	-	-	-	-	-	-	-	-	-	-	-
Harvard Gray No. 2946	9	-	-	-	-	-	-	-	-	-	-	504	-	-	-	-	-	-	-	-	-	-	-	-	-	-	-	-	-	-	-	-	-	-	-
Light Brown No. 2909	9	-	-	-	-	-	-	-	-	-	-	504	-	-	-	-	-	-	-	-	-	-	-	-	-	-	-	-	-	-	-	-	-	-	-
Mariposa Blue Decoret No. 2889	9	-	-	-	-	-	-	-	-	-	-	504	-	-	-	-	-	-	-	-	-	-	-	-	-	-	-	-	-	-	-	-	-	-	-
TL-8606 No. 43	9	-	-	-	-	-	-	-	-	-	-	504	-	-	-	-	-	-	-	-	-	-	-	-	-	-	-	-	-	-	-	-	-	-	-
TL-9465 No. 45	9	-	-	-	-	-	-	-	-	-	-	504	-	-	-	-	-	-	-	-	-	-	-	-	-	-	-	-	-	-	-	-	-	-	-
Velvet Black No. 1518	9	-	-	-	-	-	-	-	-	-	-	504	-	-	-	-	-	-	-	-	-	-	-	-	-	-	-	-	-	-	-	-	-	-	-
No. 171-A-152	9	3	-	-	-	-	-	-	-	-	-	-	-	-	-	-	-	-	-	-	-	-	-	21	-	-	-	-	-	-	-	-	-	-	-
No. 171-W-560	9	501	-	-	-	-	-	-	-	-	-	-	-	-	-	-	-	-	-	-	-	-	-	507	-	-	-	-	-	-	-	-	-	-	-
No. 172-A-1	9	-	-	-	-	-	-	-	-	-	-	13	-	-	-	-	-	-	-	-	-	-	-	-	-	-	-	-	-	-	-	-	-	-	-
No. 517-W-1	9	212	-	-	-	-	-	-	-	-	-	-	-	-	-	-	-	-	-	-	256	-	-	263	-	-	-	-	-	-	-	-	-	-	-
No. 517-W-7	9	211	-	-	-	-	-	-	-	-	-	227	-	-	-	-	-	-	-	-	-	-	-	-	-	-	-	-	-	-	-	-	-	-	-
Fused Quartz [see Silicon Dioxide (Fused)]																																			
Fused Silica [see Silicon Dioxide (Fused)]																																			
Fused Silica Glass	8	1568	1569	-	-	-	-	-	-	-	-	1571	-	-	-	-	-	-	-	-	-	-	-	1573	-	-	1574	-	-	1576	-	-	1578	-	-
Gabbro	8	-	-	-	-	-	-	-	-	-	-	1681	-	-	-	-	-	-	-	-	-	-	-	-	-	-	-	-	-	-	-	-	-	-	-
GAEC Black Epoxy Paint No. 1019	9	517	-	-	-	-	-	-	-	-	-	-	-	-	-	-	-	-	-	-	-	-	-	-	-	-	-	-	-	-	-	-	-	-	-
Gadolinium	7	-	-	-	-	-	-	-	-	-	-	204	-	-	-	-	-	-	-	-	-	-	-	-	-	-	-	-	-	207	-	-	-	-	-
Gadolinium Hexaboride	8	-	-	-	-	723	-	-	-	-	-	-	-	-	-	-	-	-	-	-	-	-	-	-	-	-	-	-	-	727	-	-	-	-	-
Gadolinium Oxide	8	-	263	-	265	267 269	-	-	-	-	-	-	-	-	-	-	-	-	-	-	-	-	-	-	-	-	-	-	-	-	-	-	-	-	-
Gadolinium Disilicide	8	-	-	-	-	1173	-	-	-	-	-	-	-	-	-	-	-	-	-	-	-	-	-	-	-	-	-	-	-	-	-	-	-	-	-
Galena	8	-	-	-	-	-	-	-	-	-	-	1199	-	-	-	-	-	-	-	-	-	-	-	-	-	-	-	-	-	-	-	-	-	-	-

MATERIAL NAME	VOLUME	EMITTANCE Total Hemispherical	EMITTANCE Total Normal	EMITTANCE Total Angular	EMITTANCE Spectral Hemispherical	EMITTANCE Spectral Normal	EMITTANCE Spectral Angular	REFLECTANCE Integrated Hemispherical	REFLECTANCE Integrated Normal	REFLECTANCE Integrated Angular	REFLECTANCE Spectral Hemispherical	REFLECTANCE Spectral Normal	REFLECTANCE Spectral Angular	REFLECTANCE Solar Hemispherical	REFLECTANCE Solar Normal	REFLECTANCE Solar Angular	ABSORPTANCE Integrated Hemispherical	ABSORPTANCE Integrated Normal	ABSORPTANCE Integrated Angular	ABSORPTANCE Spectral Hemispherical	ABSORPTANCE Spectral Normal	ABSORPTANCE Spectral Angular	ABSORPTANCE Solar Hemispherical	ABSORPTANCE Solar Normal	ABSORPTANCE Solar Angular	TRANSMITTANCE Integrated Hemispherical	TRANSMITTANCE Integrated Normal	TRANSMITTANCE Integrated Angular	TRANSMITTANCE Spectral Hemispherical	TRANSMITTANCE Spectral Normal	TRANSMITTANCE Spectral Angular	TRANSMITTANCE Solar Hemispherical	TRANSMITTANCE Solar Normal	TRANSMITTANCE Solar Angular	Absorptance to Emittance Ratio
Gallium	7	–	–	–	–	–	–	–	–	–	–	210	–	–	–	–	–	–	–	–	213	–	–	–	–	–	–	–	–	216	–	–	–	–	–
Gallium Antimonide	8	–	–	–	–	–	–	–	–	–	–	1287	–	–	–	–	–	–	–	–	–	–	–	–	–	–	–	–	–	1290	–	–	–	–	–
Gallium Arsenide	8	–	–	–	–	–	–	–	–	–	–	679	–	–	–	–	–	–	–	–	–	–	–	–	–	–	–	–	–	683	–	–	–	–	–
Gallium Arsenide Substrate Coated with Silicon Nitride	9	–	–	–	–	–	–	–	–	–	–	–	–	–	–	–	–	–	–	–	–	–	–	–	–	–	–	–	–	953	–	–	–	–	–
Gallium Nitride	8	–	–	–	–	1087	–	–	–	–	–	–	–	–	–	–	–	–	–	–	–	–	–	–	–	–	–	–	–	–	–	–	–	–	–
Gallium Phosphide	8	–	–	–	–	1092	–	–	–	–	–	1094	–	–	–	–	–	–	–	–	–	–	–	–	–	–	–	–	–	1098	–	–	–	–	–
Gallium Arsenic Phosphide	8	–	–	–	–	–	–	–	–	–	–	1107	–	–	–	–	–	–	–	–	–	–	–	–	–	–	–	–	–	–	–	–	–	–	–
Galvanized Iron	9	–	–	–	–	–	–	–	–	–	–	781	–	–	–	–	–	–	–	–	–	–	–	–	–	–	–	–	–	–	–	–	–	–	–
GaSb	8	–	–	–	–	–	–	–	–	–	–	1287	–	–	–	–	–	–	–	–	–	–	–	–	–	–	–	–	–	1290	–	–	–	–	–
G. E. 106	8	–	–	–	–	–	–	–	–	–	–	–	–	–	–	–	–	–	–	–	–	–	–	–	–	–	–	–	–	421	–	–	–	–	–
Gelatin Binder Pigmented with Silver Chloride	9	–	–	–	–	–	–	–	–	–	–	181	–	–	–	–	–	–	–	–	–	–	–	–	–	–	–	–	–	183	–	–	–	–	–
Germanium	7	–	219	–	–	222 224	–	–	–	–	–	231	–	–	–	–	–	–	–	–	–	–	–	–	–	–	–	–	–	236	240	–	–	–	–
Germanium Coating on:																																			
Calcium Fluoride Substrate	9	–	–	–	–	–	–	–	–	–	–	647	–	–	–	–	–	–	–	–	–	–	–	–	–	–	–	–	–	–	–	–	–	–	–
Lead Chloride Substrate	9	–	–	–	–	–	–	–	–	–	–	647	–	–	–	–	–	–	–	–	–	–	–	–	–	–	–	–	–	–	–	–	–	–	–
Lithium Fluoride Substrate	9	–	–	–	–	–	–	–	–	–	–	647	–	–	–	–	–	–	–	–	–	–	–	–	–	–	–	–	–	650	–	–	–	–	–
Germanium Substrate Coated with:																																			
Lead Chloride	9	–	–	–	–	–	–	–	–	–	–	912	–	–	–	–	–	–	–	–	–	–	–	–	–	–	–	–	–	–	–	–	–	–	–
Selenium	9	–	–	–	–	–	–	–	–	–	–	–	–	–	–	–	–	–	–	–	–	–	–	–	–	–	–	–	–	943	–	–	–	–	–
Germanium Monoxide	8	–	–	–	–	–	–	–	–	–	–	–	–	–	–	–	–	–	–	–	–	–	–	–	–	–	–	–	–	574	–	–	–	–	–
Germanium Dioxide	8	–	–	–	–	–	–	–	–	–	–	271	–	–	–	–	–	–	–	–	–	–	–	–	–	–	–	–	–	–	–	–	–	–	–
Germanium Telluride	8	–	–	–	–	–	–	–	–	–	–	1250	–	–	–	–	–	–	–	–	–	–	–	–	–	–	–	–	–	–	–	–	–	–	–
Glass Coating on:																																			
Aluminum Substrate	9	–	–	–	–	–	–	–	–	–	–	1035	–	–	–	–	–	–	–	–	–	–	–	1036	–	–	–	–	–	–	–	–	–	–	–
Silver Substrate	9	–	–	–	–	–	–	–	–	–	–	1035	–	–	–	–	–	–	–	–	–	–	–	–	–	–	–	–	–	–	–	–	–	–	–
Glasses (see individual glass)																																			
Glass Substrate Coated with:																																			
Aluminum	9	–	–	–	–	–	–	–	–	–	587	591	602 607	–	–	–	–	–	–	–	–	–	–	–	–	–	–	–	–	–	–	–	–	–	–
Aluminum + Magnesium	9	–	–	–	–	–	–	–	–	–	–	613	–	–	–	–	–	–	–	–	–	–	–	–	–	–	–	–	–	–	–	–	–	–	–

MATERIAL NAME	VOLUME	EMITTANCE						REFLECTANCE									ABSORPTANCE									TRANSMITTANCE									Absorptance to Emittance Ratio
		Total			Spectral			Integrated			Spectral			Solar			Integrated			Spectral			Solar			Integrated			Spectral			Solar			
		Hemispherical	Normal	Angular	Hemispherical	Normal	Angular	Hemispherical	Normal	Angular	Hemispherical	Normal	Angular	Hemispherical	Normal	Angular	Hemispherical	Normal	Angular	Hemispherical	Normal	Angular	Hemispherical	Normal	Angular	Hemispherical	Normal	Angular	Hemispherical	Normal	Angular	Hemispherical	Normal	Angular	
Glass Substrate Coated with: (continued)																																			
Antimony	9	-	-	-	-	-	-	-	-	-	-	-	614	-	-	-	-	-	-	-	-	-	-	-	-	-	-	-	-	-	-	-	-	-	-
Bismuth Oxide	9	-	-	-	-	-	-	-	-	-	-	820	-	-	-	-	-	-	-	-	-	-	-	-	-	-	-	-	-	823	-	-	-	-	-
Cadmium Arsenide	9	-	-	-	-	-	-	-	-	-	-	-	-	-	-	-	-	-	-	-	-	-	-	-	-	-	-	-	-	836	-	-	-	-	-
Cadmium Oxide	9	-	-	-	-	-	-	-	-	-	-	-	-	-	-	-	-	-	-	-	-	-	-	-	-	-	-	-	-	838	-	-	-	-	-
Calcium	9	-	-	-	-	-	-	-	-	-	-	-	-	-	-	-	-	-	-	-	-	-	-	-	-	-	-	-	-	617	-	-	-	-	-
Chromium	9	-	-	-	-	-	-	-	-	-	-	621	-	-	-	-	-	-	-	-	-	-	-	-	-	-	-	-	-	-	-	-	-	-	-
Cobalt	9	-	-	-	-	-	-	-	-	-	-	632	-	-	-	-	-	-	-	-	-	-	-	-	-	-	-	-	-	634	-	-	-	-	-
Copper	9	-	-	-	-	-	-	-	-	-	-	642	-	-	-	-	-	-	-	-	-	-	-	-	-	-	-	-	-	644	-	-	-	-	-
Copper + Tin	9	-	-	-	-	-	-	-	-	-	-	645	-	-	-	-	-	-	-	-	-	-	-	-	-	-	-	-	-	-	-	-	-	-	-
Gold	9	-	656	-	-	-	-	-	-	-	-	664	-	-	-	-	-	-	-	-	-	-	-	-	-	-	-	-	-	-	-	-	-	-	-
Gold Black	9	-	-	-	-	-	-	-	-	-	-	1175	-	-	-	-	-	-	-	-	-	-	-	-	-	-	-	-	-	1182	-	-	-	-	-
Gold + Palladium + ΣX_i	9	-	-	-	-	-	-	-	-	-	-	683	-	-	-	-	-	-	-	-	-	-	-	-	-	-	-	-	-	-	-	-	-	-	-
Indium Arsenide	9	-	-	-	-	-	-	-	-	-	-	-	-	-	-	-	-	-	-	-	-	-	-	-	-	-	-	-	-	899	-	-	-	-	-
Iridium	9	-	-	-	-	-	-	-	-	-	-	685	-	-	-	-	-	-	-	-	-	-	-	-	-	-	-	-	-	686	-	-	-	-	-
Iron	9	-	-	-	-	-	-	-	-	-	-	687	-	-	-	-	-	-	-	-	-	-	-	-	-	-	-	-	-	688	-	-	-	-	-
Lanthanum Antimonide	9	-	-	-	-	-	-	-	-	-	-	689	-	-	-	-	-	-	-	-	-	-	-	-	-	-	-	-	-	-	-	-	-	-	-
Lead Molybdenum Tetraoxide	9	-	-	-	-	-	-	-	-	-	-	913	-	-	-	-	-	-	-	-	-	-	-	-	-	-	-	-	-	-	-	-	-	-	-
Lithium Fluoride	9	-	-	-	-	-	-	-	-	-	-	1090	-	-	-	-	-	-	-	-	-	-	-	-	-	-	-	-	-	-	-	-	-	-	-
Magnesium	9	-	-	-	-	-	-	-	-	-	-	692	-	-	-	-	-	-	-	-	-	-	-	-	-	-	-	-	-	-	-	-	-	-	-
Magnesium + Aluminum	9	-	-	-	-	-	-	-	-	-	-	693	-	-	-	-	-	-	-	-	-	-	-	-	-	-	-	-	-	-	-	-	-	-	-
Magnesium Fluoride	9	-	-	-	-	-	-	-	-	-	-	1096	-	-	-	-	-	-	-	-	-	-	-	-	-	-	-	-	-	1105	-	-	-	-	-
Manganese	9	-	-	-	-	-	-	-	-	-	-	694	-	-	-	-	-	-	-	-	-	-	-	-	-	-	-	-	-	695	-	-	-	-	-
Palladium	9	-	718	-	-	-	-	-	-	-	-	720	-	-	-	-	-	-	-	-	-	-	-	-	-	-	-	-	-	721	-	-	-	-	-
Platinum	9	-	722	-	-	-	-	-	-	-	-	-	-	-	-	-	-	-	-	-	-	-	-	-	-	-	-	-	-	-	-	-	-	-	-
Polystyrene	9	-	-	-	-	-	-	-	-	-	-	1136 1137	-	-	-	-	-	-	-	-	-	-	-	-	-	-	-	-	-	-	-	-	-	-	-
Silver	9	-	-	-	-	-	-	-	-	-	-	738	743	-	-	-	-	-	-	-	-	-	-	-	-	-	-	-	-	-	752	-	-	-	-
Silver + Aluminum	9	-	-	-	-	-	-	-	-	-	-	754	-	-	-	-	-	-	-	-	-	-	-	-	-	-	-	-	-	-	-	-	-	-	-
Tellurium	9	-	-	-	-	-	-	-	-	-	-	-	971	-	-	-	-	-	-	-	-	-	-	-	-	-	-	-	-	972	-	-	-	-	-
Tin Oxide	9	-	-	-	-	-	-	-	-	-	-	-	-	-	-	-	-	-	-	-	-	-	-	-	-	-	-	-	-	979	-	-	-	-	-
Titanium	9	-	-	-	-	-	-	-	-	-	-	-	-	-	-	-	-	-	-	-	-	761	-	-	-	-	-	-	-	-	-	-	-	-	-
Uranium	9	-	-	-	-	-	-	-	-	-	-	779	-	-	-	-	-	-	-	-	-	-	-	-	-	-	-	-	-	780	-	-	-	-	-
Zinc Sulfide	9	-	-	-	-	-	-	-	-	-	-	1007	1008	-	-	-	-	-	-	-	-	-	-	-	-	-	-	-	-	-	-	-	-	-	-
Glasurit White Epoxy Paint	9	-	-	-	-	-	-	-	-	-	-	-	-	-	-	-	-	-	-	-	-	-	-	508	-	-	-	-	-	-	-	-	-	-	-
Glauberite	8	-	-	-	-	-	-	-	-	-	-	-	-	-	-	-	-	-	-	-	-	-	-	-	-	-	-	-	-	1694	-	-	-	-	-

MATERIAL NAME	VOLUME	EMITTANCE						REFLECTANCE									ABSORPTANCE									TRANSMITTANCE									Absorptance to Emittance Ratio
		Total			Spectral			Integrated			Spectral			Solar			Integrated			Spectral			Solar			Integrated			Spectral			Solar			
		Hemispherical	Normal	Angular	Hemispherical	Normal	Angular	Hemispherical	Normal	Angular	Hemispherical	Normal	Angular	Hemispherical	Normal	Angular	Hemispherical	Normal	Angular	Hemispherical	Normal	Angular	Hemispherical	Normal	Angular	Hemispherical	Normal	Angular	Hemispherical	Normal	Angular	Hemispherical	Normal	Angular	
Glidden Paints:																																			
Black Lacquer No. 131-B-190B	9	-	-	-	-	-	-	-	-	-	-	510	-	-	-	-	-	-	-	-	-	-	-	-	-	-	-	-	-	-	-	-	-	-	-
Black Lacquer No. 9099	9	-	-	-	-	-	-	-	-	-	-	510	-	-	-	-	-	-	-	-	-	-	-	-	-	-	-	-	-	-	-	-	-	-	-
Flat Black Lacquer No. 131-B-216	9	-	-	-	-	-	-	-	-	-	-	510	-	-	-	-	-	-	-	-	-	-	-	-	-	-	-	-	-	-	-	-	-	-	-
Flat White No. 2995	9	-	-	-	-	-	-	-	-	-	-	510	-	-	-	-	-	-	-	-	-	-	-	-	-	-	-	-	-	-	-	-	-	-	-
Flat White No. 5064	9	-	-	-	-	-	-	-	-	-	-	510	-	-	-	-	-	-	-	-	-	-	-	-	-	-	-	-	-	-	-	-	-	-	-
Japalac Black Enamel No. 1207	9	-	-	-	-	-	-	-	-	-	-	510	-	-	-	-	-	-	-	-	-	-	-	-	-	-	-	-	-	-	-	-	-	-	-
Japalac Flat Black No. 1208	9	-	-	-	-	-	-	-	-	-	-	510	-	-	-	-	-	-	-	-	-	-	-	-	-	-	-	-	-	-	-	-	-	-	-
Missile Black No. RGL-22818	9	-	-	-	-	-	-	-	-	-	-	510	-	-	-	-	-	-	-	-	-	-	-	-	-	-	-	-	-	-	-	-	-	-	-
Globar	8	-	-	-	-	798	-	-	-	-	-	-	-	-	-	-	-	-	-	-	808	-	-	810	-	-	-	-	-	-	-	-	-	-	-
Glyptal Paints:																																			
Black	9	513	-	-	-	-	-	-	-	-	-	515	-	-	-	-	-	-	-	-	-	-	-	-	-	-	-	-	-	-	-	-	-	-	-
Brown	9	-	-	-	-	-	-	-	-	-	-	515	-	-	-	-	-	-	-	-	-	-	-	-	-	-	-	-	-	-	-	-	-	-	-
Red	9	-	-	-	-	-	-	-	-	-	-	510	-	-	-	-	-	-	-	-	-	-	-	-	-	-	-	-	-	-	-	-	-	-	-
Glyptal (Clear) Coating on Aluminum Substrate	9	-	-	-	-	-	-	-	-	-	-	1111	-	-	-	-	-	-	-	-	-	-	-	-	-	-	-	-	-	-	-	-	-	-	-
Goddard 78-2B White	9	-	-	-	-	-	-	-	-	-	-	-	-	-	-	-	-	-	-	-	-	-	-	273	-	-	-	-	-	-	-	-	-	-	-
Goethite	8	-	-	-	-	-	-	-	-	-	-	1675	-	-	-	-	-	-	-	-	-	-	-	-	-	-	-	-	-	1661	-	-	-	-	-
Gold	7	244	248	-	-	250 254	-	-	-	-	-	258	264	-	267	-	269	-	-	-	271 273	275	-	277	-	-	-	-	-	-	-	-	-	-	-
Gold Coating on:																																			
Aluminum Substrate	9	652	-	-	-	-	-	-	-	-	-	666	-	-	-	-	-	-	-	-	-	-	-	-	-	-	-	-	-	-	-	-	-	-	-
Ceric Oxide Substrate	9	-	656	-	-	-	-	-	-	-	-	-	-	-	-	-	-	-	-	-	-	-	-	-	-	-	-	-	-	-	-	-	-	-	-
Copper Substrate	9	651	-	-	-	-	-	-	-	-	-	-	-	-	-	-	675	-	-	-	-	-	-	-	-	-	-	-	-	-	-	-	-	-	-
Dow 17 Substrate	9	-	-	-	-	660	-	-	-	-	-	-	-	-	-	-	-	-	-	-	-	-	-	-	-	-	-	-	-	-	-	-	-	-	-
Epoxy Substrate	9	-	-	-	-	-	-	-	-	-	-	672	-	-	-	-	-	-	-	-	-	-	-	-	-	-	-	-	-	-	-	-	-	-	-
Fiberglass Substrate	9	651	-	-	-	660	-	-	-	-	-	-	-	-	-	-	-	-	-	-	-	-	-	-	-	-	-	-	-	-	-	-	-	-	-
Glass Substrate	9	-	656	-	-	-	-	-	-	-	-	664	-	-	-	-	-	-	-	-	-	-	-	-	-	-	-	-	-	-	-	-	-	-	-
Inconel X Substrate	9	-	656	-	-	-	-	-	-	-	-	-	-	-	-	-	-	-	-	-	-	-	-	-	-	-	-	-	-	-	-	-	-	-	-
Magnesium Substrate	9	-	-	-	-	660	-	-	-	-	-	-	-	-	-	-	-	-	-	-	-	-	-	-	-	-	-	-	-	-	-	-	-	-	-
Molybdenum Substrate	9	651	-	-	-	-	-	-	-	-	-	-	-	-	-	-	-	-	-	-	-	-	-	-	-	-	-	-	-	-	-	-	-	-	-
Mylar Substrate	9	651	-	-	-	-	-	-	-	-	-	-	-	-	-	-	675	-	-	-	-	-	-	-	-	-	-	-	-	-	-	-	-	-	-
NBS Ceramic A418 Substrate	9	-	656	-	-	-	-	-	-	-	-	-	-	-	-	-	-	-	-	-	-	-	-	-	-	-	-	-	-	-	-	-	-	-	-
Nickel Substrate	9	652	-	-	-	-	-	-	-	-	-	-	672	-	-	-	-	-	-	-	-	-	-	-	-	-	-	-	-	-	-	-	-	-	-
Nickel Oxide Substrate	9	-	656	-	-	-	-	-	-	-	-	-	-	-	-	-	-	-	-	-	-	-	-	-	-	-	-	-	-	-	-	-	-	-	-
Polyester Substrate	9	651	-	-	-	-	-	-	-	-	-	-	-	-	-	-	-	-	-	-	-	-	-	-	-	-	-	-	-	-	-	-	-	-	-

MATERIAL NAME	VOLUME	EMITTANCE						REFLECTANCE									ABSORPTANCE									TRANSMITTANCE									Absorptance to Emittance Ratio
		Total			Spectral			Integrated			Spectral			Solar			Integrated			Spectral			Solar			Integrated			Spectral			Solar			
		Hemispherical	Normal	Angular	Hemispherical	Normal	Angular	Hemispherical	Normal	Angular	Hemispherical	Normal	Angular	Hemispherical	Normal	Angular	Hemispherical	Normal	Angular	Hemispherical	Normal	Angular	Hemispherical	Normal	Angular	Hemispherical	Normal	Angular	Hemispherical	Normal	Angular	Hemispherical	Normal	Angular	
Gold Coating on: (continued)																																			
Polyurethane Substrate	9	-	-	-	-	-	-	-	-	-	-	-	672	-	-	-	-	-	-	-	-	-	-	-	-	-	-	-	-	-	-	-	-	-	-
Quartz Substrate	9	-	-	-	-	-	-	-	-	-	-	664	-	-	-	-	-	-	-	-	678	-	-	-	-	-	-	-	-	681	-	-	-	-	-
Silicone Substrate	9	-	-	-	-	-	-	-	-	-	-	-	672	-	-	-	-	-	-	-	-	-	-	-	-	-	-	-	-	-	-	-	-	-	-
Stainless Steel Substrate	9	651	-	-	-	-	-	-	-	-	-	-	673	-	-	-	675	-	-	-	-	-	-	-	-	-	-	-	-	-	-	-	-	-	-
Titanium Substrate	9	-	-	-	-	660	-	-	-	-	-	665	-	-	-	-	-	-	-	-	-	-	-	-	-	-	-	-	-	-	-	-	-	-	-
Gold Pigment in Resin Binder	9	25	-	-	-	-	-	-	-	-	-	-	-	-	-	-	-	-	-	-	-	-	-	-	-	-	-	-	-	-	-	-	-	-	-
Gold Substrate Coated with:																																			
Aluminum Oxide	9	-	-	-	-	-	-	-	-	-	-	-	-	-	-	-	-	-	-	-	-	-	-	803	-	-	-	-	-	-	-	-	-	-	-
Iron Oxide	9	-	-	-	-	-	-	-	-	-	-	903	-	-	-	-	-	-	-	-	-	-	-	-	-	-	-	-	-	-	-	-	-	-	-
Polyester	9	1135	-	-	-	-	-	-	-	-	-	-	-	-	-	-	-	-	-	-	-	-	-	-	-	-	-	-	-	-	-	-	-	-	-
Potassium Tantalate	9	-	-	-	-	-	-	-	-	-	-	1045	-	-	-	-	-	-	-	-	-	-	-	-	-	-	-	-	-	-	-	-	-	-	-
Silicon Dioxide	9	-	-	-	-	-	-	-	-	-	-	-	1060	-	-	-	-	-	-	-	-	-	-	1068	-	-	-	-	-	-	-	-	-	-	-
Gold Black Coating on:																																			
Cellulose Nitrate Substrate	9	-	-	-	-	-	-	-	-	-	-	1175	-	-	-	-	-	-	-	-	1178	-	-	-	-	-	-	-	-	1181	-	-	-	-	-
Brass Substrate	9	-	-	-	-	-	-	-	-	-	-	1175	-	-	-	-	-	-	-	-	-	-	-	-	-	-	-	-	-	-	-	-	-	-	-
Glass Substrate	9	-	-	-	-	-	-	-	-	-	-	1175	-	-	-	-	-	-	-	-	-	-	-	-	-	-	-	-	-	1182	-	-	-	-	-
Gold Black + Copper Black Coating on Sodium Chloride Substrate	9	-	-	-	-	-	-	-	-	-	-	-	-	-	-	-	-	-	-	-	-	-	-	-	-	-	-	-	-	1185	-	-	-	-	-
Gold Black + Nickel Black Coating on Sodium Chloride Substrate	9	-	-	-	-	-	-	-	-	-	-	-	-	-	-	-	-	-	-	-	-	-	-	-	-	-	-	-	-	1187	-	-	-	-	-
Gold + Silver	7	-	-	-	-	-	-	-	-	-	-	932	-	-	-	-	-	-	-	-	-	-	-	-	-	-	-	-	-	-	-	-	-	-	-
Gold + Silver Coating on Copper Substrate	9	651	-	-	-	-	-	-	-	-	-	-	-	-	-	-	-	-	-	-	-	-	-	-	-	-	-	-	-	-	-	-	-	-	-
Gold + Silver Coating on Stainless Steel Substrate	9	651	-	-	-	-	-	-	-	-	-	-	-	-	-	-	-	-	-	-	-	-	-	-	-	-	-	-	-	-	-	-	-	-	-
Graphite(s):	8	30	31	38	40 42	44 51	57	59 61	63 65	70 73	-	-	-	-	-	-	75	-	-	-	-	-	-	76	-	-	-	-	-	-	-	-	-	-	-
Acheson Graphite	8	-	-	-	-	45	-	-	-	-	-	-	-	-	-	-	-	-	-	-	-	-	-	-	-	-	-	-	-	-	-	-	-	-	-
Pyrolytic Graphite	8	30	33	-	-	53	-	-	66	70	-	-	-	-	-	-	-	-	-	-	-	-	-	-	-	-	-	-	-	-	-	-	-	-	-
Grade 580 Graphite	8	-	33	-	-	46	-	-	-	-	-	-	-	-	-	-	-	-	-	-	-	-	-	-	-	-	-	-	-	-	-	-	-	-	-
Grade 3474D Graphite	8	-	33	-	-	46	-	-	66	-	-	-	-	-	-	-	75	-	-	-	-	-	-	-	-	-	-	-	-	-	-	-	-	-	-
Grade 7087 Graphite	8	-	34	-	-	46	-	-	66	-	-	-	-	-	-	-	-	-	-	-	-	-	-	-	-	-	-	-	-	-	-	-	-	-	-
Grade 7100 Graphite	8	-	34	-	-	47	-	-	-	-	-	-	-	-	-	-	-	-	-	-	-	-	-	-	-	-	-	-	-	-	-	-	-	-	-
Grade 8665 Graphite	8	-	-	-	-	53	-	-	-	-	-	-	-	-	-	-	75	-	-	-	-	-	-	-	-	-	-	-	-	-	-	-	-	-	-

MATERIAL NAME	VOLUME	EMITTANCE Total Hemispherical	EMITTANCE Total Normal	EMITTANCE Total Angular	EMITTANCE Spectral Hemispherical	EMITTANCE Spectral Normal	EMITTANCE Spectral Angular	REFLECTANCE Integrated Hemispherical	REFLECTANCE Integrated Normal	REFLECTANCE Integrated Angular	REFLECTANCE Spectral Hemispherical	REFLECTANCE Spectral Normal	REFLECTANCE Spectral Angular	REFLECTANCE Solar Hemispherical	REFLECTANCE Solar Normal	REFLECTANCE Solar Angular	ABSORPTANCE Integrated Hemispherical	ABSORPTANCE Integrated Normal	ABSORPTANCE Integrated Angular	ABSORPTANCE Spectral Hemispherical	ABSORPTANCE Spectral Normal	ABSORPTANCE Spectral Angular	ABSORPTANCE Solar Hemispherical	ABSORPTANCE Solar Normal	ABSORPTANCE Solar Angular	TRANSMITTANCE Integrated Hemispherical	TRANSMITTANCE Integrated Normal	TRANSMITTANCE Integrated Angular	TRANSMITTANCE Spectral Hemispherical	TRANSMITTANCE Spectral Normal	TRANSMITTANCE Spectral Angular	TRANSMITTANCE Solar Hemispherical	TRANSMITTANCE Solar Normal	TRANSMITTANCE Solar Angular	Absorptance to Emittance Ratio
Graphite(s): (continued)																																			
Grade AGKSP Graphite	8	–	–	–	41 43	–	58	–	–	–	60 62	–	70 74	–	–	–	–	–	–	–	–	–	–	–	–	–	–	–	–	–	–	–	–	–	–
Grade AGX Graphite	8	–	32	–	–	45	–	–	–	–	–	–	–	–	–	–	–	–	–	–	–	–	–	–	–	–	–	–	–	–	–	–	–	–	–
Grade ATJ Graphite	8	–	32	39	–	45	–	–	–	–	–	–	70	–	–	–	75	–	–	–	–	–	–	–	–	–	–	–	–	–	–	–	–	–	–
Grade AUC Graphite	8	–	32	–	–	45	–	–	–	–	–	–	–	–	–	–	–	–	–	–	–	–	–	–	–	–	–	–	–	–	–	–	–	–	–
Grade CEP Graphite	8	–	–	–	–	–	–	–	–	–	–	–	–	–	–	–	75	–	–	–	–	–	–	–	–	–	–	–	–	–	–	–	–	–	–
Grade GBE Graphite	8	–	32	–	–	45	–	–	–	–	–	66	–	–	–	–	–	–	–	–	–	–	–	76	–	–	–	–	–	–	–	–	–	–	–
Grade GBH Graphite	8	–	33	–	–	45	–	–	–	–	–	66	–	–	–	–	–	–	–	–	–	–	–	76	–	–	–	–	–	–	–	–	–	–	–
Grade H1LM Graphite	8	–	33	–	–	46	–	–	–	–	–	–	–	–	–	–	–	–	–	–	–	–	–	–	–	–	–	–	–	–	–	–	–	–	–
Grade H3LM Graphite	8	–	33	–	–	46	–	–	–	–	–	–	–	–	–	–	–	–	–	–	–	–	–	–	–	–	–	–	–	–	–	–	–	–	–
Grade SPK Graphite	8	–	–	–	–	–	58	–	–	–	–	–	70 74	–	–	–	–	–	–	–	–	–	–	–	–	–	–	–	–	–	–	–	–	–	–
Grade UT6 Graphite	8	–	–	–	–	–	–	–	–	–	–	64	–	–	–	–	–	–	–	–	–	–	–	–	–	–	–	–	–	–	–	–	–	–	–
Graphite, Siliconized	9	–	1325 1326	–	–	1328	–	–	–	–	–	–	–	–	–	–	–	–	–	–	–	–	–	–	–	–	–	–	–	–	–	–	–	–	–
Graphite Coating on:																																			
Aluminum Oxide Substrate	9	–	854	–	–	–	–	–	–	–	–	–	–	–	–	–	–	–	–	–	–	–	–	–	–	–	–	–	–	–	–	–	–	–	–
Brass Substrate	9	–	–	–	–	–	–	–	–	–	–	859	–	–	–	–	–	–	–	–	–	–	–	–	–	–	–	–	–	–	–	–	–	–	–
Copper Substrate	9	–	–	–	–	857	–	–	–	–	–	–	–	–	–	–	–	–	–	–	–	–	–	–	–	–	–	–	–	–	–	–	–	–	–
Silicon Dioxide Substrate	9	–	–	–	–	–	–	–	–	–	–	859	–	–	–	–	–	–	–	–	–	–	–	–	–	–	–	–	–	–	–	–	–	–	–
Tantalum Substrate	9	–	854	–	–	855	–	–	–	–	–	–	–	–	–	–	–	–	–	–	–	–	–	–	–	–	–	–	–	–	–	–	–	–	–
Graphite Substrate Coated with Silicon Carbide	9	–	945	–	–	947	–	–	–	–	–	950	–	–	–	–	–	–	–	–	–	–	–	951	–	–	–	–	–	–	–	–	–	–	–
Gypsum	8	–	–	–	–	–	–	–	–	–	–	1698	–	–	–	–	–	–	–	–	–	–	–	–	–	–	–	–	–	–	–	–	–	–	–
"H"-Film (see Polyimide)																																			
Hafnates	8	–	–	–	–	–	–	–	–	–	–	596	–	–	–	–	–	–	–	–	–	–	–	–	–	–	–	–	–	–	–	–	–	–	–
Hafnium	7	280	–	–	–	282 284	–	–	–	–	–	–	–	–	–	–	–	–	–	–	–	–	–	–	–	–	–	–	–	–	–	–	–	–	–
Hafnium Diboride	8	–	730	–	–	732	–	–	–	–	–	–	–	–	–	–	–	–	–	–	–	–	–	–	–	–	–	–	–	–	–	–	–	–	–
Hafnium Carbide	8	–	850	–	–	852	–	–	–	–	–	–	–	–	–	–	–	–	–	–	–	–	–	–	–	–	–	–	–	–	–	–	–	–	–
Hafnium Nitride	8	–	1056	–	–	1058 1060	–	–	–	–	–	–	–	–	–	–	–	–	–	–	–	–	–	–	–	–	–	–	–	–	–	–	–	–	–
Hafnium Oxide	8	273	–	–	–	–	–	–	–	–	–	275	–	–	–	–	–	–	–	–	–	–	–	–	–	–	–	–	–	–	–	–	–	–	–
Hafnium Oxide Coating on:																																			
Quartz Substrate	9	–	–	–	–	–	–	–	–	–	–	–	–	–	–	–	–	–	–	–	–	–	–	–	–	–	–	–	–	897	–	–	–	–	–

MATERIAL NAME	VOLUME	EMITTANCE						REFLECTANCE									ABSORPTANCE									TRANSMITTANCE									Absorptance to Emittance Ratio
		Total			Spectral			Integrated			Spectral			Solar			Integrated			Spectral			Solar			Integrated			Spectral			Solar			
		Hemispherical	Normal	Angular	Hemispherical	Normal	Angular	Hemispherical	Normal	Angular	Hemispherical	Normal	Angular	Hemispherical	Normal	Angular	Hemispherical	Normal	Angular	Hemispherical	Normal	Angular	Hemispherical	Normal	Angular	Hemispherical	Normal	Angular	Hemispherical	Normal	Angular	Hemispherical	Normal	Angular	
Hafnium Oxide Coating on: (continued)																																			
Tungsten Substrate	9	895	-	-	-	896	-	-	-	-	-	-	-	-	-	-	-	-	-	-	-	-	-	-	-	-	-	-	-	-	-	-	-	-	-
Hanovia Liquid Gold Coating on:																																			
Ceramic Tile Substrate	9	-	-	-	-	-	-	-	-	-	-	-	672	-	-	-	-	-	-	-	-	-	-	-	-	-	-	-	-	-	-	-	-	-	-
Glass Substrate	9	-	656	-	-	-	-	-	-	-	-	664	-	-	-	-	-	-	-	-	-	-	-	-	-	-	-	-	-	-	-	-	-	-	-
Inconel X Substrate	9	-	656	-	-	-	-	-	-	-	-	-	-	-	-	-	-	-	-	-	-	-	-	-	-	-	-	-	-	-	-	-	-	-	-
Molybdenum Substrate	9	651	-	-	-	-	-	-	-	-	-	-	-	-	-	-	-	-	-	-	-	-	-	-	-	-	-	-	-	-	-	-	-	-	-
Titanium Substrate	9	-	-	-	-	660	-	-	-	-	-	665	-	-	-	-	-	-	-	-	-	-	-	-	-	-	-	-	-	-	-	-	-	-	-
Hanovia Liquid Palladium Coating on Glass Substrate	9	-	-	-	-	-	-	-	-	-	-	683	-	-	-	-	-	-	-	-	-	-	-	-	-	-	-	-	-	-	-	-	-	-	-
Hanovia Liquid Platinum Coating on Ceramic Tile Substrate	9	-	-	-	-	-	-	-	-	-	-	-	724	-	-	-	-	-	-	-	-	-	-	-	-	-	-	-	-	-	-	-	-	-	-
Hastelloy B	7	-	1448	-	-	1452	-	-	-	-	-	1455	-	-	-	-	-	-	-	-	-	-	-	1458	-	-	-	-	-	-	-	-	-	-	-
Hastelloy C	7	-	1448	-	-	1452	-	-	-	-	-	1455	-	-	-	-	-	-	-	-	-	-	-	1458	-	-	-	-	-	-	-	-	-	-	-
Hastelloy C Coating on Stainless Steel Substrate	9	717	-	-	-	-	-	-	-	-	-	-	-	-	-	-	-	-	-	-	-	-	-	-	-	-	-	-	-	-	-	-	-	-	-
Hastelloy X	7	-	-	-	-	1364	-	-	-	-	-	1378	-	-	-	-	-	-	-	-	-	-	-	-	-	-	-	-	-	-	-	-	-	-	-
Hastelloy X Substrate Coated with Nickel + Chromium + ΣX_i	9	707	-	-	-	-	-	-	-	-	-	-	-	-	-	-	-	-	-	-	-	-	-	-	-	-	-	-	-	-	-	-	-	-	-
Haynes Alloy C	7	1444	-	-	-	-	-	-	-	-	-	-	-	-	-	-	-	-	-	-	-	-	-	-	-	-	-	-	-	-	-	-	-	-	-
Haynes Alloy No. 25	7	1140	1143	-	-	1146 1149	-	-	-	-	-	1154	-	-	-	-	-	-	-	-	-	-	-	-	-	-	-	-	-	-	-	-	-	-	-
Haynes Alloy No. 25 Substrate Coated with Iron Oxide	9	-	900	-	-	901	-	-	-	-	-	-	-	-	-	-	-	-	-	-	-	-	-	-	-	-	-	-	-	-	-	-	-	-	-
Haynes Alloy N-155	7	-	1227	-	-	1238	-	-	-	-	-	-	-	-	-	-	-	-	-	-	-	-	-	-	-	-	-	-	-	-	-	-	-	-	-
Haynes Alloy X	7	1344	-	-	-	-	-	-	-	-	-	-	-	-	-	-	-	-	-	-	-	-	-	-	-	-	-	-	-	-	-	-	-	-	-
Haynes LT-1	8	-	1356	-	-	-	-	-	-	-	-	-	-	-	-	-	-	-	-	-	-	-	-	-	-	-	-	-	-	-	-	-	-	-	-
Haynes LT-1B	8	-	1356	-	-	-	-	-	-	-	-	-	-	-	-	-	-	-	-	-	-	-	-	-	-	-	-	-	-	-	-	-	-	-	-
Haynes LT-1, LT-1B and LT-2 Cermet Coatings on Inconel Substrate	9	-	629 774	-	-	-	-	-	-	-	-	-	-	-	-	-	-	-	-	-	-	-	-	-	-	-	-	-	-	-	-	-	-	-	-
Haynes LT-2	8	-	1375	-	-	-	-	-	-	-	-	-	-	-	-	-	-	-	-	-	-	-	-	-	-	-	-	-	-	-	-	-	-	-	-
$H_3BO_3 + TiB_2$	8	-	1468	-	-	1469	-	-	-	-	-	1471	-	-	-	-	-	-	-	-	-	-	-	-	-	-	-	-	-	-	-	-	-	-	-

MATERIAL NAME	VOLUME	EMITTANCE						REFLECTANCE									ABSORPTANCE									TRANSMITTANCE									Absorptance to Emittance Ratio
		Total			Spectral			Integrated			Spectral			Solar			Integrated			Spectral			Solar			Integrated			Spectral			Solar			
		Hemispherical	Normal	Angular	Hemispherical	Normal	Angular	Hemispherical	Normal	Angular	Hemispherical	Normal	Angular	Hemispherical	Normal	Angular	Hemispherical	Normal	Angular	Hemispherical	Normal	Angular	Hemispherical	Normal	Angular	Hemispherical	Normal	Angular	Hemispherical	Normal	Angular	Hemispherical	Normal	Angular	
$H_3BO_3 + TiB_2 + TiO_2$ Powders	8	-	1515	-	-	1516	-	-	-	-	-	1518	-	-	-	-	-	-	-	-	-	-	-	-	-	-	-	-	-	-	-	-	-	-	-
Hematite	8	-	-	-	-	-	-	-	-	-	-	1677	-	-	-	-	-	-	-	-	-	-	-	-	-	-	-	-	-	-	-	-	-	-	-
Heulandite	8	-	-	-	-	-	-	-	-	-	-	-	-	-	-	-	-	-	-	-	-	-	-	-	-	-	-	-	-	1694	-	-	-	-	-
HK-31	7	-	1336	-	-	-	-	-	-	-	-	1338	-	-	-	-	-	-	-	-	-	-	-	1340	-	-	-	-	-	-	-	-	-	-	-
HK-31A, Anodized	9	-	-	-	-	-	-	-	-	-	-	1281	-	-	-	-	-	-	-	-	-	-	-	-	-	-	-	-	-	-	-	-	-	-	-
Holmium Hexaboride	8	-	-	-	-	-	-	-	-	-	-	-	-	-	-	-	-	-	-	-	-	-	-	-	-	-	-	-	-	727	-	-	-	-	-
Holmium Nitride	8	-	-	-	-	-	-	-	-	-	-	-	-	-	-	-	-	-	-	-	-	-	-	-	-	-	-	-	-	1088	-	-	-	-	-
Holmium Oxide	8	-	-	-	-	-	-	-	-	-	-	-	-	-	-	-	-	-	-	-	-	-	-	-	-	-	-	-	-	277	-	-	-	-	-
Hughes H-2	9	-	214	-	-	-	-	-	-	-	-	-	-	-	-	-	-	-	-	-	-	-	-	275 281	-	-	-	-	-	-	-	-	-	-	283
Hughes H-10	9	-	95	-	-	-	-	-	-	-	-	97	-	-	-	-	-	-	-	-	-	-	-	98 100	-	-	-	-	-	-	-	-	-	-	102
Hurwich Home Foil	7	-	4 5	-	-	-	-	-	-	-	-	-	-	-	-	-	-	42 43	-	-	-	-	-	-	-	-	-	-	-	-	-	-	-	-	-
Hydrargillite	8	-	-	-	-	-	-	-	-	-	-	-	-	-	-	-	-	-	-	-	-	-	-	-	-	-	-	-	-	1664	-	-	-	-	-
Illinois Bronze White Lacquer	9	-	-	-	-	-	-	-	-	-	-	518	-	-	-	-	-	-	-	-	-	-	-	-	-	-	-	-	-	-	-	-	-	-	-
Inconel	7	1344 1345	1351 1352	-	-	1363 1366	-	-	-	-	-	1382	-	-	-	-	-	-	-	-	-	1396	-	-	-	-	-	-	-	-	-	-	-	-	-
Inconel, Oxidized	9	-	1314	-	-	-	-	-	-	-	-	-	-	-	-	-	-	-	-	-	-	-	-	-	-	-	-	-	-	-	-	-	-	-	-
Inconel Substrate Coated with:																																			
Aluminum Oxide	9	-	788	-	-	792	-	-	-	-	-	-	-	-	-	-	-	-	-	-	-	-	-	-	-	-	-	-	-	-	-	-	-	-	-
Chromium + Aluminum Oxide + ΣX_i	9	-	629	-	-	-	-	-	-	-	-	-	-	-	-	-	-	-	-	-	-	-	-	-	-	-	-	-	-	-	-	-	-	-	-
$Cr_2O_3 + SiO_2 + \Sigma X_i$	9	-	-	-	-	885	-	-	-	-	-	-	-	-	-	-	-	-	-	-	-	-	-	-	-	-	-	-	-	-	-	-	-	-	-
Haynes LT-1, LT-1B and LT-2 Cermets	9	-	629 774	-	-	-	-	-	-	-	-	-	-	-	-	-	-	-	-	-	-	-	-	-	-	-	-	-	-	-	-	-	-	-	-
Nickel Aluminide	9	-	706	-	-	-	-	-	-	-	-	-	-	-	-	-	-	-	-	-	-	-	-	-	-	-	-	-	-	-	-	-	-	-	-
Silicon Monoxide	9	-	-	-	-	1075	-	-	-	-	-	-	-	-	-	-	-	-	-	-	-	-	-	-	-	-	-	-	-	-	-	-	-	-	-
Tungsten + Chromium + Aluminum Oxide	9	-	774	-	-	-	-	-	-	-	-	-	-	-	-	-	-	-	-	-	-	-	-	-	-	-	-	-	-	-	-	-	-	-	-
Zirconium Oxide	9	-	1011	-	-	-	-	-	-	-	-	1017	-	-	-	-	-	-	-	-	-	-	-	1021	-	-	-	-	-	-	-	-	-	-	-
Inconel 600 (see Inconel)																																			
Inconel 702	7	1345	-	-	-	1364	-	-	-	-	-	1378	-	-	-	-	-	-	-	-	-	-	-	-	-	-	-	-	-	-	-	-	-	-	-
Inconel B	7	-	1350	-	-	-	-	-	-	-	-	-	-	1393	-	-	-	-	-	-	-	-	-	1401	-	-	-	-	-	-	-	-	-	-	-

MATERIAL NAME	VOLUME	EMITTANCE						REFLECTANCE									ABSORPTANCE									TRANSMITTANCE									Absorptance to Emittance Ratio
		Total			Spectral			Integrated			Spectral			Solar			Integrated			Spectral			Solar			Integrated			Spectral			Solar			
		Hemispherical	Normal	Angular	Hemispherical	Normal	Angular	Hemispherical	Normal	Angular	Hemispherical	Normal	Angular	Hemispherical	Normal	Angular	Hemispherical	Normal	Angular	Hemispherical	Normal	Angular	Hemispherical	Normal	Angular	Hemispherical	Normal	Angular	Hemispherical	Normal	Angular	Hemispherical	Normal	Angular	
Inconel X	7	1344 1345	1351 1352 1353	-	-	1358 1365	-	-	-	-	-	1377 1379	-	-	1393	-	-	-	-	-	-	1398	-	1401 1402	-	-	-	-	-	-	-	-	-	-	-
Inconel X Substrate Coated with:																																			
Boron Carbide	9	-	-	-	-	829	-	-	-	-	-	832	-	-	-	-	-	-	-	-	-	-	-	-	-	-	-	-	-	-	-	-	-	-	-
Cobalt + Tungsten	9	-	656	-	-	636	-	-	-	-	-	639	-	-	-	-	-	-	-	-	-	-	-	-	-	-	-	-	-	-	-	-	-	-	-
Gold	9	-	656	-	-	-	-	-	-	-	-	-	-	-	-	-	-	-	-	-	-	-	-	-	-	-	-	-	-	-	-	-	-	-	-
Molybdenum Disulfide	9	-	-	-	-	-	-	-	-	-	-	930	-	-	-	-	-	-	-	-	-	-	-	-	-	-	-	-	-	-	-	-	-	-	-
Nickel + Chromium	9	-	-	-	-	709	-	-	-	-	-	712	-	-	-	-	-	-	-	-	-	-	-	-	-	-	-	-	-	-	-	-	-	-	-
Palladium	9	-	718	-	-	-	-	-	-	-	-	-	-	-	-	-	-	-	-	-	-	-	-	-	-	-	-	-	-	-	-	-	-	-	-
Rhodium	9	-	730	-	-	-	-	-	-	-	-	-	-	-	-	-	-	-	-	-	-	-	-	-	-	-	-	-	-	-	-	-	-	-	-
Tantalum Carbide	9	-	-	-	-	965	-	-	-	-	-	968	-	-	-	-	-	-	-	-	-	-	-	-	-	-	-	-	-	-	-	-	-	-	-
Tungsten	9	-	-	-	-	766	-	-	-	-	-	769	-	-	-	-	-	-	-	-	-	-	-	-	-	-	-	-	-	-	-	-	-	-	-
Tungsten + Cobalt	9	-	-	-	-	-	-	-	-	-	-	776	-	-	-	-	-	-	-	-	-	-	-	-	-	-	-	-	-	-	-	-	-	-	-
Zirconium Oxide	9	-	-	-	-	1014	-	-	-	-	-	1017	-	-	-	-	-	-	-	-	-	-	-	-	-	-	-	-	-	-	-	-	-	-	-
Inconel X-750 (see Inconel X)																																			
Indium	7	-	-	-	-	-	-	-	-	-	-	-	-	-	-	-	-	-	-	-	-	286	-	-	-	-	-	-	-	-	-	-	-	-	-
Indium Antimonide	8	-	-	-	-	-	-	-	-	-	-	1298	-	-	-	-	-	-	-	-	-	-	-	-	-	-	-	-	-	1305	-	-	-	-	-
Indium Arsenide	8	-	-	-	-	685	-	-	-	-	-	687	-	-	-	-	-	-	-	-	-	-	-	-	-	-	-	-	-	698	-	-	-	-	-
Indium Arsenide Coating on Glass Substrate	9	-	-	-	-	-	-	-	-	-	-	-	-	-	-	-	-	-	-	-	-	-	-	-	-	-	-	-	-	899	-	-	-	-	-
Diindium Trioxide	8	-	-	-	-	-	-	-	-	-	-	546	-	-	-	-	-	-	-	-	-	-	-	-	-	-	-	-	-	-	-	-	-	-	-
Indium Phosphide	8	-	-	-	-	-	-	-	-	-	-	1100	-	-	-	-	-	-	-	-	-	-	-	-	-	-	-	-	-	1103	-	-	-	-	-
$In_{0.56}Ga_{0.44}Sb$	8	-	-	-	-	-	-	-	-	-	-	1297	-	-	-	-	-	-	-	-	-	-	-	-	-	-	-	-	-	-	-	-	-	-	-
$In_{0.225}Ga_{0.775}Sb$	8	-	-	-	-	-	-	-	-	-	-	1297	-	-	-	-	-	-	-	-	-	-	-	-	-	-	-	-	-	-	-	-	-	-	-
Inor-8	7	1444	-	-	-	-	-	-	-	-	-	-	-	-	-	-	-	-	-	-	-	-	-	-	-	-	-	-	-	-	-	-	-	-	-
Iodide Hafnium	7	280	-	-	-	-	-	-	-	-	-	-	-	-	-	-	-	-	-	-	-	-	-	-	-	-	-	-	-	-	-	-	-	-	-
Iodides:																																			
Cesium Iodide	8	-	-	-	-	-	-	-	-	-	-	995	-	-	-	-	-	-	-	-	-	-	-	-	-	-	-	-	-	997	-	-	-	-	-
Copper Iodide	8	-	-	-	-	-	-	-	-	-	-	999	-	-	-	-	-	-	-	-	-	-	-	-	-	-	-	-	-	1001	-	-	-	-	-
Lead Diiodide	8	-	-	-	-	-	-	-	-	-	-	-	-	-	-	-	-	-	-	-	-	-	-	-	-	-	-	-	-	1003	-	-	-	-	-
Potassium Iodide	8	-	-	-	-	-	-	-	-	-	-	1005	-	-	-	-	-	-	-	-	-	-	-	-	-	-	-	-	-	1010	-	-	-	-	-
Rubidium Iodide	8	-	-	-	-	-	-	-	-	-	-	1014	-	-	-	-	-	-	-	-	-	-	-	-	-	-	-	-	-	1018	-	-	-	-	-
Silver Iodide	8	-	-	-	-	-	-	-	-	-	-	1022	-	-	-	-	-	-	-	-	-	-	-	-	-	-	-	-	-	1024	-	-	-	-	-
Miscellaneous Iodides	8	-	-	-	-	-	-	-	-	-	-	1026	-	-	-	-	-	-	-	-	-	-	-	-	-	-	-	-	-	1028	-	-	-	-	-

MATERIAL NAME	VOLUME	EMITTANCE Total Hemispherical	EMITTANCE Total Normal	EMITTANCE Total Angular	EMITTANCE Spectral Hemispherical	EMITTANCE Spectral Normal	EMITTANCE Spectral Angular	REFLECTANCE Integrated Hemispherical	REFLECTANCE Integrated Normal	REFLECTANCE Integrated Angular	REFLECTANCE Spectral Hemispherical	REFLECTANCE Spectral Normal	REFLECTANCE Spectral Angular	REFLECTANCE Solar Hemispherical	REFLECTANCE Solar Normal	REFLECTANCE Solar Angular	ABSORPTANCE Integrated Hemispherical	ABSORPTANCE Integrated Normal	ABSORPTANCE Integrated Angular	ABSORPTANCE Spectral Hemispherical	ABSORPTANCE Spectral Normal	ABSORPTANCE Spectral Angular	ABSORPTANCE Solar Hemispherical	ABSORPTANCE Solar Normal	ABSORPTANCE Solar Angular	TRANSMITTANCE Integrated Hemispherical	TRANSMITTANCE Integrated Normal	TRANSMITTANCE Integrated Angular	TRANSMITTANCE Spectral Hemispherical	TRANSMITTANCE Spectral Normal	TRANSMITTANCE Spectral Angular	TRANSMITTANCE Solar Hemispherical	TRANSMITTANCE Solar Normal	TRANSMITTANCE Solar Angular	Absorptance to Emittance Ratio
Iodine	8	-	-	-	-	-	-	-	-	-	-	78	-	-	-	-	-	-	-	-	-	-	-	-	-	-	-	-	-	-	-	-	-	-	-
Iridium	7	-	-	-	-	289 291	-	-	-	-	-	294	297 299	-	-	-	-	-	-	-	-	-	-	-	-	-	-	-	-	-	-	-	-	-	-
Iridium Coating on Glass Substrate	9	-	-	-	-	-	-	-	-	-	-	685	-	-	-	-	-	-	-	-	-	-	-	-	-	-	-	-	-	686	-	-	-	-	-
Iron	7	302	306	-	-	310 316	-	-	-	-	-	319 321	324	-	-	-	-	-	-	-	327 329	-	-	332	-	-	-	-	-	-	-	-	-	-	-
Iron, Oxidized	9	-	-	-	-	1297	-	-	-	-	-	1299	-	-	-	-	-	-	-	-	-	-	-	-	-	-	-	-	-	-	-	-	-	-	-
Iron Coating on Glass Substrate	9	-	-	-	-	-	-	-	-	-	-	687	-	-	-	-	-	-	-	-	-	-	-	-	-	-	-	-	-	688	-	-	-	-	-
Iron Substrate Coated with:																																			
Aluminum	9	580	-	-	-	-	-	-	-	-	-	-	-	-	-	-	-	-	-	-	-	-	-	-	-	-	-	-	-	-	-	-	-	-	-
Chromium Carbide + Cobalt	9	873	-	-	-	-	-	-	-	-	-	-	-	-	-	-	-	-	-	-	-	-	-	874	-	-	-	-	-	-	-	-	-	-	-
Molybdenum	9	696	-	-	-	-	-	-	-	-	-	-	-	-	-	-	-	-	-	-	-	-	-	697	-	-	-	-	-	-	-	-	-	-	-
Titanium Dioxide	9	981	-	-	-	-	-	-	-	-	-	-	-	-	-	-	-	-	-	-	-	-	-	-	-	-	-	-	-	-	-	-	-	-	-
Tungsten	9	764	-	-	-	-	-	-	-	-	-	-	-	-	-	-	-	-	-	-	-	-	-	772	-	-	-	-	-	-	-	-	-	-	-
Tungsten Carbide + Cobalt	9	995	-	-	-	-	-	-	-	-	-	-	-	-	-	-	-	-	-	-	-	-	-	996	-	-	-	-	-	-	-	-	-	-	-
Zinc	9	-	-	-	-	-	-	-	-	-	-	781	-	-	-	-	-	-	-	-	-	-	-	-	-	-	-	-	-	-	-	-	-	-	-
~~Iron + Carbon~~	~~7~~	-	~~935~~	-	-	-	-	-	-	-	-	-	-	-	-	-	-	-	-	-	-	-	-	-	-	-	-	-	-	-	-	-	-	-	-
Iron + Chromium	7	-	-	-	-	-	-	-	-	-	-	938	940	-	-	-	-	-	-	-	-	-	-	-	-	-	-	-	-	-	-	-	-	-	-
Iron + Chromium + ΣX_i	7	-	1178	-	-	1184 1190	-	-	-	-	-	1196	-	-	-	-	-	-	-	-	1203	-	-	1206	-	-	-	-	-	-	-	-	-	-	-
Iron + Chromium + ΣX_i, Oxidized	9	-	-	-	-	1303	-	-	-	-	-	-	-	-	-	-	-	-	-	-	-	-	-	-	-	-	-	-	-	-	-	-	-	-	-
Iron + Chromium + Nickel + ΣX_i	7	1210	1217	1231	-	1235 1242	1253 1256	-	-	-	-	1264	1283 1285	-	1287	-	1291	-	-	-	1293	1296	-	1299	-	-	-	-	-	-	-	-	-	-	-
Iron + Manganese + ΣX_i	7	1305	1307	-	-	-	-	-	-	-	-	-	-	-	-	-	-	-	-	-	-	-	-	1309	-	-	-	-	-	-	-	-	-	-	-
Iron + Nickel	7	-	-	-	-	942	-	-	-	-	-	-	-	-	-	-	-	-	-	-	-	-	-	-	-	-	-	-	-	-	-	-	-	-	-
Iron + Nickel + ΣX_i	7	-	-	-	-	-	-	-	-	-	-	1312	-	-	-	-	-	-	-	-	-	-	-	-	-	-	-	-	-	-	-	-	-	-	-
Iron + Nickel + Chromium + ΣX_i	7	1317	-	-	-	1320	-	-	-	-	-	1324	-	-	-	-	-	-	-	-	-	-	-	-	-	-	-	-	-	-	-	-	-	-	-
Iron + Nickel + Chromium + ΣX_i, Oxidized	9	-	1305	-	-	1308	-	-	-	-	-	-	-	-	-	-	-	-	-	-	-	-	-	-	-	-	-	-	-	-	-	-	-	-	-
Iron + Tungsten	7	-	-	-	-	945	-	-	-	-	-	-	-	-	-	-	-	-	-	-	-	-	-	-	-	-	-	-	-	-	-	-	-	-	-
Iron Oxide Coating on:																																			
Aluminum Substrate	9	-	-	-	-	-	-	-	-	-	-	903	-	-	-	-	-	-	-	-	-	-	-	-	-	-	-	-	-	-	-	-	-	-	-
Gold Substrate	9	-	-	-	-	-	-	-	-	-	-	903	-	-	-	-	-	-	-	-	-	-	-	-	-	-	-	-	-	-	-	-	-	-	-

MATERIAL NAME	VOLUME	EMITTANCE						REFLECTANCE									ABSORPTANCE									TRANSMITTANCE									Absorptance to Emittance Ratio
		Total			Spectral			Integrated			Spectral			Solar			Integrated			Spectral			Solar			Integrated			Spectral			Solar			
		Hemispherical	Normal	Angular	Hemispherical	Normal	Angular	Hemispherical	Normal	Angular	Hemispherical	Normal	Angular	Hemispherical	Normal	Angular	Hemispherical	Normal	Angular	Hemispherical	Normal	Angular	Hemispherical	Normal	Angular	Hemispherical	Normal	Angular	Hemispherical	Normal	Angular	Hemispherical	Normal	Angular	
Iron Oxide Coating on: (continued)																																			
Haynes Alloy 25 Substrate	9	-	900	-	-	901	-	-	-	-	-	-	-	-	-	-	-	-	-	-	-	-	-	-	-	-	-	-	-	-	-	-	-	-
Iron Oxide Pigment in Silicone Binder	9	117	119	-	-	121	-	-	-	-	-	-	-	-	-	-	-	-	-	-	-	-	-	-	-	-	-	-	-	-	-	-	-	-
Diiron Trioxide	8	-	280	-	-	282	-	-	-	-	-	284	-	-	-	-	-	-	-	-	-	-	-	-	-	-	-	-	-	-	-	-	-	-
Triiron Tetraoxide	8	-	-	-	-	282	-	-	-	-	-	-	-	-	-	-	-	-	-	-	-	-	-	-	-	-	-	-	-	-	-	-	-	-
Iron Oxide + Chromium Oxide Pigment in Base Glaze No. 3 Binder (Enamel)	9	-	466	-	-	-	-	-	-	-	-	-	-	-	-	-	-	-	-	-	-	-	-	-	-	-	-	-	-	-	-	-	-	-
Iron Oxide + Cobalt Oxide + Chromium Oxide Pigment in NBS Frit No. 332 Binder (Enamel)	9	-	466	-	-	-	-	-	-	-	-	467	-	-	-	-	-	-	-	-	-	-	-	-	-	-	-	-	-	-	-	-	-	-
Iron Oxide + Nickel Oxide Pigment in Base Glaze No. 3 Binder (Enamel)	9	-	466	-	-	-	-	-	-	-	-	-	-	-	-	-	-	-	-	-	-	-	-	-	-	-	-	-	-	-	-	-	-	-
Iron Oxide + Nickel Oxide Pigment in Base Glaze No. 3 Binder (Enamel)	9	-	466	-	-	-	-	-	-	-	-	-	-	-	-	-	-	-	-	-	-	-	-	-	-	-	-	-	-	-	-	-	-	-
Iron Oxide Substrate Coated with Magnesium Fluoride	9	1093	-	-	-	-	-	-	-	-	-	1096	-	-	-	-	-	-	-	-	-	-	-	1103	-	-	-	-	-	-	-	-	-	-
Iron Titanate Coating on:																																			
Beryllium Substrate	9	905	-	-	-	-	-	-	-	-	-	-	-	-	-	-	-	-	-	-	-	-	-	-	-	-	-	-	-	-	-	-	-	-
Nb-1Zr Substrate	9	905	-	-	-	-	-	-	-	-	-	-	-	-	-	-	-	-	-	-	-	-	-	-	-	-	-	-	-	-	-	-	-	-
Stainless Steel Substrate	9	905	-	-	-	-	-	-	-	-	-	-	-	-	-	-	-	-	-	-	-	-	-	-	-	-	-	-	-	-	-	-	-	-
Iron Titanate Pigment in Aluminum Phosphate Binder	9	123	-	-	-	-	-	-	-	-	-	-	-	-	-	-	-	-	-	-	-	-	-	-	-	-	-	-	-	-	-	-	-	-
Iron Titanate + Alumina Pigment in Aluminum Phosphate Binder	9	123	-	-	-	-	-	-	-	-	-	-	-	-	-	-	-	-	-	-	-	-	-	-	-	-	-	-	-	-	-	-	-	-
Iron Titanium Aluminum Oxide Coating on Nb-1Zr Substrate	9	909	-	-	-	911	-	-	-	-	-	-	-	-	-	-	-	-	-	-	-	-	-	-	-	-	-	-	-	-	-	-	-	-
IRC Solar Cell	8	-	88	-	-	92	-	-	-	-	-	100	-	-	-	-	-	-	-	-	-	-	-	-	-	-	-	-	-	-	-	-	-	-
Irtran I (MgF_2)	8	-	-	-	-	951	-	-	-	-	-	-	953	-	-	-	-	-	-	-	956	-	-	-	-	-	-	-	-	960	-	-	-	-	-
Irtran 2 (ZnS)	8	1214	-	-	-	1216	-	-	-	-	-	-	1223	-	-	-	-	-	-	-	1225	-	-	-	-	-	-	-	-	1228	-	-	-	-	-
Irtran 3 (CaF_2)	8	-	-	-	-	-	-	-	-	-	-	-	-	-	-	-	-	-	-	-	930	-	-	-	-	-	-	-	-	933	-	-	-	-	-
Irtran 4 (ZnSe)	8	1113	-	-	-	1115	-	-	-	-	-	1119	-	-	-	-	-	-	-	-	1122	-	-	-	-	-	-	-	-	1125	-	-	-	-	-
Irtran 5 (MgO)	8	-	-	-	-	296	-	-	-	-	-	-	-	-	-	-	-	-	-	-	-	-	-	-	-	-	-	-	-	325	-	-	-	-	-

MATERIAL NAME	VOLUME	EMITTANCE Total Hemispherical	EMITTANCE Total Normal	EMITTANCE Total Angular	EMITTANCE Spectral Hemispherical	EMITTANCE Spectral Normal	EMITTANCE Spectral Angular	REFLECTANCE Integrated Hemispherical	REFLECTANCE Integrated Normal	REFLECTANCE Integrated Angular	REFLECTANCE Spectral Hemispherical	REFLECTANCE Spectral Normal	REFLECTANCE Spectral Angular	REFLECTANCE Solar Hemispherical	REFLECTANCE Solar Normal	REFLECTANCE Solar Angular	ABSORPTANCE Integrated Hemispherical	ABSORPTANCE Integrated Normal	ABSORPTANCE Integrated Angular	ABSORPTANCE Spectral Hemispherical	ABSORPTANCE Spectral Normal	ABSORPTANCE Spectral Angular	ABSORPTANCE Solar Hemispherical	ABSORPTANCE Solar Normal	ABSORPTANCE Solar Angular	TRANSMITTANCE Integrated Hemispherical	TRANSMITTANCE Integrated Normal	TRANSMITTANCE Integrated Angular	TRANSMITTANCE Spectral Hemispherical	TRANSMITTANCE Spectral Normal	TRANSMITTANCE Spectral Angular	TRANSMITTANCE Solar Hemispherical	TRANSMITTANCE Solar Normal	TRANSMITTANCE Solar Angular	Absorptance to Emittance Ratio
Irtran 6 (CdTe)	8	1240	–	–	–	–	–	–	–	–	–	1242	–	–	–	–	–	–	–	–	–	–	–	–	–	–	–	–	–	1247	–	–	–	–	–
Japalac	9	–	–	–	–	–	–	–	–	–	–	510	–	–	–	–	–	–	–	–	–	–	–	–	–	–	–	–	–	–	–	–	–	–	–
Jersey Standard Optical Black	9	577	–	–	–	–	–	–	–	–	–	–	–	–	–	–	–	–	–	–	–	–	–	–	–	–	–	–	–	–	–	–	–	–	–
Jet Dry Black No. 78	9	–	–	–	–	–	–	–	–	–	–	519	–	–	–	–	–	–	–	–	–	–	–	–	–	–	–	–	–	–	–	–	–	–	–
K Monel	7	–	1428	–	–	–	–	–	–	–	–	–	–	–	1434	–	–	–	–	–	–	–	–	1437	–	–	–	–	–	–	–	–	–	–	–
Kaiser Foil	7	4 5	–	–	–	–	–	–	–	–	–	–	–	–	–	–	42	–	–	–	–	–	–	–	–	–	–	–	–	–	–	–	–	–	–
Kanthal	7	–	–	–	–	1192	–	–	–	–	–	–	–	–	–	–	–	–	–	–	1204	–	–	–	–	–	–	–	–	–	–	–	–	–	–
Kanthal, Oxidized	9	–	–	–	–	1303	–	–	–	–	–	–	–	–	–	–	–	–	–	–	–	–	–	–	–	–	–	–	–	–	–	–	–	–	–
Kanthal A	7	–	–	–	–	1192	–	–	–	–	–	–	–	–	–	–	–	–	–	–	–	–	–	–	–	–	–	–	–	–	–	–	–	–	–
Kaolin	8	1653	–	–	–	–	–	–	–	–	–	618 1655	–	–	–	–	–	–	–	–	–	–	–	–	–	–	–	–	–	–	–	–	–	–	–
Kapton	8	–	–	–	–	–	–	–	–	–	–	–	–	–	–	–	–	–	–	–	1714	–	–	–	–	–	–	–	–	1716	–	–	–	–	–
Kapton Coating on Aluminum Substrate	9	–	–	–	–	–	–	–	–	–	–	1037 1039	–	–	–	–	–	–	–	–	–	–	–	1041	–	–	–	–	–	–	–	–	–	–	–
KDP (see Potassium Dihydrogen Phosphate)																																			
Kemacryl Black Lacquer (M49BC12)	9	81	–	–	–	–	–	–	–	–	–	86	–	–	–	–	–	–	–	–	–	–	–	89	–	–	–	–	–	–	–	–	–	–	–
Kemacryl White Lacquer (M49WC17)	9	290	–	–	–	–	–	–	–	–	–	–	–	–	–	–	–	–	–	–	293	–	–	295	–	–	–	–	–	–	–	–	–	–	–
Kennametal K150A	8	1413	1415	–	–	–	–	–	–	–	–	–	–	–	–	–	–	–	–	–	–	–	–	–	–	–	–	–	–	–	–	–	–	–	–
Kennametal K151A	8	1413	1415	–	–	–	–	–	–	–	–	–	–	–	–	–	–	–	–	–	–	–	–	–	–	–	–	–	–	–	–	–	–	–	–
Kennametal K152B	8	1413	1415	–	–	–	–	–	–	–	–	–	–	–	–	–	–	–	–	–	–	–	–	–	–	–	–	–	–	–	–	–	–	–	–
Kennametal K153B	8	1413	1415	–	–	–	–	–	–	–	–	–	–	–	–	–	–	–	–	–	–	–	–	–	–	–	–	–	–	–	–	–	–	–	–
Kennametal K163B	8	1413	1415	–	–	–	–	–	–	–	–	–	–	–	–	–	–	–	–	–	–	–	–	–	–	–	–	–	–	–	–	–	–	–	–
Kennametal K184B	8	1413	1415	–	–	–	–	–	–	–	–	–	–	–	–	–	–	–	–	–	–	–	–	–	–	–	–	–	–	–	–	–	–	–	–
Kerpo QD Metallic Enamel No. WB-S-N 52-E-4	9	–	–	–	–	–	–	–	–	–	–	520	–	–	–	–	–	–	–	–	–	–	–	–	–	–	–	–	–	–	–	–	–	–	–
Kimbel N-51A Glass	8	–	–	–	–	1542	–	–	–	–	–	1544	–	–	–	–	–	–	–	–	–	–	–	–	–	–	–	–	–	1548	–	–	–	–	–
Kodak Black Brushing Lacquer	9	–	–	–	–	521	–	–	–	–	–	522	–	–	–	–	–	–	–	–	–	–	–	–	–	–	–	–	–	–	–	–	–	–	–

MATERIAL NAME	VOLUME	EMITTANCE						REFLECTANCE									ABSORPTANCE									TRANSMITTANCE									Absorptance to Emittance Ratio
		Total			Spectral			Integrated			Spectral			Solar			Integrated			Spectral			Solar			Integrated			Spectral			Solar			
		Hemispherical	Normal	Angular	Hemispherical	Normal	Angular	Hemispherical	Normal	Angular	Hemispherical	Normal	Angular	Hemispherical	Normal	Angular	Hemispherical	Normal	Angular	Hemispherical	Normal	Angular	Hemispherical	**Normal**	Angular	Hemispherical	Normal	Angular	Hemispherical	Normal	Angular	Hemispherical	Normal	Angular	
Konel	7	1404	-	-	-	1407	-	-	-	-	-	-	-	-	-	-	-	-	-	-	-	-	-	-	-	-	-	-	-	-	-	-	-	-	-
Kovar	7	-	-	-	-	-	-	-	-	-	-	1313	-	-	-	-	-	-	-	-	-	-	-	-	-	-	-	-	-	-	-	-	-	-	-
Kodok 515	8	-	-	-	-	-	-	-	-	-	-	-	-	-	-	-	-	-	-	-	-	-	-	521	-	-	-	-	-	-	-	-	-	-	-
KP	8	-	-	-	-	-	-	-	-	-	-	199	-	-	-	-	-	-	-	-	-	-	-	200	-	-	-	-	-	-	-	-	-	-	-
KRS-5 (see Thallium Bromide-Iodide)	8	-	-	-	-	1459	-	-	-	-	-	-	1461	-	-	-	-	-	-	-	-	-	-	-	-	-	-	-	-	1463	-	-	-	-	-
KRS-5 Substrate Coated with:																																			
Antimony Black	9	-	-	-	-	-	-	-	-	-	-	-	-	-	-	-	-	-	-	-	-	-	-	-	-	-	-	-	-	1172	-	-	-	-	-
Lithium Fluoride	9	-	-	-	-	-	-	-	-	-	-	-	-	-	-	-	-	-	-	-	-	-	-	-	-	-	-	-	-	1092	-	-	-	-	-
KRS-6 (see Thallium Bromide-Chloride)	8	-	-	-	-	-	-	-	-	-	-	-	1455	-	-	-	-	-	-	-	-	-	-	-	-	-	-	-	-	1457	-	-	-	-	-
Kry-Kote White	9	-	523	-	-	-	-	-	-	-	-	-	-	-	-	-	-	-	-	-	-	-	-	-	-	-	-	-	-	-	-	-	-	-	-
Krylon (Clear) Coating on Aluminum Substrate	9	-	1109	-	-	-	-	-	-	-	-	526	-	-	-	-	-	-	-	-	-	-	-	-	-	-	-	-	-	-	-	-	-	-	-
Krylon Paints:																																			
Black	9	-	-	-	-	524	-	-	-	-	-	526	-	-	-	-	-	-	-	-	-	-	-	-	-	-	-	-	-	-	-	-	-	-	-
Silver Aluminum	9	-	-	-	-	-	-	-	-	-	-	526	-	-	-	-	-	-	-	-	-	-	-	-	-	-	-	-	-	-	-	-	-	-	-
White	9	-	-	-	-	-	-	-	-	-	-	526	-	-	-	-	-	-	-	-	-	-	-	-	-	-	-	-	-	-	-	-	-	-	-
Yellow	9	-	-	-	-	-	-	-	-	-	-	526	-	-	-	-	-	-	-	-	-	-	-	-	-	-	-	-	-	-	-	-	-	-	-
KT Silicon Carbide	8	-	794	-	-	-	-	-	-	-	-	-	-	-	-	-	-	-	-	-	-	-	-	-	-	-	-	-	-	-	-	-	-	-	-
L34	7	-	-	-	-	-	-	-	-	-	-	1092	-	-	1095	-	-	-	-	-	-	-	-	-	-	-	-	-	-	-	-	-	-	-	-
L72	7	-	-	-	-	-	-	-	-	-	-	-	-	-	1084	-	-	-	-	-	-	-	-	1087	-	-	-	-	-	-	-	-	-	-	-
L120 Magnesium	7	-	-	-	-	-	-	-	-	-	-	-	-	-	361 362	-	-	-	-	-	-	-	-	365	-	-	-	-	-	-	-	-	-	-	-
Lacquer Binder Pigmented with Aluminum	9	-	6	-	-	-	-	-	-	-	-	-	-	-	-	-	-	-	-	-	-	-	-	-	-	-	-	-	-	-	-	-	-	-	-
Lacquer Coating on:																																			
Aluminum Substrate	9	-	1124	-	-	-	-	-	-	-	-	-	-	-	-	-	-	-	-	-	-	-	-	1127	-	-	-	-	-	-	-	-	-	-	-
Copper Substrate	9	-	-	-	-	-	-	-	-	-	-	1125	-	-	-	-	-	-	-	-	-	-	-	-	-	-	-	-	-	-	-	-	-	-	-
Dow Metal Substrate	9	-	1124	-	-	-	-	-	-	-	-	-	-	-	-	-	-	-	-	-	-	-	-	1127	-	-	-	-	-	-	-	-	-	-	-
Quartz Substrate	9	-	-	-	-	-	-	-	-	-	-	-	-	-	-	-	-	-	-	-	-	1126	-	-	-	-	-	-	-	1129	-	-	-	-	-
Steel Substrate	9	-	-	-	-	-	-	-	-	-	-	-	-	-	-	-	-	-	-	-	-	1126	-	-	-	-	-	-	-	-	-	-	-	-	-
Lacquer Substrate Coated with Aluminum	9	-	-	-	-	-	-	-	-	-	-	600	-	-	-	-	-	-	-	-	-	-	-	-	-	-	-	-	-	-	-	-	-	-	-

MATERIAL NAME	VOLUME	EMITTANCE						REFLECTANCE									ABSORPTANCE									TRANSMITTANCE									Absorptance to Emittance Ratio
		Total			Spectral			Integrated			Spectral			Solar			Integrated			Spectral			Solar			Integrated			Spectral			Solar			
		Hemispherical	Normal	Angular	Hemispherical	Normal	Angular	Hemispherical	Normal	Angular	Hemispherical	Normal	Angular	Hemispherical	Normal	Angular	Hemispherical	Normal	Angular	Hemispherical	Normal	Angular	Hemispherical	Normal	Angular	Hemispherical	Normal	Angular	Hemispherical	Normal	Angular	Hemispherical	Normal	Angular	
Laminar X-500 Coating on:																																			
Aluminum Substrate	9	-	-	-	-	-	-	-	-	-	-	1138	1140	-	-	-	-	-	-	-	-	-	-	-	-	-	-	-	-	-	-	-	-	-	-
Polyethylene Substrate	9	-	-	-	-	-	-	-	-	-	-	-	-	-	-	-	-	-	-	-	-	-	-	-	-	-	-	-	-	1143	-	-	-	-	-
Lampblack Pigment in:																																			
Decoret Binder	9	-	-	-	-	-	-	-	-	-	-	86	-	-	-	-	-	-	-	-	-	-	-	-	-	-	-	-	-	-	-	-	-	-	-
Epoxy Binder	9	-	82	-	-	84	-	-	-	-	-	86	-	-	-	-	-	-	-	-	-	-	-	89	-	-	-	-	-	-	-	-	-	-	-
Silicone Binder	9	-	82	-	-	-	-	-	-	-	-	86	-	-	-	-	-	-	-	-	-	-	-	-	-	-	-	-	-	-	-	-	-	-	-
Turpentine Binder	9	-	-	-	-	-	-	-	-	-	-	86	-	-	-	-	-	-	-	-	-	-	-	-	-	-	-	-	-	-	-	-	-	-	-
Lanthanum Antimonide Coating on Glass Substrate	9	-	-	-	-	-	-	-	-	-	-	689	-	-	-	-	-	-	-	-	-	-	-	-	-	-	-	-	-	-	-	-	-	-	-
Lanthanum Hexaboride	8	-	-	-	-	723	-	-	-	-	-	-	-	-	-	-	-	-	-	-	-	-	-	-	-	-	-	-	-	727	-	-	-	-	-
Lanthanum Trifluoride	8	-	-	-	-	-	-	-	-	-	-	-	-	-	-	-	-	-	-	-	-	-	-	-	-	-	-	-	-	994	-	-	-	-	-
Lanthanum Oxide Pigment in Potassium Silicate Binder	9	-	-	-	-	-	-	-	-	-	-	127	-	-	-	-	-	-	-	-	-	-	-	132	-	-	-	-	-	-	-	-	-	-	-
Dilanthanum Trioxide	8	-	-	-	-	-	-	-	-	-	-	546	-	-	-	-	-	-	-	-	-	-	-	-	-	-	-	-	-	-	-	-	-	-	-
Dilanthanum Trisulfide	8	-	-	-	-	1232	-	-	-	-	-	-	-	-	-	-	-	-	-	-	-	-	-	-	-	-	-	-	-	-	-	-	-	-	-
LaSb	8	-	-	-	-	-	-	-	-	-	-	1352	-	-	-	-	-	-	-	-	-	-	-	-	-	-	-	-	-	-	-	-	-	-	-
Lead	7	335	337	-	-	-	-	-	-	-	-	-	-	-	-	-	339	341	-	-	343	345	-	-	-	-	-	-	-	-	-	-	-	-	-
Lead + Tin	7	948	-	-	-	-	-	-	-	-	-	-	-	-	-	-	-	-	-	-	-	-	-	-	-	-	-	-	-	-	-	-	-	-	-
Lead + Tin Coating on Copper Substrate	9	690	-	-	-	-	-	-	-	-	-		-	-	-	-	691	-	-	-	-	-	-	-	-	-	-	-	-	-	-	-	-	-	-
Lead Dibromide	8	-	-	-	-	-	-	-	-	-	-	-	-	-	-	-	-	-	-	-	-	-	-	-	-	-	-	-	-	745	-	-	-	-	-
Lead Carbonate	8	-	-	-	-	-	-	-	-	-	-	587	-	-	-	-	-	-	-	-	-	-	-	589	-	-	-	-	-	-	-	-	-	-	-
Lead Carbonate Pigment in:																																			
Acrylic Binder	9	-	-	-	-	-	-	-	-	-	-	139	-	-	-	-	-	-	-	-	-	-	-	-	-	-	-	-	-	-	-	-	-	-	-
Acyloid A10 Binder	9	-	-	-	-	-	-	-	-	-	-	138	-	-	-	-	-	-	-	-	-	-	-	-	-	-	-	-	-	-	-	-	-	-	-
Alkyd Binder	9	-	-	-	-	-	-	-	-	-	-	139	-	-	-	-	-	-	-	-	-	-	-	-	-	-	-	-	-	-	-	-	-	-	-
Formaldehyde Binder	9	-	-	-	-	-	-	-	-	-	-	138	-	-	-	-	-	-	-	-	-	-	-	-	-	-	-	-	-	-	-	-	-	-	-
Nitrocellulose Binder	9	-	-	-	-	-	-	-	-	-	-	139	-	-	-	-	-	-	-	-	-	-	-	-	-	-	-	-	-	-	-	-	-	-	-
Silicone Binder	9	-	-	-	-	-	-	-	-	-	-	138	-	-	-	-	-	-	-	-	-	-	-	-	-	-	-	-	-	-	-	-	-	-	-
Lead Dichloride	8	-	-	-	-	-	-	-	-	-	-	-	-	-	-	-	-	-	-	-	-	-	-	-	-	-	-	-	-	908	-	-	-	-	-
Lead Chloride Coating on Germanium Substrate	9	-	-	-	-	-	-	-	-	-	-	912	-	-	-	-	-	-	-	-	-	-	-	-	-	-	-	-	-	-	-	-	-	-	-

MATERIAL NAME	VOLUME	EMITTANCE Total Hemispherical	EMITTANCE Total Normal	EMITTANCE Total Angular	EMITTANCE Spectral Hemispherical	EMITTANCE Spectral Normal	EMITTANCE Spectral Angular	REFLECTANCE Integrated Hemispherical	REFLECTANCE Integrated Normal	REFLECTANCE Integrated Angular	REFLECTANCE Spectral Hemispherical	REFLECTANCE Spectral Normal	REFLECTANCE Spectral Angular	REFLECTANCE Solar Hemispherical	REFLECTANCE Solar Normal	REFLECTANCE Solar Angular	ABSORPTANCE Integrated Hemispherical	ABSORPTANCE Integrated Normal	ABSORPTANCE Integrated Angular	ABSORPTANCE Spectral Hemispherical	ABSORPTANCE Spectral Normal	ABSORPTANCE Spectral Angular	ABSORPTANCE Solar Hemispherical	ABSORPTANCE Solar Normal	ABSORPTANCE Solar Angular	TRANSMITTANCE Integrated Hemispherical	TRANSMITTANCE Integrated Normal	TRANSMITTANCE Integrated Angular	TRANSMITTANCE Spectral Hemispherical	TRANSMITTANCE Spectral Normal	TRANSMITTANCE Spectral Angular	TRANSMITTANCE Solar Hemispherical	TRANSMITTANCE Solar Normal	TRANSMITTANCE Solar Angular	Absorptance to Emittance Ratio
Lead Chloride Substrate Coated with Germanium	9	-	-	-	-	-	-	-	-	-	-	647	-	-	-	-	-	-	-	-	-	-	-	-	-	-	-	-	-	-	-	-	-	-	-
Lead Difluoride	8	-	-	-	-	-	-	-	-	-	-	-	-	-	-	-	-	-	-	-	-	-	-	-	-	-	-	-	-	994	-	-	-	-	-
Lead Diiodide	8	-	-	-	-	-	-	-	-	-	-	-	-	-	-	-	-	-	-	-	-	-	-	-	-	-	-	-	-	1003	-	-	-	-	-
Lead Hafnate	8	-	-	-	-	-	-	-	-	-	-	597	-	-	-	-	-	-	-	-	-	-	-	-	-	-	-	-	-	-	-	-	-	-	-
Lead Molybdenum Tetraoxide Coating on:																																			
Glass Substrate	9	-	-	-	-	-	-	-	-	-	-	913	-	-	-	-	-	-	-	-	-	-	-	-	-	-	-	-	-	-	-	-	-	-	-
Potassium Bromide Substrate	9	-	-	-	-	-	-	-	-	-	-	913	-	-	-	-	-	-	-	-	-	-	-	-	-	-	-	-	-	-	-	-	-	-	-
Lead Oxide	8	-	-	-	-	-	-	-	-	-	-	546	-	-	-	-	-	-	-	-	-	-	-	-	-	-	-	-	-	-	-	-	-	-	-
Lead Selenide	8	-	-	-	-	-	-	-	-	-	-	1130	-	-	-	-	-	-	-	-	-	-	-	-	-	-	-	-	-	1133	-	-	-	-	-
Lead Sulfate	8	-	-	-	-	-	-	-	-	-	-	-	-	-	-	-	-	-	-	-	-	-	-	-	-	-	-	-	-	631	-	-	-	-	-
Lead Sulfide	8	-	-	-	-	-	-	-	-	-	-	1197	-	-	-	-	-	-	-	-	-	-	-	-	-	-	-	-	-	1204	-	-	-	-	-
Lead Sulfite	8	-	-	-	-	-	-	-	-	-	-	-	-	-	-	-	-	-	-	-	-	-	-	-	-	-	-	-	-	631	-	-	-	-	-
Lead Telluride	8	-	-	-	-	-	-	-	-	-	-	1253	-	-	-	-	-	-	-	-	-	-	-	-	-	-	-	-	-	1255	-	-	-	-	-
Lead Titanate	8	-	-	-	-	-	-	-	646	-	-	648	-	-	-	-	-	-	-	-	-	-	-	-	-	-	-	-	-	650	-	-	-	-	-
Lead Tungstate	8	-	-	-	-	-	-	-	-	-	-	666	-	-	-	-	-	-	-	-	-	-	-	-	-	-	-	-	-	-	-	-	-	-	-
Lead Zirconate	8	-	-	-	-	-	-	-	-	-	-	676	-	-	-	-	-	-	-	-	-	-	-	-	-	-	-	-	-	-	-	-	-	-	-
Leafing Aluminum Pigment in Oil Binder	9	-	-	-	-	-	-	-	-	9	-	-	-	-	-	-	-	-	-	-	-	-	-	-	-	-	-	-	-	-	-	-	-	-	-
Leafing Aluminum Pigment in Silicone Binder	9	3	7	-	-	-	-	-	-	-	-	13 17	-	-	-	-	-	-	-	-	-	-	-	21	-	-	-	-	-	-	-	-	-	-	-
Leafing Aluminum + Carbon Pigment in Silicone Binder	9	-	-	-	-	-	-	-	-	-	-	-	-	-	-	-	-	-	-	-	-	-	-	24	-	-	-	-	-	-	-	-	-	-	-
Leafing Gold Pigment in Resin Binder	9	25	-	-	-	-	-	-	-	-	-	-	-	-	-	-	-	-	-	-	-	-	-	-	-	-	-	-	-	-	-	-	-	-	-
Leonite 201-S Binder Pigmented with:																																			
Magnesium Oxide	9	-	-	-	-	-	-	-	-	-	-	161	-	-	-	-	-	-	-	-	-	-	-	-	-	-	-	-	-	-	-	-	-	-	-
Zinc Oxide	9	-	-	-	-	-	-	-	-	-	-	314	-	-	-	-	-	-	-	-	-	-	-	-	-	-	-	-	-	-	-	-	-	-	-
Zinc Sulfide	9	-	-	-	-	-	-	-	-	-	-	406	-	-	-	-	-	-	-	-	-	-	-	-	-	-	-	-	-	-	-	-	-	-	-
Limonite	8	-	-	-	-	-	-	-	-	-	-	1678	-	-	-	-	-	-	-	-	-	-	-	-	-	-	-	-	-	-	-	-	-	-	-

MATERIAL NAME	VOLUME	EMITTANCE						REFLECTANCE									ABSORPTANCE									TRANSMITTANCE									Absorptance to Emittance Ratio
		Total			Spectral			Integrated			Spectral			Solar			Integrated			Spectral			Solar			Integrated			Spectral			Solar			
		Hemispherical	Normal	Angular	Hemispherical	Normal	Angular	Hemispherical	Normal	Angular	Hemispherical	Normal	Angular	Hemispherical	Normal	Angular	Hemispherical	Normal	Angular	Hemispherical	Normal	Angular	Hemispherical	Normal	Angular	Hemispherical	Normal	Angular	Hemispherical	Normal	Angular	Hemispherical	Normal	Angular	
Liquid Platinum Coating on:																																			
Ceramic Tile Substrate	9	-	-	-	-	-	-	-	-	-	-	-	724	-	-	-	-	-	-	-	-	-	-	-	-	-	-	-	-	-	-	-	-	-	-
Glass Substrate	9	-	722	-	-	-	-	-	-	-	-	-	-	-	-	-	-	-	-	-	-	-	-	-	-	-	-	-	-	-	-	-	-	-	-
Quartz Substrate	9	-	-	-	-	-	-	-	-	-	-	-	724	-	-	-	-	-	-	-	-	-	-	-	-	-	-	-	-	-	-	-	-	-	-
Lithafrax Pigment in:																																			
Potassium Silicate Binder	9	-	-	-	-	-	-	-	-	-	-	144	-	-	-	-	-	-	-	-	-	-	-	-	-	-	-	-	-	-	-	-	-	-	-
Sodium Silicate Binder	9	142	-	-	-	-	-	-	-	-	-	-	-	-	-	-	-	-	-	-	147	-	-	150	-	-	-	-	-	-	-	-	-	-	-
Lithium Aluminum Silicate Pigment in:																																			
Potassium Silicate Binder	9	-	143	-	-	-	-	-	-	-	-	144	-	-	-	-	-	-	-	-	-	-	-	-	-	-	-	-	-	-	-	-	-	-	-
Sodium Silicate Binder	9	142	-	-	-	-	-	-	-	-	-	-	-	-	-	-	-	-	-	-	147	-	-	150	-	-	-	-	-	-	-	-	-	-	-
Lithium Borate	8	-	-	-	-	-	-	-	-	-	-	582	-	-	-	-	-	-	-	-	-	-	-	-	-	-	-	-	-	-	-	-	-	-	-
Lithium Fluoride	8	-	-	-	-	-	-	-	-	-	-	937	942	-	-	-	-	-	-	-	-	-	-	-	-	-	-	-	-	944	-	-	-	-	-
Lithium Fluoride Coating on:																																			
Aluminum Substrate	9	-	-	-	-	-	-	-	-	-	-	1090	-	-	-	-	-	-	-	-	-	-	-	-	-	-	-	-	-	-	-	-	-	-	-
Glass Substrate	9	-	-	-	-	-	-	-	-	-	-	1090	-	-	-	-	-	-	-	-	-	-	-	-	-	-	-	-	-	-	-	-	-	-	-
KRS-5 Substrate	9	-	-	-	-	-	-	-	-	-	-	-	-	-	-	-	-	-	-	-	-	-	-	-	-	-	-	-	-	1092	-	-	-	-	-
Lithium Fluoride Pigment in Sodium Silicate Binder	9	-	-	-	-	-	-	-	-	-	-	-	-	-	-	-	-	-	-	-	155	-	-	-	-	-	-	-	-	-	-	-	-	-	-
Lithium Fluoride Substrate Coated with:																																			
Germanium	9	-	-	-	-	-	-	-	-	-	-	647	-	-	-	-	-	-	-	-	-	-	-	-	-	-	-	-	-	-	-	-	-	-	-
Potassium Chloride	9	-	-	-	-	-	-	-	-	-	-	-	-	-	-	-	-	-	-	-	-	-	-	-	-	-	-	-	-	935	-	-	-	-	-
Potassium Iodide	9	-	-	-	-	-	-	-	-	-	-	936	-	-	-	-	-	-	-	-	-	-	-	-	-	-	-	-	-	-	-	-	-	-	-
Rubidium Iodide	9	-	-	-	-	-	-	-	-	-	-	940	-	-	-	-	-	-	-	-	-	-	-	-	-	-	-	-	-	-	-	-	-	-	-
Lithium Niobate	8	-	-	-	-	-	-	-	-	-	-	598	-	-	-	-	-	-	-	-	-	-	-	-	-	-	-	-	-	-	-	-	-	-	-
Lithium Yttrium Tetrafluoride	8	-	-	-	-	-	-	-	-	-	-	-	-	-	-	-	-	-	-	-	-	-	-	-	-	-	-	-	-	994	-	-	-	-	-
LMSC/Dow Corning Thermatrol Paint	9	211	-	-	-	-	-	-	-	-	-	-	-	-	-	-	-	-	-	-	257	-	-	270	-	-	-	-	-	-	-	-	-	-	-
L. O. F. 81E 19778 Glass	8	-	1560	-	-	-	-	-	-	-	-	1562	-	-	-	-	-	-	-	-	-	-	-	1563	-	-	1565	-	-	1567	-	-	-	-	-
L. O. F. PB 19195 Glass	8	-	1560	-	-	-	-	-	-	-	-	1562	-	-	-	-	-	-	-	-	-	-	-	1563	-	-	1565	-	-	1567	-	-	-	-	-
Lucite	8	-	-	-	-	-	-	-	-	-	-	1720	-	-	-	-	-	-	-	-	1722	-	-	-	-	-	-	-	-	1724	-	-	-	-	-
Ludox Binder Pigmented with SiC + Talc	9	176	-	-	-	-	-	-	-	-	-	-	-	-	-	-	-	-	-	-	-	-	-	-	-	-	-	-	-	-	-	-	-	-	-

MATERIAL NAME	VOLUME	EMITTANCE						REFLECTANCE									ABSORPTANCE									TRANSMITTANCE									Absorptance to Emittance Ratio
		Total			Spectral			Integrated			Spectral			Solar			Integrated			Spectral			Solar			Integrated			Spectral			Solar			
		Hemispherical	Normal	Angular	Hemispherical	Normal	Angular	Hemispherical	Normal	Angular	Hemispherical	Normal	Angular	Hemispherical	Normal	Angular	Hemispherical	Normal	Angular	Hemispherical	Normal	Angular	Hemispherical	Normal	Angular	Hemispherical	Normal	Angular	Hemispherical	Normal	Angular	Hemispherical	Normal	Angular	
Lunar Samples	8	–	–	–	–	–	–	–	–	–	–	–	1666	–	–	–	–	–	–	–	–	–	–	–	–	–	–	–	–	–	–	–	–	–	–
Lutetium	7	–	–	–	–	–	–	–	–	–	–	347	–	–	–	–	–	–	–	–	–	–	–	–	–	–	–	–	–	350	–	–	–	–	–
Lutetium Oxide	8	–	–	–	–	286 288	–	–	–	–	–	–	–	–	–	–	–	–	–	–	–	–	–	–	–	–	–	–	–	–	–	–	–	–	–
M252	7	–	–	–	–	1366	–	–	–	–	–	1382	–	–	–	–	–	–	–	–	–	–	–	–	–	–	–	–	–	–	–	–	–	–	–
3M Kel-F 800 Binder Pigmented with:																																			
Zinc Oxide	9	–	–	–	–	–	–	–	–	–	–	317	–	–	–	–	–	–	–	–	–	–	–	–	–	–	–	–	–	–	–	–	–	–	–
Zinc Sulfide	9	–	–	–	–	–	–	–	–	–	–	406	–	–	–	–	–	–	–	–	–	–	–	–	–	–	–	–	–	–	–	–	–	–	–
3M Kel-F 8213 Binder Pigmented with Zinc Oxide	9	–	–	–	–	–	–	–	–	–	–	314	–	–	–	–	–	–	–	–	–	–	–	–	–	–	–	–	–	–	–	–	–	–	–
3M Paints:																																			
Black Velvet	9	533	535	–	–	536	–	–	–	–	–	538	–	–	–	–	–	–	–	–	–	–	–	540	–	–	–	–	–	–	–	–	–	–	–
White Velvet	9	–	–	–	–	–	–	–	–	–	–	538	–	–	–	–	–	–	–	–	–	–	–	–	–	–	–	–	–	–	–	–	–	–	–
Magic Iron Cement Co. White Porcelain Enamel	9	–	–	–	–	–	–	–	–	–	–	528	–	–	–	–	–	–	–	–	–	–	–	–	–	–	–	–	–	–	–	–	–	–	–
Magnesium	7	353	–	–	–	–	–	–	–	–	–	356	358	–	360	–	–	–	–	–	–	–	–	364	–	–	–	–	–	367	–	–	–	–	–
Magnesium, Anodized	9	1274	–	–	–	–	–	–	–	–	–	–	–	–	–	–	–	–	–	–	–	–	–	1275	–	–	–	–	–	–	–	–	–	–	–
Magnesium, L120	7	–	–	–	–	–	–	–	–	–	–	–	–	–	361 362	–	–	–	–	–	–	–	–	365	–	–	–	–	–	–	–	–	–	–	–
Magnesium Coating on Glass Substrate	9	–	–	–	–	–	–	–	–	–	–	692	–	–	–	–	–	–	–	–	–	–	–	–	–	–	–	–	–	–	–	–	–	–	–
Magnesium Substrate Coated with:																																			
Gold	9	–	–	–	–	660	–	–	–	–	–	–	–	–	–	–	–	–	–	–	–	–	–	–	–	–	–	–	–	–	–	–	–	–	–
Silicon Dioxide	9	1049	–	–	–	–	–	–	–	–	–	–	–	–	–	–	–	–	–	–	–	–	–	1070	–	–	–	–	–	–	–	–	–	–	–
Magnesium + Aluminum	7	–	–	–	–	–	–	–	–	–	–	950	–	–	–	–	–	–	–	–	–	–	–	–	–	–	–	–	–	–	–	–	–	–	–
Magnesium + Aluminum Coating on Glass Substrate	9	–	–	–	–	–	–	–	–	–	–	693	–	–	–	–	–	–	–	–	–	–	–	–	–	–	–	–	–	–	–	–	–	–	–
Magnesium + Aluminum + ΣX_i	7	–	1327	–	–	–	–	–	–	–	–	1330	–	–	–	–	–	–	–	–	–	–	–	1334	–	–	–	–	–	–	–	–	–	–	–
Magnesium + Aluminum + ΣX_i, Anodized	9	–	–	–	–	–	–	–	–	–	–	1277	–	–	–	–	–	–	–	–	–	–	–	–	–	–	–	–	–	–	–	–	–	–	–
Magnesium + Thorium + ΣX_i	7	–	1336	–	–	–	–	–	–	–	–	1338	–	–	–	–	–	–	–	–	–	–	–	1340	–	–	–	–	–	–	–	–	–	–	–
Magnesium + Thorium + ΣX_i, Anodized	9	–	–	–	–	–	–	–	–	–	–	1281	–	–	–	–	–	–	–	–	–	–	–	–	–	–	–	–	–	–	–	–	–	–	–

MATERIAL NAME	VOLUME	EMITTANCE						REFLECTANCE									ABSORPTANCE									TRANSMITTANCE									Absorptance to Emittance Ratio
		Total			Spectral			Integrated			Spectral			Solar			Integrated			Spectral			Solar			Integrated			Spectral			Solar			
		Hemispherical	Normal	Angular	Hemispherical	Normal	Angular	Hemispherical	Normal	Angular	Hemispherical	Normal	Angular	Hemispherical	Normal	Angular	Hemispherical	Normal	Angular	Hemispherical	Normal	Angular	Hemispherical	Normal	Angular	Hemispherical	Normal	Angular	Hemispherical	Normal	Angular	Hemispherical	Normal	Angular	
Magnesium Alloys:																																			
AZ-31	7	-	1328	-	-	-	-	-	-	-	-	1332	-	-	-	-	-	-	-	-	-	-	-	1334	-	-	-	-	-	-	-	-	-	-	-
AZ-31B	7	-	-	-	-	-	-	-	-	-	-	1332	-	-	-	-	-	-	-	-	-	-	-	-	-	-	-	-	-	-	-	-	-	-	-
HK-31	7	-	1336	-	-	-	-	-	-	-	-	1338	-	-	-	-	-	-	-	-	-	-	-	1340	-	-	-	-	-	-	-	-	-	-	-
Magnesium Aluminate	8	-	-	-	-	-	-	-	-	-	-	576	-	-	-	-	-	-	-	-	-	-	-	-	-	-	-	-	-	577	-	-	-	-	-
Magnesium Aluminate Coating on:																																			
Aluminum Substrate	9	915	-	-	-	-	-	-	-	-	-	-	-	-	-	-	-	-	-	-	-	-	-	918	-	-	-	-	-	-	-	-	-	-	-
Nb-1Zr Substrate	9	915	-	-	-	917	-	-	-	-	-	-	-	-	-	-	-	-	-	-	-	-	-	-	-	-	-	-	-	-	-	-	-	-	-
Magnesium Aluminate Spinel Pigment in Potassium Silicate Binder	9	-	-	-	-	-	-	-	-	-	-	-	-	-	-	-	-	-	-	-	-	-	-	187	-	-	-	-	-	-	-	-	-	-	-
Magnesium Carbonate	8	-	-	-	-	-	-	-	590	-	-	-	-	-	-	-	-	-	-	-	-	-	-	-	-	-	-	-	-	-	-	-	-	-	-
Magnesium Carbonate Pigment in Ethylcellulose + Dow 7 Binder	9	-	-	-	-	-	-	-	156	-	-	-	-	-	-	-	-	-	-	-	-	-	-	-	-	-	-	-	-	-	-	-	-	-	-
Magnesium Difluoride	8	-	-	-	-	950	-	-	-	-	-	-	952	-	-	-	-	-	-	-	955	-	-	-	-	-	-	-	-	957	-	-	-	-	-
Magnesium Fluoride Coating on:																																			
Glass Substrate	9	-	-	-	-	-	-	-	-	-	-	1096	-	-	-	-	-	-	-	-	-	-	-	-	-	-	-	-	-	1105	-	-	-	-	-
Iron Oxide Substrate	9	1093	-	-	-	-	-	-	-	-	-	1096	-	-	-	-	-	-	-	-	-	-	-	1103	-	-	-	-	-	-	-	-	-	-	-
Platinum Substrate	9	-	-	-	-	-	-	-	-	-	-	-	1098	-	-	-	-	-	-	-	-	1101	-	-	-	-	-	-	-	1105	-	-	-	-	-
Silicon Dioxide Substrate	9	-	-	-	-	-	-	-	-	-	-	1096	1098	-	-	-	-	-	-	-	-	1101	-	-	-	-	-	-	-	1105	-	-	-	-	-
Silicon Monoxide Substrate	9	-	-	-	-	1094	-	-	-	-	-	1096	-	-	-	-	-	-	-	-	-	-	-	-	-	-	-	-	-	-	-	-	-	-	-
Magnesium Oxide	8	290	291	-	-	293 295	-	298	-	-	-	299	314	-	-	-	-	-	-	-	319	-	-	322	-	-	-	-	-	323	-	-	-	-	-
Magnesium Oxide Coating on:																																			
Aluminum Substrate	9	-	-	-	-	-	-	-	-	-	-	924	-	-	-	-	-	-	-	-	-	-	-	-	-	-	-	-	-	-	-	-	-	-	-
Black Paint Substrate	9	-	-	-	-	-	-	-	-	-	-	924	-	-	-	-	-	-	-	-	-	-	-	-	-	-	-	-	-	-	-	-	-	-	-
Nimonic 75 Substrate	9	-	919	-	-	-	-	-	-	-	-	-	-	-	-	-	-	-	-	-	-	-	-	-	-	-	-	-	-	-	-	-	-	-	-
Unknown Substrate	9	-	-	-	-	920	-	-	921	922	-	924	-	-	-	-	-	-	-	-	-	-	-	-	-	-	-	-	-	-	-	-	-	-	-
Magnesium Oxide Pigment in:																																			
Acrylic Binder	9	-	-	-	-	-	-	-	-	-	-	161	-	-	-	-	-	-	-	-	-	-	-	-	-	-	-	-	-	-	-	-	-	-	-
Acryloid Binder	9	-	-	-	-	-	-	-	-	-	-	161	-	-	-	-	-	-	-	-	-	-	-	-	-	-	-	-	-	-	-	-	-	-	-
Alkyd-Melamine Binder	9	-	-	-	-	-	-	-	-	-	-	161	-	-	-	-	-	-	-	-	-	-	-	-	-	-	-	-	-	-	-	-	-	-	-
Ethylcellulose + Dow 7 Binder	9	-	-	-	-	-	-	-	158	-	-	-	-	-	-	-	-	-	-	-	-	-	-	-	-	-	-	-	-	-	-	-	-	-	-
GE SR-112 Binder	9	-	-	-	-	-	-	-	-	-	-	-	-	-	-	-	-	-	-	-	163	-	-	-	-	-	-	-	-	-	-	-	-	-	-
Silicone Binder	9	-	-	-	-	-	-	-	-	-	-	161	-	-	-	-	-	-	-	-	-	-	-	-	-	-	-	-	-	-	-	-	-	-	-

MATERIAL NAME	VOLUME	EMITTANCE						REFLECTANCE									ABSORPTANCE									TRANSMITTANCE									Absorptance to Emittance Ratio
		Total			Spectral			Integrated			Spectral			Solar			Integrated			Spectral			Solar			Integrated			Spectral			Solar			
		Hemispherical	Normal	Angular	Hemispherical	Normal	Angular	Hemispherical	Normal	Angular	Hemispherical	Normal	Angular	Hemispherical	Normal	Angular	Hemispherical	Normal	Angular	Hemispherical	Normal	Angular	Hemispherical	Normal	Angular	Hemispherical	Normal	Angular	Hemispherical	Normal	Angular	Hemispherical	Normal	Angular	
Magnesium Oxide + Diatomaceous Earth Pigment in Silicone Binder	9	–	–	–	–	–	–	–	–	–	–	165	–	–	–	–	–	–	–	–	–	–	–	–	–	–	–	–	–	–	–	–	–	–	–
Magnesium Phosphate	8	–	–	–	–	–	–	–	–	–	–	608	–	–	–	–	–	–	–	–	–	–	–	–	–	–	–	–	–	–	–	–	–	–	–
Magnesium Silicate	8	–	–	–	–	–	–	–	–	–	–	618	–	–	–	–	–	–	–	–	–	–	–	–	–	–	–	–	–	–	–	–	–	–	–
Magnesium Silicate Pigment in Potassium Silicate Binder	9	–	–	–	–	–	–	–	–	–	–	168	–	–	–	–	–	–	–	–	–	–	–	169	–	–	–	–	–	–	–	–	–	–	–
Magnesium Trisilicate	8	–	–	–	–	–	–	–	–	–	–	618	–	–	–	–	–	–	–	–	–	–	–	–	–	–	–	–	–	–	–	–	–	–	–
Magnesium Zirconium Silicate	8	–	–	–	–	–	–	–	–	–	–	–	–	–	–	–	–	–	–	–	–	–	–	616	–	–	–	–	–	–	–	–	–	–	–
Dimagnesium Silicide	8	–	–	–	–	1173	–	–	–	–	–	–	–	–	–	–	–	–	–	–	–	–	–	–	–	–	–	–	–	–	–	–	–	–	–
Manganese Coating on Glass Substrate	9	–	–	–	–	–	–	–	–	–	–	694	–	–	–	–	–	–	–	–	–	–	–	–	–	–	–	–	–	695	–	–	–	–	–
Manganese Monoxide	8	–	–	–	–	–	–	–	–	–	–	329	–	–	–	–	–	–	–	–	–	–	–	–	–	–	–	–	–	–	–	–	–	–	–
Trimanganese Tetraoxide	8	–	–	–	–	–	–	–	–	–	–	329	–	–	–	–	–	–	–	–	–	–	–	–	–	–	–	–	–	–	–	–	–	–	–
Manganese Oxide + Cobalt Oxide Pigment in Base Glaze No. 2 Binder (Enamel)	9	–	468	–	–	–	–	–	–	–	–	–	–	–	–	–	–	–	–	–	–	–	–	–	–	–	–	–	–	–	–	–	–	–	–
Manganese Oxide + Iron Oxide Pigment in Base Glaze No. 3 Binder (Enamel)	9	–	468	–	–	–	–	–	–	–	–	–	–	–	–	–	–	–	–	–	–	–	–	–	–	–	–	–	–	–	–	–	–	–	–
Manganese Telluride	8	–	–	–	–	–	–	–	–	–	–	1256	–	–	–	–	–	–	–	–	–	–	–	–	–	–	–	–	–	–	–	–	–	–	–
Manganese Disilicide	8	–	–	–	–	1173	–	–	–	–	–	–	–	–	–	–	–	–	–	–	–	–	–	–	–	–	–	–	–	–	–	–	–	–	–
Trimanganese Silicide	8	–	–	–	–	1173	–	–	–	–	–	–	–	–	–	–	–	–	–	–	–	–	–	–	–	–	–	–	–	–	–	–	–	–	–
Manganese Sulfide	8	–	–	–	–	–	–	–	–	–	–	1234	–	–	–	–	–	–	–	–	–	–	–	–	–	–	–	–	–	–	–	–	–	–	–
Manganite	8	–	–	–	–	–	–	–	–	–	–	–	–	–	–	–	–	–	–	–	–	–	–	–	–	–	–	–	–	1664	–	–	–	–	–
McDanel AV 30	8	–	–	–	–	151	–	–	–	–	–	–	–	–	–	–	–	–	–	–	–	–	–	–	–	–	–	–	–	172	–	–	–	–	–
McDanel AP 35	8	–	–	–	–	151	–	–	–	–	–	–	–	–	–	–	–	–	–	–	–	–	–	–	–	–	–	–	–	172	–	–	–	–	–
Mealtone Hammer Finish Silver	9	–	–	–	–	–	–	–	–	–	–	529	–	–	–	–	–	–	–	–	–	–	–	–	–	–	–	–	–	–	–	–	–	–	–
Melamine Formaldehyde Coating on Quartz Substrate	9	–	–	–	–	–	–	–	–	–	–	–	–	–	–	–	–	–	–	–	–	–	–	1130	–	–	–	–	–	–	–	–	–	–	–
Dimercury Dibromide	8	–	–	–	–	–	–	–	–	–	–	747	–	–	–	–	–	–	–	–	–	–	–	–	–	–	–	–	–	–	–	–	–	–	–
Mercury Dichloride	8	–	–	–	–	–	–	–	–	–	–	908	–	–	–	–	–	–	–	–	–	–	–	–	–	–	–	–	–	–	–	–	–	–	–
Dimercury Dichloride	8	–	–	–	–	–	–	–	–	–	–	908	–	–	–	–	–	–	–	–	–	–	–	–	–	–	–	–	–	–	–	–	–	–	–

MATERIAL NAME	VOLUME	EMITTANCE						REFLECTANCE									ABSORPTANCE									TRANSMITTANCE									Absorptance to Emittance Ratio
		Total			Spectral			Integrated			Spectral			Solar			Integrated			Spectral			Solar			Integrated			Spectral			Solar			
		Hemispherical	Normal	Angular	Hemispherical	Normal	Angular	Hemispherical	Normal	Angular	Hemispherical	Normal	Angular	Hemispherical	Normal	Angular	Hemispherical	Normal	Angular	Hemispherical	Normal	Angular	Hemispherical	Normal	Angular	Hemispherical	Normal	Angular	Hemispherical	Normal	Angular	Hemispherical	Normal	Angular	
Mercury Diiodide	8	-	-	-	-	-	-	-	-	-	-	1027	-	-	-	-	-	-	-	-	-	-	-	-	-	-	-	-	-	1029	-	-	-	-	-
Dimercury Diiodide	8	-	-	-	-	-	-	-	-	-	-	1027	-	-	-	-	-	-	-	-	-	-	-	-	-	-	-	-	-	-	-	-	-	-	-
Mercury Monoxide	8	-	-	-	-	-	-	-	-	-	-	-	-	-	-	-	-	-	-	-	-	-	-	-	-	-	-	-	-	549	-	-	-	-	-
Metco XP-1103 Coating on Armco Iron Substrate	9	696	-	-	-	-	-	-	-	-	-	-	-	-	-	-	-	-	-	-	-	-	-	697	-	-	-	-	-	-	-	-	-	-	-
Metco XP-1106 Coating on Armco Iron Substrate	9	764	-	-	-	-	-	-	-	-	-	-	-	-	-	-	-	-	-	-	-	-	-	772	-	-	-	-	-	-	-	-	-	-	-
Metco XP-1109 Coating on Armco Iron Substrate	9	873	-	-	-	-	-	-	-	-	-	-	-	-	-	-	-	-	-	-	-	-	-	874	-	-	-	-	-	-	-	-	-	-	-
Metco XP-1110 Coating on Armco Iron Substrate	9	995	-	-	-	-	-	-	-	-	-	-	-	-	-	-	-	-	-	-	-	-	-	996	-	-	-	-	-	-	-	-	-	-	-
Metco XP-1114 Coating on:																																			
Niobium Substrate	9	981	-	-	-	984	-	-	-	-	-	-	-	-	-	-	-	-	-	-	-	-	-	-	-	-	-	-	-	-	-	-	-	-	-
Stainless Steel Substrate	9	981	-	-	-	-	-	-	-	-	-	-	-	-	-	-	-	-	-	-	-	-	-	-	-	-	-	-	-	-	-	-	-	-	-
Metco XP-1121 Coating on:																																			
Nb-1Zr Substrate	9	994	-	-	-	-	-	-	-	-	-	-	-	-	-	-	-	-	-	-	-	-	-	-	-	-	-	-	-	-	-	-	-	-	-
Stainless Steel Substrate	9	994	-	-	-	-	-	-	-	-	-	-	-	-	-	-	-	-	-	-	-	-	-	-	-	-	-	-	-	-	-	-	-	-	-
Mg_2Al_3	8	-	-	-	-	-	-	-	-	-	-	1310	-	-	-	-	-	-	-	-	-	-	-	-	-	-	-	-	-	-	-	-	-	-	-
Mg_3Al_2	8	-	-	-	-	-	-	-	-	-	-	1310	-	-	-	-	-	-	-	-	-	-	-	-	-	-	-	-	-	-	-	-	-	-	-
Mg_2Sn	8	-	-	-	-	-	-	-	-	-	-	1311	-	-	-	-	-	-	-	-	-	-	-	-	-	-	-	-	-	-	-	-	-	-	-
Mica	8	-	-	-	-	-	-	-	-	-	-	-	-	-	-	-	-	-	-	-	-	-	-	-	-	-	-	-	-	1694	-	-	-	-	-
Micabond, Dull Black	9	530	-	-	-	-	-	-	-	-	-	-	-	-	-	-	-	-	-	-	-	-	-	531	-	-	-	-	-	-	-	-	-	-	-
Micro-Cell "C" Pigment in Silicone Binder	9	-	-	-	-	-	-	-	-	-	-	113	-	-	-	-	-	-	-	-	-	-	-	-	-	-	-	-	-	-	-	-	-	-	-
Mild Steel	7	1305	1307	-	-	-	-	-	-	-	-	-	-	-	-	-	-	-	-	-	-	-	-	1310	-	-	-	-	-	-	-	-	-	-	-
Miscellaneous Borates	8	-	-	-	-	-	-	-	-	-	-	581	-	-	-	-	-	-	-	-	-	-	-	-	-	-	-	-	-	-	-	-	-	-	-
Miscellaneous Carbonates	8	-	-	-	-	-	-	-	-	-	-	592	-	-	-	-	-	-	-	-	-	-	-	-	-	-	-	-	-	-	-	-	-	-	-
Miscellaneous Fluorides	8	-	-	-	-	-	-	-	-	-	-	991	-	-	-	-	-	-	-	-	-	-	-	-	-	-	-	-	-	993	-	-	-	-	-
Miscellaneous Glasses	8	-	-	-	-	1643	-	-	-	-	-	1645	-	-	-	-	-	-	-	-	-	-	-	-	-	-	-	-	1647	1649	-	-	-	-	-
Miscellaneous Hafnates	8	-	-	-	-	-	-	-	-	-	-	596	-	-	-	-	-	-	-	-	-	-	-	-	-	-	-	-	-	-	-	-	-	-	-
Miscellaneous Intermetallics	8	-	-	-	-	-	-	-	-	-	-	1351	-	-	-	-	-	-	-	-	-	-	-	-	-	-	-	-	-	-	-	-	-	-	-
Miscellaneous Oxides	8	-	-	-	-	-	-	-	-	-	-	545	-	-	-	-	-	-	-	-	-	-	-	-	-	-	-	-	-	548	-	-	-	-	-

MATERIAL NAME	VOLUME	EMITTANCE Total Hemispherical	EMITTANCE Total Normal	EMITTANCE Total Angular	EMITTANCE Spectral Hemispherical	EMITTANCE Spectral Normal	EMITTANCE Spectral Angular	REFLECTANCE Integrated Hemispherical	REFLECTANCE Integrated Normal	REFLECTANCE Integrated Angular	REFLECTANCE Spectral Hemispherical	REFLECTANCE Spectral Normal	REFLECTANCE Spectral Angular	REFLECTANCE Solar Hemispherical	REFLECTANCE Solar Normal	REFLECTANCE Solar Angular	ABSORPTANCE Integrated Hemispherical	ABSORPTANCE Integrated Normal	ABSORPTANCE Integrated Angular	ABSORPTANCE Spectral Hemispherical	ABSORPTANCE Spectral Normal	ABSORPTANCE Spectral Angular	ABSORPTANCE Solar Hemispherical	ABSORPTANCE Solar Normal	ABSORPTANCE Solar Angular	TRANSMITTANCE Integrated Hemispherical	TRANSMITTANCE Integrated Normal	TRANSMITTANCE Integrated Angular	TRANSMITTANCE Spectral Hemispherical	TRANSMITTANCE Spectral Normal	TRANSMITTANCE Spectral Angular	TRANSMITTANCE Solar Hemispherical	TRANSMITTANCE Solar Normal	TRANSMITTANCE Solar Angular	Absorptance to Emittance Ratio
Miscellaneous Phosphates	8	-	-	-	-	-	-	-	-	-	-	607	-	-	-	-	-	-	-	-	-	-	-	-	-	-	-	-	-	-	-	-	-	-	-
Miscellaneous Silicates	8	-	-	-	-	-	-	-	-	-	-	617	-	-	-	-	-	-	-	-	-	-	-	-	-	-	-	-	-	621	-	-	-	-	-
Miscellaneous Zirconates	8	-	-	-	-	-	-	-	-	-	-	675	-	-	-	-	-	-	-	-	-	-	-	-	-	-	-	-	-	-	-	-	-	-	-
MnO + TiO_2 Powders	8	-	-	-	-	-	-	-	-	-	-	563	-	-	-	-	-	-	-	-	-	-	-	-	-	-	-	-	-	-	-	-	-	-	-
Mo + 0.5 Ti Alloy	7	-	954	957	-	-	-	-	-	-	-	-	-	-	-	-	-	-	-	-	-	-	-	-	-	-	-	-	-	-	-	-	-	-	-
Molochite No. 6 Pigment in Potassium Silicate Binder	9	-	-	-	-	-	-	-	-	-	-	47	-	-	-	-	-	-	-	-	-	-	-	49	-	-	-	-	-	-	-	-	-	-	-
Molochite SF Pigment in Potassium Silicate Binder	9	-	-	-	-	-	-	-	-	-	-	-	-	-	-	-	-	-	-	-	-	-	-	49	-	-	-	-	-	-	-	-	-	-	-
Molybdenum	7	376	383	-	-	387 392	-	-	-	-	-	398	402	-	-	-	-	-	-	-	404	407	-	410	-	-	-	-	-	-	-	-	-	-	-
Molybdenum, Siliconized	9	1331	1333	1335	-	1337	-	-	-	-	-	1342	-	-	-	-	-	-	-	-	1345	-	-	-	-	-	-	-	-	-	-	-	-	-	-
Molybdenum Coating on Iron Substrate	9	696	-	-	-	-	-	-	-	-	-	-	-	-	-	-	-	-	-	-	-	-	-	697	-	-	-	-	-	-	-	-	-	-	-
Molybdenum Substrate Coated with:																																			
Aluminum Oxide	9	785	788	-	-	-	-	-	-	-	-	796	-	-	-	-	-	-	-	-	-	-	-	803	-	-	-	-	-	-	-	-	-	-	-
Boron Carbide	9	827	-	-	-	-	-	-	-	-	-	-	-	-	-	-	-	-	-	-	-	-	-	-	-	-	-	-	-	-	-	-	-	-	-
Carbon	9	852	-	-	-	-	-	-	-	-	-	-	-	-	-	-	-	-	-	-	-	-	-	862	-	-	-	-	-	-	-	-	-	-	-
$Cr_2O_3 + SiO_2 + \Sigma X_i$	9	878	-	-	-	883	-	-	-	-	-	-	-	-	-	-	-	-	-	-	-	-	-	-	-	-	-	-	-	-	-	-	-	-	-
Gold	9	651	-	-	-	-	-	-	-	-	-	-	-	-	-	-	-	-	-	-	-	-	-	-	-	-	-	-	-	-	-	-	-	-	-
Molybdenum Disilicide	9	927	928	-	-	929	-	-	-	-	-	-	-	-	-	-	-	-	-	-	-	-	-	-	-	-	-	-	-	-	-	-	-	-	-
Zirconium	9	782	-	-	-	-	-	-	-	-	-	-	-	-	-	-	-	-	-	-	-	-	-	-	-	-	-	-	-	-	-	-	-	-	-
Molybdenum + Titanium	7	-	953	956	-	959	-	-	-	-	-	962	-	-	-	-	-	-	-	-	-	-	-	-	-	-	-	-	-	-	-	-	-	-	-
Molybdenum + Tungsten	7	967	-	-	-	969	-	-	-	-	-	-	-	-	-	-	-	-	-	-	-	-	-	-	-	-	-	-	-	-	-	-	-	-	-
Molybdenum Disilicide	8	-	1148	-	-	1150 1152	-	-	-	-	-	1155	-	-	-	-	-	-	-	-	-	-	-	-	-	-	-	-	-	-	-	-	-	-	-
Molybdenum Disilicide, Oxidized	9	-	-	-	-	-	-	-	-	-	-	-	-	-	-	-	-	-	-	-	-	-	-	-	-	-	-	-	-	1311	-	-	-	-	-
Molybdenum Disilicide Coating on:																																			
Bronze Substrate	9	-	928	-	-	-	-	-	-	-	-	-	-	-	-	-	-	-	-	-	-	-	-	-	-	-	-	-	-	-	-	-	-	-	-
Molybdenum Substrate	9	927	928	-	-	929	-	-	-	-	-	-	-	-	-	-	-	-	-	-	-	-	-	-	-	-	-	-	-	-	-	-	-	-	-
Molybdenum Disulfide Coating on Inconel X Substrate	9	-	-	-	-	-	-	-	-	-	-	930	-	-	-	-	-	-	-	-	-	-	-	-	-	-	-	-	-	-	-	-	-	-	-
Molybdenum Monoboride	8	-	-	-	-	692	-	-	-	-	-	-	-	-	-	-	-	-	-	-	-	-	-	-	-	-	-	-	-	-	-	-	-	-	-

MATERIAL NAME	VOLUME	EMITTANCE						REFLECTANCE									ABSORPTANCE									TRANSMITTANCE									Absorptance to Emittance Ratio
		Total			Spectral			Integrated			Spectral			Solar			Integrated			Spectral			Solar			Integrated			Spectral			Solar			
		Hemispherical	Normal	Angular	Hemispherical	Normal	Angular	Hemispherical	Normal	Angular	Hemispherical	Normal	Angular	Hemispherical	Normal	Angular	Hemispherical	Normal	Angular	Hemispherical	Normal	Angular	Hemispherical	Normal	Angular	Hemispherical	Normal	Angular	Hemispherical	Normal	Angular	Hemispherical	Normal	Angular	
Molybdenum Diboride	8	-	-	-	-	-	-	-	-	-	-	695	-	-	-	-	-	-	-	-	-	-	-	-	-	-	-	-	-	-	-	-	-	-	-
Dimolybdenum Pentaboride	8	-	-	-	-	-	-	-	-	-	-	695	-	-	-	-	-	-	-	-	-	-	-	-	-	-	-	-	-	-	-	-	-	-	-
Dimolybdenum Carbide	8	-	850	-	-	852	-	-	-	-	-	-	-	-	-	-	-	-	-	-	-	-	-	-	-	-	-	-	-	-	-	-	-	-	-
Trimolybdenum Silicide	8	-	-	-	-	1150	-	-	-	-	-	-	-	-	-	-	-	-	-	-	-	-	-	-	-	-	-	-	-	-	-	-	-	-	-
Molybdenum Trioxide	8	-	-	-	-	-	-	-	-	-	-	330	-	-	-	-	-	-	-	-	-	-	-	-	-	-	-	-	-	-	-	-	-	-	-
Pentamolybdenum Trisilicide	8	-	-	-	-	1150	-	-	-	-	-	-	-	-	-	-	-	-	-	-	-	-	-	-	-	-	-	-	-	-	-	-	-	-	-
Molybdenum Disulfide	8	-	-	-	-	-	-	-	-	-	-	1207	-	-	-	-	-	-	-	-	-	-	-	-	-	-	-	-	-	1210	-	-	-	-	-
Monel	7	1423	1428	-	-	-	-	-	-	-	-	1431	-	-	-	-	-	-	-	-	-	-	-	-	-	-	-	-	-	-	-	-	-	-	-
Monel Substrate Coated with Chromium	9	618	-	-	-	-	-	-	-	-	-	-	-	-	-	-	-	-	-	-	-	-	-	-	-	-	-	-	-	-	-	-	-	-	-
Monel 400 (see Monel)																																			
Monel K-500 (see K Monel)																																			
Monel 501 (see KR Monel)																																			
MoO_3 + MoSi Powders	8	-	1473	-	-	1474	-	-	-	-	-	1476	-	-	-	-	-	-	-	-	-	-	-	-	-	-	-	-	-	-	-	-	-	-	-
MoO_3 + $MoSi_2$ + SiO_2 Powders	8	-	1510	-	-	1511	-	-	-	-	-	1513	-	-	-	-	-	-	-	-	-	-	-	-	-	-	-	-	-	-	-	-	-	-	-
MoO_3 + Ni	8	-	-	-	-	-	-	-	-	-	-	1425	-	-	-	-	-	-	-	-	-	-	-	-	-	-	-	-	-	-	-	-	-	-	-
MoO_3 + SiO_2 Powders	8	-	-	-	-	-	-	-	-	-	-	566	-	-	-	-	-	-	-	-	-	-	-	-	-	-	-	-	-	-	-	-	-	-	-
$MoSi_2$ + MoO_3	8	-	1473	-	-	1474	-	-	-	-	-	1476	-	-	-	-	-	-	-	-	-	-	-	-	-	-	-	-	-	-	-	-	-	-	-
$MoSi_2$ + MoO_3 + SiO_2 Powders	8	-	1510	-	-	1511	-	-	-	-	-	1513	-	-	-	-	-	-	-	-	-	-	-	-	-	-	-	-	-	-	-	-	-	-	-
$MoSi_2$ + SiO_2 Powders	8	-	1478	-	-	1479	-	-	-	-	-	1482	-	-	-	-	-	-	-	-	-	-	-	-	-	-	-	-	-	-	-	-	-	-	-
Mullite	8	-	1685	-	-	1687	-	-	-	-	-	618	-	-	-	-	-	-	-	-	-	-	-	-	-	-	-	-	-	-	-	-	-	-	-
Multimet	7	1214	1227	-	-	1238	-	-	-	-	-	-	-	-	-	-	-	-	-	-	-	-	-	-	-	-	-	-	-	-	-	-	-	-	-
Mylar	8	-	-	-	-	-	-	-	-	-	-	-	-	-	-	-	-	-	-	-	1708	-	-	-	1710	-	-	-	-	1711	-	-	-	-	-
Mylar Coating on Aluminum Substrate	9	1042	-	-	-	-	-	-	-	-	-	-	-	-	-	-	-	-	-	-	-	-	-	-	-	-	-	-	-	-	-	-	-	-	-
Mylar Substrate Coated with:																																			
Aluminum	9	580	-	-	-	-	-	-	-	-	-	592	-	-	-	-	609	-	-	-	-	-	-	-	-	-	-	-	-	-	-	-	-	-	-
Gold	9	651	-	-	-	-	-	-	-	-	-	-	-	-	-	-	675	-	-	-	-	-	-	-	-	-	-	-	-	-	-	-	-	-	-
Silver	9	-	-	-	-	-	-	-	-	-	-	-	-	-	746	-	-	-	-	-	-	-	-	-	-	-	-	-	-	750	-	-	-	-	-

MATERIAL NAME	VOLUME	EMITTANCE						REFLECTANCE									ABSORPTANCE									TRANSMITTANCE									Absorptance to Emittance Ratio
		Total			Spectral			Integrated			Spectral			Solar			Integrated			Spectral			Solar			Integrated			Spectral			Solar			
		Hemispherical	Normal	Angular	Hemispherical	Normal	Angular	Hemispherical	Normal	Angular	Hemispherical	Normal	Angular	Hemispherical	Normal	Angular	Hemispherical	Normal	Angular	Hemispherical	Normal	Angular	Hemispherical	Normal	Angular	Hemispherical	Normal	Angular	Hemispherical	Normal	Angular	Hemispherical	Normal	Angular	
Natrolite	8	–	–	–	–	–	–	–	–	–	–	–	–	–	–	–	–	–	–	–	–	–	–	–	–	–	–	–	–	1694	–	–	–	–	–
$NbAl_3$	8	–	–	–	–	–	–	–	–	–	–	1322	–	–	–	–	–	–	–	–	–	–	–	–	–	–	–	–	–	–	–	–	–	–	–
NBS Coating A-418	9	–	455	–	–	459	–	–	–	–	–	–	–	–	–	–	–	–	–	–	–	–	–	–	–	–	–	–	–	–	–	–	–	–	–
NBS Ceramic A-418 Substrate Coated with Gold	9	–	656	–	–	–	–	–	–	–	–	–	–	–	–	–	–	–	–	–	–	–	–	–	–	–	–	–	–	–	–	–	–	–	–
NBS Coating N-143	9	–	445	–	–	448	–	–	–	–	–	–	–	–	–	–	–	–	–	–	–	–	–	–	–	–	–	–	–	–	–	–	–	–	–
NBS Frit No. 332 Binder Pigmented with (Enamels):																																			
Cerium Dioxide + Magnesium Oxide	9	–	–	–	–	–	–	–	–	–	–	452	–	–	–	–	–	–	–	–	–	–	–	–	–	–	–	–	–	–	–	–	–	–	–
Cerium Dioxide + Tin Oxide	9	–	445	–	–	–	–	–	–	–	–	452	–	–	–	–	–	–	–	–	–	–	–	–	–	–	–	–	–	–	–	–	–	–	–
Cerium Dioxide + Zirconium Oxide	9	–	–	–	–	–	–	–	–	–	–	452	–	–	–	–	–	–	–	–	–	–	–	–	–	–	–	–	–	–	–	–	–	–	–
Chromium Oxide	9	–	455	–	–	–	–	–	–	–	–	463	–	–	–	–	–	–	–	–	–	–	–	–	–	–	–	–	–	–	–	–	–	–	–
Chromium Oxide + Black Stain	9	–	455	–	–	–	–	–	–	–	–	463	–	–	–	–	–	–	–	–	–	–	–	–	–	–	–	–	–	–	–	–	–	–	–
$CoO \cdot Cr_2O_3$ Spinel	9	–	472	–	–	–	–	–	–	–	–	475	–	–	–	–	–	–	–	–	–	–	–	–	–	–	–	–	–	–	–	–	–	–	–
$CoO \cdot Fe_2O_2$ Spinel	9	–	472	–	–	–	–	–	–	–	–	475	–	–	–	–	–	–	–	–	–	–	–	–	–	–	–	–	–	–	–	–	–	–	–
$CoO \cdot Mn_2O_3$ Spinel	9	–	472	–	–	–	–	–	–	–	–	475	–	–	–	–	–	–	–	–	–	–	–	–	–	–	–	–	–	–	–	–	–	–	–
$Fe_2O_3 + CoO + Cr_2O_3$	9	–	466	–	–	–	–	–	–	–	–	467	–	–	–	–	–	–	–	–	–	–	–	–	–	–	–	–	–	–	–	–	–	–	–
$NiO \cdot Cr_2O_3$ Spinel	9	–	472	–	–	–	–	–	–	–	–	475	–	–	–	–	–	–	–	–	–	–	–	–	–	–	–	–	–	–	–	–	–	–	–
$NiO \cdot Fe_2O_3$ Spinel	9	–	472	–	–	–	–	–	–	–	–	475	–	–	–	–	–	–	–	–	–	–	–	–	–	–	–	–	–	–	–	–	–	–	–
Tin Oxide + Cerium Dioxide	9	–	477	–	–	–	–	–	–	–	–	478	–	–	–	–	–	–	–	–	–	–	–	–	–	–	–	–	–	–	–	–	–	–	–
Zirconium Oxide + Cerium Dioxide	9	–	479	–	–	–	–	–	–	–	–	–	–	–	–	–	–	–	–	–	–	–	–	–	–	–	–	–	–	–	–	–	–	–	–
Necoloidine Binder Pigmented with:																																			
Titanium Hydride	9	209	–	–	–	–	–	–	–	–	–	–	–	–	–	–	–	–	–	–	–	–	–	–	–	–	–	–	–	–	–	–	–	–	–
Zirconium Carbide	9	421	–	–	–	–	–	–	–	–	–	–	–	–	–	–	–	–	–	–	–	–	–	–	–	–	–	–	–	–	–	–	–	–	–
Zirconium Hydride + Iron Oxide	9	422	–	–	–	–	–	–	–	–	–	–	–	–	–	–	–	–	–	–	–	–	–	–	–	–	–	–	–	–	–	–	–	–	–
Zirconium Hydride + Zirconia	9	422	–	–	–	–	–	–	–	–	–	–	–	–	–	–	–	–	–	–	–	–	–	–	–	–	–	–	–	–	–	–	–	–	–
Neodymium Coating on Quartz Substrate	9	–	–	–	–	–	–	–	–	–	–	–	–	–	–	–	–	–	–	–	–	–	–	–	–	–	–	–	–	698	–	–	–	–	–
Neodymium Hexaboride	8	–	–	–	–	723	–	–	–	–	–	–	–	–	–	–	–	–	–	–	–	–	–	–	–	–	–	–	–	727	–	–	–	–	–
Neodymium Oxide	8	–	–	–	–	–	–	–	–	–	–	335	–	–	–	–	–	–	–	–	–	–	–	–	–	–	–	–	–	–	–	–	–	–	–
Neodymium Silicate	8	–	–	–	–	–	–	–	–	–	–	–	–	–	–	–	–	–	–	–	–	–	–	–	–	–	–	–	–	622	–	–	–	–	–
Dineodymium Trisulfide	8	–	–	–	–	1232	–	–	–	–	–	–	–	–	–	–	–	–	–	–	–	–	–	–	–	–	–	–	–	–	–	–	–	–	–
$Ni + MoO_3$	8	–	–	–	–	–	–	–	–	–	–	1425	–	–	–	–	–	–	–	–	–	–	–	–	–	–	–	–	–	–	–	–	–	–	–

MATERIAL NAME	VOLUME	EMITTANCE						REFLECTANCE									ABSORPTANCE									TRANSMITTANCE									Absorptance to Emittance Ratio
		Total			Spectral			Integrated			Spectral			Solar			Integrated			Spectral			Solar			Integrated			Spectral			Solar			
		Hemispherical	Normal	Angular	Hemispherical	Normal	Angular	Hemispherical	Normal	Angular	Hemispherical	Normal	Angular	Hemispherical	Normal	Angular	Hemispherical	Normal	Angular	Hemispherical	Normal	Angular	Hemispherical	Normal	Angular	Hemispherical	Normal	Angular	Hemispherical	Normal	Angular	Hemispherical	Normal	Angular	
Ni + TiB_2	8	-	-	-	-	-	-	-	-	-	-	1435	-	-	-	-	-	-	-	-	-	-	-	-	-	-	-	-	-	-	-	-	-	-	-
NiAl	8	-	1316	-	-	1318	-	-	-	-	-	1321	-	-	-	-	-	-	-	-	-	-	-	-	-	-	-	-	-	-	-	-	-	-	-
Ni_3Al	8	-	1316	-	-	1318	-	-	-	-	-	1321	-	-	-	-	-	-	-	-	-	-	-	-	-	-	-	-	-	-	-	-	-	-	-
Nichrome	7	1344	1353	-	-	-	-	-	-	-	-	-	-	-	-	-	-	-	-	-	-	-	-	-	-	-	-	-	-	-	-	-	-	-	-
Nickel	7	413	416	-	-	424 434	-	-	440	446	-	454	457	-	-	-	-	460	-	-	462	465	468	470	-	-	-	-	-	-	-	-	-	-	-
Nickel, Grade A	7	-	418 419 420	-	-	426 427 428 429	-	-	-	-	-	455	-	-	-	-	-	-	-	-	-	-	-	471	-	-	-	-	-	-	-	-	-	-	-
Nickel, Grade NP-3	7	-	-	-	-	435	-	-	-	-	-	-	-	-	-	-	-	-	-	-	-	-	-	-	-	-	-	-	-	-	-	-	-	-	-
Nickel, Oxidized	9	1312	-	-	-	-	-	-	-	-	-	-	-	-	-	-	-	-	-	-	-	-	-	-	-	-	-	-	-	-	-	-	-	-	-
Nickel Coating on:																																			
Copper Substrate	9	700	-	-	-	-	-	-	-	-	-	702	-	-	-	-	-	-	-	-	-	-	-	-	-	-	-	-	-	-	-	-	-	-	-
Epoxy Substrate	9	700	-	-	-	-	-	-	-	-	-	-	704	-	-	-	-	-	-	-	-	-	-	-	-	-	-	-	-	-	-	-	-	-	-
Polyurethane Substrate	9	-	-	-	-	-	-	-	-	-	-	-	704	-	-	-	-	-	-	-	-	-	-	-	-	-	-	-	-	-	-	-	-	-	-
Steel Substrate	9	700	-	-	-	-	-	-	-	-	-	-	-	-	-	-	-	-	-	-	-	-	-	-	-	-	-	-	-	-	-	-	-	-	-
Nickel Substrate Coated with:																																			
Barium + Strontium	9	616	-	-	-	-	-	-	-	-	-	-	-	-	-	-	-	-	-	-	-	-	-	-	-	-	-	-	-	-	-	-	-	-	-
Chromium	9	-	-	-	-	-	-	-	-	-	-	621	-	-	-	-	-	-	-	-	-	626	-	627	-	-	-	-	-	-	-	-	-	-	-
Copper Oxide	9	-	891	-	-	-	-	-	-	-	-	-	-	-	-	-	-	-	-	-	-	-	-	-	-	-	-	-	-	-	-	-	-	-	-
Gold	9	652	-	-	-	-	-	-	-	-	-	-	672	-	-	-	-	-	-	-	-	-	-	-	-	-	-	-	-	-	-	-	-	-	-
Silicon Dioxide	9	-	-	-	-	-	-	-	-	-	-	-	1060	-	-	-	-	-	-	-	-	-	-	-	-	-	-	-	-	-	-	-	-	-	-
Silver	9	734	-	-	-	-	-	-	-	-	-	-	-	-	-	-	747	-	-	-	-	-	-	-	-	-	-	-	-	-	-	-	-	-	-
Titanium Dioxide	9	981	-	-	-	-	-	-	-	-	-	-	-	-	-	-	-	-	-	-	-	-	-	-	-	-	-	-	-	-	-	-	-	-	-
Nickel + Chromium	7	-	-	-	-	972	-	-	-	-	-	-	-	-	-	-	-	-	-	-	-	-	-	-	-	-	-	-	-	-	-	-	-	-	-
Nickel + Chromium Coating on:																																			
Hastelloy X Substrate	9	707	-	-	-	-	-	-	-	-	-	-	-	-	-	-	-	-	-	-	-	-	-	-	-	-	-	-	-	-	-	-	-	-	-
Inconel X Substrate	9	-	-	-	-	709	-	-	-	-	-	712	-	-	-	-	-	-	-	-	-	-	-	-	-	-	-	-	-	-	-	-	-	-	-
Nickel + Chromium + ΣX_i	7	1342	1347	-	-	1356 1360	-	-	-	-	-	1374	-	-	1392	-	-	-	-	-	1395	1398	-	1400	-	-	-	-	-	-	-	-	-	-	-
Nickel + Chromium + ΣX_i, Oxidized	7	-	1314	-	-	-	-	-	-	-	-	-	-	-	-	-	-	-	-	-	-	-	-	-	-	-	-	-	-	-	-	-	-	-	-
Nickel + Cobalt Coating on Stainless Steel Substrate	9	-	715	-	-	-	-	-	-	-	-	716	-	-	-	-	-	-	-	-	-	-	-	-	-	-	-	-	-	-	-	-	-	-	-

MATERIAL NAME	VOLUME	EMITTANCE						REFLECTANCE									ABSORPTANCE									TRANSMITTANCE									Absorptance to Emittance Ratio
		Total			Spectral			Integrated			Spectral			Solar			Integrated			Spectral			Solar			Integrated			Spectral			Solar			
		Hemispherical	Normal	Angular	Hemispherical	Normal	Angular	Hemispherical	Normal	Angular	Hemispherical	Normal	Angular	Hemispherical	Normal	Angular	Hemispherical	Normal	Angular	Hemispherical	Normal	Angular	Hemispherical	Normal	Angular	Hemispherical	Normal	Angular	Hemispherical	Normal	Angular	Hemispherical	Normal	Angular	
Nickel + Cobalt + ΣX_i	7	1404	-	-	-	1406 1410	-	-	-	-	-	1416	-	-	-	-	-	-	-	-	-	-	-	-	-	-	-	-	-	-	-	-	-	-	-
Nickel + Copper + ΣX_i	7	1423	1426	-	-	-	-	-	-	-	-	1430	-	-	1433	-	-	-	-	-	-	-	-	1436	-	-	-	-	-	-	-	-	-	-	-
Nickel + Iron	7	-	-	-	-	976	-	-	-	-	-	-	-	-	-	-	-	-	-	-	-	-	-	-	-	-	-	-	-	-	-	-	-	-	-
Nickel + Iron + ΣX_i	7	-	-	-	-	1439	-	-	-	-	-	-	-	-	-	-	-	-	-	-	-	-	-	-	-	-	-	-	-	-	-	-	-	-	-
Nickel + Molybdenum + ΣX_i	7	1442	1446	-	-	1450	-	-	-	-	-	1454	-	-	-	-	-	-	-	-	-	-	-	1457	-	-	-	-	-	-	-	-	-	-	-
Nickel + Molybdenum + ΣX_i Coating on Stainless Steel Substrate	9	717	-	-	-	-	-	-	-	-	-	-	-	-	-	-	-	-	-	-	-	-	-	-	-	-	-	-	-	-	-	-	-	-	-
~~Nickel + Silver~~	~~7~~	-	~~979~~	-	-	-	-	-	-	-	-	-	-	-	-	-	-	-	-	-	-	-	-	-	-	-	-	-	-	-	-	-	-	-	-
Nickel Aluminide Coating on Inconel Substrate	9	-	706	-	-	-	-	-	-	-	-	-	-	-	-	-	-	-	-	-	-	-	-	-	-	-	-	-	-	-	-	-	-	-	-
Nickel Chromate Coating on Nb-1Zr Coating	9	932	-	-	-	-	-	-	-	-	-	-	-	-	-	-	-	-	-	-	-	-	-	-	-	-	-	-	-	-	-	-	-	-	-
Nickel Monoxides	8	-	337	-	-	339	-	-	-	-	-	342	-	-	-	-	-	-	-	-	-	-	-	-	-	-	-	-	-	-	-	-	-	-	-
Dinickel Trioxide	8	-	-	-	-	-	-	-	-	-	-	342	-	-	-	-	-	-	-	-	-	-	-	-	-	-	-	-	-	-	-	-	-	-	-
Nickel Oxide + Cobalt Oxide Pigment in Base Glaze No. 3 Binder (Enamel)	9	-	469	-	-	-	-	-	-	-	-	-	-	-	-	-	-	-	-	-	-	-	-	-	-	-	-	-	-	-	-	-	-	-	-
Nickel Oxide + Iron Oxide Pigment in Base Glaze No. 3 Binder (Enamel)	9	-	469	-	-	-	-	-	-	-	-	-	-	-	-	-	-	-	-	-	-	-	-	-	-	-	-	-	-	-	-	-	-	-	-
Nickel Oxide Substrate Coated with Gold	9	-	656	-	-	-	-	-	-	-	-	-	-	-	-	-	-	-	-	-	-	-	-	-	-	-	-	-	-	-	-	-	-	-	-
Nickel Disilicide	8	-	-	-	-	1173	-	-	-	-	-	-	-	-	-	-	-	-	-	-	-	-	-	-	-	-	-	-	-	-	-	-	-	-	-
Nimonic 75	7	1345	1350	-	-	-	-	-	-	-	-	-	-	-	-	-	-	-	-	-	-	-	-	-	-	-	-	-	-	-	-	-	-	-	-
NiO + Al_2O_3 Powders	8	-	-	-	-	556	-	-	-	-	-	-	-	-	-	-	-	-	-	-	-	-	-	-	-	-	-	-	-	-	-	-	-	-	-
NiO + NiAl + ΣX_i Cermets	8	-	1393	-	-	1394	-	-	-	-	-	1398	-	-	-	-	-	-	-	-	-	-	-	-	-	-	-	-	-	-	-	-	-	-	-
NiO · Cr_2O_3 Spinel Pigment in NBS Frit No. 332 Binder (Enamel)	9	-	472	-	-	-	-	-	-	-	-	475	-	-	-	-	-	-	-	-	-	-	-	-	-	-	-	-	-	-	-	-	-	-	-
NiO · Cr_2O_3 Spinel + SiO_2 Pigment in Aluminum Phosphate Binder	9	186	-	-	-	-	-	-	-	-	-	-	-	-	-	-	-	-	-	-	-	-	-	-	-	-	-	-	-	-	-	-	-	-	-
NiO · Fe_2O_3 Spinel Pigment in NBS Frit No. 332 Binder (Enamel)	9	-	472	-	-	-	-	-	-	-	-	475	-	-	-	-	-	-	-	-	-	-	-	-	-	-	-	-	-	-	-	-	-	-	-

MATERIAL NAME	VOLUME	EMITTANCE						REFLECTANCE									ABSORPTANCE									TRANSMITTANCE									Absorptance to Emittance Ratio
		Total			Spectral			Integrated			Spectral			Solar			Integrated			Spectral			Solar			Integrated			Spectral			Solar			
		Hemispherical	Normal	Angular	Hemispherical	Normal	Angular	Hemispherical	Normal	Angular	Hemispherical	Normal	Angular	Hemispherical	Normal	Angular	Hemispherical	Normal	Angular	Hemispherical	Normal	Angular	Hemispherical	Normal	Angular	Hemispherical	Normal	Angular	Hemispherical	Normal	Angular	Hemispherical	Normal	Angular	
Nimonic 75 Substrate Coated with:																																			
Aluminum Oxide	9	-	788	-	-	-	-	-	-	-	-	-	-	-	-	-	-	-	-	-	-	-	-	-	-	-	-	-	-	-	-	-	-	-	-
Cerium Dioxide	9	-	870	-	-	-	-	-	-	-	-	-	-	-	-	-	-	-	-	-	-	-	-	-	-	-	-	-	-	-	-	-	-	-	-
Magnesium Oxide	9	-	919	-	-	-	-	-	-	-	-	-	-	-	-	-	-	-	-	-	-	-	-	-	-	-	-	-	-	-	-	-	-	-	-
Silicon Dioxide	9	-	1052	-	-	-	-	-	-	-	-	-	-	-	-	-	-	-	-	-	-	-	-	-	-	-	-	-	-	-	-	-	-	-	-
Thorium Dioxide	9	-	973	-	-	-	-	-	-	-	-	-	-	-	-	-	-	-	-	-	-	-	-	-	-	-	-	-	-	-	-	-	-	-	-
Zirconium Oxide	9	-	1011	-	-	-	-	-	-	-	-	-	-	-	-	-	-	-	-	-	-	-	-	-	-	-	-	-	-	-	-	-	-	-	-
Niobates:																																			
Lithium Niobate	8	-	-	-	-	-	-	-	-	-	-	598	-	-	-	-	-	-	-	-	-	-	-	-	-	-	-	-	-	-	-	-	-	-	-
Niobium	7	474	480	-	-	482 486	-	-	-	-	-	492	-	-	-	-	-	-	-	-	-	497	-	-	-	-	-	-	-	-	-	-	-	-	-
Niobium Substrate Coated by Pack Cementation	9	-	1347	-	-	1348	-	-	-	-	-	1349	-	-	-	-	-	-	-	-	-	-	-	-	-	-	-	-	-	-	-	-	-	-	-
Niobium Substrate Coated with:																																			
Aluminum Oxide	9	785	-	-	-	-	-	-	-	-	-	-	-	-	-	-	-	-	-	-	-	-	-	-	-	-	-	-	-	-	-	-	-	-	-
$Cr_2O_3 + SiO_2 + \Sigma X_i$	9	878	-	-	-	885	-	-	-	-	-	-	-	-	-	-	-	-	-	-	-	-	-	-	-	-	-	-	-	-	-	-	-	-	-
Silicon Dioxide	9	1049	-	-	-	-	-	-	-	-	-	-	-	-	-	-	-	-	-	-	-	-	-	-	-	-	-	-	-	-	-	-	-	-	-
Titanium Dioxide	9	981	-	-	-	984	-	-	-	-	-	-	-	-	-	-	-	-	-	-	-	-	-	-	-	-	-	-	-	-	-	-	-	-	-
Nb-1Zr Alloy	7	990	-	-	-	-	-	-	-	-	-	-	-	-	-	-	-	-	-	-	-	-	-	-	-	-	-	-	-	-	-	-	-	-	-
Nb-1Zr Substrate Coated with:																																			
Aluminum Oxide + Aluminum Titanate	9	808	-	-	-	-	-	-	-	-	-	-	-	-	-	-	-	-	-	-	-	-	-	-	-	-	-	-	-	-	-	-	-	-	-
Barium Titanate	9	815	-	-	-	817	-	-	-	-	-	-	-	-	-	-	-	-	-	-	-	-	-	-	-	-	-	-	-	-	-	-	-	-	-
Boron	9	826	-	-	-	-	-	-	-	-	-	-	-	-	-	-	-	-	-	-	-	-	-	-	-	-	-	-	-	-	-	-	-	-	-
Calcium Titanate	9	843	-	-	-	849	-	-	-	-	-	-	-	-	-	-	-	-	-	-	-	-	-	-	-	-	-	-	-	-	-	-	-	-	-
Iron Titanate	9	905	-	-	-	-	-	-	-	-	-	-	-	-	-	-	-	-	-	-	-	-	-	-	-	-	-	-	-	-	-	-	-	-	-
Iron Titanium Aluminum Oxide	9	909	-	-	-	911	-	-	-	-	-	-	-	-	-	-	-	-	-	-	-	-	-	-	-	-	-	-	-	-	-	-	-	-	-
Magnesium Aluminate	9	915	-	-	-	917	-	-	-	-	-	-	-	-	-	-	-	-	-	-	-	-	-	-	-	-	-	-	-	-	-	-	-	-	-
Nickel Chromate	9	932	-	-	-	-	-	-	-	-	-	-	-	-	-	-	-	-	-	-	-	-	-	-	-	-	-	-	-	-	-	-	-	-	-
Strontium Titanate	9	958	-	-	-	962	-	-	-	-	-	-	-	-	-	-	-	-	-	-	-	-	-	-	-	-	-	-	-	-	-	-	-	-	-
Titanium Dioxide + Aluminum Oxide	9	994	-	-	-	-	-	-	-	-	-	-	-	-	-	-	-	-	-	-	-	-	-	-	-	-	-	-	-	-	-	-	-	-	-
Zirconium Silicate	9	1024	-	-	-	1028	-	-	-	-	-	-	-	-	-	-	-	-	-	-	-	-	-	-	-	-	-	-	-	-	-	-	-	-	-
Niobium + Molybdenum + ΣX_i	7	-	1460	-	-	-	-	-	-	-	-	-	-	-	-	-	-	-	-	-	-	-	-	-	-	-	-	-	-	-	-	-	-	-	-
Niobium + Tantalum + ΣX_i	7	1463	-	-	-	1466	-	-	-	-	-	-	-	-	-	-	-	-	-	-	-	-	-	-	-	-	-	-	-	-	-	-	-	-	-
Niobium + Tungsten	7	981	-	-	-	984	-	-	-	-	-	-	-	-	-	-	-	-	-	-	-	-	-	-	-	-	-	-	-	-	-	-	-	-	-
Niobium + Tungsten + ΣX_i	7	1469	-	-	-	-	-	-	-	-	-	-	-	-	-	-	-	-	-	-	-	-	-	-	-	-	-	-	-	-	-	-	-	-	-

MATERIAL NAME	VOLUME	EMITTANCE						REFLECTANCE									ABSORPTANCE									TRANSMITTANCE									Absorptance to Emittance Ratio
		Total			Spectral			Integrated			Spectral			Solar			Integrated			Spectral			Solar			Integrated			Spectral			Solar			
		Hemispherical	Normal	Angular	Hemispherical	Normal	Angular	Hemispherical	Normal	Angular	Hemispherical	Normal	Angular	Hemispherical	Normal	Angular	Hemispherical	Normal	Angular	Hemispherical	Normal	Angular	Hemispherical	Normal	Angular	Hemispherical	Normal	Angular	Hemispherical	Normal	Angular	Hemispherical	Normal	Angular	
Niobium + Zirconium	7	988	992	-	-	994	-	-	-	-	-	-	-	-	-	-	-	-	-	-	-	-	-	-	-	-	-	-	-	-	-	-	-	-	-
Niobium Diboride	8	697	699 701	-	-	-	-	-	-	-	-	-	-	-	-	-	-	-	-	-	-	-	-	-	-	-	-	-	-	-	-	-	-	-	-
Niobium Monocarbide	8	785	787	-	-	789	-	-	-	-	-	-	-	-	-	-	-	-	-	-	-	-	-	-	-	-	-	-	-	-	-	-	-	-	-
Niobium Nitride	8	-	-	-	-	1087	-	-	-	-	-	-	-	-	-	-	-	-	-	-	-	-	-	-	-	-	-	-	-	-	-	-	-	-	-
Diniobium Nitride	8	-	-	-	-	1087	-	-	-	-	-	-	-	-	-	-	-	-	-	-	-	-	-	-	-	-	-	-	-	-	-	-	-	-	-
Diniobium Pentoxide	8	-	-	-	-	347	-	-	-	-	-	349	-	-	-	-	-	-	-	-	-	-	-	-	-	-	-	-	-	-	-	-	-	-	-
Niobium Disilicide	8	-	-	-	-	1173	-	-	-	-	-	-	-	-	-	-	-	-	-	-	-	-	-	-	-	-	-	-	-	-	-	-	-	-	-
Nitrides:																																			
Aluminum Nitride	8	-	1030	-	-	1031 1033	-	-	-	-	-	1035	-	-	-	-	-	-	-	-	-	-	-	-	-	-	-	-	-	-	-	-	-	-	-
Boron Nitride	8	-	1037	-	-	1040 1042	-	-	-	-	-	1047	-	-	-	-	-	-	-	-	-	-	-	-	-	-	-	-	-	1054	-	-	-	-	-
Hafnium Nitride	8	-	1056	-	-	1058 1060	-	-	-	-	-	-	-	-	-	-	-	-	-	-	-	-	-	-	-	-	-	-	-	-	-	-	-	-	-
Nitrocellulose Binder Pigmented with:																																			
Aluminum	9	3	-	-	-	-	-	-	-	-	-	-	-	-	-	-	-	-	-	-	-	-	-	-	-	-	-	-	-	-	-	-	-	-	-
Carbon Black	9	-	-	-	-	-	-	-	-	-	-	86	-	-	-	-	-	-	-	-	-	-	-	-	-	-	-	-	-	-	-	-	-	-	-
Lead Carbonate	9	-	-	-	-	-	-	-	-	-	-	139	-	-	-	-	-	-	-	-	-	-	-	-	-	-	-	-	-	-	-	-	-	-	-
Titanium Dioxide	9	211	-	-	-	-	-	-	-	-	-	221	-	-	-	-	-	-	-	-	-	-	-	-	-	-	-	-	-	-	-	-	-	-	-
Titanium Pyrophosphate	9	298	-	-	-	-	-	-	-	-	-	-	-	-	-	-	-	-	-	-	-	-	-	-	-	-	-	-	-	-	-	-	-	-	-
Zirconium Oxide	9	423	-	-	-	-	-	-	-	-	-	-	-	-	-	-	-	-	-	-	-	-	-	-	-	-	-	-	-	-	-	-	-	-	-
Nitrocellulose Coating on Copper Substrate	9	1131	-	-	-	-	-	-	-	-	-	-	-	-	-	-	-	-	-	-	-	-	-	-	-	-	-	-	-	-	-	-	-	-	-
North American Aviation Paint No. 277.2.1	9	-	82	-	-	-	-	-	-	-	-	86	-	-	-	-	-	-	-	-	-	-	-	-	-	-	-	-	-	-	-	-	-	-	-
Norton A 402	8	-	-	-	-	150	-	-	-	-	-	-	-	-	-	-	-	-	-	-	-	-	-	-	-	-	-	-	-	-	-	-	-	-	-
Norton LA 603	8	-	144	-	-	147	-	-	-	-	-	160	-	-	-	-	-	-	-	-	-	-	-	168	-	-	-	-	-	-	-	-	-	-	-
Norton RA 4213	8	-	144	-	-	147	-	-	-	-	-	160	-	-	-	-	-	-	-	-	-	-	-	168	-	-	-	-	-	-	-	-	-	-	-
Norton RZ	8	-	-	-	-	534	-	-	-	-	-	-	-	-	-	-	-	-	-	-	-	-	-	-	-	-	-	-	-	-	-	-	-	-	-
Norton 5190	8	-	-	-	-	150	-	-	-	-	-	-	-	-	-	-	-	-	-	-	-	-	-	-	-	-	-	-	-	-	-	-	-	-	-
Nuclear Enterprise Paint No. 561	9	-	-	-	-	-	-	-	-	-	-	543	-	-	-	-	-	-	-	-	-	-	-	-	-	-	-	-	-	-	-	-	-	-	-
Nylon Coating on Stainless Steel Substrate	9	-	-	-	-	-	-	-	-	-	-	1132	-	-	-	-	-	-	-	-	-	-	-	-	-	-	-	-	-	-	-	-	-	-	-

MATERIAL NAME	VOLUME	EMITTANCE						REFLECTANCE									ABSORPTANCE									TRANSMITTANCE									Absorptance to Emittance Ratio
		Total			Spectral			Integrated			Spectral			Solar			Integrated			Spectral			Solar			Integrated			Spectral			Solar			
		Hemispherical	Normal	Angular	Hemispherical	Normal	Angular	Hemispherical	Normal	Angular	Hemispherical	Normal	Angular	Hemispherical	Normal	Angular	Hemispherical	Normal	Angular	Hemispherical	Normal	Angular	Hemispherical	Normal	Angular	Hemispherical	Normal	Angular	Hemispherical	Normal	Angular	Hemispherical	Normal	Angular	
Nylon Substrate Coated with Polyvinyl Chloride	9	-	-	-	-	-	-	-	-	-	-	1150	-	-	-	-	-	-	-	-	-	-	-	1152	-	-	-	-	-	-	-	-	-	-	-
Obsidian	8	-	-	-	-	-	-	-	-	-	-	1681	-	-	-	-	-	-	-	-	-	-	-	-	-	-	-	-	-	-	-	-	-	-	-
OFHC Copper	7	138	-	-	-	-	-	-	-	-	-	-	-	-	-	-	-	-	-	-	-	189	-	-	-	-	-	-	-	-	-	-	-	-	-
Oil Binder Pigmented with Aluminum	9	-	-	-	-	-	-	-	-	9	-	-	-	-	-	-	-	-	-	-	-	-	-	-	-	-	-	-	-	-	-	-	-	-	-
Oligoclase	8	-	-	-	-	-	-	-	-	-	-	-	1689	-	-	-	-	-	-	-	-	-	-	-	-	-	-	-	-	-	-	-	-	-	-
Olivene	8	-	-	-	-	-	-	-	-	-	-	-	1689	-	-	-	-	-	-	-	-	-	-	-	-	-	-	-	-	-	-	-	-	-	-
Oolitic Hematite	8	-	-	-	-	-	-	-	-	-	-	1678	-	-	-	-	-	-	-	-	-	-	-	-	-	-	-	-	-	-	-	-	-	-	-
Opal	8	-	-	-	-	-	-	-	-	-	-	-	-	-	-	-	-	-	-	-	-	-	-	-	-	-	-	-	-	421	-	-	-	-	-
Opal Glass	8	-	-	-	-	-	-	-	-	-	-	1646	-	-	-	-	-	-	-	-	-	-	-	-	-	-	-	-	-	-	-	-	-	-	-
Osmium	7	-	-	-	-	500	-	-	-	-	-	-	-	-	-	-	-	-	-	-	-	-	-	-	-	-	-	-	-	-	-	-	-	-	-
Owens-Illinois 650 Binder Pigmented with Zinc Oxide	9	-	-	-	-	-	-	-	-	-	-	322	-	-	-	-	-	-	-	-	-	-	-	377 393	-	-	-	-	-	-	-	-	-	-	-
Owens-Illinois 650 Coating on Aluminum Substrate	9	-	-	-	-	-	-	-	-	-	-	-	-	-	-	-	-	-	-	-	-	-	-	1167	-	-	-	-	-	-	-	-	-	-	-
Oxidized Armco Iron	9	-	-	-	-	1297	-	-	-	-	-	1299	-	-	-	-	-	-	-	-	-	-	-	-	-	-	-	-	-	-	-	-	-	-	-
Oxidized Iron + Chromium + ΣX_i	9	-	-	-	-	1303	-	-	-	-	-	-	-	-	-	-	-	-	-	-	-	-	-	-	-	-	-	-	-	-	-	-	-	-	-
Oxidized Iron + Nickel + Chromium + ΣX_i	9	-	1305	-	-	1308	-	-	-	-	-	-	-	-	-	-	-	-	-	-	-	-	-	-	-	-	-	-	-	-	-	-	-	-	-
Oxidized Molybdenum Silicide	9	-	-	-	-	-	-	-	-	-	-	-	-	-	-	-	-	-	-	-	-	-	-	-	-	-	-	-	-	1311	-	-	-	-	-
Oxidized Nickel	9	1312	-	-	-	-	-	-	-	-	-	-	-	-	-	-	-	-	-	-	-	-	-	-	-	-	-	-	-	-	-	-	-	-	-
Oxidized Nickel + Chromium + ΣX_i	9	-	1314	-	-	-	-	-	-	-	-	-	-	-	-	-	-	-	-	-	-	-	-	-	-	-	-	-	-	-	-	-	-	-	-
Oxidized Silicon	9	-	-	-	-	-	-	-	-	-	-	1316	1318	-	-	-	-	-	-	-	-	-	-	-	-	-	-	-	-	1320	-	-	-	-	-
Oxidized Titanium	9	-	-	-	-	-	-	-	-	-	-	1322	-	-	-	-	-	-	-	-	-	-	-	-	-	-	-	-	-	-	-	-	-	-	-
P-110 White Porcelain Enamel	9	-	-	-	-	-	-	-	-	-	-	578	-	-	-	-	-	-	-	-	-	-	-	-	-	-	-	-	-	-	-	-	-	-	-
Pack Cementation Coatings on:																																			
Molybdenum Substrate	9	-	1333	-	-	1337	-	-	-	-	-	1342	-	-	-	-	-	-	-	-	-	-	-	-	-	-	-	-	-	-	-	-	-	-	-
Niobium Substrate	9	-	1347	-	-	1348	-	-	-	-	-	1349	-	-	-	-	-	-	-	-	-	-	-	-	-	-	-	-	-	-	-	-	-	-	-
Tantalum Substrate	9	-	1353	-	-	1355	-	-	-	-	-	1358	-	-	-	-	-	-	-	-	-	-	-	-	-	-	-	-	-	-	-	-	-	-	-
Titanium Substrate	9	-	1359	-	-	1360	-	-	-	-	-	1362	-	-	-	-	-	-	-	-	-	-	-	-	-	-	-	-	-	-	-	-	-	-	-
Tungsten Substrate	9	-	1363	-	-	1364	-	-	-	-	-	1365	-	-	-	-	-	-	-	-	-	-	-	-	-	-	-	-	-	-	-	-	-	-	-

MATERIAL NAME	VOLUME	EMITTANCE						REFLECTANCE									ABSORPTANCE									TRANSMITTANCE									Absorptance to Emittance Ratio
		Total			Spectral			Integrated			Spectral			Solar			Integrated			Spectral			Solar			Integrated			Spectral			Solar			
		Hemispherical	Normal	Angular	Hemispherical	Normal	Angular	Hemispherical	Normal	Angular	Hemispherical	Normal	Angular	Hemispherical	Normal	Angular	Hemispherical	Normal	Angular	Hemispherical	Normal	Angular	Hemispherical	Normal	Angular	Hemispherical	Normal	Angular	Hemispherical	Normal	Angular	Hemispherical	Normal	Angular	
Palladium	7	502	504	-	-	507 510	-	-	-	-	-	512	-	-	-	-	-	-	-	-	515	-	-	518	-	-	-	-	-	520	-	-	-	-	-
Palladium Coating on:																																			
Glass Substrate	9	-	718	-	-	-	-	-	-	-	-	720	-	-	-	-	-	-	-	-	-	-	-	-	-	-	-	-	-	721	-	-	-	-	-
Inconel X Substrate	9	-	718	-	-	-	-	-	-	-	-	-	-	-	-	-	-	-	-	-	-	-	-	-	-	-	-	-	-	-	-	-	-	-	-
Silicon Substrate	9	-	-	-	-	-	-	-	-	-	-	720	-	-	-	-	-	-	-	-	-	-	-	-	-	-	-	-	-	-	-	-	-	-	-
Parsons' Black	9	545	546	-	-	-	-	-	-	-	-	548	-	-	-	-	-	-	-	-	-	-	-	550	-	-	-	-	-	-	-	-	-	-	-
PdIn	8	-	-	-	-	-	-	-	-	-	-	1352	-	-	-	-	-	-	-	-	-	-	-	-	-	-	-	-	-	-	-	-	-	-	-
Pedigree Red	9	-	-	-	-	-	-	-	-	-	-	551	-	-	-	-	-	-	-	-	-	-	-	-	-	-	-	-	-	-	-	-	-	-	-
Perovskite Fluorides	8	-	-	-	-	-	-	-	-	-	-	976 980 984 988	-	-	-	-	-	-	-	-	-	-	-	-	-	-	-	-	-	978 982 986 990	-	-	-	-	-
Phosphates:																																			
Aluminum Phosphate	8	-	-	-	-	-	-	-	-	-	-	602	-	-	-	-	-	-	-	-	-	-	-	-	-	-	-	-	-	-	-	-	-	-	-
Ammonium and Potassium Dihydrogen Phosphates	8	-	-	-	-	-	-	-	-	-	-	-	-	-	-	-	-	-	-	-	-	-	-	-	-	-	-	-	-	604	-	-	-	-	-
Miscellaneous Phosphates	8	-	-	-	-	-	-	-	-	-	-	607	-	-	-	-	-	-	-	-	-	-	-	-	-	-	-	-	-	-	-	-	-	-	-
Phosphate Binder Pigmented with Potassium Titanate	9	-	172	-	-	-	-	-	-	-	-	-	-	-	-	-	-	-	-	-	-	-	-	-	-	-	-	-	-	-	-	-	-	-	-
Phosphides:																																			
Gallium Phosphide	8	-	-	-	-	1092	-	-	-	-	-	1094	-	-	-	-	-	-	-	-	-	-	-	-	-	-	-	-	-	1098	-	-	-	-	-
Indium Phosphide	8	-	-	-	-	-	-	-	-	-	-	1100	-	-	-	-	-	-	-	-	-	-	-	-	-	-	-	-	1102	-	-	-	-	-	-
Miscellaneous Phosphides	8	-	-	-	-	1104	-	-	-	-	-	1106	-	-	-	-	-	-	-	-	-	-	-	-	-	-	-	-	-	-	-	-	-	-	-
Phosphor Bronze	7	-	-	-	-	-	-	-	-	-	-	1175	-	-	-	-	-	-	-	-	-	-	-	-	-	-	-	-	-	-	-	-	-	-	-
Phosphoric Acid Cement Binder Pigmented with Aluminum Oxide	9	-	-	-	-	30	-	-	-	-	-	-	-	-	-	-	-	-	-	-	-	-	-	-	-	-	-	-	-	-	-	-	-	-	-
Phthalic Alkyd Binder Pigmented with:																																			
Carbon Black	9	-	-	-	-	-	-	-	-	-	-	-	-	-	-	-	-	-	-	-	-	-	-	89	-	-	-	-	-	-	-	-	-	-	-
Carbon + Aluminum	9	91	-	-	-	-	-	-	-	-	-	-	-	-	-	-	-	-	-	-	-	-	-	93	-	-	-	-	-	-	-	-	-	-	-
Pittsburgh Flat White Enamel Undercoater LA-404	9	-	-	-	-	-	-	-	-	-	-	552	-	-	-	-	-	-	-	-	-	-	-	-	-	-	-	-	-	-	-	-	-	-	-
Platinum	7	524	529	-	-	532 536	-	-	-	-	-	544	547 549	-	-	-	-	-	-	-	551	554	-	557	-	-	-	-	-	-	-	-	-	-	-
Platinum, NBS	7	-	-	-	-	538	-	-	-	-	-	-	-	-	-	-	-	-	-	-	552	-	-	-	-	-	-	-	-	-	-	-	-	-	-

MATERIAL NAME	VOLUME	EMITTANCE						REFLECTANCE									ABSORPTANCE									TRANSMITTANCE									Absorptance to Emittance Ratio
		Total			Spectral			Integrated			Spectral			Solar			Integrated			Spectral			Solar			Integrated			Spectral			Solar			
		Hemispherical	Normal	Angular	Hemispherical	Normal	Angular	Hemispherical	Normal	Angular	Hemispherical	Normal	Angular	Hemispherical	Normal	Angular	Hemispherical	Normal	Angular	Hemispherical	Normal	Angular	Hemispherical	Normal	Angular	Hemispherical	Normal	Angular	Hemispherical	Normal	Angular	Hemispherical	Normal	Angular	
Platinum Coating on:																																			
Ceramic Tile Substrate	9	-	-	-	-	-	-	-	-	-	-	-	724	-	-	-	-	-	-	-	-	-	-	-	-	-	-	-	-	-	-	-	-	-	-
Glass Substrate	9	-	722	-	-	-	-	-	-	-	-	-	-	-	-	-	-	-	-	-	-	-	-	-	-	-	-	-	-	-	-	-	-	-	-
Quartz Substrate	9	-	-	-	-	-	-	-	-	-	-	-	724	-	-	-	-	-	-	-	-	727	-	-	-	-	-	-	-	728	-	-	-	-	-
Platinum Substrate Coated with:																																			
Cobalt	9	-	631	-	-	-	-	-	-	-	-	-	-	-	-	-	-	-	-	-	-	-	-	633	-	-	-	-	-	-	-	-	-	-	-
Magnesium Fluoride	9	-	-	-	-	-	-	-	-	-	-	-	1098	-	-	-	-	-	-	-	-	1101	-	-	-	-	-	-	-	1105	-	-	-	-	-
Potassium Bromide	9	-	-	-	-	934	-	-	-	-	-	-	-	-	-	-	-	-	-	-	-	-	-	-	-	-	-	-	-	-	-	-	-	-	-
Potassium Tantalate	9	-	-	-	-	-	-	-	-	-	-	1045	-	-	-	-	-	-	-	-	-	-	-	-	-	-	-	-	-	-	-	-	-	-	-
Silicon Dioxide	9	-	-	-	-	-	-	-	-	-	-	-	1061	-	-	-	-	-	-	-	-	1067	-	-	-	-	-	-	-	1073	-	-	-	-	-
Silicon Monoxide	9	-	-	-	-	1075	-	-	-	-	-	-	-	-	-	-	-	-	-	-	-	-	-	-	-	-	-	-	-	-	-	-	-	-	-
Platinum Black on Unknown Substrate	9	-	-	-	-	-	-	-	-	-	-	1188	-	-	-	-	-	-	-	-	-	-	-	-	-	-	-	-	-	-	-	-	-	-	-
Platinum Wire, Grade MPTU 4292-53	7	526	-	-	-	-	-	-	-	-	-	-	-	-	-	-	-	-	-	-	-	-	-	-	-	-	-	-	-	-	-	-	-	-	-
Platinum + Rhodium	7	997	1000	-	-	1004	-	-	-	-	-	-	-	-	-	-	-	-	-	-	-	-	-	-	-	-	-	-	-	-	-	-	-	-	-
Plexiglas	8	-	-	-	-	-	-	-	-	-	-	-	-	-	-	-	-	-	-	-	-	-	-	-	-	-	-	-	-	1724	-	-	-	-	-
Pliofilm Substrate Coated with Selenium	9	-	-	-	-	-	-	-	-	-	-	-	-	-	-	-	-	-	-	-	-	-	-	-	-	-	-	-	-	943	-	-	-	-	-
Polybutadiene Coating on Tin Oxide Substrate	9	-	-	-	-	-	-	-	-	-	-	-	-	-	-	-	-	-	-	-	-	-	-	-	-	-	-	-	-	1134	-	-	-	-	-
Polyester Coating on:																																			
Aluminum Substrate	9	1135	-	-	-	-	-	-	-	-	-	-	-	-	-	-	-	-	-	-	-	-	-	-	-	-	-	-	-	-	-	-	-	-	-
Gold Substrate	9	1135	-	-	-	-	-	-	-	-	-	-	-	-	-	-	-	-	-	-	-	-	-	-	-	-	-	-	-	-	-	-	-	-	-
Polyester Substrate Coated with:																																			
Aluminum	9	580	-	-	-	-	-	-	-	-	-	-	-	-	-	-	-	-	-	-	-	-	-	-	-	-	-	-	-	-	-	-	-	-	-
Gold	9	651	-	-	-	-	-	-	-	-	-	-	-	-	-	-	-	-	-	-	-	-	-	-	-	-	-	-	-	-	-	-	-	-	-
Polyethylene	8	-	-	-	-	-	-	-	-	-	-	-	-	-	-	-	-	-	-	-	-	-	-	-	-	-	-	-	-	1705	-	-	-	-	-
Polyethylene Substrate Coated with:																																			
Epoxy	9	-	-	-	-	-	-	-	-	-	-	-	-	-	-	-	-	-	-	-	-	-	-	-	-	-	-	-	-	1122	-	-	-	-	-
Polyurethane	9	-	-	-	-	-	-	-	-	-	-	-	-	-	-	-	-	-	-	-	-	-	-	-	-	-	-	-	-	1143	-	-	-	-	-
Poly(ethylene glycol - Terephthatic acid) [Mylar]	8	-	-	-	-	-	-	-	-	-	-	-	-	-	-	-	-	-	-	-	1708	-	-	-	1710	-	-	-	-	1711	-	-	-	-	-
Polyimide [Kapton]	8	-	-	-	-	-	-	-	-	-	-	-	-	-	-	-	-	-	-	-	1714	-	-	-	-	-	-	-	-	1716	-	-	-	-	-
Poly(Methyl Methacrylate)	8	-	-	-	-	-	-	-	1719	-	-	-	-	-	-	-	-	-	-	-	1721	-	-	-	-	-	-	-	-	1723	-	-	-	-	-

MATERIAL NAME	VOLUME	EMITTANCE						REFLECTANCE									ABSORPTANCE									TRANSMITTANCE									Absorptance to Emittance Ratio
		Total			Spectral			Integrated			Spectral			Solar			Integrated			Spectral			Solar			Integrated			Spectral			Solar			
		Hemispherical	Normal	Angular	Hemispherical	Normal	Angular	Hemispherical	Normal	Angular	Hemispherical	Normal	Angular	Hemispherical	Normal	Angular	Hemispherical	Normal	Angular	Hemispherical	Normal	Angular	Hemispherical	Normal	Angular	Hemispherical	Normal	Angular	Hemispherical	Normal	Angular	Hemispherical	Normal	Angular	
Polymide Fluorinated Ethylene Propylene Coating on Silver Substrate	9	1133	-	-	-	-	-	-	-	-	-	-	-	-	-	-	-	-	-	-	-	-	-	-	-	-	-	-	-	-	-	-	-	-	-
Polystyrene	8	-	-	-	-	-	-	-	-	-	-	-	1725	-	-	-	-	-	-	-	-	-	-	-	-	-	-	-	-	1728	-	-	-	-	-
Polystyrene Coating on Glass Substrate	9	-	-	-	-	-	-	-	-	-	-	-	1136 1137	-	-	-	-	-	-	-	-	-	-	-	-	-	-	-	-	-	-	-	-	-	-
Polytetrafluoroethylene [Teflon]	8	-	1730	-	-	-	-	-	-	-	-	1732	-	-	-	-	-	-	-	-	1734	-	-	-	-	-	-	-	-	1736	-	-	-	-	-
Polyurethane Coating on:																																			
Aluminum Substrate	9	-	-	-	-	-	-	-	-	-	-	1138	1140	-	-	-	-	-	-	-	-	-	-	-	-	-	-	-	-	-	-	-	-	-	-
Polyethylene Substrate	9	-	-	-	-	-	-	-	-	-	-	-	-	-	-	-	-	-	-	-	-	-	-	-	-	-	-	-	-	1143	-	-	-	-	-
Stainless Steel Substrate	9	-	-	-	-	-	-	-	-	-	-	-	1140	-	-	-	-	-	-	-	-	-	-	-	-	-	-	-	-	-	-	-	-	-	-
Polyurethane Substrate Coated with:																																			
Aluminum	9	-	-	-	-	-	-	-	-	-	-	-	602	-	-	-	-	-	-	-	-	-	-	-	-	-	-	-	-	-	-	-	-	-	-
Copper	9	-	-	-	-	-	-	-	-	-	-	-	643	-	-	-	-	-	-	-	-	-	-	-	-	-	-	-	-	-	-	-	-	-	-
Gold	9	-	-	-	-	-	-	-	-	-	-	-	672	-	-	-	-	-	-	-	-	-	-	-	-	-	-	-	-	-	-	-	-	-	-
Nickel	9	-	-	-	-	-	-	-	-	-	-	-	704	-	-	-	-	-	-	-	-	-	-	-	-	-	-	-	-	-	-	-	-	-	-
Silver	9	733	-	-	-	-	-	-	-	-	-	-	741	-	-	-	-	-	-	-	-	-	-	-	-	-	-	-	-	-	-	-	-	-	-
Polyvinyl Alcohol Binder Pigmented with Barium Sulfate	9	-	-	-	-	-	-	-	-	-	-	62	-	-	-	-	-	-	-	-	-	-	-	-	-	-	-	-	-	-	-	-	-	-	-
Polyvinyl Alcohol Coating on Fiberglass Substrate	9	-	-	-	-	-	-	-	-	-	-	1145	-	-	-	-	-	-	-	-	-	-	-	1146	-	-	-	-	-	-	-	-	-	-	-
Polyvinyl Butyral Binder Pigmented with:																																			
Titanium Dioxide	9	-	-	-	-	-	-	-	-	-	-	-	-	-	-	-	-	-	-	-	-	-	-	267	-	-	-	-	-	-	-	-	-	-	-
Zinc Sulfide	9	-	-	-	-	-	-	-	-	-	-	304	-	-	-	-	-	-	-	-	-	-	-	415	-	-	-	-	-	-	-	-	-	-	-
Polyvinyl Butyral Coating on Quartz Substrate	9	-	-	-	-	-	-	-	-	-	-	-	-	-	-	-	-	-	-	-	-	-	-	1147	-	-	-	-	-	-	-	-	-	-	-
Polyvinyl Chloride	8	-	-	-	-	-	-	-	-	-	-	1742	-	-	-	-	-	-	-	-	-	-	-	-	-	-	-	-	-	1746	-	-	-	-	-
Polyvinyl Chloride Coating on:																																			
Aluminum Substrate	9	-	-	-	-	-	-	-	-	-	-	1150	-	-	-	-	-	-	-	-	-	-	-	1154	-	-	-	-	-	-	-	-	-	-	-
Fiberglass Substrate	9	-	-	-	-	-	-	-	-	-	-	1150	-	-	-	-	-	-	-	-	-	-	-	1152	-	-	-	-	-	-	-	-	-	-	-
Nylon Substrate	9	-	-	-	-	-	-	-	-	-	-	1150	-	-	-	-	-	-	-	-	-	-	-	1152	-	-	-	-	-	-	-	-	-	-	-
Polyvinylidene Chloride	8	-	-	-	-	-	-	-	-	-	-	-	-	-	-	-	-	-	-	-	-	-	-	-	-	-	-	-	-	1746	-	-	-	-	-
Polyvinylidene Chloride Coating on Silicon Substrate	9	-	-	-	-	-	-	-	-	-	-	-	-	-	-	-	-	-	-	-	-	-	-	-	-	-	-	-	-	1155	-	-	-	-	-

MATERIAL NAME	VOLUME	EMITTANCE						REFLECTANCE									ABSORPTANCE									TRANSMITTANCE									Absorptance to Emittance Ratio
		Total			Spectral			Integrated			Spectral			Solar			Integrated			Spectral			Solar			Integrated			Spectral			Solar			
		Hemispherical	Normal	Angular	Hemispherical	Normal	Angular	Hemispherical	Normal	Angular	Hemispherical	Normal	Angular	Hemispherical	Normal	Angular	Hemispherical	Normal	Angular	Hemispherical	Normal	Angular	Hemispherical	Normal	Angular	Hemispherical	Normal	Angular	Hemispherical	Normal	Angular	Hemispherical	Normal	Angular	
Poroloy	7	–	1226	–	–	–	–	–	–	–	–	–	–	–	–	–	–	–	–	–	–	–	–	1303	–	–	–	–	–	–	–	–	–	–	–
Potassium Alum	8	–	–	–	–	–	–	–	–	–	–	–	–	–	–	–	–	–	–	–	–	–	–	–	–	–	–	–	–	1702	–	–	–	–	–
Potassium Aluminum Silicate Pigment in Sodium Silicate Binder	9	–	–	–	–	–	–	–	–	–	–	–	–	–	–	–	–	–	–	–	171	–	–	–	–	–	–	–	–	–	–	–	–	–	–
Potassium Bromide	8	–	–	–	–	749	–	–	–	751 753	–	–	–	–	–	–	–	–	–	–	–	–	–	–	–	–	–	–	–	759	–	–	–	–	–
Potassium Bromide Coating on Platinum Substrate	9	–	–	–	–	934	–	–	–	–	–	–	–	–	–	–	–	–	–	–	–	–	–	–	–	–	–	–	–	–	–	–	–	–	–
Potassium Bromide Substrate Coated with Lead Molybdenum Tetraoxide	9	–	–	–	–	–	–	–	–	–	–	913	–	–	–	–	–	–	–	–	–	–	–	–	–	–	–	–	–	–	–	–	–	–	–
Potassium Chloride	8	–	–	–	–	862	–	–	–	–	–	864	–	–	–	–	–	–	–	–	–	–	–	–	–	–	–	–	–	873	–	–	–	–	–
Potassium Chloride Coating on Lithium Fluoride Substrate	9	–	–	–	–	–	–	–	–	–	–	–	–	–	–	–	–	–	–	–	–	–	–	–	–	–	–	–	–	–	935	–	–	–	–
Potassium Cobalt Trifluoride	8	–	–	–	–	–	–	–	–	–	–	988	–	–	–	–	–	–	–	–	–	–	–	–	–	–	–	–	–	–	–	–	–	–	–
Potassium Fluoride	8	–	–	–	–	–	–	–	–	–	–	992	–	–	–	–	–	–	–	–	–	–	–	–	–	–	–	–	–	–	–	–	–	–	–
Potassium Dihydrogen Phosphate	8	–	–	–	–	–	–	–	–	–	–	–	–	–	–	–	–	–	–	–	–	–	–	–	–	–	–	–	–	604	–	–	–	–	–
Potassium Iodide	8	–	–	–	–	–	–	–	–	–	–	1005	–	–	–	–	–	–	–	–	–	–	–	–	–	–	–	–	–	1010	–	–	–	–	–
Potassium Iodide Coating on Lithium Fluoride Substrate	9	–	–	–	–	–	–	–	–	–	–	936	–	–	–	–	–	–	–	–	–	–	–	–	–	–	–	–	–	–	–	–	–	–	–
Potassium Magnesium Trifluoride	8	–	–	–	–	–	–	–	–	–	–	975	–	–	–	–	–	–	–	–	–	–	–	–	–	–	–	–	–	977	–	–	–	–	–
Potassium Manganese Trifluoride	8	–	–	–	–	–	–	–	–	–	–	988	–	–	–	–	–	–	–	–	–	–	–	–	–	–	–	–	–	–	–	–	–	–	–
Potassium Nickel Trifluoride	8	–	–	–	–	–	–	–	–	–	–	979	–	–	–	–	–	–	–	–	–	–	–	–	–	–	–	–	–	981	–	–	–	–	–
Potassium Phosphate	8	–	–	–	–	–	–	–	–	–	–	608	–	–	–	–	–	–	–	–	–	–	–	–	–	–	–	–	–	–	–	–	–	–	–
Potassium Silicate Binder Pigmented with:																																			
Aluminum Oxide	9	–	28	–	–	–	–	–	–	–	–	33 39	–	–	–	–	–	–	–	–	41	–	–	42	–	–	–	–	–	–	–	–	–	–	45
Al_2O_3 + ZnO + TiO_2	9	–	28	–	–	–	–	–	–	–	–	34	–	–	–	–	–	–	–	–	–	–	–	42	–	–	–	–	–	–	–	–	–	–	–
Aluminum Silicate	9	–	–	–	–	–	–	–	–	–	–	47	–	–	–	–	–	–	–	–	–	–	–	49	–	–	–	–	–	–	–	–	–	–	–
Antimony Oxide	9	–	–	–	–	–	–	–	–	–	–	51	–	–	–	–	54	–	–	–	56	–	–	57 60	–	–	–	–	–	–	–	–	–	–	–
Boron Nitride	9	66	–	–	–	–	–	–	–	–	–	–	–	–	–	–	–	–	–	–	–	–	–	–	–	–	–	–	–	–	–	–	–	–	–
Calcium Metasilicate	9	–	–	–	–	–	–	–	–	–	–	76	–	–	–	–	–	–	–	–	–	–	–	–	–	–	–	–	–	–	–	–	–	–	–
Cr-Co-Ni Spinel	9	–	–	–	–	–	–	–	–	–	–	–	–	–	–	–	–	–	–	–	–	–	–	187	–	–	–	–	–	–	–	–	–	–	–

MATERIAL NAME	VOLUME	EMITTANCE						REFLECTANCE									ABSORPTANCE									TRANSMITTANCE									Absorptance to Emittance Ratio
		Total			Spectral			Integrated			Spectral			Solar			Integrated			Spectral			Solar			Integrated			Spectral			Solar			
		Hemispherical	Normal	Angular	Hemispherical	Normal	Angular	Hemispherical	Normal	Angular	Hemispherical	Normal	Angular	Hemispherical	Normal	Angular	Hemispherical	Normal	Angular	Hemispherical	Normal	Angular	Hemispherical	Normal	Angular	Hemispherical	Normal	Angular	Hemispherical	Normal	Angular	Hemispherical	Normal	Angular	
Potassium Silicate Binder Pigmented with: (continued)																																			
Diatomaceous Earth	9	–	–	–	–	–	–	–	–	–	–	113	–	–	–	–	–	–	–	–	–	–	–	116	–	–	–	–	–	–	–	–	–	–	–
Lanthanum Oxide	9	–	–	–	–	–	–	–	–	–	–	127	–	–	–	–	–	–	–	–	–	–	–	132	–	–	–	–	–	–	–	–	–	–	–
Lithafrax	9	–	–	–	–	–	–	–	–	–	–	144	–	–	–	–	–	–	–	–	–	–	–	–	–	–	–	–	–	–	–	–	–	–	–
Lithium Aluminum Silicate	9	–	143	–	–	–	–	–	–	–	–	144	–	–	–	–	–	–	–	–	–	–	–	–	–	–	–	–	–	–	–	–	–	–	–
Magnesium Silicate	9	–	–	–	–	–	–	–	–	–	–	168	–	–	–	–	–	–	–	–	–	–	–	169	–	–	–	–	–	–	–	–	–	–	–
Molochite No. 6	9	–	–	–	–	–	–	–	–	–	–	47	–	–	–	–	–	–	–	–	–	–	–	49	–	–	–	–	–	–	–	–	–	–	–
Molochite SF	9	–	–	–	–	–	–	–	–	–	–	47	–	–	–	–	–	–	–	–	–	–	–	49	–	–	–	–	–	–	–	–	–	–	–
Strontium Molybdate	9	188	–	–	–	–	–	–	–	–	–	191	–	–	–	–	–	–	–	–	–	–	–	194	–	–	–	–	–	–	–	–	–	–	–
Tin Oxide	9	201	–	–	–	–	–	–	–	–	–	204	–	–	–	–	–	–	–	–	–	–	–	205 207	–	–	–	–	–	–	–	–	–	–	–
Titanium Dioxide	9	–	214	–	–	–	–	–	–	–	–	226	–	–	–	–	–	–	–	–	–	–	–	274 281	–	–	–	–	–	–	–	–	–	–	283
Titanium Dioxide + Aluminum Oxide	9	–	–	–	–	–	–	–	–	–	–	285 287	–	–	–	–	–	–	–	–	–	–	–	–	–	–	–	–	–	–	–	–	–	–	–
Zinc Oxide	9	302	304	–	–	–	–	–	–	–	306	314 355	–	–	–	–	–	360	–	–	362 365	–	–	373 392	–	–	–	–	–	–	–	–	–	–	399
Zinc Sulfide	9	–	–	–	–	–	–	–	–	–	–	405	–	–	–	–	–	–	–	–	–	–	–	418	–	–	–	–	–	–	–	–	–	–	–
Zinc Titanate	9	–	–	–	–	–	–	–	–	–	–	–	–	–	–	–	–	–	–	–	–	–	–	420	–	–	–	–	–	–	–	–	–	–	–
Zirconium Oxide	9	–	–	–	–	–	–	–	–	–	–	425	–	–	–	–	–	–	–	–	–	–	–	428	–	–	–	–	–	–	–	–	–	–	–
Zirconium Silicate	9	–	–	–	–	–	–	–	–	–	–	432	–	–	–	–	–	–	–	–	435	–	–	437 441	–	–	–	–	–	–	–	–	–	–	–
Potassium Silicate Coating on:																																			
Aluminum Substrate	9	–	–	–	–	–	–	–	–	–	–	937 1044	–	–	–	–	–	–	–	–	–	–	–	–	–	–	–	–	–	–	–	–	–	–	–
Quartz Substrate	9	–	–	–	–	–	–	–	–	–	–	–	–	–	–	–	–	–	–	–	–	–	–	938	–	–	–	–	–	–	–	–	–	–	–
Dipotassium Silicon Hexafluoride	8	–	–	–	–	–	–	–	–	–	–	992	–	–	–	–	–	–	–	–	–	–	–	–	–	–	–	–	–	–	–	–	–	–	–
Potassium Tantalate	8	–	–	–	–	–	–	–	–	–	–	633	–	–	–	–	–	–	–	–	–	–	–	–	–	–	–	–	–	–	–	–	–	–	–
Potassium Tantalate Coating on:																																			
Gold Substrate	9	–	–	–	–	–	–	–	–	–	–	1045	–	–	–	–	–	–	–	–	–	–	–	–	–	–	–	–	–	–	–	–	–	–	–
Platinum Substrate	9	–	–	–	–	–	–	–	–	–	–	1045	–	–	–	–	–	–	–	–	–	–	–	–	–	–	–	–	–	–	–	–	–	–	–
Potassium Titanate Pigment in Phosphate Binder	9	–	172	–	–	–	–	–	–	–	–	–	–	–	–	–	–	–	–	–	–	–	–	–	–	–	–	–	–	–	–	–	–	–	–
Potassium Titanate Porcelain Enamel	9	–	470	–	–	–	–	–	–	–	–	–	–	–	–	–	–	–	–	–	–	–	–	–	–	–	–	–	–	–	–	–	–	–	–
Potassium Zinc Trifluoride	8	–	–	–	–	–	–	–	–	–	–	–	–	–	–	–	–	–	–	–	–	–	–	–	–	–	–	–	–	990	–	–	–	–	–
Potomac A	7	–	–	–	–	1192	–	–	–	–	–	1198	–	–	–	–	–	–	–	–	–	–	–	–	–	–	–	–	–	–	–	–	–	–	–

MATERIAL NAME	VOLUME	EMITTANCE						REFLECTANCE									ABSORPTANCE									TRANSMITTANCE									Absorptance to Emittance Ratio
		Total			Spectral			Integrated			Spectral			Solar			Integrated			Spectral			Solar			Integrated			Spectral			Solar			
		Hemispherical	Normal	Angular	Hemispherical	Normal	Angular	Hemispherical	Normal	Angular	Hemispherical	Normal	Angular	Hemispherical	Normal	Angular	Hemispherical	Normal	Angular	Hemispherical	Normal	Angular	Hemispherical	Normal	Angular	Hemispherical	Normal	Angular	Hemispherical	Normal	Angular	Hemispherical	Normal	Angular	
Praseodymium Hexaboride	8	-	-	-	-	723	-	-	-	-	-	-	-	-	-	-	-	-	-	-	-	-	-	-	-	-	-	-	-	727	-	-	-	-	-
Praseodymium Trifluoride	8	-	-	-	-	-	-	-	-	-	-	-	-	-	-	-	-	-	-	-	-	-	-	-	-	-	-	-	-	727	-	-	-	-	-
Dipraseodymium Trisulfide	8	-	-	-	-	1232	-	-	-	-	-	-	-	-	-	-	-	-	-	-	-	-	-	-	-	-	-	-	-	-	-	-	-	-	-
PRF-6 Coating on Molybdenum Substrate	9	1331	-	-	-	929	-	-	-	-	-	-	-	-	-	-	-	-	-	-	-	-	-	-	-	-	-	-	-	-	-	-	-	-	-
Proven Flat White SP-15	9	-	-	-	-	-	-	-	-	-	-	553	-	-	-	-	-	-	-	-	-	-	-	-	-	-	-	-	-	-	-	-	-	-	-
PS-7 Binder Pigmented with:																																			
Aluminum Oxide	9	-	-	-	-	-	-	-	-	-	-	34	-	-	-	-	-	-	-	-	41	-	-	43	-	-	-	-	-	-	-	-	-	-	-
Aluminum Silicate	9	-	-	-	-	-	-	-	-	-	-	47	-	-	-	-	-	-	-	-	-	-	-	49	-	-	-	-	-	-	-	-	-	-	-
Calcium Metasilicate	9	-	-	-	-	-	-	-	-	-	-	76	-	-	-	-	-	-	-	-	-	-	-	-	-	-	-	-	-	-	-	-	-	-	-
Diatomaceous Earth	9	-	-	-	-	-	-	-	-	-	-	113	-	-	-	-	-	-	-	-	-	-	-	116	-	-	-	-	-	-	-	-	-	-	-
Lanthanum Oxide	9	-	-	-	-	-	-	-	-	-	-	127	-	-	-	-	-	-	-	-	-	-	-	-	-	-	-	-	-	-	-	-	-	-	-
Lithafrax	9	-	-	-	-	-	-	-	-	-	-	144	-	-	-	-	-	-	-	-	-	-	-	-	-	-	-	-	-	-	-	-	-	-	-
Lithium Aluminum Silicate	9	-	-	-	-	-	-	-	-	-	-	144	-	-	-	-	-	-	-	-	-	-	-	-	-	-	-	-	-	-	-	-	-	-	-
Magnesium Silicate	9	-	-	-	-	-	-	-	-	-	-	168	-	-	-	-	-	-	-	-	-	-	-	169	-	-	-	-	-	-	-	-	-	-	-
Molochite No. 6	9	-	-	-	-	-	-	-	-	-	-	47	-	-	-	-	-	-	-	-	-	-	-	49	-	-	-	-	-	-	-	-	-	-	-
Molochite SF	9	-	-	-	-	-	-	-	-	-	-	47	-	-	-	-	-	-	-	-	-	-	-	49	-	-	-	-	-	-	-	-	-	-	-
Silicon Dioxide	9	-	-	-	-	-	-	-	-	-	-	178	-	-	-	-	-	-	-	-	-	-	-	179	-	-	-	-	-	-	-	-	-	-	-
Strontium Molybdate	9	188	-	-	-	-	-	-	-	-	-	191	-	-	-	-	-	-	-	-	-	-	-	194	-	-	-	-	-	-	-	-	-	-	-
Tin Oxide	9	-	-	-	-	-	-	-	-	-	-	204	-	-	-	-	-	-	-	-	-	-	-	205	-	-	-	-	-	-	-	-	-	-	-
Titanium Dioxide	9	-	214	-	-	-	-	-	-	-	-	-	-	-	-	-	-	-	-	-	-	-	-	274 281	-	-	-	-	-	-	-	-	-	-	283
Zinc Oxide	9	-	304	-	-	-	-	-	-	-	-	314	-	-	-	-	-	-	-	-	365	-	-	370 393	-	-	-	-	-	-	-	-	-	-	399
Zinc Sulfide	9	-	-	-	-	-	-	-	-	-	-	405	-	-	-	-	-	-	-	-	-	-	-	418	-	-	-	-	-	-	-	-	-	-	-
Zirconium Oxide	9	-	-	-	-	-	-	-	-	-	-	425	-	-	-	-	-	-	-	-	-	-	-	428	-	-	-	-	-	-	-	-	-	-	-
Zirconium Silicate	9	-	-	-	-	-	-	-	-	-	-	432	-	-	-	-	-	-	-	-	-	-	-	437	-	-	-	-	-	-	-	-	-	-	-
Pt-10 Rh Alloy	7	997	-	-	-	-	-	-	-	-	-	-	-	-	-	-	-	-	-	-	-	-	-	-	-	-	-	-	-	-	-	-	-	-	-
Pt-13 Rh Alloy	7	-	1002	-	-	1005	-	-	-	-	-	-	-	-	-	-	-	-	-	-	-	-	-	-	-	-	-	-	-	-	-	-	-	-	-
PV-100	9	212	-	-	-	-	-	-	-	-	-	221	250	-	-	-	-	-	-	-	-	-	-	275	-	-	-	-	-	-	-	-	-	-	-
Pyrex 774	8	-	-	-	-	1582 1584	-	-	-	-	-	1586	1588	-	-	-	-	-	-	-	-	-	-	-	-	-	-	-	1592	-	-	-	-	-	-
Pyroceram 9606	8	1595	1597	-	-	1599 1601	-	-	-	-	-	1603	-	-	-	-	-	-	-	-	-	-	-	1606	-	-	-	-	-	-	-	-	-	-	-
Pyroceram 9608	8	1595	1597	-	-	1599 1601	-	-	-	-	-	1603	-	-	-	-	-	-	-	-	-	-	-	1606	-	-	-	-	-	1608	-	-	-	-	-

MATERIAL NAME	VOLUME	EMITTANCE						REFLECTANCE									ABSORPTANCE									TRANSMITTANCE									Absorptance to Emittance Ratio
		Total			Spectral			Integrated			Spectral			Solar			Integrated			Spectral			Solar			Integrated			Spectral			Solar			
		Hemispherical	Normal	Angular	Hemispherical	Normal	Angular	Hemispherical	Normal	Angular	Hemispherical	Normal	Angular	Hemispherical	Normal	Angular	Hemispherical	Normal	Angular	Hemispherical	Normal	Angular	Hemispherical	Normal	Angular	Hemispherical	Normal	Angular	Hemispherical	Normal	Angular	Hemispherical	Normal	Angular	
Pyromark OAO	9	-	-	-	-	-	-	-	-	-	-	244	-	-	-	-	-	-	-	-	-	-	-	-	-	-	-	-	-	-	-	-	-	-	-
Pyromark Standard Black Paint	9	117	119	-	-	123	-	-	-	-	-	-	-	-	-	-	-	-	-	-	-	-	-	-	-	-	-	-	-	-	-	-	-	-	-
Pyromark TiO_2 Paint	9	-	-	-	-	-	-	-	-	-	-	224	-	-	-	-	-	-	-	-	-	-	-	272	-	-	-	-	-	-	-	-	-	-	-
Pyroxylin Substrate Coated with Carbon	9	-	-	-	-	-	-	-	-	-	-	-	-	-	-	-	-	-	-	-	-	-	-	-	-	-	-	-	-	861	-	-	-	-	-
QMV Beryllium, Anodized	9	-	-	-	-	-	-	-	-	-	-	1265	-	-	-	-	-	-	-	-	-	-	-	-	-	-	-	-	-	-	-	-	-	-	-
Quartz, Crystal (see Silicon Dioxide, crystalline)	8	-	371	-	-	372	-	-	-	-	-	374	385 386	-	-	-	-	-	-	-	389	-	-	-	-	-	-	-	-	391	-	-	-	-	-
Quartz, Fused (see Silicon Dioxide, fused)	8	403	-	-	-	405	-	-	-	-	-	409	413	-	-	-	-	-	-	-	415	-	-	-	-	-	-	-	-	417	-	-	426	-	-
Quartz Substrate Coated with:																																			
Aluminum	9	581	-	-	-	-	-	-	-	-	-	592	602	-	-	-	-	-	-	-	-	-	-	-	-	-	-	-	-	-	-	-	-	-	-
Carbon	9	-	-	-	-	-	-	-	-	-	-	859	-	-	-	-	-	-	-	-	-	-	-	-	-	-	-	-	-	-	-	-	-	-	-
Gold	9	-	-	-	-	-	-	-	-	-	-	664	-	-	-	-	-	-	-	-	-	-	-	678	-	-	-	-	-	681	-	-	-	-	-
Hafnium Oxide	9	-	-	-	-	-	-	-	-	-	-	-	-	-	-	-	-	-	-	-	-	-	-	-	-	-	-	-	-	897	-	-	-	-	-
Lacquer	9	-	-	-	-	-	-	-	-	-	-	-	-	-	-	-	-	-	-	-	1126	-	-	-	-	-	-	-	-	1129	-	-	-	-	-
Magnesium Fluoride	9	-	-	-	-	-	-	-	-	-	-	1096	1098	-	-	-	-	-	-	-	-	1101	-	-	-	-	-	-	-	1105	-	-	-	-	-
Melamine Formaldehyde	9	-	-	-	-	-	-	-	-	-	-	-	-	-	-	-	-	-	-	-	-	-	-	1130	-	-	-	-	-	-	-	-	-	-	-
Neodymium	9	-	-	-	-	-	-	-	-	-	-	-	-	-	-	-	-	-	-	-	-	-	-	-	-	-	-	-	-	698	-	-	-	-	-
Platinum	9	-	-	-	-	-	-	-	-	-	-	-	724	-	-	-	-	-	-	-	-	727	-	-	-	-	-	-	-	728	-	-	-	-	-
Polyvinyl Butyral	9	-	-	-	-	-	-	-	-	-	-	-	-	-	-	-	-	-	-	-	-	-	-	1147	-	-	-	-	-	-	-	-	-	-	-
Potassium Silicate	9	-	-	-	-	-	-	-	-	-	-	-	-	-	-	-	-	-	-	-	-	-	-	938	-	-	-	-	-	-	-	-	-	-	-
Selenium	9	-	-	-	-	-	-	-	-	-	-	-	-	-	-	-	-	-	-	-	-	-	-	-	-	-	-	-	-	943	-	-	-	-	-
Silver	9	-	-	-	-	-	-	-	-	-	-	738	743	-	-	-	-	-	-	-	-	-	-	-	-	-	-	-	-	750	752	-	-	-	-
Thorium Dioxide	9	-	-	-	-	-	-	-	-	-	-	-	-	-	-	-	-	-	-	-	-	-	-	-	-	-	-	-	-	974	-	-	-	-	-
Zinc Selenide	9	-	-	-	-	-	-	-	-	-	-	-	1004	-	-	-	-	-	-	-	-	-	-	-	-	-	-	-	-	1005	-	-	-	-	-
Zinc Telluride	9	-	-	-	-	-	-	-	-	-	-	-	-	-	-	-	-	-	-	-	-	-	-	-	-	-	-	-	-	1009	-	-	-	-	-
Rare Earth Borides	8	-	-	-	-	722	-	-	-	-	-	-	-	-	-	-	-	-	-	-	-	-	-	-	-	-	-	-	-	726	-	-	-	-	-
René 41	7	-	1353	-	-	1365 1366	-	-	-	-	-	1380 1382	-	-	-	-	-	-	-	-	-	-	-	-	-	-	-	-	-	-	-	-	-	-	-
Resin Binder Pigmented with Gold	9	25	-	-	-	-	-	-	-	-	-	-	-	-	-	-	-	-	-	-	-	-	-	-	-	-	-	-	-	-	-	-	-	-	-
Reynolds Wrap	7	-	-	-	-	-	-	-	-	-	-	-	-	-	40	-	-	-	-	-	-	-	-	-	55	-	-	-	-	-	-	-	-	-	-
Rhenium	7	559	562	-	-	565 568	-	-	-	-	-	-	-	-	-	-	-	-	-	-	-	-	-	-	-	-	-	-	-	-	-	-	-	-	-

MATERIAL NAME	VOLUME	EMITTANCE						REFLECTANCE									ABSORPTANCE									TRANSMITTANCE									Absorptance to Emittance Ratio
		Total			Spectral			Integrated			Spectral			Solar			Integrated			Spectral			Solar			Integrated			Spectral			Solar			
		Hemispherical	Normal	Angular	Hemispherical	Normal	Angular	Hemispherical	Normal	Angular	Hemispherical	Normal	Angular	Hemispherical	Normal	Angular	Hemispherical	Normal	Angular	Hemispherical	Normal	Angular	Hemispherical	Normal	Angular	Hemispherical	Normal	Angular	Hemispherical	Normal	Angular	Hemispherical	Normal	Angular	
Rhenium Trioxide	8	-	-	-	-	-	-	-	-	-	-	546	-	-	-	-	-	-	-	-	-	-	-	-	-	-	-	-	-	-	-	-	-	-	-
Rhenium Disilicide	8	-	-	-	-	1173	-	-	-	-	-	-	-	-	-	-	-	-	-	-	-	-	-	-	-	-	-	-	-	-	-	-	-	-	-
Rhodium	7	571	573	-	-	576 579	-	-	-	-	-	581	584	-	-	-	-	-	-	-	-	587	-	589	-	-	-	-	-	-	-	-	-	-	-
Rhodium Coating on:																																			
Inconel X Substrate	9	-	730	-	-	-	-	-	-	-	-	-	-	-	-	-	-	-	-	-	-	-	-	-	-	-	-	-	-	-	-	-	-	-	-
Stainless Steel Substrate	9	729	-	-	-	-	-	-	-	-	-	-	-	-	-	-	731	-	-	-	-	-	-	-	-	-	-	-	-	-	-	-	-	-	-
Rhyolite Tuff	8	-	-	-	-	-	-	-	-	-	-	1682	-	-	-	-	-	-	-	-	-	-	-	-	-	-	-	-	-	-	-	-	-	-	-
Rocks	8	-	-	-	-	-	-	-	-	-	-	1680	-	-	-	-	-	-	-	-	-	-	-	-	-	-	-	-	-	-	-	-	-	-	-
Rokide A Coating on:																																			
Inconel Substrate	9	-	-	-	-	792	-	-	-	-	-	-	-	-	-	-	-	-	-	-	-	-	-	-	-	-	-	-	-	-	-	-	-	-	-
Molybdenum Substrate	9	-	788	-	-	-	-	-	-	-	-	796	-	-	-	-	-	-	-	-	-	-	-	803	-	-	-	-	-	-	-	-	-	-	-
Stainless Steel Substrate	9	-	788	-	-	790	-	-	-	-	-	796	-	-	-	-	-	-	-	-	-	-	-	805	-	-	-	-	-	-	-	-	-	-	-
Rokide C Coating on:																																			
Inconel Substrate	9	-	-	-	-	885	-	-	-	-	-	-	-	-	-	-	-	-	-	-	-	-	-	-	-	-	-	-	-	-	-	-	-	-	-
Molybdenum Substrate	9	878	-	-	-	883	-	-	-	-	-	-	-	-	-	-	-	-	-	-	-	-	-	-	-	-	-	-	-	-	-	-	-	-	-
Niobium Substrate	9	878	-	-	-	885	-	-	-	-	-	-	-	-	-	-	-	-	-	-	-	-	-	-	-	-	-	-	-	-	-	-	-	-	-
Stainless Steel Substrate	9	878	-	-	-	-	-	-	-	-	-	-	-	-	-	-	-	-	-	-	-	-	-	-	-	-	-	-	-	-	-	-	-	-	-
Steel Substrate	9	-	881	-	-	885	-	-	-	-	-	-	-	-	-	-	-	-	-	-	-	-	-	-	-	-	-	-	-	-	-	-	-	-	-
Titanium 6Al-4V Substrate	9	-	881	-	-	883	-	-	-	-	-	-	-	-	-	-	-	-	-	-	-	-	-	-	-	-	-	-	-	-	-	-	-	-	-
Rokide MA Coating on:																																			
Aluminum Substrate	9	915	-	-	-	-	-	-	-	-	-	-	-	-	-	-	-	-	-	-	-	-	-	918	-	-	-	-	-	-	-	-	-	-	-
Nb-1Zr Substrate	9	915	-	-	-	917	-	-	-	-	-	-	-	-	-	-	-	-	-	-	-	-	-	-	-	-	-	-	-	-	-	-	-	-	-
Rokide Z Coating on Stainless Steel Substrate	9	-	1011	-	-	-	-	-	-	-	-	1018	-	-	-	-	-	-	-	-	-	-	-	-	-	-	-	-	-	-	-	-	-	-	-
Rokide ZS Coating on:																																			
Aluminum Substrate	9	1024	-	-	-	-	-	-	-	-	-	-	-	-	-	-	-	-	-	-	-	-	-	1030	-	-	-	-	-	-	-	-	-	-	-
Nb-1Zr Substrate	9	1024	-	-	-	-	-	-	-	-	-	-	-	-	-	-	-	-	-	-	-	-	-	-	-	-	-	-	-	-	-	-	-	-	-
Stainless Steel Substrate	9	1024	1026	-	-	-	-	-	-	-	-	1029	-	-	-	-	-	-	-	-	-	-	-	-	-	-	-	-	-	-	-	-	-	-	-
RTV-11 Binder Pigmented with Zinc Oxide	9	-	-	-	-	-	-	-	-	-	-	317	-	-	-	-	-	-	-	-	-	-	-	-	-	-	-	-	-	-	-	-	-	-	-
RTV-602 Binder Pigmented with:																																			
China Clay	9	-	95	-	-	-	-	-	-	-	-	97	-	-	-	-	-	-	-	-	-	-	-	98 100	-	-	-	-	-	-	-	-	-	-	102
Titanium Dioxide	9	-	-	-	-	-	-	-	-	-	-	220 244	-	-	-	-	-	-	-	-	256	-	-	263	-	-	-	-	-	-	-	-	-	-	-

MATERIAL NAME	VOLUME	EMITTANCE						REFLECTANCE									ABSORPTANCE									TRANSMITTANCE									Absorptance to Emittance Ratio
		Total			Spectral			Integrated			Spectral			Solar			Integrated			Spectral			Solar			Integrated			Spectral			Solar			
		Hemispherical	Normal	Angular	Hemispherical	Normal	Angular	Hemispherical	Normal	Angular	Hemispherical	Normal	Angular	Hemispherical	Normal	Angular	Hemispherical	Normal	Angular	Hemispherical	Normal	Angular	Hemispherical	Normal	Angular	Hemispherical	Normal	Angular	Hemispherical	Normal	Angular	Hemispherical	Normal	Angular	
RTV-602 Binder Pigmented with: (continued)																																			
Zinc Oxide	9	-	-	-	-	-	-	-	-	-	-	314	-	-	-	-	-	-	-	-	-	-	-	373 392	-	-	-	-	-	-	-	-	-	-	-
RTV-602 Coating on Aluminum Substrate	9	-	1156	-	-	-	-	-	-	-	-	1159	-	-	-	-	-	-	-	-	-	-	-	1164 1167	-	-	-	-	-	-	-	-	-	-	1169
Rubber:	8	-	-	-	-	-	-	-	-	-	-	1739	-	-	-	-	-	-	-	-	-	-	-	-	-	-	-	-	-	-	-	-	-	-	-
PR 19-10	8	-	-	-	-	-	-	-	-	-	-	1740	-	-	-	-	-	-	-	-	-	-	-	-	-	-	-	-	-	-	-	-	-	-	-
RTV-77	8	-	-	-	-	-	-	-	-	-	-	1740	-	-	-	-	-	-	-	-	-	-	-	-	-	-	-	-	-	-	-	-	-	-	-
Rubidium Bromide	8	-	-	-	-	764	-	-	-	-	-	766	-	-	-	-	-	-	-	-	-	-	-	-	-	-	-	-	-	768	-	-	-	-	-
Rubidium Chloride	8	-	-	-	-	-	-	-	-	-	-	-	906	-	-	-	-	-	-	-	-	-	-	-	-	-	-	-	-	908	-	-	-	-	-
Rubidium Iodide	8	-	-	-	-	-	-	-	-	-	-	1014	-	-	-	-	-	-	-	-	-	-	-	-	-	-	-	-	-	1018	-	-	-	-	-
Rubidium Iodide Coating on Lithium Fluoride Substrate	9	-	-	-	-	-	-	-	-	-	-	940	-	-	-	-	-	-	-	-	-	-	-	-	-	-	-	-	-	-	-	-	-	-	-
Rubidium Manganese Trifluoride	8	-	-	-	-	-	-	-	-	-	-	983	-	-	-	-	-	-	-	-	-	-	-	-	-	-	-	-	-	985	-	-	-	-	-
Ruby	8	-	-	-	-	-	-	-	-	-	-	-	174	-	-	-	-	-	-	-	-	-	-	-	-	-	-	-	-	176	-	-	-	-	-
Rust-Oleum Paints:																																			
No. 960	9	-	-	-	-	-	-	-	-	-	-	300	-	-	-	-	-	-	-	-	-	-	-	-	-	-	-	-	-	-	-	-	-	-	-
Fire Hydrant Red No. 1210	9	-	-	-	-	-	-	-	-	-	-	555	-	-	-	-	-	-	-	-	-	-	-	-	-	-	-	-	-	-	-	-	-	-	-
Green No. 205	9	-	-	-	-	-	-	-	-	-	-	555	-	-	-	-	-	-	-	-	-	-	-	-	-	-	-	-	-	-	-	-	-	-	-
Red No. 215	9	-	-	-	-	-	-	-	-	-	-	555	-	-	-	-	-	-	-	-	-	-	-	-	-	-	-	-	-	-	-	-	-	-	-
Silver Gray No. 208	9	-	-	-	-	-	-	-	-	-	-	555	-	-	-	-	-	-	-	-	-	-	-	-	-	-	-	-	-	-	-	-	-	-	-
White No. 225	9	-	-	-	-	-	-	-	-	-	-	555	-	-	-	-	-	-	-	-	-	-	-	-	-	-	-	-	-	-	-	-	-	-	-
Ruthenium	7	-	-	-	-	591	-	-	-	-	-	-	-	-	-	-	-	-	-	-	-	-	-	-	-	-	-	-	-	-	-	-	-	-	-
Rutile	8	-	-	-	-	-	-	-	-	-	-	464 465	-	-	-	-	-	-	-	-	-	-	-	-	-	-	-	-	-	-	-	-	-	-	-
S-13	9	-	-	-	-	-	-	-	-	-	-	318 355	-	-	-	-	-	-	-	-	-	-	-	371 392	-	-	-	-	-	-	-	-	-	-	399
S-13G	9	-	304	-	-	-	-	-	-	-	-	322 355	-	-	-	-	-	-	-	-	-	-	-	382 392	-	-	-	-	-	-	-	-	-	-	399
S-13H	9	-	-	-	-	-	-	-	-	-	-	323	-	-	-	-	-	-	-	-	-	-	-	-	-	-	-	-	-	-	-	-	-	-	-
Salt Pool	8	-	-	-	-	-	-	-	-	-	-	1660	-	-	-	-	-	-	-	-	-	-	-	-	-	-	-	-	-	-	-	-	-	-	-
Samarium Hexaboride	8	-	-	-	-	723	-	-	-	-	-	-	-	-	-	-	-	-	-	-	-	-	-	-	-	-	-	-	-	-	-	-	-	-	-
Disamarium Trioxide	8	-	352	-	354	356 358	-	-	-	-	-	360	-	-	-	-	-	-	-	-	-	-	-	-	-	-	-	-	-	-	-	-	-	-	-

MATERIAL NAME	VOLUME	EMITTANCE						REFLECTANCE									ABSORPTANCE									TRANSMITTANCE									Absorptance to Emittance Ratio
		Total			Spectral			Integrated			Spectral			Solar			Integrated			Spectral			Solar			Integrated			Spectral			Solar			
		Hemispherical	Normal	Angular	Hemispherical	Normal	Angular	Hemispherical	Normal	Angular	Hemispherical	Normal	Angular	Hemispherical	Normal	Angular	Hemispherical	Normal	Angular	Hemispherical	Normal	Angular	Hemispherical	Normal	Angular	Hemispherical	Normal	Angular	Hemispherical	Normal	Angular	Hemispherical	Normal	Angular	
Samarium Silicate	8	-	-	-	-	-	-	-	-	-	-	-	-	-	-	-	-	-	-	-	-	-	-	-	-	-	-	-	-	622	-	-	-	-	-
Sand	8	-	-	-	-	-	-	-	-	-	-	1678	-	-	-	-	-	-	-	-	-	-	-	-	-	-	-	-	-	-	-	-	-	-	-
Sapphire	8	179	181	-	-	183	-	-	-	-	-	-	187	-	-	-	-	-	-	-	-	-	-	-	-	-	-	-	-	190	-	-	-	-	-
Sapphire Coating on:																																			
Aluminum Substrate	9	785	-	-	-	-	-	-	-	794	-	796	800	-	-	-	-	-	-	-	-	-	-	803	-	-	-	-	-	-	-	-	-	-	-
Gold Substrate	9	-	-	-	-	-	-	-	-	-	-	-	-	-	-	-	-	-	-	-	-	-	-	803	-	-	-	-	-	-	-	-	-	-	-
Silver Substrate	9	785	788	-	-	-	-	-	-	-	-	796	-	-	-	-	-	-	-	-	802	-	-	803	-	-	-	-	-	-	-	-	-	-	-
Stainless Steel Substrate	9	-	-	-	-	-	-	-	-	-	-	-	-	-	-	-	-	-	-	-	-	-	-	804	-	-	-	-	-	-	-	-	-	-	-
Sapphire Substrate Coated with Vanadium Oxide	9	-	-	-	-	-	-	-	-	-	-	999	-	-	-	-	-	-	-	-	-	-	-	-	-	-	-	-	-	1000	-	-	-	-	-
Scandium Diboride	8	-	-	-	-	732	-	-	-	-	-	-	-	-	-	-	-	-	-	-	-	-	-	-	-	-	-	-	-	-	-	-	-	-	-
Scandium Nitride	8	-	-	-	-	1087	-	-	-	-	-	-	-	-	-	-	-	-	-	-	-	-	-	-	-	-	-	-	-	1090	-	-	-	-	-
Scolecite	8	-	-	-	-	-	-	-	-	-	-	-	-	-	-	-	-	-	-	-	-	-	-	-	-	-	-	-	-	1694	-	-	-	-	-
Schott B & L 529516 Glass	8	-	-	-	-	-	-	-	-	-	-	-	-	-	-	-	-	-	-	-	-	-	-	-	-	-	-	-	-	1533	-	-	-	-	-
Schott BK 8 Glass	8	-	-	-	-	-	-	-	-	-	-	-	-	-	-	-	-	-	-	-	-	-	-	-	-	-	-	-	-	1533	-	-	-	-	-
Schott K5 Glass	8	-	-	-	-	-	-	-	-	-	-	-	-	-	-	-	-	-	-	-	-	-	-	-	-	-	-	-	-	1533	-	-	-	-	-
Schott K_zSF-4 Glass	8	-	-	-	-	-	-	-	-	-	-	-	-	-	-	-	-	-	-	-	-	-	-	-	-	-	-	-	-	1533	-	-	-	-	-
Schott SF-2 Glass	8	-	-	-	-	-	-	-	-	-	-	-	-	-	-	-	-	-	-	-	-	-	-	-	-	-	-	-	-	1533	-	-	-	-	-
SE 551 Binder Pigmented with Zinc Oxide	9	-	-	-	-	-	-	-	-	-	-	317	-	-	-	-	-	-	-	-	-	-	-	-	-	-	-	-	-	-	-	-	-	-	-
Selenides:																																			
Cadmium Selenide	8	-	-	-	-	-	-	-	-	-	-	1108	-	-	-	-	-	-	-	-	-	-	-	-	-	-	-	-	-	1110	-	-	-	-	-
Zinc Selenide	8	1112	-	-	-	1114	-	-	-	-	-	1117	-	-	-	-	-	-	-	-	1121	-	-	-	-	-	-	-	-	1123	-	-	-	-	-
Miscellaneous Selenides	8	-	-	-	-	-	-	-	-	-	-	1129	-	-	-	-	-	-	-	-	-	-	-	-	-	-	-	-	-	1132	-	-	-	-	-
Selenite	8	-	-	-	-	-	-	-	-	-	-	-	-	-	-	-	-	-	-	-	-	-	-	-	-	-	-	-	-	630	-	-	-	-	-
Selenium	8	-	-	-	-	-	-	-	-	-	-	80	-	-	-	-	-	-	-	-	-	-	-	-	-	-	-	-	-	84	-	-	-	-	-
Selenium Coating on:																																			
Germanium Substrate	9	-	-	-	-	-	-	-	-	-	-	-	-	-	-	-	-	-	-	-	-	-	-	-	-	-	-	-	-	943	-	-	-	-	-
Pliofilm Substrate	9	-	-	-	-	-	-	-	-	-	-	-	-	-	-	-	-	-	-	-	-	-	-	-	-	-	-	-	-	943	-	-	-	-	-
Silicon Dioxide Substrate	9	-	-	-	-	-	-	-	-	-	-	-	-	-	-	-	-	-	-	-	-	-	-	-	-	-	-	-	-	943	-	-	-	-	-
Serpentine	8	-	-	-	-	-	-	-	-	-	-	-	1689	-	-	-	-	-	-	-	-	-	-	-	-	-	-	-	-	1694	-	-	-	-	-

MATERIAL NAME	VOLUME	EMITTANCE						REFLECTANCE									ABSORPTANCE									TRANSMITTANCE									Absorptance to Emittance Ratio
		Total			Spectral			Integrated			Spectral			Solar			Integrated			Spectral			Solar			Integrated			Spectral			Solar			
		Hemispherical	Normal	Angular	Hemispherical	Normal	Angular	Hemispherical	Normal	Angular	Hemispherical	Normal	Angular	Hemispherical	Normal	Angular	Hemispherical	Normal	Angular	Hemispherical	Normal	Angular	Hemispherical	Normal	Angular	Hemispherical	Normal	Angular	Hemispherical	Normal	Angular	Hemispherical	Normal	Angular	
Sherwin-Williams Paints:																																			
Enameloid Flat Black	9	-	-	-	-	-	-	-	-	-	-	559	-	-	-	-	-	-	-	-	-	-	-	-	-	-	-	-	-	-	-	-	-	-	-
Moroon Enamel	9	-	-	-	-	557	-	-	-	-	-	-	-	-	-	-	-	-	-	-	-	-	-	-	-	-	-	-	-	-	-	-	-	-	-
Kemtone Shasta White No. 793	9	-	-	-	-	-	-	-	-	-	-	559	-	-	-	-	-	-	-	-	-	-	-	-	-	-	-	-	-	-	-	-	-	-	-
Silverbrite No. 55	9	-	6	-	-	-	-	-	-	-	-	-	-	-	-	-	-	-	-	-	-	-	-	-	-	-	-	-	-	-	-	-	-	-	-
SiC + SiO_2 Pigment in Aluminum Phosphate Binder	9	176	-	-	-	-	-	-	-	-	-	-	-	-	-	-	-	-	-	-	-	-	-	-	-	-	-	-	-	-	-	-	-	-	-
SiC + Talc Pigment in Ludox Binder	9	176	-	-	-	-	-	-	-	-	-	-	-	-	-	-	-	-	-	-	-	-	-	-	-	-	-	-	-	-	-	-	-	-	-
SiC + TiC Pigment in Dow Corning 806A Binder	9	176	-	-	-	-	-	-	-	-	-	-	-	-	-	-	-	-	-	-	-	-	-	-	-	-	-	-	-	-	-	-	-	-	-
SiC + UO_2 Pigment in Synar Binder	9	176	-	-	-	-	-	-	-	-	-	-	-	-	-	-	-	-	-	-	-	-	-	-	-	-	-	-	-	-	-	-	-	-	-
Sicon Black	9	-	-	-	-	561	-	-	-	-	-	563	-	-	-	-	-	-	-	-	-	-	-	-	-	-	-	-	-	-	-	-	-	-	-
Sicon White 7X1153	9	212	-	-	-	-	-	-	-	-	-	-	-	-	-	-	-	-	-	-	-	-	-	263	-	-	-	-	-	-	-	-	-	-	-
Silica (Crystal) [see Silicon Dioxide (Crystalline)]																																			
Silica (Fused) [see Silicon Diosice (Fused)]																																			
Silica Binder Pigmented with:																																			
Silicon Carbide	9	174	-	-	-	-	-	-	-	-	-	-	-	-	-	-	-	-	-	-	-	-	-	-	-	-	-	-	-	-	-	-	-	-	-
Titanium Carbide	9	208	-	-	-	-	-	-	-	-	-	-	-	-	-	-	-	-	-	-	-	-	-	-	-	-	-	-	-	-	-	-	-	-	-
Zirconium Oxide	9	-	-	-	-	-	-	-	-	-	-	425	-	-	-	-	-	-	-	-	-	-	-	-	-	-	-	-	-	-	-	-	-	-	-
Silica Gel	8	-	-	-	-	-	-	-	-	-	-	-	-	-	-	-	-	-	-	-	-	-	-	-	-	-	-	-	-	422	-	-	-	-	-
Silicates:																																			
Zirconium Silicate	8	-	-	-	-	610	-	-	-	-	-	612	-	-	-	-	-	-	-	-	-	-	-	615	-	-	-	-	-	-	-	-	-	-	-
Zirconium Silicate Compounds	8	-	-	-	-	-	-	-	-	-	-	-	-	-	-	-	-	-	-	-	-	-	-	616	-	-	-	-	-	-	-	-	-	-	-
Miscellaneous Silicates	8	-	-	-	-	-	-	-	-	-	-	617	-	-	-	-	-	-	-	-	-	-	-	-	-	-	-	-	-	621	-	-	-	-	-
Silicides:																																			
Boron Silicide	8	-	1134	-	-	1135	-	-	-	-	-	1137	-	-	-	-	-	-	-	-	-	-	-	-	-	-	-	-	-	-	-	-	-	-	-
Chromium Silicides	8	-	1139	-	-	1140 1142	-	-	-	-	-	1144	-	-	-	-	-	-	-	-	-	-	-	-	-	-	-	-	-	-	-	-	-	-	-
Molybdenum Silicides	8	-	1147	-	-	1149 1151	-	-	-	-	-	1154	-	-	-	-	-	-	-	-	-	-	-	-	-	-	-	-	-	-	-	-	-	-	-
Tantalum Disilicide	8	-	1156	-	-	1157 1159	-	-	-	-	-	1161	-	-	-	-	-	-	-	-	-	-	-	-	-	-	-	-	-	-	-	-	-	-	-

MATERIAL NAME	VOLUME	EMITTANCE						REFLECTANCE									ABSORPTANCE									TRANSMITTANCE									Absorptance to Emittance Ratio
		Total			Spectral			Integrated			Spectral			Solar			Integrated			Spectral			Solar			Integrated			Spectral			Solar			
		Hemispherical	Normal	Angular	Hemispherical	Normal	Angular	Hemispherical	Normal	Angular	Hemispherical	Normal	Angular	Hemispherical	Normal	Angular	Hemispherical	Normal	Angular	Hemispherical	Normal	Angular	Hemispherical	Normal	Angular	Hemispherical	Normal	Angular	Hemispherical	Normal	Angular	Hemispherical	Normal	Angular	
Silicides: (continued)																																			
Titanium Silicides	8	–	1163	–	–	1164 1166	–	–	–	–	–	1168	–	–	–	–	–	–	–	–	–	–	–	–	–	–	–	–	–	–	–	–	–	–	–
Miscellaneous Silicides	8	–	–	–	–	1172	–	–	–	–	–	1175	–	–	–	–	–	–	–	–	–	–	–	–	–	–	–	–	–	–	–	–	–	–	–
Silicon	7	–	593	–	–	595 598	–	–	–	–	–	604	611	–	–	–	–	–	–	–	614	–	–	–	–	–	–	–	–	616	–	–	–	–	–
Silicon	8	–	88	–	–	89 90	–	–	–	–	–	97	106	–	–	–	–	–	–	–	108	–	–	–	–	–	–	–	–	110	–	–	–	–	–
Silicon, Oxidized	9	–	–	–	–	–	–	–	–	–	–	1316	1318	–	–	–	–	–	–	–	–	–	–	–	–	–	–	–	–	1320	–	–	–	–	–
Silicon Coating on Aluminum Substrate	9	–	–	–	–	–	–	–	–	–	–	1046	–	–	–	–	–	–	–	–	–	–	–	1047	–	–	–	–	–	–	–	–	–	–	–
Silicon Substrate Coated with:																																			
Palladium	9	–	–	–	–	–	–	–	–	–	–	720	–	–	–	–	–	–	–	–	–	–	–	951	–	–	–	–	–	–	–	–	–	–	–
Polyvinylidene Chloride	9	–	–	–	–	–	–	–	–	–	–	–	–	–	–	–	–	–	–	–	–	–	–	–	–	–	–	–	–	1155	–	–	–	–	–
Silicon Nitride	9	–	–	–	–	–	–	–	–	–	–	–	–	–	–	–	–	–	–	–	–	–	–	–	–	–	–	–	–	953	–	–	–	–	–
Silicon Dioxide	9	–	–	–	–	–	–	–	–	–	–	–	–	–	–	–	–	–	–	–	–	–	–	–	–	–	–	–	–	1073	–	–	–	–	–
Silicon Carbide Coating on:																																			
Aluminum Oxide Substrate	9	–	–	–	–	949	–	–	–	–	–	–	–	–	–	–	–	–	–	–	–	–	–	–	–	–	–	–	–	–	–	–	–	–	–
Graphite Substrate	9	–	945	–	–	947	–	–	–	–	–	950	–	–	–	–	–	–	–	–	–	–	–	–	–	–	–	–	–	–	–	–	–	–	–
Tantalum Substrate	9	–	945	–	–	947	–	–	–	–	–	–	–	–	–	–	–	–	–	–	–	–	–	–	–	–	–	–	–	–	–	–	–	–	–
Silicon Carbide Pigment in:																																			
Aluminum Phosphate Binder	9	174	–	–	–	–	–	–	–	–	–	–	–	–	–	–	–	–	–	–	–	–	–	179	–	–	–	–	–	–	–	–	–	–	–
Dow Corning 20 Binder	9	174	–	–	–	–	–	–	–	–	–	–	–	–	–	–	–	–	–	–	–	–	–	–	–	–	–	–	–	–	–	–	–	–	–
Silica Binder	9	174	–	–	–	–	–	–	–	–	–	–	–	–	–	–	–	–	–	–	–	–	–	–	–	–	–	–	–	–	–	–	–	–	–
Synar Binder	9	174	–	–	–	–	–	–	–	–	–	–	–	–	–	–	–	–	–	–	–	–	–	–	–	–	–	–	–	–	–	–	–	–	–
Silicon Monocarbide	8	791	792	–	–	796 798	–	–	–	–	–	802	–	–	–	–	–	–	–	–	808	–	–	810	–	–	–	–	–	–	–	–	–	–	–
Silicon Dioxide Coating on:																																			
Aluminum Substrate	9	1049	1052	–	–	–	–	–	–	–	–	1054 1057	~~1060~~ 1061	–	–	–	–	–	–	–	–	–	–	1068	–	–	–	–	–	–	–	–	–	–	–
Gold Substrate	9	–	–	–	–	–	–	–	–	–	–	–	1060	–	–	–	–	–	–	–	–	–	–	1068	–	–	–	–	–	–	–	–	–	–	–
Magnesium Substrate	9	1049	–	–	–	–	–	–	–	–	–	–	–	–	–	–	–	–	–	–	–	–	–	1070	–	–	–	–	–	–	–	–	–	–	–
Nickel Substrate	9	–	–	–	–	–	–	–	–	–	–	–	1060	–	–	–	–	–	–	–	–	–	–	–	–	–	–	–	–	–	–	–	–	–	–
Nimonic 75 Substrate	9	–	1052	–	–	–	–	–	–	–	–	–	–	–	–	–	–	–	–	–	–	–	–	–	–	–	–	–	–	–	–	–	–	–	–
Niobium Substrate	9	1049	–	–	–	–	–	–	–	–	–	–	–	–	–	–	–	–	–	–	–	–	–	–	–	–	–	–	–	–	–	–	–	–	–
Platinum Substrate	9	–	–	–	–	–	–	–	–	–	–	–	1061	–	–	–	–	–	–	–	–	1067	–	–	–	–	–	–	–	1073	–	–	–	–	–

MATERIAL NAME	VOLUME	EMITTANCE						REFLECTANCE									ABSORPTANCE									TRANSMITTANCE									Absorptance to Emittance Ratio
		Total			Spectral			Integrated			Spectral			Solar			Integrated			Spectral			Solar			Integrated			Spectral			Solar			
		Hemispherical	Normal	Angular	Hemispherical	Normal	Angular	Hemispherical	Normal	Angular	Hemispherical	Normal	Angular	Hemispherical	Normal	Angular	Hemispherical	Normal	Angular	Hemispherical	Normal	Angular	Hemispherical	Normal	Angular	Hemispherical	Normal	Angular	Hemispherical	Normal	Angular	Hemispherical	Normal	Angular	
Silicon Dioxide Coating on: (continued)																																			
Silicon Substrate	9	-	-	-	-	-	-	-	-	-	-	-	-	-	-	-	-	-	-	-	-	-	-	-	-	-	-	-	-	1073	-	-	-	-	-
Silver Substrate	9	1049	1052	-	-	-	-	-	-	-	-	1054	1060	-	-	-	-	-	-	-	1065	-	-	1068 1072	-	-	-	-	-	-	-	-	-	-	-
Taylor Wire Substrate	9	1049	-	-	-	-	-	-	-	-	-	-	-	-	-	-	-	-	-	-	-	-	-	-	-	-	-	-	-	-	-	-	-	-	-
Silicon Dioxide Pigment in:																																			
Potassium Silicate Binder	9	-	-	-	-	-	-	-	-	-	-	178	-	-	-	-	-	-	-	-	-	-	-	-	-	-	-	-	-	-	-	-	-	-	-
Silicone Binder	9	-	-	-	-	-	-	-	-	-	-	178	-	-	-	-	-	-	-	-	-	-	-	-	-	-	-	-	-	-	-	-	-	-	-
Silicon Dioxide Substrate Coated with: (See Quartz)																																			
Silicon Dioxide (Crystalline)	8	-	371	-	-	372	-	-	-	-	-	374	385 387	-	-	-	-	-	-	-	389	-	-	-	-	-	-	-	-	391	-	-	-	-	-
Silicon Dioxide (Fused)	8	403	-	-	-	405	-	-	-	-	-	409	413	-	-	-	-	-	-	-	415	-	-	-	-	-	-	-	-	417	-	-	426	-	-
Silicon Monoxide	8	-	-	-	-	362	-	-	-	-	-	365	-	-	-	-	-	-	-	-	-	-	-	-	-	-	-	-	-	367	-	-	-	-	-
Silicon Monoxide Coating on:																																			
Aluminum Substrate	9	1074	-	-	-	-	-	-	-	-	-	1077	-	-	-	-	-	-	-	-	1080	-	-	-	-	-	-	-	-	-	-	-	-	-	-
Aluminum Oxide Substrate	9	-	-	-	-	-	-	-	-	-	-	1077	-	-	-	-	-	-	-	-	-	-	-	-	-	-	-	-	-	-	-	-	-	-	-
Inconel Substrate	9	-	-	-	-	1075	-	-	-	-	-	-	-	-	-	-	-	-	-	-	-	-	-	-	-	-	-	-	-	-	-	-	-	-	-
Platinum Substrate	9	-	-	-	-	1075	-	-	-	-	-	-	-	-	-	-	-	-	-	-	-	-	-	-	-	-	-	-	-	-	-	-	-	-	-
Silicon Monoxide Substrate Coated with Magnesium Fluoride	9	-	-	-	-	1094	-	-	-	-	-	1096	-	-	-	-	-	-	-	-	-	-	-	-	-	-	-	-	-	-	-	-	-	-	-
Silicon Nitride	8	-	1062	-	-	1065 1067	-	-	-	-	-	1069	-	-	-	-	-	-	-	-	-	-	-	-	-	-	-	-	-	-	-	-	-	-	-
Silicon Nitride Coating on:																																			
Gallium Arsenide Substrate	9	-	-	-	-	-	-	-	-	-	-	-	-	-	-	-	-	-	-	-	-	-	-	-	-	-	-	-	-	953	-	-	-	-	-
Silicon Substrate	9	-	-	-	-	-	-	-	-	-	-	-	-	-	-	-	-	-	-	-	-	-	-	-	-	-	-	-	-	953	-	-	-	-	-
Silicone Binder Pigmented with:																																			
Aluminum	9	3	6	-	-	-	-	-	-	-	-	-	-	-	-	-	-	-	-	-	-	-	-	-	-	-	-	-	-	-	-	-	-	-	-
Aluminum + Carbon	9	-	-	-	-	-	-	-	-	-	-	-	-	-	-	-	-	-	-	-	-	-	-	24	-	-	-	-	-	-	-	-	-	-	-
Antimony Oxide	9	-	-	-	-	-	-	-	-	-	-	51	-	-	-	-	-	-	-	-	-	-	-	-	-	-	-	-	-	-	-	-	-	-	-
Barium Titanate	9	63	-	-	-	-	-	-	-	-	-	-	-	-	-	-	-	-	-	-	-	-	-	64	-	-	-	-	-	-	-	-	-	-	-
Boron Nitride	9	-	-	-	-	-	-	-	-	-	-	68	-	-	-	-	-	-	-	-	-	-	-	-	-	-	-	-	-	-	-	-	-	-	-
Boron Nitride + Diatomaceous Earth	9	-	-	-	-	-	-	-	-	-	-	70	-	-	-	-	-	-	-	-	-	-	-	-	-	-	-	-	-	-	-	-	-	-	-
Calcium Carbonate	9	-	-	-	-	-	-	-	-	-	-	73	-	-	-	-	-	-	-	-	-	-	-	-	-	-	-	-	-	-	-	-	-	-	-
Carbon Black	9	81	82	-	-	-	-	-	-	-	-	-	-	-	-	-	-	-	-	-	-	-	-	89	-	-	-	-	-	-	-	-	-	-	-

MATERIAL NAME	VOLUME	EMITTANCE						REFLECTANCE									ABSORPTANCE									TRANSMITTANCE									Absorptance to Emittance Ratio
		Total			Spectral			Integrated			Spectral			Solar			Integrated			Spectral			Solar			Integrated			Spectral			Solar			
		Hemispherical	Normal	Angular	Hemispherical	Normal	Angular	Hemispherical	Normal	Angular	Hemispherical	Normal	Angular	Hemispherical	Normal	Angular	Hemispherical	Normal	Angular	Hemispherical	Normal	Angular	Hemispherical	Normal	Angular	Hemispherical	Normal	Angular	Hemispherical	Normal	Angular	Hemispherical	Normal	Angular	
Silicone Binder Pigmented with: (continued)																																			
China Clay	9	-	95	-	-	-	-	-	-	-	-	97	-	-	-	-	-	-	-	-	-	-	-	98 100	-	-	-	-	-	-	-	-	-	-	102
Clay + TiO_2	9	105	-	-	-	-	-	-	-	-	-	-	-	-	-	-	-	-	-	-	-	-	-	107	-	-	-	-	-	-	-	-	-	-	-
Diatomaceous Earth	9	-	-	-	-	-	-	-	-	-	-	113	-	-	-	-	-	-	-	-	-	-	-	-	-	-	-	-	-	-	-	-	-	-	-
Iron Oxide	9	117	119	-	-	121	-	-	-	-	-	-	-	-	-	-	-	-	-	-	-	-	-	-	-	-	-	-	-	-	-	-	-	-	-
Lampblack	9	-	82	-	-	-	-	-	-	-	-	86	-	-	-	-	-	-	-	-	-	-	-	-	-	-	-	-	-	-	-	-	-	-	-
Lead Carbonate	9	-	-	-	-	-	-	-	-	-	-	138	-	-	-	-	-	-	-	-	-	-	-	-	-	-	-	-	-	-	-	-	-	-	-
Leafing Aluminum	9	3	7	-	-	-	-	-	-	-	-	13 17	-	-	-	-	-	-	-	-	-	-	-	21	-	-	-	-	-	-	-	-	-	-	-
Magnesium Oxide	9	-	-	-	-	-	-	-	-	-	-	161	-	-	-	-	-	-	-	-	-	-	-	-	-	-	-	-	-	-	-	-	-	-	-
Magnesium Oxide + Diatomaceous Earth	9	-	-	-	-	-	-	-	-	-	-	165	-	-	-	-	-	-	-	-	-	-	-	-	-	-	-	-	-	-	-	-	-	-	-
Silicon Dioxide	9	-	-	-	-	-	-	-	-	-	-	178	-	-	-	-	-	-	-	-	-	-	-	-	-	-	-	-	-	-	-	-	-	-	-
Strontium Zirconate	9	198	-	-	-	-	-	-	-	-	-	-	-	-	-	-	-	-	-	-	-	-	-	199	-	-	-	-	-	-	-	-	-	-	-
Titanium Dioxide	9	211	-	-	-	-	-	-	-	-	-	220 244	-	-	-	-	-	252	-	-	255	-	-	263 281	-	-	-	-	-	-	-	-	-	-	283
Zinc Oxide	9	302	304	-	-	-	-	-	308	-	-	314 355	-	-	-	-	-	-	-	-	362	-	-	371 392	-	-	-	-	-	-	-	-	-	-	399
Zinc Sulfide	9	-	401	-	-	-	-	-	-	-	-	404	-	-	-	-	-	-	-	-	412	-	-	415	-	-	-	-	-	-	-	-	-	-	-
Zirconium Silicate	9	430	-	-	-	-	-	-	-	-	-	-	-	-	-	-	-	-	-	-	-	-	-	437	-	-	-	-	-	-	-	-	-	-	-
Silicone Coating on:																																			
Aluminum Substrate	9	-	1156	-	-	-	-	-	-	-	-	1159	-	-	-	-	-	-	-	-	-	-	-	1164 1167	-	-	-	-	-	-	-	-	-	-	1169
Dow Metal Substrate	9	-	1156	-	-	-	-	-	-	-	-	-	-	-	-	-	-	-	-	-	-	-	-	1164	-	-	-	-	-	-	-	-	-	-	-
Mild Steel Substrate	9	-	1156	-	-	-	-	-	-	-	-	-	-	-	-	-	-	-	-	-	-	-	-	1164	-	-	-	-	-	-	-	-	-	-	-
Silver Substrate	9	-	-	-	-	-	-	-	-	-	-	-	1163	-	-	-	-	-	-	-	-	-	-	-	-	-	-	-	-	-	-	-	-	-	-
Stainless Steel Substrate	9	-	1156	-	-	-	-	-	-	-	-	-	-	-	-	-	-	-	-	-	-	-	-	1164	-	-	-	-	-	-	-	-	-	-	-
Silicone Substrate Coated with:																																			
Gold	9	-	-	-	-	-	-	-	-	-	-	-	672	-	-	-	-	-	-	-	-	-	-	-	-	-	-	-	-	-	-	-	-	-	-
Silver	9	733	-	-	-	-	-	-	-	-	-	-	741	-	-	-	-	-	-	-	-	-	-	-	-	-	-	-	-	-	-	-	-	-	-
Siliconized ATJ Graphite	9	-	-	1325	-	1328	-	-	-	-	-	-	-	-	-	-	-	-	-	-	-	-	-	-	-	-	-	-	-	-	-	-	-	-	-
Siliconized Graphite	9	-	-	1325 1326	-	-	-	-	-	-	-	-	-	-	-	-	-	-	-	-	-	-	-	-	-	-	-	-	-	-	-	-	-	-	-
Siliconized Molybdenum	9	1331	1333	1335	-	1337	-	-	-	-	-	1342	-	-	-	-	-	-	-	-	1345	-	-	-	-	-	-	-	-	-	-	-	-	-	-
Silver	7	620	623	-	-	625 627	-	-	-	-	-	630	636	-	-	-	639	-	-	-	641 643	645	-	648	-	-	-	-	-	651	-	-	-	-	-

MATERIAL NAME	VOLUME	EMITTANCE						REFLECTANCE									ABSORPTANCE									TRANSMITTANCE									Absorptance to Emittance Ratio
		Total			Spectral			Integrated			Spectral			Solar			Integrated			Spectral			Solar			Integrated			Spectral			Solar			
		Hemispherical	Normal	Angular	Hemispherical	Normal	Angular	Hemispherical	Normal	Angular	Hemispherical	Normal	Angular	Hemispherical	Normal	Angular	Hemispherical	Normal	Angular	Hemispherical	Normal	Angular	Hemispherical	Normal	Angular	Hemispherical	Normal	Angular	Hemispherical	Normal	Angular	Hemispherical	Normal	Angular	
Silver Coating on:																																			
BeCu Substrate	9	–	–	–	–	–	–	–	–	–	–	–	–	–	–	–	–	–	–	–	–	–	–	–	–	–	–	–	–	–	–	–	–	–	753
Chromium Substrate	9	–	–	–	–	–	–	–	–	–	–	738	–	–	–	–	–	–	–	–	–	–	–	–	–	–	–	–	–	–	–	–	–	–	–
Copper Substrate	9	734	–	–	–	–	–	–	–	–	–	–	–	–	–	–	747	–	–	–	–	–	–	–	–	–	–	–	–	–	–	–	–	–	–
Epoxy Substrate	9	733	–	–	–	–	–	–	–	–	–	–	742	–	–	–	–	–	–	–	–	–	–	–	–	–	–	–	–	–	–	–	–	–	–
Glass Substrate	9	–	–	–	–	–	–	–	–	–	–	738	743	–	–	–	–	–	–	–	–	–	–	–	–	–	–	–	–	–	752	–	–	–	–
Mylar Substrate	9	–	–	–	–	–	–	–	–	–	–	–	–	–	746	–	–	–	–	–	–	–	–	–	–	–	–	–	–	750	–	–	–	–	–
Nickel Substrate	9	734	–	–	–	–	–	–	–	–	–	–	–	–	–	–	747	–	–	–	–	–	–	–	–	–	–	–	–	–	–	–	–	–	–
Polyurethane	9	733	–	–	–	–	–	–	–	–	–	–	–	–	–	741	–	–	–	–	–	–	–	–	–	–	–	–	–	–	–	–	–	–	–
Quartz Substrate	9	–	–	–	–	–	–	–	–	–	–	738	743	–	–	–	–	–	–	–	–	–	–	–	–	–	–	–	–	750	752	–	–	–	–
Sapphire Substrate	9	–	–	–	–	–	–	–	–	–	–	–	–	–	–	–	–	–	–	–	–	–	–	–	–	–	–	–	–	750	–	–	–	–	–
Silicone Substrate	9	733	–	–	–	–	–	–	–	–	–	741	–	–	–	–	–	–	–	–	–	–	–	–	–	–	–	–	–	–	–	–	–	–	–
Stainless Steel Substrate	9	734	736	–	–	–	–	–	–	–	–	738	742	–	–	–	747	–	–	–	–	–	–	–	–	–	–	–	–	–	–	–	–	–	–
Silver Substrate Coated with:																																			
Aluminum	9	–	–	–	–	–	–	–	–	–	–	–	–	–	–	603	–	–	–	–	–	–	–	–	–	–	–	–	–	–	–	–	–	–	–
Aluminum Oxide	9	785	788	–	–	–	–	–	–	–	–	796	–	–	–	–	–	–	–	–	802	–	–	803	–	–	–	–	–	–	–	–	–	–	–
Carbon	9	–	–	–	–	–	–	–	–	–	–	859	–	–	–	–	–	–	–	–	–	–	–	–	–	–	–	–	–	–	–	–	–	–	–
Chromium	9	–	–	–	–	–	–	–	–	–	–	621	–	–	–	–	–	–	–	–	–	–	–	–	–	–	–	–	–	–	–	–	–	–	–
Cobalt Oxide	9	–	887	–	–	–	–	–	–	–	–	–	–	–	–	–	–	–	–	–	–	–	–	–	–	–	–	–	–	–	–	–	–	–	–
Copper Oxide	9	–	891	–	–	–	–	–	–	–	–	–	–	–	–	–	–	–	–	–	–	–	–	–	–	–	–	–	–	–	–	–	–	–	–
Glass	9	–	–	–	–	–	–	–	–	–	–	1035	–	–	–	–	–	–	–	–	–	–	–	–	–	–	–	–	–	–	–	–	–	–	–
Polymide Fluorinated Ethylene Propylene	9	1133	–	–	–	–	–	–	–	–	–	–	–	–	–	–	–	–	–	–	–	–	–	–	–	–	–	–	–	–	–	–	–	–	–
Sapphire	9	785	–	–	–	–	–	–	–	–	–	–	–	–	–	–	–	–	–	–	–	–	–	–	–	–	–	–	–	–	–	–	–	–	–
Silicon Dioxide	9	1049	1052	–	–	–	–	–	–	–	–	1054	1060	–	–	–	–	–	–	–	1065	–	–	1068 1072	–	–	–	–	–	–	–	–	–	–	–
Silicone	9	–	–	–	–	–	–	–	–	–	–	–	1163	–	–	–	–	–	–	–	–	–	–	–	–	–	–	–	–	–	–	–	–	–	–
Silver Sulfide	9	954	–	–	–	–	–	–	–	–	–	–	–	–	–	–	–	–	–	–	–	–	–	955	–	–	–	–	–	–	–	–	–	–	–
Silver + Aluminum	7	–	–	–	–	–	–	–	–	–	–	1007	1009	–	–	–	–	–	–	–	–	–	–	–	–	–	–	–	–	–	–	–	–	–	–
Silver + Aluminum Coating on Glass Substrate	9	–	–	–	–	–	–	–	–	–	–	754	–	–	–	–	–	–	–	–	–	–	–	–	–	–	–	–	–	–	–	–	–	–	–
Silver + Beryllium	7	–	–	–	–	–	–	–	–	–	–	–	1012	–	–	–	–	–	–	–	–	–	–	–	–	–	–	–	–	–	–	–	–	–	–
Silver + Cadmium + ΣX_i	7	–	–	–	–	–	–	–	–	–	–	–	1472	–	–	–	–	–	–	–	–	–	–	–	–	–	–	–	–	–	–	–	–	–	–
Silver + Copper + ΣX_i	7	–	–	–	–	–	–	–	–	–	–	–	1475	–	–	–	–	–	–	–	–	–	–	–	–	–	–	–	–	–	–	–	–	–	–
Silver + Gold	7	–	–	–	–	–	–	–	–	–	–	1015	–	–	–	–	–	–	–	–	–	–	–	–	–	–	–	–	–	–	–	–	–	–	–
Silver + Silicon	7	–	–	–	–	–	–	–	–	–	–	–	1018	–	–	–	–	–	–	–	–	–	–	–	–	–	–	–	–	–	–	–	–	–	–

MATERIAL NAME	VOLUME	EMITTANCE						REFLECTANCE									ABSORPTANCE									TRANSMITTANCE									Absorptance to Emittance Ratio
		Total			Spectral			Integrated			Spectral			Solar			Integrated			Spectral			Solar			Integrated			Spectral			Solar			
		Hemispherical	Normal	Angular	Hemispherical	Normal	Angular	Hemispherical	Normal	Angular	Hemispherical	Normal	Angular	Hemispherical	Normal	Angular	Hemispherical	Normal	Angular	Hemispherical	Normal	Angular	Hemispherical	Normal	Angular	Hemispherical	Normal	Angular	Hemispherical	Normal	Angular	Hemispherical	Normal	Angular	
Silver + Zinc + ΣX_i	7	-	-	-	-	-	-	-	-	-	-	-	1478	-	-	-	-	-	-	-	-	-	-	-	-	-	-	-	-	-	-	-	-	-	-
Silver Black Coating on Unknown Substrate	9	-	-	-	-	-	-	-	-	-	-	-	-	-	-	-	-	-	-	-	-	-	-	-	-	-	-	-	-	1189	-	-	-	-	-
Silver Bromide	8	-	-	-	-	-	-	-	-	-	-	770	773	-	-	-	-	-	-	-	-	-	-	-	-	-	-	-	-	775	-	-	-	-	-
Silver Chloride	8	-	-	-	-	-	-	-	-	-	-	-	876	-	-	-	-	-	-	-	-	-	-	-	-	-	-	-	-	879	-	-	-	-	-
Silver Chloride Pigment in Gelatin Binder	9	-	-	-	-	-	-	-	-	-	-	181	-	-	-	-	-	-	-	-	-	-	-	-	-	-	-	-	-	183	-	-	-	-	-
Silver Iodide	8	-	-	-	-	-	-	-	-	-	-	1022	-	-	-	-	-	-	-	-	-	-	-	-	-	-	-	-	-	1024	-	-	-	-	-
Silver Sulfide Coating on Silver Substrate	9	954	-	-	-	-	-	-	-	-	-	-	-	-	-	-	-	-	-	-	-	-	-	955	-	-	-	-	-	-	-	-	-	-	-
SiO_2 + Al	8	-	-	-	-	-	-	-	-	-	-	1428	-	-	-	-	-	-	-	-	-	-	-	-	-	-	-	-	-	-	-	-	-	-	-
SiO_2 + Al_2O_3 Powders	8	-	-	-	-	558	-	-	-	-	-	560	-	-	-	-	-	-	-	-	-	-	-	-	-	-	-	-	-	-	-	-	-	-	-
SiO_2 + Cr Cermets	8	-	-	-	-	-	-	-	-	-	-	-	-	-	-	-	-	-	-	-	-	-	-	-	-	-	-	-	-	1401	-	-	-	-	-
SiO_2 + MoO_3 Powders	8	-	-	-	-	-	-	-	-	-	-	566	-	-	-	-	-	-	-	-	-	-	-	-	-	-	-	-	-	-	-	-	-	-	-
SiO_2 + $MoSi_2$ Powders	8	-	1478	-	-	1479	-	-	-	-	-	1482	-	-	-	-	-	-	-	-	-	-	-	-	-	-	-	-	-	-	-	-	-	-	-
SiO_2 + $MoSi_2$ + MoO_3 Powders	8	-	1510	-	-	1511	-	-	-	-	-	1513	-	-	-	-	-	-	-	-	-	-	-	-	-	-	-	-	-	-	-	-	-	-	-
SiO_2 + TiB_2 Powders	8	-	-	-	-	-	-	-	-	-	-	1484	-	-	-	-	-	-	-	-	-	-	-	-	-	-	-	-	-	-	-	-	-	-	-
Skyspar (White) A423	9	211	-	-	-	-	-	-	-	-	-	221	-	-	-	-	-	252	-	-	255	-	-	263	-	-	-	-	-	-	-	-	-	-	-
Soda Lime Glass	8	-	1609	-	-	-	-	-	-	-	-	1612	-	-	-	-	-	-	-	-	-	-	-	1614	-	-	1615	-	-	1617	-	-	-	-	-
Soda Silica Glass	8	-	-	-	-	-	-	-	-	-	-	-	-	-	-	-	-	-	-	-	-	-	-	-	-	-	-	-	-	1650	-	-	-	-	-
Sodium Aluminum Silicate Pigment in Sodium Silicate Binder	9	-	-	-	-	-	-	-	-	-	-	-	-	-	-	-	-	-	-	-	185	-	-	-	-	-	-	-	-	-	-	-	-	-	-
Sodium Borate	8	-	-	-	-	-	-	-	-	-	-	582	-	-	-	-	-	-	-	-	-	-	-	-	-	-	-	-	-	-	-	-	-	-	-
Sodium Carbonate	8	-	-	-	-	-	-	-	-	-	-	593	-	-	-	-	-	-	-	-	-	-	-	-	-	-	-	-	-	-	-	-	-	-	-
Sodium Chlorate	8	-	-	-	-	-	-	-	-	-	-	594	-	-	-	-	-	-	-	-	-	-	-	-	-	-	-	-	-	-	-	-	-	-	-
Sodium Chloride	8	881	-	-	-	883	-	-	-	-	885 886	880	893	-	-	-	-	-	-	-	-	-	-	-	-	-	-	-	-	895	-	-	-	-	-
Sodium Chloride Coating on Aluminum Substrate	9	-	-	-	-	-	-	-	-	-	-	-	1081	-	-	-	-	-	-	-	-	-	-	-	-	-	-	-	-	-	-	-	-	-	-
Sodium Chloride Substrate Coated with: Gold Black + Copper Black	9	-	-	-	-	-	-	-	-	-	-	-	-	-	-	-	-	-	-	-	-	-	-	-	-	-	-	-	-	1185	-	-	-	-	-

MATERIAL NAME	VOLUME	EMITTANCE Total Hemispherical	EMITTANCE Total Normal	EMITTANCE Total Angular	EMITTANCE Spectral Hemispherical	EMITTANCE Spectral Normal	EMITTANCE Spectral Angular	REFLECTANCE Integrated Hemispherical	REFLECTANCE Integrated Normal	REFLECTANCE Integrated Angular	REFLECTANCE Spectral Hemispherical	REFLECTANCE Spectral Normal	REFLECTANCE Spectral Angular	REFLECTANCE Solar Hemispherical	REFLECTANCE Solar Normal	REFLECTANCE Solar Angular	ABSORPTANCE Integrated Hemispherical	ABSORPTANCE Integrated Normal	ABSORPTANCE Integrated Angular	ABSORPTANCE Spectral Hemispherical	ABSORPTANCE Spectral Normal	ABSORPTANCE Spectral Angular	ABSORPTANCE Solar Hemispherical	ABSORPTANCE Solar Normal	ABSORPTANCE Solar Angular	TRANSMITTANCE Integrated Hemispherical	TRANSMITTANCE Integrated Normal	TRANSMITTANCE Integrated Angular	TRANSMITTANCE Spectral Hemispherical	TRANSMITTANCE Spectral Normal	TRANSMITTANCE Spectral Angular	TRANSMITTANCE Solar Hemispherical	TRANSMITTANCE Solar Normal	TRANSMITTANCE Solar Angular	Absorptance to Emittance Ratio
Sodium Chloride Substrate Coated with: (continued)																																			
Gold Black + Nickel Black	9	-	-	-	-	-	-	-	-	-	-	-	-	-	-	-	-	-	-	-	-	-	-	-	-	-	-	-	-	1187	-	-	-	-	-
Sodium Fluoride	8	-	-	-	-	-	-	-	-	-	-	963	-	-	-	-	-	-	-	-	-	-	-	-	-	-	-	-	-	966	-	-	-	-	-
Sodium Nickel Trifluoride	8	-	-	-	-	-	-	-	-	-	-	-	-	-	-	-	-	-	-	-	-	-	-	-	-	-	-	-	-	990	-	-	-	-	-
Sodium Nitrate	8	-	-	-	-	-	-	-	-	-	-	600	-	-	-	-	-	-	-	-	-	-	-	-	-	-	-	-	-	-	-	-	-	-	-
Sodium Phosphate	8	-	-	-	-	-	-	-	-	-	-	608	-	-	-	-	-	-	-	-	-	-	-	-	-	-	-	-	-	-	-	-	-	-	-
Sodium Tripolyphosphate	8	-	-	-	-	-	-	-	-	-	-	608	-	-	-	-	-	-	-	-	-	-	-	-	-	-	-	-	-	-	-	-	-	-	-
Sodium Salicylate Coating on MgO Pigmented Paint Substrate	9	-	-	-	-	-	-	-	-	-	-	956	-	-	-	-	-	-	-	-	-	-	-	-	-	-	-	-	-	-	-	-	-	-	-
Sodium Silicate	8	-	-	-	-	-	-	-	-	-	-	618	-	-	-	-	-	-	-	-	-	-	-	-	-	-	-	-	-	-	-	-	-	-	-
Sodium Disilicate	8	-	-	-	-	-	-	-	-	-	-	618	-	-	-	-	-	-	-	-	-	-	-	-	-	-	-	-	-	-	-	-	-	-	-
Sodium Metasilicate	8	-	-	-	-	-	-	-	-	-	-	618	-	-	-	-	-	-	-	-	-	-	-	-	-	-	-	-	-	-	-	-	-	-	-
Sodium Silicate Binder Pigmented with:																																			
Aquablack B	9	574	-	-	-	-	-	-	-	-	-	-	-	-	-	-	-	-	-	-	-	-	-	575	-	-	-	-	-	-	-	-	-	-	-
Calcium Fluoride	9	-	-	-	-	-	-	-	-	-	-	-	-	-	-	-	-	-	-	-	75	-	-	-	-	-	-	-	-	-	-	-	-	-	-
Lithafrax	9	142	-	-	-	-	-	-	-	-	-	-	-	-	-	-	-	-	-	-	147	-	-	150	-	-	-	-	-	-	-	-	-	-	-
Lithium Aluminum Silicate	9	142	-	-	-	-	-	-	-	-	-	-	-	-	-	-	-	-	-	-	147	-	-	150	-	-	-	-	-	-	-	-	-	-	-
Lithium Fluoride	9	-	-	-	-	-	-	-	-	-	-	-	-	-	-	-	-	-	-	-	155	-	-	-	-	-	-	-	-	-	-	-	-	-	-
Potassium Aluminum Silicate	9	-	-	-	-	-	-	-	-	-	-	-	-	-	-	-	-	-	-	-	171	-	-	-	-	-	-	-	-	-	-	-	-	-	-
Sodium Aluminum Silicate	9	-	-	-	-	-	-	-	-	-	-	-	-	-	-	-	-	-	-	-	185	-	-	-	-	-	-	-	-	-	-	-	-	-	-
Spodumene	9	-	-	-	-	-	-	-	-	-	-	-	-	-	-	-	-	-	-	-	-	-	-	150 153	-	-	-	-	-	-	-	-	-	-	-
Titanium Dioxide	9	-	-	-	-	-	-	-	-	-	-	-	-	-	-	-	-	-	-	-	255	-	-	-	-	-	-	-	-	-	-	-	-	-	-
Zinc Oxide	9	-	-	-	-	-	-	-	-	-	-	-	-	-	-	-	-	-	-	-	362	-	-	-	-	-	-	-	-	-	-	-	-	-	-
Zinc Sulfide	9	-	-	-	-	-	-	-	-	-	-	-	-	-	-	-	-	-	-	-	412	-	-	-	-	-	-	-	-	-	-	-	-	-	-
Zirconium	9	26	-	-	-	-	-	-	-	-	-	-	-	-	-	-	-	-	-	-	-	-	-	-	-	-	-	-	-	-	-	-	-	-	-
Zirconium Silicate	9	430	-	-	-	-	-	-	-	-	-	-	-	-	-	-	-	-	-	-	-	-	-	437	-	-	-	-	-	-	-	-	-	-	-
Sodium Tungstate	8	-	-	-	-	-	-	-	-	-	-	666	-	-	-	-	-	-	-	-	-	-	-	-	-	-	-	-	-	-	-	-	-	-	-
Sodium Vanadate	8	-	-	-	-	-	-	-	-	-	-	667	-	-	-	-	-	-	-	-	-	-	-	-	-	-	-	-	-	-	-	-	-	-	-
Solar Cell	8	-	88	-	-	92	-	-	-	-	-	100	-	-	-	-	-	-	-	-	-	-	-	-	-	-	-	-	-	-	-	-	-	-	-
Solder	7	948	-	-	-	-	-	-	-	-	-	-	-	-	-	-	-	-	-	-	-	-	-	-	-	-	-	-	-	-	-	-	-	-	-
Solder Coating on Copper Substrate	9	758	-	-	-	-	-	-	-	-	-	-	-	-	-	-	759	-	-	-	-	-	-	-	-	-	-	-	-	-	-	-	-	-	-

MATERIAL NAME	VOLUME	EMITTANCE						REFLECTANCE									ABSORPTANCE									TRANSMITTANCE									Absorptance to Emittance Ratio
		Total			Spectral			Integrated			Spectral			Solar			Integrated			Spectral			Solar			Integrated			Spectral			Solar			
		Hemispherical	Normal	Angular	Hemispherical	Normal	Angular	Hemispherical	Normal	Angular	Hemispherical	Normal	Angular	Hemispherical	Normal	Angular	Hemispherical	Normal	Angular	Hemispherical	Normal	Angular	Hemispherical	Normal	Angular	Hemispherical	Normal	Angular	Hemispherical	Normal	Angular	Hemispherical	Normal	Angular	
SP 500	8	–	–	–	–	–	–	–	–	–	–	510 511 512	–	–	–	–	–	–	–	–	–	–	–	521	–	–	–	–	–	–	–	–	–	–	–
SP 500 Pigment in:																																			
Acrylic Binder	9	–	–	–	–	–	–	–	–	–	–	314	–	–	–	–	–	–	–	–	–	–	–	–	–	–	–	–	–	–	–	–	–	–	–
Copolymer Binder	9	–	–	–	–	–	–	–	–	–	–	314	–	–	–	–	–	–	–	–	–	–	–	–	–	–	–	–	–	–	–	–	–	–	–
Copolymer Acetone Binder	9	–	–	–	–	–	–	–	–	–	–	314	–	–	–	–	–	–	–	–	–	–	–	–	–	–	–	–	–	–	–	–	–	–	–
Potassium Silicate Binder	9	–	304	–	–	–	–	–	–	–	–	314	–	–	–	–	–	360	–	–	365	–	–	370 394	–	–	–	–	–	–	–	–	–	–	–
Silicone Binder	9	–	304	–	–	–	–	–	308	–	306	314	–	–	–	–	–	–	–	–	–	–	–	370 394	–	–	–	–	–	–	–	–	–	–	–
Speculum Coating on:																																			
Glass Substrate	9	–	–	–	–	–	–	–	–	–	–	645	–	–	–	–	–	–	–	–	–	–	–	–	–	–	–	–	–	–	–	–	–	–	–
Steel Substrate	9	–	–	–	–	–	–	–	–	–	–	645 757	–	–	–	–	–	–	–	–	–	–	–	–	–	–	–	–	–	–	–	–	–	–	–
Spinel	8	–	1674	–	–	1676	–	–	–	–	–	576	–	–	–	–	–	–	–	–	–	–	–	–	–	–	–	–	–	578	–	–	–	–	–
Spinel Pigments:																																			
$CoO \cdot Cr_2O_3$ Spinel Pigment in NBS Frit No. 332 Binder (Enamel)	9	–	472	–	–	–	–	–	–	–	–	475	–	–	–	–	–	–	–	–	–	–	–	–	–	–	–	–	–	–	–	–	–	–	–
$CoO \cdot Fe_2O_3$ Spinel Pigment in NBS Frit No. 332 Binder (Enamel)	9	–	472	–	–	–	–	–	–	–	–	475	–	–	–	–	–	–	–	–	–	–	–	–	–	–	–	–	–	–	–	–	–	–	–
$CoO \cdot Mn_2O_3$ Spinel Pigment in NBS Frit No. 332 Binder (Enamel)	9	–	472	–	–	–	–	–	–	–	–	475	–	–	–	–	–	–	–	–	–	–	–	–	–	–	–	–	–	–	–	–	–	–	–
Cr–Co–Ni Spinel Pigment in Aluminum Phosphate Binder	9	186	–	–	–	–	–	–	–	–	–	–	–	–	–	–	–	–	–	–	–	–	–	–	–	–	–	–	–	–	–	–	–	–	–
Magnesium Aluminate Spinel Pigment in Potassium Silicate Binder	9	–	–	–	–	–	–	–	–	–	–	–	–	–	–	–	–	–	–	–	–	–	–	187	–	–	–	–	–	–	–	–	–	–	–
$NiO \cdot Cr_2O_3$ Spinel Pigment in NBS Frit No. 332 Binder (Enamel)	9	–	472	–	–	–	–	–	–	–	–	475	–	–	–	–	–	–	–	–	–	–	–	–	–	–	–	–	–	–	–	–	–	–	–
$NiO \cdot Cr_2O_3$ Spinel + SiO_2 Pigment in Aluminum Phosphate Binder	9	186	–	–	–	–	–	–	–	–	–	–	–	–	–	–	–	–	–	–	–	–	–	–	–	–	–	–	–	–	–	–	–	–	–
$NiO \cdot Fe_2O_3$ Spinel Pigment in NBS Frit No. 332 Binder (Enamel)	9	–	472	–	–	–	–	–	–	–	–	475	–	–	–	–	–	–	–	–	–	–	–	–	–	–	–	–	–	–	–	–	–	–	–
Spodumene Pigment in Sodium Silicate Binder	9	–	–	–	–	–	–	–	–	–	–	–	–	–	–	–	–	–	–	–	–	–	–	150 153	–	–	–	–	–	–	–	–	–	–	–

MATERIAL NAME	VOLUME	EMITTANCE						REFLECTANCE									ABSORPTANCE									TRANSMITTANCE									Absorptance to Emittance Ratio
		Total			Spectral			Integrated			Spectral			Solar			Integrated			Spectral			Solar			Integrated			Spectral			Solar			
		Hemispherical	Normal	Angular	Hemispherical	Normal	Angular	Hemispherical	Normal	Angular	Hemispherical	Normal	Angular	Hemispherical	Normal	Angular	Hemispherical	Normal	Angular	Hemispherical	Normal	Angular	Hemispherical	Normal	Angular	Hemispherical	Normal	Angular	Hemispherical	Normal	Angular	Hemispherical	Normal	Angular	
Spraint Grey No. 63	9	-	-	-	-	-	-	-	-	-	-	564	-	-	-	-	-	-	-	-	-	-	-	-	-	-	-	-	-	-	-	-	-	-	-
Sprayon Paints:																																			
High-Heat No. 324	9	-	-	-	-	-	-	-	-	-	-	566	-	-	-	-	-	-	-	-	-	-	-	-	-	-	-	-	-	-	-	-	-	-	-
Machinery Dark Grey Enamel No. 325	9	-	-	-	-	-	-	-	-	-	-	566	-	-	-	-	-	-	-	-	-	-	-	-	-	-	-	-	-	-	-	-	-	-	-
Machinery Light Grey Enamel No. 326	9	-	-	-	-	-	-	-	-	-	-	566	-	-	-	-	-	-	-	-	-	-	-	-	-	-	-	-	-	-	-	-	-	-	-
SR-111 Coating on Silver Substrate	9	-	-	-	-	-	-	-	-	-	-	-	1163	-	-	-	-	-	-	-	-	-	-	-	-	-	-	-	-	-	-	-	-	-	-
Stainless Steels:																																			
17-7 PH	7	-	1223	-	-	1237 1245	-	-	-	-	-	1266 1268	-	-	-	-	-	-	-	-	-	-	-	1302	-	-	-	-	-	-	-	-	-	-	-
18-8	7	1212 1214	1225 1226	-	-	-	-	-	-	-	-	-	-	-	-	-	-	-	-	-	-	-	-	1303	-	-	-	-	-	-	-	-	-	-	-
301	7	-	1221 1226	-	-	-	-	-	-	-	-	1269	-	-	1288	-	-	-	-	-	-	-	-	1300	-	-	-	-	-	-	-	-	-	-	-
302	7	1212 1213	-	-	-	-	-	-	-	-	-	-	-	-	-	-	1291	-	-	-	-	-	-	-	-	-	-	-	-	-	-	-	-	-	-
303	7	1212	1226	-	-	-	1254 1258 1259 1260	-	-	-	-	-	-	-	-	-	-	-	-	-	-	1297	-	-	-	-	-	-	-	-	-	-	-	-	-
304	7	1213	1227	-	-	1244	-	-	-	-	-	1270	-	-	-	-	-	-	-	-	-	-	-	-	-	-	-	-	-	-	-	-	-	-	-
304 ELC	7	1213	-	-	-	-	-	-	-	-	-	-	-	-	-	-	-	-	-	-	-	-	-	-	-	-	-	-	-	-	-	-	-	-	-
310	7	1212 1213	-	-	-	-	-	-	-	-	-	-	-	-	-	-	-	-	-	-	-	-	-	-	-	-	-	-	-	-	-	-	-	-	-
316	7	-	1221 1224	-	-	1237 1244	-	-	-	-	-	1266 1270 1271	-	-	1288	-	-	-	-	-	-	-	-	1300 1302	-	-	-	-	-	-	-	-	-	-	-
321	7	-	1224 1225 1226	-	-	1237 1238 1244 1246	-	-	-	-	-	1266 1267 1270 1272	1285	-	-	-	-	-	-	-	1294	-	-	1302 1303	-	-	-	-	-	-	-	-	-	-	-
347	7	1212	1222	-	-	-	-	-	-	-	-	-	-	-	1288	-	-	-	-	-	-	-	-	1301	-	-	-	-	-	-	-	-	-	-	-
430	7	-	-	-	-	1193	-	-	-	-	-	-	-	-	-	-	-	-	-	-	-	-	-	-	-	-	-	-	-	-	-	-	-	-	-
446	7	-	1180	-	-	1187	-	-	-	-	-	1198	-	-	-	-	-	-	-	-	-	-	-	1207	-	-	-	-	-	-	-	-	-	-	-
A286	7	-	-	-	-	1322	-	-	-	-	-	1325	-	-	-	-	-	-	-	-	-	-	-	-	-	-	-	-	-	-	-	-	-	-	-
AM 350	7	-	1225	-	-	1238 1245	-	-	-	-	-	1266 1267 1268	-	-	-	-	-	-	-	-	-	-	-	1302 1303	-	-	-	-	-	-	-	-	-	-	-
PH 15-7 Mo	7	-	1223	-	-	1237 1245	-	-	-	-	-	1266 1267	-	-	-	-	-	-	-	-	-	-	-	1301	-	-	-	-	-	-	-	-	-	-	-
SF 11	7	-	-	-	-	-	-	-	-	-	-	938	-	-	-	-	-	-	-	-	-	-	-	-	-	-	-	-	-	-	-	-	-	-	-
SF 20	7	-	-	-	-	-	-	-	-	-	-	1266	-	-	-	-	-	-	-	-	-	-	-	-	-	-	-	-	-	-	-	-	-	-	-

MATERIAL NAME	VOLUME	EMITTANCE						REFLECTANCE									ABSORPTANCE									TRANSMITTANCE									Absorptance to Emittance Ratio
		Total			Spectral			Integrated			Spectral			Solar			Integrated			Spectral			Solar			Integrated			Spectral			Solar			
		Hemispherical	Normal	Angular	Hemispherical	Normal	Angular	Hemispherical	Normal	Angular	Hemispherical	Normal	Angular	Hemispherical	Normal	Angular	Hemispherical	Normal	Angular	Hemispherical	Normal	Angular	Hemispherical	Normal	Angular	Hemispherical	Normal	Angular	Hemispherical	Normal	Angular	Hemispherical	Normal	Angular	
Stainless Steel Substrate Coated with:																																			
Aluminum	9	580	-	-	-	-	-	-	-	-	-	-	602	-	-	-	609	-	-	-	-	-	-	612	-	-	-	-	-	-	-	-	-	-	-
Aluminum Oxide	9	-	788	-	-	790 792	-	-	-	-	-	796	-	-	-	-	-	-	-	-	-	-	-	-	-	-	-	-	-	-	-	-	-	-	-
Calcium Titanate	9	843	-	-	-	-	-	-	-	-	-	-	-	-	-	-	-	-	-	-	-	-	-	-	-	-	-	-	-	-	-	-	-	-	-
Carbon Dioxide	9	-	-	-	-	-	-	-	-	-	864	-	-	-	-	-	-	-	-	-	-	-	-	-	-	-	-	-	-	-	-	-	-	-	-
Chromium	9	-	619	-	-	-	-	-	-	-	-	621	624	-	-	-	-	-	-	-	-	-	-	-	-	-	-	-	-	-	-	-	-	-	-
Chromium Black	9	1173	-	-	-	-	-	-	-	-	-	-	-	-	-	-	-	-	-	-	-	-	-	-	-	-	-	-	-	-	-	-	-	-	-
$Cr_2O_3 + SiO_2 + \Sigma X_i$	9	878	-	-	-	-	-	-	-	-	-	-	-	-	-	-	-	-	-	-	-	-	-	-	-	-	-	-	-	-	-	-	-	-	-
Cobalt	9	-	631	-	-	-	-	-	-	-	-	632	-	-	-	-	-	-	-	-	-	-	-	-	-	-	-	-	-	-	-	-	-	-	-
Copper + Tin	9	-	-	-	-	-	-	-	-	-	-	645	-	-	-	-	-	-	-	-	-	-	-	-	-	-	-	-	-	-	-	-	-	-	-
Copper Oxide	9	890	-	-	-	-	-	-	-	-	-	-	-	-	-	-	-	-	-	-	-	-	-	-	-	-	-	-	-	-	-	-	-	-	-
Gold	9	651	-	-	-	-	-	-	-	-	-	-	672	-	-	-	675	-	-	-	-	-	-	-	-	-	-	-	-	-	-	-	-	-	-
Gold + Silver	9	651	-	-	-	-	-	-	-	-	-	-	-	-	-	-	675	-	-	-	-	-	-	-	-	-	-	-	-	-	-	-	-	-	-
Iron Titanate	9	905	-	-	-	-	-	-	-	-	-	-	-	-	-	-	-	-	-	-	-	-	-	-	-	-	-	-	-	-	-	-	-	-	-
Nickel	9	700	-	-	-	-	-	-	-	-	-	-	-	-	-	-	-	-	-	-	-	-	-	-	-	-	-	-	-	-	-	-	-	-	-
Nickel + Cobalt	9	-	715	-	-	-	-	-	-	-	-	716	-	-	-	-	-	-	-	-	-	-	-	-	-	-	-	-	-	-	-	-	-	-	-
Nickel + Molybdenum + ΣX_i	9	717	-	-	-	-	-	-	-	-	-	-	-	-	-	-	-	-	-	-	-	-	-	-	-	-	-	-	-	-	-	-	-	-	-
Nylon	9	-	-	-	-	-	-	-	-	-	-	1132	-	-	-	-	-	-	-	-	-	-	-	-	-	-	-	-	-	-	-	-	-	-	-
Polyurethane	9	-	-	-	-	-	-	-	-	-	-	-	1140	-	-	-	-	-	-	-	-	-	-	-	-	-	-	-	-	-	-	-	-	-	-
Rhodium	9	729	-	-	-	-	-	-	-	-	-	-	-	-	-	-	731	-	-	-	-	-	-	-	-	-	-	-	-	-	-	-	-	-	-
Silicone	9	-	1156	-	-	-	-	-	-	-	-	-	1163	-	-	-	-	-	-	-	-	-	-	1164	-	-	-	-	-	-	-	-	-	-	-
Silver	9	734	736	-	-	-	-	-	-	-	-	738	742	-	-	-	747	-	-	-	-	-	-	-	-	-	-	-	-	-	-	-	-	-	-
Strontium Titanate	9	958	-	-	-	-	-	-	-	-	-	-	-	-	-	-	-	-	-	-	-	-	-	-	-	-	-	-	-	-	-	-	-	-	-
Tin + Copper	9	-	-	-	-	-	-	-	-	-	-	757	-	-	-	-	-	-	-	-	-	-	-	-	-	-	-	-	-	-	-	-	-	-	-
Titanium Dioxide	9	981	-	-	-	-	-	-	-	-	-	-	-	-	-	-	-	-	-	-	-	-	-	-	-	-	-	-	-	-	-	-	-	-	-
Titanium Dioxide + Aluminum Oxide	9	994	-	-	-	-	-	-	-	-	-	-	-	-	-	-	-	-	-	-	-	-	-	-	-	-	-	-	-	-	-	-	-	-	-
Zirconium Oxide	9	-	1011	-	-	-	-	-	-	-	-	1017	-	-	-	-	-	-	-	-	-	-	-	-	-	-	-	-	-	-	-	-	-	-	-
Zirconium Silicate	9	1024	1026	-	-	-	-	-	-	-	-	1029	-	-	-	-	-	-	-	-	-	-	-	-	-	-	-	-	-	-	-	-	-	-	-
Stainless Steel 347, Oxidized	9	-	1305	-	-	1308	-	-	-	-	-	-	-	-	-	-	-	-	-	-	-	-	-	-	-	-	-	-	-	-	-	-	-	-	-
Stannic Oxide (see Tin Dioxide)																																			
Stannous Telluride (see Tin Monotelluride)																																			

MATERIAL NAME	VOLUME	EMITTANCE						REFLECTANCE									ABSORPTANCE									TRANSMITTANCE									Absorptance to Emittance Ratio
		Total			Spectral			Integrated			Spectral			Solar			Integrated			Spectral			Solar			Integrated			Spectral			Solar			
		Hemispherical	Normal	Angular	Hemispherical	Normal	Angular	Hemispherical	Normal	Angular	Hemispherical	Normal	Angular	Hemispherical	Normal	Angular	Hemispherical	Normal	Angular	Hemispherical	Normal	Angular	Hemispherical	Normal	Angular	Hemispherical	Normal	Angular	Hemispherical	Normal	Angular	Hemispherical	Normal	Angular	
Steel Substrate Coated with:																																			
Aluminum Oxide	9	-	788	-	-	792	-	-	-	-	-	-	-	-	-	-	-	-	-	-	-	-	-	-	-	-	-	-	-	-	-	-	-	-	-
$Cr_2O_3 + SiO_2 + \Sigma X_i$	9	-	881	-	-	885	-	-	-	-	-	-	-	-	-	-	-	-	-	-	-	-	-	1164	-	-	-	-	-	-	-	-	-	-	-
Lacquer	9	-	-	-	-	-	-	-	-	-	-	-	-	-	-	-	-	-	-	-	1126	-	-	-	-	-	-	-	-	-	-	-	-	-	-
Silicone	9	-	1156	-	-	-	-	-	-	-	-	-	-	-	-	-	-	-	-	-	-	-	-	-	-	-	-	-	-	-	-	-	-	-	-
Zirconium Oxide	9	-	1011	-	-	1014	-	-	-	-	-	-	-	-	-	-	-	-	-	-	-	-	-	-	-	-	-	-	-	-	-	-	-	-	-
Stellite	7	-	-	-	-	-	-	-	-	-	-	1154	-	-	-	-	-	-	-	-	-	-	-	-	-	-	-	-	-	-	-	-	-	-	-
Stilbene Substrate Coated with Antimony	9	-	-	-	-	-	-	-	-	-	-	-	-	-	-	-	-	-	-	-	-	-	-	-	-	-	-	-	-	615	-	-	-	-	-
Stilbite	8	-	-	-	-	-	-	-	-	-	-	-	-	-	-	-	-	-	-	-	-	-	-	-	-	-	-	-	-	1694	-	-	-	-	-
Strontium Hexaboride	8	-	-	-	-	732	-	-	-	-	-	-	-	-	-	-	-	-	-	-	-	-	-	-	-	-	-	-	-	-	-	-	-	-	-
Strontium Fluoride	8	-	-	-	-	-	-	-	-	-	-	968	-	-	-	-	-	-	-	-	-	-	-	-	-	-	-	-	-	971	-	-	-	-	-
Strontium Hafnate	8	-	-	-	-	-	-	-	-	-	-	597	-	-	-	-	-	-	-	-	-	-	-	-	-	-	-	-	-	-	-	-	-	-	-
Strontium Molybdate Pigment in:																																			
Acrylic Binder	9	-	189	-	-	-	-	-	-	-	-	191	-	-	-	-	-	-	-	-	-	-	-	194	-	-	-	-	-	-	-	-	-	-	-
Potassium Silicate Binder	9	188	-	-	-	-	-	-	-	-	-	191	-	-	-	-	-	-	-	-	-	-	-	194	-	-	-	-	-	-	-	-	-	-	-
Strontium Oxide	8	-	-	-	-	-	-	-	-	-	-	546	-	-	-	-	-	-	-	-	-	-	-	-	-	-	-	-	-	-	-	-	-	-	-
Strontium Titanate	8	-	-	-	-	651	-	-	-	-	-	653	-	-	-	-	-	-	-	-	-	-	-	-	-	-	-	-	-	657	-	-	-	-	-
Strontium Titanate Coating on:																																			
Aluminum Substrate	9	958	-	-	-	-	-	-	-	-	-	-	-	-	-	-	-	-	-	-	-	-	-	963	-	-	-	-	-	-	-	-	-	-	-
Nb-1Zr Substrate	9	958	-	-	-	962	-	-	-	-	-	-	-	-	-	-	-	-	-	-	-	-	-	-	-	-	-	-	-	-	-	-	-	-	-
Stainless Steel Substrate	9	958	-	-	-	-	-	-	-	-	-	-	-	-	-	-	-	-	-	-	-	-	-	-	-	-	-	-	-	-	-	-	-	-	-
Strontium Titanate Pigment in Aluminum Phosphate Binder	9	197	-	-	-	-	-	-	-	-	-	-	-	-	-	-	-	-	-	-	-	-	-	-	-	-	-	-	-	-	-	-	-	-	-
Strontium Zirconate	8	-	-	-	-	-	-	-	-	-	-	676	-	-	-	-	-	-	-	-	-	-	-	-	-	-	-	-	-	-	-	-	-	-	-
Strontium Zirconate Pigment in Silicone Binder	9	198	-	-	-	-	-	-	-	-	-	-	-	-	-	-	-	-	-	-	-	-	-	199	-	-	-	-	-	-	-	-	-	-	-
Sulfates:																																			
Barium Sulfate	8	-	-	-	-	-	-	-	-	-	-	623	-	-	-	-	-	-	-	-	-	-	-	626	-	-	-	-	-	-	-	-	-	-	-
Calcium Sulfates	8	-	-	-	-	-	-	-	-	-	-	627	-	-	-	-	-	-	-	-	-	-	-	-	-	-	-	-	-	629	-	-	-	-	-
Lead Sulfate and Sulfite	8	-	-	-	-	-	-	-	-	-	-	-	-	-	-	-	-	-	-	-	-	-	-	-	-	-	-	-	-	631	-	-	-	-	-
Sulfides:																																			
Diarsenic Trisulfide	8	-	-	-	-	-	-	-	-	-	-	1177	-	-	-	-	-	-	-	-	-	-	-	-	-	-	-	-	-	1179	-	-	-	-	-
Cadmium Sulfide	8	-	1181	-	-	1183	-	-	-	-	-	1188	-	-	-	-	-	-	-	-	-	-	-	-	-	-	-	-	-	1194	-	-	-	-	-
Lead Sulfide	8	-	-	-	-	-	-	-	-	-	-	1197	-	-	-	-	-	-	-	-	-	-	-	-	-	-	-	-	-	1204	-	-	-	-	-
Molybdenum Disulfide	8	-	-	-	-	-	-	-	-	-	-	1207	-	-	-	-	-	-	-	-	-	-	-	-	-	-	-	-	-	1210	-	-	-	-	-

MATERIAL NAME	VOLUME	EMITTANCE						REFLECTANCE									ABSORPTANCE									TRANSMITTANCE									Absorptance to Emittance Ratio
		Total			Spectral			Integrated			Spectral			Solar			Integrated			Spectral			Solar			Integrated			Spectral			Solar			
		Hemispherical	Normal	Angular	Hemispherical	Normal	Angular	Hemispherical	Normal	Angular	Hemispherical	Normal	Angular	Hemispherical	Normal	Angular	Hemispherical	Normal	Angular	Hemispherical	Normal	Angular	Hemispherical	Normal	Angular	Hemispherical	Normal	Angular	Hemispherical	Normal	Angular	Hemispherical	Normal	Angular	
Sulfides: (continued)																																			
Zinc Sulfide	8	1213	-	-	-	1215	-	-	-	-	-	1217	1222	-	-	-	-	-	-	-	1224	-	-	1226	-	-	-	-	-	1227	-	-	-	-	-
Miscellaneous Sulfides	8	-	-	-	-	1231	-	-	-	-	-	1233	-	-	-	-	-	-	-	-	-	-	-	-	-	-	-	-	-	1235	-	-	-	-	-
Sulfur	8	-	-	-	-	-	-	-	-	-	-	115	-	-	-	-	-	-	-	-	-	-	-	-	-	-	-	-	-	121	-	-	-	-	-
Superlith XXXN	8	-	-	-	-	-	-	-	-	-	-	1219	-	-	-	-	-	-	-	-	-	-	-	1226	-	-	-	-	-	-	-	-	-	-	-
Superlith XXXN Pigment in Silicone Binder	8	-	-	-	-	-	-	-	-	-	-	405	-	-	-	-	-	-	-	-	-	-	-	415	-	-	-	-	-	-	-	-	-	-	-
Superpax	8	-	-	-	-	-	-	-	-	-	-	613	-	-	-	-	-	-	-	-	-	-	-	-	-	-	-	-	-	-	-	-	-	-	-
Superpax Pigment in Potassium Silicate Binder	8	-	-	-	-	-	-	-	-	-	-	432	-	-	-	-	-	-	-	-	-	-	-	437	-	-	-	-	-	-	-	-	-	-	-
SY-627-119 Coating on Stainless Steel Substrate	9	-	-	-	-	-	-	-	-	-	-	-	1140	-	-	-	-	-	-	-	-	-	-	-	-	-	-	-	-	-	-	-	-	-	-
Sylvania Phosphor Paint	9	-	-	-	-	568	-	-	-	-	-	-	-	-	-	-	-	-	-	-	-	-	-	-	-	-	-	-	-	-	-	-	-	-	-
Synar Binder Pigmented with:																																			
Boron Carbide	9	65	-	-	-	-	-	-	-	-	-	-	-	-	-	-	-	-	-	-	-	-	-	-	-	-	-	-	-	-	-	-	-	-	-
$Cr_2O_3 + Fe_3O_4 + NiO$	9	103	-	-	-	-	-	-	-	-	-	-	-	-	-	-	-	-	-	-	-	-	-	-	-	-	-	-	-	-	-	-	-	-	-
Silicon Carbide	9	174	-	-	-	-	-	-	-	-	-	-	-	-	-	-	-	-	-	-	-	-	-	-	-	-	-	-	-	-	-	-	-	-	-
$SiC + UO_2$	9	176	-	-	-	-	-	-	-	-	-	-	-	-	-	-	-	-	-	-	-	-	-	-	-	-	-	-	-	-	-	-	-	-	-
Synar Coating on Niobium Substrate	9	1049	-	-	-	-	-	-	-	-	-	-	-	-	-	-	-	-	-	-	-	-	-	-	-	-	-	-	-	-	-	-	-	-	-
T-12 Commercial Optical Material	8	-	-	-	-	-	-	-	-	-	-	-	1451	-	-	-	-	-	-	-	-	-	-	-	-	-	-	-	-	1453	-	-	-	-	-
T-40-C-C-9 Coating on Aluminum Substrate	9	-	-	-	-	-	-	-	-	-	-	1110	-	-	-	-	-	-	-	-	-	-	-	-	-	-	-	-	-	-	-	-	-	-	-
T-111	7	1481	-	-	-	-	-	-	-	-	-	-	-	-	-	-	-	-	-	-	-	-	-	-	-	-	-	-	-	-	-	-	-	-	-
$Ta + TaAl_3$	8	-	-	-	-	-	-	-	-	-	-	1433	-	-	-	-	-	-	-	-	-	-	-	-	-	-	-	-	-	-	-	-	-	-	-
$TaAl_3$	8	-	-	-	-	1330	-	-	-	-	-	1332	-	-	-	-	-	-	-	-	-	-	-	-	-	-	-	-	-	-	-	-	-	-	-
$TaAl_3 + Al$	8	-	-	-	-	-	-	-	-	-	-	1431	-	-	-	-	-	-	-	-	-	-	-	-	-	-	-	-	-	-	-	-	-	-	-
$TaAl_3 + Ta$	8	-	-	-	-	-	-	-	-	-	-	1433	-	-	-	-	-	-	-	-	-	-	-	-	-	-	-	-	-	-	-	-	-	-	-
Talc	8	-	-	-	-	-	-	-	-	-	-	-	-	-	-	-	-	-	-	-	-	-	-	-	-	-	-	-	-	1694	-	-	-	-	-
TAM-CP Coating on Stainless Steel Substrate	9	-	-	-	-	-	-	-	-	-	-	1017	-	-	-	-	-	-	-	-	-	-	-	-	-	-	-	-	-	-	-	-	-	-	-
TAM Zirconia	8	-	-	-	-	-	-	-	-	-	-	540	-	-	-	-	-	-	-	-	-	-	-	-	-	-	-	-	-	-	-	-	-	-	-
Tantalum	7	654	661	-	-	666 672	-	-	-	-	-	678	684	-	-	-	-	-	-	-	-	-	-	687	-	-	-	-	-	-	-	-	-	-	-
Tantalum, Anodized	9	1284	-	-	-	-	-	-	-	-	-	1286	-	-	-	-	-	-	-	-	-	-	-	1287	-	-	-	-	-	-	-	-	-	-	-

MATERIAL NAME	VOLUME	EMITTANCE						REFLECTANCE									ABSORPTANCE									TRANSMITTANCE									Absorptance to Emittance Ratio
		Total			Spectral			Integrated			Spectral			Solar			Integrated			Spectral			Solar			Integrated			Spectral			Solar			
		Hemispherical	Normal	Angular	Hemispherical	Normal	Angular	Hemispherical	Normal	Angular	Hemispherical	Normal	Angular	Hemispherical	Normal	Angular	Hemispherical	Normal	Angular	Hemispherical	Normal	Angular	Hemispherical	Normal	Angular	Hemispherical	Normal	Angular	Hemispherical	Normal	Angular	Hemispherical	Normal	Angular	
Tantalum Substrate Coated by Pack Cementation	9	-	1353	-	-	1355	-	-	-	-	-	1358	-	-	-	-	-	-	-	-	-	-	-	-	-	-	-	-	-	-	-	-	-	-	-
Tantalum Substrate Coated with:																																			
Cobalt Oxide	9	-	887	-	-	888	-	-	-	-	-	-	-	-	-	-	-	-	-	-	-	-	-	-	-	-	-	-	-	-	-	-	-	-	-
Graphite	9	-	854	-	-	855	-	-	-	-	-	-	-	-	-	-	-	-	-	-	-	-	-	-	-	-	-	-	-	-	-	-	-	-	-
Silicon Carbide	9	-	945	-	-	947	-	-	-	-	-	-	-	-	-	-	-	-	-	-	-	-	-	-	-	-	-	-	-	-	-	-	-	-	-
Tantalum + Tungsten	7	-	1021	-	-	1024	-	-	-	-	-	-	-	-	-	-	-	-	-	-	-	-	-	-	-	-	-	-	-	-	-	-	-	-	-
Tantalum + Tungsten + ΣX_i	7	1481	-	-	-	-	-	-	-	-	-	-	-	-	-	-	-	-	-	-	-	-	-	-	-	-	-	-	-	-	-	-	-	-	-
Tantalum Carbide Coating on Inconel X Substrate	9	-	-	-	-	965	-	-	-	-	-	968	-	-	-	-	-	-	-	-	-	-	-	-	-	-	-	-	-	-	-	-	-	-	-
Tantalates:																																			
Potassium Tantalate	8	-	-	-	-	-	-	-	-	-	-	634	-	-	-	-	-	-	-	-	-	-	-	-	-	-	-	-	-	-	-	-	-	-	-
Tantalum Carbide	8	811	813	-	-	815 817	-	-	-	-	-	-	-	-	-	-	-	-	-	-	-	-	-	-	-	-	-	-	-	-	-	-	-	-	-
Tantalum Nitride	8	-	1072	-	-	1074 1076	-	-	-	-	-	-	-	-	-	-	-	-	-	-	-	-	-	-	-	-	-	-	-	-	-	-	-	-	-
Ditantalum Nitride	8	-	-	-	-	1075	-	-	-	-	-	-	-	-	-	-	-	-	-	-	-	-	-	-	-	-	-	-	-	-	-	-	-	-	-
Tantalum Oxide	8	-	427	-	-	428	-	-	-	-	-	430	-	-	-	-	-	-	-	-	-	-	-	-	-	-	-	-	-	-	-	-	-	-	-
Tantalum Disilicide	8	-	1156	-	-	1157 1159	-	-	-	-	-	1161	-	-	-	-	-	-	-	-	-	-	-	-	-	-	-	-	-	-	-	-	-	-	-
Ta_2O_5 + $Be_{12}Ta$ Cermet	8	-	1403	-	-	1404	-	-	-	-	-	1406	-	-	-	-	-	-	-	-	-	-	-	-	-	-	-	-	-	-	-	-	-	-	-
Teflon:	8	-	1730	-	-	-	-	-	-	-	-	1732	-	-	-	-	-	-	-	-	1734	-	-	-	-	-	-	-	-	1736	-	-	-	-	-
AMS 3651	8	-	-	-	-	-	-	-	-	-	-	1733	-	-	-	-	-	-	-	-	-	-	-	-	-	-	-	-	-	-	-	-	-	-	-
BMS-8-71	8	-	-	-	-	-	-	-	-	-	-	1733	-	-	-	-	-	-	-	-	-	-	-	-	-	-	-	-	-	-	-	-	-	-	-
Teflon Binder Pigmented with:																																			
Zinc Oxide	9	-	-	-	-	-	-	-	-	-	-	317	-	-	-	-	-	-	-	-	-	-	-	-	-	-	-	-	-	-	-	-	-	-	-
Zirconium Oxide	9	-	-	-	-	-	-	-	-	-	-	425	-	-	-	-	-	-	-	-	-	-	-	-	-	-	-	-	-	-	-	-	-	-	-
Teflon Coating on:																																			
Aluminum Substrate	9	-	-	-	-	-	-	-	-	-	-	1085	-	-	-	-	-	-	-	-	-	-	-	1087	-	-	-	-	-	-	-	-	-	-	-
S-13 Substrate	9	-	-	-	-	-	-	-	-	-	-	1085	-	-	-	-	-	-	-	-	-	-	-	1087	-	-	-	-	-	-	-	-	-	-	-
Teflon Substrate Coated with Aluminum	9	-	586	-	-	-	-	-	-	-	-	-	-	-	-	-	-	-	-	-	-	-	-	610	-	-	-	-	-	-	-	-	-	-	-
Television Tube Glass	8	-	-	-	-	1644	-	-	-	-	-	1646	-	-	-	-	-	-	-	-	-	-	-	-	-	-	-	-	1648	-	-	-	-	-	-

MATERIAL NAME	VOLUME	EMITTANCE						REFLECTANCE									ABSORPTANCE									TRANSMITTANCE									Absorptance to Emittance Ratio
		Total			Spectral			Integrated			Spectral			Solar			Integrated			Spectral			Solar			Integrated			Spectral			Solar			
		Hemispherical	Normal	Angular	Hemispherical	Normal	Angular	Hemispherical	Normal	Angular	Hemispherical	Normal	Angular	Hemispherical	Normal	Angular	Hemispherical	Normal	Angular	Hemispherical	Normal	Angular	Hemispherical	Normal	Angular	Hemispherical	Normal	Angular	Hemispherical	Normal	Angular	Hemispherical	Normal	Angular	
Tellurides:																																			
Dibismuth Tritelluride	8	-	-	-	-	-	-	-	-	-	-	1237	-	-	-	-	-	-	-	-	-	-	-	-	-	-	-	-	-	-	-	-	-	-	-
Cadmium Telluride	8	1239	-	-	-	-	-	-	-	-	-	1241	-	-	-	-	-	-	-	-	-	-	-	-	-	-	-	-	-	1244	-	-	-	-	-
Germanium Telluride	8	-	-	-	-	-	-	-	-	-	-	1250	-	-	-	-	-	-	-	-	-	-	-	-	-	-	-	-	-	-	-	-	-	-	-
Lead Telluride	8	-	-	-	-	-	-	-	-	-	-	1252	-	-	-	-	-	-	-	-	-	-	-	-	-	-	-	-	-	1254	-	-	-	-	-
Manganese Telluride	8	-	-	-	-	-	-	-	-	-	-	1256	-	-	-	-	-	-	-	-	-	-	-	-	-	-	-	-	-	-	-	-	-	-	-
Tin Telluride	8	-	-	-	-	-	-	-	-	-	-	1259	-	-	-	-	-	-	-	-	-	-	-	-	-	-	-	-	-	1264	-	-	-	-	-
Zinc Telluride	8	-	-	-	-	-	-	-	-	-	-	1266	-	-	-	-	-	-	-	-	-	-	-	-	-	-	-	-	-	1268	-	-	-	-	-
Tellurium	8	-	-	-	-	-	-	-	-	-	-	123 125	128	-	-	-	-	-	-	-	-	-	-	-	-	-	-	-	-	130	136	-	-	-	-
Tellurium Coating on:																																			
Glass Substrate	9	-	-	-	-	-	-	-	-	-	-	-	971	-	-	-	-	-	-	-	-	-	-	-	-	-	-	-	-	972	-	-	-	-	-
Stilbene Substrate	9	-	-	-	-	-	-	-	-	-	-	-	971	-	-	-	-	-	-	-	-	-	-	-	-	-	-	-	-	972	-	-	-	-	-
Tellurium Dioxide	8	-	-	-	-	432	-	-	-	-	-	434	-	-	-	-	-	-	-	-	-	-	-	-	-	-	-	-	-	436	-	-	-	-	-
Terbium Hexaboride	8	-	-	-	-	723	-	-	-	-	-	-	-	-	-	-	-	-	-	-	-	-	-	-	-	-	-	-	-	-	-	-	-	-	-
Tesslar Coating on Aluminum Substrate	9	-	-	-	-	-	-	-	-	-	-	-	-	-	-	-	-	-	-	-	-	-	-	1082	-	-	-	-	-	-	-	-	-	-	-
Thallium	7	-	-	-	-	-	-	-	-	-	-	690	693	-	-	-	-	-	-	-	-	-	-	-	-	-	-	-	-	696	-	-	-	-	-
Thallium Bromide	8	-	-	-	-	-	-	-	-	-	-	778	780 782	-	-	-	-	-	-	-	-	-	-	-	-	-	-	-	-	783	-	-	-	-	-
TlBr + TlCl (KRS-6)	8	-	-	-	-	-	-	-	-	-	-	-	1455	-	-	-	-	-	-	-	-	-	-	-	-	-	-	-	-	1457	-	-	-	-	-
TlBr + TlI (KRS-5)	8	-	-	-	-	1459	-	-	-	-	-	-	1461	-	-	-	-	-	-	-	-	-	-	-	-	-	-	-	-	1463	-	-	-	-	-
Thallium Bromide Chloride	8	-	-	-	-	-	-	-	-	-	-	-	1455	-	-	-	-	-	-	-	-	-	-	-	-	-	-	-	-	1457	-	-	-	-	-
Thallium Bromide Iodide	8	-	-	-	-	1459	-	-	-	-	-	-	1461	-	-	-	-	-	-	-	-	-	-	-	-	-	-	-	-	1463	-	-	-	-	-
Thallium Chloride	8	-	-	-	-	-	-	-	-	-	-	899	901	-	-	-	-	-	-	-	-	-	-	-	-	-	-	-	-	903	-	-	-	-	-
TlCl + TlBr (KRS-6)	8	-	-	-	-	-	-	-	-	-	-	-	1455	-	-	-	-	-	-	-	-	-	-	-	-	-	-	-	-	1457	-	-	-	-	-
Thallium Iodide	8	-	-	-	-	-	-	-	-	-	-	-	-	-	-	-	-	-	-	-	-	-	-	-	-	-	-	-	-	1029	-	-	-	-	-
TlI + TlBr (KRS-5)	8	-	-	-	-	1459	-	-	-	-	-	-	1461	-	-	-	-	-	-	-	-	-	-	-	-	-	-	-	-	1463	-	-	-	-	-
Thermatrol ZA-100	9	-	-	-	-	-	-	-	-	-	-	224	-	-	-	-	-	-	-	-	-	-	-	-	-	-	-	-	-	-	-	-	-	-	-
ThO_2 + Ni Cermet	8	-	1408	-	-	-	-	-	-	-	-	-	-	-	-	-	-	-	-	-	-	-	-	-	-	-	-	-	-	-	-	-	-	-	-
Thoria (see Thorium Oxide)																																			

MATERIAL NAME	VOLUME	EMITTANCE						REFLECTANCE									ABSORPTANCE									TRANSMITTANCE									Absorptance to Emittance Ratio
		Total			Spectral			Integrated			Spectral			Solar			Integrated			Spectral			Solar			Integrated			Spectral			Solar			
		Hemispherical	Normal	Angular	Hemispherical	Normal	Angular	Hemispherical	Normal	Angular	Hemispherical	Normal	Angular	Hemispherical	Normal	Angular	Hemispherical	Normal	Angular	Hemispherical	Normal	Angular	Hemispherical	**Normal**	Angular	Hemispherical	Normal	Angular	Hemispherical	Normal	Angular	Hemispherical	Normal	Angular	
Thorium	7	-	-	-	-	699 701	-	-	-	-	-	-	-	-	-	-	-	-	-	-	-	-	-	-	-	-	-	-	-	-	-	-	-	-	-
Thorium Tetraboride	8	-	730	-	-	-	-	-	-	-	-	-	-	-	-	-	-	-	-	-	-	-	-	-	-	-	-	-	-	-	-	-	-	-	-
Thorium Monocarbide	8	-	-	-	-	852	-	-	-	-	-	-	-	-	-	-	-	-	-	-	-	-	-	-	-	-	-	-	-	-	-	-	-	-	-
Thorium Dicarbide	8	-	-	-	-	852	-	-	-	-	-	-	-	-	-	-	-	-	-	-	-	-	-	-	-	-	-	-	-	-	-	-	-	-	-
Thorium Oxide	8	-	438	-	-	440	-	-	-	-	-	442 444	-	-	-	-	-	-	-	-	-	-	-	446	-	-	-	-	-	-	-	-	-	-	-
Thorium Dioxide Coating on:																																			
Nimonic 75 Substrate	9	-	973	-	-	-	-	-	-	-	-	-	-	-	-	-	-	-	-	-	-	-	-	-	-	-	-	-	-	-	-	-	-	-	-
Quartz Substrate	9	-	-	-	-	-	-	-	-	-	-	-	-	-	-	-	-	-	-	-	-	-	-	-	-	-	-	-	-	974	-	-	-	-	-
Thulia (see Thulium Oxide)																																			
Thulium Hexaboride	8	-	-	-	-	723	-	-	-	-	-	-	-	-	-	-	-	-	-	-	-	-	-	-	-	-	-	-	-	-	-	-	-	-	-
Thulium Oxide	8	-	-	-	447 450	-	-	-	-	-	-	-	-	-	-	-	-	-	-	-	-	-	-	-	-	-	-	-	-	-	-	-	-	-	-
$Ti + TiB_2$	8	-	-	-	-	-	-	-	-	-	-	1437	-	-	-	-	-	-	-	-	-	-	-	-	-	-	-	-	-	-	-	-	-	-	-
$Ti + TiO_2$	8	-	-	-	-	-	-	-	-	-	-	1439	-	-	-	-	-	-	-	-	-	-	-	-	-	-	-	-	-	-	-	-	-	-	-
TiAl	8	-	1338	-	-	1339	-	-	-	-	-	1341	-	-	-	-	-	-	-	-	-	-	-	-	-	-	-	-	-	-	-	-	-	-	-
TiB_2 + B Powders	8	-	-	-	-	-	-	-	-	-	-	1447	-	-	-	-	-	-	-	-	-	-	-	-	-	-	-	-	-	-	-	-	-	-	-
$TiB_2 + H_3BO_3$	8	-	1468	-	-	1469	-	-	-	-	-	1471	-	-	-	-	-	-	-	-	-	-	-	-	-	-	-	-	-	-	-	-	-	-	-
TiB_2 + Ni	8	-	-	-	-	-	-	-	-	-	-	1435	-	-	-	-	-	-	-	-	-	-	-	-	-	-	-	-	-	-	-	-	-	-	-
TiB_2 + Ti Cermet	8	-	-	-	-	1410	-	-	-	-	-	-	-	-	-	-	-	-	-	-	-	-	-	-	-	-	-	-	-	-	-	-	-	-	-
TiB_2 + Ti	8	-	-	-	-	-	-	-	-	-	-	1437	-	-	-	-	-	-	-	-	-	-	-	-	-	-	-	-	-	-	-	-	-	-	-
$TiB_2 + TiO_2$ Powders	8	-	1489	-	-	1490	-	-	-	-	-	1493	-	-	-	-	-	-	-	-	-	-	-	-	-	-	-	-	-	-	-	-	-	-	-
TiC + Ni + ΣX_i Cermet	8	1412	1415	-	-	-	-	-	-	-	-	-	-	-	-	-	-	-	-	-	-	-	-	-	-	-	-	-	-	-	-	-	-	-	-
$TiCr_2$	8	-	1343	-	-	1344	-	-	-	-	-	1346	-	-	-	-	-	-	-	-	-	-	-	-	-	-	-	-	-	-	-	-	-	-	-
Tin	7	703	705	-	-	-	-	-	-	-	-	707	-	-	-	-	-	710	712	-	714	717	-	-	-	-	-	-	-	720	-	-	-	-	-
Tin Coating on Copper Substrate	9	755	-	-	-	-	-	-	-	-	-	-	-	-	-	-	756	-	-	-	-	-	-	-	-	-	-	-	-	-	-	-	-	-	-
Tin + Copper Coating on Steel Substrate	9	-	-	-	-	-	-	-	-	-	-	757	-	-	-	-	-	-	-	-	-	-	-	-	-	-	-	-	-	-	-	-	-	-	-
Tin + Indium	7	-	-	-	-	-	-	-	-	-	-	-	-	-	-	-	-	1026	-	-	-	-	-	-	-	-	-	-	-	-	-	-	-	-	-
Tin + Lead Coating on Copper Substrate	9	758	-	-	-	-	-	-	-	-	-	-	-	-	-	-	756	-	-	-	-	-	-	-	-	-	-	-	-	-	-	-	-	-	-

MATERIAL NAME	VOLUME	EMITTANCE						REFLECTANCE									ABSORPTANCE									TRANSMITTANCE									Absorptance to Emittance Ratio
		Total			Spectral			Integrated			Spectral			Solar			Integrated			Spectral			Solar			Integrated			Spectral			Solar			
		Hemispherical	Normal	Angular	Hemispherical	Normal	Angular	Hemispherical	Normal	Angular	Hemispherical	Normal	Angular	Hemispherical	Normal	Angular	Hemispherical	Normal	Angular	Hemispherical	Normal	Angular	Hemispherical	Normal	Angular	Hemispherical	Normal	Angular	Hemispherical	Normal	Angular	Hemispherical	Normal	Angular	
Tin Dioxide	8	-	-	-	-	-	-	-	-	-	-	451	-	-	-	-	-	-	-	-	-	-	-	-	-	-	-	-	-	-	-	-	-	-	-
Tin Oxide Coating on:																																			
Glass Substrate	9	-	-	-	-	-	-	-	-	-	-	-	-	-	-	-	-	-	-	-	-	-	-	-	-	-	-	-	-	979	-	-	-	-	-
Tin Oxide Substrate	9	-	-	-	-	-	-	-	-	-	-	977	-	-	-	-	-	-	-	-	-	-	-	-	-	-	-	-	-	-	-	-	-	-	-
Tin Oxide Pigment in:																																			
Aluminum Phosphate Binder	9	201	-	-	-	-	-	-	-	-	-	-	-	-	-	-	-	-	-	-	-	-	-	205	-	-	-	-	-	-	-	-	-	-	-
Potassium Silicate Binder	9	201	-	-	-	-	-	-	-	-	-	204	-	-	-	-	-	-	-	-	-	-	-	205 207	-	-	-	-	-	-	-	-	-	-	-
Tin Oxide + Cerium Dioxide Pigment in NBS Frit No. 332 Binder (Enamel)	9	-	477	-	-	-	-	-	-	-	-	478	-	-	-	-	-	-	-	-	-	-	-	-	-	-	-	-	-	-	-	-	-	-	-
Tin Sulfide	8	-	-	-	-	-	-	-	-	-	-	1233	-	-	-	-	-	-	-	-	-	-	-	-	-	-	-	-	-	-	-	-	-	-	-
Tin Monotelluride	8	-	-	-	-	-	-	-	-	-	-	1259	-	-	-	-	-	-	-	-	-	-	-	-	-	-	-	-	-	1264	-	-	-	-	-
$TiO_2 + HBO_3 + TiB_2$ Powders	8	-	1515	-	-	1516	-	-	-	-	-	1518	-	-	-	-	-	-	-	-	-	-	-	-	-	-	-	-	-	-	-	-	-	-	-
TiO_2 + MnO Powders	8	-	-	-	-	-	-	-	-	-	-	563	-	-	-	-	-	-	-	-	-	-	-	-	-	-	-	-	-	-	-	-	-	-	-
TiO_2 + Ti	8	-	-	-	-	-	-	-	-	-	-	1439	-	-	-	-	-	-	-	-	-	-	-	-	-	-	-	-	-	-	-	-	-	-	-
$TiO_2 + TiB_2$	8	-	1489	-	-	1490	-	-	-	-	-	1493	-	-	-	-	-	-	-	-	-	-	-	-	-	-	-	-	-	-	-	-	-	-	-
$TiO_2 + TiCr_2$ Cermet	8	-	1419	-	-	1420	-	-	-	-	-	1421	-	-	-	-	-	-	-	-	-	-	-	-	-	-	-	-	-	-	-	-	-	-	-
$TiB_2 + TiO_2 + H_3BO_3$ Powders	8	-	1515	-	-	1516	-	-	-	-	-	1518	-	-	-	-	-	-	-	-	-	-	-	-	-	-	-	-	-	-	-	-	-	-	-
$TiB_2 + ZrSi_2$ Powders	8	-	-	-	-	-	-	-	-	-	-	1495	-	-	-	-	-	-	-	-	-	-	-	-	-	-	-	-	-	-	-	-	-	-	-
TiPure R-900-1 Pigment in LTV-602 Binder	9	-	-	-	-	-	-	-	-	-	-	220	-	-	-	-	-	-	-	-	-	-	-	-	-	-	-	-	-	-	-	-	-	-	-
Titanates:																																			
Barium Titanate	8	-	-	-	-	-	-	-	-	-	-	635 637	-	-	-	-	-	-	-	-	-	-	-	-	-	-	-	-	-	642	-	-	-	-	-
Bismuth Titanate	8	-	-	-	-	-	-	-	-	-	-	-	-	-	-	-	-	-	-	-	-	-	-	-	-	-	-	-	-	644	-	-	-	-	-
Lead Titanate	8	-	-	-	-	-	-	646	-	-	-	647	-	-	-	-	-	-	-	-	-	-	-	-	-	-	-	-	-	649	-	-	-	-	-
Strontium Titanate	8	-	-	-	-	651	-	-	-	-	-	653	-	-	-	-	-	-	-	-	-	-	-	-	-	-	-	-	-	657	-	-	-	-	-
Zinc Titanate	8	-	-	-	-	-	-	-	-	-	-	660	-	-	-	-	-	-	-	-	-	-	-	662	-	-	-	-	-	-	-	-	-	-	-
Titania (see Titanium Dioxide)																																			
Titanium	7	723	726	-	-	729 732	735 738	-	-	-	-	744	751	-	769	-	-	-	-	-	-	771	-	-	-	-	-	-	-	773	-	-	-	-	-
Titanium, AMS 4901	7	-	727	-	-	-	-	-	-	-	-	746	-	-	-	-	-	-	-	-	-	-	-	-	-	-	-	-	-	-	-	-	-	-	-
Titanium, Oxidized	9	-	-	-	-	-	-	-	-	-	-	1322	-	-	-	-	-	-	-	-	-	-	-	-	-	-	-	-	-	-	-	-	-	-	-
Titanium, Ti-75A	7	724	727	-	-	-	-	-	-	-	-	746	-	-	-	-	-	-	-	-	-	-	-	-	-	-	-	-	-	-	-	-	-	-	-

MATERIAL NAME	VOLUME	EMITTANCE						REFLECTANCE									ABSORPTANCE									TRANSMITTANCE									Absorptance to Emittance Ratio
		Total			Spectral			Integrated			Spectral			Solar			Integrated			Spectral			Solar			Integrated			Spectral			Solar			
		Hemispherical	Normal	Angular	Hemispherical	Normal	Angular	Hemispherical	Normal	Angular	Hemispherical	Normal	Angular	Hemispherical	Normal	Angular	Hemispherical	Normal	Angular	Hemispherical	Normal	Angular	Hemispherical	Normal	Angular	Hemispherical	Normal	Angular	Hemispherical	Normal	Angular	Hemispherical	Normal	Angular	
Titanium Coating on:																																			
Aluminum Substrate	9	–	–	–	–	–	–	–	–	–	–	760	–	–	–	–	–	–	–	–	–	–	–	763	–	–	–	–	–	–	–	–	–	–	–
Brass Substrate	9	–	–	–	–	–	–	–	–	–	–	–	–	–	–	–	–	–	–	–	–	762	–	–	–	–	–	–	–	–	–	–	–	–	–
Glass Substrate	9	–	–	–	–	–	–	–	–	–	–	–	761	–	–	–	–	–	–	–	–	–	–	–	–	–	–	–	–	–	–	–	–	–	–
Titanium Substrate Coated by Pack Cementation	9	–	1359	–	–	1360	–	–	–	–	–	1362	–	–	–	–	–	–	–	–	–	–	–	–	–	–	–	–	–	–	–	–	–	–	–
Titanium Substrate Coated with Zinc Oxide	9	–	–	–	–	–	–	–	–	–	–	1003	–	–	–	–	–	–	–	–	–	–	–	–	–	–	–	–	–	–	–	–	–	–	–
Titanium + Aluminum + ΣX_i	7	1483	1486	–	–	1490	–	–	–	–	–	1497	–	–	–	–	–	–	–	–	–	–	–	1500	–	–	–	–	–	–	–	–	–	–	–
Titanium + Aluminum + ΣX_i, Anodized	9	–	–	–	–	–	–	–	–	–	–	1289	–	–	–	–	–	–	–	–	–	–	–	–	–	–	–	–	–	–	–	–	–	–	–
Titanium + Manganese	7	–	1028	–	–	1032	–	–	–	–	–	1037	–	–	–	–	–	–	–	–	–	–	–	1041	–	–	–	–	–	–	–	–	–	–	–
Titanium + Manganese + ΣX_i	7	1503	1505	–	–	–	–	–	–	–	–	–	–	–	–	–	–	–	–	–	–	–	–	–	–	–	–	–	–	–	–	–	–	–	–
Titanium + Vanadium + ΣX_i	7	–	–	–	–	–	–	–	–	–	–	1508	–	–	–	–	–	–	–	–	–	–	–	–	–	–	–	–	–	–	–	–	–	–	–
Titanium + Vanadium + ΣX_i, Anodized	9	–	–	–	–	–	–	–	–	–	–	1293	–	–	–	–	–	–	–	–	–	–	–	–	–	–	–	–	–	–	–	–	–	–	–
Titanium 6Al-4V Substrate Coated with Cr_2O_3 + SiO_2 + ΣX_i	9	–	878	–	–	883	–	–	–	–	–	–	–	–	–	–	–	–	–	–	–	–	–	–	–	–	–	–	–	–	–	–	–	–	–
Titanium Alloys:																																			
6Al-4V	7	1484	1488	–	–	1492 1493	–	–	–	–	–	1498	–	–	–	–	–	–	–	–	–	–	–	1501	–	–	–	–	–	–	–	–	–	–	–
A-110-AT	7	–	1488	–	–	1492	–	–	–	–	–	1498	–	–	–	–	–	–	–	–	–	–	–	1501	–	–	–	–	–	–	–	–	–	–	–
AMS 4908	7	–	1030	–	–	–	–	–	–	–	–	1038	–	–	–	–	–	–	–	–	–	–	–	–	–	–	–	–	–	–	–	–	–	–	–
B120 VAC	7	–	–	–	–	–	–	–	–	–	–	1509	–	–	–	–	–	–	–	–	–	–	–	–	–	–	–	–	–	–	–	–	–	–	–
C-110 M	7	–	1030	–	–	1034	–	–	–	–	–	1038	–	–	–	–	–	–	–	–	–	–	–	1041	–	–	–	–	–	–	–	–	–	–	–
RS-120	7	1503	1506	–	–	–	–	–	–	–	–	–	–	–	–	–	–	–	–	–	–	–	–	–	–	–	–	–	–	–	–	–	–	–	–
TA5E	7	1484	–	–	–	–	–	–	–	–	–	–	–	–	–	–	–	–	–	–	–	–	–	–	–	–	–	–	–	–	–	–	–	–	–
TA6V	7	1484	–	–	–	–	–	–	–	–	–	–	–	–	–	–	–	–	–	–	–	–	–	–	–	–	–	–	–	–	–	–	–	–	–
Titanium Diboride	8	–	703	–	–	705 707	–	–	–	–	–	710	–	–	–	–	–	–	–	–	–	–	–	–	–	–	–	–	–	–	–	–	–	–	–
Titanium Carbide	8	819	821	–	–	823	825	–	–	–	–	–	–	–	–	–	–	–	–	–	–	–	–	–	–	–	–	–	–	–	–	–	–	–	–
Titanium Carbide Pigment in Silica Binder	9	208	–	–	–	–	–	–	–	–	–	–	–	–	–	–	–	–	–	–	–	–	–	–	–	–	–	–	–	–	–	–	–	–	–
Titanium Nitride	8	–	1084	–	–	1086	–	–	–	–	–	–	–	–	–	–	–	–	–	–	–	–	–	–	–	–	–	–	–	–	–	–	–	–	–
Titanium Monoxide	8	–	–	–	–	–	–	–	–	–	–	454	–	–	–	–	–	–	–	–	–	–	–	–	–	–	–	–	–	–	–	–	–	–	–
Titanium Dioxide	8	–	456	–	–	458	–	–	–	–	–	461	473	–	–	–	–	–	–	–	–	–	–	475	–	–	–	–	–	476	–	–	–	–	–
Titanium Dioxide Coating on:																																			
Aluminum Substrate	9	981	–	–	–	–	–	–	–	–	–	986	989 990	–	–	–	–	–	–	–	–	–	–	991	–	–	–	–	–	–	–	–	–	992	–

MATERIAL NAME	VOLUME	EMITTANCE						REFLECTANCE									ABSORPTANCE									TRANSMITTANCE									Absorptance to Emittance Ratio
		Total			Spectral			Integrated			Spectral			Solar			Integrated			Spectral			Solar			Integrated			Spectral			Solar			
		Hemispherical	Normal	Angular	Hemispherical	Normal	Angular	Hemispherical	Normal	Angular	Hemispherical	Normal	Angular	Hemispherical	Normal	Angular	Hemispherical	Normal	Angular	Hemispherical	Normal	Angular	Hemispherical	Normal	Angular	Hemispherical	Normal	Angular	Hemispherical	Normal	Angular	Hemispherical	Normal	Angular	
Titanium Dioxide Coating on: (continued)																																		
Black Paint Substrate	9	-	-	-	-	-	-	-	-	-	-	986	-	-	-	-	-	-	-	-	-	-	-	-	-	-	-	-	-	-	-	-	-	-	-
Iron Substrate	9	981	-	-	-	-	-	-	-	-	-	-	-	-	-	-	-	-	-	-	-	-	-	-	-	-	-	-	-	-	-	-	-	-	-
Nickel Substrate	9	981	-	-	-	-	-	-	-	-	-	-	-	-	-	-	-	-	-	-	-	-	-	-	-	-	-	-	-	-	-	-	-	-	-
Niobium Substrate	9	981	-	-	-	984	-	-	-	-	-	-	-	-	-	-	-	-	-	-	-	-	-	-	-	-	-	-	-	-	-	-	-	-	-
Stainless Steel Substrate	9	981	-	-	-	-	-	-	-	-	-	-	-	-	-	-	-	-	-	-	-	-	-	-	-	-	-	-	-	-	-	-	-	-	-
Titanium Dioxide Pigment in:																																			
Acrylic Binder	9	211	-	-	-	-	-	-	-	-	-	-	-	-	-	-	-	-	-	-	-	-	-	-	-	-	-	-	-	-	-	-	-	-	-
Alkyd Binder	9	211	214	-	-	-	-	-	-	-	-	221	250	-	-	-	-	-	-	-	256	-	-	270	-	-	-	-	-	-	-	-	-	-	-
Alkyd-Melamine Binder	9	-	-	-	-	-	-	-	-	-	-	221	-	-	-	-	-	-	-	-	-	-	-	-	-	-	-	-	-	-	-	-	-	-	-
Cellulose Nitrate Binder	9	-	-	-	-	-	-	-	-	-	-	-	-	-	-	-	-	-	-	-	256	-	-	-	-	-	-	-	-	-	-	-	-	-	-
Epoxide Binder	9	211	-	-	-	-	-	-	-	-	-	-	-	-	-	-	-	-	-	-	-	-	-	-	-	-	-	-	-	-	-	-	-	-	-
Epoxy Binder	9	211	-	-	-	-	-	-	-	-	-	221	-	-	-	-	-	252	-	-	255	-	-	263 281	-	-	-	-	-	-	-	-	-	-	-
Nitrocellulose Binder	9	211	-	-	-	-	-	-	-	-	-	221	-	-	-	-	-	-	-	-	-	-	-	-	-	-	-	-	-	-	-	-	-	-	-
Polyvinyl Butyral Binder	9	-	-	-	-	-	-	-	-	-	-	-	-	-	-	-	-	-	-	-	-	-	-	267	-	-	-	-	-	-	-	-	-	-	-
Potassium Silicate Binder	9	-	214	-	-	-	-	-	-	-	-	226	-	-	-	-	-	-	-	-	-	-	-	274 281	-	-	-	-	-	-	-	-	-	-	283
Silicone Binder	9	211	-	-	-	-	-	-	-	-	-	220 244	-	-	-	-	-	-	-	-	255	-	-	263 281	-	-	-	-	-	-	-	-	-	-	283
Siloxane Binder	9	-	-	-	-	-	-	-	-	-	-	220	-	-	-	-	-	-	-	-	255	-	-	264	-	-	-	-	-	-	-	-	-	-	-
Sodium Silicate Binder	9	-	-	-	-	-	-	-	-	-	-	-	-	-	-	-	-	-	-	-	255	-	-	-	-	-	-	-	-	-	-	-	-	-	-
Titanium Dioxide + Aluminum Oxide Coating on:																																			
Nb-1Z Substrate	9	994	-	-	-	-	-	-	-	-	-	-	-	-	-	-	-	-	-	-	-	-	-	-	-	-	-	-	-	-	-	-	-	-	-
Stainless Steel Substrate	9	994	-	-	-	-	-	-	-	-	-	-	-	-	-	-	-	-	-	-	-	-	-	-	-	-	-	-	-	-	-	-	-	-	-
Titanium Dioxide + Aluminum Oxide Pigment in Potassium Silicate Binder	9	-	-	-	-	-	-	-	-	-	-	285 287	-	-	-	-	-	-	-	-	-	-	-	-	-	-	-	-	-	-	-	-	-	-	-
Titanium Dioxide + Calcium Pigment in Alkyd Binder	9	-	-	-	-	-	-	-	-	-	-	289	-	-	-	-	-	-	-	-	-	-	-	-	-	-	-	-	-	-	-	-	-	-	-
Titanium Dioxide + Talc Pigment in Acrylic Binder	9	290	-	-	-	-	-	-	-	-	-	-	-	-	-	-	-	-	-	-	293	-	-	295	-	-	-	-	-	-	-	-	-	-	-
TiO_2 + Ti	8	-	-	-	-	-	-	-	-	-	-	1439	-	-	-	-	-	-	-	-	-	-	-	-	-	-	-	-	-	-	-	-	-	-	-
TiO_2 + TiB_2 + H_3BO_3	8	-	1515	-	-	1516	-	-	-	-	-	1518	-	-	-	-	-	-	-	-	-	-	-	-	-	-	-	-	-	-	-	-	-	-	-
TiO_2 + $ZrSi_2$	8	-	-	-	-	-	-	-	-	-	-	1507	-	-	-	-	-	-	-	-	-	-	-	-	-	-	-	-	-	-	-	-	-	-	-
Titanium Hydride Pigment in Necoloidine Binder	9	209	-	-	-	-	-	-	-	-	-	-	-	-	-	-	-	-	-	-	-	-	-	-	-	-	-	-	-	-	-	-	-	-	-

MATERIAL NAME	VOLUME	EMITTANCE						REFLECTANCE									ABSORPTANCE									TRANSMITTANCE									Absorptance to Emittance Ratio
		Total			Spectral			Integrated			Spectral			Solar			Integrated			Spectral			Solar			Integrated			Spectral			Solar			
		Hemispherical	Normal	Angular	Hemispherical	Normal	Angular	Hemispherical	Normal	Angular	Hemispherical	Normal	Angular	Hemispherical	Normal	Angular	Hemispherical	Normal	Angular	Hemispherical	Normal	Angular	Hemispherical	Normal	Angular	Hemispherical	Normal	Angular	Hemispherical	Normal	Angular	Hemispherical	Normal	Angular	
Titanium Pyrophosphate Pigment in Nitrocellulose Binder	9	298	-	-	-	-	-	-	-	-	-	-	-	-	-	-	-	-	-	-	-	-	-	-	-	-	-	-	-	-	-	-	-	-	-
$TiSi_2 + Ti_5Si_3$ Powders	8	-	1498	-	-	1500	-	-	-	-	-	1504	-	-	-	-	-	-	-	-	-	-	-	-	-	-	-	-	-	-	-	-	-	-	-
$Ti_5Si_3 + TiSi_2$ Powders	8	-	1498	-	-	1500	-	-	-	-	-	1504	-	-	-	-	-	-	-	-	-	-	-	-	-	-	-	-	-	-	-	-	-	-	-
Titanium Phosphide	8	-	-	-	-	1105	-	-	-	-	-	-	-	-	-	-	-	-	-	-	-	-	-	-	-	-	-	-	-	-	-	-	-	-	-
Titanium Disilicide	8	-	1163	-	-	1165 1167	-	-	-	-	-	1169	-	-	-	-	-	-	-	-	-	-	-	-	-	-	-	-	-	-	-	-	-	-	-
Trititanium Trisilicate	8	-	-	-	-	-	-	-	-	-	-	1169	-	-	-	-	-	-	-	-	-	-	-	-	-	-	-	-	-	-	-	-	-	-	-
Pentatitanium Trisilicate	8	-	-	-	-	-	-	-	-	-	-	1169	-	-	-	-	-	-	-	-	-	-	-	-	-	-	-	-	-	-	-	-	-	-	-
Titanox AMO Pigment in Alkyd Binder	9	-	-	-	-	-	-	-	-	-	-	222	-	-	-	-	-	-	-	-	-	-	-	-	-	-	-	-	-	-	-	-	-	-	-
Titanox AMO Pigment in Silicone Binder	9	-	-	-	-	-	-	-	-	-	-	222	-	-	-	-	-	-	-	-	-	-	-	-	-	-	-	-	-	-	-	-	-	-	-
Titanox A-WO Pigment in Silicone Binder	9	-	-	-	-	-	-	-	-	-	-	222	-	-	-	-	-	-	-	-	-	-	-	-	-	-	-	-	-	-	-	-	-	-	-
Titanox C-50 Pigment in Alkyd-Melamine Binder	9	-	-	-	-	-	-	-	-	-	-	78 289	-	-	-	-	-	-	-	-	-	-	-	-	-	-	-	-	-	-	-	-	-	-	-
Titanox RA-10 Pigment in LTV-602 Binder	9	-	-	-	-	-	-	-	-	-	-	220	-	-	-	-	-	-	-	-	-	-	-	-	-	-	-	-	-	-	-	-	-	-	-
Titanox RC Pigment in Alkyd-Melamine Binder	9	-	-	-	-	-	-	-	-	-	-	78 289	-	-	-	-	-	-	-	-	-	-	-	-	-	-	-	-	-	-	-	-	-	-	-
Trial Aluminum No. 18270	9	-	6	-	-	-	-	-	-	-	-	-	-	-	-	-	-	-	-	-	-	-	-	-	-	-	-	-	-	-	-	-	-	-	-
Tungstates:																																			
Calcium Tungstates	8	-	-	-	-	-	-	-	-	-	-	-	-	-	-	-	-	-	-	-	-	-	-	-	-	-	-	-	-	663	-	-	-	-	-
Miscellaneous Tungstates	8	-	-	-	-	-	-	-	-	-	-	665	-	-	-	-	-	-	-	-	-	-	-	-	-	-	-	-	-	-	-	-	-	-	-
Tungsten	7	776	782	-	-	790 796	808 810	-	-	812	-	814 819	823	-	-	-	-	-	-	-	-	825	-	-	-	-	-	-	-	-	-	-	-	-	-
Tungsten Alloys:																																			
VR-27-VT	7	1049	-	-	-	-	-	-	-	-	-	-	-	-	-	-	-	-	-	-	-	-	-	-	-	-	-	-	-	-	-	-	-	-	-
Tungsten Coating on:																																			
Armco Iron Substrate	9	764	-	-	-	-	-	-	-	-	-	-	-	-	-	-	-	-	-	-	-	-	-	772	-	-	-	-	-	-	-	-	-	-	-
Inconel X Substrate	9	-	-	-	-	766	-	-	-	-	-	769	-	-	-	-	-	-	-	-	-	-	-	-	-	-	-	-	-	-	-	-	-	-	-
Tungsten Substrate Coated by Pack Cementation	9	-	1363	-	-	1364	-	-	-	-	-	1365	-	-	-	-	-	-	-	-	-	-	-	-	-	-	-	-	-	-	-	-	-	-	-
Tungsten Substrate Coated with:																																			
Cerium Dioxide	9	868	-	-	-	872	-	-	-	-	-	-	-	-	-	-	-	-	-	-	-	-	-	-	-	-	-	-	-	-	-	-	-	-	-
Hafnium Oxide	9	895	-	-	-	896	-	-	-	-	-	-	-	-	-	-	-	-	-	-	-	-	-	-	-	-	-	-	-	-	-	-	-	-	-
Uranium Dioxide	9	-	-	-	-	997	-	-	-	-	-	-	-	-	-	-	-	-	-	-	-	-	-	-	-	-	-	-	-	-	-	-	-	-	-

MATERIAL NAME	VOLUME	EMITTANCE						REFLECTANCE									ABSORPTANCE									TRANSMITTANCE									Absorptance to Emittance Ratio
		Total			Spectral			Integrated			Spectral			Solar			Integrated			Spectral			Solar			Integrated			Spectral			Solar			
		Hemispherical	Normal	Angular	Hemispherical	Normal	Angular	Hemispherical	Normal	Angular	Hemispherical	Normal	Angular	Hemispherical	Normal	Angular	Hemispherical	Normal	Angular	Hemispherical	Normal	Angular	Hemispherical	Normal	Angular	Hemispherical	Normal	Angular	Hemispherical	Normal	Angular	Hemispherical	Normal	Angular	
Tungsten Substrate Coated with: (continued)																																			
Vanadium Oxide	9	-	-	-	-	998	-	-	-	-	-	-	-	-	-	-	-	-	-	-	-	-	-	-	-	-	-	-	-	-	-	-	-	-	-
Yttrium Oxide	9	-	-	-	-	1001	-	-	-	-	-	-	-	-	-	-	-	-	-	-	-	-	-	-	-	-	-	-	-	-	-	-	-	-	-
Tungsten + Chromium + Aluminum Oxide Coating on Inconel Substrate	9	-	774	-	-	-	-	-	-	-	-	-	-	-	-	-	-	-	-	-	-	-	-	-	-	-	-	-	-	-	-	-	-	-	-
Tungsten + Cobalt Coating on Inconel X Substrate	9	-	-	-	-	-	-	-	-	-	-	776	-	-	-	-	-	-	-	-	-	-	-	-	-	-	-	-	-	-	-	-	-	-	-
Tungsten + Molybdenum	7	1043	-	-	-	1045	-	-	-	-	-	-	-	-	-	-	-	-	-	-	-	-	-	-	-	-	-	-	-	-	-	-	-	-	-
Tungsten + Rhenium	7	1048	-	-	-	1051	-	-	-	-	-	-	-	-	-	-	-	-	-	-	-	-	-	-	-	-	-	-	-	-	-	-	-	-	-
Tungsten Boride	8	-	730	-	-	-	-	-	-	-	-	735	-	-	-	-	-	-	-	-	-	-	-	-	-	-	-	-	-	-	-	-	-	-	-
Ditungsten Pentaboride	8	-	-	-	-	732	-	-	-	-	-	-	-	-	-	-	-	-	-	-	-	-	-	-	-	-	-	-	-	-	-	-	-	-	-
Tungsten Monocarbide	8	-	827	-	-	829 832	-	-	-	-	-	-	-	-	-	-	-	-	-	-	-	-	-	-	-	-	-	-	-	-	-	-	-	-	-
Ditungsten Carbide	8	-	-	-	-	829 832	-	-	-	-	-	-	-	-	-	-	-	-	-	-	-	-	-	-	-	-	-	-	-	-	-	-	-	-	-
Tungsten Carbide + Cobalt Coating on Armco Iron Substrate	9	995	-	-	-	-	-	-	-	-	-	-	-	-	-	-	-	-	-	-	-	-	-	996	-	-	-	-	-	-	-	-	-	-	-
Tungsten Disilicide	8	-	-	-	-	-	-	-	-	-	-	1176	-	-	-	-	-	-	-	-	-	-	-	-	-	-	-	-	-	-	-	-	-	-	-
Tritungsten Silicide	8	-	-	-	-	1173	-	-	-	-	-	-	-	-	-	-	-	-	-	-	-	-	-	-	-	-	-	-	-	-	-	-	-	-	-
Turpentine Binder Pigmented with Lampblack	9	-	-	-	-	-	-	-	-	-	-	86	-	-	-	-	-	-	-	-	-	-	-	-	-	-	-	-	-	-	-	-	-	-	-
Udimet 500	7	-	-	-	-	1365 1412	-	-	-	-	-	1381 1418	-	-	-	-	-	-	-	-	-	-	-	-	-	-	-	-	-	-	-	-	-	-	-
Ultrox Pigment in:																																			
Aluminum Phosphate Binder	9	-	-	-	-	-	-	-	-	-	-	-	-	-	-	-	-	-	-	-	435	-	-	-	-	-	-	-	-	-	-	-	-	-	-
Potassium Silicate Binder	9	-	-	-	-	-	-	-	-	-	-	-	-	-	-	-	-	-	-	-	435	-	-	-	-	-	-	-	-	-	-	-	-	-	-
Ultrox 1000 W	8	-	-	-	-	-	-	-	-	-	-	613	-	-	-	-	-	-	-	-	-	-	-	-	-	-	-	-	-	-	-	-	-	-	-
Unitane	8	-	-	-	-	-	-	-	-	-	-	465	-	-	-	-	-	-	-	-	-	-	-	-	-	-	475	-	-	-	-	-	-	-	-
Unitemp No. 41	7	-	936 1352	-	-	-	-	-	-	-	-	-	-	-	-	-	-	-	-	-	-	-	-	-	-	-	-	-	-	-	-	-	-	-	-
Uranium	7	828	-	-	-	834 838	-	-	-	-	-	-	-	-	-	-	-	-	-	-	-	-	-	-	-	-	-	-	-	-	-	-	-	-	-
Uranium Coating on Glass Substrate	9	-	-	-	-	-	-	-	-	-	-	779	-	-	-	-	-	-	-	-	-	-	-	-	-	-	-	-	-	780	-	-	-	-	-

MATERIAL NAME	VOLUME	EMITTANCE						REFLECTANCE									ABSORPTANCE									TRANSMITTANCE									Absorptance to Emittance Ratio
		Total			Spectral			Integrated			Spectral			Solar			Integrated			Spectral			Solar			Integrated			Spectral			Solar			
		Hemispherical	Normal	Angular	Hemispherical	Normal	Angular	Hemispherical	Normal	Angular	Hemispherical	Normal	Angular	Hemispherical	Normal	Angular	Hemispherical	Normal	Angular	Hemispherical	Normal	Angular	Hemispherical	Normal	Angular	Hemispherical	Normal	Angular	Hemispherical	Normal	Angular	Hemispherical	Normal	Angular	
Uranium + Niobium	7	1053	-	-	-	1056	-	-	-	-	-	-	-	-	-	-	-	-	-	-	-	-	-	-	-	-	-	-	-	-	-	-	-	-	-
Uranium + Zirconium + ΣX_i	7	1511	-	-	-	1514	-	-	-	-	-	-	-	-	-	-	-	-	-	-	-	-	-	-	-	-	-	-	-	-	-	-	-	-	-
Uranium Dodecaboride	8	-	-	-	-	732	-	-	-	-	-	-	-	-	-	-	-	-	-	-	-	-	-	-	-	-	-	-	-	-	-	-	-	-	-
Uranium Dicarbide	8	848	-	-	-	852	-	-	-	-	-	-	-	-	-	-	-	-	-	-	-	-	-	-	-	-	-	-	-	-	-	-	-	-	-
Uranium Dioxide	8	-	-	-	478	485	-	-	-	-	-	-	-	-	-	-	-	-	-	-	-	-	-	-	-	-	-	-	-	486	-	-	-	-	-
Uranium Dioxide Coating on Tungsten Substrate	9	-	-	-	-	997	-	-	-	-	-	-	-	-	-	-	-	-	-	-	-	-	-	-	-	-	-	-	-	-	-	-	-	-	-
Uranus 10	7	-	-	-	-	-	-	-	-	-	-	-	1283	-	-	-	-	-	-	-	-	-	-	-	-	-	-	-	-	-	-	-	-	-	-
U. S. Army Olive Drab Paint	9	-	-	-	-	-	-	-	-	-	-	569	-	-	-	-	-	-	-	-	-	-	-	-	-	-	-	-	-	-	-	-	-	-	-
Vanadates:																																			
Sodium Vanadate	8	-	-	-	-	-	-	-	-	-	-	667	-	-	-	-	-	-	-	-	-	-	-	-	-	-	-	-	-	-	-	-	-	-	-
Yttrium Vanadate	8	-	-	-	-	669	-	-	-	-	-	671	-	-	-	-	-	-	-	-	-	-	-	-	-	-	-	-	-	-	-	-	-	-	-
Vanadium	7	-	-	-	-	840	-	-	-	-	-	844	848	-	-	-	-	-	-	-	-	850	-	-	-	-	-	-	-	-	-	-	-	-	-
Vanadium Diboride	8	-	-	-	-	732	-	-	-	-	-	-	-	-	-	-	-	-	-	-	-	-	-	-	-	-	-	-	-	-	-	-	-	-	-
Vanadium Carbide	8	-	850	-	-	-	-	-	-	-	-	-	-	-	-	-	-	-	-	-	-	-	-	-	-	-	-	-	-	-	-	-	-	-	-
Vanadium Nitride	8	-	-	-	-	1087	-	-	-	-	-	-	-	-	-	-	-	-	-	-	-	-	-	-	-	-	-	-	-	-	-	-	-	-	-
Trivanadium Nitride	8	-	-	-	-	1087	-	-	-	-	-	-	-	-	-	-	-	-	-	-	-	-	-	-	-	-	-	-	-	-	-	-	-	-	-
Vanadium Oxide Coating on:																																			
Sapphire Substrate	9	-	-	-	-	-	-	-	-	-	-	999	-	-	-	-	-	-	-	-	-	-	-	-	-	-	-	-	-	1000	-	-	-	-	-
Tungsten Substrate	9	-	-	-	-	998	-	-	-	-	-	-	-	-	-	-	-	-	-	-	-	-	-	-	-	-	-	-	-	-	-	-	-	-	-
Divanadium Pentaoxide	8	-	-	-	-	-	-	-	-	-	-	546	-	-	-	-	-	-	-	-	-	-	-	-	-	-	-	-	-	-	-	-	-	-	-
Vanadium Disilicide	8	-	-	-	-	1173	-	-	-	-	-	-	-	-	-	-	-	-	-	-	-	-	-	-	-	-	-	-	-	-	-	-	-	-	-
Varnish Substrate Coated with Diacetyl Cellulose	9	-	-	-	-	-	-	-	-	-	-	-	-	-	-	-	-	-	-	-	-	-	-	-	-	-	-	-	-	1119	-	-	-	-	-
Vascojet 1000	7	-	-	-	-	1192	-	-	-	-	-	1199	-	-	-	-	-	-	-	-	-	-	-	-	-	-	-	-	-	-	-	-	-	-	-
Vickers F. D. P.	7	-	1221	-	-	-	-	-	-	-	-	-	-	-	-	-	-	-	-	-	-	-	-	-	-	-	-	-	-	-	-	-	-	-	-
Vinyl Plastics	8	-	-	-	-	-	-	-	-	-	-	1741	-	-	-	-	-	-	-	-	1743	-	-	-	-	-	-	-	-	1745	-	-	-	-	-
Vinyl Resins	8	-	-	-	-	-	-	-	-	-	-	1741	-	-	-	-	-	-	-	-	1743	-	-	-	-	-	-	-	-	1745	-	-	-	-	-
Vita-Var Grey Paint	9	-	-	-	-	-	-	-	-	-	-	-	-	-	-	-	-	-	-	-	570	-	-	-	-	-	-	-	-	571	-	-	-	-	-
Viton B Binder Pigmented with Zinc Oxide	9	-	-	-	-	-	-	-	-	-	-	317	-	-	-	-	-	-	-	-	-	-	-	-	-	-	-	-	-	-	-	-	-	-	-

MATERIAL NAME	VOLUME	EMITTANCE						REFLECTANCE									ABSORPTANCE									TRANSMITTANCE									Absorptance to Emittance Ratio
		Total			Spectral			Integrated			Spectral			Solar			Integrated			Spectral			Solar			Integrated			Spectral			Solar			
		Hemispherical	Normal	Angular	Hemispherical	Normal	Angular	Hemispherical	Normal	Angular	Hemispherical	Normal	Angular	Hemispherical	Normal	Angular	Hemispherical	Normal	Angular	Hemispherical	Normal	Angular	Hemispherical	Normal	Angular	Hemispherical	Normal	Angular	Hemispherical	Normal	Angular	Hemispherical	Normal	Angular	
Vitreous Silica [see Silicon Dioxide (Fused)]																																			
VM-1 Molybdenum Substrate Coated with Molybdenum Disilicide	9	927	-	-	-	-	-	-	-	-	-	-	-	-	-	-	-	-	-	-	-	-	-	-	-	-	-	-	-	-	-	-	-	-	-
Vycor 790	8	-	-	-	-	-	-	-	-	-	-	-	-	-	-	-	-	-	-	-	-	-	-	-	-	-	-	-	-	1633	-	-	-	-	-
Vycor 791	8	-	-	-	-	-	-	-	-	-	-	-	-	-	-	-	-	-	-	-	-	-	-	-	-	-	-	-	-	1633	-	-	-	-	-
VR-27-VT	7	1049	-	-	-	-	-	-	-	-	-	-	-	-	-	-	-	-	-	-	-	-	-	-	-	-	-	-	-	-	-	-	-	-	-
Wollastonite	8	-	-	-	-	-	-	-	-	-	-	618	-	-	-	-	-	-	-	-	-	-	-	-	-	-	-	-	-	-	-	-	-	-	-
Wulfenite	8	-	-	-	-	-	-	-	-	-	-	1673	-	-	-	-	-	-	-	-	-	-	-	-	-	-	-	-	-	-	-	-	-	-	-
Wulfenite Coating on Potassium Bromide Substrate	9	-	-	-	-	-	-	-	-	-	-	913	-	-	-	-	-	-	-	-	-	-	-	-	-	-	-	-	-	-	-	-	-	-	-
XX254 Pigment in Potassium Silicate Binder	9	-	-	-	-	-	-	-	-	-	-	316	-	-	-	-	-	-	-	-	-	-	-	-	-	-	-	-	-	-	-	-	-	-	-
Xylol Binder Pigmented with Acetylene Black	9	81	-	-	-	-	-	-	-	-	-	-	-	-	-	-	-	-	-	-	-	-	-	-	-	-	-	-	-	-	-	-	-	-	-
YAG (see Yttrium Aluminate)																																			
Yellow Brass	7	912	-	-	-	-	-	-	-	-	-	-	-	-	-	-	923	-	-	-	-	-	-	-	-	-	-	-	-	-	-	-	-	-	-
$Y_2O_3 + Cr_2O_3$ Powders	8	-	-	-	-	509	-	-	-	-	-	-	-	-	-	-	-	-	-	-	-	-	-	-	-	-	-	-	-	-	-	-	-	-	-
Ytterbium Hexaboride	8	-	-	-	-	723	-	-	-	-	-	-	-	-	-	-	-	-	-	-	-	-	-	-	-	-	-	-	-	-	-	-	-	-	-
Ytterbium Oxide	8	-	-	-	488	-	-	-	-	-	-	-	-	-	-	-	-	-	-	-	-	-	-	-	-	-	-	-	-	490	-	-	-	-	-
Yttrium	7	-	-	-	-	853	-	-	-	-	-	-	-	-	-	-	-	-	-	-	-	-	-	-	-	-	-	-	-	-	-	-	-	-	-
Yttrium Aluminate	8	-	-	-	-	-	-	-	-	-	-	579	-	-	-	-	-	-	-	-	-	-	-	-	-	-	-	-	-	-	-	-	-	-	-
Yttrium Aluminate Garnet (see Yttrium Aluminate)																																			
Yttrium Hexaboride	8	-	-	-	-	723	-	-	-	-	-	-	-	-	-	-	-	-	-	-	-	-	-	-	-	-	-	-	-	-	-	-	-	-	-
Yttrium Carbide	8	-	-	-	-	852	-	-	-	-	-	-	-	-	-	-	-	-	-	-	-	-	-	-	-	-	-	-	-	-	-	-	-	-	-
Yttrium Dicarbide	8	-	-	-	-	852	-	-	-	-	-	-	-	-	-	-	-	-	-	-	-	-	-	-	-	-	-	-	-	-	-	-	-	-	-
Diyttrium Tricarbide	8	-	-	-	-	852	-	-	-	-	-	-	-	-	-	-	-	-	-	-	-	-	-	-	-	-	-	-	-	-	-	-	-	-	-
Yttrium Nitride	8	-	-	-	-	-	-	-	-	-	-	-	-	-	-	-	-	-	-	-	-	-	-	-	-	-	-	-	-	1090	-	-	-	-	-
Yttrium Oxide	8	-	492	-	494	496 498	-	-	-	-	-	501	-	-	-	-	-	-	-	-	-	-	-	-	-	-	-	-	-	504	-	-	-	-	-

MATERIAL NAME	VOLUME	EMITTANCE						REFLECTANCE									ABSORPTANCE									TRANSMITTANCE									Absorptance to Emittance Ratio
		Total			Spectral			Integrated			Spectral			Solar			Integrated			Spectral			Solar			Integrated			Spectral			Solar			
		Hemispherical	Normal	Angular	Hemispherical	Normal	Angular	Hemispherical	Normal	Angular	Hemispherical	Normal	Angular	Hemispherical	Normal	Angular	Hemispherical	Normal	Angular	Hemispherical	Normal	Angular	Hemispherical	Normal	Angular	Hemispherical	Normal	Angular	Hemispherical	Normal	Angular	Hemispherical	Normal	Angular	
Yttrium Oxide Coating on Tungsten Substrate	9	-	-	-	-	1001	-	-	-	-	-	-	-	-	-	-	-	-	-	-	-	-	-	-	-	-	-	-	-	-	-	-	-	-	-
Yttrium Phosphate	8	-	-	-	-	-	-	-	-	-	-	607	-	-	-	-	-	-	-	-	-	-	-	-	-	-	-	-	-	-	-	-	-	-	-
Yttrium Vanadate	8	-	-	-	-	669	-	-	-	-	-	671	-	-	-	-	-	-	-	-	-	-	-	-	-	-	-	-	-	-	-	-	-	-	-
Z-93	9	302	304	-	-	-	-	-	-	-	-	314	-	-	-	-	-	360	-	-	-	-	-	372 392	-	-	-	-	-	-	-	-	-	-	399
Zapon Black	9	-	-	-	-	-	-	-	-	-	-	510 572	-	-	-	-	-	-	-	-	-	-	-	-	-	-	-	-	-	-	-	-	-	-	-
Zinc	7	855	857	-	-	-	-	-	-	-	-	860	-	-	-	-	864	-	-	-	866	869	-	-	-	-	-	-	-	-	-	-	-	-	-
Zinc Coating on Iron Substrate	9	-	-	-	-	-	-	-	-	-	-	781	-	-	-	-	-	-	-	-	-	-	-	-	-	-	-	-	-	-	-	-	-	-	-
Zinc + Aluminum	7	-	-	-	-	-	-	-	-	-	-	1059	-	-	-	-	-	-	-	-	-	-	-	-	-	-	-	-	-	-	-	-	-	-	-
Zinc Antimonide	8	-	-	-	-	-	-	-	-	-	-	1348	-	-	-	-	-	-	-	-	-	-	-	-	-	-	-	-	-	-	-	-	-	-	-
Zinc Black Coating on:																																			
Brass Substrate	9	-	-	-	-	-	-	-	-	-	-	1190	-	-	-	-	-	-	-	-	-	-	-	-	-	-	-	-	-	-	-	-	-	-	-
Metastyrene Substrate	9	-	-	-	-	-	-	-	-	-	-	1190	-	-	-	-	-	-	-	-	-	-	-	-	-	-	-	-	-	-	-	-	-	-	-
Pliofilm Substrate	9	-	-	-	-	-	-	-	-	-	-	-	-	-	-	-	-	-	-	-	-	-	-	-	-	-	-	-	-	1191	-	-	-	-	-
Pyroxylin Substrate	9	-	-	-	-	-	-	-	-	-	-	-	-	-	-	-	-	-	-	-	-	-	-	-	-	-	-	-	-	1191	-	-	-	-	-
Zinc Chromate Pigment in Unknown Binder	9	-	299	-	-	-	-	-	-	-	-	300	-	-	-	-	-	-	-	-	-	-	-	-	-	-	-	-	-	-	-	-	-	-	-
Zinc Diiodine Tetrasulfide	8	-	-	-	-	-	-	-	-	-	-	-	-	-	-	-	-	-	-	-	-	-	-	-	-	-	-	-	-	1236	-	-	-	-	-
Trizinc Diiodine Hexasulfide	8	-	-	-	-	-	-	-	-	-	-	-	-	-	-	-	-	-	-	-	-	-	-	-	-	-	-	-	-	1236	-	-	-	-	-
Zinc Oxide	8	506	-	-	-	-	-	-	-	-	-	507	-	-	-	-	-	-	-	-	519	-	-	521	-	-	-	-	-	522	-	-	-	-	-
Zinc Oxide Coating on Titanium Substrate	9	-	-	-	-	-	-	-	-	-	-	1003	-	-	-	-	-	-	-	-	-	-	-	-	-	-	-	-	-	-	-	-	-	-	-
Zinc Oxide Pigment in:																																			
Acrylic Binder	9	-	-	-	-	-	-	-	-	-	-	314	-	-	-	-	-	-	-	-	-	-	-	-	-	-	-	-	-	-	-	-	-	-	-
Copolymer Binder	9	-	-	-	-	-	-	-	-	-	-	314	-	-	-	-	-	-	-	-	-	-	-	-	-	-	-	-	-	-	-	-	-	-	-
Copolymer Acetone Binder	9	-	-	-	-	-	-	-	-	-	-	314	-	-	-	-	-	-	-	-	-	-	-	-	-	-	-	-	-	-	-	-	-	-	-
Potassium Silicate Binder	9	302	304	-	-	-	-	-	308	-	306	314 357	-	-	-	-	-	360	-	-	362 365	-	-	370 392	-	-	-	-	-	-	-	-	-	-	399
Silicone Binder	9	302	304	-	-	-	-	-	-	-	-	314 355	-	-	-	-	-	-	-	-	362	-	-	371 392	-	-	-	-	-	-	-	-	-	-	399
Siloxane Binder	9	-	-	-	-	-	-	-	-	-	-	316	-	-	-	-	-	-	-	-	362	-	-	392	-	-	-	-	-	-	-	-	-	-	-
Sodium Silicate Binder	9	-	-	-	-	-	-	-	-	-	-	-	-	-	-	-	-	-	-	-	362	-	-	-	-	-	-	-	-	-	-	-	-	-	-
Teflon Binder	9	-	-	-	-	-	-	-	-	-	-	317	-	-	-	-	-	-	-	-	-	-	-	-	-	-	-	-	-	-	-	-	-	-	-

MATERIAL NAME	VOLUME	EMITTANCE						REFLECTANCE									ABSORPTANCE									TRANSMITTANCE									Absorptance to Emittance Ratio
		Total			Spectral			Integrated			Spectral			Solar			Integrated			Spectral			Solar			Integrated			Spectral			Solar			
		Hemispherical	Normal	Angular	Hemispherical	Normal	Angular	Hemispherical	Normal	Angular	Hemispherical	Normal	Angular	Hemispherical	Normal	Angular	Hemispherical	Normal	Angular	Hemispherical	Normal	Angular	Hemispherical	Normal	Angular	Hemispherical	Normal	Angular	Hemispherical	Normal	Angular	Hemispherical	Normal	Angular	
Zinc Selenide	8	1112	–	–	–	1114	–	–	–	–	–	1117	–	–	–	–	–	–	–	–	1121	–	–	–	–	–	–	–	–	1123	–	–	–	–	–
Zinc Selenide Coating on Quartz Substrate	9	–	–	–	–	–	–	–	–	–	–	–	1004	–	–	–	–	–	–	–	–	–	–	–	–	–	–	–	–	1005	–	–	–	–	–
Zinc Selenide Substrate Coated with Barium Fluoride	9	–	–	–	–	–	–	–	–	–	–	810	–	–	–	–	–	–	–	–	–	–	–	–	–	–	–	–	–	812	–	–	–	–	–
Zinc Sulfide	8	1213	–	–	–	1215	–	–	–	–	–	1217	1222	–	–	–	–	–	–	–	1224	–	–	1226	–	–	–	–	–	1227	–	–	–	–	–
Zinc Sulfide Coating on Glass Substrate	9	–	–	–	–	–	–	–	–	–	–	1007	1008	–	–	–	–	–	–	–	–	–	–	–	–	–	–	–	–	–	–	–	–	–	–
Zinc Sulfide Pigment in:																																			
Acrylic Binder	9	–	–	–	–	–	–	–	–	–	–	404	–	–	–	–	–	–	–	–	–	–	–	–	–	–	–	–	–	–	–	–	–	–	–
Polyvinyl Butyral Binder	9	–	–	–	–	–	–	–	–	–	–	405	–	–	–	–	–	–	–	–	–	–	–	415	–	–	–	–	–	–	–	–	–	–	–
Potassium Silicate Binder	9	–	–	–	–	–	–	–	–	–	–	405	–	–	–	–	–	–	–	–	–	–	–	–	–	–	–	–	–	–	–	–	–	–	–
Silicone Binder	9	–	401	–	–	–	–	–	–	–	–	404	–	–	–	–	–	–	–	–	412	–	–	415	–	–	–	–	–	–	–	–	–	–	–
Sodium Silicate Binder	9	–	–	–	–	–	–	–	–	–	–	–	–	–	–	–	–	–	–	–	412	–	–	–	–	–	–	–	–	–	–	–	–	–	–
Zinc Sulfide + Clay Pigment in Silicone-Alkyd Binder	9	–	–	–	–	–	–	–	–	–	–	–	–	–	–	–	–	–	–	–	–	–	–	416	–	–	–	–	–	–	–	–	–	–	–
Zinc Telluride	8	–	–	–	–	–	–	–	–	–	–	1266	–	–	–	–	–	–	–	–	–	–	–	–	–	–	–	–	–	1268	–	–	–	–	–
Zinc Telluride Coating on Quartz Substrate	9	–	–	–	–	–	–	–	–	–	–	–	–	–	–	–	–	–	–	–	–	–	–	–	–	–	–	–	–	1009	–	–	–	–	–
Zinc Titanate	8	–	–	–	–	–	–	–	–	–	–	660	–	–	–	–	–	–	–	–	–	–	–	662	–	–	–	–	–	–	–	–	–	–	–
Zinc Titanate Pigment in Potassium Silicate Binder	9	–	–	–	–	–	–	–	–	–	–	–	–	–	–	–	–	–	–	–	–	–	–	420	–	–	–	–	–	–	–	–	–	–	–
Zinc Zirconium Silicate	8	–	–	–	–	–	–	–	–	–	–	–	–	–	–	–	–	–	–	–	–	–	–	616	–	–	–	–	–	–	–	–	–	–	–
Zircalloy 2	7	–	1520	–	–	1523	–	–	–	–	–	–	–	–	–	–	–	–	–	–	–	–	–	–	–	–	–	–	–	–	–	–	–	–	–
Zircoa (AH-C)	8	–	–	–	–	–	–	–	–	–	–	540	–	–	–	–	–	–	–	–	–	–	–	–	–	–	–	–	–	–	–	–	–	–	–
Zircon	8	–	1685	–	–	1687	–	–	–	–	–	613	–	–	–	–	–	–	–	–	–	–	–	615	–	–	–	–	–	–	–	–	–	–	–
Zirconates	8	–	–	–	–	–	–	–	–	–	–	675	–	–	–	–	–	–	–	–	–	–	–	–	–	–	–	–	–	–	–	–	–	–	–
Zirconia (see Zirconium Dioxide)																																			
Zirconium	7	–	872	–	–	875 878	–	–	–	–	–	881	–	–	–	–	–	–	–	–	–	883	–	–	–	–	–	–	–	–	–	–	–	–	–
Zirconium Coating on Molybdenum Substrate	9	782	–	–	–	–	–	–	–	–	–	–	–	–	–	–	–	–	–	–	–	–	–	–	–	–	–	–	–	–	–	–	–	–	–

MATERIAL NAME	VOLUME	EMITTANCE						REFLECTANCE									ABSORPTANCE									TRANSMITTANCE									Absorptance to Emittance Ratio
		Total			Spectral			Integrated			Spectral			Solar			Integrated			Spectral			Solar			Integrated			Spectral			Solar			
		Hemispherical	Normal	Angular	Hemispherical	Normal	Angular	Hemispherical	Normal	Angular	Hemispherical	Normal	Angular	Hemispherical	Normal	Angular	Hemispherical	Normal	Angular	Hemispherical	Normal	Angular	Hemispherical	Normal	Angular	Hemispherical	Normal	Angular	Hemispherical	Normal	Angular	Hemispherical	Normal	Angular	
Zirconium Pigment in Sodium Silicate Binder	9	26	-	-	-	-	-	-	-	-	-	-	-	-	-	-	-	-	-	-	-	-	-	-	-	-	-	-	-	-	-	-	-	-	-
Zirconium + Hafnium + ΣX_i	7	-	-	-	-	1517	-	-	-	-	-	-	-	-	-	-	-	-	-	-	-	-	-	-	-	-	-	-	-	-	-	-	-	-	-
Zirconium + Tin + ΣX_i	7	-	1519	-	-	1522	-	-	-	-	-	-	-	-	-	-	-	-	-	-	-	-	-	-	-	-	-	-	-	-	-	-	-	-	-
Zirconium + Uranium + ΣX_i	7	1525	-	-	-	1528	-	-	-	-	-	-	-	-	-	-	-	-	-	-	-	-	-	-	-	-	-	-	-	-	-	-	-	-	-
Zirconium Diboride	8	-	713	-	-	715 717	-	-	-	-	-	720	-	-	-	-	-	-	-	-	-	-	-	-	-	-	-	-	-	-	-	-	-	-	-
Zirconium Carbide	8	833	835	-	-	837 841	-	-	-	-	-	843	-	-	-	-	-	-	-	-	-	-	-	-	-	-	-	-	-	-	-	-	-	-	-
Zirconium Carbide Pigment in Necoloidine Binder	9	421	-	-	-	-	-	-	-	-	-	-	-	-	-	-	-	-	-	-	-	-	-	-	-	-	-	-	-	-	-	-	-	-	-
Zirconium Hydride + Iron Oxide Pigment in Necoloidine Binder	9	422	-	-	-	-	-	-	-	-	-	-	-	-	-	-	-	-	-	-	-	-	-	-	-	-	-	-	-	-	-	-	-	-	-
Zirconium Hydride + Zirconia Pigment in Necoloidine Binder	9	422	-	-	-	-	-	-	-	-	-	-	-	-	-	-	-	-	-	-	-	-	-	-	-	-	-	-	-	-	-	-	-	-	-
Zirconium Nitride	8	1078	-	-	-	1080 1082	-	-	-	-	-	-	-	-	-	-	-	-	-	-	-	-	-	-	-	-	-	-	-	-	-	-	-	-	-
Zirconium Dioxide	8	524	526	-	529	531 533	-	-	-	-	-	536	-	-	-	-	-	-	-	-	-	-	-	544	-	-	-	-	-	-	-	-	-	-	-
Zirconium Dioxide, Stabilized	8	-	527	-	-	532	-	-	-	-	-	538	-	-	-	-	-	-	-	-	-	-	-	544	-	-	-	-	-	-	-	-	-	-	-
Zirconium Oxide Coating on:																																			
Aluminum Substrate	9	-	-	-	-	-	-	-	-	-	-	1017	-	-	-	-	-	-	-	-	-	-	-	-	-	-	-	-	-	-	-	-	-	-	-
Inconel Substrate	9	-	1011	-	-	-	-	-	-	-	-	1017	-	-	-	-	-	-	-	-	-	-	-	1021	-	-	-	-	-	-	-	-	-	-	-
Inconel X Substrate	9	-	-	-	-	1014	-	-	-	-	-	1017	-	-	-	-	-	-	-	-	-	-	-	-	-	-	-	-	-	-	-	-	-	-	-
Mild Steel Substrate	9	-	1011	-	-	1014	-	-	-	-	-	-	-	-	-	-	-	-	-	-	-	-	-	-	-	-	-	-	-	-	-	-	-	-	-
Nimonic 75 Substrate	9	-	1011	-	-	-	-	-	-	-	-	-	-	-	-	-	-	-	-	-	-	-	-	-	-	-	-	-	-	-	-	-	-	-	-
Stainless Steel Substrate	9	-	1011	-	-	-	-	-	-	-	-	1017	-	-	-	-	-	-	-	-	-	-	-	-	-	-	-	-	-	-	-	-	-	-	-
Zirconium Oxide Pigment in:																																			
Aluminum Phosphate Binder	9	-	-	-	-	-	-	-	-	-	-	425	-	-	-	-	-	-	-	-	-	-	-	-	-	-	-	-	-	-	-	-	-	-	-
Nitrocellulose Binder	9	423	-	-	-	-	-	-	-	-	-	-	-	-	-	-	-	-	-	-	-	-	-	-	-	-	-	-	-	-	-	-	-	-	-
Potassium Silicate Binder	9	-	-	-	-	-	-	-	-	-	-	425	-	-	-	-	-	-	-	-	-	-	-	428	-	-	-	-	-	-	-	-	-	-	-
Silica Binder	9	-	-	-	-	-	-	-	-	-	-	425	-	-	-	-	-	-	-	-	-	-	-	-	-	-	-	-	-	-	-	-	-	-	-
Teflon Binder	9	-	-	-	-	-	-	-	-	-	-	425	-	-	-	-	-	-	-	-	-	-	-	-	-	-	-	-	-	-	-	-	-	-	-
Zirconium Oxide + Cerium Dioxide Pigment in NBS Frit No. 322 Binder (Enamel)	9	-	479	-	-	-	-	-	-	-	-	-	-	-	-	-	-	-	-	-	-	-	-	-	-	-	-	-	-	-	-	-	-	-	-

MATERIAL NAME	VOLUME	EMITTANCE						REFLECTANCE									ABSORPTANCE									TRANSMITTANCE									Absorptance to Emittance Ratio
		Total			Spectral			Integrated			Spectral			Solar			Integrated			Spectral			Solar			Integrated			Spectral			Solar			
		Hemispherical	Normal	Angular	Hemispherical	Normal	Angular	Hemispherical	Normal	Angular	Hemispherical	Normal	Angular	Hemispherical	Normal	Angular	Hemispherical	Normal	Angular	Hemispherical	Normal	Angular	Hemispherical	Normal	Angular	Hemispherical	Normal	Angular	Hemispherical	Normal	Angular	Hemispherical	Normal	Angular	
Zirconium Silicate	8	-	-	-	-	610	-	-	-	-	-	612	-	-	-	-	-	-	-	-	-	-	-	615	-	-	-	-	-	-	-	-	-	-	-
Zirconium Silicate Coating on:																																			
Aluminum Substrate	9	1024	-	-	-	-	-	-	-	-	-	-	-	-	-	-	-	-	-	-	-	-	-	1030	-	-	-	-	-	-	-	-	-	-	-
Nb-1Zr Substrate	9	1024	-	-	-	1028	-	-	-	-	-	-	-	-	-	-	-	-	-	-	-	-	-	-	-	-	-	-	-	-	-	-	-	-	-
Stainless Steel Substrate	9	1024	1026	-	-	-	-	-	-	-	-	1029	-	-	-	-	-	-	-	-	-	-	-	-	-	-	-	-	-	-	-	-	-	-	-
Zirconium Silicate Pigment in:																																			
Aluminum Phosphate Binder	9	-	-	-	-	-	-	-	-	-	-	-	-	-	-	-	-	-	-	-	435	-	-	-	-	-	-	-	-	-	-	-	-	-	-
Potassium Silicate Binder	9	-	-	-	-	-	-	-	-	-	-	432	-	-	-	-	-	-	-	-	435	-	-	437 441	-	-	-	-	-	-	-	-	-	-	-
Silicone Binder	9	430	-	-	-	-	-	-	-	-	-	-	-	-	-	-	-	-	-	-	-	-	-	437	-	-	-	-	-	-	-	-	-	-	-
Sodium Silicate Binder	9	430	-	-	-	-	-	-	-	-	-	-	-	-	-	-	-	-	-	-	-	-	-	437	-	-	-	-	-	-	-	-	-	-	-
Zirconium Disilicate	8	-	-	-	-	1173	-	-	-	-	-	1176	-	-	-	-	-	-	-	-	-	-	-	-	-	-	-	-	-	-	-	-	-	-	-
Zirconium Titanate Coating on Beryllium Substrate	9	1032	-	-	-	-	-	-	-	-	-	-	-	-	-	-	-	-	-	-	-	-	-	-	-	-	-	-	-	-	-	-	-	-	-
Zircopax	8	-	-	-	-	611	-	-	-	-	-	613	-	-	-	-	-	-	-	-	-	-	-	-	-	-	-	-	-	-	-	-	-	-	-
ZnSb	8	-	-	-	-	-	-	-	-	-	-	1352	-	-	-	-	-	-	-	-	-	-	-	-	-	-	-	-	-	-	-	-	-	-	-
ZrO_2 + Al	8	-	-	-	-	-	-	-	-	-	-	1442	-	-	-	-	-	-	-	-	-	-	-	-	-	-	-	-	-	-	-	-	-	-	-
$ZrSi_2$ + TiB_2 Powders	8	-	-	-	-	-	-	-	-	-	-	1495	-	-	-	-	-	-	-	-	-	-	-	-	-	-	-	-	-	-	-	-	-	-	-
$ZrSi_2$ + TiO_2 Powders	8	-	-	-	-	-	-	-	-	-	-	1507	-	-	-	-	-	-	-	-	-	-	-	-	-	-	-	-	-	-	-	-	-	-	-
ZW 40 Binder Pigmented with Zinc Sulfide	9	-	401	-	-	-	-	-	-	-	-	-	-	-	-	-	-	-	-	-	-	-	-	415	-	-	-	-	-	-	-	-	-	-	-
ZW 60 Binder Pigmented with Zinc Sulfide	9	-	401	-	-	-	-	-	-	-	-	-	-	-	-	-	-	-	-	-	-	-	-	415	-	-	-	-	-	-	-	-	-	-	-